The Aldrich Library of ^{13}C and ^1H FT NMR Spectra

EDITION I

Charles J. Pouchert
Jacqlynn Behnke

I

Printed in the United States of America

Library of Congress Catalog Card Number 92-073044
ISBN 0-941633-34-9

This book produced through programming by Professor H. W. Whitlock.

ACKNOWLEDGMENTS

It is obvious that a reference work of this size involved the efforts of more than just the two people named on its covers. Over the eight years since its conception, the efforts of a large number of people with a variety of talents contributed to the completion of this project.

Chris Lein was a major contributor to spectrum generation and data manipulation. Rick Hutson, Don Skalitzky, Tom Koresch, Payman Farid, Linda Mossburg, and Dee Williams assisted in laboratory operations. Proofreading and data organization assistance were supplied by Diana Brien, Ronnie Martin, Barb Rajzer, Tom Reed, David Swessel, Peter Vail, and Jolene Wall. Editing of ^{13}C peak tables was performed by Shirley Wing, Linda Ingalls, Kim Michaels, and Valerie Podlewski. Many other members of the Quality Control/Quality Assurance Department provided indispensable support and backup. Jerry Muccio coordinated graphics efforts. Jill Pallo helped interface the graphics and quality control contributions. Cover design was created by Toni Mendina. Tom Moertl and Joel Seewald were instrumental in bringing the data to the printed page through a maze of computer networks and hardware. Andy Smith was responsible for the program creating the indices for the collection.

PREFACE

The combination of high field superconducting magnets, computerization, and Fourier transform design has revolutionized the field of nuclear magnetic resonance spectroscopy. Through advances in instrumentation and software design the power of this technique, once the exclusive domain of the specialist, is now available to the general researcher for use in solving problems in a growing variety of applications.

Aldrich has a tradition of using state-of-the-art instrumentation to ensure the quality of its products; this publication has been produced in the spirit of that commitment. Since the first publication of the 60 MHz library of NMR spectra and the advent of higher field Fourier transform spectrometers, we began introducing 300 MHz instrumentation into our research and quality control laboratories. As with other spectral libraries that Sigma-Aldrich has published, the actual mechanisms of sample selection and preparation prior to recording the spectra are a by-product of our extensive quality control efforts.

We feel that the value of this reference work lies in the organization of the data. Combined proton and ^{13}C spectra for a compound are arranged by functional group, and by increasing complexity within a functional group. With this arrangement it is possible to make comparisons for either proton or ^{13}C spectra within a functional group, or to make a comparison of a proton to ^{13}C spectrum for a given compound.

This reference is a valuable companion to the other reference libraries available from the Sigma-Aldrich published libraries of data. These include *The Aldrich Library of NMR Spectra* (60 MHz), *The Aldrich Library of FT-IR Spectra* (both condensed and vapor phase), *The Sigma Library of FT-IR Spectra* and *The Sigma-Aldrich Handbook of Stains, Dyes and Indicators*. These publications have been created not only for our own internal use but also to satisfy the needs of our customers. They are a direct result of Sigma-Aldrich's commitment to product excellence while offering the largest inventories of research chemicals in the world.

TABLE OF CONTENTS

V

TABLE OF CONTENTS

TABLE OF CONTENTS

TABLE OF CONTENTS

TRADEMARKS AND REGISTERED TRADEMARKS

Aldrithiol™-2	Aldrich Chemical Co., Inc.
Aldrithiol™-4	Aldrich Chemical Co., Inc.
Aliquat®	Henkel Corp.
AM-ex-OL®	Aldrich Chemical Co., Inc.
Chirald®	Aldrich Chemical Co., Inc.
Colcemid®	Ciba-Geigy AG
Diazald®	Aldrich Chemical Co., Inc.
Fertilysin™	Aldrich Chemical Co., Inc.
IBM®	International Business Machines Corp.
(+)-Noe-lactol® dimer	Aldrich Chemical Co., Inc.
(-)-Noe-lactol® dimer	Aldrich Chemical Co., Inc.
PostScript®	Adobe Systems, Inc.
Triton® X-405	Union Carbide Corp.
Trolox®	Hoffmann-La Roche
WIMP™	Aldrich Chemical Co., Inc.

INTRODUCTION

Organization of the Library

This collection contains the combined proton and ^{13}C NMR spectra of nearly 12,000 compounds representing the breadth of products offered in the *Aldrich Catalog/Handbook of Fine Chemicals.*

As with previous Aldrich spectral libraries, the compounds are arranged by functionality, and by increasing complexity within a functional group. The ^{13}C spectrum is displayed jointly with the 300 MHz proton spectrum to aid the researcher in comparing the information obtained from the two nuclei, while observing trends within a family for either nucleus.

Several features have been included to assist the researcher in making spectral assignments:

1) Integration is included with the proton spectra.

2) A printout of the digital peak table for ^{13}C signals is displayed. The values in this table include, where practical, all signals for coupled heteronuclei, such as ^{13}C - ^{19}F, ^{13}C - ^{31}P, and ^{13}C - ^{14}N.

3) Signals in the peak table contain an asterisk (*) where the Attached Proton Test (APT) displays a signal of odd parity. In the majority of cases, this signal corresponds to a carbon atom in a methyl or methine group.

4) References to other Aldrich spectral libraries are listed.

5) Structures from the Aldrich files are drawn with WIMP™ and include stereochemical relationships where known.

Sample Purity and Preparation

A majority of the samples were prepared quantitatively at a concentration of 8-10% weight/volume (solid) or volume/volume (liquid) using Aldrich deuterated solvents. Every chemical used to produce the library was carefully analyzed by chromatographic techniques such as TLC, HPLC and GLC, spectroscopic techniques such as IR and UV, and other assay methods such as titrimetry and elemental analysis.

Generation of the Spectra

The proton, ^{13}C, and APT spectra were generated on a QE-300 Spectrometer with 7.05 Tesla superconducting magnet using a 5-mm ^{13}C-^{1}H dual probe. Chemical shifts were calculated in parts per million (ppm) from TMS or TSP-d_4. Typical instrument parameters for each nucleus are summarized in the table below.

Instrument Parameters

Nucleus	^{1}H	^{13}C
Frequency	300 MHz	75 MHz
Data points	32 K	32 K
Spectral width	±3,000 Hz	±10,000 Hz
Pulse angle	30°	30°
Recyle time between pulses	3 s	1-3 s
Average number of scans	32	1,024 - 4,096
Exponential line broadening	0	1 Hz
Quadrature detector	ON	ON

Processing of the Data

For both nuclei, 32 K data point free induction decays (FID's) were collected. Spectral widths encompassed the normal range of chemical shifts found for the two nuclei in organic molecules:

-5 ppm to 15 ppm for proton spectra
   (expanded if necessary, e.g. enol signals)

-25 ppm to 240 ppm for ^{13}C spectra

No exponential line broadening was applied to the proton FID; a line broadening of 1 Hz was applied to the ^{13}C FID to improve sensitivity. Although processing of the ^{13}C FID without the line broadening results in greater resolution, for consistency, all values listed in the peak table were generated under the standard conditions described. For both nuclei, the spectra were baseline-fixed after phasing.

With the sophistication of modern computer graphics software, it is easy to have a spectrum represent an artist's perception of the spectrum rather than experimental reality, and we have made a deliberate effort to exercise restraint in this regard. No chemical or solvent impurities were deleted from the spectra; only electronic noise spikes were removed. The signal intensity of TMS or TSP-d_4 used as a reference was reduced, and some artifacts of this procedure may be visible in the proton spectrum. In a few cases, the ^{13}C spectrum was scaled up relative to the solvent signals.

Display of the proton offset beyond 10 ppm was controlled by a signal-to-noise parameter in the plotting program. Therefore, broad signals from exchangeable protons in groups such as carboxylic acids are not always shown. Although some signals in the ^{13}C spectrum were too weak to be easily detected within the limitations of the display, all values listed in the peak table were verified by reproducible spectral observations.

Where protons were not present in the molecule, only the ^{13}C spectrum is presented. Similarly, if the ^{13}C spectrum was considered uninformative, only the proton spectrum is displayed.

Comments regarding the Attached Proton Test (APT)

With advances in Fourier transform pulsed programmer design in the last decade, experiments such as the APT have become routine tools in the assignment of ^{13}C signals. The pulse sequence is described in:
Patt, Steven L.; Shoolery, James N.; *Journal of Magnetic Resonance* **46**, 535-539 (1982).

The parameters used to generate the APT spectra result in methyl and methine signals with a phase opposite to those for methylene and quaternary carbon atoms. However, for carbon atoms with large ^{13}C-^{1}H coupling constants, this generalization is not necessarily true.

Representative molecules containing methine groups with large first order coupling constants include acetylenes, formate esters, and some imidazole and other similar heterocyclic aromatics. In these cases, asterisks were not used to note these methines in order to remain consistent with our objective of reporting APT experimental results as observed.

Page Composition and Typesetting
Prof. H.W. Whitlock

These books have been completely computer generated and represent a unique collaboration between colleagues from several areas of expertise. The input consisted of data sets of ^{1}H and ^{13}C spectra, text files including WIMP™ structure drawings, and ^{13}C peak chemical shifts. The output consisted of Postscript® files that were proofread after printing on 300 dpi laser printers, and sent directly to 1200 dpi typesetting machines. The spectra were rendered at their full point resolution. Point averaging was not employed as a computational shortcut. The spectra are thus as detailed as typesetting resolution permits. The program employed was written in C++, comprises twenty-eight modules totalling 148,000 bytes, and runs output on an IBM® PC or compatibles.

ABBREVIATIONS AND ACRONYMS

APT	Attached Proton Test
as	Asymmetric
bp	Boiling point in °C at 760mm pressure, unless otherwise specified
C	Celsius
ca.	Approximately
CAS	Chemical Abstracts Service
d	Dextrorotatory
d	Density of liquid compound or solid inorganic compound at 20° ±5°C relative to water at 4°C. For solid organic compounds, density is of liquid at melting-point temperature.
d.	Decompose
D	Configuration relative to D-glyceraldehyde
DOT	Department of Transportation
dpi	Dots per inch
F	Fahrenheit
fp	Freezing point
Fp	Flash point in °F (closed cup). Flash points are determined with the "Setaflash" apparatus recommended by the DOT using ASTM Procedure D3278.
FT	Fourier Transform
FT-IR	Reference to the page location of the spectrum in *The Aldrich Library of FT-IR Spectra,* Edition 1, Vol. I or II.
FW	Formula weight based on carbon mass = 12.011. Calculated values in this book include solvent or water of hydration if degree of solvation or hydration is known.
GLC	Gas-liquid chromatography
HPLC	High-performance liquid chromatography
Hz	Hertz
IR	Infrared
K	2^{10} or 1,024
l	Levorotatory

L	Configuration relative to L-glyceraldehyde
M	Molarity of solution
mp	Melting point in °C
n_D^{20}	Index of refraction for the sodium D line at 20°C (or temperature indicated)
NMR	Nuclear magnetic resonance
60MHz	Reference to the page location of the spectrum in *The Aldrich Library of NMR Spectra,* Edition II.
MHz	Megahertz
no.	Number
pp.	Pages
ppm	Parts per million
s	Second
s.	Sublime
sym	Symmetrical
temp.	Temperature
TLC	Thin-layer chromatography
TMS	Tetramethylsilane
TSP-d_4	3-(Trimethylsilyl)propionic-2,2,3,3-d_4, sodium salt
unsym	Unsymmetrical
UV	Ultraviolet
VPC	Vapor-phase chromatography
VP-FT-IR	Reference to the page location of the spectrum in *The Aldrich Library of FT-IR Spectra,* Edition 1, Vol. III.
[ ]	Italicized numbers within brackets after the chemical name denote the Chemical Abstracts Service (CAS) number
~	Approximately
=	Equals
$\cong$	Approximately equal to
<	Less than
$\leq$	Less than or equal to
>	Greater than
$\geq$	Greater than or equal to

SOLVENT ABBREVIATIONS

ACETONE-d_6	Acetone-d_6
CDCl$_3$	Chloroform-*d*
CDCl$_3$ + DMSO-d_6	A 50:50 mixture by volume of CDCl$_3$ and DMSO-d_6
CD$_2$Cl$_2$	Dichloromethane-d_2
CD$_3$CN	Acetonitrile-d_3
CD$_3$NO$_2$	Nitromethane-d_3
CD$_3$OD	Methyl-d_3 alcohol-*d*
DMSO-d_6	(Methylsulfoxide)-d_6
D$_2$O	Deuterium oxide
D$_2$O + DCl	Deuterium oxide and deuterium chloride solution
D$_2$O + NAOD	Deuterium oxide and sodium deuteroxide solution
PYRIDINE-d_5	Pyridine-d_5
TFA-*d*	Trifluoroacetic acid-*d*
THF-d_8	Tetrahydrofuran-d_8

KEY TO SPECTRAL DATA

2 — Aldrich 17,465-3   CAS [3699-66-9] — 3

1 — **Triethyl 2-phosphonopropionate, 98%** — 4

8 — CDCl₃   9 — QE-300

10 —

5
C₉H₁₉O₅P   Fp 192°F
FW 238.22
bp 144°C (12 mm)
d 1.111
n²⁰_D 1.4320

6
60 MHz: *2*, 870B
FT-IR: *1*, 918B
VP-FT-IR: *3*, 843C

7
169.71	16.
169.65	16.
62.66	16.
62.58	16.
61.36	14.
40.24*	11.
38.47*	11.

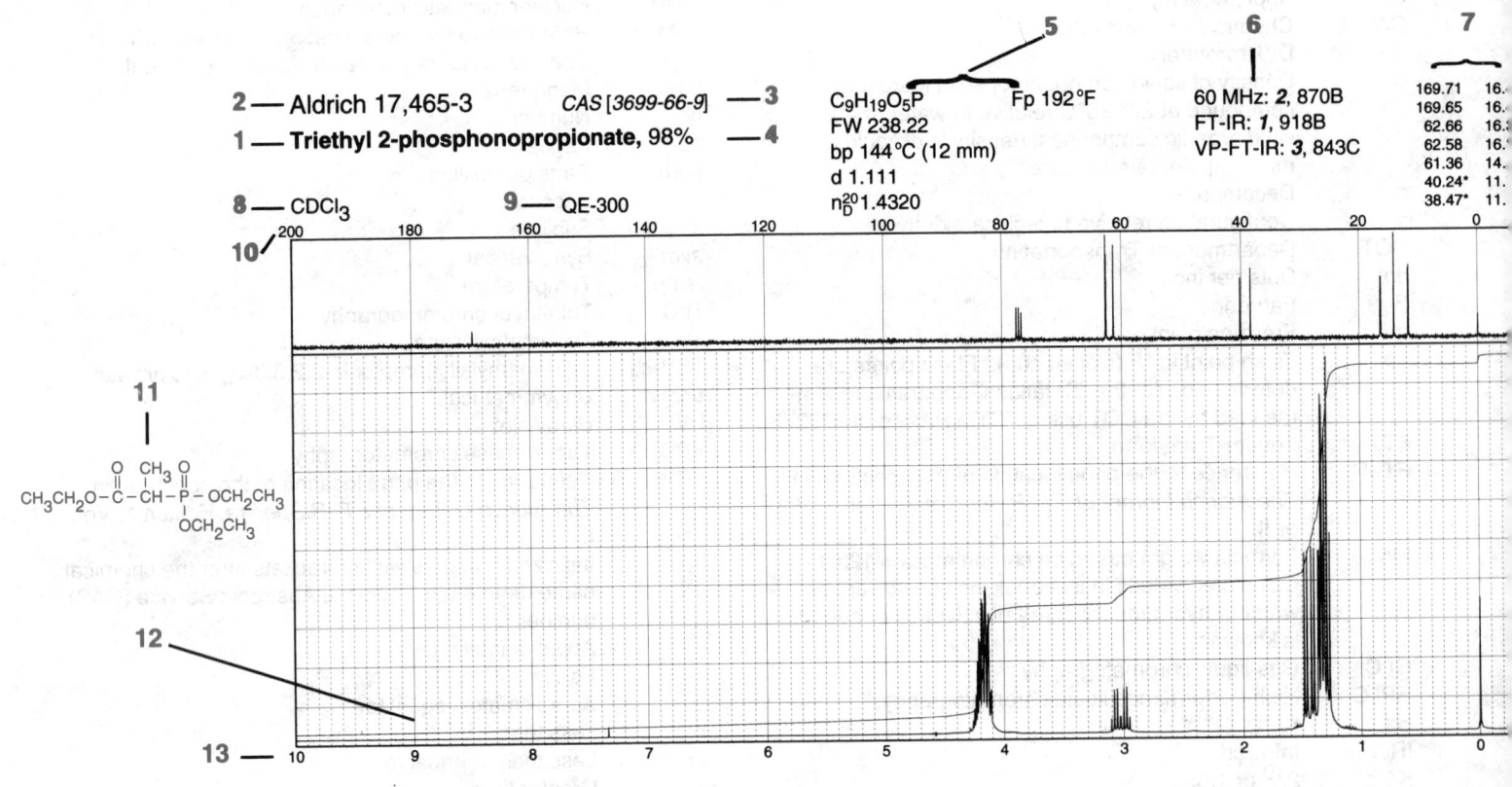

11

CH₃CH₂O–C–CH–P–OCH₂CH₃
         CH₃
         OCH₂CH₃

12

13 —

---

1. Product name
2. Aldrich catalog number
3. Chemical Abstract Service Registry Number
4. Chemical purity
5. Physical constants*, including:
   Molecular Formula
   Formula weight of product
   Boiling point (bp)
   Melting point (mp)
   Density (d)
   Index of refraction (n²⁰_D)
   Flash point (Fp)

6. Aldrich spectral references*
7. ¹³C peak listing table. Asterisk indicates ¹³C signals of odd parity.
8. Solvent
   Solvent in ¹³C spectrum shows ¹³C-D coupling. Proton spectrum shows residual protons from deuterated solvent. See Volume III, pp. 619-627.
9. Spectrometer Model
10. ¹³C scale in ppm
11. Chemical structure
12. Integral of proton spectrum
13. Proton scale in ppm

* See Abbreviations & Acronyms for additional information

Non-Aromatic Hydrocarbons

A

Aldrich 15,495-4   *CAS [109-66-0]* C_5H_{12} $n_D^{20}1.3580$ 34.24
                                                                                                        22.44
Pentane, 99+%   FW 72.15   Fp -57°F   14.07*
mp -130°C   60 MHz: *1*, 9A
bp 36°C   FT-IR: *1*, 1A

CDCl$_3$ QE-300   d 0.626   VP-FT-IR: *3*, 2A

$CH_3CH_2CH_2CH_2CH_3$

B

Aldrich 13,938-6   *CAS [110-54-3]* C_6H_{14} $n_D^{20}1.3750$ 31.68
                                                                                                        22.73
Hexane, 99+%   FW 86.18   Fp -10°F   14.12*
mp -95°C   60 MHz: *1*, 9B
bp 69°C   FT-IR: *1*, 1B

CDCl$_3$ QE-300   d 0.659   VP-FT-IR: *3*, 2B

$CH_3(CH_2)_4CH_3$

C

Aldrich 15,487-3   *CAS [142-82-5]* C_7H_{16} $n_D^{20}1.3870$ 31.98
                                                                                                        29.11
Heptane, 99%   FW 100.21   Fp 30°F   22.77
mp -91°C   60 MHz: *1*, 9D   14.12*
bp 98°C   FT-IR: *1*, 1D

CDCl$_3$ QE-300   d 0.684   VP-FT-IR: *3*, 2D

$CH_3(CH_2)_5CH_3$

ALDRICH

Non-Aromatic Hydrocarbons

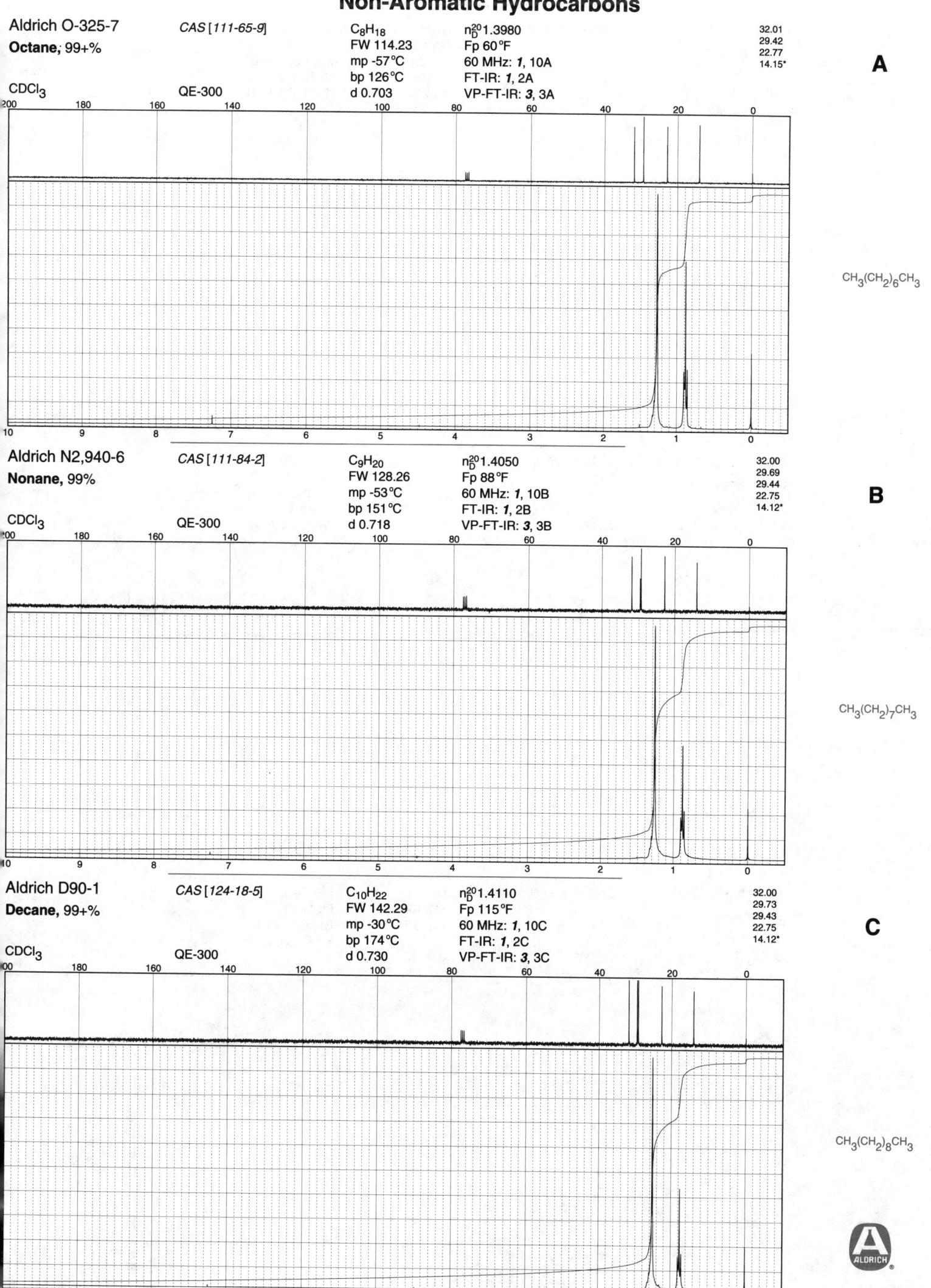

Aldrich O-325-7 CAS [111-65-9] C_8H_{18} n_D^{20} 1.3980 32.01
Octane, 99+% FW 114.23 Fp 60°F 29.42
  mp -57°C 60 MHz: **1**, 10A 22.77
CDCl₃ QE-300 bp 126°C FT-IR: **1**, 2A 14.15*
  d 0.703 VP-FT-IR: **3**, 3A

A

$CH_3(CH_2)_6CH_3$

Aldrich N2,940-6 CAS [111-84-2] C_9H_{20} n_D^{20} 1.4050 32.00
Nonane, 99% FW 128.26 Fp 88°F 29.69
  mp -53°C 60 MHz: **1**, 10B 29.44
CDCl₃ QE-300 bp 151°C FT-IR: **1**, 2B 22.75
  d 0.718 VP-FT-IR: **3**, 3B 14.12*

B

$CH_3(CH_2)_7CH_3$

Aldrich D90-1 CAS [124-18-5] $C_{10}H_{22}$ n_D^{20} 1.4110 32.00
Decane, 99+% FW 142.29 Fp 115°F 29.73
  mp -30°C 60 MHz: **1**, 10C 29.43
CDCl₃ QE-300 bp 174°C FT-IR: **1**, 2C 22.75
  d 0.730 VP-FT-IR: **3**, 3C 14.12*

C

$CH_3(CH_2)_8CH_3$

ALDRICH

Non-Aromatic Hydrocarbons

A

Aldrich U40-7   *CAS [1120-21-4]*

Undecane, 99+%

$C_{11}H_{24}$
FW 156.31
mp -26°C
bp 196°C

n_D^{20}1.4170
Fp 140°F
60 MHz: *1*, 10D
FT-IR: *1*, 2D
VP-FT-IR: *3*, 3D

32.08
29.85
29.81
29.50
22.82
14.16*

CDCl$_3$ QE-300

$CH_3(CH_2)_9CH_3$

B

Aldrich D22,110-4   *CAS [112-40-3]*

Dodecane, 99+%

$C_{12}H_{26}$
FW 170.34
mp -10°C
bp 216°C

n_D^{20}1.4220
Fp 160°F
60 MHz: *1*, 11A
FT-IR: *1*, 3A
VP-FT-IR: *3*, 4A

32.08
29.85
29.81
29.50
22.82
14.17*

CDCl$_3$ QE-300

$CH_3(CH_2)_{10}CH_3$

C

Aldrich T5,740-1   *CAS [629-50-5]*

Tridecane, 99+%

$C_{13}H_{28}$
FW 184.37
mp -5°C
bp 234°C

n_D^{20}1.4250
Fp 175°F
FT-IR: *1*, 3B
VP-FT-IR: *3*, 4B

31.98
29.76
29.72
29.42
22.74
14.12*

CDCl$_3$ QE-300

$CH_3(CH_2)_{11}CH_3$

Non-Aromatic Hydrocarbons

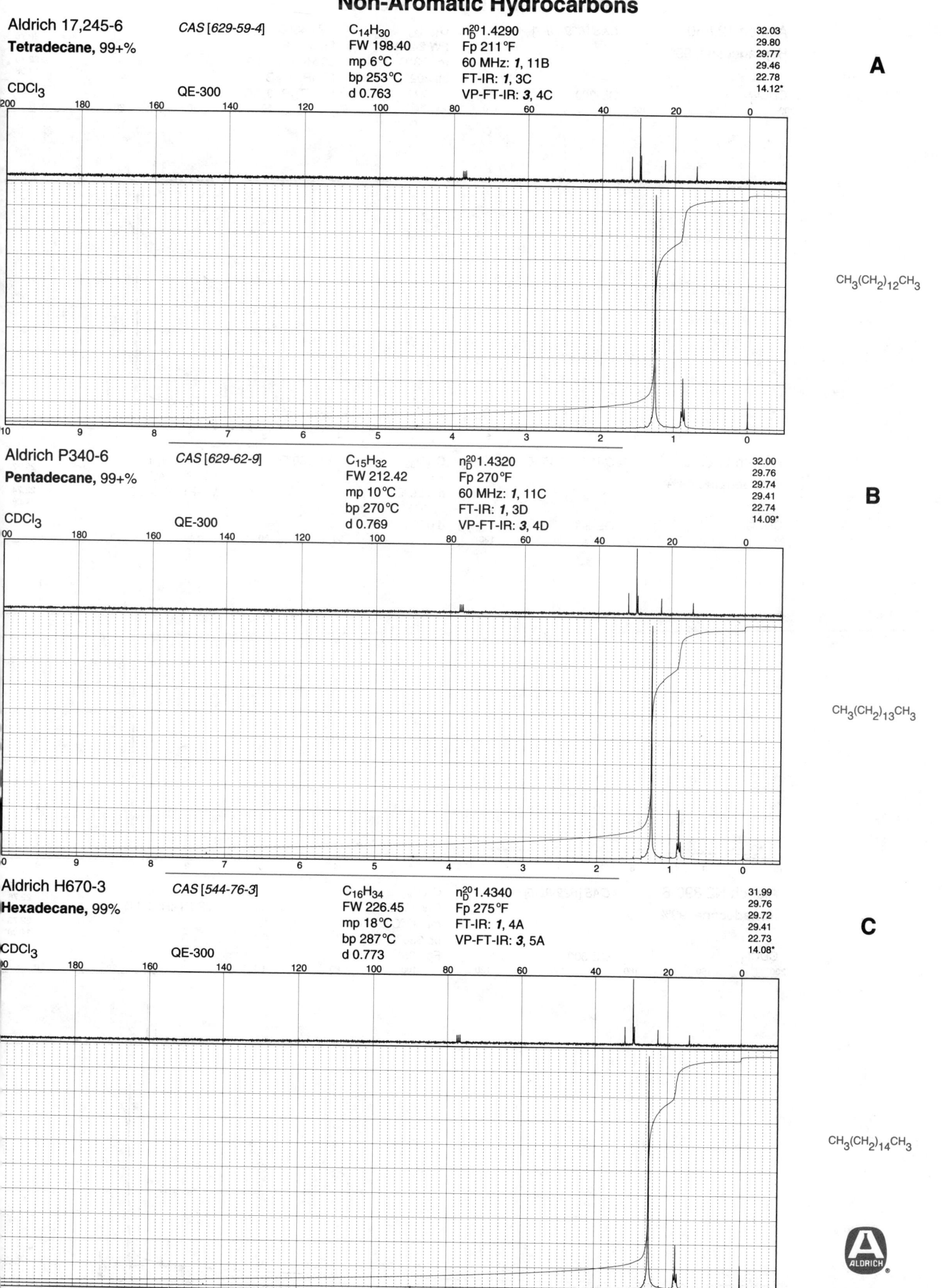

Aldrich 17,245-6 CAS [629-59-4] $C_{14}H_{30}$ $n_D^{20} 1.4290$ 32.03
Tetradecane, 99+% FW 198.40 Fp 211°F 29.80
  mp 6°C 60 MHz: *1*, 11B 29.77
  bp 253°C FT-IR: *1*, 3C 29.46
CDCl₃ QE-300 d 0.763 VP-FT-IR: *3*, 4C 22.78
  14.12*

A

$CH_3(CH_2)_{12}CH_3$

Aldrich P340-6 CAS [629-62-9] $C_{15}H_{32}$ $n_D^{20} 1.4320$ 32.00
Pentadecane, 99+% FW 212.42 Fp 270°F 29.76
  mp 10°C 60 MHz: *1*, 11C 29.74
  bp 270°C FT-IR: *1*, 3D 29.41
CDCl₃ QE-300 d 0.769 VP-FT-IR: *3*, 4D 22.74
  14.09*

B

$CH_3(CH_2)_{13}CH_3$

Aldrich H670-3 CAS [544-76-3] $C_{16}H_{34}$ $n_D^{20} 1.4340$ 31.99
Hexadecane, 99% FW 226.45 Fp 275°F 29.76
  mp 18°C FT-IR: *1*, 4A 29.72
  bp 287°C VP-FT-IR: *3*, 5A 29.41
CDCl₃ QE-300 d 0.773 22.73
  14.08*

C

$CH_3(CH_2)_{14}CH_3$

ALDRICH

Non-Aromatic Hydrocarbons

5

A

Aldrich 12,850-3   *CAS [629-78-7]* $C_{17}H_{36}$ n_D^{20} 1.4360
Heptadecane, 99%   FW 240.48   Fp 300°F
                                        mp 23°C   60 MHz: *1*, 11D
                                        bp 302°C   FT-IR: *1*, 4B
CDCl$_3$ QE-300   d 0.777   VP-FT-IR: *3*, 5B

31.99
29.76
29.41
22.73
14.08*

$CH_3(CH_2)_{15}CH_3$

B

Aldrich O-65-2   *CAS [593-45-3]* $C_{18}H_{38}$ Fp 330°F   60 MHz: *1*, 12A
Octadecane, 99%   FW 254.50   FT-IR: *1*, 4C
                                       mp 29°C   VP-FT-IR: *3*, 5C
                                       bp 317°C
CDCl$_3$ QE-300   d 0.777

32.00
29.77
29.42
22.74
14.08*

$CH_3(CH_2)_{16}CH_3$

C

Aldrich N2,890-6   *CAS [629-92-5]* $C_{19}H_{40}$ FT-IR: *1*, 4D
Nonadecane, 99%   FW 268.53   VP-FT-IR: *3*, 5D
                                        mp 33°C
                                        bp 330°C
CDCl$_3$ QE-300   Fp 335°F

31.99
29.76
29.42
22.73
14.08*

$CH_3(CH_2)_{17}CH_3$

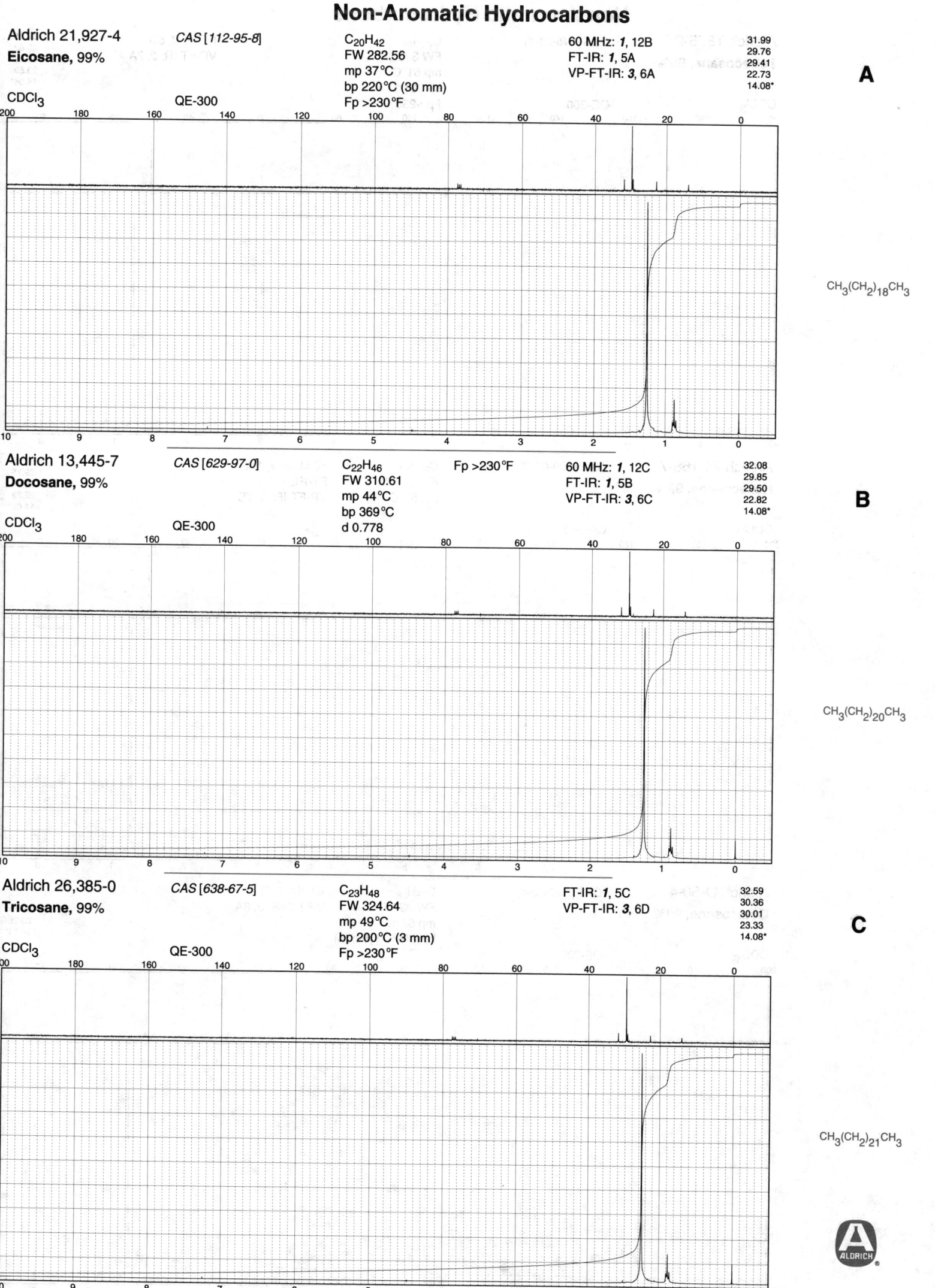

Aldrich 21,927-4   CAS [112-95-8]
Eicosane, 99%

$C_{20}H_{42}$
FW 282.56
mp 37°C
bp 220°C (30 mm)
Fp >230°F

60 MHz: *1*, 12B
FT-IR: *1*, 5A
VP-FT-IR: *3*, 6A

31.99
29.76
29.41
22.73
14.08*

A

CDCl₃   QE-300

$CH_3(CH_2)_{18}CH_3$

Aldrich 13,445-7   CAS [629-97-0]
Docosane, 99%

$C_{22}H_{46}$
FW 310.61
mp 44°C
bp 369°C
d 0.778

Fp >230°F

60 MHz: *1*, 12C
FT-IR: *1*, 5B
VP-FT-IR: *3*, 6C

32.08
29.85
29.50
22.82
14.08*

B

CDCl₃   QE-300

$CH_3(CH_2)_{20}CH_3$

Aldrich 26,385-0   CAS [638-67-5]
Tricosane, 99%

$C_{23}H_{48}$
FW 324.64
mp 49°C
bp 200°C (3 mm)
Fp >230°F

FT-IR: *1*, 5C
VP-FT-IR: *3*, 6D

32.59
30.36
30.01
23.33
14.08*

C

CDCl₃   QE-300

$CH_3(CH_2)_{21}CH_3$

ALDRICH

Non-Aromatic Hydrocarbons

A

Aldrich T875-2   CAS [646-31-1] $C_{24}H_{50}$ FT-IR: *1*, 5D

Tetracosane, 99%   FW 338.66   VP-FT-IR: *3*, 7A

mp 51 °C
bp 391 °C
Fp >230 °F

CDCl$_3$ QE-300

```
32.85
30.62
30.27
23.59
14.08*
```

$CH_3(CH_2)_{22}CH_3$

B

Aldrich 24,168-7   CAS [630-01-3] $C_{26}H_{54}$ 60 MHz: *1*, 12D

Hexacosane, 99%   FW 366.72   FT-IR: *1*, 6A

mp 57 °C   VP-FT-IR: *3*, 7C

CDCl$_3$ QE-300

```
31.98
29.75
29.40
22.72
14.08*
```

$CH_3(CH_2)_{24}CH_3$

C

Aldrich O-50-4   CAS [630-02-4] $C_{28}H_{58}$ FT-IR: *1*, 6B

Octacosane, 99%   FW 394.77   VP-FT-IR: *3*, 8A

mp 62 °C
bp 278 °C (15 mm)

CDCl$_3$ QE-300

```
32.52
30.29
29.94
23.26
14.11*
```

$CH_3(CH_2)_{26}CH_3$

Non-Aromatic Hydrocarbons

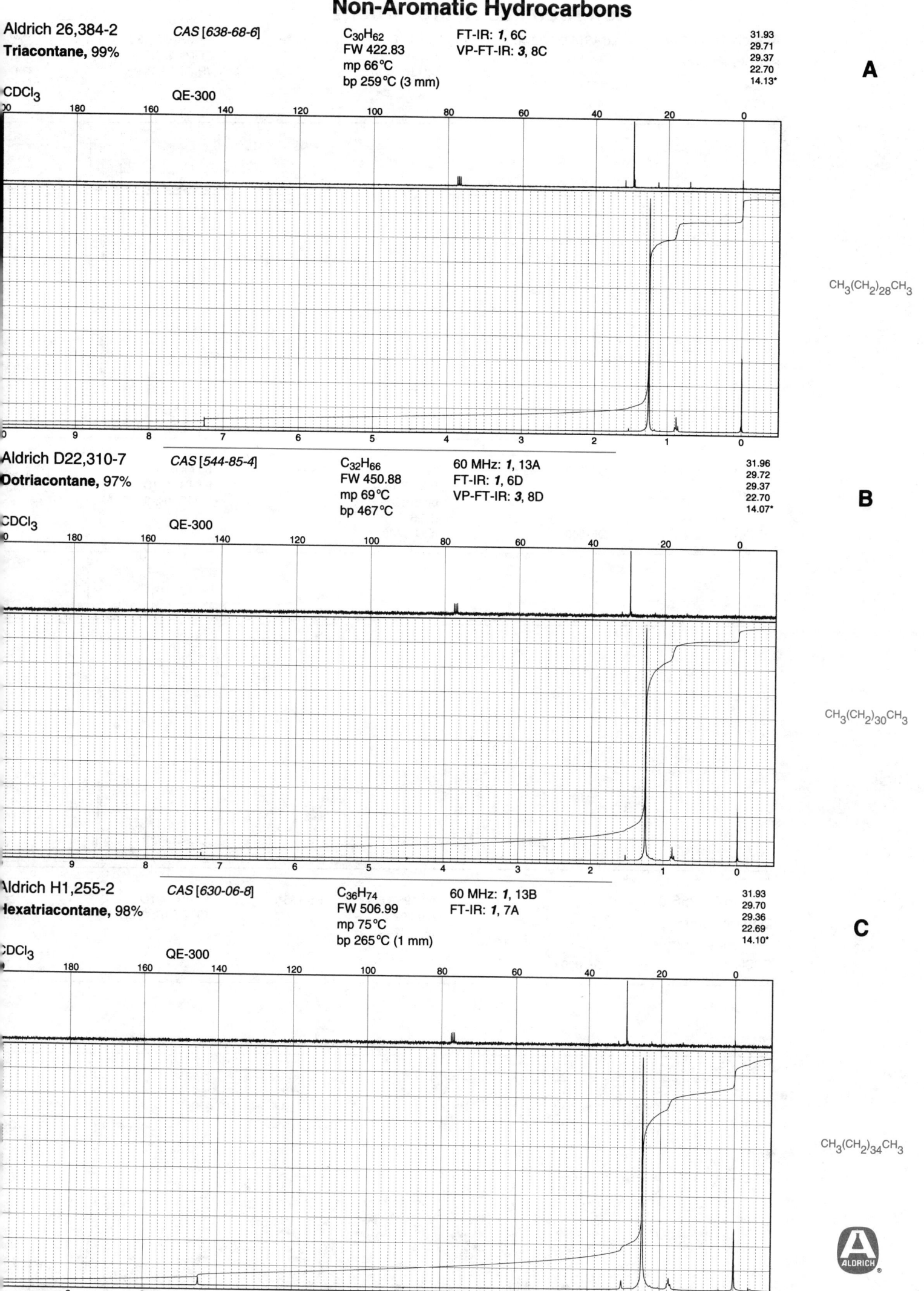

Aldrich 26,384-2   CAS [638-68-6]

Triacontane, 99%

CDCl$_3$ QE-300

$C_{30}H_{62}$
FW 422.83
mp 66 °C
bp 259 °C (3 mm)

FT-IR: *1*, 6C
VP-FT-IR: *3*, 8C

31.93
29.71
29.37
22.70
14.13*

A

$CH_3(CH_2)_{28}CH_3$

Aldrich D22,310-7   CAS [544-85-4]

Dotriacontane, 97%

CDCl$_3$ QE-300

$C_{32}H_{66}$
FW 450.88
mp 69 °C
bp 467 °C

60 MHz: *1*, 13A
FT-IR: *1*, 6D
VP-FT-IR: *3*, 8D

31.96
29.72
29.37
22.70
14.07*

B

$CH_3(CH_2)_{30}CH_3$

Aldrich H1,255-2   CAS [630-06-8]

Hexatriacontane, 98%

CDCl$_3$ QE-300

$C_{36}H_{74}$
FW 506.99
mp 75 °C
bp 265 °C (1 mm)

60 MHz: *1*, 13B
FT-IR: *1*, 7A

31.93
29.70
29.36
22.69
14.10*

C

$CH_3(CH_2)_{34}CH_3$

ALDRICH

Non-Aromatic Hydrocarbons

9

A

Aldrich 18,451-9   *CAS [8032-32-4]*   bp 45 °C   Fp -57 °F   60 MHz: *1*, 13C

Petroleum ether   d 0.640   FT-IR: *1*, 7B

n_D^{20} 1.3630   VP-FT-IR: *3*, 9A

34.23
28.93
22.64*
22.41
14.06*

CDCl$_3$ QE-300

Aldrich 15,617-5   *CAS [73513-42-5]*   C$_6$H$_{14}$ Fp -9 °F   60 MHz: *1*, 9C

Hexanes   FW 86.18   FT-IR: *1*, 1C

bp 69 °C   VP-FT-IR: *3*, 2C

d 0.670

n_D^{20} 1.3790

31.68
22.73
14.12*

B

C$_6$H$_{14}$

CDCl$_3$ QE-300

Aldrich 26,256-0   bp 195 °C   Fp 135 °F   FT-IR: *1*, 7D

Mineral spirits   d 0.752   VP-FT-IR: *3*, 9B

n_D^{20} 1.4240

29.46*
27.38*
22.76*

C

CDCl$_3$ QE-300

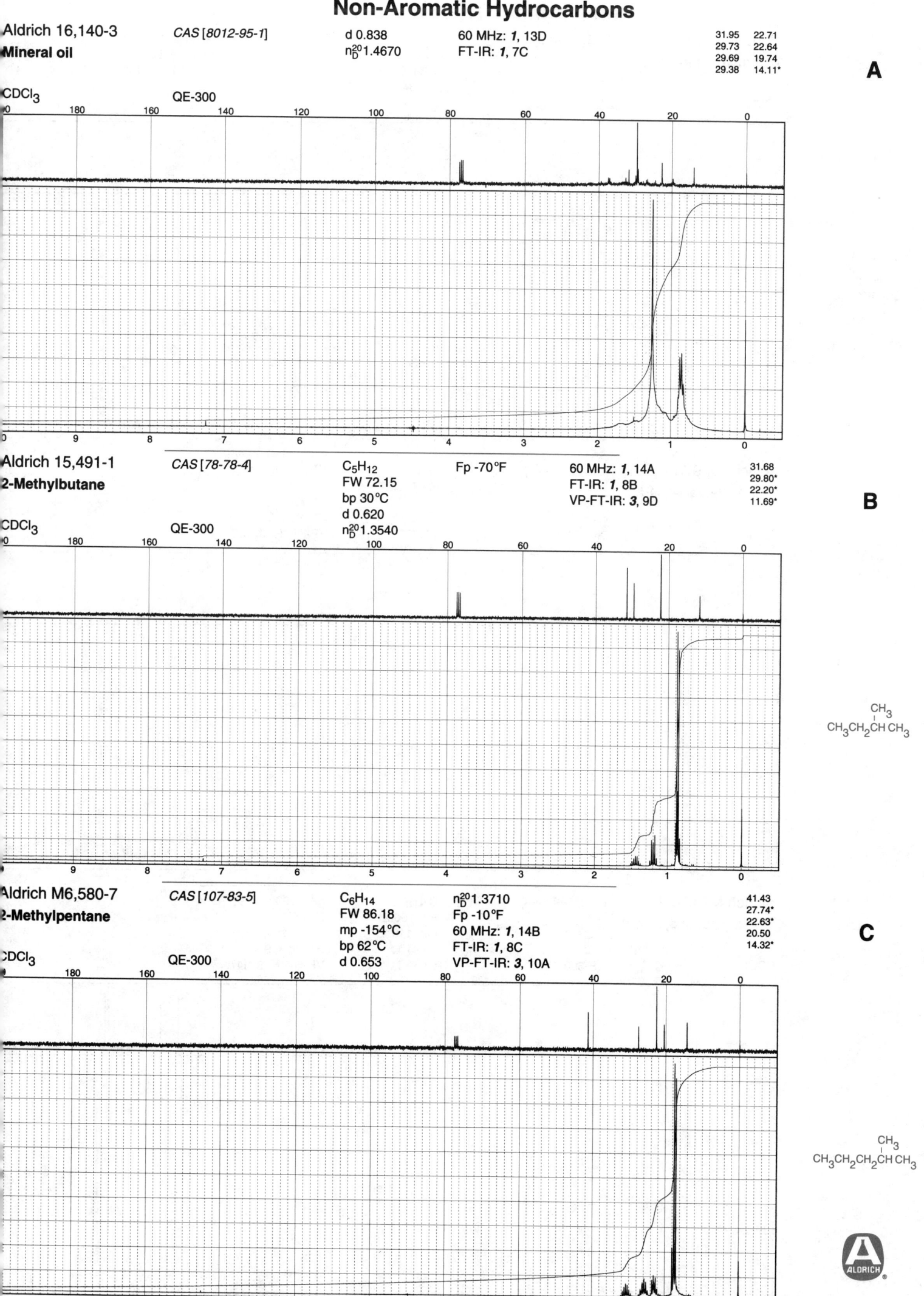

Aldrich 16,140-3   *CAS [8012-95-1]*   d 0.838   60 MHz: *1*, 13D   31.95 22.71
Mineral oil n_D^{20} 1.4670   FT-IR: *1*, 7C   29.73 22.64
                                                                              29.69 19.74
                                                                              29.38 14.11*

A

CDCl₃   QE-300

Aldrich 15,491-1   *CAS [78-78-4]*   C₅H₁₂   Fp -70°F   60 MHz: *1*, 14A   31.68
2-Methylbutane   FW 72.15   FT-IR: *1*, 8B   29.80*
                                       bp 30°C   VP-FT-IR: *3*, 9D   22.20*
                                       d 0.620   11.69*

B

CDCl₃   QE-300 n_D^{20} 1.3540

$$CH_3CH_2\underset{\underset{CH_3}{|}}{CH}CH_3$$

Aldrich M6,580-7   *CAS [107-83-5]*   C₆H₁₄ n_D^{20} 1.3710   41.43
2-Methylpentane   FW 86.18   Fp -10°F   27.74*
                                        mp -154°C   60 MHz: *1*, 14B   22.63*
                                        bp 62°C   FT-IR: *1*, 8C   20.50
                                        d 0.653   VP-FT-IR: *3*, 10A   14.32*

C

CDCl₃   QE-300

$$CH_3CH_2CH_2\underset{\underset{CH_3}{|}}{CH}CH_3$$

ALDRICH

Non-Aromatic Hydrocarbons

11

A

Aldrich M6,600-5   CAS [96-14-0]   C$_6$H$_{14}$ Fp 20°F   60 MHz: **1**, 14C   36.16*
3-Methylpentane, 99+%   FW 86.18   FT-IR: **1**, 8D   29.13
bp 64°C   VP-FT-IR: **3**, 10B   18.78*
d 0.664   11.46*
n$_D^{20}$1.3770

CDCl$_3$ QE-300

CH$_3$
|
CH$_3$CH$_2$CH CH$_2$CH$_3$

B

Aldrich M4,970-4   CAS [591-76-4]   C$_7$H$_{16}$ n$_D^{20}$1.3840   38.85
2-Methylhexane, 99%   FW 100.21   Fp 25°F   29.75
mp -118°C   60 MHz: **1**, 15A   28.05*
bp 90°C   FT-IR: **1**, 9A   23.03
d 0.679   VP-FT-IR: **3**, 10D   22.67*
14.13*

CDCl$_3$ QE-300

CH$_3$
|
CH$_3$CH$_2$CH$_2$CH$_2$CH CH$_3$

C

Aldrich M4,980-1   CAS [589-34-4]   C$_7$H$_{16}$ n$_D^{20}$1.3880   39.10
3-Methylhexane, 99%   FW 100.21   Fp 25°F   34.27*
mp -119°C   60 MHz: **1**, 15B   29.60
bp 91°C   FT-IR: **1**, 9B   20.25
d 0.687   VP-FT-IR: **3**, 11A   19.21*
14.40*
11.38*

CDCl$_3$ QE-300

CH$_3$
|
CH$_3$CH$_2$CH$_2$CH CH$_2$CH$_3$

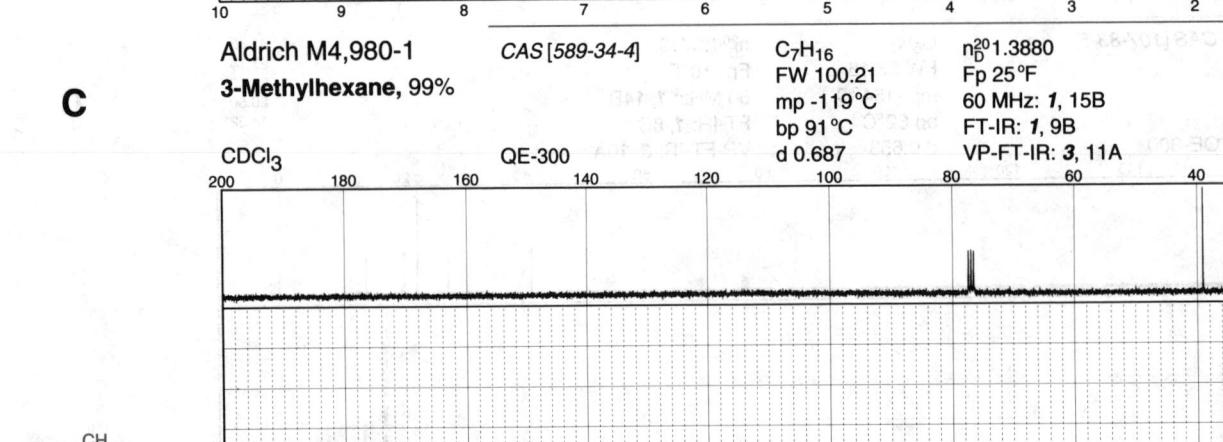

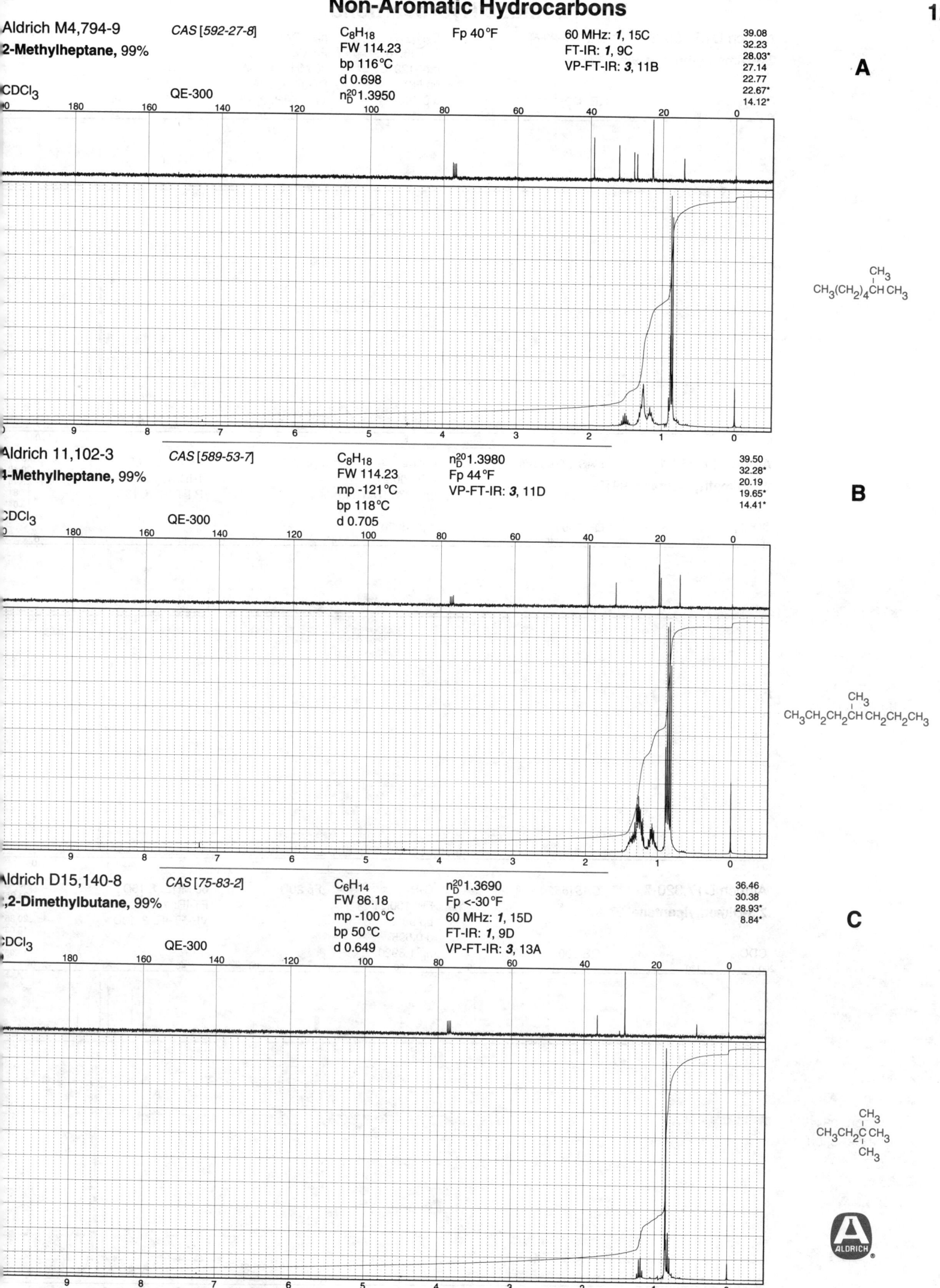

Aldrich M4,794-9   CAS [592-27-8] C_8H_{18} Fp 40 °F   60 MHz: **1**, 15C   39.08
2-Methylheptane, 99%   FW 114.23   FT-IR: **1**, 9C   32.23
bp 116 °C   VP-FT-IR: **3**, 11B   28.03*
CDCl3   QE-300   d 0.698   27.14
n_D^{20} 1.3950   22.77
22.67*
14.12*

A

$CH_3(CH_2)_4CHCH_3$
$\overset{\displaystyle CH_3}{|}$

Aldrich 11,102-3   CAS [589-53-7] C_8H_{18} n_D^{20} 1.3980   39.50
4-Methylheptane, 99%   FW 114.23   Fp 44 °F   32.28*
mp -121 °C   VP-FT-IR: **3**, 11D   20.19
bp 118 °C   19.65*
CDCl3   QE-300   d 0.705   14.41*

B

$CH_3CH_2CH_2CHCH_2CH_2CH_3$
$\overset{\displaystyle CH_3}{|}$

Aldrich D15,140-8   CAS [75-83-2] C_6H_{14} n_D^{20} 1.3690   36.46
2,2-Dimethylbutane, 99%   FW 86.18   Fp <-30 °F   30.38
mp -100 °C   60 MHz: **1**, 15D   28.93*
bp 50 °C   FT-IR: **1**, 9D   8.84*
CDCl3   QE-300   d 0.649   VP-FT-IR: **3**, 13A

C

$CH_3CH_2\underset{\displaystyle CH_3}{\overset{\displaystyle CH_3}{\underset{|}{\overset{|}{C}}}}CH_3$

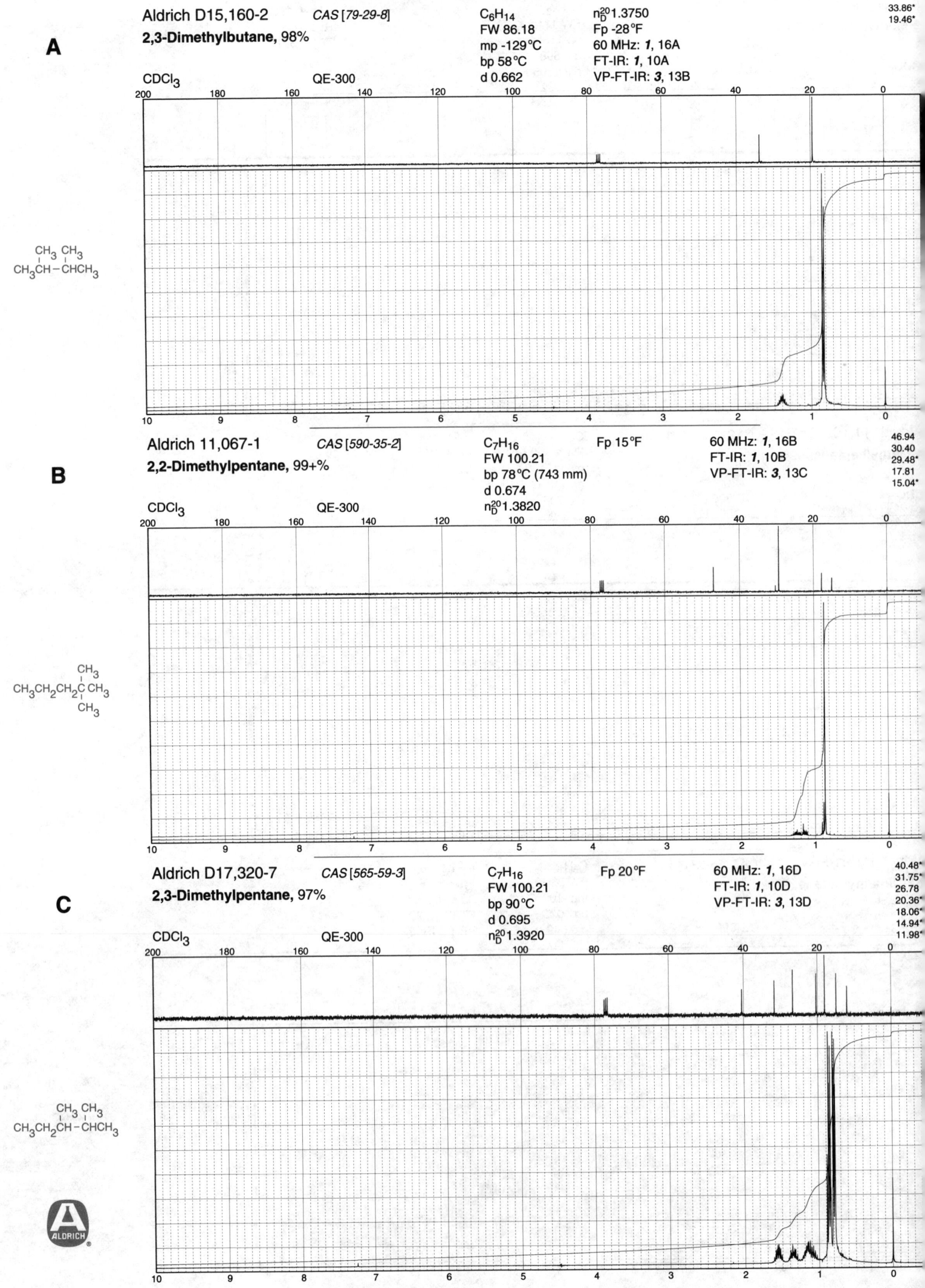

A

Aldrich D15,160-2   *CAS [79-29-8]* C_6H_{14} $n_D^{20}1.3750$ 33.86*
19.46*

2,3-Dimethylbutane, 98%   FW 86.18   Fp -28°F
mp -129°C   60 MHz: *1*, 16A
bp 58°C   FT-IR: *1*, 10A
d 0.662   VP-FT-IR: *3*, 13B

CDCl₃   QE-300

CH₃ CH₃
CH₃CH−CHCH₃

B

Aldrich 11,067-1   *CAS [590-35-2]* C_7H_{16} Fp 15°F   60 MHz: *1*, 16B   46.94
30.40

2,2-Dimethylpentane, 99+%   FW 100.21   FT-IR: *1*, 10B   29.48*
bp 78°C (743 mm)   VP-FT-IR: *3*, 13C   17.81
d 0.674   15.04*
$n_D^{20}1.3820$

CDCl₃   QE-300

CH₃
CH₃CH₂CH₂C−CH₃
CH₃

C

Aldrich D17,320-7   *CAS [565-59-3]* C_7H_{16} Fp 20°F   60 MHz: *1*, 16D   40.48*
31.75*

2,3-Dimethylpentane, 97%   FW 100.21   FT-IR: *1*, 10D   26.78
bp 90°C   VP-FT-IR: *3*, 13D   20.36*
d 0.695   18.06*
$n_D^{20}1.3920$ 14.94*
11.98*

CDCl₃   QE-300

CH₃ CH₃
CH₃CH₂CH−CHCH₃

Non-Aromatic Hydrocarbons

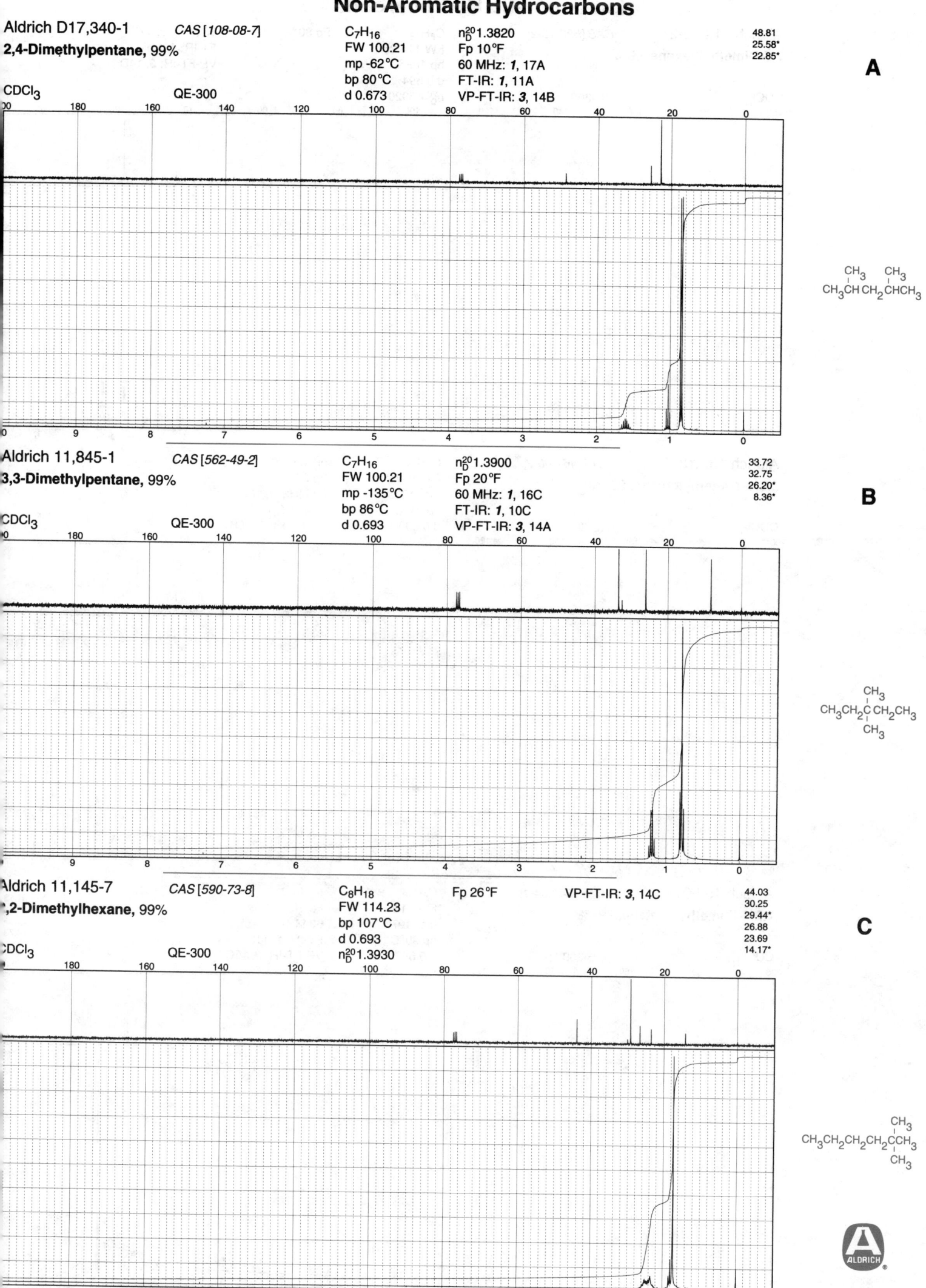

Aldrich D17,340-1   CAS [108-08-7]   C₇H₁₆   n₂₀ᴰ 1.3820   48.81
2,4-Dimethylpentane, 99%   FW 100.21   Fp 10°F   25.58*
CDCl₃   QE-300   mp -62°C   60 MHz: 1, 17A   22.85*
bp 80°C   FT-IR: 1, 11A
d 0.673   VP-FT-IR: 3, 14B

A

Aldrich 11,845-1   CAS [562-49-2]   C₇H₁₆   n₂₀ᴰ 1.3900   33.72
3,3-Dimethylpentane, 99%   FW 100.21   Fp 20°F   32.75
CDCl₃   QE-300   mp -135°C   60 MHz: 1, 16C   26.20*
bp 86°C   FT-IR: 1, 10C   8.36*
d 0.693   VP-FT-IR: 3, 14A

B

Aldrich 11,145-7   CAS [590-73-8]   C₈H₁₈   Fp 26°F   VP-FT-IR: 3, 14C   44.03
2,2-Dimethylhexane, 99%   FW 114.23   bp 107°C   30.25
CDCl₃   QE-300   d 0.693   29.44*
n₂₀ᴰ 1.3930   26.88
23.69
14.17*

C

A

Aldrich 11,058-2   *CAS [592-13-2]*   C₈H₁₈   Fp 80°F   60 MHz: *1*, 17C   36.84*
2,5-Dimethylhexane, 99%   FW 114.23   FT-IR: *1*, 11B   28.34*
                                           bp 108°C   VP-FT-IR: *3*, 14D 22.70*
                                           d 0.694
CDCl₃   QE-300 n_D^{20} 1.3920

$$CH_3CH\ CH_2\ CH_2\ CHCH_3$$
with CH₃ groups

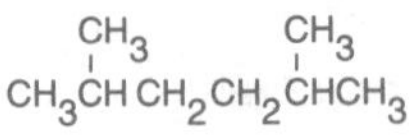

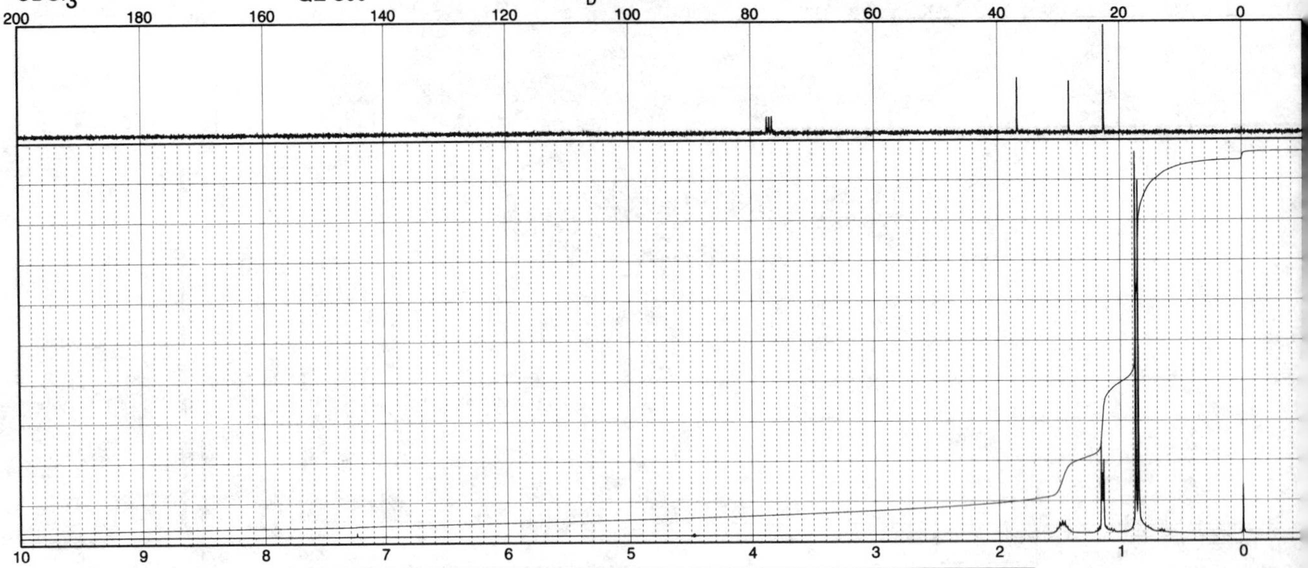

B

Aldrich 13,218-7   *CAS [464-06-2]*   C₇H₁₆ n_D^{20} 1.3890   37.70*
2,2,3-Trimethylbutane, 99+%   FW 100.21 Fp 20°F   32.70
                                          mp -13°C   60 MHz: *1*, 18C   27.11*
                                          bp 81°C   FT-IR: *1*, 11C   17.83*
CDCl₃   QE-300   d 0.690   VP-FT-IR: *3*, 15B

$$CH_3CH-C\ CH_3$$
with CH₃ groups

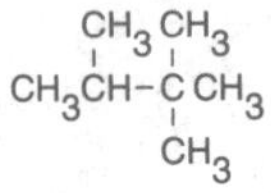

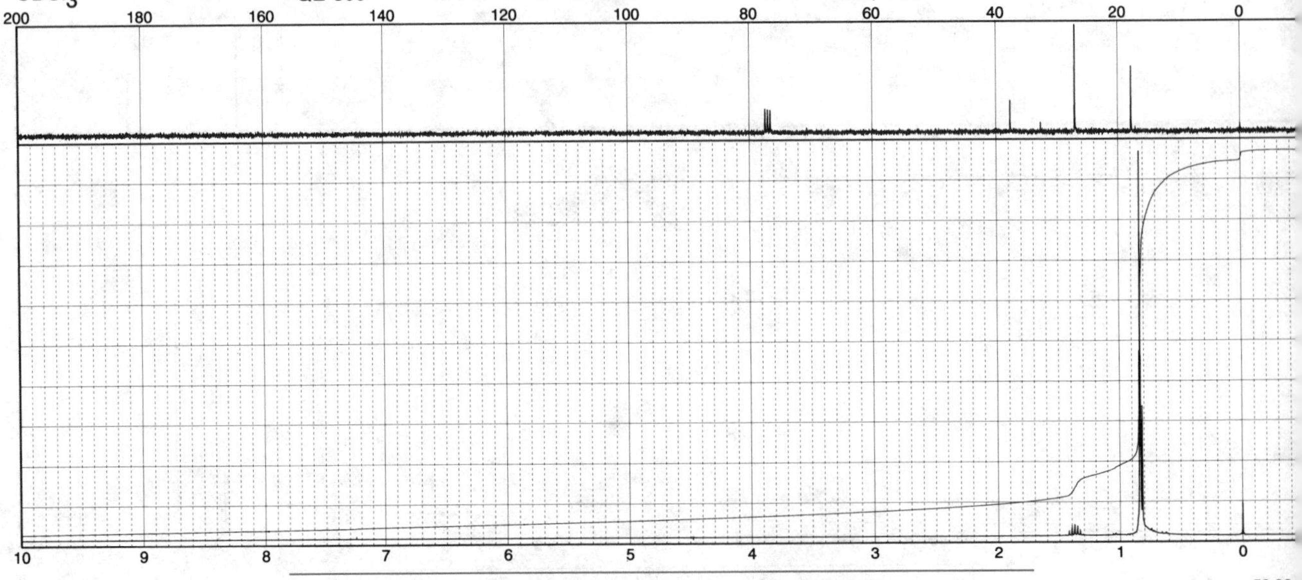

C

Aldrich 15,501-2   *CAS [540-84-1]*   C₈H₁₈ n_D^{20} 1.3910   53.33
2,2,4-Trimethylpentane, 99+%   FW 114.23 Fp 18°F   31.15
                                          mp -107°C 60 MHz: *1*, 18B   30.23*
                                          bp 99°C   FT-IR: *1*, 11D   25.56*
CDCl₃   QE-300   d 0.692   VP-FT-IR: *3*, 15C   24.79*

$$CH_3CH\ CH_2\ CCH_3$$
with CH₃ groups

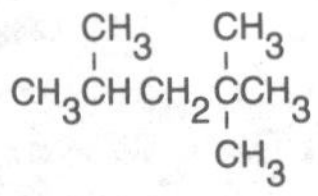

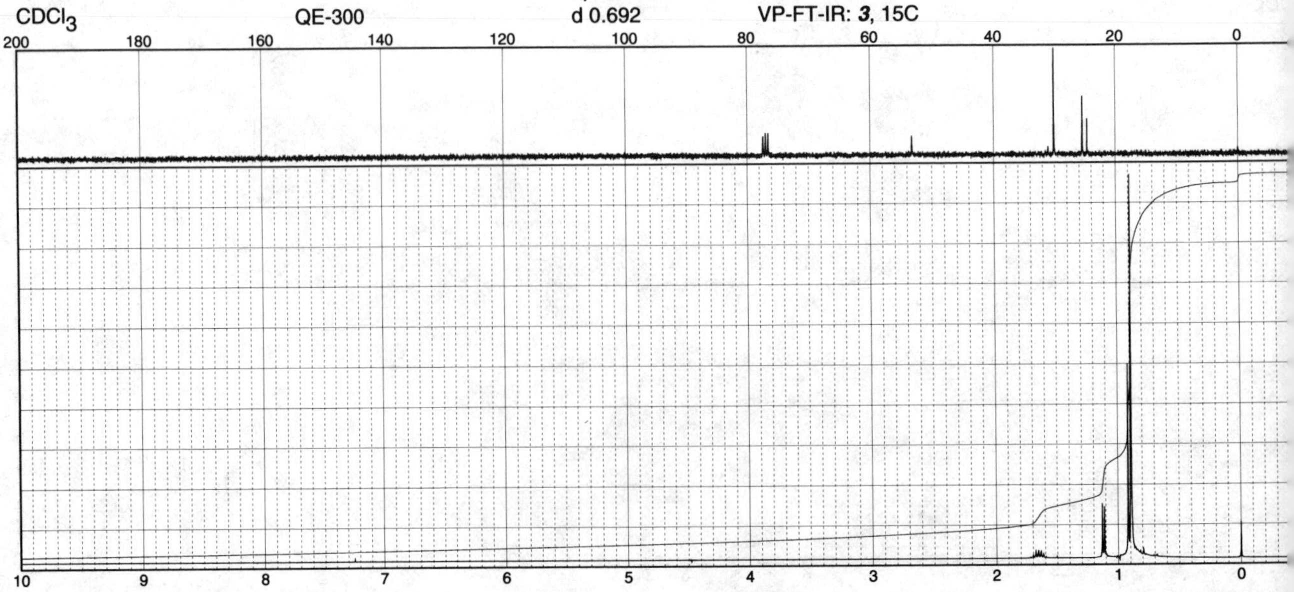

Non-Aromatic Hydrocarbons

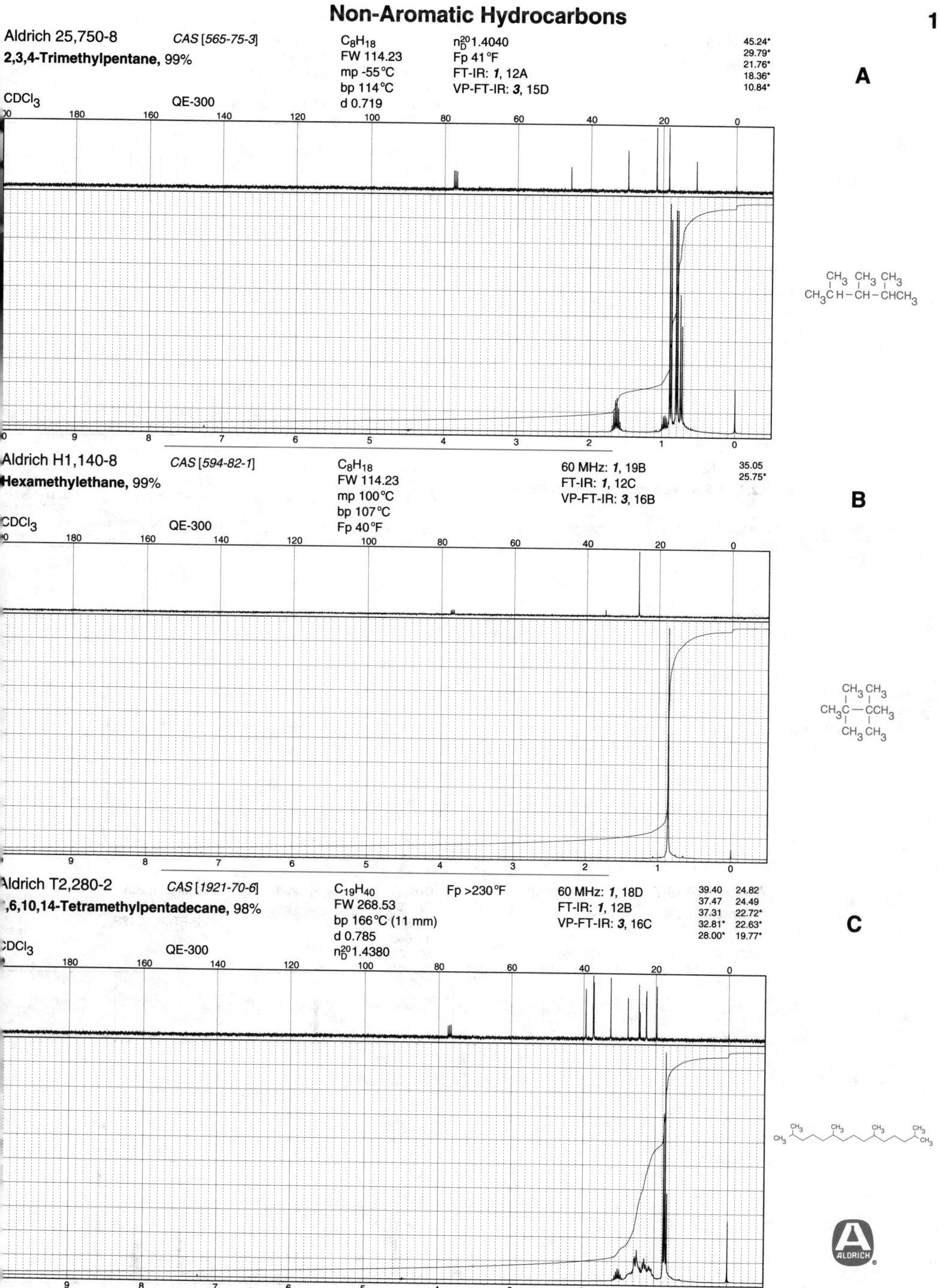

Aldrich 25,750-8   CAS [565-75-3]

2,3,4-Trimethylpentane, 99%

C₈H₁₈
FW 114.23
mp -55°C
bp 114°C
d 0.719

n_D^{20} 1.4040
Fp 41°F
FT-IR: *1*, 12A
VP-FT-IR: *3*, 15D

45.24*
29.79*
21.76*
18.36*
10.84*

A

CDCl₃   QE-300

Aldrich H1,140-8   CAS [594-82-1]

Hexamethylethane, 99%

C₈H₁₈
FW 114.23
mp 100°C
bp 107°C
Fp 40°F

60 MHz: *1*, 19B
FT-IR: *1*, 12C
VP-FT-IR: *3*, 16B

35.05
25.75*

B

CDCl₃   QE-300

Aldrich T2,280-2   CAS [1921-70-6]

2,6,10,14-Tetramethylpentadecane, 98%

C₁₉H₄₀
FW 268.53
bp 166°C (11 mm)
d 0.785
n_D^{20} 1.4380

Fp >230°F

60 MHz: *1*, 18D
FT-IR: *1*, 12B
VP-FT-IR: *3*, 16C

39.40   24.82
37.47   24.49
37.31   22.72*
32.81* 22.63*
28.00* 19.77*

C

CDCl₃   QE-300

Non-Aromatic Hydrocarbons

17

A

Aldrich 23,431-1   *CAS [111-01-3]* $C_{30}H_{62}$ n_D^{20}1.4510

Squalane, 99%

FW 422.83   Fp 424°F
mp -38°C   60 MHz: *1*, 19D
bp 176°C   FT-IR: *1*, 13A
d 0.810   VP-FT-IR: *3*, 17B

39.45	37.16	24.52
37.51	32.83*	22.71*
37.45	28.01*	22.63*
37.37	27.49	19.79*
37.25	24.83	19.74*

CDCl₃   QE-300

$$\left[CH_3CH(CH_2)_3CH(CH_2)_3CHCH_2CH_2- \right]_2$$
(with CH₃ groups)

B

Aldrich 12,851-1   *CAS [4390-04-9]* $C_{16}H_{34}$ Fp 204°F

2,2,4,4,6,8,8-Heptamethylnonane, 98%

FW 226.45   60 MHz: *1*, 19C
bp 240°C   FT-IR: *1*, 12D
d 0.793   VP-FT-IR: *3*, 17A
n_D^{20}1.4390

55.76	30.46*
54.45	29.19*
36.18	28.91*
32.39	25.84*
32.24*	25.47*

CDCl₃   QE-300

CH₃C CH₂ CHCH₂ C CH₂C CH₃ (with CH₃ groups)

C

Aldrich 24,199-7   *CAS [109-67-1]* C_5H_{10} Fp -20°F

1-Pentene, 99%

FW 70.14   60 MHz: *1*, 20A
bp 30°C   FT-IR: *1*, 13B
d 0.640   VP-FT-IR: *3*, 18B
n_D^{20}1.3720

| 138.92* |
| 114.26 |
| 35.96 |
| 22.17 |
| 13.59* |

CDCl₃   QE-300

$CH_3CH_2CH_2 CH=CH_2$

ALDRICH

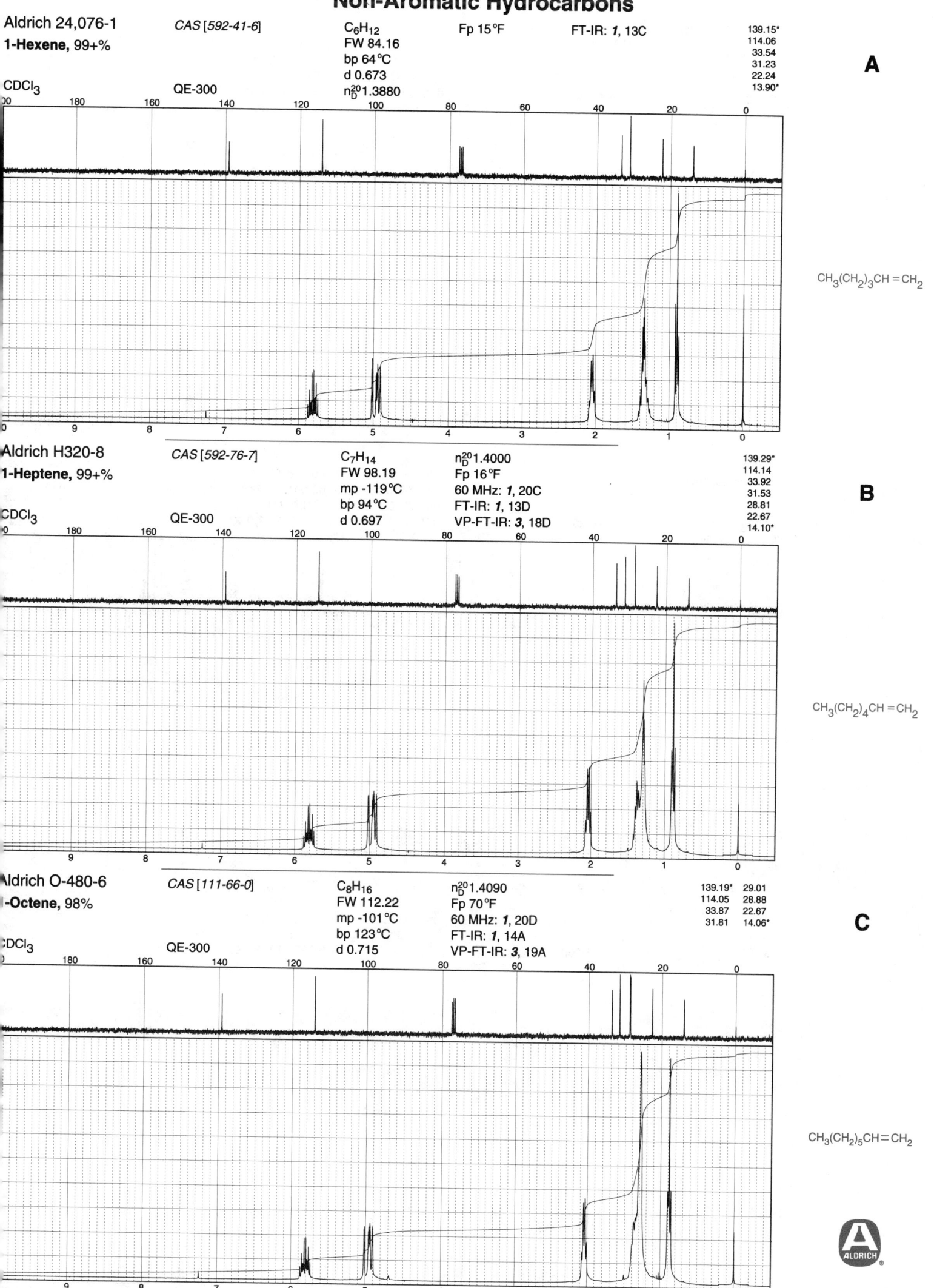

Aldrich 24,076-1   CAS [592-41-6] C_6H_{12} Fp 15°F   FT-IR: **1**, 13C
1-Hexene, 99+%
FW 84.16
bp 64°C
d 0.673
n_D^{20} 1.3880

139.15*
114.06
33.54
31.23
22.24
13.90*

A

$CH_3(CH_2)_3CH=CH_2$

CDCl₃   QE-300

Aldrich H320-8   CAS [592-76-7] C_7H_{14} n_D^{20} 1.4000
1-Heptene, 99+%
FW 98.19   Fp 16°F
mp -119°C   60 MHz: **1**, 20C
bp 94°C   FT-IR: **1**, 13D
d 0.697   VP-FT-IR: **3**, 18D

139.29*
114.14
33.92
31.53
28.81
22.67
14.10*

B

$CH_3(CH_2)_4CH=CH_2$

CDCl₃   QE-300

Aldrich O-480-6   CAS [111-66-0] C_8H_{16} n_D^{20} 1.4090
1-Octene, 98%
FW 112.22   Fp 70°F
mp -101°C   60 MHz: **1**, 20D
bp 123°C   FT-IR: **1**, 14A
d 0.715   VP-FT-IR: **3**, 19A

139.19*   29.01
114.05   28.88
33.87   22.67
31.81   14.06*

C

$CH_3(CH_2)_5CH=CH_2$

CDCl₃   QE-300

Non-Aromatic Hydrocarbons

A

Aldrich N3,040-4

1-Nonene, 99%

CAS [124-11-8]

C_9H_{18}
FW 126.24
mp -81 °C
bp 146 °C

n_D^{20} 1.4160
Fp 115 °F
60 MHz: *1*, 21A
FT-IR: *1*, 14B
VP-FT-IR: *3*, 19B

139.19*	29.18
114.04	29.05
33.86	22.71
31.91	14.06*
29.23	

CDCl$_3$

QE-300

$CH_3(CH_2)_6CH=CH_2$

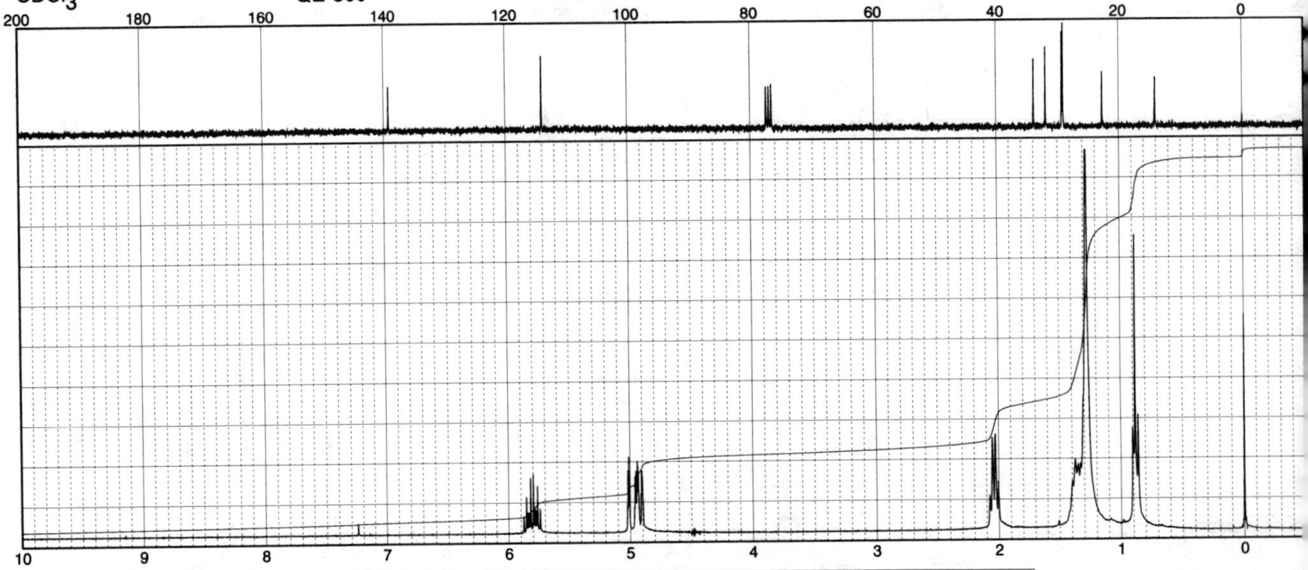

B

Aldrich D180-7

1-Decene, 94%

CAS [872-05-9]

$C_{10}H_{20}$
FW 140.27
mp -66 °C
bp 170 °C
d 0.741

n_D^{20} 1.4210
Fp 118 °F
60 MHz: *1*, 21B
FT-IR: *1*, 14C
VP-FT-IR: *3*, 19C

139.19*	29.34
114.04	29.22
33.86	29.05
31.96	22.72
29.53	14.07*

CDCl$_3$

QE-300

$CH_3(CH_2)_7CH=CH_2$

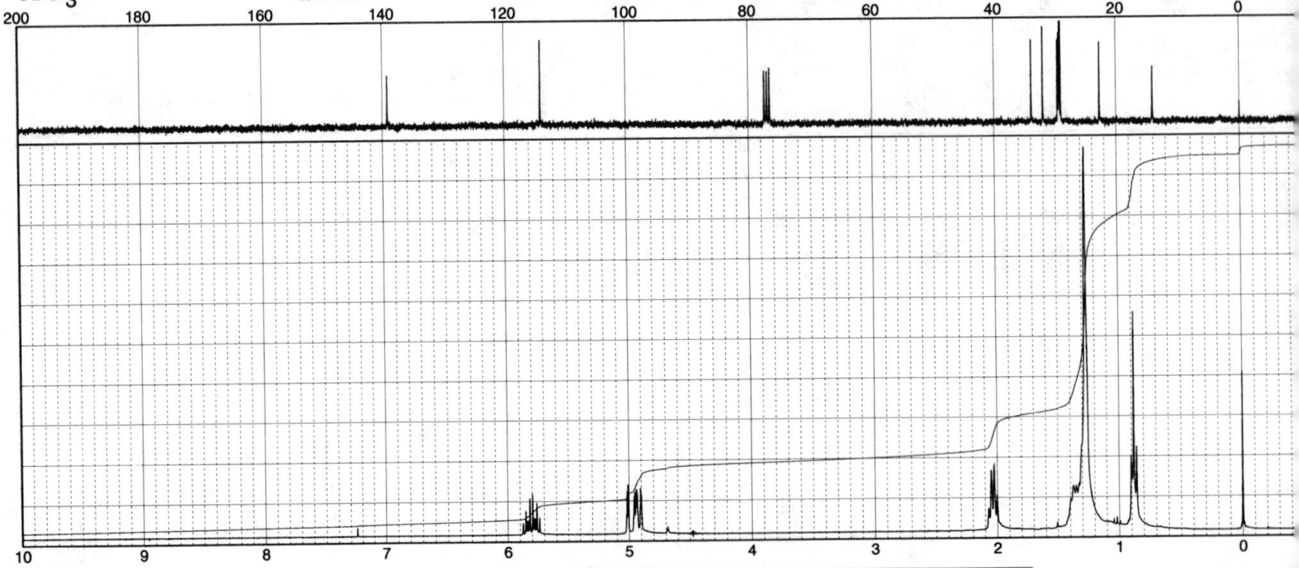

C

Aldrich 24,252-7

1-Undecene, 99%

CAS [821-95-4]

$C_{11}H_{22}$
FW 154.30
mp -49 °C
bp 193 °C
d 0.750

n_D^{20} 1.4260
Fp 145 °F
60 MHz: *1*, 21C
FT-IR: *1*, 14D
VP-FT-IR: *3*, 19D

139.19*	29.37
114.03	29.21
33.85	29.04
31.96	22.72
29.63	14.07*
29.57	

CDCl$_3$

QE-300

$CH_3(CH_2)_8CH=CH_2$

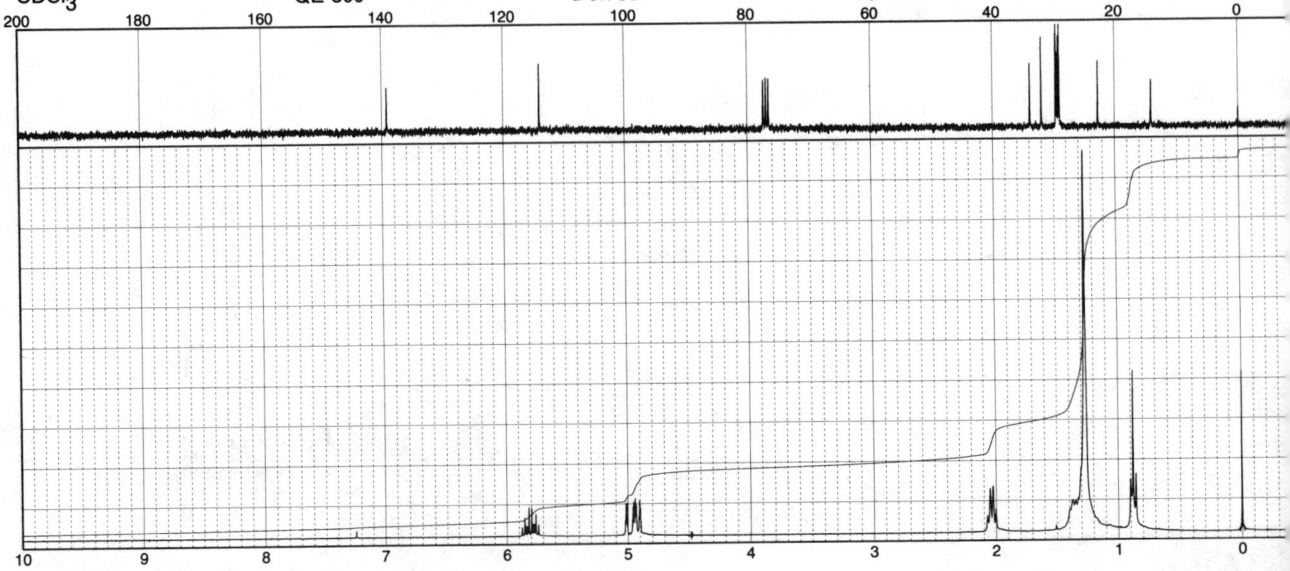

Non-Aromatic Hydrocarbons

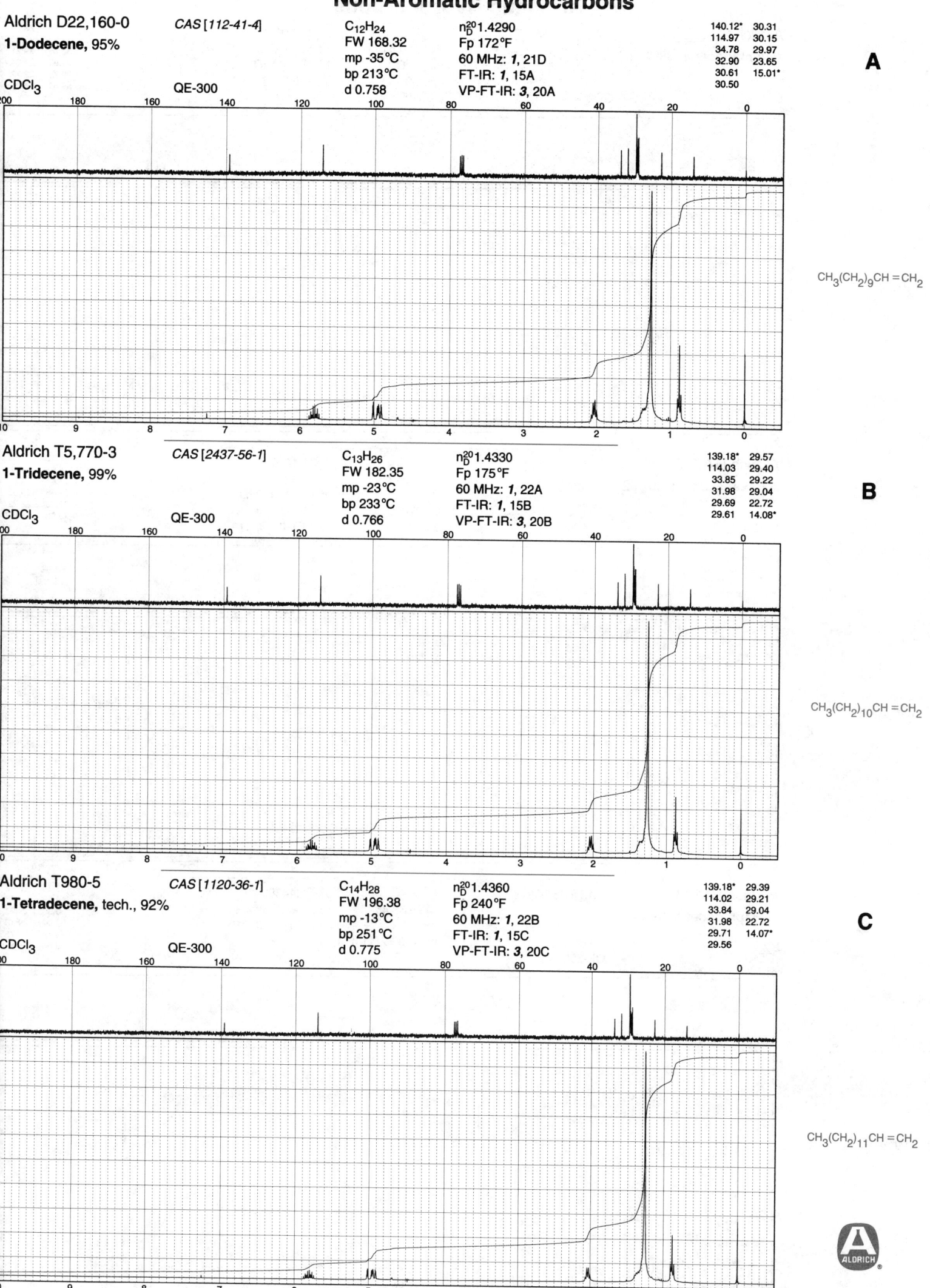

Aldrich D22,160-0   *CAS [112-41-4]*
1-Dodecene, 95%

$C_{12}H_{24}$
FW 168.32
mp -35°C
bp 213°C
d 0.758

n_D^{20} 1.4290
Fp 172°F
60 MHz: *1*, 21D
FT-IR: *1*, 15A
VP-FT-IR: *3*, 20A

140.12*	30.31
114.97	30.15
34.78	29.97
32.90	23.65
30.61	15.01*
30.50	

A

CDCl₃   QE-300

$CH_3(CH_2)_9CH=CH_2$

Aldrich T5,770-3   *CAS [2437-56-1]*
1-Tridecene, 99%

$C_{13}H_{26}$
FW 182.35
mp -23°C
bp 233°C
d 0.766

n_D^{20} 1.4330
Fp 175°F
60 MHz: *1*, 22A
FT-IR: *1*, 15B
VP-FT-IR: *3*, 20B

139.18*	29.57
114.03	29.40
33.85	29.22
31.98	29.04
29.69	22.72
29.61	14.08*

B

CDCl₃   QE-300

$CH_3(CH_2)_{10}CH=CH_2$

Aldrich T980-5   *CAS [1120-36-1]*
1-Tetradecene, tech., 92%

$C_{14}H_{28}$
FW 196.38
mp -13°C
bp 251°C
d 0.775

n_D^{20} 1.4360
Fp 240°F
60 MHz: *1*, 22B
FT-IR: *1*, 15C
VP-FT-IR: *3*, 20C

139.18*	29.39
114.02	29.21
33.84	29.04
31.98	22.72
29.71	14.07*
29.56	

C

CDCl₃   QE-300

$CH_3(CH_2)_{11}CH=CH_2$

ALDRICH

Non-Aromatic Hydrocarbons

A

Aldrich 22,288-7   *CAS [13360-61-7]* $C_{15}H_{30}$ $n_D^{20}1.4390$

1-Pentadecene, 98%   FW 210.41   Fp >230°F

mp -4°C   FT-IR: *1*, 15D

bp 269°C   VP-FT-IR: *3*, 20D

139.18*	29.56
114.03	29.40
33.85	29.21
31.98	29.04
29.72	22.72
29.60	14.08*

CDCl$_3$ QE-300

$CH_3(CH_2)_{12}CH=CH_2$

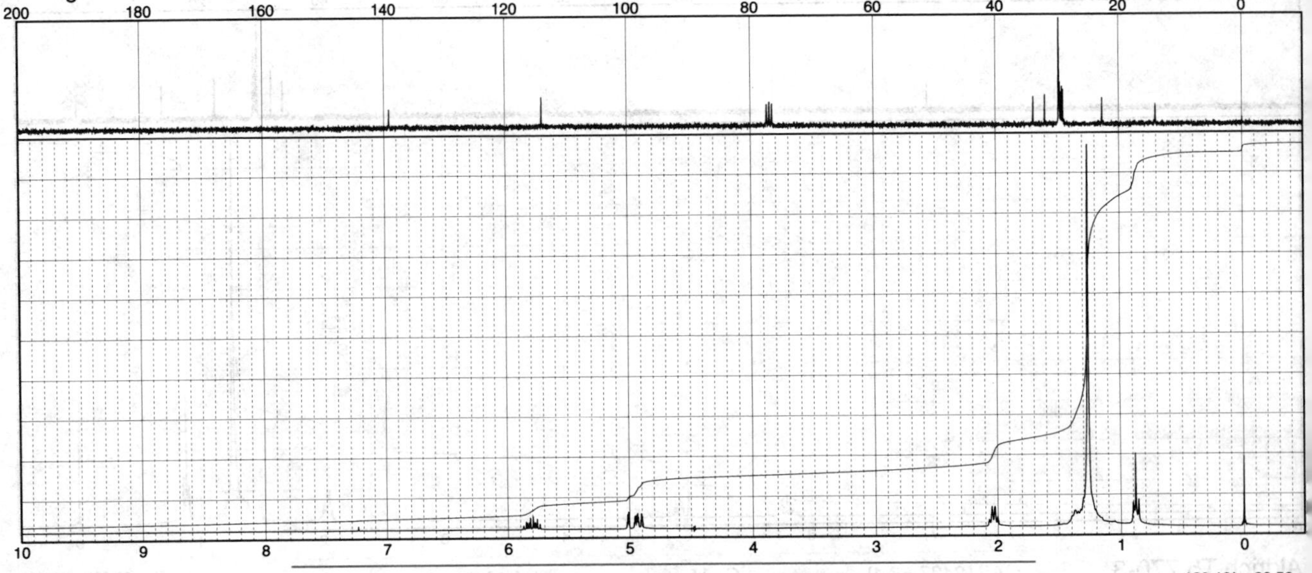

B

Aldrich H700-9   *CAS [629-73-2]* $C_{16}H_{32}$ $n_D^{20}1.4420$

1-Hexadecene, tech., 92%   FW 224.43   Fp 270°F

mp 4°C   60 MHz: *1*, 22D

bp 274°C   FT-IR: *1*, 16A

d 0.783   VP-FT-IR: *3*, 21A

139.18*	29.56
114.03	29.40
33.84	29.21
31.98	29.03
29.73	22.72
29.60	14.08*

CDCl$_3$ QE-300

$CH_3(CH_2)_{12}CH_2CH=CH_2$

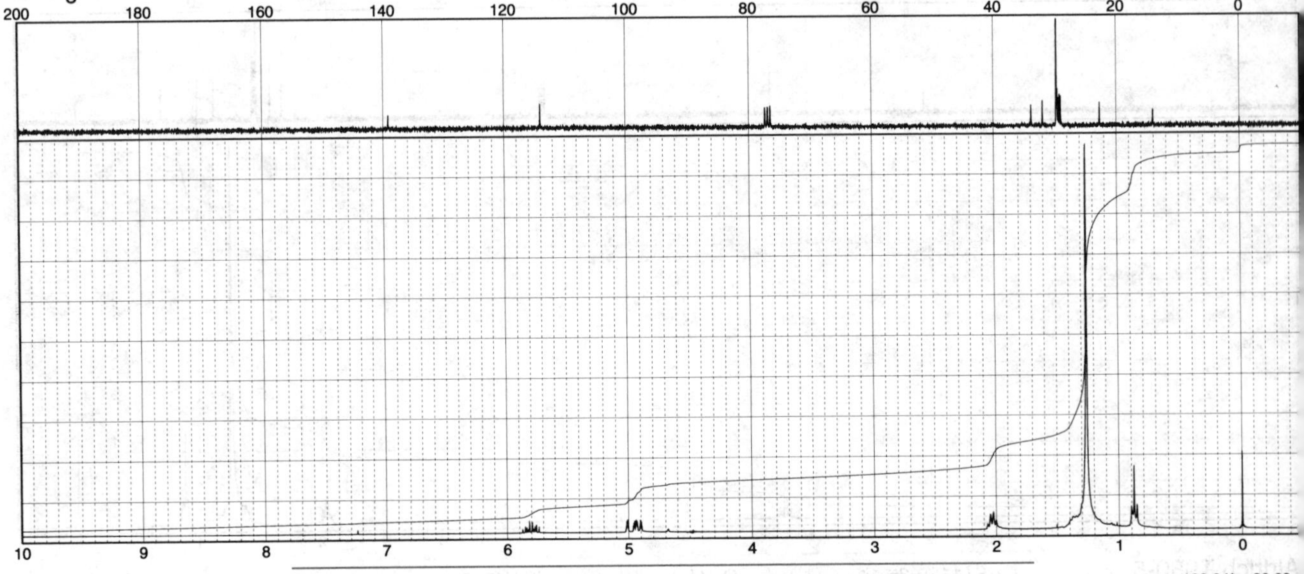

C

Aldrich H110-8   *CAS [6765-39-5]* $C_{17}H_{34}$ $n_D^{20}1.4430$

1-Heptadecene, 99%   FW 238.46   Fp >230°F

mp 11°C   VP-FT-IR: *3*, 21B

bp 158°C (11 mm)

d 0.785

139.24*	29.62
114.09	29.46
33.90	29.27
32.04	29.10
29.79	22.78
29.66	14.14*

CDCl$_3$ QE-300

$CH_3(CH_2)_{14}CH=CH_2$

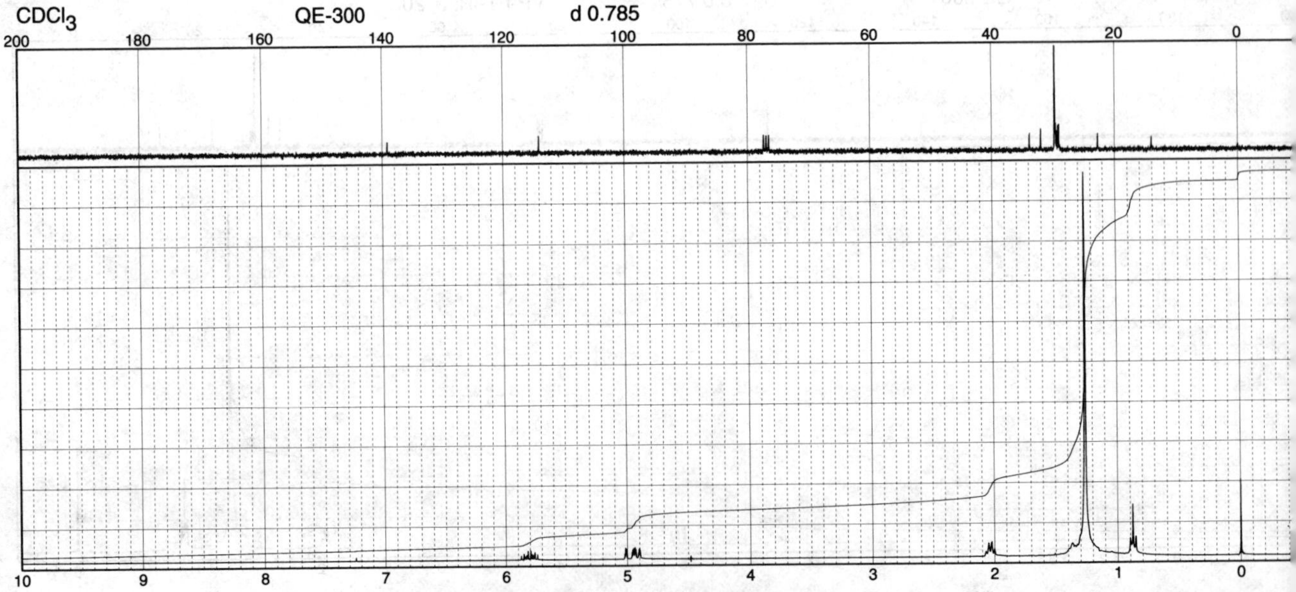

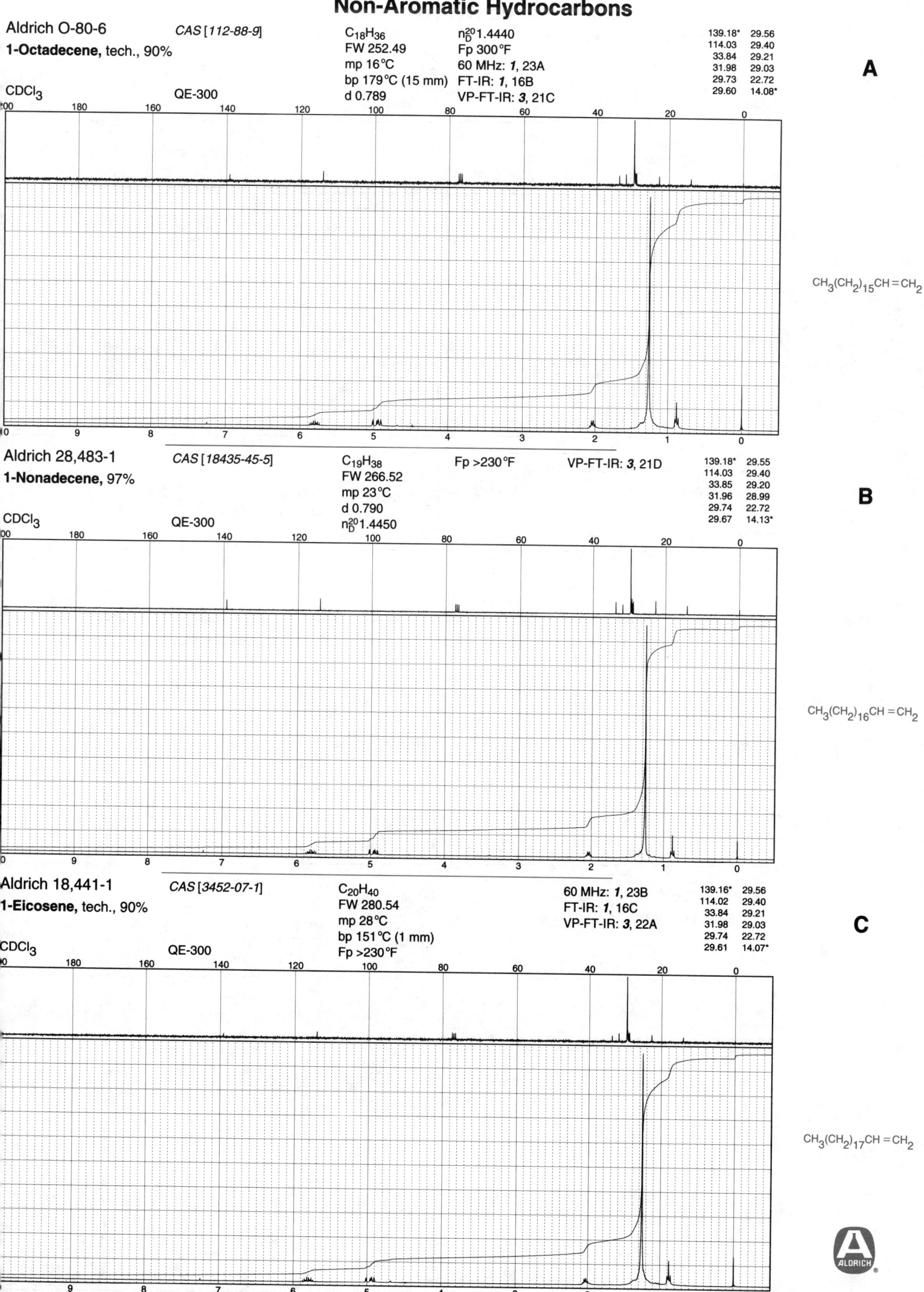

Aldrich O-80-6   CAS [112-88-9]

1-Octadecene, tech., 90%

$C_{18}H_{36}$
FW 252.49
mp 16°C
bp 179°C (15 mm)
d 0.789

n_D^{20}1.4440
Fp 300°F
60 MHz: *1*, 23A
FT-IR: *1*, 16B
VP-FT-IR: *3*, 21C

139.18*	29.56
114.03	29.40
33.84	29.21
31.98	29.03
29.73	22.72
29.60	14.08*

A

CDCl₃   QE-300

$CH_3(CH_2)_{15}CH=CH_2$

Aldrich 28,483-1   CAS [18435-45-5]

1-Nonadecene, 97%

$C_{19}H_{38}$
FW 266.52
mp 23°C
d 0.790
n_D^{20}1.4450

Fp >230°F

VP-FT-IR: *3*, 21D

139.18*	29.55
114.03	29.40
33.85	29.20
31.96	28.99
29.74	22.72
29.67	14.13*

B

CDCl₃   QE-300

$CH_3(CH_2)_{16}CH=CH_2$

Aldrich 18,441-1   CAS [3452-07-1]

1-Eicosene, tech., 90%

$C_{20}H_{40}$
FW 280.54
mp 28°C
bp 151°C (1 mm)
Fp >230°F

60 MHz: *1*, 23B
FT-IR: *1*, 16C
VP-FT-IR: *3*, 22A

139.16*	29.56
114.02	29.40
33.84	29.21
31.98	29.03
29.74	22.72
29.61	14.07*

C

CDCl₃   QE-300

$CH_3(CH_2)_{17}CH=CH_2$

ALDRICH®

Non-Aromatic Hydrocarbons

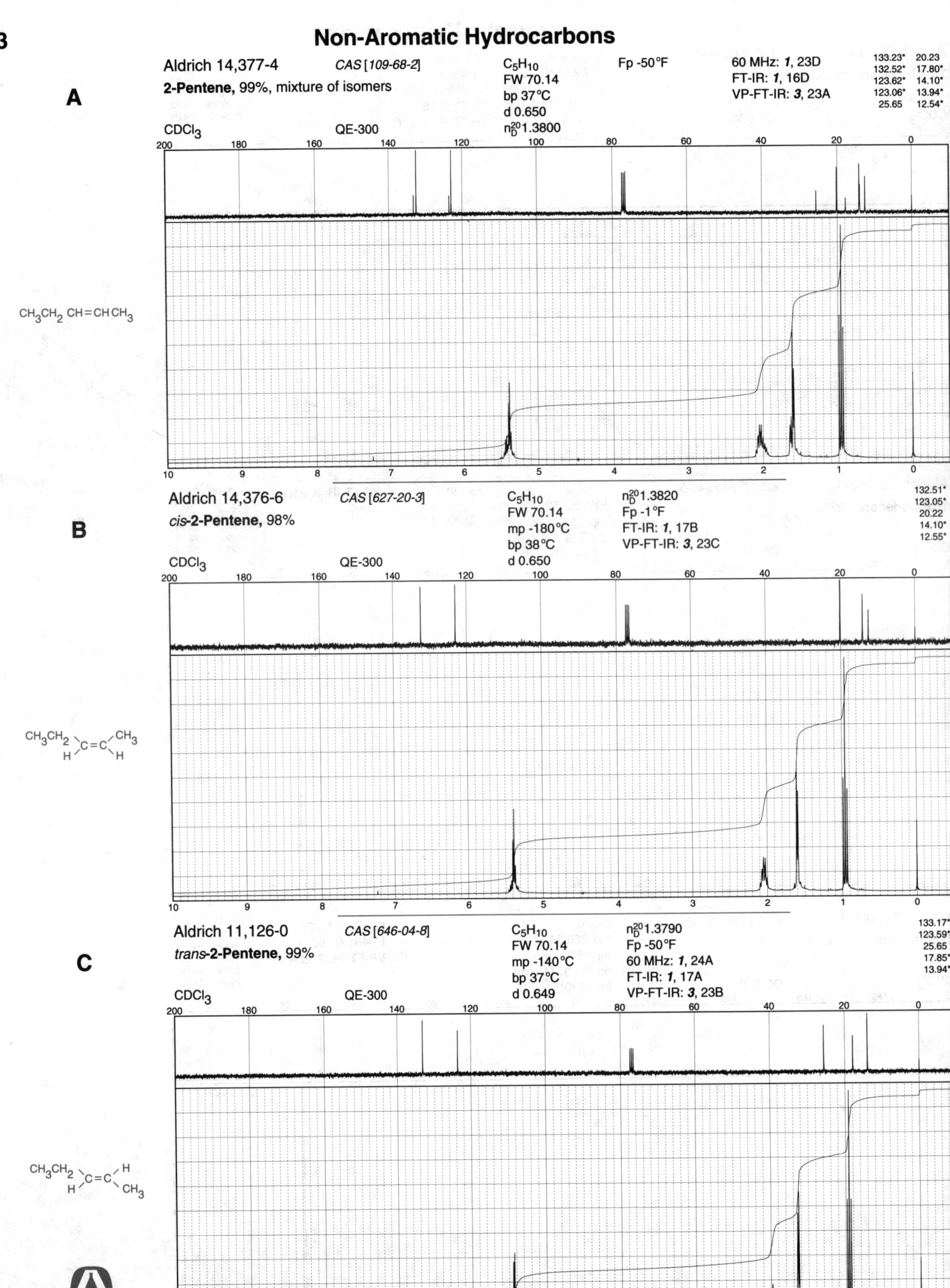

A

Aldrich 14,377-4   *CAS [109-68-2]*   C₅H₁₀   Fp -50°F   60 MHz: *1*, 23D
2-Pentene, 99%, mixture of isomers   FW 70.14   FT-IR: *1*, 16D
                                          bp 37°C   VP-FT-IR: *3*, 23A
                                          d 0.650
CDCl₃   QE-300   n²⁰_D 1.3800

133.23*	20.23
132.52*	17.80*
123.62*	14.10*
123.06*	13.94*
25.65	12.54*

$CH_3CH_2\,CH{=}CHCH_3$

B

Aldrich 14,376-6   *CAS [627-20-3]*   C₅H₁₀   n²⁰_D 1.3820
cis-**2-Pentene**, 98%   FW 70.14   Fp -1°F
                            mp -180°C   FT-IR: *1*, 17B
                            bp 38°C   VP-FT-IR: *3*, 23C
                            d 0.650
CDCl₃   QE-300

| 132.51* |
| 123.05* |
| 20.22 |
| 14.10* |
| 12.55* |

$\underset{H}{CH_3CH_2}\!{>}C{=}C\!{<}\underset{H}{CH_3}$

C

Aldrich 11,126-0   *CAS [646-04-8]*   C₅H₁₀   n²⁰_D 1.3790
trans-**2-Pentene**, 99%   FW 70.14   Fp -50°F
                              mp -140°C   60 MHz: *1*, 24A
                              bp 37°C   FT-IR: *1*, 17A
                              d 0.649   VP-FT-IR: *3*, 23B
CDCl₃   QE-300

| 133.17* |
| 123.59* |
| 25.65 |
| 17.85* |
| 13.94* |

$\underset{H}{CH_3CH_2}\!{>}C{=}C\!{<}\underset{CH_3}{H}$

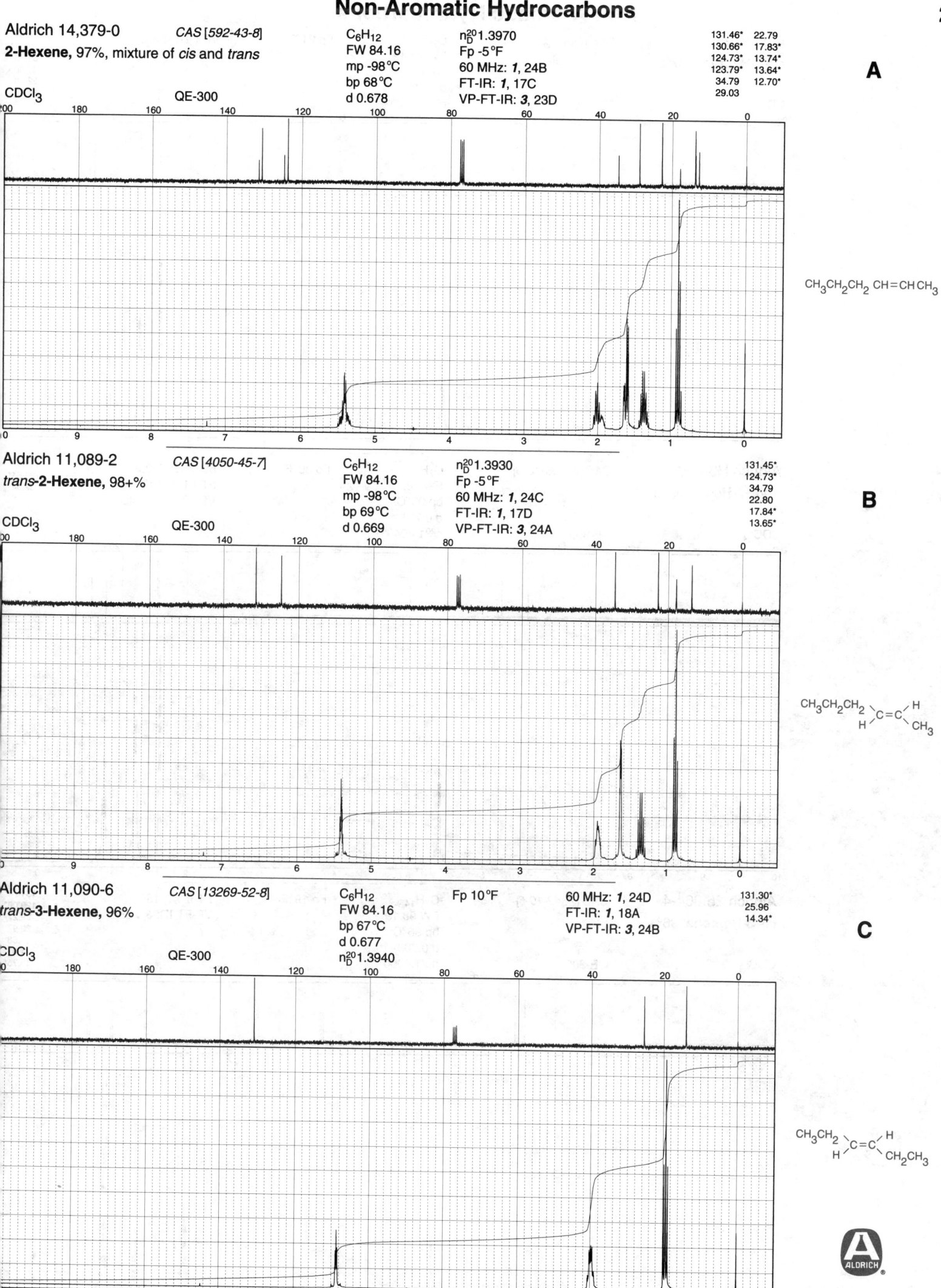

Aldrich 14,379-0   CAS [592-43-8]

2-Hexene, 97%, mixture of *cis* and *trans*

C_6H_{12}	n_D^{20} 1.3970
FW 84.16	Fp -5°F
mp -98°C	60 MHz: *1*, 24B
bp 68°C	FT-IR: *1*, 17C
d 0.678	VP-FT-IR: *3*, 23D

131.46* 22.79
130.66* 17.83*
124.73* 13.74*
123.79* 13.64*
34.79   12.70*
29.03

A

CDCl₃   QE-300

$CH_3CH_2CH_2\ CH{=}CH\ CH_3$

Aldrich 11,089-2   CAS [4050-45-7]

trans-**2-Hexene**, 98+%

C_6H_{12}	n_D^{20} 1.3930
FW 84.16	Fp -5°F
mp -98°C	60 MHz: *1*, 24C
bp 69°C	FT-IR: *1*, 17D
d 0.669	VP-FT-IR: *3*, 24A

131.45*
124.73*
34.79
22.80
17.84*
13.65*

B

CDCl₃   QE-300

Aldrich 11,090-6   CAS [13269-52-8]

trans-**3-Hexene**, 96%

C_6H_{12}	Fp 10°F
FW 84.16	
bp 67°C	60 MHz: *1*, 24D
d 0.677	FT-IR: *1*, 18A
n_D^{20} 1.3940	VP-FT-IR: *3*, 24B

131.30*
25.96
14.34*

C

CDCl₃   QE-300

Non-Aromatic Hydrocarbons

25

A

Aldrich 26,894-1
cis-2-Heptene, 97%

CAS [6443-92-1]

C₇H₁₄
FW 98.19
bp 99 °C
d 0.708
n²⁰_D 1.4070

Fp 21 °F

VP-FT-IR: *3*, 24C

CDCl₃   QE-300

130.84*
123.57*
31.83
26.60
22.39
14.00*
12.72*

$CH_3CH_2CH_2CH_2$ C=C CH_3 / H   H

B

Aldrich H341-0
trans-2-Heptene, 99+%

CAS [14686-13-6]

C₇H₁₄
FW 98.19
bp 98 °C
d 0.701
n²⁰_D 1.4040

Fp 30 °F

60 MHz: *1*, 25A
FT-IR: *1*, 18B
VP-FT-IR: *3*, 24D

131.63*
124.51*
32.34
31.88
22.28
17.91*
13.97*

CDCl₃   QE-300

$CH_3CH_2CH_2CH_2$ C=C H / H CH_3

C

Aldrich 25,864-4
cis-3-Heptene, 96%

CAS [7642-10-6]

C₇H₁₄
FW 98.19
bp 96 °C
d 0.703
n²⁰_D 1.4060

Fp 19 °F

FT-IR: *1*, 18C
VP-FT-IR: *3*, 25A

131.74*
129.08*
29.27
22.97
20.58
14.37*
13.75*

CDCl₃   QE-300

$CH_3CH_2CH_2$ C=C CH_2CH_3 / H   H

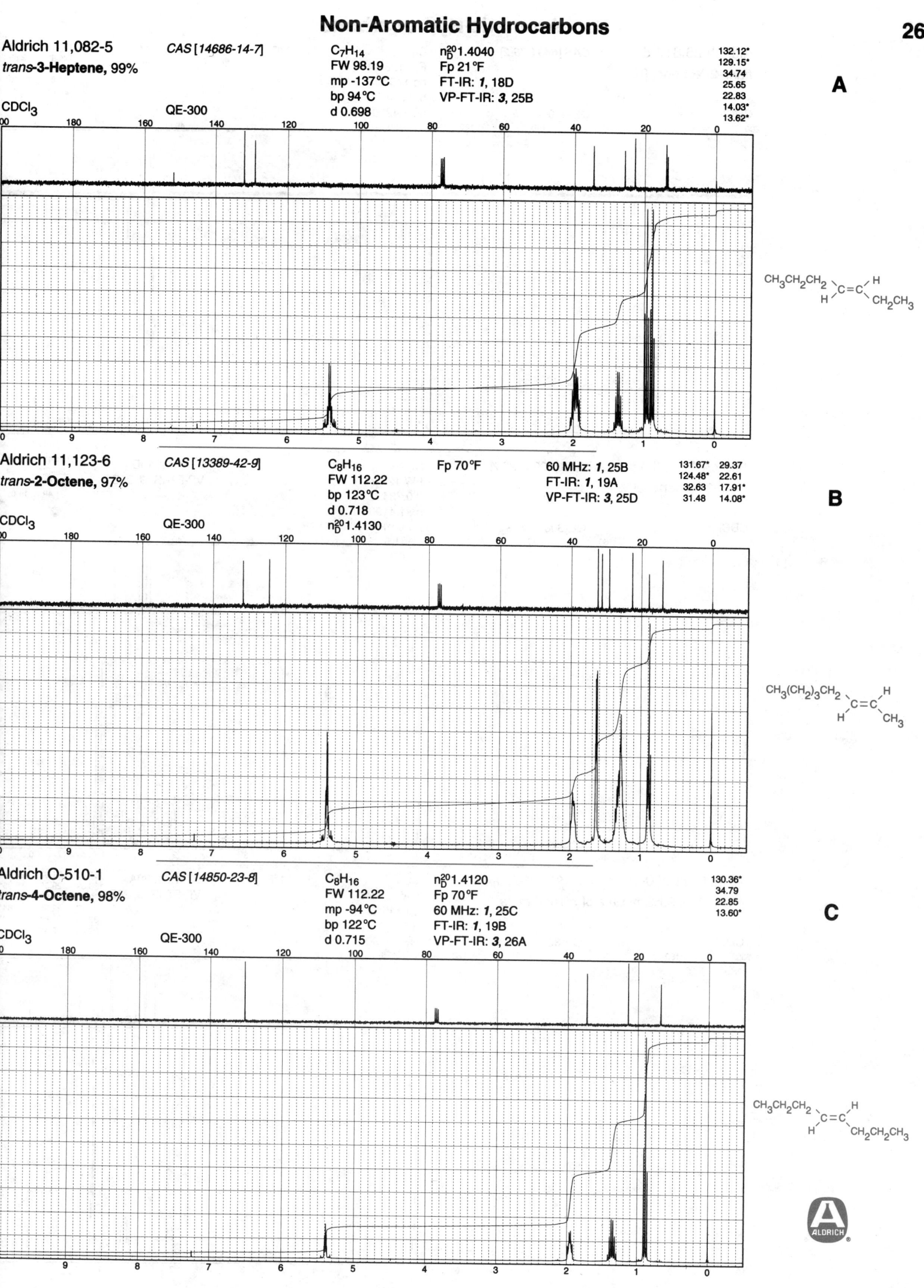

Aldrich 11,082-5
trans-**3-Heptene**, 99%

CAS [14686-14-7]

C₇H₁₄
FW 98.19
mp -137 °C
bp 94 °C
d 0.698

n_D^{20} 1.4040
Fp 21 °F
FT-IR: *1*, 18D
VP-FT-IR: *3*, 25B

132.12*
129.15*
34.74
25.65
22.83
14.03*
13.62*

A

CDCl₃   QE-300

Aldrich 11,123-6
trans-**2-Octene**, 97%

CAS [13389-42-9]

C₈H₁₆
FW 112.22
bp 123 °C
d 0.718
n_D^{20} 1.4130

Fp 70 °F

60 MHz: *1*, 25B
FT-IR: *1*, 19A
VP-FT-IR: *3*, 25D

131.67*   29.37
124.48*   22.61
32.63   17.91*
31.48   14.08*

B

CDCl₃   QE-300

Aldrich O-510-1
trans-**4-Octene**, 98%

CAS [14850-23-8]

C₈H₁₆
FW 112.22
mp -94 °C
bp 122 °C
d 0.715

n_D^{20} 1.4120
Fp 70 °F
60 MHz: *1*, 25C
FT-IR: *1*, 19B
VP-FT-IR: *3*, 26A

130.36*
34.79
22.85
13.60*

C

CDCl₃   QE-300

A

Aldrich 26,817-8   *CAS [6434-78-2]* C_9H_{18} Fp 90 °F   FT-IR: *1*, 19C

trans-2-Nonene, 99%   FW 126.24   VP-FT-IR: *3*, 26B

bp 145 °C
d 0.734
n_D^{20} 1.4200

131.67*	28.93
124.48*	22.69
32.66	17.90*
31.83	14.10*
29.65	

$CDCl_3$ QE-300

$CH_3(CH_2)_4CH_2$ C=C $\overset{H}{\underset{CH_3}{}}$

B

Aldrich 11,120-1   *CAS [20063-92-7]* C_9H_{18} FT-IR: *1*, 19D

trans-3-Nonene, 99%   FW 126.24   VP-FT-IR: *3*, 26C

d 0.734
n_D^{20} 1.4190
Fp 90 °F

131.87*	29.43
129.42*	25.64
32.58	22.61
31.49	14.02*

$CDCl_3$ QE-300

$CH_3(CH_2)_3CH_2$ C=C $\overset{H}{\underset{CH_2CH_3}{}}$

C

Aldrich N3,050-1   *CAS [2198-23-4]* C_9H_{18} Fp 81 °F   FT-IR: *1*, 20A

4-Nonene, 98%, mixture of *cis* and *trans*   FW 126.24   VP-FT-IR: *3*, 26D

bp 145 °C
d 0.732
n_D^{20} 1.4190

130.53*	31.92	22.24
130.09*	29.35	13.96*
129.61*	26.98	13.80*
34.77	22.95	13.64*
32.33	22.81	13.60*
32.04	22.40	

$CDCl_3$ QE-300

$CH_3CH_2CH_2CH_2$ CH=CH $CH_2CH_2CH_3$

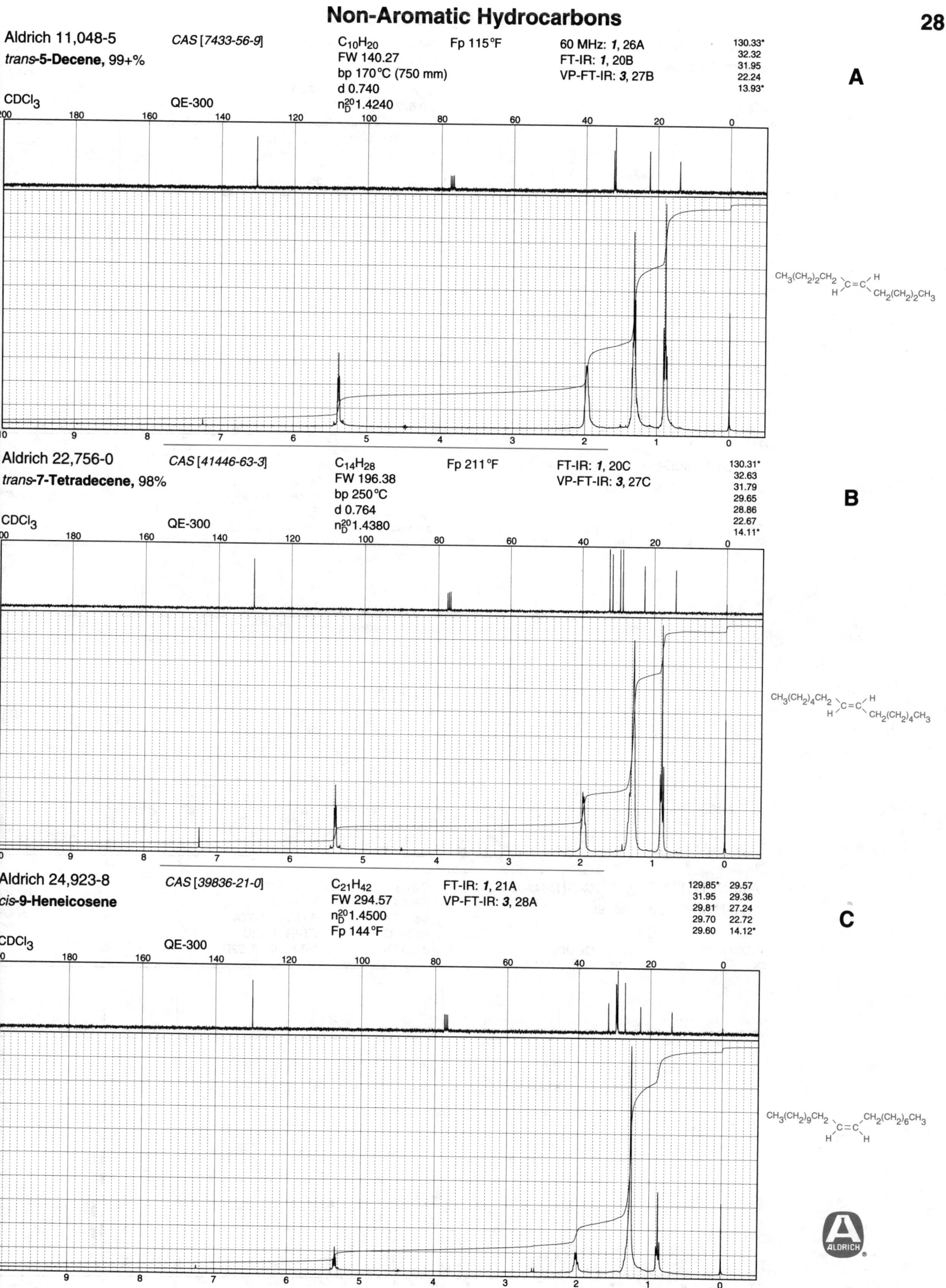

Aldrich 11,048-5 *CAS [7433-56-9]*
trans-**5-Decene**, 99+%

C₁₀H₂₀
FW 140.27
bp 170°C (750 mm)
d 0.740
n²⁰_D 1.4240

Fp 115°F

60 MHz: *1*, 26A
FT-IR: *1*, 20B
VP-FT-IR: *3*, 27B

A

130.33*
32.32
31.95
22.24
13.93*

CDCl₃   QE-300

Aldrich 22,756-0 *CAS [41446-63-3]*
trans-**7-Tetradecene**, 98%

C₁₄H₂₈
FW 196.38
bp 250°C
d 0.764
n²⁰_D 1.4380

Fp 211°F

FT-IR: *1*, 20C
VP-FT-IR: *3*, 27C

B

130.31*
32.63
31.79
29.65
28.86
22.67
14.11*

CDCl₃   QE-300

Aldrich 24,923-8 *CAS [39836-21-0]*
cis-**9-Heneicosene**

C₂₁H₄₂
FW 294.57
n²⁰_D 1.4500
Fp 144°F

FT-IR: *1*, 21A
VP-FT-IR: *3*, 28A

C

129.85* 29.57
31.95   29.36
29.81   27.24
29.70   22.72
29.60   14.12*

CDCl₃   QE-300

A

Aldrich 25,793-1   *CAS [563-45-1]*

3-Methyl-1-butene, 95%

C$_5$H$_{10}$
FW 70.14
mp -168°C
bp 20°C
d 0.627

n$_D^{20}$1.3640
Fp -70°F
FT-IR: *1*, 21C
VP-FT-IR: *3*, 28D

145.95*
111.06
32.01*
22.00*

CDCl$_3$ QE-300

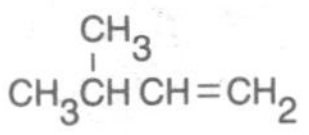

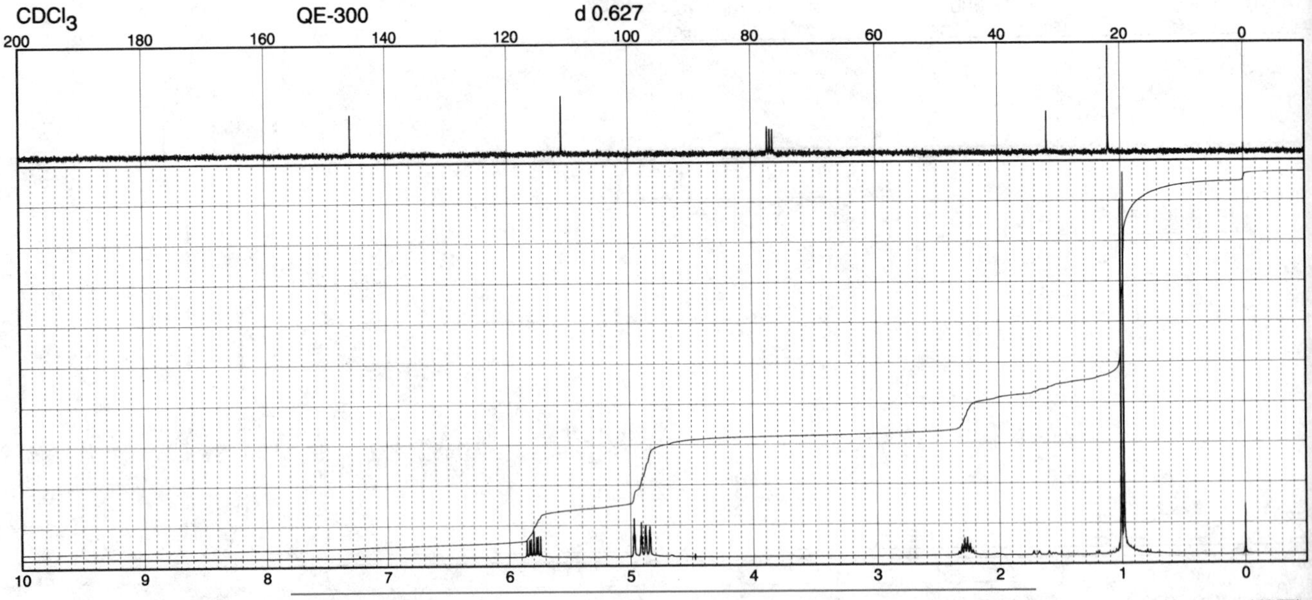

CH$_3$
CH$_3$CH CH=CH$_2$

B

Aldrich 11,905-9   *CAS [558-37-2]*

3,3-Dimethyl-1-butene, 95%

C$_6$H$_{12}$
FW 84.16
mp -115°C
bp 41°C
d 0.653

n$_D^{20}$1.3760
Fp -20°F
60 MHz: *1*, 26B
FT-IR: *1*, 25B
VP-FT-IR: *3*, 33D

149.77*
108.83
33.64
29.16*

CDCl$_3$ QE-300

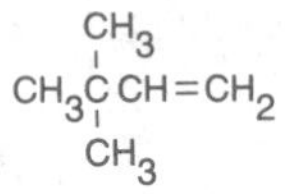

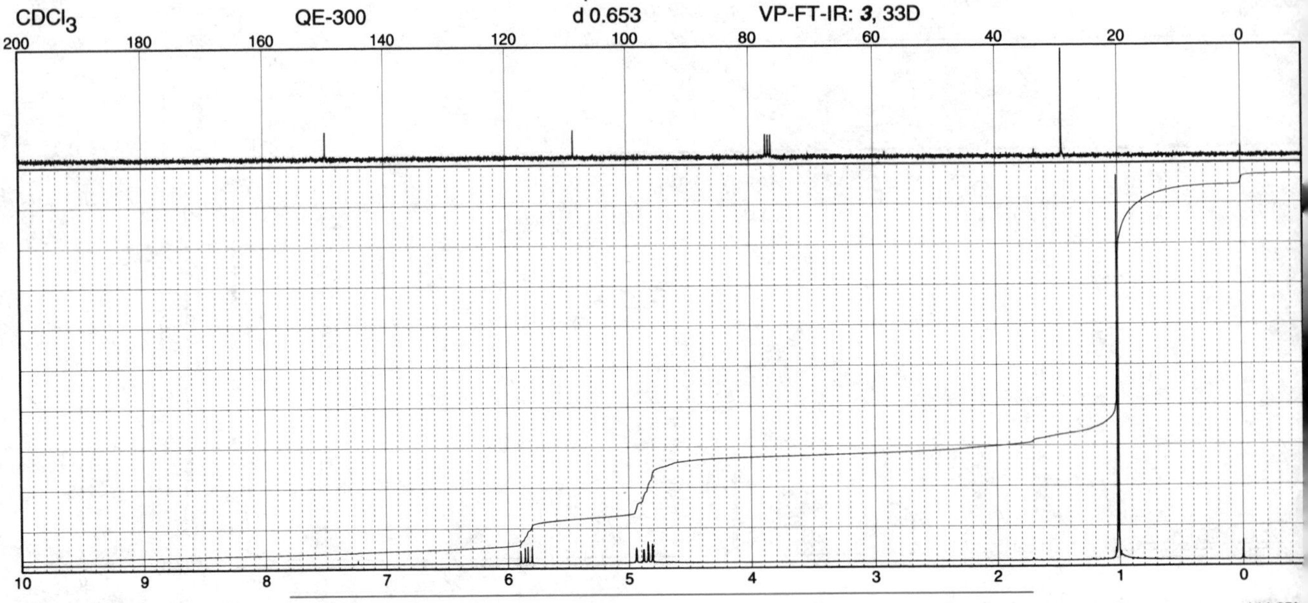

CH$_3$
CH$_3$C CH=CH$_2$
CH$_3$

C

Aldrich 11,114-7   *CAS [13643-02-2]*

(±)-3-Methyl-1-pentene, 99%

C$_6$H$_{12}$
FW 84.16
mp -154°C
bp 54°C
d 0.670

n$_D^{20}$1.3840
Fp -20°F
60 MHz: *1*, 27A
FT-IR: *1*, 22C
VP-FT-IR: *3*, 29D

144.65*
112.37
39.45*
29.39
19.72*
11.65*

CDCl$_3$ QE-300

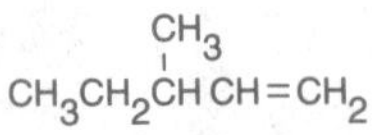

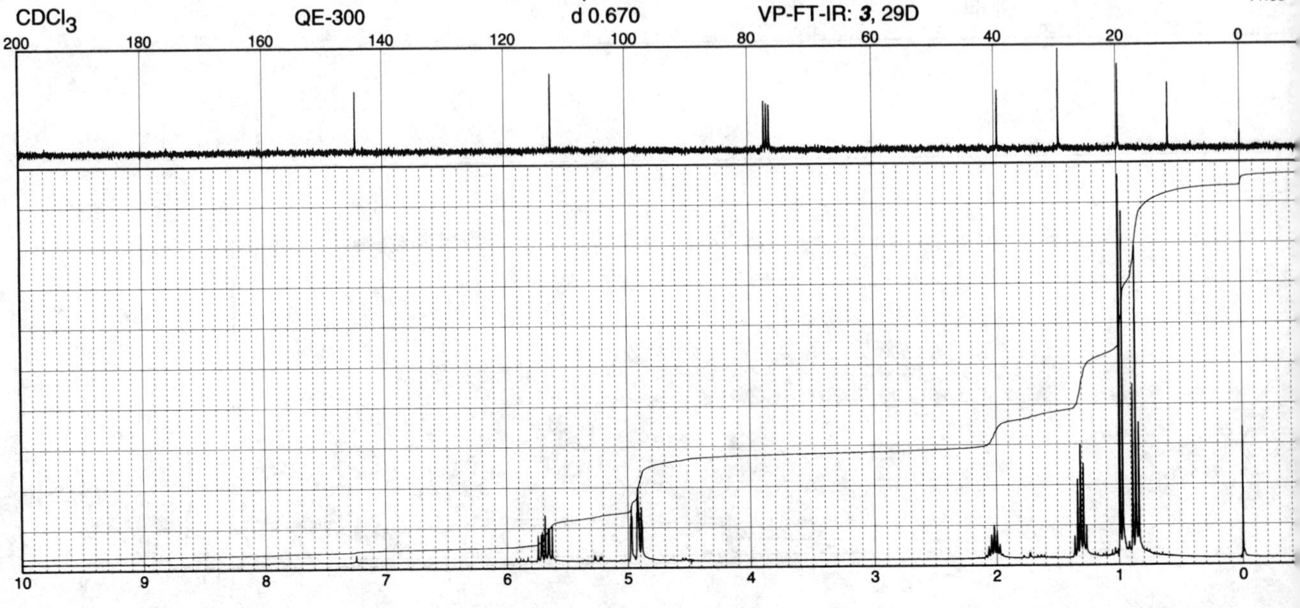

CH$_3$
CH$_3$CH$_2$CH CH=CH$_2$

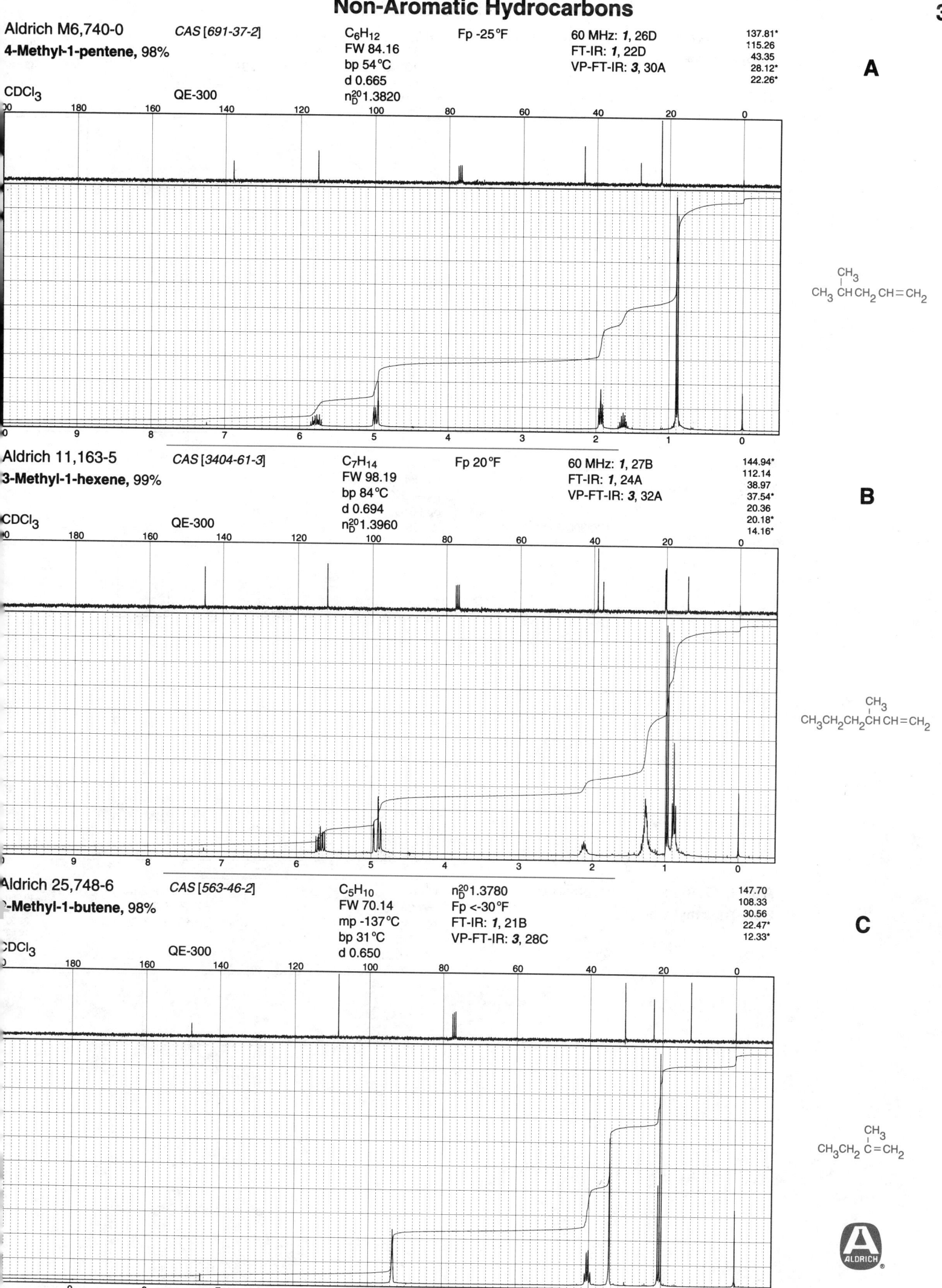

Aldrich M6,740-0   CAS [691-37-2]   C₆H₁₂   Fp -25°F   60 MHz: 1, 26D   137.81*
4-Methyl-1-pentene, 98%   FW 84.16   FT-IR: 1, 22D   115.26
    bp 54°C   VP-FT-IR: 3, 30A   43.35
CDCl₃   QE-300   d 0.665   28.12*
    n²⁰_D 1.3820   22.26*

A

Aldrich 11,163-5   CAS [3404-61-3]   C₇H₁₄   Fp 20°F   60 MHz: 1, 27B   144.94*
3-Methyl-1-hexene, 99%   FW 98.19   FT-IR: 1, 24A   112.14
    bp 84°C   VP-FT-IR: 3, 32A   38.97
CDCl₃   QE-300   d 0.694   37.54*
    n²⁰_D 1.3960   20.36
            20.18*
            14.16*

B

Aldrich 25,748-6   CAS [563-46-2]   C₅H₁₀   n²⁰_D 1.3780   147.70
2-Methyl-1-butene, 98%   FW 70.14   Fp <-30°F   108.33
    mp -137°C   FT-IR: 1, 21B   30.56
CDCl₃   QE-300   bp 31°C   VP-FT-IR: 3, 28C   22.47*
    d 0.650   12.33*

C

ALDRICH

A

Aldrich E1,470-5   CAS [760-21-4]

2-Ethyl-1-butene, 98%

C$_6$H$_{12}$

FW 84.16

mp -131°C

bp 65°C

d 0.689

n$_D^{20}$1.3960

Fp -15°F

VP-FT-IR: **3**, 29A

154.11
107.13
29.85
13.34*

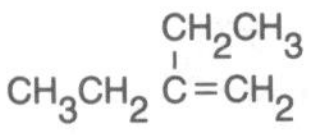

CDCl$_3$ QE-300

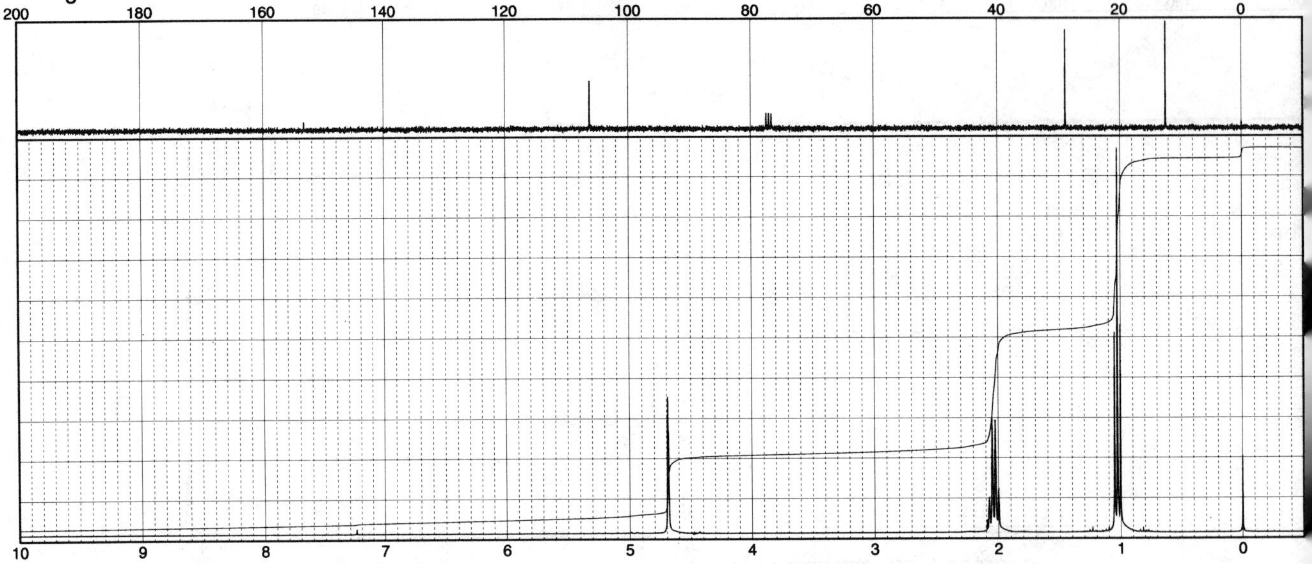

CH$_2$CH$_3$

CH$_3$CH$_2$ C=CH$_2$

B

Aldrich 19,040-3   CAS [563-78-0]

2,3-Dimethyl-1-butene, 97%

C$_6$H$_{12}$

FW 84.16

mp -79°C

bp 56°C

d 0.680

n$_D^{20}$1.3890

Fp -1°F

60 MHz: **1**, 27D

FT-IR: **1**, 22A

VP-FT-IR: **3**, 33C

151.80
107.56
35.22*
21.47*
20.13*

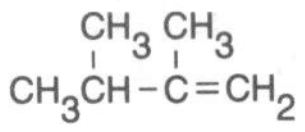

CDCl$_3$ QE-300

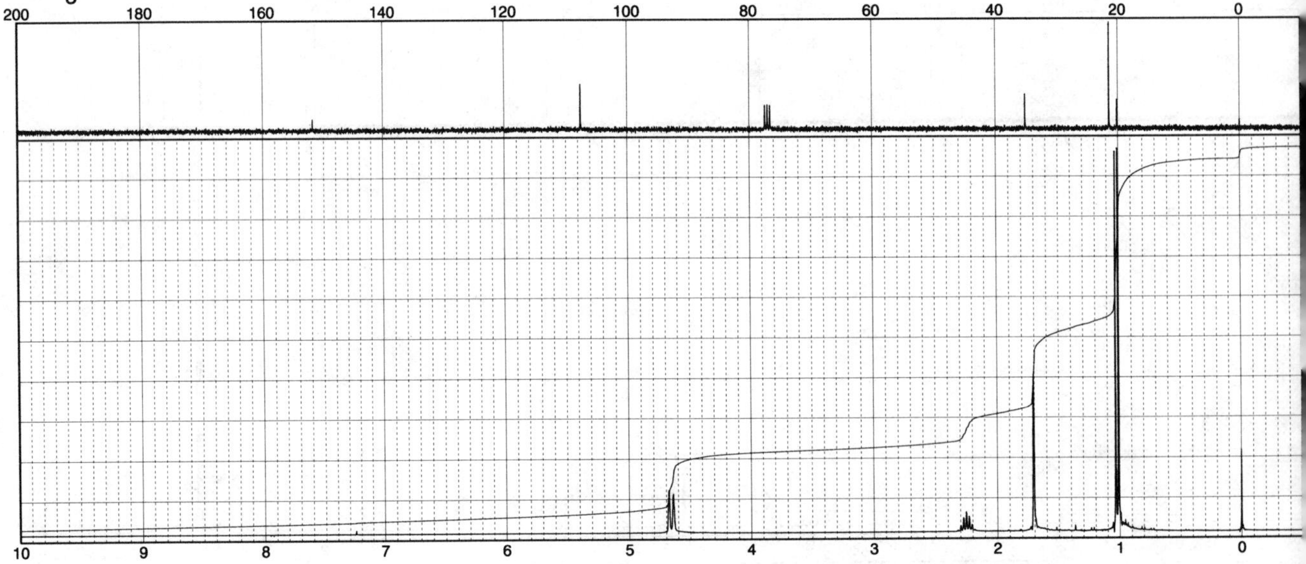

CH$_3$ CH$_3$

CH$_3$CH – C=CH$_2$

C

Aldrich T7,569-8   CAS [594-56-9]

2,3,3-Trimethyl-1-butene, 99+%

C$_7$H$_{14}$

FW 98.19

bp 79°C

d 0.705

n$_D^{20}$1.4020

Fp 1°F

60 MHz: **1**, 28A

FT-IR: **1**, 25C

VP-FT-IR: **3**, 36D

154.13
107.50
35.69
29.09*
19.53*

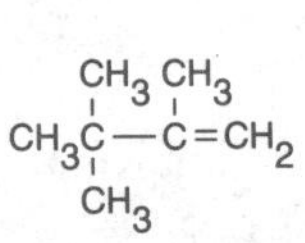

CDCl$_3$ QE-300

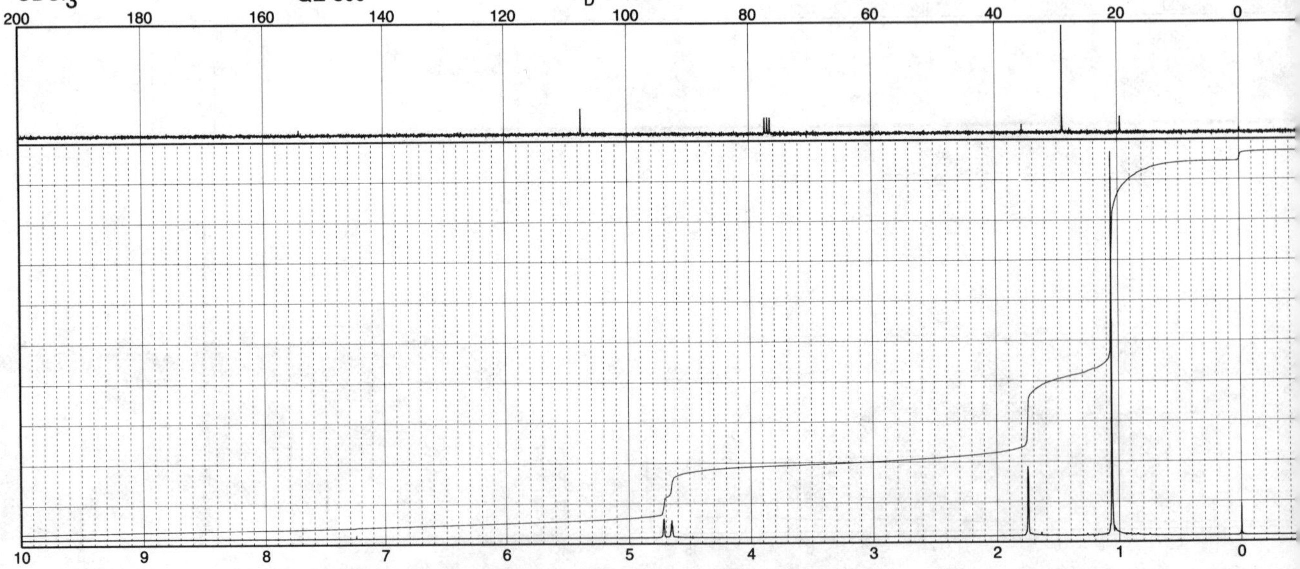

CH$_3$ CH$_3$

CH$_3$C — C=CH$_2$

CH$_3$

Non-Aromatic Hydrocarbons

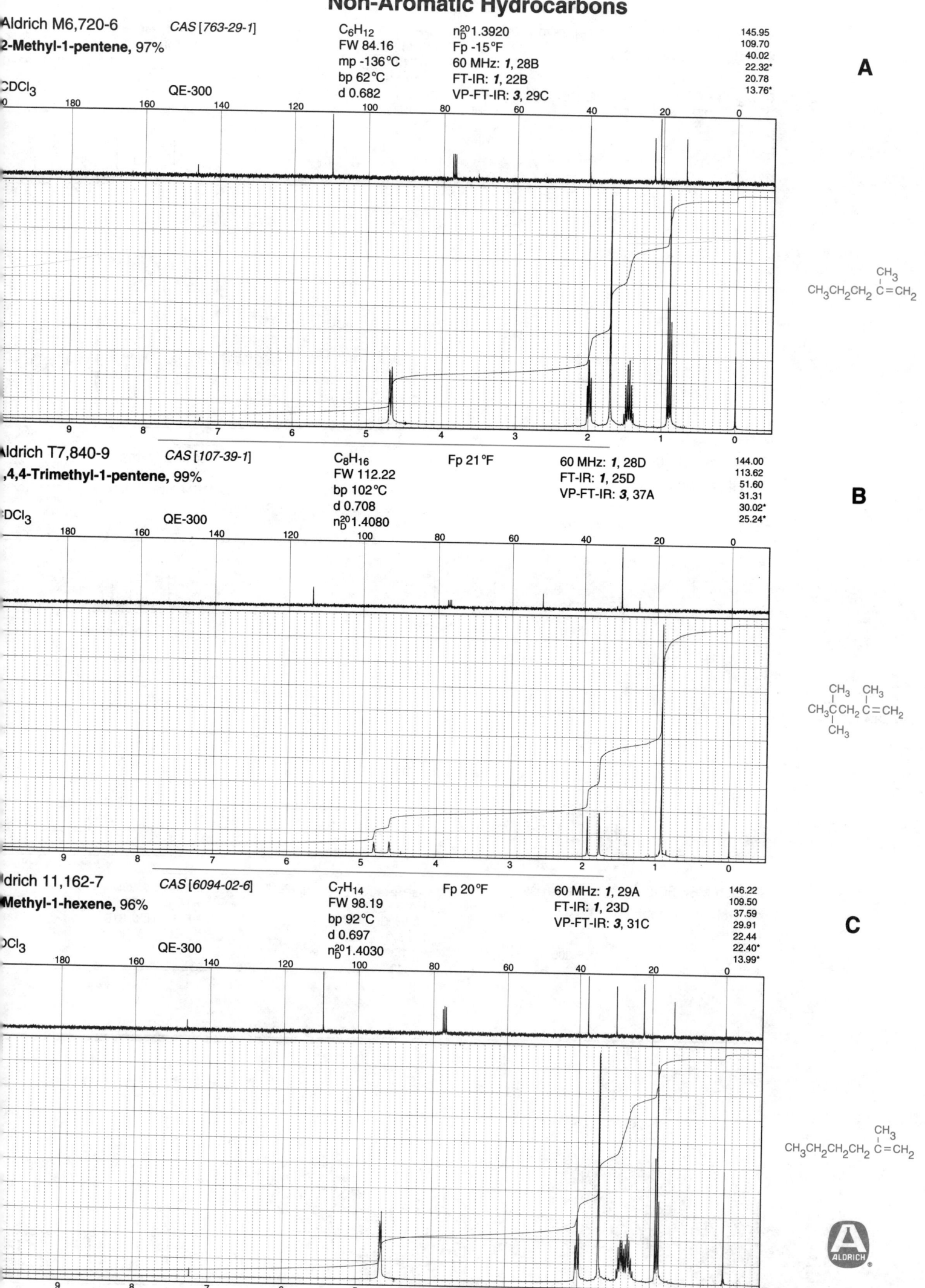

Aldrich M6,720-6   CAS [763-29-1]
2-Methyl-1-pentene, 97%

CDCl₃   QE-300

C_6H_{12}
FW 84.16
mp -136 °C
bp 62 °C
d 0.682

$n_D^{20} 1.3920$
Fp -15 °F
60 MHz: **1**, 28B
FT-IR: **1**, 22B
VP-FT-IR: **3**, 29C

145.95
109.70
40.02
22.32*
20.78
13.76*

A

$CH_3CH_2CH_2 \overset{CH_3}{\underset{}{C}} = CH_2$

Aldrich T7,840-9   CAS [107-39-1]
4,4-Trimethyl-1-pentene, 99%

CDCl₃   QE-300

C_8H_{16}
FW 112.22
bp 102 °C
d 0.708
$n_D^{20} 1.4080$

Fp 21 °F

60 MHz: **1**, 28D
FT-IR: **1**, 25D
VP-FT-IR: **3**, 37A

144.00
113.62
51.60
31.31
30.02*
25.24*

B

$CH_3\overset{CH_3}{\underset{CH_3}{C}}CH_2 \overset{CH_3}{\underset{}{C}} = CH_2$

Aldrich 11,162-7   CAS [6094-02-6]
Methyl-1-hexene, 96%

CDCl₃   QE-300

C_7H_{14}
FW 98.19
bp 92 °C
d 0.697
$n_D^{20} 1.4030$

Fp 20 °F

60 MHz: **1**, 29A
FT-IR: **1**, 23D
VP-FT-IR: **3**, 31C

146.22
109.50
37.59
29.91
22.44
22.40*
13.99*

C

$CH_3CH_2CH_2CH_2 \overset{CH_3}{\underset{}{C}} = CH_2$

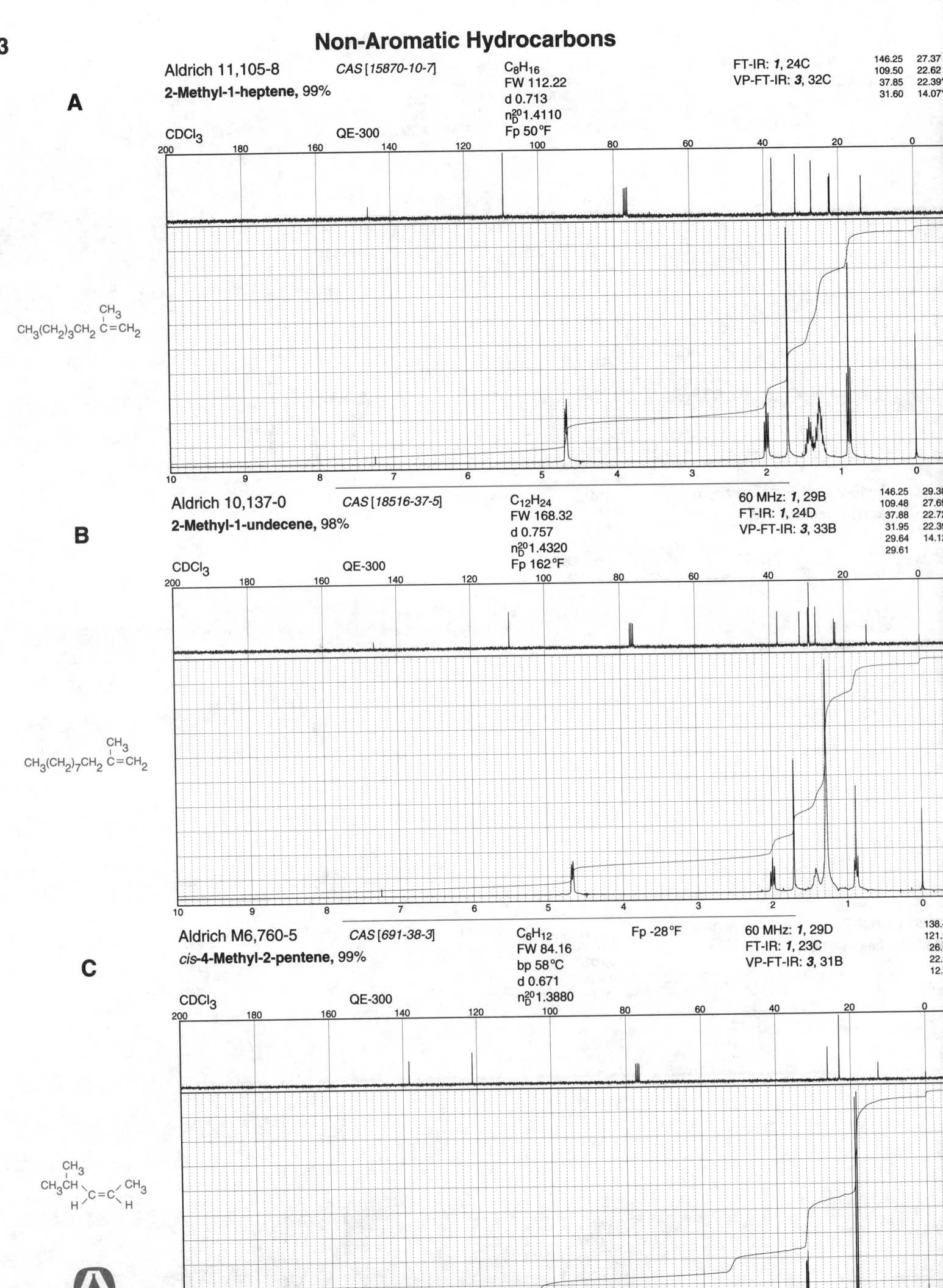

A

Aldrich 11,105-8   CAS [15870-10-7]   C$_8$H$_{16}$
2-Methyl-1-heptene, 99%   FW 112.22
d 0.713
n$_D^{20}$1.4110
Fp 50°F

CDCl$_3$ QE-300

FT-IR: **1**, 24C
VP-FT-IR: **3**, 32C

146.25   27.37
109.50   22.62
37.85   22.39*
31.60   14.07*

CH$_3$(CH$_2$)$_3$CH$_2$ C=CH$_2$ with CH$_3$

B

Aldrich 10,137-0   CAS [18516-37-5]   C$_{12}$H$_{24}$
2-Methyl-1-undecene, 98%   FW 168.32
d 0.757
n$_D^{20}$1.4320
Fp 162°F

CDCl$_3$ QE-300

60 MHz: **1**, 29B
FT-IR: **1**, 24D
VP-FT-IR: **3**, 33B

146.25   29.38
109.48   27.69
37.88   22.72
31.95   22.39
29.64   14.12
29.61

CH$_3$(CH$_2$)$_7$CH$_2$ C=CH$_2$ with CH$_3$

C

Aldrich M6,760-5   CAS [691-38-3]   C$_6$H$_{12}$ Fp -28°F
cis-4-Methyl-2-pentene, 99%   FW 84.16
bp 58°C
d 0.671
n$_D^{20}$1.3880

CDCl$_3$ QE-300

60 MHz: **1**, 29D
FT-IR: **1**, 23C
VP-FT-IR: **3**, 31B

138.4
121.2
26.0
22.9
12.7

CH$_3$CH C=C with CH$_3$, CH$_3$, H, H

ALDRICH

Non-Aromatic Hydrocarbons

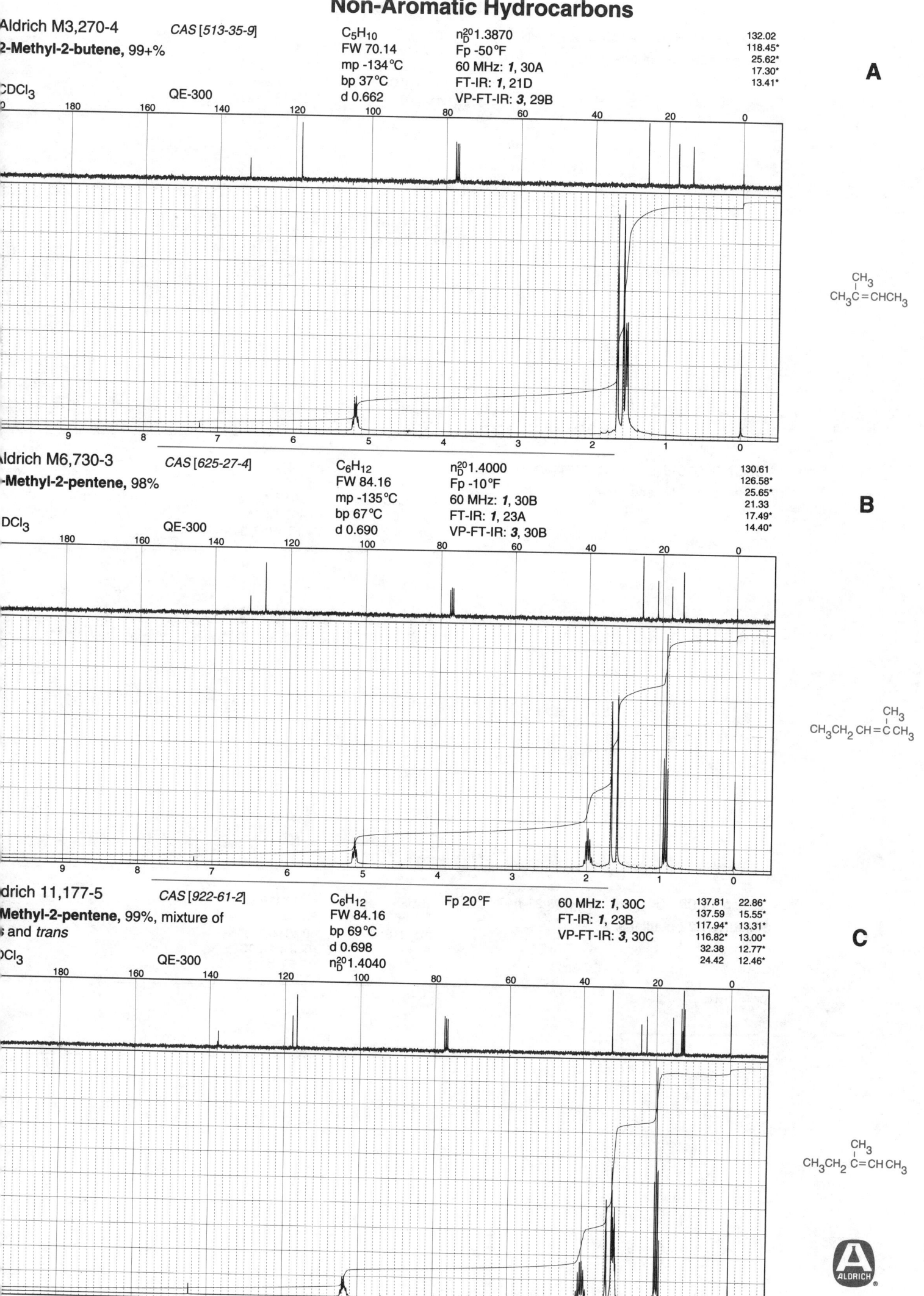

Aldrich M3,270-4   CAS [513-35-9]

2-Methyl-2-butene, 99+%

CDCl$_3$ QE-300

C_5H_{10}
FW 70.14
mp -134°C
bp 37°C
d 0.662

n_D^{20} 1.3870
Fp -50°F
60 MHz: 1, 30A
FT-IR: 1, 21D
VP-FT-IR: 3, 29B

132.02
118.45*
25.62*
17.30*
13.41*

A

CH$_3$C=CHCH$_3$ (with CH$_3$ above)

Aldrich M6,730-3   CAS [625-27-4]

Methyl-2-pentene, 98%

DCl$_3$ QE-300

C_6H_{12}
FW 84.16
mp -135°C
bp 67°C
d 0.690

n_D^{20} 1.4000
Fp -10°F
60 MHz: 1, 30B
FT-IR: 1, 23A
VP-FT-IR: 3, 30B

130.61
126.58*
25.65*
21.33
17.49*
14.40*

B

CH$_3$CH$_2$CH=CCH$_3$ (with CH$_3$ above)

drich 11,177-5   CAS [922-61-2]

Methyl-2-pentene, 99%, mixture of
s and trans

DCl$_3$ QE-300

C_6H_{12}
FW 84.16
bp 69°C
d 0.698
n_D^{20} 1.4040

Fp 20°F

60 MHz: 1, 30C
FT-IR: 1, 23B
VP-FT-IR: 3, 30C

137.81	22.86*
137.59	15.55*
117.94*	13.31*
116.82*	13.00*
32.38	12.77*
24.42	12.46*

C

CH$_3$CH$_2$C=CHCH$_3$ (with CH$_3$ above)

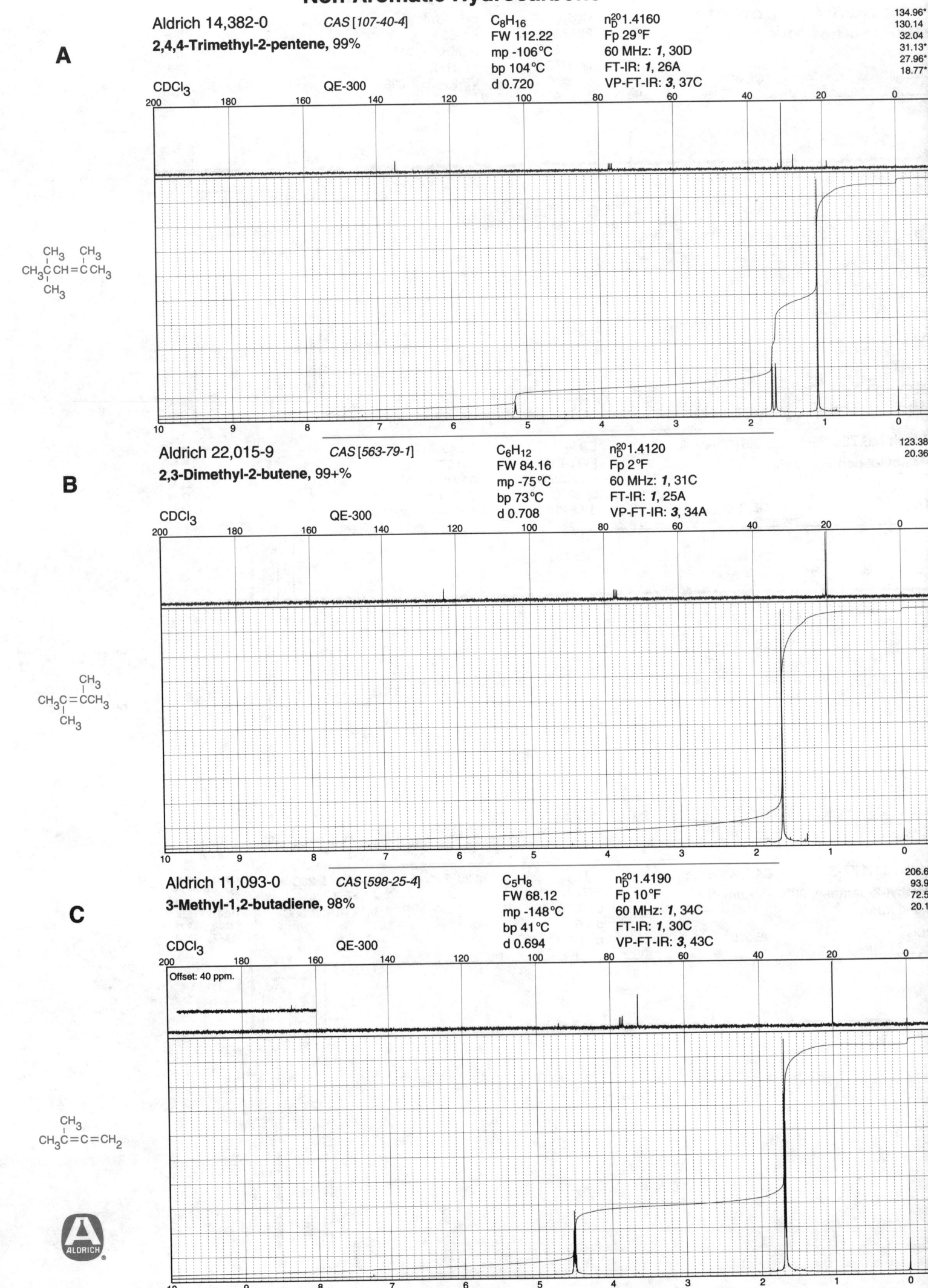

A

Aldrich 14,382-0   *CAS [107-40-4]*

2,4,4-Trimethyl-2-pentene, 99%

C_8H_{16}
FW 112.22
mp -106°C
bp 104°C
d 0.720

n_D^{20}1.4160
Fp 29°F
60 MHz: *1*, 30D
FT-IR: *1*, 26A
VP-FT-IR: *3*, 37C

134.96*
130.14
32.04
31.13*
27.96*
18.77*

CDCl$_3$ QE-300

B

Aldrich 22,015-9   *CAS [563-79-1]*

2,3-Dimethyl-2-butene, 99+%

C_6H_{12}
FW 84.16
mp -75°C
bp 73°C
d 0.708

n_D^{20}1.4120
Fp 2°F
60 MHz: *1*, 31C
FT-IR: *1*, 25A
VP-FT-IR: *3*, 34A

123.38
20.36*

CDCl$_3$ QE-300

C

Aldrich 11,093-0   *CAS [598-25-4]*

3-Methyl-1,2-butadiene, 98%

C_5H_8
FW 68.12
mp -148°C
bp 41°C
d 0.694

n_D^{20}1.4190
Fp 10°F
60 MHz: *1*, 34C
FT-IR: *1*, 30C
VP-FT-IR: *3*, 43C

206.65
93.97
72.53
20.14

CDCl$_3$ QE-300

Offset: 40 ppm.

Non-Aromatic Hydrocarbons

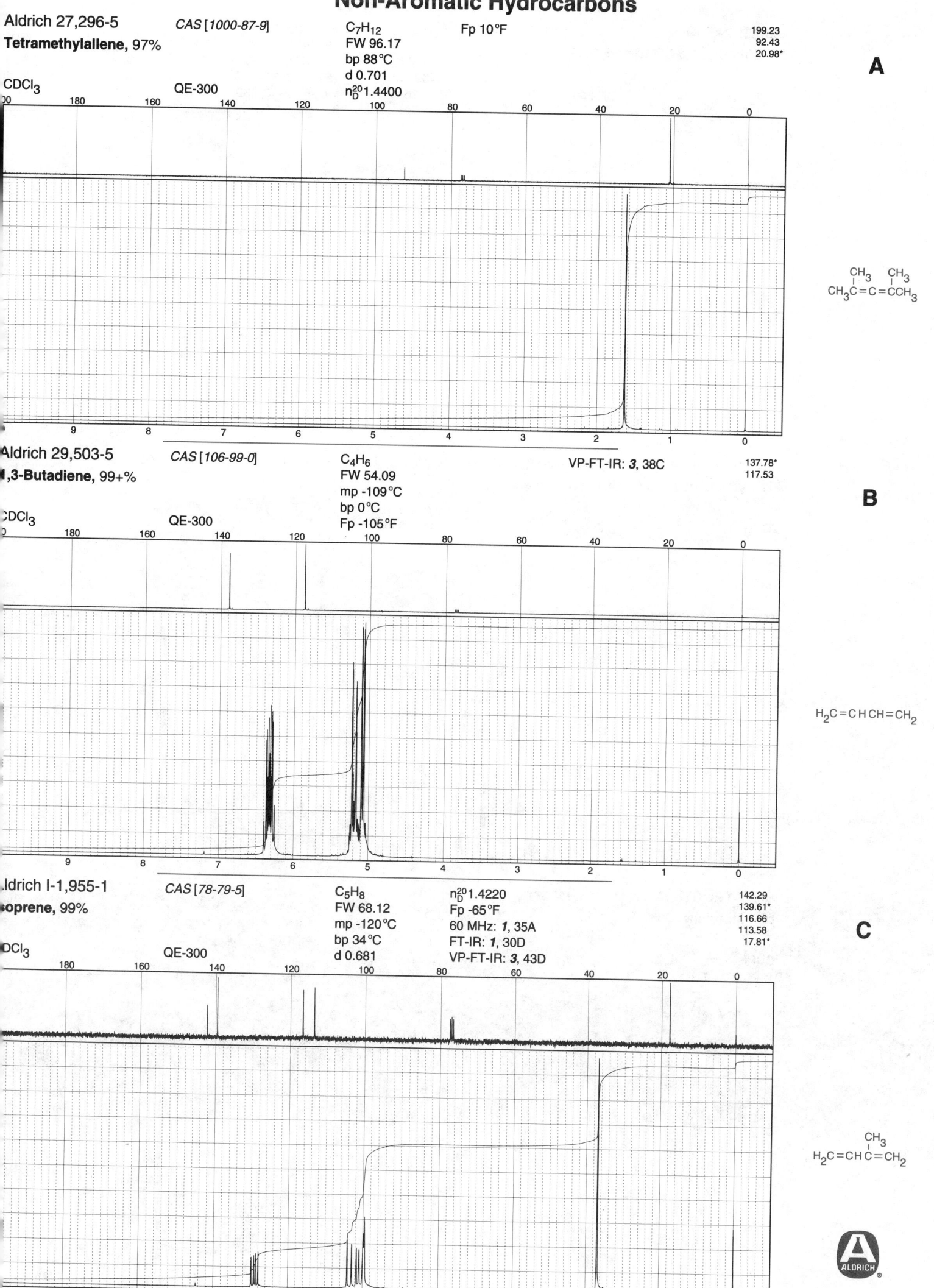

Aldrich 27,296-5   CAS [1000-87-9] C_7H_{12} Fp 10°F
Tetramethylallene, 97%   FW 96.17
bp 88°C
d 0.701
n_D^{20} 1.4400

CDCl3   QE-300

199.23
92.43
20.98*

A

CH_3 CH_3
$CH_3C=C=CCH_3$

Aldrich 29,503-5   CAS [106-99-0] C_4H_6 VP-FT-IR: **3**, 38C
1,3-Butadiene, 99+%   FW 54.09
mp -109°C
bp 0°C
Fp -105°F

137.78*
117.53

CDCl3   QE-300

B

$H_2C=CHCH=CH_2$

Aldrich I-1,955-1   CAS [78-79-5] C_5H_8 n_D^{20} 1.4220
Isoprene, 99%   FW 68.12   Fp -65°F
mp -120°C   60 MHz: **1**, 35A
bp 34°C   FT-IR: **1**, 30D
d 0.681   VP-FT-IR: **3**, 43D

DCl3   QE-300

142.29
139.61*
116.66
113.58
17.81*

C

CH_3
$H_2C=CHC=CH_2$

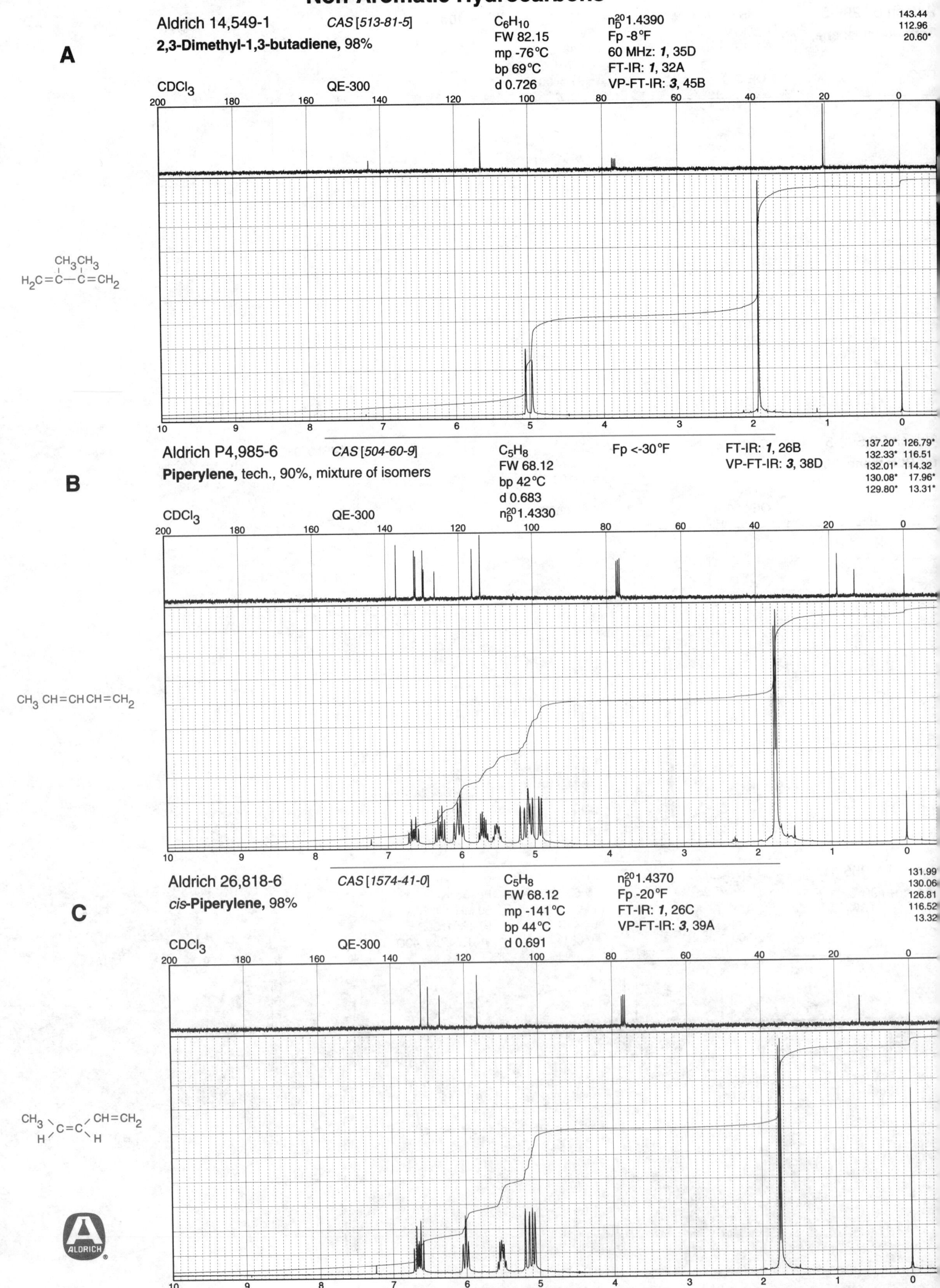

Aldrich 14,549-1   *CAS [513-81-5]*   C₆H₁₀   n²⁰_D 1.4390

2,3-Dimethyl-1,3-butadiene, 98%

FW 82.15   Fp -8°F
mp -76°C   60 MHz: *1*, 35D
bp 69°C   FT-IR: *1*, 32A
d 0.726   VP-FT-IR: *3*, 45B

143.44
112.96
20.60*

CDCl₃   QE-300

$CH_3 \quad CH_3$
$H_2C=C-C=CH_2$

Aldrich P4,985-6   *CAS [504-60-9]*   C₅H₈   Fp <-30°F   FT-IR: *1*, 26B

Piperylene, tech., 90%, mixture of isomers

FW 68.12
bp 42°C
d 0.683
n²⁰_D 1.4330

VP-FT-IR: *3*, 38D

137.20* 126.79*
132.33* 116.51
132.01* 114.32
130.08* 17.96*
129.80* 13.31*

CDCl₃   QE-300

$CH_3\ CH=CHCH=CH_2$

Aldrich 26,818-6   *CAS [1574-41-0]*   C₅H₈   n²⁰_D 1.4370

***cis*-Piperylene, 98%**

FW 68.12   Fp -20°F
mp -141°C   FT-IR: *1*, 26C
bp 44°C   VP-FT-IR: *3*, 39A
d 0.691

131.99
130.06
126.81
116.52
13.32

CDCl₃   QE-300

$CH_3 \quad CH=CH_2$
$\ \ \ \ \ C=C$
$H \quad\quad H$

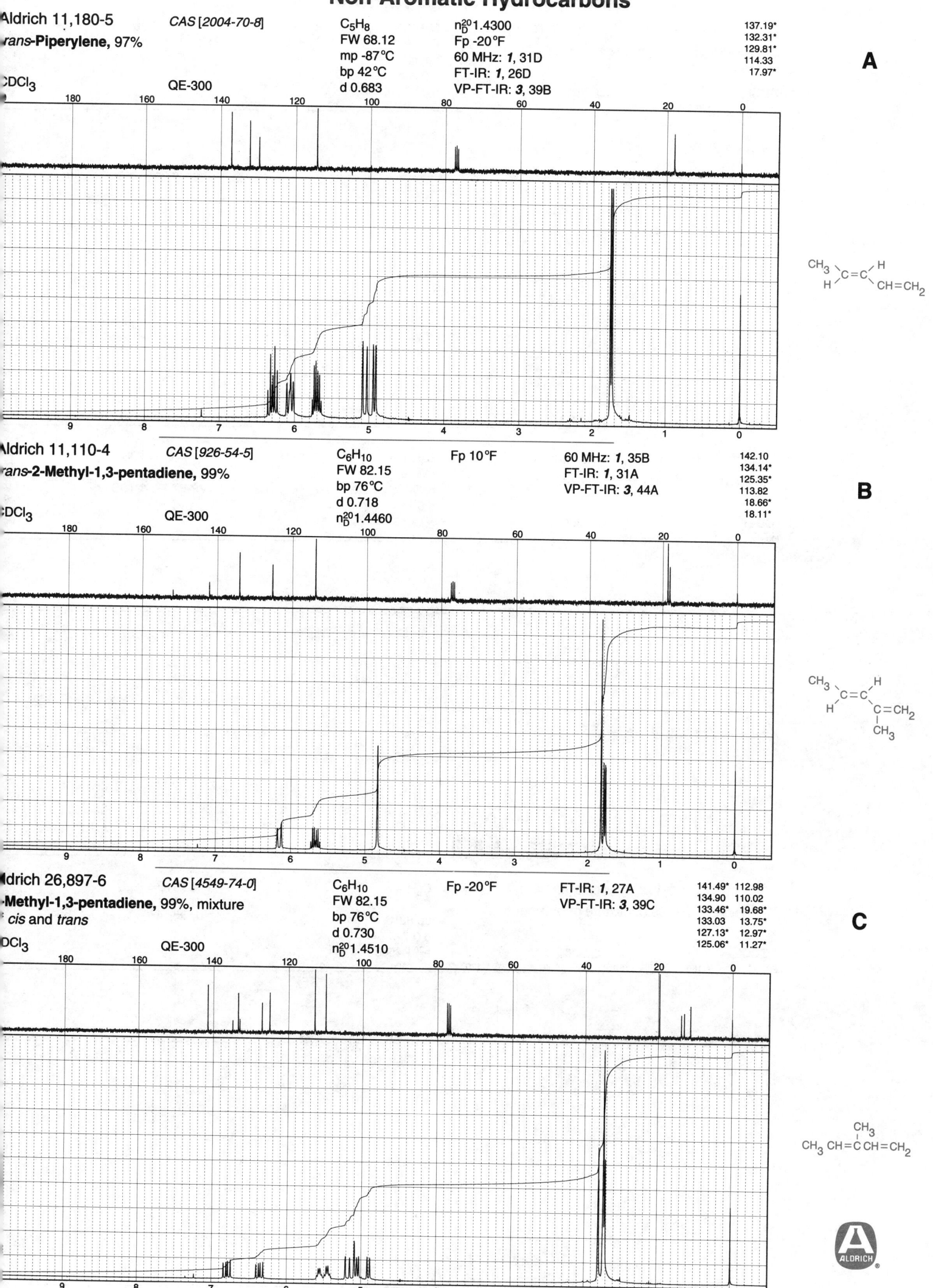

Aldrich 11,180-5   CAS [2004-70-8]
trans-**Piperylene**, 97%

C₅H₈
FW 68.12
mp -87 °C
bp 42 °C
d 0.683

n_D^{20} 1.4300
Fp -20 °F
60 MHz: *1*, 31D
FT-IR: *1*, 26D
VP-FT-IR: *3*, 39B

137.19*
132.31*
129.81*
114.33
17.97*

A

CDCl₃   QE-300

Aldrich 11,110-4   CAS [926-54-5]
trans-**2-Methyl-1,3-pentadiene**, 99%

C₆H₁₀
FW 82.15
bp 76 °C
d 0.718
n_D^{20} 1.4460

Fp 10 °F

60 MHz: *1*, 35B
FT-IR: *1*, 31A
VP-FT-IR: *3*, 44A

142.10
134.14*
125.35*
113.82
18.66*
18.11*

B

CDCl₃   QE-300

Aldrich 26,897-6   CAS [4549-74-0]
Methyl-1,3-pentadiene, 99%, mixture
of *cis* and *trans*

C₆H₁₀
FW 82.15
bp 76 °C
d 0.730
n_D^{20} 1.4510

Fp -20 °F

FT-IR: *1*, 27A
VP-FT-IR: *3*, 39C

141.49* 112.98
134.90   110.02
133.46* 19.68*
133.03   13.75*
127.13* 12.97*
125.06* 11.27*

C

CDCl₃   QE-300

A

Aldrich 12,655-1   *CAS [1000-86-8]* C_7H_{12} Fp 50°F   60 MHz: *1*, 36A

2,4-Dimethyl-1,3-pentadiene, 98+%   FW 96.17   FT-IR: *1*, 32B
bp 94°C   VP-FT-IR: *3*, 45C
d 0.744
n_D^{20}1.4410

142.27
134.02
127.10
113.56
26.92
23.70
19.43

CDCl$_3$ QE-300

$$CH_3 \underset{CH_3}{\overset{CH_3}{C}} = CHC = CH_2$$

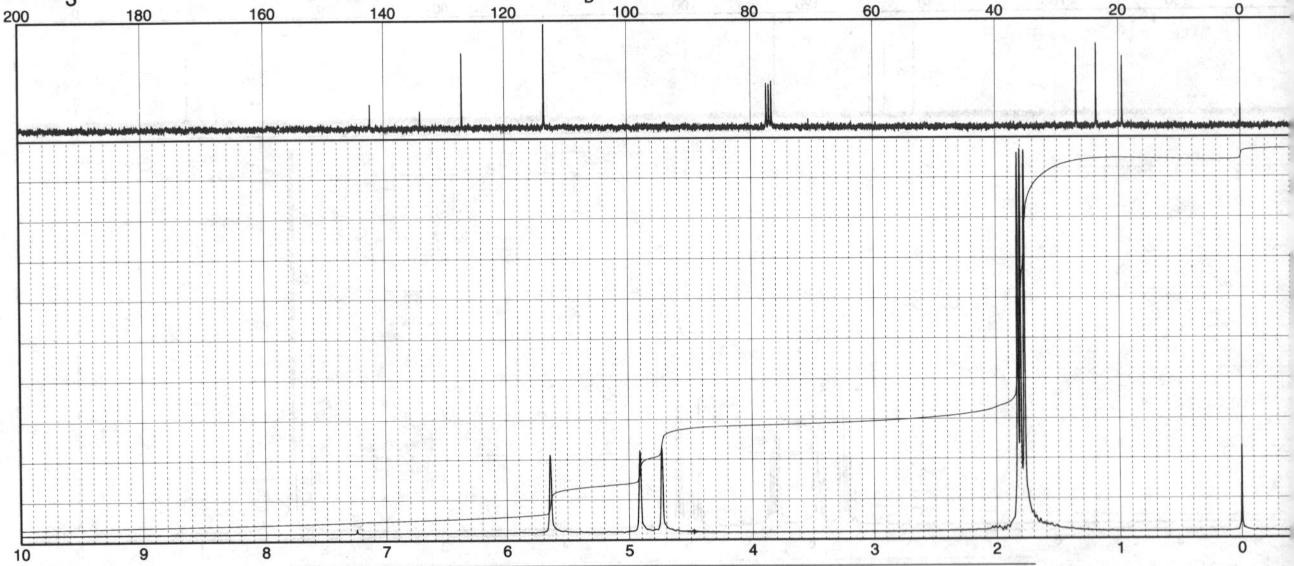

B

Aldrich 11,083-3   *CAS [592-48-3]* C_6H_{10} Fp 25°F   60 MHz: *1*, 32B

1,3-Hexadiene, 99%, predominantly *trans*   FW 82.15   FT-IR: *1*, 27D
bp 74°C   VP-FT-IR: *3*, 40C
d 0.714
n_D^{20}1.4380

137.36
136.87
129.98
114.54
25.57
13.42

CDCl$_3$ QE-300

$CH_3CH_2\ CH=CHCH=CH_2$

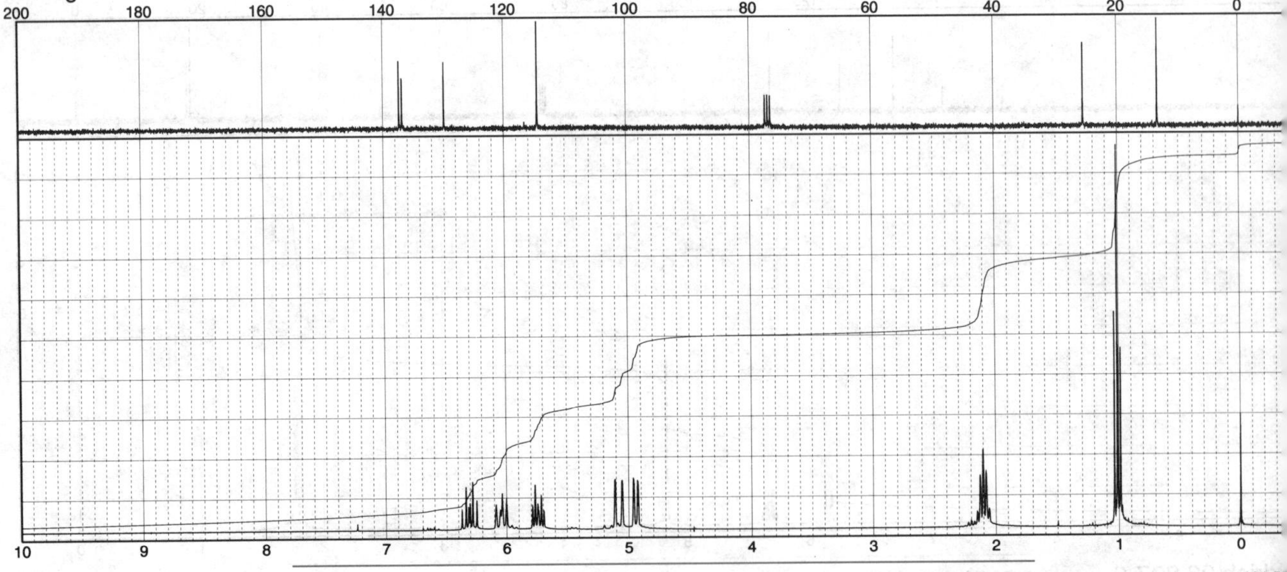

C

Aldrich 11,084-1   *CAS [592-46-1]* C_6H_{10} Fp 18°F   60 MHz: *1*, 32D

2,4-Hexadiene, 99%, mixture of isomers   FW 82.15   FT-IR: *1*, 28B
bp 82°C   VP-FT-IR: *3*, 41A
d 0.720
n_D^{20}1.4560

131.5
126.3
17.9

CDCl$_3$ QE-300

$CH_3\ CH=CHCH=CHCH_3$

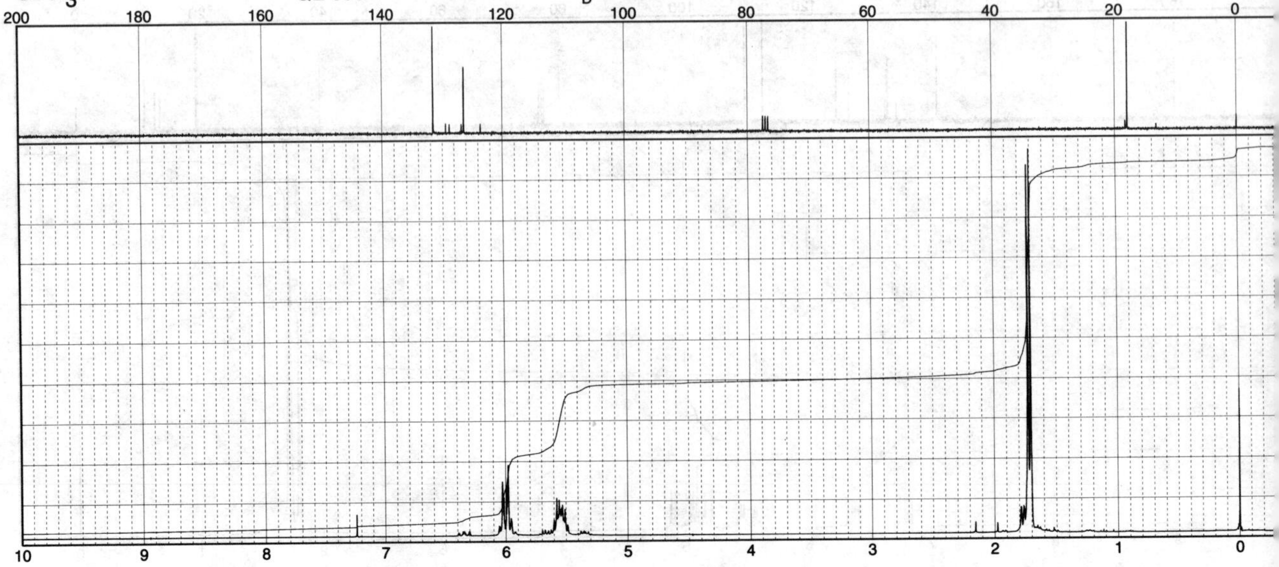

Non-Aromatic Hydrocarbons

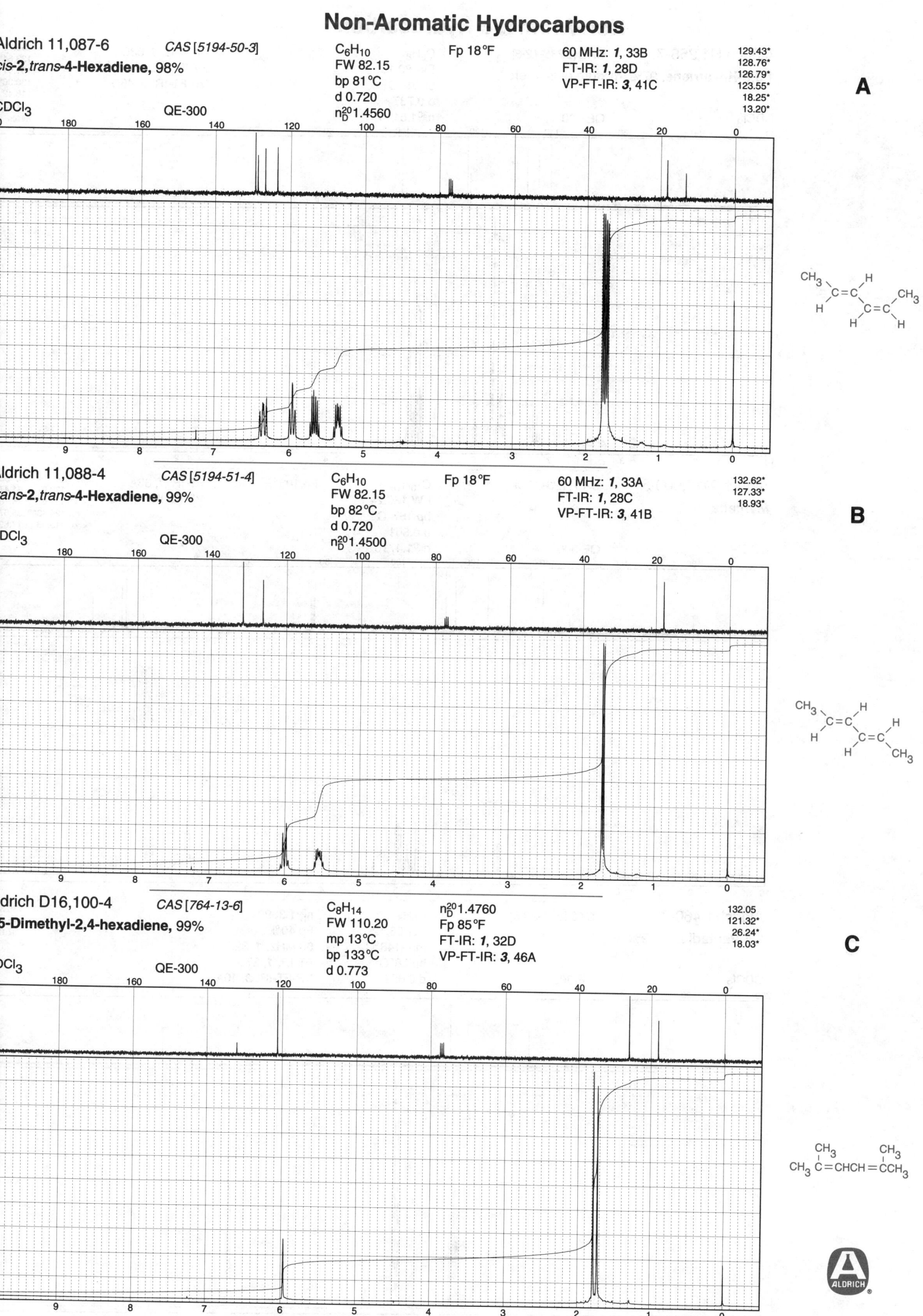

Aldrich 11,087-6   CAS [5194-50-3]

cis-2,trans-4-Hexadiene, 98%

CDCl₃   QE-300

C_6H_{10}
FW 82.15
bp 81 °C
d 0.720
n_D^{20} 1.4560

Fp 18 °F

60 MHz: **1**, 33B
FT-IR: **1**, 28D
VP-FT-IR: **3**, 41C

129.43*
128.76*
126.79*
123.55*
18.25*
13.20*

A

Aldrich 11,088-4   CAS [5194-51-4]

trans-2,trans-4-Hexadiene, 99%

CDCl₃   QE-300

C_6H_{10}
FW 82.15
bp 82 °C
d 0.720
n_D^{20} 1.4500

Fp 18 °F

60 MHz: **1**, 33A
FT-IR: **1**, 28C
VP-FT-IR: **3**, 41B

132.62*
127.33*
18.93*

B

Aldrich D16,100-4   CAS [764-13-6]

5-Dimethyl-2,4-hexadiene, 99%

CDCl₃   QE-300

C_8H_{14}
FW 110.20
mp 13 °C
bp 133 °C
d 0.773

n_D^{20} 1.4760
Fp 85 °F
FT-IR: **1**, 32D
VP-FT-IR: **3**, 46A

132.05
121.32*
26.24*
18.03*

C

A

Aldrich H1,258-7   CAS [2235-12-3] C_6H_8 Fp 101°F   60 MHz: *1*, 33D

1,3,5-Hexatriene, 97%, mixture of isomers   FW 80.13   FT-IR: *1*, 29B

bp 78°C   VP-FT-IR: *3*, 46D

d 0.737

n_D^{20} 1.5110

136.73*
133.64*
131.91*
130.36*
118.42
117.71

CDCl₃   QE-300

$H_2C=CHCH=CHCH=CH_2$

B

Aldrich M10,000-5   CAS [123-35-3] $C_{10}H_{16}$ Fp 103°F   FT-IR: *1*, 33A

Myrcene   FW 136.24   VP-FT-IR: *3*, 46C

bp 167°C

d 0.801

n_D^{20} 1.4710

146.14 112.96
138.97* 31.47
131.67 26.78
124.13* 25.67
115.56 17.69*

CDCl₃   QE-300

$H_2C=CH-\overset{\overset{\displaystyle CH_2}{\|}}{C}-CH_2CH_2CH=\overset{\overset{\displaystyle CH_3}{}}{C}CH_3$

C

Aldrich P460-7   CAS [591-93-5] C_5H_8 n_D^{20} 1.3890

1,4-Pentadiene, 99%   FW 68.12   Fp 40°F

mp -148°C   60 MHz: *1*, 32A

bp 26°C   FT-IR: *1*, 27B

d 0.659   VP-FT-IR: *3*, 40A

136.34
115.46
37.85

CDCl₃   QE-300

$H_2C=CHCH_2CH=CH_2$

Non-Aromatic Hydrocarbons

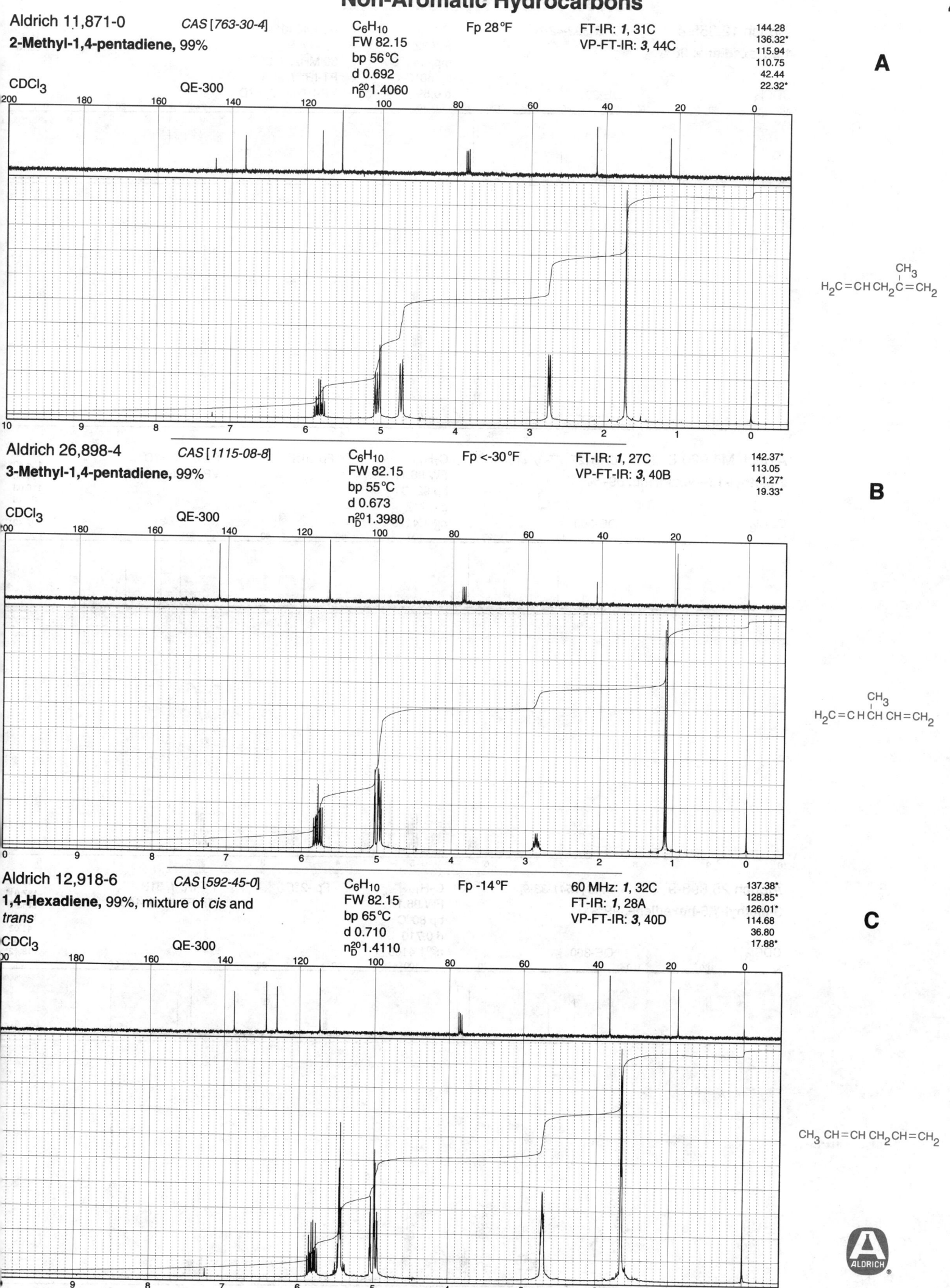

A

Aldrich 11,871-0   *CAS [763-30-4]*
2-Methyl-1,4-pentadiene, 99%

CDCl₃   QE-300

C_6H_{10}
FW 82.15
bp 56°C
d 0.692
n_D^{20} 1.4060

Fp 28°F

FT-IR: *1*, 31C
VP-FT-IR: *3*, 44C

144.28
136.32*
115.94
110.75
42.44
22.32*

$H_2C{=}CH\,CH_2\,\overset{\displaystyle CH_3}{\underset{|}{C}}{=}CH_2$

B

Aldrich 26,898-4   *CAS [1115-08-8]*
3-Methyl-1,4-pentadiene, 99%

CDCl₃   QE-300

C_6H_{10}
FW 82.15
bp 55°C
d 0.673
n_D^{20} 1.3980

Fp <-30°F

FT-IR: *1*, 27C
VP-FT-IR: *3*, 40B

142.37*
113.05
41.27*
19.33*

$H_2C{=}CH\,\overset{\displaystyle CH_3}{\underset{|}{C}}H\,CH{=}CH_2$

C

Aldrich 12,918-6   *CAS [592-45-0]*
1,4-Hexadiene, 99%, mixture of *cis* and *trans*

CDCl₃   QE-300

C_6H_{10}
FW 82.15
bp 65°C
d 0.710
n_D^{20} 1.4110

Fp -14°F

60 MHz: *1*, 32C
FT-IR: *1*, 28A
VP-FT-IR: *3*, 40D

137.38*
128.85*
126.01*
114.68
36.80
17.88*

$CH_3\,CH{=}CH\,CH_2\,CH{=}CH_2$

ALDRICH

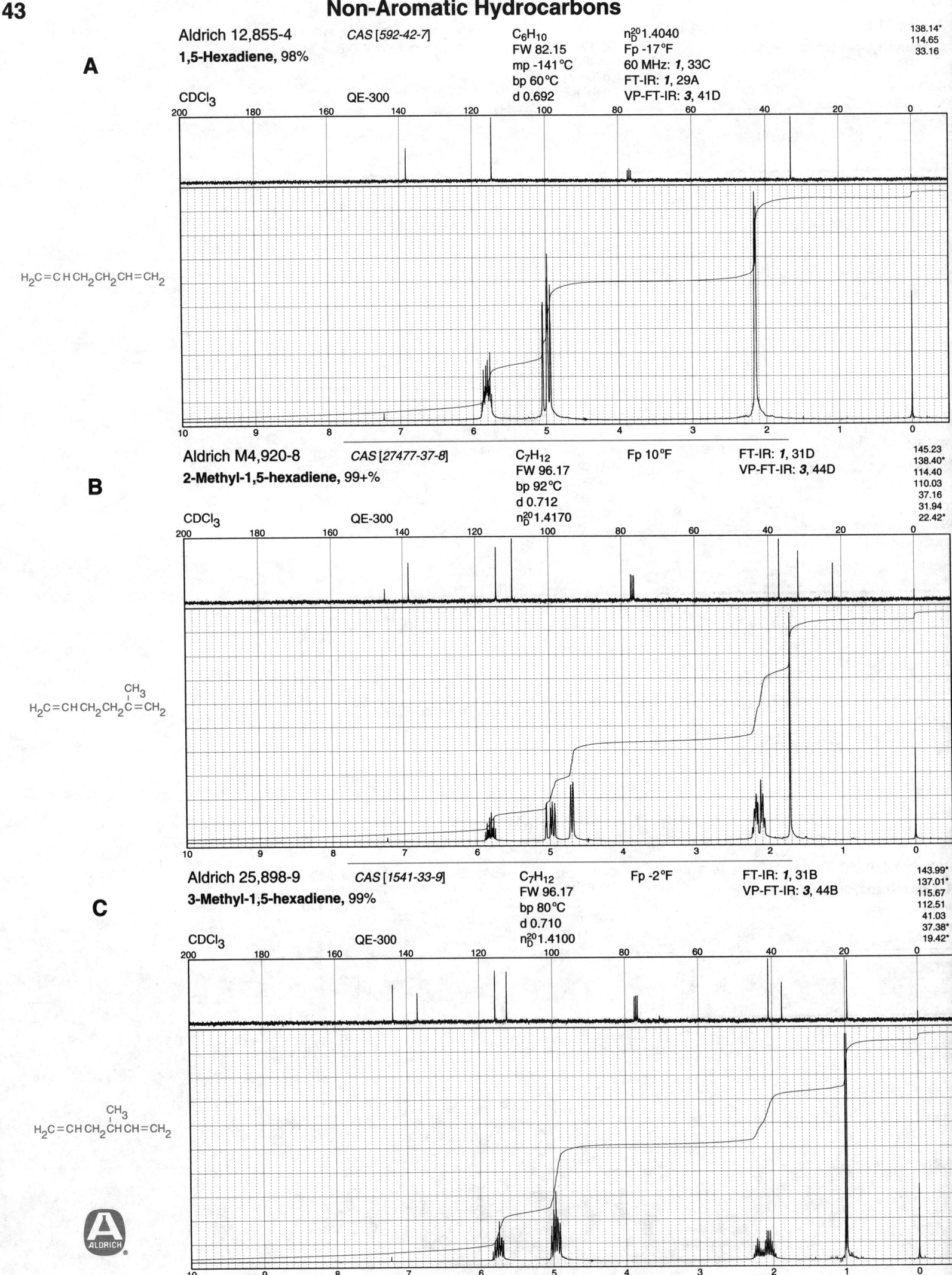

A

Aldrich 12,855-4   *CAS [592-42-7]*

1,5-Hexadiene, 98%

CDCl₃   QE-300

$H_2C=CHCH_2CH_2CH=CH_2$

C_6H_{10}
FW 82.15
mp -141 °C
bp 60 °C
d 0.692

$n_D^{20} 1.4040$
Fp -17 °F
60 MHz: *1*, 33C
FT-IR: *1*, 29A
VP-FT-IR: *3*, 41D

138.14*
114.65
33.16

B

Aldrich M4,920-8   *CAS [27477-37-8]*

2-Methyl-1,5-hexadiene, 99+%

CDCl₃   QE-300

$H_2C=CHCH_2CH_2\overset{\underset{\displaystyle |}{CH_3}}{C}=CH_2$

C_7H_{12}
FW 96.17
bp 92 °C
d 0.712
$n_D^{20} 1.4170$

Fp 10 °F

FT-IR: *1*, 31D
VP-FT-IR: *3*, 44D

145.23
138.40*
114.40
110.03
37.16
31.94
22.42*

C

Aldrich 25,898-9   *CAS [1541-33-9]*

3-Methyl-1,5-hexadiene, 99%

CDCl₃   QE-300

$H_2C=CHCH_2\overset{\underset{\displaystyle |}{CH_3}}{CH}CH=CH_2$

C_7H_{12}
FW 96.17
bp 80 °C
d 0.710
$n_D^{20} 1.4100$

Fp -2 °F

FT-IR: *1*, 31B
VP-FT-IR: *3*, 44B

143.99*
137.01*
115.67
112.51
41.03
37.38*
19.42*

ALDRICH

Non-Aromatic Hydrocarbons

Aldrich 36,224-7   CAS [627-58-7]

2,5-Dimethyl-1,5-hexadiene, 99%

C_8H_{14}
FW 110.20
mp -38°C
bp 114°C
d 0.742

n_D^{20}1.4290
Fp 45°F

145.64
109.85
36.07
22.43*

A

CDCl$_3$ QE-300

$H_2C=CCH_2CH_2C=CH_2$
| |
CH$_3$ CH$_3$

Aldrich 25,884-9   CAS [3070-53-9]

1,6-Heptadiene, 99%

C_7H_{12}
FW 96.17
mp -129°C
bp 90°C
d 0.714

n_D^{20}1.4140
Fp 14°F
FT-IR: *1*, 29C
VP-FT-IR: *3*, 42C

138.65*
114.47
33.20
28.19

B

CDCl$_3$ QE-300

$H_2C=CHCH_2CH_2CH_2CH=CH_2$

Aldrich O-250-1   CAS [3710-30-3]

1,7-Octadiene, 98%

C_8H_{14}
FW 110.20
bp 118°C
d 0.746
n_D^{20}1.4220

Fp 49°F

60 MHz: *1*, 34A
FT-IR: *1*, 29D
VP-FT-IR: *3*, 42D

138.89*
114.25
33.65
28.43

C

CDCl$_3$ QE-300

$H_2C=CH(CH_2)_4CH=CH_2$

ALDRICH

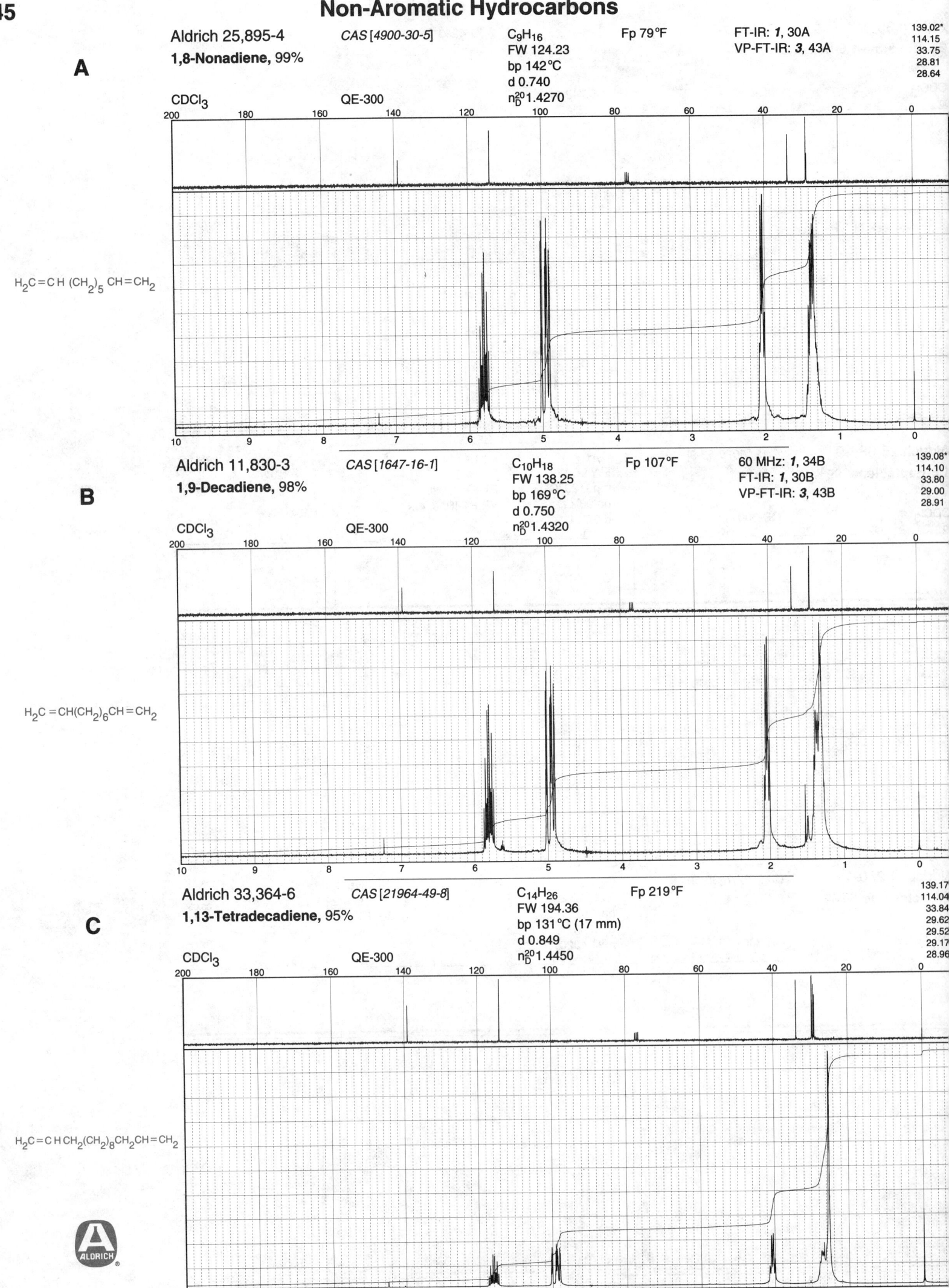

A

Aldrich 25,895-4
1,8-Nonadiene, 99%

CAS [4900-30-5]

C_9H_{16}
FW 124.23
bp 142°C
d 0.740
n_D^{20} 1.4270

Fp 79°F

FT-IR: **1**, 30A
VP-FT-IR: **3**, 43A

139.02*
114.15
33.75
28.81
28.64

CDCl₃

QE-300

$H_2C=CH(CH_2)_5CH=CH_2$

B

Aldrich 11,830-3
1,9-Decadiene, 98%

CAS [1647-16-1]

$C_{10}H_{18}$
FW 138.25
bp 169°C
d 0.750
n_D^{20} 1.4320

Fp 107°F

60 MHz: **1**, 34B
FT-IR: **1**, 30B
VP-FT-IR: **3**, 43B

139.08*
114.10
33.80
29.00
28.91

CDCl₃

QE-300

$H_2C=CH(CH_2)_6CH=CH_2$

C

Aldrich 33,364-6
1,13-Tetradecadiene, 95%

CAS [21964-49-8]

$C_{14}H_{26}$
FW 194.36
bp 131°C (17 mm)
d 0.849
n_D^{20} 1.4450

Fp 219°F

139.17
114.04
33.84
29.62
29.52
29.17
28.96

CDCl₃

QE-300

$H_2C=CHCH_2(CH_2)_8CH_2CH=CH_2$

ALDRICH

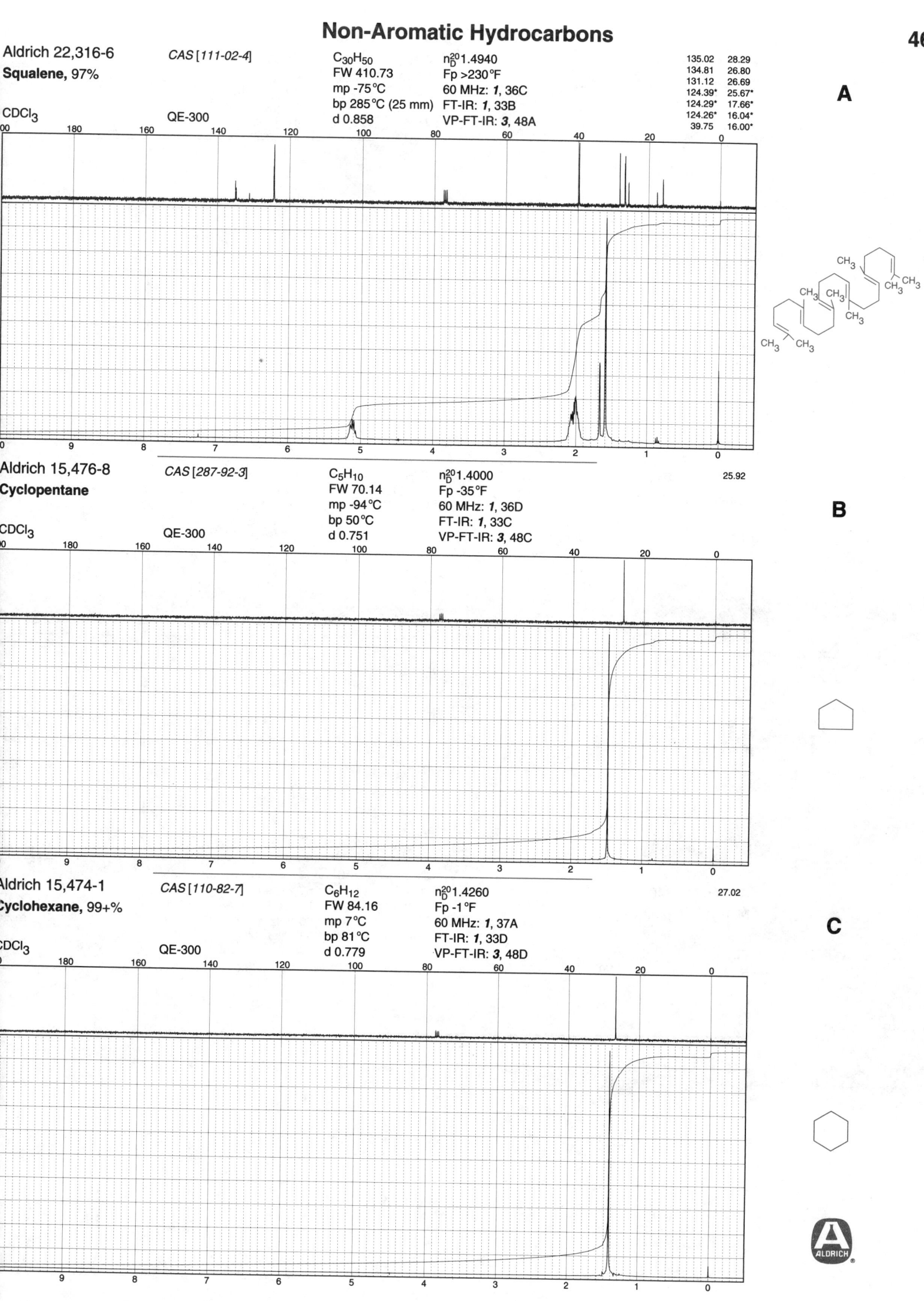

Aldrich 22,316-6
Squalene, 97%

CAS [111-02-4]

$C_{30}H_{50}$
FW 410.73
mp -75°C
bp 285°C (25 mm)
d 0.858

n_D^{20} 1.4940
Fp >230°F
60 MHz: *1*, 36C
FT-IR: *1*, 33B
VP-FT-IR: *3*, 48A

CDCl₃

QE-300

A

135.02	28.29
134.81	26.80
131.12	26.69
124.39*	25.67*
124.29*	17.66*
124.26*	16.04*
39.75	16.00*

25.92

Aldrich 15,476-8
Cyclopentane

CAS [287-92-3]

C_5H_{10}
FW 70.14
mp -94°C
bp 50°C
d 0.751

n_D^{20} 1.4000
Fp -35°F
60 MHz: *1*, 36D
FT-IR: *1*, 33C
VP-FT-IR: *3*, 48C

CDCl₃

QE-300

B

27.02

Aldrich 15,474-1
Cyclohexane, 99+%

CAS [110-82-7]

C_6H_{12}
FW 84.16
mp 7°C
bp 81°C
d 0.779

n_D^{20} 1.4260
Fp -1°F
60 MHz: *1*, 37A
FT-IR: *1*, 33D
VP-FT-IR: *3*, 48D

CDCl₃

QE-300

C

ALDRICH

Aldrich C9,840-3   *CAS [291-64-5]* C_7H_{14} n_D^{20} 1.4450

Cycloheptane, 98%

FW 98.19   Fp 43°F

mp -12°C   60 MHz: *1*, 37B

bp 119°C   FT-IR: *1*, 34A

CDCl₃   QE-300   d 0.811   VP-FT-IR: *3*, 49A

A

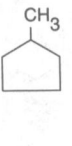

200 180 160 140 120 100 80 60 40 20 0

10 9 8 7 6 5 4 3 2 1 0

27.50

Aldrich C10,940-1   *CAS [292-64-8]* C_8H_{16} n_D^{20} 1.4580

Cyclooctane, 99+%

FW 112.22   Fp 86°F

mp 12°C   60 MHz: *1*, 37C

bp 151°C (740 mm)   FT-IR: *1*, 34B

CDCl₃   QE-300   d 0.834   VP-FT-IR: *3*, 49B

B

200 180 160 140 120 100 80 60 40 20 0

10 9 8 7 6 5 4 3 2 1 0

34.74
34.61*
25.35
20.72*

Aldrich M3,940-7   *CAS [96-37-7]* C_6H_{12} n_D^{20} 1.4090

Methylcyclopentane, 98%

FW 84.16   Fp -11°F

mp -142°C   60 MHz: *1*, 38B

bp 72°C   FT-IR: *1*, 34D

CDCl₃   QE-300   d 0.749   VP-FT-IR: *3*, 50A

C

CH₃

200 180 160 140 120 100 80 60 40 20 0

10 9 8 7 6 5 4 3 2 1 0

Non-Aromatic Hydrocarbons

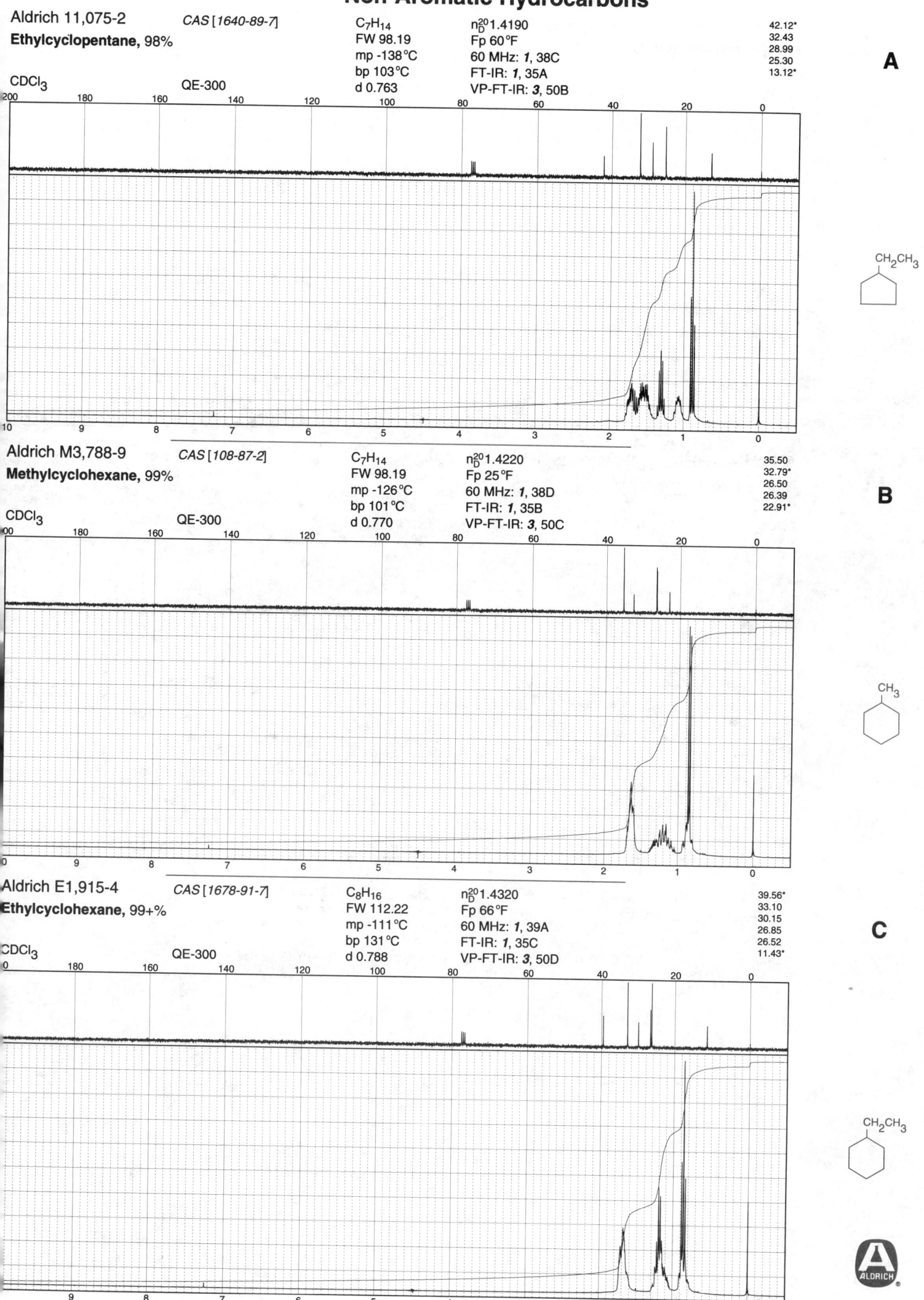

Aldrich 11,075-2   CAS [1640-89-7]
Ethylcyclopentane, 98%

C_7H_{14}
FW 98.19
mp -138°C
bp 103°C
d 0.763

n_D^{20} 1.4190
Fp 60°F
60 MHz: **1**, 38C
FT-IR: **1**, 35A
VP-FT-IR: **3**, 50B

42.12*
32.43
28.99
25.30
13.12*

A

CDCl₃   QE-300

Aldrich M3,788-9   CAS [108-87-2]
Methylcyclohexane, 99%

C_7H_{14}
FW 98.19
mp -126°C
bp 101°C
d 0.770

n_D^{20} 1.4220
Fp 25°F
60 MHz: **1**, 38D
FT-IR: **1**, 35B
VP-FT-IR: **3**, 50C

35.50
32.79*
26.50
26.39
22.91*

B

CDCl₃   QE-300

Aldrich E1,915-4   CAS [1678-91-7]
Ethylcyclohexane, 99+%

C_8H_{16}
FW 112.22
mp -111°C
bp 131°C
d 0.788

n_D^{20} 1.4320
Fp 66°F
60 MHz: **1**, 39A
FT-IR: **1**, 35C
VP-FT-IR: **3**, 50D

39.56*
33.10
30.15
26.85
26.52
11.43*

C

CDCl₃   QE-300

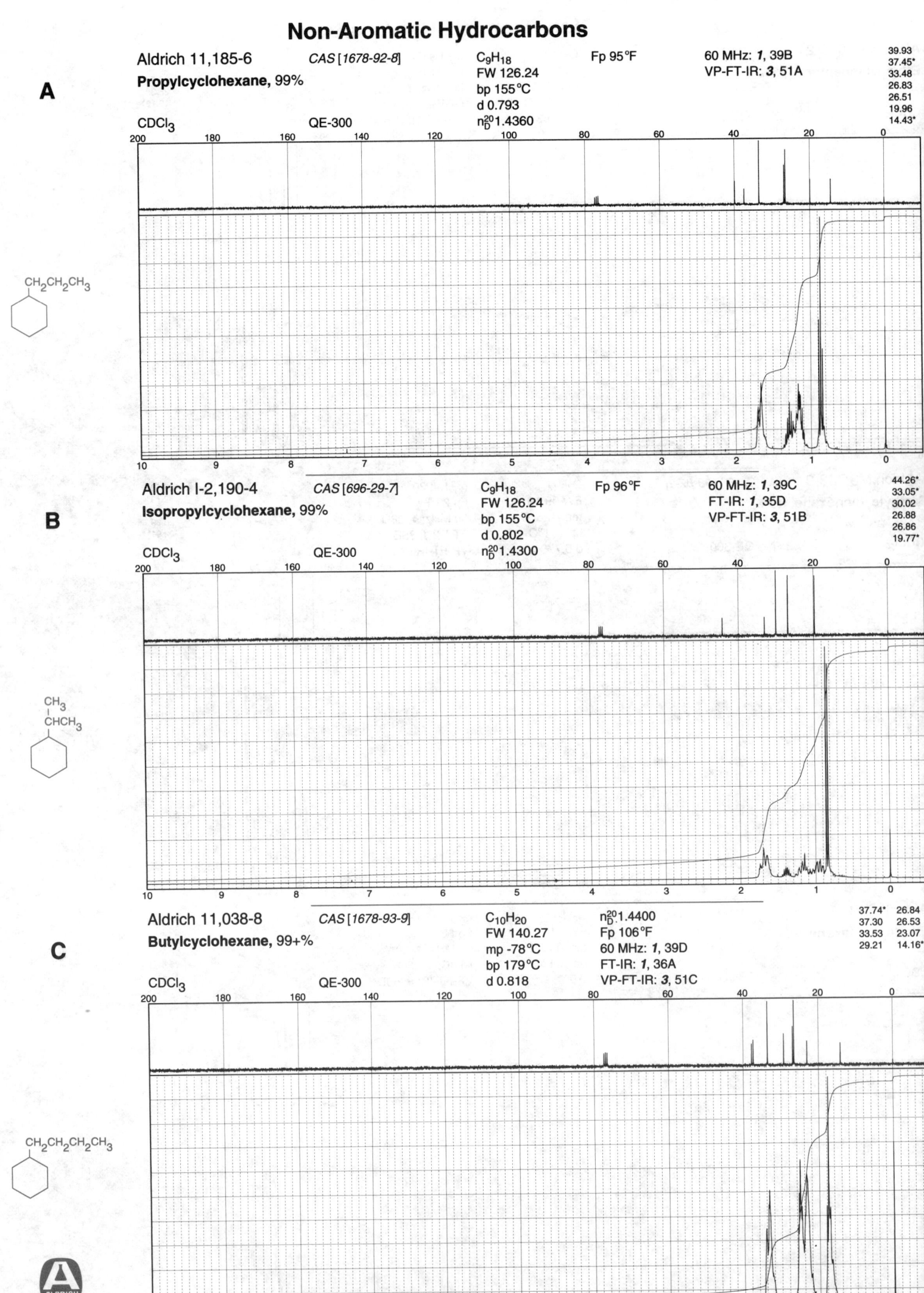

A

Aldrich 11,185-6   *CAS [1678-92-8]* C_9H_{18} Fp 95°F   60 MHz: *1*, 39B
Propylcyclohexane, 99%   FW 126.24   VP-FT-IR: *3*, 51A
                                             bp 155°C
                                             d 0.793
CDCl$_3$ QE-300   n$_D^{20}$1.4360

39.93
37.45*
33.48
26.83
26.51
19.96
14.43*

B

Aldrich I-2,190-4   *CAS [696-29-7]* C_9H_{18} Fp 96°F   60 MHz: *1*, 39C
Isopropylcyclohexane, 99%   FW 126.24   FT-IR: *1*, 35D
                                             bp 155°C   VP-FT-IR: *3*, 51B
                                             d 0.802
CDCl$_3$ QE-300   n$_D^{20}$1.4300

44.26*
33.05*
30.02
26.88
26.86
19.77*

C

Aldrich 11,038-8   *CAS [1678-93-9]* $C_{10}H_{20}$ n$_D^{20}$1.4400   60 MHz: *1*, 39D
Butylcyclohexane, 99+%   FW 140.27   Fp 106°F   FT-IR: *1*, 36A
                                             mp -78°C   VP-FT-IR: *3*, 51C
                                             bp 179°C
CDCl$_3$ QE-300   d 0.818

37.74*   26.84
37.30   26.53
33.53   23.07
29.21   14.16*

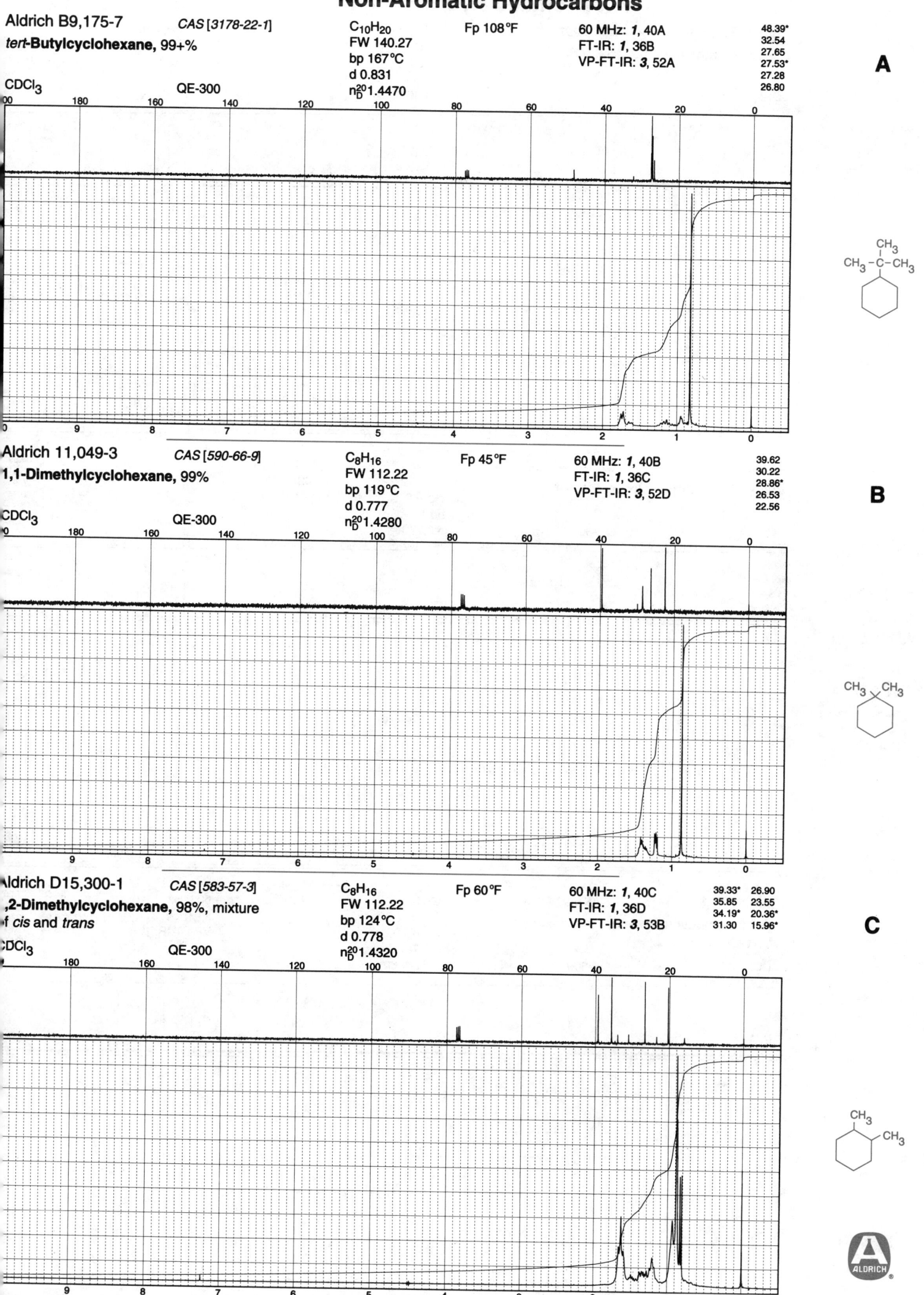

Aldrich B9,175-7 CAS [3178-22-1] C₁₀H₂₀ Fp 108°F 60 MHz: *1*, 40A 48.39*
tert-**Butylcyclohexane**, 99+% FW 140.27   FT-IR: *1*, 36B 32.54
  bp 167°C   VP-FT-IR: *3*, 52A 27.65
CDCl₃ QE-300 d 0.831   27.53*
    n²⁰_D 1.4470   27.28
          26.80

A

Aldrich 11,049-3 CAS [590-66-9] C₈H₁₆ Fp 45°F 60 MHz: *1*, 40B 39.62
1,1-Dimethylcyclohexane, 99% FW 112.22   FT-IR: *1*, 36C 30.22
  bp 119°C   VP-FT-IR: *3*, 52D 28.86*
CDCl₃ QE-300 d 0.777   26.53
    n²⁰_D 1.4280   22.56

B

Aldrich D15,300-1 CAS [583-57-3] C₈H₁₆ Fp 60°F 60 MHz: *1*, 40C 39.33* 26.90
,2-Dimethylcyclohexane, 98%, mixture FW 112.22   FT-IR: *1*, 36D 35.85 23.55
f *cis* and *trans* bp 124°C   VP-FT-IR: *3*, 53B 34.19* 20.36*
CDCl₃ QE-300 d 0.778   31.30 15.96*
    n²⁰_D 1.4320

C

A

Aldrich 29,063-7   *CAS [2207-01-4]*

cis-1,2-Dimethylcyclohexane, 99%

C8H16
FW 112.22
bp 130°C
d 0.796
nD20 1.4360

Fp 54°F

VP-FT-IR: **3**, 53C

34.20*
31.30
23.56
15.97*

CDCl3   QE-300

B

Aldrich 29,064-5   *CAS [6876-23-9]*

trans-1,2-Dimethylcyclohexane, 99%

C8H16
FW 112.22
mp -89°C
bp 124°C
d 0.770

nD20 1.4270
Fp 43°F
VP-FT-IR: **3**, 53D

39.39*
35.91
26.94
20.34*

CDCl3   QE-300

C

Aldrich 11,838-9   *CAS [591-21-9]*

1,3-Dimethylcyclohexane, 99%, mixture
of *cis* and *trans*

C8H16
FW 112.22
bp 123°C
d 0.767
nD20 1.4260

Fp 49°F

60 MHz: **1**, 40D
FT-IR: **1**, 37A
VP-FT-IR: **3**, 54A

44.55   27.20*
41.20   26.48
35.08   22.93
33.74   20.81
32.83*   20.68

CDCl3   QE-300

Non-Aromatic Hydrocarbons

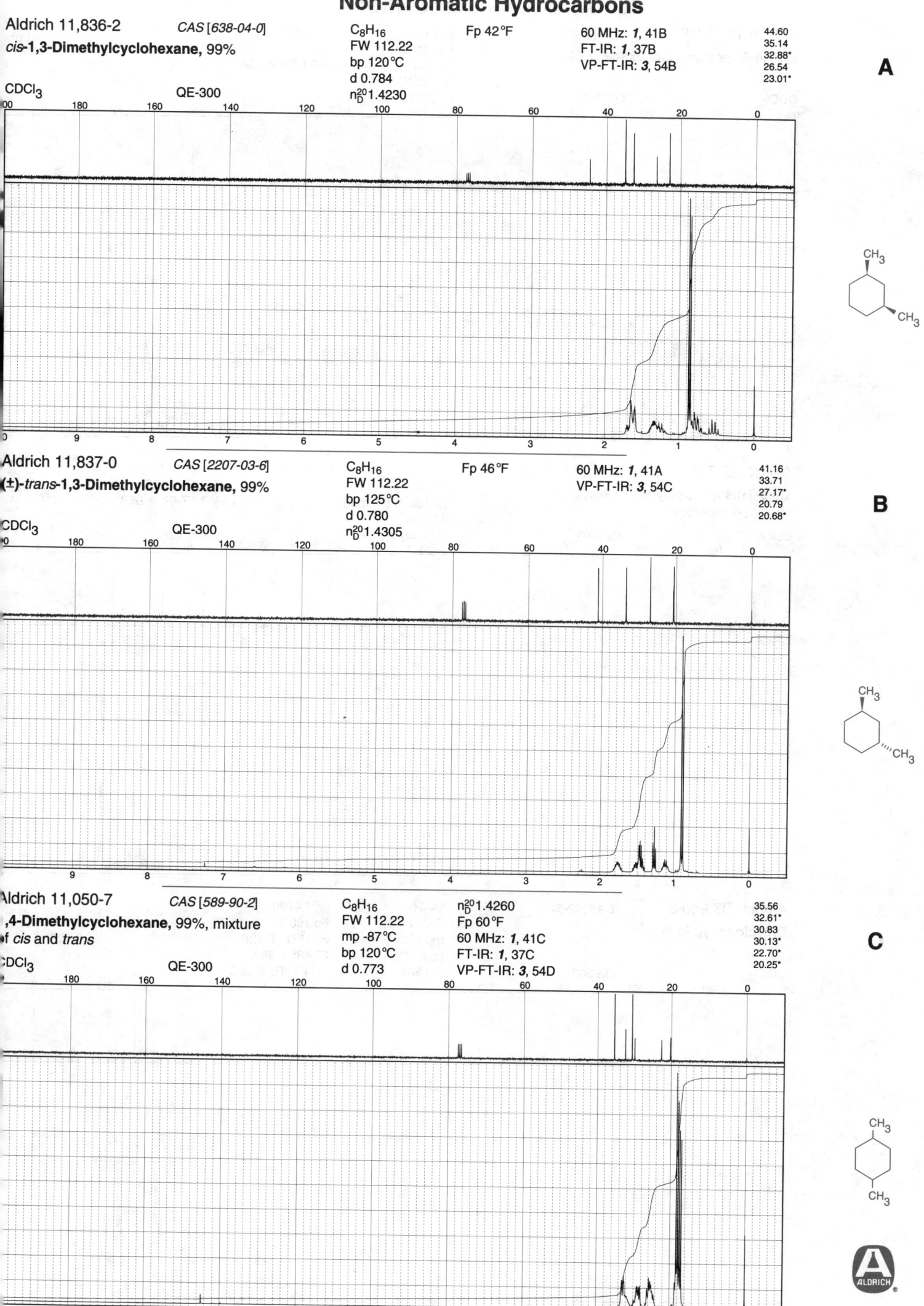

Aldrich 11,836-2   CAS [638-04-0]
cis-1,3-Dimethylcyclohexane, 99%

C_8H_{16}
FW 112.22
bp 120°C
d 0.784
n_D^{20}1.4230

Fp 42°F

60 MHz: 1, 41B
FT-IR: 1, 37B
VP-FT-IR: 3, 54B

44.60
35.14
32.88*
26.54
23.01*

CDCl₃   QE-300

A

Aldrich 11,837-0   CAS [2207-03-6]
(±)-trans-1,3-Dimethylcyclohexane, 99%

C_8H_{16}
FW 112.22
bp 125°C
d 0.780
n_D^{20}1.4305

Fp 46°F

60 MHz: 1, 41A
VP-FT-IR: 3, 54C

41.16
33.71
27.17*
20.79
20.68*

CDCl₃   QE-300

B

Aldrich 11,050-7   CAS [589-90-2]
,4-Dimethylcyclohexane, 99%, mixture
of cis and trans

C_8H_{16}
FW 112.22
mp -87°C
bp 120°C
d 0.773

n_D^{20}1.4260
Fp 60°F
60 MHz: 1, 41C
FT-IR: 1, 37C
VP-FT-IR: 3, 54D

35.56
32.61*
30.83
30.13*
22.70*
20.25*

CDCl₃   QE-300

C

A

Aldrich 11,051-5   *CAS [624-29-3]* C_8H_{16} n_D^{20} 1.4300   31.35
                                                                                      30.67*
cis-**1,4-Dimethylcyclohexane, 99%**   FW 112.22   Fp 43°F   20.85*
                                           mp -88°C   VP-FT-IR: **3**, 55A
                                           bp 125°C
CDCl₃   QE-300   d 0.783

B

Aldrich 22,794-3   *CAS [34387-60-5]* $C_{15}H_{30}$ Fp 201°F   60 MHz: **1**, 42A   44.02*
                                                                                                     33.23*
1,3,5-Triisopropylcyclohexane, 98%,   FW 210.41   FT-IR: **1**, 37D   33.03
mixture of isomers   bp 122°C (10 mm)   VP-FT-IR: **3**, 55B   19.84*
                                             d 0.810
CDCl₃   QE-300 n_D^{20} 1.4570

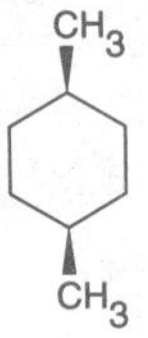

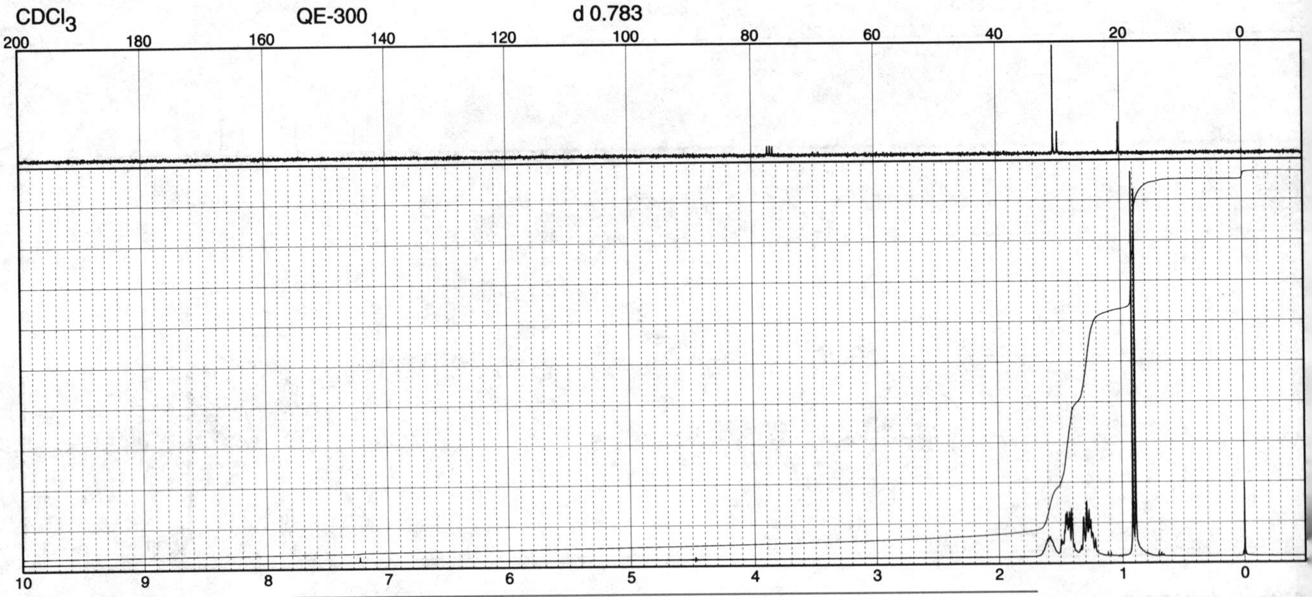

C

Aldrich D7,940-3   *CAS [92-51-3]* $C_{12}H_{22}$ n_D^{20} 1.4790   43.52*
                                                                                      30.24
Dicyclohexyl, 99%   FW 166.31   Fp 198°F   26.95
                          mp 4°C   60 MHz: **1**, 42B
                          bp 227°C   FT-IR: **1**, 38A
CDCl₃   QE-300   d 0.864   VP-FT-IR: **3**, 55C

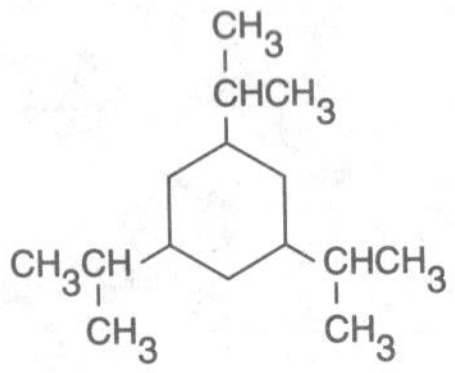

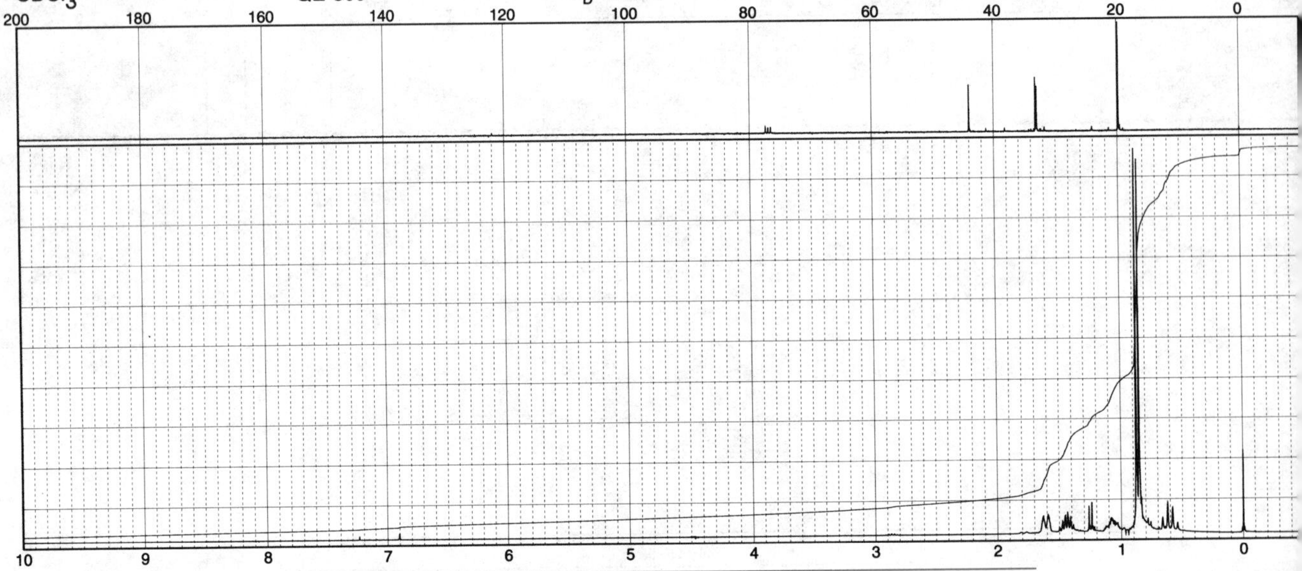

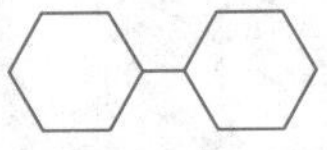

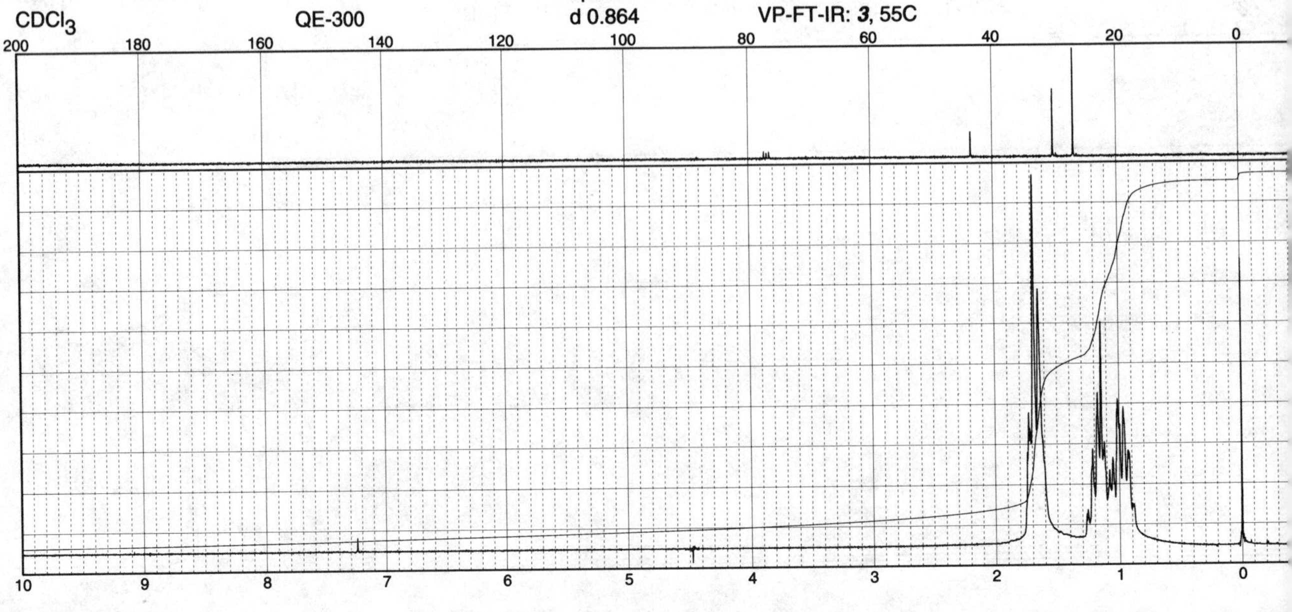

Non-Aromatic Hydrocarbons

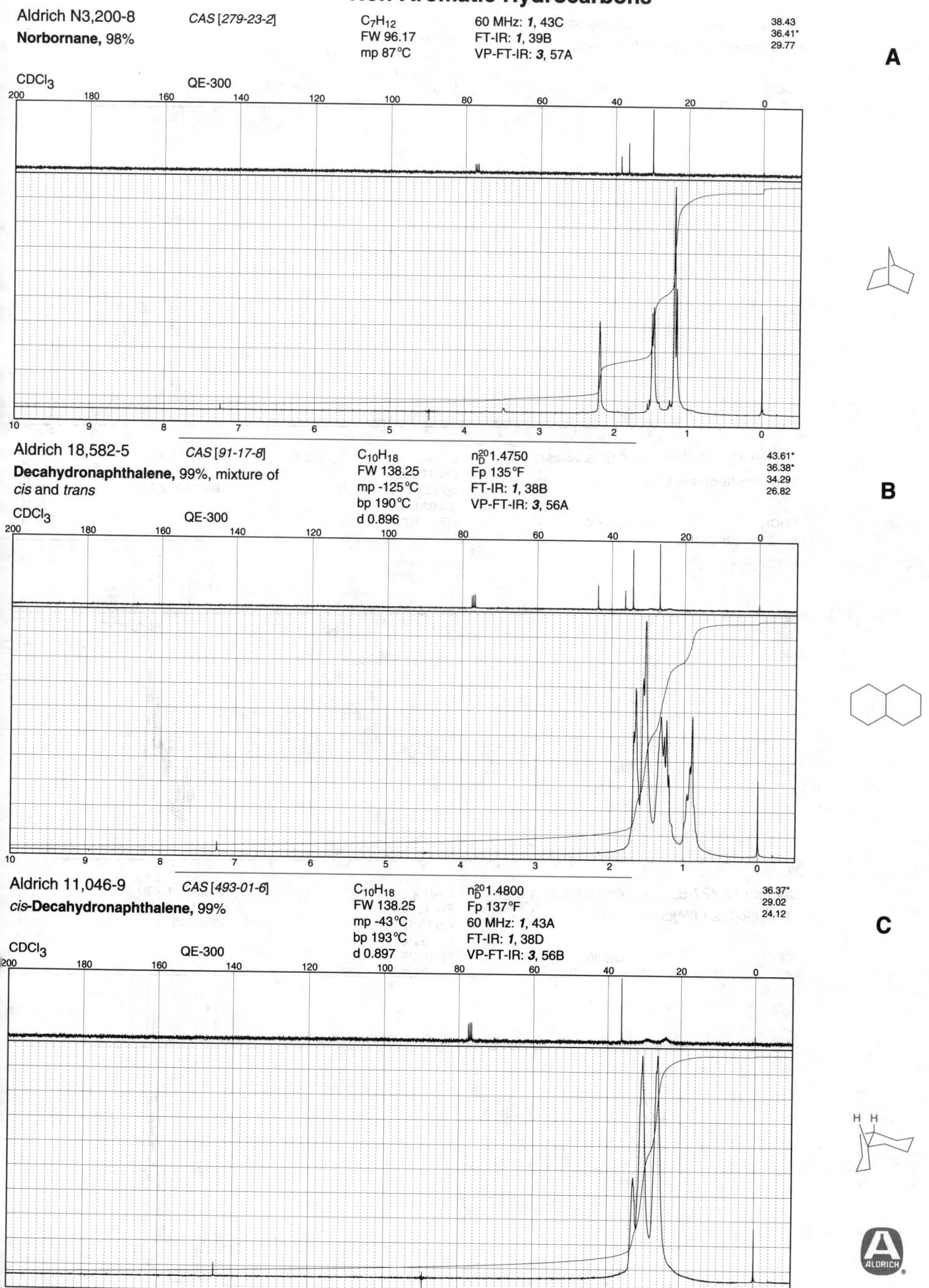

Aldrich N3,200-8   CAS [279-23-2]

Norbornane, 98%

C₇H₁₂
FW 96.17
mp 87°C

60 MHz: *1*, 43C
FT-IR: *1*, 39B
VP-FT-IR: *3*, 57A

38.43
36.41*
29.77

A

CDCl₃   QE-300

Aldrich 18,582-5   CAS [91-17-8]

Decahydronaphthalene, 99%, mixture of
cis and *trans*

C₁₀H₁₈
FW 138.25
mp -125°C
bp 190°C
d 0.896

n_D^{20} 1.4750
Fp 135°F
FT-IR: *1*, 38B
VP-FT-IR: *3*, 56A

43.61*
36.38*
34.29
26.82

B

CDCl₃   QE-300

Aldrich 11,046-9   CAS [493-01-6]

***cis*-Decahydronaphthalene, 99%**

C₁₀H₁₈
FW 138.25
mp -43°C
bp 193°C
d 0.897

n_D^{20} 1.4800
Fp 137°F
60 MHz: *1*, 43A
FT-IR: *1*, 38D
VP-FT-IR: *3*, 56B

36.37*
29.02
24.12

C

CDCl₃   QE-300

A

Aldrich 11,047-7   *CAS [493-02-7]*

trans-**Decahydronaphthalene, 99%**

$C_{10}H_{18}$
FW 138.25
mp -32°C
bp 185°C (756 mm)
d 0.870

n_D^{20} 1.4690
Fp 127°F
60 MHz: *1*, 42D
FT-IR: *1*, 38C
VP-FT-IR: *3*, 56C

43.58*
34.26
26.79

CDCl₃   QE-300
200 180 160 140 120 100 80 60 40 20 0

10 9 8 7 6 5 4 3 2 1 0

B

Aldrich 15,051-7   *CAS [5744-03-6]*

Perhydrofluorene, 97%

$C_{13}H_{22}$
FW 178.32
bp 253°C
d 0.920
n_D^{20} 1.5010

Fp 148°F

60 MHz: *1*, 43B
FT-IR: *1*, 39A
VP-FT-IR: *3*, 56D

43.04*
37.34*
35.37
29.21
24.39
24.29
22.14

CDCl₃   QE-300
200 180 160 140 120 100 80 60 40 20 0

10 9 8 7 6 5 4 3 2 1 0

C

Aldrich 16,427-5   *CAS [2825-83-4]*

Tricyclo[5.2.1.0²,⁶]decane, 98%

$C_{10}H_{16}$
FW 136.24
mp 78°C
bp 193°C
Fp 105°F

60 MHz: *1*, 43D
FT-IR: *1*, 39D
VP-FT-IR: *3*, 57C

45.53*
43.31
41.59*
28.78
26.98
23.09

CDCl₃   QE-300
200 180 160 140 120 100 80 60 40 20 0

10 9 8 7 6 5 4 3 2 1 0

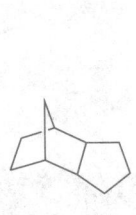

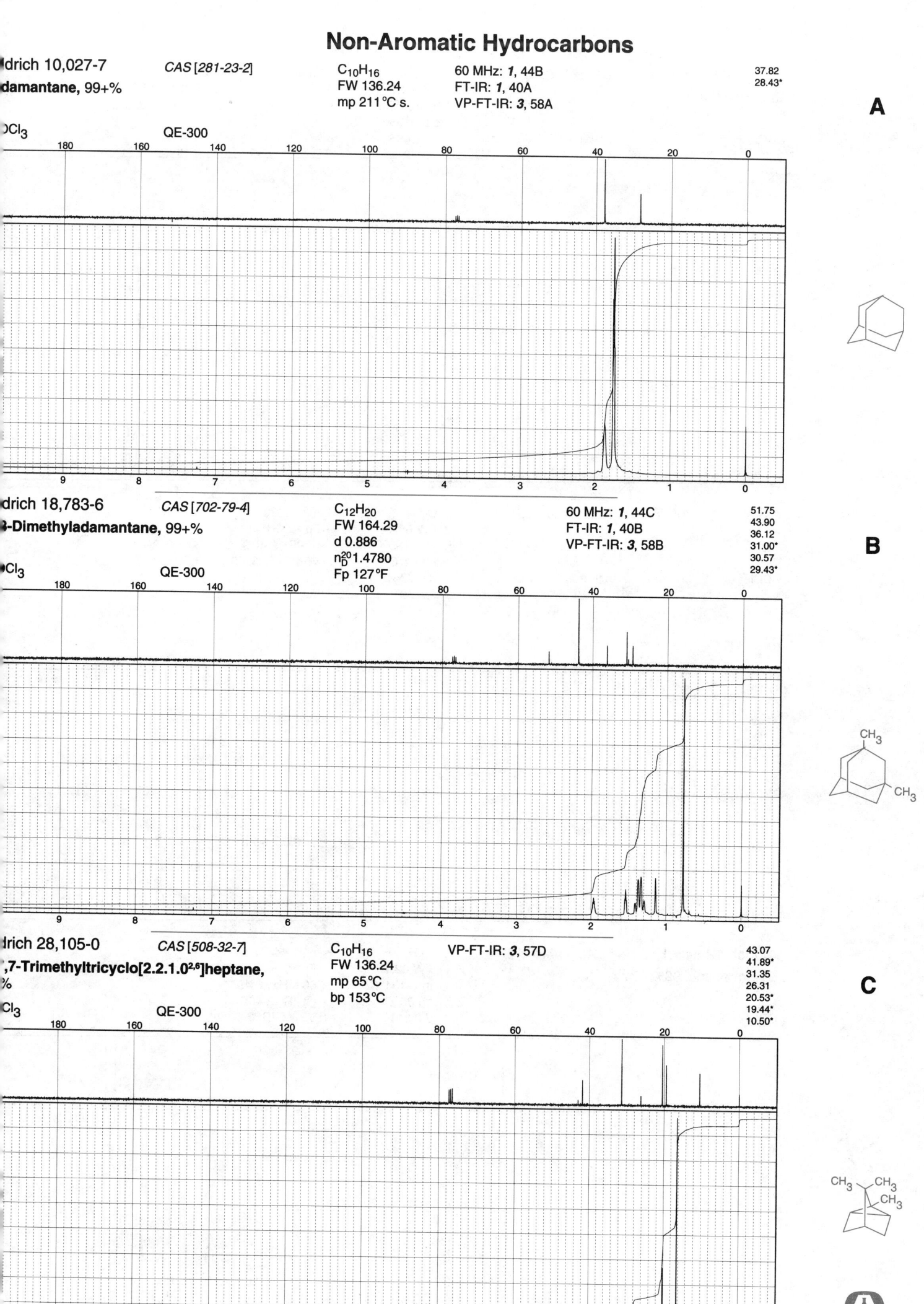

Idrich 10,027-7 *CAS [281-23-2]* **damantane, 99+%**

C₁₀H₁₆
FW 136.24
mp 211°C s.

60 MHz: *1*, 44B
FT-IR: *1*, 40A
VP-FT-IR: *3*, 58A

37.82
28.43*

A

OCl₃ QE-300

Idrich 18,783-6 *CAS [702-79-4]* **,3-Dimethyladamantane, 99+%**

C₁₂H₂₀
FW 164.29
d 0.886
n$_D^{20}$1.4780
Fp 127°F

60 MHz: *1*, 44C
FT-IR: *1*, 40B
VP-FT-IR: *3*, 58B

51.75
43.90
36.12
31.00*
30.57
29.43*

B

Cl₃ QE-300

Idrich 28,105-0 *CAS [508-32-7]* **,7-Trimethyltricyclo[2.2.1.0²,⁶]heptane, %**

C₁₀H₁₆
FW 136.24
mp 65°C
bp 153°C

VP-FT-IR: *3*, 57D

43.07
41.89*
31.35
26.31
20.53*
19.44*
10.50*

C

Cl₃ QE-300

A

Aldrich 20,108-1 *CAS [278-06-8]* C$_7$H$_8$ Fp 52°F 60 MHz: *1*, 44A 31.9
Quadricyclane, 99% FW 92.14 FT-IR: *1*, 39C 22.9
bp 108°C (740 mm) VP-FT-IR: *3*, 57B 14.6
d 0.919

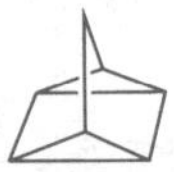

CDCl$_3$ QE-300 n$_D^{20}$1.4850

200 180 160 140 120 100 80 60 40 20 0

10   9   8   7   6   5   4   3   2   1   0

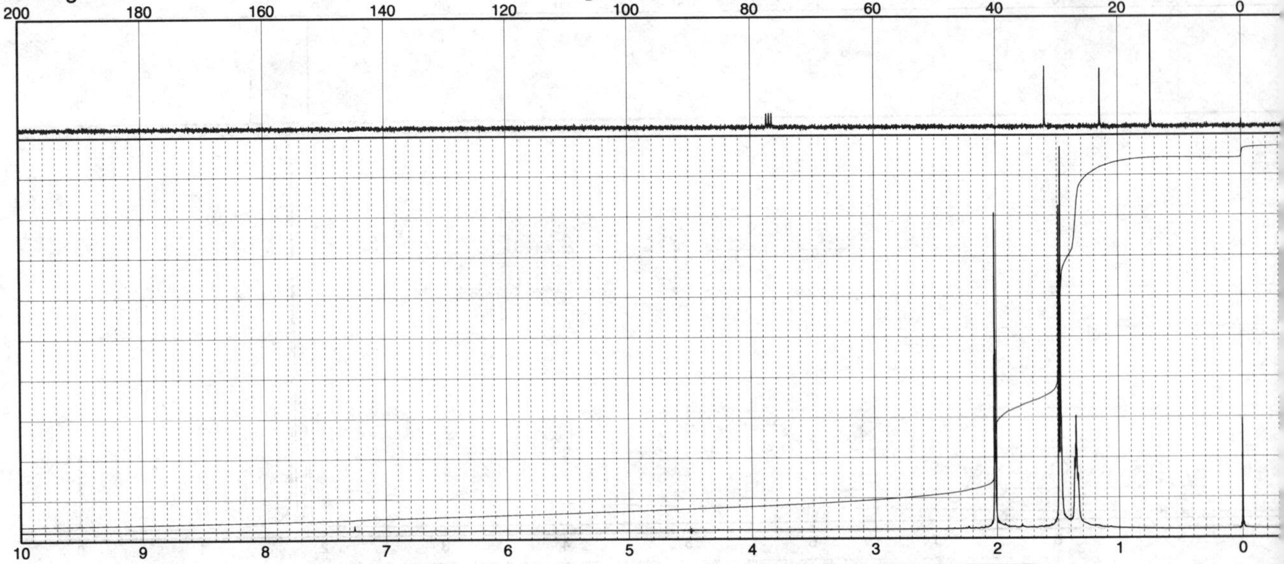

B

Aldrich C11,260-7 *CAS [142-29-0]* C$_5$H$_8$ n$_D^{20}$1.4230 130.6
Cyclopentene, 99% FW 68.12 Fp <-30°F 32.4
mp -135°C FT-IR: *1*, 40C 22.8
bp 44°C VP-FT-IR: *3*, 58C
d 0.774

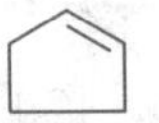

CDCl$_3$ QE-300

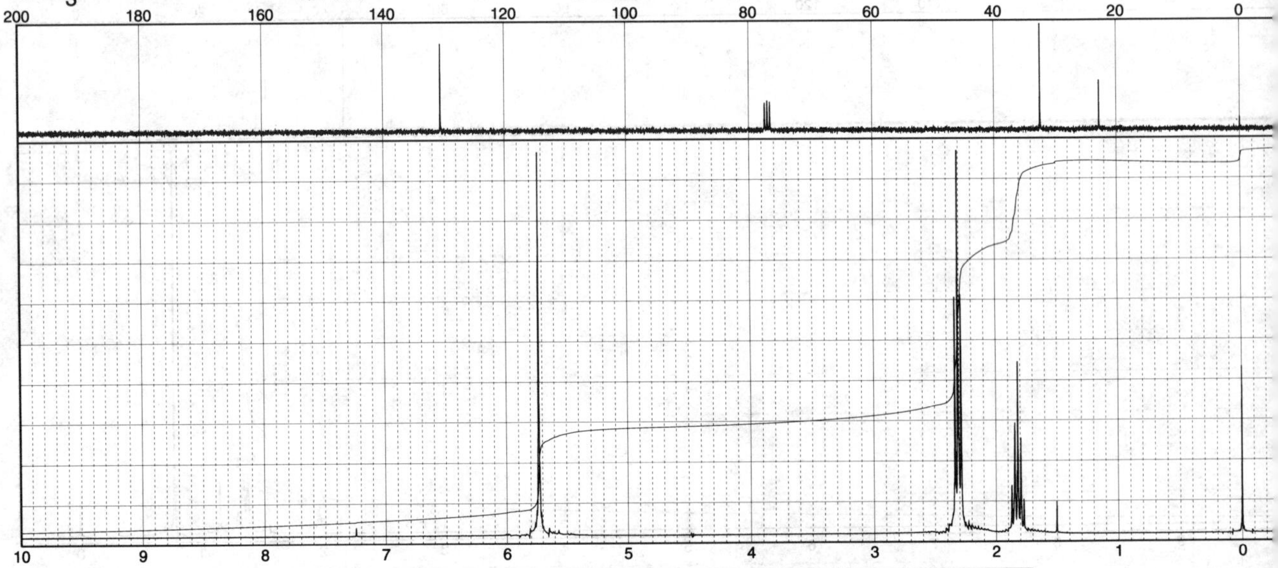

200 180 160 140 120 100 80 60 40 20 0

10   9   8   7   6   5   4   3   2   1   0

C

Aldrich 12,543-1 *CAS [110-83-8]* C$_6$H$_{10}$ n$_D^{20}$1.4460 127.
Cyclohexene, 99% FW 82.15 Fp 10°F 25
mp -104°C 60 MHz: *1*, 45A 22.
bp 83°C FT-IR: *1*, 40D
d 0.811 VP-FT-IR: *3*, 58D

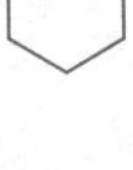

CDCl$_3$ QE-300

200 180 160 140 120 100 80 60 40 20 0

10   9   8   7   6   5   4   3   2   1   0

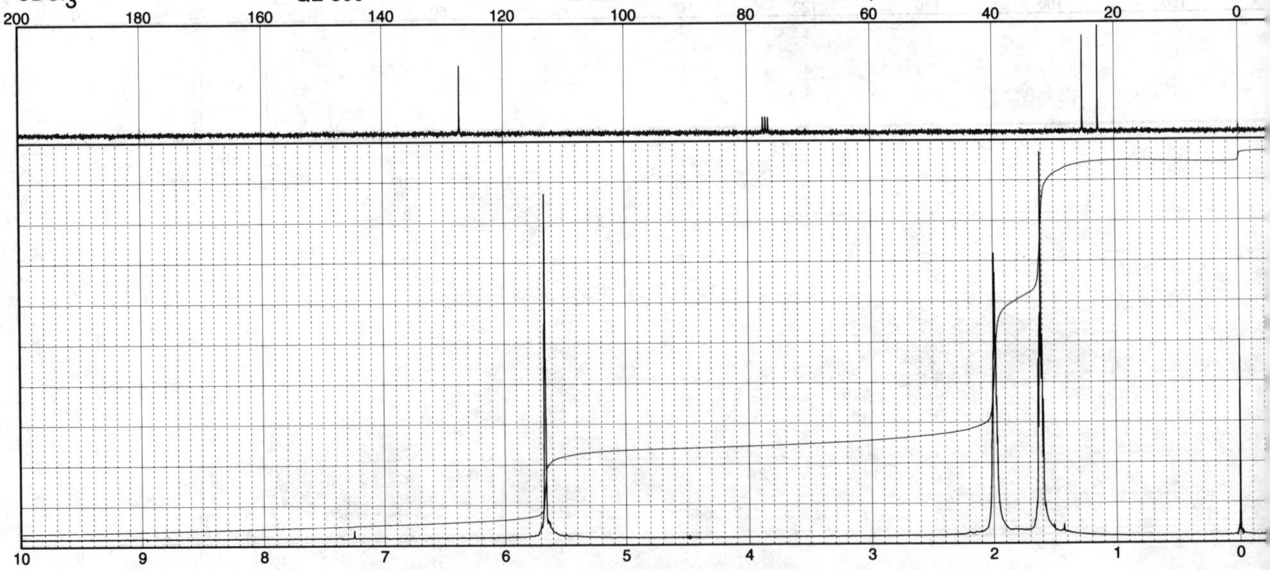

Non-Aromatic Hydrocarbons

58

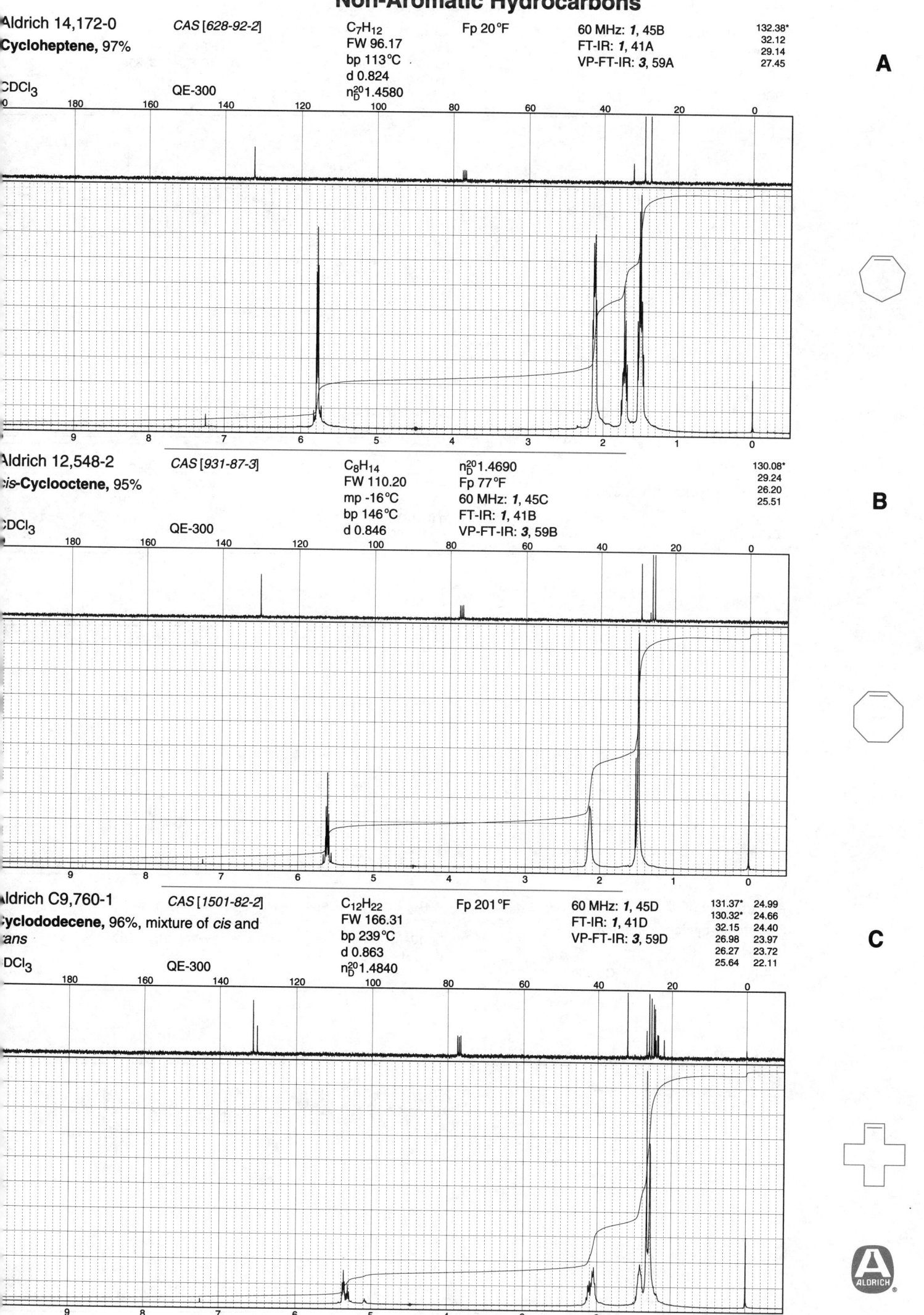

Aldrich 14,172-0 CAS [628-92-2] C7H12 Fp 20°F 60 MHz: 1, 45B 132.38*
Cycloheptene, 97% FW 96.17 FT-IR: 1, 41A 32.12
bp 113°C VP-FT-IR: 3, 59A 29.14
d 0.824 27.45
CDCl3 QE-300 n20D 1.4580

A

Aldrich 12,548-2 CAS [931-87-3] C8H14 n20D 1.4690 130.08*
cis-Cyclooctene, 95% FW 110.20 Fp 77°F 29.24
mp -16°C 60 MHz: 1, 45C 26.20
bp 146°C FT-IR: 1, 41B 25.51
CDCl3 QE-300 d 0.846 VP-FT-IR: 3, 59B

B

Aldrich C9,760-1 CAS [1501-82-2] C12H22 Fp 201°F 60 MHz: 1, 45D 131.37* 24.99
Cyclododecene, 96%, mixture of cis and FW 166.31 FT-IR: 1, 41D 130.32* 24.66
trans bp 239°C VP-FT-IR: 3, 59D 32.15 24.40
d 0.863 26.98 23.97
CDCl3 QE-300 n20D 1.4840 26.27 23.72
25.64 22.11

C

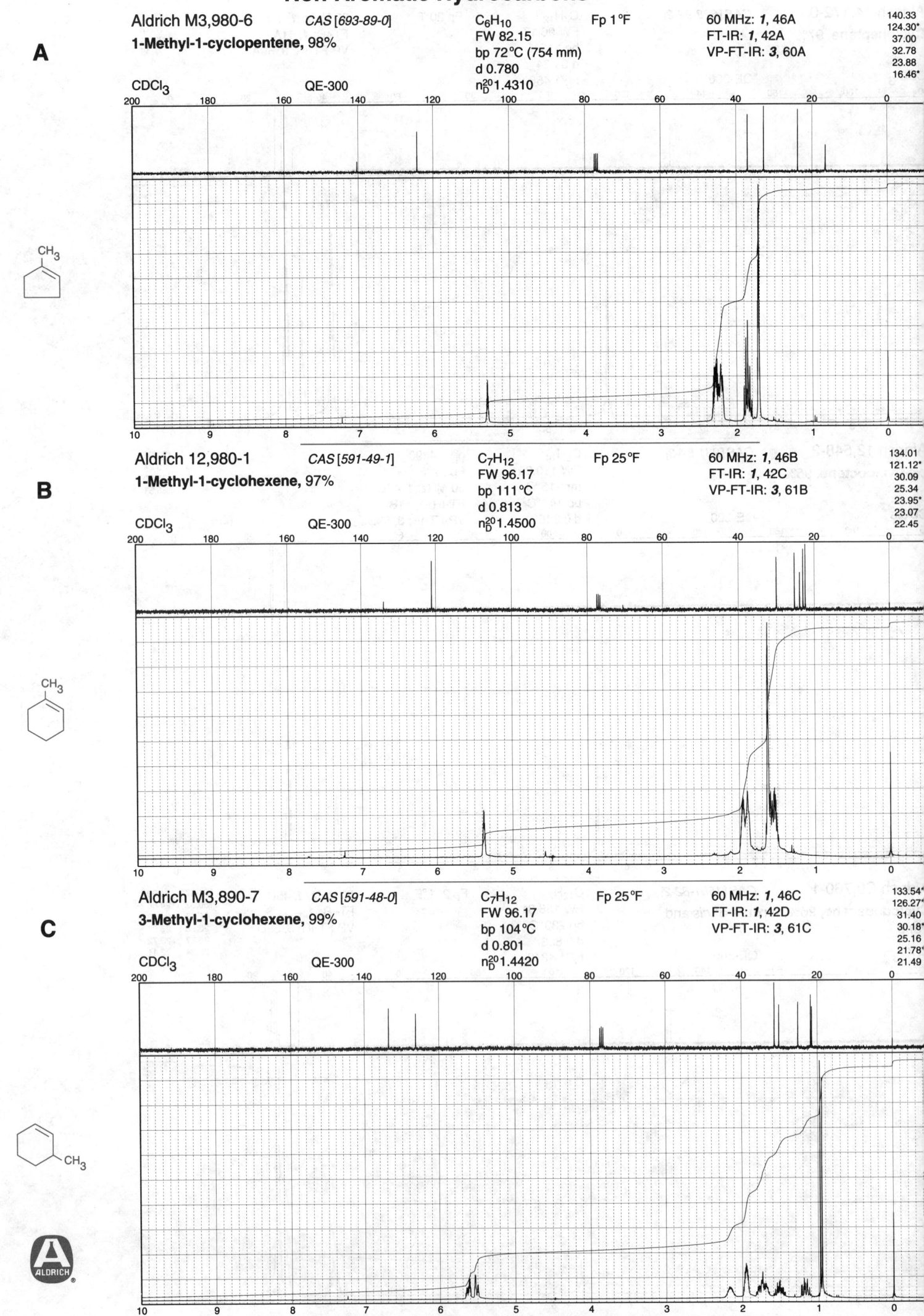

A

Aldrich M3,980-6   *CAS [693-89-0]* C_6H_{10} Fp 1 °F   60 MHz: **1**, 46A

1-Methyl-1-cyclopentene, 98%   FW 82.15   FT-IR: **1**, 42A
bp 72 °C (754 mm)   VP-FT-IR: **3**, 60A
d 0.780
$CDCl_3$ QE-300 n_D^{20} 1.4310

140.33
124.30*
37.00
32.78
23.88
16.46*

B

Aldrich 12,980-1   *CAS [591-49-1]* C_7H_{12} Fp 25 °F   60 MHz: **1**, 46B

1-Methyl-1-cyclohexene, 97%   FW 96.17   FT-IR: **1**, 42C
bp 111 °C   VP-FT-IR: **3**, 61B
d 0.813
$CDCl_3$ QE-300 n_D^{20} 1.4500

134.01
121.12*
30.09
25.34
23.95*
23.07
22.45

C

Aldrich M3,890-7   *CAS [591-48-0]* C_7H_{12} Fp 25 °F   60 MHz: **1**, 46C

3-Methyl-1-cyclohexene, 99%   FW 96.17   FT-IR: **1**, 42D
bp 104 °C   VP-FT-IR: **3**, 61C
d 0.801
$CDCl_3$ QE-300 n_D^{20} 1.4420

133.54*
126.27*
31.40
30.18*
25.16
21.78*
21.49

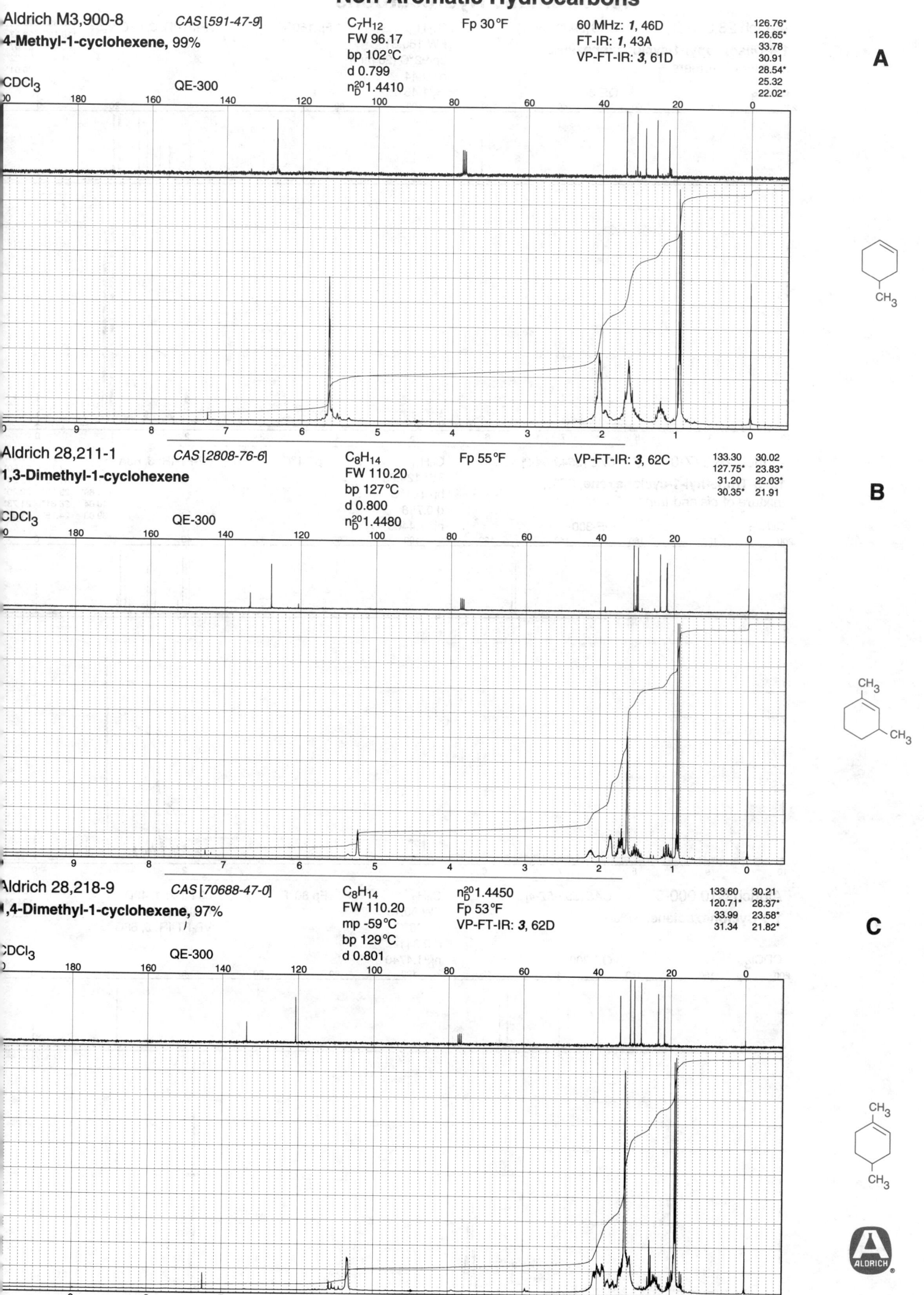

Aldrich M3,900-8 *CAS [591-47-9]*
4-Methyl-1-cyclohexene, 99%

CDCl₃ QE-300

C_7H_{12}
FW 96.17
bp 102°C
d 0.799
n_D^{20}1.4410

Fp 30°F

60 MHz: *1*, 46D
FT-IR: *1*, 43A
VP-FT-IR: *3*, 61D

126.76*
126.65*
33.78
30.91
28.54*
25.32
22.02*

A

Aldrich 28,211-1 *CAS [2808-76-6]*
1,3-Dimethyl-1-cyclohexene

CDCl₃ QE-300

C_8H_{14}
FW 110.20
bp 127°C
d 0.800
n_D^{20}1.4480

Fp 55°F

VP-FT-IR: *3*, 62C

133.30 30.02
127.75* 23.83*
31.20 22.03*
30.35* 21.91

B

Aldrich 28,218-9 *CAS [70688-47-0]*
1,4-Dimethyl-1-cyclohexene, 97%

CDCl₃ QE-300

C_8H_{14}
FW 110.20
mp -59°C
bp 129°C
d 0.801

n_D^{20}1.4450
Fp 53°F
VP-FT-IR: *3*, 62D

133.60 30.21
120.71* 28.37*
33.99 23.58*
31.34 21.82*

C

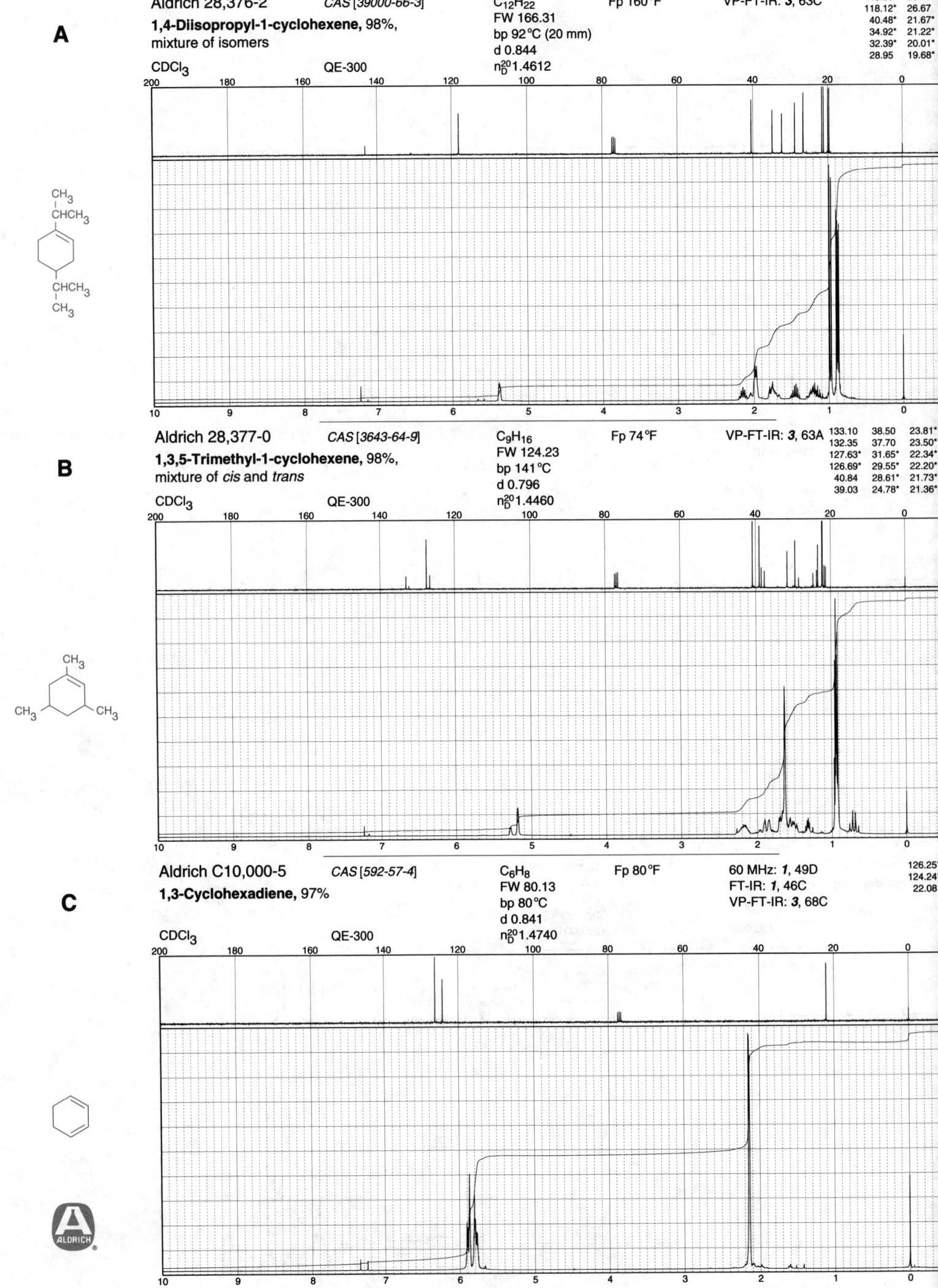

A

Aldrich 28,376-2   *CAS [39000-66-3]*

1,4-Diisopropyl-1-cyclohexene, 98%,
mixture of isomers

CDCl₃   QE-300

C₁₂H₂₂
FW 166.31
bp 92°C (20 mm)
d 0.844
n²⁰_D 1.4612

Fp 160°F

VP-FT-IR: **3**, 63C

143.36	26.73
118.12*	26.67
40.48*	21.67*
34.92*	21.22*
32.39*	20.01*
28.95	19.68*

B

Aldrich 28,377-0   *CAS [3643-64-9]*

1,3,5-Trimethyl-1-cyclohexene, 98%,
mixture of *cis* and *trans*

CDCl₃   QE-300

C₉H₁₆
FW 124.23
bp 141°C
d 0.796
n²⁰_D 1.4460

Fp 74°F

VP-FT-IR: **3**, 63A

133.10	38.50	23.81*
132.35	37.70	23.50*
127.63*	31.65*	22.34*
126.69*	29.55*	22.20*
40.84	28.61*	21.73*
39.03	24.78*	21.36*

C

Aldrich C10,000-5   *CAS [592-57-4]*

1,3-Cyclohexadiene, 97%

CDCl₃   QE-300

C₆H₈
FW 80.13
bp 80°C
d 0.841
n²⁰_D 1.4740

Fp 80°F

60 MHz: **1**, 49D
FT-IR: **1**, 46C
VP-FT-IR: **3**, 68C

126.25*
124.24*
22.08

Non-Aromatic Hydrocarbons

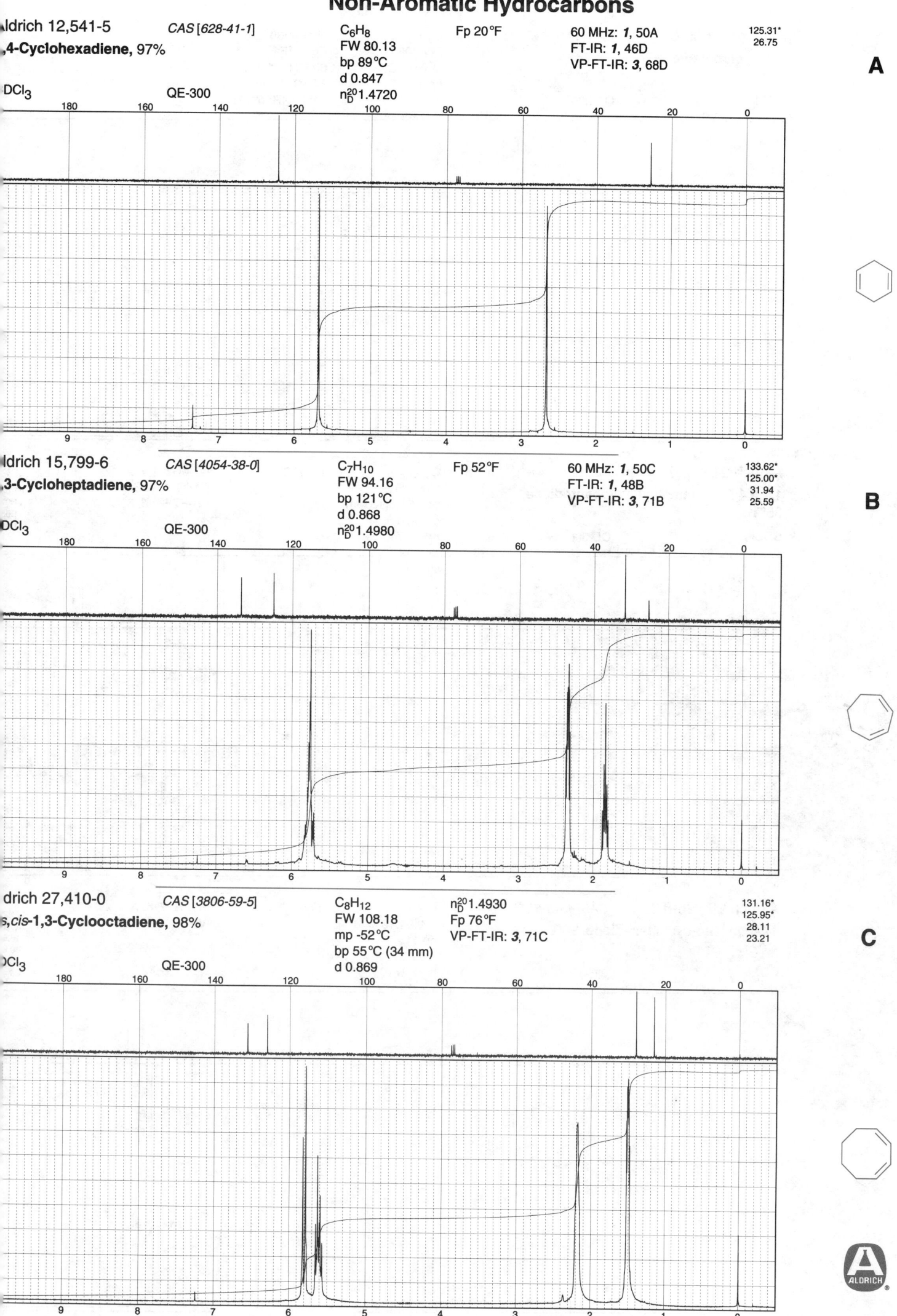

Aldrich 12,541-5   CAS [628-41-1]   C6H8   Fp 20°F   60 MHz: 1, 50A   125.31*
,4-Cyclohexadiene, 97%   FW 80.13   FT-IR: 1, 46D   26.75
bp 89°C   VP-FT-IR: 3, 68D
d 0.847
DCl3   QE-300 n_D^{20}1.4720

A

Aldrich 15,799-6   CAS [4054-38-0]   C7H10   Fp 52°F   60 MHz: 1, 50C   133.62*
,3-Cycloheptadiene, 97%   FW 94.16   FT-IR: 1, 48B   125.00*
bp 121°C   VP-FT-IR: 3, 71B   31.94
d 0.868   25.59
DCl3   QE-300 n_D^{20}1.4980

B

drich 27,410-0   CAS [3806-59-5]   C8H12 n_D^{20}1.4930   131.16*
s,cis-1,3-Cyclooctadiene, 98%   FW 108.18   Fp 76°F   125.95*
mp -52°C   VP-FT-IR: 3, 71C   28.11
bp 55°C (34 mm)   23.21
DCl3   QE-300   d 0.869

C

A

Aldrich 24,605-0   CAS [111-78-4]

1,5-Cyclooctadiene

C_8H_{12}
FW 108.18
mp -69°C
bp 150°C
d 0.882

n_D^{20} 1.4930
Fp 89°F
60 MHz: **1**, 51A
FT-IR: **1**, 48C
VP-FT-IR: **3**, 71D

128.67
28.17

CDCl₃   QE-300

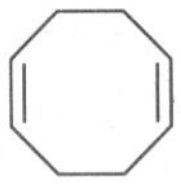

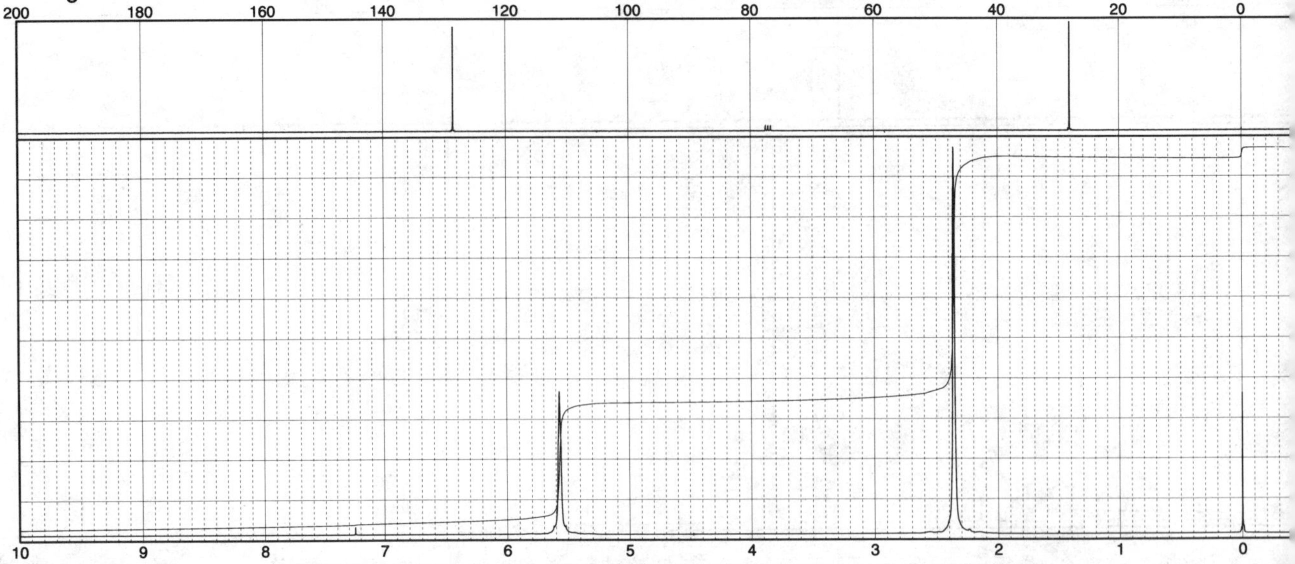

B

Aldrich 21,402-7   CAS [4045-44-7]

1,2,3,4,5-Pentamethylcyclopentadiene, 95%

$C_{10}H_{16}$
FW 136.24
bp 58°C (13 mm)
d 0.870
n_D^{20} 1.4740

Fp 112°F

60 MHz: **1**, 56B
FT-IR: **1**, 42B
VP-FT-IR: **3**, 60C

137.70
134.11
51.52
14.07
11.55
11.06

CDCl₃   QE-300

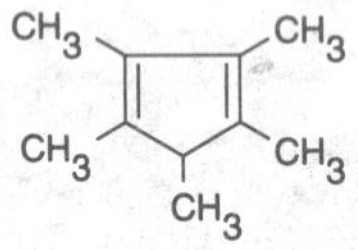

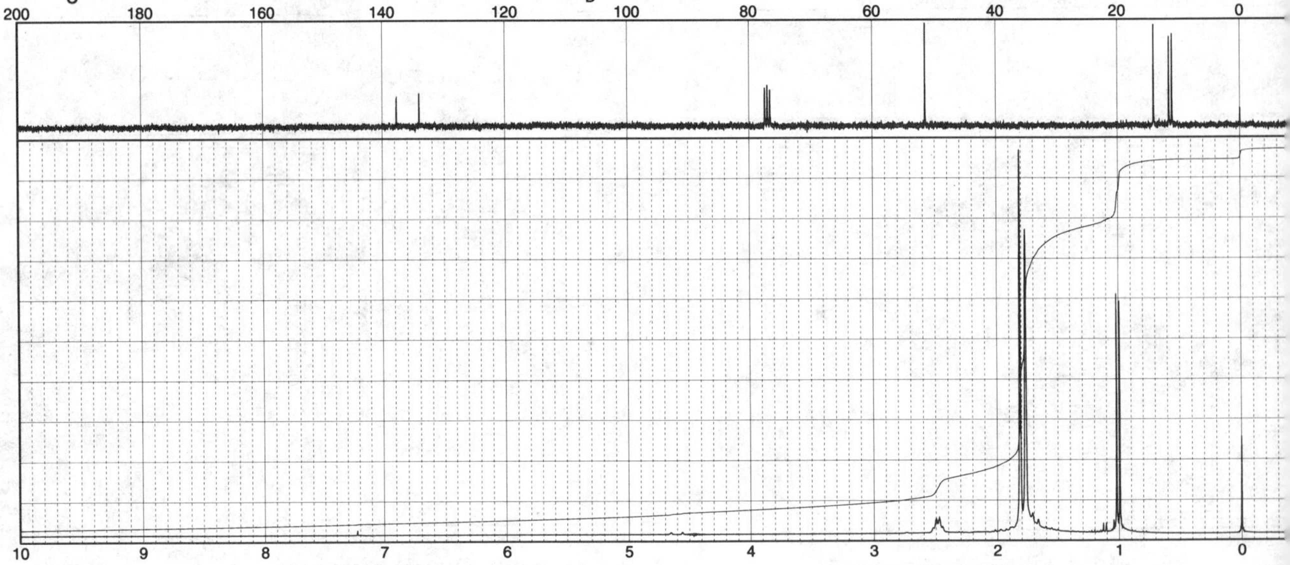

C

Aldrich 26,896-8   CAS [4313-57-9]

1-Methyl-1,4-cyclohexadiene, 99%

C_7H_{10}
FW 94.16
bp 115°C
d 0.838
n_D^{20} 1.4710

Fp 51°F

FT-IR: **1**, 47A
VP-FT-IR: **3**, 69A

131.2?
124.22
124.1?
118.5?
30.6?
26.8?
23.4?

CDCl₃   QE-300

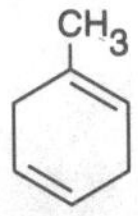

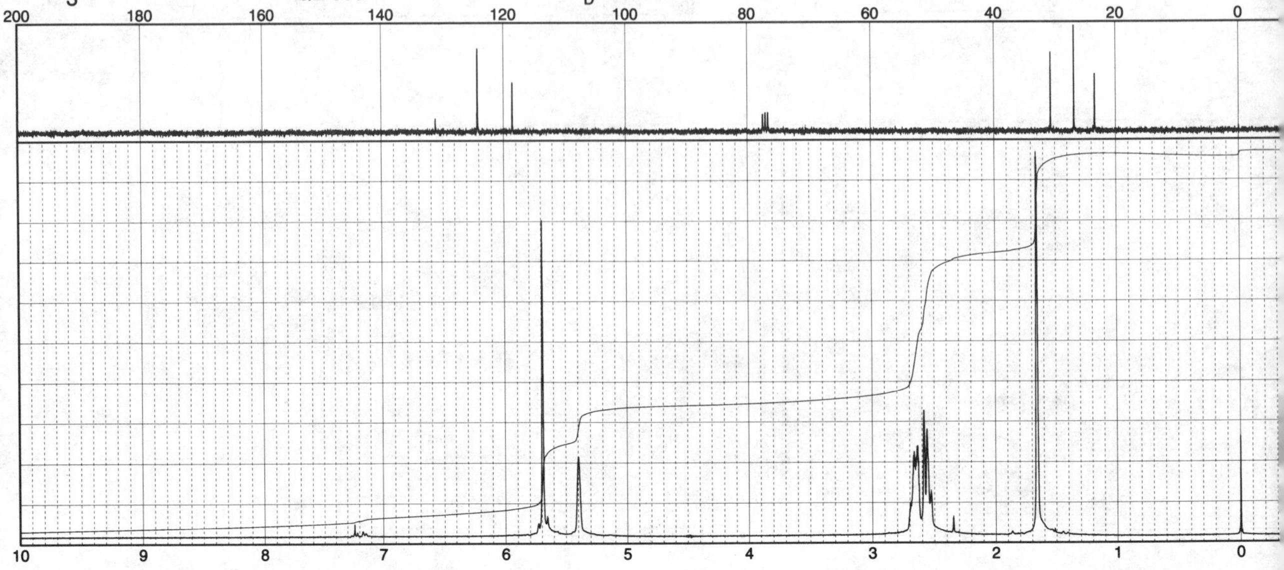

A

Aldrich 22,318-2
α-Terpinene, 85%

CAS [99-86-5]

C₁₀H₁₆
FW 136.24
bp 174°C
d 0.837
n²⁰_D 1.4780

Fp 115°F

60 MHz: *1*, 53B
FT-IR: *1*, 47B
VP-FT-IR: *3*, 69C

142.40 29.17
132.95 25.40
119.66* 22.80*
116.58* 21.27*
34.58*

CDCl₃   QE-300

B

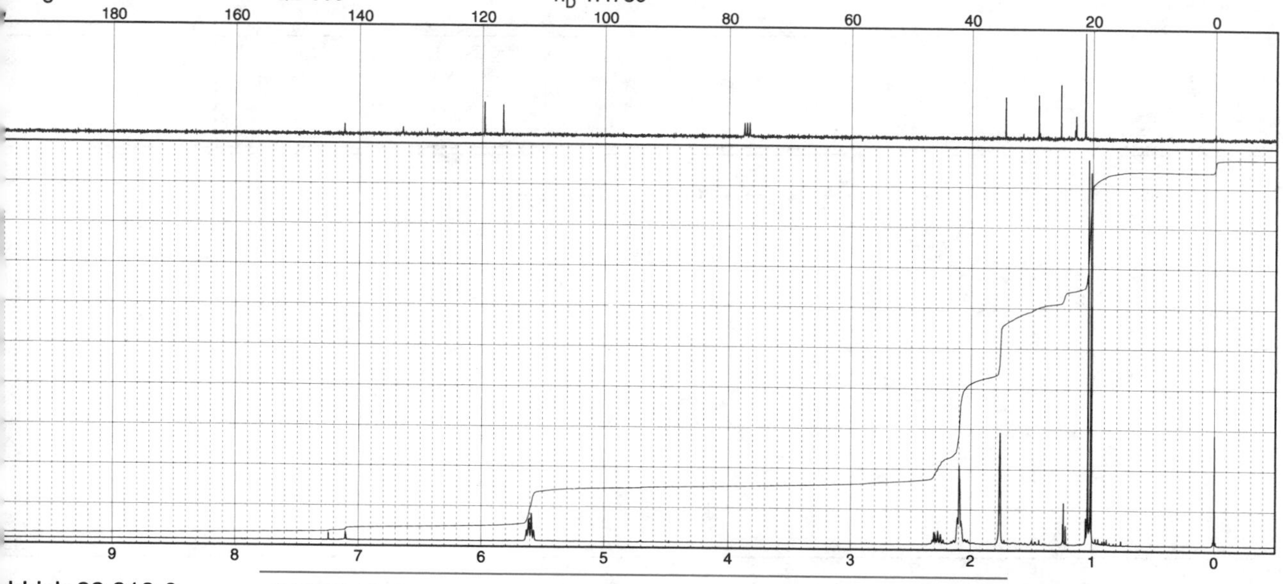

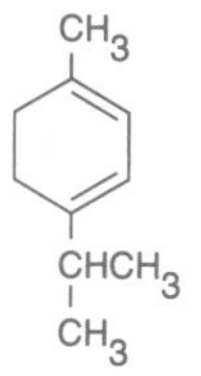

Aldrich 22,319-0
γ-Terpinene, 97%

CAS [99-85-4]

C₁₀H₁₆
FW 136.24
bp 182°C
d 0.849
n²⁰_D 1.4740

Fp 125°F

60 MHz: *1*, 53C
FT-IR: *1*, 47C
VP-FT-IR: *3*, 69D

140.59 31.66
131.21 27.59
118.86* 22.97*
115.98* 21.31*
34.57*

CDCl₃   QE-300

C

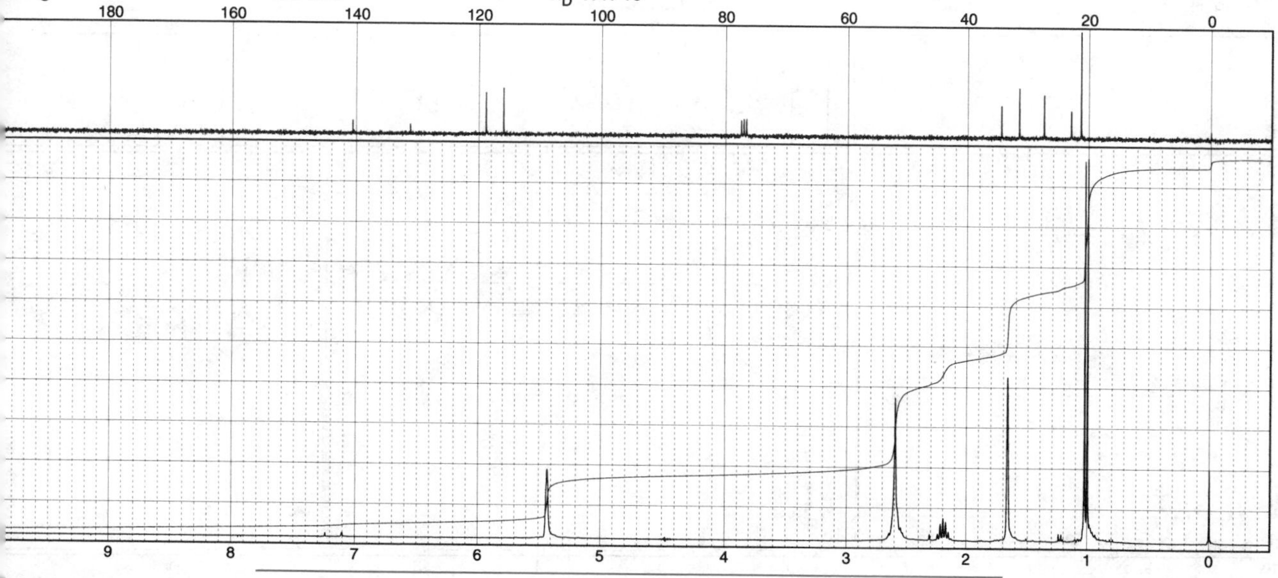

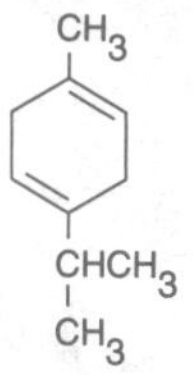

Aldrich 28,334-7
2,4,5-Tetramethyl-1,4-cyclohexadiene,
3%

CAS [26976-92-1]

C₁₀H₁₆
FW 136.24
mp 63°C

VP-FT-IR: *3*, 70A

123.40
40.01
17.97*

CDCl₃   QE-300

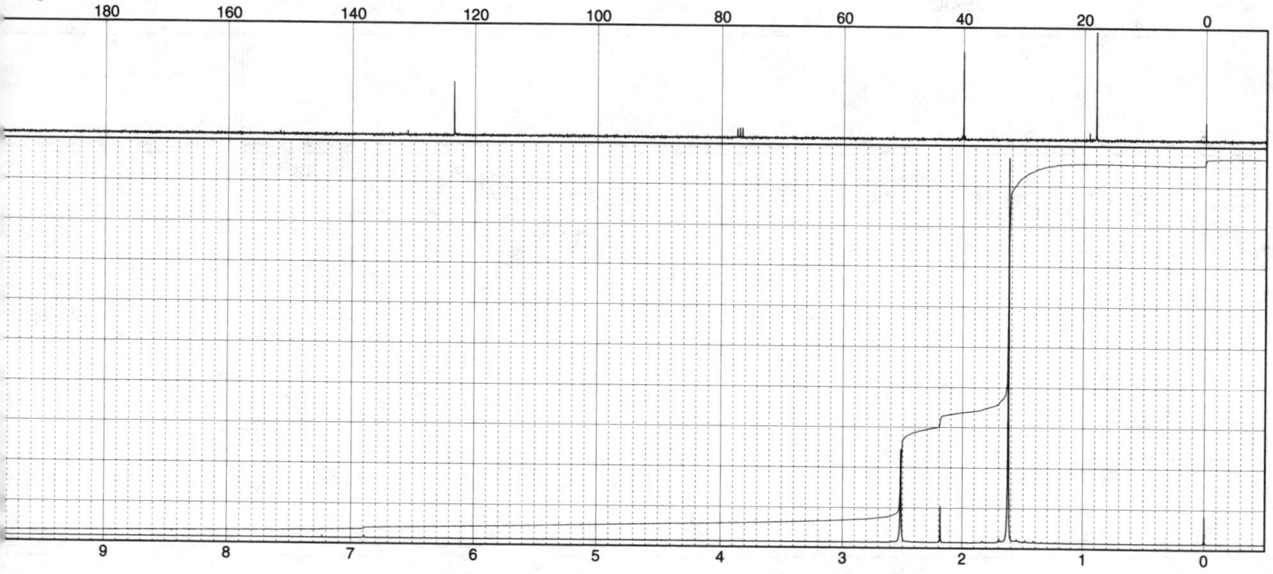

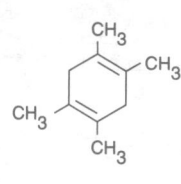

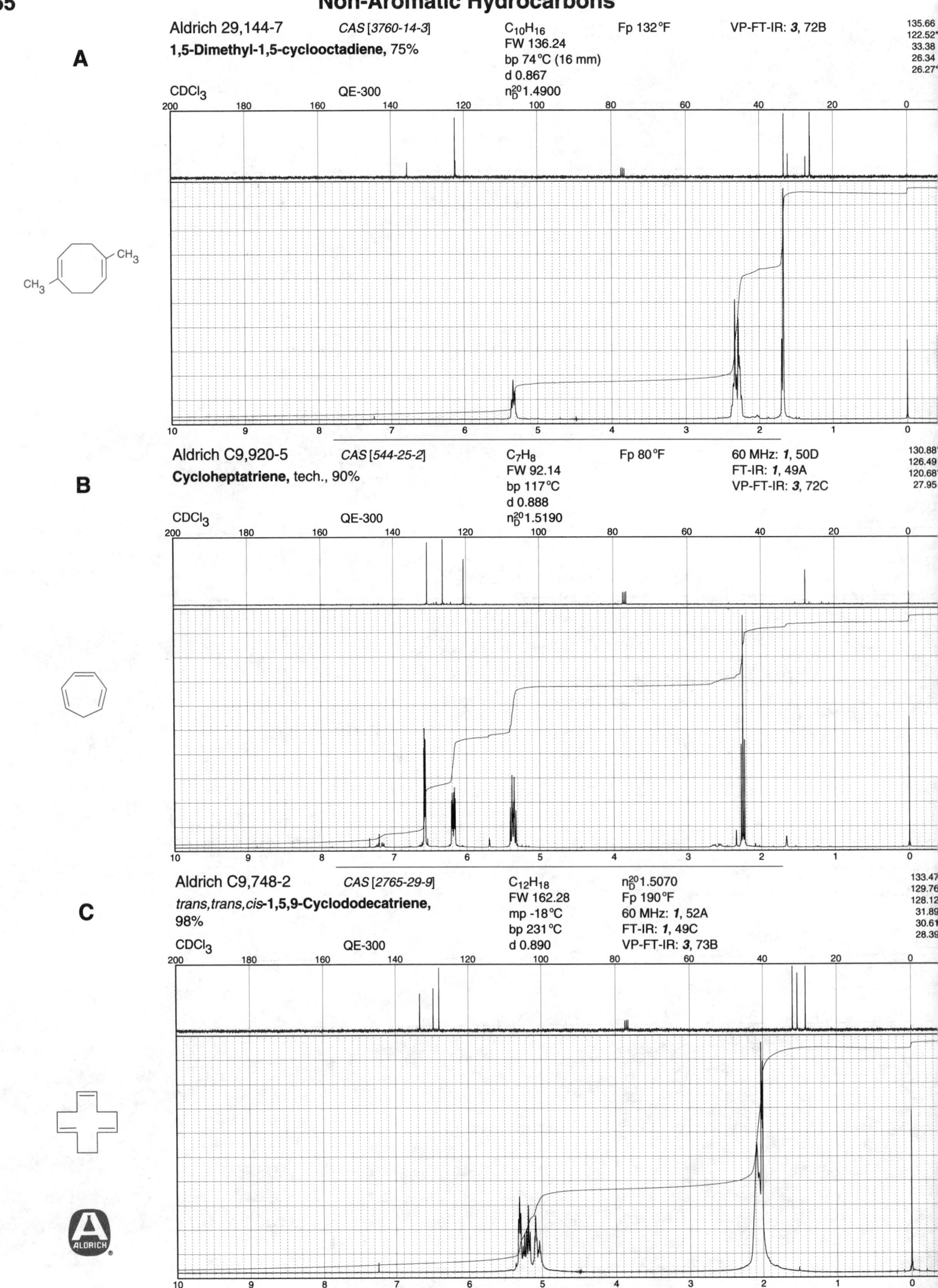

A

Aldrich 29,144-7   *CAS [3760-14-3]* $C_{10}H_{16}$ Fp 132°F   VP-FT-IR: *3*, 72B

1,5-Dimethyl-1,5-cyclooctadiene, 75%

FW 136.24
bp 74°C (16 mm)
d 0.867
n_D^{20} 1.4900

CDCl$_3$ QE-300

135.66
122.52*
33.38
26.34
26.27*

B

Aldrich C9,920-5   *CAS [544-25-2]* C_7H_8 Fp 80°F   60 MHz: *1*, 50D

Cycloheptatriene, tech., 90%

FW 92.14
bp 117°C
d 0.888
n_D^{20} 1.5190

CDCl$_3$ QE-300

FT-IR: *1*, 49A
VP-FT-IR: *3*, 72C

130.88*
126.49
120.68*
27.95

C

Aldrich C9,748-2   *CAS [2765-29-9]* $C_{12}H_{18}$ n_D^{20} 1.5070

trans,trans,cis-**1,5,9-Cyclododecatriene, 98%**

FW 162.28
mp -18°C
bp 231°C
d 0.890

Fp 190°F
60 MHz: *1*, 52A
FT-IR: *1*, 49C
VP-FT-IR: *3*, 73B

CDCl$_3$ QE-300

133.47
129.76
128.12
31.89
30.61
28.39

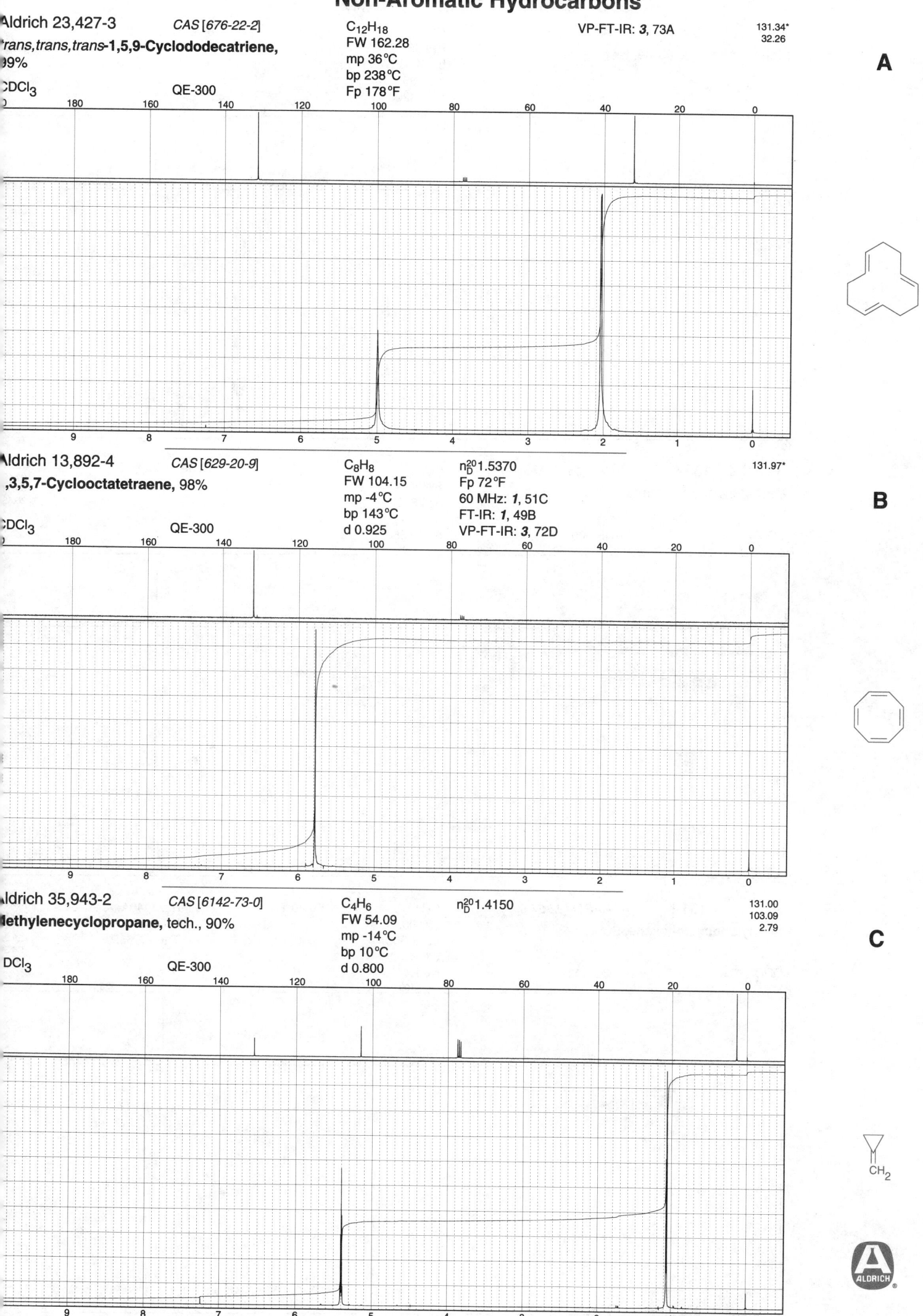

Aldrich 23,427-3   *CAS [676-22-2]* $C_{12}H_{18}$ VP-FT-IR: *3*, 73A   131.34*
trans,trans,trans-**1,5,9-Cyclododecatriene**,   FW 162.28   32.26
99%   mp 36°C
   bp 238°C
CDCl₃   QE-300   Fp 178°F

A

Aldrich 13,892-4   *CAS [629-20-9]* C_8H_8 n_D^{20}1.5370   131.97*
1,3,5,7-Cyclooctatetraene, 98%   FW 104.15   Fp 72°F
   mp -4°C   60 MHz: *1*, 51C
   bp 143°C   FT-IR: *1*, 49B
CDCl₃   QE-300   d 0.925   VP-FT-IR: *3*, 72D

B

Aldrich 35,943-2   *CAS [6142-73-0]* C_4H_6 n_D^{20}1.4150   131.00
Methylenecyclopropane, tech., 90%   FW 54.09   103.09
   mp -14°C   2.79
   bp 10°C
CDCl₃   QE-300   d 0.800

C

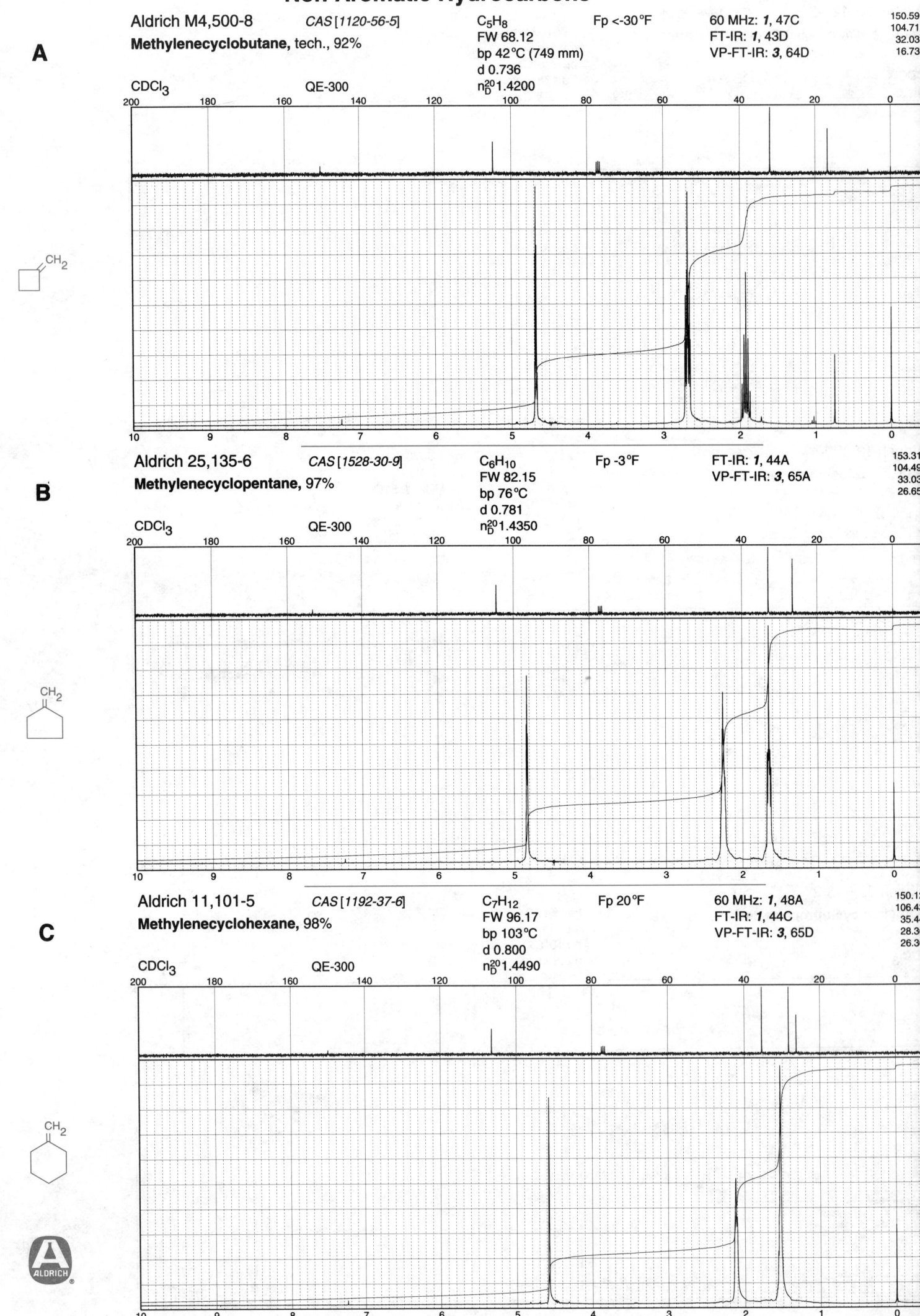

A

Aldrich M4,500-8   *CAS [1120-56-5]*   C₅H₈   Fp <-30°F   60 MHz: *1*, 47C   150.59
Methylenecyclobutane, tech., 92%   FW 68.12   FT-IR: *1*, 43D   104.71
bp 42°C (749 mm)   VP-FT-IR: *3*, 64D   32.03
d 0.736   16.73
n_D^20 1.4200

CDCl₃   QE-300

B

Aldrich 25,135-6   *CAS [1528-30-9]*   C₆H₁₀   Fp -3°F   FT-IR: *1*, 44A   153.31
Methylenecyclopentane, 97%   FW 82.15   VP-FT-IR: *3*, 65A   104.49
bp 76°C   33.03
d 0.781   26.65
n_D^20 1.4350

CDCl₃   QE-300

C

Aldrich 11,101-5   *CAS [1192-37-6]*   C₇H₁₂   Fp 20°F   60 MHz: *1*, 48A   150.12
Methylenecyclohexane, 98%   FW 96.17   FT-IR: *1*, 44C   106.43
bp 103°C   VP-FT-IR: *3*, 65D   35.44
d 0.800   28.36
n_D^20 1.4490   26.36

CDCl₃   QE-300

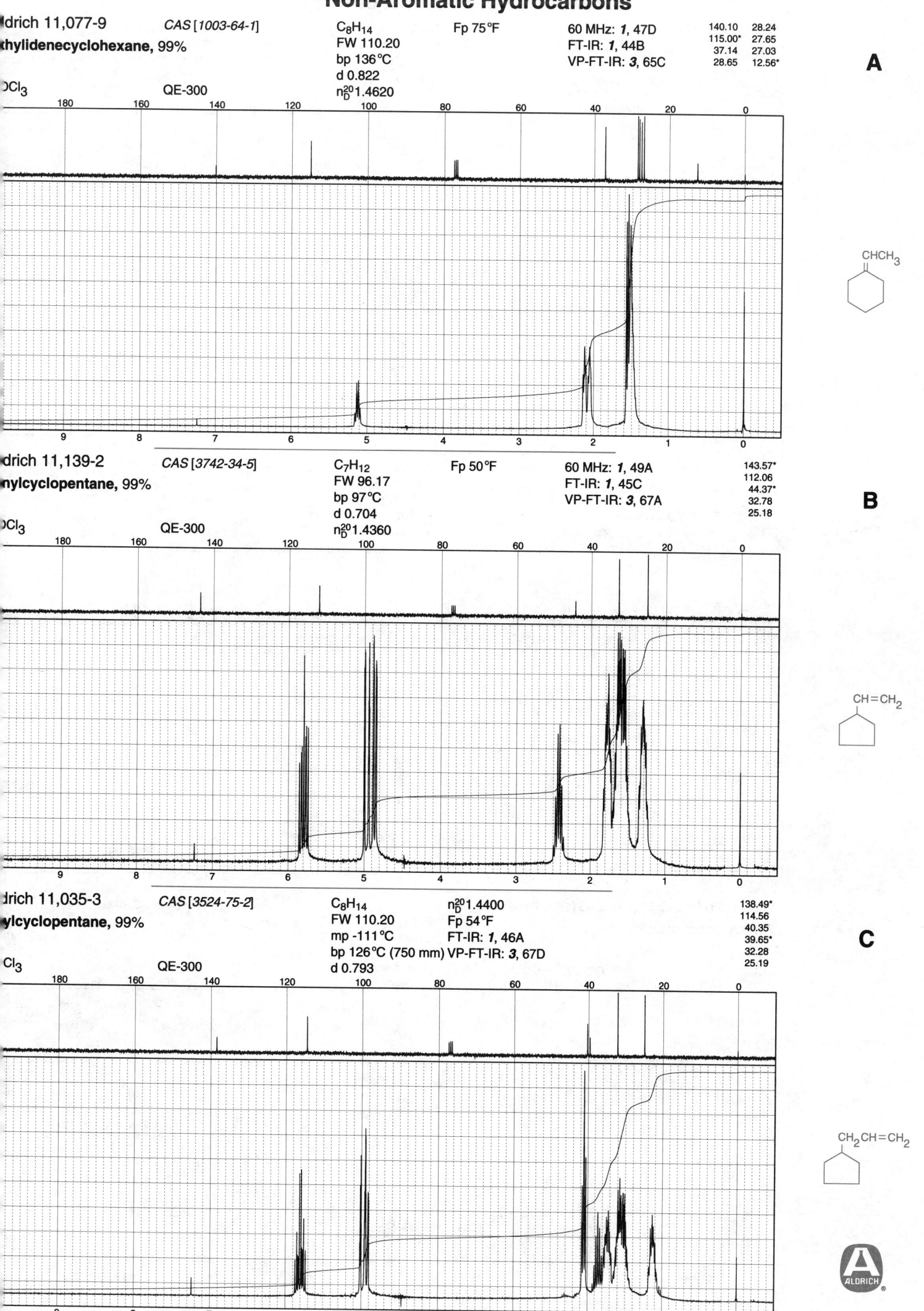

Aldrich 11,077-9   CAS [1003-64-1]
...thylidenecyclohexane, 99%
C₈H₁₄
FW 110.20
bp 136°C
d 0.822
n²⁰_D 1.4620
Fp 75°F
CDCl₃   QE-300

60 MHz: *1*, 47D
FT-IR: *1*, 44B
VP-FT-IR: *3*, 65C

140.10 28.24
115.00* 27.65
37.14 27.03
28.65 12.56*

A

Aldrich 11,139-2   CAS [3742-34-5]
...nylcyclopentane, 99%
C₇H₁₂
FW 96.17
bp 97°C
d 0.704
n²⁰_D 1.4360
Fp 50°F
CDCl₃   QE-300

60 MHz: *1*, 49A
FT-IR: *1*, 45C
VP-FT-IR: *3*, 67A

143.57*
112.06
44.37*
32.78
25.18

B

Aldrich 11,035-3   CAS [3524-75-2]
...ylcyclopentane, 99%
C₈H₁₄
FW 110.20
mp -111°C
bp 126°C (750 mm)
d 0.793
n²⁰_D 1.4400
Fp 54°F
FT-IR: *1*, 46A
VP-FT-IR: *3*, 67D
CDCl₃   QE-300

138.49*
114.56
40.35
39.65*
32.28
25.19

C

ALDRICH

A

Aldrich 11,140-6

CAS [695-12-5]

Vinylcyclohexane, 99+%

C$_8$H$_{14}$
FW 110.20
bp 128°C
d 0.805
n$_D^{20}$1.4460

Fp 70°F

60 MHz: *1*, 49B
FT-IR: *1*, 45D
VP-FT-IR: *3*, 67B

144.8
111.5
41.6
32.6
26.2
26.0

CDCl$_3$

QE-300

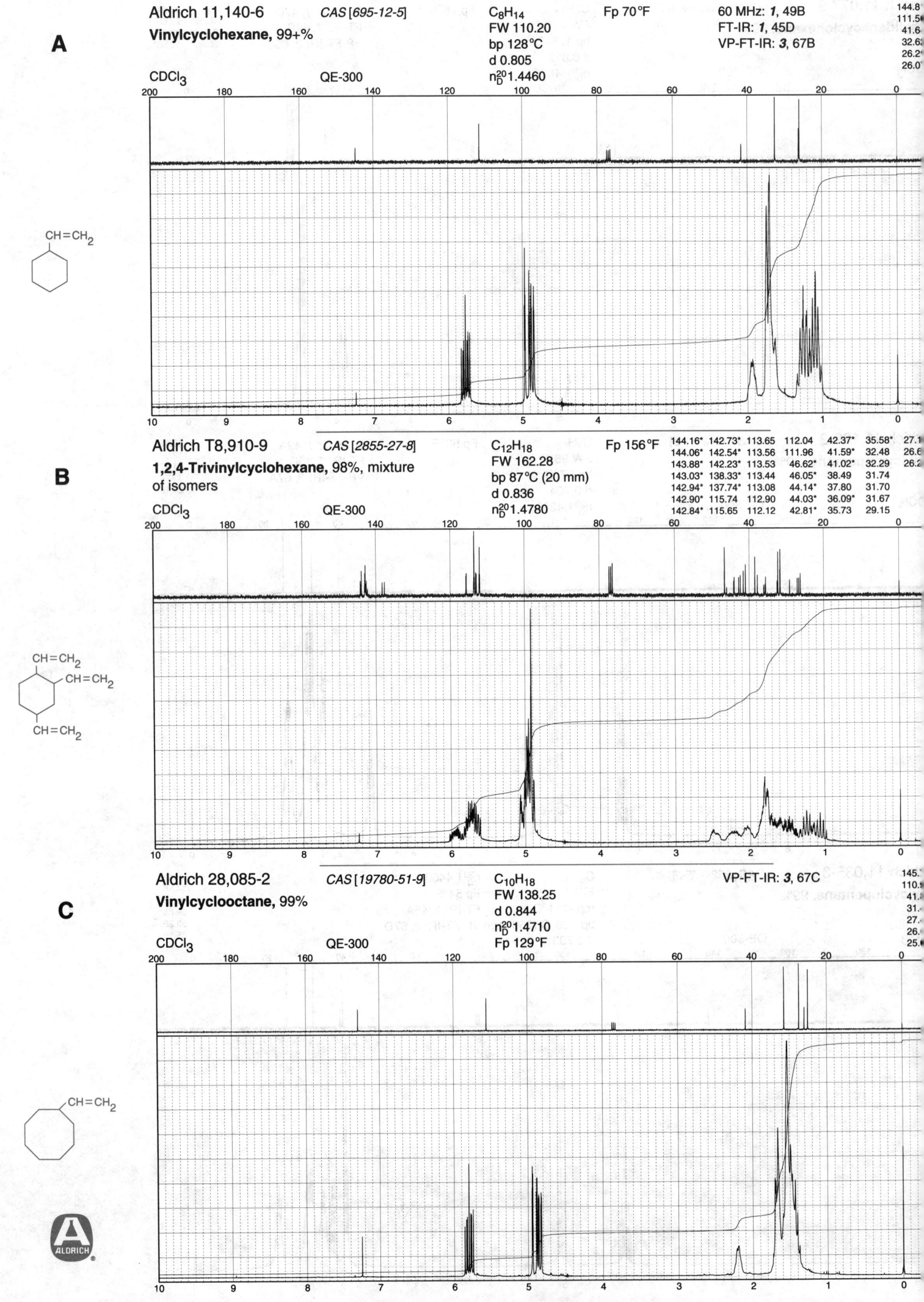

B

Aldrich T8,910-9

CAS [2855-27-8]

1,2,4-Trivinylcyclohexane, 98%, mixture of isomers

C$_{12}$H$_{18}$
FW 162.28
bp 87°C (20 mm)
d 0.836
n$_D^{20}$1.4780

Fp 156°F

144.16*	142.73*	113.65	112.04	42.37*	35.58*	27.1
144.06*	142.54*	113.56	111.96	41.59*	32.48	26.6
143.88*	142.23*	113.53	46.62*	41.02*	32.29	26.2
143.03*	138.33*	113.44	46.05*	38.49	31.74	
142.94*	137.74*	113.08	44.14*	37.80	31.70	
142.90*	115.74	112.90	44.03*	36.09*	31.67	
142.84*	115.65	112.12	42.81*	35.73	29.15	

CDCl$_3$

QE-300

C

Aldrich 28,085-2

CAS [19780-51-9]

Vinylcyclooctane, 99%

C$_{10}$H$_{18}$
FW 138.25
d 0.844
n$_D^{20}$1.4710
Fp 129°F

VP-FT-IR: *3*, 67C

145.1
110.9
41.8
31.4
27.4
26.6
25.6

CDCl$_3$

QE-300

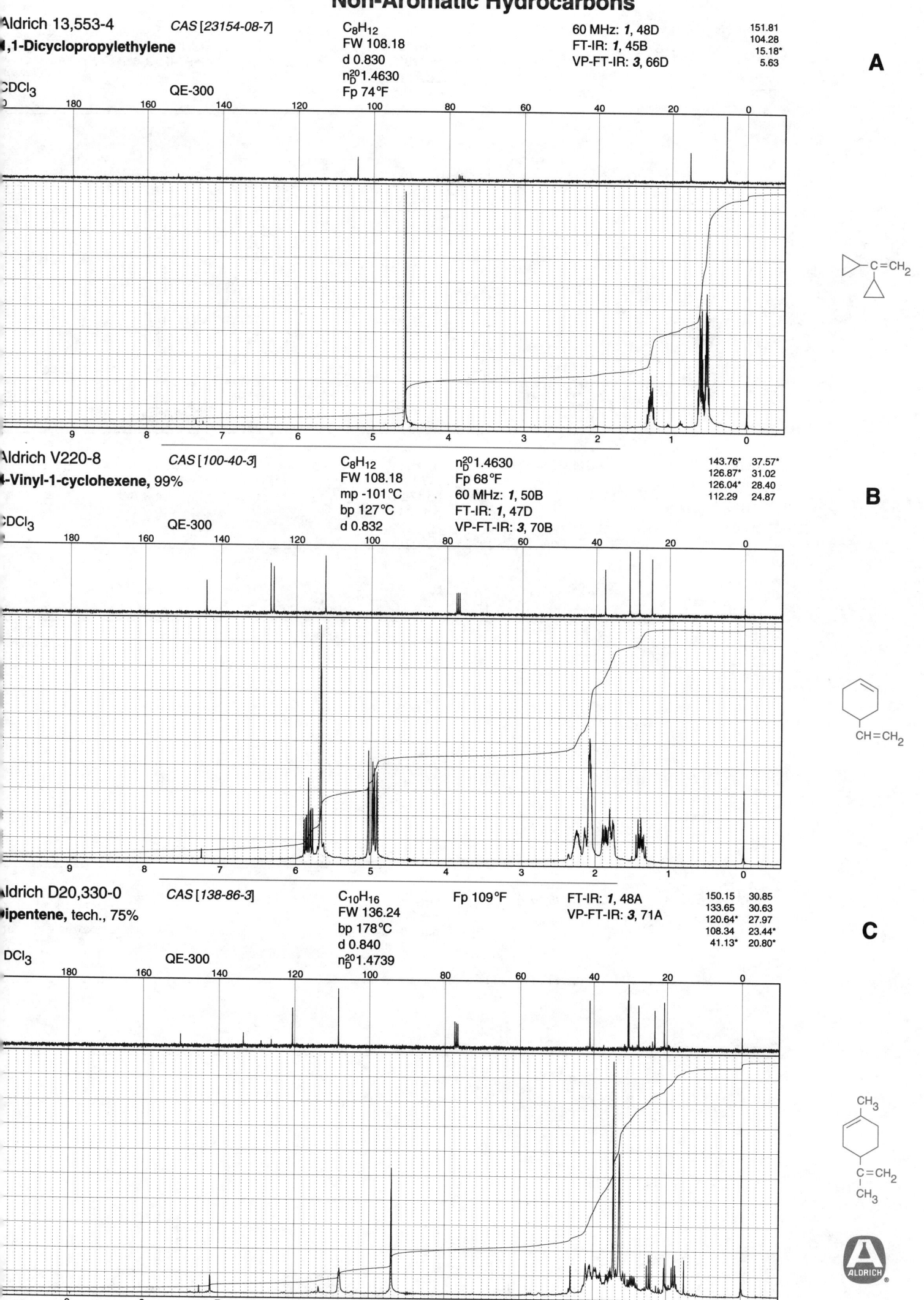

Aldrich 13,553-4   CAS [23154-08-7]

1,1-Dicyclopropylethylene

CDCl₃   QE-300

C₈H₁₂
FW 108.18
d 0.830
n_D^{20} 1.4630
Fp 74°F

60 MHz: *1*, 48D
FT-IR: *1*, 45B
VP-FT-IR: *3*, 66D

151.81
104.28
15.18*
5.63

A

Aldrich V220-8   CAS [100-40-3]

4-Vinyl-1-cyclohexene, 99%

CDCl₃   QE-300

C₈H₁₂
FW 108.18
mp -101°C
bp 127°C
d 0.832

n_D^{20} 1.4630
Fp 68°F
60 MHz: *1*, 50B
FT-IR: *1*, 47D
VP-FT-IR: *3*, 70B

143.76* 37.57*
126.87* 31.02
126.04* 28.40
112.29   24.87

B

Aldrich D20,330-0   CAS [138-86-3]

Dipentene, tech., 75%

DCl₃   QE-300

C₁₀H₁₆
FW 136.24
bp 178°C
d 0.840
n_D^{20} 1.4739

Fp 109°F

FT-IR: *1*, 48A
VP-FT-IR: *3*, 71A

150.15 30.85
133.65 30.63
120.64* 27.97
108.34 23.44*
41.13* 20.80*

C

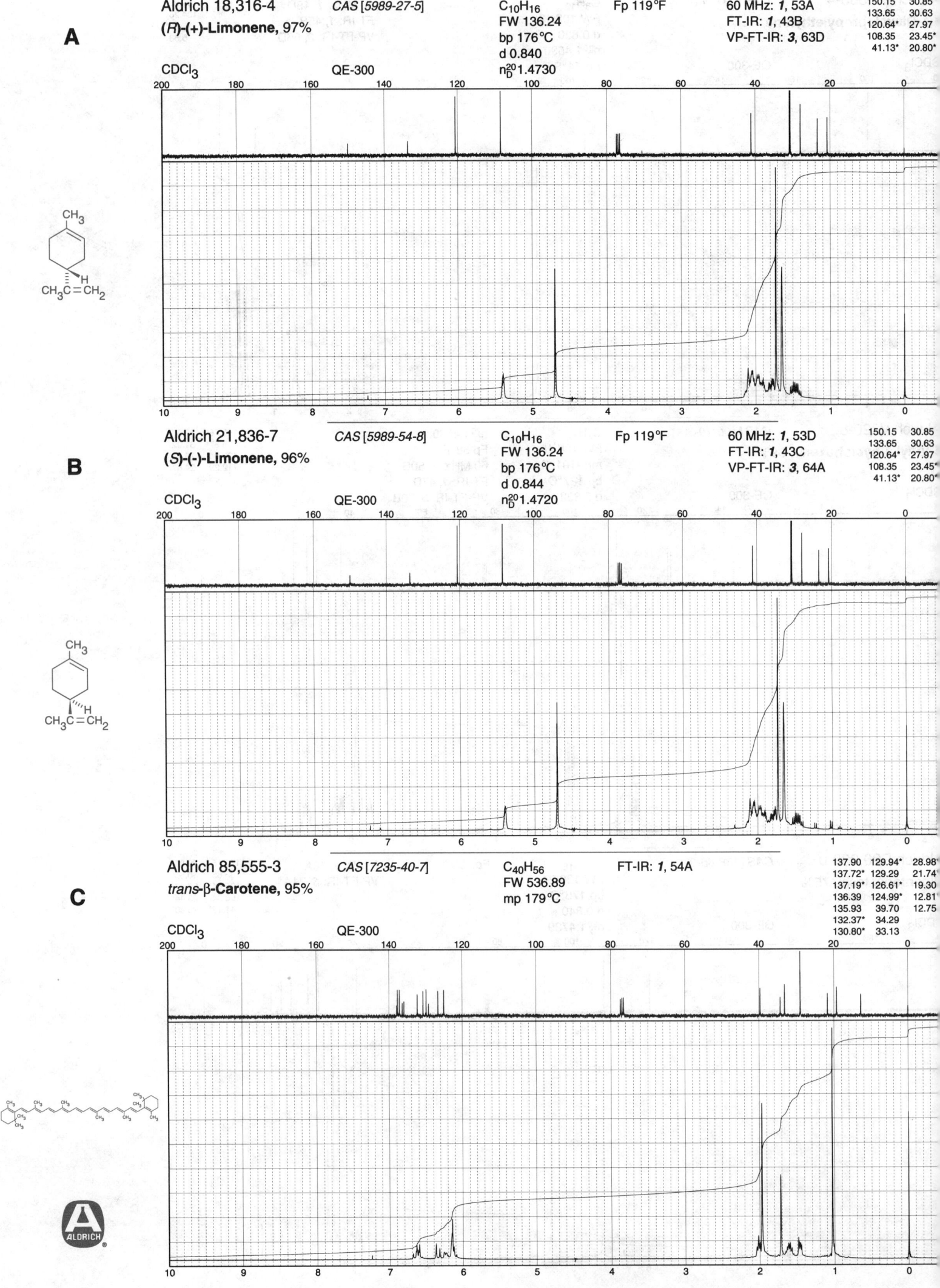

A

Aldrich 18,316-4

(R)-(+)-Limonene, 97%

CAS [5989-27-5]

$C_{10}H_{16}$
FW 136.24
bp 176°C
d 0.840
n_D^{20} 1.4730

Fp 119°F

60 MHz: *1*, 53A
FT-IR: *1*, 43B
VP-FT-IR: *3*, 63D

150.15	30.85
133.65	30.63
120.64*	27.97
108.35	23.45*
41.13*	20.80*

CDCl₃   QE-300

B

Aldrich 21,836-7

(S)-(−)-Limonene, 96%

CAS [5989-54-8]

$C_{10}H_{16}$
FW 136.24
bp 176°C
d 0.844
n_D^{20} 1.4720

Fp 119°F

60 MHz: *1*, 53D
FT-IR: *1*, 43C
VP-FT-IR: *3*, 64A

150.15	30.85
133.65	30.63
120.64*	27.97
108.35	23.45*
41.13*	20.80*

CDCl₃   QE-300

C

Aldrich 85,555-3

trans-β-**Carotene**, 95%

CAS [7235-40-7]

$C_{40}H_{56}$
FW 536.89
mp 179°C

FT-IR: *1*, 54A

137.90	129.94*	28.98*
137.72*	129.29	21.74*
137.19*	126.61*	19.30
136.39	124.99*	12.81*
135.93	39.70	12.75
132.37*	34.29	
130.80*	33.13	

CDCl₃   QE-300

Non-Aromatic Hydrocarbons

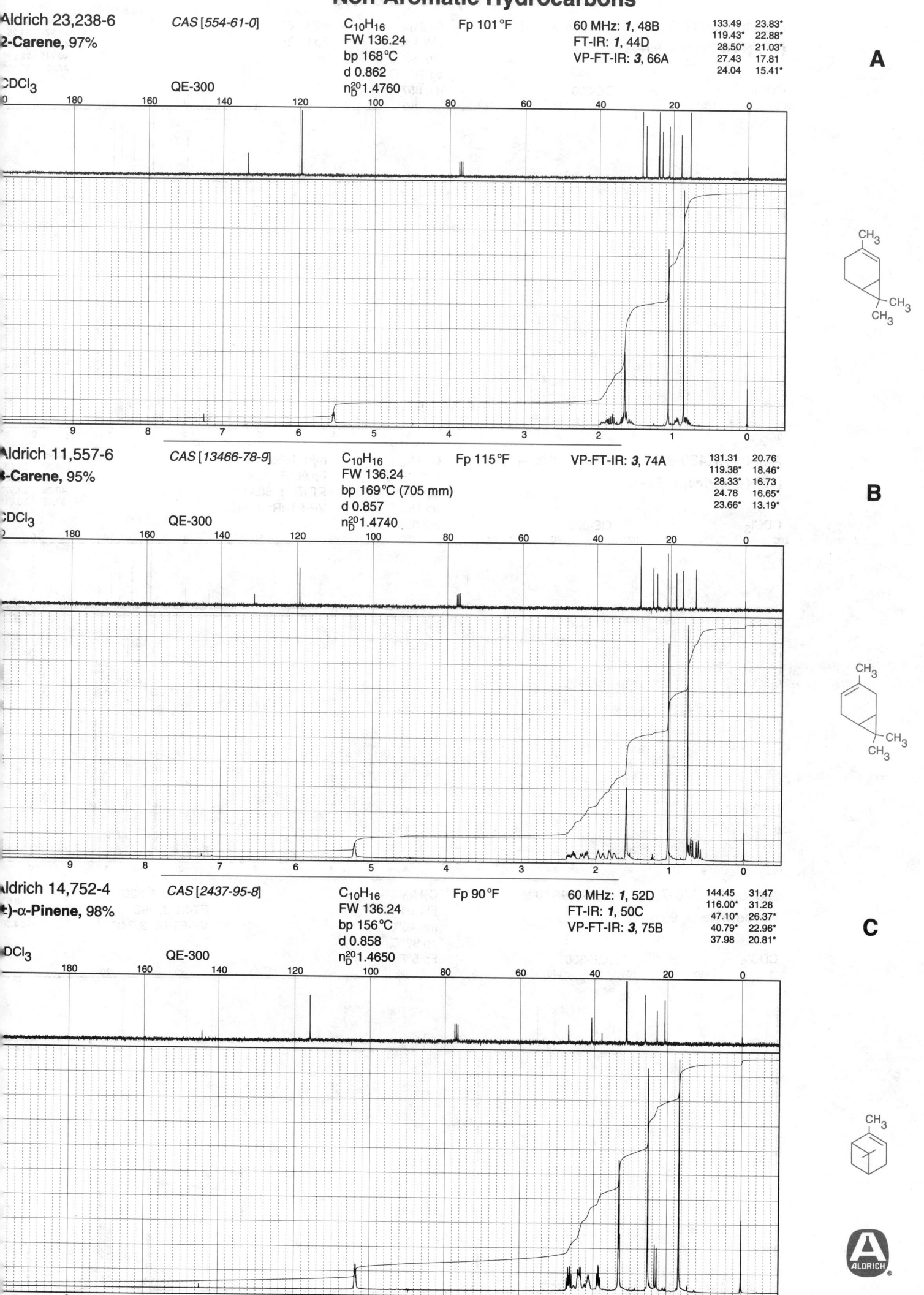

Aldrich 23,238-6

2-Carene, 97%

CAS [554-61-0]

$C_{10}H_{16}$
FW 136.24
bp 168°C
d 0.862
n_D^{20}1.4760

Fp 101°F

CDCl₃   QE-300

60 MHz: **1**, 48B
FT-IR: **1**, 44D
VP-FT-IR: **3**, 66A

133.49	23.83*
119.43*	22.88*
28.50*	21.03*
27.43	17.81
24.04	15.41*

A

Aldrich 11,557-6

3-Carene, 95%

CAS [13466-78-9]

$C_{10}H_{16}$
FW 136.24
bp 169°C (705 mm)
d 0.857
n_D^{20}1.4740

Fp 115°F

CDCl₃   QE-300

VP-FT-IR: **3**, 74A

131.31	20.76
119.38*	18.46*
28.33*	16.73
24.78	16.65*
23.66*	13.19*

B

Aldrich 14,752-4

(±)-α-Pinene, 98%

CAS [2437-95-8]

$C_{10}H_{16}$
FW 136.24
bp 156°C
d 0.858
n_D^{20}1.4650

Fp 90°F

CDCl₃   QE-300

60 MHz: **1**, 52D
FT-IR: **1**, 50C
VP-FT-IR: **3**, 75B

144.45	31.47
116.00*	31.28
47.10*	26.37*
40.79*	22.96*
37.98	20.81*

C

ALDRICH

A

Aldrich 26,807-0

(1*R*)-(+)-α-Pinene, 99+%

CAS [7785-70-8]

C₁₀H₁₆
FW 136.24
mp -62°C
bp 156°C
d 0.857

n_D^{20} 1.4660
Fp 90°F

144.44	31.45
115.97*	31.25
47.01*	26.35*
40.71*	22.99*
37.97	20.80*

CDCl₃   QE-300

B

Aldrich 27,439-9

(1*S*)-(-)-α-Pinene, 99+%

CAS [7785-26-4]

C₁₀H₁₆
FW 136.24
mp -64°C
bp 156°C
d 0.855

n_D^{20} 1.4650
Fp 90°F
FT-IR: *1*, 50A
VP-FT-IR: *3*, 74D

144.45	31.47
116.00*	31.28
47.09*	26.38*
40.78*	22.97*
37.98	20.81*

CDCl₃   QE-300

C

Aldrich N3,240-7

Norbornylene, 99%

CAS [498-66-8]

C₇H₁₀
FW 94.16
mp 45°C
bp 96°C
Fp 5°F

60 MHz: *1*, 52C
FT-IR: *1*, 49D
VP-FT-IR: *3*, 74C

135.29
48.51
41.74
24.59

CDCl₃   QE-300

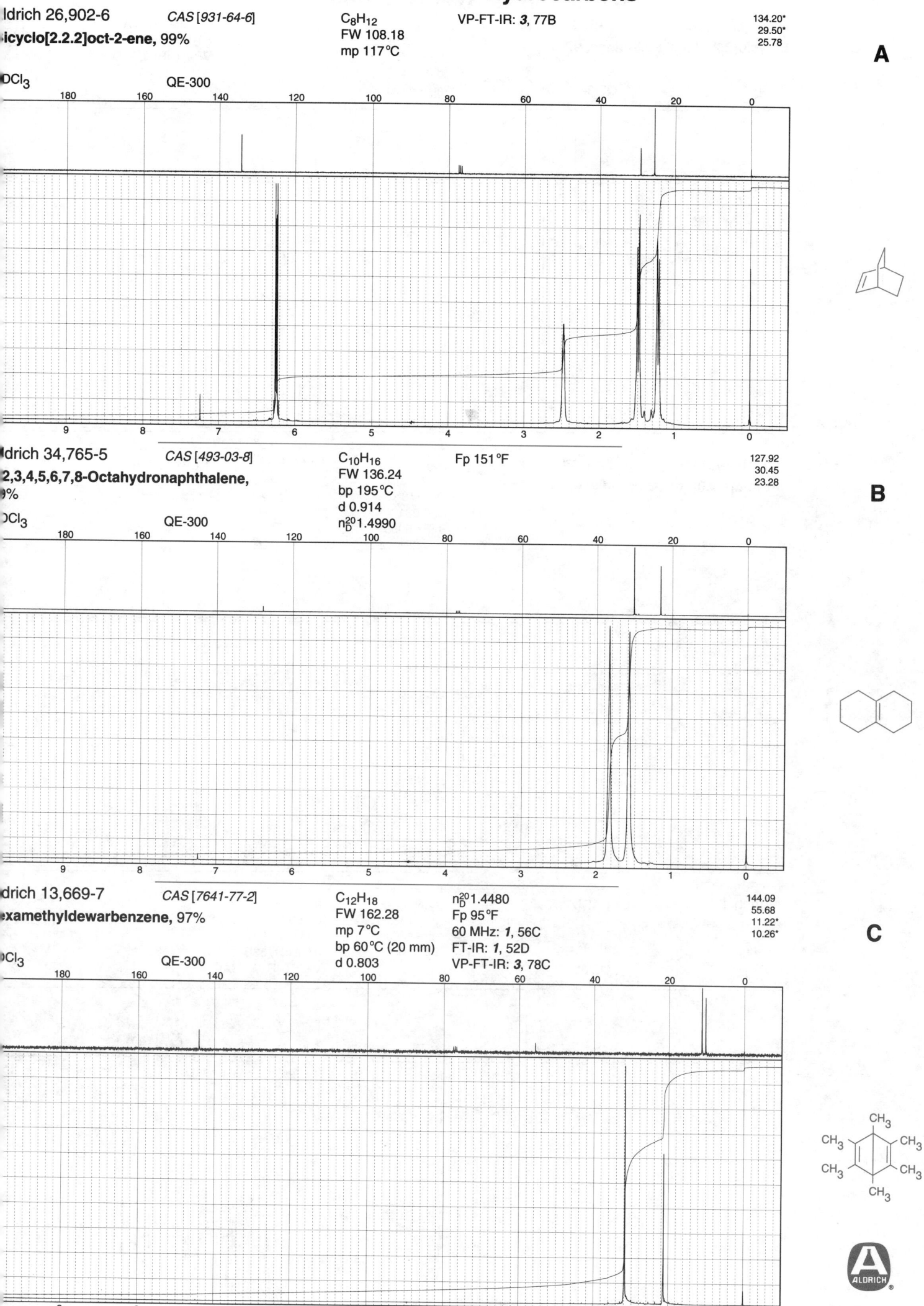

Aldrich 26,902-6   CAS [931-64-6] C_8H_{12} VP-FT-IR: **3**, 77B   134.20*
Bicyclo[2.2.2]oct-2-ene, 99%   FW 108.18   29.50*
mp 117°C   25.78

CDCl₃   QE-300

A

Aldrich 34,765-5   CAS [493-03-8] $C_{10}H_{16}$ Fp 151°F   127.92
1,2,3,4,5,6,7,8-Octahydronaphthalene,   FW 136.24   30.45
99%   bp 195°C   23.28
d 0.914
CDCl₃   QE-300 $n_D^{20}1.4990$

B

Aldrich 13,669-7   CAS [7641-77-2] $C_{12}H_{18}$ $n_D^{20}1.4480$ 144.09
Hexamethyldewarbenzene, 97%   FW 162.28   Fp 95°F   55.68
mp 7°C   60 MHz: **1**, 56C   11.22*
bp 60°C (20 mm)   FT-IR: **1**, 52D   10.26*
CDCl₃   QE-300   d 0.803   VP-FT-IR: **3**, 78C

C

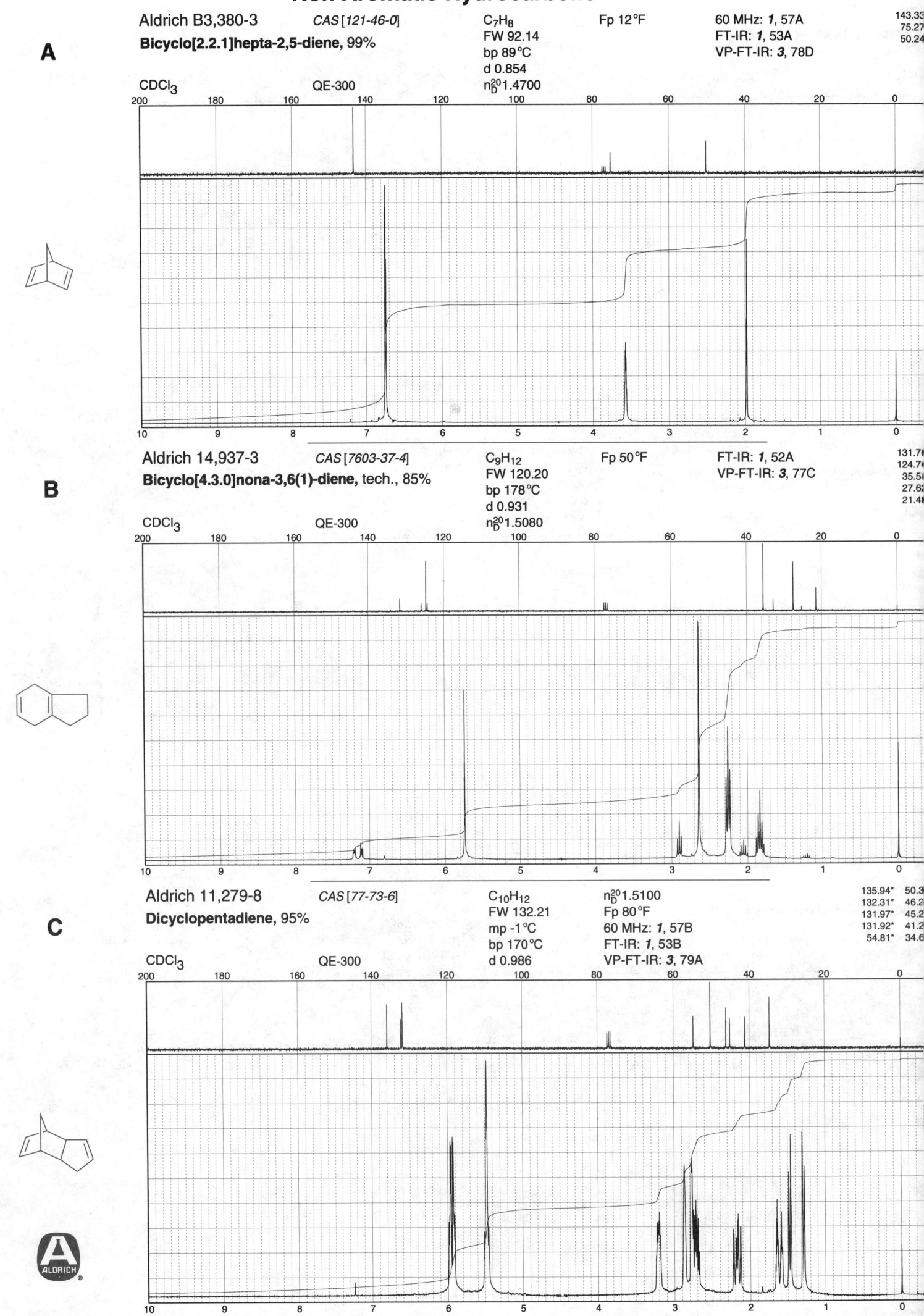

A

Aldrich B3,380-3   CAS [121-46-0] C_7H_8 Fp 12°F   60 MHz: 1, 57A   143.33
Bicyclo[2.2.1]hepta-2,5-diene, 99%   FW 92.14   FT-IR: 1, 53A   75.27
bp 89°C   VP-FT-IR: 3, 78D   50.24
d 0.854
n_D^{20}1.4700

CDCl₃   QE-300

B

Aldrich 14,937-3   CAS [7603-37-4] C_9H_{12} Fp 50°F   FT-IR: 1, 52A   131.76
Bicyclo[4.3.0]nona-3,6(1)-diene, tech., 85%   FW 120.20   VP-FT-IR: 3, 77C   124.7
bp 178°C   35.5
d 0.931   27.62
n_D^{20}1.5080   21.4

CDCl₃   QE-300

C

Aldrich 11,279-8   CAS [77-73-6] $C_{10}H_{12}$ n_D^{20}1.5100   135.94*   50.3
Dicyclopentadiene, 95%   FW 132.21   Fp 80°F   132.31*   46.2
mp -1°C   60 MHz: 1, 57B   131.97*   45.2
bp 170°C   FT-IR: 1, 53B   131.92*   41.2
d 0.986   VP-FT-IR: 3, 79A   54.81*   34.6

CDCl₃   QE-300

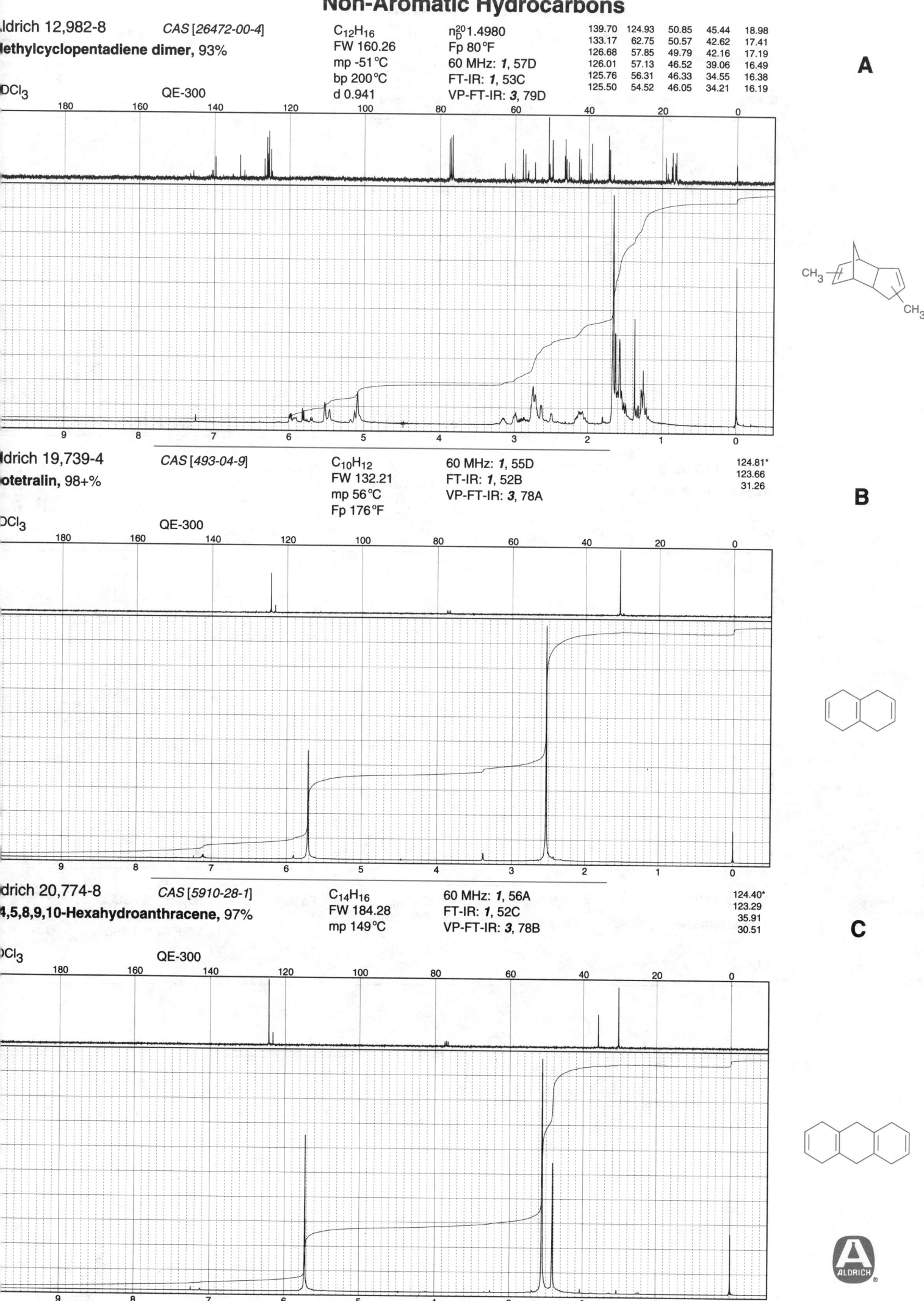

Aldrich 12,982-8 CAS [26472-00-4]
Methylcyclopentadiene dimer, 93%

$C_{12}H_{16}$
FW 160.26
mp -51°C
bp 200°C
d 0.941

n_D^{20} 1.4980
Fp 80°F
60 MHz: **1**, 57D
FT-IR: **1**, 53C
VP-FT-IR: **3**, 79D

CDCl₃ QE-300

A

139.70	124.93	50.85	45.44	18.98
133.17	62.75	50.57	42.62	17.41
126.68	57.85	49.79	42.16	17.19
126.01	57.13	46.52	39.06	16.49
125.76	56.31	46.33	34.55	16.38
125.50	54.52	46.05	34.21	16.19

Aldrich 19,739-4 CAS [493-04-9]
Isotetralin, 98+%

$C_{10}H_{12}$
FW 132.21
mp 56°C
Fp 176°F

60 MHz: **1**, 55D
FT-IR: **1**, 52B
VP-FT-IR: **3**, 78A

124.81*
123.66
31.26

CDCl₃ QE-300

B

Aldrich 20,774-8 CAS [5910-28-1]
1,4,5,8,9,10-Hexahydroanthracene, 97%

$C_{14}H_{16}$
FW 184.28
mp 149°C

60 MHz: **1**, 56A
FT-IR: **1**, 52C
VP-FT-IR: **3**, 78B

124.40*
123.29
35.91
30.51

CDCl₃ QE-300

C

Non-Aromatic Hydrocarbons

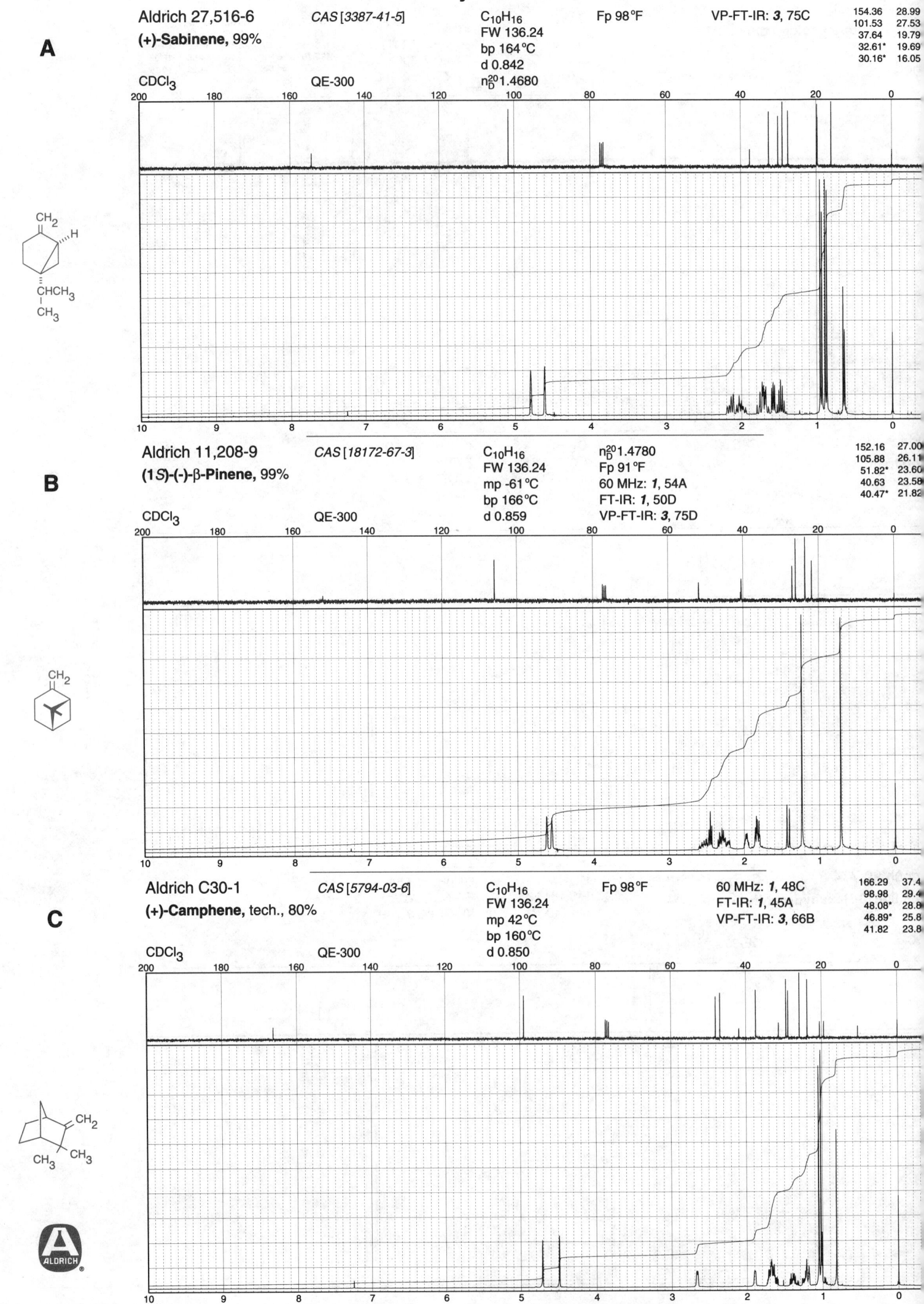

A

Aldrich 27,516-6
(+)-Sabinene, 99%

CAS [3387-41-5]

$C_{10}H_{16}$
FW 136.24
bp 164 °C
d 0.842
n_D^{20} 1.4680

Fp 98 °F

VP-FT-IR: **3**, 75C

CDCl₃   QE-300

154.36	28.99
101.53	27.53
37.64	19.79
32.61*	19.69
30.16*	16.05

B

Aldrich 11,208-9
(1S)-(-)-β-Pinene, 99%

CAS [18172-67-3]

$C_{10}H_{16}$
FW 136.24
mp -61 °C
bp 166 °C
d 0.859

n_D^{20} 1.4780
Fp 91 °F
60 MHz: **1**, 54A
FT-IR: **1**, 50D
VP-FT-IR: **3**, 75D

CDCl₃   QE-300

152.16	27.00
105.88	26.11
51.82*	23.60
40.63	23.58
40.47*	21.82

C

Aldrich C30-1
(+)-Camphene, tech., 80%

CAS [5794-03-6]

$C_{10}H_{16}$
FW 136.24
mp 42 °C
bp 160 °C
d 0.850

Fp 98 °F

60 MHz: **1**, 48C
FT-IR: **1**, 45A
VP-FT-IR: **3**, 66B

CDCl₃   QE-300

166.29	37.4
98.98	29.4
48.08*	28.8
46.89*	25.8
41.82	23.8

Non-Aromatic Hydrocarbons

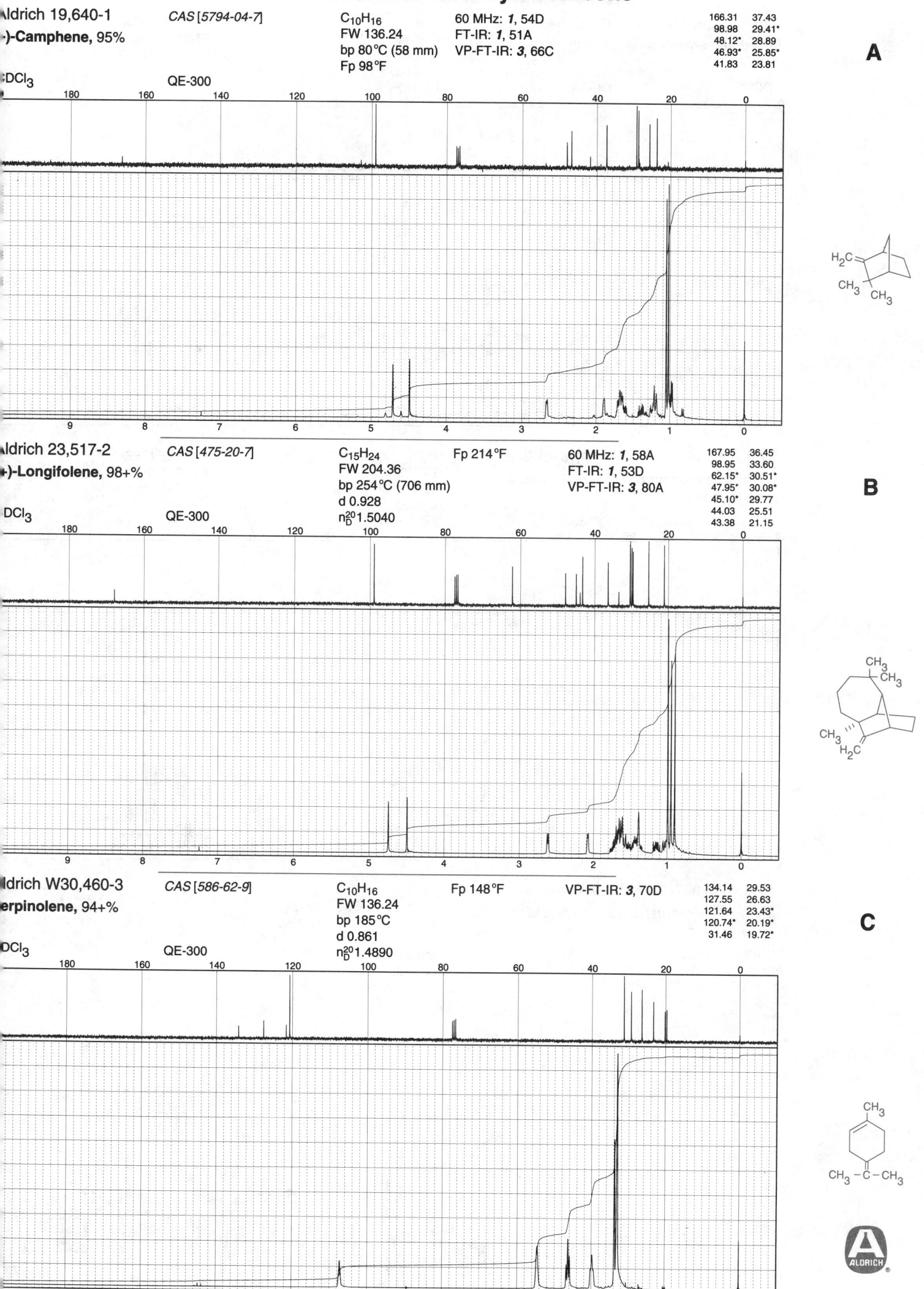

Aldrich 19,640-1 CAS [5794-04-7] C₁₀H₁₆ 60 MHz: *1*, 54D
(±)-Camphene, 95% FW 136.24 FT-IR: *1*, 51A
CDCl₃ QE-300 bp 80 °C (58 mm) VP-FT-IR: *3*, 66C
Fp 98 °F

166.31	37.43
98.98	29.41*
48.12*	28.89
46.93*	25.85*
41.83	23.81

A

Aldrich 23,517-2 CAS [475-20-7] C₁₅H₂₄ Fp 214 °F 60 MHz: *1*, 58A
(+)-Longifolene, 98+% FW 204.36 FT-IR: *1*, 53D
CDCl₃ QE-300 bp 254 °C (706 mm) VP-FT-IR: *3*, 80A
d 0.928
n²⁰_D 1.5040

167.95	36.45
98.95	33.60
62.15*	30.51*
47.95*	30.08*
45.10*	29.77
44.03	25.51
43.38	21.15

B

Aldrich W30,460-3 CAS [586-62-9] C₁₀H₁₆ Fp 148 °F VP-FT-IR: *3*, 70D
Terpinolene, 94+% FW 136.24
CDCl₃ QE-300 bp 185 °C
d 0.861
n²⁰_D 1.4890

134.14	29.53
127.55	26.63
121.64	23.43*
120.74*	20.19*
31.46	19.72*

C

A

Aldrich 12,984-4   *CAS [694-91-7]* C_8H_{10} 60 MHz: *1*, 55A   151.21   51.04

5-Methylene-2-norbornene, tech., 90%   FW 106.17   FT-IR: *1*, 51B   136.49*   50.18

d 0.981   VP-FT-IR: *3*, 76C   134.27*   42.10

n_D^{20}1.4830   103.21   33.61

CDCl$_3$ QE-300   Fp 40°F

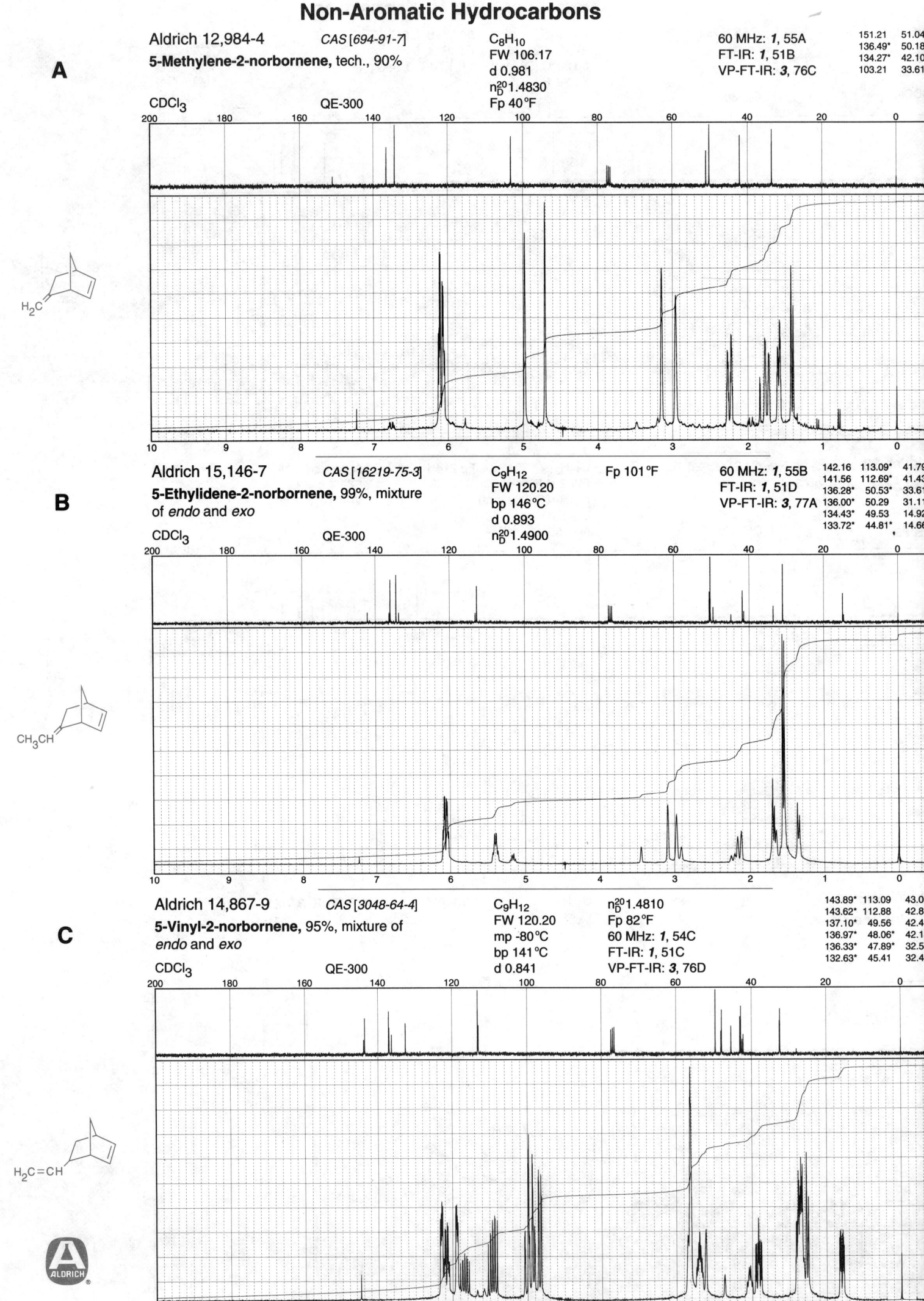

Aldrich 15,146-7   *CAS [16219-75-3]* C_9H_{12} Fp 101°F   60 MHz: *1*, 55B   142.16   113.09*   41.79

5-Ethylidene-2-norbornene, 99%, mixture   FW 120.20   FT-IR: *1*, 51D   141.56   112.69*   41.43

of *endo* and *exo*   bp 146°C   VP-FT-IR: *3*, 77A   136.28*   50.53*   33.61

d 0.893   136.00*   50.29   31.11

CDCl$_3$ QE-300 n_D^{20}1.4900   134.43*   49.53   14.92

133.72*   44.81*   14.66

B

Aldrich 14,867-9   *CAS [3048-64-4]* C_9H_{12} n_D^{20}1.4810   143.89*   113.09   43.0

5-Vinyl-2-norbornene, 95%, mixture of   FW 120.20   Fp 82°F   143.62*   112.88   42.8

endo and *exo*   mp -80°C   60 MHz: *1*, 54C   137.10*   49.56   42.4

bp 141°C   FT-IR: *1*, 51C   136.97*   48.06*   42.1

CDCl$_3$ QE-300   d 0.841   VP-FT-IR: *3*, 76D   136.33*   47.89*   32.5

132.63*   45.41   32.4

C

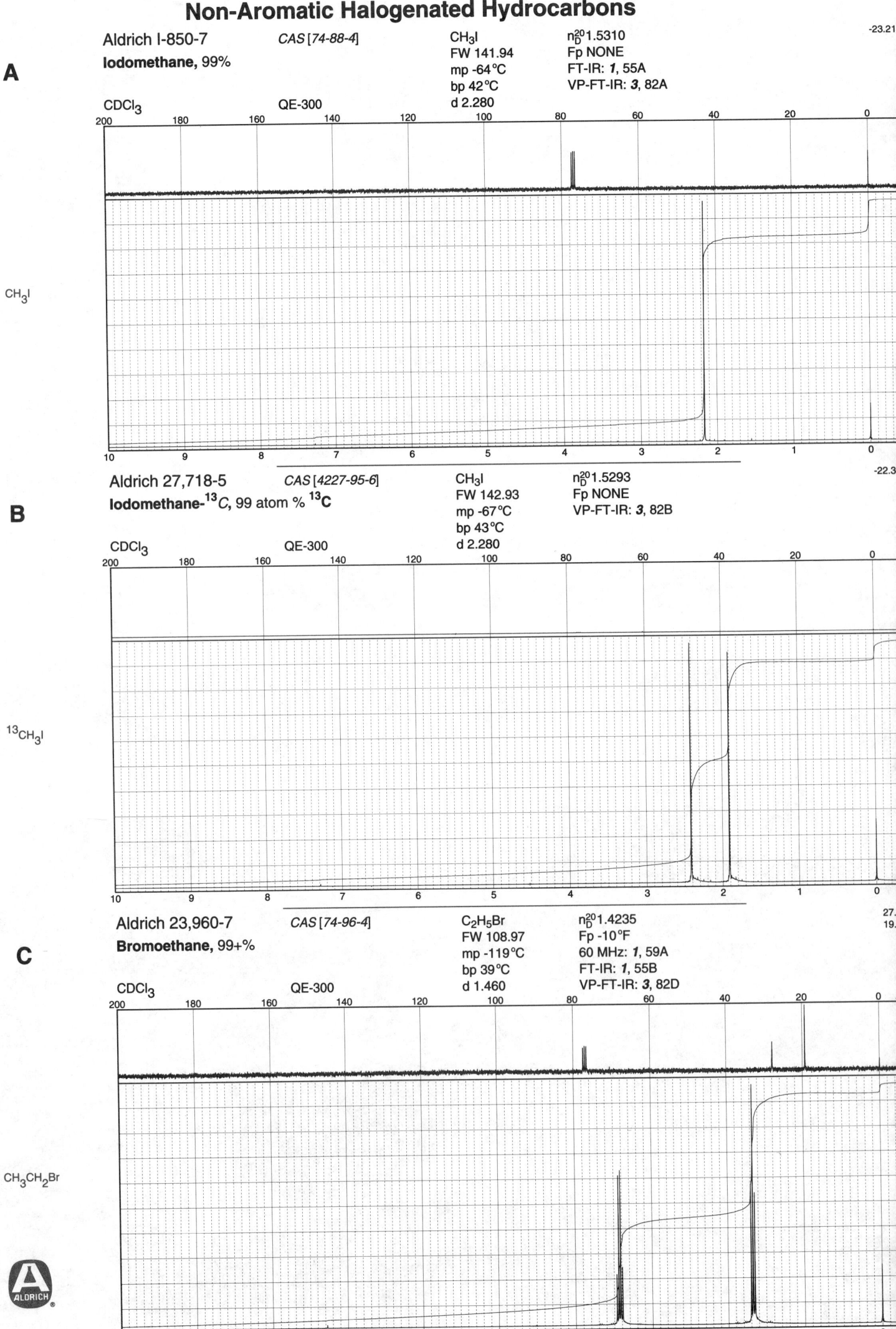

A

Aldrich I-850-7   CAS [74-88-4]   CH₃I   n²⁰_D 1.5310

Iodomethane, 99%

CH₃I
FW 141.94
mp -64°C
bp 42°C
d 2.280

Fp NONE
FT-IR: *1*, 55A
VP-FT-IR: *3*, 82A

CDCl₃   QE-300

CH₃I

-23.21

-22.3

B

Aldrich 27,718-5   CAS [4227-95-6]

Iodomethane-¹³C, 99 atom % ¹³C

CH₃I
FW 142.93
mp -67°C
bp 43°C
d 2.280

n²⁰_D 1.5293
Fp NONE
VP-FT-IR: *3*, 82B

CDCl₃   QE-300

¹³CH₃I

27.
19.

C

Aldrich 23,960-7   CAS [74-96-4]

Bromoethane, 99+%

C₂H₅Br
FW 108.97
mp -119°C
bp 39°C
d 1.460

n²⁰_D 1.4235
Fp -10°F
60 MHz: *1*, 59A
FT-IR: *1*, 55B
VP-FT-IR: *3*, 82D

CDCl₃   QE-300

CH₃CH₂Br

ALDRICH

Aldrich I-778-0

CAS [75-03-6]

C₂H₅I
FW 155.97
mp -108°C
bp 71°C
d 1.950

n_D^{20} 1.5130
Fp NONE
60 MHz: *1*, 59B
FT-IR: *1*, 55C
VP-FT-IR: *3*, 83A

20.50*
-1.02

A

Iodoethane, 99%

CDCl₃

QE-300

CH_3CH_2I

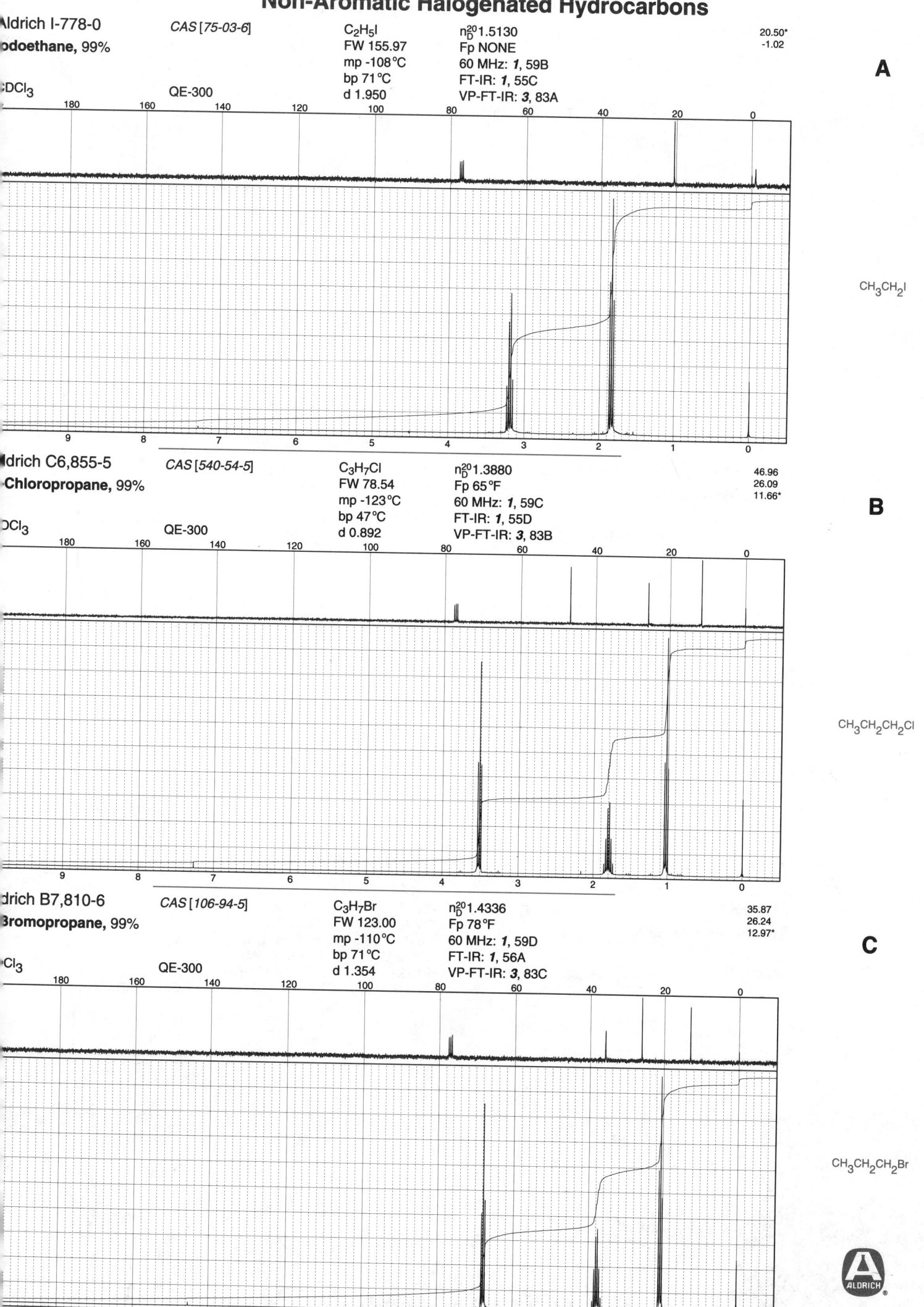

Aldrich C6,855-5

CAS [540-54-5]

C₃H₇Cl
FW 78.54
mp -123°C
bp 47°C
d 0.892

n_D^{20} 1.3880
Fp 65°F
60 MHz: *1*, 59C
FT-IR: *1*, 55D
VP-FT-IR: *3*, 83B

46.96
26.09
11.66*

B

Chloropropane, 99%

CDCl₃

QE-300

$CH_3CH_2CH_2Cl$

Aldrich B7,810-6

CAS [106-94-5]

C₃H₇Br
FW 123.00
mp -110°C
bp 71°C
d 1.354

n_D^{20} 1.4336
Fp 78°F
60 MHz: *1*, 59D
FT-IR: *1*, 56A
VP-FT-IR: *3*, 83C

35.87
26.24
12.97*

C

Bromopropane, 99%

CDCl₃

QE-300

$CH_3CH_2CH_2Br$

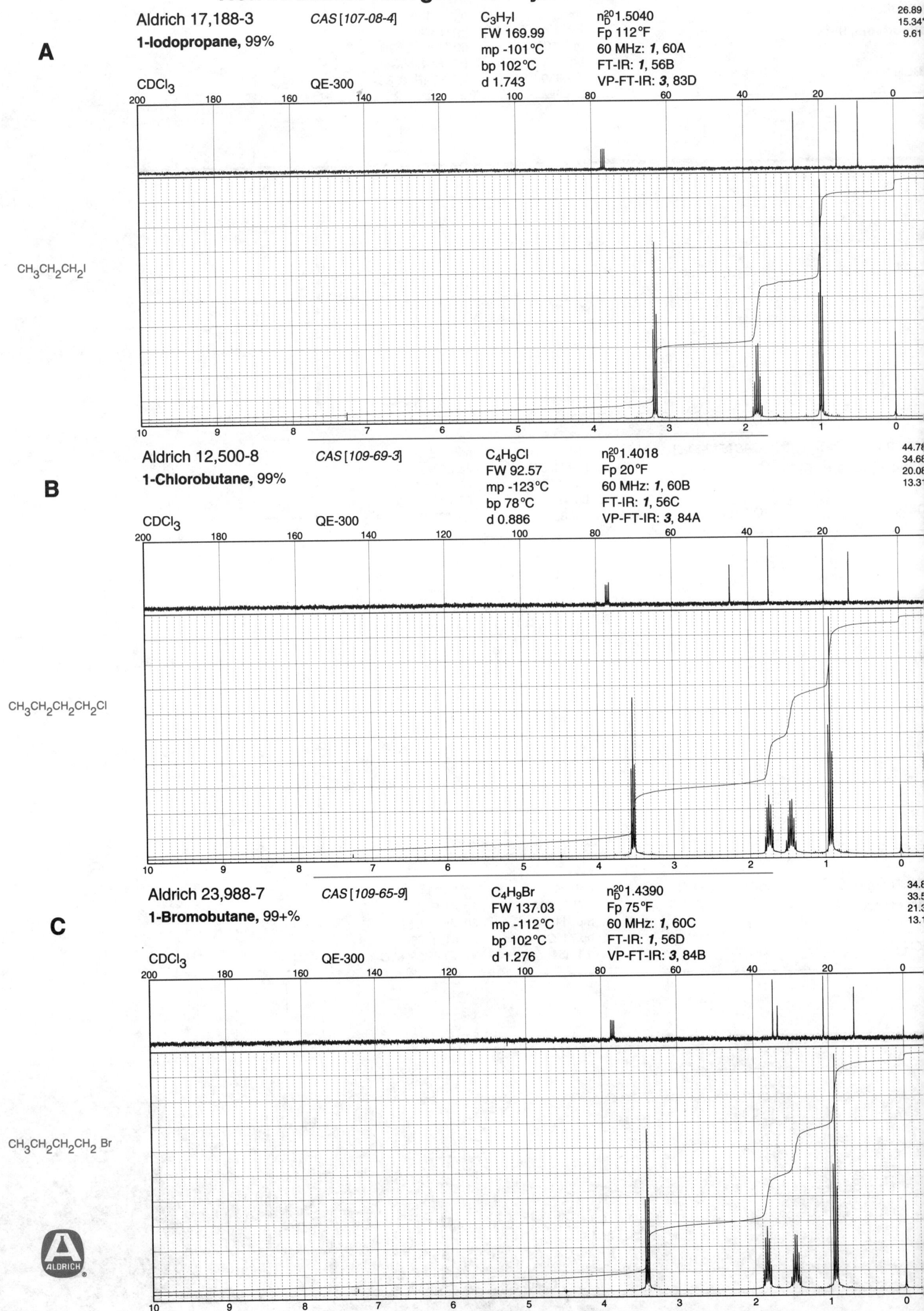

A

Aldrich 17,188-3   CAS [107-08-4]   C₃H₇I n_D^{20}1.5040
1-Iodopropane, 99%   FW 169.99   Fp 112°F
                          mp -101°C   60 MHz: *1*, 60A
                          bp 102°C   FT-IR: *1*, 56B
                          d 1.743   VP-FT-IR: *3*, 83D

26.89
15.34
9.61

CDCl₃   QE-300

CH₃CH₂CH₂I

B

Aldrich 12,500-8   CAS [109-69-3]   C₄H₉Cl n_D^{20}1.4018
1-Chlorobutane, 99%   FW 92.57   Fp 20°F
                          mp -123°C   60 MHz: *1*, 60B
                          bp 78°C   FT-IR: *1*, 56C
                          d 0.886   VP-FT-IR: *3*, 84A

44.78
34.68
20.08
13.3

CDCl₃   QE-300

CH₃CH₂CH₂CH₂Cl

C

Aldrich 23,988-7   CAS [109-65-9]   C₄H₉Br n_D^{20}1.4390
1-Bromobutane, 99+%   FW 137.03   Fp 75°F
                          mp -112°C   60 MHz: *1*, 60C
                          bp 102°C   FT-IR: *1*, 56D
                          d 1.276   VP-FT-IR: *3*, 84B

34.8
33.5
21.3
13.1

CDCl₃   QE-300

CH₃CH₂CH₂CH₂ Br

Non-Aromatic Halogenated Hydrocarbons

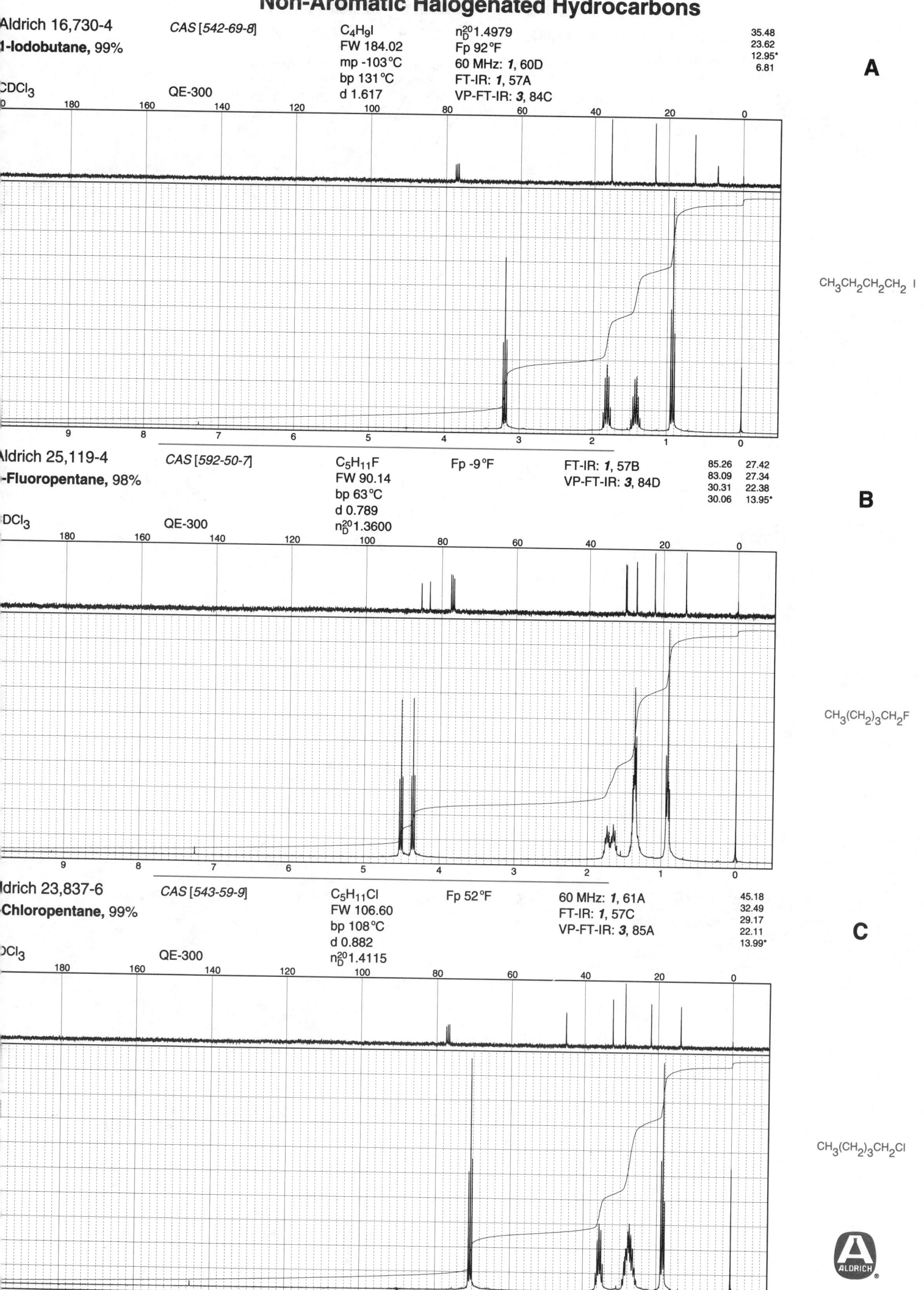

Aldrich 16,730-4   CAS [542-69-8]   C₄H₉I n_D^{20} 1.4979   35.48
1-Iodobutane, 99%   FW 184.02   Fp 92°F   23.62
mp -103°C   60 MHz: 1, 60D   12.95*
CDCl₃   QE-300   bp 131°C   FT-IR: 1, 57A   6.81
d 1.617   VP-FT-IR: 3, 84C

A

CH₃CH₂CH₂CH₂ I

Aldrich 25,119-4   CAS [592-50-7]   C₅H₁₁F   Fp -9°F   FT-IR: 1, 57B   85.26   27.42
Fluoropentane, 98%   FW 90.14   VP-FT-IR: 3, 84D   83.09   27.34
bp 63°C   30.31   22.38
DCl₃   QE-300   d 0.789   30.06   13.95*
n_D^{20} 1.3600

B

CH₃(CH₂)₃CH₂F

Aldrich 23,837-6   CAS [543-59-9]   C₅H₁₁Cl   Fp 52°F   60 MHz: 1, 61A   45.18
Chloropentane, 99%   FW 106.60   FT-IR: 1, 57C   32.49
bp 108°C   VP-FT-IR: 3, 85A   29.17
DCl₃   QE-300   d 0.882   22.11
n_D^{20} 1.4115   13.99*

C

CH₃(CH₂)₃CH₂Cl

A

Aldrich 11,781-1 *CAS [110-53-2]*

1-Bromopentane, 99%

$C_5H_{11}Br$
FW 151.05
mp -95°C
bp 130°C
d 1.218

n_D^{20} 1.4436
Fp 88°F
60 MHz: *1*, 61B
FT-IR: *1*, 57D
VP-FT-IR: *3*, 85B

33.88
32.58
30.36
21.91
13.90*

CDCl₃   QE-300

$CH_3(CH_2)_3CH_2Br$

B

Aldrich 24,194-6 *CAS [628-17-1]*

1-Iodopentane, 98%

$C_5H_{11}I$
FW 198.05
bp 155°C
d 1.517
n_D^{20} 1.4950

Fp 124°F

60 MHz: *1*, 61C
FT-IR: *1*, 58A
VP-FT-IR: *3*, 85C

33.28
32.64
21.65
13.88*
7.08

CDCl₃   QE-300

$CH_3(CH_2)_3CH_2I$

C

Aldrich 25,007-4 *CAS [373-14-8]*

1-Fluorohexane, 98%

$C_6H_{13}F$
FW 104.17
bp 93°C
d 0.800
n_D^{20} 1.3755

Fp 81°F

VP-FT-IR: *3*, 85D

85.27   24.9*
83.10   24.8*
31.52   22.6
30.60   13.9
30.34

CDCl₃   QE-300

$CH_3(CH_2)_4CH_2F$

Non-Aromatic Halogenated Hydrocarbons

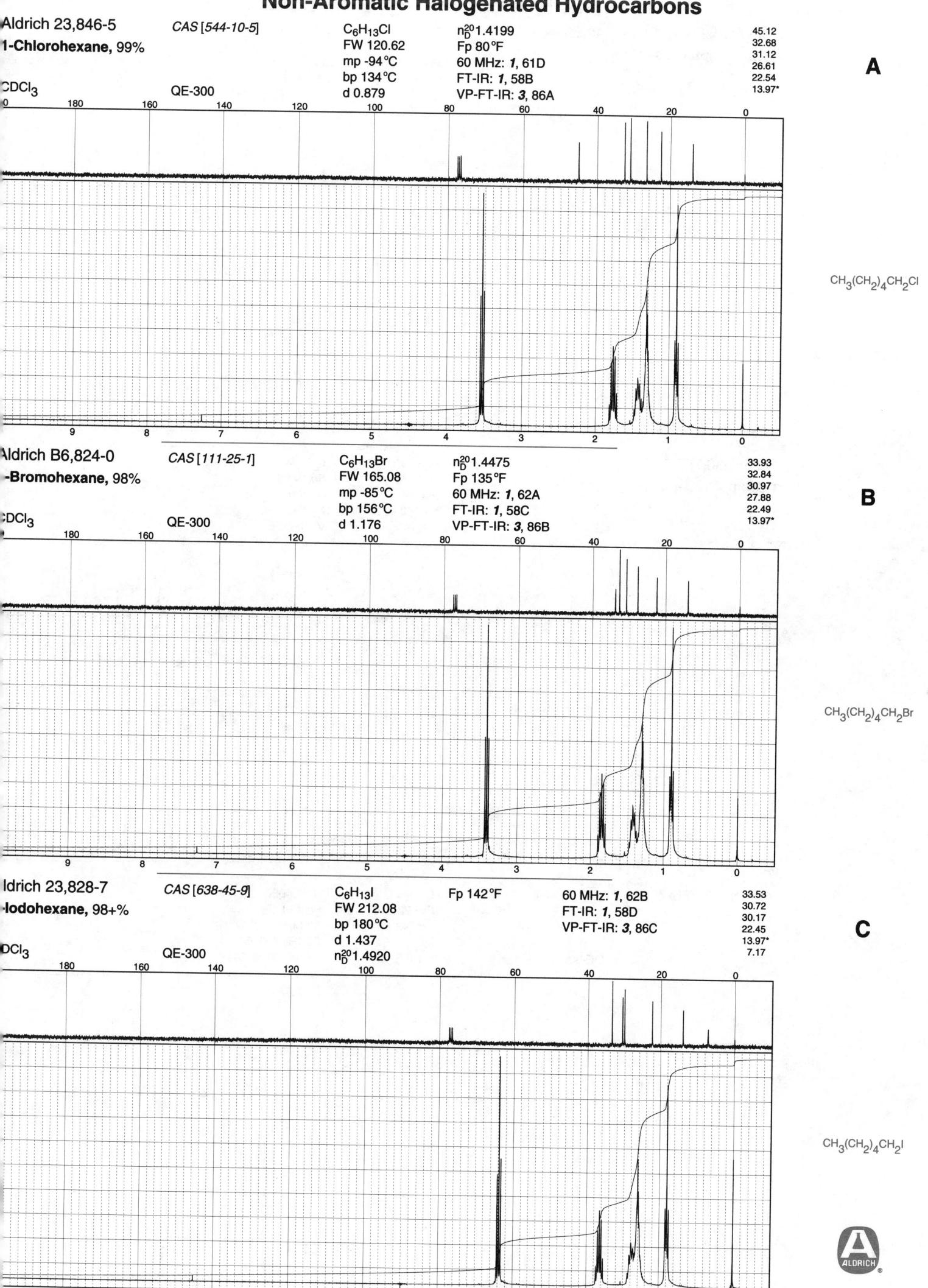

Aldrich 23,846-5
1-Chlorohexane, 99%

CAS [544-10-5]

C₆H₁₃Cl
FW 120.62
mp -94°C
bp 134°C
d 0.879

n_D^{20} 1.4199
Fp 80°F
60 MHz: *1*, 61D
FT-IR: *1*, 58B
VP-FT-IR: *3*, 86A

45.12
32.68
31.12
26.61
22.54
13.97*

A

CDCl₃   QE-300

CH₃(CH₂)₄CH₂Cl

Aldrich B6,824-0
-Bromohexane, 98%

CAS [111-25-1]

C₆H₁₃Br
FW 165.08
mp -85°C
bp 156°C
d 1.176

n_D^{20} 1.4475
Fp 135°F
60 MHz: *1*, 62A
FT-IR: *1*, 58C
VP-FT-IR: *3*, 86B

33.93
32.84
30.97
27.88
22.49
13.97*

B

CDCl₃   QE-300

CH₃(CH₂)₄CH₂Br

Idrich 23,828-7
-Iodohexane, 98+%

CAS [638-45-9]

C₆H₁₃I
FW 212.08
bp 180°C
d 1.437
n_D^{20} 1.4920

Fp 142°F

60 MHz: *1*, 62B
FT-IR: *1*, 58D
VP-FT-IR: *3*, 86C

33.53
30.72
30.17
22.45
13.97*
7.17

C

DCl₃   QE-300

CH₃(CH₂)₄CH₂I

ALDRICH

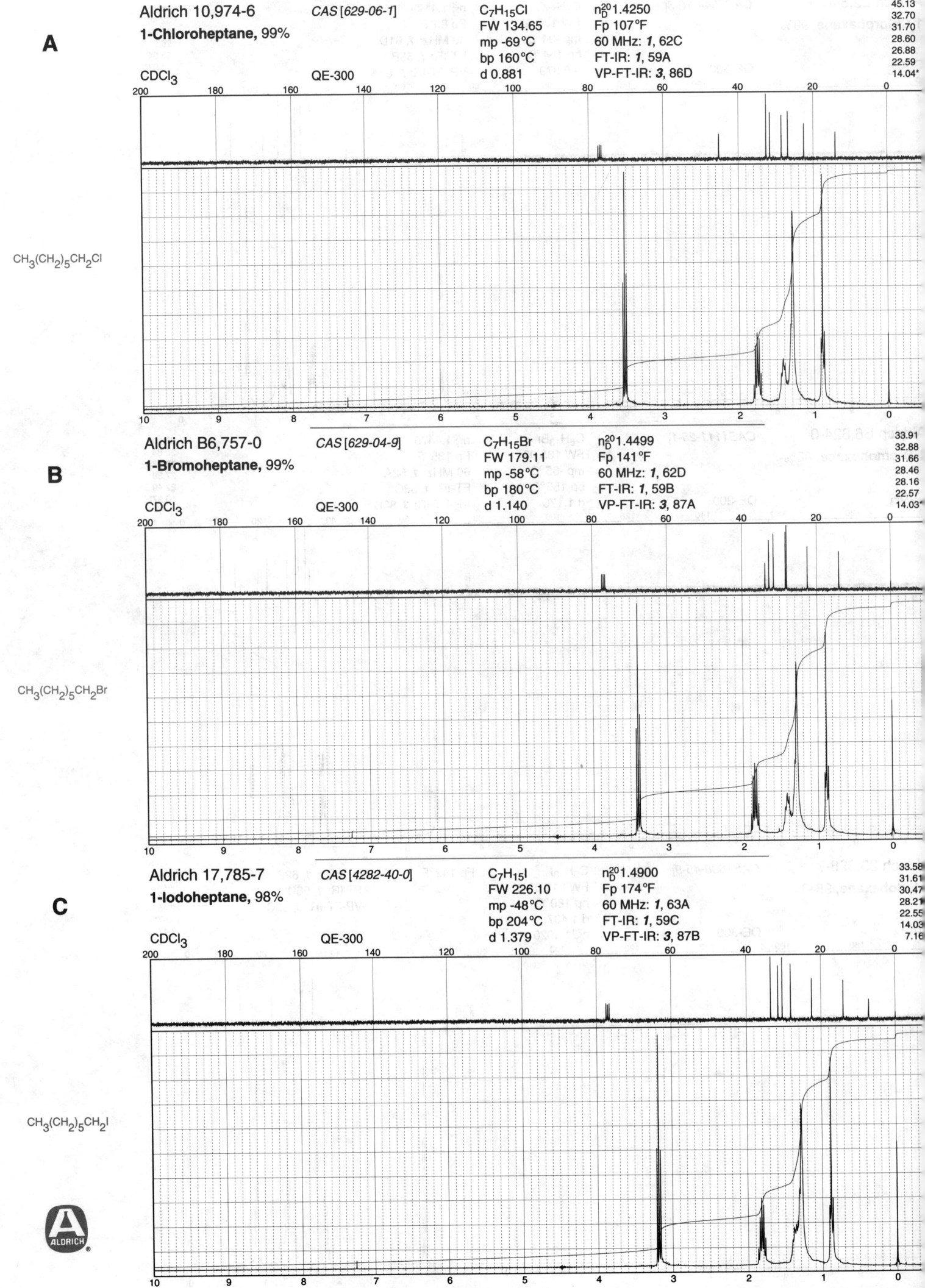

A

Aldrich 10,974-6   *CAS [629-06-1]*

1-Chloroheptane, 99%

CDCl₃   QE-300

$C_7H_{15}Cl$
FW 134.65
mp -69°C
bp 160°C
d 0.881

n_D^{20} 1.4250
Fp 107°F
60 MHz: *1*, 62C
FT-IR: *1*, 59A
VP-FT-IR: *3*, 86D

45.13
32.70
31.70
28.60
26.88
22.59
14.04*

$CH_3(CH_2)_5CH_2Cl$

B

Aldrich B6,757-0   *CAS [629-04-9]*

1-Bromoheptane, 99%

CDCl₃   QE-300

$C_7H_{15}Br$
FW 179.11
mp -58°C
bp 180°C
d 1.140

n_D^{20} 1.4499
Fp 141°F
60 MHz: *1*, 62D
FT-IR: *1*, 59B
VP-FT-IR: *3*, 87A

33.91
32.88
31.66
28.46
28.16
22.57
14.03*

$CH_3(CH_2)_5CH_2Br$

C

Aldrich 17,785-7   *CAS [4282-40-0]*

1-Iodoheptane, 98%

CDCl₃   QE-300

$C_7H_{15}I$
FW 226.10
mp -48°C
bp 204°C
d 1.379

n_D^{20} 1.4900
Fp 174°F
60 MHz: *1*, 63A
FT-IR: *1*, 59C
VP-FT-IR: *3*, 87B

33.58
31.61
30.47
28.21
22.55
14.03
7.16

$CH_3(CH_2)_5CH_2I$

Non-Aromatic Halogenated Hydrocarbons

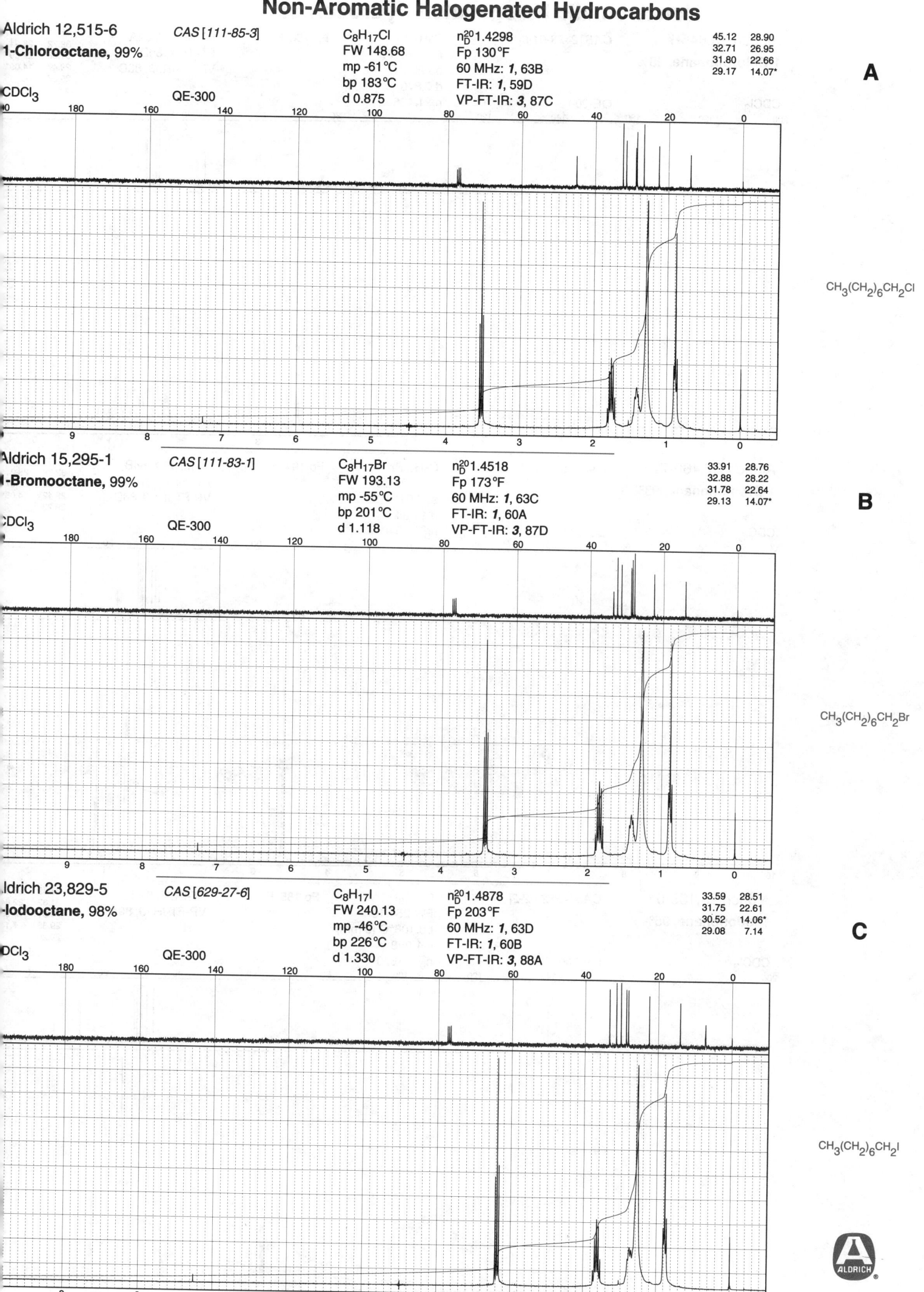

Aldrich 12,515-6   CAS [111-85-3]   C₈H₁₇Cl   n²⁰_D 1.4298   45.12 28.90
1-Chlorooctane, 99%   FW 148.68   Fp 130°F   32.71 26.95
   mp -61°C   60 MHz: 1, 63B   31.80 22.66
CDCl₃   QE-300   bp 183°C   FT-IR: 1, 59D   29.17 14.07*
   d 0.875   VP-FT-IR: 3, 87C

A

$CH_3(CH_2)_6CH_2Cl$

Aldrich 15,295-1   CAS [111-83-1]   C₈H₁₇Br   n²⁰_D 1.4518   33.91 28.76
1-Bromooctane, 99%   FW 193.13   Fp 173°F   32.88 28.22
   mp -55°C   60 MHz: 1, 63C   31.78 22.64
CDCl₃   QE-300   bp 201°C   FT-IR: 1, 60A   29.13 14.07*
   d 1.118   VP-FT-IR: 3, 87D

B

$CH_3(CH_2)_6CH_2Br$

Aldrich 23,829-5   CAS [629-27-6]   C₈H₁₇I   n²⁰_D 1.4878   33.59 28.51
1-Iodooctane, 98%   FW 240.13   Fp 203°F   31.75 22.61
   mp -46°C   60 MHz: 1, 63D   30.52 14.06*
CDCl₃   QE-300   bp 226°C   FT-IR: 1, 60B   29.08 7.14
   d 1.330   VP-FT-IR: 3, 88A

C

$CH_3(CH_2)_6CH_2I$

ALDRICH

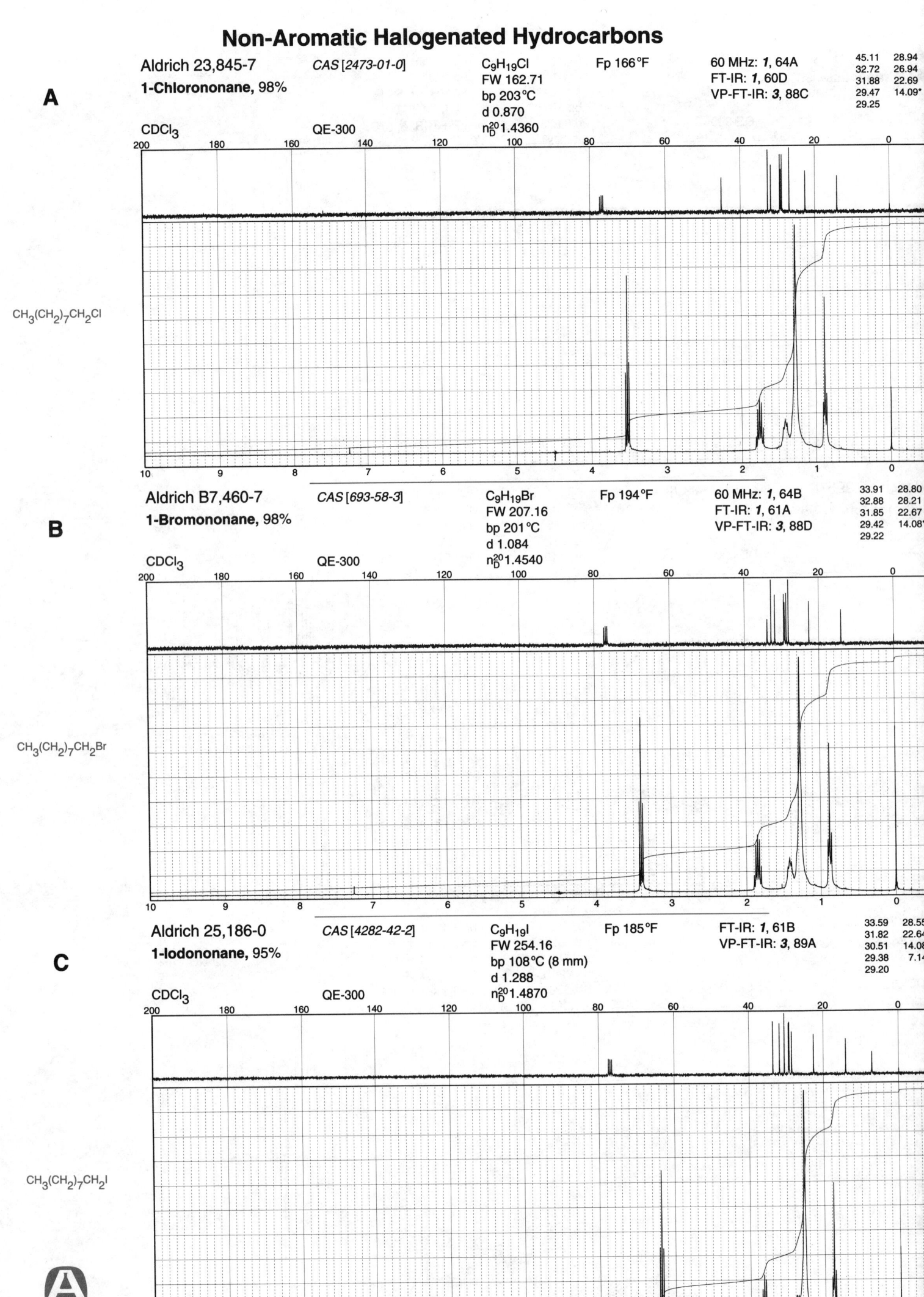

A

Aldrich 23,845-7
1-Chlorononane, 98%

CAS [2473-01-0]

C9H19Cl
FW 162.71
bp 203°C
d 0.870
n_D^{20}1.4360

Fp 166°F

60 MHz: *1*, 64A
FT-IR: *1*, 60D
VP-FT-IR: *3*, 88C

45.11	28.94
32.72	26.94
31.88	22.69
29.47	14.09*
29.25	

CDCl₃ QE-300

CH₃(CH₂)₇CH₂Cl

B

Aldrich B7,460-7
1-Bromononane, 98%

CAS [693-58-3]

C9H19Br
FW 207.16
bp 201°C
d 1.084
n_D^{20}1.4540

Fp 194°F

60 MHz: *1*, 64B
FT-IR: *1*, 61A
VP-FT-IR: *3*, 88D

33.91	28.80
32.88	28.21
31.85	22.67
29.42	14.08*
29.22	

CDCl₃ QE-300

CH₃(CH₂)₇CH₂Br

C

Aldrich 25,186-0
1-Iodononane, 95%

CAS [4282-42-2]

C9H19I
FW 254.16
bp 108°C (8 mm)
d 1.288
n_D^{20}1.4870

Fp 185°F

FT-IR: *1*, 61B
VP-FT-IR: *3*, 89A

33.59	28.55
31.82	22.64
30.51	14.08
29.38	7.14
29.20	

CDCl₃ QE-300

CH₃(CH₂)₇CH₂I

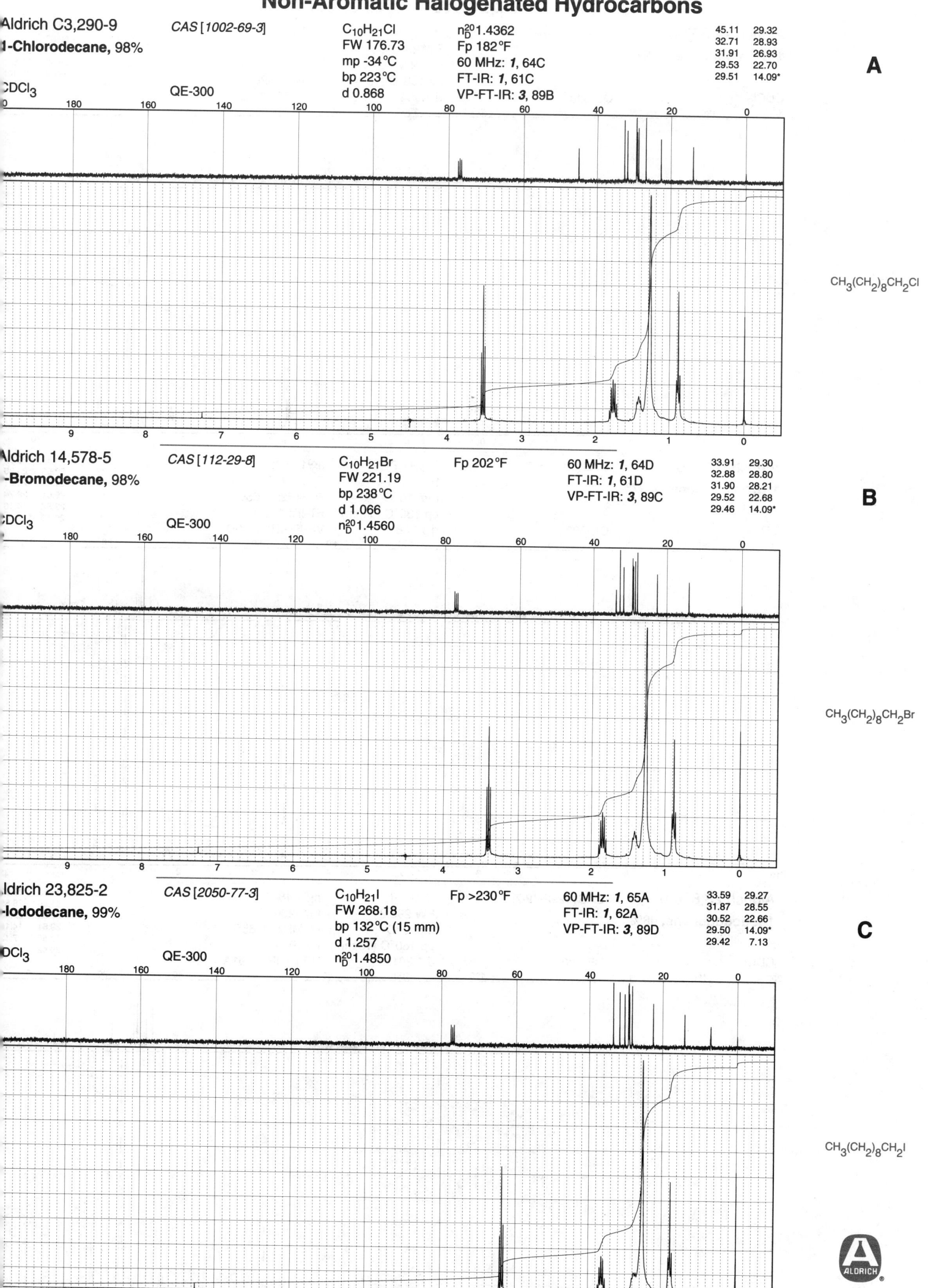

Aldrich C3,290-9
1-Chlorodecane, 98%

CAS [1002-69-3]

$C_{10}H_{21}Cl$
FW 176.73
mp -34°C
bp 223°C
d 0.868

n_D^{20} 1.4362
Fp 182°F
60 MHz: *1*, 64C
FT-IR: *1*, 61C
VP-FT-IR: *3*, 89B

45.11	29.32
32.71	28.93
31.91	26.93
29.53	22.70
29.51	14.09*

A

CDCl3

QE-300

$CH_3(CH_2)_8CH_2Cl$

Aldrich 14,578-5
1-Bromodecane, 98%

CAS [112-29-8]

$C_{10}H_{21}Br$
FW 221.19
bp 238°C
d 1.066
n_D^{20} 1.4560

Fp 202°F

60 MHz: *1*, 64D
FT-IR: *1*, 61D
VP-FT-IR: *3*, 89C

33.91	29.30
32.88	28.80
31.90	28.21
29.52	22.68
29.46	14.09*

B

CDCl3

QE-300

$CH_3(CH_2)_8CH_2Br$

Aldrich 23,825-2
1-Iododecane, 99%

CAS [2050-77-3]

$C_{10}H_{21}I$
FW 268.18
bp 132°C (15 mm)
d 1.257
n_D^{20} 1.4850

Fp >230°F

60 MHz: *1*, 65A
FT-IR: *1*, 62A
VP-FT-IR: *3*, 89D

33.59	29.27
31.87	28.55
30.52	22.66
29.50	14.09*
29.42	7.13

C

CDCl3

QE-300

$CH_3(CH_2)_8CH_2I$

A

Aldrich 24,503-8   *CAS [693-67-4]* $C_{11}H_{23}Br$ n_D^{20} 1.4563
1-Bromoundecane, 98%   FW 235.22   Fp >230°F
mp -9°C   60 MHz: *1*, 65B
bp 138°C (18 mm)   FT-IR: *1*, 62B

CDCl$_3$ QE-300   d 1.054   VP-FT-IR: *3*, 90A

33.93	29.33
32.87	28.79
31.91	28.20
29.58	22.69
29.46	14.10*

$CH_3(CH_2)_9CH_2Br$

B

Aldrich B6,555-1   *CAS [143-15-7]* $C_{12}H_{25}Br$ n_D^{20} 1.4580
1-Bromododecane, 97%   FW 249.24   Fp >230°F
mp -10°C   60 MHz: *1*, 65C
bp 135°C (6 mm)   FT-IR: *1*, 62C

CDCl$_3$ QE-300   d 1.038   VP-FT-IR: *3*, 90D

33.89	29.36
32.88	28.80
31.93	28.21
29.64	22.70
29.57	14.10*
29.46	

$CH_3(CH_2)_{10}CH_2Br$

C

Aldrich 23,826-0   *CAS [4292-19-7]* $C_{12}H_{25}I$ n_D^{20} 1.4844
1-Iodododecane, 98%   FW 296.24   Fp >230°F
mp -3°C   60 MHz: *1*, 65D
bp 160°C (15 mm)   FT-IR: *1*, 62D

CDCl$_3$ QE-300   d 1.201   VP-FT-IR: *3*, 91A

33.59	29.33
31.90	28.55
30.52	22.6
29.61	14.1
29.55	7.1
29.42	

$CH_3(CH_2)_{10}CH_2I$

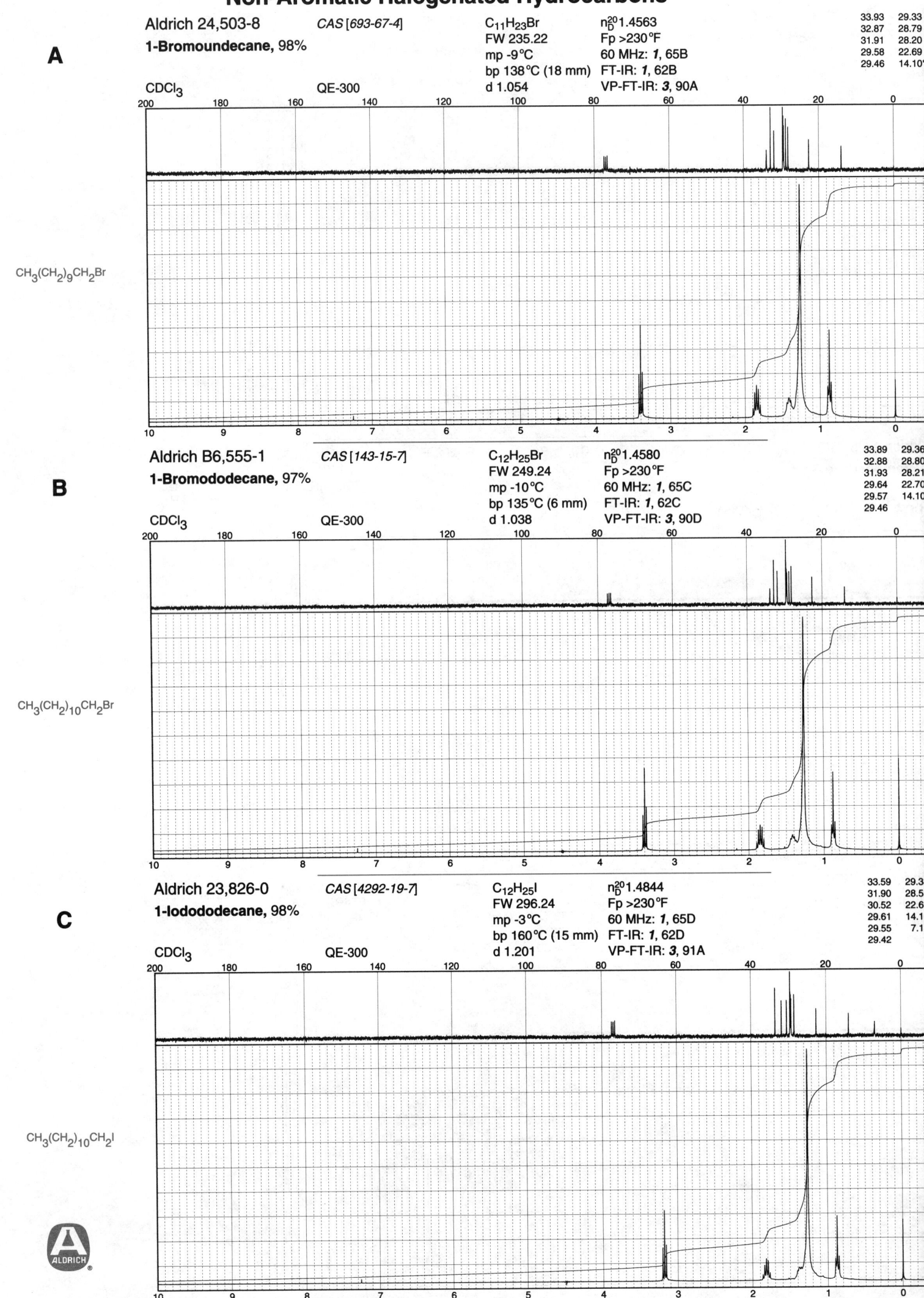

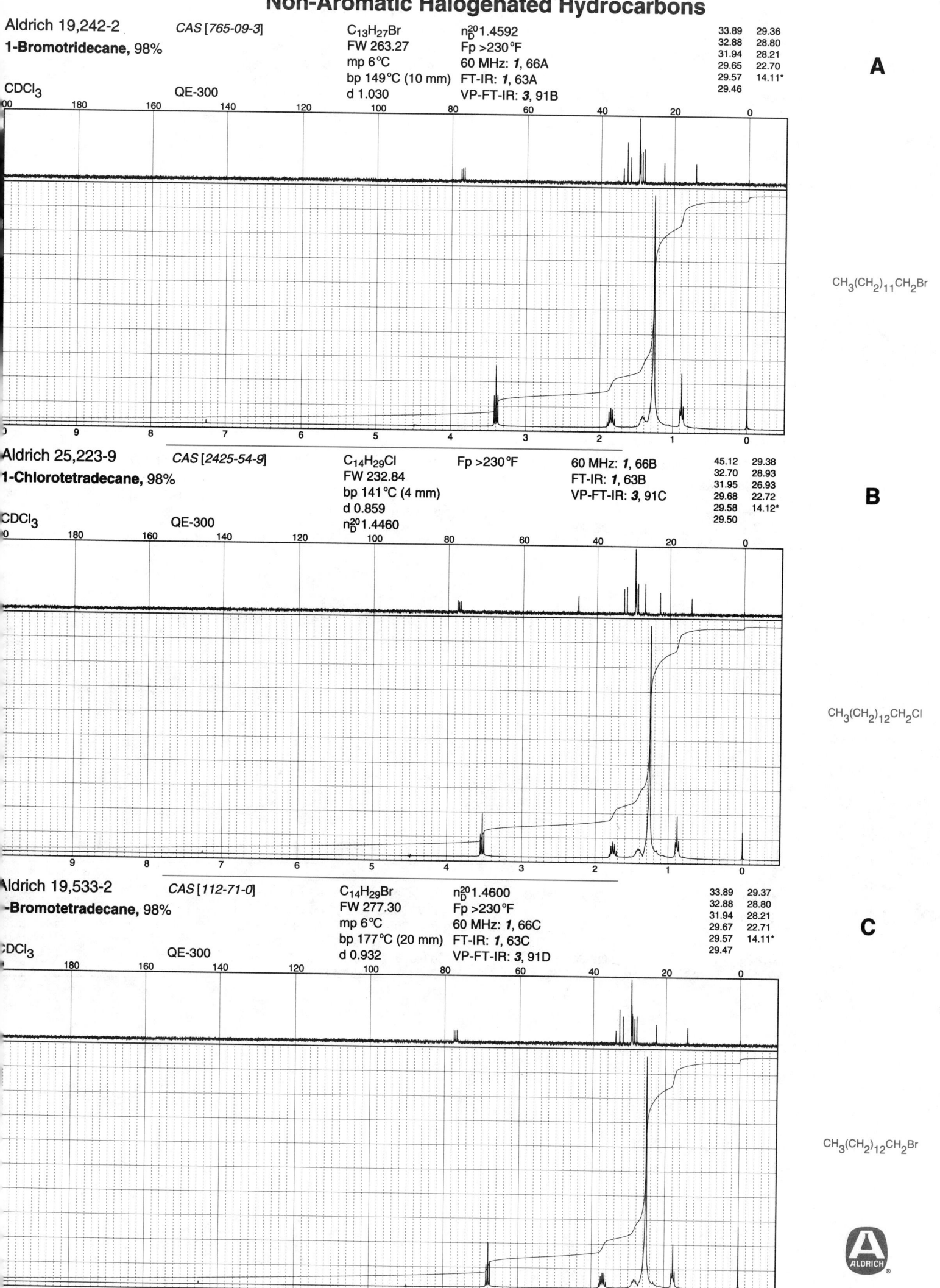

Aldrich 19,242-2   *CAS [765-09-3]*

1-Bromotridecane, 98%

$C_{13}H_{27}Br$
FW 263.27
mp 6°C
bp 149°C (10 mm)
d 1.030

n_D^{20} 1.4592
Fp >230°F
60 MHz: *1*, 66A
FT-IR: *1*, 63A
VP-FT-IR: *3*, 91B

33.89	29.36
32.88	28.80
31.94	28.21
29.65	22.70
29.57	14.11*
29.46	

CDCl₃   QE-300

A

$CH_3(CH_2)_{11}CH_2Br$

Aldrich 25,223-9   *CAS [2425-54-9]*

1-Chlorotetradecane, 98%

$C_{14}H_{29}Cl$
FW 232.84
bp 141°C (4 mm)
d 0.859
n_D^{20} 1.4460

Fp >230°F

60 MHz: *1*, 66B
FT-IR: *1*, 63B
VP-FT-IR: *3*, 91C

45.12	29.38
32.70	28.93
31.95	26.93
29.68	22.72
29.58	14.12*
29.50	

CDCl₃   QE-300

B

$CH_3(CH_2)_{12}CH_2Cl$

Aldrich 19,533-2   *CAS [112-71-0]*

1-Bromotetradecane, 98%

$C_{14}H_{29}Br$
FW 277.30
mp 6°C
bp 177°C (20 mm)
d 0.932

n_D^{20} 1.4600
Fp >230°F
60 MHz: *1*, 66C
FT-IR: *1*, 63C
VP-FT-IR: *3*, 91D

33.89	29.37
32.88	28.80
31.94	28.21
29.67	22.71
29.57	14.11*
29.47	

CDCl₃   QE-300

C

$CH_3(CH_2)_{12}CH_2Br$

ALDRICH

A

Aldrich 23,833-3   *CAS [629-72-1]* $C_{15}H_{31}Br$ n_D^{20} 1.4608
1-Bromopentadecane, 97%   FW 291.32   Fp >230 °F
    mp 18 °C   60 MHz: *1*, 66D
    bp 160 °C (5 mm)   FT-IR: *1*, 63D
    d 1.005   VP-FT-IR: *3*, 92A

32.88	29.38
31.95	28.88
29.68	28.21
29.57	22.71
29.46	14.11*

CDCl$_3$ QE-300

$CH_3(CH_2)_{13}CH_2Br$

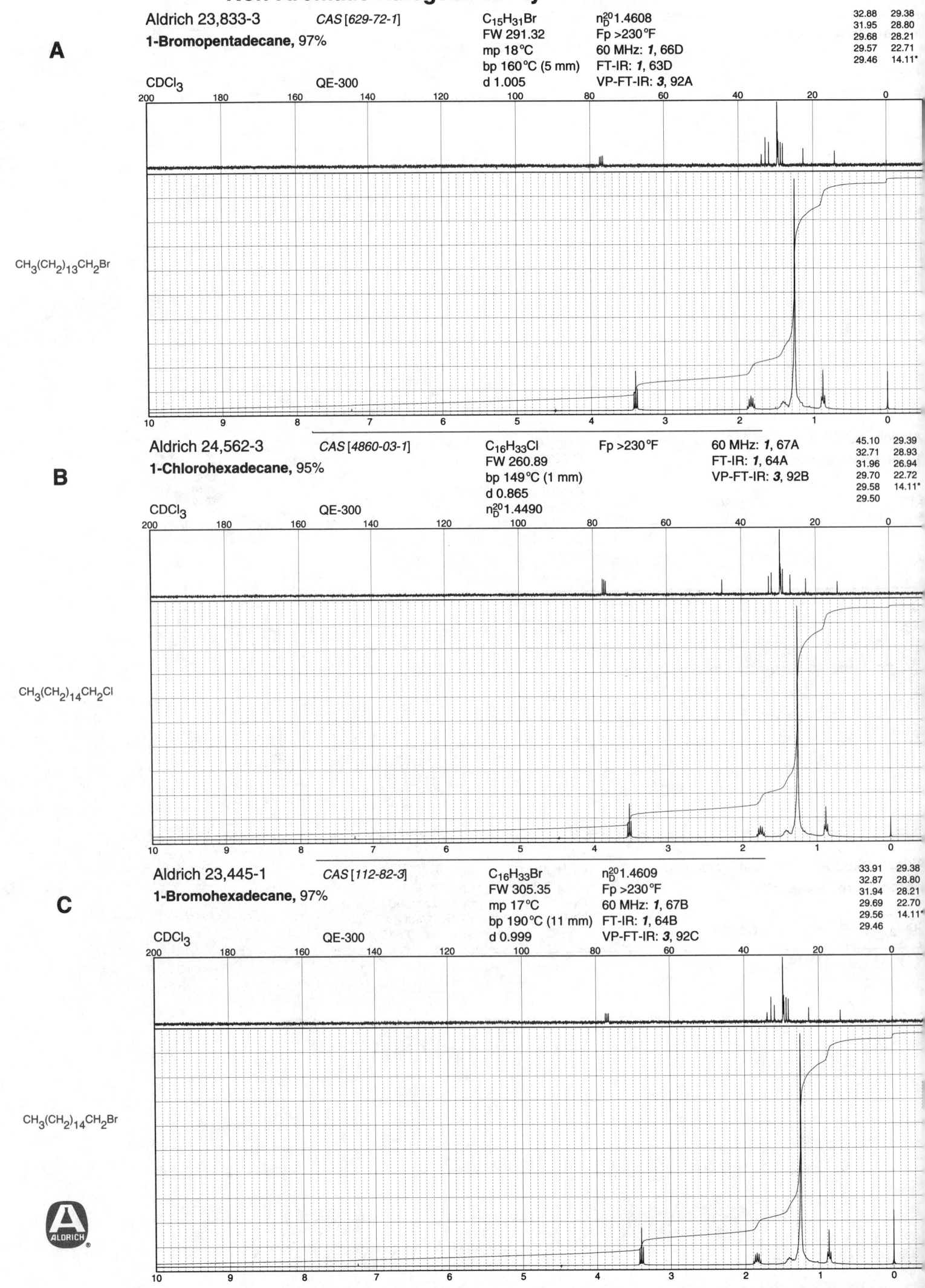

B

Aldrich 24,562-3   *CAS [4860-03-1]* $C_{16}H_{33}Cl$ Fp >230 °F   60 MHz: *1*, 67A
1-Chlorohexadecane, 95%   FW 260.89   FT-IR: *1*, 64A
    bp 149 °C (1 mm)   VP-FT-IR: *3*, 92B
    d 0.865
    n_D^{20} 1.4490

45.10	29.39
32.71	28.93
31.96	26.94
29.70	22.72
29.58	14.11*
29.50	

CDCl$_3$ QE-300

$CH_3(CH_2)_{14}CH_2Cl$

C

Aldrich 23,445-1   *CAS [112-82-3]* $C_{16}H_{33}Br$ n_D^{20} 1.4609
1-Bromohexadecane, 97%   FW 305.35   Fp >230 °F
    mp 17 °C   60 MHz: *1*, 67B
    bp 190 °C (11 mm)   FT-IR: *1*, 64B
    d 0.999   VP-FT-IR: *3*, 92C

33.91	29.38
32.87	28.80
31.94	28.21
29.69	22.70
29.56	14.11*
29.46	

CDCl$_3$ QE-300

$CH_3(CH_2)_{14}CH_2Br$

ALDRICH

Aldrich 23,827-9   CAS [544-77-4]   C16H33I   n$_D^{20}$1.4806

1-Iodohexadecane, 95%   FW 352.35   Fp >230°F

mp 22°C   FT-IR: *1*, 64C

bp 207°C (10 mm)   VP-FT-IR: *3*, 92D

d 1.121

CDCl3   QE-300

33.60	29.43
31.93	29.36
30.52	28.55
29.68	22.69
29.63	14.11*
29.55	7.12

A

CH3(CH2)14CH2I

Aldrich 23,836-8   CAS [3386-33-2]   C18H37Cl   Fp >230°F

1-Chlorooctadecane, 96%   FW 288.95   60 MHz: *1*, 67C

bp 158°C (1 mm)   FT-IR: *1*, 64D

d 0.849   VP-FT-IR: *3*, 93A

n$_D^{20}$1.4516

CDCl3   QE-300

45.09	29.39
32.71	28.93
31.96	26.94
29.72	22.71
29.58	14.11*
29.50	

B

CH3(CH2)16CH2Cl

Aldrich 19,949-4   CAS [112-89-0]   C18H37Br   Fp >230°F

1-Bromooctadecane, 96%   FW 333.41   60 MHz: *1*, 67D

mp 27°C   FT-IR: *1*, 65A

bp 215°C (12 mm)   VP-FT-IR: *3*, 93B

d 0.976

CDCl3   QE-300

33.90	29.38
32.87	28.80
31.94	28.21
29.71	22.71
29.56	14.11*
29.46	

C

CH3(CH2)16CH2Br

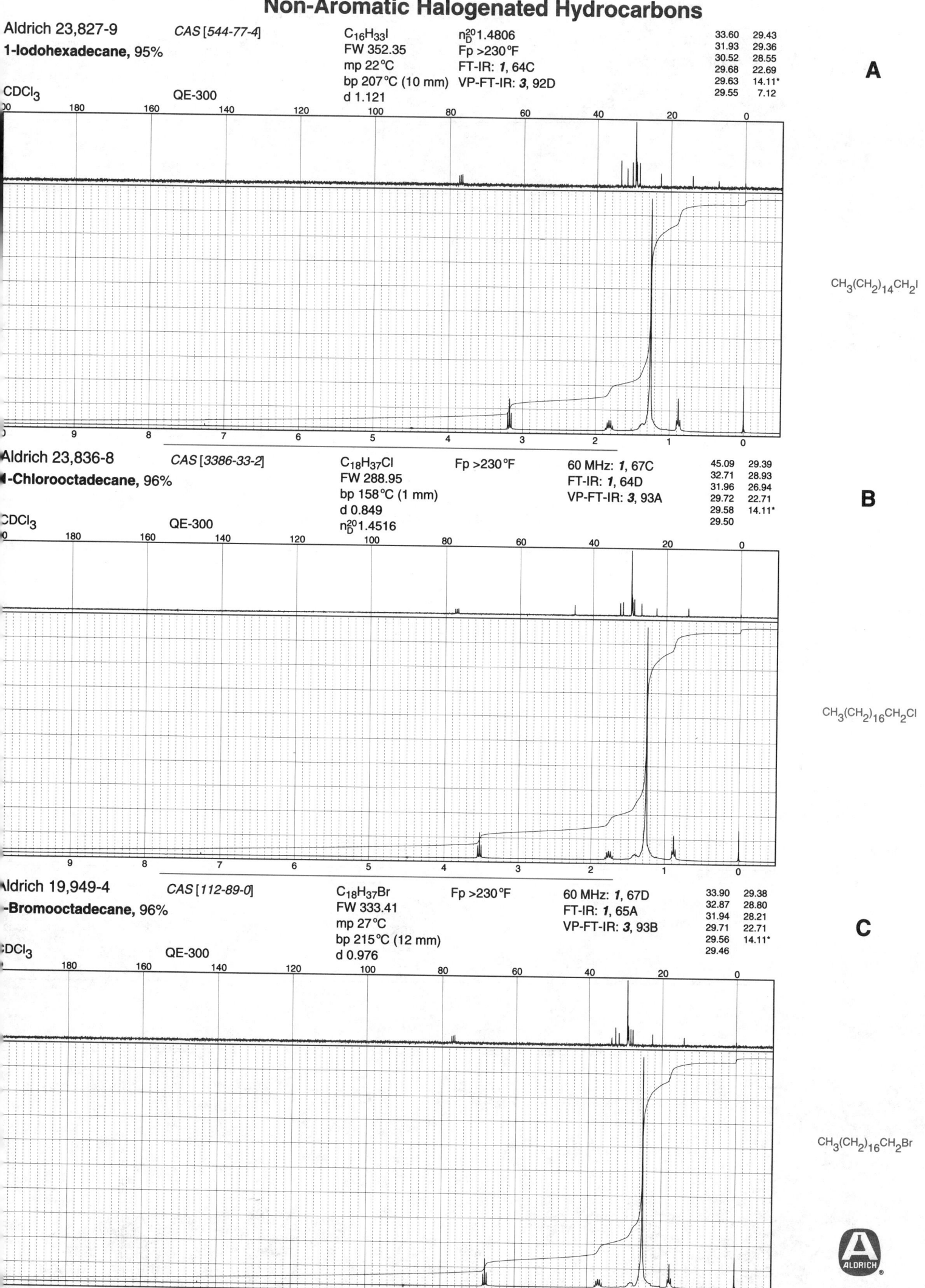

ALDRICH

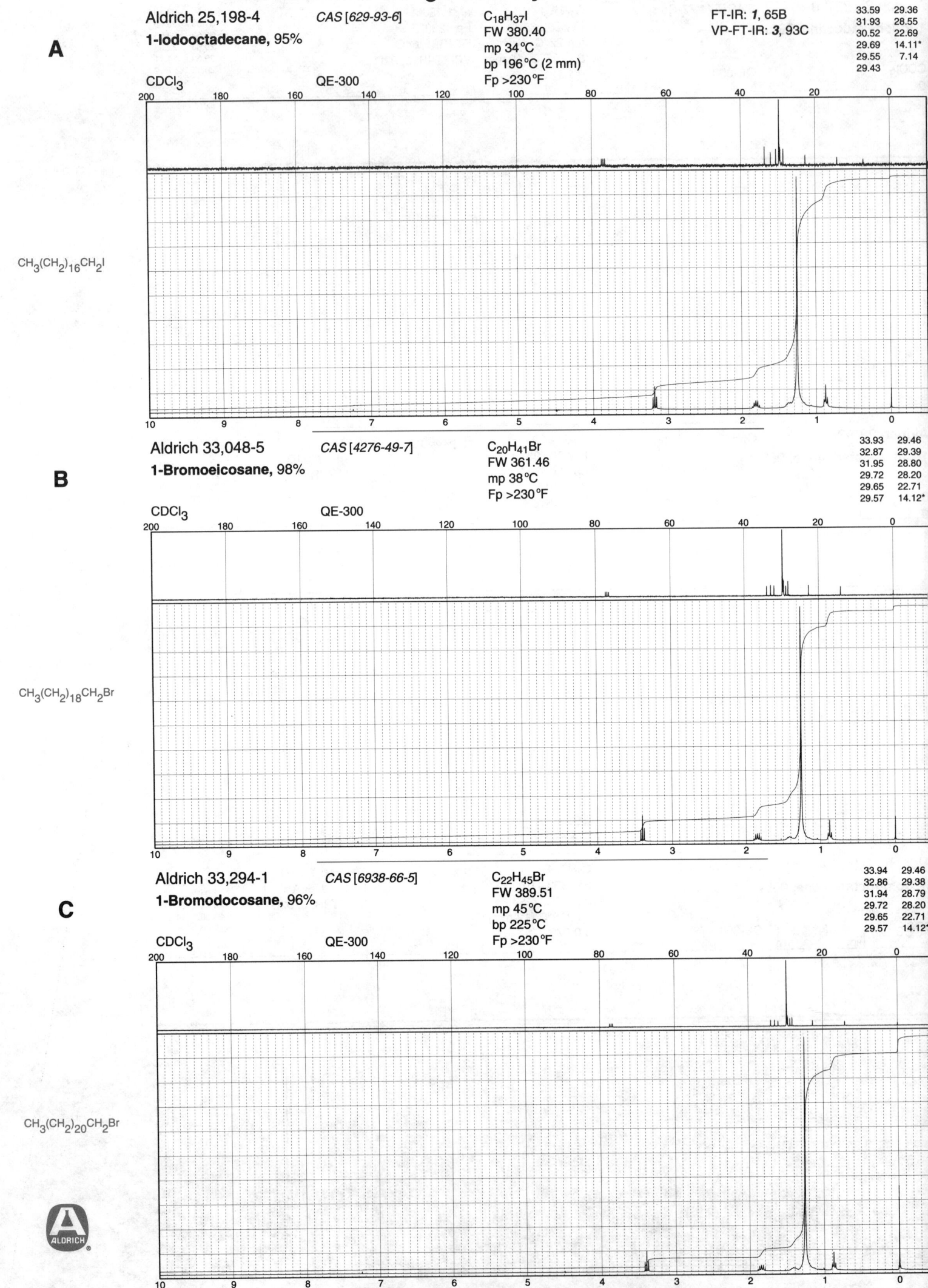

A

Aldrich 25,198-4   CAS [629-93-6]   C$_{18}$H$_{37}$I
1-Iodooctadecane, 95%   FW 380.40
mp 34 °C
bp 196 °C (2 mm)
Fp >230 °F

FT-IR: **1**, 65B
VP-FT-IR: **3**, 93C

33.59	29.36
31.93	28.55
30.52	22.69
29.69	14.11*
29.55	7.14
29.43	

CDCl$_3$ QE-300

CH$_3$(CH$_2$)$_{16}$CH$_2$I

B

Aldrich 33,048-5   CAS [4276-49-7]   C$_{20}$H$_{41}$Br
1-Bromoeicosane, 98%   FW 361.46
mp 38 °C
Fp >230 °F

33.93	29.46
32.87	29.39
31.95	28.80
29.72	28.20
29.65	22.71
29.57	14.12*

CDCl$_3$ QE-300

CH$_3$(CH$_2$)$_{18}$CH$_2$Br

C

Aldrich 33,294-1   CAS [6938-66-5]   C$_{22}$H$_{45}$Br
1-Bromodocosane, 96%   FW 389.51
mp 45 °C
bp 225 °C
Fp >230 °F

33.94	29.46
32.86	29.38
31.94	28.79
29.72	28.20
29.65	22.71
29.57	14.12*

CDCl$_3$ QE-300

CH$_3$(CH$_2$)$_{20}$CH$_2$Br

ALDRICH

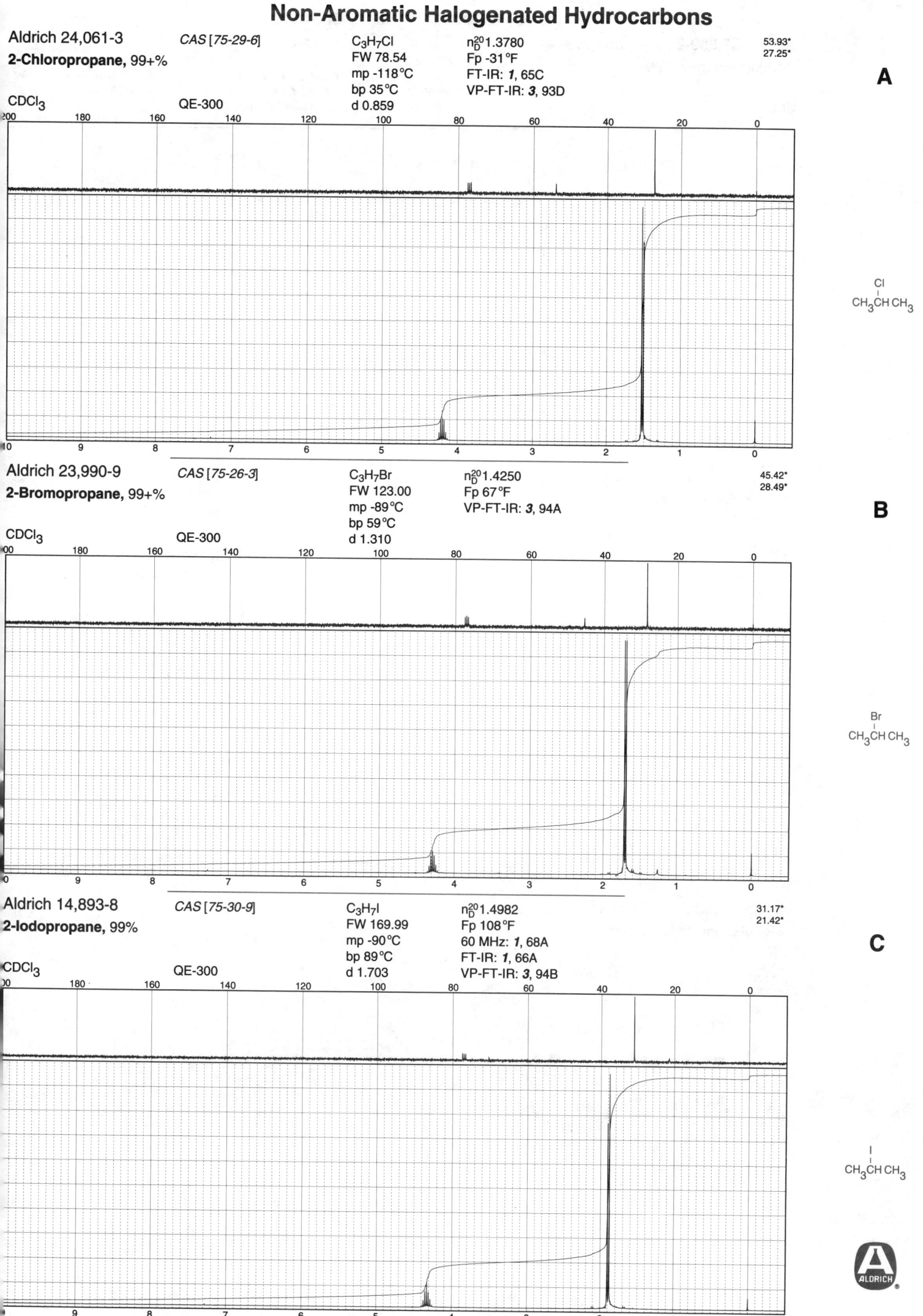

Aldrich 24,061-3   CAS [75-29-6]   C₃H₇Cl n_D^{20} 1.3780   53.93*
2-Chloropropane, 99+%   FW 78.54   Fp -31°F   27.25*
mp -118°C   FT-IR: *1*, 65C
bp 35°C   VP-FT-IR: *3*, 93D
d 0.859

CDCl₃   QE-300

A

Cl
CH₃CHCH₃

Aldrich 23,990-9   CAS [75-26-3]   C₃H₇Br n_D^{20} 1.4250   45.42*
2-Bromopropane, 99+%   FW 123.00   Fp 67°F   28.49*
mp -89°C   VP-FT-IR: *3*, 94A
bp 59°C
d 1.310

CDCl₃   QE-300

B

Br
CH₃CHCH₃

Aldrich 14,893-8   CAS [75-30-9]   C₃H₇I n_D^{20} 1.4982   31.17*
2-Iodopropane, 99%   FW 169.99   Fp 108°F   21.42*
mp -90°C   60 MHz: *1*, 68A
bp 89°C   FT-IR: *1*, 66A
d 1.703   VP-FT-IR: *3*, 94B

CDCl₃   QE-300

C

I
CH₃CHCH₃

ALDRICH

Non-Aromatic Halogenated Hydrocarbons

A

Aldrich C2,889-8   *CAS [78-86-4]*

2-Chlorobutane, 99+%

C$_4$H$_9$Cl
FW 92.57
mp -140°C
bp 69°C
d 0.873

n$_D^{20}$1.3960
Fp 5°F
60 MHz: *1*, 68B
FT-IR: *1*, 66B
VP-FT-IR: *3*, 94C

60.35*
33.38
24.90*
11.04*

CDCl$_3$ QE-300

CH$_3$CH$_2$CH CH$_3$
|
Cl

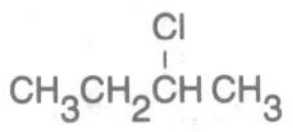

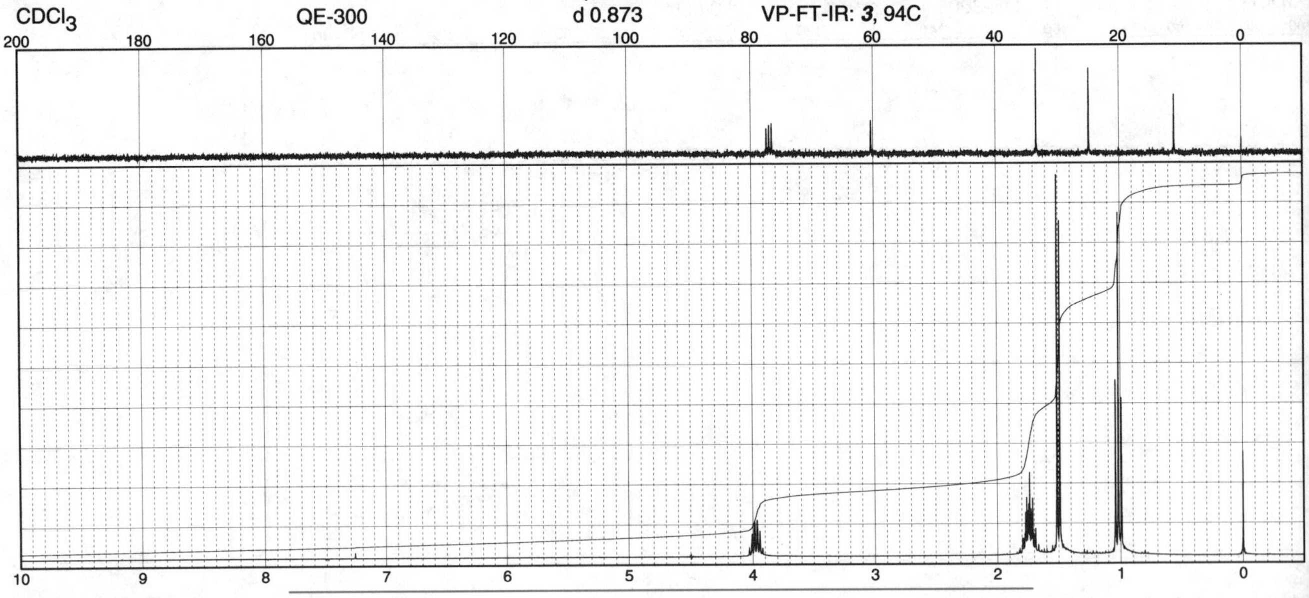

B

Aldrich B5,950-0   *CAS [78-76-2]*

2-Bromobutane, 98%

C$_4$H$_9$Br
FW 137.03
bp 91°C
d 1.255
n$_D^{20}$1.4369

Fp 70°F

60 MHz: *1*, 68C
FT-IR: *1*, 66C
VP-FT-IR: *3*, 94D

53.55*
34.26
26.08*
12.24*

CDCl$_3$ QE-300

CH$_3$CH$_2$CH CH$_3$
|
Br

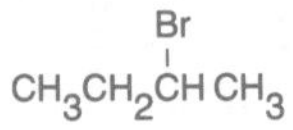

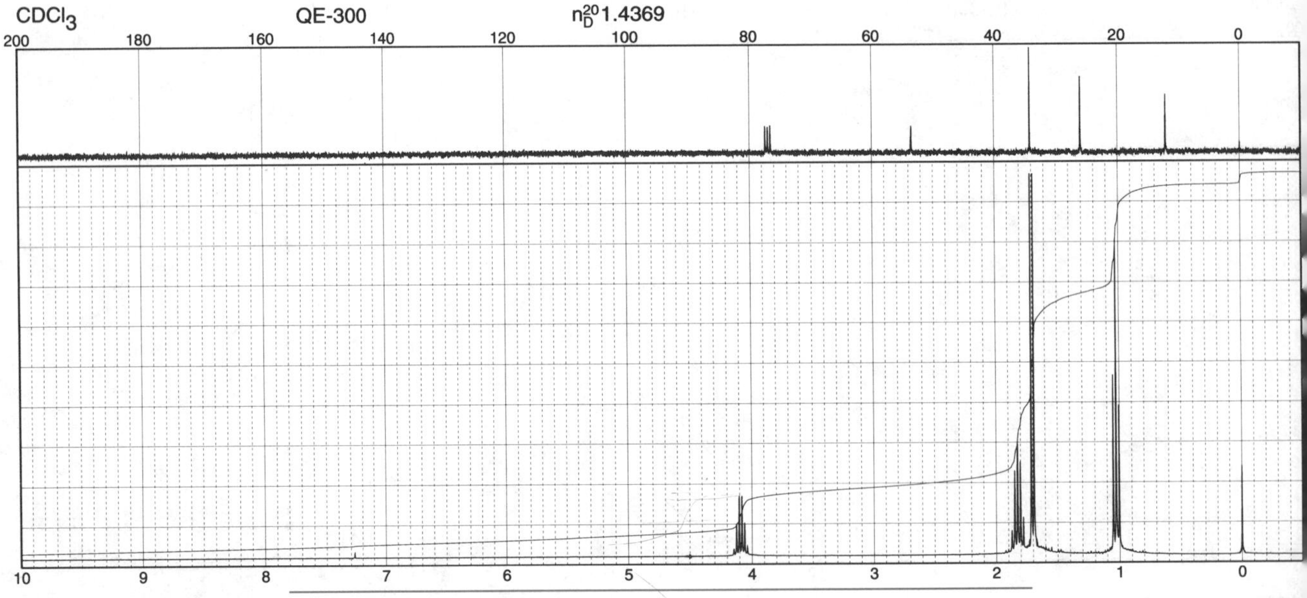

C

Aldrich 24,432-5   *CAS [513-48-4]*

2-Iodobutane, 99%

C$_4$H$_9$I
FW 184.02
mp -104°C
bp 120°C
d 1.598

n$_D^{20}$1.4991
Fp 75°F
60 MHz: *1*, 68D
FT-IR: *1*, 66D
VP-FT-IR: *3*, 95A

35.97
32.60*
28.50*
14.19*

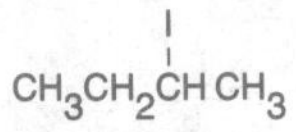

CDCl$_3$ QE-300

CH$_3$CH$_2$CH CH$_3$
|
I

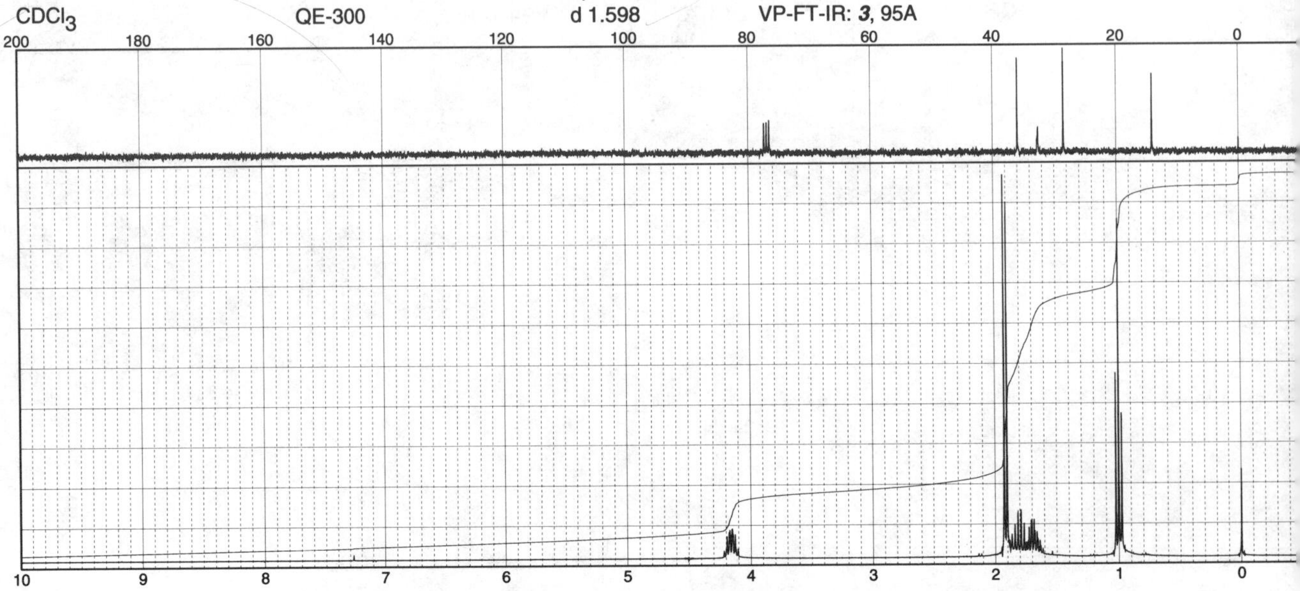

Non-Aromatic Halogenated Hydrocarbons

Aldrich 17,800-4   CAS [513-36-0]   C₄H₉Cl n_D^{20} 1.3975
1-Chloro-2-methylpropane, 98%   FW 92.57   Fp 70°F   60 MHz: 1, 69D
    mp -131°C   FT-IR: 1, 68C
CDCl₃   QE-300   bp 69°C   VP-FT-IR: 3, 95B
    d 0.883

52.48
30.99*
20.06*

A

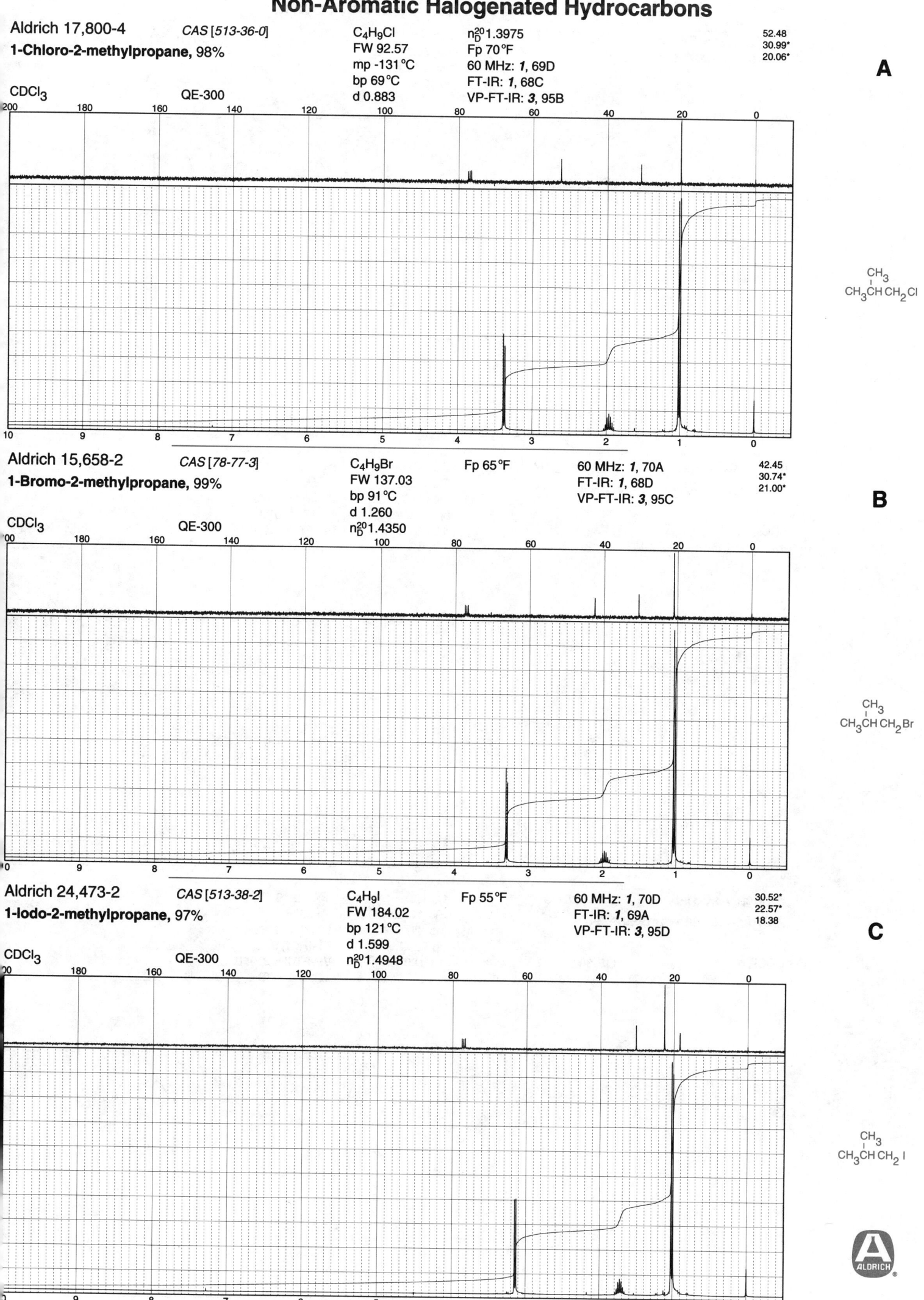

$$CH_3CHCH_2Cl \quad (CH_3)$$

Aldrich 15,658-2   CAS [78-77-3]   C₄H₉Br   Fp 65°F   60 MHz: 1, 70A
1-Bromo-2-methylpropane, 99%   FW 137.03   FT-IR: 1, 68D
    bp 91°C   VP-FT-IR: 3, 95C
    d 1.260
CDCl₃   QE-300 n_D^{20} 1.4350

42.45
30.74*
21.00*

B

$$CH_3CHCH_2Br \quad (CH_3)$$

Aldrich 24,473-2   CAS [513-38-2]   C₄H₉I   Fp 55°F   60 MHz: 1, 70D
1-Iodo-2-methylpropane, 97%   FW 184.02   FT-IR: 1, 69A
    bp 121°C   VP-FT-IR: 3, 95D
    d 1.599
CDCl₃   QE-300 n_D^{20} 1.4948

30.52*
22.57*
18.38

C

$$CH_3CHCH_2I \quad (CH_3)$$

Non-Aromatic Halogenated Hydrocarbons

A

Aldrich 17,760-1   *CAS [353-61-7]*

2-Fluoro-2-methylpropane, 97%

CDCl₃   QE-300

C₄H₉F
FW 76.11
mp -77°C
bp 12°C
Fp 10°F

FT-IR: *1*, 67A

94.75
92.60
28.73*
28.40*

CH₃
CH₃C CH₃
F

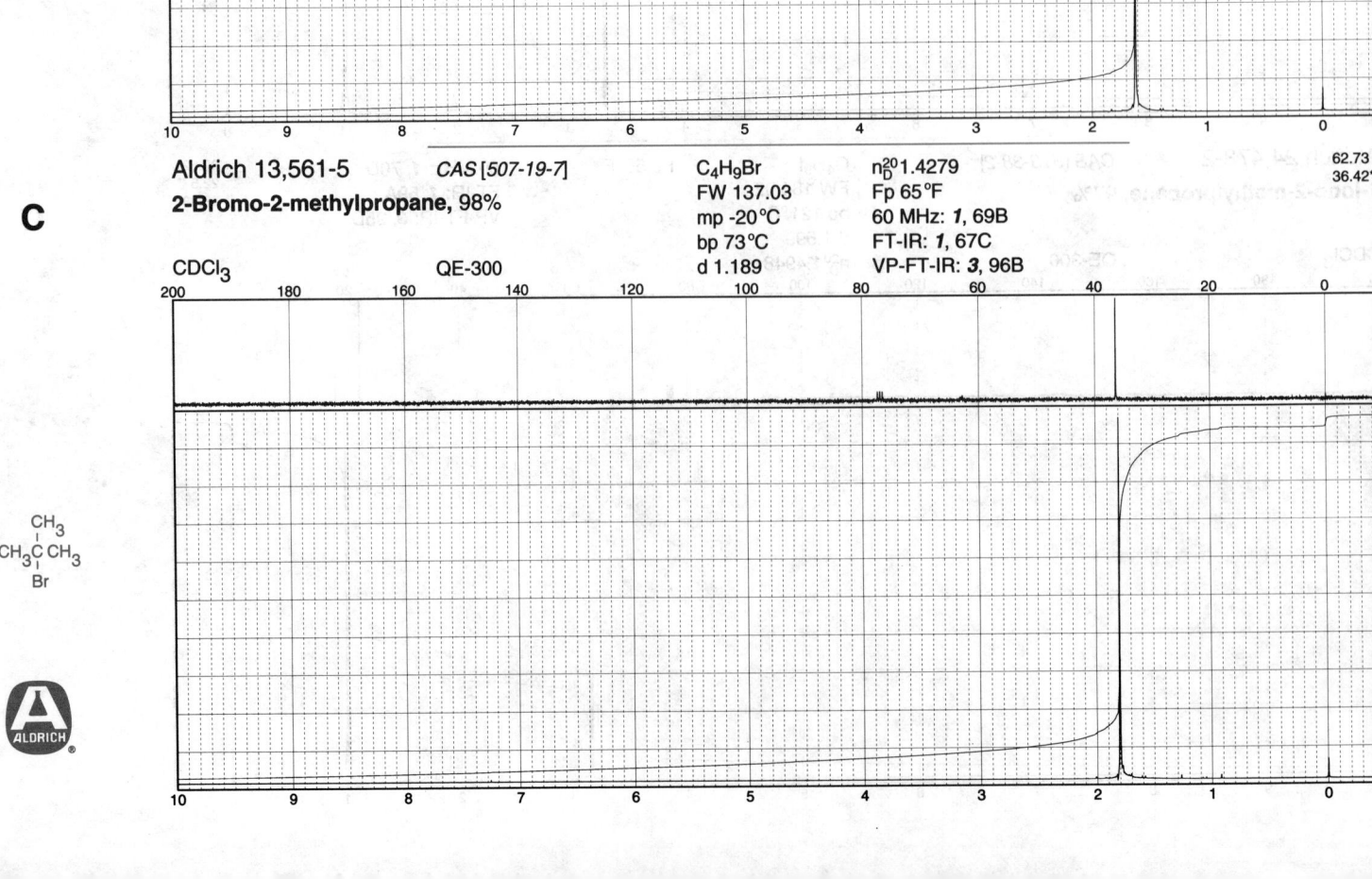

B

Aldrich C5,635-2   *CAS [507-20-0]*

2-Chloro-2-methylpropane, 99%

CDCl₃   QE-300

C₄H₉Cl
FW 92.57
mp -25°C
bp 52°C
d 0.851

n_D^{20}1.3848
Fp 65°F
60 MHz: *1*, 69A
FT-IR: *1*, 67B
VP-FT-IR: *3*, 96A

67.29
34.46*

CH₃
CH₃C CH₃
Cl

C

Aldrich 13,561-5   *CAS [507-19-7]*

2-Bromo-2-methylpropane, 98%

CDCl₃   QE-300

C₄H₉Br
FW 137.03
mp -20°C
bp 73°C
d 1.189

n_D^{20}1.4279
Fp 65°F
60 MHz: *1*, 69B
FT-IR: *1*, 67C
VP-FT-IR: *3*, 96B

62.73
36.42*

CH₃
CH₃C CH₃
Br

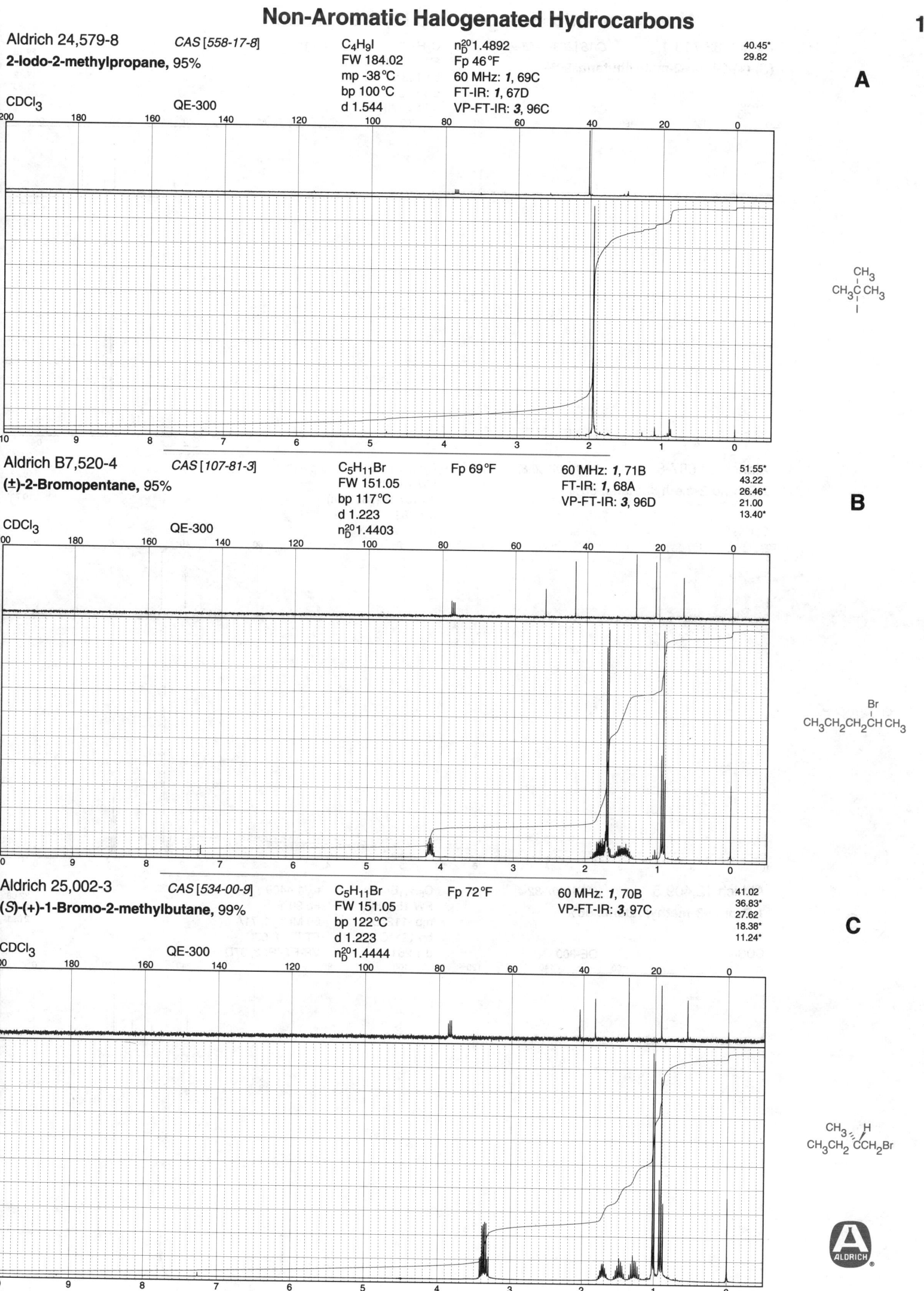

Aldrich 24,579-8   *CAS [558-17-8]*
2-Iodo-2-methylpropane, 95%

C$_4$H$_9$I
FW 184.02
mp -38°C
bp 100°C
d 1.544

n$_D^{20}$1.4892
Fp 46°F
60 MHz: *1*, 69C
FT-IR: *1*, 67D
VP-FT-IR: *3*, 96C

40.45*
29.82

CDCl$_3$ QE-300

A

Aldrich B7,520-4   *CAS [107-81-3]*
(±)-2-Bromopentane, 95%

C$_5$H$_{11}$Br
FW 151.05
bp 117°C
d 1.223
n$_D^{20}$1.4403

Fp 69°F

60 MHz: *1*, 71B
FT-IR: *1*, 68A
VP-FT-IR: *3*, 96D

51.55*
43.22
26.46*
21.00
13.40*

CDCl$_3$ QE-300

B

Aldrich 25,002-3   *CAS [534-00-9]*
(S)-(+)-1-Bromo-2-methylbutane, 99%

C$_5$H$_{11}$Br
FW 151.05
bp 122°C
d 1.223
n$_D^{20}$1.4444

Fp 72°F

60 MHz: *1*, 70B
VP-FT-IR: *3*, 97C

41.02
36.83*
27.62
18.38*
11.24*

CDCl$_3$ QE-300

C

Non-Aromatic Halogenated Hydrocarbons

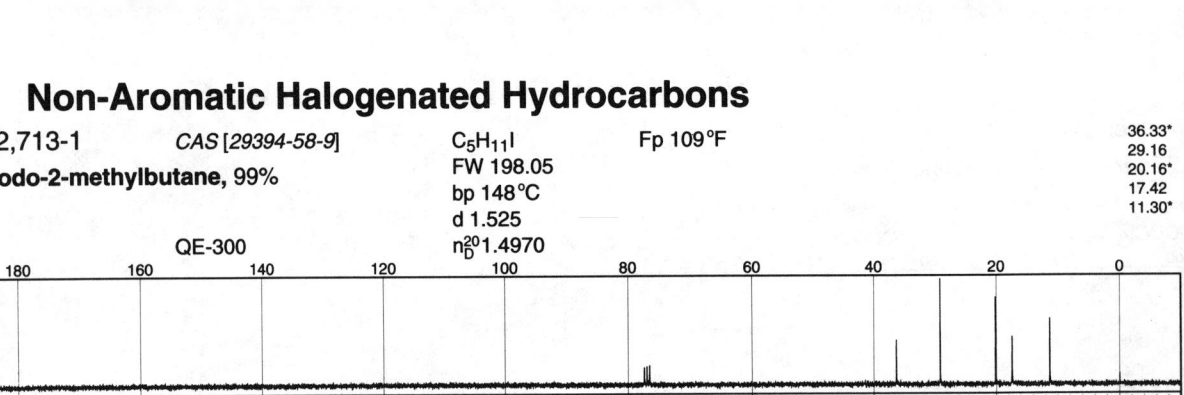

Aldrich 32,713-1   *CAS [29394-58-9]* $C_5H_{11}I$ Fp 109°F

A

(S)-(+)-1-Iodo-2-methylbutane, 99%

FW 198.05
bp 148°C
d 1.525
n_D^{20}1.4970

CDCl₃   QE-300

36.33*
29.16
20.16*
17.42
11.30*

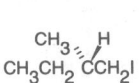

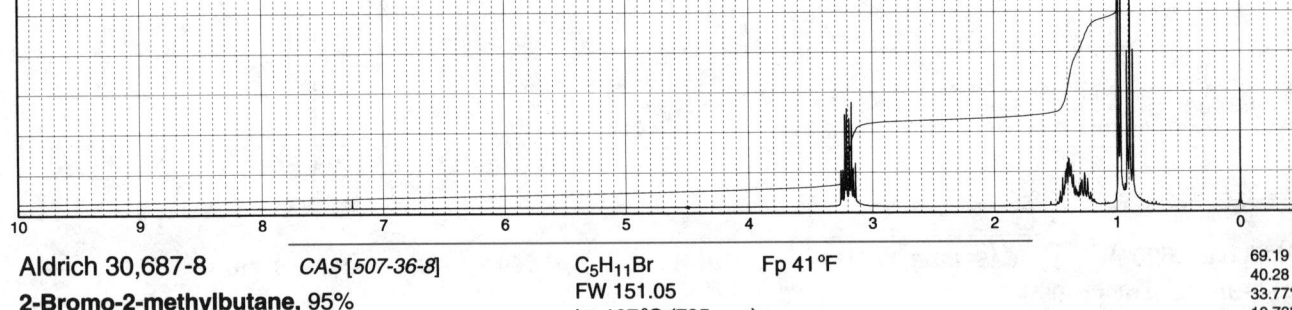

Aldrich 30,687-8   *CAS [507-36-8]* $C_5H_{11}Br$ Fp 41°F

B

2-Bromo-2-methylbutane, 95%

FW 151.05
bp 107°C (735 mm)
d 1.182
n_D^{20}1.4423

CDCl₃   QE-300

69.19
40.28
33.77*
10.73*

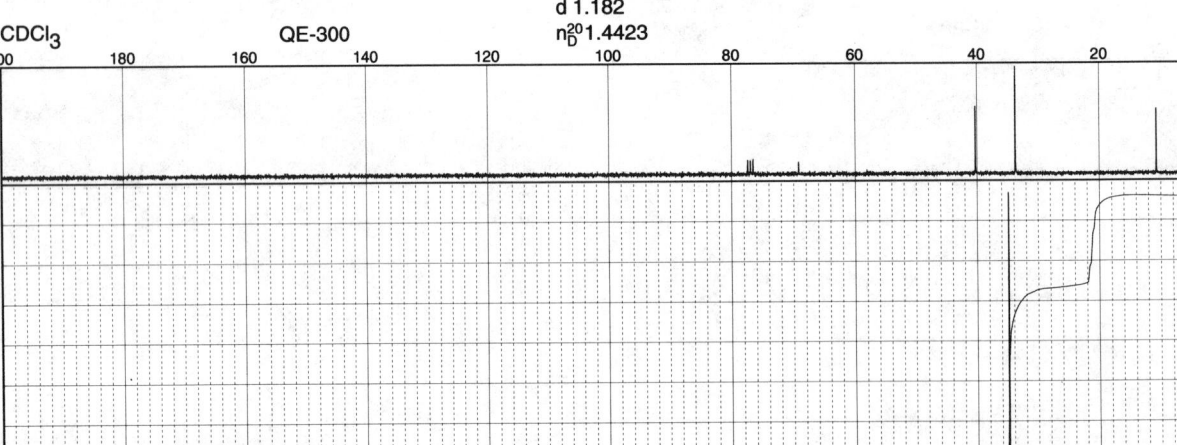

Aldrich 12,409-5   *CAS [107-82-4]* $C_5H_{11}Br$ n_D^{20}1.4409

C

1-Bromo-3-methylbutane, 96%

FW 151.05
mp -112°C
bp 121°C
d 1.261

Fp 90°F
60 MHz: *1*, 71A
FT-IR: *1*, 69C
VP-FT-IR: *3*, 97D

CDCl₃   QE-300

41.73
32.08
26.85*
21.91*

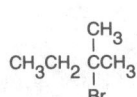

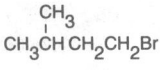

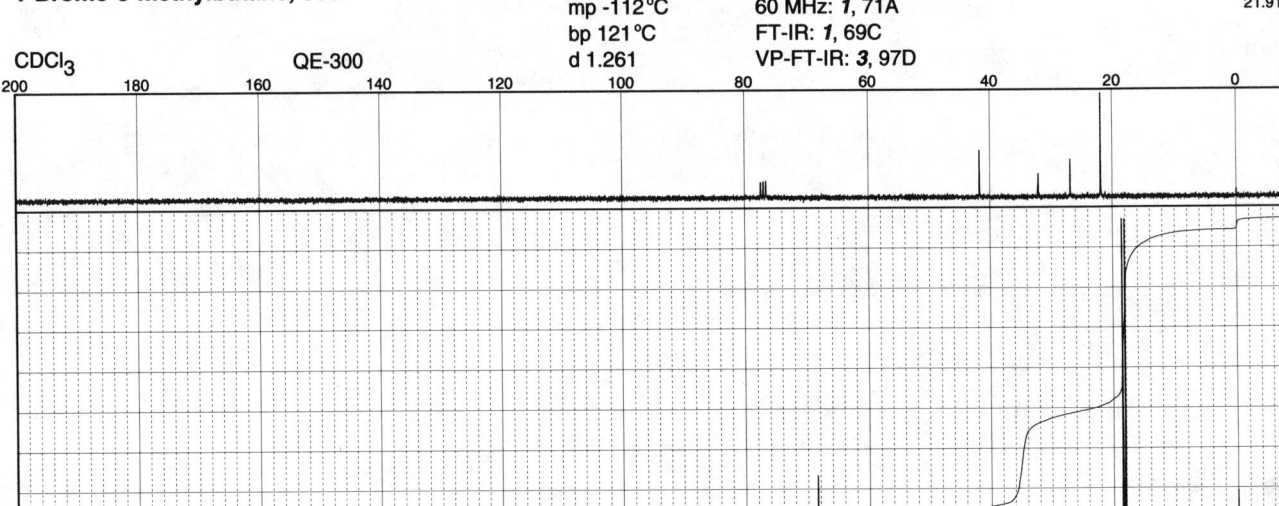

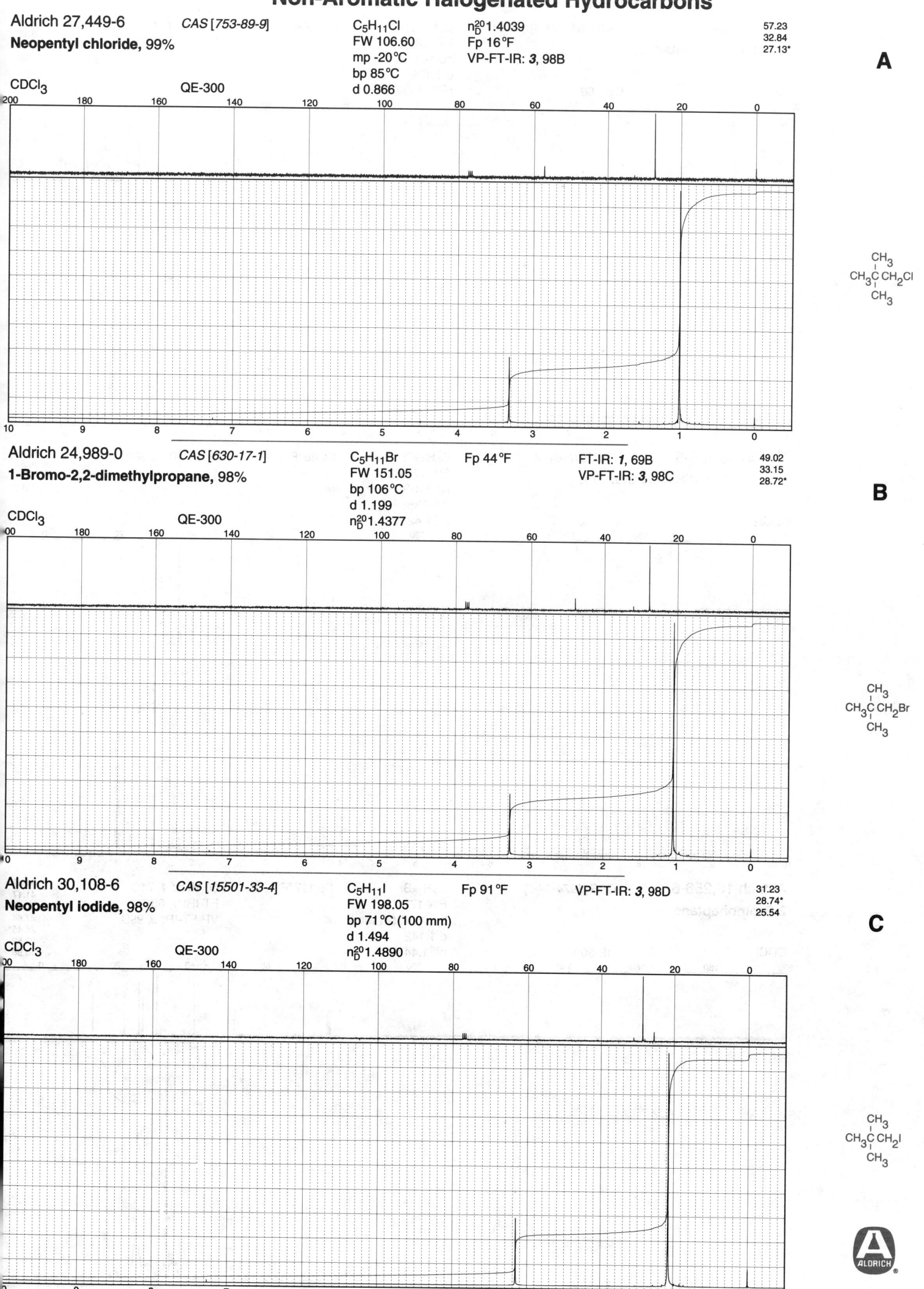

Aldrich 27,449-6   CAS [753-89-9] $C_5H_{11}Cl$ n_D^{20} 1.4039   57.23
Neopentyl chloride, 99%   FW 106.60   Fp 16°F   32.84
   mp -20°C   VP-FT-IR: *3*, 98B   27.13*
   bp 85°C
CDCl₃   QE-300   d 0.866

A

Aldrich 24,989-0   CAS [630-17-1] $C_5H_{11}Br$ Fp 44°F   FT-IR: *1*, 69B   49.02
1-Bromo-2,2-dimethylpropane, 98%   FW 151.05   VP-FT-IR: *3*, 98C   33.15
   bp 106°C   28.72*
   d 1.199
CDCl₃   QE-300 n_D^{20} 1.4377

B

Aldrich 30,108-6   CAS [15501-33-4] $C_5H_{11}I$ Fp 91°F   VP-FT-IR: *3*, 98D   31.23
Neopentyl iodide, 98%   FW 198.05   28.74*
   bp 71°C (100 mm)   25.54
   d 1.494
CDCl₃   QE-300 n_D^{20} 1.4890

C

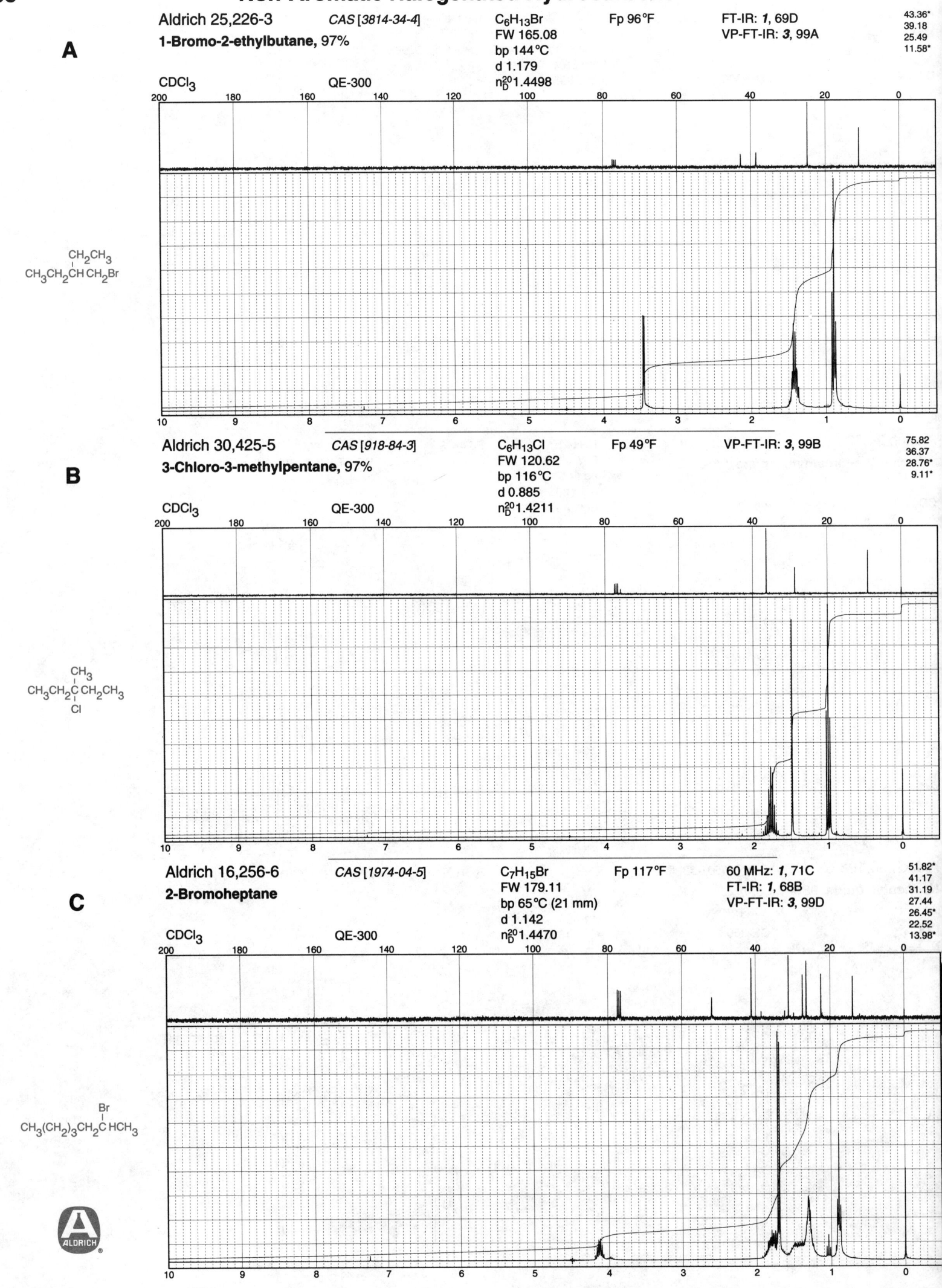

A

Aldrich 25,226-3   *CAS [3814-34-4]* $C_6H_{13}Br$ Fp 96°F   FT-IR: *1*, 69D   43.36*
1-Bromo-2-ethylbutane, 97%   FW 165.08   VP-FT-IR: *3*, 99A   39.18
bp 144°C   25.49
d 1.179   11.58*
n_D^{20}1.4498

CDCl₃   QE-300

CH₂CH₃
CH₃CH₂CH CH₂Br

B

Aldrich 30,425-5   *CAS [918-84-3]* $C_6H_{13}Cl$ Fp 49°F   VP-FT-IR: *3*, 99B   75.82
3-Chloro-3-methylpentane, 97%   FW 120.62   36.37
bp 116°C   28.76*
d 0.885   9.11*
n_D^{20}1.4211

CDCl₃   QE-300

CH₃
CH₃CH₂C CH CH₂CH₃
Cl

C

Aldrich 16,256-6   *CAS [1974-04-5]* $C_7H_{15}Br$ Fp 117°F   60 MHz: *1*, 71C   51.82*
2-Bromoheptane   FW 179.11   FT-IR: *1*, 68B   41.17
bp 65°C (21 mm)   VP-FT-IR: *3*, 99D   31.19
d 1.142   27.44
n_D^{20}1.4470   26.45*
22.52
13.98*

CDCl₃   QE-300

Br
CH₃(CH₂)₃CH₂C HCH₃

Aldrich 24,941-6   CAS [18908-66-2]   C$_8$H$_{17}$Br   Fp 157°F   FT-IR: *1*, 70A

2-Ethylhexyl bromide, 95%   FW 193.13   VP-FT-IR: *3*, 100C

bp 76°C (16 mm)

d 1.086

CDCl$_3$ QE-300   n$_D^{20}$1.4538

A

41.13*	25.22
39.01	22.86
31.93	14.03*
28.86	10.88*

CH$_2$CH$_3$

CH$_3$CH$_2$CH$_2$CH$_2$CH CH$_2$Br

Aldrich 10,849-9   CAS [13187-99-0]   C$_{12}$H$_{25}$Br   n$_D^{20}$1.4555

2-Bromododecane, tech., 85%   FW 249.24   Fp >230°F

mp -96°C   60 MHz: *1*, 72A

bp 130°C (6 mm)   FT-IR: *1*, 70B

CDCl$_3$ QE-300   d 1.020   VP-FT-IR: *3*, 101C

B

60.49*	29.59	26.44*
51.77*	29.50	22.69
41.24	29.33	14.09*
38.81	29.09	12.05*
32.18	29.02	
31.92	27.78	

Br

CH$_3$(CH$_2$)$_8$CH$_2$CH CH$_3$

Aldrich B8,230-8   CAS [59157-17-4]   C$_{13}$H$_{27}$Br   60 MHz: *1*, 72B

2-Bromotridecane, 96%   FW 263.27   FT-IR: *1*, 70C

d 0.998   VP-FT-IR: *3*, 101D

n$_D^{20}$1.4570

CDCl$_3$ QE-300   Fp >230°F

C

51.88*	29.36
41.20	29.01
31.92	27.79
29.64	26.46*
29.58	22.70
29.50	14.13*

Br

CH$_3$(CH$_2$)$_9$CH$_2$CH CH$_3$

ALDRICH®

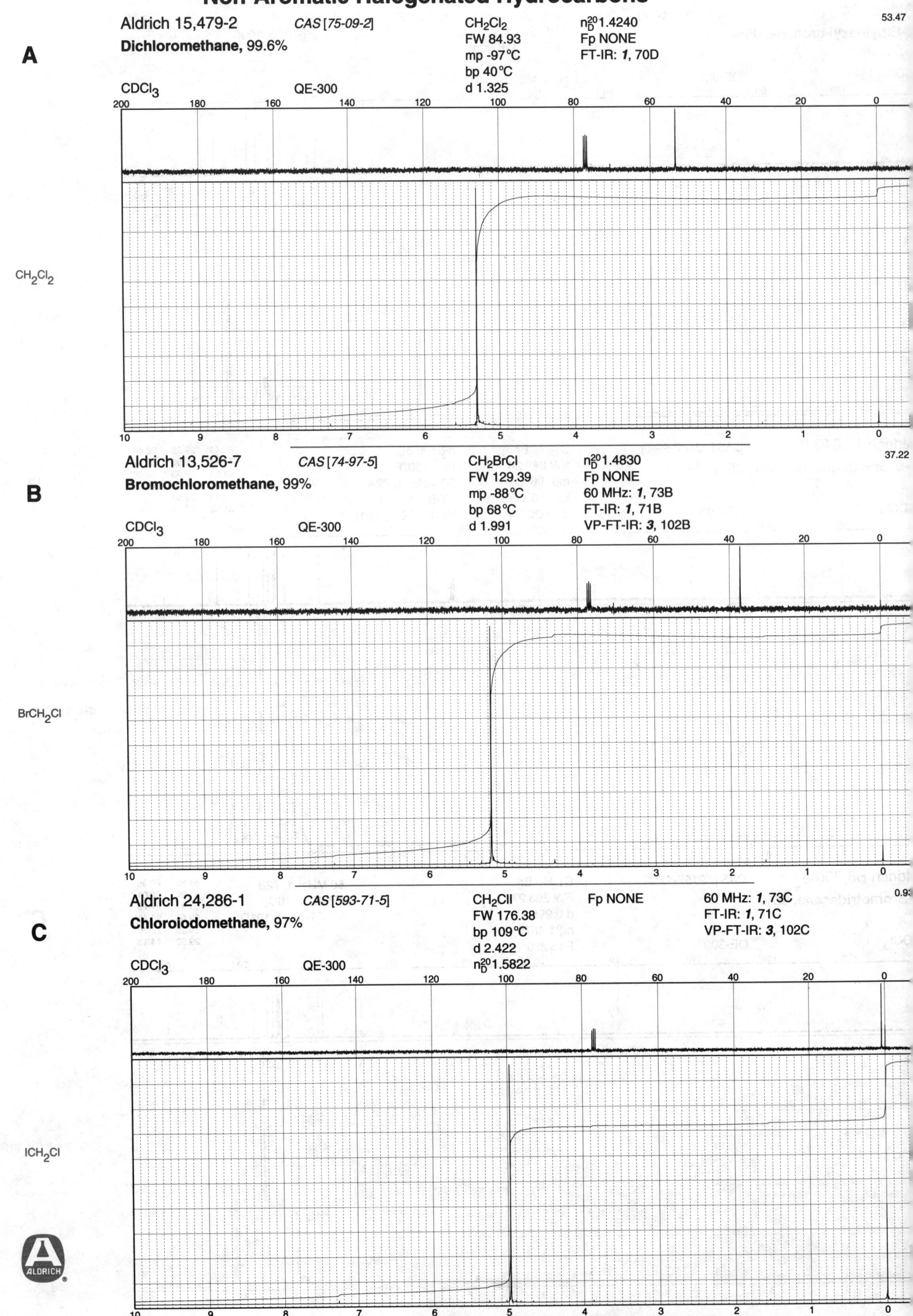

A

53.47

Aldrich 15,479-2   *CAS [75-09-2]*   CH₂Cl₂ n_D^{20}1.4240
Dichloromethane, 99.6%   FW 84.93   Fp NONE
mp -97°C   FT-IR: *1*, 70D
bp 40°C
CDCl₃   QE-300   d 1.325

CH₂Cl₂

B

37.22

Aldrich 13,526-7   *CAS [74-97-5]*   CH₂BrCl n_D^{20}1.4830
Bromochloromethane, 99%   FW 129.39   Fp NONE
mp -88°C   60 MHz: *1*, 73B
bp 68°C   FT-IR: *1*, 71B
CDCl₃   QE-300   d 1.991   VP-FT-IR: *3*, 102B

BrCH₂Cl

C

0.93

Aldrich 24,286-1   *CAS [593-71-5]*   CH₂ClI   Fp NONE   60 MHz: *1*, 73C
Chloroiodomethane, 97%   FW 176.38   FT-IR: *1*, 71C
bp 109°C   VP-FT-IR: *3*, 102C
d 2.422
CDCl₃   QE-300 n_D^{20}1.5822

ICH₂Cl

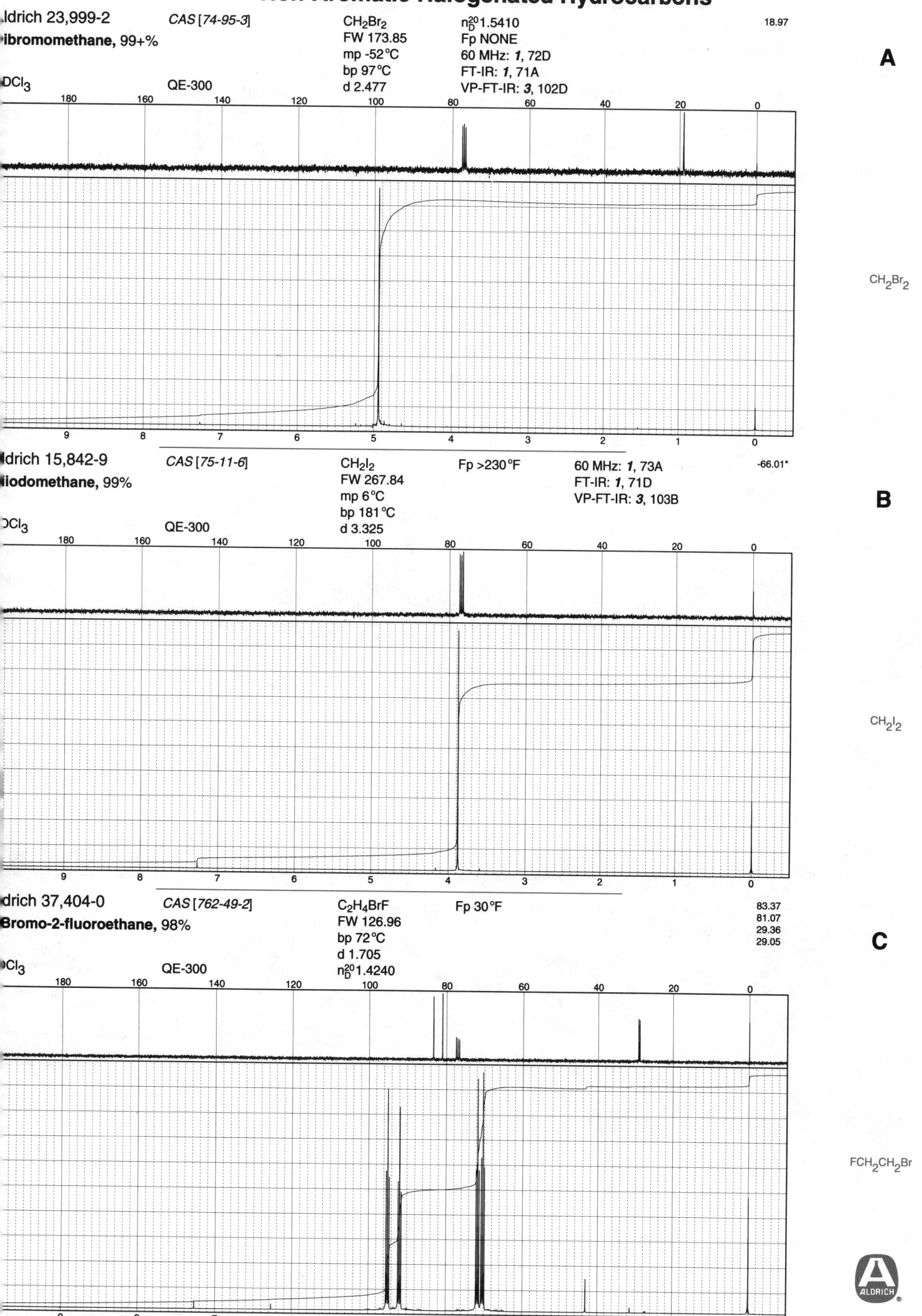

A

Aldrich 23,999-2 CAS [74-95-3] CH₂Br₂ n²⁰_D 1.5410 18.97
ibromomethane, 99+% FW 173.85 Fp NONE
DCl₃ QE-300 mp -52°C 60 MHz: *1*, 72D
bp 97°C FT-IR: *1*, 71A
d 2.477 VP-FT-IR: *3*, 102D

CH_2Br_2

B

Aldrich 15,842-9 CAS [75-11-6] CH₂I₂ Fp >230°F 60 MHz: *1*, 73A -66.01*
iodomethane, 99% FW 267.84 FT-IR: *1*, 71D
DCl₃ QE-300 mp 6°C VP-FT-IR: *3*, 103B
bp 181°C
d 3.325

CH_2I_2

C

Aldrich 37,404-0 CAS [762-49-2] C₂H₄BrF Fp 30°F 83.37
Bromo-2-fluoroethane, 98% FW 126.96 81.07
bp 72°C 29.36
DCl₃ QE-300 d 1.705 29.05
n²⁰_D 1.4240

FCH_2CH_2Br

ALDRICH

Non-Aromatic Halogenated Hydrocarbons

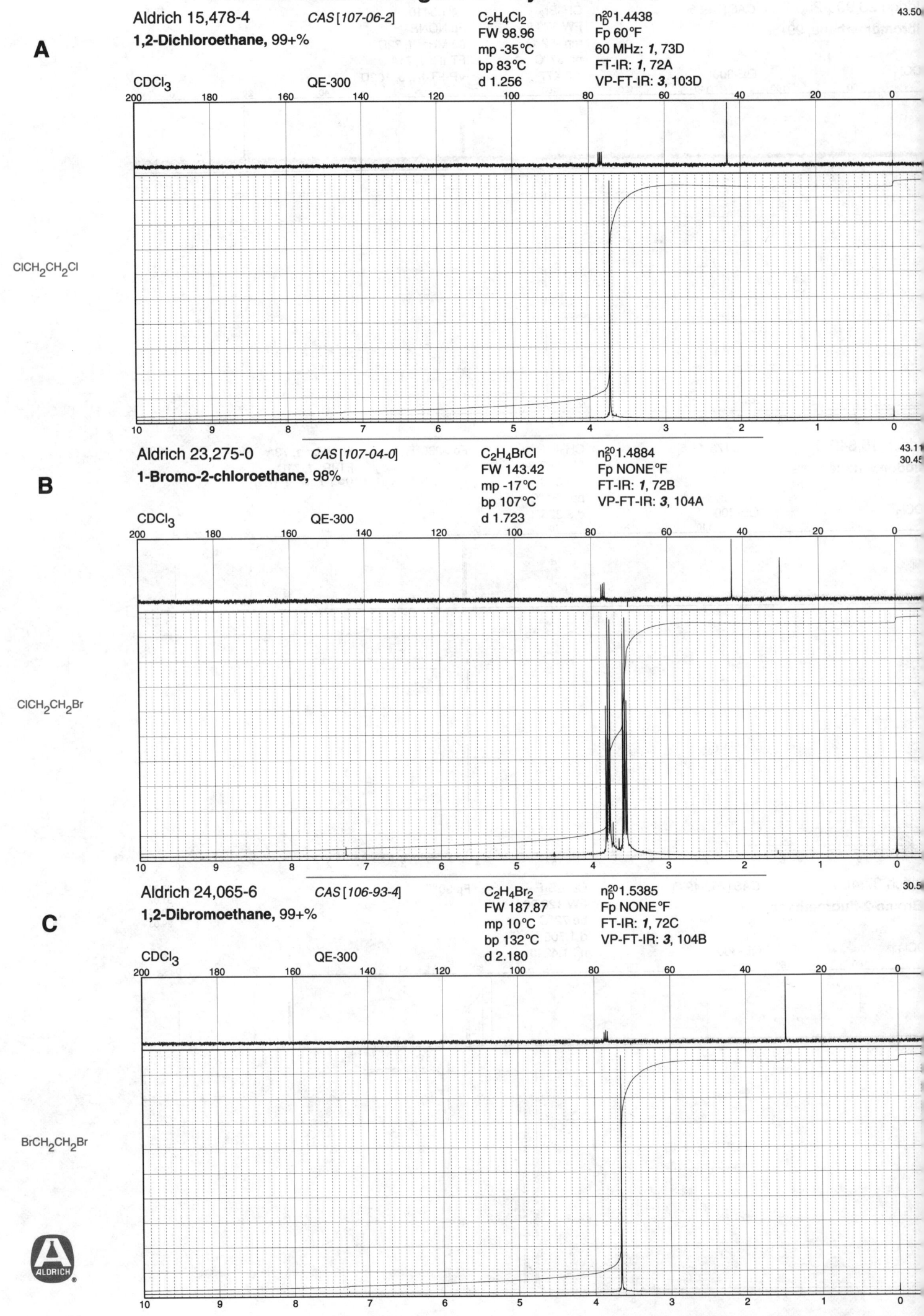

A

Aldrich 15,478-4   *CAS [107-06-2]*

1,2-Dichloroethane, 99+%

$C_2H_4Cl_2$
FW 98.96
mp -35°C
bp 83°C
d 1.256

n_D^{20} 1.4438
Fp 60°F
60 MHz: *1*, 73D
FT-IR: *1*, 72A
VP-FT-IR: *3*, 103D

43.50

CDCl₃   QE-300

ClCH₂CH₂Cl

B

Aldrich 23,275-0   *CAS [107-04-0]*

1-Bromo-2-chloroethane, 98%

C_2H_4BrCl
FW 143.42
mp -17°C
bp 107°C
d 1.723

n_D^{20} 1.4884
Fp NONE°F
FT-IR: *1*, 72B
VP-FT-IR: *3*, 104A

43.11
30.45

CDCl₃   QE-300

ClCH₂CH₂Br

C

Aldrich 24,065-6   *CAS [106-93-4]*

1,2-Dibromoethane, 99+%

$C_2H_4Br_2$
FW 187.87
mp 10°C
bp 132°C
d 2.180

n_D^{20} 1.5385
Fp NONE°F
FT-IR: *1*, 72C
VP-FT-IR: *3*, 104B

30.5

CDCl₃   QE-300

BrCH₂CH₂Br

ALDRICH

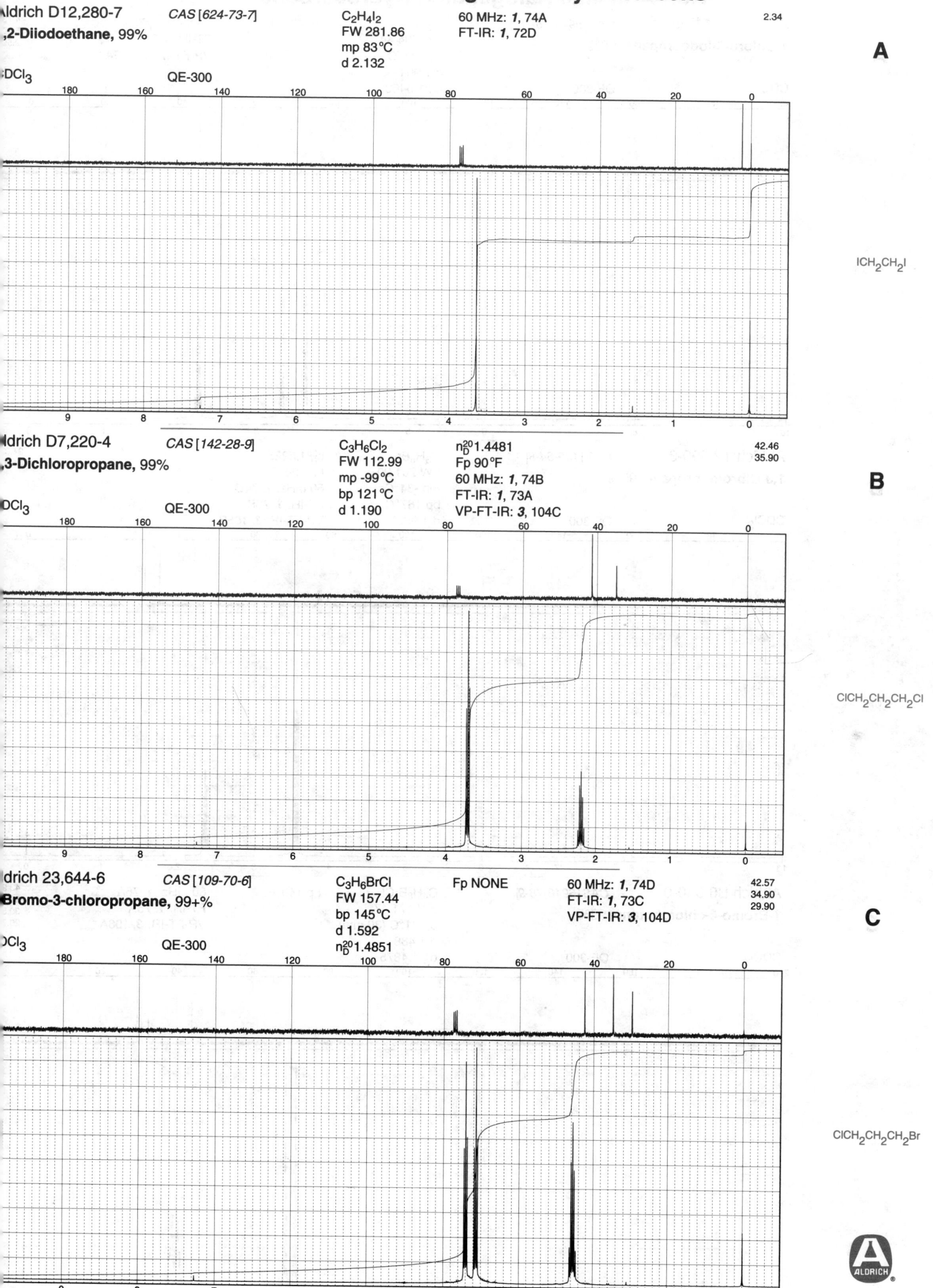

Aldrich D12,280-7 CAS [624-73-7] $C_2H_4I_2$ 60 MHz: *1*, 74A 2.34

,2-Diiodoethane, 99% FW 281.86 FT-IR: *1*, 72D

mp 83°C

d 2.132

DCl3 QE-300

A

ICH_2CH_2I

Aldrich D7,220-4 CAS [142-28-9] $C_3H_6Cl_2$ n_D^{20}1.4481 42.46

,3-Dichloropropane, 99% FW 112.99 Fp 90°F 35.90

mp -99°C 60 MHz: *1*, 74B

bp 121°C FT-IR: *1*, 73A

DCl3 QE-300 d 1.190 VP-FT-IR: *3*, 104C

B

$ClCH_2CH_2CH_2Cl$

Aldrich 23,644-6 CAS [109-70-6] C_3H_6BrCl Fp NONE 60 MHz: *1*, 74D 42.57

Bromo-3-chloropropane, 99+% FW 157.44 FT-IR: *1*, 73C 34.90

bp 145°C VP-FT-IR: *3*, 104D 29.90

d 1.592

DCl3 QE-300 n_D^{20}1.4851

C

$ClCH_2CH_2CH_2Br$

A

Aldrich 23,447-8   *CAS [6940-76-7]*   C₃H₆ClI   Fp >230°F   60 MHz: *1*, 75A

1-Chloro-3-iodopropane, 99%

C_3H_6ClI
FW 204.44
bp 171°C
d 1.904
$n_D^{20}1.5463$

Fp >230°F

60 MHz: *1*, 75A
FT-IR: *1*, 73D
VP-FT-IR: *3*, 105A

44.73
35.4
2.5

CDCl₃   QE-300

ICH₂CH₂CH₂Cl

B

Aldrich 12,590-3   *CAS [109-64-8]*

1,3-Dibromopropane, 99%

$C_3H_6Br_2$
FW 201.90
mp -34°C
bp 167°C
d 1.989

$n_D^{20}1.5225$
Fp 130°F
60 MHz: *1*, 74C
FT-IR: *1*, 73B
VP-FT-IR: *3*, 105B

34.8
31.0

CDCl₃   QE-300

BrCH₂CH₂CH₂Br

C

Aldrich B6,080-0   *CAS [6940-78-9]*   C₄H₈BrCl   Fp 140°F

1-Bromo-4-chlorobutane, 99%

C_4H_8BrCl
FW 171.47
bp 81°C (30 mm)
d 1.488
$n_D^{20}1.4875$

Fp 140°F

60 MHz: *1*, 76A
FT-IR: *1*, 75A
VP-FT-IR: *3*, 106A

43.9
32.6
30.8
29.7

CDCl₃   QE-300

ClCH₂CH₂CH₂CH₂Br

Aldrich 14,080-5   *CAS [110-52-1]*   C₄H₈Br₂ n_D^{20}1.5186   32.48 30.94

1,4-Dibromobutane, 99%   FW 215.93   Fp >230°F

   mp -20°C   60 MHz: *1*, 75D

CDCl₃   QE-300   bp 64°C (6 mm)   FT-IR: *1*, 74D

   d 1.808   VP-FT-IR: *3*, 106B

A

BrCH₂CH₂CH₂CH₂Br

Aldrich D12,260-2   *CAS [628-21-7]*   C₄H₈I₂ n_D^{20}1.6212   33.74 4.92

1,4-Diiodobutane, 99+%   FW 309.92   Fp NONE

   mp 6°C   60 MHz: *1*, 76B

CDCl₃   QE-300   bp 150°C (26 mm)   FT-IR: *1*, 75B

   d 2.350   VP-FT-IR: *3*, 106C

B

ICH₂CH₂CH₂CH₂I

Aldrich D6,960-2   *CAS [628-76-2]*   C₅H₁₀Cl₂ n_D^{20}1.4553   44.65 31.86 24.28

1,5-Dichloropentane, 99%   FW 141.04   Fp 80°F

   mp -72°C   60 MHz: *1*, 76C

CDCl₃   QE-300   bp 65°C (10 mm)   FT-IR: *1*, 75C

   d 1.106   VP-FT-IR: *3*, 106D

C

ClCH₂(CH₂)₃CH₂Cl

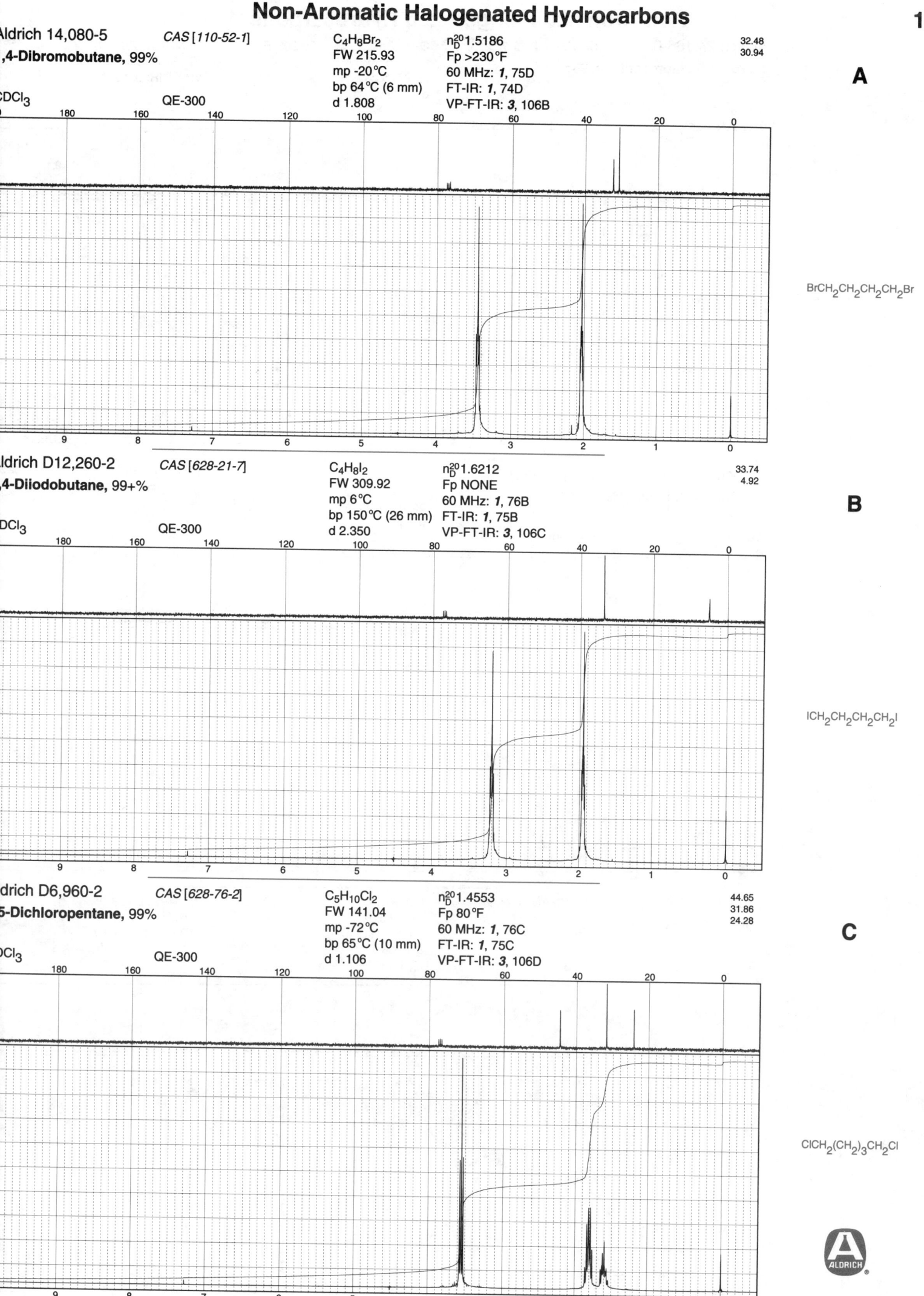

A

Aldrich 24,166-0   CAS [54512-75-3] $C_5H_{10}BrCl$ Fp 203°F   60 MHz: *1*, 77A   44.61
1-Bromo-5-chloropentane, 98%   FW 185.50   FT-IR: *1*, 75D   33.26
                                    bp 211°C   VP-FT-IR: *3*, 107A   31.99
                                    d 1.408   31.71
CDCl₃   QE-300 n_D^{20}1.4836   25.52

$ClCH_2(CH_2)_3CH_2Br$

B

Aldrich 25,213-1   CAS [628-77-3] $C_5H_{10}I_2$ Fp >230°F   FT-IR: *1*, 76B   32.27
1,5-Diiodopentane, 97%   FW 323.94   VP-FT-IR: *3*, 107C   31.31
                                    bp 102°C (3 mm)   6.22
CDCl₃   QE-300   d 2.177
                                    n_D^{20}1.6002

$ICH_2(CH_2)_3CH_2I$

C

Aldrich D6,380-9   CAS [2163-00-0] $C_6H_{12}Cl_2$ Fp 165°F   60 MHz: *1*, 77B   44.8
1,6-Dichlorohexane, 98%   FW 155.07   FT-IR: *1*, 76C   32.4
                                    bp 89°C (15 mm)   VP-FT-IR: *3*, 107D   26.1
                                    d 1.068
CDCl₃   QE-300 n_D^{20}1.4568

$ClCH_2(CH_2)_4CH_2Cl$

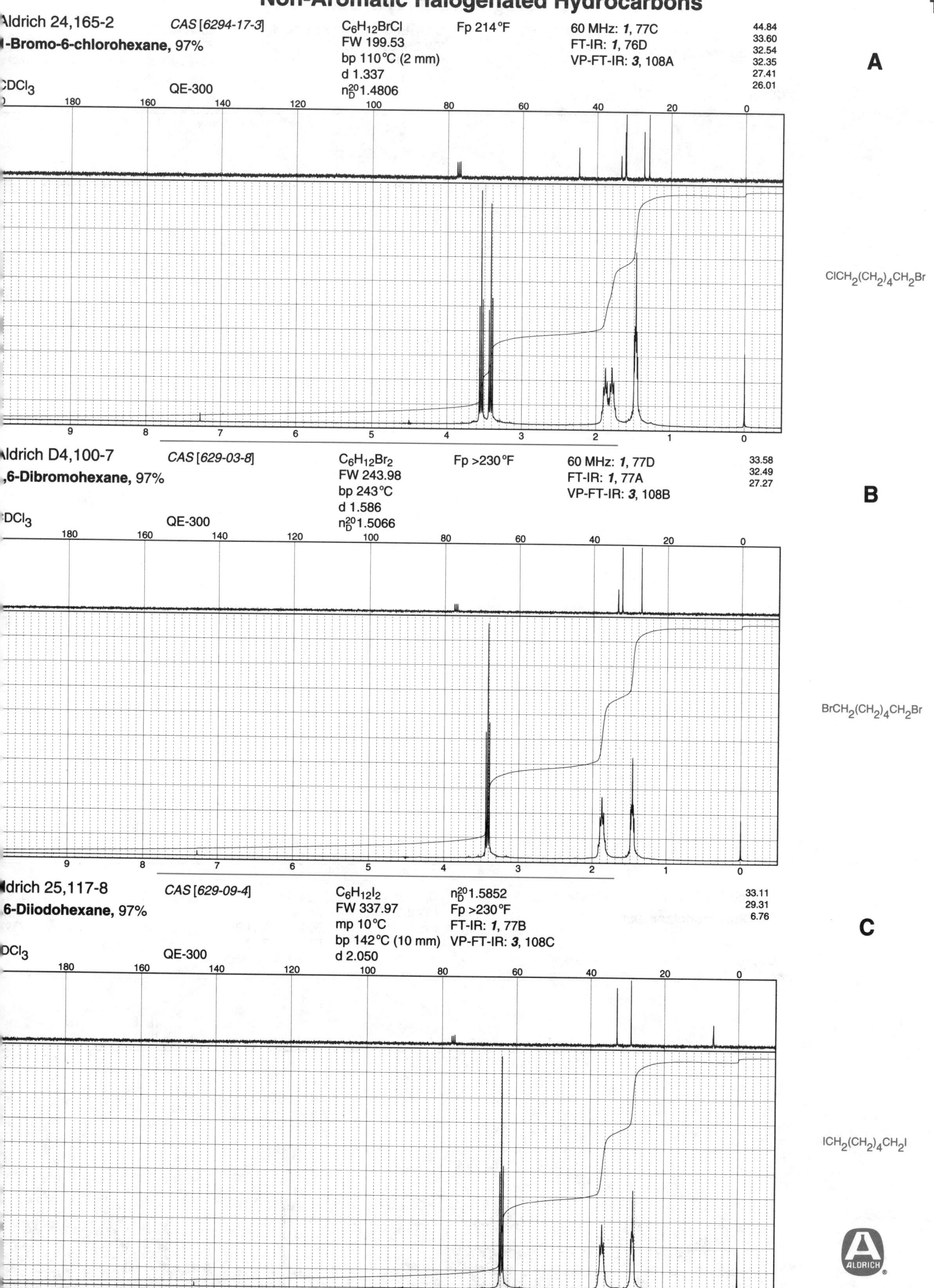

Aldrich 24,165-2   *CAS [6294-17-3]*   C6H12BrCl   Fp 214°F   60 MHz: *1*, 77C   44.84
1-Bromo-6-chlorohexane, 97%   FW 199.53   FT-IR: *1*, 76D   33.60
bp 110°C (2 mm)   VP-FT-IR: *3*, 108A   32.54
CDCl3   QE-300   d 1.337   32.35
n20D 1.4806   27.41
26.01

A

ClCH2(CH2)4CH2Br

Aldrich D4,100-7   *CAS [629-03-8]*   C6H12Br2   Fp >230°F   60 MHz: *1*, 77D   33.58
1,6-Dibromohexane, 97%   FW 243.98   FT-IR: *1*, 77A   32.49
bp 243°C   VP-FT-IR: *3*, 108B   27.27
CDCl3   QE-300   d 1.586
n20D 1.5066

B

BrCH2(CH2)4CH2Br

Aldrich 25,117-8   *CAS [629-09-4]*   C6H12I2   n20D 1.5852   33.11
1,6-Diiodohexane, 97%   FW 337.97   Fp >230°F   29.31
mp 10°C   FT-IR: *1*, 77B   6.76
bp 142°C (10 mm)   VP-FT-IR: *3*, 108C
CDCl3   QE-300   d 2.050

C

ICH2(CH2)4CH2I

Non-Aromatic Halogenated Hydrocarbons

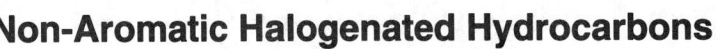

A

Aldrich 14,499-1 *CAS* [*4549-31-9*] $C_7H_{14}Br_2$ Fp >230°F 60 MHz: ***1***, 78A

1,7-Dibromoheptane, 97% FW 258.01 FT-IR: ***1***, 77C

bp 255°C VP-FT-IR: ***3***, 108D

d 1.510

n_D^{20} 1.5017

33.74
32.61
27.92
27.87

CDCl$_3$ QE-300

$BrCH_2(CH_2)_5CH_2Br$

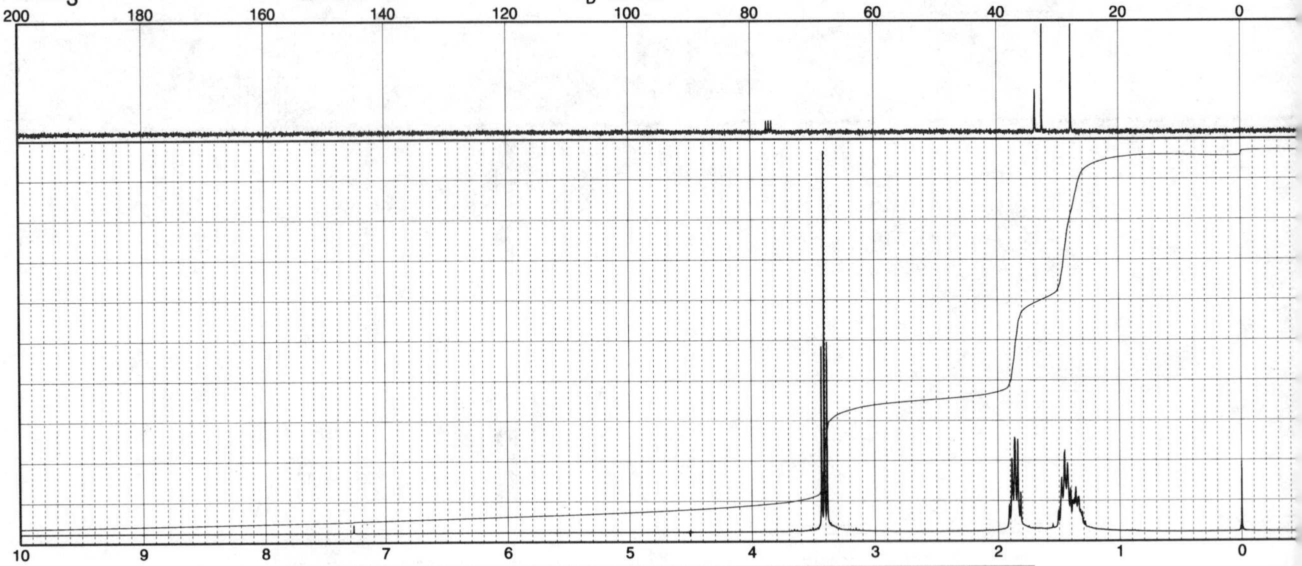

B

Aldrich 36,128-3 *CAS* [*2162-99-4*] $C_8H_{16}Cl_2$ Fp 228°F

1,8-Dichlorooctane, 98% FW 183.12

bp 116°C (11 mm)

d 1.025

n_D^{20} 1.4590

44.99
32.62
28.74
26.80

CDCl$_3$ QE-300

$ClCH_2(CH_2)_6CH_2Cl$

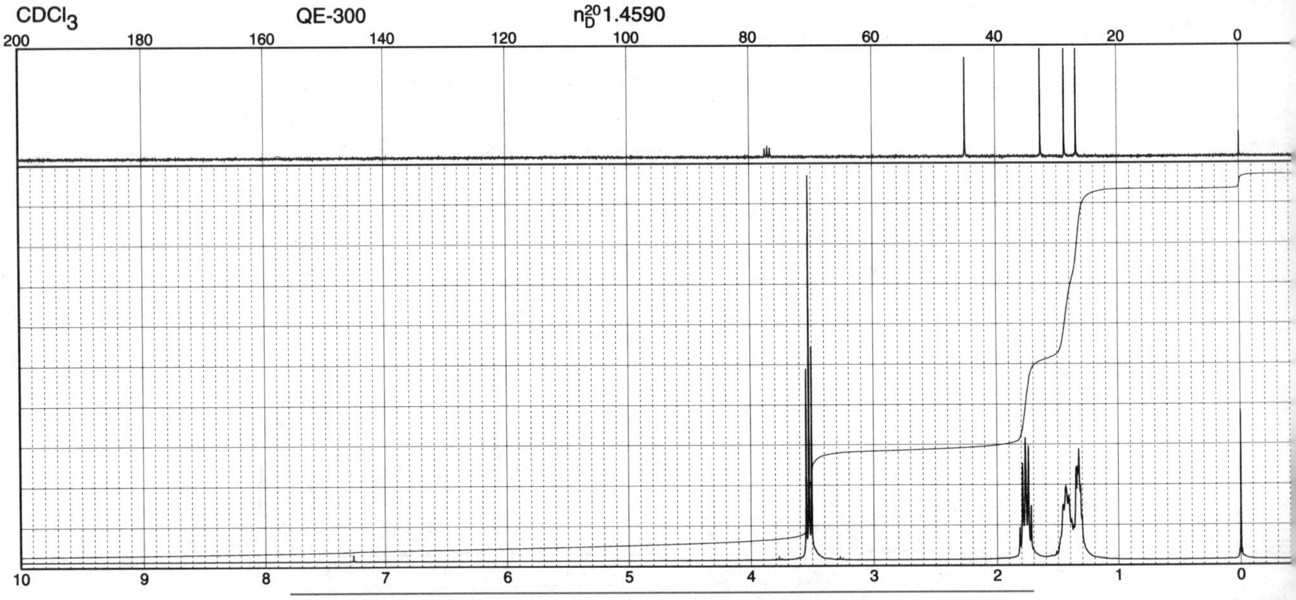

C

Aldrich D4,260-7 *CAS* [*4549-32-0*] $C_8H_{16}Br_2$ n_D^{20} 1.4981

1,8-Dibromooctane, 98% FW 272.03 Fp >230°F

mp 16°C 60 MHz: ***1***, 78B

bp 271°C FT-IR: ***1***, 77D

d 1.477 VP-FT-IR: ***3***, 109A

33.83
32.71
28.54
28.01

CDCl$_3$ QE-300

$BrCH_2(CH_2)_6CH_2Br$

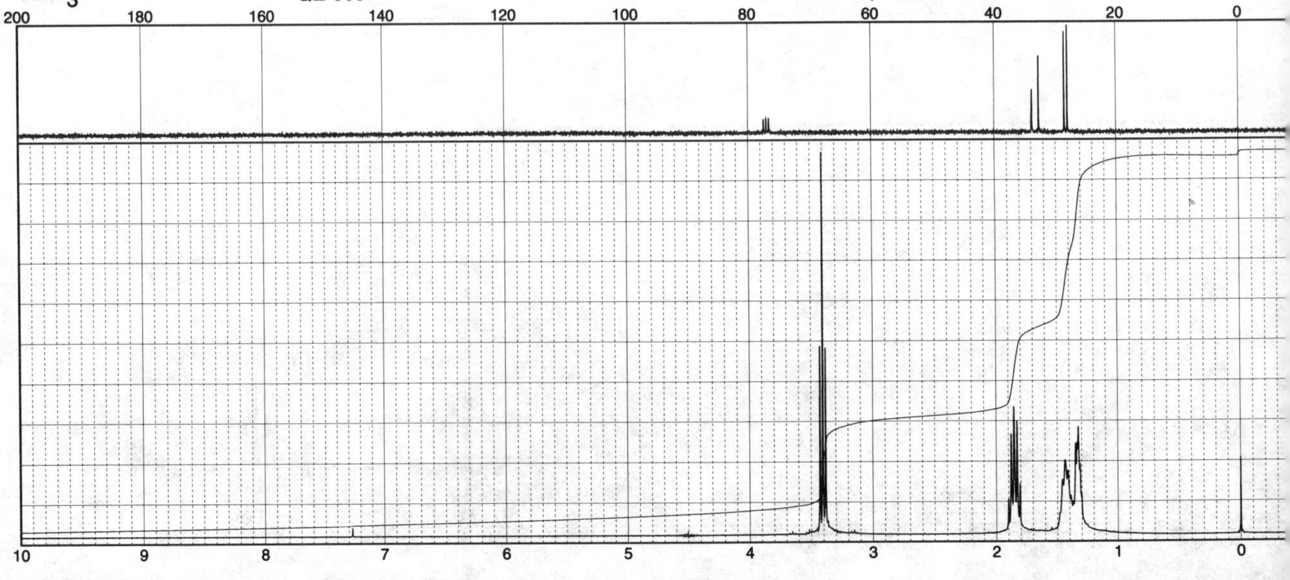

Non-Aromatic Halogenated Hydrocarbons

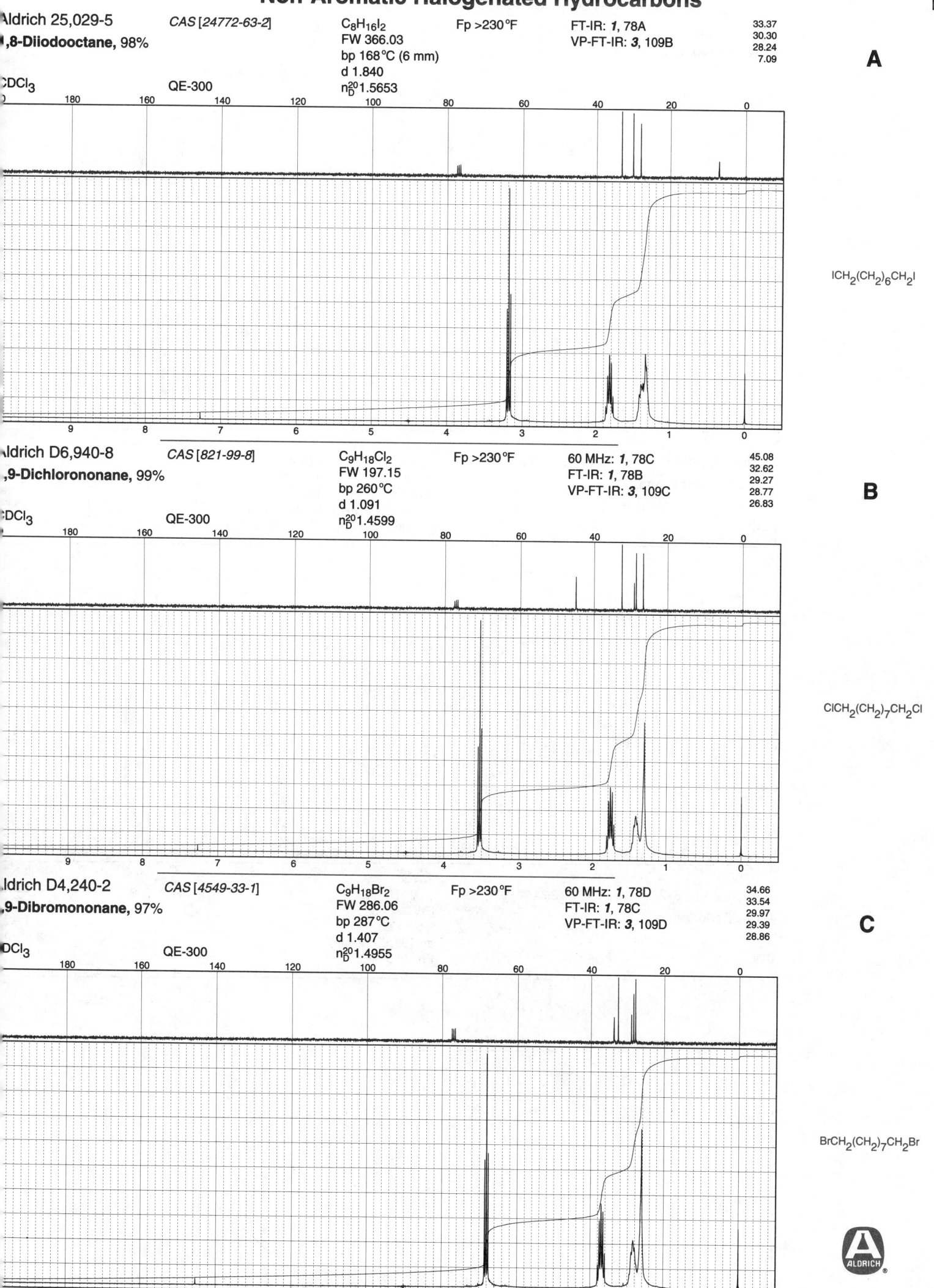

Aldrich 25,029-5   CAS [24772-63-2] $C_8H_{16}I_2$ Fp >230°F   FT-IR: **1**, 78A   33.37
,8-Diiodooctane, 98%   FW 366.03   VP-FT-IR: **3**, 109B   30.30
bp 168°C (6 mm)   28.24
CDCl₃   QE-300   d 1.840   7.09
n_D^{20} 1.5653

A

$ICH_2(CH_2)_6CH_2I$

Aldrich D6,940-8   CAS [821-99-8] $C_9H_{18}Cl_2$ Fp >230°F   60 MHz: **1**, 78C   45.08
,9-Dichlorononane, 99%   FW 197.15   FT-IR: **1**, 78B   32.62
bp 260°C   VP-FT-IR: **3**, 109C   29.27
CDCl₃   QE-300   d 1.091   28.77
n_D^{20} 1.4599   26.83

B

$ClCH_2(CH_2)_7CH_2Cl$

Aldrich D4,240-2   CAS [4549-33-1] $C_9H_{18}Br_2$ Fp >230°F   60 MHz: **1**, 78D   34.66
,9-Dibromononane, 97%   FW 286.06   FT-IR: **1**, 78C   33.54
bp 287°C   VP-FT-IR: **3**, 109D   29.97
CDCl₃   QE-300   d 1.407   29.39
n_D^{20} 1.4955   28.86

C

$BrCH_2(CH_2)_7CH_2Br$

ALDRICH

A

Aldrich 25,478-9   CAS [2162-98-3]

1,10-Dichlorodecane, 99%

$C_{10}H_{20}Cl_2$
FW 211.18
mp 16°C
bp 168°C (28 mm)
d 0.999

n_D^{20}1.4605
Fp >230°F
FT-IR: **1**, 78D
VP-FT-IR: **3**, 110A

45.14
32.68
29.38
28.88
26.91

CDCl₃   QE-300

ClCH₂(CH₂)₈CH₂Cl

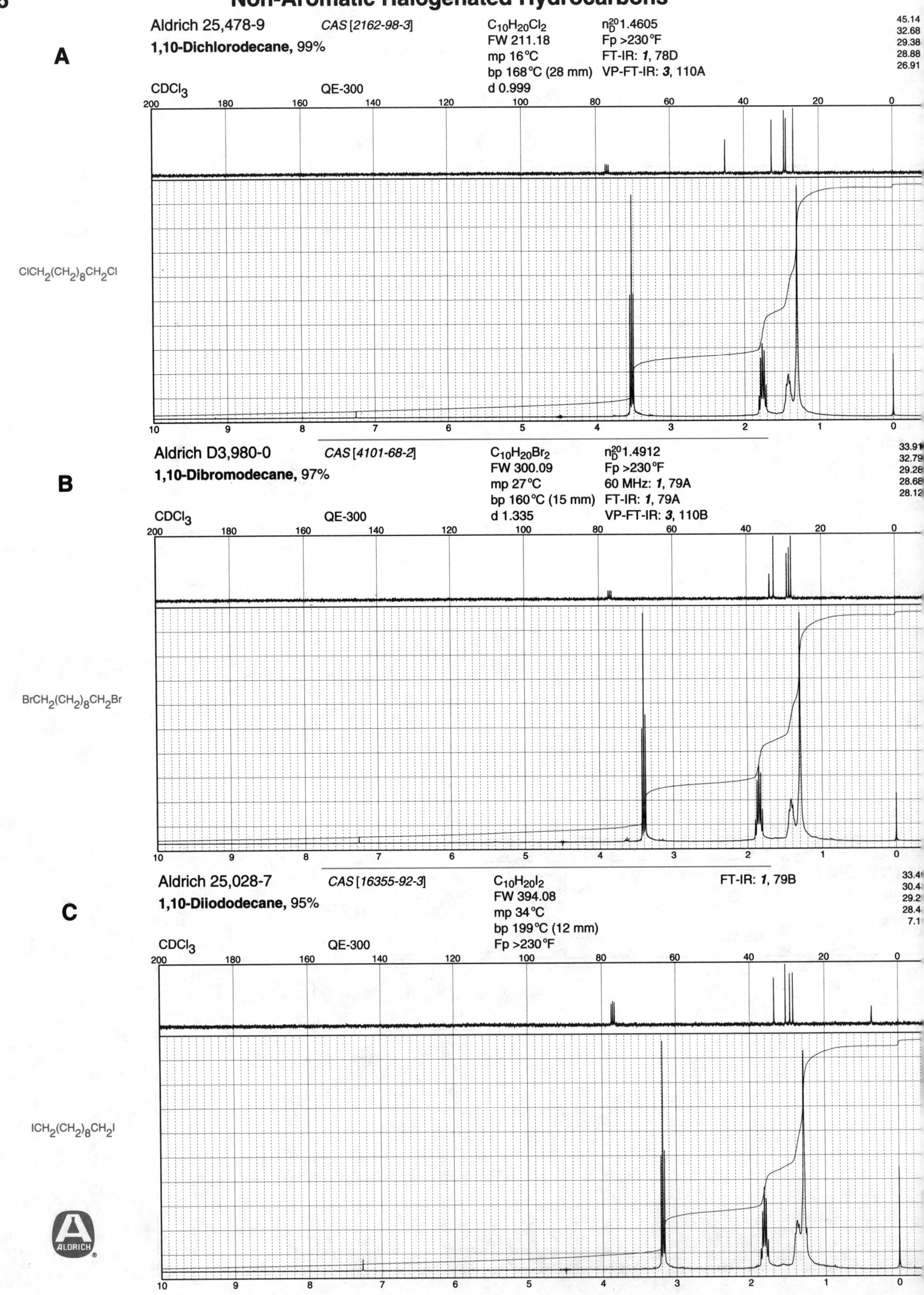

B

Aldrich D3,980-0   CAS [4101-68-2]

1,10-Dibromodecane, 97%

$C_{10}H_{20}Br_2$
FW 300.09
mp 27°C
bp 160°C (15 mm)
d 1.335

n_D^{20}1.4912
Fp >230°F
60 MHz: **1**, 79A
FT-IR: **1**, 79A
VP-FT-IR: **3**, 110B

33.91
32.79
29.28
28.68
28.12

CDCl₃   QE-300

BrCH₂(CH₂)₈CH₂Br

C

Aldrich 25,028-7   CAS [16355-92-3]

1,10-Diiododecane, 95%

$C_{10}H_{20}I_2$
FW 394.08
mp 34°C
bp 199°C (12 mm)
Fp >230°F

FT-IR: **1**, 79B

33.4
30.4
29.2
28.4
7.1

CDCl₃   QE-300

ICH₂(CH₂)₈CH₂I

Non-Aromatic Halogenated Hydrocarbons

116

Aldrich 23,246-7   CAS [16696-65-4]   C₁₁H₂₂Br₂   n₂₀ᴅ 1.4916

A

,11-Dibromoundecane, 98+%   FW 314.12   Fp >230°F

CDCl₃   QE-300   mp -11°C   60 MHz: *1*, 79B
bp 191°C (18 mm)   FT-IR: *1*, 79C
d 1.335   VP-FT-IR: *3*, 110C

34.24
33.13
29.68
29.04
28.46

BrCH₂(CH₂)₉CH₂Br

Aldrich 13,338-8   CAS [3344-70-5]   C₁₂H₂₄Br₂   60 MHz: *1*, 79C

B

,12-Dibromododecane   FW 328.14   FT-IR: *1*, 79D
mp 40°C   VP-FT-IR: *3*, 110D
CDCl₃   QE-300   bp 215°C (15 mm)
Fp >230°F

34.06
32.94
29.57
29.51
28.85
28.27

BrCH₂(CH₂)₁₀CH₂Br

Aldrich D6,155-5   CAS [75-34-3]   C₂H₄Cl₂   n₂₀ᴅ 1.4165   60 MHz: *1*, 79D

C

,1-Dichloroethane   FW 98.96   FT-IR: *1*, 80A
mp -97°C   VP-FT-IR: *3*, 111A
CDCl₃   QE-300   bp 57°C
d 1.177

69.99*
32.28*

CH₃CHCl₂

ALDRICH

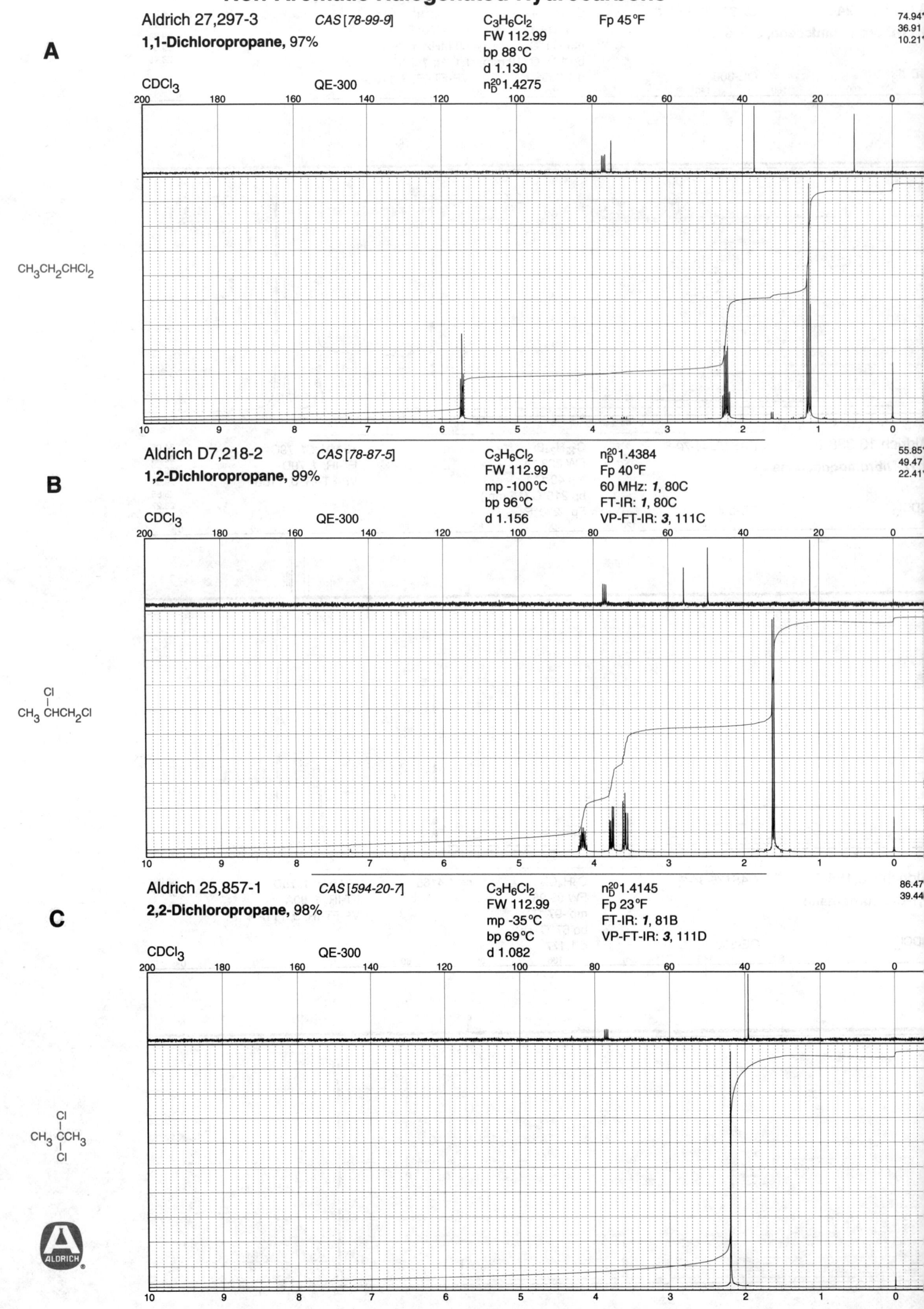

A

Aldrich 27,297-3 CAS [78-99-9] C3H6Cl2 Fp 45°F 74.94*
36.91
10.21*

1,1-Dichloropropane, 97% FW 112.99
bp 88°C
d 1.130

CDCl3 QE-300 n_D^20 1.4275

CH3CH2CHCl2

B

Aldrich D7,218-2 CAS [78-87-5] C3H6Cl2 n_D^20 1.4384 55.85*
49.47
22.41*

1,2-Dichloropropane, 99% FW 112.99 Fp 40°F
mp -100°C 60 MHz: 1, 80C
bp 96°C FT-IR: 1, 80C
d 1.156 VP-FT-IR: 3, 111C

CDCl3 QE-300

Cl
|
CH3 CHCH2Cl

C

Aldrich 25,857-1 CAS [594-20-7] C3H6Cl2 n_D^20 1.4145 86.47
39.44

2,2-Dichloropropane, 98% FW 112.99 Fp 23°F
mp -35°C FT-IR: 1, 81B
bp 69°C VP-FT-IR: 3, 111D
d 1.082

CDCl3 QE-300

Cl
|
CH3 CCH3
|
Cl

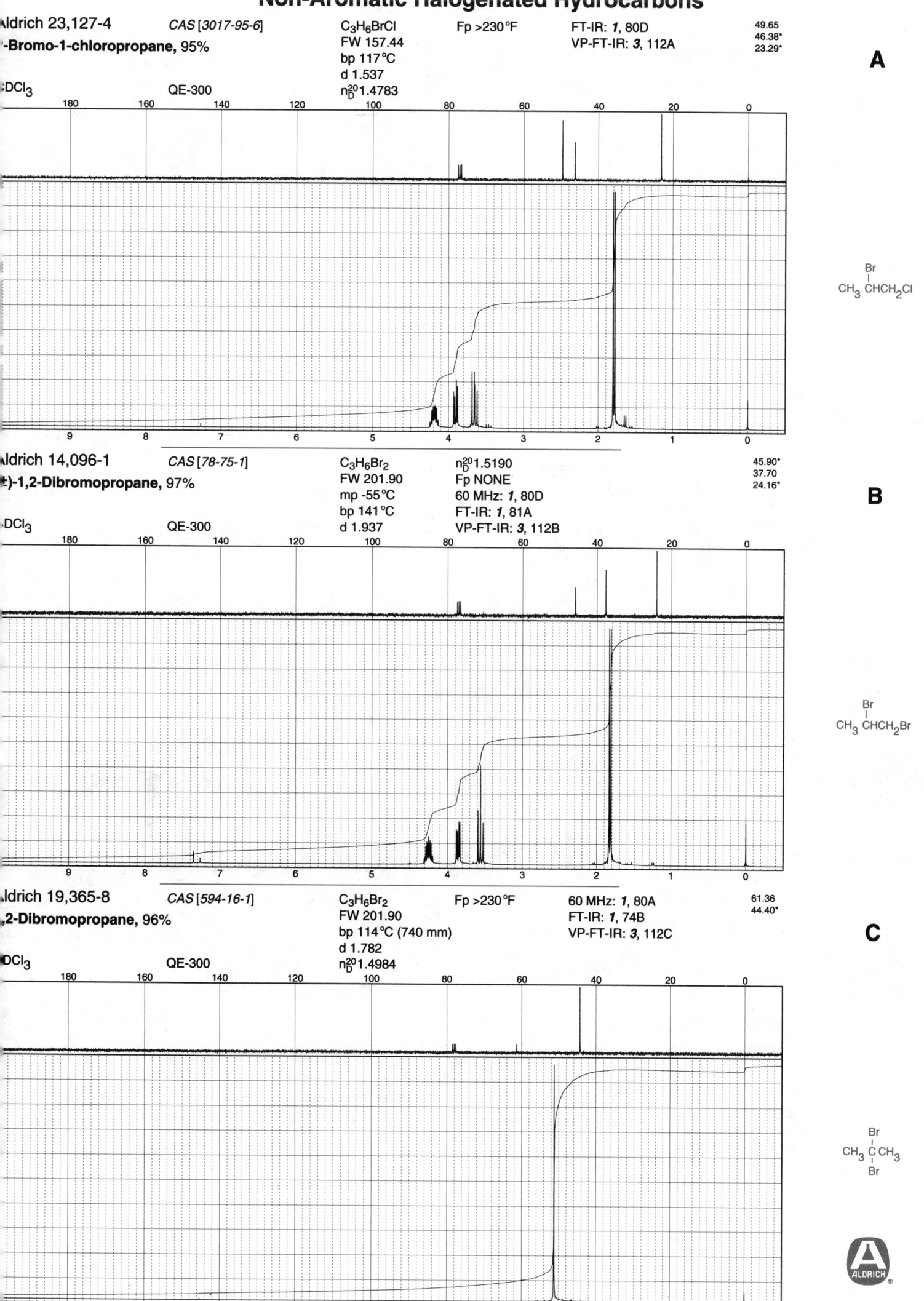

Aldrich 23,127-4   CAS [3017-95-6] C_3H_6BrCl Fp >230°F   FT-IR: **1**, 80D   49.65 / 46.38* / 23.29*
-Bromo-1-chloropropane, 95%   FW 157.44   VP-FT-IR: **3**, 112A

bp 117°C
d 1.537
n_D^{20} 1.4783

DCl₃   QE-300

A

Br
|
$CH_3 CHCH_2Cl$

Aldrich 14,096-1   CAS [78-75-1] $C_3H_6Br_2$ n_D^{20} 1.5190   45.90* / 37.70 / 24.16*
±)-1,2-Dibromopropane, 97%   FW 201.90   Fp NONE

mp -55°C   60 MHz: **1**, 80D
bp 141°C   FT-IR: **1**, 81A
d 1.937   VP-FT-IR: **3**, 112B

DCl₃   QE-300

B

Br
|
$CH_3 CHCH_2Br$

Aldrich 19,365-8   CAS [594-16-1] $C_3H_6Br_2$ Fp >230°F   60 MHz: **1**, 80A   61.36 / 44.40*
,2-Dibromopropane, 96%   FW 201.90   FT-IR: **1**, 74B

bp 114°C (740 mm)   VP-FT-IR: **3**, 112C
d 1.782
n_D^{20} 1.4984

DCl₃   QE-300

C

Br
|
$CH_3 C CH_3$
|
Br

Non-Aromatic Halogenated Hydrocarbons

A

Aldrich D5,900-3   *CAS [1190-22-3]* $C_4H_8Cl_2$ Fp 87°F   60 MHz: *1*, 81A

1,3-Dichlorobutane, 99%

FW 127.01   FT-IR: *1*, 82C

bp 134°C   VP-FT-IR: *3*, 113D

d 1.115

n_D^{20}1.4431

CDCl₃   QE-300

Cl
|
CH₃ CHCH₂CH₂Cl

54.98
42.60
41.80
25.16

B

Aldrich D5,940-2   *CAS [7581-97-7]* $C_4H_8Cl_2$ n_D^{20}1.4420

2,3-Dichlorobutane, 98%, mixture of isomers

FW 127.01   Fp 65°F

mp -80°C   60 MHz: *1*, 80B

bp 118°C   FT-IR: *1*, 80B

d 1.107   VP-FT-IR: *3*, 114A

CDCl₃   QE-300

Cl Cl
| |
CH₃CH CHCH₃

61.60
60.41
22.17
19.85

C

Aldrich 25,232-8   *CAS [533-98-2]* $C_4H_8Br_2$ Fp >230°F   FT-IR: *1*, 82A

(±)-1,2-Dibromobutane, 97%

FW 215.93   VP-FT-IR: *3*, 114B

bp 60°C (20 mm)

d 1.789

n_D^{20}1.5141

CDCl₃   QE-300

Br
|
CH₃CH₂ CHCH₂Br

54.57
35.70
29.14
11.07

ALDRICH

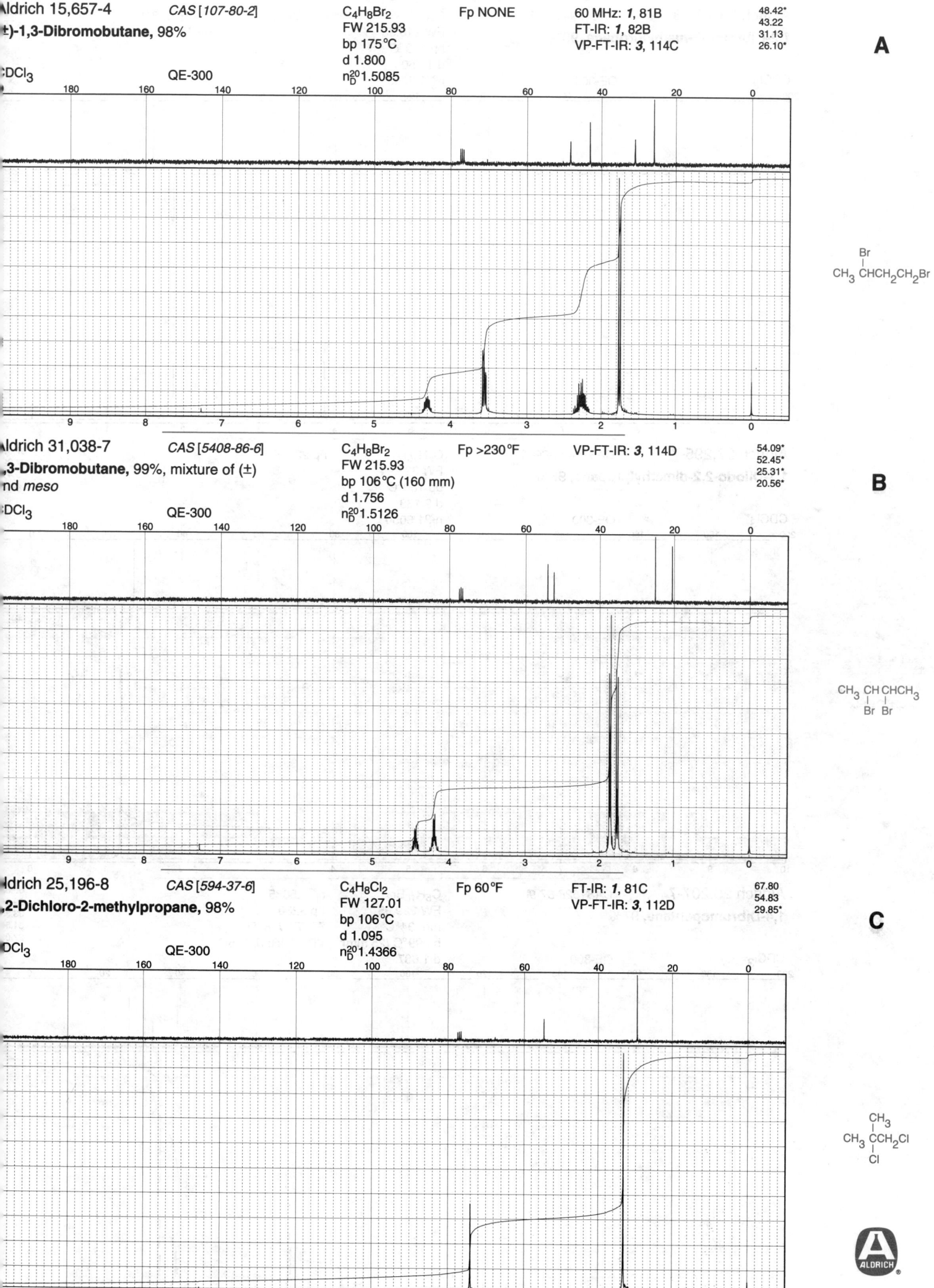

Aldrich 15,657-4   CAS [107-80-2]   C₄H₈Br₂   Fp NONE   60 MHz: 1, 81B   48.42*
(±)-1,3-Dibromobutane, 98%   FW 215.93   FT-IR: 1, 82B   43.22
bp 175°C   VP-FT-IR: 3, 114C   31.13
CDCl₃   QE-300   d 1.800   26.10*
n²⁰_D 1.5085

A

$$CH_3\,CHCH_2CH_2Br$$ with Br

Aldrich 31,038-7   CAS [5408-86-6]   C₄H₈Br₂   Fp >230°F   VP-FT-IR: 3, 114D   54.09*
2,3-Dibromobutane, 99%, mixture of (±)   FW 215.93   52.45*
and meso   bp 106°C (160 mm)   25.31*
CDCl₃   QE-300   d 1.756   20.56*
n²⁰_D 1.5126

B

$$CH_3\,CH\,CHCH_3$$ with Br Br

Aldrich 25,196-8   CAS [594-37-6]   C₄H₈Cl₂   Fp 60°F   FT-IR: 1, 81C   67.80
1,2-Dichloro-2-methylpropane, 98%   FW 127.01   VP-FT-IR: 3, 112D   54.83
bp 106°C   29.85*
CDCl₃   QE-300   d 1.095
n²⁰_D 1.4366

C

$$CH_3\,CCH_2Cl$$ with CH₃ and Cl

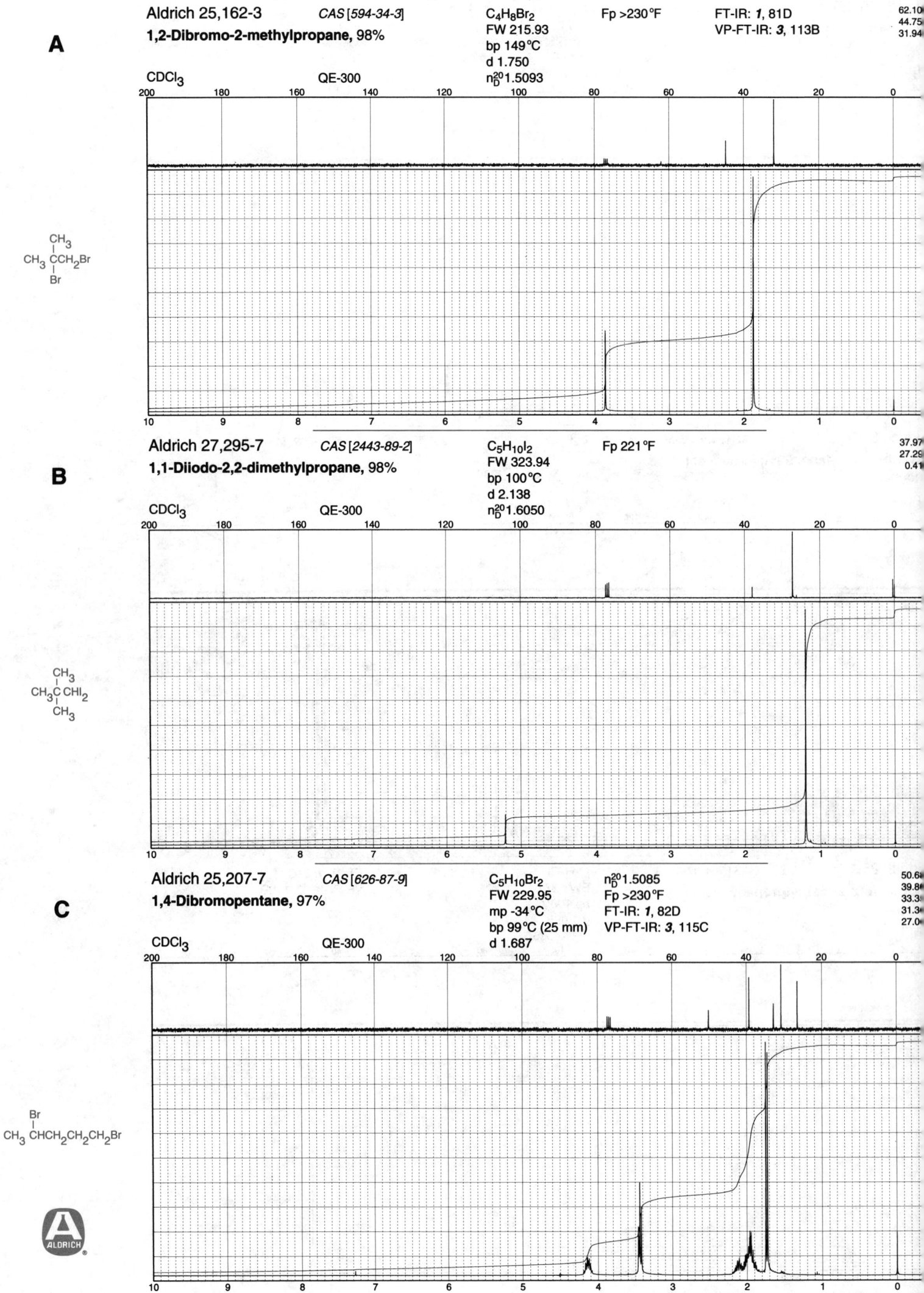

A

Aldrich 25,162-3 — *CAS [594-34-3]* — $C_4H_8Br_2$ — Fp >230 °F — FT-IR: *1*, 81D

1,2-Dibromo-2-methylpropane, 98% — FW 215.93 — VP-FT-IR: *3*, 113B

bp 149 °C

d 1.750

n_D^{20} 1.5093

62.10
44.75
31.94

CDCl$_3$ — QE-300

CH_3
|
CH_3 CCH_2Br
|
Br

B

Aldrich 27,295-7 — *CAS [2443-89-2]* — $C_5H_{10}I_2$ — Fp 221 °F

1,1-Diiodo-2,2-dimethylpropane, 98% — FW 323.94

bp 100 °C

d 2.138

n_D^{20} 1.6050

37.97
27.29
0.41

CDCl$_3$ — QE-300

CH_3
|
CH_3C CHI_2
|
CH_3

C

Aldrich 25,207-7 — *CAS [626-87-9]* — $C_5H_{10}Br_2$ — n_D^{20} 1.5085

1,4-Dibromopentane, 97% — FW 229.95 — Fp >230 °F

mp -34 °C — FT-IR: *1*, 82D

bp 99 °C (25 mm) — VP-FT-IR: *3*, 115C

d 1.687

50.6
39.8
33.3
31.3
27.0

CDCl$_3$ — QE-300

Br
|
CH_3 $CHCH_2CH_2CH_2Br$

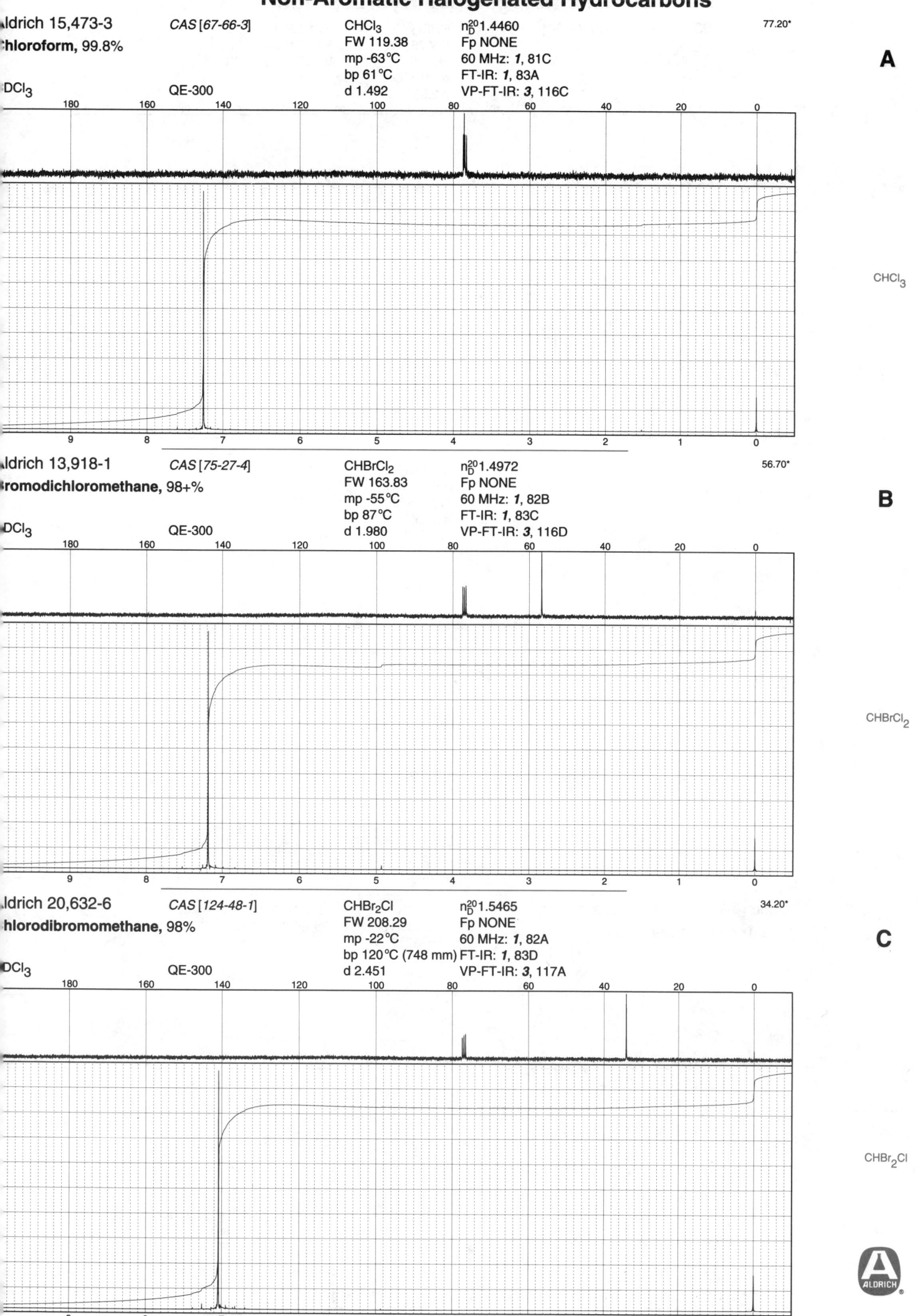

Aldrich 15,473-3   *CAS [67-66-3]*   CHCl₃   n²⁰_D 1.4460   77.20*
Chloroform, 99.8%   FW 119.38   Fp NONE   **A**
                                       mp -63°C   60 MHz: *1*, 81C
DCl₃   QE-300   bp 61°C   FT-IR: *1*, 83A
                                       d 1.492   VP-FT-IR: *3*, 116C

CHCl₃

Aldrich 13,918-1   *CAS [75-27-4]*   CHBrCl₂   n²⁰_D 1.4972   56.70*
Bromodichloromethane, 98+%   FW 163.83   Fp NONE
                                       mp -55°C   60 MHz: *1*, 82B   **B**
DCl₃   QE-300   bp 87°C   FT-IR: *1*, 83C
                                       d 1.980   VP-FT-IR: *3*, 116D

CHBrCl₂

Aldrich 20,632-6   *CAS [124-48-1]*   CHBr₂Cl   n²⁰_D 1.5465   34.20*
Chlorodibromomethane, 98%   FW 208.29   Fp NONE
                                       mp -22°C   60 MHz: *1*, 82A   **C**
                                       bp 120°C (748 mm) FT-IR: *1*, 83D
DCl₃   QE-300   d 2.451   VP-FT-IR: *3*, 117A

CHBr₂Cl

ALDRICH

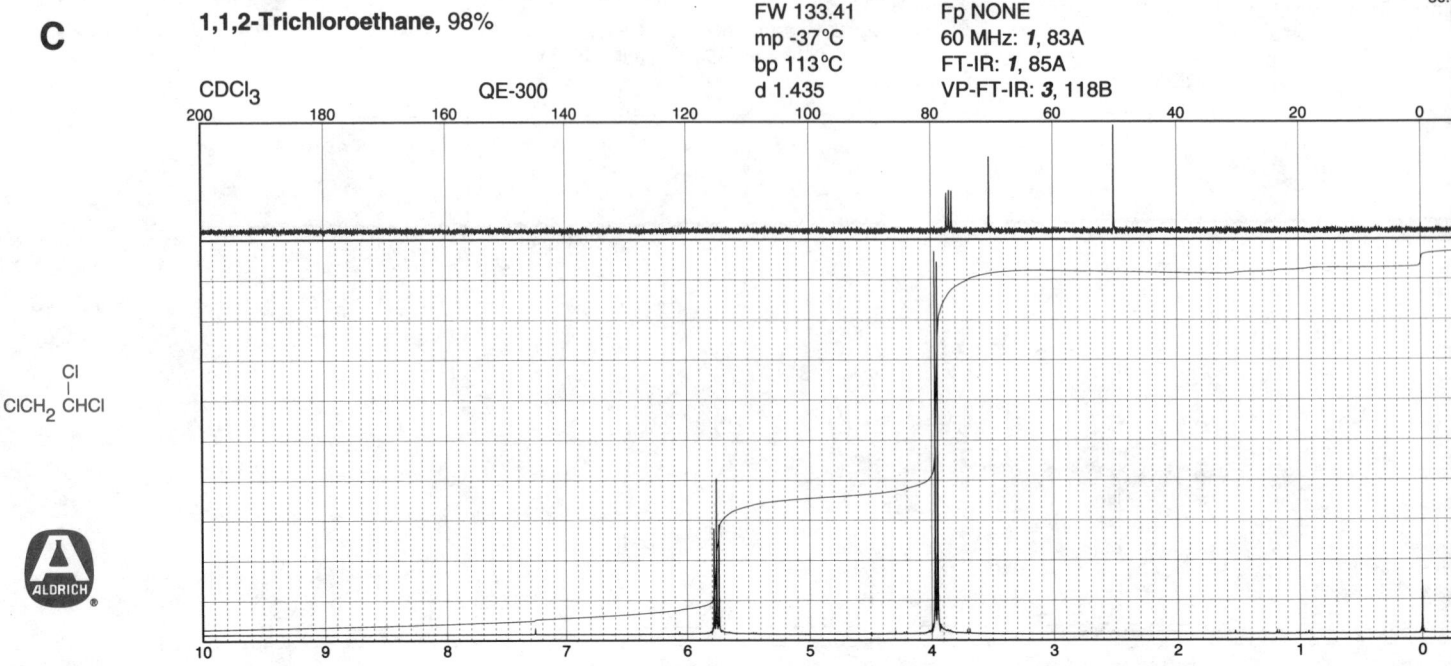

A

Aldrich 24,103-2   *CAS [75-25-2]* $CHBr_3$ $n_D^{20}1.5960$ 9.80*

Bromoform, 99+%

FW 252.75   Fp NONE

mp 8°C   VP-FT-IR: **3**, 117B

bp 151°C

d 2.894

$CDCl_3$ QE-300

$CHBr_3$

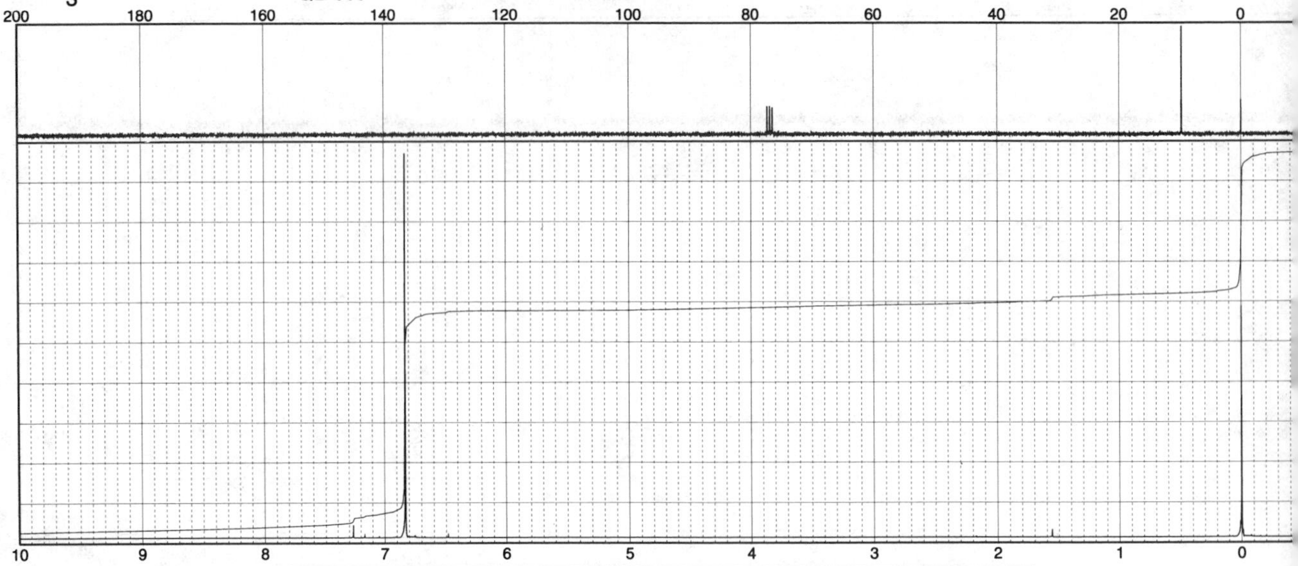

B

Aldrich 23,557-1   *CAS [71-55-6]* $C_2H_3Cl_3$ $n_D^{20}1.4366$ 95.23
45.34*

1,1,1-Trichloroethane, 99%

FW 133.41   Fp NONE

mp -35°C   60 MHz: **1**, 82D

bp 75°C   FT-IR: **1**, 84D

d 1.338   VP-FT-IR: **3**, 118A

$CDCl_3$ QE-300

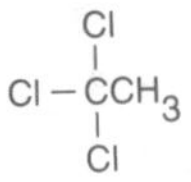

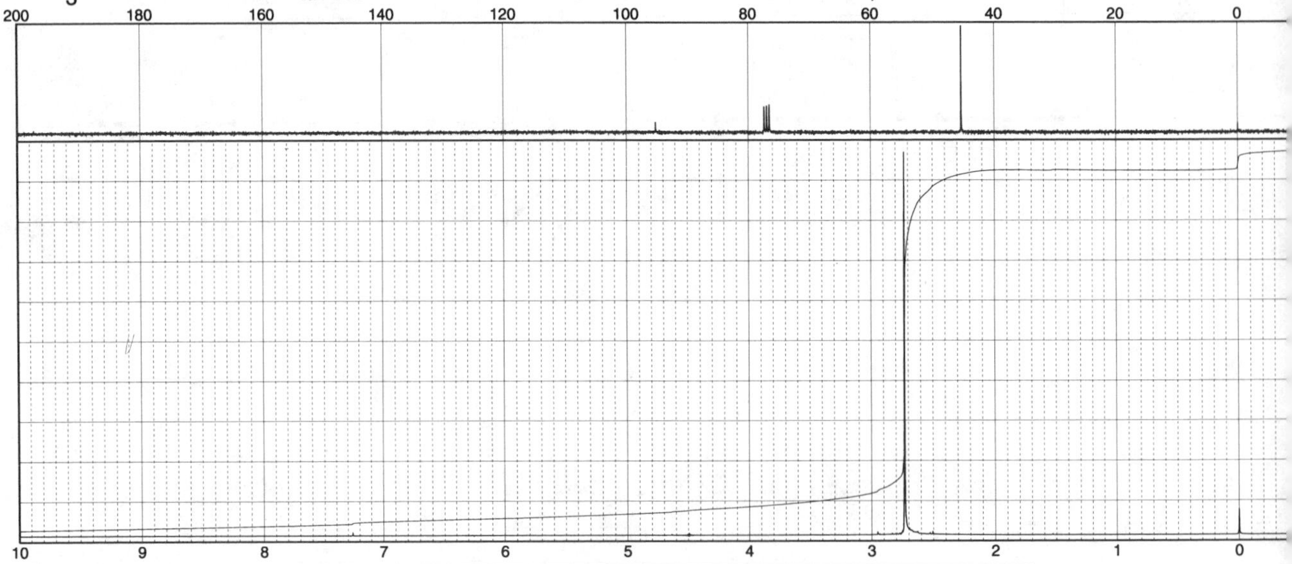

Cl
|
Cl—CCH₃
|
Cl

C

Aldrich T5,475-5   *CAS [79-00-5]* $C_2H_3Cl_3$ $n_D^{20}1.4700$ 70.74
50.45*

1,1,2-Trichloroethane, 98%

FW 133.41   Fp NONE

mp -37°C   60 MHz: **1**, 83A

bp 113°C   FT-IR: **1**, 85A

d 1.435   VP-FT-IR: **3**, 118B

$CDCl_3$ QE-300

Cl
|
$ClCH_2CHCl$

ALDRICH

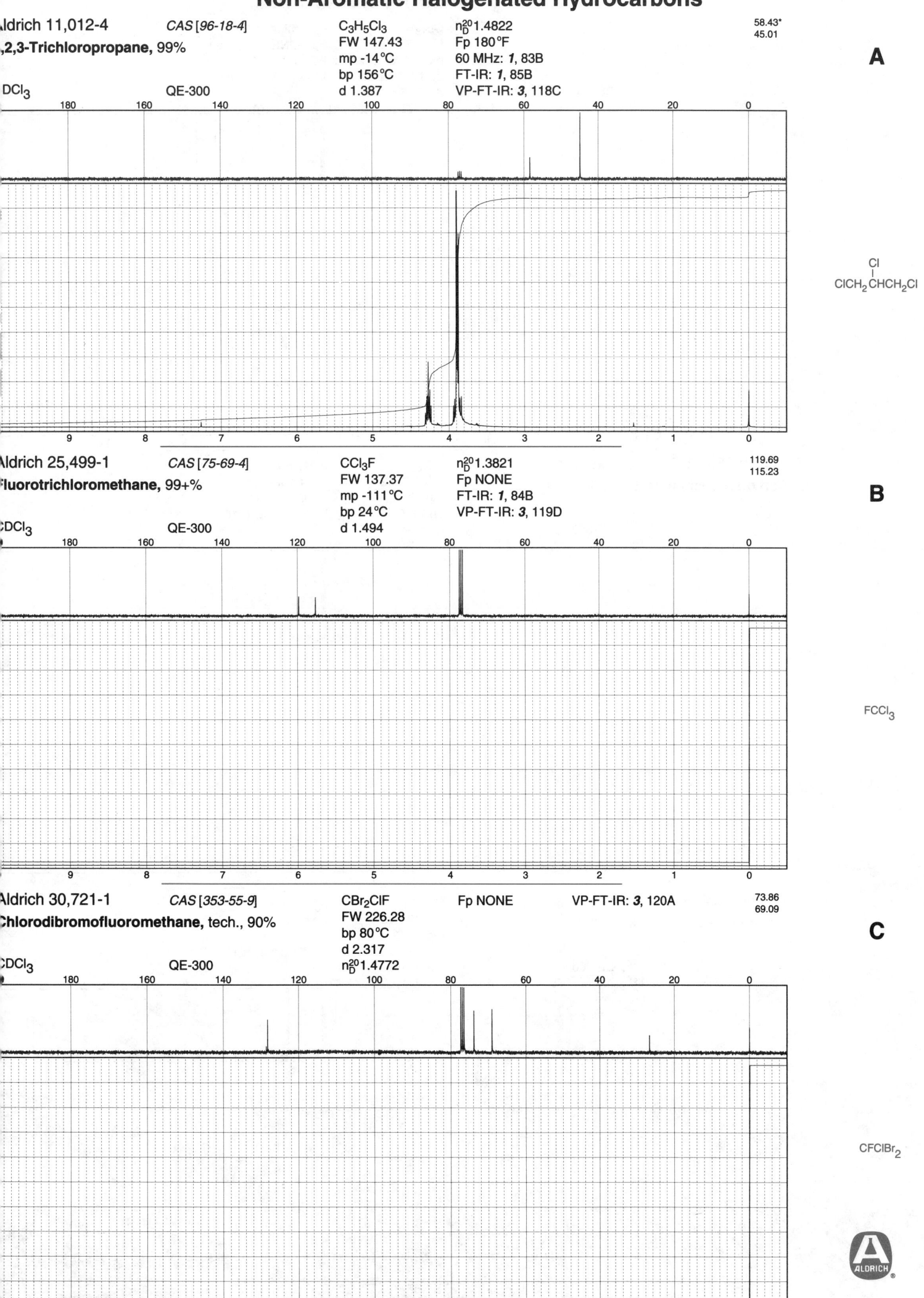

Aldrich 11,012-4   *CAS [96-18-4]* $C_3H_5Cl_3$ n_D^{20} 1.4822   58.43*
45.01

,2,3-Trichloropropane, 99%   FW 147.43   Fp 180°F
mp -14°C   60 MHz: *1*, 83B
bp 156°C   FT-IR: *1*, 85B
d 1.387   VP-FT-IR: *3*, 118C

DCl₃   QE-300

A

$$\underset{\underset{|}{Cl}}{ClCH_2CHCH_2Cl}$$

Aldrich 25,499-1   *CAS [75-69-4]* CCl_3F n_D^{20} 1.3821   119.69
115.23

Fluorotrichloromethane, 99+%   FW 137.37   Fp NONE
mp -111°C   FT-IR: *1*, 84B
bp 24°C   VP-FT-IR: *3*, 119D
d 1.494

DCl₃   QE-300

B

$FCCl_3$

Aldrich 30,721-1   *CAS [353-55-9]* CBr_2ClF Fp NONE   VP-FT-IR: *3*, 120A   73.86
69.09

Chlorodibromofluoromethane, tech., 90%   FW 226.28
bp 80°C
d 2.317

DCl₃   QE-300 n_D^{20} 1.4772

C

$CFClBr_2$

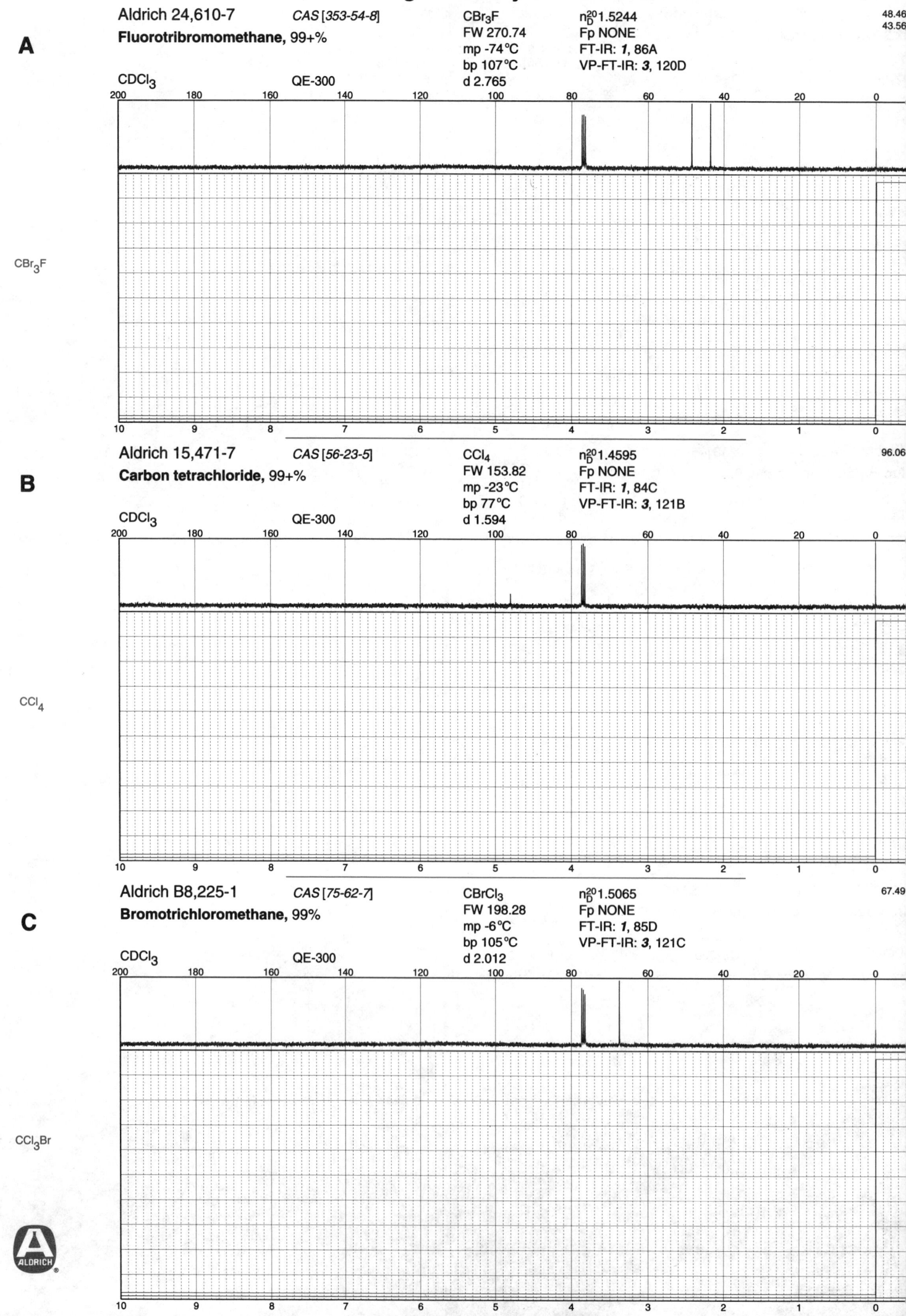

A

Aldrich 24,610-7   CAS [353-54-8]   CBr₃F n_D^{20} 1.5244   48.46

Fluorotribromomethane, 99+%   FW 270.74   Fp NONE   43.56

mp -74°C   FT-IR: *1*, 86A

bp 107°C   VP-FT-IR: *3*, 120D

d 2.765

CDCl₃   QE-300

CBr₃F

B

Aldrich 15,471-7   CAS [56-23-5]   CCl₄ n_D^{20} 1.4595   96.06

Carbon tetrachloride, 99+%   FW 153.82   Fp NONE

mp -23°C   FT-IR: *1*, 84C

bp 77°C   VP-FT-IR: *3*, 121B

d 1.594

CDCl₃   QE-300

CCl₄

C

Aldrich B8,225-1   CAS [75-62-7]   CBrCl₃ n_D^{20} 1.5065   67.49

Bromotrichloromethane, 99%   FW 198.28   Fp NONE

mp -6°C   FT-IR: *1*, 85D

bp 105°C   VP-FT-IR: *3*, 121C

d 2.012

CDCl₃   QE-300

CCl₃Br

ALDRICH

Non-Aromatic Halogenated Hydrocarbons

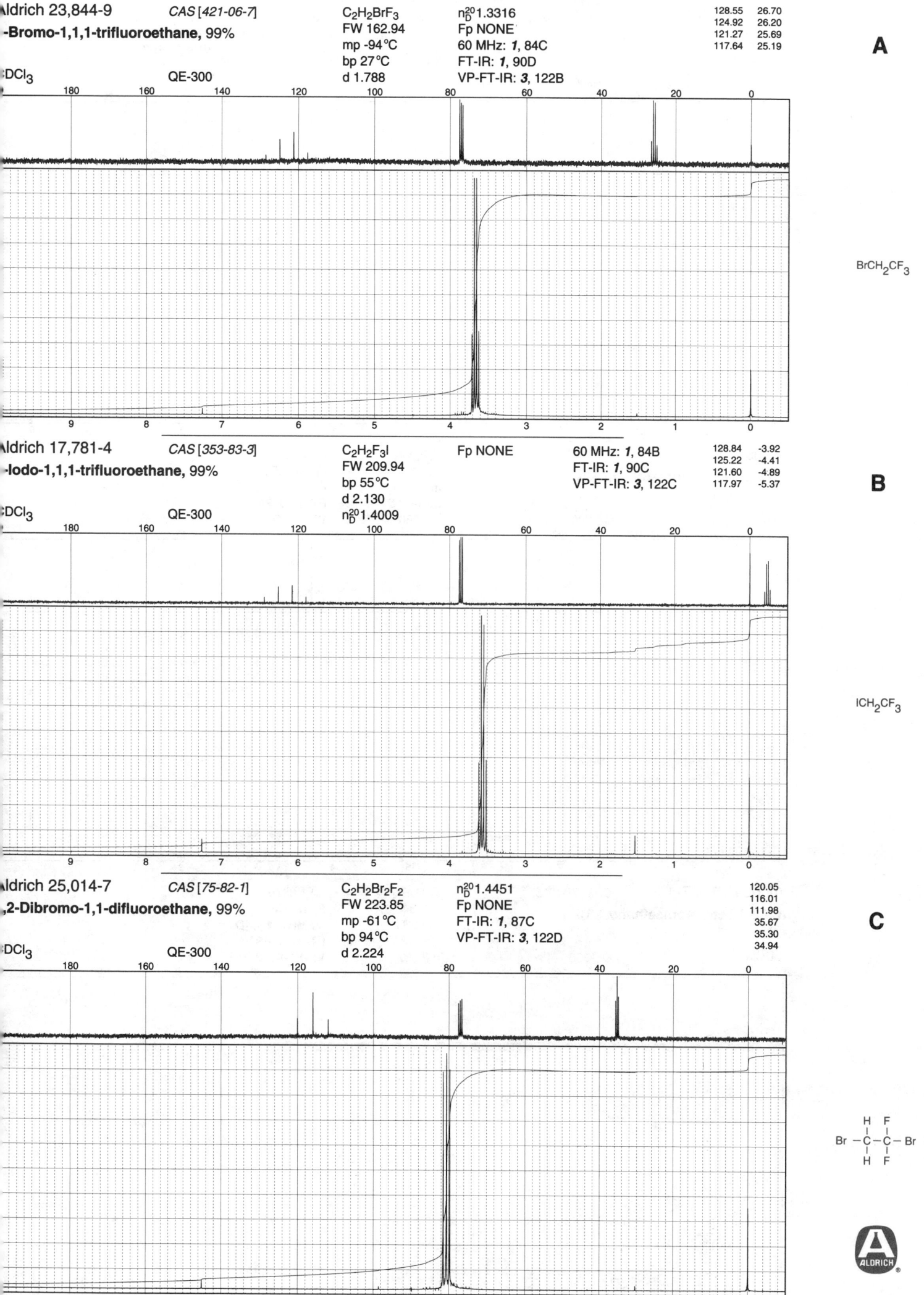

Aldrich 23,844-9   CAS [421-06-7]   C₂H₂BrF₃   n²⁰_D 1.3316
-Bromo-1,1,1-trifluoroethane, 99%   FW 162.94   Fp NONE
                                      mp -94°C   60 MHz: *1*, 84C
                                      bp 27°C   FT-IR: *1*, 90D
CDCl₃   QE-300   d 1.788   VP-FT-IR: *3*, 122B

128.55	26.70
124.92	26.20
121.27	25.69
117.64	25.19

A

BrCH₂CF₃

Aldrich 17,781-4   CAS [353-83-3]   C₂H₂F₃I   Fp NONE   60 MHz: *1*, 84B
-Iodo-1,1,1-trifluoroethane, 99%   FW 209.94   FT-IR: *1*, 90C
                                     bp 55°C   VP-FT-IR: *3*, 122C
                                     d 2.130
CDCl₃   QE-300   n²⁰_D 1.4009

128.84	-3.92
125.22	-4.41
121.60	-4.89
117.97	-5.37

B

ICH₂CF₃

Aldrich 25,014-7   CAS [75-82-1]   C₂H₂Br₂F₂   n²⁰_D 1.4451
,2-Dibromo-1,1-difluoroethane, 99%   FW 223.85   Fp NONE
                                     mp -61°C   FT-IR: *1*, 87C
                                     bp 94°C   VP-FT-IR: *3*, 122D
CDCl₃   QE-300   d 2.224

120.05
116.01
111.98
35.67
35.30
34.94

C

```
    H   F
    |   |
Br—C—C—Br
    |   |
    H   F
```

ALDRICH

A

Aldrich T720-9   *CAS [630-20-6]* $C_2H_2Cl_4$ Fp NONE   60 MHz: *1*, 83D   96.16

1,1,1,2-Tetrachloroethane, 99%   FW 167.85   FT-IR: *1*, 87A   58.85

bp 138°C   VP-FT-IR: *3*, 123A

d 1.598

n_D^{20}1.4819

CDCl$_3$ QE-300

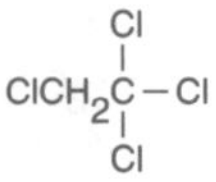

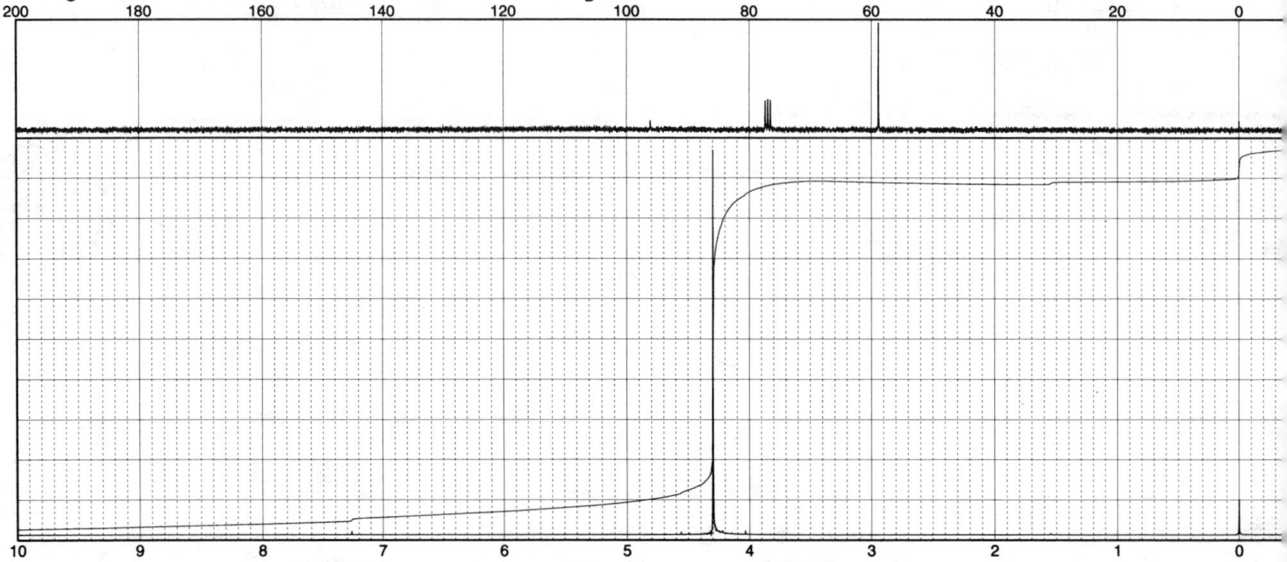

$ClCH_2C-Cl$ (structure, with Cl above and Cl below central C)

B

Aldrich 18,543-4   *CAS [79-34-5]* $C_2H_2Cl_4$ n_D^{20}1.4935   74.10

1,1,2,2-Tetrachloroethane, 99%   FW 167.85   Fp NONE

mp -43°C   60 MHz: *1*, 83C

bp 147°C   FT-IR: *1*, 86D

CDCl$_3$ QE-300   d 1.586   VP-FT-IR: *3*, 123B

$Cl_2CHCHCl_2$

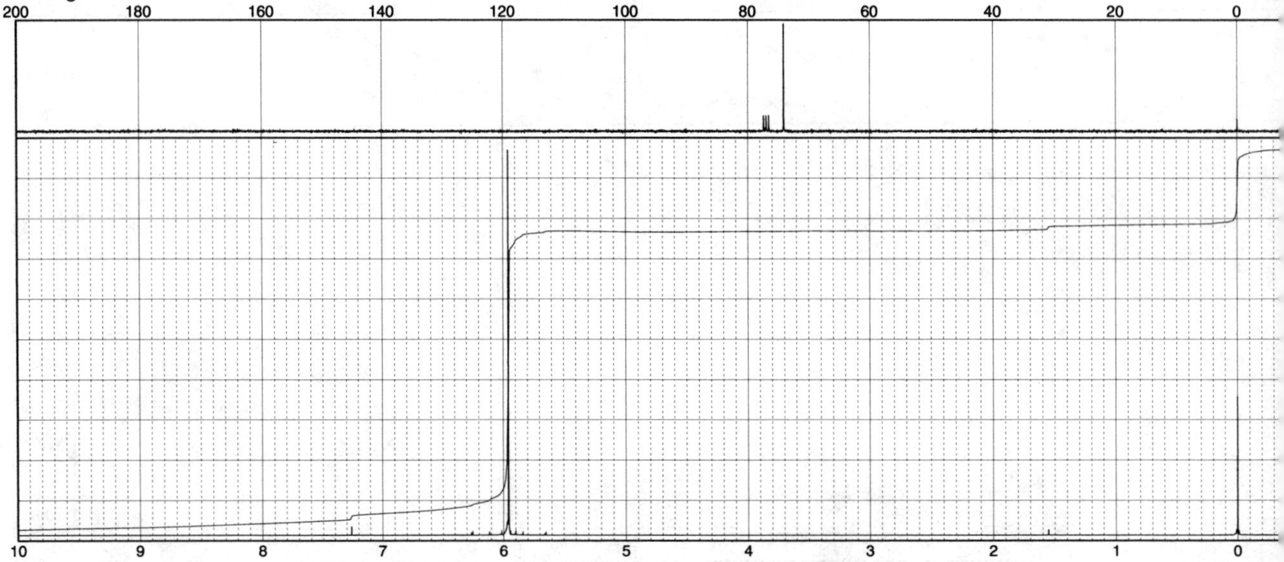

C

Aldrich 18,557-4   *CAS [79-27-6]* $C_2H_2Br_4$ n_D^{20}1.6370   46.99

1,1,2,2-Tetrabromoethane, 98%   FW 345.67   Fp NONE

mp 0°C   60 MHz: *1*, 84D

bp 119°C (15 mm)   FT-IR: *1*, 88A

CDCl$_3$ QE-300   d 2.967   VP-FT-IR: *3*, 123D

$Br_2CHCHBr_2$

ALDRICH

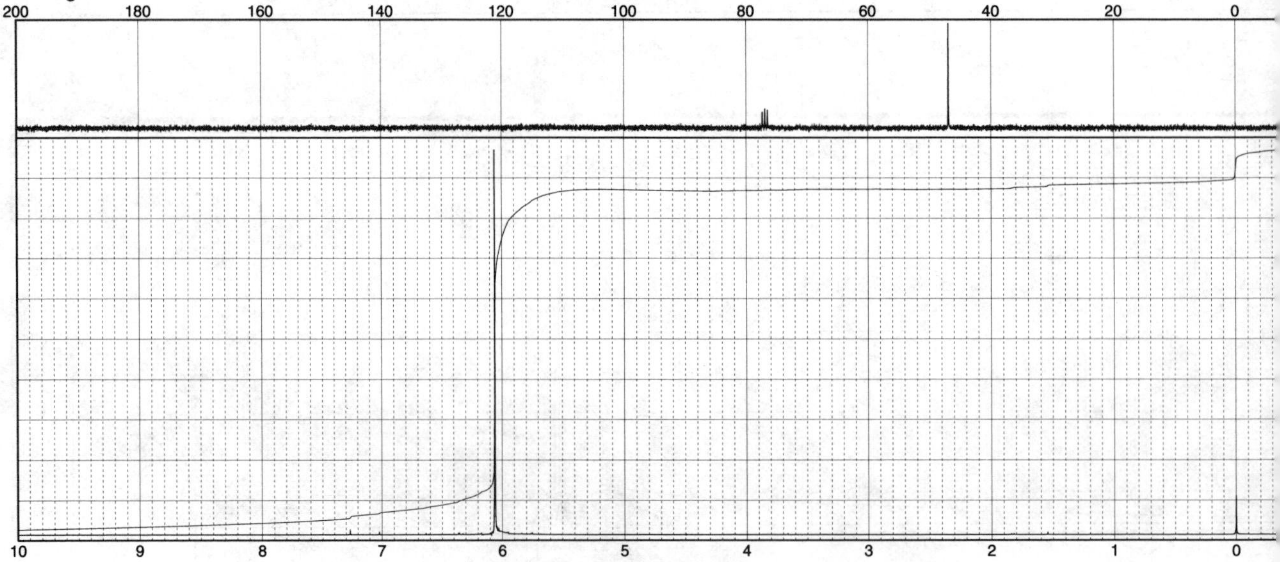

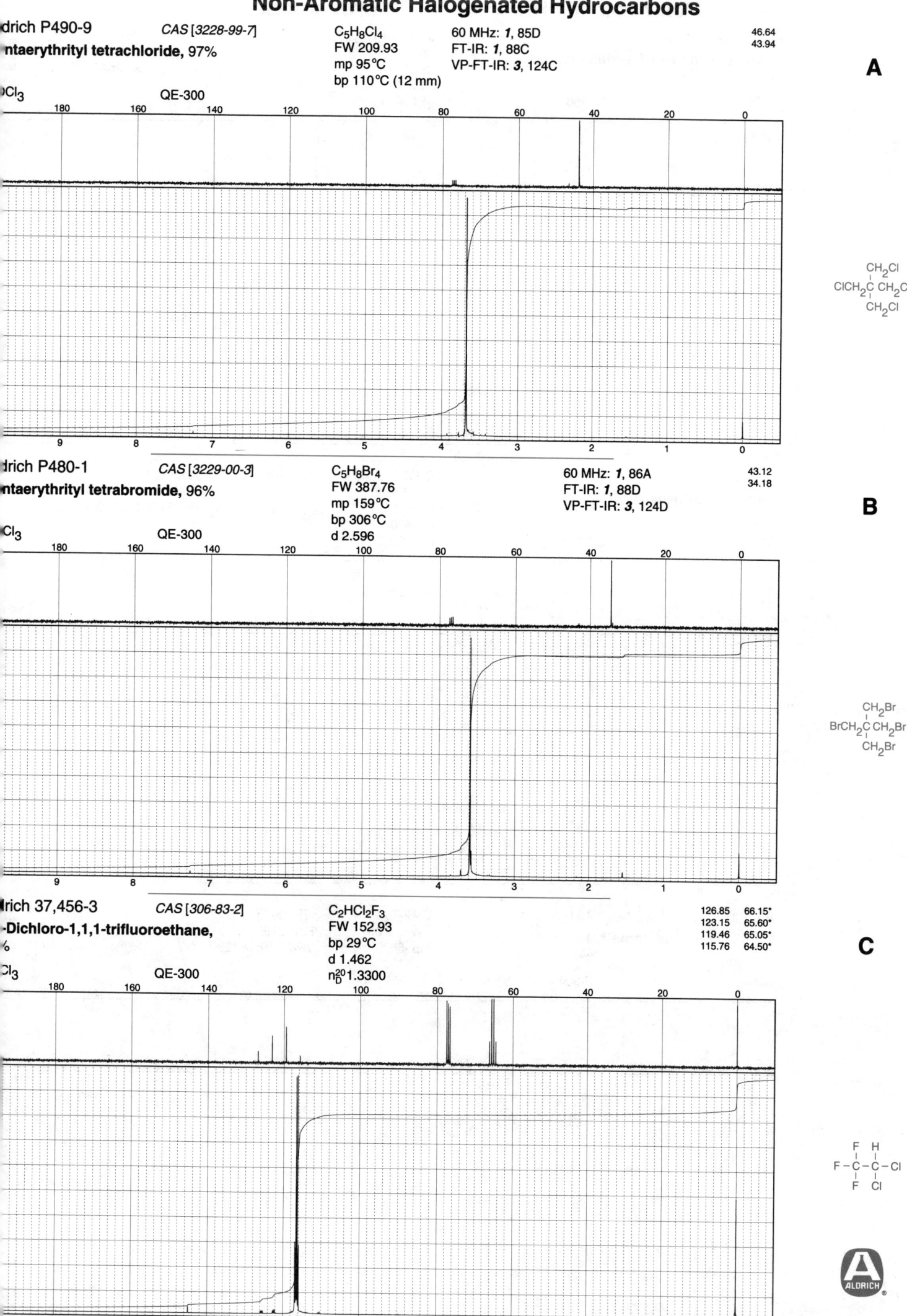

drich P490-9   CAS [3228-99-7] $C_5H_8Cl_4$ 60 MHz: **1**, 85D

ntaerythrityl tetrachloride, 97%   FW 209.93   FT-IR: **1**, 88C

mp 95 °C   VP-FT-IR: **3**, 124C

bp 110 °C (12 mm)

46.64
43.94

A

Cl_3 QE-300

CH_2Cl
$ClCH_2C\ CH_2Cl$
CH_2Cl

drich P480-1   CAS [3229-00-3] $C_5H_8Br_4$

ntaerythrityl tetrabromide, 96%   FW 387.76

mp 159 °C

bp 306 °C

d 2.596

60 MHz: **1**, 86A

FT-IR: **1**, 88D

VP-FT-IR: **3**, 124D

43.12
34.18

B

Cl_3 QE-300

CH_2Br
$BrCH_2C\ CH_2Br$
CH_2Br

drich 37,456-3   CAS [306-83-2] $C_2HCl_2F_3$

-Dichloro-1,1,1-trifluoroethane,

%

FW 152.93

bp 29 °C

d 1.462

n_D^{20} 1.3300

126.85   66.15*
123.15   65.60*
119.46   65.05*
115.76   64.50*

C

Cl_3 QE-300

F   H
F−C−C−Cl
F   Cl

ALDRICH

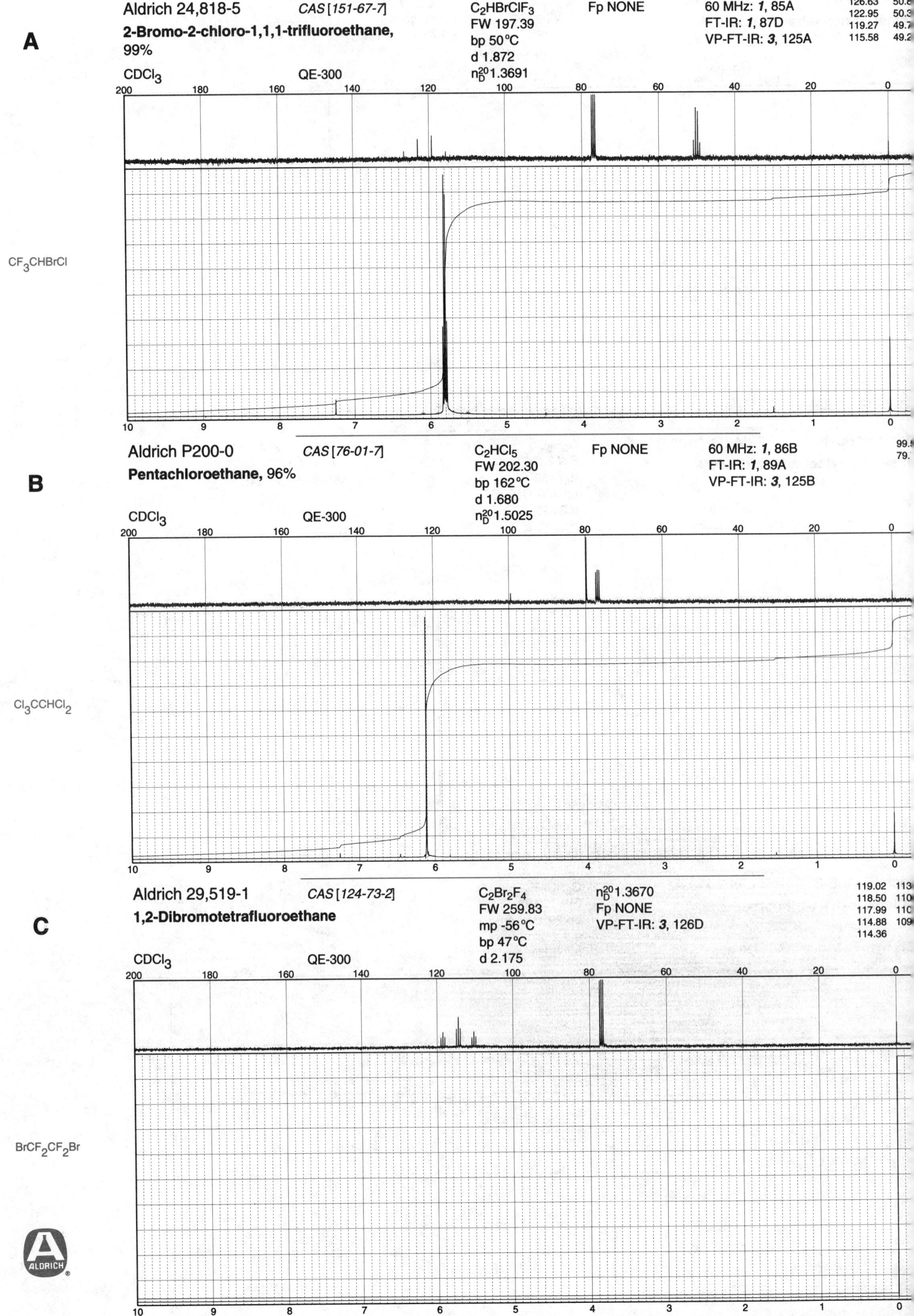

A

Aldrich 24,818-5   *CAS [151-67-7]*   C₂HBrClF₃   Fp NONE   60 MHz: *1*, 85A

2-Bromo-2-chloro-1,1,1-trifluoroethane, 99%

$C_2HBrClF_3$ FW 197.39 bp 50°C d 1.872 n_D^{20} 1.3691

FT-IR: *1*, 87D
VP-FT-IR: *3*, 125A

126.63   50.8
122.95   50.3
119.27   49.7
115.58   49.2

CDCl₃   QE-300

CF₃CHBrCl

B

Aldrich P200-0   *CAS [76-01-7]*   C₂HCl₅   Fp NONE   60 MHz: *1*, 86B

Pentachloroethane, 96%

C_2HCl_5 FW 202.30 bp 162°C d 1.680 n_D^{20} 1.5025

FT-IR: *1*, 89A
VP-FT-IR: *3*, 125B

99.9
79.

CDCl₃   QE-300

Cl₃CCHCl₂

C

Aldrich 29,519-1   *CAS [124-73-2]*   C₂Br₂F₄ n_D^{20} 1.3670

1,2-Dibromotetrafluoroethane

$C_2Br_2F_4$ FW 259.83 mp -56°C bp 47°C d 2.175

Fp NONE
VP-FT-IR: *3*, 126D

119.02   113
118.50   110
117.99   110
114.88   109
114.36

CDCl₃   QE-300

BrCF₂CF₂Br

Aldrich 13,040-0   CAS [354-58-5] $C_2Cl_3F_3$ n_D^{20} 1.3599   125.96 92.09
1,1,1-Trichlorotrifluoroethane, 99%   FW 187.38   Fp NONE   122.21 91.52
mp 14°C   FT-IR: *1*, 90A   118.46 90.96
bp 46°C   VP-FT-IR: *3*, 127A   114.71 90.39
d 1.579

A

CDCl$_3$ QE-300

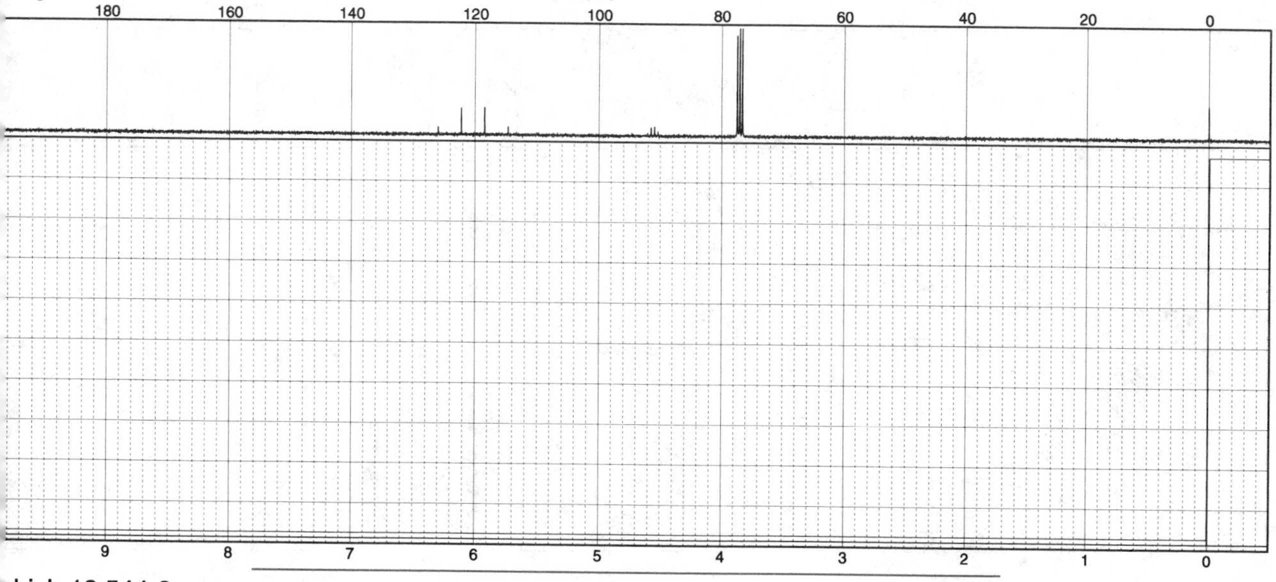

CCl_3CF_3

Aldrich 18,544-2   CAS [67-72-1] C_2Cl_6 FT-IR: *1*, 89B   105.18
Hexachloroethane, 99%   FW 236.74   VP-FT-IR: *3*, 127D
mp 193°C s.
d 2.091
Fp NONE

B

CDCl$_3$ QE-300

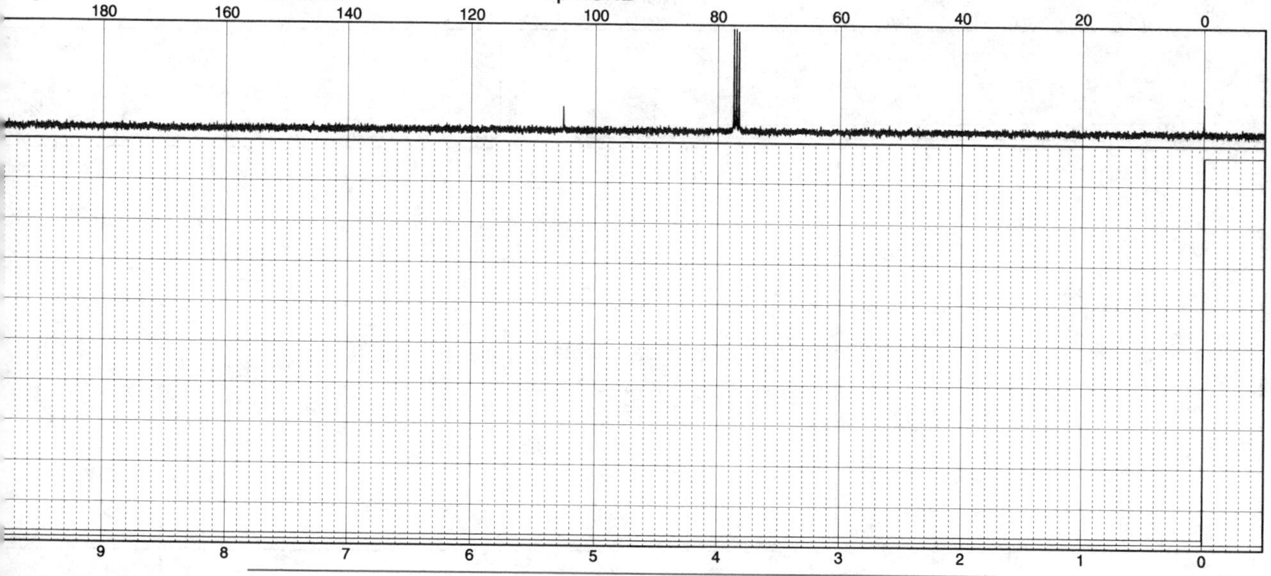

CCl_3CCl_3

Aldrich 13,339-6   CAS [630-25-1] $C_2Br_2Cl_4$ FT-IR: *1*, 89C   89.38
1,2-Dibromotetrachloroethane, 97%   FW 325.65   VP-FT-IR: *3*, 128B
mp 221°C d.
d 2.713
Fp NONE

C

CDCl$_3$ QE-300

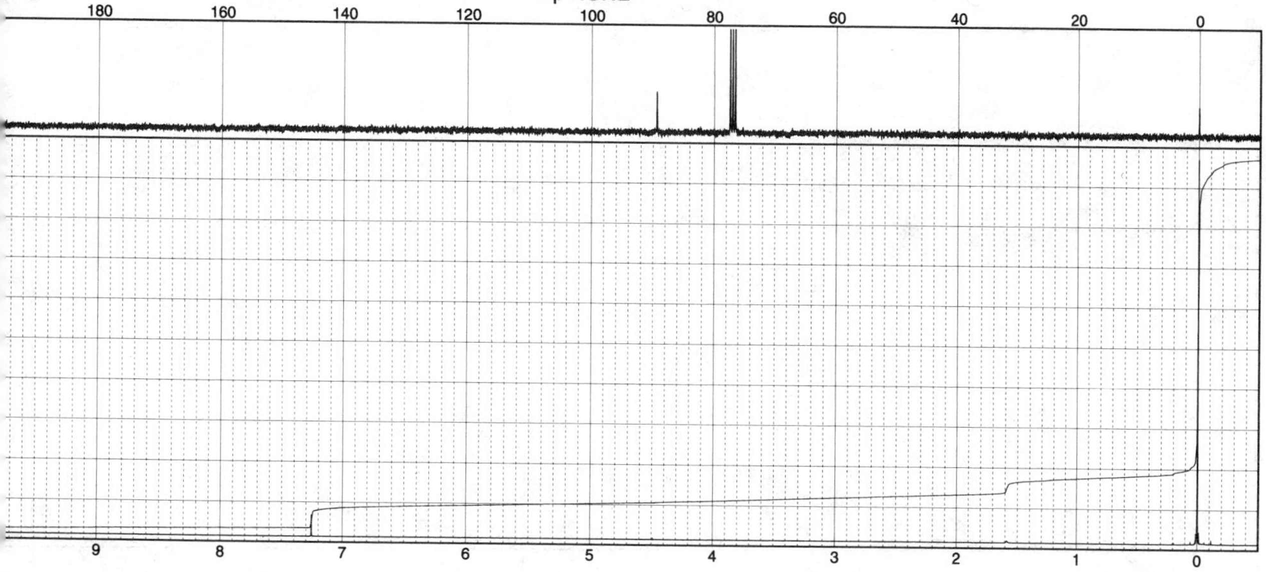

$BrCCl_2CCl_2Br$

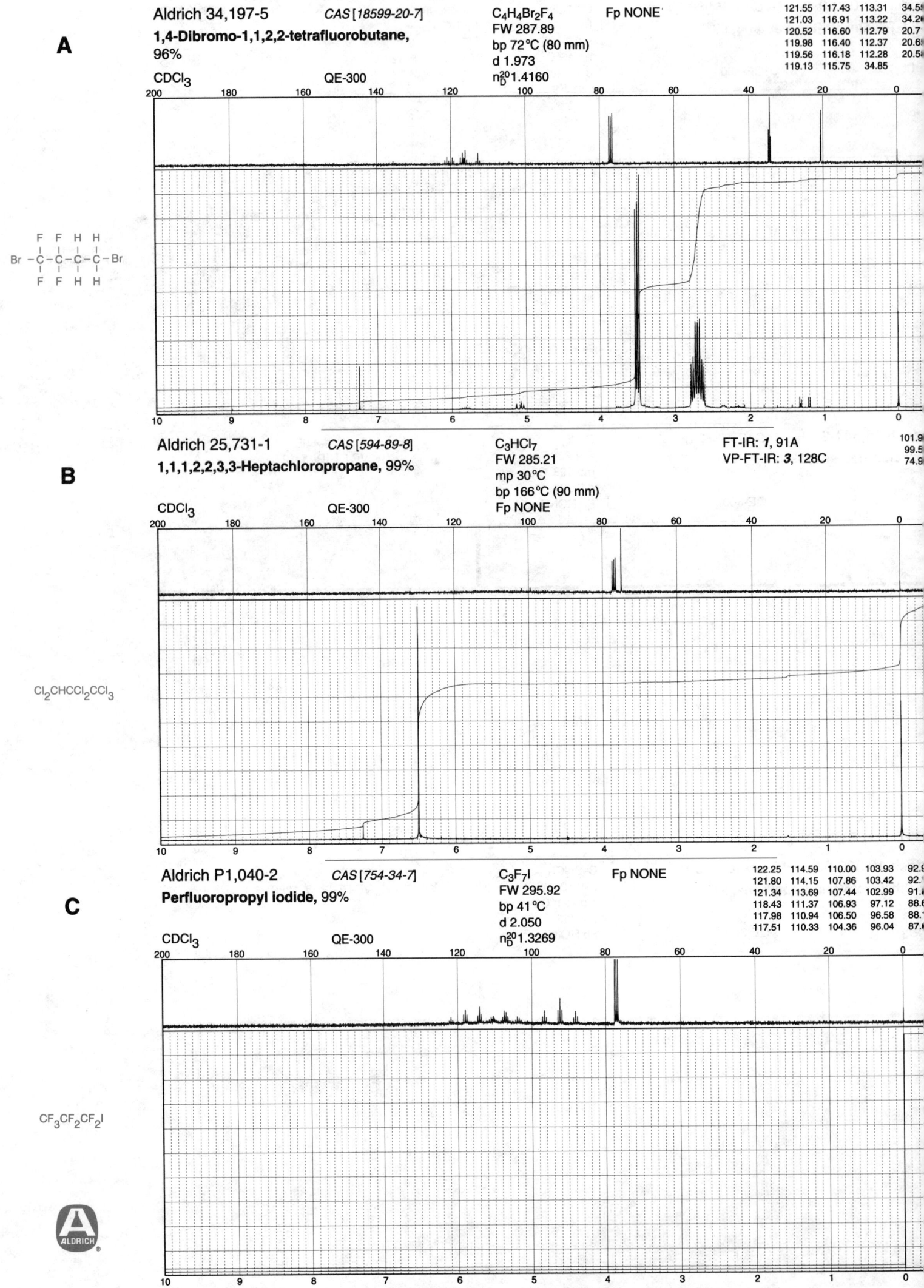

A

Aldrich 34,197-5   *CAS [18599-20-7]*   C₄H₄Br₂F₄   Fp NONE

1,4-Dibromo-1,1,2,2-tetrafluorobutane, 96%

FW 287.89
bp 72°C (80 mm)
d 1.973
n²⁰_D 1.4160

CDCl₃   QE-300

121.55	117.43	113.31	34.5
121.03	116.91	113.22	34.2
120.52	116.60	112.79	20.7
119.98	116.40	112.37	20.6
119.56	116.18	112.28	20.5
119.13	115.75	34.85	

Br—C—C—C—C—Br (F F H H / F F H H)

B

Aldrich 25,731-1   *CAS [594-89-8]*   C₃HCl₇   FT-IR: *1*, 91A

1,1,1,2,2,3,3-Heptachloropropane, 99%

FW 285.21
mp 30°C
bp 166°C (90 mm)
Fp NONE

VP-FT-IR: *3*, 128C

101.9
99.5
74.9

CDCl₃   QE-300

Cl₂CHCCl₂CCl₃

C

Aldrich P1,040-2   *CAS [754-34-7]*   C₃F₇I   Fp NONE

Perfluoropropyl iodide, 99%

FW 295.92
bp 41°C
d 2.050
n²⁰_D 1.3269

122.25	114.59	110.00	103.93	92.9
121.80	114.15	107.86	103.42	92.
121.34	113.69	107.44	102.99	91.
118.43	111.37	106.93	97.12	88.
117.98	110.94	106.50	96.58	88.
117.51	110.33	104.36	96.04	87.

CDCl₃   QE-300

CF₃CF₂CF₂I

Non-Aromatic Halogenated Hydrocarbons

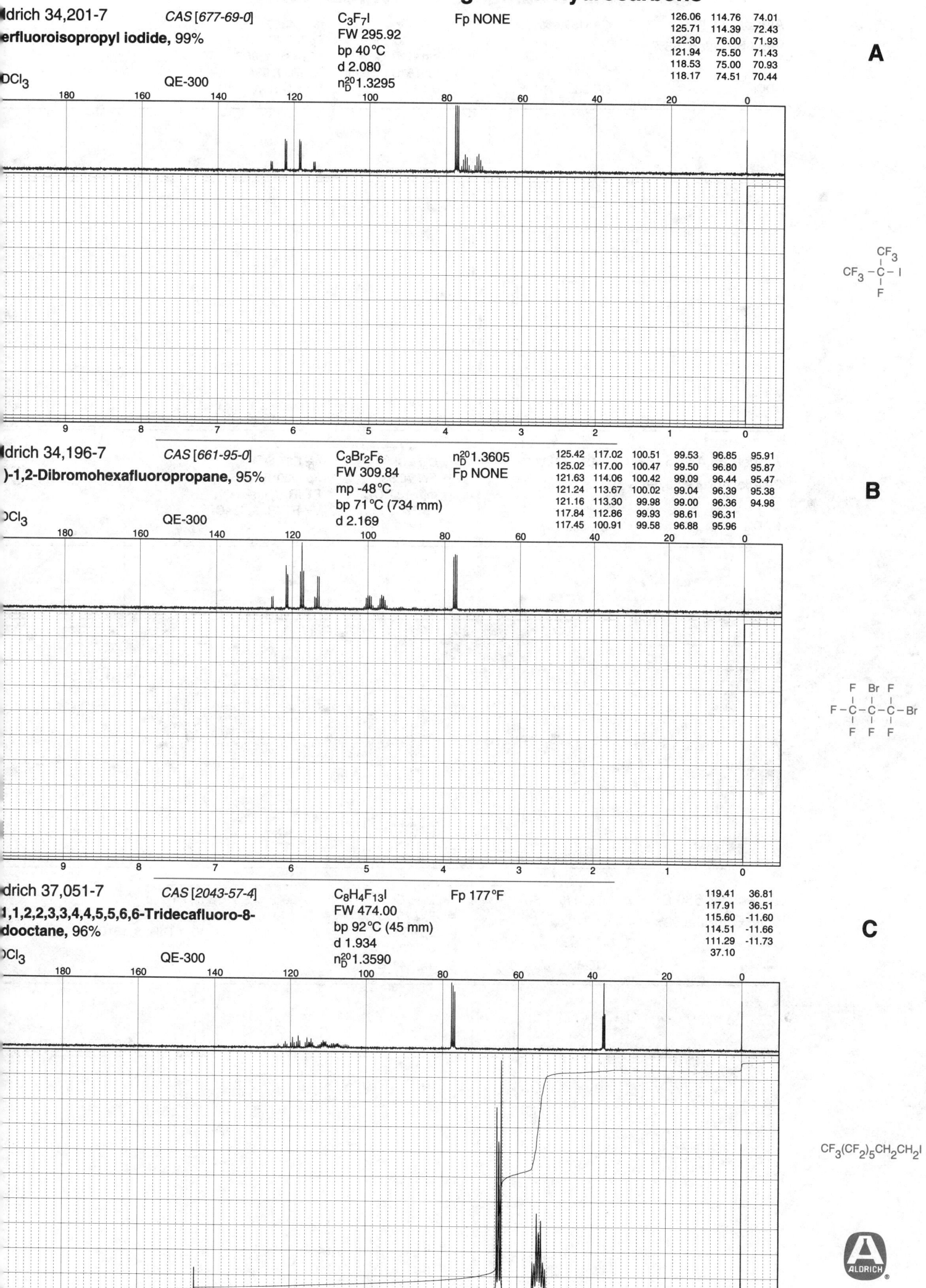

Aldrich 34,201-7   *CAS [677-69-0]*   C$_3$F$_7$I   Fp NONE

Perfluoroisopropyl iodide, 99%

FW 295.92
bp 40 °C
d 2.080
n$_D^{20}$1.3295

CDCl$_3$ QE-300

126.06	114.76	74.01
125.71	114.39	72.43
122.30	76.00	71.93
121.94	75.50	71.43
118.53	75.00	70.93
118.17	74.51	70.44

A

CF$_3$–C–I with CF$_3$ and F

Aldrich 34,196-7   *CAS [661-95-0]*   C$_3$Br$_2$F$_6$ n$_D^{20}$1.3605

(±)-1,2-Dibromohexafluoropropane, 95%

FW 309.84
mp -48 °C
bp 71 °C (734 mm)
d 2.169

Fp NONE

CDCl$_3$ QE-300

125.42	117.02	100.51	99.53	96.85	95.91
125.02	117.00	100.47	99.50	96.80	95.87
121.63	114.06	100.42	99.09	96.44	95.47
121.24	113.67	100.02	99.04	96.39	95.38
121.16	113.30	99.98	99.00	96.36	94.98
117.84	112.86	99.93	98.61	96.31	
117.45	100.91	99.58	96.88	95.96	

B

F–C–C–C–Br with F, Br, F, F, F, F

Aldrich 37,051-7   *CAS [2043-57-4]*   C$_8$H$_4$F$_{13}$I   Fp 177 °F

1,1,1,2,2,3,3,4,4,5,5,6,6-Tridecafluoro-8-iodooctane, 96%

FW 474.00
bp 92 °C (45 mm)
d 1.934
n$_D^{20}$1.3590

CDCl$_3$ QE-300

119.41	36.81
117.91	36.51
115.60	-11.60
114.51	-11.66
111.29	-11.73
37.10	

C

CF$_3$(CF$_2$)$_5$CH$_2$CH$_2$I

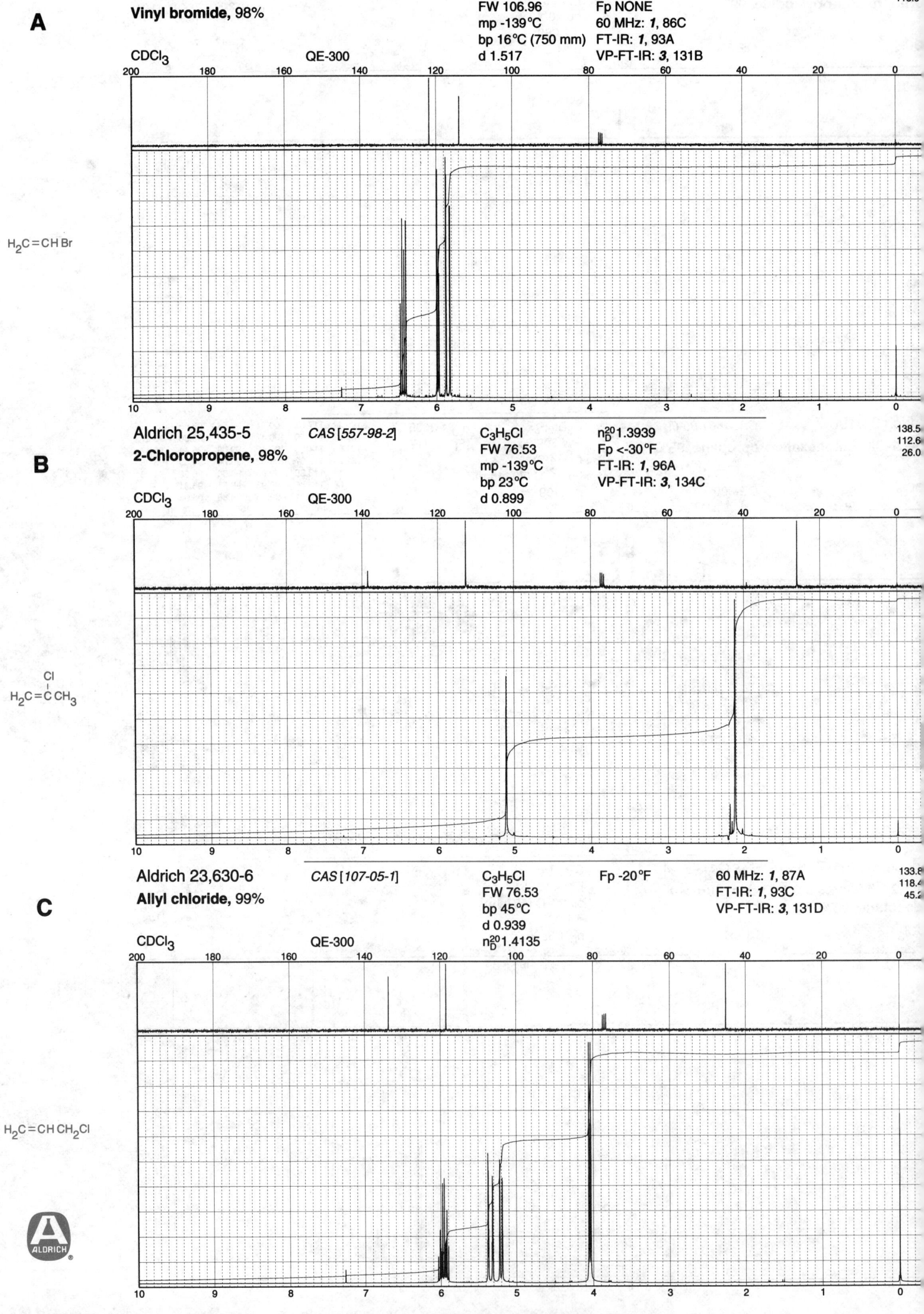

Aldrich V190-2 **Vinyl bromide, 98%** CAS [593-60-2]

C_2H_3Br
FW 106.96
mp -139°C
bp 16°C (750 mm)
d 1.517

n_D^{20} 1.4350
Fp NONE
60 MHz: *1*, 86C
FT-IR: *1*, 93A
VP-FT-IR: *3*, 131B

121.82
113.9

CDCl₃ QE-300

$H_2C=CHBr$

Aldrich 25,435-5 **2-Chloropropene, 98%** CAS [557-98-2]

C_3H_5Cl
FW 76.53
mp -139°C
bp 23°C
d 0.899

n_D^{20} 1.3939
Fp <-30°F
FT-IR: *1*, 96A
VP-FT-IR: *3*, 134C

138.5
112.6
26.0

CDCl₃ QE-300

$H_2C=C \overset{Cl}{\underset{}{CH_3}}$

Aldrich 23,630-6 **Allyl chloride, 99%** CAS [107-05-1]

C_3H_5Cl
FW 76.53
bp 45°C
d 0.939
n_D^{20} 1.4135

Fp -20°F

60 MHz: *1*, 87A
FT-IR: *1*, 93C
VP-FT-IR: *3*, 131D

133.8
118.4
45.2

CDCl₃ QE-300

$H_2C=CHCH_2Cl$

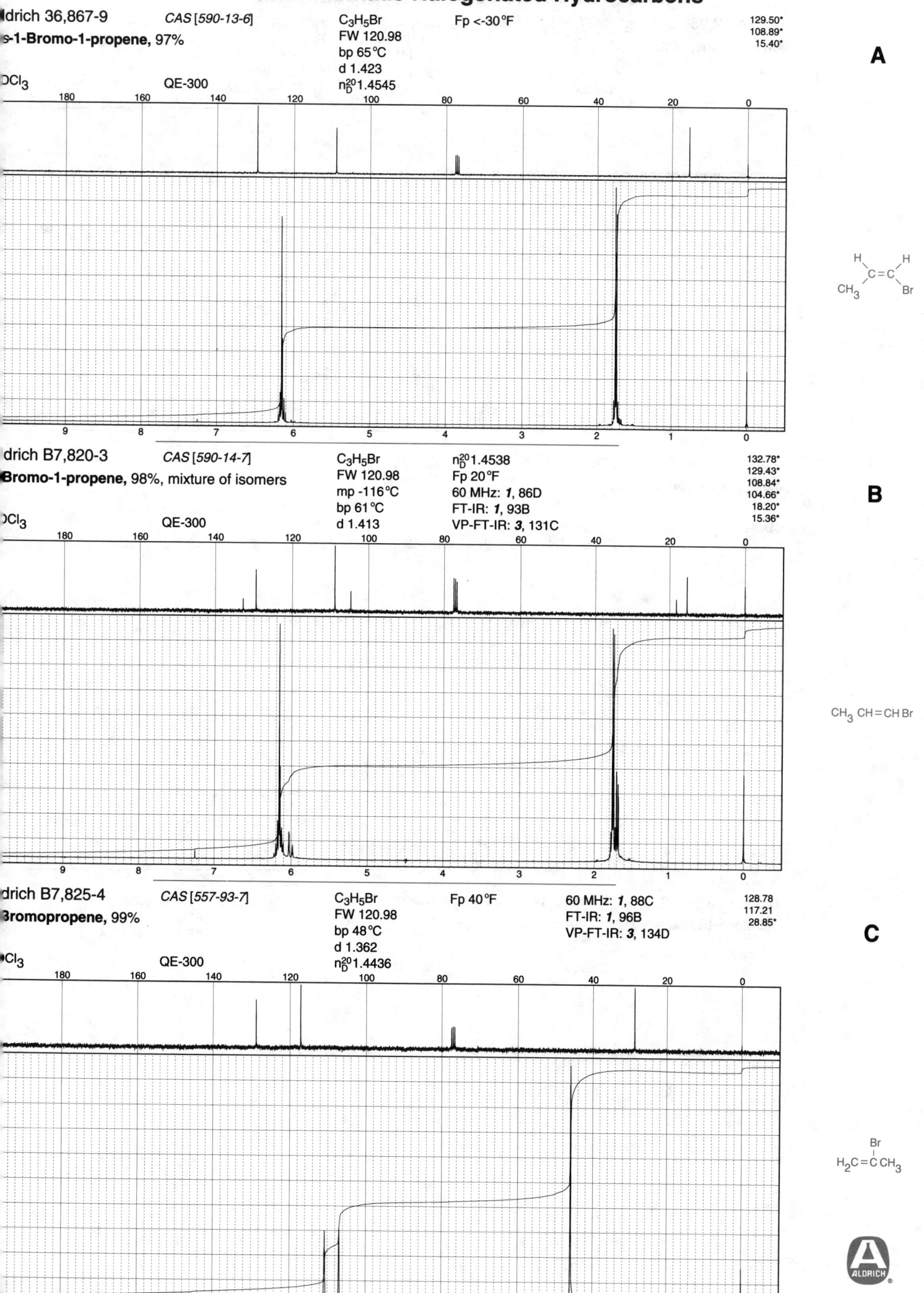

Aldrich 36,867-9   CAS [590-13-6]   C₃H₅Br   Fp <-30°F

C_3H_5Br
FW 120.98
bp 65°C
d 1.423
n_D^{20} 1.4545

cis-1-Bromo-1-propene, 97%

CDCl₃   QE-300

129.50*
108.89*
15.40*

A

Aldrich B7,820-3   CAS [590-14-7]

C_3H_5Br
FW 120.98
mp -116°C
bp 61°C
d 1.413

n_D^{20} 1.4538
Fp 20°F
60 MHz: 1, 86D
FT-IR: 1, 93B
VP-FT-IR: 3, 131C

1-Bromo-1-propene, 98%, mixture of isomers

CDCl₃   QE-300

132.78*
129.43*
108.84*
104.66*
18.20*
15.36*

B

CH₃ CH=CH Br

Aldrich B7,825-4   CAS [557-93-7]

C_3H_5Br
FW 120.98
bp 48°C
d 1.362
n_D^{20} 1.4436

Fp 40°F

60 MHz: 1, 88C
FT-IR: 1, 96B
VP-FT-IR: 3, 134D

2-Bromopropene, 99%

CDCl₃   QE-300

128.78
117.21
28.85*

C

Br
|
H₂C=C CH₃

A

Aldrich A2,958-5   *CAS [106-95-6]*   C₃H₅Br   n₂₀1.4690   134.3⁄
Allyl bromide, 99%   FW 120.98   Fp 28°F   119.2⁄
                                           mp -119°C   60 MHz: *1*, 87B   33.0
                                           bp 71°C   FT-IR: *1*, 93D
CDCl₃   QE-300   d 1.398   VP-FT-IR: *3*, 132A

200 180 160 140 120 100 80 60 40 20 0

H₂C =CHCH₂Br

10   9   8   7   6   5   4   3   2   1   0

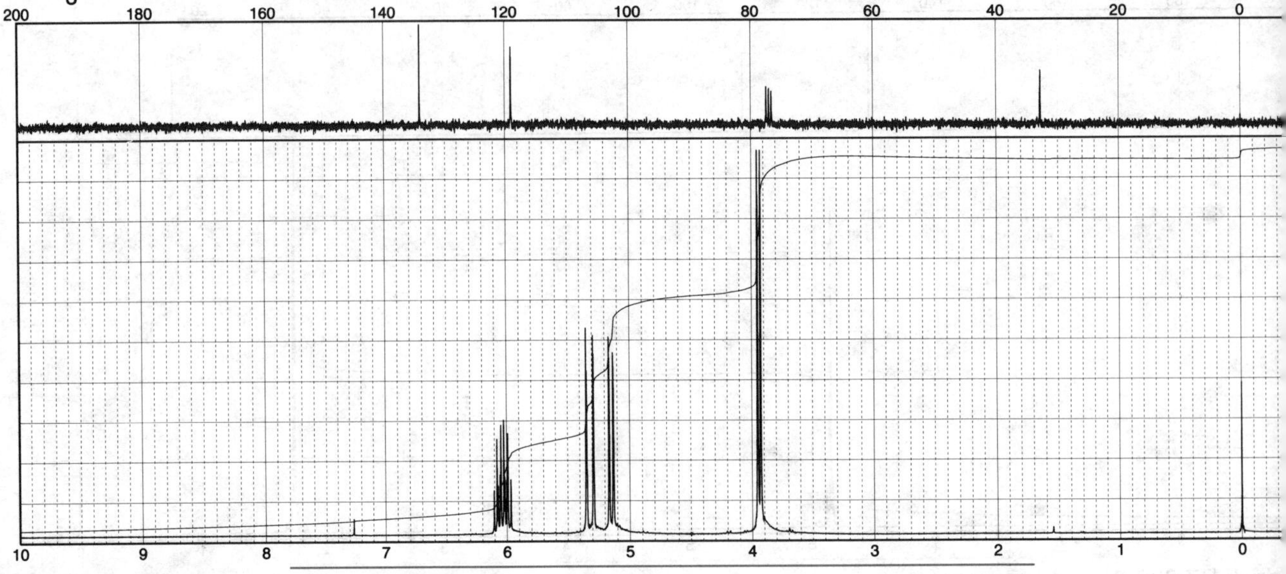

B

Aldrich 37,137-8   *CAS [677-21-4]*   C₃H₃F₃
3,3,3-Trifluoropropene, 99%   FW 96.05
                                          bp 0°C

CDCl₃   QE-300

200 180 160 140 120 100 80 60 40 20 0

CF₃ CH=CH₂

10   9   8   7   6   5   4   3   2   1   0

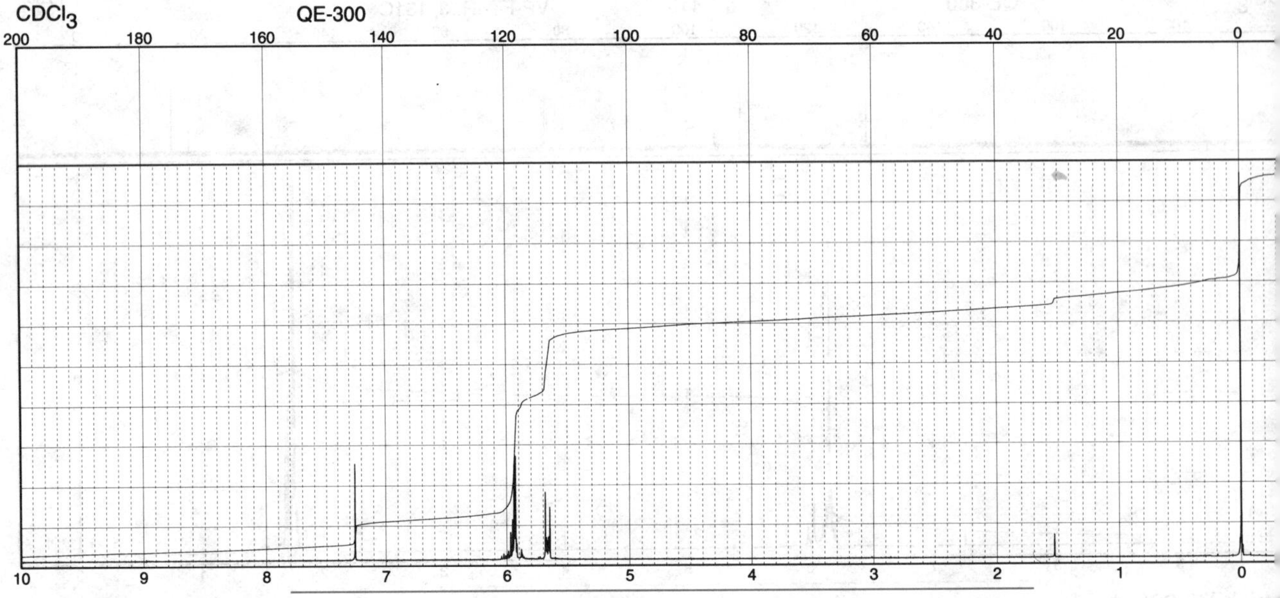

C

Aldrich C2,900-2   *CAS [563-52-0]*   C₄H₇Cl   Fp -4°F   60 MHz: *1*, 88D   139.⁄
3-Chloro-1-butene, 98%   FW 90.55   FT-IR: *1*, 96C   115.⁄
                                          bp 64°C   VP-FT-IR: *3*, 135A 57.⁄
                                          d 0.900   24.⁄
CDCl₃   QE-300   n₂₀1.4155

200 180 160 140 120 100 80 60 40 20 0

          Cl
          |
CH₃ CHCH=CH₂

10   9   8   7   6   5   4   3   2   1   0

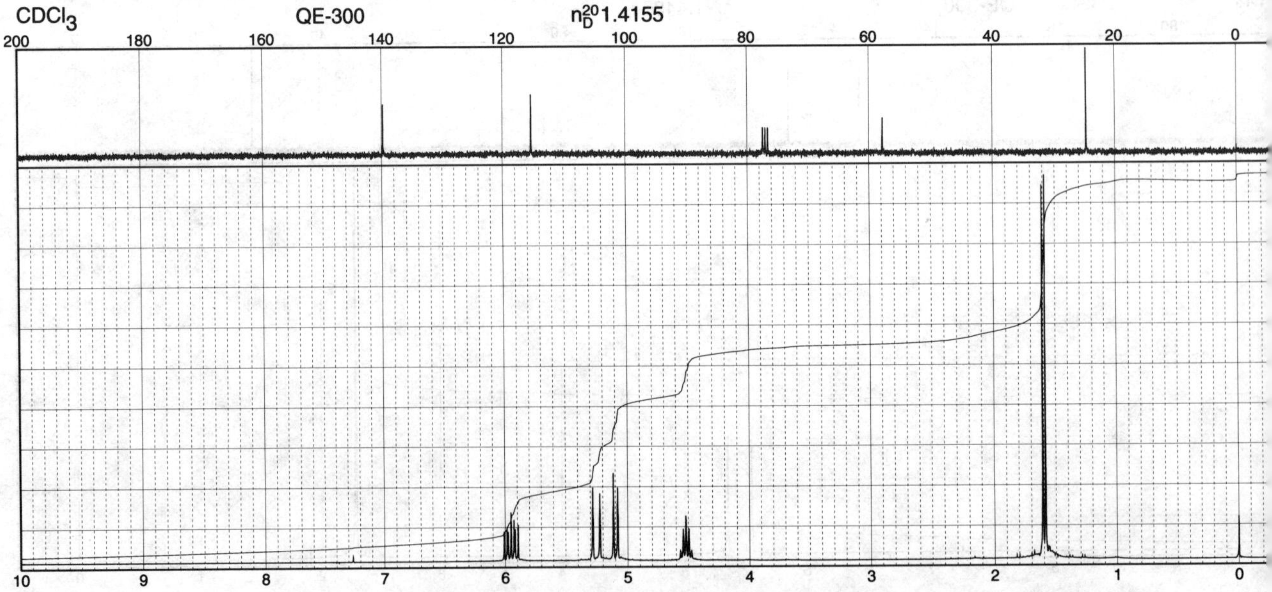

Aldrich 12,533-4   CAS [4894-61-5]   C₄H₇Cl   Fp 5°F   60 MHz: 1, 88A   130.82*
Crotyl chloride, 70%, predominantly   FW 90.55   FT-IR: 1, 95A   127.18*
trans   bp 74°C   VP-FT-IR: 3, 133C   45.35
    d 0.920   17.56*
CDCl₃   QE-300   n_D^20 1.4310

A

$CH_3 CH = CHCH_2Cl$

Aldrich 25,205-0   CAS [4461-41-0]   C₄H₇Cl   Fp -3°F   FT-IR: 1, 95C   130.89   26.05*
-Chloro-2-butene, 98%, mixture of cis   FW 90.55   VP-FT-IR: 3, 134A   129.25   20.36*
and trans   bp 65°C   122.02*   14.05*
    d 0.926   120.12*   13.98*
CDCl₃   QE-300   n_D^20 1.4205

B

$CH_3CH = CCH_3$ with Cl substituent

Aldrich 16,785-1   CAS [5162-44-7]   C₄H₇Br   Fp 49°F   60 MHz: 1, 87D   135.36*
Bromo-1-butene, 97%   FW 135.01   FT-IR: 1, 94B   117.66
    bp 99°C   VP-FT-IR: 3, 132C   37.23
    d 1.330   32.18
CDCl₃   QE-300   n_D^20 1.4625

C

$BrCH_2CH_2CH = CH_2$

A

Aldrich 21,556-2   *CAS [13294-71-8]* C_4H_7Br Fp 34°F   60 MHz: *1*, 89A   126.48
123.37
2-Bromo-2-butene, 98%, mixture of *cis*   FW 135.01   FT-IR: *1*, 95D   28.65
and *trans*   bp 86°C (740 mm)   VP-FT-IR: *3*, 134B   16.93
   d 1.328

CDCl₃   QE-300 $n_D^{20}1.4613$

CH₃CH=CCH₃ (Br)

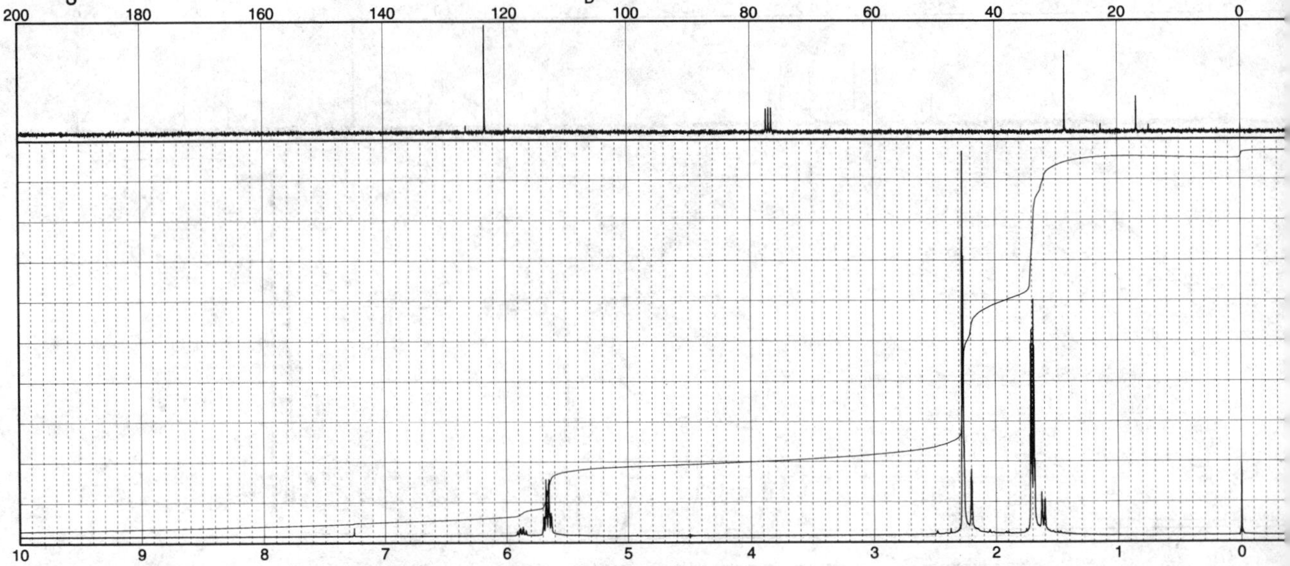

B

Aldrich C8,640-5   *CAS [29576-14-5]* C_4H_7Br Fp 52°F   60 MHz: *1*, 88B   131.25
127.5
Crotyl bromide, tech., 85%   FW 135.01   FT-IR: *1*, 95B   33.3
   bp 98°C   VP-FT-IR: *3*, 133D   17.6
   d 1.312

CDCl₃   QE-300 $n_D^{20}1.4795$

CH₃CH=CHCH₂Br

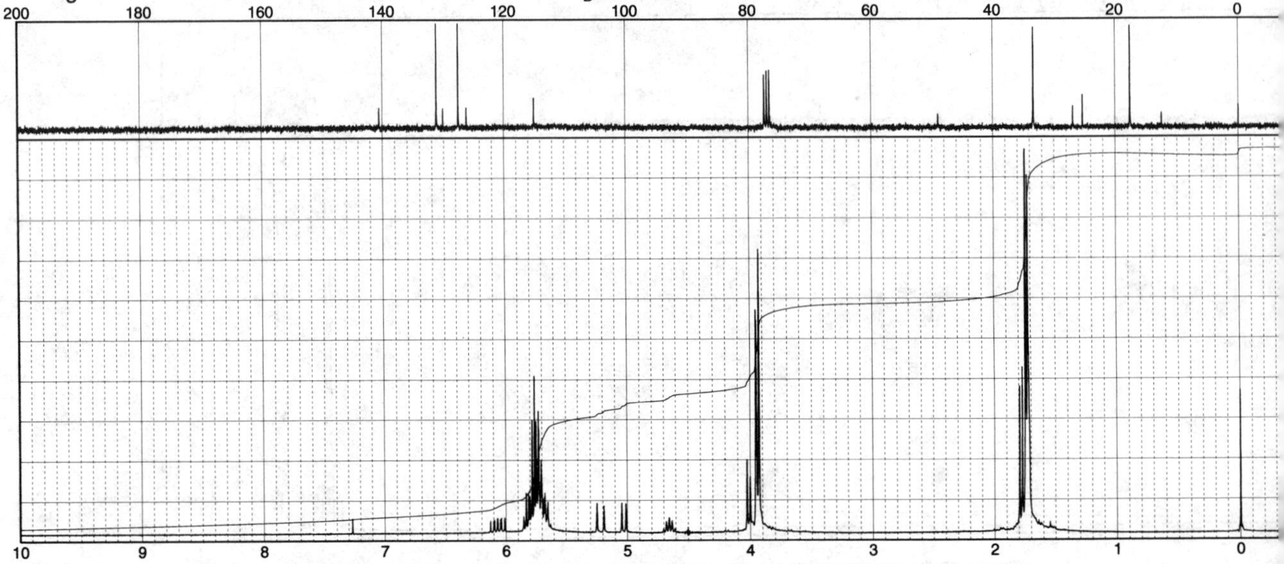

C

Aldrich 12,335-8   *CAS [513-37-1]* C_4H_7Cl Fp 30°F   60 MHz: *1*, 89D   134.8
111.4
1-Chloro-2-methylpropene, 98%   FW 90.55   FT-IR: *1*, 97A   22.8
   bp 68°C   VP-FT-IR: *3*, 135C   18.1
   d 0.920

CDCl₃   QE-300 $n_D^{20}1.4225$

CH₃C(CH₃)=CHCl

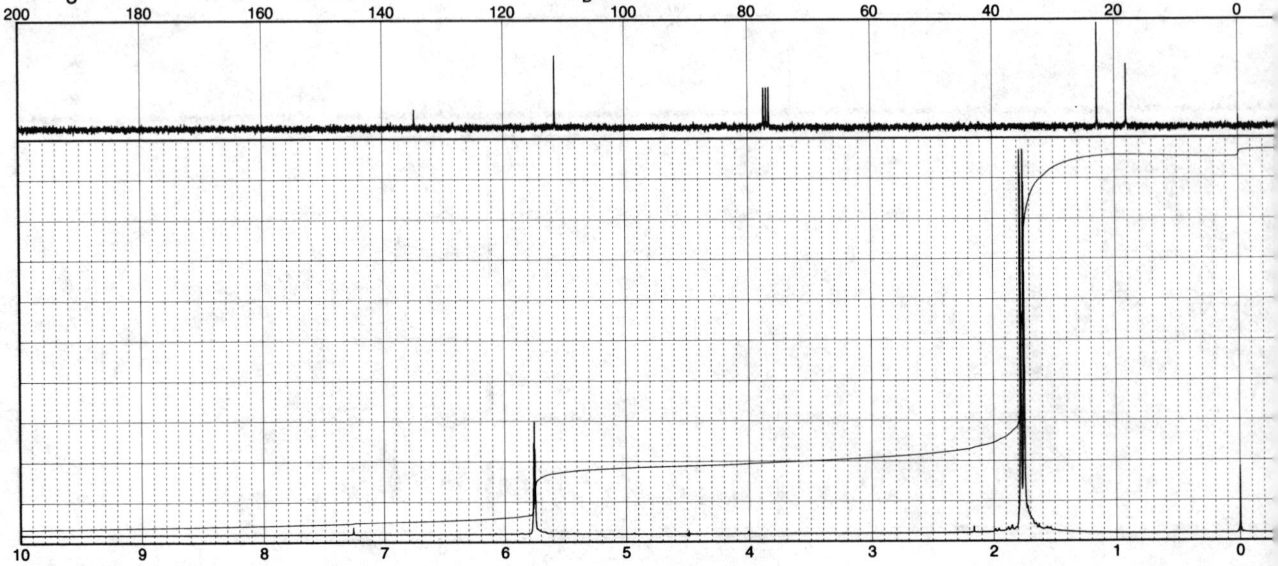

ALDRICH

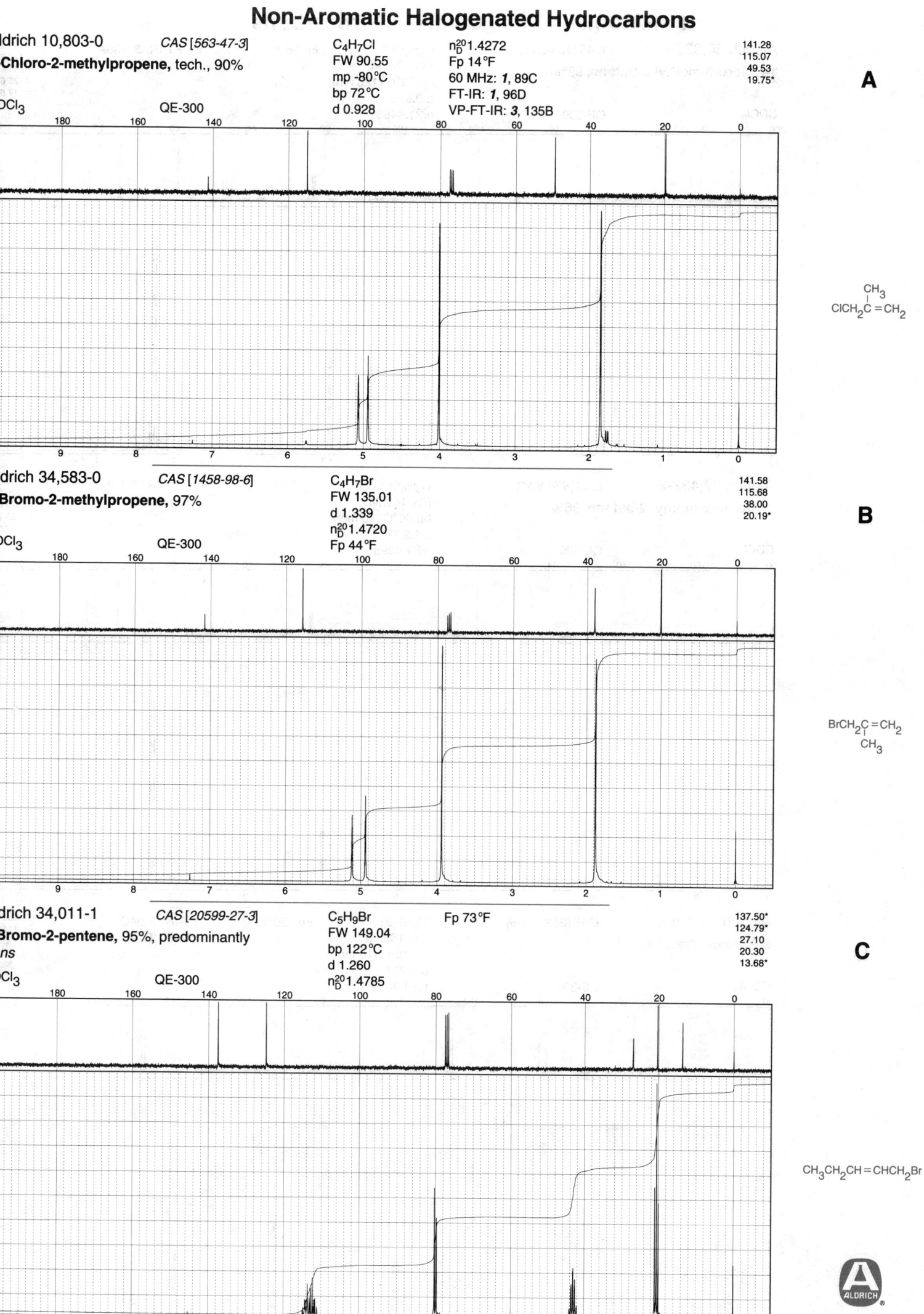

Aldrich 10,803-0 — CAS [563-47-3]

-Chloro-2-methylpropene, tech., 90%

C₄H₇Cl
FW 90.55
mp -80°C
bp 72°C
d 0.928

n$_D^{20}$1.4272
Fp 14°F
60 MHz: *1*, 89C
FT-IR: *1*, 96D
VP-FT-IR: *3*, 135B

CDCl₃   QE-300

141.28
115.07
49.53
19.75*

A

ClCH₂C=CH₂ with CH₃

Aldrich 34,583-0 — CAS [1458-98-6]

-Bromo-2-methylpropene, 97%

C₄H₇Br
FW 135.01
d 1.339
n$_D^{20}$1.4720
Fp 44°F

CDCl₃   QE-300

141.58
115.68
38.00
20.19*

B

BrCH₂C=CH₂ with CH₃

Aldrich 34,011-1 — CAS [20599-27-3]

-Bromo-2-pentene, 95%, predominantly *trans*

C₅H₉Br
FW 149.04
bp 122°C
d 1.260
n$_D^{20}$1.4785

Fp 73°F

CDCl₃   QE-300

137.50*
124.79*
27.10
20.30
13.68*

C

CH₃CH₂CH=CHCH₂Br

A

Aldrich 30,325-9   CAS [503-60-6]   C5H9Cl   Fp 56°F   VP-FT-IR: 3, 136A

1-Chloro-3-methyl-2-butene, 95%

FW 104.58
bp 59°C (120 mm)
d 0.928
n²⁰_D 1.4488

CDCl₃   QE-300

139.23
120.57
41.16
25.66
17.65

$CH_3C=CHCH_2Cl$ with CH_3

B

Aldrich 27,437-2   CAS [870-63-3]   C5H9Br   Fp 91°F   VP-FT-IR: 3, 136B

4-Bromo-2-methyl-2-butene, 96%

FW 149.04
bp 60°C (60 mm)
d 1.293
n²⁰_D 1.4898

CDCl₃   QE-300

140.0
120.7
29.6
25.7
17.5

$BrCH_2CH=CCH_3$ with CH_3

C

Aldrich 24,721-9   CAS [2695-47-8]   C6H11Br   Fp 126°F   FT-IR: 1, 94C   VP-FT-IR: 3, 133A

6-Bromo-1-hexene, 98%

FW 163.06
bp 49°C (16 mm)
d 1.217
n²⁰_D 1.4652

CDCl₃   QE-300

138.0
114.9
33.6
32.8
32.2
27.4

$BrCH_2(CH_2)_3CH=CH_2$

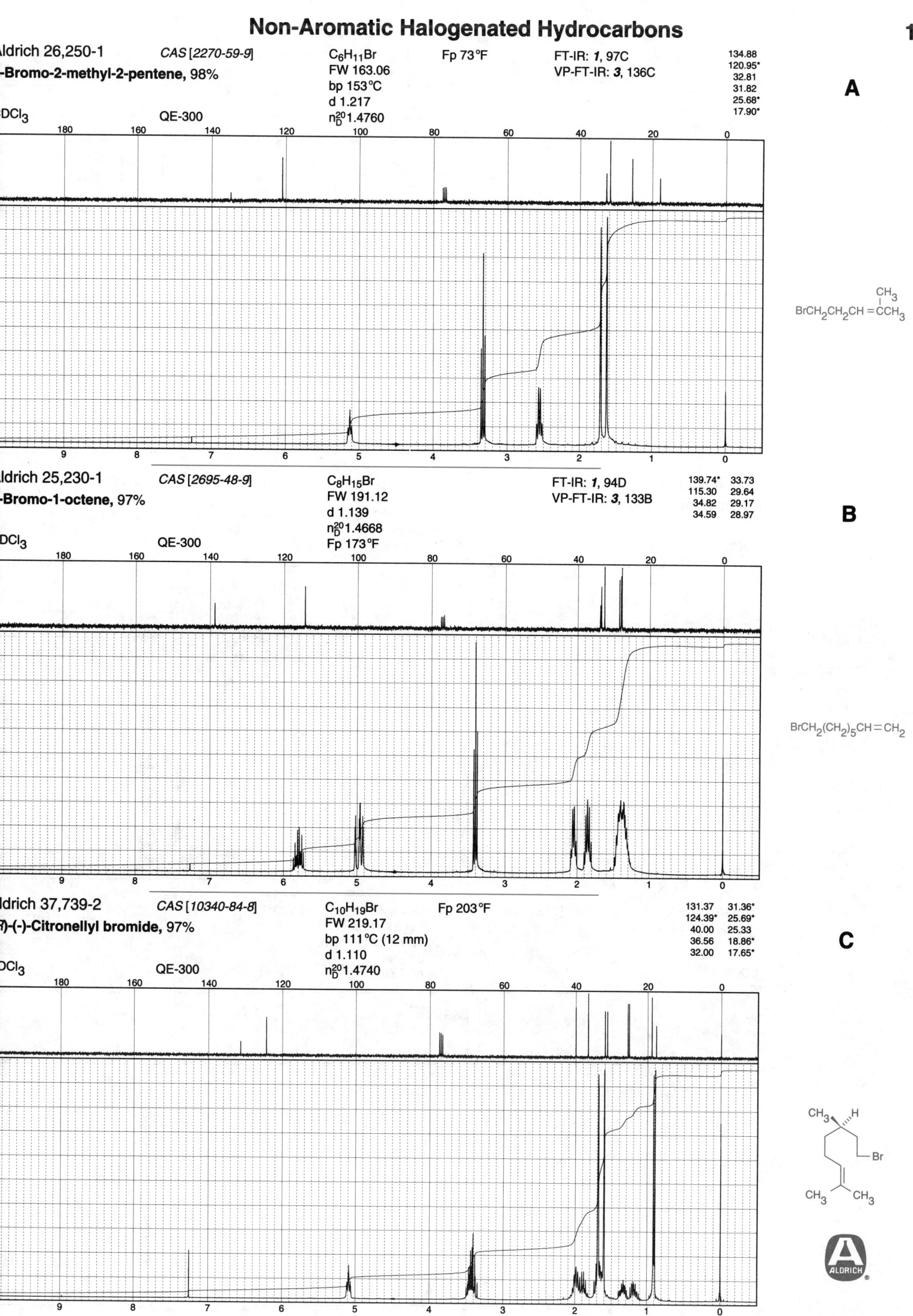

Aldrich 26,250-1 CAS [2270-59-9] C$_6$H$_{11}$Br Fp 73°F FT-IR: *1*, 97C 134.88 120.95* 32.81 31.82 25.68* 17.90*
-Bromo-2-methyl-2-pentene, 98% FW 163.06 bp 153°C d 1.217 n$_D^{20}$1.4760 VP-FT-IR: *3*, 136C

CDCl$_3$ QE-300

A

Aldrich 25,230-1 CAS [2695-48-9] C$_8$H$_{15}$Br FT-IR: *1*, 94D 139.74* 115.30 34.82 34.59 33.73 29.64 29.17 28.97
-Bromo-1-octene, 97% FW 191.12 d 1.139 n$_D^{20}$1.4668 Fp 173°F VP-FT-IR: *3*, 133B

CDCl$_3$ QE-300

B

Aldrich 37,739-2 CAS [10340-84-8] C$_{10}$H$_{19}$Br Fp 203°F 131.37 124.39* 40.00 36.56 32.00 31.36* 25.69* 25.33 18.86* 17.65*
(-)-Citronellyl bromide, 97% FW 219.17 bp 111°C (12 mm) d 1.110 n$_D^{20}$1.4740

CDCl$_3$ QE-300

C

A

Aldrich 37,771-6

(S)-(+)-Citronellyl bromide, 97%

$C_{10}H_{19}Br$
FW 219.17
bp 111°C (12 mm)
d 1.110
$n_D^{20}1.4740$

Fp 203°F

131.36	31.36
124.39*	25.70*
40.00	25.33
36.56	18.86
31.99	17.65

CDCl₃   QE-300

B

Aldrich 30,279-1   *CAS [5389-87-7]*

Geranyl chloride, 95%

$C_{10}H_{17}Cl$
FW 172.70
bp 103°C (12 mm)
d 0.931
$n_D^{20}1.4808$

Fp 194°F

VP-FT-IR: *3*, 136D

143.24	40.06
132.48	26.84
124.15*	26.25
120.89*	18.29
41.69	16.69

CDCl₃   QE-300

C

Aldrich 32,911-8   *CAS [6138-90-5]*

Geranyl bromide, 97%

$C_{10}H_{17}Br$
FW 217.16
bp 102°C (12 mm)
d 1.094
$n_D^{20}1.5031$

Fp 203°F

143.48	29.6
131.89	26.2
123.47*	25.6
120.50*	17.6
39.52	15.9

CDCl₃   QE-300

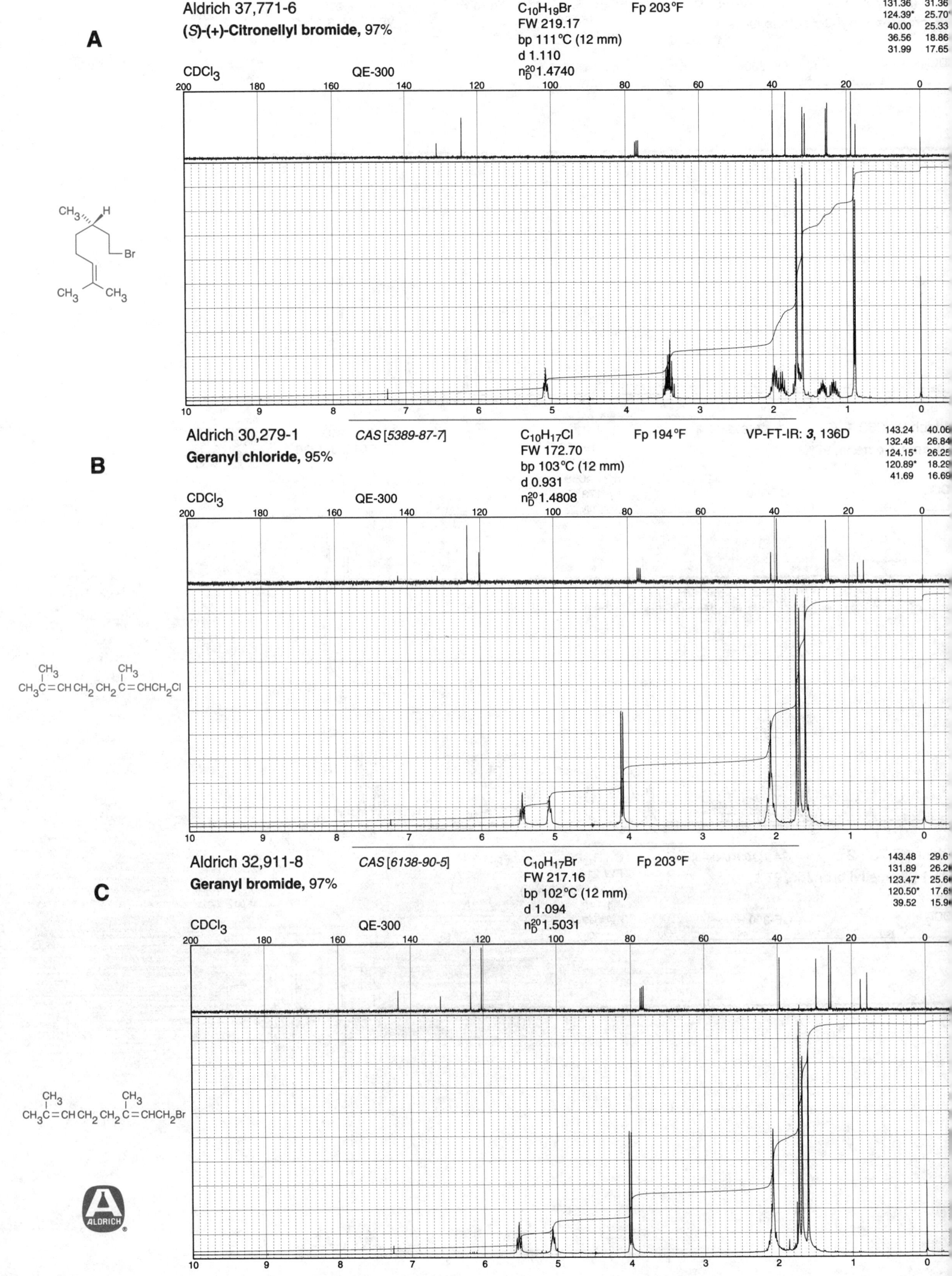

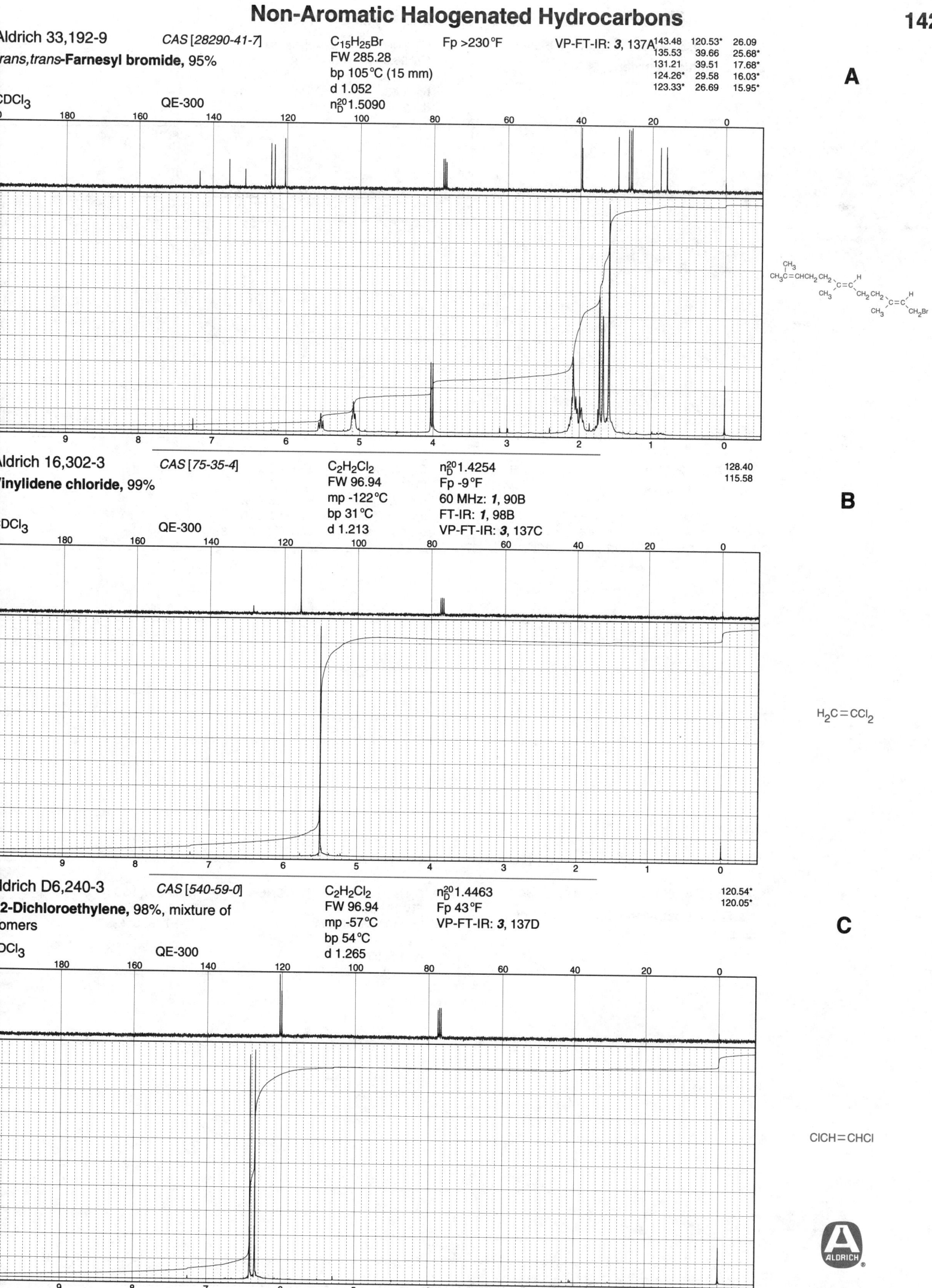

Aldrich 33,192-9   CAS [28290-41-7]   C₁₅H₂₅Br   Fp >230°F   VP-FT-IR: **3**, 137A

$C_{15}H_{25}Br$
FW 285.28
bp 105°C (15 mm)
d 1.052
n_D^{20} 1.5090

trans,trans-**Farnesyl bromide**, 95%

CDCl₃   QE-300

A

143.48	120.53*	26.09
135.53	39.66	25.68*
131.21	39.51	17.68*
124.26*	29.58	16.03*
123.33*	26.69	15.95*

Aldrich 16,302-3   CAS [75-35-4]

Vinylidene chloride, 99%

$C_2H_2Cl_2$
FW 96.94
mp -122°C
bp 31°C
d 1.213

n_D^{20} 1.4254
Fp -9°F
60 MHz: **1**, 90B
FT-IR: **1**, 98B
VP-FT-IR: **3**, 137C

128.40
115.58

CDCl₃   QE-300

$H_2C=CCl_2$

B

Aldrich D6,240-3   CAS [540-59-0]

1,2-Dichloroethylene, 98%, mixture of isomers

$C_2H_2Cl_2$
FW 96.94
mp -57°C
bp 54°C
d 1.265

n_D^{20} 1.4463
Fp 43°F
VP-FT-IR: **3**, 137D

120.54*
120.05*

CDCl₃   QE-300

ClCH=CHCl

C

A

Aldrich D6,200-4   CAS [156-59-2]   C₂H₂Cl₂   n$_D^{20}$1.4481

cis-**1,2-Dichloroethylene, 97%**

FW 96.94   Fp 43°F
mp -80°C   FT-IR: **1**, 98A
bp 60°C   VP-FT-IR: **3**, 138A
d 1.284

120.05*

CDCl₃   QE-300

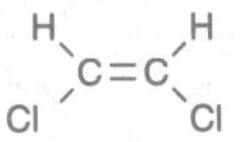

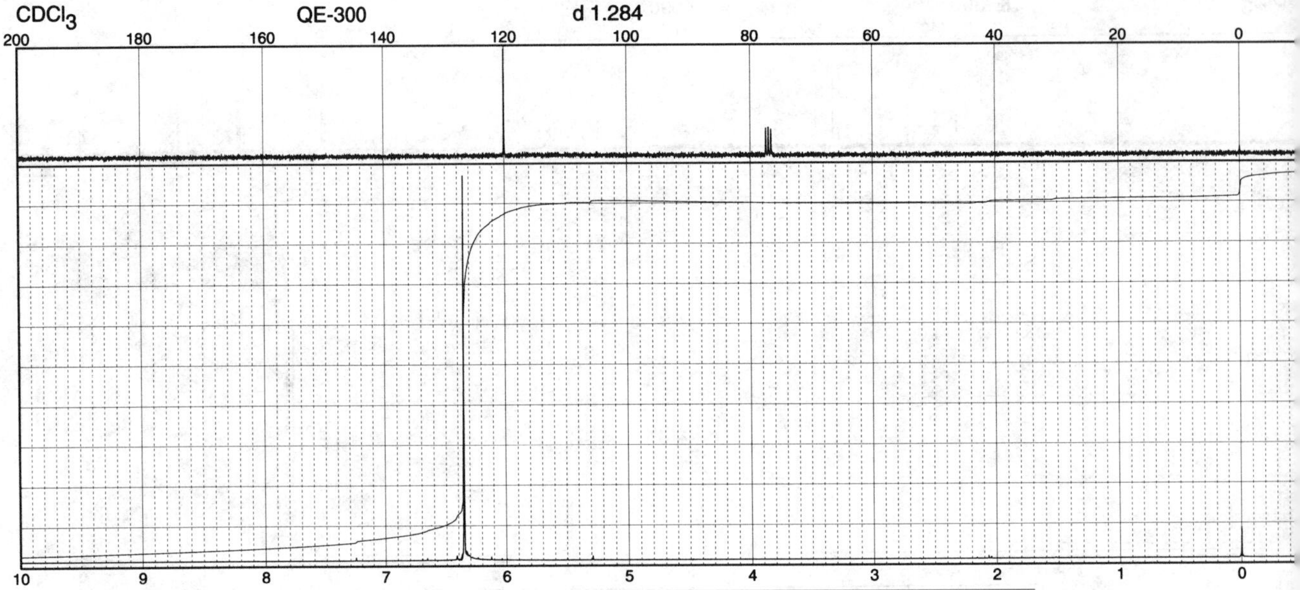

B

Aldrich D6,220-9   CAS [156-60-5]   C₂H₂Cl₂   n$_D^{20}$1.4456

trans-**1,2-Dichloroethylene, 98%**

FW 96.94   Fp 43°F
mp -50°C   60 MHz: **1**, 90A
bp 48°C   FT-IR: **1**, 97D
d 1.257   VP-FT-IR: **3**, 138B

120.55*

CDCl₃   QE-300

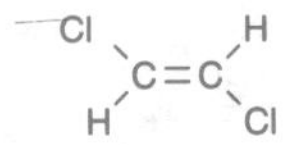

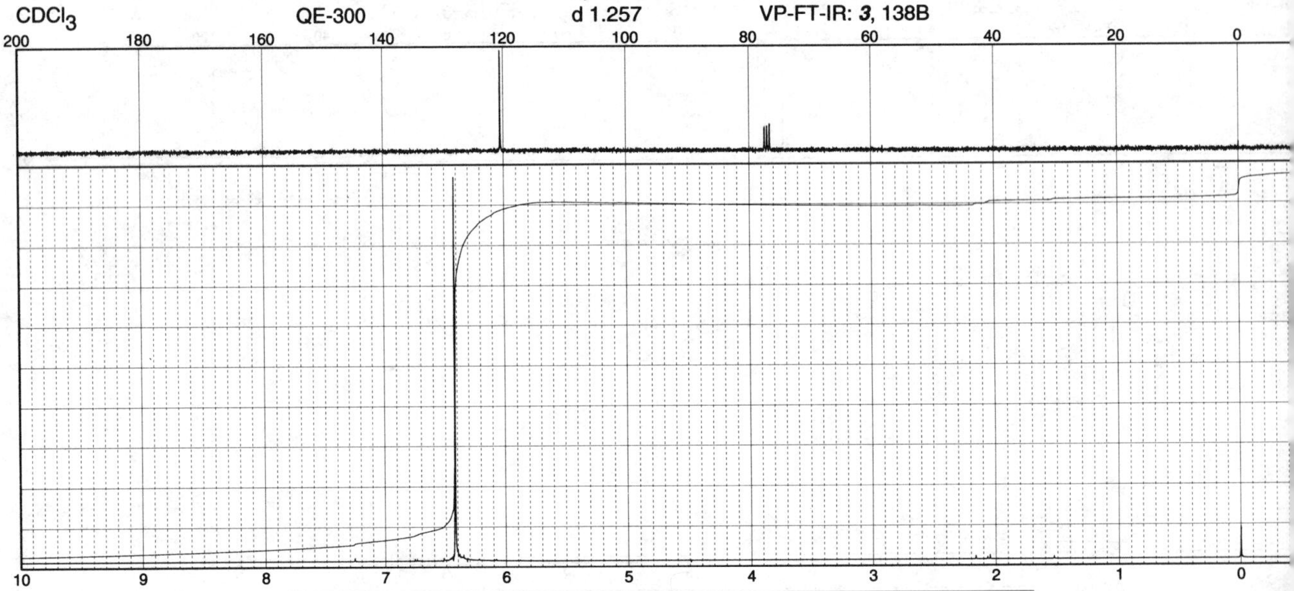

C

Aldrich D4,080-9   CAS [540-49-8]   C₂H₂Br₂   Fp NONE   60 MHz: **1**, 90C

1,2-Dibromoethylene, 98%, mixture of
cis and *trans*

FW 185.86   FT-IR: **1**, 98C
bp 110°C (754 mm)   VP-FT-IR: **3**, 138C
d 2.246
n$_D^{20}$1.5405

113.29
107.06

CDCl₃   QE-300

ALDRICH

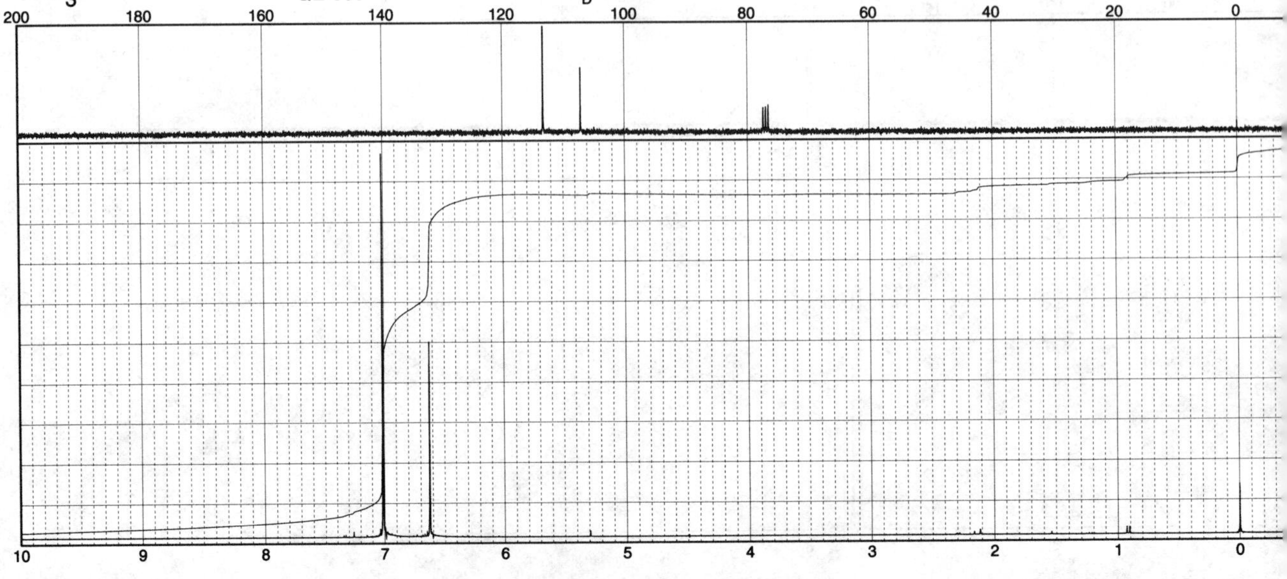

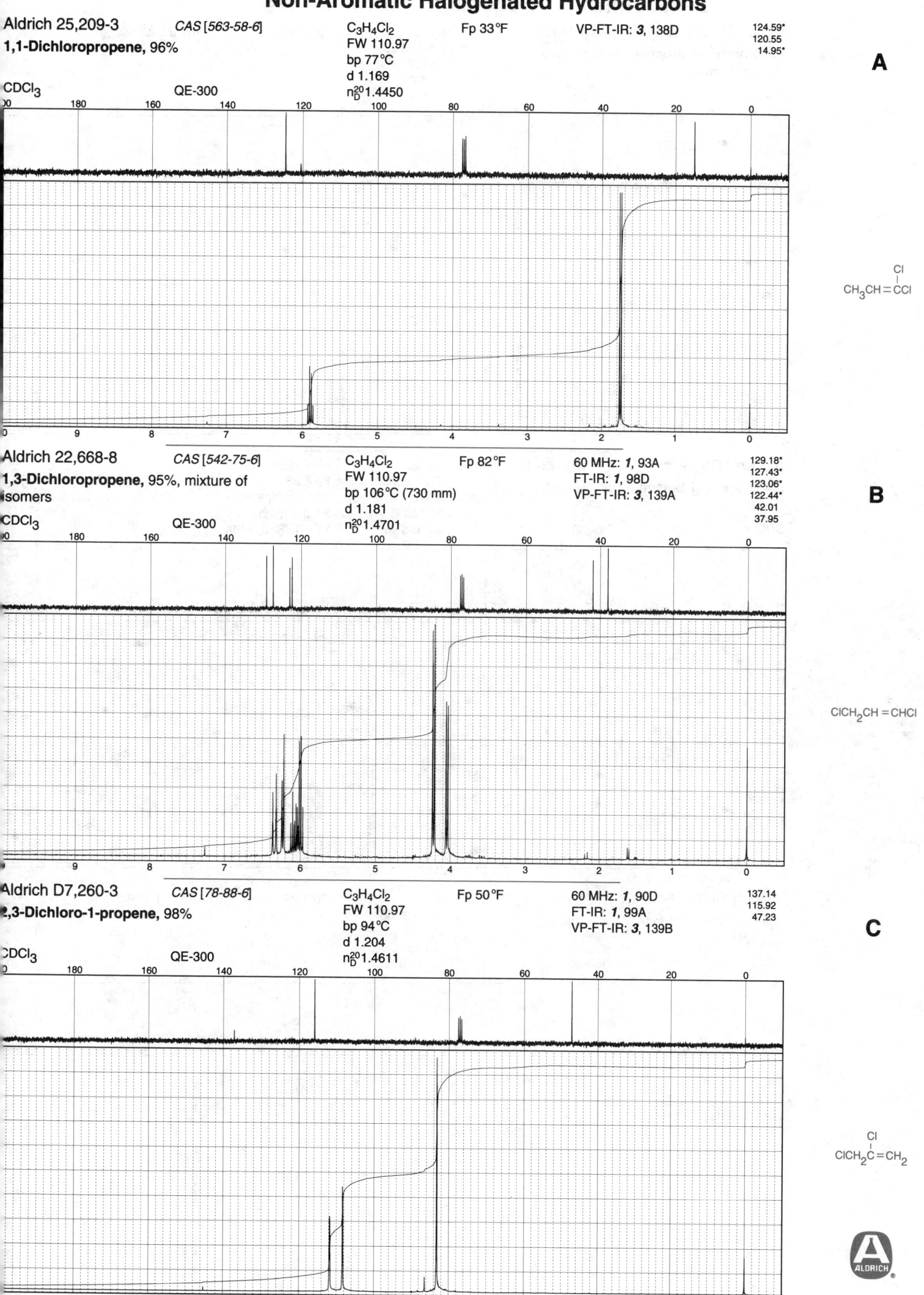

Aldrich 25,209-3   *CAS [563-58-6]* $C_3H_4Cl_2$ Fp 33°F   VP-FT-IR: *3*, 138D

1,1-Dichloropropene, 96%
FW 110.97
bp 77°C
d 1.169
n_D^{20}1.4450

CDCl$_3$ QE-300

124.59*
120.55
14.95*

A

$CH_3CH=CCl$ with Cl

Aldrich 22,668-8   *CAS [542-75-6]* $C_3H_4Cl_2$ Fp 82°F   60 MHz: *1*, 93A

1,3-Dichloropropene, 95%, mixture of isomers
FW 110.97
bp 106°C (730 mm)
d 1.181
n_D^{20}1.4701

FT-IR: *1*, 98D
VP-FT-IR: *3*, 139A

CDCl$_3$ QE-300

129.18*
127.43*
123.06*
122.44*
42.01
37.95

B

$ClCH_2CH=CHCl$

Aldrich D7,260-3   *CAS [78-88-6]* $C_3H_4Cl_2$ Fp 50°F   60 MHz: *1*, 90D

2,3-Dichloro-1-propene, 98%
FW 110.97
bp 94°C
d 1.204
n_D^{20}1.4611

FT-IR: *1*, 99A
VP-FT-IR: *3*, 139B

CDCl$_3$ QE-300

137.14
115.92
47.23

C

$ClCH_2C=CH_2$ with Cl

ALDRICH

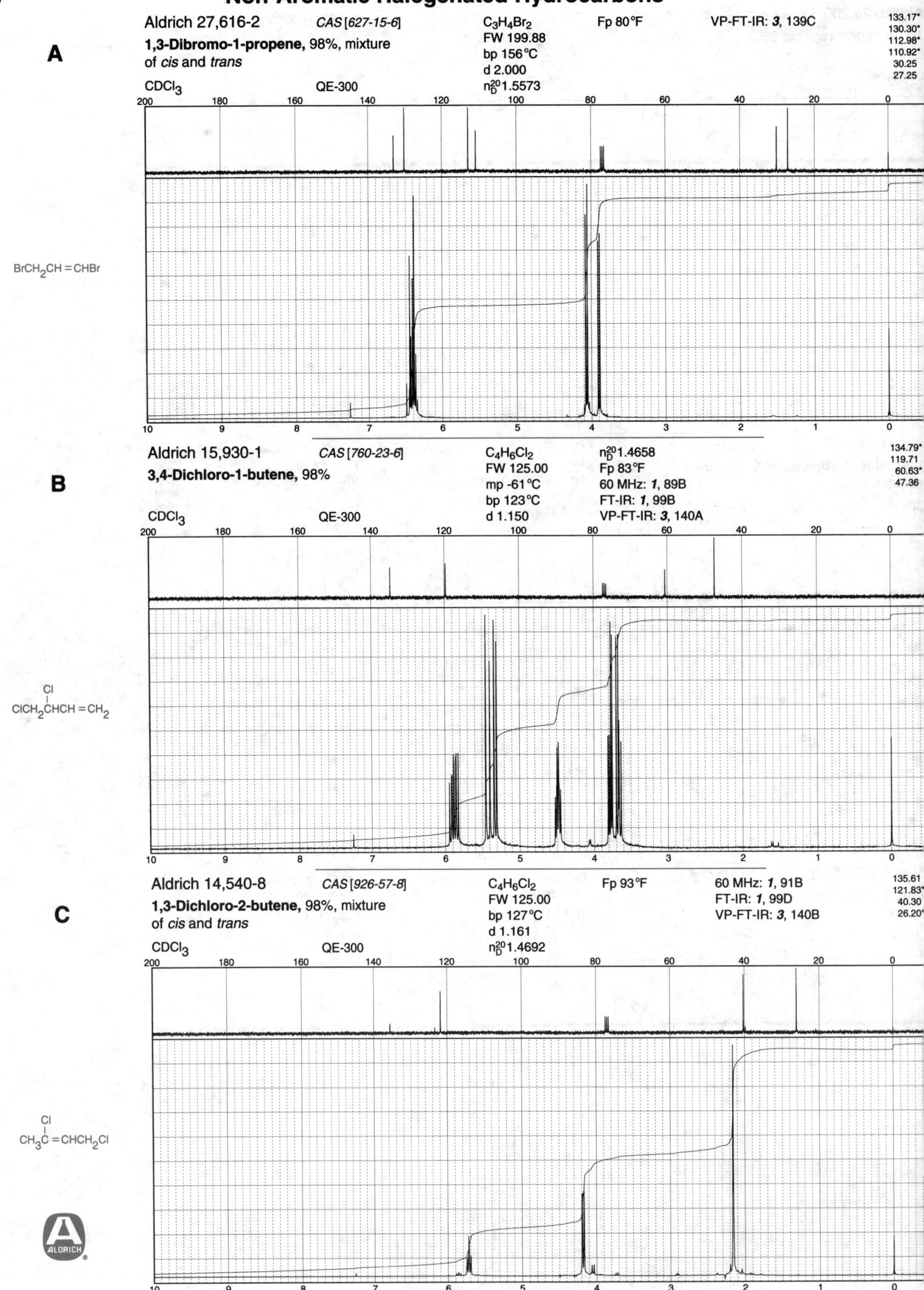

A

Aldrich 27,616-2 | *CAS [627-15-6]* | $C_3H_4Br_2$ | Fp 80°F | VP-FT-IR: **3**, 139C

1,3-Dibromo-1-propene, 98%, mixture
of *cis* and *trans*

FW 199.88
bp 156°C
d 2.000
n_D^{20} 1.5573

133.17*
130.30*
112.98*
110.92*
30.25
27.25

CDCl$_3$ QE-300

$BrCH_2CH = CHBr$

B

Aldrich 15,930-1 | *CAS [760-23-6]* | $C_4H_6Cl_2$ | n_D^{20} 1.4658 | 134.79*
119.71
60.63*
47.36

3,4-Dichloro-1-butene, 98%

FW 125.00
mp -61°C
bp 123°C
d 1.150

Fp 83°F
60 MHz: **1**, 89B
FT-IR: **1**, 99B
VP-FT-IR: **3**, 140A

CDCl$_3$ QE-300

Cl
|
$ClCH_2CHCH = CH_2$

C

Aldrich 14,540-8 | *CAS [926-57-8]* | $C_4H_6Cl_2$ | Fp 93°F | 60 MHz: **1**, 91B | 135.61
121.83*
40.30
26.20*

1,3-Dichloro-2-butene, 98%, mixture
of *cis* and *trans*

FW 125.00
bp 127°C
d 1.161
n_D^{20} 1.4692

FT-IR: **1**, 99D
VP-FT-IR: **3**, 140B

CDCl$_3$ QE-300

Cl
|
$CH_3C = CHCH_2Cl$

ALDRICH

Aldrich 15,932-8   *CAS [764-41-0]* $C_4H_6Cl_2$ Fp 139°F   FT-IR: *1*, 100B   130.00*
                                            FW 125.00   VP-FT-IR: *3*, 140D   129.53*
1,4-Dichloro-2-butene, tech., 95%, mixture   bp 74°C (40 mm)   43.54
of *cis* and *trans*   d 1.183   37.93

A

CDCl₃   QE-300 n_D^{20} 1.4874

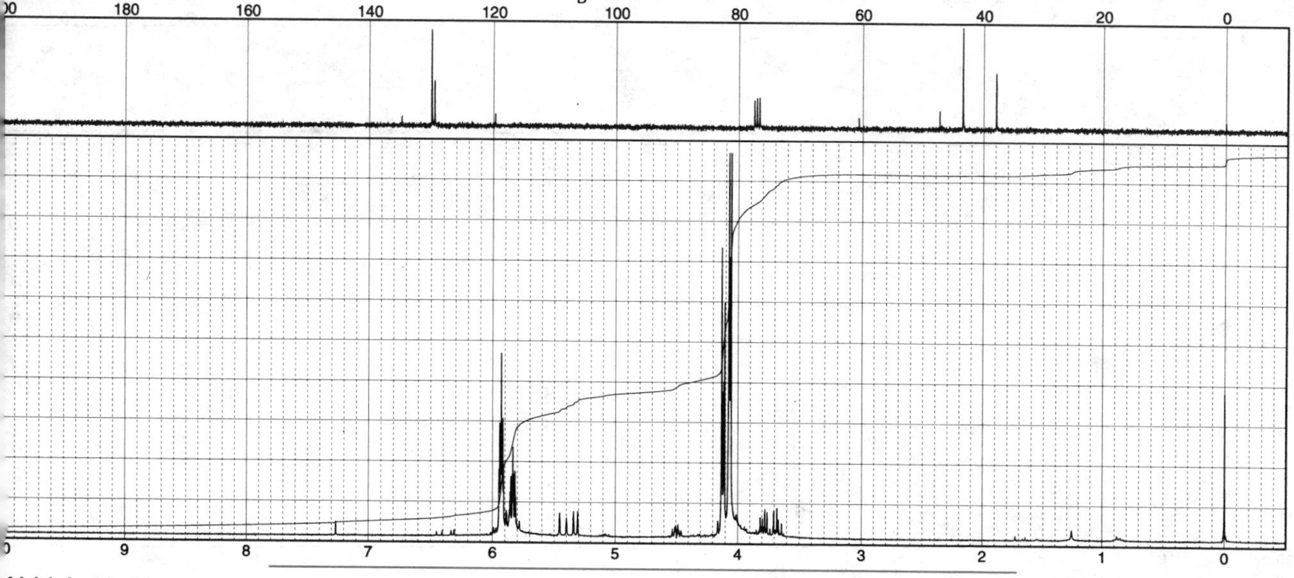

ClCH₂ CH=CH CH₂Cl

Aldrich 19,570-7   *CAS [1476-11-5]* $C_4H_6Cl_2$ n_D^{20} 1.4884   130.43*
                                             FW 125.00   Fp 132°F   38.86
cis-**1,4-Dichloro-2-butene**, 95%   mp -48°C   60 MHz: *1*, 91D
                                             bp 152°C (758 mm) FT-IR: *1*, 100C
                                             d 1.188   VP-FT-IR: *3*, 141A

B

CDCl₃   QE-300

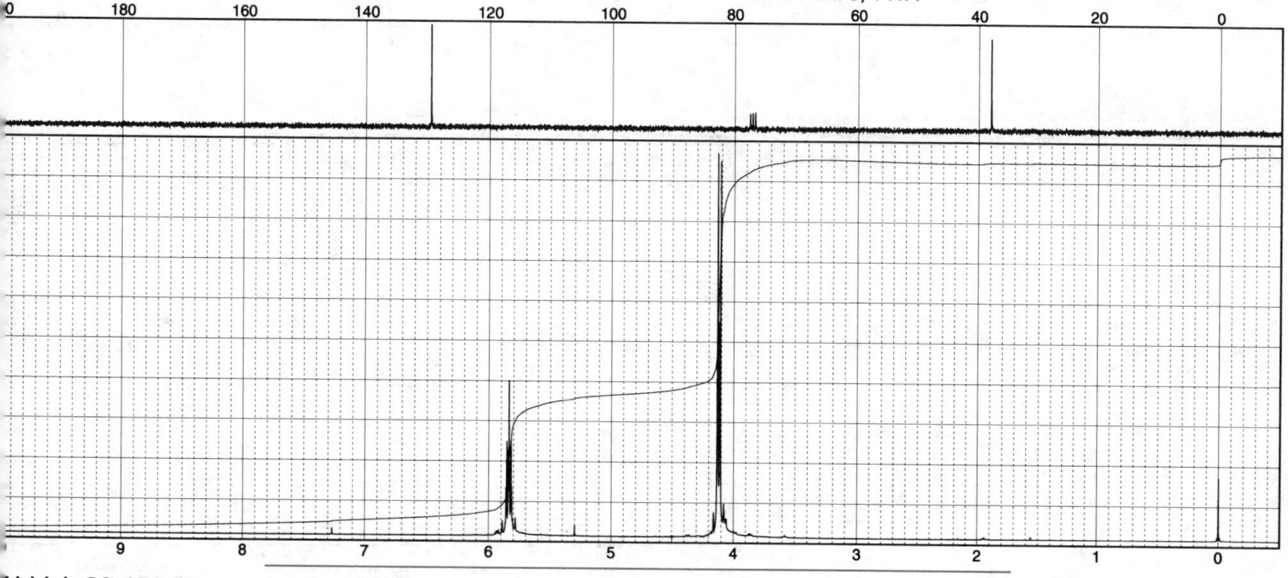

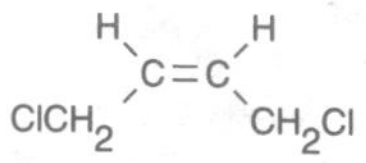

Aldrich 32,451-5   *CAS [110-57-6]* $C_4H_6Cl_2$ Fp 129°F   VP-FT-IR: *3*, 141B   129.98*
                                            FW 125.00   43.58
trans-**1,4-Dichloro-2-butene**, 98%   bp 75°C (40 mm)
                                            d 1.183

C

CDCl₃   QE-300 n_D^{20} 1.4887

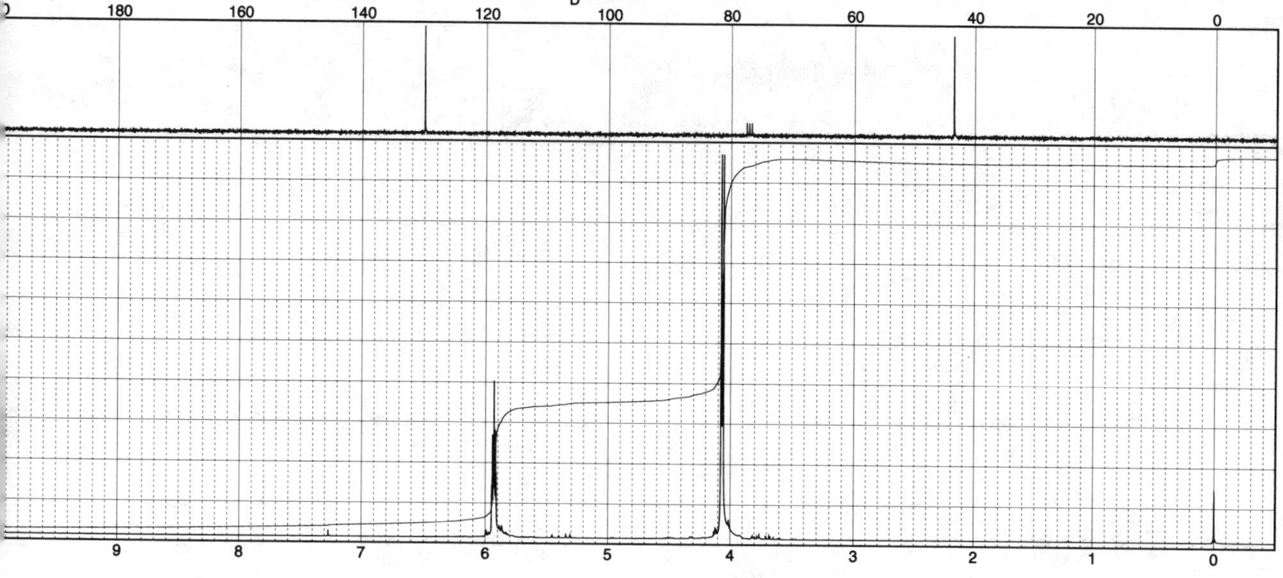

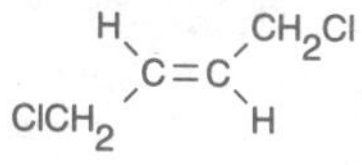

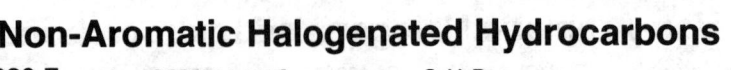

A

Aldrich D3,920-7   CAS [821-06-7] $C_4H_6Br_2$ 60 MHz: *1*, 92B   130.71*
                                      FW 213.91   FT-IR: *1*, 101A   30.83
1,4-Dibromo-2-butene, 99%, predominantly   mp 52°C   VP-FT-IR: *3*, 141C
trans   bp 205°C
                                       Fp >230°F
CDCl₃   QE-300

200 180 160 140 120 100 80 60 40 20 0

$BrCH_2\ CH=CH\,CH_2Br$

10 9 8 7 6 5 4 3 2 1 0

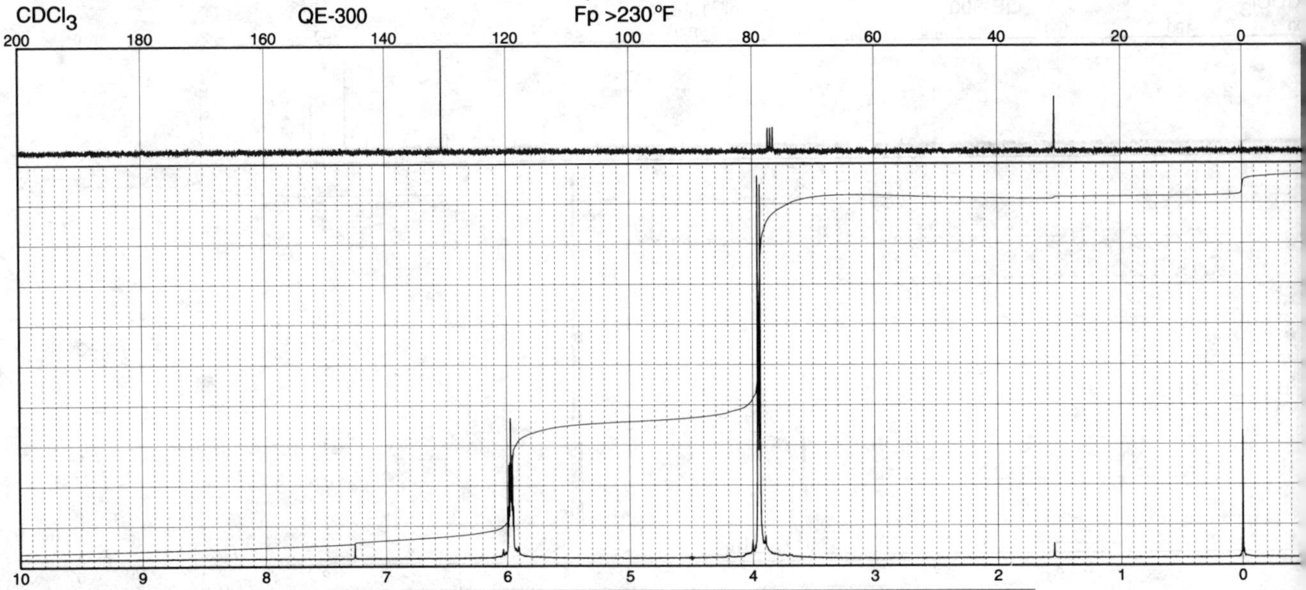

B

Aldrich C3,110-4   CAS [1871-57-4] $C_4H_6Cl_2$ n_D^{20} 1.4753   141.25
                                        FW 125.00   Fp 98°F   118.91
3-Chloro-2-chloromethyl-1-propene, 99+%   mp -14°C   60 MHz: *1*, 91C   44.72
                                        bp 138°C   FT-IR: *1*, 100A
CDCl₃   QE-300   d 1.080   VP-FT-IR: *3*, 140C

200 180 160 140 120 100 80 60 40 20 0

CH_2Cl
$ClCH_2-C=CH_2$

10 9 8 7 6 5 4 3 2 1 0

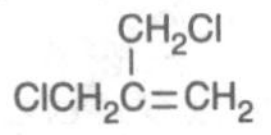

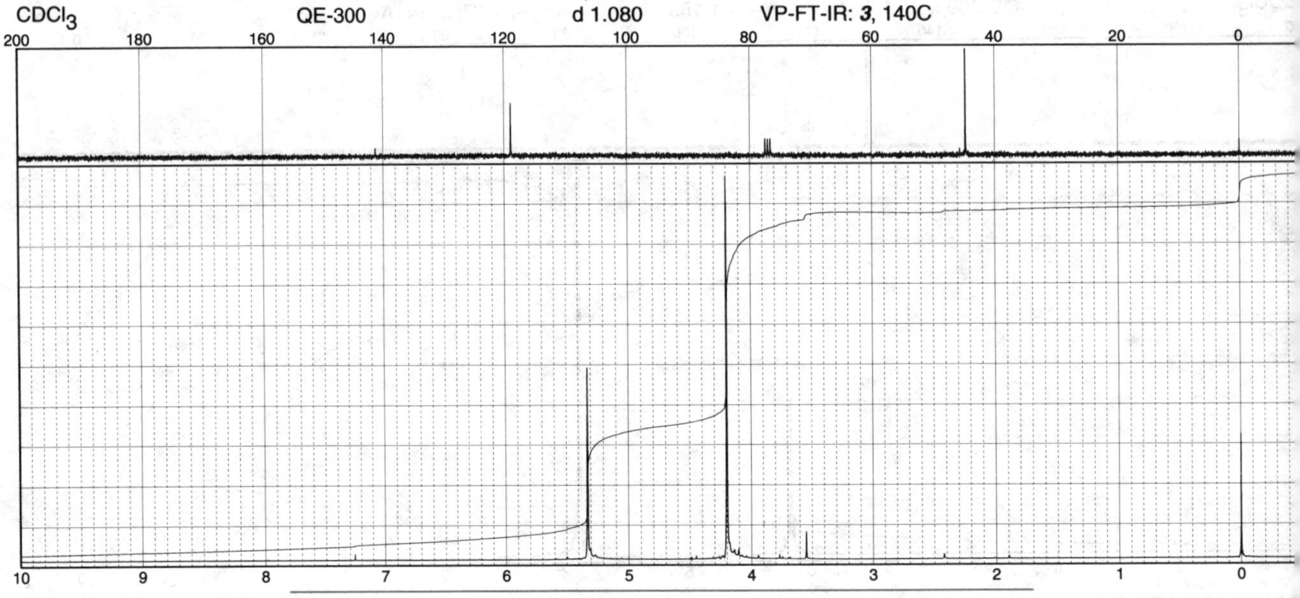

C

Aldrich 25,140-2   CAS [79-01-6] C_2HCl_3 n_D^{20} 1.4765   123.79
                                      FW 131.39   Fp NONE   116.58*
Trichloroethylene, 99+%   mp -87°C   60 MHz: *1*, 92D
                                      bp 87°C   FT-IR: *1*, 101B
CDCl₃   QE-300   d 1.462   VP-FT-IR: *3*, 141D

200 180 160 140 120 100 80 60 40 20 0

$Cl_2\ C=CH\,Cl$

10 9 8 7 6 5 4 3 2 1 0

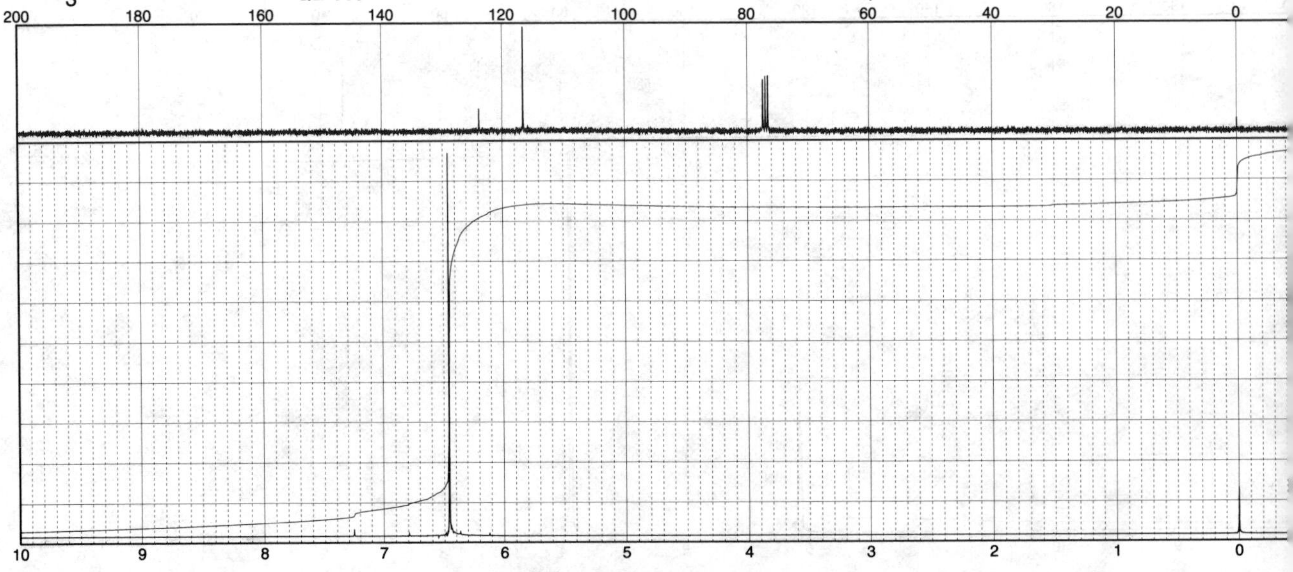

ALDRICH

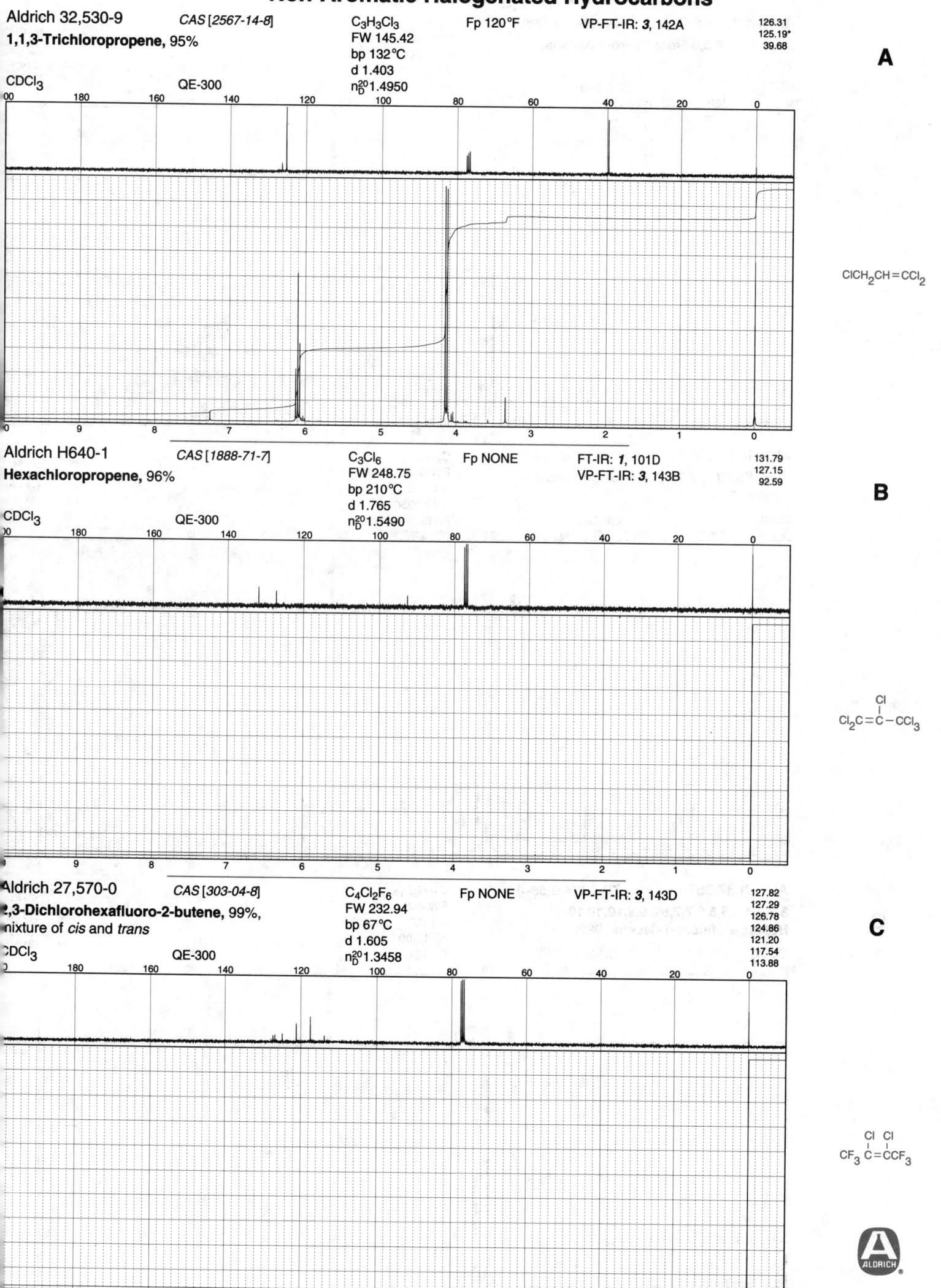

Aldrich 32,530-9   CAS [2567-14-8]

1,1,3-Trichloropropene, 95%

C$_3$H$_3$Cl$_3$
FW 145.42
bp 132°C
d 1.403
n$_D^{20}$ 1.4950

Fp 120°F   VP-FT-IR: **3**, 142A

126.31
125.19*
39.68

A

CDCl$_3$ QE-300

ClCH$_2$CH=CCl$_2$

Aldrich H640-1   CAS [1888-71-7]

Hexachloropropene, 96%

C$_3$Cl$_6$
FW 248.75
bp 210°C
d 1.765
n$_D^{20}$ 1.5490

Fp NONE   FT-IR: **1**, 101D
VP-FT-IR: **3**, 143B

131.79
127.15
92.59

B

CDCl$_3$ QE-300

Cl$_2$C=C(Cl)-CCl$_3$

Aldrich 27,570-0   CAS [303-04-8]

2,3-Dichlorohexafluoro-2-butene, 99%, mixture of cis and trans

C$_4$Cl$_2$F$_6$
FW 232.94
bp 67°C
d 1.605
n$_D^{20}$ 1.3458

Fp NONE   VP-FT-IR: **3**, 143D

127.82
127.29
126.78
124.86
121.20
117.54
113.88

C

CDCl$_3$ QE-300

CF$_3$C(Cl)=C(Cl)CF$_3$

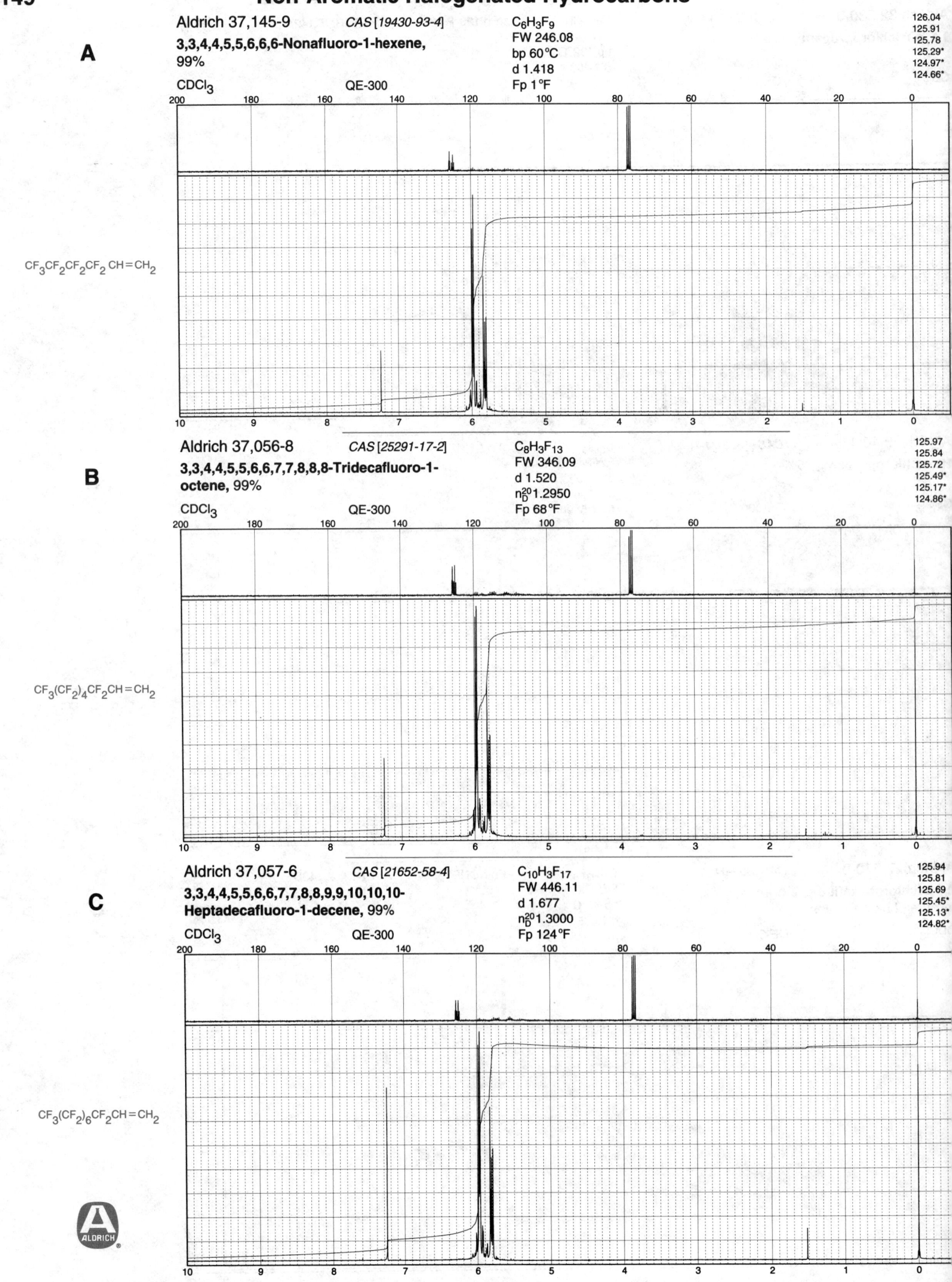

A

Aldrich 37,145-9   *CAS [19430-93-4]* $C_6H_3F_9$
FW 246.08
3,3,4,4,5,5,6,6,6-Nonafluoro-1-hexene, bp 60 °C
99%   d 1.418

CDCl$_3$ QE-300   Fp 1 °F

126.04
125.91
125.78
125.29*
124.97*
124.66*

$CF_3CF_2CF_2CF_2CH=CH_2$

B

Aldrich 37,056-8   *CAS [25291-17-2]* $C_8H_3F_{13}$
FW 346.09
3,3,4,4,5,5,6,6,7,7,8,8,8-Tridecafluoro-1- d 1.520
octene, 99% n_D^{20}1.2950

CDCl$_3$ QE-300   Fp 68 °F

125.97
125.84
125.72
125.49*
125.17*
124.86*

$CF_3(CF_2)_4CF_2CH=CH_2$

C

Aldrich 37,057-6   *CAS [21652-58-4]* $C_{10}H_3F_{17}$
FW 446.11
3,3,4,4,5,5,6,6,7,7,8,8,9,9,10,10,10- d 1.677
Heptadecafluoro-1-decene, 99% n_D^{20}1.3000

CDCl$_3$ QE-300   Fp 124 °F

125.94
125.81
125.69
125.45*
125.13*
124.82*

$CF_3(CF_2)_6CF_2CH=CH_2$

Aldrich 11,219-4   CAS [87-68-3] C_4Cl_6 n_D^{20} 1.5550   126.53
Hexachloro-1,3-butadiene, 97%   FW 260.76   Fp NONE   123.76
mp -21°C   FT-IR: *1*, 102A
bp 215°C   VP-FT-IR: *3*, 144A
d 1.665

CDCl₃   QE-300

A

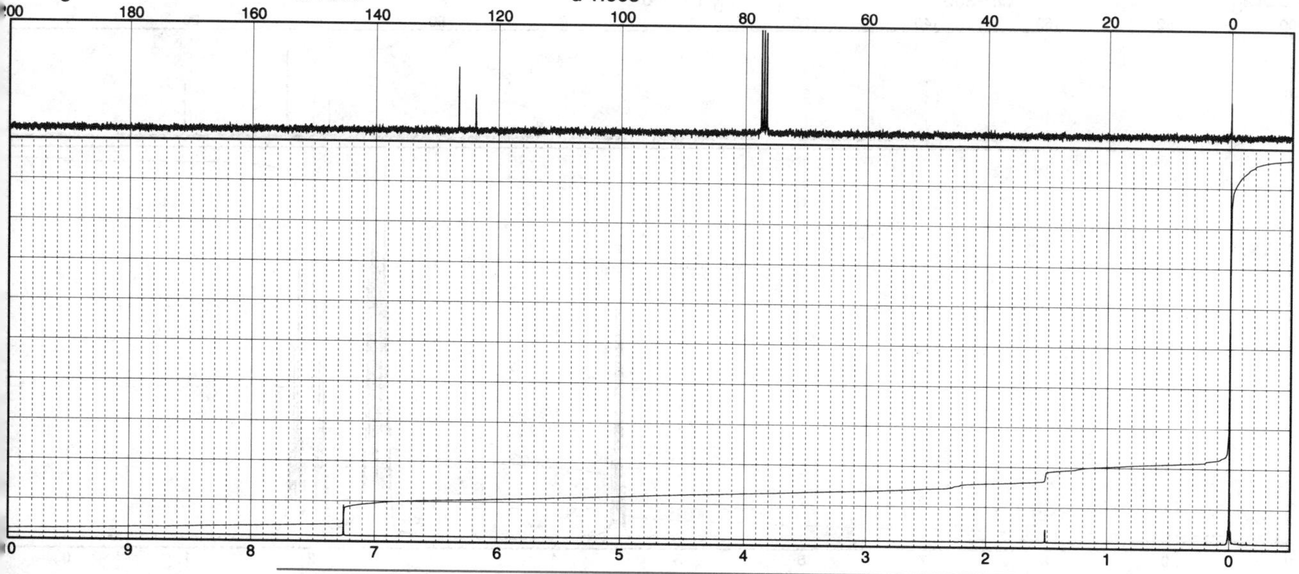

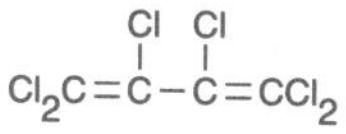

Aldrich C11,730-7   CAS [4333-56-6] C_3H_5Br Fp 20°F   14.27*
Cyclopropyl bromide, 99%   FW 120.98   60 MHz: *1*, 93D   9.05
bp 69°C   FT-IR: *1*, 102B
d 1.510   VP-FT-IR: *3*, 144B
n_D^{20} 1.4600

CDCl₃   QE-300

B

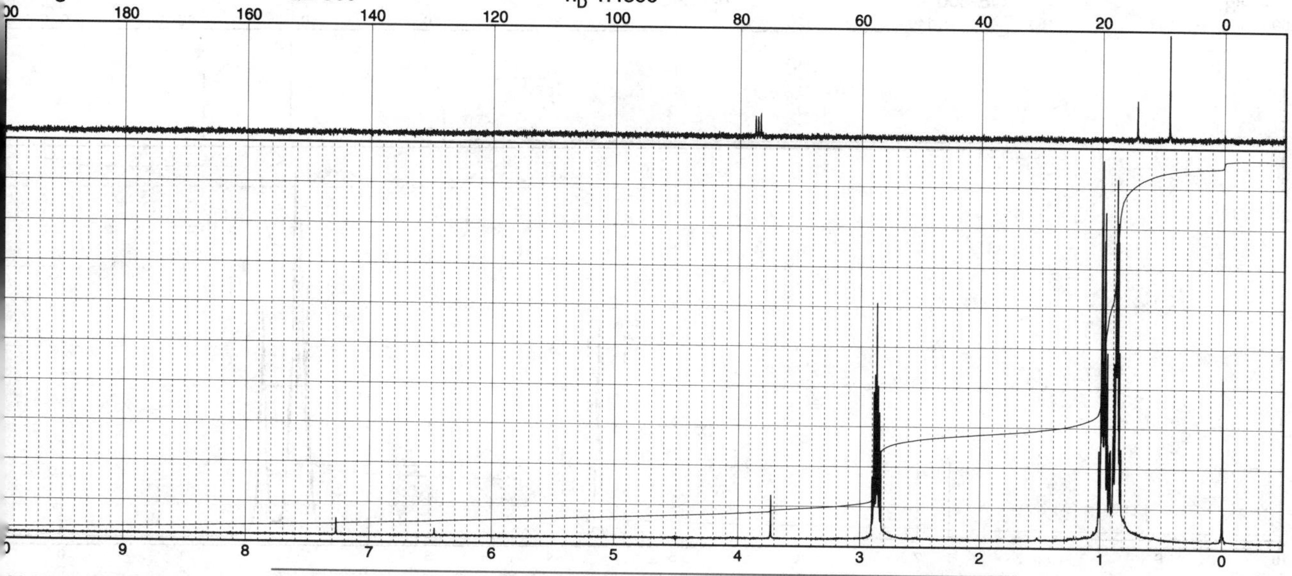

Aldrich 33,356-5   CAS [1120-57-6] C_4H_7Cl Fp 16°F   52.52*
Cyclobutyl chloride, 97%   FW 90.55   34.95
bp 83°C   16.55
d 0.991
n_D^{20} 1.4360

CDCl₃   QE-300

C

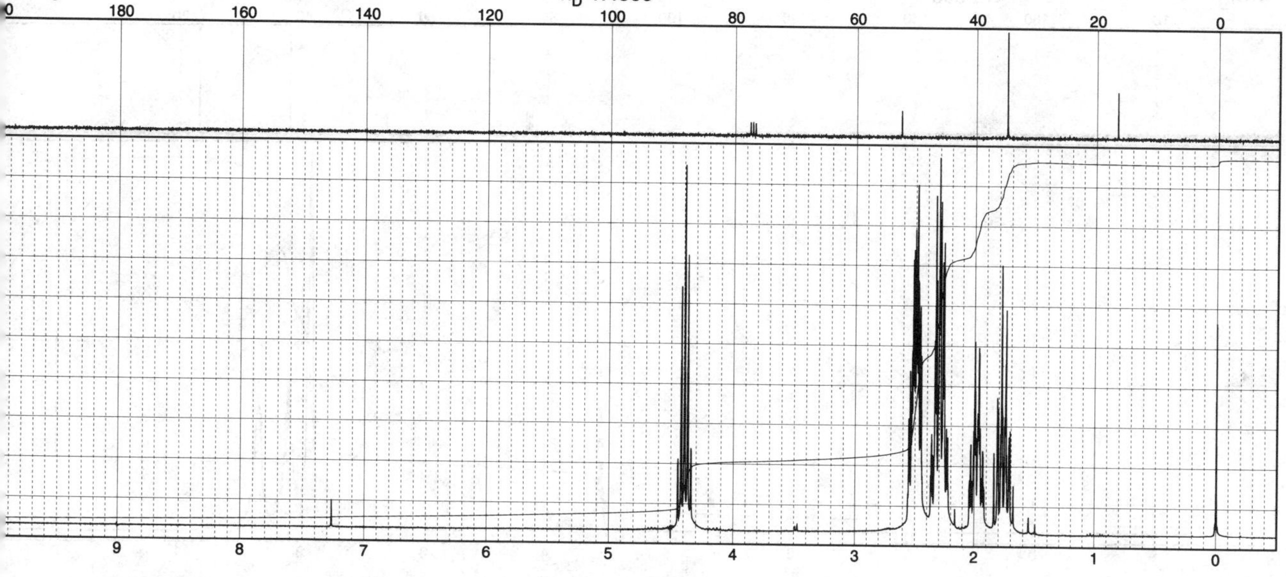

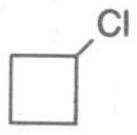

151

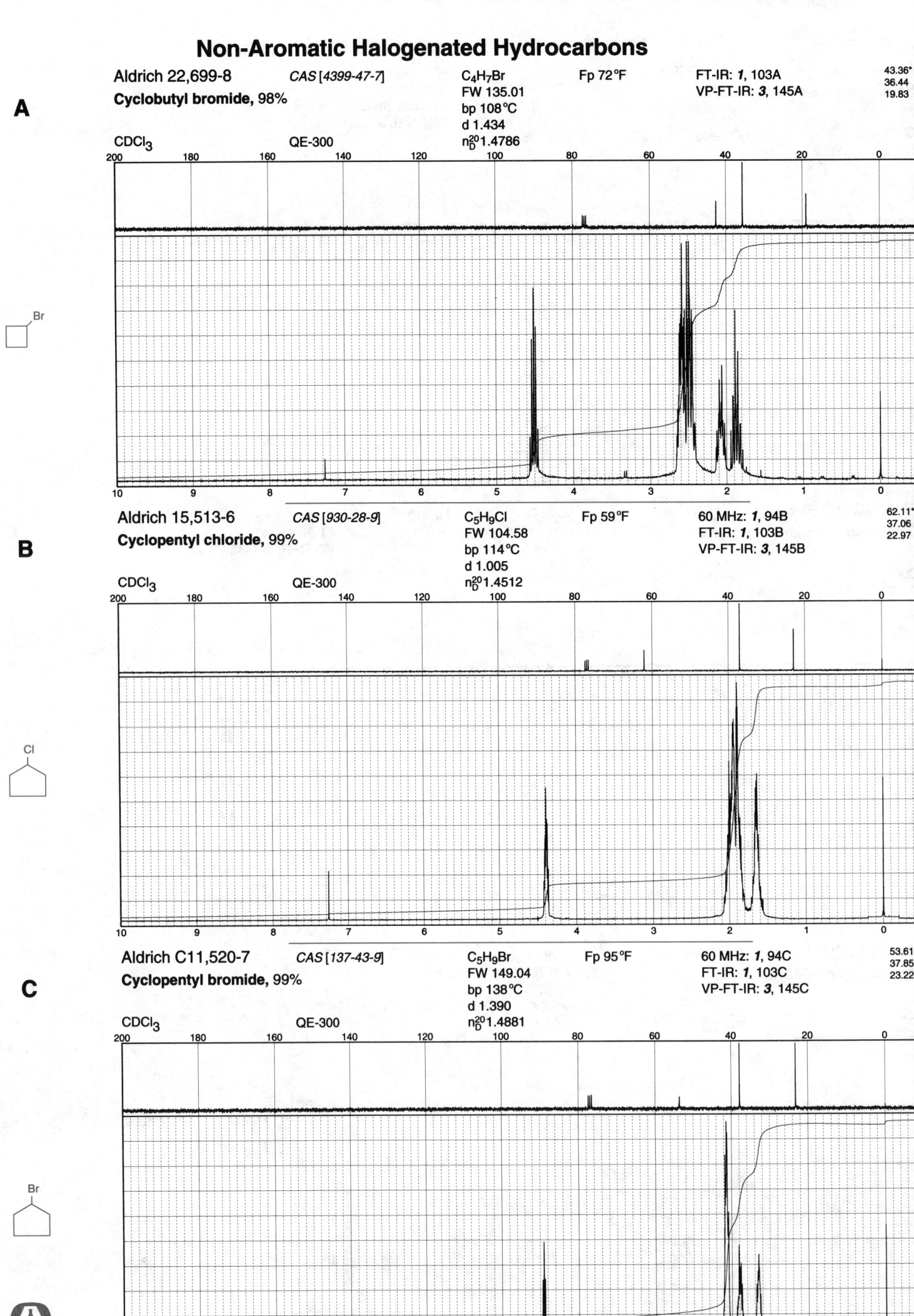

A

Aldrich 22,699-8   *CAS [4399-47-7]* C_4H_7Br Fp 72°F   FT-IR: *1*, 103A
Cyclobutyl bromide, 98%   FW 135.01   VP-FT-IR: *3*, 145A
bp 108°C
d 1.434
n_D^{20} 1.4786

CDCl_3   QE-300

43.36*
36.44
19.83

B

Aldrich 15,513-6   *CAS [930-28-9]* C_5H_9Cl Fp 59°F   60 MHz: *1*, 94B
Cyclopentyl chloride, 99%   FW 104.58   FT-IR: *1*, 103B
bp 114°C   VP-FT-IR: *3*, 145B
d 1.005
n_D^{20} 1.4512

CDCl_3   QE-300

62.11*
37.06
22.97

C

Aldrich C11,520-7   *CAS [137-43-9]* C_5H_9Br Fp 95°F   60 MHz: *1*, 94C
Cyclopentyl bromide, 99%   FW 149.04   FT-IR: *1*, 103C
bp 138°C   VP-FT-IR: *3*, 145C
d 1.390
n_D^{20} 1.4881

CDCl_3   QE-300

53.61*
37.85
23.22

ALDRICH

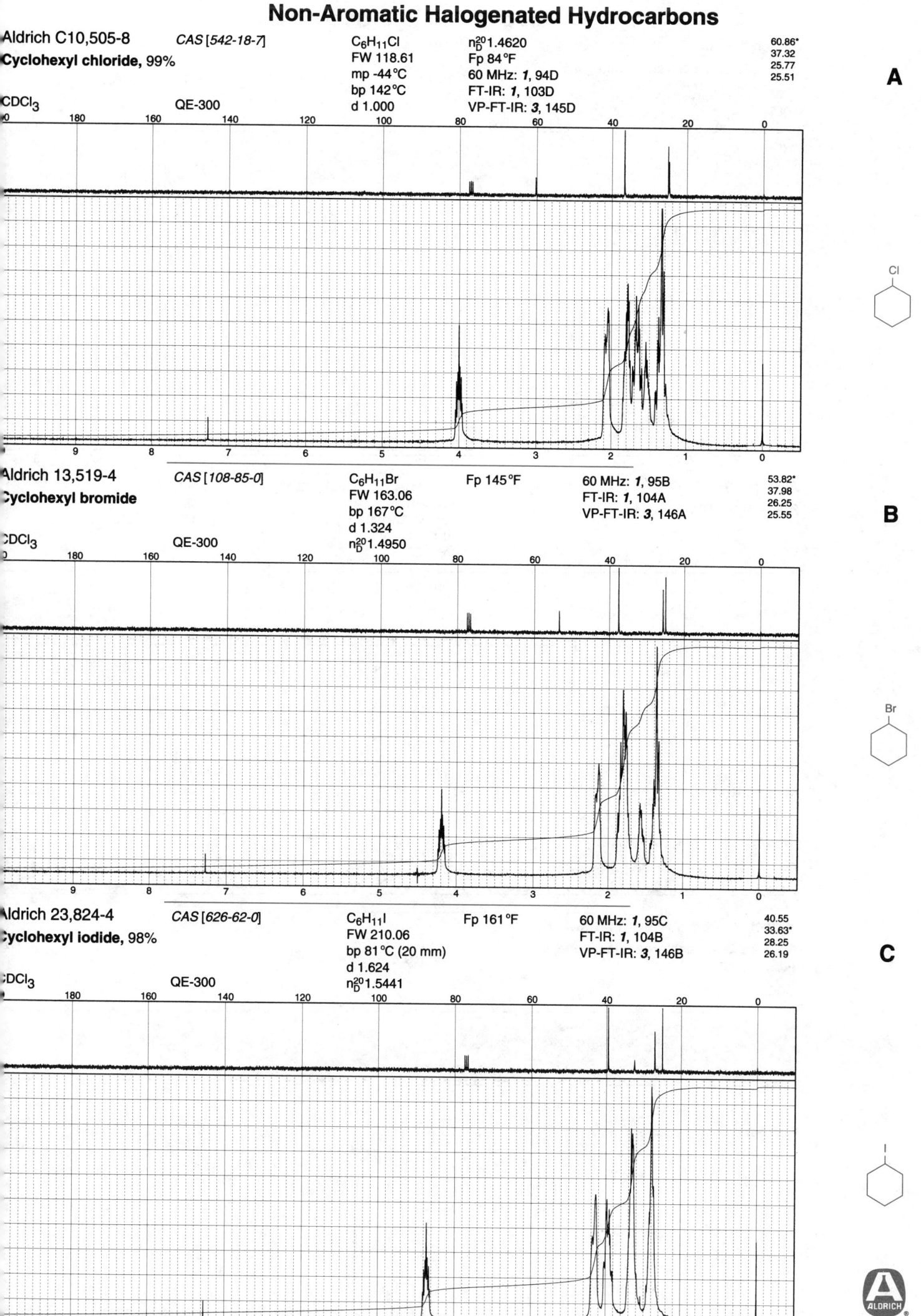

Aldrich C10,505-8　　CAS [542-18-7]　　$C_6H_{11}Cl$　　n_D^{20}1.4620　　60.86*
Cyclohexyl chloride, 99%　　FW 118.61　　Fp 84°F　　37.32
　　mp -44°C　　60 MHz: *1*, 94D　　25.77
CDCl₃　　QE-300　　bp 142°C　　FT-IR: *1*, 103D　　25.51
　　d 1.000　　VP-FT-IR: *3*, 145D

A

Aldrich 13,519-4　　CAS [108-85-0]　　$C_6H_{11}Br$　　Fp 145°F　　60 MHz: *1*, 95B　　53.82*
Cyclohexyl bromide　　FW 163.06　　FT-IR: *1*, 104A　　37.98
　　bp 167°C　　VP-FT-IR: *3*, 146A　　26.25
　　d 1.324　　25.55
CDCl₃　　QE-300　　n_D^{20}1.4950

B

Aldrich 23,824-4　　CAS [626-62-0]　　$C_6H_{11}I$　　Fp 161°F　　60 MHz: *1*, 95C　　40.55
Cyclohexyl iodide, 98%　　FW 210.06　　FT-IR: *1*, 104B　　33.63*
　　bp 81°C (20 mm)　　VP-FT-IR: *3*, 146B　　28.25
　　d 1.624　　26.19
CDCl₃　　QE-300　　n_D^{20}1.5441

C

A

Aldrich D6,000-1   *CAS [822-86-6]* $C_6H_{10}Cl_2$ Fp 151°F   60 MHz: *1*, 98A

trans-1,2-Dichlorocyclohexane, 99%

FW 153.05
bp 194°C
d 1.164
n_D^{20} 1.4917

FT-IR: *1*, 106A
VP-FT-IR: *3*, 150A

63.12*
33.43
23.06

CDCl₃   QE-300

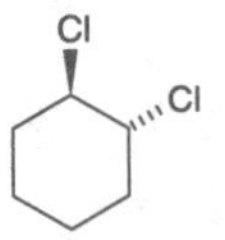

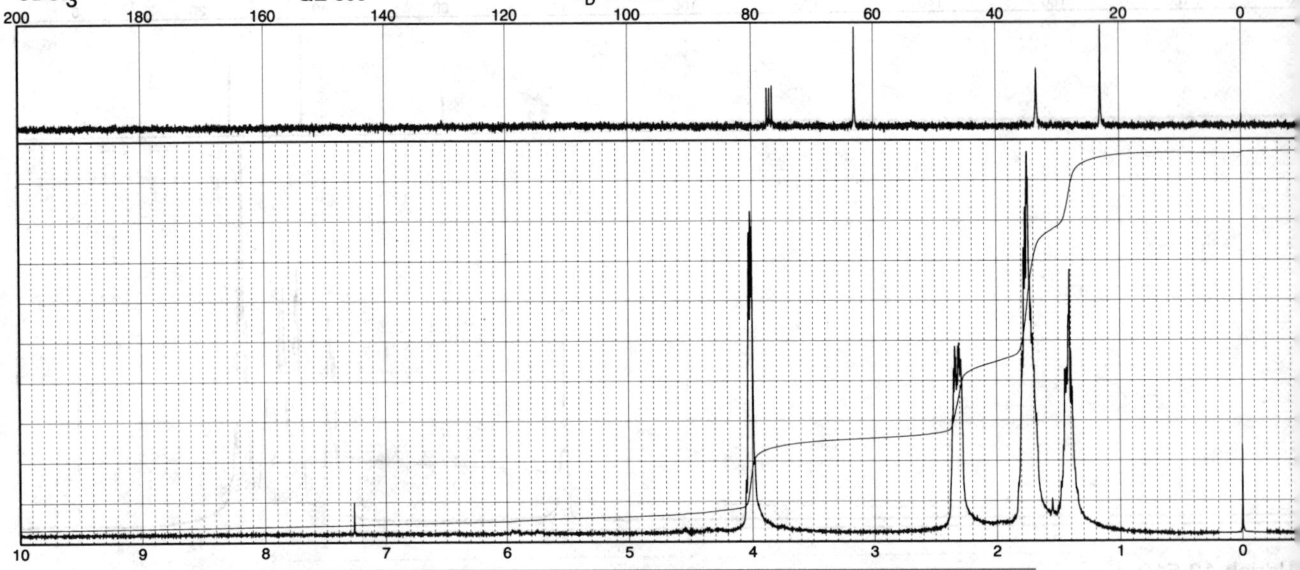

B

Aldrich D3,960-6   *CAS [7429-37-0]* $C_6H_{10}Br_2$ Fp >230°F   60 MHz: *1*, 98B

(±)-*trans*-1,2-Dibromocyclohexane, 99%

FW 241.96
bp 145°C (100 mm)
d 1.784
n_D^{20} 1.5515

FT-IR: *1*, 106B
VP-FT-IR: *3*, 150B

55.13*
31.98
22.38

CDCl₃   QE-300

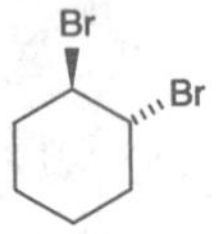

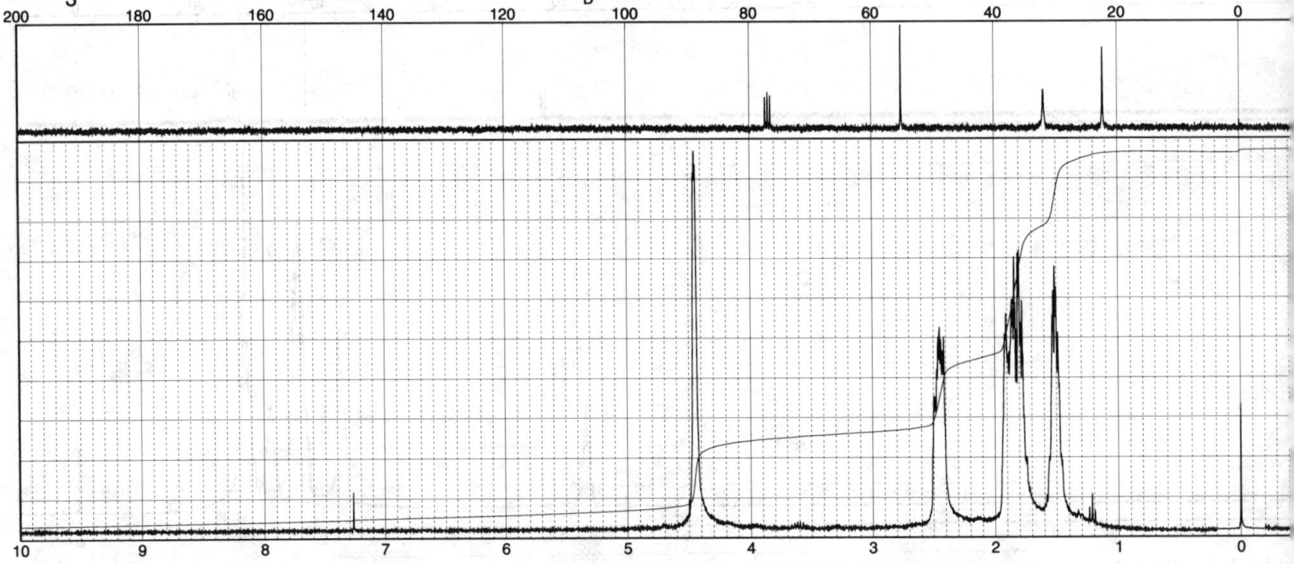

C

Aldrich 30,068-3   *CAS [319-84-6]* $C_6H_6Cl_6$

1,2,3,4,5,6-Hexachlorocyclohexane, α-isomer, 99%

FW 290.83
mp 159°C

63.88
62.63
58.72

CDCl₃   QE-300

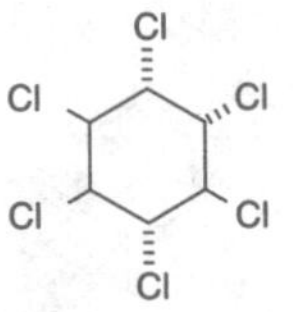

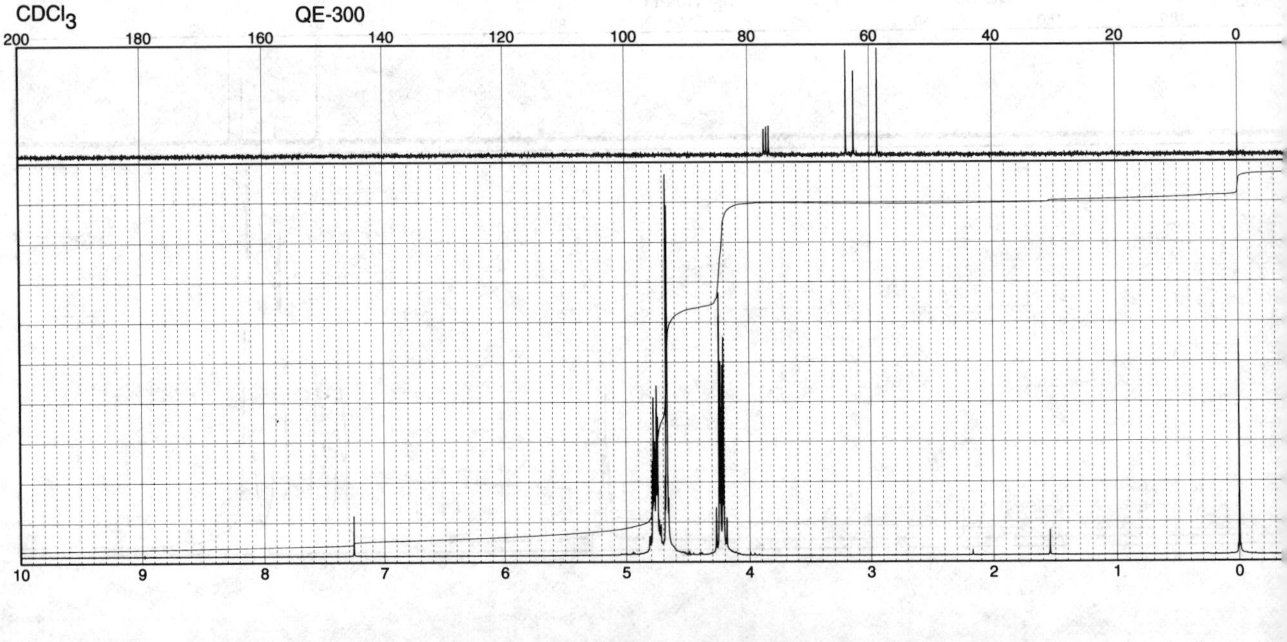

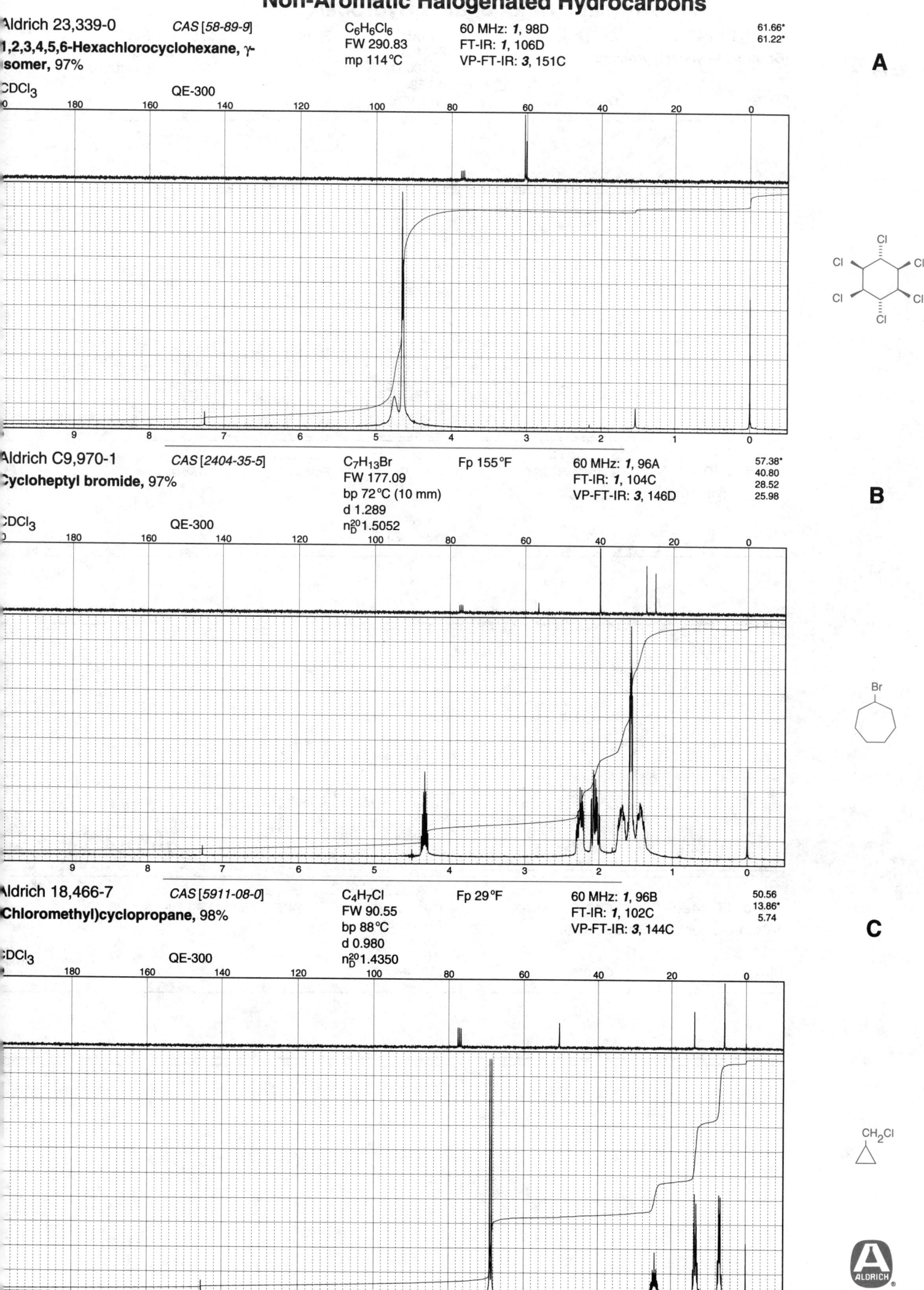

Aldrich 23,339-0   CAS [58-89-9] $C_6H_6Cl_6$ 60 MHz: *1*, 98D   61.66*
61.22*

1,2,3,4,5,6-Hexachlorocyclohexane, γ-isomer, 97%
FW 290.83   FT-IR: *1*, 106D
mp 114°C   VP-FT-IR: *3*, 151C

A

CDCl$_3$ QE-300

Aldrich C9,970-1   CAS [2404-35-5] $C_7H_{13}Br$ Fp 155°F   60 MHz: *1*, 96A   57.38*
40.80
28.52
25.98

Cycloheptyl bromide, 97%
FW 177.09   FT-IR: *1*, 104C
bp 72°C (10 mm)   VP-FT-IR: *3*, 146D
d 1.289
n$_D^{20}$1.5052

B

CDCl$_3$ QE-300

Aldrich 18,466-7   CAS [5911-08-0] C_4H_7Cl Fp 29°F   60 MHz: *1*, 96B   50.56
13.86*
5.74

(Chloromethyl)cyclopropane, 98%
FW 90.55   FT-IR: *1*, 102C
bp 88°C   VP-FT-IR: *3*, 144C
d 0.980
n$_D^{20}$1.4350

C

CDCl$_3$ QE-300

A

Aldrich 24,240-3   CAS [7051-34-5]   C₄H₇Br   Fp 107°F   60 MHz: *1*, 94A   40.26
                                                                                                        14.82*
(Bromomethyl)cyclopropane, 97%   FW 135.01   FT-IR: *1*, 102D   8.34
                                                                bp 106°C   VP-FT-IR: *3*, 144D
CDCl₃   QE-300   d 1.392
                                                                n²⁰_D 1.4570

CH₂Br

Aldrich C10,600-3   CAS [2550-36-9]   C₇H₁₃Br   Fp 135°F   60 MHz: *1*, 96C   40.70
                                                                                                        40.08*
Cyclohexylmethyl bromide, 99%   FW 177.09   FT-IR: *1*, 104D   31.75
                                                                bp 77°C (26 mm)   VP-FT-IR: *3*, 147A   26.17
                                                                                                        25.85
CDCl₃   QE-300   d 1.269
                                                                n²⁰_D 1.4920

CH₂Br

B

Aldrich 30,442-5   CAS [60192-64-5]   C₉H₁₇Cl   Fp 174°F   VP-FT-IR: *3*, 147B   45.48
                                                                                                        37.15
1-Chloro-3-cyclohexylpropane, 97%   FW 160.69   34.66
                                                                bp 78°C (5 mm)   33.29
                                                                                                        30.18
CDCl₃   QE-300   d 0.997   26.62
                                                                n²⁰_D 1.4662   26.33

C

CH₂CH₂CH₂Cl

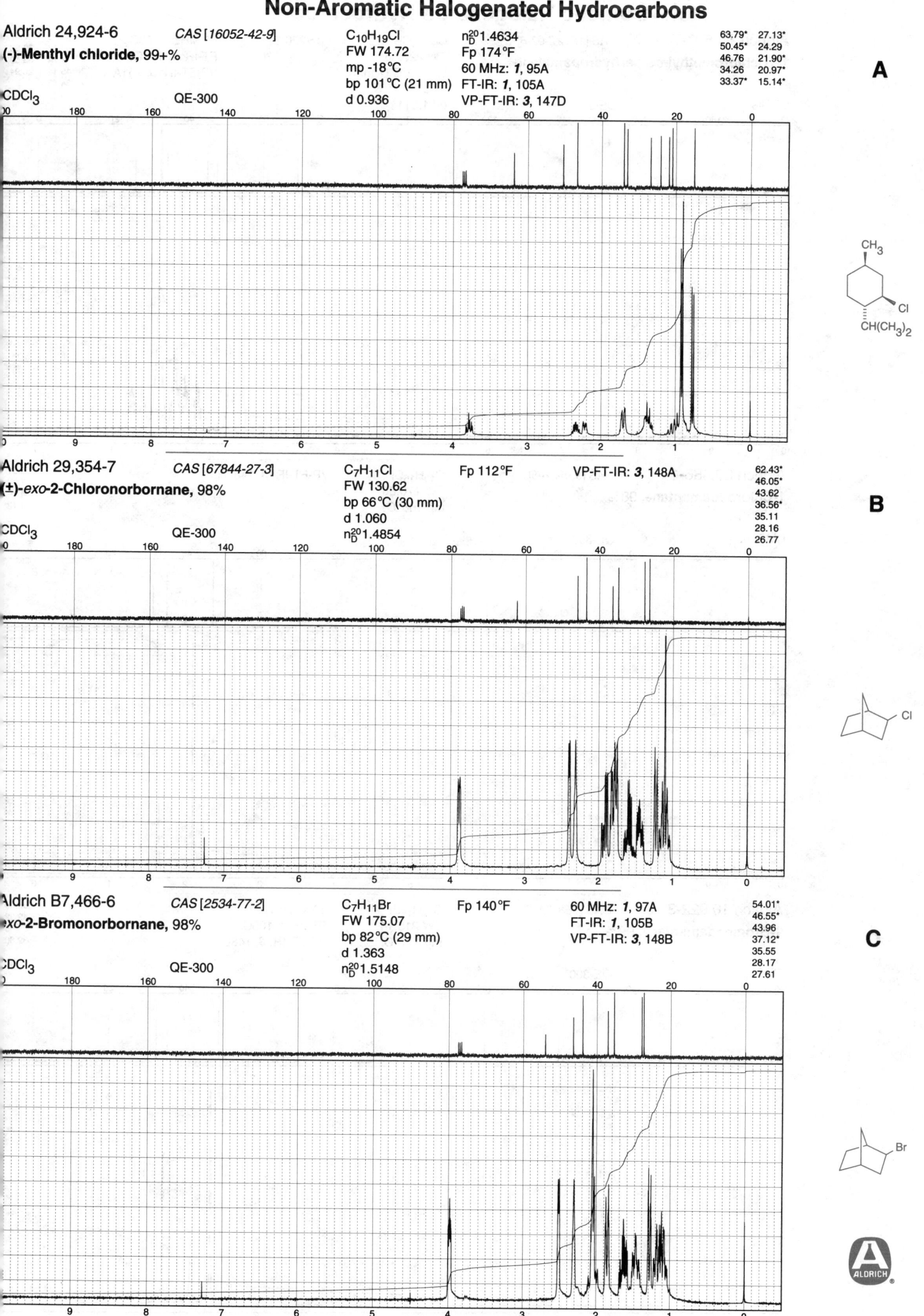

Aldrich 24,924-6   CAS [16052-42-9]   C₁₀H₁₉Cl   n²⁰_D 1.4634
(-)-Menthyl chloride, 99+%   FW 174.72   Fp 174°F   60 MHz: 1, 95A
    mp -18°C   FT-IR: 1, 105A
    bp 101°C (21 mm)   VP-FT-IR: 3, 147D
    d 0.936

63.79* 27.13*
50.45* 24.29
46.76 21.90*
34.26 20.97*
33.37* 15.14*

A

CDCl₃   QE-300

Aldrich 29,354-7   CAS [67844-27-3]   C₇H₁₁Cl   Fp 112°F   VP-FT-IR: 3, 148A
(±)-exo-2-Chloronorbornane, 98%   FW 130.62
    bp 66°C (30 mm)
    d 1.060
    n²⁰_D 1.4854

62.43*
46.05*
43.62
36.56*
35.11
28.16
26.77

B

CDCl₃   QE-300

Aldrich B7,466-6   CAS [2534-77-2]   C₇H₁₁Br   Fp 140°F   60 MHz: 1, 97A
exo-2-Bromonorbornane, 98%   FW 175.07   FT-IR: 1, 105B
    bp 82°C (29 mm)   VP-FT-IR: 3, 148B
    d 1.363
    n²⁰_D 1.5148

54.01*
46.55*
43.96
37.12*
35.55
28.17
27.61

C

CDCl₃   QE-300

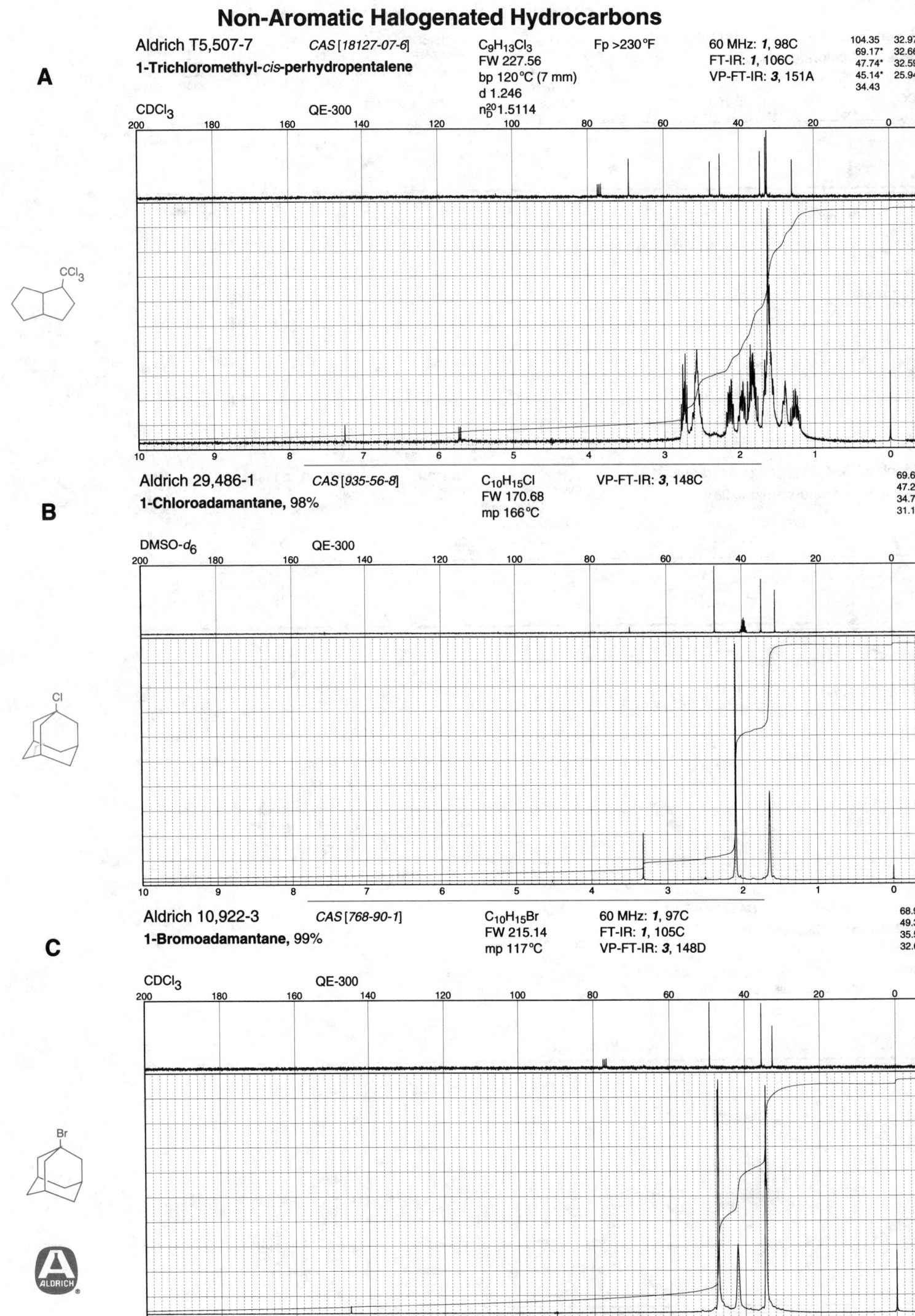

A

Aldrich T5,507-7   CAS [18127-07-6] $C_9H_{13}Cl_3$ Fp >230 °F   60 MHz: 1, 98C
1-Trichloromethyl-cis-perhydropentalene   FW 227.56   FT-IR: 1, 106C
bp 120 °C (7 mm)   VP-FT-IR: 3, 151A
d 1.246
n_D^{20} 1.5114

104.35   32.97
69.17*   32.66
47.74*   32.59
45.14*   25.94
34.43

CDCl₃   QE-300

Aldrich 29,486-1   CAS [935-56-8] $C_{10}H_{15}Cl$ VP-FT-IR: 3, 148C
1-Chloroadamantane, 98%   FW 170.68
mp 166 °C

69.61
47.20
34.79
31.12*

B

DMSO-d_6 QE-300

Aldrich 10,922-3   CAS [768-90-1] $C_{10}H_{15}Br$ 60 MHz: 1, 97C
1-Bromoadamantane, 99%   FW 215.14   FT-IR: 1, 105C
mp 117 °C   VP-FT-IR: 3, 148D

68.96
49.32
35.55
32.60

C

CDCl₃   QE-300

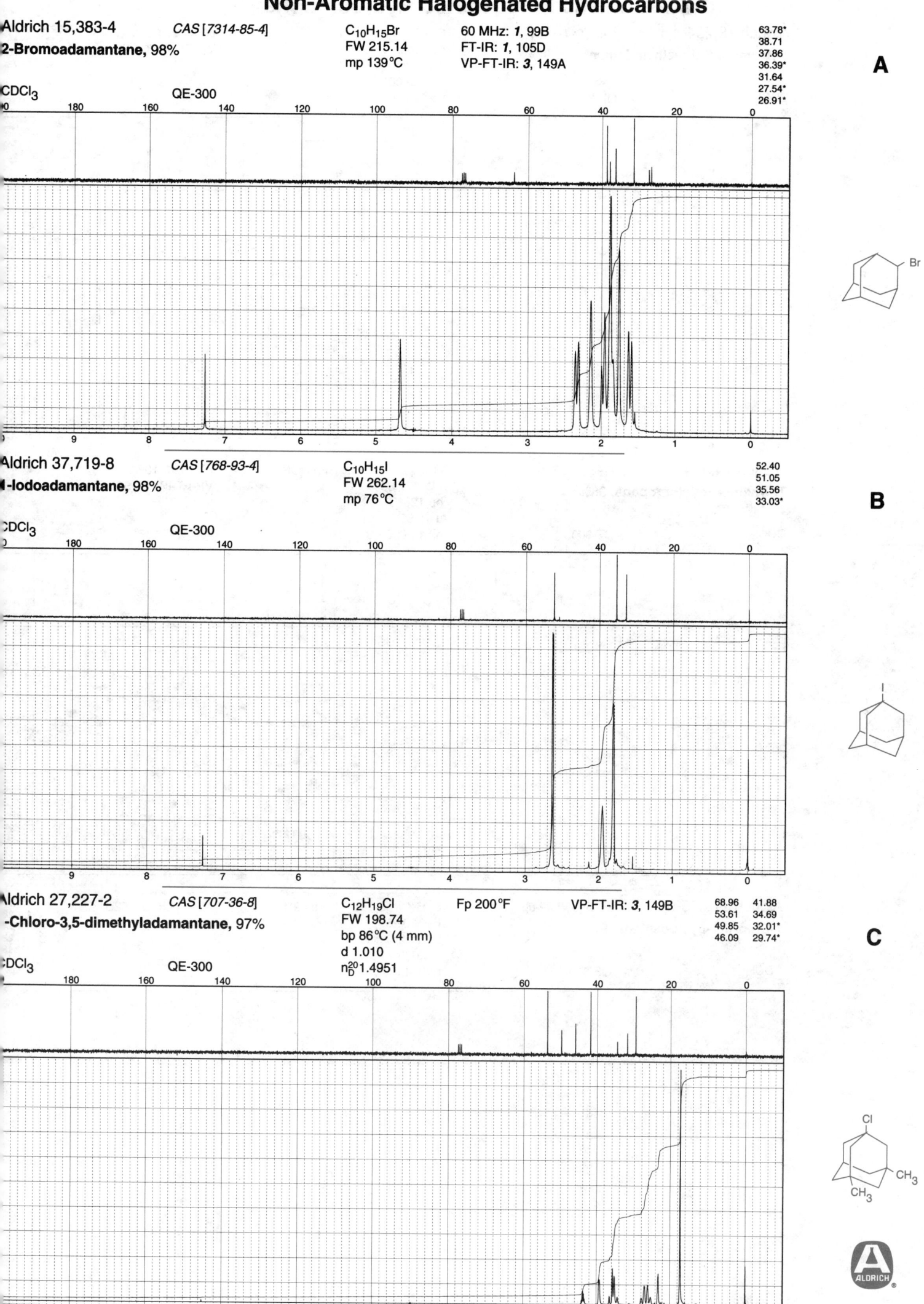

Aldrich 15,383-4   CAS [7314-85-4] $C_{10}H_{15}Br$ 60 MHz: **1**, 99B
2-Bromoadamantane, 98%   FW 215.14   FT-IR: **1**, 105D
   mp 139°C   VP-FT-IR: **3**, 149A

CDCl₃   QE-300

63.78*
38.71
37.86
36.39*
31.64
27.54*
26.91*

A

Aldrich 37,719-8   CAS [768-93-4] $C_{10}H_{15}I$
1-Iodoadamantane, 98%   FW 262.14
   mp 76°C

CDCl₃   QE-300

52.40
51.05
35.56
33.03*

B

Aldrich 27,227-2   CAS [707-36-8] $C_{12}H_{19}Cl$ Fp 200°F   VP-FT-IR: **3**, 149B
1-Chloro-3,5-dimethyladamantane, 97%   FW 198.74
   bp 86°C (4 mm)
   d 1.010
   n_D^{20} 1.4951

CDCl₃   QE-300

68.96   41.88
53.61   34.69
49.85   32.01*
46.09   29.74*

C

Non-Aromatic Halogenated Hydrocarbons

A

Aldrich 18,784-4   *CAS [941-37-7]* $C_{12}H_{19}Br$ 60 MHz: *1*, 97D

1-Bromo-3,5-dimethyladamantane, 98%

FW 243.19
d 1.224
n_D^{20}1.5200
Fp 228°F

VP-FT-IR: *3*, 149C

66.27	41.77
55.13	35.46
49.71	32.76*
47.58	29.78*

CDCl₃   QE-300

B

Aldrich 14,594-7   *CAS [6262-42-6]* C_3Cl_4 Fp NONE°F   FT-IR: *1*, 108A

Tetrachlorocyclopropene, 98%

FW 177.85
bp 128°C
d 1.450
n_D^{20}1.5063

VP-FT-IR: *3*, 153B

122.48
62.22

CDCl₃   QE-300

C

Aldrich 27,665-0   *CAS [930-29-0]* C_5H_7Cl Fp 49°F   VP-FT-IR: *3*, 152D

1-Chloro-1-cyclopentene, 97%

FW 102.57
bp 114°C
d 1.035
n_D^{20}1.4651

132.39
126.28
37.24
31.22
22.66

CDCl₃   QE-300

Aldrich H600-2   *CAS [77-47-4]* C_5Cl_6 n_D^{20} 1.5644
Hexachlorocyclopentadiene, 98%   FW 272.77   Fp NONE
mp -10°C   FT-IR: *1*, 108B
bp 239°C (753 mm)   VP-FT-IR: *3*, 154A
d 1.702

133.07
128.43
81.61

A

CDCl₃   QE-300

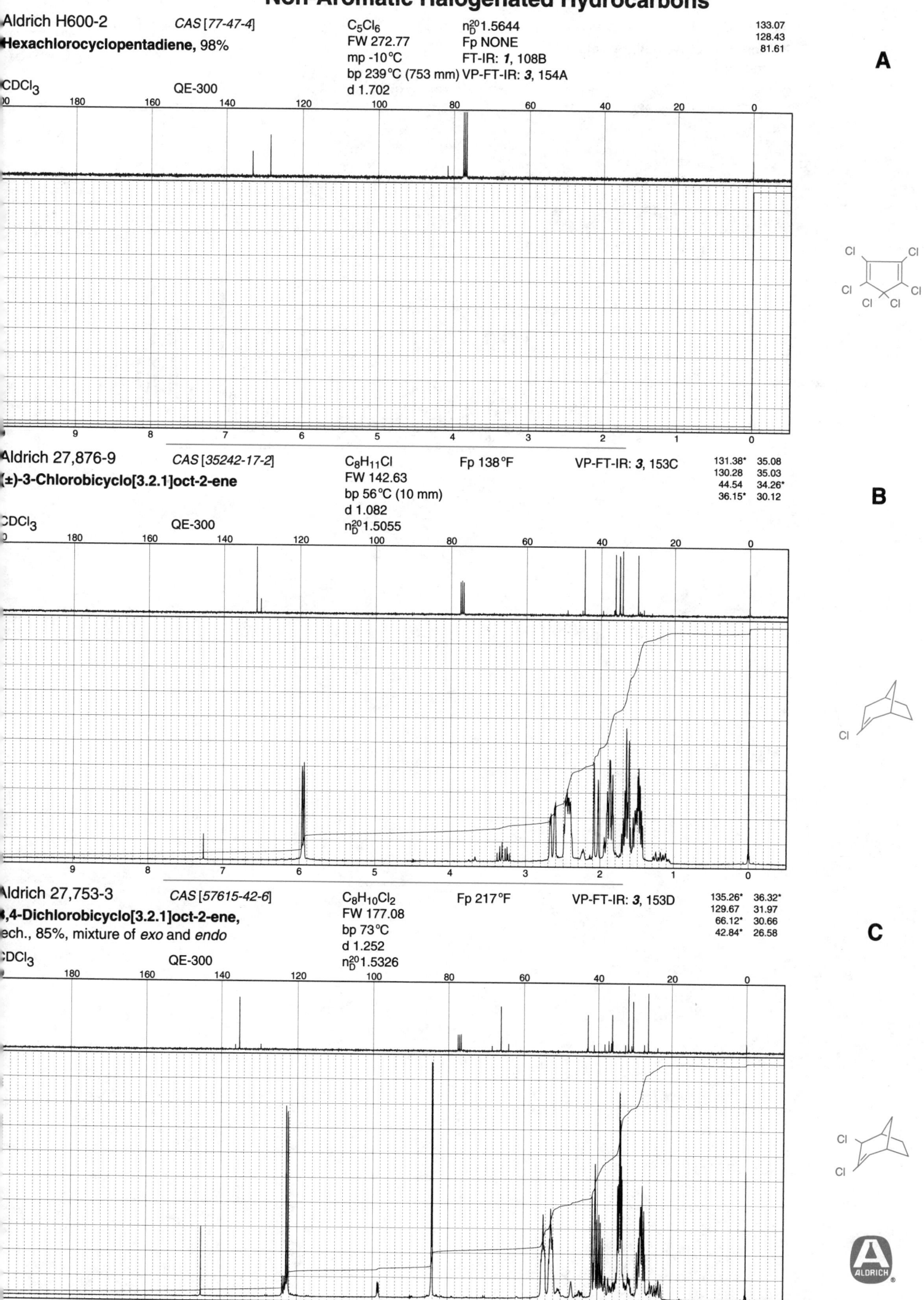

Aldrich 27,876-9   *CAS [35242-17-2]* $C_8H_{11}Cl$ Fp 138°F   VP-FT-IR: *3*, 153C
(±)-3-Chlorobicyclo[3.2.1]oct-2-ene   FW 142.63
bp 56°C (10 mm)
d 1.082
n_D^{20} 1.5055

131.38*   35.08
130.28   35.03
44.54   34.26*
36.15*   30.12

B

CDCl₃   QE-300

Aldrich 27,753-3   *CAS [57615-42-6]* $C_8H_{10}Cl_2$ Fp 217°F   VP-FT-IR: *3*, 153D
3,4-Dichlorobicyclo[3.2.1]oct-2-ene,
ech., 85%, mixture of *exo* and *endo*   FW 177.08
bp 73°C
d 1.252
n_D^{20} 1.5326

135.26*   36.32*
129.67   31.97
66.12*   30.66
42.84*   26.58

C

CDCl₃   QE-300

Non-Aromatic Halogenated Hydrocarbons

Aldrich 30,384-4   *CAS [18127-12-3]* $C_9H_{12}Cl_2$ Fp 212°F   VP-FT-IR: **3**, 150D

2-(Dichloromethylene)bicyclo[3.3.0]-octane, 95%

$CDCl_3$ QE-300

FW 191.10
bp 55°C
d 1.170
$n_D^{20}1.5210$

148.38	32.86
110.45	32.40
49.67*	31.31
45.25*	26.88
33.21	

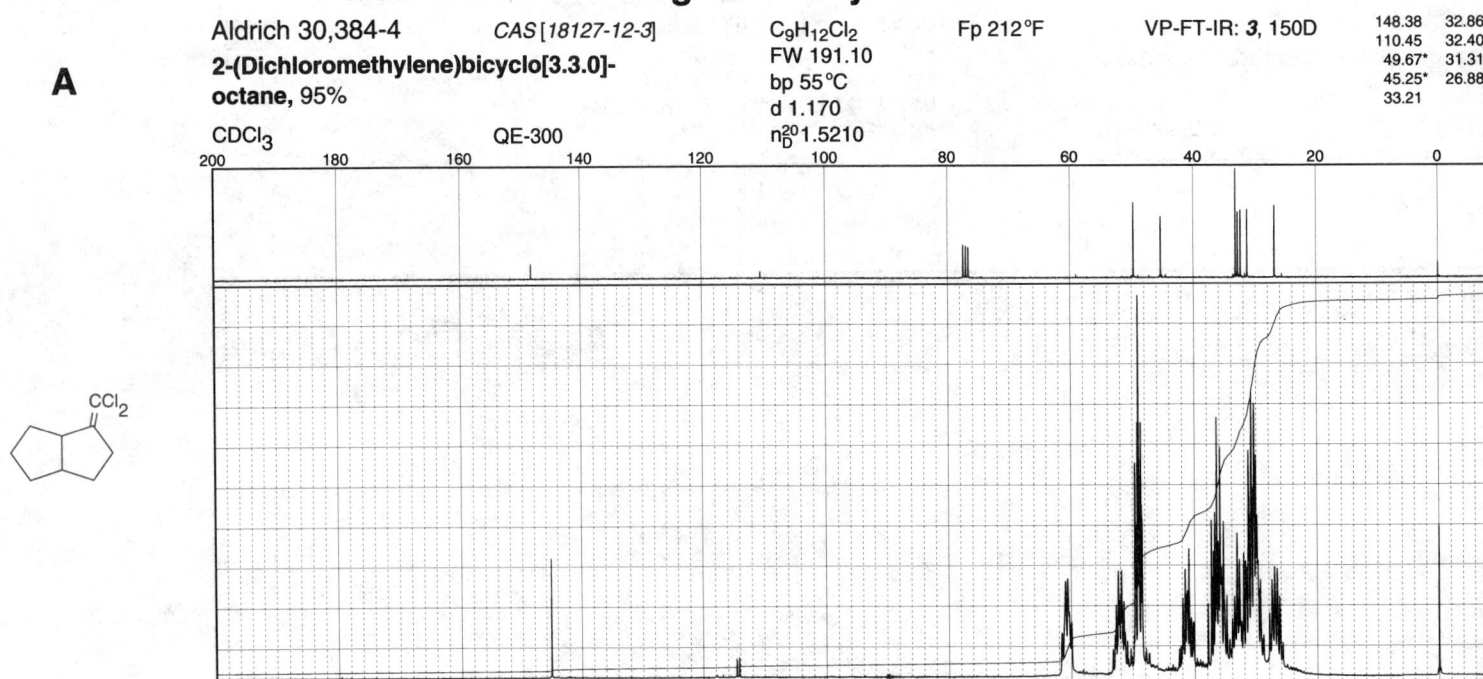

A

Aldrich 17,995-7 | CAS [67-56-1] | CH$_4$O | n$_D^{20}$ 1.3290 | 50.28*

Methyl alcohol, 99+%

FW 32.04 | Fp 52°F
mp -98°C | 60 MHz: **1**, 101A
bp 65°C | FT-IR: **1**, 109B
d 0.791 | VP-FT-IR: **3**, 155B

CDCl$_3$ | QE-300

200 180 160 140 120 100 80 60 40 20 0

CH$_3$OH

10 9 8 7 6 5 4 3 2 1 0

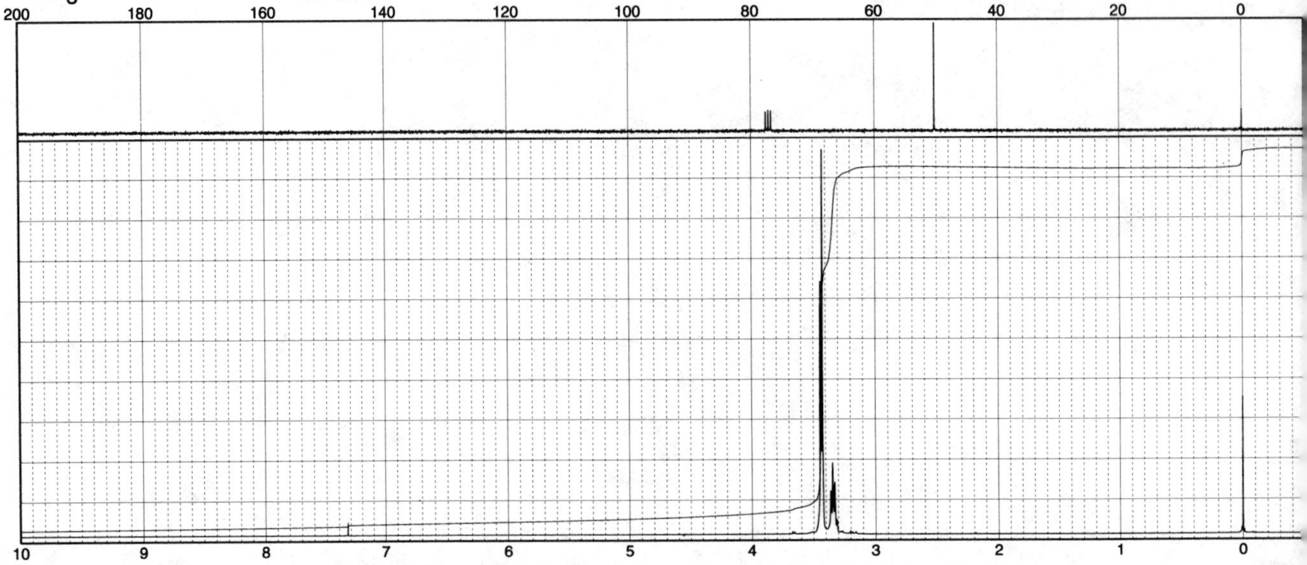

B

Aldrich 24,511-9 | CAS [64-17-5] | C$_2$H$_6$O | Fp 48°F | 60 MHz: **1**, 101B | 57.97
18.22*

Ethyl alcohol

FW 46.07 | FT-IR: **1**, 109C
bp 78°C | VP-FT-IR: **3**, 155C
d 0.785
n$_D^{20}$ 1.3600

CDCl$_3$ | QE-300

200 180 160 140 120 100 80 60 40 20 0

CH$_3$CH$_2$OH

10 9 8 7 6 5 4 3 2 1 0

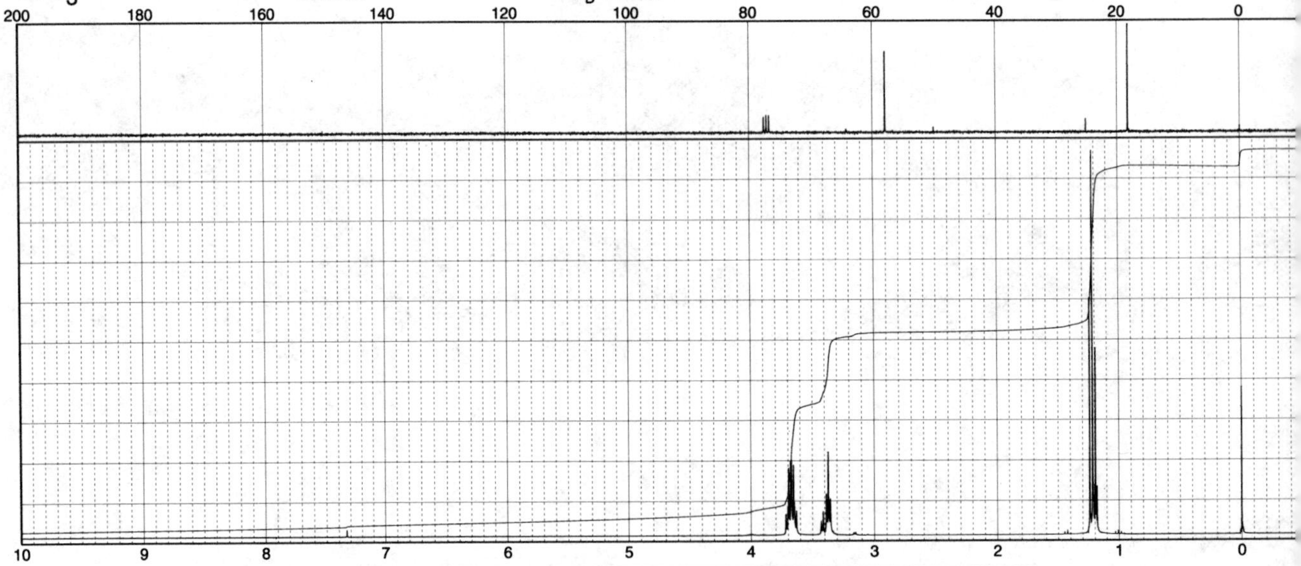

C

Aldrich 25,640-4 | CAS [71-23-8] | C$_3$H$_8$O | n$_D^{20}$ 1.3840 | 64.36
25.84
10.20*

1-Propanol, 99+%

FW 60.10 | Fp 59°F
mp -127°C | 60 MHz: **1**, 101C
bp 97°C | FT-IR: **1**, 109D
d 0.804 | VP-FT-IR: **3**, 156A

CDCl$_3$ | QE-300

200 180 160 140 120 100 80 60 40 20 0

CH$_3$CH$_2$CH$_2$OH

ALDRICH

10 9 8 7 6 5 4 3 2 1 0

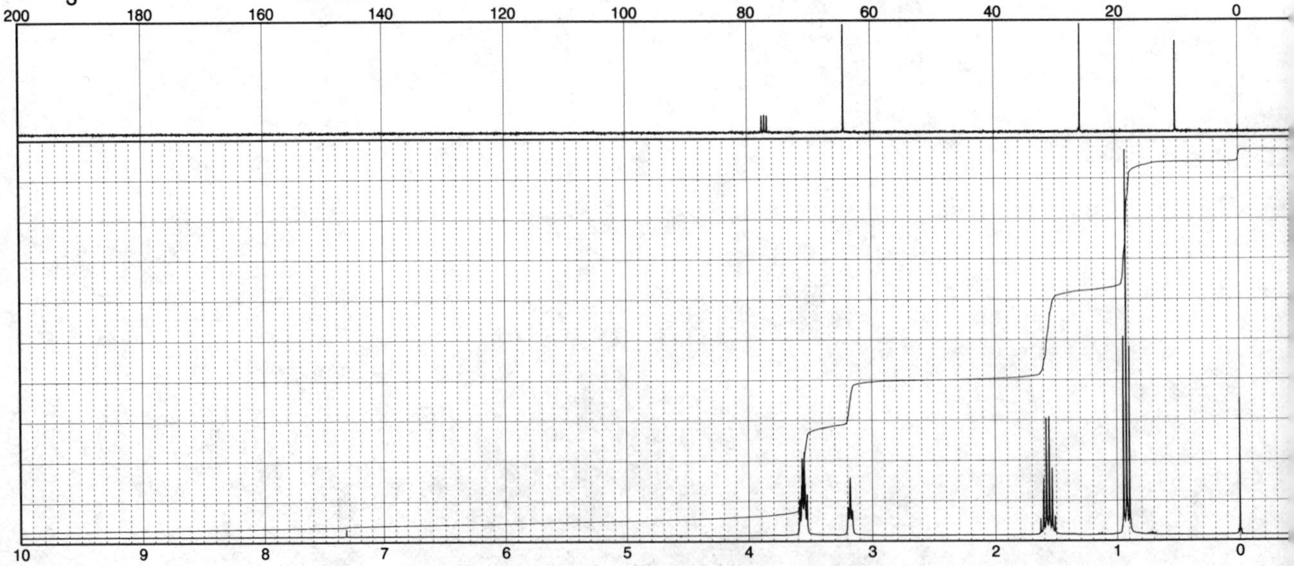

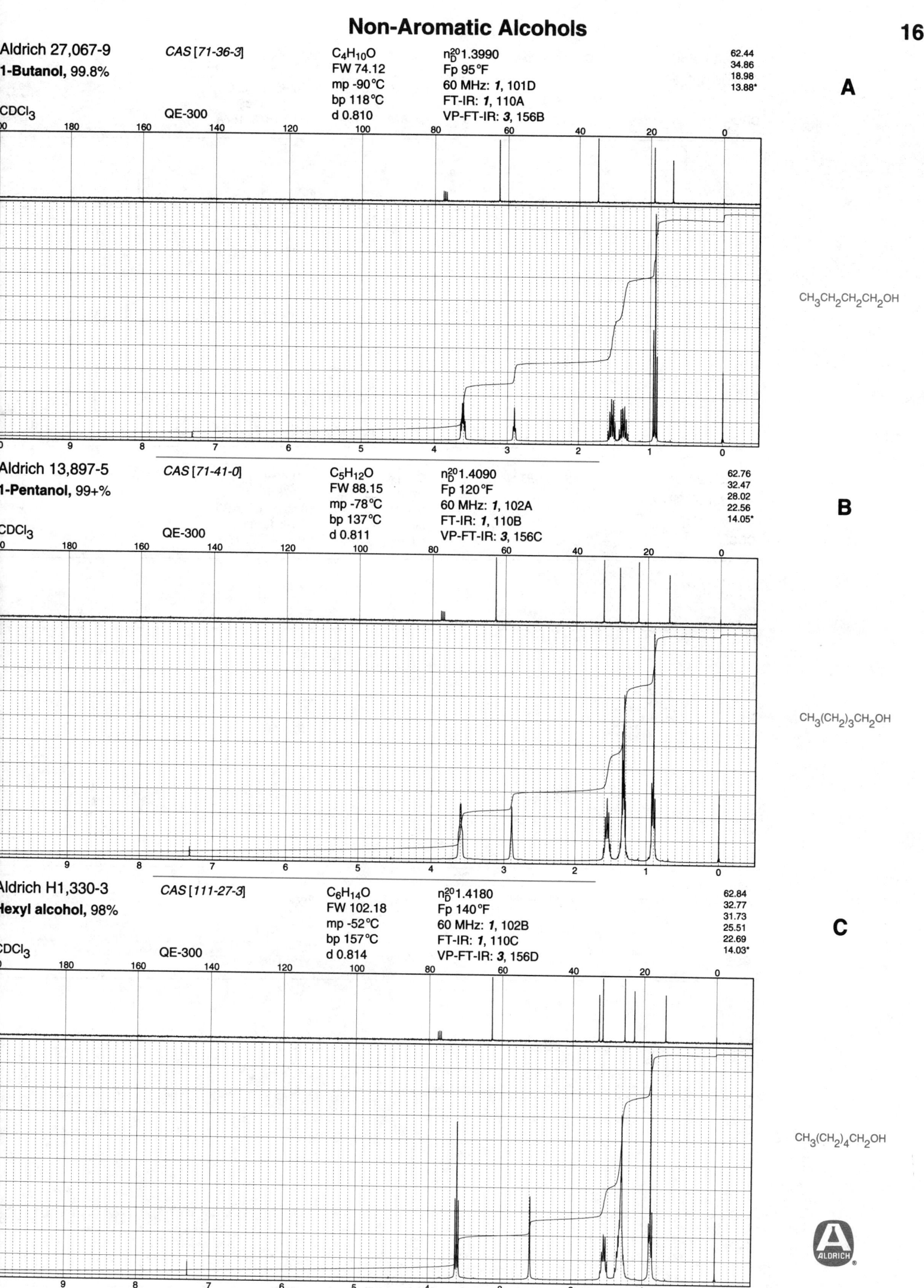

Aldrich 27,067-9 CAS [71-36-3] $C_4H_{10}O$ n_D^{20} 1.3990 62.44

1-Butanol, 99.8% FW 74.12 Fp 95°F 34.86
18.98
mp -90°C 60 MHz: *1*, 101D 13.88*

CDCl₃ QE-300 bp 118°C FT-IR: *1*, 110A
d 0.810 VP-FT-IR: *3*, 156B

A

$CH_3CH_2CH_2CH_2OH$

Aldrich 13,897-5 CAS [71-41-0] $C_5H_{12}O$ n_D^{20} 1.4090 62.76

1-Pentanol, 99+% FW 88.15 Fp 120°F 32.47
28.02
mp -78°C 60 MHz: *1*, 102A 22.56
14.05*

CDCl₃ QE-300 bp 137°C FT-IR: *1*, 110B
d 0.811 VP-FT-IR: *3*, 156C

B

$CH_3(CH_2)_3CH_2OH$

Aldrich H1,330-3 CAS [111-27-3] $C_6H_{14}O$ n_D^{20} 1.4180 62.84

Hexyl alcohol, 98% FW 102.18 Fp 140°F 32.77
31.73
mp -52°C 60 MHz: *1*, 102B 25.51
22.69
CDCl₃ QE-300 bp 157°C FT-IR: *1*, 110C 14.03*
d 0.814 VP-FT-IR: *3*, 156D

C

$CH_3(CH_2)_4CH_2OH$

A
Aldrich H280-5
1-Heptanol, 98%

CAS [111-70-6]

C$_7$H$_{16}$O
FW 116.20
mp -36°C
bp 176°C
d 0.822

n$_D^{20}$1.4240
Fp 165°F
60 MHz: *1*, 102C
FT-IR: *1*, 110D
VP-FT-IR: *3*, 157A

62.93
32.79
31.86
29.14
25.75
22.63
14.08*

CDCl$_3$ QE-300

CH$_3$(CH$_2$)$_5$CH$_2$OH

B
Aldrich 24,041-9
1-Octanol, 99+%

CAS [111-87-5]

C$_8$H$_{18}$O
FW 130.23
mp -15°C
bp 196°C
d 0.827

n$_D^{20}$1.4290
Fp 178°F
60 MHz: *1*, 102B
FT-IR: *1*, 111A
VP-FT-IR: *3*, 157B

62.97 29.31
32.81 25.80
31.85 22.68
29.44 14.09*

CDCl$_3$ QE-300

CH$_3$(CH$_2$)$_6$CH$_2$OH

C
Aldrich 13,121-0
1-Nonanol, 98%

CAS [143-08-8]

C$_9$H$_{20}$O
FW 144.26
mp -7°C
bp 215°C
d 0.827

n$_D^{20}$1.4330
Fp 168°F
60 MHz: *1*, 103A
FT-IR: *1*, 111B
VP-FT-IR: *3*, 157C

62.96 29.30
32.82 25.80
31.91 22.70
29.62 14.10*
29.49

CDCl$_3$ QE-300

CH$_3$(CH$_2$)$_7$CH$_2$OH

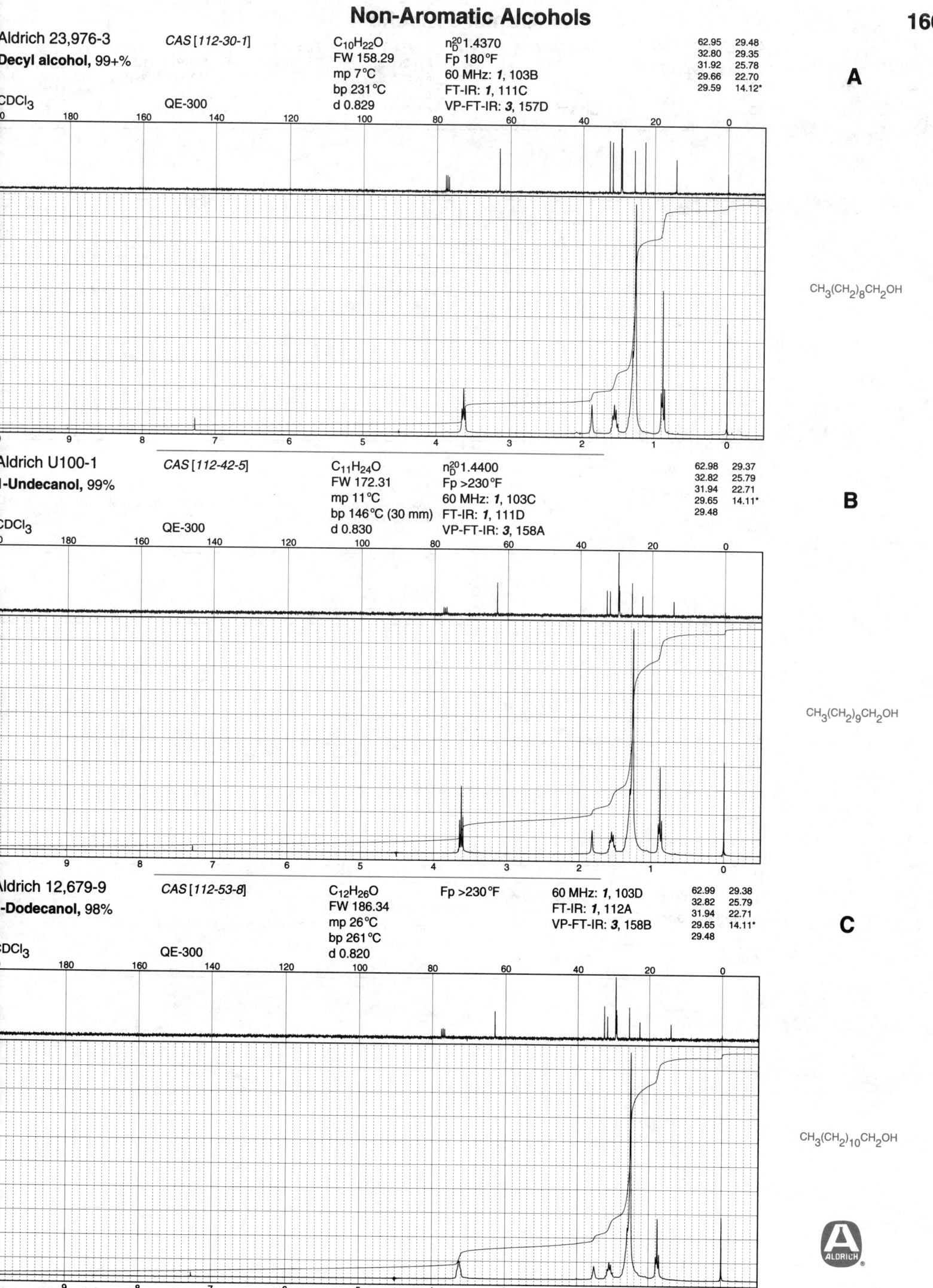

Aldrich 23,976-3
Decyl alcohol, 99+%
CDCl₃   QE-300

CAS [112-30-1]

$C_{10}H_{22}O$
FW 158.29
mp 7°C
bp 231°C
d 0.829

n_D^{20} 1.4370
Fp 180°F
60 MHz: *1*, 103B
FT-IR: *1*, 111C
VP-FT-IR: *3*, 157D

62.95   29.48
32.80   29.35
31.92   25.78
29.66   22.70
29.59   14.12*

A

$CH_3(CH_2)_8CH_2OH$

Aldrich U100-1
1-Undecanol, 99%
CDCl₃   QE-300

CAS [112-42-5]

$C_{11}H_{24}O$
FW 172.31
mp 11°C
bp 146°C (30 mm)
d 0.830

n_D^{20} 1.4400
Fp >230°F
60 MHz: *1*, 103C
FT-IR: *1*, 111D
VP-FT-IR: *3*, 158A

62.98   29.37
32.82   25.79
31.94   22.71
29.65   14.11*
29.48

B

$CH_3(CH_2)_9CH_2OH$

Aldrich 12,679-9
1-Dodecanol, 98%
CDCl₃   QE-300

CAS [112-53-8]

$C_{12}H_{26}O$
FW 186.34
mp 26°C
bp 261°C
d 0.820

Fp >230°F

60 MHz: *1*, 103D
FT-IR: *1*, 112A
VP-FT-IR: *3*, 158B

62.99   29.38
32.82   25.79
31.94   22.71
29.65   14.11*
29.48

C

$CH_3(CH_2)_{10}CH_2OH$

ALDRICH

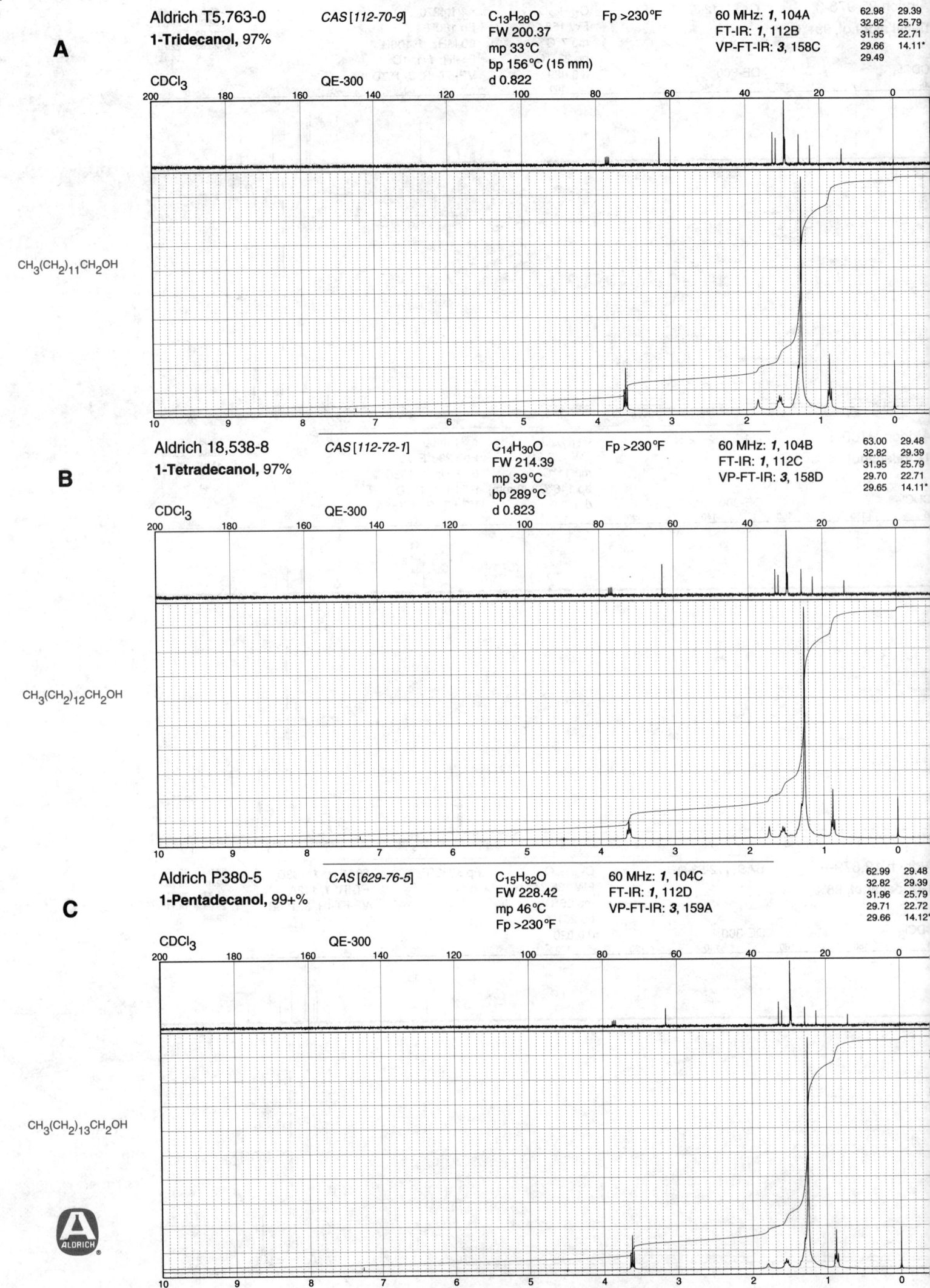

A

Aldrich T5,763-0

1-Tridecanol, 97%

CAS [112-70-9]

$C_{13}H_{28}O$
FW 200.37
mp 33°C
bp 156°C (15 mm)
d 0.822

Fp >230°F

60 MHz: *1*, 104A
FT-IR: *1*, 112B
VP-FT-IR: *3*, 158C

62.98 29.39
32.82 25.79
31.95 22.71
29.66 14.11*
29.49

CDCl₃

QE-300

$CH_3(CH_2)_{11}CH_2OH$

B

Aldrich 18,538-8

1-Tetradecanol, 97%

CAS [112-72-1]

$C_{14}H_{30}O$
FW 214.39
mp 39°C
bp 289°C
d 0.823

Fp >230°F

60 MHz: *1*, 104B
FT-IR: *1*, 112C
VP-FT-IR: *3*, 158D

63.00 29.48
32.82 29.39
31.95 25.79
29.70 22.71
29.65 14.11*

CDCl₃

QE-300

$CH_3(CH_2)_{12}CH_2OH$

C

Aldrich P380-5

1-Pentadecanol, 99+%

CAS [629-76-5]

$C_{15}H_{32}O$
FW 228.42
mp 46°C
Fp >230°F

60 MHz: *1*, 104C
FT-IR: *1*, 112D
VP-FT-IR: *3*, 159A

62.99 29.48
32.82 29.39
31.96 25.79
29.71 22.72
29.66 14.12*

CDCl₃

QE-300

$CH_3(CH_2)_{13}CH_2OH$

Non-Aromatic Alcohols

A

Aldrich H680-0   CAS [36653-82-4] $C_{16}H_{34}O$ Fp 275°F   60 MHz: *1*, 104D

1-Hexadecanol, 95%   FW 242.45   FT-IR: *1*, 113A

mp 49°C   VP-FT-IR: *3*, 159B

bp 189°C (15 mm)

CDCl₃   QE-300   d 0.818

63.00	29.48
32.83	29.39
31.96	25.79
29.72	22.72
29.66	14.11*

$CH_3(CH_2)_{14}CH_2OH$

Aldrich 24,169-5   CAS [1454-85-9] $C_{17}H_{36}O$ 60 MHz: *1*, 105A

1-Heptadecanol, 98%   FW 256.48   FT-IR: *1*, 113B

mp 57°C   VP-FT-IR: *3*, 159C

Fp >230°F

CDCl₃   QE-300

63.00	29.47
32.81	29.39
31.95	25.78
29.72	22.72
29.65	14.13*

B

$CH_3(CH_2)_{15}CH_2OH$

Aldrich 25,876-8   CAS [112-92-5] $C_{18}H_{38}O$ 60 MHz: *1*, 105B

1-Octadecanol, 99%   FW 270.50   FT-IR: *1*, 113C

mp 61°C   VP-FT-IR: *3*, 159D

bp 171°C (2 mm)

CDCl₃   QE-300

63.00	29.48
32.83	29.39
31.96	25.79
29.73	22.71
29.66	14.11*

C

$CH_3(CH_2)_{16}CH_2OH$

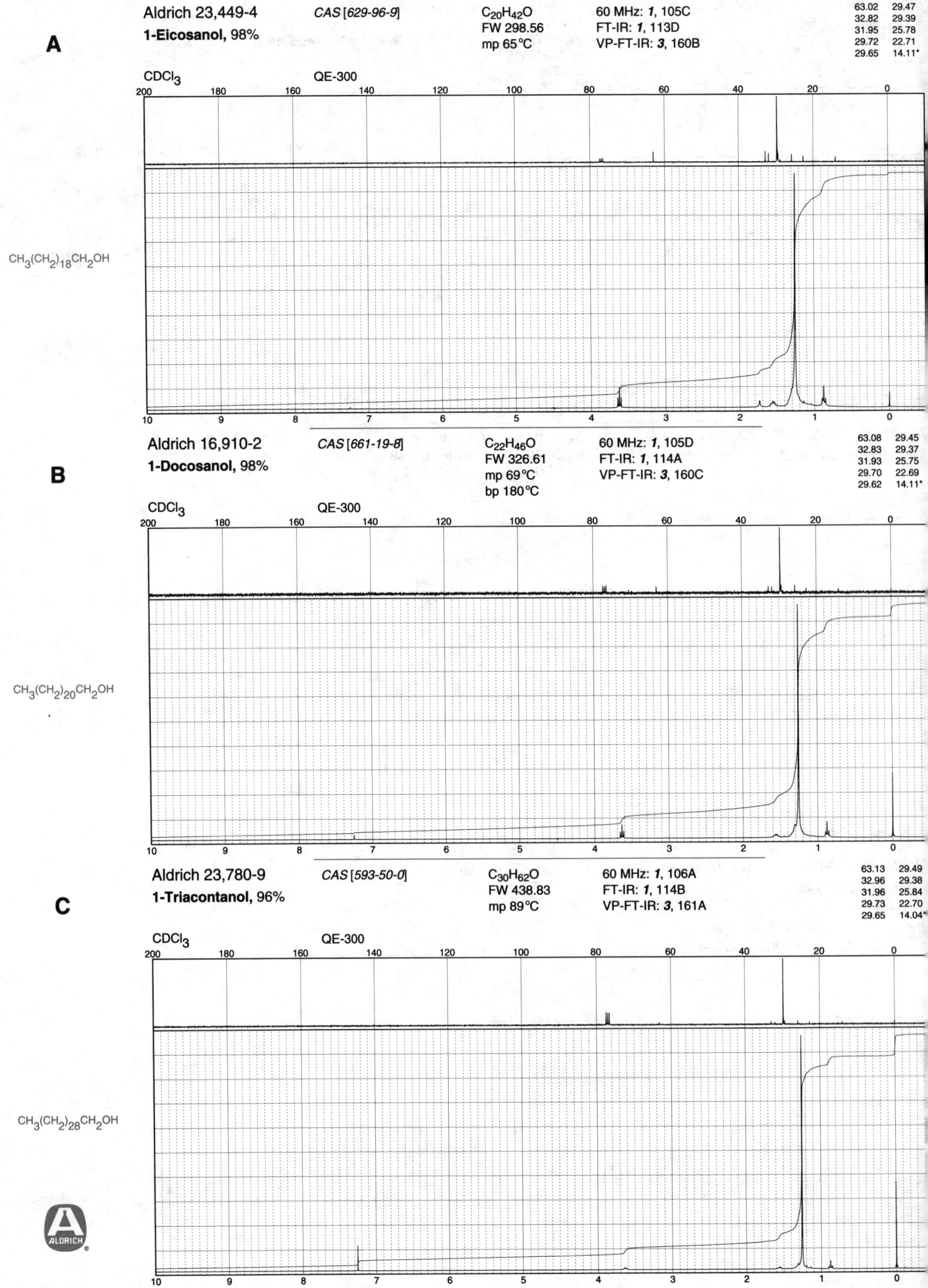

A

Aldrich 23,449-4   *CAS [629-96-9]*   C$_{20}$H$_{42}$O   60 MHz: *1*, 105C

1-Eicosanol, 98%   FW 298.56   FT-IR: *1*, 113D

mp 65°C   VP-FT-IR: *3*, 160B

63.02	29.47
32.82	29.39
31.95	25.78
29.72	22.71
29.65	14.11*

CDCl$_3$ QE-300

CH$_3$(CH$_2$)$_{18}$CH$_2$OH

B

Aldrich 16,910-2   *CAS [661-19-8]*   C$_{22}$H$_{46}$O   60 MHz: *1*, 105D

1-Docosanol, 98%   FW 326.61   FT-IR: *1*, 114A

mp 69°C   VP-FT-IR: *3*, 160C

bp 180°C

63.08	29.45
32.83	29.37
31.93	25.75
29.70	22.69
29.62	14.11*

CDCl$_3$ QE-300

CH$_3$(CH$_2$)$_{20}$CH$_2$OH

C

Aldrich 23,780-9   *CAS [593-50-0]*   C$_{30}$H$_{62}$O   60 MHz: *1*, 106A

1-Triacontanol, 96%   FW 438.83   FT-IR: *1*, 114B

mp 89°C   VP-FT-IR: *3*, 161A

63.13	29.49
32.96	29.38
31.96	25.84
29.73	22.70
29.65	14.04*

CDCl$_3$ QE-300

CH$_3$(CH$_2$)$_{28}$CH$_2$OH

ALDRICH

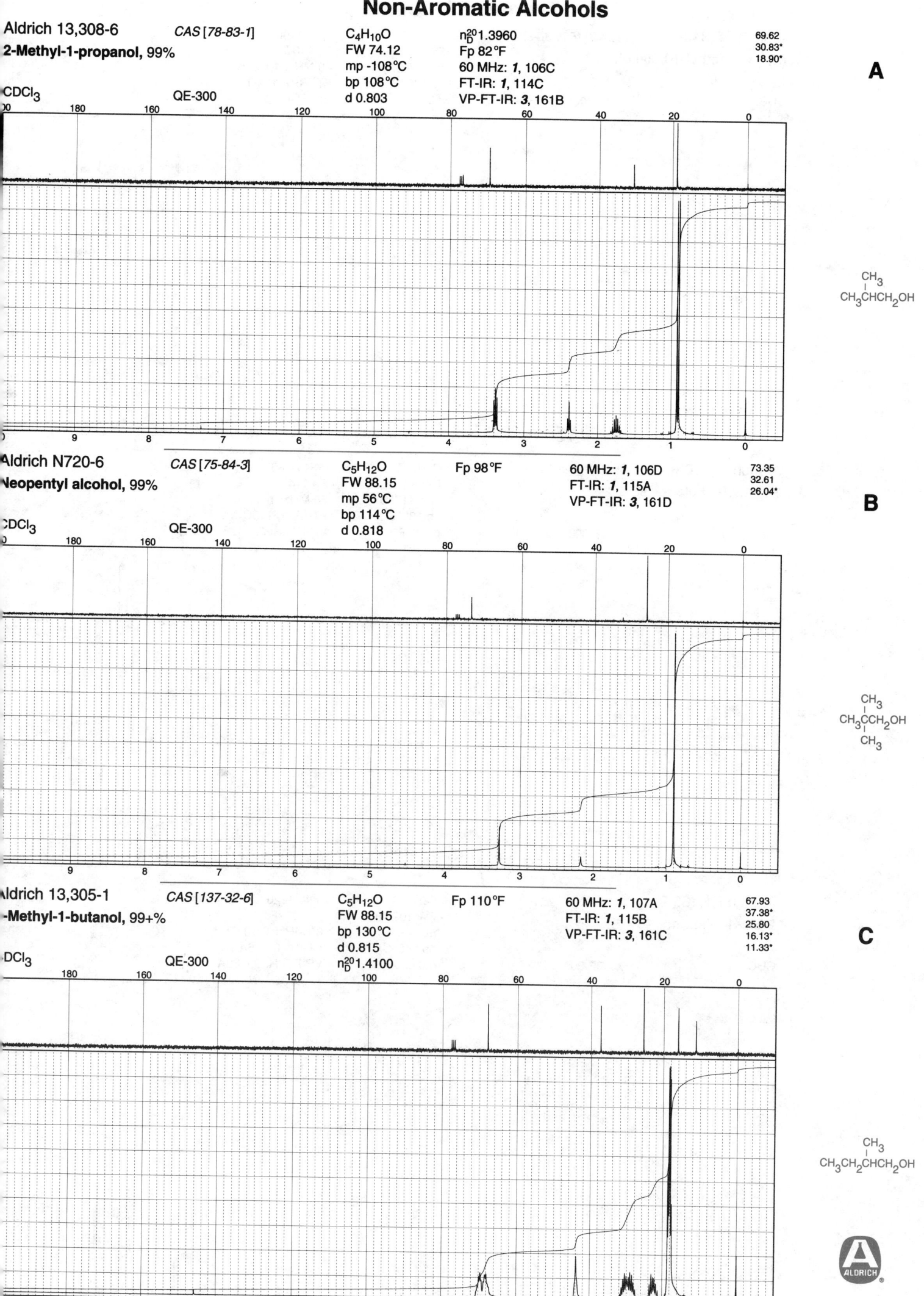

Aldrich 13,308-6   CAS [78-83-1]

2-Methyl-1-propanol, 99%

CDCl₃   QE-300

$C_4H_{10}O$
FW 74.12
mp -108°C
bp 108°C
d 0.803

$n_D^{20} 1.3960$
Fp 82°F
60 MHz: **1**, 106C
FT-IR: **1**, 114C
VP-FT-IR: **3**, 161B

69.62
30.83*
18.90*

A

CH_3CHCH_2OH with CH_3

Aldrich N720-6   CAS [75-84-3]

Neopentyl alcohol, 99%

CDCl₃   QE-300

$C_5H_{12}O$
FW 88.15
mp 56°C
bp 114°C
d 0.818

Fp 98°F

60 MHz: **1**, 106D
FT-IR: **1**, 115A
VP-FT-IR: **3**, 161D

73.35
32.61
26.04*

B

$CH_3C CH_2OH$ with CH_3 and CH_3

Aldrich 13,305-1   CAS [137-32-6]

2-Methyl-1-butanol, 99+%

CDCl₃   QE-300

$C_5H_{12}O$
FW 88.15
bp 130°C
d 0.815
$n_D^{20} 1.4100$

Fp 110°F

60 MHz: **1**, 107A
FT-IR: **1**, 115B
VP-FT-IR: **3**, 161C

67.93
37.38*
25.80
16.13*
11.33*

C

$CH_3CH_2CHCH_2OH$ with CH_3

ALDRICH

A

Aldrich A8,340-7   *CAS [1565-80-6]*

(S)-(-)-2-Methyl-1-butanol, 99%

$C_5H_{12}O$	n_D^{20} 1.4100
FW 88.15	Fp 110°F
mp -70°C	60 MHz: *1*, 106B
bp 128°C	FT-IR: *1*, 114D
d 0.811	VP-FT-IR: *3*, 162A

67.93
37.38*
25.80
16.13*
11.33*

CDCl₃   QE-300

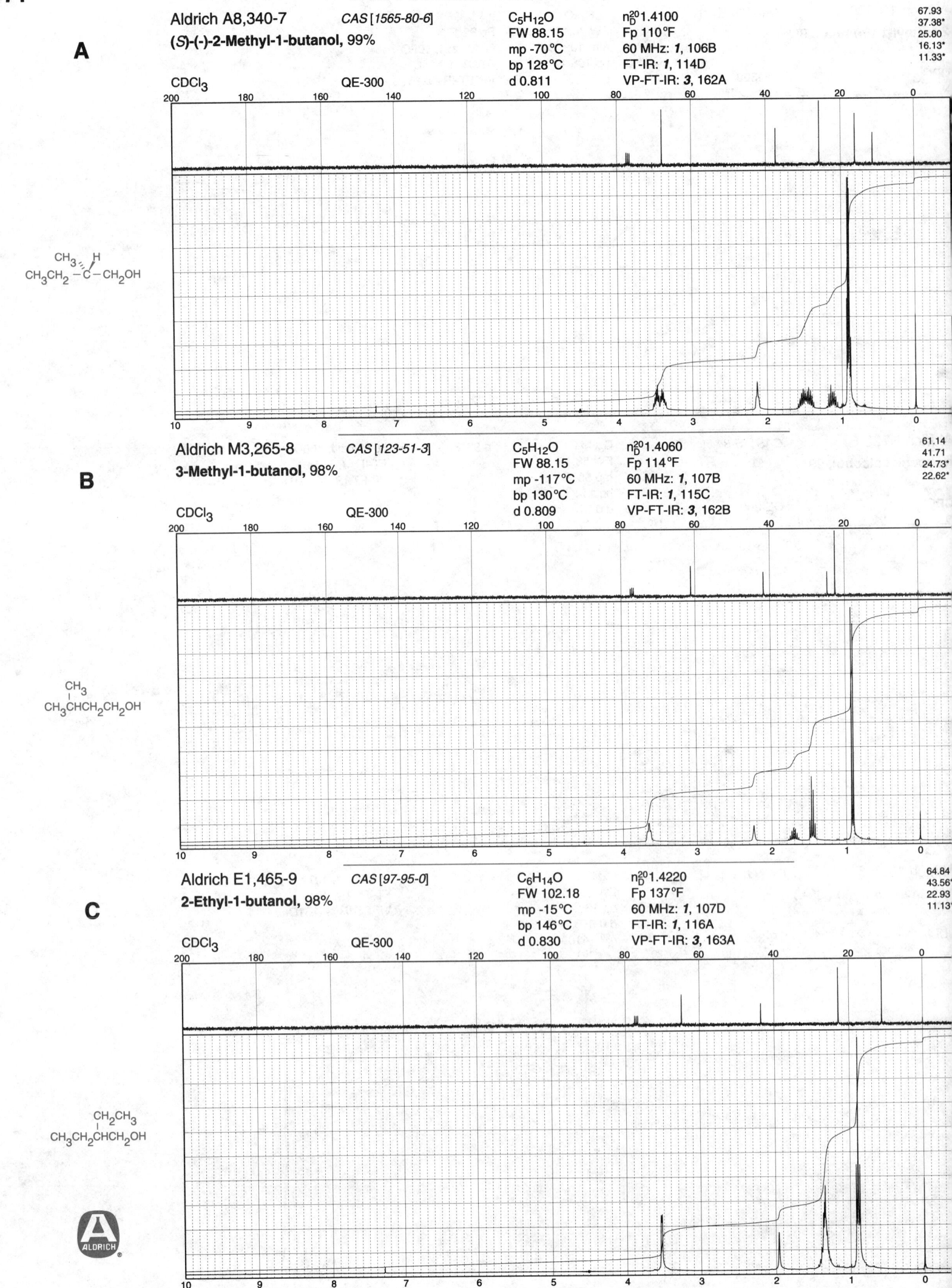

B

Aldrich M3,265-8   *CAS [123-51-3]*

3-Methyl-1-butanol, 98%

$C_5H_{12}O$	n_D^{20} 1.4060
FW 88.15	Fp 114°F
mp -117°C	60 MHz: *1*, 107B
bp 130°C	FT-IR: *1*, 115C
d 0.809	VP-FT-IR: *3*, 162B

61.14
41.71
24.73*
22.62*

CDCl₃   QE-300

C

Aldrich E1,465-9   *CAS [97-95-0]*

2-Ethyl-1-butanol, 98%

$C_6H_{14}O$	n_D^{20} 1.4220
FW 102.18	Fp 137°F
mp -15°C	60 MHz: *1*, 107D
bp 146°C	FT-IR: *1*, 116A
d 0.830	VP-FT-IR: *3*, 163A

64.84
43.56*
22.93
11.13*

CDCl₃   QE-300

Aldrich 18,310-5   CAS [624-95-3]

3,3-Dimethyl-1-butanol, 99%

CDCl₃   QE-300

$C_6H_{14}O$	n_D^{20} 1.4140
FW 102.18	Fp 118 °F
mp -60 °C	60 MHz: *1*, 107C
bp 143 °C	FT-IR: *1*, 115D
d 0.844	VP-FT-IR: *3*, 162D

59.96
46.43
29.75*
29.69

A

$CH_3CCH_2CH_2OH$ with two CH_3 groups

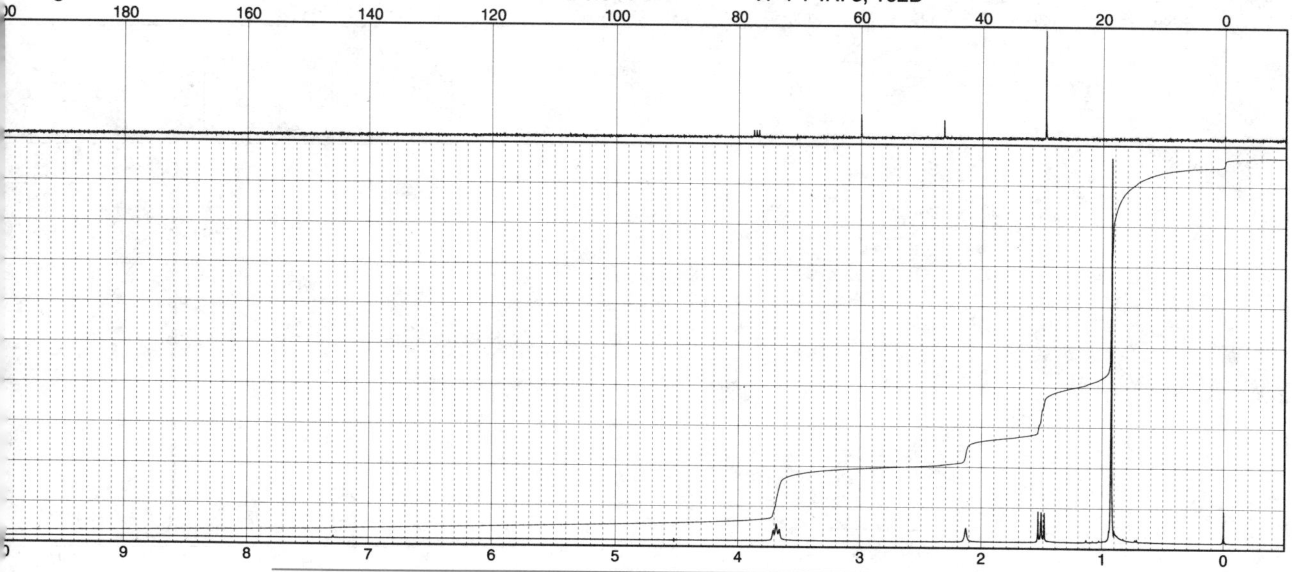

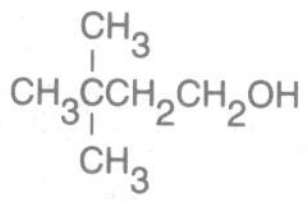

Aldrich 21,401-9   CAS [105-30-6]

2-Methyl-1-pentanol, 99%

CDCl₃   QE-300

$C_6H_{14}O$	Fp 123 °F
FW 102.18	FT-IR: *1*, 116B
bp 148 °C	VP-FT-IR: *3*, 163B
d 0.824	
n_D^{20} 1.4180	

68.25
35.50*
35.47
20.10
16.57*
14.36*

B

$CH_3CH_2CH_2CHCH_2OH$ with CH_3 group

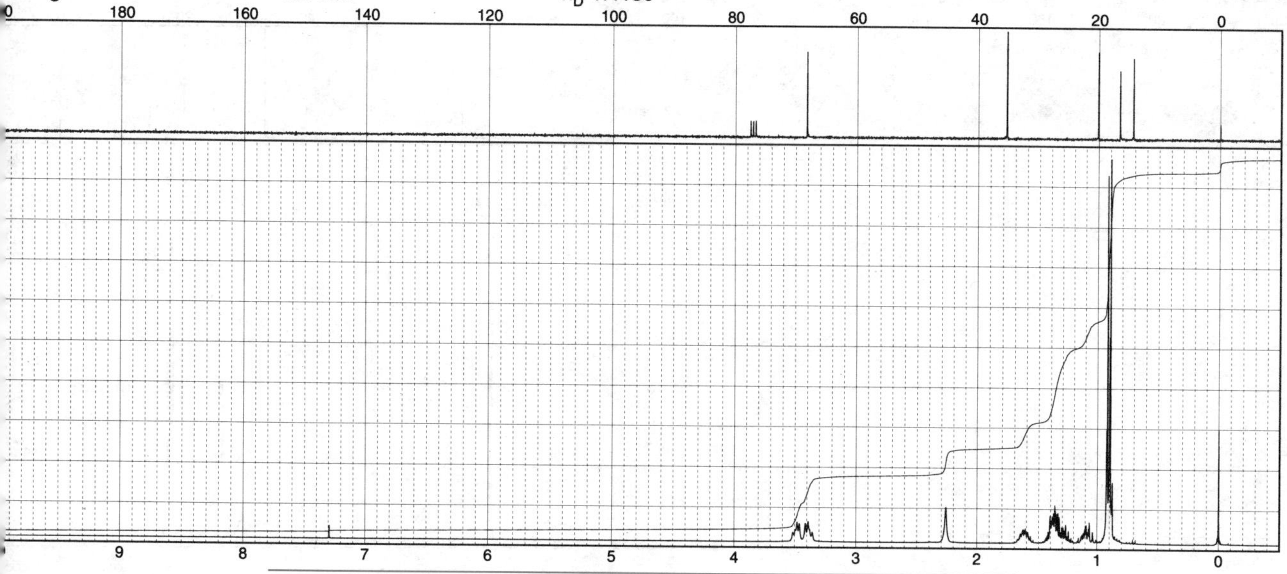

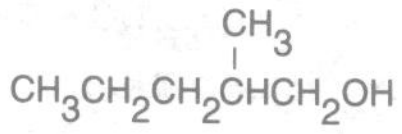

Aldrich 11,111-2   CAS [589-35-5]

3-Methyl-1-pentanol, 99%

CDCl₃   QE-300

$C_6H_{14}O$	Fp 138 °F
FW 102.18	60 MHz: *1*, 108A
bp 152 °C	FT-IR: *1*, 116C
d 0.823	VP-FT-IR: *3*, 163C
n_D^{20} 1.4180	

61.10
39.52
31.12*
29.62
19.16*
11.27*

C

$CH_3CH_2 CHCH_2CH_2OH$ with CH_3 group

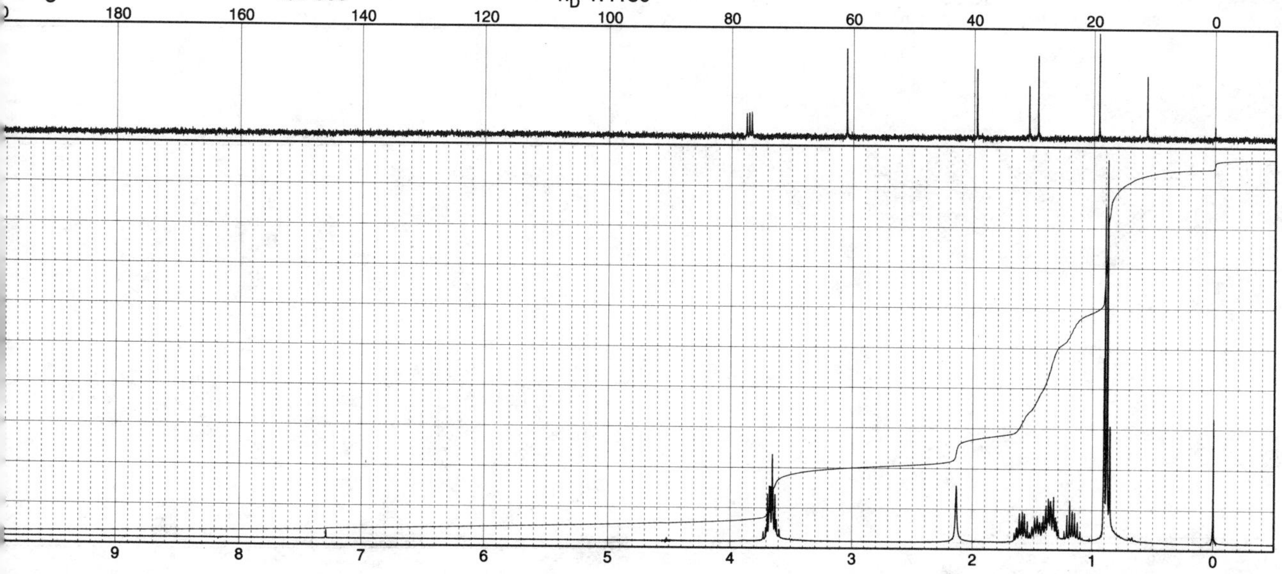

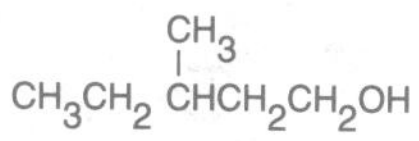

Non-Aromatic Alcohols

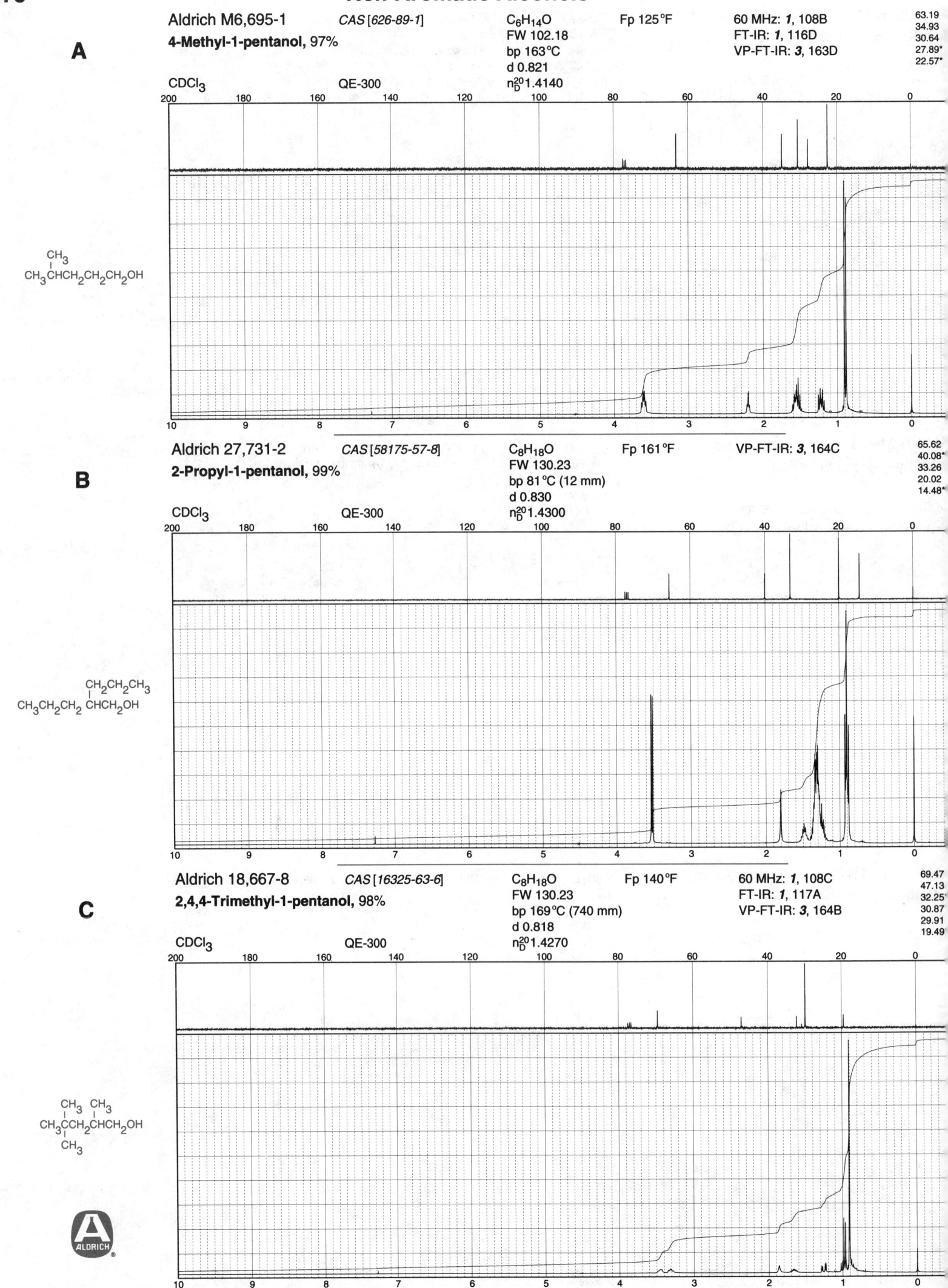

Aldrich M6,695-1 CAS [626-89-1] C₆H₁₄O Fp 125°F 60 MHz: *1*, 108B 63.19 34.93 30.64 27.89* 22.57*
$C_6H_{14}O$ FW 102.18 bp 163°C d 0.821 n_D^{20}1.4140

4-Methyl-1-pentanol, 97% FT-IR: *1*, 116D VP-FT-IR: *3*, 163D

CDCl₃ QE-300

Aldrich 27,731-2 CAS [58175-57-8] $C_8H_{18}O$ FW 130.23 bp 81°C (12 mm) d 0.830 n_D^{20}1.4300 Fp 161°F VP-FT-IR: *3*, 164C 65.62 40.08* 33.26 20.02 14.48*

2-Propyl-1-pentanol, 99%

CDCl₃ QE-300

Aldrich 18,667-8 CAS [16325-63-6] $C_8H_{18}O$ FW 130.23 bp 169°C (740 mm) d 0.818 n_D^{20}1.4270 Fp 140°F 60 MHz: *1*, 108C FT-IR: *1*, 117A VP-FT-IR: *3*, 164B 69.47 47.13 32.25 30.87 29.91 19.49

2,4,4-Trimethyl-1-pentanol, 98%

CDCl₃ QE-300

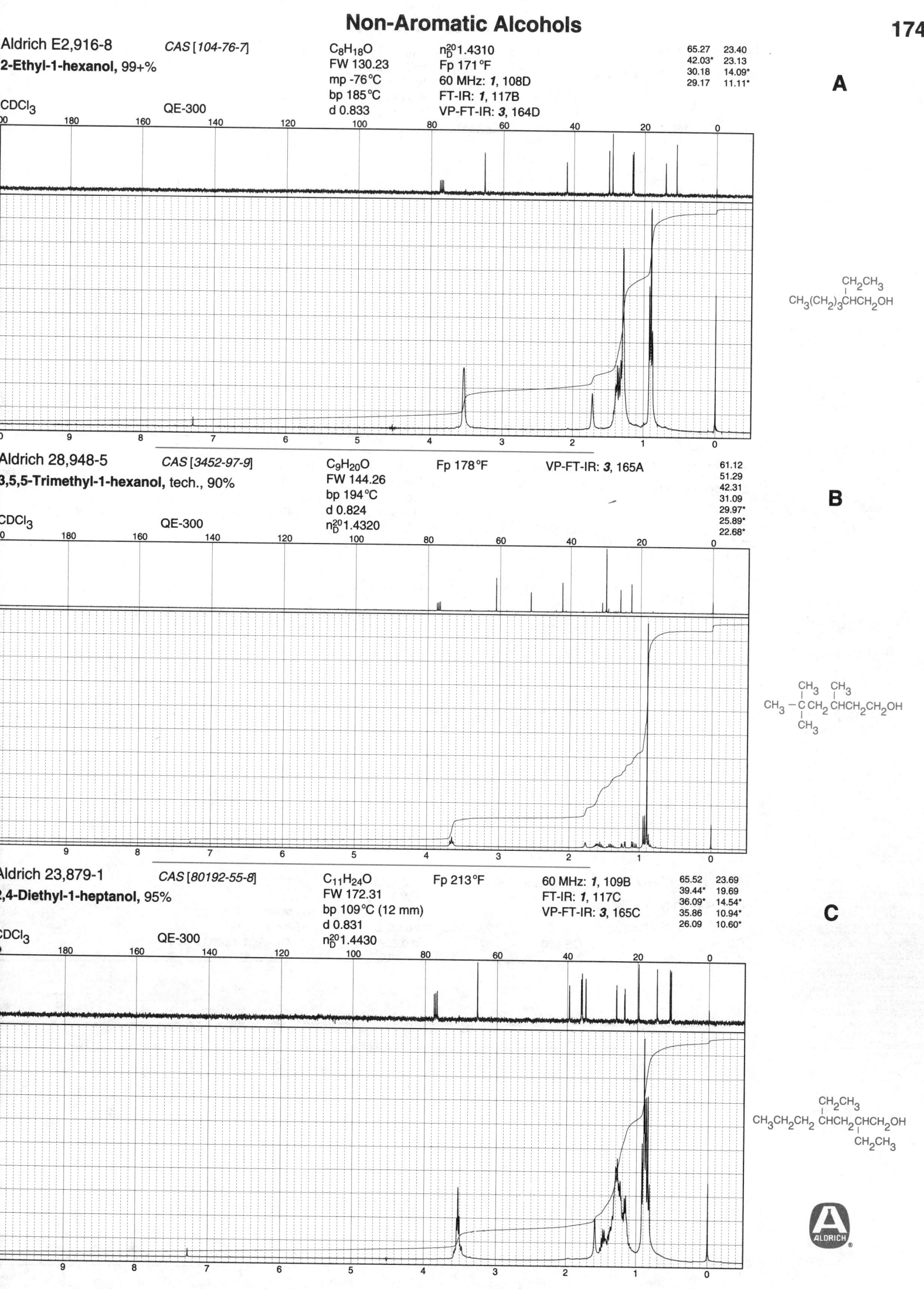

Aldrich E2,916-8 CAS [104-76-7]
2-Ethyl-1-hexanol, 99+%

$C_8H_{18}O$
FW 130.23
mp -76°C
bp 185°C
d 0.833

n_D^{20} 1.4310
Fp 171°F
60 MHz: *1*, 108D
FT-IR: *1*, 117B
VP-FT-IR: *3*, 164D

65.27	23.40
42.03*	23.13
30.18	14.09*
29.17	11.11*

A

CDCl$_3$ QE-300

CH_2CH_3
$CH_3(CH_2)_3CHCH_2OH$

Aldrich 28,948-5 CAS [3452-97-9]
3,5,5-Trimethyl-1-hexanol, tech., 90%

$C_9H_{20}O$
FW 144.26
bp 194°C
d 0.824
n_D^{20} 1.4320

Fp 178°F

VP-FT-IR: *3*, 165A

61.12
51.29
42.31
31.09
29.97*
25.89*
22.68*

B

CDCl$_3$ QE-300

CH_3 CH_3
$CH_3-C\,CH_2\,CHCH_2CH_2OH$
CH_3

Aldrich 23,879-1 CAS [80192-55-8]
2,4-Diethyl-1-heptanol, 95%

$C_{11}H_{24}O$
FW 172.31
bp 109°C (12 mm)
d 0.831
n_D^{20} 1.4430

Fp 213°F

60 MHz: *1*, 109B
FT-IR: *1*, 117C
VP-FT-IR: *3*, 165C

65.52	23.69
39.44*	19.69
36.09*	14.54*
35.86	10.94*
26.09	10.60*

C

CDCl$_3$ QE-300

CH_2CH_3
$CH_3CH_2CH_2\,CHCH_2\,CHCH_2OH$
CH_2CH_3

ALDRICH

Non-Aromatic Alcohols

175

A

Aldrich 30,577-4   *CAS [106-21-8]*   C$_{10}$H$_{22}$O   Fp 203°F   VP-FT-IR: **3**, 165D

3,7-Dimethyl-1-octanol, 99%

FW 158.29
bp 96°C (9 mm)
d 0.840
n$_D^{20}$1.4360

61.15	27.98*
40.02	24.70
39.30	22.69*
37.41	22.60*
29.56*	19.66*

CDCl$_3$ QE-300

CH$_3$CHCH$_2$CH$_2$CH$_2$CHCH$_2$CH$_2$OH (with CH$_3$ groups)

B

Aldrich 10,982-7   *CAS [67-63-0]*   C$_3$H$_8$O   n$_D^{20}$1.3770   64.18*
                                                                          25.29*

2-Propanol, 99+%

FW 60.10   Fp 60°F
mp -90°C   60 MHz: **1**, 109C
bp 83°C   FT-IR: **1**, 117D
d 0.785   VP-FT-IR: **3**, 166A

CDCl$_3$ QE-300

CH$_3$CHCH$_3$ (OH)

C

Aldrich 24,056-7   *CAS [15892-23-6]*   C$_4$H$_{10}$O   n$_D^{20}$1.3970   69.37*
                                                                              32.04
(±)-2-Butanol, 99+%   FW 74.12   Fp 80°F   22.87*
                           mp -115°C   60 MHz: **1**, 109D   9.97*
                           bp 98°C   FT-IR: **1**, 118A
                           d 0.808   VP-FT-IR: **3**, 166B

CDCl$_3$ QE-300

CH$_3$CH$_2$CHCH$_3$ (OH)

ALDRICH

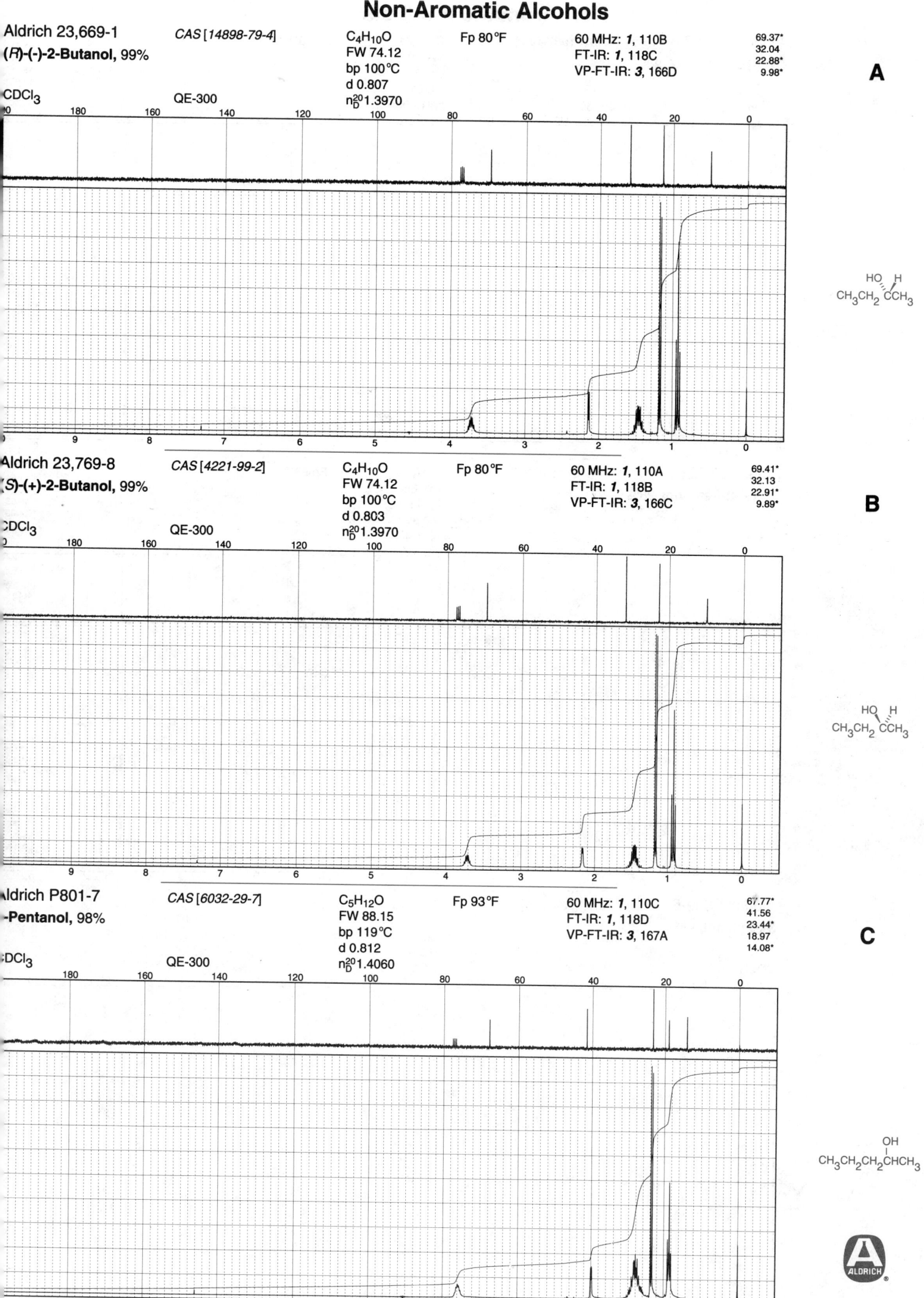

Aldrich 23,669-1　　CAS [14898-79-4]　　C₄H₁₀O　Fp 80°F　60 MHz: **1**, 110B　69.37*
(*R*)-(-)-2-Butanol, 99%　　FW 74.12　　FT-IR: **1**, 118C　32.04
　　bp 100°C　VP-FT-IR: **3**, 166D　22.88*
CDCl₃　QE-300　d 0.807　9.98*
　　n²⁰_D 1.3970

A

Aldrich 23,769-8　　CAS [4221-99-2]　　C₄H₁₀O　Fp 80°F　60 MHz: **1**, 110A　69.41*
(S)-(+)-2-Butanol, 99%　　FW 74.12　　FT-IR: **1**, 118B　32.13
　　bp 100°C　VP-FT-IR: **3**, 166C　22.91*
CDCl₃　QE-300　d 0.803　9.89*
　　n²⁰_D 1.3970

B

Aldrich P801-7　　CAS [6032-29-7]　　C₅H₁₂O　Fp 93°F　60 MHz: **1**, 110C　67.77*
-Pentanol, 98%　　FW 88.15　　FT-IR: **1**, 118D　41.56
　　bp 119°C　VP-FT-IR: **3**, 167A　23.44*
CDCl₃　QE-300　d 0.812　18.97
　　n²⁰_D 1.4060　14.08*

C

Non-Aromatic Alcohols

177

A

Aldrich 33,052-3

(*R*)-(-)-2-Pentanol, 98%

CAS [31087-44-2]

C$_5$H$_{12}$O
FW 88.15
bp 120°C
d 0.814
n$_D^{20}$1.4060

Fp 93°F

67.75*
41.51
23.42*
18.96
14.07*

CDCl$_3$ QE-300

CH$_3$CH$_2$CH$_2$CCH$_3$ (HO H)

B

Aldrich 33,051-5

(*S*)-(+)-2-Pentanol, 98%

CAS [26184-62-3]

C$_5$H$_{12}$O
FW 88.15
bp 119°C
d 0.810
n$_D^{20}$1.4060

Fp 93°F

67.77*
41.54
23.44*
18.98
14.09*

CDCl$_3$ QE-300

CH$_3$CH$_2$CH$_2$CCH$_3$ (HO H)

C

Aldrich P802-5

3-Pentanol, 98%

CAS [584-02-1]

C$_5$H$_{12}$O
FW 88.15
bp 115°C (749 mm)
d 0.815
n$_D^{20}$1.4100

Fp 105°F

60 MHz: *1*, 110D
FT-IR: *1*, 119A
VP-FT-IR: *3*, 167B

74.6*
29.74
9.81*

CDCl$_3$ QE-300

CH$_3$CH$_2$CHCH$_2$CH$_3$ (OH)

Non-Aromatic Alcohols

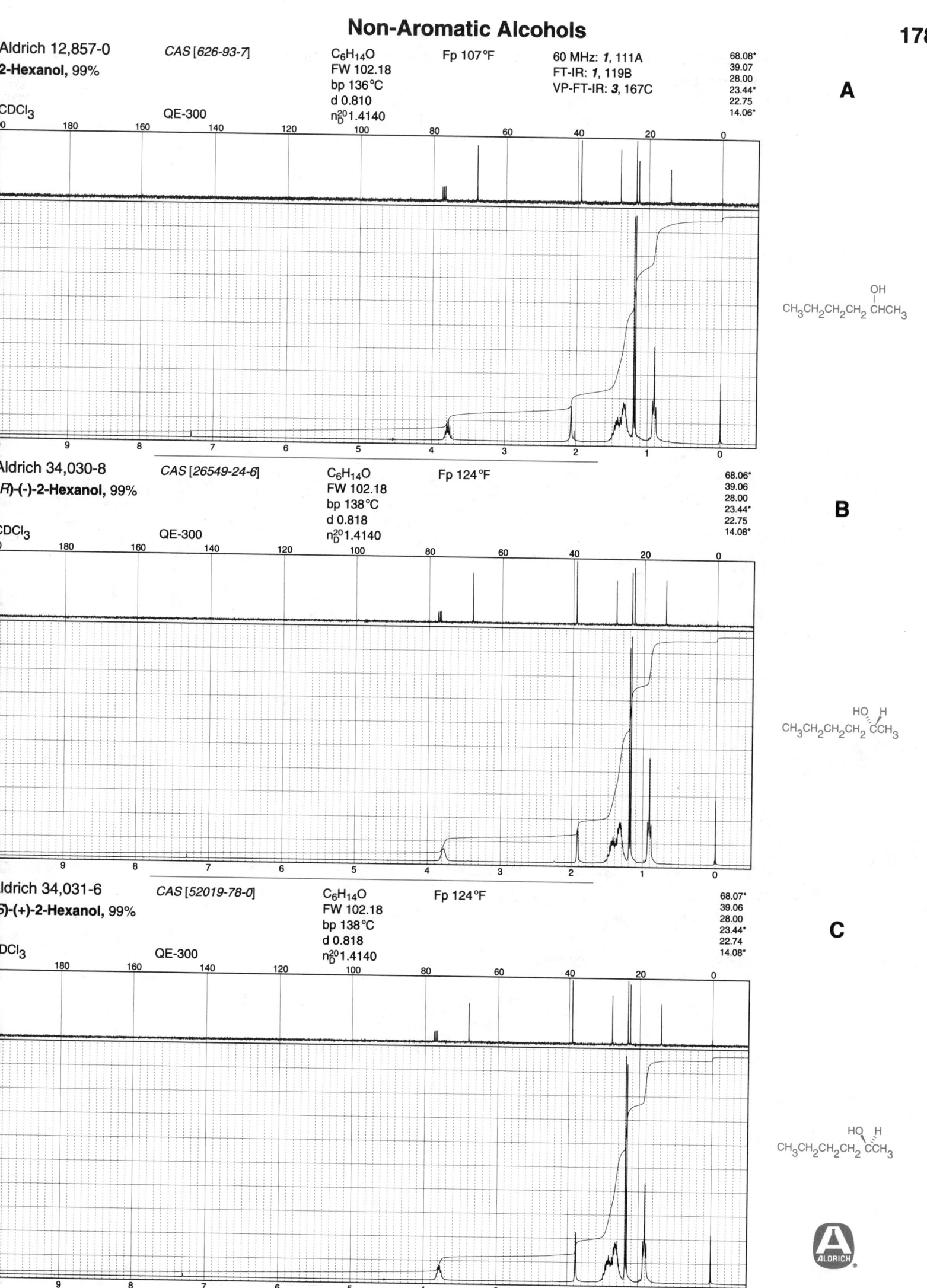

Aldrich 12,857-0
CAS [626-93-7]
2-Hexanol, 99%

CDCl₃   QE-300

$C_6H_{14}O$
FW 102.18
bp 136°C
d 0.810
n_D^{20} 1.4140

Fp 107°F

60 MHz: *1*, 111A
FT-IR: *1*, 119B
VP-FT-IR: *3*, 167C

68.08*
39.07
28.00
23.44*
22.75
14.06*

A

$CH_3CH_2CH_2CH_2\overset{\overset{\displaystyle OH}{|}}{C}HCH_3$

Aldrich 34,030-8
CAS [26549-24-6]
(R)-(-)-2-Hexanol, 99%

CDCl₃   QE-300

$C_6H_{14}O$
FW 102.18
bp 138°C
d 0.818
n_D^{20} 1.4140

Fp 124°F

68.06*
39.06
28.00
23.44*
22.75
14.08*

B

$CH_3CH_2CH_2CH_2\overset{\overset{\displaystyle HO}{}}{\underset{}{C}}\overset{\displaystyle H}{}CH_3$

Aldrich 34,031-6
CAS [52019-78-0]
(S)-(+)-2-Hexanol, 99%

DCl₃   QE-300

$C_6H_{14}O$
FW 102.18
bp 138°C
d 0.818
n_D^{20} 1.4140

Fp 124°F

68.07*
39.06
28.00
23.44*
22.74
14.08*

C

$CH_3CH_2CH_2CH_2\overset{\overset{\displaystyle HO}{}}{\underset{}{C}}\overset{\displaystyle H}{}CH_3$

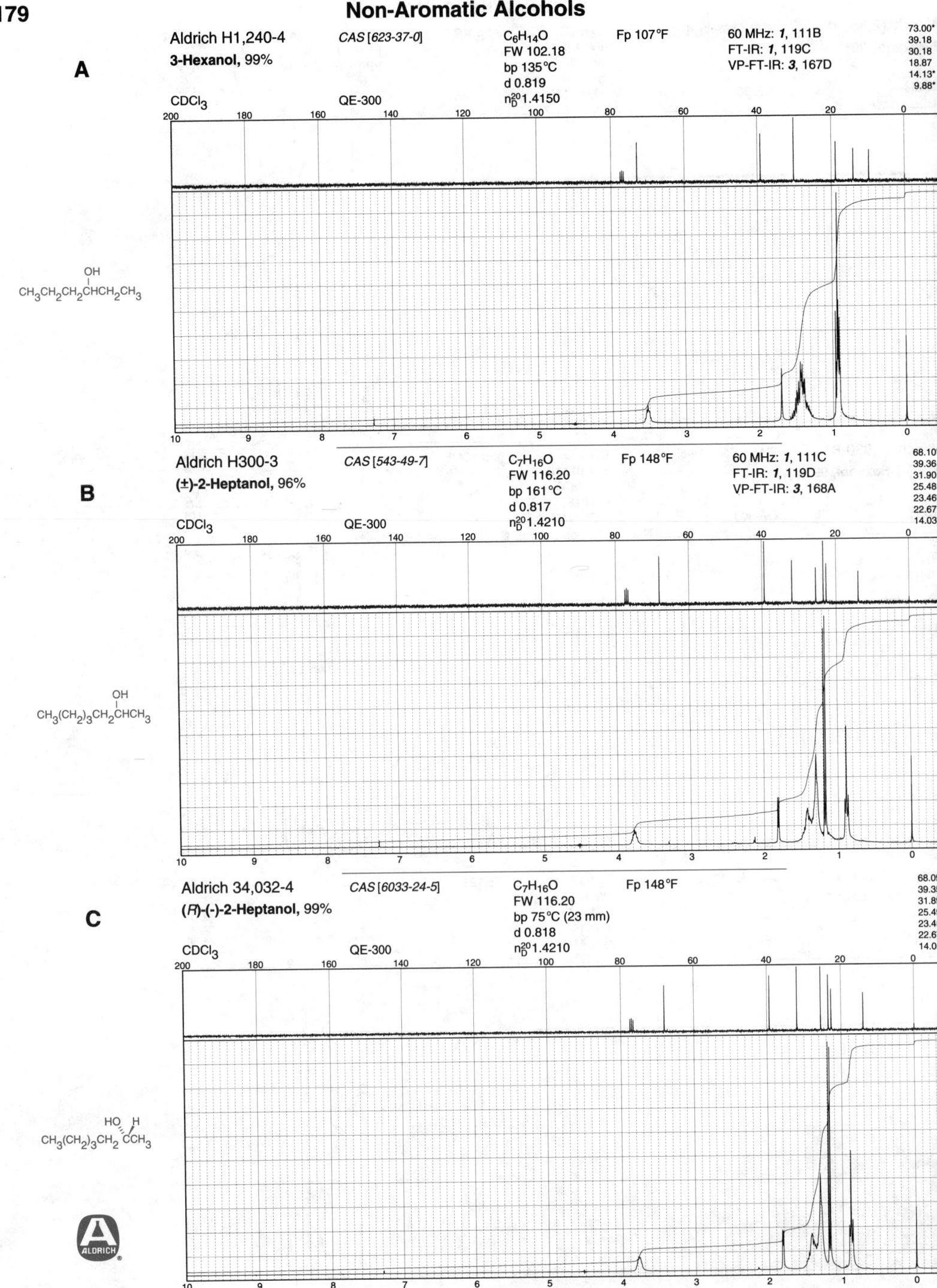

A

Aldrich H1,240-4

3-Hexanol, 99%

CAS [623-37-0]

C₆H₁₄O
FW 102.18
bp 135°C
d 0.819
n²⁰_D 1.4150

Fp 107°F

60 MHz: *1*, 111B
FT-IR: *1*, 119C
VP-FT-IR: *3*, 167D

73.00*
39.18
30.18
18.87
14.13*
9.88*

CDCl₃   QE-300

OH
|
CH₃CH₂CH₂CHCH₂CH₃

B

Aldrich H300-3

(±)-2-Heptanol, 96%

CAS [543-49-7]

C₇H₁₆O
FW 116.20
bp 161°C
d 0.817
n²⁰_D 1.4210

Fp 148°F

60 MHz: *1*, 111C
FT-IR: *1*, 119D
VP-FT-IR: *3*, 168A

68.10*
39.36
31.90
25.48
23.46*
22.67
14.03*

CDCl₃   QE-300

OH
|
CH₃(CH₂)₃CH₂CHCH₃

C

Aldrich 34,032-4

(R)-(-)-2-Heptanol, 99%

CAS [6033-24-5]

C₇H₁₆O
FW 116.20
bp 75°C (23 mm)
d 0.818
n²⁰_D 1.4210

Fp 148°F

68.09*
39.35
31.89
25.49
23.45
22.67
14.05

CDCl₃   QE-300

HO H
\ /
CH₃(CH₂)₃CH₂ CCH₃

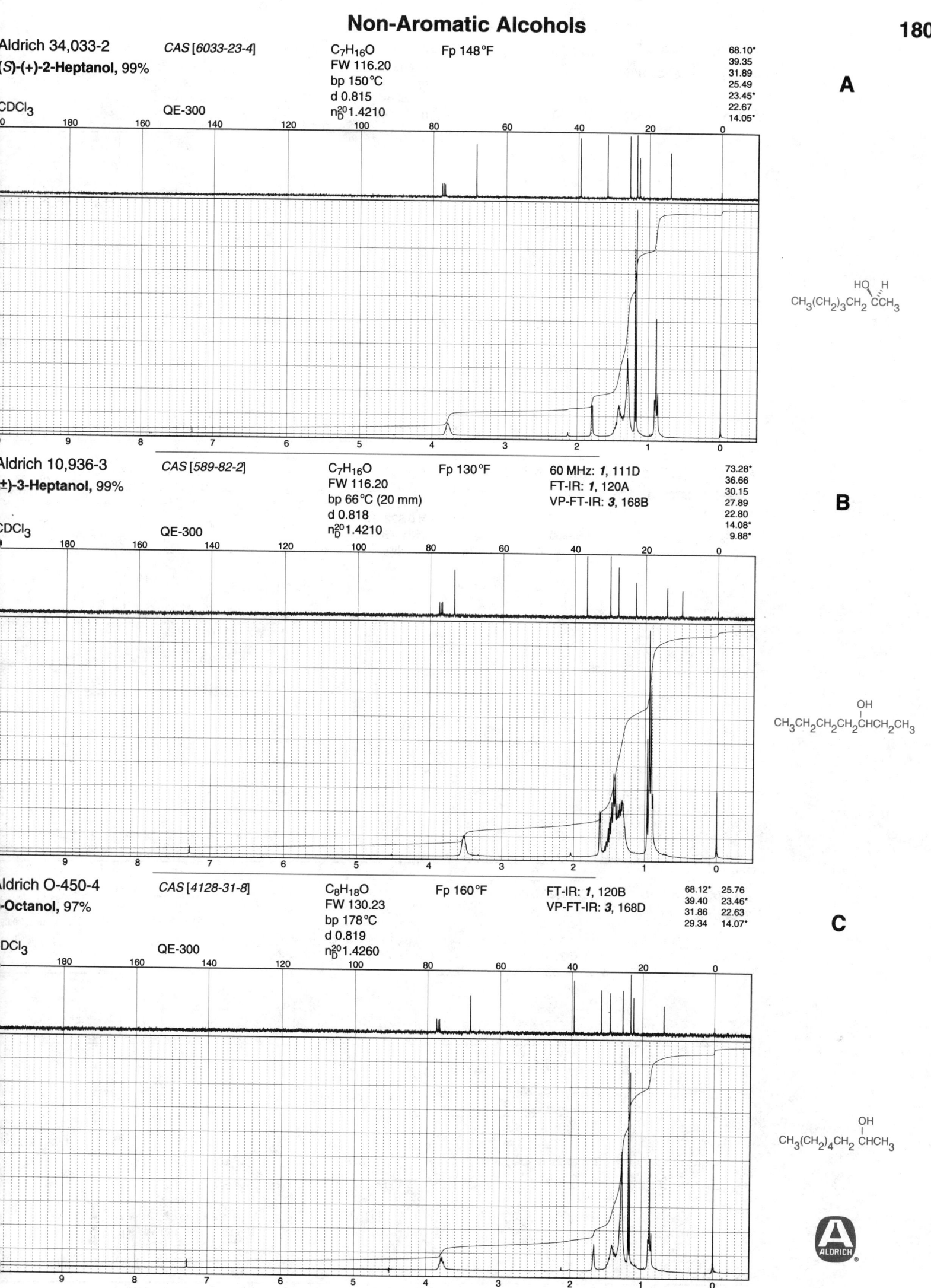

Aldrich 34,033-2   CAS [6033-23-4] $C_7H_{16}O$ Fp 148 °F

(S)-(+)-2-Heptanol, 99%

FW 116.20
bp 150 °C
d 0.815
n_D^{20} 1.4210

CDCl$_3$ QE-300

68.10*
39.35
31.89
25.49
23.45*
22.67
14.05*

A

$CH_3(CH_2)_3CH_2 \overset{HO \quad H}{\underset{}{CCH_3}}$

Aldrich 10,936-3   CAS [589-82-2] $C_7H_{16}O$ Fp 130 °F

(±)-3-Heptanol, 99%

FW 116.20
bp 66 °C (20 mm)
d 0.818
n_D^{20} 1.4210

CDCl$_3$ QE-300

60 MHz: **1**, 111D
FT-IR: **1**, 120A
VP-FT-IR: **3**, 168B

73.28*
36.66
30.15
27.89
22.80
14.08*
9.88*

B

$CH_3CH_2CH_2CH_2 \overset{OH}{\underset{}{CHCH_2CH_3}}$

Aldrich O-450-4   CAS [4128-31-8] $C_8H_{18}O$ Fp 160 °F

-Octanol, 97%

FW 130.23
bp 178 °C
d 0.819
n_D^{20} 1.4260

CDCl$_3$ QE-300

FT-IR: **1**, 120B
VP-FT-IR: **3**, 168D

68.12*   25.76
39.40   23.46*
31.86   22.63
29.34   14.07*

C

$CH_3(CH_2)_4CH_2 \overset{OH}{\underset{}{CHCH_3}}$

ALDRICH

A

Aldrich 14,799-0

(*R*)-(-)-2-Octanol, 99%

CAS [5978-70-1]

C₈H₁₈O
FW 130.23
bp 175°C
d 0.838
n²⁰_D 1.4260

Fp 160°F

60 MHz: *1*, 112C
FT-IR: *1*, 120C
VP-FT-IR: *3*, 169B

68.11*	25.77
39.40	23.46*
31.87	22.63
29.35	14.07*

CDCl₃   QE-300

B

Aldrich 14,798-2

(*S*)-(+)-2-Octanol, 99%

CAS [6169-06-8]

C₈H₁₈O
FW 130.23
bp 175°C
d 0.822
n²⁰_D 1.4260

Fp 160°F

60 MHz: *1*, 112D
FT-IR: *1*, 120D
VP-FT-IR: *3*, 169A

68.12*	25.77
39.40	23.46*
31.87	22.63
29.35	14.08*

CDCl₃   QE-300

C

Aldrich 21,840-5

3-Octanol, 99%

CAS [20296-29-1]

C₈H₁₈O
FW 130.23
bp 175°C
d 0.819
n²⁰_D 1.4260

Fp 150°F

60 MHz: *1*, 112B
FT-IR: *1*, 121A
VP-FT-IR: *3*, 169C

73.29*	25.39
36.96	22.69
31.98	14.05
30.16	9.89

CDCl₃   QE-300

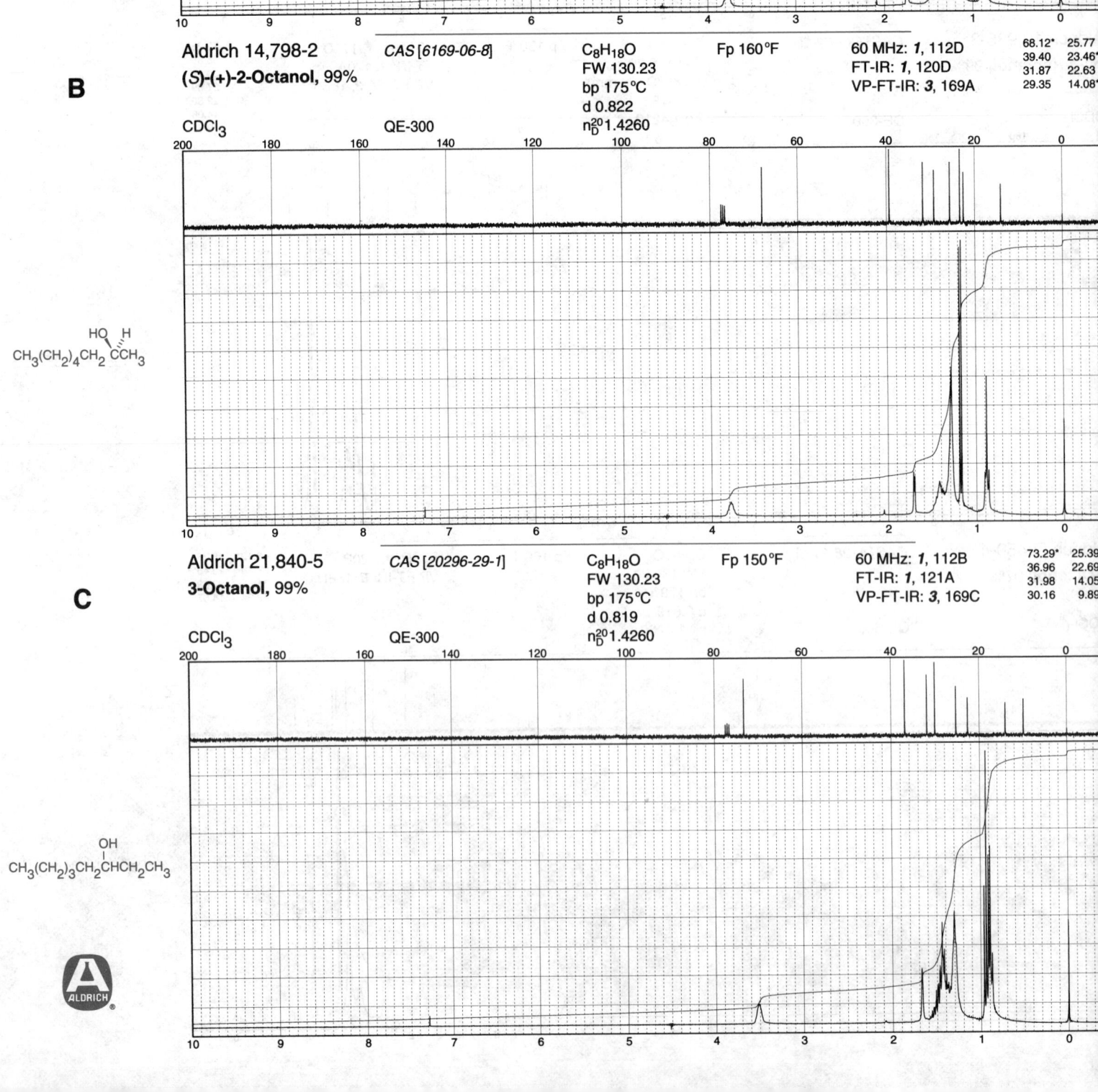

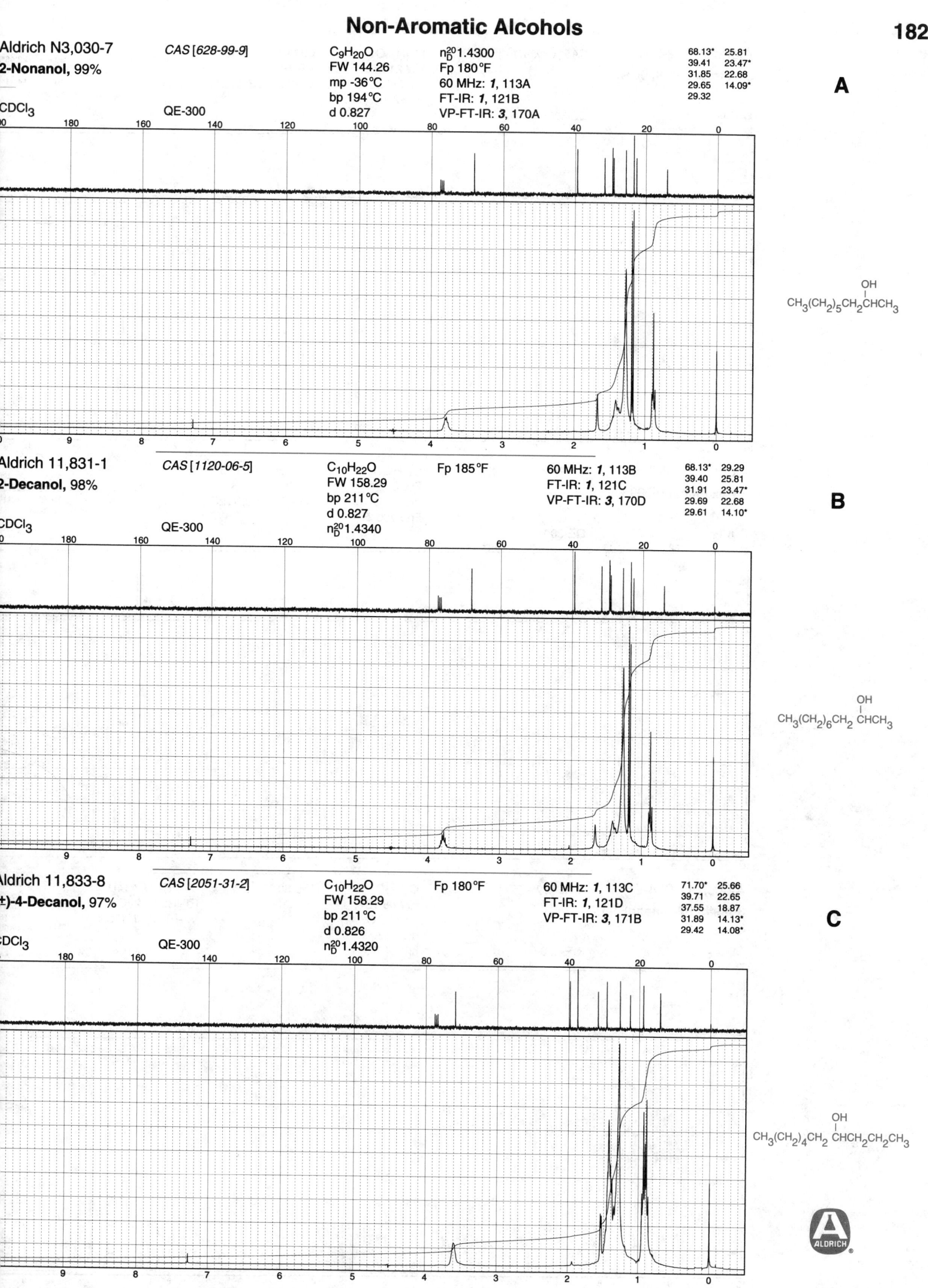

Aldrich N3,030-7
2-Nonanol, 99%
CDCl₃   QE-300

CAS [628-99-9]

C₉H₂₀O
FW 144.26
mp -36°C
bp 194°C
d 0.827

n_D^{20} 1.4300
Fp 180°F
60 MHz: *1*, 113A
FT-IR: *1*, 121B
VP-FT-IR: *3*, 170A

68.13*   25.81
39.41   23.47*
31.85   22.68
29.65   14.09*
29.32

A

$CH_3(CH_2)_5 CH_2 CHCH_3$ (OH)

Aldrich 11,831-1
2-Decanol, 98%
CDCl₃   QE-300

CAS [1120-06-5]

C₁₀H₂₂O
FW 158.29
bp 211°C
d 0.827
n_D^{20} 1.4340

Fp 185°F

60 MHz: *1*, 113B
FT-IR: *1*, 121C
VP-FT-IR: *3*, 170D

68.13*   29.29
39.40   25.81
31.91   23.47*
29.69   22.68
29.61   14.10*

B

$CH_3(CH_2)_6 CH_2 CHCH_3$ (OH)

Aldrich 11,833-8
(±)-4-Decanol, 97%
CDCl₃   QE-300

CAS [2051-31-2]

C₁₀H₂₂O
FW 158.29
bp 211°C
d 0.826
n_D^{20} 1.4320

Fp 180°F

60 MHz: *1*, 113C
FT-IR: *1*, 121D
VP-FT-IR: *3*, 171B

71.70*   25.66
39.71   22.65
37.55   18.87
31.89   14.13*
29.42   14.08*

C

$CH_3(CH_2)_4 CH_2 CHCH_2CH_2CH_3$ (OH)

Non-Aromatic Alcohols

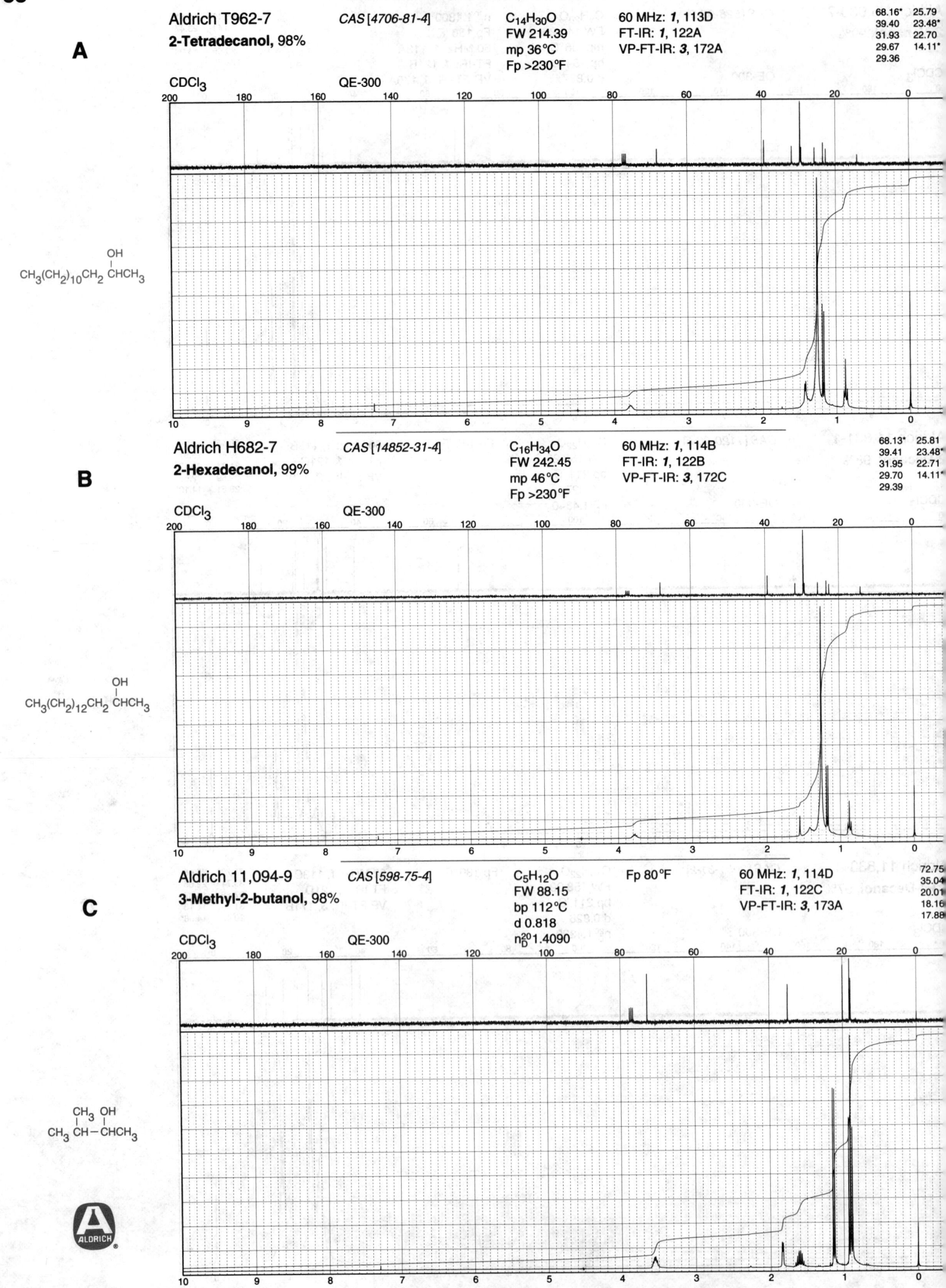

A

Aldrich T962-7

2-Tetradecanol, 98%

CAS [4706-81-4]

C₁₄H₃₀O
FW 214.39
mp 36°C
Fp >230°F

60 MHz: *1*, 113D
FT-IR: *1*, 122A
VP-FT-IR: *3*, 172A

68.16* 25.79
39.40* 23.48*
31.93 22.70
29.67 14.11*
29.36

CDCl₃   QE-300

$CH_3(CH_2)_{10}CH_2-CHCH_3$ (OH)

B

Aldrich H682-7

2-Hexadecanol, 99%

CAS [14852-31-4]

C₁₆H₃₄O
FW 242.45
mp 46°C
Fp >230°F

60 MHz: *1*, 114B
FT-IR: *1*, 122B
VP-FT-IR: *3*, 172C

68.13* 25.81
39.41 23.48*
31.95 22.71
29.70 14.11*
29.39

CDCl₃   QE-300

$CH_3(CH_2)_{12}CH_2-CHCH_3$ (OH)

C

Aldrich 11,094-9

3-Methyl-2-butanol, 98%

CAS [598-75-4]

C₅H₁₂O
FW 88.15
bp 112°C
d 0.818
n²⁰_D 1.4090

Fp 80°F

60 MHz: *1*, 114D
FT-IR: *1*, 122C
VP-FT-IR: *3*, 173A

72.75
35.04
20.01
18.16
17.88

CDCl₃   QE-300

$CH_3-CH-CHCH_3$ (CH₃)(OH)

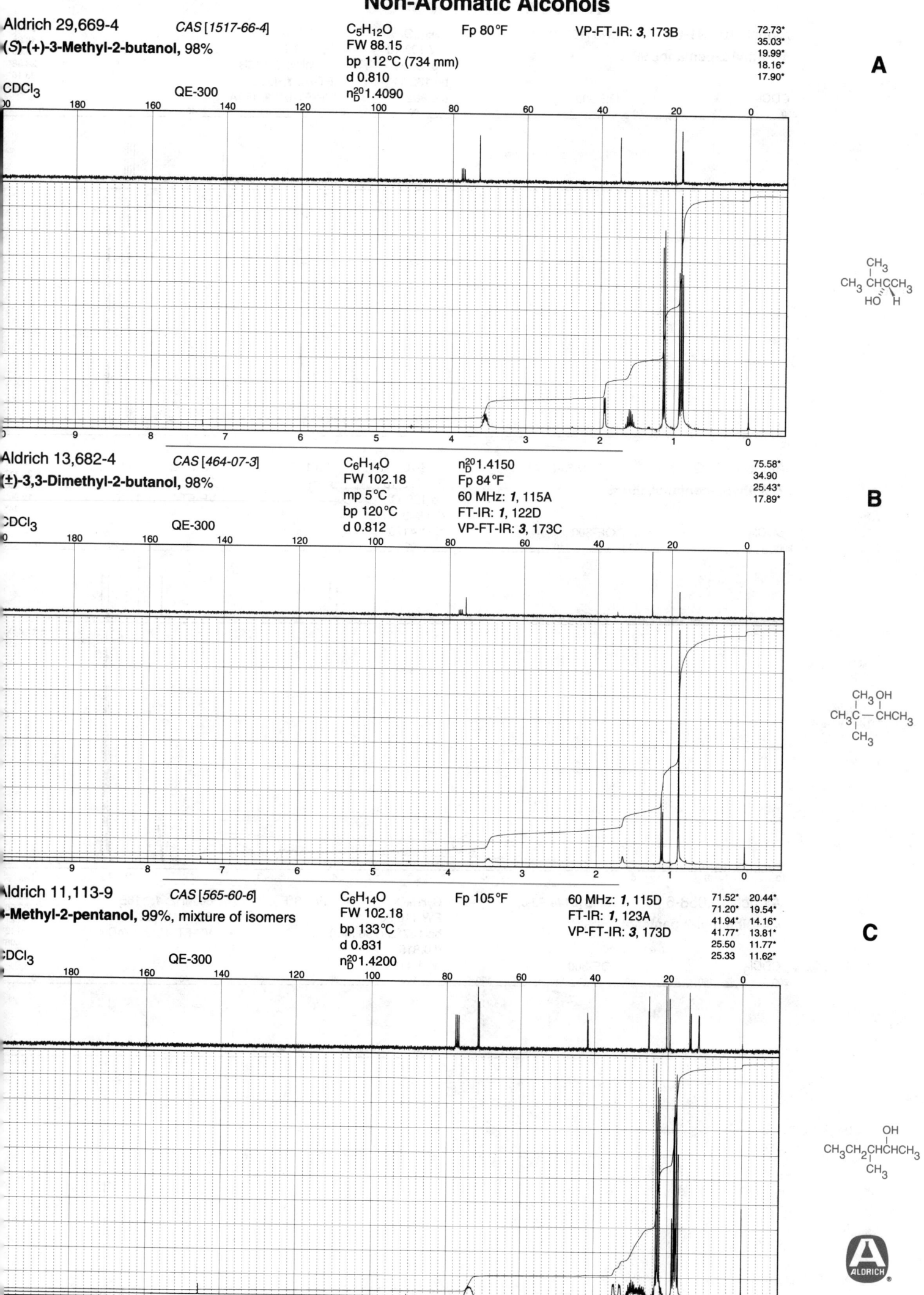

Aldrich 29,669-4   CAS [1517-66-4]   C5H12O   Fp 80°F   VP-FT-IR: 3, 173B
(S)-(+)-3-Methyl-2-butanol, 98%   FW 88.15
bp 112°C (734 mm)
d 0.810
n_D^{20} 1.4090

72.73*
35.03*
19.99*
18.16*
17.90*

A

CDCl3   QE-300

Aldrich 13,682-4   CAS [464-07-3]   C6H14O n_D^{20} 1.4150   75.58*
(±)-3,3-Dimethyl-2-butanol, 98%   FW 102.18   Fp 84°F   34.90
mp 5°C   60 MHz: 1, 115A   25.43*
bp 120°C   FT-IR: 1, 122D   17.89*
d 0.812   VP-FT-IR: 3, 173C

B

CDCl3   QE-300

Aldrich 11,113-9   CAS [565-60-6]   C6H14O   Fp 105°F   60 MHz: 1, 115D   71.52* 20.44*
4-Methyl-2-pentanol, 99%, mixture of isomers   FW 102.18   FT-IR: 1, 123A   71.20* 19.54*
bp 133°C   VP-FT-IR: 3, 173D   41.94* 14.16*
d 0.831   41.77* 13.81*
n_D^{20} 1.4200   25.50 11.77*
25.33 11.62*

C

CDCl3   QE-300

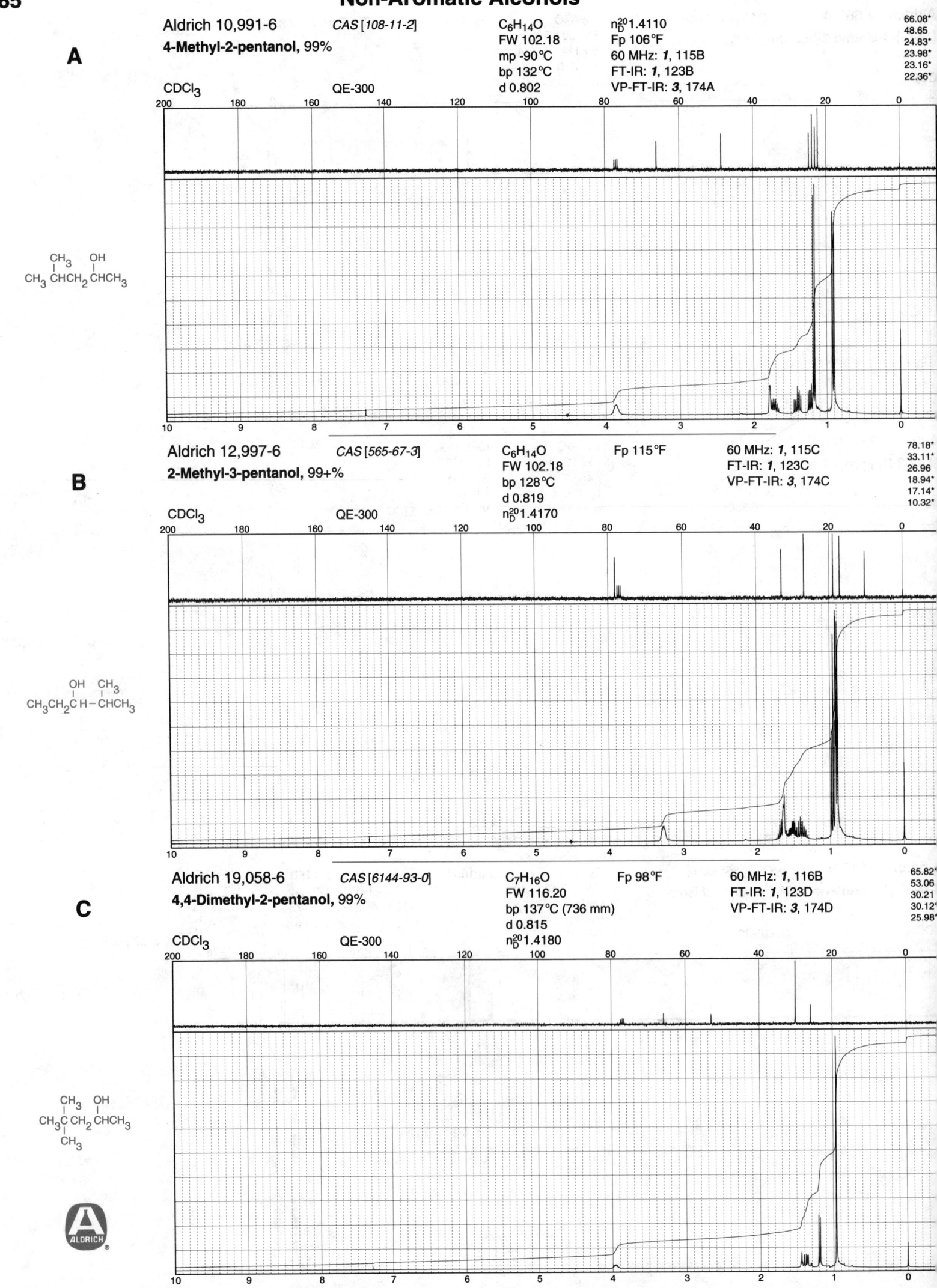

A

Aldrich 10,991-6   CAS [108-11-2]

4-Methyl-2-pentanol, 99%

$C_6H_{14}O$
FW 102.18
mp -90°C
bp 132°C
d 0.802

n_D^{20}1.4110
Fp 106°F
60 MHz: **1**, 115B
FT-IR: **1**, 123B
VP-FT-IR: **3**, 174A

66.08*
48.65
24.83*
23.98*
23.16*
22.36*

CDCl₃   QE-300

B

Aldrich 12,997-6   CAS [565-67-3]

2-Methyl-3-pentanol, 99+%

$C_6H_{14}O$
FW 102.18
bp 128°C
d 0.819
n_D^{20}1.4170

Fp 115°F

60 MHz: **1**, 115C
FT-IR: **1**, 123C
VP-FT-IR: **3**, 174C

78.18*
33.11*
26.96
18.94*
17.14*
10.32*

CDCl₃   QE-300

C

Aldrich 19,058-6   CAS [6144-93-0]

4,4-Dimethyl-2-pentanol, 99%

$C_7H_{16}O$
FW 116.20
bp 137°C (736 mm)
d 0.815
n_D^{20}1.4180

Fp 98°F

60 MHz: **1**, 116B
FT-IR: **1**, 123D
VP-FT-IR: **3**, 174D

65.82
53.06
30.21
30.12
25.98

CDCl₃   QE-300

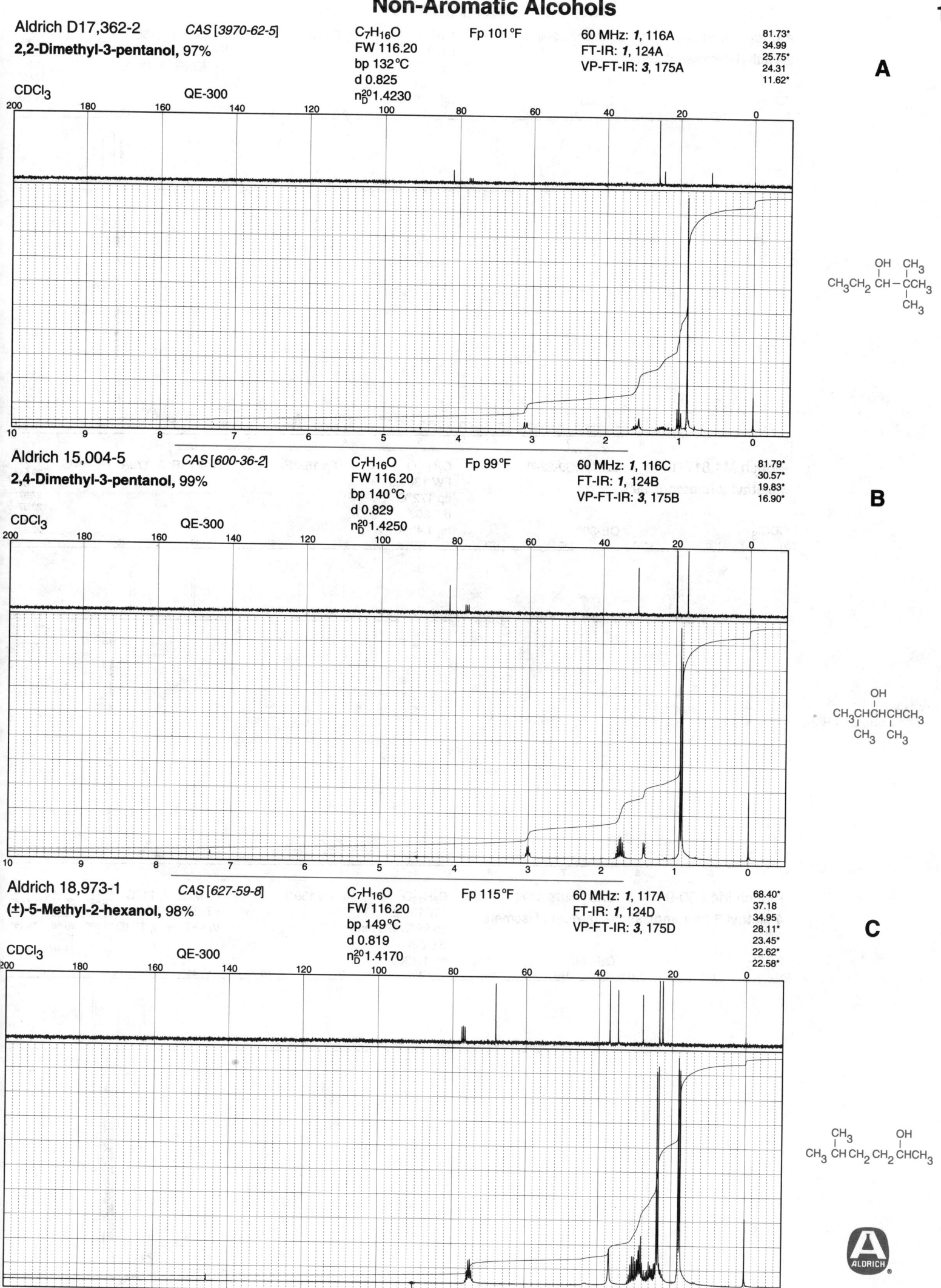

Aldrich D17,362-2 *CAS [3970-62-5]* C₇H₁₆O Fp 101°F 60 MHz: *1*, 116A

2,2-Dimethyl-3-pentanol, 97%

$C_7H_{16}O$
FW 116.20
bp 132°C
d 0.825
n_D^{20} 1.4230

Fp 101°F

60 MHz: *1*, 116A
FT-IR: *1*, 124A
VP-FT-IR: *3*, 175A

81.73*
34.99
25.75*
24.31
11.62*

CDCl₃   QE-300

A

Aldrich 15,004-5 *CAS [600-36-2]*

2,4-Dimethyl-3-pentanol, 99%

$C_7H_{16}O$
FW 116.20
bp 140°C
d 0.829
n_D^{20} 1.4250

Fp 99°F

60 MHz: *1*, 116C
FT-IR: *1*, 124B
VP-FT-IR: *3*, 175B

81.79*
30.57*
19.83*
16.90*

CDCl₃   QE-300

B

Aldrich 18,973-1 *CAS [627-59-8]*

(±)-5-Methyl-2-hexanol, 98%

$C_7H_{16}O$
FW 116.20
bp 149°C
d 0.819
n_D^{20} 1.4170

Fp 115°F

60 MHz: *1*, 117A
FT-IR: *1*, 124D
VP-FT-IR: *3*, 175D

68.40*
37.18
34.95
28.11*
23.45*
22.62*
22.58*

CDCl₃   QE-300

C

A

Aldrich M4,983-6   *CAS [617-29-8]* $C_7H_{16}O$ Fp 105°F   60 MHz: *1*, 116D

2-Methyl-3-hexanol, 98%

FW 116.20   FT-IR: *1*, 124C

bp 142°C   VP-FT-IR: *3*, 176A

d 0.821

n_D^{20} 1.4210

CDCl$_3$ QE-300

76.44*
36.39
33.54*
19.24
18.86*
17.13*
14.16*

$CH_3CH_2CH_2$ $\underset{|}{CH}$ $-$ $\underset{|}{CH}CH_3$ (with OH and CH₃ groups)

B

Aldrich M4,817-1   *CAS [4730-22-7]* $C_8H_{18}O$ Fp 153°F   VP-FT-IR: *3*, 178A

6-Methyl-2-heptanol, 99%

FW 130.23

bp 172°C

d 0.803

n_D^{20} 1.4240

CDCl$_3$ QE-300

68.11*
39.64
39.00
27.97*
23.57
23.47*
22.59*

CH_3 $\underset{|}{CH}CH_2CH_2CH_2$ $\underset{|}{CH}CH_3$ (with CH₃ and OH groups)

C

Aldrich M4,830-9   *CAS [14979-39-6]* $C_8H_{18}O$ Fp 130°F   60 MHz: *1*, 117C

4-Methyl-3-heptanol, 99+%, mixture of isomers

FW 130.23   FT-IR: *1*, 125A

bp 99°C (75 mm)   VP-FT-IR: *3*, 178D

d 0.827

n_D^{20} 1.4300

CDCl$_3$ QE-300

77.52*	27.29	14.35*
76.69*	26.24	13.56*
38.30*	20.50	10.62*
37.59*	20.46	10.45*
35.72	15.34*	
34.22	14.40*	

$CH_3CH_2CH_2$ $\underset{|}{CH}$ $-$ $\underset{|}{CH}CH_2CH_3$ (with CH₃ and OH groups)

Non-Aromatic Alcohols

Aldrich 29,297-4   *CAS [108-82-7]*   C₉H₂₀O   Fp 151 °F   VP-FT-IR: *3*, 180A

2,6-Dimethyl-4-heptanol, tech., 90%

$C_9H_{20}O$ Fp 151 °F   VP-FT-IR: *3*, 180A

FW 144.26

bp 178 °C

d 0.809

n_D^{20} 1.4230

67.99*
47.45
24.63*
23.47*
22.14*

A

CDCl₃   QE-300

$CH_3CHCH_2CHCH_2CHCH_3$ with CH_3, OH, CH_3 substituents

Aldrich B8,592-7   *CAS [75-65-0]*   C₄H₁₀O n_D^{20} 1.3870

2-Methyl-2-propanol, 99.5%

FW 74.12   Fp 52 °F

mp 25 °C   60 MHz: *1*, 117D

bp 83 °C   FT-IR: *1*, 125B

d 0.776   VP-FT-IR: *3*, 181D

69.11
31.23*

B

CDCl₃   QE-300

$CH_3 - C(CH_3)(CH_3) - OH$

Aldrich 21,312-8   *CAS [75-91-2]*   C₄H₁₀O₂   FT-IR: *1*, 125C

tert-Butyl hydroperoxide, 90%

FW 90.12   VP-FT-IR: *3*, 182A

d 0.901

n_D^{20} 1.3960

Fp 95 °F

80.92
25.78*

C

CDCl₃   QE-300

$CH_3 - C(CH_3)(CH_3) - OOH$

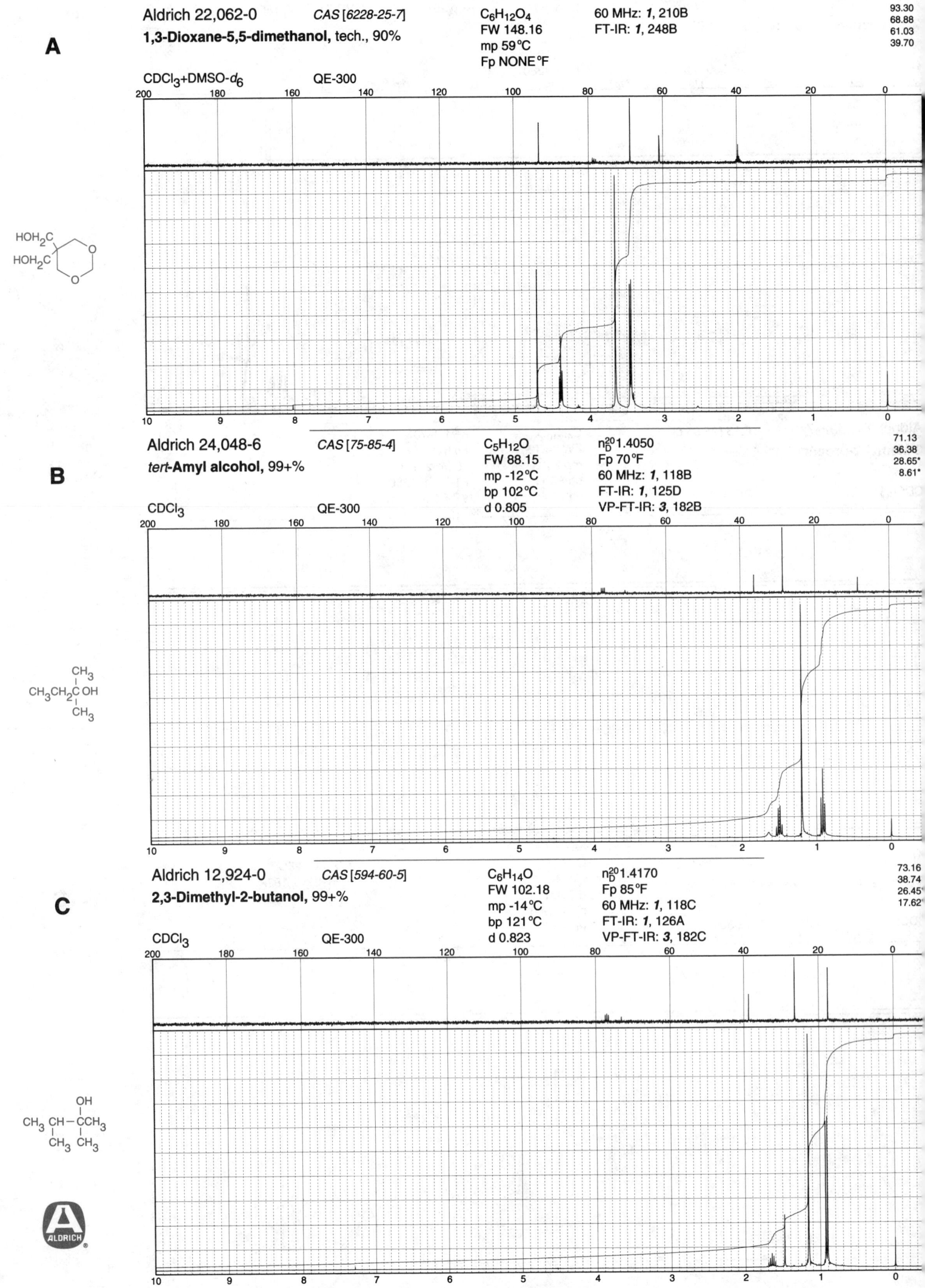

A

Aldrich 22,062-0   *CAS [6228-25-7]* $C_6H_{12}O_4$ 60 MHz: *1*, 210B

1,3-Dioxane-5,5-dimethanol, tech., 90%

FW 148.16
mp 59°C
Fp NONE °F

FT-IR: *1*, 248B

93.30
68.88
61.03
39.70

$CDCl_3+DMSO$-d_6 QE-300

B

Aldrich 24,048-6   *CAS [75-85-4]* $C_5H_{12}O$ n_D^{20}1.4050

tert-**Amyl alcohol, 99+%**

FW 88.15
mp -12°C
bp 102°C
d 0.805

Fp 70°F
60 MHz: *1*, 118B
FT-IR: *1*, 125D
VP-FT-IR: *3*, 182B

71.13
36.38
28.65*
8.61*

$CDCl_3$ QE-300

C

Aldrich 12,924-0   *CAS [594-60-5]* $C_6H_{14}O$ n_D^{20}1.4170

2,3-Dimethyl-2-butanol, 99+%

FW 102.18
mp -14°C
bp 121°C
d 0.823

Fp 85°F
60 MHz: *1*, 118C
FT-IR: *1*, 126A
VP-FT-IR: *3*, 182C

73.16
38.74
26.45*
17.62*

$CDCl_3$ QE-300

Non-Aromatic Alcohols

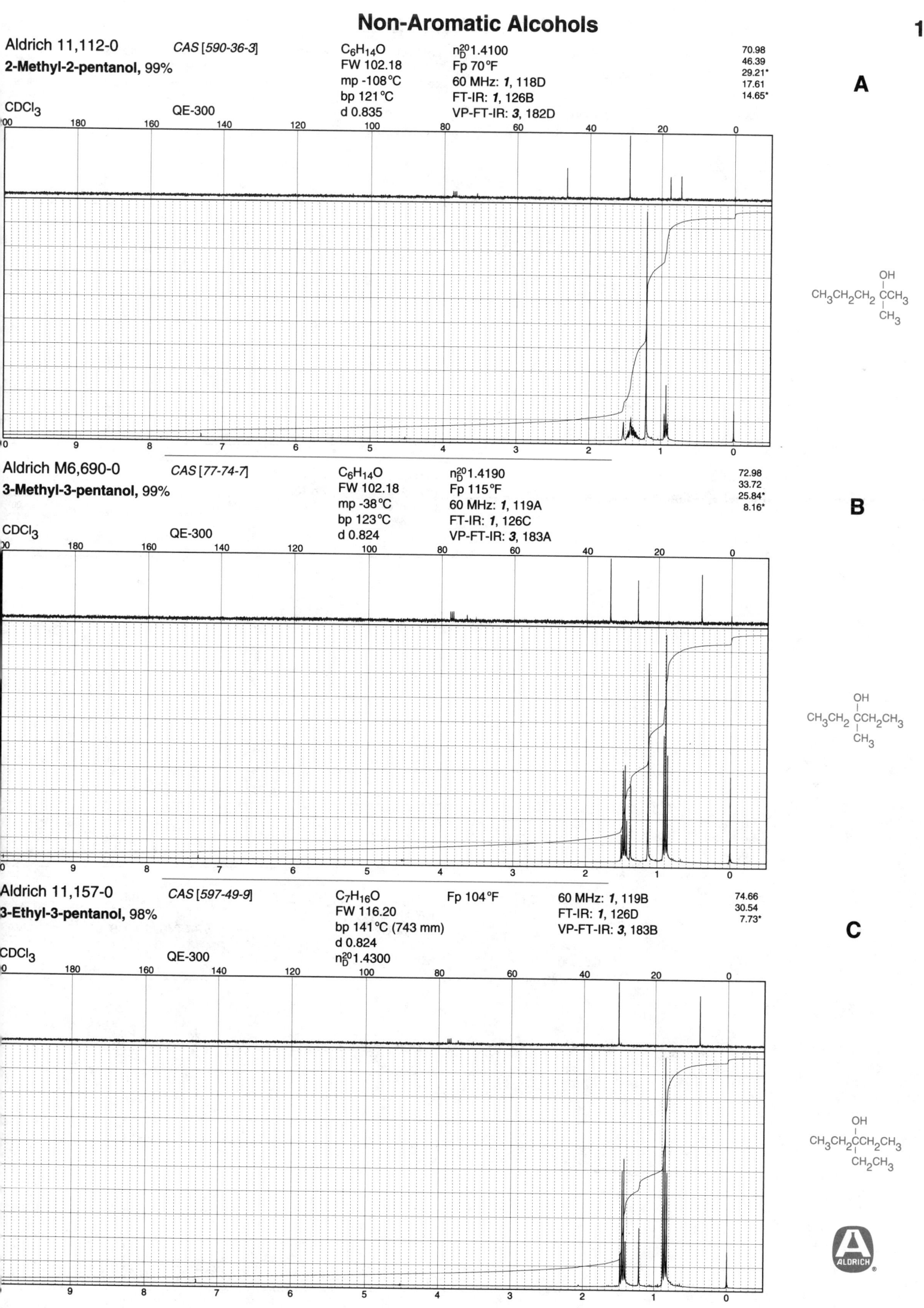

Aldrich 11,112-0   *CAS [590-36-3]*
2-Methyl-2-pentanol, 99%

CDCl₃   QE-300

C₆H₁₄O
FW 102.18
mp -108°C
bp 121°C
d 0.835

n²⁰_D 1.4100
Fp 70°F
60 MHz: *1*, 118D
FT-IR: *1*, 126B
VP-FT-IR: *3*, 182D

70.98
46.39
29.21*
17.61
14.65*

A

CH₃CH₂CH₂CCH₃ with OH and CH₃

Aldrich M6,690-0   *CAS [77-74-7]*
3-Methyl-3-pentanol, 99%

CDCl₃   QE-300

C₆H₁₄O
FW 102.18
mp -38°C
bp 123°C
d 0.824

n²⁰_D 1.4190
Fp 115°F
60 MHz: *1*, 119A
FT-IR: *1*, 126C
VP-FT-IR: *3*, 183A

72.98
33.72
25.84*
8.16*

B

CH₃CH₂CCH₂CH₃ with OH and CH₃

Aldrich 11,157-0   *CAS [597-49-9]*
3-Ethyl-3-pentanol, 98%

CDCl₃   QE-300

C₇H₁₆O
FW 116.20
bp 141°C (743 mm)
d 0.824
n²⁰_D 1.4300

Fp 104°F

60 MHz: *1*, 119B
FT-IR: *1*, 126D
VP-FT-IR: *3*, 183B

74.66
30.54
7.73*

C

CH₃CH₂CCH₂CH₃ with OH and CH₂CH₃

ALDRICH®

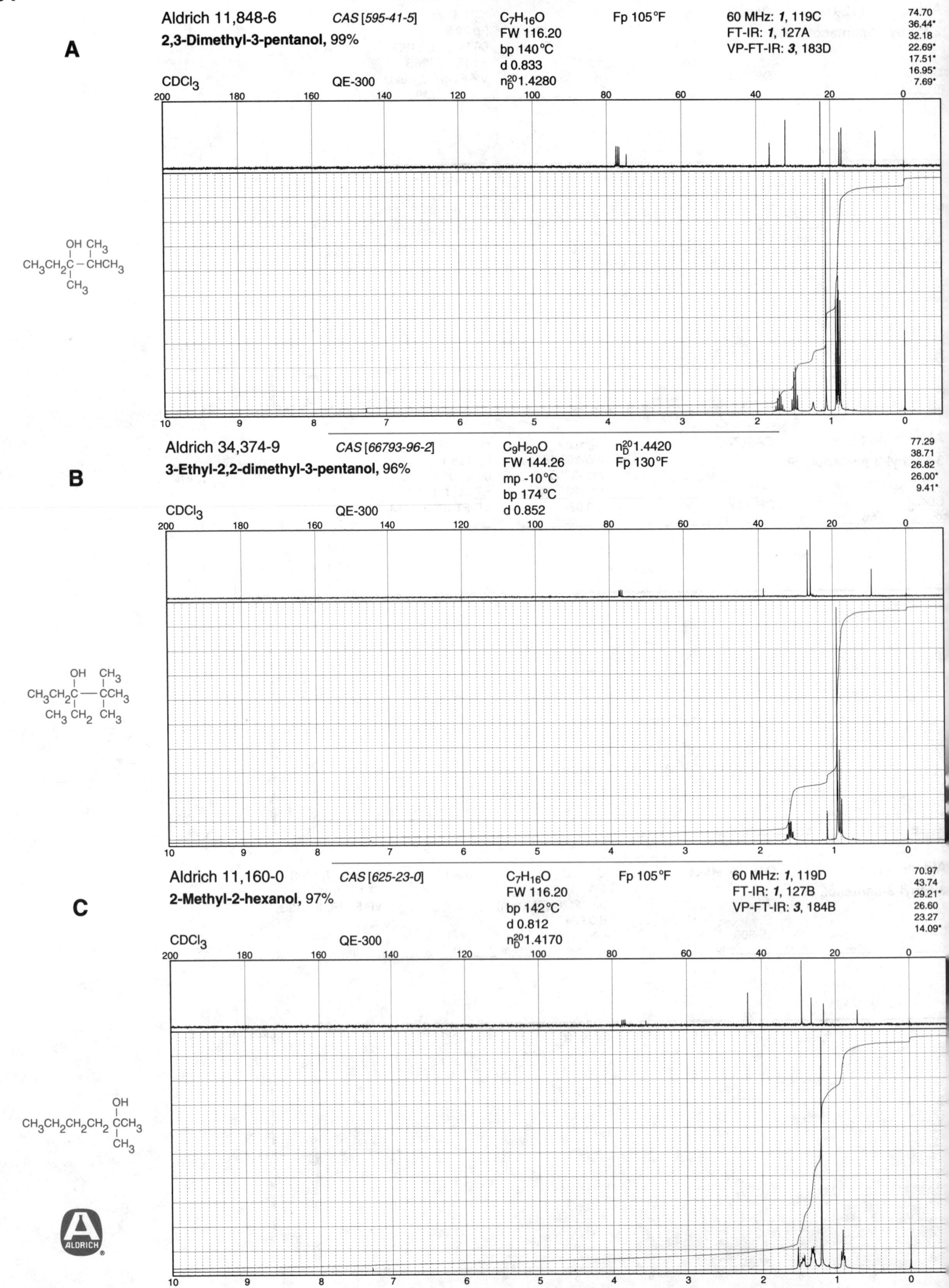

A

Aldrich 11,848-6   CAS [595-41-5] $C_7H_{16}O$ Fp 105°F   60 MHz: *1*, 119C

2,3-Dimethyl-3-pentanol, 99%
FW 116.20
bp 140°C
d 0.833
n_D^{20}1.4280

FT-IR: *1*, 127A
VP-FT-IR: *3*, 183D

74.70
36.44*
32.18
22.69*
17.51*
16.95*
7.69*

CDCl₃   QE-300

B

Aldrich 34,374-9   CAS [66793-96-2] $C_9H_{20}O$ n_D^{20}1.4420

3-Ethyl-2,2-dimethyl-3-pentanol, 96%
FW 144.26
mp -10°C
bp 174°C
d 0.852

Fp 130°F

77.29
38.71
26.82
26.00*
9.41*

CDCl₃   QE-300

C

Aldrich 11,160-0   CAS [625-23-0] $C_7H_{16}O$ Fp 105°F   60 MHz: *1*, 119D

2-Methyl-2-hexanol, 97%
FW 116.20
bp 142°C
d 0.812
n_D^{20}1.4170

FT-IR: *1*, 127B
VP-FT-IR: *3*, 184B

70.97
43.74
29.21*
26.60
23.27
14.09*

CDCl₃   QE-300

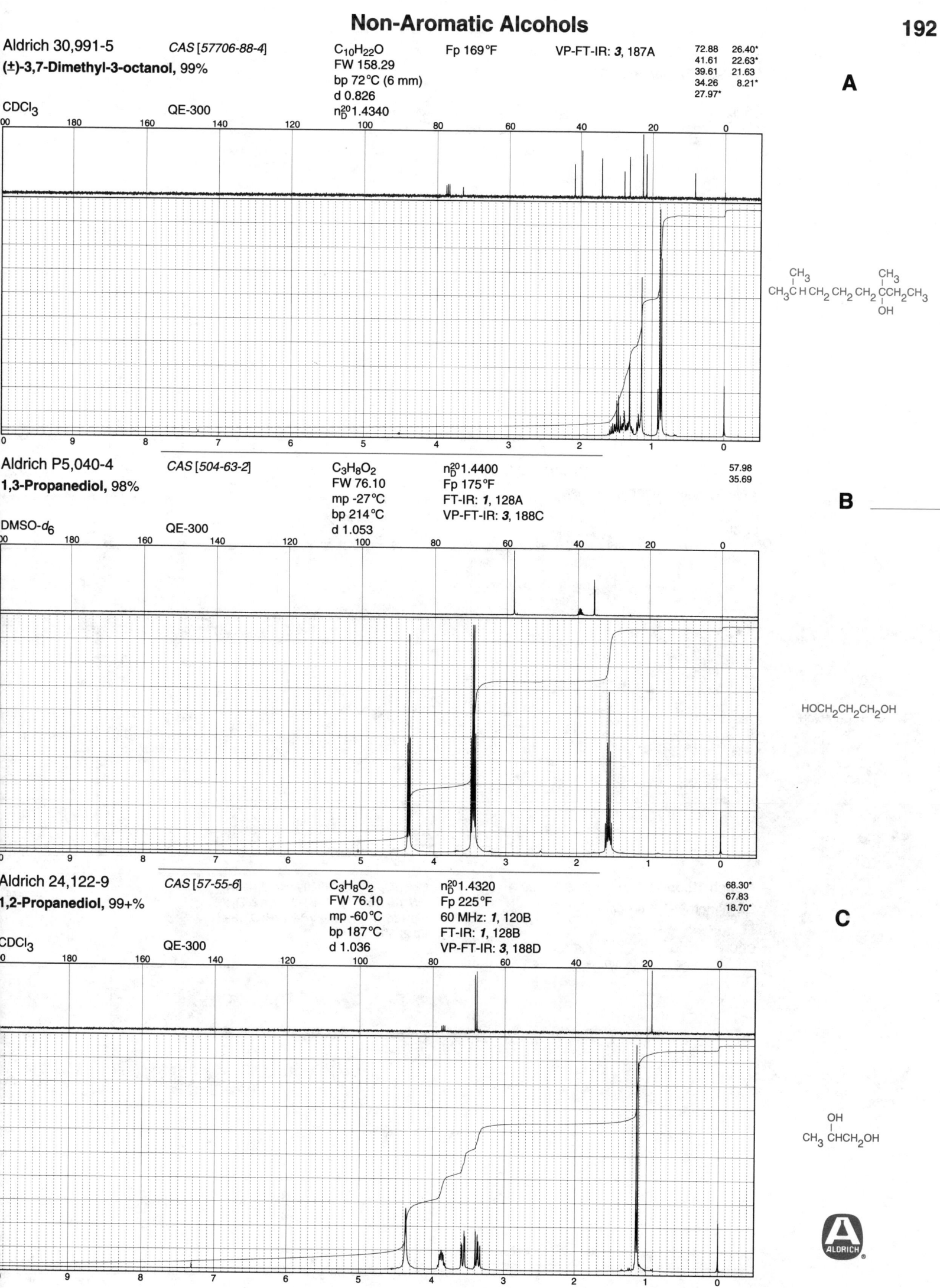

Aldrich 30,991-5   CAS [57706-88-4]
(±)-3,7-Dimethyl-3-octanol, 99%

$C_{10}H_{22}O$
FW 158.29
bp 72°C (6 mm)
d 0.826
n_D^{20} 1.4340

Fp 169°F

VP-FT-IR: 3, 187A

72.88   26.40*
41.61   22.63*
39.61   21.63
34.26   8.21*
27.97*

A

CDCl₃   QE-300

Aldrich P5,040-4   CAS [504-63-2]
1,3-Propanediol, 98%

$C_3H_8O_2$
FW 76.10
mp -27°C
bp 214°C
d 1.053

n_D^{20} 1.4400
Fp 175°F
FT-IR: 1, 128A
VP-FT-IR: 3, 188C

57.98
35.69

B

DMSO-d₆   QE-300

Aldrich 24,122-9   CAS [57-55-6]
1,2-Propanediol, 99+%

$C_3H_8O_2$
FW 76.10
mp -60°C
bp 187°C
d 1.036

n_D^{20} 1.4320
Fp 225°F
60 MHz: 1, 120B
FT-IR: 1, 128B
VP-FT-IR: 3, 188D

68.30*
67.83
18.70*

C

CDCl₃   QE-300

ALDRICH

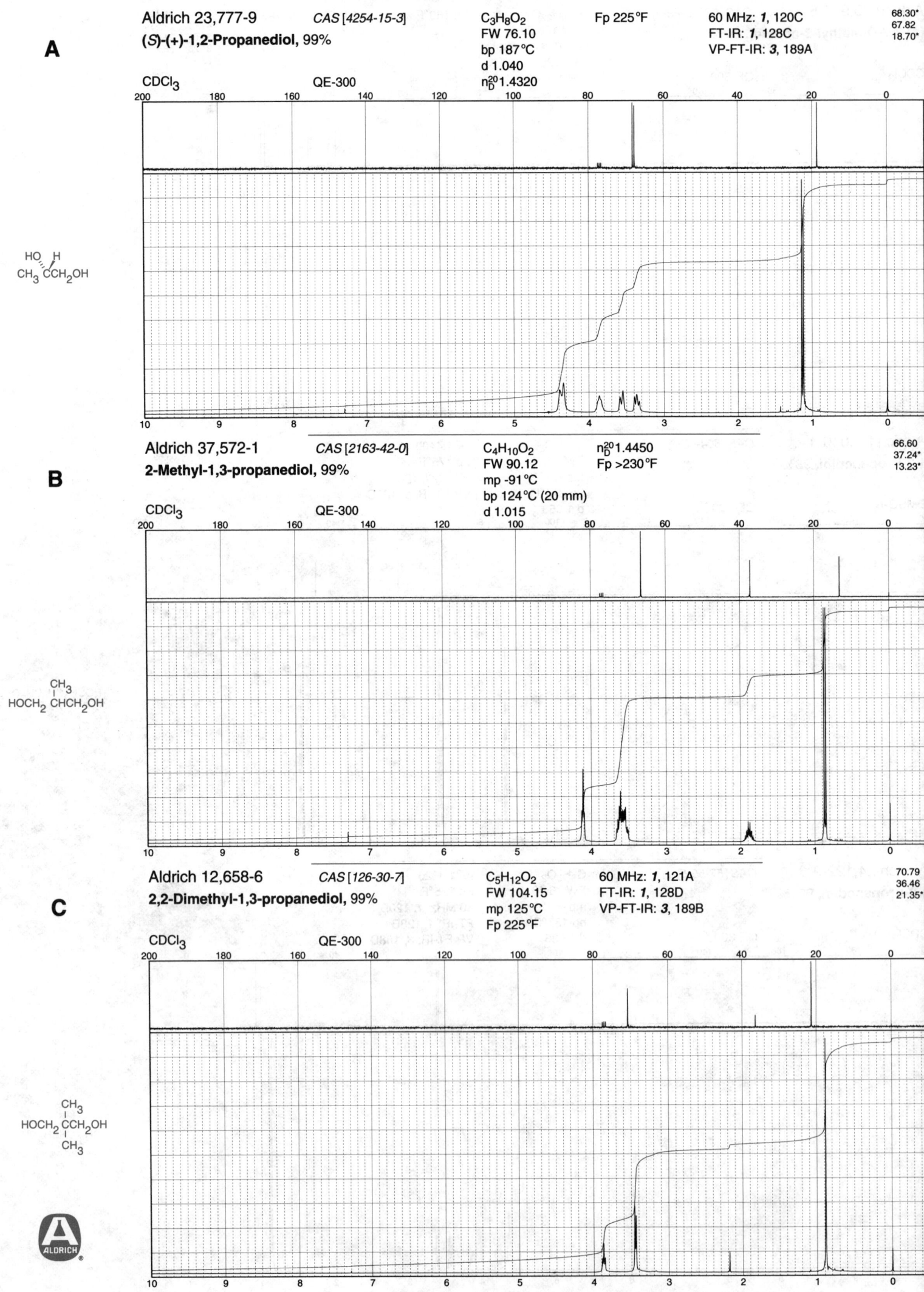

A

Aldrich 23,777-9   *CAS [4254-15-3]* $C_3H_8O_2$ Fp 225°F   60 MHz: *1*, 120C   68.30*

(S)-(+)-1,2-Propanediol, 99%   FW 76.10   FT-IR: *1*, 128C   67.82

bp 187°C   VP-FT-IR: *3*, 189A   18.70*

d 1.040

CDCl₃   QE-300 n_D^{20} 1.4320

HO H
CH₃ CCH₂OH

B

Aldrich 37,572-1   *CAS [2163-42-0]* $C_4H_{10}O_2$ n_D^{20} 1.4450   66.60

2-Methyl-1,3-propanediol, 99%   FW 90.12   Fp >230°F   37.24*

mp -91°C   13.23*

bp 124°C (20 mm)

CDCl₃   QE-300   d 1.015

CH₃
HOCH₂ CHCH₂OH

C

Aldrich 12,658-6   *CAS [126-30-7]* $C_5H_{12}O_2$ 60 MHz: *1*, 121A   70.79

2,2-Dimethyl-1,3-propanediol, 99%   FW 104.15   FT-IR: *1*, 128D   36.46

mp 125°C   VP-FT-IR: *3*, 189B   21.35*

Fp 225°F

CDCl₃   QE-300

CH₃
HOCH₂ CCH₂OH
CH₃

ALDRICH

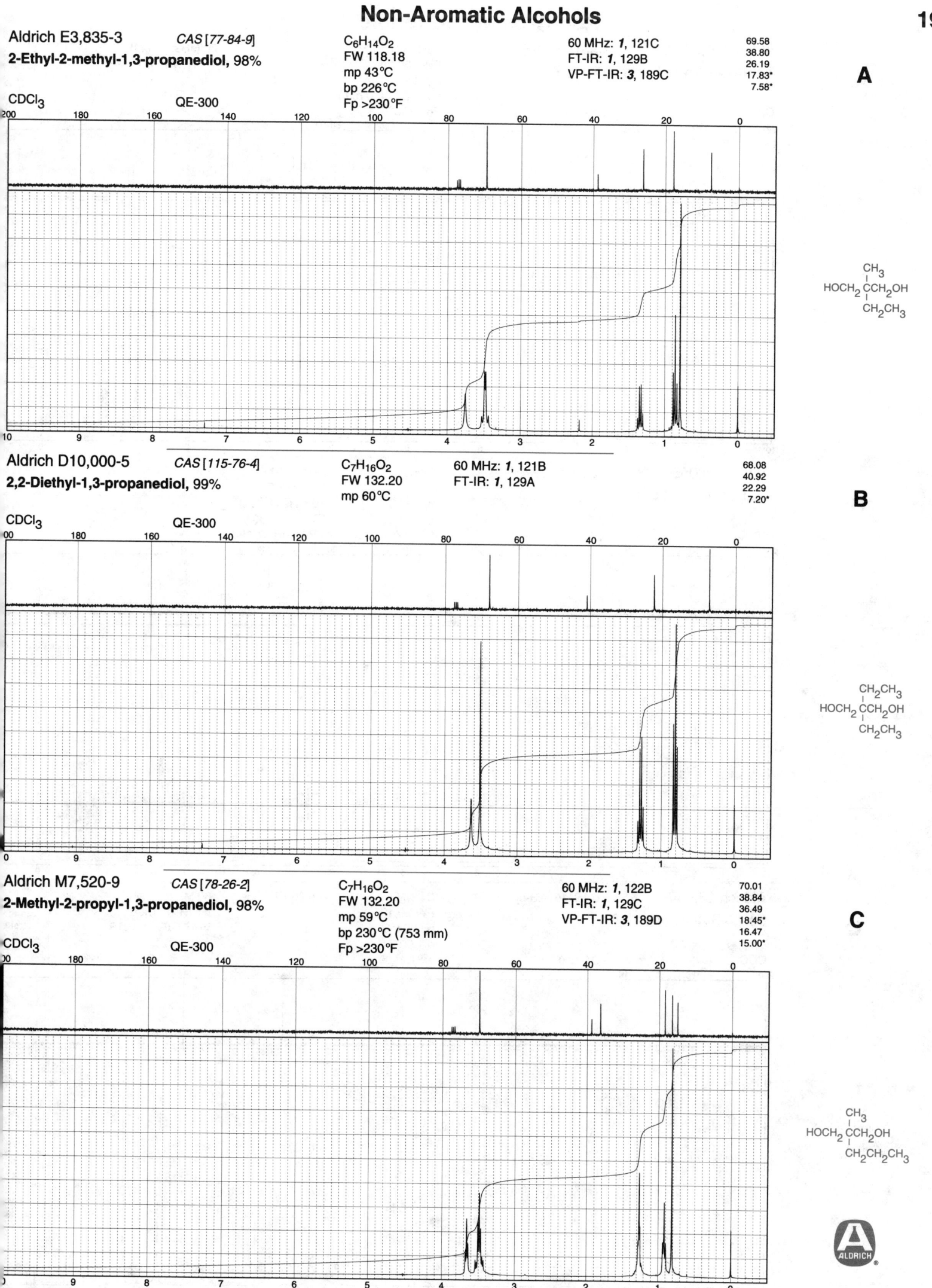

Aldrich E3,835-3   *CAS [77-84-9]*
2-Ethyl-2-methyl-1,3-propanediol, 98%

$C_6H_{14}O_2$
FW 118.18
mp 43°C
bp 226°C
Fp >230°F

60 MHz: *1*, 121C
FT-IR: *1*, 129B
VP-FT-IR: *3*, 189C

69.58
38.80
26.19
17.83*
7.58*

A

CDCl₃   QE-300

HOCH₂CCH₂OH with CH₃ and CH₂CH₃

Aldrich D10,000-5   *CAS [115-76-4]*
2,2-Diethyl-1,3-propanediol, 99%

$C_7H_{16}O_2$
FW 132.20
mp 60°C

60 MHz: *1*, 121B
FT-IR: *1*, 129A

68.08
40.92
22.29
7.20*

B

CDCl₃   QE-300

HOCH₂CCH₂OH with CH₂CH₃ and CH₂CH₃

Aldrich M7,520-9   *CAS [78-26-2]*
2-Methyl-2-propyl-1,3-propanediol, 98%

$C_7H_{16}O_2$
FW 132.20
mp 59°C
bp 230°C (753 mm)
Fp >230°F

60 MHz: *1*, 122B
FT-IR: *1*, 129C
VP-FT-IR: *3*, 189D

70.01
38.84
36.49
18.45*
16.47
15.00*

C

CDCl₃   QE-300

HOCH₂CCH₂OH with CH₃ and CH₂CH₂CH₃

ALDRICH

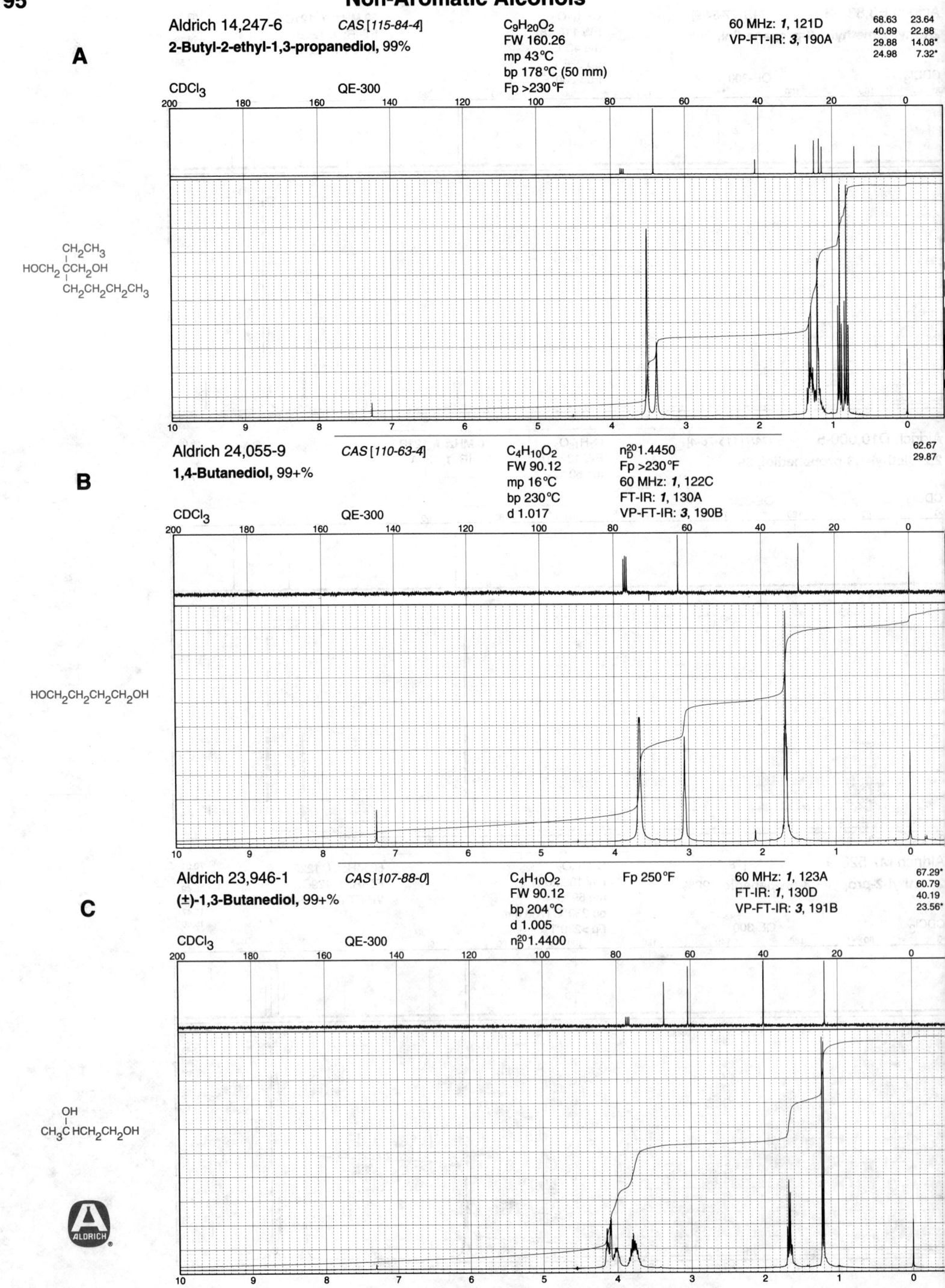

A

Aldrich 14,247-6   *CAS [115-84-4]*   C₉H₂₀O₂   60 MHz: *1*, 121D

2-Butyl-2-ethyl-1,3-propanediol, 99%   FW 160.26   VP-FT-IR: *3*, 190A

68.63   23.64
40.89   22.88
29.88   14.08*
24.98   7.32*

mp 43 °C
bp 178 °C (50 mm)
Fp >230 °F

CDCl₃   QE-300

$$CH_2CH_3$$
$$HOCH_2\ CCH_2OH$$
$$CH_2CH_2CH_2CH_3$$

B

Aldrich 24,055-9   *CAS [110-63-4]*   C₄H₁₀O₂   n²⁰_D 1.4450   62.67
29.87

1,4-Butanediol, 99+%   FW 90.12   Fp >230 °F

mp 16 °C   60 MHz: *1*, 122C
bp 230 °C   FT-IR: *1*, 130A
d 1.017   VP-FT-IR: *3*, 190B

CDCl₃   QE-300

$$HOCH_2CH_2CH_2CH_2OH$$

C

Aldrich 23,946-1   *CAS [107-88-0]*   C₄H₁₀O₂   Fp 250 °F   60 MHz: *1*, 123A

67.29*
60.79
40.19
23.56*

(±)-1,3-Butanediol, 99+%   FW 90.12   FT-IR: *1*, 130D

bp 204 °C   VP-FT-IR: *3*, 191B
d 1.005
n²⁰_D 1.4400

CDCl₃   QE-300

$$OH$$
$$CH_3CHCH_2CH_2OH$$

ALDRICH

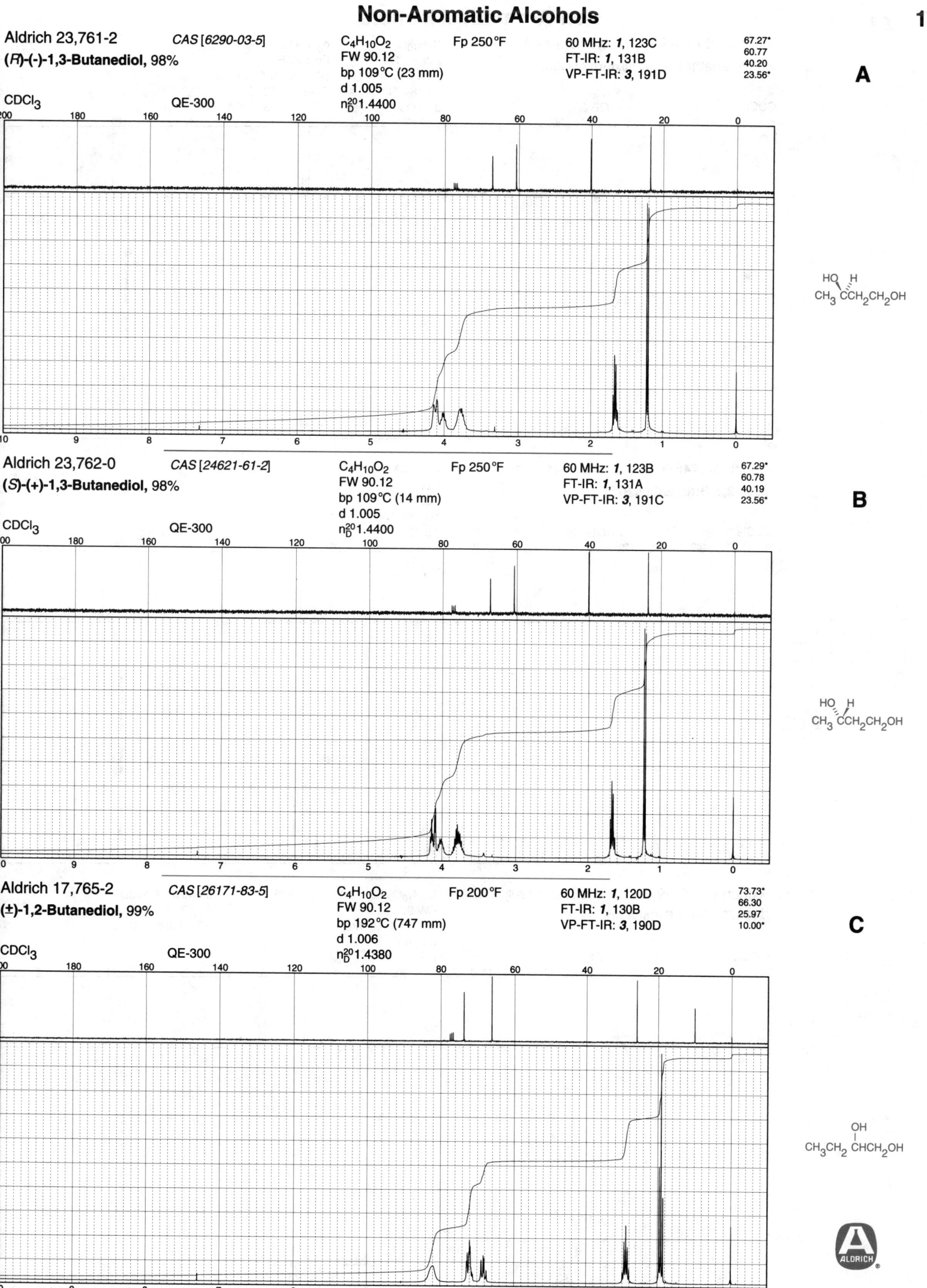

Aldrich 23,761-2   CAS [6290-03-5]   C₄H₁₀O₂   Fp 250°F   60 MHz: 1, 123C   67.27*
(R)-(-)-1,3-Butanediol, 98%   FW 90.12   FT-IR: 1, 131B   60.77
                                              bp 109°C (23 mm)   VP-FT-IR: 3, 191D   40.20
                                              d 1.005   23.56*
CDCl₃   QE-300   n₂₀ᴅ 1.4400

A

Aldrich 23,762-0   CAS [24621-61-2]   C₄H₁₀O₂   Fp 250°F   60 MHz: 1, 123B   67.29*
(S)-(+)-1,3-Butanediol, 98%   FW 90.12   FT-IR: 1, 131A   60.78
                                              bp 109°C (14 mm)   VP-FT-IR: 3, 191C   40.19
                                              d 1.005   23.56*
CDCl₃   QE-300   n₂₀ᴅ 1.4400

B

Aldrich 17,765-2   CAS [26171-83-5]   C₄H₁₀O₂   Fp 200°F   60 MHz: 1, 120D   73.73*
(±)-1,2-Butanediol, 99%   FW 90.12   FT-IR: 1, 130B   66.30
                                              bp 192°C (747 mm)   VP-FT-IR: 3, 190D   25.97
                                              d 1.006   10.00*
CDCl₃   QE-300   n₂₀ᴅ 1.4380

C

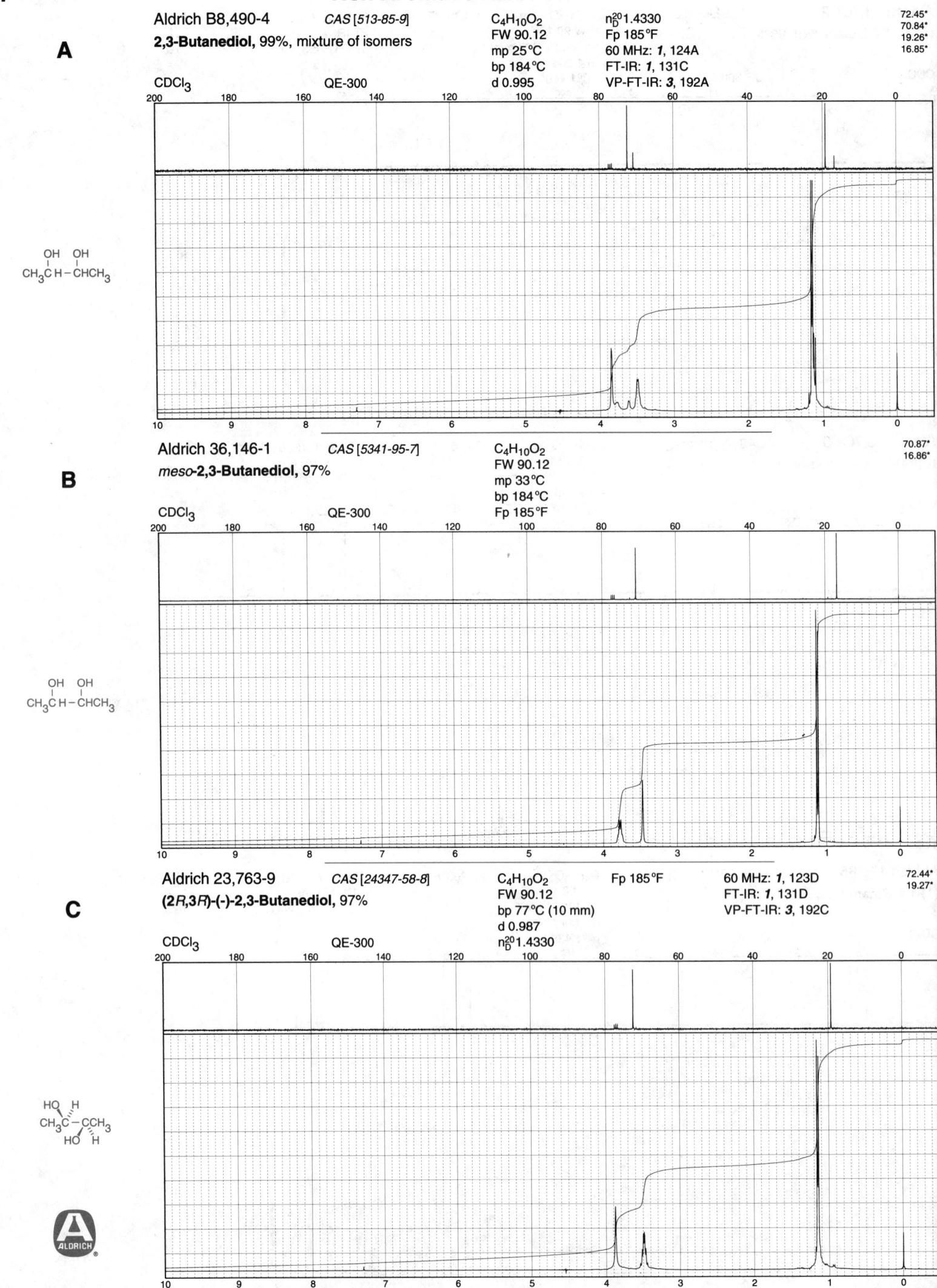

A

Aldrich B8,490-4   *CAS [513-85-9]*

2,3-Butanediol, 99%, mixture of isomers

$C_4H_{10}O_2$
FW 90.12
mp 25°C
bp 184°C
d 0.995

n_D^{20}1.4330
Fp 185°F
60 MHz: *1*, 124A
FT-IR: *1*, 131C
VP-FT-IR: *3*, 192A

72.45*
70.84*
19.26*
16.85*

CDCl$_3$ QE-300

B

Aldrich 36,146-1   *CAS [5341-95-7]*

meso-**2,3-Butanediol**, 97%

$C_4H_{10}O_2$
FW 90.12
mp 33°C
bp 184°C
Fp 185°F

70.87*
16.86*

CDCl$_3$ QE-300

C

Aldrich 23,763-9   *CAS [24347-58-8]*

(2R,3R)-(-)-2,3-Butanediol, 97%

$C_4H_{10}O_2$
FW 90.12
bp 77°C (10 mm)
d 0.987
n_D^{20}1.4330

Fp 185°F

60 MHz: *1*, 123D
FT-IR: *1*, 131D
VP-FT-IR: *3*, 192C

72.44*
19.27*

CDCl$_3$ QE-300

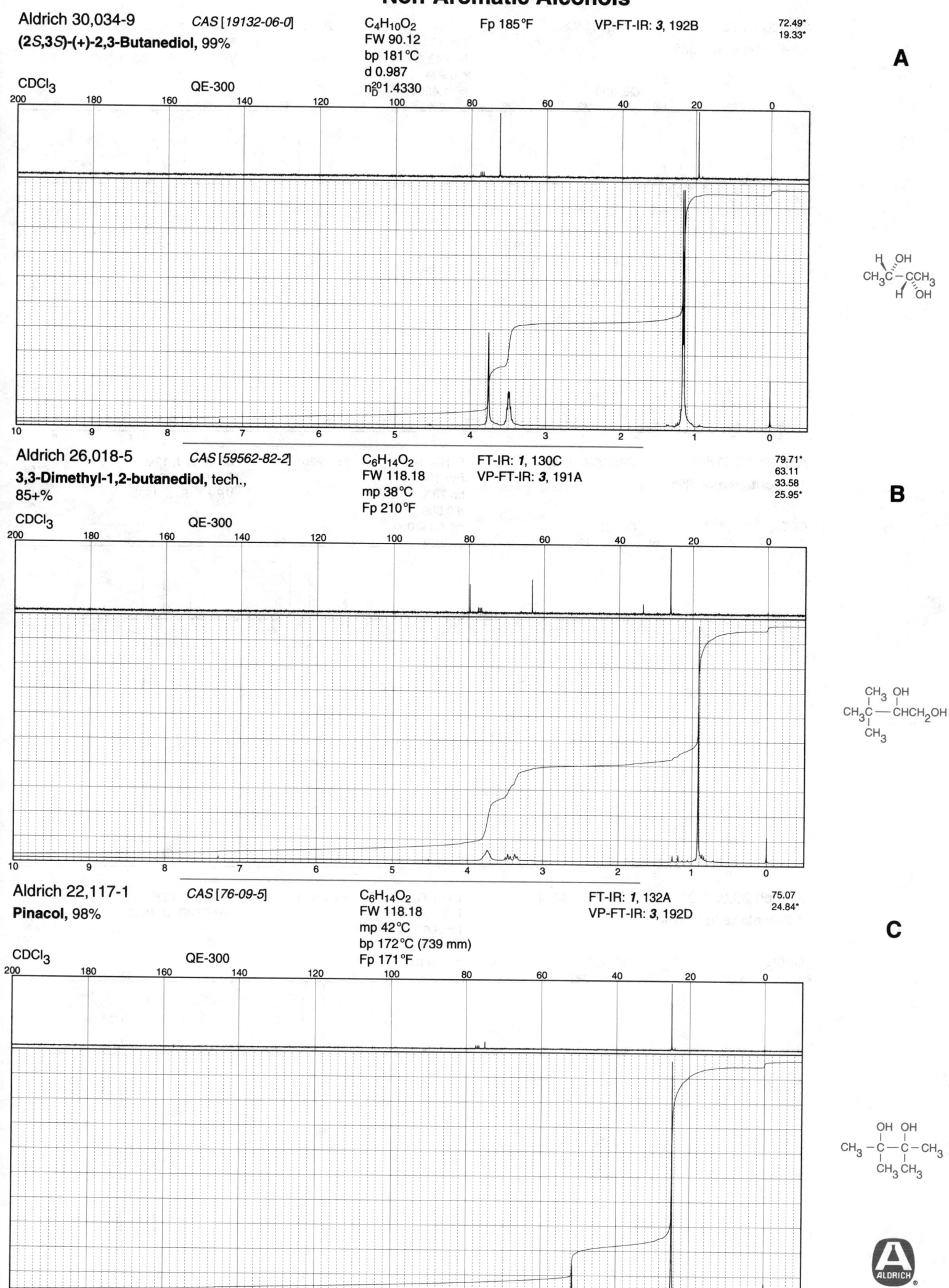

Aldrich 30,034-9   CAS [19132-06-0]   C₄H₁₀O₂
(2S,3S)-(+)-2,3-Butanediol, 99%

$C_4H_{10}O_2$
FW 90.12
bp 181°C
d 0.987
n_D^{20} 1.4330

Fp 185°F   VP-FT-IR: **3**, 192B

72.49*
19.33*

A

CDCl₃   QE-300

Aldrich 26,018-5   CAS [59562-82-2]
3,3-Dimethyl-1,2-butanediol, tech., 85+%

$C_6H_{14}O_2$
FW 118.18
mp 38°C
Fp 210°F

FT-IR: **1**, 130C
VP-FT-IR: **3**, 191A

79.71*
63.11
33.58
25.95*

B

CDCl₃   QE-300

Aldrich 22,117-1   CAS [76-09-5]
Pinacol, 98%

$C_6H_{14}O_2$
FW 118.18
mp 42°C
bp 172°C (739 mm)
Fp 171°F

FT-IR: **1**, 132A
VP-FT-IR: **3**, 192D

75.07
24.84*

C

CDCl₃   QE-300

Non-Aromatic Alcohols

A

Aldrich P770-3

1,5-Pentanediol, 96%

CAS [111-29-5]

$C_5H_{12}O_2$
FW 104.15
bp 242°C
d 0.994
n_D^{20} 1.4500

Fp 265°F

60 MHz: *1*, 124D
FT-IR: *1*, 132B
VP-FT-IR: *3*, 193A

64.34
33.79
24.18

D_2O QE-300

$HOCH_2(CH_2)_3CH_2OH$

B

Aldrich 19,418-2

1,4-Pentanediol, 99%

CAS [626-95-9]

$C_5H_{12}O_2$
FW 104.15
bp 73°C
d 0.986
n_D^{20} 1.4470

Fp >230°F

60 MHz: *1*, 124C
FT-IR: *1*, 133A
VP-FT-IR: *3*, 193B

67.80*
62.68
36.23
29.11
23.48*

$CDCl_3$ QE-300

OH
|
$CH_3 CHCH_2CH_2CH_2OH$

C

Aldrich 26,028-2

1,2-Pentanediol, 98%

CAS [5343-92-0]

$C_5H_{12}O_2$
FW 104.15
bp 206°C
d 0.971
n_D^{20} 1.4400

Fp 220°F

FT-IR: *1*, 133B
VP-FT-IR: *3*, 193C

72.08*
66.72
35.19
18.80
14.07*

$CDCl_3$ QE-300

OH
|
$CH_3CH_2CH_2 CHCH_2OH$

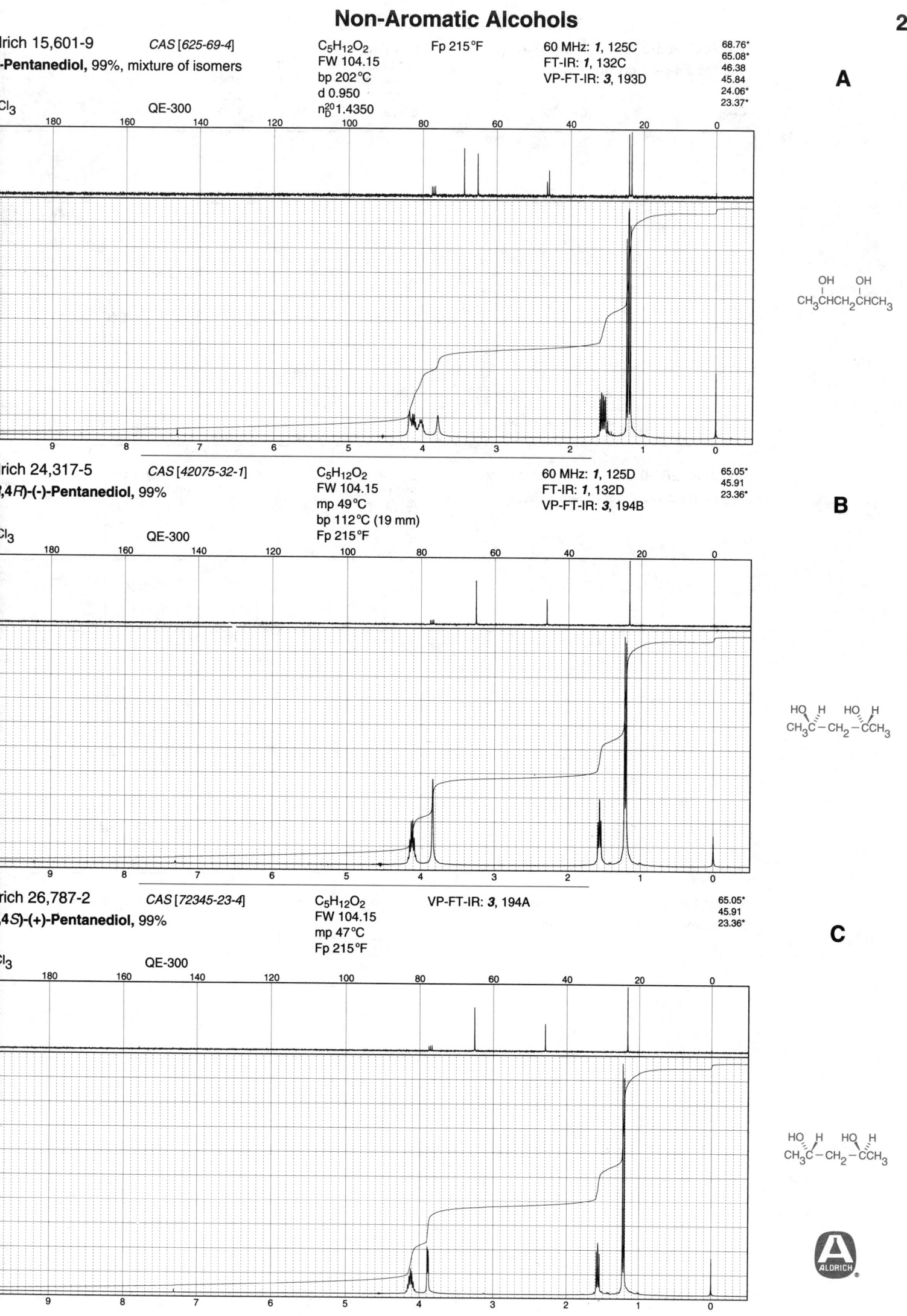

Aldrich 15,601-9   *CAS [625-69-4]* $C_5H_{12}O_2$ Fp 215°F   60 MHz: *1*, 125C   68.76*

-Pentanediol, 99%, mixture of isomers   FW 104.15   FT-IR: *1*, 132C   65.08*

bp 202°C   VP-FT-IR: *3*, 193D   46.38

Cl3   QE-300   d 0.950   45.84

n_D^{20} 1.4350   24.06*

23.37*

A

OH   OH

$CH_3CHCH_2CHCH_3$

Aldrich 24,317-5   *CAS [42075-32-1]* $C_5H_{12}O_2$ 60 MHz: *1*, 125D   65.05*

,4R)-(-)-Pentanediol, 99%   FW 104.15   FT-IR: *1*, 132D   45.91

mp 49°C   VP-FT-IR: *3*, 194B   23.36*

bp 112°C (19 mm)

Cl3   QE-300   Fp 215°F

B

HO H   HO H

$CH_3C - CH_2 - CCH_3$

Aldrich 26,787-2   *CAS [72345-23-4]* $C_5H_{12}O_2$ VP-FT-IR: *3*, 194A   65.05*

,4S)-(+)-Pentanediol, 99%   FW 104.15   45.91

mp 47°C   23.36*

Fp 215°F

C

HO H   HO H

$CH_3C - CH_2 - CCH_3$

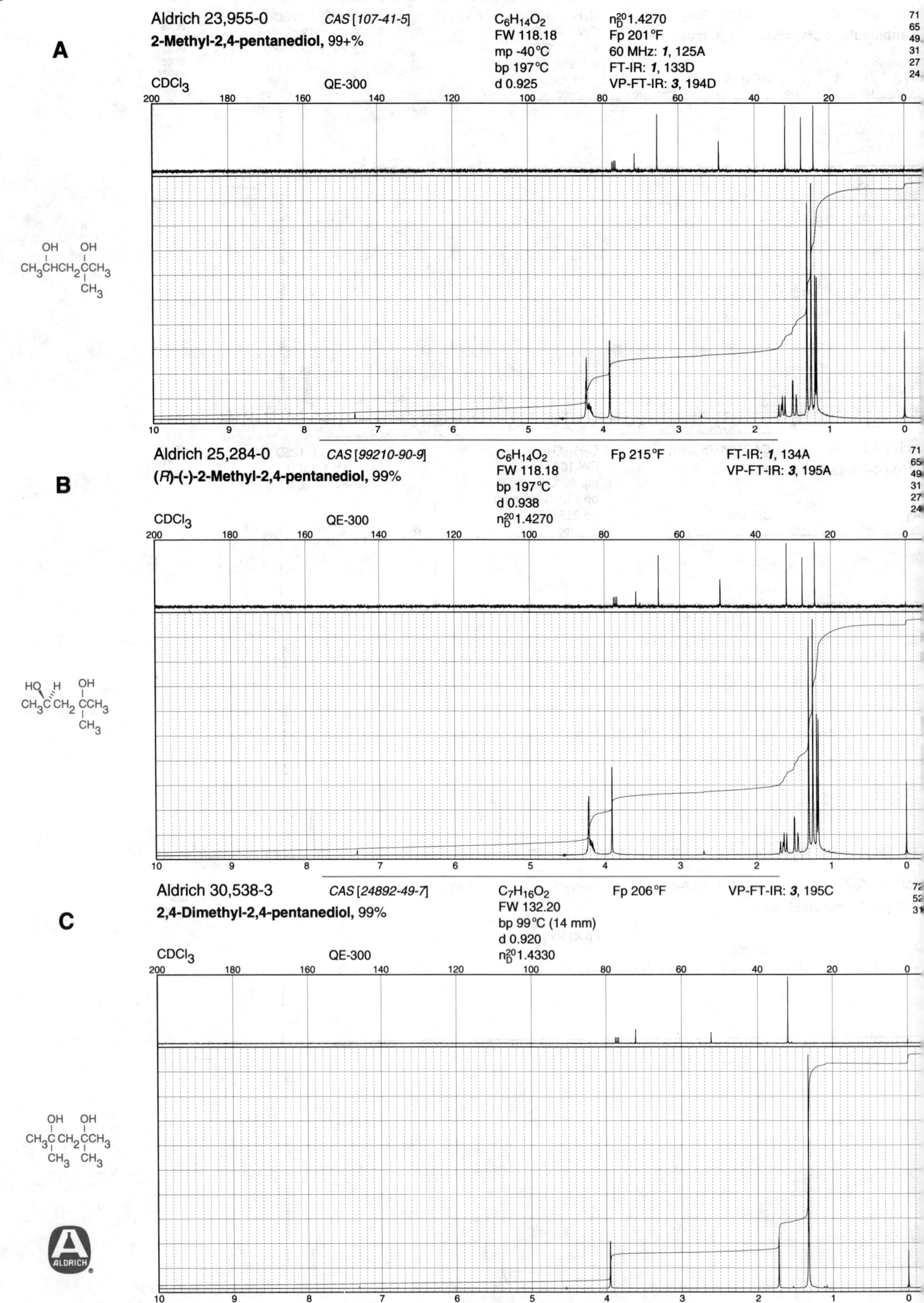

A

Aldrich 23,955-0   *CAS [107-41-5]* $C_6H_{14}O_2$ $n_D^{20} 1.4270$

2-Methyl-2,4-pentanediol, 99+%   FW 118.18   Fp 201°F

mp -40°C   60 MHz: *1*, 125A

bp 197°C   FT-IR: *1*, 133D

CDCl₃   QE-300   d 0.925   VP-FT-IR: *3*, 194D

71
65
49
31
27
24.

B

Aldrich 25,284-0   *CAS [99210-90-9]* $C_6H_{14}O_2$ Fp 215°F   FT-IR: *1*, 134A

(R)-(-)-2-Methyl-2,4-pentanediol, 99%   FW 118.18   VP-FT-IR: *3*, 195A

bp 197°C

d 0.938

CDCl₃   QE-300 $n_D^{20} 1.4270$

71
65
49
31
27
24

C

Aldrich 30,538-3   *CAS [24892-49-7]* $C_7H_{16}O_2$ Fp 206°F   VP-FT-IR: *3*, 195C

2,4-Dimethyl-2,4-pentanediol, 99%   FW 132.20

bp 99°C (14 mm)

d 0.920

CDCl₃   QE-300 $n_D^{20} 1.4330$

72
52
31

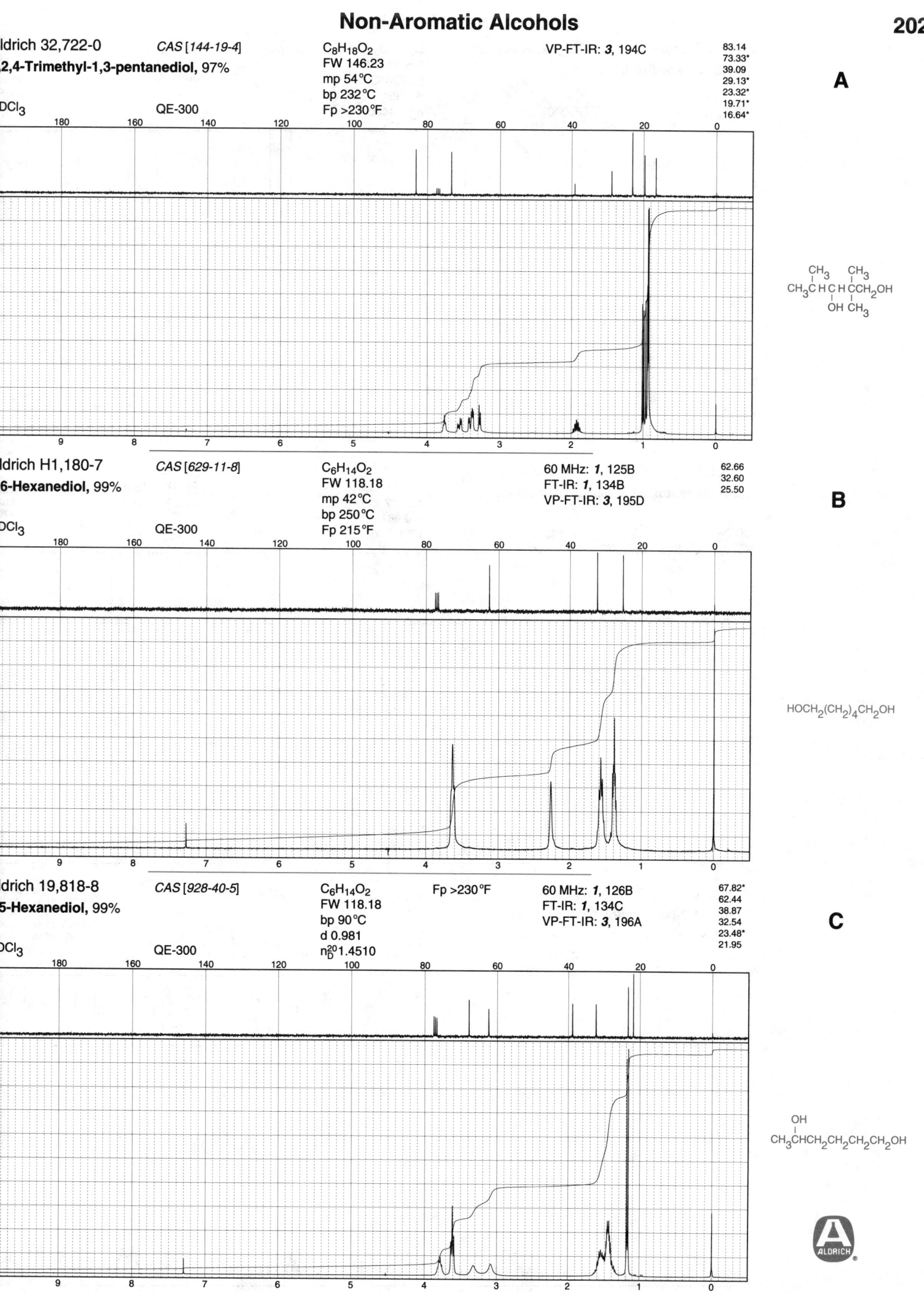

Aldrich 32,722-0   CAS [144-19-4]   C₈H₁₈O₂   VP-FT-IR: 3, 194C

2,4-Trimethyl-1,3-pentanediol, 97%

FW 146.23
mp 54°C
bp 232°C
Fp >230°F

DCl₃   QE-300

83.14
73.33*
39.09
29.13*
23.32*
19.71*
16.64*

A

Aldrich H1,180-7   CAS [629-11-8]   C₆H₁₄O₂   60 MHz: 1, 125B

1,6-Hexanediol, 99%

FW 118.18
mp 42°C
bp 250°C
Fp 215°F

DCl₃   QE-300

FT-IR: 1, 134B
VP-FT-IR: 3, 195D

62.66
32.60
25.50

B

Aldrich 19,818-8   CAS [928-40-5]   C₆H₁₄O₂   Fp >230°F   60 MHz: 1, 126B

1,5-Hexanediol, 99%

FW 118.18
bp 90°C
d 0.981
n²⁰_D 1.4510

DCl₃   QE-300

FT-IR: 1, 134C
VP-FT-IR: 3, 196A

67.82*
62.44
38.87
32.54
23.48*
21.95

C

A

Aldrich 21,369-1   CAS [6920-22-5]

(±)-1,2-Hexanediol, 98%

$C_6H_{14}O_2$
FW 118.18
bp 224°C
d 0.951
n_D^{20} 1.4420

Fp >230°F

60 MHz: 1, 126A
FT-IR: 1, 133C
VP-FT-IR: 3, 196B

72.38
66.74
32.80
27.80
22.74
13.98

CDCl₃   QE-300

$CH_3CH_2CH_2CH_2\overset{\displaystyle OH}{\underset{\displaystyle |}{CH}}CH_2OH$

B

Aldrich H1,190-4   CAS [2935-44-6]

2,5-Hexanediol, 99%, mixture of isomers

$C_6H_{14}O_2$
FW 118.18
bp 217°C
d 0.961
n_D^{20} 1.4470

Fp 215°F

60 MHz: 1, 126C
FT-IR: 1, 134D
VP-FT-IR: 3, 196C

68.20
67.69
36.01
35.02
23.70
23.35

CDCl₃   QE-300

$CH_3\overset{\displaystyle OH}{\underset{\displaystyle |}{CH}}CH_2CH_2\overset{\displaystyle OH}{\underset{\displaystyle |}{CH}}CH_3$

C

Aldrich E2,912-5   CAS [94-96-2]

2-Ethyl-1,3-hexanediol, 97%, mixture of isomers

$C_8H_{18}O_2$
FW 146.23
mp -40°C
bp 245°C
d 0.933

n_D^{20} 1.4510
Fp 265°F
60 MHz: 1, 126D
FT-IR: 1, 135A
VP-FT-IR: 3, 196D

75.17* 45.90* 18.95
74.83* 37.84   18.18
64.24   35.33   14.11
63.65   21.49   12.32
46.11* 19.53   11.76

CDCl₃   QE-300

$CH_3CH_2CH_2\overset{\displaystyle OH}{\underset{\displaystyle |}{CH}}\underset{\displaystyle \underset{\displaystyle CH_2CH_3}{|}}{CH}CH_2OH$

A

Aldrich 14,361-8   *CAS [110-03-2]* $C_8H_{18}O_2$ 60 MHz: *1*, 127A   70.52
2,5-Dimethyl-2,5-hexanediol, 97%   FW 146.23   FT-IR: *1*, 135B   37.72
mp 88°C   VP-FT-IR: *3*, 197A   29.39*
bp 215°C
Fp 260°F

CDCl₃   QE-300

OH   OH

CH₃CCH₂CH₂CCH₃

CH₃   CH₃

B

Aldrich H220-1   *CAS [629-30-1]* $C_7H_{16}O_2$ n_D^{20} 1.4550   62.73
1,7-Heptanediol, 95%   FW 132.20   Fp >230°F   32.57
mp 18°C   60 MHz: *1*, 127B   29.15
bp 259°C   FT-IR: *1*, 135C   25.69
CDCl₃   QE-300   d 0.951   VP-FT-IR: *3*, 197B

HOCH₂(CH₂)₅CH₂OH

C

Aldrich O-330-3   *CAS [629-41-4]* $C_8H_{18}O_2$ 60 MHz: *1*, 127D   62.94
1,8-Octanediol, 98%   FW 146.23   FT-IR: *1*, 135D   32.74
mp 60°C   VP-FT-IR: *3*, 197D   29.36
bp 172°C (20 mm)   25.67

CDCl₃   QE-300

HOCH₂(CH₂)₆CH₂OH

ALDRICH

Non-Aromatic Alcohols

A

Aldrich 21,370-5
1,2-Octanediol, 98+%

CAS [1117-86-8]

$C_8H_{18}O_2$
FW 146.23
mp 37°C
bp 132°C (10 mm)
Fp >230°F

60 MHz: *1*, 127C
FT-IR: *1*, 136A
VP-FT-IR: *3*, 198A

72.49*   29.39
66.84   25.61
33.31   22.61
31.82   13.99

CDCl₃   QE-300

$$CH_3(CH_2)_4CH_2\underset{\underset{OH}{|}}{C}HCH_2OH$$

B

Aldrich N2,960-0
1,9-Nonanediol, 98%

CAS [3937-56-2]

$C_9H_{20}O_2$
FW 160.26
mp 48°C
bp 177°C (15 mm)
Fp >230°F

60 MHz: *1*, 128A
FT-IR: *1*, 136B
VP-FT-IR: *3*, 198B

62.65
32.65
29.55
29.36
25.73

CDCl₃   QE-300

$$HOCH_2(CH_2)_7CH_2OH$$

C

Aldrich D120-3
1,10-Decanediol, 98%

CAS [112-47-0]

$C_{10}H_{22}O_2$
FW 174.28

60 MHz: *1*, 128B

63.02
32.87
29.51
29.42
25.79

CDCl₃   QE-300

$$HOCH_2(CH_2)_8CH_2OH$$

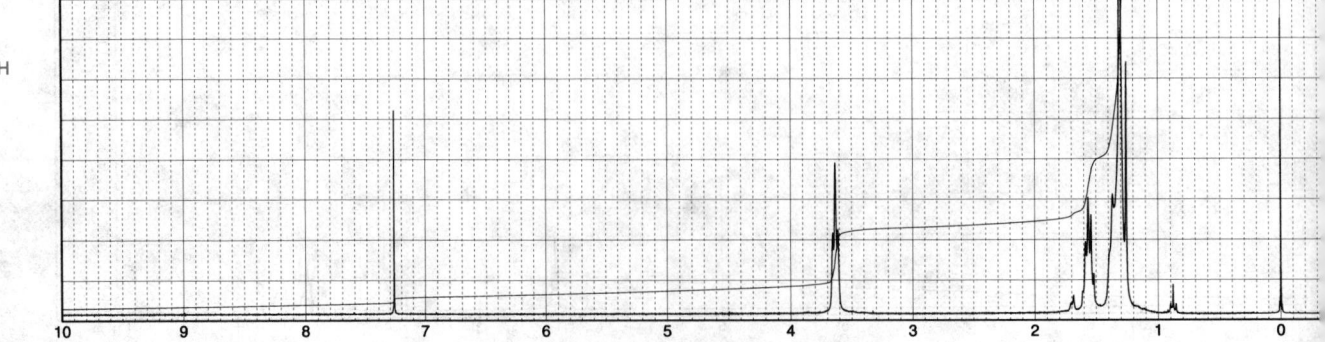

ALDRICH

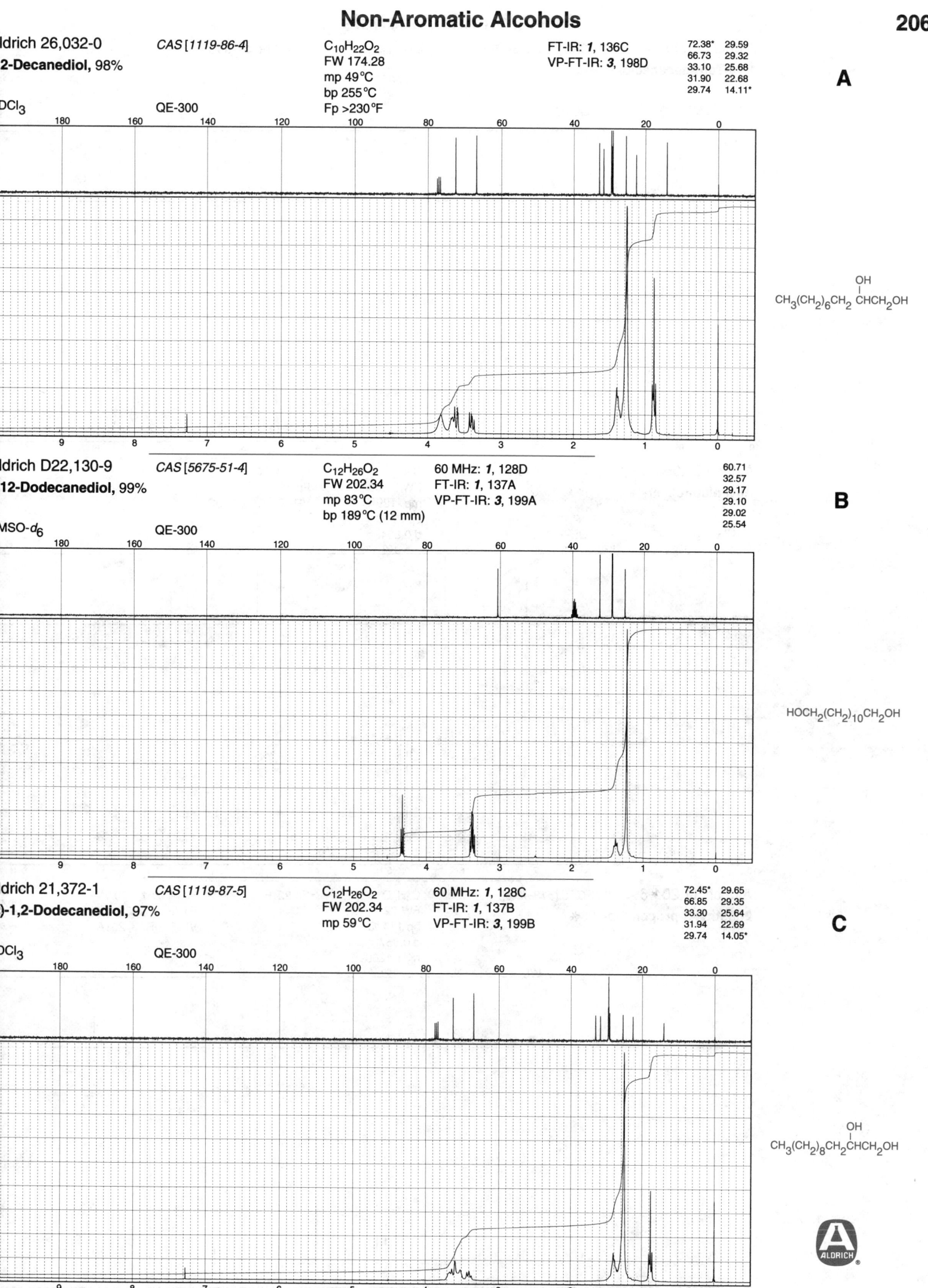

Aldrich 26,032-0   CAS [1119-86-4]   C₁₀H₂₂O₂   FT-IR: *1*, 136C
1,2-Decanediol, 98%   FW 174.28   VP-FT-IR: *3*, 198D
mp 49°C
bp 255°C
Fp >230°F

CDCl₃   QE-300

72.38*   29.59
66.73   29.32
33.10   25.68
31.90   22.68
29.74   14.11*

A

$CH_3(CH_2)_6CH_2CHCH_2OH$ with OH

Aldrich D22,130-9   CAS [5675-51-4]   C₁₂H₂₆O₂   60 MHz: *1*, 128D
1,12-Dodecanediol, 99%   FW 202.34   FT-IR: *1*, 137A
mp 83°C   VP-FT-IR: *3*, 199A
bp 189°C (12 mm)

DMSO-d₆   QE-300

60.71
32.57
29.17
29.10
29.02
25.54

B

$HOCH_2(CH_2)_{10}CH_2OH$

Aldrich 21,372-1   CAS [1119-87-5]   C₁₂H₂₆O₂   60 MHz: *1*, 128C
( )-1,2-Dodecanediol, 97%   FW 202.34   FT-IR: *1*, 137B
mp 59°C   VP-FT-IR: *3*, 199B

CDCl₃   QE-300

72.45*   29.65
66.85   29.35
33.30   25.64
31.94   22.69
29.74   14.05*

C

$CH_3(CH_2)_8CH_2CHCH_2OH$ with OH

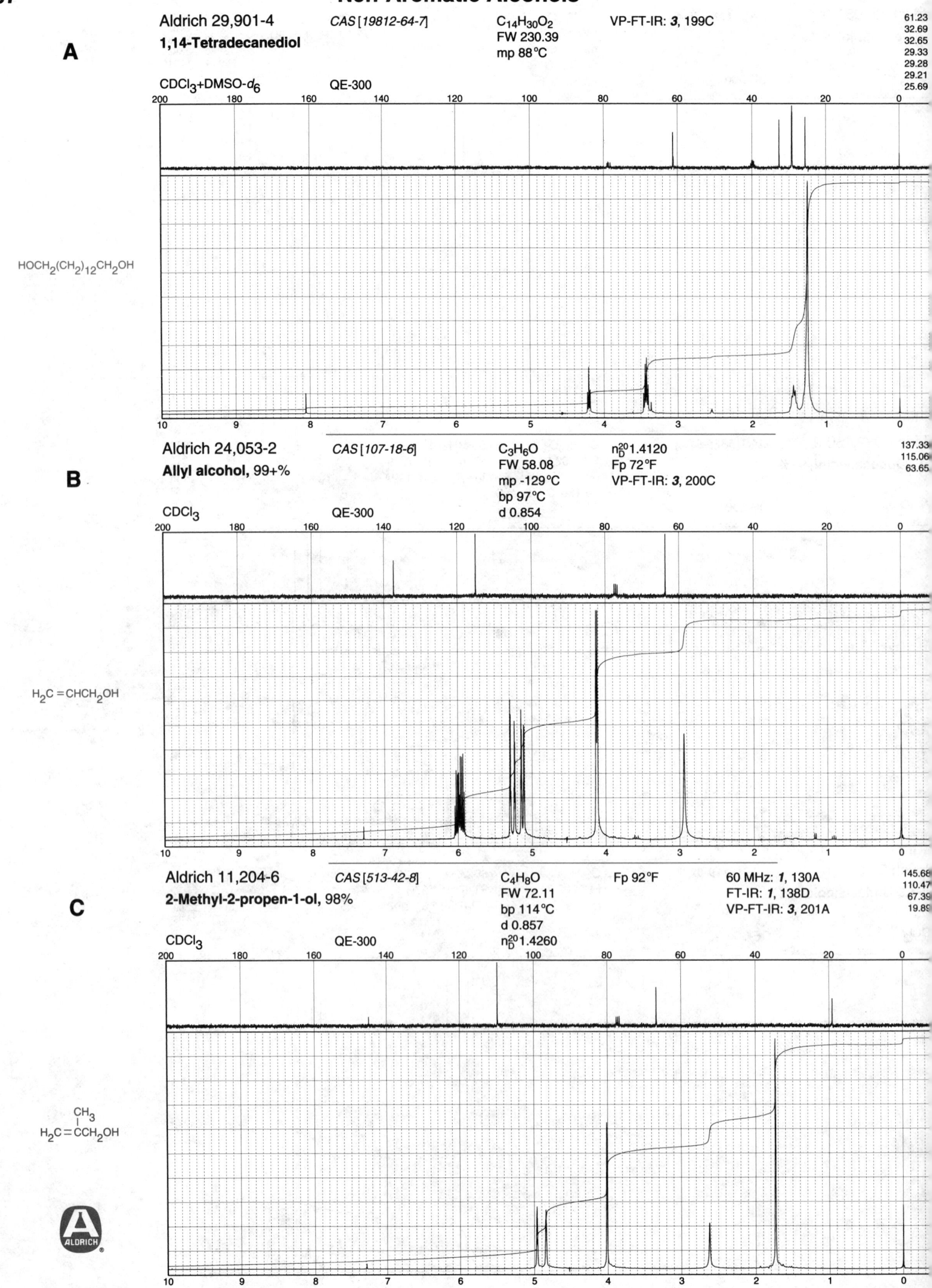

A

Aldrich 29,901-4
1,14-Tetradecanediol

CAS [19812-64-7]

$C_{14}H_{30}O_2$
FW 230.39
mp 88°C

VP-FT-IR: *3*, 199C

61.23
32.69
32.65
29.33
29.28
29.21
25.69

CDCl$_3$+DMSO-d_6 QE-300

HOCH$_2$(CH$_2$)$_{12}$CH$_2$OH

B

Aldrich 24,053-2
Allyl alcohol, 99+%

CAS [107-18-6]

C_3H_6O
FW 58.08
mp -129°C
bp 97°C
d 0.854

n_D^{20}1.4120
Fp 72°F
VP-FT-IR: *3*, 200C

137.33
115.06
63.65

CDCl$_3$ QE-300

H$_2$C=CHCH$_2$OH

C

Aldrich 11,204-6
2-Methyl-2-propen-1-ol, 98%

CAS [513-42-8]

C_4H_8O
FW 72.11
bp 114°C
d 0.857
n_D^{20}1.4260

Fp 92°F

60 MHz: *1*, 130A
FT-IR: *1*, 138D
VP-FT-IR: *3*, 201A

145.68
110.47
67.39
19.89

CDCl$_3$ QE-300

CH$_3$
|
H$_2$C=CCH$_2$OH

Non-Aromatic Alcohols

208

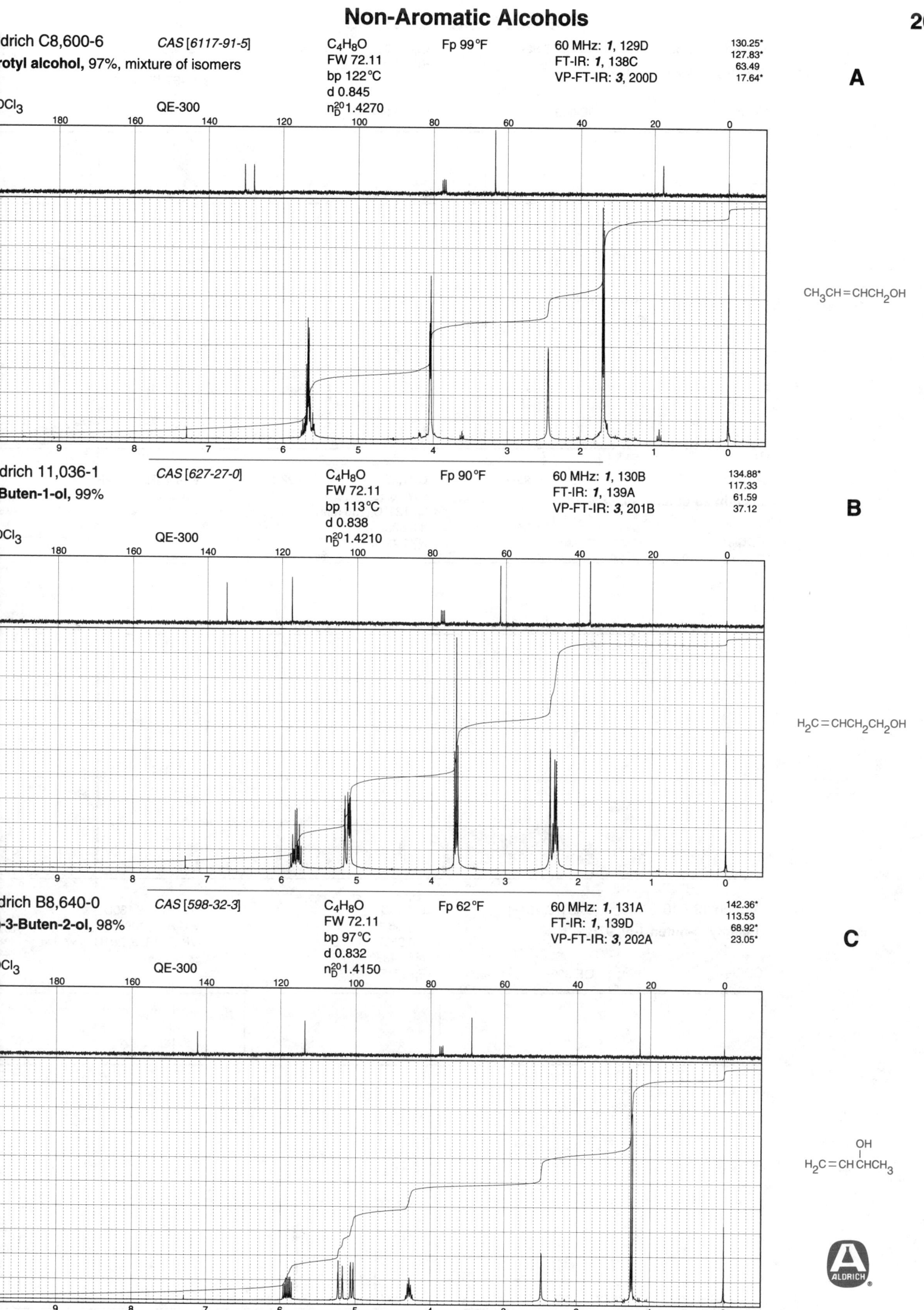

Aldrich C8,600-6   CAS [6117-91-5]   C4H8O   Fp 99°F   60 MHz: *1*, 129D   130.25*
rotyl alcohol, 97%, mixture of isomers   FW 72.11   FT-IR: *1*, 138C   127.83*
bp 122°C   VP-FT-IR: *3*, 200D   63.49
DCl3   QE-300   d 0.845   n_D^20 1.4270   17.64*

A

CH3CH=CHCH2OH

Aldrich 11,036-1   CAS [627-27-0]   C4H8O   Fp 90°F   60 MHz: *1*, 130B   134.88*
-Buten-1-ol, 99%   FW 72.11   FT-IR: *1*, 139A   117.33
bp 113°C   VP-FT-IR: *3*, 201B   61.59
DCl3   QE-300   d 0.838   n_D^20 1.4210   37.12

B

H2C=CHCH2CH2OH

Aldrich B8,640-0   CAS [598-32-3]   C4H8O   Fp 62°F   60 MHz: *1*, 131A   142.36*
)-3-Buten-2-ol, 98%   FW 72.11   FT-IR: *1*, 139D   113.53
bp 97°C   VP-FT-IR: *3*, 202A   68.92*
DCl3   QE-300   d 0.832   n_D^20 1.4150   23.05*

C

OH
H2C=CHCHCH3

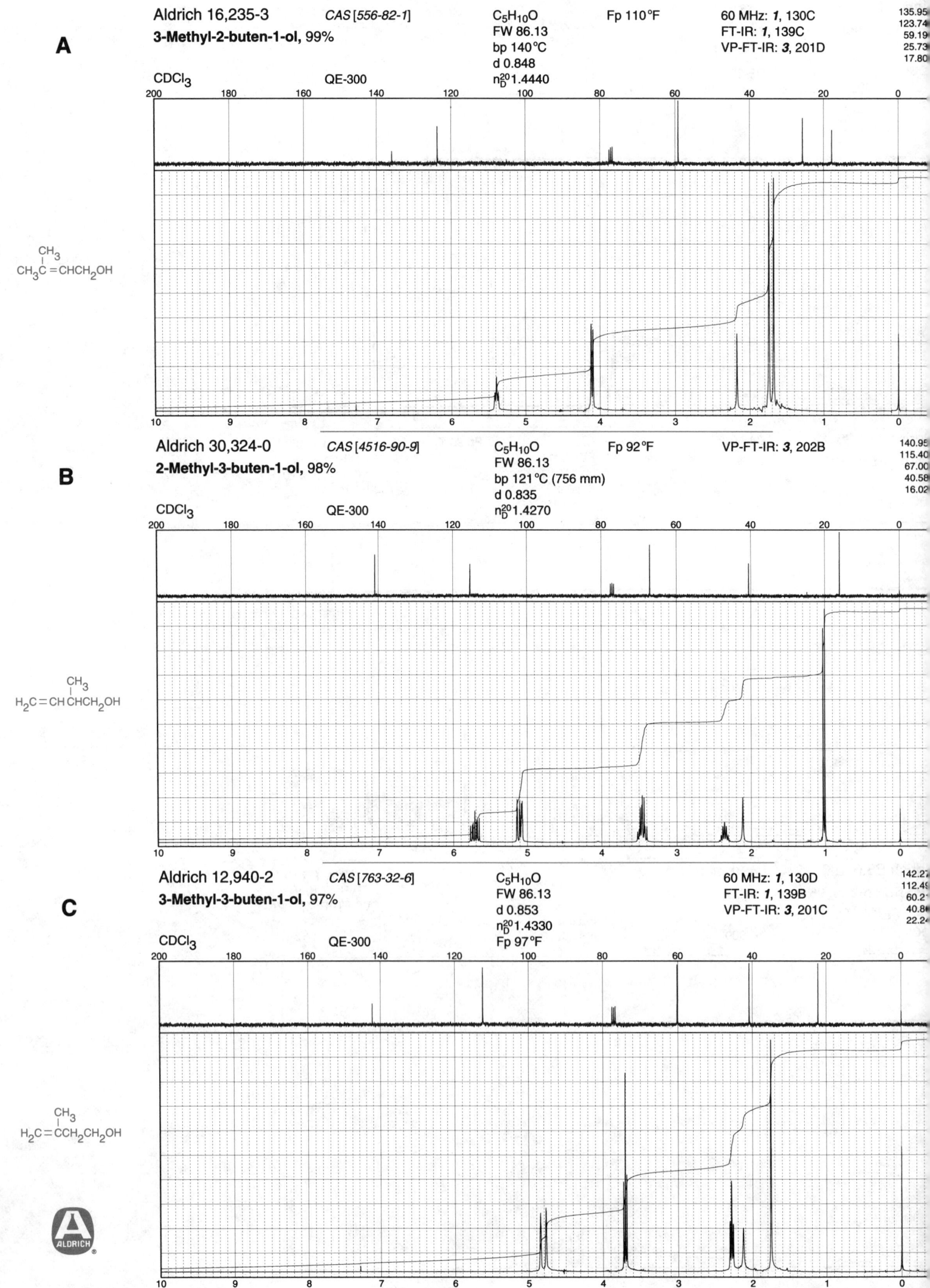

A

Aldrich 16,235-3   CAS [556-82-1]   C₅H₁₀O   Fp 110°F   60 MHz: **1**, 130C   135.95
3-Methyl-2-buten-1-ol, 99%   FW 86.13   FT-IR: **1**, 139C   123.74
bp 140°C   VP-FT-IR: **3**, 201D   59.19
d 0.848   25.73
n²⁰_D 1.4440   17.80

CDCl₃   QE-300

$$CH_3C=CHCH_2OH$$
with CH_3 above

B

Aldrich 30,324-0   CAS [4516-90-9]   C₅H₁₀O   Fp 92°F   VP-FT-IR: **3**, 202B   140.95
2-Methyl-3-buten-1-ol, 98%   FW 86.13   115.40
bp 121°C (756 mm)   67.00
d 0.835   40.58
n²⁰_D 1.4270   16.02

CDCl₃   QE-300

$$H_2C=CHCH CH_2OH$$
with CH_3 above

C

Aldrich 12,940-2   CAS [763-32-6]   C₅H₁₀O   60 MHz: **1**, 130D   142.27
3-Methyl-3-buten-1-ol, 97%   FW 86.13   FT-IR: **1**, 139B   112.49
d 0.853   VP-FT-IR: **3**, 201C   60.21
n²⁰_D 1.4330   40.86
Fp 97°F   22.24

CDCl₃   QE-300

$$H_2C=CCH_2CH_2OH$$
with CH_3 above

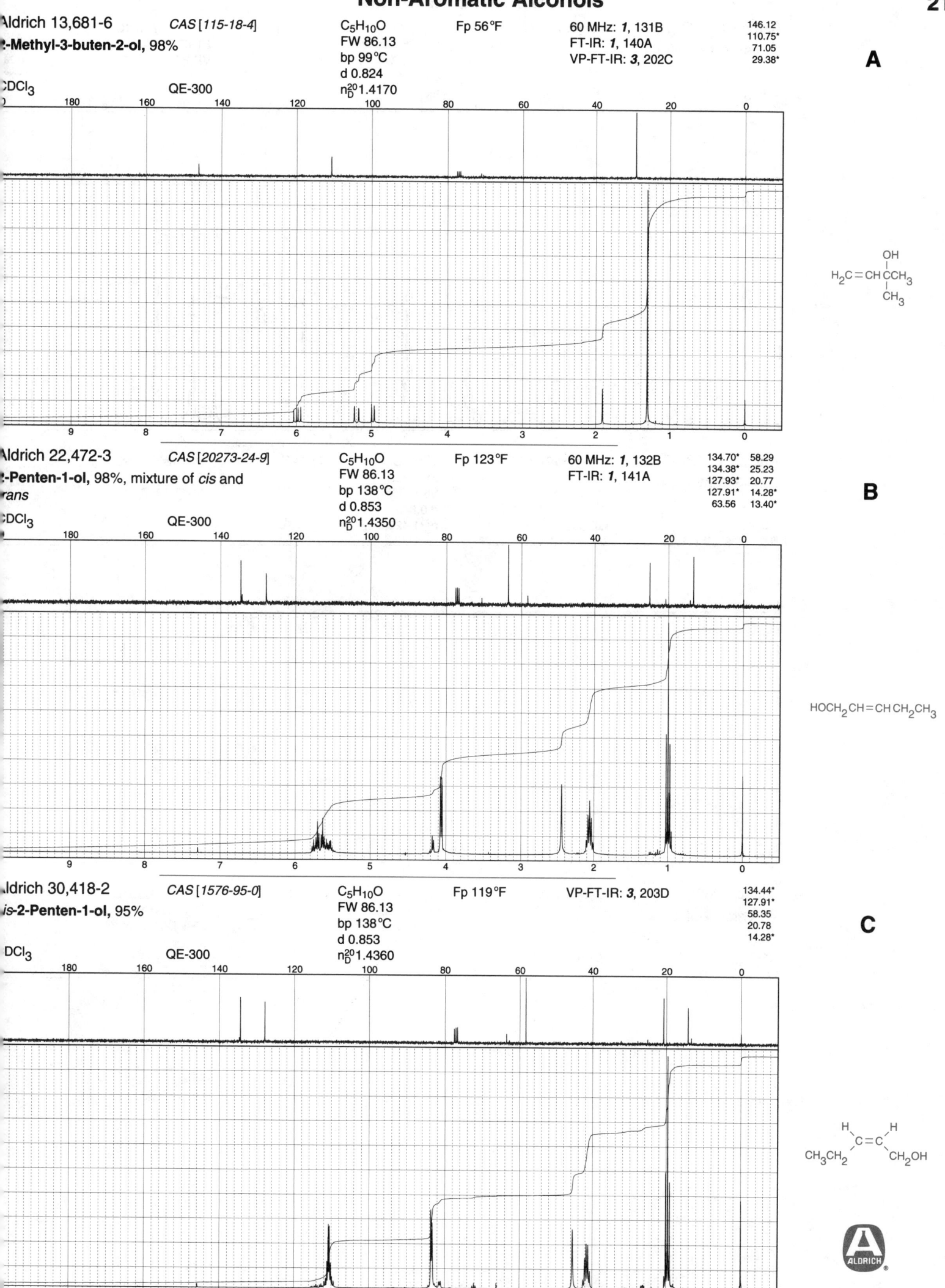

Aldrich 13,681-6   *CAS [115-18-4]* $C_5H_{10}O$ Fp 56°F   60 MHz: *1*, 131B   146.12
110.75*
71.05
29.38*
2-Methyl-3-buten-2-ol, 98%   FW 86.13   FT-IR: *1*, 140A
bp 99°C   VP-FT-IR: *3*, 202C
d 0.824
CDCl₃   QE-300   n²⁰_D 1.4170

A

Aldrich 22,472-3   *CAS [20273-24-9]* $C_5H_{10}O$ Fp 123°F   60 MHz: *1*, 132B   134.70*   58.29
134.38*   25.23
127.93*   20.77
127.91*   14.28*
63.56   13.40*
2-Penten-1-ol, 98%, mixture of *cis* and   FW 86.13   FT-IR: *1*, 141A
trans   bp 138°C
d 0.853
CDCl₃   QE-300   n²⁰_D 1.4350

B

Aldrich 30,418-2   *CAS [1576-95-0]* $C_5H_{10}O$ Fp 119°F   VP-FT-IR: *3*, 203D   134.44*
127.91*
58.35
20.78
14.28*
cis-2-Penten-1-ol, 95%   FW 86.13
bp 138°C
d 0.853
CDCl₃   QE-300   n²⁰_D 1.4360

C

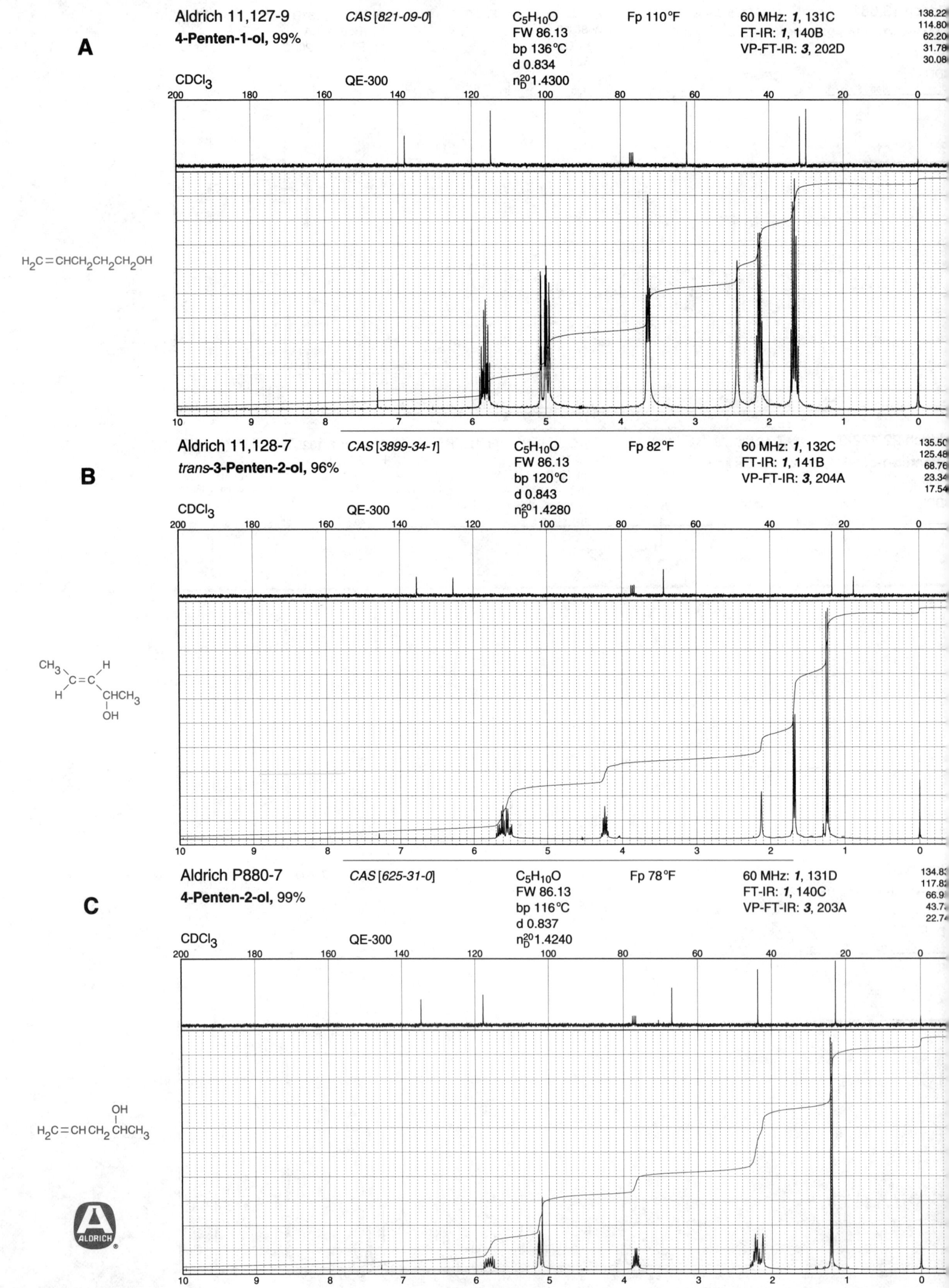

A

Aldrich 11,127-9

4-Penten-1-ol, 99%

CAS [821-09-0]

$C_5H_{10}O$
FW 86.13
bp 136°C
d 0.834
n_D^{20}1.4300

Fp 110°F

60 MHz: *1*, 131C
FT-IR: *1*, 140B
VP-FT-IR: *3*, 202D

138.22
114.80
62.20
31.78
30.08

CDCl₃

QE-300

$H_2C=CHCH_2CH_2CH_2OH$

B

Aldrich 11,128-7

trans-**3-Penten-2-ol,** 96%

CAS [3899-34-1]

$C_5H_{10}O$
FW 86.13
bp 120°C
d 0.843
n_D^{20}1.4280

Fp 82°F

60 MHz: *1*, 132C
FT-IR: *1*, 141B
VP-FT-IR: *3*, 204A

135.50
125.48
68.76
23.34
17.54

CDCl₃

QE-300

C

Aldrich P880-7

4-Penten-2-ol, 99%

CAS [625-31-0]

$C_5H_{10}O$
FW 86.13
bp 116°C
d 0.837
n_D^{20}1.4240

Fp 78°F

60 MHz: *1*, 131D
FT-IR: *1*, 140C
VP-FT-IR: *3*, 203A

134.83
117.82
66.9
43.7
22.74

CDCl₃

QE-300

$H_2C=CHCH_2CHCH_3$

Non-Aromatic Alcohols

212

Irich P860-2   CAS [616-25-1] $C_5H_{10}O$ Fp 77°F   60 MHz: **1**, 132A

Penten-3-ol, 99%

FW 86.13   FT-IR: **1**, 140D

bp 115°C   VP-FT-IR: **3**, 203B

Cl_3 QE-300   d 0.839

$n_D^{20} 1.4240$

140.97*
114.68
74.51*
29.83
9.62*

A

OH
$CH_3CH_2CHCH=CH_2$

Irich 31,120-0   CAS [763-89-3] $C_6H_{12}O$ Fp 146°F   VP-FT-IR: **3**, 204D

Methyl-3-penten-1-ol, 97%

FW 100.16

bp 157°C

Cl_3 QE-300   d 0.858

$n_D^{20} 1.4450$

134.81
120.12*
62.46
31.66
25.81*
17.86*

B

CH_3
$CH_3C=CHCH_2CH_2OH$

Irich 27,415-1   CAS [918-85-4] $C_6H_{12}O$ Fp 78°F   VP-FT-IR: **3**, 205A

3-Methyl-1-penten-3-ol, 99%

FW 100.16

bp 118°C

Cl_3 QE-300   d 0.838

$n_D^{20} 1.4280$

144.89*
111.70
73.45
34.83
27.09*
8.18*

C

OH
$CH_3CH_2CCH=CH_2$
CH_3

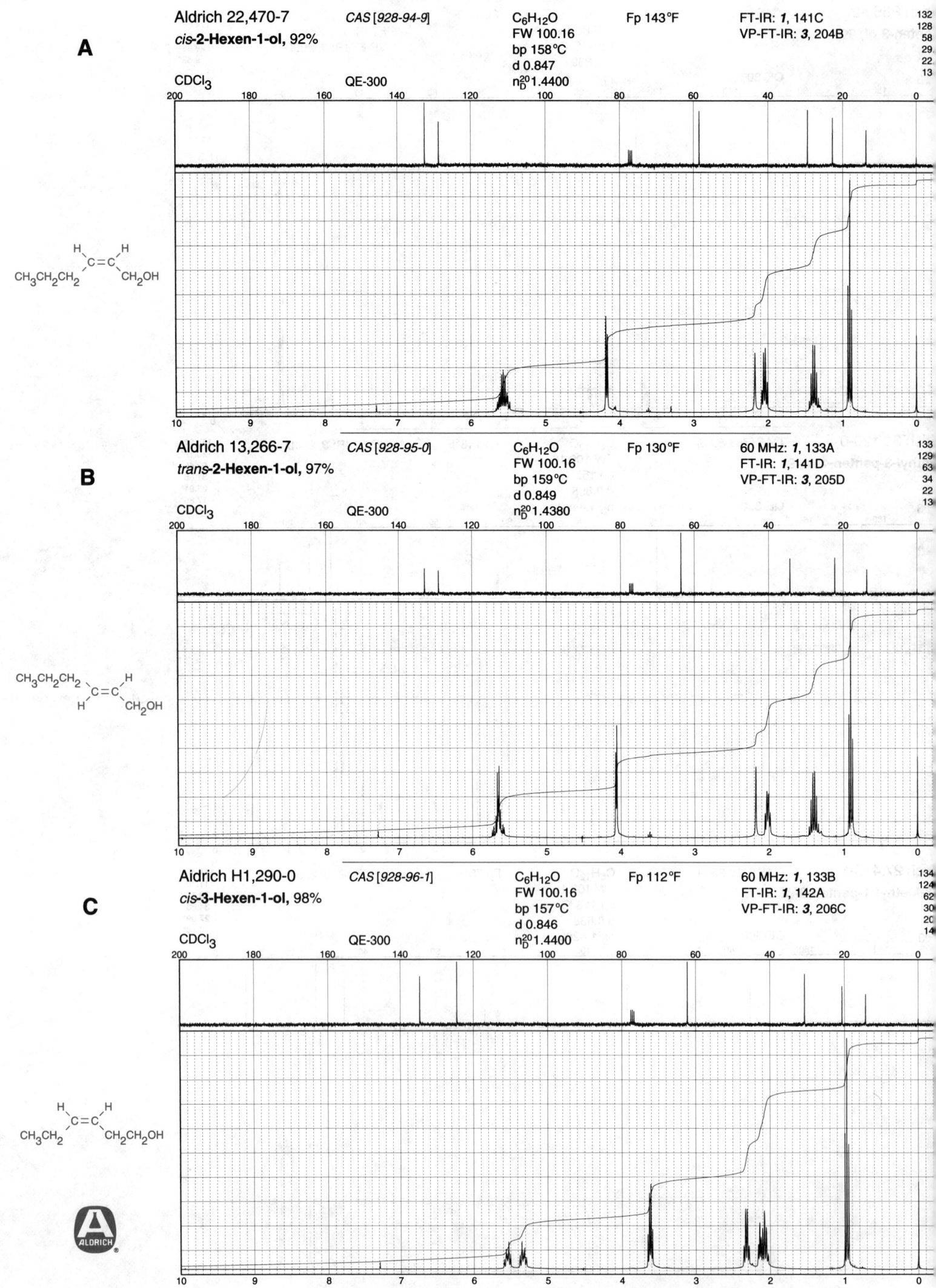

A

Aldrich 22,470-7
cis-**2-Hexen-1-ol,** 92%

CAS [928-94-9]

$C_6H_{12}O$
FW 100.16
bp 158°C
d 0.847
n_D^{20}1.4400

Fp 143°F

FT-IR: *1*, 141C
VP-FT-IR: *3*, 204B

132
128
58
29
22
13

CDCl₃

QE-300

B

Aldrich 13,266-7
trans-**2-Hexen-1-ol,** 97%

CAS [928-95-0]

$C_6H_{12}O$
FW 100.16
bp 159°C
d 0.849
n_D^{20}1.4380

Fp 130°F

60 MHz: *1*, 133A
FT-IR: *1*, 141D
VP-FT-IR: *3*, 205D

133
129
63
34
22
13

CDCl₃

QE-300

C

Aidrich H1,290-0
cis-**3-Hexen-1-ol,** 98%

CAS [928-96-1]

$C_6H_{12}O$
FW 100.16
bp 157°C
d 0.846
n_D^{20}1.4400

Fp 112°F

60 MHz: *1*, 133B
FT-IR: *1*, 142A
VP-FT-IR: *3*, 206C

134
124
62
30
20
14

CDCl₃

QE-300

Non-Aromatic Alcohols

214

Aldrich 22,471-5   CAS [928-97-2]   C₆H₁₂O   Fp 138°F   60 MHz: 1, 133C   135.50*
trans-3-Hexen-1-ol, 98%   FW 100.16   FT-IR: 1, 142B   124.81*
                                       bp 62°C (12 mm)   VP-FT-IR: 3, 206D 62.08
DCl₃   QE-300   d 0.817   35.96
                                       n²⁰_D 1.4390   25.67
                                                                            13.79*

A

$HOCH_2CH_2 \,C=C\, H / H / CH_2CH_3$

Aldrich 23,760-4   CAS [928-92-7]   C₆H₁₂O   Fp 142°F   60 MHz: 1, 133D   130.68*
-Hexen-1-ol, 97%, predominantly trans   FW 100.16   FT-IR: 1, 142C   125.34*
                                       bp 160°C   VP-FT-IR: 3, 207A 62.36
DCl₃   QE-300   d 0.851   32.44
                                       n²⁰_D 1.4390   28.90
                                                                           17.86*

B

$CH_3CH=CHCH_2CH_2CH_2OH$

Aldrich 23,032-4   CAS [821-41-0]   C₆H₁₂O   Fp 117°F   60 MHz: 1, 134A   138.63*
-Hexen-1-ol, 99%   FW 100.16   FT-IR: 1, 142D   114.54
                                       bp 79°C (25 mm)   VP-FT-IR: 3, 207B 62.61
DCl₃   QE-300   d 0.834   33.49
                                       n²⁰_D 1.4350   32.16
                                                                           25.06

C

$H_2C=CHCH_2CH_2CH_2CH_2OH$

A

Aldrich H1,285-4

1-Hexen-3-ol, 98%

CAS [4798-44-1]

$C_6H_{12}O$
FW 100.16
bp 135°C
d 0.834
n_D^{20} 1.4280

Fp 95°F

60 MHz: *1*, 134B
FT-IR: *1*, 143A
VP-FT-IR: *3*, 207C

141.36*
114.37
72.95*
39.19
18.57
13.98*

CDCl₃   QE-300

$CH_3CH_2CH_2\ \overset{OH}{\underset{|}{CH}}CH=CH_2$

B

Aldrich 19,587-1

(±)-6-Methyl-5-hepten-2-ol, 99%

CAS [4630-06-2]

$C_8H_{16}O$
FW 128.22
bp 78°C (14 mm)
d 0.844
n_D^{20} 1.4480

Fp 154°F

FT-IR: *1*, 143B
VP-FT-IR: *3*, 209D

131.89   25.69
124.08* 24.51
67.84*   23.44
39.25   17.65

CDCl₃   QE-300

$CH_3\overset{CH_3}{\underset{|}{C}}=CHCH_2\ CH_2\ \overset{OH}{\underset{|}{CH}}CH_3$

C

Aldrich O-528-4

1-Octen-3-ol, 98%

CAS [3391-86-4]

$C_8H_{16}O$
FW 128.22
bp 85°C (25 mm)
d 0.830
n_D^{20} 1.4370

Fp 142°F

60 MHz: *1*, 134C
FT-IR: *1*, 143C
VP-FT-IR: *3*, 210C

141.36* 31.79
114.40   25.03
73.24*   22.62
37.03   14.02

CDCl₃   QE-300

$CH_3(CH_2)_3CH_2\ \overset{OH}{\underset{|}{CH}}CH=CH_2$

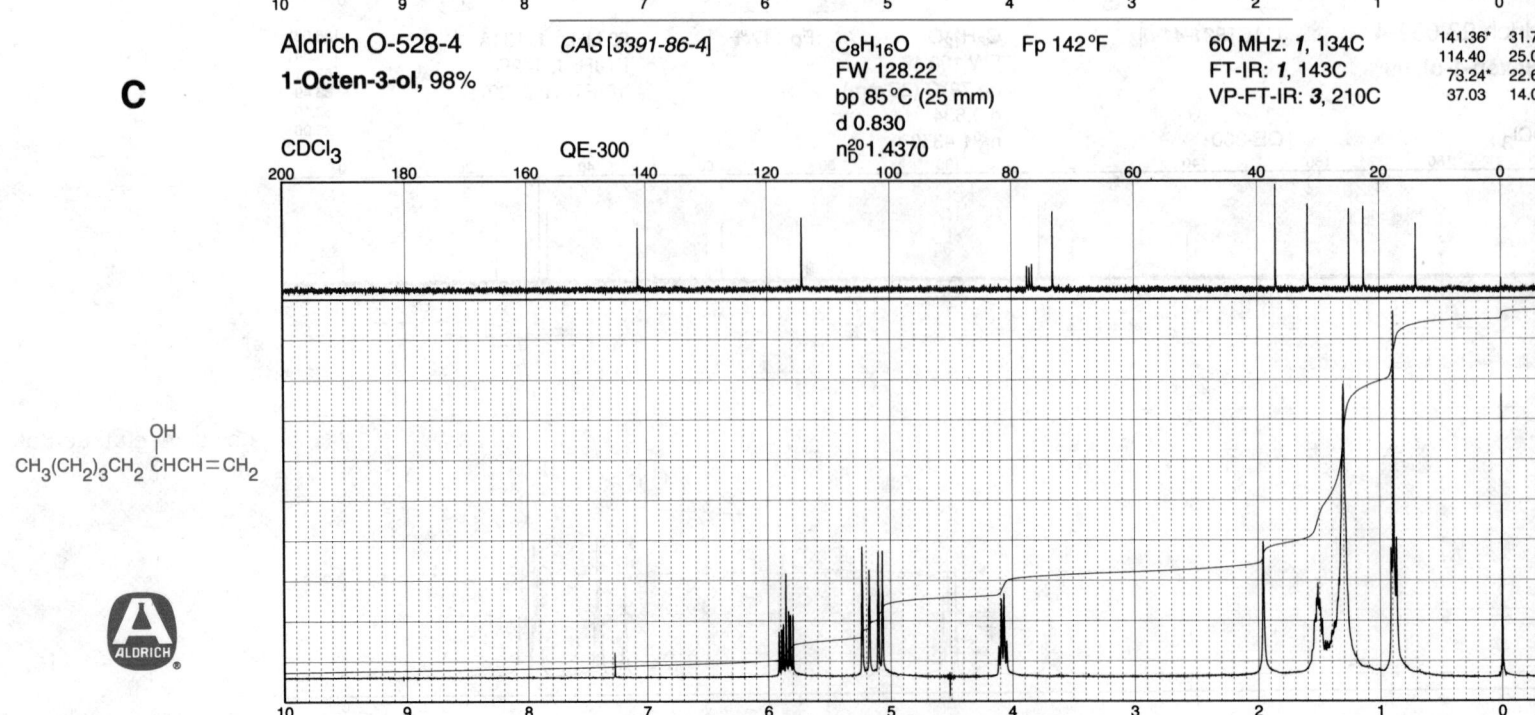

Aldrich W50,521-8   CAS [53907-72-5]   C$_8$H$_{16}$O
7-Octen-4-ol, 99+%   FW 128.22
                                       d 0.830
                                       n$_D^{20}$1.4380
CDCl$_3$ QE-300   Fp 152°F

138.64*	36.48
114.63	30.09
71.13*	18.83
39.67	14.12*

A

H$_2$C=CHCH$_2$CH$_2$CHCH$_2$CH$_2$CH$_3$ (OH)

Aldrich C8,320-1   CAS [106-22-9]   C$_{10}$H$_{20}$O   Fp 209°F   60 MHz: 1, 135B
β-Citronellol, 95%   FW 156.27   FT-IR: 1, 145C
                                      bp 222°C   VP-FT-IR: 3, 213A
CDCl$_3$ QE-300   d 0.857
                                     n$_D^{20}$1.4560

131.16	29.22*
124.70*	25.69*
61.09	25.48
39.91	19.53*
37.25	17.63*

B

CH$_3$C=CHCH$_2$CH$_2$CHCH$_2$CH$_2$OH (CH$_3$, CH$_3$)

Aldrich 30,346-1   CAS [1117-61-9]   C$_{10}$H$_{20}$O   Fp 213°F   VP-FT-IR: 3, 213C
(R)-(+)-β-Citronellol, 98%   FW 156.27
                                     bp 113°C (12 mm)
CDCl$_3$ QE-300   d 0.857
                                     n$_D^{20}$1.4560

131.15	29.23*
124.70*	25.69*
61.07	25.48
39.90	19.53*
37.25	17.63*

C

CH$_3$C=CHCH$_2$CH$_2$CCH$_2$CH$_2$OH (CH$_3$, CH$_3$, H)

ALDRICH

Non-Aromatic Alcohols

A

Aldrich 30,348-8
(*S*)-(-)-β-Citronellol, 99%

CAS [*7540-51-4*]

$C_{10}H_{20}O$
FW 156.27
bp 226°C
d 0.856
n_D^{20} 1.4560

Fp 210°F

VP-FT-IR: ***3***, 213D

131.15	29.25
124.72*	25.68
61.08	25.49
39.93	19.54
37.26	17.62

CDCl3   QE-300

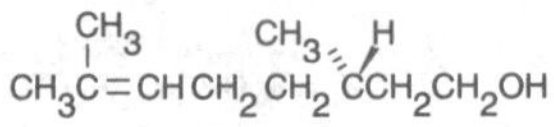

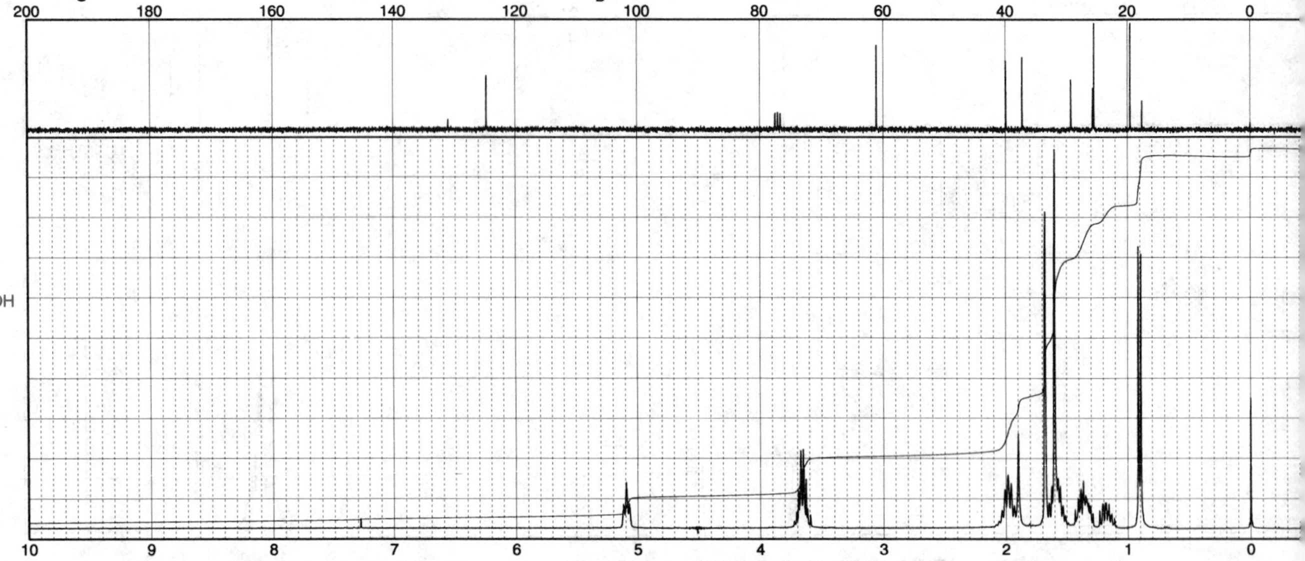

B

Aldrich 19,642-8
Dihydromyrcenol, 99%

CAS [*18479-58-8*]

$C_{10}H_{20}O$
FW 156.27
bp 84°C (10 mm)
d 0.784
n_D^{20} 1.4430

Fp 170°F

60 MHz: ***1***, 135D
FT-IR: ***1***, 143D
VP-FT-IR: ***3***, 211B

144.69*	37.22
112.40	29.29
70.96	22.00
44.10	20.19
37.73*	

CDCl3   QE-300

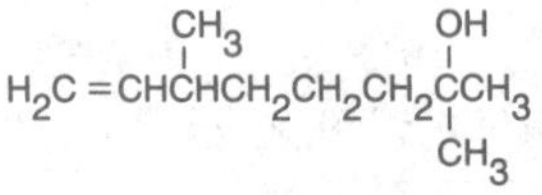

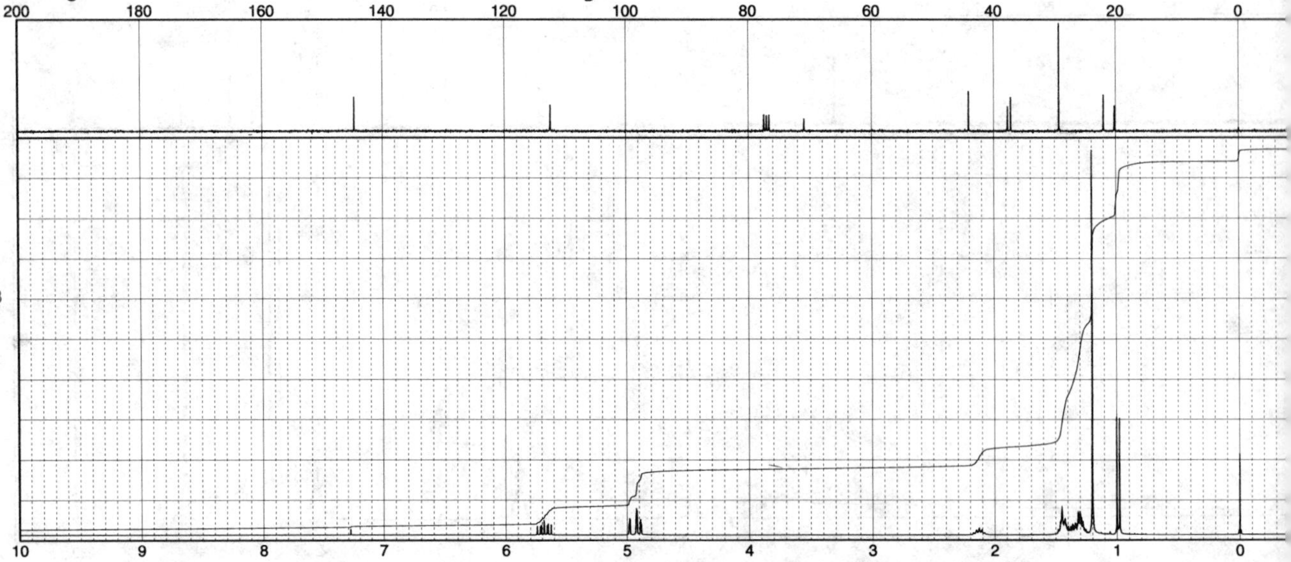

C

Aldrich 11,835-4
9-Decen-1-ol, 97%

CAS [*13019-22-2*]

$C_{10}H_{20}O$
FW 156.27
bp 236°C
d 0.876
n_D^{20} 1.4480

Fp 210°F

60 MHz: ***1***, 134D
FT-IR: ***1***, 145B
VP-FT-IR: ***3***, 212D

139.11*	29.46
114.09	29.40
62.93	29.07
33.79	28.93
32.79	25.76

CDCl3   QE-300

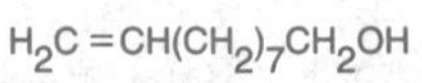

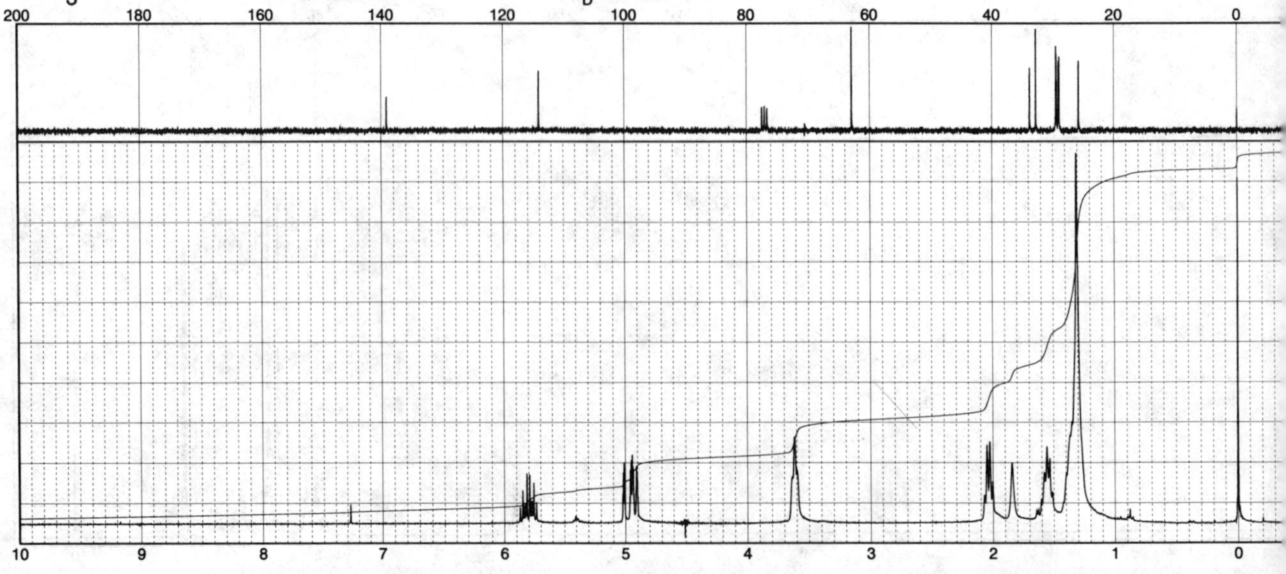

Non-Aromatic Alcohols

A

Aldrich U200-8 CAS [112-43-6]
-Undecylenyl alcohol, 99%

C₁₁H₂₂O
FW 170.30
bp 133°C (15 mm)
d 0.850
n²⁰_D 1.4500

Fp 200°F

60 MHz: *1*, 136A
FT-IR: *1*, 145D
VP-FT-IR: *3*, 214A

139.14*	29.57
114.07	29.44
62.94	29.13
33.81	28.95
32.80	25.77

DCl₃   QE-300

H₂C=CH(CH₂)₈CH₂OH

B

Aldrich 24,899-1 CAS [20056-92-2]
is-7-Dodecen-1-ol, 98%

C₁₂H₂₄O
FW 184.32
n²⁰_D 1.4550
Fp 142°F

FT-IR: *1*, 144A
VP-FT-IR: *3*, 214B

129.96*	29.09
129.65*	27.13
62.93	26.93
32.78	25.67
31.97	22.36
29.73	13.99*

DCl₃   QE-300

CH₃CH₂CH₂CH₂ H\C=C/H CH₂(CH₂)₅OH

C

Aldrich 24,900-9 CAS [40642-43-1]
is-7-Tetradecen-1-ol, 95%

C₁₄H₂₈O
FW 212.38
n²⁰_D 1.4575
Fp 142°F

FT-IR: *1*, 144B
VP-FT-IR: *3*, 214C

130.02*	29.00
129.64*	27.24
62.99	27.14
32.79	25.67
31.80	22.67
29.73	14.09*
29.09	

DCl₃   QE-300

CH₃(CH₂)₄CH₂ H\C=C/H CH₂(CH₂)₅OH

ALDRICH®

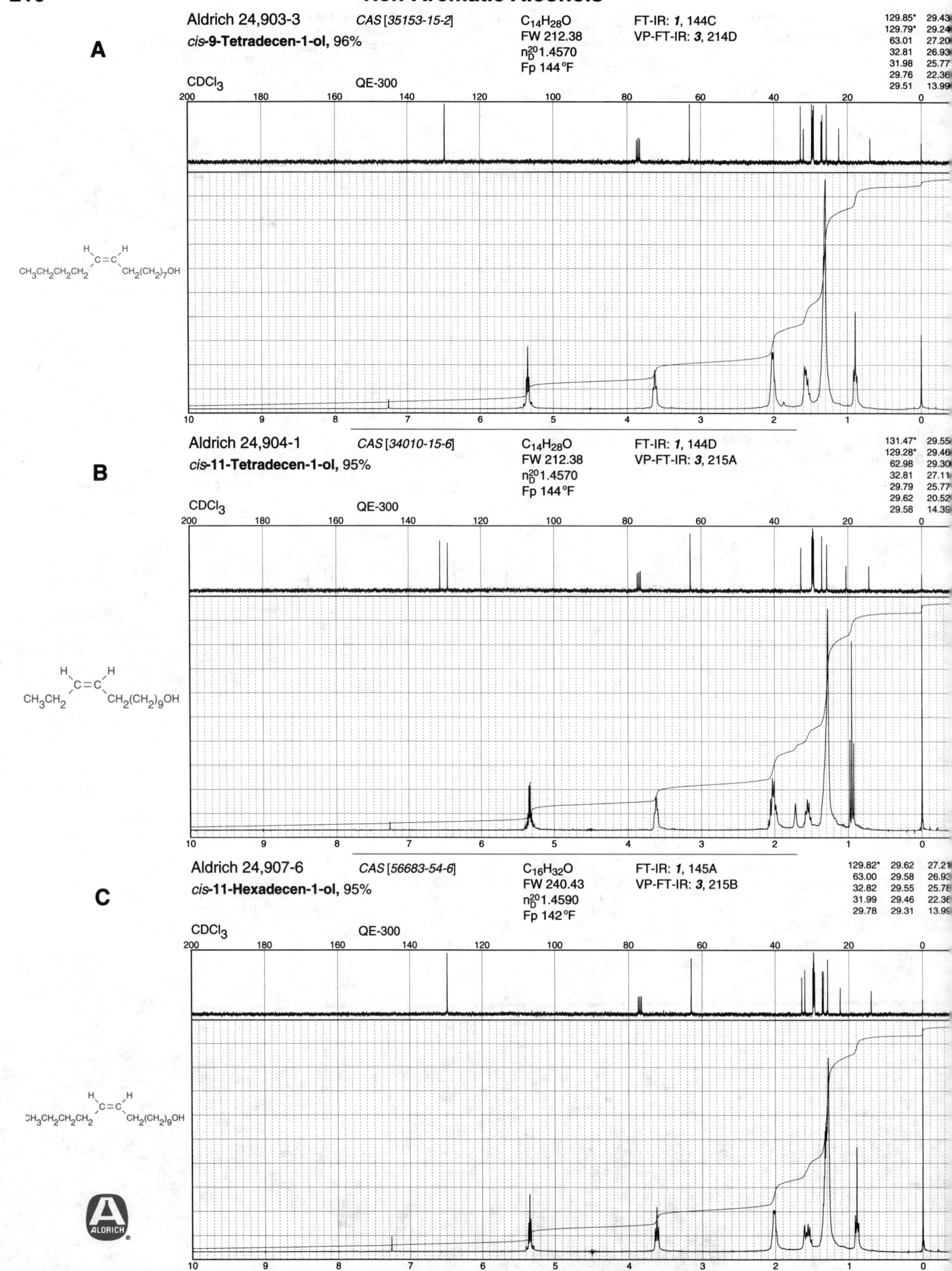

A

Aldrich 24,903-3

CAS [35153-15-2]

cis-9-Tetradecen-1-ol, 96%

$C_{14}H_{28}O$
FW 212.38
n_D^{20} 1.4570
Fp 144°F

FT-IR: *1*, 144C
VP-FT-IR: *3*, 214D

129.85*	29.43
129.79*	29.24
63.01	27.20
32.81	26.93
31.98	25.77
29.76	22.36
29.51	13.99

CDCl₃   QE-300

B

Aldrich 24,904-1

CAS [34010-15-6]

cis-11-Tetradecen-1-ol, 95%

$C_{14}H_{28}O$
FW 212.38
n_D^{20} 1.4570
Fp 144°F

FT-IR: *1*, 144D
VP-FT-IR: *3*, 215A

131.47*	29.55
129.28*	29.46
62.98	29.30
32.81	27.11
29.79	25.77
29.62	20.52
29.58	14.39

CDCl₃   QE-300

C

Aldrich 24,907-6

CAS [56683-54-6]

cis-11-Hexadecen-1-ol, 95%

$C_{16}H_{32}O$
FW 240.43
n_D^{20} 1.4590
Fp 142°F

FT-IR: *1*, 145A
VP-FT-IR: *3*, 215B

129.82*	29.62	27.21
63.00	29.58	26.93
32.82	29.55	25.78
31.99	29.46	22.36
29.78	29.31	13.99

CDCl₃   QE-300

Non-Aromatic Alcohols

Aldrich 13,991-2   CAS [7541-49-3]   C$_{20}$H$_{40}$O   Fp >230°F

Phytol, 97%, mixture of isomers

FW 296.54
bp 203°C (10 mm)
d 0.850
n$_D^{20}$ 1.4630

CDCl$_3$ QE-300

140.44	39.36	36.76	25.15	19.71*
140.11	37.42	36.66	24.80	19.68*
123.94*	37.37	32.77*	24.46	16.17*
123.08*	37.31	32.67*	23.44*	
59.35	37.28	32.18	22.72*	
59.06	36.93	27.97*	22.62*	
39.87	36.84	25.70	19.74*	

A

$$H-\left[-CH_2CH(CH_3)CH_2CH_2-\right]_3-CH_2C(CH_3)=CHCH_2OH$$

Aldrich O-760-0   CAS [143-28-2]   C$_{18}$H$_{36}$O   Fp >230°F

Oleyl alcohol, tech., 65%

FW 268.49
bp 207°C (13 mm)
d 0.849
n$_D^{20}$ 1.4600

CDCl$_3$ QE-300

60 MHz: **1**, 136B
FT-IR: **1**, 146A
VP-FT-IR: **3**, 215C

130.34*	29.69	29.17
130.28*	29.65	27.23
32.82	29.54	25.78
31.93	29.44	22.70
29.78	29.34	14.11*
29.71	29.26	

B

$$CH_3(CH_2)_7CH=CH(CH_2)_7CH_2OH$$

Aldrich 33,951-2   CAS [3513-81-3]   C$_4$H$_8$O$_2$ Fp >230°F

2-Methylene-1,3-propanediol, 99%

FW 88.11
bp 94°C (2 mm)
d 1.081
n$_D^{20}$ 1.4750

CDCl$_3$ + DMSO-d_6 QE-300

| 149.62 |
| 108.21 |
| 62.23 |

C

$$HOCH_2-\underset{\underset{CH_2}{\|}}{C}-CH_2OH$$

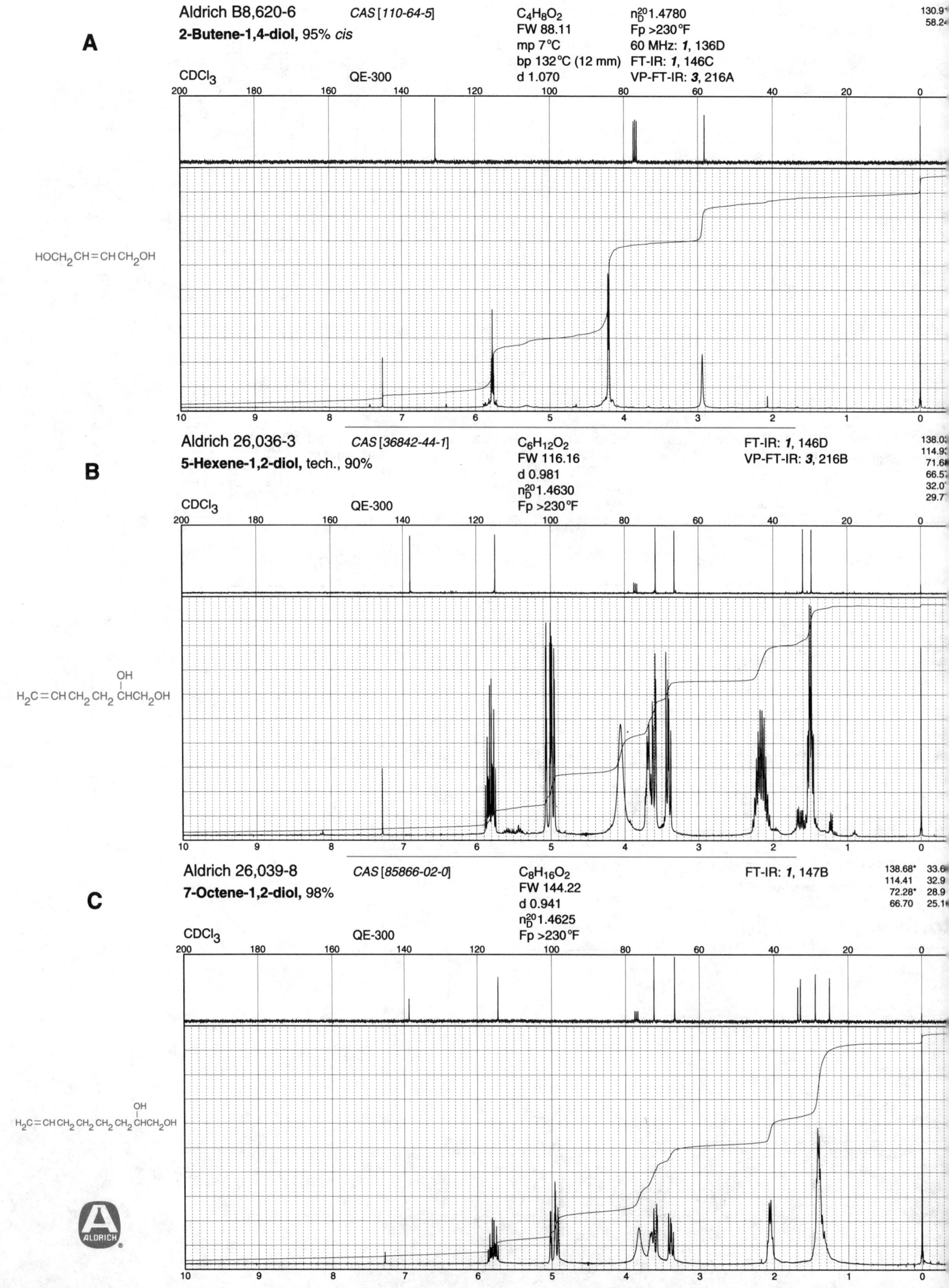

A

Aldrich B8,620-6   CAS [110-64-5] $C_4H_8O_2$ n_D^{20}1.4780   130.91
2-Butene-1,4-diol, 95% *cis*   FW 88.11   Fp >230°F   58.24
mp 7°C   60 MHz: *1*, 136D
bp 132°C (12 mm)   FT-IR: *1*, 146C
d 1.070   VP-FT-IR: *3*, 216A

CDCl₃   QE-300

$HOCH_2CH=CHCH_2OH$

Aldrich 26,036-3   CAS [36842-44-1] $C_6H_{12}O_2$ FT-IR: *1*, 146D   138.03
5-Hexene-1,2-diol, tech., 90%   FW 116.16   VP-FT-IR: *3*, 216B   114.93
d 0.981   71.6
n_D^{20}1.4630   66.5
Fp >230°F   32.0
   29.7

B

CDCl₃   QE-300

OH
|
$H_2C=CHCH_2CH_2$ CHCH₂OH

Aldrich 26,039-8   CAS [85866-02-0] $C_8H_{16}O_2$ FT-IR: *1*, 147B   138.68*   33.6
7-Octene-1,2-diol, 98%   FW 144.22   114.41   32.9
d 0.941   72.28*   28.9
n_D^{20}1.4625   66.70   25.1
Fp >230°F

C

CDCl₃   QE-300

OH
|
$H_2C=CHCH_2 CH_2 CH_2$ CHCH₂OH

Non-Aromatic Alcohols

222

Aldrich 32,466-3   CAS [922-65-6]   C₅H₈O   Fp 86°F   VP-FT-IR: 3, 217A
1,4-Pentadien-3-ol, 99%   FW 84.12   bp 116°C   d 0.865   n₂₀D 1.4460

139.15*
115.20
73.91*

A

CDCl₃   QE-300

$H_2C=CHCHCH=CH_2$ with OH

Aldrich 18,305-9   CAS [17102-64-6]   C₆H₁₀O   n₂₀D 1.5000
trans,trans-2,4-Hexadien-1-ol, 97%   FW 98.15   Fp 162°F
mp 32°C   60 MHz: 1, 137B
bp 80°C (12 mm)   FT-IR: 1, 147A
d 0.871   VP-FT-IR: 3, 216C

131.72*
130.79*
129.82*
129.25*
63.24
18.06*

B

CDCl₃   QE-300

Aldrich 11,855-9   CAS [924-41-4]   C₆H₁₀O   60 MHz: 1, 137C
1,5-Hexadien-3-ol, 95%   FW 98.15   FT-IR: 1, 147C
d 0.878   VP-FT-IR: 3, 217B
n₂₀D 1.4480   Fp 85°F

140.28*
134.06*
118.13
114.71
71.85*
41.65

C

CDCl₃   QE-300

$H_2C=CHCH_2CHCH=CH_2$ with OH

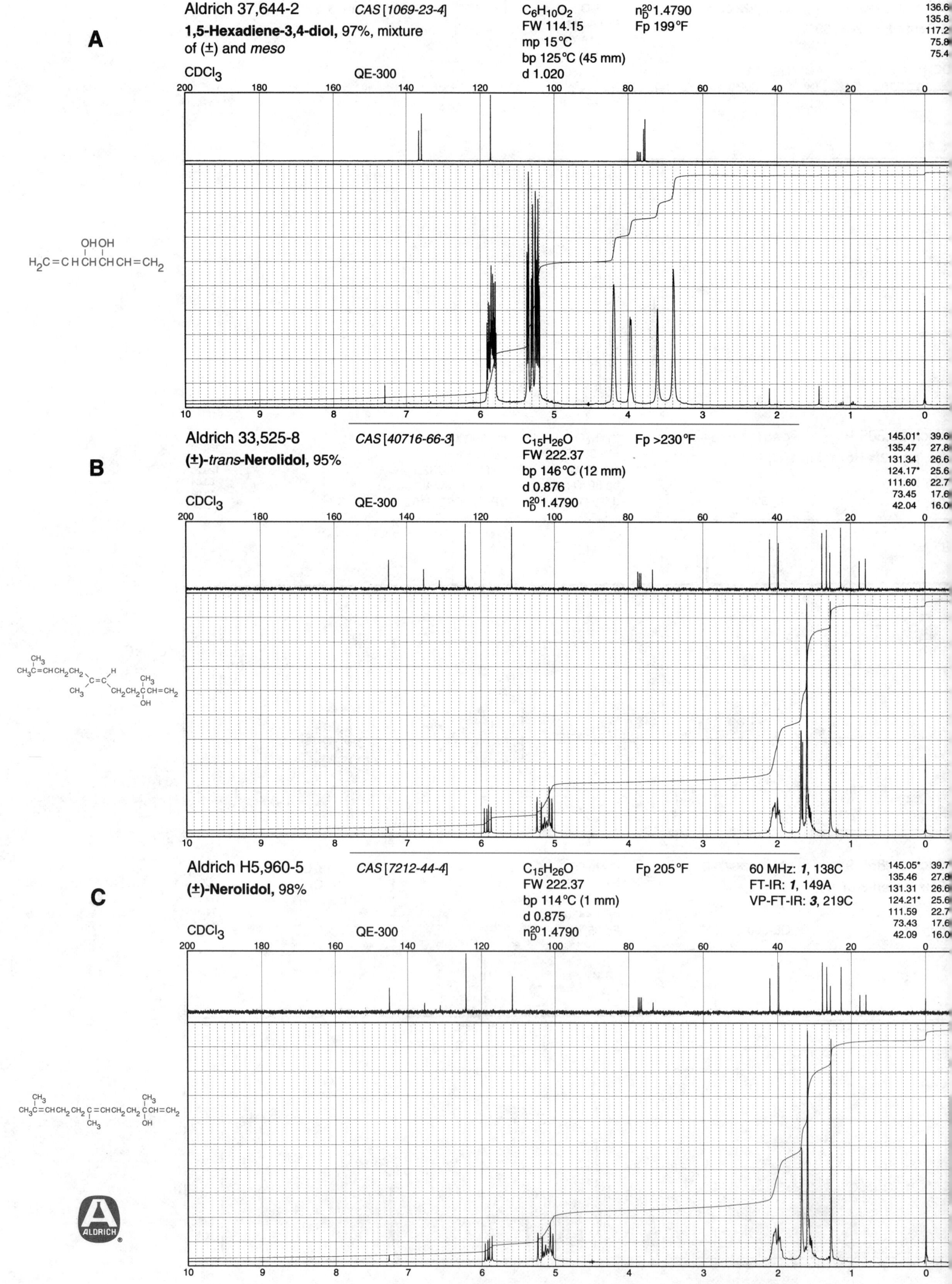

A

Aldrich 37,644-2   CAS [1069-23-4] $C_6H_{10}O_2$ $n_D^{20}1.4790$
1,5-Hexadiene-3,4-diol, 97%, mixture   FW 114.15   Fp 199°F
of (±) and *meso*   mp 15°C
                                            bp 125°C (45 mm)
CDCl$_3$ QE-300   d 1.020

136.6
135.8
117.2
75.8
75.4

B

Aldrich 33,525-8   CAS [40716-66-3] $C_{15}H_{26}O$ Fp >230°F
(±)-*trans*-Nerolidol, 95%   FW 222.37
                                    bp 146°C (12 mm)
                                    d 0.876
CDCl$_3$ QE-300 $n_D^{20}1.4790$

145.01*   39.6
135.47   27.8
131.34   26.6
124.17*   25.6
111.60   22.7
73.45   17.6
42.04   16.0

C

Aldrich H5,960-5   CAS [7212-44-4] $C_{15}H_{26}O$ Fp 205°F   60 MHz: *1*, 138C
(±)-Nerolidol, 98%   FW 222.37   FT-IR: *1*, 149A
                          bp 114°C (1 mm)   VP-FT-IR: *3*, 219C
                          d 0.875
CDCl$_3$ QE-300 $n_D^{20}1.4790$

145.05*   39.7
135.46   27.8
131.31   26.6
124.21*   25.6
111.59   22.7
73.43   17.6
42.09   16.0

ALDRICH

Aldrich 11,158-9 CAS [2883-45-6] C$_7$H$_{12}$O Fp 104°F 60 MHz: *1*, 137D

1,6-Heptadien-4-ol, 97% FW 112.17 FT-IR: *1*, 147D
bp 151°C VP-FT-IR: *3*, 217D
CDCl$_3$ QE-300 d 0.864
n$_D^{20}$ 1.4500

134.64*	
117.89	
69.85*	
41.23	

A

$H_2C=CH\,CH_2\,CHCH_2\,CH=CH_2$ (with OH)

Aldrich 23,876-7 CAS [80192-56-9] C$_9$H$_{16}$O Fp 173°F 60 MHz: *1*, 138A

4-Dimethyl-2,6-heptadien-1-ol, 96%, FW 140.23 FT-IR: *1*, 148A
mixture of isomers bp 86°C (10 mm) VP-FT-IR: *3*, 218A
DCl$_3$ QE-300 d 1.351
n$_D^{20}$ 1.4640

137.13*	41.66
133.55	32.05*
131.77*	20.43*
115.55	13.86*
68.81	

B

$HOCH_2\,C=CH\,CHCH_2\,CH=CH_2$ (with CH$_3$, CH$_3$)

Aldrich 26,890-9 CAS [106-25-2] C$_{10}$H$_{18}$O Fp 170°F VP-FT-IR: *3*, 219A

rol, 97% FW 154.25
bp 104°C (9 mm)
Cl$_3$ QE-300 d 0.876
n$_D^{20}$ 1.4750

139.69	31.99
132.29	26.57
124.48*	25.65*
123.82*	23.41*
58.93	17.64*

C

$CH_3C=CHCH_2\,CH_2\,C=CHCH_2OH$ (with CH$_3$, CH$_3$)

ALDRICH®

Non-Aromatic Alcohols

A

Aldrich 16,333-3   CAS [106-24-1] $C_{10}H_{18}O$ Fp 170°F   60 MHz: *1*, 135A

Geraniol, 98%

FW 154.25   FT-IR: *1*, 148D

bp 230°C   VP-FT-IR: *3*, 219B

d 0.889

n_D^{20} 1.4760

139.48	39.56
131.64	26.42
123.89*	25.65*
123.42*	17.67*
59.30	16.25*

CDCl₃   QE-300

$CH_3C=CHCH_2CH_2C=CHCH_2OH$ (with two CH_3 groups)

B

Aldrich L260-2   CAS [78-70-6] $C_{10}H_{18}O$ Fp 169°F   60 MHz: *1*, 138B

(±)-Linalool, 97%

FW 154.25   FT-IR: *1*, 148C

bp 196°C (720 mm)   VP-FT-IR: *3*, 218D

d 0.870

n_D^{20} 1.4620

145.03*	42.1
131.79	27.85
124.32*	25.68
111.61	22.8
73.41	17.6

CDCl₃   QE-300

$CH_3C=CHCH_2CH_2C(OH)CH=CH_2$ (with CH_3 groups)

C

Aldrich 24,917-3   CAS [57002-06-9] $C_{12}H_{22}O$ FT-IR: *1*, 148B

***trans*-8,*trans*-10-Dodecadien-1-ol, 98%**

FW 182.31   VP-FT-IR: *3*, 218B

mp 32°C

Fp 144°F

131.98*	32.5
131.68*	29.3
130.25*	29.3
126.59*	29.
62.91	25.
32.75	17.

CDCl₃   QE-300

$CH_3C=C$ ... $C=C$ ... $CH_2(CH_2)_6OH$

ALDRICH

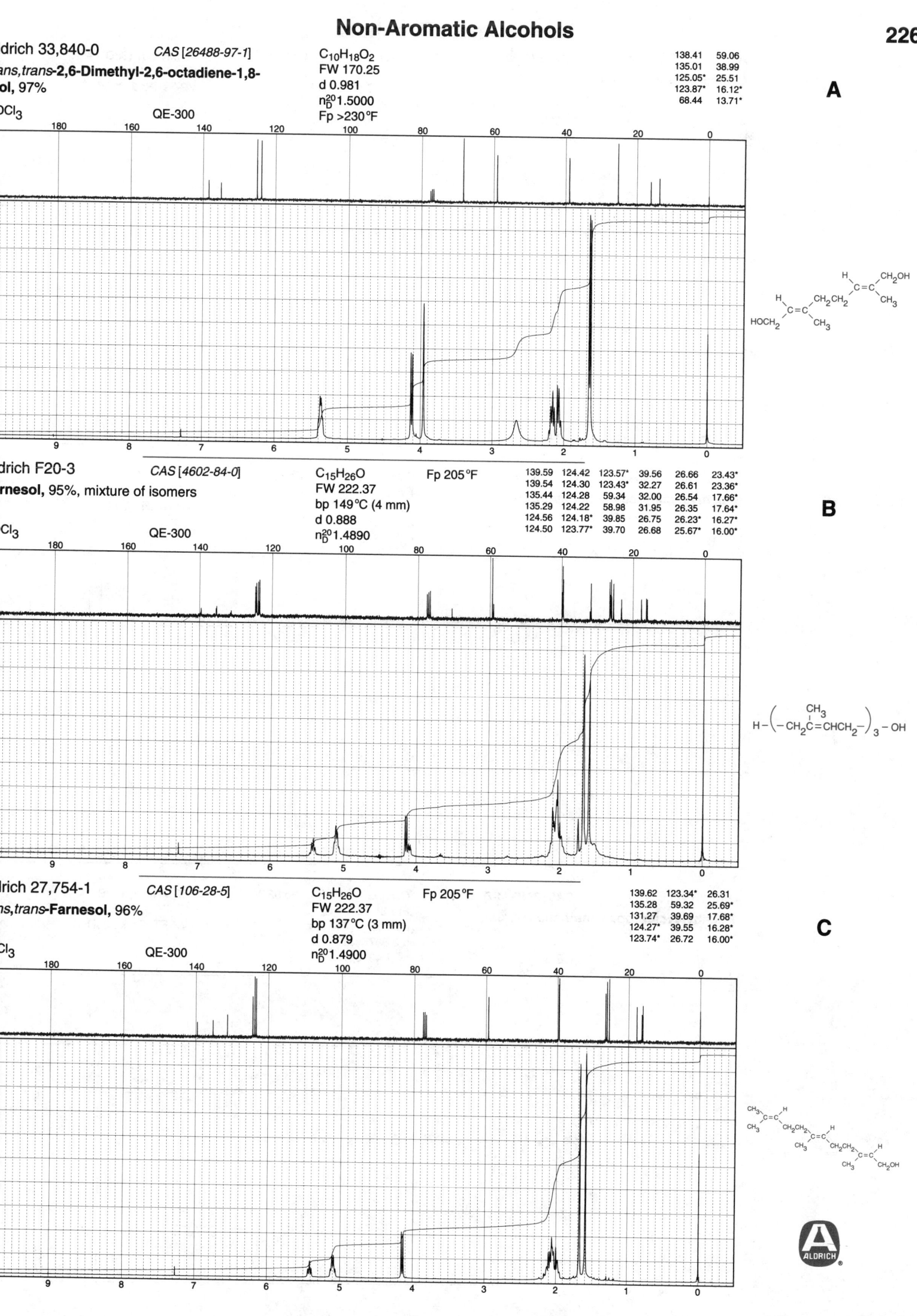

A

Aldrich 33,840-0 CAS [26488-97-1]

trans,trans-2,6-Dimethyl-2,6-octadiene-1,8-diol, 97%

CDCl₃ QE-300

C₁₀H₁₈O₂
FW 170.25
d 0.981
n²⁰_D 1.5000
Fp >230°F

138.41	59.06
135.01	38.99
125.05*	25.51
123.87*	16.12*
68.44	13.71*

B

Aldrich F20-3 CAS [4602-84-0]

Farnesol, 95%, mixture of isomers

CDCl₃ QE-300

C₁₅H₂₆O
FW 222.37
bp 149°C (4 mm)
d 0.888
n²⁰_D 1.4890

Fp 205°F

139.59	124.42	123.57*	39.56	26.66	23.43*
139.54	124.30	123.43*	32.27	26.61	23.36*
135.44	124.28	59.34	32.00	26.54	17.66*
135.29	124.22	58.98	31.95	26.35	17.64*
124.56	124.18*		39.85	26.75	16.27*
124.50	123.77*		39.70	26.68	25.67*	16.00*

C

Aldrich 27,754-1 CAS [106-28-5]

trans,trans-Farnesol, 96%

CDCl₃ QE-300

C₁₅H₂₆O
FW 222.37
bp 137°C (3 mm)
d 0.879
n²⁰_D 1.4900

Fp 205°F

139.62	123.34*	26.31
135.28	59.32	25.69*
131.27	39.69	17.68*
124.27*	39.55	16.28*
123.74*	26.72	16.00*

Non-Aromatic Alcohols

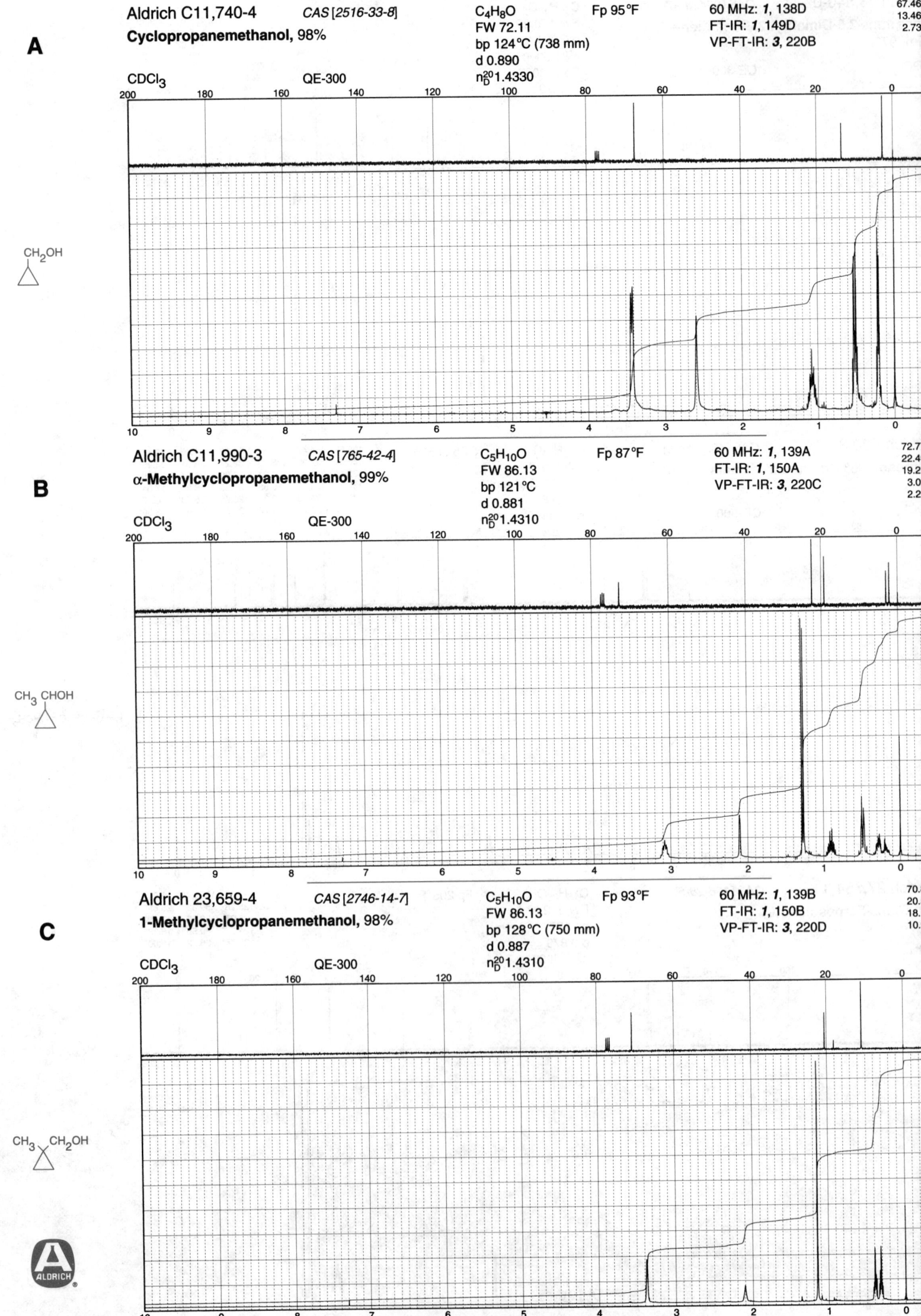

A

Aldrich C11,740-4   *CAS [2516-33-8]* C_4H_8O Fp 95°F   60 MHz: *1*, 138D   67.46

Cyclopropanemethanol, 98%   FW 72.11   FT-IR: *1*, 149D   13.46

bp 124°C (738 mm)   VP-FT-IR: *3*, 220B   2.73

d 0.890

n_D^{20} 1.4330

CDCl₃   QE-300

CH₂OH

B

Aldrich C11,990-3   *CAS [765-42-4]* $C_5H_{10}O$ Fp 87°F   60 MHz: *1*, 139A   72.77

α-Methylcyclopropanemethanol, 99%   FW 86.13   FT-IR: *1*, 150A   22.4

bp 121°C   VP-FT-IR: *3*, 220C   19.2

d 0.881   3.0

n_D^{20} 1.4310   2.2

CDCl₃   QE-300

CH₃ CHOH

C

Aldrich 23,659-4   *CAS [2746-14-7]* $C_5H_{10}O$ Fp 93°F   60 MHz: *1*, 139B   70.

1-Methylcyclopropanemethanol, 98%   FW 86.13   FT-IR: *1*, 150B   20.

bp 128°C (750 mm)   VP-FT-IR: *3*, 220D   18.

d 0.887   10.

n_D^{20} 1.4310

CDCl₃   QE-300

CH₃ CH₂OH

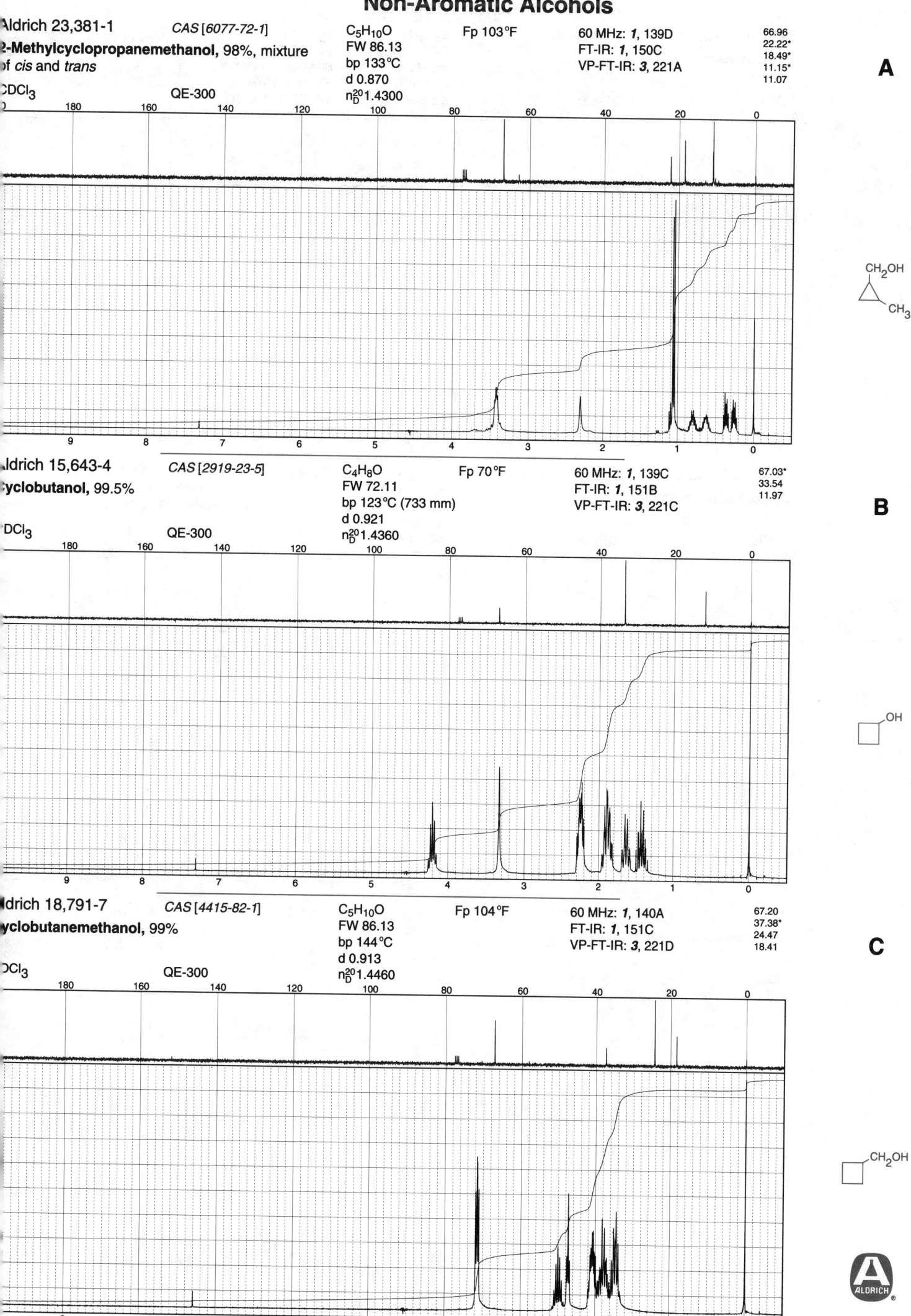

Aldrich 23,381-1   CAS [6077-72-1]   C5H10O   Fp 103°F   60 MHz: 1, 139D   66.96
2-Methylcyclopropanemethanol, 98%, mixture   FW 86.13   FT-IR: 1, 150C   22.22*
of cis and trans   bp 133°C   VP-FT-IR: 3, 221A   18.49*
   d 0.870   11.15*
CDCl3   QE-300   n²⁰_D 1.4300   11.07

A

Aldrich 15,643-4   CAS [2919-23-5]   C4H8O   Fp 70°F   60 MHz: 1, 139C   67.03*
Cyclobutanol, 99.5%   FW 72.11   FT-IR: 1, 151B   33.54
   bp 123°C (733 mm)   VP-FT-IR: 3, 221C   11.97
   d 0.921
CDCl3   QE-300   n²⁰_D 1.4360

B

Aldrich 18,791-7   CAS [4415-82-1]   C5H10O   Fp 104°F   60 MHz: 1, 140A   67.20
Cyclobutanemethanol, 99%   FW 86.13   FT-IR: 1, 151C   37.38*
   bp 144°C   VP-FT-IR: 3, 221D   24.47
   d 0.913   18.41
CDCl3   QE-300   n²⁰_D 1.4460

C

Non-Aromatic Alcohols

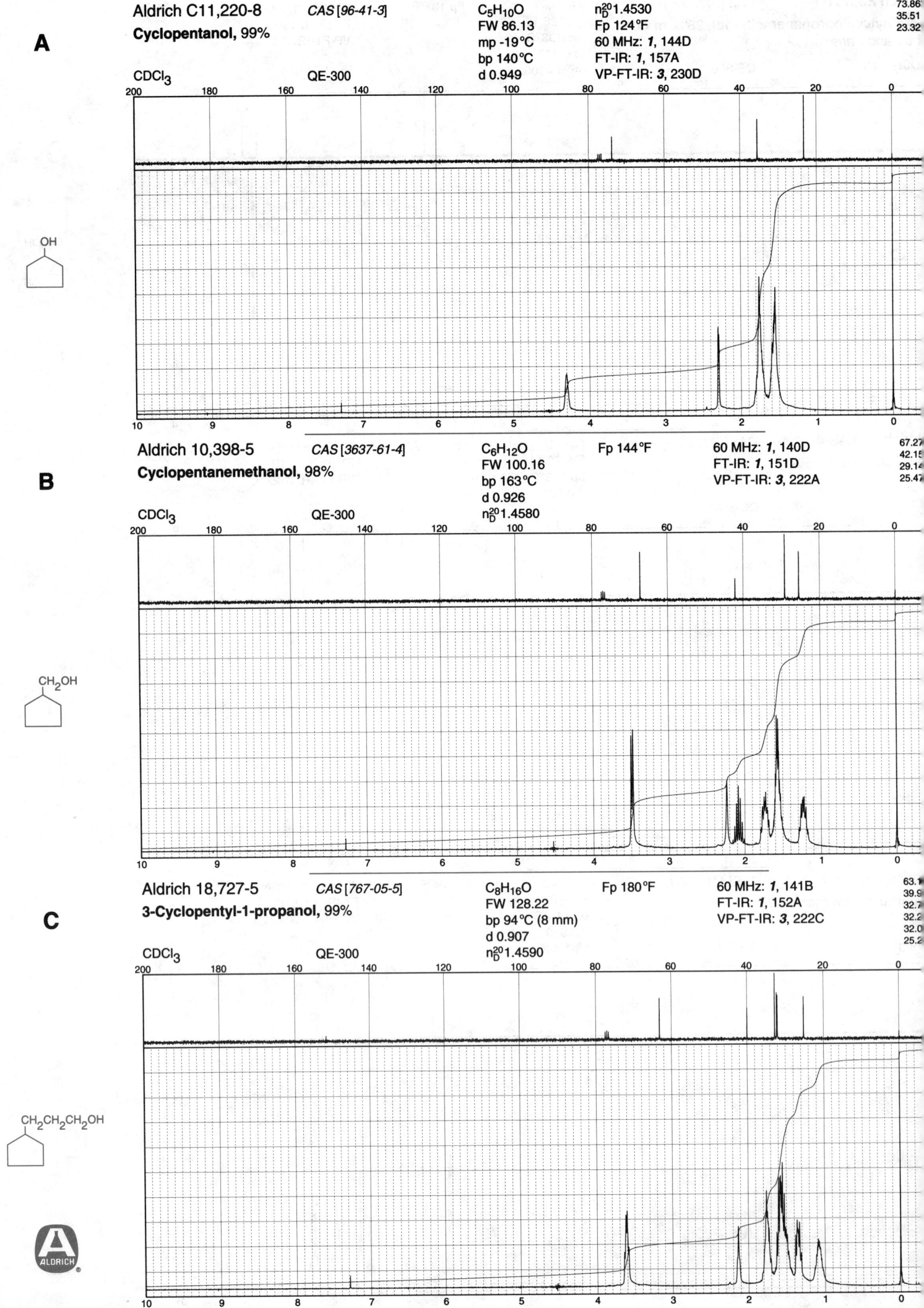

A

Aldrich C11,220-8   *CAS [96-41-3]* $C_5H_{10}O$ n_D^{20} 1.4530   73.86
Cyclopentanol, 99%   FW 86.13   Fp 124°F   35.51
mp -19°C   60 MHz: *1*, 144D   23.32
bp 140°C   FT-IR: *1*, 157A
CDCl₃   QE-300   d 0.949   VP-FT-IR: *3*, 230D

B

Aldrich 10,398-5   *CAS [3637-61-4]* $C_6H_{12}O$ Fp 144°F   60 MHz: *1*, 140D   67.27
Cyclopentanemethanol, 98%   FW 100.16   FT-IR: *1*, 151D   42.15
bp 163°C   VP-FT-IR: *3*, 222A   29.14
d 0.926   25.47
CDCl₃   QE-300 n_D^{20} 1.4580

C

Aldrich 18,727-5   *CAS [767-05-5]* $C_8H_{16}O$ Fp 180°F   60 MHz: *1*, 141B   63.1
3-Cyclopentyl-1-propanol, 99%   FW 128.22   FT-IR: *1*, 152A   39.9
bp 94°C (8 mm)   VP-FT-IR: *3*, 222C   32.7
d 0.907   32.2
CDCl₃   QE-300 n_D^{20} 1.4590   32.0
25.2

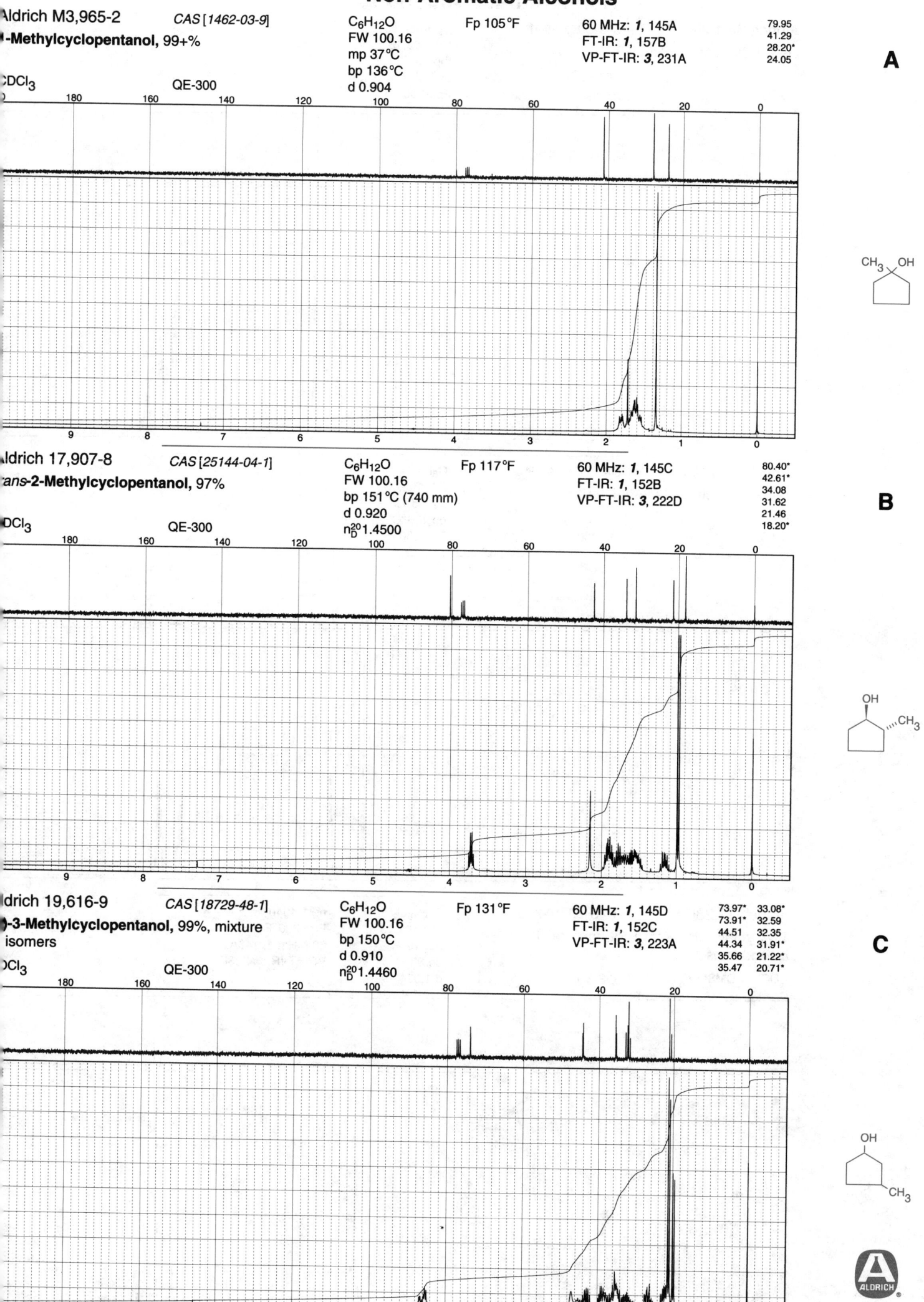

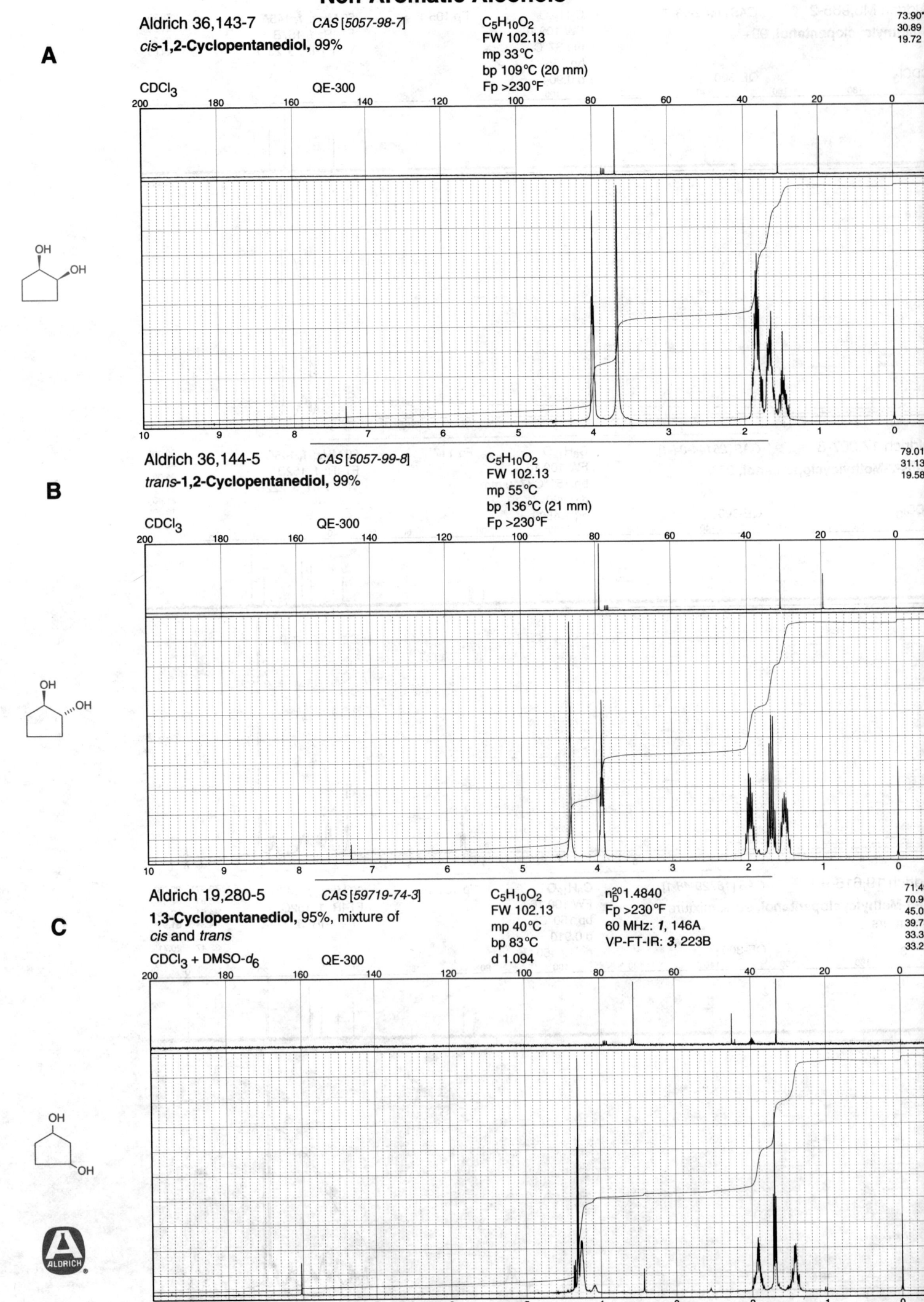

A

Aldrich 36,143-7   CAS [5057-98-7] $C_5H_{10}O_2$
cis-**1,2-Cyclopentanediol, 99%**   FW 102.13
                                                  mp 33°C
                                                  bp 109°C (20 mm)
                                                  Fp >230°F

CDCl$_3$ QE-300   73.90*
                                                  30.89
                                                  19.72

B

Aldrich 36,144-5   CAS [5057-99-8] $C_5H_{10}O_2$
trans-**1,2-Cyclopentanediol, 99%**   FW 102.13
                                                  mp 55°C
                                                  bp 136°C (21 mm)
                                                  Fp >230°F

CDCl$_3$ QE-300   79.01*
                                                  31.13
                                                  19.58

C

Aldrich 19,280-5   CAS [59719-74-3] $C_5H_{10}O_2$ n_D^{20}1.4840
1,3-Cyclopentanediol, 95%, mixture of   FW 102.13   Fp >230°F
cis and trans   mp 40°C   60 MHz: 1, 146A
                                                 bp 83°C   VP-FT-IR: 3, 223B
CDCl$_3$ + DMSO-d_6 QE-300   d 1.094

71.4*
70.9*
45.02
39.73
33.3*
33.2*

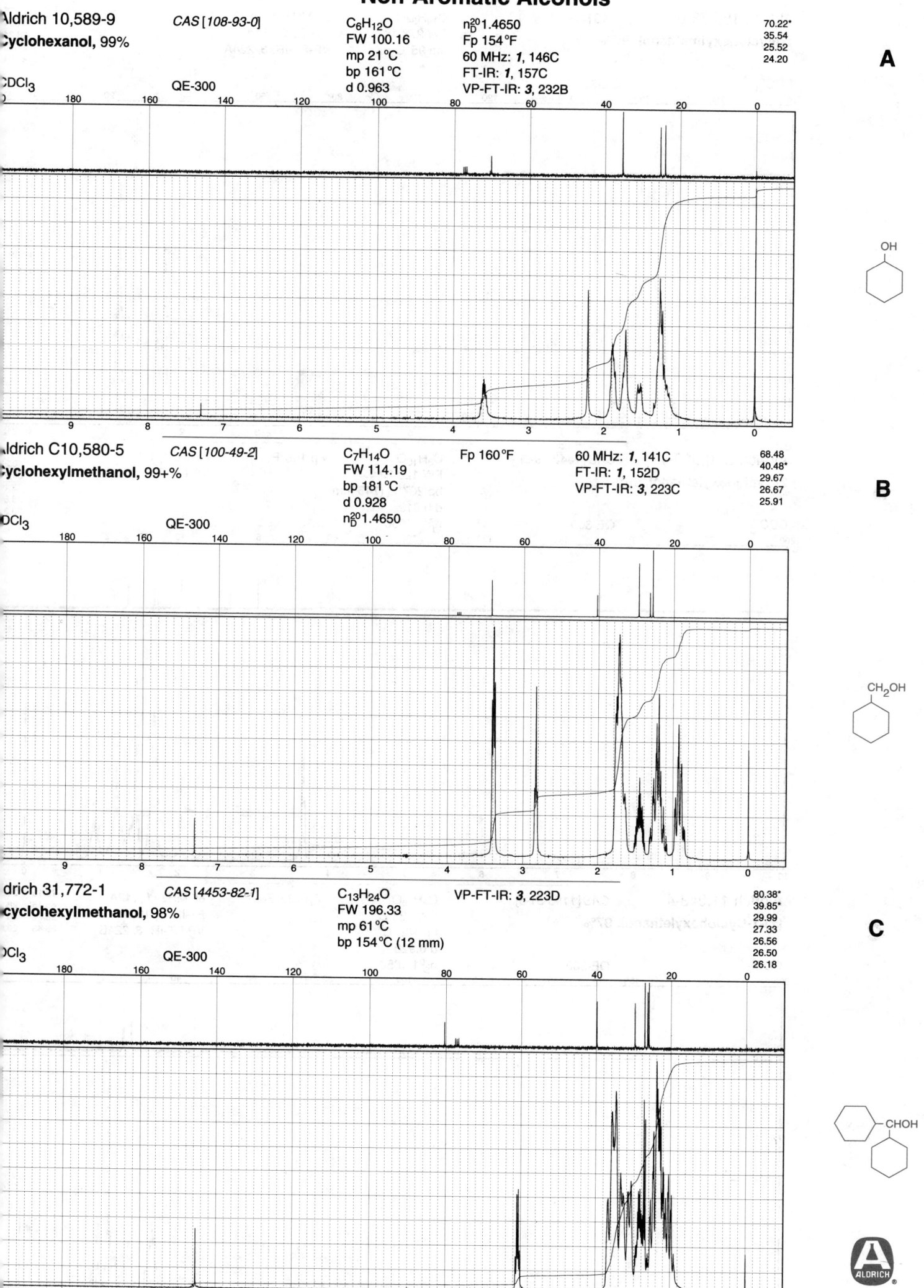

Aldrich 10,589-9 *CAS [108-93-0]*
Cyclohexanol, 99%

CDCl₃   QE-300

$C_6H_{12}O$
FW 100.16
mp 21 °C
bp 161 °C
d 0.963

n_D^{20} 1.4650
Fp 154 °F
60 MHz: *1*, 146C
FT-IR: *1*, 157C
VP-FT-IR: *3*, 232B

70.22*
35.54
25.52
24.20

A

Aldrich C10,580-5 *CAS [100-49-2]*
Cyclohexylmethanol, 99+%

CDCl₃   QE-300

$C_7H_{14}O$
FW 114.19
bp 181 °C
d 0.928
n_D^{20} 1.4650

Fp 160 °F

60 MHz: *1*, 141C
FT-IR: *1*, 152D
VP-FT-IR: *3*, 223C

68.48
40.48*
29.67
26.67
25.91

B

Aldrich 31,772-1 *CAS [4453-82-1]*
Dicyclohexylmethanol, 98%

CDCl₃   QE-300

$C_{13}H_{24}O$
FW 196.33
mp 61 °C
bp 154 °C (12 mm)

VP-FT-IR: *3*, 223D

80.38*
39.85*
29.99
27.33
26.56
26.50
26.18

C

ALDRICH®

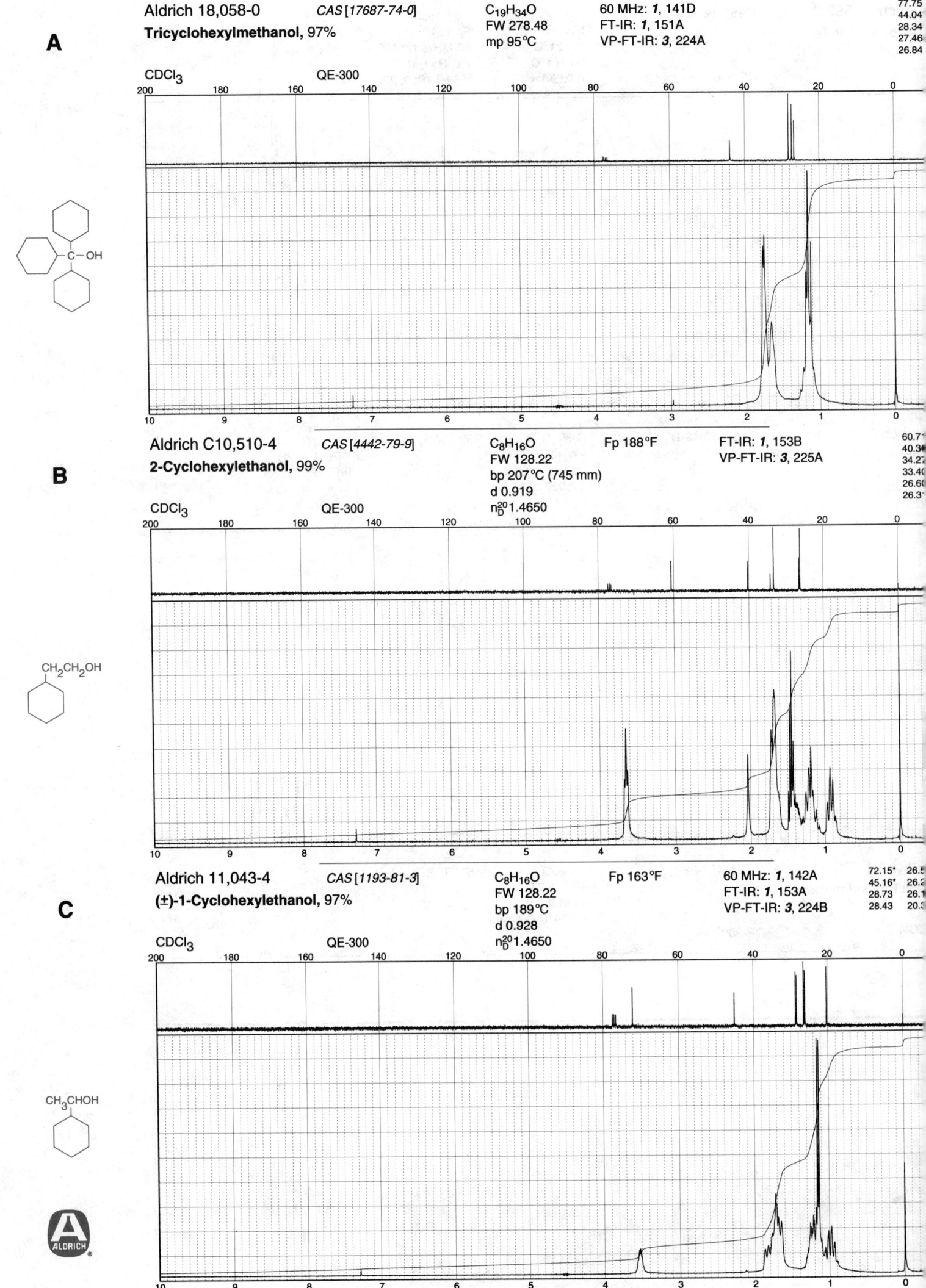

A Aldrich 18,058-0   *CAS [17687-74-0]* $C_{19}H_{34}O$ 60 MHz: *1*, 141D
Tricyclohexylmethanol, 97%

FW 278.48   FT-IR: *1*, 151A
mp 95°C   VP-FT-IR: *3*, 224A

77.75
44.04
28.34
27.46
26.84

CDCl₃   QE-300

B Aldrich C10,510-4   *CAS [4442-79-9]* $C_8H_{16}O$ Fp 188°F   FT-IR: *1*, 153B
2-Cyclohexylethanol, 99%

FW 128.22   VP-FT-IR: *3*, 225A
bp 207°C (745 mm)
d 0.919
n_D^{20}1.4650

60.71
40.30
34.27
33.40
26.60
26.3

CDCl₃   QE-300

C Aldrich 11,043-4   *CAS [1193-81-3]* $C_8H_{16}O$ Fp 163°F   60 MHz: *1*, 142A
(±)-1-Cyclohexylethanol, 97%

FW 128.22   FT-IR: *1*, 153A
bp 189°C   VP-FT-IR: *3*, 224B
d 0.928
n_D^{20}1.4650

72.15*
45.16*
28.73
28.43

26.5
26.2
26.1
20.3

CDCl₃   QE-300

Non-Aromatic Alcohols

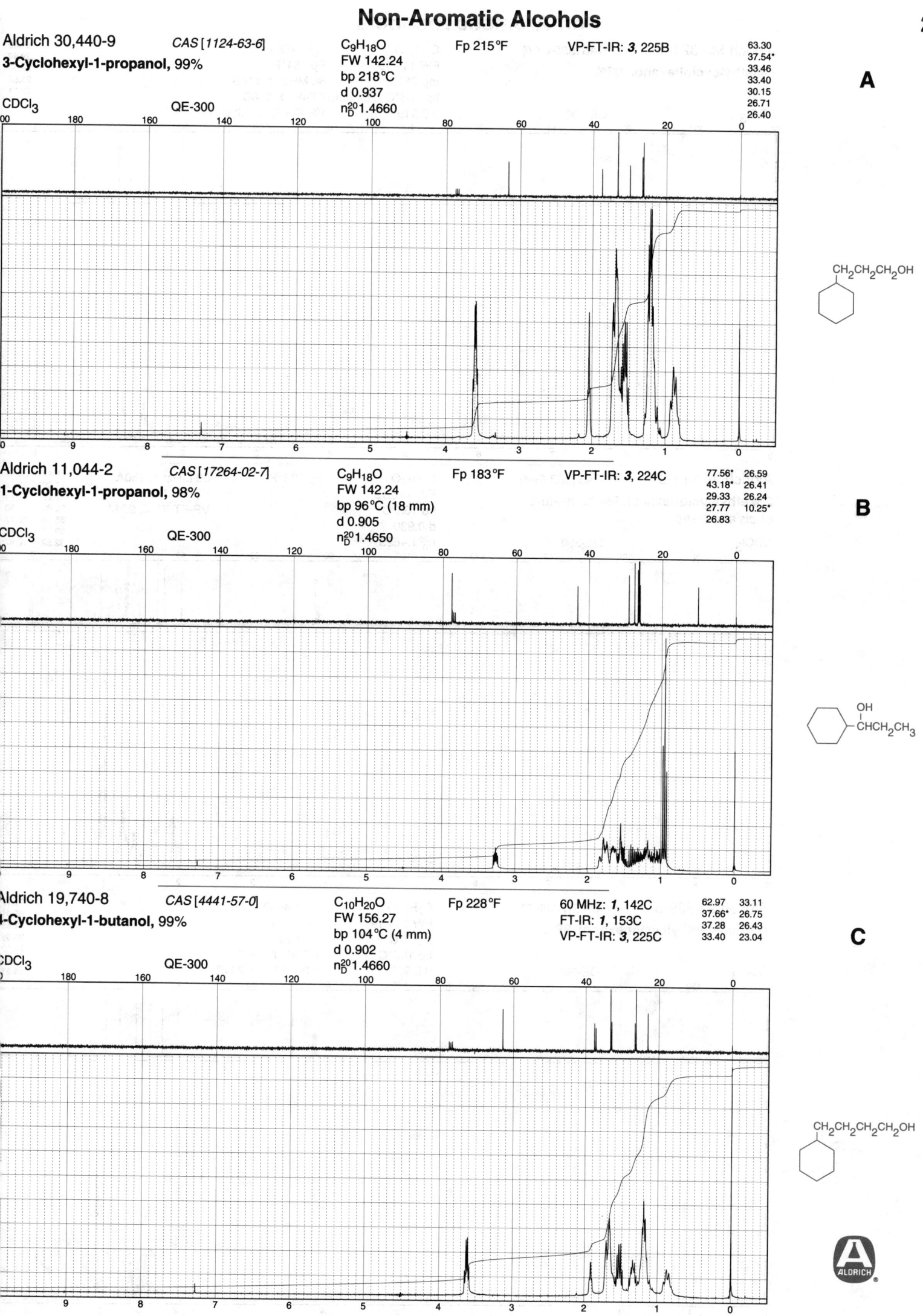

Aldrich 30,440-9 CAS [1124-63-6]
3-Cyclohexyl-1-propanol, 99%
C₉H₁₈O
FW 142.24
bp 218°C
d 0.937
n²⁰_D 1.4660
Fp 215°F
VP-FT-IR: **3**, 225B
CDCl₃ QE-300

A

Aldrich 11,044-2 CAS [17264-02-7]
1-Cyclohexyl-1-propanol, 98%
C₉H₁₈O
FW 142.24
bp 96°C (18 mm)
d 0.905
n²⁰_D 1.4650
Fp 183°F
VP-FT-IR: **3**, 224C
CDCl₃ QE-300

B

Aldrich 19,740-8 CAS [4441-57-0]
4-Cyclohexyl-1-butanol, 99%
C₁₀H₂₀O
FW 156.27
bp 104°C (4 mm)
d 0.902
n²⁰_D 1.4660
Fp 228°F
60 MHz: **1**, 142C
FT-IR: **1**, 153C
VP-FT-IR: **3**, 225C
CDCl₃ QE-300

C

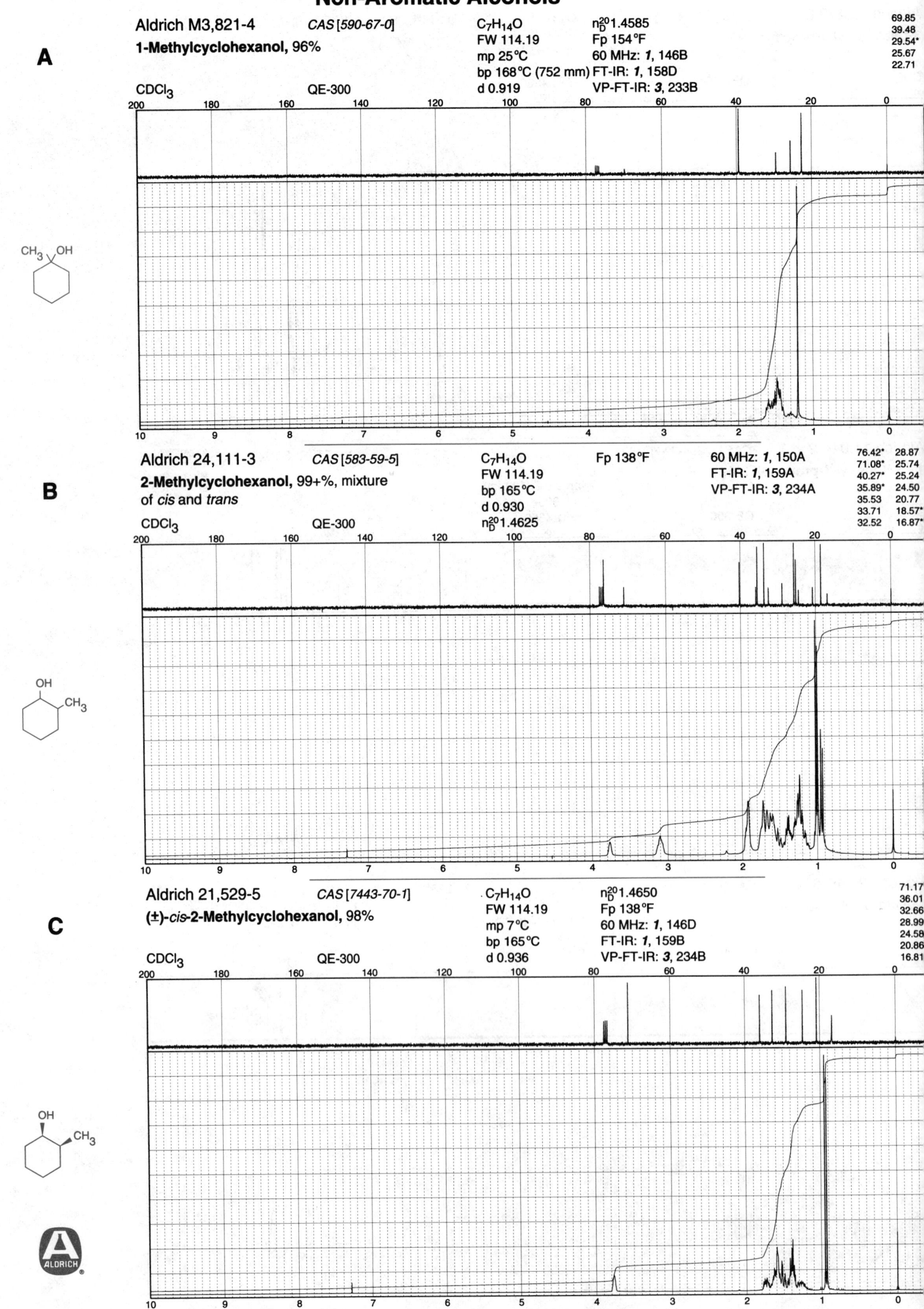

A

Aldrich M3,821-4   CAS [590-67-0]

1-Methylcyclohexanol, 96%

$C_7H_{14}O$
FW 114.19
mp 25°C
bp 168°C (752 mm)

n_D^{20} 1.4585
Fp 154°F
60 MHz: *1*, 146B
FT-IR: *1*, 158D

CDCl₃   QE-300   d 0.919   VP-FT-IR: *3*, 233B

69.85
39.48
29.54*
25.67
22.71

B

Aldrich 24,111-3   CAS [583-59-5]

2-Methylcyclohexanol, 99+%, mixture of *cis* and *trans*

$C_7H_{14}O$
FW 114.19
bp 165°C
d 0.930
n_D^{20} 1.4625

Fp 138°F

60 MHz: *1*, 150A
FT-IR: *1*, 159A
VP-FT-IR: *3*, 234A

CDCl₃   QE-300

76.42* 28.87
71.08* 25.74
40.27* 25.24
35.89* 24.50
35.53   20.77
33.71   18.57*
32.52   16.87*

C

Aldrich 21,529-5   CAS [7443-70-1]

(±)-*cis*-2-Methylcyclohexanol, 98%

$C_7H_{14}O$
FW 114.19
mp 7°C
bp 165°C
d 0.936

n_D^{20} 1.4650
Fp 138°F
60 MHz: *1*, 146D
FT-IR: *1*, 159B

CDCl₃   QE-300   VP-FT-IR: *3*, 234B

71.17
36.01
32.66
28.99
24.58
20.86
16.81

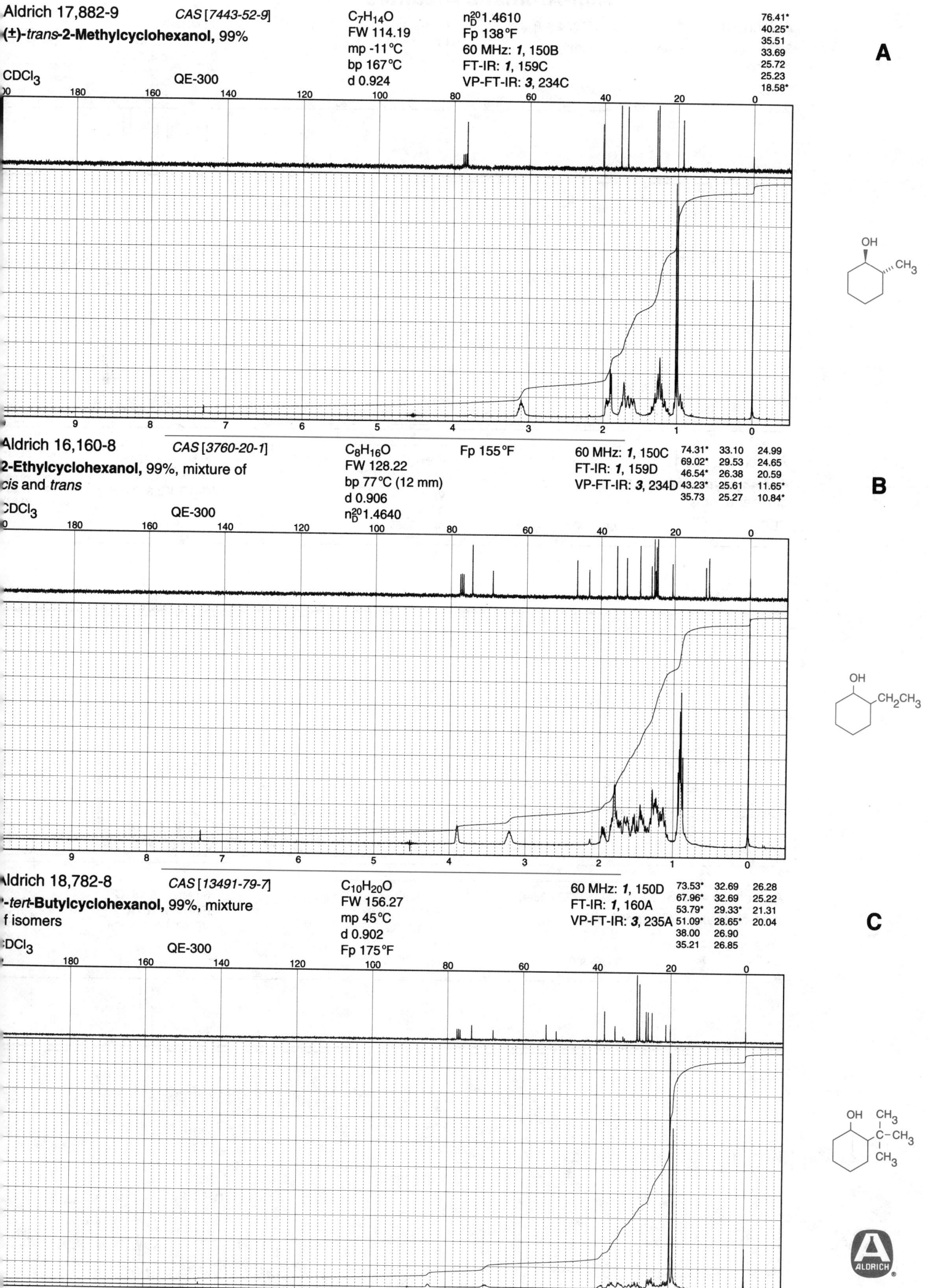

A

Aldrich 17,882-9   CAS [7443-52-9]

(±)-trans-2-Methylcyclohexanol, 99%

CDCl₃   QE-300

$C_7H_{14}O$
FW 114.19
mp -11°C
bp 167°C
d 0.924

n_D^{20} 1.4610
Fp 138°F
60 MHz: *1*, 150B
FT-IR: *1*, 159C
VP-FT-IR: *3*, 234C

76.41*
40.25*
35.51
33.69
25.72
25.23
18.58*

B

Aldrich 16,160-8   CAS [3760-20-1]

2-Ethylcyclohexanol, 99%, mixture of
cis and trans

CDCl₃   QE-300

$C_8H_{16}O$
FW 128.22
bp 77°C (12 mm)
d 0.906
n_D^{20} 1.4640

Fp 155°F

60 MHz: *1*, 150C	74.31*	33.10	24.99
FT-IR: *1*, 159D	69.02*	29.53	24.65
VP-FT-IR: *3*, 234D	46.54*	26.38	20.59
	43.23*	25.61	11.65*
	35.73	25.27	10.84*

C

Aldrich 18,782-8   CAS [13491-79-7]

2-tert-Butylcyclohexanol, 99%, mixture
of isomers

CDCl₃   QE-300

$C_{10}H_{20}O$
FW 156.27
mp 45°C
d 0.902
Fp 175°F

60 MHz: *1*, 150D	73.53*	32.69	26.28
FT-IR: *1*, 160A	67.96*	32.69	25.22
VP-FT-IR: *3*, 235A	53.79*	29.33*	21.31
	51.09*	28.65*	20.04
	38.00	26.90	
	35.21	26.85	

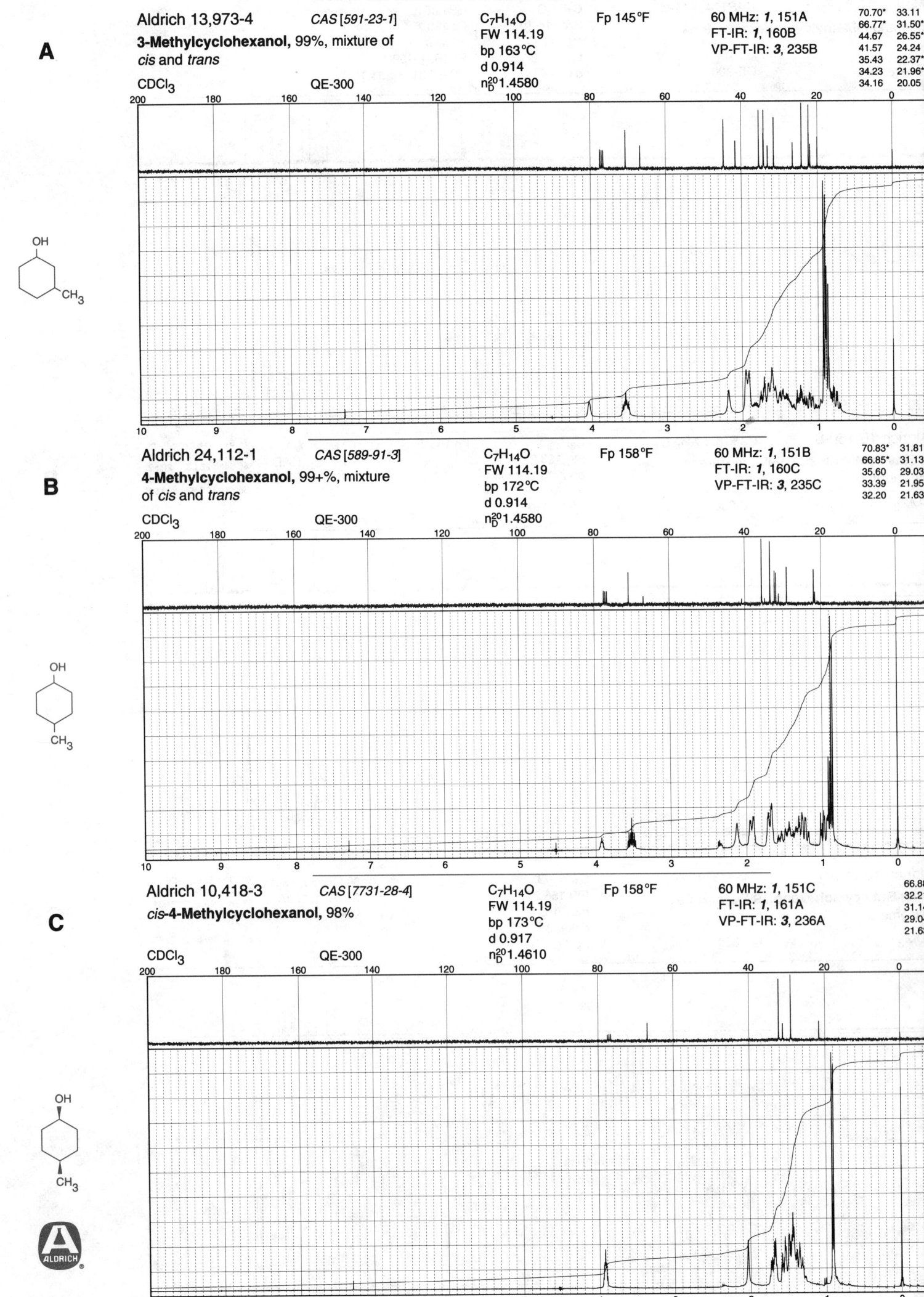

A

Aldrich 13,973-4   *CAS [591-23-1]*   C$_7$H$_{14}$O   Fp 145°F   60 MHz: *1*, 151A

3-Methylcyclohexanol, 99%, mixture of   FW 114.19   FT-IR: *1*, 160B
cis and *trans*   bp 163°C   VP-FT-IR: *3*, 235B
    d 0.914
CDCl$_3$ QE-300   n$_D^{20}$1.4580

70.70*	33.11
66.77*	31.50*
44.67	26.55*
41.57	24.24
35.43	22.37*
34.23	21.96*
34.16	20.05

B

Aldrich 24,112-1   *CAS [589-91-3]*   C$_7$H$_{14}$O   Fp 158°F   60 MHz: *1*, 151B

4-Methylcyclohexanol, 99+%, mixture   FW 114.19   FT-IR: *1*, 160C
of *cis* and *trans*   bp 172°C   VP-FT-IR: *3*, 235C
    d 0.914
CDCl$_3$ QE-300   n$_D^{20}$1.4580

70.83*	31.81*
66.85*	31.13*
35.60	29.03
33.39	21.95*
32.20	21.63*

C

Aldrich 10,418-3   *CAS [7731-28-4]*   C$_7$H$_{14}$O   Fp 158°F   60 MHz: *1*, 151C

***cis*-4-Methylcyclohexanol,** 98%   FW 114.19   FT-IR: *1*, 161A
    bp 173°C   VP-FT-IR: *3*, 236A
    d 0.917
CDCl$_3$ QE-300   n$_D^{20}$1.4610

| 66.88 |
| 32.21 |
| 31.14 |
| 29.04 |
| 21.63 |

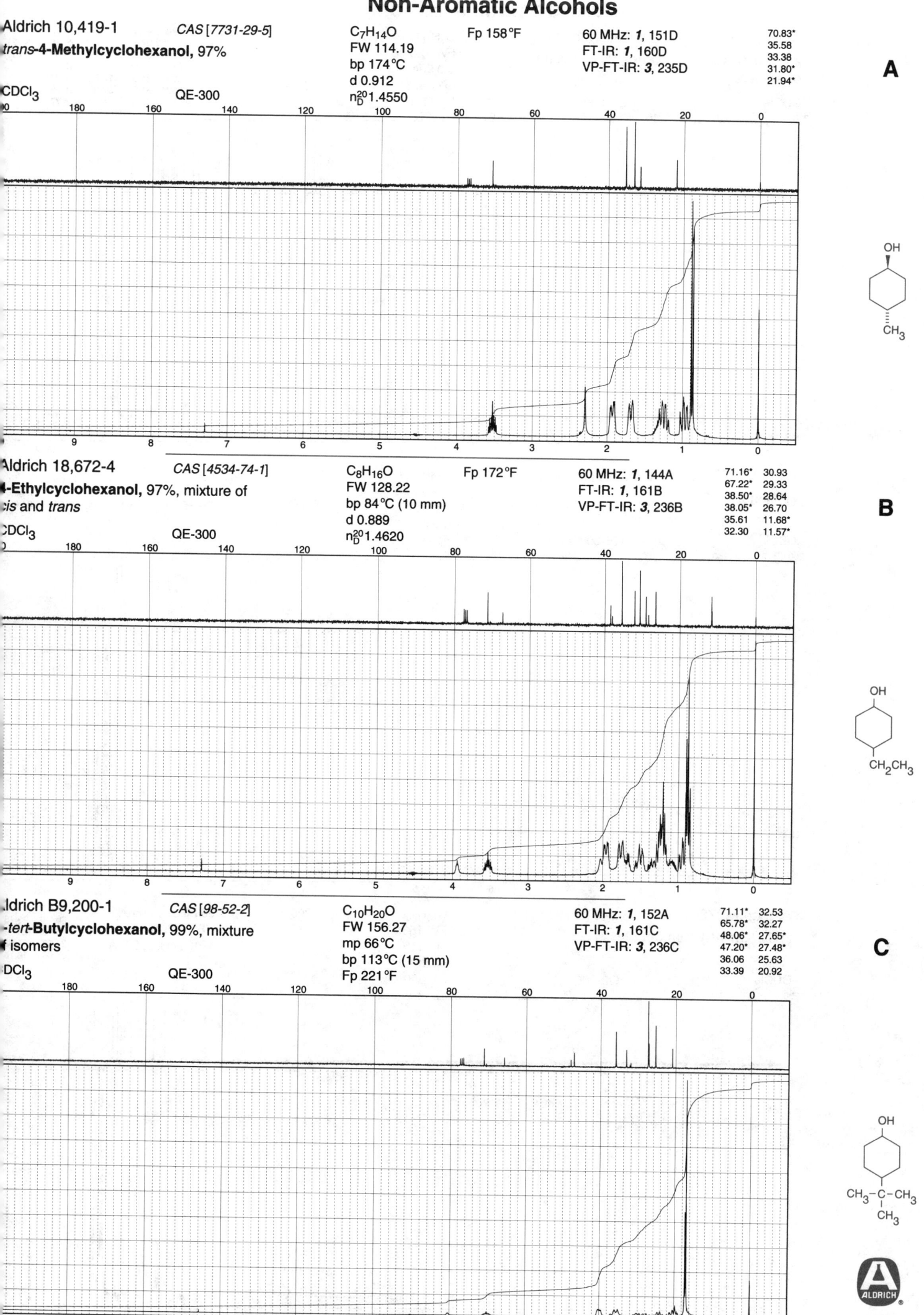

Aldrich 10,419-1   *CAS [7731-29-5]*   C₇H₁₄O   Fp 158°F   60 MHz: *1*, 151D   70.83*
trans-4-Methylcyclohexanol, 97%   FW 114.19   FT-IR: *1*, 160D   35.58
    bp 174°C   VP-FT-IR: *3*, 235D   33.38
CDCl₃   QE-300   d 0.912   31.80*
    n²⁰_D 1.4550   21.94*

A

Aldrich 18,672-4   *CAS [4534-74-1]*   C₈H₁₆O   Fp 172°F   60 MHz: *1*, 144A   71.16* 30.93
4-Ethylcyclohexanol, 97%, mixture of   FW 128.22   FT-IR: *1*, 161B   67.22* 29.33
cis and *trans*   bp 84°C (10 mm)   VP-FT-IR: *3*, 236B   38.50* 28.64
CDCl₃   QE-300   d 0.889   38.05* 26.70
    n²⁰_D 1.4620   35.61 11.68*
    32.30 11.57*

B

Aldrich B9,200-1   *CAS [98-52-2]*   C₁₀H₂₀O   60 MHz: *1*, 152A   71.11* 32.53
4-*tert*-Butylcyclohexanol, 99%, mixture   FW 156.27   FT-IR: *1*, 161C   65.78* 32.27
of isomers   mp 66°C   VP-FT-IR: *3*, 236C   48.06* 27.65*
CDCl₃   QE-300   bp 113°C (15 mm)   47.20* 27.48*
    Fp 221°F   36.06 25.63
    33.39 20.92

C

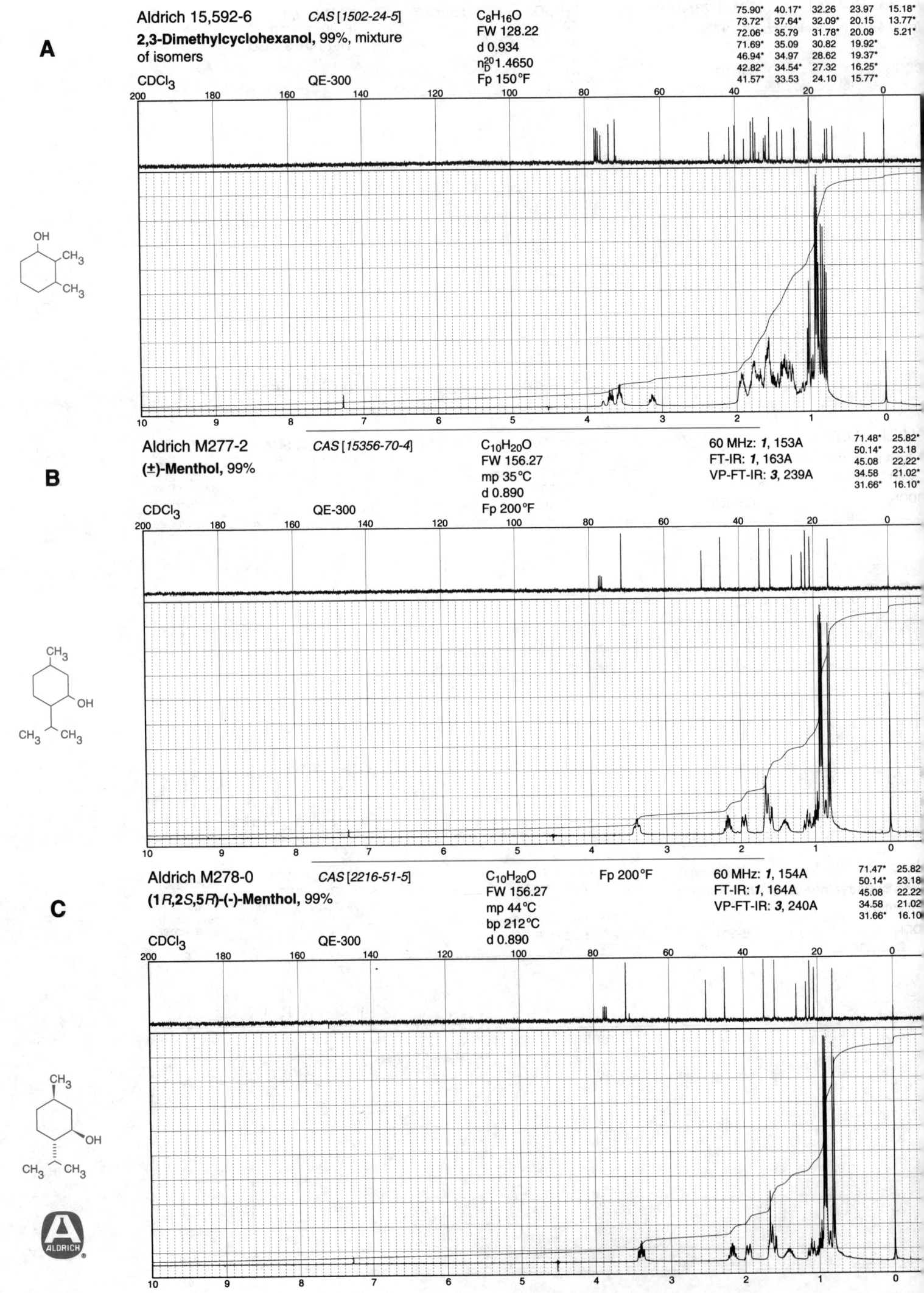

A

Aldrich 15,592-6   *CAS [1502-24-5]*   C8H16O

2,3-Dimethylcyclohexanol, 99%, mixture
of isomers

FW 128.22
d 0.934
n_D^20 1.4650
Fp 150°F

75.90*	40.17*	32.26	23.97	15.18*
73.72*	37.64*	32.09*	20.15	13.77*
72.06*	35.79	31.78*	20.09	5.21*
71.69*	35.09	30.82	19.92*	
46.94*	34.97	28.62	19.37*	
42.82*	34.54*	27.32	16.25*	
41.57*	33.53	24.10	15.77*	

CDCl3   QE-300

B

Aldrich M277-2   *CAS [15356-70-4]*   C10H20O

(±)-Menthol, 99%

FW 156.27
mp 35°C
d 0.890
Fp 200°F

60 MHz: *1*, 153A
FT-IR: *1*, 163A
VP-FT-IR: *3*, 239A

71.48*	25.82*
50.14*	23.18
45.08	22.22*
34.58	21.02*
31.66*	16.10*

CDCl3   QE-300

C

Aldrich M278-0   *CAS [2216-51-5]*   C10H20O   Fp 200°F

(1R,2S,5R)-(-)-Menthol, 99%

FW 156.27
mp 44°C
bp 212°C
d 0.890

60 MHz: *1*, 154A
FT-IR: *1*, 164A
VP-FT-IR: *3*, 240A

71.47*	25.82
50.14*	23.18
45.08	22.22
34.58	21.02
31.66*	16.10

CDCl3   QE-300

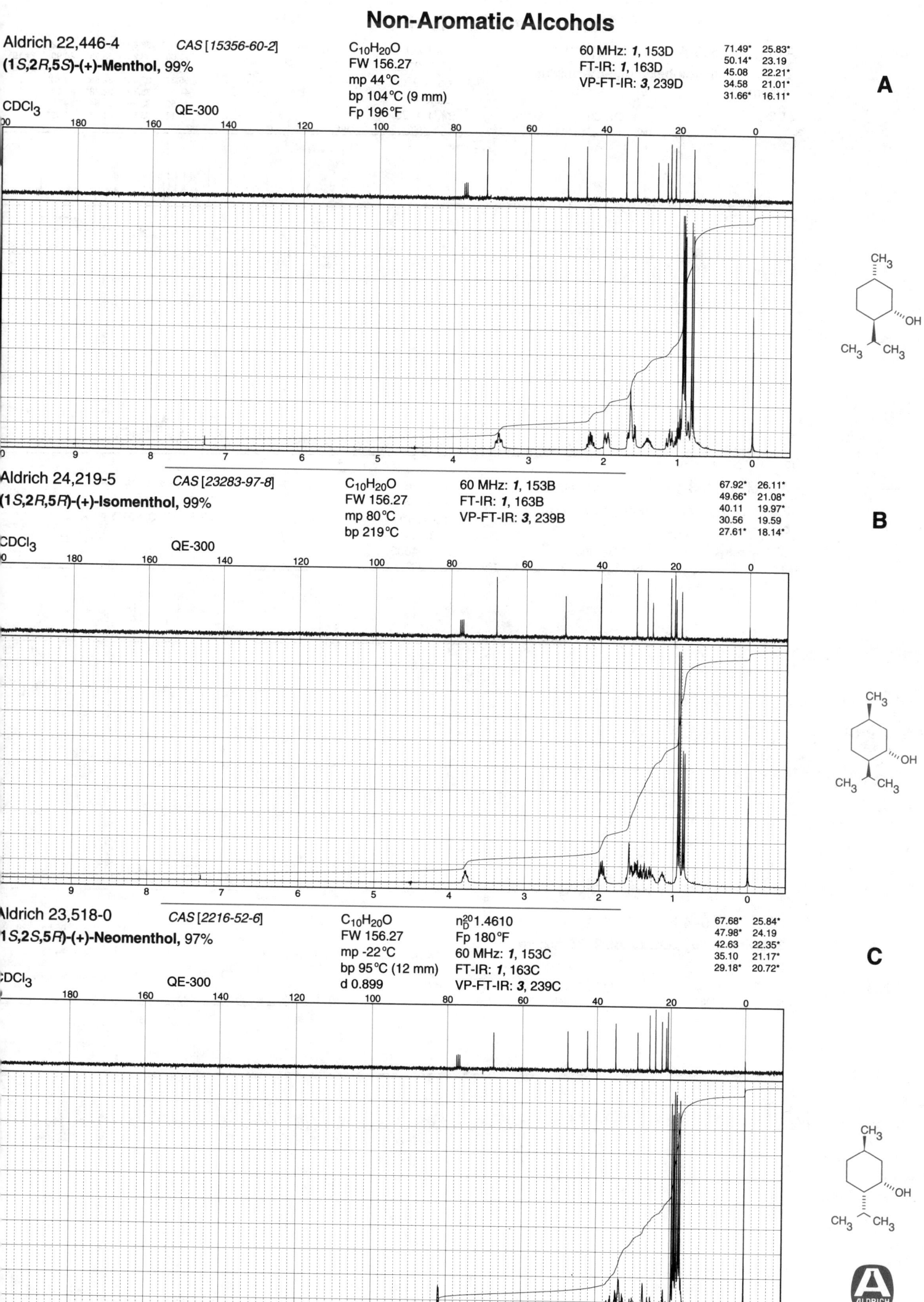

Aldrich 22,446-4   CAS [15356-60-2]   C₁₀H₂₀O
(1S,2R,5S)-(+)-Menthol, 99%

$C_{10}H_{20}O$
FW 156.27
mp 44°C
bp 104°C (9 mm)
Fp 196°F

60 MHz: 1, 153D
FT-IR: 1, 163D
VP-FT-IR: 3, 239D

71.49* 25.83*
50.14* 23.19
45.08   22.21*
34.58   21.01*
31.66* 16.11*

CDCl₃   QE-300

A

Aldrich 24,219-5   CAS [23283-97-8]

(1S,2R,5R)-(+)-Isomenthol, 99%

$C_{10}H_{20}O$
FW 156.27
mp 80°C
bp 219°C

60 MHz: 1, 153B
FT-IR: 1, 163B
VP-FT-IR: 3, 239B

67.92* 26.11*
49.66* 21.08*
40.11   19.97*
30.56   19.59
27.61* 18.14*

CDCl₃   QE-300

B

Aldrich 23,518-0   CAS [2216-52-6]

(1S,2S,5R)-(+)-Neomenthol, 97%

$C_{10}H_{20}O$
FW 156.27
mp -22°C
bp 95°C (12 mm)
d 0.899

n_D^{20} 1.4610
Fp 180°F
60 MHz: 1, 153C
FT-IR: 1, 163C
VP-FT-IR: 3, 239C

67.68* 25.84*
47.98* 24.19
42.63   22.35*
35.10   21.17*
29.18* 20.72*

CDCl₃   QE-300

C

ALDRICH

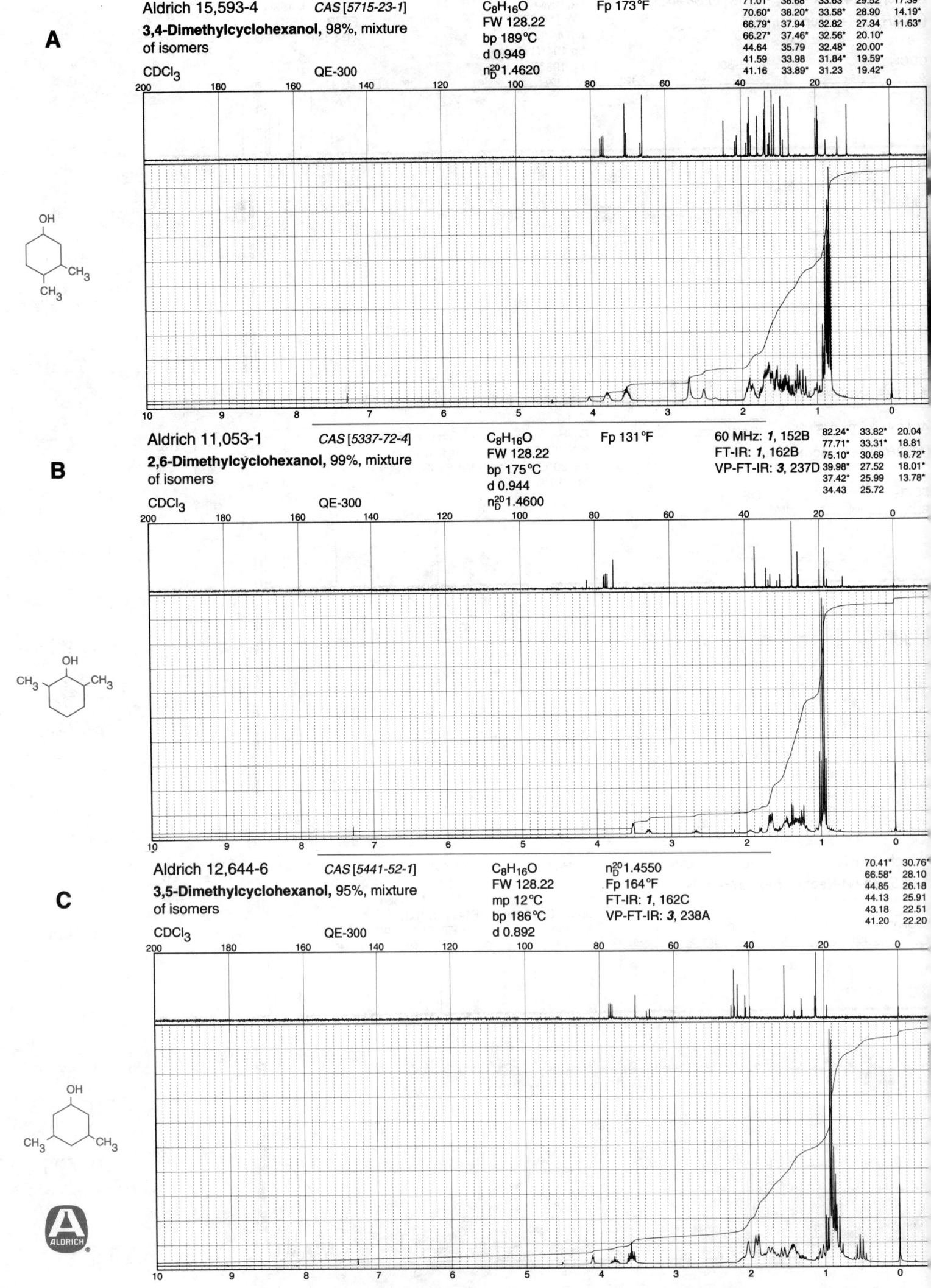

A

Aldrich 15,593-4   *CAS [5715-23-1]*

3,4-Dimethylcyclohexanol, 98%, mixture of isomers

CDCl₃   QE-300

C₈H₁₆O
FW 128.22
bp 189°C
d 0.949
n²⁰_D 1.4620

Fp 173°F

71.01*	38.68*	33.63	29.52	17.39*
70.60*	38.20*	33.58*	28.90	14.19*
66.79*	37.94	32.82	27.34	11.63*
66.27*	37.46*	32.56*	20.10*	
44.64	35.79	32.48*	20.00*	
41.59	33.98	31.84*	19.59*	
41.16	33.89*	31.23	19.42*	

B

Aldrich 11,053-1   *CAS [5337-72-4]*

2,6-Dimethylcyclohexanol, 99%, mixture of isomers

CDCl₃   QE-300

C₈H₁₆O
FW 128.22
bp 175°C
d 0.944
n²⁰_D 1.4600

Fp 131°F

60 MHz: *1*, 152B
FT-IR: *1*, 162B
VP-FT-IR: *3*, 237D

82.24*	33.82*	20.04
77.71*	33.31*	18.81
75.10*	30.69	18.72*
39.98*	27.52	18.01*
37.42*	25.99	13.78*
34.43	25.72	

C

Aldrich 12,644-6   *CAS [5441-52-1]*

3,5-Dimethylcyclohexanol, 95%, mixture of isomers

CDCl₃   QE-300

C₈H₁₆O
FW 128.22
mp 12°C
bp 186°C
d 0.892

n²⁰_D 1.4550
Fp 164°F
FT-IR: *1*, 162C
VP-FT-IR: *3*, 238A

70.41*	30.76*	
66.58*	28.10	
44.85	26.18	
44.13	25.91	
43.18	22.51	
41.20	22.20	

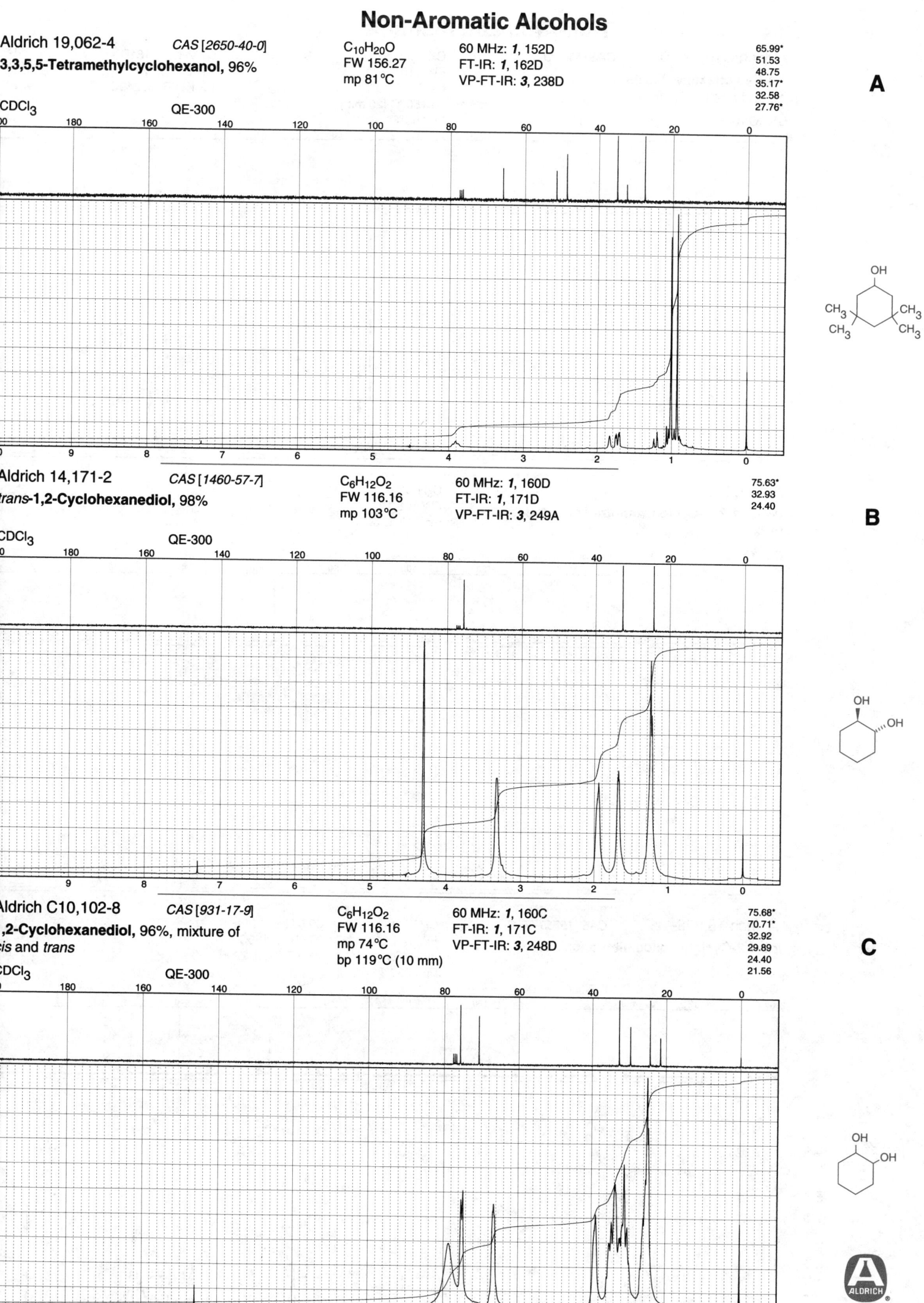

Aldrich 19,062-4   CAS [2650-40-0]
3,3,5,5-Tetramethylcyclohexanol, 96%

C$_{10}$H$_{20}$O
FW 156.27
mp 81°C

60 MHz: *1*, 152D
FT-IR: *1*, 162D
VP-FT-IR: *3*, 238D

65.99*
51.53
48.75
35.17*
32.58
27.76*

A

CDCl$_3$ QE-300

Aldrich 14,171-2   CAS [1460-57-7]
trans-1,2-Cyclohexanediol, 98%

C$_6$H$_{12}$O$_2$
FW 116.16
mp 103°C

60 MHz: *1*, 160D
FT-IR: *1*, 171D
VP-FT-IR: *3*, 249A

75.63*
32.93
24.40

B

CDCl$_3$ QE-300

Aldrich C10,102-8   CAS [931-17-9]
**1,2-Cyclohexanediol, 96%, mixture of
cis and trans**

C$_6$H$_{12}$O$_2$
FW 116.16
mp 74°C
bp 119°C (10 mm)

60 MHz: *1*, 160C
FT-IR: *1*, 171C
VP-FT-IR: *3*, 248D

75.68*
70.71*
32.92
29.89
24.40
21.56

C

CDCl$_3$ QE-300

ALDRICH

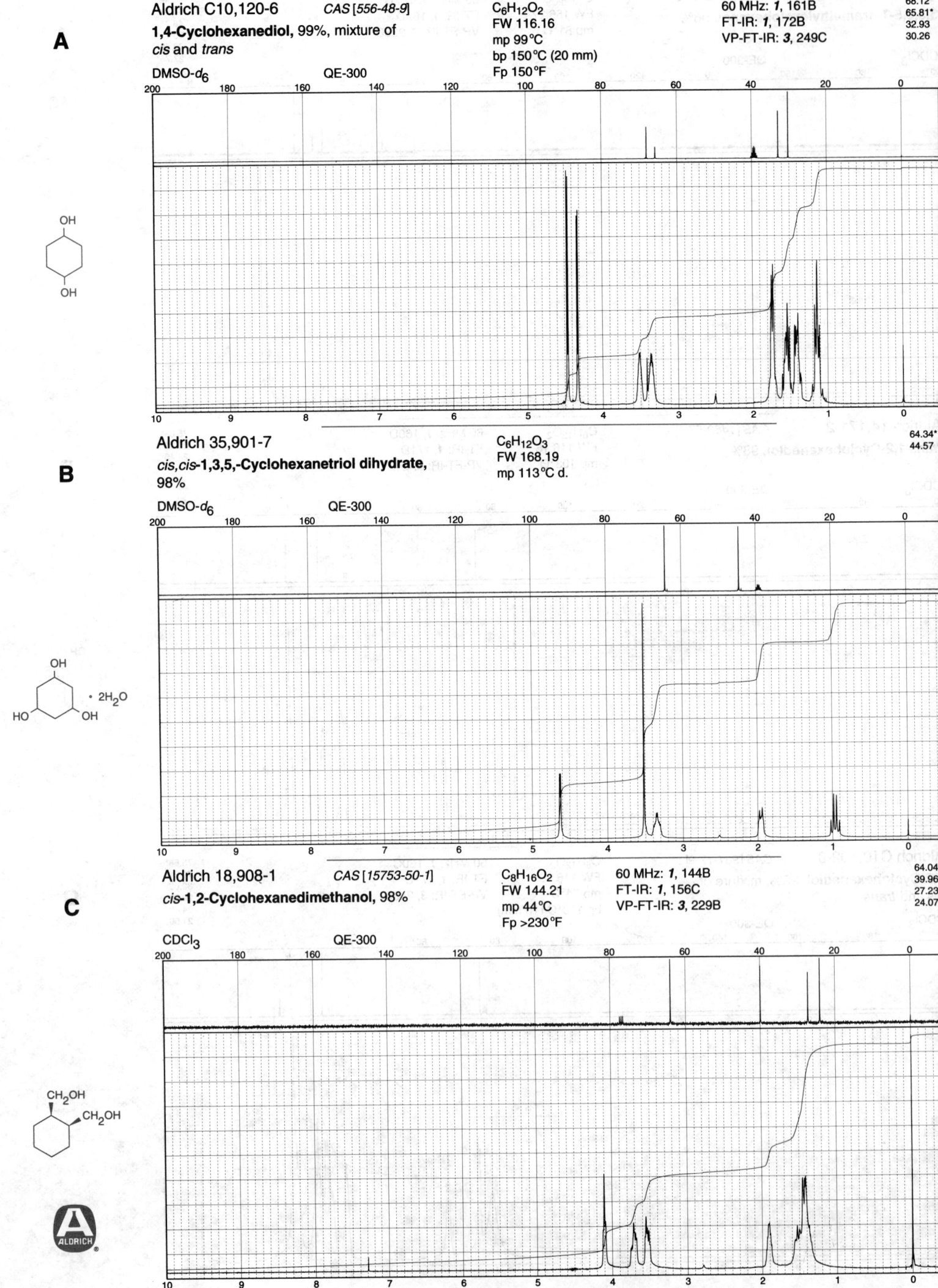

A

Aldrich C10,120-6   CAS [556-48-9] $C_6H_{12}O_2$
FW 116.16

1,4-Cyclohexanediol, 99%, mixture of
cis and *trans*

mp 99°C
bp 150°C (20 mm)
Fp 150°F

60 MHz: *1*, 161B
FT-IR: *1*, 172B
VP-FT-IR: *3*, 249C

68.12*
65.81*
32.93
30.26

DMSO-d_6 QE-300

B

Aldrich 35,901-7 $C_6H_{12}O_3$
FW 168.19

cis,cis-**1,3,5,-Cyclohexanetriol dihydrate,**
98%

mp 113°C d.

64.34*
44.57

DMSO-d_6 QE-300

C

Aldrich 18,908-1   CAS [15753-50-1] $C_8H_{16}O_2$
FW 144.21

cis-**1,2-Cyclohexanedimethanol,** 98%

mp 44°C
Fp >230°F

60 MHz: *1*, 144B
FT-IR: *1*, 156C
VP-FT-IR: *3*, 229B

64.04
39.96*
27.23
24.07

CDCl$_3$ QE-300

Non-Aromatic Alcohols

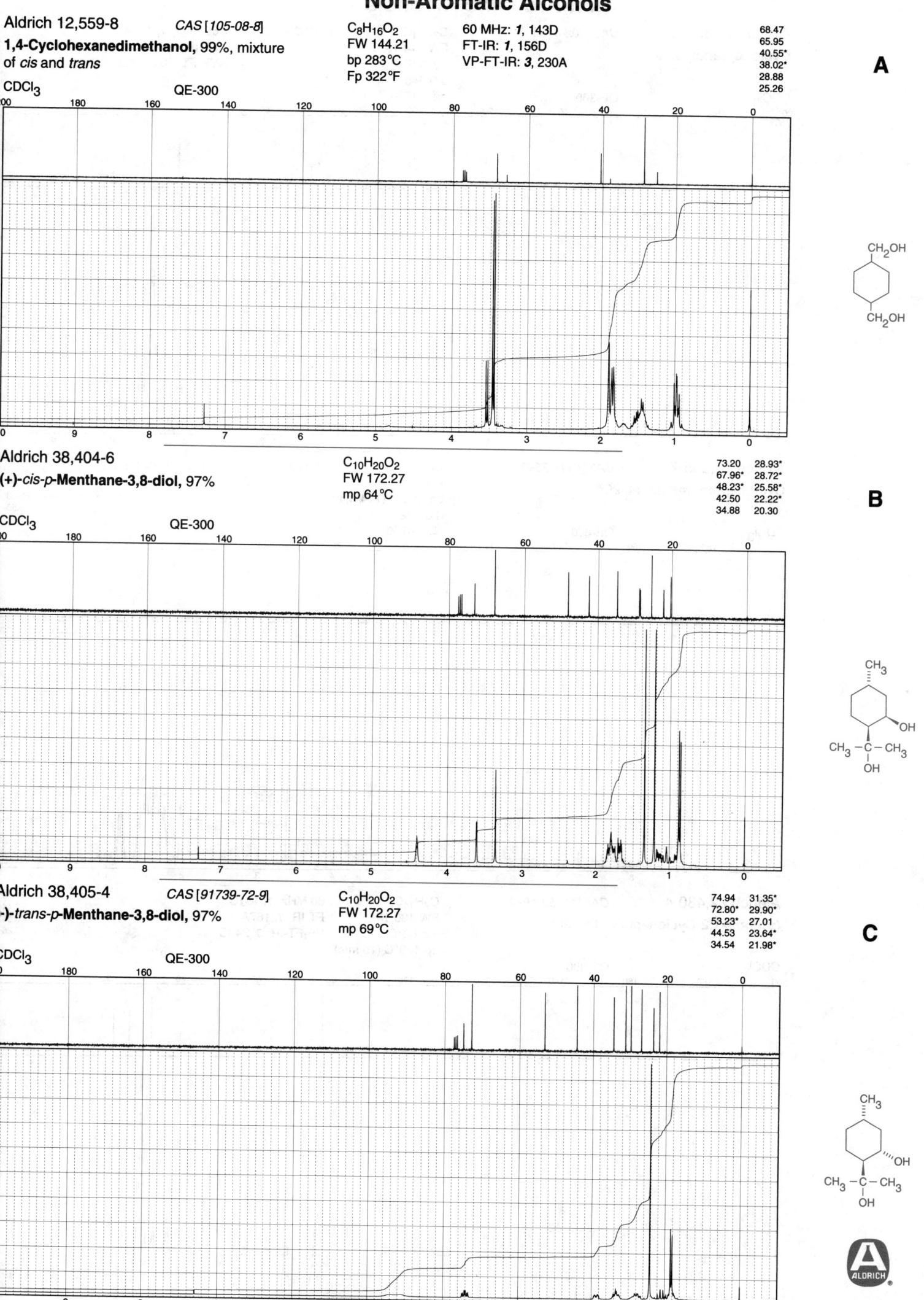

Aldrich 12,559-8   CAS [105-08-8] $C_8H_{16}O_2$ 60 MHz: 1, 143D   FW 144.21   FT-IR: 1, 156D

1,4-Cyclohexanedimethanol, 99%, mixture of cis and trans   bp 283°C   VP-FT-IR: 3, 230A   Fp 322°F

CDCl₃   QE-300

68.47 / 65.95 / 40.55* / 38.02* / 28.88 / 25.26

A

Aldrich 38,404-6 $C_{10}H_{20}O_2$ FW 172.27   mp 64°C

(+)-cis-p-Menthane-3,8-diol, 97%

CDCl₃   QE-300

73.20 / 67.96* / 48.23* / 42.50 / 34.88 / 28.93* / 28.72* / 25.58* / 22.22* / 20.30

B

Aldrich 38,405-4   CAS [91739-72-9] $C_{10}H_{20}O_2$ FW 172.27   mp 69°C

(-)-trans-p-Menthane-3,8-diol, 97%

CDCl₃   QE-300

74.94 / 72.80* / 53.23* / 44.53 / 34.54 / 31.35* / 29.90* / 27.01 / 23.64* / 21.98*

C

ALDRICH

Non-Aromatic Alcohols

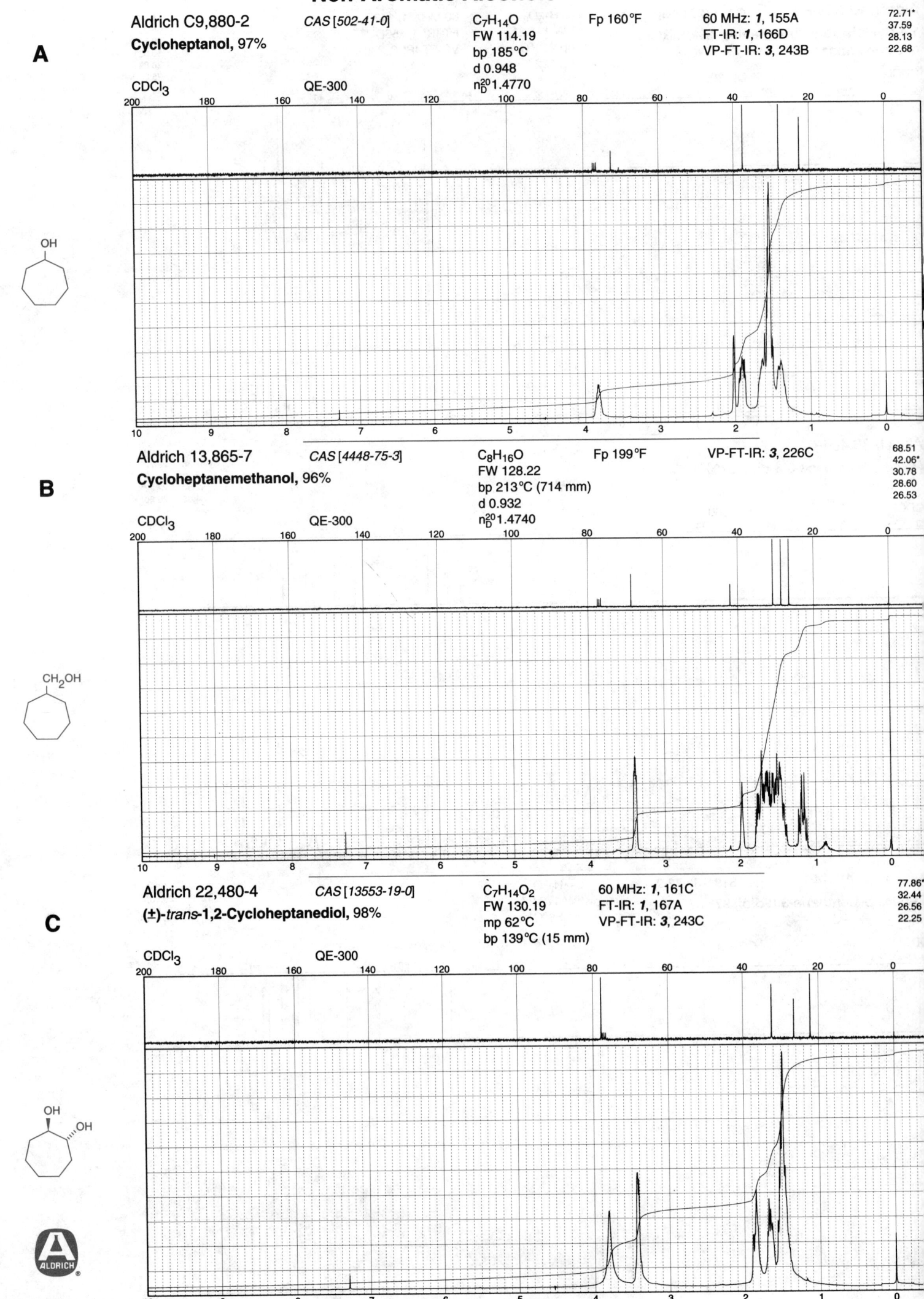

A

Aldrich C9,880-2   CAS [502-41-0] $C_7H_{14}O$ Fp 160°F   60 MHz: *1*, 155A   72.71*
Cycloheptanol, 97%   FW 114.19   FT-IR: *1*, 166D   37.59
                                      bp 185°C   VP-FT-IR: *3*, 243B 28.13
                                      d 0.948   22.68
CDCl₃   QE-300 n_D^{20}1.4770

B

Aldrich 13,865-7   CAS [4448-75-3] $C_8H_{16}O$ Fp 199°F   VP-FT-IR: *3*, 226C   68.51
Cycloheptanemethanol, 96%   FW 128.22   42.06*
                                       bp 213°C (714 mm)   30.78
                                       d 0.932   28.60
                                       n_D^{20}1.4740   26.53
CDCl₃   QE-300

C

Aldrich 22,480-4   CAS [13553-19-0] $C_7H_{14}O_2$ 60 MHz: *1*, 161C   77.86*
(±)-*trans*-1,2-Cycloheptanediol, 98%   FW 130.19   FT-IR: *1*, 167A   32.44
                                            mp 62°C   VP-FT-IR: *3*, 243C   26.56
                                            bp 139°C (15 mm)   22.25
CDCl₃   QE-300

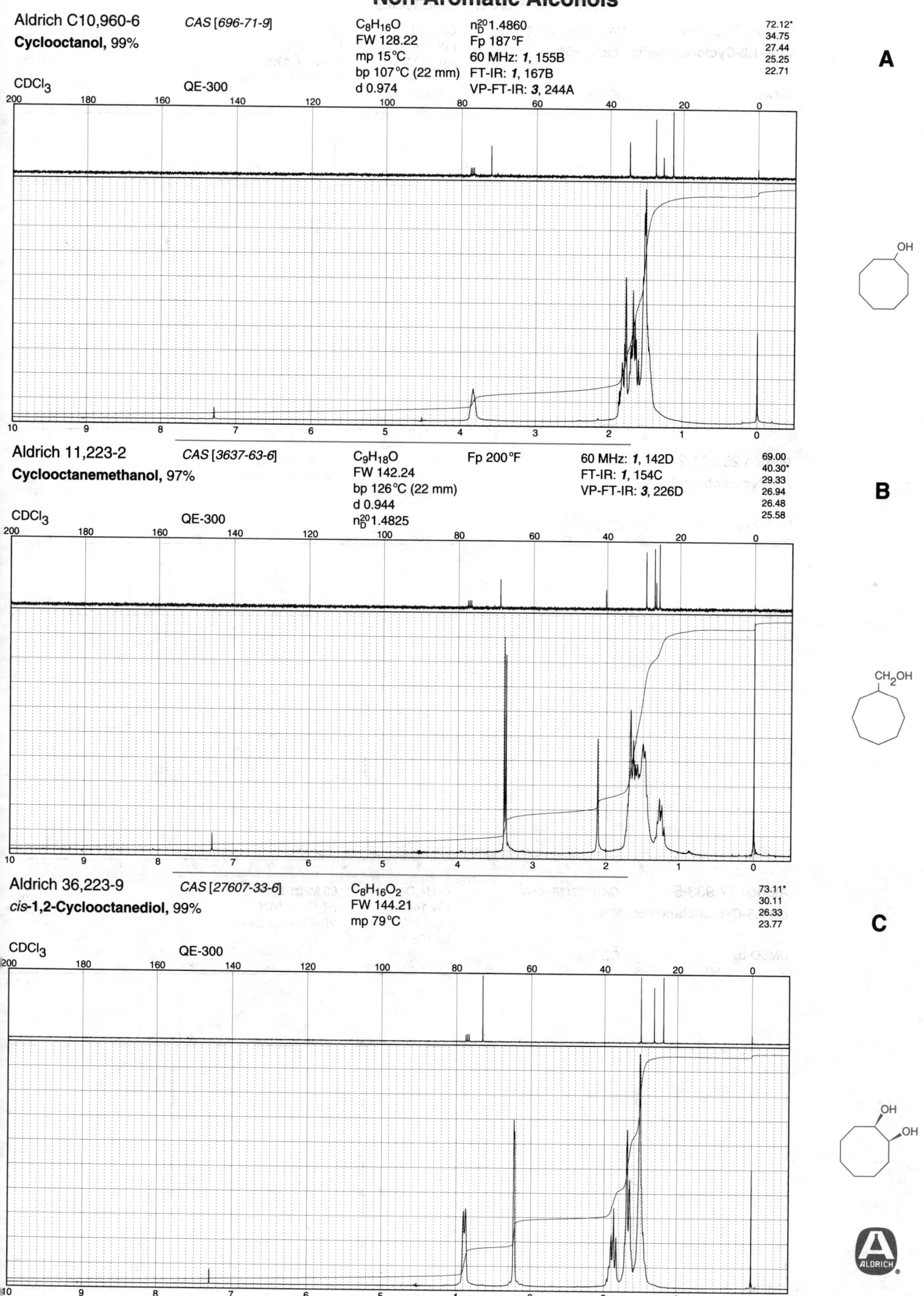

Aldrich C10,960-6   *CAS [696-71-9]*   C₈H₁₆O   n²⁰_D 1.4860
Cyclooctanol, 99%

Aldrich C10,960-6   CAS [696-71-9] $C_8H_{16}O$ n_D^{20} 1.4860
Cyclooctanol, 99%   FW 128.22   Fp 187°F
mp 15°C   60 MHz: *1*, 155B
CDCl₃   QE-300   bp 107°C (22 mm)   FT-IR: *1*, 167B
d 0.974   VP-FT-IR: *3*, 244A

72.12*
34.75
27.44
25.25
22.71

A

Aldrich 11,223-2   CAS [3637-63-6] $C_9H_{18}O$ Fp 200°F   60 MHz: *1*, 142D
Cyclooctanemethanol, 97%   FW 142.24   FT-IR: *1*, 154C
bp 126°C (22 mm)   VP-FT-IR: *3*, 226D
d 0.944
CDCl₃   QE-300 n_D^{20} 1.4825

69.00
40.30*
29.33
26.94
26.48
25.58

B

Aldrich 36,223-9   CAS [27607-33-6] $C_8H_{16}O_2$
cis-1,2-**Cyclooctanediol**, 99%   FW 144.21
mp 79°C

73.11*
30.11
26.33
23.77

CDCl₃   QE-300

C

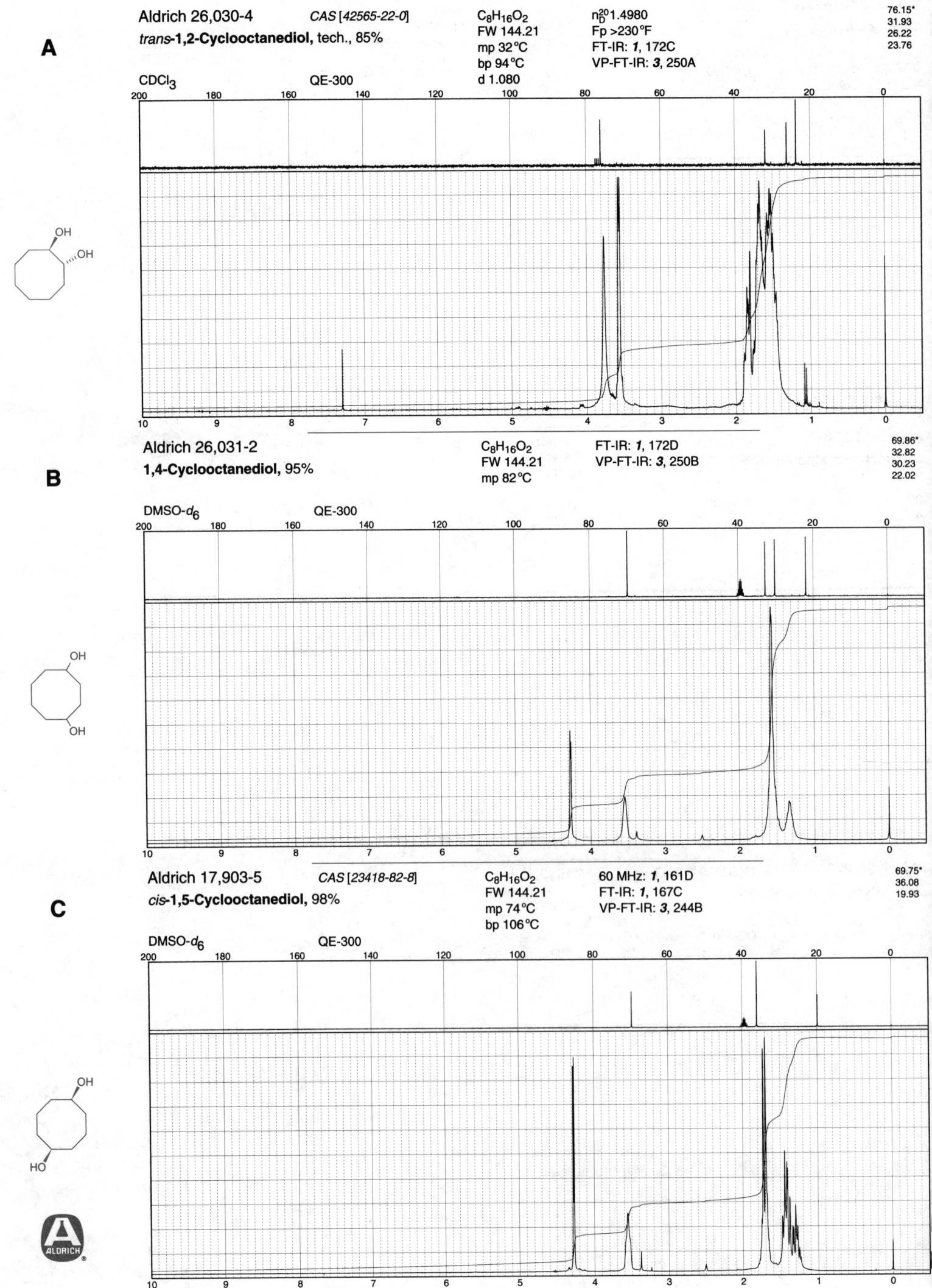

A Aldrich 26,030-4   CAS [42565-22-0] $C_8H_{16}O_2$ n_D^{20} 1.4980   76.15*

trans-**1,2-Cyclooctanediol**, tech., 85%   FW 144.21   Fp >230°F   31.93

mp 32°C   FT-IR: *1*, 172C   26.22

bp 94°C   VP-FT-IR: *3*, 250A   23.76

d 1.080

CDCl₃   QE-300

B Aldrich 26,031-2 $C_8H_{16}O_2$ FT-IR: *1*, 172D   69.86*

1,4-Cyclooctanediol, 95%   FW 144.21   VP-FT-IR: *3*, 250B   32.82

mp 82°C   30.23

22.02

DMSO-*d₆*   QE-300

C Aldrich 17,903-5   CAS [23418-82-8] $C_8H_{16}O_2$ 60 MHz: *1*, 161D   69.75*

cis-**1,5-Cyclooctanediol**, 98%   FW 144.21   FT-IR: *1*, 167C   36.08

mp 74°C   VP-FT-IR: *3*, 244B   19.93

bp 106°C

DMSO-*d₆*   QE-300

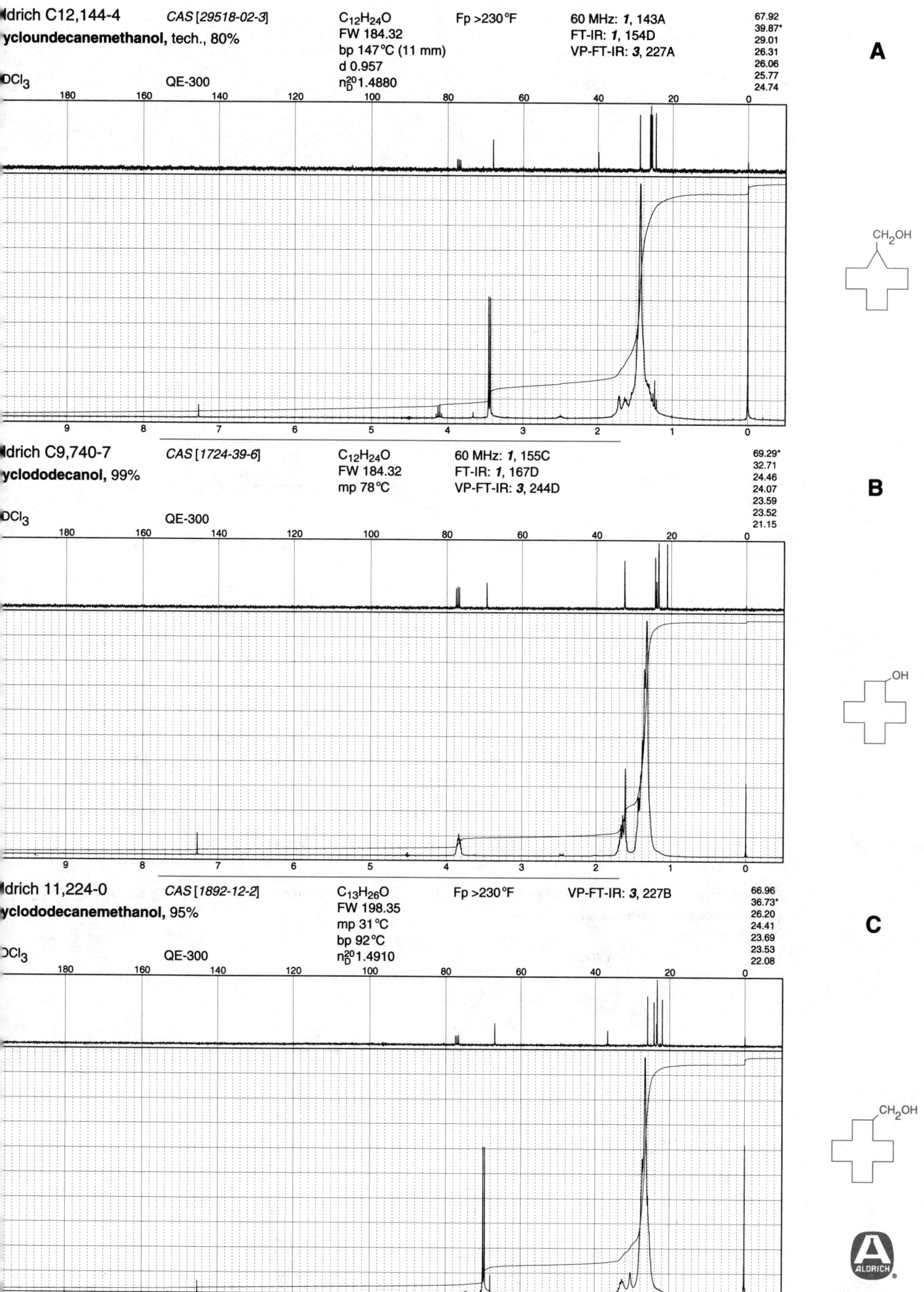

A

Aldrich C12,144-4   CAS [29518-02-3]

Cycloundecanemethanol, tech., 80%

CDCl₃   QE-300

$C_{12}H_{24}O$
FW 184.32
bp 147°C (11 mm)
d 0.957
n_D^{20}1.4880

Fp >230°F

60 MHz: *1*, 143A
FT-IR: *1*, 154D
VP-FT-IR: *3*, 227A

67.92
39.87*
29.01
26.31
26.06
25.77
24.74

CH₂OH

B

Aldrich C9,740-7   CAS [1724-39-6]

Cyclododecanol, 99%

CDCl₃   QE-300

$C_{12}H_{24}O$
FW 184.32
mp 78°C

60 MHz: *1*, 155C
FT-IR: *1*, 167D
VP-FT-IR: *3*, 244D

69.29*
32.71
24.46
24.07
23.59
23.52
21.15

OH

C

Aldrich 11,224-0   CAS [1892-12-2]

Cyclododecanemethanol, 95%

CDCl₃   QE-300

$C_{13}H_{26}O$
FW 198.35
mp 31°C
bp 92°C
n_D^{20}1.4910

Fp >230°F

VP-FT-IR: *3*, 227B

66.96
36.73*
26.20
24.41
23.69
23.53
22.08

CH₂OH

ALDRICH

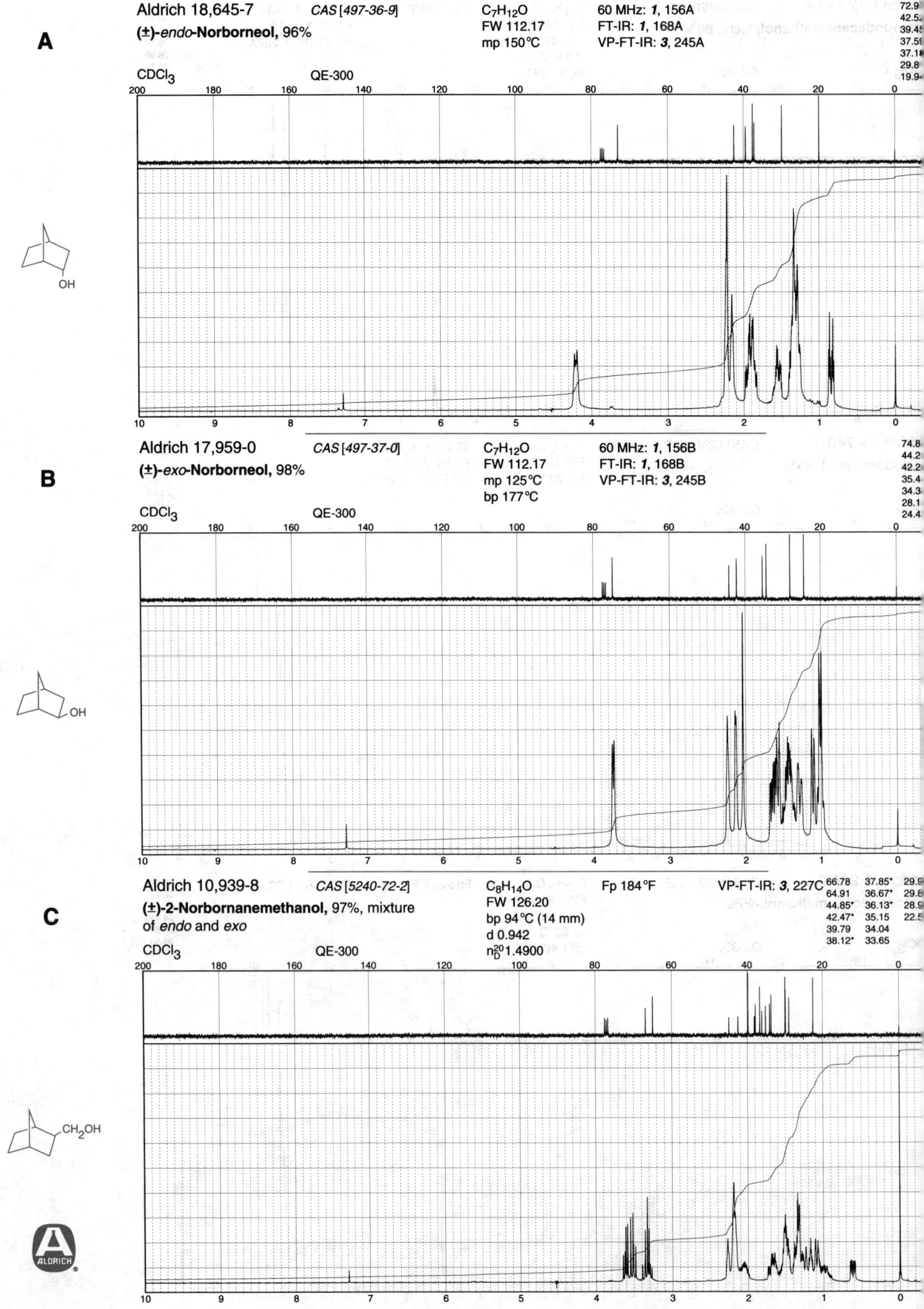

A

Aldrich 18,645-7   CAS [497-36-9]   C₇H₁₂O   60 MHz: **1**, 156A
(±)-endo-Norborneol, 96%   FW 112.17   FT-IR: **1**, 168A
                                         mp 150°C   VP-FT-IR: **3**, 245A

CDCl₃   QE-300

72.9
42.5
39.4
37.5
37.1
29.8
19.9

B

Aldrich 17,959-0   CAS [497-37-0]   C₇H₁₂O   60 MHz: **1**, 156B
(±)-exo-Norborneol, 98%   FW 112.17   FT-IR: **1**, 168B
                                         mp 125°C   VP-FT-IR: **3**, 245B
                                         bp 177°C

CDCl₃   QE-300

74.8
44.2
42.2
35.4
34.3
28.1
24.4

C

Aldrich 10,939-8   CAS [5240-72-2]   C₈H₁₄O   Fp 184°F   VP-FT-IR: **3**, 227C
(±)-2-Norbornanemethanol, 97%, mixture   FW 126.20
of endo and exo   bp 94°C (14 mm)
                                         d 0.942
CDCl₃   QE-300   n²⁰_D 1.4900

66.78	37.85*	29.9
64.91	36.67*	29.8
44.85*	36.13*	28.9
42.47*	35.15	22.5
39.79	34.04	
38.12*	33.65	

Aldrich 13,057-5   *CAS [6968-75-8]*

(±)-3-Methyl-2-norbornanemethanol, 93%, mixture of isomers

$C_9H_{16}O$
FW 140.23
bp 215°C
d 0.959
n_D^{20}1.4840

Fp 195°F

CDCl3   QE-300

60 MHz: *1*, 156C
FT-IR: *1*, 155C
VP-FT-IR: *3*, 228B

66.41	40.80*	30.07
64.12	39.40*	21.74*
53.52*	38.71*	21.71
52.55*	38.45*	16.51*
43.35*	37.31	
41.73*	36.46	

A

Aldrich 19,644-4   *CAS [2217-02-9]*

(1R)-endo-(+)-Fenchyl alcohol, 97%

$C_{10}H_{18}O$
FW 154.25
mp 42°C
Fp 165°F

CDCl3   QE-300

60 MHz: *1*, 157C
FT-IR: *1*, 169D
VP-FT-IR: *3*, 246C

85.20*	30.80*
49.23	26.16
48.14*	25.23
41.11	20.16*
39.19	19.44*

B

Aldrich W34,910-0   *CAS [18368-91-7]*

2-Ethylfenchol, 97+%

$C_{12}H_{22}O$
FW 182.31
bp 105°C (15 mm)
d 0.956
n_D^{20}1.4800

Fp 192°F

CDCl3   QE-300

80.87	27.80
52.61	27.49*
50.22*	24.95
44.29	22.70*
41.11	18.13*
30.81	9.52*

C

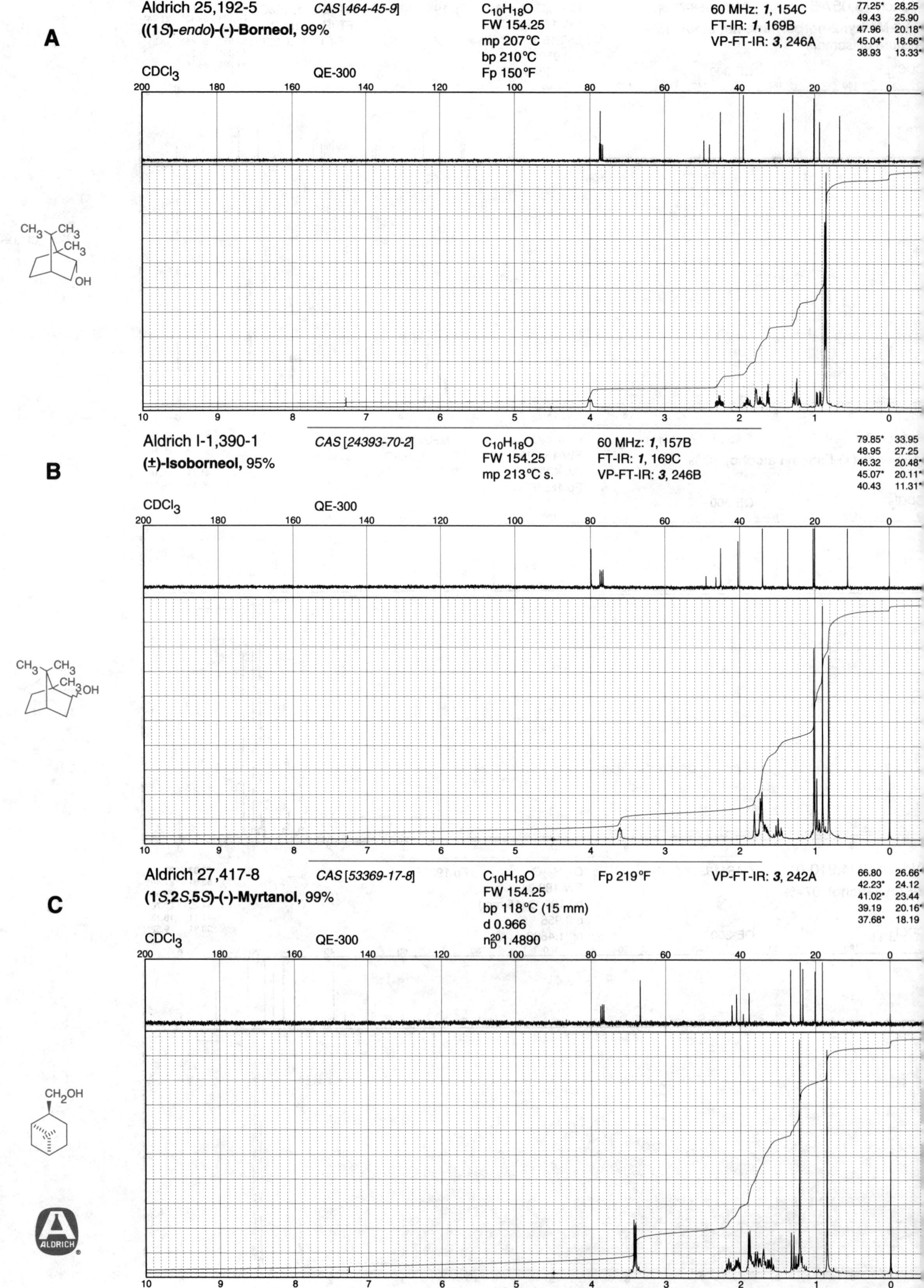

A

Aldrich 25,192-5   *CAS [464-45-9]* $C_{10}H_{18}O$ 60 MHz: *1*, 154C   77.25* 28.25
((1*S*)-*endo*)-(-)-**Borneol**, 99%   FW 154.25   FT-IR: *1*, 169B   49.43 25.90
mp 207°C   VP-FT-IR: *3*, 246A   47.96 20.18*
bp 210°C   45.04* 18.66*
Fp 150°F   38.93 13.33*

CDCl₃   QE-300

B

Aldrich I-1,390-1   *CAS [24393-70-2]* $C_{10}H_{18}O$ 60 MHz: *1*, 157B   79.85* 33.95
(±)-**Isoborneol**, 95%   FW 154.25   FT-IR: *1*, 169C   48.95 27.25
mp 213°C s.   VP-FT-IR: *3*, 246B   46.32 20.48*
45.07* 20.11*
40.43 11.31*

CDCl₃   QE-300

C

Aldrich 27,417-8   *CAS [53369-17-8]* $C_{10}H_{18}O$ Fp 219°F   VP-FT-IR: *3*, 242A   66.80 26.66*
(1*S*,2*S*,5*S*)-(-)-**Myrtanol**, 99%   FW 154.25   42.23* 24.12
bp 118°C (15 mm)   41.02* 23.44
d 0.966   39.19 20.16*
n₂₀D 1.4890   37.68* 18.19

CDCl₃   QE-300

Aldrich B4,590-9   CAS [26896-48-0]   C₁₂H₂₀O₂   FT-IR: *1*, 173B

$C_{12}H_{20}O_2$
FW 196.29
n_D^{20} 1.5280
Fp >230 °F

4,8-Bis(hydroxymethyl)tricyclo[5.2.1.0²,⁶]-
decane, 98%, mixture of isomers

CDCl₃   QE-300

A

67.92	47.96*	43.01*	40.30*	32.73	27.80
67.27	45.85*	42.59*	39.34	32.22	27.61
66.87	45.59*	42.48*	39.23*	30.72	27.32
66.58	45.36*	41.60*	37.80*	30.65	25.27
66.53	45.11*	41.53*	37.56*	29.04	24.62
50.20*	44.74*	40.50*	37.01*	28.37	
48.82*	43.73*	40.39	36.30*	27.98	

Aldrich 18,322-9   CAS [51152-11-5]

$C_{10}H_{18}O$
FW 154.25
mp 40 °C
bp 217 °C
Fp 200 °F

(±)-Isopinocampheol, 98%

CDCl₃   QE-300

60 MHz: *1*, 157D
FT-IR: *1*, 165D
VP-FT-IR: *3*, 242B

71.56*	38.15
47.89*	34.36
47.70*	27.69*
41.81*	23.67*
39.02	20.73*

B

Aldrich 22,190-2   CAS [25465-65-0]

$C_{10}H_{18}O$
FW 154.25
mp 52 °C
bp 219 °C
Fp 200 °F

(1R,2R,3R,5S)-(−)-Isopinocampheol, 98%

CDCl₃   QE-300

60 MHz: *1*, 158A
FT-IR: *1*, 166A
VP-FT-IR: *3*, 242C

71.55*	38.16
47.86*	34.36
47.68*	27.68*
41.79*	23.68*
39.01	20.73*

C

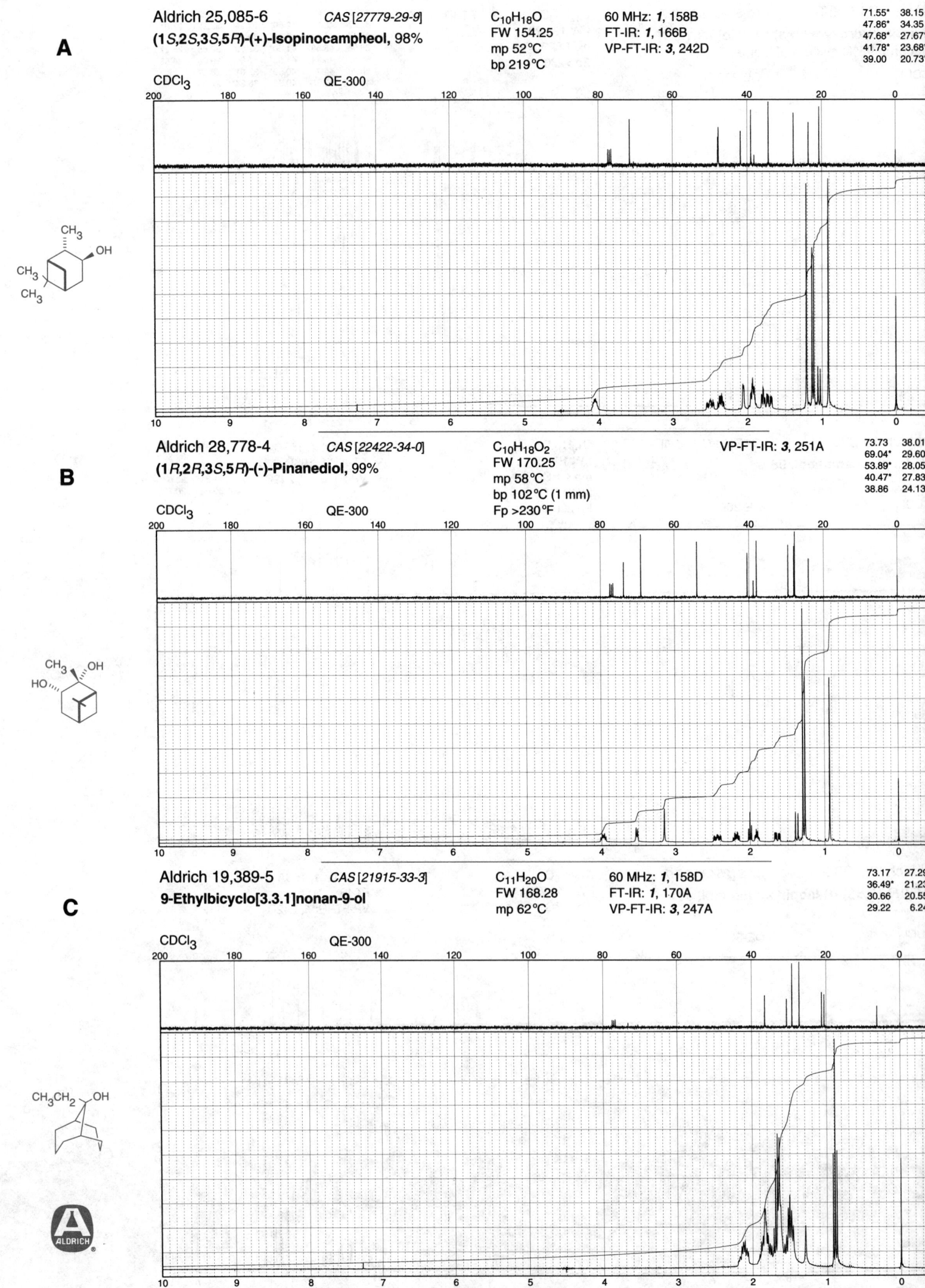

A

Aldrich 25,085-6   *CAS [27779-29-9]*

(1*S*,2*S*,3*S*,5*R*)-(+)-Isopinocampheol, 98%

C₁₀H₁₈O
FW 154.25
mp 52°C
bp 219°C

60 MHz: *1*, 158B
FT-IR: *1*, 166B
VP-FT-IR: *3*, 242D

71.55*	38.15
47.86*	34.35
47.68*	27.67*
41.78*	23.68*
39.00	20.73*

CDCl₃   QE-300

B

Aldrich 28,778-4   *CAS [22422-34-0]*

(1*R*,2*R*,3*S*,5*R*)-(-)-Pinanediol, 99%

C₁₀H₁₈O₂
FW 170.25
mp 58°C
bp 102°C (1 mm)
Fp >230°F

VP-FT-IR: *3*, 251A

73.73	38.01
69.04*	29.60*
53.89*	28.05
40.47*	27.83*
38.86	24.13

CDCl₃   QE-300

C

Aldrich 19,389-5   *CAS [21915-33-3]*

9-Ethylbicyclo[3.3.1]nonan-9-ol

C₁₁H₂₀O
FW 168.28
mp 62°C

60 MHz: *1*, 158D
FT-IR: *1*, 170A
VP-FT-IR: *3*, 247A

73.17	27.29
36.49*	21.23
30.66	20.55
29.22	6.24

CDCl₃   QE-300

Aldrich 19,390-9   *CAS [21915-40-2]* $C_{15}H_{26}O$ 60 MHz: *1*, 159A
-Cyclohexylbicyclo[3.3.1]nonan-9-ol   FW 222.37   FT-IR: *1*, 170B
mp 111°C

73.93	26.90
39.72*	26.65
33.66*	24.67
28.46	21.14
27.59	20.37

A

CDCl₃   QE-300

Aldrich 15,698-1   *CAS [36159-47-4]* $C_{10}H_{18}O$ 60 MHz: *1*, 160A
cis-Decahydro-1-naphthol, 99%   FW 154.25   VP-FT-IR: *3*, 247D
mp 91°C
bp 238°C

73.50*	26.35
42.97*	24.46
35.80*	24.38
31.83	21.59
29.37	18.97

B

CDCl₃   QE-300

Aldrich D30-8   *CAS [825-51-4]* $C_{10}H_{18}O$ Fp >230°F
Decahydro-2-naphthol, 98%, mixture of   FW 154.25
isomers   bp 109°C (14 mm)
d 0.996
CDCl₃   QE-300   n²⁰_D 1.5000

71.62*	40.40	34.73*	30.54	26.58
70.75*	36.48*	33.84	30.02	26.33
66.94*	35.84	33.80	28.09	25.90
43.27	35.56	33.30	27.55	24.90
42.36*	35.44*	32.04	26.71	22.19
41.24*	34.96*	31.77	26.66	21.08

C

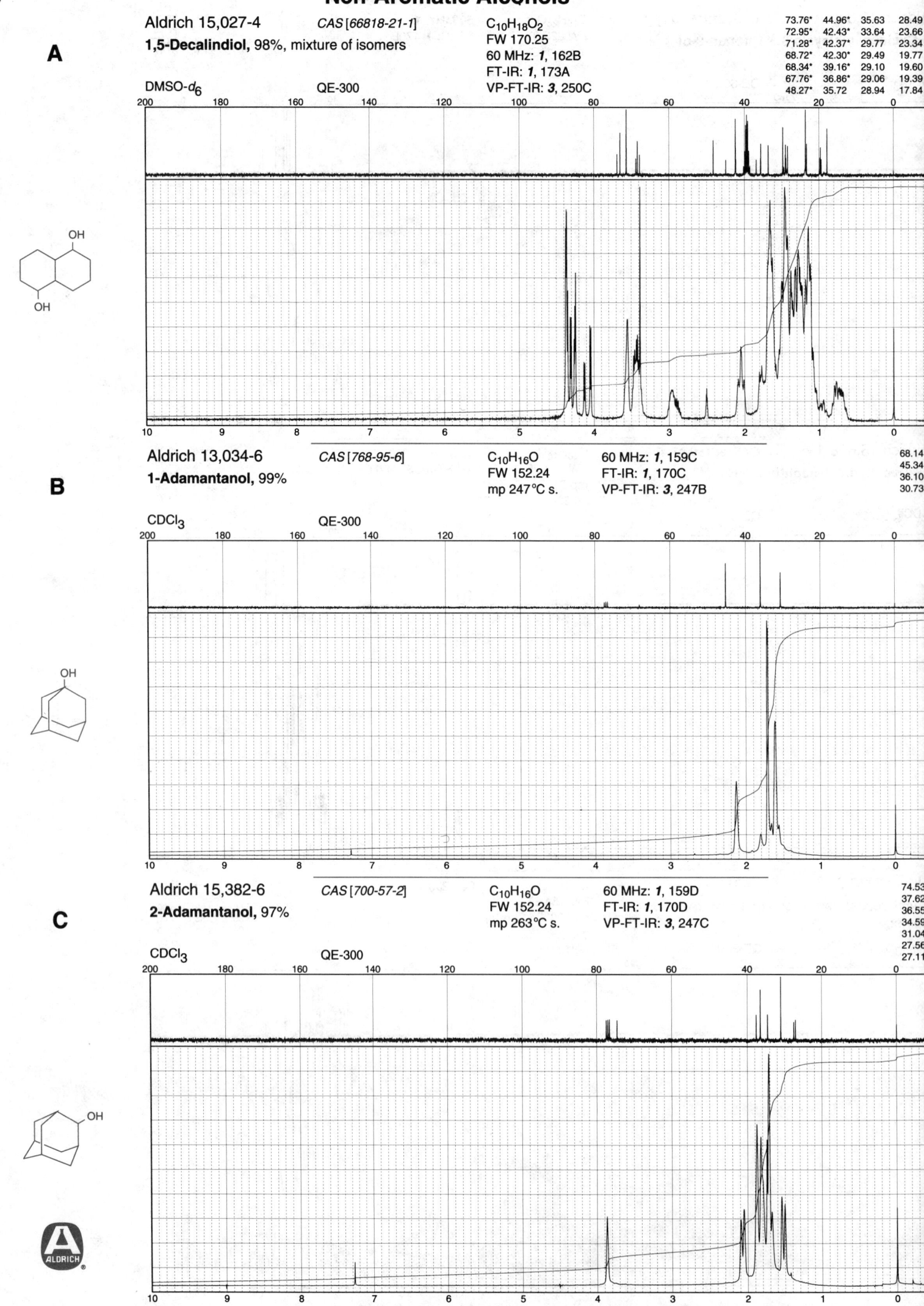

A

Aldrich 15,027-4   *CAS [66818-21-1]* $C_{10}H_{18}O_2$
1,5-Decalindiol, 98%, mixture of isomers   FW 170.25
60 MHz: *1*, 162B
FT-IR: *1*, 173A
DMSO-d_6 QE-300   VP-FT-IR: *3*, 250C

73.76*	44.96*	35.63	28.49
72.95*	42.43*	33.64	23.66
71.28*	42.37*	29.77	23.34
68.72*	42.30*	29.49	19.77
68.34*	39.16*	29.10	19.60
67.76*	36.86*	29.06	19.39
48.27*	35.72	28.94	17.84

B

Aldrich 13,034-6   *CAS [768-95-6]* $C_{10}H_{16}O$ 60 MHz: *1*, 159C
1-Adamantanol, 99%   FW 152.24   FT-IR: *1*, 170C
mp 247°C s.   VP-FT-IR: *3*, 247B

| 68.14 |
| 45.34 |
| 36.10 |
| 30.73* |

CDCl₃   QE-300

C

Aldrich 15,382-6   *CAS [700-57-2]* $C_{10}H_{16}O$ 60 MHz: *1*, 159D
2-Adamantanol, 97%   FW 152.24   FT-IR: *1*, 170D
mp 263°C s.   VP-FT-IR: *3*, 247C

| 74.53 |
| 37.62 |
| 36.55 |
| 34.59 |
| 31.04 |
| 27.56 |
| 27.11 |

CDCl₃   QE-300

ALDRICH

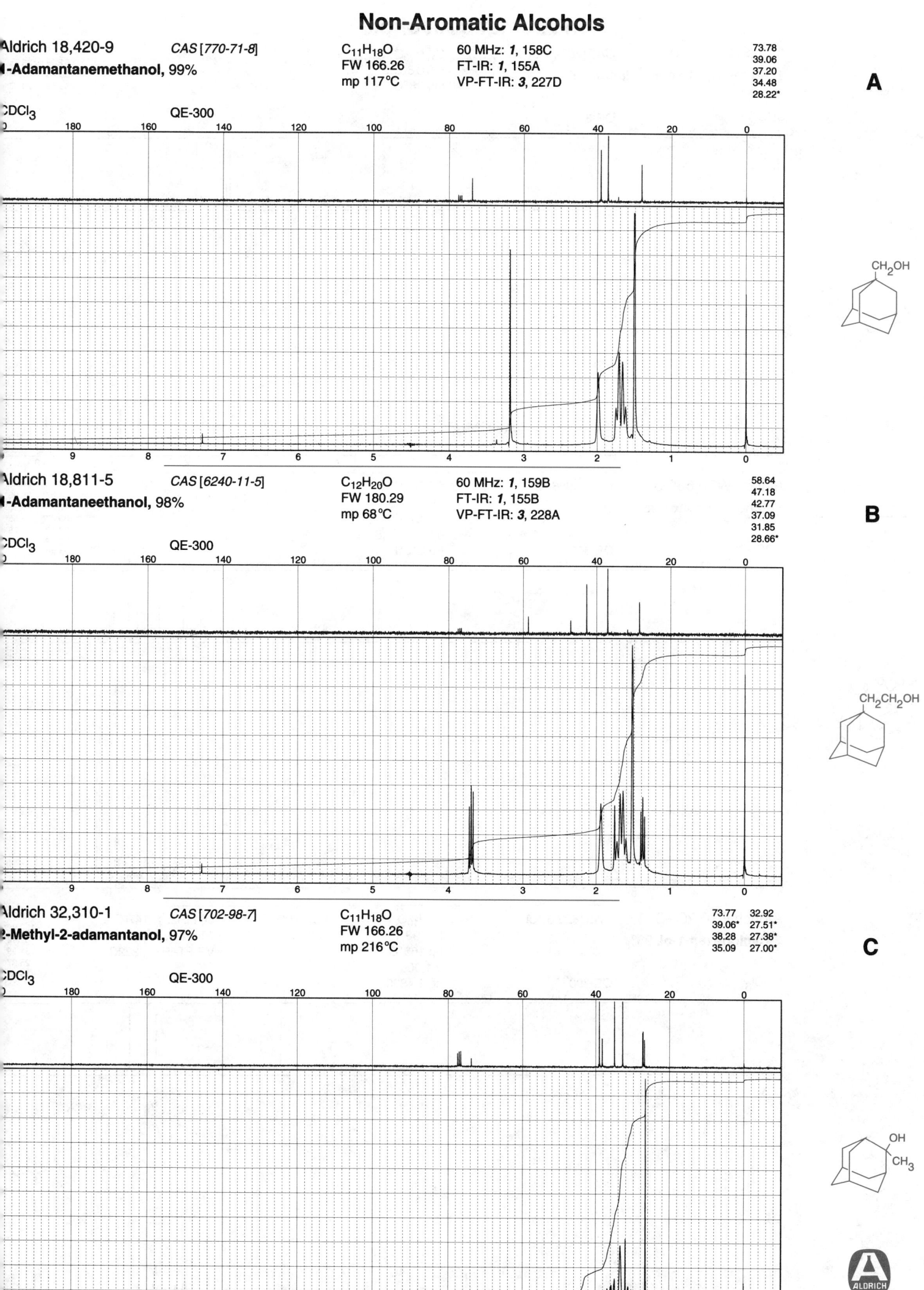

Aldrich 18,420-9   *CAS [770-71-8]*
1-Adamantanemethanol, 99%

C₁₁H₁₈O
FW 166.26
mp 117°C

60 MHz: *1*, 158C
FT-IR: *1*, 155A
VP-FT-IR: *3*, 227D

73.78
39.06
37.20
34.48
28.22*

A

CDCl₃   QE-300

Aldrich 18,811-5   *CAS [6240-11-5]*
1-Adamantaneethanol, 98%

C₁₂H₂₀O
FW 180.29
mp 68°C

60 MHz: *1*, 159B
FT-IR: *1*, 155B
VP-FT-IR: *3*, 228A

58.64
47.18
42.77
37.09
31.85
28.66*

B

CDCl₃   QE-300

Aldrich 32,310-1   *CAS [702-98-7]*
2-Methyl-2-adamantanol, 97%

C₁₁H₁₈O
FW 166.26
mp 216°C

73.77   32.92
39.06* 27.51*
38.28   27.38*
35.09   27.00*

C

CDCl₃   QE-300

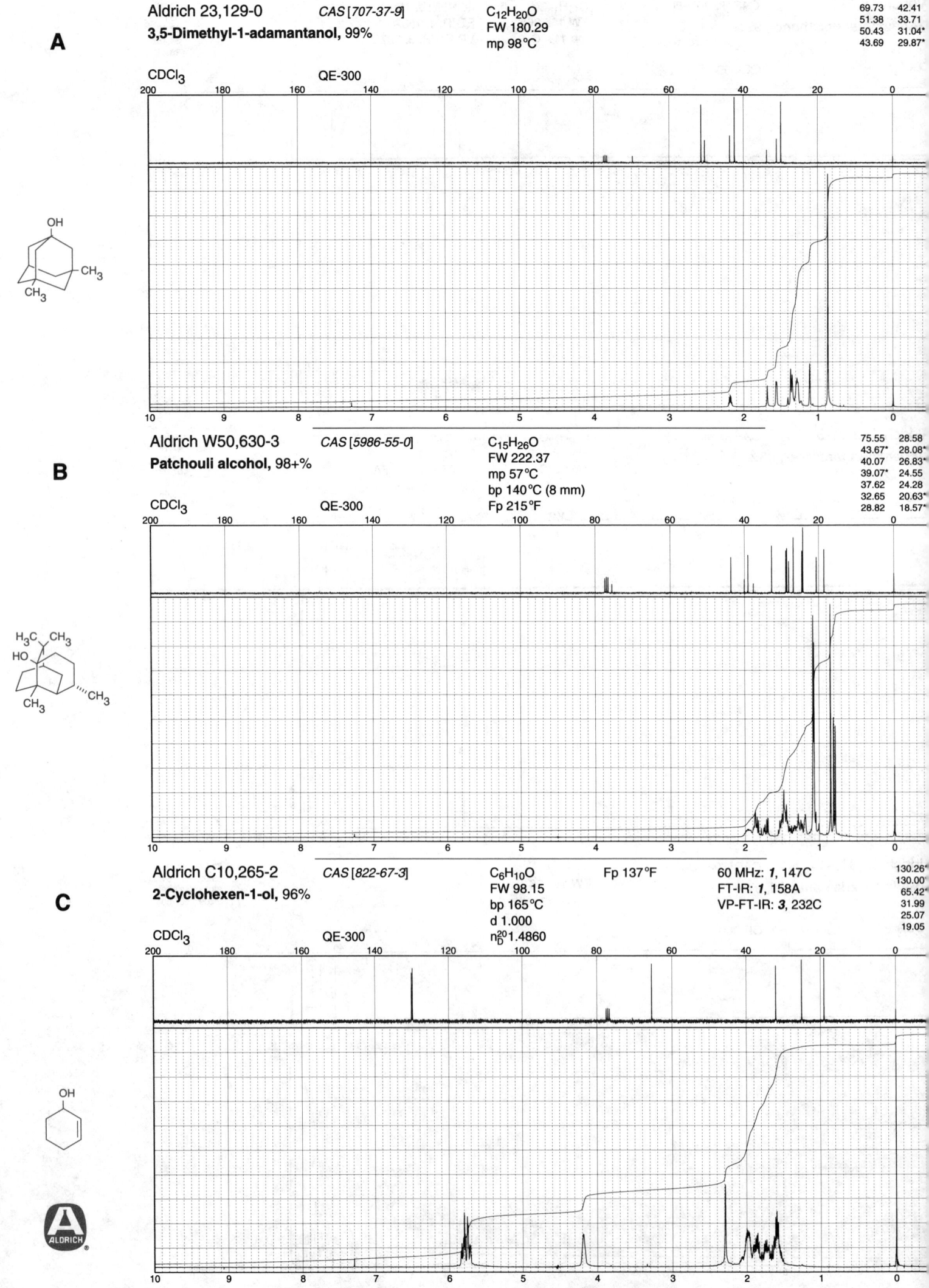

A

Aldrich 23,129-0

3,5-Dimethyl-1-adamantanol, 99%

CAS [707-37-9]

$C_{12}H_{20}O$
FW 180.29
mp 98°C

69.73 42.41
51.38 33.71
50.43 31.04*
43.69 29.87*

CDCl₃   QE-300

B

Aldrich W50,630-3

Patchouli alcohol, 98+%

CAS [5986-55-0]

$C_{15}H_{26}O$
FW 222.37
mp 57°C
bp 140°C (8 mm)
Fp 215°F

75.55 28.58
43.67* 28.08*
40.07 26.83*
39.07* 24.55
37.62 24.28
32.65 20.63*
28.82 18.57*

CDCl₃   QE-300

C

Aldrich C10,265-2

2-Cyclohexen-1-ol, 96%

CAS [822-67-3]

$C_6H_{10}O$
FW 98.15
bp 165°C
d 1.000
n²⁰_D 1.4860

Fp 137°F

60 MHz: *1*, 147C
FT-IR: *1*, 158A
VP-FT-IR: *3*, 232C

130.26*
130.00
65.42*
31.99
25.07
19.05

CDCl₃   QE-300

Non-Aromatic Alcohols

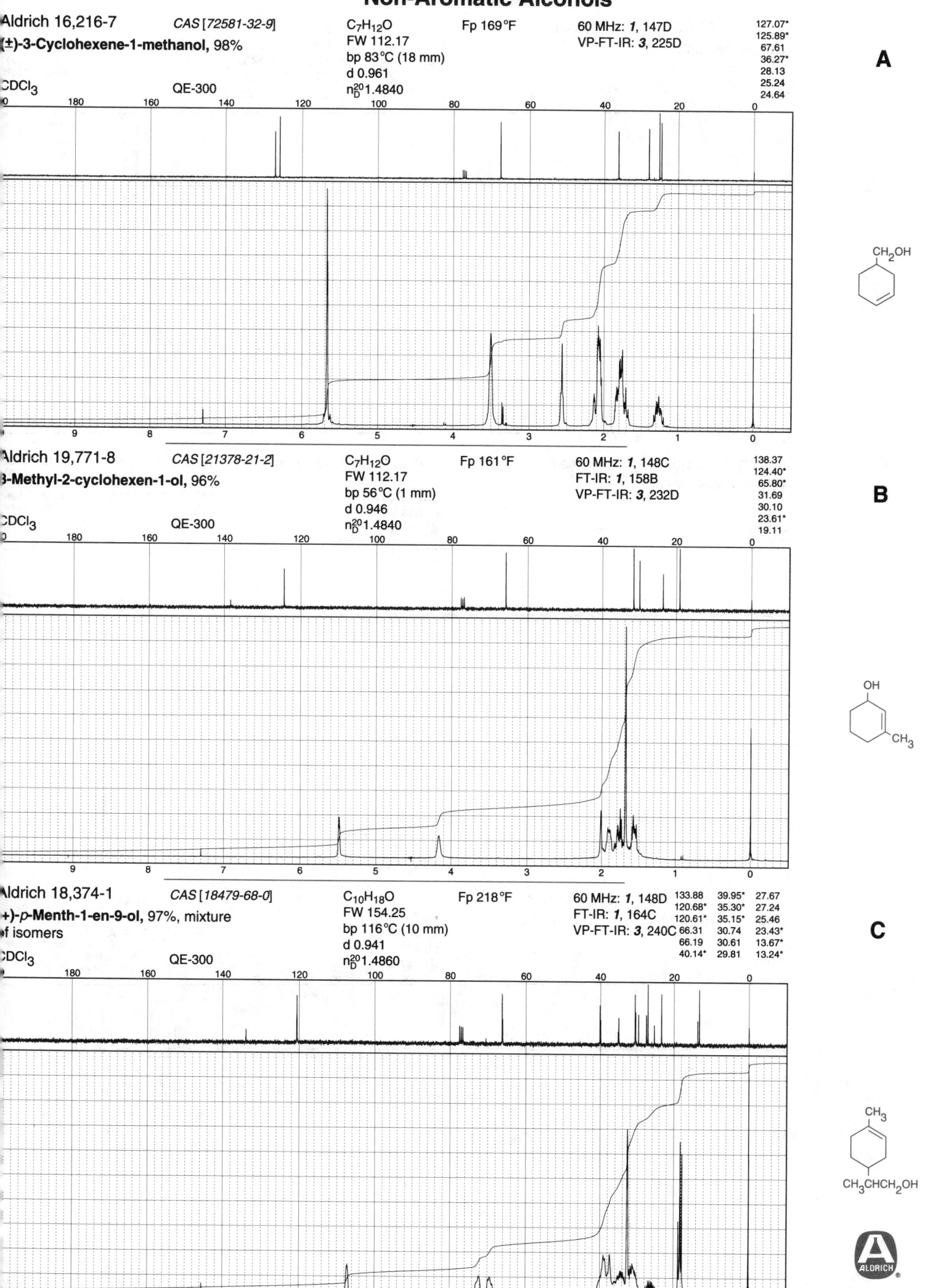

Aldrich 16,216-7   *CAS [72581-32-9]* $C_7H_{12}O$ Fp 169°F   60 MHz: *1*, 147D

(±)-3-Cyclohexene-1-methanol, 98%   FW 112.17   VP-FT-IR: *3*, 225D

bp 83°C (18 mm)
d 0.961
CDCl₃   QE-300 n_D^{20} 1.4840

127.07*
125.89*
67.61
36.27*
28.13
25.24
24.64

A

Aldrich 19,771-8   *CAS [21378-21-2]* $C_7H_{12}O$ Fp 161°F   60 MHz: *1*, 148C

3-Methyl-2-cyclohexen-1-ol, 96%   FW 112.17   FT-IR: *1*, 158B

bp 56°C (1 mm)   VP-FT-IR: *3*, 232D
d 0.946
CDCl₃   QE-300 n_D^{20} 1.4840

138.37
124.40*
65.80*
31.69
30.10
23.61*
19.11

B

Aldrich 18,374-1   *CAS [18479-68-0]* $C_{10}H_{18}O$ Fp 218°F   60 MHz: *1*, 148D

(+)-p-Menth-1-en-9-ol, 97%, mixture of isomers   FW 154.25   FT-IR: *1*, 164C

bp 116°C (10 mm)   VP-FT-IR: *3*, 240C
d 0.941
CDCl₃   QE-300 n_D^{20} 1.4860

133.88   39.95*   27.67
120.68*   35.30*   27.24
120.61*   35.15*   25.46
66.31   30.74   23.43*
66.19   30.61   13.67*
40.14*   29.81   13.24*

C

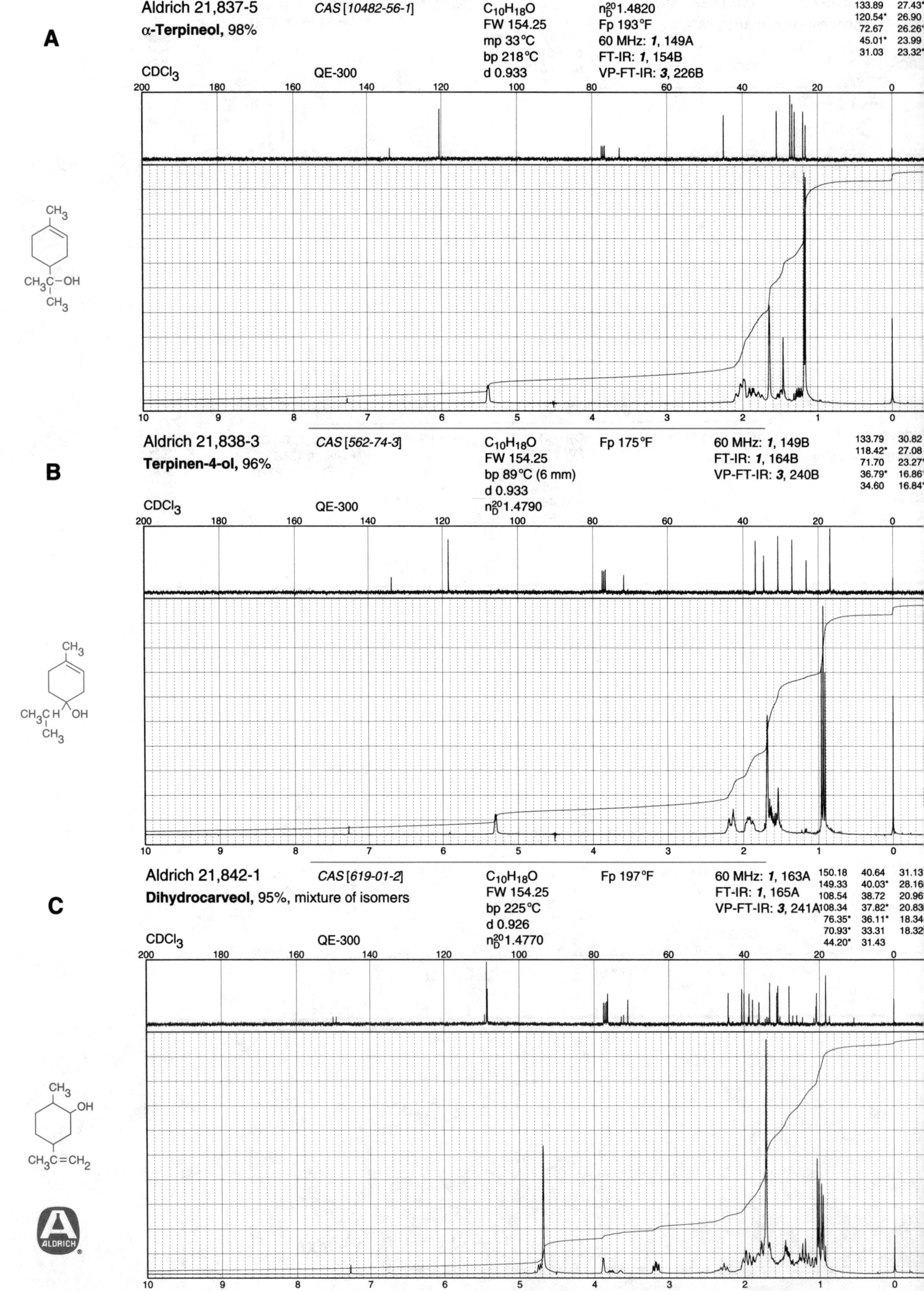

Aldrich 21,837-5
α-Terpineol, 98%

A

CAS [10482-56-1]

CDCl₃   QE-300

$C_{10}H_{18}O$
FW 154.25
mp 33°C
bp 218°C
d 0.933

n_D^{20} 1.4820
Fp 193°F
60 MHz: *1*, 149A
FT-IR: *1*, 154B
VP-FT-IR: *3*, 226B

133.89	27.43*
120.54*	26.90
72.67	26.26*
45.01*	23.99
31.03	23.32*

Aldrich 21,838-3
Terpinen-4-ol, 96%

B

CAS [562-74-3]

CDCl₃   QE-300

$C_{10}H_{18}O$
FW 154.25
bp 89°C (6 mm)
d 0.933
n_D^{20} 1.4790

Fp 175°F

60 MHz: *1*, 149B
FT-IR: *1*, 164B
VP-FT-IR: *3*, 240B

133.79	30.82
118.42*	27.08
71.70	23.27*
36.79*	16.86*
34.60	16.84*

Aldrich 21,842-1
Dihydrocarveol, 95%, mixture of isomers

C

CAS [619-01-2]

CDCl₃   QE-300

$C_{10}H_{18}O$
FW 154.25
bp 225°C
d 0.926
n_D^{20} 1.4770

Fp 197°F

60 MHz: *1*, 163A
FT-IR: *1*, 165A
VP-FT-IR: *3*, 241A

150.18	40.64	31.13
149.33	40.03*	28.16
108.54	38.72	20.96
108.34	37.82*	20.83
76.35*	36.11*	18.34
70.93*	33.31	18.32
44.20*	31.43	

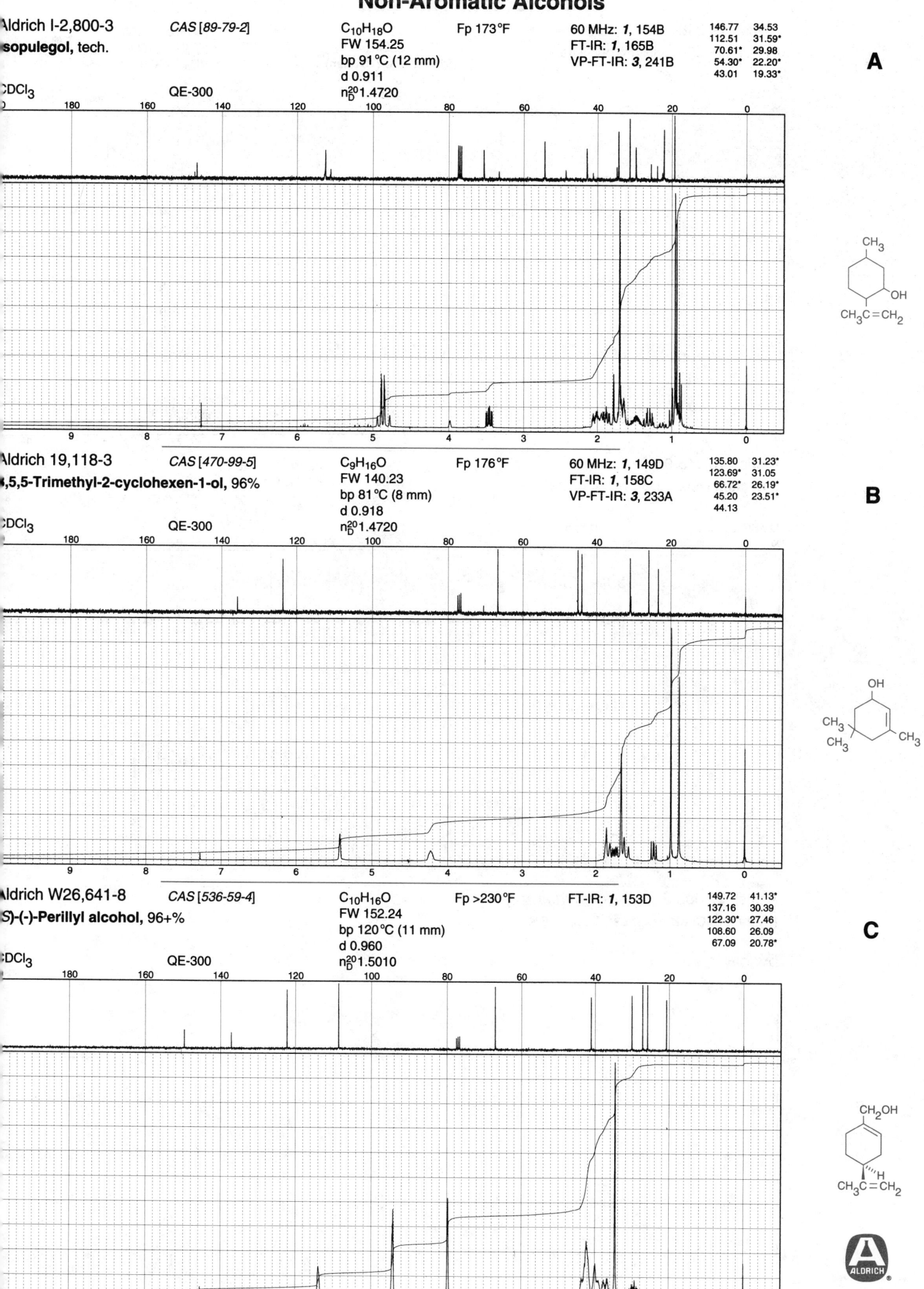

Aldrich I-2,800-3 CAS [89-79-2] C₁₀H₁₈O Fp 173°F 60 MHz: *1*, 154B
Isopulegol, tech. FW 154.25 FT-IR: *1*, 165B
bp 91°C (12 mm) VP-FT-IR: *3*, 241B
d 0.911
CDCl₃ QE-300 n²⁰_D 1.4720

146.77	34.53
112.51	31.59*
70.61*	29.98
54.30*	22.20*
43.01	19.33*

A

Aldrich 19,118-3 CAS [470-99-5] C₉H₁₆O Fp 176°F 60 MHz: *1*, 149D
3,5,5-Trimethyl-2-cyclohexen-1-ol, 96% FW 140.23 FT-IR: *1*, 158C
bp 81°C (8 mm) VP-FT-IR: *3*, 233A
d 0.918
CDCl₃ QE-300 n²⁰_D 1.4720

135.80	31.23*
123.69*	31.05
66.72*	26.19*
45.20	23.51*
44.13	

B

Aldrich W26,641-8 CAS [536-59-4] C₁₀H₁₆O Fp >230°F FT-IR: *1*, 153D
(S)-(-)-Perillyl alcohol, 96+% FW 152.24
bp 120°C (11 mm)
d 0.960
CDCl₃ QE-300 n²⁰_D 1.5010

149.72	41.13*
137.16	30.39
122.30*	27.46
108.60	26.09
67.09	20.78*

C

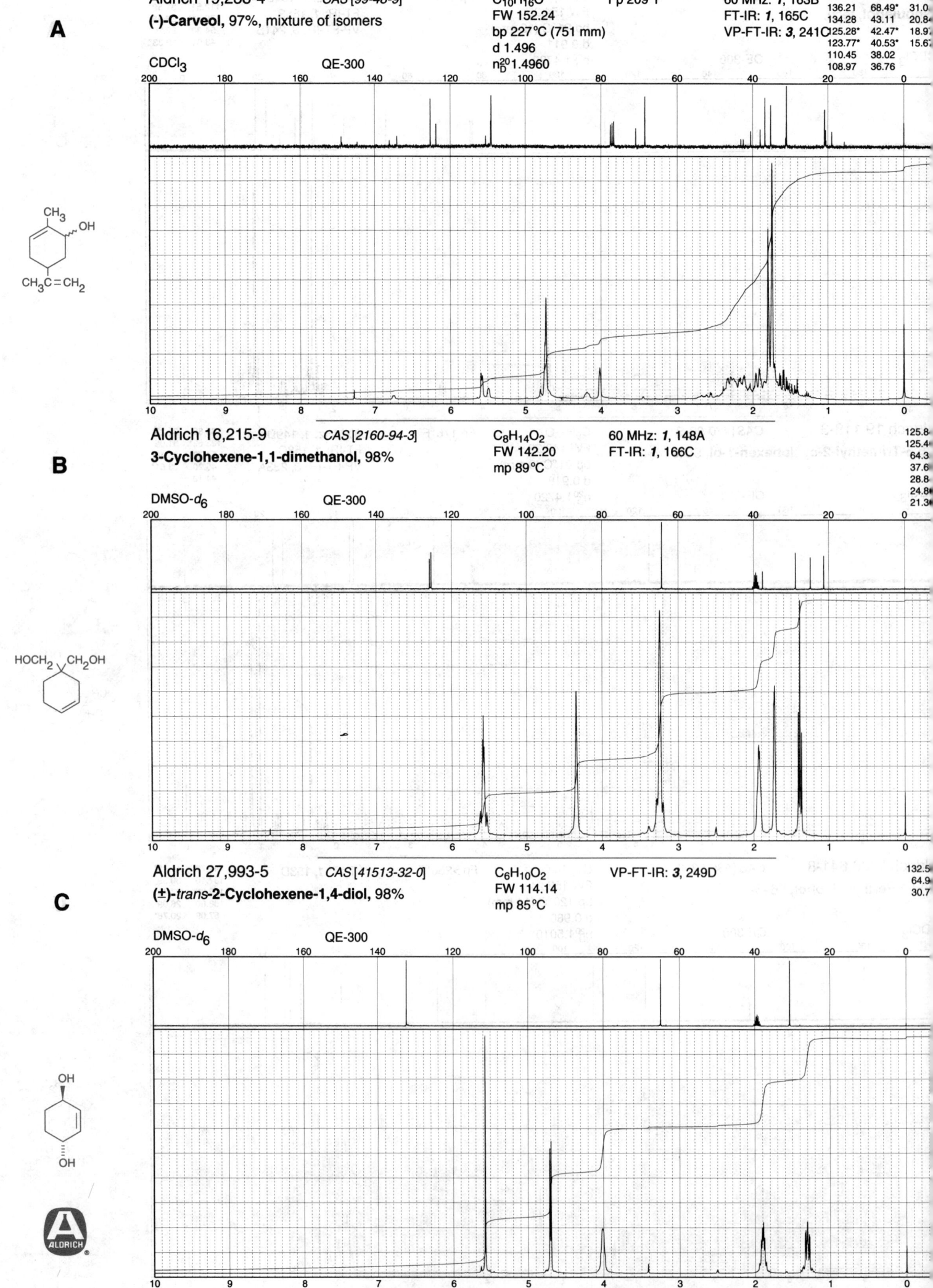

A

Aldrich 19,238-4   *CAS [99-48-9]*

(-)-Carveol, 97%, mixture of isomers

$C_{10}H_{16}O$
FW 152.24
bp 227°C (751 mm)
d 1.496
n_D^{20} 1.4960

Fp 209°F

60 MHz: *1*, 163B
FT-IR: *1*, 165C
VP-FT-IR: *3*, 241C

149.13   70.86*   35.2
136.21   68.49*   31.0
134.28   43.11   20.8
125.28* 42.47*   18.9
123.77* 40.53*   15.6
110.45   38.02
108.97   36.76

CDCl₃   QE-300

B

Aldrich 16,215-9   *CAS [2160-94-3]*

3-Cyclohexene-1,1-dimethanol, 98%
FW 142.20
mp 89°C

$C_8H_{14}O_2$

60 MHz: *1*, 148A
FT-IR: *1*, 166C

125.8
125.4
64.3
37.6
28.8
24.8
21.3

DMSO-*d₆*   QE-300

C

Aldrich 27,993-5   *CAS [41513-32-0]*

(±)-*trans*-2-Cyclohexene-1,4-diol, 98%
FW 114.14
mp 85°C

$C_6H_{10}O_2$

VP-FT-IR: *3*, 249D

132.5
64.9
30.7

DMSO-*d₆*   QE-300

Aldrich 24,777-4   CAS [32226-54-3] $C_{10}H_{18}O_2$ FT-IR: **1**, 164D

(±)-*trans*-*p*-**Menth-6-ene-2,8-diol**, 99%   FW 170.25   VP-FT-IR: **3**, 240D

mp 131°C
bp 271°C

134.65	32.96
123.93*	27.27*
70.49	26.82
67.06*	26.75*
38.20*	20.98*

A

CDCl$_3$+DMSO-d_6 QE-300

Aldrich 30,152-3   CAS [17793-95-2] $C_6H_8O_2$

cis-**3,5-Cyclohexadiene-1,2-diol**, 94%   FW 112.13

mp 59°C

130.19*
123.48*
66.79*

B

CDCl$_3$+DMSO-d_6 QE-300

Aldrich 22,302-6   CAS [68-26-8] $C_{20}H_{30}O$ 60 MHz: **1**, 163D

all-trans-**Retinol**   FW 286.46   FT-IR: **1**, 149C

mp 62°C

137.77	129.18	28.94*
137.58*	126.69*	21.70*
136.78	125.10*	19.27
136.26*	59.41	12.71*
136.07	39.62	12.62*
130.02*	34.24	
129.97*	33.04	

C

CDCl$_3$ QE-300

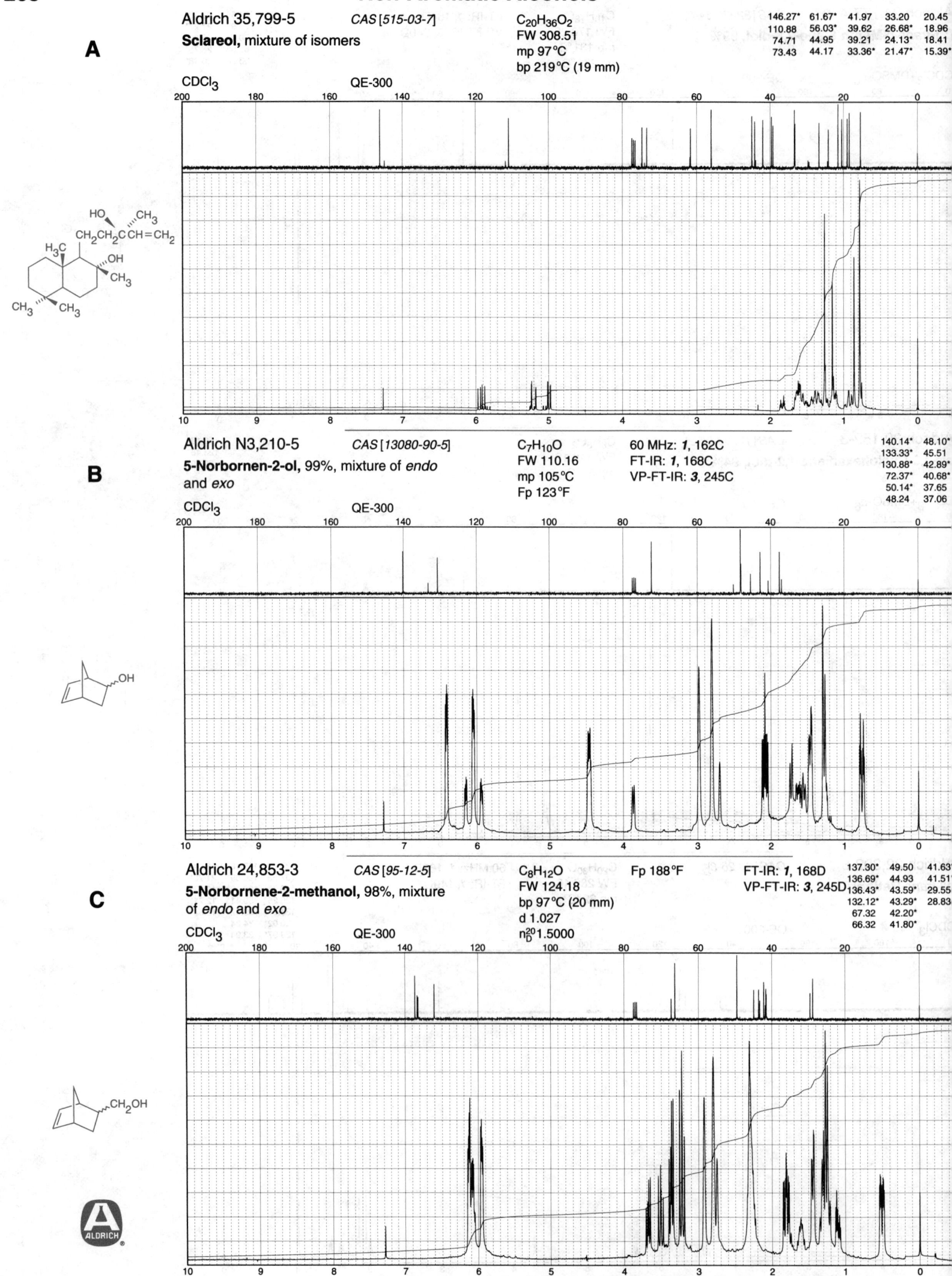

A Aldrich 35,799-5   *CAS [515-03-7]*   C_{20}H_{36}O_2
Sclareol, mixture of isomers   FW 308.51   mp 97°C   bp 219°C (19 mm)

146.27*	61.67*	41.97	33.20	20.45
110.88	56.03*	39.62	26.68*	18.96
74.71	44.95	39.21	24.13*	18.41
73.43	44.17	33.36*	21.47*	15.39*

CDCl₃   QE-300

B Aldrich N3,210-5   *CAS [13080-90-5]*   C_7H_{10}O   60 MHz: *1*, 162C
5-Norbornen-2-ol, 99%, mixture of *endo*   FW 110.16   FT-IR: *1*, 168C
and *exo*   mp 105°C   VP-FT-IR: *3*, 245C
   Fp 123°F

140.14*	48.10*
133.33*	45.51
130.88*	42.89*
72.37*	40.68*
50.14*	37.65
48.24	37.06

CDCl₃   QE-300

C Aldrich 24,853-3   *CAS [95-12-5]*   C_8H_{12}O   Fp 188°F   FT-IR: *1*, 168D
5-Norbornene-2-methanol, 98%, mixture   FW 124.18   VP-FT-IR: *3*, 245D
of *endo* and *exo*   bp 97°C (20 mm)
   d 1.027
   n_D^{20} 1.5000

137.30*	49.50	41.63*
136.69*	44.93	41.51*
136.43*	43.59*	29.55
132.12*	43.29*	28.83
67.32	42.20*	
66.32	41.80*	

CDCl₃   QE-300

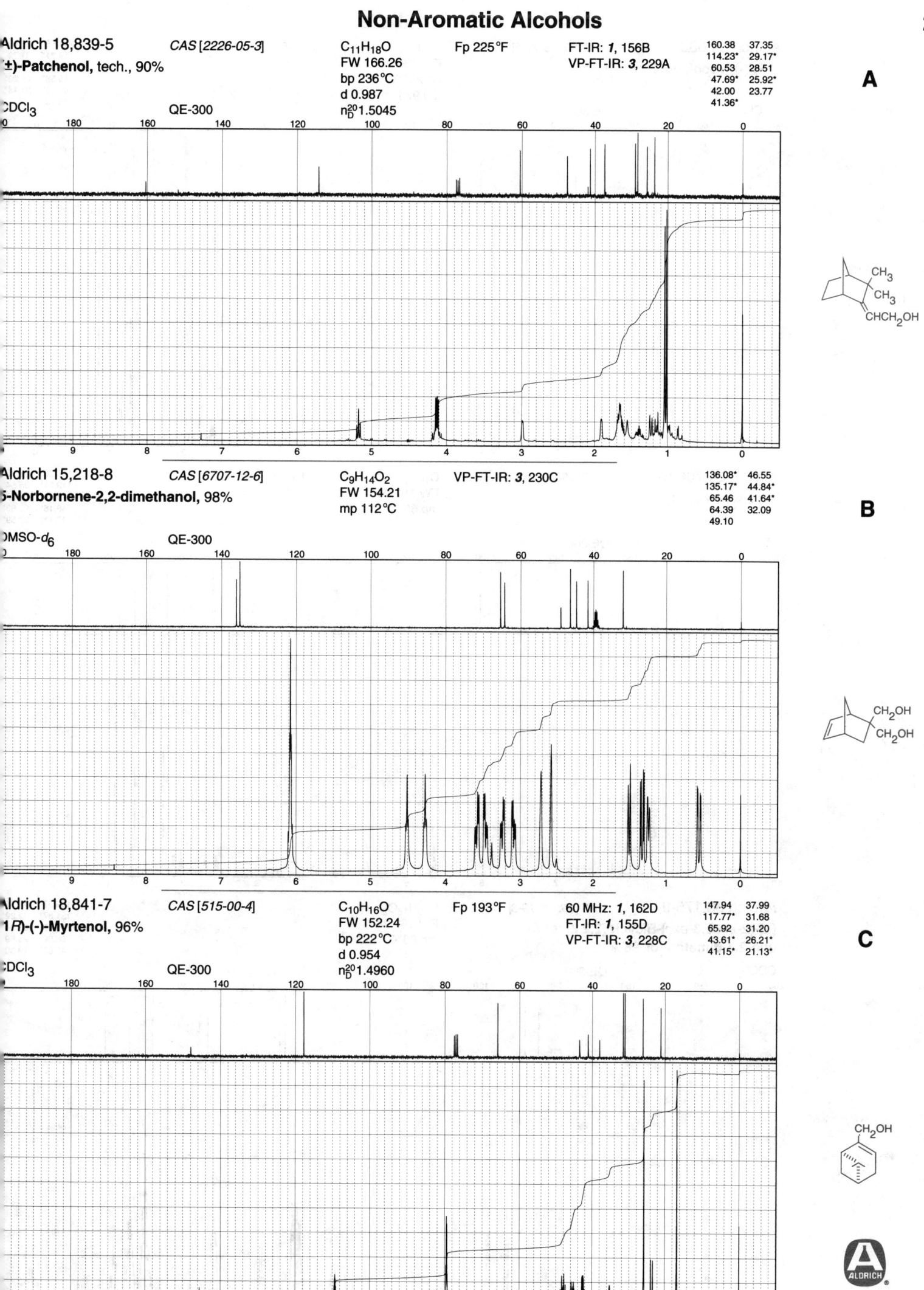

Aldrich 18,839-5 CAS [2226-05-3] $C_{11}H_{18}O$ Fp 225°F FT-IR: **1**, 156B
(±)-**Patchenol**, tech., 90% FW 166.26 VP-FT-IR: **3**, 229A
bp 236°C
d 0.987
CDCl₃ QE-300 n_D^{20} 1.5045

160.38 37.35
114.23* 29.17*
60.53 28.51
47.69* 25.92*
42.00 23.77
41.36*

A

Aldrich 15,218-8 CAS [6707-12-6] $C_9H_{14}O_2$ VP-FT-IR: **3**, 230C
5-**Norbornene-2,2-dimethanol**, 98% FW 154.21
mp 112°C
DMSO-d_6 QE-300

136.08* 46.55
135.17* 44.84*
65.46 41.64*
64.39 32.09
49.10

B

Aldrich 18,841-7 CAS [515-00-4] $C_{10}H_{16}O$ Fp 193°F 60 MHz: **1**, 162D
(1R)-(-)-**Myrtenol**, 96% FW 152.24 FT-IR: **1**, 155D
bp 222°C VP-FT-IR: **3**, 228C
d 0.954
CDCl₃ QE-300 n_D^{20} 1.4960

147.94 37.99
117.77* 31.68
65.92 31.20
43.61* 26.21*
41.15* 21.13*

C

ALDRICH

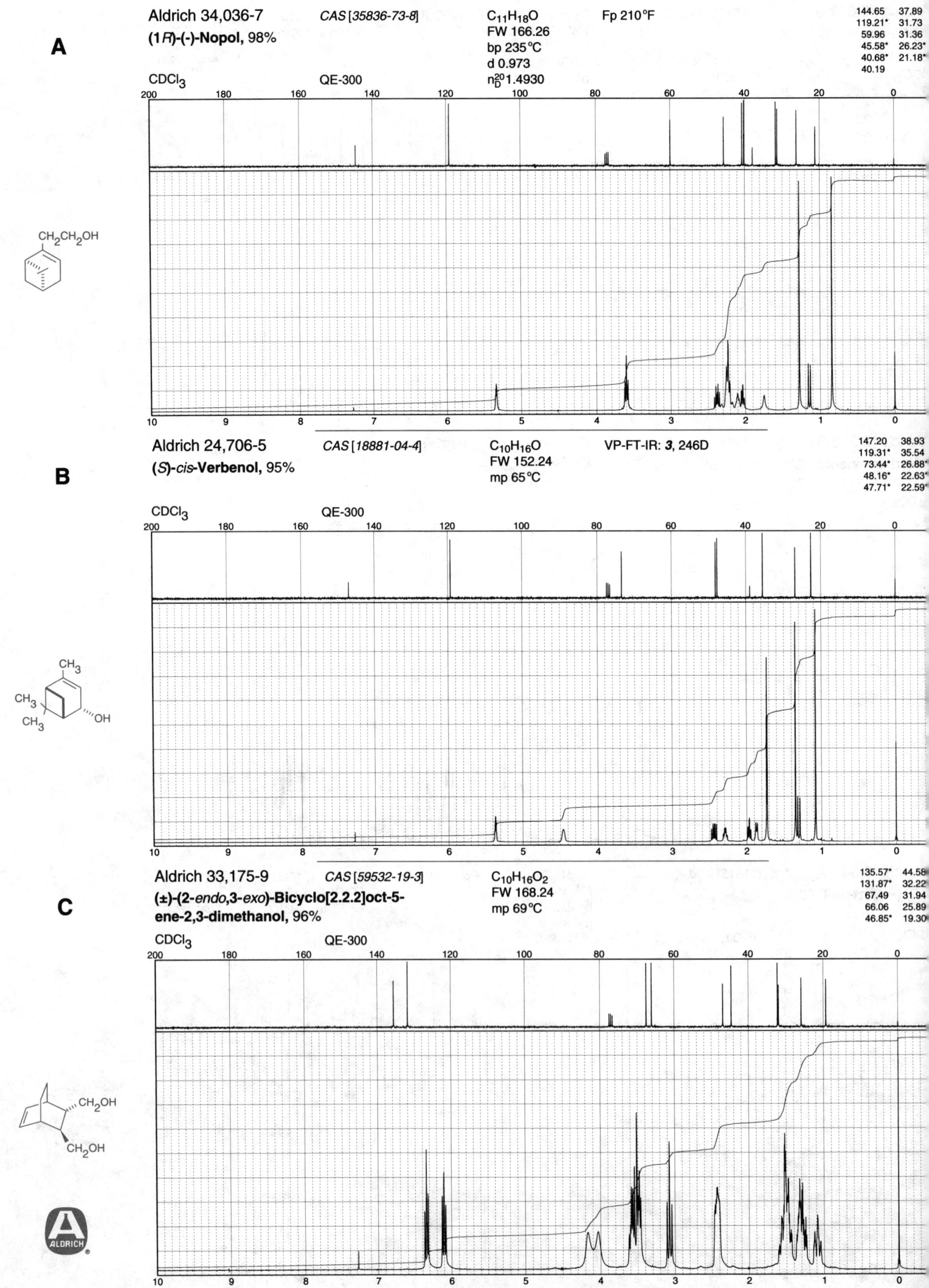

A

Aldrich 34,036-7

(1*R*)-(-)-**Nopol**, 98%

CAS [35836-73-8]

$C_{11}H_{18}O$
FW 166.26
bp 235°C
d 0.973
n_D^{20} 1.4930

Fp 210°F

144.65	37.89
119.21*	31.73
59.96	31.36
45.58*	26.23*
40.68*	21.18*
40.19	

CDCl₃   QE-300

B

Aldrich 24,706-5

(*S*)-*cis*-**Verbenol**, 95%

CAS [18881-04-4]

$C_{10}H_{16}O$
FW 152.24
mp 65°C

VP-FT-IR: **3**, 246D

147.20	38.93
119.31*	35.54
73.44*	26.88*
48.16*	22.63*
47.71*	22.59*

CDCl₃   QE-300

C

Aldrich 33,175-9

(±)-(2-*endo*,3-*exo*)-**Bicyclo[2.2.2]oct-5-ene-2,3-dimethanol**, 96%

CAS [59532-19-3]

$C_{10}H_{16}O_2$
FW 168.24
mp 69°C

135.57*	44.58
131.87*	32.22
67.49	31.94
66.06	25.89
46.85*	19.30

CDCl₃   QE-300

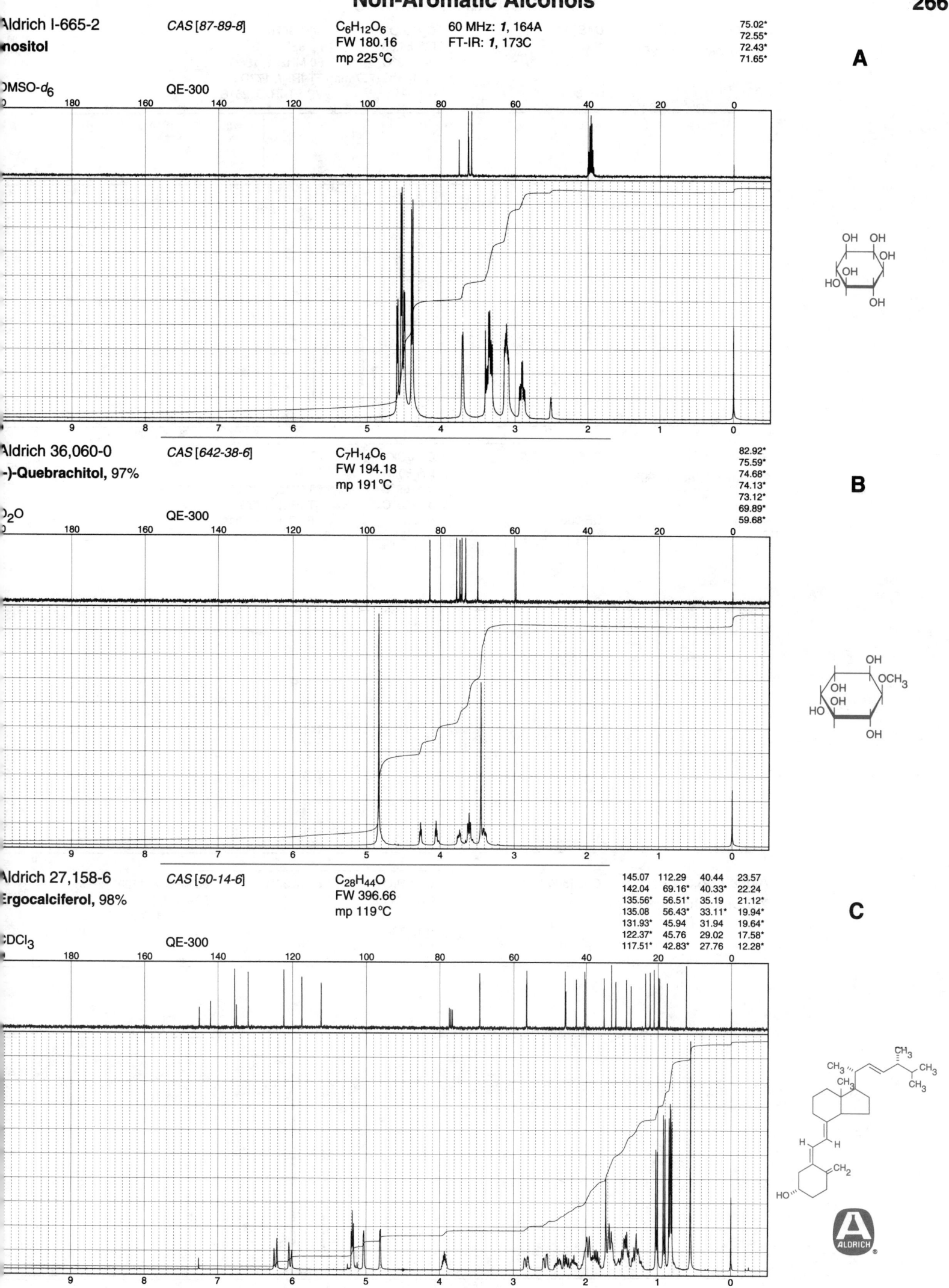

Aldrich I-665-2   *CAS [87-89-8]* $C_6H_{12}O_6$ 60 MHz: *1*, 164A

nositol   FW 180.16   FT-IR: *1*, 173C

mp 225°C

DMSO-d_6 QE-300

75.02*
72.55*
72.43*
71.65*

A

Aldrich 36,060-0   *CAS [642-38-6]* $C_7H_{14}O_6$

-)-Quebrachitol, 97%   FW 194.18

mp 191°C

D$_2$O   QE-300

82.92*
75.59*
74.68*
74.13*
73.12*
69.89*
59.68*

B

Aldrich 27,158-6   *CAS [50-14-6]* $C_{28}H_{44}O$

Ergocalciferol, 98%   FW 396.66

mp 119°C

CDCl$_3$ QE-300

145.07	112.29	40.44	23.57
142.04	69.16*	40.33*	22.24
135.56*	56.51*	35.19	21.12*
135.08	56.43*	33.11*	19.94*
131.93*	45.94	31.94	19.64*
122.37*	45.76	29.02	17.58*
117.51*	42.83*	27.76	12.28*

C

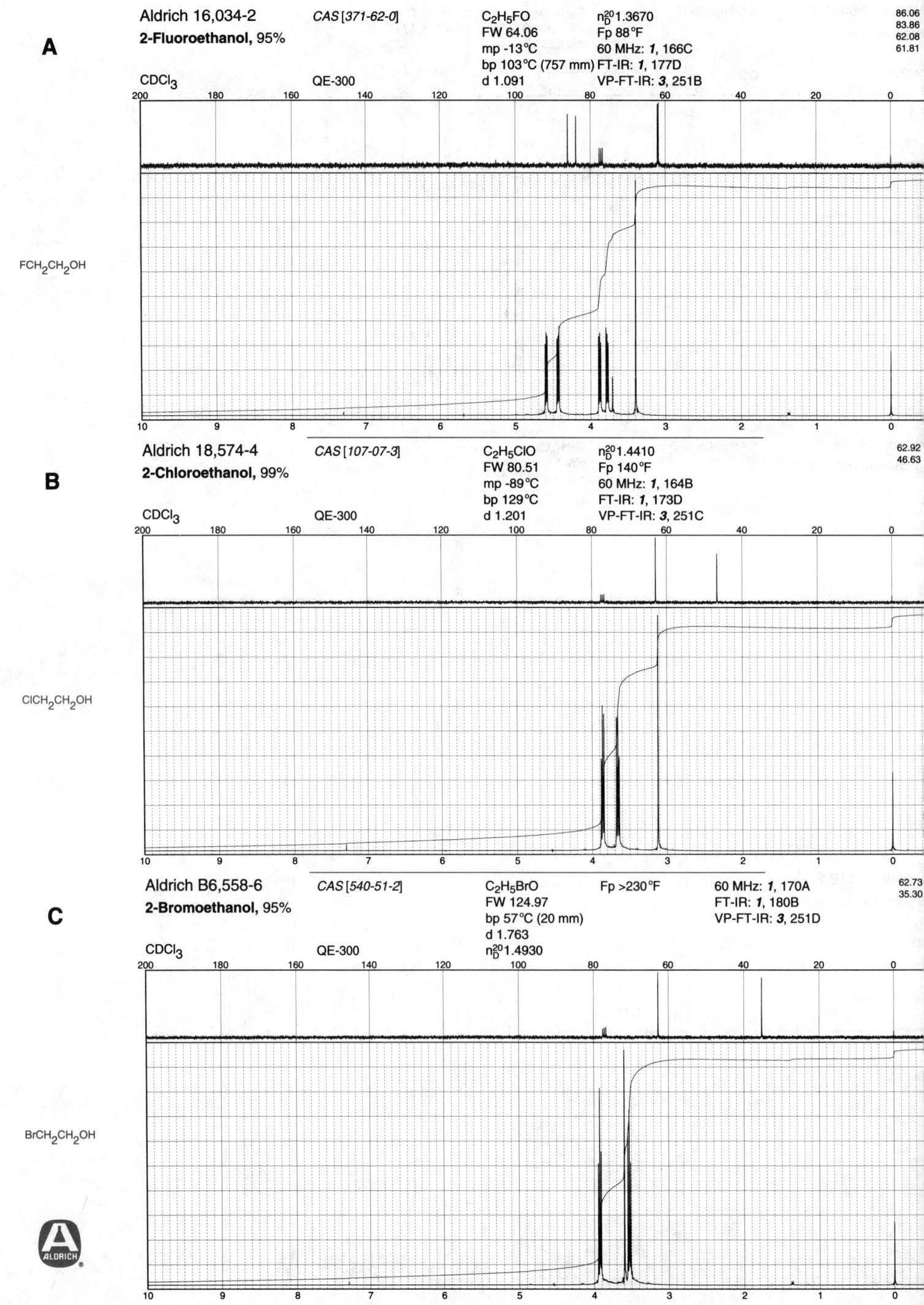

A

Aldrich 16,034-2   CAS [371-62-0] C_2H_5FO n_D^{20} 1.3670   86.06 83.86
2-Fluoroethanol, 95%   FW 64.06   Fp 88°F   62.08 61.81
mp -13°C   60 MHz: **1**, 166C
bp 103°C (757 mm)   FT-IR: **1**, 177D
CDCl$_3$ QE-300   d 1.091   VP-FT-IR: **3**, 251B

FCH_2CH_2OH

B

Aldrich 18,574-4   CAS [107-07-3] C_2H_5ClO n_D^{20} 1.4410   62.92 46.63
2-Chloroethanol, 99%   FW 80.51   Fp 140°F
mp -89°C   60 MHz: **1**, 164B
bp 129°C   FT-IR: **1**, 173D
CDCl$_3$ QE-300   d 1.201   VP-FT-IR: **3**, 251C

$ClCH_2CH_2OH$

C

Aldrich B6,558-6   CAS [540-51-2] C_2H_5BrO Fp >230°F   60 MHz: **1**, 170A   62.73 35.30
2-Bromoethanol, 95%   FW 124.97   FT-IR: **1**, 180B
bp 57°C (20 mm)   VP-FT-IR: **3**, 251D
d 1.763
CDCl$_3$ QE-300 n_D^{20} 1.4930

$BrCH_2CH_2OH$

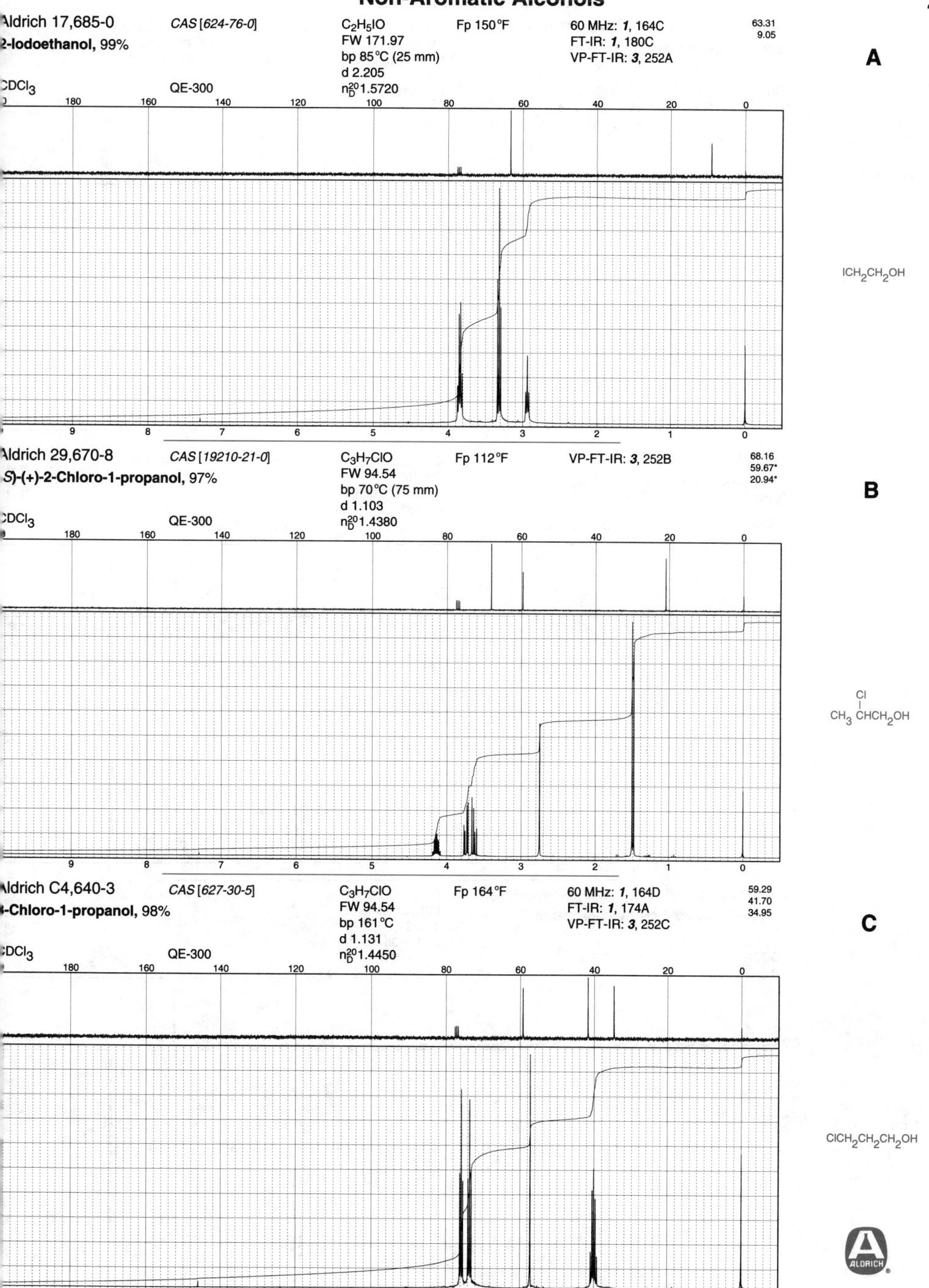

Aldrich 17,685-0   *CAS [624-76-0]* C_2H_5IO Fp 150°F   60 MHz: *1*, 164C   63.31 9.05

2-Iodoethanol, 99%   FW 171.97   FT-IR: *1*, 180C

bp 85°C (25 mm)   VP-FT-IR: *3*, 252A

CDCl₃   QE-300   d 2.205

n_D^{20}1.5720

A

ICH₂CH₂OH

Aldrich 29,670-8   *CAS [19210-21-0]* C_3H_7ClO Fp 112°F   VP-FT-IR: *3*, 252B   68.16 59.67* 20.94*

(S)-(+)-2-Chloro-1-propanol, 97%   FW 94.54

bp 70°C (75 mm)

CDCl₃   QE-300   d 1.103

n_D^{20}1.4380

B

CH₃ CHCH₂OH (with Cl)

Aldrich C4,640-3   *CAS [627-30-5]* C_3H_7ClO Fp 164°F   60 MHz: *1*, 164D   59.29 41.70 34.95

3-Chloro-1-propanol, 98%   FW 94.54   FT-IR: *1*, 174A

bp 161°C   VP-FT-IR: *3*, 252C

CDCl₃   QE-300   d 1.131

n_D^{20}1.4450

C

ClCH₂CH₂CH₂OH

ALDRICH

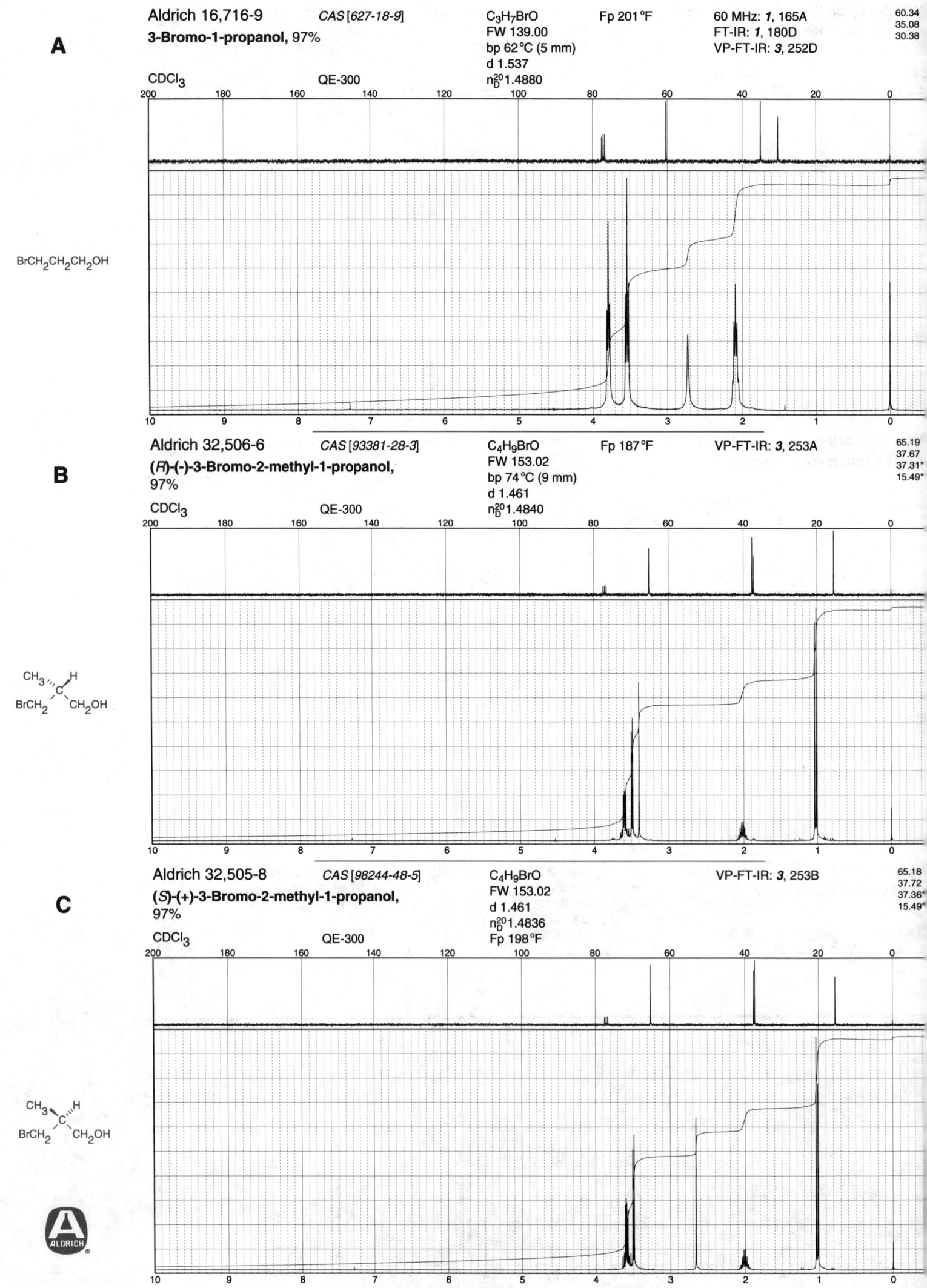

A

Aldrich 16,716-9   CAS [627-18-9]   C₃H₇BrO   Fp 201°F   60 MHz: 1, 165A   60.34
3-Bromo-1-propanol, 97%   FW 139.00   FT-IR: 1, 180D   35.08
bp 62°C (5 mm)   VP-FT-IR: 3, 252D   30.38
d 1.537
n_D^{20} 1.4880

CDCl₃   QE-300

BrCH₂CH₂CH₂OH

B

Aldrich 32,506-6   CAS [93381-28-3]   C₄H₉BrO   Fp 187°F   VP-FT-IR: 3, 253A   65.19
(R)-(-)-3-Bromo-2-methyl-1-propanol,   FW 153.02   37.67
97%   bp 74°C (9 mm)   37.31*
d 1.461   15.49*
CDCl₃   QE-300 n_D^{20} 1.4840

CH₃ ''''' H
    C
BrCH₂   CH₂OH

C

Aldrich 32,505-8   CAS [98244-48-5]   C₄H₉BrO   VP-FT-IR: 3, 253B   65.18
(S)-(+)-3-Bromo-2-methyl-1-propanol,   FW 153.02   37.72
97%   d 1.461   37.36*
n_D^{20} 1.4836   15.49*
CDCl₃   QE-300   Fp 198°F

CH₃ ,,,, H
    C
BrCH₂   CH₂OH

ALDRICH

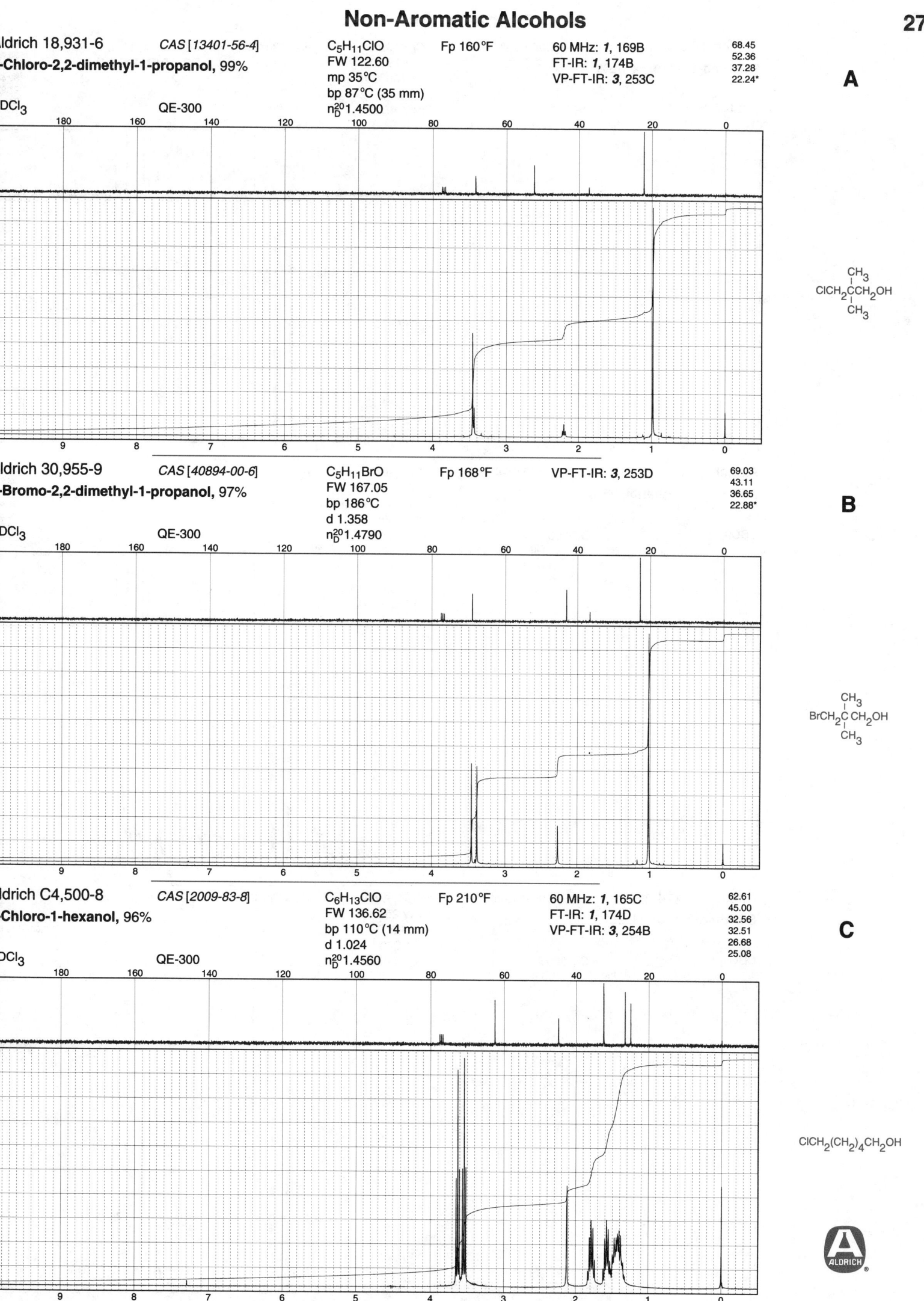

Aldrich 18,931-6   *CAS [13401-56-4]*   C₅H₁₁ClO   Fp 160°F   60 MHz: *1*, 169B   68.45
3-Chloro-2,2-dimethyl-1-propanol, 99%   FW 122.60   FT-IR: *1*, 174B   52.36
mp 35°C   VP-FT-IR: *3*, 253C   37.28
bp 87°C (35 mm)   22.24*
CDCl₃   QE-300   n²⁰_D 1.4500

A

$ClCH_2CCH_2OH$ with CH_3 / CH_3

Aldrich 30,955-9   *CAS [40894-00-6]*   C₅H₁₁BrO   Fp 168°F   VP-FT-IR: *3*, 253D   69.03
3-Bromo-2,2-dimethyl-1-propanol, 97%   FW 167.05   43.11
bp 186°C   36.65
d 1.358   22.88*
CDCl₃   QE-300   n²⁰_D 1.4790

B

$BrCH_2CCH_2OH$ with CH_3 / CH_3

Aldrich C4,500-8   *CAS [2009-83-8]*   C₆H₁₃ClO   Fp 210°F   60 MHz: *1*, 165C   62.61
6-Chloro-1-hexanol, 96%   FW 136.62   FT-IR: *1*, 174D   45.00
bp 110°C (14 mm)   VP-FT-IR: *3*, 254B   32.56
d 1.024   32.51
n²⁰_D 1.4560   26.68
CDCl₃   QE-300   25.08

C

$ClCH_2(CH_2)_4CH_2OH$

ALDRICH

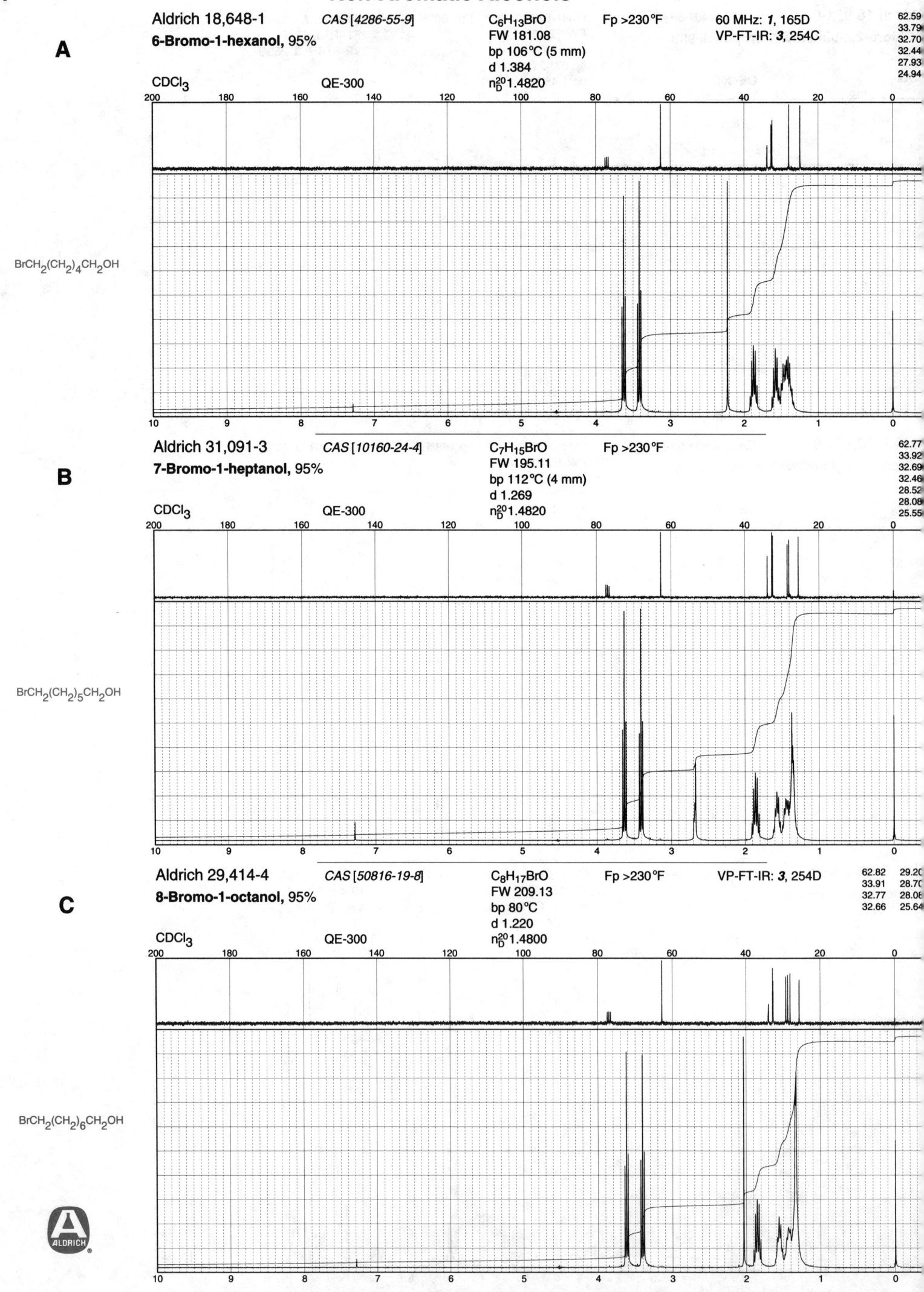

A

Aldrich 18,648-1 *CAS [4286-55-9]* C$_6$H$_{13}$BrO Fp >230°F 60 MHz: *1*, 165D
6-Bromo-1-hexanol, 95% FW 181.08 VP-FT-IR: *3*, 254C
bp 106°C (5 mm)
CDCl$_3$ QE-300 d 1.384
n$_D^{20}$1.4820

62.59
33.79
32.70
32.44
27.93
24.94

BrCH$_2$(CH$_2$)$_4$CH$_2$OH

B

Aldrich 31,091-3 *CAS [10160-24-4]* C$_7$H$_{15}$BrO Fp >230°F
7-Bromo-1-heptanol, 95% FW 195.11
bp 112°C (4 mm)
CDCl$_3$ QE-300 d 1.269
n$_D^{20}$1.4820

62.77
33.92
32.69
32.46
28.52
28.08
25.55

BrCH$_2$(CH$_2$)$_5$CH$_2$OH

C

Aldrich 29,414-4 *CAS [50816-19-8]* C$_8$H$_{17}$BrO Fp >230°F VP-FT-IR: *3*, 254D
8-Bromo-1-octanol, 95% FW 209.13
bp 80°C
CDCl$_3$ QE-300 d 1.220
n$_D^{20}$1.4800

62.82 29.20
33.91 28.70
32.77 28.08
32.66 25.64

BrCH$_2$(CH$_2$)$_6$CH$_2$OH

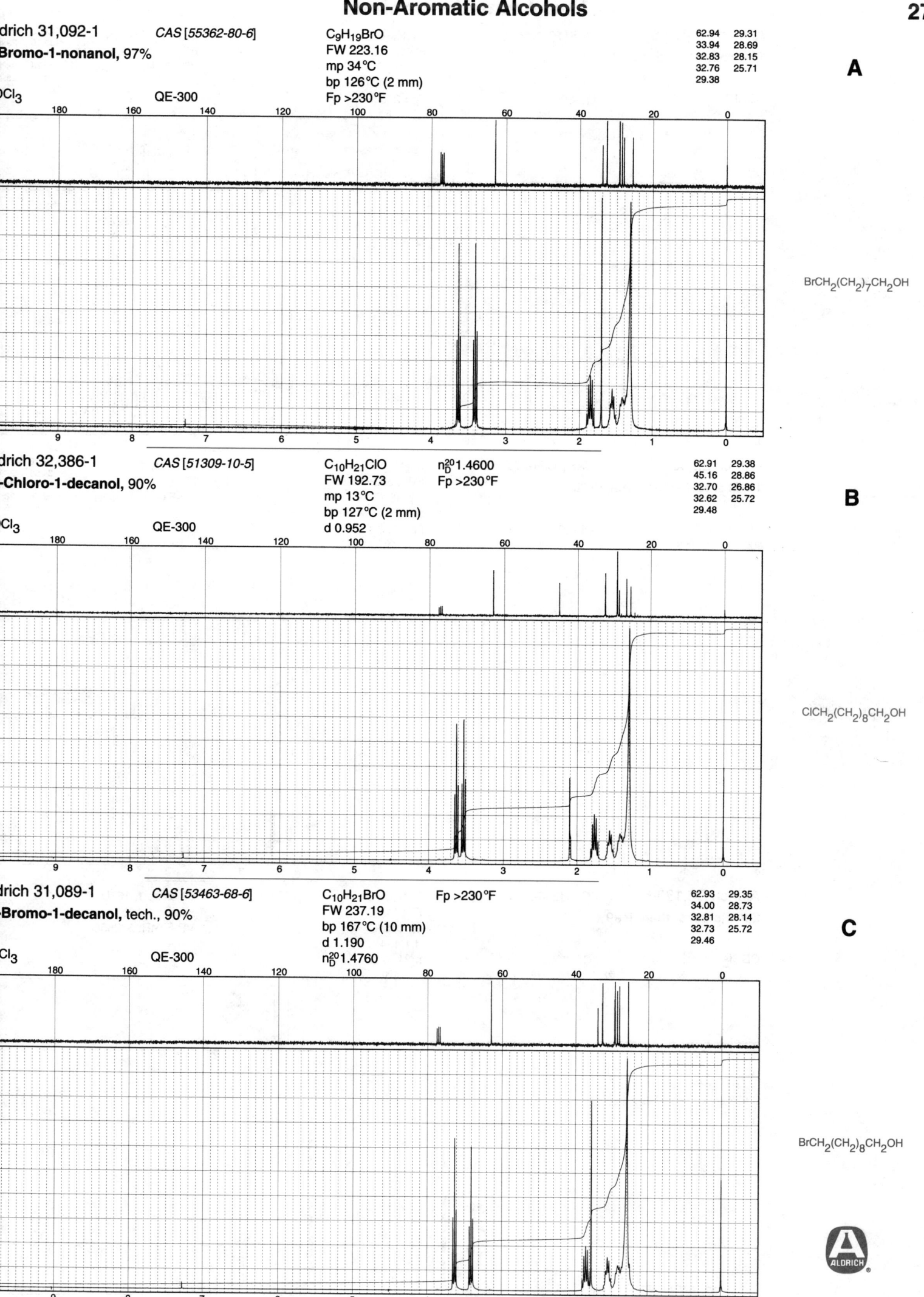

Aldrich 31,092-1 CAS [55362-80-6]

-Bromo-1-nonanol, 97%

CDCl₃ QE-300

C₉H₁₉BrO
FW 223.16
mp 34°C
bp 126°C (2 mm)
Fp >230°F

62.94	29.31
33.94	28.69
32.83	28.15
32.76	25.71
29.38	

A

$BrCH_2(CH_2)_7CH_2OH$

Aldrich 32,386-1 CAS [51309-10-5]

-Chloro-1-decanol, 90%

CDCl₃ QE-300

C₁₀H₂₁ClO n_D^{20} 1.4600
FW 192.73 Fp >230°F
mp 13°C
bp 127°C (2 mm)
d 0.952

62.91	29.38
45.16	28.86
32.70	26.86
32.62	25.72
29.48	

B

$ClCH_2(CH_2)_8CH_2OH$

Aldrich 31,089-1 CAS [53463-68-6]

-Bromo-1-decanol, tech., 90%

CDCl₃ QE-300

C₁₀H₂₁BrO Fp >230°F
FW 237.19
bp 167°C (10 mm)
d 1.190
n_D^{20} 1.4760

62.93	29.35
34.00	28.73
32.81	28.14
32.73	25.72
29.46	

C

$BrCH_2(CH_2)_8CH_2OH$

ALDRICH

Non-Aromatic Alcohols

A

Aldrich 18,413-6   *CAS [1611-56-9]*

11-Bromo-1-undecanol, 98%

C₁₁H₂₃BrO
FW 251.21
mp 48°C
bp 168°C (1 mm)
Fp >230°F

60 MHz: *1*, 166A
FT-IR: *1*, 181B
VP-FT-IR: *3*, 255A

62.96   29.4
33.95   29.4
32.83   28.7
32.78   28.1
29.53   25.7

CDCl₃   QE-300

BrCH₂(CH₂)₉CH₂OH

B

Aldrich 22,467-7   *CAS [3344-77-2]*

12-Bromo-1-dodecanol, 99%

C₁₂H₂₅BrO
FW 265.24
mp 35°C
bp 155°C (4 mm)
Fp >230°F

60 MHz: *1*, 166B
FT-IR: *1*, 181C
VP-FT-IR: *3*, 255B

62.96   29.5
34.05   29.4
32.82   28.7
32.76   28.1
29.57   25.7

CDCl₃   QE-300

BrCH₂(CH₂)₁₀CH₂OH

C

Aldrich D6,180-6   *CAS [598-38-9]*

2,2-Dichloroethanol, 99%

C₂H₄Cl₂O
FW 114.96
bp 146°C
d 1.404
n²⁰_D 1.4730

Fp 173°F

60 MHz: *1*, 167D
FT-IR: *1*, 175A
VP-FT-IR: *3*, 255C

72.6
68.8

CDCl₃   QE-300

Cl₂CHCH₂OH

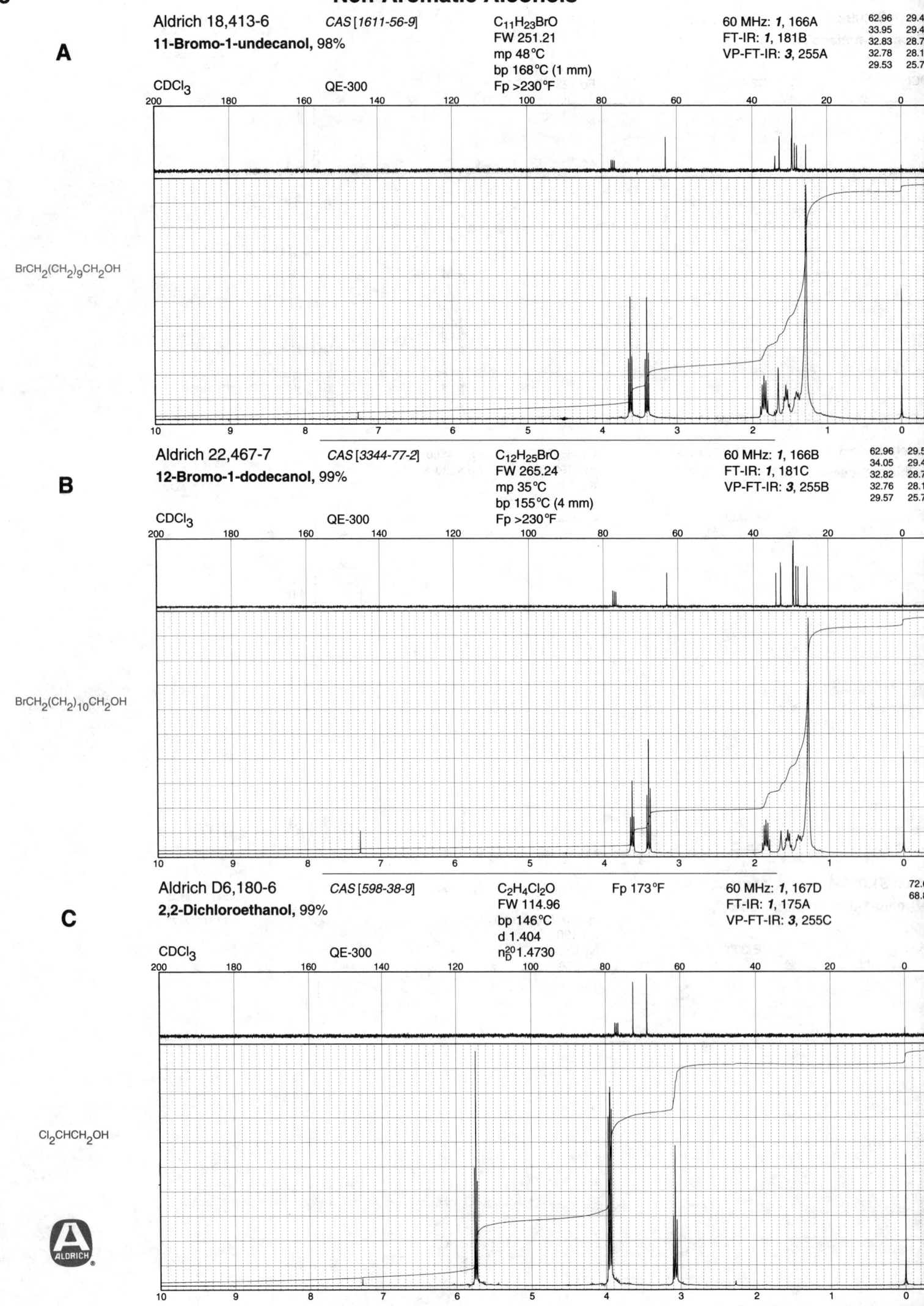

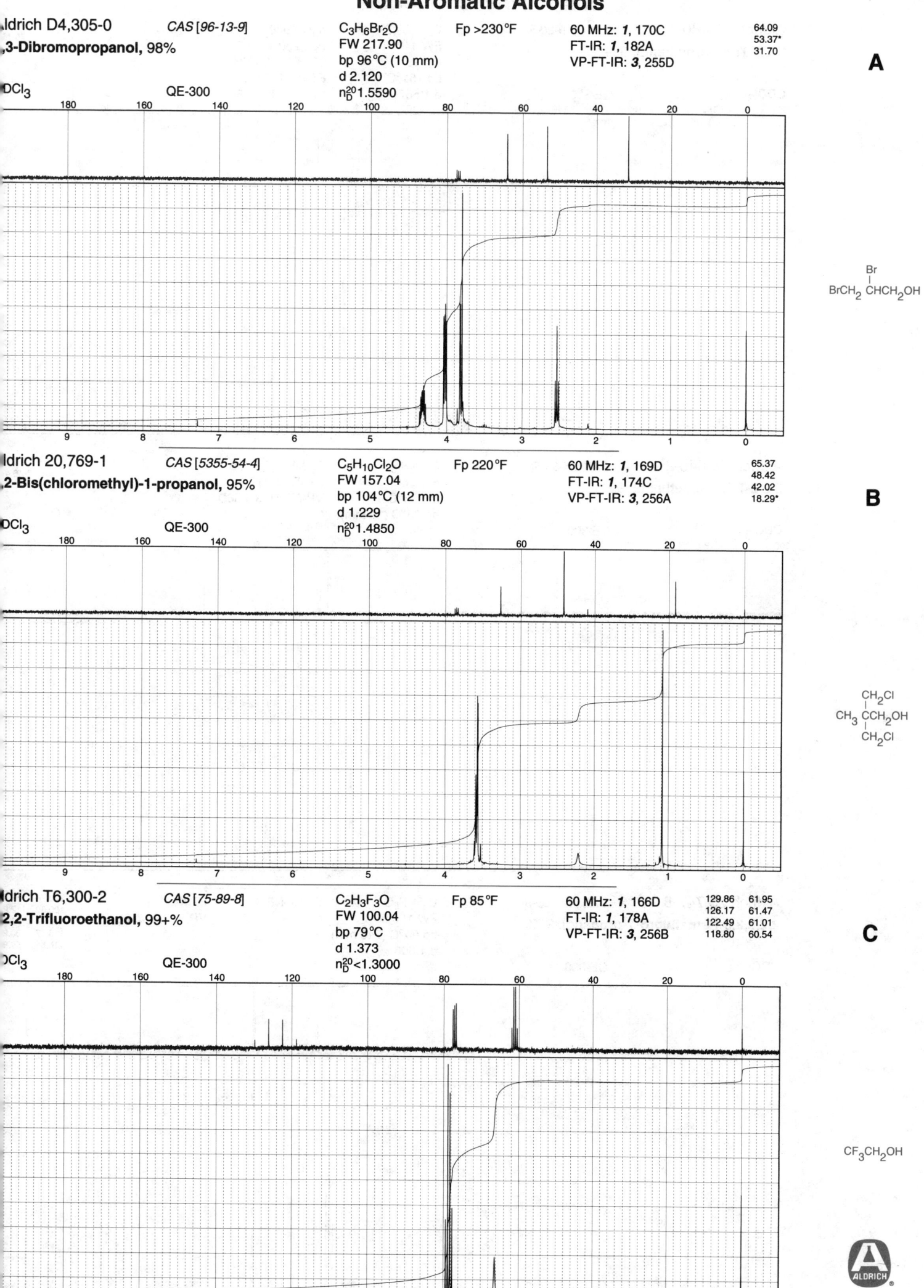

Aldrich D4,305-0   CAS [96-13-9]   C₃H₆Br₂O   Fp >230°F   60 MHz: **1**, 170C   64.09

,3-Dibromopropanol, 98%   FW 217.90   FT-IR: **1**, 182A   53.37*

bp 96°C (10 mm)   VP-FT-IR: **3**, 255D   31.70

d 2.120

DCl₃   QE-300   n²⁰_D 1.5590

A

$BrCH_2\ CHCH_2OH$ (Br above CH)

Aldrich 20,769-1   CAS [5355-54-4]   C₅H₁₀Cl₂O   Fp 220°F   60 MHz: **1**, 169D   65.37

,2-Bis(chloromethyl)-1-propanol, 95%   FW 157.04   FT-IR: **1**, 174C   48.42

bp 104°C (12 mm)   VP-FT-IR: **3**, 256A   42.02

d 1.229   18.29*

DCl₃   QE-300   n²⁰_D 1.4850

B

$CH_3\ CCH_2OH$ with CH_2Cl above and CH_2Cl below

Aldrich T6,300-2   CAS [75-89-8]   C₂H₃F₃O   Fp 85°F   60 MHz: **1**, 166D   129.86   61.95

,2,2-Trifluoroethanol, 99+%   FW 100.04   FT-IR: **1**, 178A   126.17   61.47

bp 79°C   VP-FT-IR: **3**, 256B   122.49   61.01

d 1.373   118.80   60.54

DCl₃   QE-300   n²⁰_D <1.3000

C

CF_3CH_2OH

ALDRICH

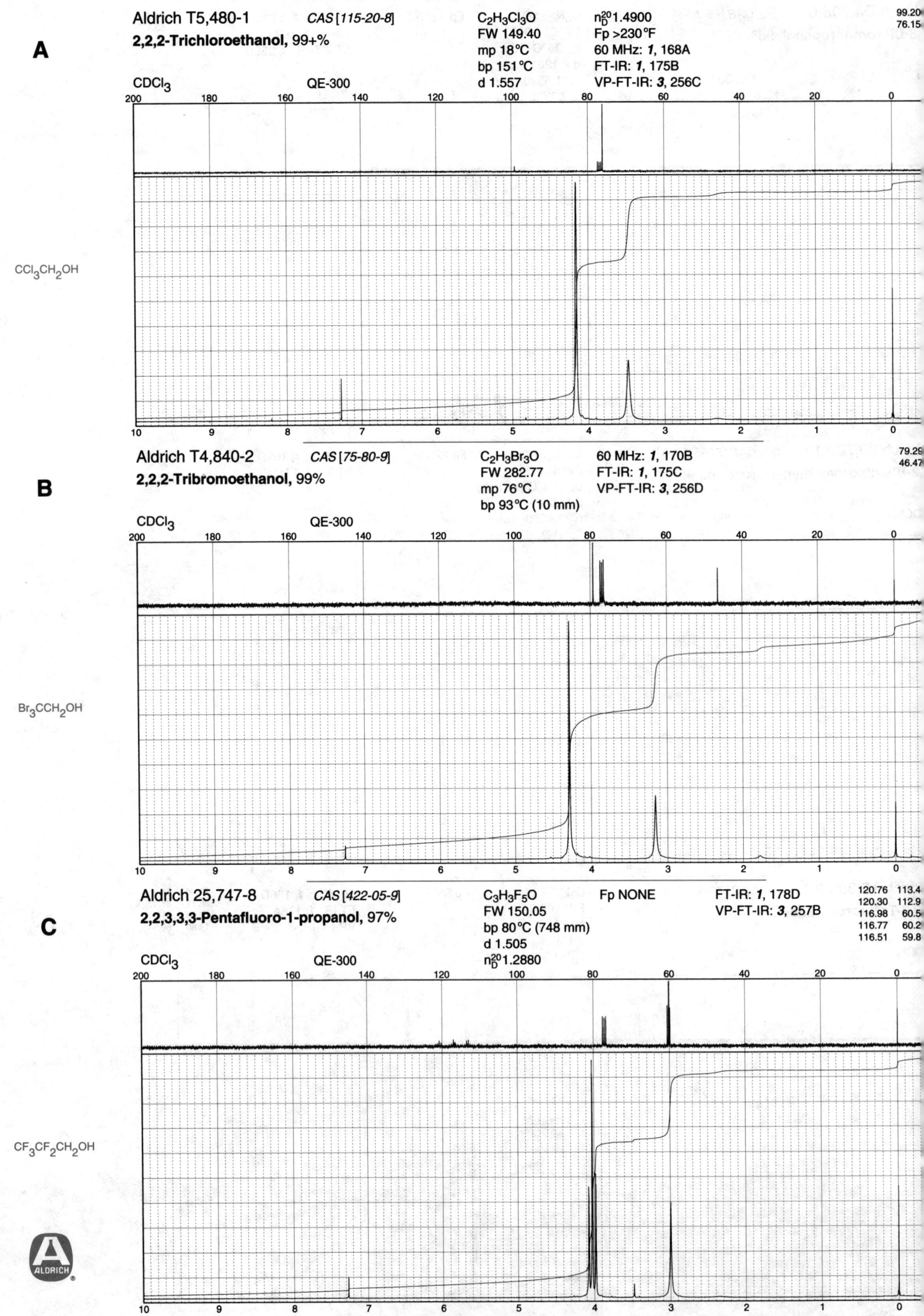

A

Aldrich T5,480-1   *CAS* [115-20-8]   C$_2$H$_3$Cl$_3$O   n$_D^{20}$1.4900   99.20
2,2,2-Trichloroethanol, 99+%   FW 149.40   Fp >230°F   76.15
mp 18°C   60 MHz: *1*, 168A
bp 151°C   FT-IR: *1*, 175B
CDCl$_3$ QE-300   d 1.557   VP-FT-IR: *3*, 256C

CCl$_3$CH$_2$OH

B

Aldrich T4,840-2   *CAS* [75-80-9]   C$_2$H$_3$Br$_3$O   60 MHz: *1*, 170B   79.29
2,2,2-Tribromoethanol, 99%   FW 282.77   FT-IR: *1*, 175C   46.47
mp 76°C   VP-FT-IR: *3*, 256D
bp 93°C (10 mm)
CDCl$_3$ QE-300

Br$_3$CCH$_2$OH

C

Aldrich 25,747-8   *CAS* [422-05-9]   C$_3$H$_3$F$_5$O   Fp NONE   FT-IR: *1*, 178D   120.76 113.4
2,2,3,3,3-Pentafluoro-1-propanol, 97%   FW 150.05   VP-FT-IR: *3*, 257B   120.30 112.9
bp 80°C (748 mm)   116.98 60.5
d 1.505   116.77 60.2
CDCl$_3$ QE-300   n$_D^{20}$1.2880   116.51 59.8

CF$_3$CF$_2$CH$_2$OH

Aldrich 37,195-5   CAS [382-31-0]

,2,3,4,4,4-Hexafluoro-1-butanol, 95%

DCl₃   QE-300

$C_4H_4F_6O$
FW 182.07
bp 114°C (740 mm)
d 1.557
n_D^{20}1.3120

Fp 125°F

126.62	118.82	114.15	84.80*	83.04*	81.29*
126.29	117.49	113.82	84.69*	82.69*	61.71
122.89	117.44	86.08*	84.33*	82.58*	61.36
122.55	117.16	85.72*	84.23*	82.22*	61.28
120.78	117.11	85.62*	83.87*	82.12*	60.93
120.45	115.43	85.26*	83.51*	81.76*	
119.16	115.09	85.15*	83.14*	81.65*	

A

$CF_3CHCF_2CH_2OH$ (with F on the CH)

Aldrich H160-4   CAS [375-01-9]

,2,3,3,4,4,4-Heptafluoro-1-butanol,
8%

DCl₃   QE-300

$C_4H_3F_7O$
FW 200.06
bp 97°C
d 1.600
n_D^{20}1.3000

Fp 77°F

FT-IR: *1*, 179B
VP-FT-IR: *3*, 257C

120.12
119.67
115.86
114.95
60.68
60.33
59.99

B

$CF_3CF_2CF_2CH_2OH$

Aldrich 26,943-3   CAS [355-80-6]

,2,3,3,4,4,5,5-Octafluoro-1-pentanol,
8%

DCl₃   QE-300

$C_5H_4F_8O$
FW 232.07
bp 142°C
d 1.667
n_D^{20}1.3180

Fp 168°F

119.93	115.70	111.23	108.82*	105.47*
119.53	115.25	111.17	108.72	105.06*
119.12	112.78	110.90	108.41*	104.65*
116.56	112.17	110.82	108.27	61.04
116.15	111.76	110.47	108.00*	60.70
115.75	111.35	110.41	107.34	60.36

C

A

Aldrich 37,053-3   *CAS [647-42-7]*

3,3,4,4,5,5,6,6,7,7,8,8,8-Tridecafluoro-1-octanol, 97%

CDCl₃   QE-300

$C_8H_5F_{13}O$
FW 364.11
bp 92°C (28 mm)
d 1.651
n_D^{20}1.3150

123.32	119.07	115.50	114.25	111.90	110.75	108.72	107.48	55.2
122.88	118.95	115.26	114.23	111.77	110.66	108.63	107.24	55.2
122.33	118.54	115.17	113.81	111.66	110.24	108.34	107.09	34.3
121.92	118.12	115.10	112.72	111.63	109.67	108.20	106.66	34.1
121.49	116.14	115.06	112.33	111.47	109.59	108.11	105.57	33.8
119.95	115.70	114.74	112.21	111.21	109.15	107.90	105.06	
119.51	115.58	114.64	112.08	111.06	109.07	107.69	55.33	

CF₃(CF₂)₅CH₂CH₂OH

B

Aldrich 17,552-8   *CAS [127-00-4]*

1-Chloro-2-propanol, 97%

CDCl₃   QE-300

C_3H_7ClO
FW 94.54
bp 127°C
d 1.115
n_D^{20}1.4375

60 MHz: *1*, 168C
FT-IR: *1*, 176A
VP-FT-IR: *3*, 258A

67.6
51.3
20.2

OH
|
CH₃CHCH₂Cl

C

Aldrich 23,843-0   *CAS [19686-73-8]*

1-Bromo-2-propanol, tech., 70%

CDCl₃   QE-300

C_3H_7BrO
FW 139.00
bp 147°C
d 1.530
n_D^{20}1.4810

Fp 130°F

FT-IR: *1*, 176B
VP-FT-IR: *3*, 258B

67.2
41.1
21.0

OH
|
CH₃CHCH₂Br

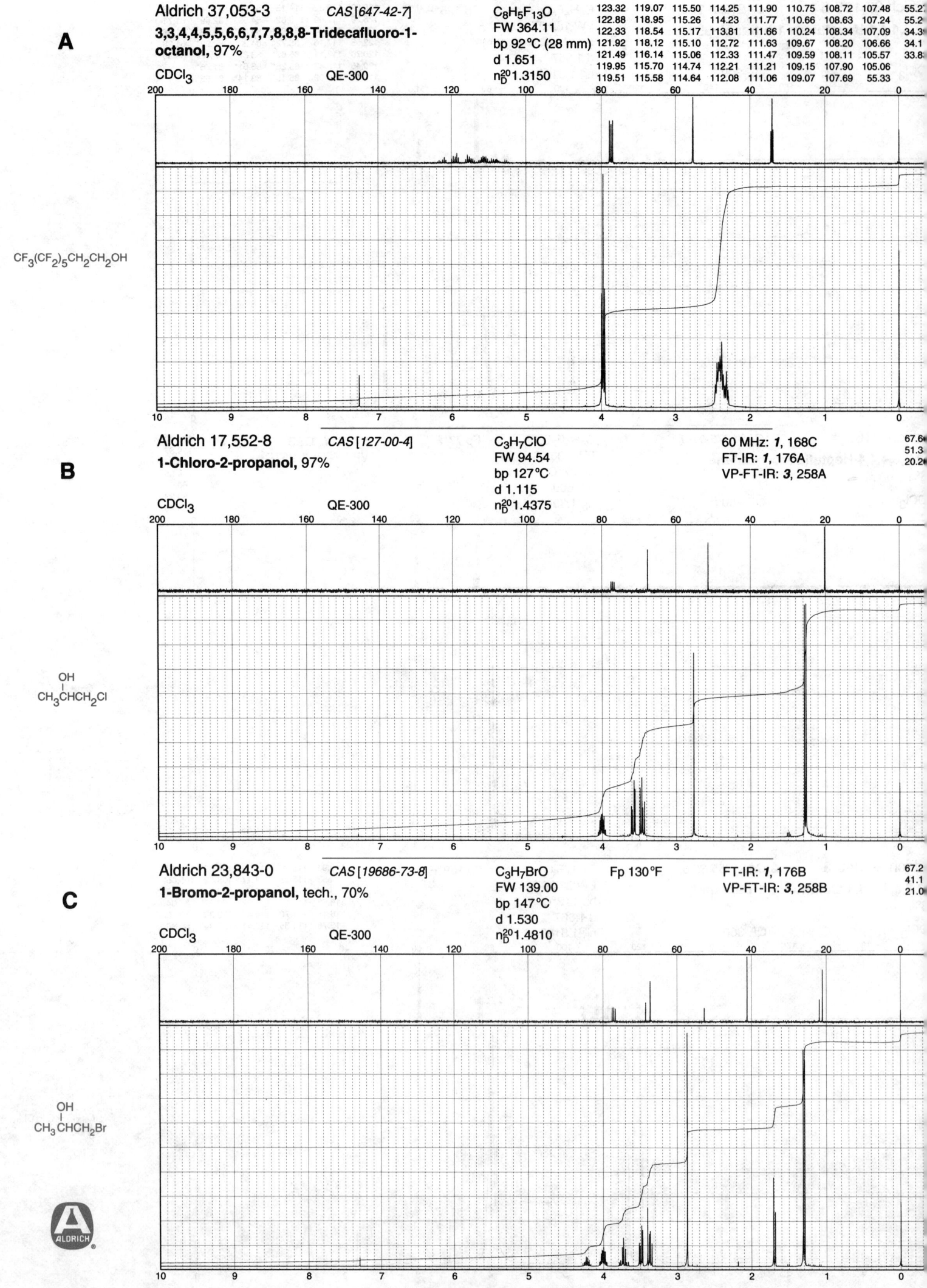

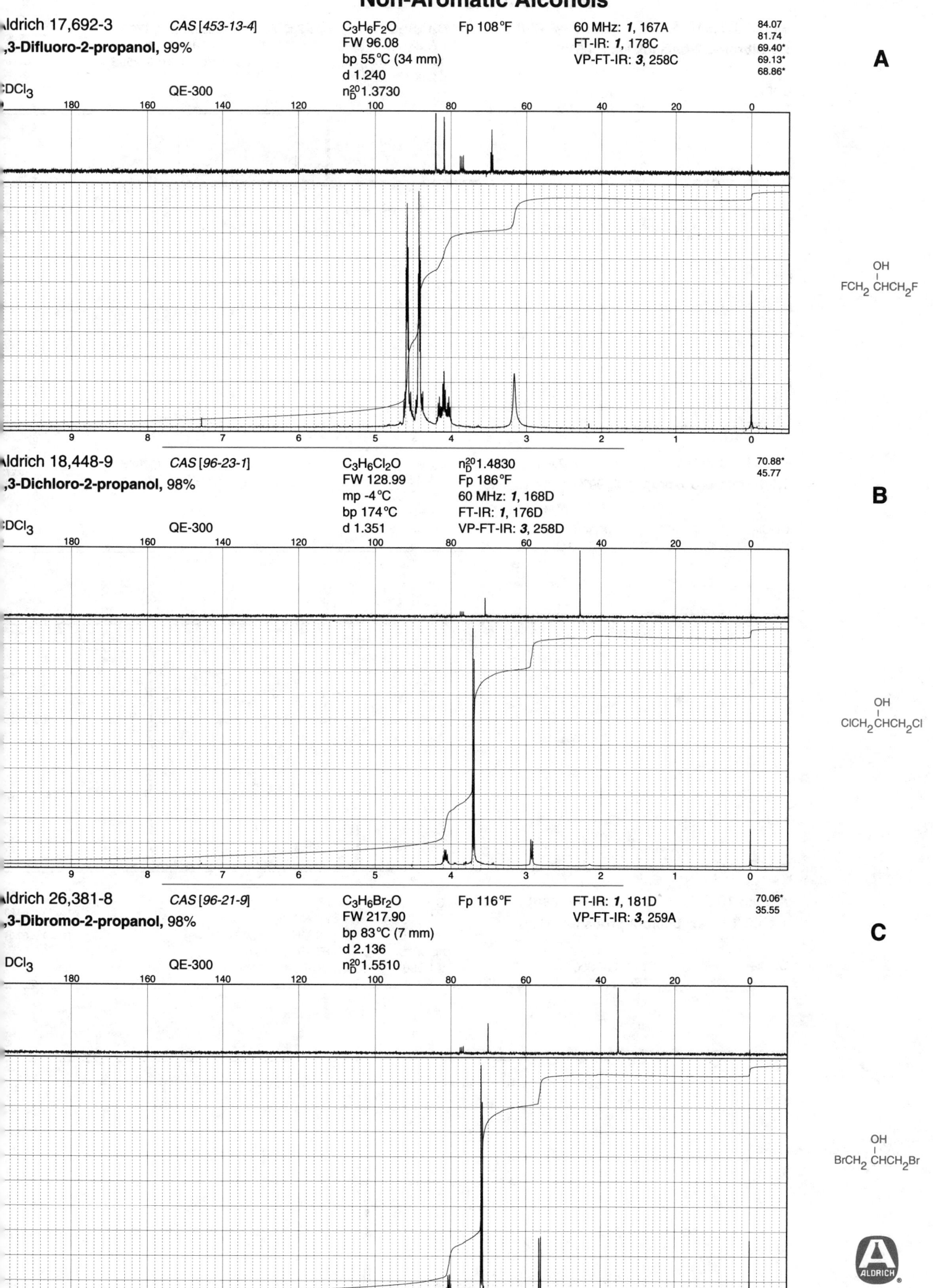

Aldrich 17,692-3   *CAS [453-13-4]*   C₃H₆F₂O   Fp 108°F   60 MHz: *1*, 167A
,3-Difluoro-2-propanol, 99%   FW 96.08   FT-IR: *1*, 178C
    bp 55°C (34 mm)   VP-FT-IR: *3*, 258C
    d 1.240
CDCl₃   QE-300   n²⁰_D 1.3730

$$C_3H_6F_2O$$

84.07
81.74
69.40*
69.13*
68.86*

A

$$FCH_2-\overset{\displaystyle OH}{\underset{\displaystyle |}{C}}HCH_2F$$

Aldrich 18,448-9   *CAS [96-23-1]*   C₃H₆Cl₂O   n²⁰_D 1.4830
,3-Dichloro-2-propanol, 98%   FW 128.99   Fp 186°F
    mp -4°C   60 MHz: *1*, 168D
    bp 174°C   FT-IR: *1*, 176D
CDCl₃   QE-300   d 1.351   VP-FT-IR: *3*, 258D

70.88*
45.77

B

$$ClCH_2-\overset{\displaystyle OH}{\underset{\displaystyle |}{C}}HCH_2Cl$$

Aldrich 26,381-8   *CAS [96-21-9]*   C₃H₆Br₂O   Fp 116°F   FT-IR: *1*, 181D
,3-Dibromo-2-propanol, 98%   FW 217.90   VP-FT-IR: *3*, 259A
    bp 83°C (7 mm)
    d 2.136
CDCl₃   QE-300   n²⁰_D 1.5510

70.06*
35.55

C

$$BrCH_2-\overset{\displaystyle OH}{\underset{\displaystyle |}{C}}HCH_2Br$$

ALDRICH

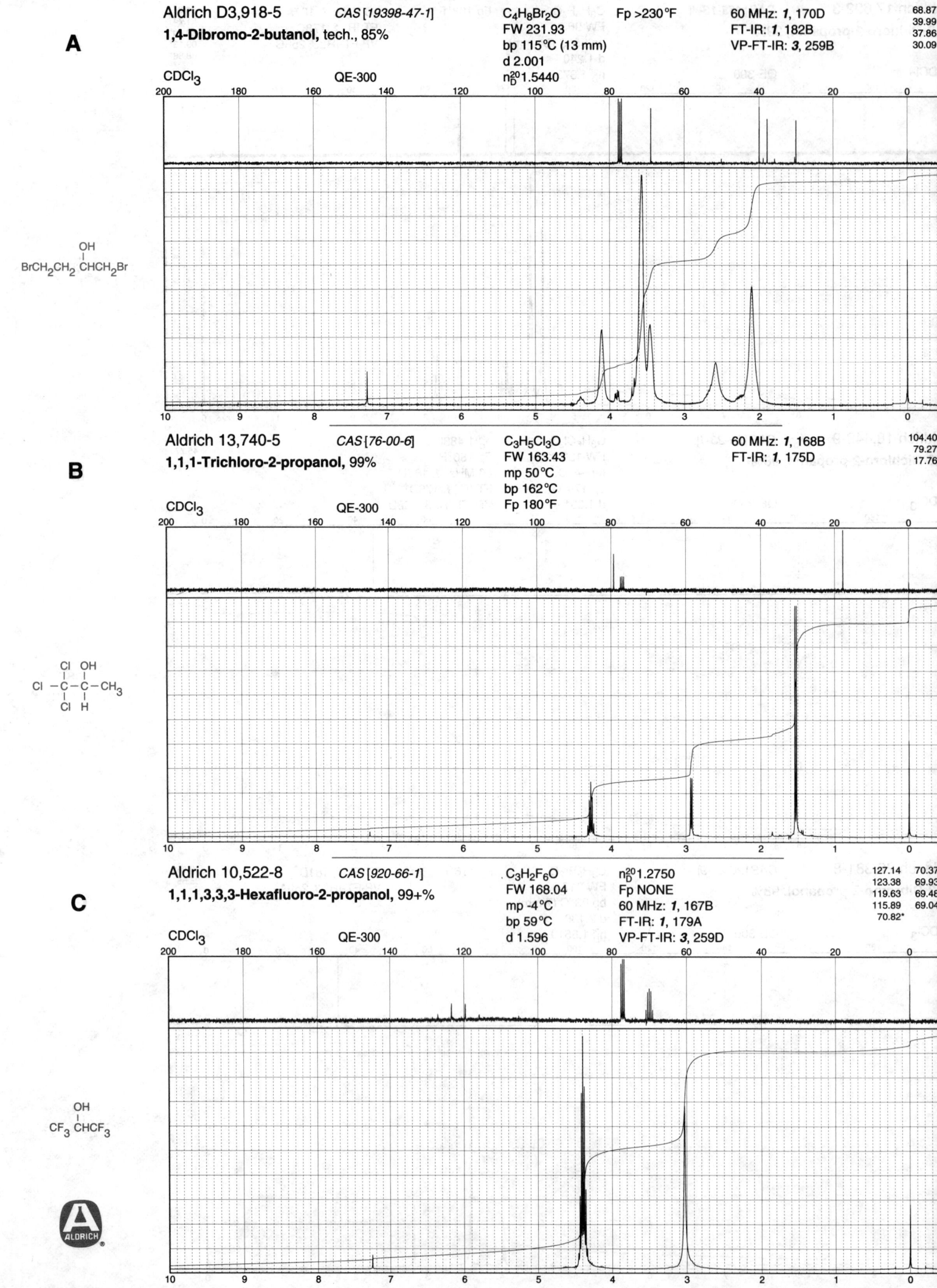

A

Aldrich D3,918-5   *CAS [19398-47-1]* $C_4H_8Br_2O$ Fp >230°F

1,4-Dibromo-2-butanol, tech., 85%   FW 231.93

bp 115°C (13 mm)

d 2.001

n_D^{20} 1.5440

60 MHz: *1*, 170D

FT-IR: *1*, 182B

VP-FT-IR: *3*, 259B

68.87*
39.99
37.86
30.09

CDCl₃   QE-300

BrCH₂CH₂ CHCH₂Br   (OH)

B

Aldrich 13,740-5   *CAS [76-00-6]* $C_3H_5Cl_3O$

1,1,1-Trichloro-2-propanol, 99%   FW 163.43

mp 50°C

bp 162°C

Fp 180°F

60 MHz: *1*, 168B

FT-IR: *1*, 175D

104.40
79.27*
17.76*

CDCl₃   QE-300

Cl–C–C–CH₃   (Cl, OH, Cl, H)

C

Aldrich 10,522-8   *CAS [920-66-1]* $C_3H_2F_6O$ n_D^{20} 1.2750

1,1,1,3,3,3-Hexafluoro-2-propanol, 99+%   FW 168.04   Fp NONE

mp -4°C   60 MHz: *1*, 167B

bp 59°C   FT-IR: *1*, 179A

d 1.596   VP-FT-IR: *3*, 259D

127.14   70.37
123.38   69.93
119.63   69.48
115.89   69.04
70.82*

CDCl₃   QE-300

CF₃ CHCF₃   (OH)

ALDRICH

Non-Aromatic Alcohols

280

A

Aldrich 11,205-4 CAS [6001-64-5]

,1,1-Trichloro-2-methyl-2-propanol
ydrate, 98%

C₄H₇Cl₃O
FW 177.46
mp 78°C

60 MHz: *1*, 169C
FT-IR: *1*, 177A

109.70
81.80
25.40*

DCl₃   QE-300

$CH_3\overset{OH}{\underset{CH_3}{C}}CCl_3$ · xH₂O

B

Aldrich M8,225-6 CAS [29553-26-2]

-Methyl-3,3,4,4-tetrafluoro-2-butanol

C₅H₈F₄O
FW 160.11
bp 117°C
d 1.283
n²⁰_D 1.3524

Fp 165°F

60 MHz: *1*, 167C
FT-IR: *1*, 178B
VP-FT-IR: *3*, 260B

116.68	112.40*	105.76*
116.68	109.98*	72.66
116.37*	109.53*	72.34
116.05	109.09*	23.43*
113.30*	106.67*	
112.85*	106.22*	

DCl₃   QE-300

$F_2CHCF_2\overset{OH}{\underset{CH_3}{C}}CH_3$

C

Aldrich 33,102-3 CAS [2378-02-1]

erfluoro-*tert*-butyl alcohol, 99%

C₄HF₉O
FW 236.04
bp 45°C
d 1.693
n²⁰_D 1.3000

Fp NONE

VP-FT-IR: *3*, 260C

126.15
122.32
118.49
114.65

DCl₃   QE-300

$F_3C-\overset{CF_3}{\underset{CF_3}{C}}-OH$

ALDRICH

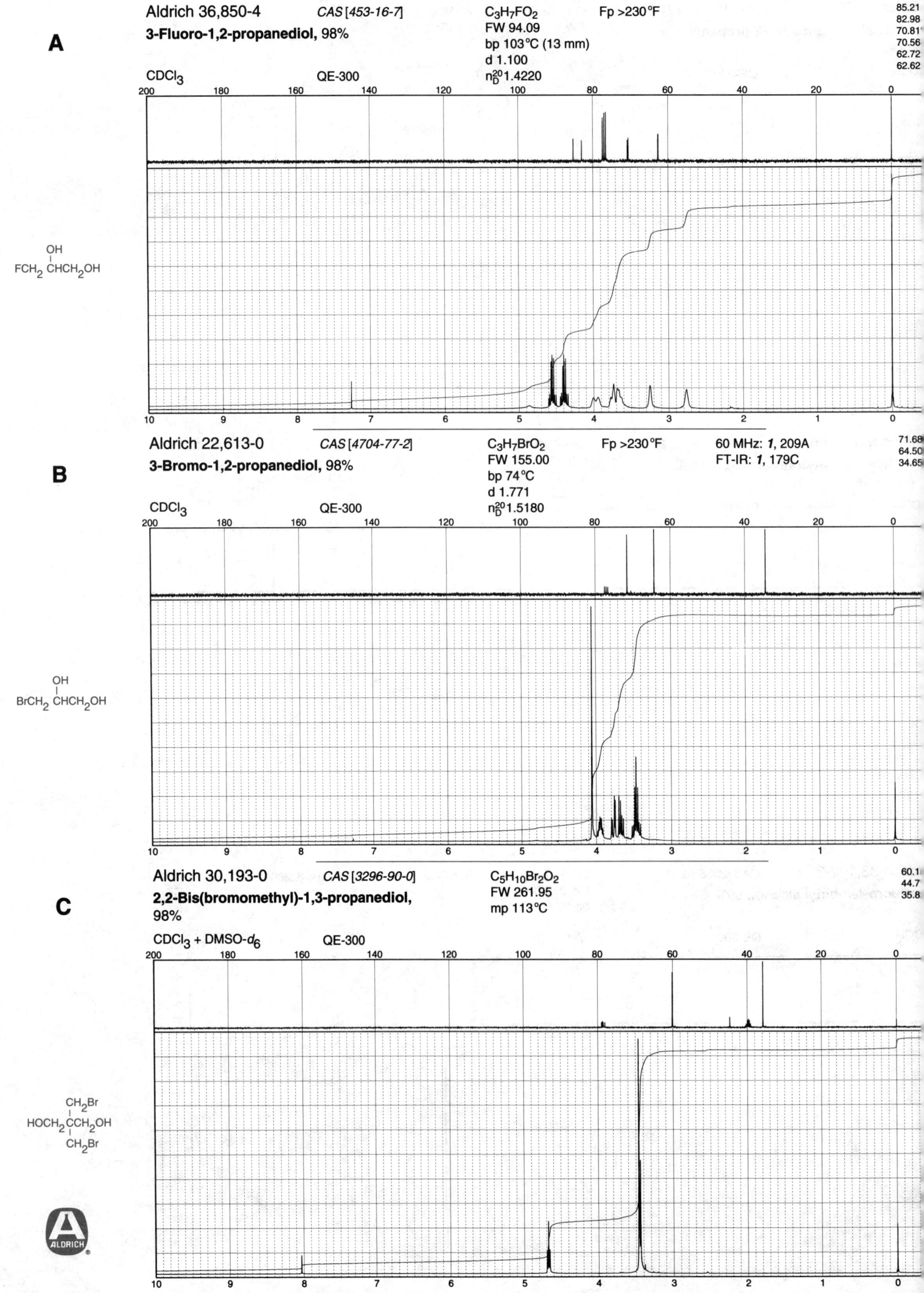

A

Aldrich 36,850-4   *CAS [453-16-7]*   C₃H₇FO₂   Fp >230°F

3-Fluoro-1,2-propanediol, 98%

$C_3H_7FO_2$
FW 94.09
bp 103°C (13 mm)
d 1.100
$n_D^{20}1.4220$

CDCl₃   QE-300

85.21
82.98
70.81
70.56
62.72
62.62

$FCH_2\overset{\overset{\textstyle OH}{|}}{C}HCH_2OH$

B

Aldrich 22,613-0   *CAS [4704-77-2]*   C₃H₇BrO₂   Fp >230°F   60 MHz: *1*, 209A

3-Bromo-1,2-propanediol, 98%

$C_3H_7BrO_2$
FW 155.00
bp 74°C
d 1.771
$n_D^{20}1.5180$

FT-IR: *1*, 179C

CDCl₃   QE-300

71.68
64.50
34.65

$BrCH_2\overset{\overset{\textstyle OH}{|}}{C}HCH_2OH$

C

Aldrich 30,193-0   *CAS [3296-90-0]*   C₅H₁₀Br₂O₂

2,2-Bis(bromomethyl)-1,3-propanediol, 98%

$C_5H_{10}Br_2O_2$
FW 261.95
mp 113°C

CDCl₃ + DMSO-*d₆*   QE-300

60.1
44.7
35.8

$HOCH_2\overset{\overset{\textstyle CH_2Br}{|}}{\underset{\underset{\textstyle CH_2Br}{|}}{C}}CH_2OH$

ALDRICH

Non-Aromatic Alcohols

Aldrich 30,104-3   *CAS [1947-58-6]*   C$_4$H$_8$Br$_2$O$_2$ 64.30
55.49*

(±)-2,3-Dibromo-1,4-butanediol, 99%   FW 247.93
mp 89°C
bp 149°C (1 mm)

A

CDCl$_3$+DMSO-d_6 QE-300

HOCH$_2$CH(Br)CH(Br)CH$_2$OH

Aldrich 23,757-4   *CAS [299-70-7]*   C$_4$H$_8$Br$_2$O$_2$ FT-IR: *1*, 182C   70.88*
34.54

(±)-1,4-Dibromo-2,3-butanediol, 95%   FW 247.93
mp 83°C

B

CDCl$_3$+DMSO-d_6 QE-300

BrCH$_2$CH(OH)CH(OH)CH$_2$Br

Aldrich 14,370-7   *CAS [3234-02-4]*   C$_4$H$_6$Br$_2$O$_2$ 60 MHz: *1*, 171A   122.69
65.72

trans-2,3-Dibromo-2-butene-1,4-diol,   FW 245.91   FT-IR: *1*, 182D
7%   mp 113°C

C

CDCl$_3$+DMSO-d_6 QE-300

Br(HOCH$_2$)C=C(Br)CH$_2$OH

ALDRICH

Non-Aromatic Alcohols

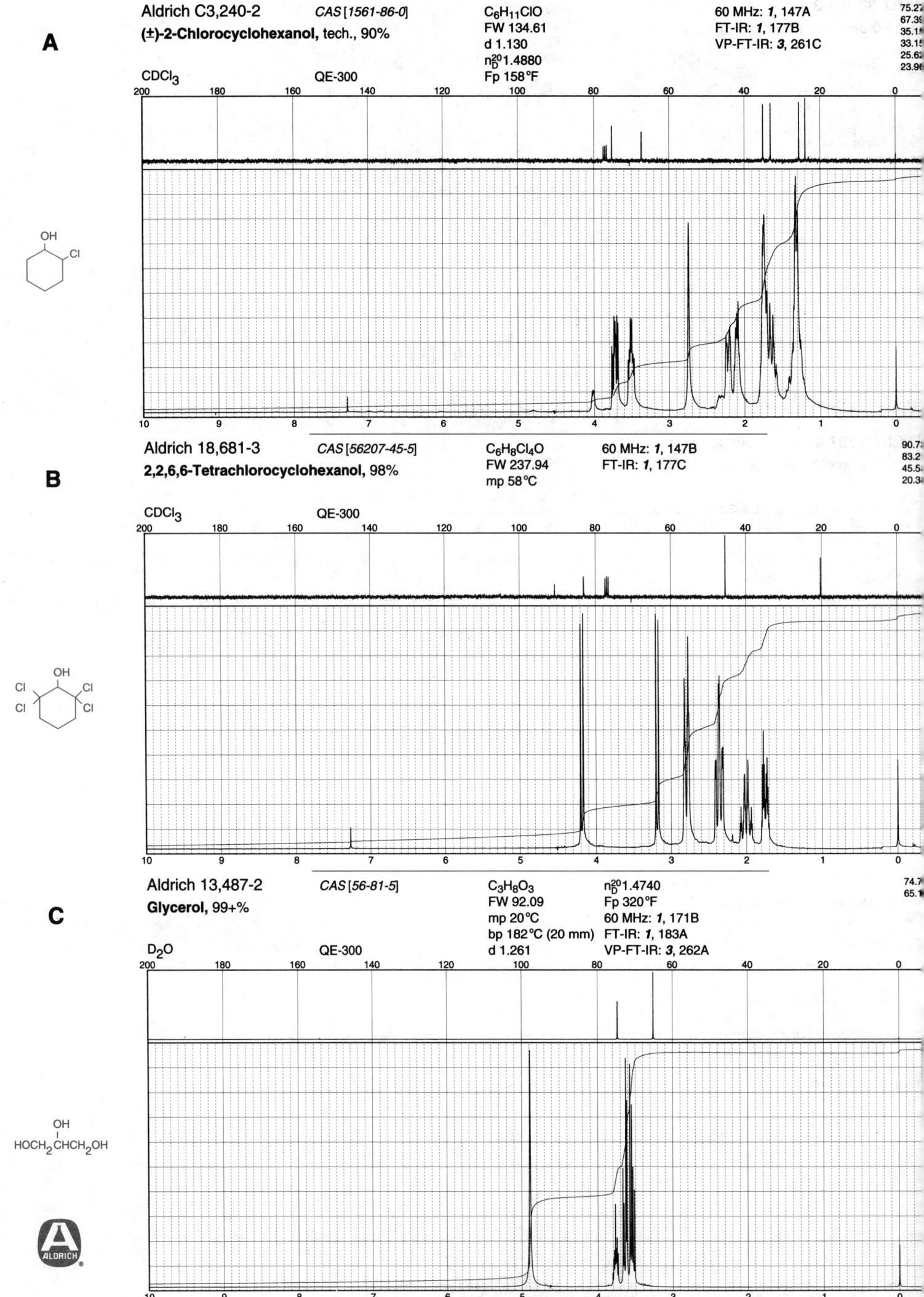

A

Aldrich C3,240-2   *CAS [1561-86-0]*   C6H11ClO   60 MHz: *1*, 147A
(±)-2-Chlorocyclohexanol, tech., 90%   FW 134.61   FT-IR: *1*, 177B
    d 1.130   VP-FT-IR: *3*, 261C
    n_D^{20}1.4880
    Fp 158°F

CDCl3   QE-300

75.27
67.39
35.1?
33.15
25.6?
23.9?

B

Aldrich 18,681-3   *CAS [56207-45-5]*   C6H8Cl4O   60 MHz: *1*, 147B
2,2,6,6-Tetrachlorocyclohexanol, 98%   FW 237.94   FT-IR: *1*, 177C
    mp 58°C

CDCl3   QE-300

90.7?
83.2?
45.5?
20.3?

C

Aldrich 13,487-2   *CAS [56-81-5]*   C3H8O3 n_D^{20}1.4740
Glycerol, 99+%   FW 92.09   Fp 320°F
    mp 20°C   60 MHz: *1*, 171B
    bp 182°C (20 mm)   FT-IR: *1*, 183A
    d 1.261   VP-FT-IR: *3*, 262A

D2O   QE-300

74.7?
65.1?

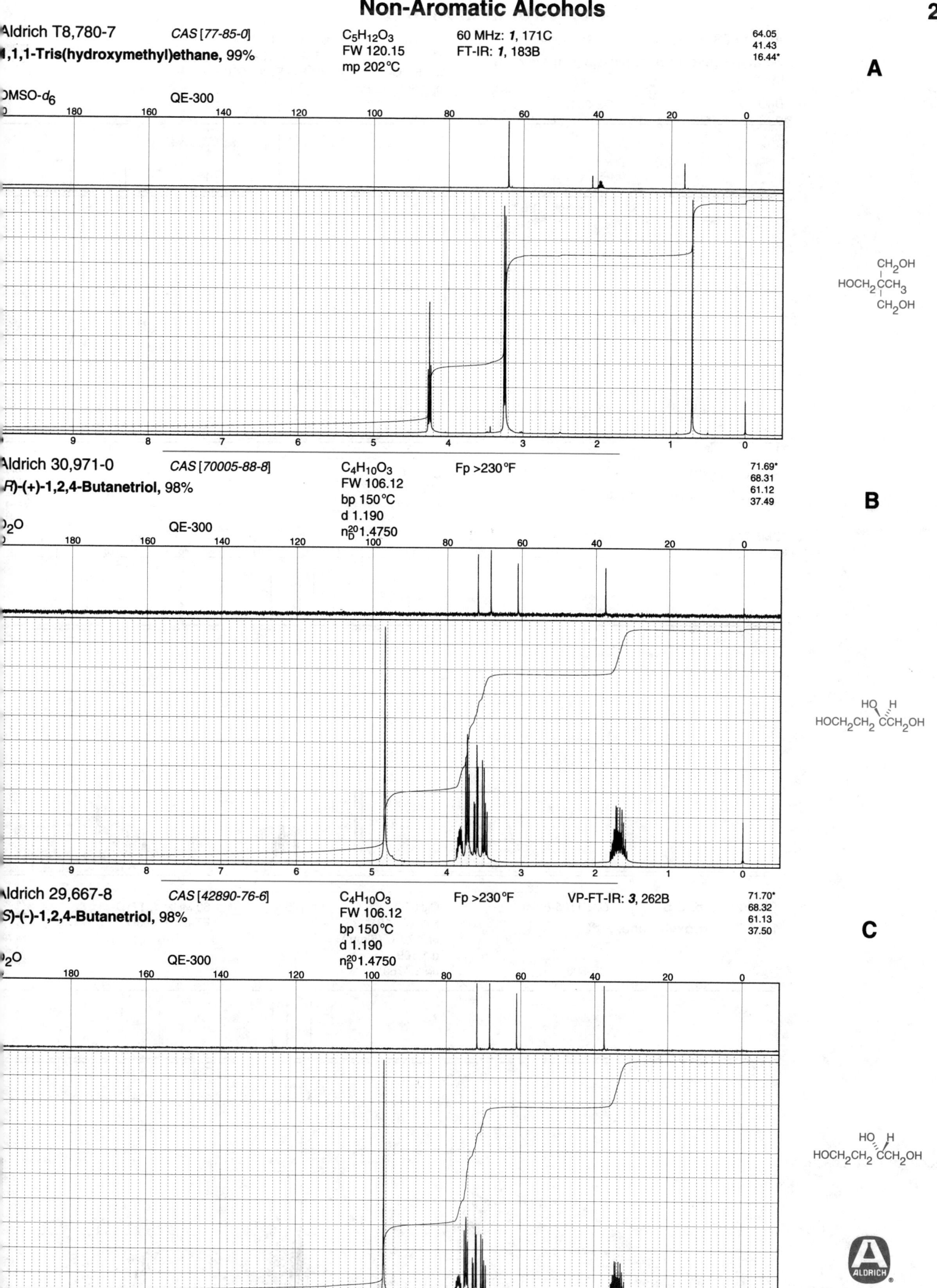

Aldrich T8,780-7 *CAS [77-85-0]* C₅H₁₂O₃ 60 MHz: *1*, 171C
1,1,1-Tris(hydroxymethyl)ethane, 99% FW 120.15 FT-IR: *1*, 183B
mp 202°C

64.05
41.43
16.44*

A

DMSO-d₆ QE-300

CH₂OH
HOCH₂CCH₃
CH₂OH

Aldrich 30,971-0 *CAS [70005-88-8]* C₄H₁₀O₃ Fp >230°F
(R)-(+)-1,2,4-Butanetriol, 98% FW 106.12
bp 150°C
d 1.190
n²⁰D 1.4750

71.69*
68.31
61.12
37.49

B

D₂O QE-300

HO H
HOCH₂CH₂ CCH₂OH

Aldrich 29,667-8 *CAS [42890-76-6]* C₄H₁₀O₃ Fp >230°F VP-FT-IR: *3*, 262B
(S)-(-)-1,2,4-Butanetriol, 98% FW 106.12
bp 150°C
d 1.190
n²⁰D 1.4750

71.70*
68.32
61.13
37.50

C

D₂O QE-300

HO H
HOCH₂CH₂ CCH₂OH

ALDRICH

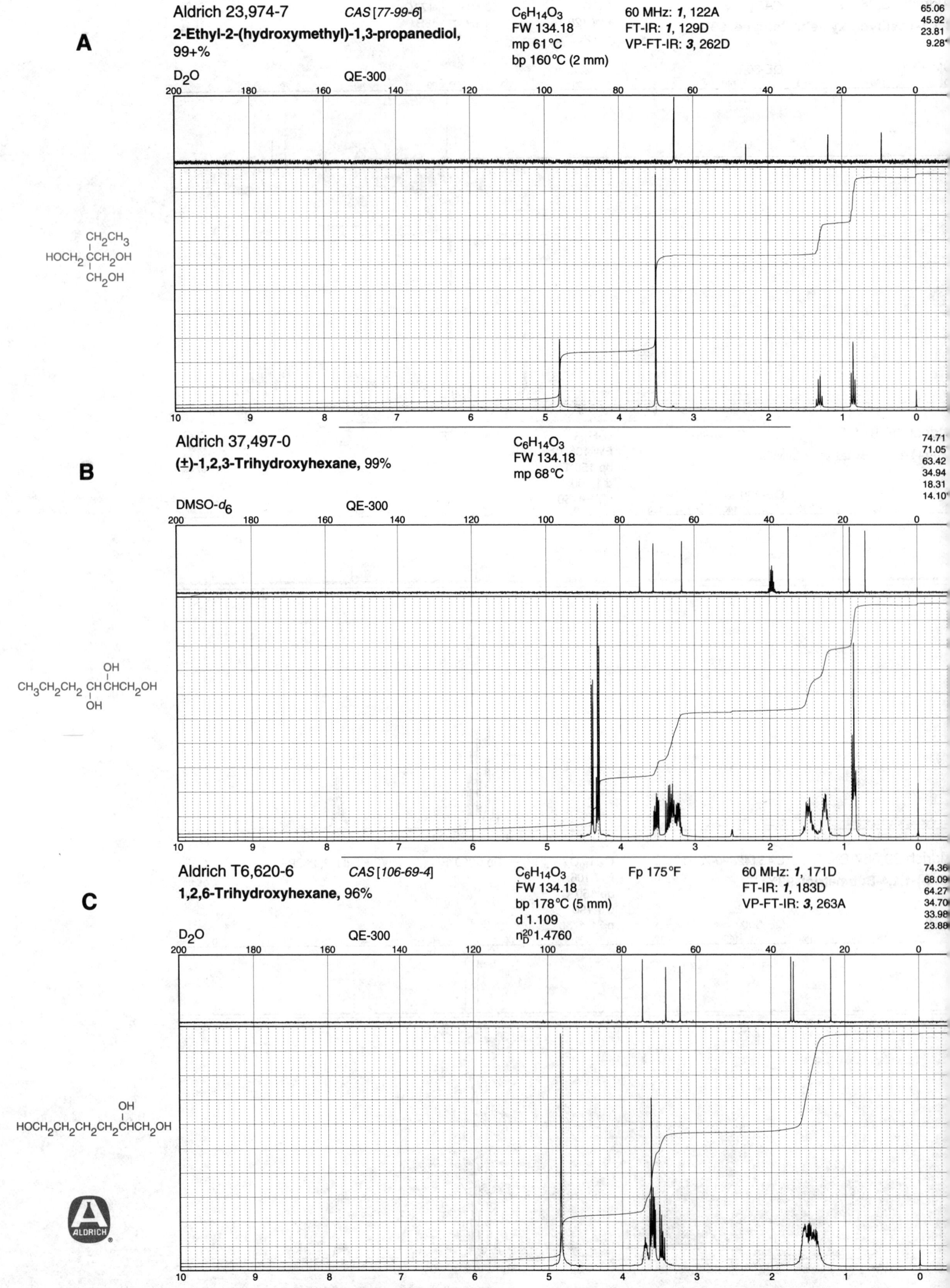

A

Aldrich 23,974-7   *CAS [77-99-6]*

2-Ethyl-2-(hydroxymethyl)-1,3-propanediol, 99+%

$C_6H_{14}O_3$
FW 134.18
mp 61°C
bp 160°C (2 mm)

60 MHz: *1*, 122A
FT-IR: *1*, 129D
VP-FT-IR: *3*, 262D

65.06
45.92
23.81
9.28

D$_2$O   QE-300

CH_2CH_3
$HOCH_2$ CCH_2OH
CH_2OH

B

Aldrich 37,497-0

(±)-1,2,3-Trihydroxyhexane, 99%

$C_6H_{14}O_3$
FW 134.18
mp 68°C

74.71
71.05
63.42
34.94
18.31
14.10

DMSO-d_6 QE-300

$CH_3CH_2CH_2$ $CHCHCH_2OH$
OH
OH

C

Aldrich T6,620-6   *CAS [106-69-4]*

1,2,6-Trihydroxyhexane, 96%

$C_6H_{14}O_3$
FW 134.18
bp 178°C (5 mm)
d 1.109
n$_D^{20}$1.4760

Fp 175°F

60 MHz: *1*, 171D
FT-IR: *1*, 183D
VP-FT-IR: *3*, 263A

74.36
68.09
64.27
34.70
33.98
23.88

D$_2$O   QE-300

OH
$HOCH_2CH_2CH_2CH_2CHCH_2OH$

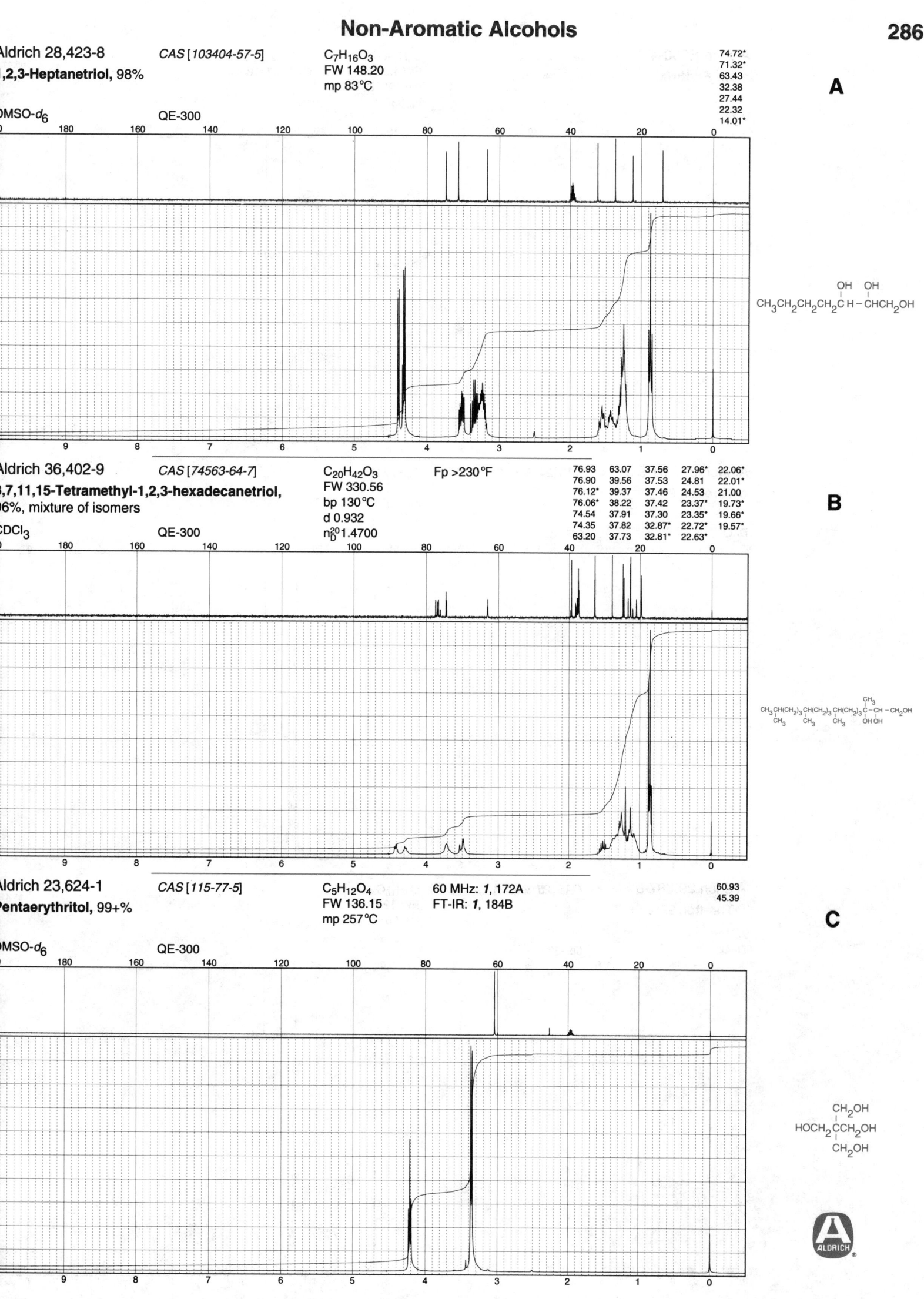

Aldrich 28,423-8

1,2,3-Heptanetriol, 98%

CAS [103404-57-5]

$C_7H_{16}O_3$
FW 148.20
mp 83°C

DMSO-d_6

QE-300

74.72*
71.32*
63.43
32.38
27.44
22.32
14.01*

A

$CH_3CH_2CH_2CH_2CH$ OH $-$ OH $CHCH_2OH$

Aldrich 36,402-9

3,7,11,15-Tetramethyl-1,2,3-hexadecanetriol, 96%, mixture of isomers

CAS [74563-64-7]

$C_{20}H_{42}O_3$
FW 330.56
bp 130°C
d 0.932
n$_D^{20}$ 1.4700

Fp >230°F

CDCl$_3$

QE-300

76.93	63.07	37.56	27.96*	22.06*
76.90	39.56	37.53	24.81	22.01*
76.12*	39.37	37.46	24.53	21.00
76.06*	38.22	37.42	23.37*	19.73*
74.54	37.91	37.30	23.35*	19.66*
74.35	37.82	32.87*	22.72*	19.57*
63.20	37.73	32.81*	22.63*	

B

$CH_3CH(CH_2)_3 CH(CH_2)_3 CH(CH_2)_3 C-CH-CH_2OH$
$CH_3 \quad CH_3 \quad CH_3 \quad OH OH$ (with CH_3 above)

Aldrich 23,624-1

Pentaerythritol, 99+%

CAS [115-77-5]

$C_5H_{12}O_4$
FW 136.15
mp 257°C

60 MHz: *1*, 172A
FT-IR: *1*, 184B

60.93
45.39

DMSO-d_6

QE-300

C

CH_2OH
$HOCH_2CCH_2OH$
CH_2OH

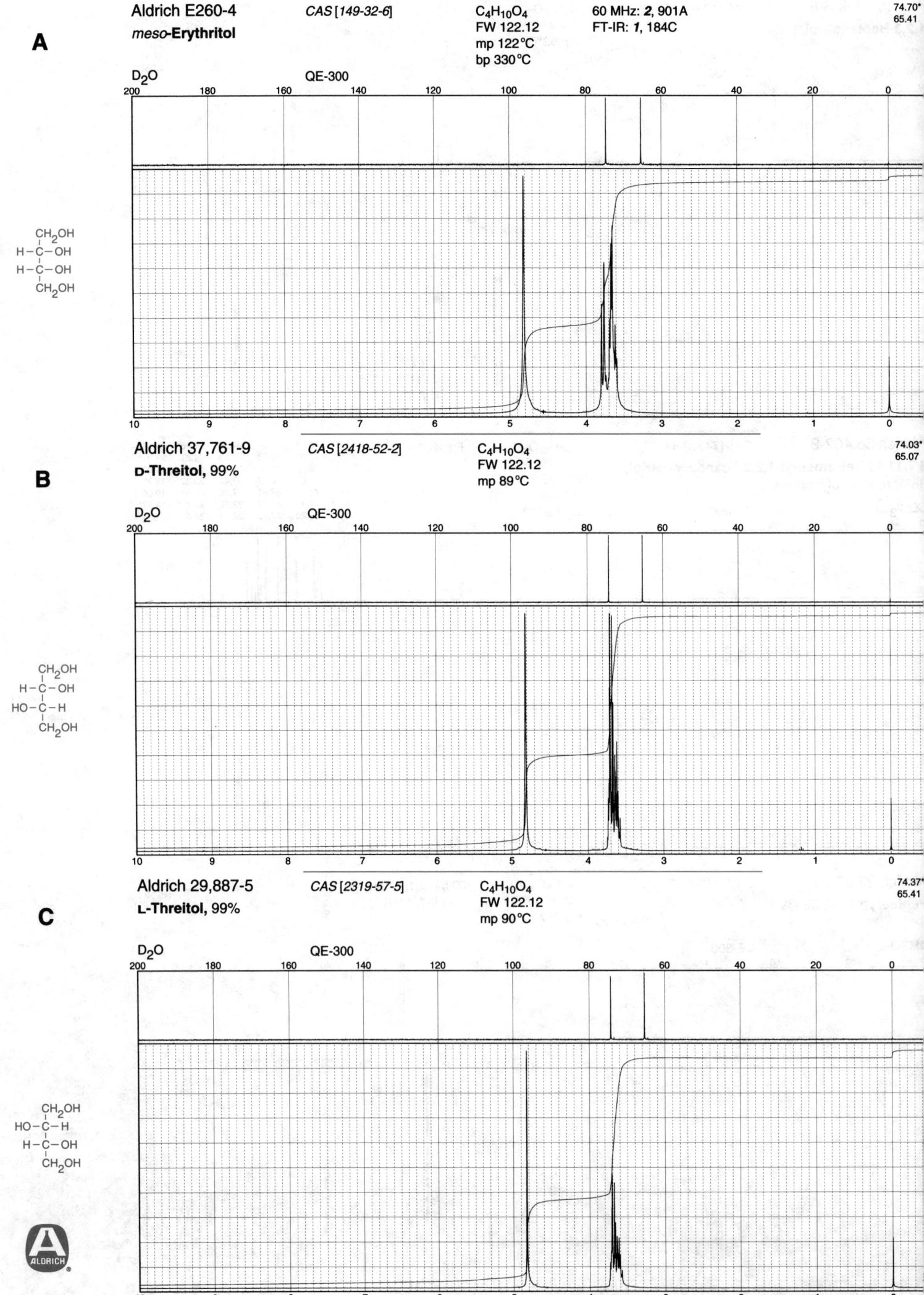

A

Aldrich E260-4
meso-**Erythritol**

CAS [149-32-6]

$C_4H_{10}O_4$
FW 122.12
mp 122°C
bp 330°C

60 MHz: *2*, 901A
FT-IR: *1*, 184C

74.70*
65.41

D_2O QE-300

CH₂OH
H–C–OH
H–C–OH
CH₂OH

B

Aldrich 37,761-9
D-Threitol, 99%

CAS [2418-52-2]

$C_4H_{10}O_4$
FW 122.12
mp 89°C

74.03*
65.07

D_2O QE-300

CH₂OH
H–C–OH
HO–C–H
CH₂OH

C

Aldrich 29,887-5
L-Threitol, 99%

CAS [2319-57-5]

$C_4H_{10}O_4$
FW 122.12
mp 90°C

74.37*
65.41

D_2O QE-300

CH₂OH
HO–C–H
H–C–OH
CH₂OH

ALDRICH

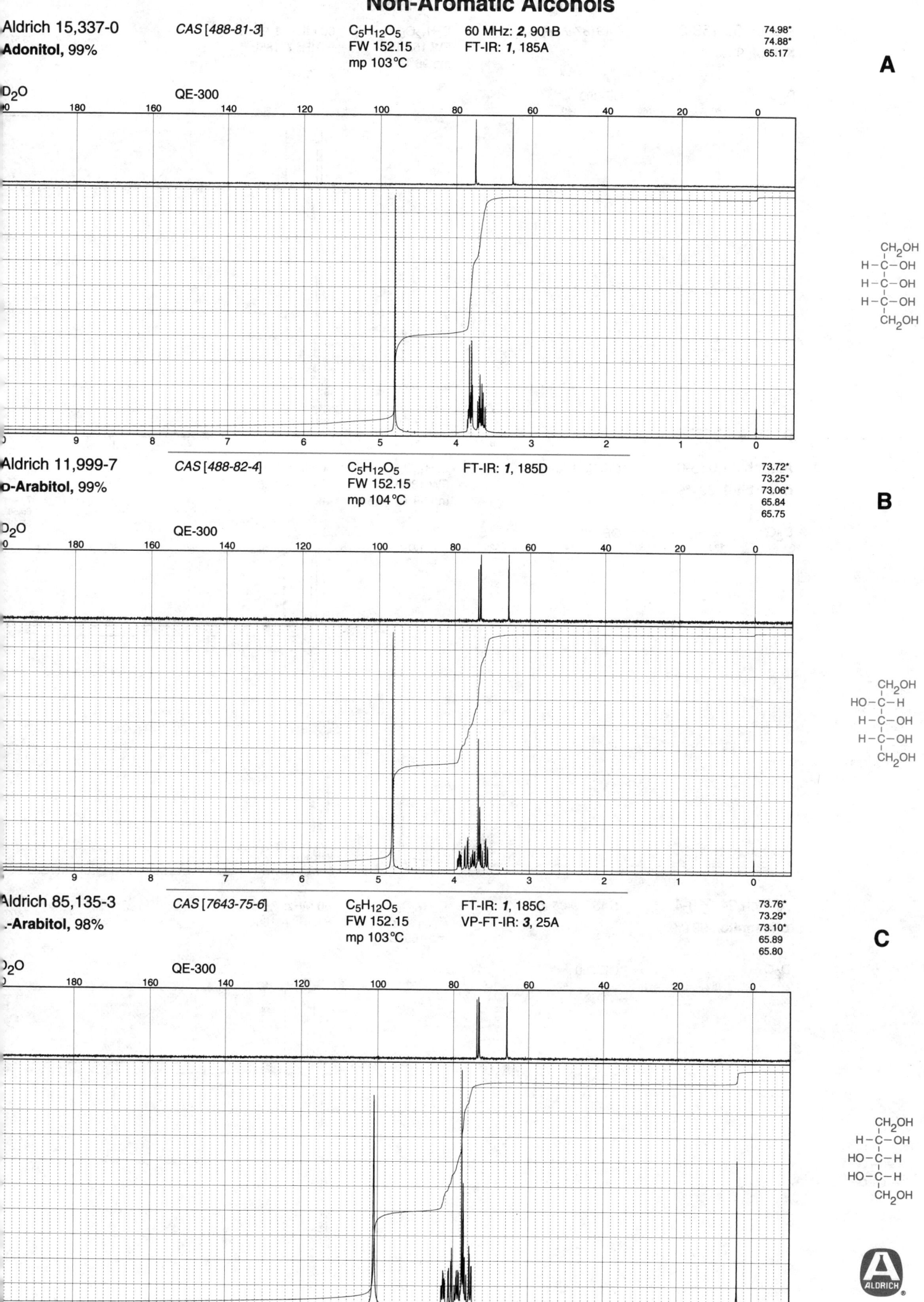

Aldrich 15,337-0
Adonitol, 99%
CAS [488-81-3]
$C_5H_{12}O_5$
FW 152.15
mp 103°C
60 MHz: 2, 901B
FT-IR: 1, 185A
74.98*
74.88*
65.17

A

D_2O
QE-300

Aldrich 11,999-7
D-Arabitol, 99%
CAS [488-82-4]
$C_5H_{12}O_5$
FW 152.15
mp 104°C
FT-IR: 1, 185D
73.72*
73.25*
73.06*
65.84
65.75

B

D_2O
QE-300

Aldrich 85,135-3
L-Arabitol, 98%
CAS [7643-75-6]
$C_5H_{12}O_5$
FW 152.15
mp 103°C
FT-IR: 1, 185C
VP-FT-IR: 3, 25A
73.76*
73.29*
73.10*
65.89
65.80

C

D_2O
QE-300

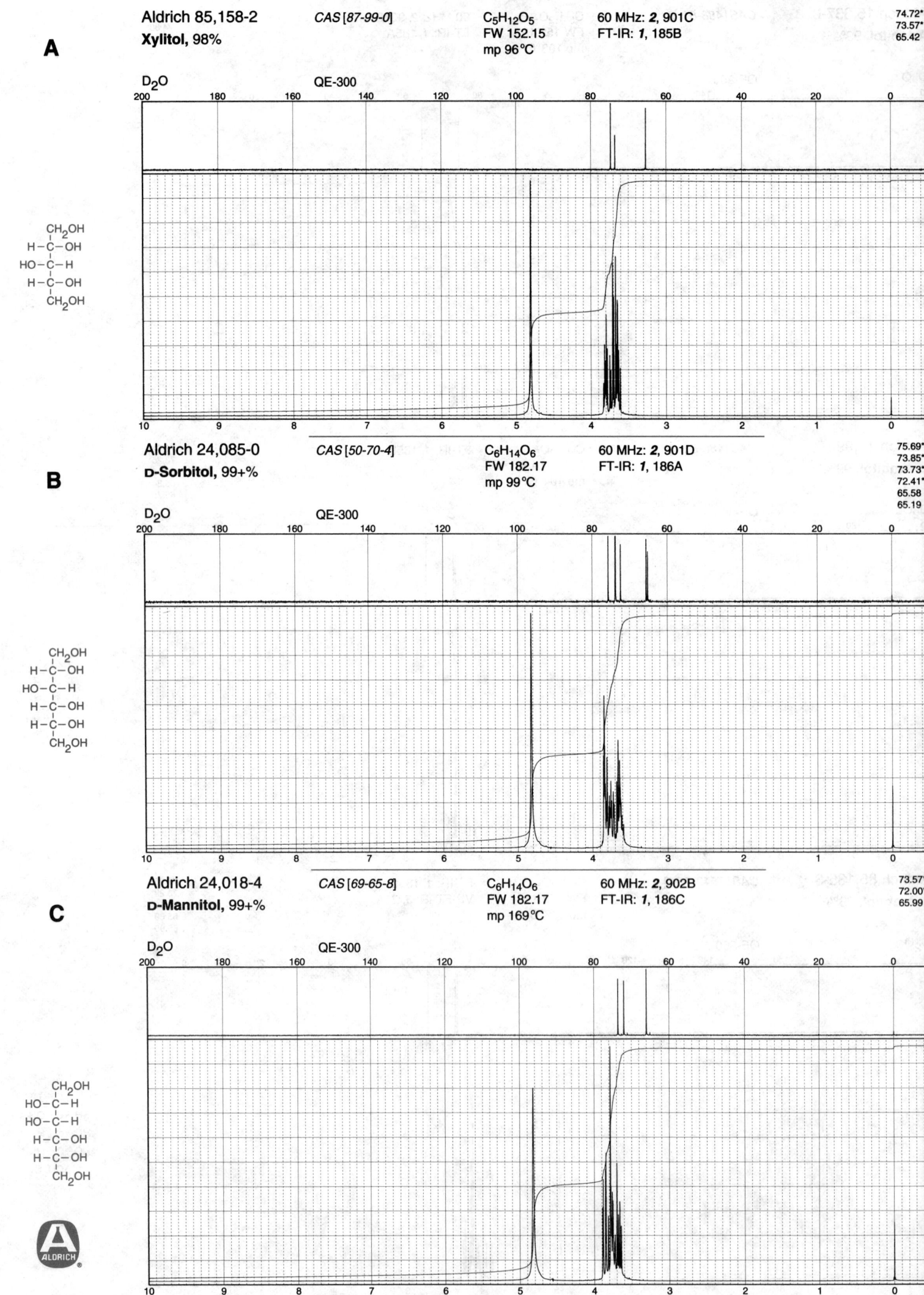

A

Aldrich 85,158-2

Xylitol, 98%

CAS [87-99-0]

$C_5H_{12}O_5$
FW 152.15
mp 96°C

60 MHz: **2**, 901C
FT-IR: **1**, 185B

74.72*
73.57*
65.42

B

Aldrich 24,085-0

D-Sorbitol, 99+%

CAS [50-70-4]

$C_6H_{14}O_6$
FW 182.17
mp 99°C

60 MHz: **2**, 901D
FT-IR: **1**, 186A

75.69*
73.85*
73.73*
72.41*
65.58
65.19

C

Aldrich 24,018-4

D-Mannitol, 99+%

CAS [69-65-8]

$C_6H_{14}O_6$
FW 182.17
mp 169°C

60 MHz: **2**, 902B
FT-IR: **1**, 186C

73.57*
72.00*
65.99

Non-Aromatic Alcohols

Aldrich 30,227-9   CAS [488-45-9]   C6H14O6   71.82*
L-Iditol, 99%   FW 182.17   70.85*
mp 79°C   62.69

A

DMSO-d6   QE-300

CH2OH
H–C–OH
HO–C–H
H–C–OH
HO–C–H
CH2OH

Aldrich D22,315-8   CAS [608-66-2]   C6H14O6   60 MHz: *2*, 902D   72.98*
Dulcitol, 97%   FW 182.17   FT-IR: *1*, 186D   72.17*
mp 190°C   66.06
bp 278°C (1 mm)

B

D2O   QE-300

CH2OH
H–C–OH
HO–C–H
HO–C–H
H–C–OH
CH2OH

Aldrich 37,467-9   CAS [4306-35-8]   C9H18O6   107.82   66.70
1,2-O-Isopropylidene-D-mannitol, 97%   FW 222.24   74.98*   63.58
mp 160°C   70.74*   26.71*
69.94*   25.42*

C

DMSO-d6   QE-300

CH2OH
HO–C–H
HO–C–H
H–C–OH
H–C–O   CH3
              \\ /
               C
              / \\
CH2O–   CH3

290

Non-Aromatic Alcohols

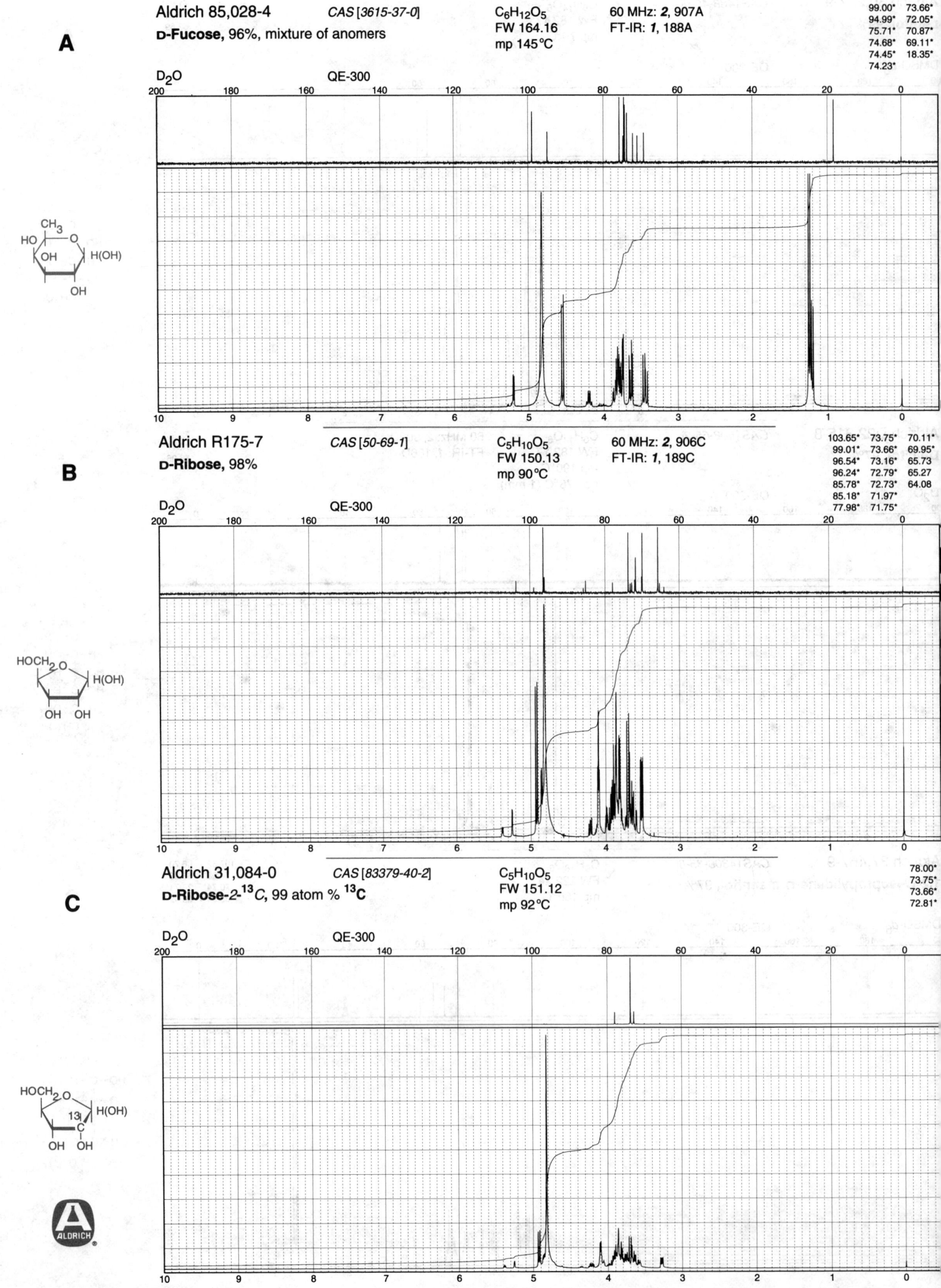

A

Aldrich 85,028-4   *CAS [3615-37-0]* $C_6H_{12}O_5$ 60 MHz: *2*, 907A

D-Fucose, 96%, mixture of anomers   FW 164.16   FT-IR: *1*, 188A   mp 145°C

99.00*	73.66*
94.99*	72.05*
75.71*	70.87*
74.68*	69.11*
74.45*	18.35*
74.23*	

D₂O   QE-300

B

Aldrich R175-7   *CAS [50-69-1]* $C_5H_{10}O_5$ 60 MHz: *2*, 906C

D-Ribose, 98%   FW 150.13   FT-IR: *1*, 189C   mp 90°C

103.65*	73.75*	70.11*
99.01*	73.66*	69.95*
96.54*	73.16*	65.73
96.24*	72.79*	65.27
85.78*	72.73*	64.08
85.18*	71.97*	
77.98*	71.75*	

D₂O   QE-300

C

Aldrich 31,084-0   *CAS [83379-40-2]* $C_5H_{10}O_5$

D-Ribose-2-^{13}C, 99 atom % ^{13}C FW 151.12   mp 92°C

78.00*	
73.75*	
73.66*	
72.81*	

D₂O   QE-300

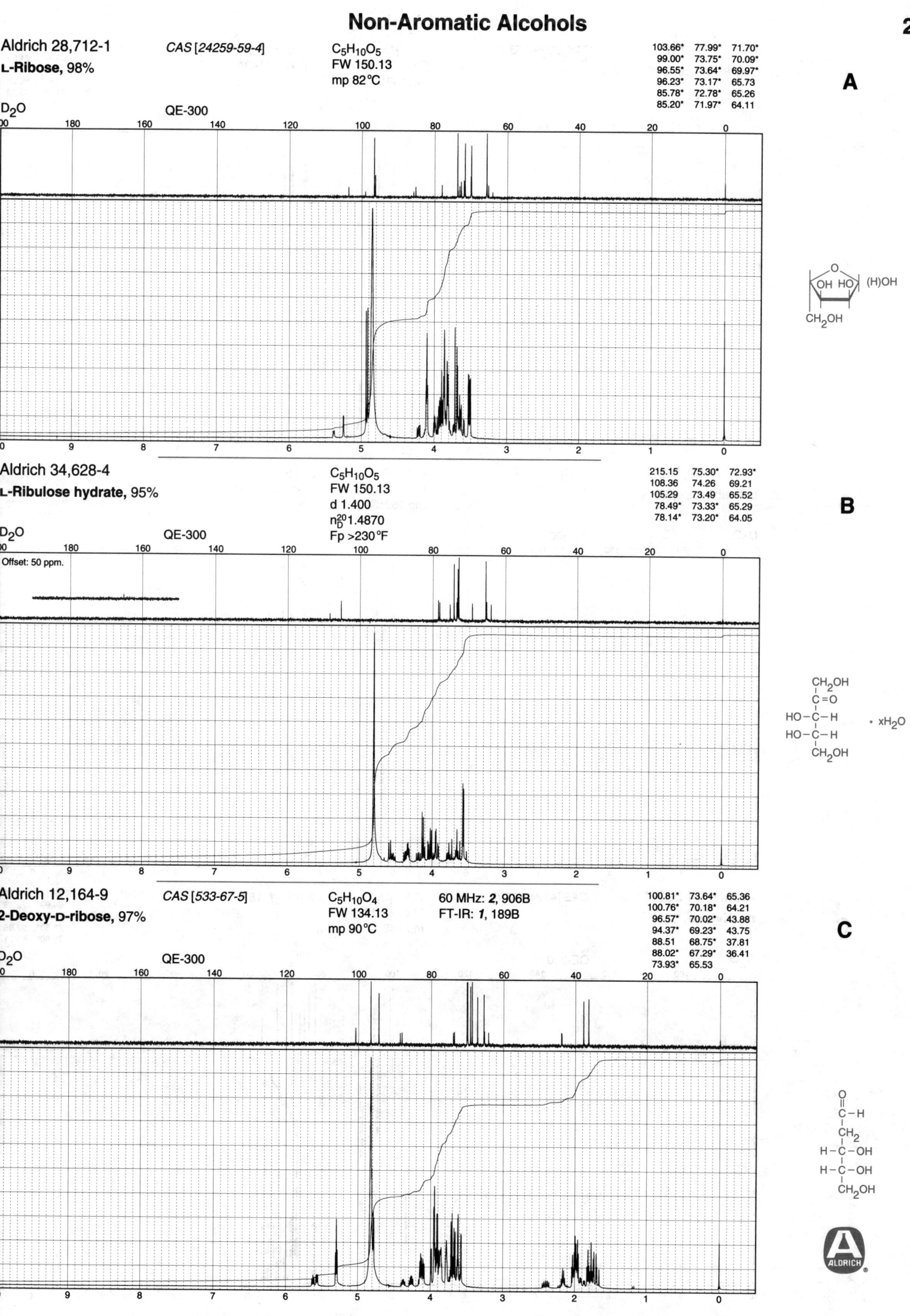

C₅H₁₀O₅
FW 150.13
mp 82 °C

103.66*	77.99*	71.70*
99.00*	73.75*	70.09*
96.55*	73.64*	69.97*
96.23*	73.17*	65.73
85.78*	72.78*	65.26
85.20*	71.97*	64.11

A

Aldrich 34,628-4

L-Ribulose hydrate, 95%

D₂O   QE-300

Offset: 50 ppm.

C₅H₁₀O₅
FW 150.13
d 1.400
n²⁰_D 1.4870
Fp >230 °F

215.15	75.30*	72.93*
108.36	74.26	69.21
105.29	73.49	65.52
78.49*	73.33*	65.29
78.14*	73.20*	64.05

B

Aldrich 12,164-9   *CAS [533-67-5]*

2-Deoxy-D-ribose, 97%

D₂O   QE-300

C₅H₁₀O₄
FW 134.13
mp 90 °C

60 MHz: **2**, 906B
FT-IR: **1**, 189B

100.81*	73.64*	65.36
100.76*	70.18*	64.21
96.57*	70.02*	43.88
94.37*	69.23*	43.75
88.51	68.75*	37.81
88.02*	67.29*	36.41
73.93*	65.53	

C

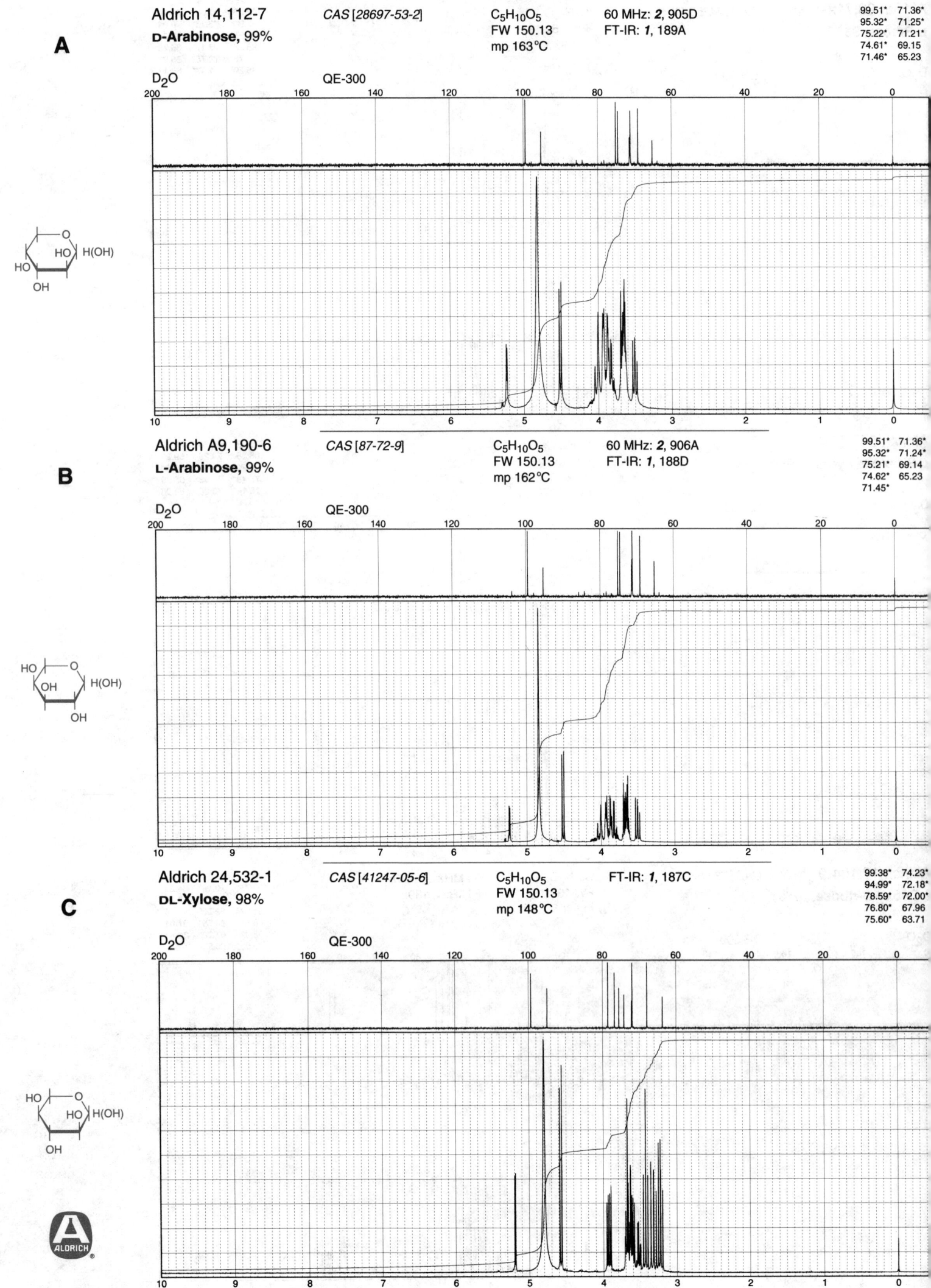

A

Aldrich 14,112-7   *CAS [28697-53-2]*   C$_5$H$_{10}$O$_5$ 60 MHz: *2*, 905D   99.51* 71.36*
D-Arabinose, 99%   FW 150.13   FT-IR: *1*, 189A   95.32* 71.25*
                                           mp 163°C   75.22* 71.21*
                                                                           74.61* 69.15
                                                                           71.46* 65.23

D$_2$O   QE-300

B

Aldrich A9,190-6   *CAS [87-72-9]*   C$_5$H$_{10}$O$_5$ 60 MHz: *2*, 906A   99.51* 71.36*
L-Arabinose, 99%   FW 150.13   FT-IR: *1*, 188D   95.32* 71.24*
                                       mp 162°C   75.21* 69.14
                                                                       74.62* 65.23
                                                                       71.45*

D$_2$O   QE-300

C

Aldrich 24,532-1   *CAS [41247-05-6]*   C$_5$H$_{10}$O$_5$ FT-IR: *1*, 187C   99.38* 74.23*
DL-Xylose, 98%   FW 150.13   94.99* 72.18*
                                          mp 148°C   78.59* 72.00*
                                                                             76.80* 67.96
                                                                             75.60* 63.71

D$_2$O   QE-300

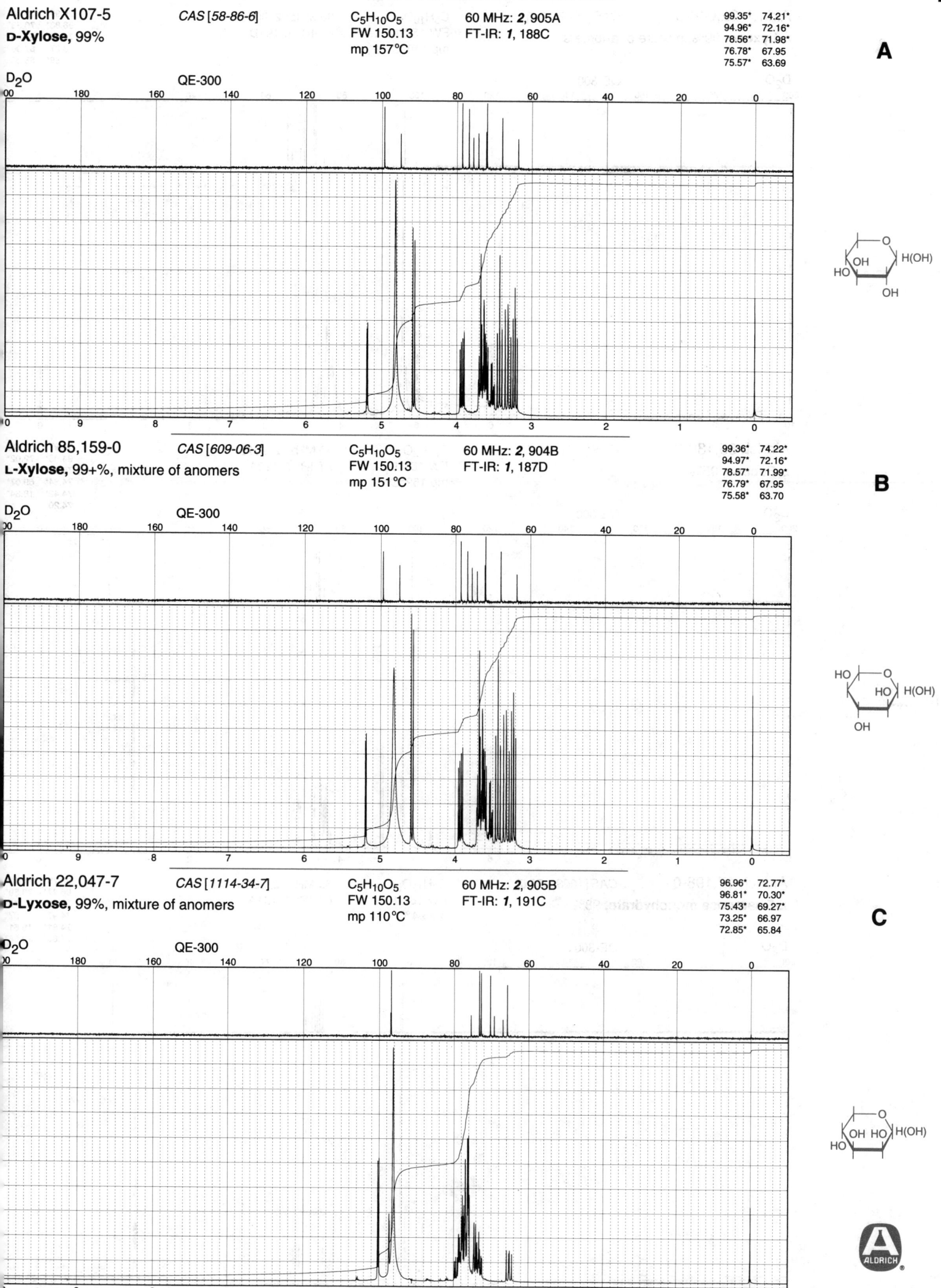

Aldrich X107-5 *CAS [58-86-6]* $C_5H_{10}O_5$ 60 MHz: *2*, 905A
D-Xylose, 99% FW 150.13 FT-IR: *1*, 188C
mp 157°C

99.35*	74.21*
94.96*	72.16*
78.56*	71.98*
76.78*	67.95
75.57*	63.69

A

D_2O QE-300

Aldrich 85,159-0 *CAS [609-06-3]* $C_5H_{10}O_5$ 60 MHz: *2*, 904B
L-Xylose, 99+%, mixture of anomers FW 150.13 FT-IR: *1*, 187D
mp 151°C

99.36*	74.22*
94.97*	72.16*
78.57*	71.99*
76.79*	67.95
75.58*	63.70

B

D_2O QE-300

Aldrich 22,047-7 *CAS [1114-34-7]* $C_5H_{10}O_5$ 60 MHz: *2*, 905B
D-Lyxose, 99%, mixture of anomers FW 150.13 FT-IR: *1*, 191C
mp 110°C

96.96*	72.77*
96.81*	70.30*
75.43*	69.27*
73.25*	66.97
72.85*	65.84

C

D_2O QE-300

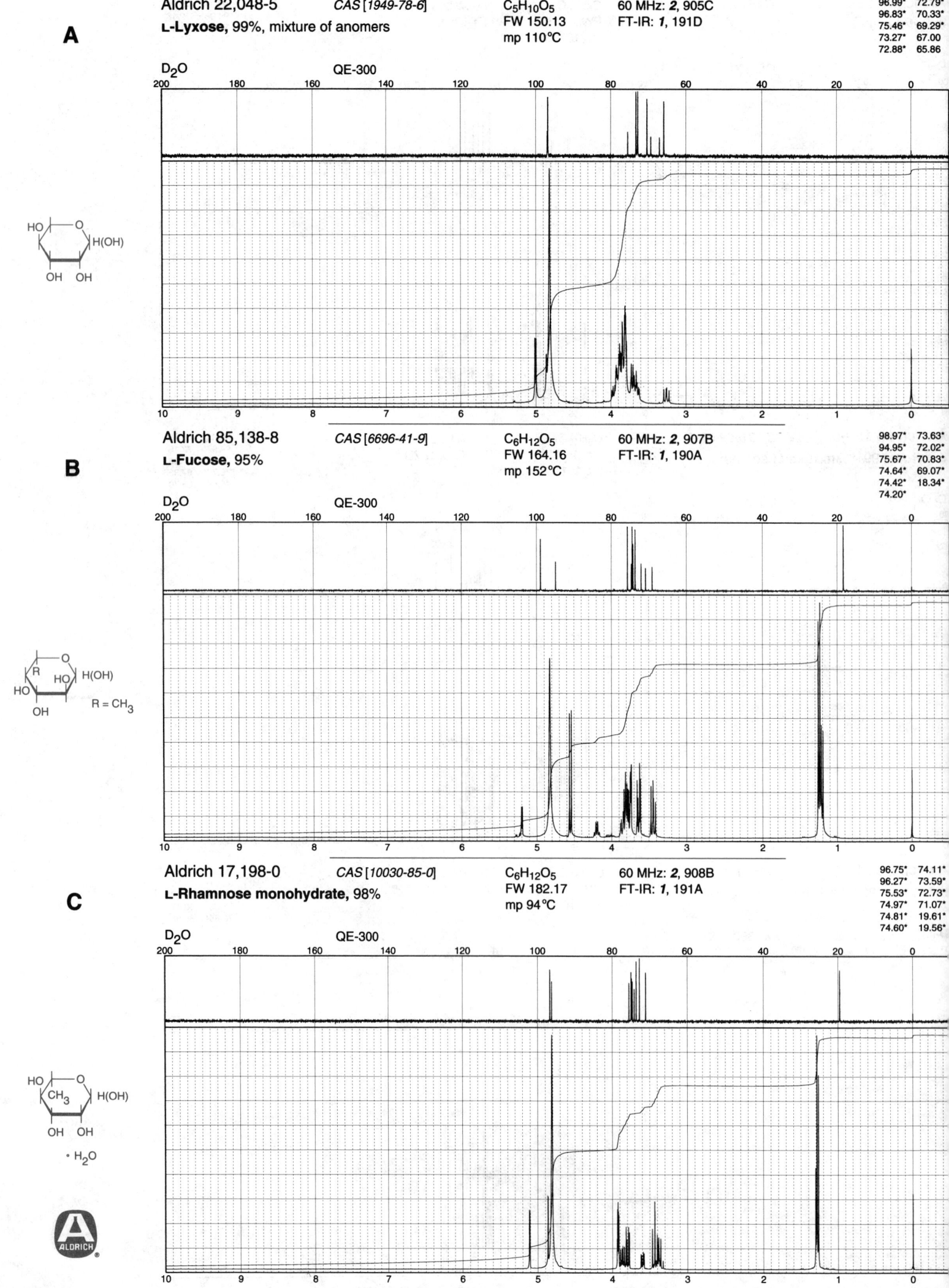

A

Aldrich 22,048-5   *CAS [1949-78-6]*   C₅H₁₀O₅   60 MHz: *2*, 905C
L-Lyxose, 99%, mixture of anomers   FW 150.13   FT-IR: *1*, 191D
mp 110°C

96.99*	72.79*
96.83*	70.33*
75.46*	69.29*
73.27*	67.00
72.88*	65.86

D₂O   QE-300

B

Aldrich 85,138-8   *CAS [6696-41-9]*   C₆H₁₂O₅   60 MHz: *2*, 907B
L-Fucose, 95%   FW 164.16   FT-IR: *1*, 190A
mp 152°C

98.97*	73.63*
94.95*	72.02*
75.67*	70.83*
74.64*	69.07*
74.42*	18.34*
74.20*	

D₂O   QE-300

R = CH₃

C

Aldrich 17,198-0   *CAS [10030-85-0]*   C₆H₁₂O₅   60 MHz: *2*, 908B
L-Rhamnose monohydrate, 98%   FW 182.17   FT-IR: *1*, 191A
mp 94°C

96.75*	74.11*
96.27*	73.59*
75.53*	72.73*
74.97*	71.07*
74.81*	19.61*
74.60*	19.56*

D₂O   QE-300

• H₂O

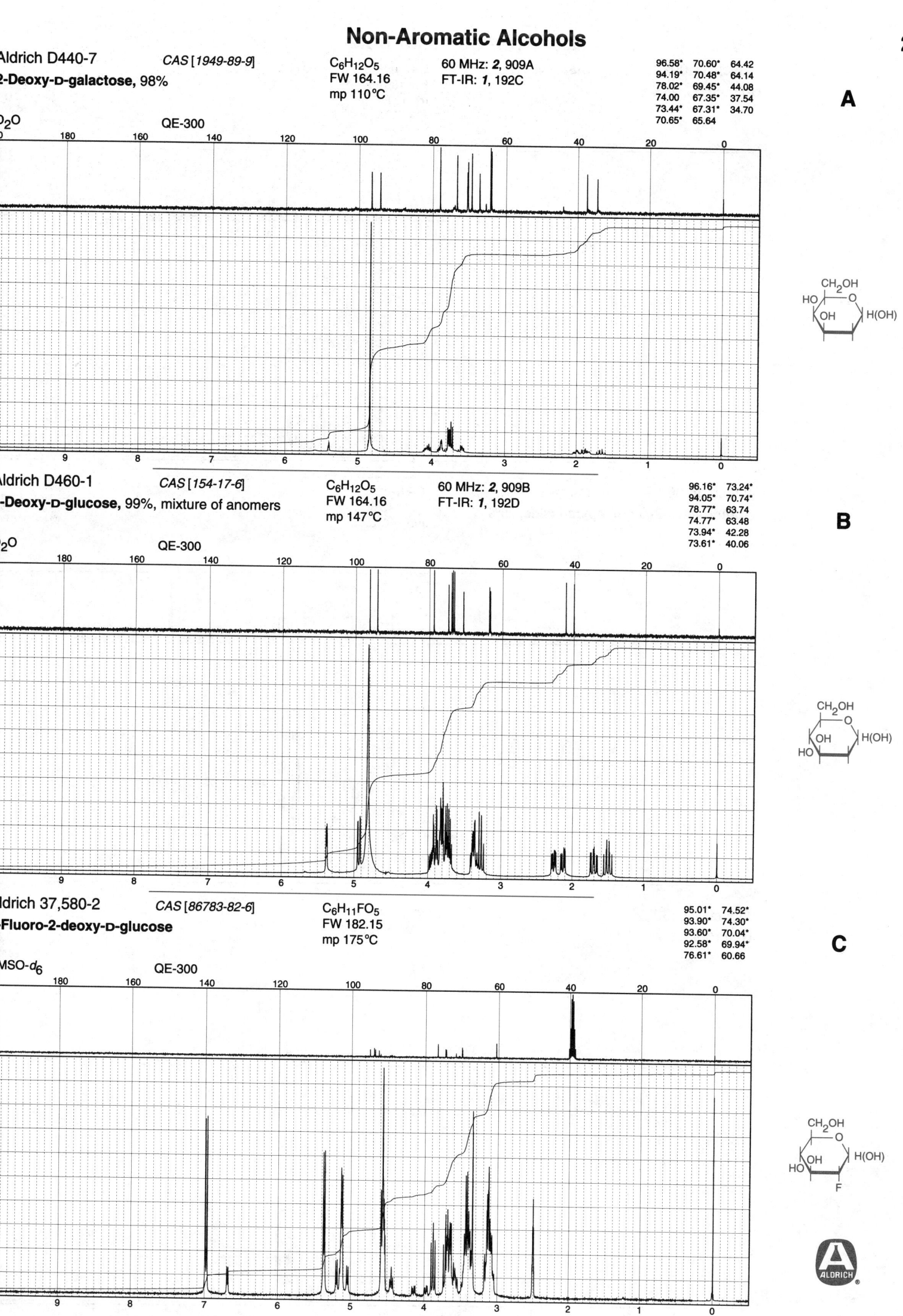

Aldrich D440-7   *CAS [1949-89-9]*
2-Deoxy-D-galactose, 98%

C₆H₁₂O₅
FW 164.16
mp 110°C

60 MHz: *2*, 909A
FT-IR: *1*, 192C

96.58*	70.60*	64.42
94.19*	70.48*	64.14
78.02*	69.45*	44.08
74.00	67.35*	37.54
73.44*	67.31*	34.70
70.65*	65.64	

A

D₂O   QE-300

Aldrich D460-1   *CAS [154-17-6]*
2-Deoxy-D-glucose, 99%, mixture of anomers

C₆H₁₂O₅
FW 164.16
mp 147°C

60 MHz: *2*, 909B
FT-IR: *1*, 192D

96.16*	73.24*
94.05*	70.74*
78.77*	63.74
74.77*	63.48
73.94*	42.28
73.61*	40.06

B

D₂O   QE-300

Aldrich 37,580-2   *CAS [86783-82-6]*
2-Fluoro-2-deoxy-D-glucose

C₆H₁₁FO₅
FW 182.15
mp 175°C

95.01*	74.52*
93.90*	74.30*
93.60*	70.04*
92.58*	69.94*
76.61*	60.66

C

DMSO-*d₆*   QE-300

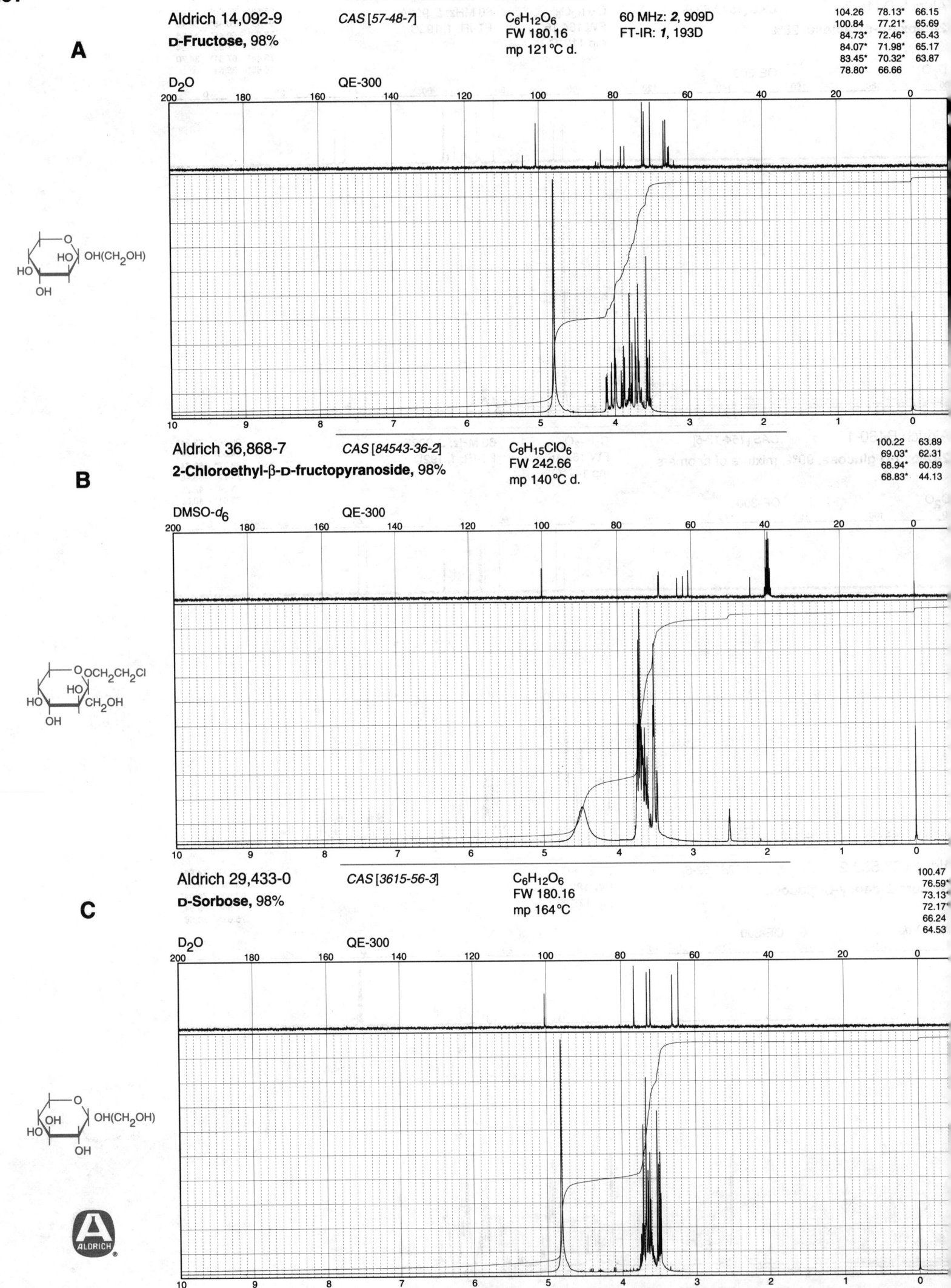

A

Aldrich 14,092-9
D-Fructose, 98%

CAS [57-48-7]

C6H12O6
FW 180.16
mp 121°C d.

60 MHz: *2*, 909D
FT-IR: *1*, 193D

104.26	78.13*	66.15
100.84	77.21*	65.69
84.73*	72.46*	65.43
84.07*	71.98*	65.17
83.45*	70.32*	63.87
78.80*	66.66	

D2O   QE-300

B

Aldrich 36,868-7
2-Chloroethyl-β-D-fructopyranoside, 98%

CAS [84543-36-2]

C8H15ClO6
FW 242.66
mp 140°C d.

100.22	63.89
69.03*	62.31
68.94*	60.89
68.83*	44.13

DMSO-*d6*   QE-300

C

Aldrich 29,433-0
D-Sorbose, 98%

CAS [3615-56-3]

C6H12O6
FW 180.16
mp 164°C

100.47
76.59*
73.13*
72.17*
66.24
64.53

D2O   QE-300

ALDRICH

Non-Aromatic Alcohols

Aldrich 85,156-6

CAS [87-79-6]

L-Sorbose, 98%

$C_6H_{12}O_6$
FW 180.16
mp 172°C

60 MHz: *2*, 909C
FT-IR: *1*, 193C

100.50
76.61*
73.15*
72.20*
66.26
64.55

A

D₂O

QE-300

Aldrich 11,259-3

CAS [59-23-4]

D-Galactose, 97%

$C_6H_{12}O_6$
FW 180.16
mp 169°C

60 MHz: *2*, 908C
FT-IR: *1*, 192A

99.23* 72.10*
95.06* 71.96*
77.93* 71.53*
75.58* 71.14*
74.66* 63.98
73.26* 63.78

B

D₂O

QE-300

Aldrich 28,592-7

CAS [15572-79-9]

L-Galactose, 98%, mixture of anomers

$C_6H_{12}O_6$
FW 180.16
mp 170°C

99.20* 73.24* 71.51*
95.03* 72.93* 71.11*
77.91* 72.11* 66.02
75.55* 72.08* 63.96
74.64* 71.93* 63.76

C

D₂O

QE-300

R = CH₂OH

ALDRICH

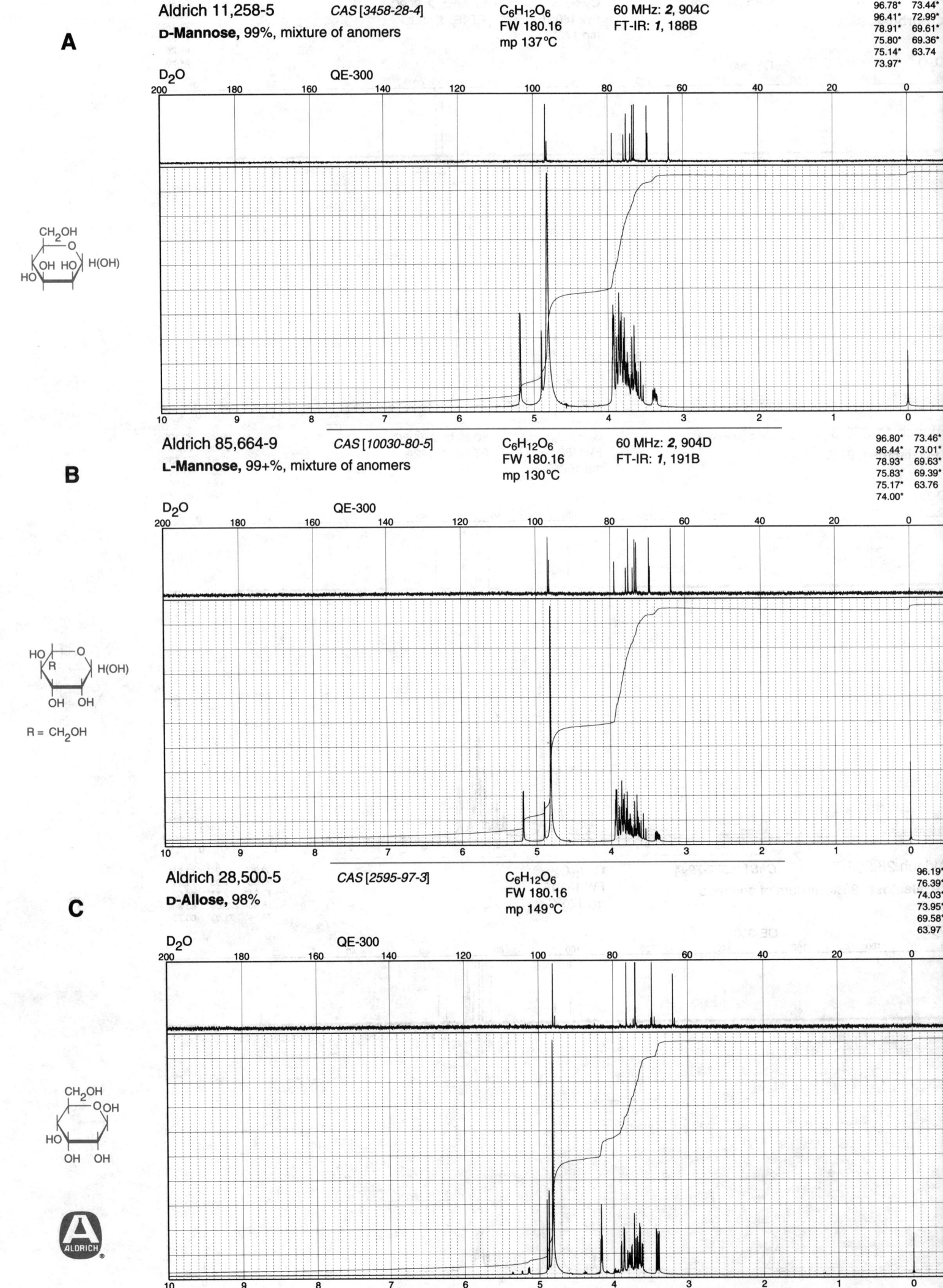

A

Aldrich 11,258-5   *CAS [3458-28-4]* $C_6H_{12}O_6$ 60 MHz: **2**, 904C

D-Mannose, 99%, mixture of anomers   FW 180.16   FT-IR: **1**, 188B

mp 137°C

96.78*	73.44*
96.41*	72.99*
78.91*	69.61*
75.80*	69.36*
75.14*	63.74
73.97*	

D₂O   QE-300

B

Aldrich 85,664-9   *CAS [10030-80-5]* $C_6H_{12}O_6$ 60 MHz: **2**, 904D

L-Mannose, 99+%, mixture of anomers   FW 180.16   FT-IR: **1**, 191B

mp 130°C

96.80*	73.46*
96.44*	73.01*
78.93*	69.63*
75.83*	69.39*
75.17*	63.76
74.00*	

R = CH₂OH

D₂O   QE-300

C

Aldrich 28,500-5   *CAS [2595-97-3]* $C_6H_{12}O_6$

D-Allose, 98%   FW 180.16

mp 149°C

96.19*
76.39*
74.03*
73.95*
69.58*
63.97

D₂O   QE-300

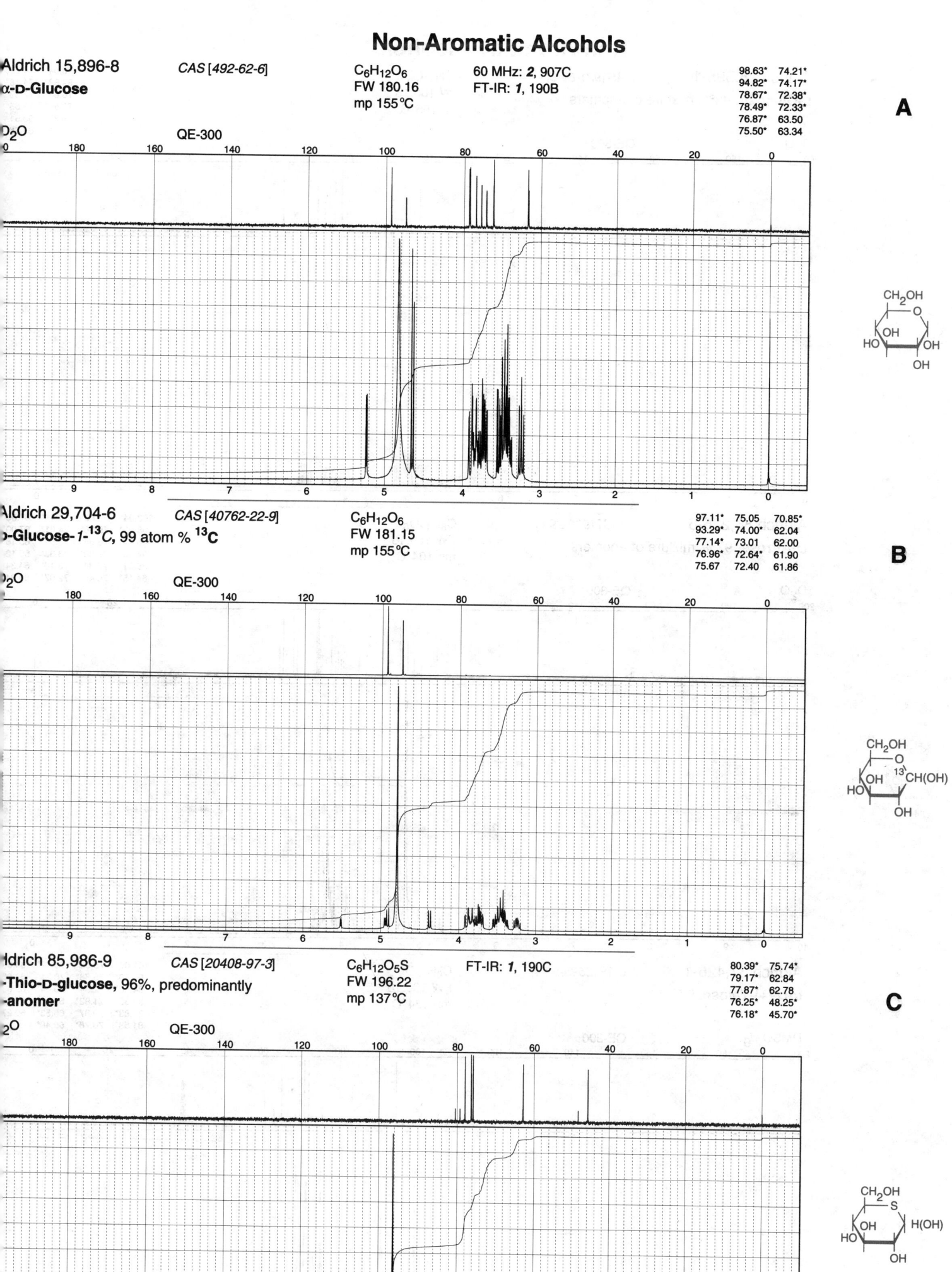

Aldrich 15,896-8   *CAS [492-62-6]* $C_6H_{12}O_6$ 60 MHz: **2**, 907C

α-D-Glucose   FW 180.16   FT-IR: **1**, 190B

mp 155 °C

D₂O   QE-300

98.63*	74.21*
94.82*	74.17*
78.67*	72.38*
78.49*	72.33*
76.87*	63.50
75.50*	63.34

A

Aldrich 29,704-6   *CAS [40762-22-9]* $C_6H_{12}O_6$

D-Glucose-1-^{13}C, 99 atom % ^{13}C FW 181.15

mp 155 °C

D₂O   QE-300

97.11*	75.05	70.85*
93.29*	74.00*	62.04
77.14*	73.01	62.00
76.96*	72.64*	61.90
75.67	72.40	61.86

B

Aldrich 85,986-9   *CAS [20408-97-3]* $C_6H_{12}O_5S$ FT-IR: **1**, 190C

5-Thio-D-glucose, 96%, predominantly   FW 196.22

α-anomer   mp 137 °C

D₂O   QE-300

80.39*	75.74*	
79.17*	62.84	
77.87*	62.78	
76.25*	48.25*	
76.18*	45.70*	

C

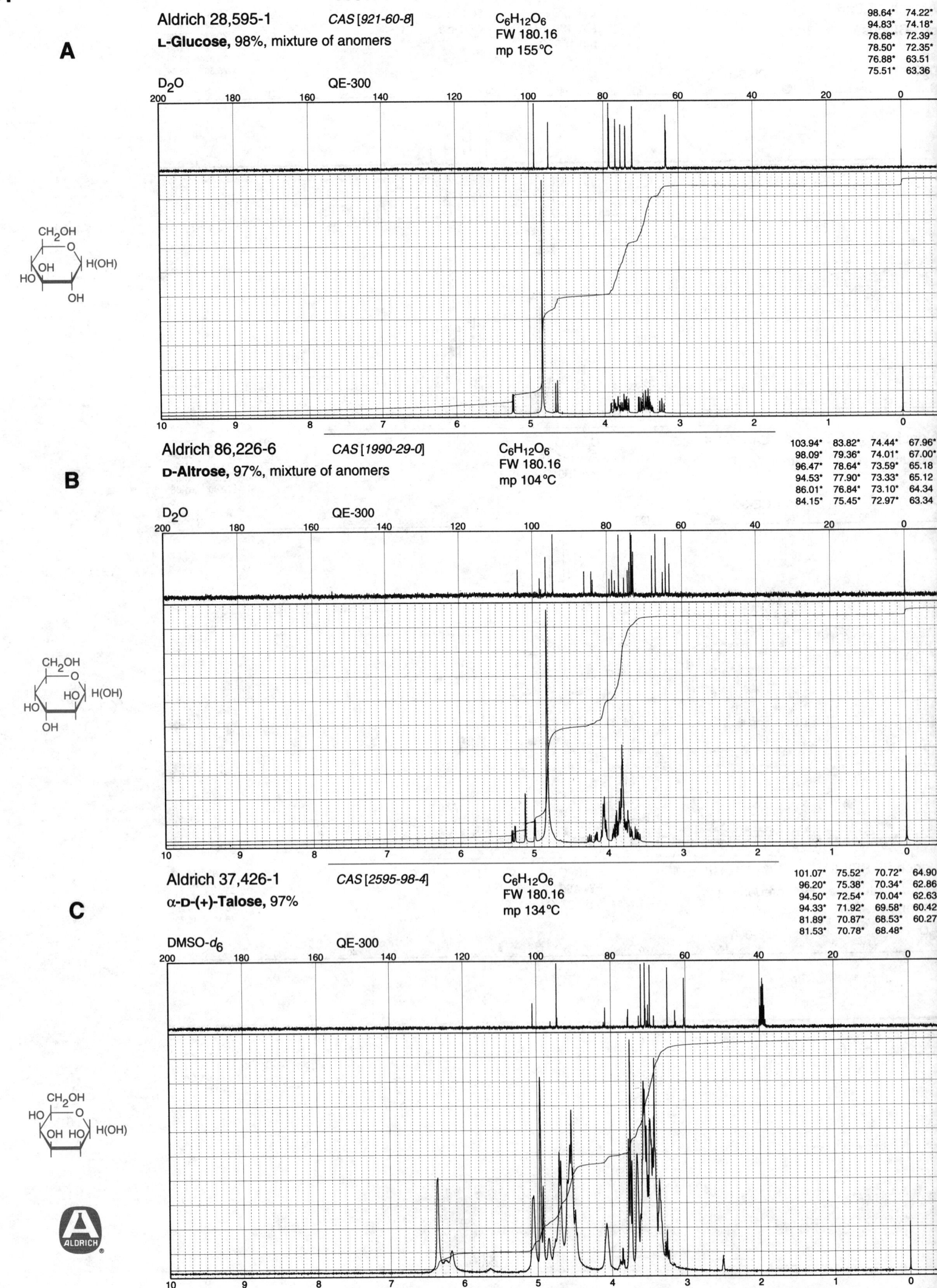

Aldrich 28,595-1   *CAS [921-60-8]*   C₆H₁₂O₆

L-Glucose, 98%, mixture of anomers

$C_6H_{12}O_6$
FW 180.16
mp 155°C

98.64*	74.22*
94.83*	74.18*
78.68*	72.39*
78.50*	72.35*
76.88*	63.51
75.51*	63.36

D₂O   QE-300

B

Aldrich 86,226-6   *CAS [1990-29-0]*

D-Altrose, 97%, mixture of anomers

$C_6H_{12}O_6$
FW 180.16
mp 104°C

103.94*	83.82*	74.44*	67.96*
98.09*	79.36*	74.01*	67.00*
96.47*	78.64*	73.59*	65.18
94.53*	77.90*	73.33*	65.12
86.01*	76.84*	73.10*	64.34
84.15*	75.45*	72.97*	63.34

D₂O   QE-300

C

Aldrich 37,426-1   *CAS [2595-98-4]*

α-D-(+)-Talose, 97%

$C_6H_{12}O_6$
FW 180.16
mp 134°C

101.07*	75.52*	70.72*	64.90
96.20*	75.38*	70.34*	62.86
94.50*	72.54*	70.04*	62.63
94.33*	71.92*	69.58*	60.42
81.89*	70.87*	68.53*	60.27
81.53*	70.78*	68.48*	

DMSO-*d₆*   QE-300

Aldrich 37,729-5   CAS [87-81-0] $C_6H_{12}O_6$ FT-IR: 1, 1142D

D-(-)-Tagatose, 96%, mixture of anomers   FW 180.16
mp 132°C

DMSO-d_6 QE-300

104.77	77.42*	64.29
102.66	71.83*	63.94
98.60	71.16*	62.97*
97.56	70.39*	62.58
79.84*	69.55*	59.69
78.94*	66.40*	

A

Aldrich 86,052-2   CAS [5328-63-2] $C_6H_{12}O_5$

Methyl-β-D-arabinopyranoside, 99%   FW 164.16
FT-IR: 1, 193A

DMSO-d_6 QE-300

| 101.93* |
| 70.37* |
| 69.94* |
| 69.59* |
| 64.19 |
| 56.14* |

B

Aldrich 29,269-9   CAS [14703-09-4] $C_6H_{12}O_5$

Methyl-α-D-xylopyranoside, 99%   FW 164.16
mp 92°C

DMSO-d_6 QE-300

| 100.00* |
| 73.28* |
| 71.78* |
| 69.82* |
| 61.54 |
| 54.42* |

C

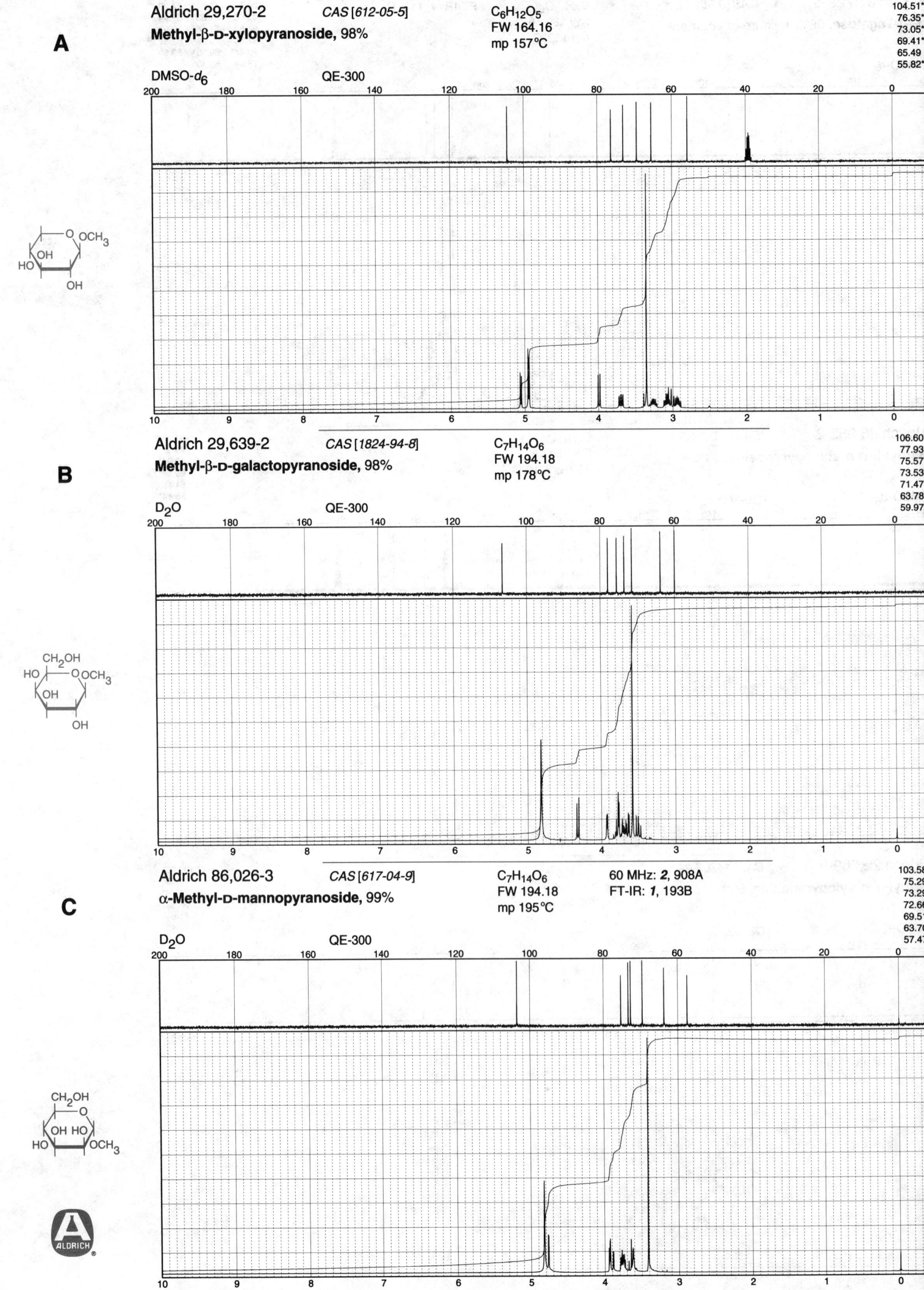

A

Aldrich 29,270-2   CAS [612-05-5] $C_6H_{12}O_5$
FW 164.16
mp 157°C

Methyl-β-D-xylopyranoside, 98%

104.51*
76.35*
73.05*
69.41*
65.49
55.82*

DMSO-d_6 QE-300

B

Aldrich 29,639-2   CAS [1824-94-8] $C_7H_{14}O_6$
FW 194.18
mp 178°C

Methyl-β-D-galactopyranoside, 98%

106.60*
77.93*
75.57*
73.53*
71.47
63.78
59.97*

D_2O QE-300

C

Aldrich 86,026-3   CAS [617-04-9] $C_7H_{14}O_6$ 60 MHz: **2**, 908A
FW 194.18   FT-IR: **1**, 193B
mp 195°C

α-Methyl-D-mannopyranoside, 99%

103.58
75.29
73.29
72.66
69.51
63.70
57.47

D_2O QE-300

Non-Aromatic Alcohols

304

Aldrich 17,044-5   *CAS [97-30-3]*
Methyl-α-D-glucopyranoside, 99%
$C_7H_{14}O_6$
FW 194.18
mp 170°C
60 MHz: *2*, 908D
FT-IR: *1*, 192B
D₂O   QE-300

102.04*
75.87*
74.36*
74.00*
72.35*
63.35
57.81*

A

Aldrich 29,268-0   *CAS [7000-27-3]*
Methyl-β-D-glucopyranoside hemihydrate, 99%
$C_7H_{14}O_6$
FW 203.19
mp 112°C
DMSO-d_6 QE-300

103.68*
76.59*
76.43*
73.18*
69.84*
60.87
55.81*

B

· $^1/_2$ H₂O

Aldrich 86,193-6   *CAS [29836-26-8]*
-O-Octyl-β-D-glucopyranoside, 98%
$C_{14}H_{28}O_6$
FW 292.38
D₂O   QE-300

105.05* 34.35
78.40* 31.95
78.27* 31.87
75.53* 28.28
72.74 25.08
71.82* 16.27*
63.22

C

ALDRICH

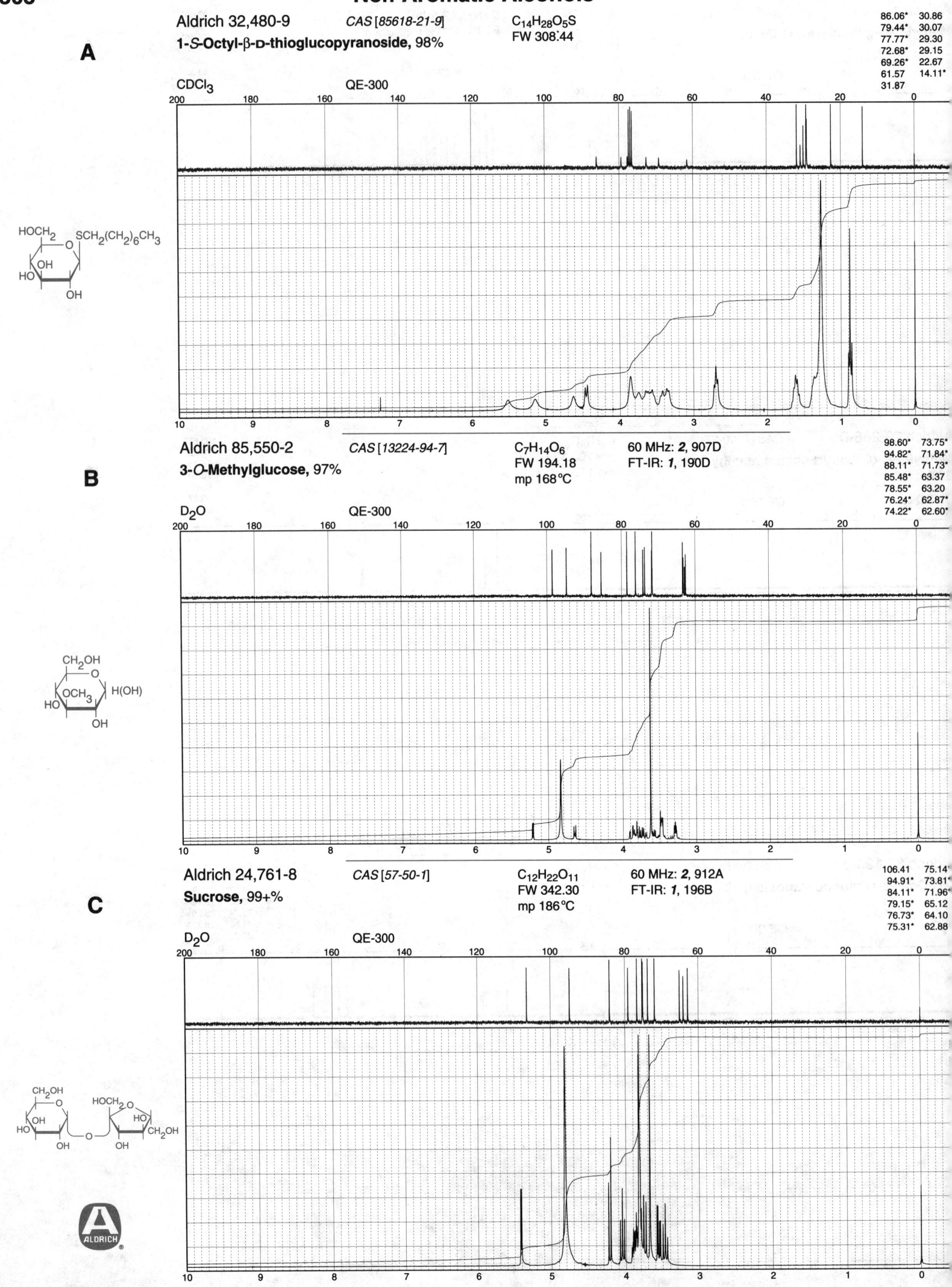

A

Aldrich 32,480-9   CAS [85618-21-9] $C_{14}H_{28}O_5S$
FW 308.44

1-S-Octyl-β-D-thioglucopyranoside, 98%

86.06*	30.86
79.44*	30.07
77.77*	29.30
72.68*	29.15
69.26*	22.67
61.57	14.11*
31.87	

CDCl₃   QE-300

B

Aldrich 85,550-2   CAS [13224-94-7] $C_7H_{14}O_6$ 60 MHz: 2, 907D
FW 194.18   FT-IR: 1, 190D
mp 168 °C

3-O-Methylglucose, 97%

98.60*	73.75*
94.82*	71.84*
88.11*	71.73*
85.48*	63.37
78.55*	63.20
76.24*	62.87*
74.22*	62.60*

D₂O   QE-300

C

Aldrich 24,761-8   CAS [57-50-1] $C_{12}H_{22}O_{11}$ 60 MHz: 2, 912A
FW 342.30   FT-IR: 1, 196B
mp 186 °C

Sucrose, 99+%

106.41	75.14
94.91*	73.81*
84.11*	71.96*
79.15*	65.12
76.73*	64.10
75.31*	62.88

D₂O   QE-300

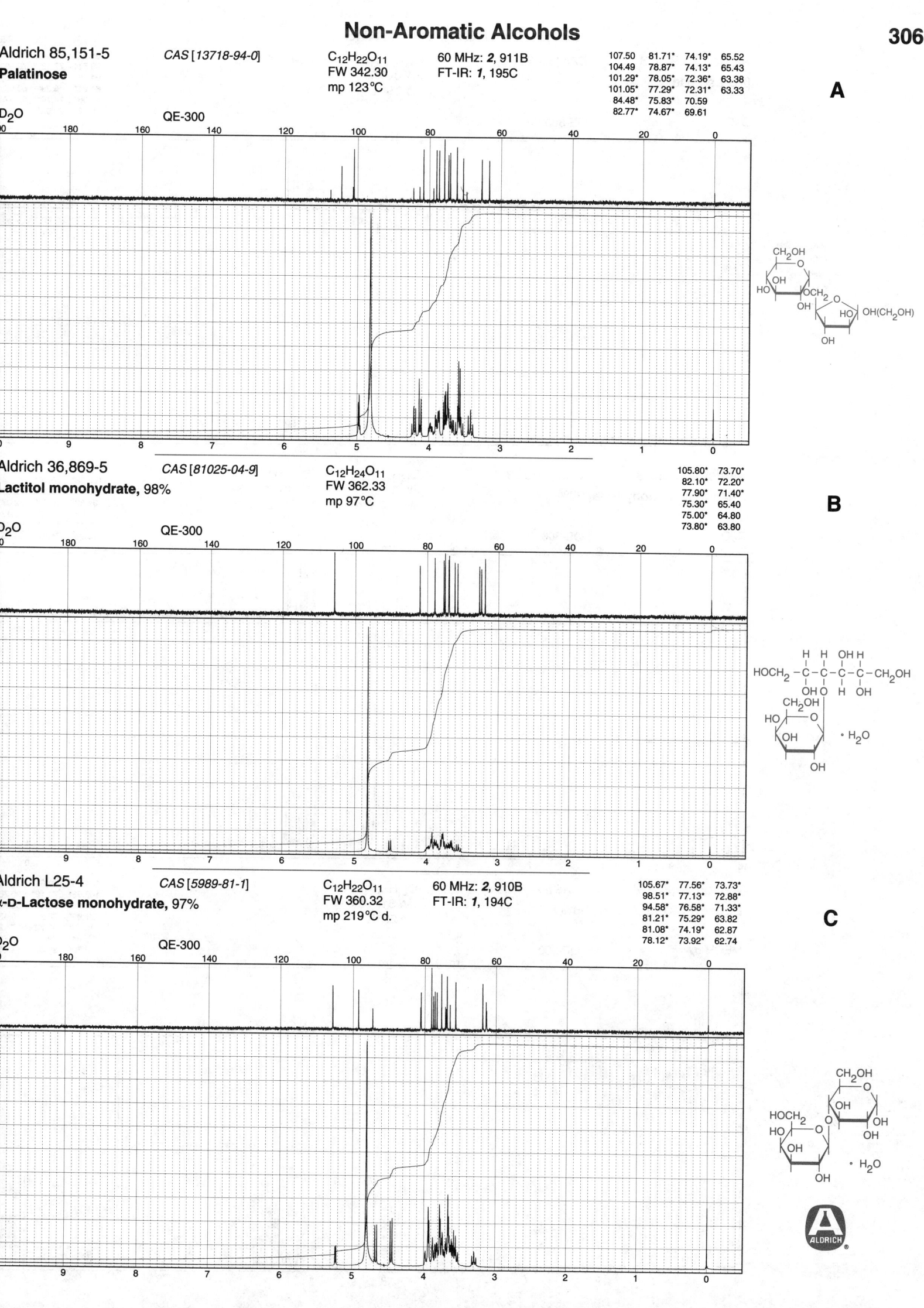

Aldrich 85,151-5   CAS [13718-94-0]   C₁₂H₂₂O₁₁   60 MHz: **2**, 911B
Palatinose   FW 342.30   FT-IR: **1**, 195C
mp 123°C

107.50	81.71*	74.19*	65.52
104.49	78.87*	74.13*	65.43
101.29*	78.05*	72.36*	63.38
101.05*	77.29*	72.31*	63.33
84.48*	75.83*	70.59	
82.77*	74.67*	69.61	

A

D₂O   QE-300

Aldrich 36,869-5   CAS [81025-04-9]   C₁₂H₂₄O₁₁
Lactitol monohydrate, 98%   FW 362.33
mp 97°C

105.80*	73.70*
82.10*	72.20*
77.90*	71.40*
75.30*	65.40*
75.00*	64.80*
73.80*	63.80*

B

D₂O   QE-300

Aldrich L25-4   CAS [5989-81-1]   C₁₂H₂₂O₁₁   60 MHz: **2**, 910B
α-D-Lactose monohydrate, 97%   FW 360.32   FT-IR: **1**, 194C
mp 219°C d.

105.67*	77.56*	73.73*
98.51*	77.13*	72.88*
94.58*	76.58*	71.33*
81.21*	75.29*	63.82
81.08*	74.19*	62.87
78.12*	73.92*	62.74

C

D₂O   QE-300

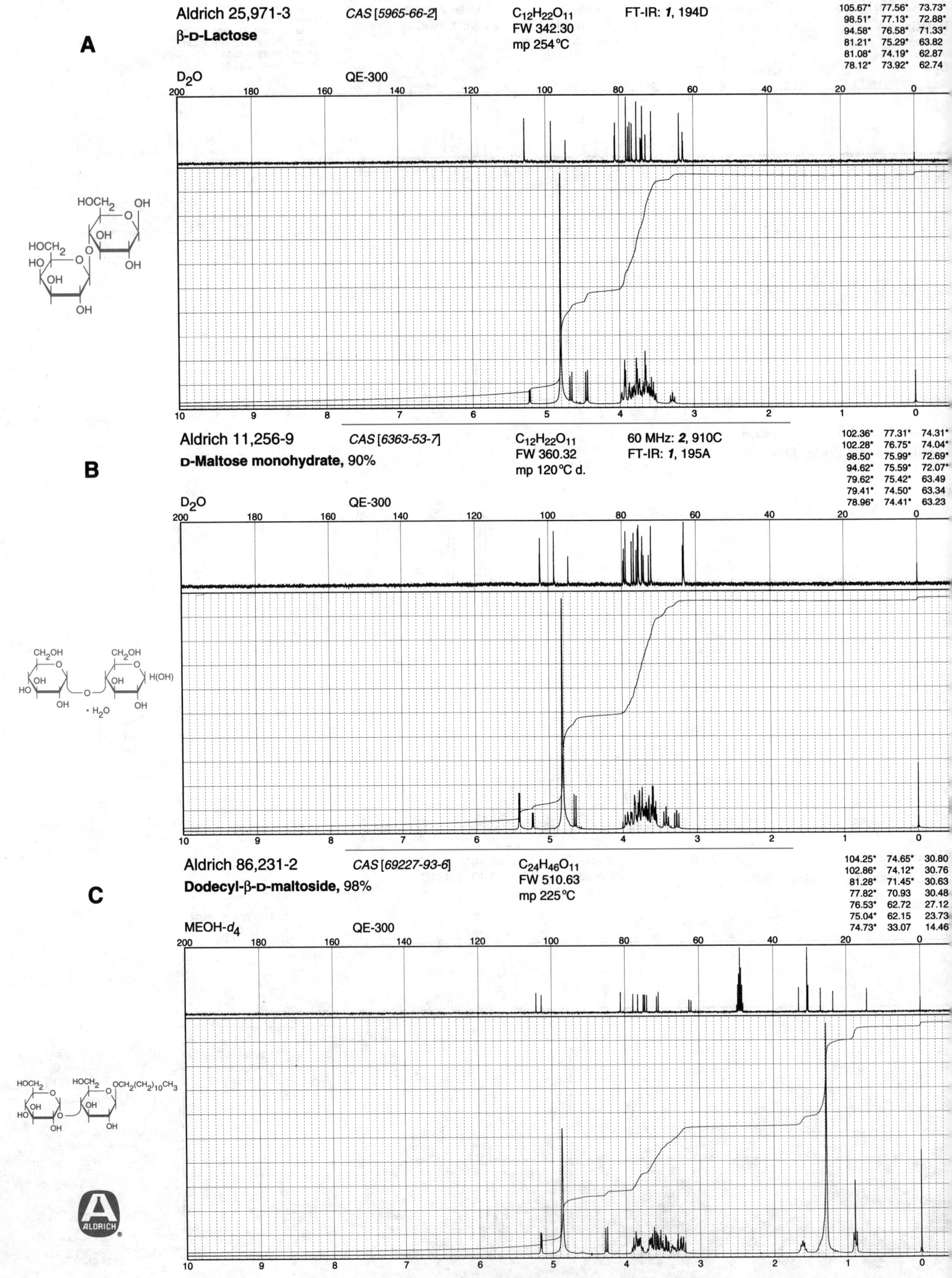

A

Aldrich 25,971-3   *CAS [5965-66-2]* $C_{12}H_{22}O_{11}$ FT-IR: *1*, 194D
β-D-Lactose   FW 342.30
   mp 254°C

105.67*	77.56*	73.73*
98.51*	77.13*	72.88*
94.58*	76.58*	71.33*
81.21*	75.29*	63.82
81.08*	74.19*	62.87
78.12*	73.92*	62.74

D_2O QE-300

B

Aldrich 11,256-9   *CAS [6363-53-7]* $C_{12}H_{22}O_{11}$ 60 MHz: *2*, 910C
D-Maltose monohydrate, 90%   FW 360.32   FT-IR: *1*, 195A
   mp 120°C d.

102.36*	77.31*	74.31*
102.28*	76.75*	74.04*
98.50*	75.99*	72.69*
94.62*	75.59*	72.07*
79.62*	75.42*	63.49
79.41*	74.50*	63.34
78.96*	74.41*	63.23

D_2O QE-300

C

Aldrich 86,231-2   *CAS [69227-93-6]* $C_{24}H_{46}O_{11}$
Dodecyl-β-D-maltoside, 98%   FW 510.63
   mp 225°C

104.25*	74.65*	30.80
102.86*	74.12*	30.76
81.28*	71.45*	30.63
77.82*	70.93	30.48
76.53*	62.72	27.12
75.04*	62.15	23.73
74.73*	33.07	14.46

MEOH-d_4 QE-300

ALDRICH

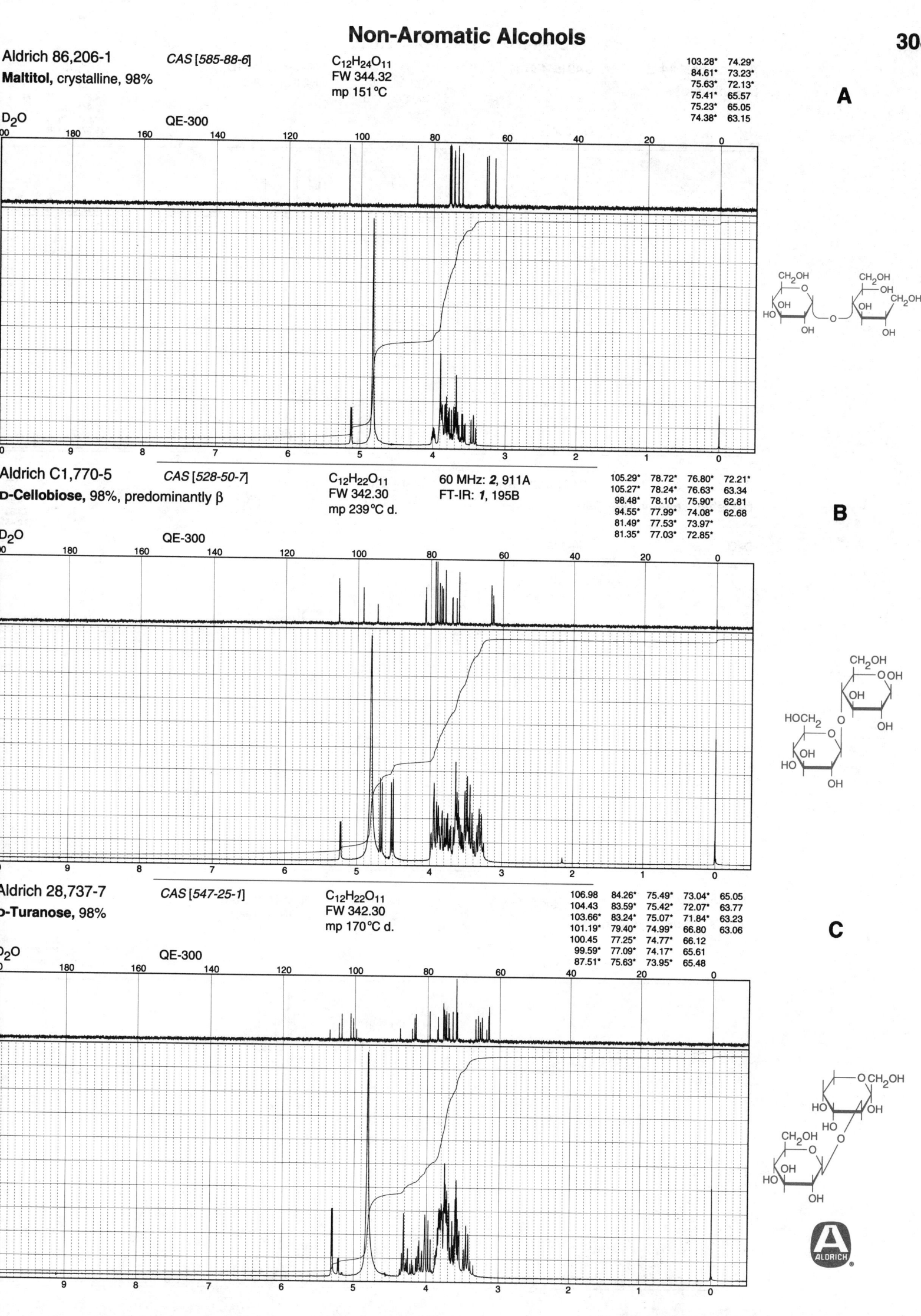

Aldrich 86,206-1 CAS [585-88-6] $C_{12}H_{24}O_{11}$ FW 344.32 mp 151°C

Maltitol, crystalline, 98%

103.28*	74.29*
84.61*	73.23*
75.63*	72.13*
75.41*	65.57
75.23*	65.05
74.38*	63.15

D_2O QE-300   **A**

Aldrich C1,770-5 CAS [528-50-7] $C_{12}H_{22}O_{11}$ FW 342.30 mp 239°C d.

D-Cellobiose, 98%, predominantly β

60 MHz: **2**, 911A FT-IR: **1**, 195B

105.29*	78.72*	76.80*	72.21*
105.27*	78.24*	76.63*	63.34
98.48*	78.10*	75.90*	62.81
94.55*	77.99*	74.08*	62.68
81.49*	77.53*	73.97*	
81.35*	77.03*	72.85*	

D_2O QE-300   **B**

Aldrich 28,737-7 CAS [547-25-1] $C_{12}H_{22}O_{11}$ FW 342.30 mp 170°C d.

D-Turanose, 98%

106.98	84.26*	75.49*	73.04*	65.05
104.43	83.59*	75.42*	72.07*	63.77
103.66*	83.24*	75.07*	71.84*	63.23
101.19*	79.40*	74.99*	66.80	63.06
100.45	77.25*	74.77*	66.12	
99.59*	77.09*	74.17*	65.61	
87.51*	75.63*	73.95*	65.48	

D_2O QE-300   **C**

Non-Aromatic Alcohols

A

Aldrich 85,144-2

CAS [554-91-6]

β-Gentiobiose

$C_{12}H_{22}O_{11}$
FW 342.30
mp 209°C d.

60 MHz: *2*, 911D
FT-IR: *1*, 195D

105.40*	77.61*	72.20*
105.38*	76.76*	72.16*
98.70*	75.82*	71.54
94.86*	75.40*	71.38
78.64*	74.15*	63.47
78.61*	73.17*	
78.37*	72.35*	

D_2O　　QE-300

B

Aldrich 18,835-2

CAS [6138-23-4]

D-Trehalose dihydrate, 99%

$C_{12}H_{22}O_{11}$
FW 378.33
mp 98°C

60 MHz: *2*, 912C
FT-IR: *1*, 196A

95.99*	
75.30*	
74.94*	
73.83*	
72.48*	
63.32	

D_2O　　QE-300

· 2H₂O

C

Aldrich M269-1

CAS [585-99-9]

α-D-Melibiose hydrate, 99%

$C_{12}H_{22}O_{11}$
FW 360.32
mp 182°C d.

60 MHz: *2*, 912B
FT-IR: *1*, 196C

100.95*	75.71*	71.97*
100.91*	74.19*	71.23*
98.82*	73.71*	68.67
94.94*	72.85*	68.57
78.65*	72.33*	63.88
77.11*	72.24*	
76.81*	72.19*	

D_2O　　QE-300

· xH₂O

ALDRICH

Non-Aromatic Alcohols

310

Aldrich 85,149-3
Maltotriose hydrate, 95%

$C_{18}H_{32}O_{16}$
FW 504.44
mp 134°C

60 MHz: *2*, 912D
FT-IR: *1*, 196D

102.49*	79.63*	75.93*	74.02*
102.27*	79.47*	75.60*	73.91*
102.18*	78.91*	75.44*	72.66*
98.50*	77.27*	74.48*	72.05*
94.63*	76.72*	74.31*	63.41
79.86*	76.06*	74.22*	63.21

A

D_2O QE-300

Aldrich 20,667-9 *CAS [17629-30-0]*
D-Raffinose pentahydrate, 98%

$C_{18}H_{32}O_{16}$
FW 594.52
mp 81°C

60 MHz: *2*, 913C
FT-IR: *1*, 197B

103.84*	72.65*	68.68*
98.91*	71.30*	68.38*
91.44*	71.09*	66.52
82.17*	70.76*	62.03
76.80*	70.09*	60.42
74.03*	69.31*	

B

DMSO-*d*₆ QE-300

Aldrich 85,178-7 *CAS [10094-58-3]*
Stachyose tetrahydrate, 98%

$C_{24}H_{42}O_{21}$
FW 738.64
mp 110°C d.

60 MHz: *2*, 913D
FT-IR: *1*, 197C

106.58	76.78*	72.16*	69.27
101.14*	75.51*	72.14*	68.64
100.80*	74.07*	72.03*	65.27
94.89*	73.78*	71.59*	64.21
84.13*	73.74*	71.24*	63.95
79.12*	72.29*	71.08*	

C

D₂O QE-300

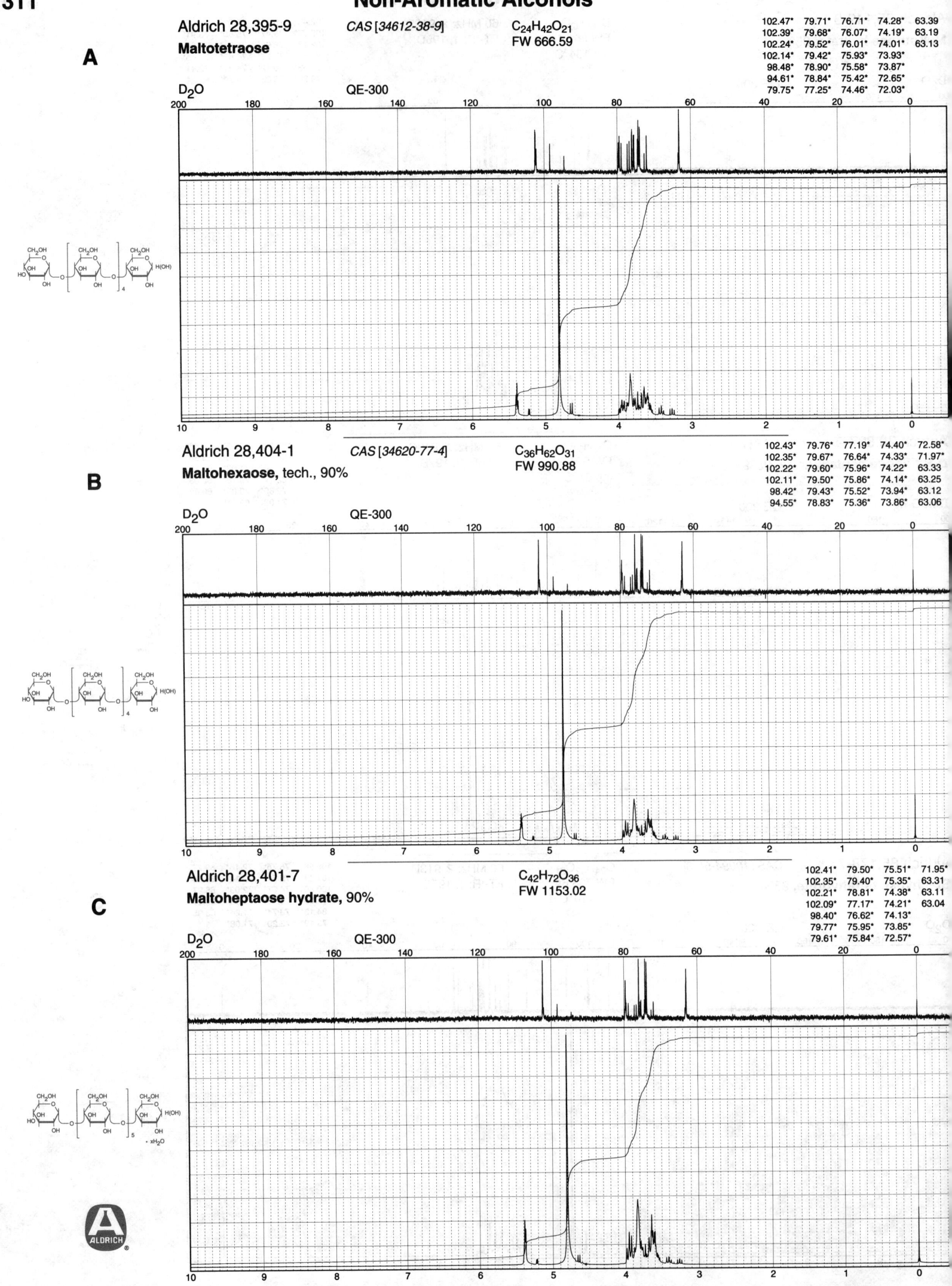

A

Aldrich 28,395-9

Maltotetraose

CAS [34612-38-9]

$C_{24}H_{42}O_{21}$
FW 666.59

102.47*	79.71*	76.71*	74.28*	63.39
102.39*	79.68*	76.07*	74.19*	63.19
102.24*	79.52*	76.01*	74.01*	63.13
102.14*	79.42*	75.93*	73.93*	
98.48*	78.90*	75.58*	73.87*	
94.61*	78.84*	75.42*	72.65*	
79.75*	77.25*	74.46*	72.03*	

D_2O QE-300

B

Aldrich 28,404-1

Maltohexaose, tech., 90%

CAS [34620-77-4]

$C_{36}H_{62}O_{31}$
FW 990.88

102.43*	79.76*	77.19*	74.40*	72.58*
102.35*	79.67*	76.64*	74.33*	71.97*
102.22*	79.60*	75.96*	74.22*	63.33
102.11*	79.50*	75.86*	74.14*	63.25
98.42*	79.43*	75.52*	73.94*	63.12
94.55*	78.83*	75.36*	73.86*	63.06

D_2O QE-300

C

Aldrich 28,401-7

Maltoheptaose hydrate, 90%

$C_{42}H_{72}O_{36}$
FW 1153.02

102.41*	79.50*	75.51*	71.95*
102.35*	79.40*	75.35*	63.31
102.21*	78.81*	74.38*	63.11
102.09*	77.17*	74.21*	63.04
98.40*	76.62*	74.13*	
79.77*	75.95*	73.85*	
79.61*	75.84*	72.57*	

D_2O QE-300

Non-Aromatic Alcohols

Aldrich 85,608-8   *CAS [68168-23-0]* $C_{42}H_{70}O_{35}$ 60 MHz: **2**, 914A

β-Cyclodextrin   FW 1135.01   FT-IR: **1**, 198A

mp 260°C d.

DMSO-d_6 QE-300

```
101.62*
81.23*
72.75*
72.10*
71.73*
59.62
```

A

Aldrich 33,261-5

Methyl-β-cyclodextrin   mp 181°C

CDCl$_3$ QE-300

B

Aldrich 27,953-6   *CAS [51166-71-3]* $C_{56}H_{98}O_{35}$

Heptakis(2,6-di-O-methyl)-β-cyclodextrin   FW 1331.40

```
101.35*  70.96
83.65*   70.40*
82.14*   60.31*
73.28*   58.98*
```

DCl$_3$ QE-300

C

Non-Aromatic Alcohols

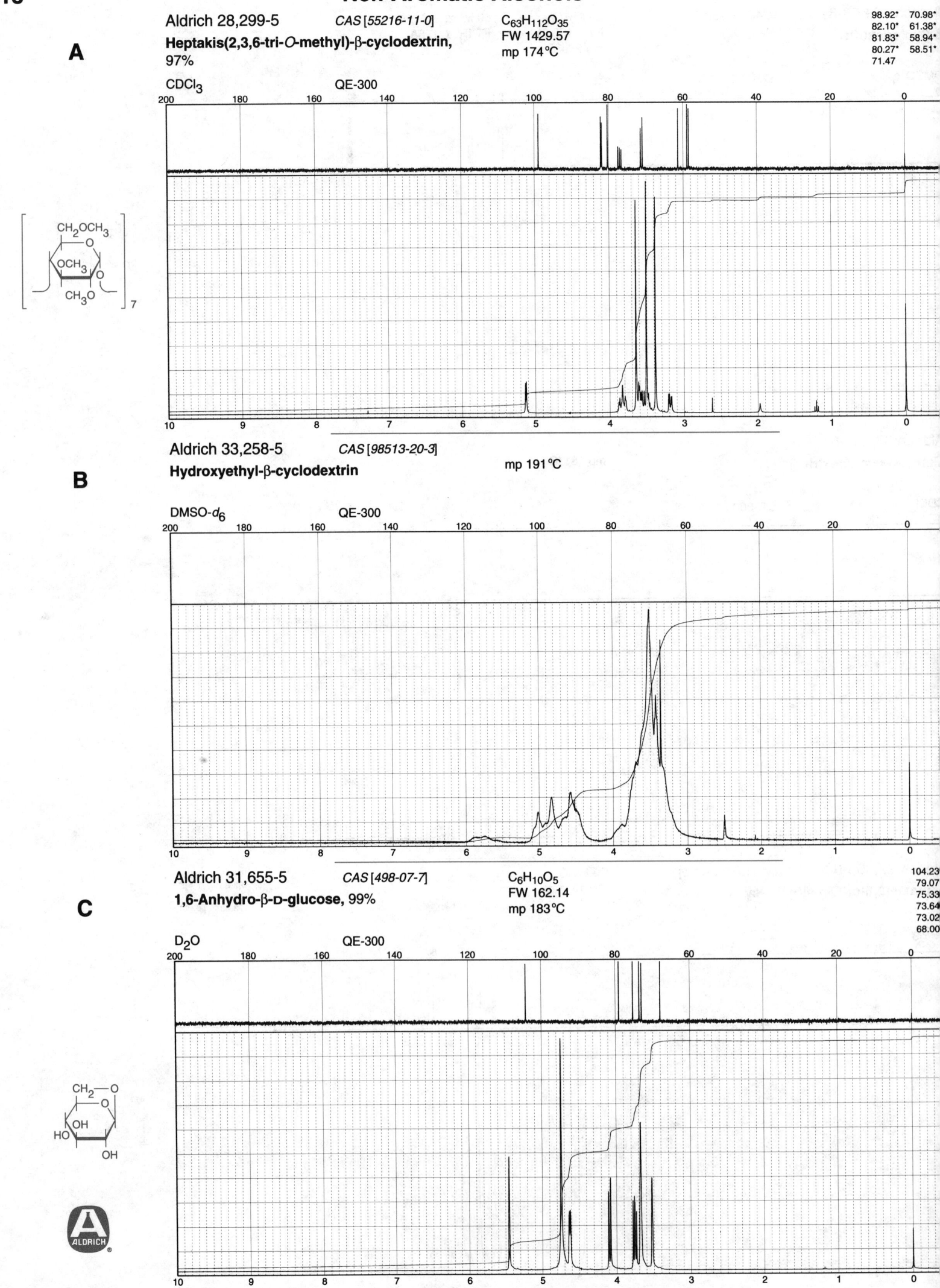

A

Aldrich 28,299-5   *CAS [55216-11-0]* $C_{63}H_{112}O_{35}$
FW 1429.57
Heptakis(2,3,6-tri-O-methyl)-β-cyclodextrin,
97%
mp 174°C

98.92* 70.98*
82.10* 61.38*
81.83* 58.94*
80.27* 58.51*
71.47

CDCl₃   QE-300

B

Aldrich 33,258-5   *CAS [98513-20-3]*
Hydroxyethyl-β-cyclodextrin   mp 191°C

DMSO-*d₆*   QE-300

C

Aldrich 31,655-5   *CAS [498-07-7]* $C_6H_{10}O_5$
FW 162.14
1,6-Anhydro-β-D-glucose, 99%
mp 183°C

104.23
79.07
75.33
73.64
73.02
68.00

D₂O   QE-300

Aldrich E3,275-4   *CAS [13224-99-2]*   C$_8$H$_{14}$O$_6$ 60 MHz: *2*, 914C

4,6-*O*-Ethylidene-α-D-glucose   FW 206.19   FT-IR: *1*, 200B
mp 169°C

102.71*	82.28*	70.59
102.65*	77.80*	70.22
99.37*	75.31*	68.73*
95.67*	74.87*	64.96*
82.79*	72.58*	22.03*

A

D$_2$O   QE-300

180 160 140 120 100 80 60 40 20 0

9 8 7 6 5 4 3 2 1 0

Aldrich D760-0   *CAS [582-52-5]*   C$_{12}$H$_{20}$O$_6$ 60 MHz: *2*, 914B

Diacetone-D-glucose, 98%   FW 260.29   FT-IR: *1*, 200A
mp 111°C

111.76	73.17*
109.55	67.58
105.18*	26.82*
85.07*	26.78*
81.16*	26.16*
74.89*	25.15*

B

CDCl$_3$ QE-300

180 160 140 120 100 80 60 40 20 0

9 8 7 6 5 4 3 2 1 0

Aldrich I-2,290-0   *CAS [18549-40-1]*   C$_9$H$_{16}$O$_6$ 60 MHz: *2*, 914D

1,2-*O*-Isopropylidene-D-glucofuranose,   FW 220.22   FT-IR: *1*, 200C
98%   mp 160°C

115.33	71.06*
107.38*	66.17
87.08*	28.26*
82.41*	27.83*
76.25*	

C

D$_2$O   QE-300

180 160 140 120 100 80 60 40 20 0

9 8 7 6 5 4 3 2 1 0

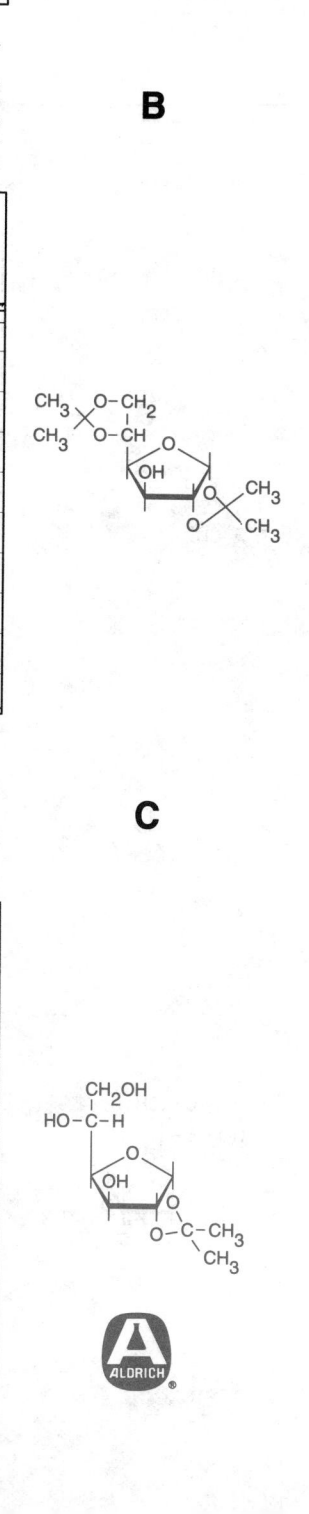

Non-Aromatic Alcohols

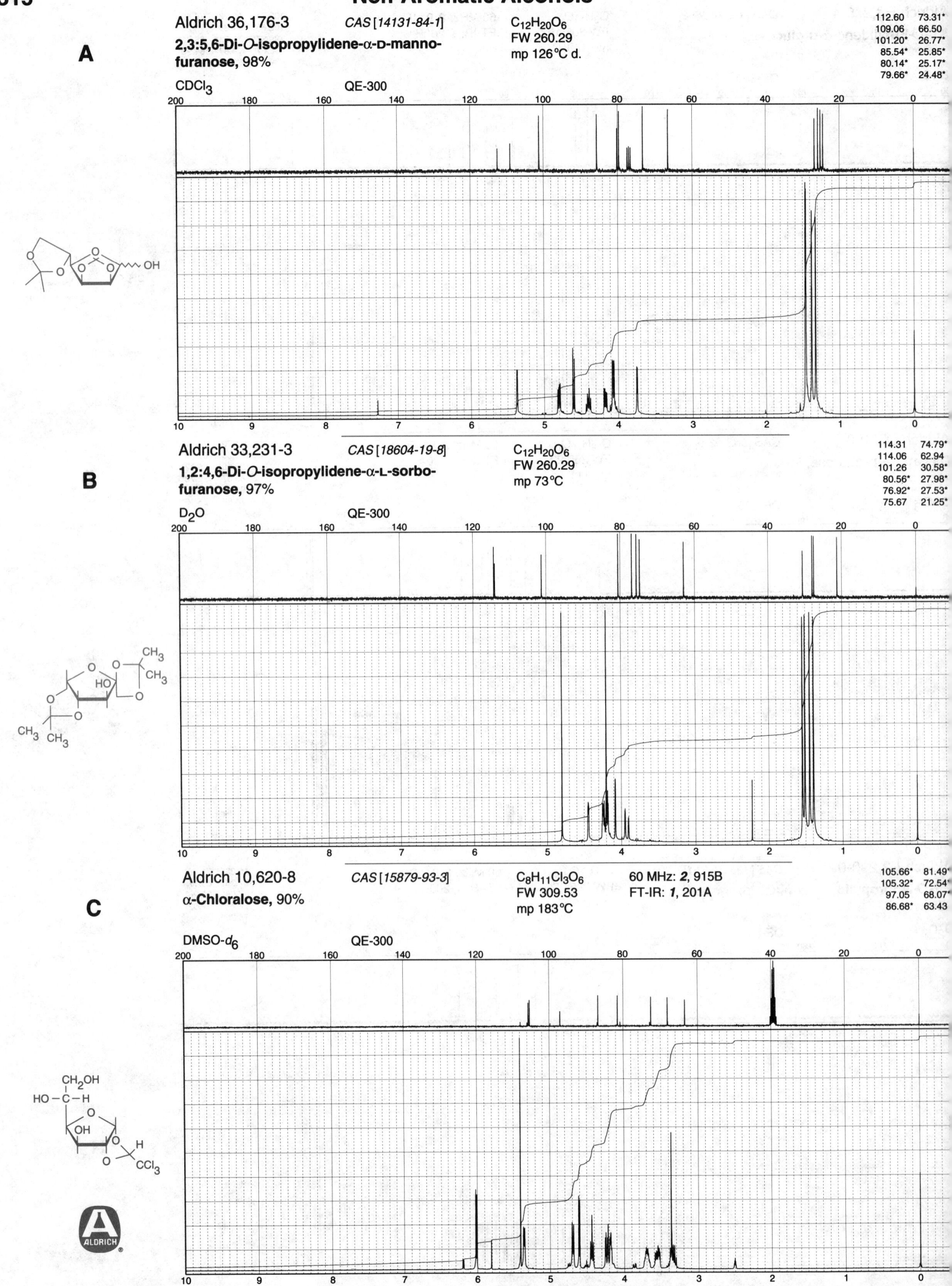

A

Aldrich 36,176-3   *CAS [14131-84-1]*   C₁₂H₂₀O₆
2,3:5,6-Di-O-isopropylidene-α-D-manno-
furanose, 98%

$C_{12}H_{20}O_6$
FW 260.29
mp 126°C d.

112.60	73.31*
109.06	66.50
101.20*	26.77*
85.54*	25.85*
80.14*	25.17*
79.66*	24.48*

CDCl₃   QE-300

B

Aldrich 33,231-3   *CAS [18604-19-8]*
1,2:4,6-Di-O-isopropylidene-α-L-sorbo-
furanose, 97%

$C_{12}H_{20}O_6$
FW 260.29
mp 73°C

114.31	74.79*
114.06	62.94
101.26	30.58*
80.56*	27.98*
76.92*	27.53*
75.67	21.25*

D₂O   QE-300

C

Aldrich 10,620-8   *CAS [15879-93-3]*   C₈H₁₁Cl₃O₆   60 MHz: **2**, 915B
α-Chloralose, 90%

$C_8H_{11}Cl_3O_6$
FW 309.53
mp 183°C

FT-IR: **1**, 201A

105.66*	81.49*
105.32*	72.54*
97.05	68.07*
86.68*	63.43

DMSO-d₆   QE-300

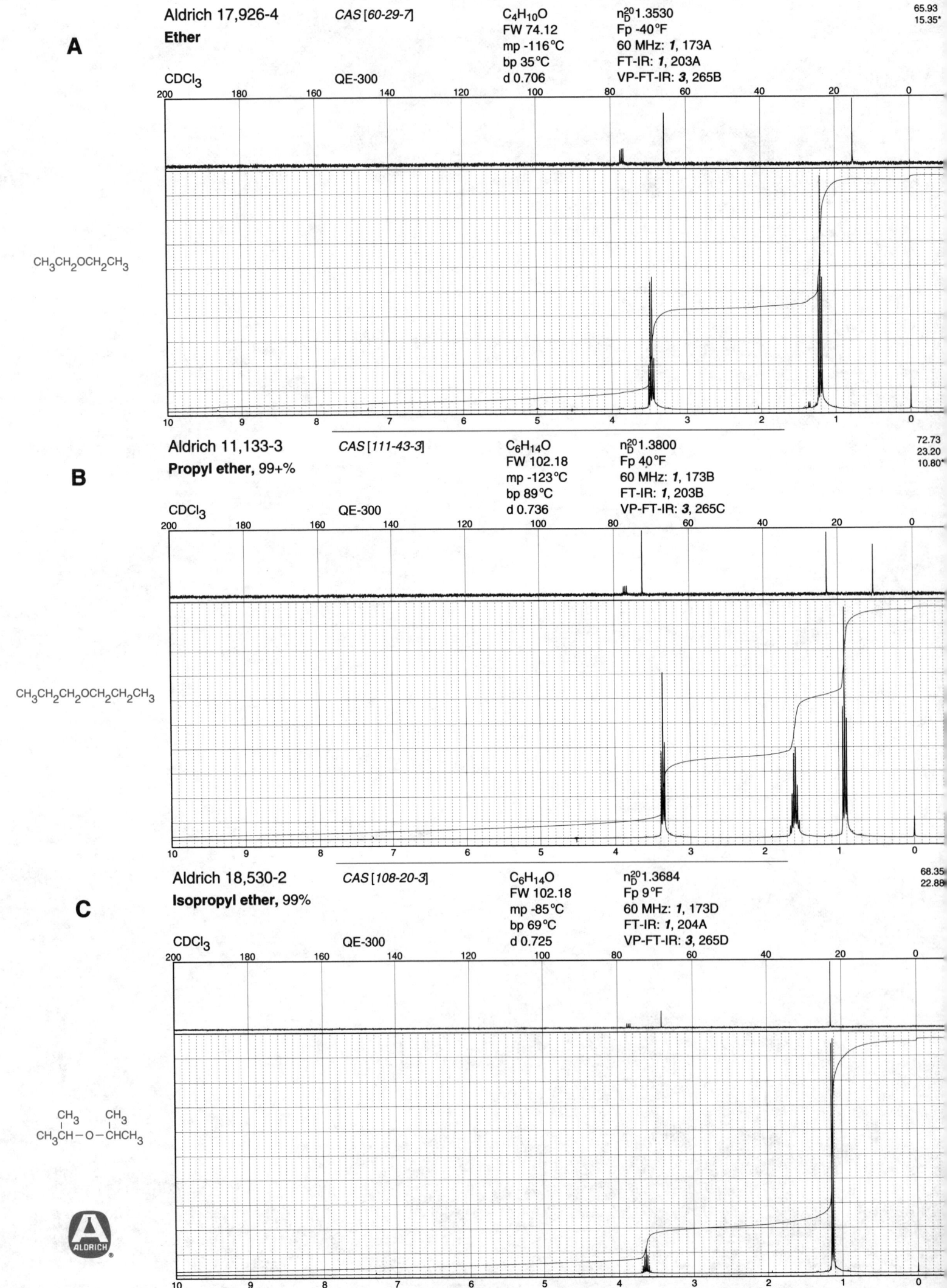

A

Aldrich 17,926-4   *CAS [60-29-7]*   C₄H₁₀O   n²⁰D 1.3530   65.93
Ether   FW 74.12   Fp -40°F   15.35*
mp -116°C   60 MHz: *1*, 173A
bp 35°C   FT-IR: *1*, 203A
CDCl₃   QE-300   d 0.706   VP-FT-IR: *3*, 265B

CH₃CH₂OCH₂CH₃

B

Aldrich 11,133-3   *CAS [111-43-3]*   C₆H₁₄O   n²⁰D 1.3800   72.73
Propyl ether, 99+%   FW 102.18   Fp 40°F   23.20
mp -123°C   60 MHz: *1*, 173B   10.80*
bp 89°C   FT-IR: *1*, 203B
CDCl₃   QE-300   d 0.736   VP-FT-IR: *3*, 265C

CH₃CH₂CH₂OCH₂CH₂CH₃

C

Aldrich 18,530-2   *CAS [108-20-3]*   C₆H₁₄O   n²⁰D 1.3684   68.35
Isopropyl ether, 99%   FW 102.18   Fp 9°F   22.88
mp -85°C   60 MHz: *1*, 173D
bp 69°C   FT-IR: *1*, 204A
CDCl₃   QE-300   d 0.725   VP-FT-IR: *3*, 265D

CH₃CH—O—CHCH₃
   | |
  CH₃   CH₃

Non-Aromatic Ethers, Acetals, and Epoxides

Aldrich 23,537-7
Butyl methyl ether, 99%

CAS [628-28-4]

C₅H₁₂O
FW 88.15
mp -115°C
bp 71°C
d 0.744

$C_5H_{12}O$
FW 88.15
mp -115°C
bp 71°C
d 0.744

n_D^{20}1.3733
Fp 14°F
60 MHz: **1**, 173C
FT-IR: **1**, 203C
VP-FT-IR: **3**, 266A

72.67
58.50*
31.79
19.35
13.92*

A

CDCl₃

QE-300

$CH_3CH_2CH_2CH_2OCH_3$

Aldrich 26,892-5
Butyl ethyl ether, 99%

CAS [628-81-9]

$C_6H_{14}O$
FW 102.18
mp -124°C
bp 92°C
d 0.750

n_D^{20}1.3818
Fp 22°F
FT-IR: **1**, 203D
VP-FT-IR: **3**, 266B

70.54
66.12
32.02
19.49
15.32*
14.00*

B

CDCl₃

QE-300

$CH_3CH_2CH_2CH_2OCH_2CH_3$

Aldrich 24,040-0
Butyl ether, 99+%

CAS [142-96-1]

$C_8H_{18}O$
FW 130.23
mp -98°C
bp 143°C
d 0.764

n_D^{20}1.3985
Fp 77°F
60 MHz: **1**, 174B
FT-IR: **1**, 204C
VP-FT-IR: **3**, 266C

70.66
31.96
19.44
13.94*

C

CDCl₃

QE-300

$CH_3CH_2CH_2CH_2OCH_2CH_2CH_2CH_3$

ALDRICH

A

Aldrich 27,599-9   CAS [628-55-7]   C$_8$H$_{18}$O   Fp 48°F   VP-FT-IR: **3**, 267C   77.94
28.47*
19.41*

Isobutyl ether, 99%   FW 130.23
bp 123°C
d 0.760

CDCl$_3$ QE-300   n$_D^{20}$1.3888

CH$_3$ CH$_3$
CH$_3$CHCH$_2$OCH$_2$CHCH$_3$

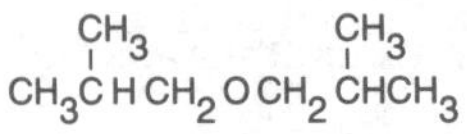

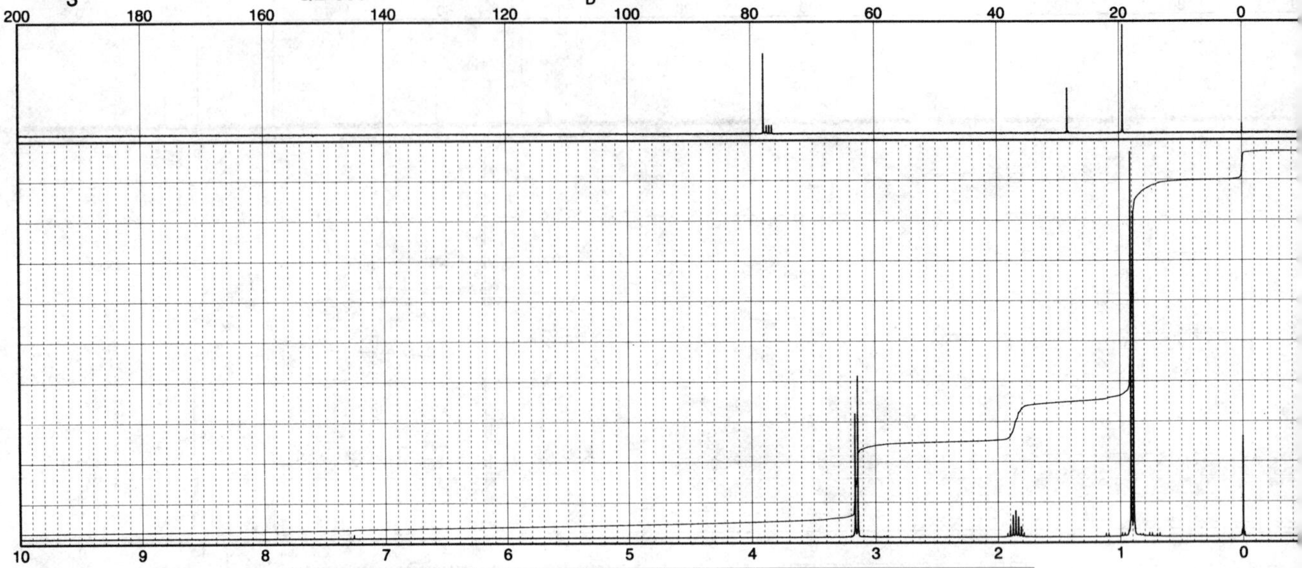

B

Aldrich 17,978-7   CAS [1634-04-4]   C$_5$H$_{12}$O   Fp 14°F   60 MHz: **1**, 174A   72.74
49.43*
26.98*

tert-Butyl methyl ether, 98%   FW 88.15
bp 55°C
d 0.740   VP-FT-IR: **3**, 266D

CDCl$_3$ QE-300   n$_D^{20}$1.3685

      CH$_3$
CH$_3$–C–OCH$_3$
      CH$_3$

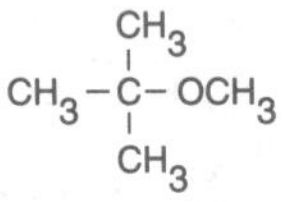

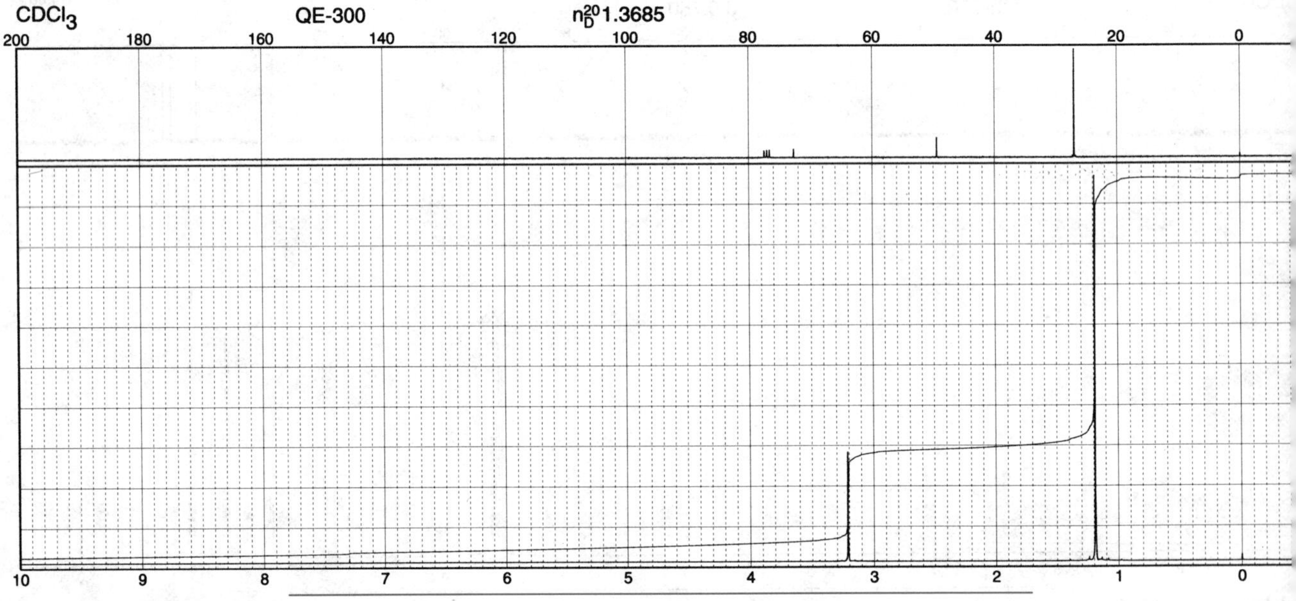

C

Aldrich 25,389-8   CAS [637-92-3]   C$_6$H$_{14}$O   n$_D^{20}$1.3756   72.51
56.75*
27.63
16.33

tert-Butyl ethyl ether, 99%   FW 102.18
mp -97°C
bp 73°C
d 0.742   Fp -3°F
FT-IR: **1**, 204B
VP-FT-IR: **3**, 267A

CDCl$_3$ QE-300

      CH$_3$
CH$_3$–C–OCH$_2$CH$_3$
      CH$_3$

ALDRICH

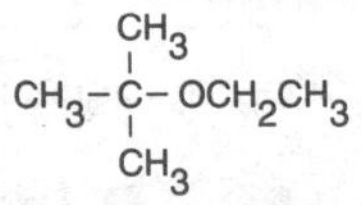

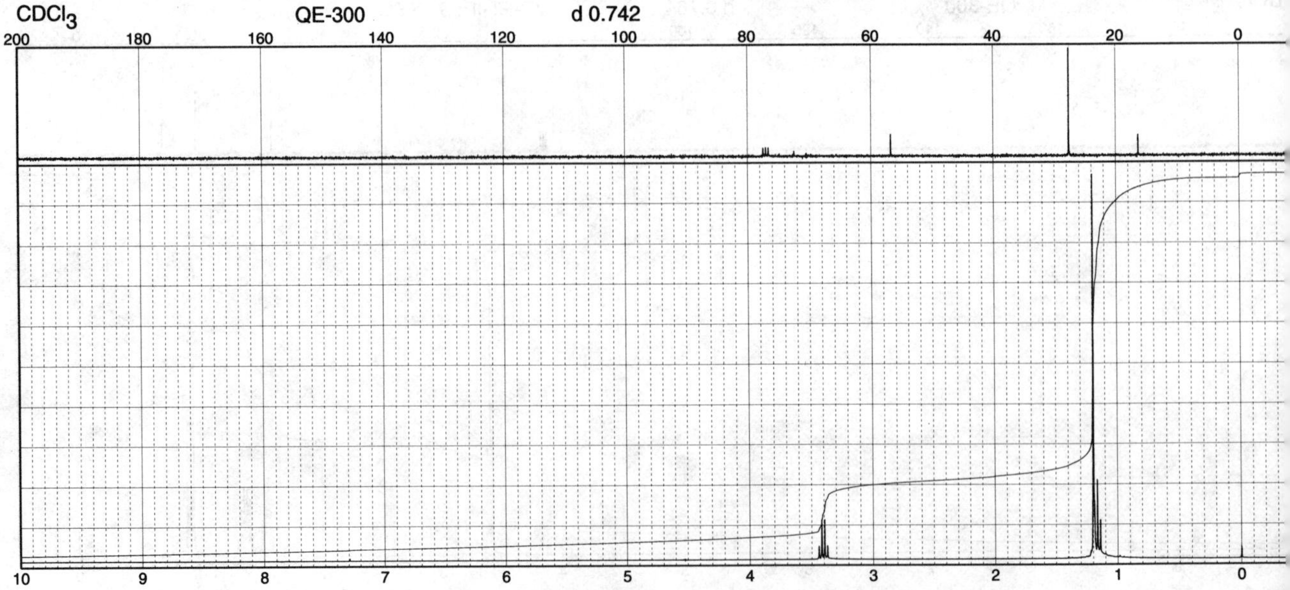

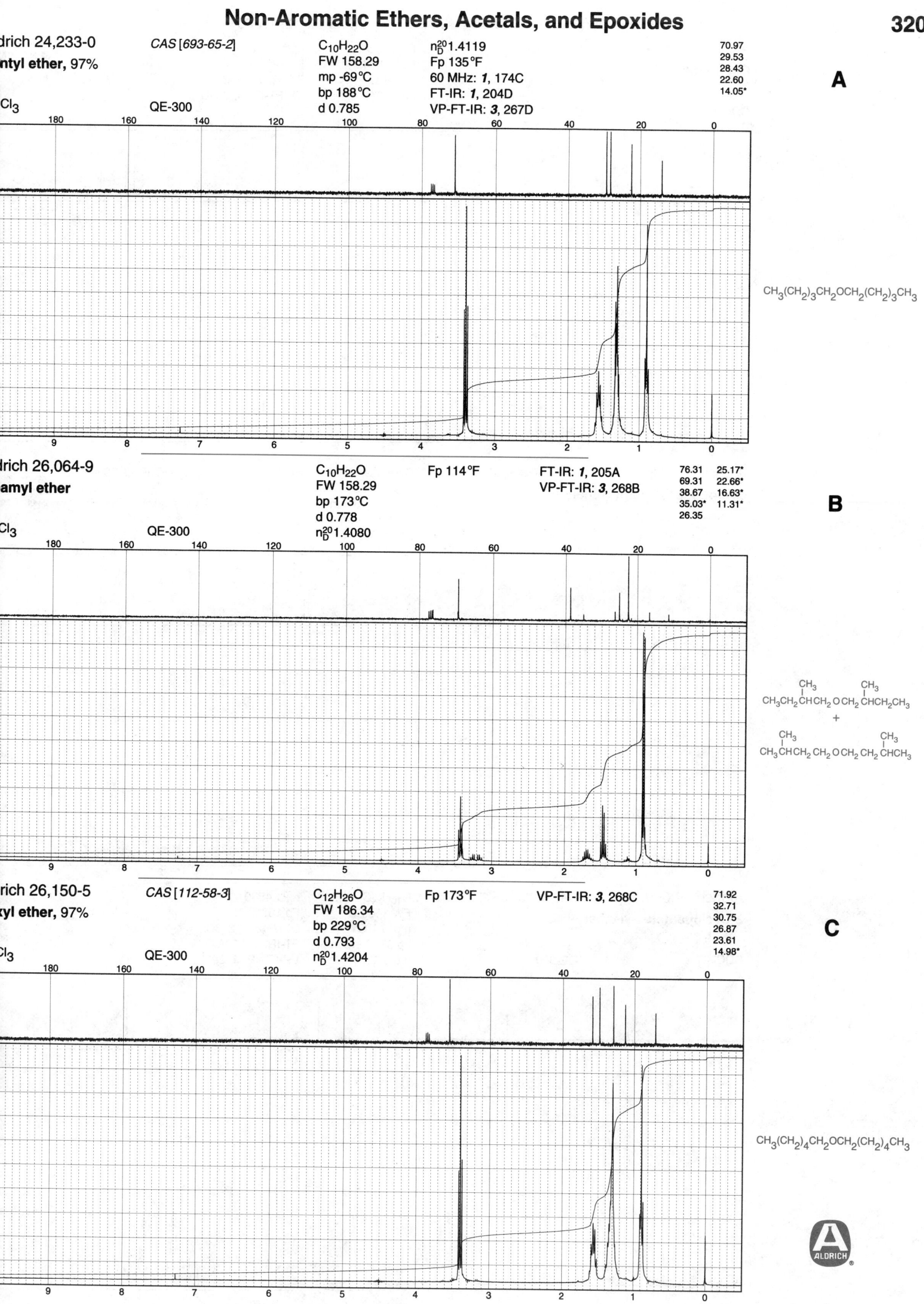

Aldrich 24,233-0 CAS [693-65-2] C₁₀H₂₂O n²⁰_D 1.4119 **A**
...ntyl ether, 97% FW 158.29 Fp 135°F
...Cl₃ QE-300 mp -69°C 60 MHz: 1, 174C
bp 188°C FT-IR: 1, 204D
d 0.785 VP-FT-IR: 3, 267D

70.97
29.53
28.43
22.60
14.05*

$CH_3(CH_2)_3CH_2OCH_2(CH_2)_3CH_3$

Aldrich 26,064-9 C₁₀H₂₂O Fp 114°F FT-IR: 1, 205A **B**
...amyl ether FW 158.29 VP-FT-IR: 3, 268B
...Cl₃ QE-300 bp 173°C
d 0.778
n²⁰_D 1.4080

76.31 25.17*
69.31 22.66*
38.67 16.63*
35.03* 11.31*
26.35

Aldrich 26,150-5 CAS [112-58-3] C₁₂H₂₆O Fp 173°F VP-FT-IR: 3, 268C **C**
...xyl ether, 97% FW 186.34
...Cl₃ QE-300 bp 229°C
d 0.793
n²⁰_D 1.4204

71.92
32.71
30.75
26.87
23.61
14.98*

$CH_3(CH_2)_4CH_2OCH_2(CH_2)_4CH_3$

A

Aldrich 24,959-9   CAS [629-82-3]   C₁₆H₃₄O   n₂₀D 1.4318   71.75   30.
$C_{16}H_{34}O$ n_D^{20} 1.4318
Octyl ether, 96%   FW 242.45   Fp >230°F   32.65   27.
mp -8°C   FT-IR: **1**, 205B   30.61   23.
bp 287°C   VP-FT-IR: **3**, 268D   30.29   14.

CDCl₃   QE-300   d 0.806

$CH_3(CH_2)_6CH_2OCH_2(CH_2)_6CH_3$

B

Aldrich 32,513-9   CAS [2986-54-1] $C_{13}H_{26}O$ Fp 225°F   VP-FT-IR: **3**, 299C   78.76*   24.
Cyclododecyl methyl ether, 98%   FW 198.35   56.13*   23.
bp 89°C (3 mm)   28.34   23.
d 0.908   24.86   20.

CDCl₃   QE-300 n_D^{20} 1.4730

—OCH₃

C

Aldrich 14,869-5   CAS [15131-55-2] $C_{10}H_{14}O$ n_D^{20} 1.4890   134
2-Cyclopenten-1-yl ether, 98%   FW 150.22   Fp 167°F   131
mp -74°C   60 MHz: **1**, 213C   83
bp 87°C (12 mm)   FT-IR: **1**, 252A   31
d 0.972   VP-FT-IR: **3**, 299D   30

CDCl₃   QE-300

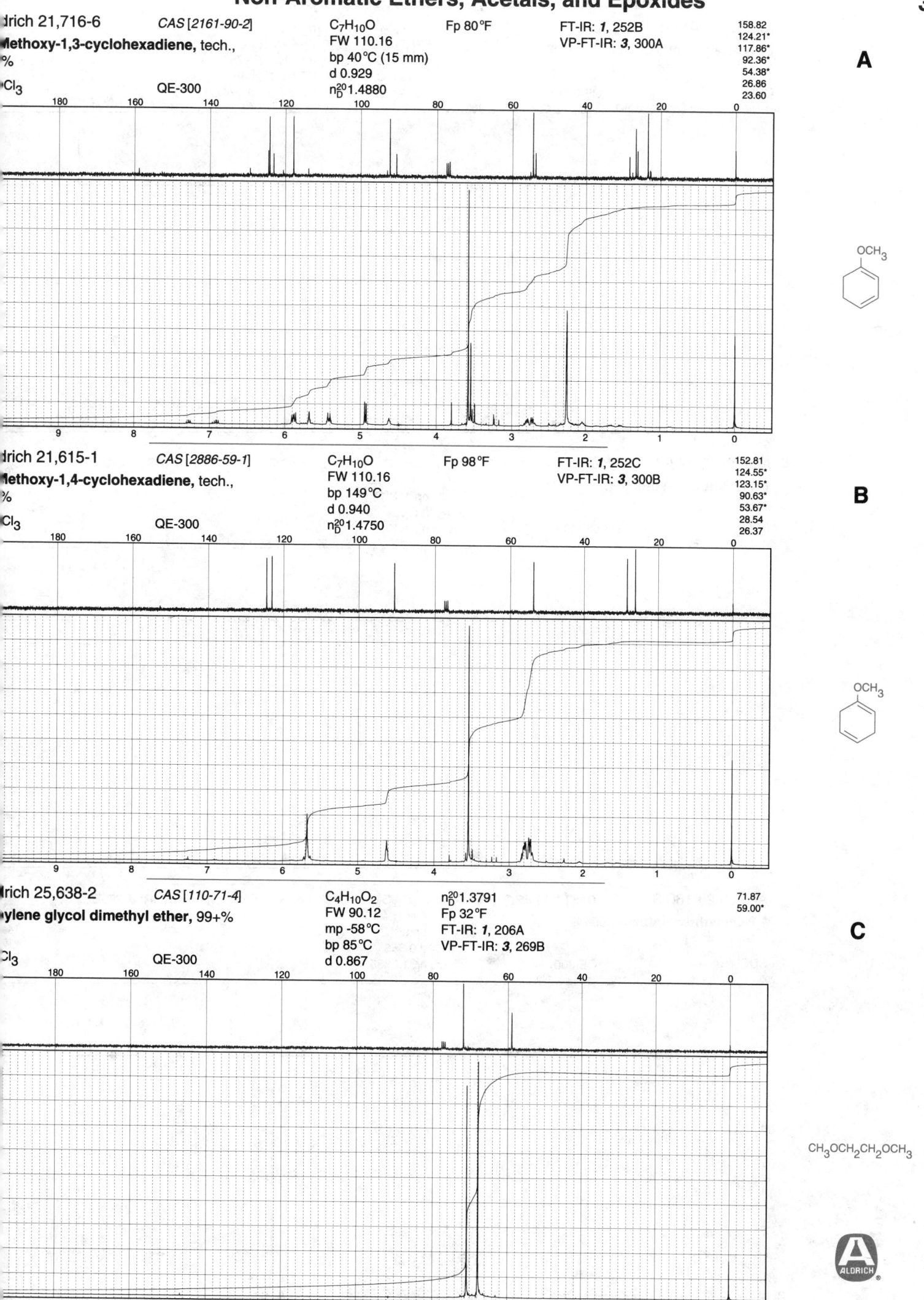

drich 21,716-6 CAS [2161-90-2] C₇H₁₀O Fp 80°F FT-IR: 1, 252B VP-FT-IR: 3, 300A

Methoxy-1,3-cyclohexadiene, tech.,

FW 110.16
bp 40°C (15 mm)
d 0.929
n₂₀ 1.4880

A

drich 21,615-1 CAS [2886-59-1] C₇H₁₀O Fp 98°F FT-IR: 1, 252C VP-FT-IR: 3, 300B

Methoxy-1,4-cyclohexadiene, tech.,

FW 110.16
bp 149°C
d 0.940
n₂₀ 1.4750

B

drich 25,638-2 CAS [110-71-4] C₄H₁₀O₂ n₂₀ 1.3791

ylene glycol dimethyl ether, 99+%

FW 90.12
mp -58°C
bp 85°C
d 0.867
Fp 32°F
FT-IR: 1, 206A
VP-FT-IR: 3, 269B

C

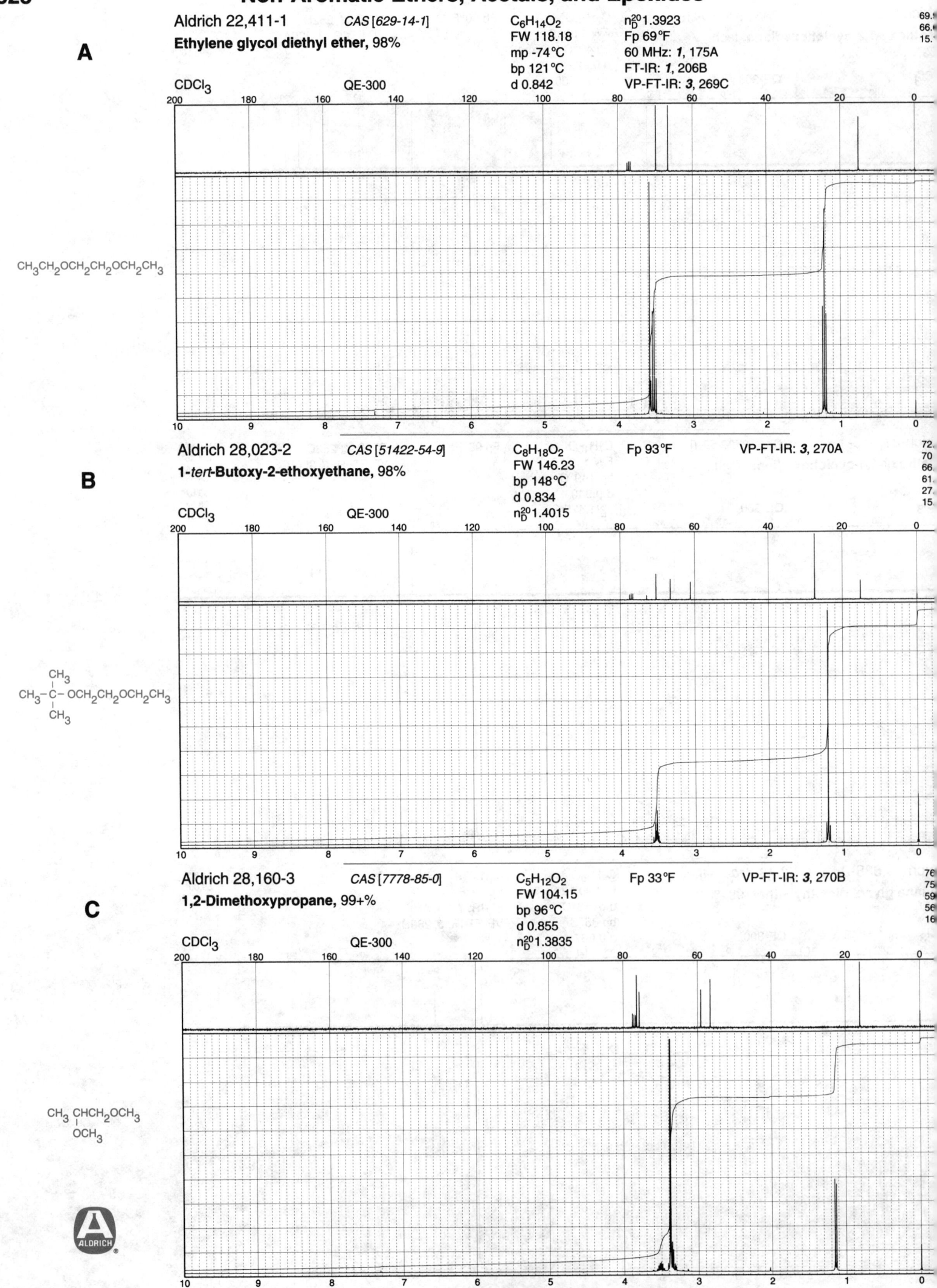

A

Aldrich 22,411-1   *CAS [629-14-1]*   C_6H_14O_2   n_D^20 1.3923
Ethylene glycol diethyl ether, 98%   FW 118.18   Fp 69°F
mp -74°C   60 MHz: *1*, 175A
bp 121°C   FT-IR: *1*, 206B
CDCl_3   QE-300   d 0.842   VP-FT-IR: *3*, 269C

$CH_3CH_2OCH_2CH_2OCH_2CH_3$

B

Aldrich 28,023-2   *CAS [51422-54-9]*   C_8H_18O_2   Fp 93°F   VP-FT-IR: *3*, 270A
1-*tert*-Butoxy-2-ethoxyethane, 98%   FW 146.23
bp 148°C
d 0.834
CDCl_3   QE-300   n_D^20 1.4015

$CH_3\!-\!\underset{\underset{CH_3}{|}}{\overset{\overset{CH_3}{|}}{C}}\!-\!OCH_2CH_2OCH_2CH_3$

C

Aldrich 28,160-3   *CAS [7778-85-0]*   C_5H_12O_2   Fp 33°F   VP-FT-IR: *3*, 270B
1,2-Dimethoxypropane, 99+%   FW 104.15
bp 96°C
d 0.855
CDCl_3   QE-300   n_D^20 1.3835

$CH_3\ \underset{\underset{OCH_3}{|}}{CHCH_2OCH_3}$

Non-Aromatic Ethers, Acetals, and Epoxides

Aldrich 25,639-0   CAS [111-96-6] $C_6H_{14}O_3$ n_D^{20}1.4080

Methoxyethyl ether, 99+%

FW 134.18   Fp 158°F

mp -64°C   60 MHz: *1*, 175B

bp 162°C   FT-IR: *1*, 206C

$CDCl_3$ QE-300   d 0.937   VP-FT-IR: *3*, 270D

71.96
70.57
58.97*

A

$CH_3OCH_2CH_2OCH_2CH_2OCH_3$

Aldrich E465-8   CAS [112-36-7] $C_8H_{18}O_3$ Fp 160°F

Ethoxyethyl ether, 98+%

FW 162.23   60 MHz: *1*, 175C

bp 185°C   FT-IR: *1*, 206D

d 0.909   VP-FT-IR: *3*, 271A

$CDCl_3$ QE-300 n_D^{20}1.4120

70.68
69.83
66.60
15.16*

B

$CH_3CH_2OCH_2CH_2OCH_2CH_2OCH_2CH_3$

Aldrich 20,562-1   CAS [112-73-2] $C_{12}H_{26}O_3$ n_D^{20}1.4240

Diethylene glycol dibutyl ether, 99+%

FW 218.34   Fp 118°F

mp -60°C   60 MHz: *1*, 175D

bp 256°C   FT-IR: *1*, 207A

$CDCl_3$ QE-300   d 0.885   VP-FT-IR: *3*, 271D

71.19
70.70
70.14
31.79
19.31
13.90*

C

$(CH_3CH_2CH_2CH_2OCH_2CH_2)_2O$

A

Aldrich 28,044-5   CAS [52788-79-1] $C_9H_{20}O_3$ VP-FT-IR: 3, 271B

72.9
71.9
71.1
70.5
61.1
58.9
27.4

Diethylene glycol tert-butyl methyl ether, 97%

FW 176.26
d 0.909
$n_D^{20}1.4160$
Fp 163°F

CDCl$_3$ QE-300

$CH_3OCH_2CH_2OCH_2CH_2O-\overset{\underset{\displaystyle CH_3}{\displaystyle CH_3}}{\underset{}{C}}-CH_3$

B

Aldrich T5,980-3   CAS [112-49-2] $C_8H_{18}O_4$ $n_D^{20}1.4230$

71.9
70.6
70.5
58.9

Triethylene glycol dimethyl ether, 99%

FW 178.23
mp -45°C
bp 216°C
d 0.986

Fp 231°F
60 MHz: 1, 176A
FT-IR: 1, 207B
VP-FT-IR: 3, 272A

CDCl$_3$ QE-300

$CH_3(OCH_2CH_2)_3OCH_3$

C

Aldrich 24,116-4   CAS [143-24-8] $C_{10}H_{22}O_5$ $n_D^{20}1.4330$

72.
71.
71.
59.

Tetraethylene glycol dimethyl ether, 99+%

FW 222.28
mp -30°C
bp 276°C
d 1.009

Fp 285°F
60 MHz: 1, 176C
FT-IR: 1, 207C
VP-FT-IR: 3, 272B

CDCl$_3$ QE-300

$CH_3(OCH_2CH_2)_4OCH_3$

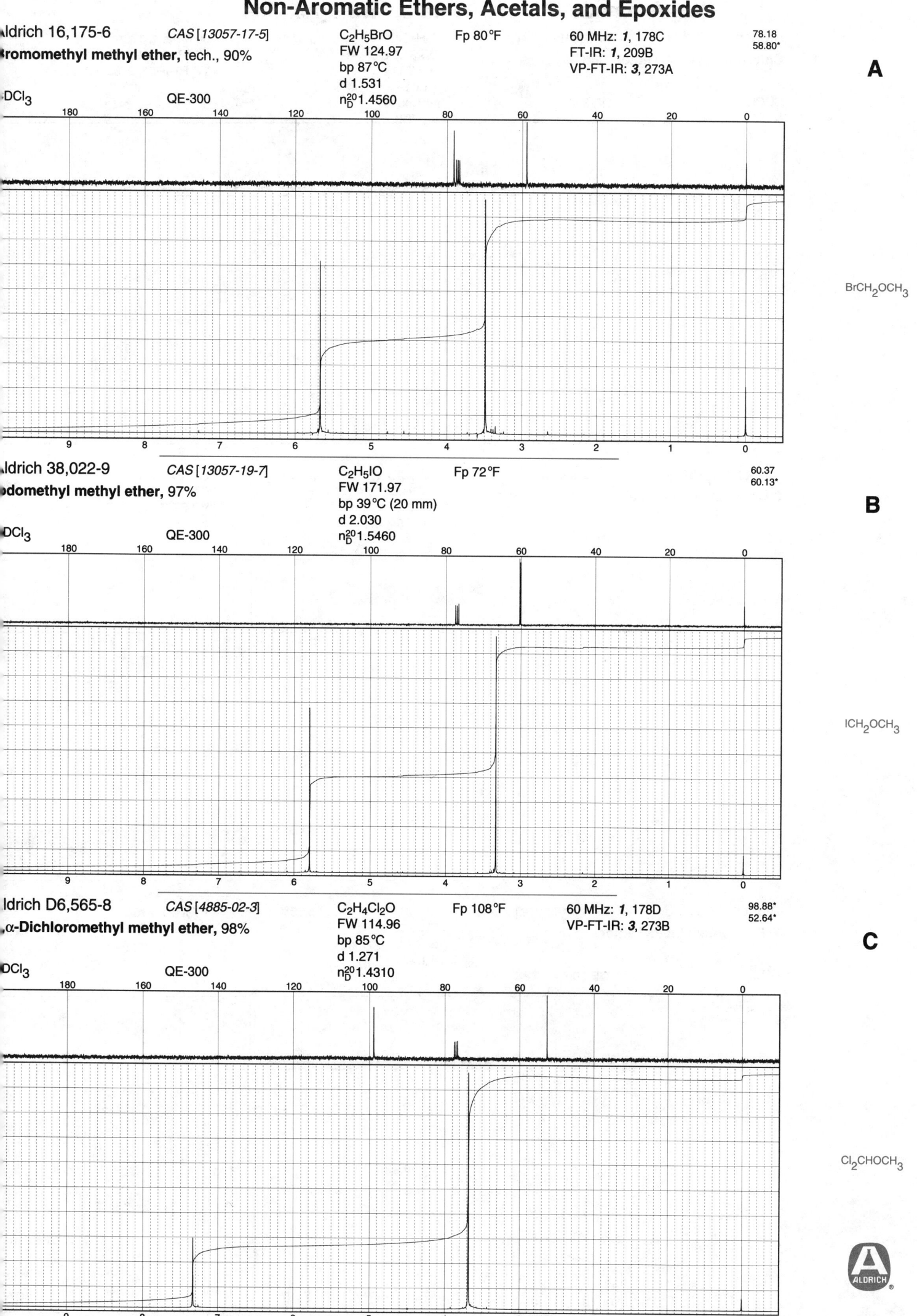

Aldrich 16,175-6   CAS [13057-17-5]   C₂H₅BrO   Fp 80 °F   60 MHz: *1*, 178C   78.18
Bromomethyl methyl ether, tech., 90%   FW 124.97   FT-IR: *1*, 209B   58.80*
bp 87 °C   VP-FT-IR: *3*, 273A
d 1.531
n_D^{20} 1.4560

CDCl₃   QE-300

A

BrCH₂OCH₃

Aldrich 38,022-9   CAS [13057-19-7]   C₂H₅IO   Fp 72 °F   60.37
Iodomethyl methyl ether, 97%   FW 171.97   60.13*
bp 39 °C (20 mm)
d 2.030
n_D^{20} 1.5460

CDCl₃   QE-300

B

ICH₂OCH₃

Aldrich D6,565-8   CAS [4885-02-3]   C₂H₄Cl₂O   Fp 108 °F   60 MHz: *1*, 178D   98.88*
α-Dichloromethyl methyl ether, 98%   FW 114.96   VP-FT-IR: *3*, 273B   52.64*
bp 85 °C
d 1.271
n_D^{20} 1.4310

CDCl₃   QE-300

C

Cl₂CHOCH₃

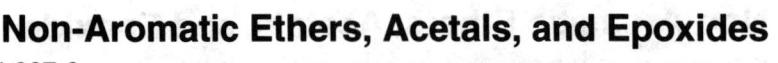

Non-Aromatic Ethers, Acetals, and Epoxides

A

Aldrich 14,267-0   *CAS [3188-13-4]* C_3H_7ClO Fp 67°F   60 MHz: *1*, 179A   82.97

Chloromethyl ethyl ether, 96%   FW 94.54   FT-IR: *1*, 209C   66.07

bp 82°C   VP-FT-IR: *3*, 273C   14.42

d 1.019

n_D^{20} 1.4040

CDCl$_3$ QE-300

ClCH$_2$OCH$_2$CH$_3$

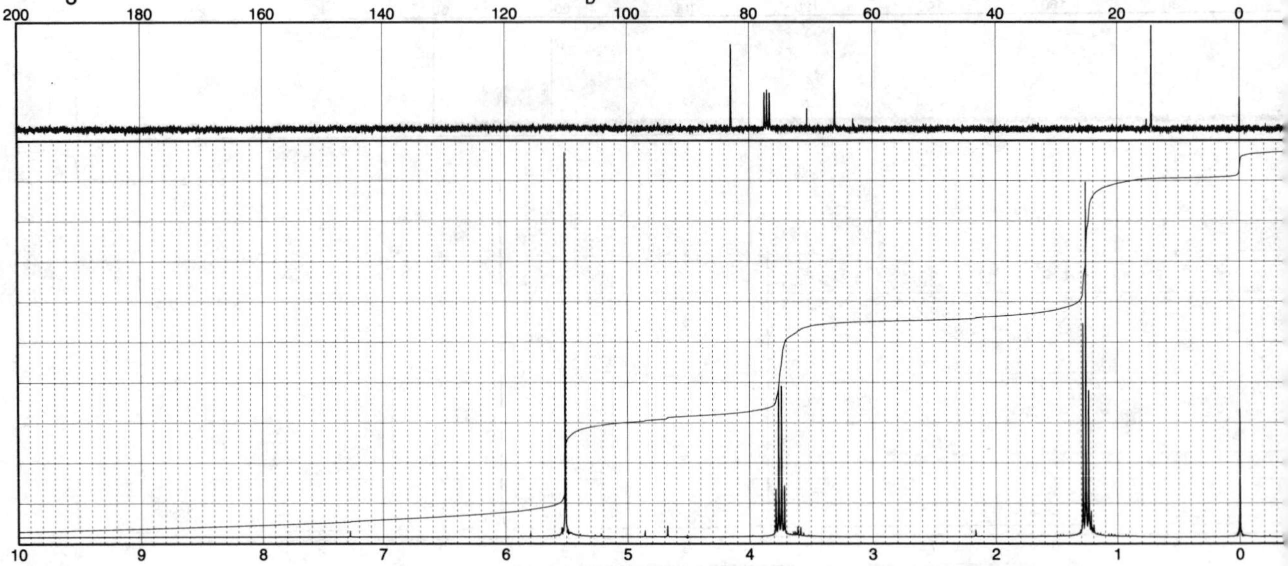

B

Aldrich 24,234-9   *CAS [627-42-9]* C_3H_7ClO Fp 59°F   60 MHz: *1*, 179B   72.58

2-Chloroethyl methyl ether, 98%   FW 94.54   FT-IR: *1*, 209D   58.8#

bp 90°C   VP-FT-IR: *3*, 274B   42.7.

d 1.035

n_D^{20} 1.4090

CDCl$_3$ QE-300

ClCH$_2$CH$_2$OCH$_3$

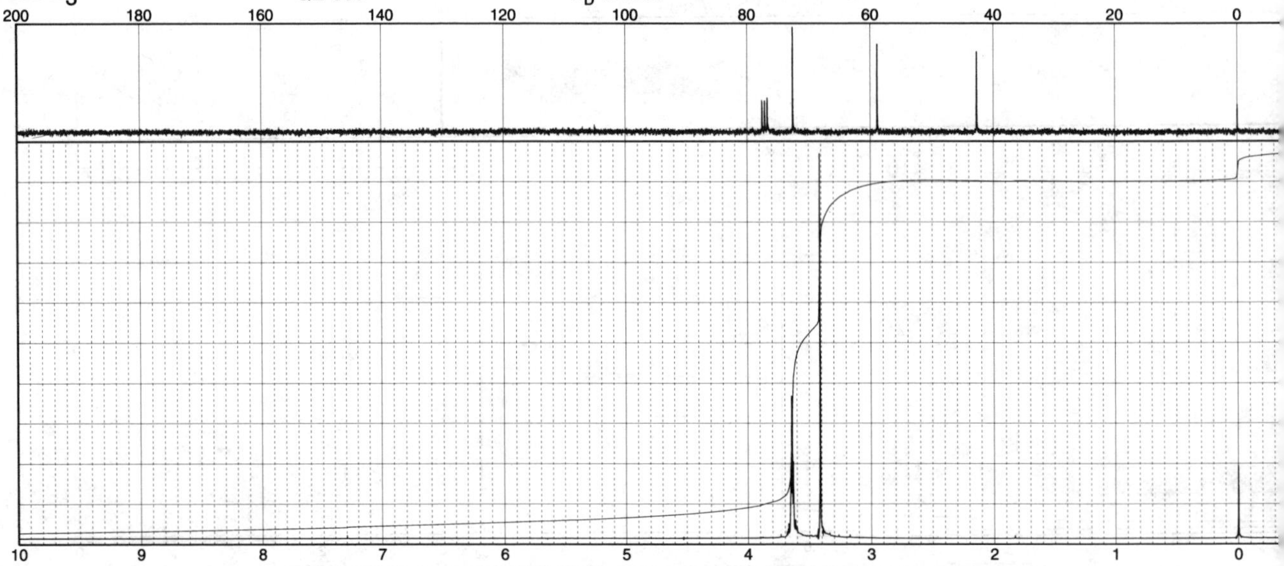

C

Aldrich 23,815-5   *CAS [6482-24-2]* C_3H_7BrO Fp 83°F   FT-IR: *1*, 210A   72.3

2-Bromoethyl methyl ether   FW 139.00   VP-FT-IR: *3*, 274C   58.7

bp 41°C (66 mm)   30.2

d 1.479

n_D^{20} 1.4470

CDCl$_3$ QE-300

BrCH$_2$CH$_2$OCH$_3$

ALDRICH

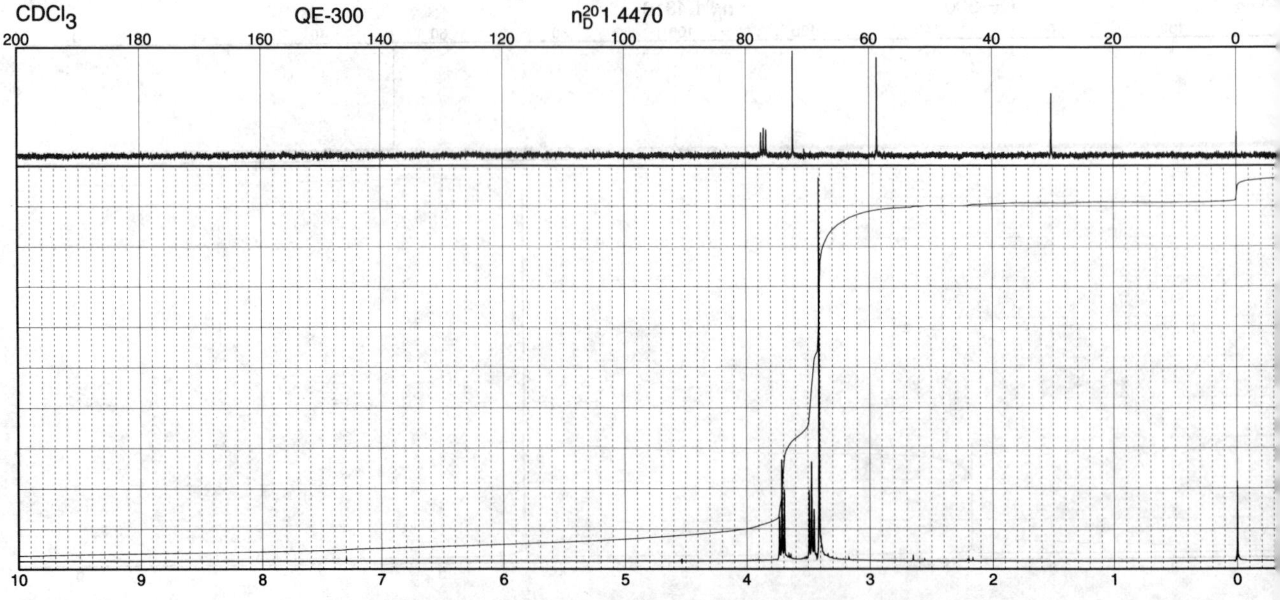

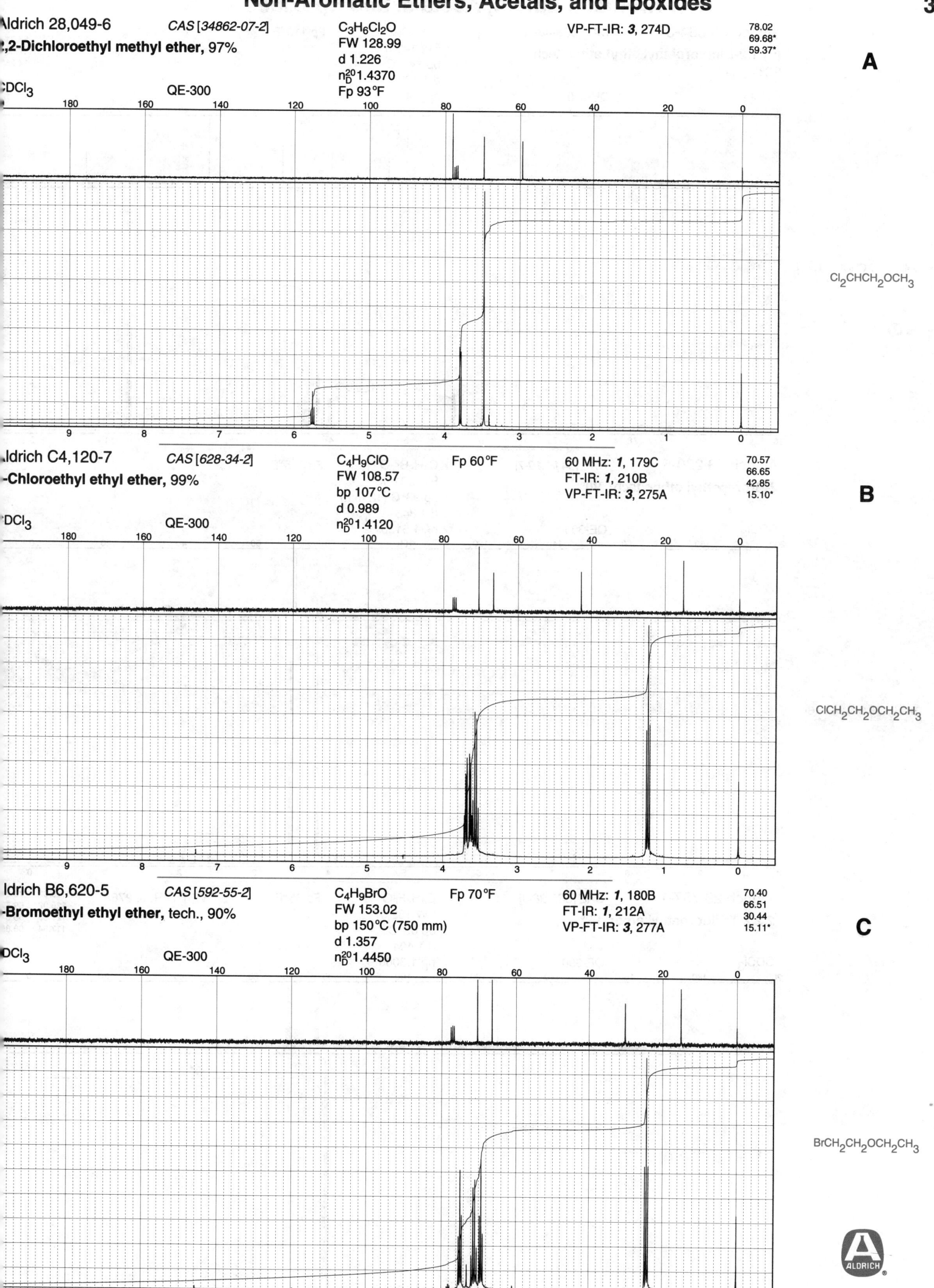

Aldrich 28,049-6   CAS [34862-07-2]   C₃H₆Cl₂O   VP-FT-IR: 3, 274D   78.02
2,2-Dichloroethyl methyl ether, 97%   FW 128.99   69.68*
    d 1.226   59.37*
CDCl₃   QE-300   n²⁰_D 1.4370   **A**
    Fp 93°F

Cl₂CHCH₂OCH₃

Aldrich C4,120-7   CAS [628-34-2]   C₄H₉ClO   Fp 60°F   60 MHz: 1, 179C   70.57
2-Chloroethyl ethyl ether, 99%   FW 108.57   FT-IR: 1, 210B   66.65
    bp 107°C   VP-FT-IR: 3, 275A   42.85
CDCl₃   QE-300   d 0.989   15.10*
    n²⁰_D 1.4120   **B**

ClCH₂CH₂OCH₂CH₃

Aldrich B6,620-5   CAS [592-55-2]   C₄H₉BrO   Fp 70°F   60 MHz: 1, 180B   70.40
2-Bromoethyl ethyl ether, tech., 90%   FW 153.02   FT-IR: 1, 212A   66.51
    bp 150°C (750 mm)   VP-FT-IR: 3, 277A   30.44
CDCl₃   QE-300   d 1.357   15.11*
    n²⁰_D 1.4450   **C**

BrCH₂CH₂OCH₂CH₃

ALDRICH

Non-Aromatic Ethers, Acetals, and Epoxides

A

Aldrich 35,884-3   *CAS [623-46-1]* $C_4H_8Cl_2O$ Fp 110°F

(±)-1,2-Dichloroethyl ethyl ether, tech., 90%

FW 143.01
bp 143°C
d 1.167
n_D^{20} 1.4430

95.14
66.71
47.06
14.26

CDCl$_3$ QE-300

$CH_3CH_2O\,CHCH_2Cl$
             |
             Cl

B

Aldrich 38,220-5   *CAS [5414-19-7]* $C_4H_8Br_2O$ Fp 185°F

2-Bromoethyl ether, tech., 90%

FW 231.93
bp 93°C (12 mm)
d 1.845
n_D^{20} 1.5130

70.94
30.07

CDCl$_3$ QE-300

$BrCH_2CH_2OCH_2CH_2Br$

C

Aldrich 28,757-1   *CAS [333-36-8]* $C_4H_4F_6O$ Fp 35°F   VP-FT-IR: *3*, 276A

2,2,2-Trifluoroethyl ether, 99%

FW 182.07
bp 63°C
d 1.404
n_D^{20} 1.3000

129.02   70.2
125.32   69.7
121.63   69.3
117.94   68.8

CDCl$_3$ QE-300

$CF_3CH_2OCH_2CF_3$

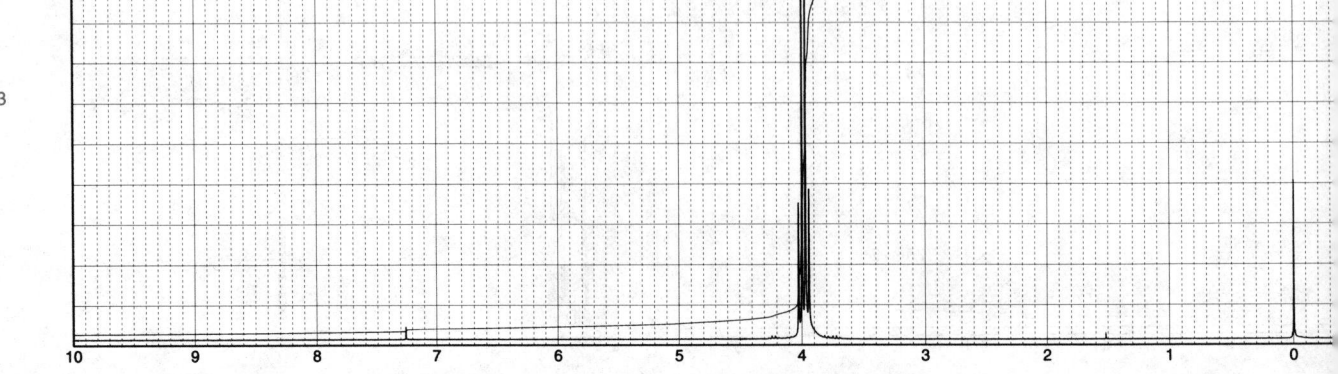

Non-Aromatic Ethers, Acetals, and Epoxides

330

Aldrich 29,451-9   CAS [24566-90-3]
Chloromethyl octyl ether, 95%

C9H19ClO
FW 178.70
d 0.924
n20D 1.4370
Fp 192°F

CDCl3   QE-300

VP-FT-IR: 3, 273D

83.30   28.96
70.71   25.99
31.82   22.67
29.29   14.07*
29.22

A

ClCH2OCH2(CH2)6CH3

Aldrich 29,452-7   CAS [96384-68-8]
Bromomethyl octyl ether, 95%

C9H19BrO
FW 223.16
d 1.120
n20D 1.4590
Fp 227°F

CDCl3   QE-300

VP-FT-IR: 3, 274A

77.01   28.62
71.89   25.92
31.78   22.65
29.24   14.09*
29.18

B

BrCH2OCH2(CH2)6CH3

Aldrich 18,078-5   CAS [54149-17-6]
1-Bromo-2-(2-methoxyethoxy)ethane, 97%

C5H11BrO2
FW 183.05
d 1.347
n20D 1.4550
Fp 165°F

CDCl3   QE-300

60 MHz: 1, 180D
FT-IR: 1, 210D
VP-FT-IR: 3, 275C

71.87
71.22
70.42
59.03*
30.13

C

CH3OCH2CH2OCH2CH2Br

A

Aldrich C4,113-4   *CAS [111-44-4]* $C_4H_8Cl_2O$ n_D^{20} 1.4560
2-Chloroethyl ether, 99%   FW 143.01   Fp 131°F
mp -47°C   60 MHz: *1*, 179D
bp 66°C (15 mm)   FT-IR: *1*, 211A
d 1.220   VP-FT-IR: *3*, 275D

71.3
42.6

CDCl₃   QE-300

$ClCH_2CH_2OCH_2CH_2Cl$

B

Aldrich 24,162-8   *CAS [112-26-5]* $C_6H_{12}Cl_2O_2$ Fp 250°F
1,2-Bis(2-chloroethoxy)ethane, 97%   FW 187.07
bp 235°C
d 1.197
n_D^{20} 1.4610

60 MHz: *1*, 180A
FT-IR: *1*, 211B
VP-FT-IR: *3*, 276B

71.3
70.6
42.7

CDCl₃   QE-300

$ClCH_2CH_2OCH_2CH_2OCH_2CH_2Cl$

C

Aldrich 33,343-3   *CAS [36839-55-1]* $C_6H_{12}I_2O_2$
1,2-Bis-(2-iodoethoxy)ethane, 96%   FW 369.97
d 2.028
n_D^{20} 1.5720
Fp >230°F

71.8
70.0
3.0

CDCl₃   QE-300

$ICH_2CH_2OCH_2CH_2OCH_2CH_2I$

ALDRICH

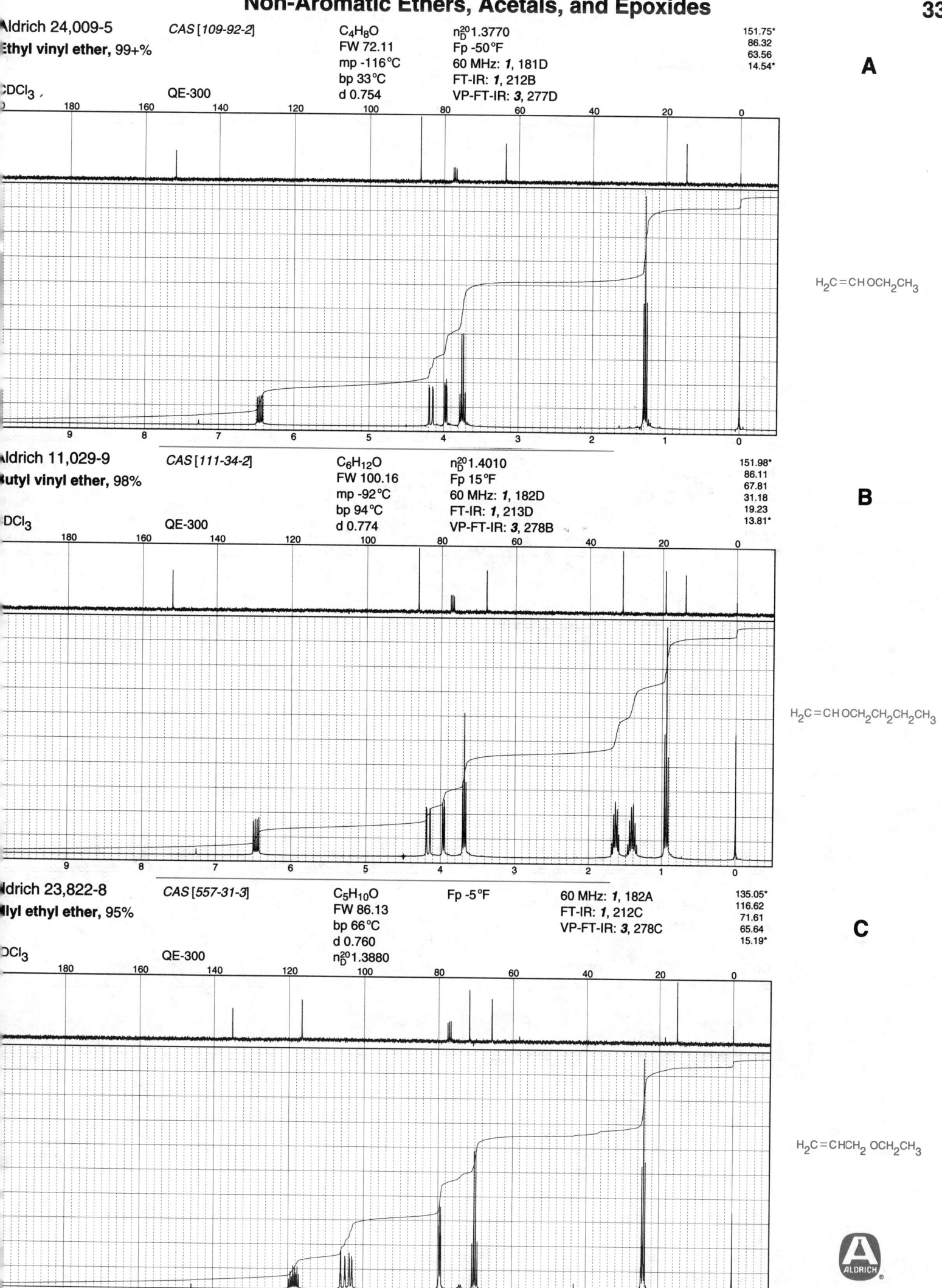

Aldrich 24,009-5 CAS [109-92-2] C₄H₈O n_D^{20} 1.3770 151.75*
Ethyl vinyl ether, 99+% FW 72.11 Fp -50°F 86.32
63.56
mp -116°C 60 MHz: *1*, 181D 14.54*
bp 33°C FT-IR: *1*, 212B
CDCl₃ QE-300 d 0.754 VP-FT-IR: *3*, 277D

A

$H_2C=CHOCH_2CH_3$

Aldrich 11,029-9 CAS [111-34-2] C₆H₁₂O n_D^{20} 1.4010 151.98*
Butyl vinyl ether, 98% FW 100.16 Fp 15°F 86.11
67.81
mp -92°C 60 MHz: *1*, 182D 31.18
19.23
bp 94°C FT-IR: *1*, 213D 13.81*
CDCl₃ QE-300 d 0.774 VP-FT-IR: *3*, 278B

B

$H_2C=CHOCH_2CH_2CH_2CH_3$

Aldrich 23,822-8 CAS [557-31-3] C₅H₁₀O Fp -5°F 60 MHz: *1*, 182A 135.05*
Allyl ethyl ether, 95% FW 86.13 FT-IR: *1*, 212C 116.62
71.61
bp 66°C VP-FT-IR: *3*, 278C 65.64
d 0.760 15.19*
CDCl₃ QE-300 n_D^{20} 1.3880

C

$H_2C=CHCH_2OCH_2CH_3$

A

Aldrich 28,192-1   *CAS [1471-03-0]*   C₆H₁₂O   Fp 23°F   VP-FT-IR: *3*, 278D

Allyl propyl ether, 95%

C_6H_12O
FW 100.16
bp 91°C
d 0.767
n_D^20 1.3990

135.06*
116.59
72.07
71.75
22.95
10.59*

CDCl₃   QE-300

$H_2C=CHCH_2\ OCH_2CH_2CH_3$

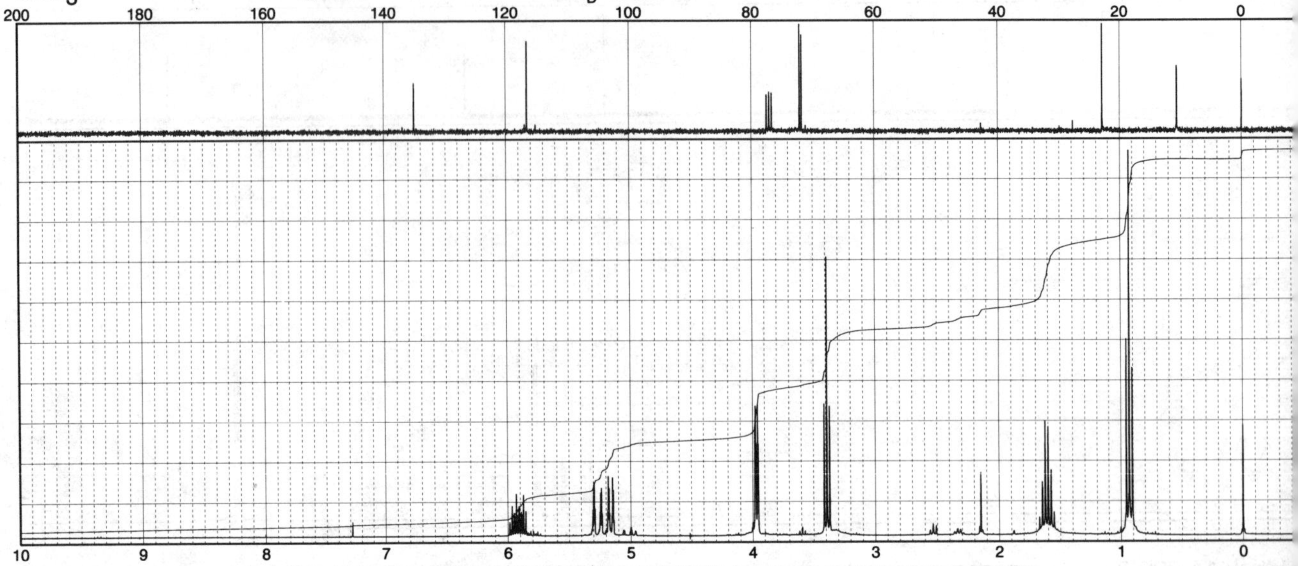

B

Aldrich 36,260-3   *CAS [3739-64-8]*   C₇H₁₄O

Allyl butyl ether, 98%

FW 114.19
d 0.783
n_D^20 1.4060
Fp 58°F

135.14
116.51
71.79
70.18
31.90
19.40
13.93

CDCl₃   QE-300

$H_2C=CHCH_2OCH_2CH_2CH_2CH_3$

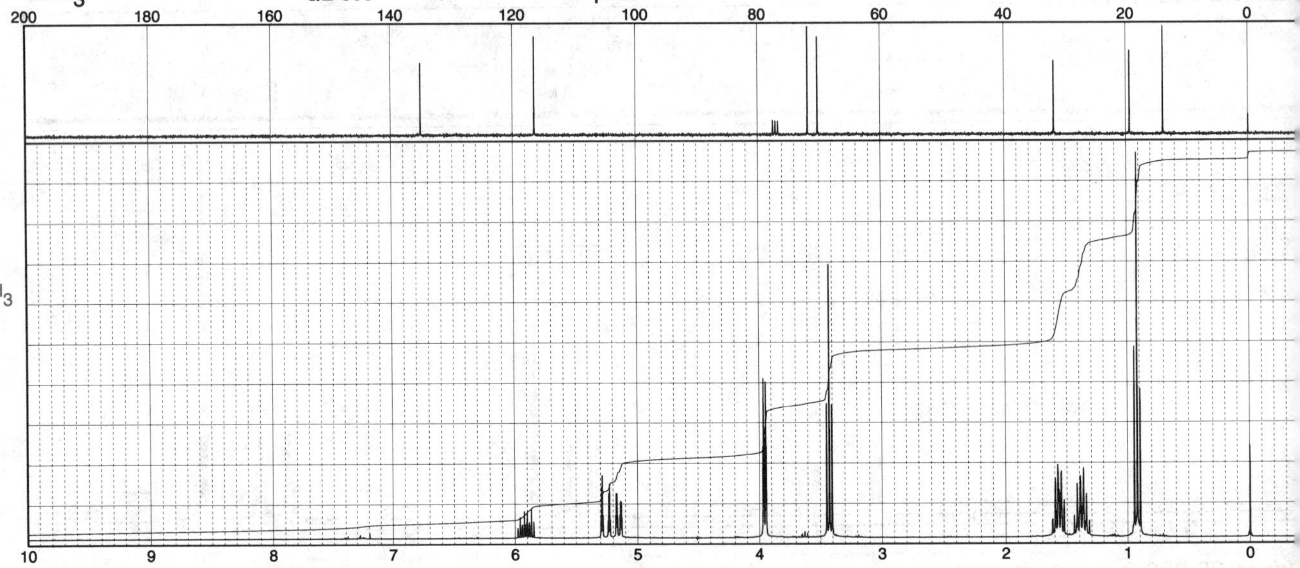

C

Aldrich 25,947-0   *CAS [557-40-4]*   C₆H₁₀O   Fp 20°F   FT-IR: *1*, 212D

Allyl ether, 99%

FW 98.15   VP-FT-IR: *3*, 279B
bp 95°C
d 0.803
n_D^20 1.4160

134.73
116.85
71.07

CDCl₃   QE-300

$H_2C=CHCH_2\ OCH_2CH=CH_2$

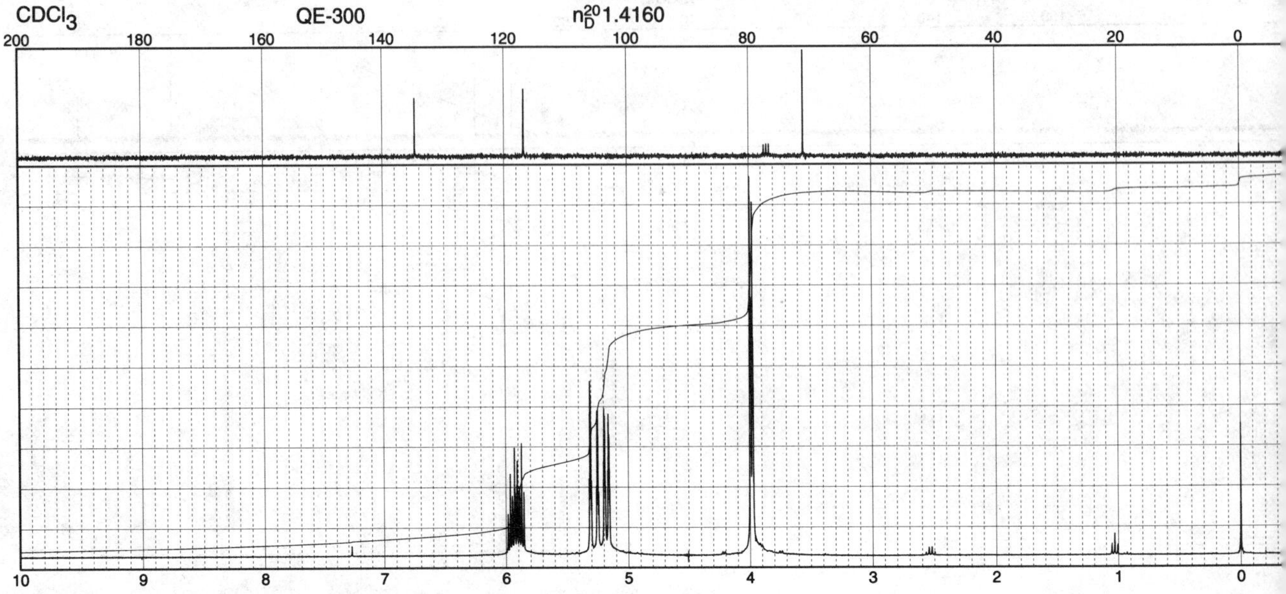

Aldrich 17,464-5 *CAS [116-11-0]* C_4H_8O Fp -21°F 60 MHz: *1*, 181B 160.66
2-Methoxypropene, 97% FW 72.11 FT-IR: *1*, 213A 80.69
bp 35°C VP-FT-IR: *3*, 279C 54.70*
CDCl3 QE-300 d 0.753 20.78*
n_D^{20} 1.3820

A

$$CH_3 \underset{\underset{OCH_3}{|}}{C} = CH_2$$

Aldrich 36,404-5 *CAS [928-55-2]* $C_5H_{10}O$ Fp -2°F 146.41* 64.59
Ethyl 1-propenyl ether, 99%, mixture FW 86.13 145.28* 15.29*
of *cis* and *trans* bp 72°C 100.90 14.84*
d 0.778 98.48 12.58*
CDCl3 QE-300 n_D^{20} 1.3980 67.40 9.20*

B

$$CH_3CH=CHOCH_2CH_3$$

Aldrich M1,200-2 *CAS [3036-66-6]* C_5H_8O Fp 22°F 60 MHz: *1*, 183A 151.73*
1-Methoxy-1,3-butadiene, 99%, mixture FW 84.12 VP-FT-IR: *3*, 280B 133.32*
of isomers bp 91°C 111.51
d 0.830 106.31*
CDCl3 QE-300 n_D^{20} 1.4640 56.33*

C

$$H_2C=CH-CH=CHOCH_3$$

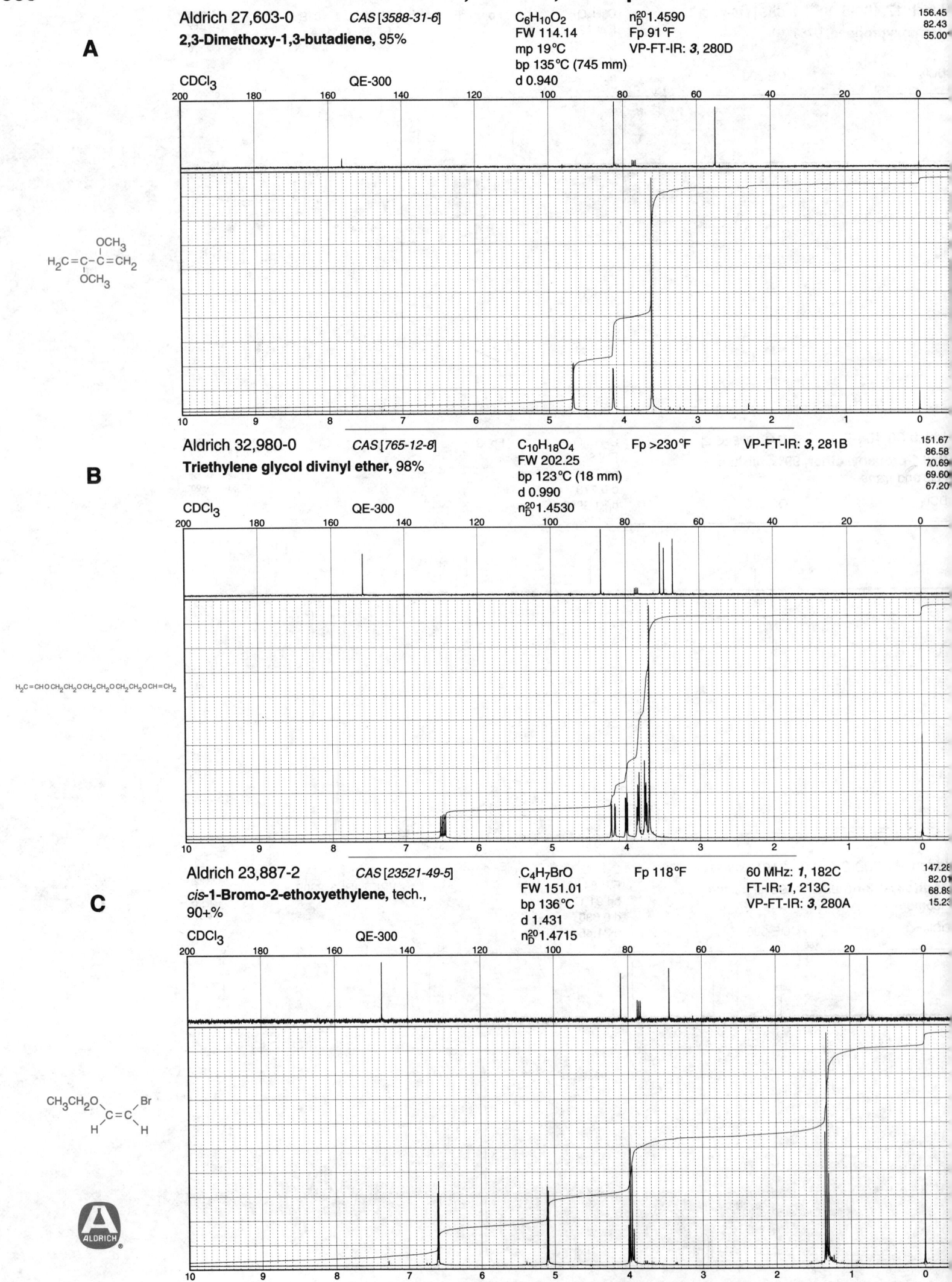

A Aldrich 27,603-0   CAS [3588-31-6] $C_6H_{10}O_2$ n_D^{20} 1.4590   156.45
2,3-Dimethoxy-1,3-butadiene, 95%   FW 114.14   Fp 91°F   82.43
mp 19°C   VP-FT-IR: *3*, 280D   55.00
bp 135°C (745 mm)
d 0.940

CDCl₃   QE-300

B Aldrich 32,980-0   CAS [765-12-8] $C_{10}H_{18}O_4$ Fp >230°F   VP-FT-IR: *3*, 281B   151.67
Triethylene glycol divinyl ether, 98%   FW 202.25   86.58
bp 123°C (18 mm)   70.69
d 0.990   69.60
CDCl₃   QE-300 n_D^{20} 1.4530   67.20

H₂C=CHOCH₂CH₂OCH₂CH₂OCH₂CH₂OCH=CH₂

C Aldrich 23,887-2   CAS [23521-49-5]   C₄H₇BrO   Fp 118°F   60 MHz: *1*, 182C   147.28
cis-1-Bromo-2-ethoxyethylene, tech.,   FW 151.01   FT-IR: *1*, 213C   82.01
90+%   bp 136°C   VP-FT-IR: *3*, 280A   68.89
d 1.431   15.23
CDCl₃   QE-300 n_D^{20} 1.4715

Non-Aromatic Ethers, Acetals, and Epoxides

Aldrich 10,998-3   CAS [110-75-8]   C₄H₇ClO   Fp 61°F   60 MHz: *1*, 182B   151.10*
-Chloroethyl vinyl ether, 99%   FW 106.55   FT-IR: *1*, 213B   87.35
bp 109°C   VP-FT-IR: *3*, 279D   67.81
d 1.048   41.84
n²⁰_D 1.4380

A

CDCl₃   QE-300

$H_2C=CHOCH_2CH_2Cl$

Aldrich 37,194-7   CAS [1428-33-7]   C₅H₆F₄O   Fp 23°F
llyl 1,1,2,2-tetrafluoroethyl ether,   FW 158.10
9%   bp 76°C
d 1.200
n²⁰_D 1.3270

131.47*	117.49	111.24*	104.60*
121.41	117.12	110.67*	104.04*
121.03	114.31	108.48*	65.34
120.66	113.94	107.92*	65.26
119.10	113.56	107.36*	65.18
117.86	111.80*	105.16*	

B

CDCl₃   QE-300

$H_2C=CHCH_2OCF_2CHF_2$

Aldrich 15,620-5   CAS [109-86-4]   C₃H₈O₂   n²⁰_D 1.4020   73.99
-Methoxyethanol, 99+%   FW 76.10   Fp 115°F   61.56
mp -85°C   FT-IR: *1*, 222D   58.83*
bp 125°C   VP-FT-IR: *3*, 293A
d 0.965

C

CDCl₃   QE-300

$CH_3OCH_2CH_2OH$

ALDRICH

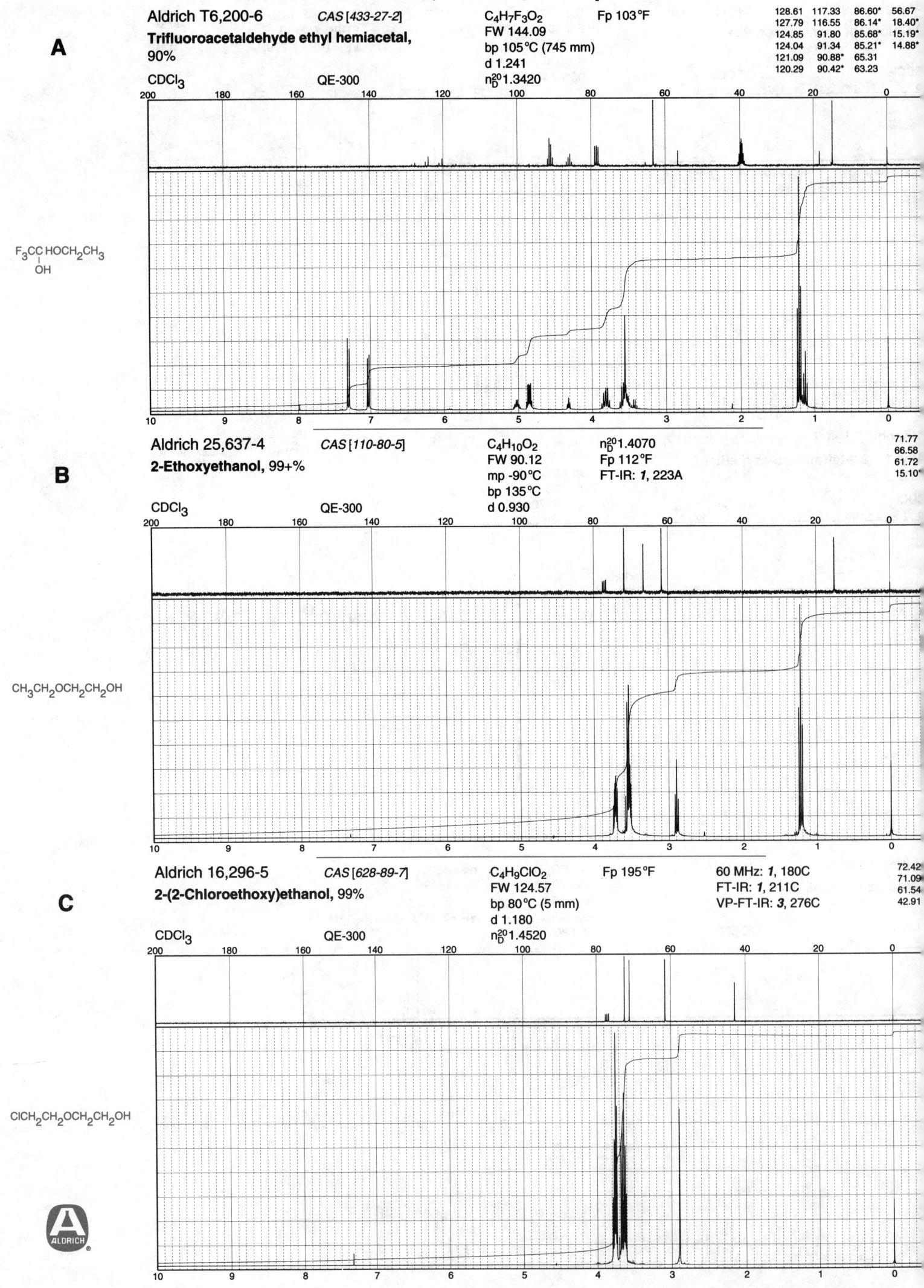

A

Aldrich T6,200-6   CAS [433-27-2] $C_4H_7F_3O_2$ Fp 103°F

Trifluoroacetaldehyde ethyl hemiacetal, 90%

FW 144.09
bp 105°C (745 mm)
d 1.241
n_D^{20} 1.3420

CDCl₃   QE-300

128.61	117.33	86.60*	56.67
127.79	116.55	86.14*	18.40*
124.85	91.80	85.68*	15.19*
124.04	91.34	85.21*	14.88*
121.09	90.88*	65.31	
120.29	90.42*	63.23	

$F_3CC\,HOCH_2CH_3$
    |
    OH

B

Aldrich 25,637-4   CAS [110-80-5] $C_4H_{10}O_2$ n_D^{20} 1.4070

2-Ethoxyethanol, 99+%

FW 90.12   Fp 112°F
mp -90°C   FT-IR: 1, 223A
bp 135°C
d 0.930

CDCl₃   QE-300

71.77
66.58
61.72
15.10*

$CH_3CH_2OCH_2CH_2OH$

C

Aldrich 16,296-5   CAS [628-89-7] $C_4H_9ClO_2$ Fp 195°F   60 MHz: 1, 180C

2-(2-Chloroethoxy)ethanol, 99%

FW 124.57   FT-IR: 1, 211C
bp 80°C (5 mm)   VP-FT-IR: 3, 276C
d 1.180
n_D^{20} 1.4520

CDCl₃   QE-300

72.42
71.09
61.54
42.91

$ClCH_2CH_2OCH_2CH_2OH$

ALDRICH

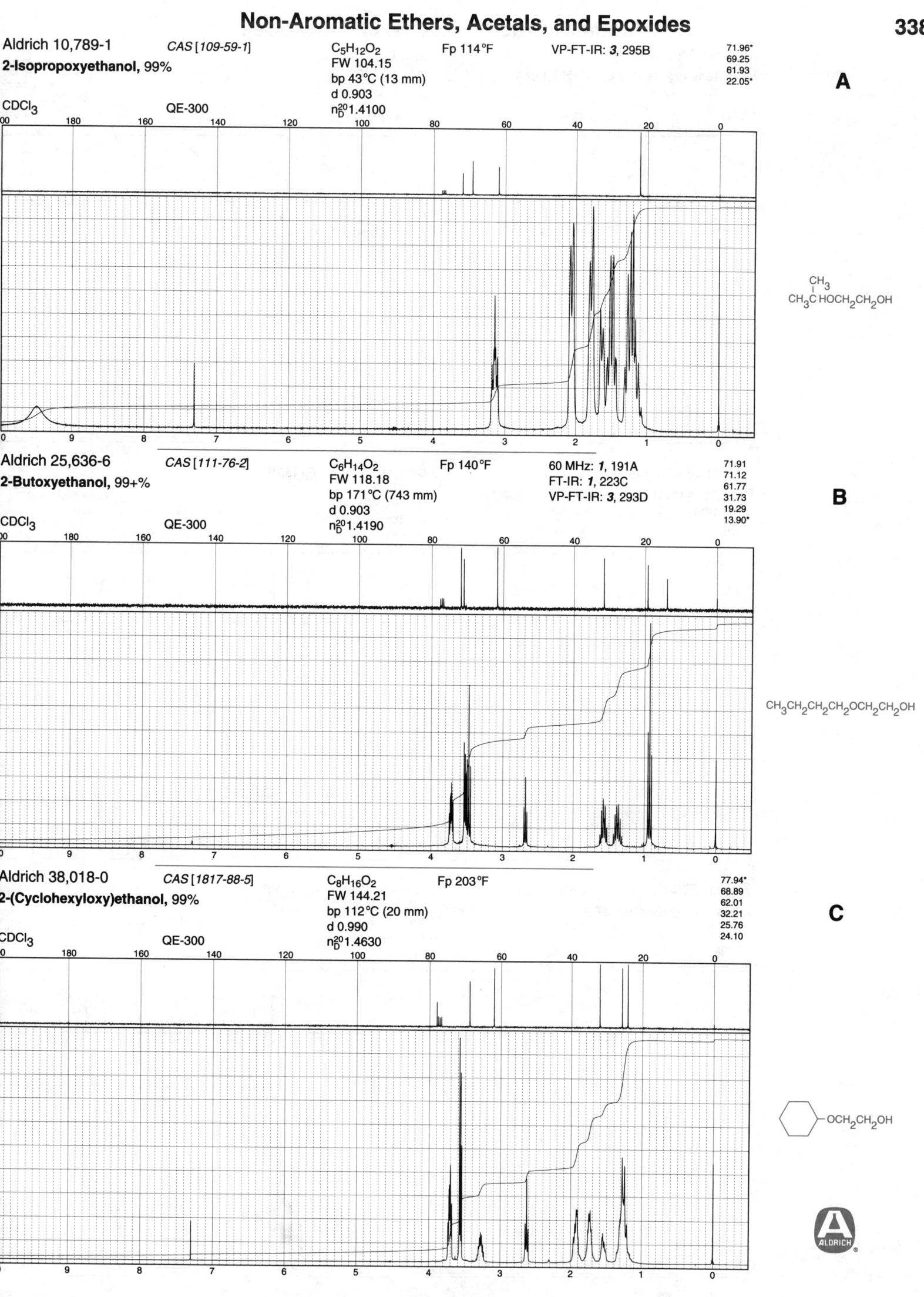

Aldrich 10,789-1   CAS [109-59-1]   C₅H₁₂O₂   Fp 114°F   VP-FT-IR: 3, 295B
$C_5H_{12}O_2$
2-Isopropoxyethanol, 99%   FW 104.15
bp 43°C (13 mm)
d 0.903
n_D^{20} 1.4100

CDCl₃   QE-300

71.96*
69.25
61.93
22.05*

A

Aldrich 25,636-6   CAS [111-76-2]   C₆H₁₄O₂   Fp 140°F   60 MHz: 1, 191A
2-Butoxyethanol, 99+%   FW 118.18   FT-IR: 1, 223C
bp 171°C (743 mm)   VP-FT-IR: 3, 293D
d 0.903
n_D^{20} 1.4190

CDCl₃   QE-300

71.91
71.12
61.77
31.73
19.29
13.90*

B

Aldrich 38,018-0   CAS [1817-88-5]   C₈H₁₆O₂   Fp 203°F
2-(Cyclohexyloxy)ethanol, 99%   FW 144.21
bp 112°C (20 mm)
d 0.990
n_D^{20} 1.4630

CDCl₃   QE-300

77.94*
68.89
62.01
32.21
25.76
24.10

C

A

Aldrich 30,287-2

Propylene glycol monomethyl ether

$C_4H_{10}O_2$

79.23
67.09*
59.78*
19.60*

CDCl$_3$ QE-300

200 180 160 140 120 100 80 60 40 20 0

$CH_3OC_3H_6OH$

10 9 8 7 6 5 4 3 2 1 0

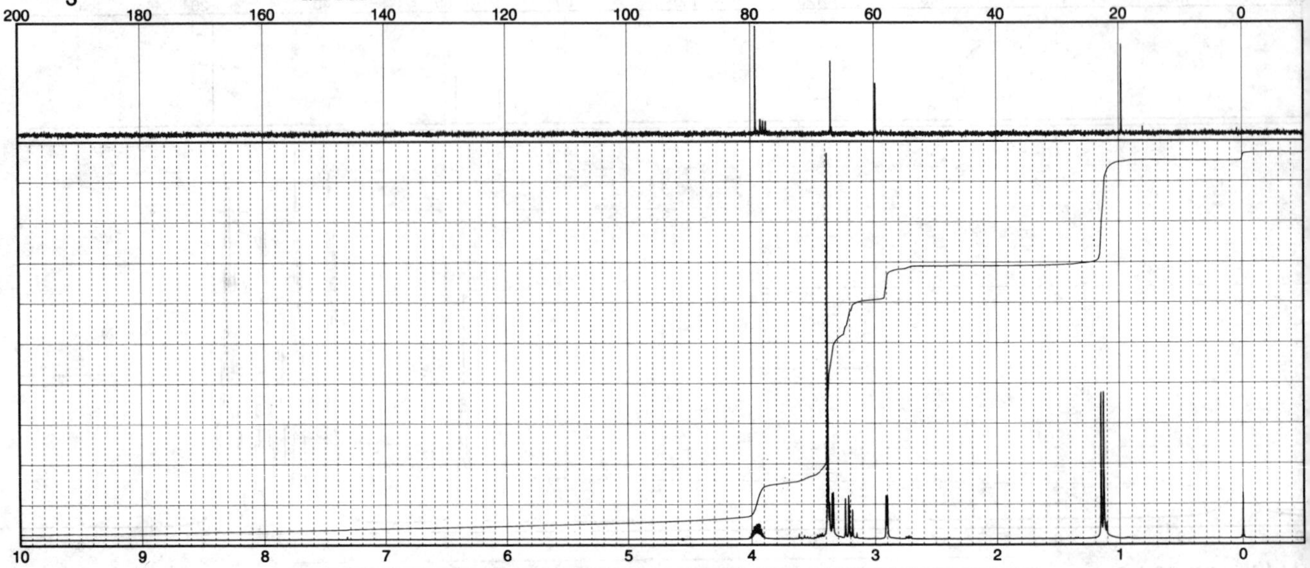

B

Aldrich 38,818-1

Propylene glycol butyl ether, 99%, mixture of isomers

$C_7H_{16}O_2$
FW 132.20
bp 72°C (20 mm)
d 0.885
n$_D^{20}$1.4160

Fp 138°F

76.36
71.15
66.37*
31.73
19.31
18.68*
13.91*

CDCl$_3$ QE-300

200 180 160 140 120 100 80 60 40 20 0

$CH_3(CH_2)_3OC_3H_6OH$

10 9 8 7 6 5 4 3 2 1 0

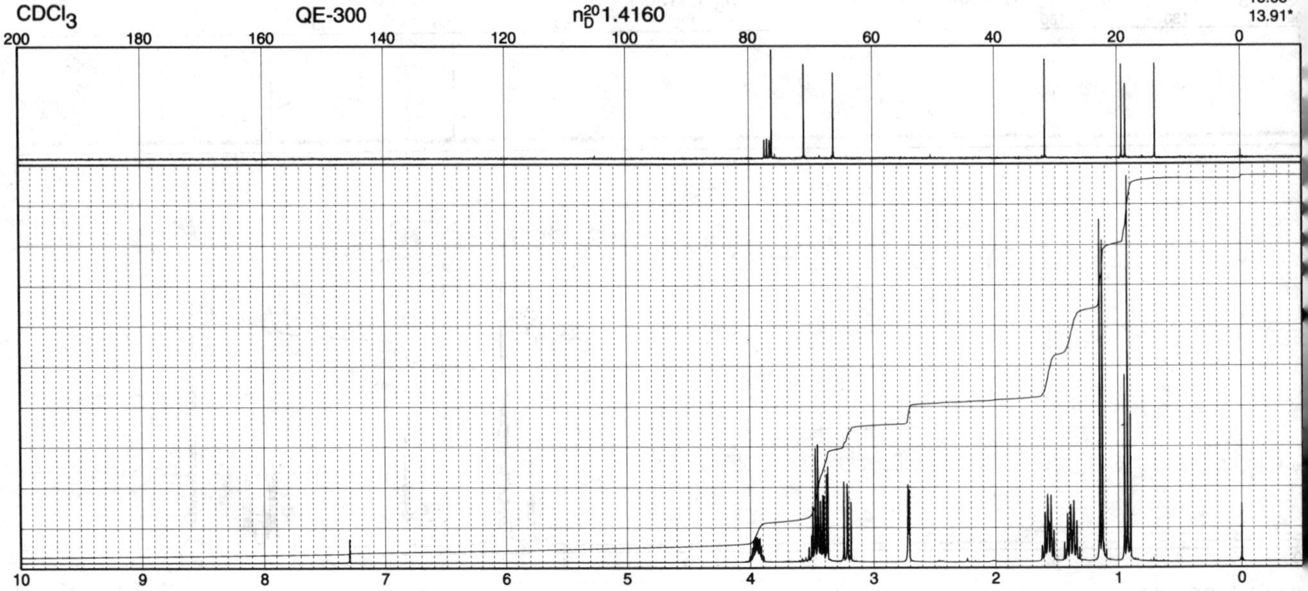

C

Aldrich E740-1   *CAS [111-35-3]*

3-Ethoxy-1-propanol, 97%

$C_5H_{12}O_2$
FW 104.15
bp 161°C
d 0.904
n$_D^{20}$1.4170

Fp 130°F

60 MHz: *1*, 191B
FT-IR: *1*, 223D
VP-FT-IR: *3*, 294A

69.49
66.50
61.60
32.17
15.18*

CDCl$_3$ QE-300

200 180 160 140 120 100 80 60 40 20 0

$CH_3CH_2OCH_2CH_2CH_2OH$

10 9 8 7 6 5 4 3 2 1 0

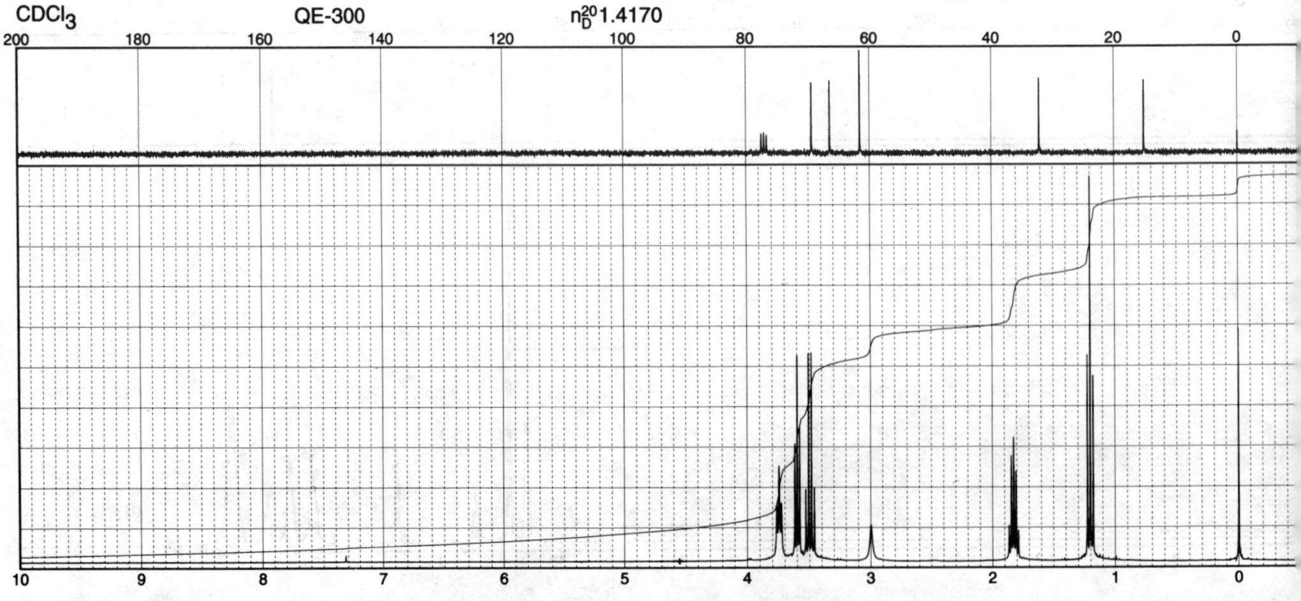

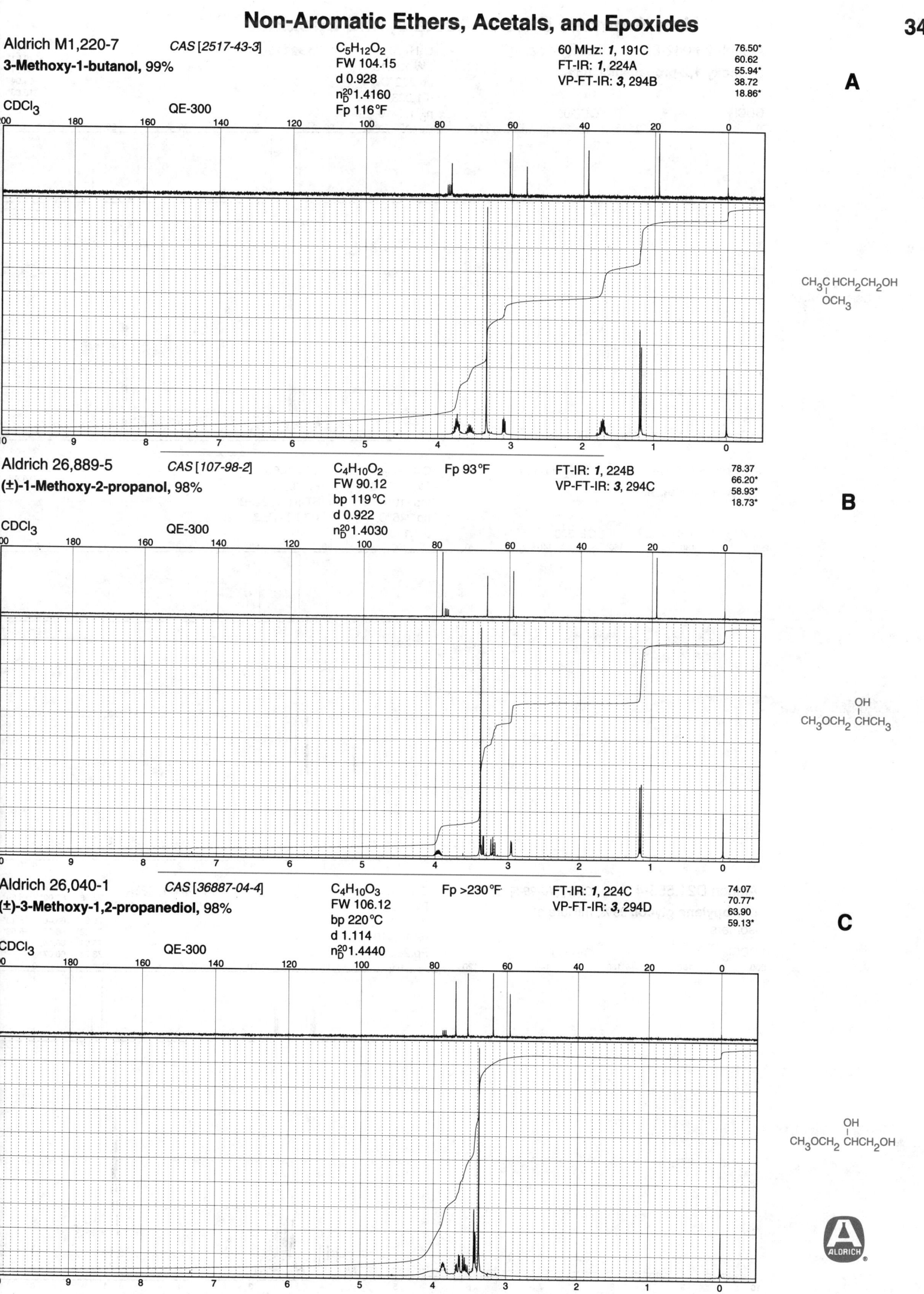

Aldrich M1,220-7   *CAS [2517-43-3]* $C_5H_{12}O_2$
3-Methoxy-1-butanol, 99%   FW 104.15
d 0.928
n_D^{20}1.4160
Fp 116°F

60 MHz: *1*, 191C
FT-IR: *1*, 224A
VP-FT-IR: *3*, 294B

76.50*
60.62
55.94*
38.72
18.86*

A

CDCl$_3$ QE-300

$CH_3CHCH_2CH_2OH$
    OCH_3

Aldrich 26,889-5   *CAS [107-98-2]* $C_4H_{10}O_2$ Fp 93°F
(±)-1-Methoxy-2-propanol, 98%   FW 90.12
bp 119°C
d 0.922
n_D^{20}1.4030

FT-IR: *1*, 224B
VP-FT-IR: *3*, 294C

78.37
66.20*
58.93*
18.73*

B

CDCl$_3$ QE-300

$CH_3OCH_2CHCH_3$
          OH

Aldrich 26,040-1   *CAS [36887-04-4]* $C_4H_{10}O_3$ Fp >230°F
(±)-3-Methoxy-1,2-propanediol, 98%   FW 106.12
bp 220°C
d 1.114
n_D^{20}1.4440

FT-IR: *1*, 224C
VP-FT-IR: *3*, 294D

74.07
70.77*
63.90
59.13*

C

CDCl$_3$ QE-300

$CH_3OCH_2CHCH_2OH$
          OH

ALDRICH

A

Aldrich 26,042-8   CAS [1874-62-0] $C_5H_{12}O_3$ Fp >230°F   FT-IR: **1**, 224D

3-Ethoxy-1,2-propanediol, 98%   FW 120.15   VP-FT-IR: **3**, 295A

bp 222°C
d 1.063
n_D^{20} 1.4410

71.96
70.81*
66.91
64.04
15.03*

CDCl$_3$ QE-300

$CH_3CH_2OCH_2$ $\overset{OH}{\underset{|}{CH}}CH_2OH$

B

Aldrich H2,645-6   CAS [111-46-6] $C_4H_{10}O_3$ n_D^{20} 1.4460

Diethylene glycol, 99%   FW 106.12   Fp 290°F

mp -10°C   FT-IR: **1**, 223B
bp 245°C   VP-FT-IR: **3**, 293C
d 1.118

72.36
61.48

CDCl$_3$ QE-300

$HOCH_2CH_2OCH_2CH_2OH$

C

Aldrich D21,555-4   CAS [110-98-5] $C_6H_{14}O_3$ FT-IR: **1**, 226A

Dipropylene glycol, 99%, mixture of   FW 134.18   VP-FT-IR: **3**, 296D
isomers   d 1.023
n_D^{20} 1.4410
Fp 280°F

77.26*	72.72*	18.74*
77.06	67.02*	18.69*
76.66	66.55	17.81*
76.27*	66.46*	16.13*
76.22*	66.37	15.93*
75.01	66.14	15.71*
73.74	66.07*	

CDCl$_3$ QE-300

$HO-C_3H_6-O-C_3H_6-OH$

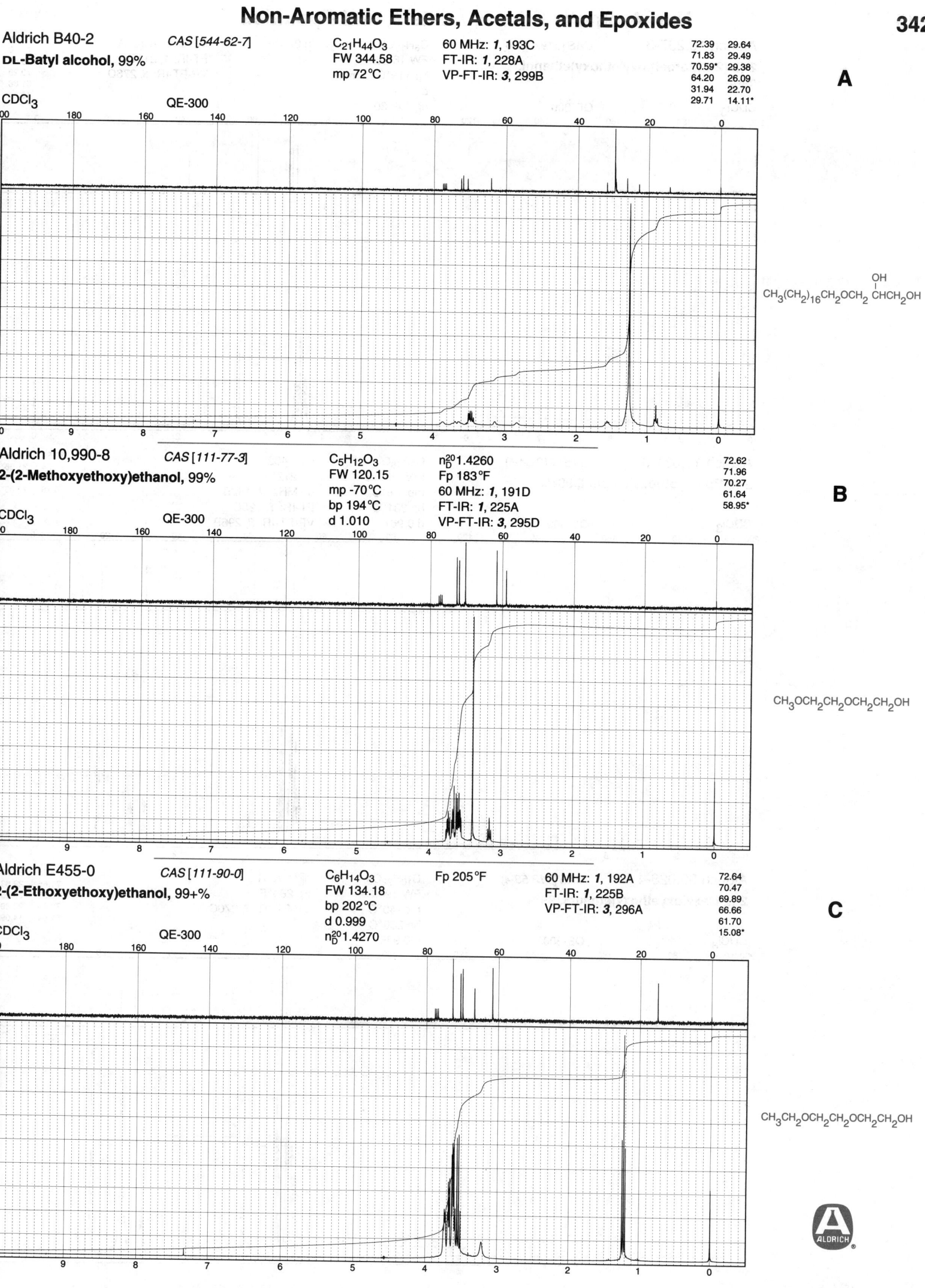

Aldrich B40-2　　　*CAS [544-62-7]*

DL-Batyl alcohol, 99%

$C_{21}H_{44}O_3$
FW 344.58
mp 72°C

60 MHz: *1*, 193C
FT-IR: *1*, 228A
VP-FT-IR: *3*, 299B

72.39　29.64
71.83　29.49
70.59*　29.38
64.20　26.09
31.94　22.70
29.71　14.11*

A

$CDCl_3$　　QE-300

$CH_3(CH_2)_{16}CH_2OCH_2\ CHCH_2OH$ with OH

Aldrich 10,990-8　　*CAS [111-77-3]*

2-(2-Methoxyethoxy)ethanol, 99%

$C_5H_{12}O_3$
FW 120.15
mp -70°C
bp 194°C
d 1.010

n_D^{20} 1.4260
Fp 183°F
60 MHz: *1*, 191D
FT-IR: *1*, 225A
VP-FT-IR: *3*, 295D

72.62
71.96
70.27
61.64
58.95*

B

$CDCl_3$　　QE-300

$CH_3OCH_2CH_2OCH_2CH_2OH$

Aldrich E455-0　　*CAS [111-90-0]*

2-(2-Ethoxyethoxy)ethanol, 99+%

$C_6H_{14}O_3$
FW 134.18
bp 202°C
d 0.999
n_D^{20} 1.4270

Fp 205°F

60 MHz: *1*, 192A
FT-IR: *1*, 225B
VP-FT-IR: *3*, 296A

72.64
70.47
69.89
66.66
61.70
15.08*

C

$CDCl_3$　　QE-300

$CH_3CH_2OCH_2CH_2OCH_2CH_2OH$

ALDRICH

A

Aldrich 16,297-3   *CAS [5197-62-6]* $C_6H_{13}ClO_3$ Fp 225°F   60 MHz: *1*, 181A

72.55
71.30
70.62
70.29
61.66
42.68

2-(2-(2-Chloroethoxy)ethoxy)ethanol, 98%

FW 168.62
bp 119°C (5 mm)
d 1.160
n_D^{20}1.4580

FT-IR: *1*, 211D
VP-FT-IR: *3*, 276D

CDCl₃   QE-300

$ClCH_2CH_2OCH_2CH_2OCH_2CH_2OH$

B

Aldrich 11,031-0   *CAS [112-34-5]* $C_8H_{18}O_3$ n_D^{20}1.4320

72.58 61.75
71.24 31.66
70.46 19.26
70.16 13.89*

2-(2-Butoxyethoxy)ethanol, 99+%

FW 162.23
mp -68°C
bp 231°C
d 0.967

Fp 212°F
60 MHz: *1*, 192B
FT-IR: *1*, 225C
VP-FT-IR: *3*, 296B

CDCl₃   QE-300

$CH_3CH_2CH_2CH_2OCH_2CH_2OCH_2CH_2OH$

C

Aldrich 30,028-4   *CAS [112-59-4]* $C_{10}H_{22}O_3$ n_D^{20}1.4381

72.65 31.75
71.65 29.63
70.55 25.83
70.23 22.69
61.84 14.09*

2-(2-Hexyloxyethoxy)ethanol, 99%

FW 190.29
mp -40°C
bp 260°C
d 0.935

Fp 284°F
VP-FT-IR: *3*, 270C

CDCl₃   QE-300

$CH_3CH_2(CH_2)_3CH_2OCH_2CH_2OCH_2CH_2OH$

ALDRICH

Aldrich 37,495-4   CAS [3055-93-4]

Diethylene glycol dodecyl ether, 97%

$C_{16}H_{34}O_3$

FW 274.45

mp 165°C

bp 169°C

d 0.904

n_D^{20}1.4480

Fp >230°F

CDCl$_3$ QE-300

72.56	29.61
71.59	29.48
70.47	29.36
70.15	26.08
61.78	22.69
31.93	14.11*
29.65	

A

$CH_3(CH_2)_{10}CH_2(OCH_2CH_2)_2OH$

Aldrich 28,328-2   CAS [34590-94-8]

Dipropylene glycol methyl ether, 97%, mixture of isomers

$C_7H_{16}O_3$

FW 148.20

bp 91°C (12 mm)

d 0.938

n_D^{20}1.4220

Fp 166°F

CDCl$_3$ QE-300

77.29	75.89*	67.04*	18.58*
77.09	75.73	66.38*	18.47*
76.73	75.05	66.12*	18.31*
76.65	74.86	65.82*	17.12*
75.98*	74.57	59.04*	16.82*
75.96*	74.42*	56.55*	16.09*

B

$CH_3OC_3H_6OC_3H_6OH$

Aldrich 38,813-0   CAS [29911-28-2]

Dipropylene glycol butyl ether, 99%, mixture of isomers

$C_{10}H_{22}O_3$

FW 190.29

bp 103°C (10 mm)

d 0.913

n_D^{20}1.4260

Fp 205°F

CDCl$_3$ QE-300

76.51*	67.23*	18.14*
75.99	65.66*	17.33*
74.93	31.66	16.97*
74.83	19.31	13.89*
74.48*	19.26	
71.26	18.46*	

C

$CH_3(CH_2)_3OC_3H_6OC_3H_6OH$

ALDRICH

A

Aldrich T5,945-5   *CAS [112-27-6]*

Triethylene glycol, 99%

CDCl$_3$ QE-300

$C_6H_{14}O_4$
FW 150.17
mp -7°C
bp 285°C
d 1.125

n_D^{20} 1.4550
Fp 330°F
60 MHz: *1*, 192C
FT-IR: *1*, 226B
VP-FT-IR: *3*, 297B

72.69
70.28
61.45

HOCH$_2$CH$_2$(OCH$_2$CH$_2$)$_2$OH

B

Aldrich 18,759-3   *CAS [24800-44-0]*

Tripropylene glycol, 97%, mixture of isomers

CDCl$_3$ QE-300

$C_9H_{20}O_4$
FW 192.26
bp 273°C
d 1.021
n_D^{20} 1.4440

Fp >230°F

77.21	75.64*	74.65*	66.15*	18.53*
77.10	75.59*	74.38	66.08*	18.35*
77.07	75.26	74.34	65.87*	16.92*
77.02	75.12	67.01*	65.76*	16.86*
75.88*	75.09	66.97*	18.65*	16.59*
75.79*	74.85	66.35*	18.55*	

HO[CH$_2$CH$_2$CH$_2$O]$_2$CH$_2$CH$_2$CH$_2$OH

C

Aldrich 30,286-4   *CAS [20324-33-8]*

Tripropylene glycol monomethyl ether, 98%, mixture of isomers

CDCl$_3$ QE-300

$C_{10}H_{22}O_4$
FW 206.29
bp 100°C (2 mm)
d 0.968
n_D^{20} 1.4300

Fp >230°F

76.69*	75.06*	74.45	65.70*	18.20*
76.43	74.95	73.34	65.57*	17.23*
75.92*	74.86	73.29	59.09*	17.14*
75.49	74.70	73.18	59.06*	16.95*
75.20*	74.58	67.19*	18.50*	16.91
75.14*	74.52	67.12*	18.25*	16.87*

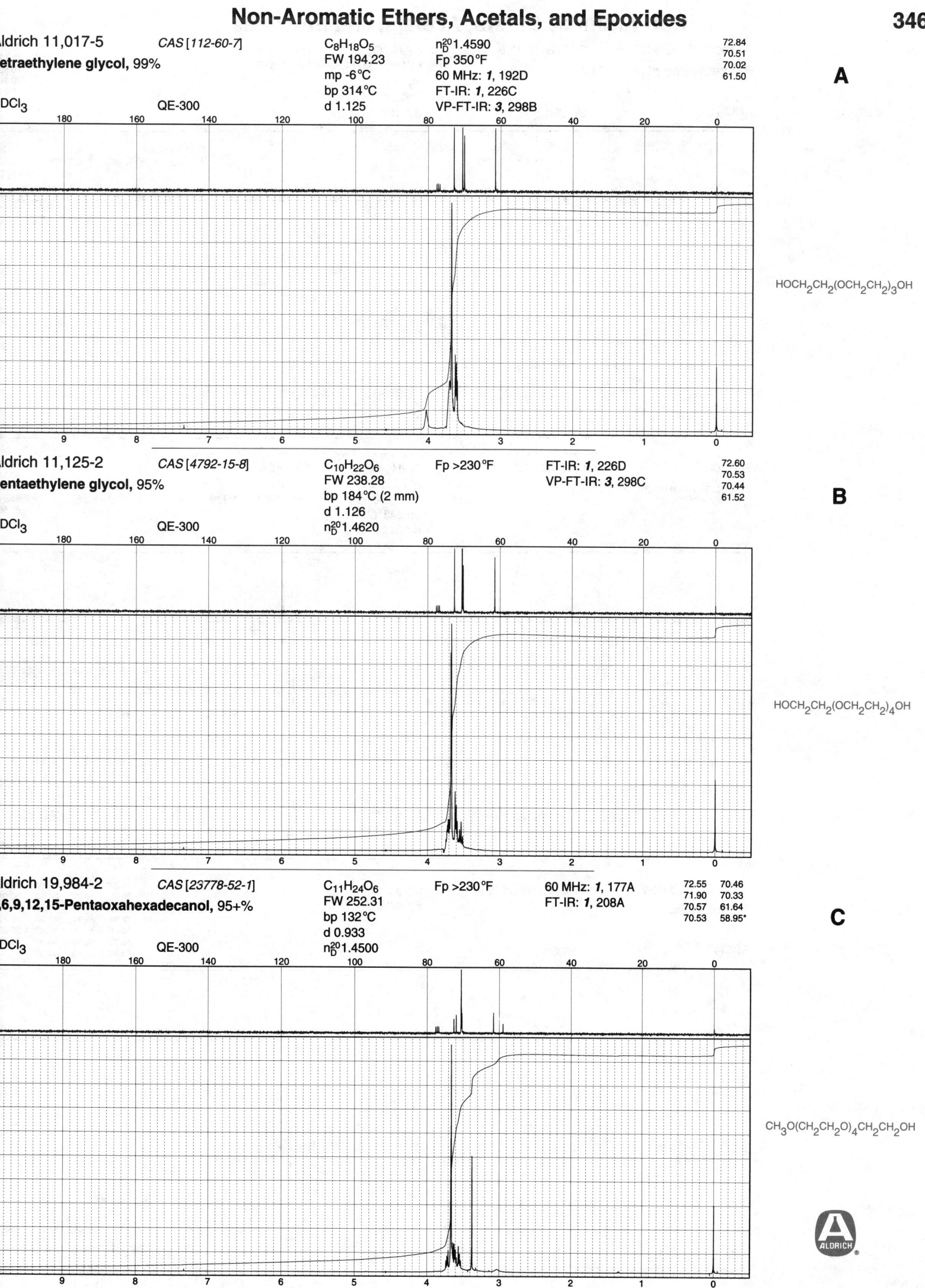

Aldrich 11,017-5   CAS [112-60-7]
Tetraethylene glycol, 99%

$C_8H_{18}O_5$
FW 194.23
mp -6°C
bp 314°C
d 1.125

n_D^{20} 1.4590
Fp 350°F
60 MHz: **1**, 192D
FT-IR: **1**, 226C
VP-FT-IR: **3**, 298B

72.84
70.51
70.02
61.50

A

CDCl₃   QE-300

$HOCH_2CH_2(OCH_2CH_2)_3OH$

Aldrich 11,125-2   CAS [4792-15-8]
Pentaethylene glycol, 95%

$C_{10}H_{22}O_6$
FW 238.28
bp 184°C (2 mm)
d 1.126
n_D^{20} 1.4620

Fp >230°F

FT-IR: **1**, 226D
VP-FT-IR: **3**, 298C

72.60
70.53
70.44
61.52

B

CDCl₃   QE-300

$HOCH_2CH_2(OCH_2CH_2)_4OH$

Aldrich 19,984-2   CAS [23778-52-1]
2,6,9,12,15-Pentaoxahexadecanol, 95+%

$C_{11}H_{24}O_6$
FW 252.31
bp 132°C
d 0.933
n_D^{20} 1.4500

Fp >230°F

60 MHz: **1**, 177A
FT-IR: **1**, 208A

72.55 70.46
71.90 70.33
70.57 61.64
70.53 58.95*

C

CDCl₃   QE-300

$CH_3O(CH_2CH_2O)_4CH_2CH_2OH$

ALDRICH

A

Aldrich 25,926-8   *CAS [2615-15-8]*

Hexaethylene glycol, 97%

$C_{12}H_{26}O_7$
FW 282.34
mp 6°C
bp 217°C (4 mm)
d 1.127

n_D^{20} 1.4640
Fp >230°F
FT-IR: *1*, 227A
VP-FT-IR: *3*, 298D

72.60
70.57
70.53
70.46
70.24
61.62

CDCl₃   QE-300

$HOCH_2CH_2(OCH_2CH_2)_5OH$

B

Aldrich 32,905-3   *CAS [5617-32-3]*

Heptaethylene glycol, 97%

$C_{14}H_{30}O_8$
FW 326.39
mp 11°C
bp 243°C
d 1.122

n_D^{20} 1.4640
Fp >230°F
VP-FT-IR: *3*, 299A

72.56
70.54
70.48
70.24
61.56

CDCl₃   QE-300

$HOCH_2CH_2(OCH_2CH_2)_6OH$

C

Aldrich 30,312-7   *CAS [92046-34-9]*

Triton®X-405

mp 49°C
Fp >230°F

78.87*	70.50	67.04	32.23
76.55	70.27	61.63	30.61
73.31*	70.08	51.49	26.80
72.49	69.97	51.30	26.73
70.86	69.56	48.95*	25.23
70.79	69.44	36.90	21.10
70.60	67.16	32.68	

CDCl₃   QE-300

x = 8(avg.)

ALDRICH®

Non-Aromatic Ethers, Acetals, and Epoxides

Aldrich D13,465-1   CAS [109-87-5]

Dimethoxymethane, 99%

CDCl₃   QE-300

C₃H₈O₂
FW 76.10
mp -105 °C
bp 42 °C
d 0.860

n_D^{20} 1.3540
Fp 1 °F
60 MHz: 1, 174D
FT-IR: 1, 205C
VP-FT-IR: 3, 281D

97.52
55.02*

A

$CH_3OCH_2OCH_3$

Aldrich 16,642-1   CAS [462-95-3]

Diethoxymethane, 99%

CDCl₃   QE-300

C₅H₁₂O₂
FW 104.15
bp 88 °C
d 0.839
n_D^{20} 1.3730

Fp 22 °F   VP-FT-IR: 3, 282A

94.87
63.11
15.21*

B

$CH_3CH_2OCH_2OCH_2CH_3$

Aldrich 18,623-6   CAS [534-15-6]

Acetaldehyde dimethyl acetal, 99%

CDCl₃   QE-300

C₄H₁₀O₂
FW 90.12
bp 64 °C
d 0.852
n_D^{20} 1.3660

Fp 1 °F

60 MHz: 1, 183B
FT-IR: 1, 214A
VP-FT-IR: 3, 282B

101.07*
52.31*
18.76*

C

$$CH_3-\underset{\underset{OCH_3}{|}}{\overset{\overset{OCH_3}{|}}{CH}}$$

A

Aldrich A90-2   *CAS [105-57-7]* $C_6H_{14}O_2$ Fp -6°F   60 MHz: *1*, 183C   99.43
Acetal, 99%   FW 118.18   FT-IR: *1*, 214B   60.63
                                                  bp 102°C   VP-FT-IR: *3*, 282C   19.98
                                                  d 0.831   15.35
CDCl₃   QE-300 n_D^{20}1.3810

$$\begin{array}{c} OCH_2CH_3 \\ | \\ CH_3CH \\ | \\ OCH_2CH_3 \end{array}$$

B

Aldrich 17,695-8   *CAS [4744-08-5]* $C_7H_{16}O_2$ Fp 55°F   60 MHz: *1*, 183D   104.11
Propionaldehyde diethyl acetal, 97%   FW 132.20   FT-IR: *1*, 215B   60.91
                                                    bp 123°C   VP-FT-IR: *3*, 283C   26.66
                                                    d 0.815   15.37
CDCl₃   QE-300 n_D^{20}1.3890   9.03

$$\begin{array}{c} CH_3CH_2CHOCH_2CH_3 \\ | \\ OCH_2CH_3 \end{array}$$

C

Aldrich 19,054-3   *CAS [6044-68-4]* $C_5H_{10}O_2$ Fp 27°F   60 MHz: *1*, 187D   134.60
Acrolein dimethyl acetal, 98%   FW 102.13   FT-IR: *1*, 220B   118.71
                                                    bp 90°C   VP-FT-IR: *3*, 289D   102.98
                                                    d 0.862   52.62
CDCl₃   QE-300 n_D^{20}1.3950

$$\begin{array}{c} OCH_3 \\ | \\ H_2C=CHCH \\ | \\ OCH_3 \end{array}$$

Aldrich A2,400-1   CAS [3054-95-3]   C$_7$H$_{14}$O$_2$ Fp 40°F   60 MHz: *1*, 188A

Acrolein diethyl acetal, 96%   FW 130.19   FT-IR: *1*, 220C

bp 125°C   VP-FT-IR: *3*, 290A

d 0.854

CDCl$_3$ QE-300   n$_D^{20}$1.4010

135.52*
117.98
101.57*
61.01
15.24*

A

H$_2$C=CHCHOCH$_2$CH$_3$
$\quad\quad$|
$\quad\quad$OCH$_2$CH$_3$

Aldrich 30,724-6   CAS [10602-36-5]   C$_8$H$_{16}$O$_2$ Fp 93°F   VP-FT-IR: *3*, 290B

3-Butenal diethyl acetal, 97%   FW 144.21

bp 67°C (50 mm)

d 0.851

CDCl$_3$ QE-300   n$_D^{20}$1.4080

133.49*
117.16
102.37*
61.14
38.48
15.29*

B

H$_2$C=CHCH$_2$CHOCH$_2$CH$_3$
$\quad\quad\quad\quad$|
$\quad\quad\quad\quad$OCH$_2$CH$_3$

Aldrich 27,742-8   CAS [54306-00-2]   C$_{10}$H$_{20}$O$_2$ Fp 145°F   VP-FT-IR: *3*, 283D

2-Hexenal diethyl acetal, 98%, predominantly
trans

FW 172.27

bp 97°C (35 mm)

d 0.848

CDCl$_3$ QE-300   n$_D^{20}$1.4210

134.78* 34.17
127.49* 22.04
101.94* 15.26*
60.87   13.72*

C

CH$_3$CH$_2$CH$_2$CH=CHCHOCH$_2$CH$_3$
$\quad\quad\quad\quad\quad\quad\quad$|
$\quad\quad\quad\quad\quad\quad\quad$OCH$_2CH_3$

Non-Aromatic Ethers, Acetals, and Epoxides

351

Aldrich 30,375-5   CAS [7549-37-3] $C_{12}H_{22}O_2$ Fp 180°F   VP-FT-IR: **3**, 284A
Citral dimethyl acetal, mixture of *cis* and *trans*
FW 198.31
bp 106°C (10 mm)
d 0.890
n_D^{20}1.4540
CDCl$_3$ QE-300

142.11, 141.95, 131.87, 131.71, 123.80*, 122.63*, 121.73*
100.45*, 100.21*, 52.30*, 52.12*, 39.43, 32.75, 26.53
26.28, 25.66*, 23.21*, 17.66*, 16.99*

Aldrich 36,820-2   CAS [6068-62-8] $C_8H_{16}O_4$ Fp 169°F
Fumaraldehyde bis(dimethyl acetal), 95%
FW 176.21
bp 102°C (15 mm)
d 1.010
n_D^{20}1.4280
CDCl$_3$ QE-300

130.91*, 101.89*, 52.64*

Aldrich D13,680-8   CAS [77-76-9] $C_5H_{12}O_2$ Fp 12°F   60 MHz: **1**, 187C
2,2-Dimethoxypropane, 98%
FW 104.15
bp 83°C
d 0.847
n_D^{20}1.3780
CDCl$_3$ QE-300
FT-IR: **1**, 219D
VP-FT-IR: **3**, 289A

99.87, 48.30*, 23.93*

Non-Aromatic Ethers, Acetals, and Epoxides

Aldrich 25,375-8   CAS [126-84-1] $C_7H_{16}O_2$ Fp 46°F   FT-IR: 1, 220A   99.65
2,2-Diethoxypropane, 94%   FW 132.21   VP-FT-IR: 3, 289B   55.85
   bp 45°C (60 mm)   25.05*
   d 0.820   15.54*
CDCl₃   QE-300 n_D^{20} 1.3890

A

$$CH_3\underset{\underset{OCH_2CH_3}{|}}{\overset{\overset{OCH_2CH_3}{|}}{C}}CH_3$$

Aldrich 16,403-8   CAS [24332-20-5] $C_5H_{12}O_3$ Fp 74°F   60 MHz: 1, 184A   102.54*
Methoxyacetaldehyde dimethyl acetal,   FW 120.15   FT-IR: 1, 214D   72.25
98%   bp 58°C (56 mm)   VP-FT-IR: 3, 283A   59.31*
   d 0.932   53.85*
CDCl₃   QE-300 n_D^{20} 1.3920

B

$$CH_3OCH_2\underset{\underset{OCH_3}{|}}{C}HOCH_3$$

Aldrich 14,930-6   CAS [4819-75-4] $C_7H_{16}O_3$ Fp 95°F   60 MHz: 1, 184B   100.89*
Methoxyacetaldehyde diethyl acetal,   FW 148.20   FT-IR: 1, 215A   73.24
97%   bp 146°C   VP-FT-IR: 3, 283B   62.15
   d 0.911   59.31*
CDCl₃   QE-300 n_D^{20} 1.3990   15.35*

C

$$CH_3OCH_2\underset{\underset{OCH_2CH_3}{|}}{C}HOCH_2CH_3$$

ALDRICH

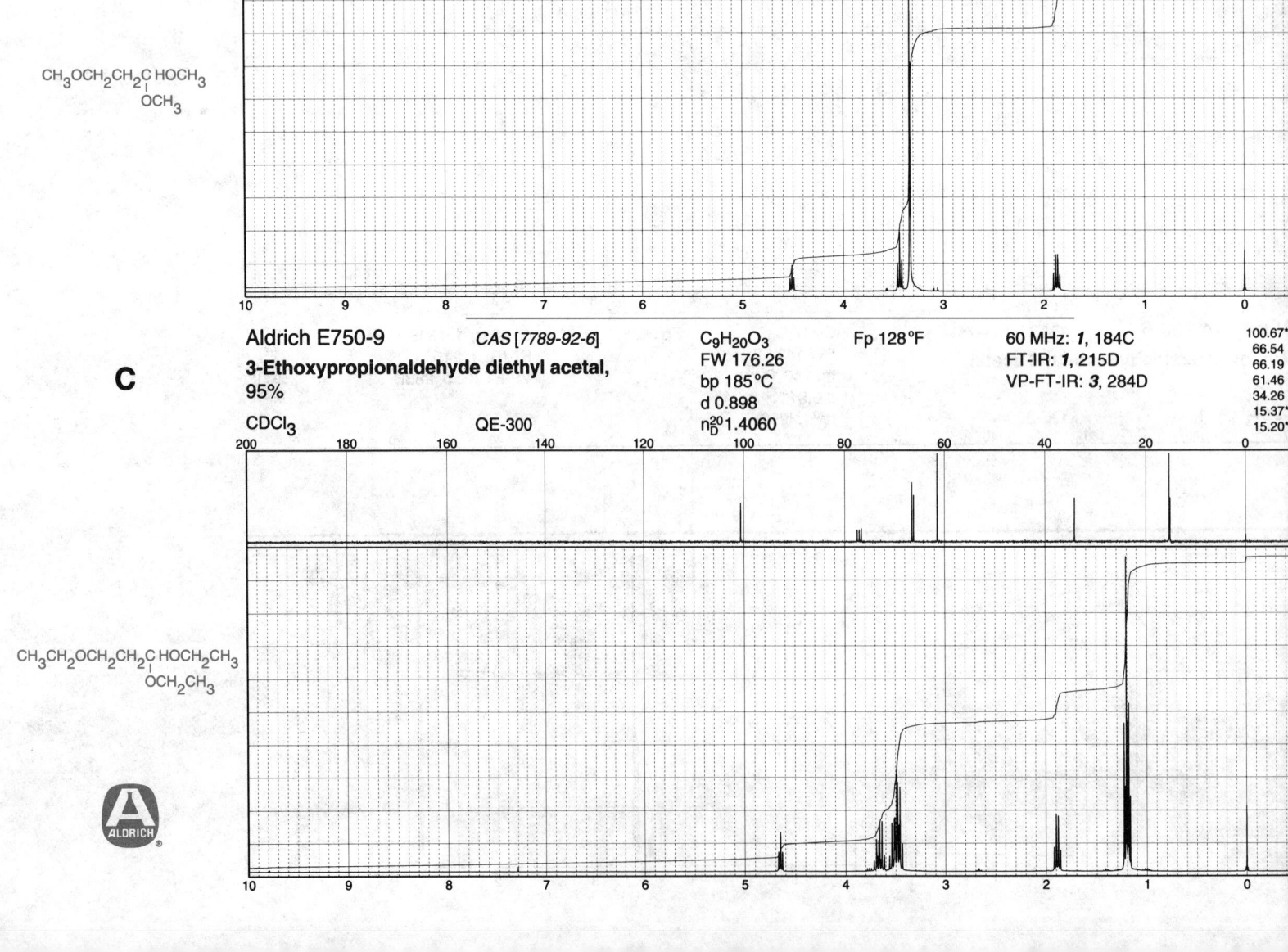

A

Aldrich 28,340-1   *CAS [4819-77-6]* $C_8H_{18}O_3$ VP-FT-IR: *3*, 284B

Ethoxyacetaldehyde diethyl acetal, 99%   FW 162.23
bp 165°C
d 0.896
n_D^{20}1.4009

101.19*
71.22
66.91
62.17
15.36*
15.13*

CDCl₃   QE-300

$CH_3CH_2OCH_2CH(OCH_2CH_3)_2$

B

Aldrich 24,470-8   *CAS [14315-97-0]* $C_6H_{14}O_3$ Fp 105°F   FT-IR: *1*, 215C

1,1,3-Trimethoxypropane, 99%   FW 134.18   VP-FT-IR: *3*, 284C
bp 46°C (17 mm)
d 0.942
n_D^{20}1.4000

102.27*
68.62
58.64*
53.06*
33.06

CDCl₃   QE-300

$CH_3OCH_2CH_2CH(OCH_3)OCH_3$

C

Aldrich E750-9   *CAS [7789-92-6]* $C_9H_{20}O_3$ Fp 128°F   60 MHz: *1*, 184C

3-Ethoxypropionaldehyde diethyl acetal,   FW 176.26   FT-IR: *1*, 215D
95%   bp 185°C   VP-FT-IR: *3*, 284D
d 0.898
n_D^{20}1.4060

100.67*
66.54
66.19
61.46
34.26
15.37*
15.20*

CDCl₃   QE-300

$CH_3CH_2OCH_2CH_2CH(OCH_2CH_3)OCH_2CH_3$

Aldrich M1,300-9   *CAS [10138-89-3]*

3-Methoxybutyraldehyde dimethyl acetal,
98%

CDCl$_3$ QE-300

$C_7H_{16}O_3$
FW 148.20
bp 157°C
d 0.921
n$_D^{20}$1.4030

Fp 117°F

60 MHz: *1*, 184D
FT-IR: *1*, 216A
VP-FT-IR: *3*, 285A

102.32*
73.48*
56.00*
53.10*
52.86*
39.98
19.25*

A

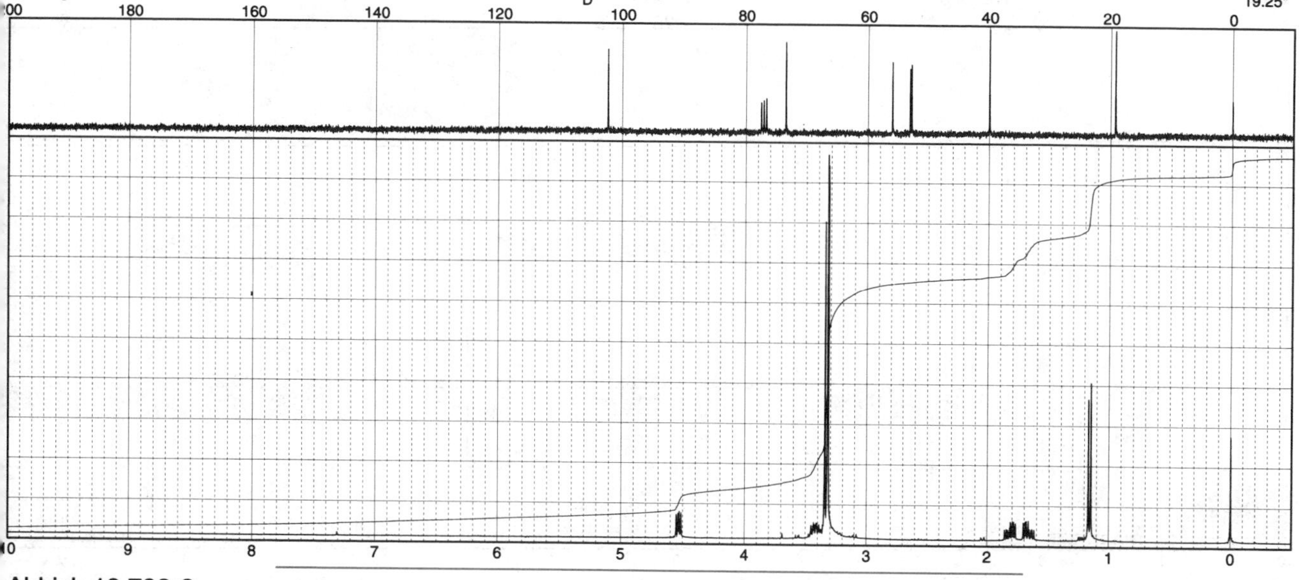

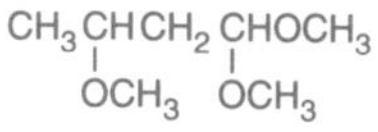

CH$_3$CHCH$_2$CHOCH$_3$
  | |
  OCH$_3$ OCH$_3$

Aldrich 19,788-2   *CAS [6607-66-5]*

1,3,3-Trimethoxybutane, 97%

CDCl$_3$ QE-300

$C_7H_{16}O_3$
FW 148.20
bp 62°C (20 mm)
d 0.940
n$_D^{20}$1.4100

Fp 114°F

60 MHz: *1*, 190B
FT-IR: *1*, 216B
VP-FT-IR: *3*, 285B

100.48
68.90
58.59*
48.01*
36.34
21.58*

B

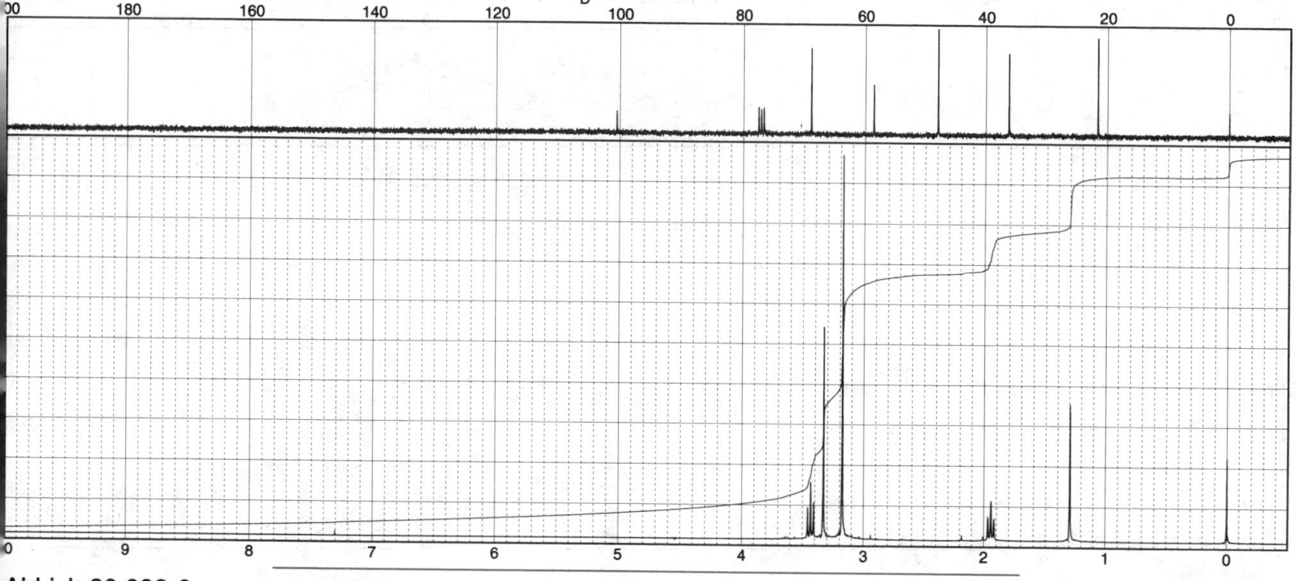

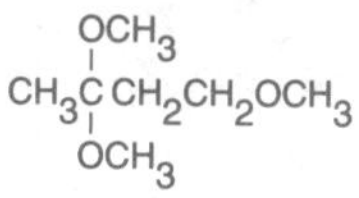

            OCH$_3$
             |
CH$_3$C CH$_2$CH$_2$OCH$_3$
             |
            OCH$_3$

Aldrich 20,098-0   *CAS [2207-27-4]*

5,5-Dimethoxy-1,2,3,4-tetrachlorocyclopentadiene,
96%

CDCl$_3$ QE-300

$C_7H_6Cl_4O_2$
FW 263.94
bp 109°C (11 mm)
d 1.501
n$_D^{20}$1.5250

Fp 152°F

60 MHz: *1*, 213D
FT-IR: *1*, 249D
VP-FT-IR: *3*, 331A

129.35
128.52
104.69
51.83*

C

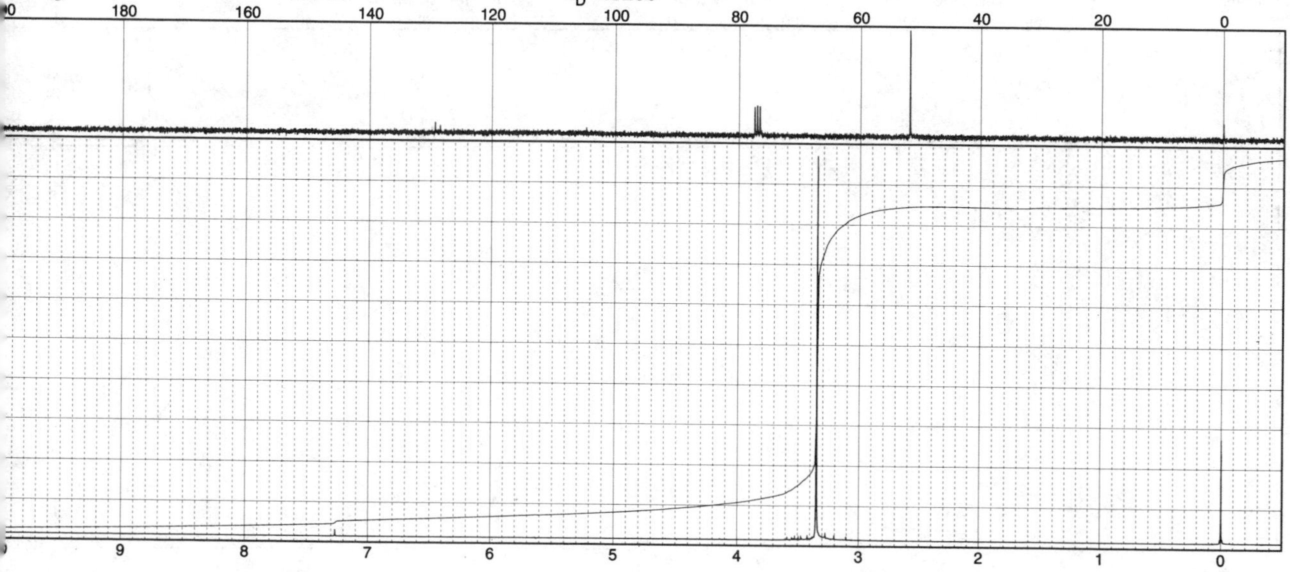

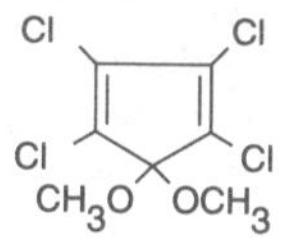

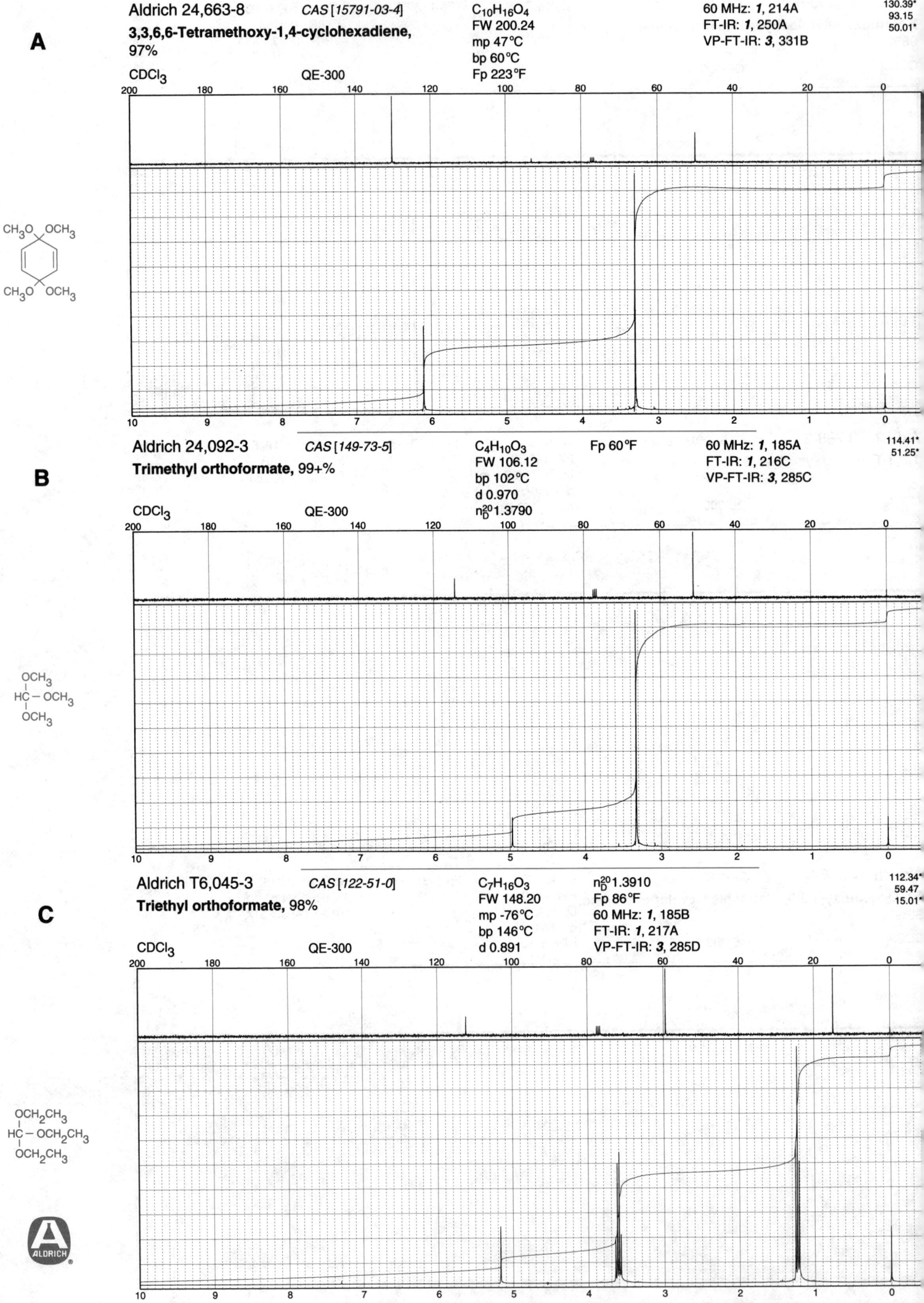

A

Aldrich 24,663-8   *CAS [15791-03-4]*

3,3,6,6-Tetramethoxy-1,4-cyclohexadiene, 97%

$C_{10}H_{16}O_4$
FW 200.24
mp 47°C
bp 60°C
Fp 223°F

60 MHz: *1*, 214A
FT-IR: *1*, 250A
VP-FT-IR: *3*, 331B

130.39*
93.15
50.01*

CDCl₃   QE-300

B

Aldrich 24,092-3   *CAS [149-73-5]*

Trimethyl orthoformate, 99+%

$C_4H_{10}O_3$
FW 106.12
bp 102°C
d 0.970
n_D^{20}1.3790

Fp 60°F

60 MHz: *1*, 185A
FT-IR: *1*, 216C
VP-FT-IR: *3*, 285C

114.41*
51.25*

CDCl₃   QE-300

C

Aldrich T6,045-3   *CAS [122-51-0]*

Triethyl orthoformate, 98%

$C_7H_{16}O_3$
FW 148.20
mp -76°C
bp 146°C
d 0.891

n_D^{20}1.3910
Fp 86°F

60 MHz: *1*, 185B
FT-IR: *1*, 217A
VP-FT-IR: *3*, 285D

112.34*
59.47
15.01*

CDCl₃   QE-300

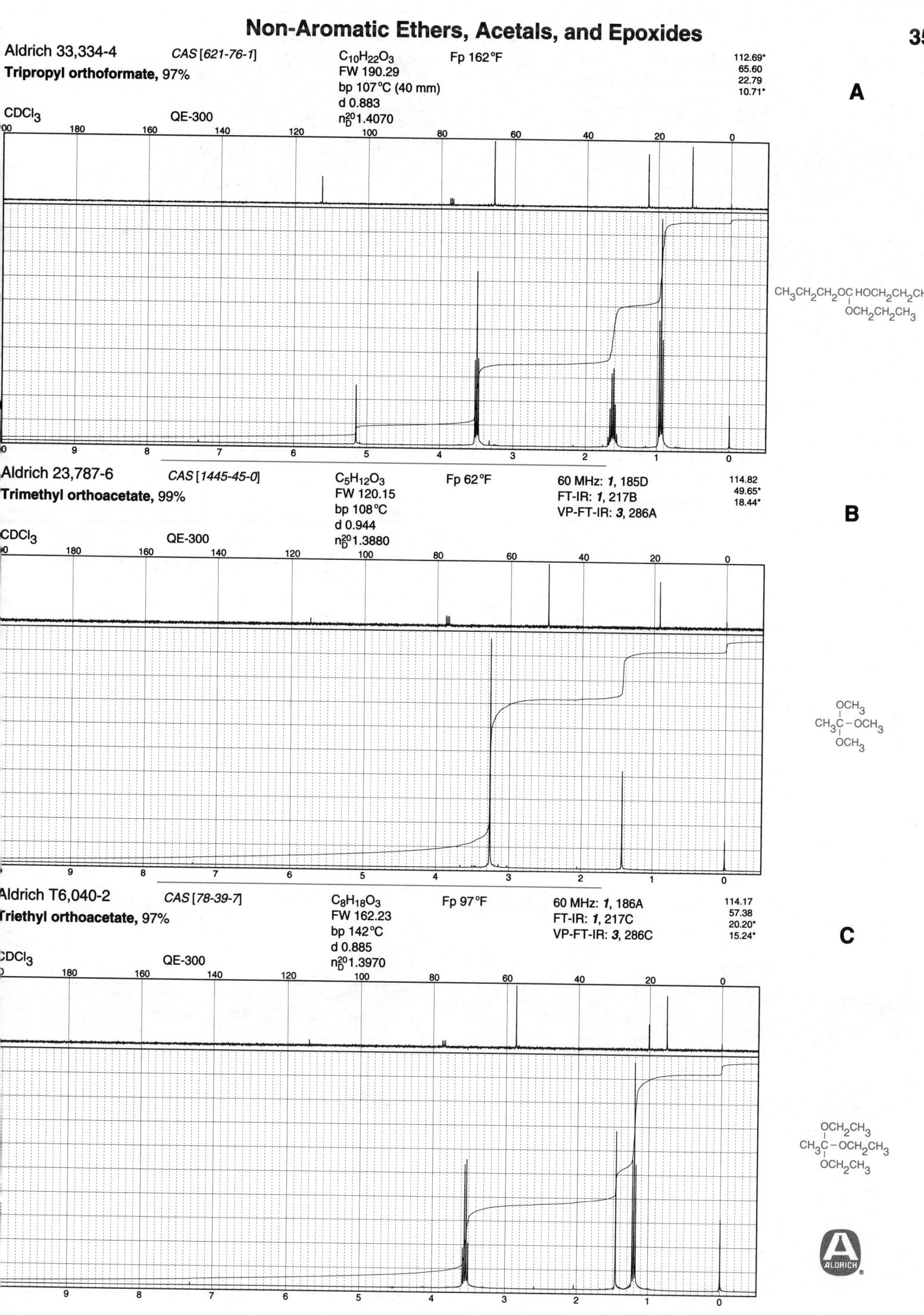

Aldrich 33,334-4   CAS [621-76-1] $C_{10}H_{22}O_3$ Fp 162°F
Tripropyl orthoformate, 97%   FW 190.29
bp 107°C (40 mm)
d 0.883
n_D^{20} 1.4070

112.69*
65.60
22.79
10.71*

A

CDCl₃   QE-300

$CH_3CH_2CH_2OC\ HOCH_2CH_2CH_3$
$OCH_2CH_2CH_3$

Aldrich 23,787-6   CAS [1445-45-0] $C_5H_{12}O_3$ Fp 62°F   60 MHz: **1**, 185D   114.82
Trimethyl orthoacetate, 99%   FW 120.15   FT-IR: **1**, 217B   49.65*
bp 108°C   VP-FT-IR: **3**, 286A   18.44*
d 0.944
n_D^{20} 1.3880

B

CDCl₃   QE-300

OCH_3
CH_3C-OCH_3
OCH_3

Aldrich T6,040-2   CAS [78-39-7] $C_8H_{18}O_3$ Fp 97°F   60 MHz: **1**, 186A   114.17
Triethyl orthoacetate, 97%   FW 162.23   FT-IR: **1**, 217C   57.38
bp 142°C   VP-FT-IR: **3**, 286C   20.20*
d 0.885   15.24*
n_D^{20} 1.3970

C

CDCl₃   QE-300

OCH_2CH_3
$CH_3C-OCH_2CH_3$
OCH_2CH_3

ALDRICH

A

Aldrich T6,060-7 *CAS [115-80-0]* $C_9H_{20}O_3$ FW 176.26 bp 158°C d 0.876 n_D^{20}1.4020 Fp 140°F FT-IR: ***1***, 217D VP-FT-IR: ***3***, 286D

Triethyl orthopropionate, 97%

115.61
56.96
24.76
15.17*
7.31*

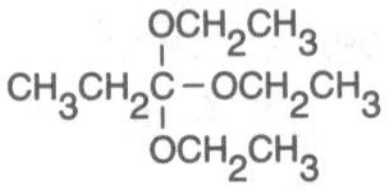

CDCl3 QE-300

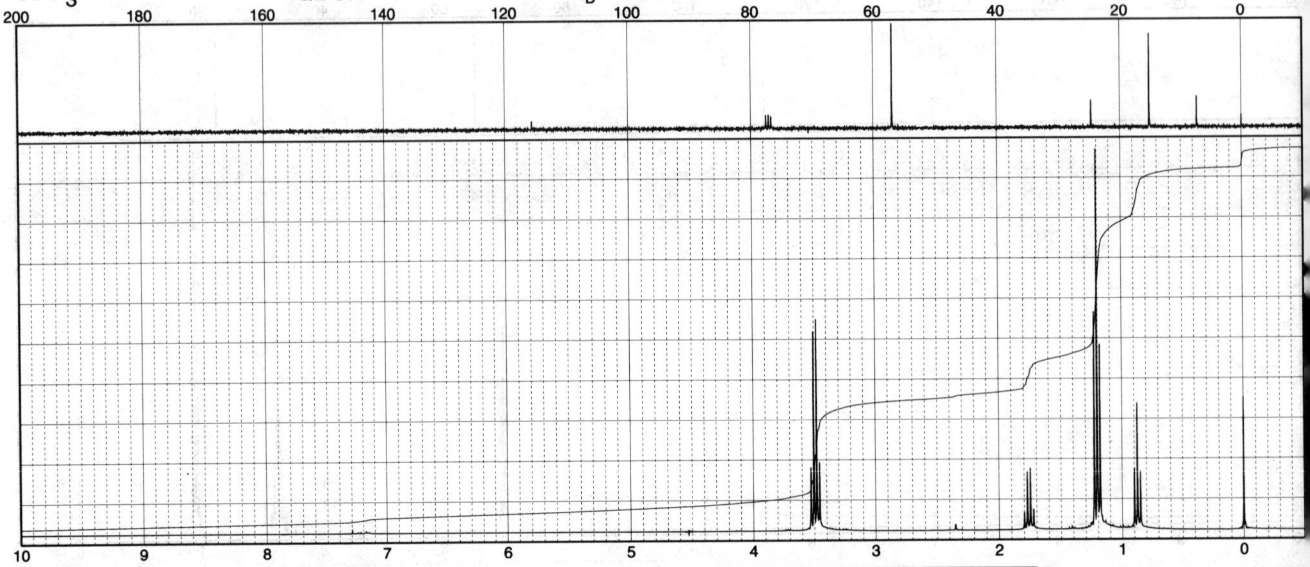

B

Aldrich 25,450-9 *CAS [43083-12-1]* $C_7H_{16}O_3$ FW 148.20 bp 146°C d 0.926 n_D^{20}1.4040 Fp 95°F FT-IR: ***1***, 218A VP-FT-IR: ***3***, 287A

Trimethyl orthobutyrate, 97%

115.79
49.32*
32.67
16.24
14.01*

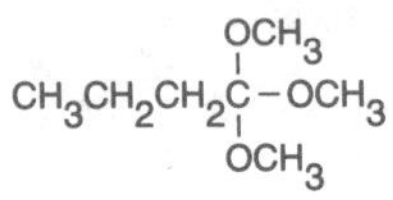

CDCl3 QE-300

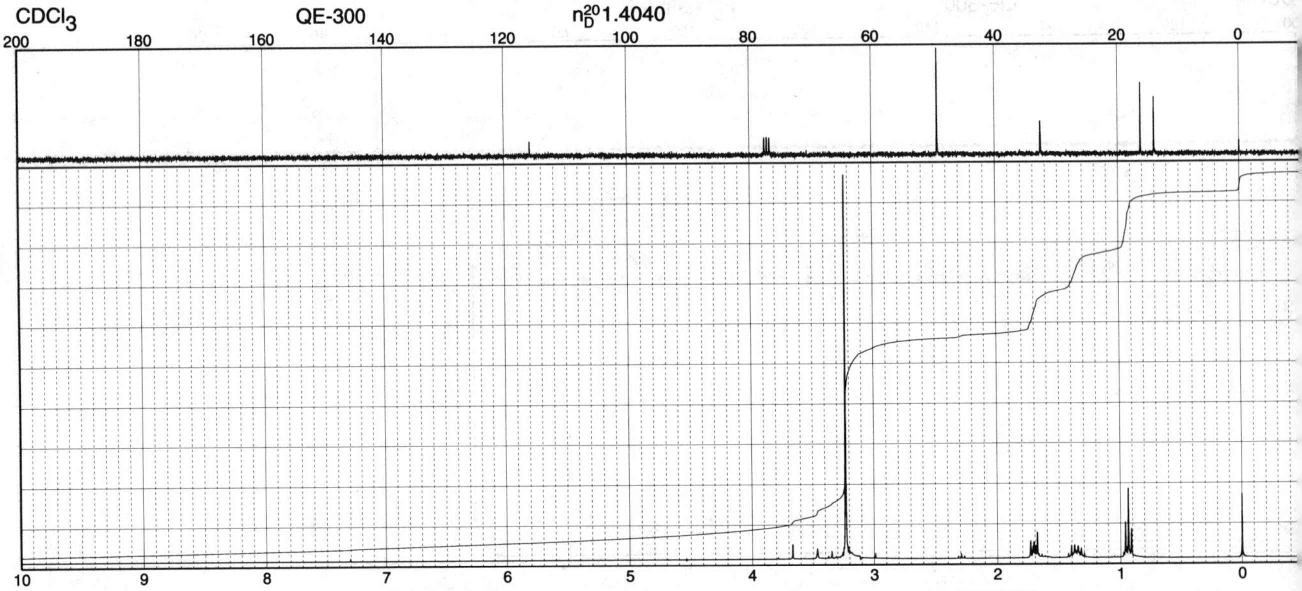

C

Aldrich 25,451-7 *CAS [13820-09-2]* $C_8H_{18}O_3$ FW 162.23 bp 165°C d 0.941 n_D^{20}1.4100 Fp 107°F FT-IR: ***1***, 218B VP-FT-IR: ***3***, 287B

Trimethyl orthovalerate, 97%

115.87
49.31*
30.09
24.95
22.60
14.03*

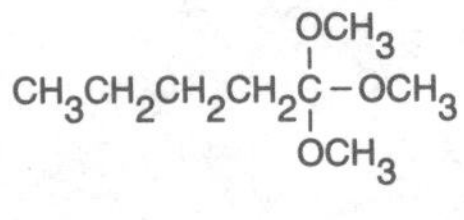

CDCl3 QE-300

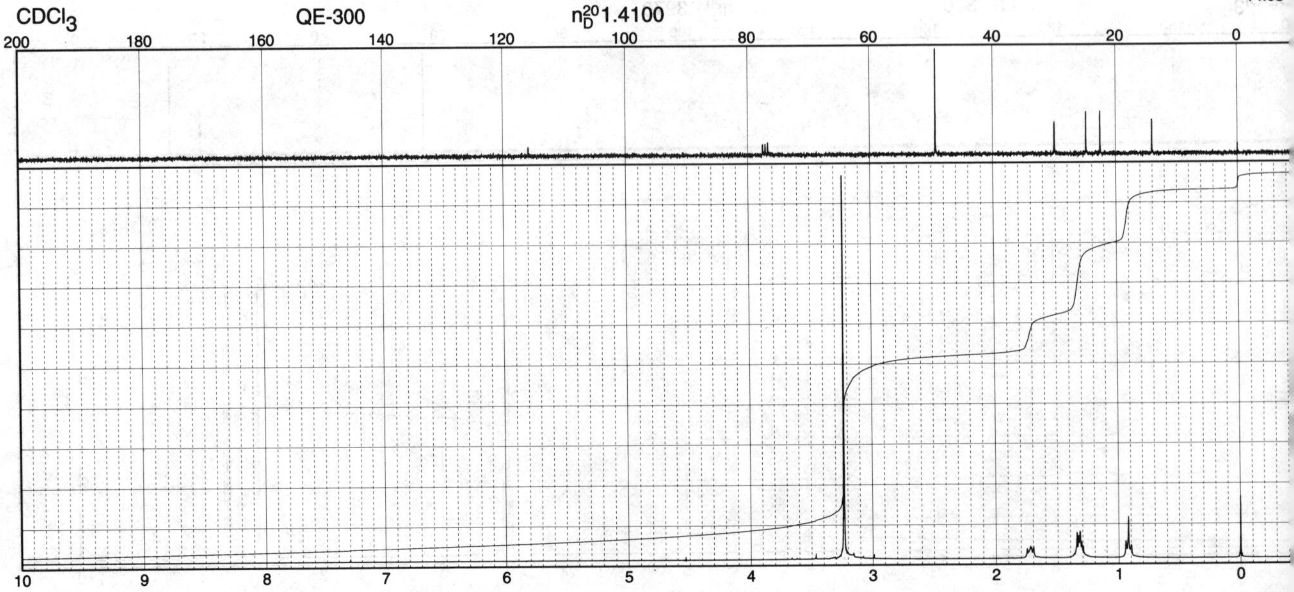

ALDRICH®

Non-Aromatic Ethers, Acetals, and Epoxides

Aldrich 13,262-4 CAS [1850-14-2] $C_5H_{12}O_4$ n_D^{20} 1.3850 120.78
Tetramethyl orthocarbonate, 99% FW 136.15 Fp 44 °F 50.32*
mp -5 °C 60 MHz: *1*, 186B
CDCl₃ QE-300 bp 114 °C FT-IR: *1*, 218C
d 1.023 VP-FT-IR: *3*, 287C

A

$$CH_3O-\underset{\underset{OCH_3}{|}}{\overset{\overset{OCH_3}{|}}{C}}-OCH_3$$

Aldrich 16,362-7 CAS [78-09-1] $C_9H_{20}O_4$ Fp 127 °F 119.63
Tetraethyl orthocarbonate, 97% FW 192.26 60 MHz: *1*, 186C 58.27
bp 159 °C FT-IR: *1*, 218D 14.79*
CDCl₃ QE-300 d 0.919 VP-FT-IR: *3*, 287D
n_D^{20} 1.3930

B

$$CH_3CH_2O-\underset{\underset{OCH_2CH_3}{|}}{\overset{\overset{OCH_2CH_3}{|}}{C}}-OCH_2CH_3$$

Aldrich 26,054-1 CAS [16646-44-9] $C_{14}H_{22}O_4$ Fp >230 °F FT-IR: *1*, 219C 134.34*
Glyoxal bis(diallyl acetal), 95% FW 254.33 VP-FT-IR: *3*, 288D 116.92
bp 158 °C (25 mm) 101.18*
d 1.001 68.44
CDCl₃ QE-300 n_D^{20} 1.4556

C

$$H_2C=CHCH_2O \quad OCH_2CH=CH_2$$
$$\underset{\underset{H_2C=CHCH_2O \quad OCH_2CH=CH_2}{}}{CHCH}$$

A

Aldrich 10,838-3   *CAS [102-52-3]* $C_7H_{16}O_4$ Fp 130°F   60 MHz: *1*, 186D

Malonaldehyde bis(dimethyl acetal),   FW 164.20   FT-IR: *1*, 219A

99%   bp 183°C   VP-FT-IR: *3*, 288B

d 0.997

CDCl$_3$ QE-300 n_D^{20} 1.4070

101.73*
52.96
36.16*

$$CH_3O \quad OCH_3$$
$$HC\ CH_2\ CH$$
$$CH_3O \quad OCH_3$$

B

Aldrich 12,960-7   *CAS [122-31-6]* $C_{11}H_{24}O_4$ Fp 108°F   60 MHz: *1*, 187A

Malonaldehyde bis(diethyl acetal), 98%   FW 220.31   FT-IR: *1*, 219B

bp 220°C   VP-FT-IR: *3*, 288C

d 0.919

CDCl$_3$ QE-300 n_D^{20} 1.4100

100.24*
61.31
38.17
15.35*

$$CH_3CH_2O \quad OCH_2CH_3$$
$$HCCH_2CH$$
$$CH_3CH_2O \quad OCH_2CH_3$$

C

Aldrich C1,940-6   *CAS [97-97-2]* $C_4H_9ClO_2$ Fp 84°F   60 MHz: *1*, 188C

Chloroacetaldehyde dimethyl acetal,   FW 124.57   FT-IR: *1*, 220D

99%   bp 129°C   VP-FT-IR: *3*, 290D

d 1.094

CDCl$_3$ QE-300 n_D^{20} 1.4150

103.27*
53.96*
42.94

$$OCH_3$$
$$ClCH_2CH$$
$$OCH_3$$

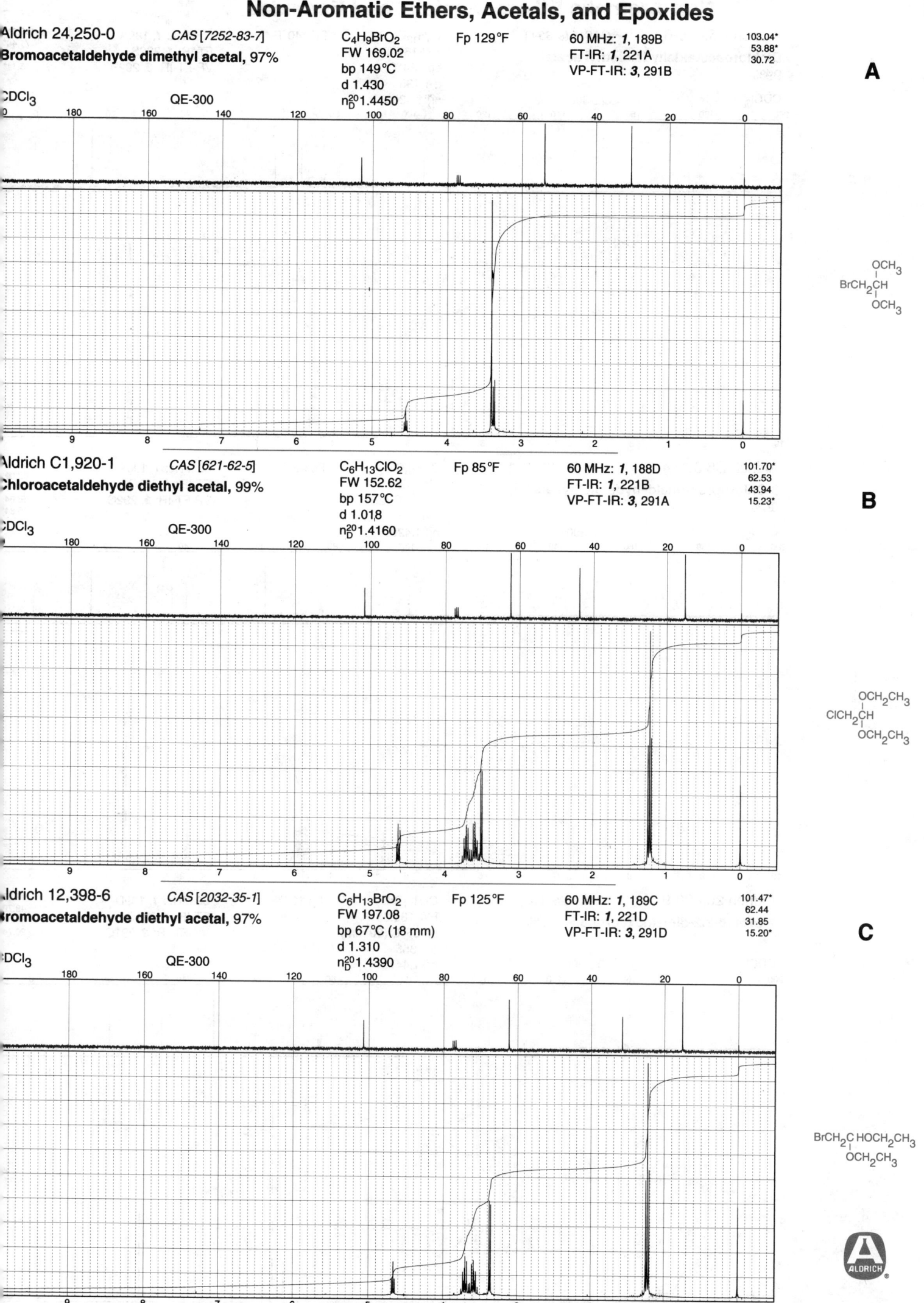

A

Aldrich 24,250-0   CAS [7252-83-7] $C_4H_9BrO_2$ Fp 129°F   60 MHz: **1**, 189B   103.04*
Bromoacetaldehyde dimethyl acetal, 97%   FW 169.02   FT-IR: **1**, 221A   53.88*
bp 149°C   VP-FT-IR: **3**, 291B   30.72
CDCl₃   QE-300   d 1.430
n₂₀D 1.4450

$$BrCH_2CH \begin{array}{c} OCH_3 \\ | \\ OCH_3 \end{array}$$

B

Aldrich C1,920-1   CAS [621-62-5] $C_6H_{13}ClO_2$ Fp 85°F   60 MHz: **1**, 188D   101.70*
Chloroacetaldehyde diethyl acetal, 99%   FW 152.62   FT-IR: **1**, 221B   62.53
bp 157°C   VP-FT-IR: **3**, 291A   43.94
CDCl₃   QE-300   d 1.018   15.23*
n₂₀D 1.4160

$$ClCH_2CH \begin{array}{c} OCH_2CH_3 \\ | \\ OCH_2CH_3 \end{array}$$

C

Aldrich 12,398-6   CAS [2032-35-1] $C_6H_{13}BrO_2$ Fp 125°F   60 MHz: **1**, 189C   101.47*
Bromoacetaldehyde diethyl acetal, 97%   FW 197.08   FT-IR: **1**, 221D   62.44
bp 67°C (18 mm)   VP-FT-IR: **3**, 291D   31.85
CDCl₃   QE-300   d 1.310   15.20*
n₂₀D 1.4390

$$BrCH_2CHOCH_2CH_3 \\ | \\ OCH_2CH_3$$

Non-Aromatic Ethers, Acetals, and Epoxides

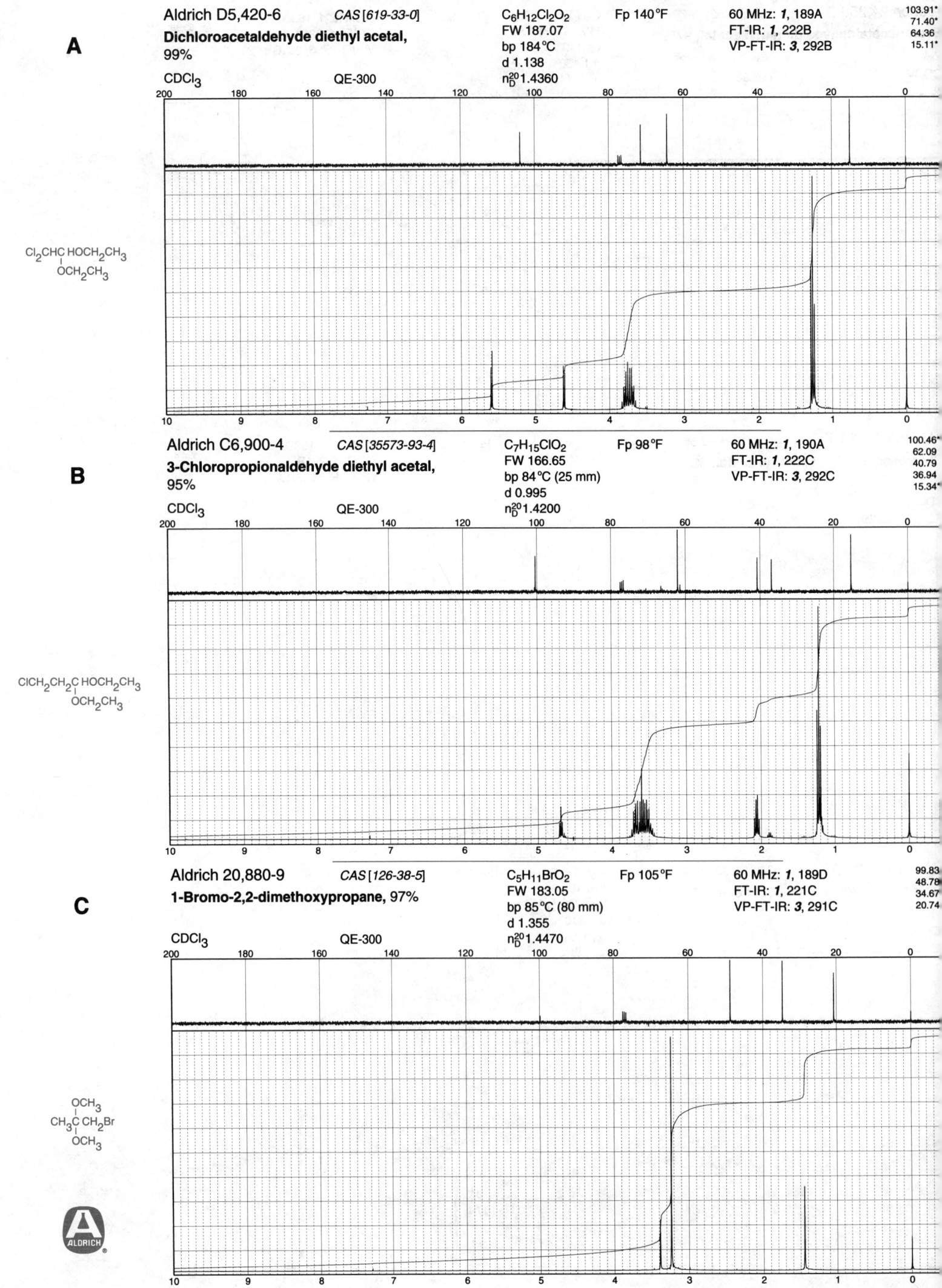

A

Aldrich D5,420-6   CAS [619-33-0] $C_6H_{12}Cl_2O_2$ Fp 140°F   60 MHz: **1**, 189A   103.91*
Dichloroacetaldehyde diethyl acetal, 99%   FW 187.07   FT-IR: **1**, 222B   71.40*
bp 184°C   VP-FT-IR: **3**, 292B   64.36
d 1.138   15.11*
CDCl₃   QE-300 n_D^{20} 1.4360

$Cl_2CHCHOCH_2CH_3$
OCH_2CH_3

B

Aldrich C6,900-4   CAS [35573-93-4] $C_7H_{15}ClO_2$ Fp 98°F   60 MHz: **1**, 190A   100.46*
3-Chloropropionaldehyde diethyl acetal, 95%   FW 166.65   FT-IR: **1**, 222C   62.09
bp 84°C (25 mm)   VP-FT-IR: **3**, 292C   40.79
d 0.995   36.94
CDCl₃   QE-300 n_D^{20} 1.4200   15.34*

$ClCH_2CH_2CHOCH_2CH_3$
OCH_2CH_3

C

Aldrich 20,880-9   CAS [126-38-5] $C_5H_{11}BrO_2$ Fp 105°F   60 MHz: **1**, 189D   99.83
1-Bromo-2,2-dimethoxypropane, 97%   FW 183.05   FT-IR: **1**, 221C   48.78
bp 85°C (80 mm)   VP-FT-IR: **3**, 291C   34.67
d 1.355   20.74
CDCl₃   QE-300 n_D^{20} 1.4470

OCH_3
CH_3CCH_2Br
OCH_3

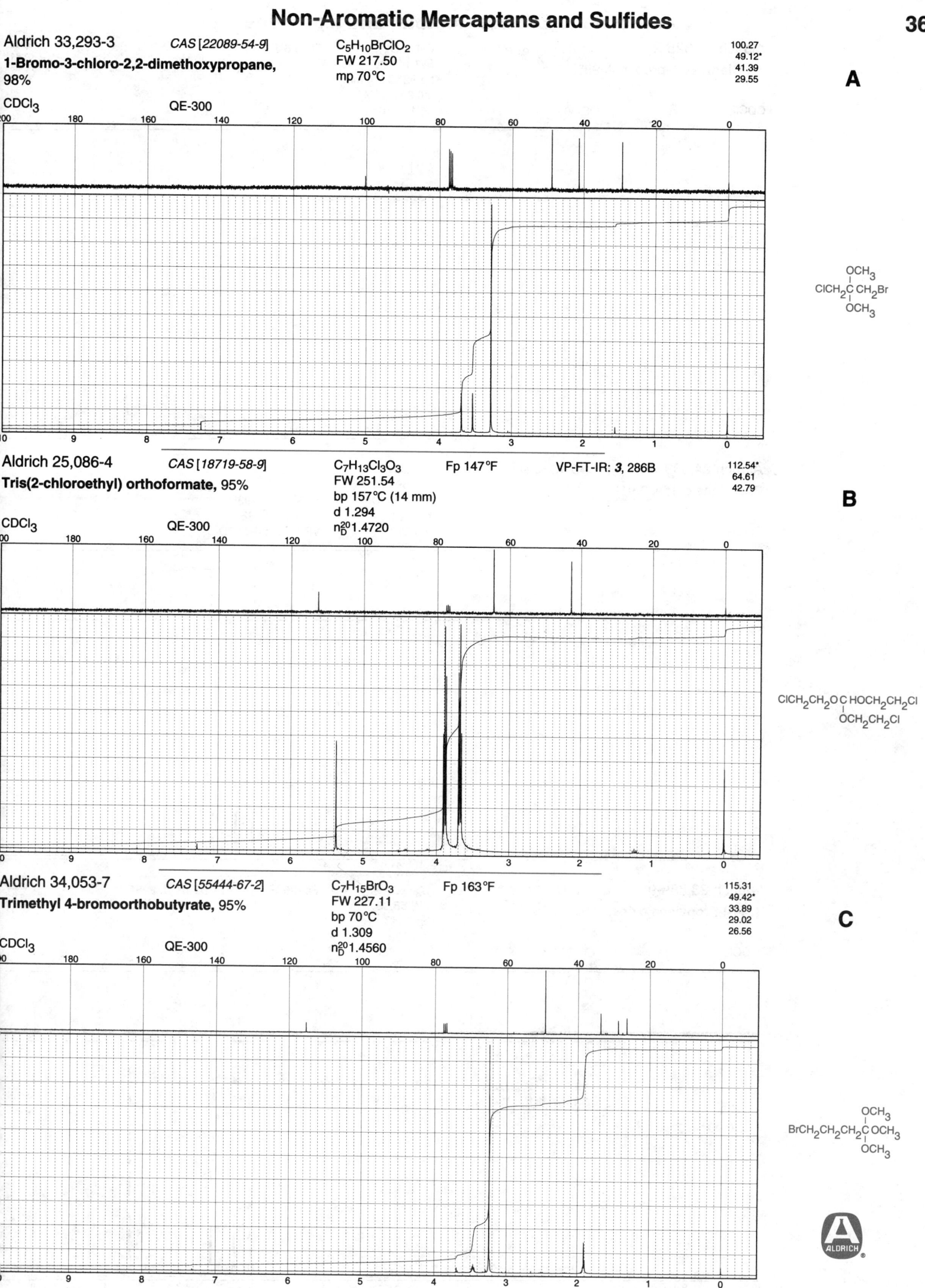

Aldrich 33,293-3   CAS [22089-54-9] $C_5H_{10}BrClO_2$ 100.27
1-Bromo-3-chloro-2,2-dimethoxypropane,   FW 217.50   49.12*
98%   mp 70°C   41.39
    29.55

A

CDCl₃   QE-300

OCH₃
|
ClCH₂C CH₂Br
|
OCH₃

Aldrich 25,086-4   CAS [18719-58-9] $C_7H_{13}Cl_3O_3$ Fp 147°F   VP-FT-IR: **3**, 286B   112.54*
Tris(2-chloroethyl) orthoformate, 95%   FW 251.54   64.61
    bp 157°C (14 mm)   42.79
    d 1.294
    n$_D^{20}$1.4720

B

CDCl₃   QE-300

ClCH₂CH₂O C HOCH₂CH₂Cl
|
OCH₂CH₂Cl

Aldrich 34,053-7   CAS [55444-67-2] $C_7H_{15}BrO_3$ Fp 163°F   115.31
Trimethyl 4-bromoorthobutyrate, 95%   FW 227.11   49.42*
    bp 70°C   33.89
    d 1.309   29.02
    n$_D^{20}$1.4560   26.56

C

CDCl₃   QE-300

OCH₃
|
BrCH₂CH₂CH₂C OCH₃
|
OCH₃

ALDRICH

A

Aldrich 27,325-2   *CAS [16777-87-0]* $C_7H_{16}O_3$ Fp 196°F

3,3-Diethoxy-1-propanol, 98%

FW 148.20
bp 92°C (14 mm)
d 0.941
n_D^{20}1.4210

102.45*
61.84
59.05
35.90
15.33*

CDCl$_3$ QE-300

CH_3CH_2O
$\quad\quad\quad\quad$ CHCH$_2$CH$_2$OH
CH_3CH_2O

B

Aldrich 24,039-7   *CAS [75-56-9]* C_3H_6O n_D^{20}1.3660

Propylene oxide, 99+%

FW 58.08
mp -112°C
bp 34°C
d 0.830

Fp -35°F
60 MHz: *1*, 193D
FT-IR: *1*, 228B
VP-FT-IR: *3*, 300D

48.22*
48.00
18.03*

CDCl$_3$ QE-300

C

Aldrich 23,889-9   *CAS [16088-62-3]* C_3H_6O Fp -35°F   FT-IR: *1*, 228C

(S)-(-)-Propylene oxide, 99%

FW 58.08
bp 35°C
d 0.829
n_D^{20}1.3660

VP-FT-IR: *3*, 301A

48.22*
48.01
18.03*

CDCl$_3$ QE-300

ALDRICH

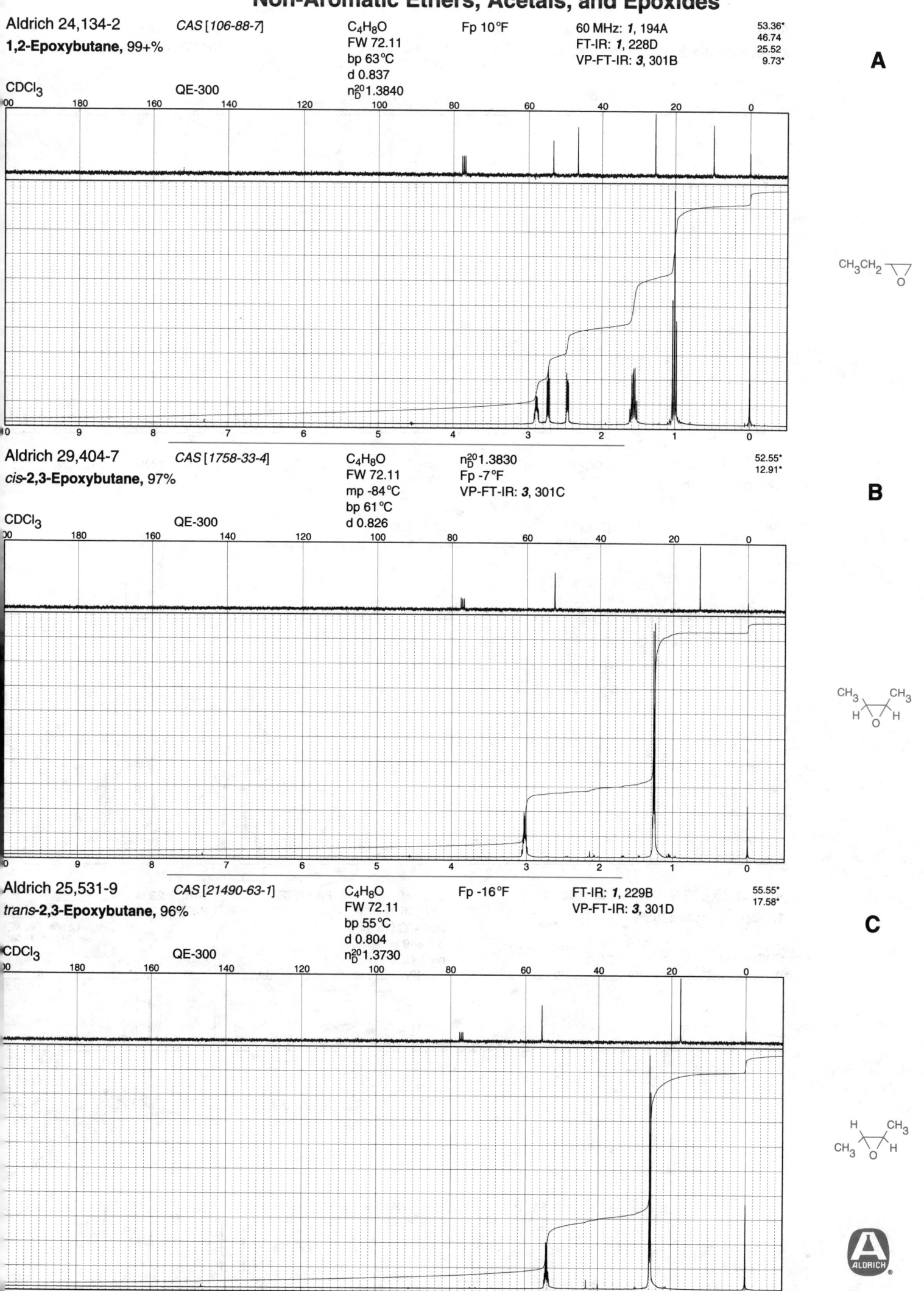

Aldrich 24,134-2

1,2-Epoxybutane, 99+%

CAS [106-88-7]

C_4H_8O
FW 72.11
bp 63°C
d 0.837
n_D^{20} 1.3840

Fp 10°F

60 MHz: *1*, 194A
FT-IR: *1*, 228D
VP-FT-IR: *3*, 301B

53.36*
46.74
25.52
9.73*

A

$CDCl_3$

QE-300

CH_3CH_2 ⏜O

Aldrich 29,404-7

cis-**2,3-Epoxybutane,** 97%

CAS [1758-33-4]

C_4H_8O
FW 72.11
mp -84°C
bp 61°C
d 0.826

n_D^{20} 1.3830
Fp -7°F
VP-FT-IR: *3*, 301C

52.55*
12.91*

B

$CDCl_3$

QE-300

CH_3 ⏜ CH_3
H ^{O} H

Aldrich 25,531-9

trans-**2,3-Epoxybutane,** 96%

CAS [21490-63-1]

C_4H_8O
FW 72.11
bp 55°C
d 0.804
n_D^{20} 1.3730

Fp -16°F

FT-IR: *1*, 229B
VP-FT-IR: *3*, 301D

55.55*
17.58*

C

$CDCl_3$

QE-300

H ^{CH_3}
CH_3 $_{O}$ H

ALDRICH

A

Aldrich 26,019-3

3,3-Dimethyl-1,2-epoxybutane, 98%

CAS [2245-30-9]

$C_6H_{12}O$
FW 100.16
bp 86°C
d 0.820
n_D^{20} 1.3995

FT-IR: *1*, 229C
VP-FT-IR: *3*, 302A

60.10*
44.05
30.57
25.57*

CDCl₃   QE-300

Structure: $CH_3-C(CH_3)(CH_3)-CH-CH_2$ with epoxide O

B

Aldrich 26,025-8

1,2-Epoxyoctane, 97%

CAS [2984-50-1]

$C_8H_{16}O$
FW 128.22
bp 63°C (17 mm)
d 0.839
n_D^{20} 1.4200

Fp 99°F

FT-IR: *1*, 229D
VP-FT-IR: *3*, 302C

52.36*	29.15
47.06	25.97
32.54	22.59
31.80	14.05*

CDCl₃   QE-300

Structure: $CH_3(CH_2)_4CH_2$ with epoxide

C

Aldrich 26,033-9

1,2-Epoxydecane, 95%

CAS [2404-44-6]

$C_{10}H_{20}O$
FW 156.27
bp 94°C (15 mm)
d 0.840
n_D^{20} 1.4290

Fp 173°F

FT-IR: *1*, 230A
VP-FT-IR: *3*, 302D

52.35*	29.49
47.06	29.25
32.54	26.01
31.88	22.69
29.55	14.09*

CDCl₃   QE-300

Structure: $CH_3(CH_2)_6CH_2$ with epoxide

ALDRICH

Aldrich 26,020-7   CAS [2855-19-8] $C_{12}H_{24}O$ Fp 221°F   FT-IR: **1**, 230B

1,2-Epoxydodecane, 98%   FW 184.32   VP-FT-IR: **3**, 303A

bp 125°C (15 mm)
d 0.844
n_D^{20} 1.4360

52.35*	29.49
47.06	29.36
32.54	26.00
31.94	22.71
29.59	14.10*

A

CDCl$_3$ QE-300

$CH_3(CH_2)_8CH_2$ (epoxide)

Aldrich 26,026-6   CAS [3234-28-4] $C_{14}H_{28}O$ Fp >230°F   FT-IR: **1**, 230C

1,2-Epoxytetradecane, tech., 85%   FW 212.38   VP-FT-IR: **3**, 303B

bp 96°C
d 0.845
n_D^{20} 1.4408

52.35*	29.48
47.06	29.38
32.53	26.00
31.95	22.71
29.67	14.10*
29.59	

B

CDCl$_3$ QE-300

$CH_3(CH_2)_{10}CH_2$ (epoxide)

Aldrich 26,021-5   CAS [7320-37-8] $C_{16}H_{32}O$ n_D^{20} 1.4460   FT-IR: **1**, 230D

1,2-Epoxyhexadecane, tech., 85%   FW 240.43   Fp 200°F   VP-FT-IR: **3**, 303C

mp 22°C
bp 178°C (12 mm)
d 0.846

52.36*	29.48
47.07	29.39
32.53	26.00
31.95	22.71
29.69	14.11*
29.59	

C

CDCl$_3$ QE-300

$CH_3(CH_2)_{12}CH_2$ (epoxide)

ALDRICH

A

Aldrich 26,023-1   *CAS [7390-81-0]*

1,2-Epoxyoctadecane, tech., 85%

$C_{18}H_{36}O$
FW 268.49
mp 34°C
bp 137°C
Fp >230°F

FT-IR: *1*, 231B
VP-FT-IR: *3*, 303D

52.33*	29.50
47.03	29.39
32.54	26.00
31.96	22.71
29.72	14.10*
29.60	

CDCl₃   QE-300

$CH_3(CH_2)_{14}CH_2 \triangle_O$

B

Aldrich 24,921-1   *CAS [29804-22-6]*

(±)-*cis*-7,8-Epoxy-2-methyloctadecane, 96%, synthetic

$C_{19}H_{38}O$
FW 282.51
Fp 142°F

FT-IR: *1*, 231A
VP-FT-IR: *3*, 304A

57.20*	27.34
38.93	26.89
31.92	26.63
29.58	22.69
29.34	22.62*
27.90*	14.10*
27.87	

CDCl₃   QE-300

$CH_3(CH_2)_8CH_2 \triangle_O CH_2 CH_2 CH_2 CH_2 CHCH_3$
CH₃

C

Aldrich 12,757-4   *CAS [930-22-3]*

Butadiene monoxide, 98%

C_4H_6O
FW 70.09
bp 66°C
d 0.870
n_D^{20} 1.4170

Fp -58°F

60 MHz: *1*, 195B
FT-IR: *1*, 233A
VP-FT-IR: *3*, 310B

135.95*	
119.39	
52.43	
48.64*	

CDCl₃   QE-300

$H_2C=CH \triangle_O$

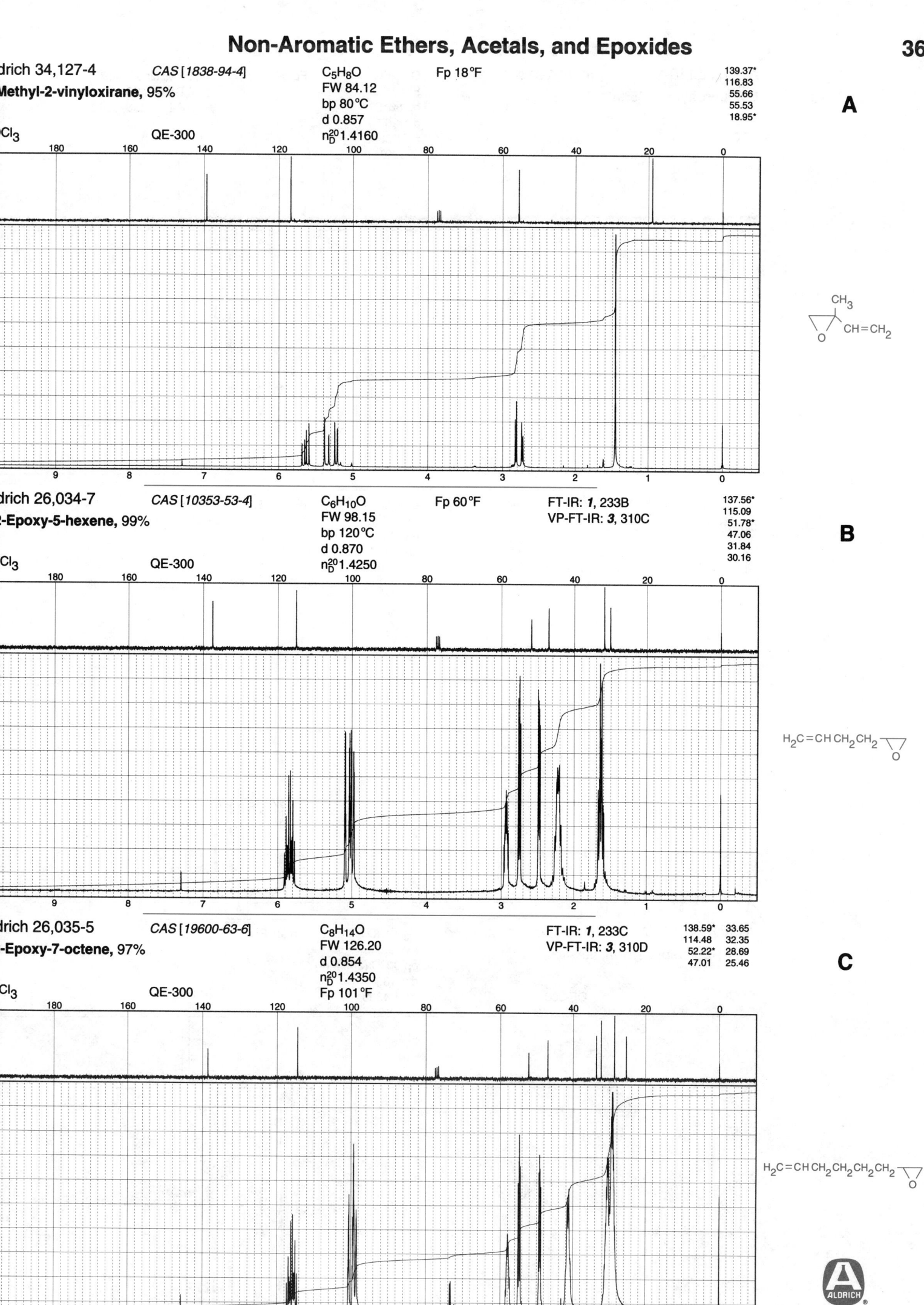

Aldrich 34,127-4   CAS [1838-94-4]   C₅H₈O   Fp 18°F
Methyl-2-vinyloxirane, 95%   FW 84.12
   bp 80°C
   d 0.857
CDCl₃   QE-300   n²⁰_D 1.4160
139.37*
116.83
55.66
55.53
18.95*
A

Aldrich 26,034-7   CAS [10353-53-4]   C₆H₁₀O   Fp 60°F   FT-IR: 1, 233B
2-Epoxy-5-hexene, 99%   FW 98.15   VP-FT-IR: 3, 310C
   bp 120°C
   d 0.870
CDCl₃   QE-300   n²⁰_D 1.4250
137.56*
115.09
51.78*
47.06
31.84
30.16
B

Aldrich 26,035-5   CAS [19600-63-6]   C₈H₁₄O   FT-IR: 1, 233C
2-Epoxy-7-octene, 97%   FW 126.20   VP-FT-IR: 3, 310D
   d 0.854
   n²⁰_D 1.4350
CDCl₃   QE-300   Fp 101°F
138.59*   33.65
114.48   32.35
52.22*   28.69
47.01   25.46
C

A

Aldrich E110-1   *CAS [503-09-3]*   C₃H₅FO   Fp 40°F   60 MHz: *1*, 194B
Epifluorohydrin, 98%   FW 76.07   FT-IR: *1*, 231C
                        bp 86°C   VP-FT-IR: *3*, 304B
                        d 1.067
CDCl₃   QE-300   n₂₀D 1.3680

84.8
82.5
50.0
49.6
43.6
43.5

FCH_2 (epoxide)

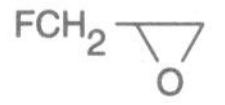

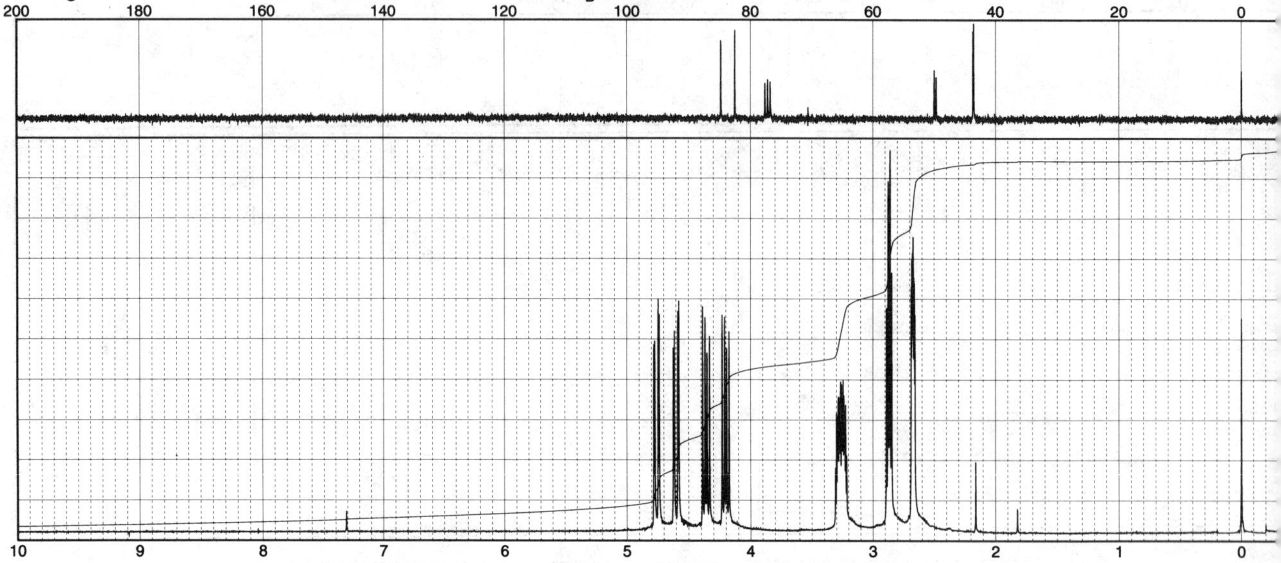

B

Aldrich 24,069-9   *CAS [106-89-8]*   C₃H₅ClO   n₂₀D 1.4380   51.2
Epichlorohydrin, 99+%   FW 92.53   Fp 93°F   46.8
                        mp -57°C   FT-IR: *1*, 231D   45.1
                        bp 116°C   VP-FT-IR: *3*, 304C
CDCl₃   QE-300   d 1.183

$ClCH_2$ (epoxide)

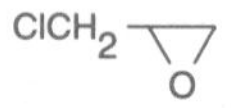

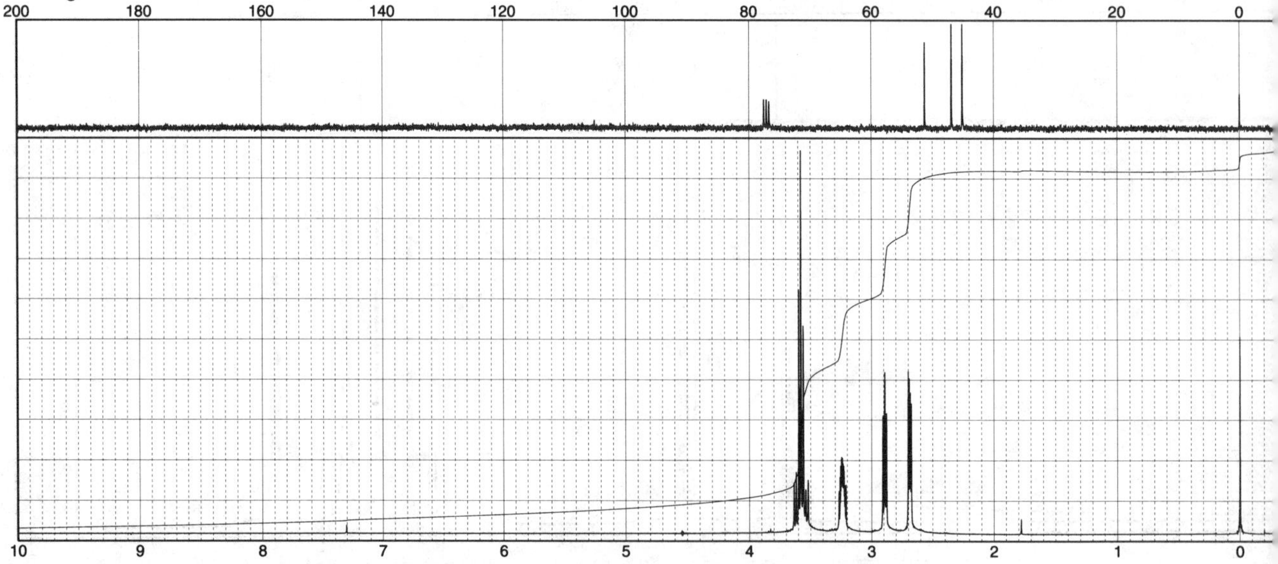

C

Aldrich 36,157-7   *CAS [51594-55-9]*   C₃H₅ClO   Fp 93°F   51.2
(R)-(-)-Epichlorohydrin, 99%   FW 92.53   46.8
                        bp 93°C (360 mm)   45.1
                        d 1.183
CDCl₃   QE-300   n₂₀D 1.4380

(epoxide) H / CH_2Cl

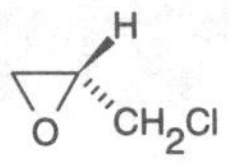

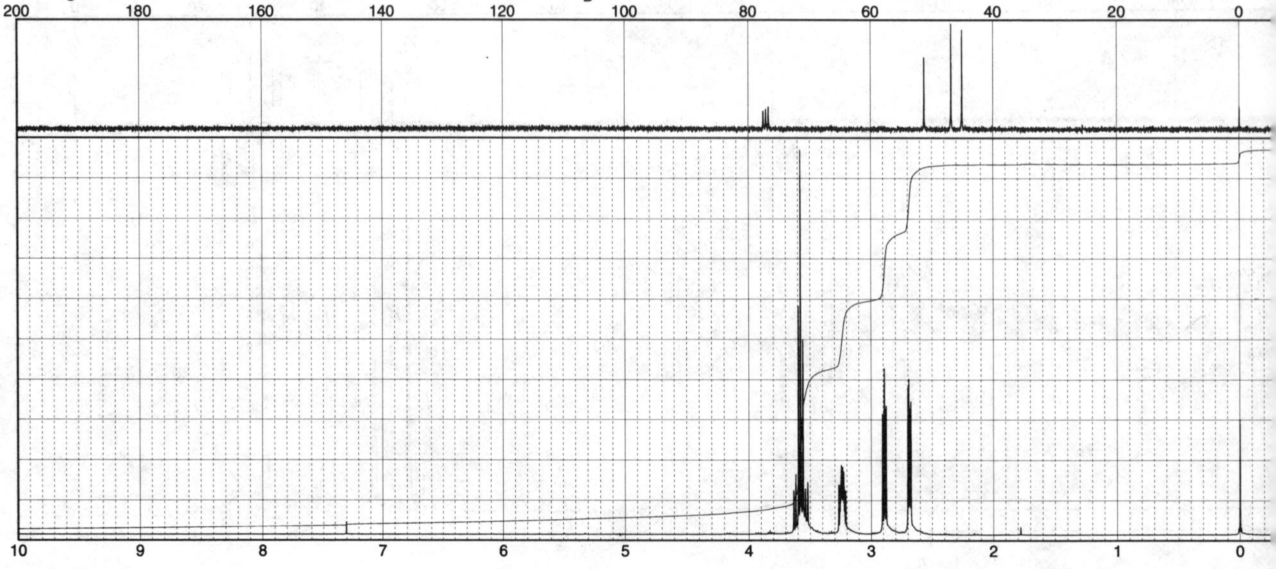

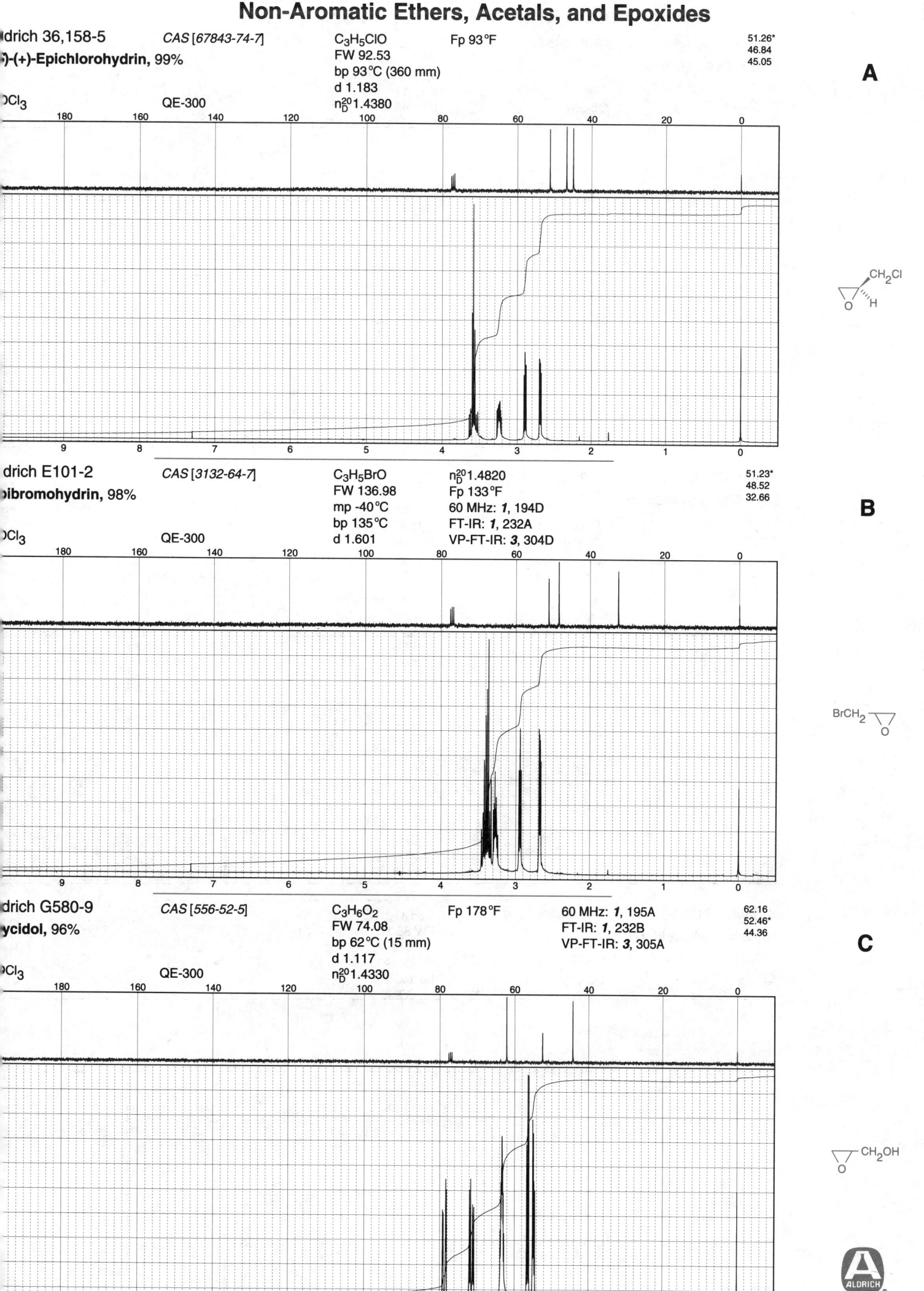

Aldrich 36,158-5   CAS [67843-74-7] C_3H_5ClO Fp 93°F

(S)-(+)-Epichlorohydrin, 99%

FW 92.53
bp 93°C (360 mm)
d 1.183
n_D^{20} 1.4380

CDCl₃   QE-300

51.26*
46.84
45.05

A

Aldrich E101-2   CAS [3132-64-7] C_3H_5BrO n_D^{20} 1.4820

Epibromohydrin, 98%

FW 136.98
mp -40°C
bp 135°C
d 1.601

Fp 133°F
60 MHz: 1, 194D
FT-IR: 1, 232A
VP-FT-IR: 3, 304D

CDCl₃   QE-300

51.23*
48.52
32.66

B

Aldrich G580-9   CAS [556-52-5] $C_3H_6O_2$ Fp 178°F

Glycidol, 96%

FW 74.08
bp 62°C (15 mm)
d 1.117
n_D^{20} 1.4330

60 MHz: 1, 195A
FT-IR: 1, 232B
VP-FT-IR: 3, 305A

CDCl₃   QE-300

62.16
52.46*
44.36

C

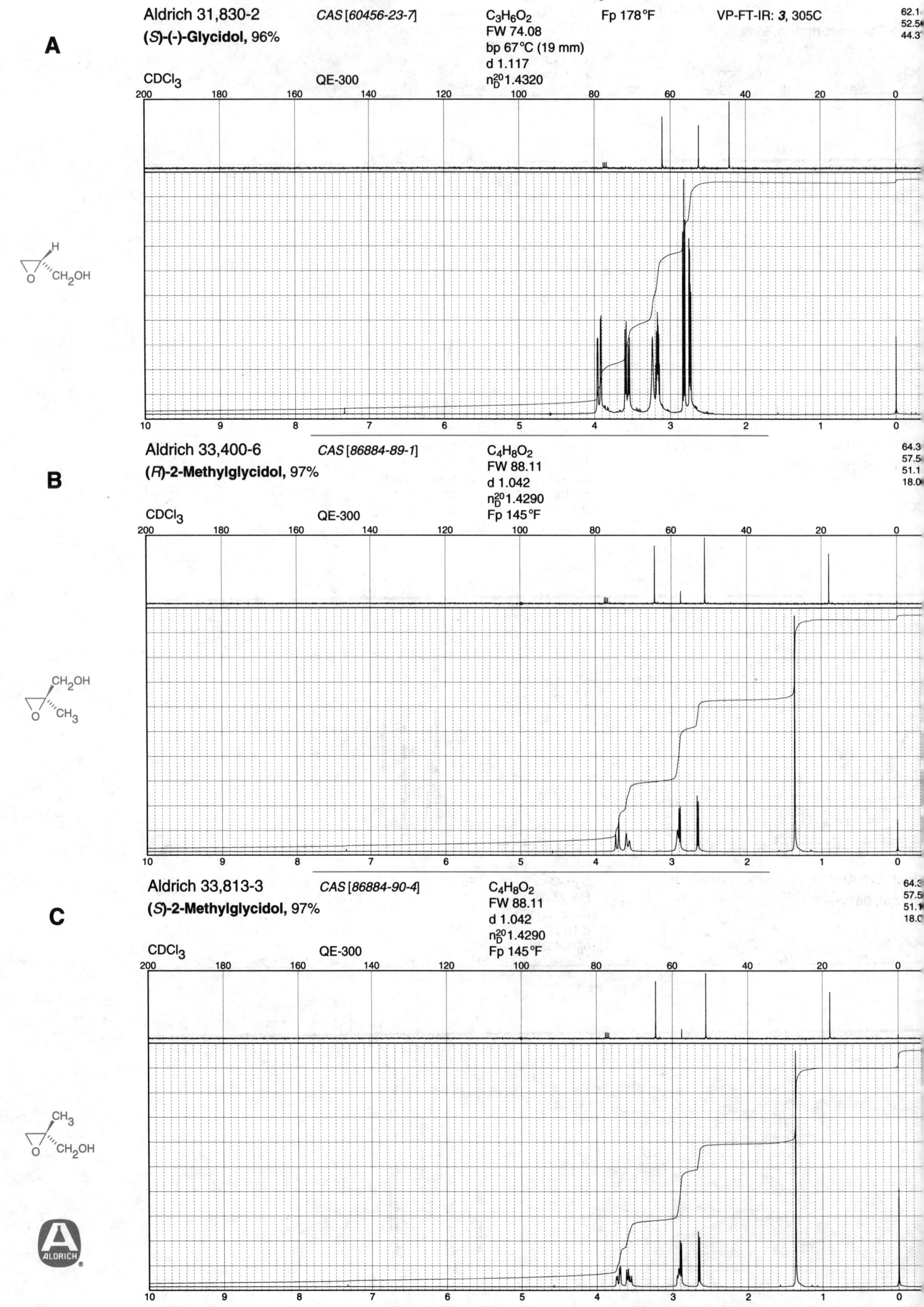

A

Aldrich 31,830-2   *CAS [60456-23-7]* $C_3H_6O_2$ Fp 178°F   VP-FT-IR: **3**, 305C

(*S*)-(-)-**Glycidol**, 96%   FW 74.08

bp 67°C (19 mm)

d 1.117

n_D^{20} 1.4320

CDCl$_3$ QE-300

B

Aldrich 33,400-6   *CAS [86884-89-1]* $C_4H_8O_2$

(*R*)-2-**Methylglycidol**, 97%   FW 88.11

d 1.042

n_D^{20} 1.4290

CDCl$_3$ QE-300   Fp 145°F

C

Aldrich 33,813-3   *CAS [86884-90-4]* $C_4H_8O_2$

(*S*)-2-**Methylglycidol**, 97%   FW 88.11

d 1.042

n_D^{20} 1.4290

CDCl$_3$ QE-300   Fp 145°F

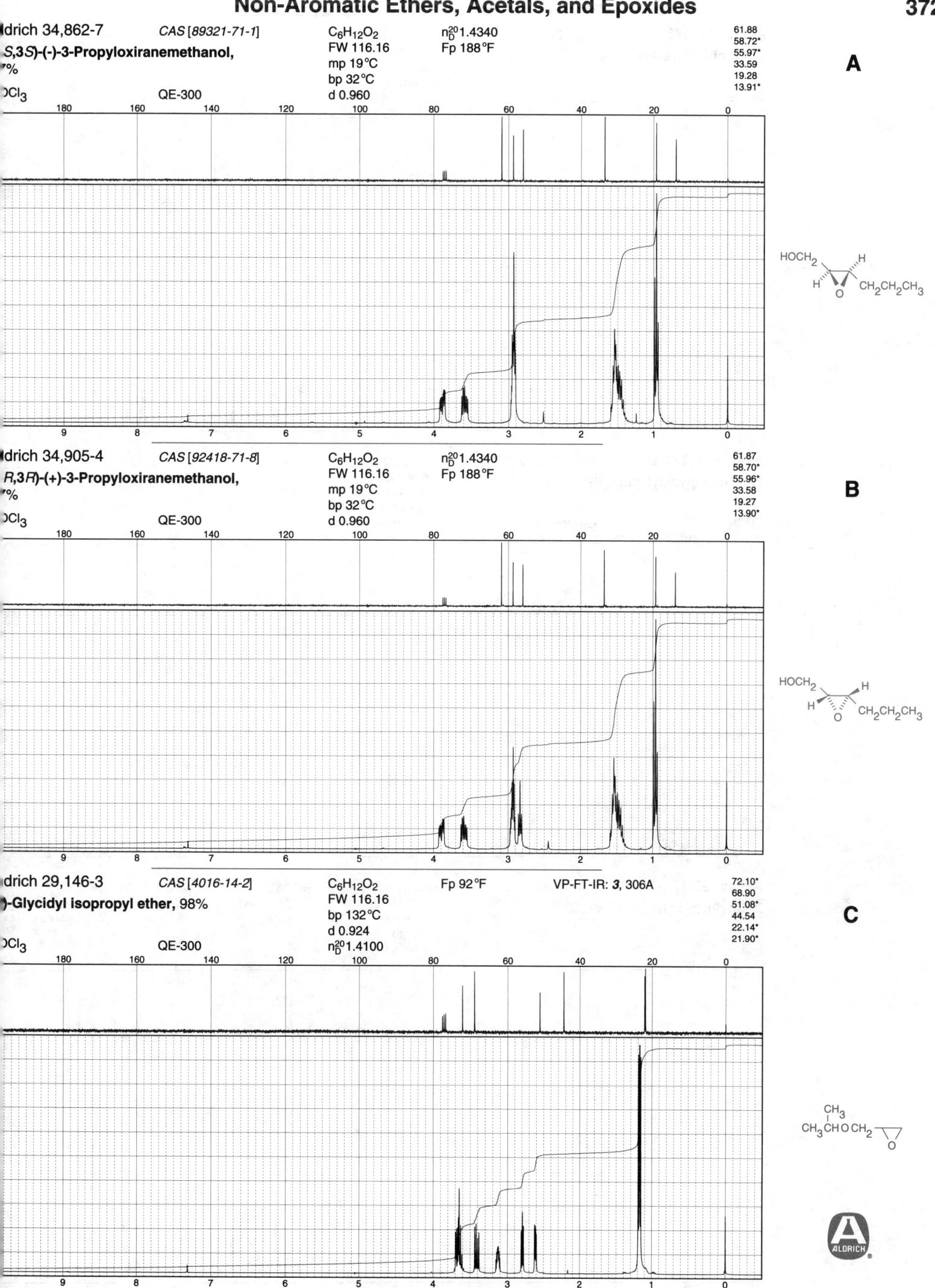

Aldrich 34,862-7   CAS [89321-71-1] $C_6H_{12}O_2$ n_D^{20}1.4340

(2S,3S)-(-)-3-Propyloxiranemethanol, 97%

FW 116.16   Fp 188°F

mp 19°C

bp 32°C

CDCl₃   QE-300   d 0.960

61.88 / 58.72* / 55.97* / 33.59 / 19.28 / 13.91*

A

HOCH₂ / H / CH₂CH₂CH₃ / O

Aldrich 34,905-4   CAS [92418-71-8] $C_6H_{12}O_2$ n_D^{20}1.4340

(2R,3R)-(+)-3-Propyloxiranemethanol, 97%

FW 116.16   Fp 188°F

mp 19°C

bp 32°C

CDCl₃   QE-300   d 0.960

61.87 / 58.70* / 55.96* / 33.58 / 19.27 / 13.90*

B

HOCH₂ / H / CH₂CH₂CH₃ / O

Aldrich 29,146-3   CAS [4016-14-2] $C_6H_{12}O_2$ Fp 92°F   VP-FT-IR: **3**, 306A

(±)-Glycidyl isopropyl ether, 98%

FW 116.16

bp 132°C

d 0.924

CDCl₃   QE-300 n_D^{20}1.4100

72.10* / 68.90 / 51.08* / 44.54 / 22.14* / 21.90*

C

CH₃ / CH₃CHOCH₂ / O

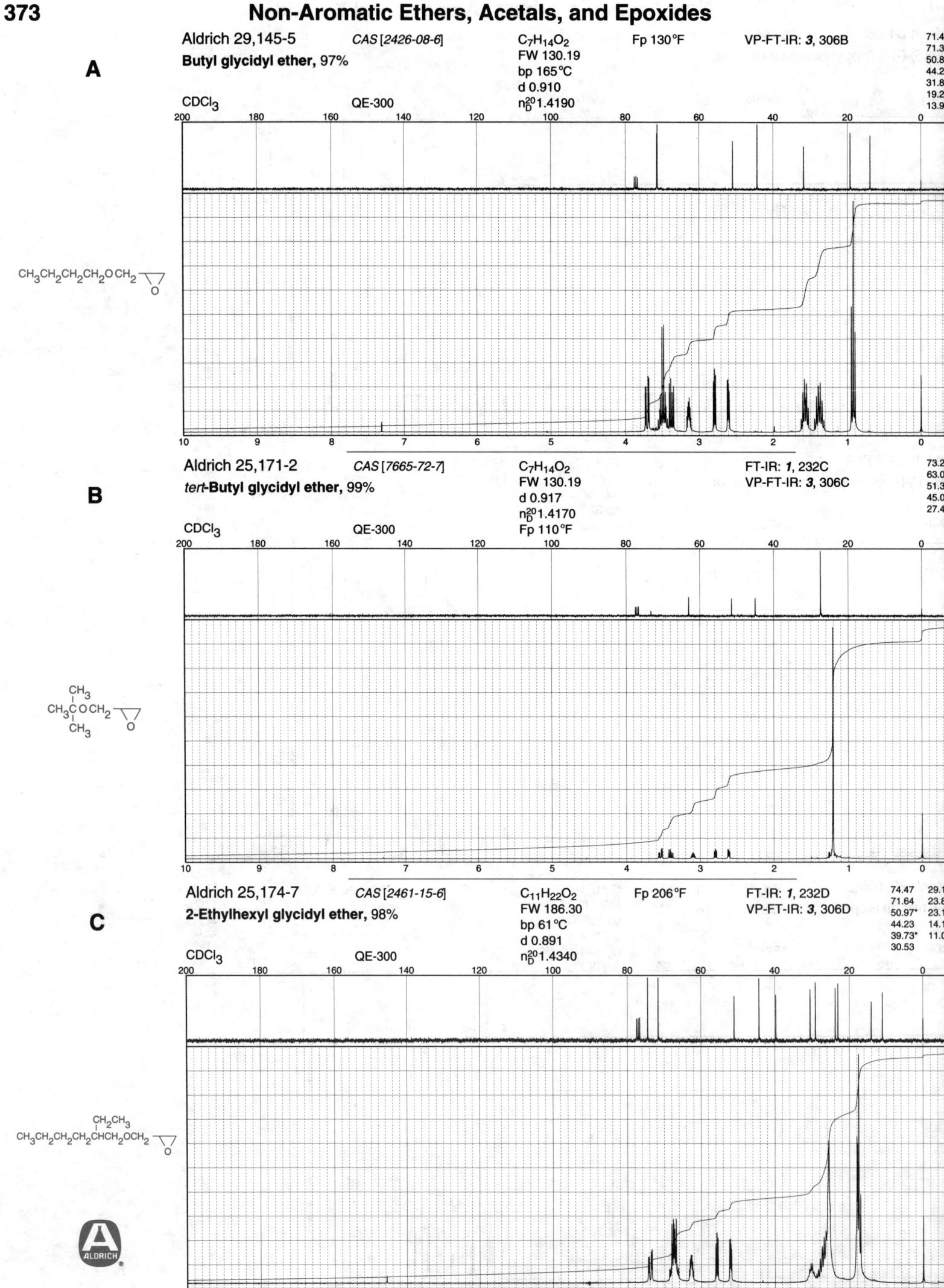

A

Aldrich 29,145-5   CAS [2426-08-6] $C_7H_{14}O_2$ Fp 130°F   VP-FT-IR: *3*, 306B
Butyl glycidyl ether, 97%   FW 130.19
bp 165°C
d 0.910
n_D^{20} 1.4190

71.4
71.3
50.8
44.2
31.8
19.2
13.9

CDCl₃   QE-300

$CH_3CH_2CH_2CH_2OCH_2$

B

Aldrich 25,171-2   CAS [7665-72-7] $C_7H_{14}O_2$ FT-IR: *1*, 232C
tert-**Butyl glycidyl ether, 99%**   FW 130.19   VP-FT-IR: *3*, 306C
d 0.917
n_D^{20} 1.4170
Fp 110°F

73.2
63.0
51.3
45.0
27.4

CDCl₃   QE-300

CH_3
CH_3COCH_2
CH_3

C

Aldrich 25,174-7   CAS [2461-15-6] $C_{11}H_{22}O_2$ Fp 206°F   FT-IR: *1*, 232D
2-Ethylhexyl glycidyl ether, 98%   FW 186.30   VP-FT-IR: *3*, 306D
bp 61°C
d 0.891
n_D^{20} 1.4340

74.47   29.1
71.64   23.8
50.97* 23.3
44.23   14.1
39.73* 11.0
30.53

CDCl₃   QE-300

CH_2CH_3
$CH_3CH_2CH_2CH_2CHCH_2OCH_2$

Aldrich A3,260-8   *CAS [106-92-3]* $C_6H_{10}O_2$ Fp 135°F   FT-IR: *1*, 233D

Allyl glycidyl ether, 99+%   FW 114.14   VP-FT-IR: *3*, 307B

bp 154°C

d 0.962

CDCl₃   QE-300 n_D^{20} 1.4330

134.40*
117.18
72.22
70.77
50.75*
44.26

A

$H_2C=CHCH_2OCH_2$ —△O

Aldrich 20,253-3   *CAS [298-18-0]* $C_4H_6O_2$ n_D^{20} 1.4340

(±)-1,3-Butadiene diepoxide, 97%   FW 86.09   Fp 114°F

mp 3°C   FT-IR: *1*, 234A

bp 57°C (25 mm)   VP-FT-IR: *3*, 308B

CDCl₃   QE-300   d 1.113

51.16*
44.19

B

Aldrich 13,956-4   *CAS [2426-07-5]* $C_8H_{14}O_2$ Fp 208°F   60 MHz: *1*, 195C

1,2,7,8-Diepoxyoctane, 97%   FW 142.20   FT-IR: *1*, 234B

bp 240°C   VP-FT-IR: *3*, 307D

d 0.997

CDCl₃   QE-300 n_D^{20} 1.4450

52.10*
46.96
32.40
25.82

C

△O — $CH_2CH_2CH_2CH_2$ — △O

A

Aldrich E2,720-3   CAS [2224-15-9] $C_8H_{14}O_4$ Fp >230°F   VP-FT-IR: *3*, 308C

Ethylene glycol diglycidyl ether, tech., 50%

FW 174.20
bp 112°C (4 mm)
d 1.118
n_D^{20}1.4630

CDCl₃   QE-300

71.97
70.67
50.77
44.17

△—CH₂OCH₂CH₂OCH₂—▽
O   O

B

Aldrich 33,803-6   CAS [17557-23-2] $C_{11}H_{20}O_4$ Fp >230°F

(±)-2,2-Dimethyl-1,3-propanediol diglycidyl ether, 97%

FW 216.28
bp 114°C (1 mm)
d 1.046
n_D^{20}1.4480

CDCl₃   QE-300

77.22
71.91
71.89
50.93
44.12
36.40
22.06

CH₃
H₂C—CH CH₂O CH₂C CH₂O CH₂CH—CH₂
O   CH₃   O

C

Aldrich 22,089-2   CAS [2425-79-8] $C_{10}H_{18}O_4$ Fp >230°F   VP-FT-IR: *3*, 308D

1,4-Butanediol diglycidyl ether, 95%

FW 202.25
bp 158°C (11 mm)
d 1.049
n_D^{20}1.4530

CDCl₃   QE-300

71.40
71.20
50.82
44.20
26.34

▽—CH₂OCH₂CH₂CH₂CH₂OCH₂—▽
O   O

Aldrich 20,562-1   CAS [112-73-2]

Diethylene glycol dibutyl ether, 99+%

$C_{12}H_{26}O_3$
FW 218.34
mp -60°C
bp 256°C
d 0.885

n_D^{20}1.4240
Fp 118°F
60 MHz: **1**, 175D
FT-IR: **1**, 207A
VP-FT-IR: **3**, 271D

71.19
70.70
70.14
31.79
19.31
13.90*

A

CDCl₃   QE-300

$(CH_3CH_2CH_2CH_2OCH_2CH_2)_2O$

Aldrich 17,518-8   CAS [285-67-6]

Cyclopentene oxide, 98%

C_5H_8O
FW 84.12
bp 102°C
d 0.964
n_D^{20}1.4340

Fp 50°F

60 MHz: **1**, 196A
FT-IR: **1**, 234D
VP-FT-IR: **3**, 309A

57.14*
27.17
18.26

B

CDCl₃   QE-300

Aldrich C10,250-4   CAS [286-20-4]

Cyclohexene oxide, 98%

$C_6H_{10}O$
FW 98.15
bp 130°C
d 0.970
n_D^{20}1.4520

Fp 81°F

60 MHz: **1**, 196C
FT-IR: **1**, 235A
VP-FT-IR: **3**, 309B

52.07*
24.48
19.47

C

CDCl₃   QE-300

A

Aldrich C11,050-7   *CAS [286-62-4]* $C_8H_{14}O$ 60 MHz: *1*, 198B   55.58
Cyclooctene oxide, 99%   FW 126.20   FT-IR: *1*, 235B   26.59
                                mp 55°C   VP-FT-IR: *3*, 309C   26.33
                                bp 55°C (5 mm)   25.62

CDCl₃   QE-300   Fp 133°F

B

Aldrich C9,720-2   *CAS [286-99-7]* $C_{12}H_{22}O$ 60 MHz: *1*, 197C   59.91* 25.31
Cyclododecane epoxide, 95%, mixture   FW 182.31   FT-IR: *1*, 236C   58.09* 24.29
of isomers   d 0.939   VP-FT-IR: *3*, 311D   31.51   24.06
                                n_D^{20} 1.4800   26.81   23.64
                                                                    26.17   22.74
CDCl₃   QE-300   Fp >230°F   25.80

C

Aldrich 21,830-8   *CAS [1686-14-2]* $C_{10}H_{16}O$ Fp 150°F   60 MHz: *1*, 198D   60.23   27.64
α-Pinene oxide, 98%   FW 152.24   FT-IR: *1*, 237A   56.85* 26.72
                                bp 103°C (50 mm)   VP-FT-IR: *3*, 312D   45.12* 25.87
                                d 0.964   40.52   22.41
                                n_D^{20} 1.4690   39.77* 20.18

CDCl₃   QE-300

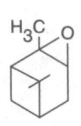

Aldrich 21,831-6   CAS [6931-54-0]   C₁₀H₁₆O   Fp 151°F

β-Pinene oxide, 90%

FW 152.24
bp 99°C (27 mm)
d 0.976
n²⁰_D 1.4770

CDCl₃   QE-300

60 MHz: *1*, 199A
FT-IR: *1*, 237B

61.40	26.12*
56.39	25.20
48.94*	23.58
40.73	22.33
40.18*	21.19*

A

Aldrich 11,780-3   CAS [3146-39-2]   C₇H₁₀O

exo-2,3-Epoxynorbornane, 98%

FW 110.16
mp 123°C
Fp 50°F

CDCl₃   QE-300

60 MHz: *1*, 196B
FT-IR: *1*, 235D
VP-FT-IR: *3*, 310A

| 51.36* |
| 36.60* |
| 26.21 |
| 25.09 |

B

Aldrich 16,046-6   CAS [27035-39-8]   C₈H₁₂O₂   Fp 221°F

1,2,5,6-Diepoxycyclooctane, 96%

FW 140.18
bp 70°C
d 1.138
n²⁰_D 1.4960

CDCl₃   QE-300

60 MHz: *1*, 198C
FT-IR: *1*, 235C
VP-FT-IR: *3*, 309D

| 55.99* |
| 22.02 |

C

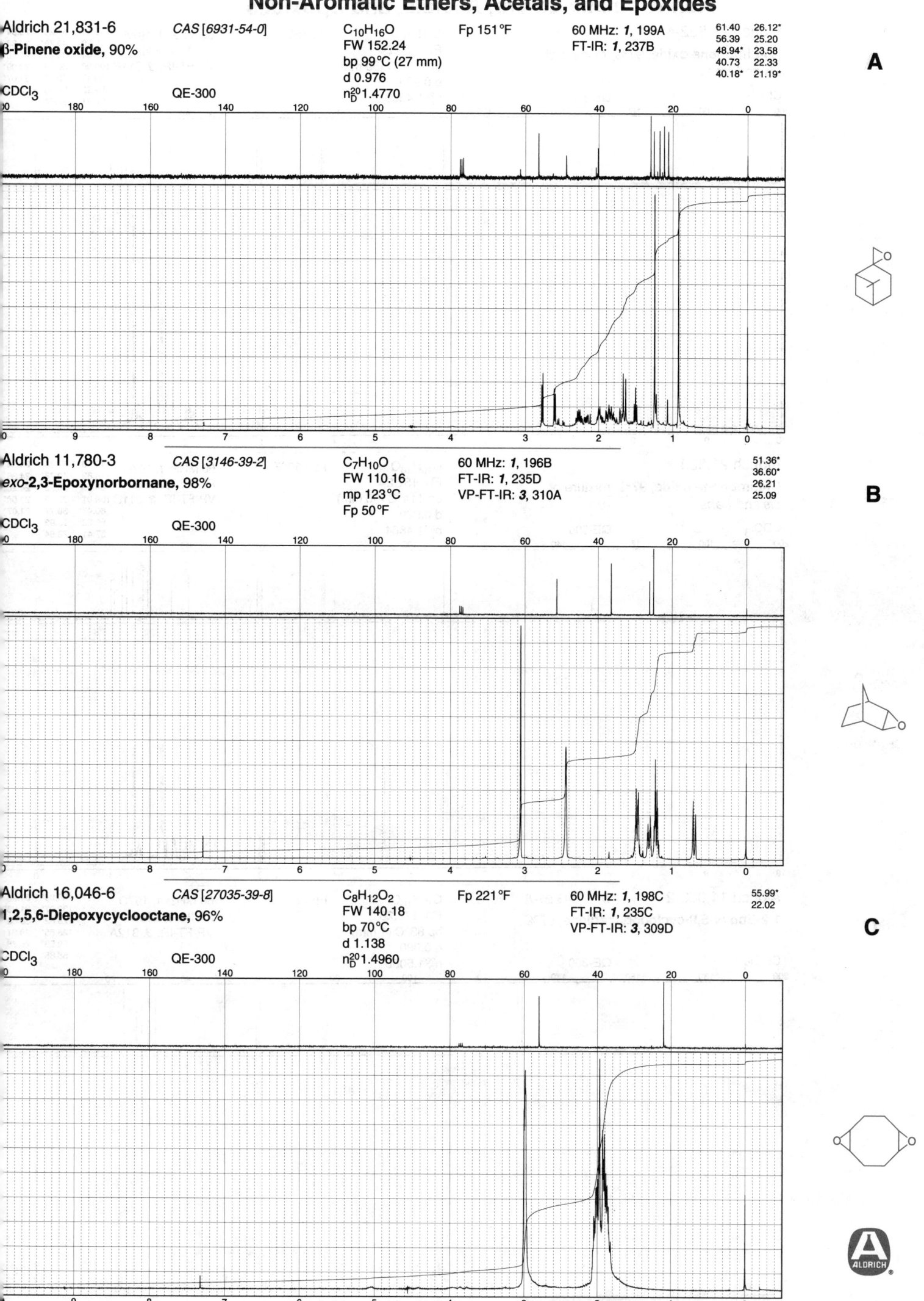

ALDRICH

A

Aldrich 21,832-4

(+)-Limonene oxide, 97%, mixture of *cis* and *trans*

CDCl₃   QE-300

C₁₀H₁₆O   Fp 150 °F
FW 152.24
bp 114 °C (50 mm)
d 0.929
n²⁰_D 1.4660

60 MHz: *1*, 197A
FT-IR: *1*, 236A
VP-FT-IR: *3*, 311B

149.11	57.25	25.94
148.90	40.78*	24.38
109.04	36.23*	24.26*
109.00	30.79	23.09*
60.47*	30.76	21.07*
59.22*	29.93	20.21*
57.41	28.64	

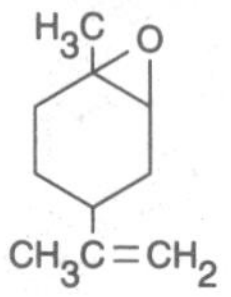

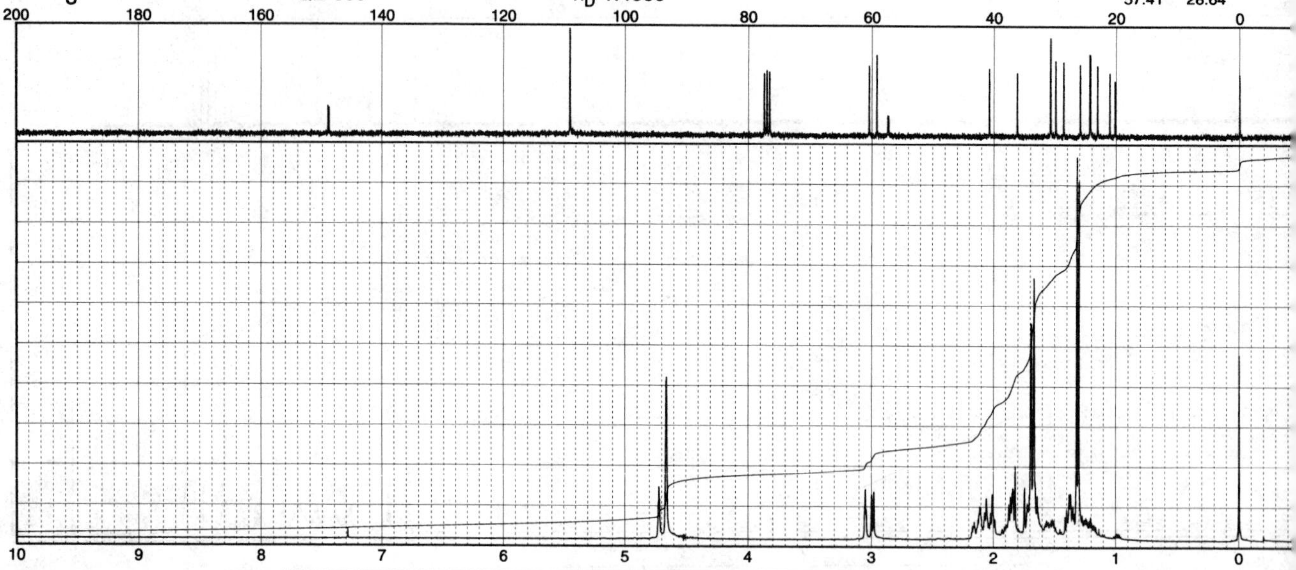

B

Aldrich 21,833-2

(−)-Limonene oxide, 97%, mixture of *cis* and *trans*

CDCl₃   QE-300

C₁₀H₁₆O   Fp 150 °F
FW 152.24
bp 114 °C (50 mm)
d 0.929
n²⁰_D 1.4664

60 MHz: *1*, 197B
FT-IR: *1*, 236B
VP-FT-IR: *3*, 311C

149.10	57.25	25.94
148.89	40.78*	24.38
109.05	36.23*	24.26*
109.01	30.79	23.08*
60.47*	30.77	21.07*
59.22*	29.93	20.21*
57.41	28.64	

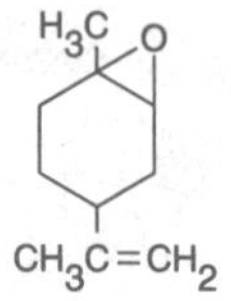

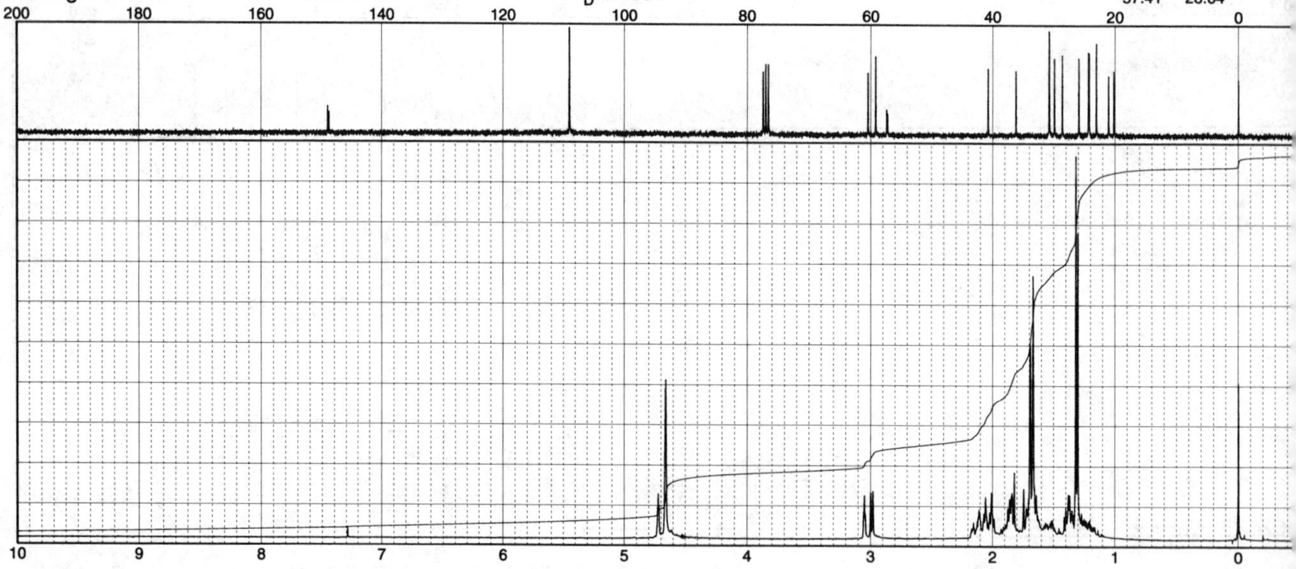

C

Aldrich 14,932-2   *CAS [943-93-1]*

1,2-Epoxy-5,9-cyclododecadiene, 97%

CDCl₃   QE-300

C₁₂H₁₈O   Fp >230 °F
FW 178.28
bp 83 °C (1 mm)
d 0.980
n²⁰_D 1.5050

60 MHz: *1*, 197D
FT-IR: *1*, 236D
VP-FT-IR: *3*, 312A

130.09*	31.98
130.07*	31.93
129.82*	30.03
128.85*	28.31
59.56*	26.86
58.85*	23.54

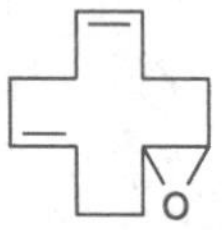

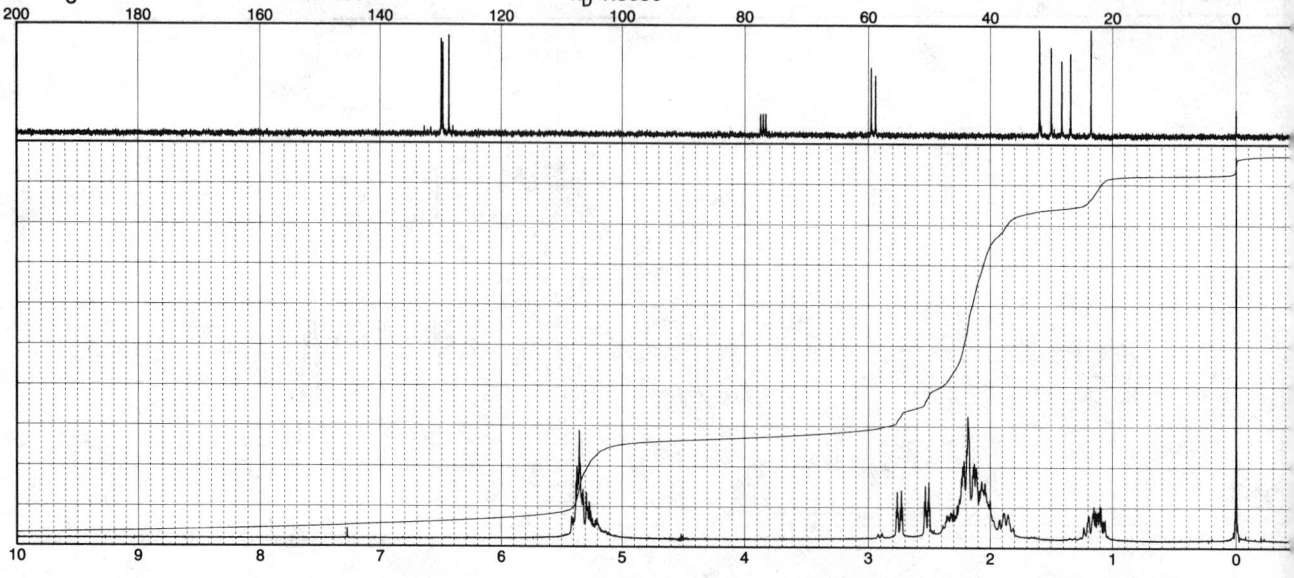

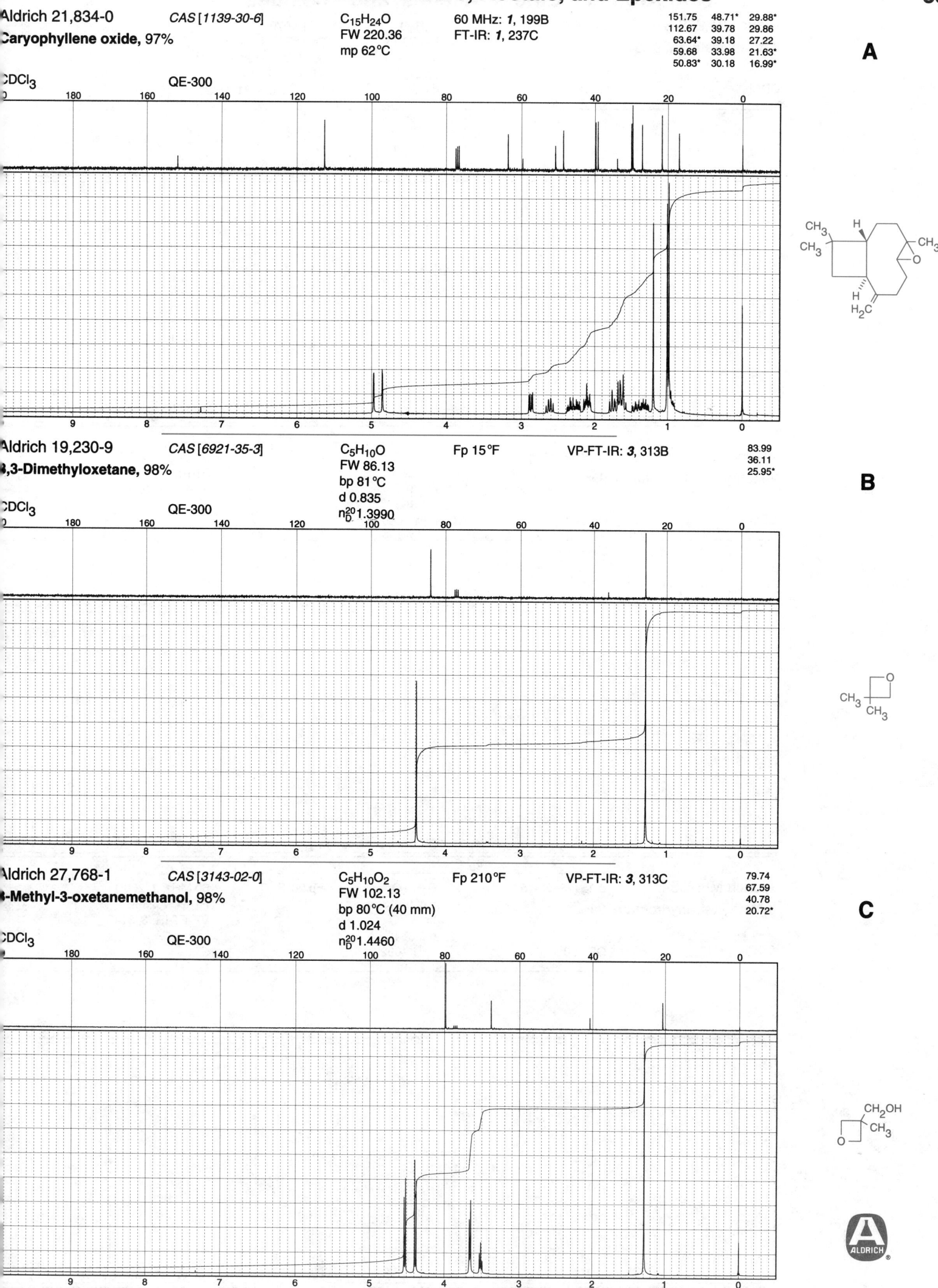

Aldrich 21,834-0   CAS [1139-30-6]
Caryophyllene oxide, 97%

$C_{15}H_{24}O$
FW 220.36
mp 62°C

60 MHz: *1*, 199B
FT-IR: *1*, 237C

151.75	48.71*	29.88*
112.67	39.78	29.86
63.64*	39.18	27.22
59.68	33.98	21.63*
50.83*	30.18	16.99*

A

CDCl₃   QE-300

Aldrich 19,230-9   CAS [6921-35-3]
3,3-Dimethyloxetane, 98%

$C_5H_{10}O$
FW 86.13
bp 81°C
d 0.835
n_D^{20} 1.3990

Fp 15°F

VP-FT-IR: *3*, 313B

83.99
36.11
25.95*

B

CDCl₃   QE-300

Aldrich 27,768-1   CAS [3143-02-0]
3-Methyl-3-oxetanemethanol, 98%

$C_5H_{10}O_2$
FW 102.13
bp 80°C (40 mm)
d 1.024
n_D^{20} 1.4460

Fp 210°F

VP-FT-IR: *3*, 313C

79.74
67.59
40.78
20.72*

C

CDCl₃   QE-300

ALDRICH

A

Aldrich 24,288-8   *CAS [109-99-9]*

Tetrahydrofuran, 99.5+%

C_4H_8O

FW 72.11

mp -108°C

bp 67°C

d 0.886

n_D^{20} 1.4070

Fp 1°F

60 MHz: *1*, 199D

FT-IR: *1*, 238A

VP-FT-IR: *3*, 313D

67.96
25.69

CDCl₃   QE-300

Aldrich 15,581-0   *CAS [96-47-9]*

2-Methyltetrahydrofuran, 99%

$C_5H_{10}O$

FW 86.13

bp 79°C

d 0.860

n_D^{20} 1.4060

Fp 12°F

60 MHz: *1*, 200B

FT-IR: *1*, 241C

VP-FT-IR: *3*, 314A

75.21
67.70
33.13
25.94
20.97

B

CDCl₃   QE-300

Aldrich M8,235-3   *CAS [13423-15-9]*

3-Methyltetrahydrofuran, 99%

$C_5H_{10}O$

FW 86.13

bp 87°C

d 0.864

n_D^{20} 1.4092

Fp 20°F

60 MHz: *1*, 200D

FT-IR: *1*, 238B

VP-FT-IR: *3*, 314B

74.90
67.94
34.32
33.65
17.77

C

CDCl₃   QE-300

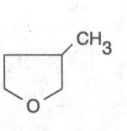

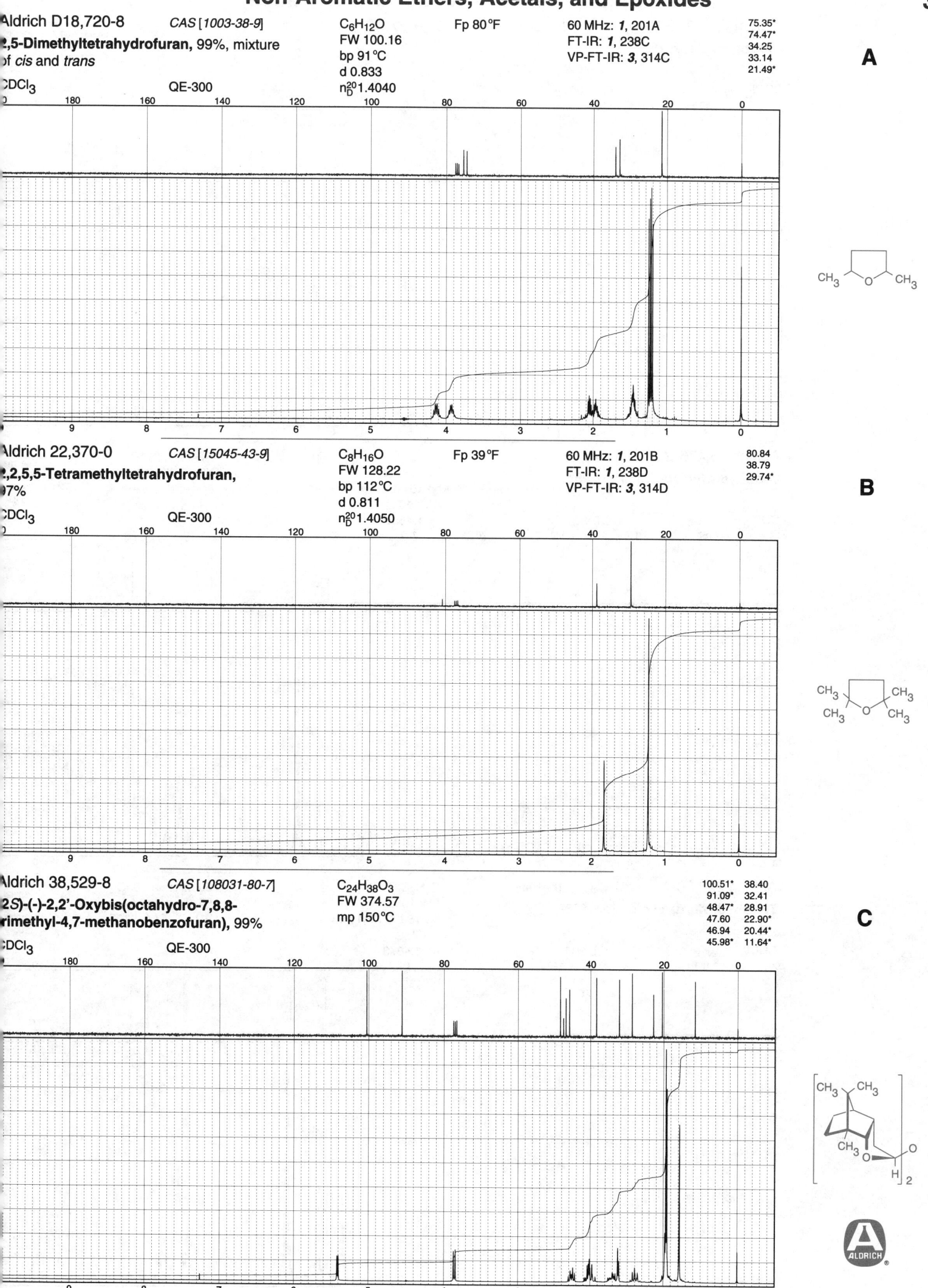

Aldrich D18,720-8   CAS [1003-38-9]   C₆H₁₂O   Fp 80°F   60 MHz: *1*, 201A

2,5-Dimethyltetrahydrofuran, 99%, mixture of cis and trans

FW 100.16
bp 91°C
d 0.833
n²⁰_D 1.4040

FT-IR: *1*, 238C
VP-FT-IR: *3*, 314C

75.35*
74.47*
34.25
33.14
21.49*

A

CDCl₃   QE-300

Aldrich 22,370-0   CAS [15045-43-9]   C₈H₁₆O   Fp 39°F   60 MHz: *1*, 201B

2,2,5,5-Tetramethyltetrahydrofuran, 97%

FW 128.22
bp 112°C
d 0.811
n²⁰_D 1.4050

FT-IR: *1*, 238D
VP-FT-IR: *3*, 314D

80.84
38.79
29.74*

B

CDCl₃   QE-300

Aldrich 38,529-8   CAS [108031-80-7]   C₂₄H₃₈O₃

(2S)-(-)-2,2'-Oxybis(octahydro-7,8,8-trimethyl-4,7-methanobenzofuran), 99%

FW 374.57
mp 150°C

100.51* 38.40
91.09*   32.41
48.47*   28.91
47.60   22.90*
46.94   20.44*
45.98*   11.64*

C

CDCl₃   QE-300

ALDRICH

A

Aldrich W34,710-8   CAS [6790-58-5] $C_{16}H_{28}O$ VP-FT-IR: **3**, 315A

(-)-Ambroxide, 99+%   FW 236.40
mp 75°C

79.84	39.95	22.63
64.93	39.74	21.13*
60.11*	36.18	20.66
57.25*	33.58*	18.41
42.44	33.05	15.04*

CDCl₃   QE-300

B

Aldrich 25,476-2   CAS [3003-84-7] C_5H_9ClO Fp 118°F   FT-IR: **1**, 241D

Tetrahydrofurfuryl chloride, 98%   FW 120.58
bp 151°C
d 1.110
n₂₀ᴅ 1.4550

VP-FT-IR: **3**, 319A

78.44*
68.78
46.94
29.37
25.82

CDCl₃   QE-300

C

Aldrich 22,750-1   CAS [1192-30-9] C_5H_9BrO 60 MHz: **1**, 200C

Tetrahydrofurfuryl bromide, 98%   FW 165.04
bp 169°C (744 mm)
d 1.452
n₂₀ᴅ 1.4873

FT-IR: **1**, 242A
VP-FT-IR: **3**, 319B

78.19*
68.82
35.68
30.39
25.91

CDCl₃   QE-300

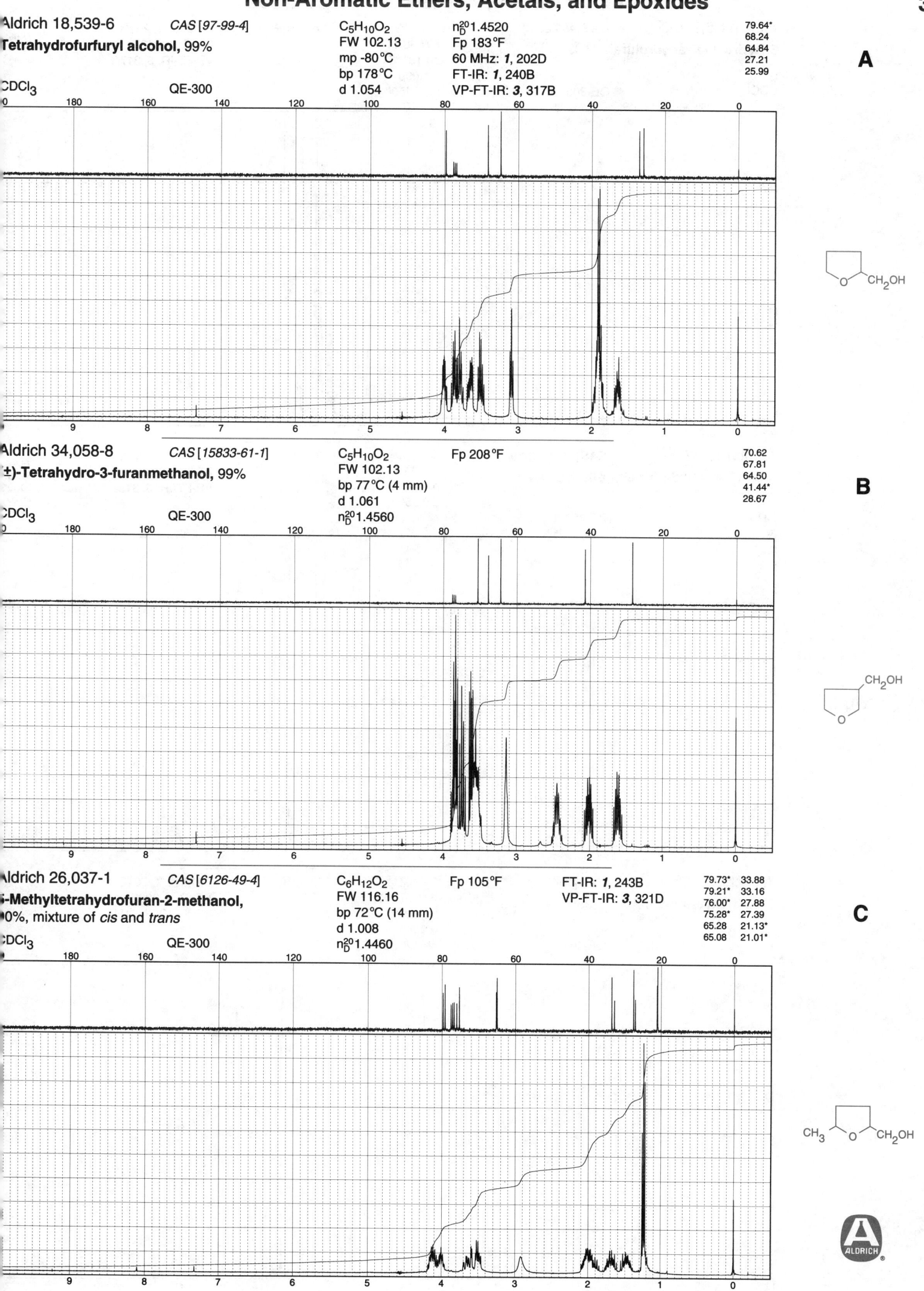

Aldrich 18,539-6   CAS [97-99-4]   C5H10O2   n20D 1.4520   79.64*
Tetrahydrofurfuryl alcohol, 99%   FW 102.13   Fp 183°F   68.24
    mp -80°C   60 MHz: 1, 202D   64.84
    bp 178°C   FT-IR: 1, 240B   27.21
    d 1.054   VP-FT-IR: 3, 317B   25.99
CDCl3   QE-300

A

Aldrich 34,058-8   CAS [15833-61-1]   C5H10O2   Fp 208°F   70.62
(±)-Tetrahydro-3-furanmethanol, 99%   FW 102.13   67.81
    bp 77°C (4 mm)   64.50
    d 1.061   41.44*
CDCl3   QE-300   n20D 1.4560   28.67

B

Aldrich 26,037-1   CAS [6126-49-4]   C6H12O2   Fp 105°F   FT-IR: 1, 243B   79.73* 33.88
5-Methyltetrahydrofuran-2-methanol,   FW 116.16   VP-FT-IR: 3, 321D   79.21* 33.16
90%, mixture of cis and trans   bp 72°C (14 mm)   76.00* 27.88
    d 1.008   75.28* 27.39
CDCl3   QE-300   n20D 1.4460   65.28 21.13*
    65.08 21.01*

C

Non-Aromatic Ethers, Acetals, and Epoxides

A

Aldrich H5,910-9   *CAS [453-20-3]* $C_4H_8O_2$ Fp 178 °F   60 MHz: *1*, 202C   75.49
3-Hydroxytetrahydrofuran, 99%   FW 88.11   FT-IR: *1*, 240A   71.66*
bp 181 °C   VP-FT-IR: *3*, 317A   66.76
d 1.090   35.48
n_D^{20} 1.4500

CDCl₃   QE-300

B

Aldrich 13,420-1   *CAS [19354-27-9]* $C_6H_{12}O_2$ Fp 92 °F   60 MHz: *1*, 202A   77.73*
Methyl tetrahydrofurfuryl ether, 98%   FW 116.16   FT-IR: *1*, 239A   75.34
bp 140 °C   VP-FT-IR: *3*, 315B   68.26
d 0.952   59.19*
n_D^{20} 1.4260   28.00
25.66

CDCl₃   QE-300

C

Aldrich 20,991-0   *CAS [13436-45-8]* $C_5H_{10}O_2$ Fp 46 °F   60 MHz: *1*, 201C   104.97*
2-Methoxytetrahydrofuran, 99%   FW 102.13   FT-IR: *1*, 239B   66.87
bp 106 °C   VP-FT-IR: *3*, 315D   54.48*
d 0.972   32.34
n_D^{20} 1.4119   23.48

CDCl₃   QE-300

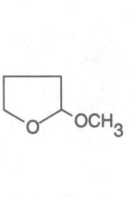

A

Aldrich 20,992-9 CAS [13436-46-9]

(±)-2-Ethoxytetrahydrofuran, 99%

$C_6H_{12}O_2$
FW 116.16
bp 171 °C
d 0.908
n_D^{20} 1.4140

Fp 61 °F

60 MHz: *1*, 201D
FT-IR: *1*, 239C
VP-FT-IR: *3*, 316A

103.63*
66.78
62.57
32.41
23.58
15.28*

CDCl₃   QE-300

B

Aldrich D13,710-3 CAS [696-59-3]

**2,5-Dimethoxytetrahydrofuran, 98%, mixture
of isomers**

$C_6H_{12}O_3$
FW 132.16
bp 146 °C
d 1.020
n_D^{20} 1.4180

Fp 95 °F

60 MHz: *1*, 202B
FT-IR: *1*, 239D
VP-FT-IR: *3*, 316B

106.09*
105.38*
55.28*
54.92*
31.00
30.08

CDCl₃   QE-300

C

Aldrich 36,234-4 CAS [3320-90-9]

**2,5-Diethoxytetrahydrofuran, 97%, mixture
of isomers**

$C_8H_{16}O_3$
FW 160.22
bp 171 °C
d 0.967
n_D^{20} 1.4190

Fp 134 °F

104.54* 31.14
104.02* 30.27
63.12   15.22*
63.07   15.20*

CDCl₃   QE-300

A

Aldrich 34,092-8   *CAS [4358-64-9]* $C_4H_8O_3$
1,4-Anhydroerythritol, 97%   FW 104.11
d 1.270
n_D^{20} 1.4770
Fp >230°F

CDCl$_3$ QE-300

72.60
71.30*

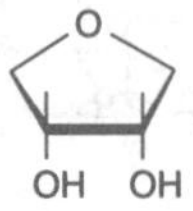

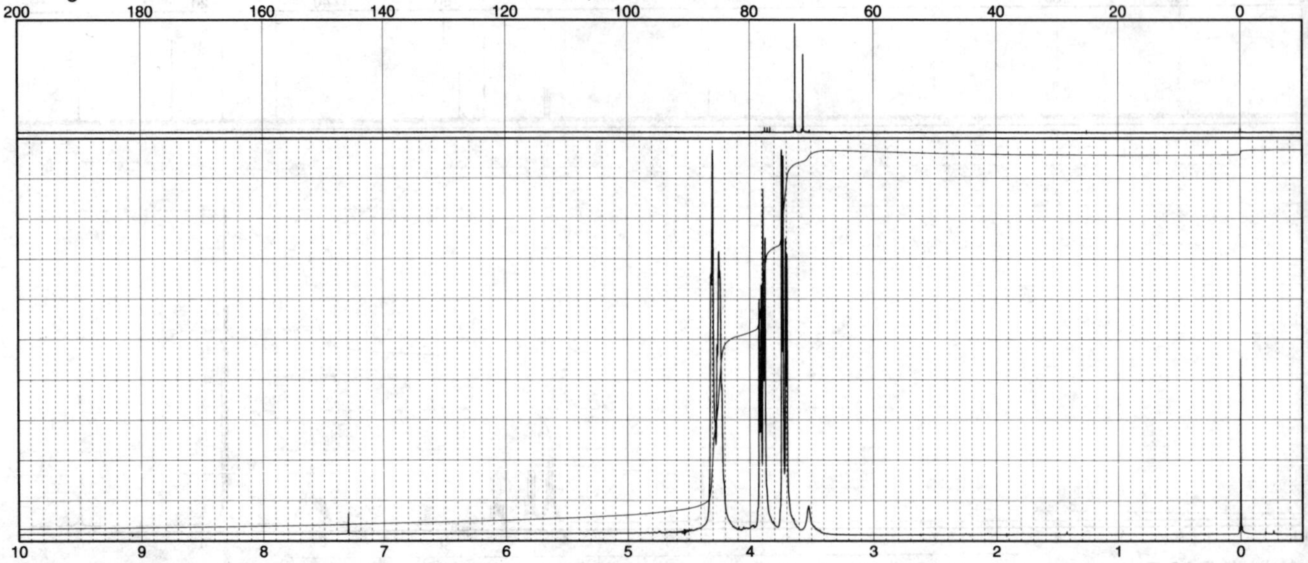

B

Aldrich 34,093-6   *CAS [22554-74-1]* $C_4H_8O_3$
1,4-Anhydro-L-threitol, 98%   FW 104.11
mp 62°C
bp 120°C (16 mm)

DMSO-d_6 QE-300

76.29*
72.90

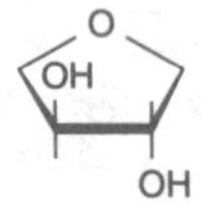

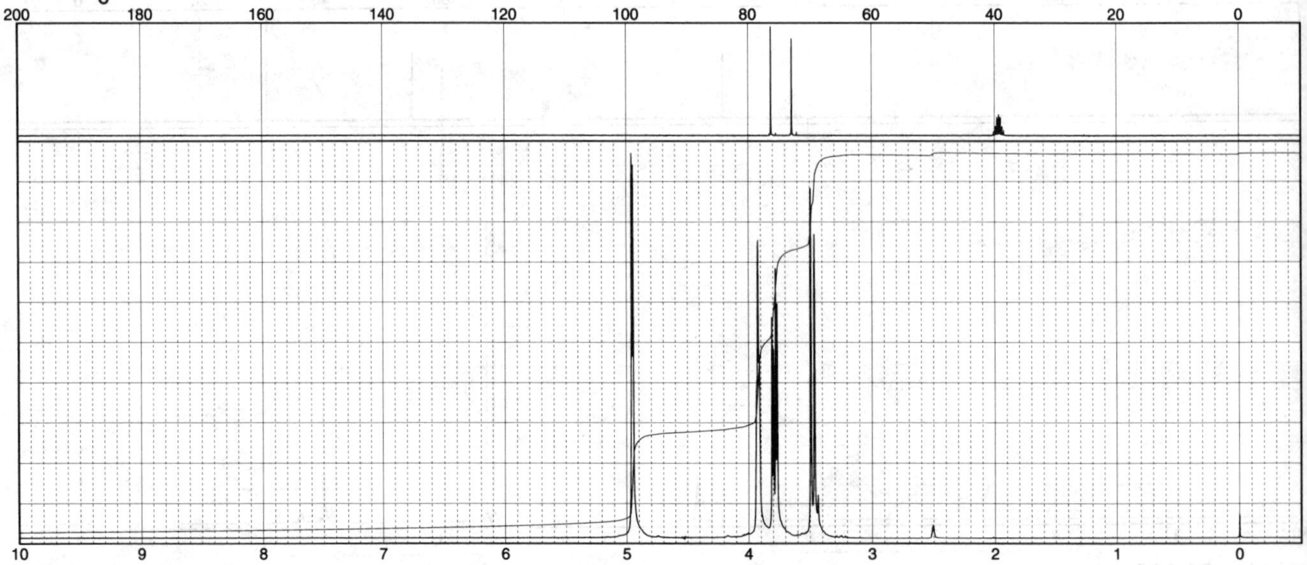

C

Aldrich 30,226-0   *CAS [41107-82-8]* $C_6H_{12}O_5$
2,5-Anhydro-D-mannitol, 99%   FW 164.16
mp 102°C

CDCl$_3$+DMSO-d_6 QE-300

84.17*
77.64*
62.11

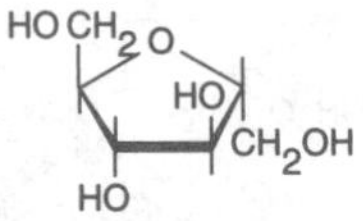

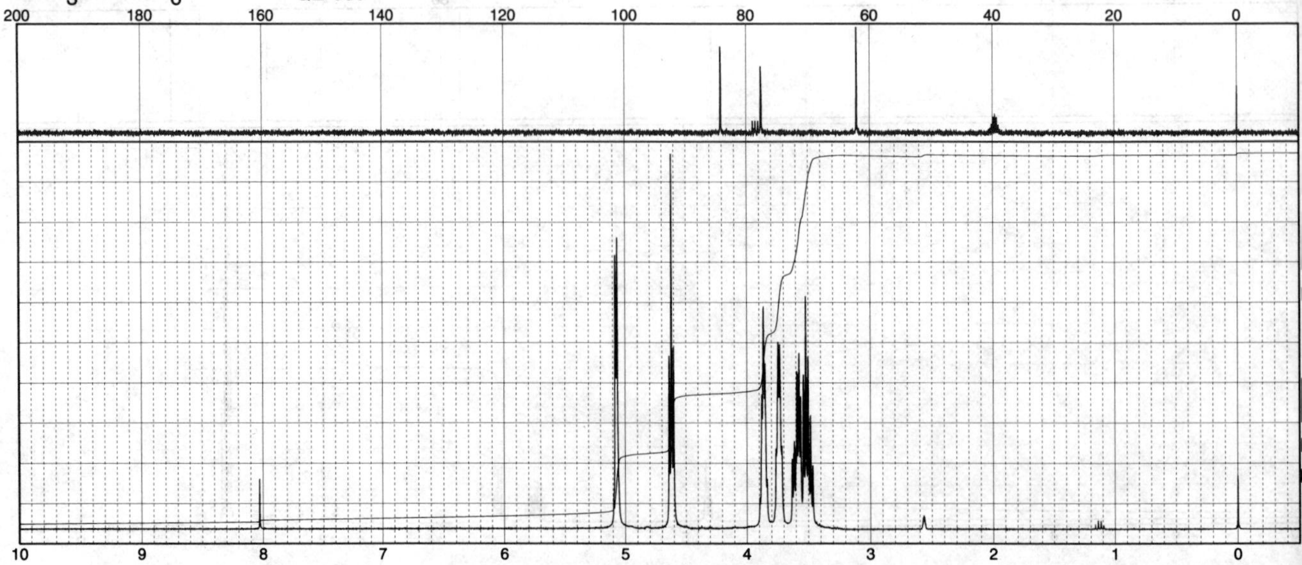

Aldrich 34,904-6
CAS [60134-26-1]
1-*O*-Methyl-2-deoxy-D-ribose, 97%

$C_6H_{12}O_4$
FW 148.16
d 1.224
n_D^{20} 1.4700
Fp >230°F

105.42*	72.36*	63.58	54.91*
105.30*	71.79*	62.70	41.96
98.86*	68.02*	60.29	41.31
98.55*	67.37*	55.57	34.77
87.08*	66.59*	55.29*	33.56
86.76*	64.82*	55.09*	

CDCl₃   QE-300

A

Aldrich 20,001-8
CAS [1191-99-7]
2,3-Dihydrofuran, 99%

C_4H_6O
FW 70.09
bp 55°C
d 0.927
n_D^{20} 1.4230

Fp -12°F

60 MHz: *1*, 203C
FT-IR: *1*, 241A
VP-FT-IR: *3*, 318C

145.75*
99.50*
69.51
29.15

CDCl₃   QE-300

B

Aldrich 25,317-0
CAS [1708-29-8]
2,5-Dihydrofuran, 97%

C_4H_6O
FW 70.09
bp 67°C
d 0.927
n_D^{20} 1.4320

Fp 2°F

FT-IR: *1*, 240D
VP-FT-IR: *3*, 318B

126.18*
75.36

CDCl₃   QE-300

C

A

Aldrich D10,580-5   *CAS [1487-15-6]*

4,5-Dihydro-2-methylfuran, 97%

C$_5$H$_8$O
FW 84.12
bp 82°C
d 0.922
n$_D^{20}$1.4300

Fp 10°F

60 MHz: *1*, 203D
FT-IR: *1*, 241B
VP-FT-IR: *3*, 318D

154.87
94.43*
69.94
30.31
13.35*

CDCl$_3$ QE-300

B

Aldrich D13,410-4   *CAS [332-77-4]*

2,5-Dimethoxy-2,5-dihydrofuran, 99%,
mixture of *cis* and *trans*

C$_6$H$_{10}$O$_3$
FW 130.14
bp 161°C
d 1.073
n$_D^{20}$1.4340

Fp 117°F

60 MHz: *1*, 203A
FT-IR: *1*, 240C
VP-FT-IR: *3*, 317D

131.73*
131.24*
108.55*
107.11*
54.17*
54.10*

CDCl$_3$ QE-300

C

Aldrich T1,440-0   *CAS [142-68-7]*

Tetrahydropyran, 99%

C$_5$H$_{10}$O
FW 86.13
mp -45°C
bp 88°C
d 0.881

n$_D^{20}$1.4200
Fp 4°F
60 MHz: *1*, 204A
FT-IR: *1*, 242C
VP-FT-IR: *3*, 320A

68.69
26.65
23.52

CDCl$_3$ QE-300

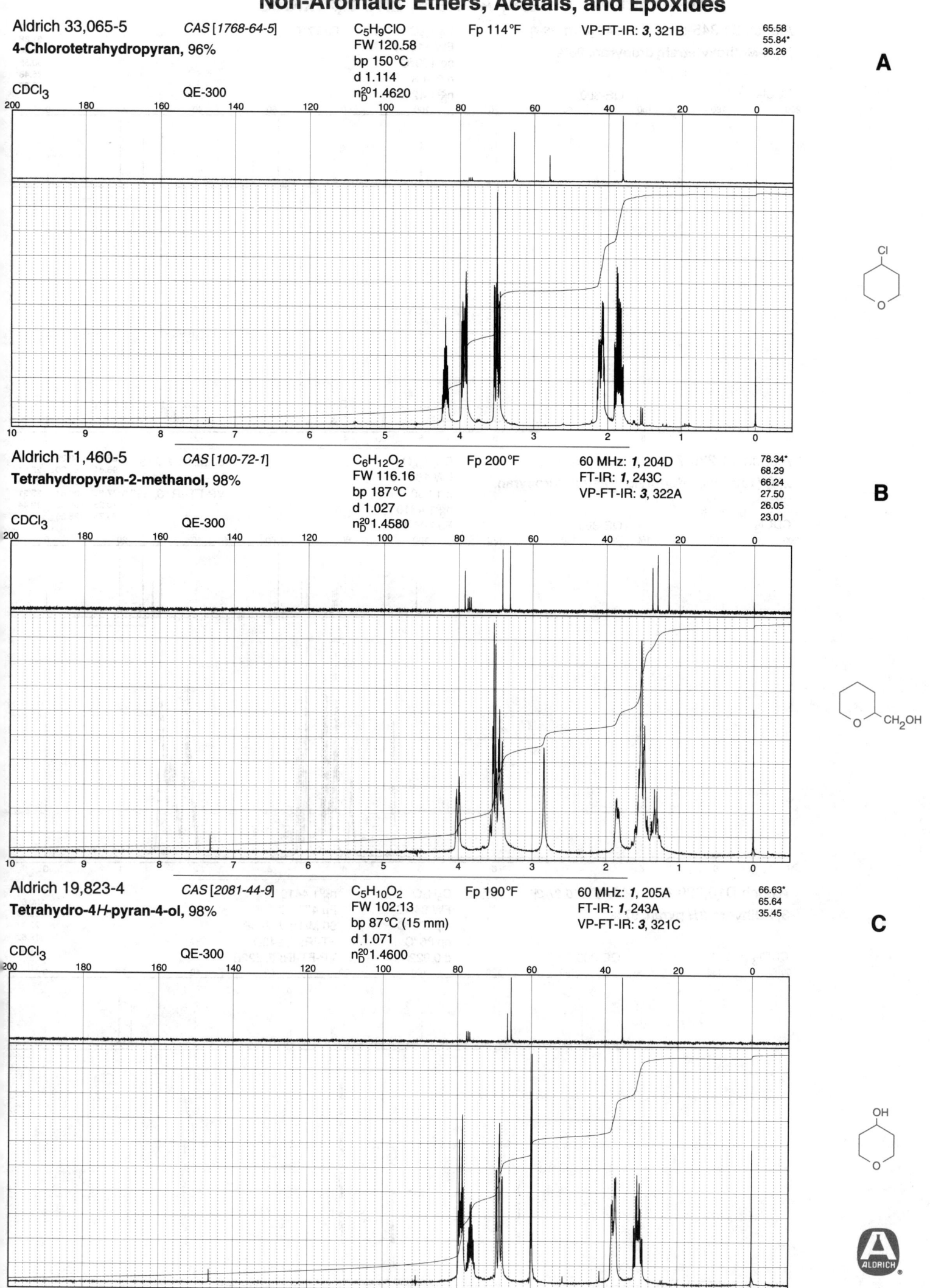

Aldrich 33,065-5   *CAS [1768-64-5]*   C₅H₉ClO   Fp 114°F   VP-FT-IR: **3**, 321B
4-Chlorotetrahydropyran, 96%   FW 120.58
bp 150°C
d 1.114
CDCl₃   QE-300   n²⁰_D 1.4620

65.58
55.84*
36.26

A

Aldrich T1,460-5   *CAS [100-72-1]*   C₆H₁₂O₂   Fp 200°F   60 MHz: **1**, 204D
Tetrahydropyran-2-methanol, 98%   FW 116.16   FT-IR: **1**, 243C
bp 187°C   VP-FT-IR: **3**, 322A
d 1.027
CDCl₃   QE-300   n²⁰_D 1.4580

78.34*
68.29
66.24
27.50
26.05
23.01

B

Aldrich 19,823-4   *CAS [2081-44-9]*   C₅H₁₀O₂   Fp 190°F   60 MHz: **1**, 205A
Tetrahydro-4H-pyran-4-ol, 98%   FW 102.13   FT-IR: **1**, 243A
bp 87°C (15 mm)   VP-FT-IR: **3**, 321C
d 1.071
CDCl₃   QE-300   n²⁰_D 1.4600

66.63*
65.64
35.45

C

A

Aldrich 34,345-5   *CAS [6581-66-4]*   C₆H₁₂O₂   Fp 77°F

(±)-2-Methoxytetrahydropyran, 98%

$C_6H_{12}O_2$
FW 116.16
bp 129°C
d 0.963
n_D^{20} 1.4250

99.77*
61.88
54.94*
30.55
25.48
19.37

CDCl₃   QE-300

B

Aldrich T1,280-7   *CAS [710-14-5]*   C₁₀H₁₈O₃

2(3)-(Tetrahydrofurfuryloxy)tetrahydropyran, 98%

$C_{10}H_{18}O_3$
FW 186.25
d 1.030
n_D^{20} 1.4610
Fp 208°F

60 MHz: *1*, 213B
FT-IR: *1*, 251C
VP-FT-IR: *3*, 333A

98.99*   68.39   28.13
98.81*   68.30   25.81
78.04*   62.09   25.60
77.68*   30.61   25.51
70.25   30.56   19.44
69.76   28.35

CDCl₃   QE-300

C

Aldrich D10,620-8   *CAS [110-87-2]*

3,4-Dihydro-2H-pyran, 97%

C_5H_8O
FW 84.12
mp -70°C
bp 86°C
d 0.922

n_D^{20} 1.4410
Fp 4°F
60 MHz: *1*, 205B
FT-IR: *1*, 243D
VP-FT-IR: *3*, 322B

144.11*
100.65*
65.78
22.87
19.62

CDCl₃   QE-300

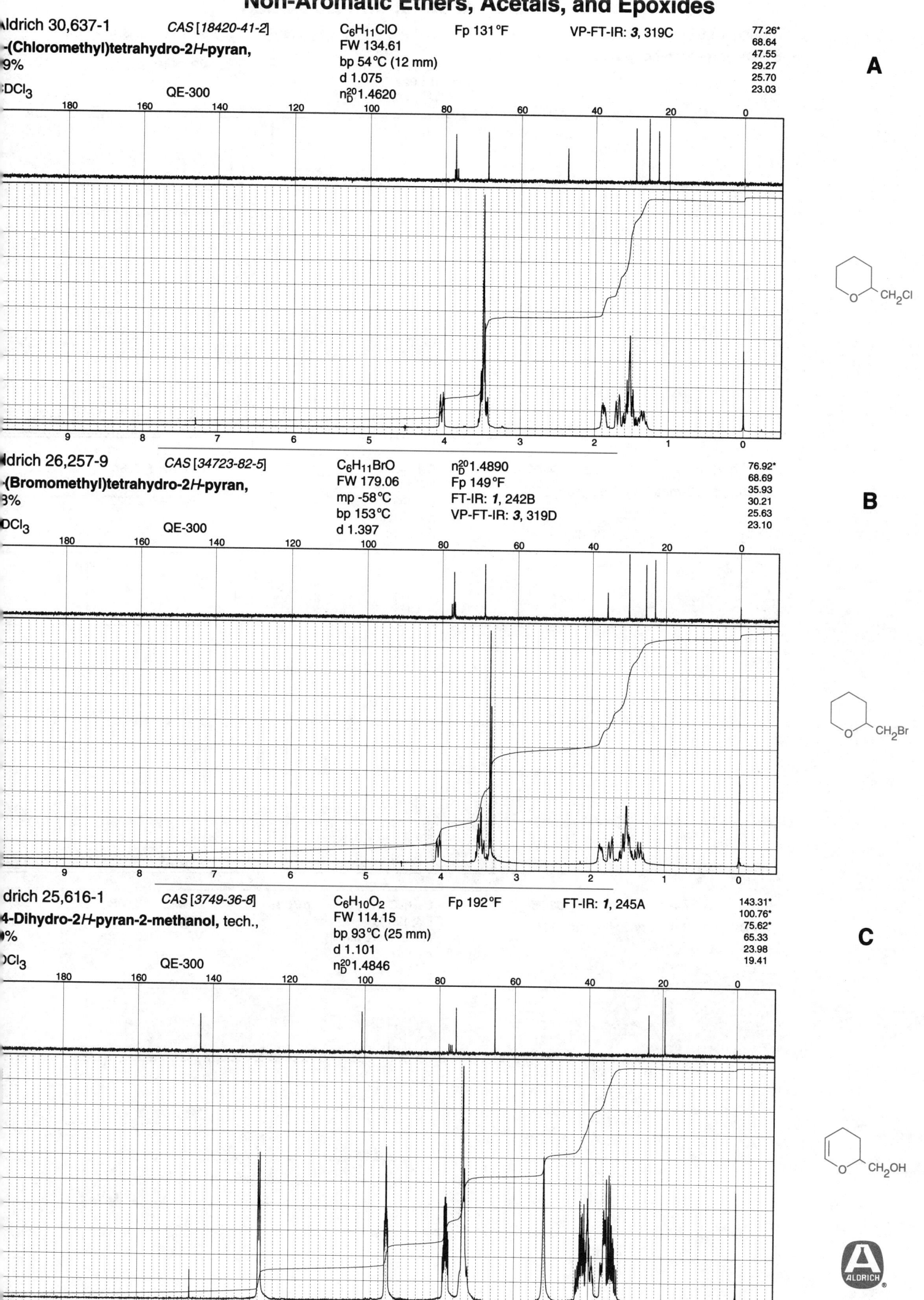

Aldrich 30,637-1   CAS [18420-41-2]

-(Chloromethyl)tetrahydro-2H-pyran,

9%

DCl₃   QE-300

$C_6H_{11}ClO$
FW 134.61
bp 54 °C (12 mm)
d 1.075
n_D^{20} 1.4620

Fp 131 °F   VP-FT-IR: **3**, 319C

77.26*
68.64
47.55
29.27
25.70
23.03

A

Aldrich 26,257-9   CAS [34723-82-5]

-(Bromomethyl)tetrahydro-2H-pyran,

8%

DCl₃   QE-300

$C_6H_{11}BrO$
FW 179.06
mp -58 °C
bp 153 °C
d 1.397

n_D^{20} 1.4890
Fp 149 °F
FT-IR: **1**, 242B
VP-FT-IR: **3**, 319D

76.92*
68.69
35.93
30.21
25.63
23.10

B

drich 25,616-1   CAS [3749-36-8]

4-Dihydro-2H-pyran-2-methanol, tech.,

9%

DCl₃   QE-300

$C_6H_{10}O_2$
FW 114.15
bp 93 °C (25 mm)
d 1.101
n_D^{20} 1.4846

Fp 192 °F   FT-IR: **1**, 245A

143.31*
100.76*
75.62*
65.33
23.98
19.41

C

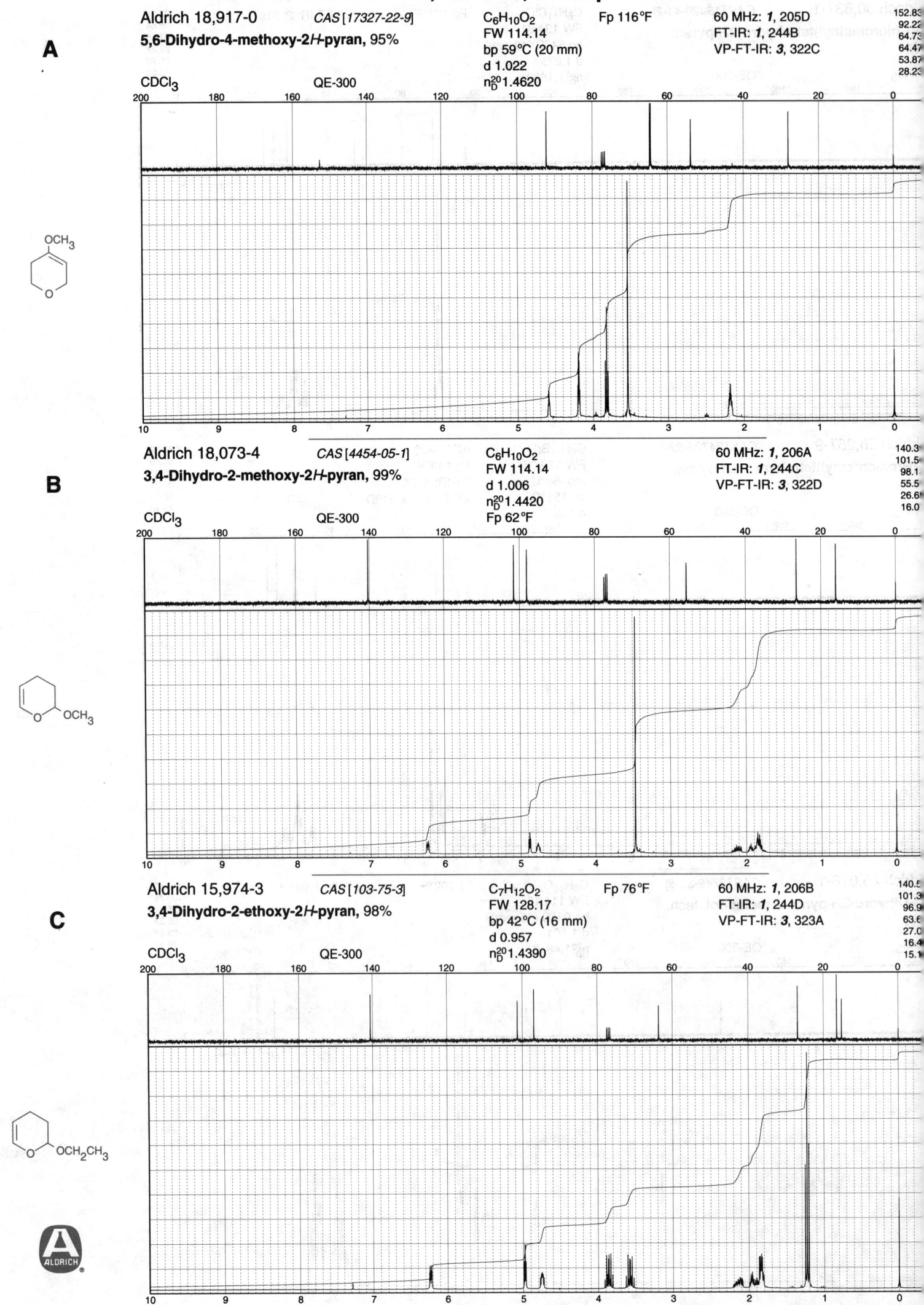

A

Aldrich 18,917-0   *CAS [17327-22-9]*

5,6-Dihydro-4-methoxy-2H-pyran, 95%

$C_6H_{10}O_2$
FW 114.14
bp 59°C (20 mm)
d 1.022
n_D^{20} 1.4620

Fp 116°F

60 MHz: *1*, 205D
FT-IR: *1*, 244B
VP-FT-IR: *3*, 322C

152.83
92.22
64.73
64.47
53.87
28.23

CDCl₃   QE-300

B

Aldrich 18,073-4   *CAS [4454-05-1]*

3,4-Dihydro-2-methoxy-2H-pyran, 99%

$C_6H_{10}O_2$
FW 114.14
d 1.006
n_D^{20} 1.4420
Fp 62°F

60 MHz: *1*, 206A
FT-IR: *1*, 244C
VP-FT-IR: *3*, 322D

140.3
101.5
98.1
55.5
26.6
16.0

CDCl₃   QE-300

C

Aldrich 15,974-3   *CAS [103-75-3]*

3,4-Dihydro-2-ethoxy-2H-pyran, 98%

$C_7H_{12}O_2$
FW 128.17
bp 42°C (16 mm)
d 0.957
n_D^{20} 1.4390

Fp 76°F

60 MHz: *1*, 206B
FT-IR: *1*, 244D
VP-FT-IR: *3*, 323A

140.5
101.3
96.9
63.6
27.0
16.4
15.1

CDCl₃   QE-300

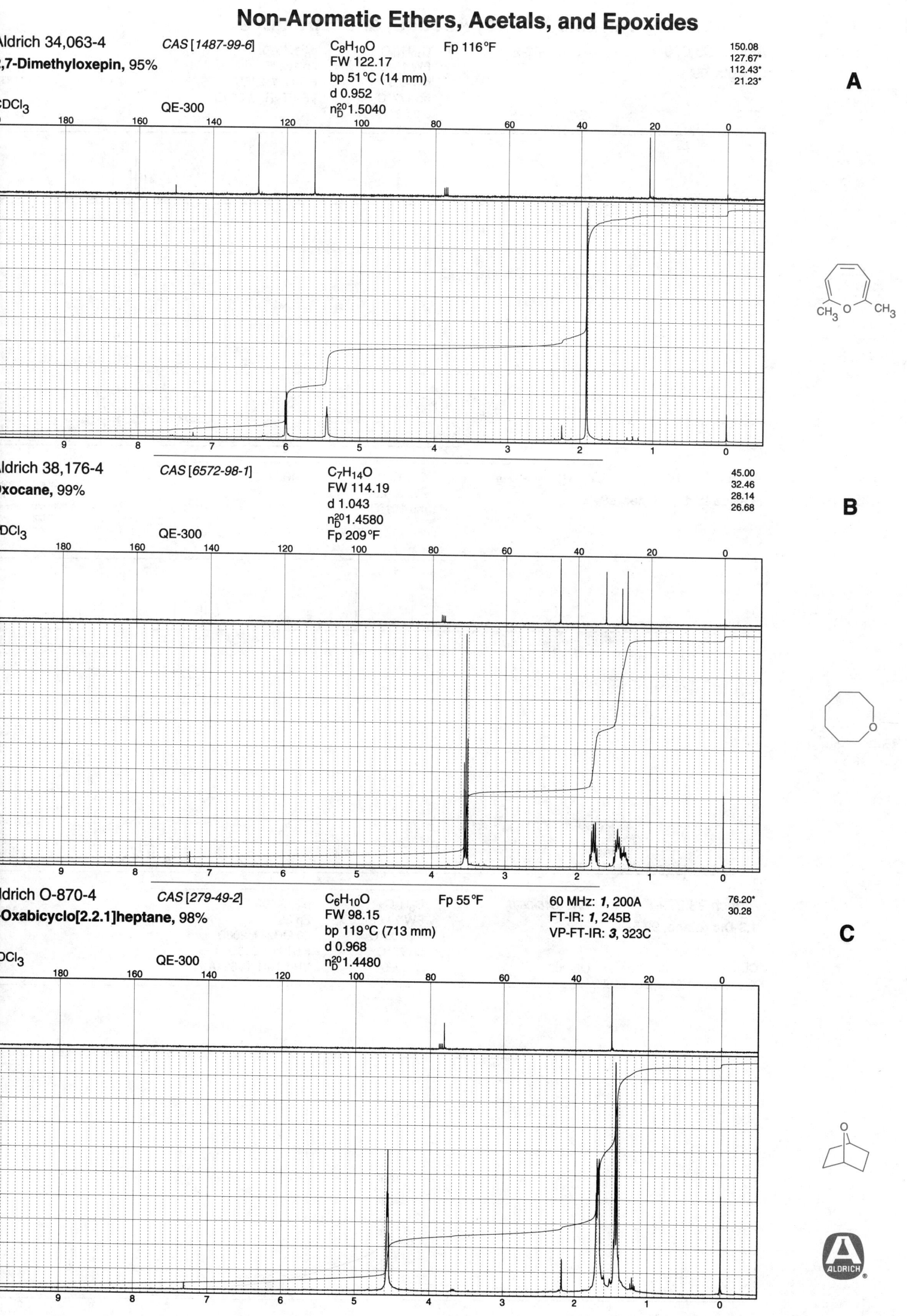

Aldrich 34,063-4   CAS [1487-99-6] $C_8H_{10}O$ Fp 116°F
,7-Dimethyloxepin, 95%   FW 122.17
bp 51°C (14 mm)
d 0.952
n_D^{20} 1.5040

150.08
127.67*
112.43*
21.23*

A

DCl₃   QE-300

Aldrich 38,176-4   CAS [6572-98-1] $C_7H_{14}O$
xocane, 99%   FW 114.19
d 1.043
n_D^{20} 1.4580
Fp 209°F

45.00
32.46
28.14
26.68

B

DCl₃   QE-300

Aldrich O-870-4   CAS [279-49-2] $C_6H_{10}O$ Fp 55°F   60 MHz: *1*, 200A
-Oxabicyclo[2.2.1]heptane, 98%   FW 98.15   FT-IR: *1*, 245B
bp 119°C (713 mm)   VP-FT-IR: *3*, 323C
d 0.968
n_D^{20} 1.4480

76.20*
30.28

C

DCl₃   QE-300

A

Aldrich C8,060-1   *CAS [470-82-6]* $C_{10}H_{18}O$ $n_D^{20}1.4570$ 73.58
Cineole, 99%   FW 154.25   Fp 122°F   69.74
    mp 2°C   FT-IR: *1*, 245C   32.96*
    bp 177°C   VP-FT-IR: *3*, 323D   31.53
CDCl₃   QE-300   d 0.921   28.90*
    27.59*
    22.86

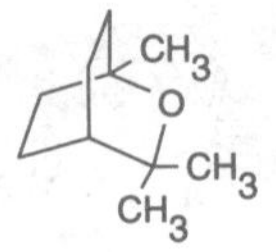

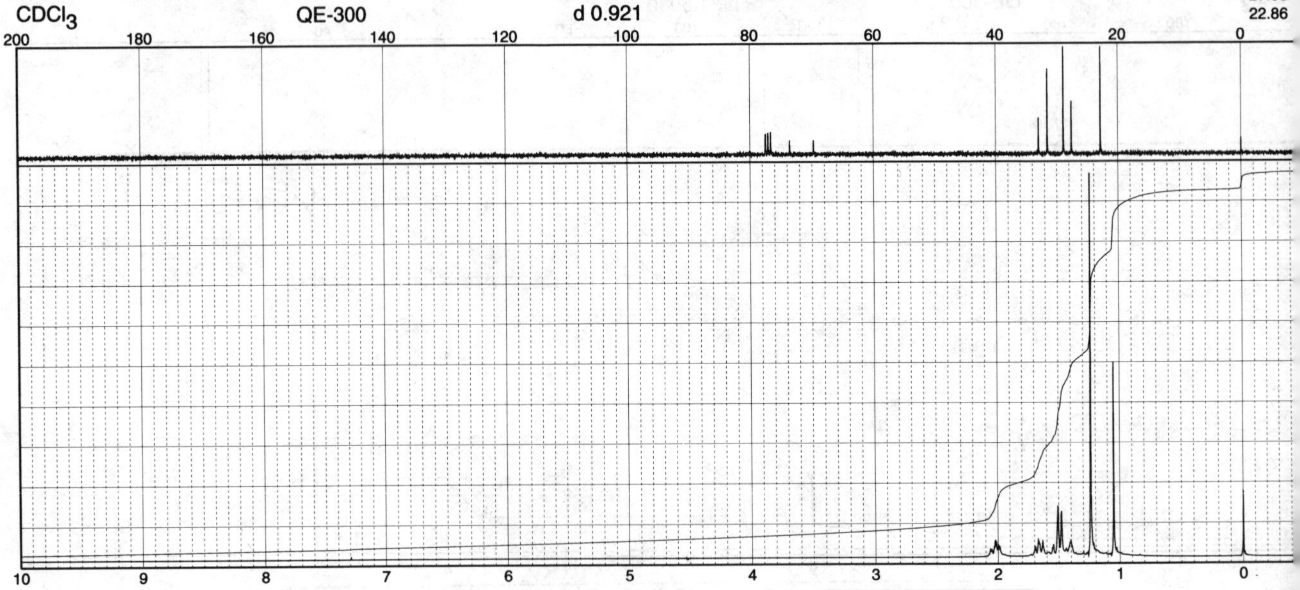

B

Aldrich 25,482-7   *CAS [87248-50-8]* $C_{24}H_{38}O_3$ FT-IR: *1*, 251D   104.71* 32.63
(+)-Noe-lactol® dimer, 99+%   FW 374.57   89.61* 26.58
    mp 151°C   52.19 20.86*
    48.83 20.54
    47.60* 18.82
CDCl₃   QE-300   40.29* 14.79*

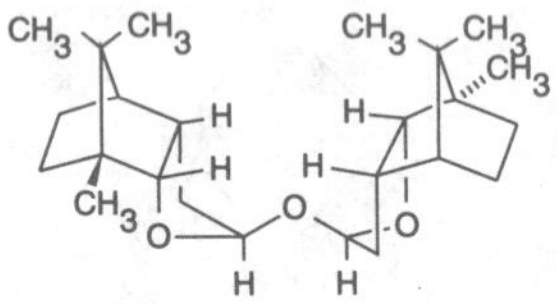

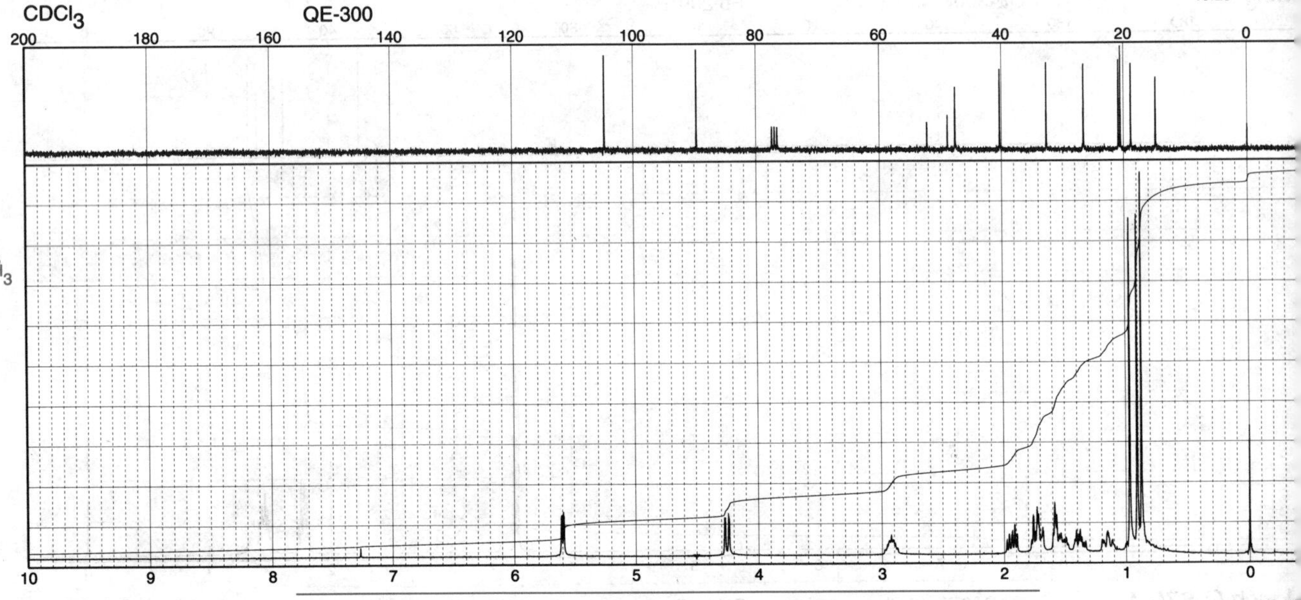

C

Aldrich 23,979-8   *CAS [646-06-0]* $C_3H_6O_2$ $n_D^{20}1.4000$ 95.07
1,3-Dioxolane, 99.5+%   FW 74.08   Fp 35°F   64.57
    mp -95°C   60 MHz: *1*, 206D
    bp 75°C   FT-IR: *1*, 245D
CDCl₃   QE-300   d 1.060   VP-FT-IR: *3*, 324A

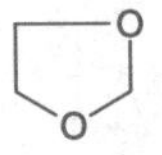

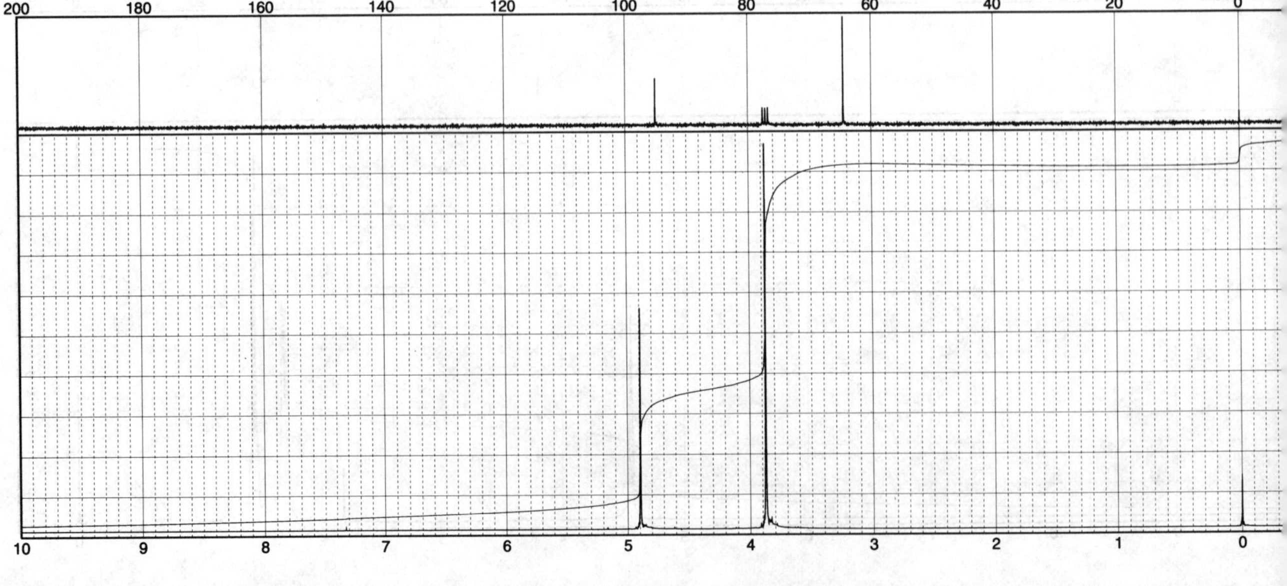

Aldrich D15,562-4 CAS [2916-31-6] $C_5H_{10}O_2$ Fp 35°F VP-FT-IR: *3*, 324C

108.40
64.46
25.62*

,2-Dimethyl-1,3-dioxolane, 99% FW 102.13
bp 93°C
d 0.926
n_D^{20} 1.3980

DCl₃ QE-300

A

Aldrich 34,104-5 CAS [126-39-6] $C_6H_{12}O_2$ Fp 55°F

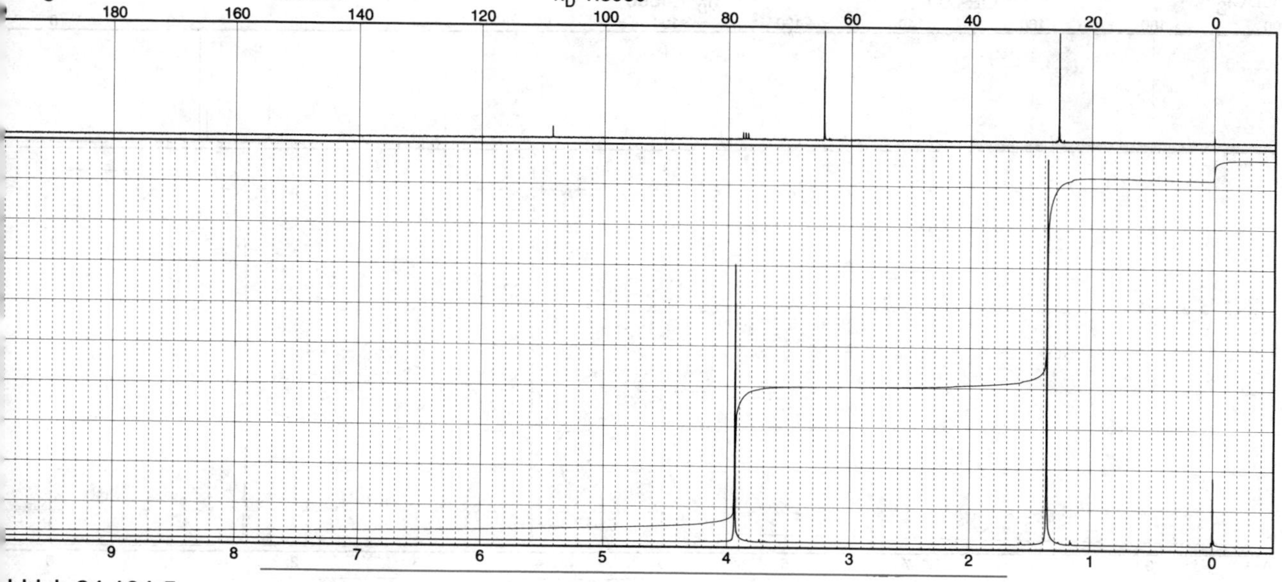

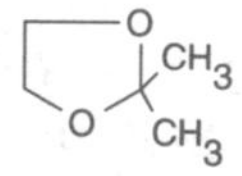

110.42
64.68
31.96
23.29*
8.34*

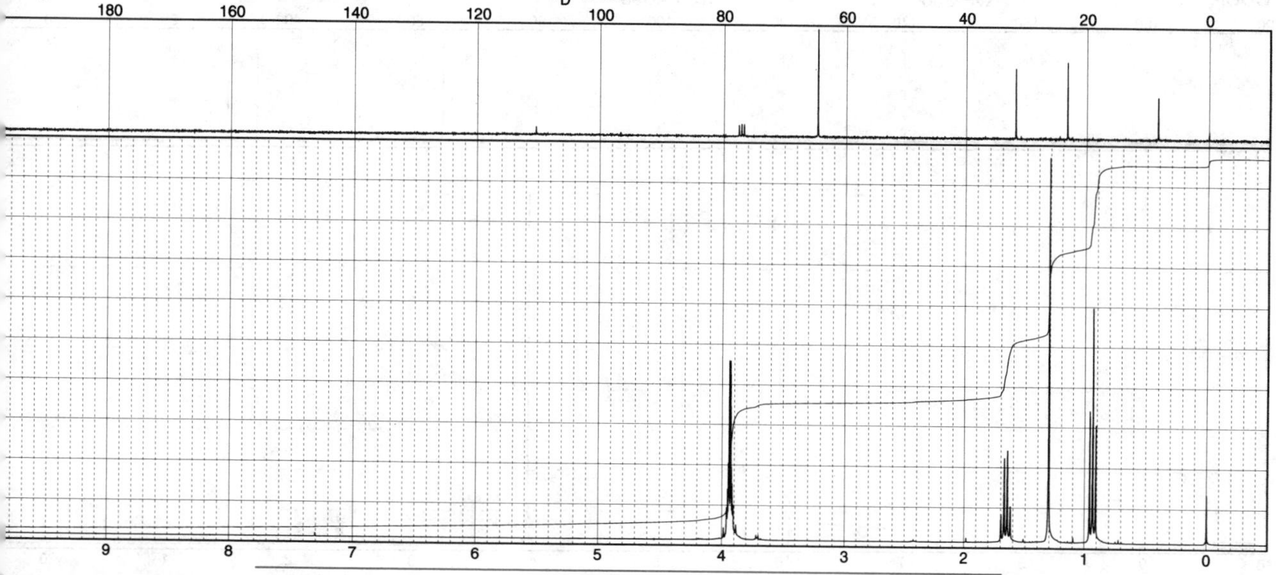

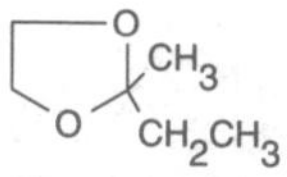

-Ethyl-2-methyl-1,3-dioxolane, 99% FW 116.16
bp 117°C
d 0.929
n_D^{20} 1.4090

DCl₃ QE-300

B

Aldrich 36,537-8 CAS [176-32-9] $C_7H_{12}O_2$ Fp 120°F

118.46
64.20
35.92
23.56

yclopentanone ethylene ketal, 98% FW 128.17
bp 55°C (35 mm)
d 1.030
n_D^{20} 1.4480

DCl₃ QE-300

C

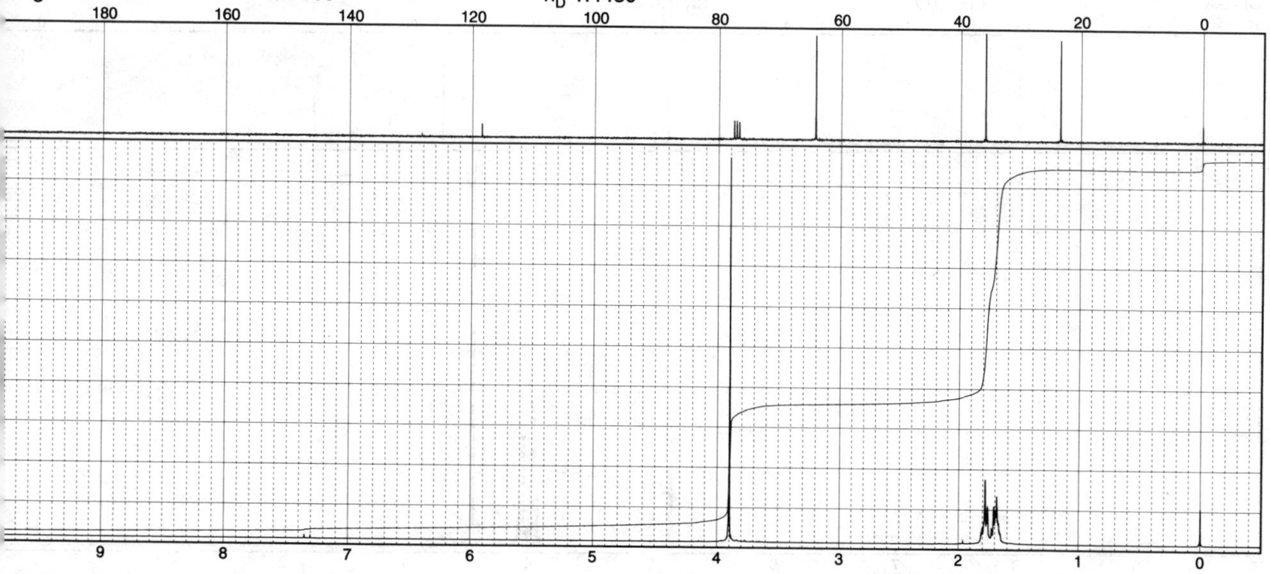

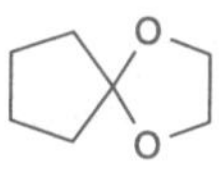

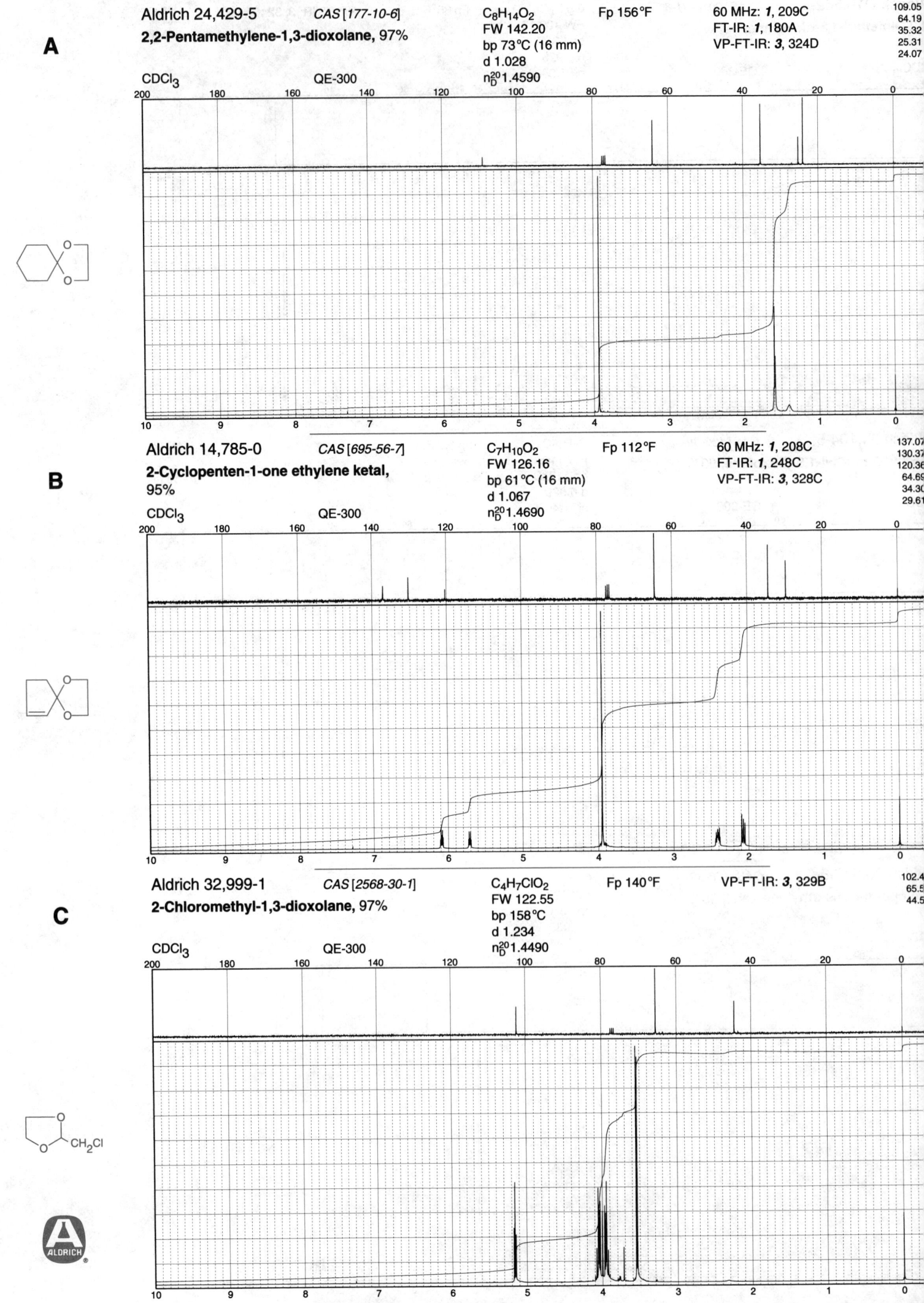

A

Aldrich 24,429-5   CAS [177-10-6]
2,2-Pentamethylene-1,3-dioxolane, 97%

$C_8H_{14}O_2$
FW 142.20
bp 73°C (16 mm)
d 1.028
n_D^{20}1.4590

Fp 156°F

60 MHz: *1*, 209C
FT-IR: *1*, 180A
VP-FT-IR: *3*, 324D

109.05
64.19
35.32
25.31
24.07

CDCl$_3$ QE-300

B

Aldrich 14,785-0   CAS [695-56-7]
2-Cyclopenten-1-one ethylene ketal, 95%

$C_7H_{10}O_2$
FW 126.16
bp 61°C (16 mm)
d 1.067
n_D^{20}1.4690

Fp 112°F

60 MHz: *1*, 208C
FT-IR: *1*, 248C
VP-FT-IR: *3*, 328C

137.07
130.37
120.36
64.69
34.30
29.61

CDCl$_3$ QE-300

C

Aldrich 32,999-1   CAS [2568-30-1]
2-Chloromethyl-1,3-dioxolane, 97%

$C_4H_7ClO_2$
FW 122.55
bp 158°C
d 1.234
n_D^{20}1.4490

Fp 140°F

VP-FT-IR: *3*, 329B

102.4
65.5
44.5

CDCl$_3$ QE-300

Aldrich 22,612-2   *CAS [4360-63-8]*

2-Bromomethyl-1,3-dioxolane, 96%

C$_4$H$_7$BrO$_2$
FW 167.01
bp 81°C (27 mm)
d 1.613
n$_D^{20}$1.4820

Fp 145°F

FT-IR: *1*, 246A
VP-FT-IR: *3*, 325A

102.02*
65.66
32.50

A

CDCl$_3$ QE-300

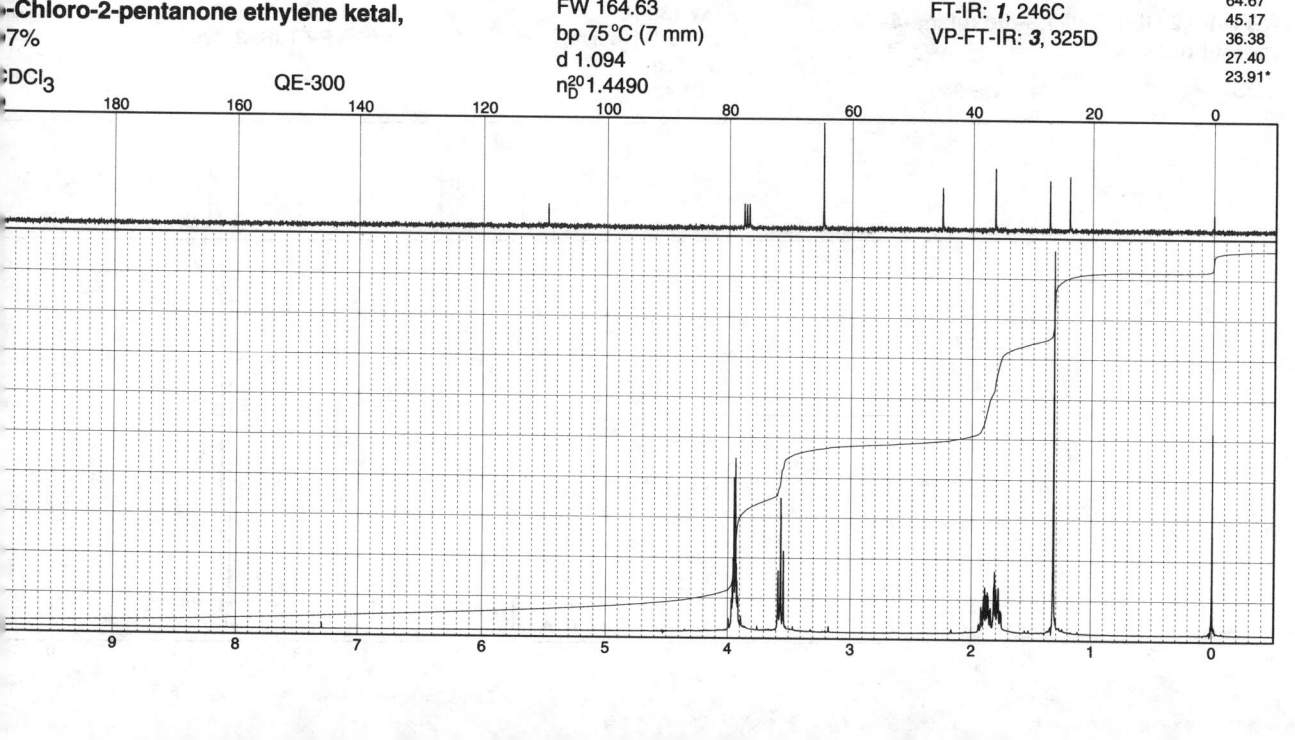

Aldrich 23,099-5   *CAS [18742-02-4]*

2-(2-Bromoethyl)-1,3-dioxolane, 96%

C$_5$H$_9$BrO$_2$
FW 181.04
bp 69°C (8 mm)
d 1.542
n$_D^{20}$1.4790

Fp 185°F

60 MHz: *1*, 209B
FT-IR: *1*, 179D
VP-FT-IR: *3*, 325B

102.58*
64.96
37.16
27.13

B

CDCl$_3$ QE-300

Aldrich 15,155-6   *CAS [5978-08-5]*

-Chloro-2-pentanone ethylene ketal, 7%

C$_7$H$_{13}$ClO$_2$
FW 164.63
bp 75°C (7 mm)
d 1.094
n$_D^{20}$1.4490

Fp 163°F

60 MHz: *1*, 207A
FT-IR: *1*, 246C
VP-FT-IR: *3*, 325D

109.55
64.67
45.17
36.38
27.40
23.91*

C

CDCl$_3$ QE-300

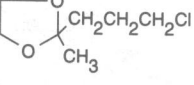

Non-Aromatic Ethers, Acetals, and Epoxides

A

Aldrich 12,269-6   *CAS* [100-79-8] $C_6H_{12}O_3$ Fp 176°F   60 MHz: *1*, 207D

Solketal, 98%   FW 132.16   FT-IR: *1*, 247B

bp 189°C   VP-FT-IR: *3*, 327B

d 1.063

CDCl₃   QE-300 n_D^{20}1.4340

109.37
76.25*
65.81
63.03
26.71*
25.29*

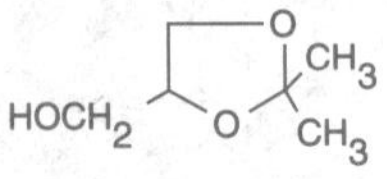

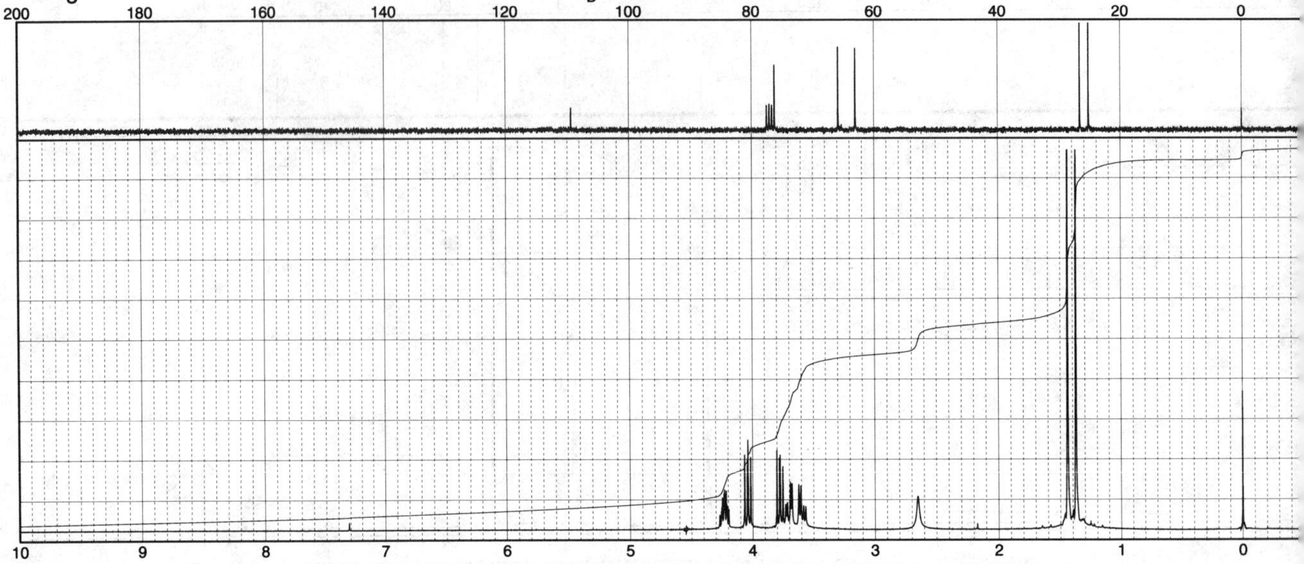

B

Aldrich 24,180-6   *CAS* [14347-78-5] $C_6H_{12}O_3$ Fp 176°F   60 MHz: *1*, 207C

(R)-(-)-2,2-Dimethyl-1,3-dioxolane-4-methanol, 98%   FW 132.16   FT-IR: *1*, 247A

bp 73°C (8 mm)   VP-FT-IR: *3*, 326B

d 1.062

CDCl₃   QE-300 n_D^{20}1.4340

109.38
76.28*
65.83
63.05
26.72*
25.30*

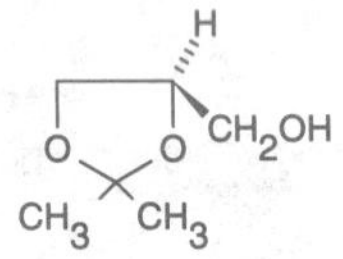

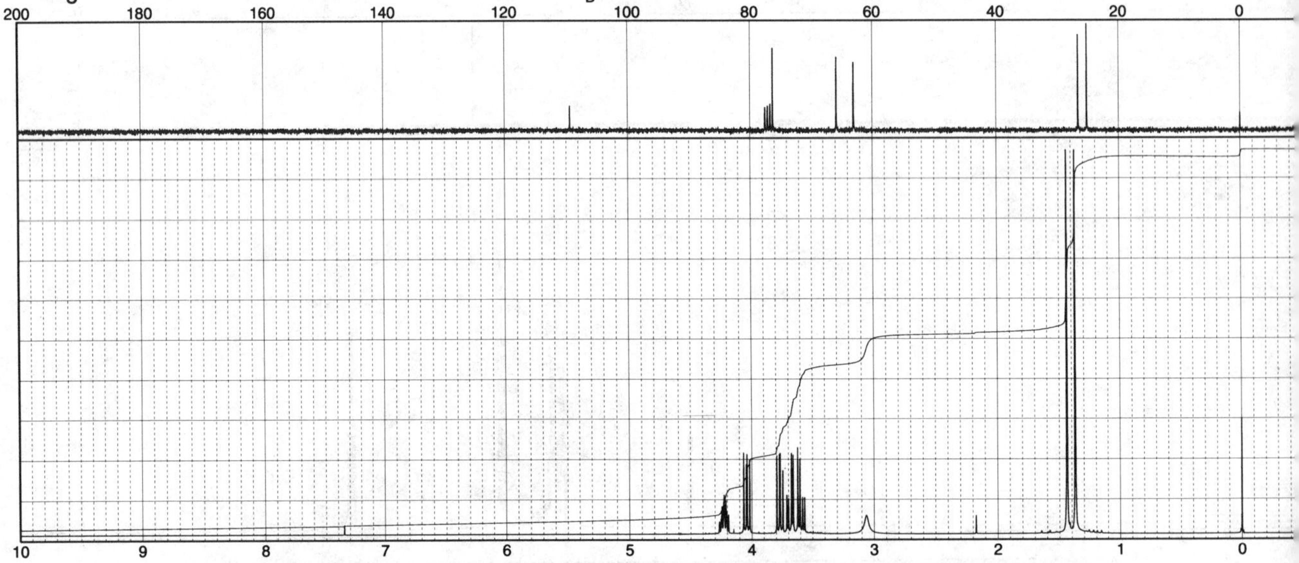

C

Aldrich 23,774-4   *CAS* [22323-82-6] $C_6H_{12}O_3$ Fp 176°F   60 MHz: *1*, 207B

(S)-(+)-2,2-Dimethyl-1,3-dioxolane-4-methanol, 98%   FW 132.16   FT-IR: *1*, 246D

bp 83°C (14 mm)   VP-FT-IR: *3*, 326A

d 1.070

CDCl₃   QE-300 n_D^{20}1.4340

109.37
76.26*
65.80
63.03
26.71*
25.29*

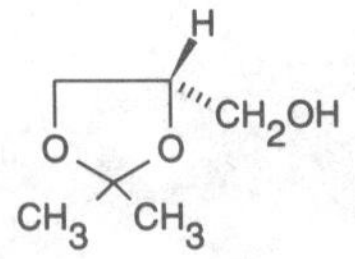

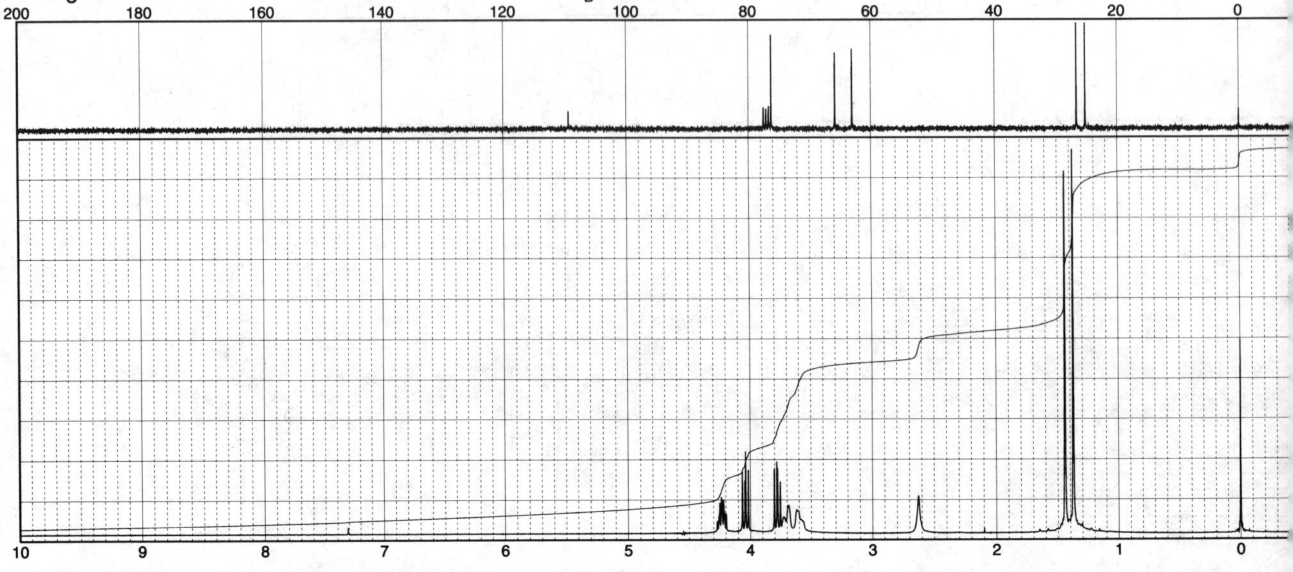

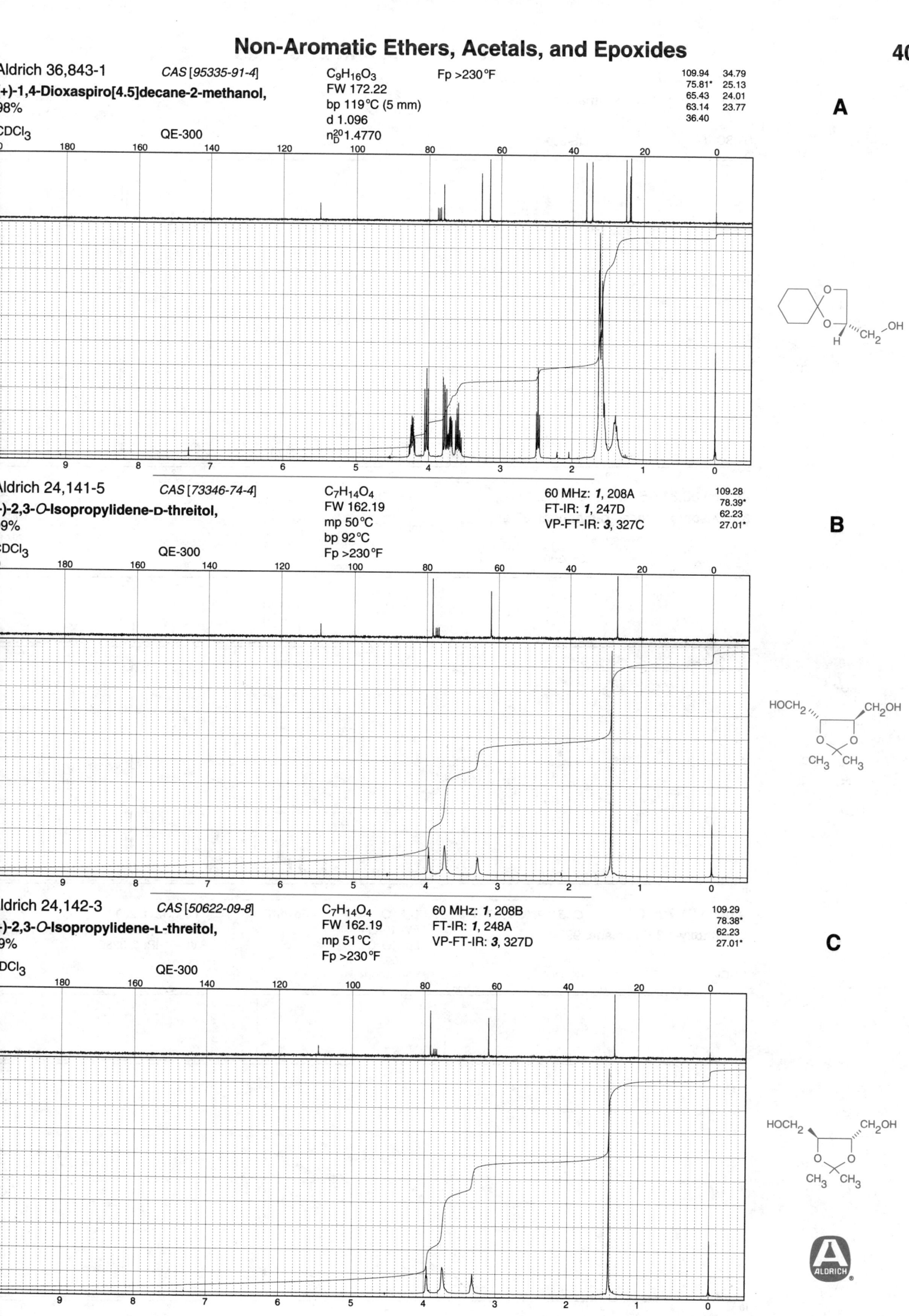

Aldrich 36,843-1   CAS [95335-91-4]   C₉H₁₆O₃   Fp >230°F

(+)-1,4-Dioxaspiro[4.5]decane-2-methanol, 98%

CDCl₃   QE-300

FW 172.22
bp 119°C (5 mm)
d 1.096
n²⁰_D 1.4770

109.94 34.79
75.81* 25.13
65.43   24.01
63.14   23.77
36.40

A

Aldrich 24,141-5   CAS [73346-74-4]   C₇H₁₄O₄

(-)-2,3-O-Isopropylidene-D-threitol, 99%

CDCl₃   QE-300

FW 162.19
mp 50°C
bp 92°C
Fp >230°F

60 MHz: 1, 208A
FT-IR: 1, 247D
VP-FT-IR: 3, 327C

109.28
78.39*
62.23
27.01*

B

Aldrich 24,142-3   CAS [50622-09-8]   C₇H₁₄O₄

(+)-2,3-O-Isopropylidene-L-threitol, 99%

CDCl₃   QE-300

FW 162.19
mp 51°C
Fp >230°F

60 MHz: 1, 208B
FT-IR: 1, 248A
VP-FT-IR: 3, 327D

109.29
78.38*
62.23
27.01*

C

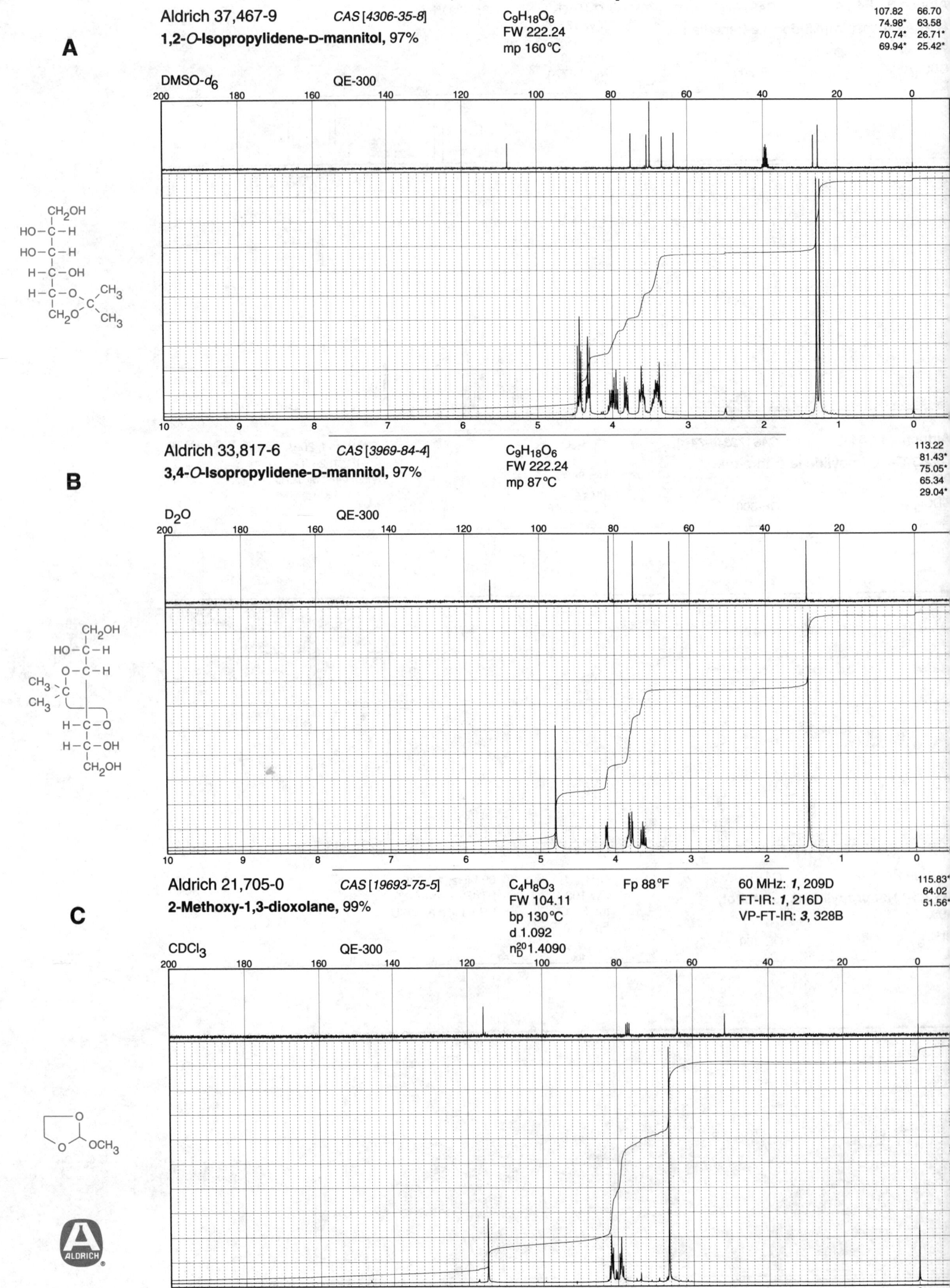

Aldrich 37,467-9   *CAS [4306-35-8]*   C9H18O6
1,2-O-Isopropylidene-D-mannitol, 97%   FW 222.24
mp 160°C

107.82 66.70
74.98* 63.58
70.74* 26.71*
69.94* 25.42*

A

DMSO-d6   QE-300

Aldrich 33,817-6   *CAS [3969-84-4]*   C9H18O6
3,4-O-Isopropylidene-D-mannitol, 97%   FW 222.24
mp 87°C

113.22
81.43*
75.05*
65.34
29.04*

B

D2O   QE-300

Aldrich 21,705-0   *CAS [19693-75-5]*   C4H8O3   Fp 88°F   60 MHz: *1*, 209D
2-Methoxy-1,3-dioxolane, 99%   FW 104.11   FT-IR: *1*, 216D
bp 130°C   VP-FT-IR: *3*, 328B
d 1.092
n_D^{20}1.4090

115.83*
64.02
51.56*

C

CDCl3   QE-300

Non-Aromatic Ethers, Acetals, and Epoxides

402

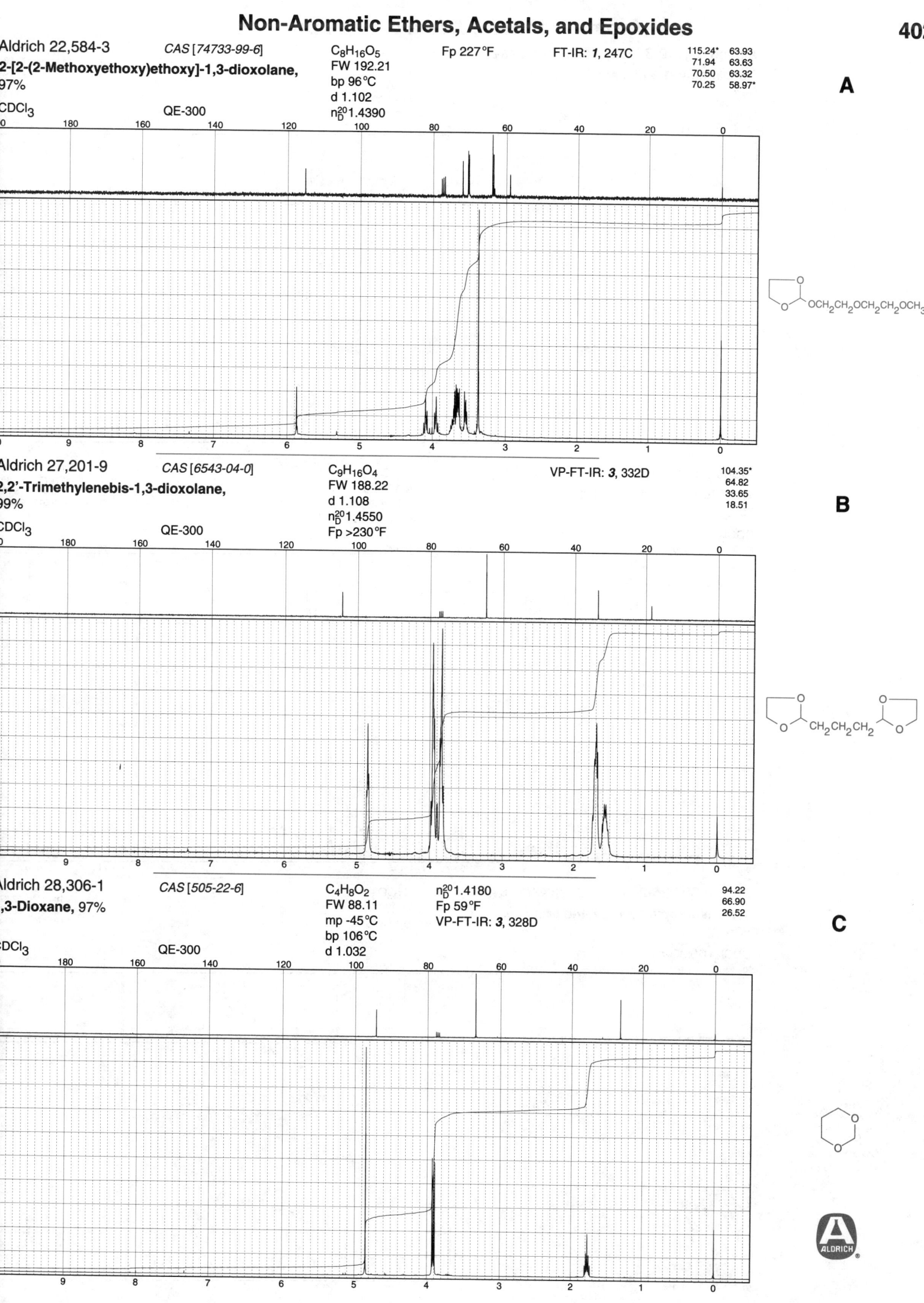

Aldrich 22,584-3   CAS [74733-99-6]   C₈H₁₆O₅   Fp 227°F   FT-IR: 1, 247C   115.24* 63.93 / 71.94 63.63 / 70.50 63.32 / 70.25 58.97*

2-[2-(2-Methoxyethoxy)ethoxy]-1,3-dioxolane, 97%

CDCl₃   QE-300   FW 192.21   bp 96°C   d 1.102   n²⁰_D 1.4390

A

Aldrich 27,201-9   CAS [6543-04-0]   C₉H₁₆O₄   VP-FT-IR: 3, 332D   104.35* / 64.82 / 33.65 / 18.51

2,2'-Trimethylenebis-1,3-dioxolane, 99%

CDCl₃   QE-300   FW 188.22   d 1.108   n²⁰_D 1.4550   Fp >230°F

B

Aldrich 28,306-1   CAS [505-22-6]   C₄H₈O₂   n²⁰_D 1.4180   94.22 / 66.90 / 26.52

1,3-Dioxane, 97%

CDCl₃   QE-300   FW 88.11   mp -45°C   bp 106°C   d 1.032   Fp 59°F   VP-FT-IR: 3, 328D

C

A

Aldrich 37,140-8 *CAS [116141-68-5]* C$_8$H$_{16}$O$_3$
FW 160.22
5,5-Dimethyl-1,3-dioxane-2-ethanol,
99% d 1.031
n$_D^{20}$1.4500

CDCl$_3$ QE-300 Fp 225°F

101.49*
77.08
58.53
36.79
30.13
22.90*
21.77*

B

Aldrich 29,636-8 *CAS [20031-21-4]* C$_8$H$_{14}$O$_5$
FW 190.20
1,2-O-Isopropylidene-D-xylofuranose,
99% mp 70°C
bp 113°C

CDCl$_3$ QE-300

111.77 76.52*
104.77* 60.85
85.53* 26.73*
79.01* 26.16*

C

Aldrich 30,225-2 *CAS [29747-91-9]* C$_9$H$_{16}$O$_6$
FW 220.22
1,2-O-Isopropylidene-β-L-idofuranose,
99% mp 114°C

CDCl$_3$+DMSO-d$_6$ QE-300

110.45 70.61*
104.11* 62.87
85.37* 26.66
81.16* 26.14*
74.43*

Non-Aromatic Ethers, Acetals, and Epoxides

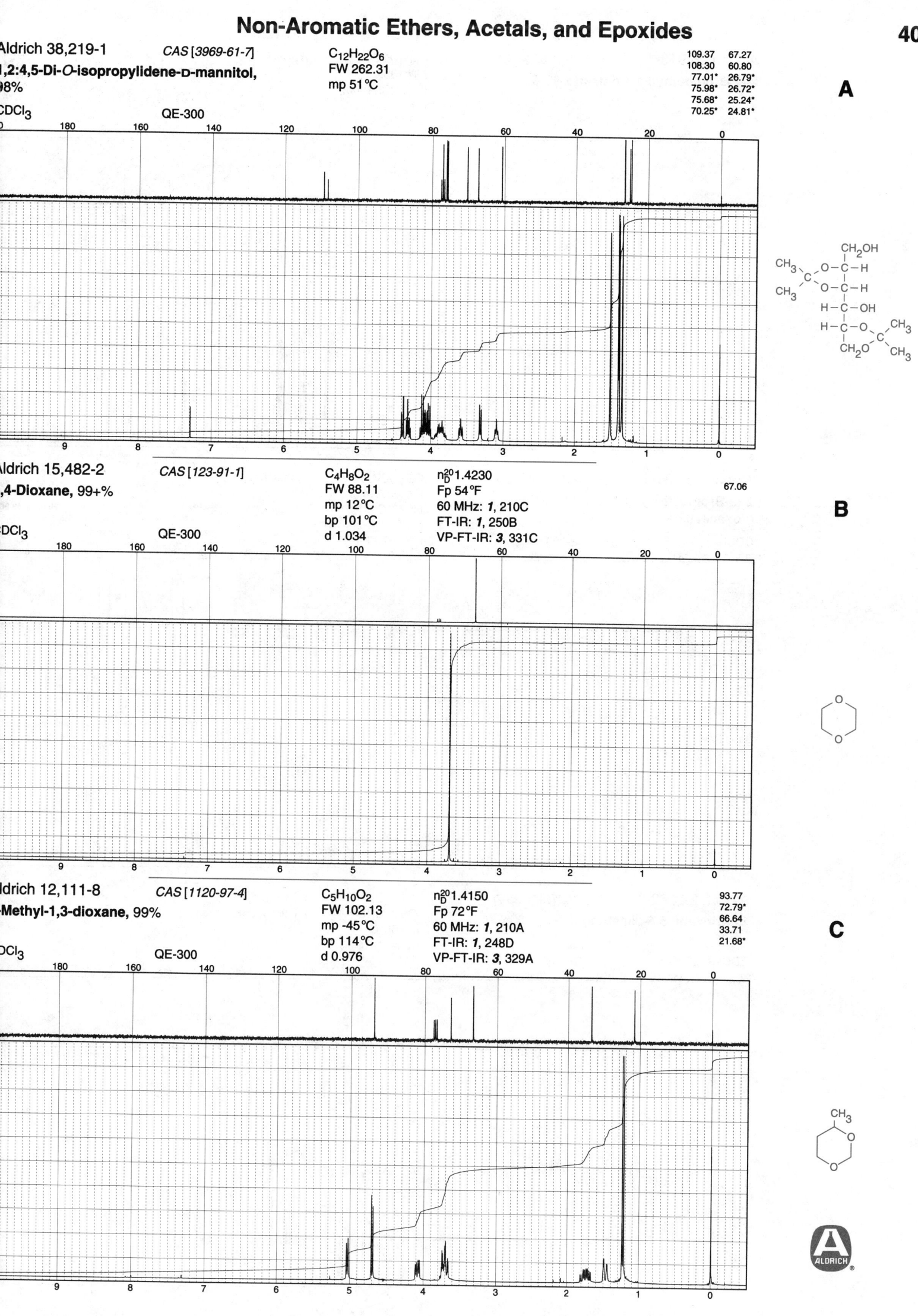

Aldrich 38,219-1   CAS [3969-61-7] $C_{12}H_{22}O_6$ 109.37 67.27
1,2:4,5-Di-O-isopropylidene-D-mannitol, 98%   FW 262.31   mp 51°C   108.30 60.80
CDCl₃   QE-300   77.01* 26.79*
75.98* 26.72*
75.68* 25.24*
70.25* 24.81*

A

Aldrich 15,482-2   CAS [123-91-1] $C_4H_8O_2$ n_D^{20}1.4230   67.06
1,4-Dioxane, 99+%   FW 88.11   Fp 54°F
mp 12°C   60 MHz: *1*, 210C
CDCl₃   QE-300   bp 101°C   FT-IR: *1*, 250B
d 1.034   VP-FT-IR: *3*, 331C

B

Aldrich 12,111-8   CAS [1120-97-4] $C_5H_{10}O_2$ n_D^{20}1.4150   93.77
2-Methyl-1,3-dioxane, 99%   FW 102.13   Fp 72°F   72.79*
mp -45°C   60 MHz: *1*, 210A   66.64
CDCl₃   QE-300   bp 114°C   FT-IR: *1*, 248D   33.71
d 0.976   VP-FT-IR: *3*, 329A   21.68*

C

A

Aldrich 26,913-1   CAS [33884-43-4] $C_6H_{11}BrO_2$ Fp 206 °F   FT-IR: **1**, 246B
2-(2-Bromoethyl)-1,3-dioxane, 98%   FW 195.06   VP-FT-IR: **3**, 325C
bp 69 °C (2 mm)
d 1.431
n_D^{20} 1.4810

CDCl₃   QE-300

100.05*
66.84
38.13
27.71
25.78

B

Aldrich 36,731-1   CAS [87842-52-2] $C_9H_{17}BrO_2$ Fp 204 °F
2-(2-Bromoethyl)-2,5,5-trimethyl-1,3-dioxane, 95%   FW 237.14
bp 73 °C
d 1.260
n_D^{20} 1.4700

CDCl₃   QE-300

98.29   27.19
70.35   22.79
42.94   22.33
29.81   19.90

C

Aldrich 22,062-0   CAS [6228-25-7] $C_6H_{12}O_4$ 60 MHz: **1**, 210B
1,3-Dioxane-5,5-dimethanol, tech., 90%   FW 148.16   FT-IR: **1**, 248B
mp 59 °C
Fp NONE

CDCl₃+DMSO-d₆   QE-300

93.3*
68.8*
61.0
39.7*

Non-Aromatic Ethers, Acetals, and Epoxides

Aldrich 31,033-6   CAS [5417-32-3]   C₅H₈O₂   Fp 79°F   VP-FT-IR: 3, 332A

$C_5H_8O_2$

cis-4,7-Dihydro-1,3-dioxepin, 97%

FW 100.12
bp 126°C
d 1.049
n_D^{20} 1.4570

CDCl₃   QE-300

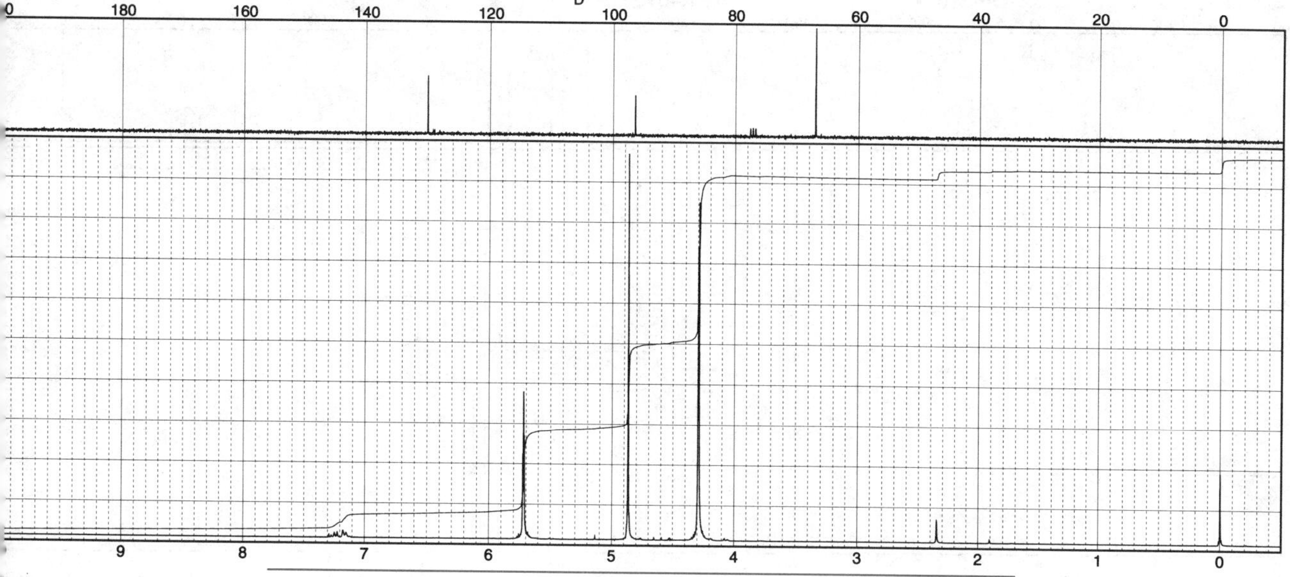

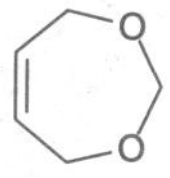

130.04*
96.40
66.99

A

Aldrich 32,926-6   CAS [641-74-7] $C_6H_{10}O_4$

Isomannide, 95%

FW 146.14
mp 83°C

CDCl₃   QE-300

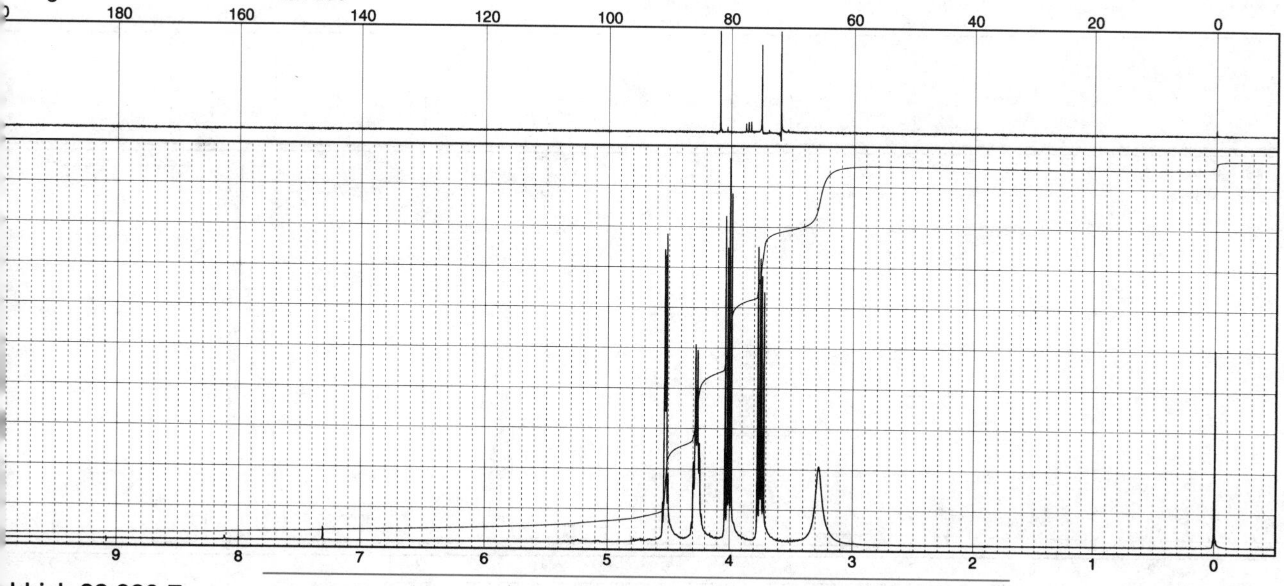

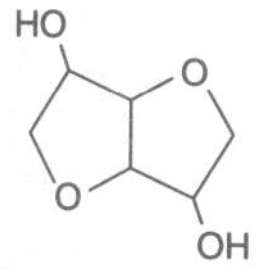

81.78*
74.94
71.82*

B

Aldrich 32,920-7   CAS [652-67-5] $C_6H_{10}O_4$

Isosorbide, 98%

FW 146.14
mp 62°C

CDCl₃ + DMSO-d_6 QE-300

87.81*
81.31*
75.81*
75.40
72.29*
71.34

C

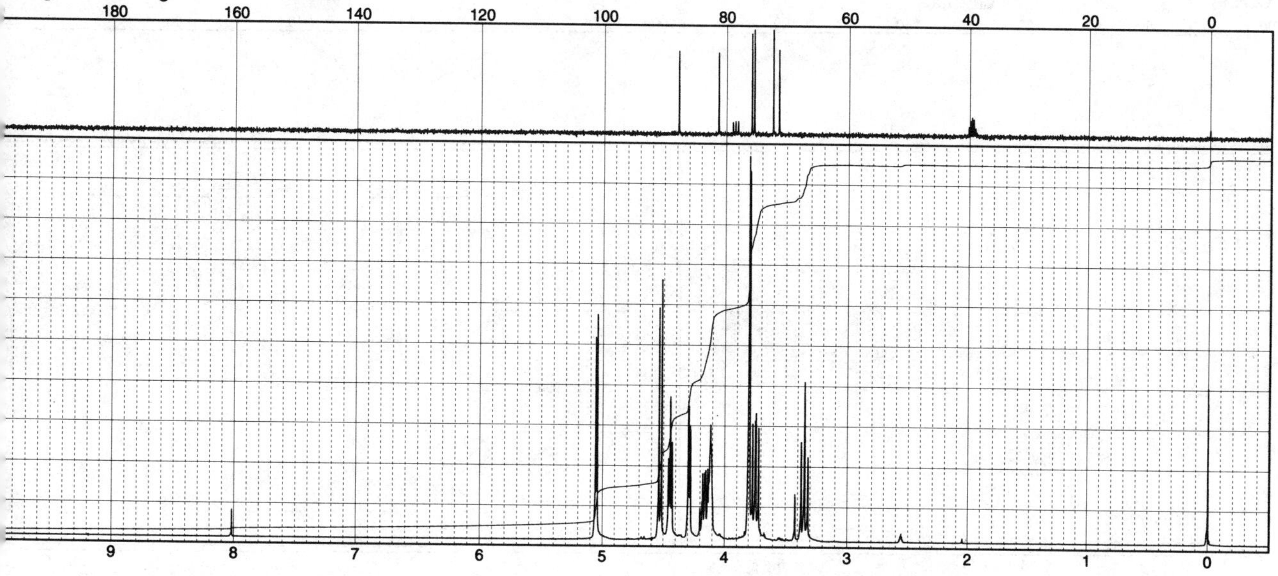

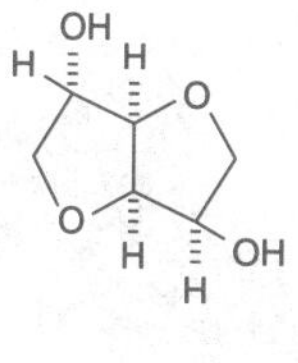

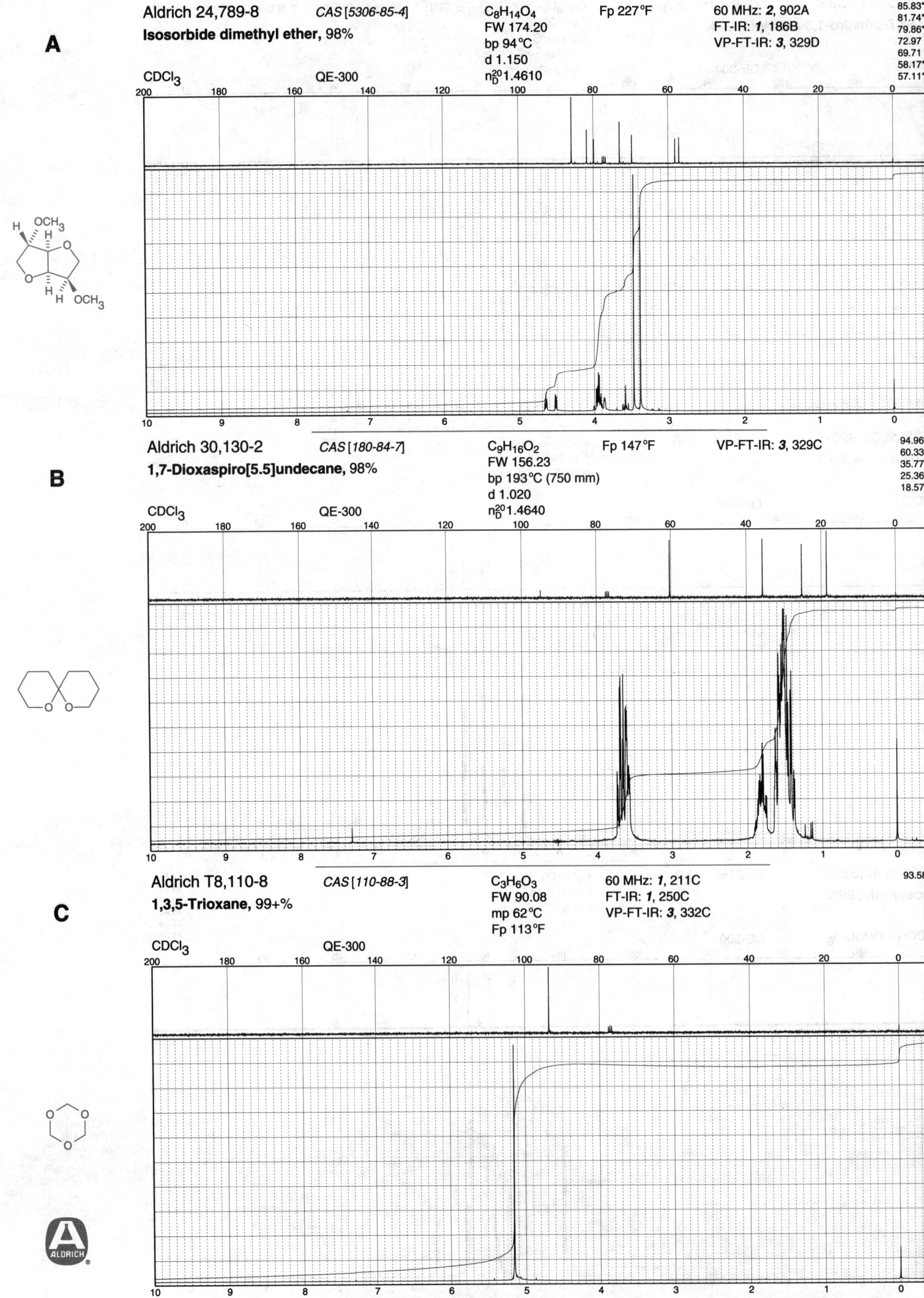

A

Aldrich 24,789-8   *CAS [5306-85-4]*   C₈H₁₄O₄   Fp 227°F   60 MHz: *2*, 902A

Isosorbide dimethyl ether, 98%
FW 174.20   FT-IR: *1*, 186B
bp 94°C   VP-FT-IR: *3*, 329D
d 1.150
n₂₀D 1.4610

CDCl₃   QE-300

85.83*
81.74*
79.86*
72.97
69.71
58.17*
57.11*

B

Aldrich 30,130-2   *CAS [180-84-7]*   C₉H₁₆O₂   Fp 147°F   VP-FT-IR: *3*, 329C

1,7-Dioxaspiro[5.5]undecane, 98%
FW 156.23
bp 193°C (750 mm)
d 1.020
n₂₀D 1.4640

CDCl₃   QE-300

94.96
60.33
35.77
25.36
18.57

93.58

C

Aldrich T8,110-8   *CAS [110-88-3]*   C₃H₆O₃   60 MHz: *1*, 211C

1,3,5-Trioxane, 99+%
FW 90.08   FT-IR: *1*, 250C
mp 62°C   VP-FT-IR: *3*, 332C
Fp 113°F

CDCl₃   QE-300

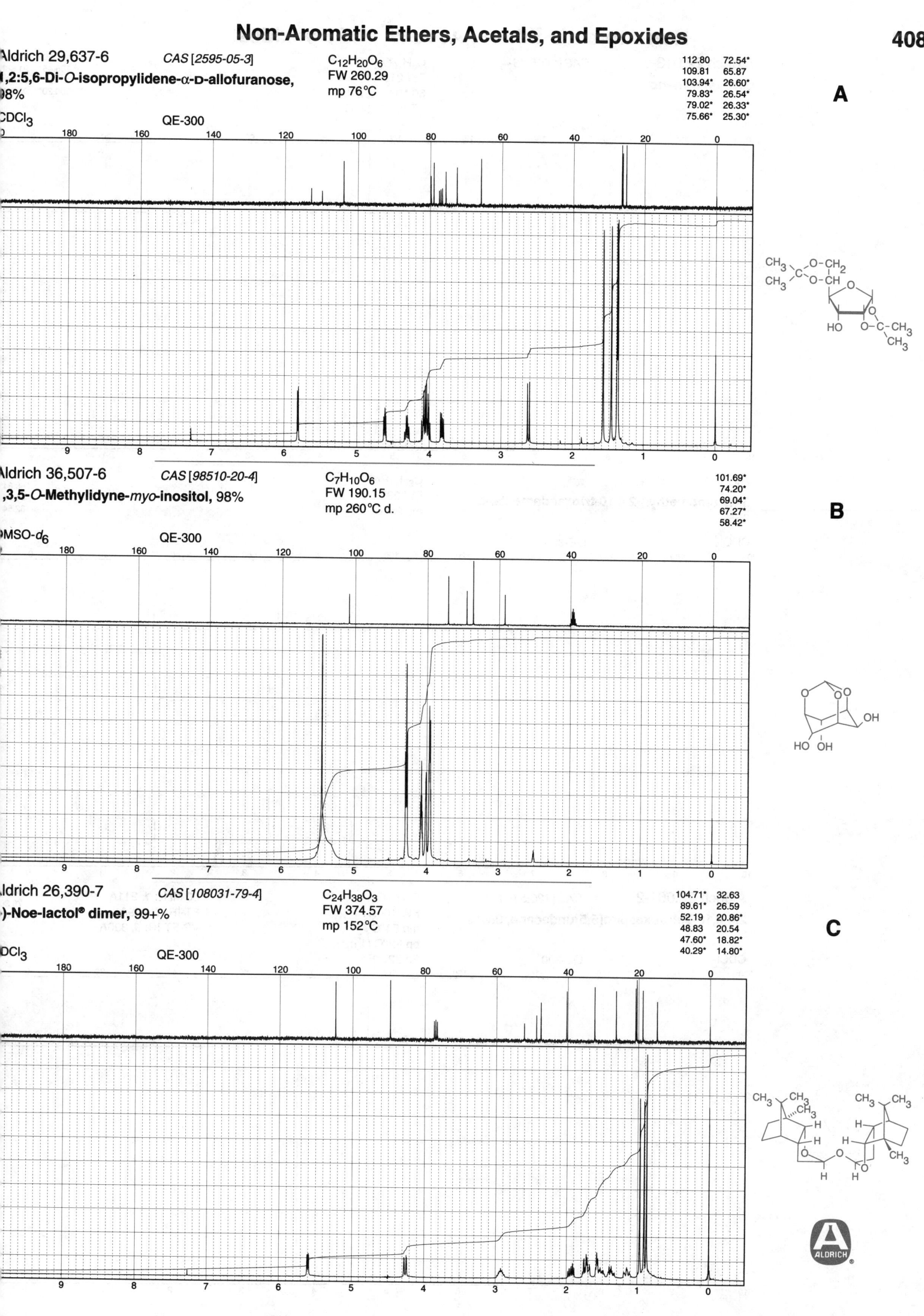

Aldrich 29,637-6 CAS [2595-05-3] $C_{12}H_{20}O_6$ FW 260.29 mp 76°C

1,2:5,6-Di-*O*-isopropylidene-α-D-allofuranose, 98%

CDCl$_3$ QE-300

112.80	72.54*
109.81	65.87
103.94*	26.60*
79.83*	26.54*
79.02*	26.33*
75.66*	25.30*

A

Aldrich 36,507-6 CAS [98510-20-4] $C_7H_{10}O_6$ FW 190.15 mp 260°C d.

1,3,5-*O*-Methylidyne-*myo*-inositol, 98%

DMSO-d_6 QE-300

101.69*
74.20*
69.04*
67.27*
58.42*

B

Aldrich 26,390-7 CAS [108031-79-4] $C_{24}H_{38}O_3$ FW 374.57 mp 152°C

(-)-Noe-lactol® dimer, 99+%

CDCl$_3$ QE-300

104.71*	32.63
89.61*	26.59
52.19	20.86*
48.83	20.54
47.60*	18.82*
40.29*	14.80*

C

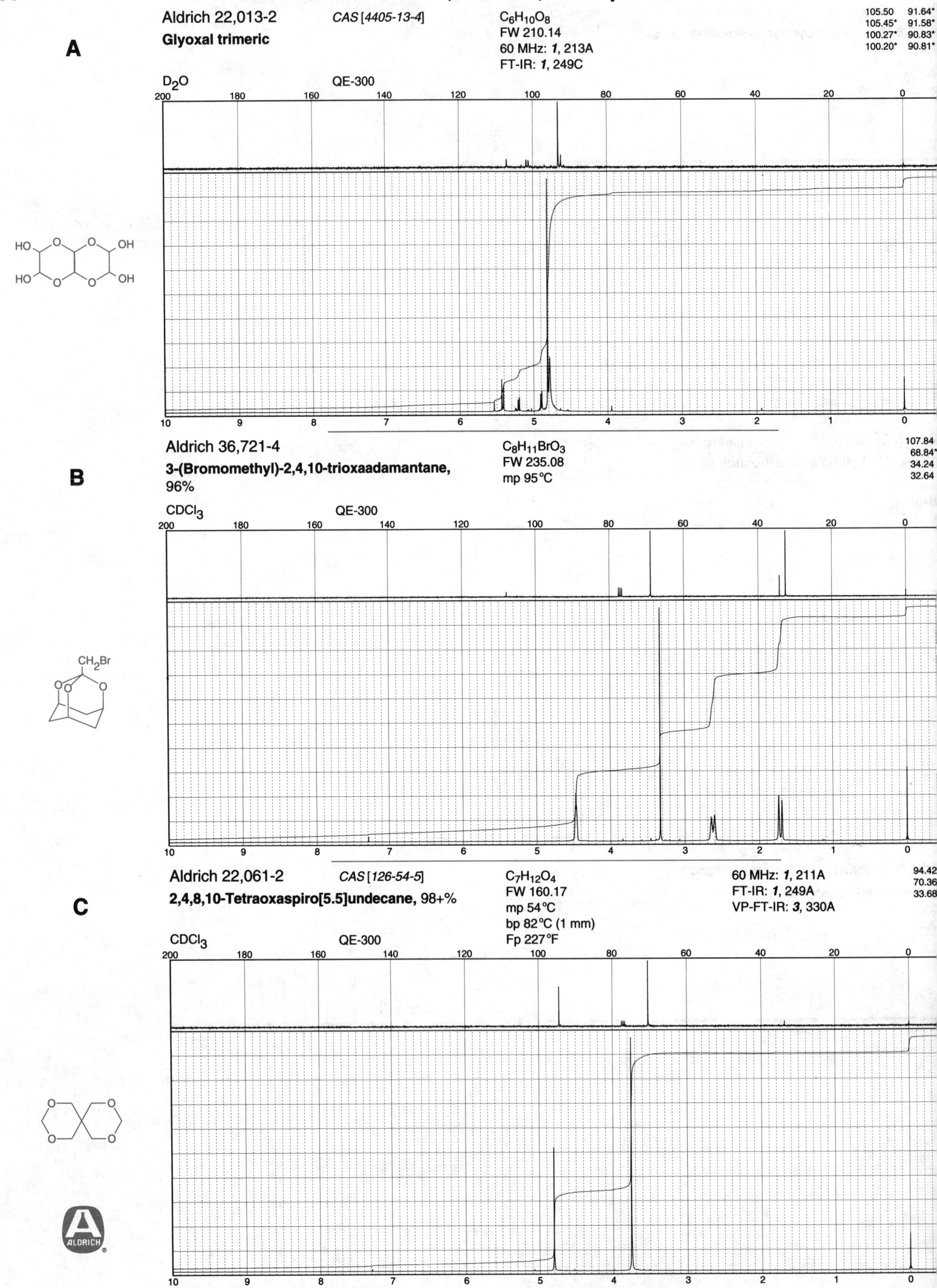

A

Aldrich 22,013-2   CAS [4405-13-4] $C_6H_{10}O_8$
FW 210.14
60 MHz: *1*, 213A
FT-IR: *1*, 249C

Glyoxal trimeric

105.50   91.64*
105.45* 91.58*
100.27* 90.83*
100.20* 90.81*

D$_2$O   QE-300

B

Aldrich 36,721-4

3-(Bromomethyl)-2,4,10-trioxaadamantane, 96%

$C_8H_{11}BrO_3$
FW 235.08
mp 95°C

107.84
68.84*
34.24
32.64

CDCl$_3$ QE-300

CH$_2$Br

C

Aldrich 22,061-2   CAS [126-54-5] $C_7H_{12}O_4$
FW 160.17
mp 54°C
bp 82°C (1 mm)
Fp 227°F

2,4,8,10-Tetraoxaspiro[5.5]undecane, 98+%

60 MHz: *1*, 211A
FT-IR: *1*, 249A
VP-FT-IR: *3*, 330A

94.42
70.36
33.68

CDCl$_3$ QE-300

Non-Aromatic Ethers, Acetals, and Epoxides

410

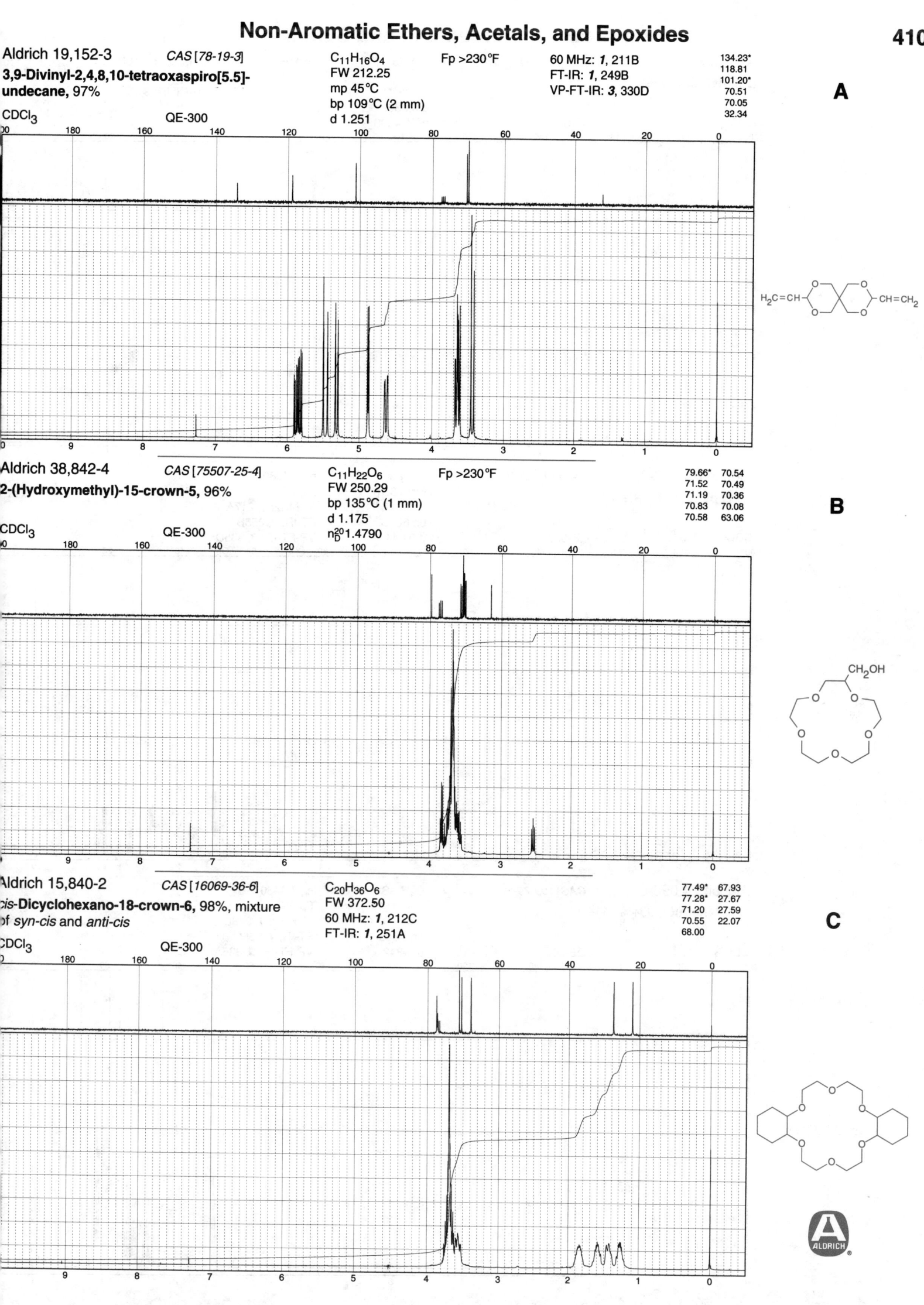

Aldrich 19,152-3   CAS [78-19-3] $C_{11}H_{16}O_4$ Fp >230°F   60 MHz: 1, 211B   134.23*
3,9-Divinyl-2,4,8,10-tetraoxaspiro[5.5]-undecane, 97%   FW 212.25   FT-IR: 1, 249B   118.81
mp 45°C   VP-FT-IR: 3, 330D   101.20*
CDCl₃   QE-300   bp 109°C (2 mm)   70.51
d 1.251   70.05
32.34

A

Aldrich 38,842-4   CAS [75507-25-4] $C_{11}H_{22}O_6$ Fp >230°F   79.66* 70.54
2-(Hydroxymethyl)-15-crown-5, 96%   FW 250.29   71.52 70.49
bp 135°C (1 mm)   71.19 70.36
CDCl₃   QE-300   d 1.175   70.83 70.08
n_D^{20}1.4790   70.58 63.06

B

Aldrich 15,840-2   CAS [16069-36-6] $C_{20}H_{36}O_6$ 77.49* 67.93
cis-Dicyclohexano-18-crown-6, 98%, mixture   FW 372.50   77.28* 27.67
of syn-cis and anti-cis   60 MHz: 1, 212C   71.20 27.59
CDCl₃   QE-300   FT-IR: 1, 251A   70.55 22.07
68.00

C

A

Aldrich E370-8   *CAS [75-08-1]* C_2H_6S Fp 1°F   60 MHz: *1*, 216D   19.66*
Ethanethiol, 97%   FW 62.13   FT-IR: *1*, 255A   19.08
bp 35°C   VP-FT-IR: *3*, 335C
d 0.839
n_D^{20} 1.4306

CDCl$_3$ QE-300

CH_3CH_2SH

B

Aldrich P5,075-7   *CAS [107-03-9]* C_3H_8S n_D^{20} 1.4380   27.26
1-Propanethiol, 99%   FW 76.16   Fp -5°F   26.67
mp -113°C   60 MHz: *1*, 217A   12.95*
bp 68°C   FT-IR: *1*, 255B
d 0.841   VP-FT-IR: *3*, 335D

CDCl$_3$ QE-300

$CH_3CH_2CH_2SH$

C

Aldrich 24,096-6   *CAS [109-79-5]* $C_4H_{10}S$ n_D^{20} 1.4430   36.12
1-Butanethiol, 99+%   FW 90.19   Fp 55°F   24.30
mp -116°C   60 MHz: *1*, 218D   21.47
bp 98°C   FT-IR: *1*, 255C   13.50*
d 0.842   VP-FT-IR: *3*, 336A

CDCl$_3$ QE-300

$CH_3CH_2CH_2CH_2SH$

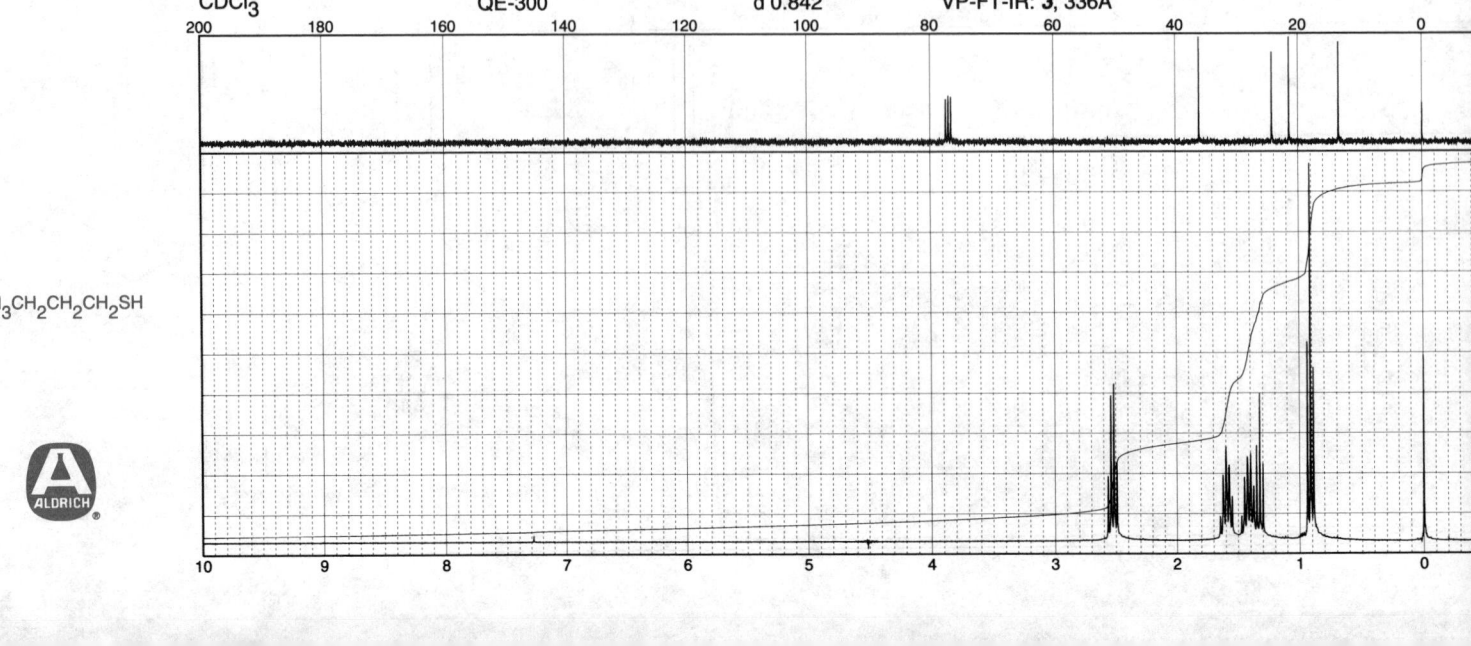

ALDRICH

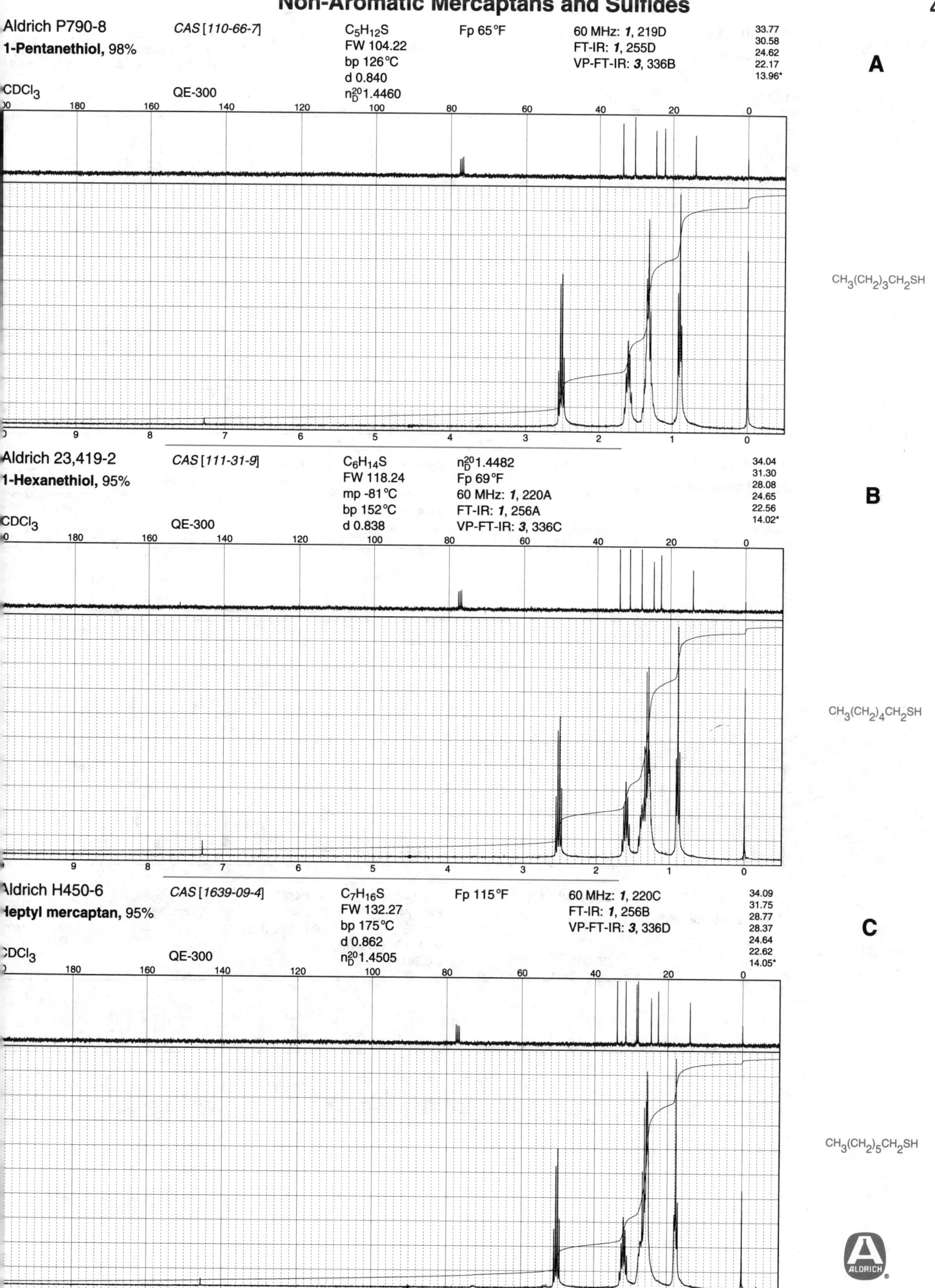

Aldrich P790-8 — CAS [110-66-7] — C₅H₁₂S — Fp 65°F — 60 MHz: *1*, 219D
1-Pentanethiol, 98%

$C_5H_{12}S$
FW 104.22
bp 126°C
d 0.840
n_D^{20} 1.4460

FT-IR: *1*, 255D
VP-FT-IR: *3*, 336B

33.77
30.58
24.62
22.17
13.96*

A

CDCl₃   QE-300

$CH_3(CH_2)_3CH_2SH$

Aldrich 23,419-2 — CAS [111-31-9]
1-Hexanethiol, 95%

$C_6H_{14}S$
FW 118.24
mp -81°C
bp 152°C
d 0.838

n_D^{20} 1.4482
Fp 69°F
60 MHz: *1*, 220A
FT-IR: *1*, 256A
VP-FT-IR: *3*, 336C

34.04
31.30
28.08
24.65
22.56
14.02*

B

CDCl₃   QE-300

$CH_3(CH_2)_4CH_2SH$

Aldrich H450-6 — CAS [1639-09-4] — C₇H₁₆S — Fp 115°F
Heptyl mercaptan, 95%

$C_7H_{16}S$
FW 132.27
bp 175°C
d 0.862
n_D^{20} 1.4505

60 MHz: *1*, 220C
FT-IR: *1*, 256B
VP-FT-IR: *3*, 336D

34.09
31.75
28.77
28.37
24.64
22.62
14.05*

C

CDCl₃   QE-300

$CH_3(CH_2)_5CH_2SH$

ALDRICH

Non-Aromatic Mercaptans and Sulfides

A

Aldrich 13,124-5   CAS [111-88-6] $C_8H_{18}S$ Fp 156°F   60 MHz: **1**, 220D

1-Octanethiol, 97+%   FW 146.30   FT-IR: **1**, 256C

bp 199°C   VP-FT-IR: **3**, 337A

CDCl₃   QE-300   d 0.843

n_D^{20} 1.4525

34.09	28.42
31.84	24.65
29.21	22.66
29.07	14.07*

$CH_3(CH_2)_6CH_2SH$

B

Aldrich N3,140-0   CAS [1455-21-6] $C_9H_{20}S$ Fp 174°F   FT-IR: **1**, 256D

Nonyl mercaptan, 98%   FW 160.32   VP-FT-IR: **3**, 337B

bp 220°C

CDCl₃   QE-300   d 0.842

n_D^{20} 1.4548

34.09	28.41
31.88	24.64
29.50	22.67
29.27	14.08*
29.11	

$CH_3(CH_2)_7CH_2SH$

C

Aldrich D160-2   CAS [143-10-2] $C_{10}H_{22}S$ n_D^{20} 1.4565

1-Decanethiol, 96%   FW 174.35   Fp 209°F

mp -26°C   VP-FT-IR: **3**, 337C

bp 114°C (13 mm)

CDCl₃   QE-300   d 0.841

34.09	28.42
31.92	24.65
29.56	22.70
29.33	14.09*
29.11	

$CH_3(CH_2)_8CH_2SH$

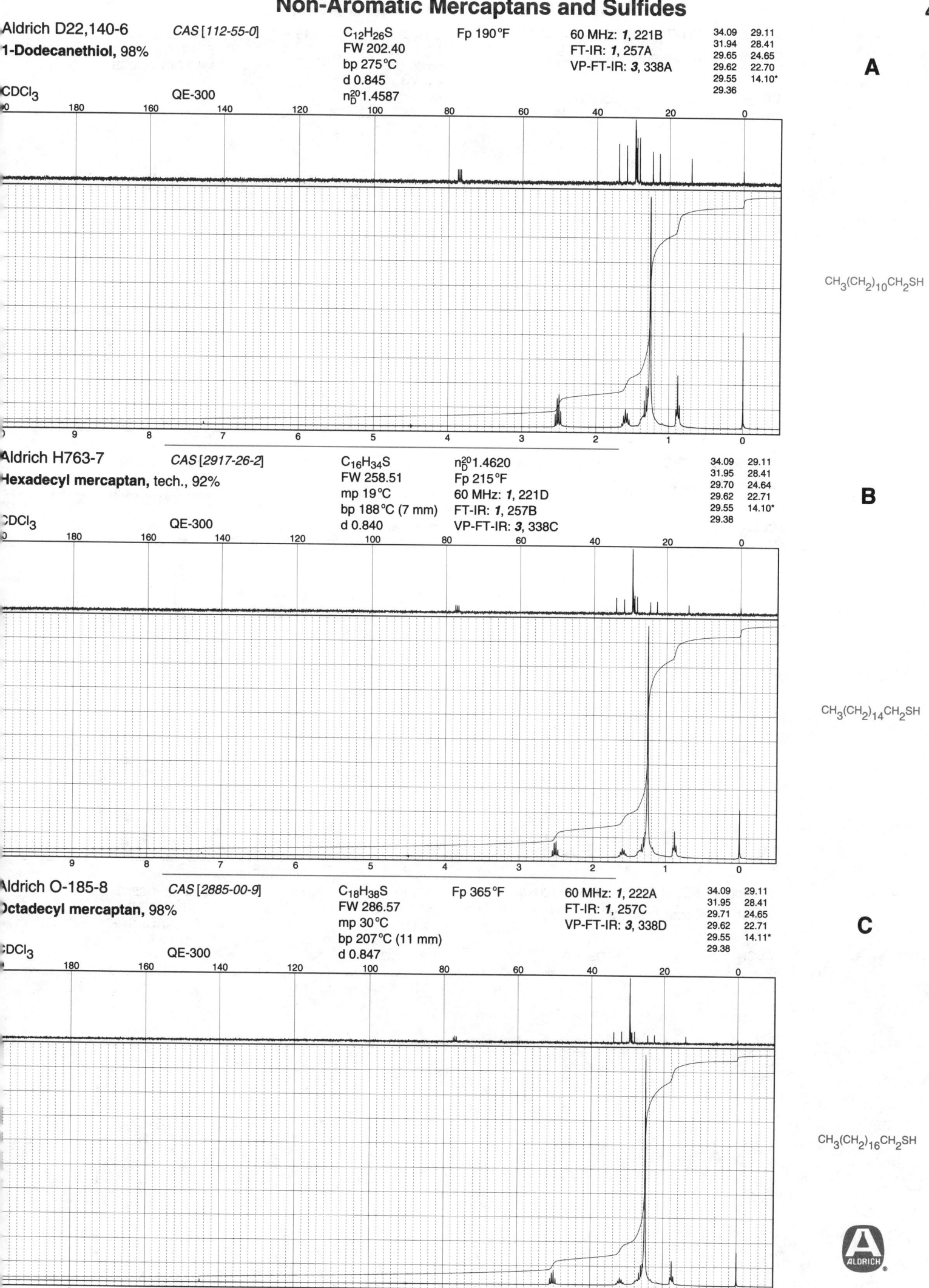

Aldrich D22,140-6   CAS [112-55-0] $C_{12}H_{26}S$ Fp 190°F
1-Dodecanethiol, 98%
FW 202.40
bp 275°C
d 0.845
n_D^{20} 1.4587

CDCl₃   QE-300

60 MHz: *1*, 221B
FT-IR: *1*, 257A
VP-FT-IR: *3*, 338A

34.09	29.11
31.94	28.41
29.65	24.65
29.62	22.70
29.55	14.10*
29.36	

A

$CH_3(CH_2)_{10}CH_2SH$

Aldrich H763-7   CAS [2917-26-2] $C_{16}H_{34}S$ n_D^{20} 1.4620
Hexadecyl mercaptan, tech., 92%
FW 258.51   Fp 215°F
mp 19°C
bp 188°C (7 mm)   FT-IR: *1*, 257B
d 0.840   VP-FT-IR: *3*, 338C

CDCl₃   QE-300   60 MHz: *1*, 221D

34.09	29.11
31.95	28.41
29.70	24.64
29.62	22.71
29.55	14.10*
29.38	

B

$CH_3(CH_2)_{14}CH_2SH$

Aldrich O-185-8   CAS [2885-00-9] $C_{18}H_{38}S$ Fp 365°F
Octadecyl mercaptan, 98%
FW 286.57
mp 30°C   FT-IR: *1*, 257C
bp 207°C (11 mm)   VP-FT-IR: *3*, 338D
d 0.847

CDCl₃   QE-300   60 MHz: *1*, 222A

34.09	29.11
31.95	28.41
29.71	24.65
29.62	22.71
29.55	14.11*
29.38	

C

$CH_3(CH_2)_{16}CH_2SH$

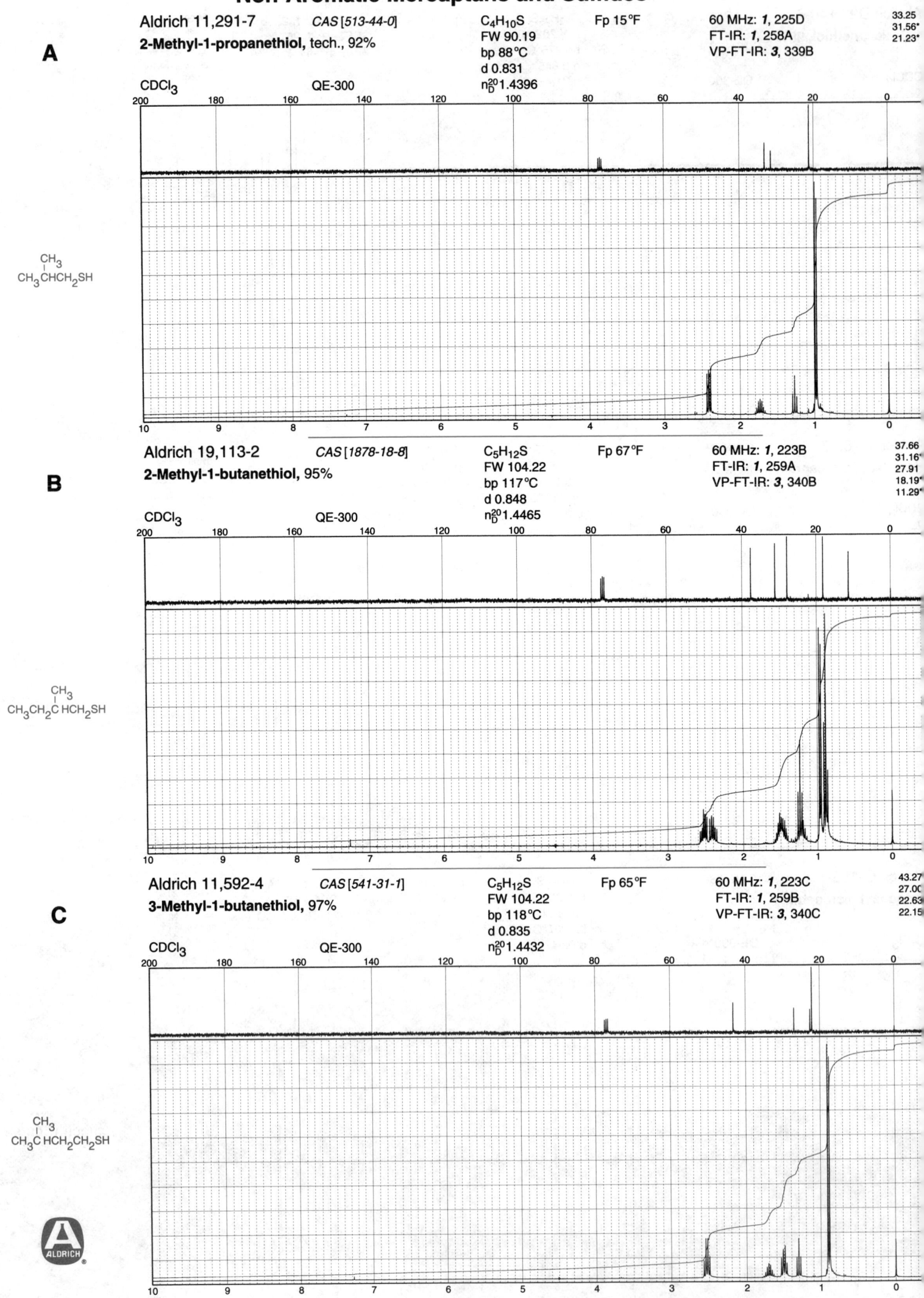

Aldrich 11,291-7   *CAS [513-44-0]* $C_4H_{10}S$ Fp 15°F   60 MHz: *1*, 225D   33.25
31.56*
2-Methyl-1-propanethiol, tech., 92%   FW 90.19   FT-IR: *1*, 258A   21.23*
bp 88°C   VP-FT-IR: *3*, 339B
d 0.831
n_D^{20}1.4396

A

CH_3
CH_3CHCH_2SH

CDCl$_3$ QE-300

Aldrich 19,113-2   *CAS [1878-18-8]* $C_5H_{12}S$ Fp 67°F   60 MHz: *1*, 223B   37.66
31.16*
2-Methyl-1-butanethiol, 95%   FW 104.22   FT-IR: *1*, 259A   27.91
bp 117°C   VP-FT-IR: *3*, 340B   18.19*
d 0.848   11.29*
n_D^{20}1.4465

B

CH_3
$CH_3CH_2CHCH_2SH$

CDCl$_3$ QE-300

Aldrich 11,592-4   *CAS [541-31-1]* $C_5H_{12}S$ Fp 65°F   60 MHz: *1*, 223C   43.27*
27.00
3-Methyl-1-butanethiol, 97%   FW 104.22   FT-IR: *1*, 259B   22.63
bp 118°C   VP-FT-IR: *3*, 340C   22.15
d 0.835
n_D^{20}1.4432

C

CH_3
$CH_3CHCH_2CH_2SH$

CDCl$_3$ QE-300

ALDRICH

Non-Aromatic Mercaptans and Sulfides

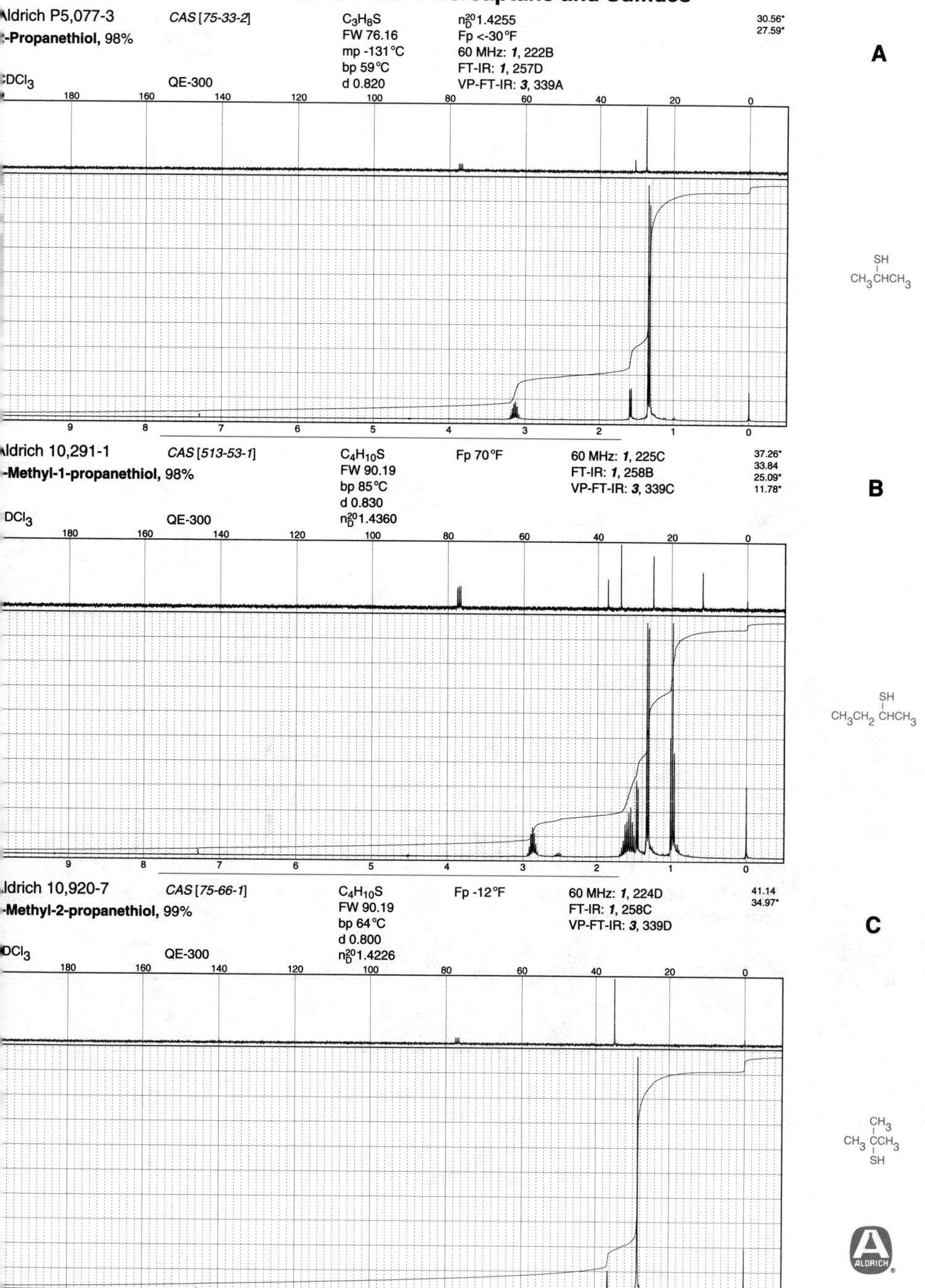

Aldrich P5,077-3   CAS [75-33-2] C_3H_8S n_D^{20} 1.4255   30.56* 27.59*
2-Propanethiol, 98%   FW 76.16   Fp <-30°F
CDCl₃   QE-300   mp -131°C   60 MHz: 1, 222B
bp 59°C   FT-IR: 1, 257D
d 0.820   VP-FT-IR: 3, 339A

A

Aldrich 10,291-1   CAS [513-53-1] $C_4H_{10}S$ Fp 70°F   37.26* 33.84 25.09* 11.78*
2-Methyl-1-propanethiol, 98%   FW 90.19
bp 85°C   60 MHz: 1, 225C
CDCl₃   QE-300   d 0.830   FT-IR: 1, 258B
n_D^{20} 1.4360   VP-FT-IR: 3, 339C

B

Aldrich 10,920-7   CAS [75-66-1] $C_4H_{10}S$ Fp -12°F   41.14 34.97*
2-Methyl-2-propanethiol, 99%   FW 90.19
bp 64°C   60 MHz: 1, 224D
CDCl₃   QE-300   d 0.800   FT-IR: 1, 258C
n_D^{20} 1.4226   VP-FT-IR: 3, 339D

C

A

Aldrich 13,409-0   *CAS [1679-09-0]* $C_5H_{12}S$ Fp 30°F   60 MHz: *1*, 223D

2-Methyl-2-butanethiol, tech., 90%   FW 104.22   FT-IR: *1*, 258D

bp 102°C   VP-FT-IR: *3*, 340A

d 0.842

n_D^{20}1.4385

45.06
39.13
32.27
9.65

CDCl₃   QE-300

$CH_3CH_2 \overset{\overset{\displaystyle CH_3}{|}}{\underset{\underset{\displaystyle SH}{|}}{C}}CH_3$

B

Aldrich 12,280-7   *CAS [870-23-5]* C_3H_6S Fp 70°F   60 MHz: *1*, 230D

Allyl mercaptan, tech., 80%   FW 74.15   FT-IR: *1*, 259C

bp 68°C   VP-FT-IR: *3*, 341A

d 0.930

n_D^{20}1.4765

137.25
115.42
27.57

CDCl₃   QE-300

$H_2C=CHCH_2SH$

C

Aldrich 31,970-8   *CAS [1679-07-8]* $C_5H_{10}S$ Fp 77°F   VP-FT-IR: *3*, 341B

Cyclopentyl mercaptan, 98%   FW 102.20

bp 130°C (745 mm)

d 0.955

n_D^{20}1.4902

38.5
37.5
24.5

CDCl₃   QE-300

Non-Aromatic Mercaptans and Sulfides

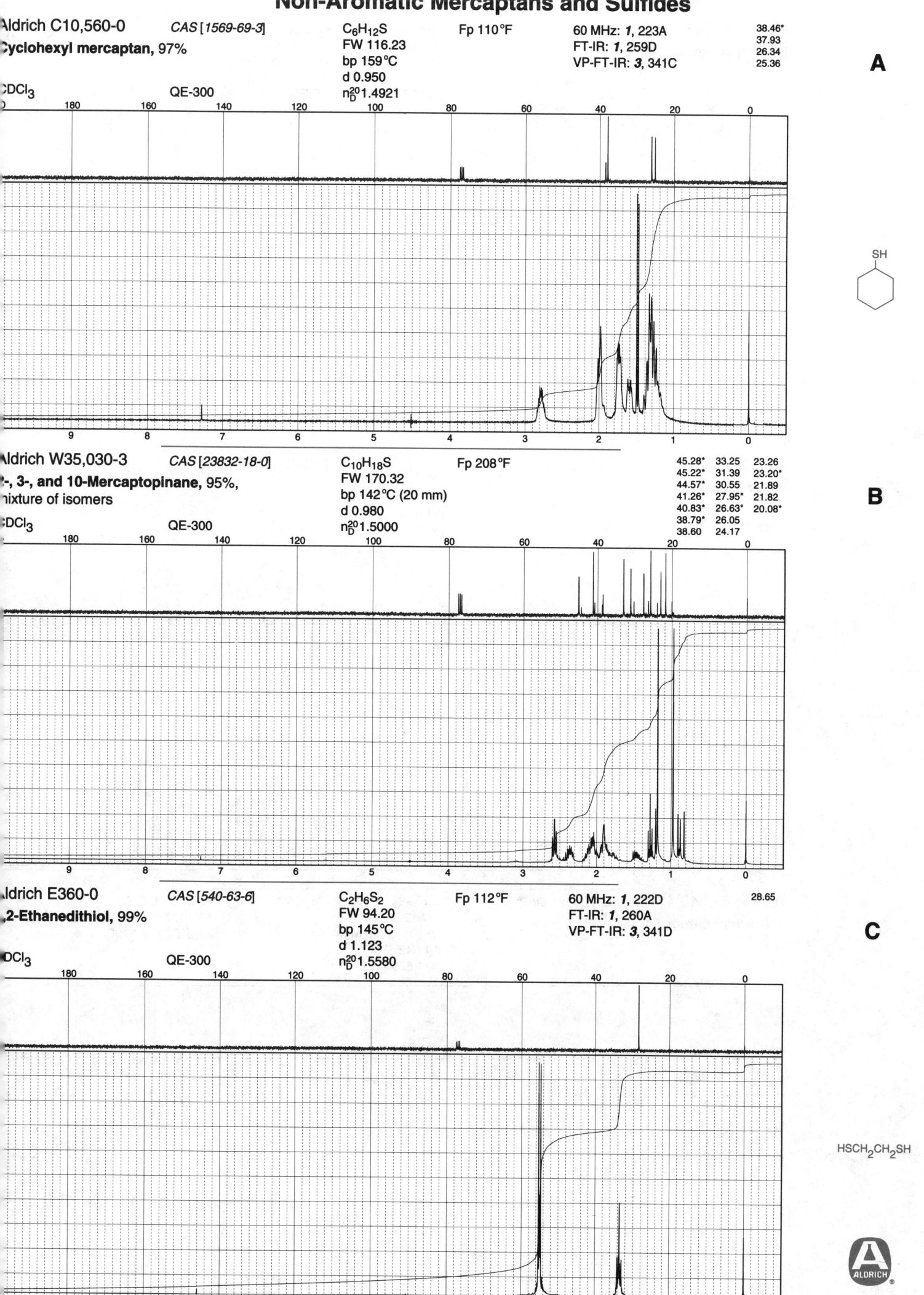

Aldrich C10,560-0   *CAS [1569-69-3]* $C_6H_{12}S$ Fp 110°F   60 MHz: *1*, 223A

Cyclohexyl mercaptan, 97%   FW 116.23   FT-IR: *1*, 259D   VP-FT-IR: *3*, 341C

bp 159°C

d 0.950

CDCl₃   QE-300 n_D^{20}1.4921

38.46*
37.93
26.34
25.36

A

Aldrich W35,030-3   *CAS [23832-18-0]* $C_{10}H_{18}S$ Fp 208°F

2-, 3-, and 10-Mercaptopinane, 95%,

mixture of isomers

FW 170.32

bp 142°C (20 mm)

d 0.980

CDCl₃   QE-300 n_D^{20}1.5000

45.28*	33.25	23.26
45.22*	31.39	23.20*
44.57*	30.55	21.89
41.26*	27.95*	21.82
40.83*	26.63*	20.08*
38.79*	26.05	
38.60	24.17	

B

Aldrich E360-0   *CAS [540-63-6]* $C_2H_6S_2$ Fp 112°F   60 MHz: *1*, 222D

1,2-Ethanedithiol, 99%   FW 94.20   FT-IR: *1*, 260A   VP-FT-IR: *3*, 341D

bp 145°C

d 1.123

CDCl₃   QE-300 n_D^{20}1.5580

28.65

C

HSCH₂CH₂SH

Non-Aromatic Mercaptans and Sulfides

A

Aldrich P5,060-9

1,3-Propanedithiol, 99%

CAS [109-80-8]

$C_3H_8S_2$
FW 108.23
mp -79°C
bp 169°C
d 1.078

n_D^{20} 1.5405
Fp 138°F
60 MHz: *1*, 226A
FT-IR: *1*, 260B
VP-FT-IR: *3*, 342A

37.24
22.83

CDCl₃

QE-300

HSCH₂CH₂CH₂SH

B

Aldrich P5,055-2

1,2-Propanedithiol, 97%

CAS [814-67-5]

$C_3H_8S_2$
FW 108.23
bp 152°C
d 1.068
n_D^{20} 1.5310

Fp 95°F

60 MHz: *1*, 226B
FT-IR: *1*, 260C

38.28
35.45
23.20

CDCl₃

QE-300

SH
|
CH₃CHCH₂SH

C

Aldrich B8,540-4

1,4-Butanedithiol, 94%

CAS [1191-08-8]

$C_4H_{10}S_2$
FW 122.25
bp 106°C (30 mm)
d 1.042
n_D^{20} 1.5290

Fp 158°F

60 MHz: *1*, 226C
FT-IR: *1*, 260D
VP-FT-IR: *3*, 342B

32.50
24.05

CDCl₃

QE-300

HSCH₂CH₂CH₂CH₂SH

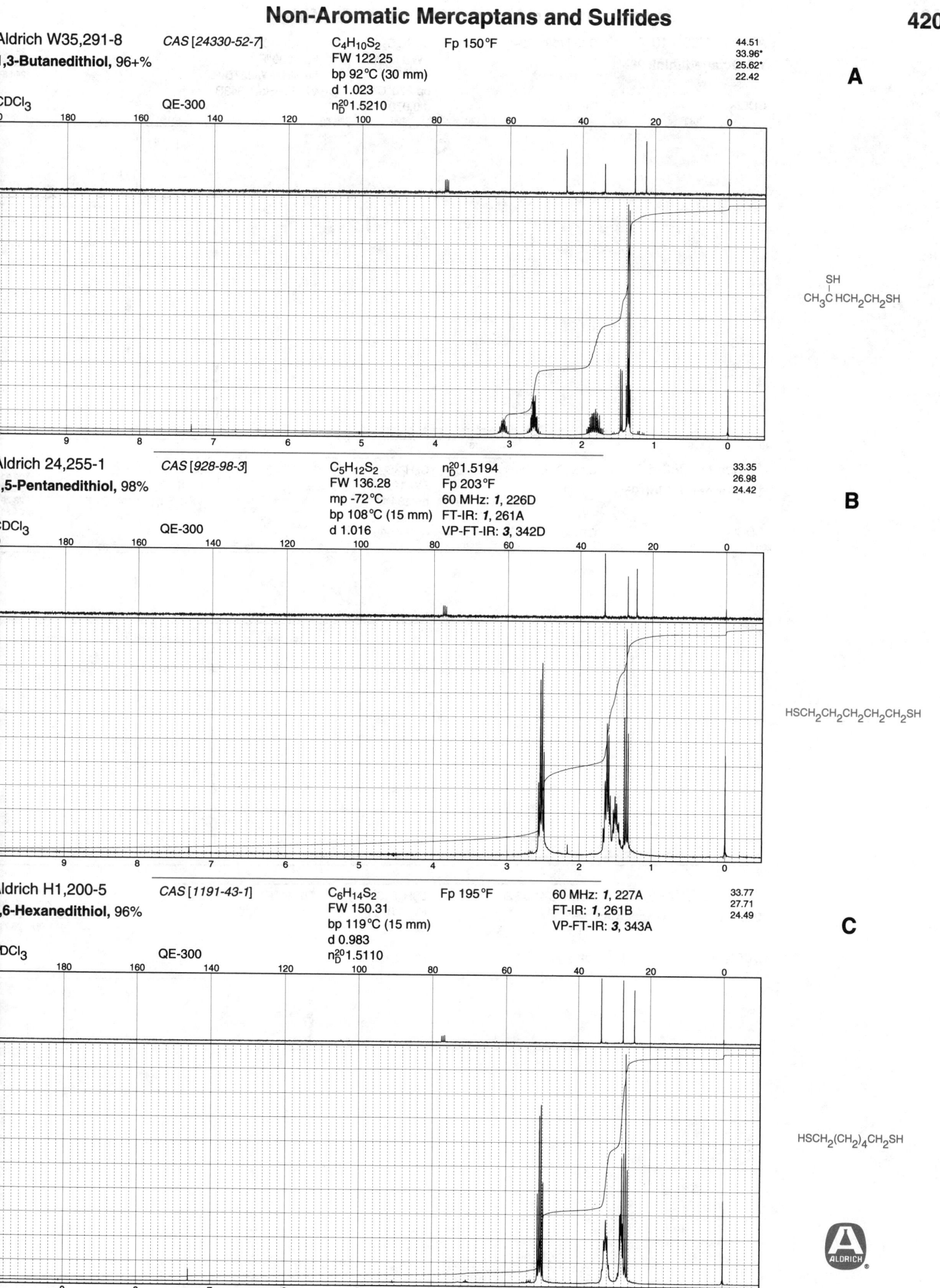

Aldrich W35,291-8

CAS [24330-52-7]

C₄H₁₀S₂
FW 122.25
bp 92°C (30 mm)
d 1.023
n²⁰_D 1.5210

Fp 150°F

44.51
33.96*
25.62*
22.42

A

1,3-Butanedithiol, 96+%

CDCl₃

QE-300

SH
|
CH₃CHCH₂CH₂SH

Aldrich 24,255-1

CAS [928-98-3]

C₅H₁₂S₂
FW 136.28
mp -72°C
bp 108°C (15 mm)
d 1.016

n²⁰_D 1.5194
Fp 203°F
60 MHz: 1, 226D
FT-IR: 1, 261A
VP-FT-IR: 3, 342D

33.35
26.98
24.42

B

1,5-Pentanedithiol, 98%

CDCl₃

QE-300

HSCH₂CH₂CH₂CH₂CH₂SH

Aldrich H1,200-5

CAS [1191-43-1]

C₆H₁₄S₂
FW 150.31
bp 119°C (15 mm)
d 0.983
n²⁰_D 1.5110

Fp 195°F

60 MHz: 1, 227A
FT-IR: 1, 261B
VP-FT-IR: 3, 343A

33.77
27.71
24.49

C

1,6-Hexanedithiol, 96%

CDCl₃

QE-300

HSCH₂(CH₂)₄CH₂SH

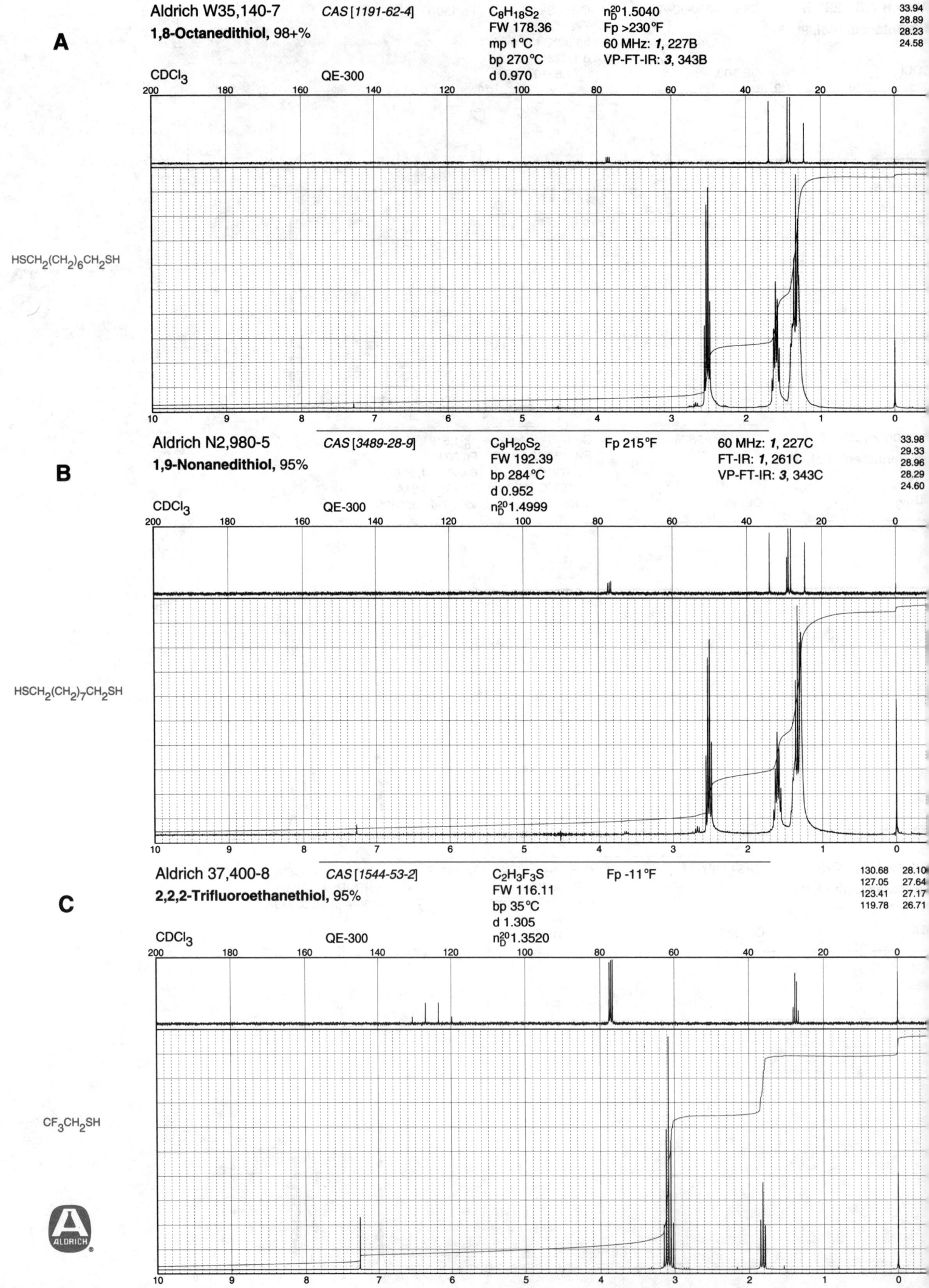

A

Aldrich W35,140-7   *CAS [1191-62-4]* $C_8H_{18}S_2$ n_D^{20} 1.5040
1,8-Octanedithiol, 98+%   FW 178.36   Fp >230°F
                                                          mp 1°C   60 MHz: *1*, 227B
                                                          bp 270°C   VP-FT-IR: *3*, 343B
                                                          d 0.970

CDCl₃   QE-300

$HSCH_2(CH_2)_6CH_2SH$

33.94
28.89
28.23
24.58

B

Aldrich N2,980-5   *CAS [3489-28-9]* $C_9H_{20}S_2$ Fp 215°F   60 MHz: *1*, 227C
1,9-Nonanedithiol, 95%   FW 192.39   FT-IR: *1*, 261C
                                                          bp 284°C   VP-FT-IR: *3*, 343C
                                                          d 0.952
                                                          n_D^{20} 1.4999

CDCl₃   QE-300

$HSCH_2(CH_2)_7CH_2SH$

33.98
29.33
28.96
28.29
24.60

C

Aldrich 37,400-8   *CAS [1544-53-2]* $C_2H_3F_3S$ Fp -11°F
2,2,2-Trifluoroethanethiol, 95%   FW 116.11
                                                          bp 35°C
                                                          d 1.305
                                                          n_D^{20} 1.3520

CDCl₃   QE-300

CF_3CH_2SH

130.68   28.10
127.05   27.64
123.41   27.17
119.78   26.71

Aldrich C6,860-1 CAS [17481-19-5] C3H7ClS Fp 110°F 42.86
3-Chloro-1-propanethiol, 98% FW 110.61 35.90
bp 146°C 21.56
d 1.136
CDCl3 QE-300 n_D^20 1.4921

A

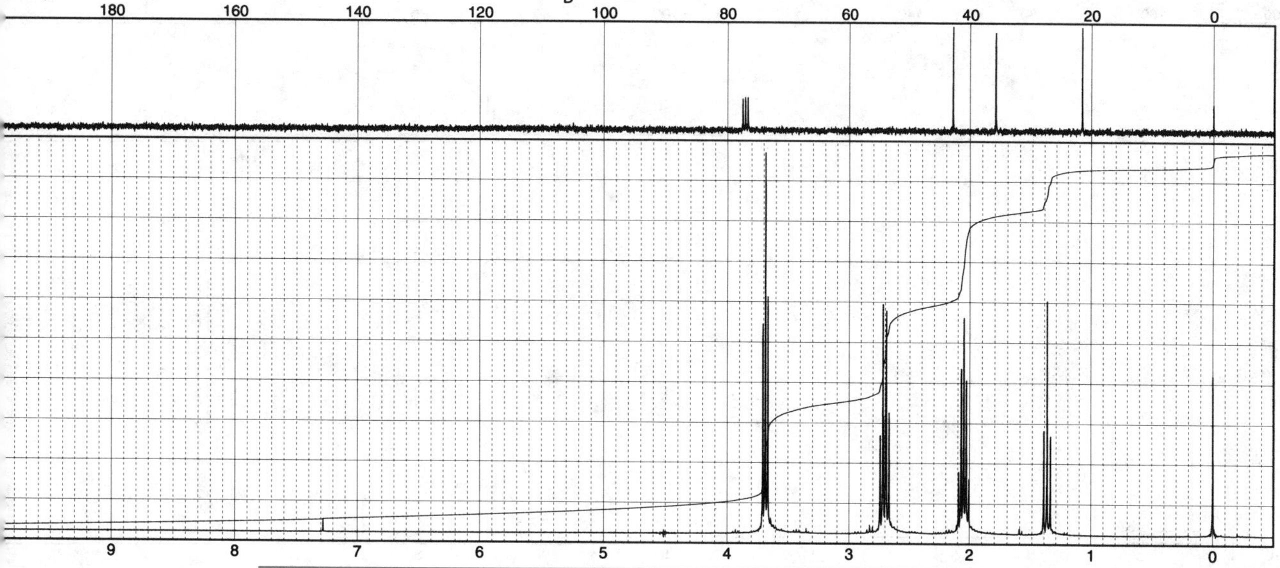

ClCH2CH2CH2SH

Aldrich M370-1 CAS [60-24-2] C2H6OS Fp 165°F 60 MHz: 1, 228B 63.95
2-Mercaptoethanol, 98% FW 78.13 FT-IR: 1, 262B 27.54
bp 157°C VP-FT-IR: 3, 344C
d 1.114
CDCl3 QE-300 n_D^20 1.5006

B

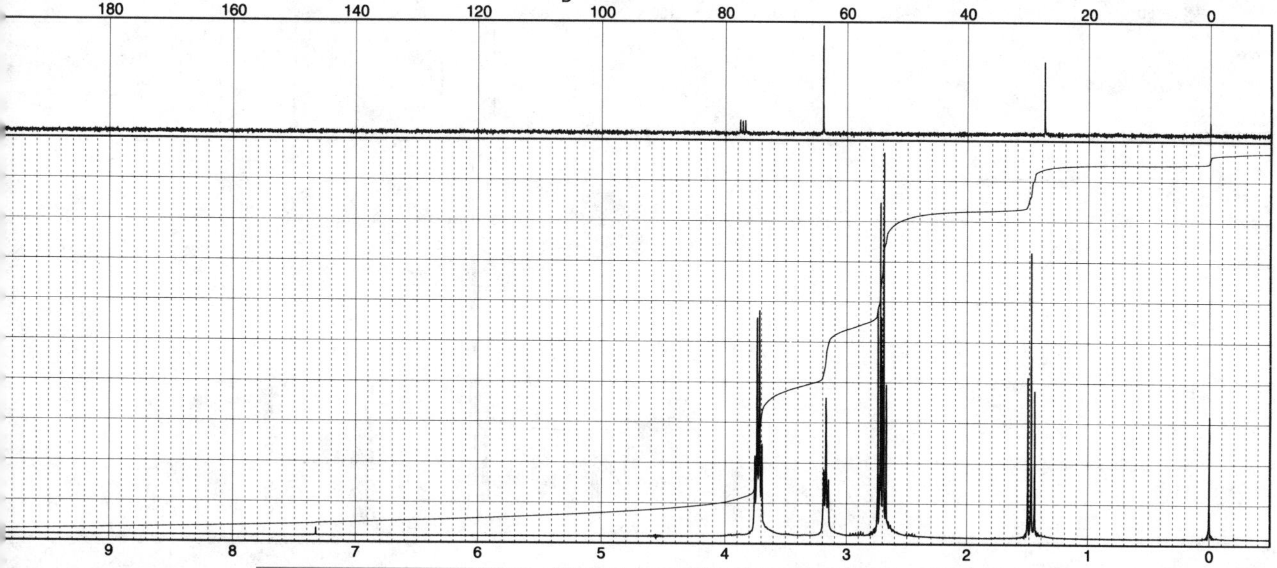

HSCH2CH2OH

Aldrich 32,837-5 CAS [1068-47-9] C3H8OS Fp 145°F VP-FT-IR: 3, 344D 68.49*
1-Mercapto-2-propanol, 95% FW 92.16 33.46
bp 59°C (17 mm) 21.62*
d 1.048
CDCl3 QE-300 n_D^20 1.4860

C

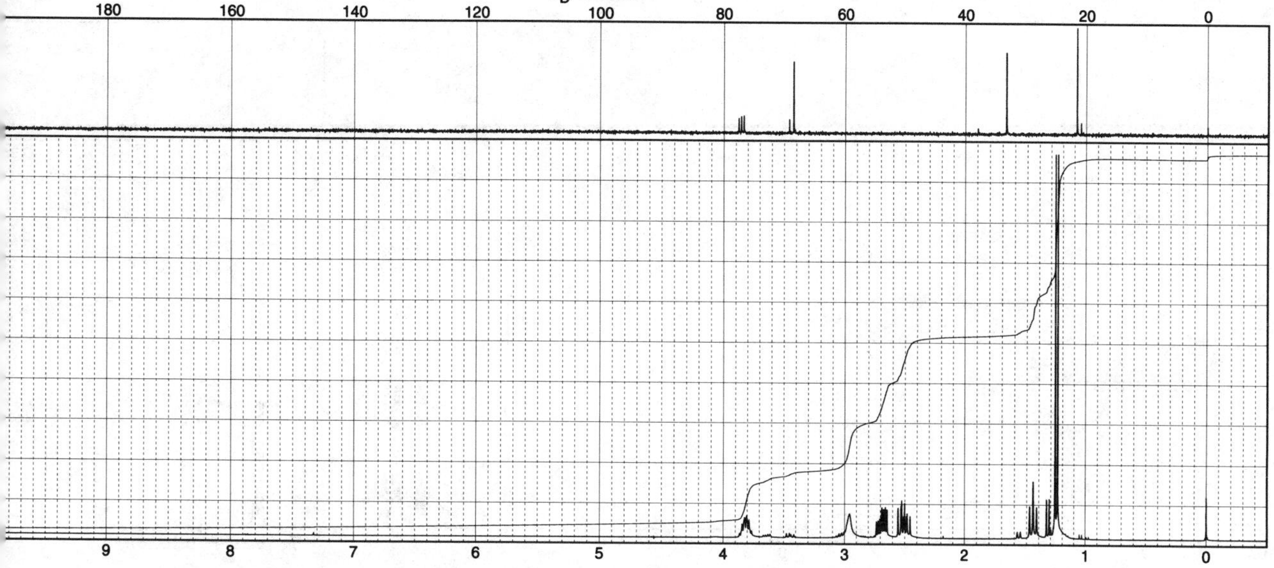

OH
|
CH3CHCH2SH

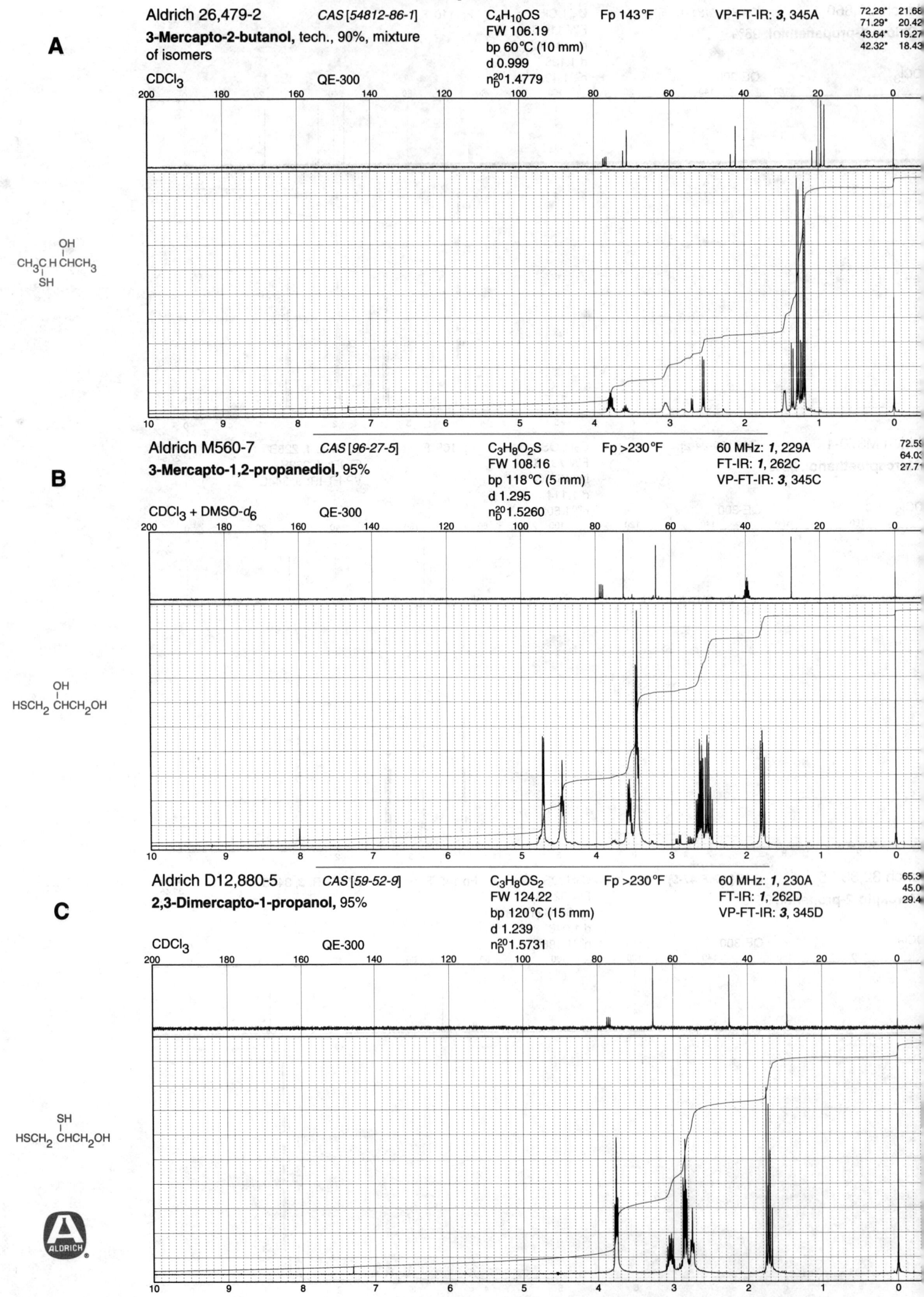

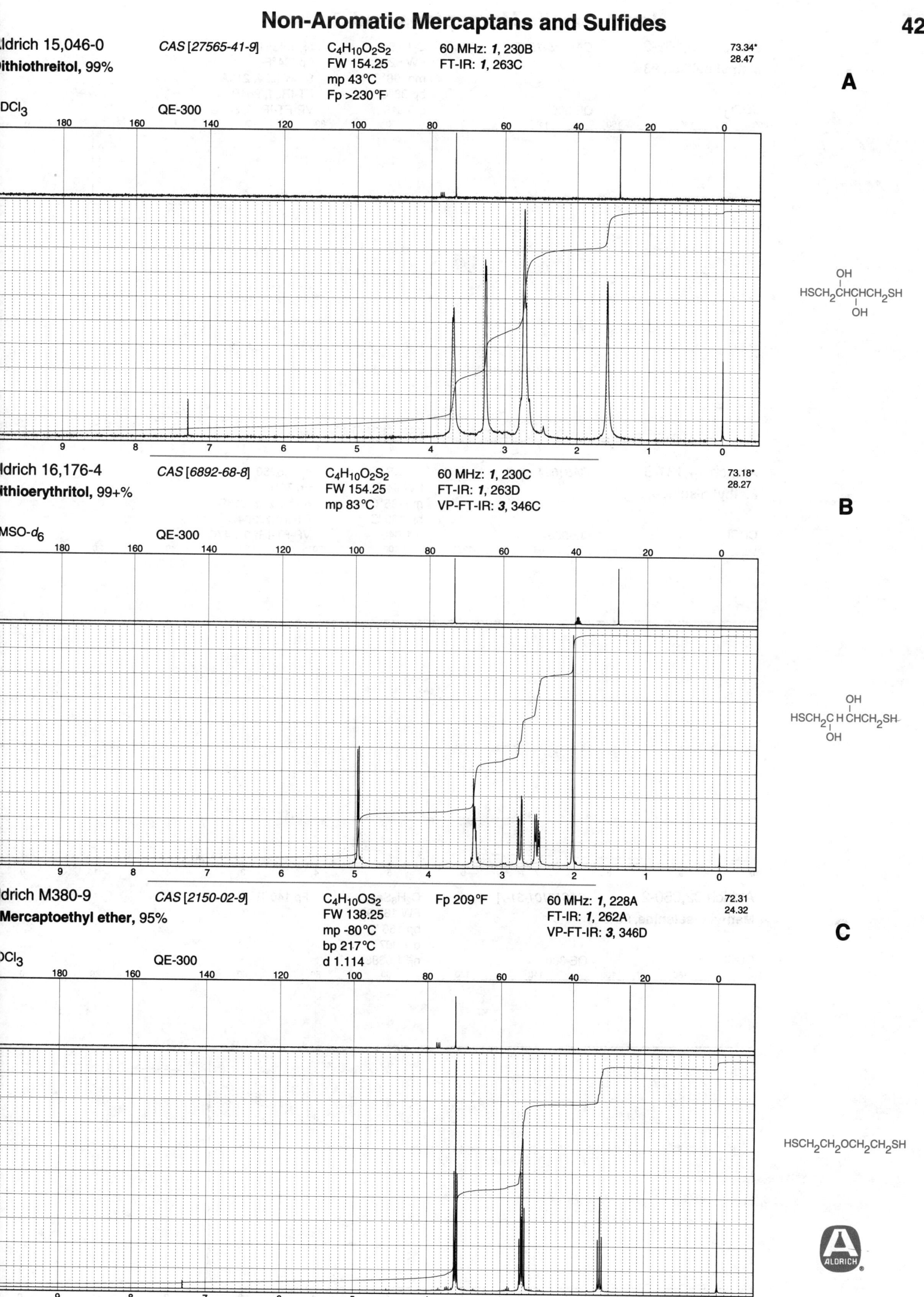

Aldrich 15,046-0   CAS [27565-41-9] $C_4H_{10}O_2S_2$ 60 MHz: 1, 230B   73.34*
Dithiothreitol, 99%   FW 154.25   FT-IR: 1, 263C   28.47
mp 43°C
Fp >230°F

CDCl₃   QE-300

A

Aldrich 16,176-4   CAS [6892-68-8] $C_4H_{10}O_2S_2$ 60 MHz: 1, 230C   73.18*
Dithioerythritol, 99+%   FW 154.25   FT-IR: 1, 263D   28.27
mp 83°C   VP-FT-IR: 3, 346C

DMSO-d_6 QE-300

B

Aldrich M380-9   CAS [2150-02-9] $C_4H_{10}OS_2$ Fp 209°F   60 MHz: 1, 228A   72.31
Mercaptoethyl ether, 95%   FW 138.25   FT-IR: 1, 262A   24.32
mp -80°C   VP-FT-IR: 3, 346D
bp 217°C
CDCl₃   QE-300   d 1.114

C

ALDRICH

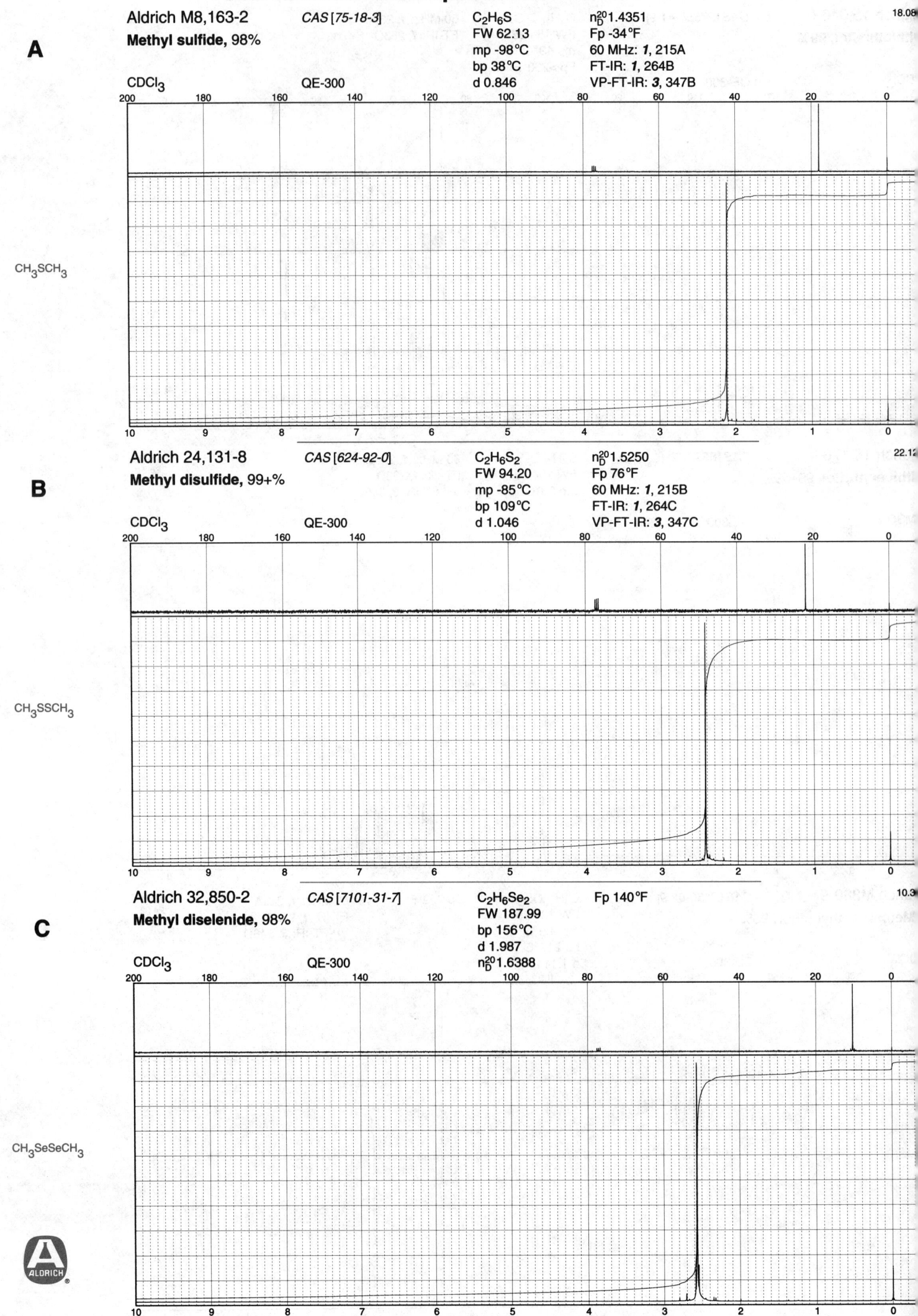

A

Aldrich M8,163-2   *CAS [75-18-3]*   C₂H₆S   n²⁰_D 1.4351   18.08

Methyl sulfide, 98%

C_2H_6S
FW 62.13
mp -98°C
bp 38°C
d 0.846

$n_D^{20} 1.4351$
Fp -34°F
60 MHz: *1*, 215A
FT-IR: *1*, 264B
VP-FT-IR: *3*, 347B

CDCl₃   QE-300

CH₃SCH₃

B

Aldrich 24,131-8   *CAS [624-92-0]*   C₂H₆S₂   n²⁰_D 1.5250   22.12

Methyl disulfide, 99+%

$C_2H_6S_2$
FW 94.20
mp -85°C
bp 109°C
d 1.046

$n_D^{20} 1.5250$
Fp 76°F
60 MHz: *1*, 215B
FT-IR: *1*, 264C
VP-FT-IR: *3*, 347C

CDCl₃   QE-300

CH₃SSCH₃

C

Aldrich 32,850-2   *CAS [7101-31-7]*   C₂H₆Se₂   Fp 140°F   10.3

Methyl diselenide, 98%

$C_2H_6Se_2$
FW 187.99
bp 156°C
d 1.987
$n_D^{20} 1.6388$

CDCl₃   QE-300

CH₃SeSeCH₃

Aldrich 22,633-5   *CAS [1618-26-4]*   C₃H₈S₂   Fp 111°F   60 MHz: *1*, 215C   40.09
Bis(methylthio)methane, 99%   FW 108.23   FT-IR: *1*, 264D   14.29*
    bp 147°C   VP-FT-IR: *3*, 347D
CDCl₃   QE-300   d 1.059
    n²⁰_D 1.5340

A

$CH_3SCH_2SCH_3$

Aldrich 25,506-8   *CAS [5418-86-0]*   C₄H₁₀S₃   n²⁰_D 1.5767   59.17*
Tris(methylthio)methane, 99%   FW 154.32   Fp 204°F   14.75*
    mp 16°C   FT-IR: *1*, 265A
CDCl₃   QE-300   bp 102°C (15 mm)   VP-FT-IR: *3*, 348A
    d 1.160

B

$CH_3SC\ HSCH_3$
$\quad\ |$
$\quad SCH_3$

Aldrich 23,831-7   *CAS [624-89-5]*   C₃H₈S   n²⁰_D 1.4392   28.05
Ethyl methyl sulfide, 99%   FW 76.16   Fp 5°F   15.03*
    mp -106°C   60 MHz: *1*, 217B   14.37*
    bp 67°C   FT-IR: *1*, 265C
CDCl₃   QE-300   d 0.842   VP-FT-IR: *3*, 348B

C

$CH_3CH_2SCH_3$

Non-Aromatic Mercaptans and Sulfides

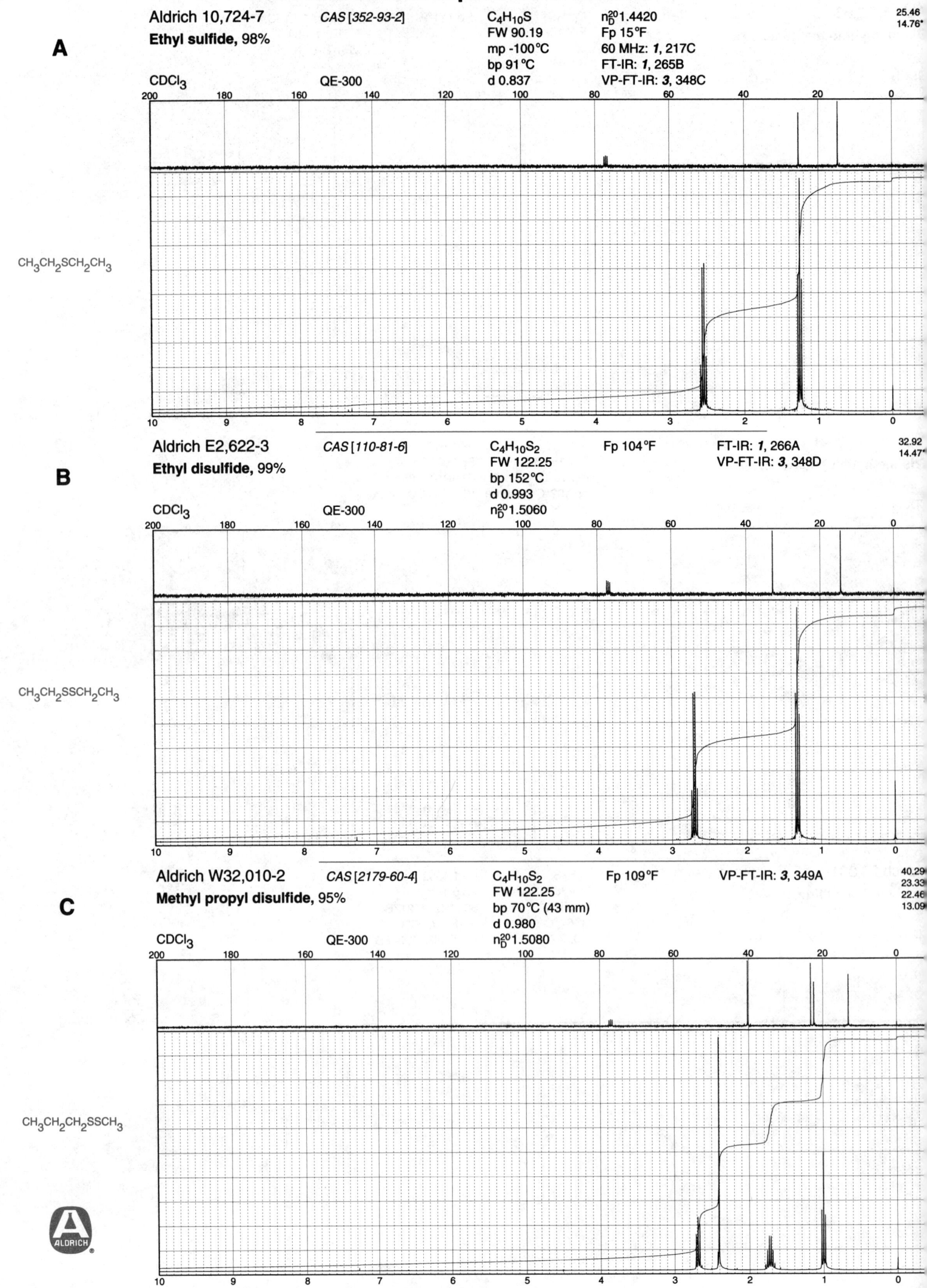

A

Aldrich 10,724-7   *CAS [352-93-2]* $C_4H_{10}S$ $n_D^{20}1.4420$ 25.46
Ethyl sulfide, 98%   FW 90.19   Fp 15°F   14.76*
mp -100°C   60 MHz: *1*, 217C
bp 91°C   FT-IR: *1*, 265B

CDCl$_3$ QE-300   d 0.837   VP-FT-IR: *3*, 348C

$CH_3CH_2SCH_2CH_3$

B

Aldrich E2,622-3   *CAS [110-81-6]* $C_4H_{10}S_2$ Fp 104°F   FT-IR: *1*, 266A   32.92
Ethyl disulfide, 99%   FW 122.25   VP-FT-IR: *3*, 348D   14.47*
bp 152°C
d 0.993

CDCl$_3$ QE-300 $n_D^{20}1.5060$

$CH_3CH_2SSCH_2CH_3$

C

Aldrich W32,010-2   *CAS [2179-60-4]* $C_4H_{10}S_2$ Fp 109°F   VP-FT-IR: *3*, 349A   40.29
Methyl propyl disulfide, 95%   FW 122.25   23.33
bp 70°C (43 mm)   22.46
d 0.980   13.09

CDCl$_3$ QE-300 $n_D^{20}1.5080$

$CH_3CH_2CH_2SSCH_3$

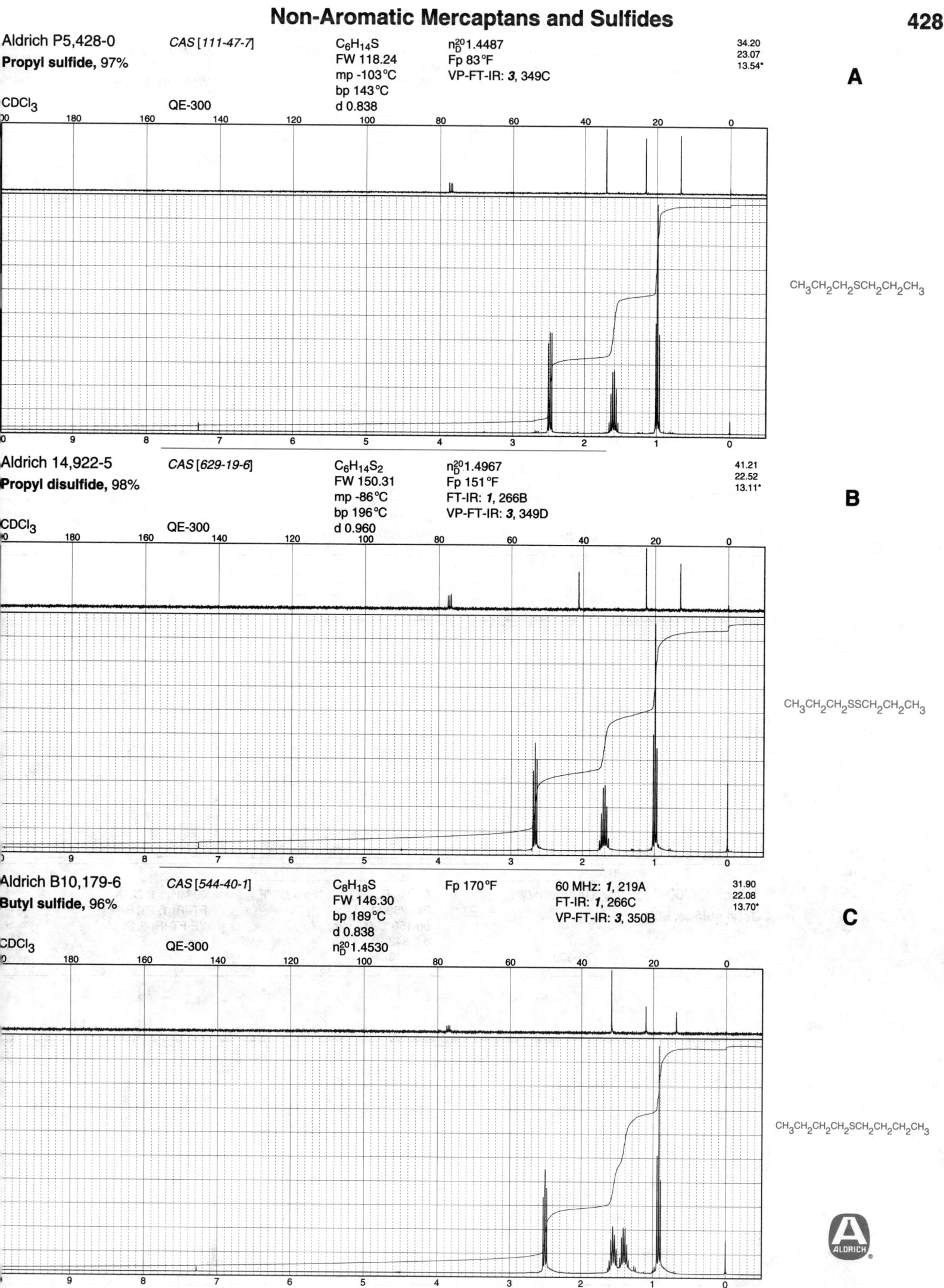

Aldrich P5,428-0   CAS [111-47-7] $C_6H_{14}S$ n_D^{20} 1.4487   34.20
Propyl sulfide, 97%   FW 118.24   Fp 83°F   23.07
   mp -103°C   VP-FT-IR: **3**, 349C   13.54*
   bp 143°C
$CDCl_3$ QE-300   d 0.838

A

$CH_3CH_2CH_2SCH_2CH_2CH_3$

Aldrich 14,922-5   CAS [629-19-6] $C_6H_{14}S_2$ n_D^{20} 1.4967   41.21
Propyl disulfide, 98%   FW 150.31   Fp 151°F   22.52
   mp -86°C   FT-IR: **1**, 266B   13.11*
   bp 196°C   VP-FT-IR: **3**, 349D
$CDCl_3$ QE-300   d 0.960

B

$CH_3CH_2CH_2SSCH_2CH_2CH_3$

Aldrich B10,179-6   CAS [544-40-1] $C_8H_{18}S$ Fp 170°F   60 MHz: **1**, 219A   31.90
Butyl sulfide, 96%   FW 146.30   FT-IR: **1**, 266C   22.08
   bp 189°C   VP-FT-IR: **3**, 350B   13.70*
   d 0.838
$CDCl_3$ QE-300 n_D^{20} 1.4530

C

$CH_3CH_2CH_2CH_2SCH_2CH_2CH_2CH_3$

Non-Aromatic Mercaptans and Sulfides

A

Aldrich B9,398-9   *CAS [629-45-8]*   C₈H₁₈S₂   Fp 200°F   60 MHz: *1*, 219B

Butyl disulfide, 98%

$C_8H_{18}S_2$
FW 178.36
bp 231°C
d 0.938
n_D^{20}1.4925

FT-IR: *1*, 266D
VP-FT-IR: *3*, 350C

38.94
31.35
21.68
13.67*

CDCl₃   QE-300

$(CH_3CH_2CH_2CH_2S)_2$

B

Aldrich H1,460-1   *CAS [6294-31-1]*   C₁₂H₂₆S   Fp >230°F   60 MHz: *1*, 220B

Hexyl sulfide, 95%

$C_{12}H_{26}S$
FW 202.40
bp 230°C
d 0.849
n_D^{20}1.4587

FT-IR: *1*, 267A
VP-FT-IR: *3*, 350D

32.26
31.51
29.76
28.67
22.59
14.02*

CDCl₃   QE-300

$CH_3(CH_2)_4CH_2SCH_2(CH_2)_4CH_3$

C

Aldrich O-700-7   *CAS [2690-08-6]*   C₁₆H₃₄S   Fp >230°F   60 MHz: *1*, 221A

Octyl sulfide, 96%

$C_{16}H_{34}S$
FW 258.51
bp 180°C (10 mm)
d 0.842
n_D^{20}1.4624

FT-IR: *1*, 267B
VP-FT-IR: *3*, 351A

32.24
31.85
29.78
29.24
28.99
22.67
14.08*

CDCl₃   QE-300

$CH_3(CH_2)_6CH_2SCH_2(CH_2)_6CH_3$

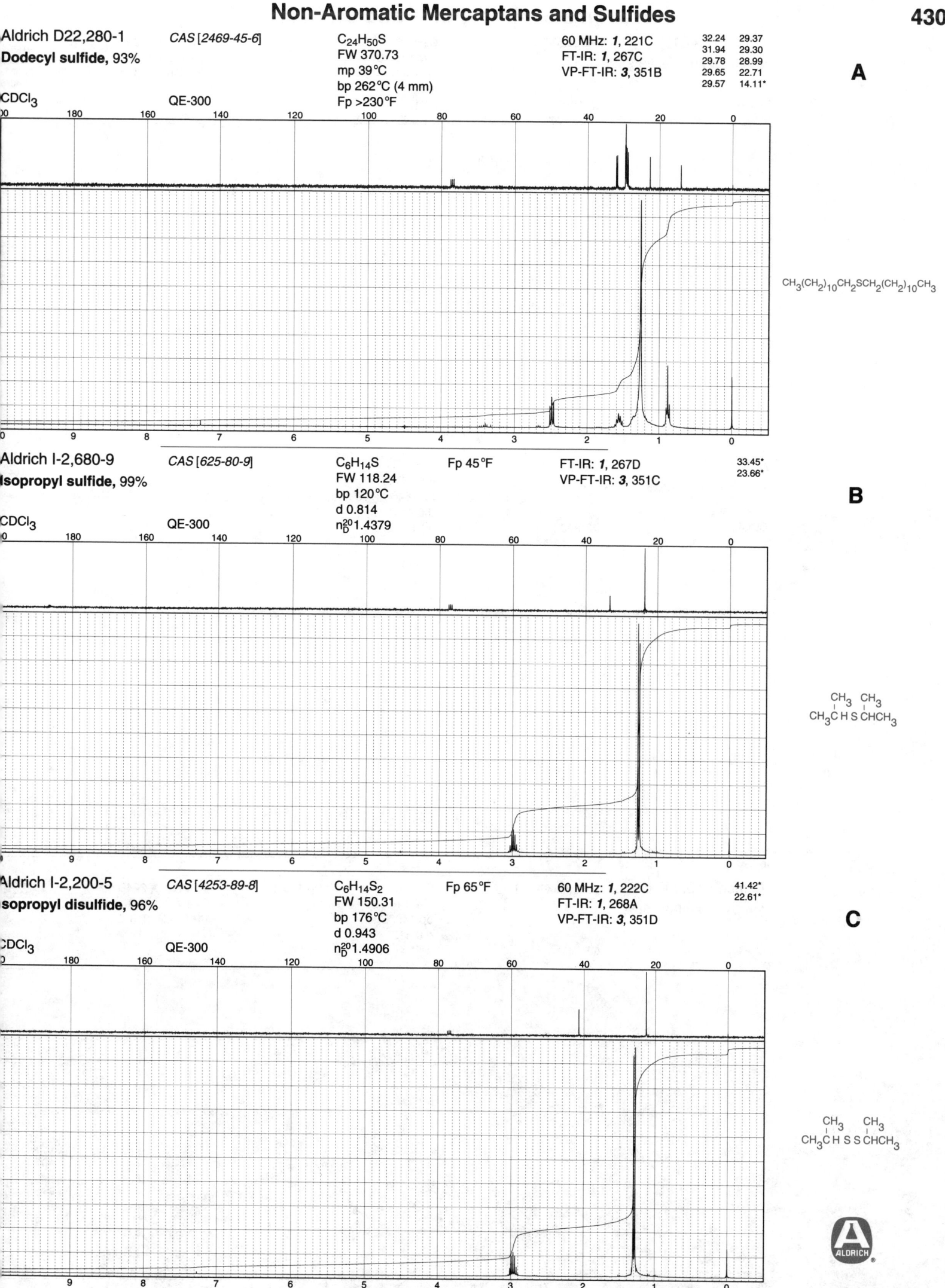

Aldrich D22,280-1   CAS [2469-45-6]   C₂₄H₅₀S
Dodecyl sulfide, 93%

$C_{24}H_{50}S$
FW 370.73
mp 39°C
bp 262°C (4 mm)
Fp >230°F

60 MHz: *1*, 221C
FT-IR: *1*, 267C
VP-FT-IR: *3*, 351B

32.24 29.37
31.94 29.30
29.78 28.99
29.65 22.71
29.57 14.11*

A

CDCl₃   QE-300

$CH_3(CH_2)_{10}CH_2SCH_2(CH_2)_{10}CH_3$

Aldrich I-2,680-9   CAS [625-80-9] $C_6H_{14}S$ Fp 45°F
Isopropyl sulfide, 99%

$C_6H_{14}S$
FW 118.24
bp 120°C
d 0.814
n₂₀D 1.4379

FT-IR: *1*, 267D
VP-FT-IR: *3*, 351C

33.45*
23.66*

B

CDCl₃   QE-300

Aldrich I-2,200-5   CAS [4253-89-8] $C_6H_{14}S_2$ Fp 65°F
Isopropyl disulfide, 96%

$C_6H_{14}S_2$
FW 150.31
bp 176°C
d 0.943
n₂₀D 1.4906

60 MHz: *1*, 222C
FT-IR: *1*, 268A
VP-FT-IR: *3*, 351D

41.42*
22.61*

C

CDCl₃   QE-300

ALDRICH

A

Aldrich 12,431-1   CAS [626-26-6] $C_8H_{18}S$ Fp 103°F   60 MHz: *1*, 225B

sec-Butyl sulfide, 98%   FW 146.30   FT-IR: *1*, 268B

bp 165°C   VP-FT-IR: *3*, 352A

d 0.839

n_D^{20} 1.4500

CDCl₃   QE-300

40.29*
30.25
30.15
21.32*
11.46*
11.30*

CH₃CH₂CHS CHCH₂CH₃

B

Aldrich 20,985-6   CAS [5943-30-6] $C_8H_{18}S_2$ Fp >230°F   FT-IR: *1*, 268C

sec-Butyl disulfide, 99+%   FW 178.36   VP-FT-IR: *3*, 352B

bp 164°C (739 mm)

d 0.950

n_D^{20} 1.4927

CDCl₃   QE-300

48.28*
29.04
20.10*
11.49*

CH₃CH₂CHS — SCHCH₂CH₃

C

Aldrich B10,200-8   CAS [107-47-1] $C_8H_{18}S$ Fp 120°F   60 MHz: *1*, 224B

tert-Butyl sulfide, 98%   FW 146.30   FT-IR: *1*, 268D

bp 149°C   VP-FT-IR: *3*, 352D

d 0.815

n_D^{20} 1.4506

CDCl₃   QE-300

45.50
33.15*

CH₃ − C − S − C − CH₃ (with CH₃ groups)

Non-Aromatic Mercaptans and Sulfides

Aldrich 24,754-5   CAS [110-06-5]   C$_8$H$_{18}$S$_2$ Fp 144°F   60 MHz: *1*, 224C   46.06
tert-Butyl disulfide, 97%   FW 178.36   FT-IR: *1*, 269A   30.56*
bp 201°C
d 0.923   VP-FT-IR: *3*, 353A

A

CDCl$_3$ QE-300   n$_D^{20}$1.4892

$$CH_3-\overset{\underset{|}{CH_3}}{\overset{|}{C}}-S-S-\overset{\underset{|}{CH_3}}{\overset{|}{C}}-CH_3$$

Aldrich A3,420-1   CAS [10152-76-8]   C$_4$H$_8$S   Fp 65°F   60 MHz: *1*, 231A   133.98*
Allyl methyl sulfide, 98%   FW 88.17   FT-IR: *1*, 269B   116.81
bp 92°C   36.84
d 0.803   VP-FT-IR: *3*, 353C   14.26*

B

CDCl$_3$ QE-300   n$_D^{20}$1.4714

$$H_2C=CHCH_2\,SCH_3$$

Aldrich A3,580-1   CAS [592-88-1]   C$_6$H$_{10}$S   Fp 115°F   60 MHz: *1*, 231B   134.21*
Allyl sulfide, 97%   FW 114.21   FT-IR: *1*, 269C   117.03
bp 138°C   33.33
d 0.887   VP-FT-IR: *3*, 353D

C

CDCl$_3$ QE-300   n$_D^{20}$1.4889

$$CH_2CH=CH_2$$
$$|$$
$$S$$
$$|$$
$$CH_2CH=CH_2$$

Non-Aromatic Mercaptans and Sulfides

A

Aldrich 12,825-2 | *CAS* [*420-12-2*] | C_2H_4S | Fp 50°F | 60 MHz: ***1***, 231C | 17.97

Ethylene sulfide, 98%

FW 60.12
bp 56°C
d 1.010
n_D^{20} 1.4935

FT-IR: ***1***, 269D
VP-FT-IR: ***3***, 354C

CDCl3 | QE-300

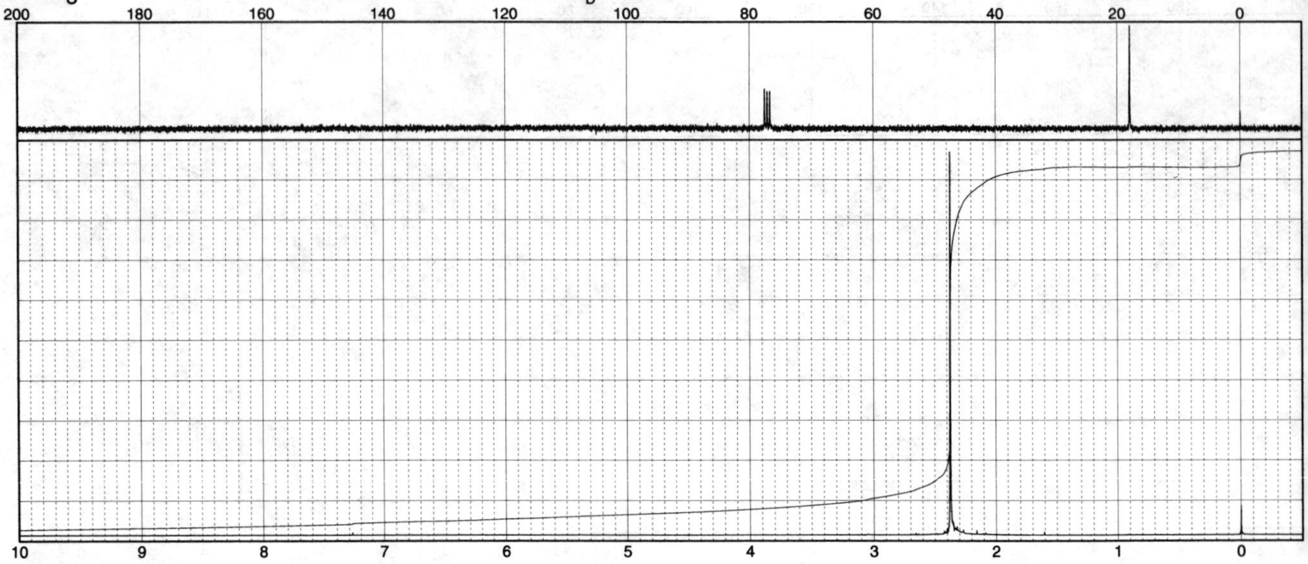

B

Aldrich P5,320-9 | *CAS* [*1072-43-1*] | C_3H_6S | Fp 50°F | 60 MHz: ***1***, 231D | 30.32* 26.96 21.89*

Propylene sulfide, 98%

FW 74.15
bp 74°C
d 0.946
n_D^{20} 1.4760

FT-IR: ***1***, 270A
VP-FT-IR: ***3***, 354D

CDCl3 | QE-300

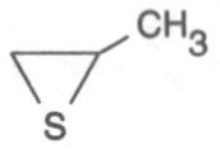

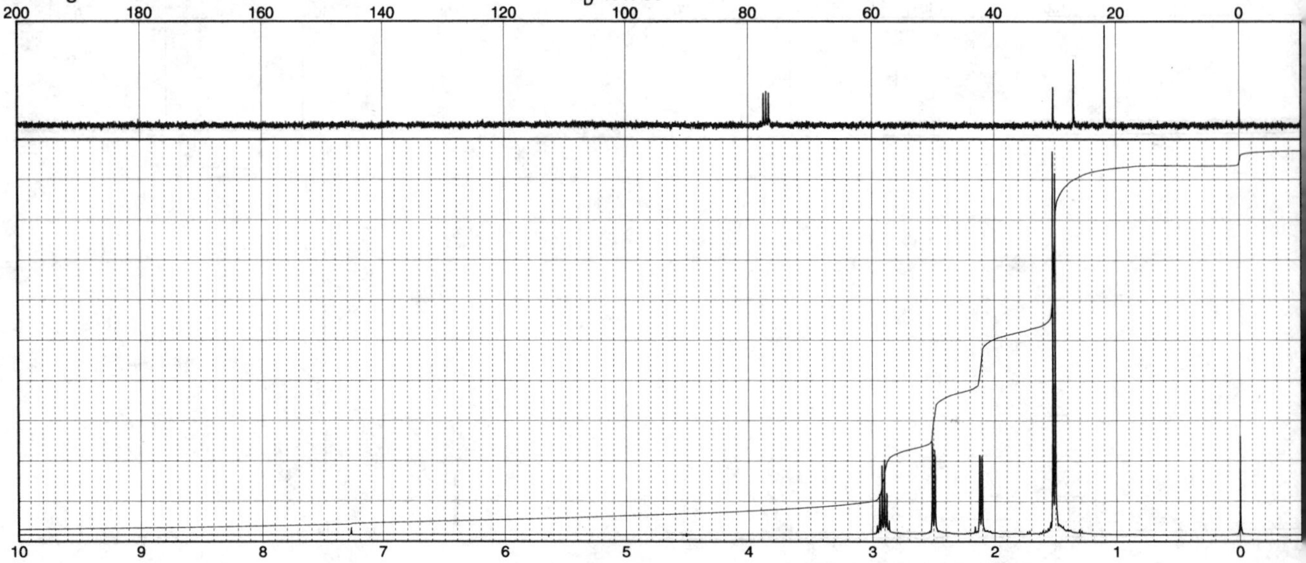

C

Aldrich 18,894-8 | *CAS* [*287-27-4*] | C_3H_6S | Fp 27°F | 60 MHz: ***1***, 232C | 27.97 26.02

Trimethylene sulfide, 97%

FW 74.15
bp 94°C
d 1.028
n_D^{20} 1.5055

FT-IR: ***1***, 270D
VP-FT-IR: ***3***, 355D

CDCl3 | QE-300

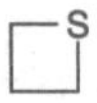

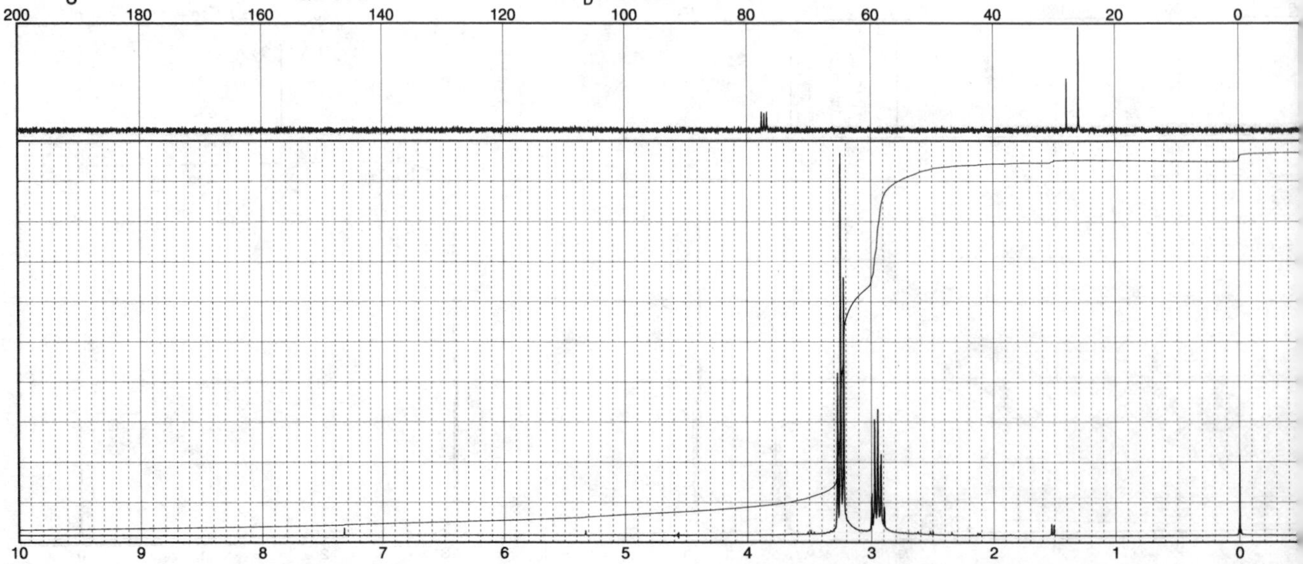

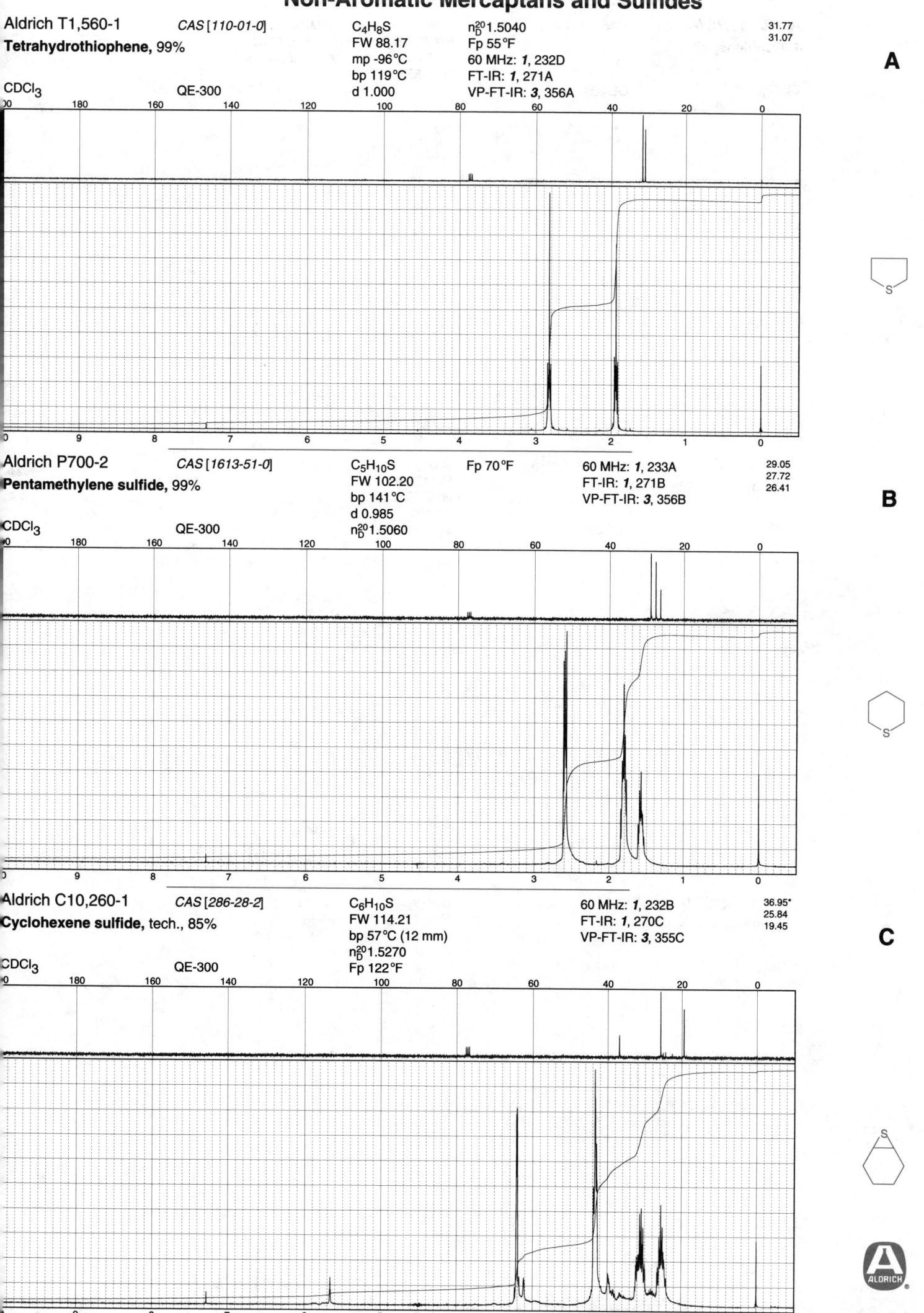

Aldrich T1,560-1   CAS [110-01-0]   C₄H₈S   n²⁰_D 1.5040   31.77
Tetrahydrothiophene, 99%   FW 88.17   Fp 55°F   31.07
mp -96°C   60 MHz: *1*, 232D
bp 119°C   FT-IR: *1*, 271A
CDCl₃   QE-300   d 1.000   VP-FT-IR: *3*, 356A

A

Aldrich P700-2   CAS [1613-51-0]   C₅H₁₀S   Fp 70°F   60 MHz: *1*, 233A   29.05
Pentamethylene sulfide, 99%   FW 102.20   FT-IR: *1*, 271B   27.72
bp 141°C   VP-FT-IR: *3*, 356B   26.41
d 0.985
CDCl₃   QE-300   n²⁰_D 1.5060

B

Aldrich C10,260-1   CAS [286-28-2]   C₆H₁₀S   60 MHz: *1*, 232B   36.95*
Cyclohexene sulfide, tech., 85%   FW 114.21   FT-IR: *1*, 270C   25.84
bp 57°C (12 mm)   VP-FT-IR: *3*, 355C   19.45
n²⁰_D 1.5270
CDCl₃   QE-300   Fp 122°F

C

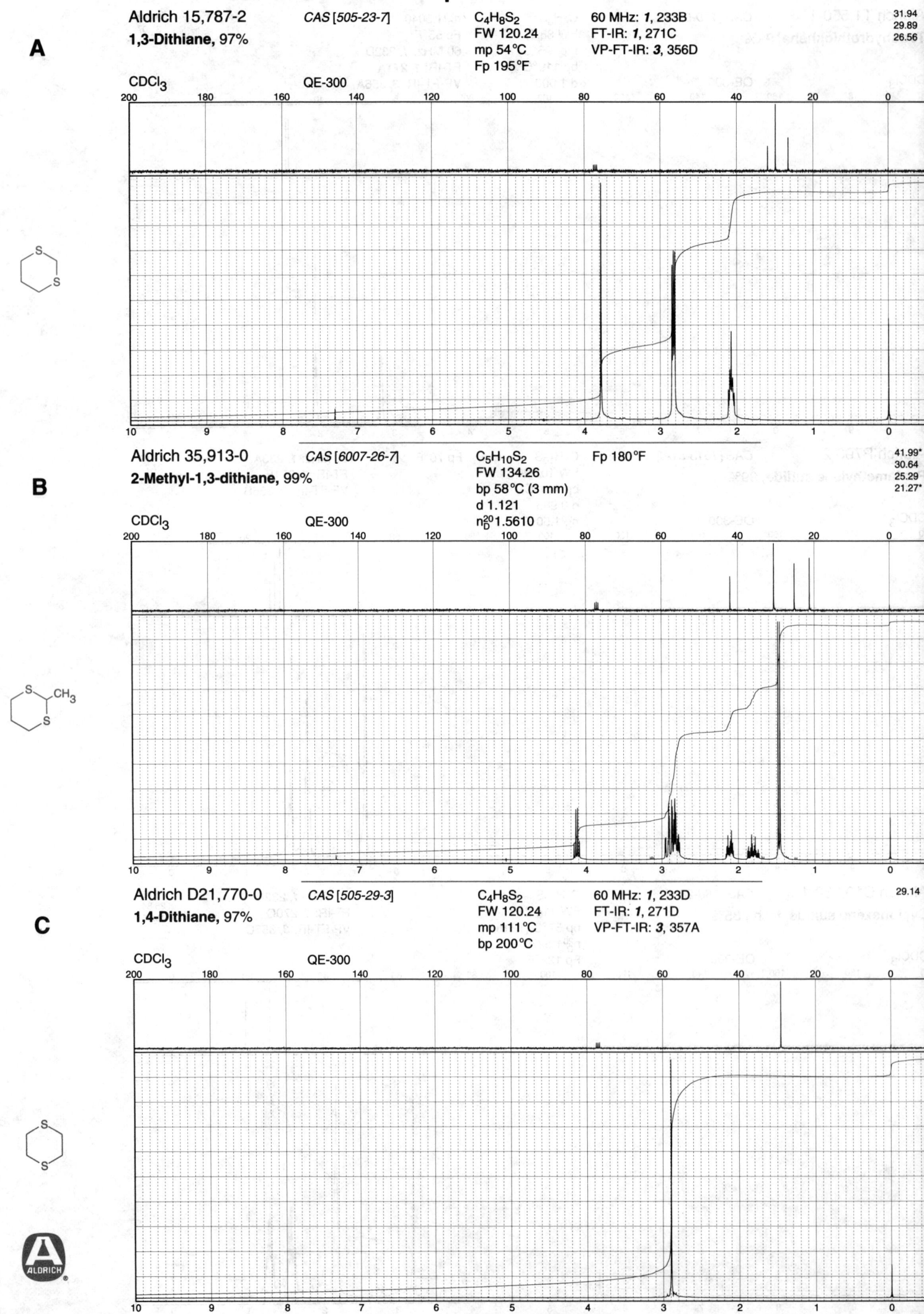

A

Aldrich 15,787-2

1,3-Dithiane, 97%

CAS [505-23-7]

$C_4H_8S_2$
FW 120.24
mp 54°C
Fp 195°F

60 MHz: *1*, 233B
FT-IR: *1*, 271C
VP-FT-IR: *3*, 356D

31.94
29.89
26.56

CDCl₃   QE-300

B

Aldrich 35,913-0

2-Methyl-1,3-dithiane, 99%

CAS [6007-26-7]

$C_5H_{10}S_2$
FW 134.26
bp 58°C (3 mm)
d 1.121
n_D^{20}1.5610

Fp 180°F

41.99*
30.64
25.29
21.27*

CDCl₃   QE-300

C

Aldrich D21,770-0

1,4-Dithiane, 97%

CAS [505-29-3]

$C_4H_8S_2$
FW 120.24
mp 111°C
bp 200°C

60 MHz: *1*, 233D
FT-IR: *1*, 271D
VP-FT-IR: *3*, 357A

29.14

CDCl₃   QE-300

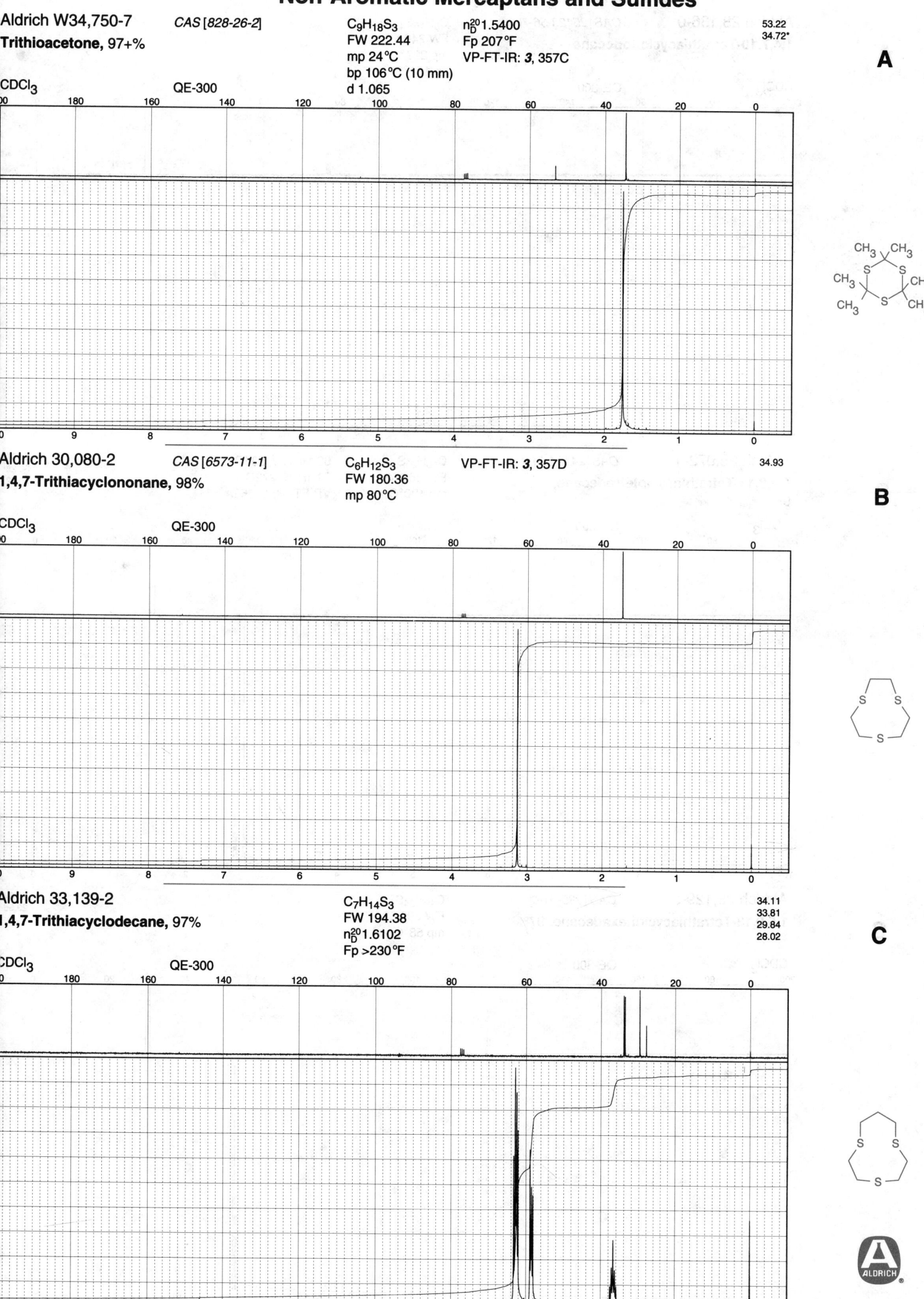

Aldrich W34,750-7   CAS [828-26-2]   C_9H_18S_3 n_D^{20}1.5400   53.22

Trithioacetone, 97+%   FW 222.44   Fp 207°F   34.72*

mp 24°C   VP-FT-IR: *3*, 357C

bp 106°C (10 mm)

CDCl_3   QE-300   d 1.065

A

Aldrich 30,080-2   CAS [6573-11-1]   C_6H_12S_3   VP-FT-IR: *3*, 357D   34.93

1,4,7-Trithiacyclononane, 98%   FW 180.36

mp 80°C

CDCl_3   QE-300

B

Aldrich 33,139-2   C_7H_14S_3   34.11

1,4,7-Trithiacyclodecane, 97%   FW 194.38   33.81

n_D^{20}1.6102   29.84

Fp >230°F   28.02

CDCl_3   QE-300

C

ALDRICH

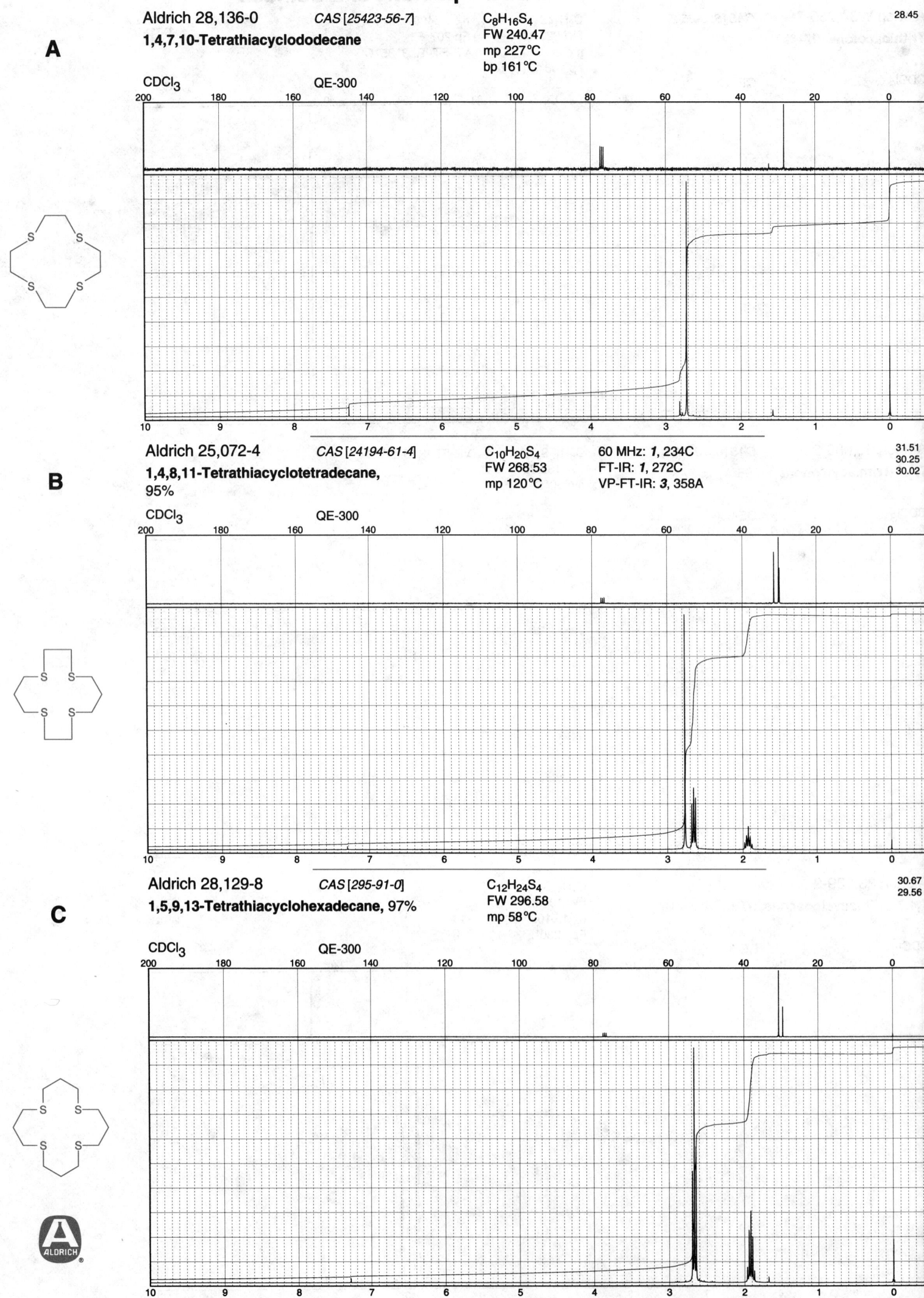

Aldrich 28,136-0   CAS [25423-56-7] $C_8H_{16}S_4$
FW 240.47
mp 227°C
bp 161°C

1,4,7,10-Tetrathiacyclododecane

28.45

A

CDCl₃   QE-300

Aldrich 25,072-4   CAS [24194-61-4] $C_{10}H_{20}S_4$ 60 MHz: *1*, 234C

1,4,8,11-Tetrathiacyclotetradecane,
95%

FW 268.53   FT-IR: *1*, 272C
mp 120°C   VP-FT-IR: *3*, 358A

31.51
30.25
30.02

B

CDCl₃   QE-300

Aldrich 28,129-8   CAS [295-91-0] $C_{12}H_{24}S_4$
FW 296.58
mp 58°C

1,5,9,13-Tetrathiacyclohexadecane, 97%

30.67
29.56

C

CDCl₃   QE-300

Non-Aromatic Mercaptans and Sulfides

438

Aldrich 28,134-4   *CAS [36338-04-2]* $C_{10}H_{20}S_5$ FW 300.59   mp 125°C   32.80

1,4,7,10,13-Pentathiacyclopentadecane

A

CDCl$_3$ QE-300

Aldrich 28,127-1   *CAS [296-41-3]* $C_{12}H_{24}S_6$ FW 360.71   mp 93°C   32.65

1,4,7,10,13,16-Hexathiacyclooctadecane

B

CDCl$_3$ QE-300

Aldrich 28,137-9 $C_{16}H_{32}S_8$ FW 480.95   mp 112°C   32.46

1,4,7,10,13,16,19,22-Octathiacyclotetracosane, 98%

C

CDCl$_3$ QE-300

439

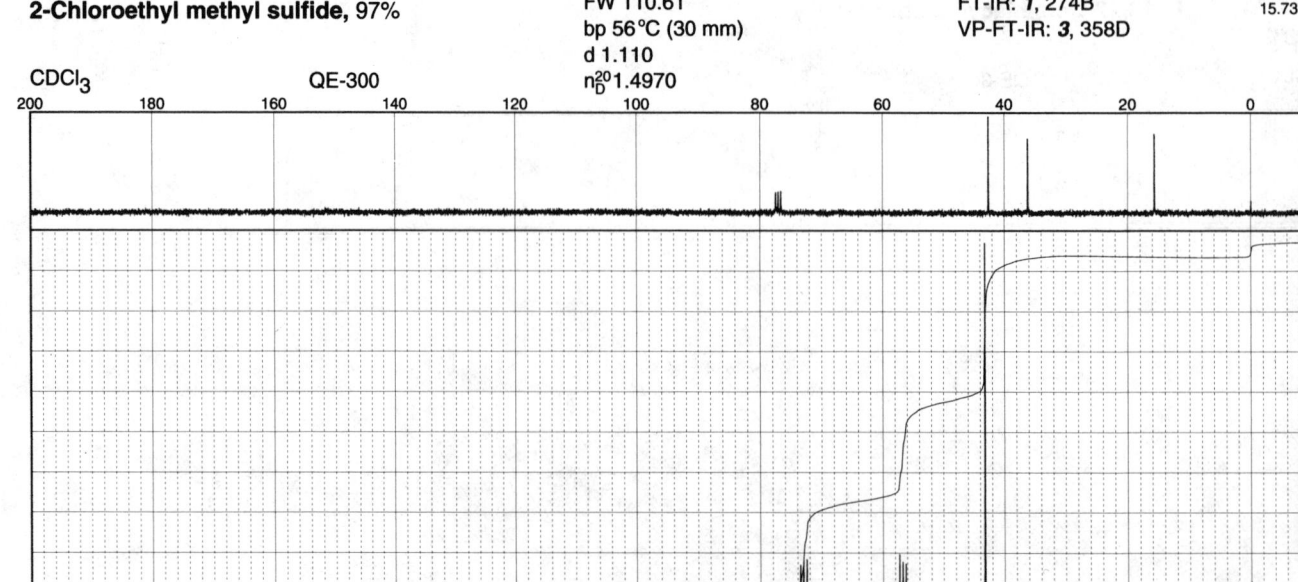

A

Aldrich 32,628-3   *CAS [28948-54-1]*   C_{20}H_{40}S_8
FW 537.05
mp 71°C

1,4,8,11,15,18,22,25-Octathiacyclooctacosane, 95%

32.22
30.82
29.48

CDCl_3   QE-300

B

Aldrich C5,400-7   *CAS [2373-51-5]*   C_2H_5ClS   Fp 63°F   60 MHz: *1*, 216A
FW 96.58   FT-IR: *1*, 274A
Chloromethyl methyl sulfide, 95%   bp 105°C   VP-FT-IR: *3*, 358C
d 1.153
n_D^{20} 1.4980

51.93
15.13*

CDCl_3   QE-300

CH_3SCH_2Cl

C

Aldrich 24,263-2   *CAS [542-81-4]*   C_3H_7ClS   Fp 108°F   60 MHz: *1*, 216B
FW 110.61   FT-IR: *1*, 274B
2-Chloroethyl methyl sulfide, 97%   bp 56°C (30 mm)   VP-FT-IR: *3*, 358D
d 1.110
n_D^{20} 1.4970

42.65
36.22
15.73*

CDCl_3   QE-300

ClCH_2CH_2SCH_3

ALDRICH

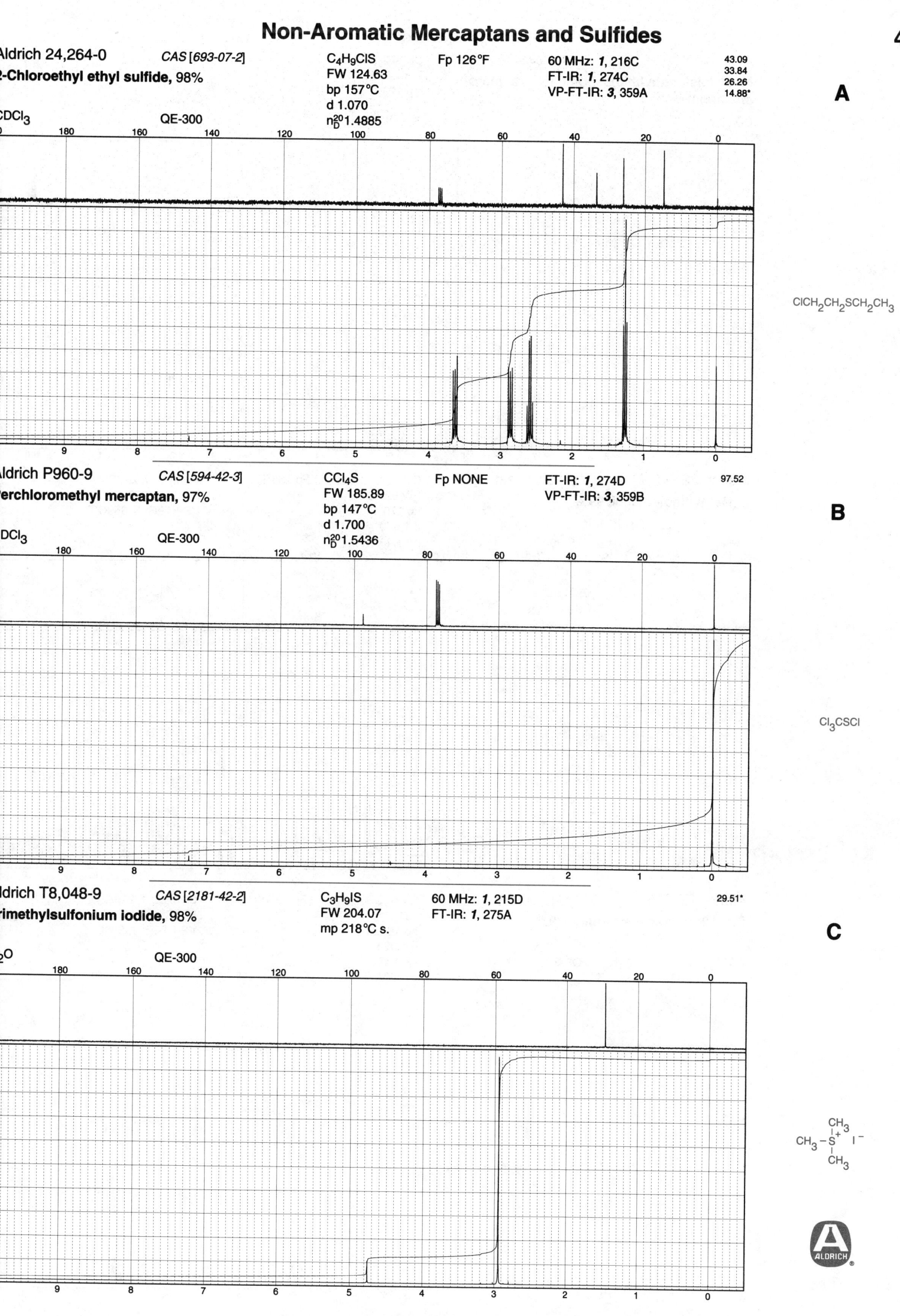

A

Aldrich 24,264-0   CAS [693-07-2]   C₄H₉ClS   Fp 126°F   60 MHz: *1*, 216C   43.09
2-Chloroethyl ethyl sulfide, 98%   FW 124.63   FT-IR: *1*, 274C   33.84
                                                 bp 157°C   VP-FT-IR: *3*, 359A 26.26
CDCl₃   QE-300   d 1.070   14.88*
                                                 n²⁰_D 1.4885

ClCH₂CH₂SCH₂CH₃

B

Aldrich P960-9   CAS [594-42-3]   CCl₄S   Fp NONE   FT-IR: *1*, 274D   97.52
Perchloromethyl mercaptan, 97%   FW 185.89   VP-FT-IR: *3*, 359B
                                  bp 147°C
CDCl₃   QE-300   d 1.700
                                  n²⁰_D 1.5436

Cl₃CSCl

C

Aldrich T8,048-9   CAS [2181-42-2]   C₃H₉IS   60 MHz: *1*, 215D   29.51*
Trimethylsulfonium iodide, 98%   FW 204.07   FT-IR: *1*, 275A
                                  mp 218°C s.
₂O   QE-300

CH₃
|
CH₃—S⁺—I⁻
|
CH₃

ALDRICH

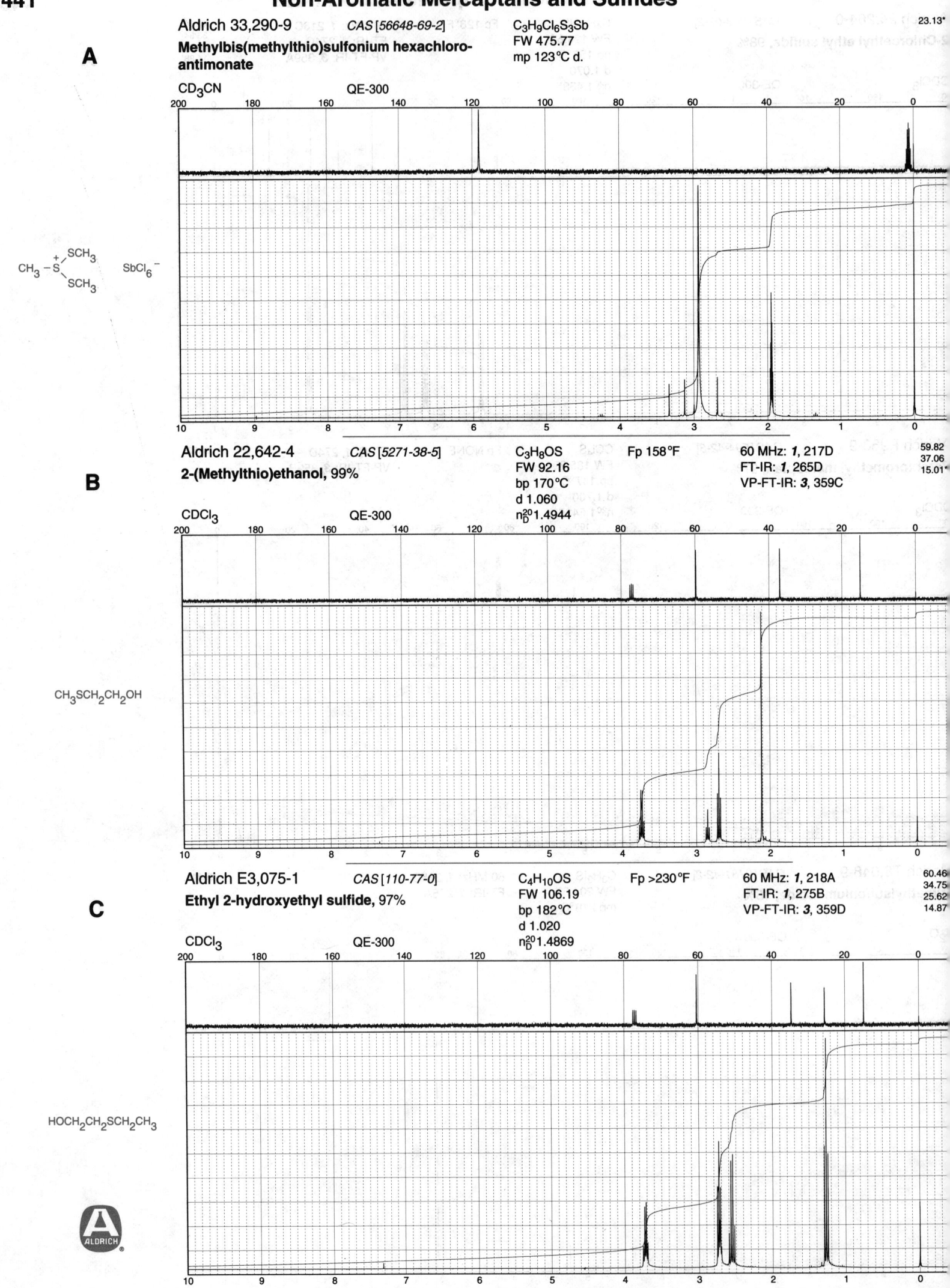

A

Aldrich 33,290-9   CAS [56648-69-2]   C₃H₉Cl₆S₃Sb
FW 475.77
mp 123°C d.

$C_3H_9Cl_6S_3Sb$
FW 475.77
mp 123°C d.

Methylbis(methylthio)sulfonium hexachloro-antimonate

CD₃CN   QE-300

23.13*

$CH_3-\overset{+}{\underset{SCH_3}{S}}-SCH_3 \quad SbCl_6^-$

B

Aldrich 22,642-4   CAS [5271-38-5]   C₃H₈OS   Fp 158°F
FW 92.16
bp 170°C
d 1.060
n²⁰_D 1.4944

60 MHz: *1*, 217D
FT-IR: *1*, 265D
VP-FT-IR: *3*, 359C

59.82
37.06
15.01*

2-(Methylthio)ethanol, 99%

CDCl₃   QE-300

$CH_3SCH_2CH_2OH$

C

Aldrich E3,075-1   CAS [110-77-0]   C₄H₁₀OS   Fp >230°F
FW 106.19
bp 182°C
d 1.020
n²⁰_D 1.4869

60 MHz: *1*, 218A
FT-IR: *1*, 275B
VP-FT-IR: *3*, 359D

60.46
34.75
25.62
14.87

Ethyl 2-hydroxyethyl sulfide, 97%

CDCl₃   QE-300

$HOCH_2CH_2SCH_2CH_3$

ALDRICH

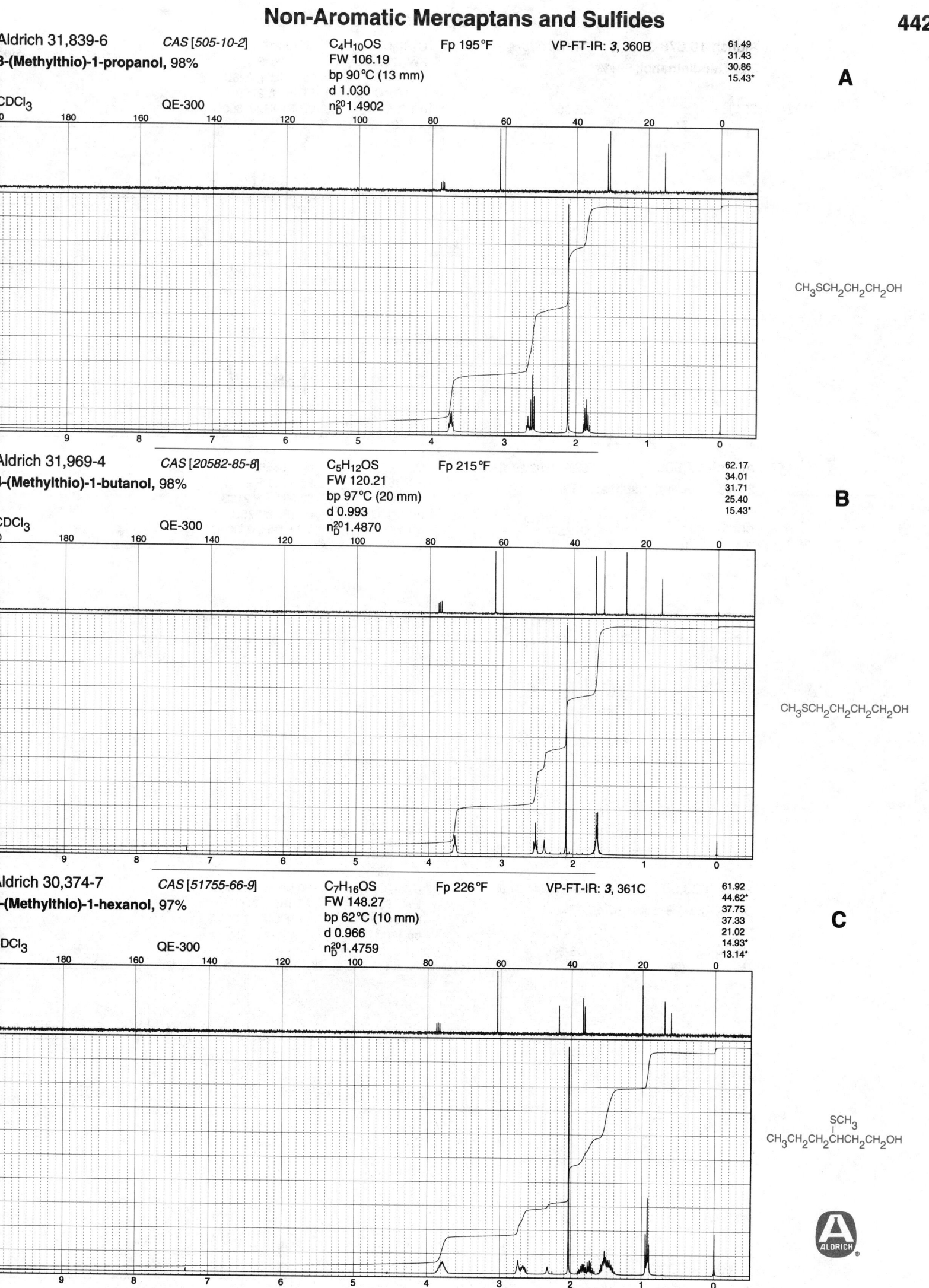

Aldrich 31,839-6   CAS [505-10-2]   C₄H₁₀OS   Fp 195°F   VP-FT-IR: *3*, 360B

3-(Methylthio)-1-propanol, 98%

CDCl₃   QE-300

$C_4H_{10}OS$
FW 106.19
bp 90°C (13 mm)
d 1.030
n_D^{20} 1.4902

61.49
31.43
30.86
15.43*

A

$CH_3SCH_2CH_2CH_2OH$

Aldrich 31,969-4   CAS [20582-85-8]   C₅H₁₂OS   Fp 215°F

4-(Methylthio)-1-butanol, 98%

CDCl₃   QE-300

$C_5H_{12}OS$
FW 120.21
bp 97°C (20 mm)
d 0.993
n_D^{20} 1.4870

62.17
34.01
31.71
25.40
15.43*

B

$CH_3SCH_2CH_2CH_2CH_2OH$

Aldrich 30,374-7   CAS [51755-66-9]   C₇H₁₆OS   Fp 226°F   VP-FT-IR: *3*, 361C

3-(Methylthio)-1-hexanol, 97%

CDCl₃   QE-300

$C_7H_{16}OS$
FW 148.27
bp 62°C (10 mm)
d 0.966
n_D^{20} 1.4759

61.92
44.62*
37.75
37.33
21.02
14.93*
13.14*

C

$CH_3CH_2CH_2CHCH_2CH_2OH$
with SCH₃ branch

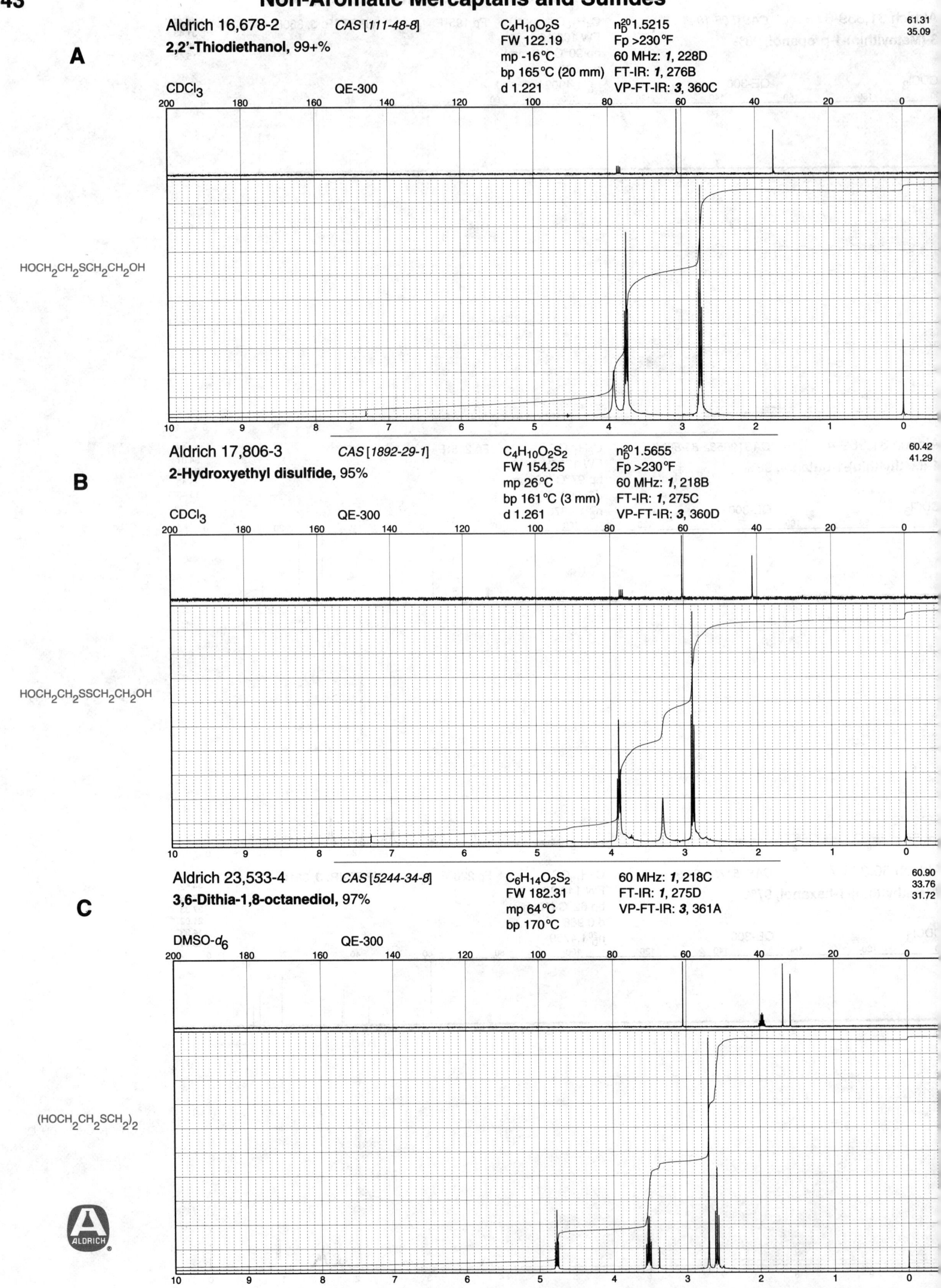

A

Aldrich 16,678-2   CAS [111-48-8]   C₄H₁₀O₂S n_D^{20}1.5215   61.31
35.09
2,2'-Thiodiethanol, 99+%   FW 122.19   Fp >230°F
mp -16°C   60 MHz: *1*, 228D
bp 165°C (20 mm)   FT-IR: *1*, 276B
d 1.221   VP-FT-IR: *3*, 360C

CDCl₃   QE-300

HOCH₂CH₂SCH₂CH₂OH

B

Aldrich 17,806-3   CAS [1892-29-1]   C₄H₁₀O₂S₂ n_D^{20}1.5655   60.42
41.29
2-Hydroxyethyl disulfide, 95%   FW 154.25   Fp >230°F
mp 26°C   60 MHz: *1*, 218B
bp 161°C (3 mm)   FT-IR: *1*, 275C
d 1.261   VP-FT-IR: *3*, 360D

CDCl₃   QE-300

HOCH₂CH₂SSCH₂CH₂OH

C

Aldrich 23,533-4   CAS [5244-34-8]   C₆H₁₄O₂S₂   60 MHz: *1*, 218C   60.90
33.76
3,6-Dithia-1,8-octanediol, 97%   FW 182.31   FT-IR: *1*, 275D   31.72
mp 64°C   VP-FT-IR: *3*, 361A
bp 170°C

DMSO-d_6 QE-300

(HOCH₂CH₂SCH₂)₂

Non-Aromatic Mercaptans and Sulfides

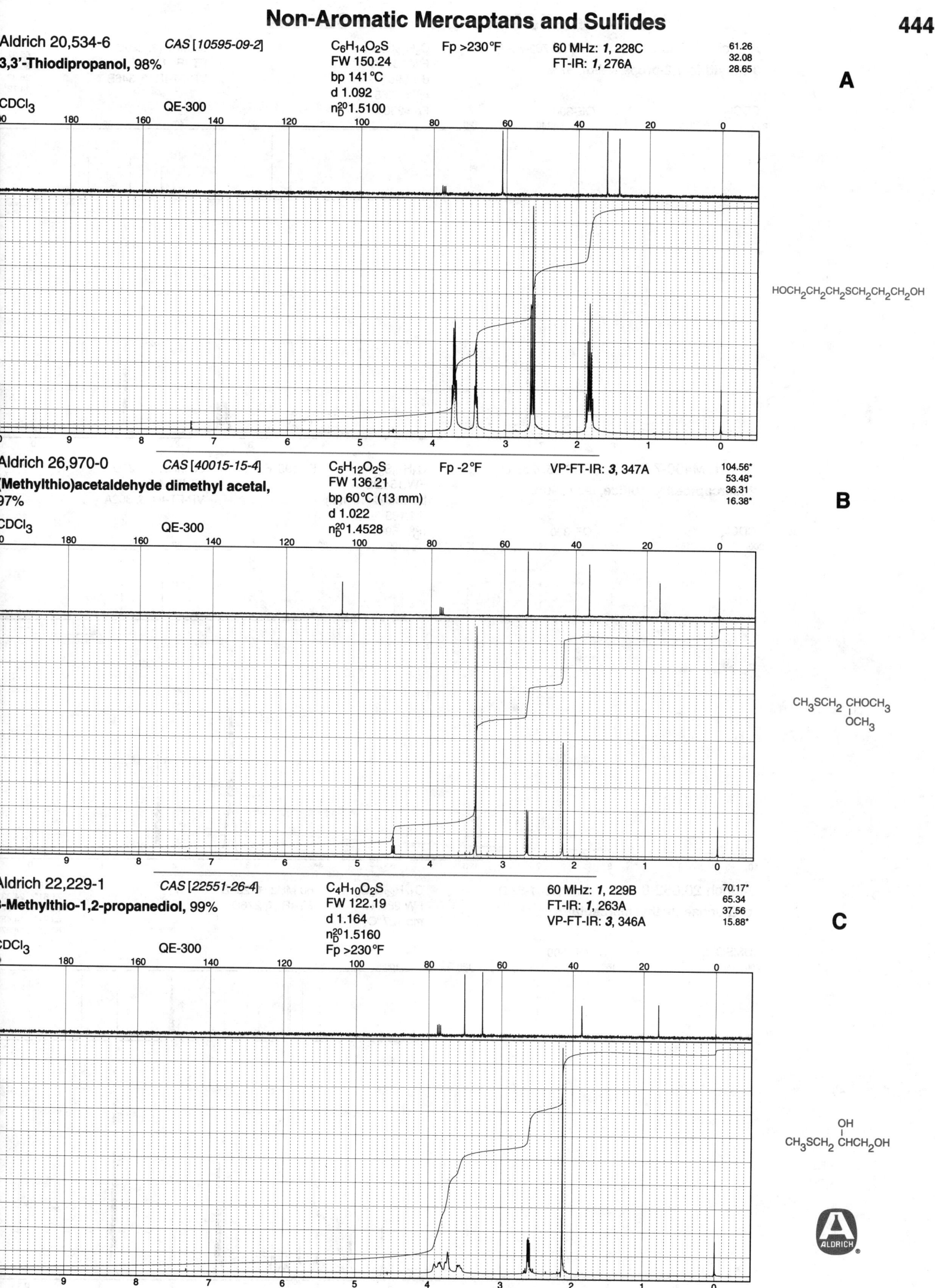

Aldrich 20,534-6   CAS [10595-09-2] $C_6H_{14}O_2S$ Fp >230°F   60 MHz: 1, 228C   61.26
3,3'-Thiodipropanol, 98%   FW 150.24   FT-IR: 1, 276A   32.08
   bp 141°C   28.65
CDCl$_3$ QE-300   d 1.092
   n$_D^{20}$1.5100

A

$HOCH_2CH_2CH_2SCH_2CH_2CH_2OH$

Aldrich 26,970-0   CAS [40015-15-4] $C_5H_{12}O_2S$ Fp -2°F   VP-FT-IR: 3, 347A   104.56*
(Methylthio)acetaldehyde dimethyl acetal,   FW 136.21   53.48*
97%   bp 60°C (13 mm)   36.31
CDCl$_3$ QE-300   d 1.022   16.38*
   n$_D^{20}$1.4528

B

$CH_3SCH_2\ CHOCH_3$
$\qquad\qquad\ |$
$\qquad\qquad OCH_3$

Aldrich 22,229-1   CAS [22551-26-4] $C_4H_{10}O_2S$ 60 MHz: 1, 229B   70.17*
3-Methylthio-1,2-propanediol, 99%   FW 122.19   FT-IR: 1, 263A   65.34
   d 1.164   VP-FT-IR: 3, 346A   37.56
CDCl$_3$ QE-300   n$_D^{20}$1.5160   15.88*
   Fp >230°F

C

$\qquad\qquad\quad OH$
$\qquad\qquad\quad |$
$CH_3SCH_2\ CHCH_2OH$

Non-Aromatic Mercaptans and Sulfides

A

Aldrich 22,230-5   *CAS [60763-78-2]* $C_5H_{12}O_2S$ 60 MHz: *1*, 229C   70.54*
3-Ethylthio-1,2-propanediol, 97%   FW 136.21   FT-IR: *1*, 263B   65.37
    d 1.095   VP-FT-IR: *3*, 346B   35.09
    n_D^{20}1.5065   26.38
    Fp >230°F   14.79*

CDCl$_3$ QE-300

$CH_3CH_2SCH_2$ CHCH$_2$OH (OH)

B

Aldrich M400-7   *CAS [3570-55-6]* $C_4H_{10}S_3$ Fp 195°F   60 MHz: *1*, 227D   36.02
2-Mercaptoethyl sulfide, tech., 90%   FW 154.32   FT-IR: *1*, 261D   24.77
    bp 136°C (10 mm)   VP-FT-IR: *3*, 362A
    d 1.183
    n_D^{20}1.5961

CDCl$_3$ QE-300

$HSCH_2CH_2SCH_2CH_2SH$

C

Aldrich 20,662-8   *CAS [1941-52-2]* $C_{10}H_{22}O_5S_2$ 60 MHz: *1*, 235B   75.18* 53.84*
D-Glucose diethyl mercaptal   FW 286.41   FT-IR: *1*, 276D   71.81* 24.33
    mp 127°C   71.28* 24.27
    69.71* 14.48*
    63.25 14.37*

DMSO-d_6 QE-300

CH_3CH_2S , SCH_2CH_3
CH
H–C–OH
HO–C–H
H–C–OH
H–C–OH
CH_2OH

ALDRICH

Non-Aromatic Mercaptans and Sulfides

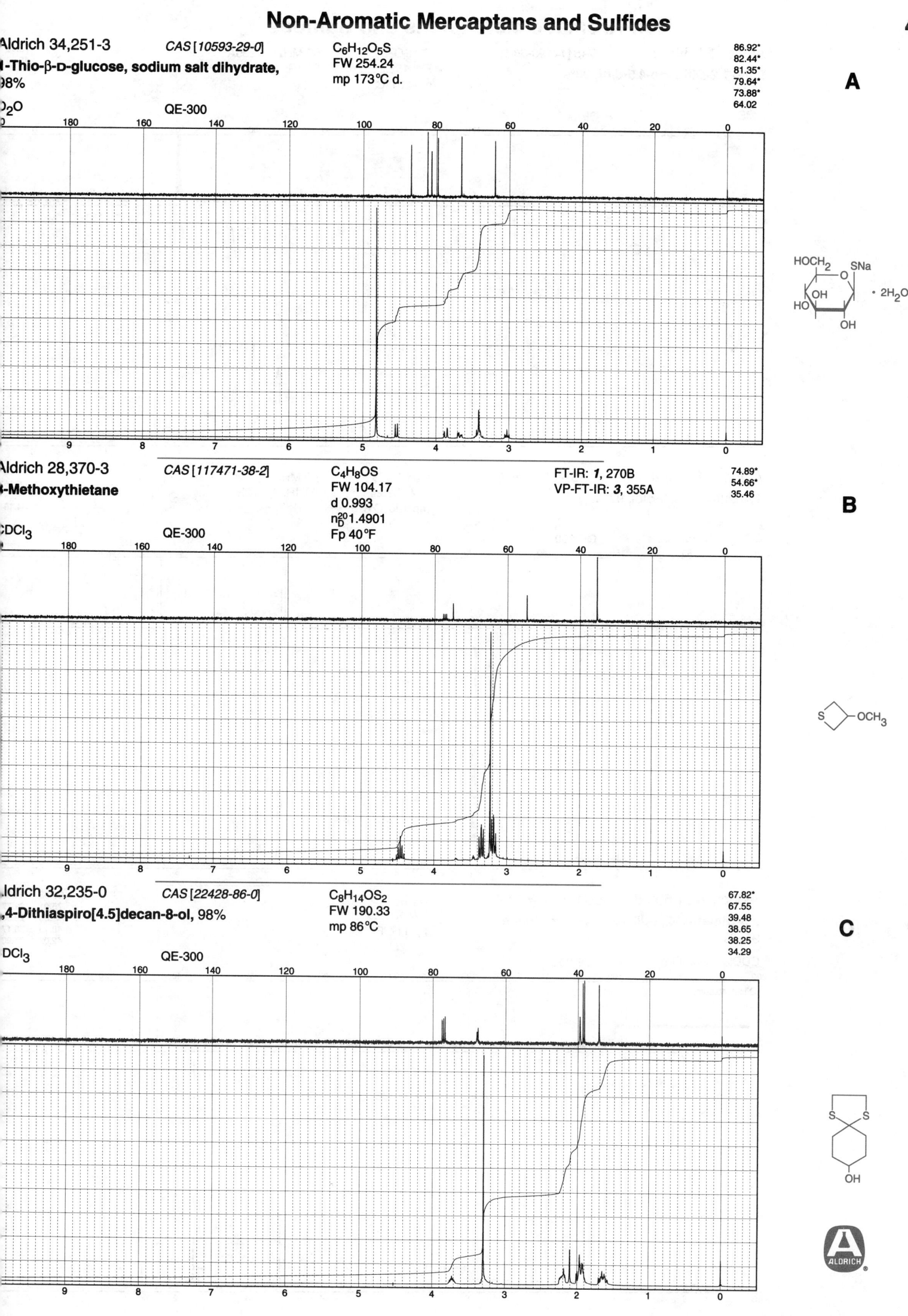

Aldrich 34,251-3 *CAS [10593-29-0]* C₆H₁₂O₅S
1-Thio-β-D-glucose, sodium salt dihydrate, 98% FW 254.24 mp 173°C d.

D₂O QE-300

86.92*
82.44*
81.35*
79.64*
73.88*
64.02

A

Aldrich 28,370-3 *CAS [117471-38-2]* C₄H₈OS
3-Methoxythietane FW 104.17
d 0.993
n²⁰_D 1.4901
Fp 40°F

CDCl₃ QE-300

FT-IR: *1*, 270B
VP-FT-IR: *3*, 355A

74.89*
54.66*
35.46

B

Aldrich 32,235-0 *CAS [22428-86-0]* C₈H₁₄OS₂
1,4-Dithiaspiro[4.5]decan-8-ol, 98% FW 190.33
mp 86°C

DCl₃ QE-300

67.82*
67.55
39.48
38.65
38.25
34.29

C

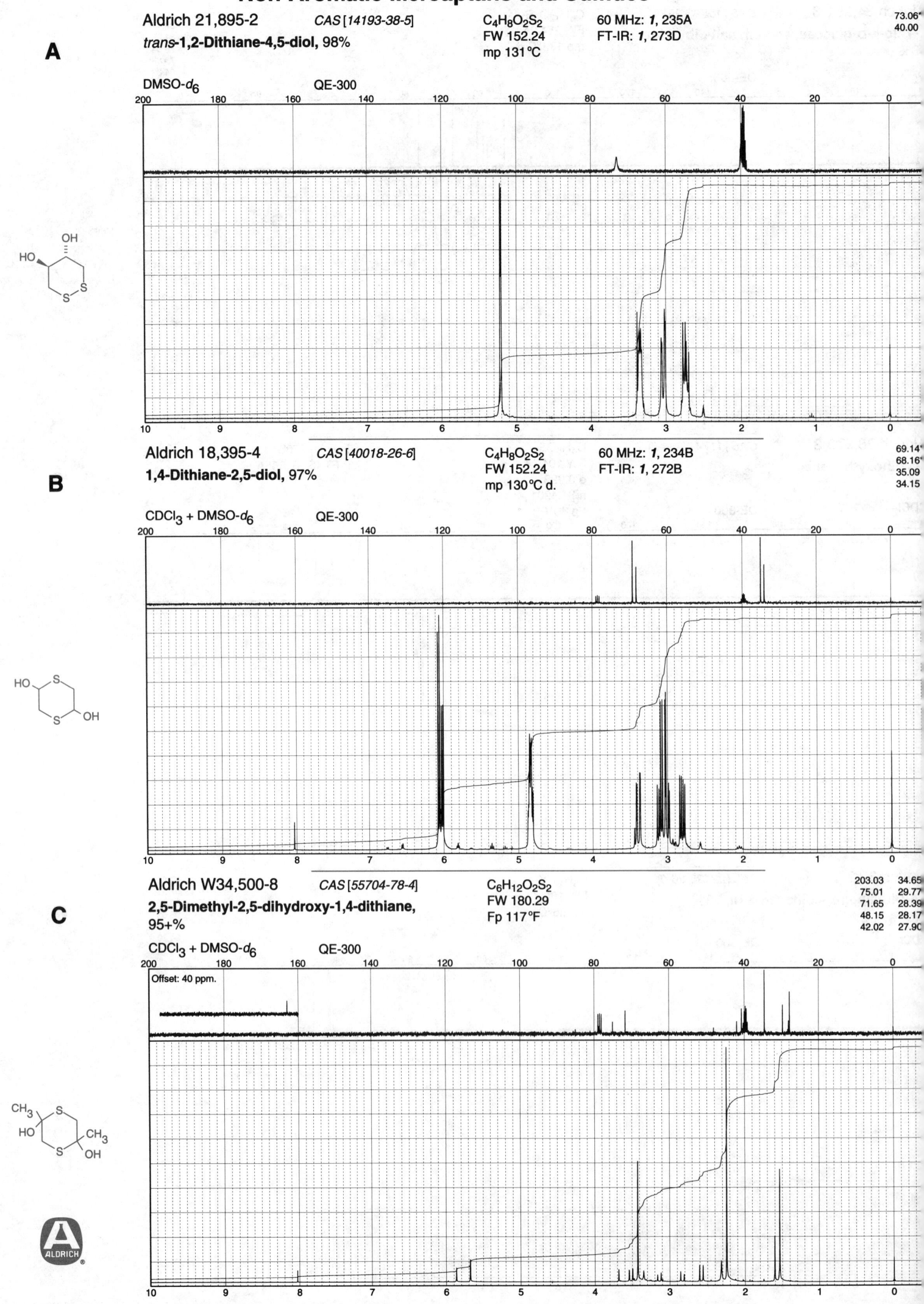

A

Aldrich 21,895-2   *CAS [14193-38-5]* $C_4H_8O_2S_2$ 60 MHz: *1*, 235A   73.06*
                                          FW 152.24   FT-IR: *1*, 273D   40.00
trans-1,2-Dithiane-4,5-diol, 98%   mp 131°C

DMSO-d_6 QE-300

B

Aldrich 18,395-4   *CAS [40018-26-6]* $C_4H_8O_2S_2$ 60 MHz: *1*, 234B   69.14*
                                          FW 152.24   FT-IR: *1*, 272B   68.16*
1,4-Dithiane-2,5-diol, 97%   mp 130°C d.   35.09
                                                                               34.15

CDCl$_3$ + DMSO-d_6 QE-300

C

Aldrich W34,500-8   *CAS [55704-78-4]* $C_6H_{12}O_2S_2$ 203.03   34.65
                                           FW 180.29   75.01   29.77
2,5-Dimethyl-2,5-dihydroxy-1,4-dithiane,   Fp 117°F   71.65   28.39
95+%   48.15   28.17
                                                             42.02   27.90

CDCl$_3$ + DMSO-d_6 QE-300

Offset: 40 ppm.

Non-Aromatic Mercaptans and Sulfides

448

Aldrich 25,073-2   CAS [86944-00-5] $C_6H_{12}OS_2$ Fp >230°F   60 MHz: *1*, 234A
69.85*
1,5-Dithiacyclooctan-3-ol, 94%
FW 164.29   FT-IR: *1*, 272A
35.97
bp 147°C (2 mm)   VP-FT-IR: *3*, 357B
31.71
d 1.272
30.29
CDCl₃   QE-300 n_D^{20}1.5970

A

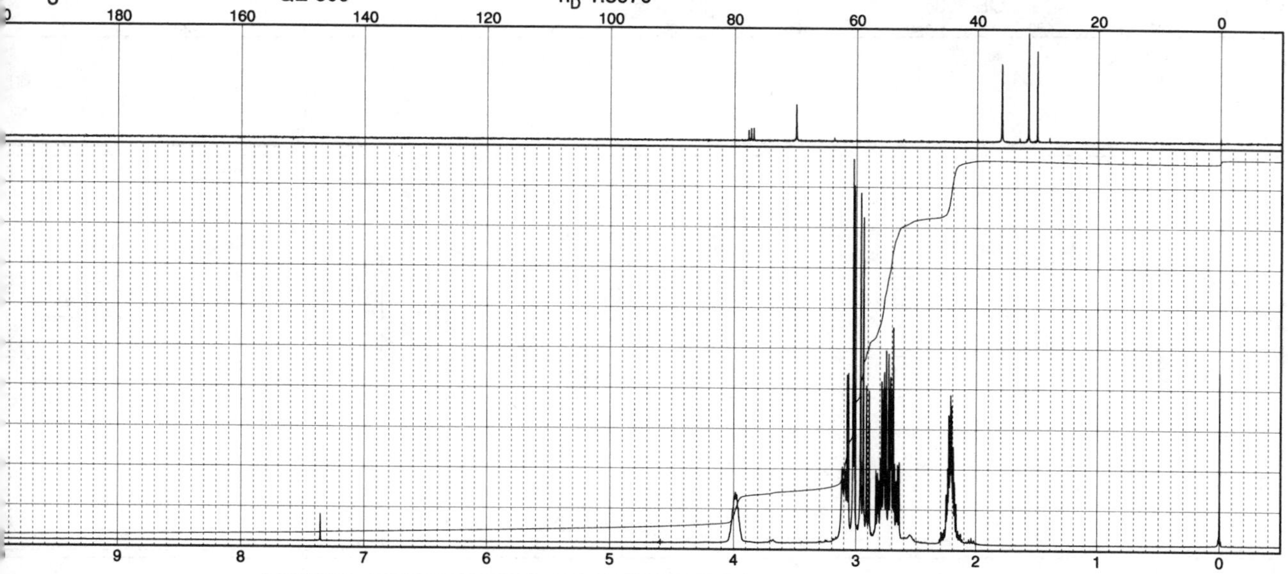

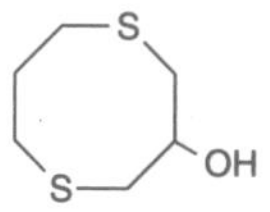

Aldrich 25,823-7 $C_{12}H_{24}O_2S_4$ FT-IR: *1*, 272D
69.23*
1,5,9,13-Tetrathiacyclohexadecane-3,11-
FW 328.58
38.09
diol, 96%, mixture of *cis* and *trans*
mp 90°C
31.07
29.70
CDCl₃   QE-300

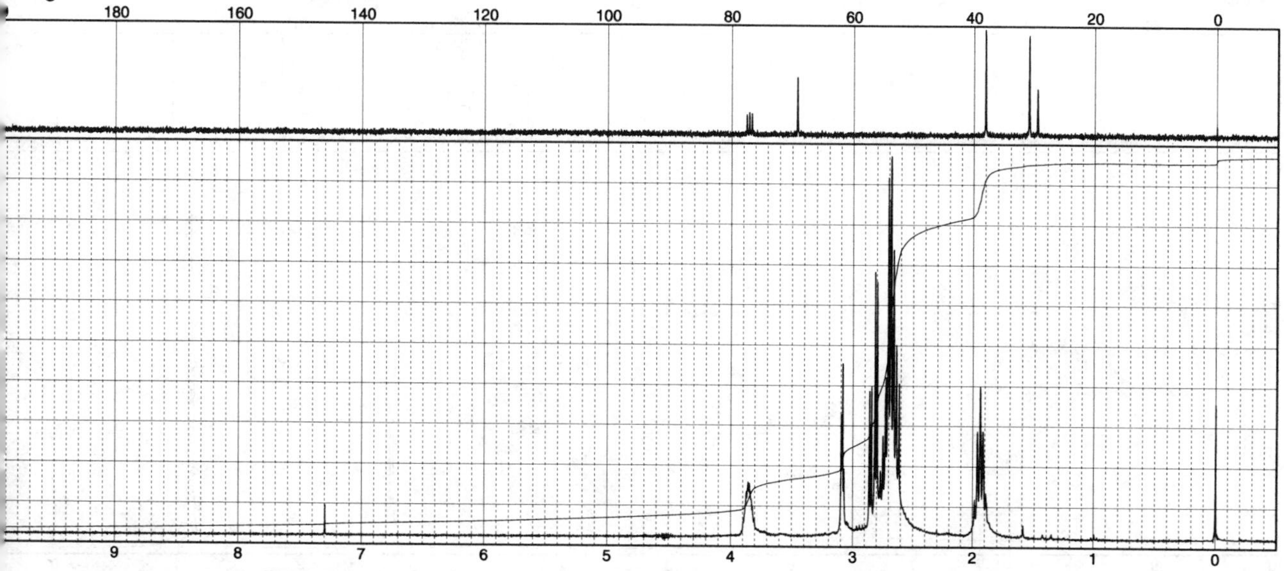

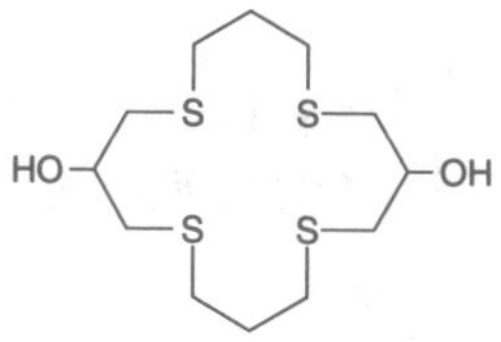

B

Aldrich 26,842-9 $C_{18}H_{36}O_3S_6$ FT-IR: *1*, 273A
69.16*
1,5,9,13,17,21-Hexathiacyclotetracosane-3,11,19-
FW 492.87
38.15
triol
Fp >230°F
31.12
29.30
CDCl₃   QE-300

C

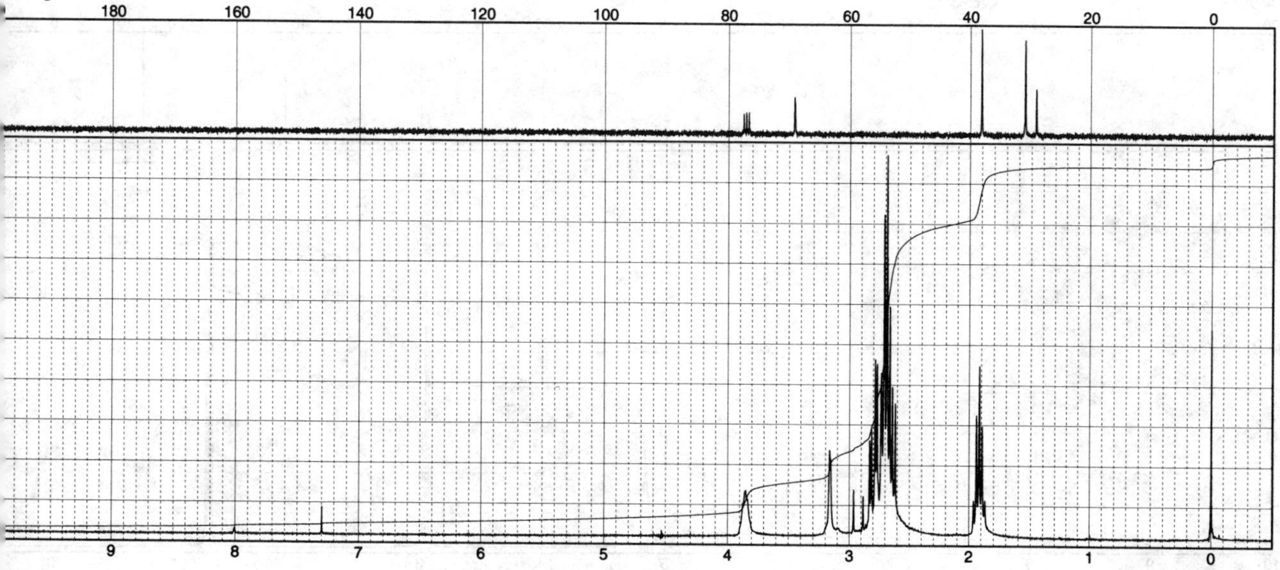

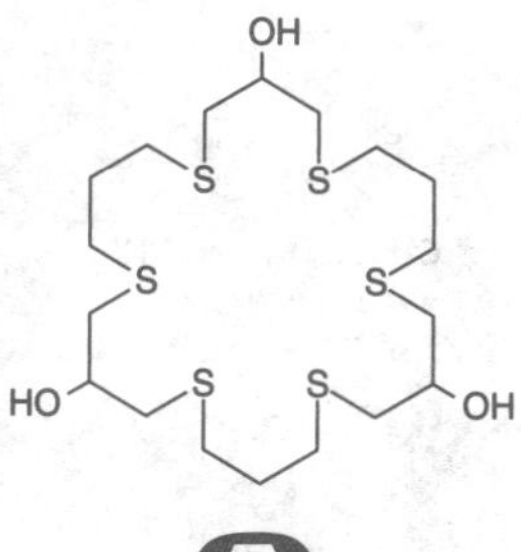

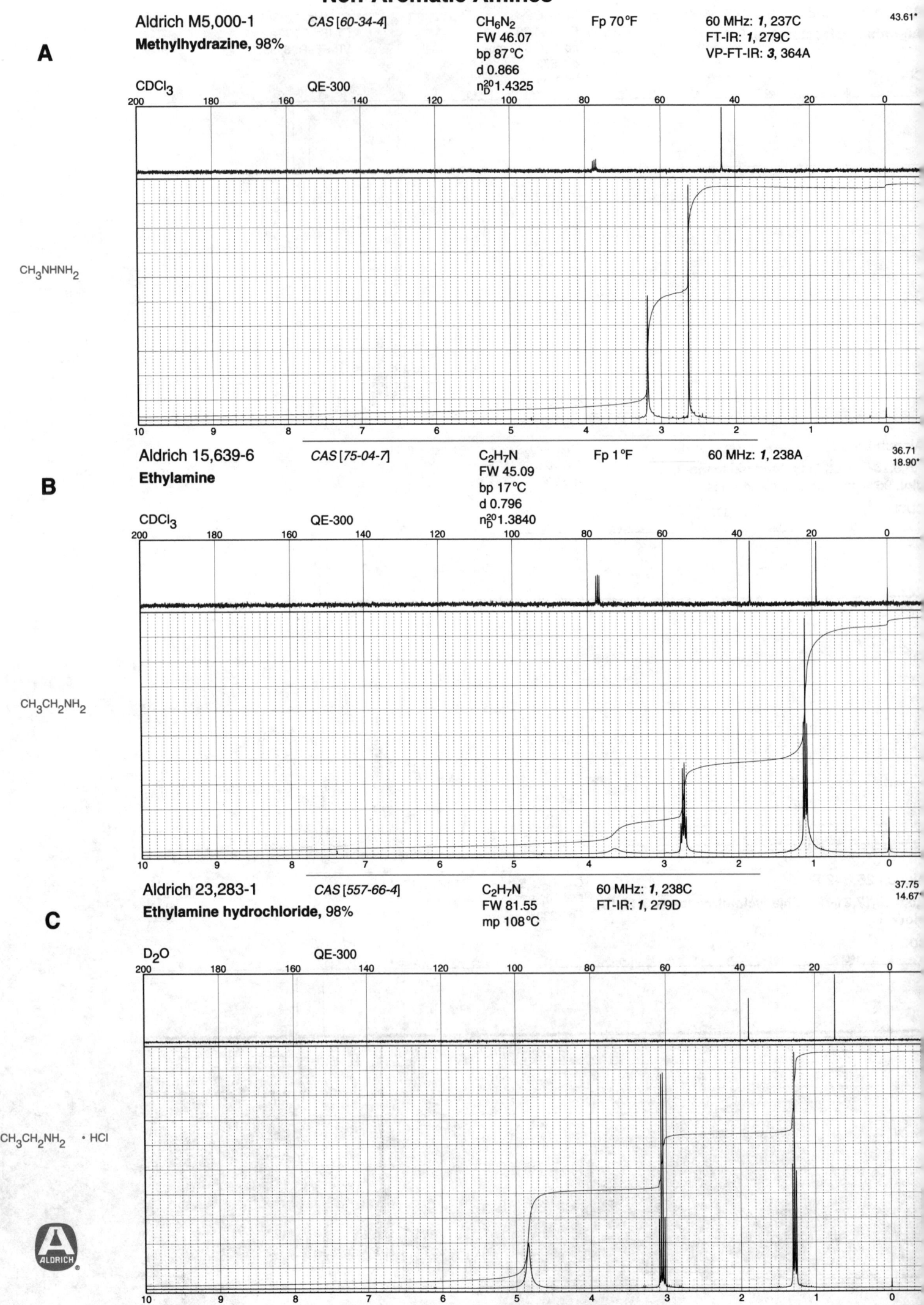

A

Aldrich M5,000-1   *CAS [60-34-4]*   CH$_6$N$_2$ Fp 70°F   60 MHz: **1**, 237C   43.61*
Methylhydrazine, 98%   FW 46.07   FT-IR: **1**, 279C
bp 87°C   VP-FT-IR: **3**, 364A
d 0.866
n$_D^{20}$1.4325

CDCl$_3$ QE-300

CH$_3$NHNH$_2$

Aldrich 15,639-6   *CAS [75-04-7]*   C$_2$H$_7$N   Fp 1°F   60 MHz: **1**, 238A   36.71
Ethylamine   FW 45.09   18.90*
bp 17°C
d 0.796
n$_D^{20}$1.3840

B

CDCl$_3$ QE-300

CH$_3$CH$_2$NH$_2$

Aldrich 23,283-1   *CAS [557-66-4]*   C$_2$H$_7$N   60 MHz: **1**, 238C   37.75
Ethylamine hydrochloride, 98%   FW 81.55   FT-IR: **1**, 279D   14.67*
mp 108°C

C

D$_2$O   QE-300

CH$_3$CH$_2$NH$_2$ • HCl

ALDRICH

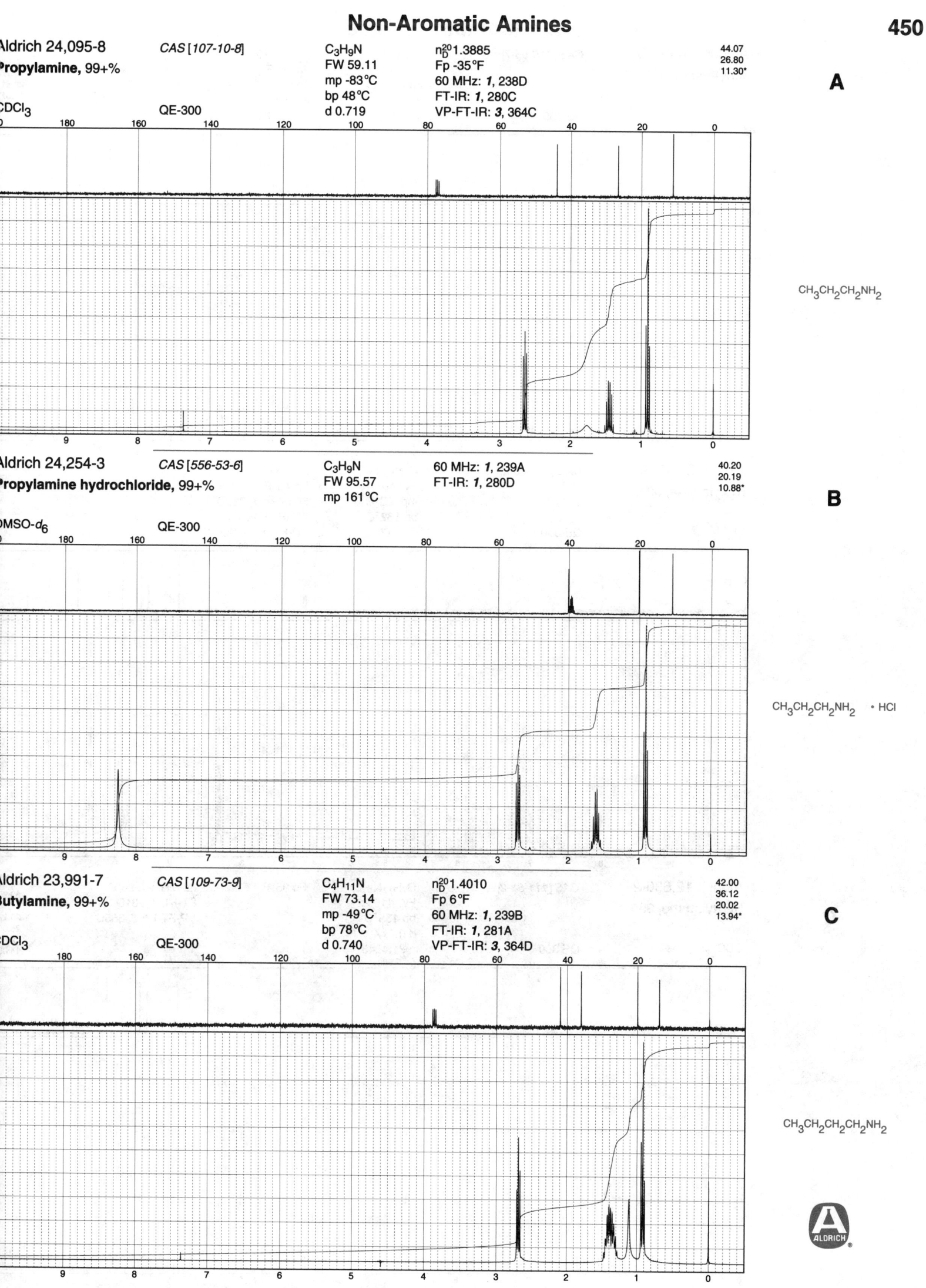

Aldrich 24,095-8   CAS [107-10-8]

Propylamine, 99+%

C$_3$H$_9$N
FW 59.11
mp -83°C
bp 48°C
d 0.719

n$_D^{20}$1.3885
Fp -35°F
60 MHz: **1**, 238D
FT-IR: **1**, 280C
VP-FT-IR: **3**, 364C

44.07
26.80
11.30*

A

CDCl$_3$ QE-300

CH$_3$CH$_2$CH$_2$NH$_2$

Aldrich 24,254-3   CAS [556-53-6]

Propylamine hydrochloride, 99+%

C$_3$H$_9$N
FW 95.57
mp 161°C

60 MHz: **1**, 239A
FT-IR: **1**, 280D

40.20
20.19
10.88*

B

DMSO-d_6 QE-300

CH$_3$CH$_2$CH$_2$NH$_2$ · HCl

Aldrich 23,991-7   CAS [109-73-9]

Butylamine, 99+%

C$_4$H$_{11}$N
FW 73.14
mp -49°C
bp 78°C
d 0.740

n$_D^{20}$1.4010
Fp 6°F
60 MHz: **1**, 239B
FT-IR: **1**, 281A
VP-FT-IR: **3**, 364D

42.00
36.12
20.02
13.94*

C

CDCl$_3$ QE-300

CH$_3$CH$_2$CH$_2$CH$_2$NH$_2$

Non-Aromatic Amines

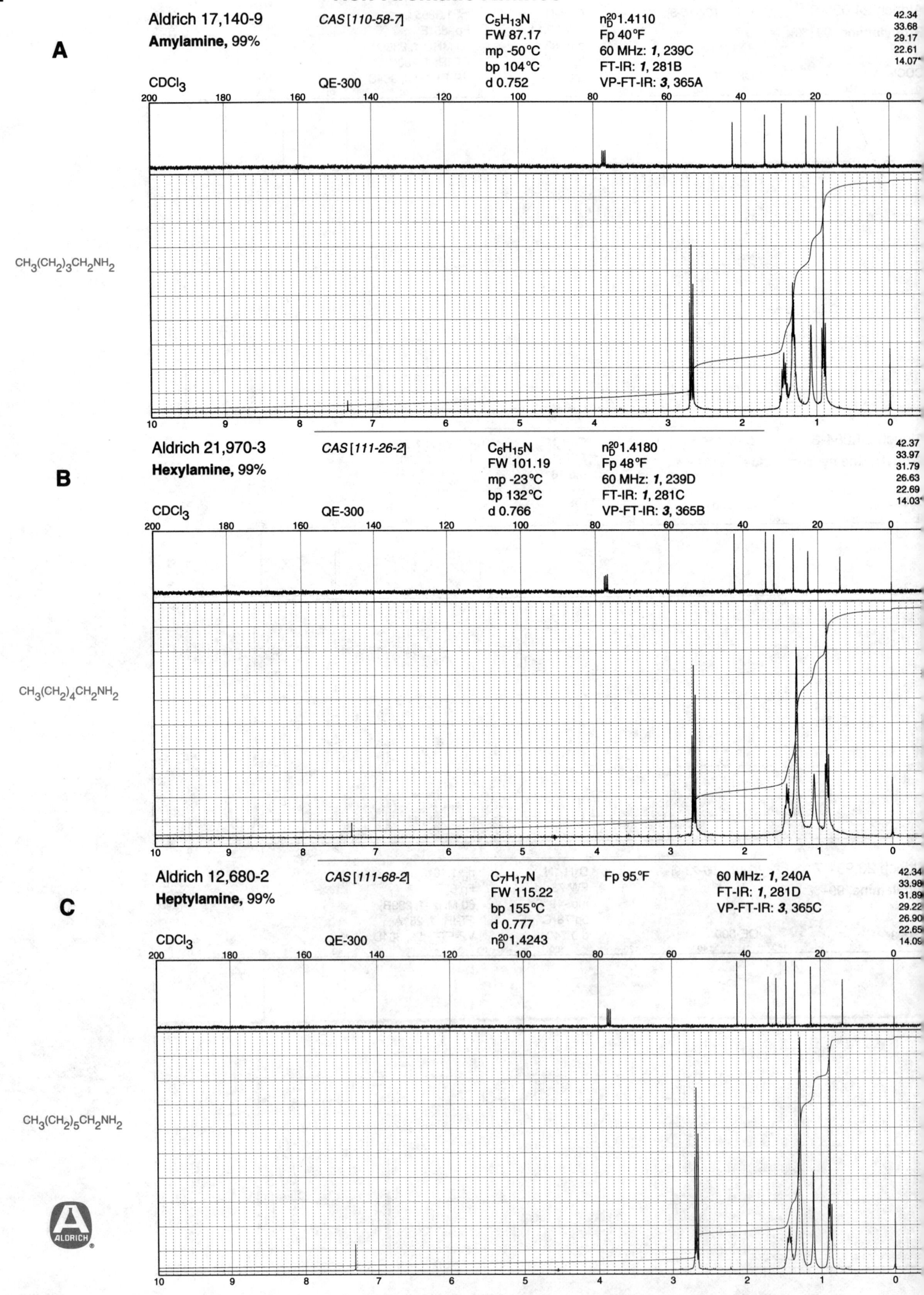

A

Aldrich 17,140-9
Amylamine, 99%

CAS [110-58-7]

C$_5$H$_{13}$N
FW 87.17
mp -50°C
bp 104°C
d 0.752

n$_D^{20}$1.4110
Fp 40°F
60 MHz: **1**, 239C
FT-IR: **1**, 281B
VP-FT-IR: **3**, 365A

42.34
33.68
29.17
22.61
14.07*

CDCl$_3$ QE-300

CH$_3$(CH$_2$)$_3$CH$_2$NH$_2$

B

Aldrich 21,970-3
Hexylamine, 99%

CAS [111-26-2]

C$_6$H$_{15}$N
FW 101.19
mp -23°C
bp 132°C
d 0.766

n$_D^{20}$1.4180
Fp 48°F
60 MHz: **1**, 239D
FT-IR: **1**, 281C
VP-FT-IR: **3**, 365B

42.37
33.97
31.79
26.63
22.69
14.03*

CDCl$_3$ QE-300

CH$_3$(CH$_2$)$_4$CH$_2$NH$_2$

C

Aldrich 12,680-2
Heptylamine, 99%

CAS [111-68-2]

C$_7$H$_{17}$N
FW 115.22
bp 155°C
d 0.777
n$_D^{20}$1.4243

Fp 95°F

60 MHz: **1**, 240A
FT-IR: **1**, 281D
VP-FT-IR: **3**, 365C

42.34
33.98
31.89
29.22
26.90
22.65
14.09

CDCl$_3$ QE-300

CH$_3$(CH$_2$)$_5$CH$_2$NH$_2$

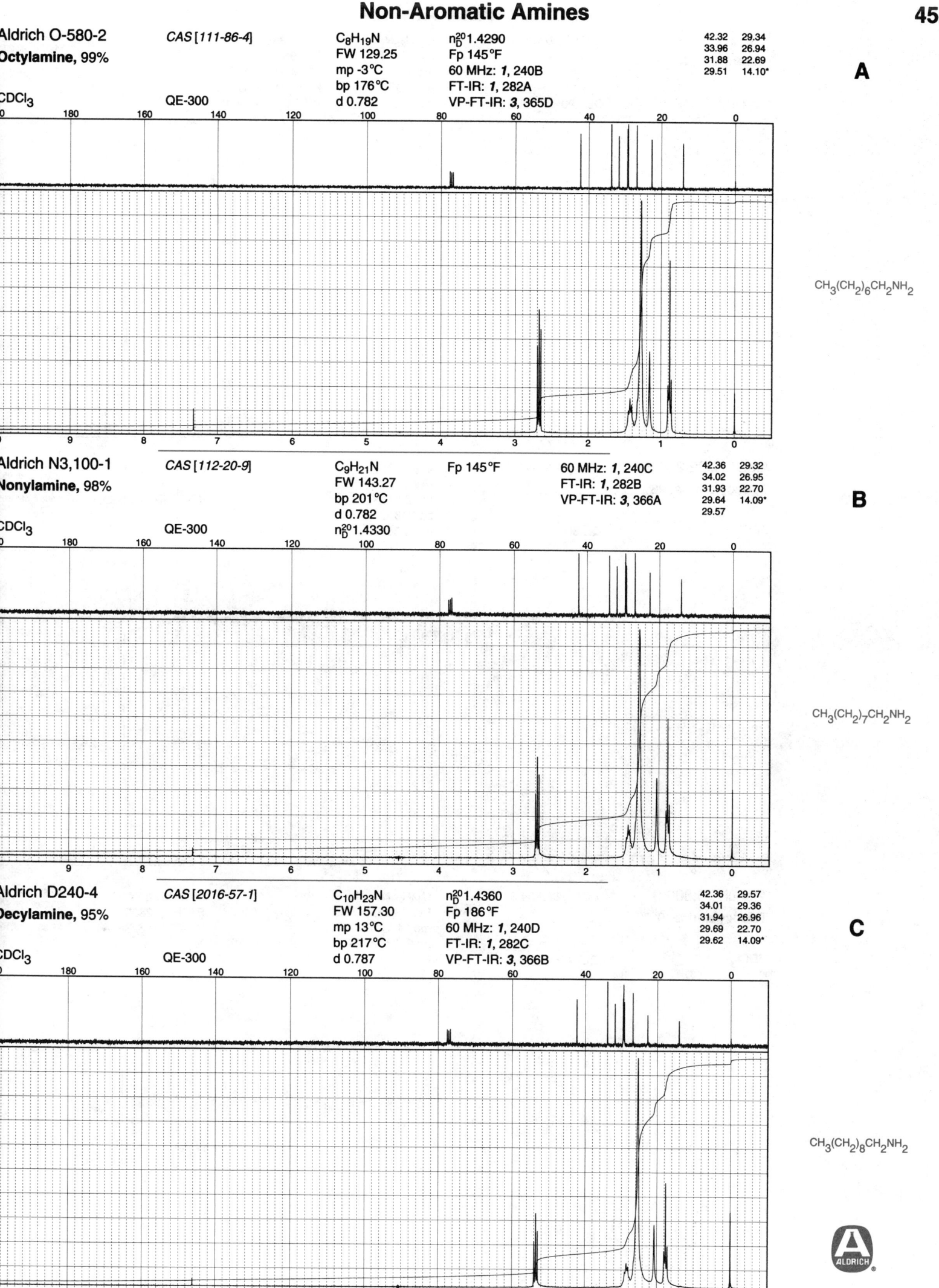

A

Aldrich O-580-2   *CAS [111-86-4]*   C₈H₁₉N
Octylamine, 99%   FW 129.25

CDCl₃   QE-300

42.32	29.34
33.96	26.94
31.88	22.69
29.51	14.10*

n_D^{20} 1.4290
Fp 145°F
mp -3°C   60 MHz: *1*, 240B
bp 176°C   FT-IR: *1*, 282A
d 0.782   VP-FT-IR: *3*, 365D

$CH_3(CH_2)_6CH_2NH_2$

B

Aldrich N3,100-1   *CAS [112-20-9]*   C₉H₂₁N   Fp 145°F   60 MHz: *1*, 240C
Nonylamine, 98%   FW 143.27   FT-IR: *1*, 282B
                                        bp 201°C   VP-FT-IR: *3*, 366A
CDCl₃   QE-300   d 0.782
                                        n_D^{20} 1.4330

42.36	29.32
34.02	26.95
31.93	22.70
29.64	14.09*
29.57	

$CH_3(CH_2)_7CH_2NH_2$

C

Aldrich D240-4   *CAS [2016-57-1]*   C₁₀H₂₃N
Decylamine, 95%   FW 157.30

CDCl₃   QE-300

42.36	29.57
34.01	29.36
31.94	26.96
29.69	22.70
29.62	14.09*

n_D^{20} 1.4360
Fp 186°F
mp 13°C   60 MHz: *1*, 240D
bp 217°C   FT-IR: *1*, 282C
d 0.787   VP-FT-IR: *3*, 366B

$CH_3(CH_2)_8CH_2NH_2$

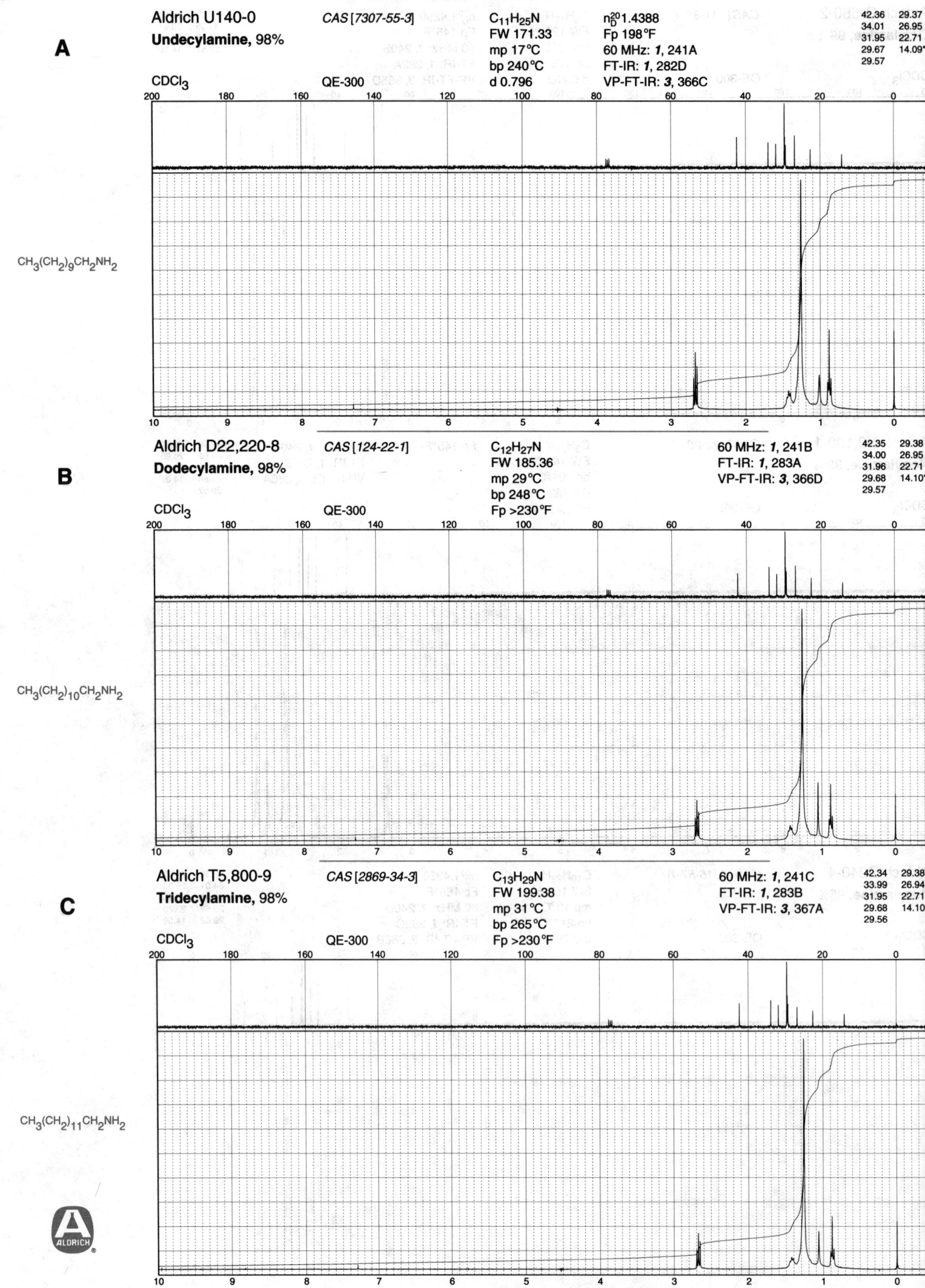

A

Aldrich U140-0

Undecylamine, 98%

CAS [7307-55-3]

$C_{11}H_{25}N$
FW 171.33
mp 17°C
bp 240°C
d 0.796

n_D^{20} 1.4388
Fp 198°F
60 MHz: *1*, 241A
FT-IR: *1*, 282D
VP-FT-IR: *3*, 366C

42.36 29.37
34.01 26.95
31.95 22.71
29.67 14.09*
29.57

CDCl₃ QE-300

$CH_3(CH_2)_9CH_2NH_2$

B

Aldrich D22,220-8

Dodecylamine, 98%

CAS [124-22-1]

$C_{12}H_{27}N$
FW 185.36
mp 29°C
bp 248°C
Fp >230°F

60 MHz: *1*, 241B
FT-IR: *1*, 283A
VP-FT-IR: *3*, 366D

42.35 29.38
34.00 26.95
31.96 22.71
29.68 14.10*
29.57

CDCl₃ QE-300

$CH_3(CH_2)_{10}CH_2NH_2$

C

Aldrich T5,800-9

Tridecylamine, 98%

CAS [2869-34-3]

$C_{13}H_{29}N$
FW 199.38
mp 31°C
bp 265°C
Fp >230°F

60 MHz: *1*, 241C
FT-IR: *1*, 283B
VP-FT-IR: *3*, 367A

42.34 29.38
33.99 26.94
31.95 22.71
29.68 14.10
29.56

CDCl₃ QE-300

$CH_3(CH_2)_{11}CH_2NH_2$

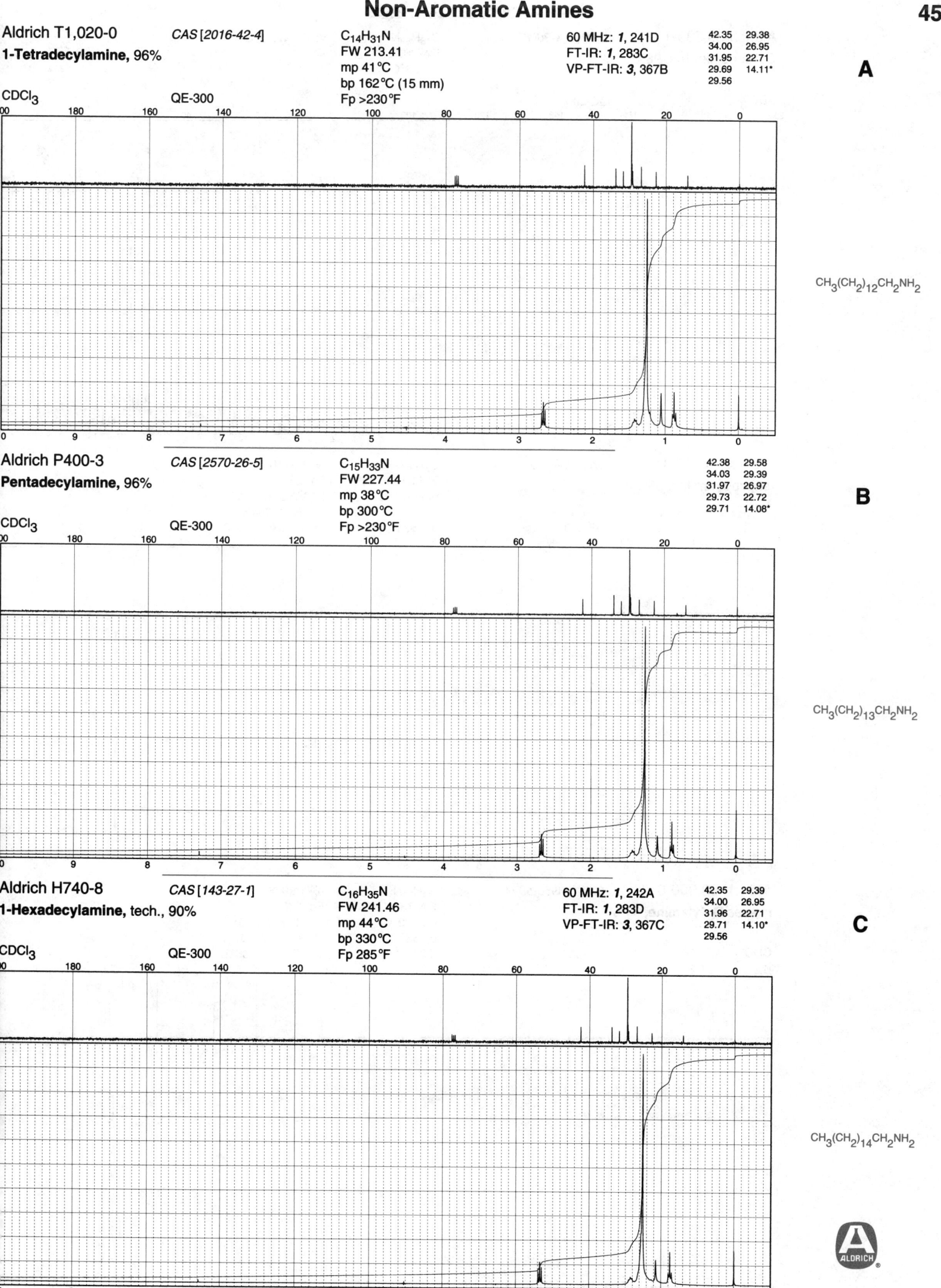

Aldrich T1,020-0   CAS [2016-42-4]

1-Tetradecylamine, 96%

$C_{14}H_{31}N$
FW 213.41
mp 41°C
bp 162°C (15 mm)
Fp >230°F

60 MHz: *1*, 241D
FT-IR: *1*, 283C
VP-FT-IR: *3*, 367B

42.35	29.38
34.00	26.95
31.95	22.71
29.69	14.11*
29.56	

A

CDCl₃   QE-300

$CH_3(CH_2)_{12}CH_2NH_2$

Aldrich P400-3   CAS [2570-26-5]

Pentadecylamine, 96%

$C_{15}H_{33}N$
FW 227.44
mp 38°C
bp 300°C
Fp >230°F

42.38	29.58
34.03	29.39
31.97	26.97
29.73	22.72
29.71	14.08*

B

CDCl₃   QE-300

$CH_3(CH_2)_{13}CH_2NH_2$

Aldrich H740-8   CAS [143-27-1]

1-Hexadecylamine, tech., 90%

$C_{16}H_{35}N$
FW 241.46
mp 44°C
bp 330°C
Fp 285°F

60 MHz: *1*, 242A
FT-IR: *1*, 283D
VP-FT-IR: *3*, 367C

42.35	29.39
34.00	26.95
31.96	22.71
29.71	14.10*
29.56	

C

CDCl₃   QE-300

$CH_3(CH_2)_{14}CH_2NH_2$

Non-Aromatic Amines

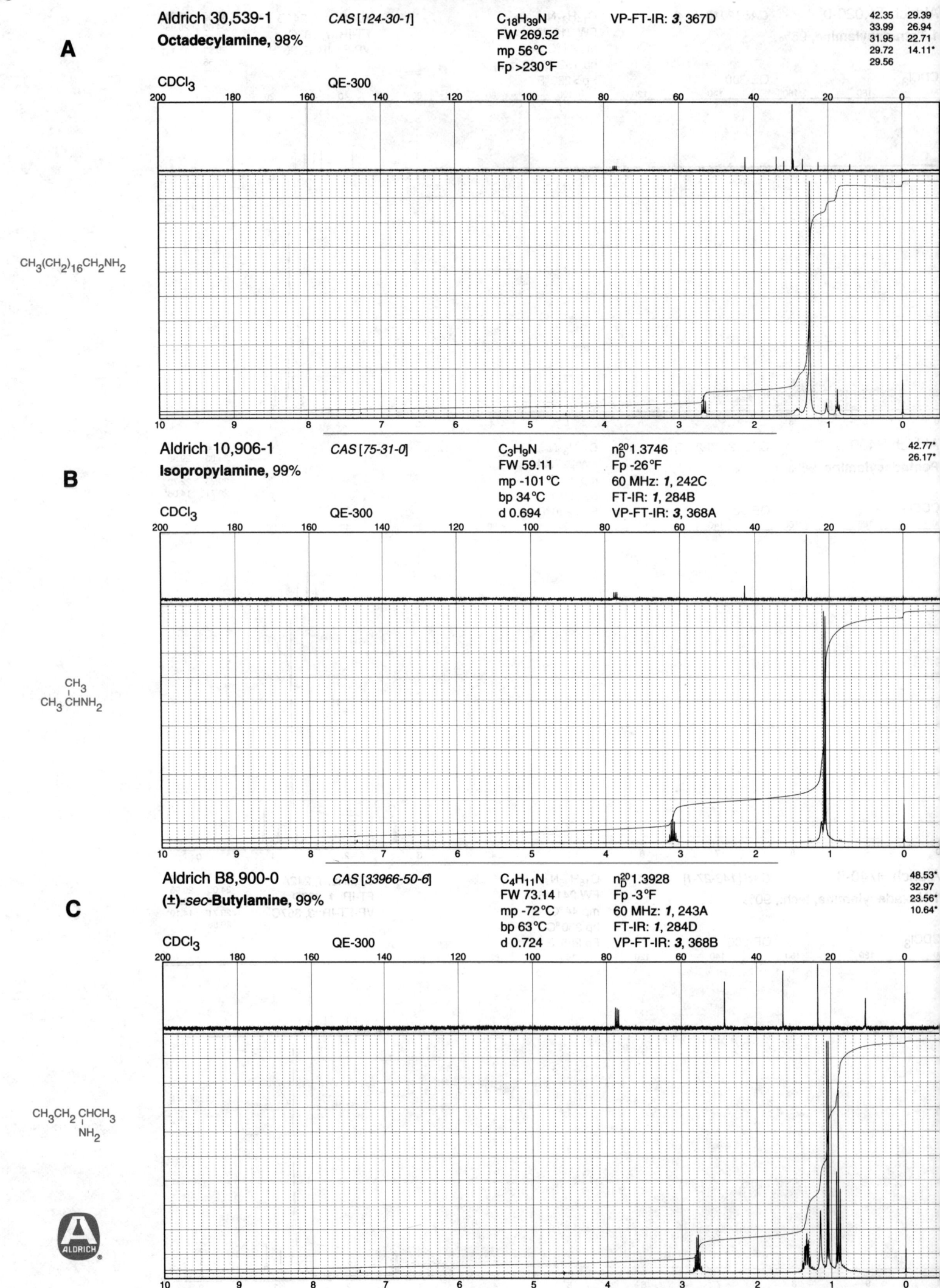

A

Aldrich 30,539-1
Octadecylamine, 98%

CAS [124-30-1]

$C_{18}H_{39}N$
FW 269.52
mp 56°C
Fp >230°F

VP-FT-IR: **3**, 367D

42.35	29.39
33.99	26.94
31.95	22.71
29.72	14.11*
29.56	

CDCl₃

QE-300

$CH_3(CH_2)_{16}CH_2NH_2$

B

Aldrich 10,906-1
Isopropylamine, 99%

CAS [75-31-0]

C_3H_9N
FW 59.11
mp -101°C
bp 34°C
d 0.694

n_D^{20} 1.3746
Fp -26°F
60 MHz: **1**, 242C
FT-IR: **1**, 284B
VP-FT-IR: **3**, 368A

| 42.77* |
| 26.17* |

CDCl₃

QE-300

$CH_3\,CHNH_2$ with CH_3

C

Aldrich B8,900-0
(±)-sec-Butylamine, 99%

CAS [33966-50-6]

$C_4H_{11}N$
FW 73.14
mp -72°C
bp 63°C
d 0.724

n_D^{20} 1.3928
Fp -3°F
60 MHz: **1**, 243A
FT-IR: **1**, 284D
VP-FT-IR: **3**, 368B

| 48.53* |
| 32.97 |
| 23.56* |
| 10.64* |

CDCl₃

QE-300

$CH_3CH_2\,CHCH_3$ with NH_2

ALDRICH

Non-Aromatic Amines

Aldrich 29,665-1   CAS [513-49-5] $C_4H_{11}N$ Fp -3°F   VP-FT-IR: **3**, 368C

48.48*
32.88
23.47*
10.69*

(S)-(+)-sec-**Butylamine**, 99%   FW 73.14
bp 63°C
d 0.731

A

CDCl₃   QE-300 n_D^{20}1.3930

$CH_3CH_2 \overset{\cdot}{\underset{H_2N}{C}} \overset{\cdot}{\underset{H}{C}}H_3$

Aldrich 29,664-3   CAS [13250-12-9] $C_4H_{11}N$ Fp -3°F   VP-FT-IR: **3**, 368D

48.50*
32.94
23.54*
10.66*

(R)-(-)-sec-**Butylamine**, 99%   FW 73.14
bp 63°C
d 0.720

B

CDCl₃   QE-300 n_D^{20}1.3936

$CH_3CH_2 \overset{\cdot}{\underset{H_2N}{C}} \overset{\cdot}{\underset{H}{C}}H_3$

Aldrich B8,920-5   CAS [75-64-9] $C_4H_{11}N$ n_D^{20}1.3790

47.33
32.58*

tert-**Butylamine**, 98%   FW 73.14   Fp 16°F
mp -67°C   60 MHz: **1**, 244A
bp 46°C   FT-IR: **1**, 285C
d 0.696   VP-FT-IR: **3**, 369B

C

CDCl₃   QE-300

$CH_3 - \overset{\underset{|}{CH_3}}{\underset{\underset{|}{CH_3}}{C}} - NH_2$

ALDRICH

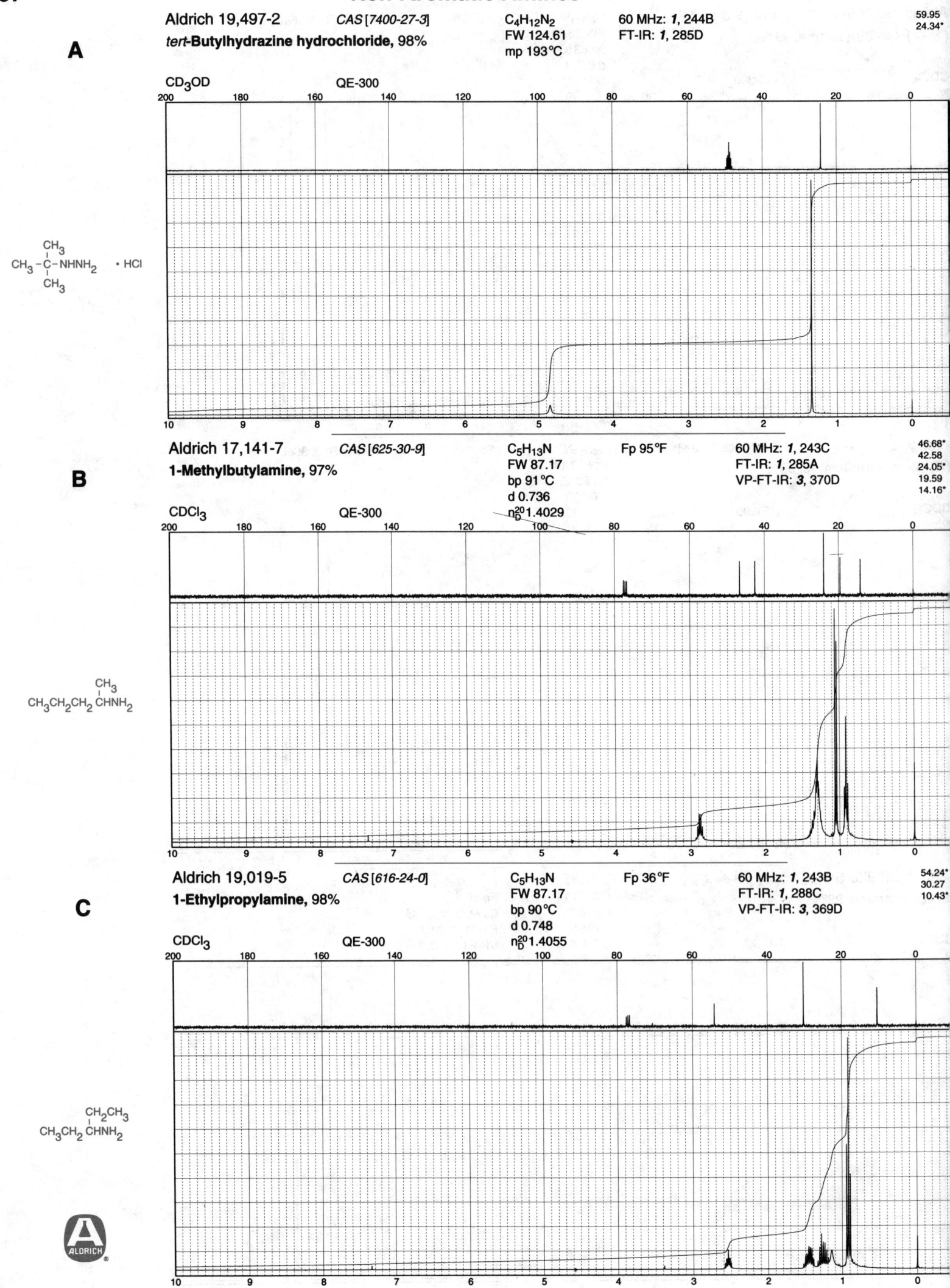

A

Aldrich 19,497-2   *CAS [7400-27-3]*
tert-Butylhydrazine hydrochloride, 98%

$C_4H_{12}N_2$
FW 124.61
mp 193°C

60 MHz: *1*, 244B
FT-IR: *1*, 285D

59.95
24.34*

CD₃OD   QE-300

$CH_3-\underset{\underset{CH_3}{|}}{\overset{\overset{CH_3}{|}}{C}}-NHNH_2$ • HCl

B

Aldrich 17,141-7   *CAS [625-30-9]*
1-Methylbutylamine, 97%

$C_5H_{13}N$
FW 87.17
bp 91°C
d 0.736
n_D^{20}1.4029

Fp 95°F

60 MHz: *1*, 243C
FT-IR: *1*, 285A
VP-FT-IR: *3*, 370D

46.68*
42.58
24.05*
19.59
14.16*

CDCl₃   QE-300

$CH_3CH_2CH_2\underset{\overset{|}{CH}}{\overset{\overset{CH_3}{|}}{C}}NH_2$

C

Aldrich 19,019-5   *CAS [616-24-0]*
1-Ethylpropylamine, 98%

$C_5H_{13}N$
FW 87.17
bp 90°C
d 0.748
n_D^{20}1.4055

Fp 36°F

60 MHz: *1*, 243B
FT-IR: *1*, 288C
VP-FT-IR: *3*, 369D

54.24*
30.27
10.43*

CDCl₃   QE-300

$CH_3CH_2\underset{\overset{|}{CH}}{\overset{\overset{CH_2CH_3}{|}}{C}}NH_2$

ALDRICH

Aldrich 24,140-7   *CAS [20626-52-2]* $C_5H_{13}N$ Fp 38°F   60 MHz: *1*, 245B

(S)-(-)-2-Methylbutylamine, 95%   FW 87.17   FT-IR: *1*, 287B   VP-FT-IR: *3*, 371B

bp 43°C (12 mm)

d 0.738

CDCl₃   QE-300 n_D^{20} 1.4126

48.15
38.08*
26.86
17.03*
11.37*

A

CH_3CH_2 CCH_2NH_2 (with CH₃ and H substituents)

Aldrich 22,052-3   *CAS [96-15-1]* $C_5H_{13}N$ Fp 38°F   60 MHz: *1*, 244D

-Methylbutylamine, 97+%   FW 87.17   FT-IR: *1*, 287A   VP-FT-IR: *3*, 371A

bp 96°C

d 0.738

CDCl₃   QE-300 n_D^{20} 1.4116

48.16
38.10*
26.86
17.03*
11.37*

B

CH_3CH_2 $CHCH_2NH_2$ (with CH₃ substituent)

Aldrich 12,681-0   *CAS [107-85-7]* $C_5H_{13}N$ Fp 65°F   60 MHz: *1*, 245A

oamylamine, 99%   FW 87.17   FT-IR: *1*, 286D   VP-FT-IR: *3*, 370C

bp 96°C

d 0.751

DCl₃   QE-300 n_D^{20} 1.4089

43.19
40.36
25.69*
22.64*

C

$CH_3CHCH_2CH_2NH_2$ (with CH₃ substituent)

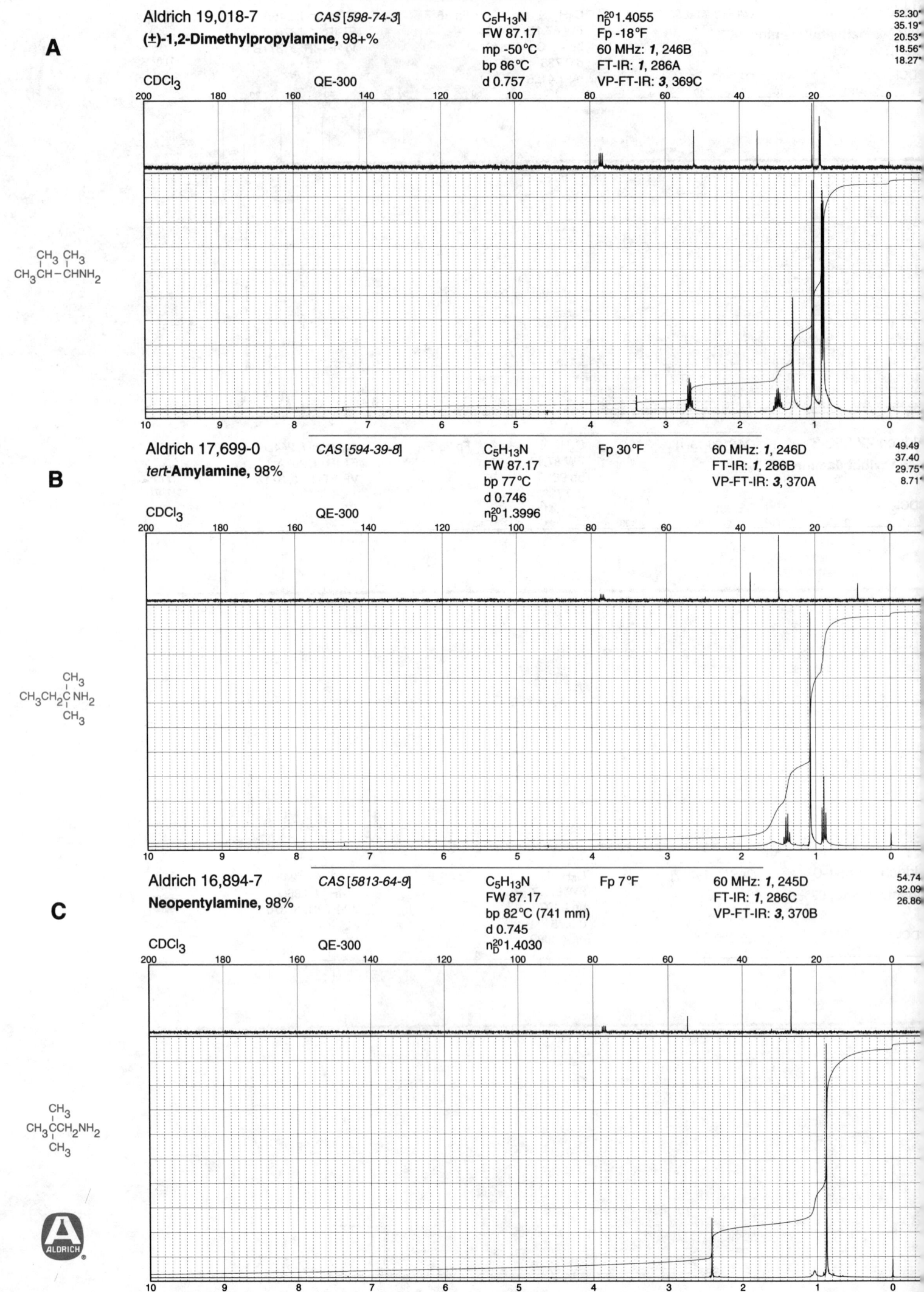

A

Aldrich 19,018-7   *CAS [598-74-3]*

(±)-1,2-Dimethylpropylamine, 98+%

CDCl₃   QE-300

$C_5H_{13}N$
FW 87.17
mp -50°C
bp 86°C
d 0.757

n_D^{20} 1.4055
Fp -18°F
60 MHz: *1*, 246B
FT-IR: *1*, 286A
VP-FT-IR: *3*, 369C

52.30*
35.19*
20.53*
18.56*
18.27*

$CH_3CH-CHNH_2$ (with CH₃ CH₃ above)

B

Aldrich 17,699-0   *CAS [594-39-8]*

tert-Amylamine, 98%

CDCl₃   QE-300

$C_5H_{13}N$
FW 87.17
bp 77°C
d 0.746
n_D^{20} 1.3996

Fp 30°F

60 MHz: *1*, 246D
FT-IR: *1*, 286B
VP-FT-IR: *3*, 370A

49.49
37.40
29.75*
8.71*

$CH_3CH_2C\,NH_2$ (with CH₃ above and CH₃ below)

C

Aldrich 16,894-7   *CAS [5813-64-9]*

Neopentylamine, 98%

CDCl₃   QE-300

$C_5H_{13}N$
FW 87.17
bp 82°C (741 mm)
d 0.745
n_D^{20} 1.4030

Fp 7°F

60 MHz: *1*, 245D
FT-IR: *1*, 286C
VP-FT-IR: *3*, 370B

54.74
32.09
26.86

$CH_3CCH_2NH_2$ (with CH₃ above and CH₃ below)

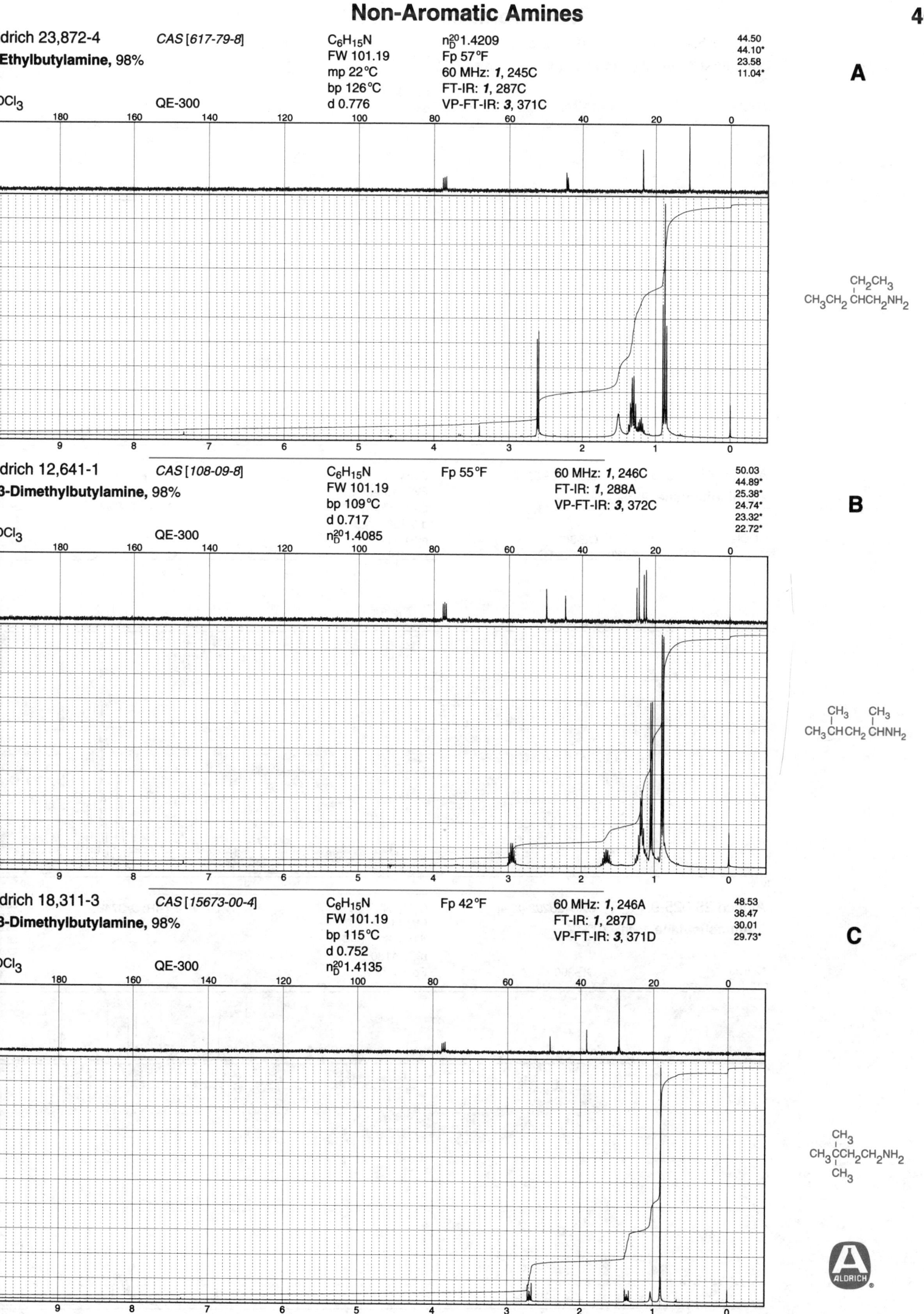

Aldrich 23,872-4   CAS [617-79-8]
-Ethylbutylamine, 98%

C₆H₁₅N
FW 101.19
mp 22°C
bp 126°C
d 0.776

n_D^{20}1.4209
Fp 57°F
60 MHz: *1*, 245C
FT-IR: *1*, 287C
VP-FT-IR: *3*, 371C

44.50
44.10*
23.58
11.04*

A

DCl₃   QE-300

CH₂CH₃
CH₃CH₂CHCH₂NH₂

Aldrich 12,641-1   CAS [108-09-8]
,3-Dimethylbutylamine, 98%

C₆H₁₅N
FW 101.19
bp 109°C
d 0.717
n_D^{20}1.4085

Fp 55°F

60 MHz: *1*, 246C
FT-IR: *1*, 288A
VP-FT-IR: *3*, 372C

50.03
44.89*
25.38*
24.74*
23.32*
22.72*

B

DCl₃   QE-300

CH₃   CH₃
CH₃CHCH₂ CHNH₂

Aldrich 18,311-3   CAS [15673-00-4]
,3-Dimethylbutylamine, 98%

C₆H₁₅N
FW 101.19
bp 115°C
d 0.752
n_D^{20}1.4135

Fp 42°F

60 MHz: *1*, 246A
FT-IR: *1*, 287D
VP-FT-IR: *3*, 371D

48.53
38.47
30.01
29.73*

C

DCl₃   QE-300

CH₃
CH₃CCH₂CH₂NH₂
CH₃

ALDRICH

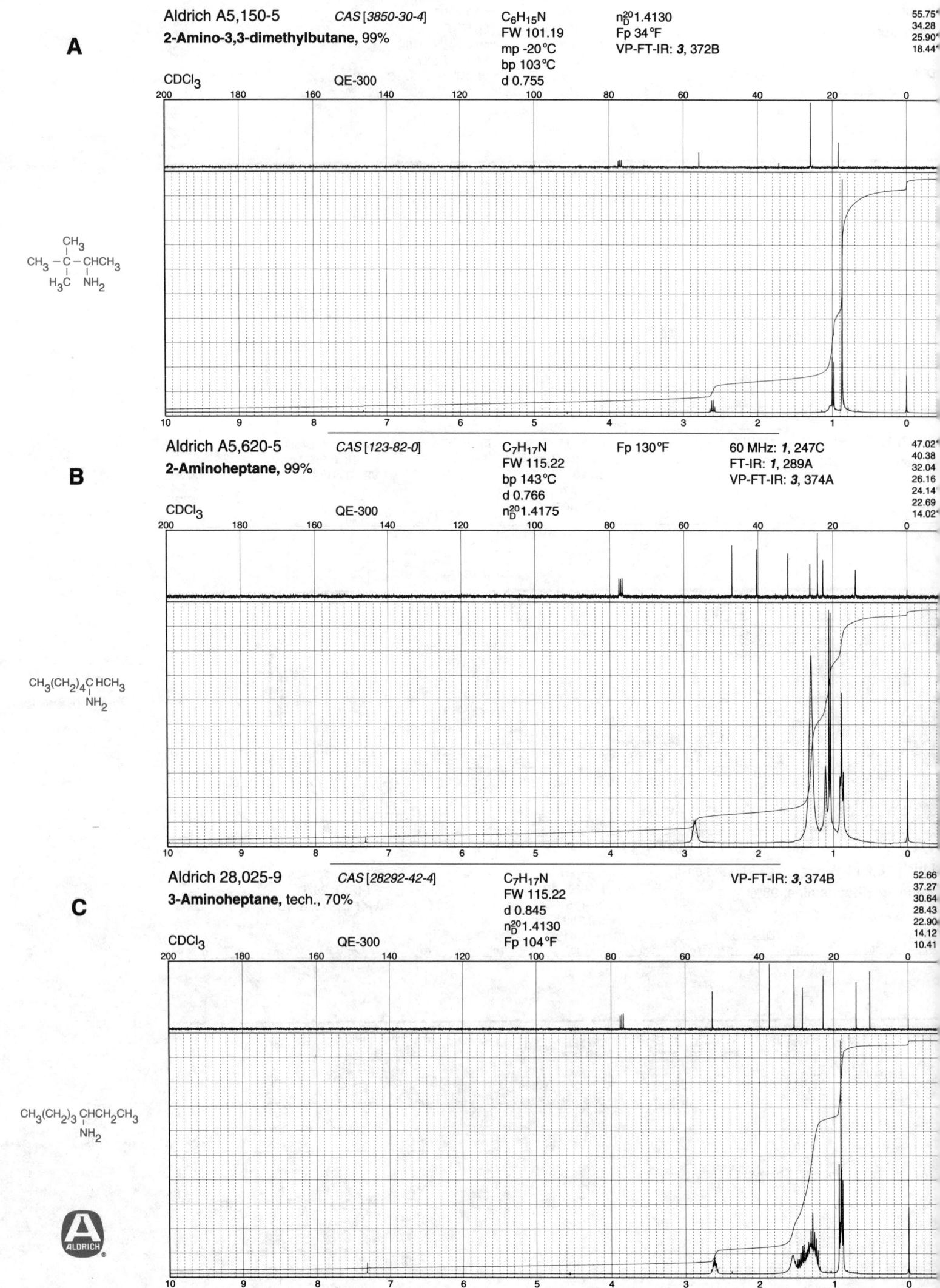

A

Aldrich A5,150-5   *CAS [3850-30-4]*   C₆H₁₅N   n²⁰_D 1.4130

2-Amino-3,3-dimethylbutane, 99%

FW 101.19   Fp 34 °F   VP-FT-IR: **3**, 372B
mp -20 °C
bp 103 °C
d 0.755

CDCl₃   QE-300

55.75*
34.28
25.90*
18.44*

B

Aldrich A5,620-5   *CAS [123-82-0]*   C₇H₁₇N   Fp 130 °F   60 MHz: **1**, 247C

2-Aminoheptane, 99%

FW 115.22   FT-IR: **1**, 289A
bp 143 °C   VP-FT-IR: **3**, 374A
d 0.766
n²⁰_D 1.4175

CDCl₃   QE-300

47.02*
40.38
32.04
26.16
24.14
22.69
14.02*

C

Aldrich 28,025-9   *CAS [28292-42-4]*   C₇H₁₇N   VP-FT-IR: **3**, 374B

3-Aminoheptane, tech., 70%

FW 115.22
d 0.845
n²⁰_D 1.4130
Fp 104 °F

CDCl₃   QE-300

52.66
37.27
30.64
28.43
22.90
14.12
10.41

Non-Aromatic Amines

Aldrich 18,398-9   CAS [693-16-3] $C_8H_{19}N$ Fp 123°F   60 MHz: **1**, 247D

-Methylheptylamine, 99%   FW 129.25   FT-IR: **1**, 289B

bp 165°C   VP-FT-IR: **3**, 374C

d 0.771

CDCl₃   QE-300 n_D^{20} 1.4235

47.00*	26.46
40.40	24.12*
31.92	22.65
29.47	14.06*

A

$CH_3(CH_2)_5 \overset{\overset{\displaystyle CH_3}{|}}{CH}NH_2$

Aldrich E2,950-8   CAS [104-75-6] $C_8H_{19}N$ n_D^{20} 1.4300

±)-2-Ethylhexylamine, 98%   FW 129.25   Fp 126°F

mp -76°C   VP-FT-IR: **3**, 373C

bp 169°C

CDCl₃   QE-300   d 0.789

44.90	23.94
42.51*	23.16
30.87	14.11*
29.12	11.01*

B

$CH_3(CH_2)_3 \overset{\overset{\displaystyle CH_2CH_3}{|}}{CH}CH_2NH_2$

Aldrich D16,129-2   CAS [543-82-8] $C_8H_{19}N$ Fp 120°F   60 MHz: **1**, 247B

,5-Dimethylhexylamine, 99%   FW 129.25   FT-IR: **1**, 288D

bp 155°C   VP-FT-IR: **3**, 373D

d 0.767

DCl₃   QE-300 n_D^{20} 1.4215

47.00*
40.62
39.14
27.98*
24.22
24.12*
22.61*

C

$CH_3 \overset{\overset{\displaystyle CH_3}{|}}{CH}(CH_2)_3 \overset{\overset{\displaystyle CH_3}{|}}{CH}NH_2$

A

Aldrich O-600-0   *CAS [107-45-9]*   C$_8$H$_{19}$N   Fp 90°F   60 MHz: *1*, 247A   57.19
tert-Octylamine, 95%   FW 129.25   FT-IR: *1*, 288B   51.21
                                       bp 140°C   VP-FT-IR: *3*, 372D 32.85
                                       d 0.805   31.75
                                       n$_D^{20}$1.4240   31.67

CDCl$_3$ QE-300

```
        CH3      CH3
         |        |
CH3 - C - CH2 - C - NH2
         |        |
        CH3      CH3
```

45.77

B

Aldrich 24,072-9   *CAS [107-15-3]*   C$_2$H$_8$N$_2$ n$_D^{20}$1.4565
Ethylenediamine, 99+%   FW 60.10   Fp 93°F
                                         mp 9°C   60 MHz: *1*, 248A
                                         bp 118°C   FT-IR: *1*, 289C
                                         d 0.899   VP-FT-IR: *3*, 374D

CDCl$_3$ QE-300

H$_2$NCH$_2$CH$_2$NH$_2$

39.30

C

Aldrich 19,580-4   *CAS [333-18-6]*   C$_2$H$_8$N$_2$ FT-IR: *1*, 289D
Ethylenediamine dihydrochloride, 98%   FW 133.02
                                         mp 300°C

D$_2$O   QE-300

H$_2$NCH$_2$CH$_2$NH$_2$ • 2HCl

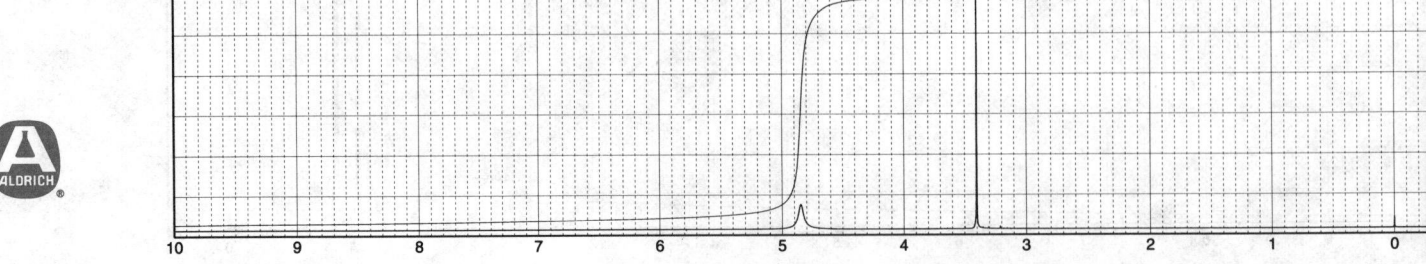

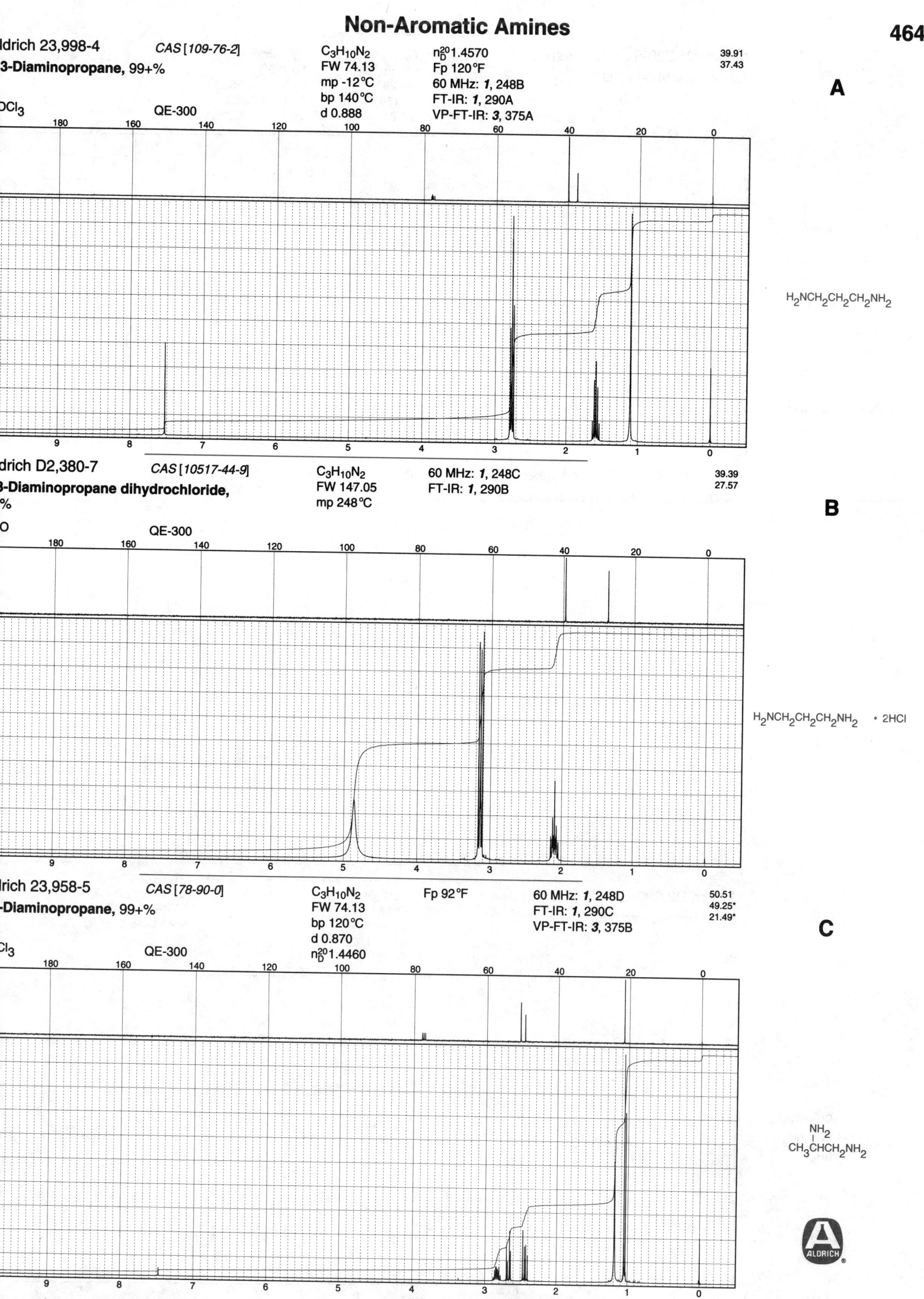

Aldrich 23,998-4   CAS [109-76-2]
,3-Diaminopropane, 99+%

C₃H₁₀N₂
FW 74.13
mp -12°C
bp 140°C
d 0.888

n_D^{20} 1.4570
Fp 120°F
60 MHz: *1*, 248B
FT-IR: *1*, 290A
VP-FT-IR: *3*, 375A

39.91
37.43

A

CDCl₃   QE-300

$H_2NCH_2CH_2CH_2NH_2$

Aldrich D2,380-7   CAS [10517-44-9]
3-Diaminopropane dihydrochloride,
%

C₃H₁₀N₂
FW 147.05
mp 248°C

60 MHz: *1*, 248C
FT-IR: *1*, 290B

39.39
27.57

B

₂O   QE-300

$H_2NCH_2CH_2CH_2NH_2$ • 2HCl

Aldrich 23,958-5   CAS [78-90-0]
-Diaminopropane, 99+%

C₃H₁₀N₂
FW 74.13
bp 120°C
d 0.870
n_D^{20} 1.4460

Fp 92°F

60 MHz: *1*, 248D
FT-IR: *1*, 290C
VP-FT-IR: *3*, 375B

50.51
49.25*
21.49*

C

Cl₃   QE-300

$CH_3CHCH_2NH_2$
|
NH_2

ALDRICH

A

Aldrich D1,320-8   *CAS [110-60-1]*   C$_4$H$_{12}$N$_2$ n$_D^{20}$1.4569   42.1(
1,4-Diaminobutane, 99%   FW 88.15   Fp 125°F   31.2(
mp 28°C   60 MHz: *1*, 250B
bp 159°C   FT-IR: *1*, 291D
d 0.877   VP-FT-IR: *3*, 376A

CDCl$_3$ QE-300

H$_2$NCH$_2$CH$_2$CH$_2$CH$_2$NH$_2$

B

Aldrich 23,400-1   *CAS [333-93-7]*   C$_4$H$_{12}$N$_2$ 60 MHz: *1*, 250C   41.6
1,4-Diaminobutane dihydrochloride, 97%   FW 161.08   FT-IR: *1*, 292A   26.6
mp 280°C d.

D$_2$O   QE-300

H$_2$NCH$_2$CH$_2$CH$_2$CH$_2$NH$_2$ • 2HCl

C

Aldrich D2,080-8   *CAS [811-93-8]*   C$_4$H$_{12}$N$_2$ 60 MHz: *1*, 249A   54
1,2-Diamino-2-methylpropane, 99%   FW 88.15   FT-IR: *1*, 290D   50
d 0.841   VP-FT-IR: *3*, 375C   28
n$_D^{20}$1.4410
Fp 75°F

CDCl$_3$ QE-300

CH$_3$
|
CH$_3$CCH$_2$NH$_2$
|
NH$_2$

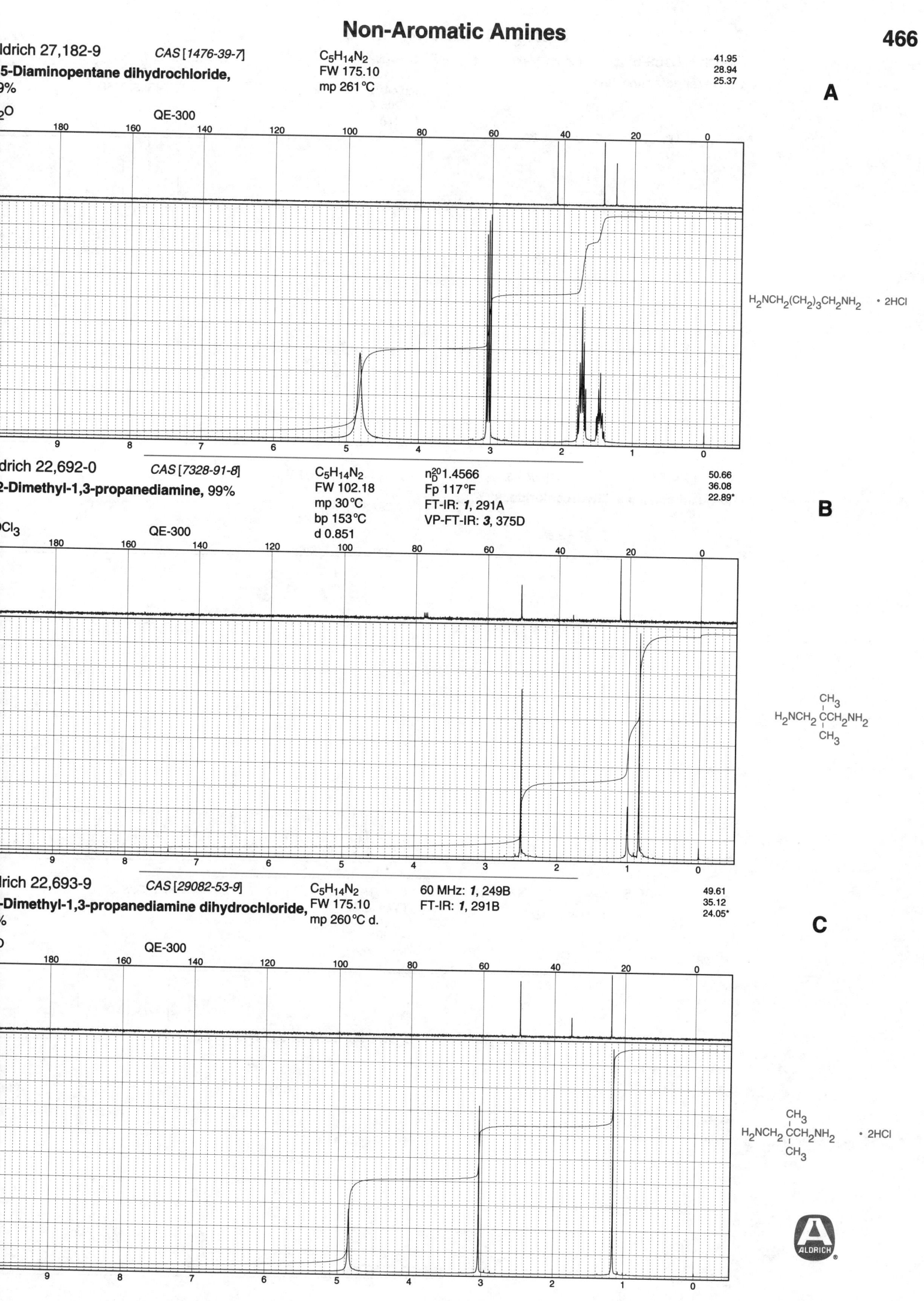

A

Aldrich 27,182-9 CAS [1476-39-7] C₅H₁₄N₂ $C_5H_{14}N_2$ FW 175.10 mp 261°C

41.95
28.94
25.37

1,5-Diaminopentane dihydrochloride, 99%

H₂O QE-300

$H_2NCH_2(CH_2)_3CH_2NH_2$ · 2HCl

B

Aldrich 22,692-0 CAS [7328-91-8] $C_5H_{14}N_2$ FW 102.18 mp 30°C bp 153°C d 0.851

n_D^{20} 1.4566
Fp 117°F
FT-IR: **1**, 291A
VP-FT-IR: **3**, 375D

50.66
36.08
22.89*

2,2-Dimethyl-1,3-propanediamine, 99%

CDCl₃ QE-300

$H_2NCH_2\overset{CH_3}{\underset{CH_3}{C}}CH_2NH_2$

C

Aldrich 22,693-9 CAS [29082-53-9] $C_5H_{14}N_2$ FW 175.10 mp 260°C d.

60 MHz: **1**, 249B
FT-IR: **1**, 291B

49.61
35.12
24.05*

2,2-Dimethyl-1,3-propanediamine dihydrochloride, %

D₂O QE-300

$H_2NCH_2\overset{CH_3}{\underset{CH_3}{C}}CH_2NH_2$ · 2HCl

ALDRICH

Non-Aromatic Amines

A

Aldrich H1,169-6   CAS [124-09-4] $C_6H_{16}N_2$ 60 MHz: *1*, 251A   42.26
1,6-Hexanediamine, 98%   FW 116.21   FT-IR: *1*, 292C   33.93
                                          mp 44°C   VP-FT-IR: *3*, 376D   26.83
                                          bp 205°C
                                          Fp 178°F

CDCl$_3$ QE-300

$H_2NCH_2(CH_2)_4CH_2NH_2$

B

Aldrich 24,773-1   CAS [6055-52-3] $C_6H_{16}N_2$ FT-IR: *1*, 292D   42.13
1,6-Hexanediamine dihydrochloride, 99%   FW 189.13   29.12
                                          mp 257°C   27.76

D$_2$O   QE-300

$H_2NCH_2(CH_2)_4CH_2NH_2$ · 2HCl

C

Aldrich 32,966-5   CAS [15520-10-2] $C_6H_{16}N_2$ Fp 177°F   VP-FT-IR: *3*, 376C   48.4
2-Methyl-1,5-pentanediamine, 99%   FW 116.21   42.5
                                          bp 193°C   36.
                                          d 0.860   31.4
                                          n_D^{20} 1.4590   31.2
                                                                                                 17.

CDCl$_3$ QE-300

$H_2NCH_2CH_2CH_2\overset{CH_3}{\underset{|}{CH}}CH_2NH_2$

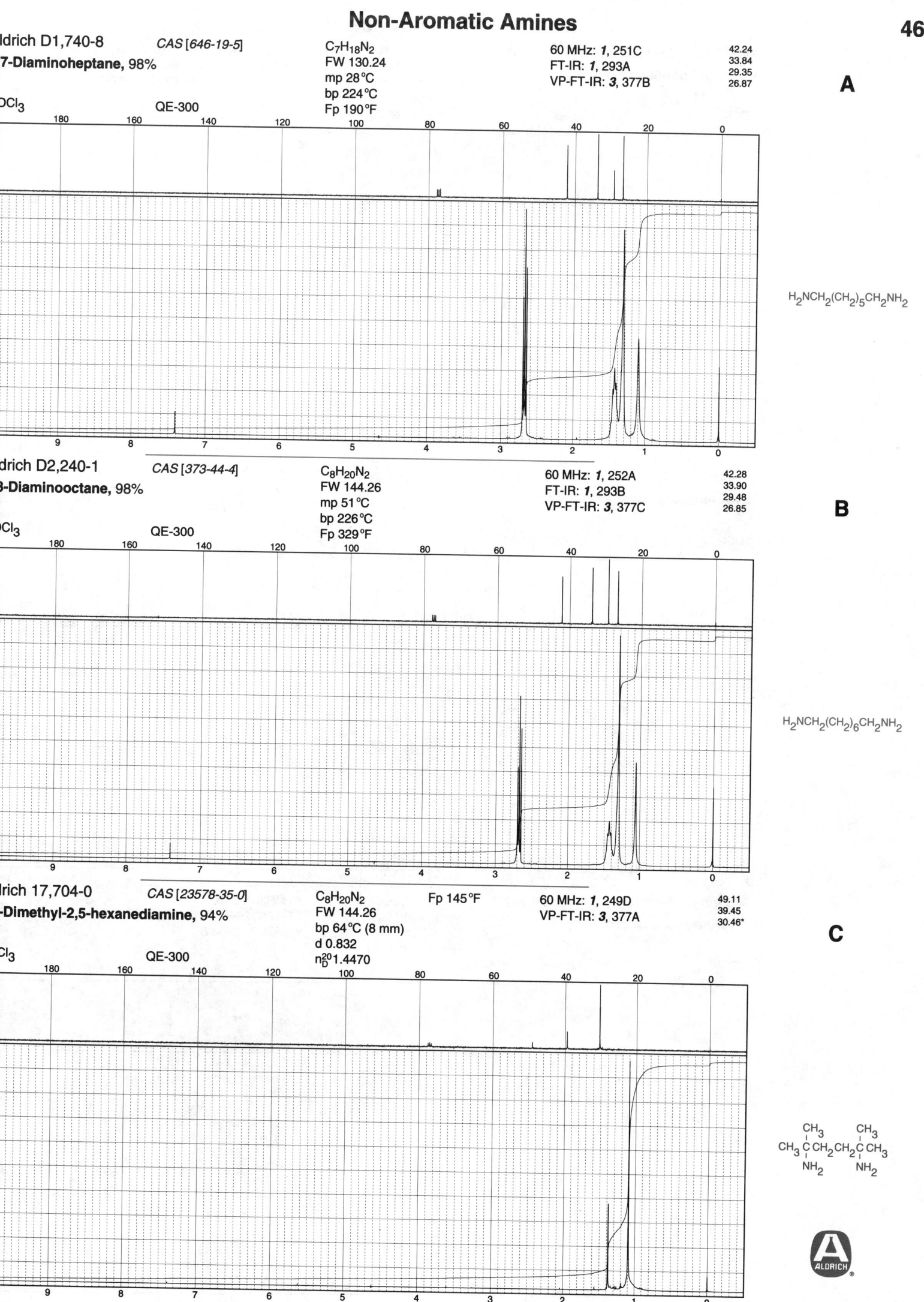

Aldrich D1,740-8   CAS [646-19-5]   C$_7$H$_{18}$N$_2$ 60 MHz: *1*, 251C   42.24
1,7-Diaminoheptane, 98%   FW 130.24   FT-IR: *1*, 293A   33.84
    mp 28°C   VP-FT-IR: *3*, 377B   29.35
CDCl$_3$ QE-300   bp 224°C   26.87
    Fp 190°F

A

H$_2$NCH$_2$(CH$_2$)$_5$CH$_2$NH$_2$

Aldrich D2,240-1   CAS [373-44-4]   C$_8$H$_{20}$N$_2$ 60 MHz: *1*, 252A   42.28
1,8-Diaminooctane, 98%   FW 144.26   FT-IR: *1*, 293B   33.90
    mp 51°C   VP-FT-IR: *3*, 377C   29.48
CDCl$_3$ QE-300   bp 226°C   26.85
    Fp 329°F

B

H$_2$NCH$_2$(CH$_2$)$_6$CH$_2$NH$_2$

Aldrich 17,704-0   CAS [23578-35-0]   C$_8$H$_{20}$N$_2$ Fp 145°F   60 MHz: *1*, 249D   49.11
2,5-Dimethyl-2,5-hexanediamine, 94%   FW 144.26   VP-FT-IR: *3*, 377A   39.45
    bp 64°C (8 mm)   30.46*
CDCl$_3$ QE-300   d 0.832
    n$_D^{20}$1.4470

C

CH$_3$ CH$_3$
CH$_3$ C CH$_2$CH$_2$ C CH$_3$
NH$_2$ NH$_2$

ALDRICH

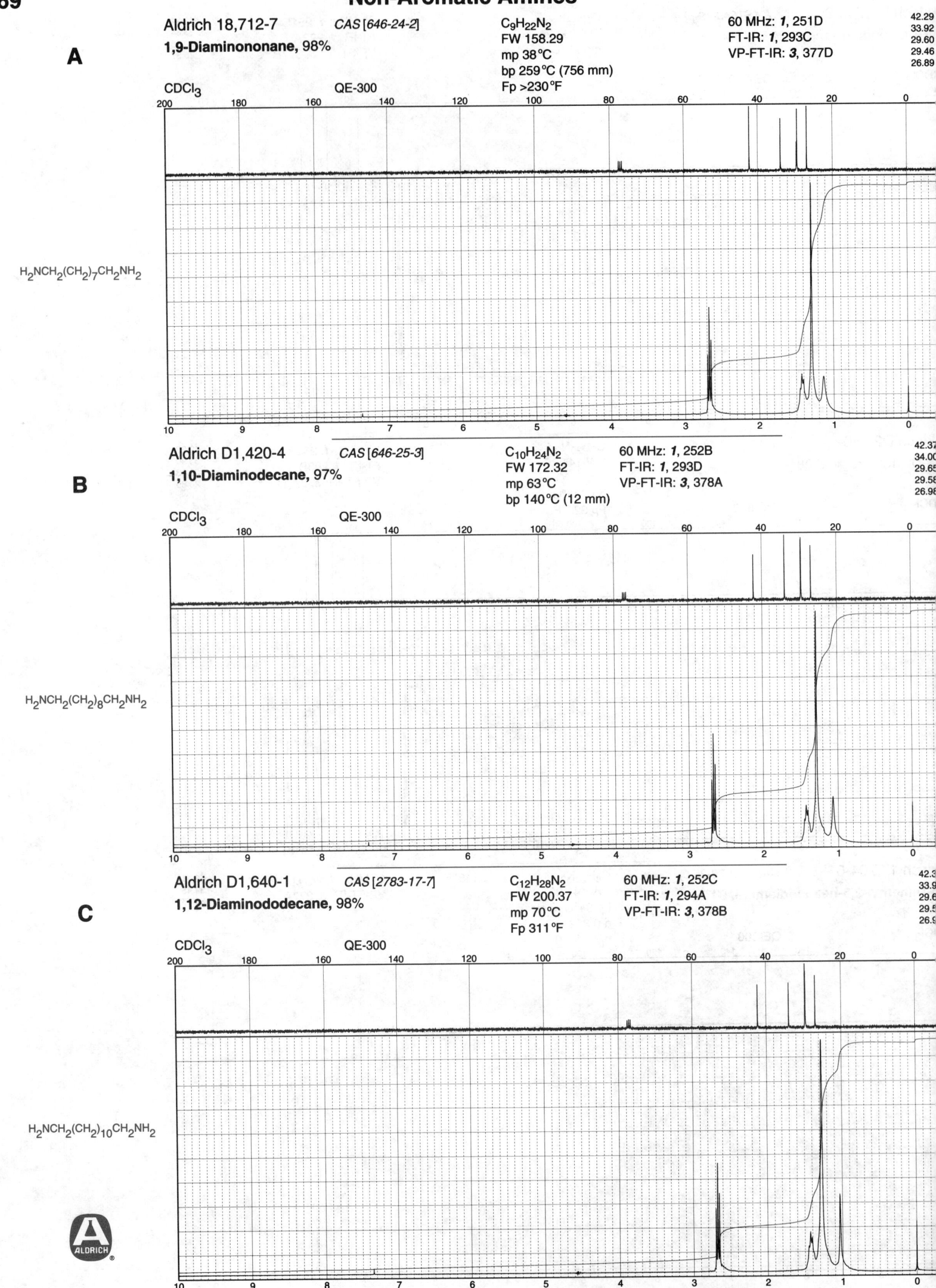

A

Aldrich 18,712-7   *CAS [646-24-2]*   C$_9$H$_{22}$N$_2$
FW 158.29
mp 38°C
bp 259°C (756 mm)
Fp >230°F

60 MHz: *1*, 251D
FT-IR: *1*, 293C
VP-FT-IR: *3*, 377D

42.29
33.92
29.60
29.46
26.89

1,9-Diaminononane, 98%

CDCl$_3$ QE-300

H$_2$NCH$_2$(CH$_2$)$_7$CH$_2$NH$_2$

B

Aldrich D1,420-4   *CAS [646-25-3]*   C$_{10}$H$_{24}$N$_2$
FW 172.32
mp 63°C
bp 140°C (12 mm)

60 MHz: *1*, 252B
FT-IR: *1*, 293D
VP-FT-IR: *3*, 378A

42.37
34.00
29.65
29.5
26.98

1,10-Diaminodecane, 97%

CDCl$_3$ QE-300

H$_2$NCH$_2$(CH$_2$)$_8$CH$_2$NH$_2$

C

Aldrich D1,640-1   *CAS [2783-17-7]*   C$_{12}$H$_{28}$N$_2$
FW 200.37
mp 70°C
Fp 311°F

60 MHz: *1*, 252C
FT-IR: *1*, 294A
VP-FT-IR: *3*, 378B

42.3
33.9
29.6
29.5
26.9

1,12-Diaminododecane, 98%

CDCl$_3$ QE-300

H$_2$NCH$_2$(CH$_2$)$_{10}$CH$_2$NH$_2$

ALDRICH

Non-Aromatic Amines

470

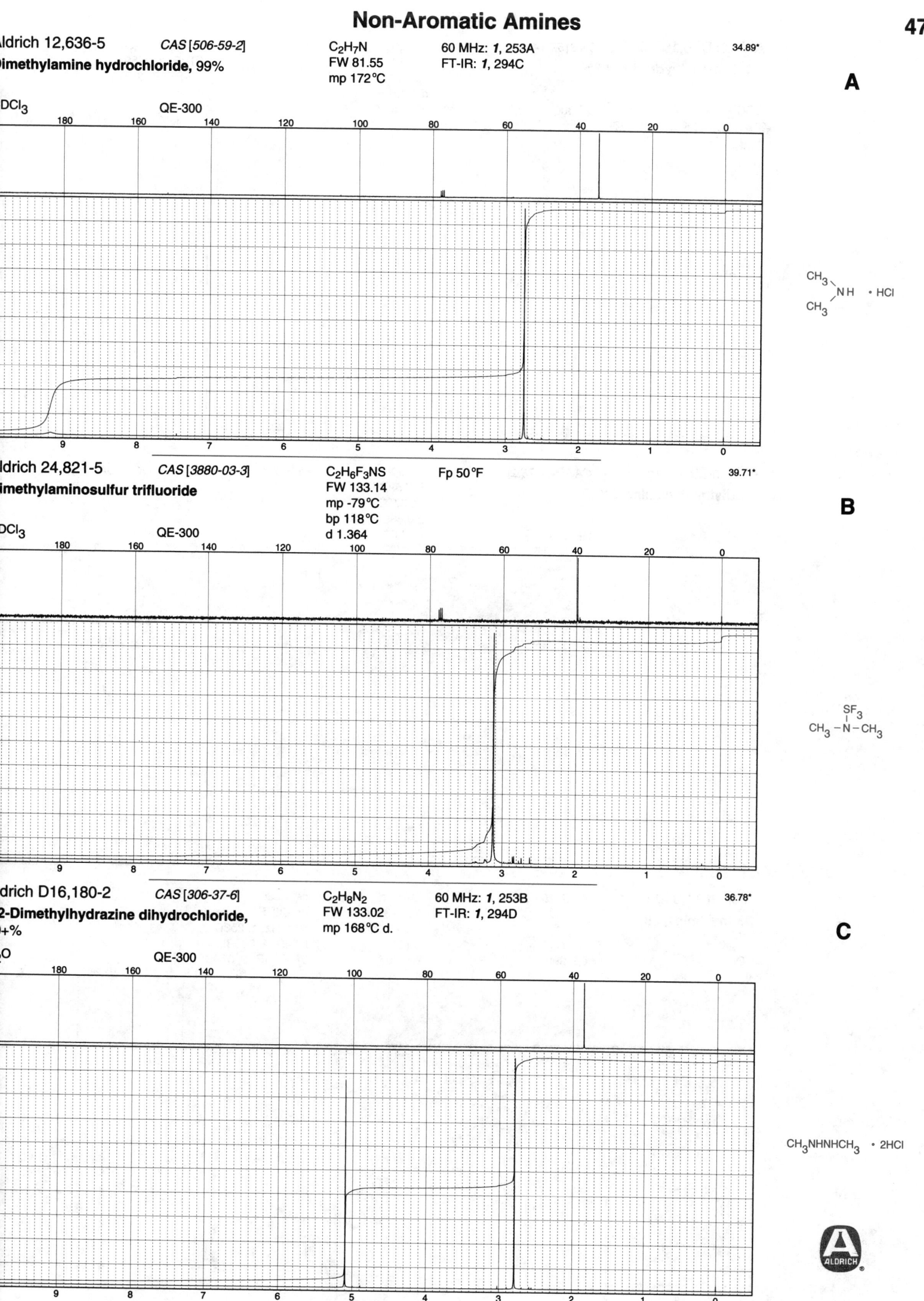

Aldrich 12,636-5   CAS [506-59-2]   C₂H₇N   60 MHz: 1, 253A   34.89*
Dimethylamine hydrochloride, 99%   FW 81.55   FT-IR: 1, 294C
mp 172°C

CDCl₃   QE-300

A

CH₃
      N H   • HCl
CH₃

Aldrich 24,821-5   CAS [3880-03-3]   C₂H₆F₃NS   Fp 50°F   39.71*
Dimethylaminosulfur trifluoride   FW 133.14
mp -79°C
bp 118°C
CDCl₃   QE-300   d 1.364

B

SF₃
CH₃ — N — CH₃

Aldrich D16,180-2   CAS [306-37-6]   C₂H₈N₂   60 MHz: 1, 253B   36.78*
2-Dimethylhydrazine dihydrochloride,   FW 133.02   FT-IR: 1, 294D
99+%   mp 168°C d.

D₂O   QE-300

C

CH₃NHNHCH₃   • 2HCl

ALDRICH

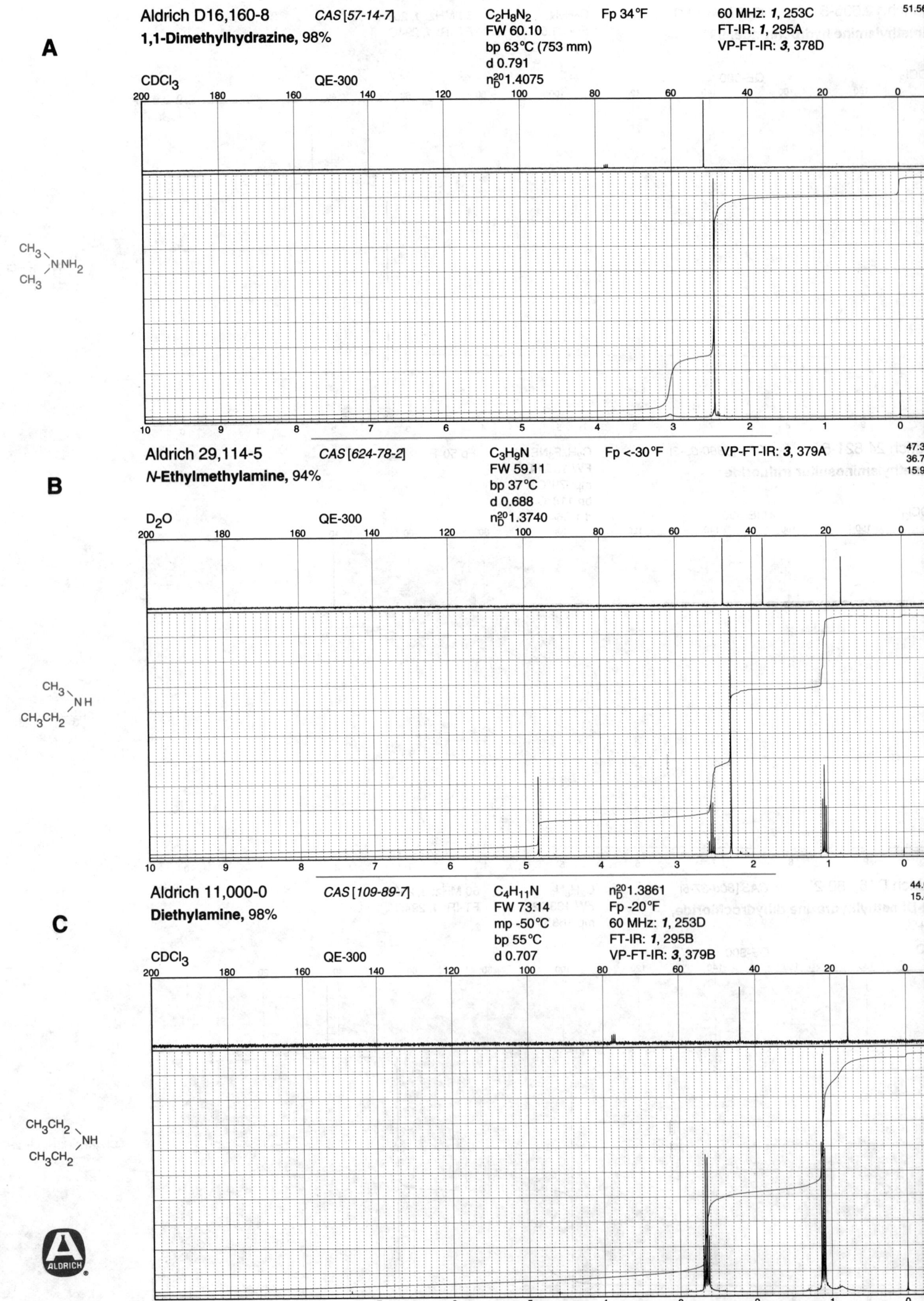

A

Aldrich D16,160-8   CAS [57-14-7]   C₂H₈N₂   Fp 34°F   60 MHz: *1*, 253C   51.56

1,1-Dimethylhydrazine, 98%   FW 60.10   FT-IR: *1*, 295A
bp 63°C (753 mm)   VP-FT-IR: *3*, 378D
d 0.791
n_D^{20} 1.4075

CDCl₃   QE-300

B

Aldrich 29,114-5   CAS [624-78-2]   C₃H₉N   Fp <-30°F   VP-FT-IR: *3*, 379A   47.30
36.7
N-Ethylmethylamine, 94%   FW 59.11   15.9
bp 37°C
d 0.688
n_D^{20} 1.3740

D₂O   QE-300

C

Aldrich 11,000-0   CAS [109-89-7]   C₄H₁₁N n_D^{20} 1.3861   44.0
15.4
Diethylamine, 98%   FW 73.14   Fp -20°F
mp -50°C   60 MHz: *1*, 253D
bp 55°C   FT-IR: *1*, 295B
d 0.707   VP-FT-IR: *3*, 379B

CDCl₃   QE-300

Aldrich 12,774-4   CAS [660-68-4]   C₄H₁₁N   60 MHz: *1*, 254A   42.49

Diethylamine hydrochloride, 99%   FW 109.60   FT-IR: *1*, 295C   11.29*

mp 229°C

A

CDCl₃   QE-300

CH_3CH_2
$\quad\quad\quad NH \cdot HCl$
CH_3CH_2

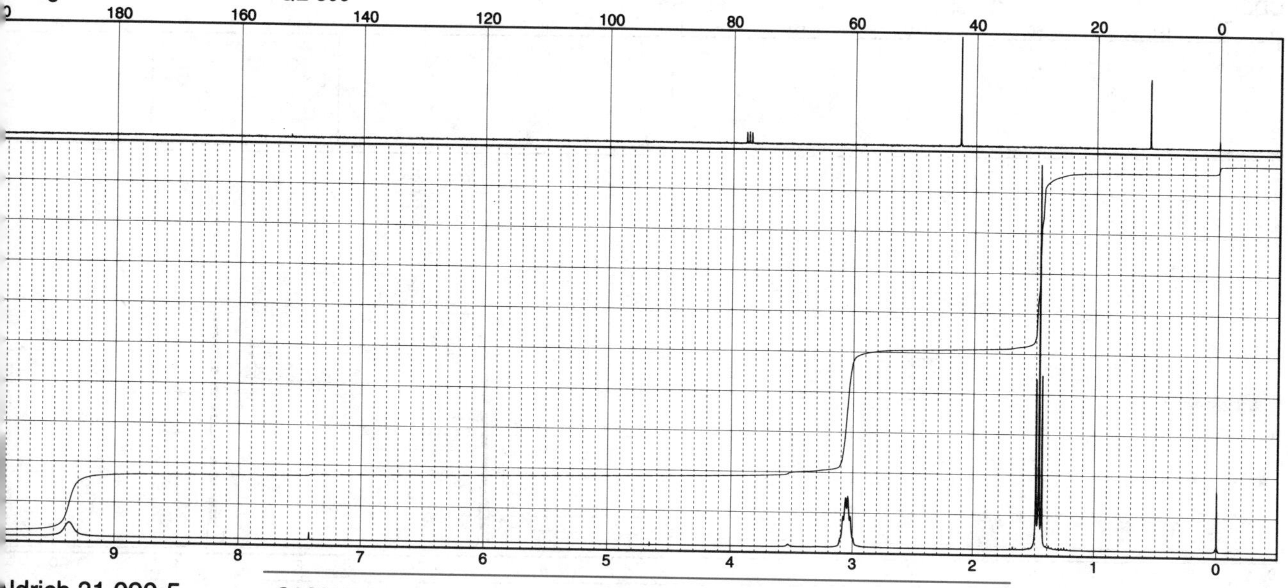

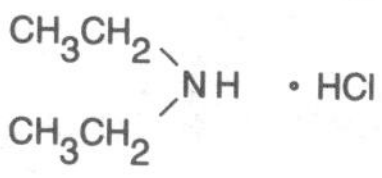

Aldrich 31,090-5   CAS [6274-12-0]   C₄H₁₁N   42.72

Diethylamine hydrobromide, 98%   FW 154.06   11.29*

mp 219°C

B

CDCl₃   QE-300

CH_3CH_2
$\quad\quad\quad NH \cdot HBr$
CH_3CH_2

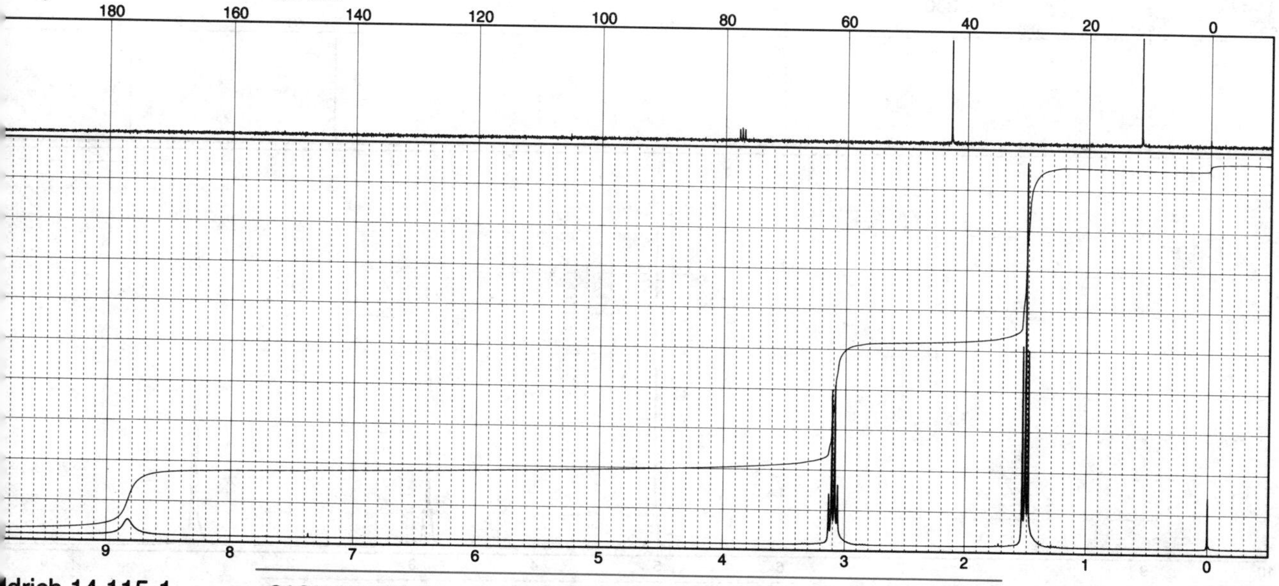

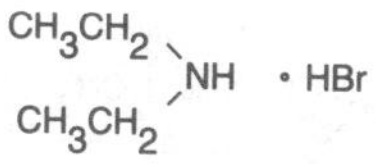

Aldrich 14,115-1   CAS [68109-72-8]   C₄H₁₁N   60 MHz: *1*, 254B   45.06

Diethylamine phosphate, 99%   FW 171.14   FT-IR: *1*, 295D   13.43*

mp 159°C

C

D₂O   QE-300

CH_3CH_2
$\quad\quad\quad NH \cdot H_3PO_4$
CH_3CH_2

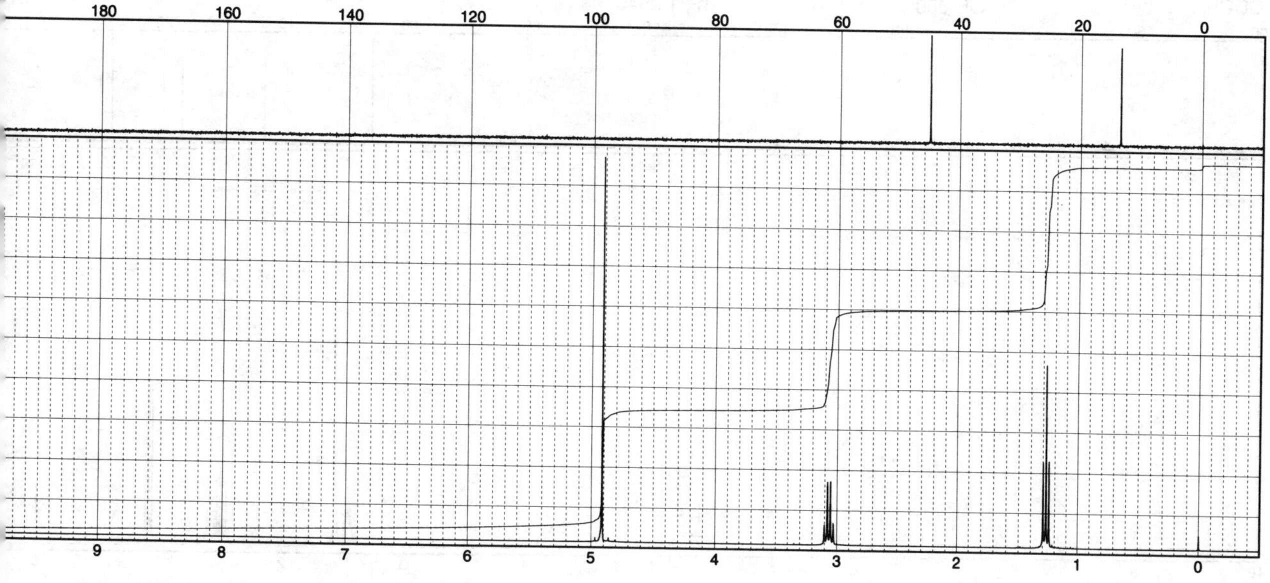

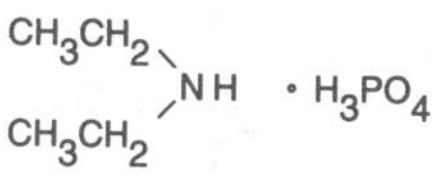

A

Aldrich 23,525-3   *CAS [38078-09-0]* $C_4H_{10}F_3NS$ 60 MHz: *1*, 254C   45.65
Diethylaminosulfur trifluoride   FW 161.19   FT-IR: *1*, 296A   15.18*
bp 31°C (3 mm)
d 1.220
Fp 75°F

CDCl$_3$ QE-300

$$CH_3CH_2-\underset{\underset{SF_3}{|}}{N}-CH_2CH_3$$

B

Aldrich 22,917-2   *CAS [7699-31-2]* $C_4H_{12}N_2$ FT-IR: *1*, 280A   46.00
1,2-Diethylhydrazine dihydrochloride   FW 161.08   12.65
mp 169°C d.

D$_2$O   QE-300

$$CH_3CH_2NHNHCH_2CH_3 \cdot 2HCl$$

C

Aldrich 30,813-7   *CAS [627-35-0]* $C_4H_{11}N$ Fp <-30°F   VP-FT-IR: *3*, 379C   54.0
N-Methylpropylamine, 96%   FW 73.14   36.4
bp 62°C   23.0
d 0.713   11.7
n_D^{20}1.3940

CDCl$_3$ QE-300

$$CH_3CH_2CH_2NHCH_3$$

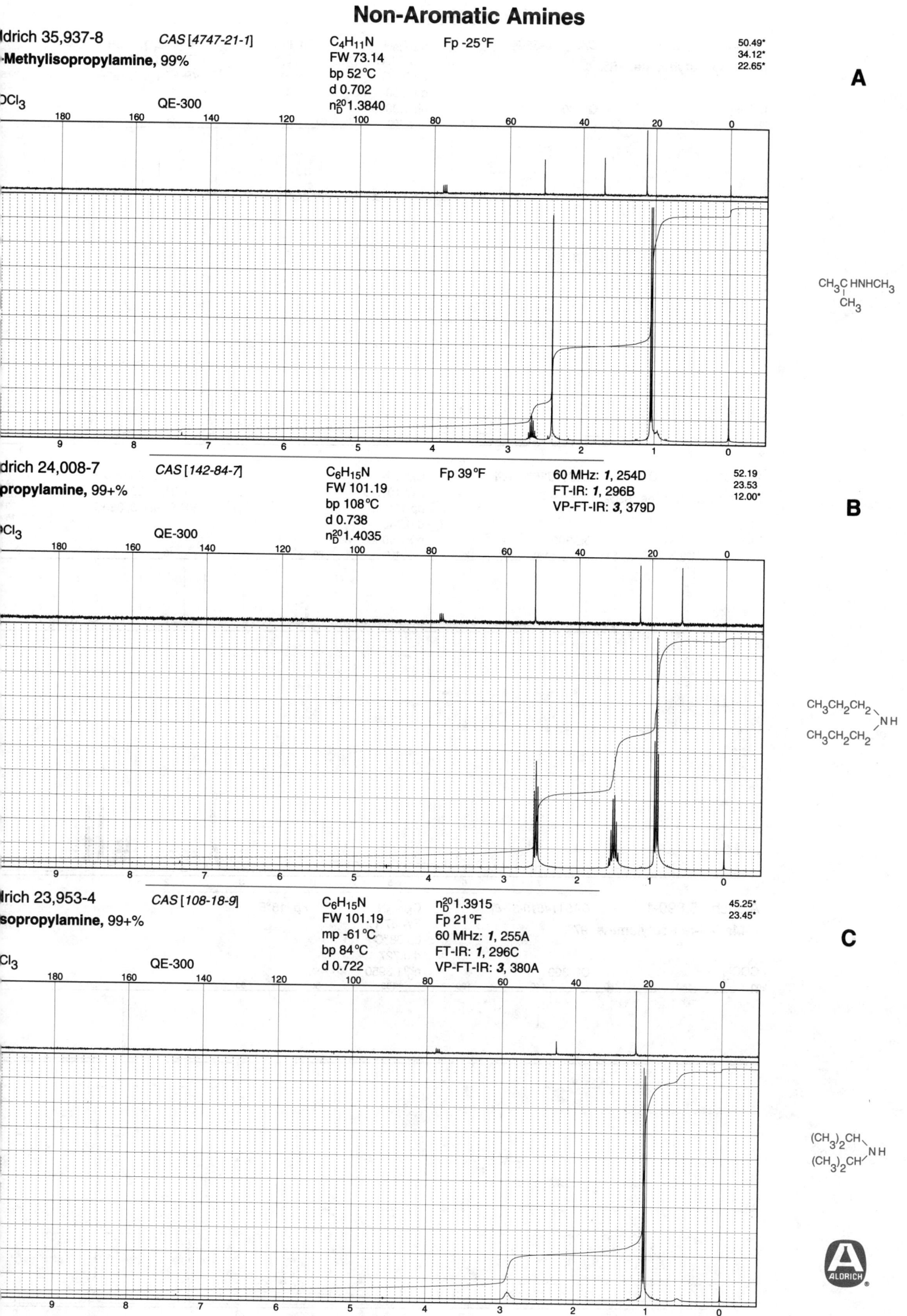

Aldrich 35,937-8   CAS [4747-21-1] $C_4H_{11}N$ Fp -25°F
-Methylisopropylamine, 99%
FW 73.14
bp 52°C
d 0.702
n_D^{20} 1.3840

$CH_3CHNHCH_3$
CH_3

50.49*
34.12*
22.65*

A

DCl₃   QE-300

180 160 140 120 100 80 60 40 20 0

9 8 7 6 5 4 3 2 1 0

Aldrich 24,008-7   CAS [142-84-7] $C_6H_{15}N$ Fp 39°F   60 MHz: **1**, 254D
propylamine, 99+%
FW 101.19
bp 108°C
d 0.738
n_D^{20} 1.4035

FT-IR: **1**, 296B
VP-FT-IR: **3**, 379D

52.19
23.53
12.00*

B

DCl₃   QE-300

180 160 140 120 100 80 60 40 20 0

$CH_3CH_2CH_2$
         NH
$CH_3CH_2CH_2$

9 8 7 6 5 4 3 2 1 0

Aldrich 23,953-4   CAS [108-18-9] $C_6H_{15}N$ n_D^{20} 1.3915
isopropylamine, 99+%
FW 101.19
mp -61°C
bp 84°C
d 0.722

Fp 21°F
60 MHz: **1**, 255A
FT-IR: **1**, 296C
VP-FT-IR: **3**, 380A

45.25*
23.45*

C

Cl₃   QE-300

180 160 140 120 100 80 60 40 20 0

$(CH_3)_2CH$
         NH
$(CH_3)_2CH$

9 8 7 6 5 4 3 2 1 0

ALDRICH

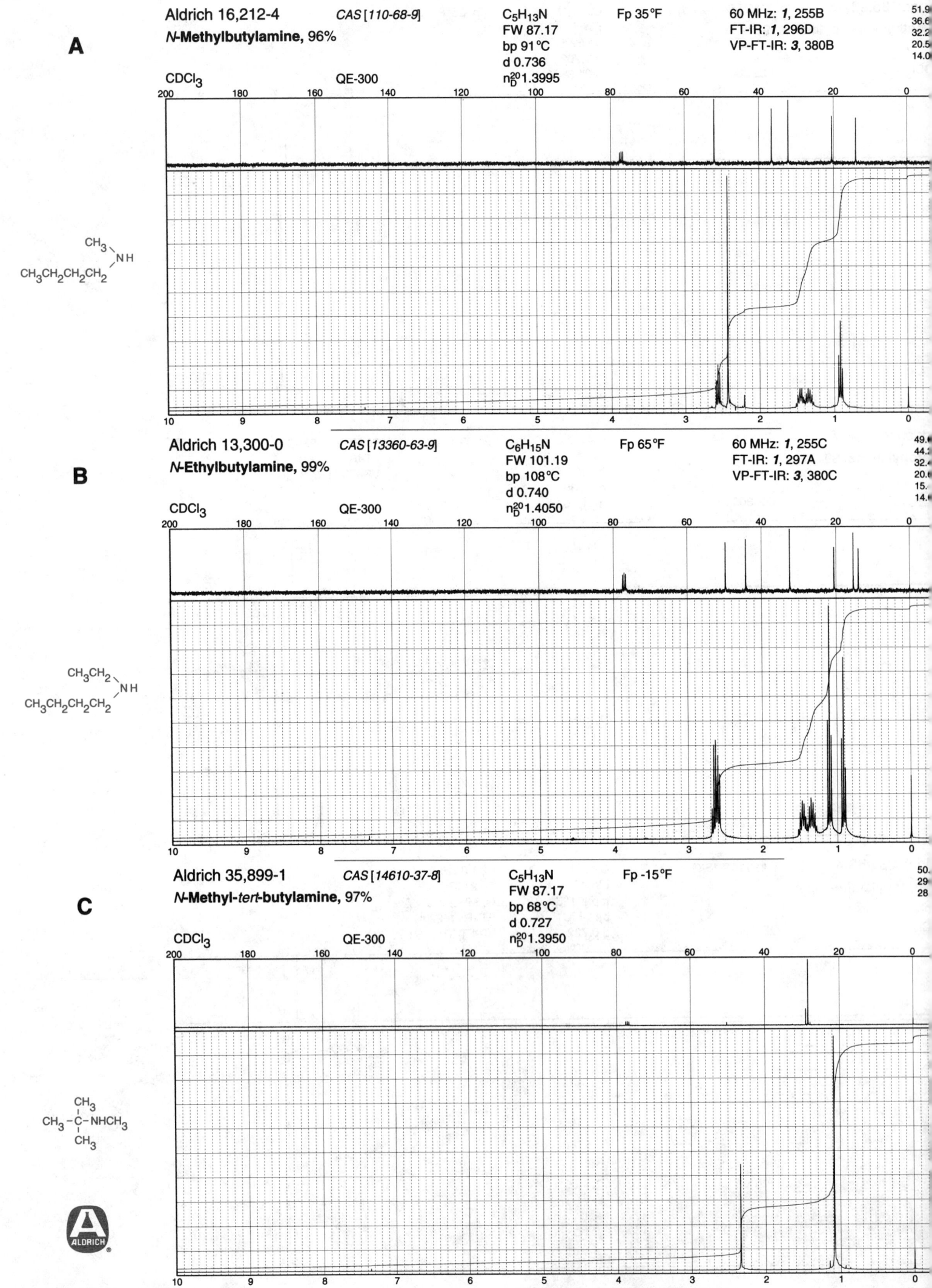

A

Aldrich 16,212-4

N-Methylbutylamine, 96%

CAS [110-68-9]

C₅H₁₃N
FW 87.17
bp 91 °C
d 0.736
n²⁰_D 1.3995

Fp 35 °F

60 MHz: *1*, 255B
FT-IR: *1*, 296D
VP-FT-IR: *3*, 380B

51.9
36.6
32.2
20.5
14.0

CDCl₃   QE-300

CH₃
    NH
CH₃CH₂CH₂CH₂

B

Aldrich 13,300-0

N-Ethylbutylamine, 99%

CAS [13360-63-9]

C₆H₁₅N
FW 101.19
bp 108 °C
d 0.740
n²⁰_D 1.4050

Fp 65 °F

60 MHz: *1*, 255C
FT-IR: *1*, 297A
VP-FT-IR: *3*, 380C

49.
44.
32.4
20.
15.
14.

CDCl₃   QE-300

CH₃CH₂
       NH
CH₃CH₂CH₂CH₂

C

Aldrich 35,899-1

N-Methyl-*tert*-butylamine, 97%

CAS [14610-37-8]

C₅H₁₃N
FW 87.17
bp 68 °C
d 0.727
n²⁰_D 1.3950

Fp -15 °F

50.
29
28

CDCl₃   QE-300

CH₃
    |
CH₃ – C – NHCH₃
    |
CH₃

Aldrich 37,414-8 CAS [7515-80-2] C$_7$H$_{17}$N Fp 26°F

N-tert-Butylisopropylamine, 99%

FW 115.22
bp 98°C
d 0.727
n$_D^{20}$1.3980

A

50.87
42.89*
30.01*
26.37*

CDCl$_3$ QE-300

CH$_3$—C NH CHCH$_3$ with CH$_3$ groups

Aldrich 24,000-1 CAS [111-92-2] C$_8$H$_{19}$N n$_D^{20}$1.4170

Dibutylamine, 99+%

FW 129.25
mp -62°C
bp 159°C
d 0.767

Fp 106°F
60 MHz: 1, 255D
FT-IR: 1, 225B
VP-FT-IR: 3, 380D

B

49.91
32.48
20.60
14.04*

CDCl$_3$ QE-300

CH$_3$CH$_2$CH$_2$CH$_2$ \
                                   NH
CH$_3$CH$_2$CH$_2$CH$_2$ /

Aldrich 30,735-1 CAS [4444-67-1] C$_8$H$_{19}$N Fp 69°F VP-FT-IR: 3, 381A

N-sec-butylamine, 99%, mixture of (±) and meso

FW 129.25
bp 135°C
d 0.753
n$_D^{20}$1.4100

C

51.33*
30.38
29.76
20.76*
20.33*
10.42*
10.19*

CDCl$_3$ QE-300

CH$_3$ CH$_3$
CH$_3$CH$_2$ C H NH CHCH$_2$CH$_3$

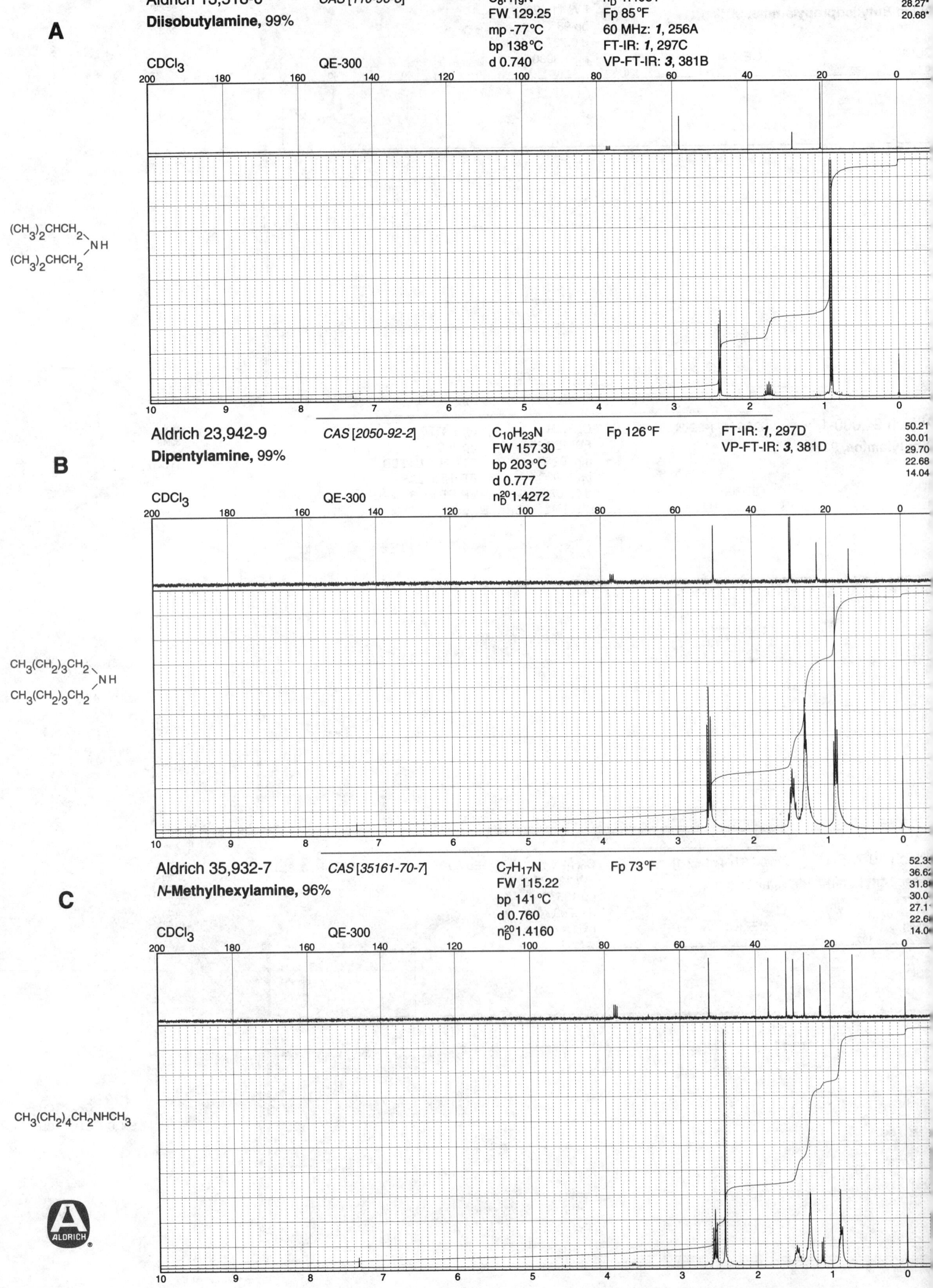

A

Aldrich 13,518-6   *CAS [110-96-3]* $C_8H_{19}N$ $n_D^{20}1.4081$ 58.22
Diisobutylamine, 99%   FW 129.25   Fp 85°F   28.27*
mp -77°C   60 MHz: *1*, 256A   20.68*
bp 138°C   FT-IR: *1*, 297C
CDCl₃   QE-300   d 0.740   VP-FT-IR: *3*, 381B

$(CH_3)_2CHCH_2$
$\qquad$ NH
$(CH_3)_2CHCH_2$

B

Aldrich 23,942-9   *CAS [2050-92-2]* $C_{10}H_{23}N$ Fp 126°F   FT-IR: *1*, 297D   50.21
Dipentylamine, 99%   FW 157.30   VP-FT-IR: *3*, 381D   30.01
bp 203°C   29.70
d 0.777   22.68
CDCl₃   QE-300 $n_D^{20}1.4272$ 14.04

$CH_3(CH_2)_3CH_2$
$\qquad$ NH
$CH_3(CH_2)_3CH_2$

C

Aldrich 35,932-7   *CAS [35161-70-7]* $C_7H_{17}N$ Fp 73°F   52.3
***N*-Methylhexylamine, 96%**   FW 115.22   36.6
bp 141°C   31.8
d 0.760   30.0
CDCl₃   QE-300 $n_D^{20}1.4160$ 27.1
22.6
14.0

$CH_3(CH_2)_4CH_2NHCH_3$

Aldrich 13,120-2   CAS [143-16-8]   C$_{12}$H$_{27}$N   Fp 203°F   60 MHz: **1**, 256C

Dihexylamine, 97%

FW 185.36   FT-IR: **1**, 298A

bp 194°C   VP-FT-IR: **3**, 382A

CDCl$_3$ QE-300   d 0.795

n$_D^{20}$1.4320

50.25
31.88
30.30
27.17
22.67
14.04*

A

$$CH_3(CH_2)_4CH_2 \diagdown$$
$$\quad\quad\quad\quad NH$$
$$CH_3(CH_2)_4CH_2 \diagup$$

Aldrich D20,114-6   CAS [1120-48-5]   C$_{16}$H$_{35}$N   n$_D^{20}$1.4432

Dioctylamine, 98%

FW 241.46   Fp >230°F

mp 15°C   60 MHz: **1**, 256D

bp 298°C   FT-IR: **1**, 298B

CDCl$_3$ QE-300   d 0.799   VP-FT-IR: **3**, 382C

51.03   30.10
32.67   28.28
31.12   23.48
30.39   14.86*

B

$$CH_3(CH_2)_6CH_2 \diagdown$$
$$\quad\quad\quad\quad NH$$
$$CH_3(CH_2)_6CH_2 \diagup$$

Aldrich 29,285-0   CAS [106-20-7]   C$_{16}$H$_{35}$N   Fp >230°F   VP-FT-IR: **3**, 382B

Bis(2-ethylhexyl)amine, 99%

FW 241.46

bp 123°C (5 mm)

d 0.805

CDCl$_3$ QE-300   n$_D^{20}$1.4425

53.61   24.62
39.38* 23.18
31.49   14.13*
29.05   10.91*

C

$$CH_2CH_3$$
$$CH_3(CH_2)_3 CHCH_2 \diagdown$$
$$\quad\quad\quad\quad\quad\quad NH$$
$$CH_3(CH_2)_3 CHCH_2 \diagup$$
$$\quad\quad\quad CH_2CH_3$$

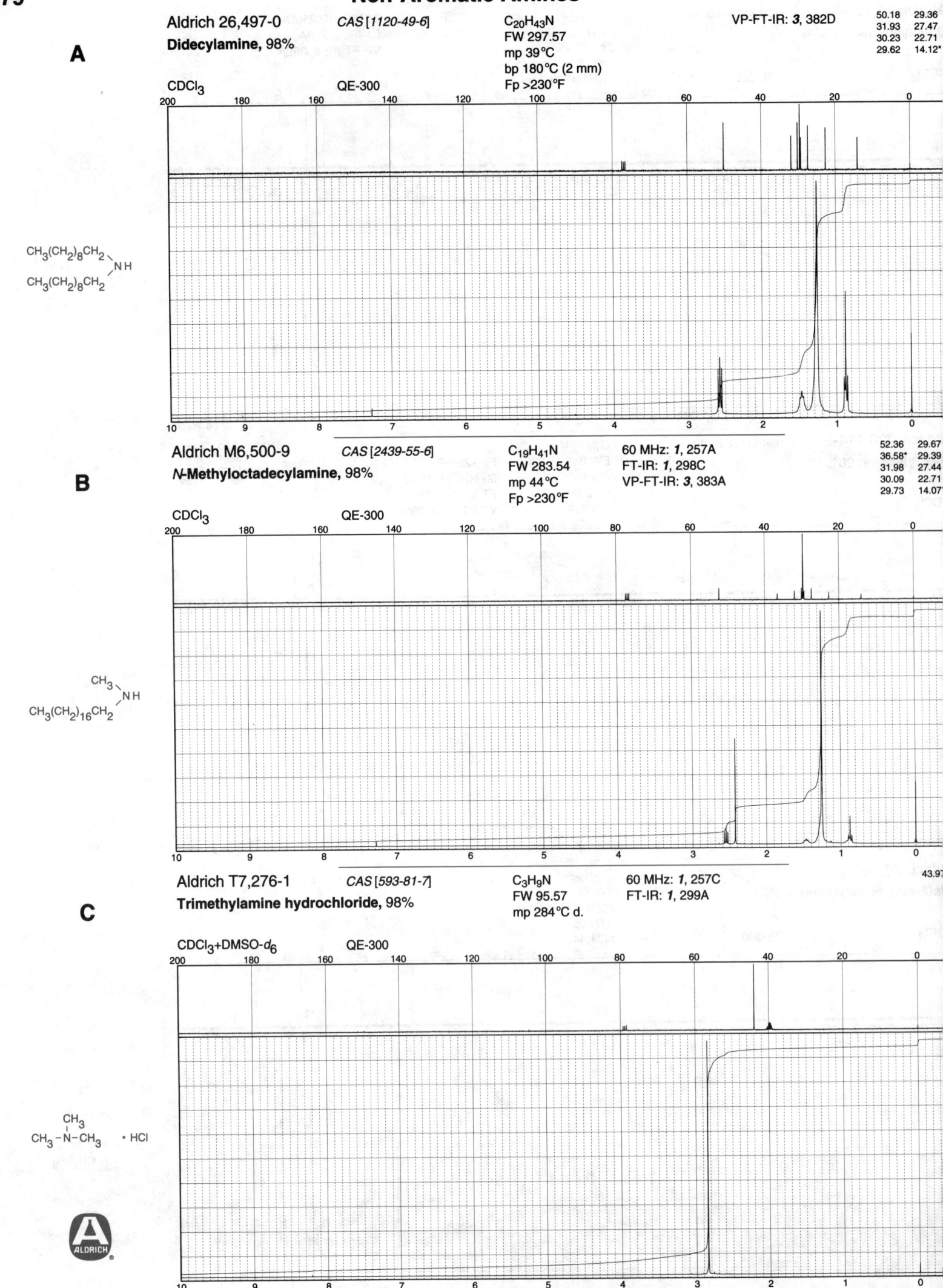

Aldrich 26,497-0 — CAS [1120-49-6] — C$_{20}$H$_{43}$N — FW 297.57 — mp 39°C — bp 180°C (2 mm) — Fp >230°F — VP-FT-IR: **3**, 382D

Didecylamine, 98%

50.18 29.36
31.93 27.47
30.23 22.71
29.62 14.12*

CDCl$_3$ QE-300

CH$_3$(CH$_2$)$_8$CH$_2$
CH$_3$(CH$_2$)$_8$CH$_2$ —NH

Aldrich M6,500-9 — CAS [2439-55-6] — C$_{19}$H$_{41}$N — FW 283.54 — mp 44°C — Fp >230°F — 60 MHz: **1**, 257A — FT-IR: **1**, 298C — VP-FT-IR: **3**, 383A

N-Methyloctadecylamine, 98%

52.36 29.67
36.58* 29.39
31.98 27.44
30.09 22.71
29.73 14.07

CDCl$_3$ QE-300

CH$_3$
CH$_3$(CH$_2$)$_{16}$CH$_2$ —NH

Aldrich T7,276-1 — CAS [593-81-7] — C$_3$H$_9$N — FW 95.57 — mp 284°C d. — 60 MHz: **1**, 257C — FT-IR: **1**, 299A

Trimethylamine hydrochloride, 98%

43.97

CDCl$_3$+DMSO-d_6 QE-300

CH$_3$
CH$_3$—N—CH$_3$ • HCl

A

Aldrich 13,587-9   CAS [3162-58-1] $C_3H_9NO_3S$ 60 MHz: *2*, 803D   48.24*
Sulfur trioxide trimethylamine complex   FW 139.17   FT-IR: *1*, 903B   44.22*
mp 232°C d.

DMSO-d_6 QE-300

$$CH_3-\underset{\underset{CH_3}{|}}{\overset{\overset{CH_3}{|}}{N}}-CH_3 \quad \cdot SO_3$$

B

Aldrich 17,686-9   CAS [62637-93-8] C_3H_9NO 60 MHz: *1*, 257D   62.31*
Trimethylamine *N*-oxide dihydrate, 98%   FW 111.14   FT-IR: *1*, 299B
mp 99°C

D_2O QE-300

$$CH_3-\underset{\underset{CH_3}{|}}{\overset{\overset{CH_3}{|}}{N}}\rightarrow O \quad \cdot 2H_2O$$

C

Aldrich 23,935-6   CAS [598-56-1] $C_4H_{11}N$ n_D^{20} 1.3720   53.56
***N,N*-Dimethylethylamine, 99%**   FW 73.14   Fp -33°F   45.05*
mp -140°C   60 MHz: *1*, 258A   12.89*
bp 37°C   FT-IR: *1*, 299C
d 0.675   VP-FT-IR: *3*, 383C

CDCl$_3$ QE-300

$$CH_3CH_2-\underset{\underset{CH_3}{|}}{\overset{\overset{CH_3}{|}}{N}}-CH_3$$

ALDRICH

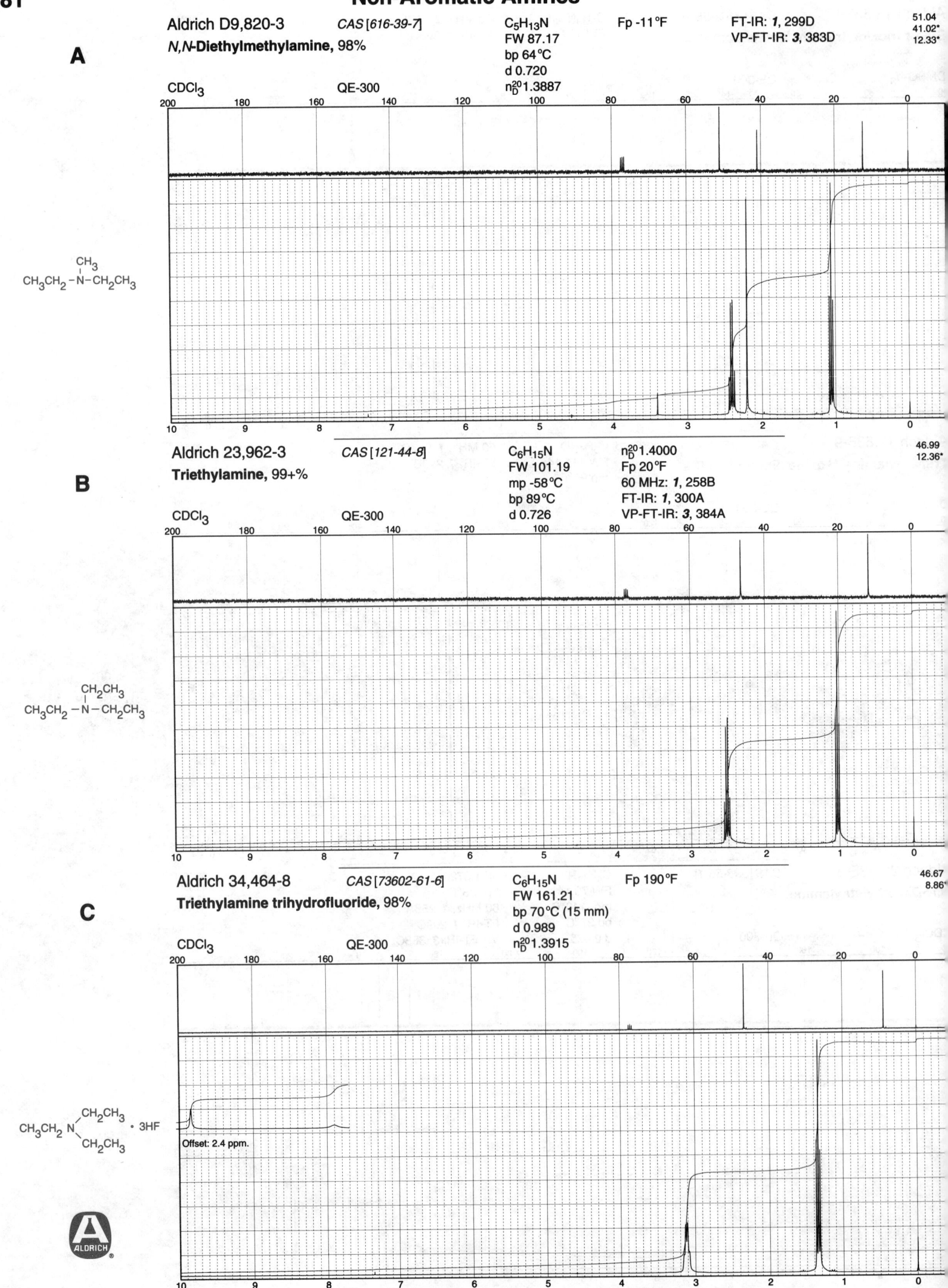

A

Aldrich D9,820-3   *CAS [616-39-7]*   C₅H₁₃N   Fp -11°F   FT-IR: *1*, 299D   51.04
N,N-Diethylmethylamine, 98%   FW 87.17   VP-FT-IR: *3*, 383D   41.02*
bp 64°C   12.33*
d 0.720
n²⁰_D 1.3887

CDCl₃   QE-300

CH₃CH₂–N–CH₂CH₃
       |
       CH₃

B

Aldrich 23,962-3   *CAS [121-44-8]*   C₆H₁₅N   n²⁰_D 1.4000   46.99
Triethylamine, 99+%   FW 101.19   Fp 20°F   12.36*
mp -58°C   60 MHz: *1*, 258B
bp 89°C   FT-IR: *1*, 300A
d 0.726   VP-FT-IR: *3*, 384A

CDCl₃   QE-300

          CH₂CH₃
          |
CH₃CH₂ – N – CH₂CH₃

C

Aldrich 34,464-8   *CAS [73602-61-6]*   C₆H₁₅N   Fp 190°F   46.67
Triethylamine trihydrofluoride, 98%   FW 161.21   8.86*
bp 70°C (15 mm)
d 0.989
n²⁰_D 1.3915

CDCl₃   QE-300

                CH₂CH₃
                |
CH₃CH₂ N   · 3HF
                |
                CH₂CH₃

Offset: 2.4 ppm.

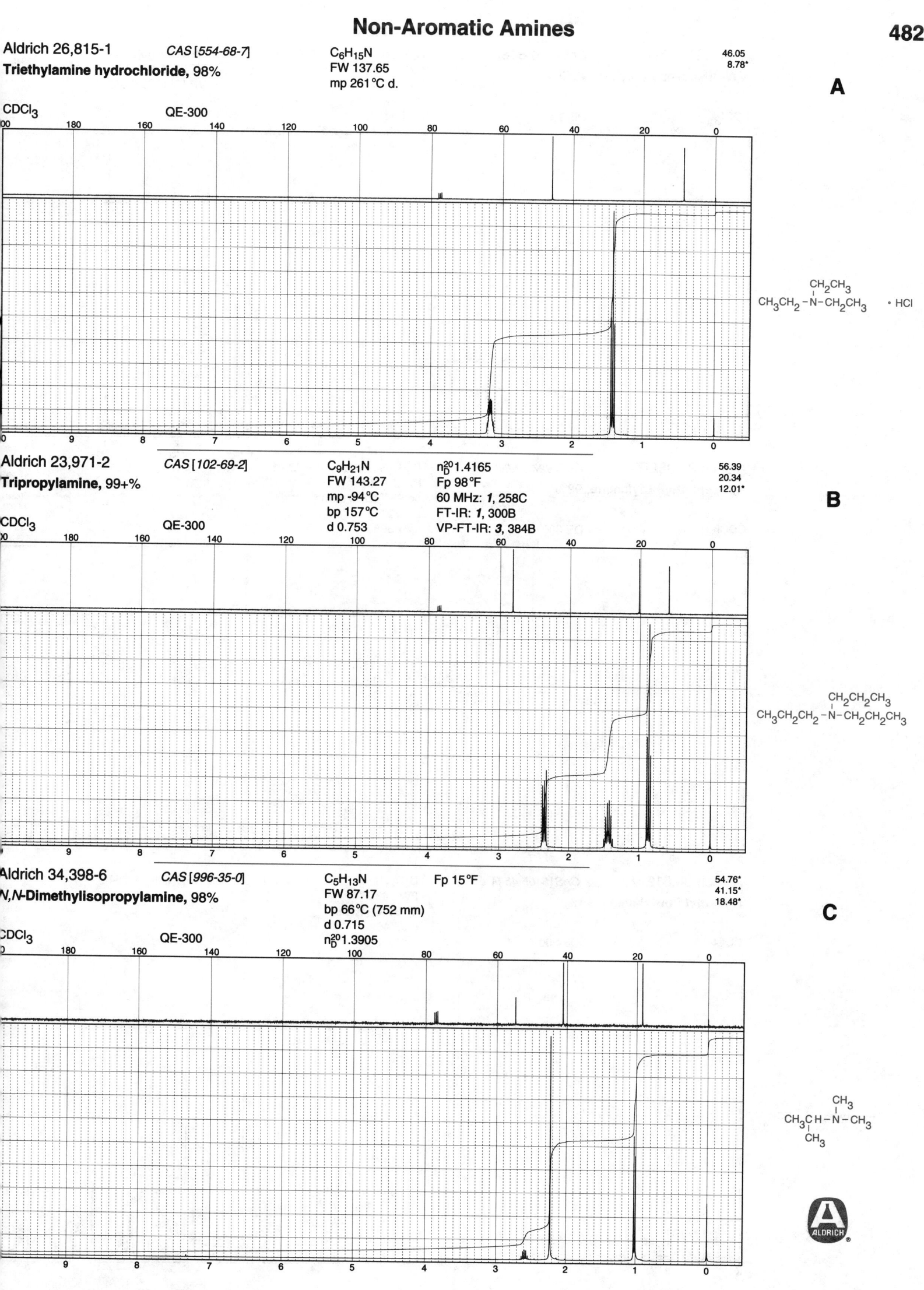

Aldrich 26,815-1 *CAS [554-68-7]*
Triethylamine hydrochloride, 98%

C₆H₁₅N
FW 137.65
mp 261 °C d.

46.05
8.78*

A

CDCl₃ QE-300

CH₃CH₂–N–CH₂CH₃ • HCl
with CH₂CH₃

Aldrich 23,971-2 *CAS [102-69-2]*
Tripropylamine, 99+%

C₉H₂₁N
FW 143.27
mp -94 °C
bp 157 °C
d 0.753

n$_D^{20}$1.4165
Fp 98 °F
60 MHz: *1*, 258C
FT-IR: *1*, 300B
VP-FT-IR: *3*, 384B

56.39
20.34
12.01*

B

CDCl₃ QE-300

CH₃CH₂CH₂–N–CH₂CH₂CH₃
with CH₂CH₂CH₃

Aldrich 34,398-6 *CAS [996-35-0]*
N,N-Dimethylisopropylamine, 98%

C₅H₁₃N
FW 87.17
bp 66 °C (752 mm)
d 0.715
n$_D^{20}$1.3905

Fp 15 °F

54.76*
41.15*
18.48*

C

CDCl₃ QE-300

CH₃CH–N–CH₃
with CH₃ and CH₃

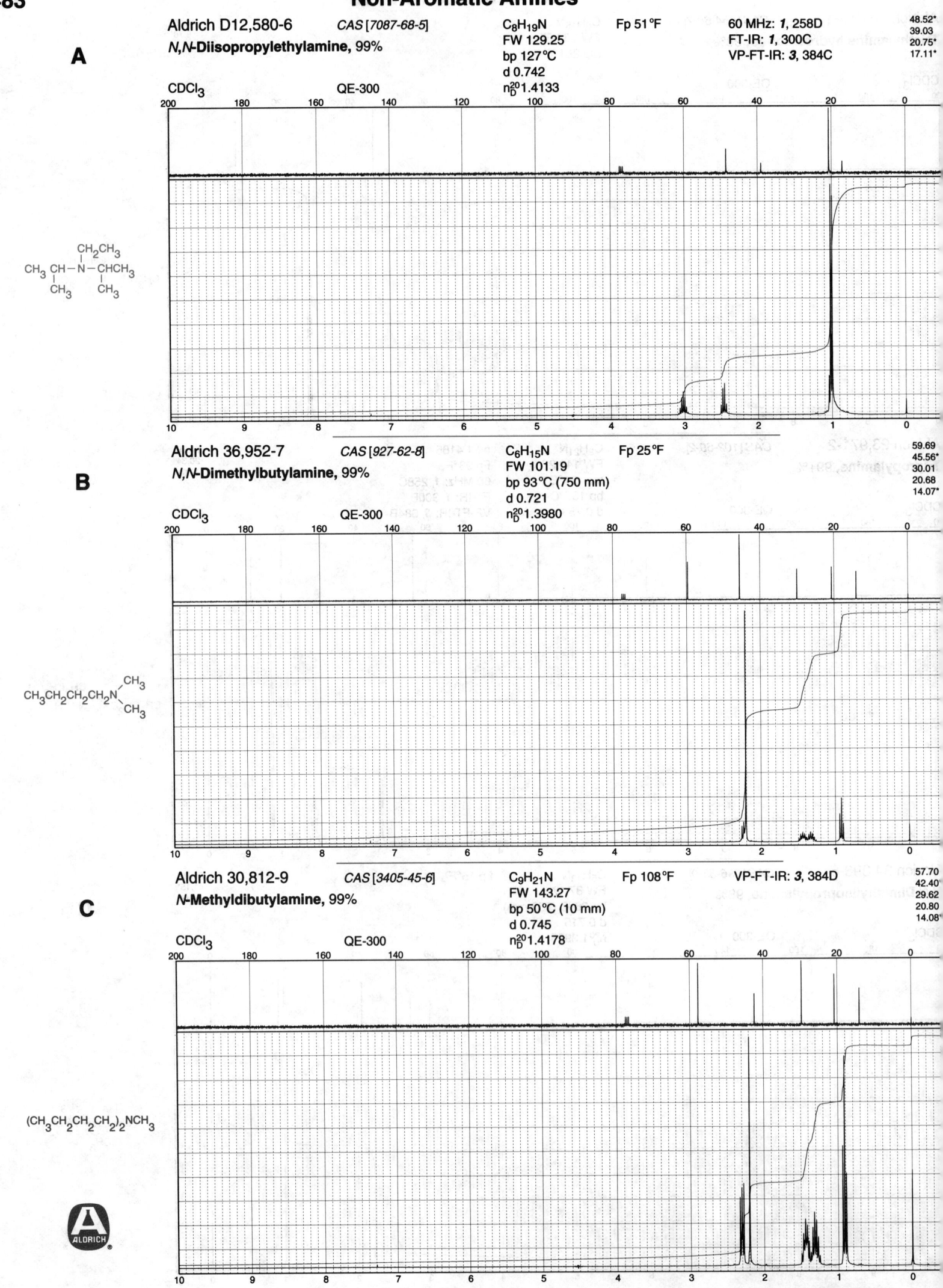

A

Aldrich D12,580-6   CAS [7087-68-5]   C₈H₁₉N   Fp 51°F   60 MHz: 1, 258D

$C_8H_{19}N$

N,N-Diisopropylethylamine, 99%   FW 129.25   FT-IR: 1, 300C

bp 127°C   VP-FT-IR: 3, 384C

d 0.742

$n_D^{20} 1.4133$

CDCl₃   QE-300

48.52*
39.03
20.75*
17.11*

B

Aldrich 36,952-7   CAS [927-62-8]   C₆H₁₅N   Fp 25°F

$C_6H_{15}N$

N,N-Dimethylbutylamine, 99%   FW 101.19

bp 93°C (750 mm)

d 0.721

$n_D^{20} 1.3980$

CDCl₃   QE-300

59.69
45.56*
30.01
20.68
14.07*

C

Aldrich 30,812-9   CAS [3405-45-6]   C₉H₂₁N   Fp 108°F   VP-FT-IR: 3, 384D

$C_9H_{21}N$

N-Methyldibutylamine, 99%   FW 143.27

bp 50°C (10 mm)

d 0.745

$n_D^{20} 1.4178$

CDCl₃   QE-300

57.70
42.40
29.62
20.80
14.08*

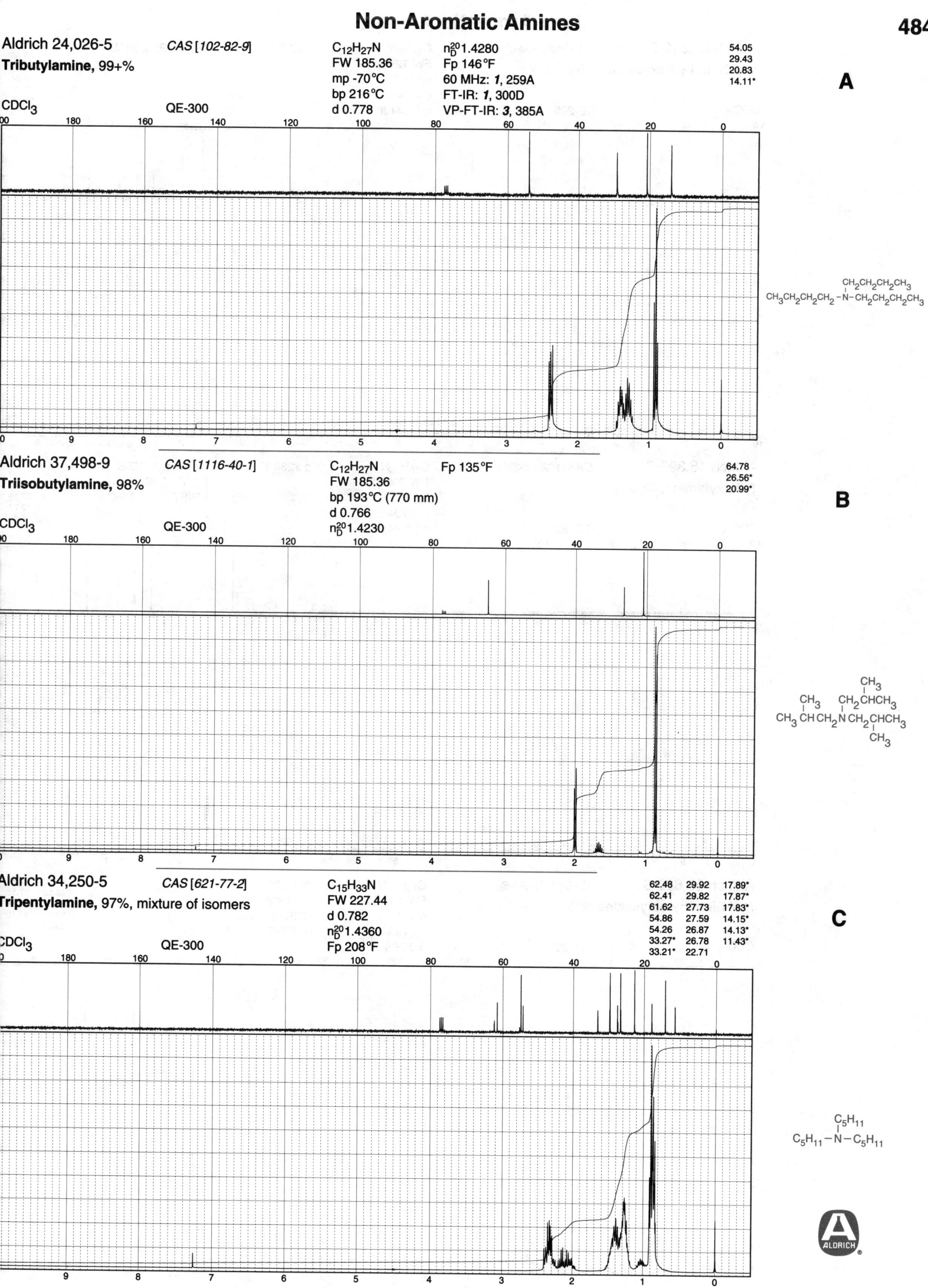

Aldrich 24,026-5   *CAS [102-82-9]*
Tributylamine, 99+%

$C_{12}H_{27}N$
FW 185.36
mp -70°C
bp 216°C
d 0.778

n_D^{20} 1.4280
Fp 146°F
60 MHz: *1*, 259A
FT-IR: *1*, 300D
VP-FT-IR: *3*, 385A

54.05
29.43
20.83
14.11*

A

CDCl₃   QE-300

$CH_3CH_2CH_2CH_2-N-CH_2CH_2CH_2CH_3$
with $CH_2CH_2CH_2CH_3$

Aldrich 37,498-9   *CAS [1116-40-1]*
Triisobutylamine, 98%

$C_{12}H_{27}N$
FW 185.36
bp 193°C (770 mm)
d 0.766
n_D^{20} 1.4230

Fp 135°F

64.78
26.56*
20.99*

B

CDCl₃   QE-300

Aldrich 34,250-5   *CAS [621-77-2]*
Tripentylamine, 97%, mixture of isomers

$C_{15}H_{33}N$
FW 227.44
d 0.782
n_D^{20} 1.4360
Fp 208°F

62.48	29.92	17.89*
62.41	29.82	17.87*
61.62	27.73	17.83*
54.86	27.59	14.15*
54.26	26.87	14.13*
33.27*	26.78	11.43*
33.21*	22.71	

C

CDCl₃   QE-300

$C_5H_{11}-N-C_5H_{11}$
with C_5H_{11}

A

Aldrich 30,810-2   CAS [4385-04-0] $C_8H_{19}N$ Fp 92°F   VP-FT-IR: 3, 385B

N,N-Dimethylhexylamine, 99%   FW 129.25

bp 148°C

d 0.744

$CDCl_3$ QE-300 n_D^{20} 1.4136

60.03
45.55*
31.91
27.84
27.25
22.68
14.08*

$CH_3(CH_2)_4CH_2$—N(CH_3)(CH_3)

B

Aldrich 18,399-7   CAS [102-86-3] $C_{18}H_{39}N$ Fp >230°F   60 MHz: 1, 259B

Trihexylamine, 96%   FW 269.52   FT-IR: 1, 301A

bp 264°C   VP-FT-IR: 3, 385C

d 0.794

$CDCl_3$ QE-300 n_D^{20} 1.4415

54.39
31.93
27.40
27.19
22.72
14.06*

$CH_3(CH_2)_4CH_2$—N—$CH_2(CH_2)_4CH_3$ with $CH_2(CH_2)_4CH_3$

C

Aldrich 25,622-6   CAS [7378-99-6] $C_{10}H_{23}N$ n_D^{20} 1.4243

N,N-Dimethyloctylamine, 95%   FW 157.30   Fp 149°F

mp -57°C   FT-IR: 1, 301C

bp 195°C   VP-FT-IR: 3, 385D

d 0.765

$CDCl_3$ QE-300

60.02   27.87
45.55*   27.56
31.88   22.69
29.62   14.10*
29.31

$CH_3(CH_2)_6CH_2$—N(CH_3)(CH_3)

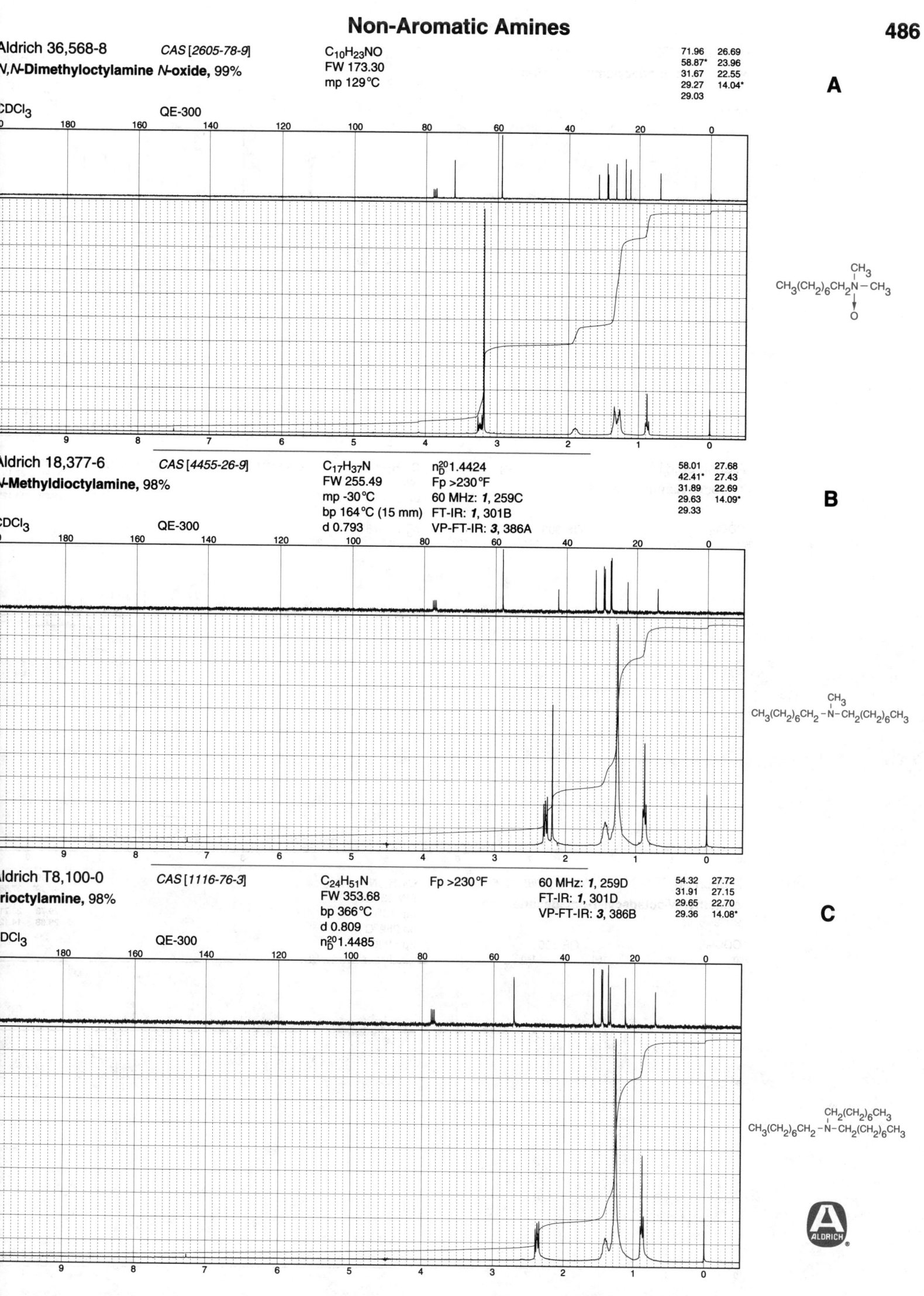

Aldrich 36,568-8   CAS [2605-78-9]
N,N-Dimethyloctylamine N-oxide, 99%

C₁₀H₂₃NO
FW 173.30
mp 129°C

71.96	26.69
58.87*	23.96
31.67	22.55
29.27	14.04*
29.03	

A

CDCl₃   QE-300

Aldrich 18,377-6   CAS [4455-26-9]
N-Methyldioctylamine, 98%

C₁₇H₃₇N
FW 255.49
mp -30°C
bp 164°C (15 mm)
d 0.793

n²⁰_D 1.4424
Fp >230°F
60 MHz: 1, 259C
FT-IR: 1, 301B
VP-FT-IR: 3, 386A

58.01	27.68
42.41*	27.43
31.89	22.69
29.63	14.09*
29.33	

B

CDCl₃   QE-300

Aldrich T8,100-0   CAS [1116-76-3]
Trioctylamine, 98%

C₂₄H₅₁N
FW 353.68
bp 366°C
d 0.809
n²⁰_D 1.4485

Fp >230°F

60 MHz: 1, 259D
FT-IR: 1, 301D
VP-FT-IR: 3, 386B

54.32	27.72
31.91	27.15
29.65	22.70
29.36	14.08*

C

CDCl₃   QE-300

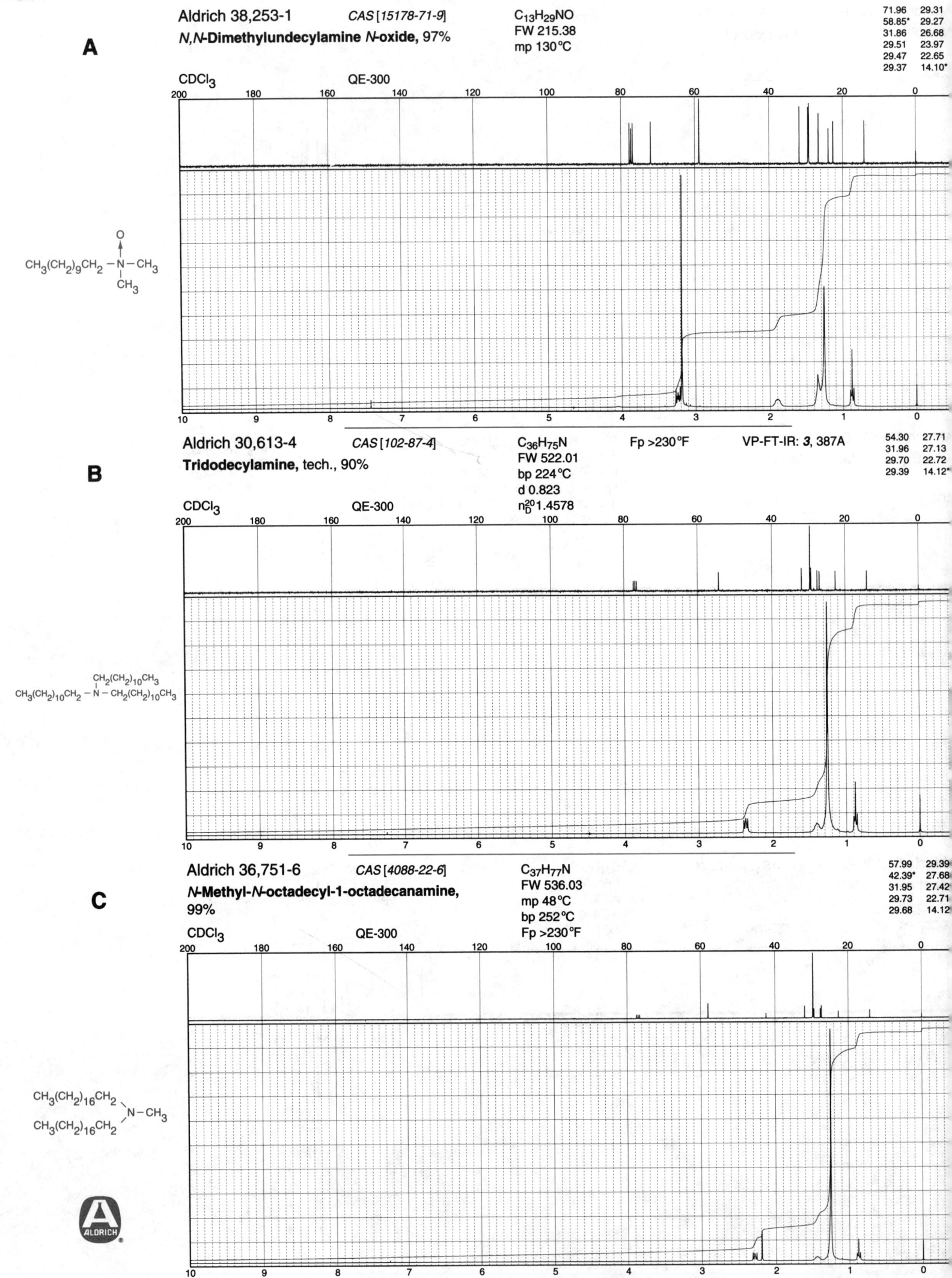

A

Aldrich 38,253-1   CAS [15178-71-9]   C₁₃H₂₉NO
N,N-Dimethylundecylamine *N*-oxide, 97%

$C_{13}H_{29}NO$
FW 215.38
mp 130°C

71.96	29.31
58.85*	29.27
31.86	26.68
29.51	23.97
29.47	22.65
29.37	14.10*

CDCl₃   QE-300

CH₃(CH₂)₉CH₂—N—CH₃ (with O above N, CH₃ below N)

B

Aldrich 30,613-4   CAS [102-87-4]   C₃₆H₇₅N   Fp >230°F   VP-FT-IR: *3*, 387A
Tridodecylamine, tech., 90%

$C_{36}H_{75}N$
FW 522.01
bp 224°C
d 0.823
n_D^{20} 1.4578

54.30	27.71
31.96	27.13
29.70	22.72
29.39	14.12*

CDCl₃   QE-300

CH₃(CH₂)₁₀CH₂—N—CH₂(CH₂)₁₀CH₃ (with CH₂(CH₂)₁₀CH₃ above N)

C

Aldrich 36,751-6   CAS [4088-22-6]   C₃₇H₇₇N
N-Methyl-*N*-octadecyl-1-octadecanamine, 99%

$C_{37}H_{77}N$
FW 536.03
mp 48°C
bp 252°C
Fp >230°F

57.99	29.39
42.39*	27.68
31.95	27.42
29.73	22.71
29.68	14.12

CDCl₃   QE-300

CH₃(CH₂)₁₆CH₂—N—CH₃ (with CH₃(CH₂)₁₆CH₂ as second chain)

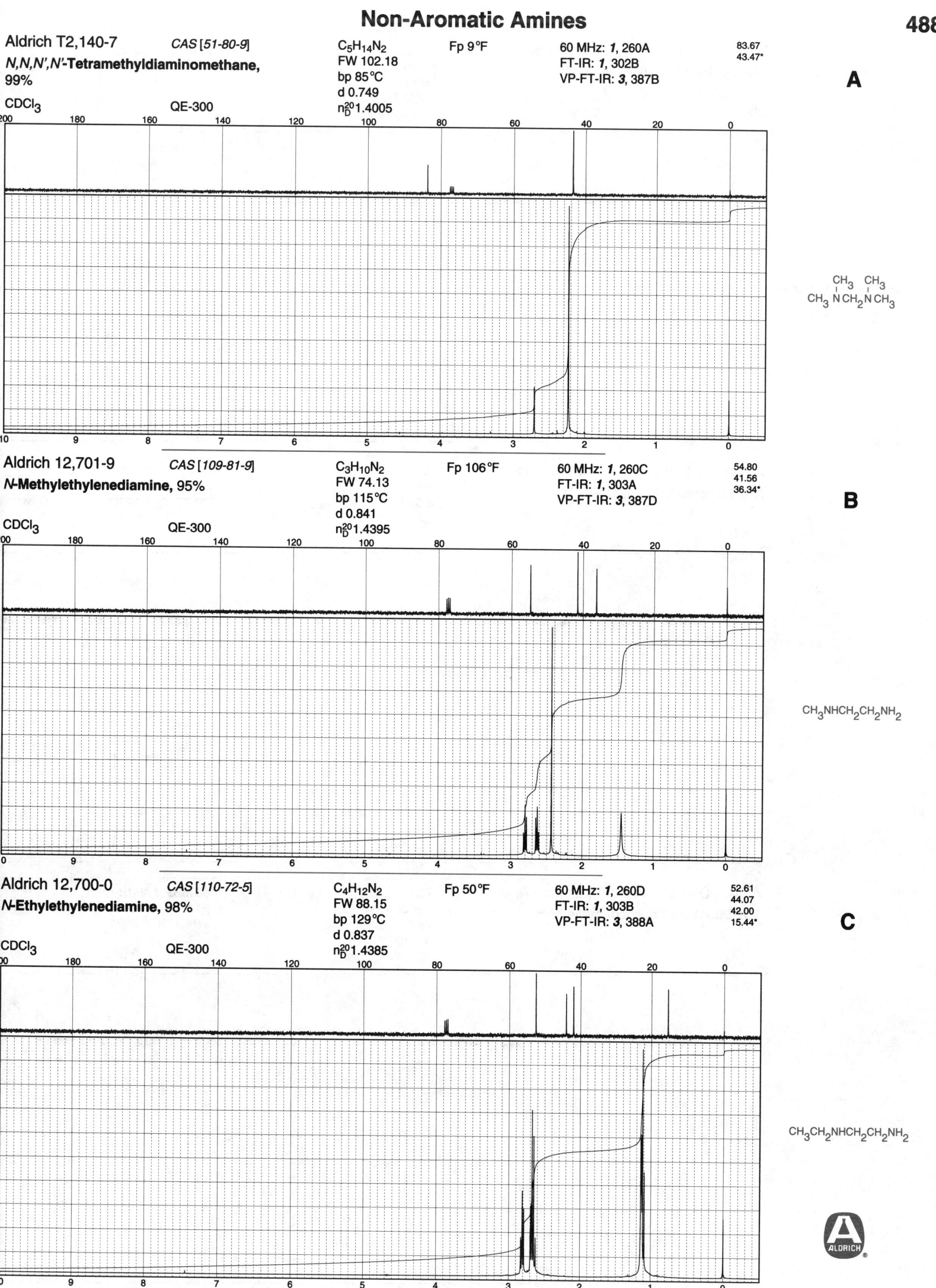

Aldrich T2,140-7   *CAS [51-80-9]*
N,N,N',N'-**Tetramethyldiaminomethane,**
99%

$C_5H_{14}N_2$
FW 102.18
bp 85°C
d 0.749
n_D^{20} 1.4005

Fp 9°F

60 MHz: *1*, 260A
FT-IR: *1*, 302B
VP-FT-IR: *3*, 387B

83.67
43.47*

A

CDCl₃   QE-300

Aldrich 12,701-9   *CAS [109-81-9]*
N-**Methylethylenediamine,** 95%

$C_3H_{10}N_2$
FW 74.13
bp 115°C
d 0.841
n_D^{20} 1.4395

Fp 106°F

60 MHz: *1*, 260C
FT-IR: *1*, 303A
VP-FT-IR: *3*, 387D

54.80
41.56
36.34*

B

CDCl₃   QE-300

Aldrich 12,700-0   *CAS [110-72-5]*
N-**Ethylethylenediamine,** 98%

$C_4H_{12}N_2$
FW 88.15
bp 129°C
d 0.837
n_D^{20} 1.4385

Fp 50°F

60 MHz: *1*, 260D
FT-IR: *1*, 303B
VP-FT-IR: *3*, 388A

52.61
44.07
42.00
15.44*

C

CDCl₃   QE-300

A

Aldrich 30,814-5   *CAS [111-39-7]* $C_5H_{14}N_2$ Fp 59°F   VP-FT-IR: **3**, 388B

N-Propylethylenediamine, 99%   FW 102.18
bp 149°C
d 0.829
n_D^{20} 1.4404

52.67
51.85
41.94
23.34
11.79*

CDCl$_3$ QE-300

$H_2NCH_2CH_2NHCH_2CH_2CH_3$

B

Aldrich I-2,210-2   *CAS [19522-67-9]* $C_5H_{14}N_2$ 60 MHz: **1**, 263B

N-Isopropylethylenediamine, 98%   FW 102.18
d 0.819
n_D^{20} 1.4369
Fp 43°F

FT-IR: **1**, 305C
VP-FT-IR: **3**, 388C

50.24
48.68*
42.30
23.13*

CDCl$_3$ QE-300

CH$_3$
|
CH$_3$ CHNHCH$_2$CH$_2$NH$_2$

C

Aldrich D15,800-3   *CAS [108-00-9]* $C_4H_{12}N_2$ Fp 75°F   60 MHz: **1**, 261B

N,N-Dimethylethylenediamine, 95%   FW 88.15
bp 105°C
d 0.803
n_D^{20} 1.4260

FT-IR: **1**, 303D
VP-FT-IR: **3**, 389A

62.68
45.54*
39.83

CDCl$_3$ QE-300

CH$_3$
|
CH$_3$ NCH$_2$CH$_2$NH$_2$

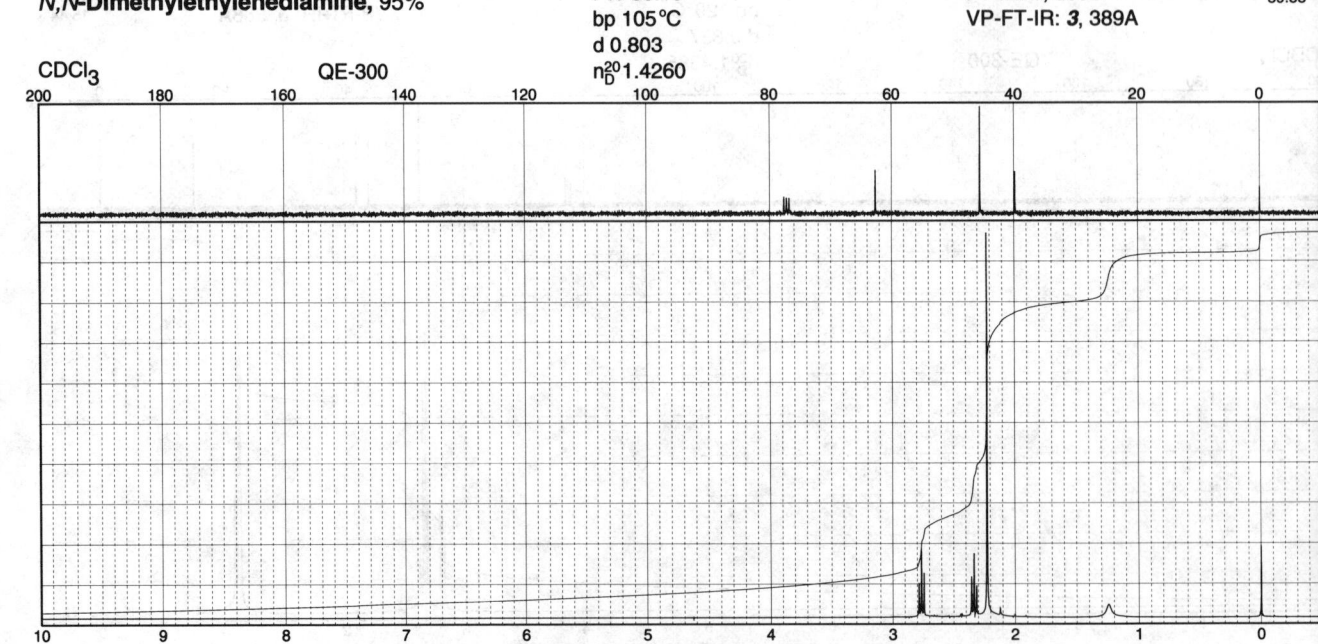

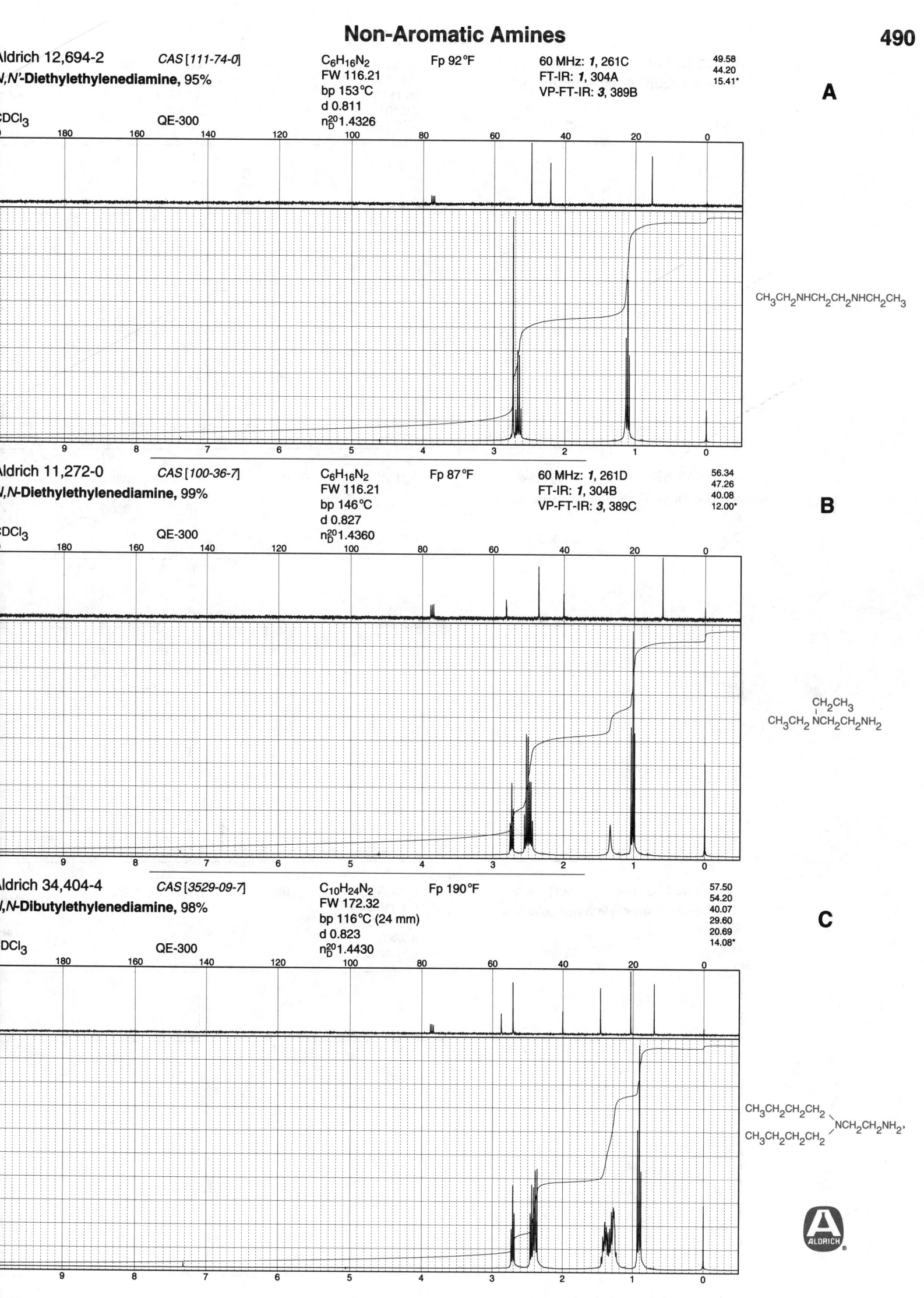

A

Aldrich 12,694-2   CAS [111-74-0]

N,N'-Diethylethylenediamine, 95%

$C_6H_{16}N_2$ Fp 92°F

FW 116.21

bp 153°C

d 0.811

n_D^{20} 1.4326

CDCl$_3$ QE-300

60 MHz: *1*, 261C

FT-IR: *1*, 304A

VP-FT-IR: *3*, 389B

49.58
44.20
15.41*

$CH_3CH_2NHCH_2CH_2NHCH_2CH_3$

B

Aldrich 11,272-0   CAS [100-36-7]

N,N-Diethylethylenediamine, 99%

$C_6H_{16}N_2$ Fp 87°F

FW 116.21

bp 146°C

d 0.827

n_D^{20} 1.4360

CDCl$_3$ QE-300

60 MHz: *1*, 261D

FT-IR: *1*, 304B

VP-FT-IR: *3*, 389C

56.34
47.26
40.08
12.00*

$CH_3CH_2\overset{\displaystyle CH_2CH_3}{\underset{\displaystyle |}{N}}CH_2CH_2NH_2$

C

Aldrich 34,404-4   CAS [3529-09-7]

N,N-Dibutylethylenediamine, 98%

$C_{10}H_{24}N_2$ Fp 190°F

FW 172.32

bp 116°C (24 mm)

d 0.823

n_D^{20} 1.4430

CDCl$_3$ QE-300

57.50
54.20
40.07
29.60
20.69
14.08*

$CH_3CH_2CH_2CH_2$
$CH_3CH_2CH_2CH_2$ $NCH_2CH_2NH_2$,

ALDRICH

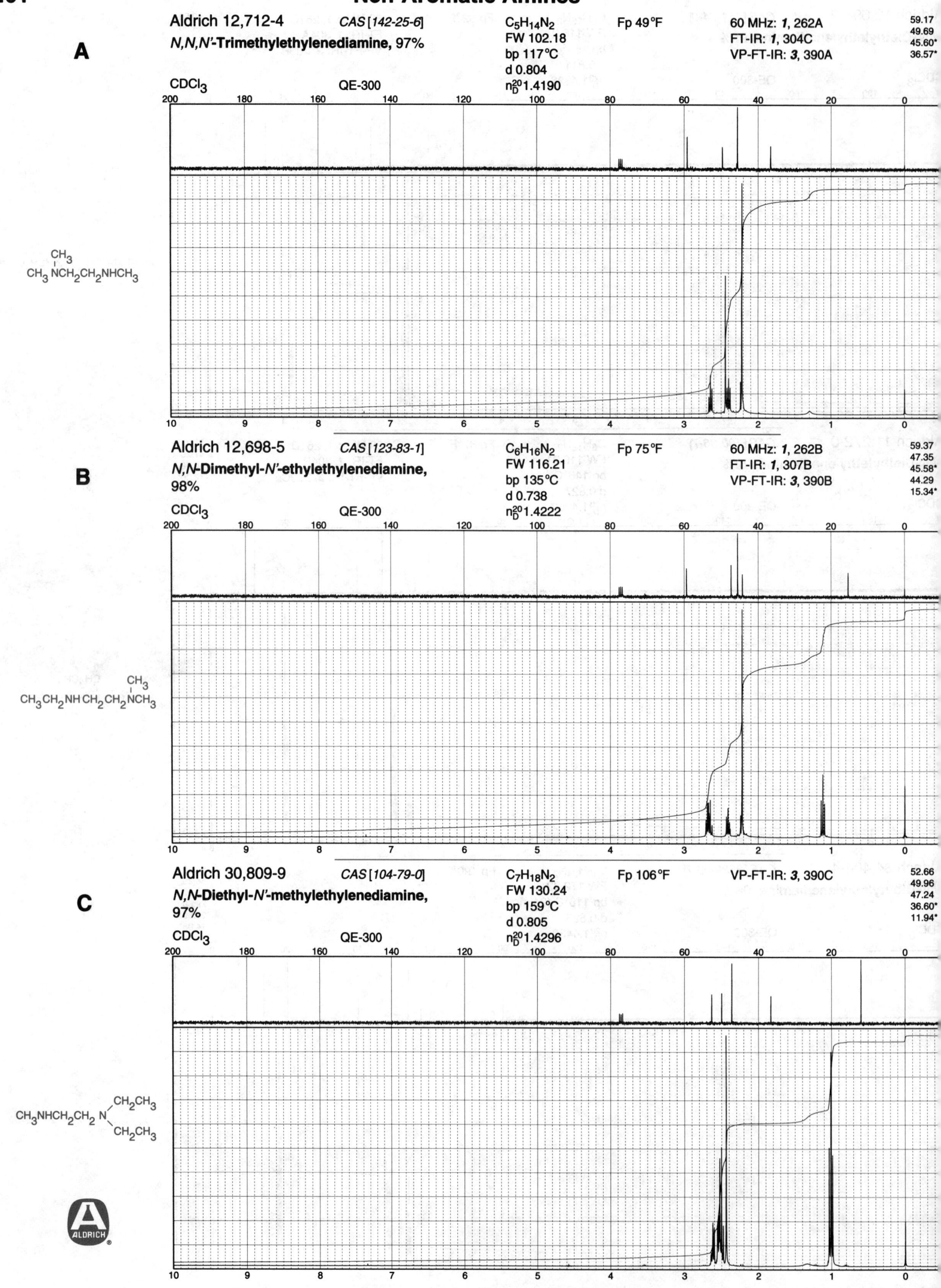

A

Aldrich 12,712-4   CAS [142-25-6]   C₅H₁₄N₂   Fp 49°F   60 MHz: *1*, 262A

$C_5H_{14}N_2$

N,N,N′-Trimethylethylenediamine, 97%   FW 102.18   FT-IR: *1*, 304C

bp 117°C   VP-FT-IR: *3*, 390A

d 0.804

CDCl₃   QE-300 n_D^{20}1.4190

59.17
49.69
45.60*
36.57*

CH_3
$CH_3\,NCH_2CH_2NHCH_3$

B

Aldrich 12,698-5   CAS [123-83-1]   C₆H₁₆N₂   Fp 75°F   60 MHz: *1*, 262B

N,N-Dimethyl-*N′*-ethylethylenediamine,   FW 116.21   FT-IR: *1*, 307B

98%   bp 135°C   VP-FT-IR: *3*, 390B

d 0.738

CDCl₃   QE-300 n_D^{20}1.4222

59.37
47.35
45.58*
44.29
15.34*

CH_3
$CH_3CH_2\,NHCH_2CH_2\,NCH_3$

C

Aldrich 30,809-9   CAS [104-79-0]   C₇H₁₈N₂   Fp 106°F   VP-FT-IR: *3*, 390C

N,N-Diethyl-*N′*-methylethylenediamine,   FW 130.24

97%   bp 159°C

d 0.805

CDCl₃   QE-300 n_D^{20}1.4296

52.66
49.96
47.24
36.60*
11.94*

CH_2CH_3
$CH_3NHCH_2CH_2\,N$
CH_2CH_3

ALDRICH

Aldrich 12,711-6   *CAS [105-04-4]* $C_8H_{20}N_2$ Fp 90 °F   60 MHz: *1*, 262C   52.91
N,N,N'-Triethylethylenediamine, 98%   FW 144.26   FT-IR: *1*, 304D   47.67
47.27
bp 55 °C (13 mm)   VP-FT-IR: *3*, 390D   44.27
CDCl₃   QE-300   d 0.804   15.37*
n_D^{20} 1.4311   11.94*

A

$$CH_3CH_2 \overset{\overset{\displaystyle CH_2CH_3}{|}}{N} CH_2CH_2NHCH_2CH_3$$

Aldrich T2,250-0   *CAS [110-18-9]* $C_6H_{16}N_2$ n_D^{20} 1.4179   57.70
N,N,N',N'-Tetramethylethylenediamine, 99%   FW 116.21   Fp 50 °F   45.81*
mp -55 °C   60 MHz: *1*, 262D
bp 121 °C   FT-IR: *1*, 305A
CDCl₃   QE-300   d 0.770   VP-FT-IR: *3*, 391A

B

$$CH_3 \overset{\overset{\displaystyle CH_3}{|}}{N} CH_2CH_2 \overset{\overset{\displaystyle CH_3}{|}}{N} CH_3$$

Aldrich 12,707-8   *CAS [150-77-6]* $C_{10}H_{24}N_2$ Fp 138 °F   60 MHz: *1*, 263A   51.50
N,N,N',N'-Tetraethylethylenediamine, 98%   FW 172.32   FT-IR: *1*, 305B   47.67
bp 191 °C   VP-FT-IR: *3*, 391B   11.93*
CDCl₃   QE-300   d 0.808
n_D^{20} 1.4343

C

$$CH_3CH_2 \overset{\overset{\displaystyle CH_2CH_3}{|}}{N} CH_2CH_2 \overset{\overset{\displaystyle CH_2CH_3}{|}}{N} CH_2CH_3$$

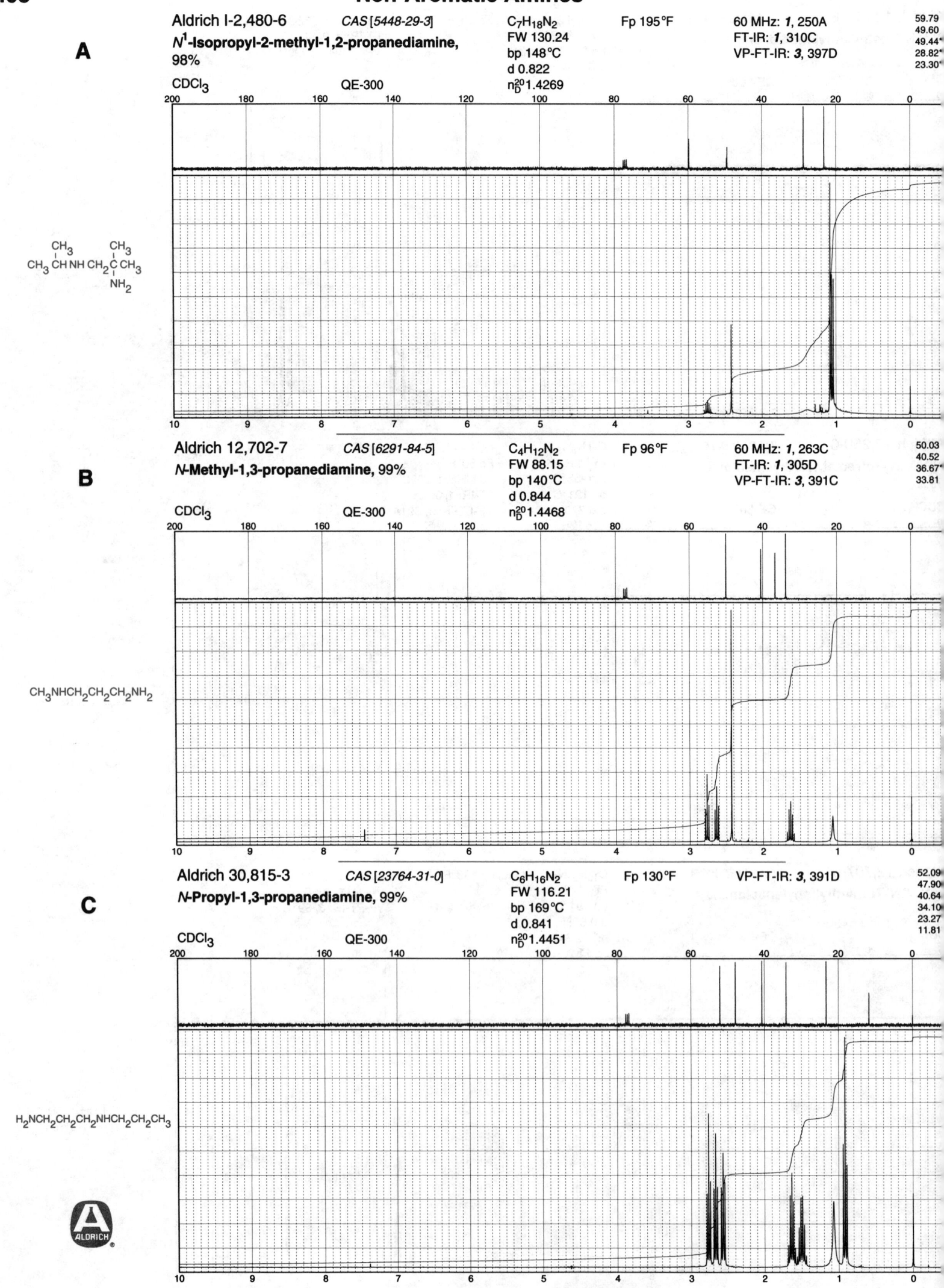

A

Aldrich I-2,480-6   *CAS [5448-29-3]* $C_7H_{18}N_2$ Fp 195°F   60 MHz: *1*, 250A

N^1-Isopropyl-2-methyl-1,2-propanediamine, 98%   FW 130.24   FT-IR: *1*, 310C

      bp 148°C   VP-FT-IR: *3*, 397D

CDCl$_3$ QE-300   d 0.822

      n_D^{20}1.4269

59.79
49.60
49.44*
28.82*
23.30*

B

Aldrich 12,702-7   *CAS [6291-84-5]* $C_4H_{12}N_2$ Fp 96°F   60 MHz: *1*, 263C

N-Methyl-1,3-propanediamine, 99%   FW 88.15   FT-IR: *1*, 305D

      bp 140°C   VP-FT-IR: *3*, 391C

CDCl$_3$ QE-300   d 0.844

      n_D^{20}1.4468

50.03
40.52
36.67*
33.81

CH$_3$NHCH$_2$CH$_2$CH$_2$NH$_2$

C

Aldrich 30,815-3   *CAS [23764-31-0]* $C_6H_{16}N_2$ Fp 130°F   VP-FT-IR: *3*, 391D

N-Propyl-1,3-propanediamine, 99%   FW 116.21

      bp 169°C

CDCl$_3$ QE-300   d 0.841

      n_D^{20}1.4451

52.09
47.90
40.64
34.10
23.27
11.81

H$_2$NCH$_2$CH$_2$CH$_2$NHCH$_2$CH$_2$CH$_3$

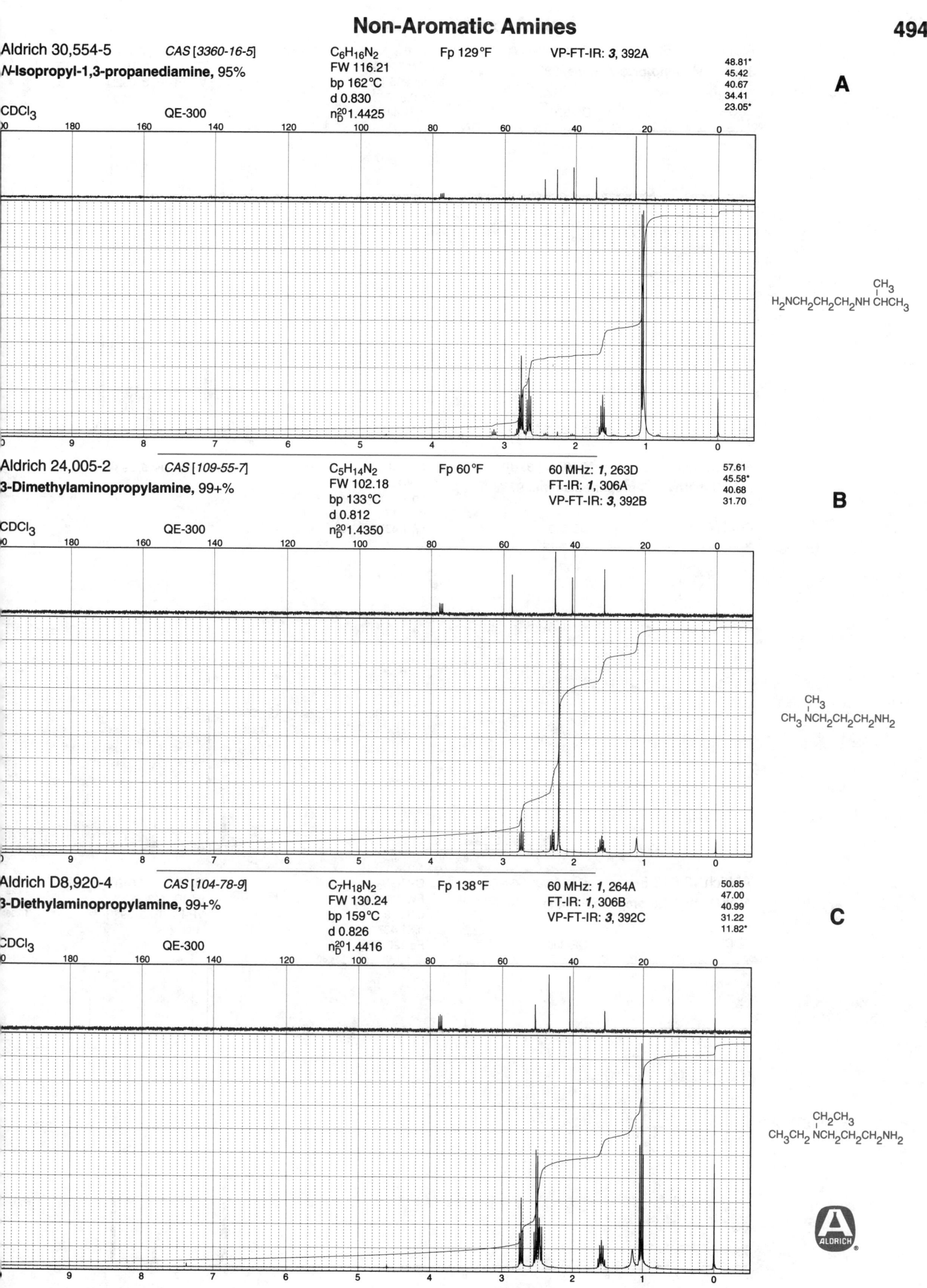

Aldrich 30,554-5   CAS [3360-16-5] $C_6H_{16}N_2$ Fp 129°F   VP-FT-IR: *3*, 392A

N-Isopropyl-1,3-propanediamine, 95%   FW 116.21
bp 162°C
d 0.830

CDCl₃   QE-300 n_D^{20}1.4425

48.81*
45.42
40.67
34.41
23.05*

A

$H_2NCH_2CH_2CH_2NH\,CHCH_3$ with CH_3

Aldrich 24,005-2   CAS [109-55-7] $C_5H_{14}N_2$ Fp 60°F   60 MHz: *1*, 263D

3-Dimethylaminopropylamine, 99+%   FW 102.18   FT-IR: *1*, 306A
bp 133°C   VP-FT-IR: *3*, 392B
d 0.812

CDCl₃   QE-300 n_D^{20}1.4350

57.61
45.58*
40.68
31.70

B

$CH_3\,NCH_2CH_2CH_2NH_2$ with CH_3

Aldrich D8,920-4   CAS [104-78-9] $C_7H_{18}N_2$ Fp 138°F   60 MHz: *1*, 264A

3-Diethylaminopropylamine, 99+%   FW 130.24   FT-IR: *1*, 306B
bp 159°C   VP-FT-IR: *3*, 392C
d 0.826

CDCl₃   QE-300 n_D^{20}1.4416

50.85
47.00
40.99
31.22
11.82*

C

$CH_3CH_2\,NCH_2CH_2CH_2NH_2$ with CH_2CH_3

A

Aldrich D4,560-6 *CAS [102-83-0]* $C_{11}H_{26}N_2$ Fp 219°F 60 MHz: *1*, 266A

3-(Dibutylamino)propylamine, 98% FW 186.34 FT-IR: *1*, 308B
bp 205°C VP-FT-IR: *3*, 392D
d 0.827

CDCl$_3$ QE-300 n_D^{20}1.4463

54.08
52.16
41.00
31.36
29.47
20.78
14.08*

$$CH_3CH_2CH_2CH_2 \overset{\displaystyle CH_2CH_2CH_2CH_3}{\underset{\displaystyle |}{N}} CH_2CH_2CH_2NH_2$$

B

Aldrich 30,811-0 *CAS [111-33-1]* $C_5H_{14}N_2$ Fp 69°F VP-FT-IR: *3*, 393A

N,N'-Dimethyl-1,3-propanediamine, 97% FW 102.18
bp 145°C
d 0.817

CDCl$_3$ QE-300 n_D^{20}1.4375

50.62
36.63*
30.17

$$CH_3NHCH_2CH_2CH_2NHCH_3$$

C

Aldrich 13,842-8 *CAS [10061-68-4]* $C_7H_{18}N_2$ 60 MHz: *1*, 264B

N,N'-Diethyl-1,3-propanediamine, 97% FW 130.24 FT-IR: *1*, 306C
d 0.819 VP-FT-IR: *3*, 393B
n_D^{20}1.4374
Fp 123°F

CDCl$_3$ QE-300

48.37
44.25
30.70
15.38*

$$CH_3CH_2NHCH_2CH_2CH_2NHCH_2CH_3$$

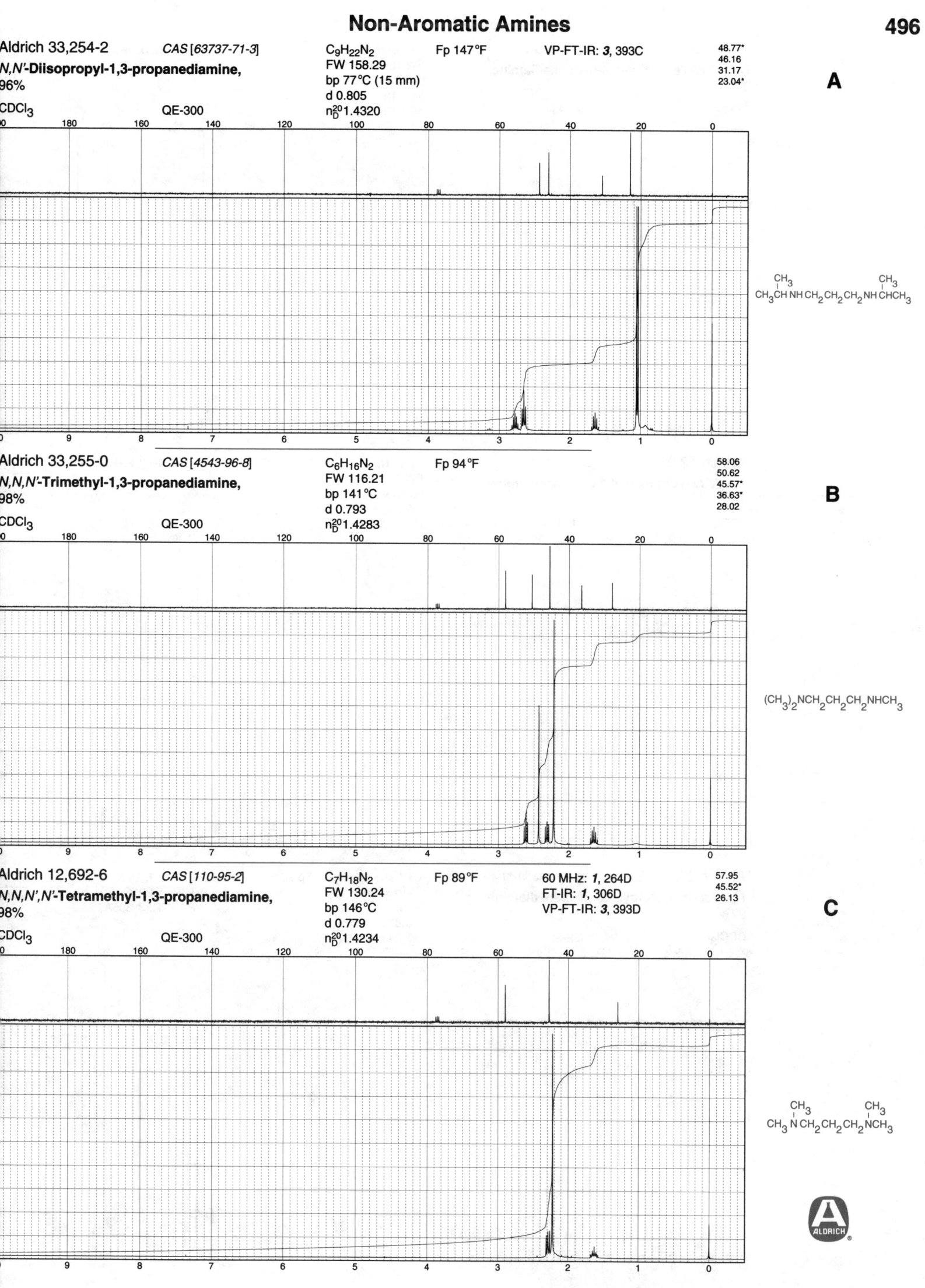

Aldrich 33,254-2   *CAS [63737-71-3]* $C_9H_{22}N_2$ Fp 147°F   VP-FT-IR: **3**, 393C

N,N'-Diisopropyl-1,3-propanediamine, 96%

FW 158.29
bp 77°C (15 mm)
d 0.805
n_D^{20} 1.4320

CDCl$_3$ QE-300

48.77*
46.16
31.17
23.04*

A

$CH_3CH\,NH\,CH_2CH_2CH_2\,NH\,CHCH_3$

Aldrich 33,255-0   *CAS [4543-96-8]* $C_6H_{16}N_2$ Fp 94°F

N,N,N'-Trimethyl-1,3-propanediamine, 98%

FW 116.21
bp 141°C
d 0.793
n_D^{20} 1.4283

CDCl$_3$ QE-300

58.06
50.62
45.57*
36.63*
28.02

B

$(CH_3)_2NCH_2CH_2CH_2NHCH_3$

Aldrich 12,692-6   *CAS [110-95-2]* $C_7H_{18}N_2$ Fp 89°F   60 MHz: **1**, 264D

N,N,N',N'-Tetramethyl-1,3-propanediamine, 98%

FW 130.24
bp 146°C
d 0.779
n_D^{20} 1.4234

FT-IR: **1**, 306D
VP-FT-IR: **3**, 393D

CDCl$_3$ QE-300

57.95
45.52*
26.13

C

$CH_3\,N\,CH_2CH_2CH_2\,N\,CH_3$

ALDRICH

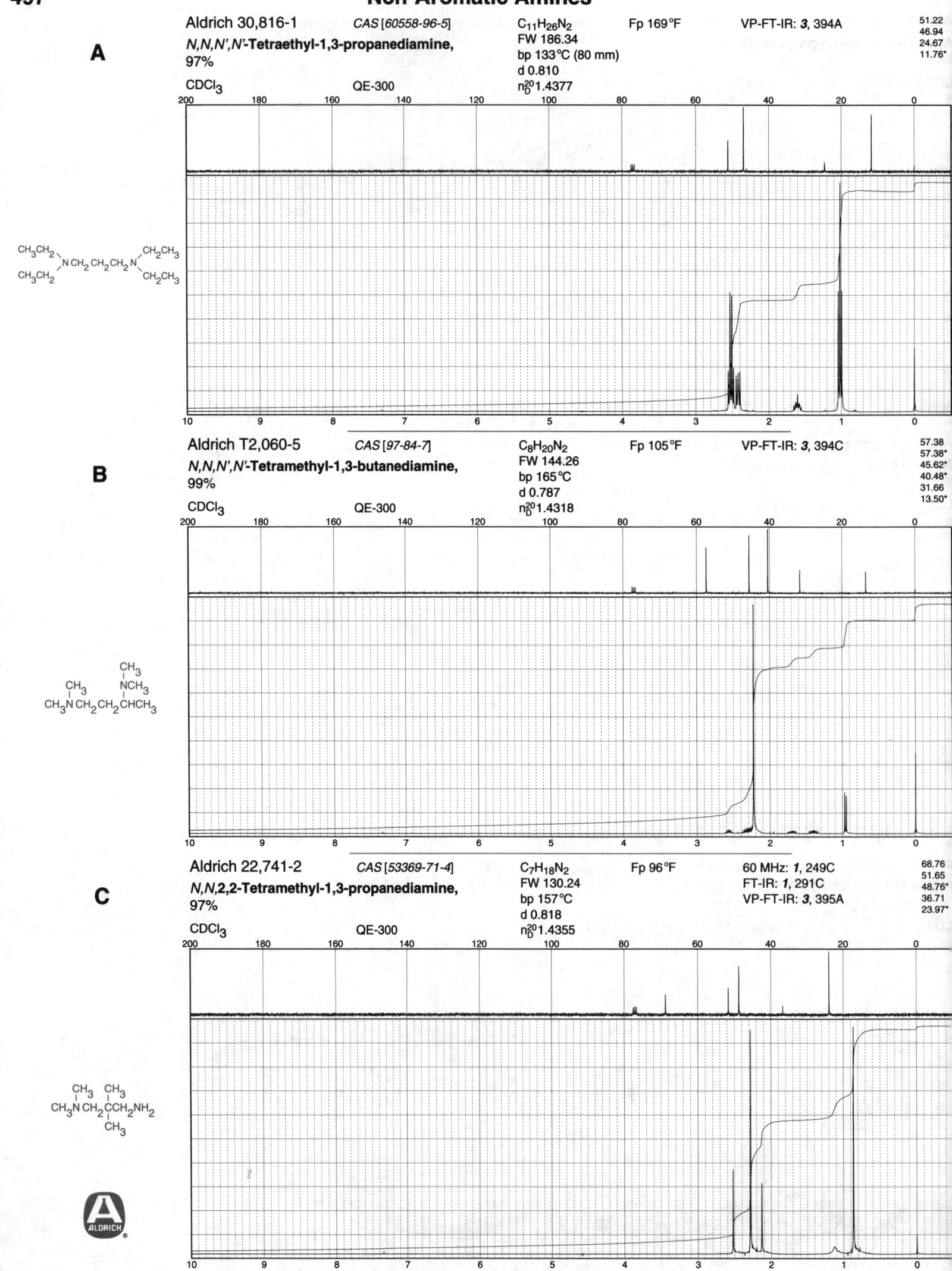

A

Aldrich 30,816-1   CAS [60558-96-5] $C_{11}H_{26}N_2$ Fp 169°F   VP-FT-IR: **3**, 394A

N,N,N',N'-Tetraethyl-1,3-propanediamine, 97%

FW 186.34
bp 133°C (80 mm)
d 0.810
n_D^{20} 1.4377

CDCl₃   QE-300

51.22
46.94
24.67
11.76*

B

Aldrich T2,060-5   CAS [97-84-7] $C_8H_{20}N_2$ Fp 105°F   VP-FT-IR: **3**, 394C

N,N,N',N'-Tetramethyl-1,3-butanediamine, 99%

FW 144.26
bp 165°C
d 0.787
n_D^{20} 1.4318

CDCl₃   QE-300

57.38
57.38*
45.62*
40.48*
31.66
13.50*

C

Aldrich 22,741-2   CAS [53369-71-4] $C_7H_{18}N_2$ Fp 96°F   60 MHz: **1**, 249C

$N,N,2,2$-Tetramethyl-1,3-propanediamine, 97%

FW 130.24
bp 157°C
d 0.818
n_D^{20} 1.4355

CDCl₃   QE-300

FT-IR: **1**, 291C
VP-FT-IR: **3**, 395A

68.76
51.65
48.76*
36.71
23.97*

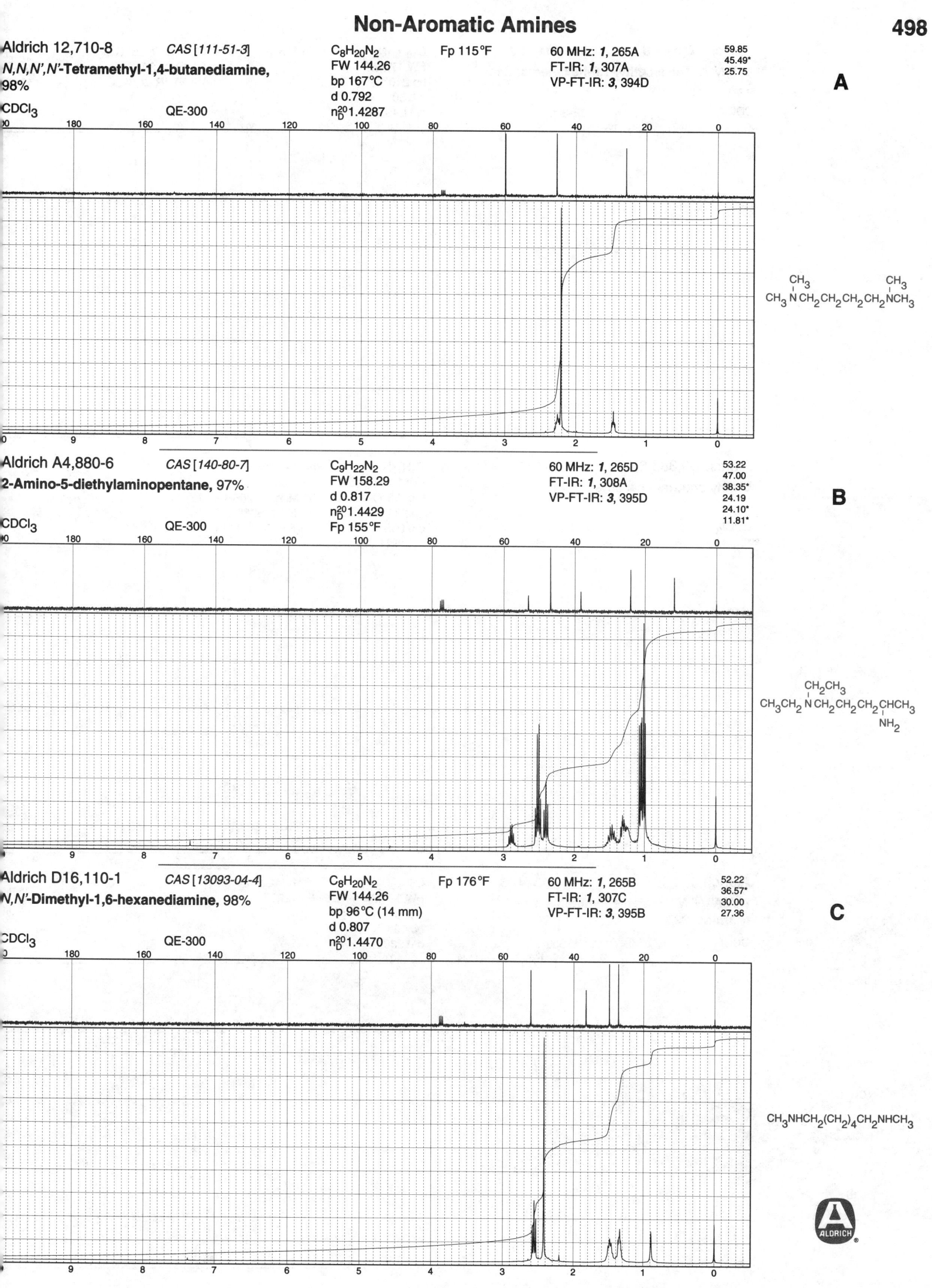

Aldrich 12,710-8　　*CAS [111-51-3]*　　$C_8H_{20}N_2$　　Fp 115°F　　60 MHz: *1*, 265A　　59.85

N,N,N',N'-Tetramethyl-1,4-butanediamine, 98%

FW 144.26　　FT-IR: *1*, 307A　　45.49*
bp 167°C　　VP-FT-IR: *3*, 394D　　25.75
d 0.792
n_D^{20}1.4287

CDCl₃　　QE-300

Aldrich A4,880-6　　*CAS [140-80-7]*　　$C_9H_{22}N_2$　　60 MHz: *1*, 265D　　53.22

2-Amino-5-diethylaminopentane, 97%

FW 158.29　　FT-IR: *1*, 308A　　47.00
d 0.817　　VP-FT-IR: *3*, 395D　　38.35*
n_D^{20}1.4429　　24.19
Fp 155°F　　24.10*
11.81*

CDCl₃　　QE-300

Aldrich D16,110-1　　*CAS [13093-04-4]*　　$C_8H_{20}N_2$　　Fp 176°F　　60 MHz: *1*, 265B　　52.22

N,N'-Dimethyl-1,6-hexanediamine, 98%

FW 144.26　　FT-IR: *1*, 307C　　36.57*
bp 96°C (14 mm)　　VP-FT-IR: *3*, 395B　　30.00
d 0.807　　27.36
n_D^{20}1.4470

CDCl₃　　QE-300

A

B

C

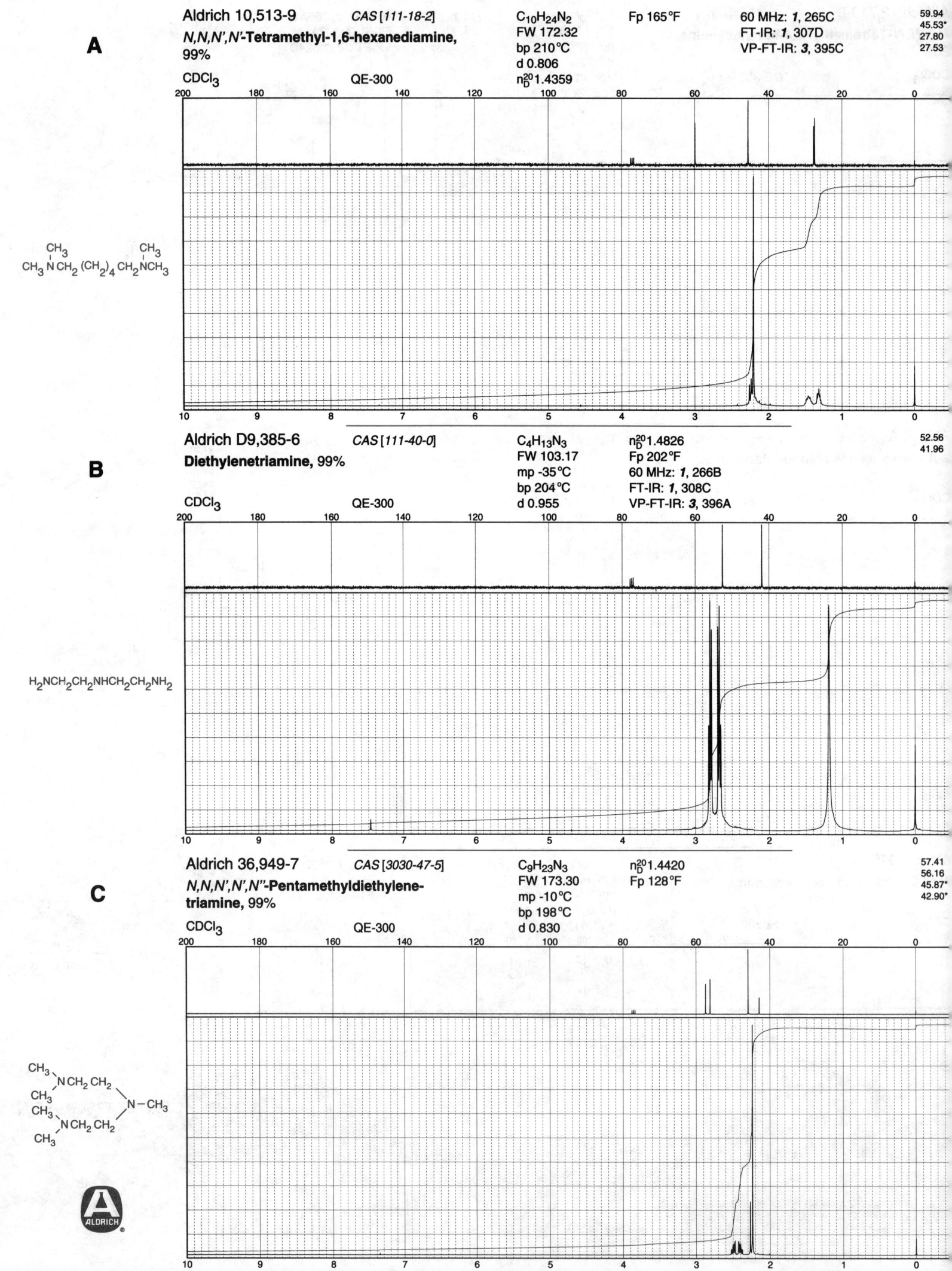

A

Aldrich 10,513-9   CAS [111-18-2] $C_{10}H_{24}N_2$ Fp 165°F   60 MHz: **1**, 265C   59.94
N,N,N',N'-Tetramethyl-1,6-hexanediamine,   FW 172.32   FT-IR: **1**, 307D   45.53*
99%   bp 210°C   VP-FT-IR: **3**, 395C   27.80
CDCl₃   QE-300   d 0.806   27.53
n$_D^{20}$1.4359

CH_3 $\quad$ CH_3
CH_3 N CH_2 $(CH_2)_4$ CH_2 N CH_3

B

Aldrich D9,385-6   CAS [111-40-0] $C_4H_{13}N_3$ n$_D^{20}$1.4826   52.56
Diethylenetriamine, 99%   FW 103.17   Fp 202°F   41.96
mp -35°C   60 MHz: **1**, 266B
CDCl₃   QE-300   bp 204°C   FT-IR: **1**, 308C
d 0.955   VP-FT-IR: **3**, 396A

$H_2NCH_2CH_2NHCH_2CH_2NH_2$

C

Aldrich 36,949-7   CAS [3030-47-5] $C_9H_{23}N_3$ n$_D^{20}$1.4420   57.41
N,N,N',N',N''-Pentamethyldiethylene-   FW 173.30   Fp 128°F   56.16
triamine, 99%   mp -10°C   45.87*
bp 198°C   42.90*
CDCl₃   QE-300   d 0.830

CH_3
$\quad$ N CH_2 CH_2
CH_3
CH_3 $\quad$ N $-$ CH_3
CH_3
$\quad$ N CH_2 CH_2
CH_3

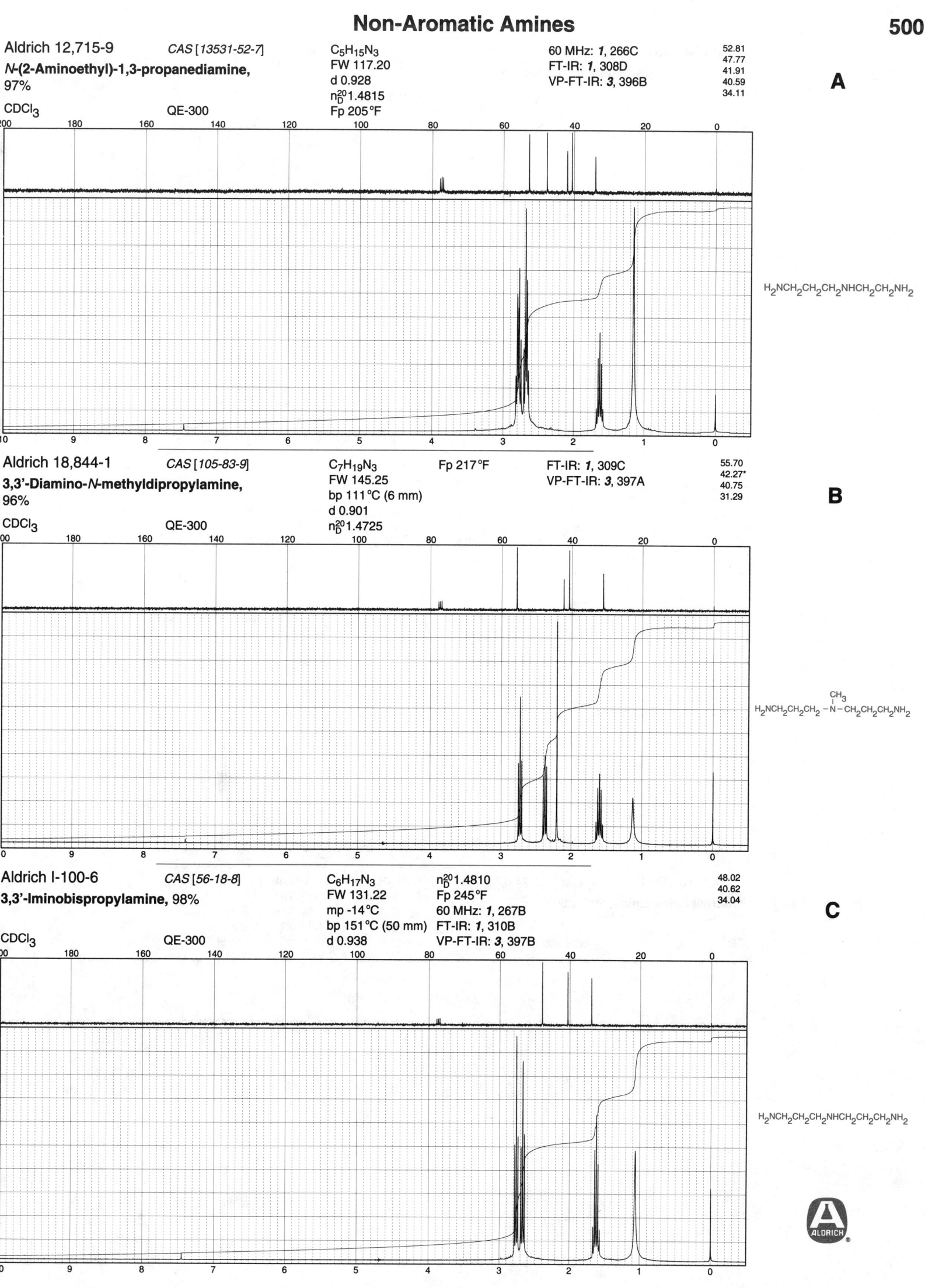

Aldrich 12,715-9 CAS [13531-52-7]

N-(2-Aminoethyl)-1,3-propanediamine, 97%

CDCl₃   QE-300

$C_5H_{15}N_3$
FW 117.20
d 0.928
n_D^{20} 1.4815
Fp 205 °F

60 MHz: *1*, 266C
FT-IR: *1*, 308D
VP-FT-IR: *3*, 396B

52.81
47.77
41.91
40.59
34.11

A

$H_2NCH_2CH_2CH_2NHCH_2CH_2NH_2$

Aldrich 18,844-1 CAS [105-83-9]

3,3'-Diamino-N-methyldipropylamine, 96%

CDCl₃   QE-300

$C_7H_{19}N_3$
FW 145.25
bp 111 °C (6 mm)
d 0.901
n_D^{20} 1.4725

Fp 217 °F

FT-IR: *1*, 309C
VP-FT-IR: *3*, 397A

55.70
42.27*
40.75
31.29

B

$H_2NCH_2CH_2CH_2 - N - CH_2CH_2CH_2NH_2$ with CH₃ on N

Aldrich I-100-6 CAS [56-18-8]

3,3'-Iminobispropylamine, 98%

CDCl₃   QE-300

$C_6H_{17}N_3$
FW 131.22
mp -14 °C
bp 151 °C (50 mm)
d 0.938

n_D^{20} 1.4810
Fp 245 °F
60 MHz: *1*, 267B
FT-IR: *1*, 310B
VP-FT-IR: *3*, 397B

48.02
40.62
34.04

C

$H_2NCH_2CH_2CH_2NHCH_2CH_2CH_2NH_2$

ALDRICH

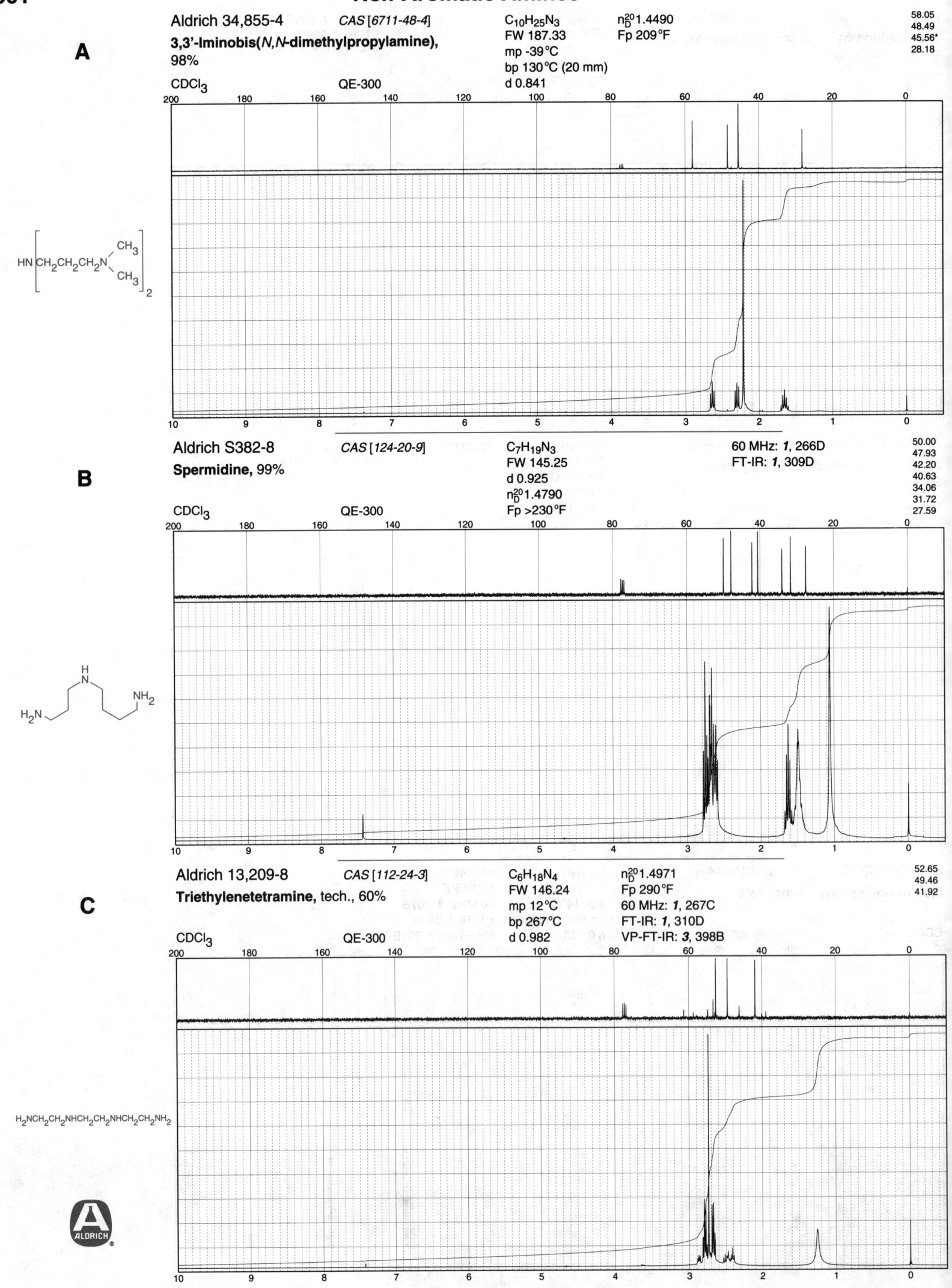

A

Aldrich 34,855-4 CAS [6711-48-4]

3,3'-Iminobis(N,N-dimethylpropylamine), 98%

$C_{10}H_{25}N_3$
FW 187.33
mp -39°C
bp 130°C (20 mm)
d 0.841

n_D^{20} 1.4490
Fp 209°F

58.05
48.49
45.56*
28.18

CDCl$_3$ QE-300

200 180 160 140 120 100 80 60 40 20 0

10 9 8 7 6 5 4 3 2 1 0

B

Aldrich S382-8 CAS [124-20-9]

Spermidine, 99%

$C_7H_{19}N_3$
FW 145.25
d 0.925
n_D^{20} 1.4790
Fp >230°F

60 MHz: **1**, 266D
FT-IR: **1**, 309D

50.00
47.93
42.20
40.63
34.06
31.72
27.59

CDCl$_3$ QE-300

200 180 160 140 120 100 80 60 40 20 0

10 9 8 7 6 5 4 3 2 1 0

C

Aldrich 13,209-8 CAS [112-24-3]

Triethylenetetramine, tech., 60%

$C_6H_{18}N_4$
FW 146.24
mp 12°C
bp 267°C
d 0.982

n_D^{20} 1.4971
Fp 290°F
60 MHz: **1**, 267C
FT-IR: **1**, 310D
VP-FT-IR: **3**, 398B

52.65
49.46
41.92

CDCl$_3$ QE-300

200 180 160 140 120 100 80 60 40 20 0

$H_2NCH_2CH_2NHCH_2CH_2NHCH_2CH_2NH_2$

10 9 8 7 6 5 4 3 2 1 0

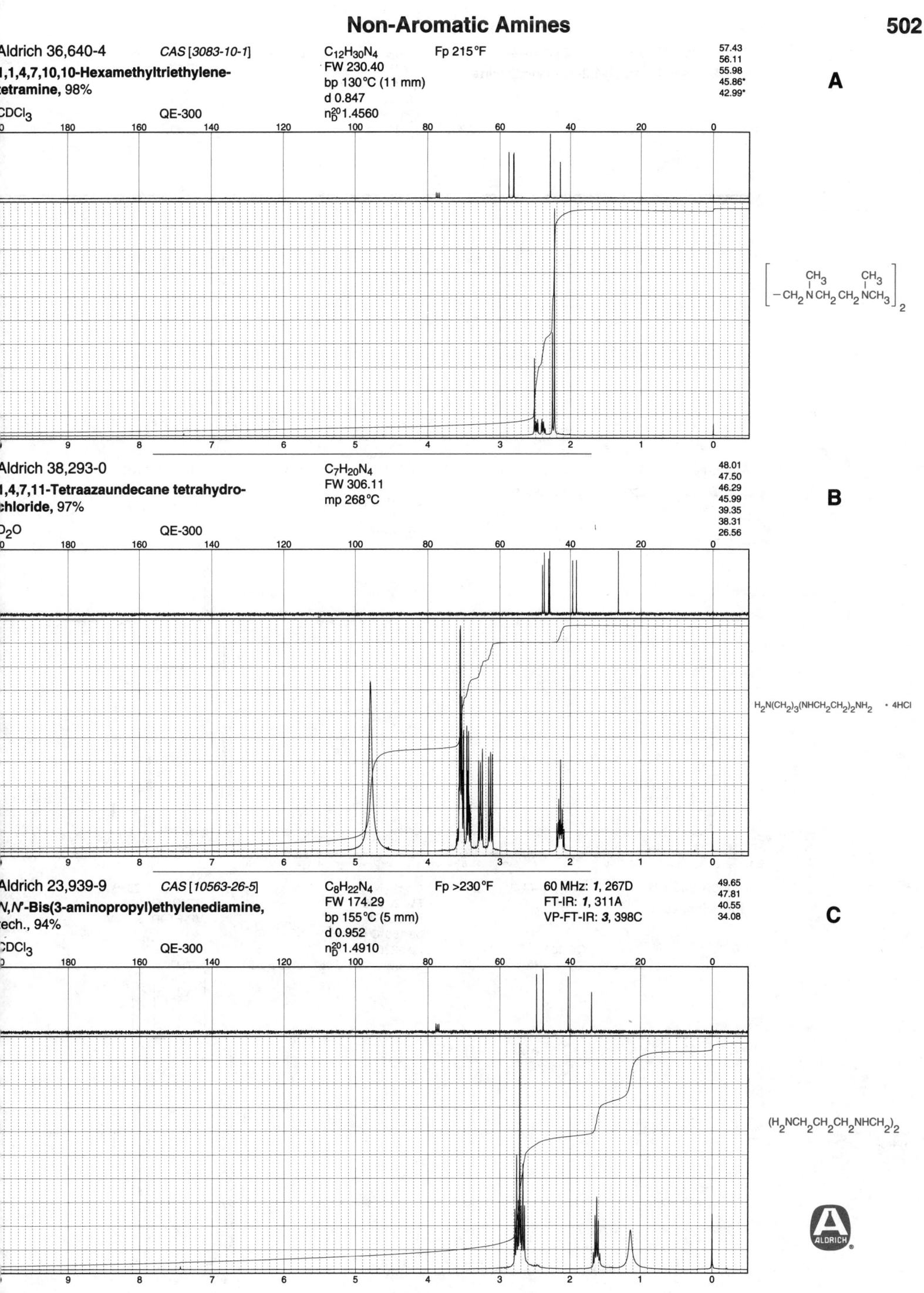

Aldrich 36,640-4   *CAS [3083-10-1]* $C_{12}H_{30}N_4$ Fp 215°F

1,1,4,7,10,10-Hexamethyltriethylene-
tetramine, 98%

CDCl$_3$ QE-300

FW 230.40
bp 130°C (11 mm)
d 0.847
n_D^{20} 1.4560

57.43
56.11
55.98
45.86*
42.99*

A

Aldrich 38,293-0

1,4,7,11-Tetraazaundecane tetrahydro-
chloride, 97%

D$_2$O   QE-300

$C_7H_{20}N_4$
FW 306.11
mp 268°C

48.01
47.50
46.29
45.99
39.35
38.31
26.56

B

$H_2N(CH_2)_3(NHCH_2CH_2)_2NH_2 \cdot 4HCl$

Aldrich 23,939-9   *CAS [10563-26-5]* $C_8H_{22}N_4$ Fp >230°F   60 MHz: *1*, 267D

N,N'-Bis(3-aminopropyl)ethylenediamine,
tech., 94%

CDCl$_3$ QE-300

FW 174.29
bp 155°C (5 mm)
d 0.952
n_D^{20} 1.4910

FT-IR: *1*, 311A
VP-FT-IR: *3*, 398C

49.65
47.81
40.55
34.08

C

$(H_2NCH_2CH_2CH_2NHCH_2)_2$

A

Aldrich 25,913-6   *CAS [4741-99-5]* $C_7H_{20}N_4$ Fp >230°F   FT-IR: *1*, 309A

N,N'-Bis(2-aminoethyl)-1,3-propanediamine

FW 160.27   VP-FT-IR: *3*, 396C

bp 144°C (8 mm)

d 0.960

n_D^{20} 1.4960

52.77
48.32
41.89
30.66

CDCl₃   QE-300

B

Aldrich 25,914-4   *CAS [4605-14-5]* $C_9H_{24}N_4$ Fp >230°F   FT-IR: *1*, 309B

N,N'-Bis(3-aminopropyl)-1,3-propane-diamine

FW 188.32   VP-FT-IR: *3*, 396D

bp 101°C (1 mm)

d 0.920

n_D^{20} 1.4915

48.66
47.96
40.61
34.03
30.51

CDCl₃   QE-300

$(H_2NCH_2CH_2CH_2NHCH_2)_2CH_2$

C

Aldrich S383-6   *CAS [71-44-3]* $C_{10}H_{26}N_4$ 60 MHz: *1*, 268B

Spermine, 98%

FW 202.35   FT-IR: *1*, 312B

mp 29°C   VP-FT-IR: *3*, 399A

bp 150°C (5 mm)

Fp >230°F

50.02
47.90
40.60
34.02
28.02

CDCl₃   QE-300

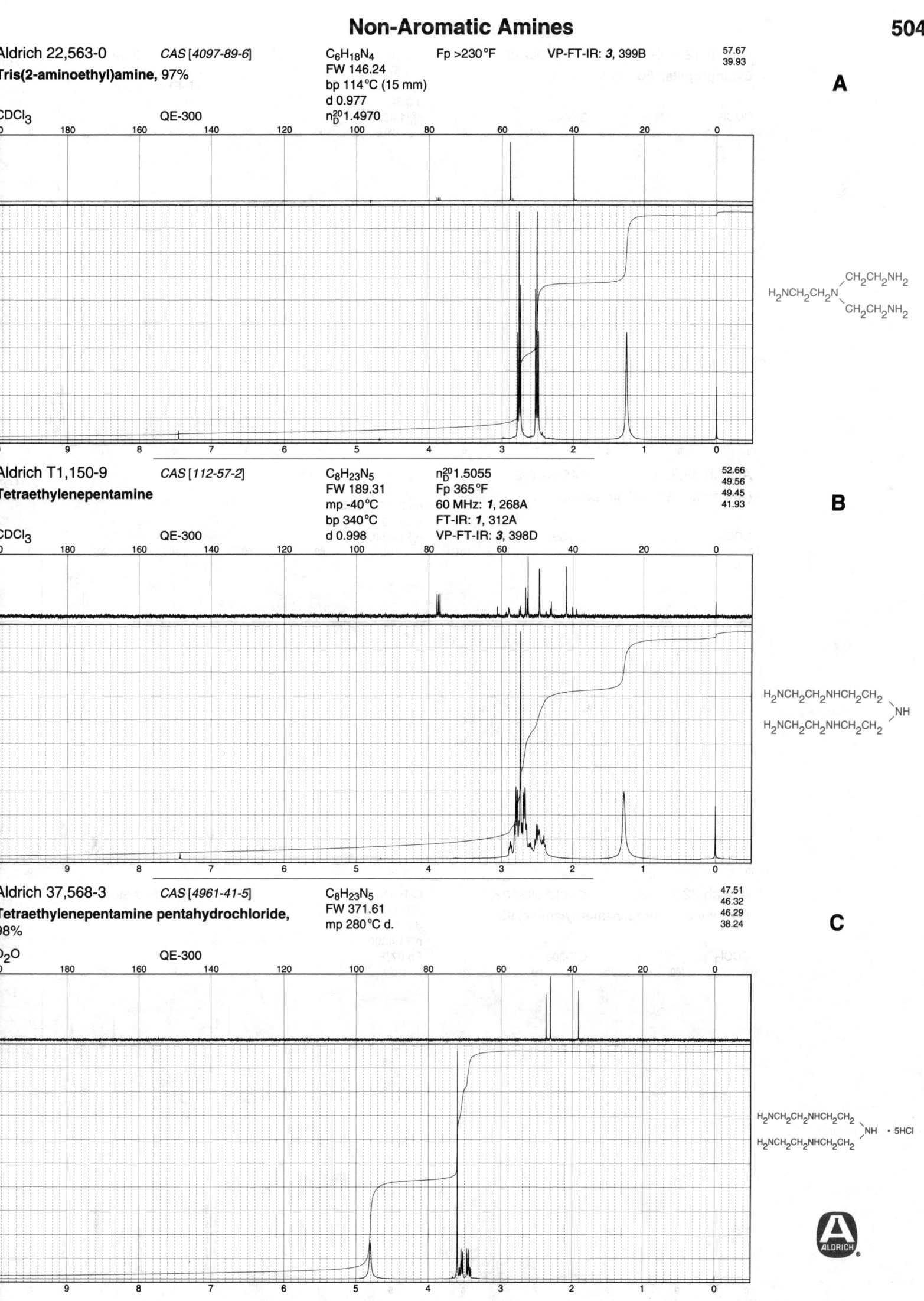

Aldrich 22,563-0   CAS [4097-89-6]   C₆H₁₈N₄   Fp >230°F   VP-FT-IR: **3**, 399B   57.67 / 39.93

Tris(2-aminoethyl)amine, 97%

FW 146.24
bp 114°C (15 mm)
d 0.977
nᴅ²⁰ 1.4970

CDCl₃   QE-300

A

Aldrich T1,150-9   CAS [112-57-2]   C₈H₂₃N₅   nᴅ²⁰ 1.5055   52.66 / 49.56 / 49.45 / 41.93

Tetraethylenepentamine

FW 189.31
mp -40°C
bp 340°C
d 0.998

60 MHz: **1**, 268A
FT-IR: **1**, 312A
VP-FT-IR: **3**, 398D

CDCl₃   QE-300

B

Aldrich 37,568-3   CAS [4961-41-5]   C₈H₂₃N₅   47.51 / 46.32 / 46.29 / 38.24

Tetraethylenepentamine pentahydrochloride, 98%

FW 371.61
mp 280°C d.

D₂O   QE-300

C

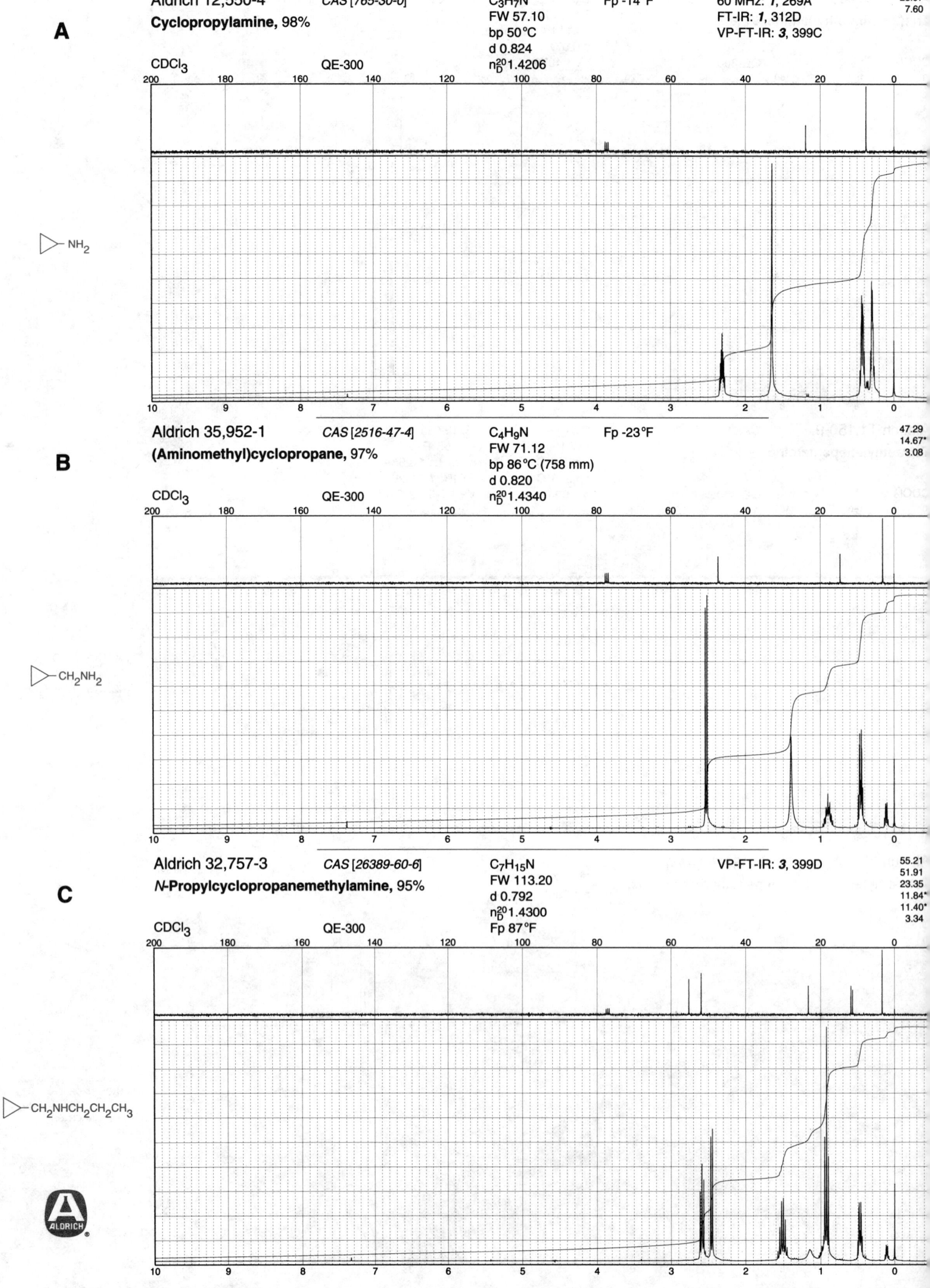

A

Aldrich 12,550-4　　*CAS [765-30-0]*　　C₃H₇N　　Fp -14°F　　60 MHz: **1**, 269A

Cyclopropylamine, 98%　　FW 57.10　　　　FT-IR: **1**, 312D

bp 50°C　　VP-FT-IR: **3**, 399C

d 0.824

n²⁰_D 1.4206

CDCl₃　　QE-300

23.97*
7.60

B

Aldrich 35,952-1　　*CAS [2516-47-4]*　　C₄H₉N　　Fp -23°F

(Aminomethyl)cyclopropane, 97%　　FW 71.12

bp 86°C (758 mm)

d 0.820

n²⁰_D 1.4340

CDCl₃　　QE-300

47.29
14.67*
3.08

C

Aldrich 32,757-3　　*CAS [26389-60-6]*　　C₇H₁₅N　　VP-FT-IR: **3**, 399D

***N*-Propylcyclopropanemethylamine, 95%**　　FW 113.20

d 0.792

n²⁰_D 1.4300

Fp 87°F

CDCl₃　　QE-300

55.21
51.91
23.35
11.84*
11.40*
3.34

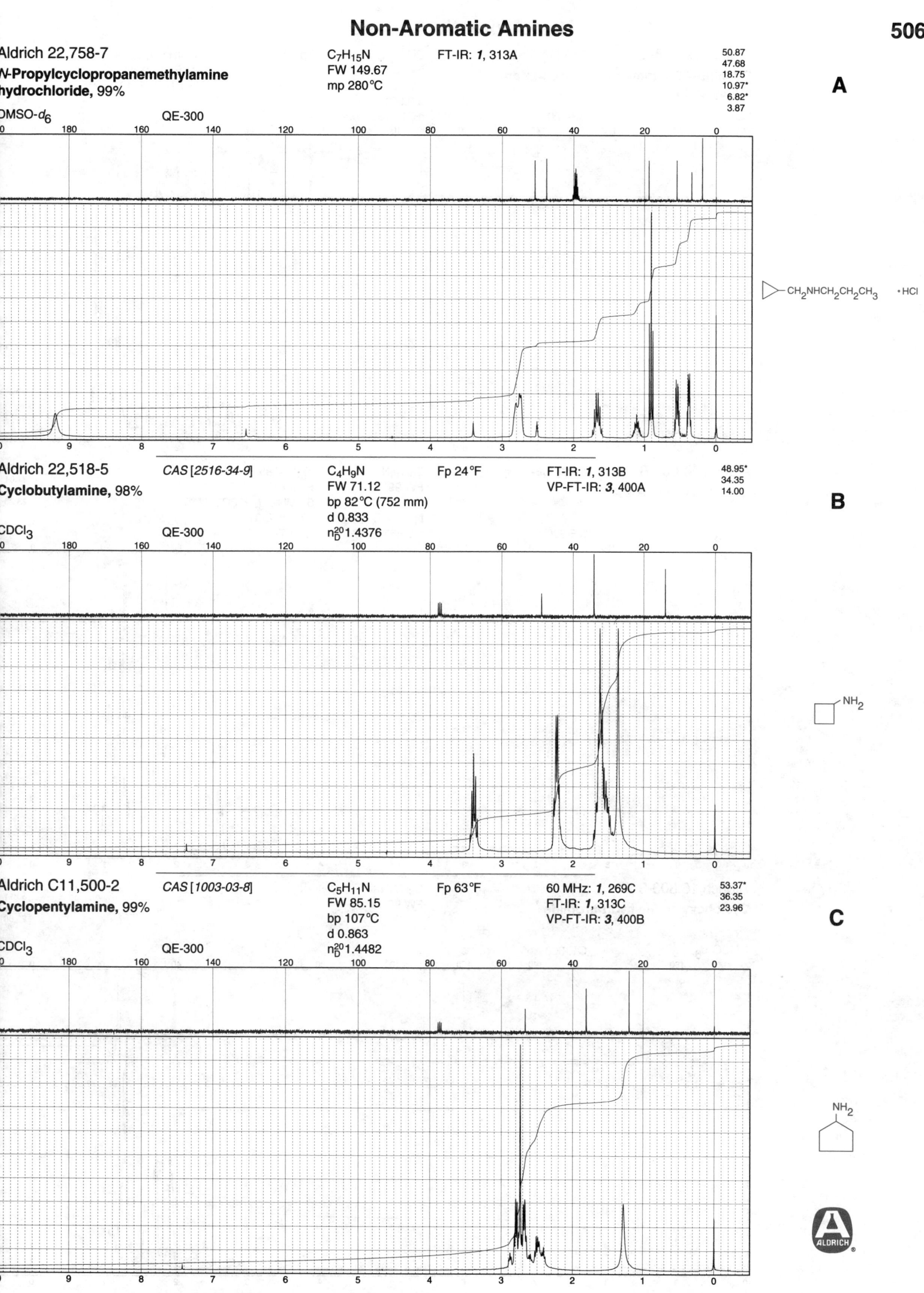

Aldrich 22,758-7
N-Propylcyclopropanemethylamine
hydrochloride, 99%
DMSO-d_6 QE-300

$C_7H_{15}N$ FT-IR: *1*, 313A
FW 149.67
mp 280°C

50.87
47.68
18.75
10.97*
6.82*
3.87

A

$\triangleright$—CH$_2$NHCH$_2$CH$_2$CH$_3$ •HCl

Aldrich 22,518-5   CAS [2516-34-9]
Cyclobutylamine, 98%
CDCl$_3$ QE-300

C_4H_9N Fp 24°F   FT-IR: *1*, 313B
FW 71.12   VP-FT-IR: *3*, 400A
bp 82°C (752 mm)
d 0.833
n_D^{20} 1.4376

48.95*
34.35
14.00

B

Aldrich C11,500-2   CAS [1003-03-8]
Cyclopentylamine, 99%
CDCl$_3$ QE-300

$C_5H_{11}N$ Fp 63°F   60 MHz: *1*, 269C
FW 85.15   FT-IR: *1*, 313C
bp 107°C   VP-FT-IR: *3*, 400B
d 0.863
n_D^{20} 1.4482

53.37*
36.35
23.96

C

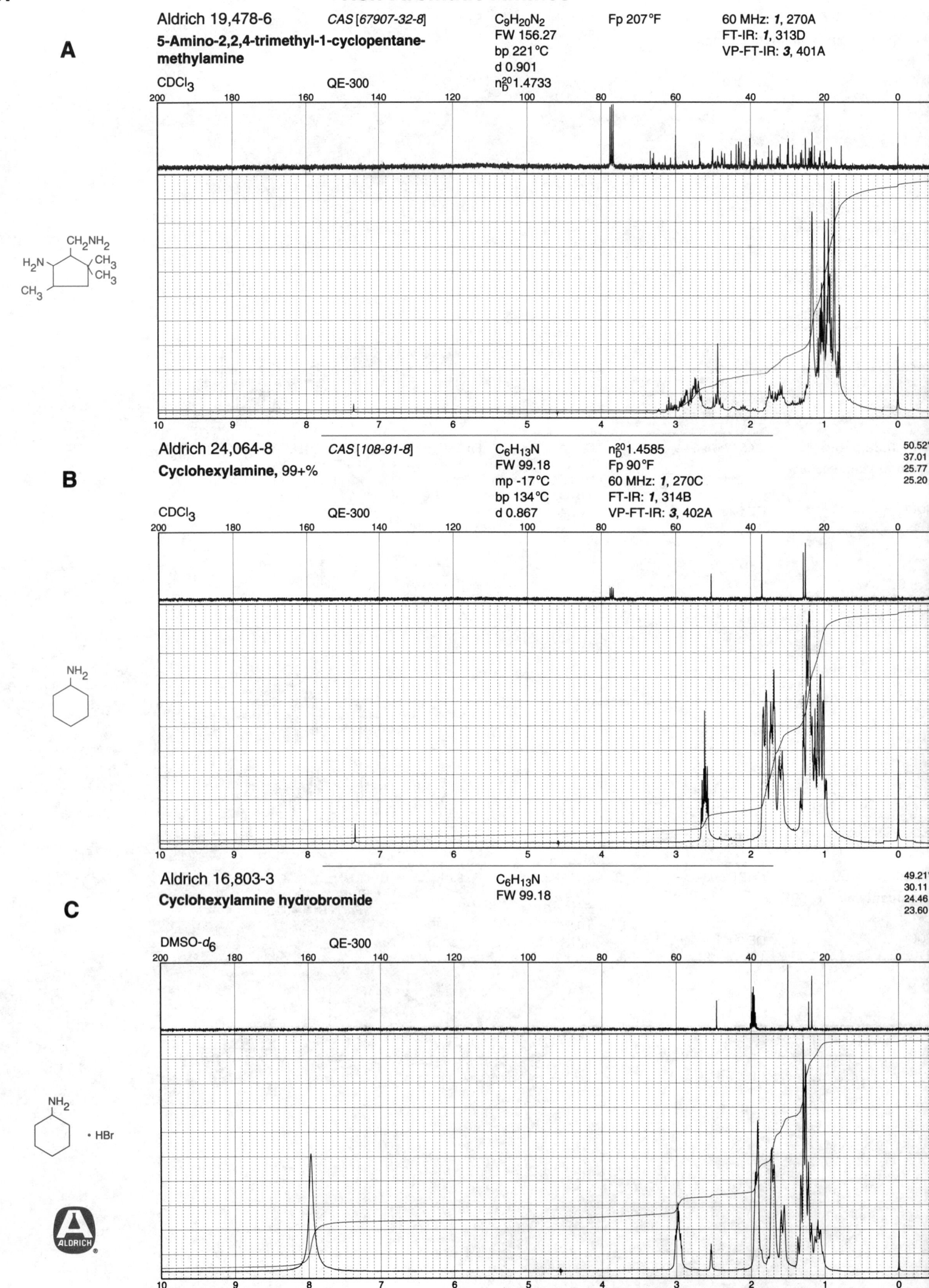

A

Aldrich 19,478-6   *CAS [67907-32-8]* $C_9H_{20}N_2$ Fp 207°F   60 MHz: *1*, 270A
5-Amino-2,2,4-trimethyl-1-cyclopentane-   FW 156.27   FT-IR: *1*, 313D
methylamine   bp 221°C   VP-FT-IR: *3*, 401A
d 0.901
CDCl₃   QE-300 n_D^{20}1.4733

B

Aldrich 24,064-8   *CAS [108-91-8]* $C_6H_{13}N$ n_D^{20}1.4585   50.52*
Cyclohexylamine, 99+%   FW 99.18   Fp 90°F   37.01
mp -17°C   60 MHz: *1*, 270C   25.77
bp 134°C   FT-IR: *1*, 314B   25.20
d 0.867   VP-FT-IR: *3*, 402A
CDCl₃   QE-300

C

Aldrich 16,803-3 $C_6H_{13}N$ 49.21*
Cyclohexylamine hydrobromide   FW 99.18   30.11
24.46
23.60
DMSO-d_6 QE-300

Aldrich 10,332-2   *CAS [100-60-7]* $C_7H_{15}N$ Fp 85°F   60 MHz: *1*, 272B   58.59*
                                                                                              33.62*
N-Methylcyclohexylamine, 99%   FW 113.20   FT-IR: *1*, 314D   33.30
                                         bp 149°C   VP-FT-IR: *3*, 402B   26.30
                                         d 0.868   25.05

A

CDCl₃   QE-300 n_D^{20}1.4560

NHCH₃

Aldrich 37,450-4   *CAS [51609-06-4]* $C_{10}H_{21}N$ Fp 126°F   51.21*
                                                                                   50.87
N-tert-Butylcyclohexylamine, 98%   FW 155.29   37.11
                                          bp 173°C   30.07*
                                          d 0.834   25.89
                                                                                   25.74

B

CDCl₃   QE-300 n_D^{20}1.4480

CH₃
|
NH—C—CH₃
|
CH₃

Aldrich 18,584-1   *CAS [101-83-7]* $C_{12}H_{23}N$ n_D^{20}1.4842   53.21*
                                                                                   34.52
Dicyclohexylamine, 99%   FW 181.32   Fp 205°F   26.29
                                          mp -2°C   60 MHz: *1*, 272C   25.36
                                          bp 256°C   FT-IR: *1*, 315A
                                          d 0.910   VP-FT-IR: *3*, 402D

C

CDCl₃   QE-300

NH

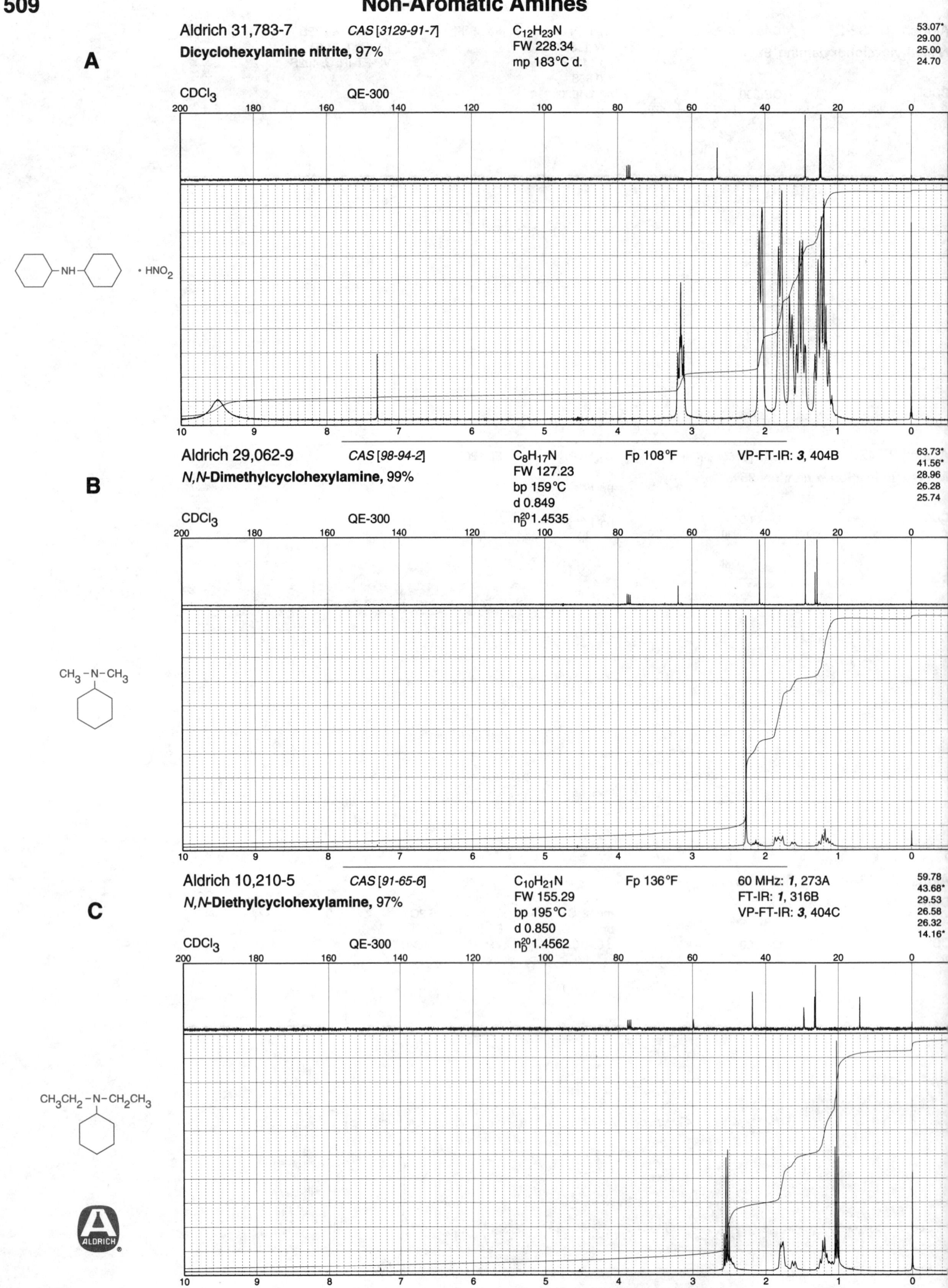

A

Dicyclohexylamine nitrite, 97%

$C_{12}H_{23}N$
FW 228.34
mp 183°C d.

53.07*
29.00
25.00
24.70

CDCl₃   QE-300

Aldrich 29,062-9   CAS [98-94-2]

B

N,N-Dimethylcyclohexylamine, 99%

$C_8H_{17}N$
FW 127.23
bp 159°C
d 0.849
n_D^{20}1.4535

Fp 108°F

VP-FT-IR: **3**, 404B

63.73*
41.56*
28.96
26.28
25.74

CDCl₃   QE-300

Aldrich 10,210-5   CAS [91-65-6]

C

N,N-Diethylcyclohexylamine, 97%

$C_{10}H_{21}N$
FW 155.29
bp 195°C
d 0.850
n_D^{20}1.4562

Fp 136°F

60 MHz: **1**, 273A
FT-IR: **1**, 316B
VP-FT-IR: **3**, 404C

59.78
43.68*
29.53
26.58
26.32
14.16*

CDCl₃   QE-300

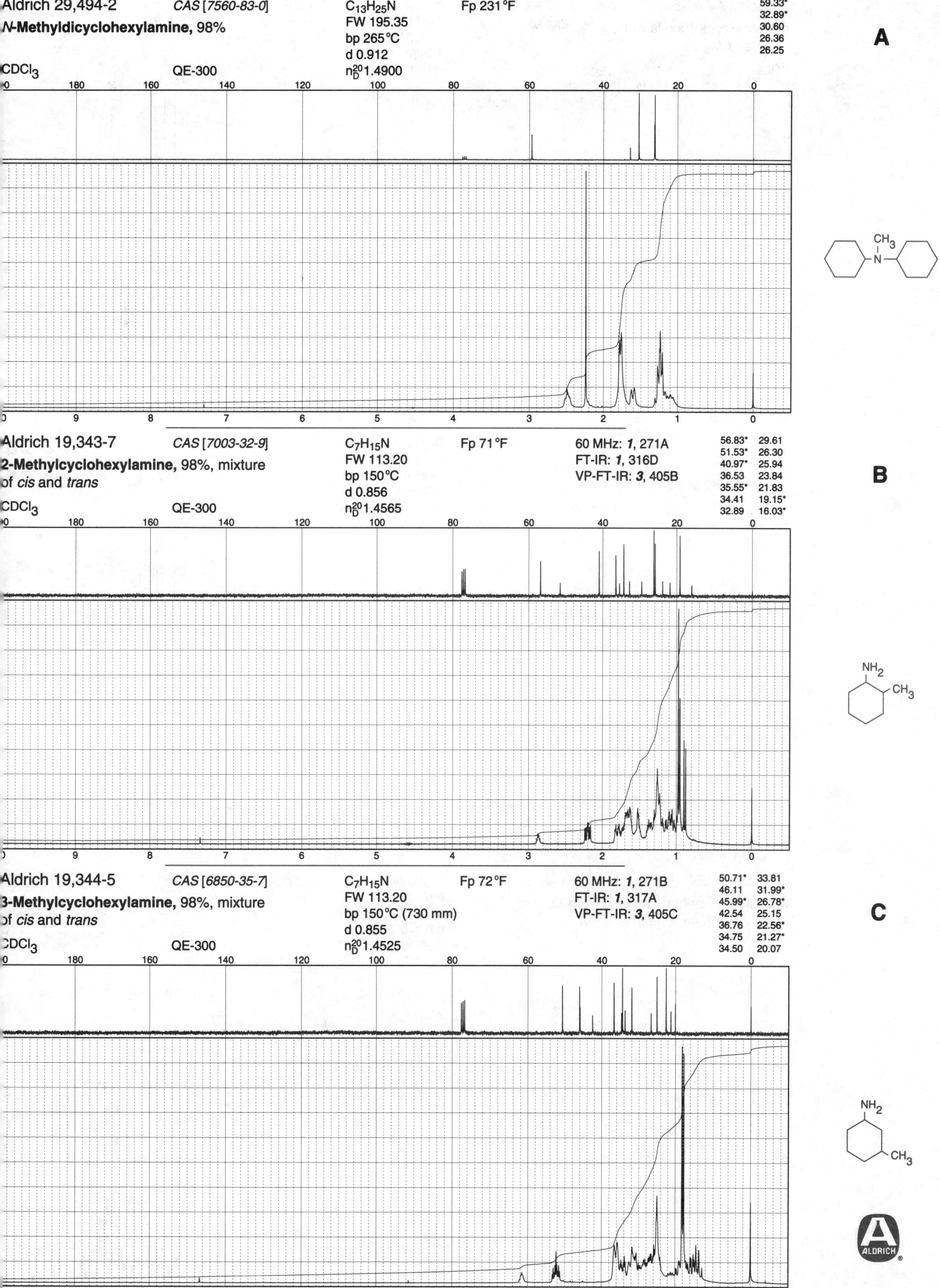

Aldrich 29,494-2   *CAS [7560-83-0]*
N-Methyldicyclohexylamine, 98%

CDCl₃   QE-300

C₁₃H₂₅N
FW 195.35
bp 265°C
d 0.912
n²⁰_D 1.4900

Fp 231°F

59.33*
32.89*
30.60
26.36
26.25

A

Aldrich 19,343-7   *CAS [7003-32-9]*
2-Methylcyclohexylamine, 98%, mixture
of *cis* and *trans*

CDCl₃   QE-300

C₇H₁₅N
FW 113.20
bp 150°C
d 0.856
n²⁰_D 1.4565

Fp 71°F

60 MHz: *1*, 271A
FT-IR: *1*, 316D
VP-FT-IR: *3*, 405B

56.83* 29.61
51.53* 26.30
40.97* 25.94
36.53 23.84
35.55* 21.83
34.41 19.15*
32.89 16.03*

B

Aldrich 19,344-5   *CAS [6850-35-7]*
3-Methylcyclohexylamine, 98%, mixture
of *cis* and *trans*

CDCl₃   QE-300

C₇H₁₅N
FW 113.20
bp 150°C (730 mm)
d 0.855
n²⁰_D 1.4525

Fp 72°F

60 MHz: *1*, 271B
FT-IR: *1*, 317A
VP-FT-IR: *3*, 405C

50.71* 33.81
46.11 31.99*
45.99* 26.78*
42.54 25.15
36.76 22.56*
34.75 21.27*
34.50 20.07

C

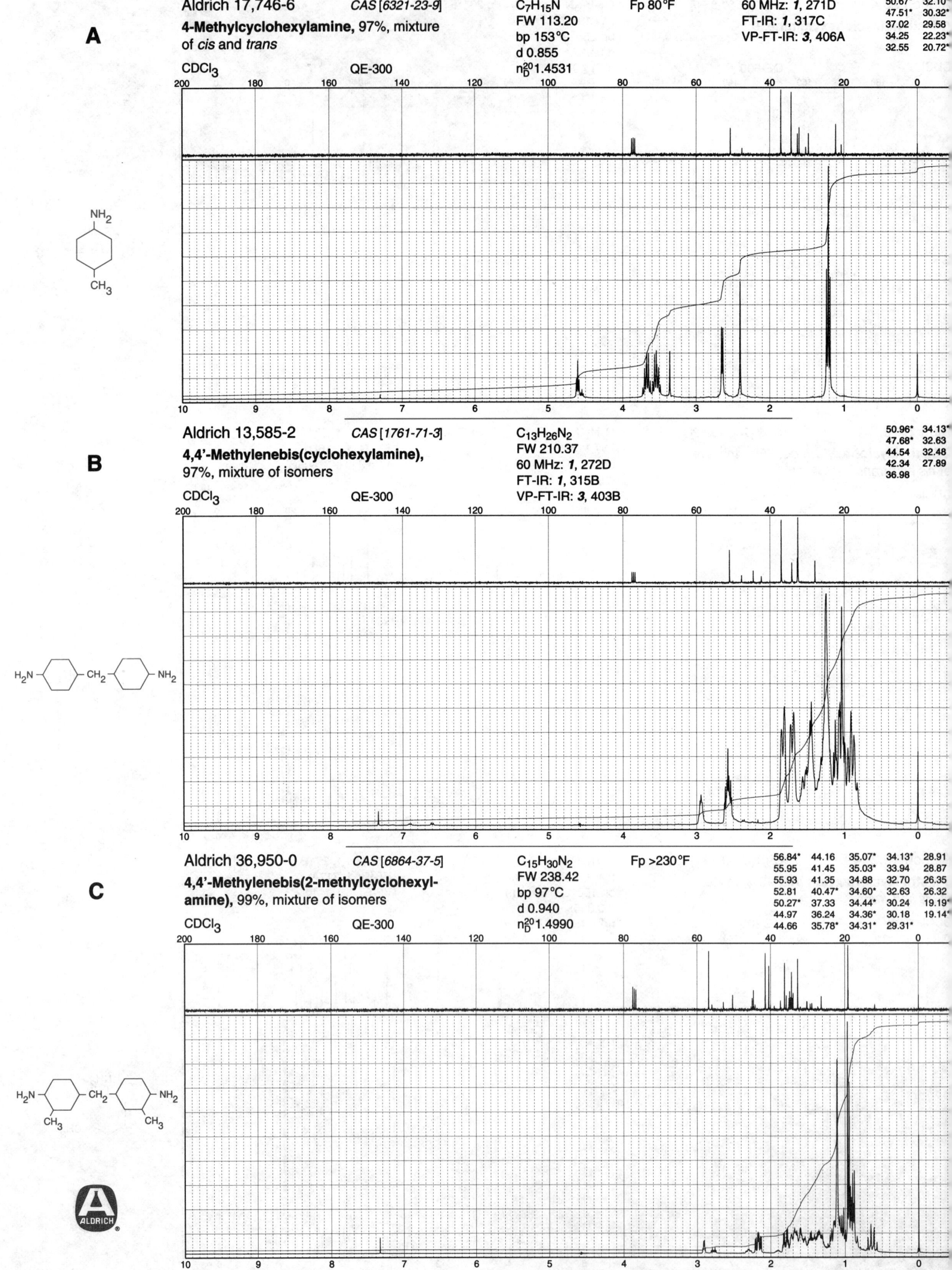

Aldrich 17,746-6 — CAS [6321-23-9]

4-Methylcyclohexylamine, 97%, mixture of *cis* and *trans*

$C_7H_{15}N$
FW 113.20
bp 153°C
d 0.855
n_D^{20} 1.4531

Fp 80°F

60 MHz: **1**, 271D
FT-IR: **1**, 317C
VP-FT-IR: **3**, 406A

CDCl₃   QE-300

50.67*	32.10*
47.51*	30.32*
37.02	29.58
34.25	22.23*
32.55	20.72*

Aldrich 13,585-2 — CAS [1761-71-3]

4,4'-Methylenebis(cyclohexylamine), 97%, mixture of isomers

$C_{13}H_{26}N_2$
FW 210.37
60 MHz: **1**, 272D
FT-IR: **1**, 315B
VP-FT-IR: **3**, 403B

CDCl₃   QE-300

50.96*	34.13*
47.68*	32.63
44.54	32.48
42.34	27.89
36.98	

Aldrich 36,950-0 — CAS [6864-37-5]

4,4'-Methylenebis(2-methylcyclohexyl-amine), 99%, mixture of isomers

$C_{15}H_{30}N_2$
FW 238.42
bp 97°C
d 0.940
n_D^{20} 1.4990

Fp >230°F

CDCl₃   QE-300

56.84*	44.16	35.07*	34.13*	28.91
55.95	41.45	35.03*	33.94	28.87
55.93	41.35	34.88	32.70	26.35
52.81	40.47*	34.60*	32.63	26.32
50.27*	37.33	34.44*	30.24	19.19*
44.97	36.24	34.36*	30.18	19.14*
44.66	35.78*	34.31*	29.31*	

Aldrich 13,255-1   CAS [694-83-7]

1,2-Diaminocyclohexane, 99%, mixture
of cis and trans

CDCl₃   QE-300

$C_6H_{14}N_2$
FW 114.19
bp 93°C (18 mm)
d 0.931
n_D^{20} 1.4900

Fp 167°F

FT-IR: 1, 315C
VP-FT-IR: 3, 404A

57.87*
52.05*
35.71
31.22
25.56
22.02

A

Aldrich 34,671-3   CAS [21436-03-3]

(S,2S)-(+)-1,2-Diaminocyclohexane,
8%

CDCl₃   QE-300

$C_6H_{14}N_2$
FW 114.19
mp 44°C
bp 107°C (40 mm)
Fp 169°F

57.75*
35.58
25.49

B

Aldrich 34,672-1   CAS [20439-47-8]

(R,2R)-(-)-1,2-Diaminocyclohexane,
8%

CDCl₃   QE-300

$C_6H_{14}N_2$
FW 114.19
Fp 169°F

57.77*
35.59
25.50

C

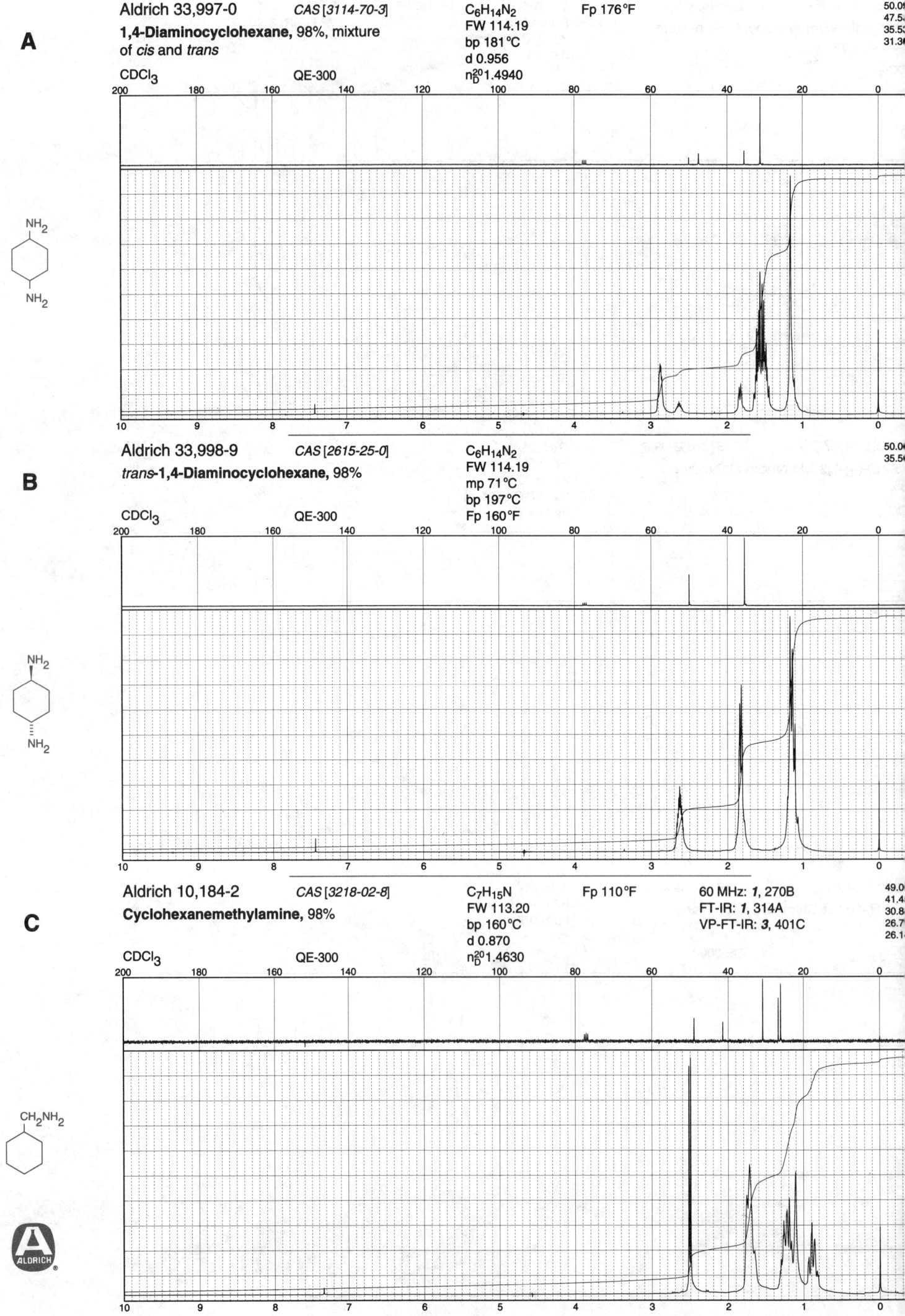

A

Aldrich 33,997-0   CAS [3114-70-3] $C_6H_{14}N_2$ Fp 176°F

1,4-Diaminocyclohexane, 98%, mixture of *cis* and *trans*

FW 114.19
bp 181°C
d 0.956
n_D^{20}1.4940

CDCl₃   QE-300

B

Aldrich 33,998-9   CAS [2615-25-0] $C_6H_{14}N_2$

***trans*-1,4-Diaminocyclohexane, 98%**

FW 114.19
mp 71°C
bp 197°C
Fp 160°F

CDCl₃   QE-300

C

Aldrich 10,184-2   CAS [3218-02-8] $C_7H_{15}N$ Fp 110°F   60 MHz: *1*, 270B

Cyclohexanemethylamine, 98%

FW 113.20
bp 160°C
d 0.870
n_D^{20}1.4630

FT-IR: *1*, 314A
VP-FT-IR: *3*, 401C

CDCl₃   QE-300

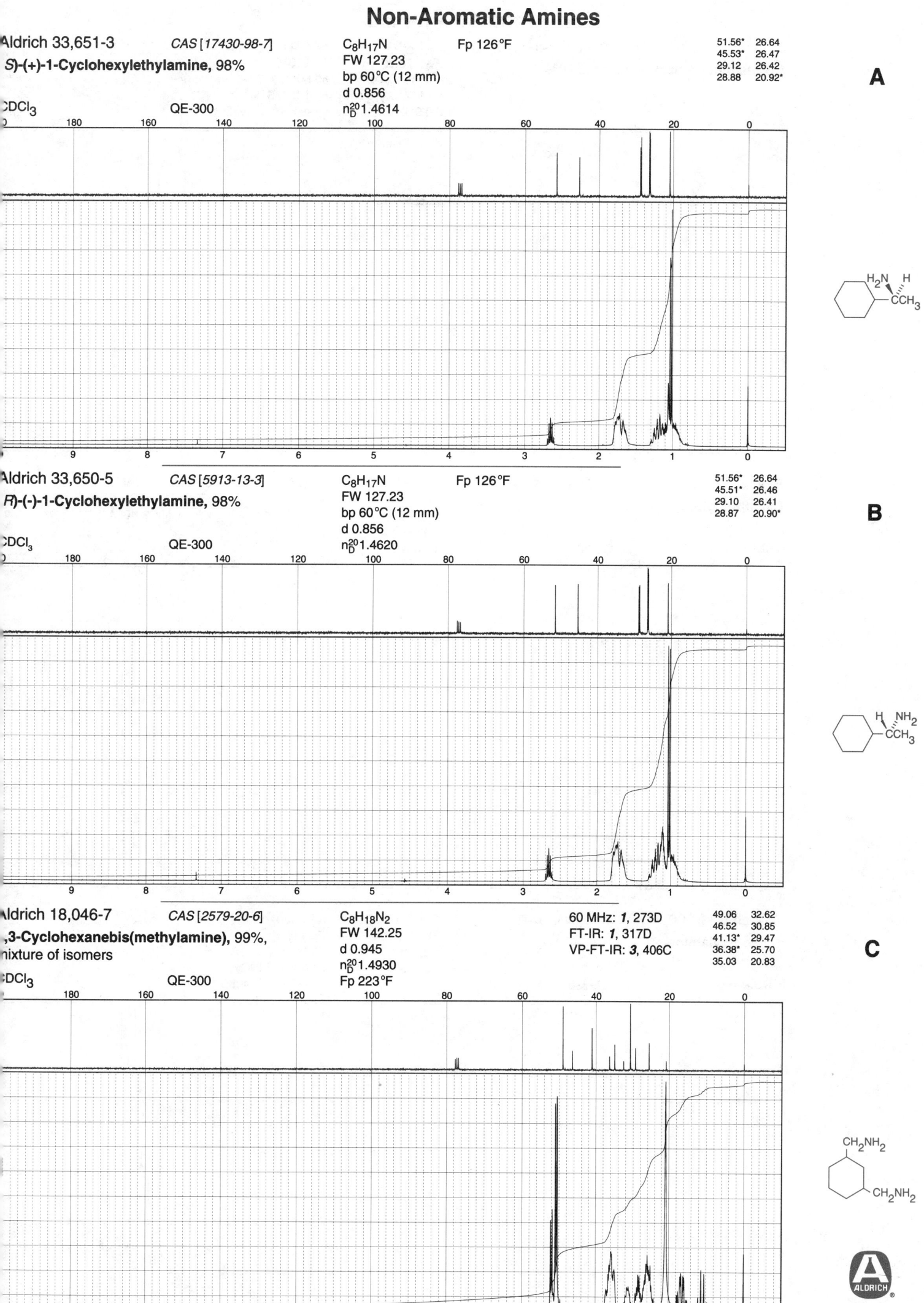

Aldrich 33,651-3   *CAS [17430-98-7]* $C_8H_{17}N$ Fp 126°F

S)-(+)-1-Cyclohexylethylamine, 98%

FW 127.23
bp 60°C (12 mm)
d 0.856
n_D^{20} 1.4614

CDCl$_3$ QE-300

51.56*	26.64
45.53*	26.47
29.12	26.42
28.88	20.92*

A

Aldrich 33,650-5   *CAS [5913-13-3]* $C_8H_{17}N$ Fp 126°F

R)-(-)-1-Cyclohexylethylamine, 98%

FW 127.23
bp 60°C (12 mm)
d 0.856
n_D^{20} 1.4620

CDCl$_3$ QE-300

51.56*	26.64
45.51*	26.46
29.10	26.41
28.87	20.90*

B

Aldrich 18,046-7   *CAS [2579-20-6]* $C_8H_{18}N_2$

,3-Cyclohexanebis(methylamine), 99%,
mixture of isomers

FW 142.25
d 0.945
n_D^{20} 1.4930
Fp 223°F

CDCl$_3$ QE-300

60 MHz: *1*, 273D
FT-IR: *1*, 317D
VP-FT-IR: *3*, 406C

49.06	32.62
46.52	30.85
41.13*	29.47
36.38*	25.70
35.03	20.83

C

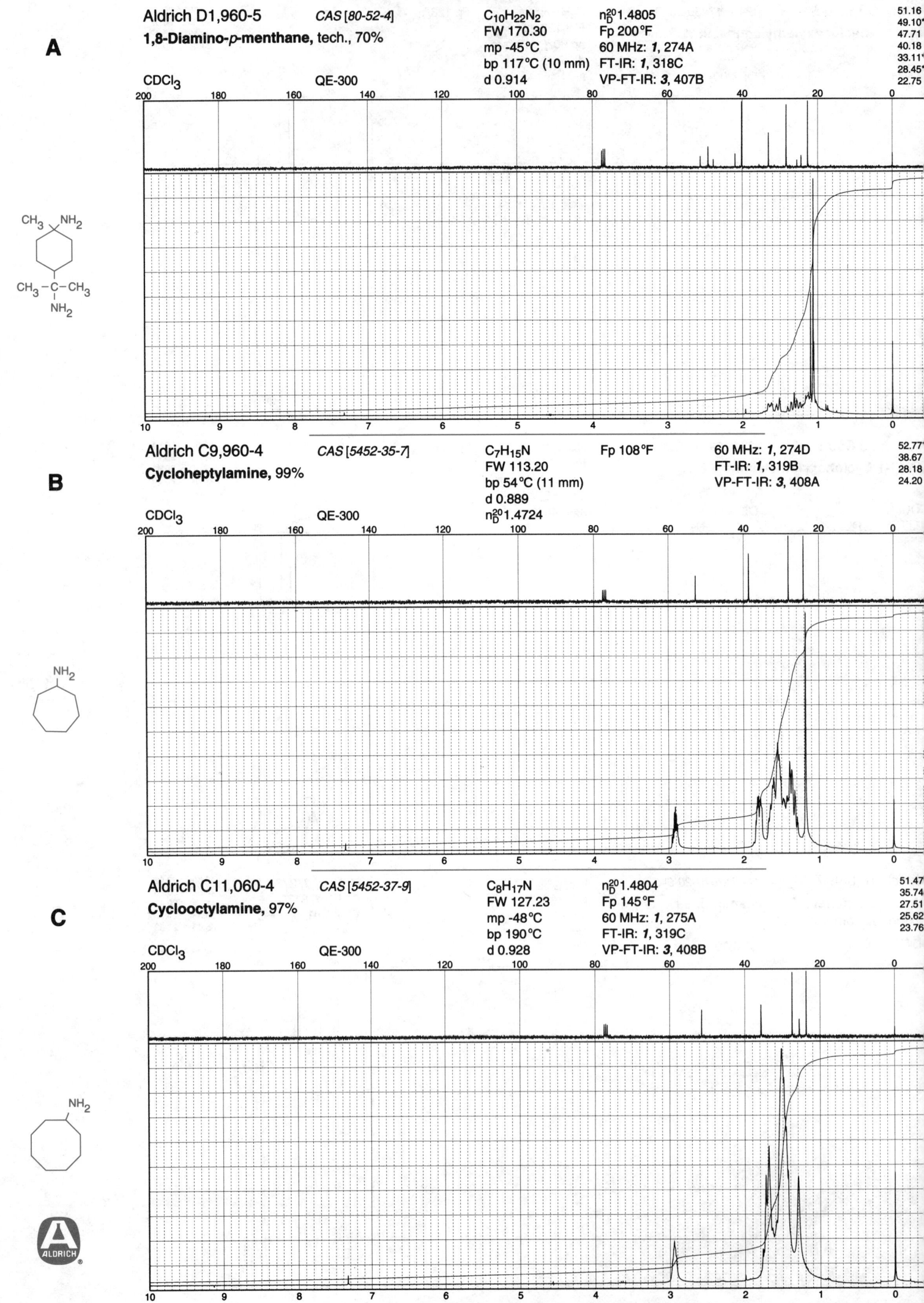

A

Aldrich D1,960-5   *CAS [80-52-4]* $C_{10}H_{22}N_2$ n_D^{20} 1.4805

1,8-Diamino-*p*-menthane, tech., 70%

FW 170.30   Fp 200°F
mp -45°C   60 MHz: *1*, 274A
bp 117°C (10 mm)   FT-IR: *1*, 318C
d 0.914   VP-FT-IR: *3*, 407B

51.16
49.10*
47.71
40.18
33.11*
28.45*
22.75

CDCl₃   QE-300

B

Aldrich C9,960-4   *CAS [5452-35-7]* $C_7H_{15}N$ Fp 108°F

Cycloheptylamine, 99%

FW 113.20   60 MHz: *1*, 274D
bp 54°C (11 mm)   FT-IR: *1*, 319B
d 0.889   VP-FT-IR: *3*, 408A
n_D^{20} 1.4724

52.77*
38.67
28.18
24.20

CDCl₃   QE-300

C

Aldrich C11,060-4   *CAS [5452-37-9]* $C_8H_{17}N$ n_D^{20} 1.4804

Cyclooctylamine, 97%

FW 127.23   Fp 145°F
mp -48°C   60 MHz: *1*, 275A
bp 190°C   FT-IR: *1*, 319C
d 0.928   VP-FT-IR: *3*, 408B

51.47
35.74
27.51
25.62
23.76

CDCl₃   QE-300

Aldrich 12,979-8   CAS [80789-66-8]   C$_{11}$H$_{23}$N   Fp 218°F   60 MHz: *1*, 275C   58.81* 25.86

N-Methylcyclodecylamine, 98%   FW 169.31   FT-IR: *1*, 319D   34.36* 25.56

bp 105°C (8 mm)   VP-FT-IR: *3*, 408C   30.45 24.59

CDCl$_3$ QE-300   d 0.982   30.30 22.82

n$_D^{20}$1.4832

A

Aldrich C9,780-6   CAS [1502-03-0]   C$_{12}$H$_{25}$N   60 MHz: *1*, 275B   47.77*

Cyclododecylamine, 98%   FW 183.34   FT-IR: *1*, 320A   33.46

mp 29°C   VP-FT-IR: *3*, 408D   24.38

bp 123°C (7 mm)   23.89

CDCl$_3$ QE-300   Fp 250°F   23.55

23.51

21.48

B

Aldrich 17,960-4   CAS [7242-92-4]   C$_7$H$_{13}$N   Fp 95°F   60 MHz: *1*, 275D   54.97*

(±)-exo-2-Aminonorbornane, 99%   FW 111.19   FT-IR: *1*, 320B   45.39*

bp 49°C (10 mm)   VP-FT-IR: *3*, 409B   42.34

d 0.938   35.99*

DCl$_3$ QE-300   n$_D^{20}$1.4807   34.23

28.24

26.77

C

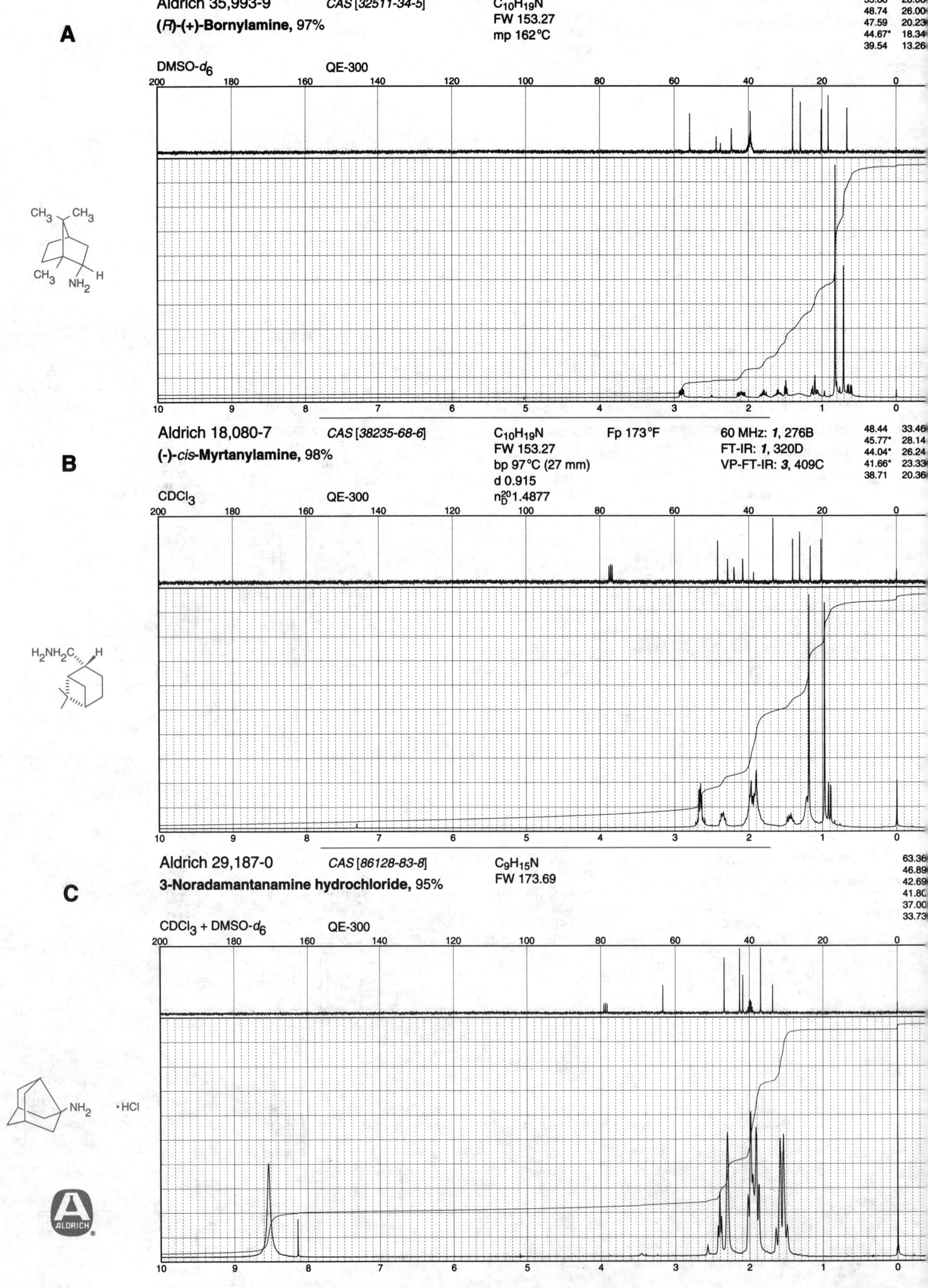

A

Aldrich 35,993-9   CAS [32511-34-5]   C₁₀H₁₉N
(R)-(+)-Bornylamine, 97%   FW 153.27
                                            mp 162°C

55.88* 28.08
48.74   26.00
47.59   20.23
44.67* 18.34
39.54   13.26

DMSO-d₆   QE-300

B

Aldrich 18,080-7   CAS [38235-68-6]   C₁₀H₁₉N   Fp 173°F   60 MHz: 1, 276B
(-)-cis-Myrtanylamine, 98%   FW 153.27   FT-IR: 1, 320D
                                           bp 97°C (27 mm)   VP-FT-IR: 3, 409C
                                           d 0.915
                                           n²⁰_D 1.4877

48.44   33.46
45.77* 28.14
44.04* 26.24
41.66* 23.33
38.71   20.36

CDCl₃   QE-300

C

Aldrich 29,187-0   CAS [86128-83-8]   C₉H₁₅N
3-Noradamantanamine hydrochloride, 95%   FW 173.69

63.36
46.89
42.69
41.80
37.00
33.73

CDCl₃ + DMSO-d₆   QE-300

Aldrich 18,037-8   CAS [17768-41-1]   C₁₁H₁₉N   Fp 198°F   60 MHz: 1, 277B   FT-IR: 1, 322A   VP-FT-IR: 3, 410A
1-Adamantanemethylamine, 98%   FW 165.28   bp 84°C   d 0.933   n²⁰_D 1.5137
CDCl₃   QE-300

55.06
40.05
37.30
33.91
28.48*

A

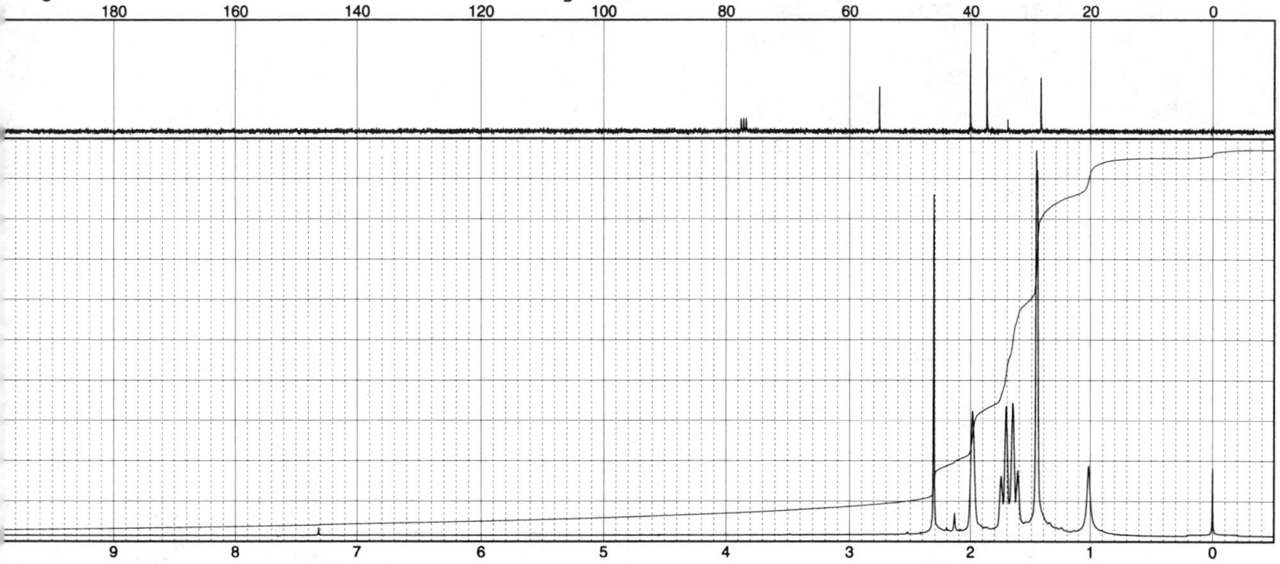

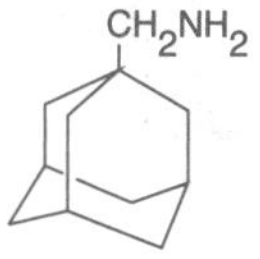

Aldrich 34,081-2   CAS [26562-81-2]   C₁₀H₁₈N₂   FW 239.19
1,3-Adamantanediamine dihydrochloride, 98%
D₂O   QE-300

55.39
45.08
40.70
35.35
31.43*

B

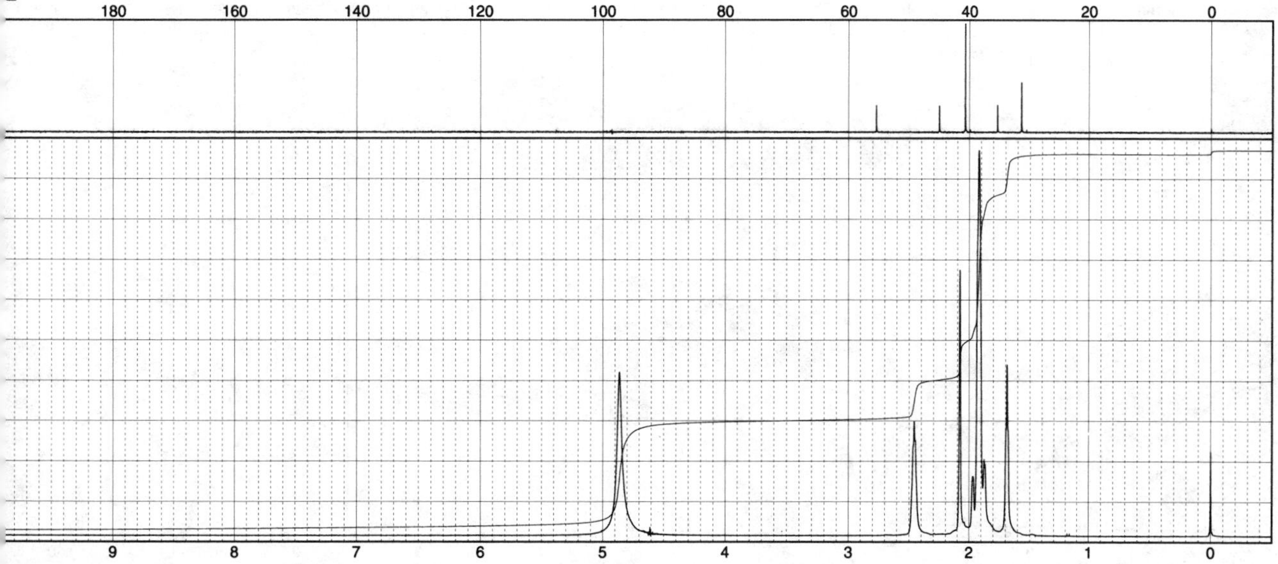

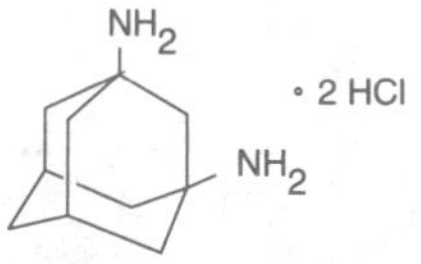

Aldrich 24,107-5   CAS [107-11-9]   C₃H₇N   FW 57.10   mp -88°C   bp 53°C   d 0.761   n²⁰_D 1.4205   Fp -20°F   60 MHz: 1, 277D   FT-IR: 1, 322B   VP-FT-IR: 3, 410C
Allylamine, 99+%
CDCl₃   QE-300

139.82*
113.51
44.77

C

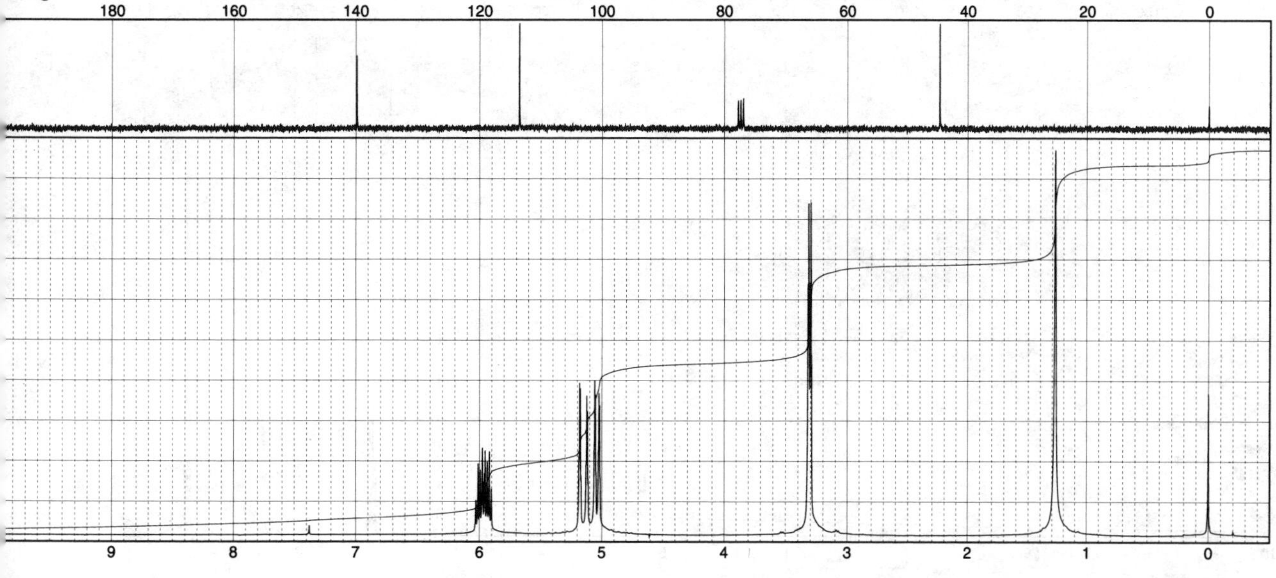

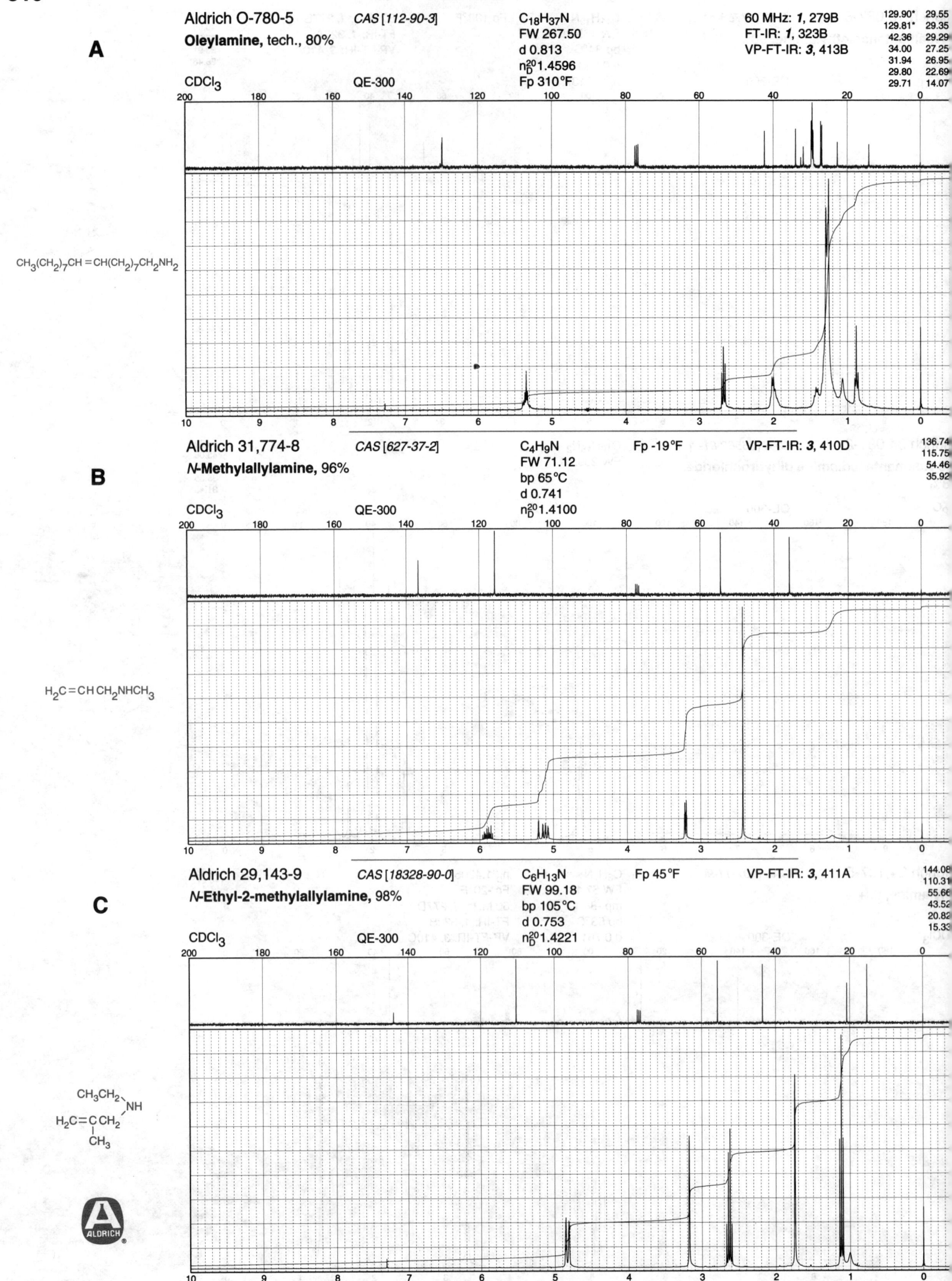

Aldrich O-780-5
Oleylamine, tech., 80%

CAS [112-90-3]

C₁₈H₃₇N
FW 267.50
d 0.813
n²⁰_D 1.4596
Fp 310°F

60 MHz: 1, 279B
FT-IR: 1, 323B
VP-FT-IR: 3, 413B

CDCl₃   QE-300

$CH_3(CH_2)_7CH=CH(CH_2)_7CH_2NH_2$

Aldrich 31,774-8
N-Methylallylamine, 96%

CAS [627-37-2]

C₄H₉N
FW 71.12
bp 65°C
d 0.741
n²⁰_D 1.4100

Fp -19°F

VP-FT-IR: 3, 410D

CDCl₃   QE-300

$H_2C=CHCH_2NHCH_3$

Aldrich 29,143-9
N-Ethyl-2-methylallylamine, 98%

CAS [18328-90-0]

C₆H₁₃N
FW 99.18
bp 105°C
d 0.753
n²⁰_D 1.4221

Fp 45°F

VP-FT-IR: 3, 411A

CDCl₃   QE-300

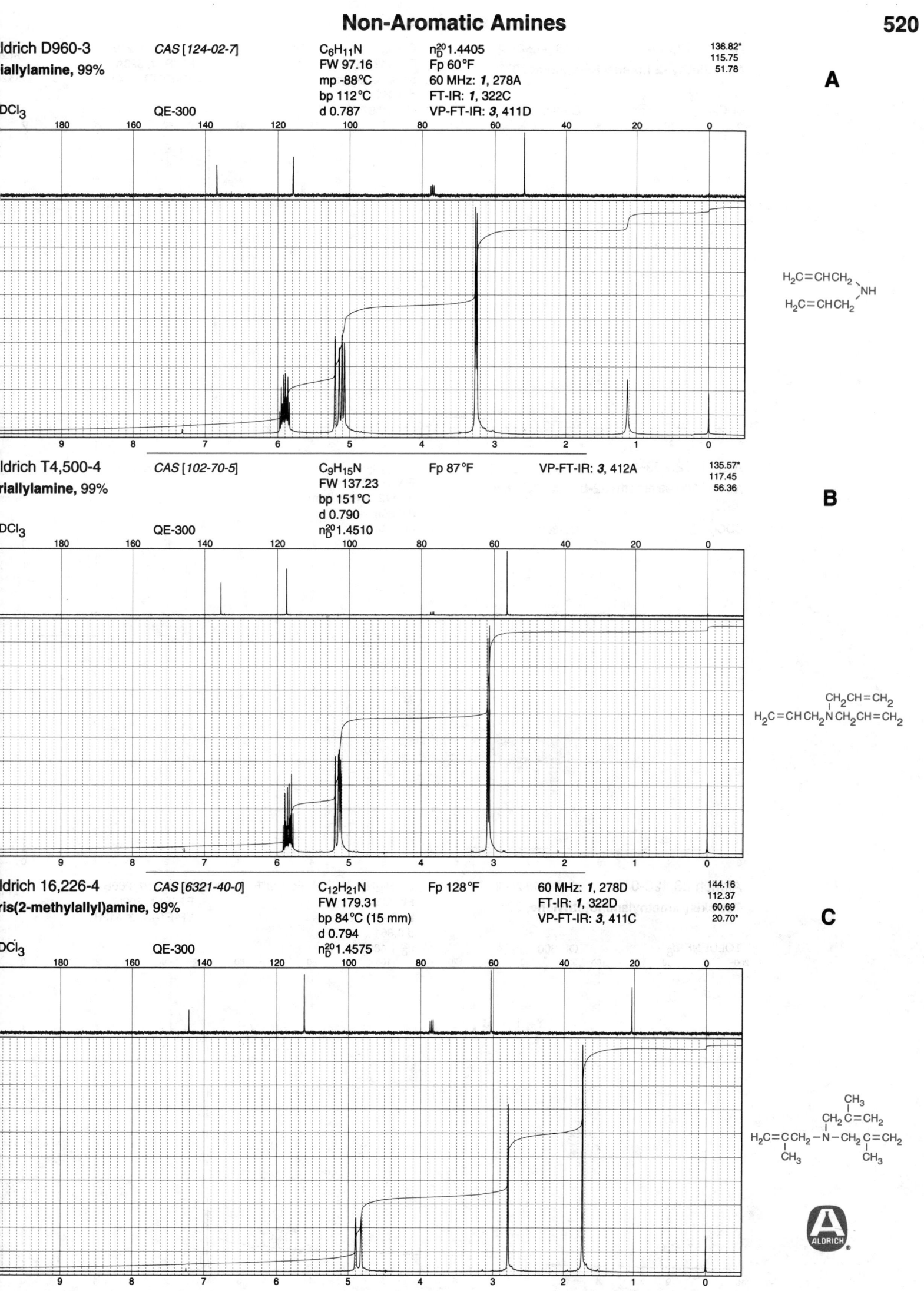

Aldrich D960-3   CAS [124-02-7]   C₆H₁₁N n_D^{20} 1.4405   136.82*
Diallylamine, 99%   FW 97.16   Fp 60°F   115.75
    mp -88°C   60 MHz: 1, 278A   51.78
CDCl₃   QE-300   bp 112°C   FT-IR: 1, 322C
    d 0.787   VP-FT-IR: 3, 411D

A

Aldrich T4,500-4   CAS [102-70-5]   C₉H₁₅N   Fp 87°F   VP-FT-IR: 3, 412A   135.57*
Triallylamine, 99%   FW 137.23   117.45
    bp 151°C   56.36
CDCl₃   QE-300   d 0.790
    n_D^{20} 1.4510

B

Aldrich 16,226-4   CAS [6321-40-0]   C₁₂H₂₁N   Fp 128°F   60 MHz: 1, 278D   144.16
Tris(2-methylallyl)amine, 99%   FW 179.31   FT-IR: 1, 322D   112.37
    bp 84°C (15 mm)   VP-FT-IR: 3, 411C   60.69
CDCl₃   QE-300   d 0.794   20.70*
    n_D^{20} 1.4575

C

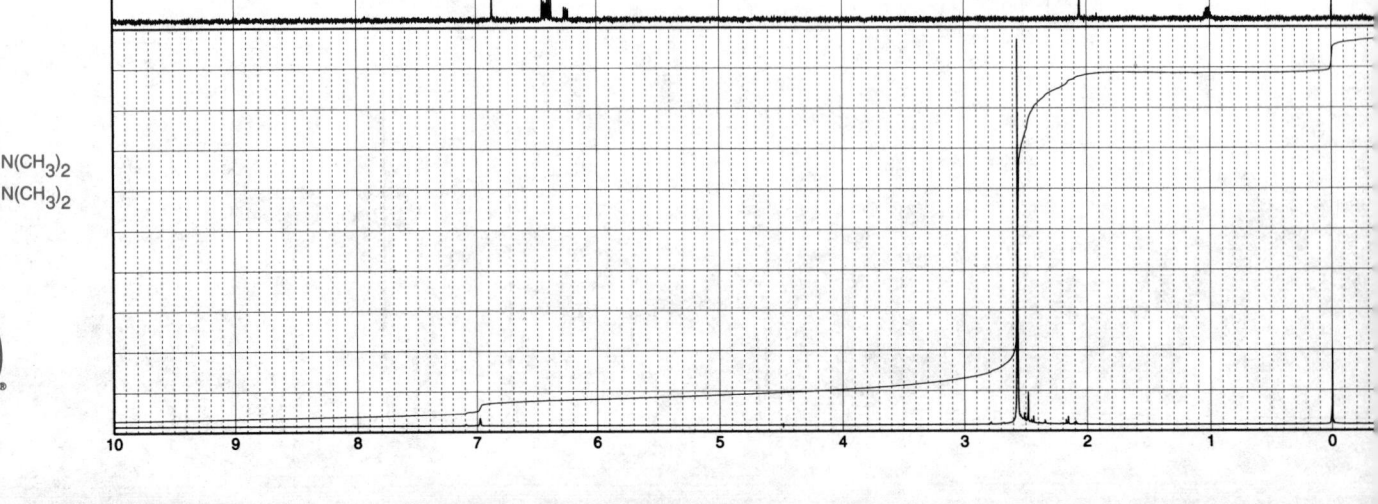

A

Aldrich 12,684-5   *CAS [112-21-0]*

N,N'-Diethyl-2-butene-1,4-diamine, 97%

C₈H₁₈N₂
FW 142.25
bp 82°C (8 mm)
d 0.841
n_D^20 1.4587

Fp 165°F

60 MHz: *1*, 279A
FT-IR: *1*, 323A
VP-FT-IR: *3*, 412C

130.27
51.28
43.59
15.28

CDCl₃   QE-300

CH₃CH₂NHCH₂CH=CHCH₂NHCH₂CH₃

B

Aldrich 12,709-4   *CAS [4559-79-9]*

N,N,N',N'-Tetramethyl-2-butene-1,4-diamine, 95%

C₈H₁₈N₂
FW 142.25
bp 172°C (735 mm)
d 0.808
n_D^20 1.4416

Fp 123°F

VP-FT-IR: *3*, 412D

130.72
61.61
45.17

CDCl₃   QE-300

CH₃NCH₂CH=CHCH₂NCH₃ with CH₃ groups

C

Aldrich 23,423-0   *CAS [996-70-3]*

Tetrakis(dimethylamino)ethylene, 97%

C₁₀H₂₄N₄
FW 200.23
bp 59°C
d 0.861
n_D^20 1.4800

Fp 128°F

60 MHz: *1*, 260B
FT-IR: *1*, 302C
VP-FT-IR: *3*, 412B

137.40
41.18

TOLUENE-d₈   QE-300

(CH₃)₂N   C=C   N(CH₃)₂
(CH₃)₂N   N(CH₃)₂

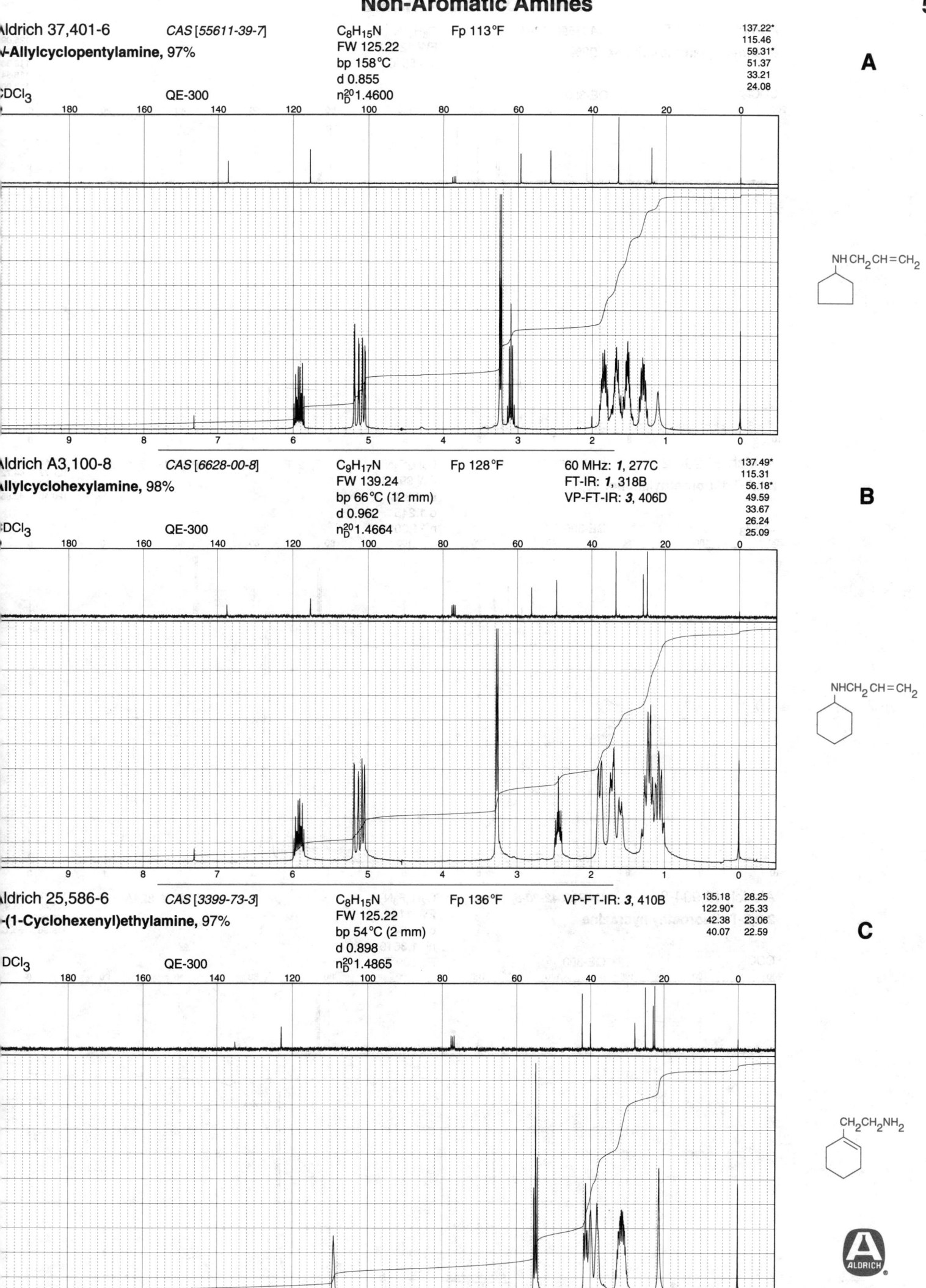

A

Aldrich 37,401-6   CAS [55611-39-7]   C₈H₁₅N   Fp 113°F
N-Allylcyclopentylamine, 97%
FW 125.22
bp 158°C
d 0.855
n²⁰_D 1.4600

CDCl₃   QE-300

137.22*
115.46
59.31*
51.37
33.21
24.08

B

Aldrich A3,100-8   CAS [6628-00-8]   C₉H₁₇N   Fp 128°F
Allylcyclohexylamine, 98%
FW 139.24
bp 66°C (12 mm)
d 0.962
n²⁰_D 1.4664

CDCl₃   QE-300

60 MHz: 1, 277C
FT-IR: 1, 318B
VP-FT-IR: 3, 406D

137.49*
115.31
56.18*
49.59
33.67
26.24
25.09

C

Aldrich 25,586-6   CAS [3399-73-3]   C₈H₁₅N   Fp 136°F
-(1-Cyclohexenyl)ethylamine, 97%
FW 125.22
bp 54°C (2 mm)
d 0.898
n²⁰_D 1.4865

CDCl₃   QE-300

VP-FT-IR: 3, 410B

135.18   28.25
122.90*   25.33
42.38   23.06
40.07   22.59

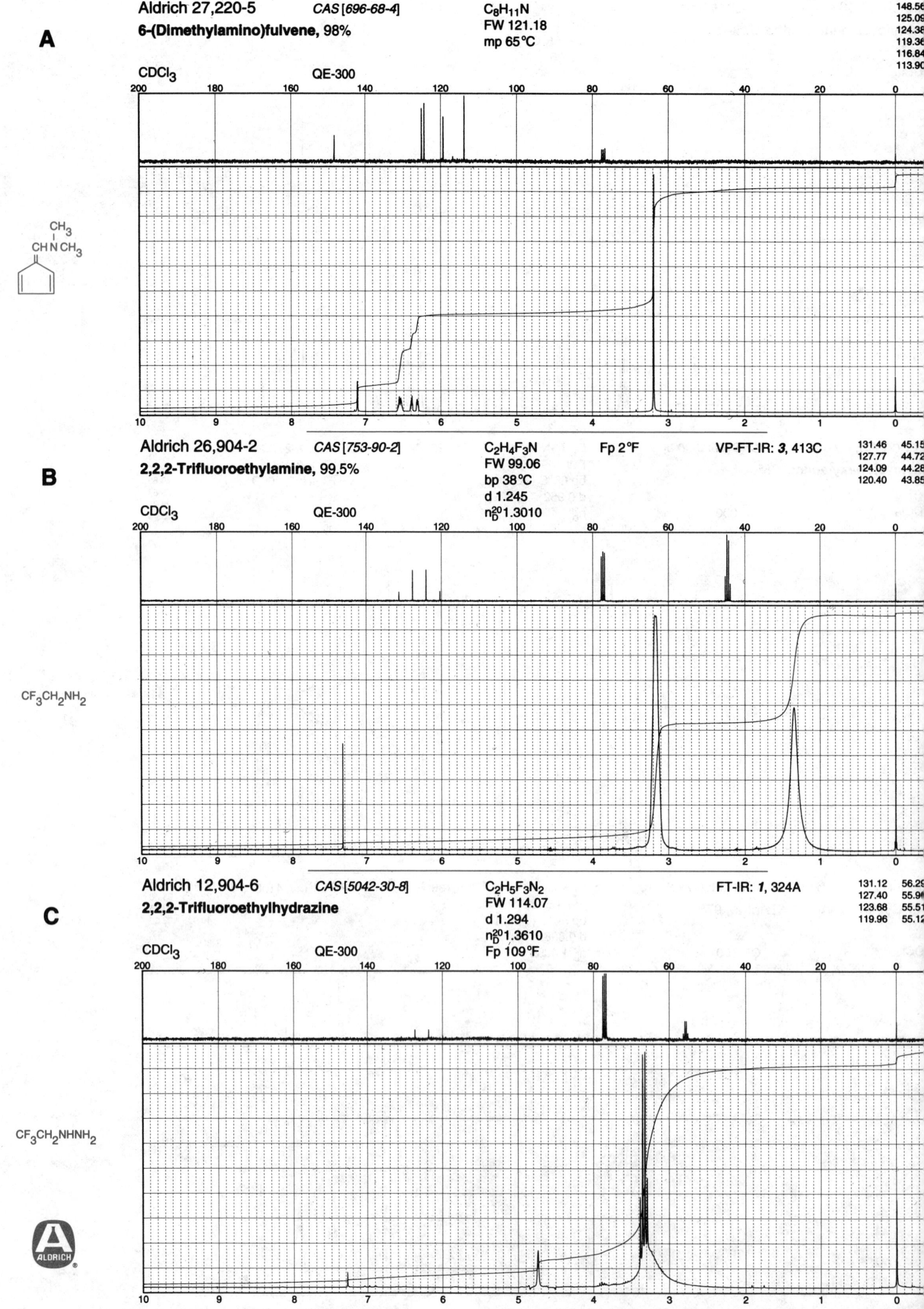

A

Aldrich 27,220-5   *CAS [696-68-4]* $C_8H_{11}N$
6-(Dimethylamino)fulvene, 98%   FW 121.18
mp 65°C

148.56
125.09
124.38
119.36
116.84
113.90

CDCl$_3$ QE-300

B

Aldrich 26,904-2   *CAS [753-90-2]* $C_2H_4F_3N$ Fp 2°F   VP-FT-IR: *3*, 413C
2,2,2-Trifluoroethylamine, 99.5%   FW 99.06
bp 38°C
d 1.245
n_D^{20}1.3010

131.46 45.15
127.77 44.72
124.09 44.28
120.40 43.85

CDCl$_3$ QE-300

$CF_3CH_2NH_2$

C

Aldrich 12,904-6   *CAS [5042-30-8]* $C_2H_5F_3N_2$ FT-IR: *1*, 324A
2,2,2-Trifluoroethylhydrazine   FW 114.07
d 1.294
n_D^{20}1.3610
Fp 109°F

131.12 56.29
127.40 55.96
123.68 55.51
119.96 55.12

CDCl$_3$ QE-300

$CF_3CH_2NHNH_2$

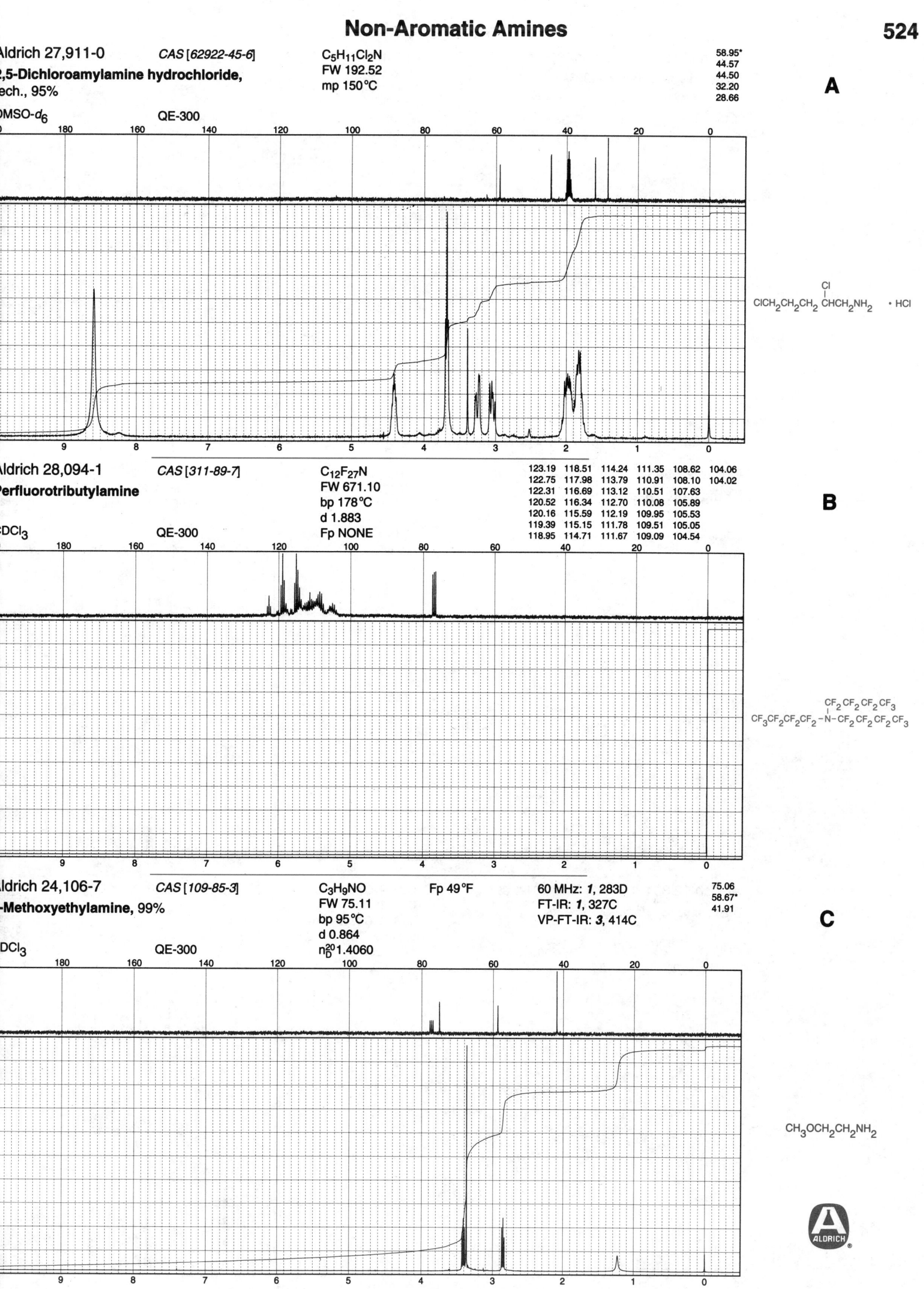

A

Aldrich 27,911-0   *CAS [62922-45-6]* $C_5H_{11}Cl_2N$
2,5-Dichloroamylamine hydrochloride,   FW 192.52
tech., 95%   mp 150°C

58.95*
44.57
44.50
32.20
28.66

DMSO-d_6 QE-300

$ClCH_2CH_2CH_2$—$\overset{Cl}{\underset{|}{CH}}CH_2NH_2$ · HCl

B

Aldrich 28,094-1   *CAS [311-89-7]* $C_{12}F_{27}N$
Perfluorotributylamine   FW 671.10
bp 178°C
d 1.883
Fp NONE

CDCl₃   QE-300

123.19	118.51	114.24	111.35	108.62	104.06
122.75	117.98	113.79	110.91	108.10	104.02
122.31	116.69	113.12	110.51	107.63	
120.52	116.34	112.70	110.08	105.89	
120.16	115.59	112.19	109.95	105.53	
119.39	115.15	111.78	109.51	105.05	
118.95	114.71	111.67	109.09	104.54	

$CF_3CF_2CF_2CF_2$—N—$\overset{CF_2CF_2CF_2CF_3}{\underset{CF_2CF_2CF_2CF_3}{|}}$

C

Aldrich 24,106-7   *CAS [109-85-3]* C_3H_9NO Fp 49°F   60 MHz: **1**, 283D
2-Methoxyethylamine, 99%   FW 75.11   FT-IR: **1**, 327C
bp 95°C   VP-FT-IR: **3**, 414C
d 0.864
n_D^{20} 1.4060

75.06
58.67*
41.91

CDCl₃   QE-300

$CH_3OCH_2CH_2NH_2$

ALDRICH

A

Aldrich M2,500-7   CAS [5332-73-0] $C_4H_{11}NO$ Fp 73°F   VP-FT-IR: **3**, 415A

3-Methoxypropylamine, 99%

FW 89.14
bp 118°C (733 mm)
d 0.874
n_D^{20} 1.4175

CDCl₃   QE-300

70.85
58.61*
39.59
33.55

$CH_3OCH_2CH_2CH_2NH_2$

B

Aldrich 23,943-7   CAS [6291-85-6] $C_5H_{13}NO$ Fp 91°F   60 MHz: **1**, 284B

3-Ethoxypropylamine, 99%

FW 103.17
bp 137°C
d 0.861
n_D^{20} 1.4178

FT-IR: **1**, 327D
VP-FT-IR: **3**, 415B

CDCl₃   QE-300

68.71
66.18
39.78
33.82
15.22*

$CH_3CH_2OCH_2CH_2CH_2NH_2$

C

Aldrich 12,354-4   CAS [16499-88-0] $C_7H_{17}NO$ Fp 146°F   VP-FT-IR: **3**, 415C

3-Butoxypropylamine, 99%

FW 131.22
bp 170°C (756 mm)
d 0.853
n_D^{20} 1.4260

CDCl₃   QE-300

70.76
68.93
39.78
33.68
31.85
19.39
13.94

$CH_3CH_2CH_2CH_2OCH_2CH_2CH_2NH_2$

Non-Aromatic Amines

Aldrich A6,100-4   *CAS [37143-54-7]*   C4H11NO   Fp 48°F   VP-FT-IR: **3**, 415D

(±)-2-Amino-1-methoxypropane, 95%   FW 89.14
bp 93°C (743 mm)
d 0.845
n_D^{20} 1.4065

79.81
58.83*
46.40*
19.90*

A

CDCl3   QE-300

CH3CHCH2OCH3
|
NH2

Aldrich 12,353-6   *CAS [2906-12-9]*   C6H15NO   Fp 103°F   VP-FT-IR: **3**, 416A

3-Isopropoxypropylamine, 99%   FW 117.19
bp 79°C (85 mm)
d 0.845
n_D^{20} 1.4195

71.39*
66.16
39.83
34.10
22.14*

B

CDCl3   QE-300

CH3
|
CH3CHOCH2CH2CH2NH2

Aldrich 22,744-7   *CAS [7300-34-7]*   C10H24N2O2   Fp >230°F   60 MHz: **1**, 306C

4,9-Dioxa-1,12-dodecanediamine, 97%   FW 204.32   FT-IR: **1**, 350C
bp 135°C (4 mm)   VP-FT-IR: **3**, 439C
d 0.962
n_D^{20} 1.4609

70.73
68.91
39.74
33.70
26.54

C

CDCl3   QE-300

(H2NCH2CH2CH2OCH2CH2)2

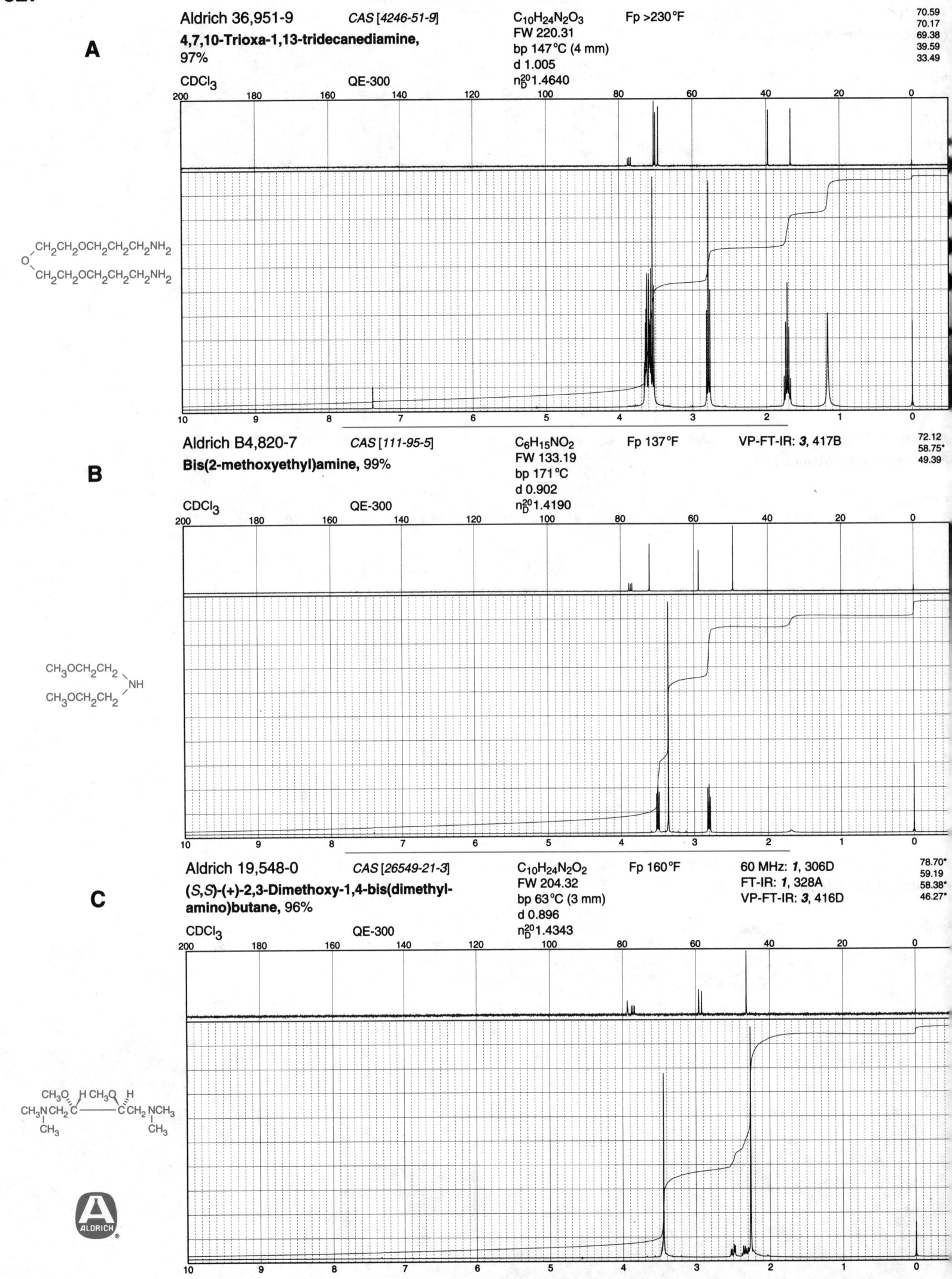

A

Aldrich 36,951-9   *CAS [4246-51-9]*   C₁₀H₂₄N₂O₃   Fp >230°F

4,7,10-Trioxa-1,13-tridecanediamine, 97%

FW 220.31
bp 147°C (4 mm)
d 1.005
n²⁰_D 1.4640

70.59
70.17
69.38
39.59
33.49

CDCl₃   QE-300

B

Aldrich B4,820-7   *CAS [111-95-5]*   C₆H₁₅NO₂   Fp 137°F   VP-FT-IR: *3*, 417B

Bis(2-methoxyethyl)amine, 99%

FW 133.19
bp 171°C
d 0.902
n²⁰_D 1.4190

72.12
58.75*
49.39

CDCl₃   QE-300

C

Aldrich 19,548-0   *CAS [26549-21-3]*   C₁₀H₂₄N₂O₂   Fp 160°F   60 MHz: *1*, 306D

(S,S)-(+)-2,3-Dimethoxy-1,4-bis(dimethyl-amino)butane, 96%

FW 204.32
bp 63°C (3 mm)
d 0.896
n²⁰_D 1.4343

FT-IR: *1*, 328A
VP-FT-IR: *3*, 416D

78.70*
59.19
58.38*
46.27*

CDCl₃   QE-300

Aldrich 21,296-2   *CAS [26549-22-4]*   C$_{10}$H$_{24}$N$_2$O$_2$ Fp 160°F   60 MHz: *1*, 307A   78.71*

(R,R)-(-)-2,3-Dimethoxy-1,4-bis(dimethyl-amino)butane, 95%   FW 204.32   FT-IR: *1*, 328B   59.20
58.38*

bp 63°C (3 mm)   VP-FT-IR: *3*, 417A   46.27*

A

d 0.896

CDCl$_3$ QE-300   n$_D^{20}$1.4345

CH$_3$O   OCH$_3$

CH$_3$NCH$_2$CHCHCH$_2$NCH$_3$
  | |
 CH$_3$ CH$_3$

Aldrich 13,191-1   *CAS [4795-29-3]*   C$_5$H$_{11}$NO   Fp 114°F   60 MHz: *1*, 285B   80.84*
67.91

Tetrahydrofurfurylamine, 97%   FW 101.15   FT-IR: *1*, 328C   46.59

B

bp 154°C (744 mm)   VP-FT-IR: *3*, 417D   28.64
25.96

d 0.980

CDCl$_3$ QE-300   n$_D^{20}$1.4560

[tetrahydrofuran ring]—CH$_2$NH$_2$

Aldrich 14,073-2   *CAS [4637-24-5]*   C$_5$H$_{13}$NO$_2$ Fp 45°F   60 MHz: *1*, 285C   113.18*
53.20*

N,N-Dimethylformamide dimethyl acetal, 94%   FW 119.16   FT-IR: *1*, 328D   37.52*

bp 103°C (720 mm)   VP-FT-IR: *3*, 418D

C

d 0.897

CDCl$_3$ QE-300   n$_D^{20}$1.3972

CH$_3$O   CH$_3$
       \ /
  HCN
       / \
CH$_3$O   CH$_3$

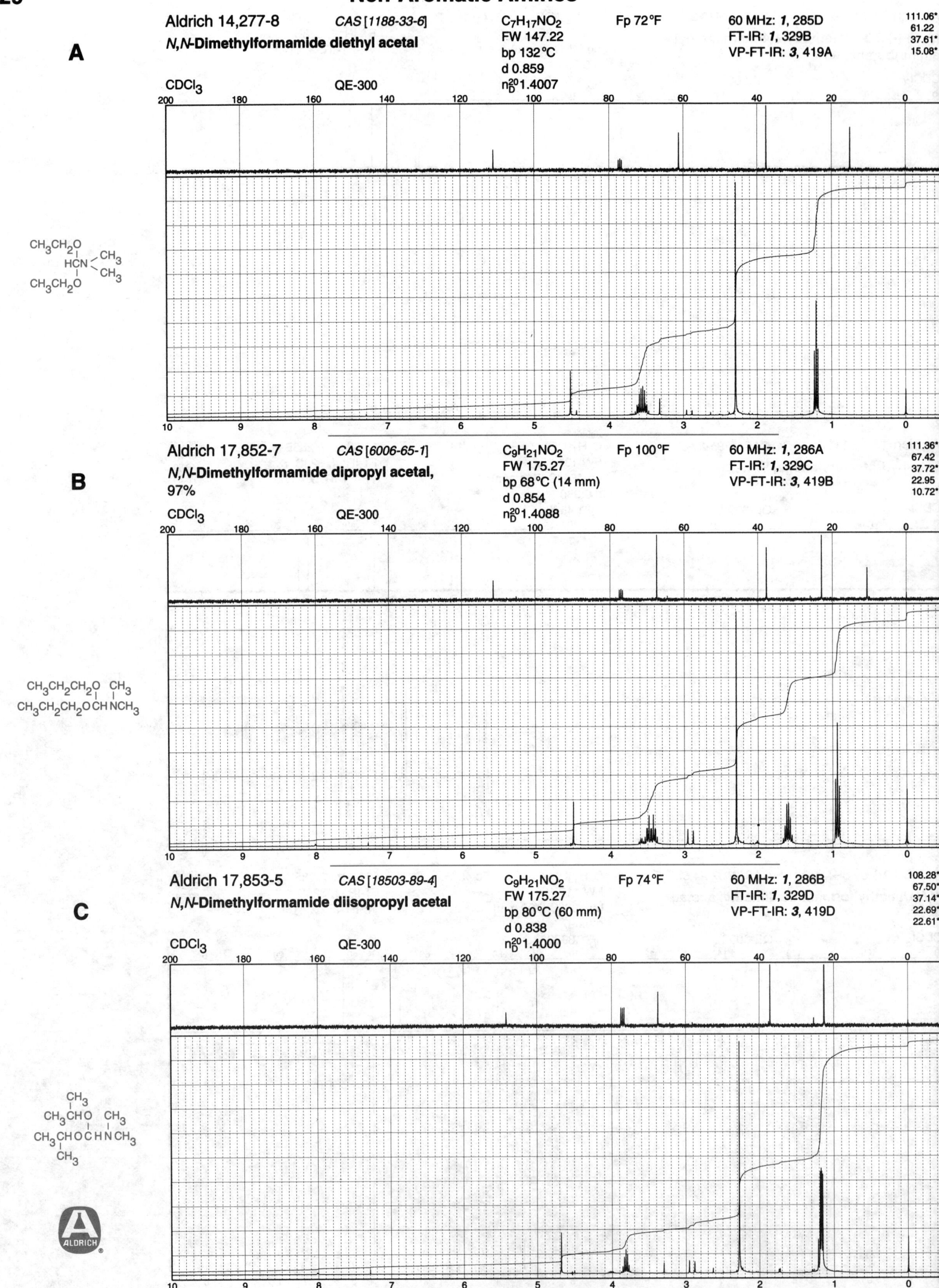

A

Aldrich 14,277-8   *CAS [1188-33-6]* $C_7H_{17}NO_2$ Fp 72°F   60 MHz: *1*, 285D   111.06*
N,N-Dimethylformamide diethyl acetal   FW 147.22   FT-IR: *1*, 329B   61.22
    bp 132°C   VP-FT-IR: *3*, 419A   37.61*
    d 0.859   15.08*
    n_D^{20}1.4007

CDCl₃   QE-300

CH_3CH_2O CH_3
      HCN
CH_3CH_2O CH_3

B

Aldrich 17,852-7   *CAS [6006-65-1]* $C_9H_{21}NO_2$ Fp 100°F   60 MHz: *1*, 286A   111.36*
N,N-Dimethylformamide dipropyl acetal,   FW 175.27   FT-IR: *1*, 329C   67.42
97%   bp 68°C (14 mm)   VP-FT-IR: *3*, 419B   37.72*
    d 0.854   22.95
    n_D^{20}1.4088   10.72*

CDCl₃   QE-300

$CH_3CH_2CH_2O$ CH_3
$CH_3CH_2CH_2O CH NCH_3$

C

Aldrich 17,853-5   *CAS [18503-89-4]* $C_9H_{21}NO_2$ Fp 74°F   60 MHz: *1*, 286B   108.28*
N,N-Dimethylformamide diisopropyl acetal   FW 175.27   FT-IR: *1*, 329D   67.50*
    bp 80°C (60 mm)   VP-FT-IR: *3*, 419D   37.14*
    d 0.838   22.69*
    n_D^{20}1.4000   22.61*

CDCl₃   QE-300

        CH_3
CH_3CHO CH_3
$CH_3CHOCHNCH_3$
        CH_3

Aldrich 35,880-0   *CAS [36805-97-7]*   C₁₁H₂₅NO₂   Fp 93°F

$C_{11}H_{25}NO_2$ FW 203.33   bp 57°C (8 mm)   d 0.848 n_D^{20} 1.4130

N,N-Dimethylformamide di-*tert*-butyl acetal, tech., 90+%

CDCl₃   QE-300

101.72*
72.96
36.48*
28.85*

A

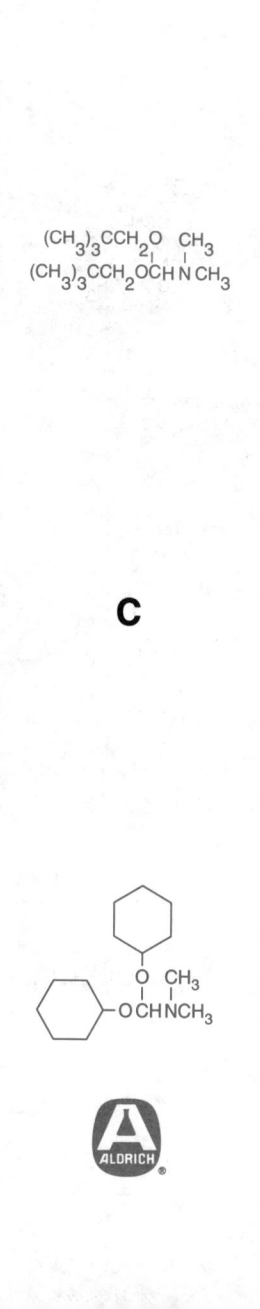

Aldrich 14,024-4   *CAS [4909-78-8]*

N,N-Dimethylformamide dineopentyl acetal, 99%

C₁₃H₂₉NO₂   Fp 125°F   60 MHz: *1*, 286D
FW 231.38   FT-IR: *1*, 330A
bp 86°C (10 mm)   VP-FT-IR: *3*, 420A
d 0.829
n_D^{20} 1.4117

111.74*
75.37
37.99*
31.96
26.80*

CDCl₃   QE-300

B

Aldrich 14,278-6   *CAS [2016-05-9]*

N,N-Dimethylformamide dicyclohexyl acetal

C₁₅H₂₉NO₂   Fp 195°F   60 MHz: *1*, 287A
FW 255.40   FT-IR: *1*, 330B
bp 116°C
d 0.958
n_D^{20} 1.4678

107.84*
73.16*
37.30*
32.90
32.69
25.89
24.22

CDCl₃   QE-300

C

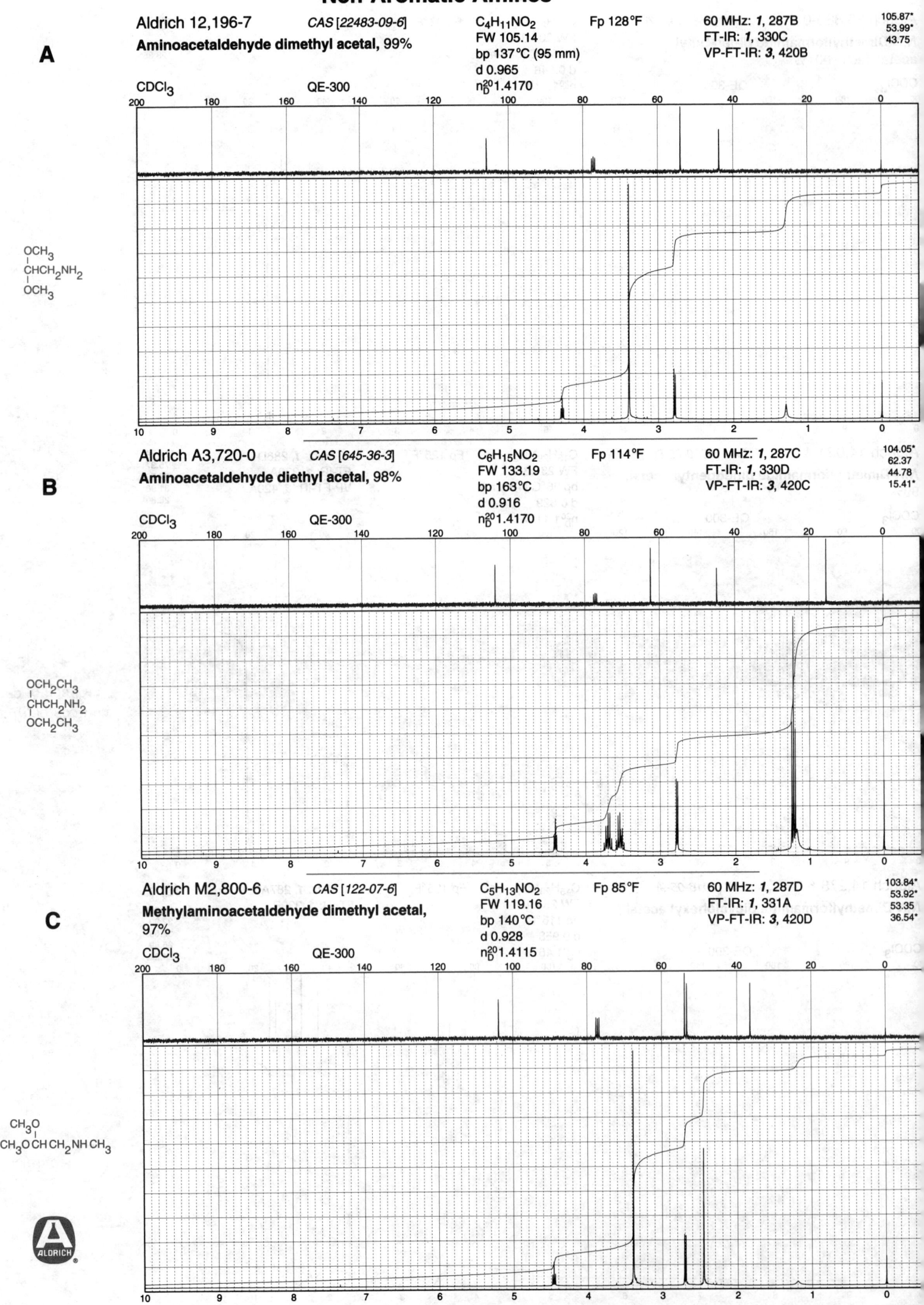

A

Aldrich 12,196-7   *CAS [22483-09-6]* $C_4H_{11}NO_2$ Fp 128°F   60 MHz: *1*, 287B   105.87*
Aminoacetaldehyde dimethyl acetal, 99%   FW 105.14   FT-IR: *1*, 330C   53.99*
bp 137°C (95 mm)   VP-FT-IR: *3*, 420B   43.75
d 0.965
n_D^{20}1.4170

CDCl$_3$ QE-300

OCH$_3$
|
CHCH$_2$NH$_2$
|
OCH$_3$

B

Aldrich A3,720-0   *CAS [645-36-3]* $C_6H_{15}NO_2$ Fp 114°F   60 MHz: *1*, 287C   104.05*
Aminoacetaldehyde diethyl acetal, 98%   FW 133.19   FT-IR: *1*, 330D   62.37
bp 163°C   VP-FT-IR: *3*, 420C   44.78
d 0.916   15.41*
n_D^{20}1.4170

CDCl$_3$ QE-300

OCH$_2$CH$_3$
|
CHCH$_2$NH$_2$
|
OCH$_2$CH$_3$

C

Aldrich M2,800-6   *CAS [122-07-6]* $C_5H_{13}NO_2$ Fp 85°F   60 MHz: *1*, 287D   103.84*
Methylaminoacetaldehyde dimethyl acetal,   FW 119.16   FT-IR: *1*, 331A   53.92*
97%   bp 140°C   VP-FT-IR: *3*, 420D   53.35
d 0.928   36.54*
n_D^{20}1.4115

CDCl$_3$ QE-300

CH$_3$O
|
CH$_3$OCHCH$_2$NHCH$_3$

ALDRICH

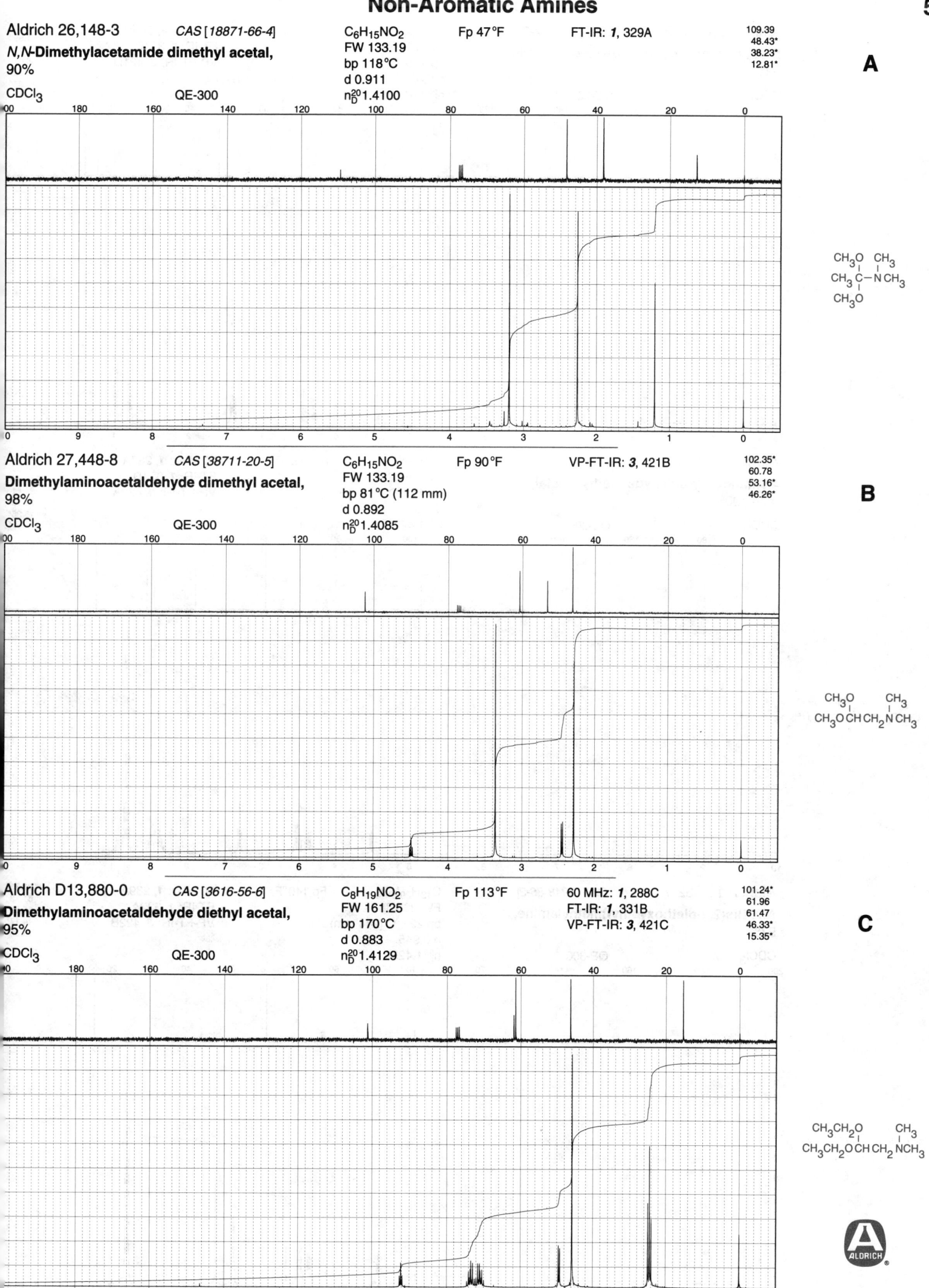

Aldrich 26,148-3   *CAS [18871-66-4]*

N,N-Dimethylacetamide dimethyl acetal,
90%

CDCl₃   QE-300

C₆H₁₅NO₂
FW 133.19
bp 118°C
d 0.911
n²⁰_D 1.4100

Fp 47°F

FT-IR: *1*, 329A

109.39
48.43*
38.23*
12.81*

A

Aldrich 27,448-8   *CAS [38711-20-5]*

Dimethylaminoacetaldehyde dimethyl acetal,
98%

CDCl₃   QE-300

C₆H₁₅NO₂
FW 133.19
bp 81°C (112 mm)
d 0.892
n²⁰_D 1.4085

Fp 90°F

VP-FT-IR: *3*, 421B

102.35*
60.78
53.16*
46.26*

B

Aldrich D13,880-0   *CAS [3616-56-6]*

Dimethylaminoacetaldehyde diethyl acetal,
95%

CDCl₃   QE-300

C₈H₁₉NO₂
FW 161.25
bp 170°C
d 0.883
n²⁰_D 1.4129

Fp 113°F

60 MHz: *1*, 288C
FT-IR: *1*, 331B
VP-FT-IR: *3*, 421C

101.24*
61.96
61.47
46.33*
15.35*

C

A

Aldrich D8,580-2   *CAS [3616-57-7]*

Diethylaminoacetaldehyde diethyl acetal,
99%

CDCl₃   QE-300

C₁₀H₂₃NO₂
FW 189.30
bp 195°C
d 0.850
n²⁰_D 1.4189

Fp 149°F

60 MHz: *1*, 288D
FT-IR: *1*, 331C
VP-FT-IR: *3*, 421D

102.62*
61.99
56.17
48.07
15.39*
11.96*

CH₃CH₂O   CH₂CH₃
CH₃CH₂OCHCH₂NCH₂CH₃

B

Aldrich A4,415-0   *CAS [6346-09-4]*

4-Aminobutyraldehyde diethyl acetal,
tech., 90%

CDCl₃   QE-300

C₈H₁₉NO₂
FW 161.25
bp 196°C
d 0.933
n²⁰_D 1.4275

Fp 145°F

60 MHz: *1*, 288A
FT-IR: *1*, 331D
VP-FT-IR: *3*, 422A

102.88*
61.02
42.14
31.11
29.13
15.36*

OCH₂CH₃
CH₃CH₂OCHCH₂CH₂CH₂NH₂

C

Aldrich 11,582-7   *CAS [6948-86-3]*

N,N-Bis(2,2-diethoxyethyl)methylamine,
97%

CDCl₃   QE-300

C₁₃H₂₉NO₄
FW 263.38
bp 221°C (244 mm)
d 0.945
n²⁰_D 1.4259

Fp 140°F

60 MHz: *1*, 289A
FT-IR: *1*, 332A
VP-FT-IR: *3*, 422B

101.71*
61.73
60.56
44.17*
15.38*

CH₃
(CH₃CH₂O)₂CHCH₂−N−CH₂CH(OCH₂CH₃)₂

Aldrich 27,275-2   *CAS [42548-78-7]*

N-Ethylhydroxylamine hydrochloride, 97%

C₂H₇NO
FW 97.55
mp 36°C
Fp >230°F

C_2H_7NO
FW 97.55
mp 36°C
Fp >230°F

45.95
8.81*

A

CDCl₃   QE-300

$CH_3CH_2NHOH \cdot HCl$

Aldrich D9,720-7   *CAS [3710-84-7]*

N,N-Diethylhydroxylamine, 99%

$C_4H_{11}NO$
FW 89.14
mp -26°C
bp 128°C
d 0.867

n_D^{20} 1.4195
Fp 113°F
60 MHz: *1*, 290C
FT-IR: *1*, 333A
VP-FT-IR: *3*, 423C

53.99
12.10*

B

CDCl₃   QE-300

$$CH_3CH_2 \overset{\displaystyle OH}{\underset{\displaystyle |}{N}} CH_2CH_3$$

Aldrich 34,006-5   *CAS [39684-28-1]*

O-(tert-Butyl)hydroxylamine hydrochloride, 99%

$C_4H_{11}NO$
FW 125.60
mp 159°C

81.65
26.01*

C

DMSO-d₆   QE-300

Offset: 1.1 ppm.

$$H_2NOC\overset{\displaystyle CH_3}{\underset{\displaystyle CH_3}{\overset{|}{\underset{|}{C}}}}CH_3 \cdot HCl$$

ALDRICH

A

Aldrich 25,456-8
O-Allylhydroxylamine hydrochloride hydrate

C$_3$H$_7$NO
FW 109.56
mp 175°C d.

FT-IR: *1*, 333C

130.61*
121.04
74.39

DMSO-d_6 QE-300

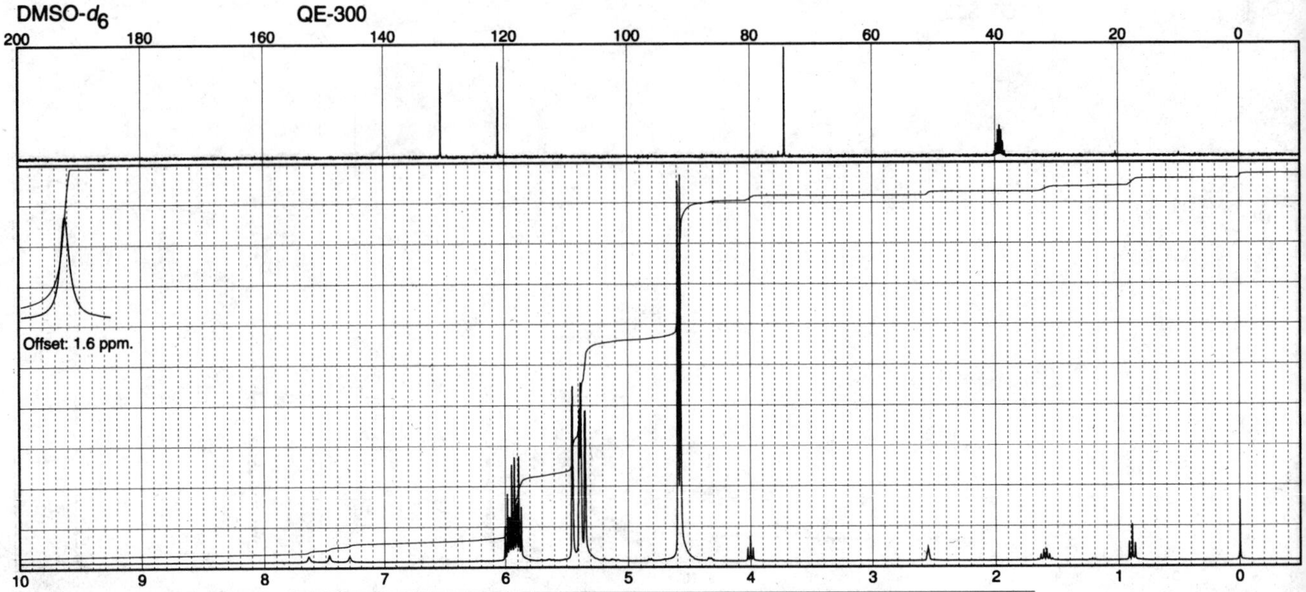

H$_2$NOCH$_2$CH=CH$_2$ · HCl   · xH$_2$O

Offset: 1.6 ppm.

B

Aldrich 11,016-7   CAS [141-43-5]
Ethanolamine, 99+%

C$_2$H$_7$NO
FW 61.08
mp 11°C
bp 170°C
d 1.012

n$_D^{20}$1.4540
Fp 200°F
60 MHz: *1*, 291B
FT-IR: *1*, 334A
VP-FT-IR: *3*, 423D

63.37
44.01

CDCl$_3$ QE-300

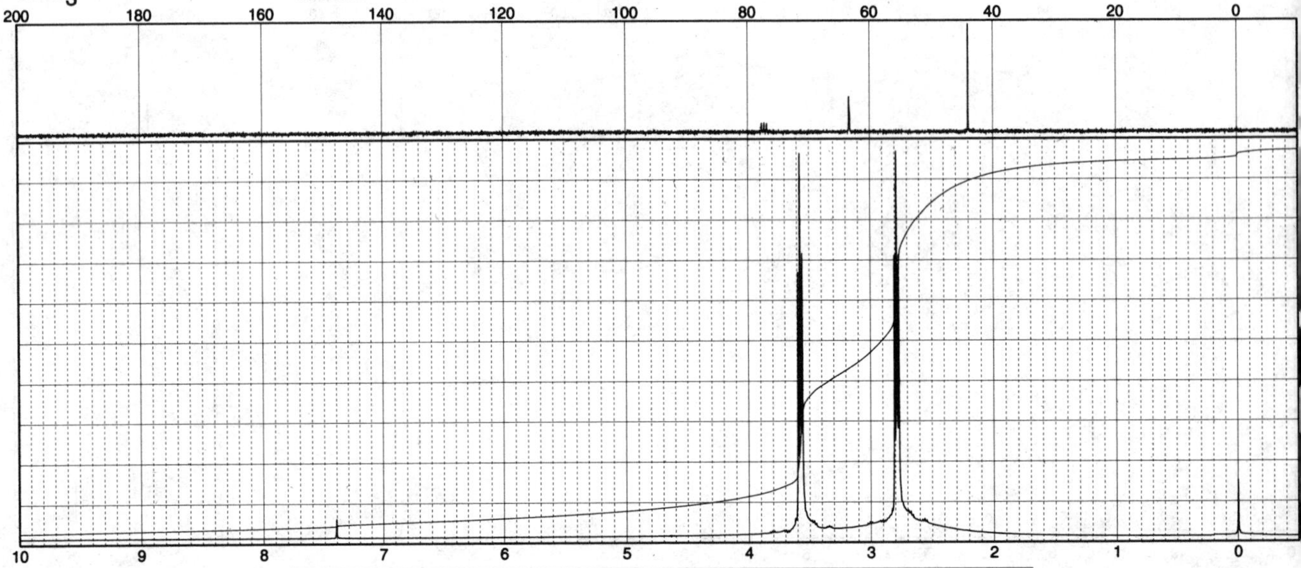

HOCH$_2$CH$_2$NH$_2$

C

Aldrich 12,862-7   CAS [109-84-2]
2-Hydroxyethylhydrazine, 97%

C$_2$H$_8$N$_2$O
FW 76.10
bp 219°C (754 mm)
d 1.119
n$_D^{20}$1.4930

Fp 165°F

60 MHz: *1*, 291C
FT-IR: *1*, 334C

59.69
56.96

CDCl$_3$ QE-300

HOCH$_2$CH$_2$NHNH$_2$

ALDRICH

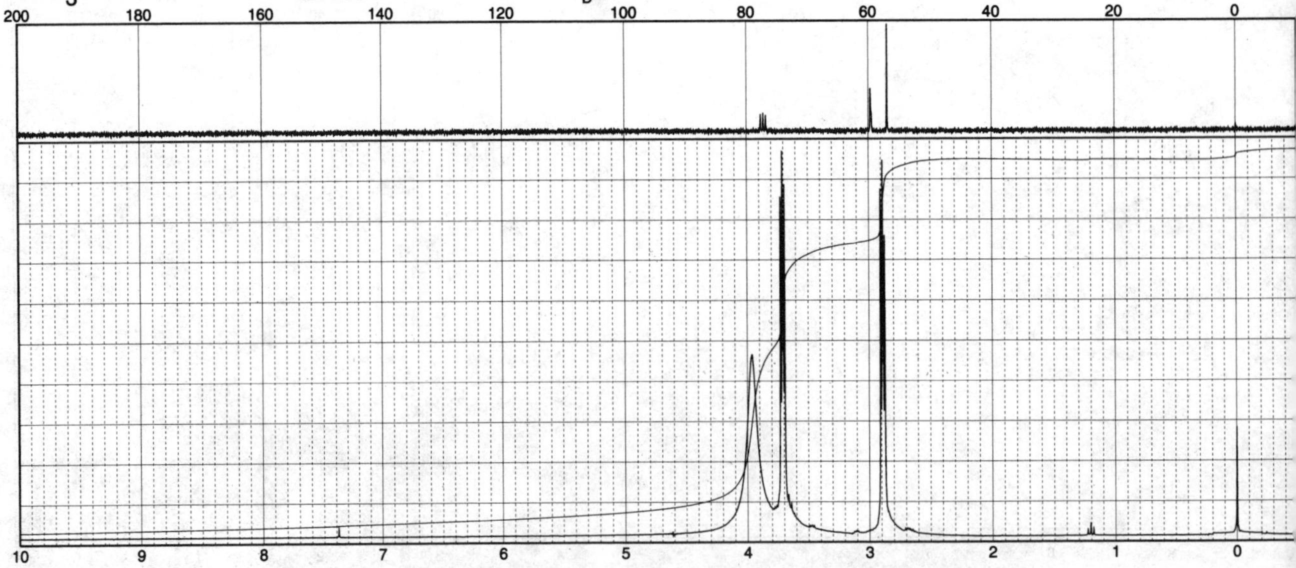

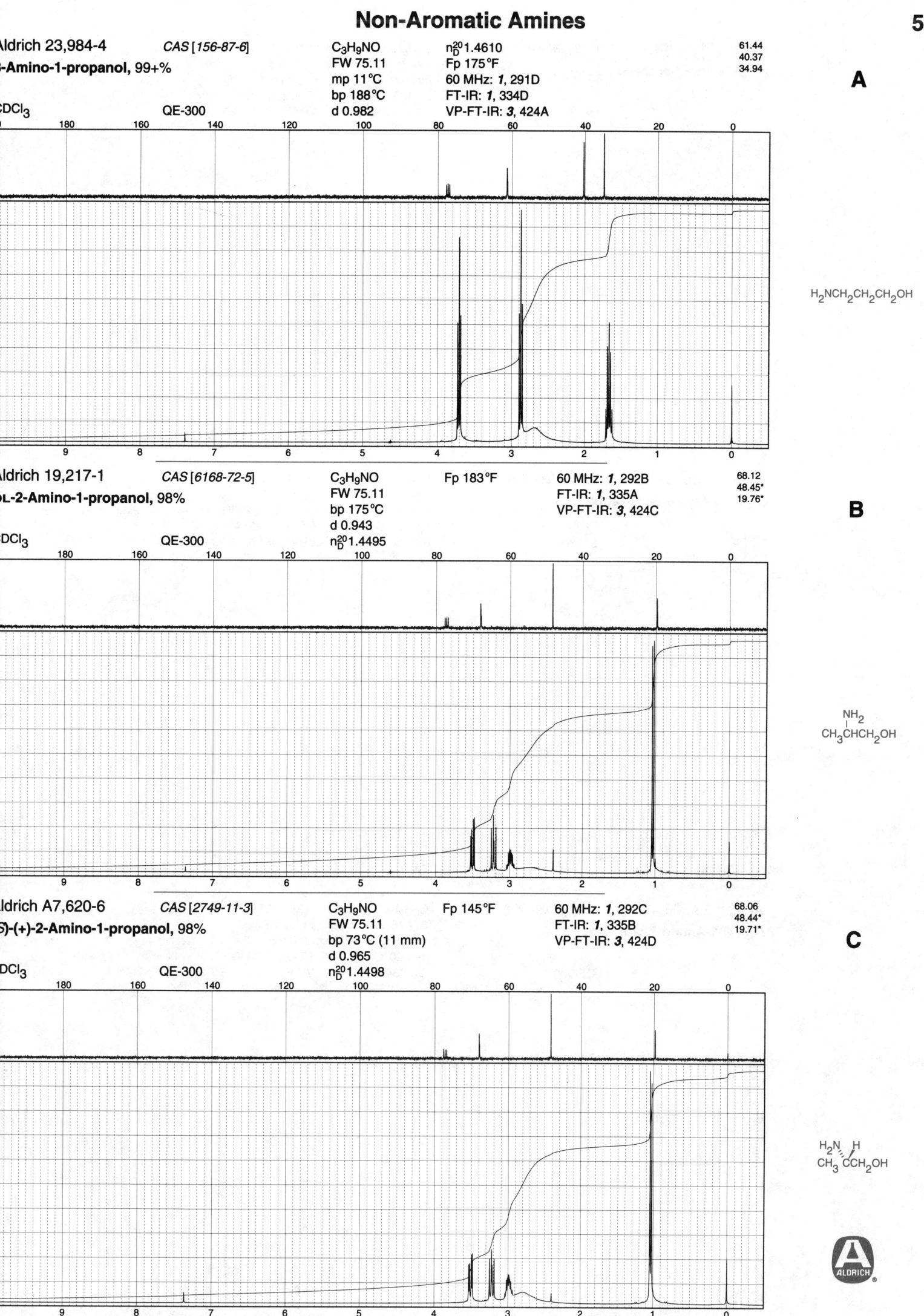

Aldrich 23,984-4   *CAS [156-87-6]*   C₃H₉NO   n²⁰_D 1.4610

3-Amino-1-propanol, 99+%

FW 75.11   Fp 175°F

mp 11°C   60 MHz: *1*, 291D

bp 188°C   FT-IR: *1*, 334D

d 0.982   VP-FT-IR: *3*, 424A

CDCl₃   QE-300

A

$H_2NCH_2CH_2CH_2OH$

Aldrich 19,217-1   *CAS [6168-72-5]*   C₃H₉NO   Fp 183°F

DL-2-Amino-1-propanol, 98%

FW 75.11   FT-IR: *1*, 335A

bp 175°C   VP-FT-IR: *3*, 424C

d 0.943

n²⁰_D 1.4495

CDCl₃   QE-300   60 MHz: *1*, 292B

B

CH_3CHCH_2OH with NH_2

Aldrich A7,620-6   *CAS [2749-11-3]*   C₃H₉NO   Fp 145°F

(S)-(+)-2-Amino-1-propanol, 98%

FW 75.11   60 MHz: *1*, 292C

bp 73°C (11 mm)   FT-IR: *1*, 335B

d 0.965   VP-FT-IR: *3*, 424D

n²⁰_D 1.4498

CDCl₃   QE-300

C

H_2N H
CH_3 CCH_2OH

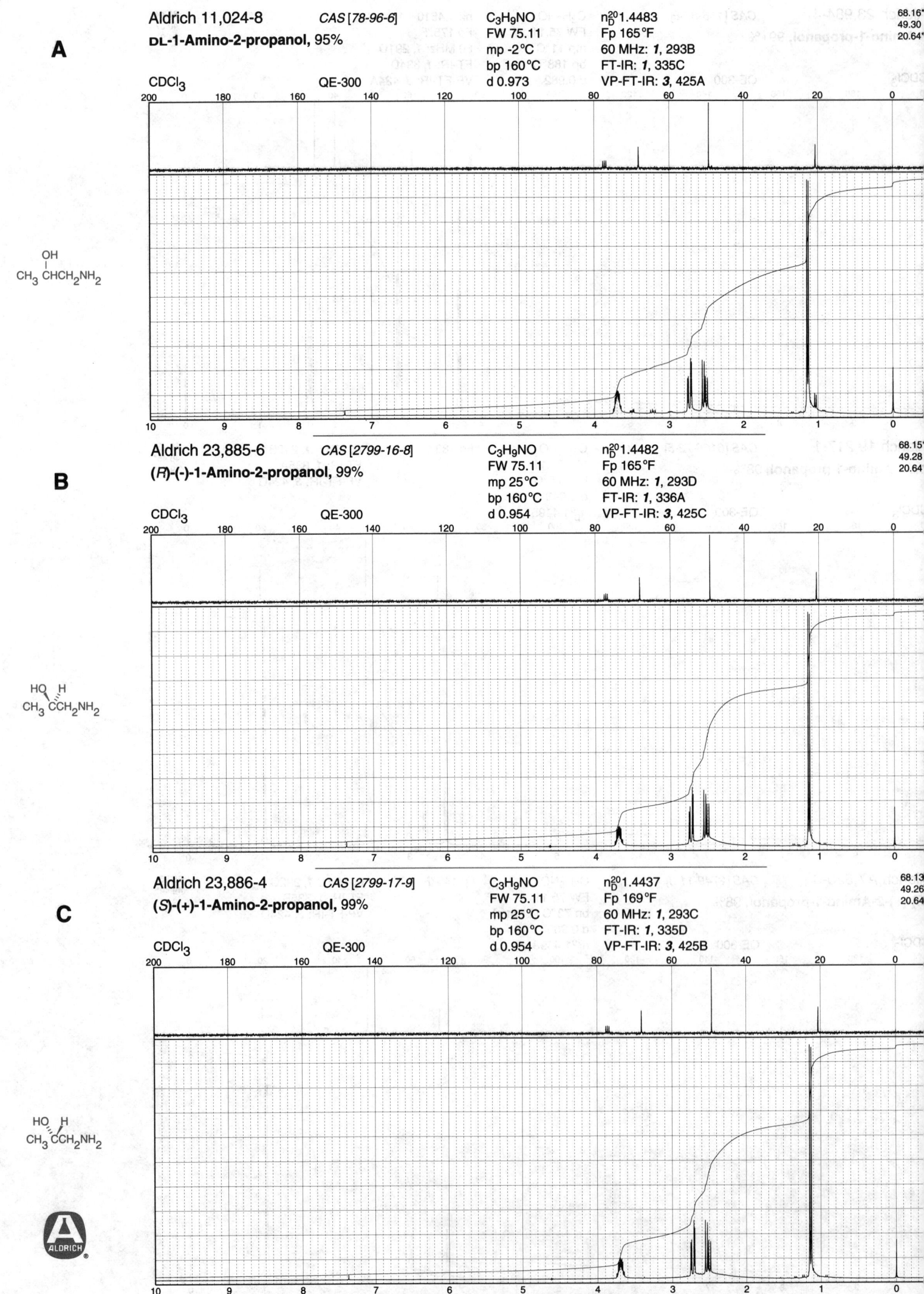

Aldrich 11,024-8   *CAS [78-96-6]* C_3H_9NO n_D^{20} 1.4483   68.16*
DL-1-Amino-2-propanol, 95%   FW 75.11   Fp 165°F   49.30
    mp -2°C   60 MHz: *1*, 293B   20.64*
    bp 160°C   FT-IR: *1*, 335C
CDCl₃   QE-300   d 0.973   VP-FT-IR: *3*, 425A

OH
CH₃ CHCH₂NH₂

Aldrich 23,885-6   *CAS [2799-16-8]* C_3H_9NO n_D^{20} 1.4482   68.15*
(R)-(-)-1-Amino-2-propanol, 99%   FW 75.11   Fp 165°F   49.28
    mp 25°C   60 MHz: *1*, 293D   20.64*
    bp 160°C   FT-IR: *1*, 336A
CDCl₃   QE-300   d 0.954   VP-FT-IR: *3*, 425C

HO H
CH₃ CCH₂NH₂

Aldrich 23,886-4   *CAS [2799-17-9]* C_3H_9NO n_D^{20} 1.4437   68.13*
(S)-(+)-1-Amino-2-propanol, 99%   FW 75.11   Fp 169°F   49.26
    mp 25°C   60 MHz: *1*, 293C   20.64*
    bp 160°C   FT-IR: *1*, 335D
CDCl₃   QE-300   d 0.954   VP-FT-IR: *3*, 425B

HO H
CH₃ CCH₂NH₂

Non-Aromatic Amines

A

Aldrich 17,833-0 CAS [13325-10-5]
4-Amino-1-butanol, 98%

C₄H₁₁NO
FW 89.14
bp 206°C
d 0.967
n_D^{20} 1.4610

Fp 226°F

60 MHz: *1*, 292A
FT-IR: *1*, 338D
VP-FT-IR: *3*, 429A

62.07
41.83
31.16
30.62

CDCl₃   QE-300

H₂NCH₂CH₂CH₂CH₂OH

B

Aldrich 13,251-9 CAS [5856-63-3]
(R)-(-)-2-Amino-1-butanol

C₄H₁₁NO
FW 89.14
mp -2°C
bp 173°C
d 0.947

n_D^{20} 1.4525
Fp 180°F
60 MHz: *1*, 295A
FT-IR: *1*, 338A
VP-FT-IR: *3*, 428A

66.33
54.49*
27.12
10.48*

CDCl₃   QE-300

CH₃CH₂CCH₂OH
H₂N H

C

Aldrich 13,252-7 CAS [5856-62-2]
(S)-(+)-2-Amino-1-butanol, 98+%

C₄H₁₁NO
FW 89.14
bp 173°C
d 0.944
n_D^{20} 1.4521

Fp 175°F

60 MHz: *1*, 292D
FT-IR: *1*, 337D
VP-FT-IR: *3*, 427D

66.29
54.47*
27.07
10.49*

CDCl₃   QE-300

CH₃CH₂CCH₂OH
H₂N H

ALDRICH

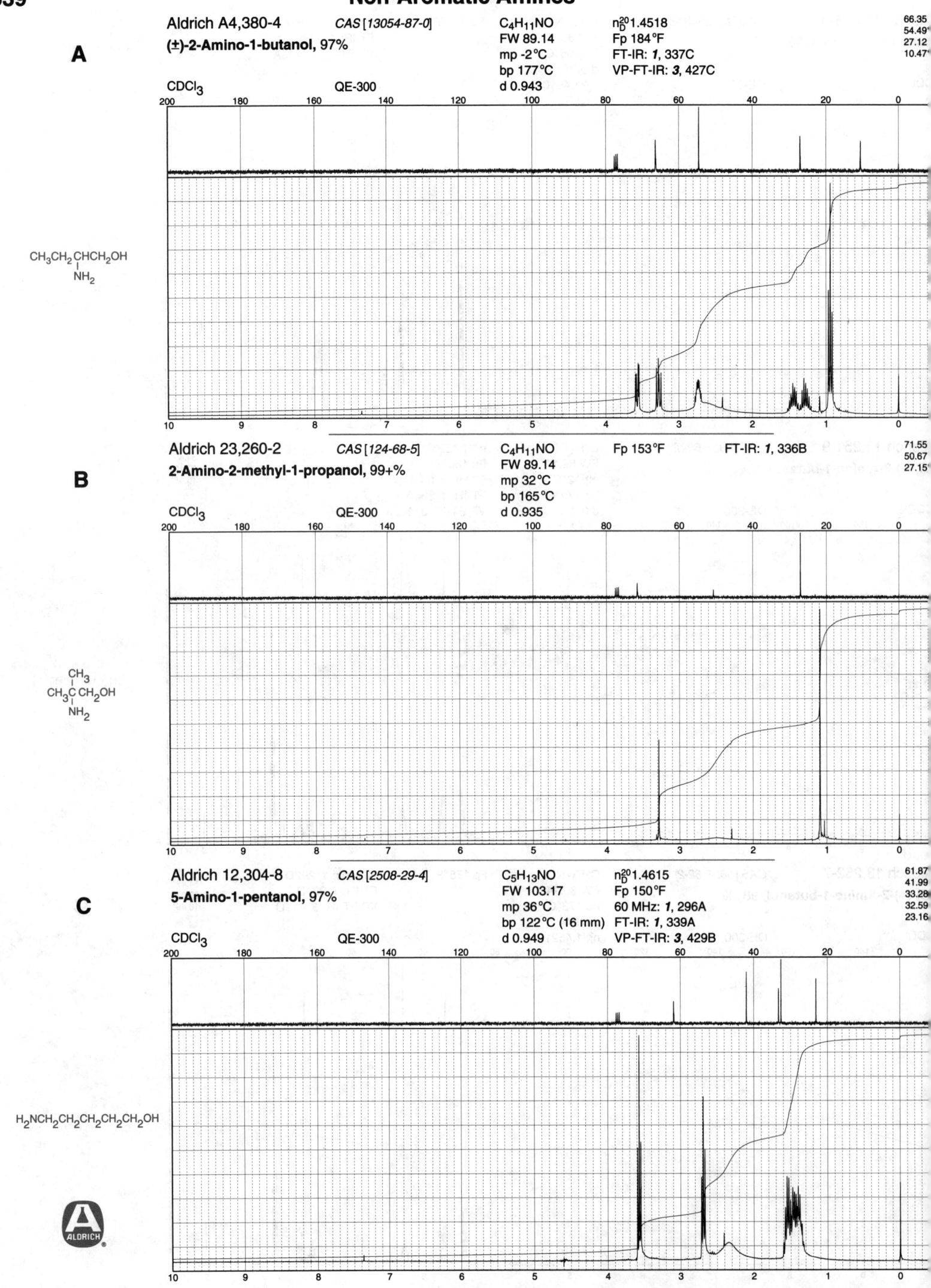

A

Aldrich A4,380-4   *CAS [13054-87-0]*   C₄H₁₁NO   n²⁰_D 1.4518   66.35
(±)-2-Amino-1-butanol, 97%   FW 89.14   Fp 184°F   54.49
    mp -2°C   FT-IR: *1*, 337C   27.12
    bp 177°C   VP-FT-IR: *3*, 427C   10.47
    d 0.943

CDCl₃   QE-300

CH₃CH₂CHCH₂OH
       |
       NH₂

B

Aldrich 23,260-2   *CAS [124-68-5]*   C₄H₁₁NO   Fp 153°F   FT-IR: *1*, 336B   71.55
2-Amino-2-methyl-1-propanol, 99+%   FW 89.14   50.67
    mp 32°C   27.15
    bp 165°C
    d 0.935

CDCl₃   QE-300

CH₃
 |
CH₃C CH₂OH
 |
 NH₂

C

Aldrich 12,304-8   *CAS [2508-29-4]*   C₅H₁₃NO   n²⁰_D 1.4615   61.87
5-Amino-1-pentanol, 97%   FW 103.17   Fp 150°F   41.99
    mp 36°C   60 MHz: *1*, 296A   33.28
    bp 122°C (16 mm)   FT-IR: *1*, 339A   32.59
    d 0.949   VP-FT-IR: *3*, 429B   23.16

CDCl₃   QE-300

H₂NCH₂CH₂CH₂CH₂CH₂OH

Non-Aromatic Amines

540

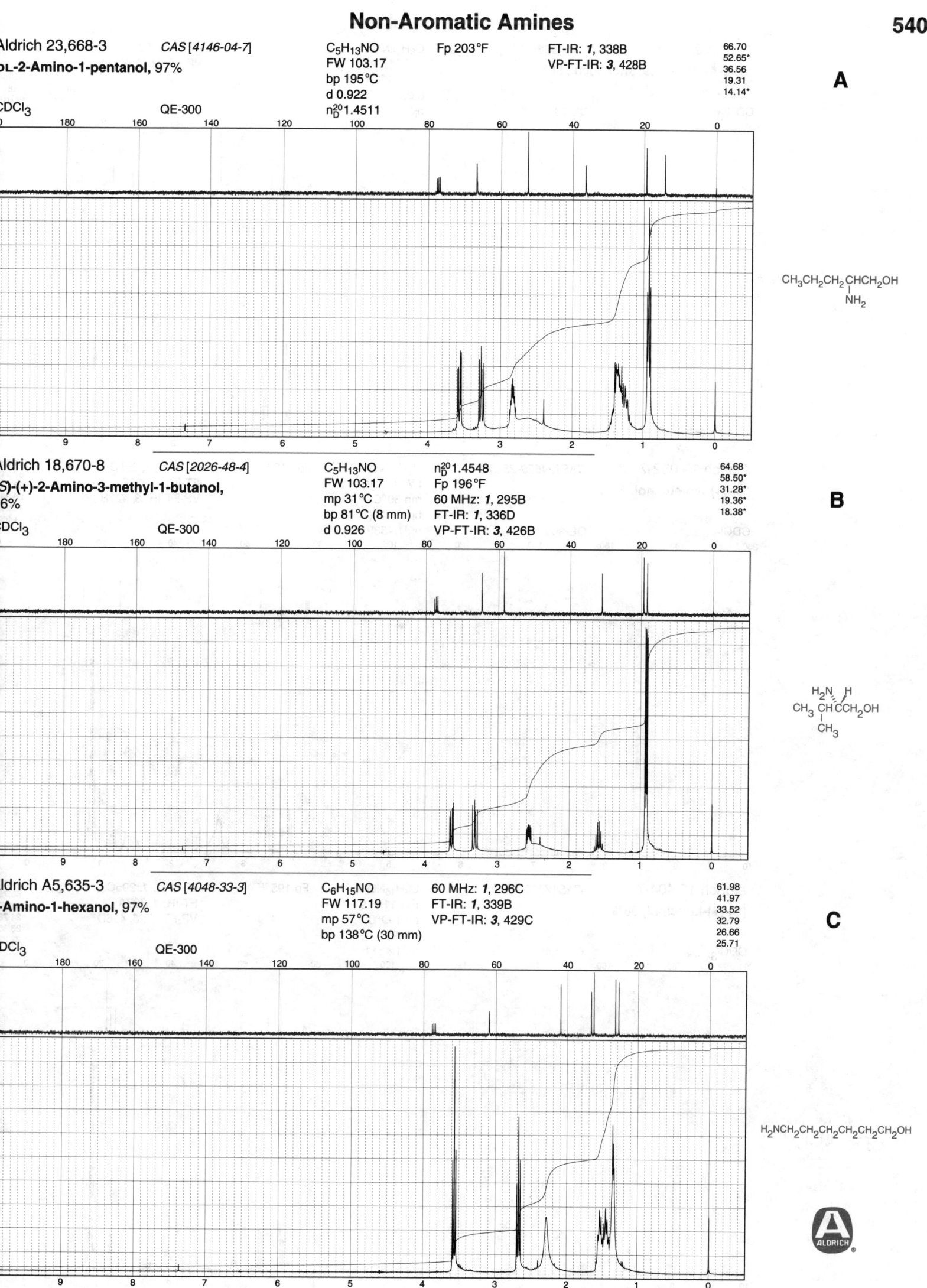

Aldrich 23,668-3   CAS [4146-04-7]   C$_5$H$_{13}$NO   Fp 203°F   FT-IR: *1*, 338B   66.70
DL-2-Amino-1-pentanol, 97%   FW 103.17   VP-FT-IR: *3*, 428B   52.65*
bp 195°C   36.56
CDCl$_3$ QE-300   d 0.922   19.31
n$_D^{20}$1.4511   14.14*

A

CH$_3$CH$_2$CH$_2$CHCH$_2$OH
NH$_2$

Aldrich 18,670-8   CAS [2026-48-4]   C$_5$H$_{13}$NO   n$_D^{20}$1.4548   64.68
(S)-(+)-2-Amino-3-methyl-1-butanol,   FW 103.17   Fp 196°F   58.50*
96%   mp 31°C   60 MHz: *1*, 295B   31.28*
bp 81°C (8 mm)   FT-IR: *1*, 336D   19.36*
CDCl$_3$ QE-300   d 0.926   VP-FT-IR: *3*, 426B   18.38*

B

H$_2$N   H
CH$_3$ CHCCH$_2$OH
CH$_3$

Aldrich A5,635-3   CAS [4048-33-3]   C$_6$H$_{15}$NO   60 MHz: *1*, 296C   61.98
6-Amino-1-hexanol, 97%   FW 117.19   FT-IR: *1*, 339B   41.97
mp 57°C   VP-FT-IR: *3*, 429C   33.52
bp 138°C (30 mm)   32.79
CDCl$_3$ QE-300   26.66
25.71

C

H$_2$NCH$_2$CH$_2$CH$_2$CH$_2$CH$_2$CH$_2$OH

ALDRICH

A

Aldrich 23,767-1   CAS [5665-74-7] $C_6H_{15}NO$ Fp 212°F   FT-IR: **1**, 338C

DL-2-Amino-1-hexanol, tech., 90%   FW 117.19   VP-FT-IR: **3**, 428C

bp 191°C (740 mm)

d 0.912

CDCl₃   QE-300 n_D^{20}1.4522

66.62
52.88*
34.01
28.35
22.81
14.01*

$CH_3CH_2CH_2CH_2 \underset{NH_2}{CH} CH_2OH$

B

Aldrich 19,052-7   CAS [24629-25-2] $C_6H_{15}NO$ Fp 213°F   60 MHz: **1**, 295D

(S)-(+)-Isoleucinol, 97%   FW 117.19   FT-IR: **1**, 337B

mp 30°C   VP-FT-IR: **3**, 427B

bp 97°C (14 mm)

CDCl₃   QE-300 n_D^{20}1.4589

64.36
57.11*
38.33*
25.39
15.20*
11.37*

$CH_3CH_2 \underset{\underset{H_2N}{|}}{\overset{\overset{CH_3}{|}}{CH}} CHCH_2OH$

C

Aldrich 18,404-7   CAS [7533-40-6] $C_6H_{15}NO$ Fp 195°F   60 MHz: **1**, 295C

(S)-(+)-Leucinol, 98%   FW 117.19   FT-IR: **1**, 337A

bp 199°C   VP-FT-IR: **3**, 426D

d 0.917

CDCl₃   QE-300 n_D^{20}1.4511

67.10
50.78
43.70
24.78
23.35
22.22

$CH_3 \underset{CH_3}{\overset{}{CH}} CH_2 \overset{\overset{H_2NH}{|}}{CCH_2OH}$

Non-Aromatic Amines

Aldrich 22,257-7   *CAS [5456-63-3]* $C_6H_{13}NO$ 60 MHz: *1*, 271C

trans-2-Aminocyclohexanol hydrochloride, 99%

FW 151.64   FT-IR: *1*, 315D

mp 174°C

DMSO-d_6 QE-300

69.91*	
55.64*	
33.57	
28.42	
23.53	
23.42	

A

Aldrich 19,479-4   *CAS [15647-11-7]* $C_{10}H_{21}NO$ n_D^{20} 1.4904

3-Aminomethyl-3,5,5-trimethylcyclohexanol, 99%, mixture of *cis* and *trans*

FW 171.29   Fp >230°F

mp 46°C   FT-IR: *1*, 319A

bp 265°C   VP-FT-IR: *3*, 407C

d 0.969

CDCl$_3$ QE-300

65.42*	47.25	35.20*
65.02*	46.45	32.32
57.39	45.48	29.47*
49.93	44.49	28.18*
49.14	37.11	27.79*
48.90	36.83	23.63*

B

Aldrich A5,405-9   *CAS [929-06-6]* $C_4H_{11}NO_2$ FT-IR: *1*, 339C

2-(2-Aminoethoxy)ethanol, 98%

FW 105.14   VP-FT-IR: *3*, 429D

bp 221°C

d 1.048

CDCl$_3$ QE-300

73.02	
72.69	
61.35	
41.73	

C

$H_2NCH_2CH_2OCH_2CH_2OH$

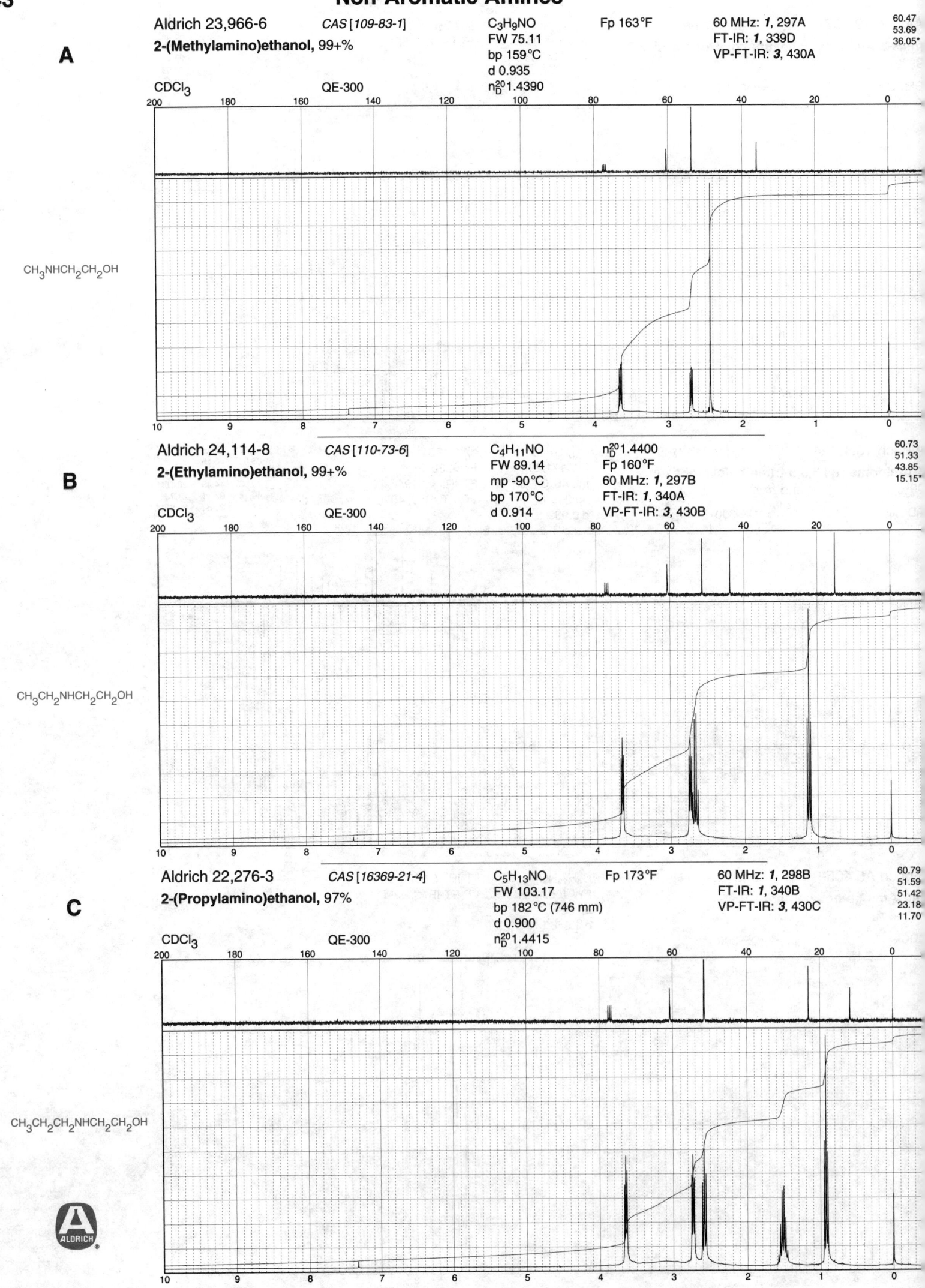

A

Aldrich 23,966-6   *CAS [109-83-1]* C_3H_9NO Fp 163°F   60 MHz: *1*, 297A   60.47
2-(Methylamino)ethanol, 99+%   FW 75.11   FT-IR: *1*, 339D   53.69
                                                bp 159°C   VP-FT-IR: *3*, 430A   36.05*
                                                d 0.935
CDCl$_3$ QE-300   n$_D^{20}$1.4390

$CH_3NHCH_2CH_2OH$

B

Aldrich 24,114-8   *CAS [110-73-6]* $C_4H_{11}NO$ n$_D^{20}$1.4400   60.73
2-(Ethylamino)ethanol, 99+%   FW 89.14   Fp 160°F   51.33
                                             mp -90°C   60 MHz: *1*, 297B 43.85
                                             bp 170°C   FT-IR: *1*, 340A   15.15*
CDCl$_3$ QE-300   d 0.914   VP-FT-IR: *3*, 430B

$CH_3CH_2NHCH_2CH_2OH$

C

Aldrich 22,276-3   *CAS [16369-21-4]* $C_5H_{13}NO$ Fp 173°F   60 MHz: *1*, 298B   60.79
2-(Propylamino)ethanol, 97%   FW 103.17   FT-IR: *1*, 340B   51.59
                                              bp 182°C (746 mm)   VP-FT-IR: *3*, 430C   51.42
                                              d 0.900   23.18
CDCl$_3$ QE-300   n$_D^{20}$1.4415   11.70

$CH_3CH_2CH_2NHCH_2CH_2OH$

Aldrich B8,960-4   *CAS [4620-70-6]*

2-(*tert*-Butylamino)ethanol, 99+%

C$_6$H$_{15}$NO
FW 117.19
mp 44°C
bp 91°C (25 mm)
Fp 156°F

60 MHz: *1*, 297C
FT-IR: *1*, 340C
VP-FT-IR: *3*, 430D

61.49
50.20
44.09
29.00*

A

CDCl$_3$ QE-300

$$CH_3$$
$$CH_3-C-NHCH_2CH_2OH$$
$$CH_3$$

Aldrich D8,330-3   *CAS [111-42-2]*

Diethanolamine, 99%

C$_4$H$_{11}$NO$_2$
FW 105.14
mp 29°C
bp 217°C (150 mm)
d 1.097

n_D^{20} 1.4770
Fp 280°F
60 MHz: *1*, 297D
FT-IR: *1*, 340D
VP-FT-IR: *3*, 431B

60.71
51.29

B

CDCl$_3$ QE-300

$$HOCH_2CH_2$$
$$\qquad\qquad NH$$
$$HOCH_2CH_2$$

Aldrich 13,301-9   *CAS [110-97-4]*

Diisopropanolamine, 95%

C$_6$H$_{15}$NO$_2$
FW 133.19
mp 45°C
bp 250°C (745 mm)
d 1.004

Fp 260°F

FT-IR: *1*, 341C
VP-FT-IR: *3*, 431C

66.13*
65.85*
57.26
56.88
20.98*
20.91*

C

CDCl$_3$ QE-300

$$OH$$
$$CH_3CHCH_2$$
$$\qquad\qquad NH$$
$$CH_3CHCH_2$$
$$OH$$

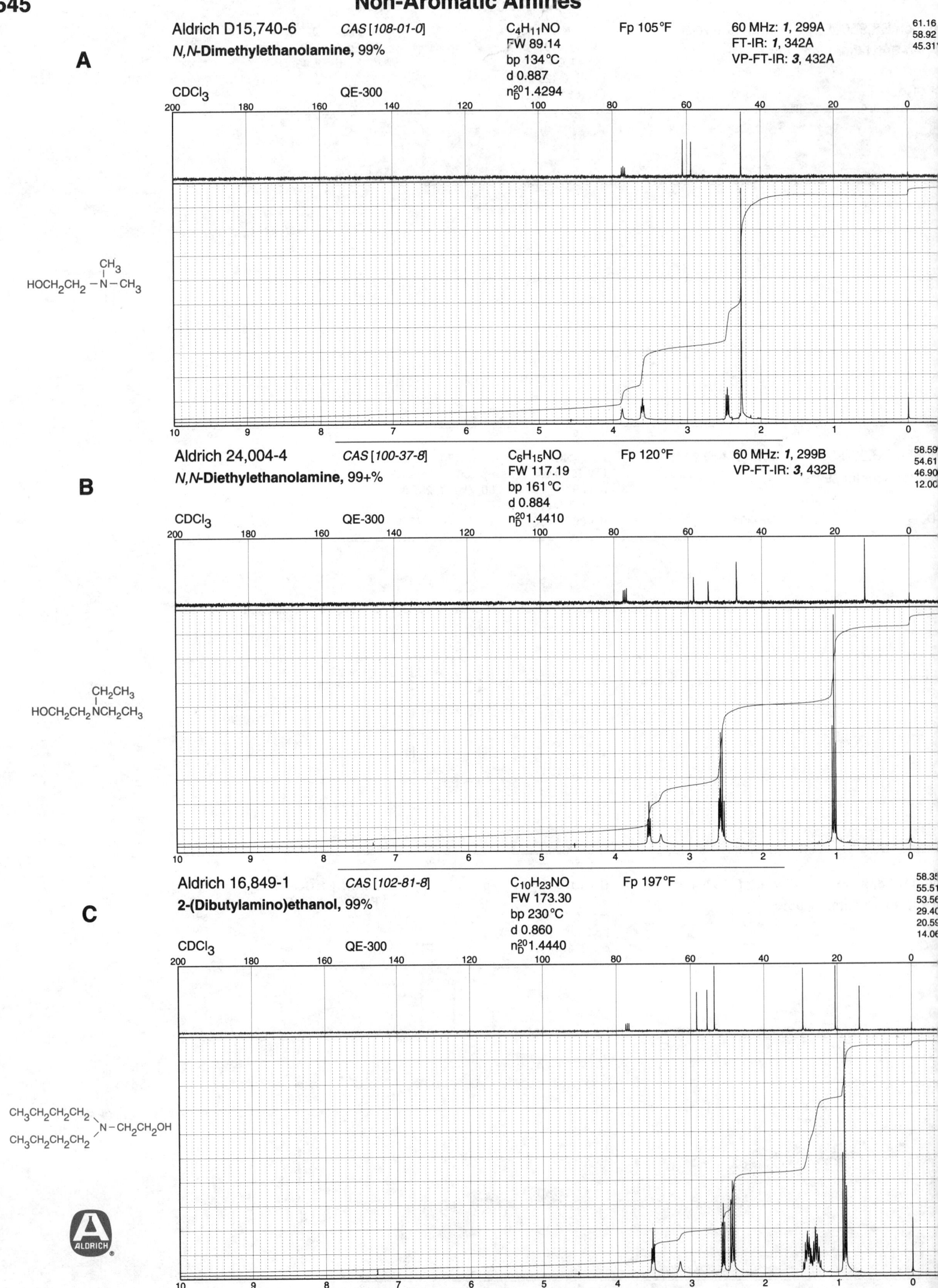

A

Aldrich D15,740-6   *CAS [108-01-0]*   C₄H₁₁NO   Fp 105°F   60 MHz: *1*, 299A   61.16

N,N-Dimethylethanolamine, 99%   FW 89.14   FT-IR: *1*, 342A   58.92
bp 134°C   VP-FT-IR: *3*, 432A   45.31
d 0.887
n²⁰_D 1.4294

CDCl₃   QE-300

HOCH₂CH₂—N(CH₃)CH₃

B

Aldrich 24,004-4   *CAS [100-37-8]*   C₆H₁₅NO   Fp 120°F   60 MHz: *1*, 299B   58.59

N,N-Diethylethanolamine, 99+%   FW 117.19   VP-FT-IR: *3*, 432B   54.61
bp 161°C   46.90
12.00

CDCl₃   QE-300   n²⁰_D 1.4410

HOCH₂CH₂NCH₂CH₃(CH₂CH₃)

C

Aldrich 16,849-1   *CAS [102-81-8]*   C₁₀H₂₃NO   Fp 197°F   58.35

2-(Dibutylamino)ethanol, 99%   FW 173.30   55.51
bp 230°C   53.56
d 0.860   29.40
n²⁰_D 1.4440   20.59
14.06

CDCl₃   QE-300

CH₃CH₂CH₂CH₂—N(CH₃CH₂CH₂CH₂)—CH₂CH₂OH

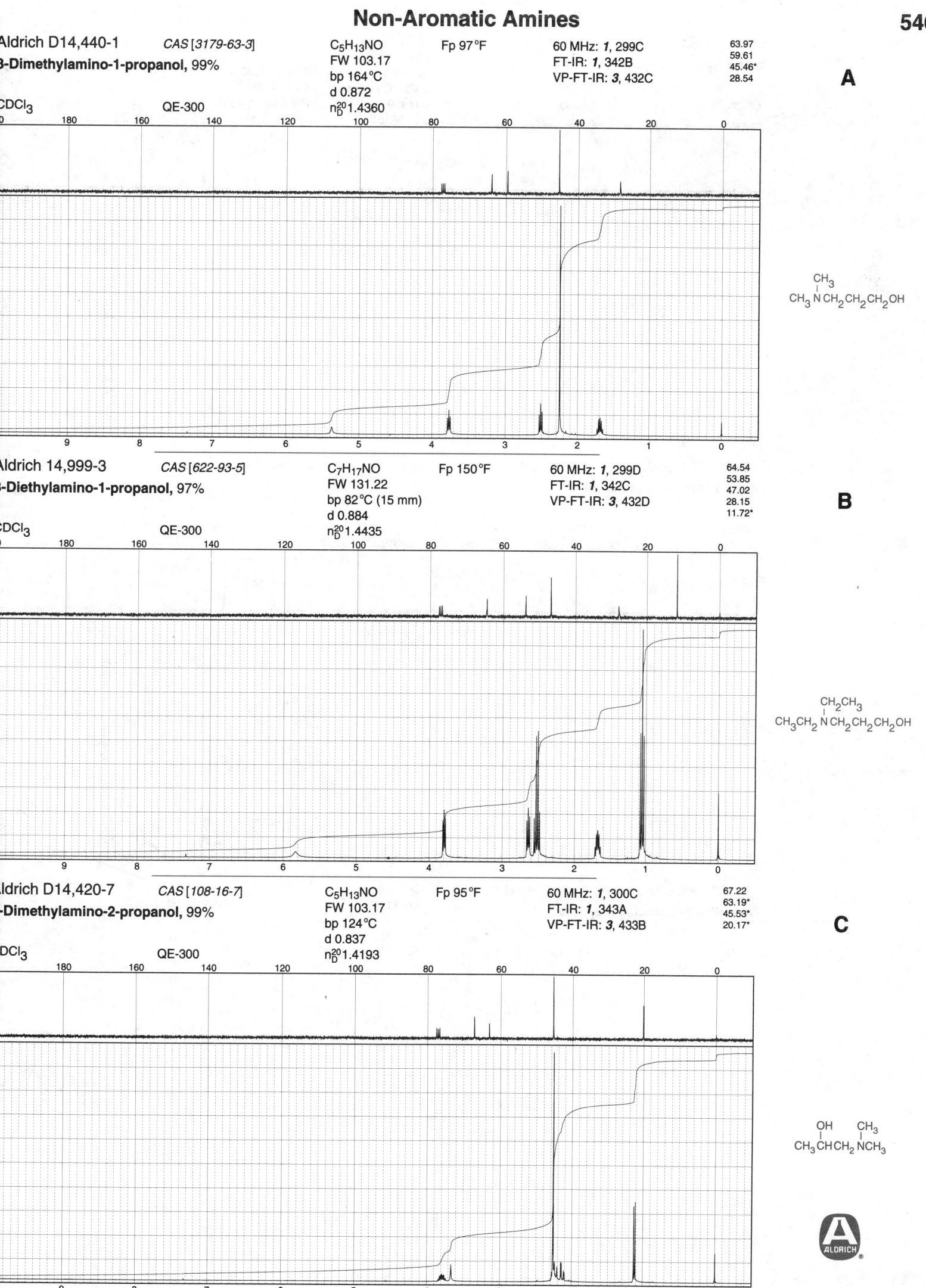

Aldrich D14,440-1   CAS [3179-63-3]
3-Dimethylamino-1-propanol, 99%

C₅H₁₃NO
FW 103.17
bp 164°C
d 0.872
nᴅ²⁰ 1.4360

Fp 97°F

CDCl₃   QE-300

60 MHz: *1*, 299C
FT-IR: *1*, 342B
VP-FT-IR: *3*, 432C

63.97
59.61
45.46*
28.54

A

CH₃
|
CH₃ N CH₂ CH₂ CH₂ OH

Aldrich 14,999-3   CAS [622-93-5]
3-Diethylamino-1-propanol, 97%

C₇H₁₇NO
FW 131.22
bp 82°C (15 mm)
nᴅ²⁰ 1.4435

Fp 150°F

CDCl₃   QE-300

60 MHz: *1*, 299D
FT-IR: *1*, 342C
VP-FT-IR: *3*, 432D

64.54
53.85
47.02
28.15
11.72*

B

CH₂CH₃
|
CH₃CH₂ N CH₂ CH₂ CH₂ OH

Aldrich D14,420-7   CAS [108-16-7]
1-Dimethylamino-2-propanol, 99%

C₅H₁₃NO
FW 103.17
bp 124°C
d 0.837
nᴅ²⁰ 1.4193

Fp 95°F

CDCl₃   QE-300

60 MHz: *1*, 300C
FT-IR: *1*, 343A
VP-FT-IR: *3*, 433B

67.22
63.19*
45.53*
20.17*

C

OH   CH₃
| |
CH₃ CHCH₂ N CH₃

Non-Aromatic Amines

A

Aldrich D8,848-8 CAS [4402-32-8]
1-Diethylamino-2-propanol, 97%

C₇H₁₇NO
FW 131.22
mp 14°C
bp 57°C (13 mm)
d 0.889

n²⁰_D 1.4255
Fp 92°F
60 MHz: *1*, 300D
FT-IR: *1*, 343B
VP-FT-IR: *3*, 433C

63.01*
61.37
47.26
20.02*
12.14*

CDCl₃ QE-300

OH   CH₂CH₃
CH₃CHCH₂ NCH₂CH₃

B

Aldrich 19,055-1 CAS [5412-69-1]
5-Diethylamino-2-pentanol, 97%

C₉H₂₁NO
FW 159.27
d 0.866

60 MHz: *1*, 296B
FT-IR: *1*, 342D
VP-FT-IR: *3*, 433A

67.34*
53.80
46.16
39.47
24.96
23.80*
10.76*

CDCl₃ QE-300

CH₂CH₃   OH
CH₃CH₂ N CH₂CH₂CH₂CH CH₃

C

Aldrich M4,220-3 CAS [105-59-9]
N-Methyldiethanolamine, 99%

C₅H₁₃NO₂
FW 119.16
bp 247°C
d 1.038
n²⁰_D 1.4685

Fp 260°F

60 MHz: *1*, 303B
FT-IR: *1*, 347A
VP-FT-IR: *3*, 437B

59.58
59.18
42.17*

CDCl₃ QE-300

CH₃
HOCH₂CH₂ NCH₂CH₂OH

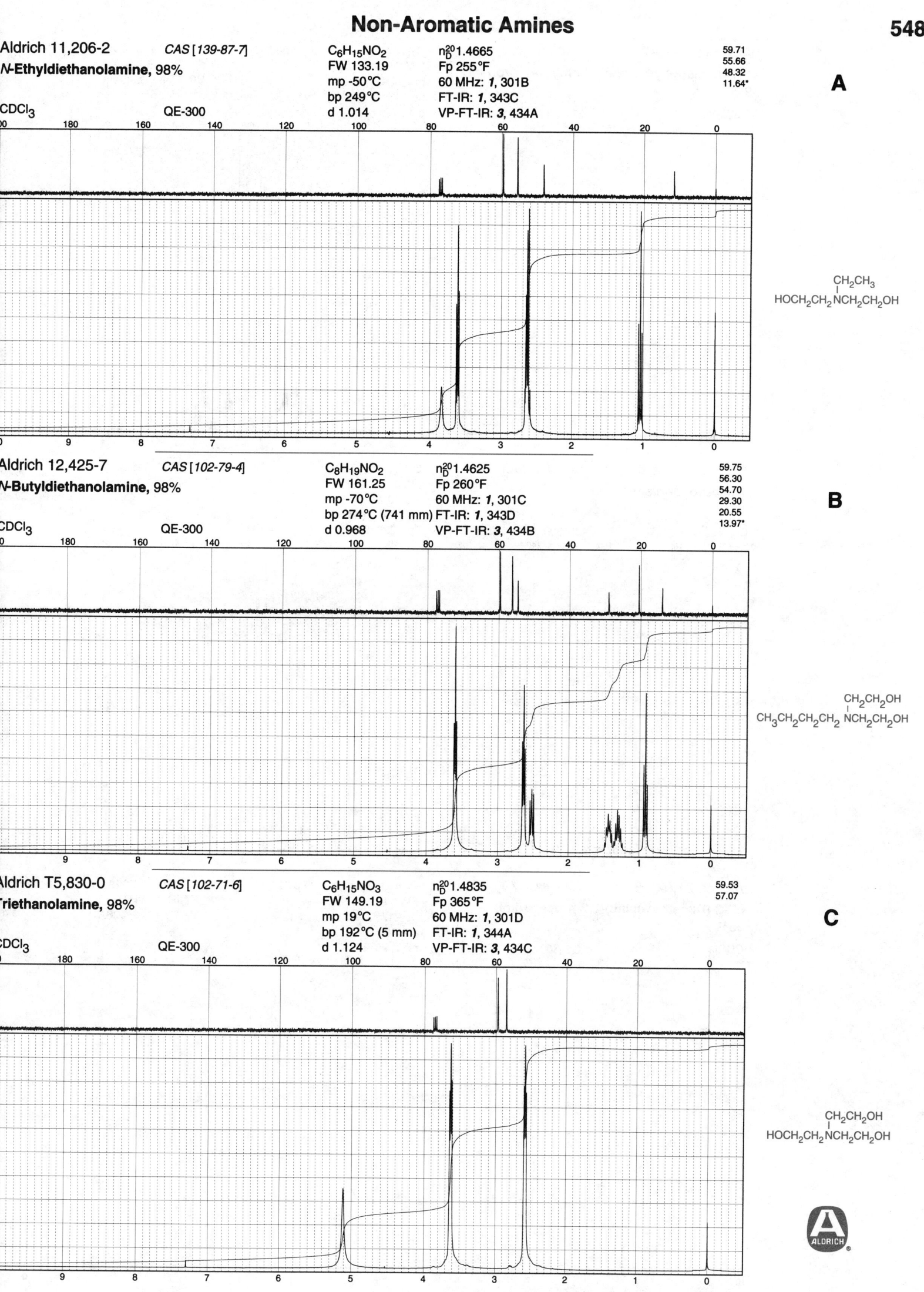

Aldrich 11,206-2   CAS [139-87-7]
N-Ethyldiethanolamine, 98%

C$_6$H$_{15}$NO$_2$
FW 133.19
mp -50°C
bp 249°C
d 1.014

n$_D^{20}$ 1.4665
Fp 255°F
60 MHz: *1*, 301B
FT-IR: *1*, 343C
VP-FT-IR: *3*, 434A

59.71
55.66
48.32
11.64*

A

CDCl$_3$ QE-300

HOCH$_2$CH$_2$NCH$_2$CH$_2$OH
CH$_2$CH$_3$

Aldrich 12,425-7   CAS [102-79-4]
N-Butyldiethanolamine, 98%

C$_8$H$_{19}$NO$_2$
FW 161.25
mp -70°C
bp 274°C (741 mm)
d 0.968

n$_D^{20}$ 1.4625
Fp 260°F
60 MHz: *1*, 301C
FT-IR: *1*, 343D
VP-FT-IR: *3*, 434B

59.75
56.30
54.70
29.30
20.55
13.97*

B

CDCl$_3$ QE-300

CH$_3$CH$_2$CH$_2$CH$_2$NCH$_2$CH$_2$OH
CH$_2$CH$_2$OH

Aldrich T5,830-0   CAS [102-71-6]
Triethanolamine, 98%

C$_6$H$_{15}$NO$_3$
FW 149.19
mp 19°C
bp 192°C (5 mm)
d 1.124

n$_D^{20}$ 1.4835
Fp 365°F
60 MHz: *1*, 301D
FT-IR: *1*, 344A
VP-FT-IR: *3*, 434C

59.53
57.07

C

CDCl$_3$ QE-300

HOCH$_2$CH$_2$NCH$_2$CH$_2$OH
CH$_2$CH$_2$OH

A

Aldrich 23,375-7   *CAS [6712-98-7]* $C_7H_{17}NO_3$ Fp >230°F   60 MHz: *1*, 302B

1-[*N,N*-Bis(2-hydroxyethyl)amino]-2-propanol,   FW 163.22   FT-IR: *1*, 344C

94%   mp 34°C   VP-FT-IR: *3*, 434D

CDCl₃   QE-300   bp 145°C   d 1.079

64.45*	57.53
63.47	20.07*
59.55	

Structure: CH₃CHCH₂NCH₂CH₂OH with OH and CH₂CH₂OH groups

B

Aldrich 25,474-6   *CAS [122-20-3]* $C_9H_{21}NO_3$ FT-IR: *1*, 345B

Triisopropanolamine, 95%   FW 191.27   VP-FT-IR: *3*, 435C

mp 50°C

bp 190°C (23 mm)

CDCl₃   QE-300   Fp >230°F

67.65*	63.12
66.50	20.49*
65.92	20.35*
65.50*	20.24*
63.71*	

Structure: CH₃CHCH₂NCH₂CHCH₃ with OH groups and CH₂CHCH₃/OH branch

C

Aldrich 21,021-8   *CAS [623-57-4]* $C_5H_{13}NO_2$ Fp 221°F   60 MHz: *1*, 300A

(±)-3-(Dimethylamino)-1,2-propanediol,   FW 119.16   FT-IR: *1*, 345D

98%   bp 217°C   VP-FT-IR: *3*, 436A

CDCl₃   QE-300   d 1.004

n_D^{20} 1.4609

| 68.21* |
| 65.23 |
| 62.13 |
| 45.75* |

Structure: CH₃NCH₂CHCH₂OH with CH₃ and OH groups

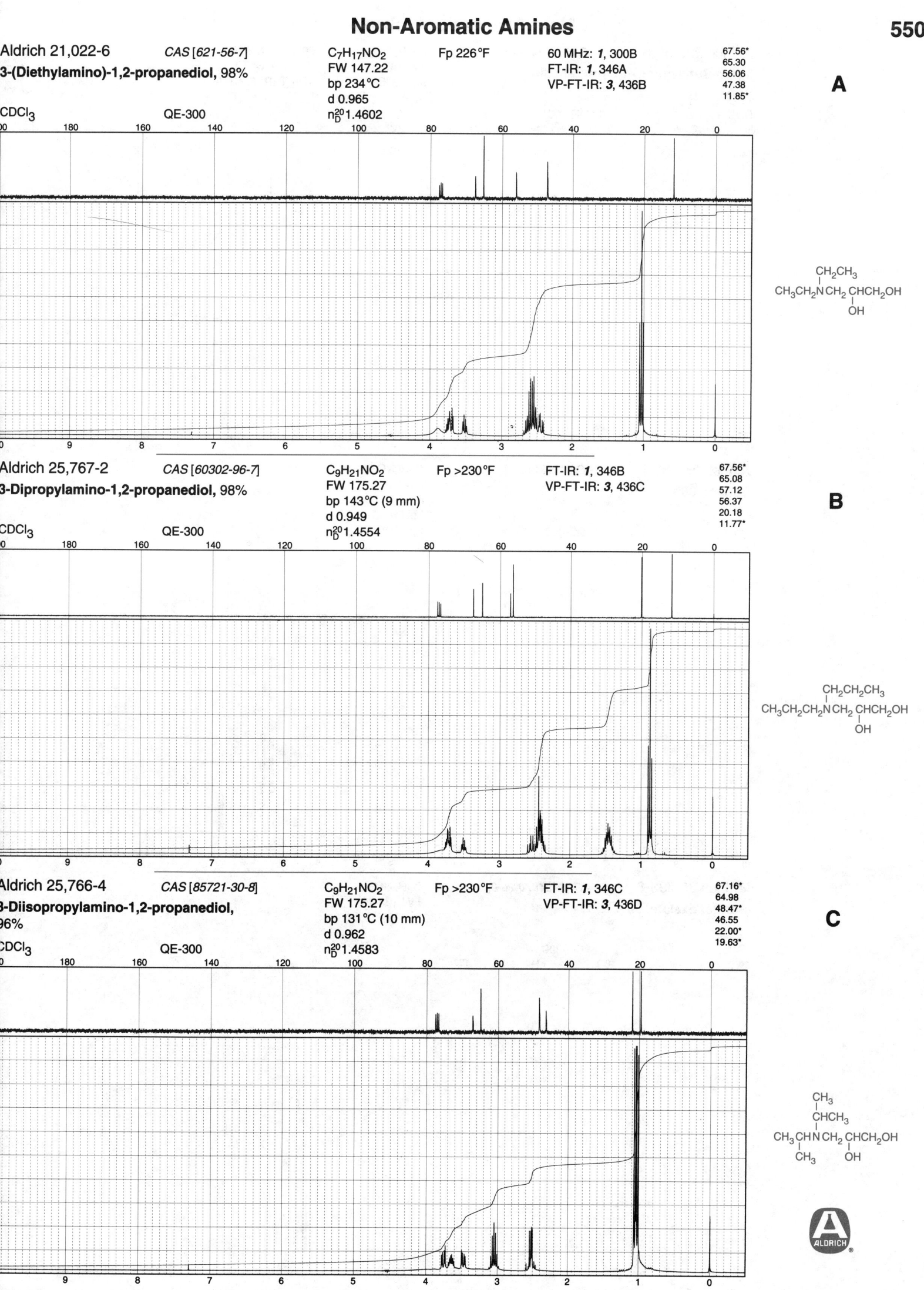

Aldrich 21,022-6   CAS [621-56-7]   C₇H₁₇NO₂   Fp 226°F   60 MHz: 1, 300B

3-(Diethylamino)-1,2-propanediol, 98%

FW 147.22   FT-IR: 1, 346A
bp 234°C   VP-FT-IR: 3, 436B
d 0.965
n²⁰_D 1.4602

CDCl₃   QE-300

A

67.56*
65.30
56.06
47.38
11.85*

Aldrich 25,767-2   CAS [60302-96-7]   C₉H₂₁NO₂   Fp >230°F   FT-IR: 1, 346B

3-Dipropylamino-1,2-propanediol, 98%

FW 175.27   VP-FT-IR: 3, 436C
bp 143°C (9 mm)
d 0.949
n²⁰_D 1.4554

CDCl₃   QE-300

B

67.56*
65.08
57.12
56.37
20.18
11.77*

Aldrich 25,766-4   CAS [85721-30-8]   C₉H₂₁NO₂   Fp >230°F   FT-IR: 1, 346C

3-Diisopropylamino-1,2-propanediol, 96%

FW 175.27   VP-FT-IR: 3, 436D
bp 131°C (10 mm)
d 0.962
n²⁰_D 1.4583

CDCl₃   QE-300

C

67.16*
64.98
48.47*
46.55
22.00*
19.63*

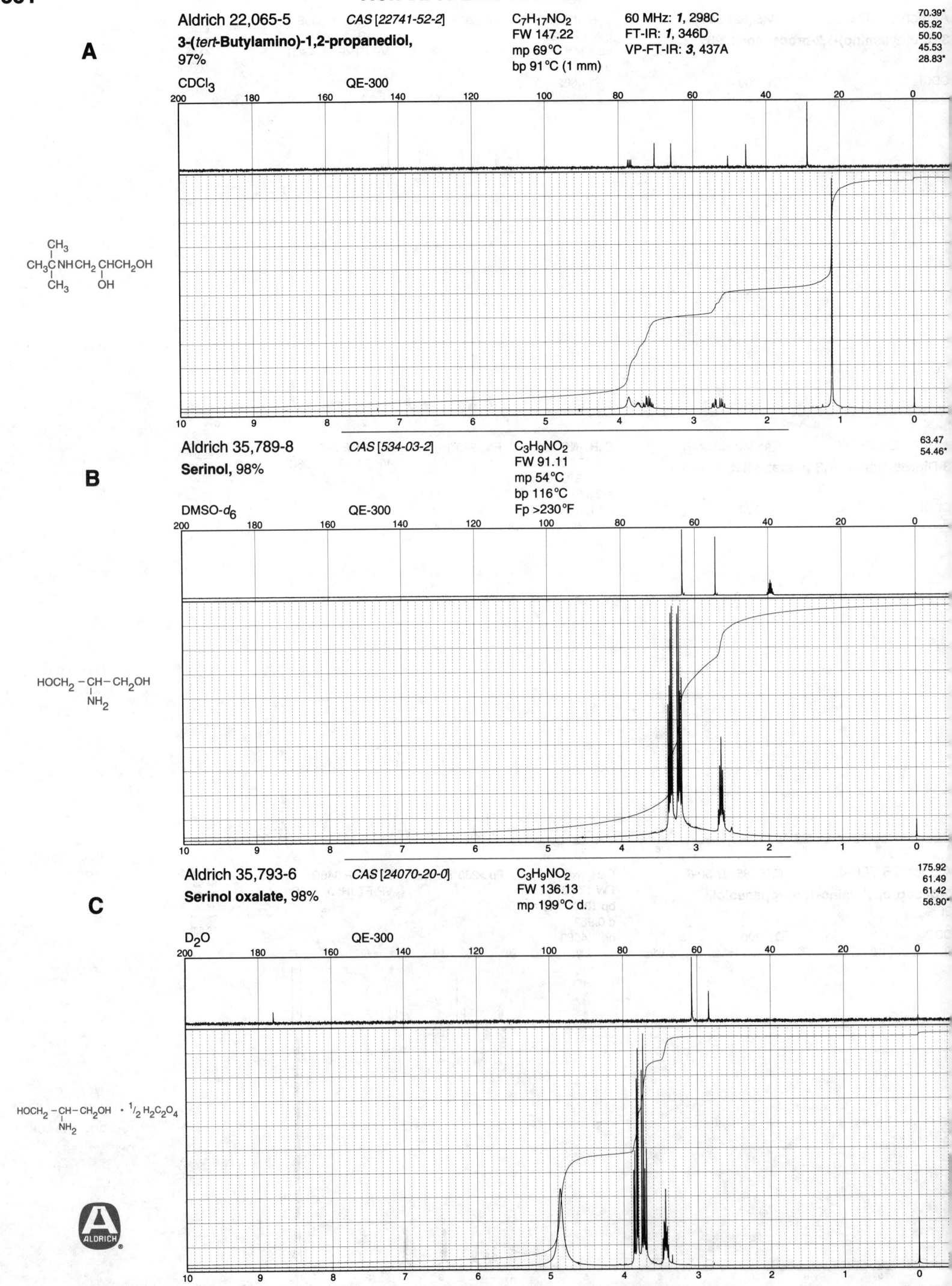

Aldrich 22,065-5   CAS [22741-52-2]   C₇H₁₇NO₂
3-(*tert*-Butylamino)-1,2-propanediol,   FW 147.22   60 MHz: **1**, 298C
97%   mp 69°C   FT-IR: **1**, 346D
    bp 91°C (1 mm)   VP-FT-IR: **3**, 437A

70.39*
65.92
50.50
45.53
28.83*

CDCl₃   QE-300

Aldrich 35,789-8   CAS [534-03-2]   C₃H₉NO₂
Serinol, 98%   FW 91.11
    mp 54°C
    bp 116°C
    Fp >230°F

63.47
54.46*

DMSO-d_6 QE-300

Aldrich 35,793-6   CAS [24070-20-0]   C₃H₉NO₂
Serinol oxalate, 98%   FW 136.13
    mp 199°C d.

175.92
61.49
61.42
56.90*

D₂O   QE-300

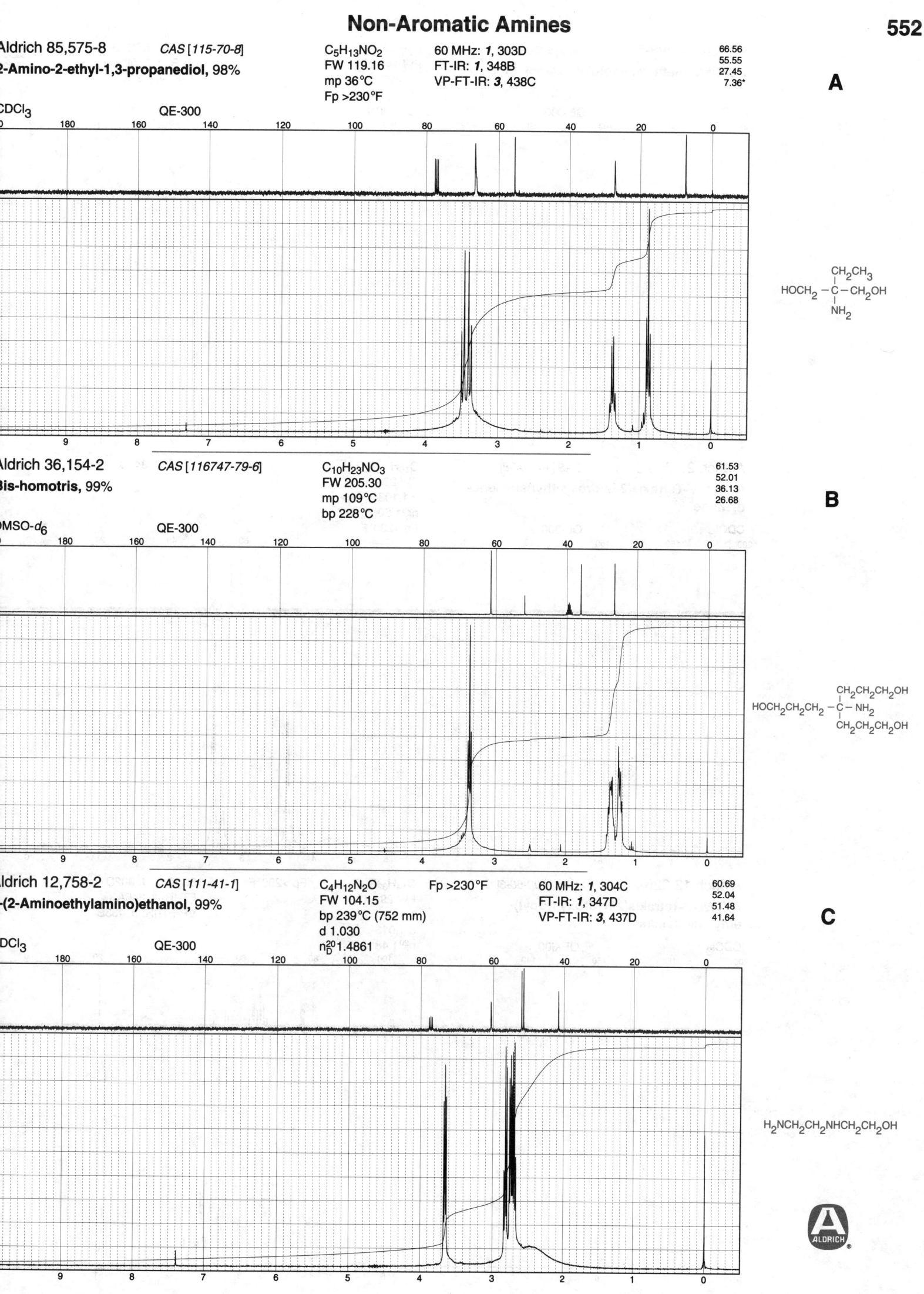

Aldrich 85,575-8   CAS [115-70-8]
2-Amino-2-ethyl-1,3-propanediol, 98%
$C_5H_{13}NO_2$
FW 119.16
mp 36°C
Fp >230°F
60 MHz: *1*, 303D
FT-IR: *1*, 348B
VP-FT-IR: *3*, 438C

66.56
55.55
27.45
7.36*

CDCl₃   QE-300

Aldrich 36,154-2   CAS [116747-79-6]
Bis-homotris, 99%
$C_{10}H_{23}NO_3$
FW 205.30
mp 109°C
bp 228°C

61.53
52.01
36.13
26.68

DMSO-d_6 QE-300

Aldrich 12,758-2   CAS [111-41-1]
2-(2-Aminoethylamino)ethanol, 99%
$C_4H_{12}N_2O$
FW 104.15
bp 239°C (752 mm)
d 1.030
n_D^{20}1.4861
Fp >230°F
60 MHz: *1*, 304C
FT-IR: *1*, 347D
VP-FT-IR: *3*, 437D

60.69
52.04
51.48
41.64

CDCl₃   QE-300

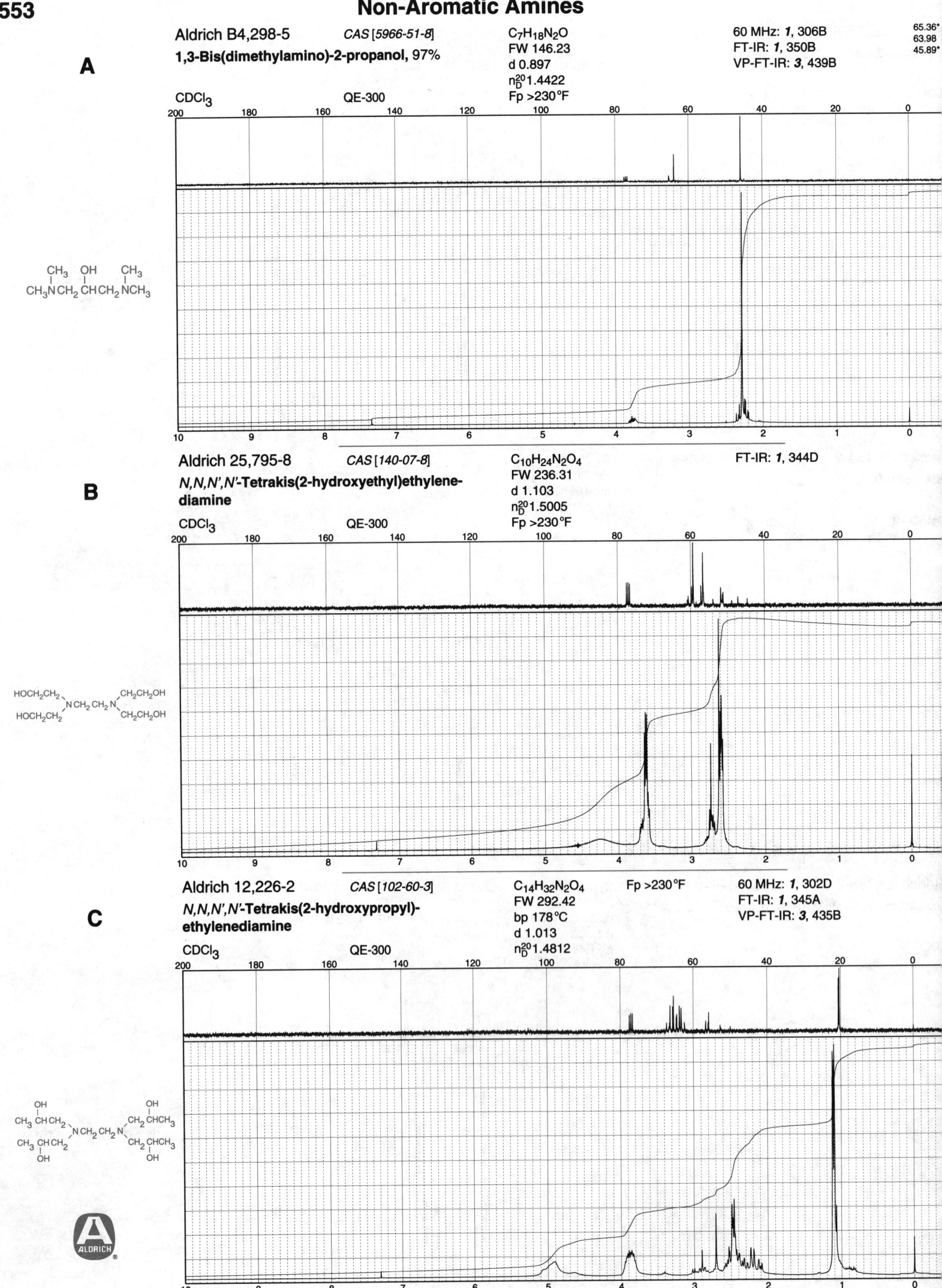

A

Aldrich B4,298-5   CAS [5966-51-8] $C_7H_{18}N_2O$ 60 MHz: *1*, 306B   65.36*
1,3-Bis(dimethylamino)-2-propanol, 97%   FW 146.23   FT-IR: *1*, 350B   63.98
    d 0.897   VP-FT-IR: *3*, 439B   45.89*
    n_D^{20} 1.4422
    Fp >230°F

CDCl₃   QE-300

B

Aldrich 25,795-8   CAS [140-07-8] $C_{10}H_{24}N_2O_4$ FT-IR: *1*, 344D
N,N,N',N'-Tetrakis(2-hydroxyethyl)ethylene-   FW 236.31
diamine   d 1.103
    n_D^{20} 1.5005
CDCl₃   QE-300   Fp >230°F

C

Aldrich 12,226-2   CAS [102-60-3] $C_{14}H_{32}N_2O_4$ Fp >230°F   60 MHz: *1*, 302D
N,N,N',N'-Tetrakis(2-hydroxypropyl)-   FW 292.42   FT-IR: *1*, 345A
ethylenediamine   bp 178°C   VP-FT-IR: *3*, 435B
    d 1.013
CDCl₃   QE-300 n_D^{20} 1.4812

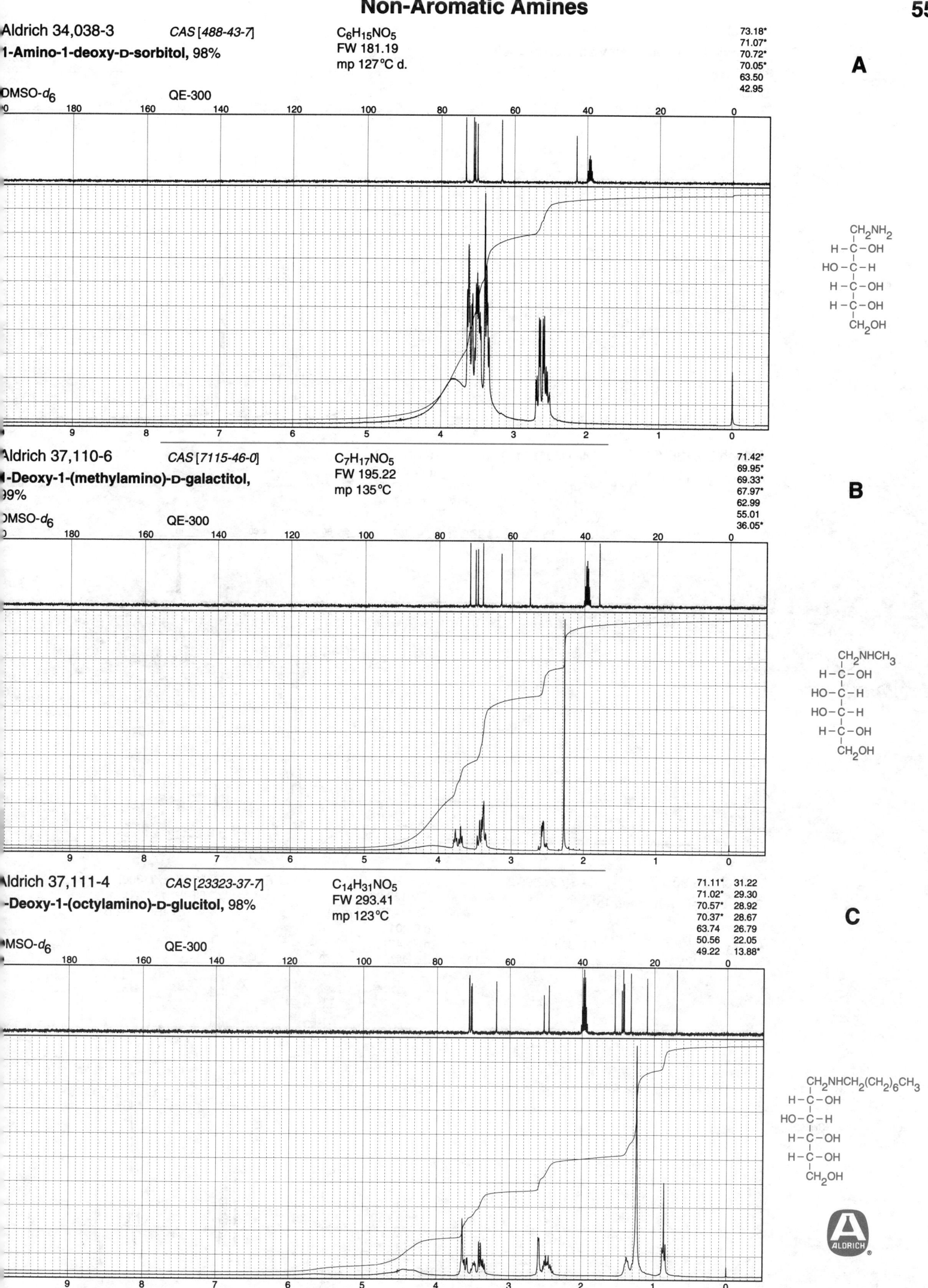

Aldrich 34,038-3   CAS [488-43-7]
1-Amino-1-deoxy-D-sorbitol, 98%
DMSO-d_6 QE-300
$C_6H_{15}NO_5$
FW 181.19
mp 127°C d.

73.18*
71.07*
70.72*
70.05*
63.50
42.95

A

Aldrich 37,110-6   CAS [7115-46-0]
1-Deoxy-1-(methylamino)-D-galactitol, 99%
DMSO-d_6 QE-300
$C_7H_{17}NO_5$
FW 195.22
mp 135°C

71.42*
69.95*
69.33*
67.97*
62.99
55.01
36.05*

B

Aldrich 37,111-4   CAS [23323-37-7]
1-Deoxy-1-(octylamino)-D-glucitol, 98%
DMSO-d_6 QE-300
$C_{14}H_{31}NO_5$
FW 293.41
mp 123°C

71.11*   31.22
71.02*   29.30
70.57*   28.92
70.37*   28.67
63.74   26.79
50.56   22.05
49.22   13.88*

C

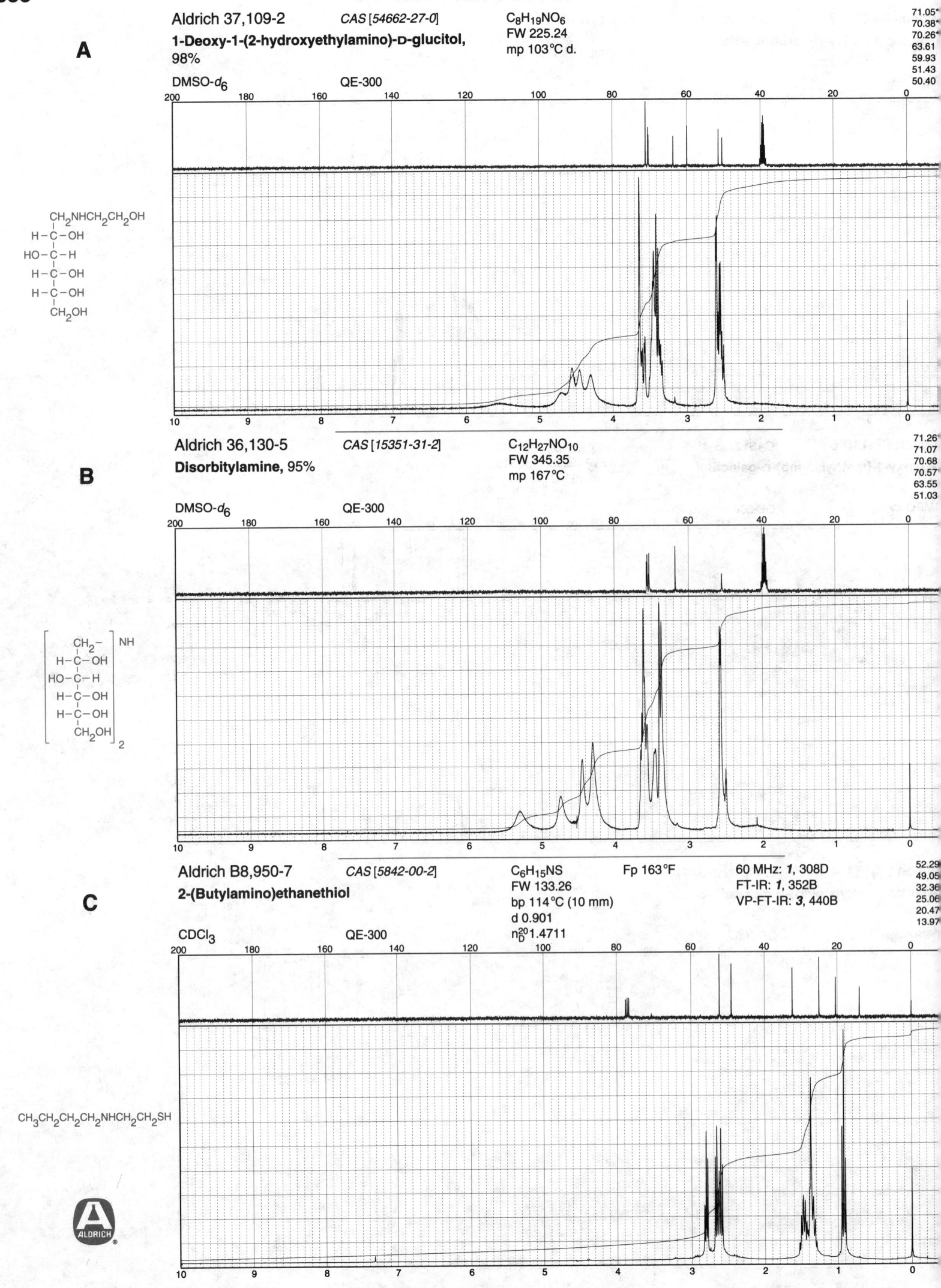

A Aldrich 37,109-2   *CAS [54662-27-0]* $C_8H_{19}NO_6$ FW 225.24   mp 103°C d.

1-Deoxy-1-(2-hydroxyethylamino)-D-glucitol, 98%

DMSO-d_6 QE-300

71.05*
70.38*
70.26*
63.61
59.93
51.43
50.40

B Aldrich 36,130-5   *CAS [15351-31-2]* $C_{12}H_{27}NO_{10}$ FW 345.35   mp 167°C

Disorbitylamine, 95%

DMSO-d_6 QE-300

71.26*
71.07
70.68
70.57*
63.55
51.03

C Aldrich B8,950-7   *CAS [5842-00-2]* $C_6H_{15}NS$ Fp 163°F   60 MHz: *1*, 308D

2-(Butylamino)ethanethiol

FW 133.26   FT-IR: *1*, 352B
bp 114°C (10 mm)   VP-FT-IR: *3*, 440B
d 0.901
n_D^{20} 1.4711

CDCl$_3$ QE-300

52.29
49.05
32.36
25.06
20.47
13.97*

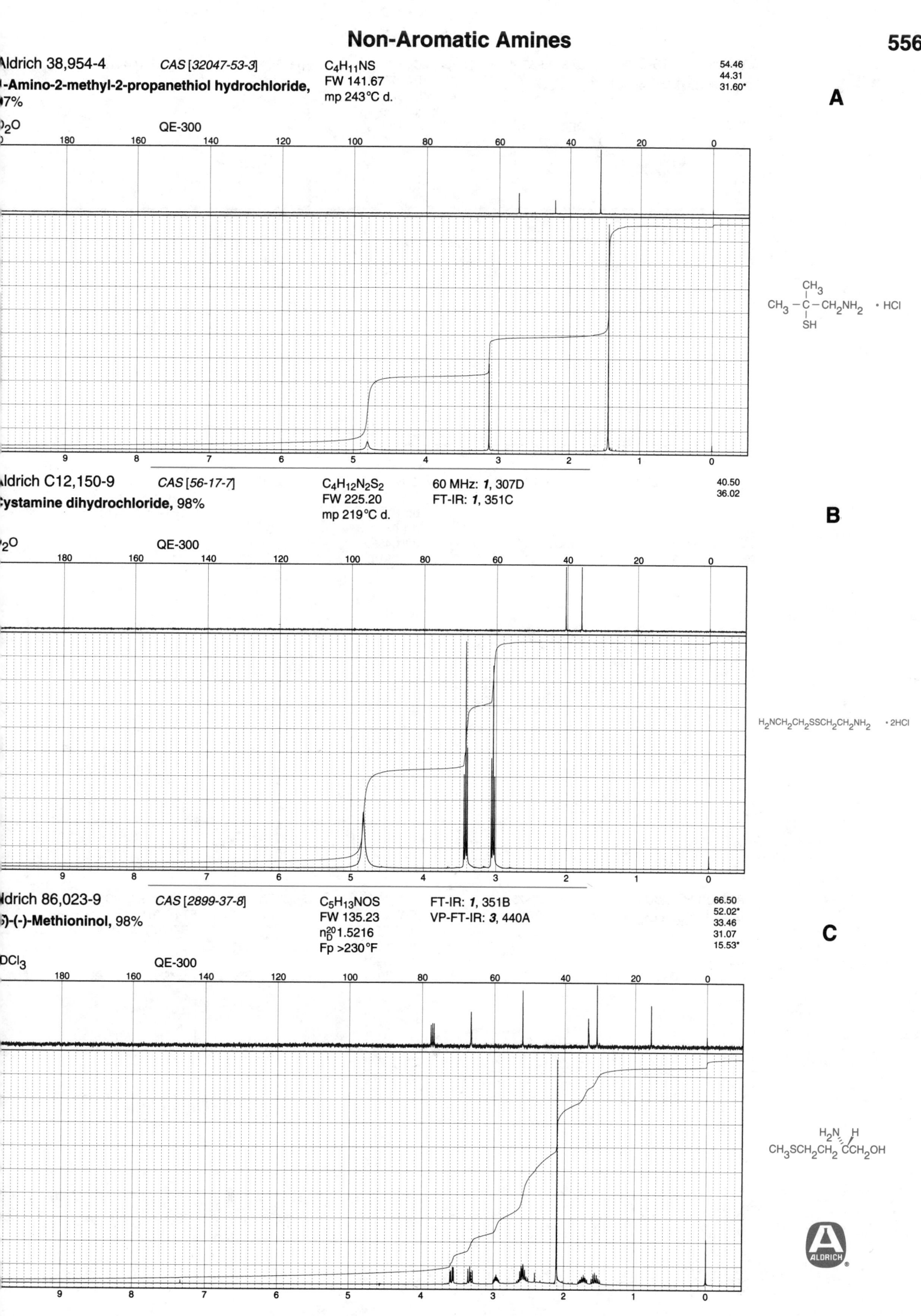

Aldrich 38,954-4   *CAS [32047-53-3]* $C_4H_{11}NS$ FW 141.67   mp 243°C d.

1-Amino-2-methyl-2-propanethiol hydrochloride, 97%

D_2O QE-300

54.46
44.31
31.60*

A

Aldrich C12,150-9   *CAS [56-17-7]* $C_4H_{12}N_2S_2$ FW 225.20   mp 219°C d.   60 MHz: *1*, 307D   FT-IR: *1*, 351C

Cystamine dihydrochloride, 98%

D_2O QE-300

40.50
36.02

B

Aldrich 86,023-9   *CAS [2899-37-8]* $C_5H_{13}NOS$ FW 135.23 n_D^{20} 1.5216   Fp >230°F   FT-IR: *1*, 351B   VP-FT-IR: *3*, 440A

(S)-(-)-Methioninol, 98%

$CDCl_3$ QE-300

66.50
52.02*
33.46
31.07
15.53*

C

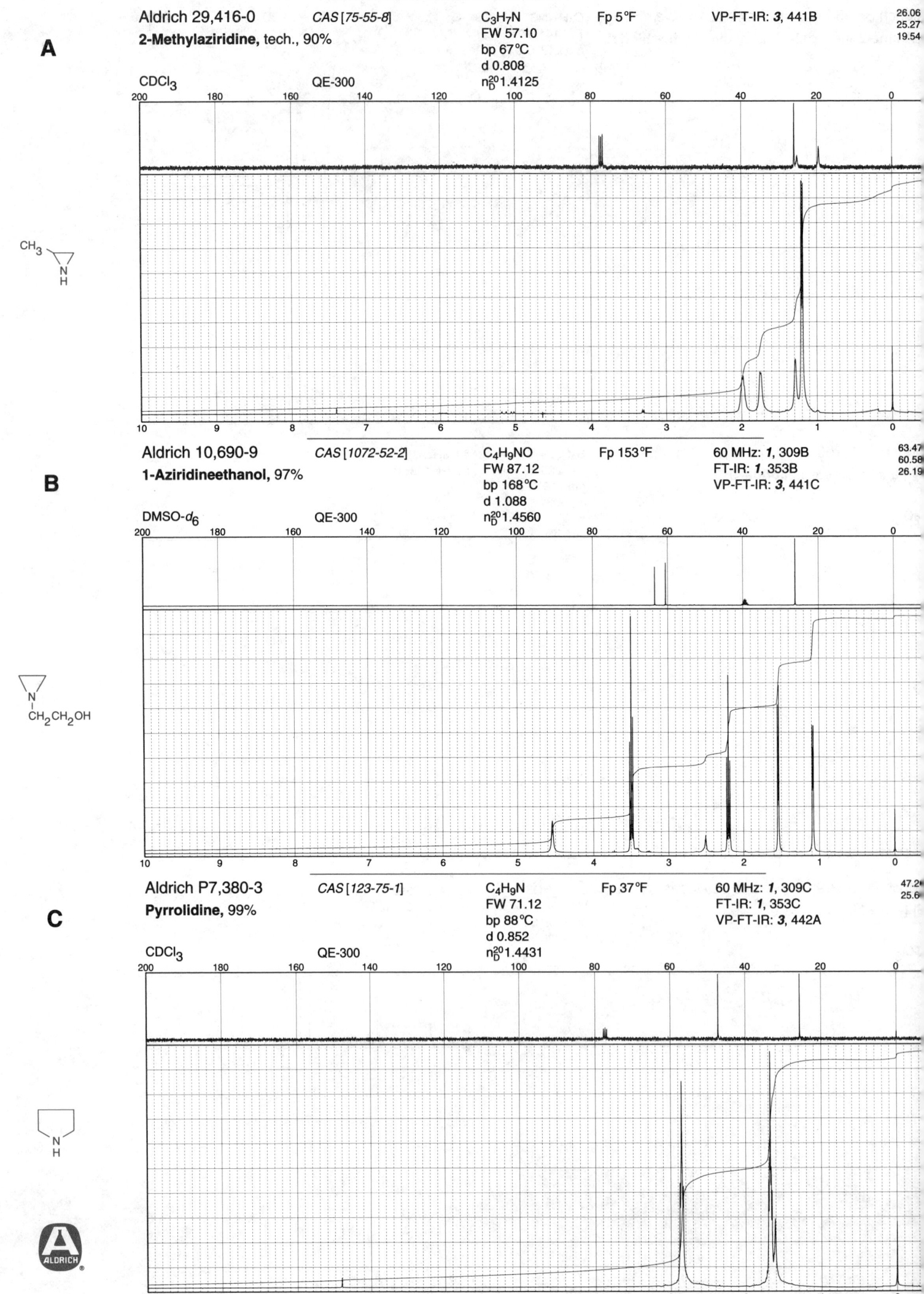

A

Aldrich 29,416-0   *CAS [75-55-8]*   C$_3$H$_7$N   Fp 5°F   VP-FT-IR: *3*, 441B   26.06
2-Methylaziridine, tech., 90%   FW 57.10   25.27
bp 67°C   19.54
d 0.808
CDCl$_3$ QE-300   n$_D^{20}$1.4125

B

Aldrich 10,690-9   *CAS [1072-52-2]*   C$_4$H$_9$NO   Fp 153°F   60 MHz: *1*, 309B   63.47
1-Aziridineethanol, 97%   FW 87.12   FT-IR: *1*, 353B   60.58
bp 168°C   VP-FT-IR: *3*, 441C   26.19
d 1.088
DMSO-d$_6$ QE-300   n$_D^{20}$1.4560

C

Aldrich P7,380-3   *CAS [123-75-1]*   C$_4$H$_9$N   Fp 37°F   60 MHz: *1*, 309C   47.2
Pyrrolidine, 99%   FW 71.12   FT-IR: *1*, 353C   25.6
bp 88°C   VP-FT-IR: *3*, 442A
d 0.852
CDCl$_3$ QE-300   n$_D^{20}$1.4431

A

Aldrich M7,920-4   CAS [120-94-5] $C_5H_{11}N$ Fp -7°F   60 MHz: *1*, 309D

1-Methylpyrrolidine, 97%   FW 85.15   FT-IR: *1*, 354B

bp 81°C   VP-FT-IR: *3*, 443A

CDCl₃   QE-300   d 0.819 n_D^{20} 1.4247

56.34
42.19*
24.14

B

Aldrich D18,380-6   CAS [3378-71-0] $C_6H_{13}N$ Fp 45°F   60 MHz: *1*, 310B

2,5-Dimethylpyrrolidine, tech., 93%,   FW 99.18   FT-IR: *1*, 354D

mixture of *cis* and *trans*   bp 106°C   VP-FT-IR: *3*, 443D

CDCl₃   QE-300   d 0.810 n_D^{20} 1.4299

54.93*
53.27*
34.81
33.79
22.27*
21.53*

C

Aldrich P7,400-1   CAS [1125-99-1] $C_{10}H_{17}N$ Fp 103°F   60 MHz: *1*, 348D

1-Pyrrolidino-1-cyclohexene, 97%   FW 151.25   FT-IR: *1*, 355B

bp 115°C (15 mm)   VP-FT-IR: *3*, 444C

CDCl₃   QE-300   d 0.940 n_D^{20} 1.5217

143.36
93.58*
47.46
27.59
24.57
23.44
23.08

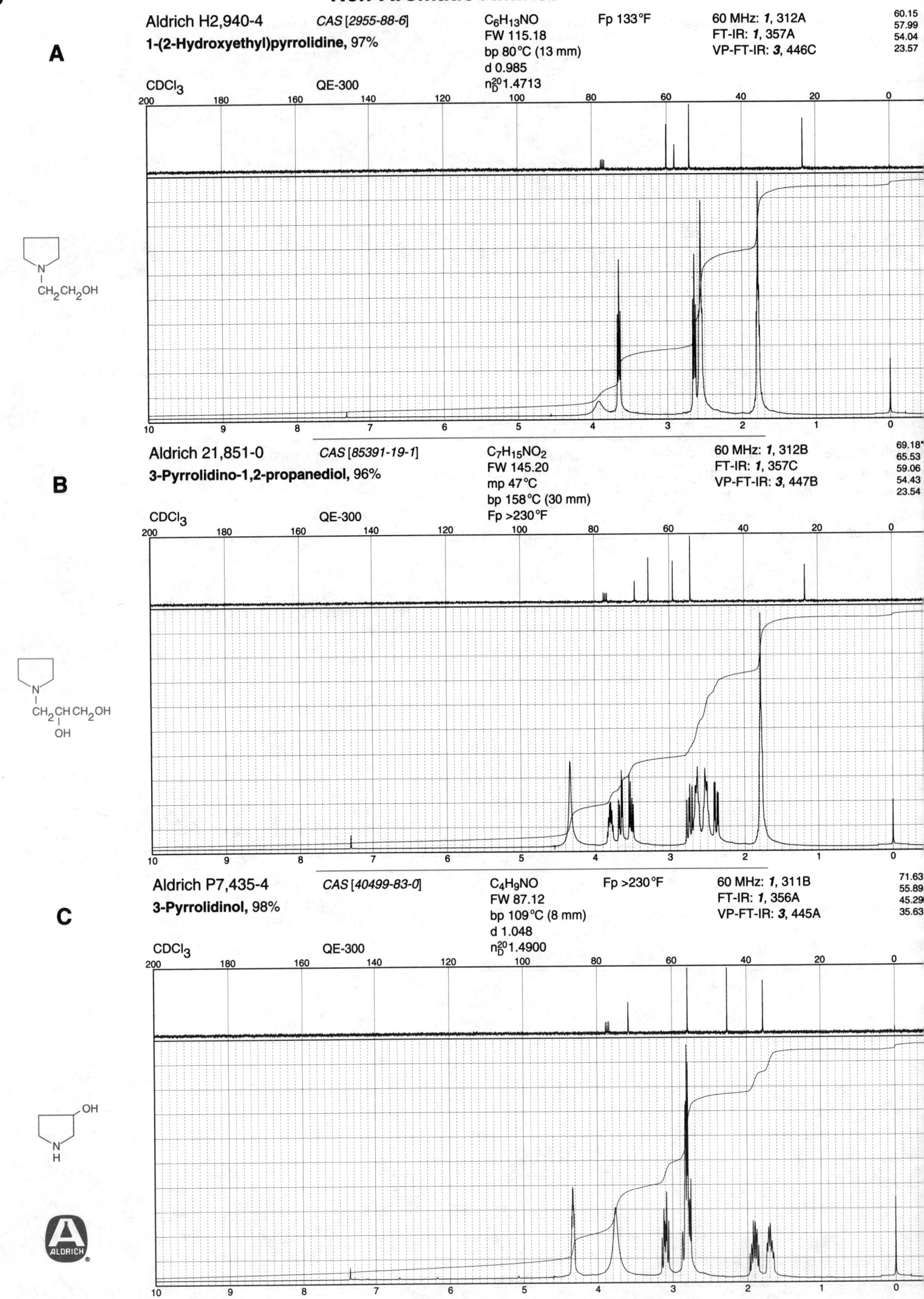

A

Aldrich H2,940-4   *CAS [2955-88-6]*

1-(2-Hydroxyethyl)pyrrolidine, 97%

$C_6H_{13}NO$
FW 115.18
bp 80°C (13 mm)
d 0.985
n_D^{20}1.4713

Fp 133°F

60 MHz: *1*, 312A
FT-IR: *1*, 357A
VP-FT-IR: *3*, 446C

60.15
57.99
54.04
23.57

CDCl₃   QE-300

B

Aldrich 21,851-0   *CAS [85391-19-1]*

3-Pyrrolidino-1,2-propanediol, 96%

$C_7H_{15}NO_2$
FW 145.20
mp 47°C
bp 158°C (30 mm)
Fp >230°F

60 MHz: *1*, 312B
FT-IR: *1*, 357C
VP-FT-IR: *3*, 447B

69.18*
65.53
59.06
54.43
23.54

CDCl₃   QE-300

C

Aldrich P7,435-4   *CAS [40499-83-0]*

3-Pyrrolidinol, 98%

C_4H_9NO
FW 87.12
bp 109°C (8 mm)
d 1.048
n_D^{20}1.4900

Fp >230°F

60 MHz: *1*, 311B
FT-IR: *1*, 356A
VP-FT-IR: *3*, 445A

71.63
55.89
45.29
35.63

CDCl₃   QE-300

Aldrich M7,950-6 CAS [13220-33-2] C5H11NO Fp 159°F 60 MHz: *1*, 311C 71.34*
65.21
-Methyl-3-pyrrolidinol, 97% FW 101.15 FT-IR: *1*, 356B 54.88
bp 51°C (1 mm) VP-FT-IR: *3*, 445B 42.07*
d 0.921 35.70

DCl3 QE-300 n20D 1.4640

A

Aldrich E4,735-2 CAS [30727-14-1] C6H13NO Fp 165°F 60 MHz: *1*, 311D 70.81*
62.86
-Ethyl-3-pyrrolidinol, 97% FW 115.18 FT-IR: *1*, 356C 52.36
bp 60°C (1 mm) VP-FT-IR: *3*, 445C 50.05
d 0.967 34.92
13.57*

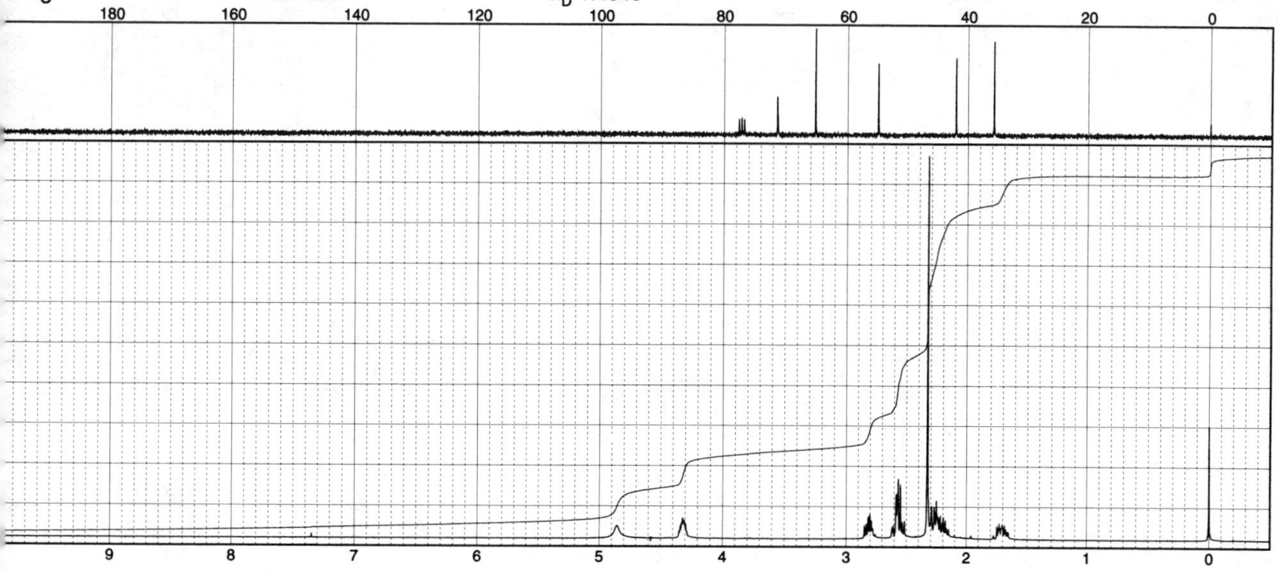

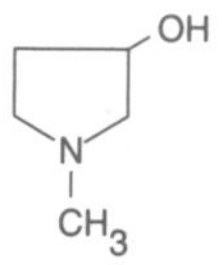

DCl3 QE-300 n20D 1.4670

B

Aldrich 18,651-1 CAS [23356-96-9] C5H11NO Fp 187°F 60 MHz: *1*, 312D 64.81
59.95*
-(+)-2-Pyrrolidinemethanol, 99% FW 101.15 FT-IR: *1*, 356D 46.45
bp 75°C (2 mm) VP-FT-IR: *3*, 445D 27.65
d 1.025 25.87

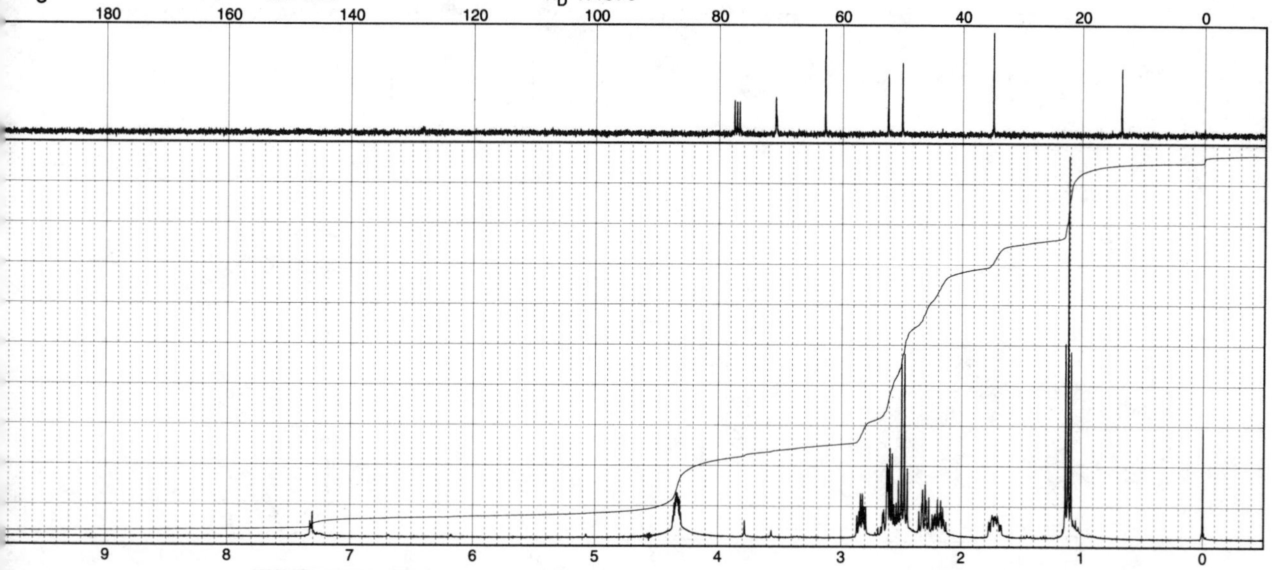

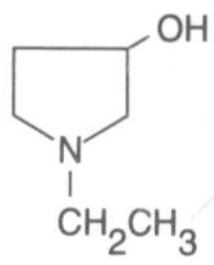

DCl3 QE-300 n20D 1.4853

C

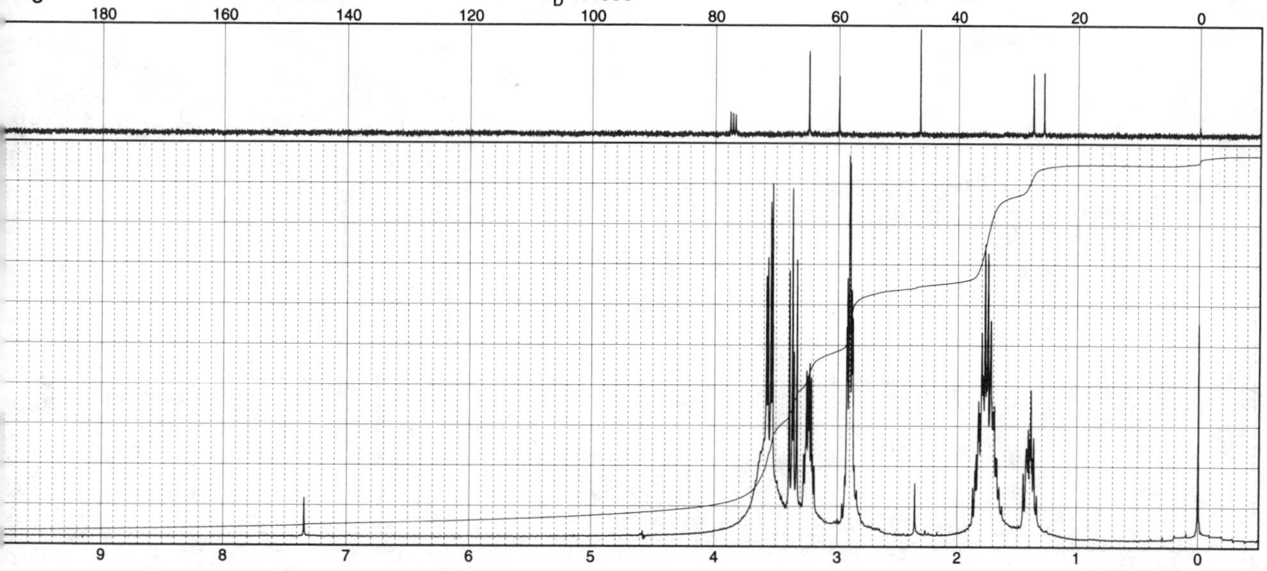

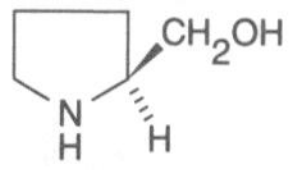

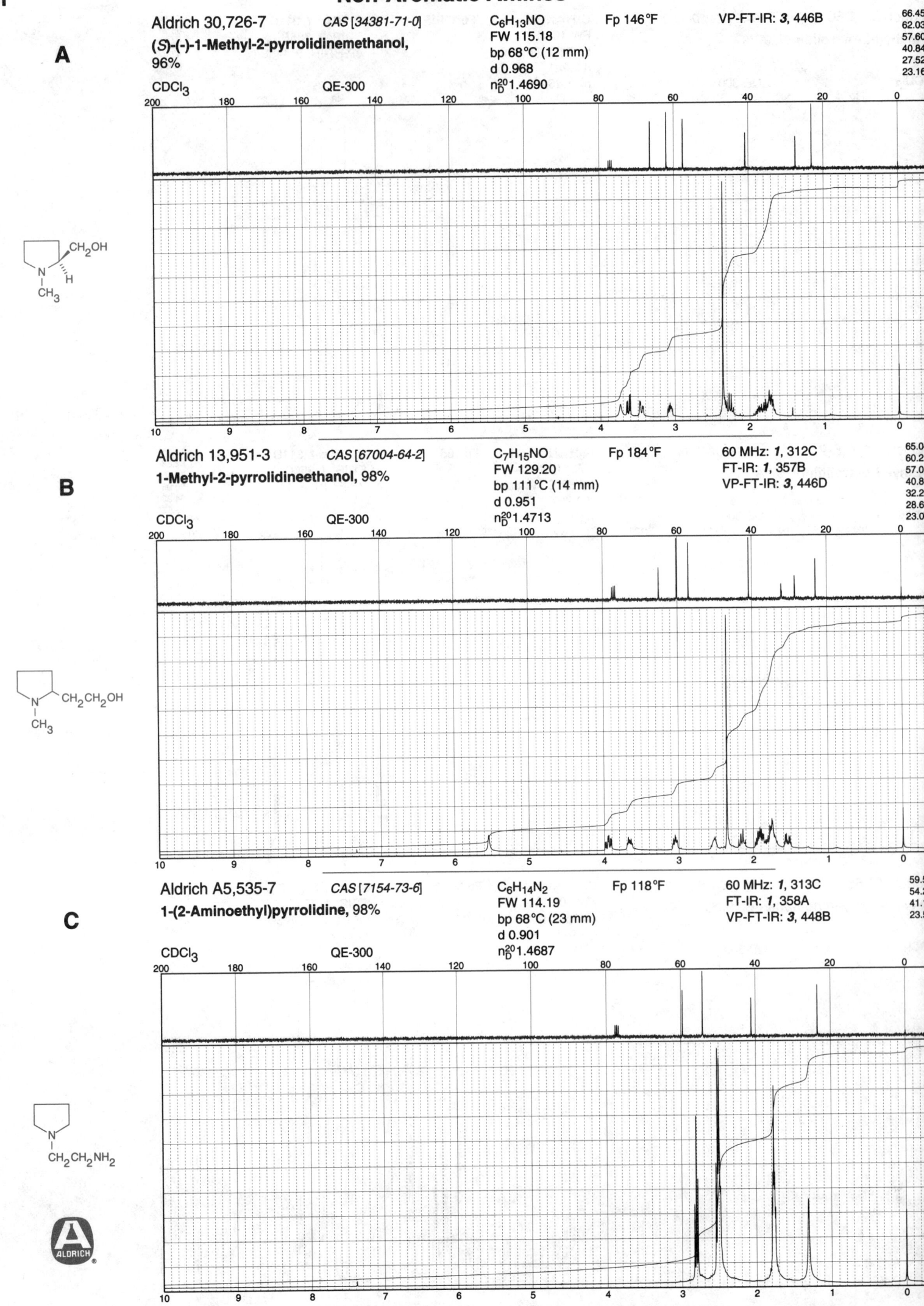

A

Aldrich 30,726-7 *CAS [34381-71-0]* C₆H₁₃NO Fp 146°F VP-FT-IR: *3*, 446B

(S)-(-)-1-Methyl-2-pyrrolidinemethanol, 96%

FW 115.18
bp 68°C (12 mm)
d 0.968
n$_D^{20}$1.4690

CDCl₃   QE-300

66.45
62.03
57.60
40.84
27.52
23.16

B

Aldrich 13,951-3 *CAS [67004-64-2]* C₇H₁₅NO Fp 184°F 60 MHz: *1*, 312C

1-Methyl-2-pyrrolidineethanol, 98%

FW 129.20
bp 111°C (14 mm)
d 0.951
n$_D^{20}$1.4713

FT-IR: *1*, 357B
VP-FT-IR: *3*, 446D

CDCl₃   QE-300

65.0
60.2
57.0
40.8
32.2
28.6
23.0

C

Aldrich A5,535-7 *CAS [7154-73-6]* C₆H₁₄N₂ Fp 118°F 60 MHz: *1*, 313C

1-(2-Aminoethyl)pyrrolidine, 98%

FW 114.19
bp 68°C (23 mm)
d 0.901
n$_D^{20}$1.4687

FT-IR: *1*, 358A
VP-FT-IR: *3*, 448B

CDCl₃   QE-300

59.5
54.2
41.1
23.5

Non-Aromatic Amines

Aldrich 18,326-1   CAS [26116-12-1]   C₇H₁₆N₂   Fp 141°F   60 MHz: **1**, 269D
-(Aminomethyl)-1-ethylpyrrolidine,   FW 128.22   FT-IR: **1**, 355C
8%   bp 59°C (16 mm)   VP-FT-IR: **3**, 444D
DCl₃   QE-300   d 0.887
n₂₀D 1.4665

66.16*
53.88
48.72
44.83
28.27
22.84
13.98*

A

Aldrich 13,950-5   CAS [51387-90-7]   C₇H₁₆N₂   60 MHz: **1**, 311A
(2-Aminoethyl)-1-methylpyrrolidine,   FW 128.22   FT-IR: **1**, 368D
8%   d 0.885   VP-FT-IR: **3**, 448C
DCl₃   QE-300   n₂₀D 1.4684
Fp 149°F

64.32*
57.21
40.49*
39.86
37.98
30.80
22.00

B

Aldrich 32,445-0   CAS [51207-66-0]   C₉H₁₈N₂   Fp 190°F   VP-FT-IR: **3**, 448D
(+)-1-(2-Pyrrolidinylmethyl)pyrrolidine,   FW 154.26
%   bp 100°C (2 mm)
DCl₃   QE-300   d 0.946
n₂₀D 1.4871

62.18
57.42*
54.58
46.14
30.15
25.08
23.45

C

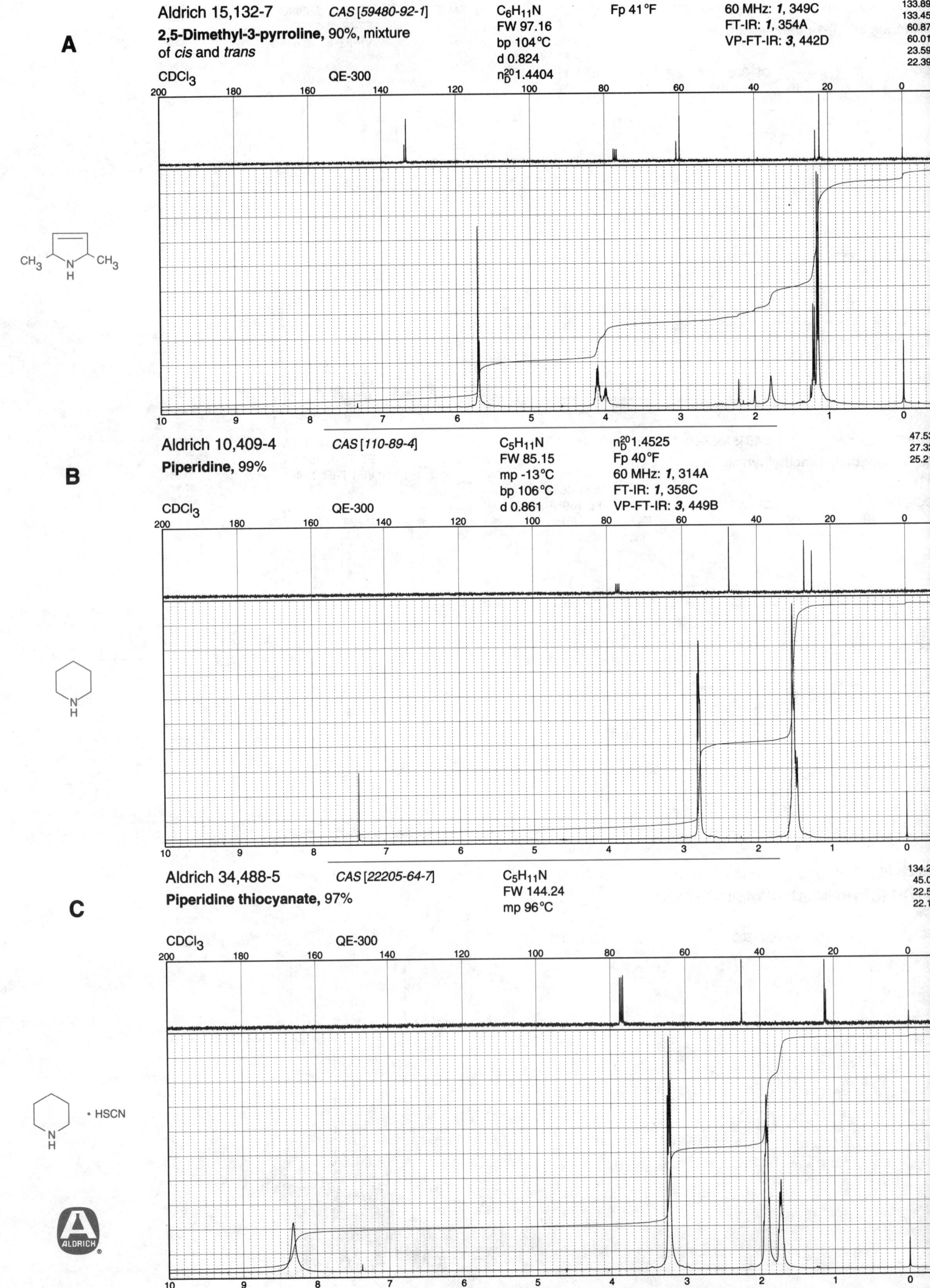

A

Aldrich 15,132-7   *CAS [59480-92-1]*   C$_6$H$_{11}$N   Fp 41°F   60 MHz: *1*, 349C
2,5-Dimethyl-3-pyrroline, 90%, mixture   FW 97.16   FT-IR: *1*, 354A
of *cis* and *trans*   bp 104°C   VP-FT-IR: *3*, 442D
d 0.824
CDCl$_3$ QE-300   n$_D^{20}$1.4404

133.89
133.45
60.87
60.01
23.59
22.39

CH$_3$ — N — CH$_3$
        H

B

Aldrich 10,409-4   *CAS [110-89-4]*   C$_5$H$_{11}$N   n$_D^{20}$1.4525   47.53
Piperidine, 99%   FW 85.15   Fp 40°F   27.32
mp -13°C   60 MHz: *1*, 314A   25.21
bp 106°C   FT-IR: *1*, 358C
d 0.861   VP-FT-IR: *3*, 449B
CDCl$_3$ QE-300

N
H

C

Aldrich 34,488-5   *CAS [22205-64-7]*   C$_5$H$_{11}$N   134.2
Piperidine thiocyanate, 97%   FW 144.24   45.0
mp 96°C   22.5
22.1
CDCl$_3$ QE-300

N   · HSCN
H

A

Aldrich M7,260-9   *CAS [626-67-5]* $C_6H_{13}N$ Fp 38°F

-Methylpiperidine, 99%

FW 99.18
bp 107°C
d 0.816
n_D^{20} 1.4378

DCl$_3$ QE-300

60 MHz: *1*, 314D
FT-IR: *1*, 360B
VP-FT-IR: *3*, 450C

56.62
46.94*
26.12
23.87

B

Aldrich 17,020-8   *CAS [880-09-1]* $C_{11}H_{22}N_2$ Fp 197°F

ipiperidinomethane, 98%

FW 182.31
bp 123°C (15 mm)
d 0.915
n_D^{20} 1.4820

DCl$_3$ QE-300

FT-IR: *1*, 363A
VP-FT-IR: *3*, 455B

82.78
53.11
26.04
25.03

C

Aldrich 19,223-6   *CAS [1932-04-3]* $C_{12}H_{24}N_2$ n_D^{20} 1.4876

2-Dipiperidinoethane, 98%

FW 196.34
mp -1°C
bp 265°C
d 0.916

Fp >230°F
60 MHz: *1*, 315B
FT-IR: *1*, 363B
VP-FT-IR: *3*, 455C

DCl$_3$ QE-300

56.90
55.14
26.07
24.46

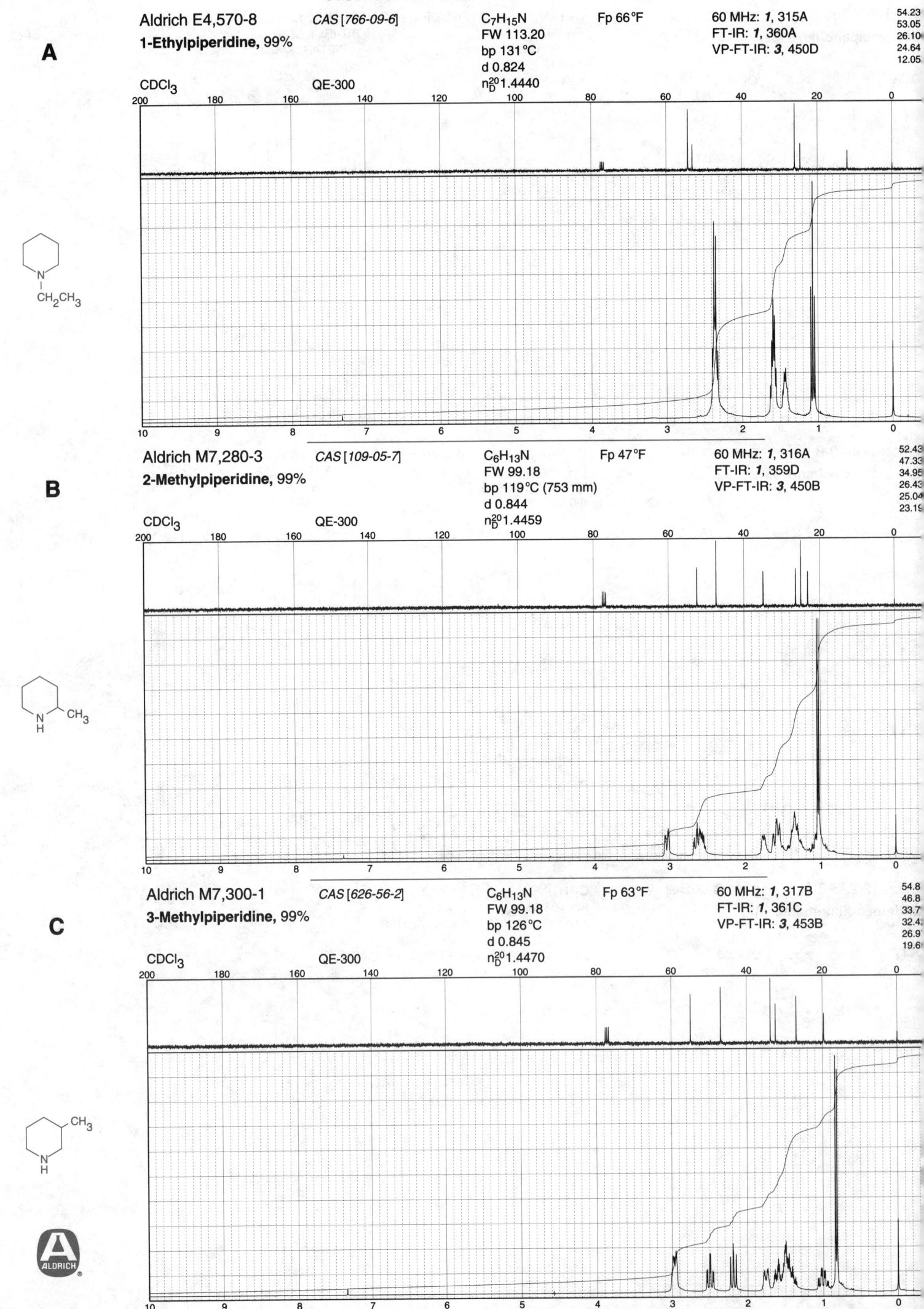

A

Aldrich E4,570-8   CAS [766-09-6] $C_7H_{15}N$ Fp 66°F   60 MHz: *1*, 315A   54.23
1-Ethylpiperidine, 99%   FW 113.20   FT-IR: *1*, 360A   53.05
bp 131°C   VP-FT-IR: *3*, 450D   26.10
d 0.824   24.64
n_D^{20}1.4440   12.05

CDCl$_3$ QE-300

B

Aldrich M7,280-3   CAS [109-05-7] $C_6H_{13}N$ Fp 47°F   60 MHz: *1*, 316A   52.43
2-Methylpiperidine, 99%   FW 99.18   FT-IR: *1*, 359D   47.33
bp 119°C (753 mm)   VP-FT-IR: *3*, 450B   34.95
d 0.844   26.43
n_D^{20}1.4459   25.04
23.19

CDCl$_3$ QE-300

C

Aldrich M7,300-1   CAS [626-56-2] $C_6H_{13}N$ Fp 63°F   60 MHz: *1*, 317B   54.8
3-Methylpiperidine, 99%   FW 99.18   FT-IR: *1*, 361C   46.8
bp 126°C   VP-FT-IR: *3*, 453B   33.7
d 0.845   32.4
n_D^{20}1.4470   26.9
19.6

CDCl$_3$ QE-300

A

Aldrich 19,225-2   CAS [68922-17-8]

1,1'-Methylenebis(3-methylpiperidine), 98%

C$_{13}$H$_{26}$N$_2$
FW 210.37
bp 160°C (50 mm)
d 0.887
n$_D^{20}$ 1.4734

Fp >230°F

60 MHz: 1, 315C
FT-IR: 1, 362D
VP-FT-IR: 3, 455A

DCl$_3$ QE-300

82.34
60.58
52.61
33.64
31.05*
25.52
19.72*

B

Aldrich M7,320-6   CAS [626-58-4]

Methylpiperidine, 96%

C$_6$H$_{13}$N
FW 99.18
bp 124°C
d 0.838
n$_D^{20}$ 1.4458

Fp 45°F

60 MHz: 1, 317C
FT-IR: 1, 361D
VP-FT-IR: 3, 453C

DCl$_3$ QE-300

46.93
35.80
31.41*
22.60*

C

Aldrich 15,005-3   CAS [4897-50-1]

Piperidinopiperidine, tech., 95%

C$_{10}$H$_{20}$N$_2$
FW 168.28
mp 65°C

FT-IR: 1, 359A
VP-FT-IR: 3, 449D

DCl$_3$ QE-300

63.04*
50.07
46.65
29.65
26.52
24.93

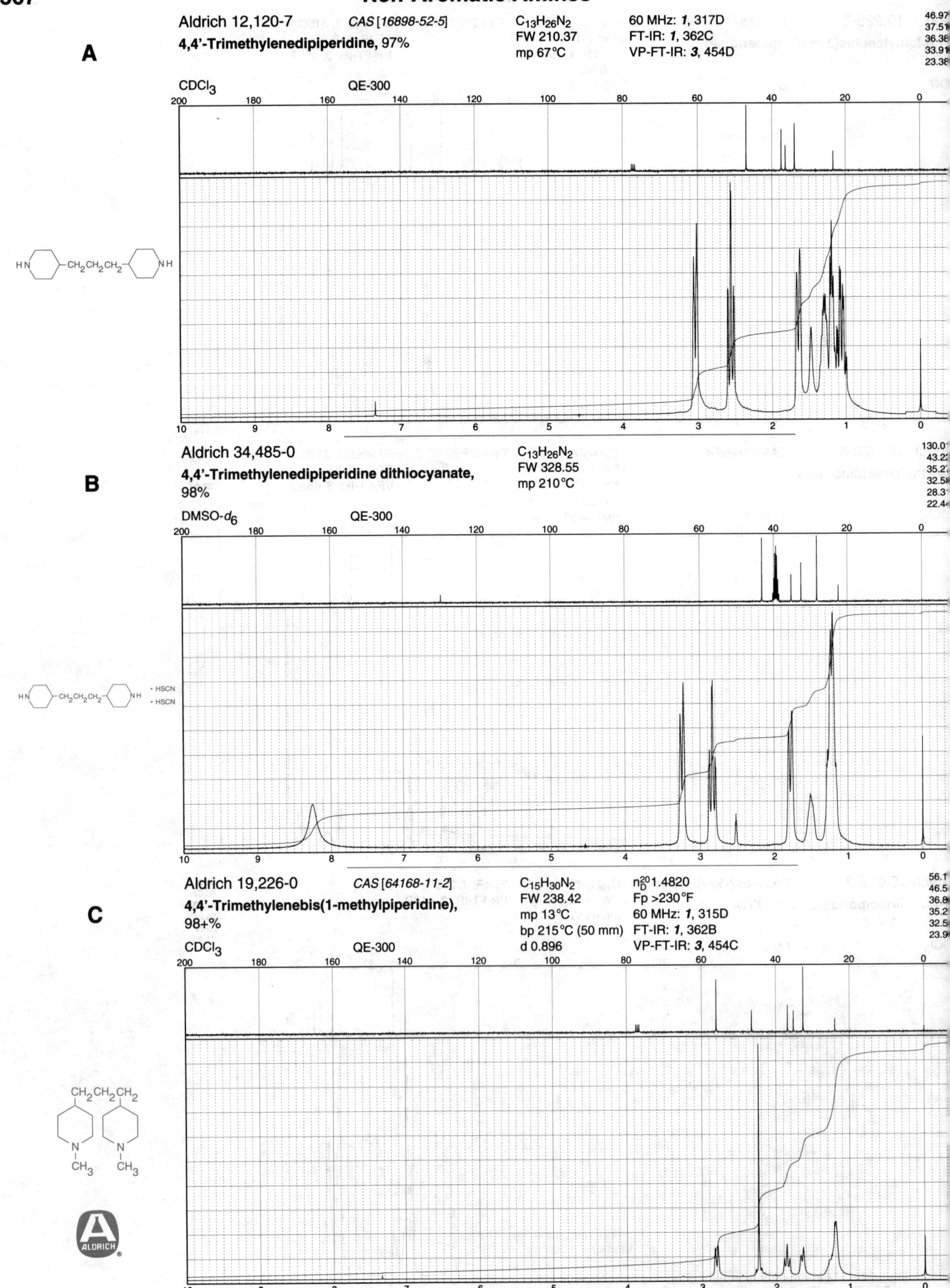

A

Aldrich 12,120-7   *CAS [16898-52-5]*

4,4'-Trimethylenedipiperidine, 97%

$C_{13}H_{26}N_2$
FW 210.37
mp 67°C

60 MHz: *1*, 317D
FT-IR: *1*, 362C
VP-FT-IR: *3*, 454D

46.97
37.51
36.38
33.91
23.38

CDCl₃   QE-300

B

Aldrich 34,485-0

4,4'-Trimethylenedipiperidine dithiocyanate, 98%

$C_{13}H_{26}N_2$
FW 328.55
mp 210°C

130.0
43.2
35.2
32.5
28.3
22.4

DMSO-d_6 QE-300

C

Aldrich 19,226-0   *CAS [64168-11-2]*

4,4'-Trimethylenebis(1-methylpiperidine), 98+%

$C_{15}H_{30}N_2$
FW 238.42
mp 13°C
bp 215°C (50 mm)
d 0.896

n_D^{20}1.4820
Fp >230°F
60 MHz: *1*, 315D
FT-IR: *1*, 362B
VP-FT-IR: *3*, 454C

56.1
46.5
36.8
35.2
32.5
23.9

CDCl₃   QE-300

Non-Aromatic Amines

Aldrich 31,631-8   CAS [1193-12-0]   C₇H₁₅N   Fp 90°F   VP-FT-IR: 3, 452A

3,3-Dimethylpiperidine, 97%

	FW 113.20
	bp 137°C
	d 0.829

CDCl₃   QE-300   n$_D^{20}$1.4476

58.87
47.04
37.98
29.90
26.77*
23.35

A

Aldrich D18,030-0   CAS [766-17-6]   C₇H₁₅N   Fp 53°F   60 MHz: 1, 316B

cis-2,6-Dimethylpiperidine, 99%

	FW 113.20
	bp 128°C
	d 0.840

CDCl₃   QE-300   n$_D^{20}$1.4394

FT-IR: 1, 360C
VP-FT-IR: 3, 451D

52.53*
34.22
25.01
23.13*

B

Aldrich 18,610-4   CAS [35794-11-7]   C₇H₁₅N   Fp 91°F   60 MHz: 1, 316C

3,5-Dimethylpiperidine, 96%, mixture
of cis and trans

	FW 113.20
	bp 144°C
	d 0.853

CDCl₃   QE-300   n$_D^{20}$1.4454

FT-IR: 1, 360D
VP-FT-IR: 3, 452B

54.41   32.75*
53.59   27.54*
42.97   19.63*
39.64   18.62*

C

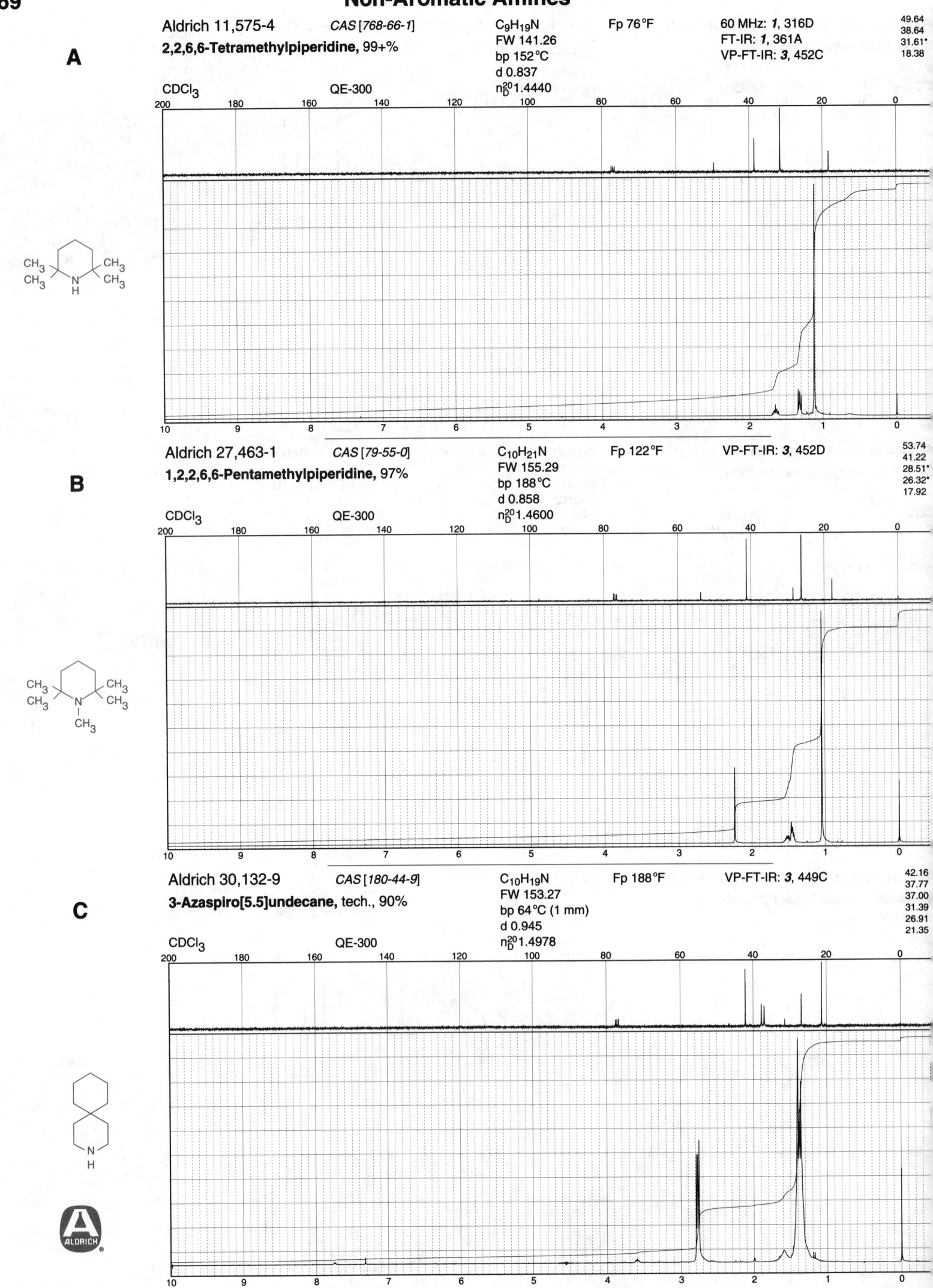

A

Aldrich 11,575-4   *CAS [768-66-1]*   C$_9$H$_{19}$N   Fp 76°F   60 MHz: *1*, 316D

2,2,6,6-Tetramethylpiperidine, 99+%   FW 141.26   FT-IR: *1*, 361A

bp 152°C   VP-FT-IR: *3*, 452C

d 0.837

n$_D^{20}$1.4440

49.64
38.64
31.61*
18.38

CDCl$_3$ QE-300

B

Aldrich 27,463-1   *CAS [79-55-0]*   C$_{10}$H$_{21}$N   Fp 122°F   VP-FT-IR: *3*, 452D

1,2,2,6,6-Pentamethylpiperidine, 97%   FW 155.29

bp 188°C

d 0.858

n$_D^{20}$1.4600

53.74
41.22
28.51*
26.32*
17.92

CDCl$_3$ QE-300

C

Aldrich 30,132-9   *CAS [180-44-9]*   C$_{10}$H$_{19}$N   Fp 188°F   VP-FT-IR: *3*, 449C

3-Azaspiro[5.5]undecane, tech., 90%   FW 153.27

bp 64°C (1 mm)

d 0.945

n$_D^{20}$1.4978

42.16
37.77
37.00
31.39
26.91
21.35

CDCl$_3$ QE-300

Aldrich P4,620-2   *CAS [3616-58-8]*

1-Piperidineacetaldehyde diethyl acetal

$C_{11}H_{23}NO_2$ Fp 177°F

FW 201.31

bp 220°C

d 0.915

n_D^{20} 1.4430

60 MHz: *1*, 319C

FT-IR: *1*, 364C

VP-FT-IR: *3*, 455D

101.47*
61.83
61.65
55.26
26.08
24.32
15.39*

CDCl₃   QE-300

A

Aldrich 17,836-5   *CAS [177-11-7]*

1,4-Dioxa-8-azaspiro[4.5]decane, 98%

$C_7H_{13}NO_2$ Fp 179°F

FW 143.19

bp 109°C (26 mm)

d 1.117

n_D^{20} 1.4819

60 MHz: *1*, 322D

FT-IR: *1*, 368A

VP-FT-IR: *3*, 459D

107.44
64.16
44.61
36.52

CDCl₃   QE-300

B

Aldrich 21,849-9   *CAS [4847-93-2]*

3-Piperidino-1,2-propanediol, 96%

$C_8H_{17}NO_2$

FW 159.23

mp 79°C

FT-IR: *1*, 364D

VP-FT-IR: *3*, 456C

67.09*
65.37
61.52
54.97
26.06
24.19

CDCl₃   QE-300

C

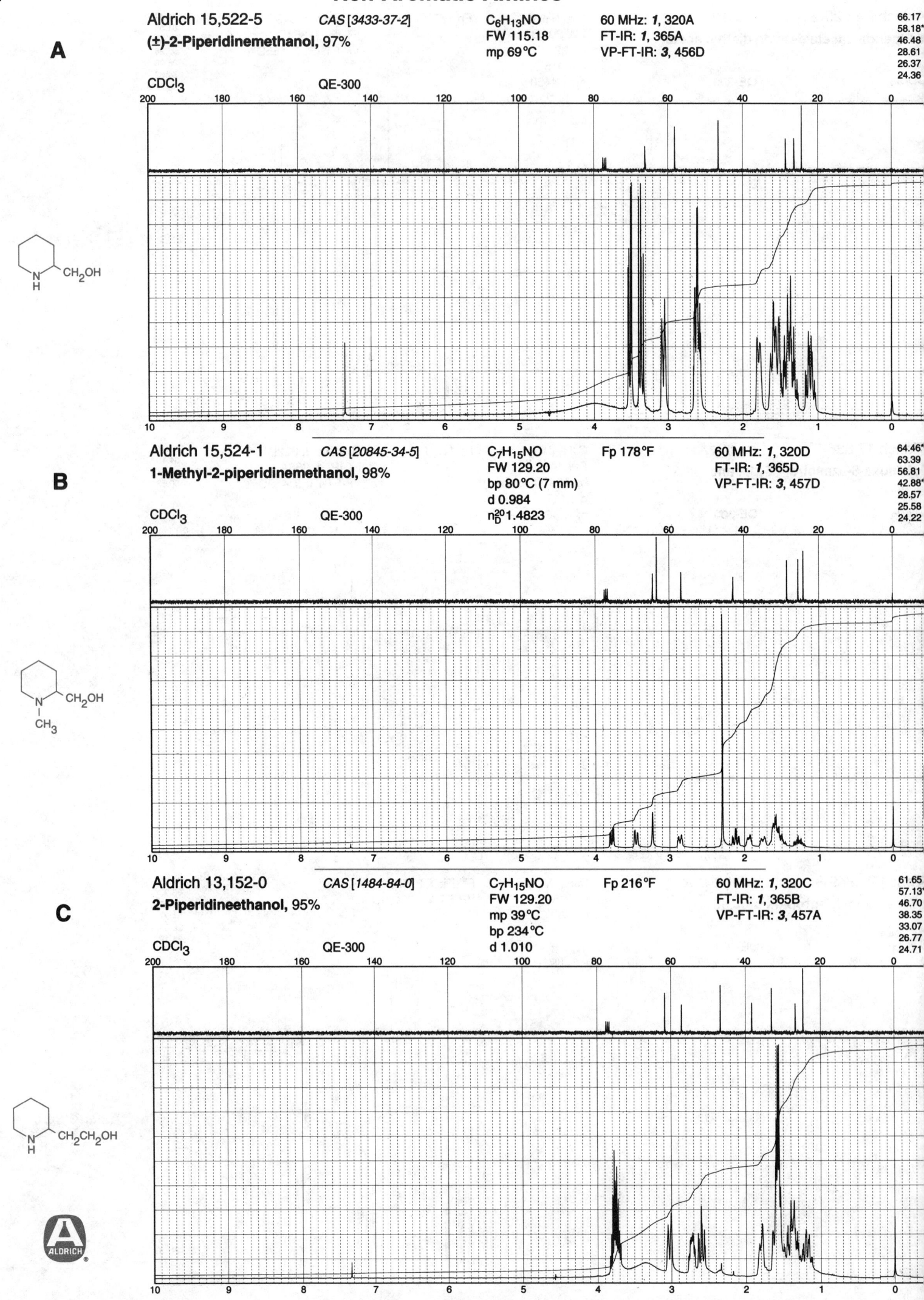

A

Aldrich 15,522-5 CAS [3433-37-2]

(±)-2-Piperidinemethanol, 97%

$C_6H_{13}NO$
FW 115.18
mp 69°C

60 MHz: *1*, 320A
FT-IR: *1*, 365A
VP-FT-IR: *3*, 456D

66.17
58.18*
46.48
28.61
26.37
24.36

CDCl$_3$ QE-300

B

Aldrich 15,524-1 CAS [20845-34-5]

1-Methyl-2-piperidinemethanol, 98%

$C_7H_{15}NO$
FW 129.20
bp 80°C (7 mm)
d 0.984
n$_D^{20}$1.4823

Fp 178°F

60 MHz: *1*, 320D
FT-IR: *1*, 365D
VP-FT-IR: *3*, 457D

64.46*
63.39
56.81
42.88*
28.57
25.58
24.22

CDCl$_3$ QE-300

C

Aldrich 13,152-0 CAS [1484-84-0]

2-Piperidineethanol, 95%

$C_7H_{15}NO$
FW 129.20
mp 39°C
bp 234°C
d 1.010

Fp 216°F

60 MHz: *1*, 320C
FT-IR: *1*, 365B
VP-FT-IR: *3*, 457A

61.65
57.13*
46.70
38.35
33.07
26.77
24.71

CDCl$_3$ QE-300

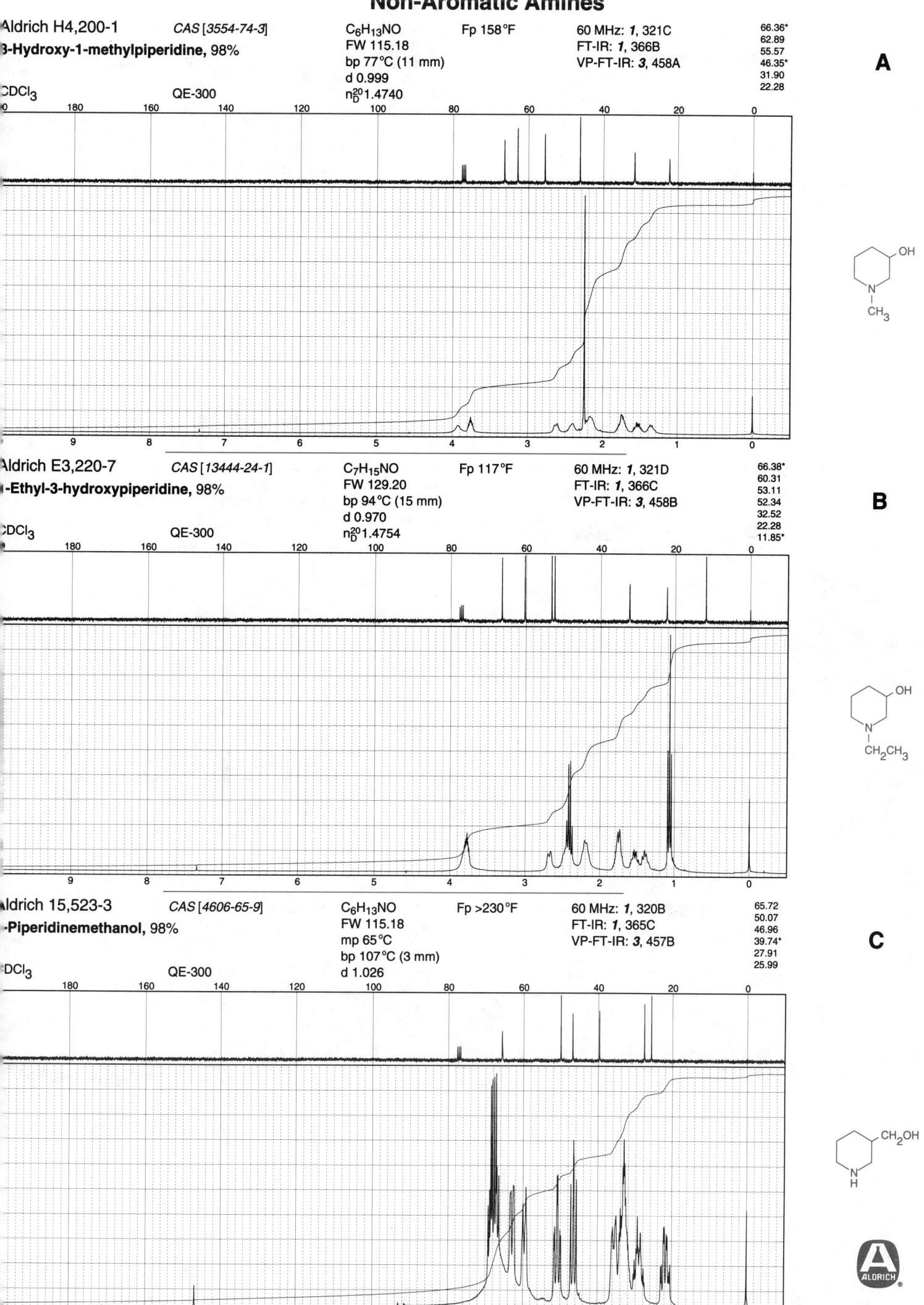

Aldrich H4,200-1   CAS [3554-74-3]

3-Hydroxy-1-methylpiperidine, 98%

CDCl₃   QE-300

$C_6H_{13}NO$
FW 115.18
bp 77°C (11 mm)
d 0.999
n_D^{20}1.4740

Fp 158°F

60 MHz: *1*, 321C
FT-IR: *1*, 366B
VP-FT-IR: *3*, 458A

66.36*
62.89
55.57
46.35*
31.90
22.28

A

Aldrich E3,220-7   CAS [13444-24-1]

1-Ethyl-3-hydroxypiperidine, 98%

CDCl₃   QE-300

$C_7H_{15}NO$
FW 129.20
bp 94°C (15 mm)
d 0.970
n_D^{20}1.4754

Fp 117°F

60 MHz: *1*, 321D
FT-IR: *1*, 366C
VP-FT-IR: *3*, 458B

66.38*
60.31
53.11
52.34
32.52
22.28
11.85*

B

Aldrich 15,523-3   CAS [4606-65-9]

3-Piperidinemethanol, 98%

CDCl₃   QE-300

$C_6H_{13}NO$
FW 115.18
mp 65°C
bp 107°C (3 mm)
d 1.026

Fp >230°F

60 MHz: *1*, 320B
FT-IR: *1*, 365C
VP-FT-IR: *3*, 457B

65.72
50.07
46.96
39.74*
27.91
25.99

C

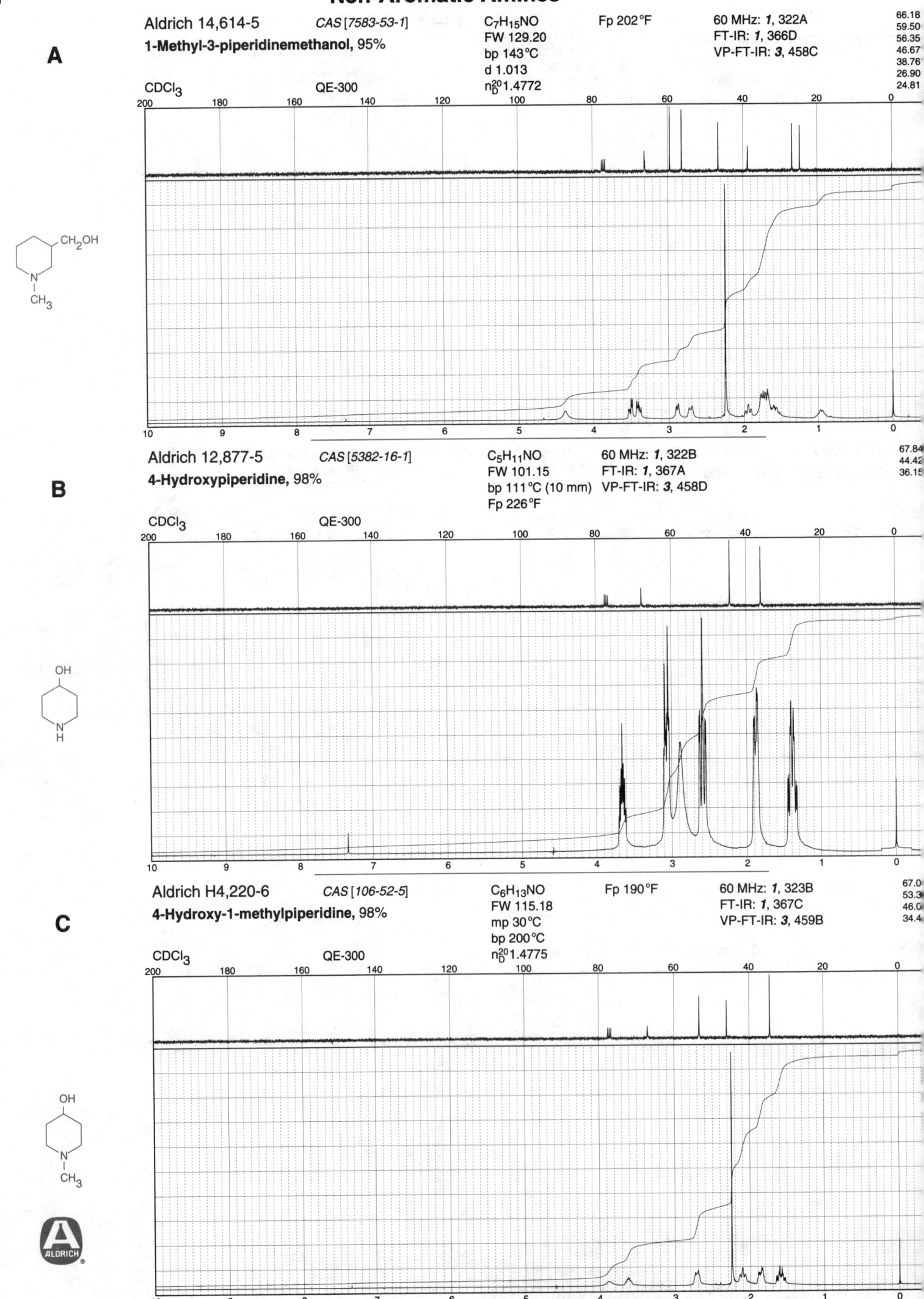

A

Aldrich 14,614-5   *CAS [7583-53-1]*

1-Methyl-3-piperidinemethanol, 95%

$C_7H_{15}NO$
FW 129.20
bp 143°C
d 1.013
n_D^{20} 1.4772

Fp 202°F

60 MHz: *1*, 322A
FT-IR: *1*, 366D
VP-FT-IR: *3*, 458C

66.18
59.50
56.35
46.67
38.76
26.90
24.81

$CDCl_3$ QE-300

B

Aldrich 12,877-5   *CAS [5382-16-1]*

4-Hydroxypiperidine, 98%

$C_5H_{11}NO$
FW 101.15
bp 111°C (10 mm)
Fp 226°F

60 MHz: *1*, 322B
FT-IR: *1*, 367A
VP-FT-IR: *3*, 458D

67.84
44.42
36.15

$CDCl_3$ QE-300

C

Aldrich H4,220-6   *CAS [106-52-5]*

4-Hydroxy-1-methylpiperidine, 98%

$C_6H_{13}NO$
FW 115.18
mp 30°C
bp 200°C
n_D^{20} 1.4775

Fp 190°F

60 MHz: *1*, 323B
FT-IR: *1*, 367C
VP-FT-IR: *3*, 459B

67.0
53.3
46.0
34.4

$CDCl_3$ QE-300

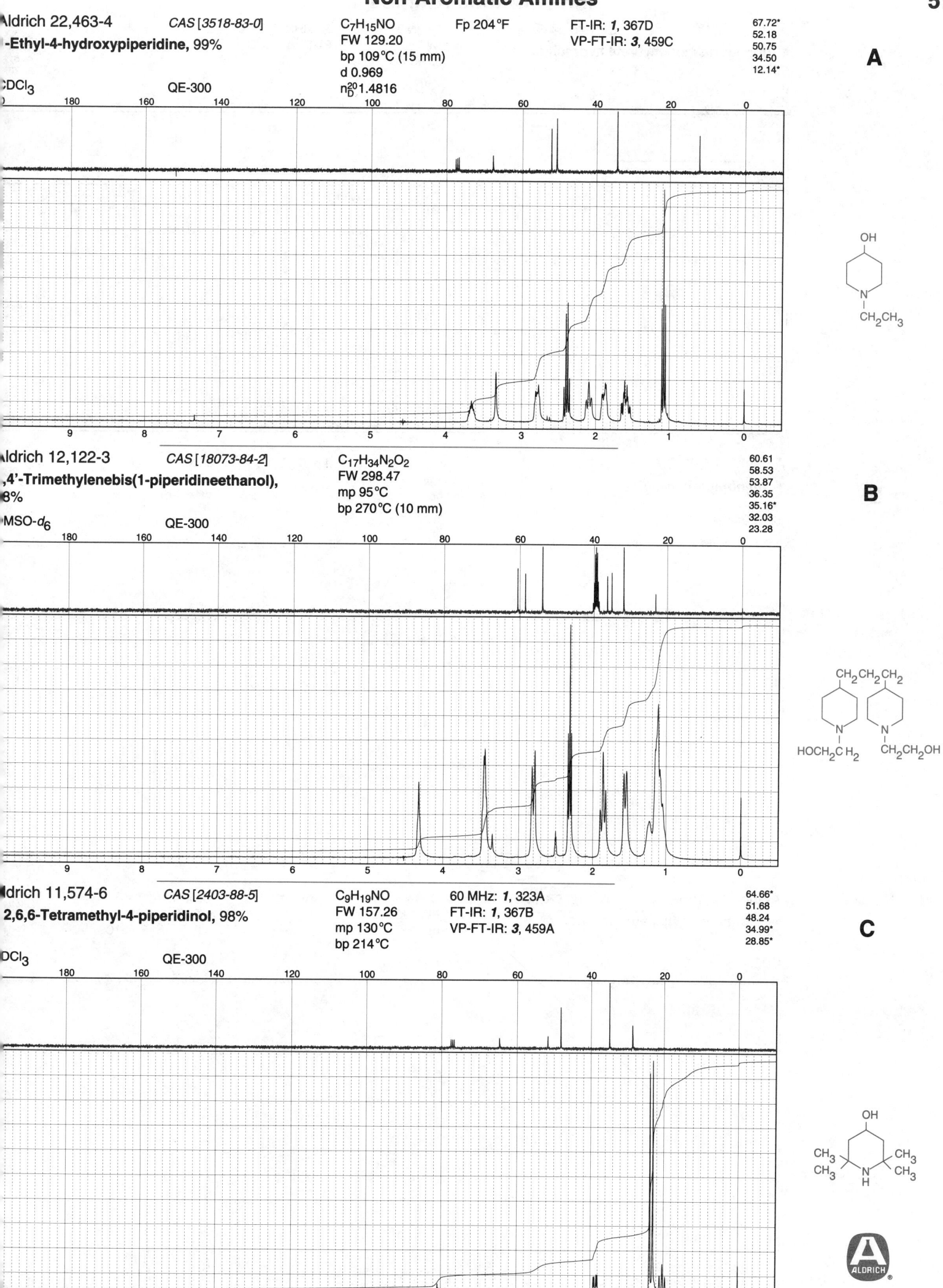

Aldrich 22,463-4　　CAS [3518-83-0]
-Ethyl-4-hydroxypiperidine, 99%

CDCl₃　　QE-300

C₇H₁₅NO　　Fp 204°F
FW 129.20
bp 109°C (15 mm)
d 0.969
n₂₀ᴰ 1.4816

FT-IR: *1*, 367D
VP-FT-IR: *3*, 459C

67.72*
52.18
50.75
34.50
12.14*

A

Aldrich 12,122-3　　CAS [18073-84-2]
,4'-Trimethylenebis(1-piperidineethanol),
%

MSO-d₆　　QE-300

C₁₇H₃₄N₂O₂
FW 298.47
mp 95°C
bp 270°C (10 mm)

60.61
58.53
53.87
36.35
35.16*
32.03
23.28

B

Aldrich 11,574-6　　CAS [2403-88-5]
2,6,6-Tetramethyl-4-piperidinol, 98%

DCl₃　　QE-300

C₉H₁₉NO
FW 157.26
mp 130°C
bp 214°C

60 MHz: *1*, 323A
FT-IR: *1*, 367B
VP-FT-IR: *3*, 459A

64.66*
51.68
48.24
34.99*
28.85*

C

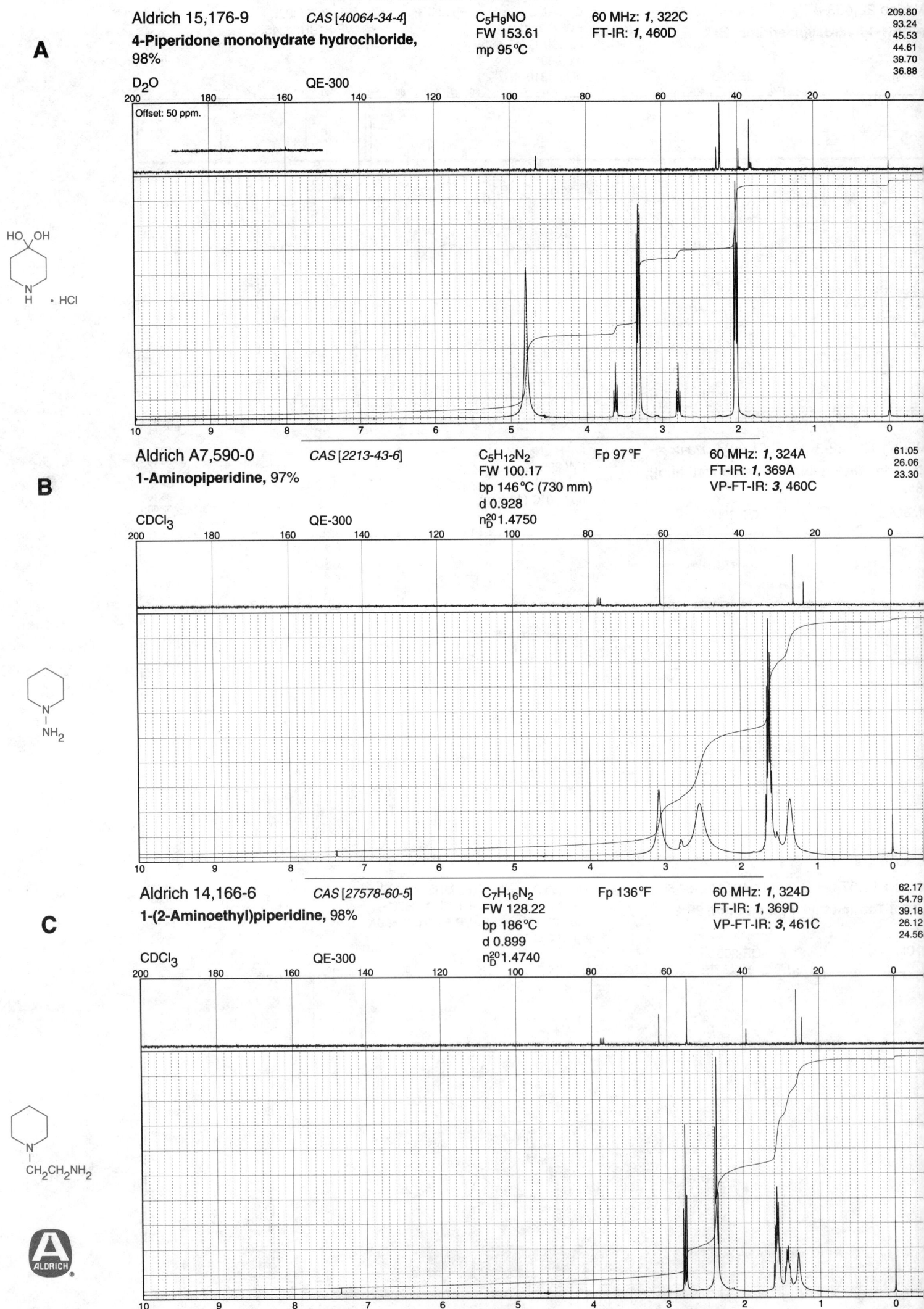

A

Aldrich 15,176-9   CAS [40064-34-4] C_5H_9NO 60 MHz: *1*, 322C

4-Piperidone monohydrate hydrochloride,   FW 153.61   FT-IR: *1*, 460D

98%   mp 95°C

209.80
93.24
45.53
44.61
39.70
36.88

D_2O QE-300

Offset: 50 ppm.

B

Aldrich A7,590-0   CAS [2213-43-6] $C_5H_{12}N_2$ Fp 97°F   60 MHz: *1*, 324A

1-Aminopiperidine, 97%   FW 100.17   FT-IR: *1*, 369A

bp 146°C (730 mm)   VP-FT-IR: *3*, 460C

d 0.928

n_D^{20} 1.4750

61.05
26.06
23.30

$CDCl_3$ QE-300

C

Aldrich 14,166-6   CAS [27578-60-5] $C_7H_{16}N_2$ Fp 136°F   60 MHz: *1*, 324D

1-(2-Aminoethyl)piperidine, 98%   FW 128.22   FT-IR: *1*, 369D

bp 186°C   VP-FT-IR: *3*, 461C

d 0.899

n_D^{20} 1.4740

62.17
54.79
39.18
26.12
24.56

$CDCl_3$ QE-300

A

Aldrich 18,611-2   CAS [25560-00-3]   C₉H₂₀N₂   Fp 191 °F

1-(3-Aminopropyl)-2-pipecoline, 96%

FW 156.27
bp 97 °C (15 mm)
d 0.889

CDCl₃   QE-300 n_D^{20} 1.4765

60 MHz: *1*, 325A
FT-IR: *1*, 370A
VP-FT-IR: *3*, 462A

55.99* 29.84
52.14   26.27
51.83   24.02
41.10   19.01*
34.77

B

Aldrich 12,293-9   CAS [6789-94-2]   C₇H₁₆N₂   Fp 141 °F

3-Amino-1-ethylpiperidine, 97%

FW 128.22
bp 155 °C
d 0.923

CDCl₃   QE-300 n_D^{20} 1.4715

60 MHz: *1*, 325C
FT-IR: *1*, 370D
VP-FT-IR: *3*, 462D

62.50
53.26
52.46
48.16*
34.48
23.93
12.01*

C

Aldrich 22,140-6   CAS [73579-08-5]   C₇H₁₆N₂

1-Methyl-4-(methylamino)piperidine, 98%

FW 128.22
d 0.882
n_D^{20} 1.4672
Fp 132 °F

CDCl₃   QE-300

FT-IR: *1*, 370C
VP-FT-IR: *3*, 462C

56.17*
54.63
46.27*
33.52*
32.53

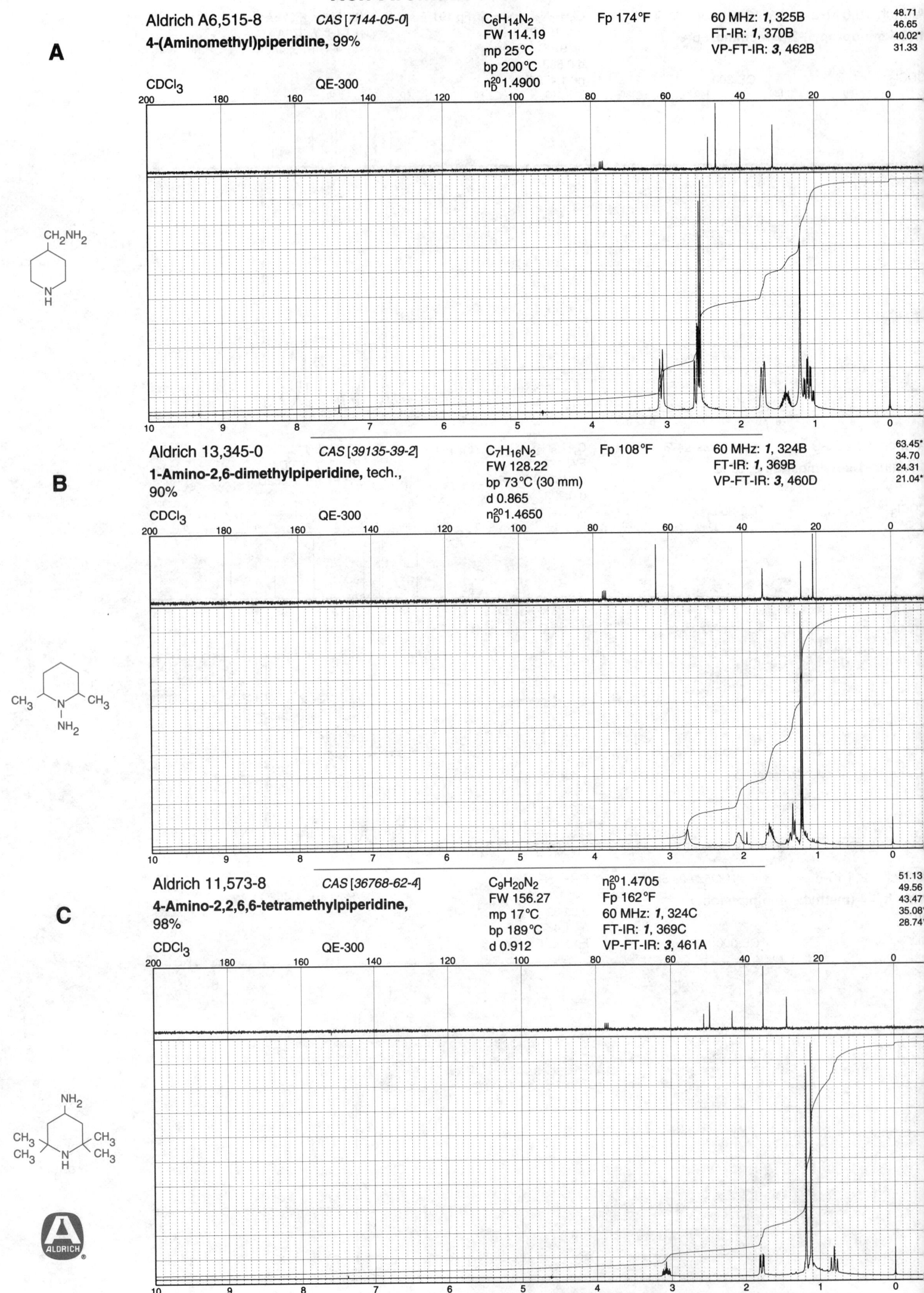

A

Aldrich A6,515-8   *CAS [7144-05-0]* $C_6H_{14}N_2$ Fp 174°F   60 MHz: *1*, 325B

4-(Aminomethyl)piperidine, 99%   FW 114.19   FT-IR: *1*, 370B

CDCl₃   QE-300   mp 25°C   VP-FT-IR: *3*, 462B
bp 200°C
$n_D^{20}1.4900$

48.71
46.65
40.02*
31.33

B

Aldrich 13,345-0   *CAS [39135-39-2]* $C_7H_{16}N_2$ Fp 108°F   60 MHz: *1*, 324B

1-Amino-2,6-dimethylpiperidine, tech.,   FW 128.22   FT-IR: *1*, 369B
90%   bp 73°C (30 mm)   VP-FT-IR: *3*, 460D
d 0.865

CDCl₃   QE-300 $n_D^{20}1.4650$

63.45*
34.70
24.31
21.04*

C

Aldrich 11,573-8   *CAS [36768-62-4]* $C_9H_{20}N_2$ $n_D^{20}1.4705$

4-Amino-2,2,6,6-tetramethylpiperidine,   FW 156.27   Fp 162°F
98%   mp 17°C   60 MHz: *1*, 324C
bp 189°C   FT-IR: *1*, 369C

CDCl₃   QE-300   d 0.912   VP-FT-IR: *3*, 461A

51.13
49.56
43.47
35.08*
28.74*

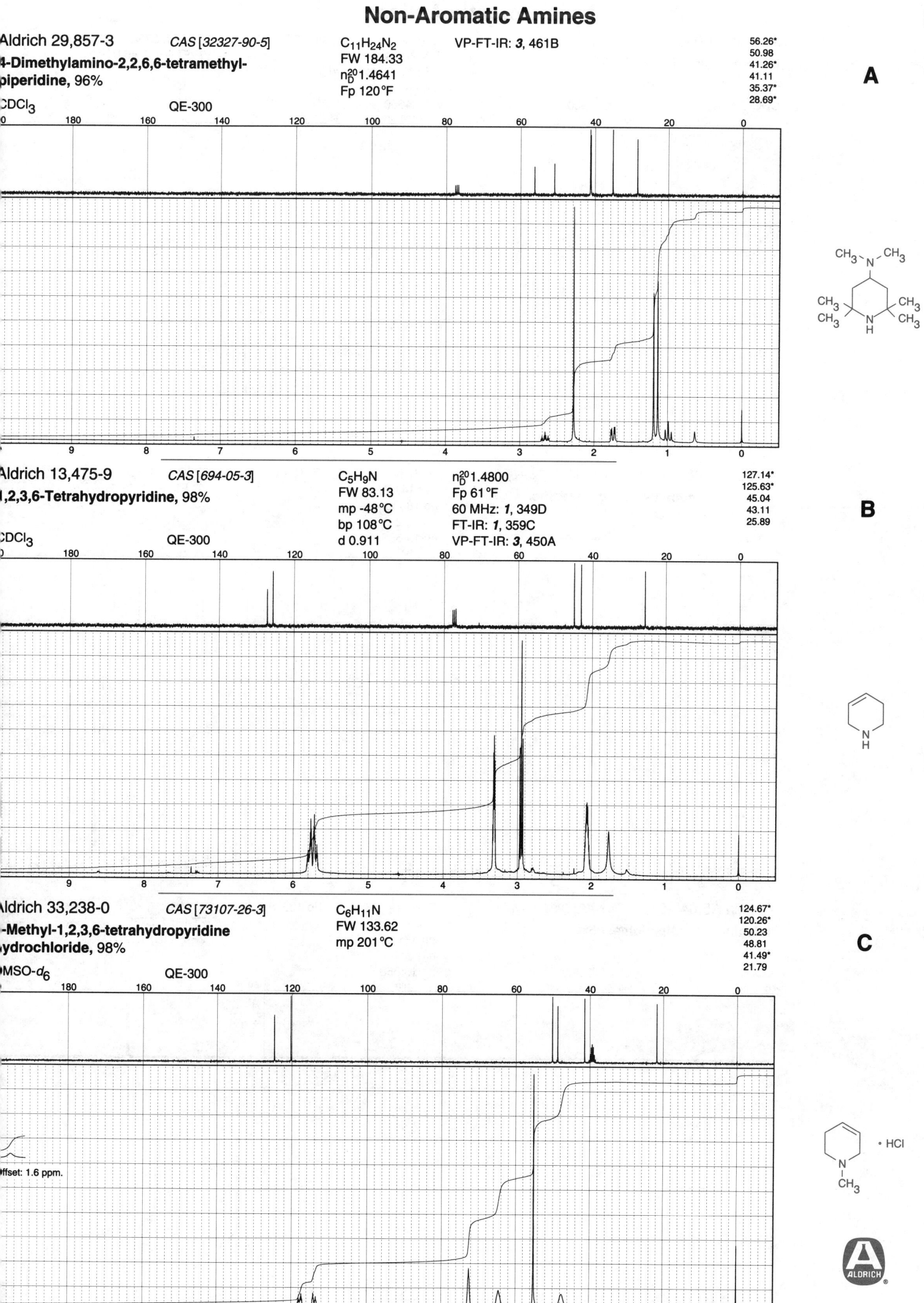

Aldrich 29,857-3   CAS [32327-90-5]   C$_{11}$H$_{24}$N$_2$ VP-FT-IR: **3**, 461B

4-Dimethylamino-2,2,6,6-tetramethyl-piperidine, 96%

FW 184.33
n$_D^{20}$1.4641
Fp 120°F

56.26*
50.98
41.26*
41.11
35.37*
28.69*

A

CDCl$_3$ QE-300

Aldrich 13,475-9   CAS [694-05-3]   C$_5$H$_9$N   n$_D^{20}$1.4800

1,2,3,6-Tetrahydropyridine, 98%

FW 83.13
mp -48°C
bp 108°C
d 0.911

Fp 61°F
60 MHz: **1**, 349D
FT-IR: **1**, 359C
VP-FT-IR: **3**, 450A

127.14*
125.63*
45.04
43.11
25.89

B

CDCl$_3$ QE-300

Aldrich 33,238-0   CAS [73107-26-3]   C$_6$H$_{11}$N

1-Methyl-1,2,3,6-tetrahydropyridine hydrochloride, 98%

FW 133.62
mp 201°C

124.67*
120.26*
50.23
48.81
41.49*
21.79

C

DMSO-d$_6$ QE-300

Offset: 1.6 ppm.

· HCl

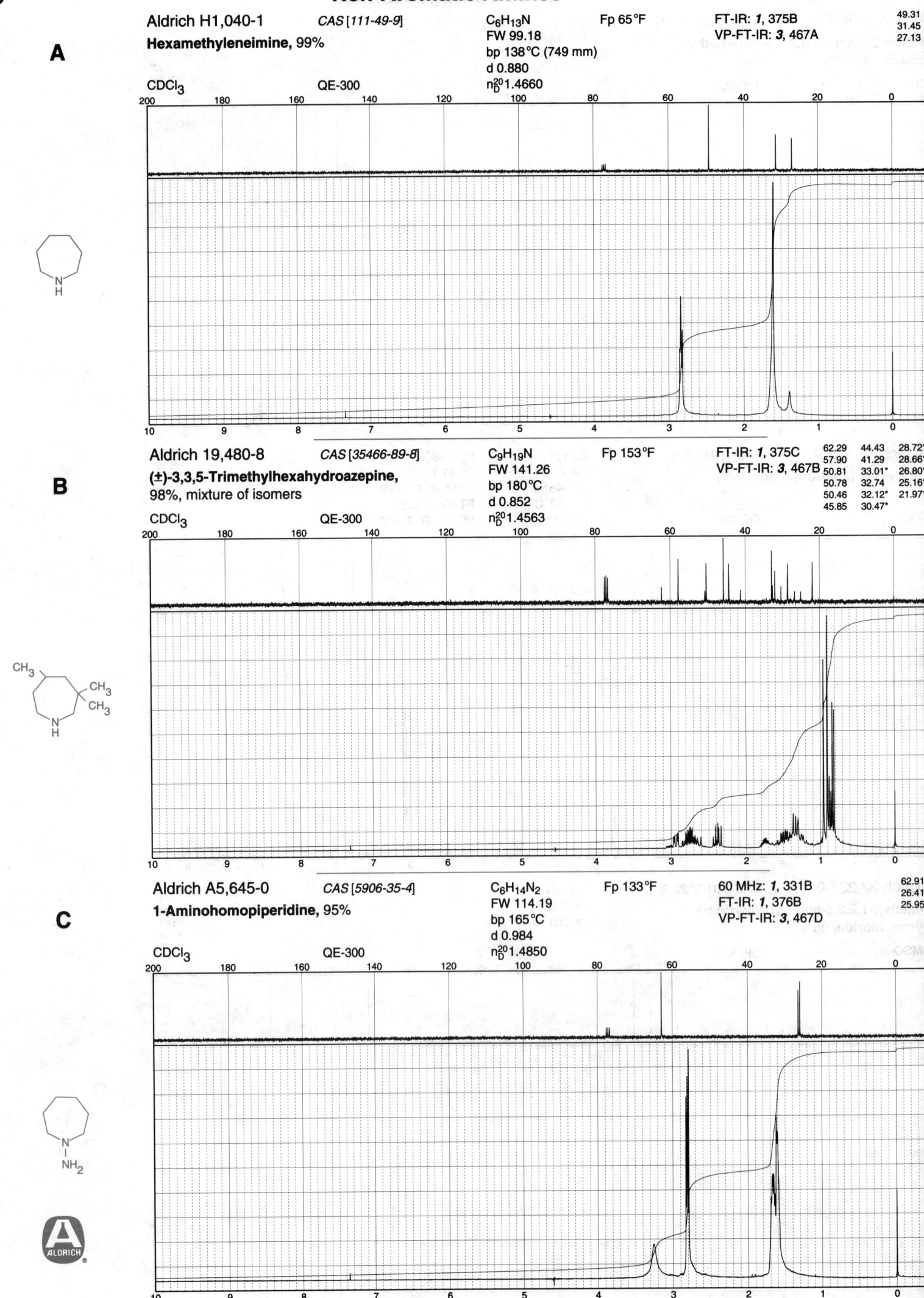

A

Aldrich H1,040-1   CAS [111-49-9]   C₆H₁₃N   Fp 65°F   FT-IR: *1*, 375B   49.31
Hexamethyleneimine, 99%   FW 99.18   VP-FT-IR: *3*, 467A   31.45
                                                                    27.13
bp 138°C (749 mm)
d 0.880
CDCl₃   QE-300   n₂₀D 1.4660

B

Aldrich 19,480-8   CAS [35466-89-8]   C₉H₁₉N   Fp 153°F   FT-IR: *1*, 375C   62.29 44.43 28.72*
(±)-3,3,5-Trimethylhexahydroazepine,   FW 141.26   VP-FT-IR: *3*, 467B   57.90 41.29 28.66*
98%, mixture of isomers   bp 180°C   50.81 33.01* 26.80*
                                                                    50.78 32.74   25.16*
d 0.852   50.46 32.12* 21.97*
CDCl₃   QE-300   n₂₀D 1.4563   45.85 30.47*

C

Aldrich A5,645-0   CAS [5906-35-4]   C₆H₁₄N₂   Fp 133°F   60 MHz: *1*, 331B   62.91
1-Aminohomopiperidine, 95%   FW 114.19   FT-IR: *1*, 376B   26.41
                                                                    25.95
bp 165°C   VP-FT-IR: *3*, 467D
d 0.984
CDCl₃   QE-300   n₂₀D 1.4850

Non-Aromatic Amines

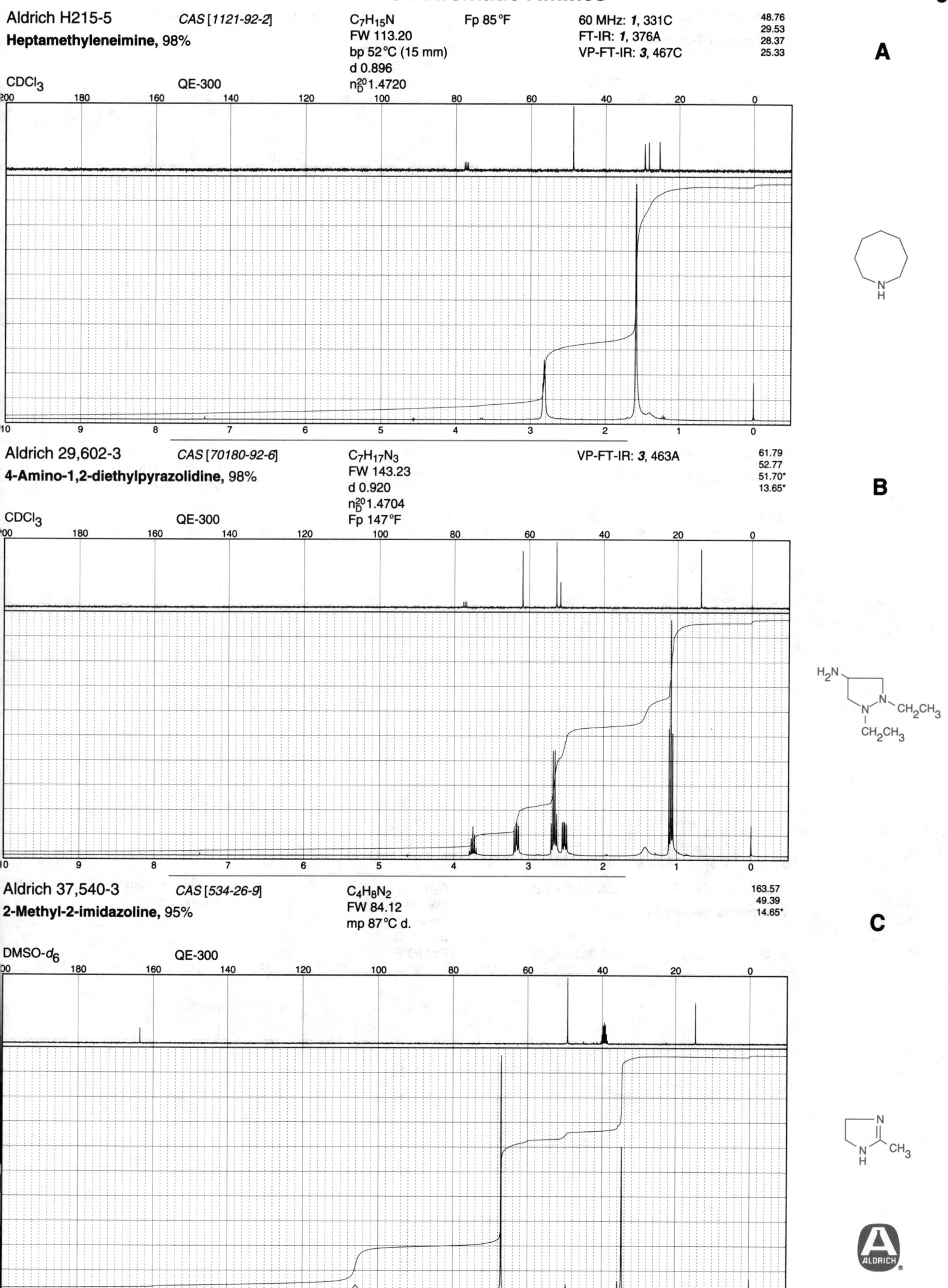

Aldrich H215-5 CAS [1121-92-2] C₇H₁₅N Fp 85°F 60 MHz: **1**, 331C 48.76
Heptamethyleneimine, 98% FW 113.20 FT-IR: **1**, 376A 29.53
bp 52°C (15 mm) VP-FT-IR: **3**, 467C 28.37
d 0.896 25.33
n$_D^{20}$1.4720

CDCl₃ QE-300

A

Aldrich 29,602-3 CAS [70180-92-6] C₇H₁₇N₃ VP-FT-IR: **3**, 463A 61.79
4-Amino-1,2-diethylpyrazolidine, 98% FW 143.23 52.77
d 0.920 51.70*
n$_D^{20}$1.4704 13.65*
Fp 147°F

CDCl₃ QE-300

B

Aldrich 37,540-3 CAS [534-26-9] C₄H₈N₂ 163.57
2-Methyl-2-imidazoline, 95% FW 84.12 49.39
mp 87°C d. 14.65*

DMSO-d₆ QE-300

C

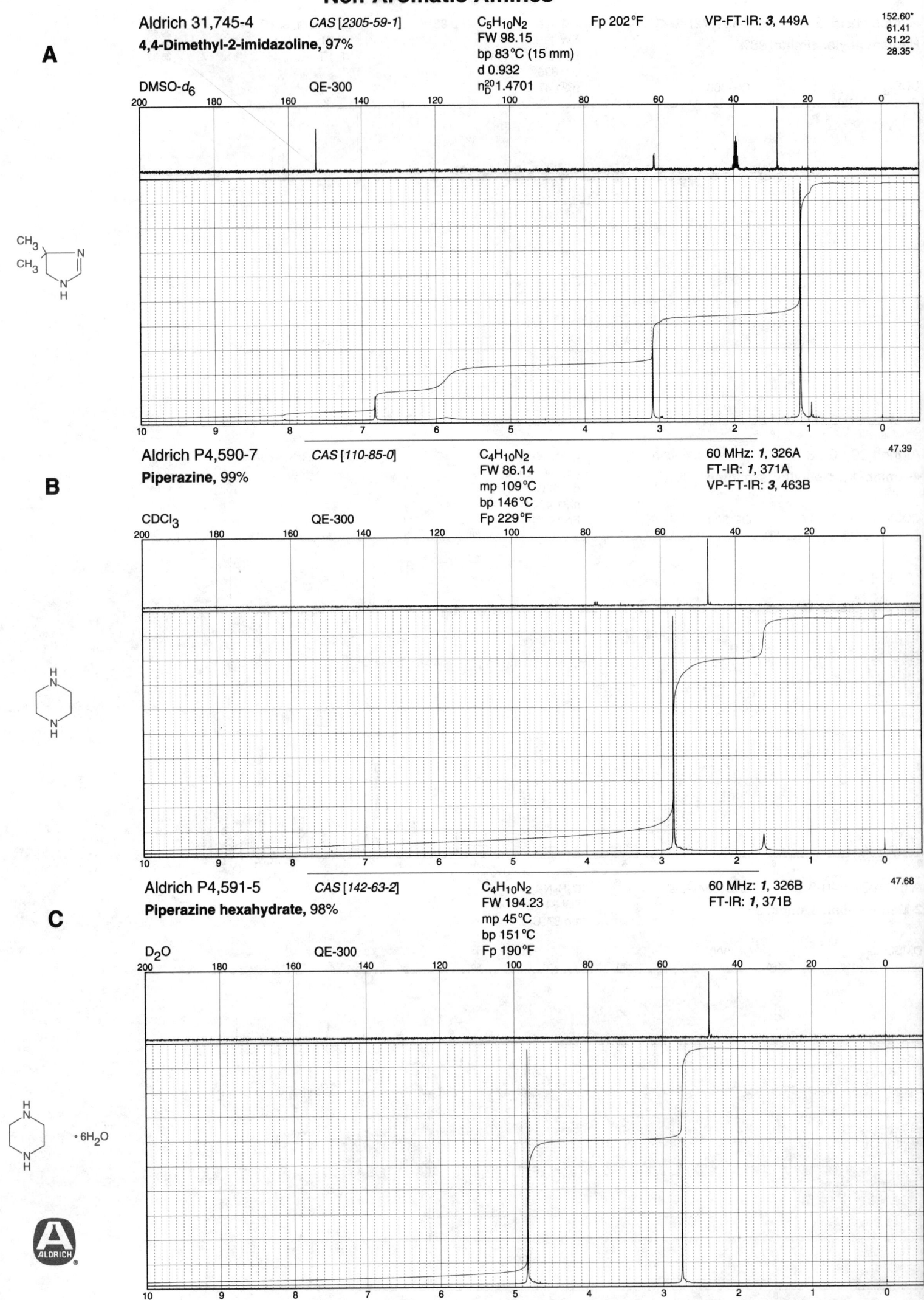

A

Aldrich 31,745-4   *CAS [2305-59-1]*   C$_5$H$_{10}$N$_2$ Fp 202°F   VP-FT-IR: *3*, 449A   152.60*
4,4-Dimethyl-2-imidazoline, 97%   FW 98.15   61.41
bp 83°C (15 mm)   61.22
d 0.932   28.35*
n$_D^{20}$1.4701

DMSO-*d*$_6$ QE-300

47.39

B

Aldrich P4,590-7   *CAS [110-85-0]*   C$_4$H$_{10}$N$_2$ 60 MHz: *1*, 326A
Piperazine, 99%   FW 86.14   FT-IR: *1*, 371A
mp 109°C   VP-FT-IR: *3*, 463B
bp 146°C
Fp 229°F

CDCl$_3$ QE-300

47.68

C

Aldrich P4,591-5   *CAS [142-63-2]*   C$_4$H$_{10}$N$_2$ 60 MHz: *1*, 326B
Piperazine hexahydrate, 98%   FW 194.23   FT-IR: *1*, 371B
mp 45°C
bp 151°C
Fp 190°F

D$_2$O   QE-300

• 6H$_2$O

ALDRICH

Aldrich 13,000-1   *CAS [109-01-3]* $C_5H_{12}N_2$ Fp 108°F   60 MHz: *1*, 326C   56.50
46.71*
46.13

1-Methylpiperazine, 99%   FW 100.17   FT-IR: *1*, 371C
bp 138°C   VP-FT-IR: *3*, 463C

CDCl$_3$ QE-300   bp 138°C
d 0.903
n_D^{20}1.4655

A

Aldrich D17,930-2   *CAS [106-58-1]* $C_6H_{14}N_2$ Fp 65°F   60 MHz: *1*, 327A   55.17
46.04*

1,4-Dimethylpiperazine, 98%   FW 114.19   FT-IR: *1*, 371D
bp 132°C (750 mm)   VP-FT-IR: *3*, 463D
d 0.844

CDCl$_3$ QE-300 n_D^{20}1.4463

B

Aldrich D17,980-9   *CAS [108-49-6]* $C_6H_{14}N_2$ 60 MHz: *1*, 327D   53.26
52.17*
19.91*

2,6-Dimethylpiperazine, 97%, predominantly   FW 114.19   FT-IR: *1*, 372B
cis   mp 110°C   VP-FT-IR: *3*, 464B
bp 162°C

CDCl$_3$ QE-300   Fp 113°F

C

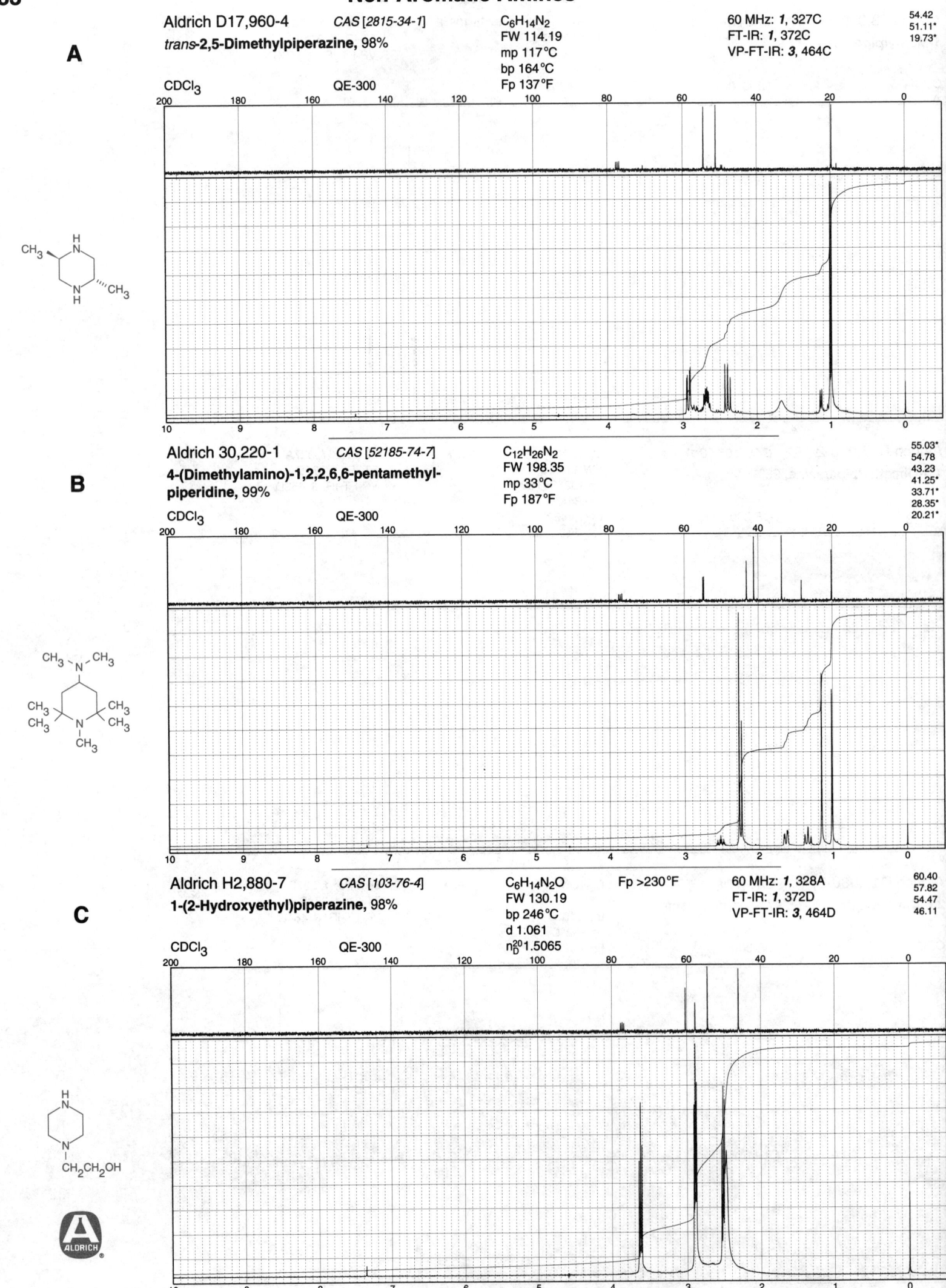

A

Aldrich D17,960-4   *CAS [2815-34-1]*

trans-2,5-Dimethylpiperazine, 98%

$C_6H_{14}N_2$
FW 114.19
mp 117°C
bp 164°C
Fp 137°F

60 MHz: *1*, 327C
FT-IR: *1*, 372C
VP-FT-IR: *3*, 464C

54.42
51.11*
19.73*

CDCl₃   QE-300

B

Aldrich 30,220-1   *CAS [52185-74-7]*

**4-(Dimethylamino)-1,2,2,6,6-pentamethyl-
piperidine, 99%**

$C_{12}H_{26}N_2$
FW 198.35
mp 33°C
Fp 187°F

55.03*
54.78
43.23
41.25*
33.71*
28.35*
20.21*

CDCl₃   QE-300

C

Aldrich H2,880-7   *CAS [103-76-4]*

1-(2-Hydroxyethyl)piperazine, 98%

$C_6H_{14}N_2O$
FW 130.19
bp 246°C
d 1.061
n_D^{20}1.5065

Fp >230°F

60 MHz: *1*, 328A
FT-IR: *1*, 372D
VP-FT-IR: *3*, 464D

60.40
57.82
54.47
46.11

CDCl₃   QE-300

A

Aldrich 19,359-3   CAS [5317-32-8]

1-Piperazinepropanol, 97%

CDCl₃   QE-300

$C_7H_{16}N_2O$
FW 144.22
mp 51°C
bp 130°C (5 mm)
Fp >230°F

60 MHz: *1*, 328C
FT-IR: *1*, 373A
VP-FT-IR: *3*, 465A

63.81
58.87
54.60
46.09
27.39

B

Aldrich 23,871-6   CAS [5317-33-9]

4-Methyl-1-piperazinepropanol, 97%

CDCl₃   QE-300

$C_8H_{18}N_2O$
FW 158.25
mp 29°C
bp 121°C (9 mm)
Fp >230°F

60 MHz: *1*, 328D
FT-IR: *1*, 373B
VP-FT-IR: *3*, 465C

64.00
58.31
55.10
53.23
45.89*
27.52

C

Aldrich B4,540-2   CAS [122-96-3]

1,4-Bis(2-hydroxyethyl)piperazine, 99%

DMSO-d₆   QE-300

$C_8H_{18}N_2O_2$
FW 174.24
mp 135°C
bp 218°C (50 mm)

60 MHz: *1*, 328B
FT-IR: *1*, 373C
VP-FT-IR: *3*, 466A

60.22
58.37
53.13

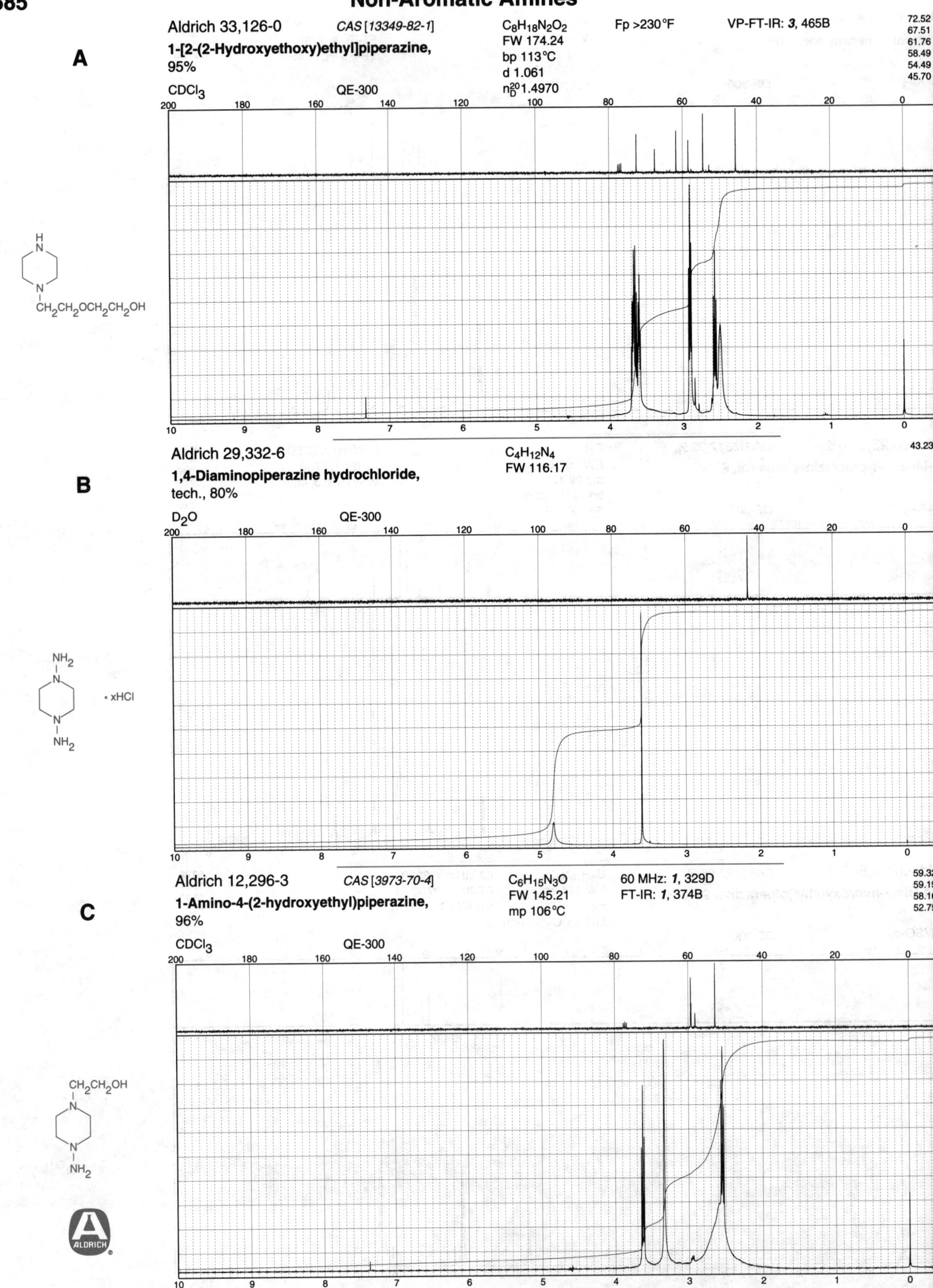

A

Aldrich 33,126-0   *CAS [13349-82-1]*   C$_8$H$_{18}$N$_2$O$_2$ Fp >230°F   VP-FT-IR: *3*, 465B

1-[2-(2-Hydroxyethoxy)ethyl]piperazine,
95%

FW 174.24
bp 113°C
d 1.061
n$_D^{20}$1.4970

72.52
67.51
61.76
58.49
54.49
45.70

CDCl$_3$ QE-300

CH$_2$CH$_2$OCH$_2$CH$_2$OH

43.23

Aldrich 29,332-6   C$_4$H$_{12}$N$_4$ FW 116.17

B

1,4-Diaminopiperazine hydrochloride,
tech., 80%

D$_2$O   QE-300

• xHCl

Aldrich 12,296-3   *CAS [3973-70-4]*   C$_6$H$_{15}$N$_3$O   60 MHz: *1*, 329D

C

1-Amino-4-(2-hydroxyethyl)piperazine,
96%

FW 145.21
mp 106°C
FT-IR: *1*, 374B

59.32
59.19
58.16
52.75

CDCl$_3$ QE-300

CH$_2$CH$_2$OH

NH$_2$

Non-Aromatic Amines

586

Aldrich A5,520-9   CAS [140-31-8]   C₆H₁₅N₃   Fp 200°F   60 MHz: *1*, 330C   61.95
-(2-Aminoethyl)piperazine, 99%   FW 129.21   FT-IR: *1*, 374D   54.80
bp 220°C   VP-FT-IR: *3*, 466B   46.23
CDCl₃   QE-300   d 0.985   38.75
n²⁰_D 1.5000

A

Aldrich 23,948-8   CAS [7209-38-3]   C₁₀H₂₄N₄   n²⁰_D 1.5020   55.78
,4-Bis(3-aminopropyl)piperazine, 99+%   FW 200.33   Fp 325°F   52.89
mp 15°C   60 MHz: *1*, 330D   40.02
bp 151°C (2 mm)   FT-IR: *1*, 375A   30.42
MSO-d₆   QE-300   d 0.973   VP-FT-IR: *3*, 466D

B

Aldrich 29,333-4   CAS [1606-49-1]   C₄H₈N₂   Fp >230°F   VP-FT-IR: *3*, 477A   147.52*
,4,5,6-Tetrahydropyrimidine, 97%   FW 84.12   40.54
bp 89°C (1 mm)   21.13
MSO-d₆   QE-300   d 1.024
n²⁰_D 1.5194

C

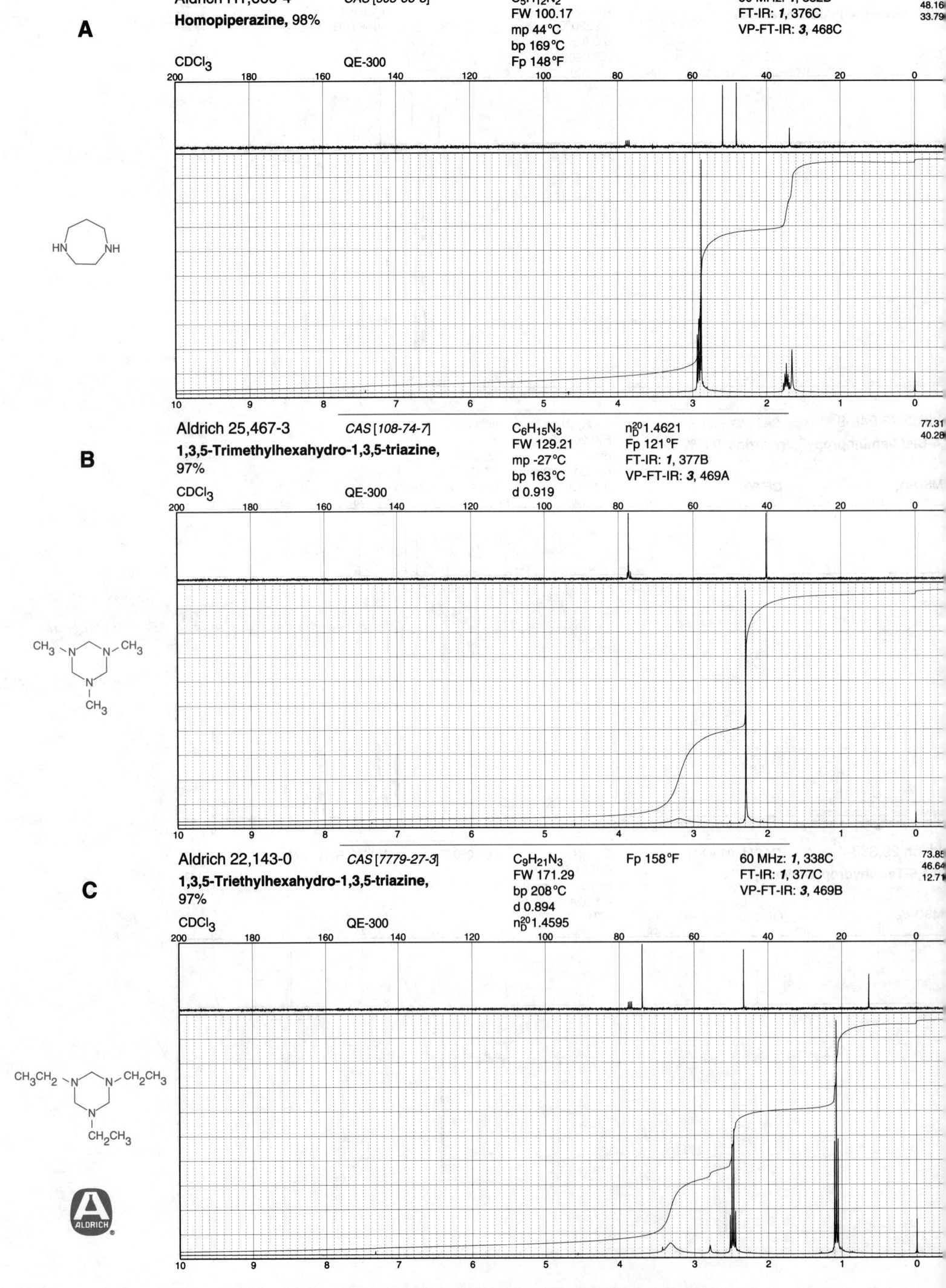

Aldrich H1,660-4   CAS [505-66-8] $C_5H_{12}N_2$ 60 MHz: *1*, 332B   51.91 48.16 33.79

Homopiperazine, 98%   FW 100.17   FT-IR: *1*, 376C

mp 44°C   VP-FT-IR: *3*, 468C

bp 169°C

CDCl$_3$ QE-300   Fp 148°F

Aldrich 25,467-3   CAS [108-74-7] $C_6H_{15}N_3$ n_D^{20}1.4621   77.31 40.28

1,3,5-Trimethylhexahydro-1,3,5-triazine, 97%   FW 129.21   Fp 121°F

mp -27°C   FT-IR: *1*, 377B

bp 163°C   VP-FT-IR: *3*, 469A

CDCl$_3$ QE-300   d 0.919

Aldrich 22,143-0   CAS [7779-27-3] $C_9H_{21}N_3$ Fp 158°F   60 MHz: *1*, 338C   73.85 46.64 12.71

1,3,5-Triethylhexahydro-1,3,5-triazine, 97%   FW 171.29   FT-IR: *1*, 377C

bp 208°C   VP-FT-IR: *3*, 469B

d 0.894

CDCl$_3$ QE-300 n_D^{20}1.4595

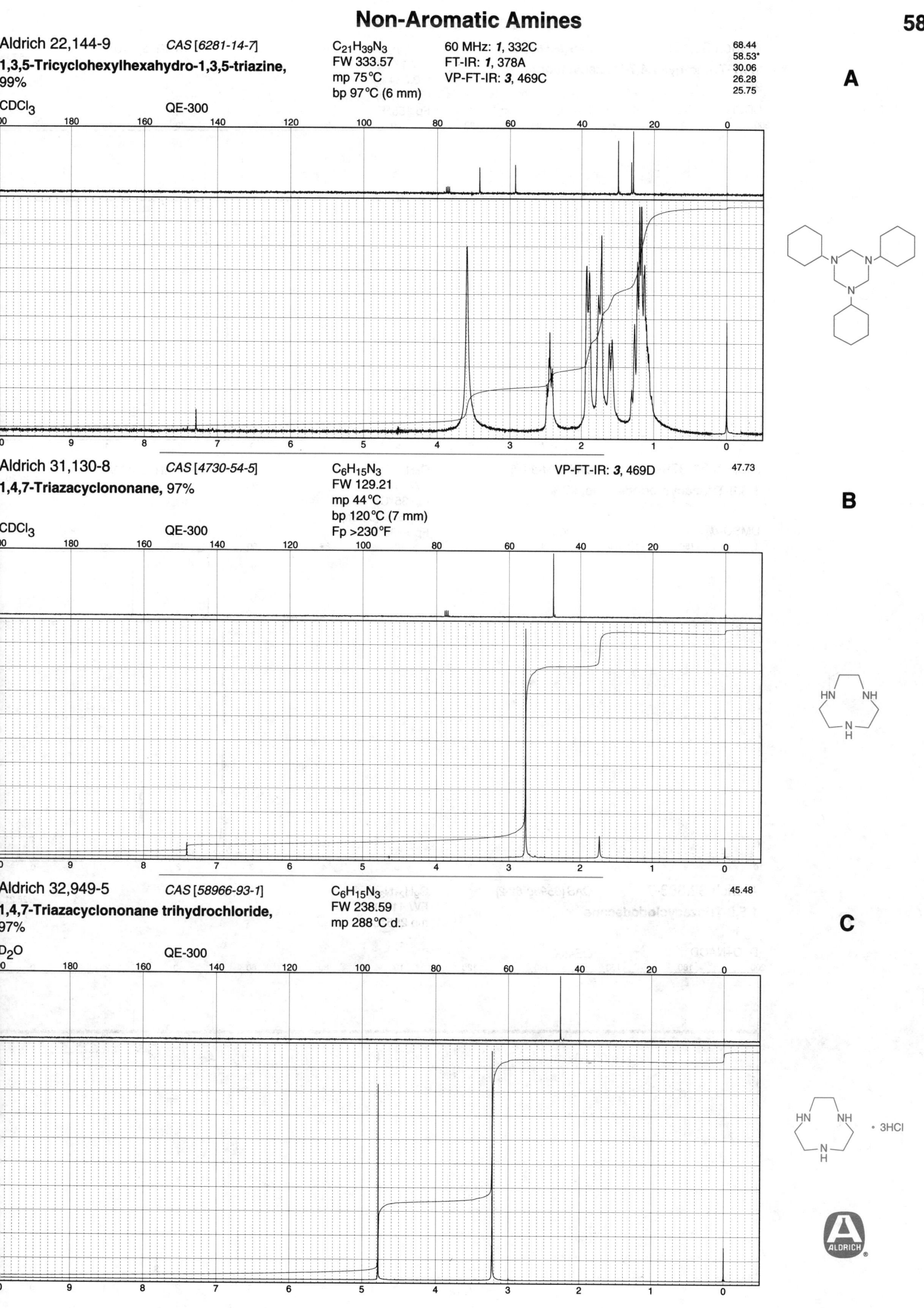

Aldrich 22,144-9   CAS [6281-14-7] $C_{21}H_{39}N_3$ 60 MHz: **1**, 332C   68.44
1,3,5-Tricyclohexylhexahydro-1,3,5-triazine,   FW 333.57   FT-IR: **1**, 378A   58.53*
99%   mp 75°C   VP-FT-IR: **3**, 469C   30.06
    bp 97°C (6 mm)   26.28
    25.75

A

CDCl₃   QE-300

Aldrich 31,130-8   CAS [4730-54-5] $C_6H_{15}N_3$ VP-FT-IR: **3**, 469D   47.73
1,4,7-Triazacyclononane, 97%   FW 129.21
    mp 44°C
    bp 120°C (7 mm)
    Fp >230°F

B

CDCl₃   QE-300

Aldrich 32,949-5   CAS [58966-93-1] $C_6H_{15}N_3$ 45.48
1,4,7-Triazacyclononane trihydrochloride,   FW 238.59
97%   mp 288°C d.

C

D₂O   QE-300

· 3HCl

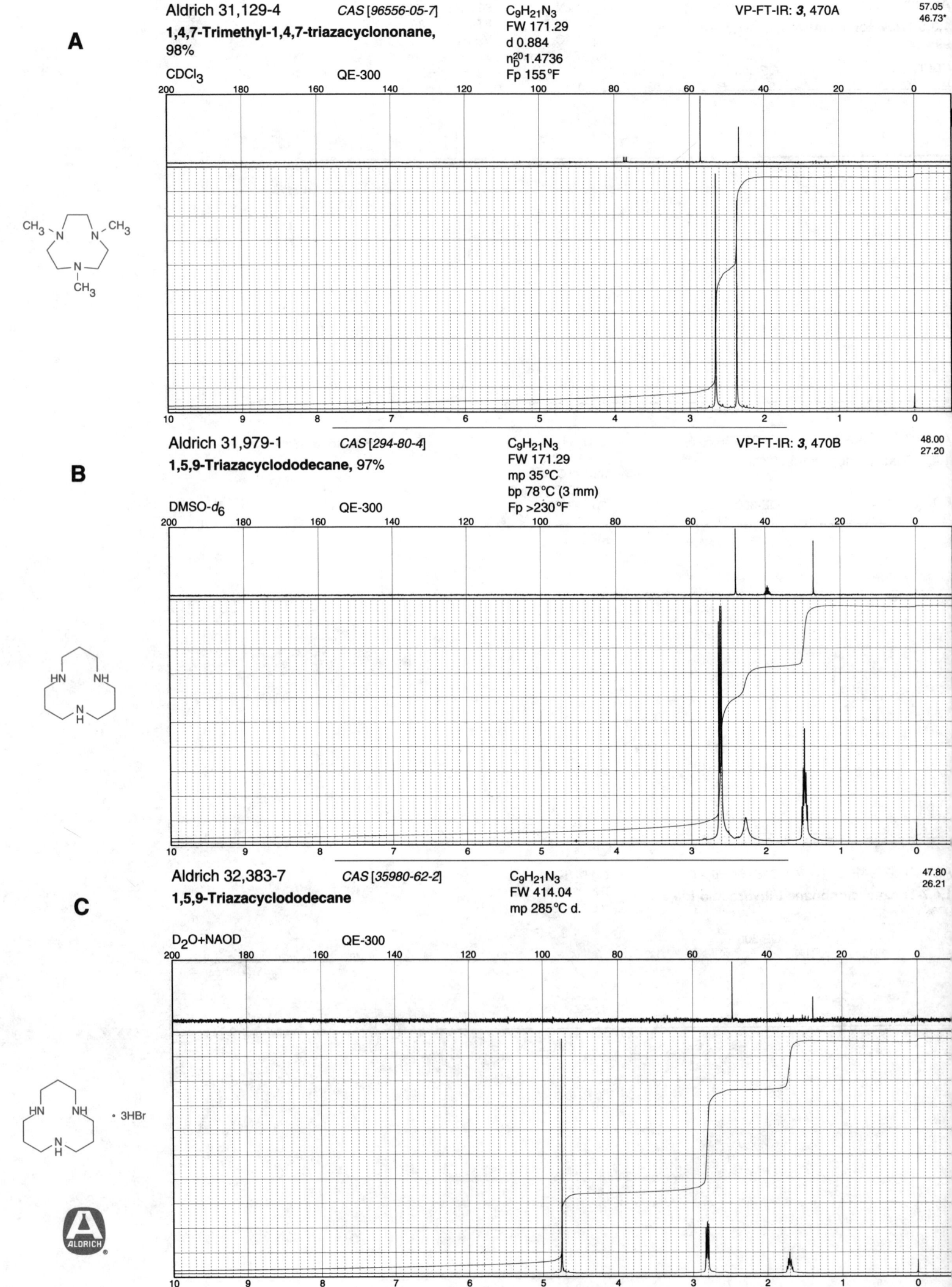

Aldrich 31,129-4   CAS [96556-05-7] $C_9H_{21}N_3$ VP-FT-IR: **3**, 470A   57.05 / 46.73*

A

1,4,7-Trimethyl-1,4,7-triazacyclononane, 98%

FW 171.29
d 0.884
n_D^{20} 1.4736
Fp 155°F

CDCl$_3$ QE-300

Aldrich 31,979-1   CAS [294-80-4] $C_9H_{21}N_3$ VP-FT-IR: **3**, 470B   48.00 / 27.20

B

1,5,9-Triazacyclododecane, 97%

FW 171.29
mp 35°C
bp 78°C (3 mm)
Fp >230°F

DMSO-d_6 QE-300

Aldrich 32,383-7   CAS [35980-62-2] $C_9H_{21}N_3$ 47.80 / 26.21

C

1,5,9-Triazacyclododecane

FW 414.04
mp 285°C d.

· 3HBr

D$_2$O+NAOD   QE-300

Aldrich 37,781-3

1,5,9-Trimethyl-1,5,9-triazacyclododecane, 95%

CDCl₃   QE-300

$C_{12}H_{27}N_3$
FW 213.37
bp 59 °C
d 0.905
n_D^{20} 1.4830

Fp 227 °F

52.08
43.18*
22.19

A

Aldrich 33,965-2   *CAS [294-90-6]*

Cyclen, 97%

CDCl₃   QE-300

$C_8H_{20}N_4$
FW 172.28
mp 112 °C

46.27

B

Aldrich 25,916-0   *CAS [295-37-4]*

1,4,8,11-Tetraazacyclotetradecane, 98%

CDCl₃   QE-300

$C_{10}H_{24}N_4$
FW 200.33
mp 185 °C

FT-IR: *1*, 378C
VP-FT-IR: *3*, 470C

50.57
49.37
29.47

C

A

Aldrich 25,915-2   *CAS [15439-16-4]* $C_{11}H_{26}N_4$ FT-IR: *1*, 378D   50.94
50.23
1,4,8,12-Tetraazacyclopentadecane, 98%   FW 214.36   VP-FT-IR: *3*, 470D   49.82
49.44
mp 101 °C   29.79
29.34

CDCl$_3$ QE-300

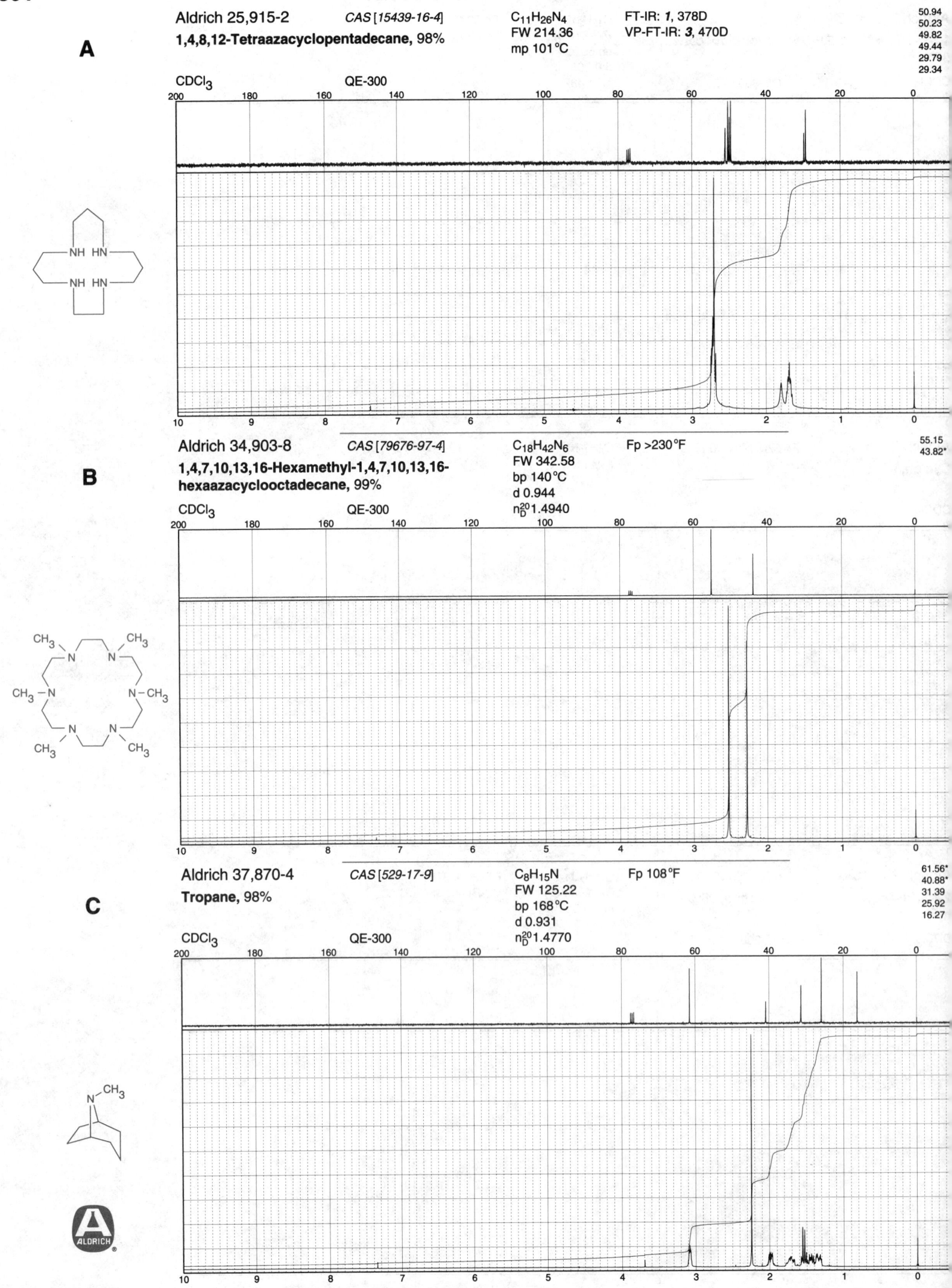

B

Aldrich 34,903-8   *CAS [79676-97-4]* $C_{18}H_{42}N_6$ Fp >230°F   55.15
43.82*
1,4,7,10,13,16-Hexamethyl-1,4,7,10,13,16-   FW 342.58
hexaazacyclooctadecane, 99%   bp 140 °C
d 0.944
CDCl$_3$ QE-300   n$_D^{20}$1.4940

C

Aldrich 37,870-4   *CAS [529-17-9]* $C_8H_{15}N$ Fp 108°F   61.56*
40.88*
Tropane, 98%   FW 125.22   31.39
bp 168 °C   25.92
d 0.931   16.27
CDCl$_3$ QE-300   n$_D^{20}$1.4770

Aldrich 19,760-2   *CAS [100-76-5]* $C_7H_{13}N$ FT-IR: *1*, 379A

Quinuclidine, 97%

FW 111.19
mp 159°C s.

47.89
26.86
20.86*

A

$CDCl_3$ QE-300

Aldrich Q187-5   *CAS [1619-34-7]* $C_7H_{13}NO$ 60 MHz: *1*, 333D

3-Quinuclidinol, 97%

FW 127.19
mp 222°C

FT-IR: *1*, 379D

67.19*
57.99
47.36
46.31
28.39*
24.82
18.92

B

$CDCl_3$ QE-300

Aldrich M4,612-8 $C_8H_{11}NO$ 60 MHz: *1*, 410B

2-Methylene-3-quinuclidinone hydrate hydrochloride, 97%

FW 173.66
mp 263°C d.

FT-IR: *1*, 462C

95.47   35.26
63.12   35.18*
57.54   22.32*
51.87   22.22

C

D_2O QE-300

· $2H_2O$
· HCl

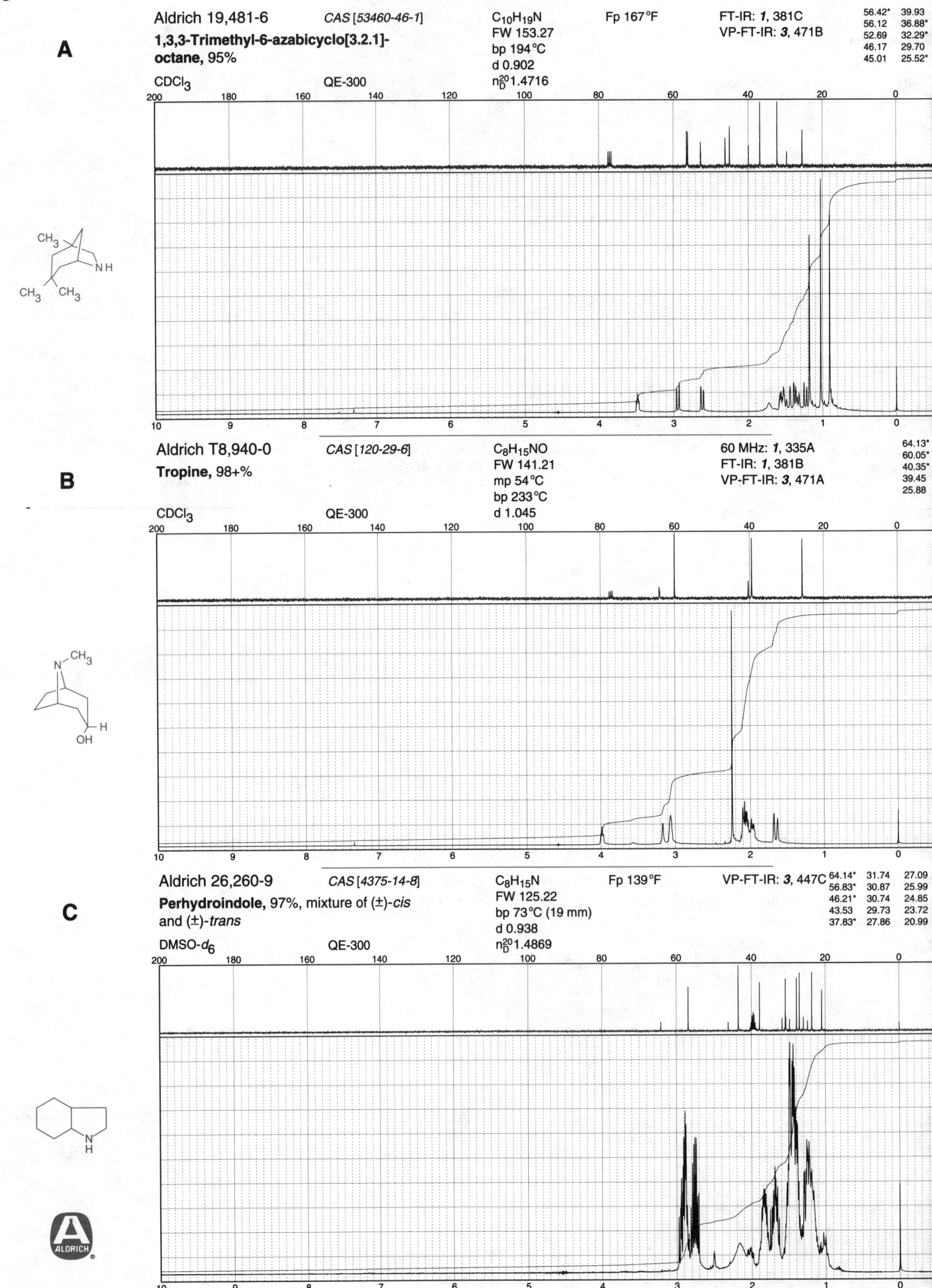

A

Aldrich 19,481-6   *CAS [53460-46-1]* $C_{10}H_{19}N$ Fp 167°F   FT-IR: *1*, 381C   56.42* 39.93
1,3,3-Trimethyl-6-azabicyclo[3.2.1]-   FW 153.27   VP-FT-IR: *3*, 471B   56.12 36.88*
octane, 95%   bp 194°C   52.69 32.29*
d 0.902   46.17 29.70
$CDCl_3$ QE-300 $n_D^{20}1.4716$ 45.01 25.52*

B

Aldrich T8,940-0   *CAS [120-29-6]* $C_8H_{15}NO$ 60 MHz: *1*, 335A   64.13*
Tropine, 98+%   FW 141.21   FT-IR: *1*, 381B   60.05*
mp 54°C   VP-FT-IR: *3*, 471A   40.35*
bp 233°C   39.45
$CDCl_3$ QE-300   d 1.045   25.88

C

Aldrich 26,260-9   *CAS [4375-14-8]* $C_8H_{15}N$ Fp 139°F   VP-FT-IR: *3*, 447C 64.14* 31.74 27.09
Perhydroindole, 97%, mixture of (±)-*cis*   FW 125.22   56.83* 30.87 25.99
and (±)-*trans*   bp 73°C (19 mm)   46.21* 30.74 24.85
d 0.938   43.53 29.73 23.72
$DMSO$-d_6 QE-300 $n_D^{20}1.4869$ 37.83* 27.86 20.99

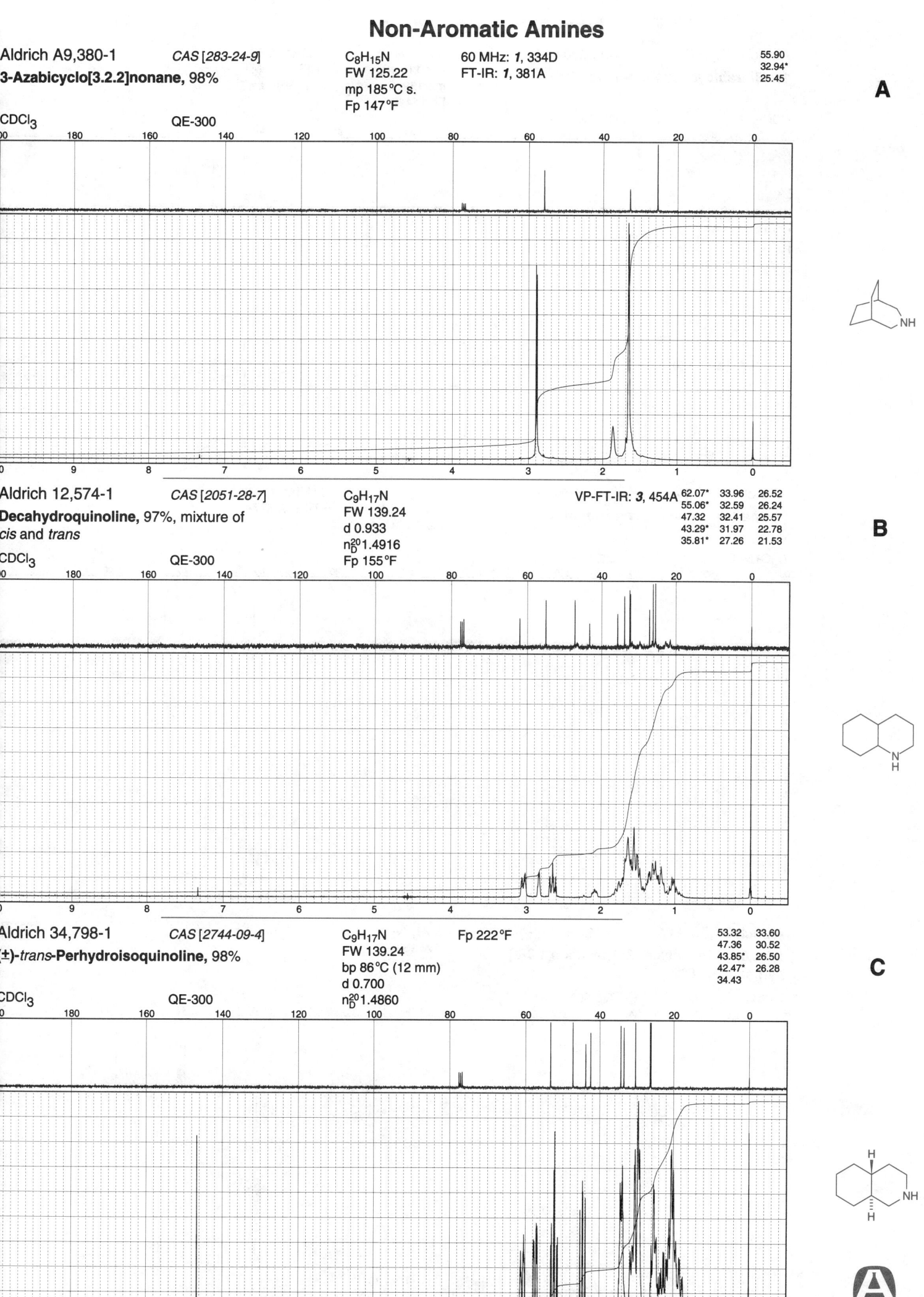

Aldrich A9,380-1   CAS [283-24-9]
3-Azabicyclo[3.2.2]nonane, 98%

$C_8H_{15}N$
FW 125.22
mp 185°C s.
Fp 147°F

60 MHz: *1*, 334D
FT-IR: *1*, 381A

55.90
32.94*
25.45

A

CDCl₃   QE-300

Aldrich 12,574-1   CAS [2051-28-7]
Decahydroquinoline, 97%, mixture of
cis and trans

$C_9H_{17}N$
FW 139.24
d 0.933
$n_D^{20}1.4916$
Fp 155°F

VP-FT-IR: *3*, 454A

62.07*	33.96	26.52
55.06*	32.59	26.24
47.32	32.41	25.57
43.29*	31.97	22.78
35.81*	27.26	21.53

B

CDCl₃   QE-300

Aldrich 34,798-1   CAS [2744-09-4]
(±)-trans-Perhydroisoquinoline, 98%

$C_9H_{17}N$
FW 139.24
bp 86°C (12 mm)
d 0.700
$n_D^{20}1.4860$

Fp 222°F

53.32   33.60
47.36   30.52
43.85*   26.50
42.47*   26.28
34.43

C

CDCl₃   QE-300

ALDRICH

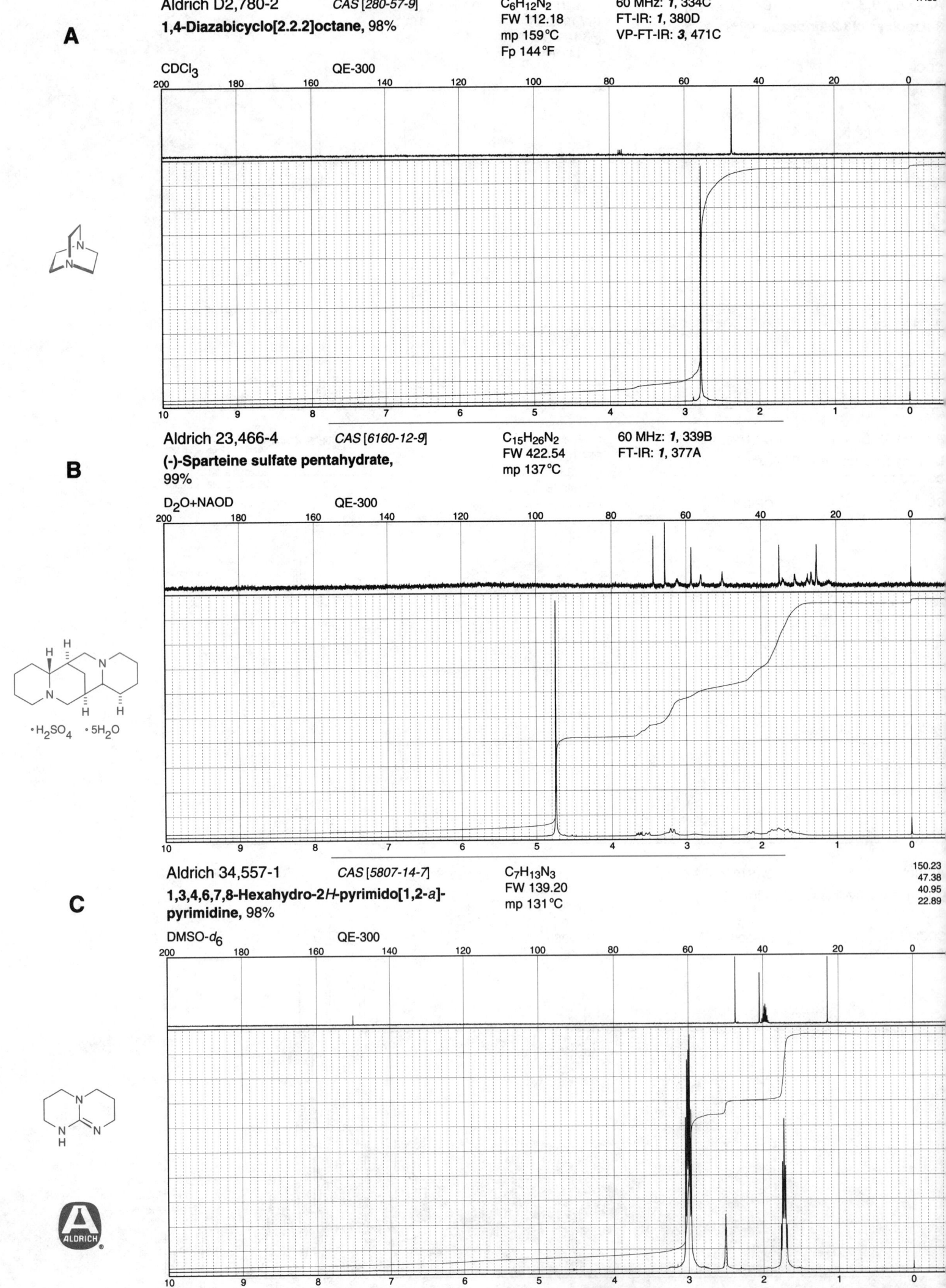

A

Aldrich D2,780-2   *CAS [280-57-9]*   C₆H₁₂N₂   60 MHz: *1*, 334C
1,4-Diazabicyclo[2.2.2]octane, 98%   FW 112.18   FT-IR: *1*, 380D
    mp 159°C   VP-FT-IR: *3*, 471C
    Fp 144°F

CDCl₃   QE-300

B

Aldrich 23,466-4   *CAS [6160-12-9]*   C₁₅H₂₆N₂   60 MHz: *1*, 339B
(-)-Sparteine sulfate pentahydrate,   FW 422.54   FT-IR: *1*, 377A
99%   mp 137°C

D₂O+NAOD   QE-300

•H₂SO₄   •5H₂O

C

Aldrich 34,557-1   *CAS [5807-14-7]*   C₇H₁₃N₃   150.23
    FW 139.20   47.38
1,3,4,6,7,8-Hexahydro-2*H*-pyrimido[1,2-*a*]-   mp 131°C   40.95
pyrimidine, 98%   22.89

DMSO-*d*₆   QE-300

Aldrich 35,950-5   *CAS [84030-20-6]*   C₈H₁₅N₃   Fp >230°F

A

1,3,4,6,7,8-Hexahydro-1-methyl-2*H*-pyrimido-
[1,2-a]pyrimidine, 98%

$C_8H_{15}N_3$
FW 153.23
bp 77°C
d 1.067
n_D^{20} 1.5380

151.99   43.82
48.47   37.17*
48.44   23.08
48.18   22.96

CDCl₃   QE-300

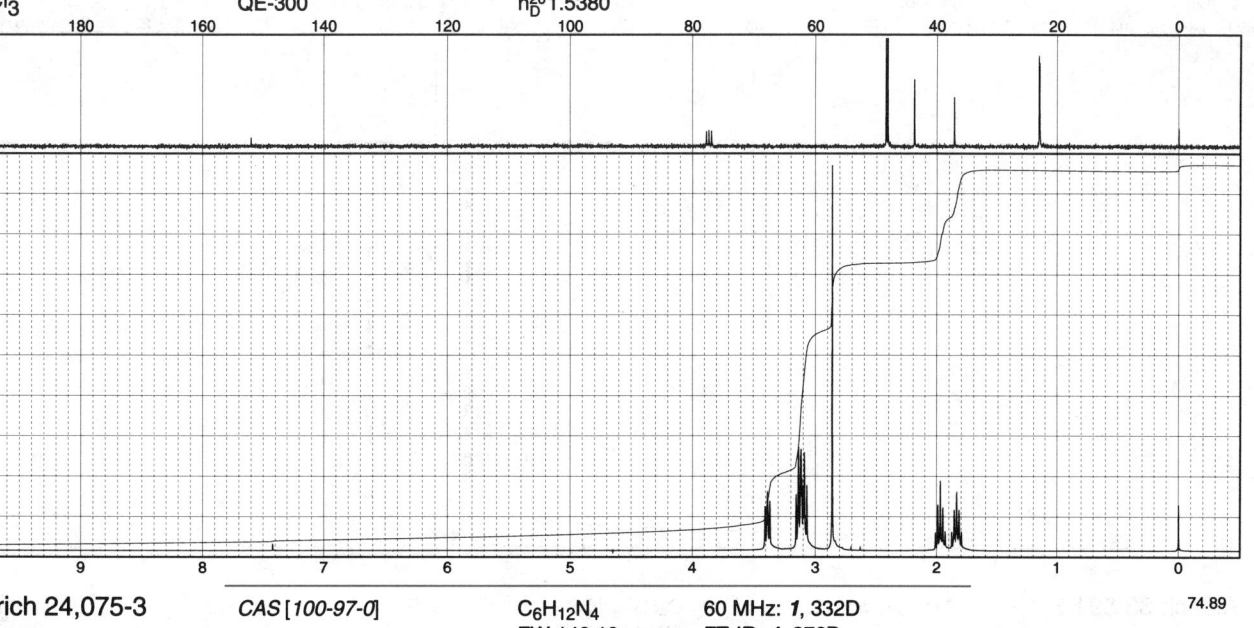
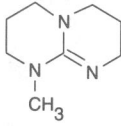

74.89

Aldrich 24,075-3   *CAS [100-97-0]*   C₆H₁₂N₄   60 MHz: *1*, 332D

B

Hexamethylenetetramine, 99+%

$C_6H_{12}N_4$
FW 140.19
mp 280°C s.
Fp 482°F

FT-IR: *1*, 378B
VP-FT-IR: *3*, 471D

CDCl₃   QE-300

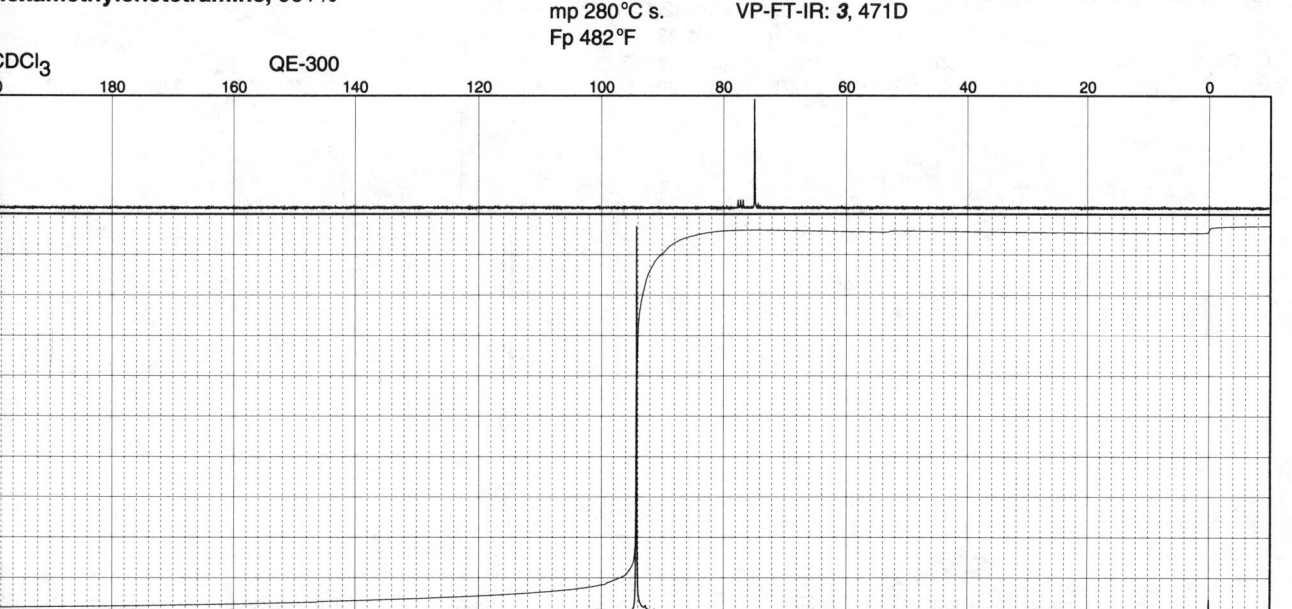
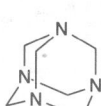

Aldrich 17,874-8   *CAS [1772-43-6]*   C₆H₁₁NO   Fp 55°F   FT-IR: *1*, 385B

C

2,4,4-Trimethyl-2-oxazoline, 99%

$C_6H_{11}NO$
FW 113.16
bp 113°C
d 0.887
n_D^{20} 1.4213

VP-FT-IR: *3*, 476C

162.77
79.18
67.06
28.41*
13.95*

CDCl₃   QE-300

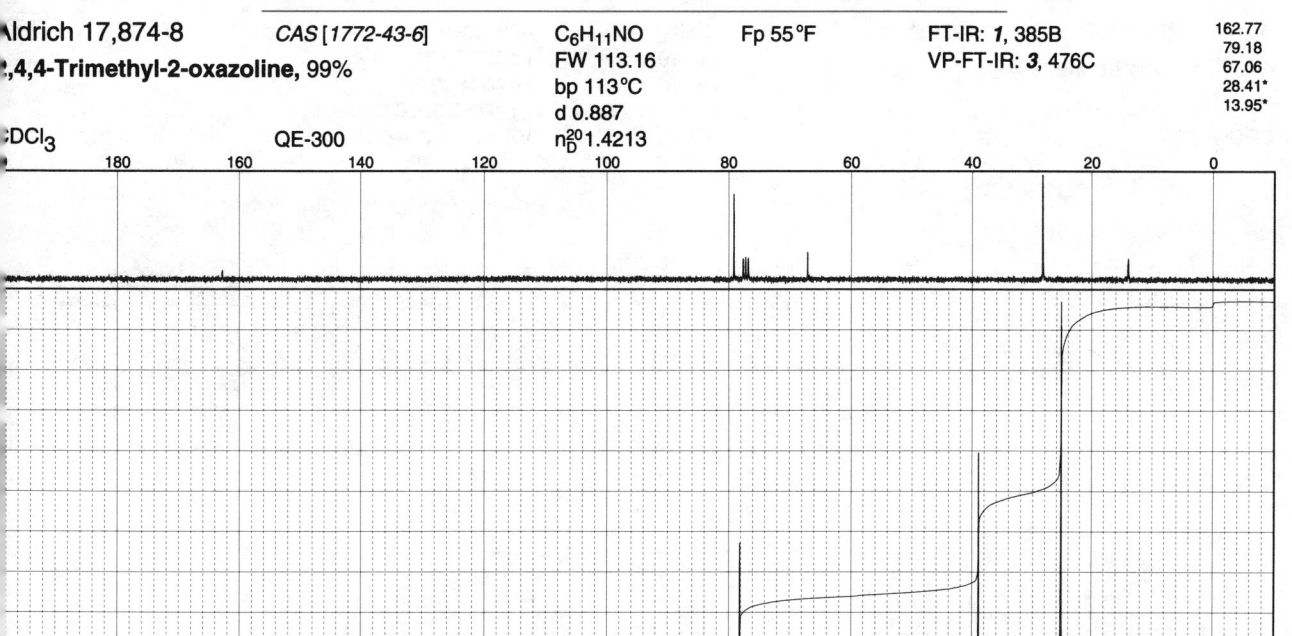
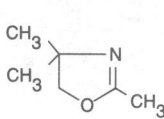

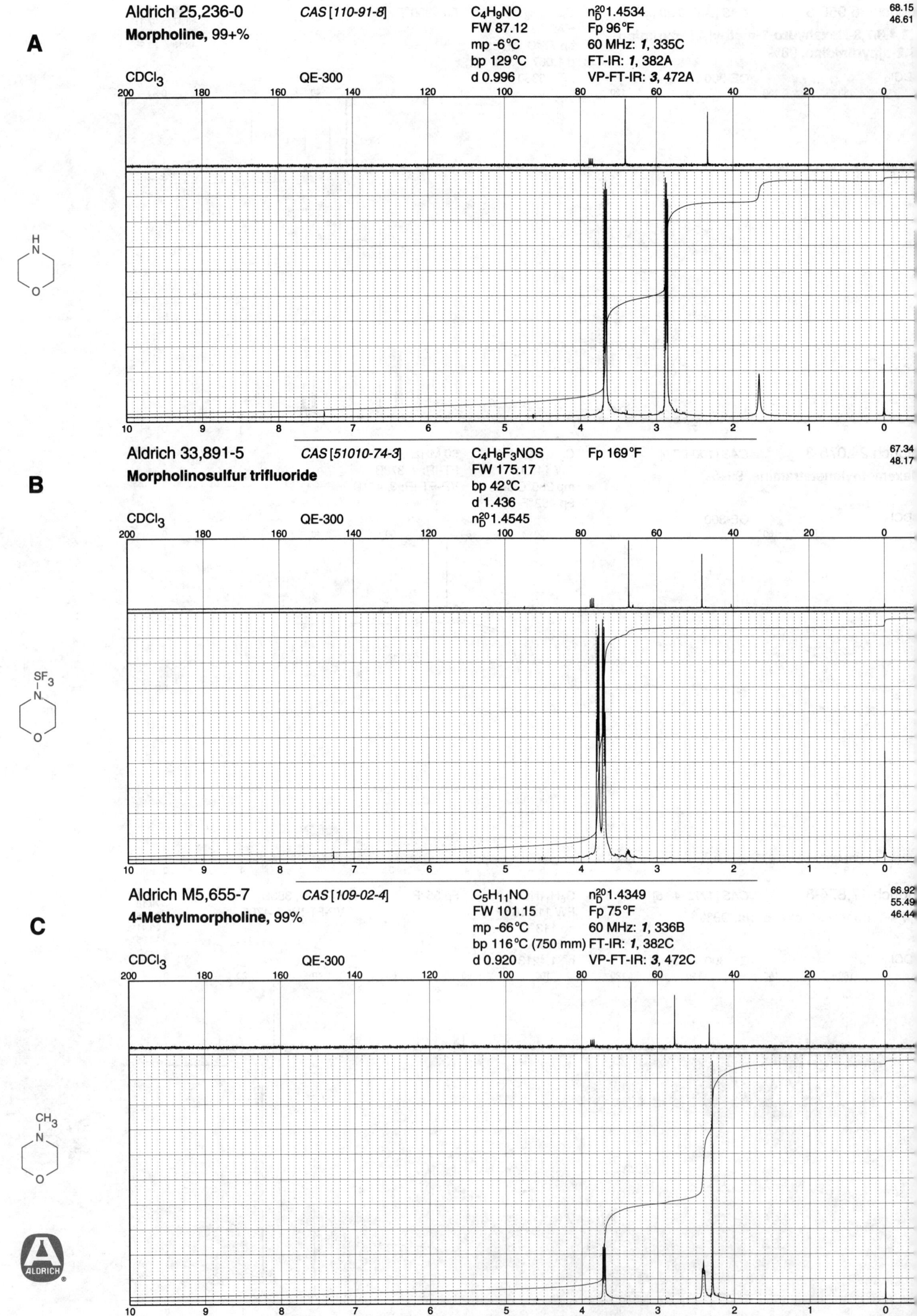

A

Aldrich 25,236-0
Morpholine, 99+%

CAS [110-91-8]

C_4H_9NO
FW 87.12
mp -6°C
bp 129°C
d 0.996

n_D^{20}1.4534
Fp 96°F
60 MHz: **1**, 335C
FT-IR: **1**, 382A
VP-FT-IR: **3**, 472A

68.15
46.61

CDCl₃   QE-300

B

Aldrich 33,891-5
Morpholinosulfur trifluoride

CAS [51010-74-3]

$C_4H_8F_3NOS$
FW 175.17
bp 42°C
d 1.436
n_D^{20}1.4545

Fp 169°F

67.34
48.17

CDCl₃   QE-300

C

Aldrich M5,655-7
4-Methylmorpholine, 99%

CAS [109-02-4]

$C_5H_{11}NO$
FW 101.15
mp -66°C
bp 116°C (750 mm)
d 0.920

n_D^{20}1.4349
Fp 75°F
60 MHz: **1**, 336B
FT-IR: **1**, 382C
VP-FT-IR: **3**, 472C

66.92
55.49
46.44

CDCl₃   QE-300

A

Aldrich 23,952-6   *CAS [100-74-3]*

-Ethylmorpholine, 99+%

CDCl₃   QE-300

C₆H₁₃NO	n²⁰_D 1.4415
FW 115.18	Fp 82°F
mp -63°C	60 MHz: *1*, 336C
bp 139°C	FT-IR: *1*, 382D
d 0.905	VP-FT-IR: *3*, 472D

67.03
53.42
52.76
11.70*

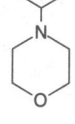

B

Aldrich 12,652-7   *CAS [141-91-3]*

**6-Dimethylmorpholine, 97%, mixture
f isomers**

DCl₃   QE-300

C₆H₁₃NO	n²⁰_D 1.4470
FW 115.18	Fp 120°F
mp -85°C	60 MHz: *1*, 336A
bp 147°C	FT-IR: *1*, 382B
d 0.935	VP-FT-IR: *3*, 472B

73.10*
66.50*
52.16
51.27
19.15*
17.61*

C

Aldrich C11,490-1   *CAS [936-52-7]*

-(1-Cyclopenten-1-yl)morpholine, 98%

DCl₃   QE-300

C₉H₁₅NO	Fp 140°F
FW 153.23	FT-IR: *1*, 383A
bp 106°C (12 mm)	VP-FT-IR: *3*, 473B
d 0.957	
n²⁰_D 1.5118	

151.72
98.25*
66.68
49.11
31.38
30.33
22.54

A

Aldrich M8,780-0   *CAS [670-80-4]*   C$_{10}$H$_{17}$NO   Fp 155°F   60 MHz: *1*, 349A   145.49   26.90

1-Morpholino-1-cyclohexene, 98%   FW 167.25   FT-IR: *1*, 383B   100.43*   24.48

bp 119°C (10 mm)   VP-FT-IR: *3*, 473C   67.03   23.30

d 0.995   48.56   22.84

n$_D^{20}$1.5146

CDCl$_3$ QE-300

B

Aldrich 38,557-3   *CAS [7182-08-3]*   C$_{11}$H$_{19}$NO   Fp 201°F   154.23   31.64

1-Morpholino-1-cycloheptene, 95%   FW 181.28   107.00*   28.03

bp 108°C (4 mm)   67.03   26.78

d 0.994   49.94   26.35

n$_D^{20}$1.5090   32.66

CDCl$_3$ QE-300

C

Aldrich H2,820-3   *CAS [622-40-2]*   C$_6$H$_{13}$NO$_2$ Fp 211°F   60 MHz: *1*, 337A   66.95

4-(2-Hydroxyethyl)morpholine, 99%   FW 131.18   FT-IR: *1*, 383D   60.14

bp 227°C (757 mm)   VP-FT-IR: *3*, 473D   57.79

d 1.083   53.53

n$_D^{20}$1.4760

CDCl$_3$ QE-300

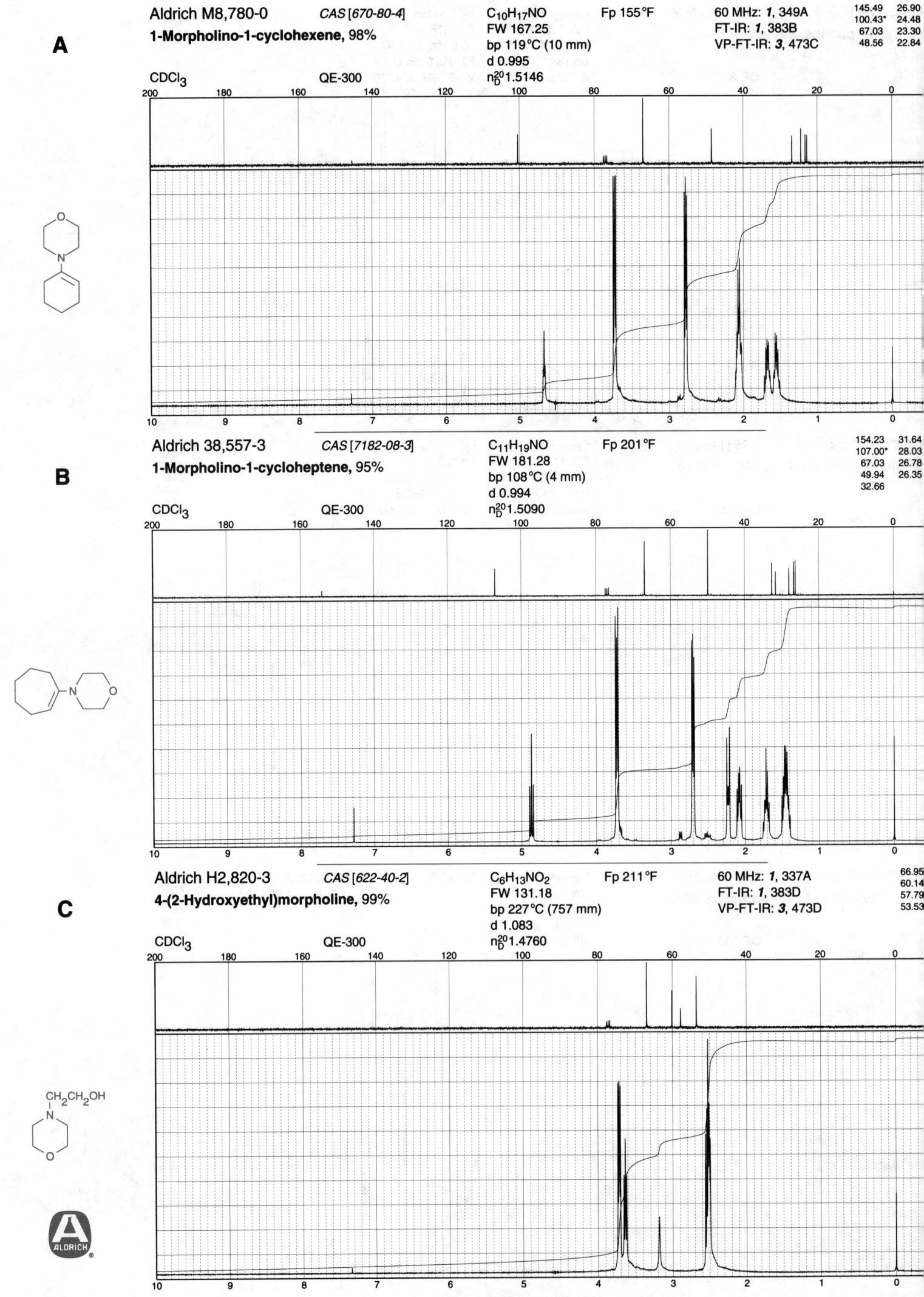

Aldrich 21,848-0   *CAS [6425-32-7]* $C_7H_{15}NO_3$ Fp >230°F   60 MHz: *1*, 337B   67.09*
3-Morpholino-1,2-propanediol, 96+%   FW 161.20   FT-IR: *1*, 384A   66.90
                                          mp 38°C   VP-FT-IR: *3*, 474A   64.94
                                          bp 191°C (30 mm)   61.06
CDCl₃   QE-300   d 1.157   53.89

A

Aldrich A6,630-8   *CAS [4319-49-7]* $C_4H_{10}N_2O$ Fp 137°F   60 MHz: *1*, 337D   66.73
4-Aminomorpholine, 99%   FW 102.14   FT-IR: *1*, 384B   59.94
                              bp 168°C   VP-FT-IR: *3*, 474C
                              d 1.059
CDCl₃   QE-300   n²⁰_D 1.4772

B

Aldrich A5,500-4   *CAS [2038-03-1]* $C_6H_{14}N_2O$ Fp 347°F   60 MHz: *1*, 338A   66.99
4-(2-Aminoethyl)morpholine, 99%   FW 130.19   FT-IR: *1*, 384C   61.65
                                       bp 205°C   VP-FT-IR: *3*, 474D   53.77
                                       d 0.992   38.56
CDCl₃   QE-300   n²⁰_D 1.4755

C

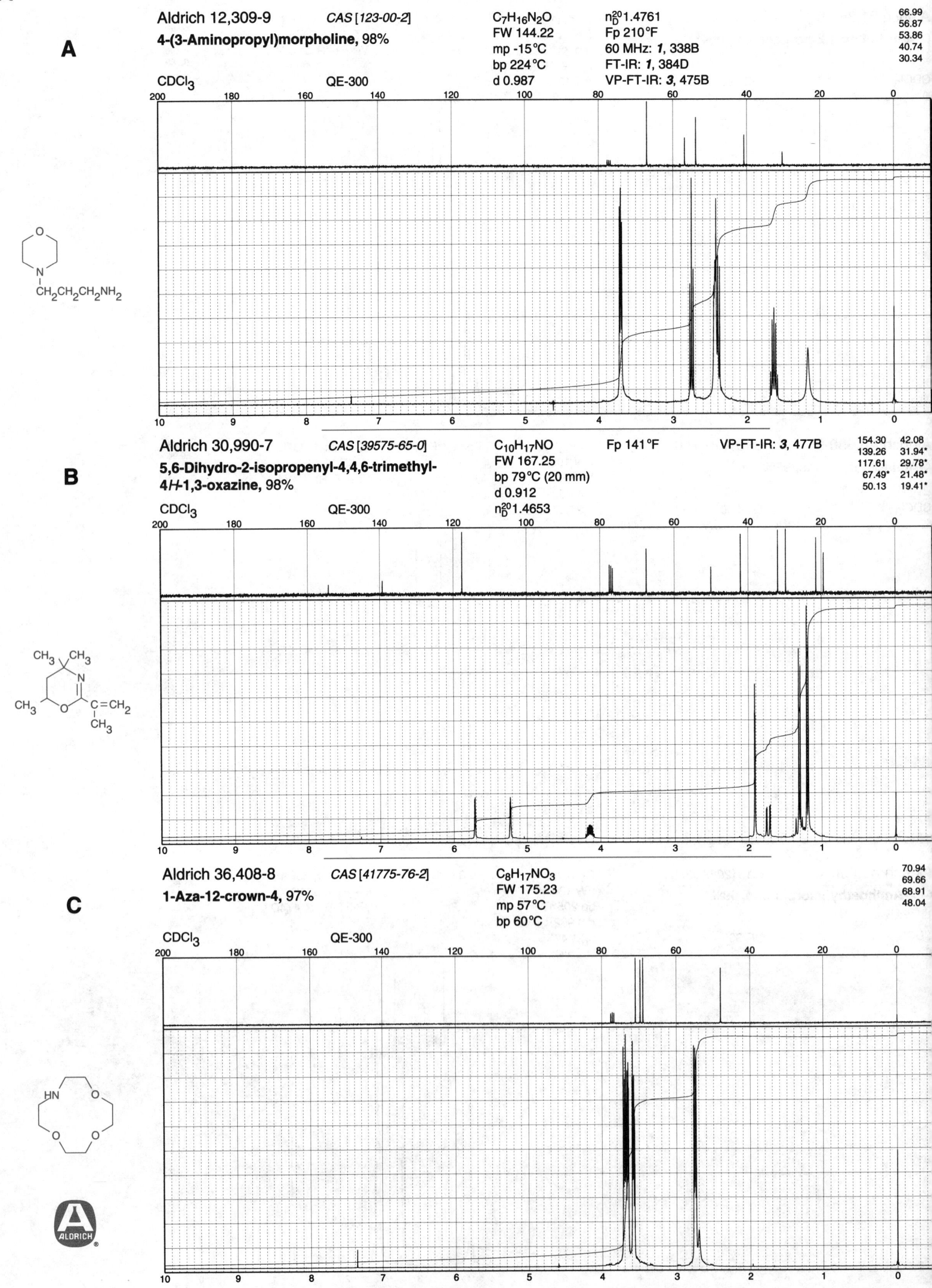

A

Aldrich 12,309-9   *CAS [123-00-2]*   C$_7$H$_{16}$N$_2$O   n$_D^{20}$1.4761

4-(3-Aminopropyl)morpholine, 98%   FW 144.22   Fp 210°F

mp -15°C   60 MHz: *1*, 338B

bp 224°C   FT-IR: *1*, 384D

CDCl$_3$ QE-300   d 0.987   VP-FT-IR: *3*, 475B

66.99
56.87
53.86
40.74
30.34

B

Aldrich 30,990-7   *CAS [39575-65-0]*   C$_{10}$H$_{17}$NO   Fp 141°F   VP-FT-IR: *3*, 477B

5,6-Dihydro-2-isopropenyl-4,4,6-trimethyl-   FW 167.25

4H-1,3-oxazine, 98%   bp 79°C (20 mm)

d 0.912

CDCl$_3$ QE-300   n$_D^{20}$1.4653

154.30   42.08
139.26   31.94*
117.61   29.78*
67.49*   21.48*
50.13   19.41*

C

Aldrich 36,408-8   *CAS [41775-76-2]*   C$_8$H$_{17}$NO$_3$

1-Aza-12-crown-4, 97%   FW 175.23

mp 57°C

bp 60°C

CDCl$_3$ QE-300

70.94
69.66
68.91
48.04

Aldrich 36,409-6   CAS [66943-05-3]

1-Aza-15-crown-5, 97%

CDCl₃   QE-300

$C_{10}H_{21}NO_4$
FW 219.28
mp 36°C
Fp >230°F

70.33
70.00
69.80
49.11

A

Aldrich 36,411-8   CAS [33941-15-0]

1-Aza-18-crown-6, 97%

CDCl₃   QE-300

$C_{12}H_{25}NO_5$
FW 263.34
mp 48°C
Fp >230°F

70.73
70.49
70.43
70.38
70.28
49.26

B

Aldrich 30,732-7   CAS [31249-95-3]

1,4,10-Trioxa-7,13-diazacyclopentadecane, 99%

CDCl₃   QE-300

$C_{10}H_{22}N_2O_3$
FW 218.30
mp 90°C

VP-FT-IR: **3**, 475D

69.86
69.73
69.36
49.09
48.86

C

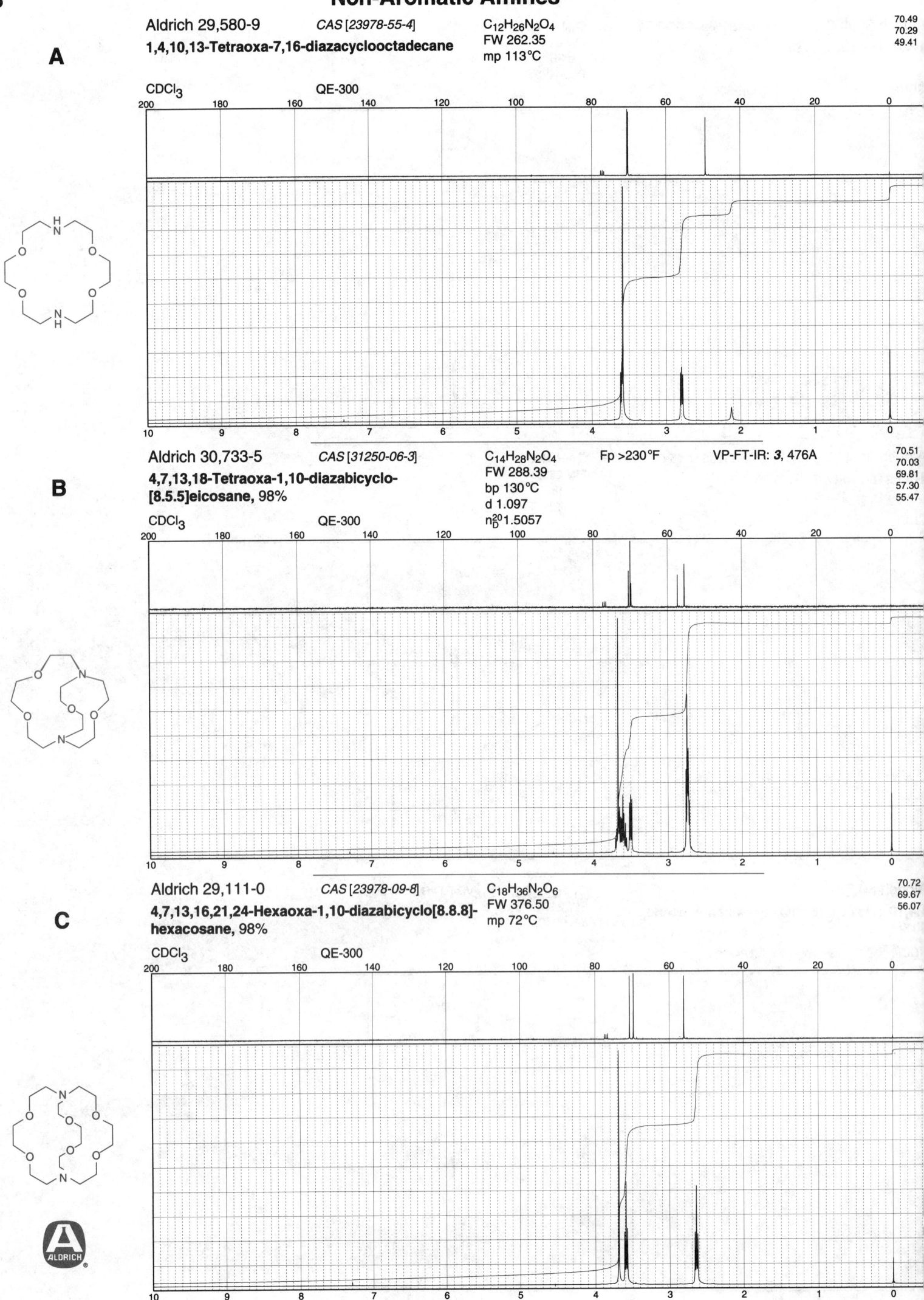

A

Aldrich 29,580-9   *CAS [23978-55-4]* $C_{12}H_{26}N_2O_4$
1,4,10,13-Tetraoxa-7,16-diazacyclooctadecane   FW 262.35
mp 113°C

70.49
70.29
49.41

CDCl$_3$ QE-300

B

Aldrich 30,733-5   *CAS [31250-06-3]* $C_{14}H_{28}N_2O_4$ Fp >230°F   VP-FT-IR: *3*, 476A
4,7,13,18-Tetraoxa-1,10-diazabicyclo-   FW 288.39
[8.5.5]eicosane, 98%   bp 130°C
d 1.097
n_D^{20} 1.5057

70.51
70.03
69.81
57.30
55.47

CDCl$_3$ QE-300

C

Aldrich 29,111-0   *CAS [23978-09-8]* $C_{18}H_{36}N_2O_6$
4,7,13,16,21,24-Hexaoxa-1,10-diazabicyclo[8.8.8]-   FW 376.50
hexacosane, 98%   mp 72°C

70.72
69.67
56.07

CDCl$_3$ QE-300

ALDRICH

Aldrich 14,969-1   *CAS [504-78-9]* C_3H_7NS Fp 133°F   60 MHz: *1*, 313D

Thiazolidine, 95%   FW 89.16   FT-IR: *1*, 352D

bp 74°C (25 mm)   VP-FT-IR: *3*, 440D

d 1.131

CDCl₃   QE-300 n_D^{20} 1.5508

55.39
52.78
33.95

A

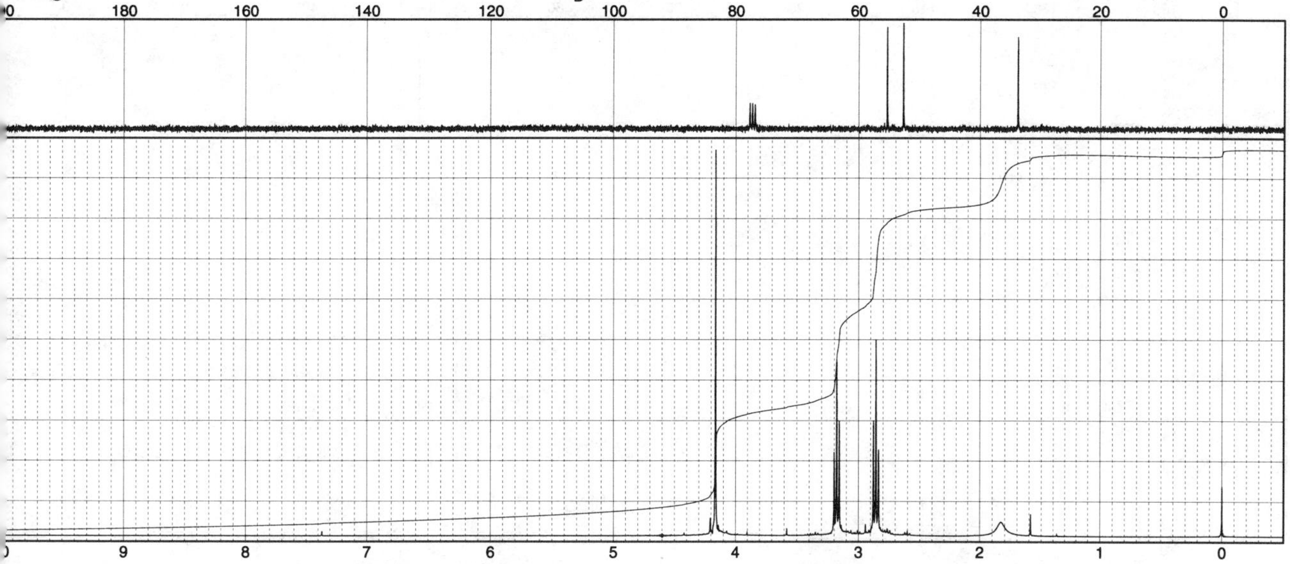

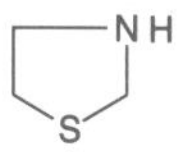

Aldrich 19,627-4   *CAS [123-90-0]* C_4H_9NS Fp 140°F   60 MHz: *1*, 335D

Thiomorpholine, 98%   FW 103.19   FT-IR: *1*, 353A

bp 169°C   VP-FT-IR: *3*, 441A

d 1.026

CDCl₃   QE-300 n_D^{20} 1.5384

47.91
28.36

B

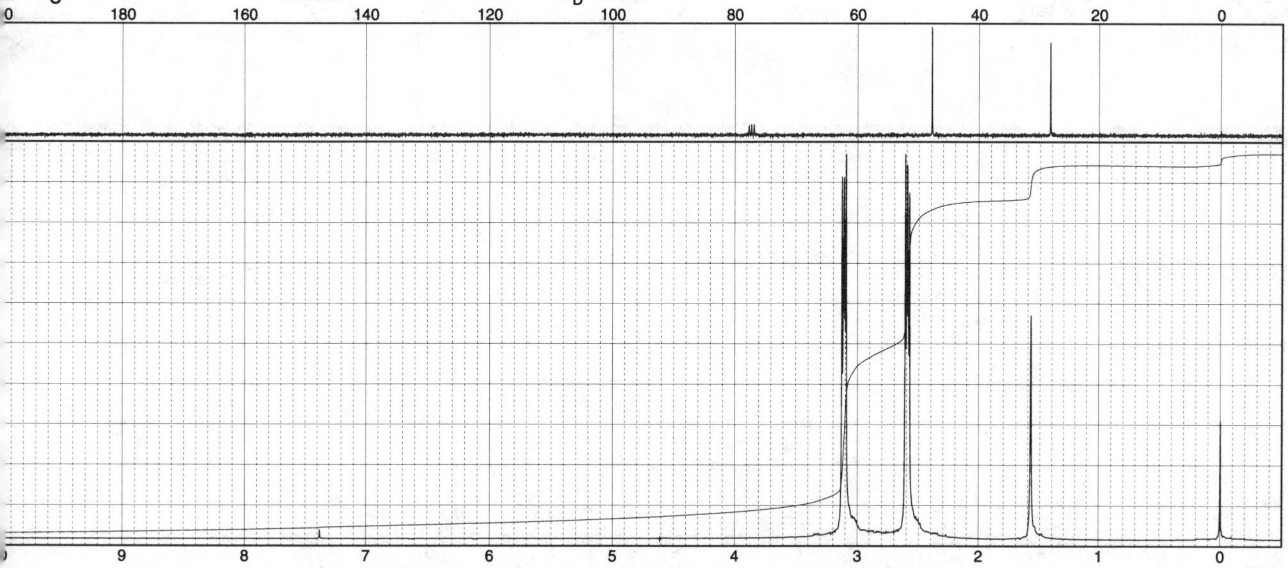

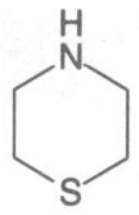

Aldrich 10,721-2   *CAS [17787-40-5]* $C_4H_{12}FN$ 60 MHz: *1*, 341C

Tetramethylammonium fluoride tetrahydrate, 98%   FW 165.21   FT-IR: *1*, 385C

mp 41°C

Fp >230°F

D₂O   QE-300

58.04*
57.99*
57.94*

C

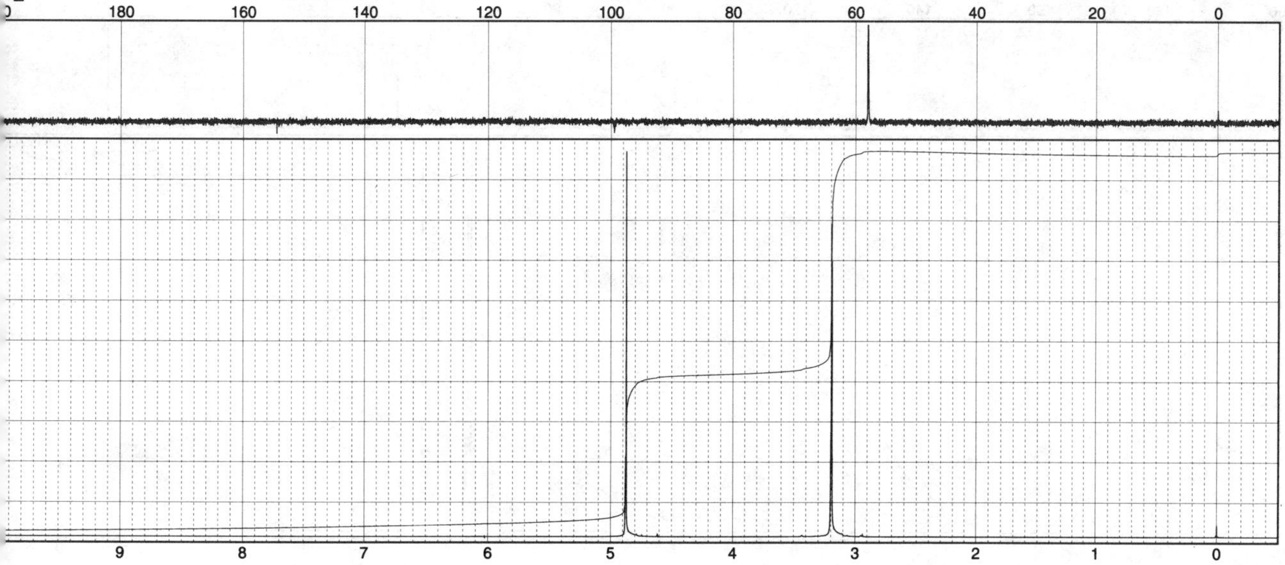

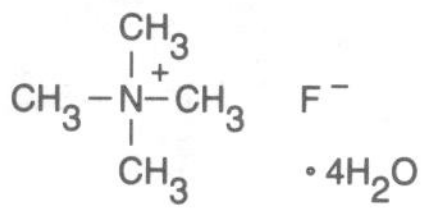

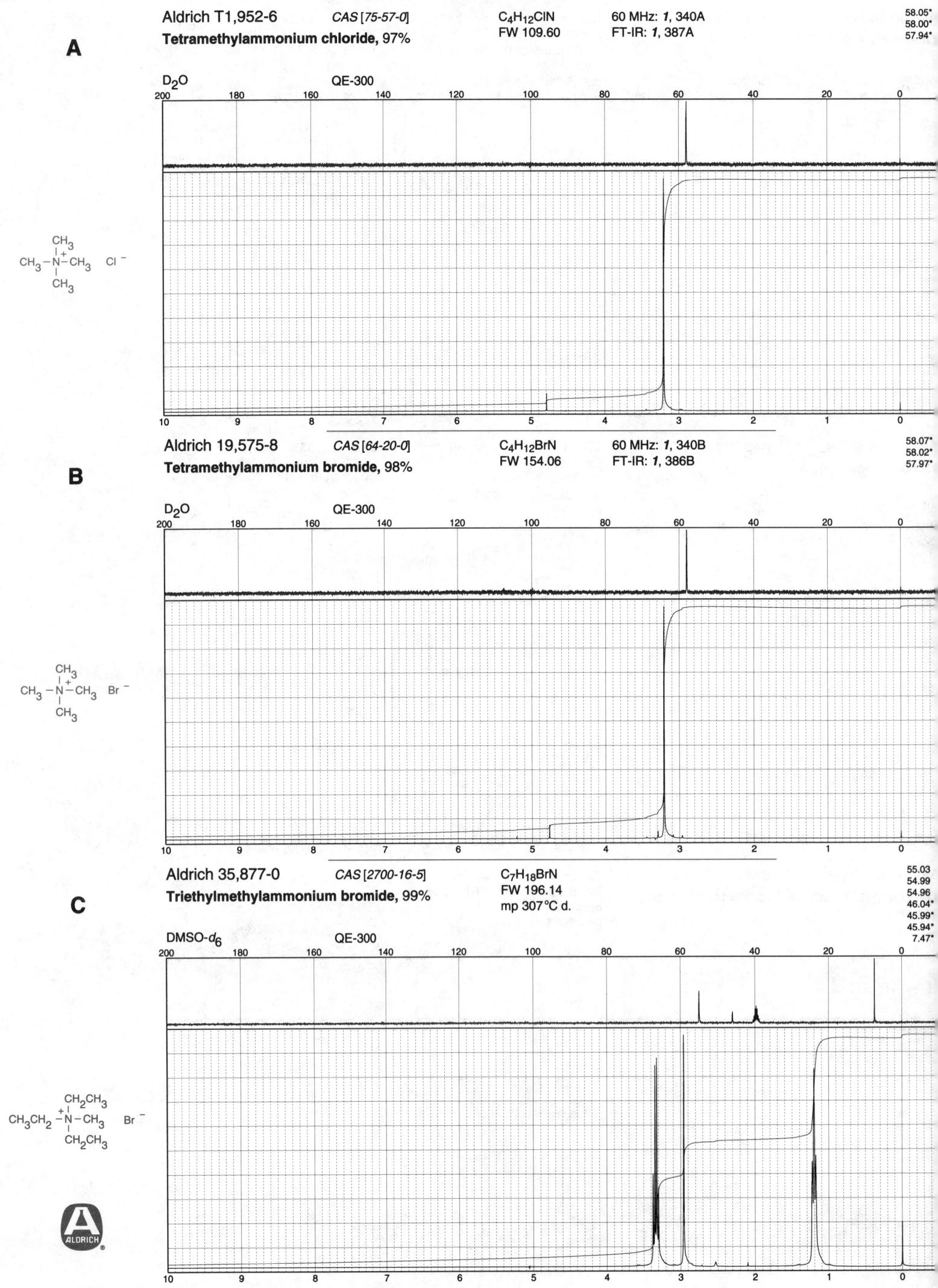

Aldrich T1,952-6   *CAS [75-57-0]* $C_4H_{12}ClN$ 60 MHz: *1*, 340A

Tetramethylammonium chloride, 97%   FW 109.60   FT-IR: *1*, 387A

A

D$_2$O   QE-300

58.05*
58.00*
57.94*

Aldrich 19,575-8   *CAS [64-20-0]* $C_4H_{12}BrN$ 60 MHz: *1*, 340B

Tetramethylammonium bromide, 98%   FW 154.06   FT-IR: *1*, 386B

B

D$_2$O   QE-300

58.07*
58.02*
57.97*

Aldrich 35,877-0   *CAS [2700-16-5]* $C_7H_{18}BrN$

Triethylmethylammonium bromide, 99%   FW 196.14

mp 307°C d.

C

DMSO-d_6 QE-300

55.03
54.99
54.96
46.04*
45.99*
45.94*
7.47*

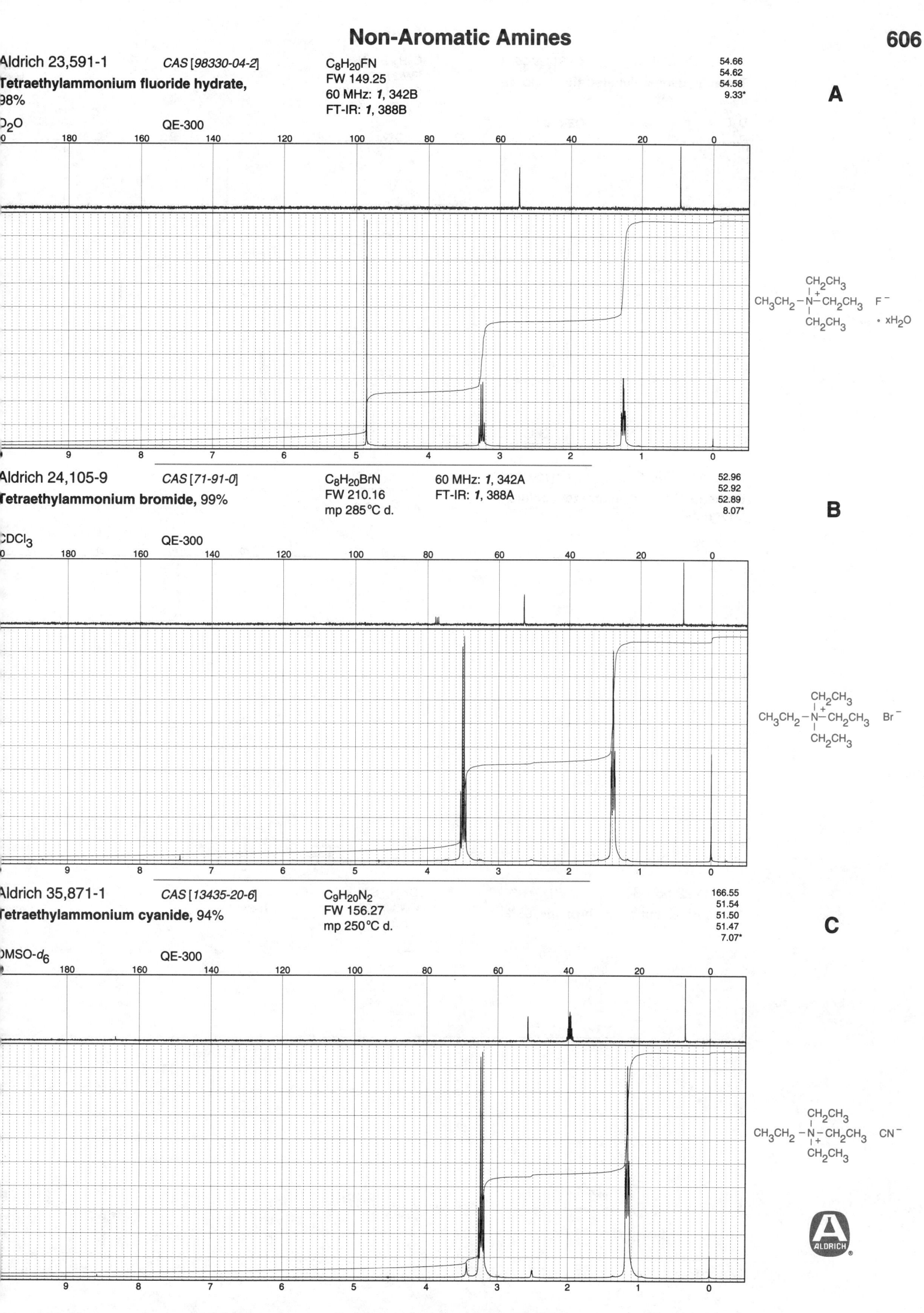

Aldrich 23,591-1   CAS [98330-04-2]

Tetraethylammonium fluoride hydrate, 98%

C$_8$H$_{20}$FN
FW 149.25
60 MHz: 1, 342B
FT-IR: 1, 388B

54.66
54.62
54.58
9.33*

A

D$_2$O   QE-300

Aldrich 24,105-9   CAS [71-91-0]

Tetraethylammonium bromide, 99%

C$_8$H$_{20}$BrN
FW 210.16
mp 285°C d.
60 MHz: 1, 342A
FT-IR: 1, 388A

52.96
52.92
52.89
8.07*

B

CDCl$_3$ QE-300

Aldrich 35,871-1   CAS [13435-20-6]

Tetraethylammonium cyanide, 94%

C$_9$H$_{20}$N$_2$
FW 156.27
mp 250°C d.

166.55
51.54
51.50
51.47
7.07*

C

DMSO-d_6 QE-300

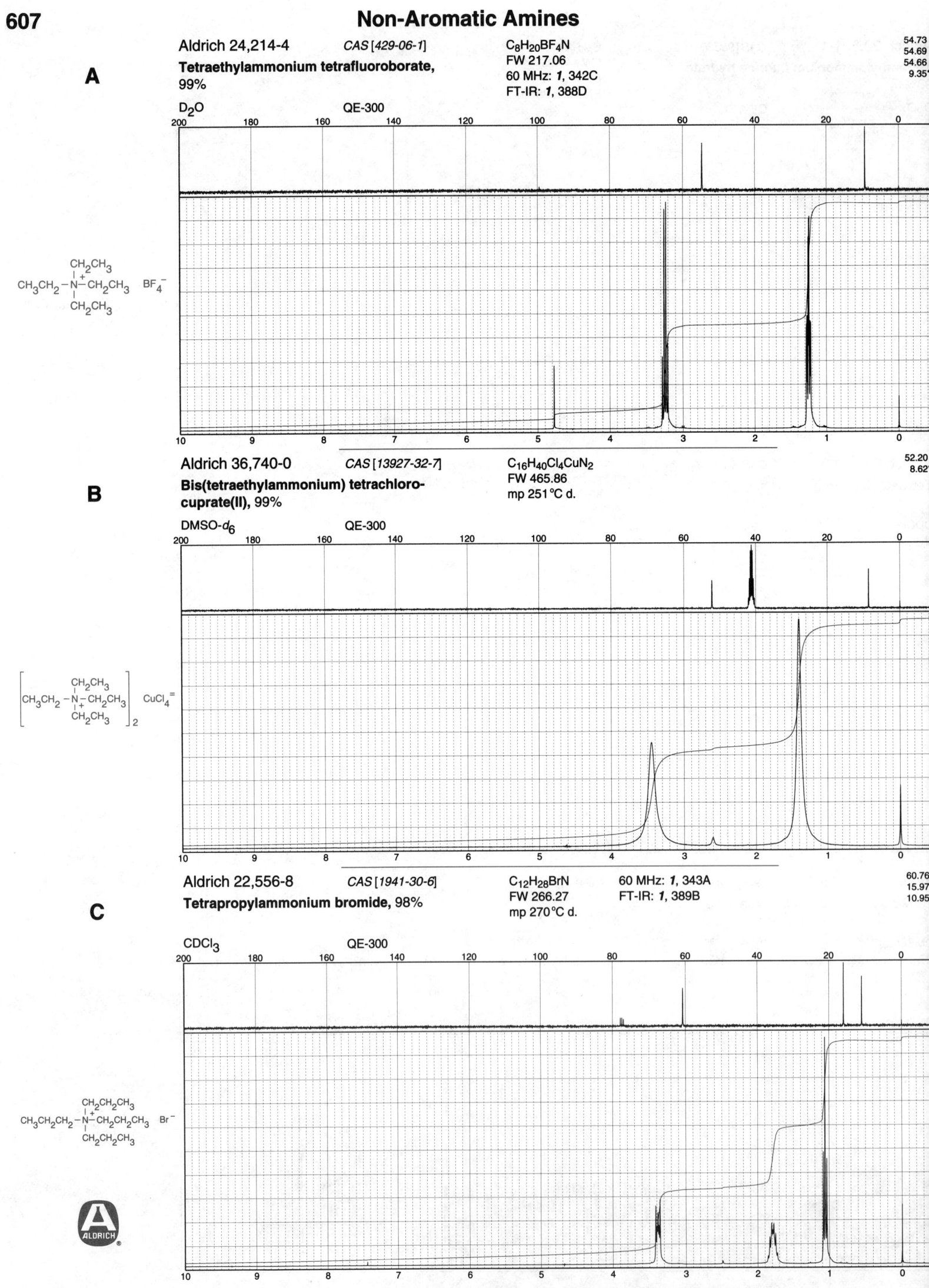

A

Aldrich 24,214-4   CAS [429-06-1]

Tetraethylammonium tetrafluoroborate,
99%

$C_8H_{20}BF_4N$
FW 217.06
60 MHz: *1*, 342C
FT-IR: *1*, 388D

54.73
54.69
54.66
9.35*

D₂O   QE-300

B

Aldrich 36,740-0   CAS [13927-32-7]

**Bis(tetraethylammonium) tetrachloro-
cuprate(II), 99%**

$C_{16}H_{40}Cl_4CuN_2$
FW 465.86
mp 251 °C d.

52.20
8.62*

DMSO-d_6 QE-300

C

Aldrich 22,556-8   CAS [1941-30-6]

Tetrapropylammonium bromide, 98%

$C_{12}H_{28}BrN$
FW 266.27
mp 270 °C d.

60 MHz: *1*, 343A
FT-IR: *1*, 389B

60.76
15.97
10.95

CDCl₃   QE-300

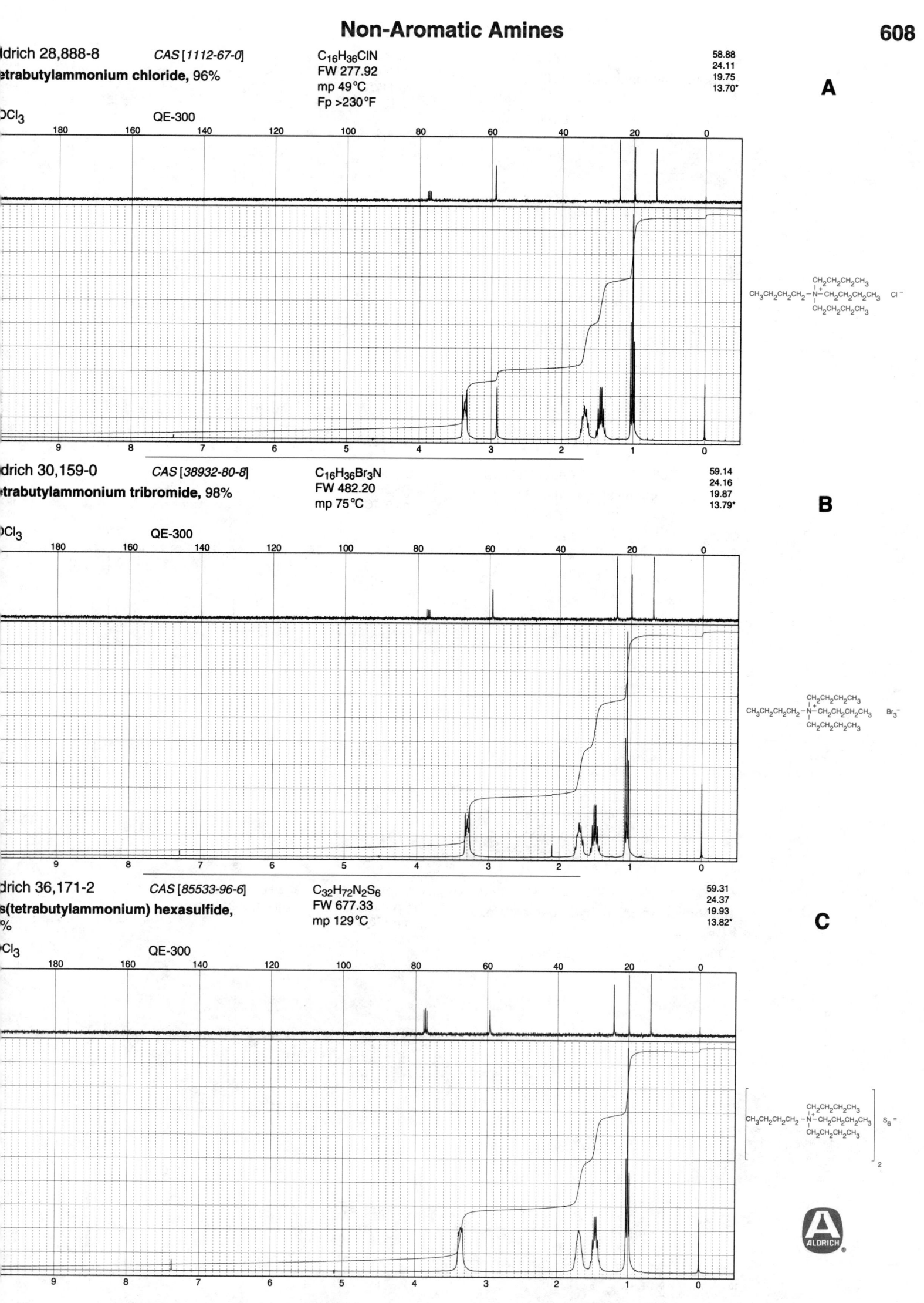

A

Aldrich 28,888-8 CAS [1112-67-0]
tetrabutylammonium chloride, 96%

$C_{16}H_{36}ClN$
FW 277.92
mp 49°C
Fp >230°F

58.88
24.11
19.75
13.70*

DCl₃ QE-300

B

Aldrich 30,159-0 CAS [38932-80-8]
tetrabutylammonium tribromide, 98%

$C_{16}H_{36}Br_3N$
FW 482.20
mp 75°C

59.14
24.16
19.87
13.79*

DCl₃ QE-300

C

Aldrich 36,171-2 CAS [85533-96-6]
bis(tetrabutylammonium) hexasulfide, %

$C_{32}H_{72}N_2S_6$
FW 677.33
mp 129°C

59.31
24.37
19.93
13.82*

Cl₃ QE-300

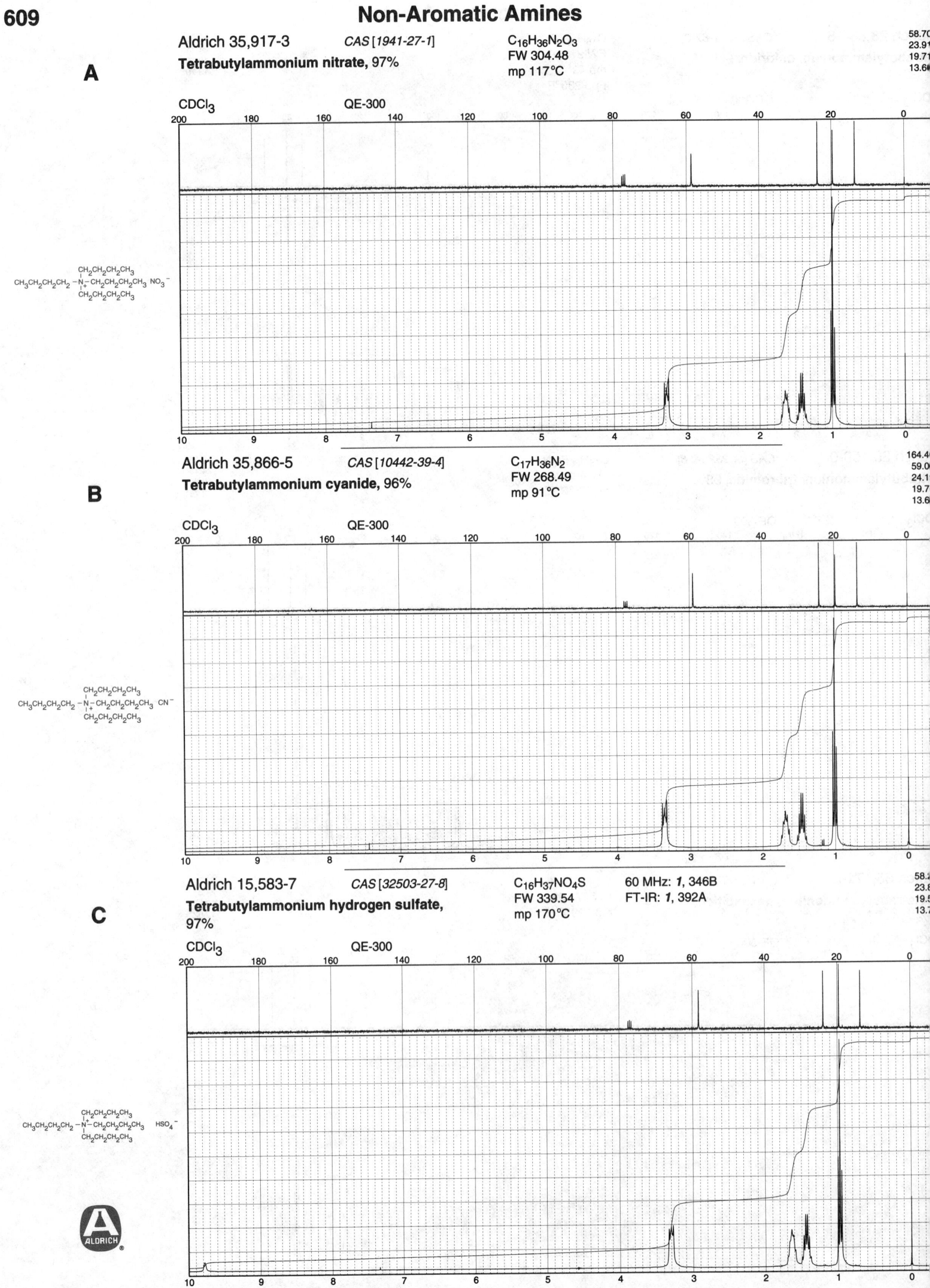

A

Aldrich 35,917-3   *CAS [1941-27-1]*   C₁₆H₃₆N₂O₃
Tetrabutylammonium nitrate, 97%   FW 304.48
mp 117°C

58.70
23.91
19.71
13.6

CDCl₃   QE-300

B

Aldrich 35,866-5   *CAS [10442-39-4]*   C₁₇H₃₆N₂
Tetrabutylammonium cyanide, 96%   FW 268.49
mp 91°C

164.4
59.0
24.1
19.7
13.6

CDCl₃   QE-300

C

Aldrich 15,583-7   *CAS [32503-27-8]*   C₁₆H₃₇NO₄S   60 MHz: *1*, 346B
Tetrabutylammonium hydrogen sulfate,   FW 339.54   FT-IR: *1*, 392A
97%   mp 170°C

58.2
23.8
19.5
13.7

CDCl₃   QE-300

Aldrich 28,102-6   *CAS [3109-63-5]* $C_{16}H_{36}F_6NP$

Tetrabutylammonium hexafluorophosphate,

%

$C_{16}H_{36}F_6NP$
FW 387.44
mp 245°C

58.42
23.65
19.51
13.48*

A

DCl₃   QE-300

$CH_3CH_2CH_2CH_2 - \overset{+}{N} \langle \begin{array}{l} CH_2CH_2CH_2CH_3 \\ CH_2CH_2CH_2CH_3 \\ CH_2CH_2CH_2CH_3 \end{array} \quad PF_6^-$

Aldrich 21,796-4   *CAS [429-42-5]* $C_{16}H_{36}BF_4N$

Tetrabutylammonium tetrafluoroborate,

%

$C_{16}H_{36}BF_4N$
FW 329.28
mp 161°C

60 MHz: *1*, 346A
FT-IR: *1*, 391D

58.38
23.72
19.56
13.55*

B

DCl₃   QE-300

$CH_3CH_2CH_2CH_2 - \overset{+}{N} \langle \begin{array}{l} CH_2CH_2CH_2CH_3 \\ CH_2CH_2CH_2CH_3 \\ CH_2CH_2CH_2CH_3 \end{array} \quad BF_4^-$

Aldrich 34,214-9   *CAS [18129-78-7]* $C_{32}H_{72}Cl_6N_2Pt$

Tetrabutylammonium hexachloroplatinate(IV)

$C_{32}H_{72}Cl_6N_2Pt$
FW 892.75
mp 225°C d.

57.39
22.97
19.10
13.41*

C

SO-*d*₆   QE-300

$\left[CH_3CH_2CH_2CH_2 - \overset{+}{N} \langle \begin{array}{l} CH_2CH_2CH_2CH_3 \\ CH_2CH_2CH_2CH_3 \\ CH_2CH_2CH_2CH_3 \end{array} \right]_2 \quad PtCl_6^-$

Non-Aromatic Amines

A

Aldrich 35,920-3   *CAS [2498-20-6]*
Tetrapentylammonium iodide, 98%

$C_{20}H_{44}IN$
FW 425.49
mp 138°C

59.38
28.43
22.24
22.17
13.84

CDCl₃   QE-300

B

Aldrich 38,083-0   *CAS [13028-71-2]*
Triethylhexylammonium bromide, 99%

$C_{12}H_{28}BrN$
FW 266.27
mp 116°C

57.49	22.3
53.50	22.0
31.18	13.8
26.08	8.1

CDCl₃   QE-300

C

Aldrich 36,961-6   *CAS [32503-34-7]*
Tetrahexylammonium hydrogen sulfate, 98%

$C_{24}H_{53}NO_4S$
FW 451.76
mp 99°C

57.6
30.5
25.3
21.8
20.9
13.7

DMSO-d_6 QE-300

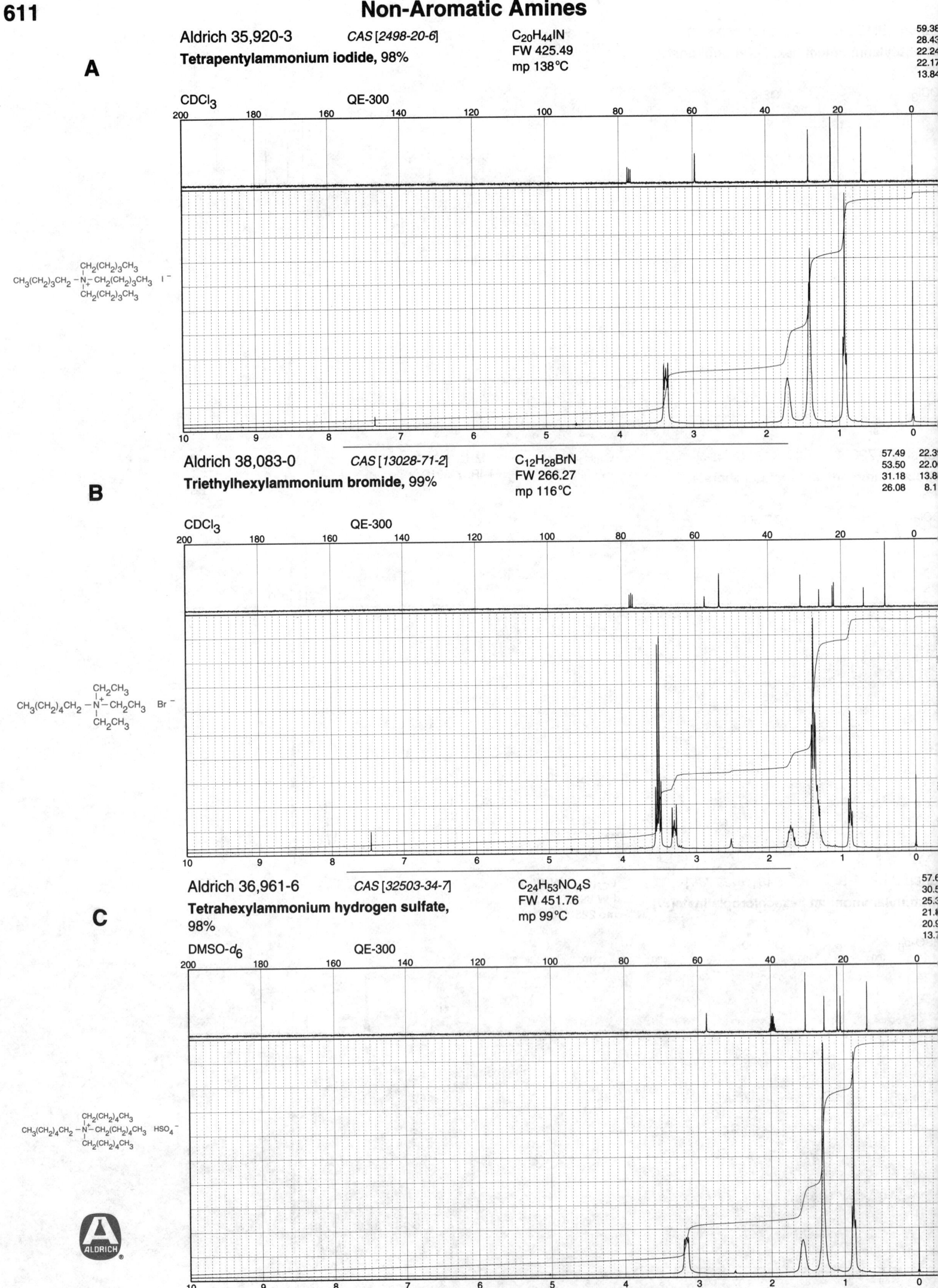

Aldrich 36,571-8   CAS [35675-80-0]

Methyltrioctylammonium bromide, 97%

CDCl₃   QE-300

61.62	26.35
48.96*	22.57
31.64	22.52
29.14	14.05*
29.04	

A

C₂₅H₅₄BrN
FW 448.63
mp 39 °C
Fp >230 °F

$$CH_3-\overset{CH_2(CH_2)_6CH_3}{\underset{CH_2(CH_2)_6CH_3}{\overset{|}{\underset{|}{N^+}}}}-CH_2(CH_2)_6CH_3 \quad Br^-$$

Aldrich 37,566-7   CAS [6243-39-6]

Aliquat® 336 nitrate

CDCl₃   QE-300

62.89	31.64	29.33	25.81
61.61	29.61	29.25	22.65
48.41*	29.58	29.15	22.58
32.85	29.49	29.09	22.34
31.89	29.43	29.03	14.09*
31.84	29.39	26.33	14.05*

B

C₂₅H₅₄N₂O₃
FW 430.72
d 0.880
n²⁰_D 1.4630
Fp >230 °F

$$CH_3(CH_2)_6CH_2-\overset{CH_3}{\underset{CH_2(CH_2)_6CH_3}{\overset{|}{\underset{|}{N^+}}}}-CH_2(CH_2)_6CH_3 \quad NO_3^-$$

Aldrich 29,801-8   CAS [2390-68-3]

decyldimethylammonium bromide,
wt. % solution in ethyl alcohol/water (15:5)

CDCl₃   QE-300

64.50	29.72
51.80*	26.77
32.32	23.31
29.91	23.13
29.89	14.56*

C

C₂₂H₄₈BrN
FW 406.54
d 0.944
n²⁰_D 1.4614
Fp 95 °F

$$CH_3(CH_2)_6CH_2-\overset{CH_3}{\underset{CH_3}{\overset{|}{\underset{|}{N^+}}}}-CH_2(CH_2)_8CH_3 \quad Br^-$$

Non-Aromatic Amines

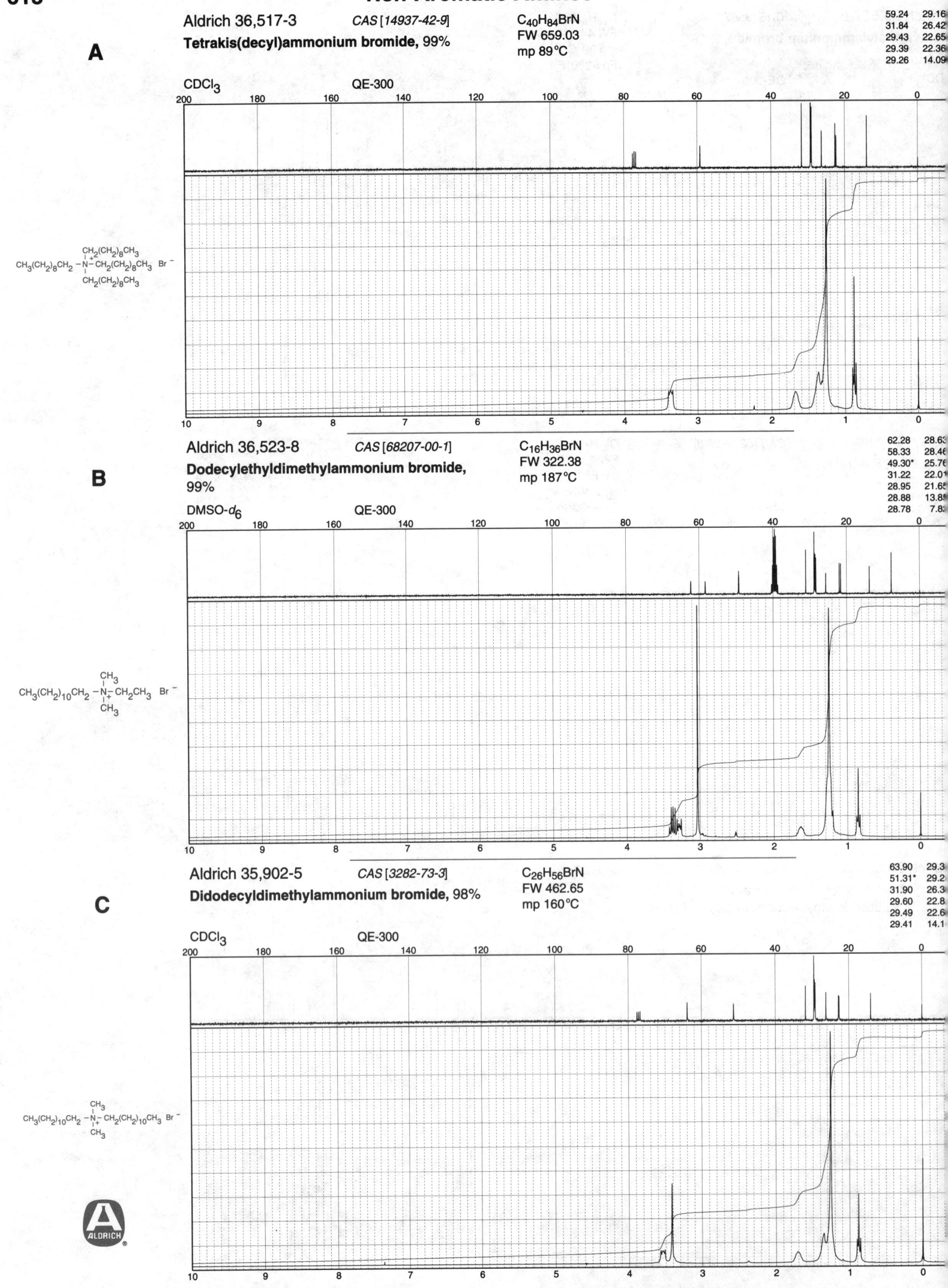

A

Aldrich 36,517-3 *CAS [14937-42-9]* C₄₀H₈₄BrN
Tetrakis(decyl)ammonium bromide, 99% FW 659.03
mp 89°C

59.24	29.16
31.84	26.42
29.43	22.65
29.39	22.36
29.26	14.09

CDCl₃   QE-300

B

Aldrich 36,523-8 *CAS [68207-00-1]* C₁₆H₃₆BrN
Dodecylethyldimethylammonium bromide, 99% FW 322.38
mp 187°C

62.28	28.63
58.33	28.46
49.30*	25.76
31.22	22.01
28.95	21.65
28.88	13.85
28.78	7.82

DMSO-d₆   QE-300

C

Aldrich 35,902-5 *CAS [3282-73-3]* C₂₆H₅₆BrN
Didodecyldimethylammonium bromide, 98% FW 462.65
mp 160°C

63.90	29.3
51.31*	29.2
31.90	26.3
29.60	22.8
29.49	22.6
29.41	14.1

CDCl₃   QE-300

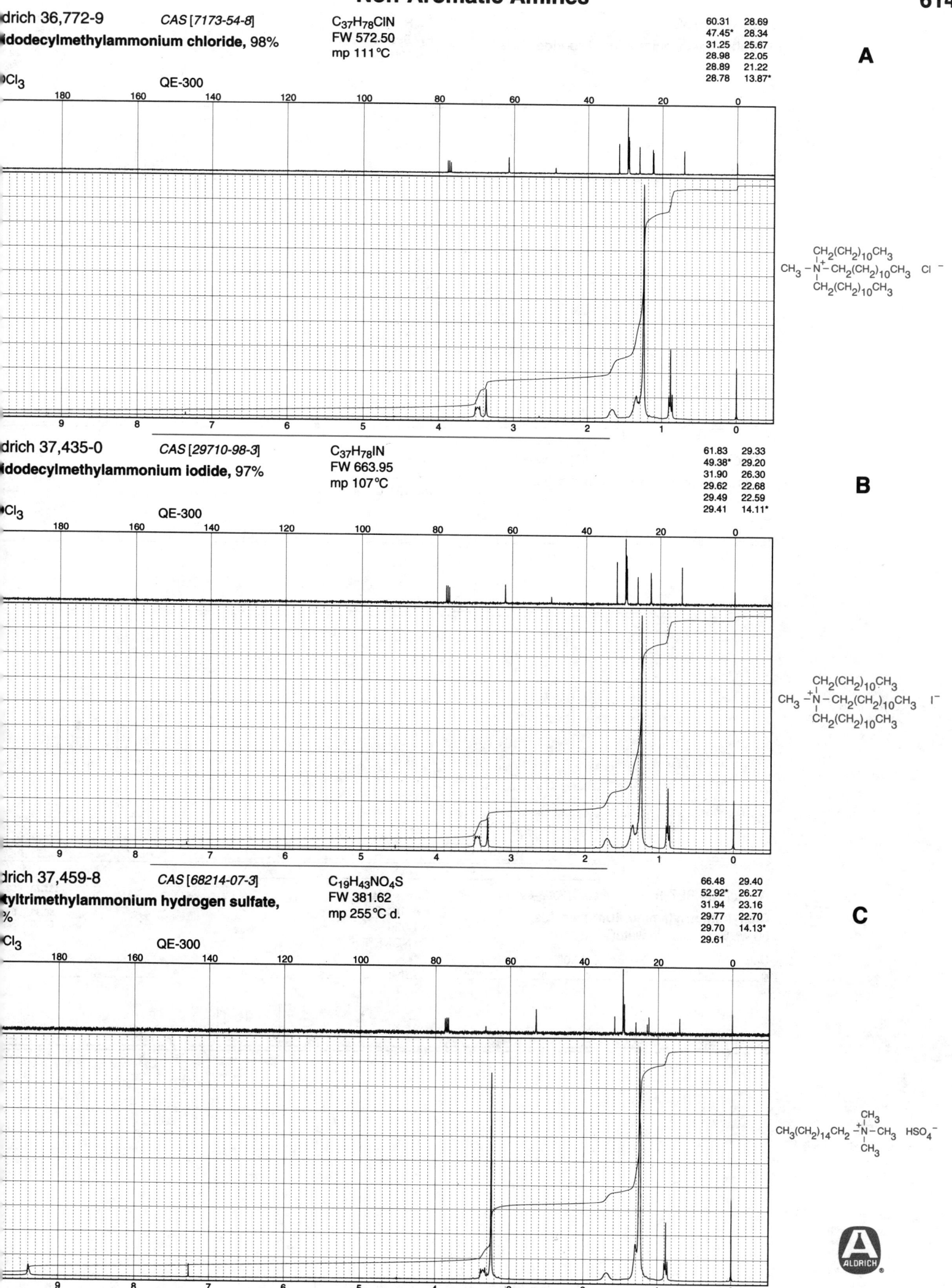

ldrich 36,772-9   CAS [7173-54-8]   C₃₇H₇₈ClN   FW 572.50   mp 111°C

A

60.31	28.69
47.45*	28.34
31.25	25.67
28.98	22.05
28.89	21.22
28.78	13.87*

Cl₃   QE-300

Idodecylmethylammonium chloride, 98%

ldrich 37,435-0   CAS [29710-98-3]   C₃₇H₇₈IN   FW 663.95   mp 107°C

B

61.83	29.33
49.38*	29.20
31.90	26.30
29.62	22.68
29.49	22.59
29.41	14.11*

Cl₃   QE-300

Idodecylmethylammonium iodide, 97%

ldrich 37,459-8   CAS [68214-07-3]   C₁₉H₄₃NO₄S   FW 381.62   mp 255°C d.

C

66.48	29.40
52.92*	26.27
31.94	23.16
29.77	22.70
29.70	14.13*
29.61	

Cl₃   QE-300

tyltrimethylammonium hydrogen sulfate, %

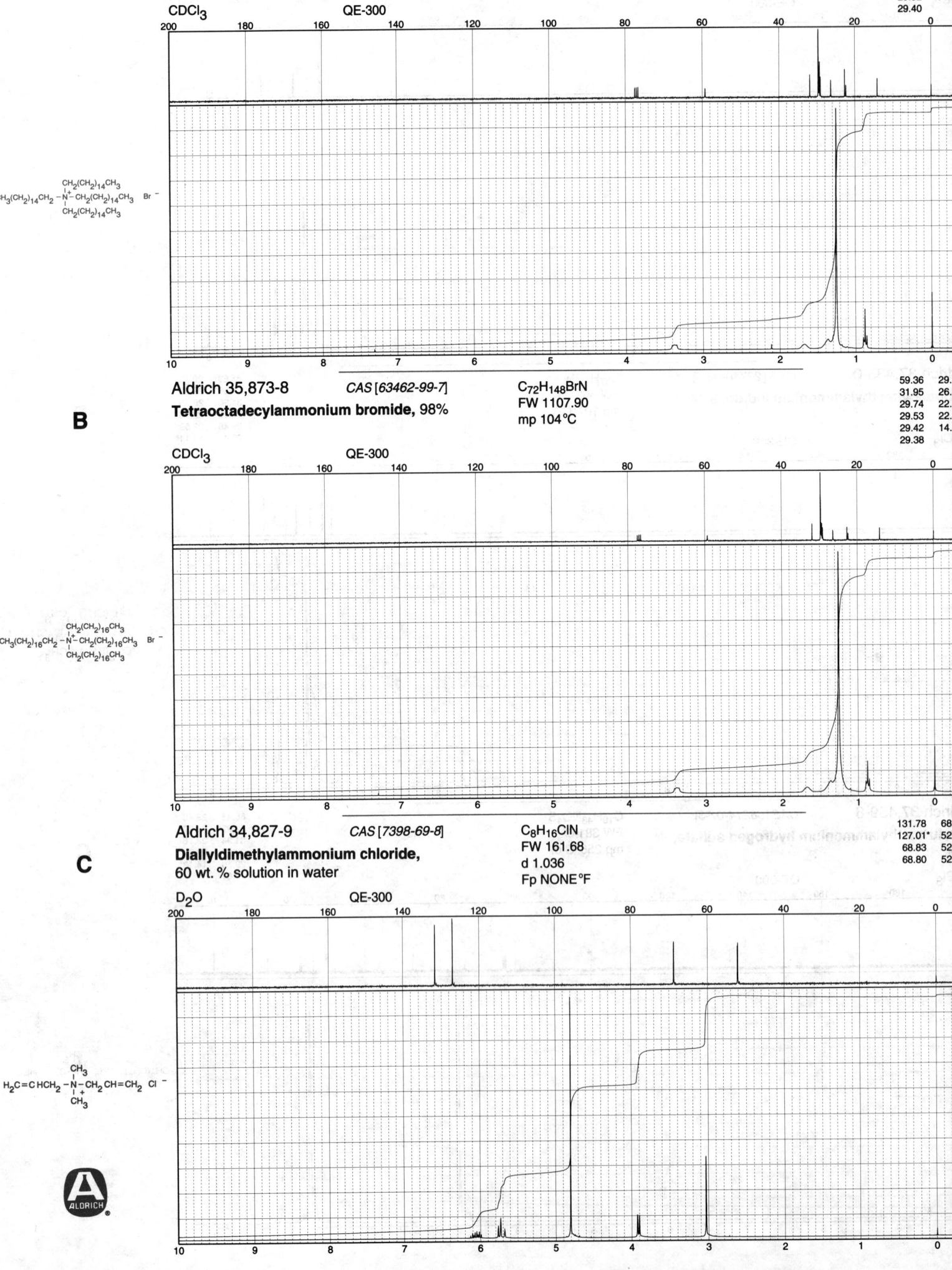

A

Aldrich 36,752-4

Tetrahexadecylammonium bromide, 98%

C₆₄H₁₃₂BrN
FW 995.68
mp 100 °C

$C_{64}H_{132}BrN$

59.28	29.3
31.94	29.1
29.72	26.4
29.69	22.7
29.66	22.3
29.52	14.1
29.40	

CDCl₃   QE-300

B

Aldrich 35,873-8   CAS [63462-99-7]

Tetraoctadecylammonium bromide, 98%

C₇₂H₁₄₈BrN
FW 1107.90
mp 104 °C

$C_{72}H_{148}BrN$

59.36	29.1
31.95	26.4
29.74	22.7
29.53	22.4
29.42	14.1
29.38	

CDCl₃   QE-300

C

Aldrich 34,827-9   CAS [7398-69-8]

Diallyldimethylammonium chloride,
60 wt. % solution in water

C₈H₁₆ClN
FW 161.68
d 1.036
Fp NONE °F

$C_8H_{16}ClN$

131.78	68.7
127.01*	52.2
68.83	52.1
68.80	52.

D₂O   QE-300

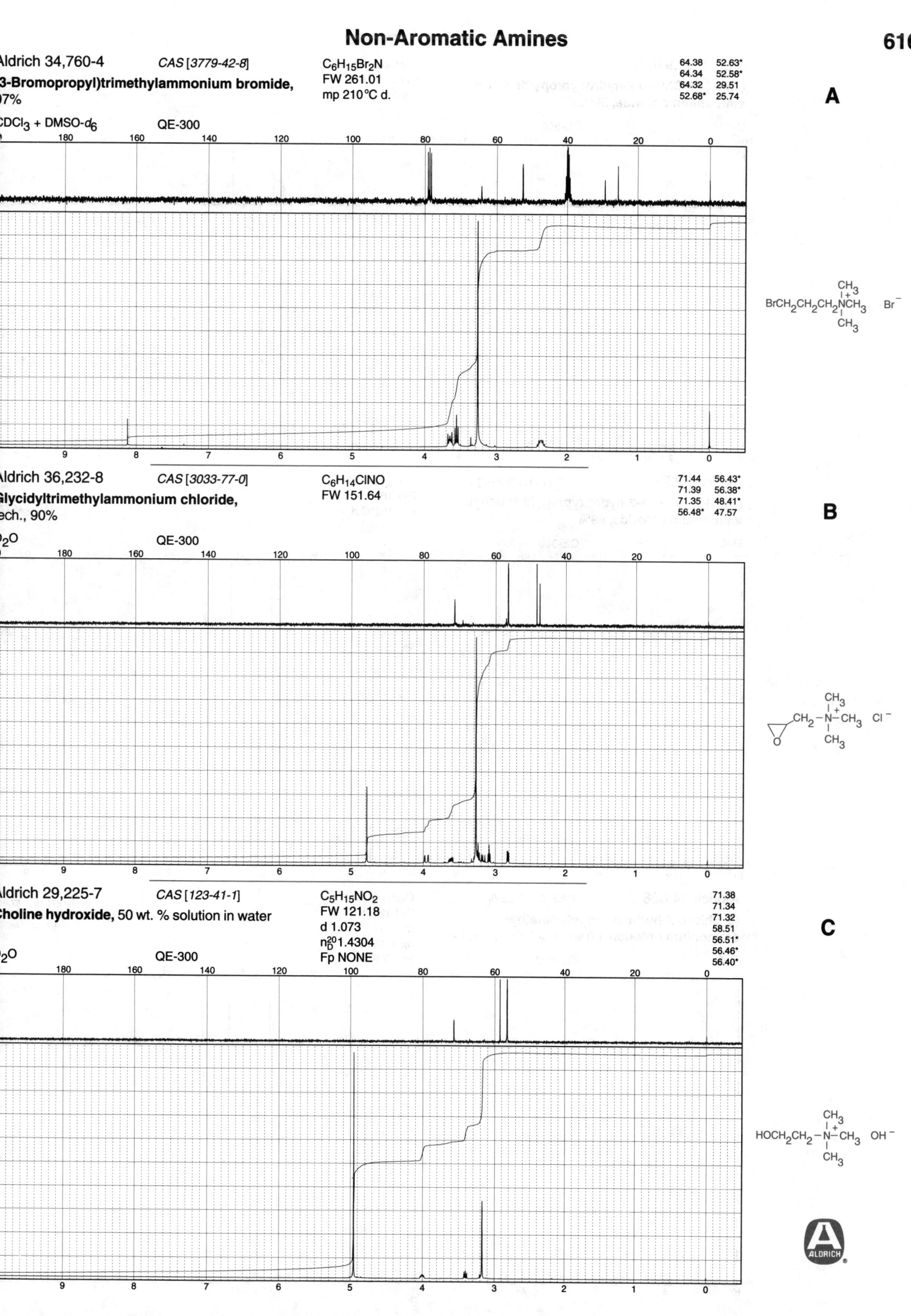

Aldrich 34,760-4   CAS [3779-42-8]   C₆H₁₅Br₂N

(3-Bromopropyl)trimethylammonium bromide, 97%

$C_6H_{15}Br_2N$
FW 261.01
mp 210°C d.

64.38	52.63*
64.34	52.58*
64.32	29.51
52.68*	25.74

CDCl₃ + DMSO-d₆   QE-300

A

Aldrich 36,232-8   CAS [3033-77-0]

Glycidyltrimethylammonium chloride, tech., 90%

$C_6H_{14}ClNO$
FW 151.64

71.44	56.43*
71.39	56.38*
71.35	48.41*
56.48*	47.57

D₂O   QE-300

B

Aldrich 29,225-7   CAS [123-41-1]

Choline hydroxide, 50 wt. % solution in water

$C_5H_{15}NO_2$
FW 121.18
d 1.073
n_D^{20} 1.4304
Fp NONE

71.38
71.34
71.32
58.51
56.51*
56.46*
56.40*

D₂O   QE-300

C

A

Aldrich 32,916-9

(*R*)-(+)-(3-Chloro-2-hydroxypropyl)trimethyl-
ammonium chloride, 99%

$C_6H_{15}Cl_2NO$
FW 188.10
mp 217°C d.

70.99
68.26*
56.98*
49.76

D$_2$O   QE-300

HO H CH$_3$
| | | +
ClCH$_2$CCH$_2$N—CH$_3$ Cl$^-$
           |
           CH$_3$

B

Aldrich 32,917-7   *CAS [101396-91-2]*

(*S*)-(-)-(3-Chloro-2-hydroxypropyl)trimethyl-
ammonium chloride, 99%

$C_6H_{15}Cl_2NO$
FW 188.10
mp 210°C d.

70.97
68.24*
56.96*
49.77

D$_2$O   QE-300

HO H CH$_3$
| | | +
ClCH$_2$C CH$_2$N—CH$_3$ Cl$^-$
           |
           CH$_3$

C

Aldrich 34,828-7   *CAS [3327-22-8]*

(3-Chloro-2-hydroxypropyl)trimethyl-
ammonium chloride, 60 wt. % solution in water

$C_6H_{15}Cl_2NO$
FW 188.10
d 1.154
n_D^{20}1.4541
Fp >230°F

71.20 57.09
71.16 57.05
71.14 57.00
68.29* 49.72

D$_2$O   QE-300

OH   CH$_3$
| |
ClCH$_2$CHCH$_2$—N—CH$_3$ Cl$^-$
         | +
         CH$_3$

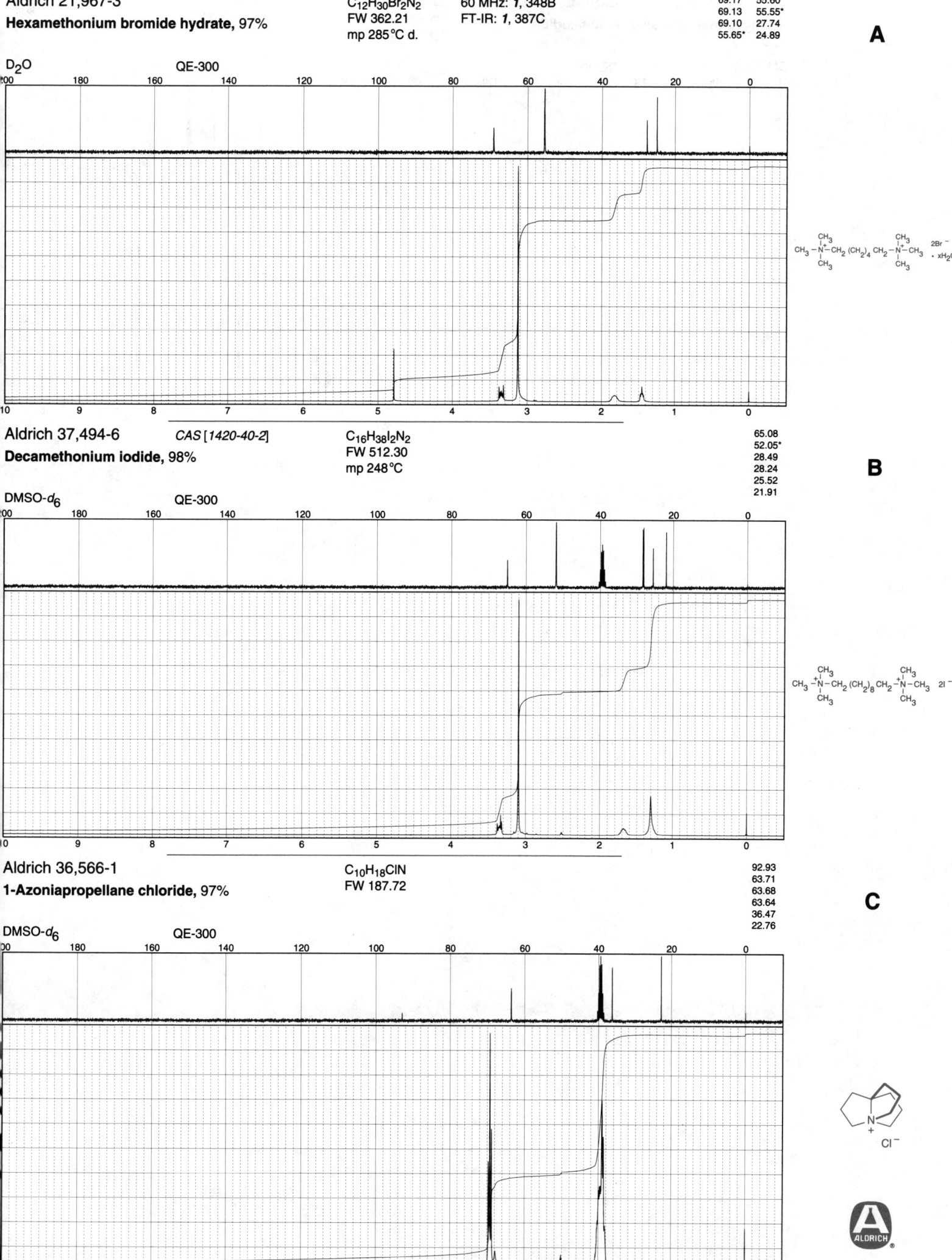

Aldrich 21,967-3
Hexamethonium bromide hydrate, 97%

$C_{12}H_{30}Br_2N_2$
FW 362.21
mp 285°C d.

60 MHz: *1*, 348B
FT-IR: *1*, 387C

69.17 55.60*
69.13 55.55*
69.10 27.74
55.65* 24.89

A

D₂O   QE-300

Aldrich 37,494-6   *CAS [1420-40-2]*
Decamethonium iodide, 98%

$C_{16}H_{38}I_2N_2$
FW 512.30
mp 248°C

65.08
52.05*
28.49
28.24
25.52
21.91

B

DMSO-*d₆*   QE-300

Aldrich 36,566-1
1-Azoniapropellane chloride, 97%

$C_{10}H_{18}ClN$
FW 187.72

92.93
63.71
63.68
63.64
36.47
22.76

C

DMSO-*d₆*   QE-300

Non-Aromatic Amines

A

Aldrich 32,249-0   *CAS [54654-71-6]*   C₆H₁₂INS

3-Ethyl-2-methyl-2-thiazolinium iodide, 98%

FW 257.14
mp 192°C

188.39
58.69
46.27
28.85
17.19*
11.86*

DMSO-*d₆*   QE-300

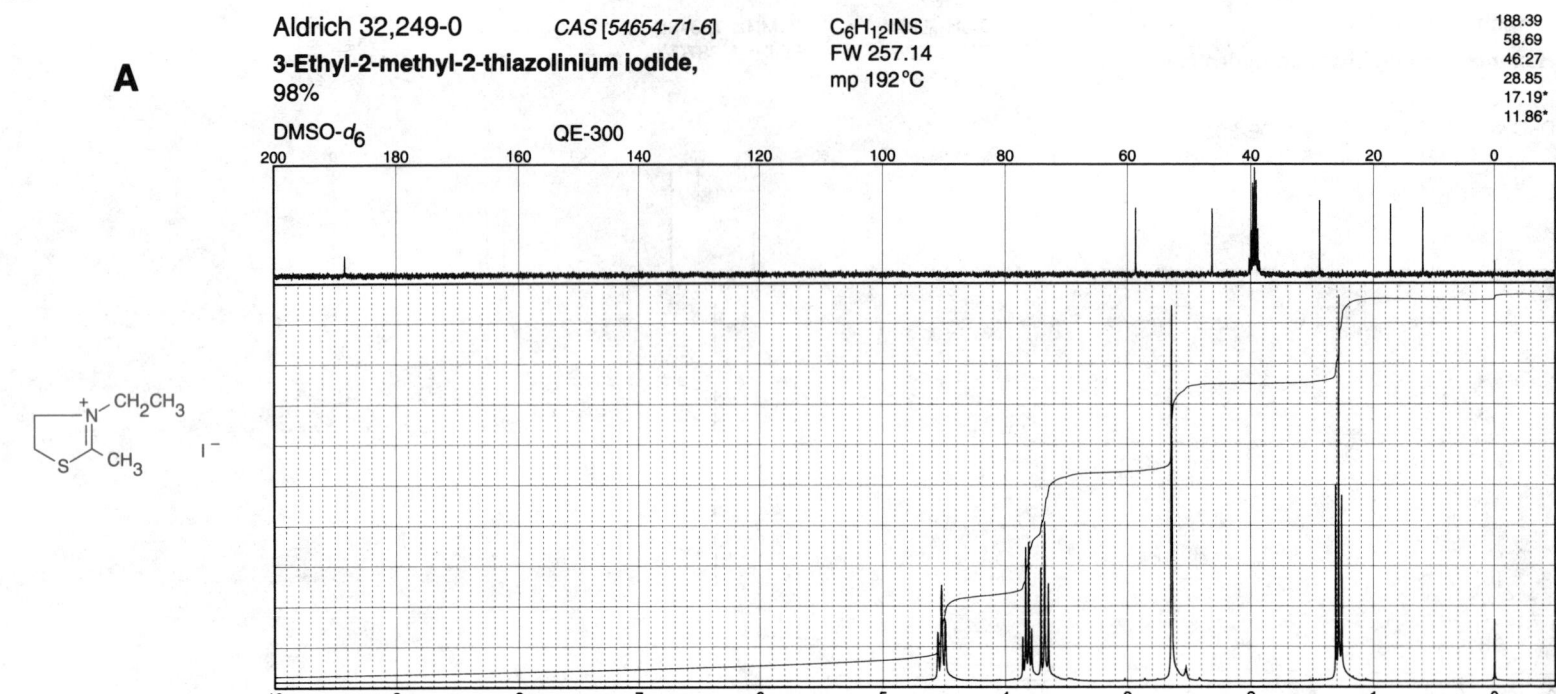

Non-Aromatic Nitro and Nitroso Compounds

A

Aldrich 22,663-7
Butyl nitrite, 95%
CAS [544-16-1]
C$_4$H$_9$NO$_2$
FW 103.12
bp 78°C
d 0.882
n$_D^{20}$1.3770
Fp 8°F
VP-FT-IR: **3**, 479A

68.57
31.59
19.65
14.09*

CDCl$_3$
QE-300

CH$_3$CH$_2$CH$_2$CH$_2$ONO

B

Aldrich 32,719-0
Isobutyl nitrite, 95%
CAS [542-56-3]
C$_4$H$_9$NO$_2$
FW 103.12
bp 67°C
d 0.870
n$_D^{20}$1.3730
Fp -6°F
VP-FT-IR: **3**, 479B

74.67
28.36*
19.10*

CDCl$_3$
QE-300

CH$_3$CHCH$_2$ONO
CH$_3$

C

Aldrich 15,049-5
Isoamyl nitrite, 97%
CAS [110-46-3]
C$_5$H$_{11}$NO$_2$
FW 117.15
bp 99°C
d 0.872
n$_D^{20}$1.3860
Fp 50°F
60 MHz: **1**, 351A
FT-IR: **1**, 397A
VP-FT-IR: **3**, 479D

66.83
37.79
25.13
22.44*

CDCl$_3$
QE-300

CH$_3$CHCH$_2$CH$_2$ONO
CH$_3$

ALDRICH

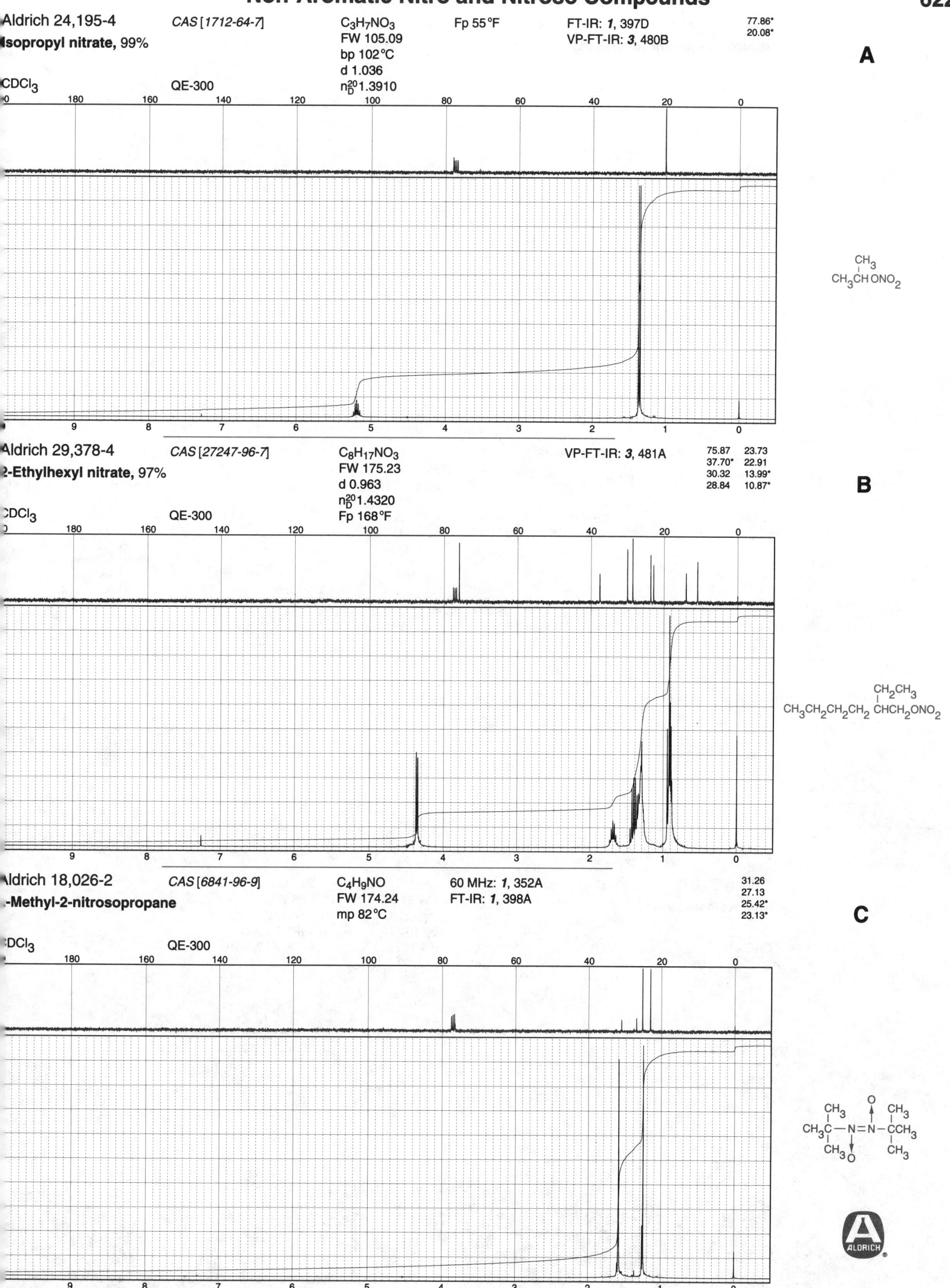

Aldrich 24,195-4

Isopropyl nitrate, 99%

CAS [1712-64-7]

$C_3H_7NO_3$
FW 105.09
bp 102°C
d 1.036
n_D^{20} 1.3910

Fp 55°F

FT-IR: *1*, 397D
VP-FT-IR: *3*, 480B

77.86*
20.08*

A

CDCl₃   QE-300

Aldrich 29,378-4

2-Ethylhexyl nitrate, 97%

CAS [27247-96-7]

$C_8H_{17}NO_3$
FW 175.23
d 0.963
n_D^{20} 1.4320
Fp 168°F

VP-FT-IR: *3*, 481A

75.87   23.73
37.70* 22.91
30.32   13.99*
28.84   10.87*

B

CDCl₃   QE-300

Aldrich 18,026-2

2-Methyl-2-nitrosopropane

CAS [6841-96-9]

C_4H_9NO
FW 174.24
mp 82°C

60 MHz: *1*, 352A
FT-IR: *1*, 398A

31.26
27.13
25.42*
23.13*

C

CDCl₃   QE-300

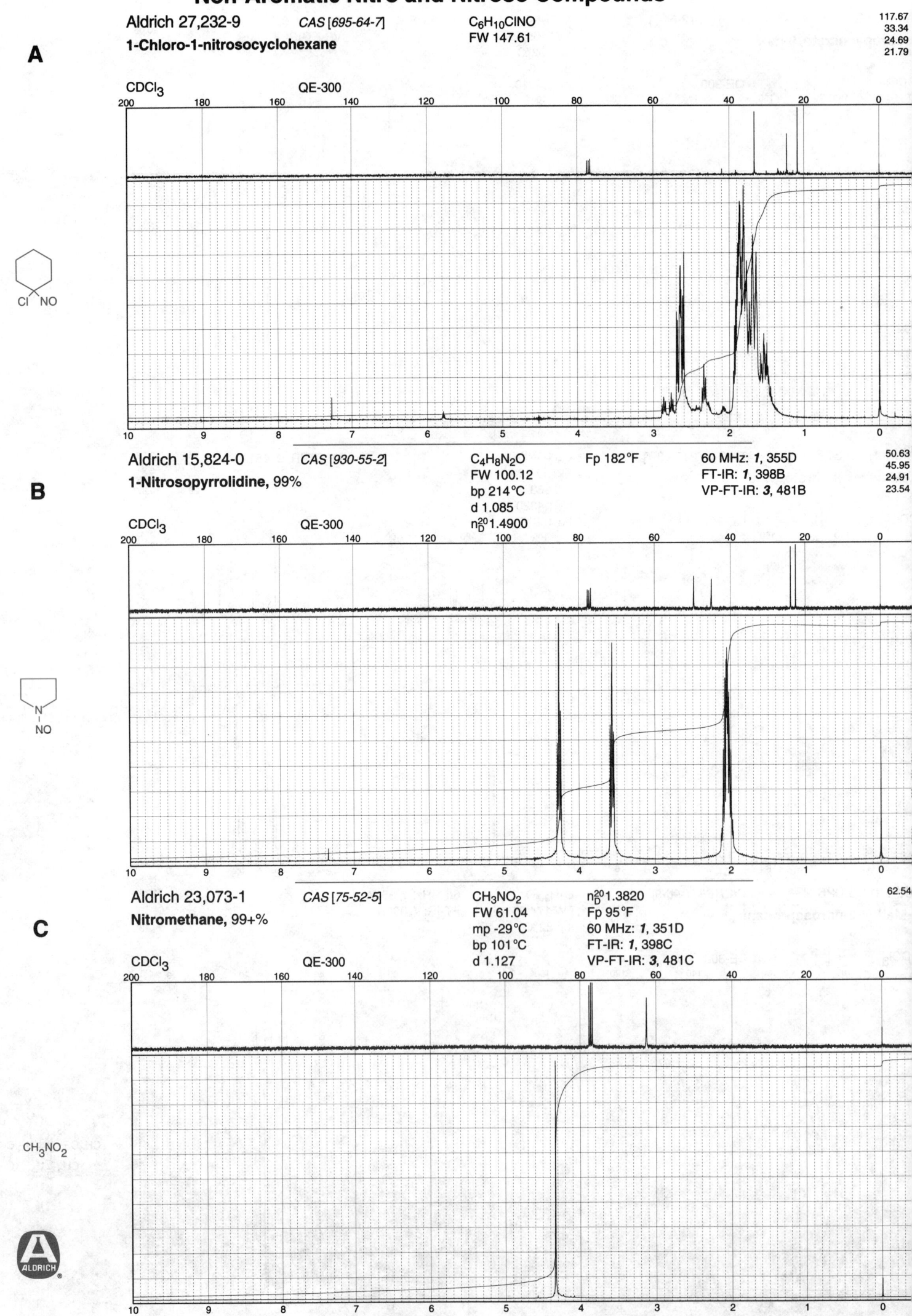

A

Aldrich 27,232-9   *CAS [695-64-7]* $C_6H_{10}ClNO$
FW 147.61

1-Chloro-1-nitrosocyclohexane

117.67
33.34
24.69
21.79

CDCl₃   QE-300

B

Aldrich 15,824-0   *CAS [930-55-2]* $C_4H_8N_2O$ Fp 182°F   60 MHz: *1*, 355D

1-Nitrosopyrrolidine, 99%   FW 100.12   FT-IR: *1*, 398B
bp 214°C   VP-FT-IR: *3*, 481B
d 1.085
$n_D^{20} 1.4900$

50.63
45.95
24.91
23.54

CDCl₃   QE-300

C

Aldrich 23,073-1   *CAS [75-52-5]* CH_3NO_2 $n_D^{20} 1.3820$

Nitromethane, 99+%   FW 61.04   Fp 95°F
mp -29°C   60 MHz: *1*, 351D
bp 101°C   FT-IR: *1*, 398C
d 1.127   VP-FT-IR: *3*, 481C

62.54

CDCl₃   QE-300

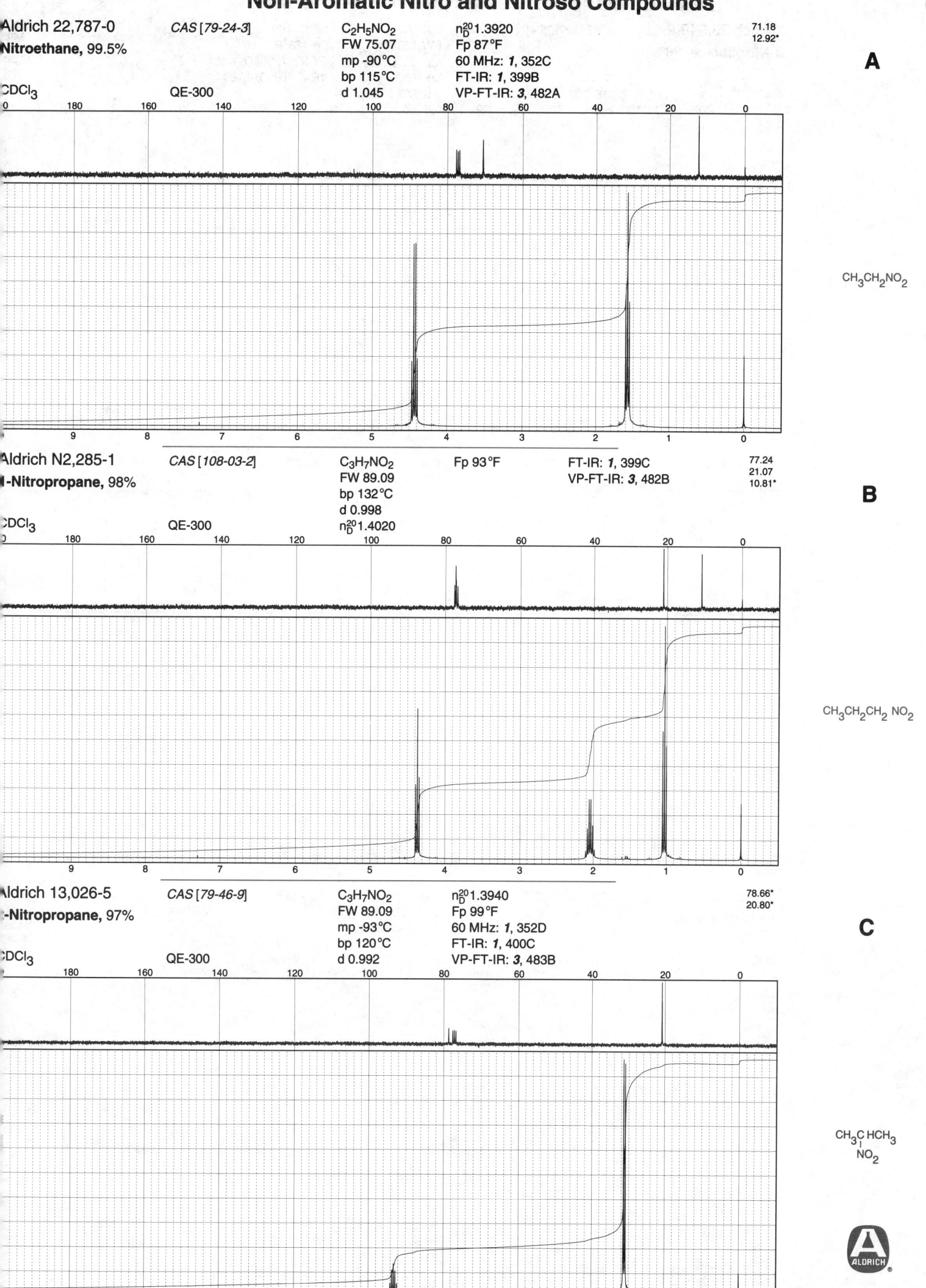

Aldrich 22,787-0
Nitroethane, 99.5%
CDCl₃   QE-300

CAS [79-24-3]

$C_2H_5NO_2$
FW 75.07
mp -90°C
bp 115°C
d 1.045

n_D^{20}1.3920
Fp 87°F
60 MHz: 1, 352C
FT-IR: 1, 399B
VP-FT-IR: 3, 482A

71.18
12.92*

A

$CH_3CH_2NO_2$

Aldrich N2,285-1
-Nitropropane, 98%
CDCl₃   QE-300

CAS [108-03-2]

$C_3H_7NO_2$
FW 89.09
bp 132°C
d 0.998
n_D^{20}1.4020

Fp 93°F

FT-IR: 1, 399C
VP-FT-IR: 3, 482B

77.24
21.07
10.81*

B

$CH_3CH_2CH_2\ NO_2$

Aldrich 13,026-5
-Nitropropane, 97%
CDCl₃   QE-300

CAS [79-46-9]

$C_3H_7NO_2$
FW 89.09
mp -93°C
bp 120°C
d 0.992

n_D^{20}1.3940
Fp 99°F
60 MHz: 1, 352D
FT-IR: 1, 400C
VP-FT-IR: 3, 483B

78.66*
20.80*

C

CH_3CHCH_3
NO_2

ALDRICH

A

Aldrich 25,948-9   CAS [627-05-4]   C4H9NO2   n_D^20 1.4100

1-Nitrobutane, 98%

FW 103.12   Fp 118°F
mp -81°C   FT-IR: *1*, 399D
bp 153°C   VP-FT-IR: *3*, 482C
d 0.973

75.50
29.40
19.61
13.28*

CDCl3   QE-300

$CH_3CH_2CH_2CH_2 NO_2$

B

Aldrich 18,902-2   CAS [594-70-7]   C4H9NO2   Fp 67°F   60 MHz: *1*, 352B

2-Methyl-2-nitropropane, 99%

FW 103.12   FT-IR: *1*, 400D
bp 127°C   VP-FT-IR: *3*, 483C
d 0.950
n_D^20 1.4000

85.07
27.87*

CDCl3   QE-300

$CH_3-\underset{\underset{NO_2}{|}}{\overset{\overset{CH_3}{|}}{C}}-CH_3$

C

Aldrich 25,950-0   CAS [628-05-7]   C5H11NO2   Fp 139°F   FT-IR: *1*, 400A

1-Nitropentane, 97%

FW 117.15   VP-FT-IR: *3*, 482D
bp 76°C (23 mm)
d 0.952
n_D^20 1.4170

76.61
29.25
28.01
22.86
14.55*

CDCl3   QE-300

$CH_3CH_2CH_2CH_2CH_2NO_2$

Aldrich 25,949-7   CAS [646-14-0] $C_6H_{13}NO_2$ Fp 164°F   FT-IR: *1*, 400B

Nitrohexane, 98%

FW 131.18   VP-FT-IR: *3*, 483A

bp 92°C (24 mm)

d 0.940

CDCl₃   QE-300 n_D^{20} 1.4230

75.77
31.05
27.44
25.98
22.35
13.84*

A

$CH_3(CH_2)_4CH_2NO_2$

Aldrich 26,971-9   CAS [2562-38-1] $C_5H_9NO_2$ Fp 153°F   FT-IR: *1*, 401A

Nitrocyclopentane, 99%

FW 115.13   VP-FT-IR: *3*, 483D

bp 180°C

d 1.086

CDCl₃   QE-300 n_D^{20} 1.4540

86.84*
32.50
24.72

B

Aldrich N1,660-6   CAS [1122-60-7] $C_6H_{11}NO_2$ n_D^{20} 1.4620

Nitrocyclohexane, 97%

FW 129.16   Fp 166°F

mp -34°C   60 MHz: *1*, 355B

bp 206°C   FT-IR: *1*, 401B

CDCl₃   QE-300   d 1.061   VP-FT-IR: *3*, 484A

84.63*
30.98
24.84
24.15

C

A

Aldrich 21,953-3   *CAS [2562-37-0]* $C_6H_9NO_2$ Fp 175°F   60 MHz: *1*, 355C

1-Nitro-1-cyclohexene, 99%   FW 127.15   FT-IR: *1*, 401C

bp 67°C (1 mm)   VP-FT-IR: *3*, 484B

d 1.127

CDCl₃   QE-300 n_D^{20} 1.5050

149.87
133.98
24.85
24.04
21.93
20.81

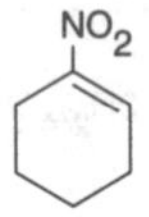

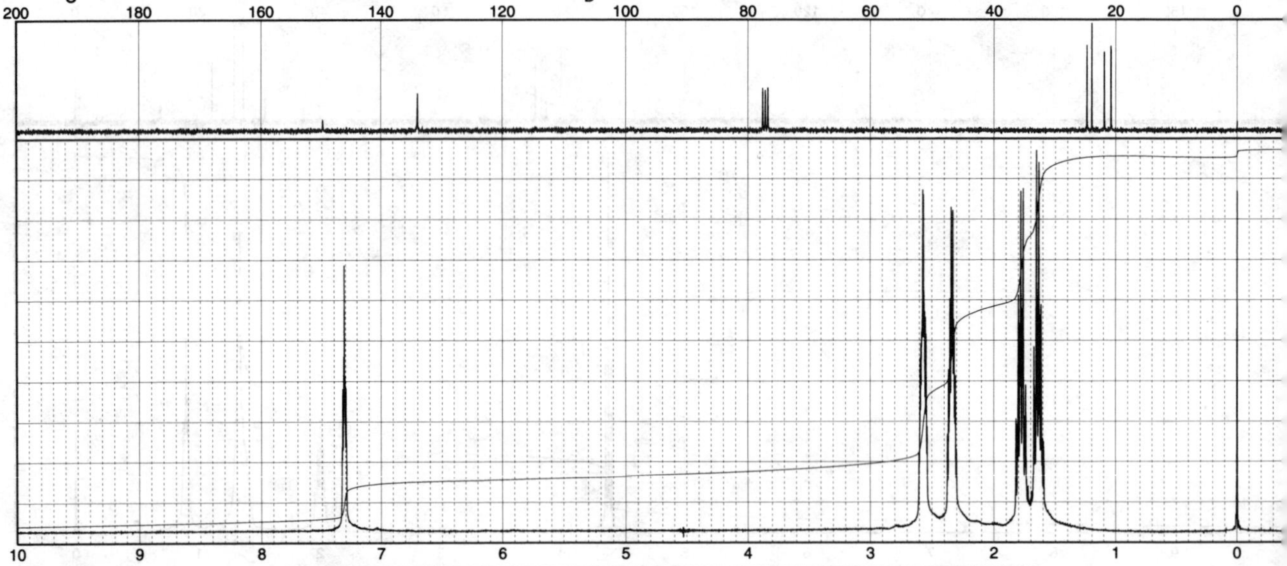

B

Aldrich 25,585-8   *CAS [563-70-2]* CH_2BrNO_2 Fp >230°F   FT-IR: *1*, 399A

Bromonitromethane, tech., 90%   FW 139.94   VP-FT-IR: *3*, 484C

bp 147°C (750 mm)

d 2.007

CDCl₃   QE-300 n_D^{20} 1.4960

60.82

$BrCH_2NO_2$

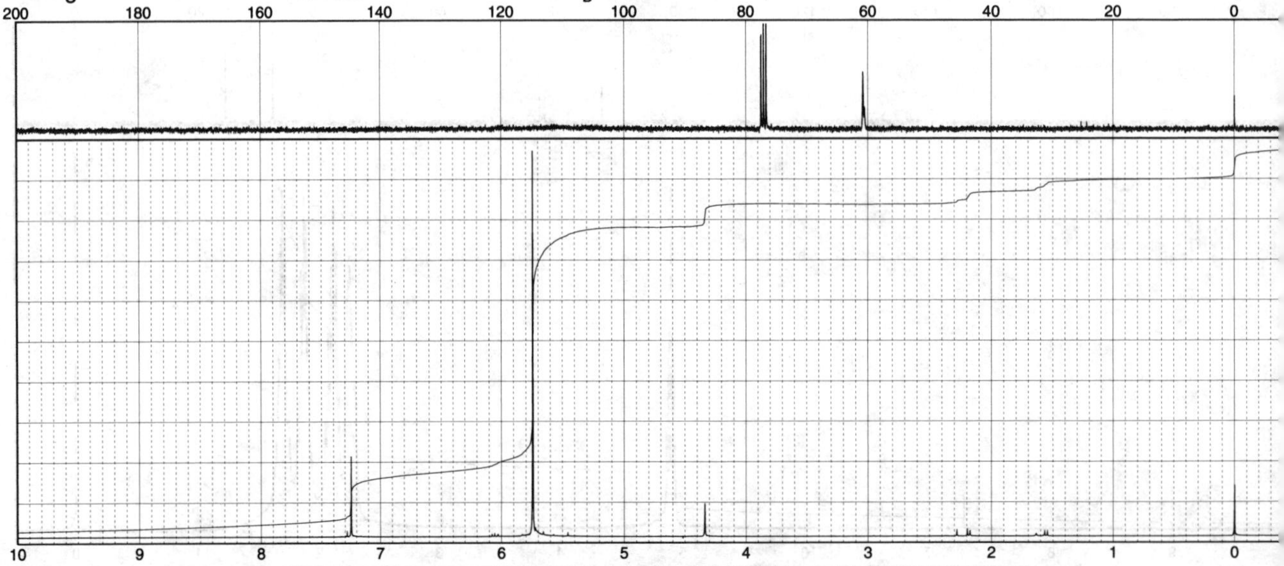

C

Aldrich B7,430-5   *CAS [5447-97-2]* $C_3H_6BrNO_2$ Fp 142°F   FT-IR: *1*, 403B

2-Bromo-2-nitropropane, 97%   FW 168.00   VP-FT-IR: *3*, 484D

bp 48°C (15 mm)

d 1.582

CDCl₃   QE-300 n_D^{20} 1.4640

88.92
32.34

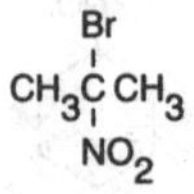

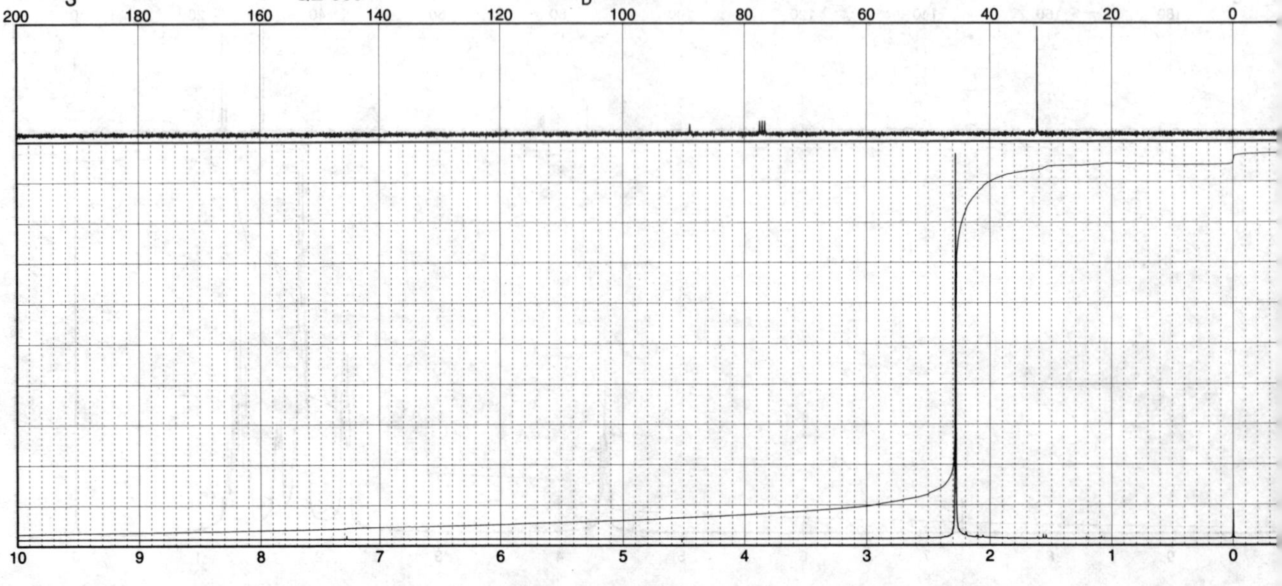

Br
|
CH₃-C-CH₃
|
NO₂

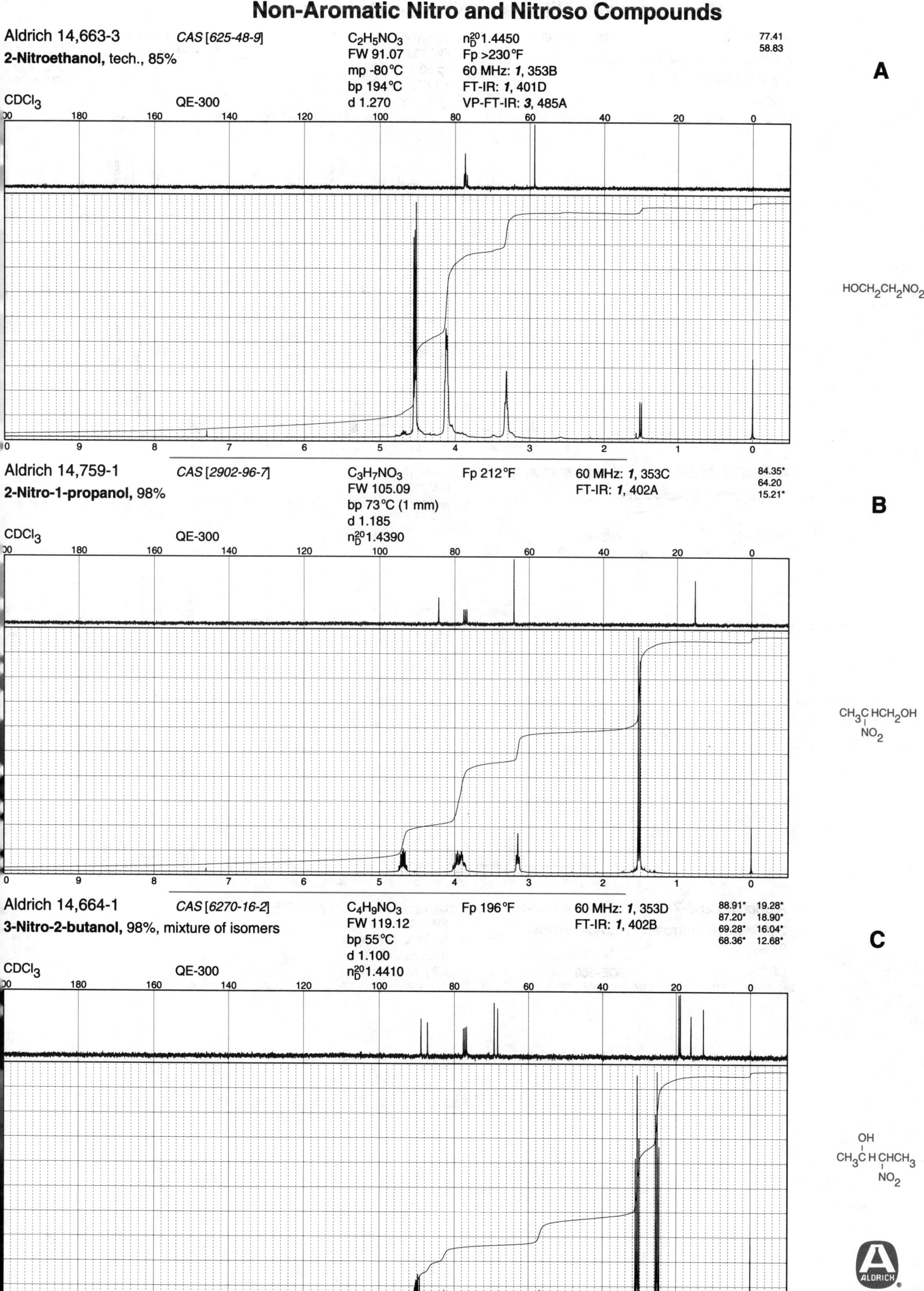

A

Aldrich 14,663-3   *CAS [625-48-9]* $C_2H_5NO_3$ $n_D^{20}1.4450$ 77.41
2-Nitroethanol, tech., 85%   FW 91.07   Fp >230°F   58.83
mp -80°C   60 MHz: *1*, 353B
bp 194°C   FT-IR: *1*, 401D
CDCl₃   QE-300   d 1.270   VP-FT-IR: *3*, 485A

A

$HOCH_2CH_2NO_2$

Aldrich 14,759-1   *CAS [2902-96-7]* $C_3H_7NO_3$ Fp 212°F   60 MHz: *1*, 353C   84.35*
2-Nitro-1-propanol, 98%   FW 105.09   FT-IR: *1*, 402A   64.20
bp 73°C (1 mm)   15.21*
d 1.185
CDCl₃   QE-300 $n_D^{20}1.4390$

B

CH_3CHCH_2OH
NO_2

Aldrich 14,664-1   *CAS [6270-16-2]* $C_4H_9NO_3$ Fp 196°F   60 MHz: *1*, 353D   88.91*   19.28*
3-Nitro-2-butanol, 98%, mixture of isomers   FW 119.12   FT-IR: *1*, 402B   87.20*   18.90*
bp 55°C   69.28*   16.04*
d 1.100   68.36*   12.68*
CDCl₃   QE-300 $n_D^{20}1.4410$

C

OH
$CH_3CHCHCH_3$
NO_2

Non-Aromatic Nitro and Nitroso Compounds

629

A

Aldrich 15,970-0   CAS [5447-99-4] $C_5H_{11}NO_3$ Fp 195°F   60 MHz: *1*, 354A   95.72* 21.97
3-Nitro-2-pentanol, 97%, mixture of   FW 133.15   FT-IR: *1*, 402C   94.69* 19.75*
(±)-*threo* and (±)-*erythro*   bp 60°C   68.34* 19.14*
    d 1.075   68.26* 10.46*
CDCl₃   QE-300 n_D^{20}1.4430   23.80 10.10*

Aldrich 36,153-4   CAS [116747-80-9] $C_{10}H_{21}NO_5$ 94.41
Nitromethanetrispropanol   FW 235.28   60.23
    mp 62°C   31.73
    26.56

B

DMSO-d_6 QE-300

Aldrich 36,286-7   CAS [57620-56-1] $C_8H_{15}NO_4$ Fp >230°F   103.52*
2-(3-Methyl-3-nitrobutyl)-1,3-dioxolane,   FW 189.21   87.75
97%   bp 99°C (1 mm)   65.01
    d 1.130   34.74
CDCl₃   QE-300 n_D^{20}1.4540   28.79
    25.81*

C

Aldrich 30,237-6   *CAS [81104-52-1]*

1,3-Dimorpholino-2-nitropropane, 98%

$C_{11}H_{21}N_3O_4$
FW 259.31
mp 122°C

82.56*
66.22
58.71
53.25

A

CDCl$_3$ + DMSO-d_6 QE-300

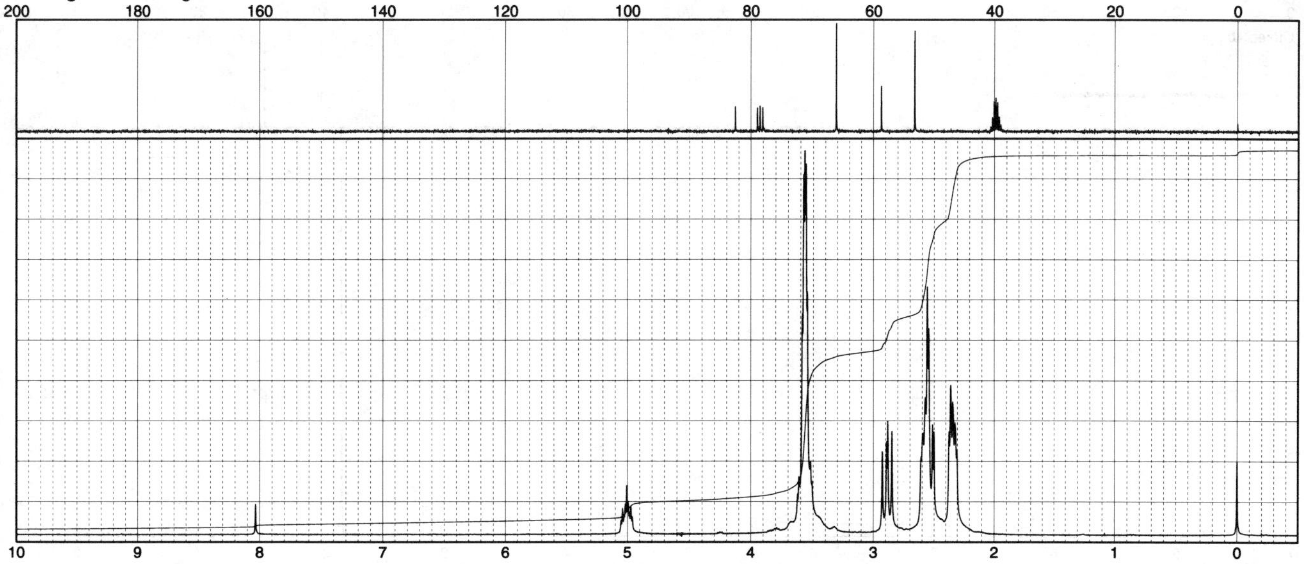

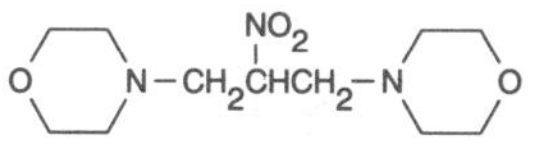

Aldrich 15,634-5   *CAS [3964-18-9]*

2,3-Dimethyl-2,3-dinitrobutane, 98%

$C_6H_{12}N_2O_4$
FW 176.17
mp 215°C d.

60 MHz: *1*, 353A
FT-IR: *1*, 402D

91.54*
23.10*

B

CDCl$_3$ QE-300

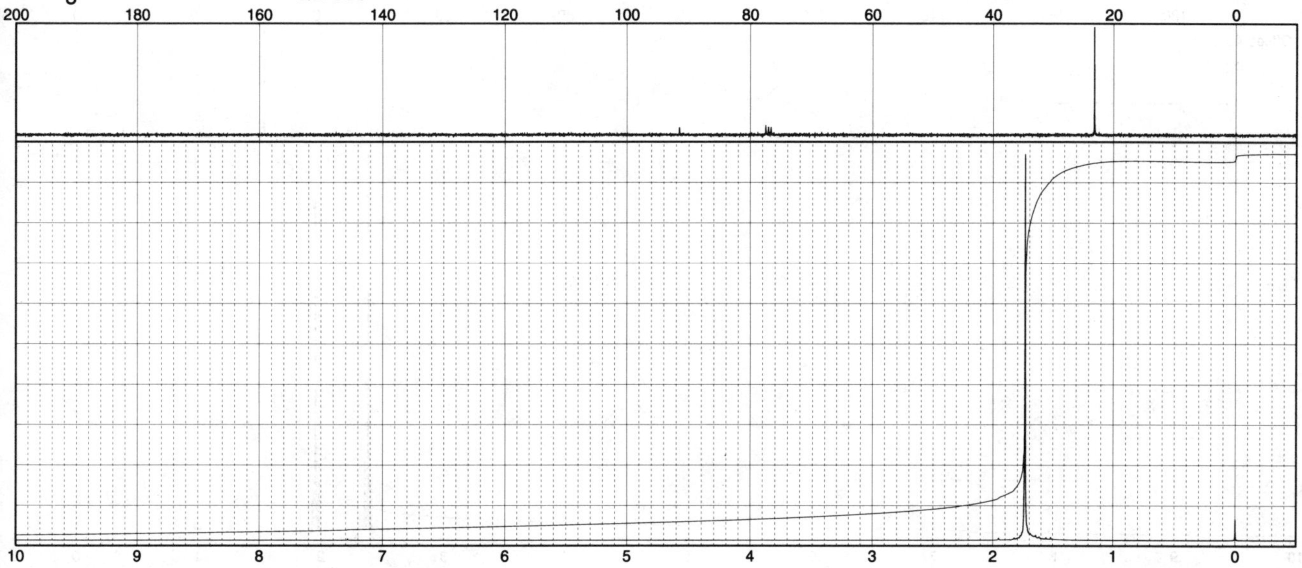

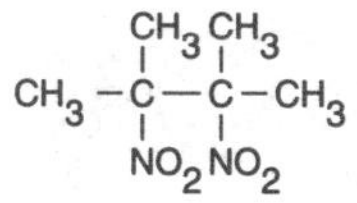

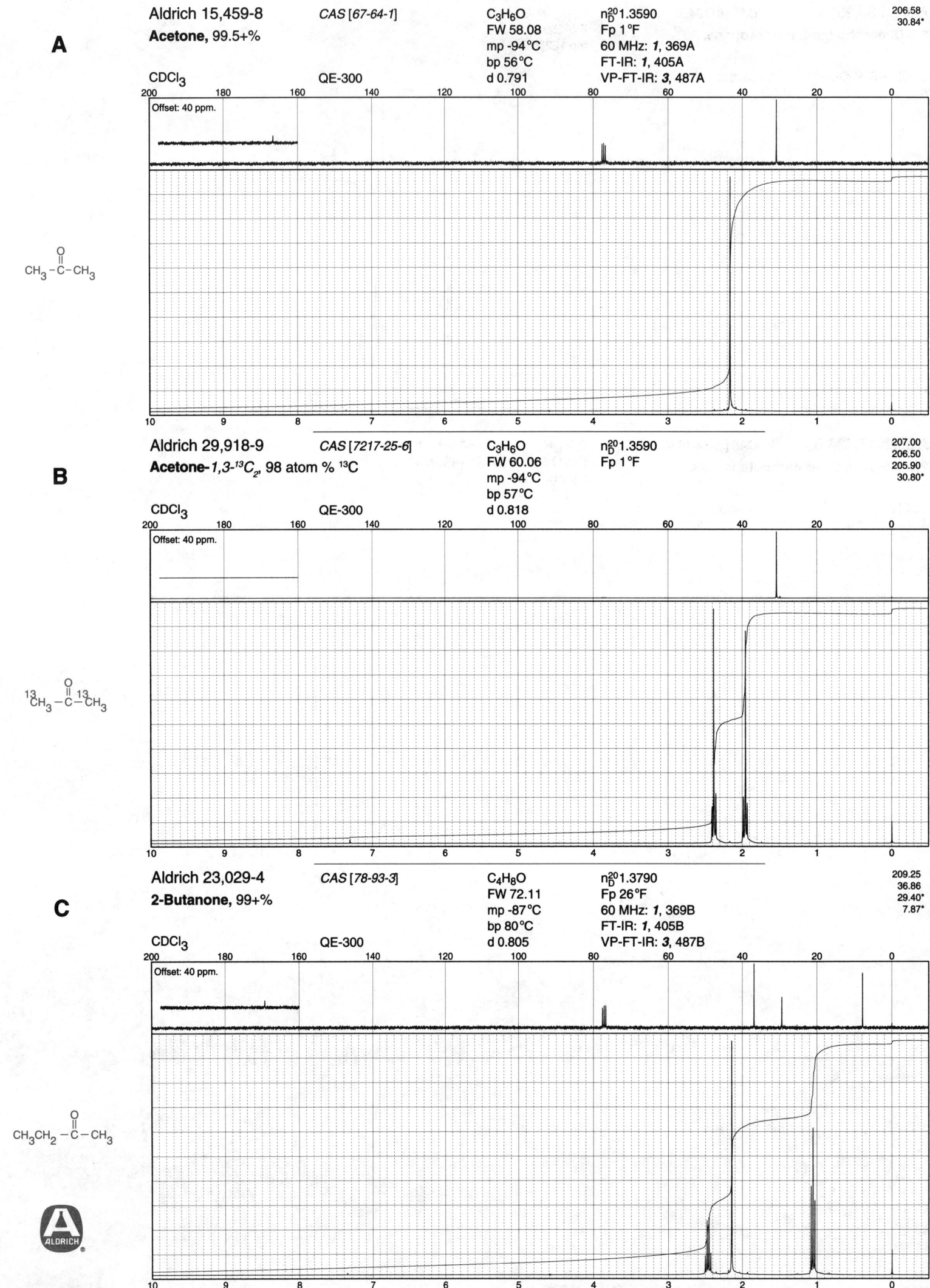

A

Aldrich 15,459-8
Acetone, 99.5+%

CAS [67-64-1]

C_3H_6O
FW 58.08
mp -94°C
bp 56°C
d 0.791

n_D^{20} 1.3590
Fp 1°F
60 MHz: *1*, 369A
FT-IR: *1*, 405A
VP-FT-IR: *3*, 487A

206.58
30.84*

CDCl₃

QE-300

Offset: 40 ppm.

$CH_3-\overset{O}{\underset{}{C}}-CH_3$

B

Aldrich 29,918-9
Acetone-*1,3-¹³C₂,* 98 atom % ¹³C

CAS [7217-25-6]

C_3H_6O
FW 60.06
mp -94°C
bp 57°C
d 0.818

n_D^{20} 1.3590
Fp 1°F

207.00
206.50
205.90
30.80*

CDCl₃

QE-300

Offset: 40 ppm.

$^{13}CH_3-\overset{O}{\underset{}{C}}-^{13}CH_3$

C

Aldrich 23,029-4
2-Butanone, 99+%

CAS [78-93-3]

C_4H_8O
FW 72.11
mp -87°C
bp 80°C
d 0.805

n_D^{20} 1.3790
Fp 26°F
60 MHz: *1*, 369B
FT-IR: *1*, 405B
VP-FT-IR: *3*, 487B

209.25
36.86
29.40*
7.87*

CDCl₃

QE-300

Offset: 40 ppm.

$CH_3CH_2-\overset{O}{\underset{}{C}}-CH_3$

ALDRICH

A

Irich 23,861-9   CAS [563-80-4]   C$_5$H$_{10}$O   n$_D^{20}$1.3880   212.47
Methyl-2-butanone, 99%   FW 86.13   Fp 43°F   41.65*
mp -92°C   60 MHz: **1**, 369D   27.43*
Cl$_3$ QE-300   bp 95°C   FT-IR: **1**, 406B   18.15*
d 0.805   VP-FT-IR: **3**, 487D

et: 40 ppm.

CH$_3$ O
CH$_3$CH — C — CH$_3$

B

Irich P4560-5   CAS [75-97-8]   C$_6$H$_{12}$O   Fp 75°F   60 MHz: **1**, 370A   214.21
acolone, 98%   FW 100.16   FT-IR: **1**, 405D   44.30
bp 106°C   VP-FT-IR: **3**, 487C   26.37*
Cl$_3$ QE-300   d 0.801   24.69*
n$_D^{20}$1.3960

et: 40 ppm.

CH$_3$ O
CH$_3$ — C — C — CH$_3$
CH$_3$

C

rich P810-6   CAS [107-87-9]   C$_5$H$_{10}$O   n$_D^{20}$1.3900   208.85
entanone, 97%   FW 86.13   Fp 45°F   45.68
mp -78°C   60 MHz: **1**, 369C   29.79*
Cl$_3$ QE-300   bp 101°C   FT-IR: **1**, 406A   17.35
d 0.812   VP-FT-IR: **3**, 488A   13.68*

et: 40 ppm.

O
CH$_3$CH$_2$CH$_2$ — C — CH$_3$

ALDRICH

A

Aldrich 24,304-3

3-Pentanone, 96%

CAS [96-22-0]

$C_5H_{10}O$
FW 86.13
mp -40°C
bp 102°C
d 0.814

n_D^{20}1.3920
Fp 55°F
60 MHz: *1*, 370B
FT-IR: *1*, 406C
VP-FT-IR: *3*, 488B

211
35.
7

CDCl$_3$

QE-300

Offset: 40 ppm.

$CH_3CH_2-\overset{\overset{\displaystyle O}{\|}}{C}-CH_2CH_3$

B

Aldrich M6,700-1

3-Methyl-2-pentanone, 99%

CAS [565-61-7]

$C_6H_{12}O$
FW 100.16
bp 118°C (758 mm)
d 0.815
n_D^{20}1.4000

Fp 54°F

60 MHz: *1*, 370C
FT-IR: *1*, 406D
VP-FT-IR: *3*, 488C

212
48
27
25
15
11

CDCl$_3$

QE-300

Offset: 40 ppm.

$CH_3CH_2\underset{\underset{\displaystyle CH_3}{|}}{CH}-\overset{\overset{\displaystyle O}{\|}}{C}-CH_3$

C

Aldrich 24,289-6

4-Methyl-2-pentanone, 99+%

CAS [108-10-1]

$C_6H_{12}O$
FW 100.16
mp -80°C
bp 118°C
d 0.800

n_D^{20}1.3960
Fp 56°F
60 MHz: *1*, 370D
FT-IR: *1*, 407A
VP-FT-IR: *3*, 489A

208
52
3
24
22

CDCl$_3$

QE-300

Offset: 40 ppm.

$CH_3\underset{\underset{\displaystyle CH_3}{|}}{\overset{\overset{\displaystyle CH_3}{|}}{CH}}CH_2-\overset{\overset{\displaystyle O}{\|}}{C}-CH_3$

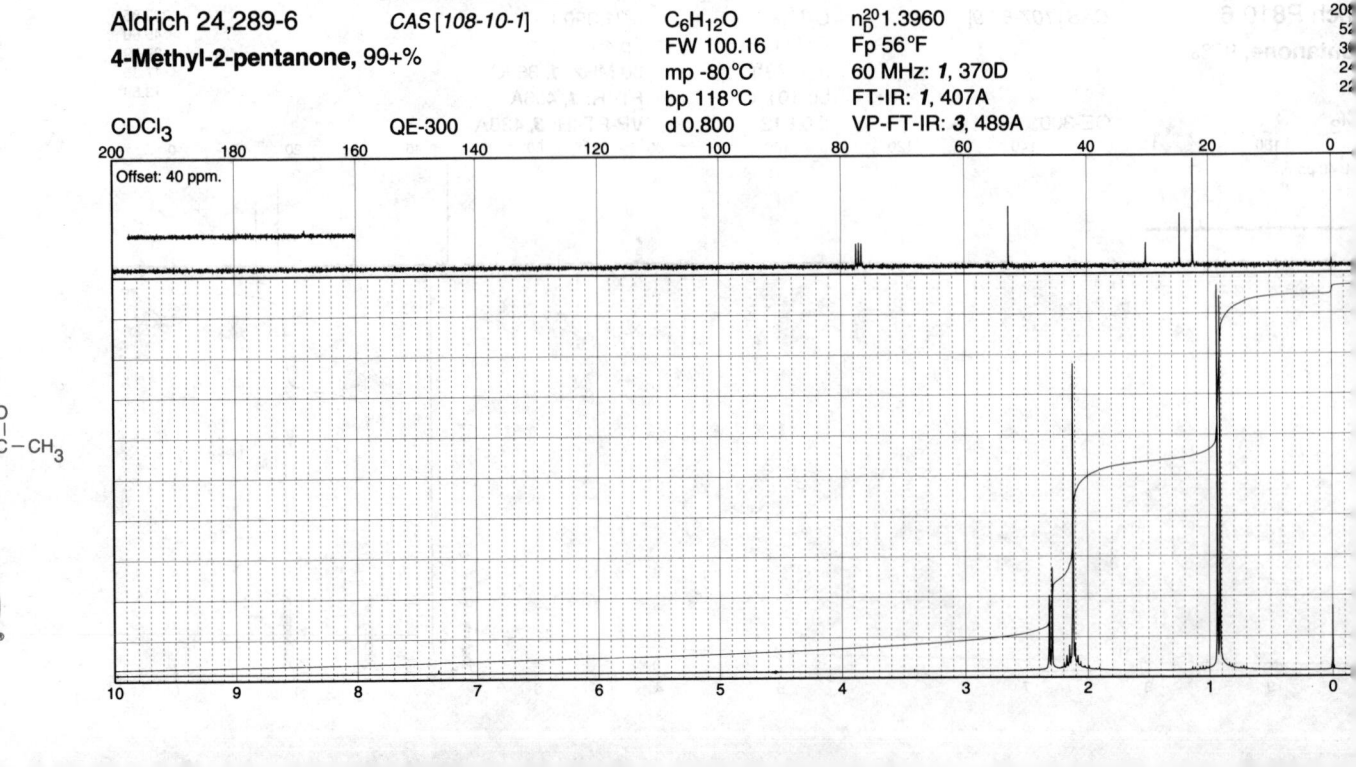

Aldrich 10,870-7　　*CAS [565-69-5]*

-Methyl-3-pentanone, 97%

A

$C_6H_{12}O$
FW 100.16
bp 113°C
d 0.811
n_D^{20} 1.3970

Fp 57°F

FT-IR: *1*, 407B
VP-FT-IR: *3*, 489B

215.19
40.58*
33.40
18.38*
7.89*

DCl₃　　QE-300

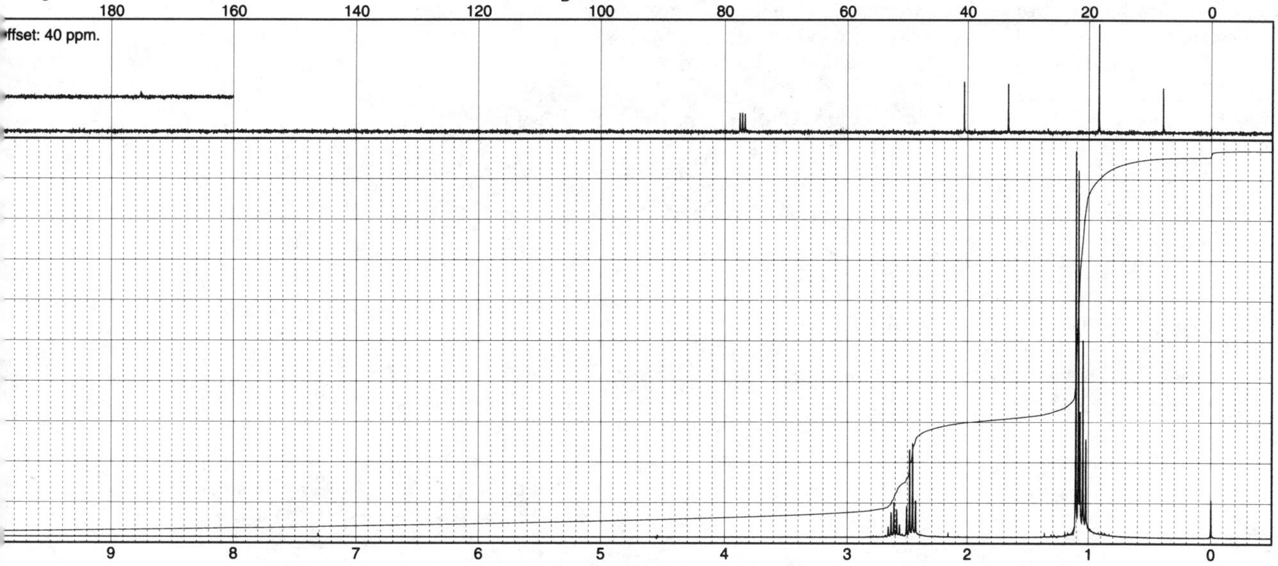

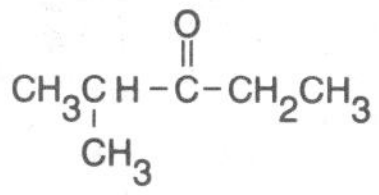

ffset: 40 ppm.

180　160　140　120　100　80　60　40　20　0

9　8　7　6　5　4　3　2　1　0

$$CH_3CH-C-CH_2CH_3$$
with O above and CH₃ below

Aldrich 13,687-5　　*CAS [590-50-1]*

4-Dimethyl-2-pentanone, 99%

B

$C_7H_{14}O$
FW 114.19
bp 128°C
d 0.809
n_D^{20} 1.4040

Fp 66°F

60 MHz: *1*, 371B
FT-IR: *1*, 407D
VP-FT-IR: *3*, 489D

208.38
56.00
32.30*
30.91
29.77*

DCl₃　　QE-300

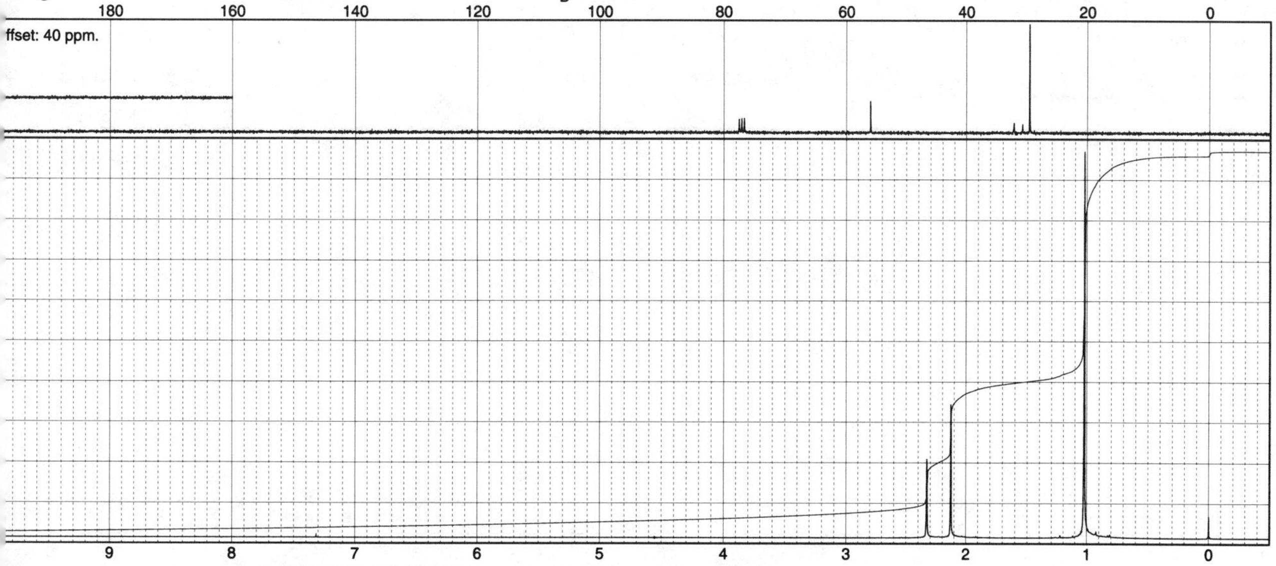

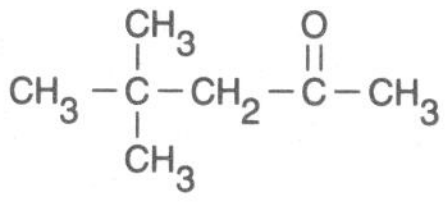

ffset: 40 ppm.

180　160　140　120　100　80　60　40　20　0

9　8　7　6　5　4　3　2　1　0

$$CH_3-C-CH_2-C-CH_3$$
with CH₃ and CH₃ on left carbon, O on right carbon

Aldrich 13,686-7　　*CAS [565-80-0]*

4-Dimethyl-3-pentanone, 98%

C

$C_7H_{14}O$
FW 114.19
mp -80°C
bp 124°C
d 0.806

n_D^{20} 1.4000
Fp 60°F
60 MHz: *1*, 371A
FT-IR: *1*, 407C
VP-FT-IR: *3*, 489C

218.35
38.84*
18.54*

DCl₃　　QE-300

ffset: 40 ppm.

180　160　140　120　100　80　60　40　20　0

9　8　7　6　5　4　3　2　1　0

$$CH-C-CH$$
with CH₃/CH₃ on left, O above center, CH₃/CH₃ on right

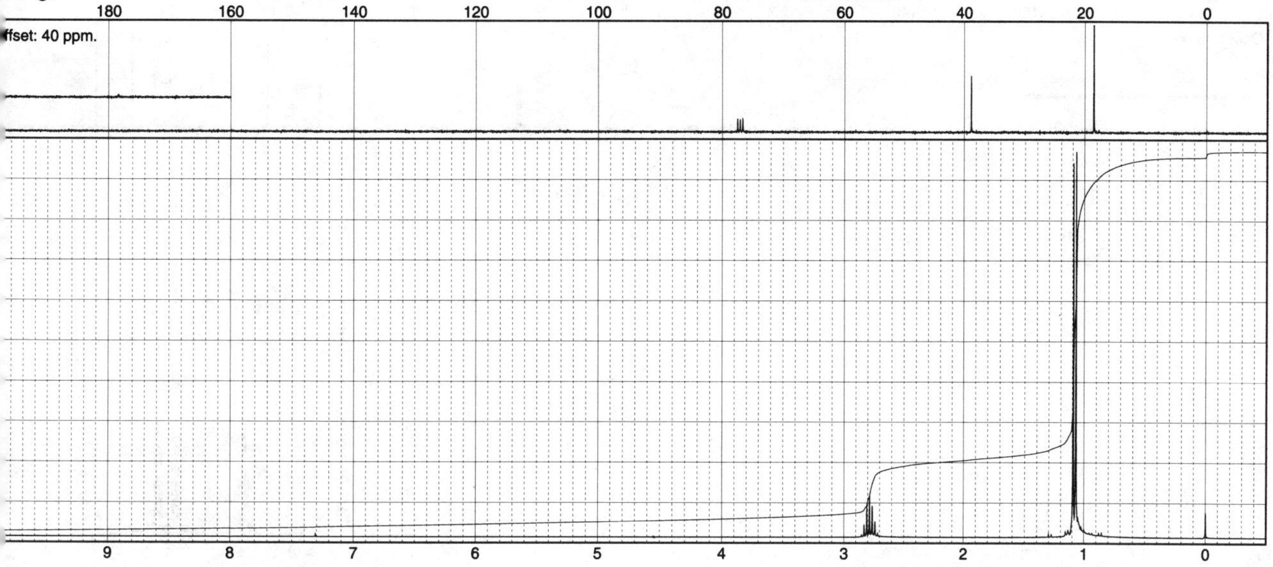

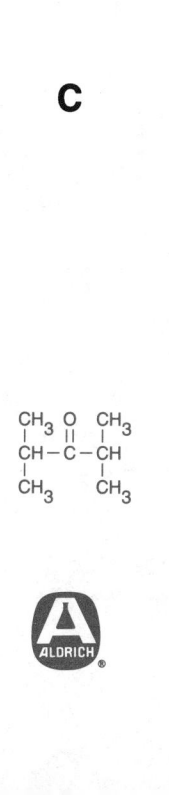

Non-Aromatic Ketones

635

A	

Aldrich 28,261-8   CAS [815-24-7] $C_9H_{18}O$ Fp 91°F   VP-FT-IR: **3**, 490A

2,2,4,4-Tetramethyl-3-pentanone, 99%

FW 142.24
bp 153°C
d 0.824
n_D^{20} 1.4190

218.75
45.67
28.46

CDCl₃   QE-300

Offset: 40 ppm.

Aldrich 10,300-4   CAS [591-78-6] $C_6H_{12}O$ n_D^{20} 1.4000

B

2-Hexanone, 99+%

FW 100.16
mp -57°C
bp 127°C
d 0.812

Fp 95°F
60 MHz: **1**, 371C
FT-IR: **1**, 408A
VP-FT-IR: **3**, 490B

208.88
43.50
29.75
26.05
22.34
13.81

CDCl₃   QE-300

Offset: 40 ppm.

Aldrich 10,302-0   CAS [589-38-8] $C_6H_{12}O$ Fp 95°F   60 MHz: **1**, 371D

C

3-Hexanone, 98%

FW 100.16
bp 123°C
d 0.815
n_D^{20} 1.4000

FT-IR: **1**, 408B
VP-FT-IR: **3**, 490C

211.55
44.34
35.89
17.43
13.79
7.85

CDCl₃   QE-300

Offset: 40 ppm.

Aldrich 11,025-6 *CAS [110-12-3]* $C_7H_{14}O$ Fp 106°F 60 MHz: *1*, 372A

5-Methyl-2-hexanone, 99% FW 114.19 FT-IR: *1*, 408C VP-FT-IR: *3*, 490D

bp 145°C
d 0.888
n_D^{20}1.4070

CDCl₃ QE-300

208.79
41.82
32.83
29.68*
27.76*
22.32*

A

Offset: 40 ppm.

$$CH_3CHCH_2CH_2 - \overset{\overset{\displaystyle O}{\|}}{C} - CH_3$$
(with CH₃ branch)

Aldrich 10,794-8 *CAS [7379-12-6]* $C_7H_{14}O$ Fp 75°F VP-FT-IR: *3*, 491A

2-Methyl-3-hexanone, 99% FW 114.19
bp 132°C
d 0.825
n_D^{20}1.4060

CDCl₃ QE-300

214.77
42.26
40.80*
18.23*
17.24
13.81*

B

Offset: 40 ppm.

$$CH_3CH_2CH_2 - \overset{\overset{\displaystyle O}{\|}}{C} - CHCH_3$$
(with CH₃ branch)

Aldrich 12,336-6 *CAS [110-43-0]* $C_7H_{14}O$ n_D^{20}1.4080

2-Heptanone, 98% FW 114.19 Fp 117°F

mp -35°C 60 MHz: *1*, 372B
bp 150°C FT-IR: *1*, 408D
d 0.820 VP-FT-IR: *3*, 492B

CDCl₃ QE-300

208.77
43.78
31.45
29.71*
23.66
22.46
13.86*

C

Offset: 40 ppm.

$$CH_3(CH_2)_3CH_2 - \overset{\overset{\displaystyle O}{\|}}{C} - CH_3$$

ALDRICH

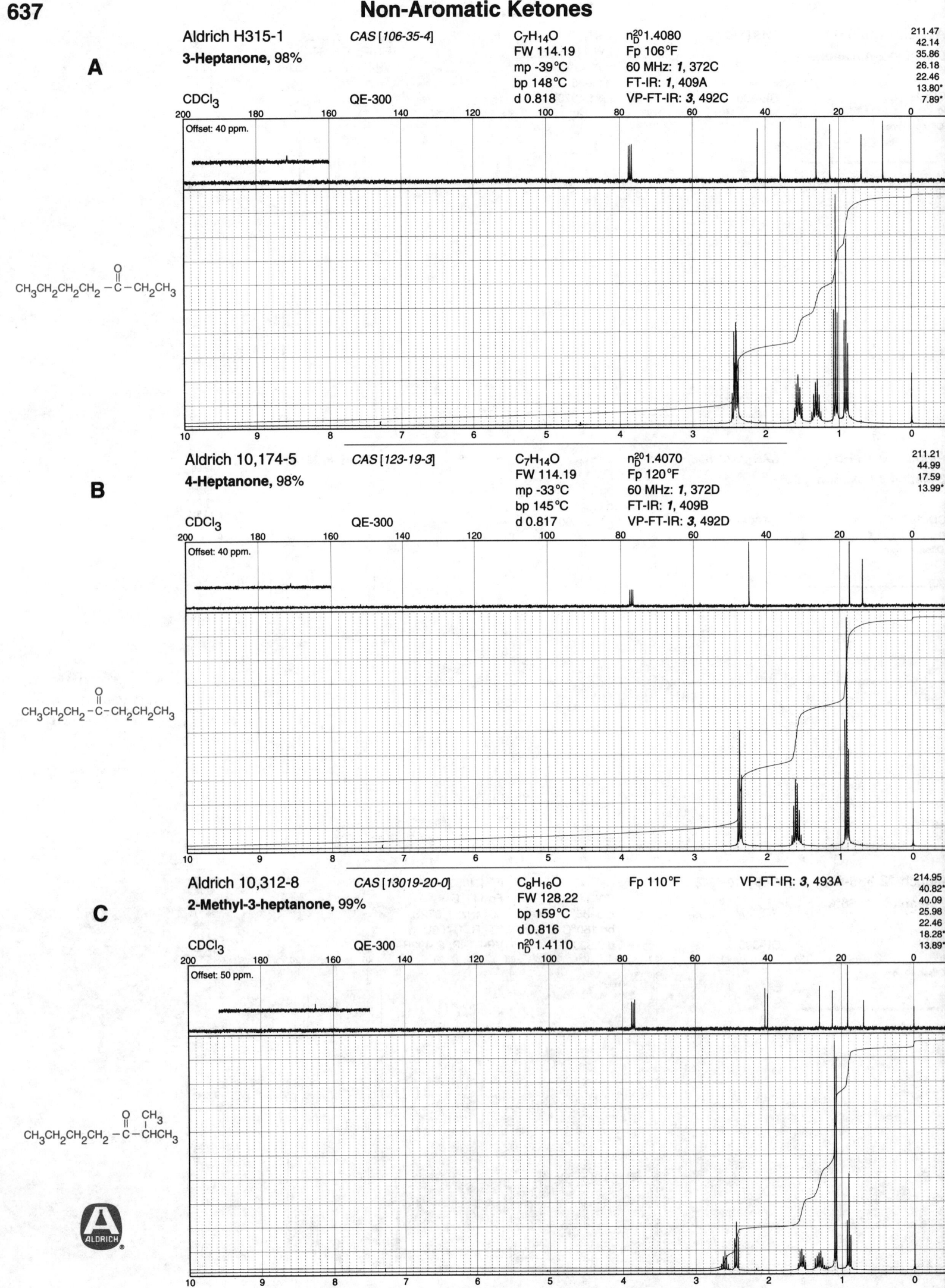

A

Aldrich H315-1
3-Heptanone, 98%

CAS [106-35-4]

$C_7H_{14}O$
FW 114.19
mp -39°C
bp 148°C

n_D^{20} 1.4080
Fp 106°F
60 MHz: *1*, 372C
FT-IR: *1*, 409A
VP-FT-IR: *3*, 492C

211.47
42.14
35.86
26.18
22.46
13.80*
7.89*

CDCl₃   QE-300   d 0.818

Offset: 40 ppm.

$CH_3CH_2CH_2CH_2 - \overset{O}{\underset{||}{C}} - CH_2CH_3$

B

Aldrich 10,174-5
4-Heptanone, 98%

CAS [123-19-3]

$C_7H_{14}O$
FW 114.19
mp -33°C
bp 145°C

n_D^{20} 1.4070
Fp 120°F
60 MHz: *1*, 372D
FT-IR: *1*, 409B
VP-FT-IR: *3*, 492D

211.21
44.99
17.59
13.99*

CDCl₃   QE-300

Offset: 40 ppm.

$CH_3CH_2CH_2 - \overset{O}{\underset{||}{C}} - CH_2CH_2CH_3$

C

Aldrich 10,312-8
2-Methyl-3-heptanone, 99%

CAS [13019-20-0]

$C_8H_{16}O$
FW 128.22
bp 159°C
d 0.816
n_D^{20} 1.4110

Fp 110°F   VP-FT-IR: *3*, 493A

214.95
40.82*
40.09
25.98
22.46
18.28*
13.89*

CDCl₃   QE-300

Offset: 50 ppm.

$CH_3CH_2CH_2CH_2 - \overset{O}{\underset{||}{C}} - \overset{CH_3}{\underset{|}{C}HCH_3}$

Non-Aromatic Ketones

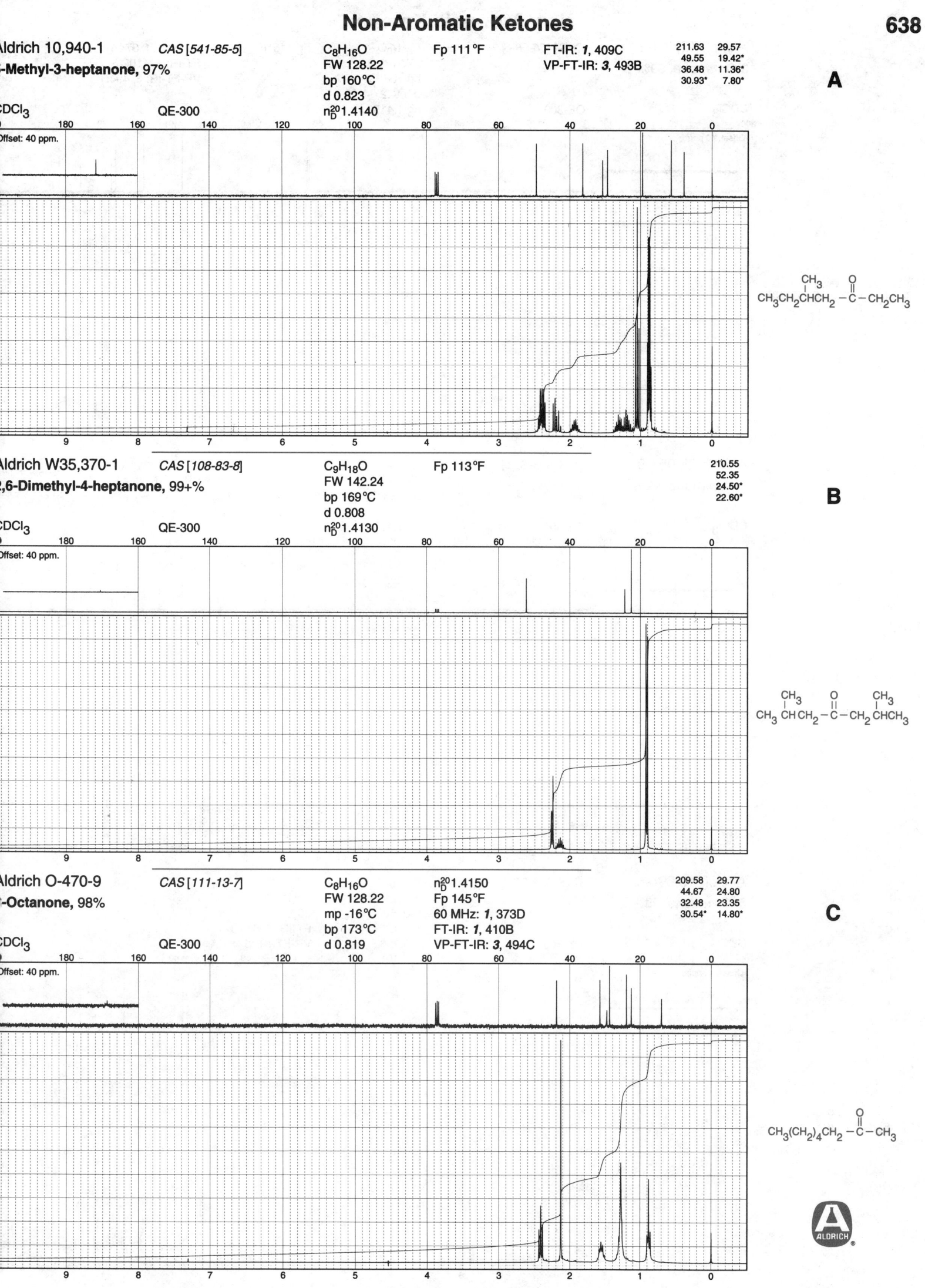

A

Aldrich 10,940-1   CAS [541-85-5]
5-Methyl-3-heptanone, 97%

$C_8H_{16}O$
FW 128.22
bp 160°C
d 0.823
n_D^{20}1.4140

Fp 111°F

FT-IR: *1*, 409C
VP-FT-IR: *3*, 493B

CDCl₃   QE-300

Offset: 40 ppm.

211.63	29.57
49.55	19.42*
36.48	11.36*
30.93*	7.80*

B

Aldrich W35,370-1   CAS [108-83-8]
2,6-Dimethyl-4-heptanone, 99+%

$C_9H_{18}O$
FW 142.24
bp 169°C
d 0.808
n_D^{20}1.4130

Fp 113°F

CDCl₃   QE-300

Offset: 40 ppm.

210.55
52.35
24.50*
22.60*

C

Aldrich O-470-9   CAS [111-13-7]
2-Octanone, 98%

$C_8H_{16}O$
FW 128.22
mp -16°C
bp 173°C
d 0.819

n_D^{20}1.4150
Fp 145°F
60 MHz: *1*, 373D
FT-IR: *1*, 410B
VP-FT-IR: *3*, 494C

CDCl₃   QE-300

Offset: 40 ppm.

209.58	29.77
44.67	24.80
32.48	23.35
30.54*	14.80*

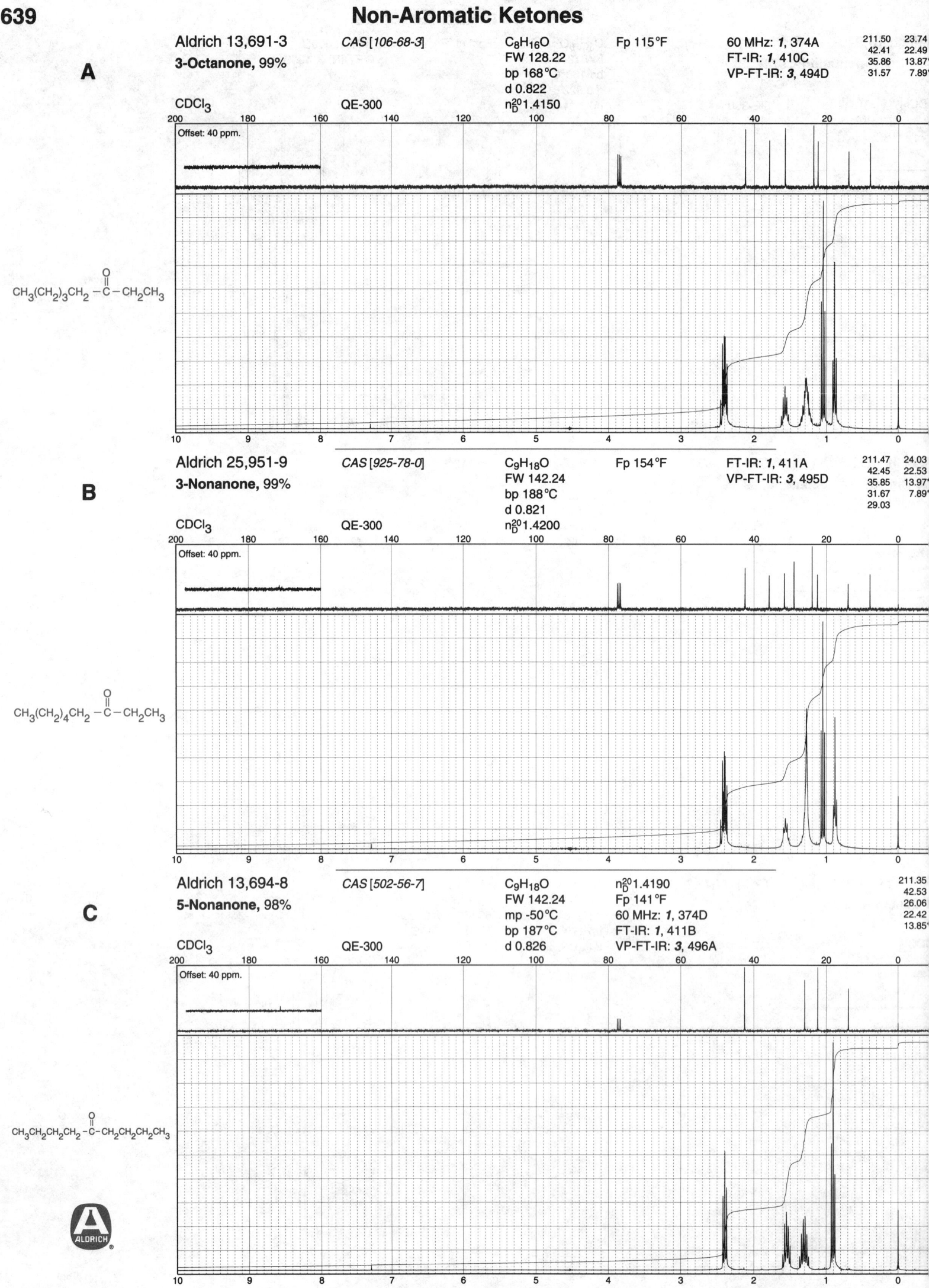

A

Aldrich 13,691-3
3-Octanone, 99%

CAS [106-68-3]

C$_8$H$_{16}$O
FW 128.22
bp 168°C
d 0.822
n$_D^{20}$1.4150

Fp 115°F

60 MHz: **1**, 374A
FT-IR: **1**, 410C
VP-FT-IR: **3**, 494D

211.50   23.74
42.41   22.49
35.86   13.87
31.57   7.89

CDCl$_3$ QE-300

CH$_3$(CH$_2$)$_3$CH$_2$ —C—CH$_2$CH$_3$

B

Aldrich 25,951-9
3-Nonanone, 99%

CAS [925-78-0]

C$_9$H$_{18}$O
FW 142.24
bp 188°C
d 0.821
n$_D^{20}$1.4200

Fp 154°F

FT-IR: **1**, 411A
VP-FT-IR: **3**, 495D

211.47   24.03
42.45   22.53
35.85   13.97
31.67   7.89
29.03

CDCl$_3$ QE-300

CH$_3$(CH$_2$)$_4$CH$_2$ —C—CH$_2$CH$_3$

C

Aldrich 13,694-8
5-Nonanone, 98%

CAS [502-56-7]

C$_9$H$_{18}$O
FW 142.24
mp -50°C
bp 187°C
d 0.826

n$_D^{20}$1.4190
Fp 141°F
60 MHz: **1**, 374D
FT-IR: **1**, 411B
VP-FT-IR: **3**, 496A

211.35
42.53
26.06
22.42
13.85

CDCl$_3$ QE-300

CH$_3$CH$_2$CH$_2$CH$_2$ —C—CH$_2$CH$_2$CH$_2$CH$_3$

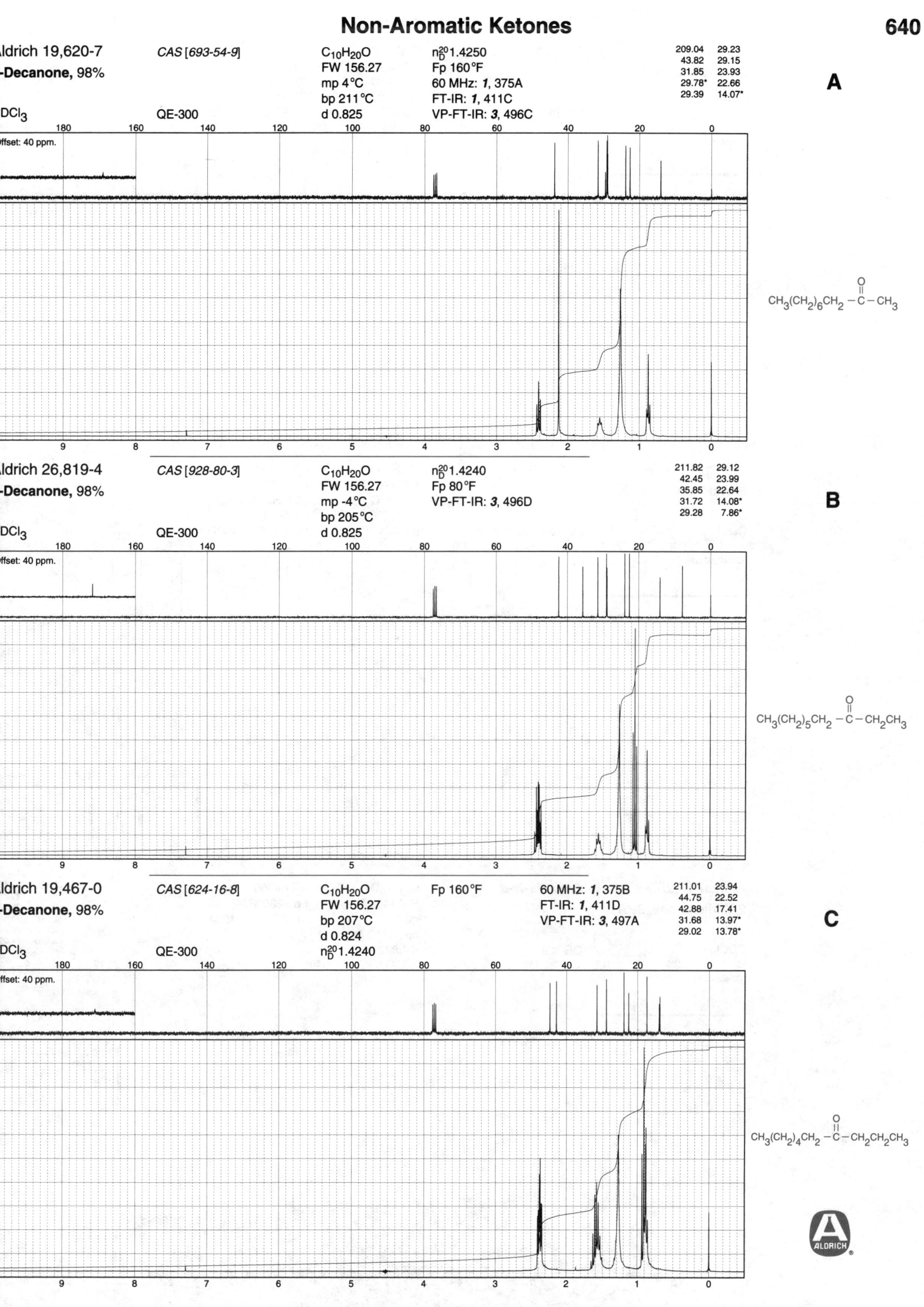

Aldrich 19,620-7

CAS [693-54-9]

2-Decanone, 98%

CDCl₃   QE-300

$C_{10}H_{20}O$
FW 156.27
mp 4°C
bp 211°C
d 0.825

n_D^{20} 1.4250
Fp 160°F
60 MHz: **1**, 375A
FT-IR: **1**, 411C
VP-FT-IR: **3**, 496C

209.04	29.23
43.82	29.15
31.85	23.93
29.78*	22.66
29.39	14.07*

A

Offset: 40 ppm.

$CH_3(CH_2)_6CH_2 - \overset{\overset{\text{O}}{\|}}{C} - CH_3$

Aldrich 26,819-4

CAS [928-80-3]

3-Decanone, 98%

CDCl₃   QE-300

$C_{10}H_{20}O$
FW 156.27
mp -4°C
bp 205°C
d 0.825

n_D^{20} 1.4240
Fp 80°F
VP-FT-IR: **3**, 496D

211.82	29.12
42.45	23.99
35.85	22.64
31.72	14.08*
29.28	7.86*

B

Offset: 40 ppm.

$CH_3(CH_2)_5CH_2 - \overset{\overset{\text{O}}{\|}}{C} - CH_2CH_3$

Aldrich 19,467-0

CAS [624-16-8]

4-Decanone, 98%

CDCl₃   QE-300

$C_{10}H_{20}O$
FW 156.27
bp 207°C
d 0.824
n_D^{20} 1.4240

Fp 160°F
60 MHz: **1**, 375B
FT-IR: **1**, 411D
VP-FT-IR: **3**, 497A

211.01	23.94
44.75	22.52
42.88	17.41
31.68	13.97*
29.02	13.78*

C

Offset: 40 ppm.

$CH_3(CH_2)_4CH_2 - \overset{\overset{\text{O}}{\|}}{C} - CH_2CH_2CH_3$

A

Aldrich U130-3

2-Undecanone, 99%

CAS [112-12-9]

$C_{11}H_{22}O$
FW 170.30
mp 12°C
bp 232°C
d 0.825

n_D^{20} 1.4300
Fp 192°F
60 MHz: *1*, 375C
FT-IR: *1*, 412A
VP-FT-IR: *3*, 497B

208.84	29.27
43.83	23.97
31.90	22.68
29.72*	14.05*
29.45	

CDCl₃   QE-300

Offset: 40 ppm.

$CH_3(CH_2)_7CH_2-\overset{\displaystyle O}{\overset{\|}{C}}-CH_3$

B

Aldrich 13,699-9

6-Undecanone, 97%

CAS [927-49-1]

$C_{11}H_{22}O$
FW 170.30
mp 15°C
bp 228°C
d 0.831

n_D^{20} 1.4270
Fp 191°F
60 MHz: *1*, 375D
FT-IR: *1*, 412B
VP-FT-IR: *3*, 497D

211.37
42.79
31.51
23.62
22.49
13.90*

CDCl₃   QE-300

Offset: 40 ppm.

$CH_3(CH_2)_3CH_2-\overset{\displaystyle O}{\overset{\|}{C}}-CH_2(CH_2)_3CH_3$

C

Aldrich 17,283-9

2-Tridecanone, 99%

CAS [593-08-8]

$C_{13}H_{26}O$
FW 198.35
mp 30°C
bp 134°C (10 mm)
d 0.822

n_D^{20} 1.4350
Fp >230°F
VP-FT-IR: *3*, 498A

208.97	29.43
43.82	29.35
31.94	29.23
29.76*	23.93
29.63	22.70
29.50	14.09*

CDCl₃   QE-300

Offset: 40 ppm.

$CH_3(CH_2)_9CH_2-\overset{\displaystyle O}{\overset{\|}{C}}-CH_3$

A

ldrich D10,420-5
ihexyl ketone, 97%

CAS [462-18-0]

C13H26O
FW 198.35
mp 31°C
bp 264°C
d 0.825

Fp >230°F

60 MHz: 1, 376A
FT-IR: 1, 412C
VP-FT-IR: 3, 498B

211.14
42.85
31.67
29.02
23.97
22.52
13.97*

DCl3　　QE-300

Offset: 40 ppm.

CH3(CH2)4CH2–C(=O)–CH2(CH2)4CH3

B

ldrich 15,838-0
-Pentadecanone, 99%

CAS [818-23-5]

C15H30O
FW 226.40
mp 42°C
bp 178°C
Fp >230°F

FT-IR: 1, 412D
VP-FT-IR: 3, 498D

211.40　29.11
42.83　23.95
31.72　22.63
29.28　14.05*

DCl3　　QE-300

Offset: 40 ppm.

CH3(CH2)5CH2–C(=O)–CH2(CH2)5CH3

C

ldrich 37,418-0
-Hexadecanone, 98%

CAS [18787-63-8]

C16H32O
FW 240.43
mp 45°C
bp 139°C (2 mm)
Fp >230°F

209.12　29.43
43.82　29.38
31.95　29.21
29.81*　23.90
29.67　22.71
29.64　14.12*
29.50

DCl3　　QE-300

Offset: 40 ppm.

CH3(CH2)12CH2–C(=O)–CH3

ALDRICH

Non-Aromatic Ketones

A

Aldrich 15,990-5   *CAS [540-08-9]* $C_{17}H_{34}O$ 60 MHz: *1*, 376C

9-Heptadecanone, 94%   FW 254.46   FT-IR: *1*, 413A

mp 52°C

Fp >230°F

211.41	29.17
42.83	23.95
31.86	22.67
29.41	14.07
29.33	

CDCl$_3$ QE-300

200 180 160 140 120 100 80 60 40 20 0

Offset: 40 ppm.

$CH_3(CH_2)_6CH_2-\overset{\overset{\displaystyle O}{\|}}{C}-CH_2(CH_2)_6CH_3$

10 9 8 7 6 5 4 3 2 1 0

B

Aldrich 10,366-7   *CAS [504-57-4]* $C_{19}H_{38}O$ 60 MHz: *1*, 376B

10-Nonadecanone, 99%   FW 282.51   FT-IR: *1*, 413C

mp 56°C   VP-FT-IR: *3*, 499B

Fp >230°F

211.76	29.74
43.27	24.41
32.35	23.13
29.91	14.51
29.78	

CDCl$_3$ QE-300

200 180 160 140 120 100 80 60 40 20 0

Offset: 40 ppm.

$CH_3(CH_2)_7CH_2-\overset{\overset{\displaystyle O}{\|}}{C}-CH_2(CH_2)_7CH_3$

10 9 8 7 6 5 4 3 2 1 0

C

Aldrich 26,954-9   *CAS [78-94-4]* C_4H_6O Fp 20°F   60 MHz: *1*, 377B

Methyl vinyl ketone, 99%   FW 70.09   FT-IR: *1*, 413D

bp 37°C (145 mm)   VP-FT-IR: *3*, 499C

d 0.842

n_D^{20}1.4110

198.6
137.4
128.7
26.3

CDCl$_3$ QE-300

200 180 160 140 120 100 80 60 40 20 0

$CH_3-\overset{\overset{\displaystyle O}{\|}}{C}-CH=CH_2$

10 9 8 7 6 5 4 3 2 1 0

Non-Aromatic Ketones

Aldrich E5,130-9 CAS [1629-58-9] C5H8O Fp 20°F 60 MHz: *1*, 377C 201.01
ethyl vinyl ketone, 97%   FW 84.12   FT-IR: *1*, 414A 136.41*
     bp 38°C (60 mm)   VP-FT-IR: *3*, 499D 127.44
     d 0.845   32.88
CDCl3 QE-300 n²⁰_D 1.4200   7.91*
offset: 40 ppm.

A

$$CH_3CH_2-\overset{\overset{\displaystyle O}{\|}}{C}-CH=CH_2$$

Aldrich 14,501-7 CAS [625-33-2] C5H8O Fp 70°F 60 MHz: *1*, 377D 197.97
3-Penten-2-one, 65%   FW 84.12   FT-IR: *1*, 414B 143.12*
     bp 123°C   VP-FT-IR: *3*, 500A 133.00*
     d 0.862   26.73*
CDCl3 QE-300 n²⁰_D 1.4370   18.08*

B

$$CH_3CH=CH-\overset{\overset{\displaystyle O}{\|}}{C}-CH_3$$

Aldrich M785-5 CAS [141-79-7] C6H10O n²⁰_D 1.4450 198.23
mesityl oxide, 98%   FW 98.15 Fp 87°F 154.56
     mp -53°C 60 MHz: *1*, 378B 124.30*
     bp 129°C FT-IR: *1*, 414D 31.49*
CDCl3 QE-300 d 0.858 VP-FT-IR: *3*, 500B 27.48*
          20.55*

C

$$CH_3-\overset{\overset{\displaystyle O}{\|}}{C}-CH=\underset{\underset{\displaystyle CH_3}{|}}{C}CH_3$$

A

Aldrich H1,300-1   CAS [109-49-9] $C_6H_{10}O$ Fp 75 °F   FT-IR: *1*, 415B

5-Hexen-2-one, 99%   FW 98.15   VP-FT-IR: *3*, 500D

bp 129 °C

d 0.847

$CDCl_3$ QE-300 n_D^{20} 1.4190

Offset: 40 ppm.

$H_2C=CHCH_2CH_2-\overset{\displaystyle O}{\overset{\displaystyle \|}{C}}-CH_3$

B

Aldrich 26,473-3   CAS [2497-21-4] $C_6H_{10}O$ Fp 94 °F   VP-FT-IR: *3*, 501A

4-Hexen-3-one, 98%, predominantly *trans*   FW 98.15

bp 136 °C

d 0.858

$CDCl_3$ QE-300 n_D^{20} 1.4400

Offset: 40 ppm.

$CH_3CH=CH-\overset{\displaystyle O}{\overset{\displaystyle \|}{C}}-CH_2CH_3$

C

Aldrich 18,760-7   CAS [5166-53-0] $C_7H_{12}O$ FT-IR: *1*, 415A

5-Methyl-3-hexen-2-one, tech., 80%   FW 112.17   VP-FT-IR: *3*, 500C

d 0.850

n_D^{20} 1.4400

Fp 118 °F

$CDCl_3$ QE-300

$\underset{\displaystyle CH_3CHCH=CH-\overset{O}{\overset{\|}{C}}-CH_3}{\overset{\displaystyle CH_3}{|}}$

A

Aldrich 36,447-9   CAS [3240-09-3] $C_7H_{12}O$ Fp 108°F

5-Methyl-5-hexen-2-one, 99%

FW 112.17
bp 149°C
d 0.865
n_D^{20} 1.4310

CDCl$_3$ QE-300

208.11
144.33
110.12
41.86
31.52
29.84*
22.60*

Offset: 40 ppm.

$H_2C=C(CH_3)-CH_2CH_2-C(=O)-CH_3$

B

Aldrich M4,880-5   CAS [110-93-0] $C_8H_{14}O$ Fp 123°F

6-Methyl-5-hepten-2-one, 99%

FW 126.20
bp 73°C (18 mm)
d 0.855
n_D^{20} 1.4390

CDCl$_3$ QE-300

60 MHz: 1, 378C
FT-IR: 1, 415D
VP-FT-IR: 3, 501B

208.36   29.79*
132.62   25.61*
122.78*   22.67
43.77   17.61*

Offset: 40 ppm.

$CH_3-C(CH_3)=CHCH_2CH_2-C(=O)-CH_3$

C

Aldrich 26,253-6   CAS [14309-57-0] $C_9H_{16}O$ Fp 179°F

3-Nonen-2-one, 95%

FW 140.23
bp 85°C (12 mm)
d 0.848
n_D^{20} 1.4490

CDCl$_3$ QE-300

FT-IR: 1, 415C
VP-FT-IR: 3, 501C

198.35   27.90
148.30*   26.86*
131.36*   22.49
32.47   13.95*
31.45

$CH_3(CH_2)_3CH_2CH=CH-C(=O)-CH_3$

Non-Aromatic Ketones

A

Aldrich 14,923-3
Phorone, 97%

CAS [504-20-1]

$C_9H_{14}O$
FW 138.21
mp 27°C
bp 199°C

n_D^{20} 1.4970
Fp 175°F
60 MHz: **1**, 378D
FT-IR: **1**, 416A
VP-FT-IR: **3**, 502A

191.34
153.86
126.23*
27.64*
20.49*

CDCl$_3$

QE-300

d 0.885

Structure: $CH_3C=CH-C-CH=CCH_3$ with CH_3 groups and O

B

Aldrich 33,001-9
Nerylacetone, 97%

CAS [3879-26-3]

$C_{13}H_{22}O$
FW 194.32
bp 80°C
d 0.868
n_D^{20} 1.4670

Fp 225°F

208.61   29.90*
136.39   26.50
131.58   25.71*
124.12*   23.35*
123.30*   22.28
43.98   17.64*
31.86

CDCl$_3$

QE-300

Offset: 40 ppm.

Structure: $CH_3-C-CH_2CH_2-C=C$ ... $CH_2CH_2-C=CCH_3$ with CH_3 and H groups

C

Aldrich 32,867-7
Geranylacetone, 96%

CAS [3796-70-1]

$C_{13}H_{22}O$
FW 194.32
bp 124°C (10 mm)
d 0.873
n_D^{20} 1.4670

Fp >230°F

60 MHz: **1**, 379A
FT-IR: **1**, 416B
VP-FT-IR: **3**, 502B

208.69   29.92*
136.28   26.61
131.32   25.68*
124.15*   22.48
122.50*   17.67
43.74   15.96*
39.64

CDCl$_3$

QE-300

Offset: 40 ppm.

ALDRICH

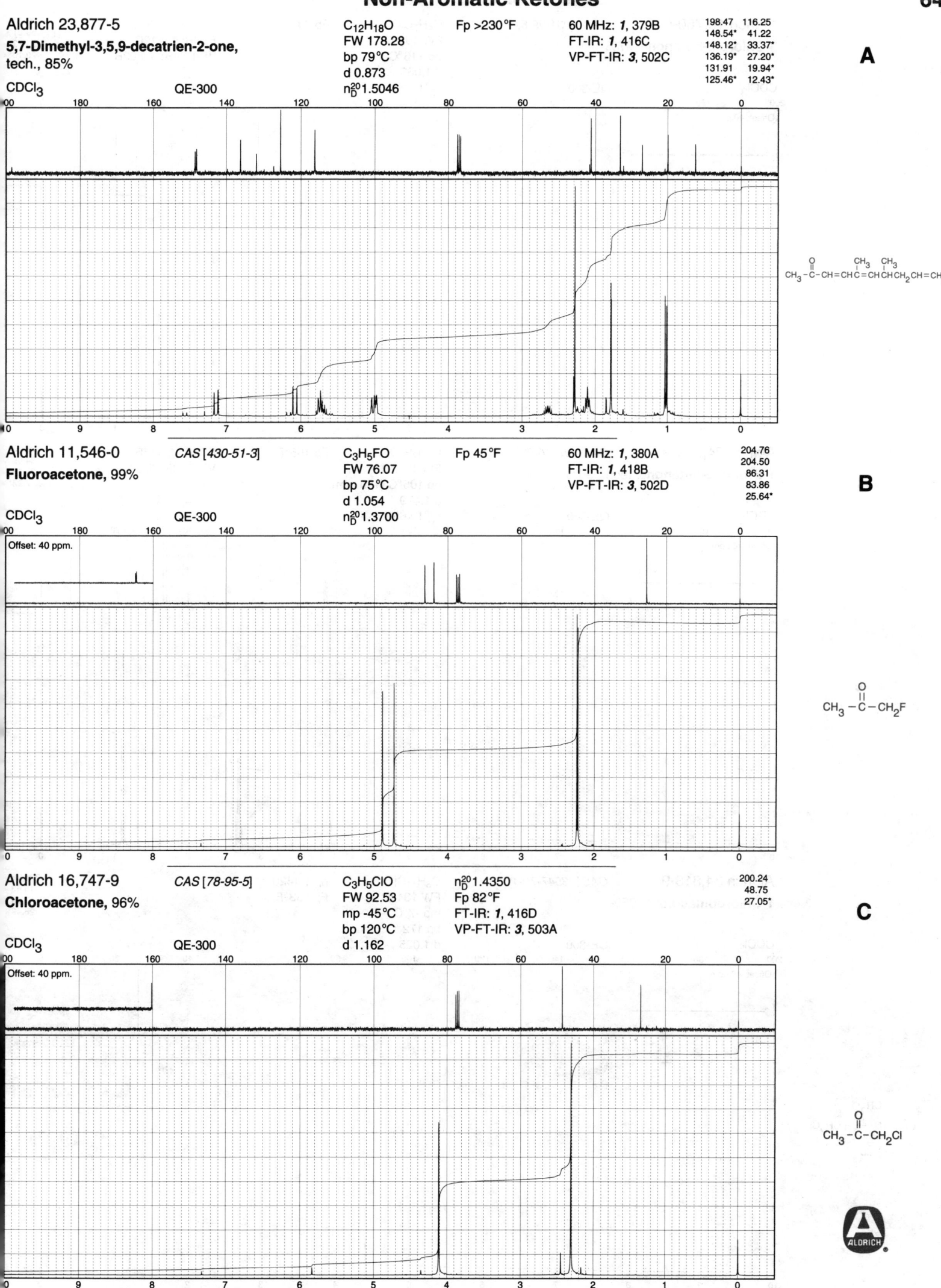

Aldrich 23,877-5

5,7-Dimethyl-3,5,9-decatrien-2-one,
tech., 85%

C₁₂H₁₈O Fp >230 °F
FW 178.28
bp 79 °C
d 0.873
n²⁰_D 1.5046

60 MHz: *1*, 379B
FT-IR: *1*, 416C
VP-FT-IR: *3*, 502C

198.47 116.25
148.54* 41.22
148.12* 33.37*
136.19* 27.20*
131.91 19.94*
125.46* 12.43*

CDCl₃   QE-300

A

Aldrich 11,546-0   *CAS [430-51-3]*

Fluoroacetone, 99%

C₃H₅FO Fp 45 °F
FW 76.07
bp 75 °C
d 1.054
n²⁰_D 1.3700

60 MHz: *1*, 380A
FT-IR: *1*, 418B
VP-FT-IR: *3*, 502D

204.76
204.50
86.31
83.86
25.64*

CDCl₃   QE-300

Offset: 40 ppm.

B

Aldrich 16,747-9   *CAS [78-95-5]*

Chloroacetone, 96%

C₃H₅ClO n²⁰_D 1.4350
FW 92.53 Fp 82 °F
mp -45 °C FT-IR: *1*, 416D
bp 120 °C VP-FT-IR: *3*, 503A
d 1.162

200.24
48.75
27.05*

CDCl₃   QE-300

Offset: 40 ppm.

C

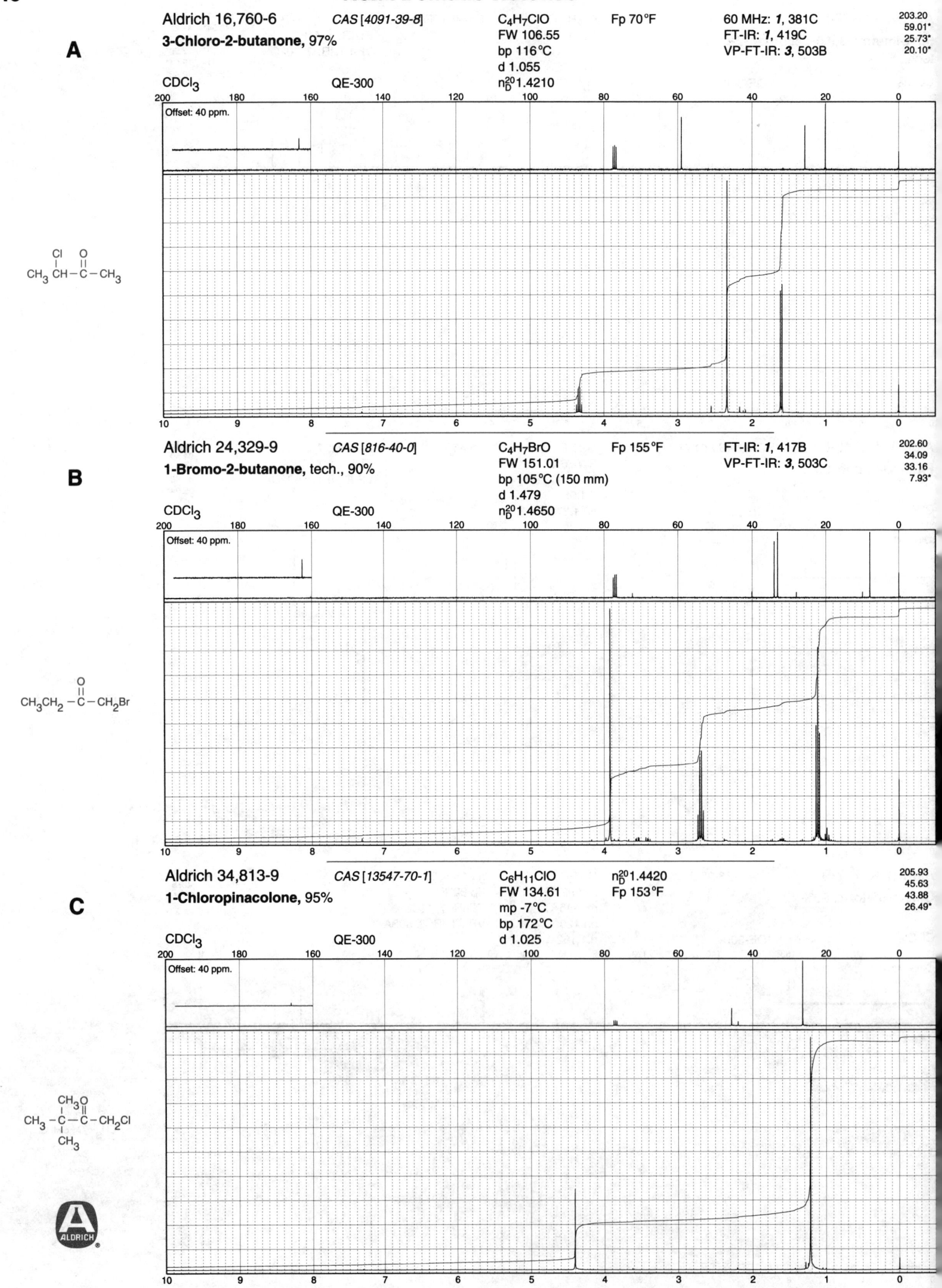

A

Aldrich 16,760-6   *CAS [4091-39-8]*   C₄H₇ClO   Fp 70°F   60 MHz: *1*, 381C   203.20
3-Chloro-2-butanone, 97%   FW 106.55   FT-IR: *1*, 419C   59.01*
bp 116°C   VP-FT-IR: *3*, 503B   25.73*
d 1.055   20.10*
CDCl₃   QE-300   n²⁰_D 1.4210

CH₃ CH–C–CH₃ (Cl, O)

B

Aldrich 24,329-9   *CAS [816-40-0]*   C₄H₇BrO   Fp 155°F   FT-IR: *1*, 417B   202.60
1-Bromo-2-butanone, tech., 90%   FW 151.01   VP-FT-IR: *3*, 503C   34.09
bp 105°C (150 mm)   33.16
d 1.479   7.93*
CDCl₃   QE-300   n²⁰_D 1.4650

CH₃CH₂–C–CH₂Br (O)

C

Aldrich 34,813-9   *CAS [13547-70-1]*   C₆H₁₁ClO   n²⁰_D 1.4420   205.93
1-Chloropinacolone, 95%   FW 134.61   Fp 153°F   45.63
mp -7°C   43.88
bp 172°C   26.49*
CDCl₃   QE-300   d 1.025

CH₃–C–C–CH₂Cl (CH₃, CH₃, O)

A

Aldrich 37,805-4   *CAS [3884-71-7]*   C$_5$H$_9$BrO   Fp 169°F

-Bromo-2-pentanone, tech., 85%

FW 165.04
bp 88°C (25 mm)
d 1.448
n$_D^{20}$1.4840

CDCl$_3$ QE-300

Offset: 40 ppm.

207.26
41.41
33.32
30.08*
26.35

BrCH$_2$CH$_2$CH$_2$ –C–CH$_3$

B

Aldrich 22,714-5   *CAS [513-88-2]*   C$_3$H$_4$Cl$_2$O   Fp 76°F   60 MHz: *1*, 380C

,1-Dichloroacetone, 98%

FW 126.97
bp 118°C
d 1.327
n$_D^{20}$1.4460

CDCl$_3$ QE-300

FT-IR: *1*, 418D
VP-FT-IR: *3*, 504D

194.50
70.08*
22.31*

CH$_3$ –C–CHCl$_2$

C

Aldrich 16,854-8   *CAS [534-07-6]*   C$_3$H$_4$Cl$_2$O   Fp 193°F   60 MHz: *1*, 380D

,3-Dichloroacetone, 95+%

FW 126.97
mp 40°C
bp 173°C
d 1.383

CDCl$_3$ QE-300

FT-IR: *1*, 417A
VP-FT-IR: *3*, 505A

194.89
46.17

ClCH$_2$ –C–CH$_2$Cl

ALDRICH

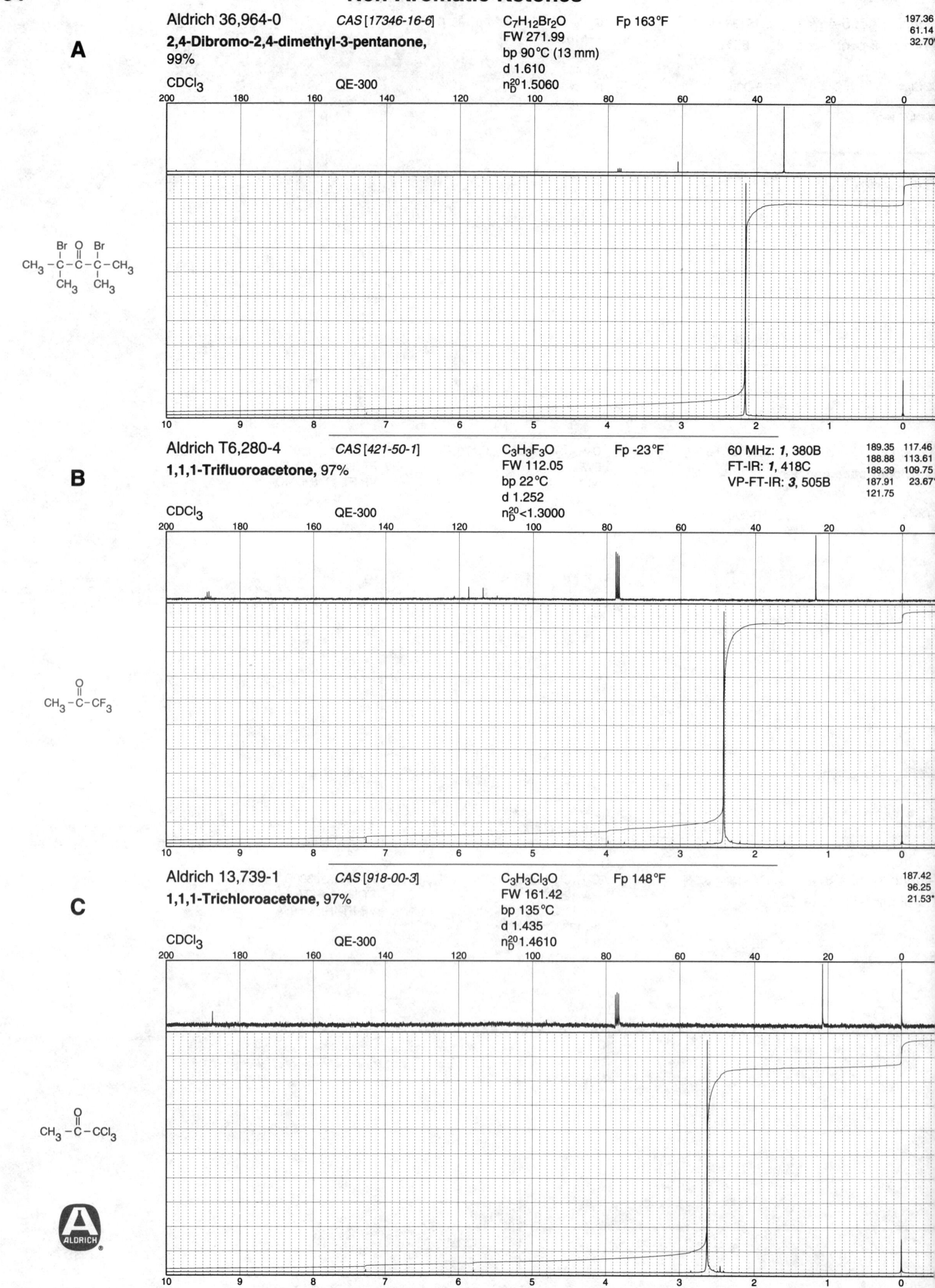

A

Aldrich 36,964-0   CAS [17346-16-6]   C₇H₁₂Br₂O   Fp 163°F

2,4-Dibromo-2,4-dimethyl-3-pentanone, 99%

FW 271.99
bp 90°C (13 mm)
d 1.610
n²⁰_D 1.5060

CDCl₃   QE-300

197.36
61.14
32.70*

B

Aldrich T6,280-4   CAS [421-50-1]   C₃H₃F₃O   Fp -23°F

1,1,1-Trifluoroacetone, 97%

FW 112.05
bp 22°C
d 1.252
n²⁰_D <1.3000

CDCl₃   QE-300

60 MHz: 1, 380B
FT-IR: 1, 418C
VP-FT-IR: 3, 505B

189.35 117.46
188.88 113.61
188.39 109.75
187.91 23.67*
121.75

C

Aldrich 13,739-1   CAS [918-00-3]   C₃H₃Cl₃O   Fp 148°F

1,1,1-Trichloroacetone, 97%

FW 161.42
bp 135°C
d 1.435
n²⁰_D 1.4610

CDCl₃   QE-300

187.42
96.25
21.53*

Aldrich 24,664-6   CAS [921-03-9] $C_3H_3Cl_3O$ $n_D^{20}1.4892$ 188.72
                                      FW 161.42   Fp 175°F   67.75*
1,3-Trichloroacetone, tech., 85%   mp 14°C   VP-FT-IR: *3*, 505C   42.78

CDCl₃   QE-300   bp 172°C
                   d 1.508

A

$Cl_2CH-\overset{\displaystyle O}{\underset{\displaystyle \|}{C}}-CH_2Cl$

Aldrich 37,405-9   CAS [431-35-6] $C_3H_2BrF_3O$ Fp 41°F   183.88 116.89
                                      FW 190.95   183.39 113.03
3-Bromo-1,1,1-trifluoroacetone, 96%   bp 87°C (743 mm)   182.91 109.16
                                      d 1.839   182.42 27.25
CDCl₃   QE-300 $n_D^{20}1.3760$ 120.76

B

$BrCH_2-\overset{\displaystyle O}{\underset{\displaystyle \|}{C}}-CF_3$

Aldrich 13,860-6   CAS [1768-31-6] C_3HCl_5O Fp NONE   60 MHz: *1*, 381B   179.42
Pentachloroacetone, tech., 85%   FW 230.31   FT-IR: *1*, 419B   92.47
                                      bp 192°C   VP-FT-IR: *3*, 505D   61.83*
CDCl₃   QE-300   d 1.690
                   $n_D^{20}1.4967$

C

$Cl-\overset{\displaystyle Cl}{\underset{\displaystyle Cl}{C}}-\overset{\displaystyle O}{\underset{\displaystyle \|}{C}}-\overset{\displaystyle H}{\underset{\displaystyle Cl}{C}}-Cl$

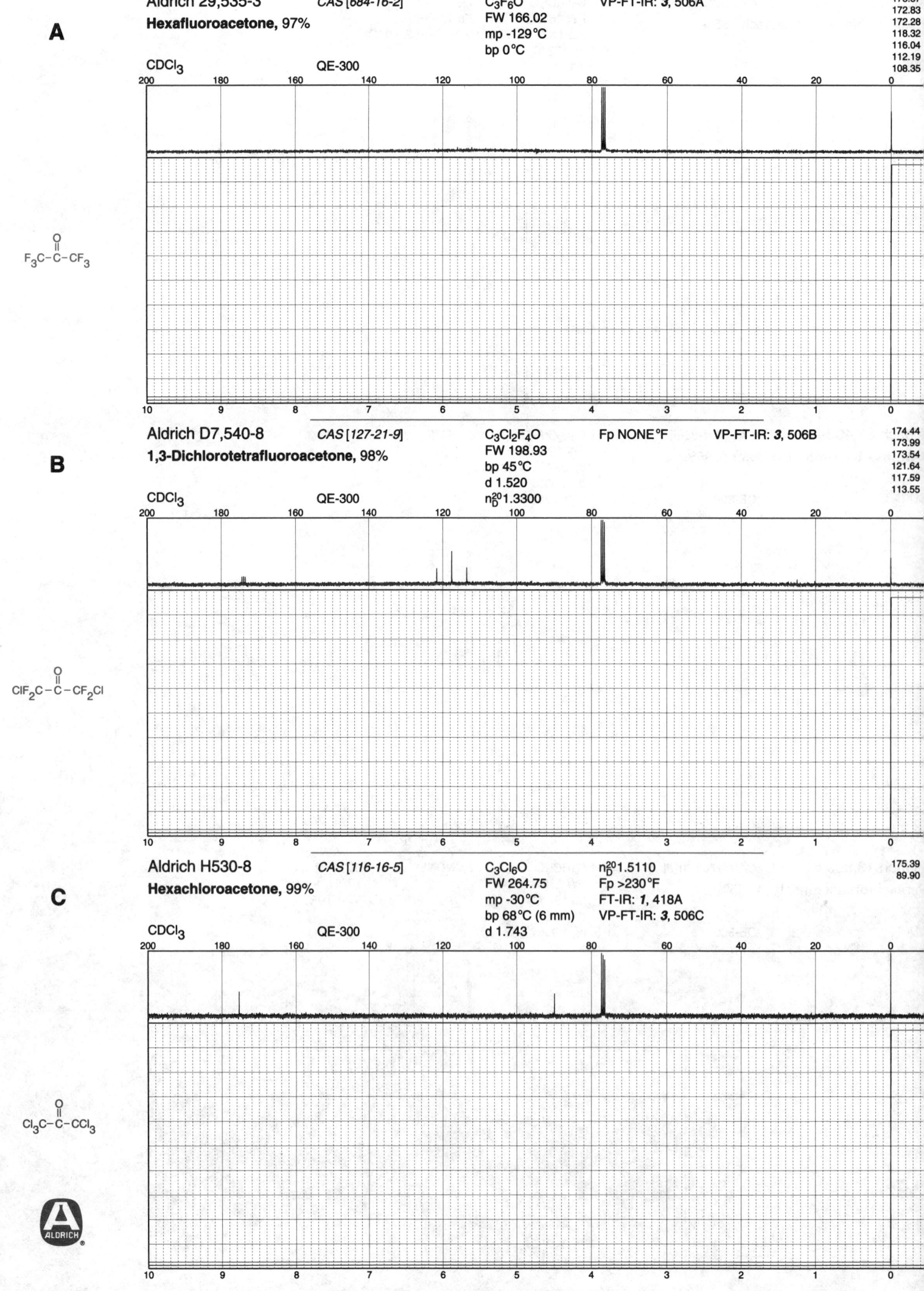

A

Aldrich 29,535-3   CAS [684-16-2]   C$_3$F$_6$O   VP-FT-IR: **3**, 506A

Hexafluoroacetone, 97%

CDCl$_3$ QE-300

FW 166.02
mp -129°C
bp 0°C

173.37
172.83
172.28
118.32
116.04
112.19
108.35

F$_3$C–C–CF$_3$ (O)

B

Aldrich D7,540-8   CAS [127-21-9]   C$_3$Cl$_2$F$_4$O   Fp NONE°F   VP-FT-IR: **3**, 506B

1,3-Dichlorotetrafluoroacetone, 98%

CDCl$_3$ QE-300

FW 198.93
bp 45°C
d 1.520
n$_D^{20}$1.3300

174.44
173.99
173.54
121.64
117.59
113.55

ClF$_2$C–C–CF$_2$Cl (O)

C

Aldrich H530-8   CAS [116-16-5]   C$_3$Cl$_6$O   n$_D^{20}$1.5110

Hexachloroacetone, 99%

CDCl$_3$ QE-300

FW 264.75
mp -30°C
bp 68°C (6 mm)
d 1.743

Fp >230°F
FT-IR: **1**, 418A
VP-FT-IR: **3**, 506C

175.39
89.90

Cl$_3$C–C–CCl$_3$ (O)

ALDRICH

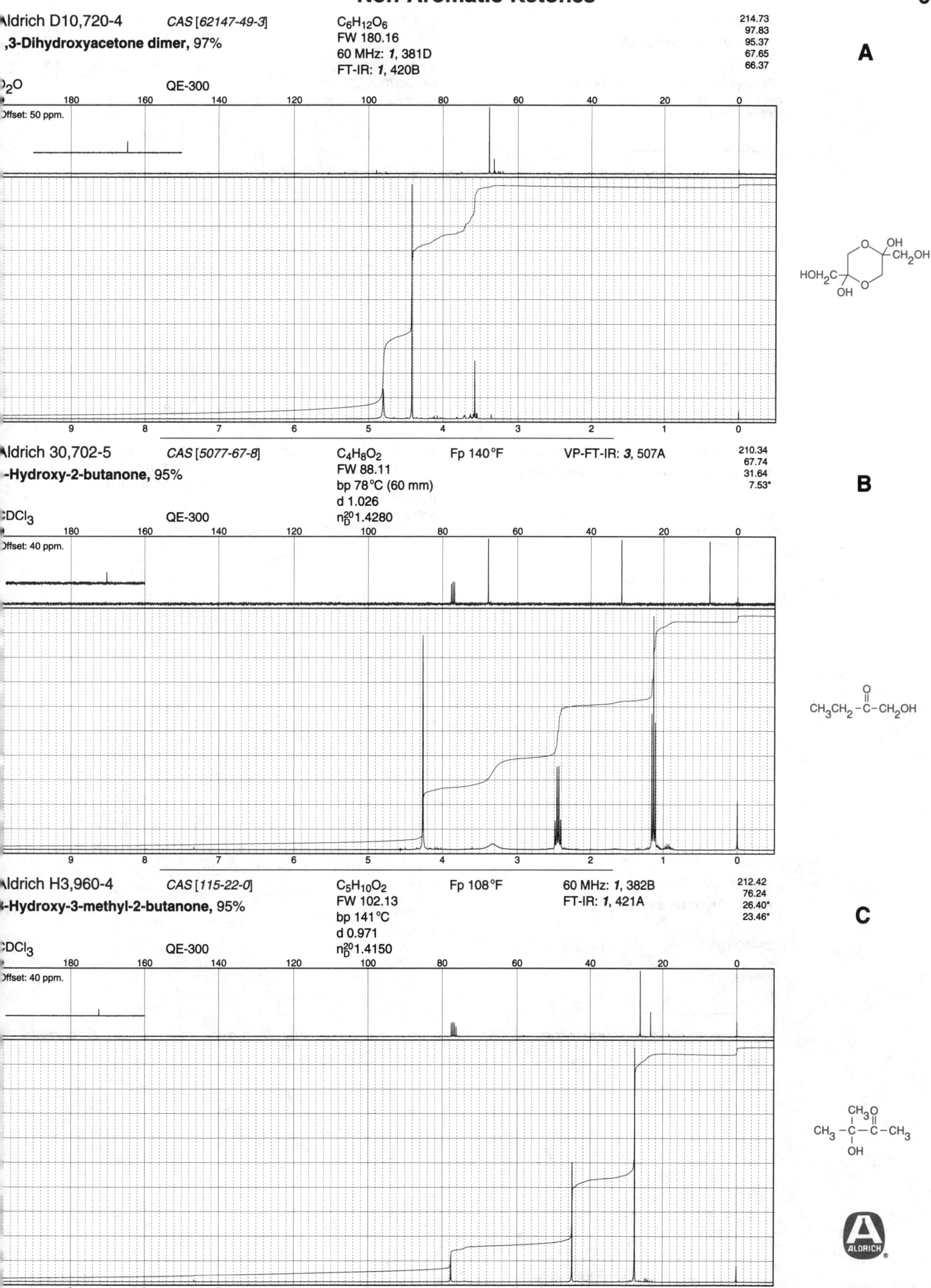

Aldrich D10,720-4   *CAS [62147-49-3]* $C_6H_{12}O_6$
,3-Dihydroxyacetone dimer, 97%   FW 180.16
60 MHz: *1*, 381D
FT-IR: *1*, 420B

A

214.73
97.83
95.37
67.65
66.37

O_2O QE-300

Offset: 50 ppm.

Aldrich 30,702-5   *CAS [5077-67-8]* $C_4H_8O_2$ Fp 140°F   VP-FT-IR: *3*, 507A
-Hydroxy-2-butanone, 95%   FW 88.11
bp 78°C (60 mm)
d 1.026
n_D^{20} 1.4280

B

210.34
67.74
31.64
7.53*

$CDCl_3$ QE-300

Offset: 40 ppm.

$CH_3CH_2-\overset{\overset{\displaystyle O}{\|}}{C}-CH_2OH$

Aldrich H3,960-4   *CAS [115-22-0]* $C_5H_{10}O_2$ Fp 108°F   60 MHz: *1*, 382B
-Hydroxy-3-methyl-2-butanone, 95%   FW 102.13   FT-IR: *1*, 421A
bp 141°C
d 0.971
n_D^{20} 1.4150

C

212.42
76.24
26.40*
23.46*

$CDCl_3$ QE-300

Offset: 40 ppm.

$CH_3-\overset{\overset{\displaystyle CH_3}{|}}{\underset{\underset{\displaystyle OH}{|}}{C}}-\overset{\overset{\displaystyle O}{\|}}{C}-CH_3$

ALDRICH

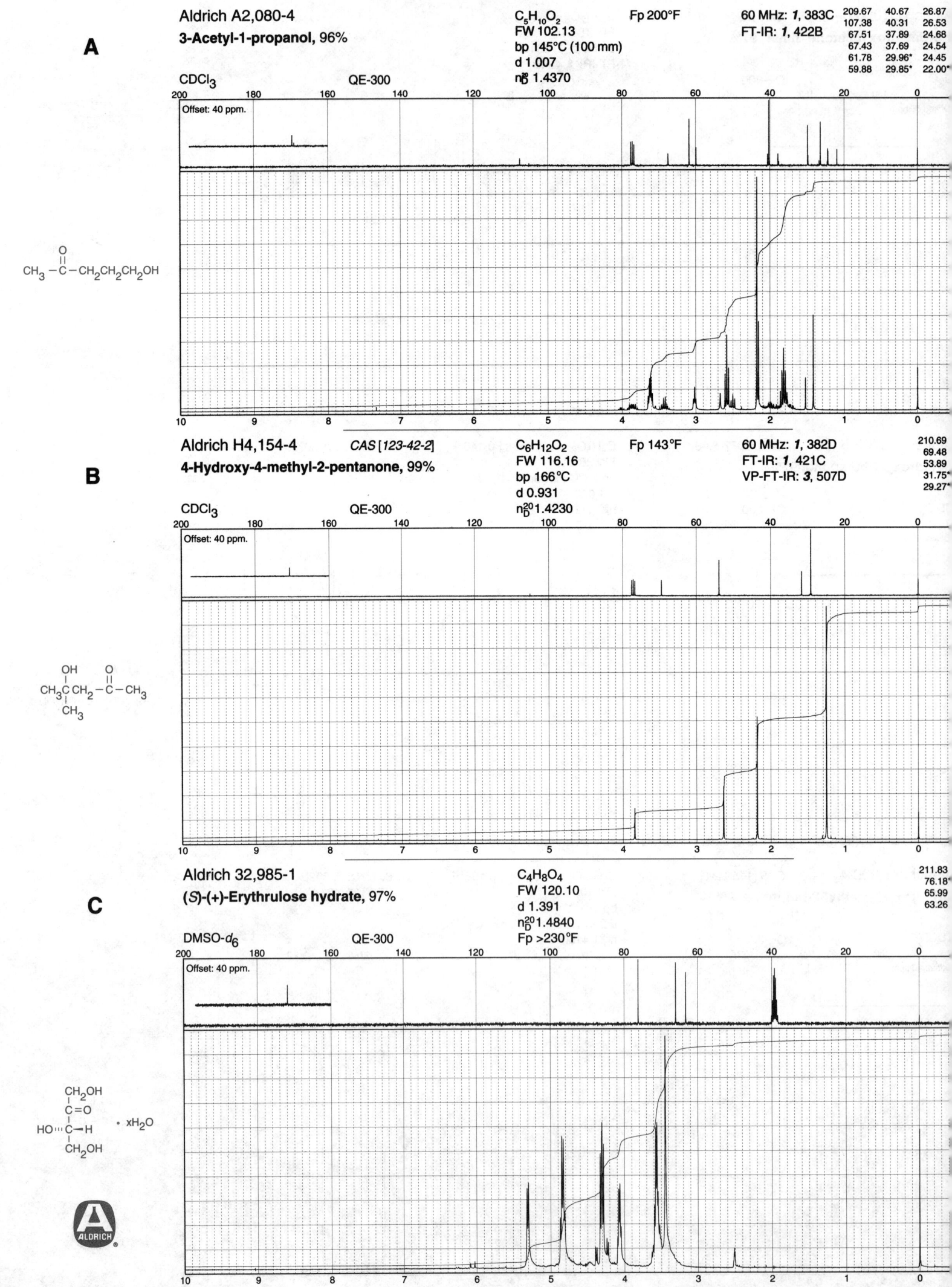

A

Aldrich A2,080-4

3-Acetyl-1-propanol, 96%

$C_5H_{10}O_2$
FW 102.13
bp 145°C (100 mm)
d 1.007
n_D^{20} 1.4370

Fp 200°F

CDCl$_3$ QE-300   Offset: 40 ppm.

60 MHz: *1*, 383C
FT-IR: *1*, 422B

209.67	40.67	26.87
107.38	40.31	26.53
67.51	37.89	24.68
67.43	37.69	24.54
61.78	29.96*	24.45
59.88	29.85*	22.00*

CH$_3$ – C(=O) – CH$_2$CH$_2$CH$_2$OH

B

Aldrich H4,154-4   CAS [123-42-2]

4-Hydroxy-4-methyl-2-pentanone, 99%

$C_6H_{12}O_2$
FW 116.16
bp 166°C
d 0.931
n_D^{20} 1.4230

Fp 143°F

CDCl$_3$ QE-300   Offset: 40 ppm.

60 MHz: *1*, 382D
FT-IR: *1*, 421C
VP-FT-IR: *3*, 507D

210.69
69.48
53.89
31.75*
29.27*

CH$_3$C(OH)(CH$_3$)CH$_2$ – C(=O) – CH$_3$

C

Aldrich 32,985-1

(S)-(+)-Erythrulose hydrate, 97%

$C_4H_8O_4$
FW 120.10
d 1.391
n_D^{20} 1.4840
Fp >230°F

DMSO-d_6 QE-300   Offset: 40 ppm.

211.83
76.18*
65.99
63.26

CH$_2$OH – C(=O) – HO···C–H – CH$_2$OH · xH$_2$O

ALDRICH

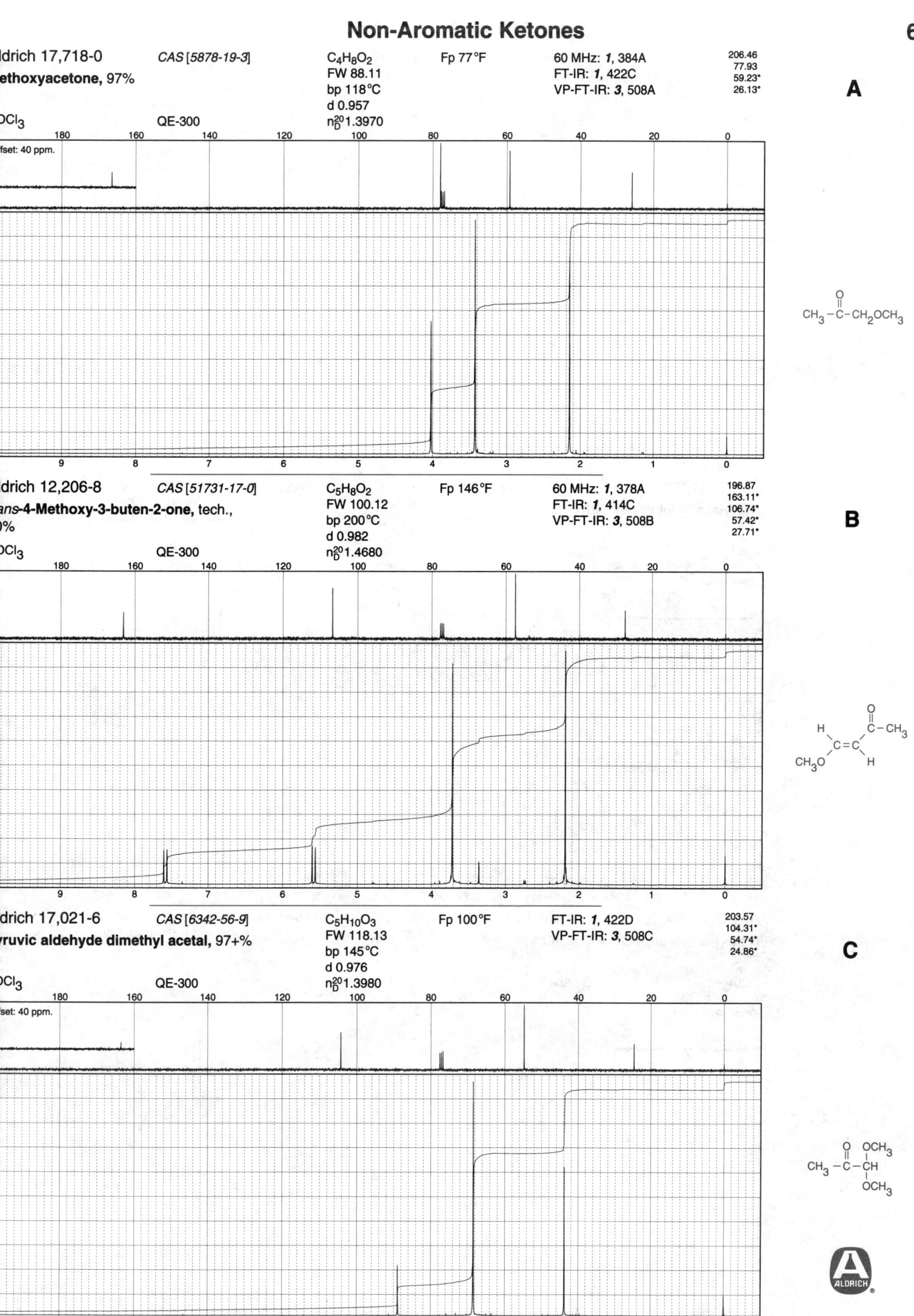

Aldrich 17,718-0   CAS [5878-19-3] $C_4H_8O_2$ Fp 77°F   60 MHz: **1**, 384A   206.46
Methoxyacetone, 97%   FW 88.11   FT-IR: **1**, 422C   77.93
bp 118°C   VP-FT-IR: **3**, 508A   59.23*
d 0.957   26.13*
$CDCl_3$ QE-300 n_D^{20} 1.3970

offset: 40 ppm.

A

$CH_3\text{--}\overset{\displaystyle O}{\overset{\|}{C}}\text{--}CH_2OCH_3$

Aldrich 12,206-8   CAS [51731-17-0] $C_5H_8O_2$ Fp 146°F   60 MHz: **1**, 378A   196.87
trans-4-Methoxy-3-buten-2-one, tech.,   FW 100.12   FT-IR: **1**, 414C   163.11*
90%   bp 200°C   VP-FT-IR: **3**, 508B   106.74*
d 0.982   57.42*
$CDCl_3$ QE-300 n_D^{20} 1.4680   27.71*

B

Aldrich 17,021-6   CAS [6342-56-9] $C_5H_{10}O_3$ Fp 100°F   FT-IR: **1**, 422D   203.57
Pyruvic aldehyde dimethyl acetal, 97+%   FW 118.13   VP-FT-IR: **3**, 508C   104.31*
bp 145°C   54.74*
d 0.976   24.86*
$CDCl_3$ QE-300 n_D^{20} 1.3980

offset: 40 ppm.

C

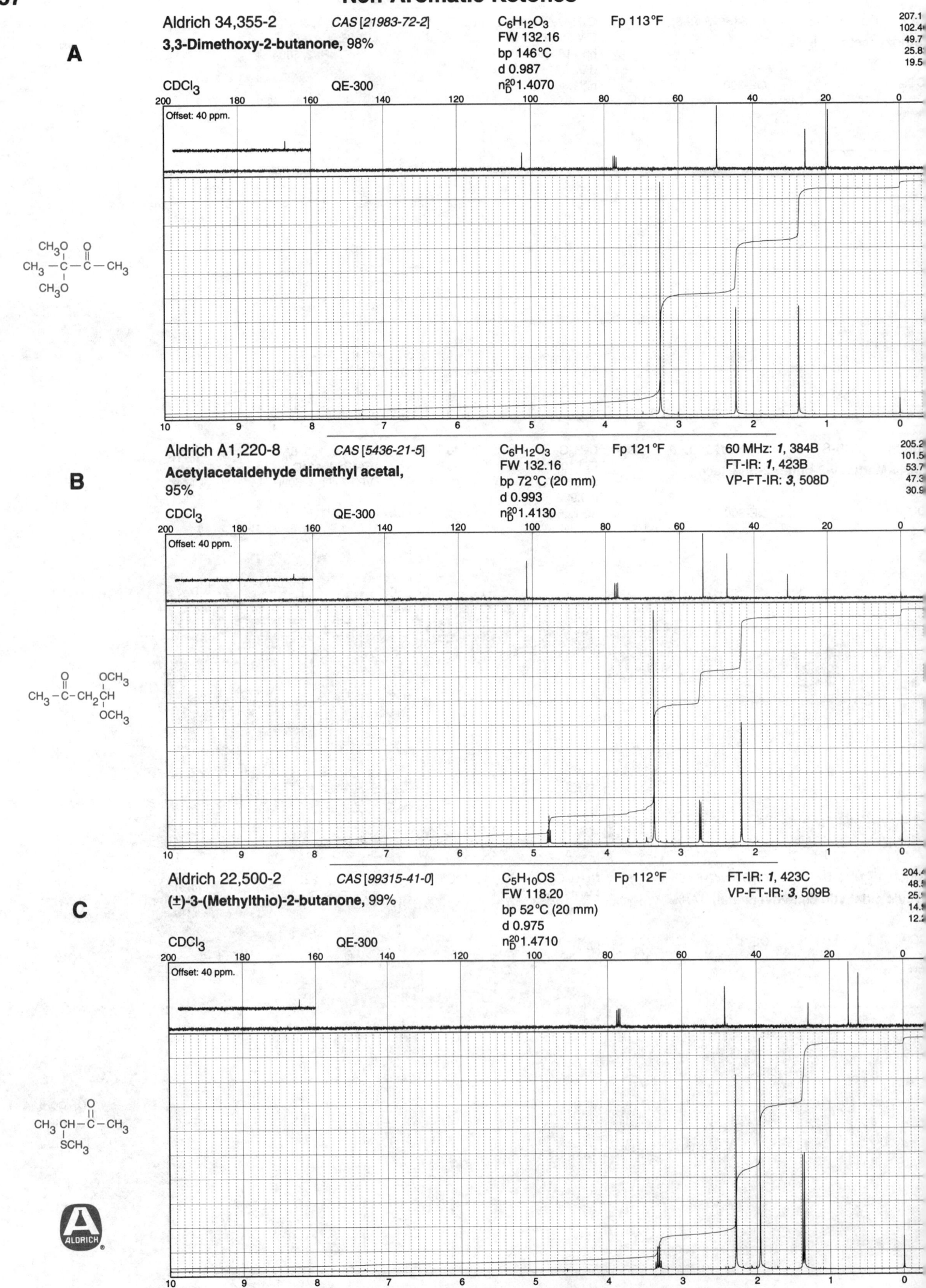

A

Aldrich 34,355-2   CAS [21983-72-2]   C6H12O3   Fp 113°F   207.1
102.4
3,3-Dimethoxy-2-butanone, 98%   FW 132.16   49.7
bp 146°C   25.8
d 0.987   19.5

CDCl3   QE-300   nD20 1.4070

Offset: 40 ppm.

Aldrich A1,220-8   CAS [5436-21-5]   C6H12O3   Fp 121°F   60 MHz: 1, 384B   205.2
101.5
Acetylacetaldehyde dimethyl acetal,   FW 132.16   FT-IR: 1, 423B   53.7
95%   bp 72°C (20 mm)   VP-FT-IR: 3, 508D   47.3
d 0.993   30.9

B

CDCl3   QE-300   nD20 1.4130

Offset: 40 ppm.

Aldrich 22,500-2   CAS [99315-41-0]   C5H10OS   Fp 112°F   FT-IR: 1, 423C   204.4
48.5
(±)-3-(Methylthio)-2-butanone, 99%   FW 118.20   VP-FT-IR: 3, 509B   25.9
bp 52°C (20 mm)   14.9
d 0.975   12.2

C

CDCl3   QE-300   nD20 1.4710

Offset: 40 ppm.

Aldrich W33,750-1   *CAS [34047-39-7]*   C$_5$H$_{10}$OS   Fp 162°F   VP-FT-IR: **3**, 509C

206.78
43.25
30.04*
27.86
15.72*

A

-Methylthio-2-butanone, 97+%   FW 118.20
bp 106°C (55 mm)
d 1.003
n$_D^{20}$1.4730

DCl$_3$ QE-300

Offset: 40 ppm.

$$CH_3SCH_2CH_2-\overset{\overset{\displaystyle O}{\|}}{C}-CH_3$$

Aldrich 10,769-7   *CAS [15364-56-4]*   C$_5$H$_{11}$NO   Fp 86°F   60 MHz: **1**, 385A
FT-IR: **1**, 458D
VP-FT-IR: **3**, 544C

206.72
69.54
45.75*
27.49*

B

Dimethylamino)acetone, 99%   FW 101.15
bp 120°C
d 0.883
n$_D^{20}$1.4130

DCl$_3$ QE-300

Offset: 40 ppm.

$$CH_3NCH_2-\overset{\overset{\displaystyle O}{\|}}{C}-CH_3$$
$$\underset{}{CH_3}$$

Aldrich D8,600-0   *CAS [1620-14-0]*   C$_7$H$_{15}$NO   Fp 101°F   60 MHz: **1**, 385D
FT-IR: **1**, 459A
VP-FT-IR: **3**, 544D

209.21
64.11
48.24
27.49*
12.09*

C

Methylaminoacetone, 96%   FW 129.20
bp 64°C (16 mm)
d 0.832
n$_D^{20}$1.4250

DCl$_3$ QE-300

Offset: 40 ppm.

$$CH_3CH_2NCH_2-\overset{\overset{\displaystyle O}{\|}}{C}-CH_3$$
$$\underset{CH_2CH_3}{}$$

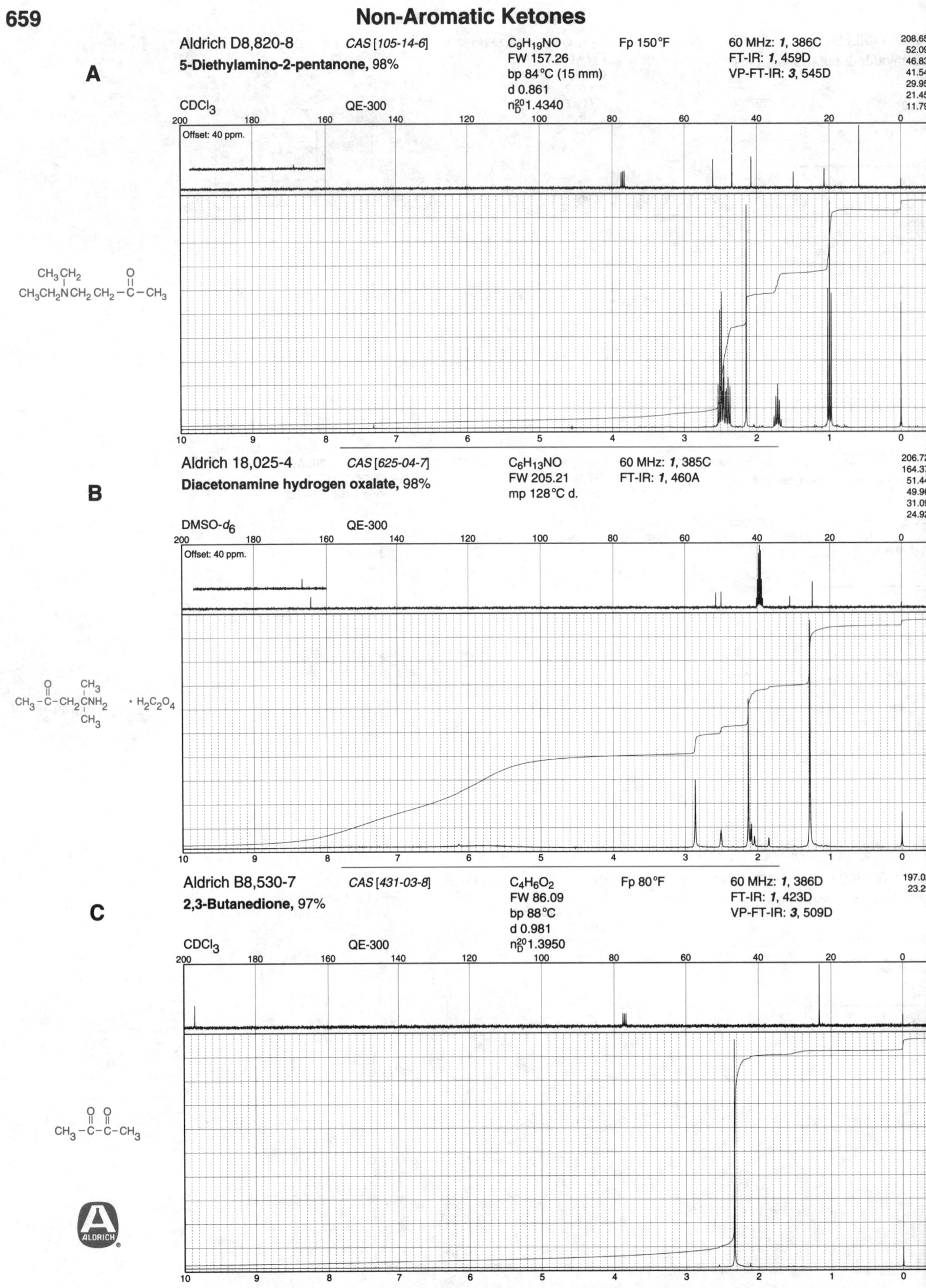

A

Aldrich D8,820-8   *CAS [105-14-6]*

5-Diethylamino-2-pentanone, 98%

$C_9H_{19}NO$
FW 157.26
bp 84°C (15 mm)
d 0.861
n_D^{20}1.4340

Fp 150°F

60 MHz: *1*, 386C
FT-IR: *1*, 459D
VP-FT-IR: *3*, 545D

208.65
52.09
46.83
41.54
29.95
21.45
11.79

CDCl₃   QE-300

Offset: 40 ppm.

B

Aldrich 18,025-4   *CAS [625-04-7]*

Diacetonamine hydrogen oxalate, 98%

$C_6H_{13}NO$
FW 205.21
mp 128°C d.

60 MHz: *1*, 385C
FT-IR: *1*, 460A

206.72
164.37
51.44
49.96
31.09
24.92

DMSO-*d₆*   QE-300

Offset: 40 ppm.

C

Aldrich B8,530-7   *CAS [431-03-8]*

2,3-Butanedione, 97%

$C_4H_6O_2$
FW 86.09
bp 88°C
d 0.981
n_D^{20}1.3950

Fp 80°F

60 MHz: *1*, 386D
FT-IR: *1*, 423D
VP-FT-IR: *3*, 509D

197.02
23.29

CDCl₃   QE-300

ALDRICH

Aldrich D3,916-9   CAS [6305-43-7] $C_4H_4Br_2O_2$ 60 MHz: *1*, 387C   187.87
                                        FW 243.89   FT-IR: *1*, 424B   28.41
4-Dibromo-2,3-butanedione, 99%   mp 118°C

A

DCl₃   QE-300

180   160   140   120   100   80   60   40   20   0

$BrCH_2-\overset{O}{\overset{\|}{C}}-\overset{O}{\overset{\|}{C}}-CH_2Br$

9   8   7   6   5   4   3   2   1   0

Aldrich 24,196-2   CAS [600-14-6] $C_5H_8O_2$ n_D^{20} 1.4040   199.78
                                      FW 100.12   Fp 66°F   197.45
3-Pentanedione, 97%   mp -52°C   60 MHz: *1*, 387A   29.21
                                      bp 111°C   FT-IR: *1*, 424A   23.70*
DCl₃   QE-300   d 0.957   VP-FT-IR: *3*, 510A   6.95*

180   160   140   120   100   80   60   40   20   0

B

$CH_3CH_2-\overset{O}{\overset{\|}{C}}-\overset{O}{\overset{\|}{C}}-CH_3$

9   8   7   6   5   4   3   2   1   0

Aldrich 30,693-2   CAS [4437-51-8] $C_6H_{10}O_2$ Fp 88°F   VP-FT-IR: *3*, 510C   200.26
                                       FW 114.15   29.61
4-Hexanedione, 95%   bp 124°C   6.93*
                                       d 0.939
DCl₃   QE-300 n_D^{20} 1.4100

fset: 40 ppm.
180   160   140   120   100   80   60   40   20   0

C

$CH_3CH_2-\overset{O}{\overset{\|}{C}}-\overset{O}{\overset{\|}{C}}-CH_2CH_3$

9   8   7   6   5   4   3   2   1   0

ALDRICH

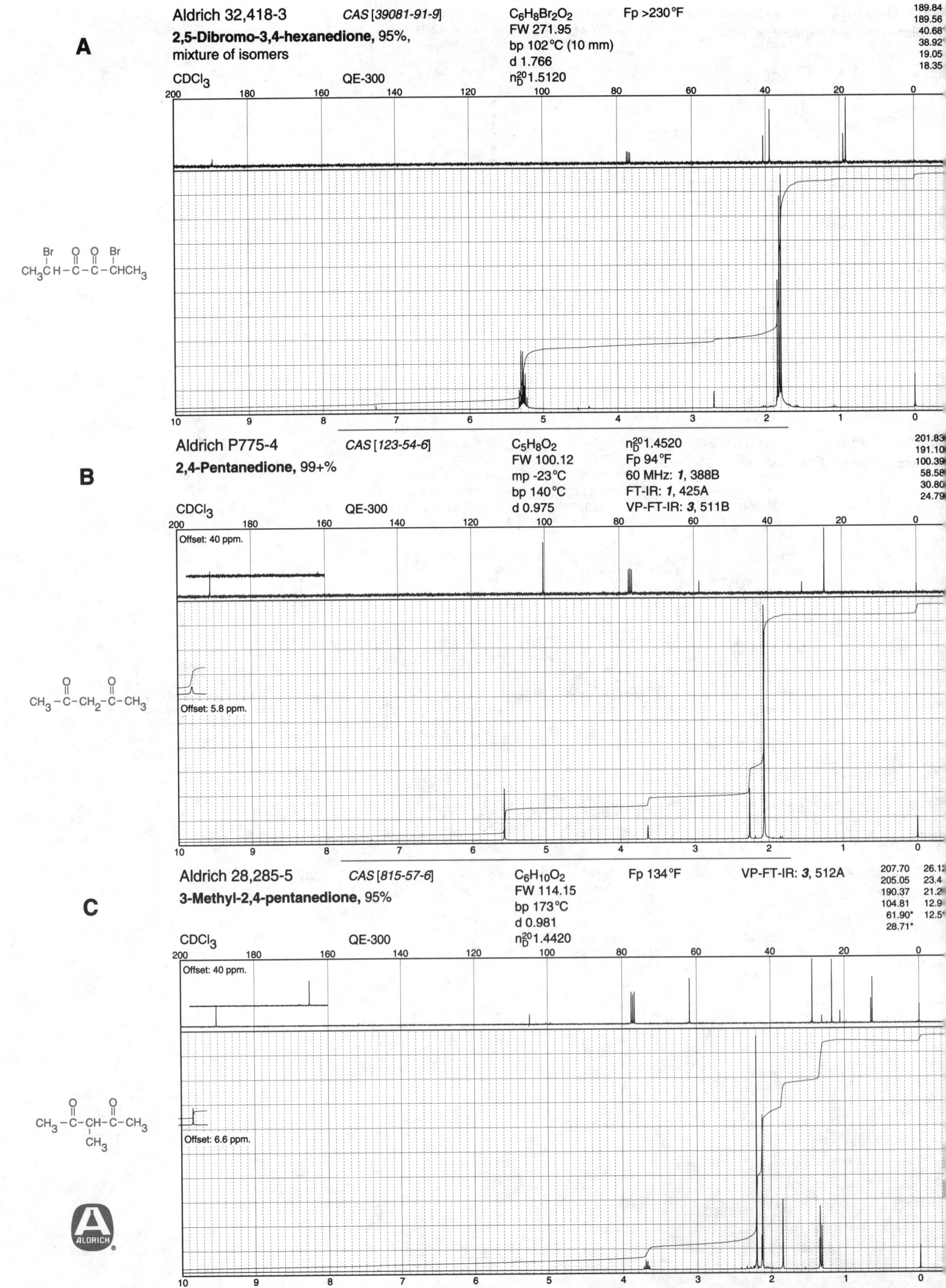

A

Aldrich 32,418-3   CAS [39081-91-9] $C_6H_8Br_2O_2$ Fp >230°F

2,5-Dibromo-3,4-hexanedione, 95%,
mixture of isomers

FW 271.95
bp 102°C (10 mm)
d 1.766
n_D^{20} 1.5120

189.84
189.56
40.68
38.92
19.05
18.35

CDCl₃   QE-300

Structure A: CH₃CHBr—CO—CO—CHBrCH₃

B

Aldrich P775-4   CAS [123-54-6]

2,4-Pentanedione, 99+%

$C_5H_8O_2$
FW 100.12
mp -23°C
bp 140°C
d 0.975

n_D^{20} 1.4520
Fp 94°F
60 MHz: 1, 388B
FT-IR: 1, 425A
VP-FT-IR: 3, 511B

201.83
191.10
100.39
58.58
30.80
24.79

CDCl₃   QE-300

Offset: 40 ppm.

Offset: 5.8 ppm.

Structure B: CH₃—CO—CH₂—CO—CH₃

C

Aldrich 28,285-5   CAS [815-57-6]

3-Methyl-2,4-pentanedione, 95%

$C_6H_{10}O_2$
FW 114.15
bp 173°C
d 0.981
n_D^{20} 1.4420

Fp 134°F   VP-FT-IR: 3, 512A

207.70   26.1
205.05   23.4
190.37   21.2
104.81   12.9
61.90*   12.5
28.71*

CDCl₃   QE-300

Offset: 40 ppm.

Offset: 6.6 ppm.

Structure C: CH₃—CO—CH—CO—CH₃ with CH₃ below

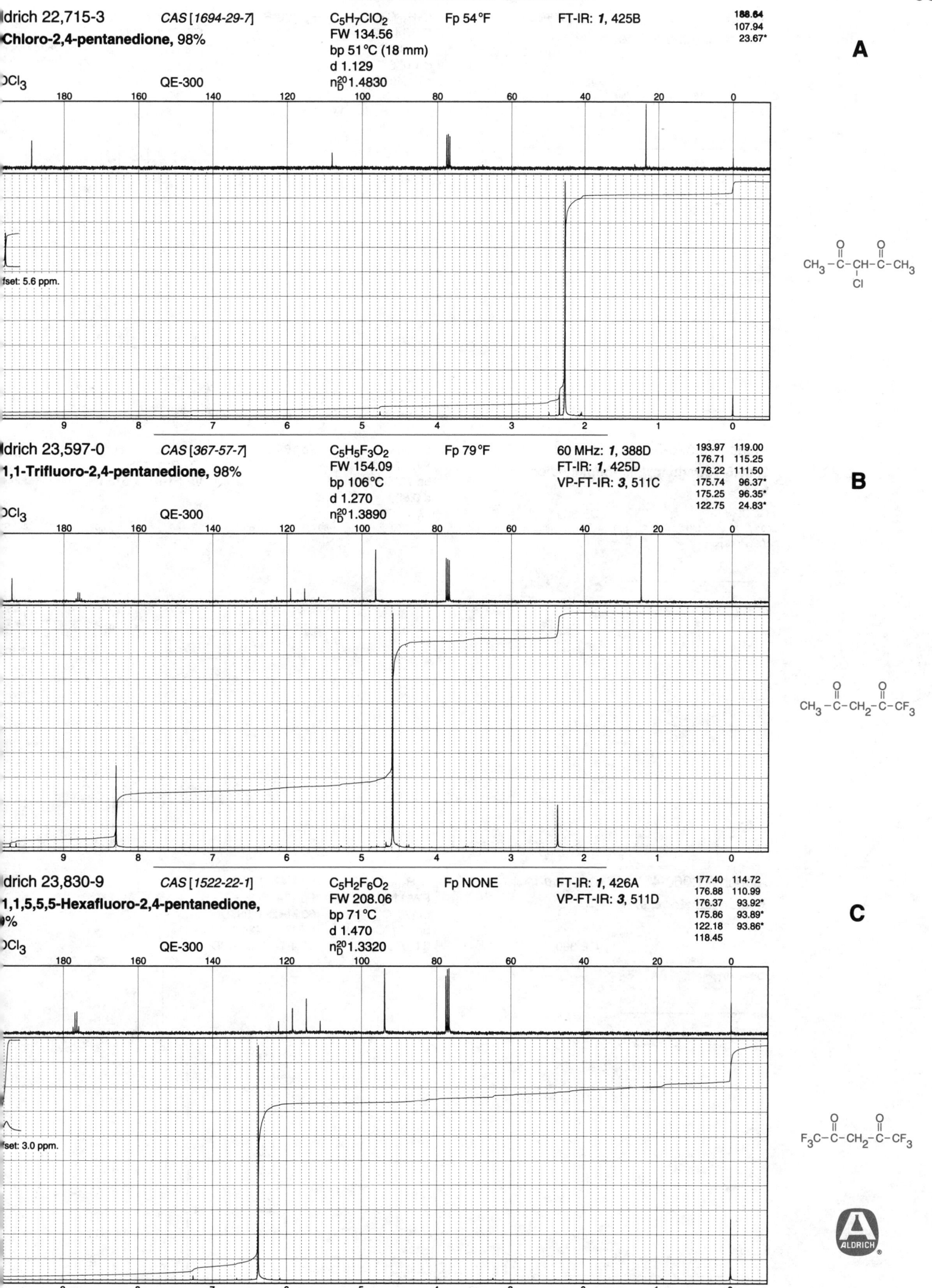

A

ldrich 22,715-3   *CAS [1694-29-7]* $C_5H_7ClO_2$ Fp 54°F   FT-IR: *1*, 425B   **188.64** 107.94 23.67*

Chloro-2,4-pentanedione, 98%

FW 134.56
bp 51°C (18 mm)
d 1.129
n_D^{20}1.4830

DCl₃   QE-300

fset: 5.6 ppm.

$CH_3-\overset{O}{\underset{\|}{C}}-\underset{\underset{Cl}{|}}{CH}-\overset{O}{\underset{\|}{C}}-CH_3$

B

ldrich 23,597-0   *CAS [367-57-7]* $C_5H_5F_3O_2$ Fp 79°F   60 MHz: *1*, 388D

1,1,1-Trifluoro-2,4-pentanedione, 98%

FW 154.09
bp 106°C
d 1.270
n_D^{20}1.3890

DCl₃   QE-300

193.97	119.00
176.71	115.25
176.22	111.50
175.74	96.37*
175.25	96.35*
122.75	24.83*

FT-IR: *1*, 425D
VP-FT-IR: *3*, 511C

$CH_3-\overset{O}{\underset{\|}{C}}-CH_2-\overset{O}{\underset{\|}{C}}-CF_3$

C

ldrich 23,830-9   *CAS [1522-22-1]* $C_5H_2F_6O_2$ Fp NONE   FT-IR: *1*, 426A

1,1,1,5,5,5-Hexafluoro-2,4-pentanedione, 9%

FW 208.06
bp 71°C
d 1.470
n_D^{20}1.3320

DCl₃   QE-300

177.40	114.72
176.88	110.99
176.37	93.92*
175.86	93.89*
122.18	93.86*
118.45	

VP-FT-IR: *3*, 511D

fset: 3.0 ppm.

$F_3C-\overset{O}{\underset{\|}{C}}-CH_2-\overset{O}{\underset{\|}{C}}-CF_3$

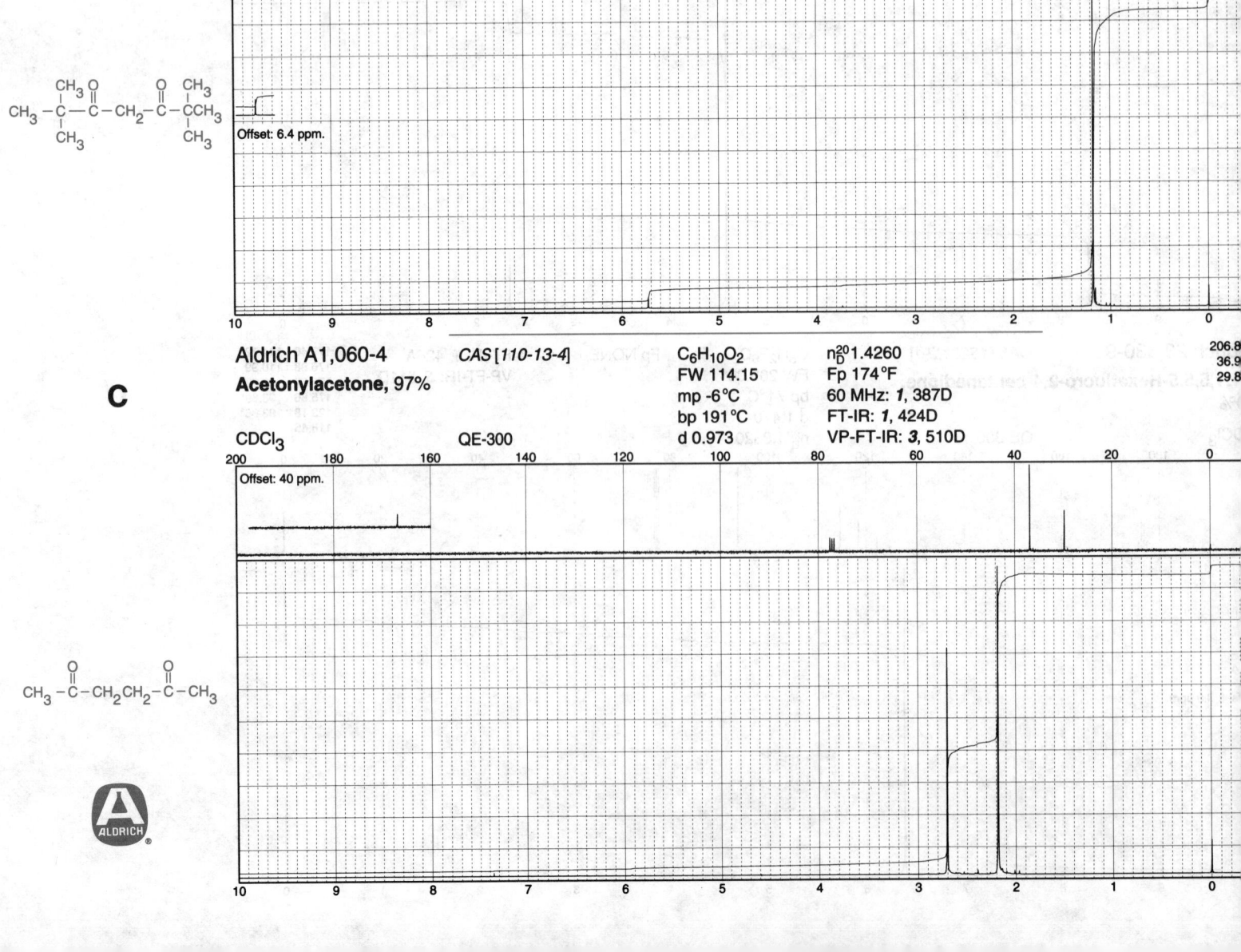

A

Aldrich 14,738-9   *CAS [815-68-9]*   C₇H₁₀O₃   Fp 190°F   60 MHz: *1*, 392A   200.80 100.42
Triacetylmethane, 97%   FW 142.15   FT-IR: *1*, 428D   192.92 32.70
bp 204°C   191.17 24.99
d 1.066   176.16 24.81
CDCl₃   QE-300 n_D^{20}1.4870   118.97 20.68

Offset: 40 ppm.

Offset: 7.4 ppm.

B

Aldrich 15,575-6   *CAS [1118-71-4]*   C₁₁H₂₀O₂   Fp 153°F   60 MHz: *1*, 388C   201.3
2,2,6,6-Tetramethyl-3,5-heptanedione,   FW 184.28   FT-IR: *1*, 425C   90.6
95%   bp 73°C (6 mm)   VP-FT-IR: *3*, 513A   39.4
d 0.883   27.4
CDCl₃   QE-300 n_D^{20}1.4590

Offset: 40 ppm.

Offset: 6.4 ppm.

C

Aldrich A1,060-4   *CAS [110-13-4]*   C₆H₁₀O₂ n_D^{20}1.4260   206.8
Acetonylacetone, 97%   FW 114.15   Fp 174°F   36.9
mp -6°C   60 MHz: *1*, 387D   29.8
bp 191°C   FT-IR: *1*, 424D
d 0.973   VP-FT-IR: *3*, 510D
CDCl₃   QE-300

Offset: 40 ppm.

ALDRICH

Non-Aromatic Ketones

664

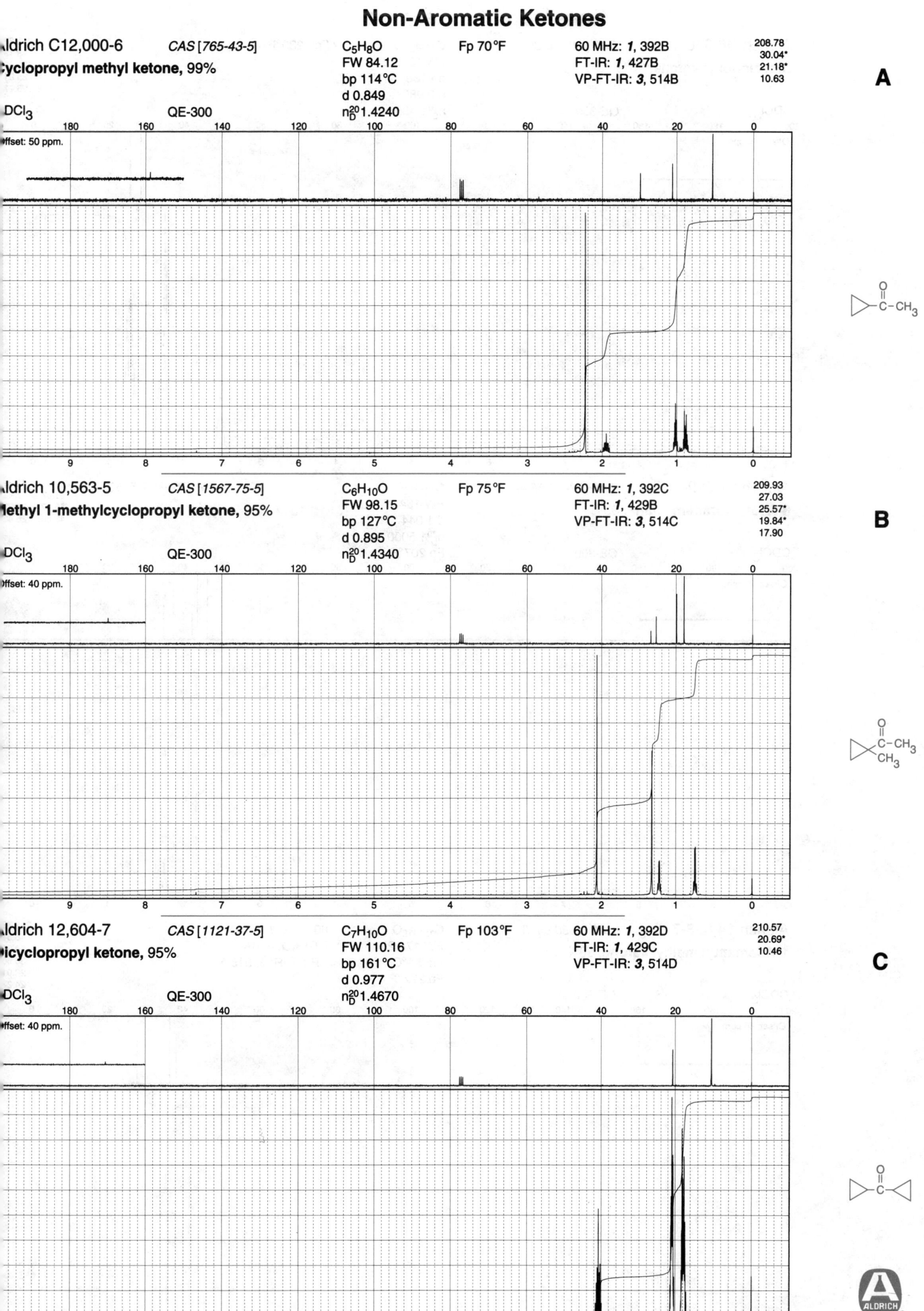

Aldrich C12,000-6   CAS [765-43-5]   C5H8O   Fp 70°F   60 MHz: 1, 392B   208.78 / 30.04* / 21.18* / 10.63   **A**
Cyclopropyl methyl ketone, 99%   FW 84.12   FT-IR: 1, 427B
CDCl3   QE-300   bp 114°C   d 0.849   n20D 1.4240   VP-FT-IR: 3, 514B
Offset: 50 ppm.

Aldrich 10,563-5   CAS [1567-75-5]   C6H10O   Fp 75°F   60 MHz: 1, 392C   209.93 / 27.03 / 25.57* / 19.84* / 17.90   **B**
Methyl 1-methylcyclopropyl ketone, 95%   FW 98.15   FT-IR: 1, 429B
CDCl3   QE-300   bp 127°C   d 0.895   n20D 1.4340   VP-FT-IR: 3, 514C
Offset: 40 ppm.

Aldrich 12,604-7   CAS [1121-37-5]   C7H10O   Fp 103°F   60 MHz: 1, 392D   210.57 / 20.69* / 10.46   **C**
Dicyclopropyl ketone, 95%   FW 110.16   FT-IR: 1, 429C
CDCl3   QE-300   bp 161°C   d 0.977   n20D 1.4670   VP-FT-IR: 3, 514D
Offset: 40 ppm.

ALDRICH

Non-Aromatic Ketones

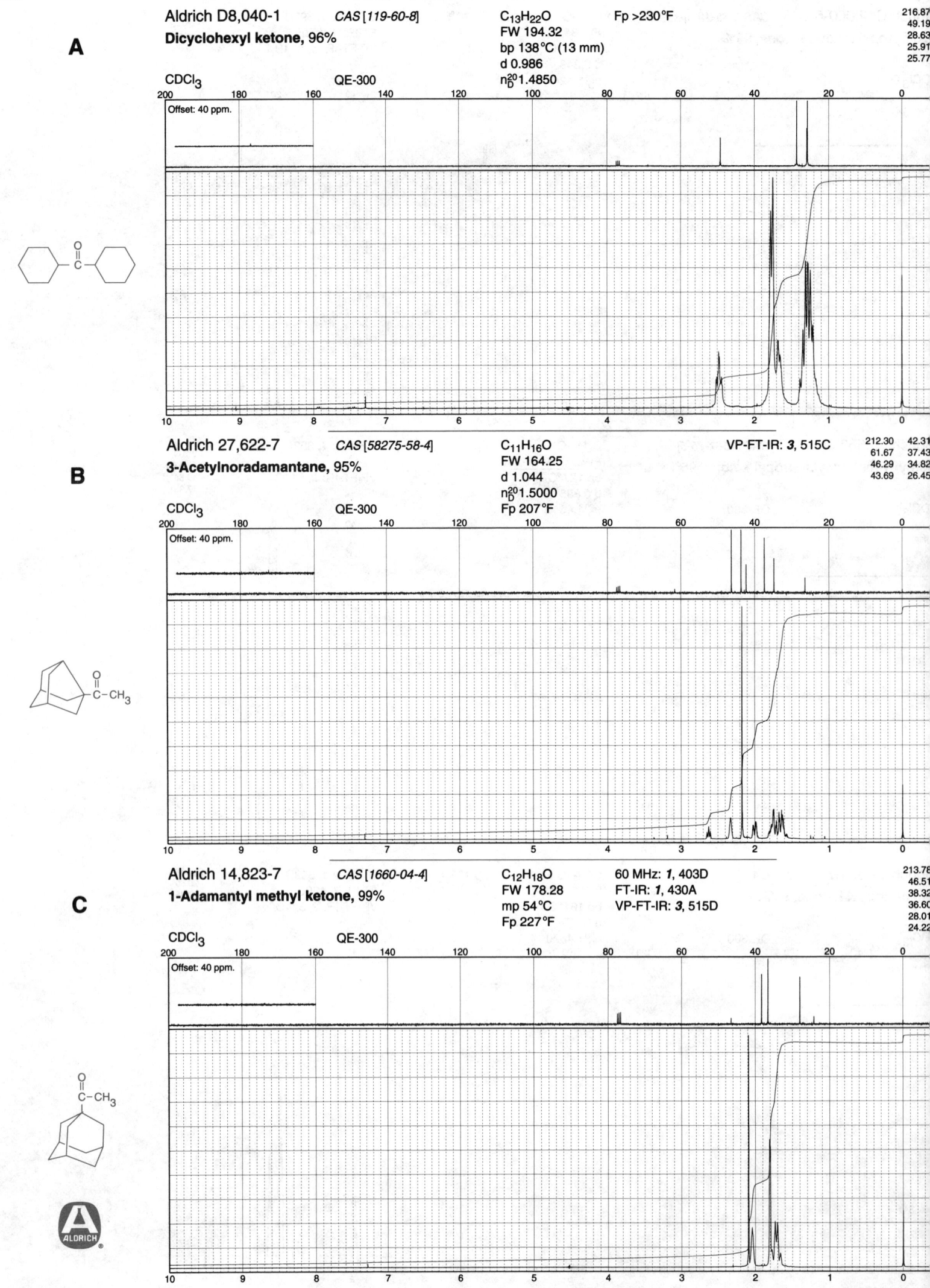

A

Aldrich D8,040-1   *CAS [119-60-8]* $C_{13}H_{22}O$ Fp >230°F

Dicyclohexyl ketone, 96%   FW 194.32
bp 138°C (13 mm)
d 0.986
n_D^{20} 1.4850

CDCl₃   QE-300

Offset: 40 ppm.

216.87
49.19
28.63
25.91
25.77

B

Aldrich 27,622-7   *CAS [58275-58-4]* $C_{11}H_{16}O$ VP-FT-IR: *3*, 515C

3-Acetylnoradamantane, 95%   FW 164.25
d 1.044
n_D^{20} 1.5000
Fp 207°F

CDCl₃   QE-300

Offset: 40 ppm.

212.30   42.31
61.67   37.43
46.29   34.82
43.69   26.45

C

Aldrich 14,823-7   *CAS [1660-04-4]* $C_{12}H_{18}O$ 60 MHz: *1*, 403D

1-Adamantyl methyl ketone, 99%   FW 178.28   FT-IR: *1*, 430A
mp 54°C   VP-FT-IR: *3*, 515D
Fp 227°F

CDCl₃   QE-300

Offset: 40 ppm.

213.78
46.51
38.32
36.60
28.01
24.22

ALDRICH

A

Aldrich 14,929-2   CAS [5122-82-7]   C₁₂H₁₇BrO   60 MHz: *1*, 404A
$C_{12}H_{17}BrO$
Adamantyl bromomethyl ketone, 98%   FW 257.18   FT-IR: *1*, 430B
mp 78°C

205.32
46.61
38.55
36.36
31.67
27.87*

CDCl₃   QE-300
offset: 40 ppm.

B

Aldrich C9,600-1   CAS [1191-95-3] C_4H_6O Fp 50°F   60 MHz: *1*, 393A
Cyclobutanone, 99%   FW 70.09   FT-IR: *1*, 430C
bp 99°C   VP-FT-IR: *3*, 516A
d 0.938
CDCl₃   QE-300 n_D^{20} 1.4210
offset: 40 ppm.

208.90
47.73
9.74

C

Aldrich C11,240-2   CAS [120-92-3] C_5H_8O n_D^{20} 1.4370
Cyclopentanone, 99+%   FW 84.12   Fp 87°F
mp -51°C   60 MHz: *1*, 393B
bp 131°C   FT-IR: *1*, 430D
d 0.951   VP-FT-IR: *3*, 516B
CDCl₃   QE-300
offset: 40 ppm.

220.16
38.30
23.24

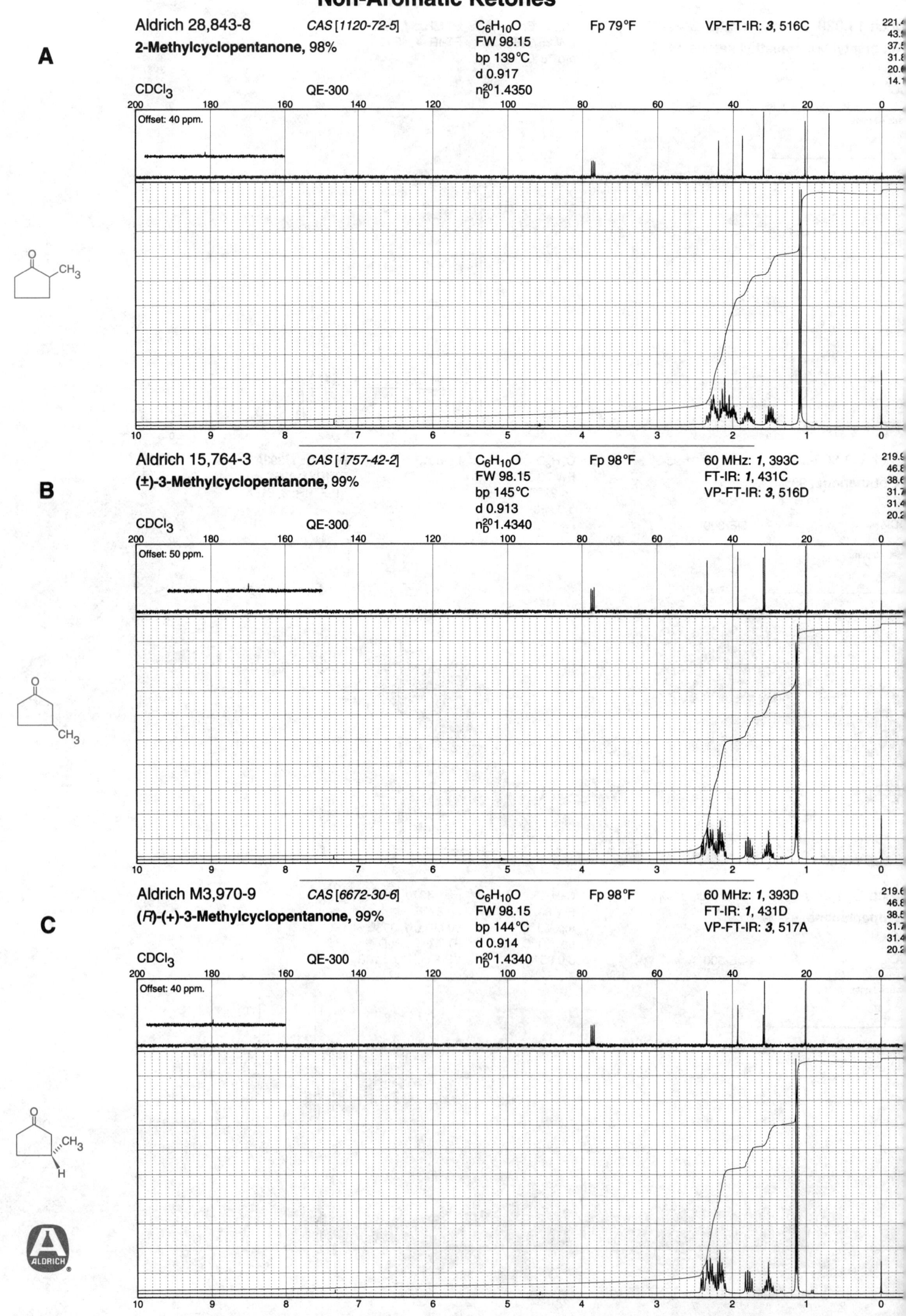

A

Aldrich 28,843-8   *CAS* [*1120-72-5*] $C_6H_{10}O$ Fp 79°F   VP-FT-IR: **3**, 516C

2-Methylcyclopentanone, 98%

FW 98.15
bp 139°C
d 0.917
n_D^{20} 1.4350

221.4
43.5
37.5
31.8
20.6
14.1

$CDCl_3$ QE-300

Offset: 40 ppm.

B

Aldrich 15,764-3   *CAS* [*1757-42-2*] $C_6H_{10}O$ Fp 98°F   60 MHz: **1**, 393C

(±)-3-Methylcyclopentanone, 99%

FW 98.15
bp 145°C
d 0.913
n_D^{20} 1.4340

FT-IR: **1**, 431C
VP-FT-IR: **3**, 516D

219.9
46.8
38.6
31.7
31.4
20.2

$CDCl_3$ QE-300

Offset: 50 ppm.

C

Aldrich M3,970-9   *CAS* [*6672-30-6*] $C_6H_{10}O$ Fp 98°F   60 MHz: **1**, 393D

(R)-(+)-3-Methylcyclopentanone, 99%

FW 98.15
bp 144°C
d 0.914
n_D^{20} 1.4340

FT-IR: **1**, 431D
VP-FT-IR: **3**, 517A

219.6
46.8
38.5
31.7
31.4
20.2

$CDCl_3$ QE-300

Offset: 40 ppm.

ALDRICH

Aldrich 37,147-5   *CAS [4541-32-6]* $C_7H_{12}O$ Fp 91°F

2,2-Dimethylcyclopentanone, 96+%

FW 112.17
bp 144°C
d 0.894
n_D^{20} 1.4330

CDCl₃   QE-300

Offset: 40 ppm.

A

223.63
44.87
38.47
37.09
23.73*
18.66

Aldrich 26,893-3   *CAS [1121-33-1]* $C_7H_{12}O$ Fp 107°F   FT-IR: *1*, 432A

(±)-2,4-Dimethylcyclopentanone, 99%,
mixture of *cis* and *trans*

FW 112.17
bp 153°C
d 0.895
n_D^{20} 1.4310

VP-FT-IR: *3*, 517B

CDCl₃   QE-300

Offset: 40 ppm.

B

220.81	29.62*
46.20	28.06*
46.05	20.79*
45.68*	20.32*
41.59*	15.35*
40.90	14.02*
39.13	

Aldrich C3,280-1   *CAS [694-28-0]* C_5H_7ClO Fp 171°F   60 MHz: *1*, 394B

2-Chlorocyclopentanone, 98%

FW 118.56
bp 73°C (12 mm)
d 1.185
n_D^{20} 1.4740

FT-IR: *1*, 446B
VP-FT-IR: *3*, 534C

CDCl₃   QE-300

Offset: 40 ppm.

C

210.71
58.40*
35.06
33.49
19.28

ALDRICH

Non-Aromatic Ketones

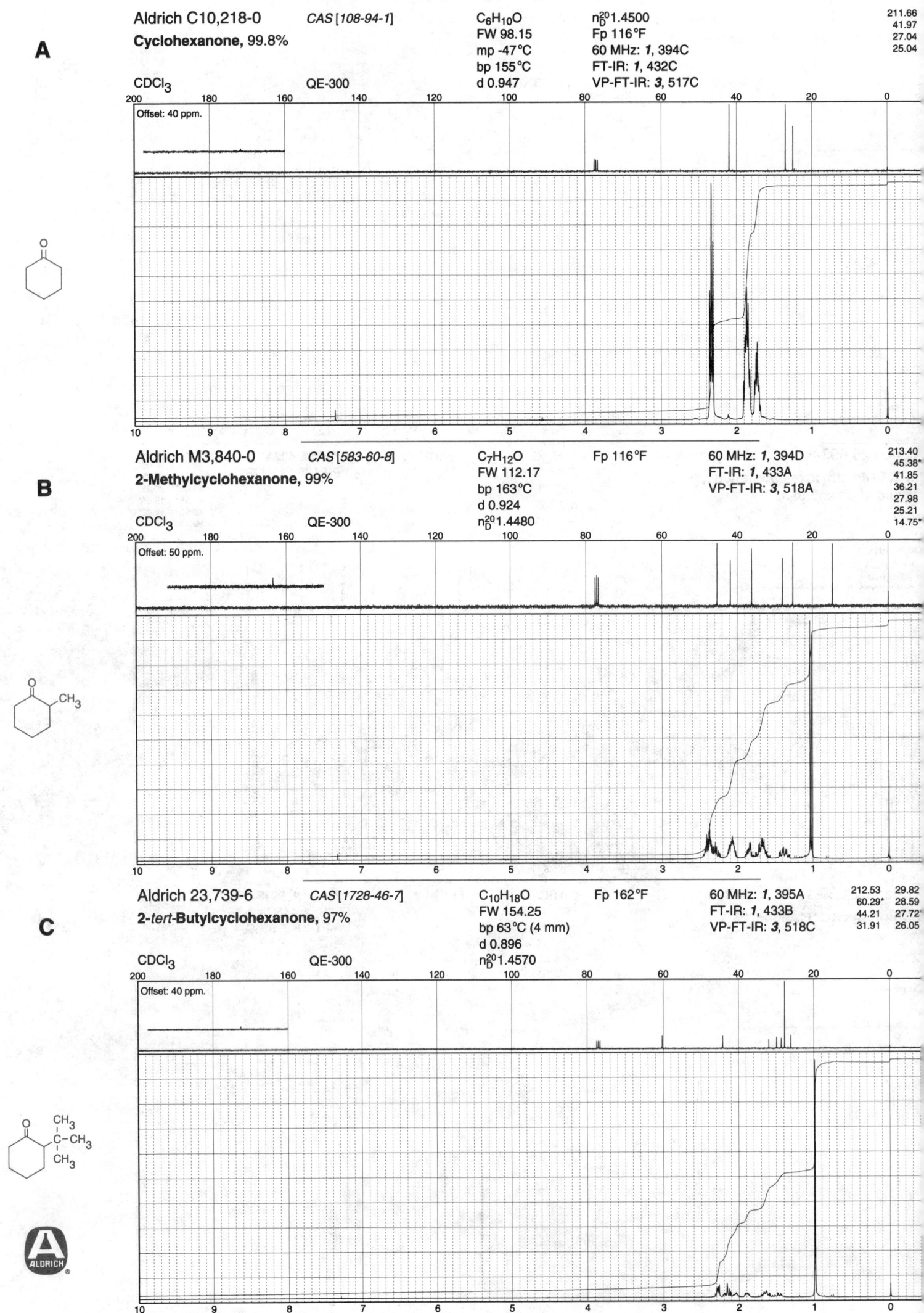

A

Aldrich C10,218-0

Cyclohexanone, 99.8%

CDCl$_3$

QE-300

CAS [108-94-1]

C$_6$H$_{10}$O
FW 98.15
mp -47°C
bp 155°C
d 0.947

n$_D^{20}$1.4500
Fp 116°F
60 MHz: **1**, 394C
FT-IR: **1**, 432C
VP-FT-IR: **3**, 517C

211.66
41.97
27.04
25.04

Offset: 40 ppm.

B

Aldrich M3,840-0

2-Methylcyclohexanone, 99%

CDCl$_3$

QE-300

CAS [583-60-8]

C$_7$H$_{12}$O
FW 112.17
bp 163°C
d 0.924
n$_D^{20}$1.4480

Fp 116°F

60 MHz: **1**, 394D
FT-IR: **1**, 433A
VP-FT-IR: **3**, 518A

213.40
45.38*
41.85
36.21
27.98
25.21
14.75*

Offset: 50 ppm.

C

Aldrich 23,739-6

2-tert-Butylcyclohexanone, 97%

CDCl$_3$

QE-300

CAS [1728-46-7]

C$_{10}$H$_{18}$O
FW 154.25
bp 63°C (4 mm)
d 0.896
n$_D^{20}$1.4570

Fp 162°F

60 MHz: **1**, 395A
FT-IR: **1**, 433B
VP-FT-IR: **3**, 518C

212.53   29.82
60.29*   28.59
44.21   27.72*
31.91   26.05

Offset: 40 ppm.

A

Aldrich M3,860-5   CAS [591-24-2]   C₇H₁₂O   Fp 125°F   60 MHz: *1*, 395B   211.76
(±)-3-Methylcyclohexanone, 97%   FW 112.17   FT-IR: *1*, 433C   49.97
bp 170°C   VP-FT-IR: *3*, 518D   41.10
d 0.914   34.19*
CDCl₃   QE-300   n²⁰_D 1.4450   33.28
25.30
22.06*

Offset: 40 ppm.

B

Aldrich M3,858-3   CAS [13368-65-5]   C₇H₁₂O   Fp 125°F   60 MHz: *1*, 395C   211.62
(R)-(+)-3-Methylcyclohexanone, 98%   FW 112.17   FT-IR: *1*, 433D   49.96
bp 169°C   VP-FT-IR: *3*, 519A   41.09
d 0.916   34.18*
CDCl₃   QE-300   n²⁰_D 1.4460   33.28
25.30
22.05*

Offset: 40 ppm.

C

Aldrich 17,361-4   CAS [589-92-4]   C₇H₁₂O   Fp 105°F   60 MHz: *1*, 395D   211.82
4-Methylcyclohexanone, 99%   FW 112.17   FT-IR: *1*, 434A   40.80
bp 170°C   VP-FT-IR: *3*, 519B   34.80
d 0.914   31.17*
CDCl₃   QE-300   n²⁰_D 1.4460   20.99*

Offset: 40 ppm.

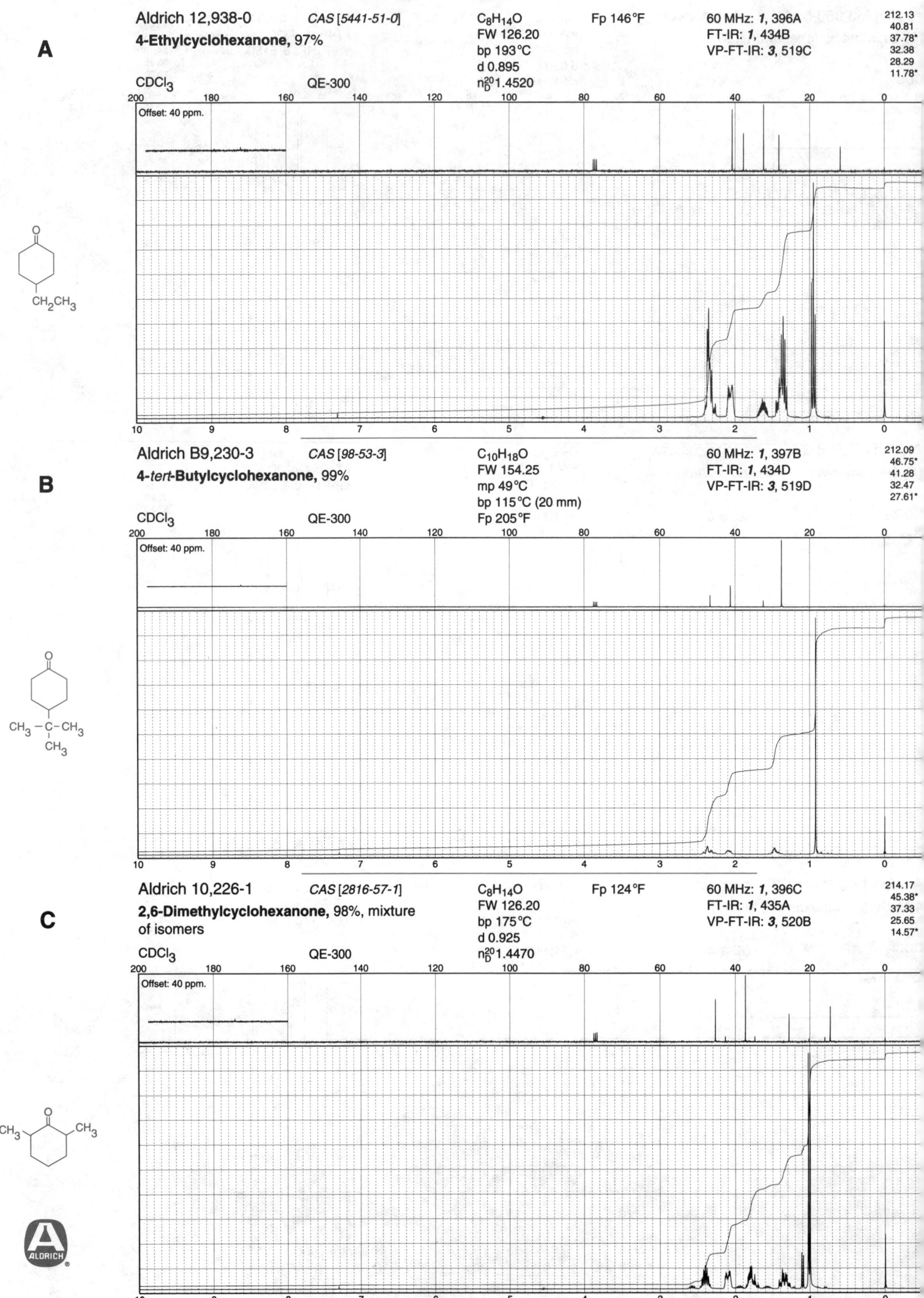

A

Aldrich 12,938-0   *CAS [5441-51-0]* $C_8H_{14}O$ Fp 146°F   60 MHz: *1*, 396A

4-Ethylcyclohexanone, 97%   FW 126.20   FT-IR: *1*, 434B

bp 193°C   VP-FT-IR: *3*, 519C

d 0.895

$CDCl_3$ QE-300 n_D^{20}1.4520

212.13
40.81
37.78*
32.38
28.29
11.78*

Offset: 40 ppm.

B

Aldrich B9,230-3   *CAS [98-53-3]* $C_{10}H_{18}O$ 60 MHz: *1*, 397B

4-*tert*-Butylcyclohexanone, 99%   FW 154.25   FT-IR: *1*, 434D

mp 49°C   VP-FT-IR: *3*, 519D

bp 115°C (20 mm)

Fp 205°F

$CDCl_3$ QE-300

212.09
46.75*
41.28
32.47
27.61*

Offset: 40 ppm.

C

Aldrich 10,226-1   *CAS [2816-57-1]* $C_8H_{14}O$ Fp 124°F   60 MHz: *1*, 396C

2,6-Dimethylcyclohexanone, 98%, mixture   FW 126.20   FT-IR: *1*, 435A

of isomers   bp 175°C   VP-FT-IR: *3*, 520B

d 0.925

$CDCl_3$ QE-300 n_D^{20}1.4470

214.17
45.38*
37.33
25.65
14.57*

Offset: 40 ppm.

Aldrich 21,823-5   *CAS [14073-97-3]* $C_{10}H_{18}O$ Fp 163°F   60 MHz: *1*, 397D

(-)-Menthone, 90%   FW 154.25   FT-IR: *1*, 436A

bp 209°C   VP-FT-IR: *3*, 521C

d 0.893

$CDCl_3$ QE-300 n_D^{20} 1.4500

Offset: 40 ppm.

212.07	27.92
55.91*	25.96*
50.87	22.27*
35.46*	21.19*
33.97	18.72*

A

Aldrich T7,573-6   *CAS [2408-37-9]* $C_9H_{16}O$ Fp 125°F   60 MHz: *1*, 396D

2,2,6-Trimethylcyclohexanone, 98%   FW 140.23   FT-IR: *1*, 435B

bp 179°C   VP-FT-IR: *3*, 520D

d 0.904

$CDCl_3$ QE-300 n_D^{20} 1.4470

Offset: 40 ppm.

216.97	25.67*
45.22	25.34*
41.87	21.60
40.79*	14.99*
36.80	

B

Aldrich T2,110-5   *CAS [14376-79-5]* $C_{10}H_{18}O$ 60 MHz: *1*, 397A

3,3,5,5-Tetramethylcyclohexanone, 98%   FW 154.25   FT-IR: *1*, 435C

d 0.881   VP-FT-IR: *3*, 521B

n_D^{20} 1.4510

Fp 164°F

$CDCl_3$ QE-300

Offset: 40 ppm.

| 211.98 |
| 53.94 |
| 51.65 |
| 36.07 |
| 31.38* |

C

Non-Aromatic Ketones

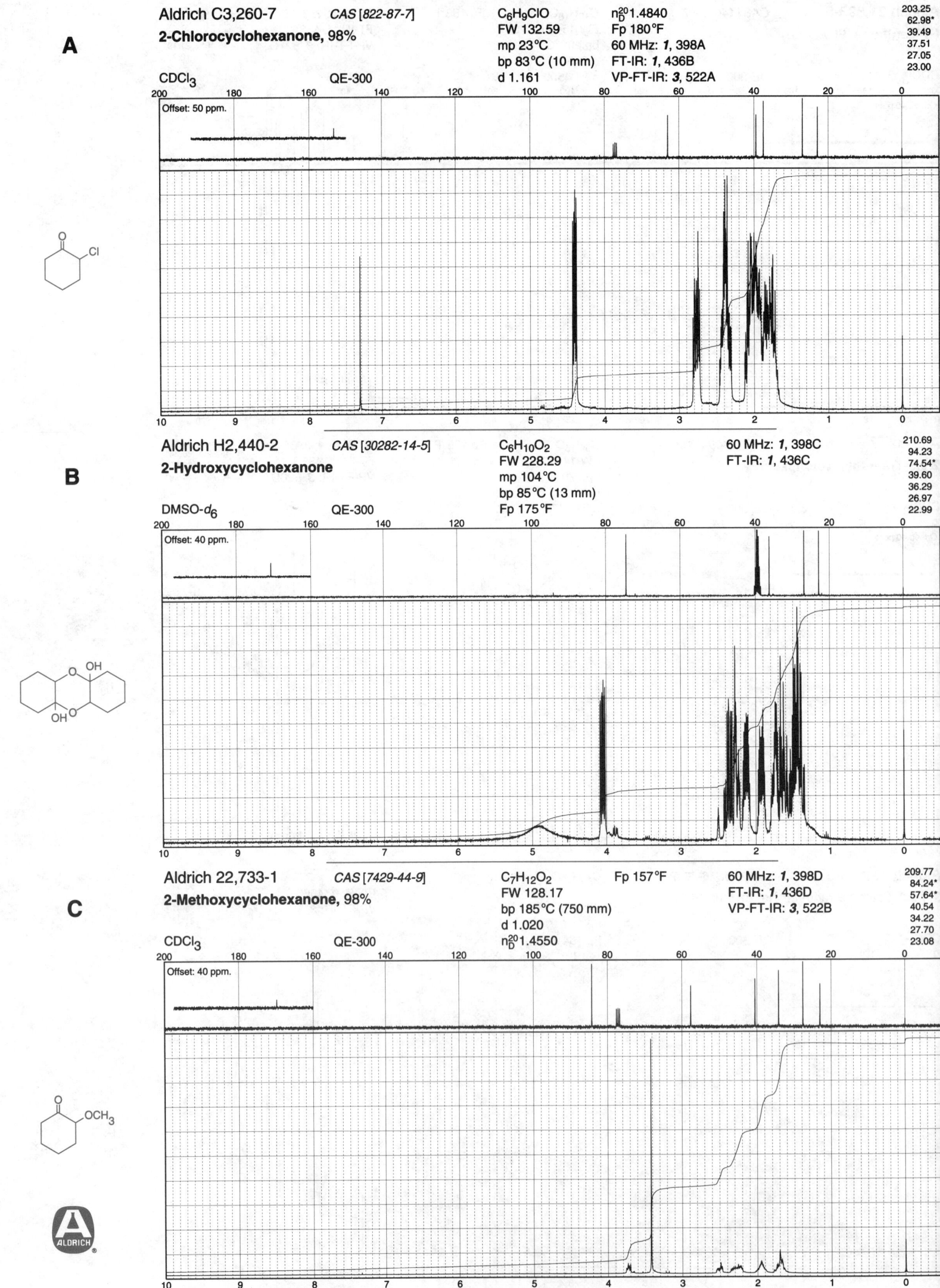

Aldrich C3,260-7   CAS [822-87-7]

2-Chlorocyclohexanone, 98%

C_6H_9ClO
FW 132.59
mp 23°C
bp 83°C (10 mm)
d 1.161

n_D^{20}1.4840
Fp 180°F
60 MHz: **1**, 398A
FT-IR: **1**, 436B
VP-FT-IR: **3**, 522A

203.25
62.98*
39.49
37.51
27.05
23.00

CDCl$_3$ QE-300

Offset: 50 ppm.

Aldrich H2,440-2   CAS [30282-14-5]

2-Hydroxycyclohexanone

$C_6H_{10}O_2$
FW 228.29
mp 104°C
bp 85°C (13 mm)
Fp 175°F

60 MHz: **1**, 398C
FT-IR: **1**, 436C

210.69
94.23
74.54*
39.60
36.29
26.97
22.99

DMSO-d_6 QE-300

Offset: 40 ppm.

Aldrich 22,733-1   CAS [7429-44-9]

2-Methoxycyclohexanone, 98%

$C_7H_{12}O_2$
FW 128.17
bp 185°C (750 mm)
d 1.020
n_D^{20}1.4550

Fp 157°F

60 MHz: **1**, 398D
FT-IR: **1**, 436D
VP-FT-IR: **3**, 522B

209.77
84.24*
57.64*
40.54
34.22
27.70
23.08

CDCl$_3$ QE-300

Offset: 40 ppm.

ALDRICH

Non-Aromatic Ketones

Aldrich W31,770-5   CAS [38462-22-5]   C₁₀H₁₈OS   Fp 227°F

p-Mentha-8-thiol-3-one, 94+%, mixture
of cis and trans isomers

FW 186.32
bp 56°C
d 1.000
n²⁰_D 1.4950

210.62	36.20*	30.41*
61.03*	34.27	30.28*
60.92*	32.19*	29.28
52.03	31.03*	25.06
49.98	30.90*	22.26*
44.27	30.56*	18.63*

CDCl₃   QE-300

Offset: 40 ppm.

A

Aldrich 22,448-0   CAS [4883-67-4]   C₆H₉NO₃   60 MHz: 1, 399B

2-Nitrocyclohexanone, 99%

FW 143.14
mp 41°C
Fp >230°F

FT-IR: 1, 463C

198.68	29.95
172.77	26.47
124.62	23.81
91.94*	22.58
40.77	21.97
31.65	21.16

CDCl₃   QE-300

Offset: 4.1 ppm.

B

Aldrich C9,900-0   CAS [502-42-1]   C₇H₁₂O   Fp 132°F

Cycloheptanone, 99%

FW 112.17
bp 179°C
d 0.951
n²⁰_D 1.4610

60 MHz: 1, 399C
FT-IR: 1, 437B
VP-FT-IR: 3, 522D

214.93
43.87
30.45
24.39

CDCl₃   QE-300

Offset: 40 ppm.

C

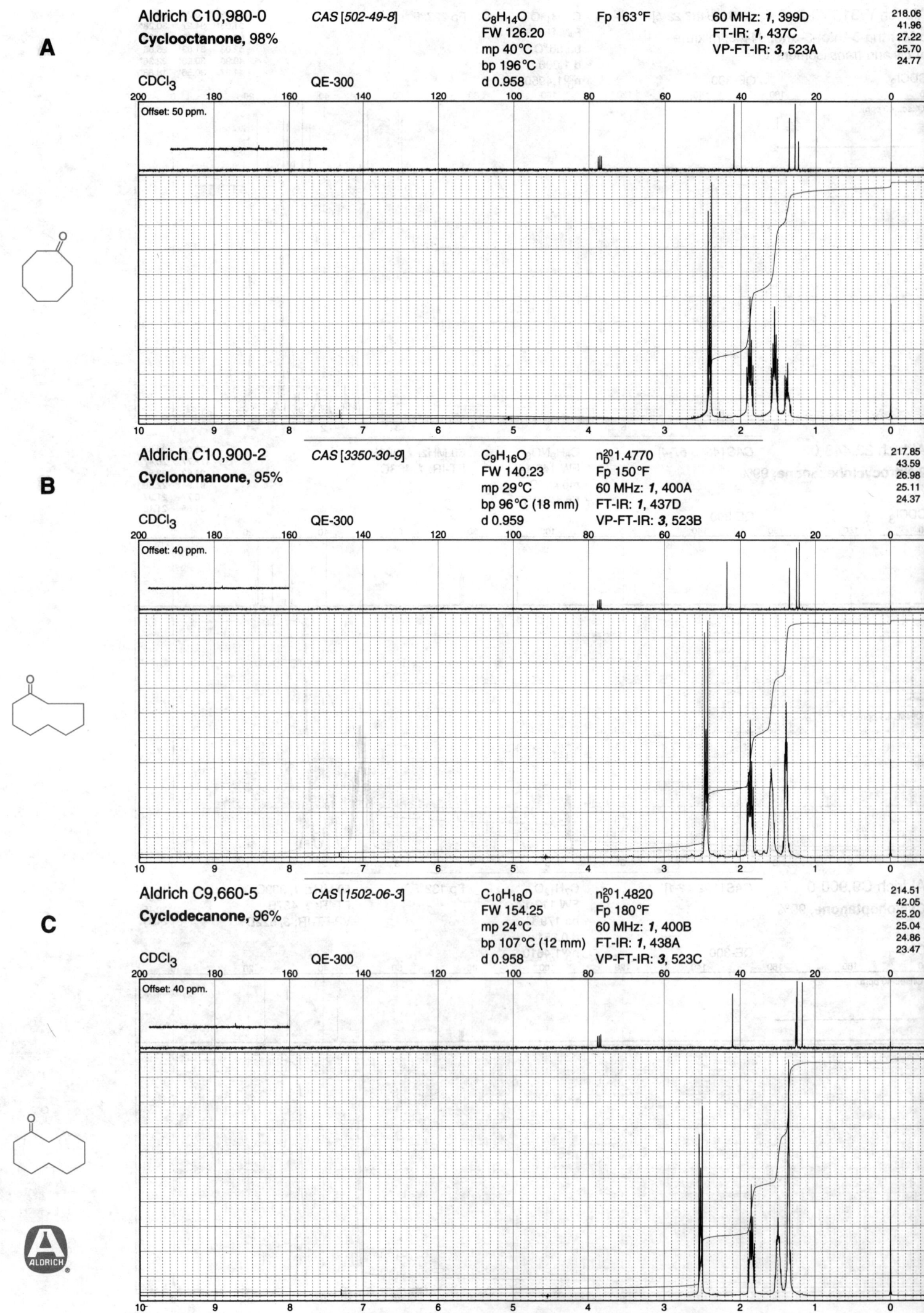

A

Aldrich C10,980-0
Cyclooctanone, 98%

CAS [502-49-8]

C$_8$H$_{14}$O
FW 126.20
mp 40°C
bp 196°C
d 0.958

Fp 163°F

60 MHz: **1**, 399D
FT-IR: **1**, 437C
VP-FT-IR: **3**, 523A

218.06
41.96
27.22
25.70
24.77

CDCl$_3$ QE-300 Offset: 50 ppm.

B

Aldrich C10,900-2
Cyclononanone, 95%

CAS [3350-30-9]

C$_9$H$_{16}$O
FW 140.23
mp 29°C
bp 96°C (18 mm)
d 0.959

n$_D^{20}$1.4770
Fp 150°F
60 MHz: **1**, 400A
FT-IR: **1**, 437D
VP-FT-IR: **3**, 523B

217.85
43.59
26.98
25.11
24.37

CDCl$_3$ QE-300 Offset: 40 ppm.

C

Aldrich C9,660-5
Cyclodecanone, 96%

CAS [1502-06-3]

C$_{10}$H$_{18}$O
FW 154.25
mp 24°C
bp 107°C (12 mm)
d 0.958

n$_D^{20}$1.4820
Fp 180°F
60 MHz: **1**, 400B
FT-IR: **1**, 438A
VP-FT-IR: **3**, 523C

214.51
42.05
25.20
25.04
24.86
23.47

CDCl$_3$ QE-300 Offset: 40 ppm.

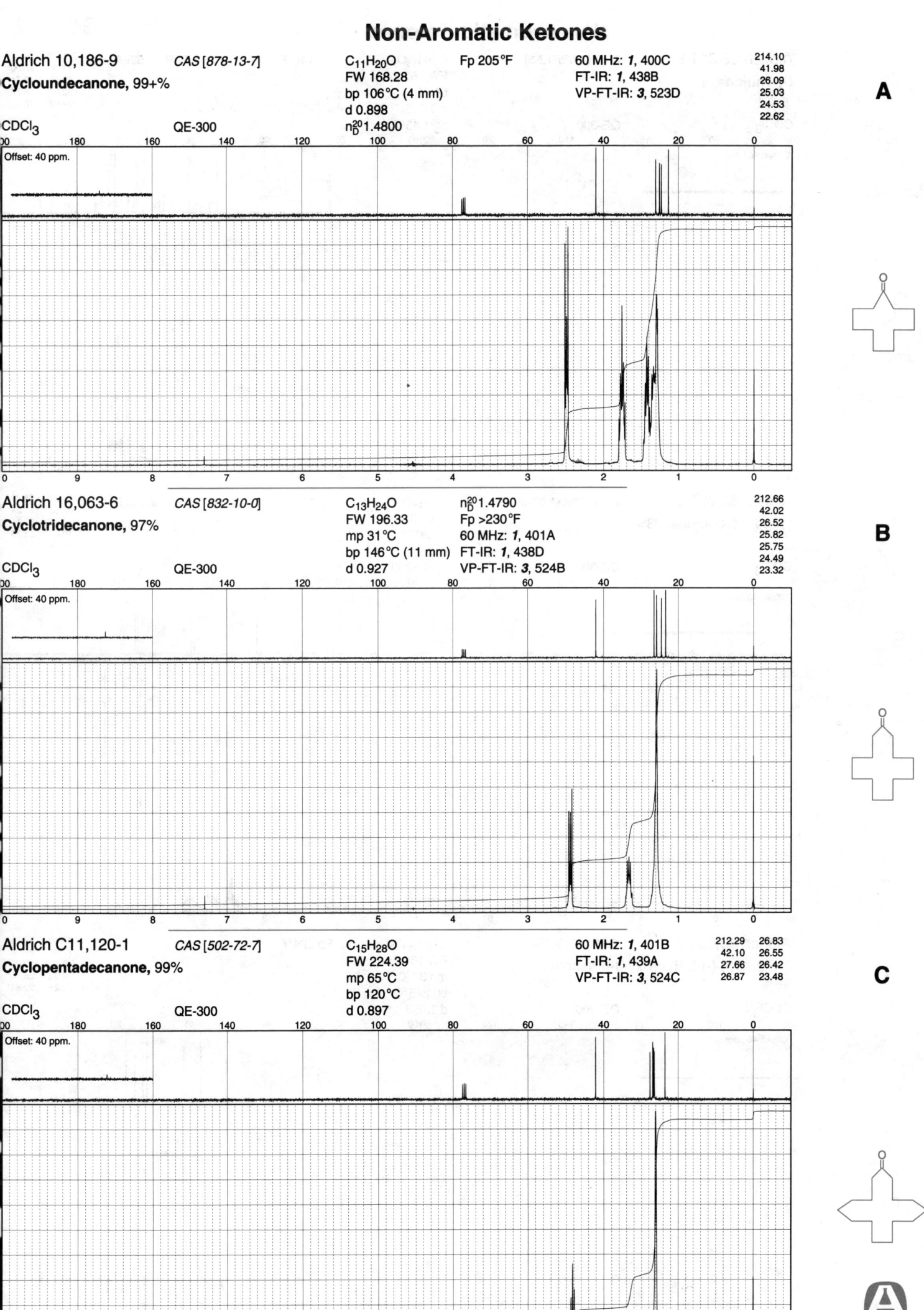

Aldrich 10,186-9   CAS [878-13-7]
Cycloundecanone, 99+%

$C_{11}H_{20}O$ Fp 205°F
FW 168.28
bp 106°C (4 mm)
d 0.898
n_D^{20} 1.4800

60 MHz: **1**, 400C
FT-IR: **1**, 438B
VP-FT-IR: **3**, 523D

214.10
41.98
26.09
25.03
24.53
22.62

CDCl₃   QE-300

Offset: 40 ppm.

Aldrich 16,063-6   CAS [832-10-0]
Cyclotridecanone, 97%

$C_{13}H_{24}O$ n_D^{20} 1.4790
FW 196.33   Fp >230°F
mp 31°C
bp 146°C (11 mm)   FT-IR: **1**, 438D
d 0.927   VP-FT-IR: **3**, 524B

60 MHz: **1**, 401A

212.66
42.02
26.52
25.82
25.75
24.49
23.32

CDCl₃   QE-300

Offset: 40 ppm.

Aldrich C11,120-1   CAS [502-72-7]
Cyclopentadecanone, 99%

$C_{15}H_{28}O$
FW 224.39
mp 65°C
bp 120°C
d 0.897

60 MHz: **1**, 401B
FT-IR: **1**, 439A
VP-FT-IR: **3**, 524C

212.29   26.83
42.10   26.55
27.66   26.42
26.87   23.48

CDCl₃   QE-300

Offset: 40 ppm.

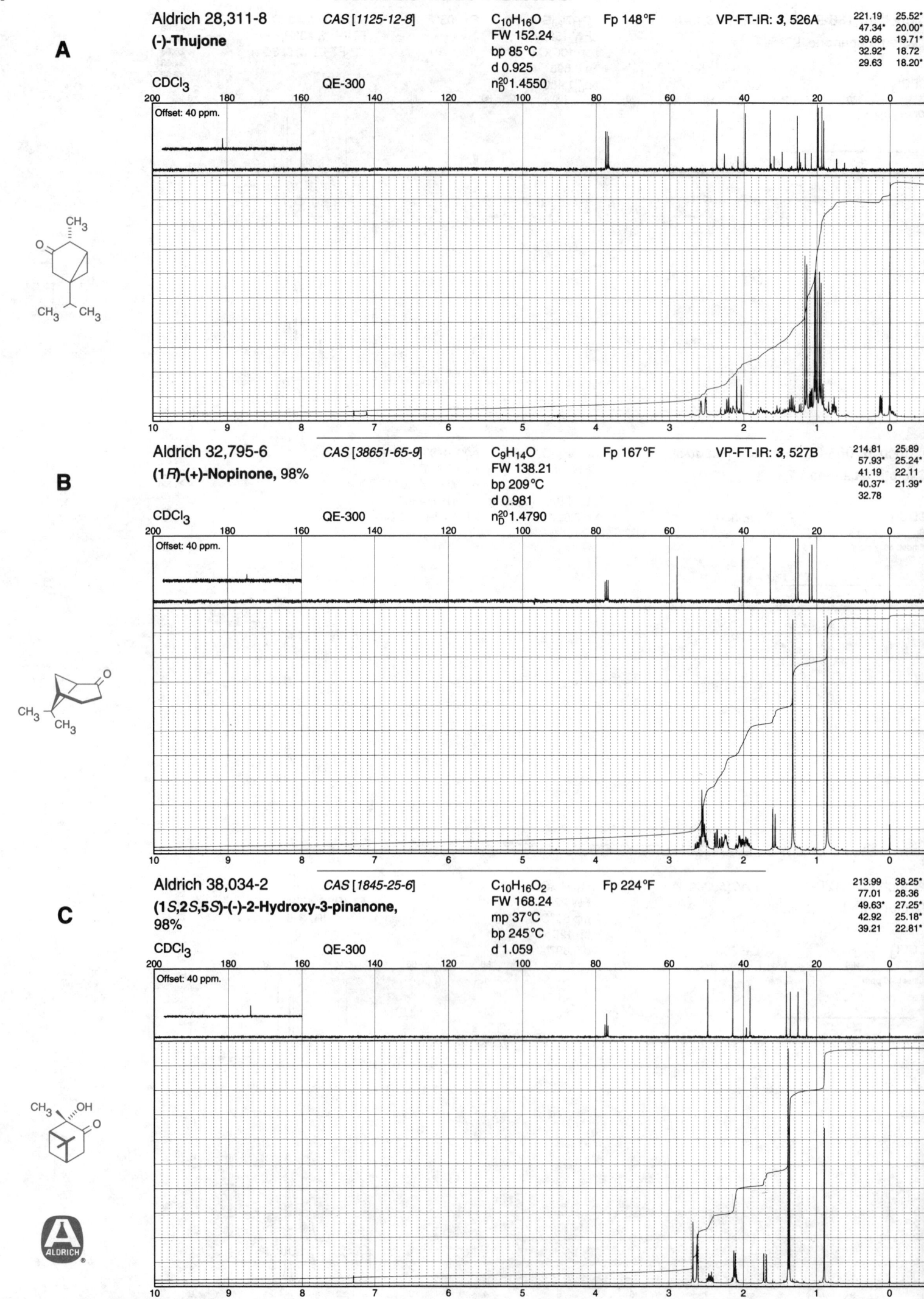

A

Aldrich 28,311-8

(-)-Thujone

CDCl₃   QE-300

CAS [1125-12-8]

$C_{10}H_{16}O$
FW 152.24
bp 85°C
d 0.925
n_D^{20}1.4550

Fp 148°F

VP-FT-IR: **3**, 526A

221.19	25.52*
47.34*	20.00*
39.66	19.71*
32.92*	18.72
29.63	18.20*

Offset: 40 ppm.

B

Aldrich 32,795-6

(1R)-(+)-Nopinone, 98%

CDCl₃   QE-300

CAS [38651-65-9]

$C_9H_{14}O$
FW 138.21
bp 209°C
d 0.981
n_D^{20}1.4790

Fp 167°F

VP-FT-IR: **3**, 527B

214.81	25.89
57.93*	25.24*
41.19	22.11
40.37*	21.39*
32.78	

Offset: 40 ppm.

C

Aldrich 38,034-2

(1S,2S,5S)-(-)-2-Hydroxy-3-pinanone, 98%

CDCl₃   QE-300

CAS [1845-25-6]

$C_{10}H_{16}O_2$
FW 168.24
mp 37°C
bp 245°C
d 1.059

Fp 224°F

213.99	38.25*
77.01	28.36
49.63*	27.25*
42.92	25.18*
39.21	22.81*

Offset: 40 ppm.

ALDRICH

Aldrich N3,260-1

CAS [22270-13-9]

$C_7H_{10}O$
FW 110.16
mp 95°C
bp 170°C
Fp 93°F

60 MHz: **1**, 402C
FT-IR: **1**, 440B
VP-FT-IR: **3**, 526B

217.73
49.84*
45.23
37.69
35.34*
27.23
24.21

(±)-Norcamphor, 98%

CDCl₃

QE-300

Offset: 40 ppm.

A

Aldrich 19,643-6

CAS [7787-20-4]

$C_{10}H_{16}O$
FW 152.24
mp 5°C
bp 193°C
d 0.948

n_D^{20} 1.4620
Fp 127°F
60 MHz: **1**, 406A
FT-IR: **1**, 442B
VP-FT-IR: **3**, 527C

222.94 31.89
54.11   24.98
47.36   23.37*
45.40* 21.69*
41.69   14.60*

(1R)-(-)-Fenchone, 98+%

CDCl₃

QE-300

Offset: 40 ppm.

B

Aldrich 14,807-5

CAS [21368-68-3]

$C_{10}H_{16}O$
FW 152.24
mp 176°C
bp 204°C
Fp 148°F

60 MHz: **1**, 404D
FT-IR: **1**, 441D
VP-FT-IR: **3**, 527D

219.21 29.95
57.64   27.07
46.75   19.77*
43.28   19.15*
43.09* 9.23*

(±)-Camphor, 96%

CDCl₃

QE-300

Offset: 40 ppm.

C

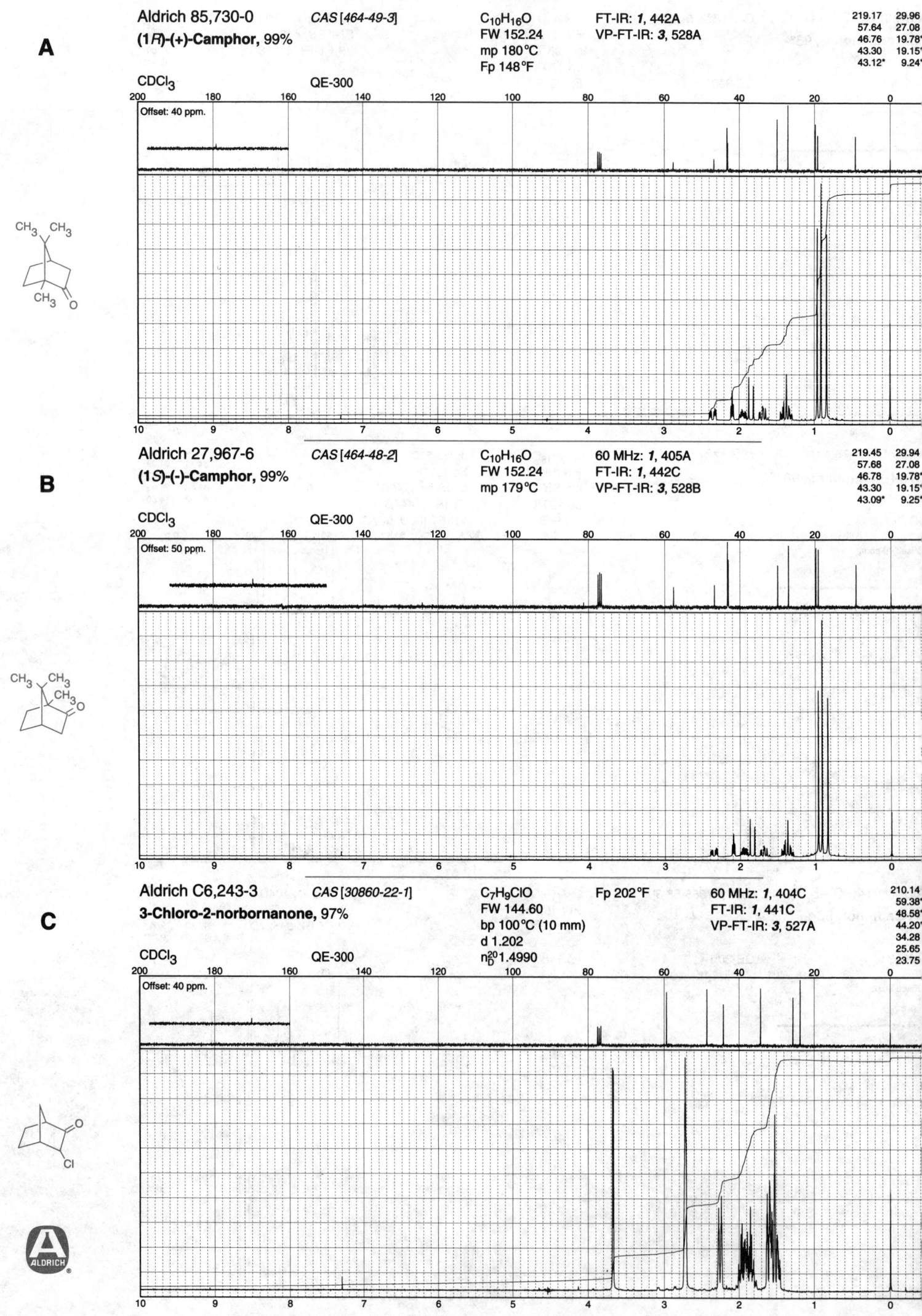

A

Aldrich 85,730-0   CAS [464-49-3] $C_{10}H_{16}O$ FT-IR: **1**, 442A

(1*R*)-(+)-Camphor, 99%   FW 152.24   VP-FT-IR: **3**, 528A

mp 180°C

Fp 148°F

219.17   29.96
57.64   27.08
46.76   19.78
43.30   19.15
43.12*   9.24*

CDCl₃   QE-300

Offset: 40 ppm.

B

Aldrich 27,967-6   CAS [464-48-2] $C_{10}H_{16}O$ 60 MHz: **1**, 405A

(1*S*)-(-)-Camphor, 99%   FW 152.24   FT-IR: **1**, 442C

mp 179°C   VP-FT-IR: **3**, 528B

219.45   29.94
57.68   27.08
46.78   19.78
43.30   19.15
43.09*   9.25*

CDCl₃   QE-300

Offset: 50 ppm.

C

Aldrich C6,243-3   CAS [30860-22-1] C_7H_9ClO Fp 202°F   60 MHz: **1**, 404C

3-Chloro-2-norbornanone, 97%   FW 144.60   FT-IR: **1**, 441C

bp 100°C (10 mm)   VP-FT-IR: **3**, 527A

d 1.202

n²⁰_D 1.4990

210.14
59.38*
48.58*
44.20*
34.28
25.65
23.75

CDCl₃   QE-300

Offset: 40 ppm.

Non-Aromatic Ketones

Aldrich 14,716-8   CAS [10293-06-8]   C₁₀H₁₅BrO   60 MHz: *1*, 405D

[(1*R*)-*endo*]-(+)-3-Bromocamphor, 98%

$C_{10}H_{15}BrO$

FW 231.14   FT-IR: *1*, 442D

mp 77°C   VP-FT-IR: *3*, 528C

212.15	30.52
57.55	22.43
53.83*	19.96*
49.63*	19.86*
45.85	9.59*

A

CDCl₃   QE-300

Offset: 40 ppm.

Aldrich 31,103-0   CAS [64474-54-0]

[(1*S*)-*endo*]-(-)-3-Bromocamphor, 98%

$C_{10}H_{15}BrO$

FW 231.14

mp 76°C

212.23	30.50
57.54	22.41
53.83*	19.95*
49.59*	19.86*
45.84	9.60*

B

CDCl₃   QE-300

Offset: 40 ppm.

Aldrich 30,642-8   CAS [10293-10-4]

3,9-Dibromo-(+)-camphor, 98%

$C_{10}H_{14}Br_2O$

FW 310.04

mp 158°C

210.48	38.96
58.24	30.13
52.30*	21.85
50.06	16.96*
47.60*	9.91*

C

CDCl₃   QE-300

Offset: 40 ppm.

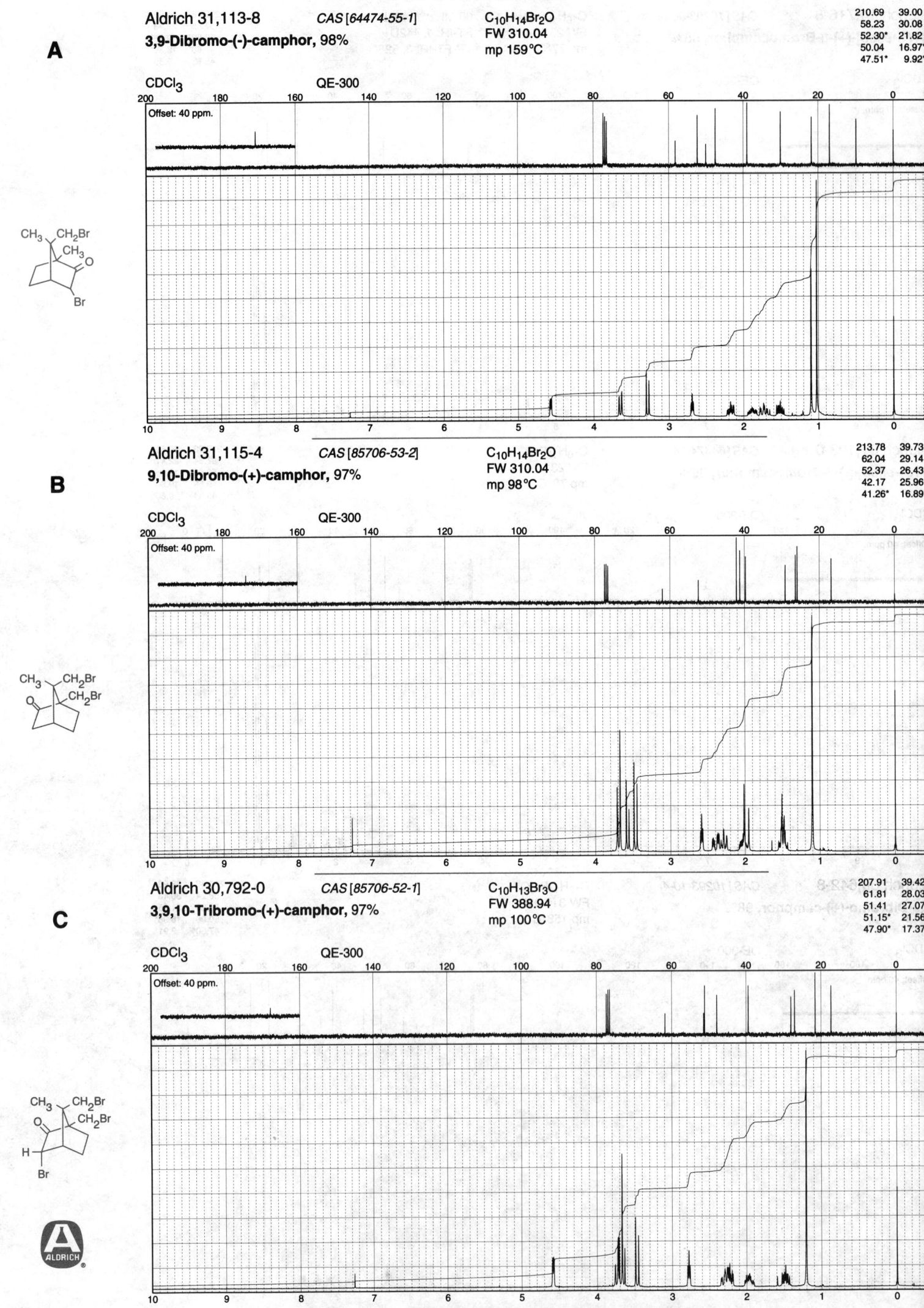

Aldrich 31,113-8   CAS [64474-55-1] $C_{10}H_{14}Br_2O$

A

3,9-Dibromo-(-)-camphor, 98%   FW 310.04   mp 159°C

210.69	39.00
58.23	30.08
52.30*	21.82
50.04	16.97*
47.51*	9.92*

CDCl₃   QE-300

Offset: 40 ppm.

Aldrich 31,115-4   CAS [85706-53-2] $C_{10}H_{14}Br_2O$

B

9,10-Dibromo-(+)-camphor, 97%   FW 310.04   mp 98°C

213.78	39.73
62.04	29.14
52.37	26.43
42.17	25.96
41.26*	16.89*

CDCl₃   QE-300

Offset: 40 ppm.

Aldrich 30,792-0   CAS [85706-52-1] $C_{10}H_{13}Br_3O$

C

3,9,10-Tribromo-(+)-camphor, 97%   FW 388.94   mp 100°C

207.91	39.42
61.81	28.03
51.41	27.07
51.15*	21.56
47.90*	17.37

CDCl₃   QE-300

Offset: 40 ppm.

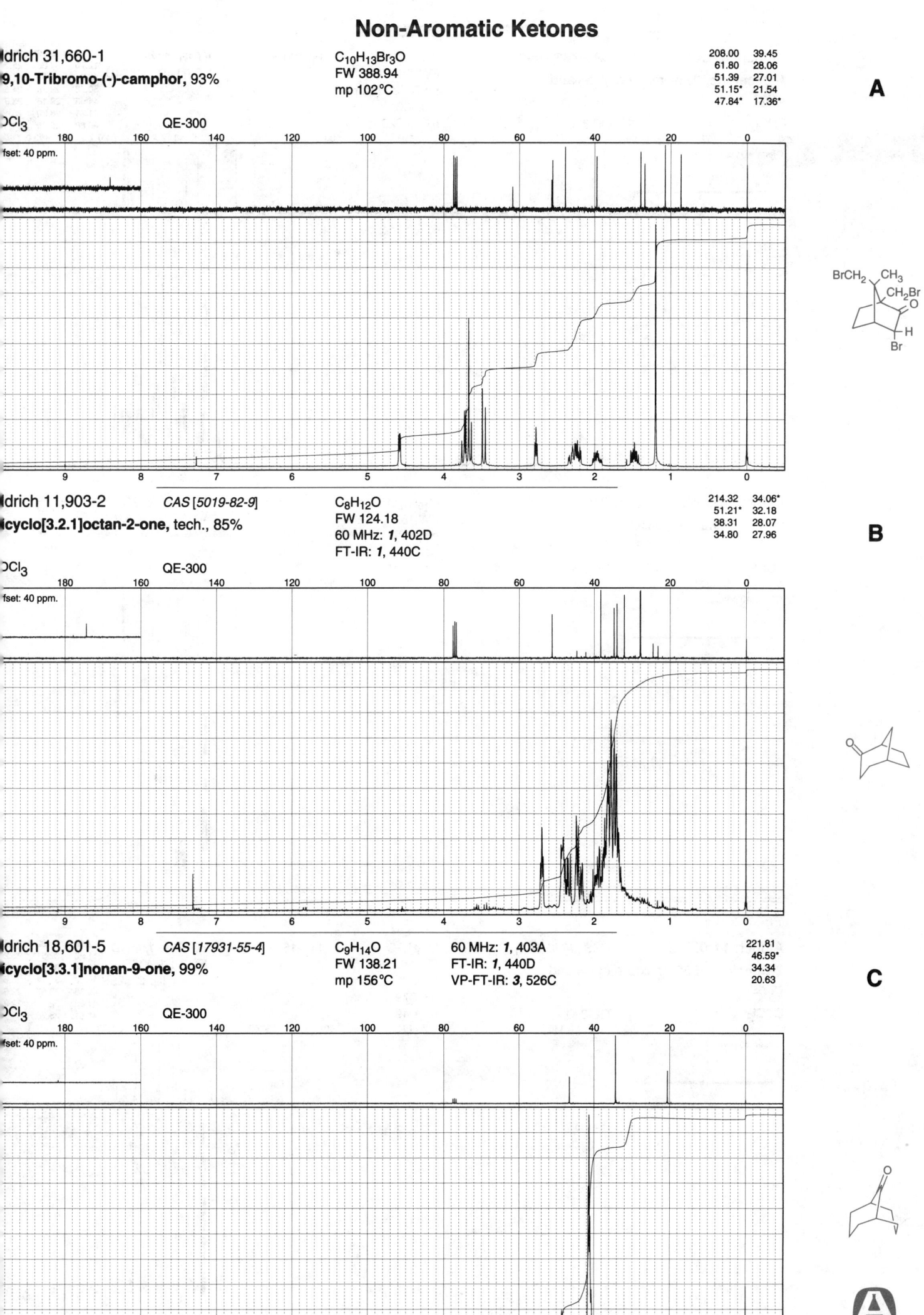

A

Idrich 31,660-1

9,10-Tribromo-(-)-camphor, 93%

$C_{10}H_{13}Br_3O$
FW 388.94
mp 102°C

208.00	39.45
61.80	28.06
51.39	27.01
51.15*	21.54
47.84*	17.36*

DCl₃   QE-300

fset: 40 ppm.

BrCH₂ CH₃
CH₂Br
O
H
Br

B

Idrich 11,903-2   CAS [5019-82-9]

cyclo[3.2.1]octan-2-one, tech., 85%

$C_8H_{12}O$
FW 124.18
60 MHz: 1, 402D
FT-IR: 1, 440C

214.32	34.06*
51.21*	32.18
38.31	28.07
34.80	27.96

DCl₃   QE-300

fset: 40 ppm.

C

Idrich 18,601-5   CAS [17931-55-4]

cyclo[3.3.1]nonan-9-one, 99%

$C_9H_{14}O$
FW 138.21
mp 156°C

60 MHz: 1, 403A
FT-IR: 1, 440D
VP-FT-IR: 3, 526C

221.81	
46.59*	
34.34	
20.63	

DCl₃   QE-300

fset: 40 ppm.

ALDRICH

Non-Aromatic Ketones

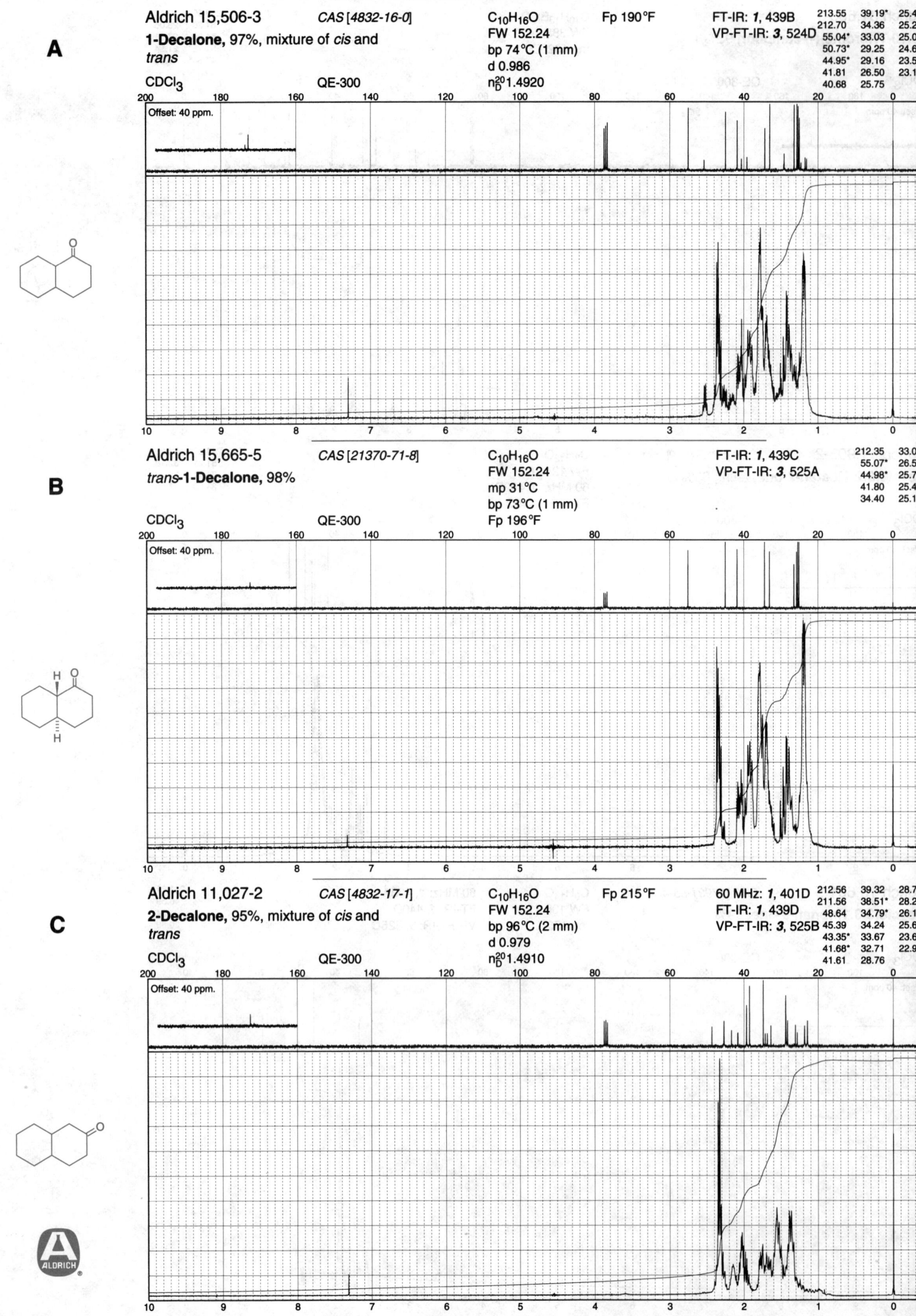

A

Aldrich 15,506-3   CAS [4832-16-0] $C_{10}H_{16}O$ Fp 190°F   FT-IR: *1*, 439B
1-Decalone, 97%, mixture of *cis* and   FW 152.24   VP-FT-IR: *3*, 524D
trans   bp 74°C (1 mm)
   d 0.986
CDCl₃   QE-300 n_D^{20}1.4920

213.55	39.19*	25.4
212.70	34.36	25.2
55.04*	33.03	25.0
50.73*	29.25	24.6
44.95*	29.16	23.5
41.81	26.50	23.1
40.68	25.75	

Offset: 40 ppm.

B

Aldrich 15,665-5   CAS [21370-71-8] $C_{10}H_{16}O$ FT-IR: *1*, 439C
trans-**1-Decalone**, 98%   FW 152.24   VP-FT-IR: *3*, 525A
   mp 31°C
   bp 73°C (1 mm)
CDCl₃   QE-300   Fp 196°F

212.35	33.0
55.07*	26.5
44.98*	25.7
41.80	25.4
34.40	25.1

Offset: 40 ppm.

C

Aldrich 11,027-2   CAS [4832-17-1] $C_{10}H_{16}O$ Fp 215°F   60 MHz: *1*, 401D
2-Decalone, 95%, mixture of *cis* and   FW 152.24   FT-IR: *1*, 439D
trans   bp 96°C (2 mm)   VP-FT-IR: *3*, 525B
   d 0.979
CDCl₃   QE-300 n_D^{20}1.4910

212.56	39.32	28.7
211.56	38.51*	28.2
48.64	34.79*	26.1
45.39	34.24	25.6
43.35*	33.67	23.6
41.68*	32.71	22.9
41.61	28.76	

Offset: 40 ppm.

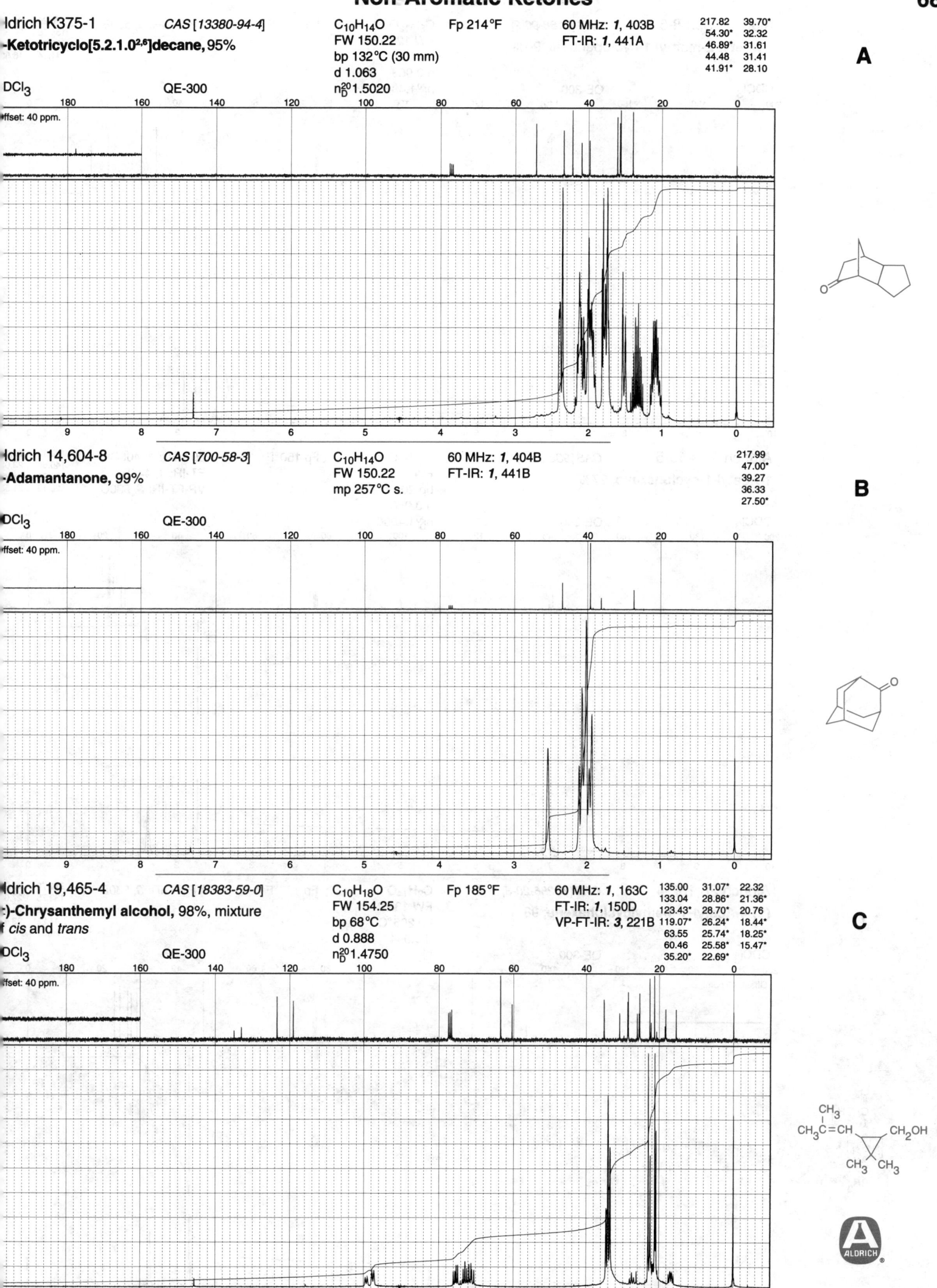

Idrich K375-1   *CAS [13380-94-4]* $C_{10}H_{14}O$ Fp 214°F   60 MHz: *1*, 403B

-Ketotricyclo[5.2.1.0²,⁶]decane, 95%

FW 150.22   FT-IR: *1*, 441A

bp 132°C (30 mm)

d 1.063

DCl₃   QE-300   n²⁰_D 1.5020

A

217.82	39.70*
54.30*	32.32
46.89*	31.61
44.48	31.41
41.91*	28.10

ffset: 40 ppm.

Idrich 14,604-8   *CAS [700-58-3]* $C_{10}H_{14}O$ 60 MHz: *1*, 404B

-Adamantanone, 99%

FW 150.22   FT-IR: *1*, 441B

mp 257°C s.

DCl₃   QE-300

B

217.99
47.00*
39.27
36.33
27.50*

ffset: 40 ppm.

Idrich 19,465-4   *CAS [18383-59-0]* $C_{10}H_{18}O$ Fp 185°F   60 MHz: *1*, 163C

)-Chrysanthemyl alcohol, 98%, mixture

FW 154.25   FT-IR: *1*, 150D

f *cis* and *trans*

bp 68°C   VP-FT-IR: *3*, 221B

DCl₃   QE-300   n²⁰_D 1.4750

d 0.888

C

135.00	31.07*	22.32
133.04	28.86*	21.36*
123.43*	28.70*	20.76
119.07*	26.24*	18.44*
63.55	25.74*	18.25*
60.46	25.58*	15.47*
35.20*	22.69*	

ffset: 40 ppm.

CH₃—C=CH, CH₂OH structure labels

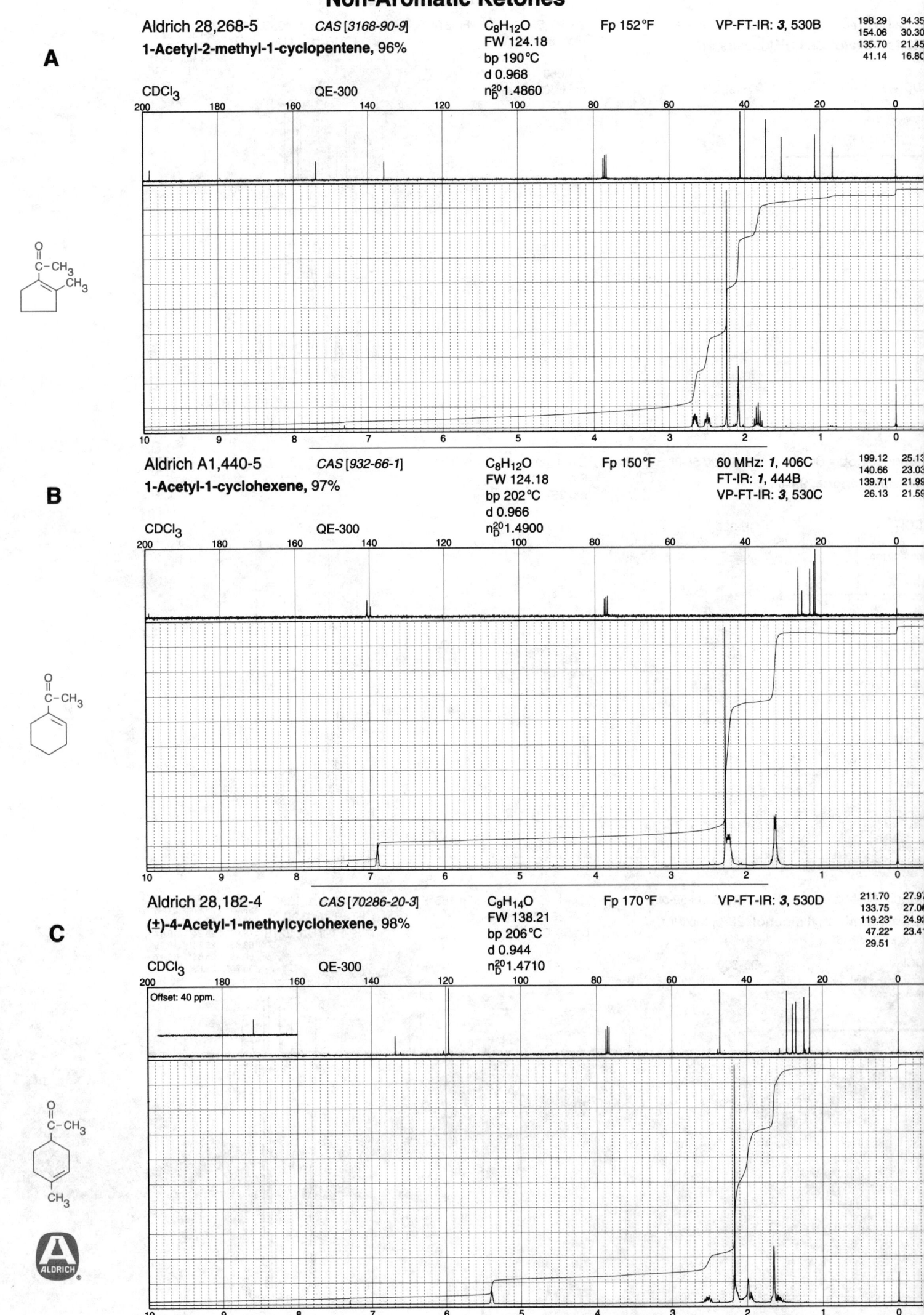

A

Aldrich 28,268-5   *CAS [3168-90-9]*   C$_8$H$_{12}$O   Fp 152°F   VP-FT-IR: **3**, 530B

198.29	34.35
154.06	30.30
135.70	21.45
41.14	16.80

1-Acetyl-2-methyl-1-cyclopentene, 96%

FW 124.18
bp 190°C
d 0.968
n$_D^{20}$1.4860

CDCl$_3$ QE-300

B

Aldrich A1,440-5   *CAS [932-66-1]*   C$_8$H$_{12}$O   Fp 150°F   60 MHz: **1**, 406C

199.12	25.13
140.66	23.03
139.71*	21.99
26.13	21.59

1-Acetyl-1-cyclohexene, 97%

FW 124.18   FT-IR: **1**, 444B
bp 202°C   VP-FT-IR: **3**, 530C
d 0.966
n$_D^{20}$1.4900

CDCl$_3$ QE-300

C

Aldrich 28,182-4   *CAS [70286-20-3]*   C$_9$H$_{14}$O   Fp 170°F   VP-FT-IR: **3**, 530D

211.70	27.97
133.75	27.06
119.23*	24.92
47.22*	23.4
29.51	

(±)-4-Acetyl-1-methylcyclohexene, 98%

FW 138.21
bp 206°C
d 0.944
n$_D^{20}$1.4710

CDCl$_3$ QE-300

Offset: 40 ppm.

Non-Aromatic Ketones

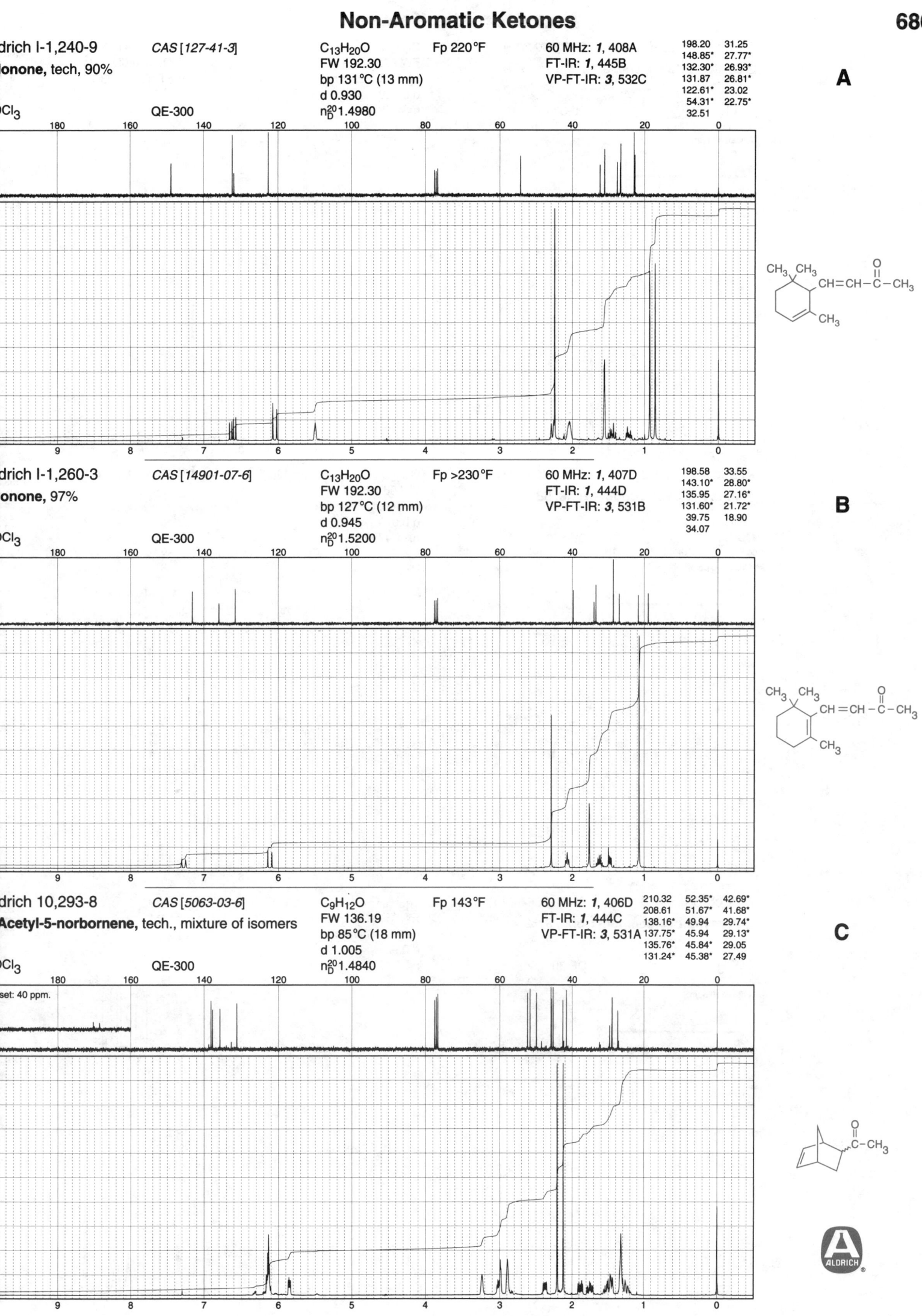

drich I-1,240-9   CAS [127-41-3]   C₁₃H₂₀O   Fp 220°F   60 MHz: *1*, 408A

lonone, tech, 90%

FW 192.30
bp 131°C (13 mm)
d 0.930
n₂₀D 1.4980

FT-IR: *1*, 445B
VP-FT-IR: *3*, 532C

198.20	31.25
148.85*	27.77*
132.30*	26.93*
131.87	26.81*
122.61*	23.02
54.31*	22.75*
32.51	

A

QE-300

)Cl₃

drich I-1,260-3   CAS [14901-07-6]   C₁₃H₂₀O   Fp >230°F   60 MHz: *1*, 407D

lonone, 97%

FW 192.30
bp 127°C (12 mm)
d 0.945
n₂₀D 1.5200

FT-IR: *1*, 444D
VP-FT-IR: *3*, 531B

198.58	33.55
143.10*	28.80*
135.95	27.16*
131.60*	21.72*
39.75	18.90
34.07	

B

QE-300

)Cl₃

drich 10,293-8   CAS [5063-03-6]   C₉H₁₂O   Fp 143°F   60 MHz: *1*, 406D

Acetyl-5-norbornene, tech., mixture of isomers

FW 136.19
bp 85°C (18 mm)
d 1.005
n₂₀D 1.4840

FT-IR: *1*, 444C
VP-FT-IR: *3*, 531A

210.32	52.35*	42.69*
208.61	51.67*	41.68*
138.16*	49.94	29.74*
137.75*	45.94	29.13*
135.76*	45.84*	29.05
131.24*	45.38*	27.49

C

fset: 40 ppm.

QE-300

)Cl₃

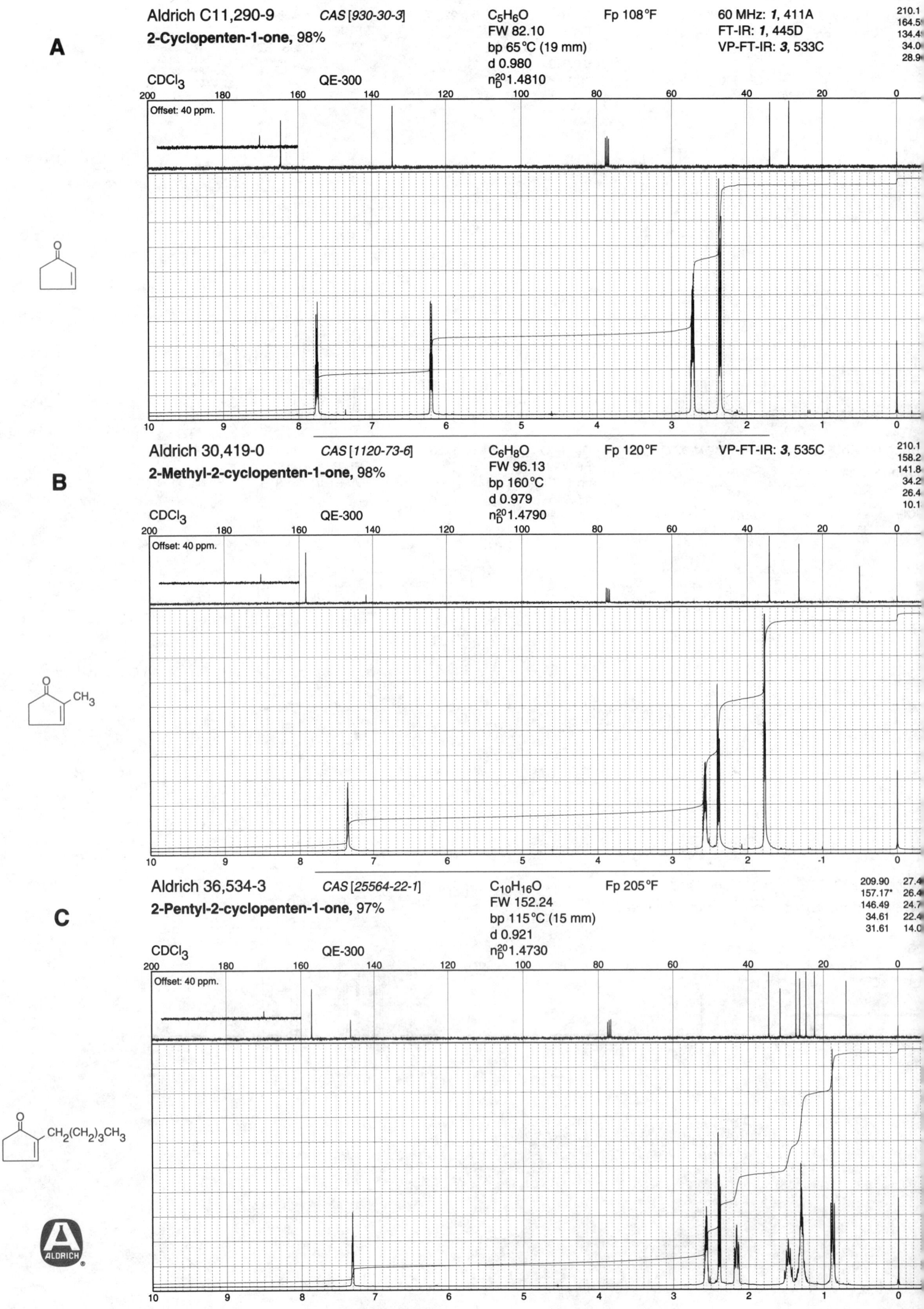

A

Aldrich C11,290-9   *CAS [930-30-3]*   C$_5$H$_6$O   Fp 108°F   60 MHz: *1*, 411A
2-Cyclopenten-1-one, 98%   FW 82.10   FT-IR: *1*, 445D
bp 65°C (19 mm)   VP-FT-IR: *3*, 533C
d 0.980
n$_D^{20}$1.4810

CDCl$_3$ QE-300

Offset: 40 ppm.

210.1
164.5
134.4
34.0
28.9

B

Aldrich 30,419-0   *CAS [1120-73-6]*   C$_6$H$_8$O   Fp 120°F   VP-FT-IR: *3*, 535C
2-Methyl-2-cyclopenten-1-one, 98%   FW 96.13
bp 160°C
d 0.979
n$_D^{20}$1.4790

CDCl$_3$ QE-300

Offset: 40 ppm.

210.1
158.2
141.8
34.2
26.4
10.1

C

Aldrich 36,534-3   *CAS [25564-22-1]*   C$_{10}$H$_{16}$O   Fp 205°F
2-Pentyl-2-cyclopenten-1-one, 97%   FW 152.24
bp 115°C (15 mm)
d 0.921
n$_D^{20}$1.4730

CDCl$_3$ QE-300

Offset: 40 ppm.

209.90   27.4
157.17*   26.4
146.49   24.7
34.61   22.4
31.61   14.0

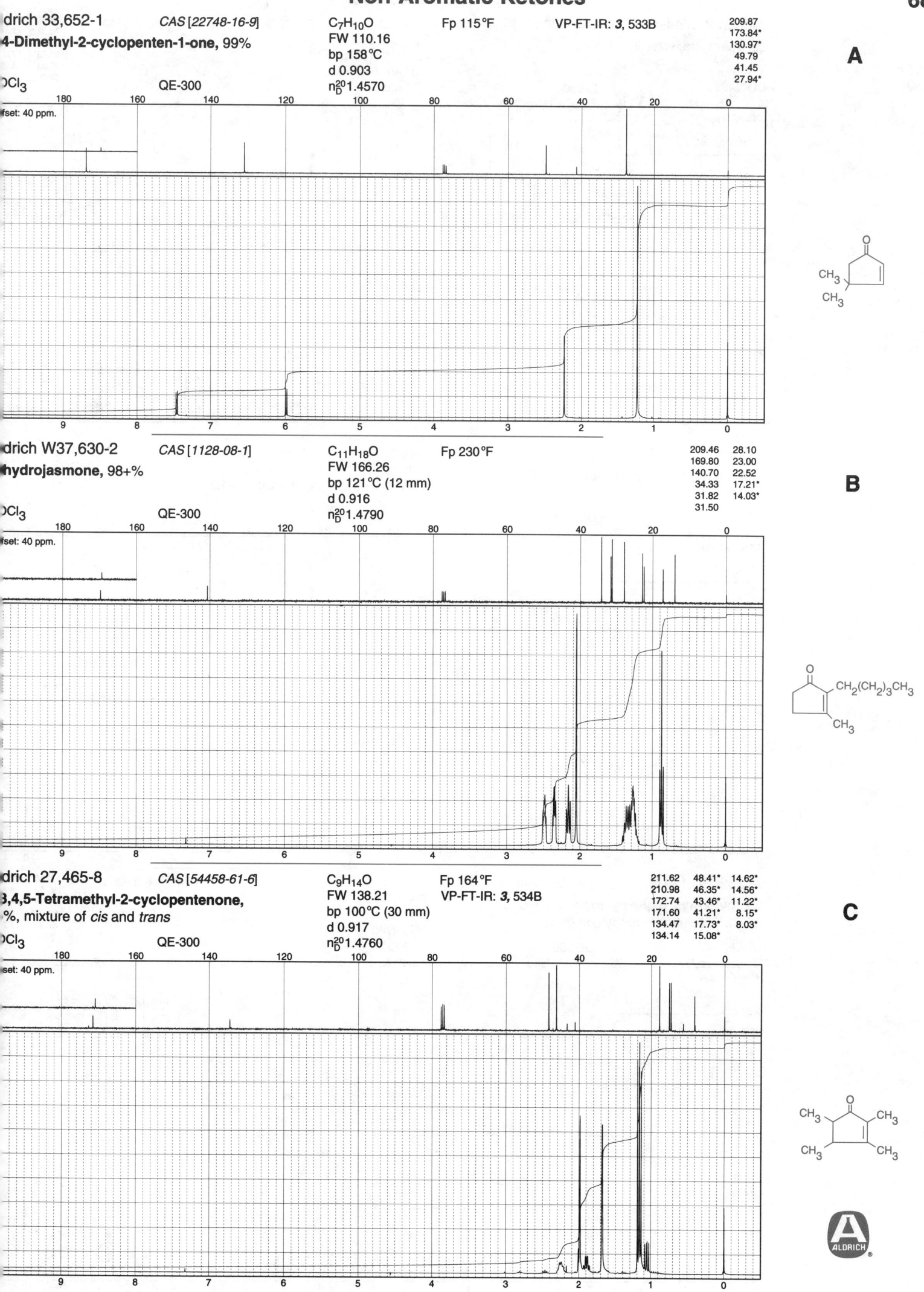

Aldrich 33,652-1 CAS [22748-16-9] C₇H₁₀O Fp 115°F VP-FT-IR: 3, 533B
4-Dimethyl-2-cyclopenten-1-one, 99% FW 110.16 bp 158°C d 0.903 n²⁰_D 1.4570

$C_7H_{10}O$ FW 110.16 bp 158°C d 0.903 n_D^{20} 1.4570

209.87
173.84*
130.97*
49.79
41.45
27.94*

A

CDCl₃ QE-300
offset: 40 ppm.

Aldrich W37,630-2 CAS [1128-08-1] C₁₁H₁₈O Fp 230°F
hydrojasmone, 98+%

$C_{11}H_{18}O$ FW 166.26 bp 121°C (12 mm) d 0.916 n_D^{20} 1.4790

209.46 28.10
169.80 23.00
140.70 22.52
34.33 17.21*
31.82 14.03*
31.50

B

CDCl₃ QE-300
offset: 40 ppm.

Aldrich 27,465-8 CAS [54458-61-6] C₉H₁₄O Fp 164°F VP-FT-IR: 3, 534B
3,4,5-Tetramethyl-2-cyclopentenone,
%, mixture of cis and trans

$C_9H_{14}O$ FW 138.21 bp 100°C (30 mm) d 0.917 n_D^{20} 1.4760

211.62 48.41* 14.62*
210.98 46.35* 14.56*
172.74 43.46* 11.22*
171.60 41.21* 8.15*
134.47 17.73* 8.03*
134.14 15.08*

C

CDCl₃ QE-300
offset: 40 ppm.

ALDRICH

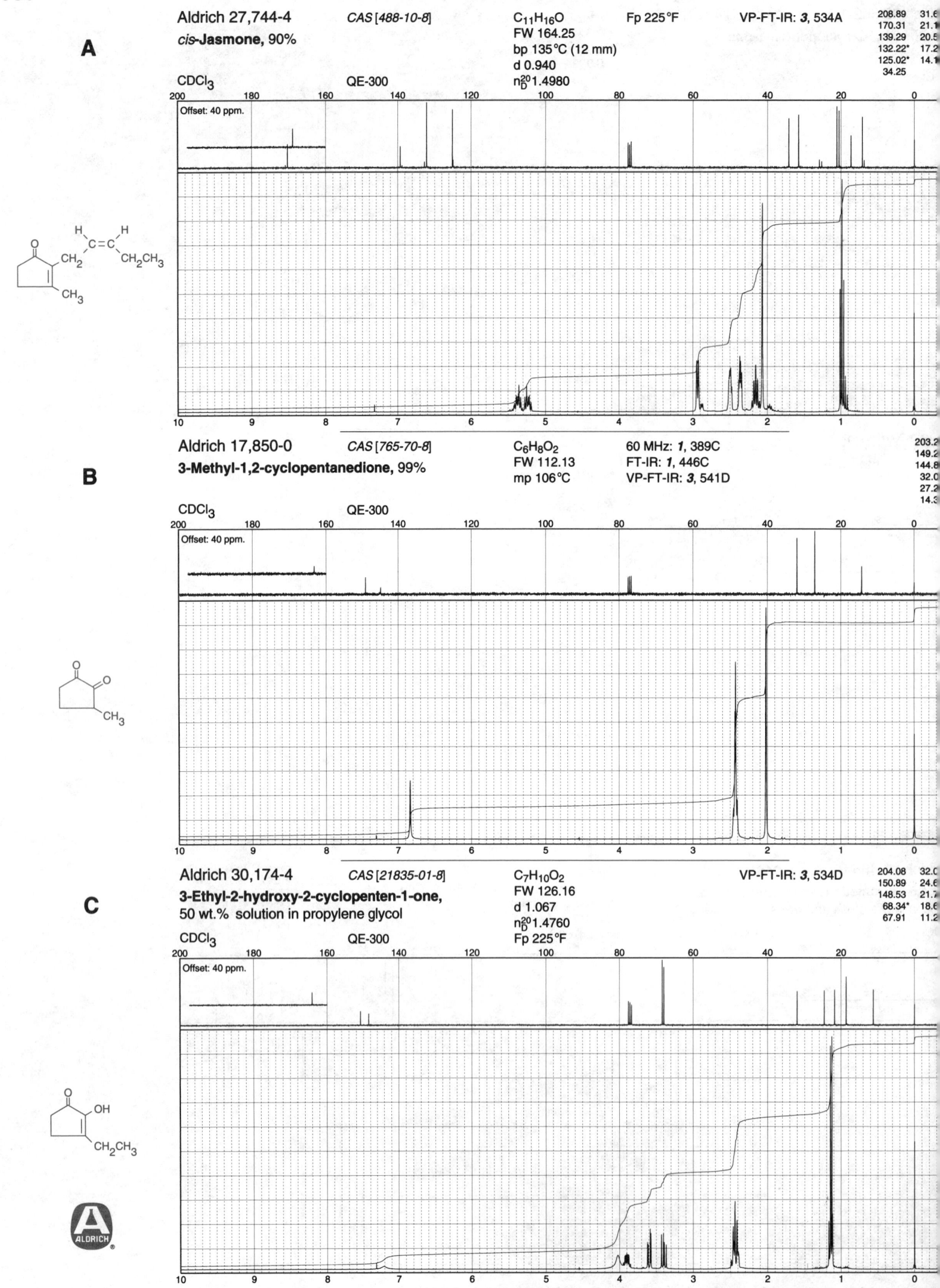

A

Aldrich 27,744-4

cis-Jasmone, 90%

CAS [488-10-8]

C₁₁H₁₆O
FW 164.25
bp 135°C (12 mm)
d 0.940
n²⁰_D 1.4980

Fp 225°F

VP-FT-IR: *3*, 534A

208.89   31.6
170.31   21.1
139.29   20.5
132.22*   17.2
125.02*   14.1
34.25

CDCl₃   QE-300

Offset: 40 ppm.

B

Aldrich 17,850-0

3-Methyl-1,2-cyclopentanedione, 99%

CAS [765-70-8]

C₆H₈O₂
FW 112.13
mp 106°C

60 MHz: *1*, 389C
FT-IR: *1*, 446C
VP-FT-IR: *3*, 541D

203.2
149.2
144.8
32.0
27.2
14.3

CDCl₃   QE-300

Offset: 40 ppm.

C

Aldrich 30,174-4

3-Ethyl-2-hydroxy-2-cyclopenten-1-one,
50 wt.% solution in propylene glycol

CAS [21835-01-8]

C₇H₁₀O₂
FW 126.16
d 1.067
n²⁰_D 1.4760
Fp 225°F

VP-FT-IR: *3*, 534D

204.08   32.0
150.89   24.6
148.53   21.7
68.34*   18.6
67.91   11.2

CDCl₃   QE-300

Offset: 40 ppm.

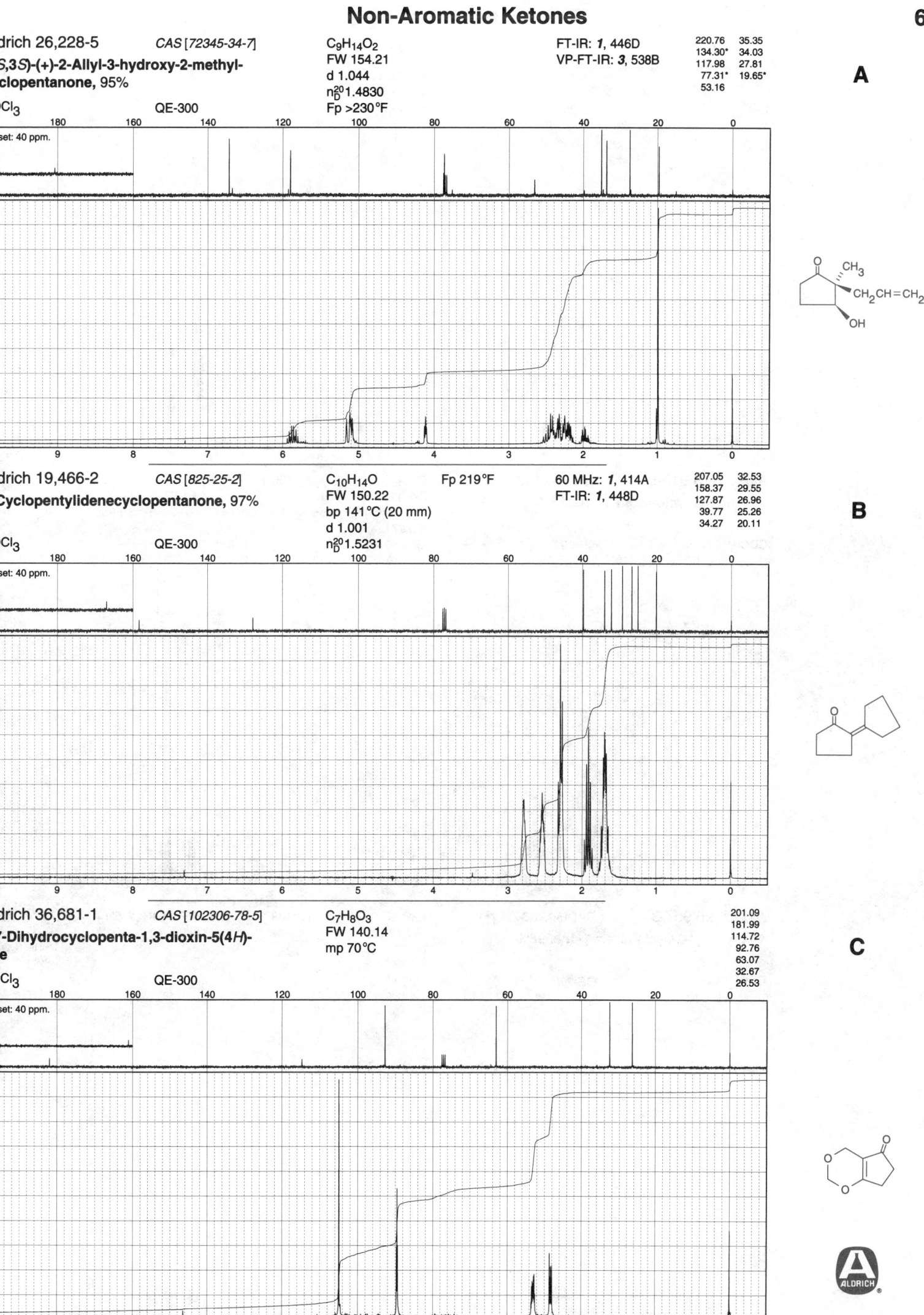

Aldrich 26,228-5   CAS [72345-34-7]   C9H14O2   FT-IR: 1, 446D   220.76 35.35
(2S,3S)-(+)-2-Allyl-3-hydroxy-2-methyl-   FW 154.21   VP-FT-IR: 3, 538B   134.30* 34.03
cyclopentanone, 95%   d 1.044   117.98 27.81
CDCl3   QE-300   n20D 1.4830   77.31* 19.65*
Fp >230°F   53.16

A

offset: 40 ppm.

Aldrich 19,466-2   CAS [825-25-2]   C10H14O   Fp 219°F   60 MHz: 1, 414A   207.05 32.53
2-Cyclopentylidenecyclopentanone, 97%   FW 150.22   FT-IR: 1, 448D   158.37 29.55
bp 141°C (20 mm)   127.87 26.96
CDCl3   QE-300   d 1.001   39.77   25.26
n20D 1.5231   34.27   20.11

B

offset: 40 ppm.

Aldrich 36,681-1   CAS [102306-78-5]   C7H8O3   201.09
6,7-Dihydrocyclopenta-1,3-dioxin-5(4H)-   FW 140.14   181.99
one   mp 70°C   114.72
CDCl3   QE-300   92.76
63.07
32.67
26.53

C

offset: 40 ppm.

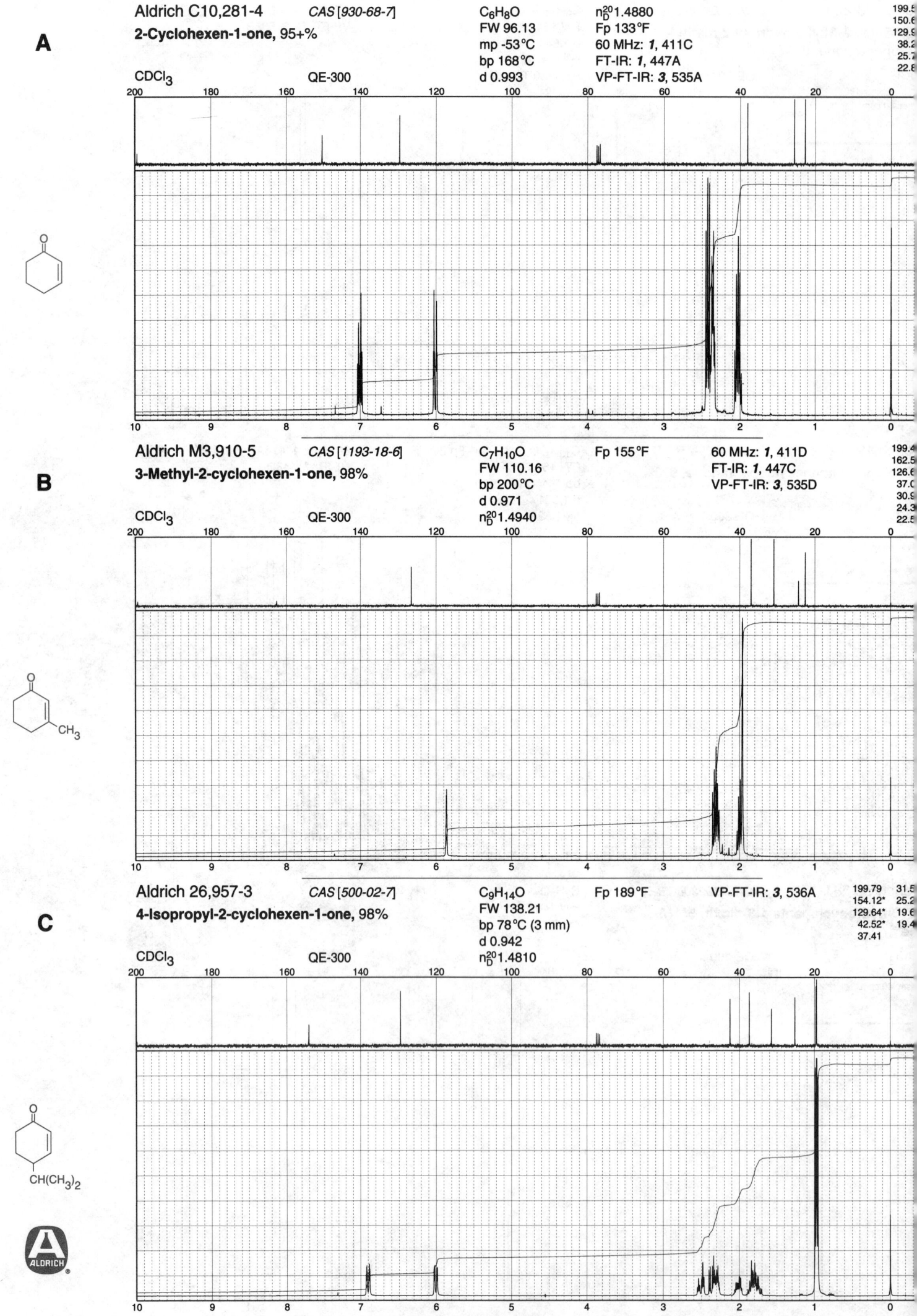

A

Aldrich C10,281-4   *CAS [930-68-7]* C_6H_8O n_D^{20} 1.4880   199.5
2-Cyclohexen-1-one, 95+%   FW 96.13   Fp 133°F   150.6
                                 mp -53°C   60 MHz: *1*, 411C   129.9
                                 bp 168°C   FT-IR: *1*, 447A   38.2
                                             VP-FT-IR: *3*, 535A   25.7
CDCl₃   QE-300   d 0.993   22.8

B

Aldrich M3,910-5   *CAS [1193-18-6]* $C_7H_{10}O$ Fp 155°F   60 MHz: *1*, 411D   199.4
3-Methyl-2-cyclohexen-1-one, 98%   FW 110.16   FT-IR: *1*, 447C   162.5
                                    bp 200°C   VP-FT-IR: *3*, 535D   126.6
                                    d 0.971   37.0
CDCl₃   QE-300   n₂₀D 1.4940   30.9
                                                               24.3
                                                               22.5

C

Aldrich 26,957-3   *CAS [500-02-7]* $C_9H_{14}O$ Fp 189°F   VP-FT-IR: *3*, 536A   199.79   31.5
4-Isopropyl-2-cyclohexen-1-one, 98%   FW 138.21   154.12*   25.2
                                       bp 78°C (3 mm)   129.64*   19.6
                                       d 0.942   42.52*   19.4
CDCl₃   QE-300   n₂₀D 1.4810   37.41

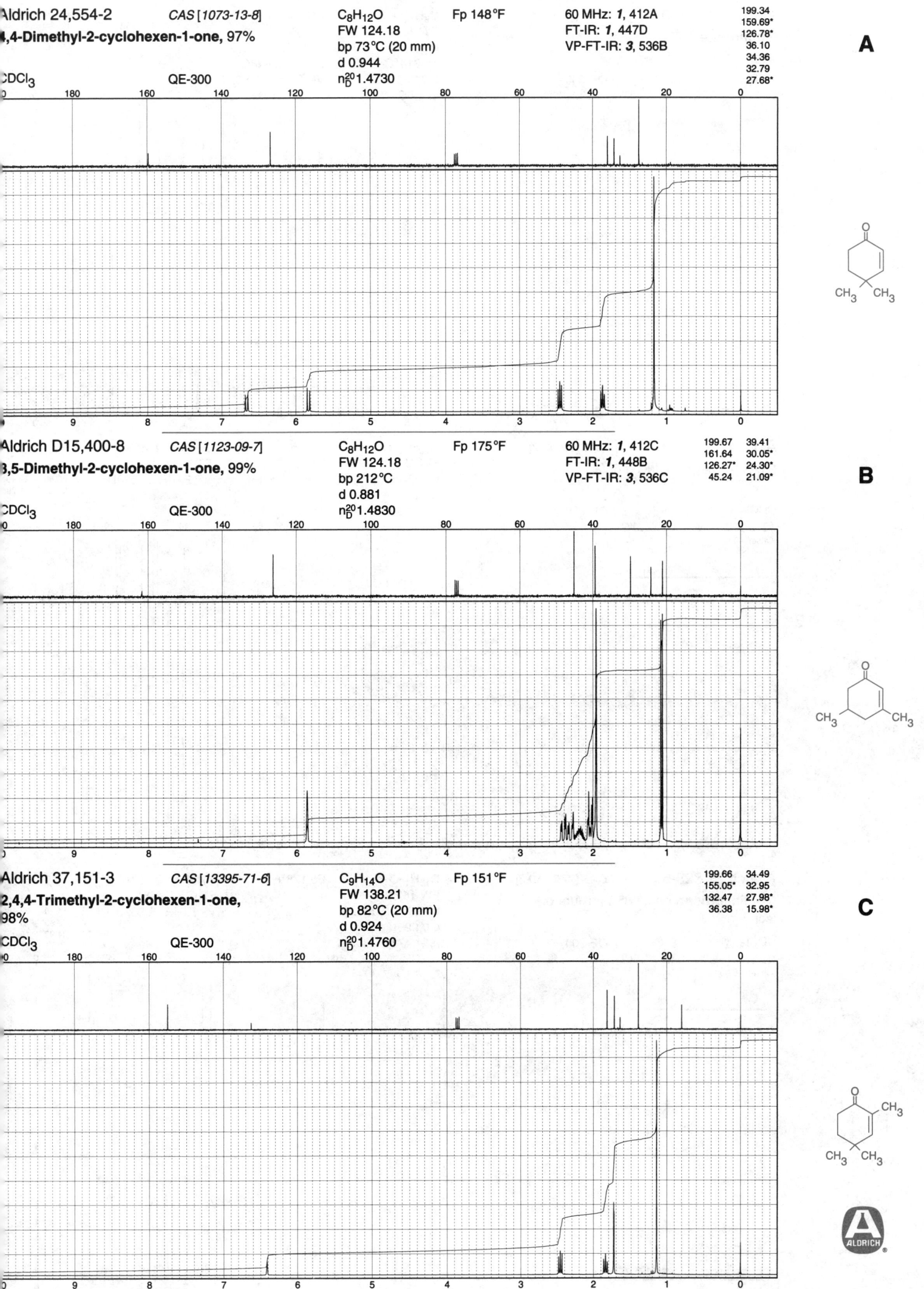

Aldrich 24,554-2   CAS [1073-13-8]

4,4-Dimethyl-2-cyclohexen-1-one, 97%

CDCl₃   QE-300

C₈H₁₂O   Fp 148°F
FW 124.18
bp 73°C (20 mm)
d 0.944
n²⁰_D 1.4730

60 MHz: 1, 412A
FT-IR: 1, 447D
VP-FT-IR: 3, 536B

199.34
159.69*
126.78*
36.10
34.36
32.79
27.68*

A

Aldrich D15,400-8   CAS [1123-09-7]

3,5-Dimethyl-2-cyclohexen-1-one, 99%

CDCl₃   QE-300

C₈H₁₂O   Fp 175°F
FW 124.18
bp 212°C
d 0.881
n²⁰_D 1.4830

60 MHz: 1, 412C
FT-IR: 1, 448B
VP-FT-IR: 3, 536C

199.67   39.41
161.64   30.05*
126.27* 24.30*
45.24   21.09*

B

Aldrich 37,151-3   CAS [13395-71-6]

2,4,4-Trimethyl-2-cyclohexen-1-one,
98%

CDCl₃   QE-300

C₉H₁₄O   Fp 151°F
FW 138.21
bp 82°C (20 mm)
d 0.924
n²⁰_D 1.4760

199.66   34.49
155.05* 32.95
132.47   27.98*
36.38   15.98*

C

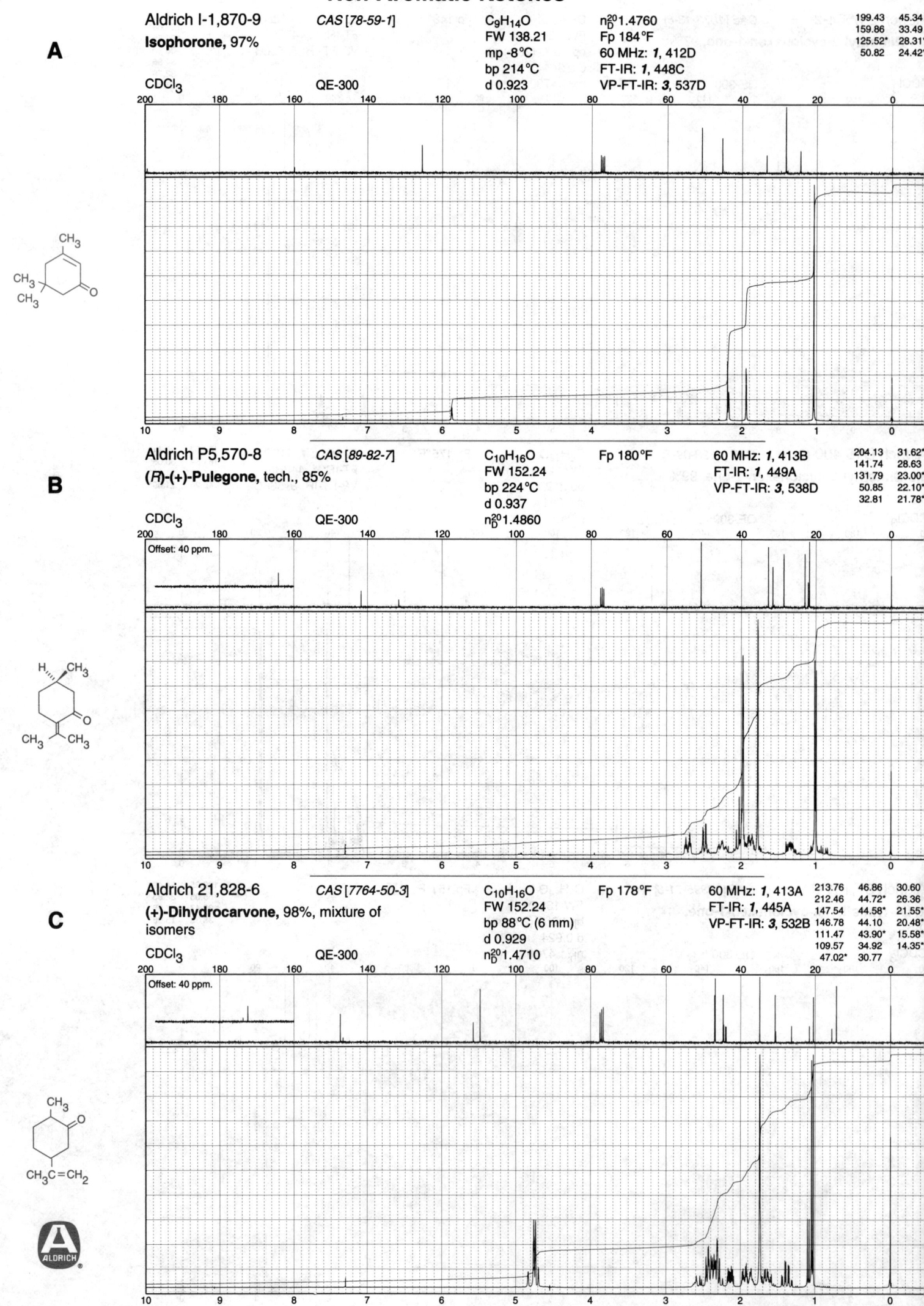

A

Aldrich I-1,870-9   *CAS [78-59-1]* $C_9H_{14}O$ n_D^{20} 1.4760

Isophorone, 97%   FW 138.21   Fp 184°F   60 MHz: *1*, 412D

mp -8°C   FT-IR: *1*, 448C

CDCl₃   QE-300   bp 214°C   VP-FT-IR: *3*, 537D

d 0.923

199.43	45.34
159.86	33.49
125.52*	28.31*
50.82	24.42*

B

Aldrich P5,570-8   *CAS [89-82-7]* $C_{10}H_{16}O$ Fp 180°F   60 MHz: *1*, 413B

(*R*)-(+)-Pulegone, tech., 85%   FW 152.24   FT-IR: *1*, 449A

bp 224°C   VP-FT-IR: *3*, 538D

d 0.937

CDCl₃   QE-300 n_D^{20} 1.4860

Offset: 40 ppm.

204.13	31.62*
141.74	28.63
131.79	23.00*
50.85	22.10*
32.81	21.78*

C

Aldrich 21,828-6   *CAS [7764-50-3]* $C_{10}H_{16}O$ Fp 178°F   60 MHz: *1*, 413A

(+)-Dihydrocarvone, 98%, mixture of   FW 152.24   FT-IR: *1*, 445A

isomers   bp 88°C (6 mm)   VP-FT-IR: *3*, 532B

d 0.929

CDCl₃   QE-300 n_D^{20} 1.4710

Offset: 40 ppm.

213.76	46.86	30.60
212.46	44.72*	26.36
147.54	44.58*	21.55*
146.78	44.10	20.48*
111.47	43.90*	15.58*
109.57	34.92	14.35*
47.02*	30.77	

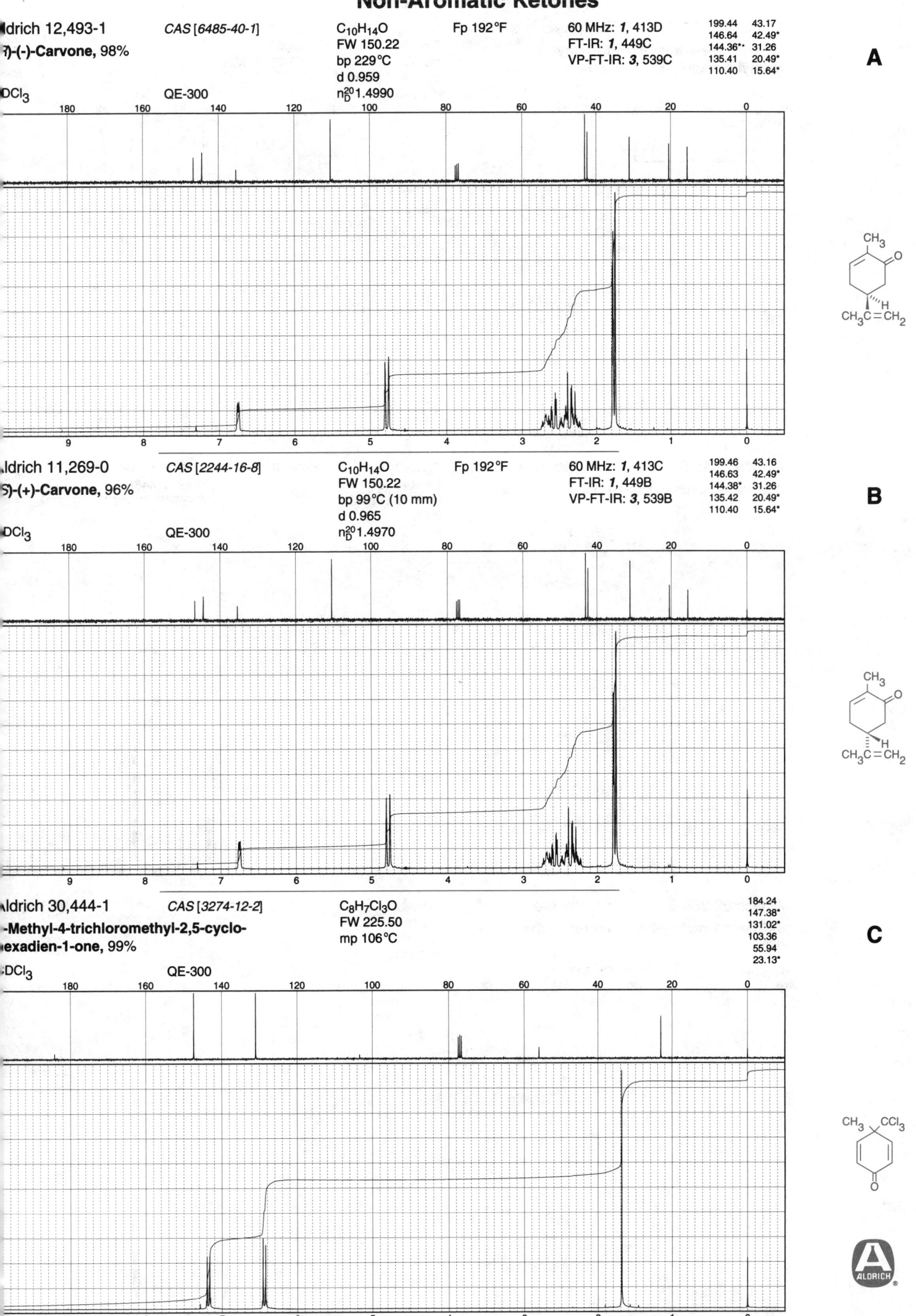

Aldrich 12,493-1   CAS [6485-40-1]   C₁₀H₁₄O   Fp 192°F   60 MHz: 1, 413D
(R)-(-)-Carvone, 98%   FW 150.22   FT-IR: 1, 449C
    bp 229°C   VP-FT-IR: 3, 539C
    d 0.959
CDCl₃   QE-300   n²⁰_D 1.4990

199.44 43.17
146.64 42.49*
144.36** 31.26
135.41 20.49*
110.40 15.64*

A

Aldrich 11,269-0   CAS [2244-16-8]   C₁₀H₁₄O   Fp 192°F   60 MHz: 1, 413C
(S)-(+)-Carvone, 96%   FW 150.22   FT-IR: 1, 449B
    bp 99°C (10 mm)   VP-FT-IR: 3, 539B
    d 0.965
CDCl₃   QE-300   n²⁰_D 1.4970

199.46 43.16
146.63 42.49*
144.38* 31.26
135.42 20.49*
110.40 15.64*

B

Aldrich 30,444-1   CAS [3274-12-2]   C₈H₇Cl₃O
4-Methyl-4-trichloromethyl-2,5-cyclo-   FW 225.50
hexadien-1-one, 99%   mp 106°C
CDCl₃   QE-300

184.24
147.38*
131.02*
103.36
55.94
23.13*

C

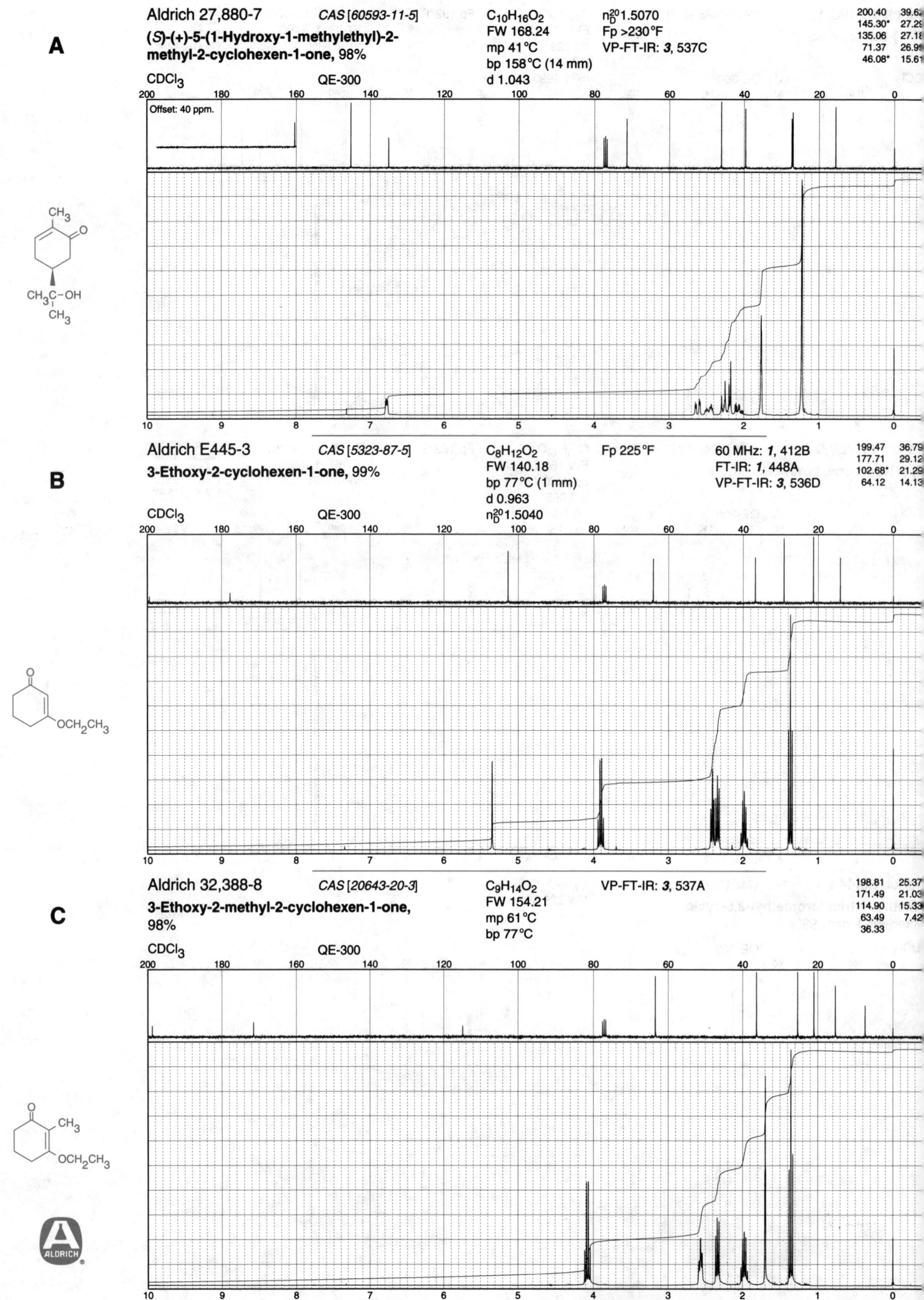

A

Aldrich 27,880-7   *CAS [60593-11-5]*

(S)-(+)-5-(1-Hydroxy-1-methylethyl)-2-methyl-2-cyclohexen-1-one, 98%

CDCl$_3$ QE-300

$C_{10}H_{16}O_2$
FW 168.24
mp 41 °C
bp 158 °C (14 mm)
d 1.043

n_D^{20} 1.5070
Fp >230 °F
VP-FT-IR: **3**, 537C

200.40	39.6
145.30*	27.29
135.06	27.18
71.37	26.9
46.08*	15.61

Offset: 40 ppm.

B

Aldrich E445-3   *CAS [5323-87-5]*

3-Ethoxy-2-cyclohexen-1-one, 99%

CDCl$_3$ QE-300

$C_8H_{12}O_2$
FW 140.18
bp 77 °C (1 mm)
d 0.963
n_D^{20} 1.5040

Fp 225 °F

60 MHz: **1**, 412B
FT-IR: **1**, 448A
VP-FT-IR: **3**, 536D

199.47	36.79
177.71	29.12
102.68*	21.29
64.12	14.13

C

Aldrich 32,388-8   *CAS [20643-20-3]*

3-Ethoxy-2-methyl-2-cyclohexen-1-one, 98%

CDCl$_3$ QE-300

$C_9H_{14}O_2$
FW 154.21
mp 61 °C
bp 77 °C

VP-FT-IR: **3**, 537A

198.81	25.37
171.49	21.03
114.90	15.33
63.49	7.42
36.33	

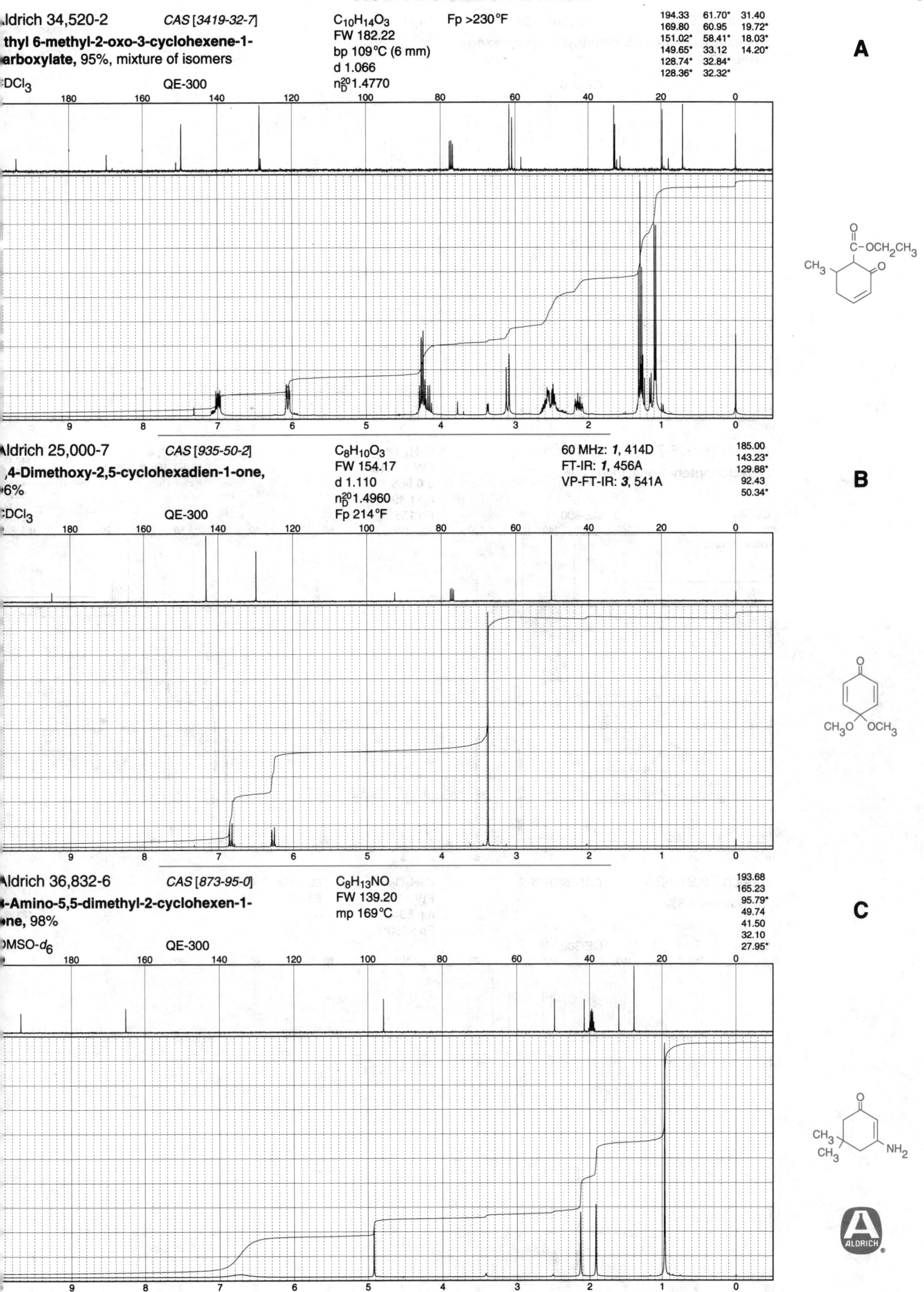

Aldrich 34,520-2   CAS [3419-32-7]

thyl 6-methyl-2-oxo-3-cyclohexene-1-
arboxylate, 95%, mixture of isomers

DCl₃   QE-300

$C_{10}H_{14}O_3$
FW 182.22
bp 109°C (6 mm)
d 1.066
n_D^{20} 1.4770

Fp >230°F

A

194.33	61.70*	31.40
169.80	60.95	19.72*
151.02*	58.41*	18.03*
149.65*	33.12	14.20*
128.74*	32.84*	
128.36*	32.32*	

Aldrich 25,000-7   CAS [935-50-2]

,4-Dimethoxy-2,5-cyclohexadien-1-one,
6%

DCl₃   QE-300

$C_8H_{10}O_3$
FW 154.17
d 1.110
n_D^{20} 1.4960
Fp 214°F

60 MHz: *1*, 414D
FT-IR: *1*, 456A
VP-FT-IR: *3*, 541A

B

| 185.00 |
| 143.23* |
| 129.88* |
| 92.43 |
| 50.34* |

Aldrich 36,832-6   CAS [873-95-0]

-Amino-5,5-dimethyl-2-cyclohexen-1-
ne, 98%

MSO-d₆   QE-300

$C_8H_{13}NO$
FW 139.20
mp 169°C

C

| 193.68 |
| 165.23 |
| 95.79* |
| 49.74 |
| 41.50 |
| 32.10 |
| 27.95* |

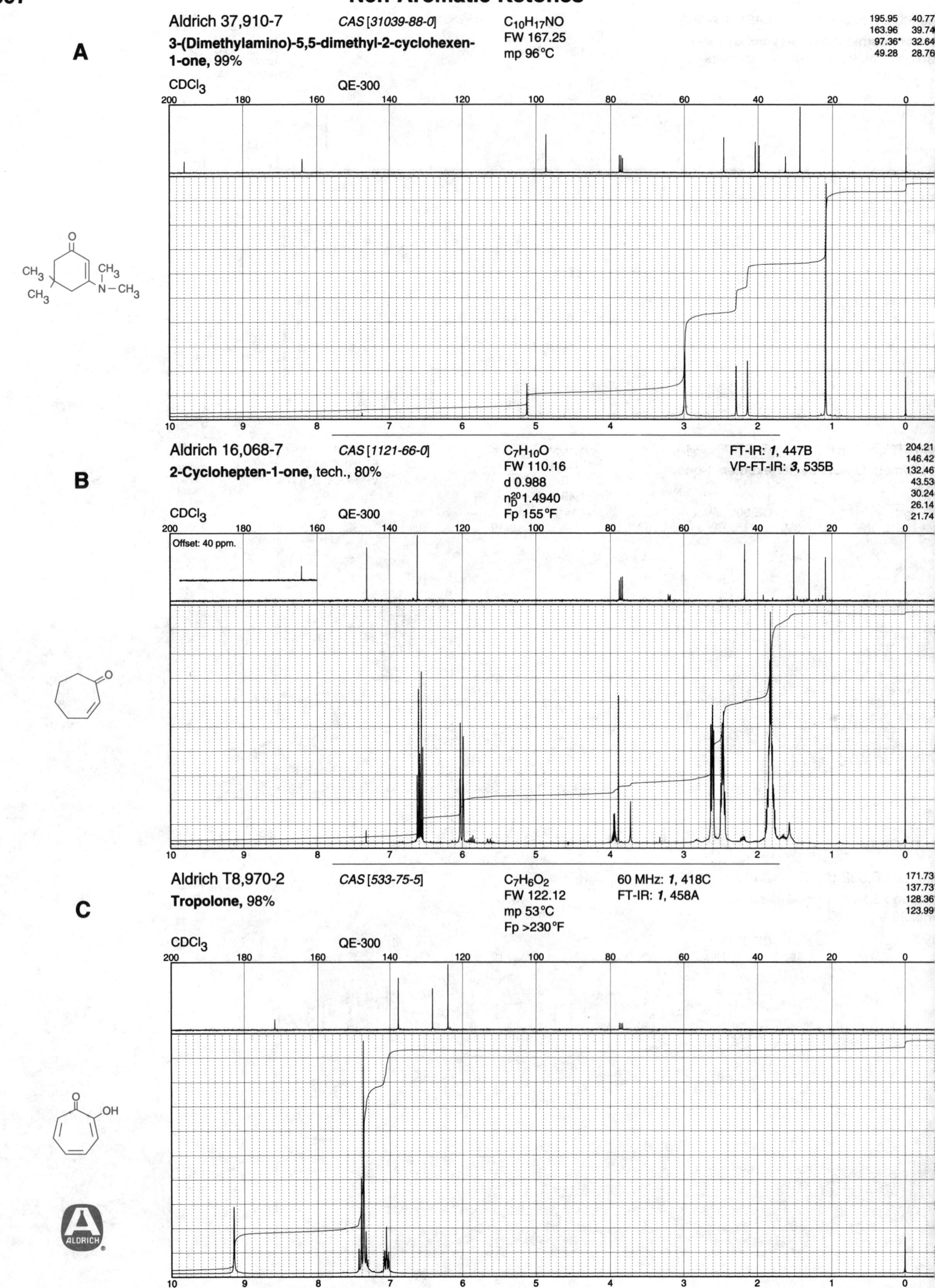

A

Aldrich 37,910-7   *CAS [31039-88-0]*   C_{10}H_{17}NO
3-(Dimethylamino)-5,5-dimethyl-2-cyclohexen-1-one, 99%   FW 167.25   mp 96°C

CDCl_{3}   QE-300

195.95   40.77
163.96   39.74
97.36*   32.64
49.28   28.76

B

Aldrich 16,068-7   *CAS [1121-66-0]*   C_{7}H_{10}O   FT-IR: *1*, 447B
2-Cyclohepten-1-one, tech., 80%   FW 110.16   VP-FT-IR: *3*, 535B
d 0.988
n_{D}^{20} 1.4940
Fp 155°F

CDCl_{3}   QE-300

Offset: 40 ppm.

204.21
146.42
132.46
43.53
30.24
26.14
21.74

C

Aldrich T8,970-2   *CAS [533-75-5]*   C_{7}H_{6}O_{2}   60 MHz: *1*, 418C
Tropolone, 98%   FW 122.12   FT-IR: *1*, 458A
mp 53°C
Fp >230°F

171.73
137.73
128.36
123.99

CDCl_{3}   QE-300

Non-Aromatic Ketones

A

Idrich 30,967-2   CAS [3100-36-5]

Cyclohexadecen-1-one, 98%, mixture
cis and *trans*

DCl_3 QE-300
fset: 40 ppm.

212.33	40.64	28.20	26.99
131.13*	40.35	28.02	26.71
130.85*	31.95	27.90	26.58
130.00*	31.57	27.83	24.04
129.89*	29.01	27.77	23.62
42.88	28.90	27.59	23.29
42.31	28.58	27.29	

$C_{16}H_{28}O$
FW 236.40
bp 194°C (19 mm)
n_D^{20} 1.4890
Fp >230°F

B

Idrich 21,825-1   CAS [1196-01-6]

(S)-(-)-Verbenone, 94%

DCl_3 QE-300
fset: 40 ppm.

$C_{10}H_{14}O$ Fp 185°F
FW 150.22
bp 228°C
d 0.974
n_D^{20} 1.4960

60 MHz: *1*, 406B
FT-IR: *1*, 444A
VP-FT-IR: *3*, 530A

203.54	49.81*
169.80	40.75
121.17*	26.59*
57.67*	23.45*
53.85	22.04*

C

Idrich M4,605-5   CAS [5597-27-3]

Methylene-2-norbornanone, 97%

DCl_3 QE-300
fset: 40 ppm.

$C_8H_{10}O$ Fp 131°F
FW 122.17
bp 70°C (11 mm)
d 0.997
n_D^{20} 1.4930

FT-IR: *1*, 449D
VP-FT-IR: *3*, 540A

205.77	42.43*
149.96	36.80
111.68	28.04
49.10*	23.60

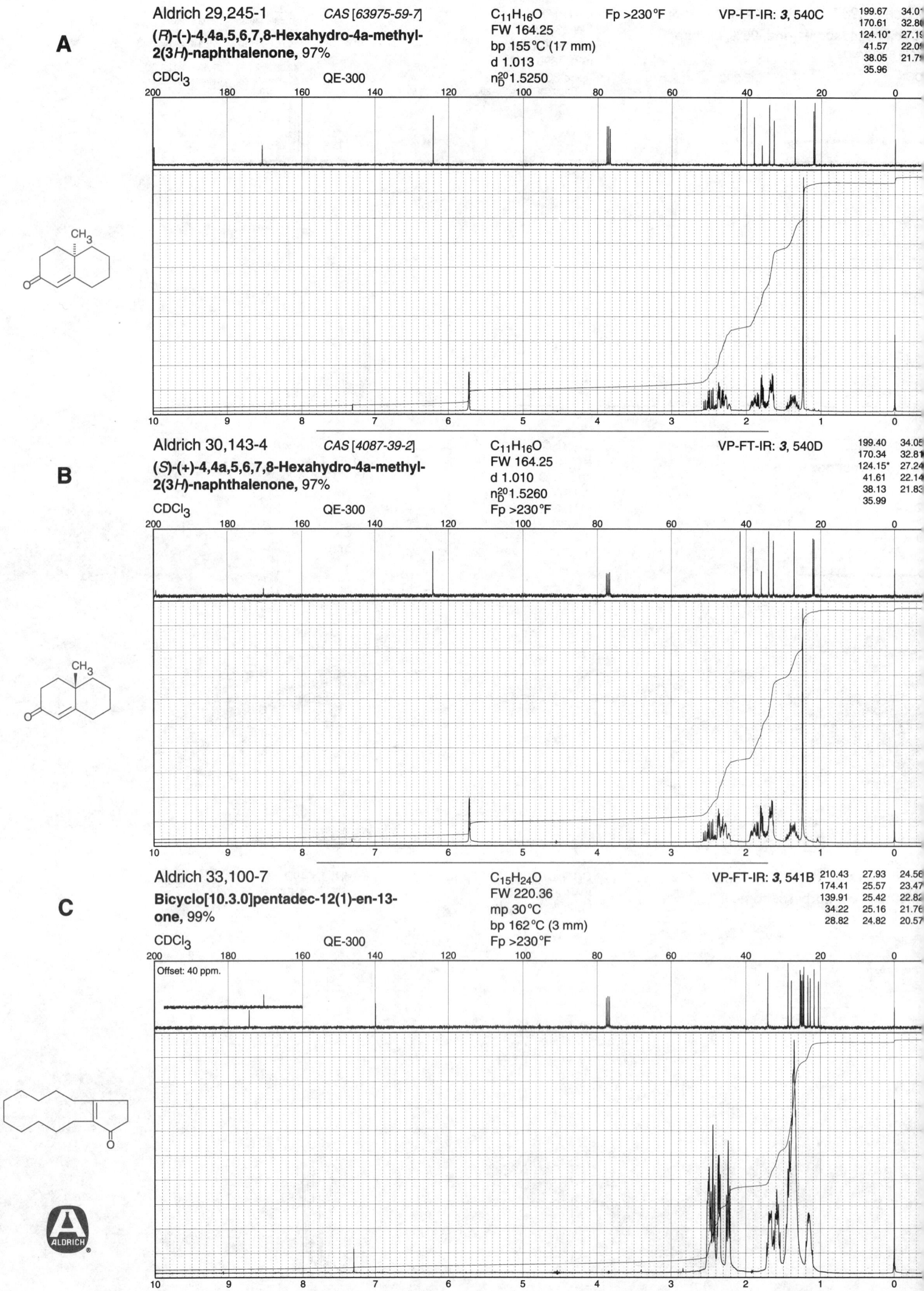

A Aldrich 29,245-1   *CAS [63975-59-7]*   C$_{11}$H$_{16}$O   Fp >230°F   VP-FT-IR: **3**, 540C

(R)-(-)-4,4a,5,6,7,8-Hexahydro-4a-methyl-2(3H)-naphthalenone, 97%

FW 164.25
bp 155°C (17 mm)
d 1.013
n$_D^{20}$1.5250

CDCl$_3$ QE-300

199.67	34.0
170.61	32.8
124.10*	27.19
41.57	22.0
38.05	21.7
35.96	

B Aldrich 30,143-4   *CAS [4087-39-2]*   C$_{11}$H$_{16}$O   VP-FT-IR: **3**, 540D

(S)-(+)-4,4a,5,6,7,8-Hexahydro-4a-methyl-2(3H)-naphthalenone, 97%

FW 164.25
d 1.010
n$_D^{20}$1.5260
Fp >230°F

CDCl$_3$ QE-300

199.40	34.05
170.34	32.81
124.15*	27.24
41.61	22.14
38.13	21.83
35.99	

C Aldrich 33,100-7   C$_{15}$H$_{24}$O   VP-FT-IR: **3**, 541B

Bicyclo[10.3.0]pentadec-12(1)-en-13-one, 99%

FW 220.36
mp 30°C
bp 162°C (3 mm)
Fp >230°F

CDCl$_3$ QE-300

210.43	27.93	24.56
174.41	25.57	23.47
139.91	25.42	22.82
34.22	25.16	21.76
28.82	24.82	20.57

Offset: 40 ppm.

A

drich 16,200-0   CAS [52962-99-9]

syn-**Methoxymethyl-5-norbornen-2-one**,
ch., 90%

$C_9H_{12}O_2$
FW 152.19
d 1.068
n_D^{20} 1.4820
Fp 197°F

60 MHz: *1*, 407C
FT-IR: *1*, 443D
VP-FT-IR: *3*, 529D

213.24	58.84*
140.18*	58.18*
126.89*	42.07*
69.42	38.77
61.71*	

OCl₃   QE-300

set: 40 ppm.

CH_3OCH_2

B

drich 17,977-9   CAS [1670-46-8]   Fp 163°F

Acetylcyclopentanone, 98%

$C_7H_{10}O_2$
FW 126.16
bp 74°C (8 mm)
d 1.043
n_D^{20} 1.4900

60 MHz: *1*, 390A
FT-IR: *1*, 428C
VP-FT-IR: *3*, 514A

212.81	36.93
205.11	30.21*
202.35	25.80
175.59	25.16
109.82	20.68*
62.68*	20.28
38.75	

OCl₃   QE-300

set: 40 ppm.

C

drich 17,976-0   CAS [874-23-7]   Fp 175°F

Acetylcyclohexanone, 97%

$C_8H_{12}O_2$
FW 140.18
bp 112°C (18 mm)
d 1.078
n_D^{20} 1.5090

60 MHz: *1*, 391C
FT-IR: *1*, 428B
VP-FT-IR: *3*, 513D

198.63	24.80*
182.12	24.37
107.02*	22.90
31.20	21.73

OCl₃   QE-300

set: 6.1 ppm.

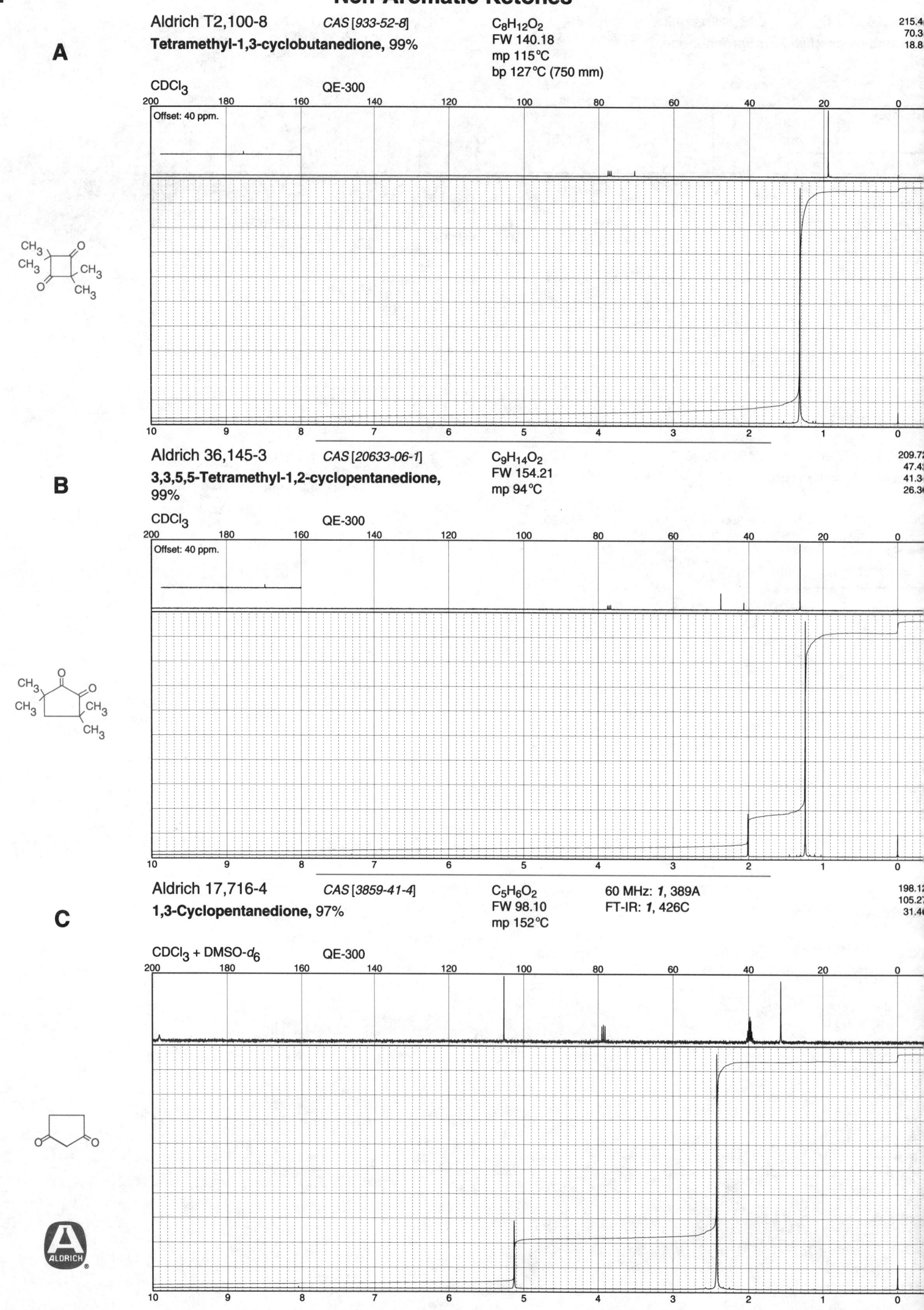

A

Aldrich T2,100-8   CAS [933-52-8]   C₈H₁₂O₂
Tetramethyl-1,3-cyclobutanedione, 99%
FW 140.18
mp 115°C
bp 127°C (750 mm)

215.4
70.3
18.8

CDCl₃   QE-300

Offset: 40 ppm.

B

Aldrich 36,145-3   CAS [20633-06-1]   C₉H₁₄O₂
3,3,5,5-Tetramethyl-1,2-cyclopentanedione, 99%
FW 154.21
mp 94°C

209.72
47.4
41.3
26.3

CDCl₃   QE-300

Offset: 40 ppm.

C

Aldrich 17,716-4   CAS [3859-41-4]   C₅H₆O₂   60 MHz: 1, 389A
1,3-Cyclopentanedione, 97%
FW 98.10   FT-IR: 1, 426C
mp 152°C

198.12
105.27
31.4

CDCl₃ + DMSO-d₆   QE-300

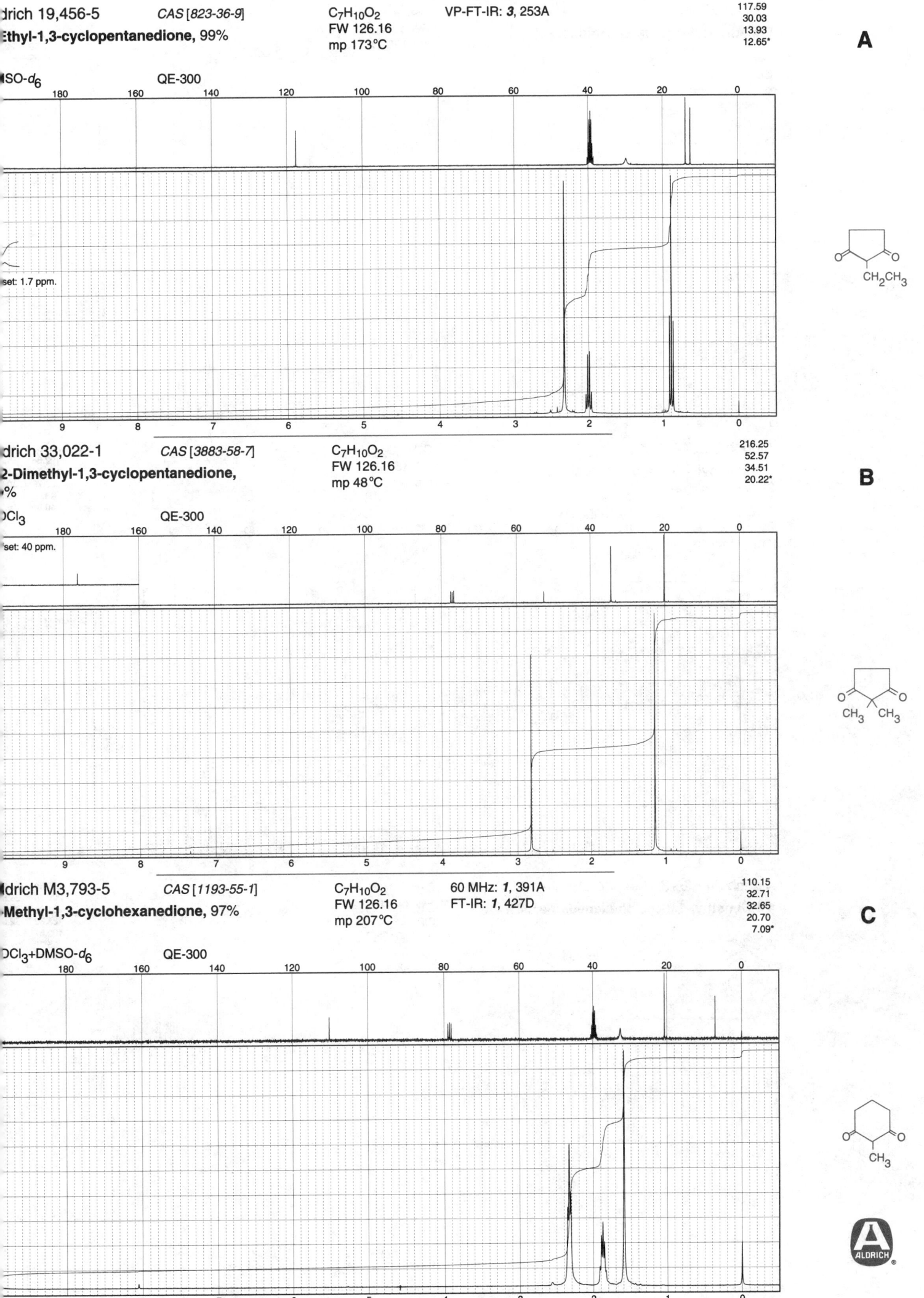

Aldrich 19,456-5   CAS [823-36-9]   C$_7$H$_{10}$O$_2$ VP-FT-IR: **3**, 253A   117.59 / 30.03 / 13.93 / 12.65*
Ethyl-1,3-cyclopentanedione, 99%   FW 126.16   mp 173°C

A

DMSO-d$_6$ QE-300   set: 1.7 ppm.

Aldrich 33,022-1   CAS [3883-58-7]   C$_7$H$_{10}$O$_2$ 216.25 / 52.57 / 34.51 / 20.22*
2-Dimethyl-1,3-cyclopentanedione,   FW 126.16   mp 48°C
%

B

CDCl$_3$ QE-300   set: 40 ppm.

Aldrich M3,793-5   CAS [1193-55-1]   C$_7$H$_{10}$O$_2$ 60 MHz: **1**, 391A   110.15 / 32.71 / 32.65 / 20.70 / 7.09*
Methyl-1,3-cyclohexanedione, 97%   FW 126.16   FT-IR: **1**, 427D
mp 207°C

C

CDCl$_3$+DMSO-d$_6$ QE-300

ALDRICH

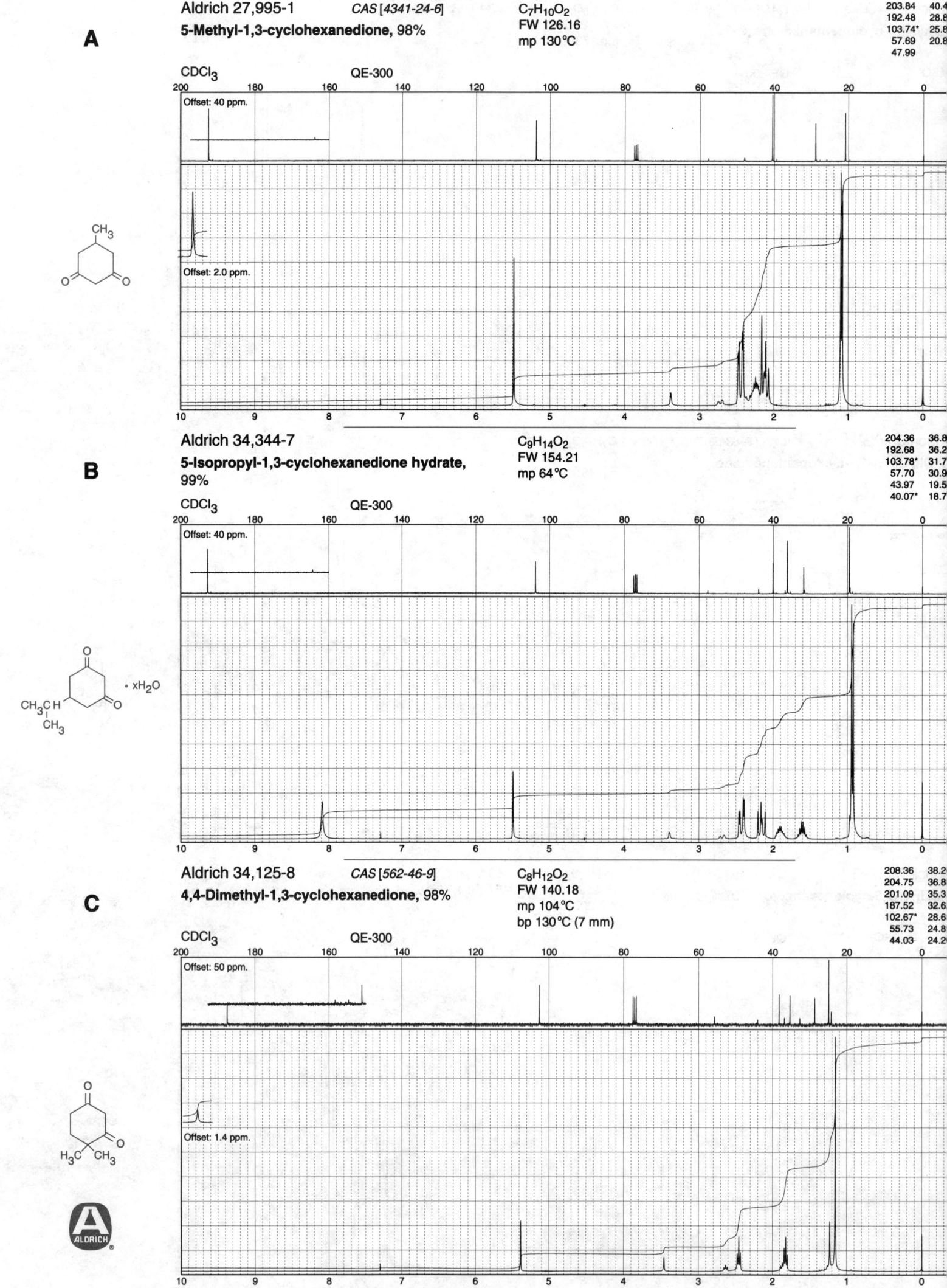

A

Aldrich 27,995-1   CAS [4341-24-6] $C_7H_{10}O_2$
5-Methyl-1,3-cyclohexanedione, 98%   FW 126.16
mp 130°C

203.84   40.4
192.48   28.8
103.74*   25.8
57.69   20.8
47.99

CDCl₃   QE-300

Offset: 40 ppm.

Offset: 2.0 ppm.

B

Aldrich 34,344-7 $C_9H_{14}O_2$
5-Isopropyl-1,3-cyclohexanedione hydrate, 99%   FW 154.21
mp 64°C

204.36   36.8
192.68   36.2
103.78*   31.7
57.70   30.9
43.97   19.5
40.07*   18.7

CDCl₃   QE-300

Offset: 40 ppm.

C

Aldrich 34,125-8   CAS [562-46-9] $C_8H_{12}O_2$
4,4-Dimethyl-1,3-cyclohexanedione, 98%   FW 140.18
mp 104°C
bp 130°C (7 mm)

208.36   38.2
204.75   36.8
201.09   35.3
187.52   32.6
102.67*   28.6
55.73   24.8
44.03   24.2

CDCl₃   QE-300

Offset: 50 ppm.

Offset: 1.4 ppm.

Aldrich D15,330-3   CAS [126-81-8] $C_8H_{12}O_2$ 60 MHz: *1*, 391B   102.66*
46.27
32.12
28.12*

5,5-Dimethyl-1,3-cyclohexanedione, 95%   FW 140.18   FT-IR: *1*, 428A

mp 150°C

A

CDCl$_3$+DMSO-d_6 QE-300

Aldrich 28,672-9   CAS [7298-89-7] $C_8H_{11}ClO_2$ 105.71
46.66
31.40
27.49*

2-Chloro-5,5-dimethyl-1,3-cyclohexane-   FW 174.63
dione, 98%   mp 163°C

B

DMSO-d_6 QE-300

Aldrich 12,489-3   CAS [10373-78-1] $C_{10}H_{14}O_2$ 60 MHz: *1*, 405C   204.67   29.96
202.67   22.28
58.63   21.05*
58.01* 17.40*
42.56   8.75*

(±)-Camphorquinone, 99%   FW 166.22   FT-IR: *1*, 450D

mp 199°C

C

CDCl$_3$ QE-300

offset: 40 ppm.

R = CH$_3$ R' = H   or
R = H   R' = CH$_3$

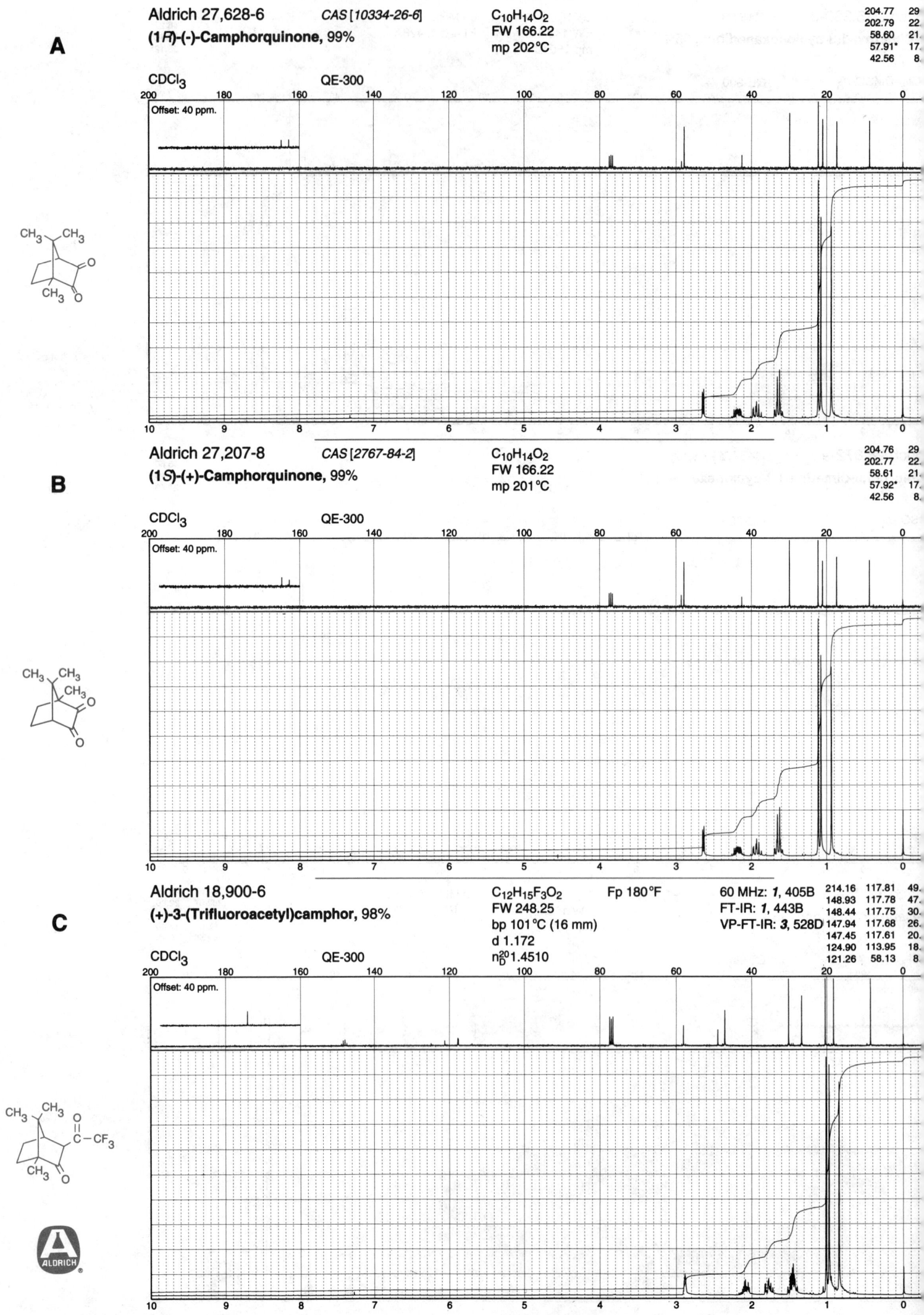

A

Aldrich 27,628-6   *CAS [10334-26-6]* $C_{10}H_{14}O_2$
(1*R*)-(-)-Camphorquinone, 99%   FW 166.22
mp 202°C

204.77 29
202.79 22.
58.60   21.
57.91* 17.
42.56   8.

CDCl$_3$ QE-300

Offset: 40 ppm.

B

Aldrich 27,207-8   *CAS [2767-84-2]* $C_{10}H_{14}O_2$
(1*S*)-(+)-Camphorquinone, 99%   FW 166.22
mp 201°C

204.76 29
202.77 22.
58.61   21.
57.92* 17.
42.56   8.

CDCl$_3$ QE-300

Offset: 40 ppm.

C

Aldrich 18,900-6 $C_{12}H_{15}F_3O_2$ Fp 180°F
(+)-3-(Trifluoroacetyl)camphor, 98%   FW 248.25
bp 101°C (16 mm)
d 1.172
n$_D^{20}$1.4510

60 MHz: *1*, 405B   214.16 117.81 49.
FT-IR: *1*, 443B   148.93 117.78 47.
VP-FT-IR: *3*, 528D   148.44 117.75 30.
                      147.94 117.68 26.
                      147.45 117.61 20.
                      124.90 113.95 18.
                      121.26 58.13   8

CDCl$_3$ QE-300

Offset: 40 ppm.

Aldrich 29,834-4

-3-(Trifluoroacetyl)camphor, 98%

$C_{12}H_{15}F_3O_2$
FW 248.25
d 1.172
n_D^{20} 1.4510
Fp 180 °F

Cl₃   QE-300

set: 40 ppm.

214.05	121.33	47.29*
149.08	117.49	30.29
148.58	117.68	26.77
148.08	114.03	20.47*
147.59	58.18	18.30*
124.98	49.03	8.54*

A

Aldrich 19,593-6   CAS [51800-99-8]

Heptafluorobutyryl-(+)-camphor, 96%

$C_{14}H_{15}F_7O_2$ Fp 201 °F
FW 348.26
bp 65 °C
d 1.218
n_D^{20} 1.4210

Cl₃   QE-300

set: 40 ppm.

213.91	115.81	47.57*
148.87	114.57	30.19
148.51	111.15	26.76
148.49	109.05	20.39*
148.14	107.71	18.30*
120.58	58.09	8.55*
119.61	49.08	

B

Aldrich 29,833-6   CAS [115224-00-5]

Heptafluorobutyryl-(-)-camphor, 98%

$C_{14}H_{15}F_7O_2$
FW 348.26
d 1.218
n_D^{20} 1.4220
Fp 201 °F

Cl₃   QE-300

set: 50 ppm.

214.13	119.80	58.22	30.30
149.09	116.00	49.20	26.85
148.73	114.74	47.76*	20.43*
148.70	111.33	47.73*	18.32*
148.35	109.24	47.70*	8.57*
120.76	107.90		

C

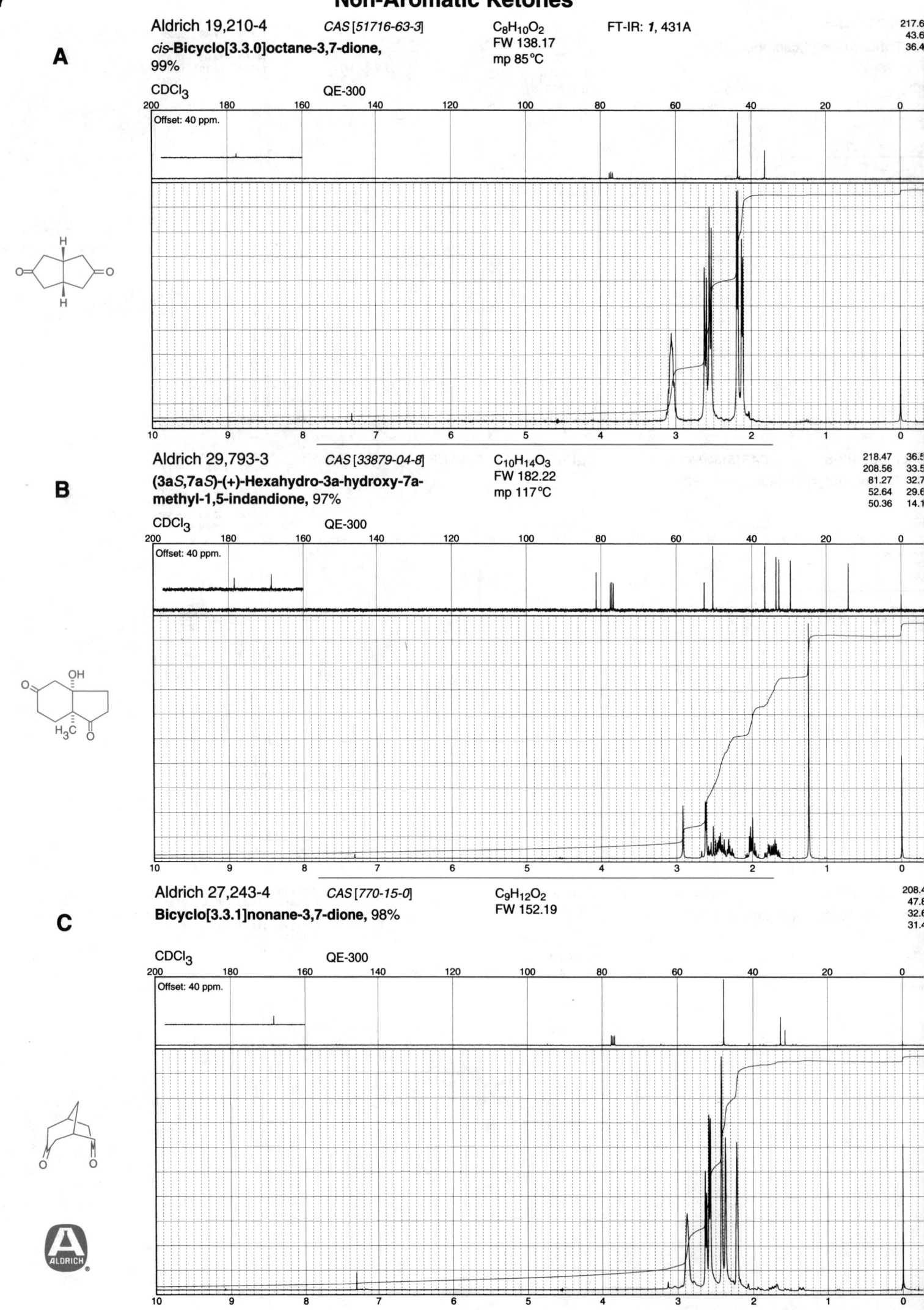

A

Aldrich 19,210-4   *CAS [51716-63-3]*   C₈H₁₀O₂   FT-IR: *1*, 431A
$C_8H_{10}O_2$
FW 138.17
mp 85°C

cis-Bicyclo[3.3.0]octane-3,7-dione, 99%

217.6
43.6
36.4

CDCl₃   QE-300

Offset: 40 ppm.

B

Aldrich 29,793-3   *CAS [33879-04-8]* $C_{10}H_{14}O_3$
FW 182.22
mp 117°C

(3a*S*,7a*S*)-(+)-Hexahydro-3a-hydroxy-7a-methyl-1,5-indandione, 97%

218.47   36.5
208.56   33.5
81.27   32.7
52.64   29.6
50.36   14.1

CDCl₃   QE-300

Offset: 40 ppm.

C

Aldrich 27,243-4   *CAS [770-15-0]* $C_9H_{12}O_2$
FW 152.19

Bicyclo[3.3.1]nonane-3,7-dione, 98%

208.4
47.8
32.6
31.4

CDCl₃   QE-300

Offset: 40 ppm.

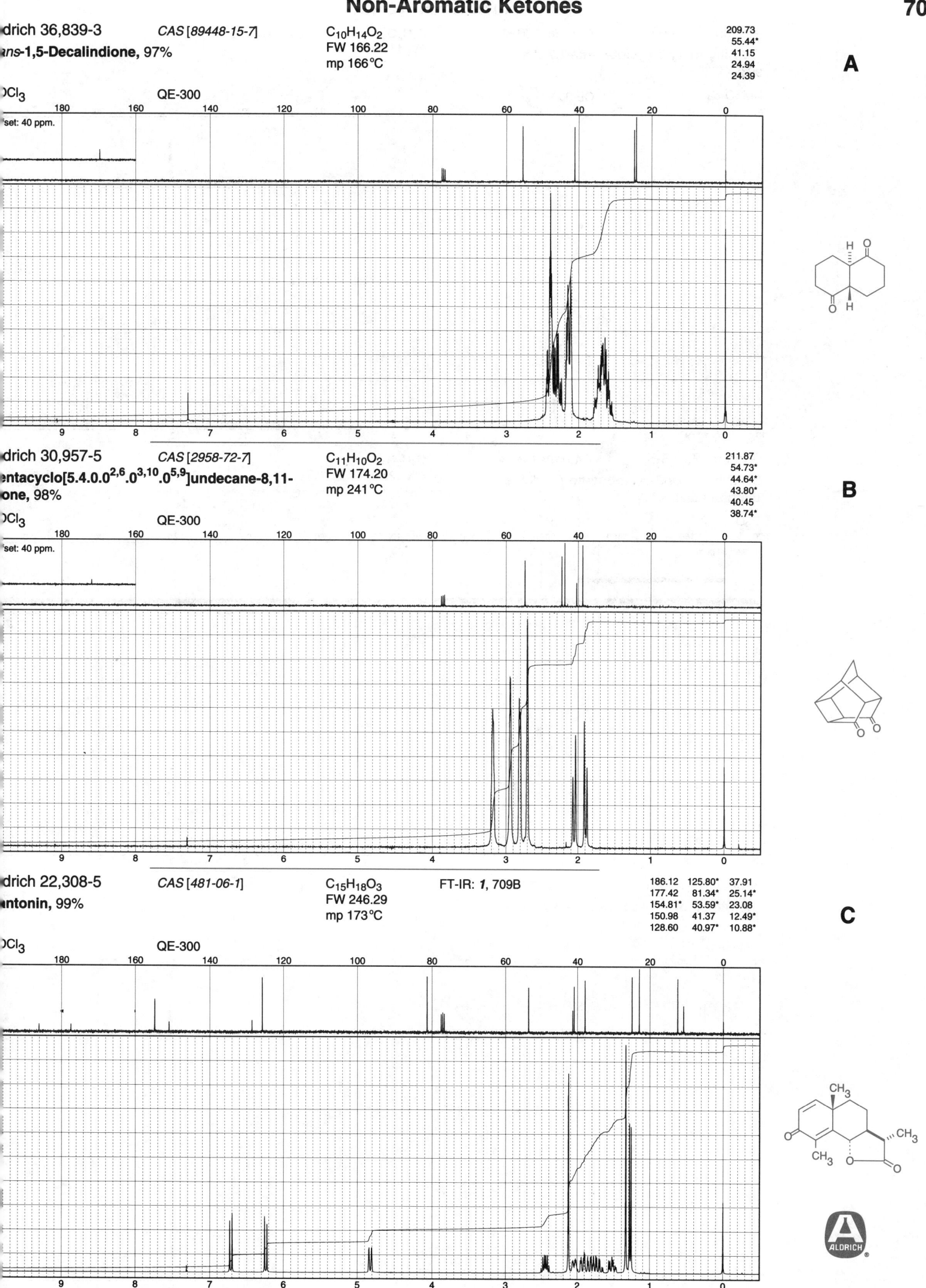

A

Aldrich 36,839-3   CAS [89448-15-7] $C_{10}H_{14}O_2$ FW 166.22   mp 166°C

trans-1,5-Decalindione, 97%

CDCl$_3$ QE-300   offset: 40 ppm.

209.73
55.44*
41.15
24.94
24.39

B

Aldrich 30,957-5   CAS [2958-72-7] $C_{11}H_{10}O_2$ FW 174.20   mp 241°C

Pentacyclo[5.4.0.0^{2,6}.0^{3,10}.0^{5,9}]undecane-8,11-dione, 98%

CDCl$_3$ QE-300   offset: 40 ppm.

211.87
54.73*
44.64*
43.80*
40.45
38.74*

C

Aldrich 22,308-5   CAS [481-06-1] $C_{15}H_{18}O_3$ FW 246.29   mp 173°C   FT-IR: 1, 709B

Santonin, 99%

CDCl$_3$ QE-300

186.12	125.80*
177.42	81.34*
154.81*	53.59*
150.98	41.37
128.60	40.97*

37.91
25.14*
23.08
12.49*
10.88*

ALDRICH

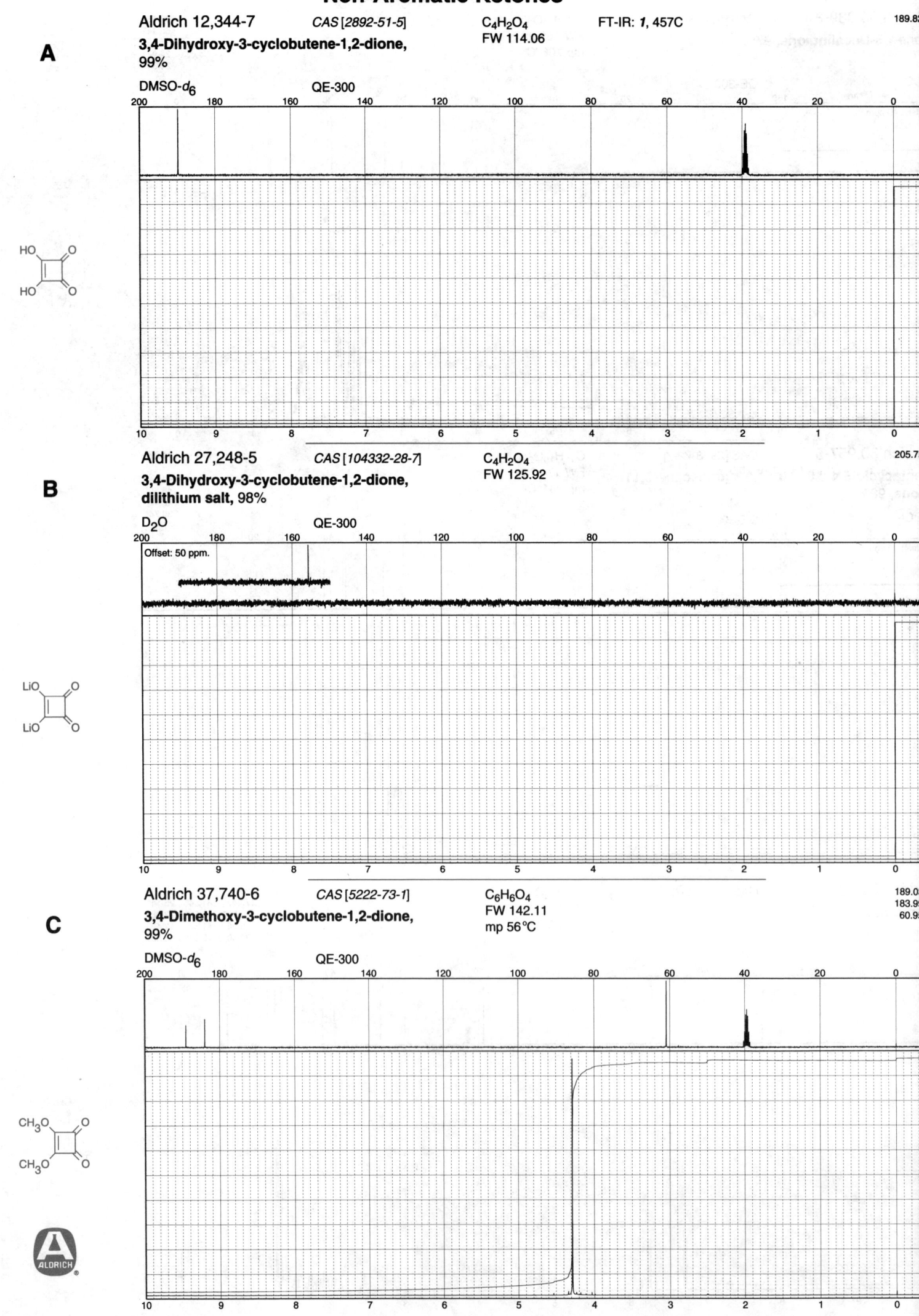

A

Aldrich 12,344-7   *CAS [2892-51-5]* $C_4H_2O_4$ FT-IR: *1*, 457C   189.82

3,4-Dihydroxy-3-cyclobutene-1,2-dione, 99%   FW 114.06

DMSO-d_6 QE-300

B

Aldrich 27,248-5   *CAS [104332-28-7]* $C_4H_2O_4$ 205.75

3,4-Dihydroxy-3-cyclobutene-1,2-dione, dilithium salt, 98%   FW 125.92

D_2O QE-300

Offset: 50 ppm.

C

Aldrich 37,740-6   *CAS [5222-73-1]* $C_6H_6O_4$ 189.03
                                                        183.99
3,4-Dimethoxy-3-cyclobutene-1,2-dione, 99%   FW 142.11   60.95
                                                   mp 56°C

DMSO-d_6 QE-300

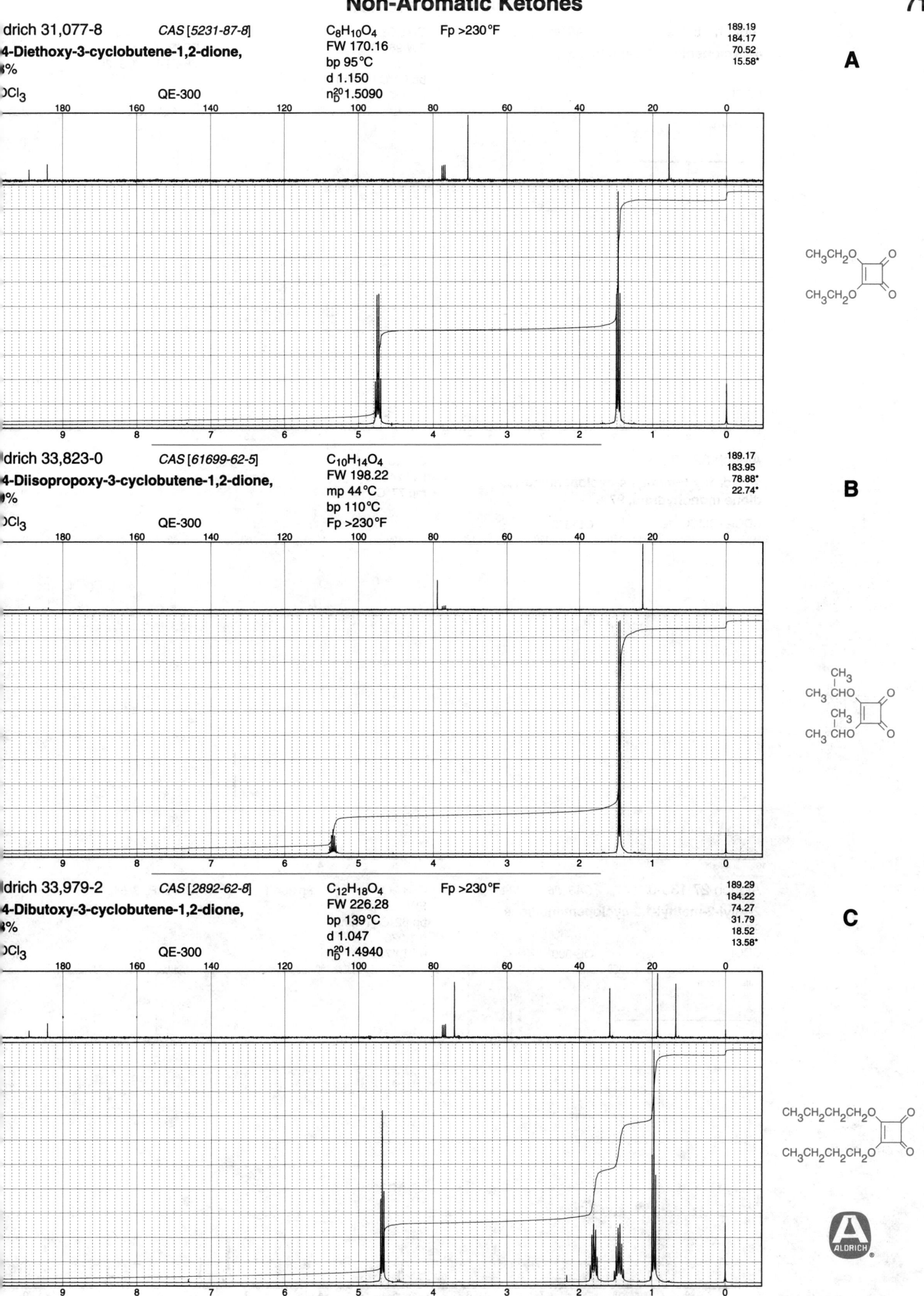

Aldrich 31,077-8   CAS [5231-87-8]   C$_8$H$_{10}$O$_4$ Fp >230°F

4-Diethoxy-3-cyclobutene-1,2-dione, %

CDCl$_3$ QE-300   FW 170.16   bp 95°C   d 1.150   n$_D^{20}$1.5090

189.19
184.17
70.52
15.58*

A

CH$_3$CH$_2$O

CH$_3$CH$_2$O

Aldrich 33,823-0   CAS [61699-62-5]   C$_{10}$H$_{14}$O$_4$

4-Diisopropoxy-3-cyclobutene-1,2-dione, %

CDCl$_3$ QE-300   FW 198.22   mp 44°C   bp 110°C   Fp >230°F

189.17
183.95
78.88*
22.74*

B

CH$_3$ CH$_3$ CHO

CH$_3$ CH$_3$ CHO

Aldrich 33,979-2   CAS [2892-62-8]   C$_{12}$H$_{18}$O$_4$ Fp >230°F

4-Dibutoxy-3-cyclobutene-1,2-dione, %

CDCl$_3$ QE-300   FW 226.28   bp 139°C   d 1.047   n$_D^{20}$1.4940

189.29
184.22
74.27
31.79
18.52
13.58*

C

CH$_3$CH$_2$CH$_2$CH$_2$O

CH$_3$CH$_2$CH$_2$CH$_2$O

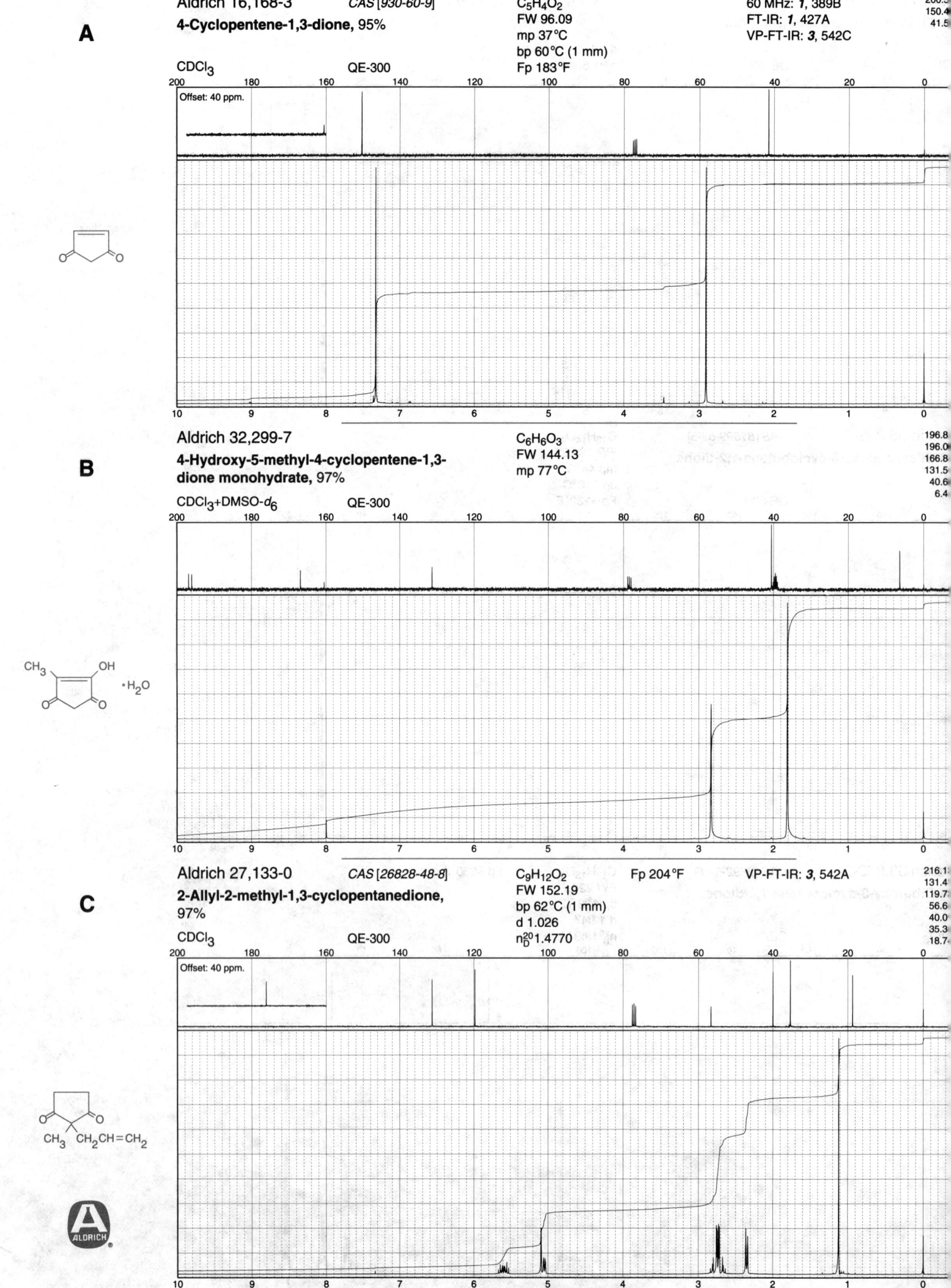

A

Aldrich 16,168-3   *CAS [930-60-9]*

4-Cyclopentene-1,3-dione, 95%

$C_5H_4O_2$
FW 96.09
mp 37°C
bp 60°C (1 mm)
Fp 183°F

60 MHz: *1*, 389B
FT-IR: *1*, 427A
VP-FT-IR: *3*, 542C

200.5
150.4
41.5

CDCl$_3$ QE-300

Offset: 40 ppm.

B

Aldrich 32,299-7

**4-Hydroxy-5-methyl-4-cyclopentene-1,3-
dione monohydrate, 97%**

$C_6H_6O_3$
FW 144.13
mp 77°C

196.8
196.0
166.8
131.5
40.6
6.4

CDCl$_3$+DMSO-*d*$_6$ QE-300

C

Aldrich 27,133-0   *CAS [26828-48-8]*

**2-Allyl-2-methyl-1,3-cyclopentanedione,
97%**

$C_9H_{12}O_2$
FW 152.19
bp 62°C (1 mm)
d 1.026
n$_D^{20}$1.4770

Fp 204°F   VP-FT-IR: *3*, 542A

216.1
131.4
119.7
56.6
40.0
35.3
18.7

CDCl$_3$ QE-300

Offset: 40 ppm.

A

Aldrich 32,951-7   CAS [1125-21-9] $C_9H_{12}O_2$ Fp 205°F   VP-FT-IR: **3**, 542D

2,6-Trimethyl-2-cyclohexene-1,4-dione,
%

Cl_3 QE-300
FW 152.19
mp 27°C
bp 93°C (11 mm)
n_D^{20}1.4910

203.25   51.79
197.51   45.09
148.79   26.05*
136.96* 16.76*

set: 40 ppm.

B

Aldrich T660-1   CAS [2435-53-2] $C_6Cl_4O_2$ FT-IR: **1**, 454B

trachloro-1,2-benzoquinone, 97%
FW 245.88
mp 128°C

168.88
143.82
132.00

Cl_3 QE-300

C

Aldrich 15,147-5   CAS [2435-54-3] $C_6Br_4O_2$ FT-IR: **1**, 454C

trabromo-1,2-benzoquinone, 98%
FW 423.70
mp 150°C

168.65
142.18
128.24

Cl_3 QE-300

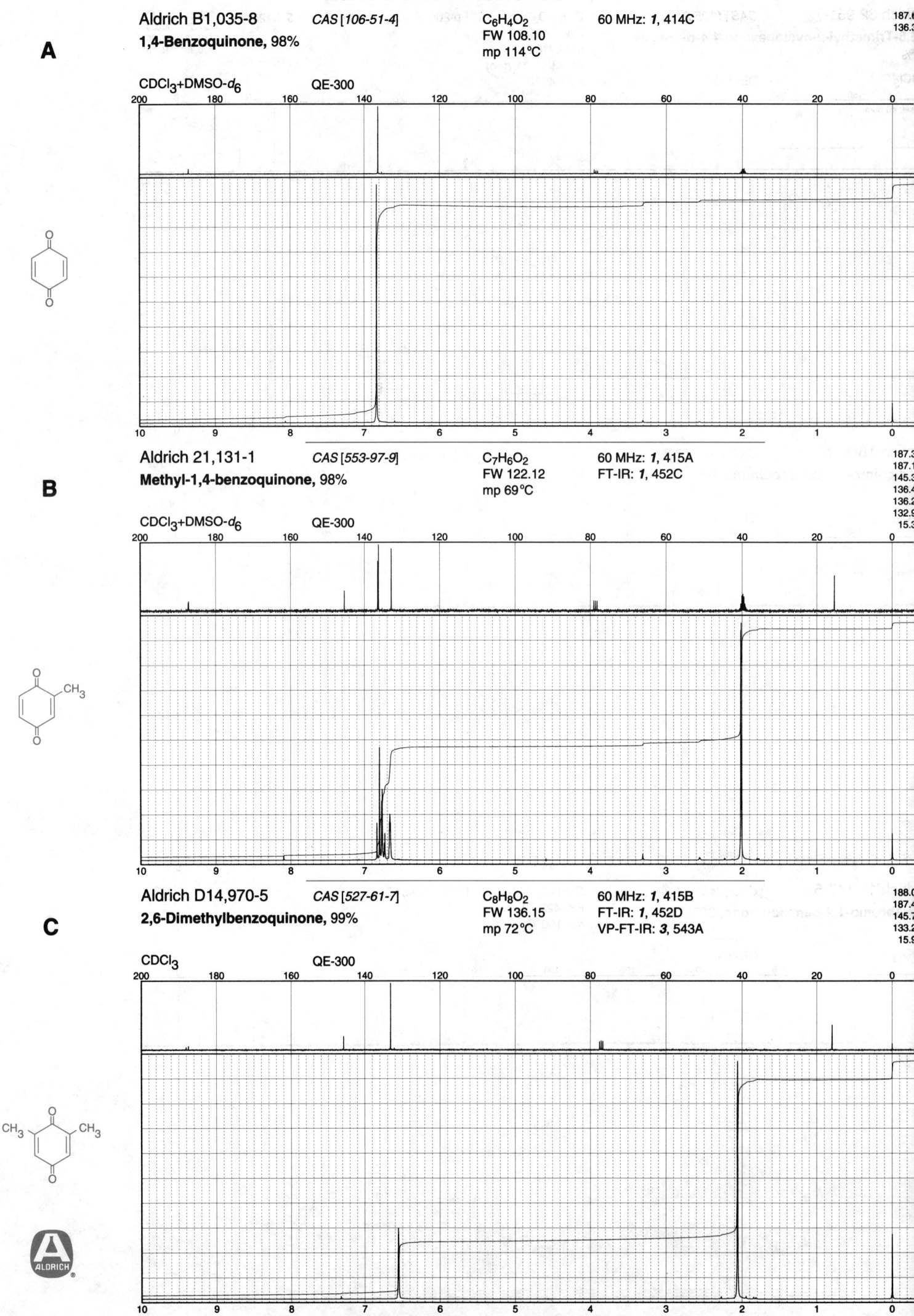

A

Aldrich B1,035-8   *CAS [106-51-4]* $C_6H_4O_2$ 60 MHz: *1*, 414C
1,4-Benzoquinone, 98%   FW 108.10
mp 114°C

$CDCl_3$+DMSO-d_6 QE-300

187.6
136.3

B

Aldrich 21,131-1   *CAS [553-97-9]* $C_7H_6O_2$ 60 MHz: *1*, 415A
Methyl-1,4-benzoquinone, 98%   FW 122.12   FT-IR: *1*, 452C
mp 69°C

$CDCl_3$+DMSO-d_6 QE-300

187.3
187.1
145.3
136.4
136.2
132.9
15.3

C

Aldrich D14,970-5   *CAS [527-61-7]* $C_8H_8O_2$ 60 MHz: *1*, 415B
2,6-Dimethylbenzoquinone, 99%   FW 136.15   FT-IR: *1*, 452D
mp 72°C   VP-FT-IR: *3*, 543A

$CDCl_3$ QE-300

188.0
187.4
145.7
133.2
15.9

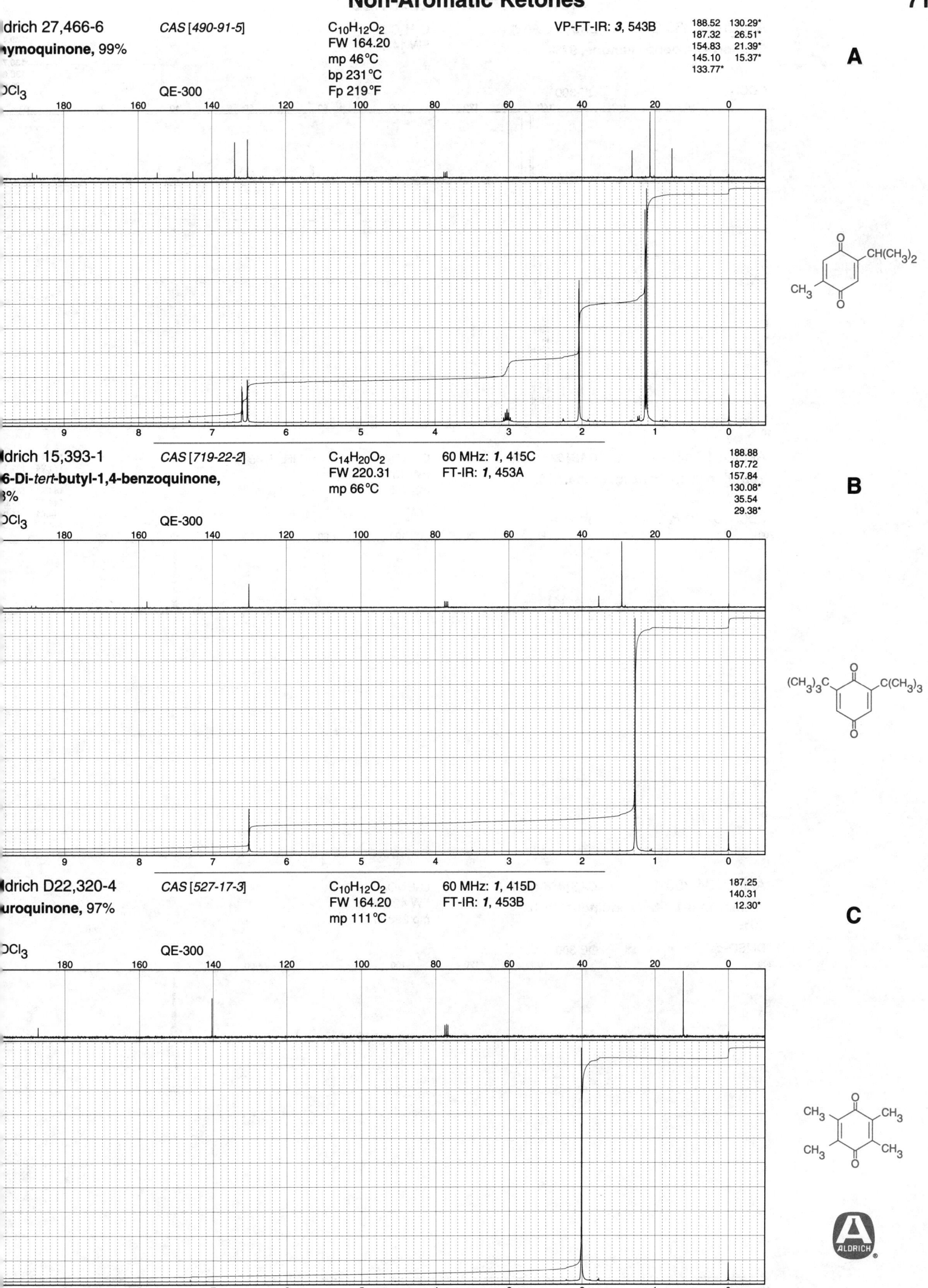

Aldrich 27,466-6 CAS [490-91-5] C₁₀H₁₂O₂ VP-FT-IR: 3, 543B
Thymoquinone, 99% FW 164.20
mp 46°C
bp 231°C
Fp 219°F
CDCl₃ QE-300

188.52 130.29*
187.32 26.51*
154.83 21.39*
145.10 15.37*
133.77*

A

Aldrich 15,393-1 CAS [719-22-2] C₁₄H₂₀O₂ 60 MHz: 1, 415C
2,6-Di-tert-butyl-1,4-benzoquinone, FW 220.31 FT-IR: 1, 453A
98% mp 66°C
CDCl₃ QE-300

188.88
187.72
157.84
130.08*
35.54
29.38*

B

Aldrich D22,320-4 CAS [527-17-3] C₁₀H₁₂O₂ 60 MHz: 1, 415D
Duroquinone, 97% FW 164.20 FT-IR: 1, 453B
mp 111°C
CDCl₃ QE-300

187.25
140.31
12.30*

C

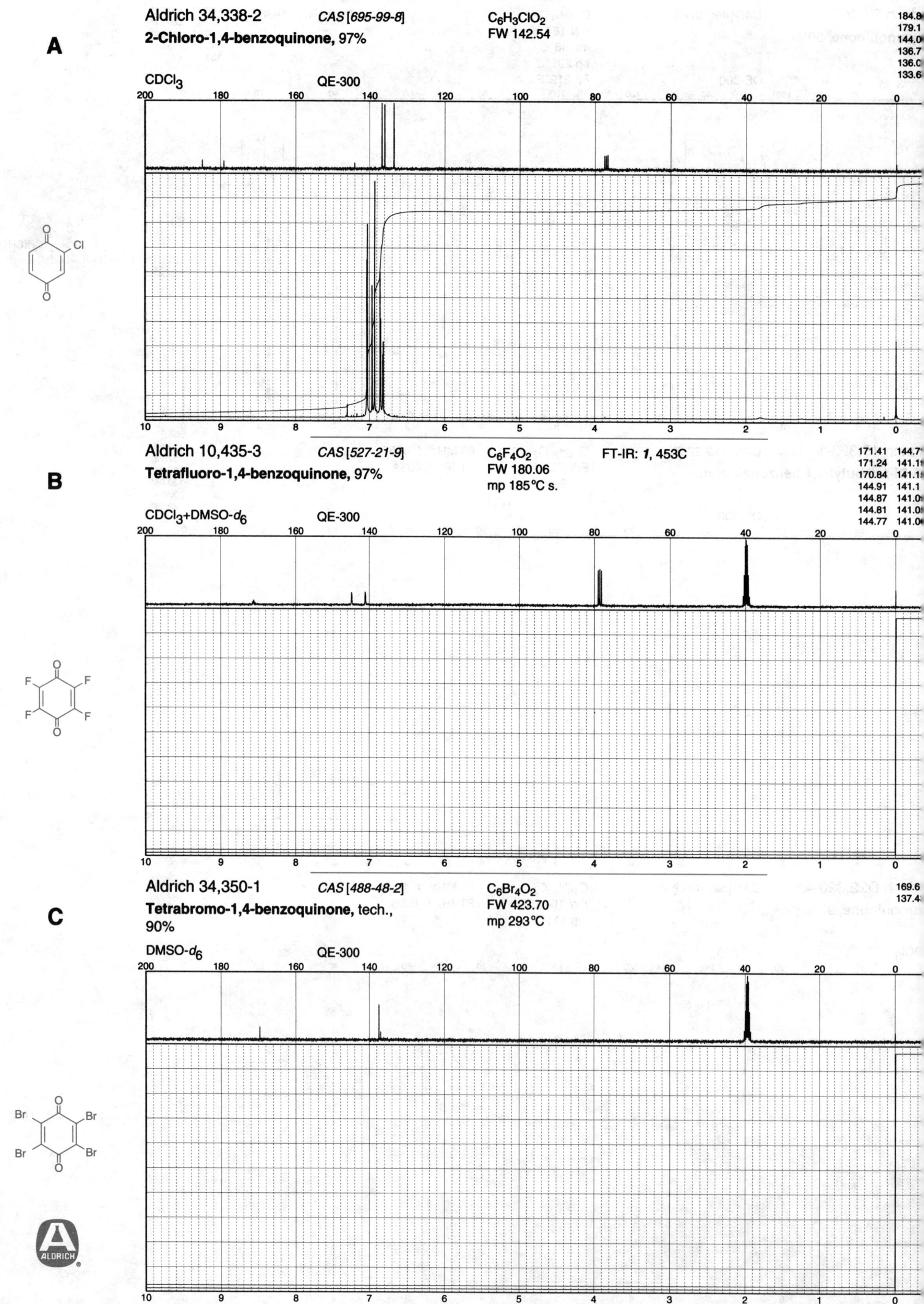

A

Aldrich 34,338-2   *CAS [695-99-8]* $C_6H_3ClO_2$
2-Chloro-1,4-benzoquinone, 97%   FW 142.54

184.8
179.1
144.0
136.7
136.0
133.6

CDCl₃   QE-300

B

Aldrich 10,435-3   *CAS [527-21-9]* $C_6F_4O_2$ FT-IR: *1*, 453C
Tetrafluoro-1,4-benzoquinone, 97%   FW 180.06
mp 185°C s.

171.41 144.7
171.24 141.1
170.84 141.1
144.91 141.1
144.87 141.0
144.81 141.0
144.77 141.0

CDCl₃+DMSO-*d₆*   QE-300

C

Aldrich 34,350-1   *CAS [488-48-2]* $C_6Br_4O_2$
Tetrabromo-1,4-benzoquinone, tech.,   FW 423.70
90%   mp 293°C

169.6
137.4

DMSO-*d₆*   QE-300

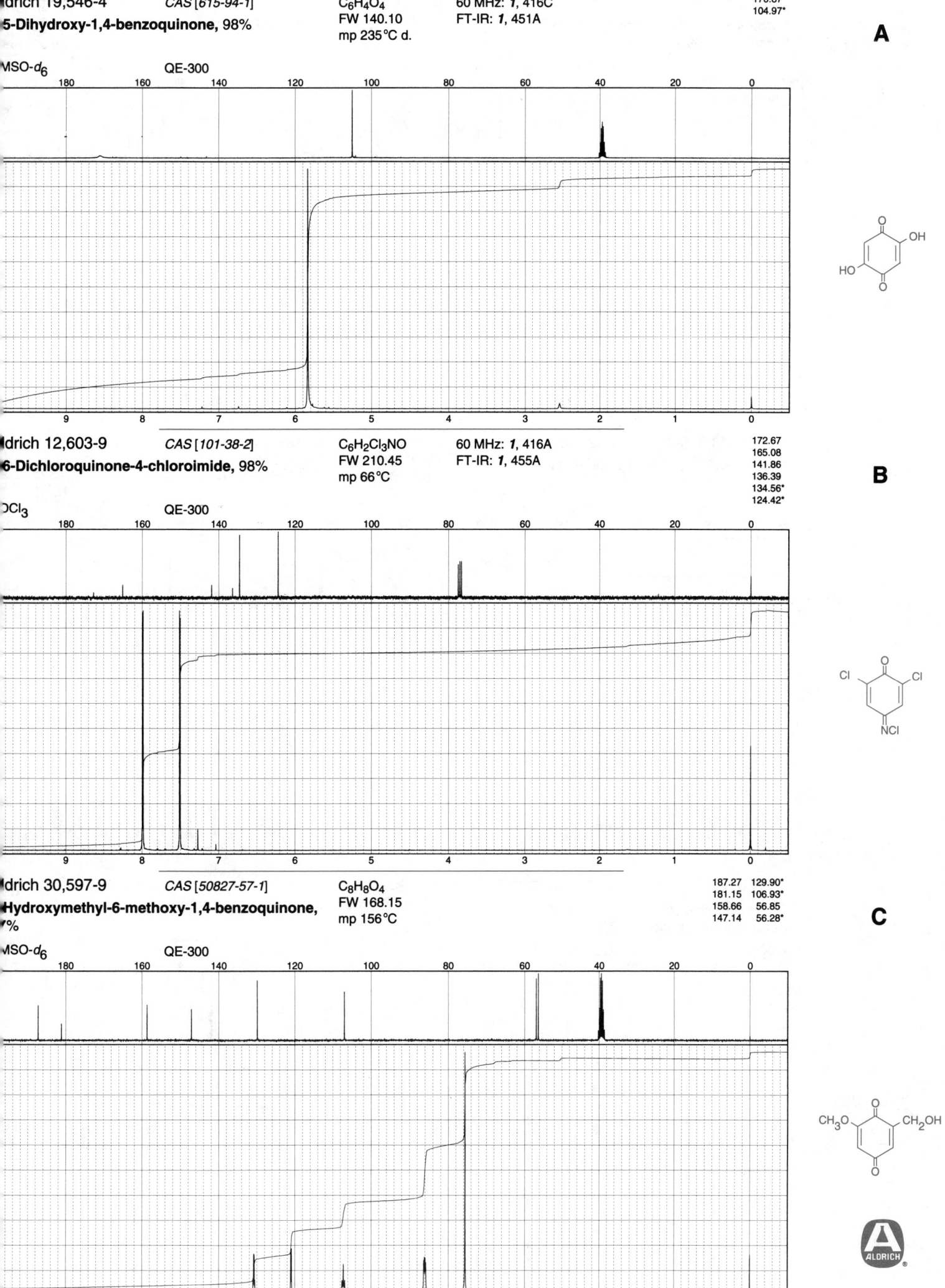

A

Idrich 19,546-4   *CAS [615-94-1]*
5-Dihydroxy-1,4-benzoquinone, 98%

$C_6H_4O_4$
FW 140.10
mp 235°C d.

60 MHz: *1*, 416C
FT-IR: *1*, 451A

170.87
104.97*

MSO-d_6 QE-300

B

Idrich 12,603-9   *CAS [101-38-2]*
6-Dichloroquinone-4-chloroimide, 98%

$C_6H_2Cl_3NO$
FW 210.45
mp 66°C

60 MHz: *1*, 416A
FT-IR: *1*, 455A

172.67
165.08
141.86
136.39
134.56*
124.42*

OCl₃   QE-300

C

Idrich 30,597-9   *CAS [50827-57-1]*
Hydroxymethyl-6-methoxy-1,4-benzoquinone,
%

$C_8H_8O_4$
FW 168.15
mp 156°C

187.27   129.90*
181.15   106.93*
158.66   56.85
147.14   56.28*

MSO-d_6 QE-300

ALDRICH

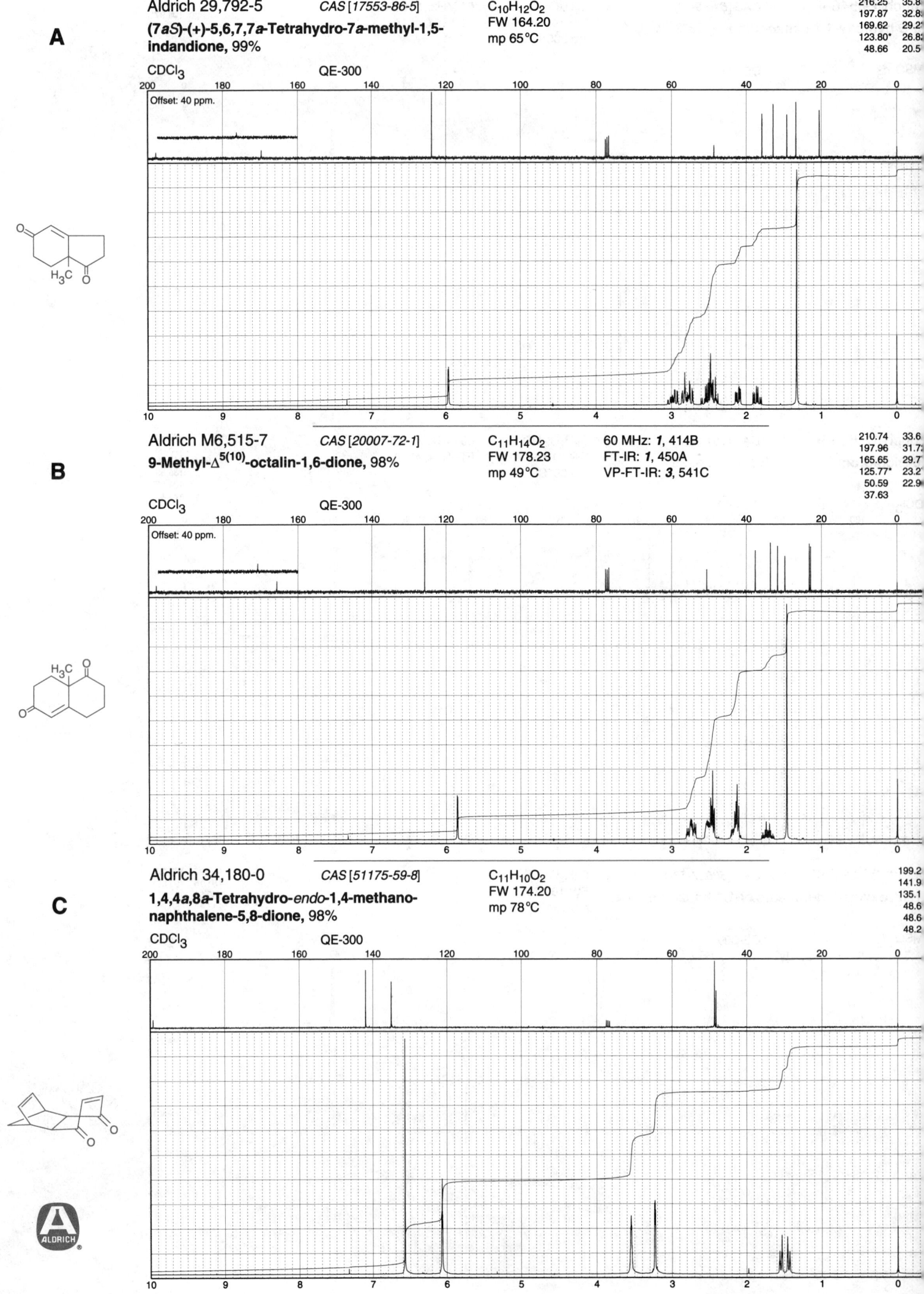

A

Aldrich 29,792-5   CAS [17553-86-5] $C_{10}H_{12}O_2$ FW 164.20 mp 65°C

(7aS)-(+)-5,6,7,7a-Tetrahydro-7a-methyl-1,5-indandione, 99%

216.25   35.8
197.87   32.8
169.62   29.2
123.80*   26.8
48.66   20.5

CDCl₃   QE-300

Offset: 40 ppm.

B

Aldrich M6,515-7   CAS [20007-72-1] $C_{11}H_{14}O_2$ FW 178.23 mp 49°C

9-Methyl-Δ⁵⁽¹⁰⁾-octalin-1,6-dione, 98%

60 MHz: *1*, 414B
FT-IR: *1*, 450A
VP-FT-IR: *3*, 541C

210.74   33.6
197.96   31.7
165.96   29.7
125.77*   23.2
50.59   22.9
37.63

CDCl₃   QE-300

Offset: 40 ppm.

C

Aldrich 34,180-0   CAS [51175-59-8] $C_{11}H_{10}O_2$ FW 174.20 mp 78°C

1,4,4a,8a-Tetrahydro-endo-1,4-methano-naphthalene-5,8-dione, 98%

199.2
141.9
135.1
48.6
48.6
48.2

CDCl₃   QE-300

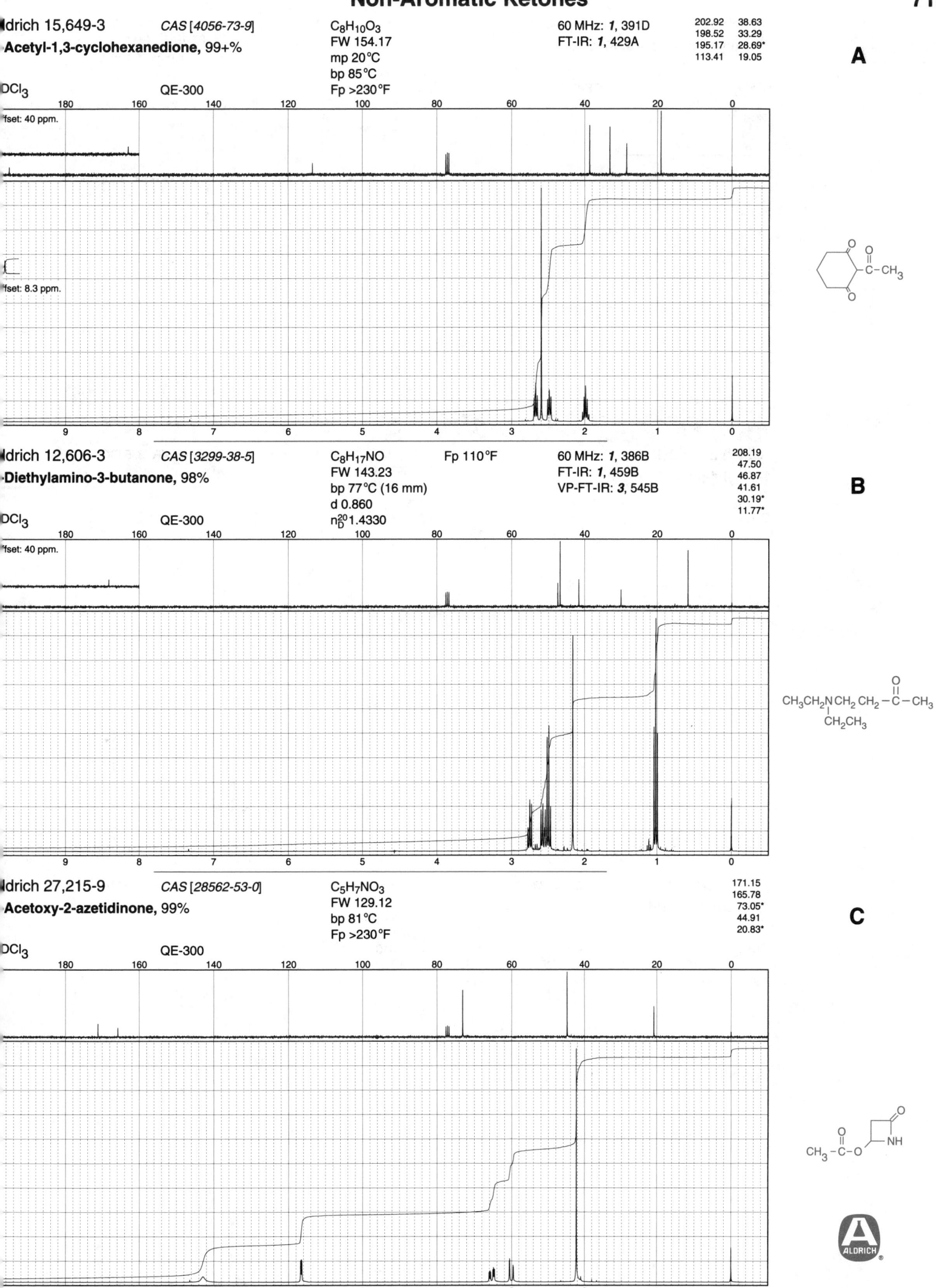

Aldrich 15,649-3   CAS [4056-73-9]
Acetyl-1,3-cyclohexanedione, 99+%

C₈H₁₀O₃
FW 154.17
mp 20°C
bp 85°C
Fp >230°F

60 MHz: *1*, 391D
FT-IR: *1*, 429A

202.92   38.63
198.52   33.29
195.17   28.69*
113.41   19.05

CDCl₃   QE-300

offset: 40 ppm.

offset: 8.3 ppm.

A

Aldrich 12,606-3   CAS [3299-38-5]
Diethylamino-3-butanone, 98%

C₈H₁₇NO   Fp 110°F
FW 143.23
bp 77°C (16 mm)
d 0.860
n²⁰_D 1.4330

60 MHz: *1*, 386B
FT-IR: *1*, 459B
VP-FT-IR: *3*, 545B

208.19
47.50
46.87
41.61
30.19*
11.77*

CDCl₃   QE-300

offset: 40 ppm.

$CH_3CH_2NCH_2CH_2-C-CH_3$
     |
   CH_2CH_3

B

Aldrich 27,215-9   CAS [28562-53-0]
Acetoxy-2-azetidinone, 99%

C₅H₇NO₃
FW 129.12
bp 81°C
Fp >230°F

171.15
165.78
73.05*
44.91
20.83*

CDCl₃   QE-300

CH_3-C-O

C

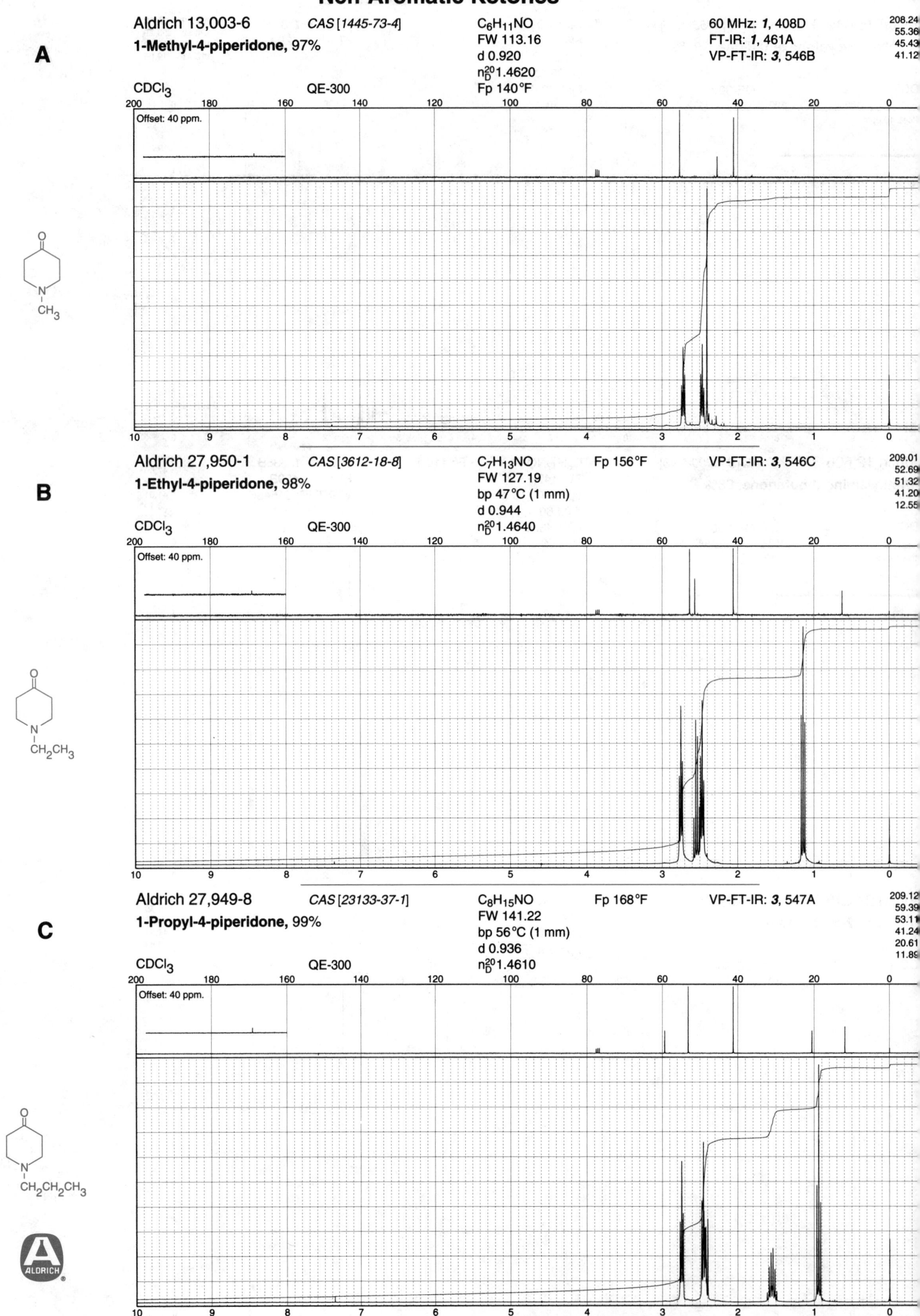

A

Aldrich 13,003-6   CAS [1445-73-4]   C$_6$H$_{11}$NO   60 MHz: *1*, 408D   208.24
1-Methyl-4-piperidone, 97%   FW 113.16   FT-IR: *1*, 461A   55.36
                                                           d 0.920   VP-FT-IR: *3*, 546B   45.43
                                                           n$_D^{20}$1.4620   41.12
                                                           Fp 140°F

CDCl$_3$ QE-300

Offset: 40 ppm.

B

Aldrich 27,950-1   CAS [3612-18-8]   C$_7$H$_{13}$NO   Fp 156°F   VP-FT-IR: *3*, 546C   209.01
1-Ethyl-4-piperidone, 98%   FW 127.19   52.69
                                                           bp 47°C (1 mm)   51.32
                                                           d 0.944   41.20
                                                           n$_D^{20}$1.4640   12.55

CDCl$_3$ QE-300

Offset: 40 ppm.

C

Aldrich 27,949-8   CAS [23133-37-1]   C$_8$H$_{15}$NO   Fp 168°F   VP-FT-IR: *3*, 547A   209.12
1-Propyl-4-piperidone, 99%   FW 141.22   59.39
                                                           bp 56°C (1 mm)   53.11
                                                           d 0.936   41.24
                                                           n$_D^{20}$1.4610   20.61
                                                                                                      11.89

CDCl$_3$ QE-300

Offset: 40 ppm.

Idrich 24,775-8   *CAS [5355-68-0]* $C_8H_{15}NO$ VP-FT-IR: **3**, 546D

-Isopropyl-4-piperidone, 99%

FW 141.22
bp 101 °C (27 mm)
d 0.950
n_D^{20} 1.4634

DCl₃   QE-300

209.37
54.03*
48.35
41.85
18.48*

A

ffset: 40 ppm.

Idrich Q190-5   *CAS [1193-65-3]* $C_7H_{11}NO$

-Quinuclidinone hydrochloride, 97%

FW 161.63
60 MHz: **1**, 410A
FT-IR: **1**, 462B

209.74   48.78
95.28   39.84*
63.47   34.76*
61.92   22.83
49.53   21.74

B

₂O   QE-300

D₂O

ffset: 40 ppm.

• HCl

Idrich 24,548-8   *CAS [10581-38-1]* $C_9H_{17}NO$ 60 MHz: **1**, 409B

,2,6,6-Tetramethyl-4-piperidone monohydrate,
9%

FW 173.26
mp 62 °C
bp 104 °C (18 mm)

FT-IR: **1**, 461C

210.88
55.33
54.08
31.96*

C

DCl₃   QE-300

ffset: 40 ppm.

• H₂O

CH₃   CH₃
CH₃   N   CH₃
    H

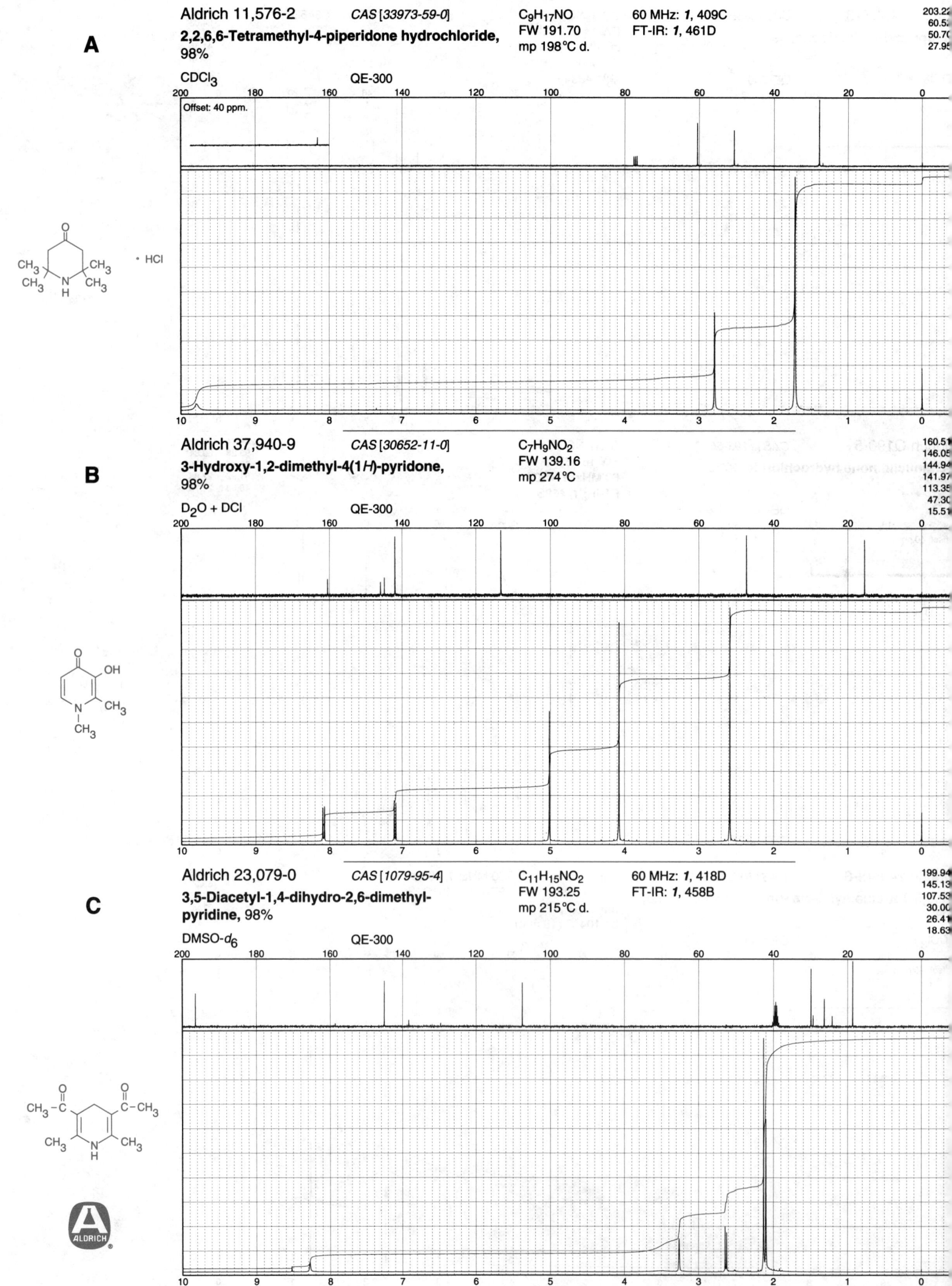

A

Aldrich 11,576-2   CAS [33973-59-0]   C₉H₁₇NO   60 MHz: *1*, 409C   203.22
2,2,6,6-Tetramethyl-4-piperidone hydrochloride, 98%   FW 191.70   FT-IR: *1*, 461D   60.52
mp 198°C d.   50.70
27.95

CDCl₃   QE-300
Offset: 40 ppm.

B

Aldrich 37,940-9   CAS [30652-11-0]   C₇H₉NO₂   160.51
3-Hydroxy-1,2-dimethyl-4(1H)-pyridone, 98%   FW 139.16   146.05
mp 274°C   144.94
141.97
113.35
47.30
15.51

D₂O + DCl   QE-300

C

Aldrich 23,079-0   CAS [1079-95-4]   C₁₁H₁₅NO₂   60 MHz: *1*, 418D   199.94
3,5-Diacetyl-1,4-dihydro-2,6-dimethyl-pyridine, 98%   FW 193.25   FT-IR: *1*, 458B   145.13
mp 215°C d.   107.53
30.00
26.41
18.63

DMSO-d₆   QE-300

Non-Aromatic Ketones

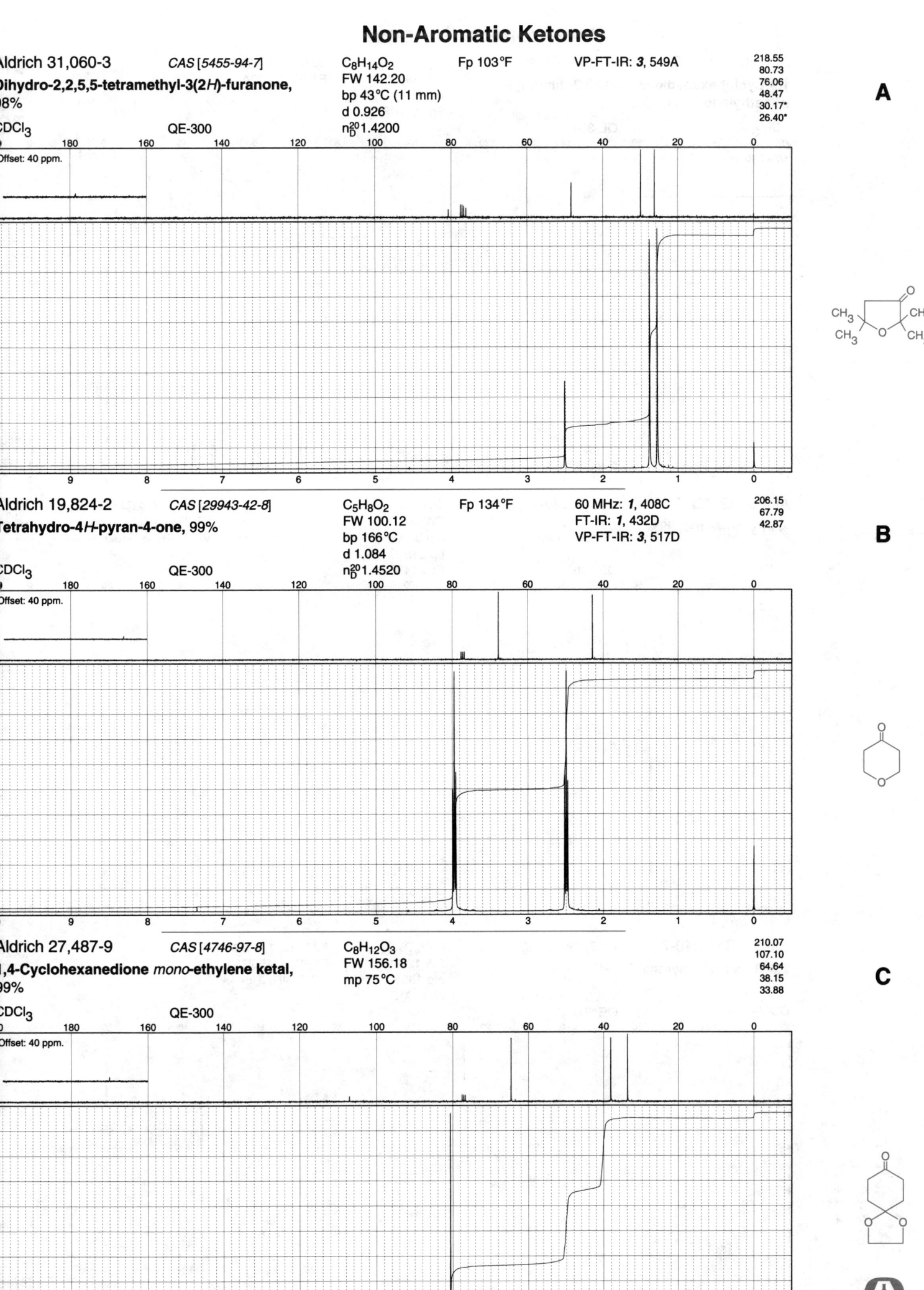

Aldrich 31,060-3 — CAS [5455-94-7]
Dihydro-2,2,5,5-tetramethyl-3(2H)-furanone, 98%
CDCl₃   QE-300
$C_8H_{14}O_2$
FW 142.20
bp 43°C (11 mm)
d 0.926
n_D^{20} 1.4200
Fp 103°F
VP-FT-IR: *3*, 549A
Offset: 40 ppm.

218.55
80.73
76.06
48.47
30.17*
26.40*

A

Aldrich 19,824-2 — CAS [29943-42-8]
Tetrahydro-4H-pyran-4-one, 99%
CDCl₃   QE-300
$C_5H_8O_2$
FW 100.12
bp 166°C
d 1.084
n_D^{20} 1.4520
Fp 134°F
60 MHz: *1*, 408C
FT-IR: *1*, 432D
VP-FT-IR: *3*, 517D
Offset: 40 ppm.

206.15
67.79
42.87

B

Aldrich 27,487-9 — CAS [4746-97-8]
1,4-Cyclohexanedione mono-ethylene ketal, 99%
CDCl₃   QE-300
$C_8H_{12}O_3$
FW 156.18
mp 75°C
Offset: 40 ppm.

210.07
107.10
64.64
38.15
33.88

C

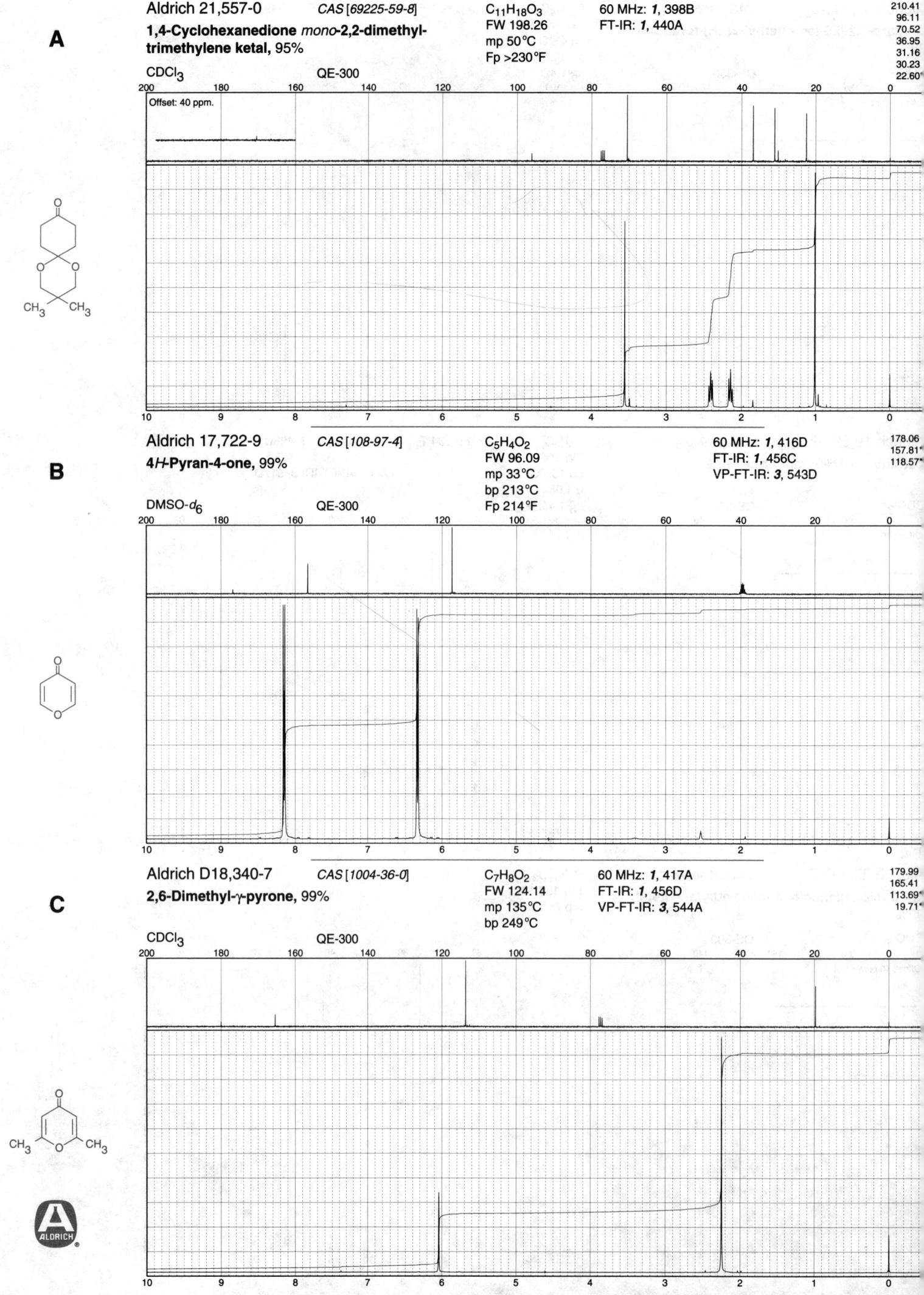

A

Aldrich 21,557-0   CAS [69225-59-8]

1,4-Cyclohexanedione mono-2,2-dimethyl-trimethylene ketal, 95%

$C_{11}H_{18}O_3$
FW 198.26
mp 50°C
Fp >230°F

60 MHz: *1*, 398B
FT-IR: *1*, 440A

210.41
96.11
70.52
36.95
31.16
30.23
22.60*

CDCl₃   QE-300

Offset: 40 ppm.

B

Aldrich 17,722-9   CAS [108-97-4]

4H-Pyran-4-one, 99%

$C_5H_4O_2$
FW 96.09
mp 33°C
bp 213°C
Fp 214°F

60 MHz: *1*, 416D
FT-IR: *1*, 456C
VP-FT-IR: *3*, 543D

178.06
157.81
118.57*

DMSO-d₆   QE-300

C

Aldrich D18,340-7   CAS [1004-36-0]

2,6-Dimethyl-γ-pyrone, 99%

$C_7H_8O_2$
FW 124.14
mp 135°C
bp 249°C

60 MHz: *1*, 417A
FT-IR: *1*, 456D
VP-FT-IR: *3*, 544A

179.99
165.41
113.69*
19.71*

CDCl₃   QE-300

Idrich H4,340-7 *CAS [118-71-8]* C₆H₆O₃ 60 MHz: *1*, 417B
-Hydroxy-2-methyl-4-pyrone, 99% FW 126.11 FT-IR: *1*, 457A
mp 162°C

DCl₃+DMSO-d₆ QE-300

172.62
154.06*
148.99
143.14
113.42*
13.99*

A

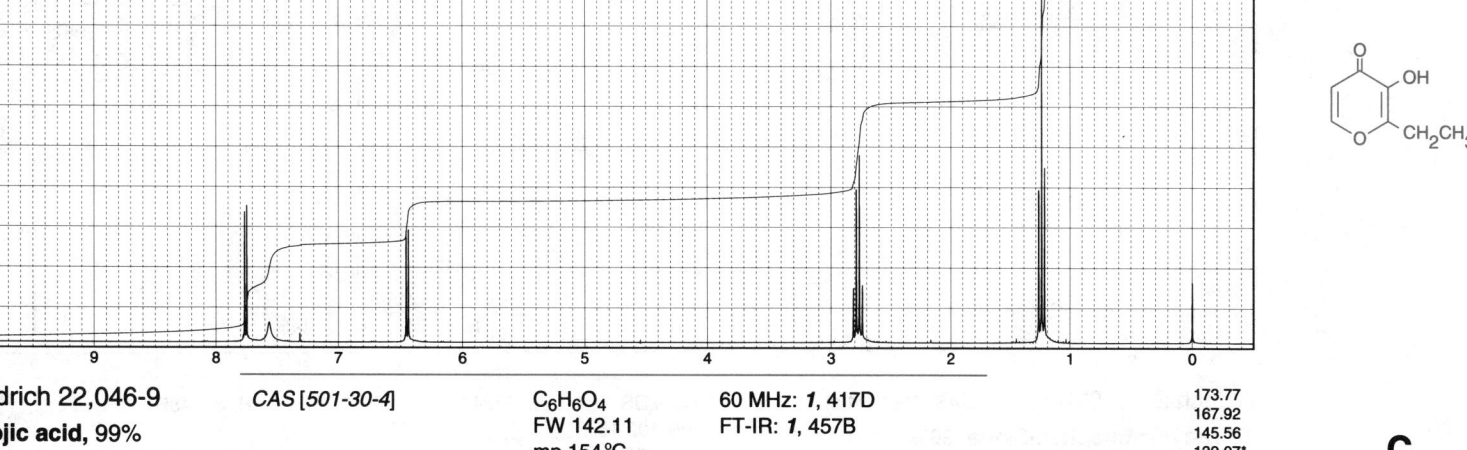

Idrich W34,870-8 *CAS [4940-11-8]* C₇H₈O₃ VP-FT-IR: *3*, 544B
thyl maltol, 99+% FW 140.14
mp 90°C

DCl₃ QE-300

173.32
154.19*
153.56
142.60
113.09*
21.71
10.78*

B

Idrich 22,046-9 *CAS [501-30-4]* C₆H₆O₄ 60 MHz: *1*, 417D
ojic acid, 99% FW 142.11 FT-IR: *1*, 457B
mp 154°C

MSO-d₆ QE-300

173.77
167.92
145.56
139.07*
109.73*
59.42

C

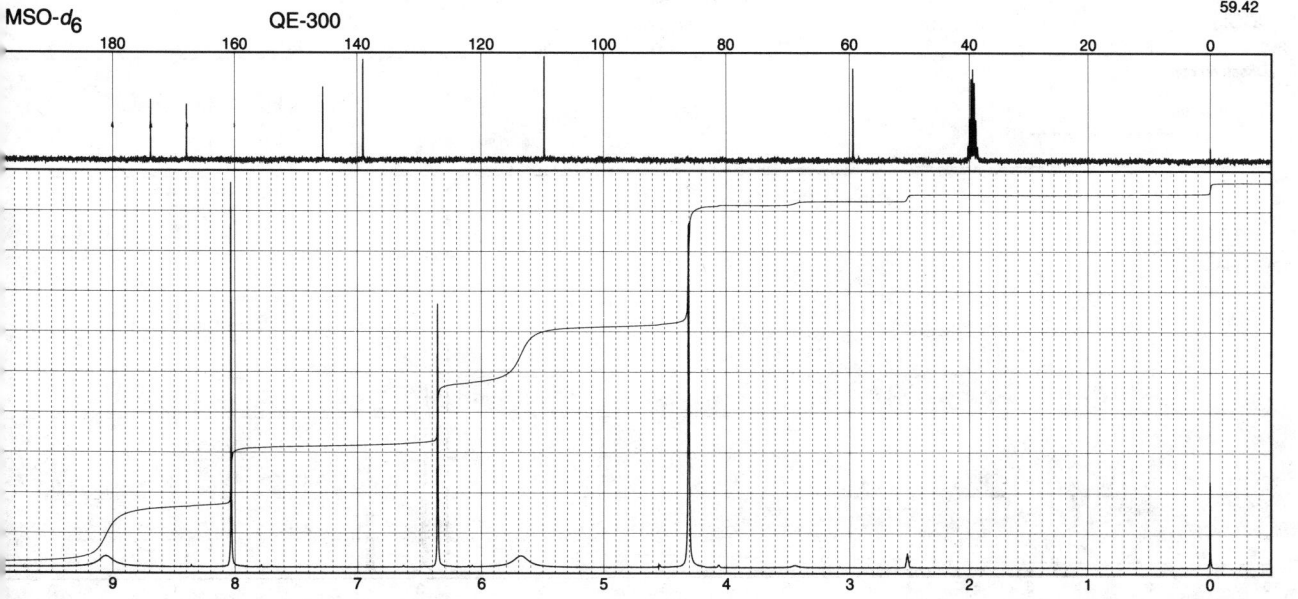

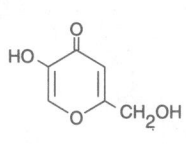

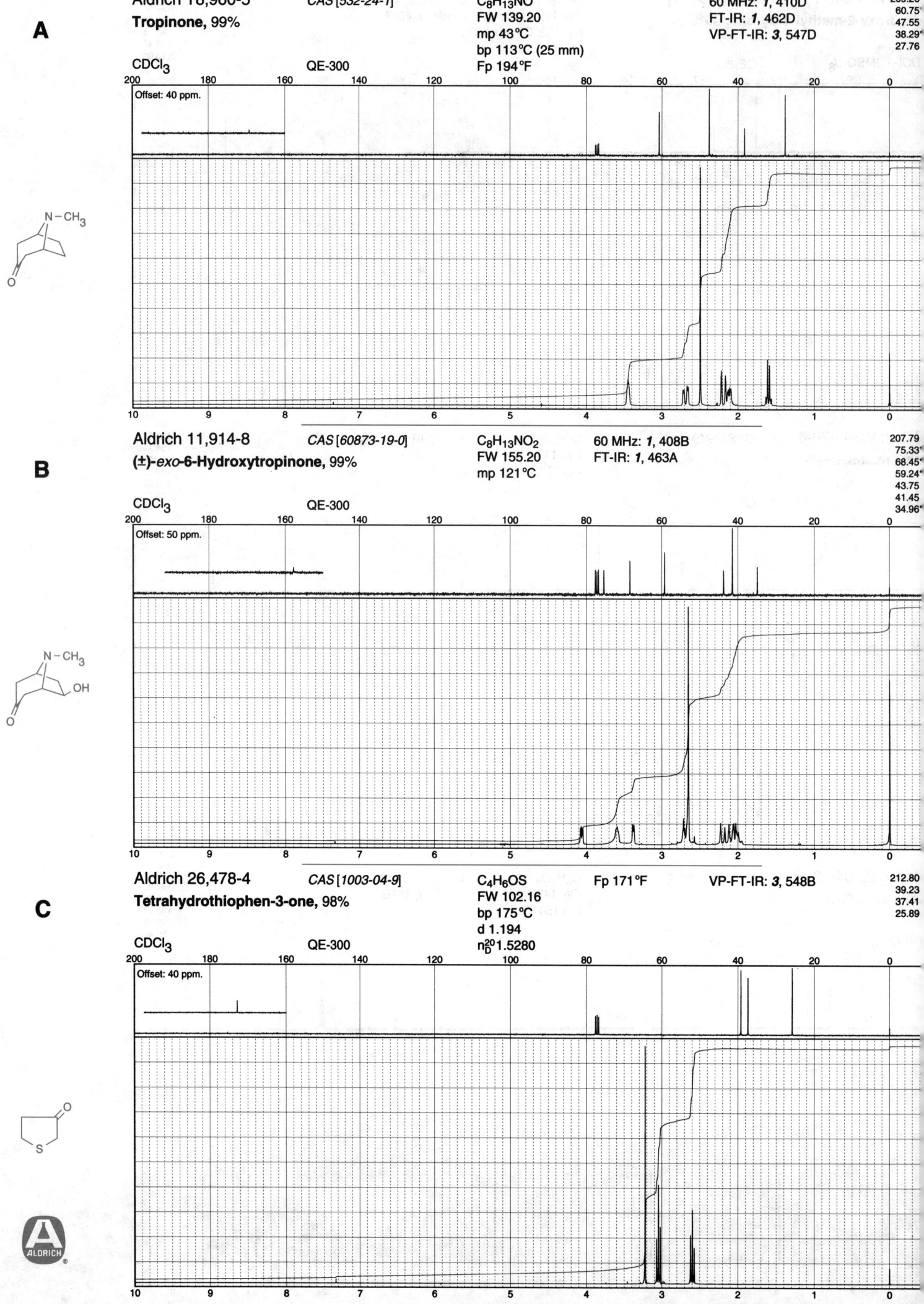

A

Aldrich T8,960-5   *CAS [532-24-1]* $C_8H_{13}NO$
Tropinone, 99%   FW 139.20
mp 43°C
bp 113°C (25 mm)
Fp 194°F

60 MHz: *1*, 410D
FT-IR: *1*, 462D
VP-FT-IR: *3*, 547D

209.26
60.75*
47.55
38.29*
27.76

CDCl₃   QE-300
Offset: 40 ppm.

B

Aldrich 11,914-8   *CAS [60873-19-0]* $C_8H_{13}NO_2$
(±)-*exo*-6-Hydroxytropinone, 99%   FW 155.20
mp 121°C

60 MHz: *1*, 408B
FT-IR: *1*, 463A

207.79
75.33*
68.45*
59.24*
43.75
41.45
34.96*

CDCl₃   QE-300
Offset: 50 ppm.

C

Aldrich 26,478-4   *CAS [1003-04-9]* C_4H_6OS Fp 171°F   VP-FT-IR: *3*, 548B
Tetrahydrothiophen-3-one, 98%   FW 102.16
bp 175°C
d 1.194
nD²⁰ 1.5280

212.80
39.23
37.41
25.89

CDCl₃   QE-300
Offset: 40 ppm.

Non-Aromatic Ketones

726

Aldrich W35,120-2   CAS [13679-85-1]   C₅H₈OS   Fp 160°F

-Methyltetrahydrothiophen-3-one, 97+%   FW 116.18

    bp 82°C (28 mm)
    d 1.119
    n²⁰_D 1.5070

CDCl₃   QE-300

180 160 140 120 100 80 60 40 20 0

Offset: 40 ppm.

213.82
46.01*
38.62
23.44
16.33*

A

9 8 7 6 5 4 3 2 1 0

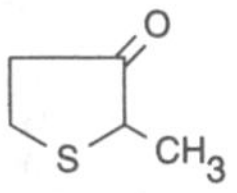

Aldrich 15,516-0   CAS [1072-72-6]   C₅H₈OS   60 MHz: 1, 418B

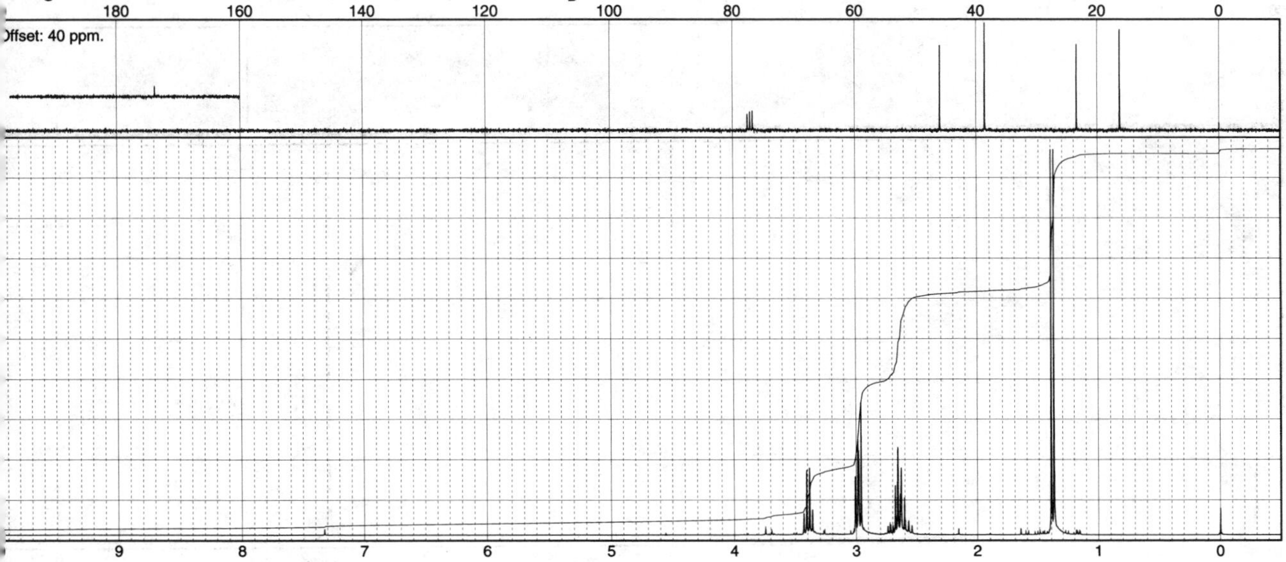

Tetrahydrothiopyran-4-one, 99%   FW 116.18   FT-IR: 1, 463B

    mp 63°C   VP-FT-IR: 3, 548C

CDCl₃   QE-300

180 160 140 120 100 80 60 40 20 0

Offset: 40 ppm.

207.95
43.99
30.00

B

9 8 7 6 5 4 3 2 1 0

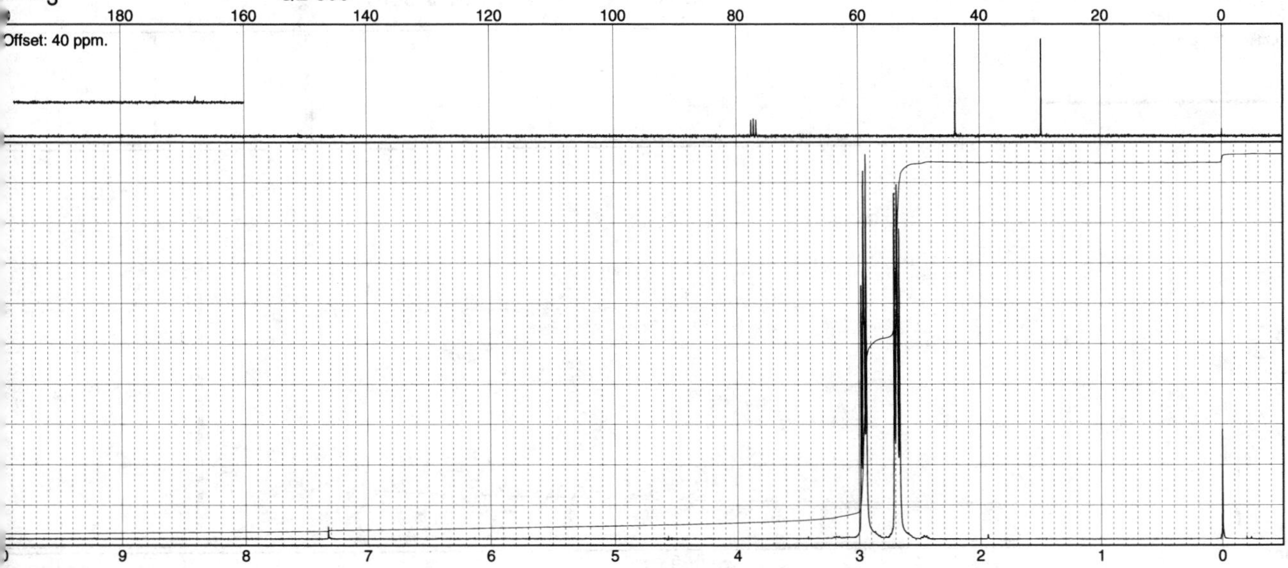

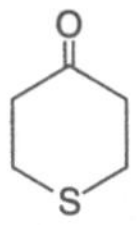

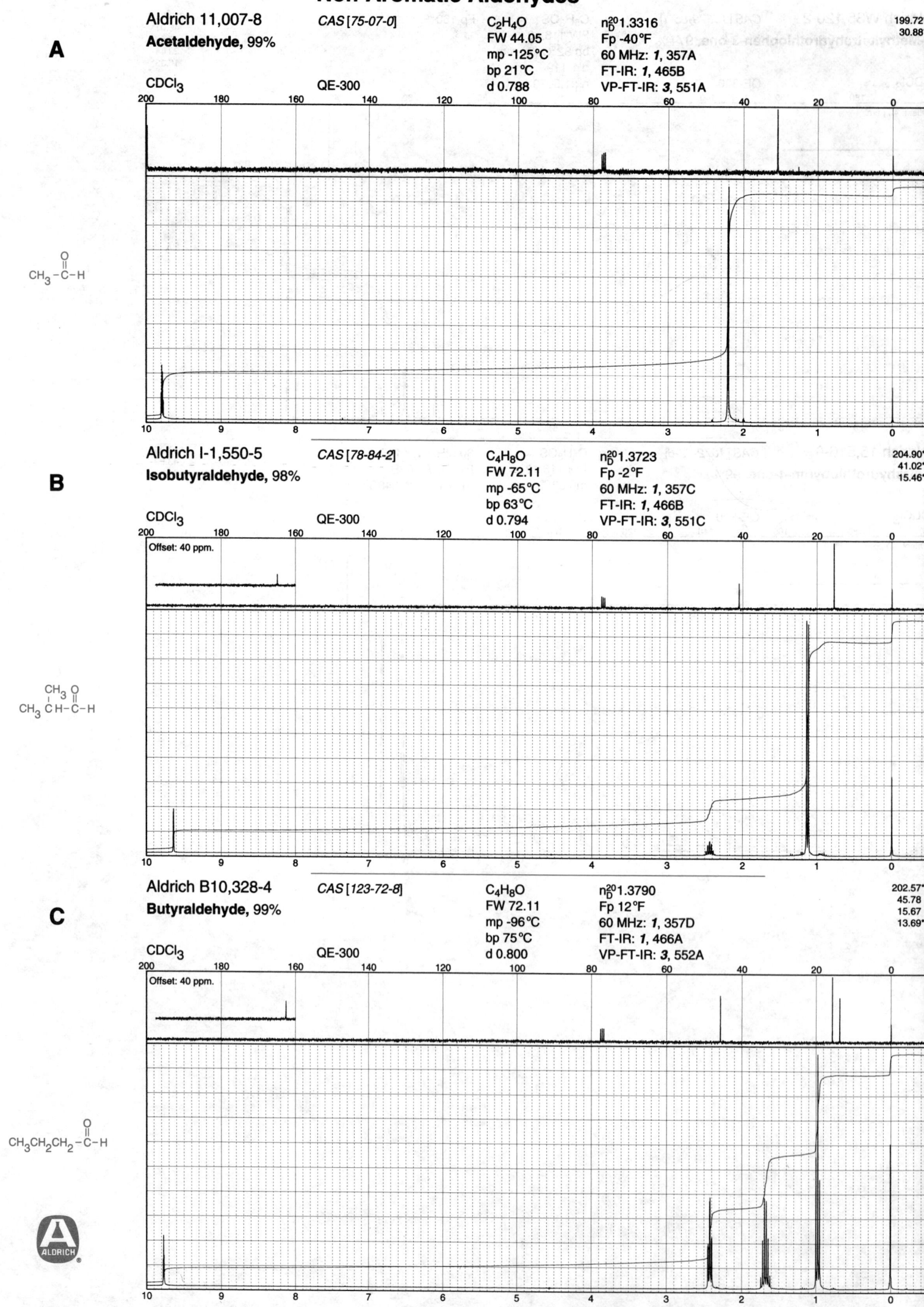

A

Aldrich 11,007-8
Acetaldehyde, 99%

CAS [75-07-0]

C_2H_4O
FW 44.05
mp -125°C
bp 21°C
d 0.788

n_D^{20} 1.3316
Fp -40°F
60 MHz: *1*, 357A
FT-IR: *1*, 465B
VP-FT-IR: *3*, 551A

199.72*
30.88*

CDCl₃

QE-300

$CH_3-\overset{O}{\overset{\|}{C}}-H$

B

Aldrich I-1,550-5
Isobutyraldehyde, 98%

CAS [78-84-2]

C_4H_8O
FW 72.11
mp -65°C
bp 63°C
d 0.794

n_D^{20} 1.3723
Fp -2°F
60 MHz: *1*, 357C
FT-IR: *1*, 466B
VP-FT-IR: *3*, 551C

204.90*
41.02*
15.46*

CDCl₃

QE-300

Offset: 40 ppm.

$\begin{array}{c} CH_3 \\ | \\ CH_3 \end{array} CH-\overset{O}{\overset{\|}{C}}-H$

C

Aldrich B10,328-4
Butyraldehyde, 99%

CAS [123-72-8]

C_4H_8O
FW 72.11
mp -96°C
bp 75°C
d 0.800

n_D^{20} 1.3790
Fp 12°F
60 MHz: *1*, 357D
FT-IR: *1*, 466A
VP-FT-IR: *3*, 552A

202.57*
45.78
15.67
13.69*

CDCl₃

QE-300

Offset: 40 ppm.

$CH_3CH_2CH_2-\overset{O}{\overset{\|}{C}}-H$

ALDRICH

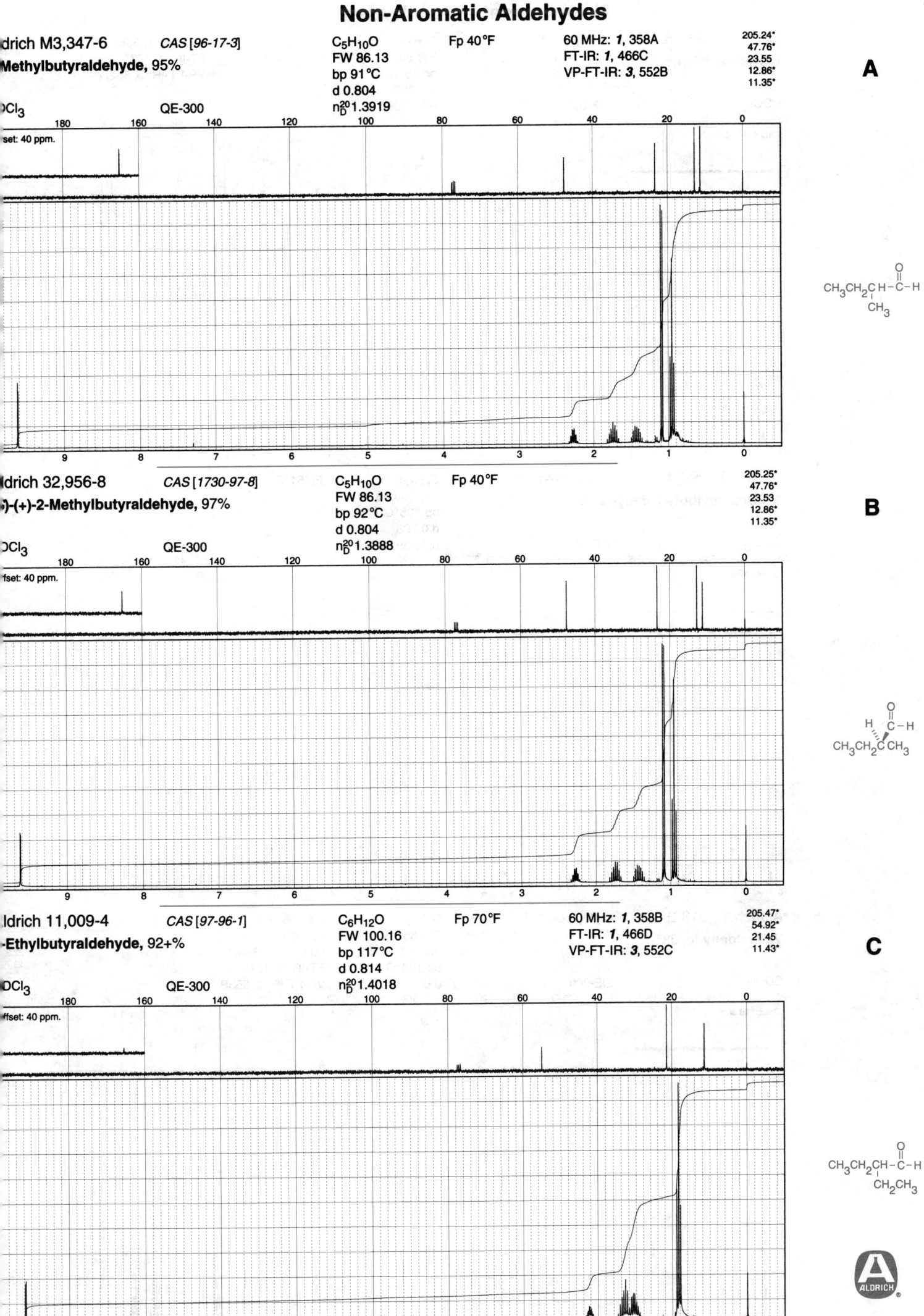

Aldrich M3,347-6   CAS [96-17-3]
Methylbutyraldehyde, 95%

$C_5H_{10}O$
FW 86.13
bp 91 °C
d 0.804
n_D^{20} 1.3919

Fp 40 °F

60 MHz: **1**, 358A
FT-IR: **1**, 466C
VP-FT-IR: **3**, 552B

205.24*
47.76*
23.55
12.86*
11.35*

A

OCl₃   QE-300
offset: 40 ppm.

$CH_3CH_2CH-C-H$ with O double bond and CH_3

Aldrich 32,956-8   CAS [1730-97-8]
(S)-(+)-2-Methylbutyraldehyde, 97%

$C_5H_{10}O$
FW 86.13
bp 92 °C
d 0.804
n_D^{20} 1.3888

Fp 40 °F

205.25*
47.76*
12.86*
11.35*

B

OCl₃   QE-300
offset: 40 ppm.

Aldrich 11,009-4   CAS [97-96-1]
Ethylbutyraldehyde, 92+%

$C_6H_{12}O$
FW 100.16
bp 117 °C
d 0.814
n_D^{20} 1.4018

Fp 70 °F

60 MHz: **1**, 358B
FT-IR: **1**, 466D
VP-FT-IR: **3**, 552C

205.47*
54.92*
21.45
11.43*

C

OCl₃   QE-300
offset: 40 ppm.

$CH_3CH_2CH-C-H$ with O double bond and CH_2CH_3

ALDRICH

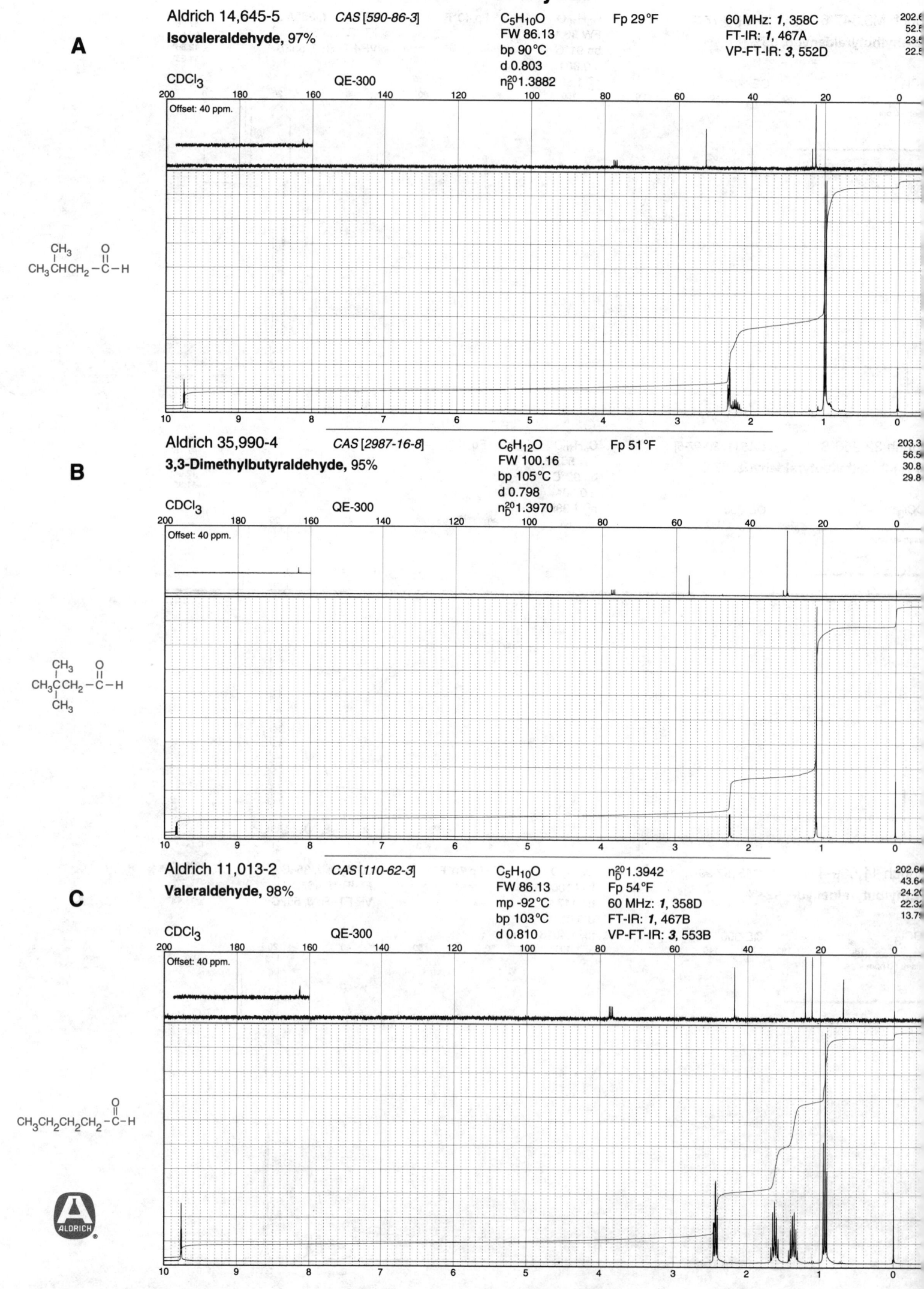

A

Aldrich 14,645-5

Isovaleraldehyde, 97%

CAS [590-86-3]

$C_5H_{10}O$
FW 86.13
bp 90°C
d 0.803
n_D^{20} 1.3882

Fp 29°F

60 MHz: **1**, 358C
FT-IR: **1**, 467A
VP-FT-IR: **3**, 552D

202.6
52.5
23.5
22.5

CDCl₃   QE-300

Offset: 40 ppm.

B

Aldrich 35,990-4

3,3-Dimethylbutyraldehyde, 95%

CAS [2987-16-8]

$C_6H_{12}O$
FW 100.16
bp 105°C
d 0.798
n_D^{20} 1.3970

Fp 51°F

203.3
56.5
30.8
29.8

CDCl₃   QE-300

Offset: 40 ppm.

C

Aldrich 11,013-2

Valeraldehyde, 98%

CAS [110-62-3]

$C_5H_{10}O$
FW 86.13
mp -92°C
bp 103°C
d 0.810

n_D^{20} 1.3942
Fp 54°F
60 MHz: **1**, 358D
FT-IR: **1**, 467B
VP-FT-IR: **3**, 553B

202.6
43.64
24.20
22.32
13.7

CDCl₃   QE-300

Offset: 40 ppm.

ALDRICH

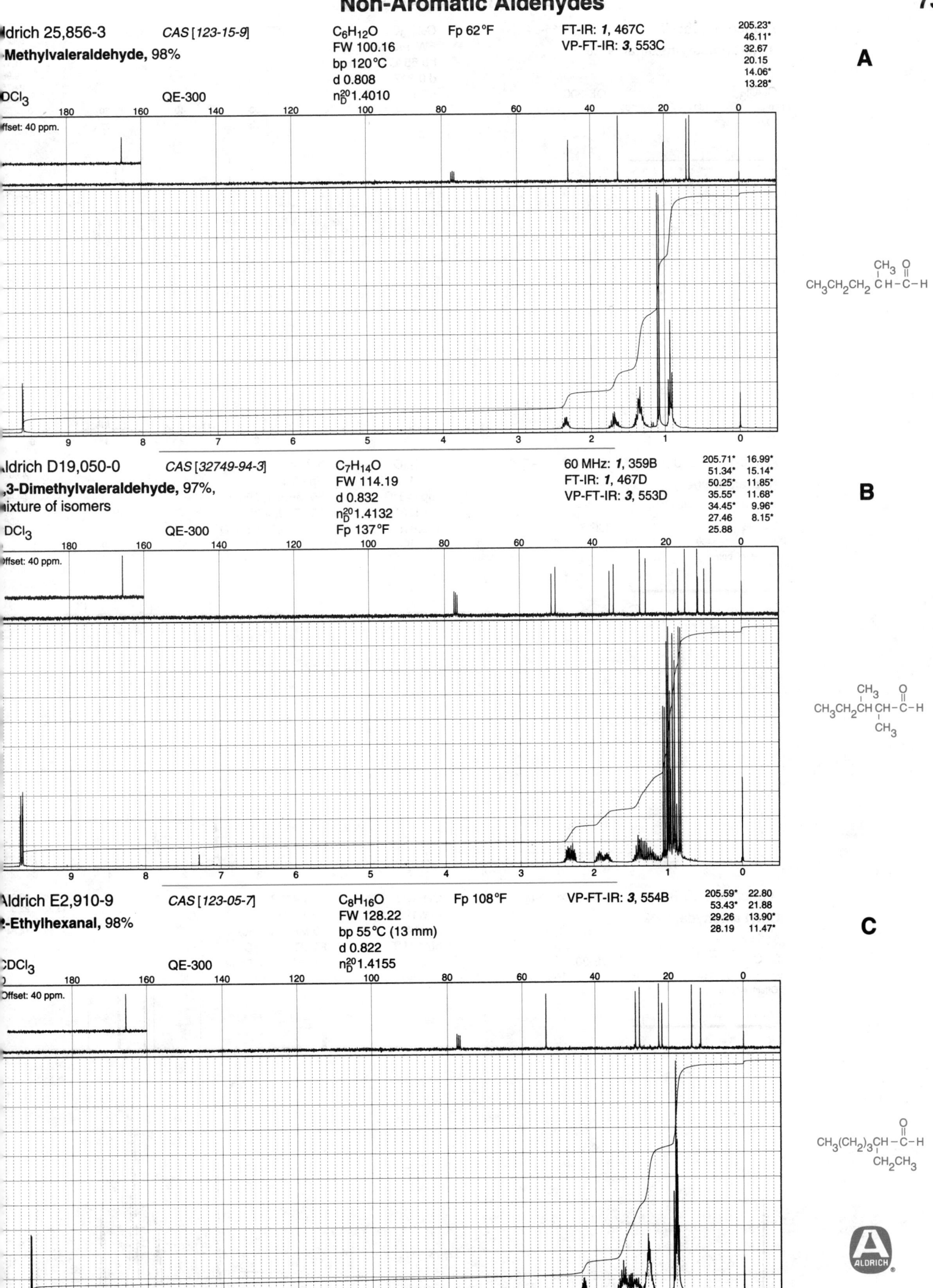

Aldrich 25,856-3   CAS [123-15-9] $C_6H_{12}O$ Fp 62°F   FT-IR: *1*, 467C
Methylvaleraldehyde, 98%   FW 100.16   VP-FT-IR: *3*, 553C
CDCl₃   QE-300   bp 120°C
    d 0.808
    n_D^{20}1.4010
Offset: 40 ppm.

205.23*
46.11*
32.67
20.15
14.06*
13.28*

A

Aldrich D19,050-0   CAS [32749-94-3] $C_7H_{14}O$ 60 MHz: *1*, 359B
3-Dimethylvaleraldehyde, 97%,   FW 114.19   FT-IR: *1*, 467D
mixture of isomers   d 0.832   VP-FT-IR: *3*, 553D
CDCl₃   QE-300 n_D^{20}1.4132
    Fp 137°F
Offset: 40 ppm.

205.71* 16.99*
51.34* 15.14*
50.25* 11.85*
35.55* 11.68*
34.45* 9.96*
27.46 8.15*
25.88

B

Aldrich E2,910-9   CAS [123-05-7] $C_8H_{16}O$ Fp 108°F   VP-FT-IR: *3*, 554B
2-Ethylhexanal, 98%   FW 128.22
    bp 55°C (13 mm)
CDCl₃   QE-300   d 0.822
    n_D^{20}1.4155
Offset: 40 ppm.

205.59* 22.80
53.43* 21.88
29.26 13.90*
28.19 11.47*

C

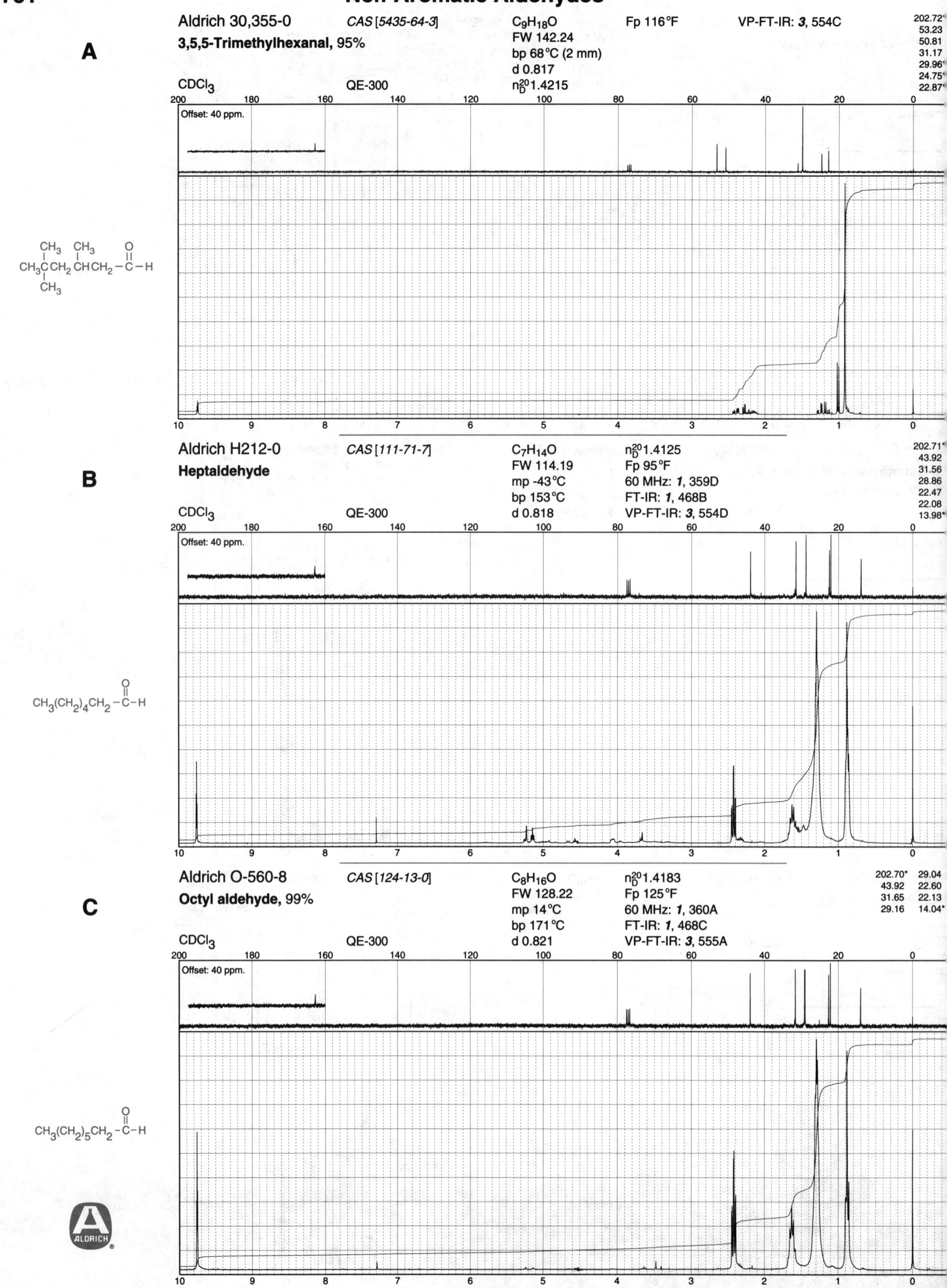

A

Aldrich 30,355-0
3,5,5-Trimethylhexanal, 95%

CAS [5435-64-3]

$C_9H_{18}O$
FW 142.24
bp 68°C (2 mm)
d 0.817
n_D^{20} 1.4215

Fp 116°F

VP-FT-IR: **3**, 554C

202.72*
53.23
50.81
31.17
29.96*
24.75*
22.87*

CDCl₃

QE-300

Offset: 40 ppm.

B

Aldrich H212-0
Heptaldehyde

CAS [111-71-7]

$C_7H_{14}O$
FW 114.19
mp -43°C
bp 153°C
d 0.818

n_D^{20} 1.4125
Fp 95°F
60 MHz: **1**, 359D
FT-IR: **1**, 468B
VP-FT-IR: **3**, 554D

202.71*
43.92
31.56
28.86
22.47
22.08
13.98*

CDCl₃

QE-300

Offset: 40 ppm.

$CH_3(CH_2)_4CH_2-\overset{\overset{\displaystyle O}{\|}}{C}-H$

C

Aldrich O-560-8
Octyl aldehyde, 99%

CAS [124-13-0]

$C_8H_{16}O$
FW 128.22
mp 14°C
bp 171°C
d 0.821

n_D^{20} 1.4183
Fp 125°F
60 MHz: **1**, 360A
FT-IR: **1**, 468C
VP-FT-IR: **3**, 555A

202.70*   29.04
43.92   22.60
31.65   22.13
29.16   14.04*

CDCl₃

QE-300

Offset: 40 ppm.

$CH_3(CH_2)_5CH_2-\overset{\overset{\displaystyle O}{\|}}{C}-H$

A

Aldrich N3,080-3 CAS [124-19-6] C9H18O Fp 147°F 60 MHz: 1, 360B
nonyl aldehyde, 95% FW 142.24 FT-IR: 1, 468D
bp 93°C (23 mm) VP-FT-IR: 3, 555B
d 0.827
CDCl3 QE-300 n$_D^{20}$ 1.4240
offset: 40 ppm.

202.69*	29.12
43.93	22.65
31.82	22.13
29.35	14.07*
29.20	

$CH_3(CH_2)_6CH_2-\overset{O}{\overset{\|}{C}}-H$

B

Aldrich 12,577-6 CAS [112-31-2] C10H20O Fp 186°F 60 MHz: 1, 360C
decyl aldehyde, 95% FW 156.27 FT-IR: 1, 469A
bp 208°C VP-FT-IR: 3, 555C
d 0.830
CDCl3 QE-300 n$_D^{20}$ 1.4280
offset: 40 ppm.

202.78*	29.32
43.98	29.25
31.93	22.73
29.64	22.17
29.45	14.15*

$CH_3(CH_2)_7CH_2-\overset{O}{\overset{\|}{C}}-H$

C

Aldrich U220-2 CAS [112-44-7] C11H22O Fp 205°F FT-IR: 1, 469B
undecylic aldehyde, 97% FW 170.30 VP-FT-IR: 3, 555D
bp 112°C (5 mm)
d 0.825
CDCl3 QE-300 n$_D^{20}$ 1.4322
offset: 40 ppm.

202.68*	29.32
43.92	29.20
31.90	22.69
29.56	22.12
29.45	14.10*
29.38	

$CH_3(CH_2)_8CH_2-\overset{O}{\overset{\|}{C}}-H$

ALDRICH

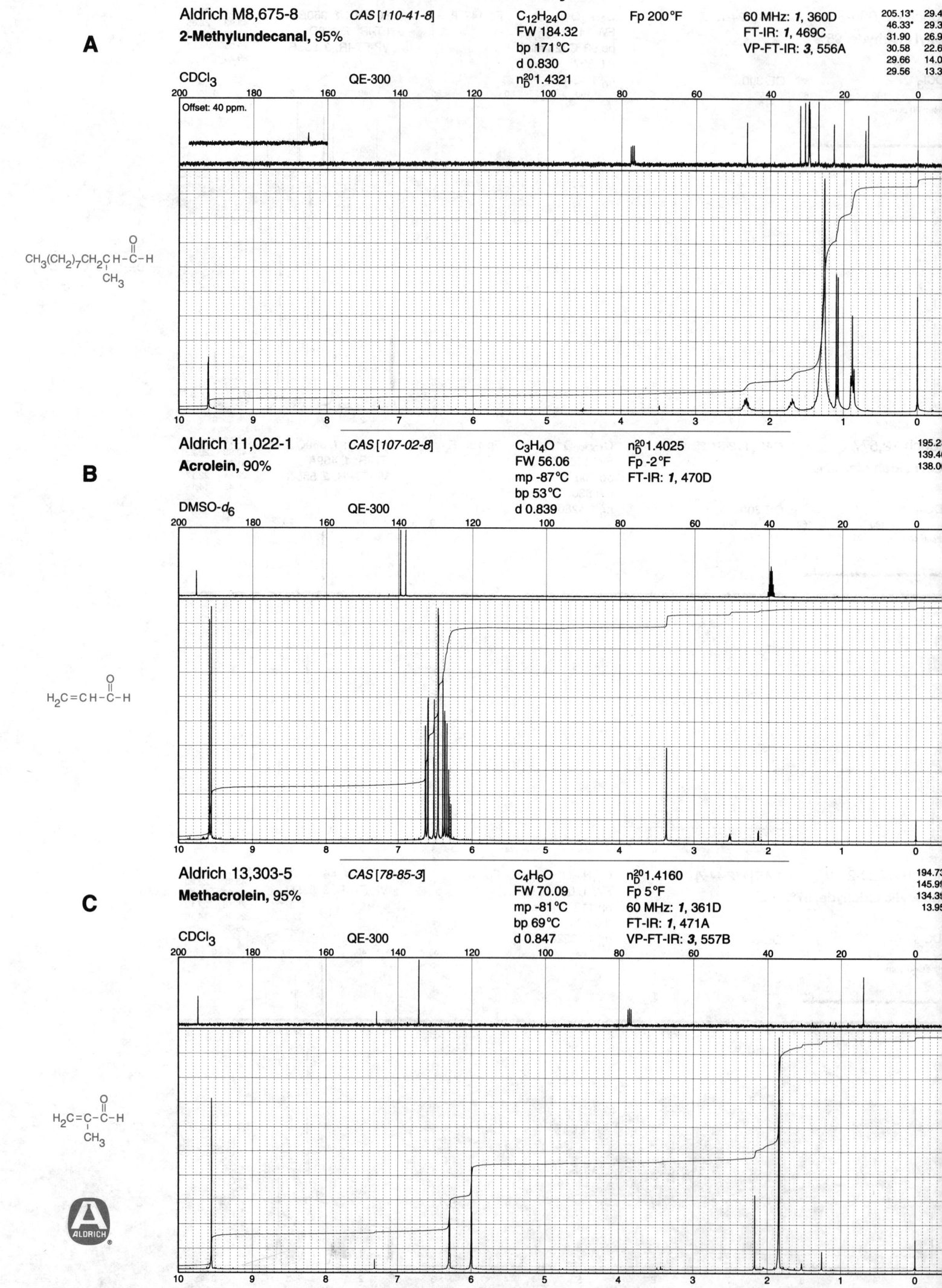

A

Aldrich M8,675-8   *CAS* [*110-41-8*]

2-Methylundecanal, 95%

$C_{12}H_{24}O$
FW 184.32
bp 171 °C
d 0.830
n_D^{20} 1.4321

Fp 200 °F

60 MHz: *1*, 360D
FT-IR: *1*, 469C
VP-FT-IR: *3*, 556A

205.13* 29.4
46.33* 29.3
31.90 26.9
30.58 22.6
29.66 14.0
29.56 13.3

CDCl₃

QE-300

Offset: 40 ppm.

$CH_3(CH_2)_7CH_2CH-C-H$ with CH_3 and O

B

Aldrich 11,022-1   *CAS* [*107-02-8*]

Acrolein, 90%

C_3H_4O
FW 56.06
mp -87 °C
bp 53 °C
d 0.839

n_D^{20} 1.4025
Fp -2 °F
FT-IR: *1*, 470D

195.2
139.4
138.0

DMSO-d_6

QE-300

$H_2C=CH-C-H$ with O

C

Aldrich 13,303-5   *CAS* [*78-85-3*]

Methacrolein, 95%

C_4H_6O
FW 70.09
mp -81 °C
bp 69 °C
d 0.847

n_D^{20} 1.4160
Fp 5 °F
60 MHz: *1*, 361D
FT-IR: *1*, 471A
VP-FT-IR: *3*, 557B

194.73
145.95
134.3
13.95

CDCl₃

QE-300

$H_2C=C-C-H$ with CH_3 and O

Non-Aromatic Aldehydes

ldrich 25,614-5 CAS [922-63-4] C$_5$H$_8$O Fp 34°F FT-IR: 1, 471C 194.60*
Ethylacrolein FW 84.12 VP-FT-IR: 3, 557C 151.77
bp 93°C 132.99
d 0.859 20.91
Cl$_3$ QE-300 n$_D^{20}$1.4271 11.92*

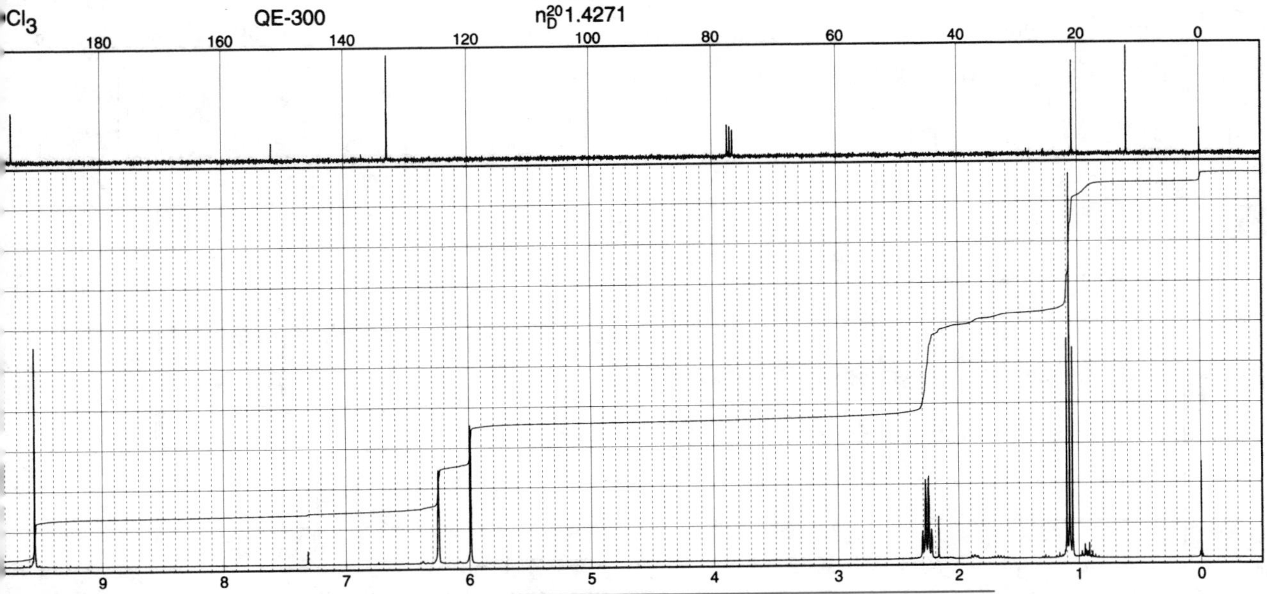

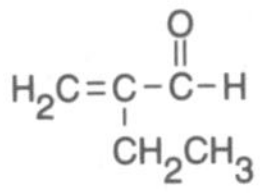

A

ldrich 25,613-7 CAS [1070-66-2] C$_7$H$_{12}$O Fp 92°F FT-IR: 1, 472B 194.62*
Butylacrolein, 95% FW 112.17 VP-FT-IR: 3, 558B 150.50
bp 139°C 133.70
d 0.843 29.95
27.52
Cl$_3$ QE-300 n$_D^{20}$1.4348 22.39
13.82*

B

ldrich 19,261-9 CAS [497-03-0] C$_5$H$_8$O Fp 65°F VP-FT-IR: 3, 558D 195.05*
ans-2-Methyl-2-butenal, 98% FW 84.12 149.52*
bp 118°C (752 mm) 140.39
d 0.871 14.83*
8.86*
Cl$_3$ QE-300 n$_D^{20}$1.4480

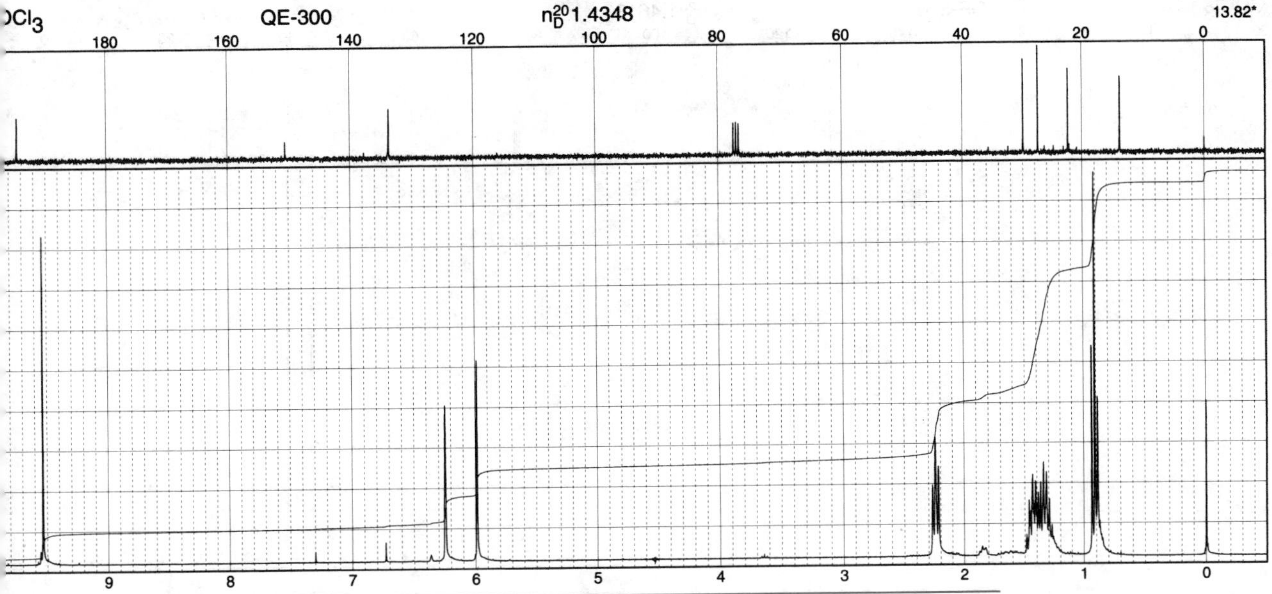

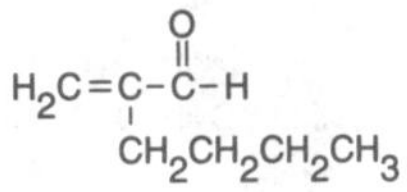

C

Offset: 0.4 ppm.

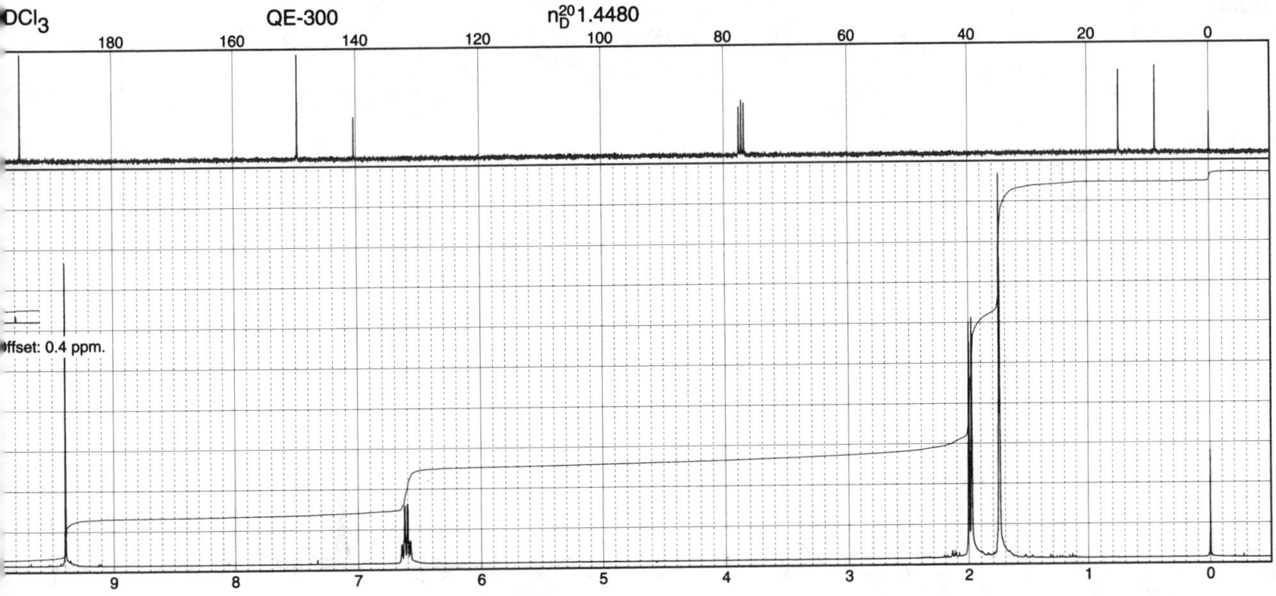

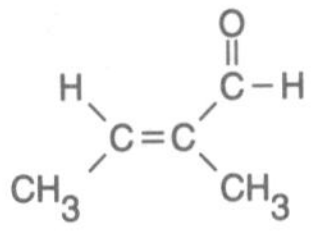

Non-Aromatic Aldehydes

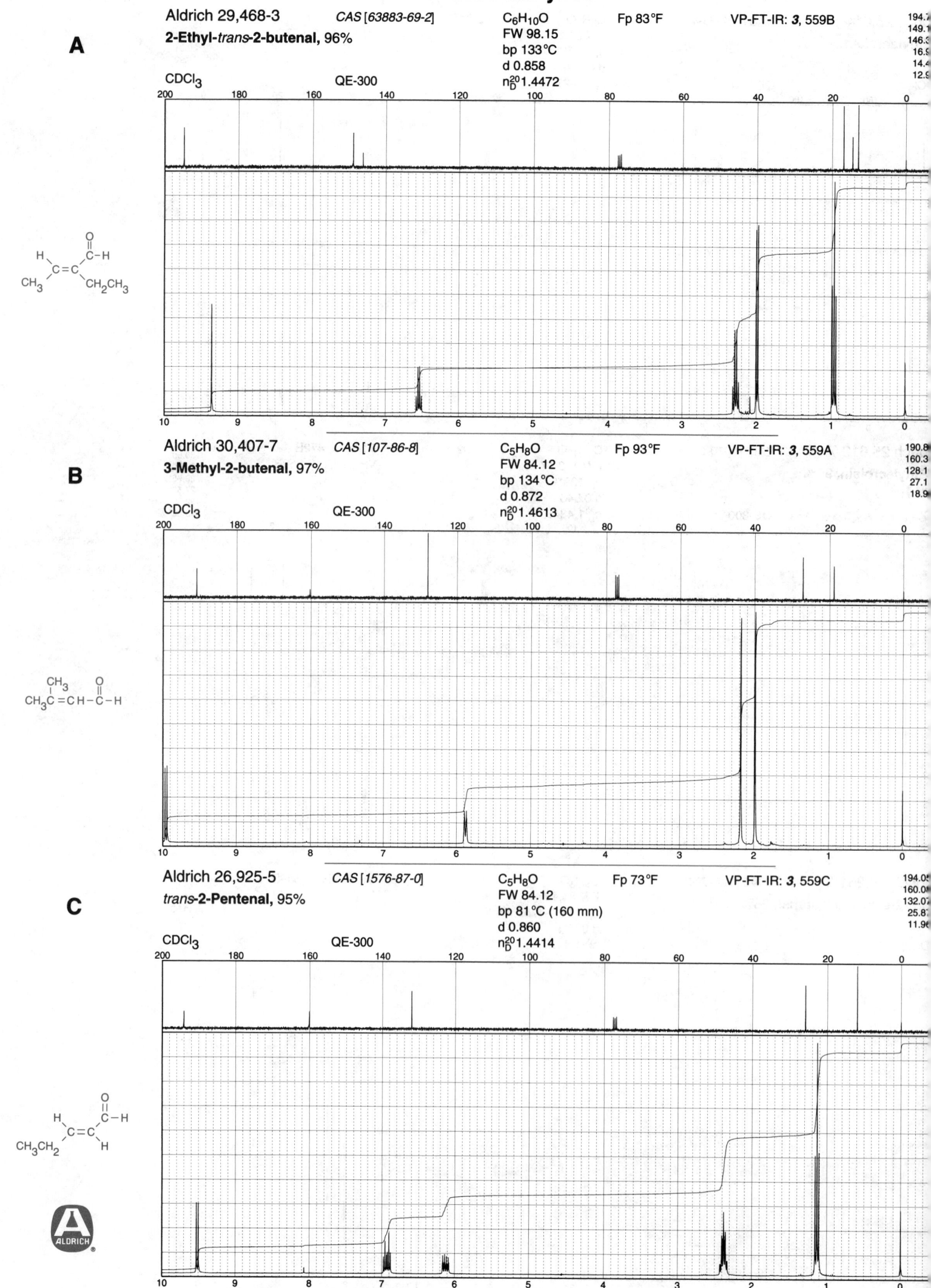

A

Aldrich 29,468-3   *CAS [63883-69-2]* $C_6H_{10}O$ Fp 83°F   VP-FT-IR: **3**, 559B

2-Ethyl-*trans*-2-butenal, 96%

FW 98.15
bp 133°C
d 0.858
n_D^{20}1.4472

CDCl₃   QE-300

194.7
149.1
146.3
16.9
14.4
12.9

B

Aldrich 30,407-7   *CAS [107-86-8]* C_5H_8O Fp 93°F   VP-FT-IR: **3**, 559A

3-Methyl-2-butenal, 97%

FW 84.12
bp 134°C
d 0.872
n_D^{20}1.4613

CDCl₃   QE-300

190.8
160.3
128.1
27.1
18.9

C

Aldrich 26,925-5   *CAS [1576-87-0]* C_5H_8O Fp 73°F   VP-FT-IR: **3**, 559C

***trans*-2-Pentenal, 95%**

FW 84.12
bp 81°C (160 mm)
d 0.860
n_D^{20}1.4414

CDCl₃   QE-300

194.05
160.0
132.07
25.87
11.96

Non-Aromatic Aldehydes

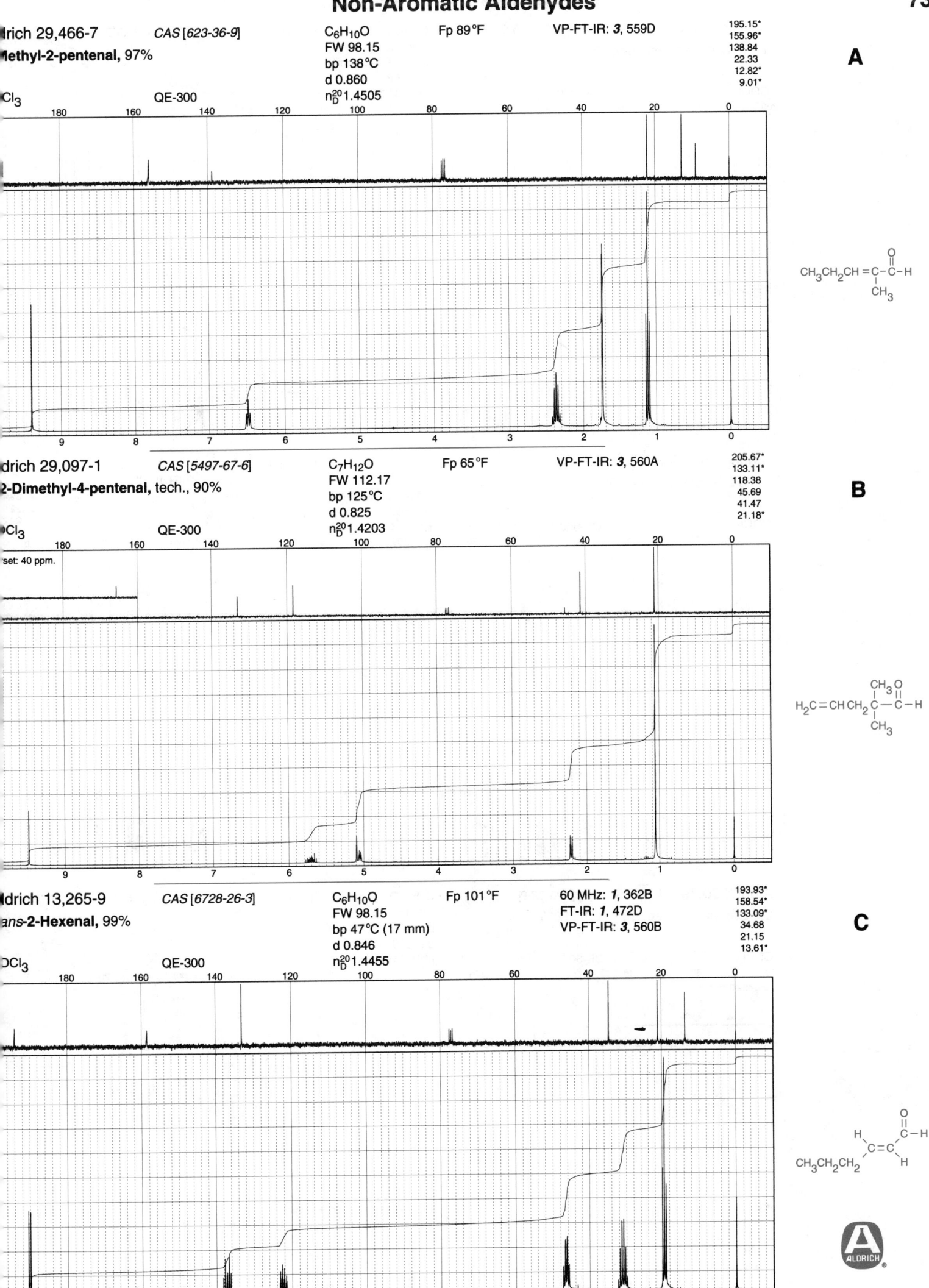

Aldrich 29,466-7   CAS [623-36-9]   C₆H₁₀O   Fp 89°F   VP-FT-IR: 3, 559D
Methyl-2-pentenal, 97%
FW 98.15
bp 138°C
d 0.860
n²⁰_D 1.4505
CDCl₃   QE-300

Aldrich 29,097-1   CAS [5497-67-6]   C₇H₁₂O   Fp 65°F   VP-FT-IR: 3, 560A
2-Dimethyl-4-pentenal, tech., 90%
FW 112.17
bp 125°C
d 0.825
n²⁰_D 1.4203
CDCl₃   QE-300

Aldrich 13,265-9   CAS [6728-26-3]   C₆H₁₀O   Fp 101°F   60 MHz: 1, 362B; FT-IR: 1, 472D; VP-FT-IR: 3, 560B
trans-2-Hexenal, 99%
FW 98.15
bp 47°C (17 mm)
d 0.846
n²⁰_D 1.4455
CDCl₃   QE-300

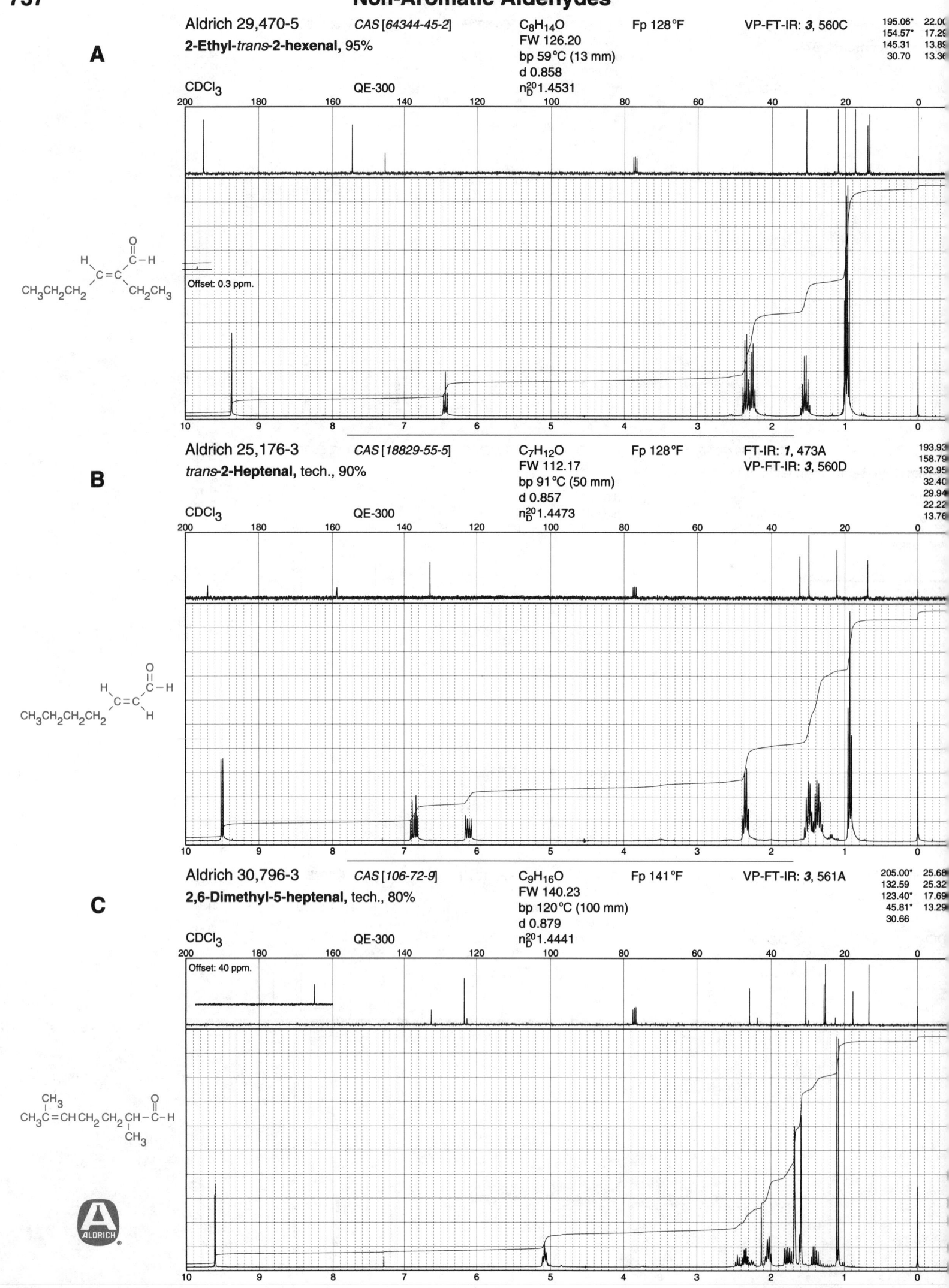

A

Aldrich 29,470-5   CAS [64344-45-2] $C_8H_{14}O$ Fp 128 °F   VP-FT-IR: **3**, 560C

2-Ethyl-*trans*-2-hexenal, 95%

FW 126.20
bp 59 °C (13 mm)
d 0.858
n_D^{20} 1.4531

CDCl₃   QE-300

195.06*	22.0
154.57*	17.29
145.31	13.89
30.70	13.36

Offset: 0.3 ppm.

B

Aldrich 25,176-3   CAS [18829-55-5] $C_7H_{12}O$ Fp 128 °F   FT-IR: **1**, 473A

***trans*-2-Heptenal, tech., 90%**

FW 112.17
bp 91 °C (50 mm)
d 0.857
n_D^{20} 1.4473

CDCl₃   QE-300

| 193.93 |
| 158.79 |
| 132.95 |
| 32.40 |
| 29.94 |
| 22.22 |
| 13.76 |

VP-FT-IR: **3**, 560D

C

Aldrich 30,796-3   CAS [106-72-9] $C_9H_{16}O$ Fp 141 °F   VP-FT-IR: **3**, 561A

2,6-Dimethyl-5-heptenal, tech., 80%

FW 140.23
bp 120 °C (100 mm)
d 0.879
n_D^{20} 1.4441

CDCl₃   QE-300

205.00*	25.68
132.59	25.32
123.40*	17.69
45.81*	13.29
30.66	

Offset: 40 ppm.

Non-Aromatic Aldehydes

Aldrich 26,995-6   *CAS [2548-87-0]*   C₈H₁₄O   Fp 150°F   VP-FT-IR: **3**, 561C

trans-2-Octenal, tech., 94%   FW 126.20

bp 85°C (19 mm)

d 0.846

n²⁰_D 1.4500

CDCl₃   QE-300

193.94*	31.30
158.82*	27.53
132.94*	22.39
32.68	13.91*

A

Aldrich 34,364-1   *CAS [2385-77-5]*   C₁₀H₁₈O   Fp 168°F

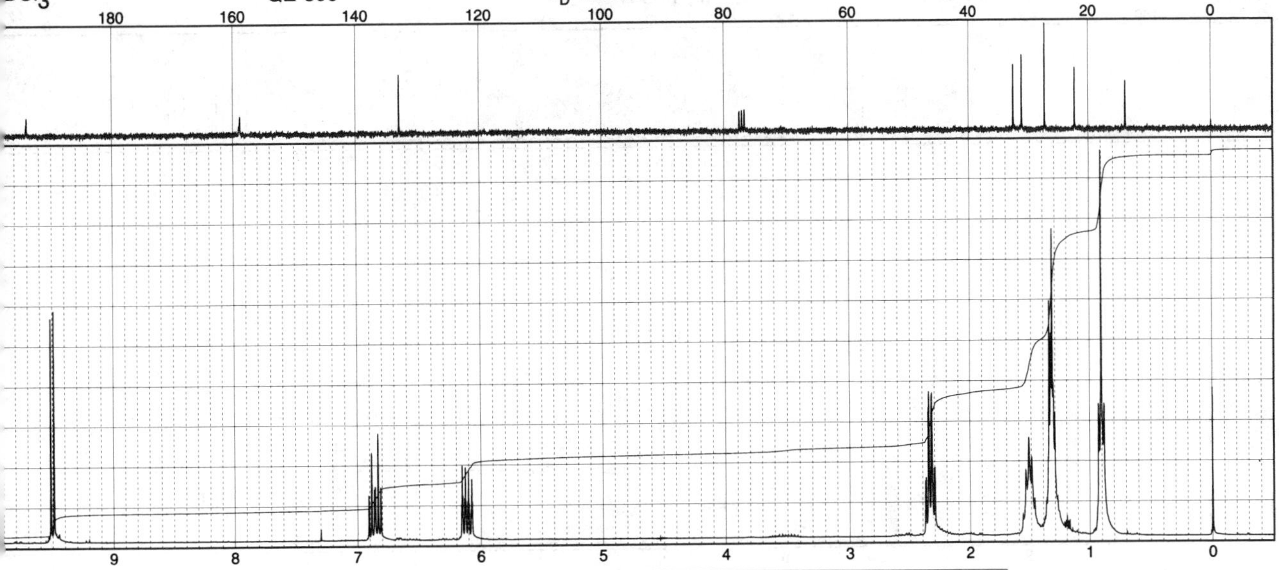

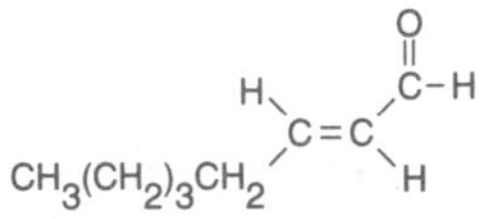

(R)-(+)-Citronellal, 96%   FW 154.25

bp 207°C

d 0.851

n²⁰_D 1.4480

CDCl₃   QE-300

Offset: 40 ppm.

202.87*	27.76*
131.67	25.70*
124.00*	25.39
50.99	19.86*
36.93	17.65*

B

Aldrich 37,375-3   *CAS [5949-05-3]*   C₁₀H₁₈O

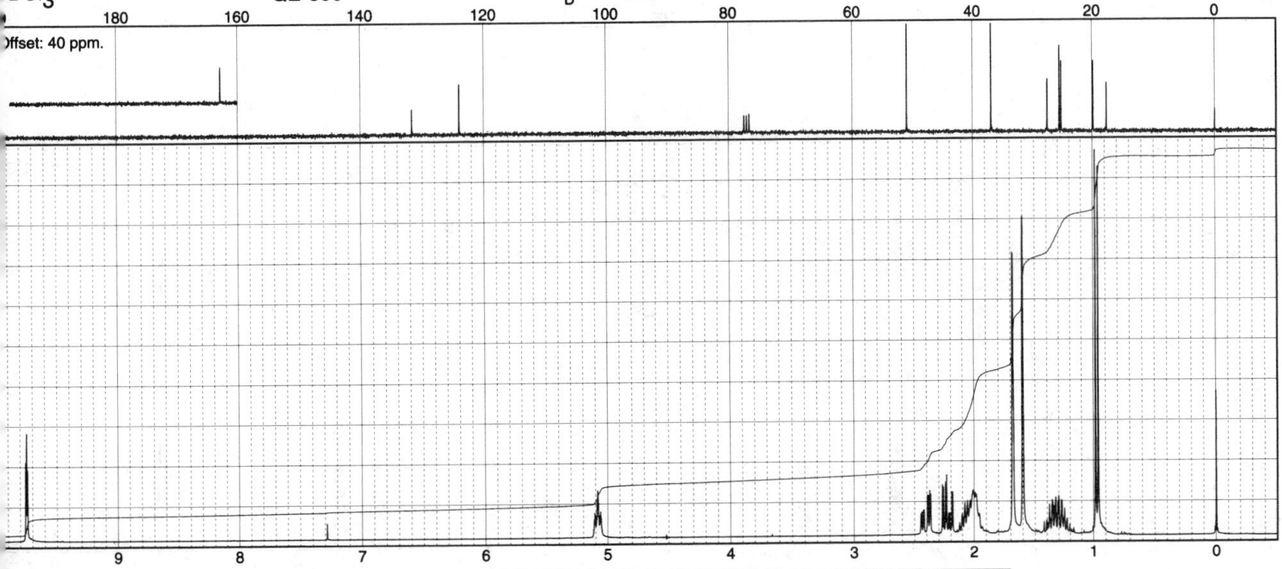

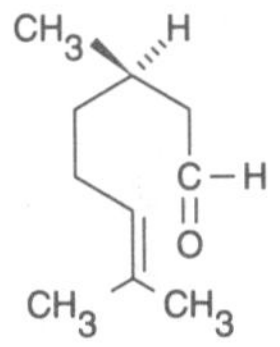

(S)-(-)-Citronellal, 96%   FW 154.25

d 0.851

n²⁰_D 1.4460

Fp 168°F

CDCl₃   QE-300

Offset: 40 ppm.

202.82*	27.78*
131.68	25.69*
124.01*	25.40
51.00	19.87*
36.95	17.65*

C

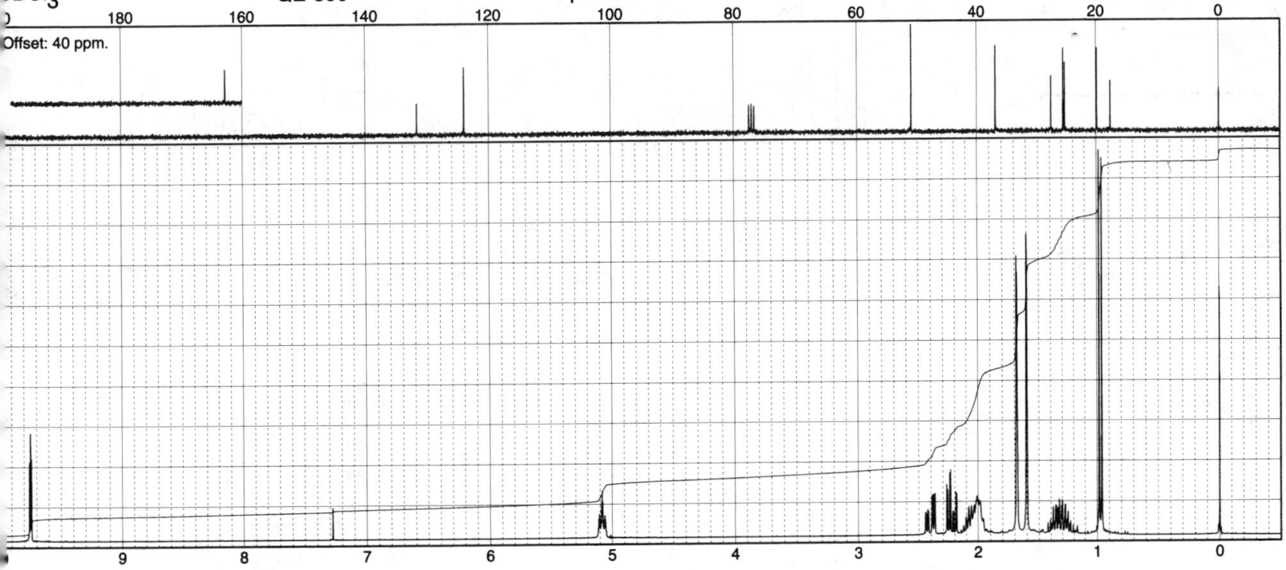

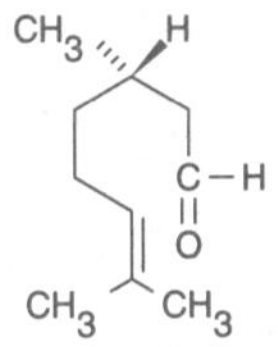

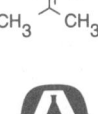

A

Aldrich 25,565-3

trans-2-Nonenal, 97%

CAS [18829-56-6]

$C_9H_{16}O$
FW 140.23
bp 89°C (12 mm)
d 0.846
n_D^{20}1.4531

Fp 184°F

FT-IR: **1**, 473B
VP-FT-IR: **3**, 562A

194.08*   28.8
159.03*   27.8
132.91*   22.5
32.74   14.0
31.55

CDCl₃   QE-300

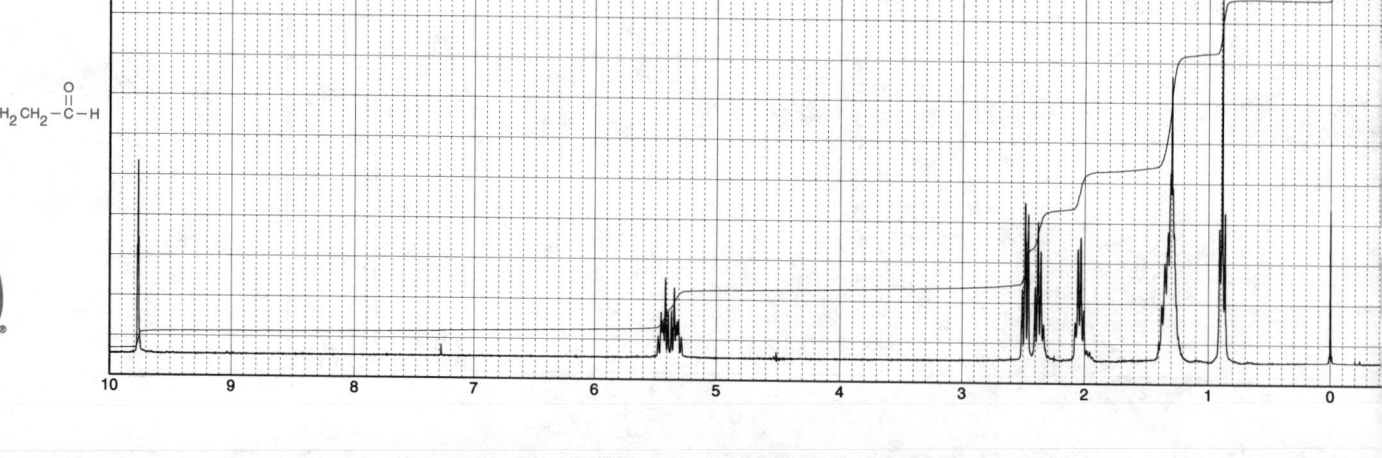

B

Aldrich W23,660-8

trans-2-Decenal, 92+%

CAS [3913-71-1]

$C_{10}H_{18}O$
FW 154.25
bp 79°C (3 mm)
d 0.841
n_D^{20}1.4540

Fp 205°F

VP-FT-IR: **3**, 562B

193.88*   29.1
158.75*   29.0
132.99*   27.9
32.72   22.6
31.72   14.0

CDCl₃   QE-300

C

Aldrich 30,406-9

cis-4-Decenal, 98%

CAS [21662-09-9]

$C_{10}H_{18}O$
FW 154.25
bp 79°C (10 mm)
d 0.857
n_D^{20}1.4439

Fp 179°F

VP-FT-IR: **3**, 562D

201.91*   29.26
131.69*   27.21
127.03*   22.57
43.86   20.14
31.52   14.03

CDCl₃   QE-300

Offset: 40 ppm.

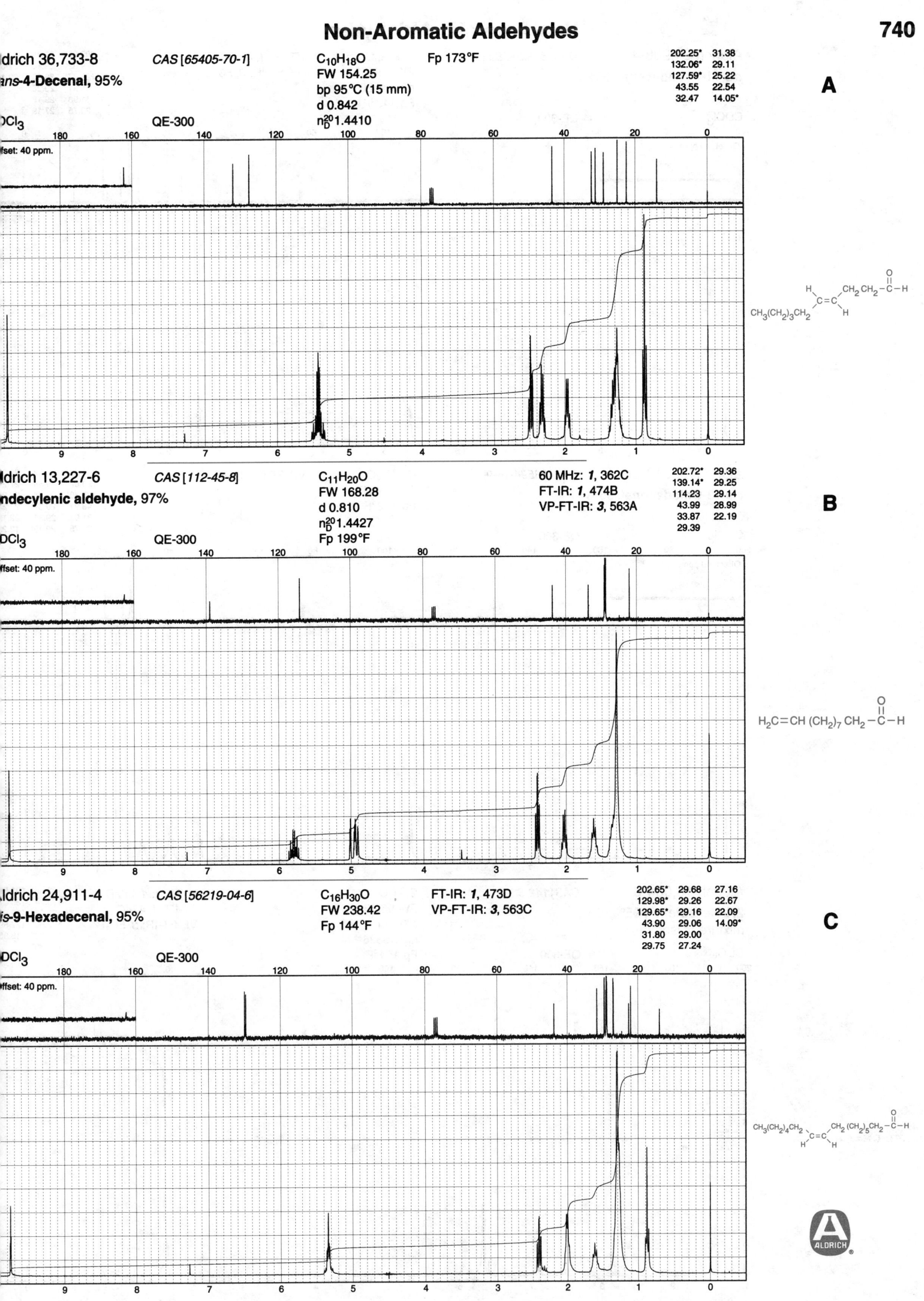

A

Aldrich 36,733-8   CAS [65405-70-1]   C₁₀H₁₈O   Fp 173°F

trans-4-Decenal, 95%

FW 154.25
bp 95°C (15 mm)
d 0.842
n²⁰_D 1.4410

CDCl₃   QE-300

offset: 40 ppm.

202.25*	31.38
132.06*	29.11
127.59*	25.22
43.55	22.54
32.47	14.05*

B

Aldrich 13,227-6   CAS [112-45-8]   C₁₁H₂₀O

undecylenic aldehyde, 97%

FW 168.28
d 0.810
n²⁰_D 1.4427
Fp 199°F

CDCl₃   QE-300

offset: 40 ppm.

60 MHz: *1*, 362C
FT-IR: *1*, 474B
VP-FT-IR: *3*, 563A

202.72*	29.36
139.14*	29.25
114.23	29.14
43.99	28.99
33.87	22.19
29.39	

C

Aldrich 24,911-4   CAS [56219-04-6]   C₁₆H₃₀O

cis-9-Hexadecenal, 95%

FW 238.42
Fp 144°F

FT-IR: *1*, 473D
VP-FT-IR: *3*, 563C

CDCl₃   QE-300

offset: 40 ppm.

202.65*	29.68	27.16
129.98*	29.26	22.67
129.65*	29.16	22.09
43.90	29.06	14.09*
31.80	29.00	
29.75	27.24	

Non-Aromatic Aldehydes

A

Aldrich 24,908-4
CAS [53939-28-9]

cis-11-Hexadecenal, 95%

C$_{16}$H$_{30}$O
FW 238.42
n$_D^{20}$1.4526
Fp 142°F

FT-IR: **1**, 474A
VP-FT-IR: **3**, 563D

202.66*	29.46	26.9
129.82*	29.40	22.3
129.77*	29.35	22.1
43.91	29.26	13.9
31.98	29.18	
29.75	27.19	

CDCl$_3$

QE-300

Offset: 40 ppm.

CH$_3$CH$_2$CH$_2$CH$_2$ C=C CH$_2$ (CH$_2$)$_7$ C=O H

B

Aldrich 24,913-0
CAS [58594-45-9]

cis-13-Octadecenal, 95%

C$_{18}$H$_{34}$O
FW 266.47
Fp 142°F

VP-FT-IR: **3**, 564A

202.70*	29.58	29.0
129.82*	29.55	27.2
129.80*	29.44	26.9
43.91	29.37	22.3
31.99	29.31	22.1
29.78	29.19	13.9

CDCl$_3$

QE-300

Offset: 40 ppm.

CH$_3$CH$_2$CH$_2$CH$_2$ C=C CH$_2$ (CH$_2$)$_9$ CH$_2$ C=O H

C

Aldrich 18,034-3
CAS [142-83-6]

2,4-Hexadienal, 95%,
mixture of isomers

C$_6$H$_8$O
FW 96.13
d 0.871
n$_D^{20}$1.5386
Fp 154°F

60 MHz: **1**, 362D
FT-IR: **1**, 474C
VP-FT-IR: **3**, 564B

193.86*	131.5
193.74*	130.0
152.63*	129.8
152.50*	127.5
141.90*	18.85
141.80*	

CDCl$_3$

QE-300

CH$_3$ CH=CHCH=CH—C=O H

Offset: 0.4 ppm.

Non-Aromatic Aldehydes

742

A

Aldrich 18,054-8 CAS [4313-03-5]
ans,trans-2,4-Heptadienal, tech.,
%
DCl₃ QE-300

C₇H₁₀O
FW 110.16
bp 84°C
d 0.881
n²⁰_D 1.5325

Fp 150°F

60 MHz: **1**, 363A
FT-IR: **1**, 474D
VP-FT-IR: **3**, 564C

193.78*
152.84*
148.51*
130.02*
127.67*
26.23
12.69*

B

Aldrich 23,878-3
4-Dimethyl-2,6-heptadienal, 96%, mixture
f isomers
DCl₃ QE-300

C₉H₁₄O
FW 138.21
bp 47°C (2 mm)
d 0.870
n²⁰_D 1.4664

Fp 148°F

60 MHz: **1**, 363B
FT-IR: **1**, 475A
VP-FT-IR: **3**, 564D

195.28* 135.55* 33.42*
191.29* 117.01 31.08*
159.24* 116.76 21.01*
159.12* 41.47 19.35*
138.08 40.66 9.39*

Offset: 0.3 ppm.

C

Aldrich 23,880-5 CAS [85136-07-8]
4-Diethyl-2,6-heptadienal, tech.,
0%, mixture of isomers
DCl₃ QE-300

C₁₁H₁₈O
FW 166.27
bp 91°C (12 mm)
d 0.862
n²⁰_D 1.4676

Fp 187°F

60 MHz: **1**, 363C
FT-IR: **1**, 475B
VP-FT-IR: **3**, 565A

195.04* 39.10
158.06* 27.43
145.24 17.72
135.63* 13.52*
116.66 11.82*
40.53*

Offset: 0.2 ppm.

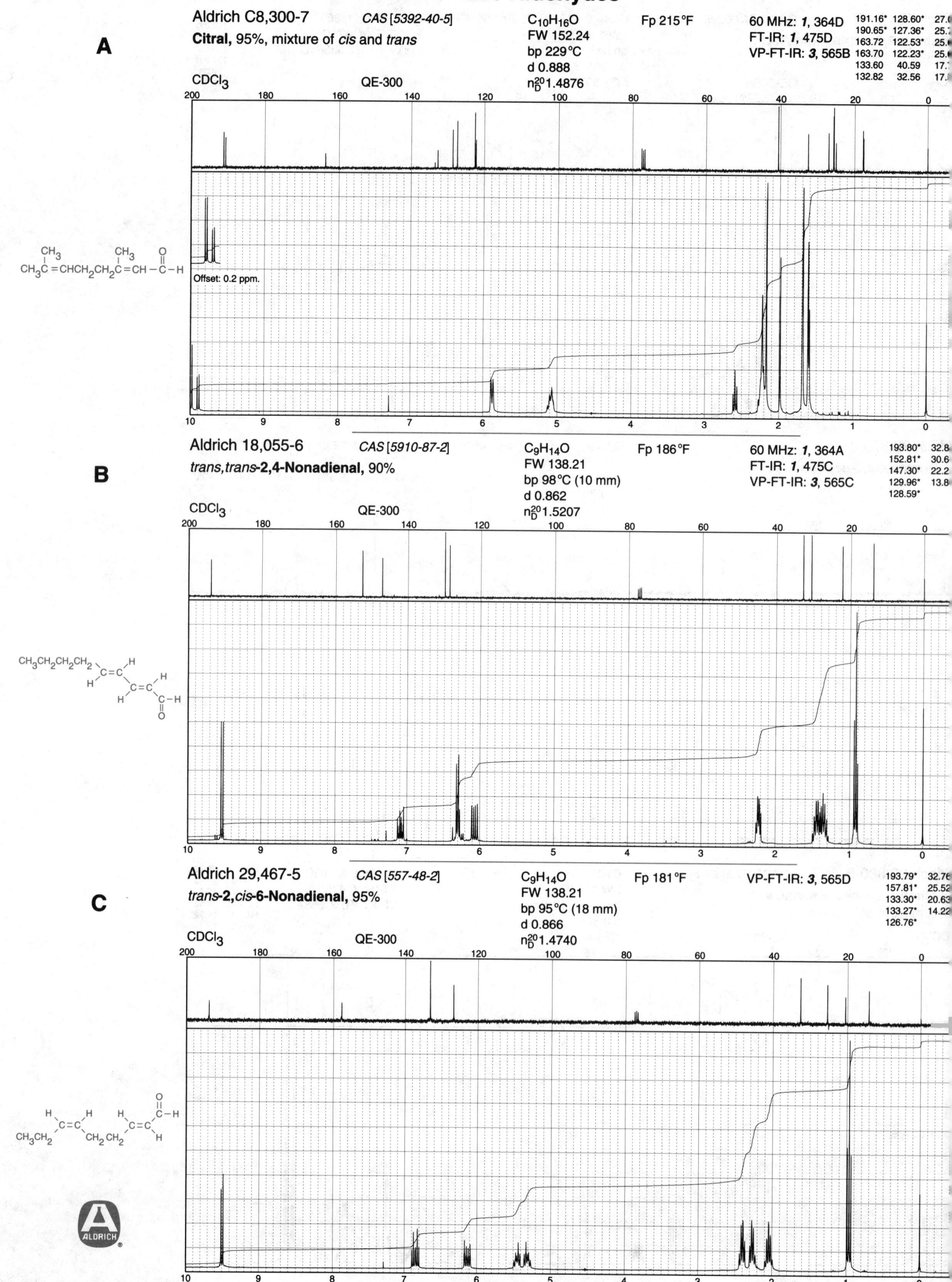

A

Aldrich C8,300-7 CAS [5392-40-5] $C_{10}H_{16}O$ Fp 215°F

Citral, 95%, mixture of *cis* and *trans*

FW 152.24
bp 229°C
d 0.888
n_D^{20} 1.4876

60 MHz: **1**, 364D
FT-IR: **1**, 475D
VP-FT-IR: **3**, 565B

191.16*	128.60*	27.0
190.65*	127.36*	25.2
163.72	122.53*	25.6
163.70	122.23*	25.6
133.60	40.59	17.7
132.82	32.56	17.7

CDCl₃ QE-300

Offset: 0.2 ppm.

B

Aldrich 18,055-6 CAS [5910-87-2] $C_9H_{14}O$ Fp 186°F

trans,trans-2,4-Nonadienal, 90%

FW 138.21
bp 98°C (10 mm)
d 0.862
n_D^{20} 1.5207

60 MHz: **1**, 364A
FT-IR: **1**, 475C
VP-FT-IR: **3**, 565C

193.80*	32.8
152.81*	30.6
147.30*	22.2
129.96*	13.8
128.59*	

CDCl₃ QE-300

C

Aldrich 29,467-5 CAS [557-48-2] $C_9H_{14}O$ Fp 181°F

trans-2,cis-6-Nonadienal, 95%

FW 138.21
bp 95°C (18 mm)
d 0.866
n_D^{20} 1.4740

VP-FT-IR: **3**, 565D

193.79*	32.76
157.81*	25.52
133.30*	20.63
133.27*	14.22
126.76*	

CDCl₃ QE-300

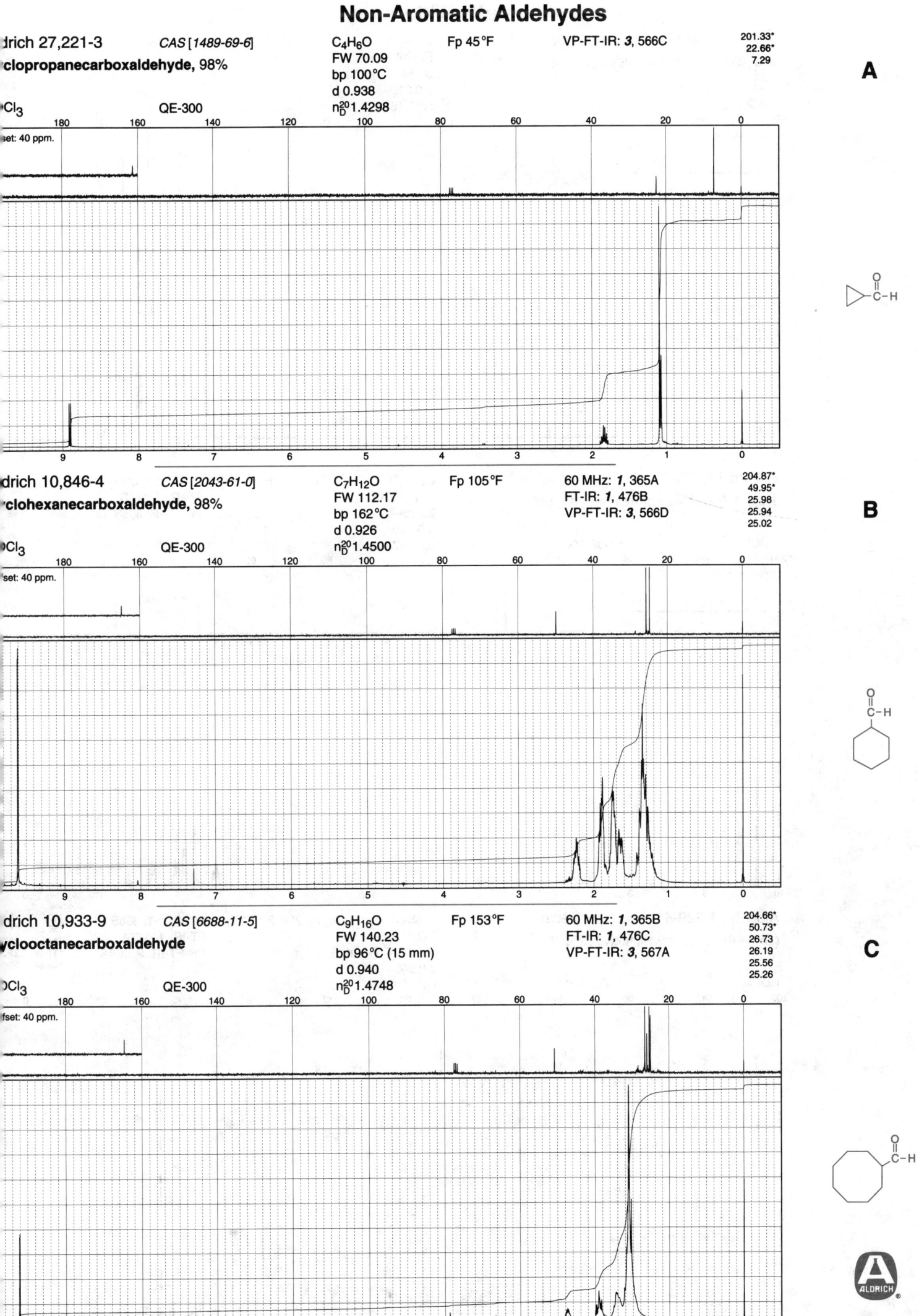

drich 27,221-3   CAS *[1489-69-6]*   C$_4$H$_6$O   Fp 45°F   VP-FT-IR: *3*, 566C   201.33*
22.66*
7.29

A

clopropanecarboxaldehyde, 98%   FW 70.09
bp 100°C
d 0.938
n$_D^{20}$1.4298

Cl$_3$ QE-300

set: 40 ppm.

drich 10,846-4   CAS *[2043-61-0]*   C$_7$H$_{12}$O   Fp 105°F   60 MHz: *1*, 365A   204.87*
49.95*
25.98
25.94
25.02

B

clohexanecarboxaldehyde, 98%   FW 112.17   FT-IR: *1*, 476B
bp 162°C   VP-FT-IR: *3*, 566D
d 0.926
n$_D^{20}$1.4500

Cl$_3$ QE-300

set: 40 ppm.

drich 10,933-9   CAS *[6688-11-5]*   C$_9$H$_{16}$O   Fp 153°F   60 MHz: *1*, 365B   204.66*
50.73*
26.73
26.19
25.56
25.26

C

yclooctanecarboxaldehyde   FW 140.23   FT-IR: *1*, 476C
bp 96°C (15 mm)   VP-FT-IR: *3*, 567A
d 0.940
n$_D^{20}$1.4748

Cl$_3$ QE-300

set: 40 ppm.

ALDRICH

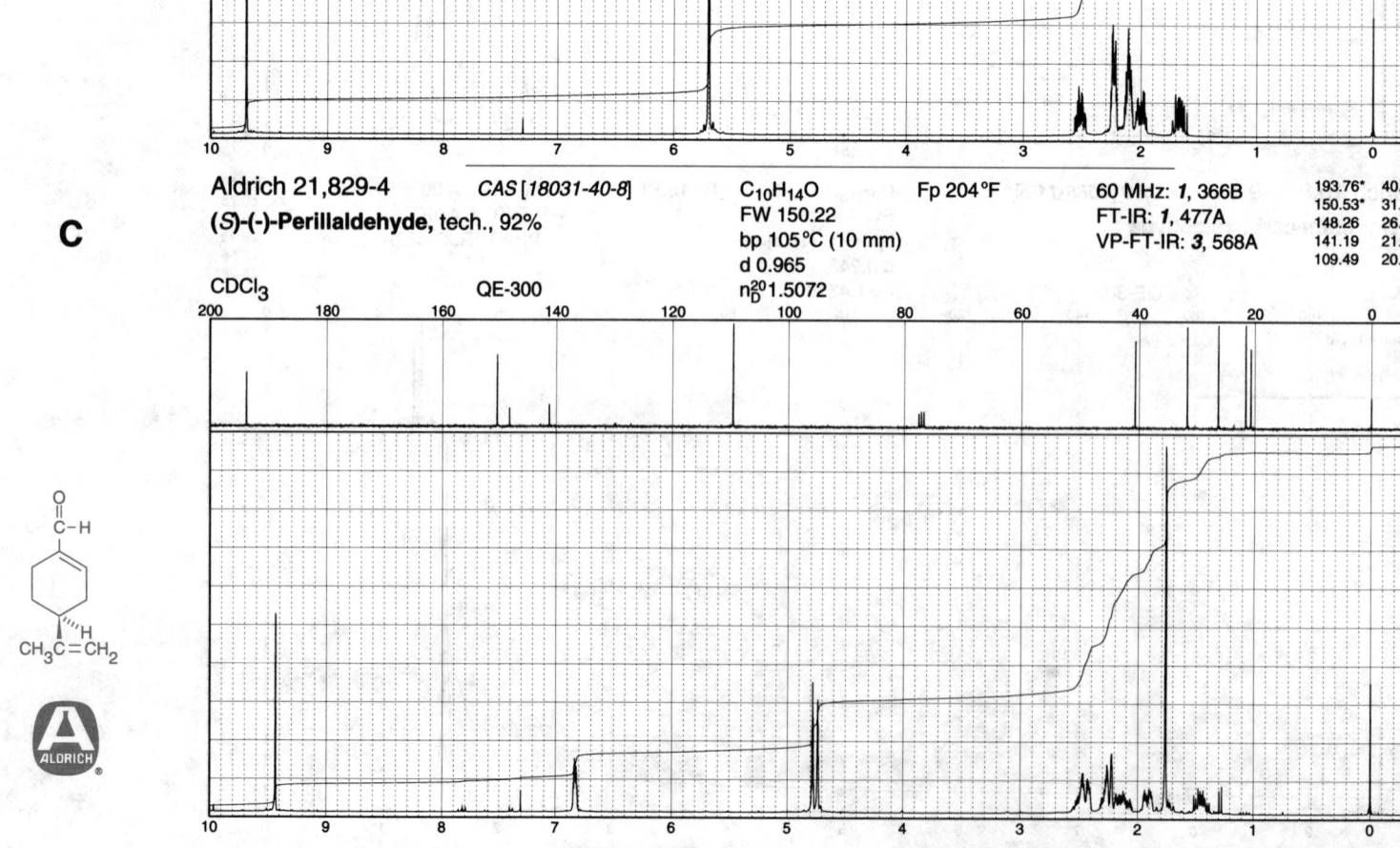

A

Aldrich 30,441-7   *CAS [4361-28-8]*

3-Cyclohexylpropionaldehyde, 98%

$C_9H_{16}O$
FW 140.23
bp 88°C (18 mm)
d 0.919
n_D^{20}1.4613

Fp 174°F

VP-FT-IR: *3*, 567B

202.
41.
37.
33.
29.
26.
26.

$CDCl_3$ QE-300

Offset: 40 ppm.

B

Aldrich T1,220-3   *CAS [100-50-5]*

1,2,3,6-Tetrahydrobenzaldehyde, 99%

$C_7H_{10}O$
FW 110.16
bp 164°C
d 0.940
n_D^{20}1.4745

Fp 135°F

60 MHz: *1*, 365C
FT-IR: *1*, 476D
VP-FT-IR: *3*, 567D

204.
127.
124.7
46.0
24.3
23.7
22.0

$CDCl_3$ QE-300

Offset: 40 ppm.

C

Aldrich 21,829-4   *CAS [18031-40-8]*

(S)-(-)-Perillaldehyde, tech., 92%

$C_{10}H_{14}O$
FW 150.22
bp 105°C (10 mm)
d 0.965
n_D^{20}1.5072

Fp 204°F

60 MHz: *1*, 366B
FT-IR: *1*, 477A
VP-FT-IR: *3*, 568A

193.76*   40.6
150.53*   31.7
148.26   26.3
141.19   21.5
109.49   20.7

$CDCl_3$ QE-300

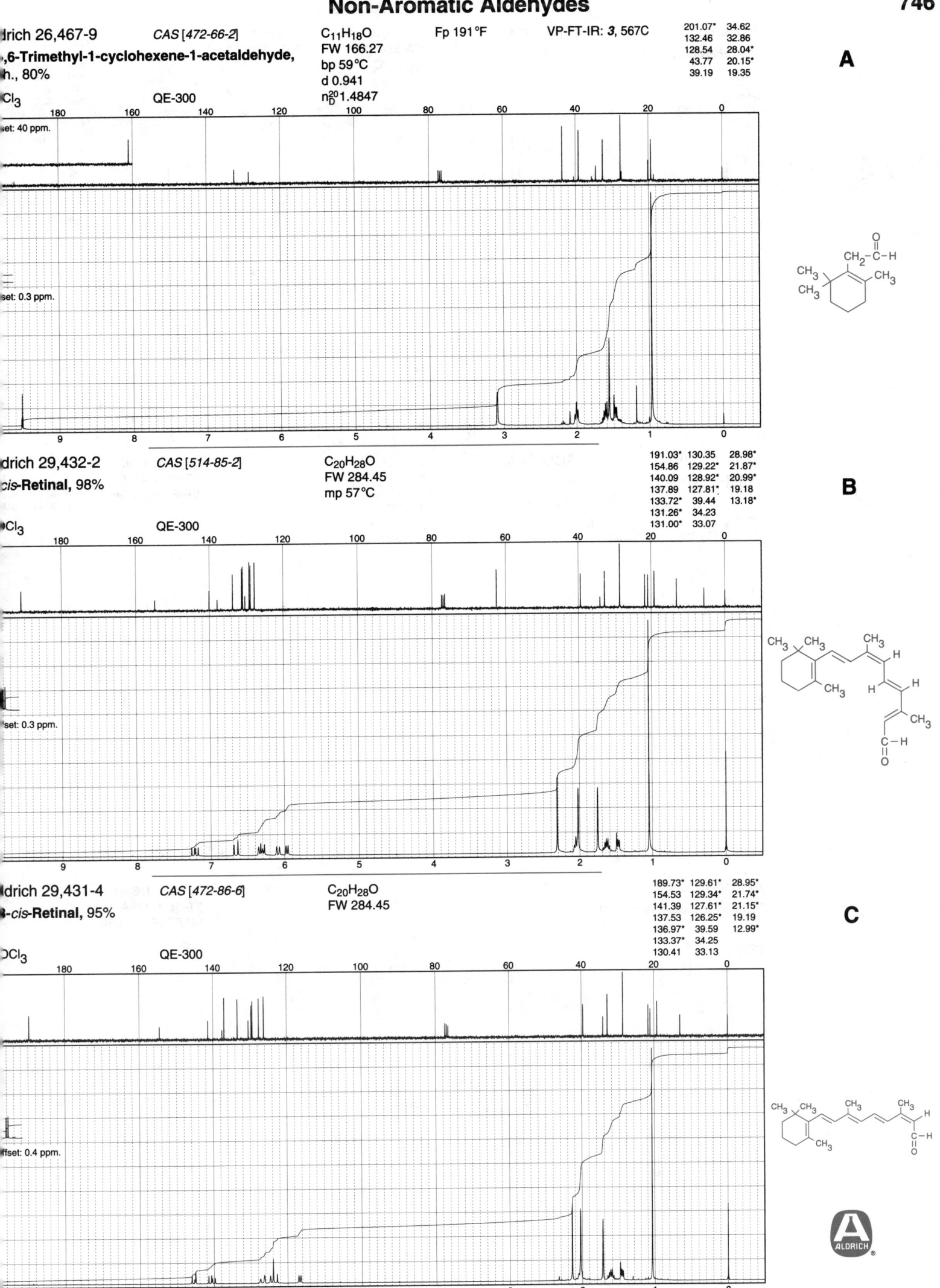

Idrich 26,467-9   *CAS [472-66-2]*   C₁₁H₁₈O   Fp 191 °F   VP-FT-IR: **3**, 567C

,6-Trimethyl-1-cyclohexene-1-acetaldehyde,
h., 80%

FW 166.27
bp 59 °C
d 0.941
n²⁰_D 1.4847

Cl₃   QE-300

201.07*	34.62
132.46	32.86
128.54	28.04*
43.77	20.15*
39.19	19.35

A

set: 40 ppm.

set: 0.3 ppm.

drich 29,432-2   *CAS [514-85-2]*   C₂₀H₂₈O   FW 284.45   mp 57 °C

cis-Retinal, 98%

Cl₃   QE-300

191.03*	130.35	28.98*
154.86	129.22*	21.87*
140.09	128.92*	20.99*
137.89	127.81*	19.18
133.72*	39.44	13.18*
131.26*	34.23	
131.00*	33.07	

B

set: 0.3 ppm.

ldrich 29,431-4   *CAS [472-86-6]*   C₂₀H₂₈O   FW 284.45

-cis-Retinal, 95%

Cl₃   QE-300

189.73*	129.61*	28.95*
154.53	129.34*	21.74*
141.39	127.61*	21.15*
137.53	126.25*	19.19
136.97*	39.59	12.99*
133.37*	34.25	
130.41	33.13	

C

offset: 0.4 ppm.

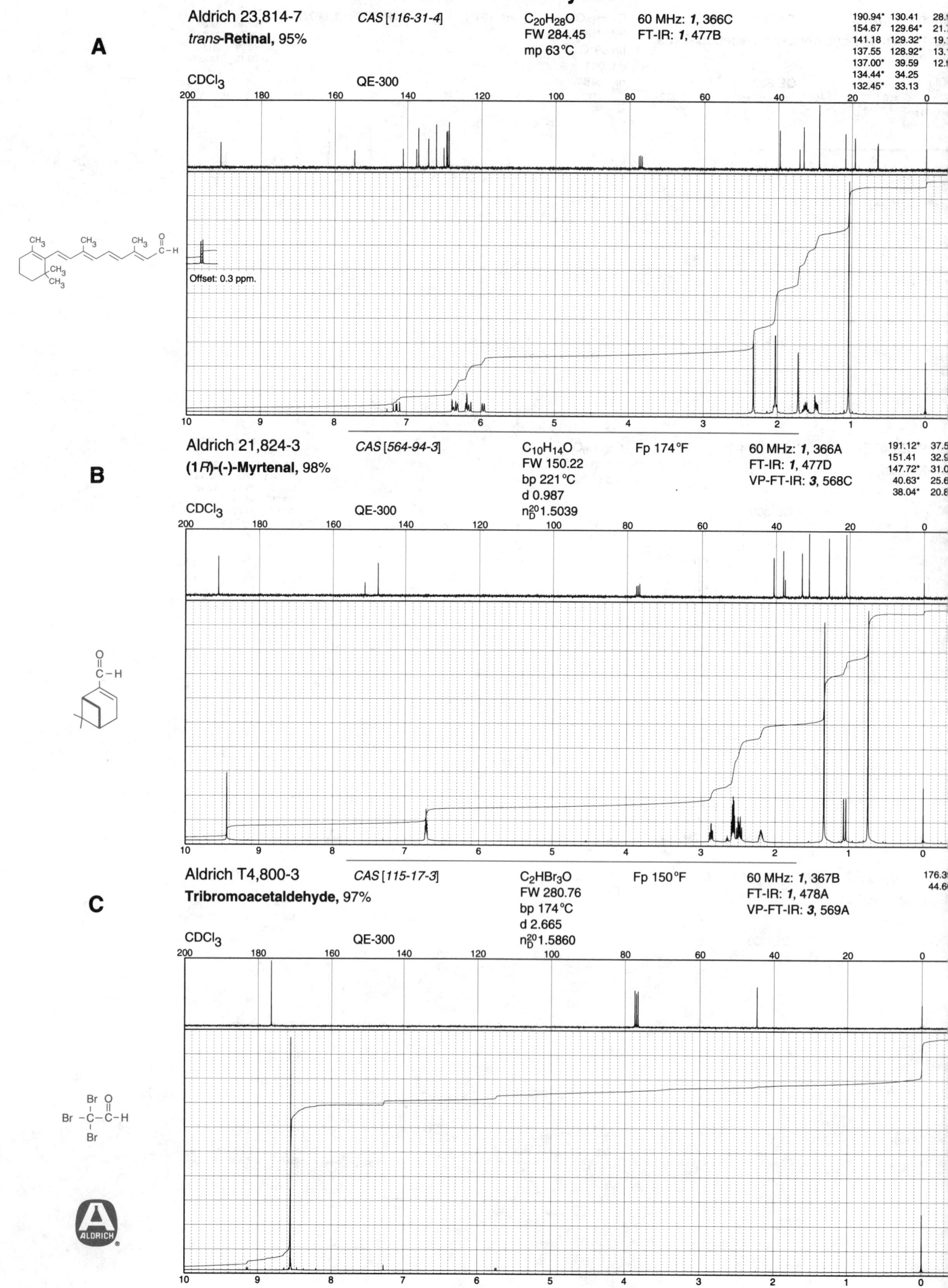

A

Aldrich 23,814-7

trans-Retinal, 95%

CAS [116-31-4]

$C_{20}H_{28}O$
FW 284.45
mp 63°C

60 MHz: *1*, 366C
FT-IR: *1*, 477B

190.94*	130.41
154.67	129.64*
141.18	129.32*
137.55	128.92*
137.00*	39.59
134.44*	34.25
132.45*	33.13

28.
21.
19.
13.
12.

CDCl₃

QE-300

Offset: 0.3 ppm.

B

Aldrich 21,824-3

(1*R*)-(-)-Myrtenal, 98%

CAS [564-94-3]

$C_{10}H_{14}O$
FW 150.22
bp 221°C
d 0.987
n_D^{20} 1.5039

Fp 174°F

60 MHz: *1*, 366A
FT-IR: *1*, 477D
VP-FT-IR: *3*, 568C

191.12*	37.5
151.41	32.9
147.72*	31.0
40.63*	25.6
38.04*	20.8

CDCl₃

QE-300

C

Aldrich T4,800-3

Tribromoacetaldehyde, 97%

CAS [115-17-3]

C_2HBr_3O
FW 280.76
bp 174°C
d 2.665
n_D^{20} 1.5860

Fp 150°F

60 MHz: *1*, 367B
FT-IR: *1*, 478A
VP-FT-IR: *3*, 569A

176.3
44.6

CDCl₃

QE-300

rich G680-5   CAS [23147-58-2]   C₂H₄O₂
colaldehyde   FW 120.10
                                     FT-IR: 1, 478D

SO-d₆   QE-300

A

104.33*	71.08
102.21*	68.21
94.32*	66.36
94.11*	64.23
90.22*	63.12
89.05*	62.11
71.15	

drich 27,746-0   CAS [3268-49-3]   C₄H₈OS   Fp 142°F   VP-FT-IR: 3, 571A
Methylthio)propionaldehyde   FW 104.17
                                      bp 166°C
                                      d 1.041
                                      n_D^20 1.4830

Cl₃   QE-300
set: 40 ppm.

B

| 200.54* |
| 43.19 |
| 26.40 |
| 15.55* |

$CH_3SCH_2CH_2-\overset{O}{\overset{\|}{C}}-H$

drich W33,740-4   CAS [16630-52-7]   C₅H₁₀OS   Fp 144°F   VP-FT-IR: 3, 570D
Methylthio)butanal, 96+%   FW 118.20
                                      bp 63°C (10 mm)
                                      d 1.001
                                      n_D^20 1.4760

Cl₃   QE-300
set: 40 ppm.

C

| 200.47* |
| 49.84 |
| 35.17* |
| 20.93* |
| 13.18* |

$CH_3\underset{SCH_3}{C}HCH_2-\overset{O}{\overset{\|}{C}}-H$

A

Aldrich 30,583-9   *CAS [927-63-9]* C_5H_9NO Fp >230°F   VP-FT-IR: **3**, 572A

3-(Dimethylamino)acrolein, 95%   FW 99.13

bp 272°C

d 0.990

n_D^{20} 1.5844

CDCl₃   QE-300

188
160
101
44
37

B

Aldrich 12,846-5   *CAS [107-22-2]* $C_2H_2O_2$ FT-IR: **1**, 480D

Glyoxal   FW 58.04

d 1.265

n_D^{20} 1.4087

Fp NONE

D₂O   QE-300

105.
100.
99.
91.
90.

C

Aldrich 29,211-7   *CAS [61020-06-2]* $C_6H_{11}NO_2$ Fp 219°F   VP-FT-IR: **3**, 571C

2-Methoxy-1-pyrrolidinecarboxaldehyde,

97%, mixture of *cis* and *trans*   FW 129.16

bp 136°C (48 mm)

d 1.078

n_D^{20} 1.4720

CDCl₃   QE-300

162.44* 45.0
161.19* 42.5
89.57* 31.7
85.41* 22.0
56.46* 21.3
54.27*

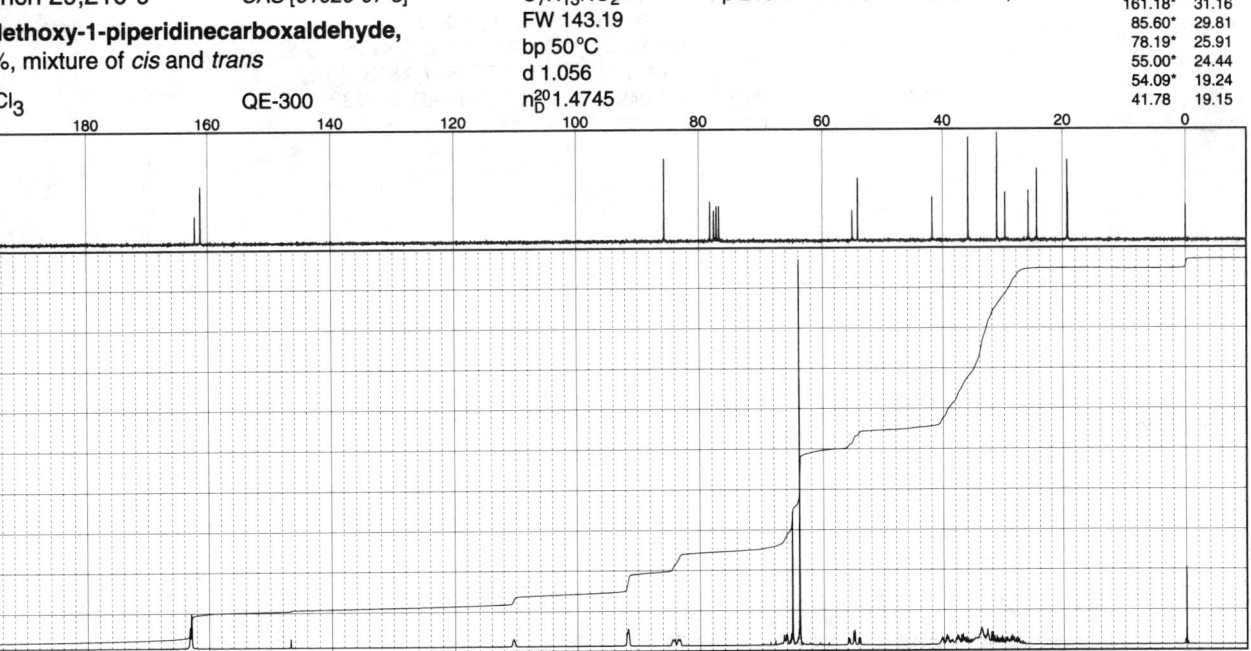

drich 29,210-9   *CAS [61020-07-3]*

Methoxy-1-piperidinecarboxaldehyde,
%, mixture of *cis* and *trans*

Cl₃   QE-300

$C_7H_{13}NO_2$
FW 143.19
bp 50 °C
d 1.056
n_D^{20} 1.4745

Fp 210 °F

VP-FT-IR: *3*, 571D

A

162.09*	35.85
161.18*	31.16
85.60*	29.81
78.19*	25.91
55.00*	24.44
54.09*	19.24
41.78	19.15

Non-Aromatic Carboxylic Acids

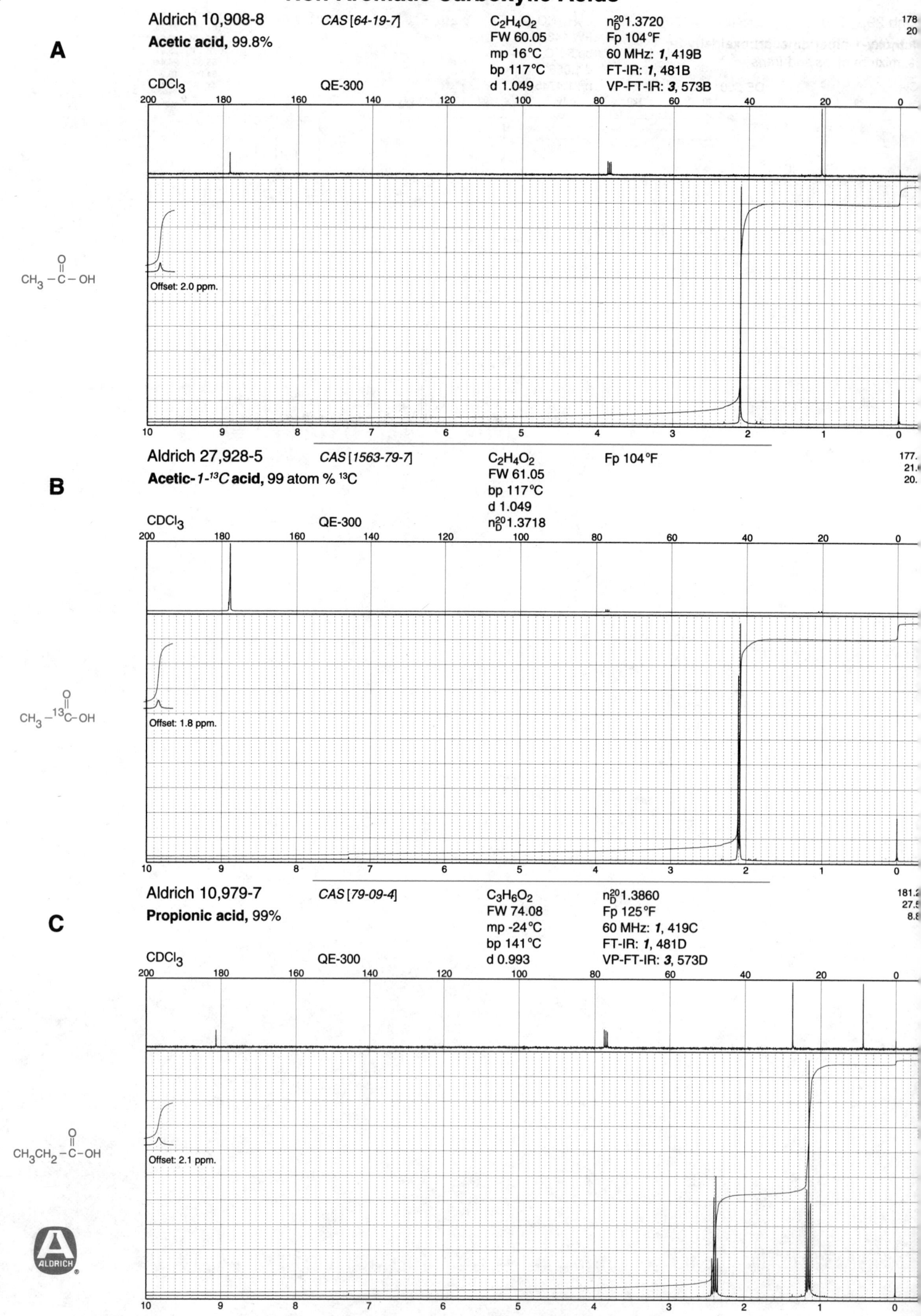

A

Aldrich 10,908-8
Acetic acid, 99.8%

CAS [64-19-7]

C$_2$H$_4$O$_2$
FW 60.05
mp 16°C
bp 117°C
d 1.049

n$_D^{20}$1.3720
Fp 104°F
60 MHz: *1*, 419B
FT-IR: *1*, 481B
VP-FT-IR: *3*, 573B

178
20

CDCl$_3$ QE-300

CH$_3$—C—OH

Offset: 2.0 ppm.

B

Aldrich 27,928-5
Acetic-1-^{13}C acid, 99 atom % ^{13}C

CAS [1563-79-7]

C$_2$H$_4$O$_2$
FW 61.05
bp 117°C
d 1.049
n$_D^{20}$1.3718

Fp 104°F

177.
21.
20.

CDCl$_3$ QE-300

CH$_3$—^{13}C—OH

Offset: 1.8 ppm.

C

Aldrich 10,979-7
Propionic acid, 99%

CAS [79-09-4]

C$_3$H$_6$O$_2$
FW 74.08
mp -24°C
bp 141°C
d 0.993

n$_D^{20}$1.3860
Fp 125°F
60 MHz: *1*, 419C
FT-IR: *1*, 481D
VP-FT-IR: *3*, 573D

181.
27.
8.8

CDCl$_3$ QE-300

CH$_3$CH$_2$—C—OH

Offset: 2.1 ppm.

ALDRICH

Aldrich 28,244-8   *CAS [6212-69-7]* $C_3H_6O_2$ n_D^{20} 1.3864

Propionic-*1-13C* acid, 99 atom % ^{13}C

FW 75.07   Fp 125°F

mp -24°C

bp 141°C

CDCl₃   QE-300   d 0.993

180.93
27.86
27.13
8.85*

A

Offset: 2.2 ppm.

$CH_3CH_2 - {}^{13}C{-}OH$ (with C=O)

Aldrich B10,350-0   *CAS [107-92-6]* $C_4H_8O_2$ n_D^{20} 1.3969

Butyric acid, 99+%

FW 88.11   Fp 170°F

mp -6°C   60 MHz: *1*, 419D

bp 162°C   FT-IR: *1*, 482A

CDCl₃   QE-300   d 0.964   VP-FT-IR: *3*, 574A

180.55
36.03
18.19
13.60*

B

Offset: 2.0 ppm.

$CH_3CH_2CH_2 - C{-}OH$ (with C=O)

Aldrich 11,014-0   *CAS [109-52-4]* $C_5H_{10}O_2$ n_D^{20} 1.4076

Valeric acid, 99%

FW 102.13   Fp 192°F

mp -19°C   60 MHz: *1*, 420A

bp 185°C   FT-IR: *1*, 482B

CDCl₃   QE-300   d 0.939   VP-FT-IR: *3*, 574B

180.71
33.89
26.76
22.21
13.69*

C

Offset: 2.0 ppm.

$CH_3CH_2CH_2CH_2 - C{-}OH$ (with C=O)

ALDRICH®

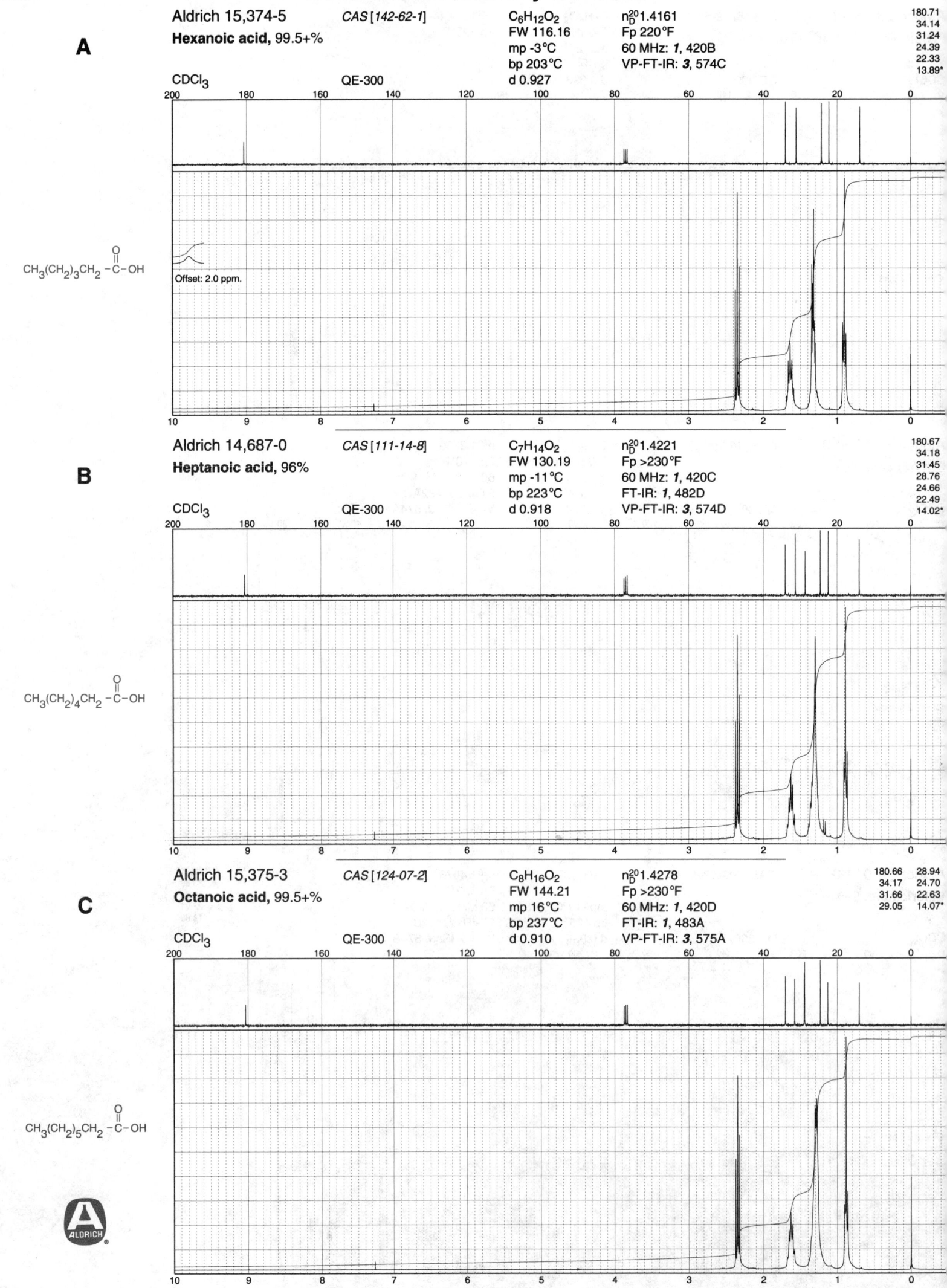

A

Aldrich 15,374-5
Hexanoic acid, 99.5+%

CAS [142-62-1]

$C_6H_{12}O_2$
FW 116.16
mp -3°C
bp 203°C
d 0.927

n_D^{20} 1.4161
Fp 220°F
60 MHz: *1*, 420B
VP-FT-IR: *3*, 574C

180.71
34.14
31.24
24.39
22.33
13.89*

CDCl₃

QE-300

$CH_3(CH_2)_3CH_2 - \overset{\overset{O}{\|}}{C} - OH$

Offset: 2.0 ppm.

B

Aldrich 14,687-0
Heptanoic acid, 96%

CAS [111-14-8]

$C_7H_{14}O_2$
FW 130.19
mp -11°C
bp 223°C
d 0.918

n_D^{20} 1.4221
Fp >230°F
60 MHz: *1*, 420C
FT-IR: *1*, 482D
VP-FT-IR: *3*, 574D

180.67
34.18
31.45
28.76
24.66
22.49
14.02*

CDCl₃

QE-300

$CH_3(CH_2)_4CH_2 - \overset{\overset{O}{\|}}{C} - OH$

C

Aldrich 15,375-3
Octanoic acid, 99.5+%

CAS [124-07-2]

$C_8H_{16}O_2$
FW 144.21
mp 16°C
bp 237°C
d 0.910

n_D^{20} 1.4278
Fp >230°F
60 MHz: *1*, 420D
FT-IR: *1*, 483A
VP-FT-IR: *3*, 575A

180.66   28.94
34.17   24.70
31.66   22.63
29.05   14.07*

CDCl₃

QE-300

$CH_3(CH_2)_5CH_2 - \overset{\overset{O}{\|}}{C} - OH$

Non-Aromatic Carboxylic Acids

Aldrich N2,990-2   *CAS [112-05-0]*

Nonanoic acid, 96%

$C_9H_{18}O_2$	$n_D^{20}1.4319$
FW 158.24	Fp 212°F
mp 9°C	60 MHz: *1*, 421A
bp 254°C	FT-IR: *1*, 483B
d 0.906	VP-FT-IR: *3*, 575C

180.65   29.09
34.17   24.69
31.83   22.67
29.23   14.10*
29.12

A

CDCl$_3$ QE-300

$CH_3(CH_2)_6CH_2-\overset{\displaystyle O}{\overset{\|}{C}}-OH$

Aldrich 15,376-1   *CAS [334-48-5]*

Decanoic acid, 99+%

$C_{10}H_{20}O_2$	Fp >230°F
FW 172.27	60 MHz: *1*, 421B
mp 32°C	FT-IR: *1*, 483C
bp 269°C	VP-FT-IR: *3*, 575D
d 0.893	

180.64   29.08
34.17   24.69
31.88   22.69
29.42   14.11*
29.28

B

CDCl$_3$ QE-300

$CH_3(CH_2)_7CH_2-\overset{\displaystyle O}{\overset{\|}{C}}-OH$

Aldrich 17,147-6   *CAS [112-37-8]*

Undecanoic acid, 99%

$C_{11}H_{22}O_2$	60 MHz: *1*, 421C
FW 186.30	FT-IR: *1*, 483D
mp 29°C	VP-FT-IR: *3*, 576A
bp 228°C (160 mm)	
Fp >230°F	

180.59   29.27
34.16   29.09
31.91   24.69
29.57   22.70
29.46   14.12*
29.33

C

CDCl$_3$ QE-300

$CH_3(CH_2)_8CH_2-\overset{\displaystyle O}{\overset{\|}{C}}-OH$

ALDRICH®

Non-Aromatic Carboxylic Acids

A

Aldrich 15,378-8
Lauric acid, 99.5+%

CAS [143-07-7]

$C_{12}H_{24}O_2$
FW 200.32
mp 45°C
bp 225°C (100 mm)
d 0.883

Fp >230°F

60 MHz: *1*, 421D
FT-IR: *1*, 484A
VP-FT-IR: *3*, 576B

CDCl₃ QE-300

180.23	29.24
34.08	29.06
31.91	24.68
29.59	22.69
29.43	14.12*
29.33	

$CH_3(CH_2)_9CH_2-C(=O)-OH$

B

Aldrich T5,760-6
Tridecanoic acid, 98%

CAS [638-53-9]

$C_{13}H_{26}O_2$
FW 214.35
mp 42°C
bp 236°C (100 mm)
Fp >230°F

60 MHz: *1*, 422A
FT-IR: *1*, 484B
VP-FT-IR: *3*, 576C

CDCl₃ QE-300

180.35	29.37
34.15	29.27
31.95	29.11
29.66	24.72
29.63	22.71
29.47	14.09*

$CH_3(CH_2)_{10}CH_2-C(=O)-OH$

C

Aldrich 15,379-6
Myristic acid, 99.5+%

CAS [544-63-8]

$C_{14}H_{28}O_2$
FW 228.38
mp 55°C
bp 250°C (100 mm)
Fp >230°F

60 MHz: *1*, 422B
FT-IR: *1*, 484C
VP-FT-IR: *3*, 576D

CDCl₃ QE-300

180.59	29.39
34.16	29.27
31.95	29.09
29.67	24.69
29.62	22.72
29.46	14.13*

$CH_3(CH_2)_{11}CH_2-C(=O)-OH$

Aldrich P360-0   *CAS [1002-84-2]* $C_{15}H_{30}O_2$ 60 MHz: *1*, 422C

Pentadecanoic acid, 99+%   FW 242.40   FT-IR: *1*, 484D

mp 52°C   VP-FT-IR: *3*, 577A

bp 257°C (100 mm)

Fp >230°F

180.55	29.39
34.15	29.27
31.95	29.08
29.68	24.69
29.62	22.72
29.46	14.13*

A

CDCl$_3$ QE-300

$CH_3(CH_2)_{12}CH_2-\overset{\displaystyle O}{\overset{\|}{C}}-OH$

Aldrich 25,872-5   *CAS [57-10-3]* $C_{16}H_{32}O_2$ VP-FT-IR: *3*, 577B

Palmitic acid, 99%   FW 256.43

mp 63°C

180.50	29.39
34.14	29.27
31.95	29.09
29.70	24.69
29.62	22.72
29.46	14.12*

B

CDCl$_3$ QE-300

$CH_3(CH_2)_{13}CH_2-\overset{\displaystyle O}{\overset{\|}{C}}-OH$

Aldrich 29,212-5   *CAS [57677-53-9]* $C_{16}H_{32}O_2$

Palmitic-1-^{13}C acid, 99 atom % ^{13}C FW 257.42

mp 64°C

179.80	29.40
34.43	29.28
33.70	29.16
31.98	29.11
29.72	24.75
29.64	22.72
29.48	14.08*

C

CDCl$_3$ QE-300

$CH_3(CH_2)_{13}CH_2-{}^{13}\overset{\displaystyle O}{\overset{\|}{C}}-OH$

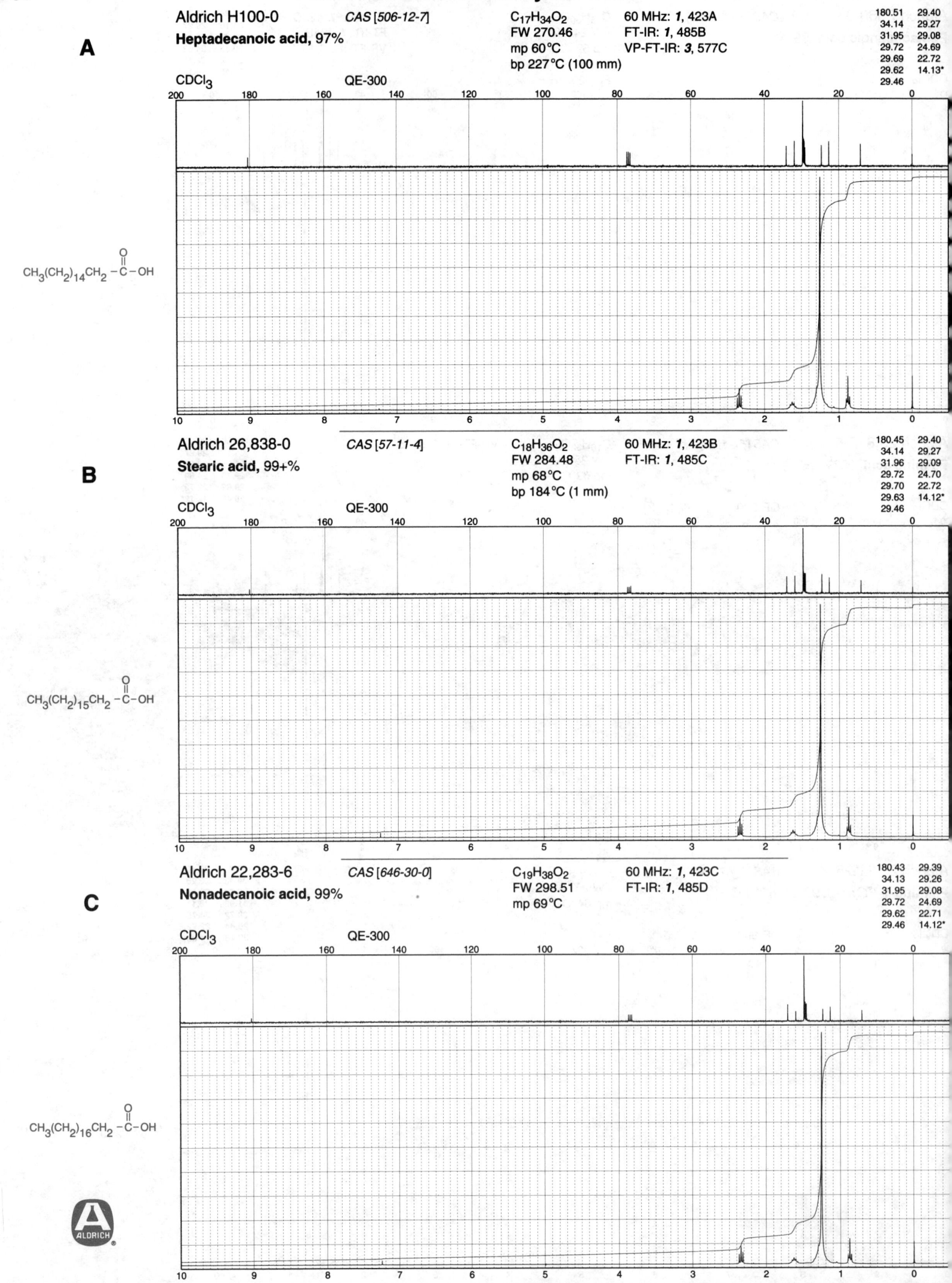

A

Aldrich H100-0   CAS [506-12-7]   C$_{17}$H$_{34}$O$_2$ 60 MHz: **1**, 423A
Heptadecanoic acid, 97%   FW 270.46   FT-IR: **1**, 485B
     mp 60°C   VP-FT-IR: **3**, 577C
     bp 227°C (100 mm)

180.51 29.40
34.14   29.27
31.95   29.08
29.72   24.69
29.69   22.72
29.62   14.13*
29.46

CDCl$_3$ QE-300

CH$_3$(CH$_2$)$_{14}$CH$_2$–C–OH

B

Aldrich 26,838-0   CAS [57-11-4]   C$_{18}$H$_{36}$O$_2$ 60 MHz: **1**, 423B
Stearic acid, 99+%   FW 284.48   FT-IR: **1**, 485C
     mp 68°C
     bp 184°C (1 mm)

180.45 29.40
34.14   29.27
31.96   29.09
29.72   24.70
29.70   22.72
29.63   14.12*
29.46

CDCl$_3$ QE-300

CH$_3$(CH$_2$)$_{15}$CH$_2$–C–OH

C

Aldrich 22,283-6   CAS [646-30-0]   C$_{19}$H$_{38}$O$_2$ 60 MHz: **1**, 423C
Nonadecanoic acid, 99%   FW 298.51   FT-IR: **1**, 485D
     mp 69°C

180.43 29.39
34.13   29.26
31.95   29.08
29.72   24.69
29.62   22.71
29.46   14.12*

CDCl$_3$ QE-300

CH$_3$(CH$_2$)$_{16}$CH$_2$–C–OH

ALDRICH

Aldrich E23-1
Eicosanoic acid, 99%

CAS [506-30-9]

$C_{20}H_{40}O_2$
FW 312.54
mp 75°C

60 MHz: *1*, 423D
FT-IR: *1*, 486A

180.05	29.37
34.06	29.25
31.94	29.08
29.70	24.69
29.60	22.70
29.44	14.11*

A

CDCl₃   QE-300

$CH_3(CH_2)_{17}CH_2 - \overset{\text{O}}{\overset{\|}{C}} - OH$

Aldrich 21,966-5
Heneicosanoic acid, 99%

CAS [2363-71-5]

$C_{21}H_{42}O_2$
FW 326.57
mp 75°C

60 MHz: *1*, 424A
FT-IR: *1*, 486B

174.69	28.96
33.86	28.83
31.53	24.65
29.29	22.31
29.15	13.98*
29.00	

B

CDCl₃ + DMSO-*d₆*   QE-300

$CH_3(CH_2)_{18}CH_2 - \overset{\text{O}}{\overset{\|}{C}} - OH$

Aldrich 21,694-1
Docosanoic acid, 99%

CAS [112-85-6]

$C_{22}H_{44}O_2$
FW 340.60
mp 81°C

FT-IR: *1*, 486C

174.61	28.90
33.86	28.79
31.49	24.64
29.23	22.26
29.11	13.94*
28.96	

C

CDCl₃ + DMSO-*d₆*   QE-300

$CH_3(CH_2)_{19}CH_2 - \overset{\text{O}}{\overset{\|}{C}} - OH$

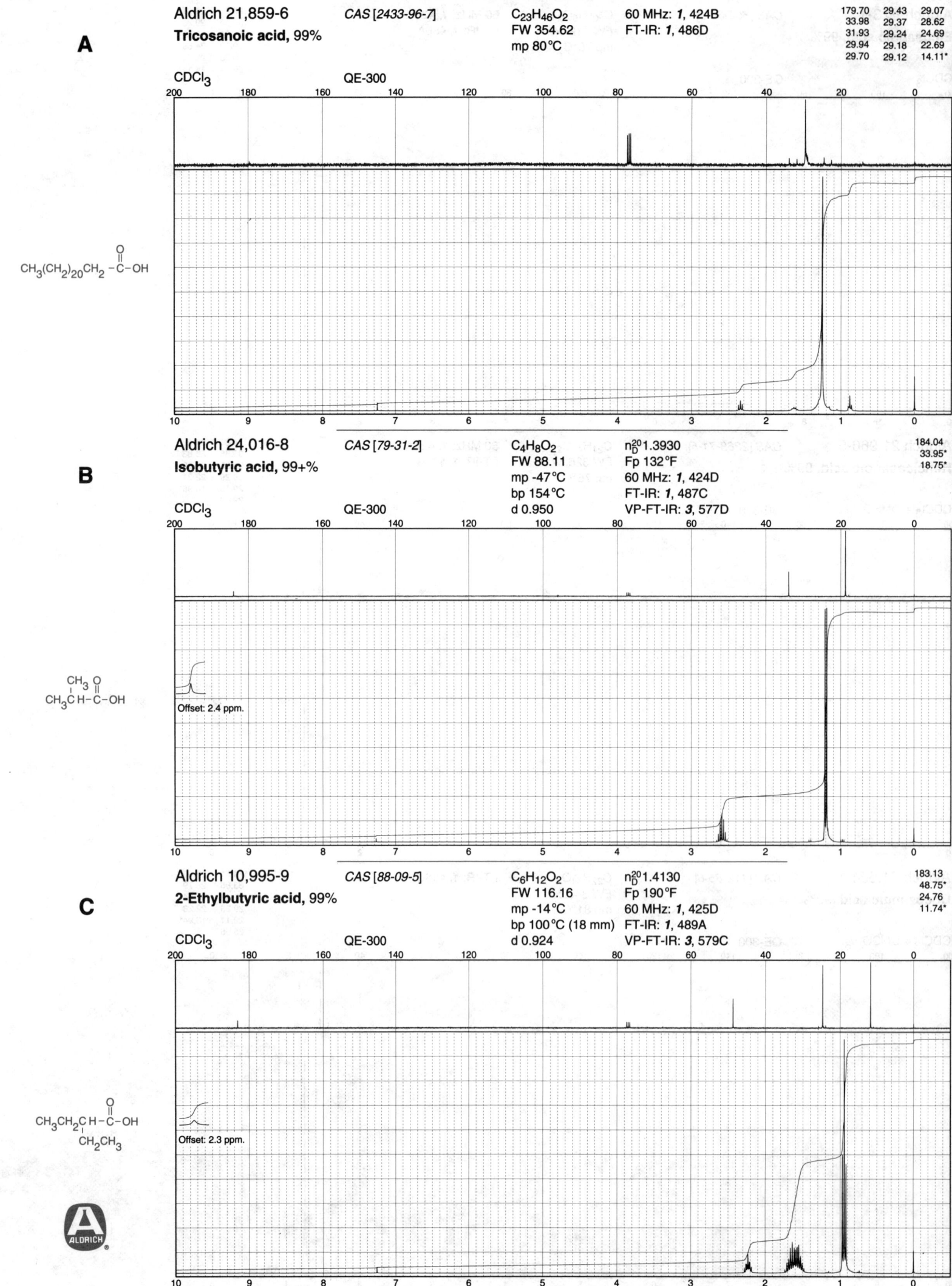

A

Aldrich 21,859-6   CAS [2433-96-7]   C₂₃H₄₆O₂   60 MHz: *1*, 424B
Tricosanoic acid, 99%   FW 354.62   FT-IR: *1*, 486D
                                         mp 80°C

179.70	29.43	29.07
33.98	29.37	28.62
31.93	29.24	24.69
29.94	29.18	22.69
29.70	29.12	14.11*

CDCl₃   QE-300

CH₃(CH₂)₂₀CH₂ –C–OH (with O double bond)

B

Aldrich 24,016-8   CAS [79-31-2]   C₄H₈O₂   n²⁰_D 1.3930
Isobutyric acid, 99+%   FW 88.11   Fp 132°F
                                       mp -47°C   60 MHz: *1*, 424D
                                       bp 154°C   FT-IR: *1*, 487C
                                       d 0.950   VP-FT-IR: *3*, 577D

| 184.04 |
| 33.95* |
| 18.75* |

CDCl₃   QE-300

CH₃CH–C–OH (with CH₃ and O double bond)

Offset: 2.4 ppm.

C

Aldrich 10,995-9   CAS [88-09-5]   C₆H₁₂O₂   n²⁰_D 1.4130
2-Ethylbutyric acid, 99%   FW 116.16   Fp 190°F
                                      mp -14°C   60 MHz: *1*, 425D
                                      bp 100°C (18 mm) FT-IR: *1*, 489A
                                      d 0.924   VP-FT-IR: *3*, 579C

| 183.13 |
| 48.75* |
| 24.76 |
| 11.74* |

CDCl₃   QE-300

CH₃CH₂CH–C–OH (with CH₂CH₃ and O double bond)

Offset: 2.3 ppm.

ALDRICH

Non-Aromatic Carboxylic Acids

Aldrich W26,950-6 *CAS [600-07-7]* $C_5H_{10}O_2$ Fp 165°F 60 MHz: *1*, 425C
2-Methylbutyric acid, 98+% FW 102.13 FT-IR: *1*, 488B
bp 177°C VP-FT-IR: *3*, 578C
d 0.936
CDCl$_3$ QE-300 n_D^{20}1.4050

183.63
40.97*
26.56
16.36*
11.55*

A

Offset: 2.3 ppm.

$CH_3CH_2CH-C-OH$ with CH_3 and O

Aldrich 24,552-6 *CAS [1730-91-2]* $C_5H_{10}O_2$ Fp 165°F FT-IR: *1*, 488C
(S)-(+)-2-Methylbutyric acid, 98% FW 102.13 VP-FT-IR: *3*, 578D
bp 78°C (15 mm)
d 0.938
CDCl$_3$ QE-300 n_D^{20}1.4051

183.64
40.97*
26.56
16.36*
11.55*

B

Offset: 2.3 ppm.

$CH_3CH_2-C-C-OH$

Aldrich W31,020-4 *CAS [503-74-2]* $C_5H_{10}O_2$ n_D^{20}1.4040
Isovaleric acid, 99+% FW 102.13 Fp 159°F
mp -29°C 60 MHz: *1*, 425B
bp 176°C FT-IR: *1*, 488A
d 0.929 VP-FT-IR: *3*, 578B
CDCl$_3$ QE-300

179.98
43.23
25.50*
22.36*

C

Offset: 2.2 ppm.

CH_3CHCH_2-C-OH with CH_3 and O

ALDRICH

A

Aldrich D15,260-9   CAS [595-37-9]

2,2-Dimethylbutyric acid, 96%

$C_6H_{12}O_2$
FW 116.16
bp 95°C (5 mm)
d 0.928
n_D^{20} 1.4154

Fp 175°F

VP-FT-IR: **3**, 579A

185.19
42.51
33.15
24.42*
9.20*

CDCl₃   QE-300

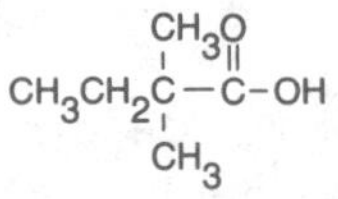

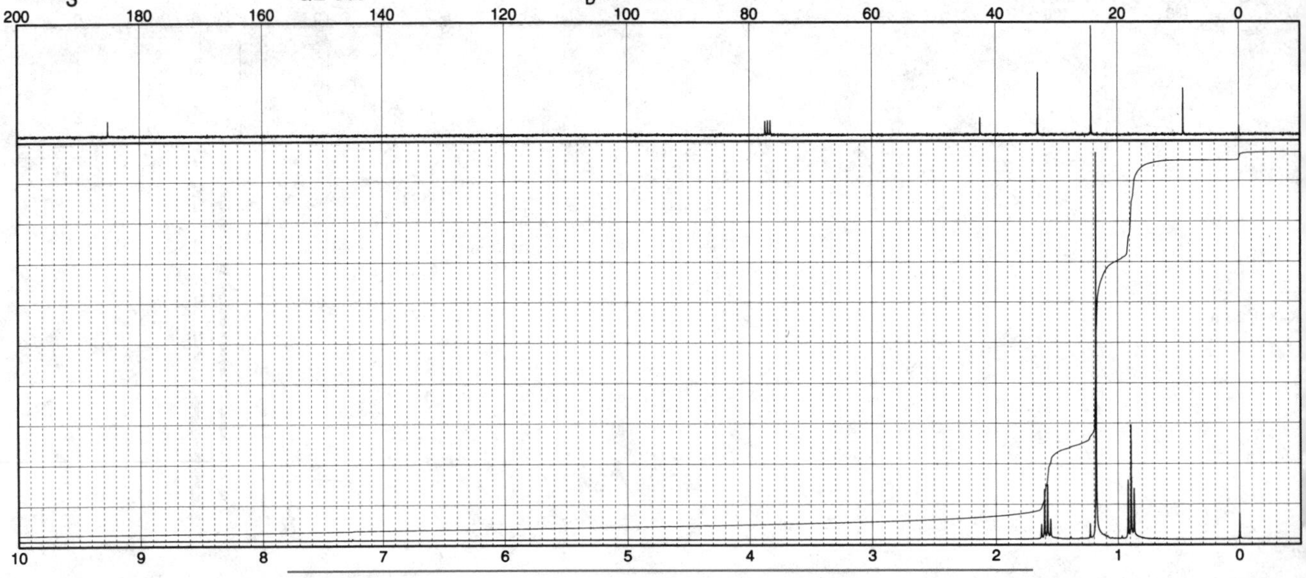

$CH_3CH_2\overset{\overset{\displaystyle CH_3}{|}}{\underset{\underset{\displaystyle CH_3}{|}}{C}}\overset{\displaystyle O}{\overset{\|}{C}}-OH$

B

Aldrich B8,840-3   CAS [1070-83-3]

***tert*-Butylacetic acid, 98%**

$C_6H_{12}O_2$
FW 116.16
mp -11°C
bp 188°C
d 0.912

n_D^{20} 1.4100
Fp 192°F
60 MHz: **1**, 426A
FT-IR: **1**, 488D
VP-FT-IR: **3**, 579B

179.29
47.82
30.62
29.57*

CDCl₃   QE-300

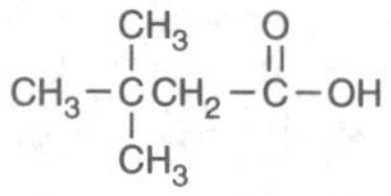

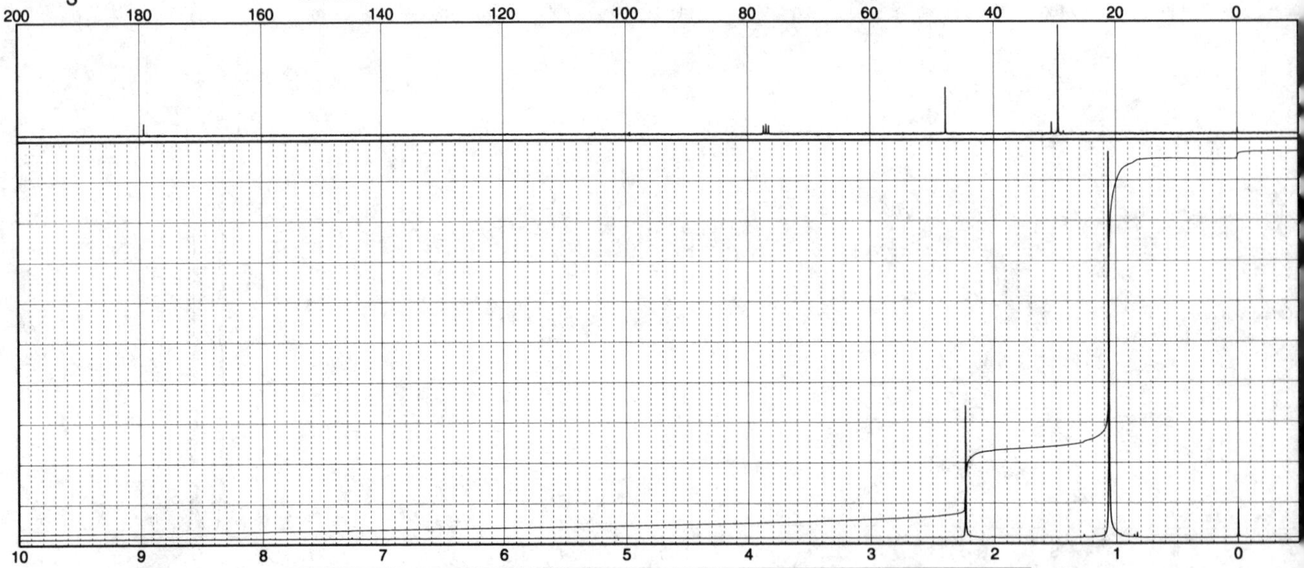

$CH_3-\overset{\overset{\displaystyle CH_3}{|}}{\underset{\underset{\displaystyle CH_3}{|}}{C}}-CH_2-\overset{\displaystyle O}{\overset{\|}{C}}-OH$

C

Aldrich 10,987-8   CAS [97-61-0]

(±)-2-Methylvaleric acid, 98%

$C_6H_{12}O_2$
FW 116.16
bp 197°C
d 0.931
n_D^{20} 1.4140

Fp 196°F

VP-FT-IR: **3**, 579D

183.81
39.23*
35.70
20.35
16.80*
13.92*

CDCl₃   QE-300

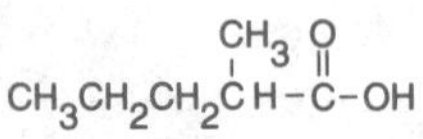

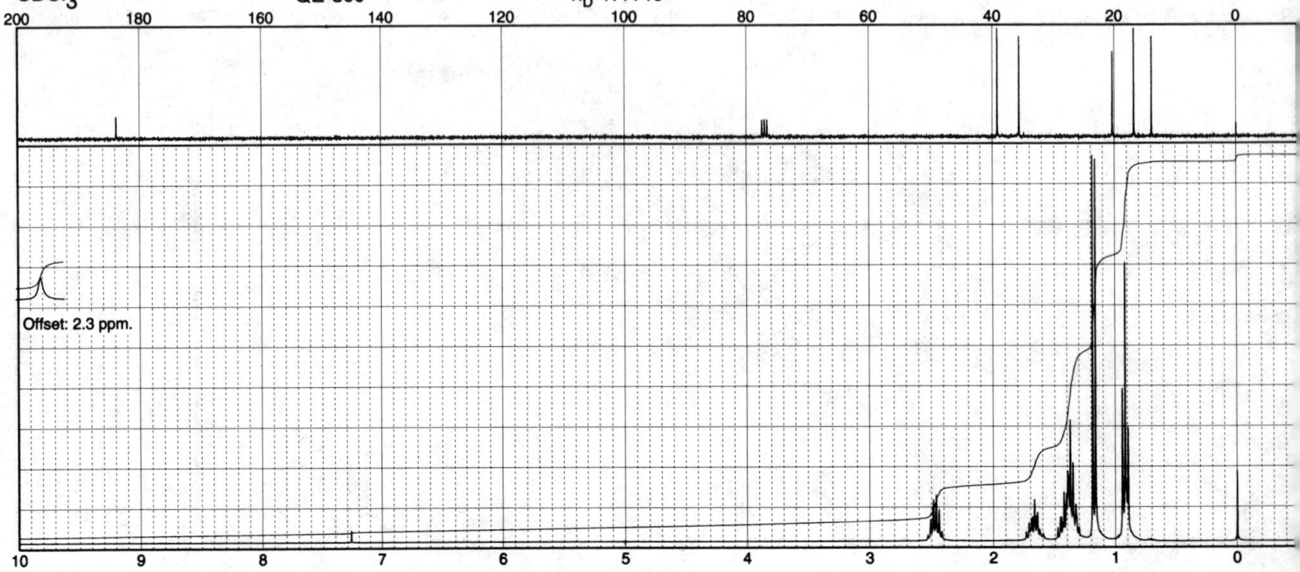

$CH_3CH_2CH_2\overset{\overset{\displaystyle CH_3}{|}}{C}H-\overset{\displaystyle O}{\overset{\|}{C}}-OH$

Offset: 2.3 ppm.

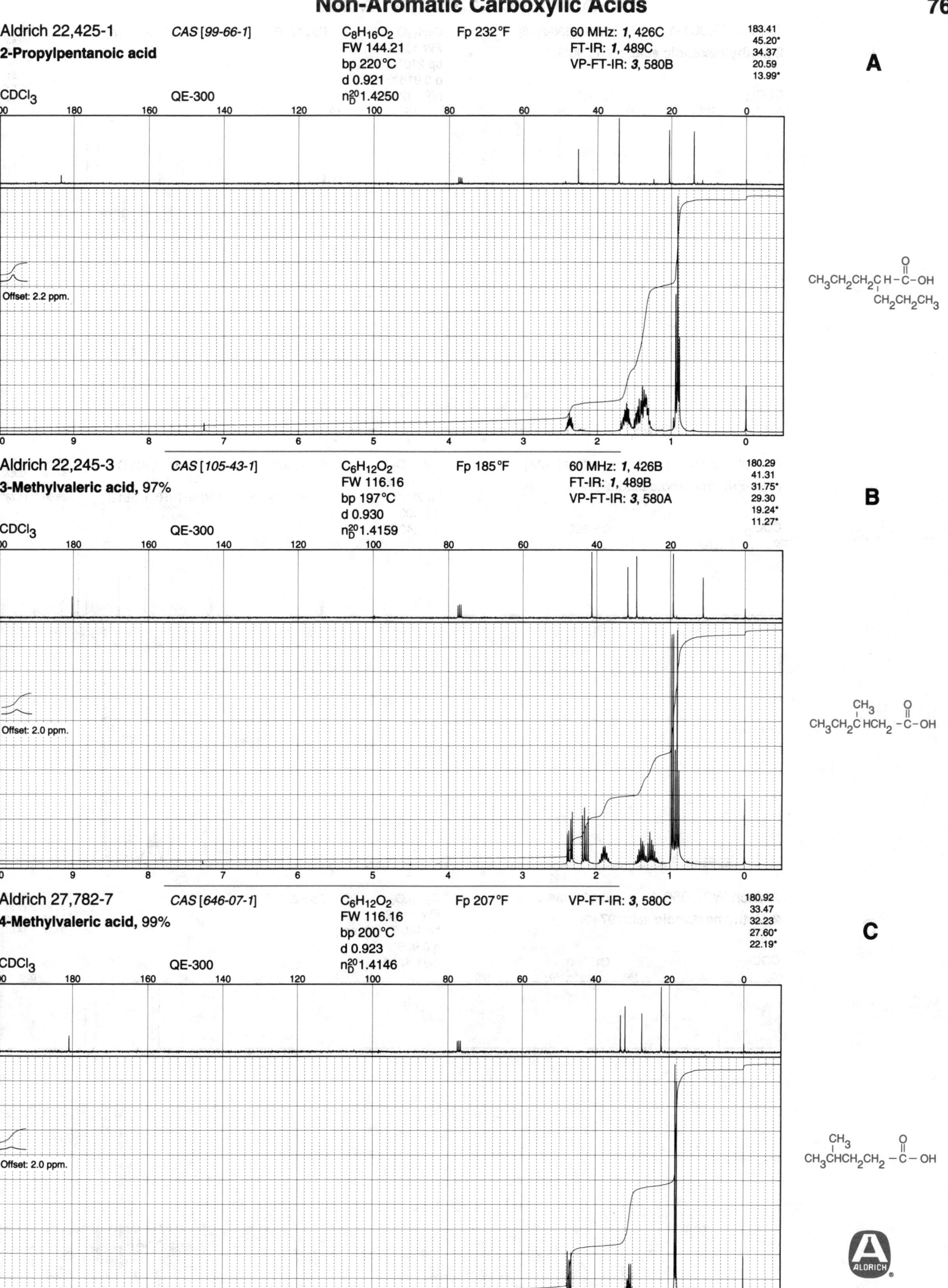

Aldrich 22,425-1   CAS [99-66-1]   C₈H₁₆O₂   Fp 232°F   60 MHz: *1*, 426C   183.41 / 45.20* / 34.37 / 20.59 / 13.99*
2-Propylpentanoic acid   FW 144.21   FT-IR: *1*, 489C   **A**
bp 220°C   VP-FT-IR: *3*, 580B
CDCl₃   QE-300   d 0.921 n_D^{20} 1.4250

Offset: 2.2 ppm.

Aldrich 22,245-3   CAS [105-43-1]   C₆H₁₂O₂   Fp 185°F   60 MHz: *1*, 426B   180.29 / 41.31 / 31.75* / 29.30 / 19.24* / 11.27*
3-Methylvaleric acid, 97%   FW 116.16   FT-IR: *1*, 489B   **B**
bp 197°C   VP-FT-IR: *3*, 580A
CDCl₃   QE-300   d 0.930 n_D^{20} 1.4159

Offset: 2.0 ppm.

Aldrich 27,782-7   CAS [646-07-1]   C₆H₁₂O₂   Fp 207°F   VP-FT-IR: *3*, 580C   180.92 / 33.47 / 32.23 / 27.60* / 22.19*
4-Methylvaleric acid, 99%   FW 116.16   **C**
bp 200°C
CDCl₃   QE-300   d 0.923 n_D^{20} 1.4146

Offset: 2.0 ppm.

A

Aldrich 28,001-1   *CAS [4536-23-6]* $C_7H_{14}O_2$ Fp 222°F   VP-FT-IR: *3*, 581A

2-Methylhexanoic acid, tech., 90%

FW 130.19
bp 210°C
d 0.918
$n_D^{20} 1.4201$

183.81
39.45*
33.26
29.34
22.61
16.83*
13.94*

$CDCl_3$ QE-300

$CH_3CH_2CH_2CH_2\overset{\overset{\displaystyle CH_3}{|}}{C}H-\overset{\overset{\displaystyle O}{\|}}{C}-OH$

B

Aldrich E2,914-1   *CAS [149-57-5]* $C_8H_{16}O_2$ Fp >230°F   60 MHz: *1*, 426D

(±)-2-Ethylhexanoic acid, 99%

FW 144.21
bp 228°C
d 0.903
$n_D^{20} 1.4250$

FT-IR: *1*, 489D
VP-FT-IR: *3*, 581B

183.36   25.21
47.19*   22.67
31.50   13.93*
29.56   11.78*

$CDCl_3$ QE-300

Offset: 2.3 ppm.

$CH_3CH_2CH_2CH_2\overset{\displaystyle CH}{\underset{\displaystyle CH_2CH_3}{|}}-\overset{\overset{\displaystyle O}{\|}}{C}-OH$

C

Aldrich W27,060-1   *CAS [1188-02-9]* $C_8H_{16}O_2$ Fp >230°F

2-Methylheptanoic acid, 97+%

FW 144.21
bp 140°C (30 mm)
d 0.906
$n_D^{20} 1.4250$

183.76   26.84
39.47*   22.51
33.52   16.82*
31.72   14.01*

$CDCl_3$ QE-300

Offset: 2.4 ppm.

$CH_3(CH_2)_3CH_2\overset{\overset{\displaystyle CH_3}{|}}{C}H-\overset{\overset{\displaystyle O}{\|}}{C}-OH$

Non-Aromatic Carboxylic Acids

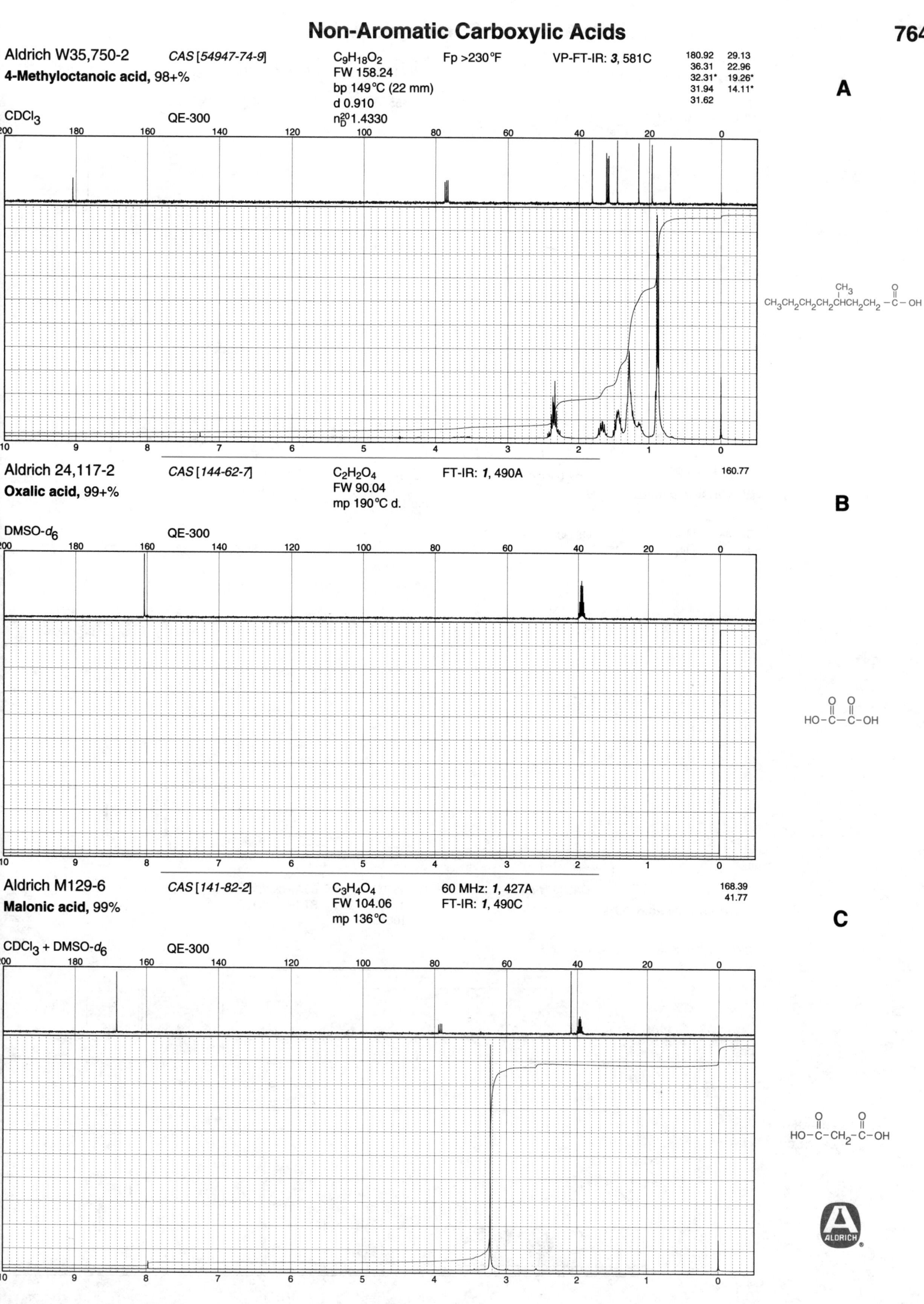

Aldrich W35,750-2   *CAS [54947-74-9]* $C_9H_{18}O_2$ Fp >230°F   VP-FT-IR: **3**, 581C

4-Methyloctanoic acid, 98+%

FW 158.24
bp 149°C (22 mm)
d 0.910
n_D^{20}1.4330

180.92	29.13
36.31	22.96
32.31*	19.26*
31.94	14.11*
31.62	

A

CDCl3   QE-300

$CH_3CH_2CH_2CH_2CHCH_2CH_2-C-OH$ (with CH_3 branch and O on carbonyl)

Aldrich 24,117-2   *CAS [144-62-7]* $C_2H_2O_4$ FT-IR: **1**, 490A   160.77

Oxalic acid, 99+%

FW 90.04
mp 190°C d.

B

DMSO-d_6 QE-300

$HO-C-C-OH$ (with two O double bonds)

Aldrich M129-6   *CAS [141-82-2]* $C_3H_4O_4$ 60 MHz: **1**, 427A   168.39
                                                                           41.77

Malonic acid, 99%

FW 104.06
mp 136°C
FT-IR: **1**, 490C

C

CDCl3 + DMSO-d_6 QE-300

$HO-C-CH_2-C-OH$ (with two O double bonds)

ALDRICH

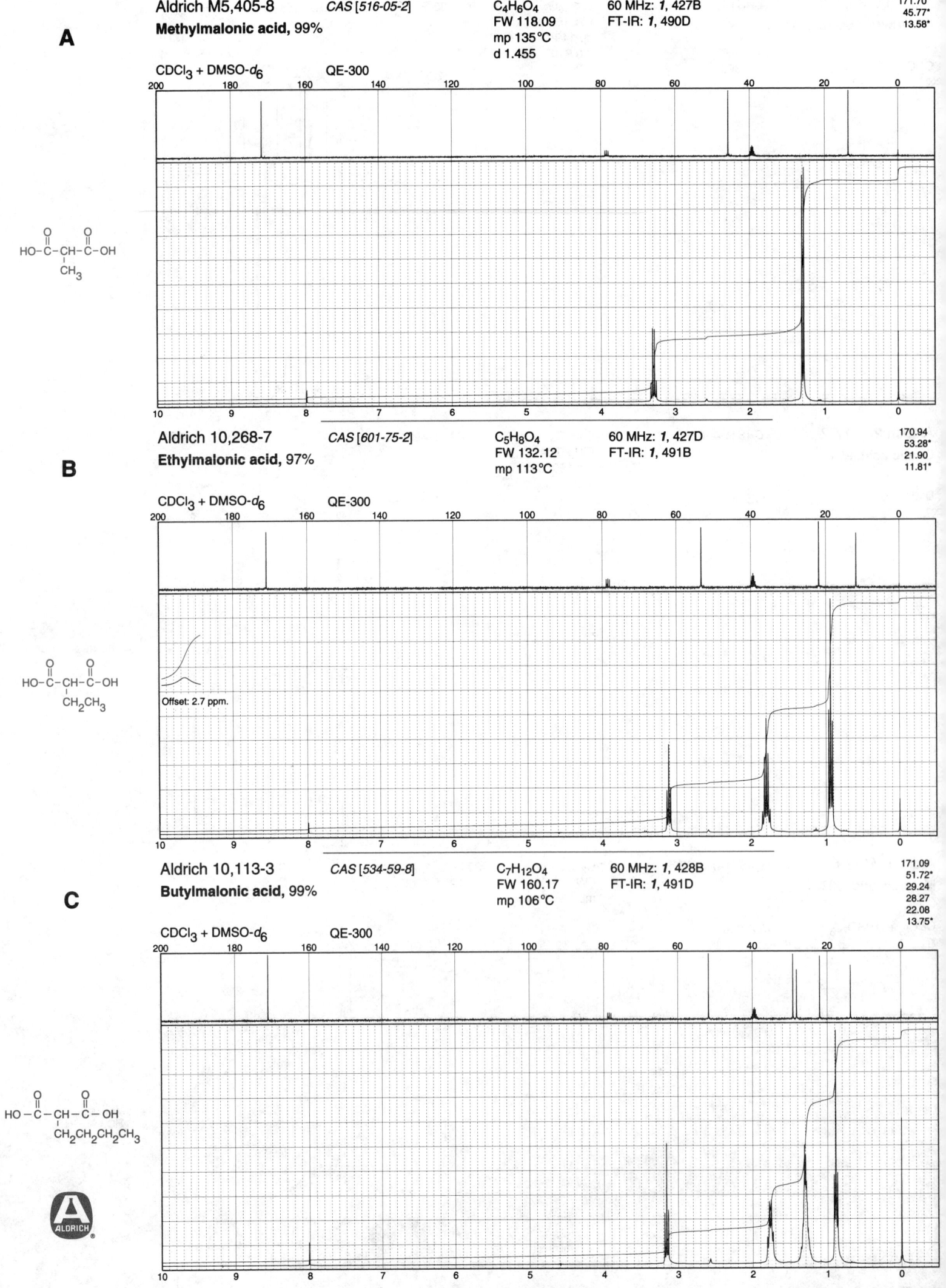

A

Aldrich M5,405-8
Methylmalonic acid, 99%

CAS [516-05-2]

C4H6O4
FW 118.09
mp 135°C
d 1.455

60 MHz: *1*, 427B
FT-IR: *1*, 490D

171.70
45.77*
13.58*

CDCl3 + DMSO-d6   QE-300

B

Aldrich 10,268-7
Ethylmalonic acid, 97%

CAS [601-75-2]

C5H8O4
FW 132.12
mp 113°C

60 MHz: *1*, 427D
FT-IR: *1*, 491B

170.94
53.28*
21.90
11.81*

CDCl3 + DMSO-d6   QE-300

Offset: 2.7 ppm.

C

Aldrich 10,113-3
Butylmalonic acid, 99%

CAS [534-59-8]

C7H12O4
FW 160.17
mp 106°C

60 MHz: *1*, 428B
FT-IR: *1*, 491D

171.09
51.72*
29.24
28.27
22.08
13.75*

CDCl3 + DMSO-d6   QE-300

Aldrich D16,800-9 *CAS [595-46-0]*

Dimethylmalonic acid, 98%

$C_5H_8O_4$
FW 132.12
mp 192°C d.

60 MHz: *1*, 427C
FT-IR: *1*, 491A

174.01
48.77
22.56*

A

DMSO-d_6 QE-300

Offset: 2.8 ppm.

$$\underset{\underset{CH_3}{|}}{HO-\overset{O}{\overset{||}{C}}-\overset{CH_3}{\underset{|}{C}}-\overset{O}{\overset{||}{C}}-OH}$$

Aldrich 23,968-2 *CAS [110-15-6]*

Succinic acid, 99+%

$C_4H_6O_4$
FW 118.09
mp 188°C

60 MHz: *1*, 428C
FT-IR: *1*, 492A

180.00
31.82

B

D$_2$O   QE-300

$$HO-\overset{O}{\overset{||}{C}}-CH_2-CH_2-\overset{O}{\overset{||}{C}}-OH$$

Aldrich M8,120-9 *CAS [498-21-5]*

Methylsuccinic acid, 99%

$C_5H_8O_4$
FW 132.12
mp 118°C

60 MHz: *1*, 428D
FT-IR: *1*, 492B

176.33
172.91
37.12
35.12*
16.76*

C

DMSO-d_6 QE-300

Offset: 2.3 ppm.

$$HO-\overset{O}{\overset{||}{C}}-\overset{CH_3}{\underset{|}{C}}H\,CH_2-\overset{O}{\overset{||}{C}}-OH$$

ALDRICH®

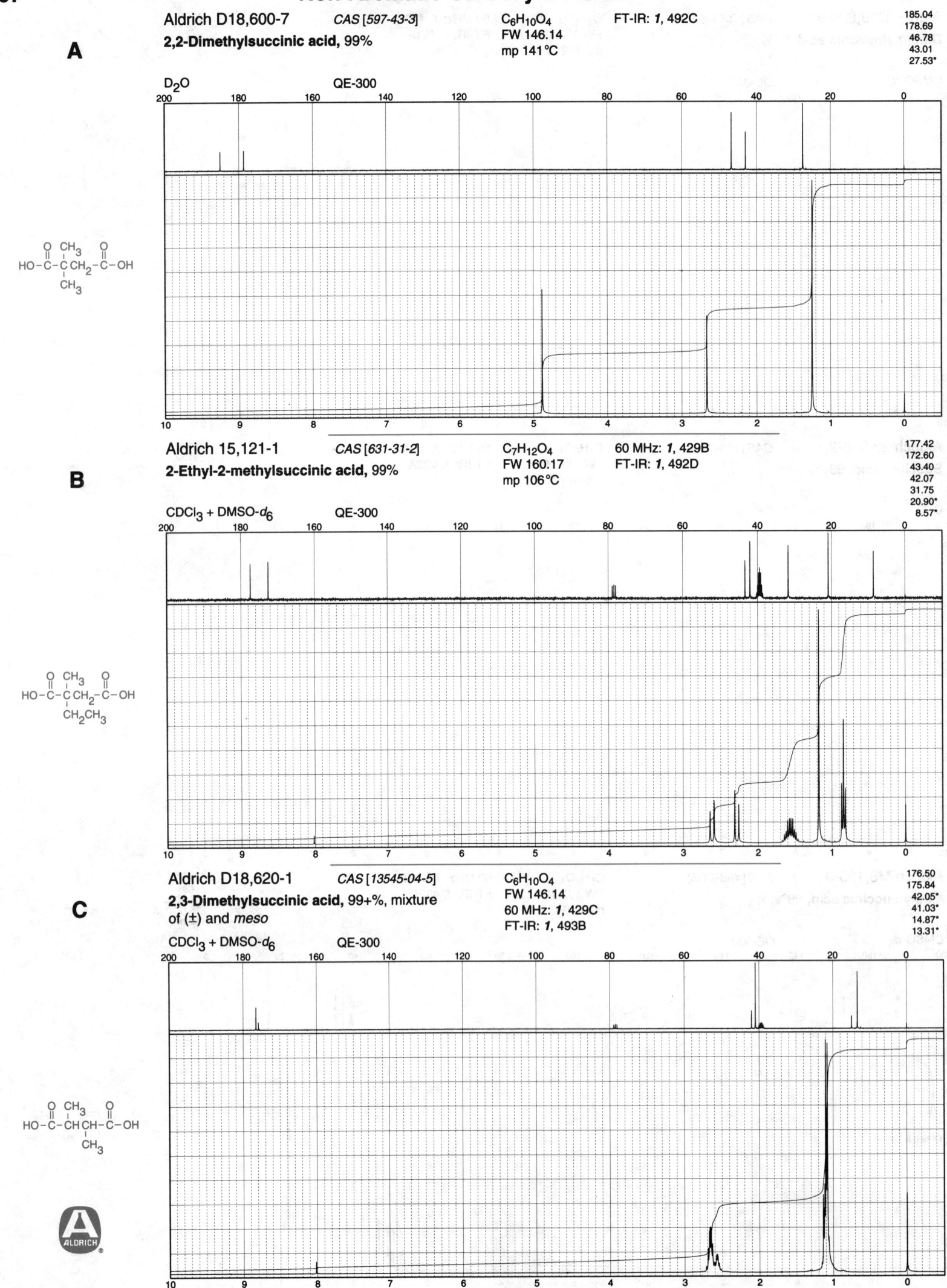

A

Aldrich D18,600-7 CAS [597-43-3] $C_6H_{10}O_4$ FT-IR: 1, 492C
2,2-Dimethylsuccinic acid, 99% FW 146.14
mp 141°C

185.04
178.69
46.78
43.01
27.53*

D₂O QE-300

B

Aldrich 15,121-1 CAS [631-31-2] $C_7H_{12}O_4$ 60 MHz: 1, 429B
2-Ethyl-2-methylsuccinic acid, 99% FW 160.17 FT-IR: 1, 492D
mp 106°C

177.42
172.60
43.40
42.07
31.75
20.90*
8.57*

CDCl₃ + DMSO-d₆ QE-300

C

Aldrich D18,620-1 CAS [13545-04-5] $C_6H_{10}O_4$ FW 146.14
2,3-Dimethylsuccinic acid, 99+%, mixture 60 MHz: 1, 429C
of (±) and meso FT-IR: 1, 493B

176.50
175.84
42.05*
41.03*
14.87*
13.31*

CDCl₃ + DMSO-d₆ QE-300

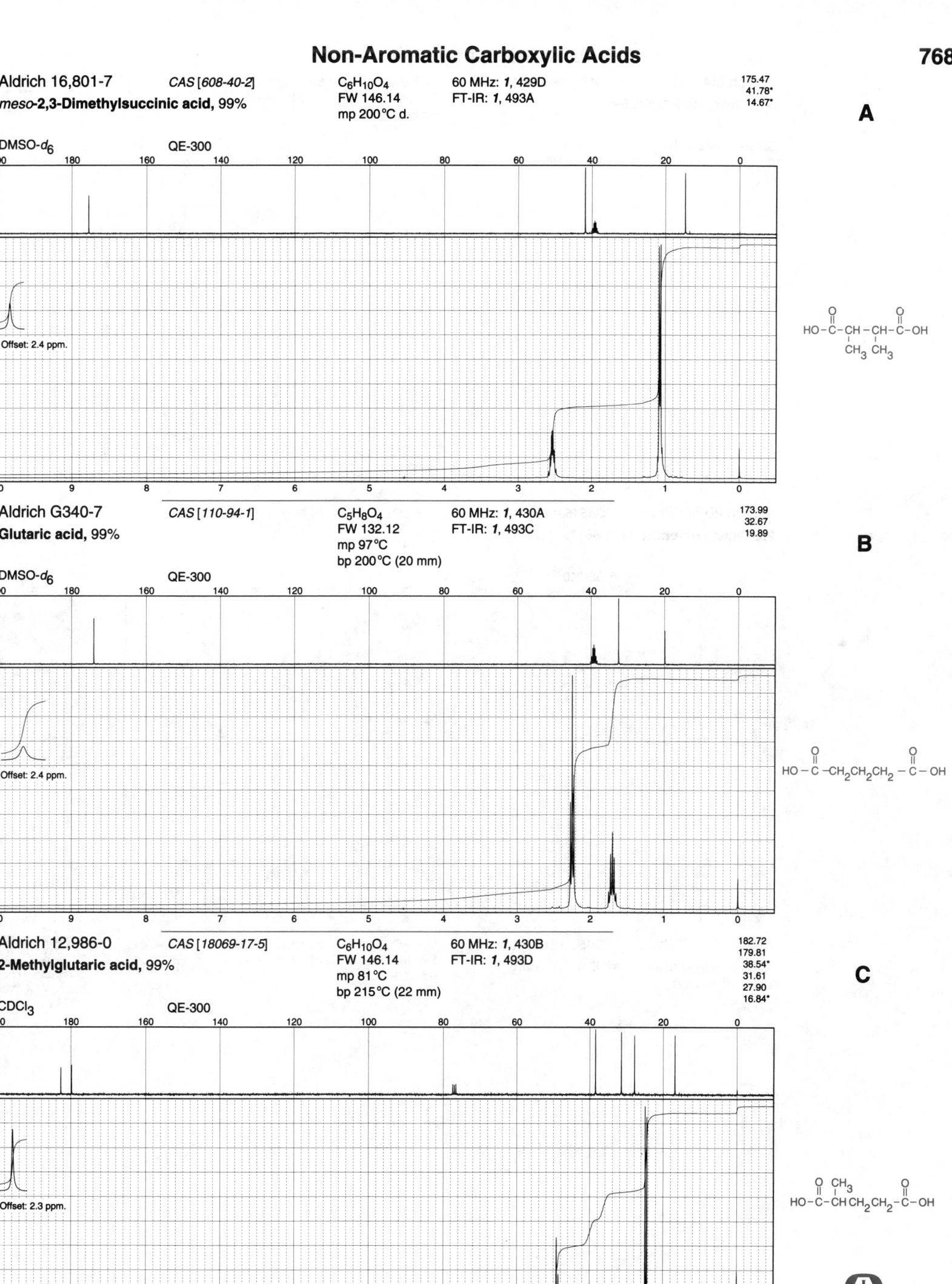

Aldrich 16,801-7   CAS [608-40-2]   C₆H₁₀O₄   60 MHz: 1, 429D

$C_6H_{10}O_4$

meso-2,3-Dimethylsuccinic acid, 99%   FW 146.14   FT-IR: 1, 493A

mp 200°C d.

175.47
41.78*
14.67*

A

DMSO-d_6 QE-300

Offset: 2.4 ppm.

Aldrich G340-7   CAS [110-94-1] $C_5H_8O_4$ 60 MHz: 1, 430A

Glutaric acid, 99%   FW 132.12   FT-IR: 1, 493C

mp 97°C

bp 200°C (20 mm)

173.99
32.67
19.89

B

DMSO-d_6 QE-300

Offset: 2.4 ppm.

Aldrich 12,986-0   CAS [18069-17-5] $C_6H_{10}O_4$ 60 MHz: 1, 430B

2-Methylglutaric acid, 99%   FW 146.14   FT-IR: 1, 493D

mp 81°C

bp 215°C (22 mm)

182.72
179.81
38.54*
31.61
27.90
16.84*

C

CDCl₃   QE-300

Offset: 2.3 ppm.

Non-Aromatic Carboxylic Acids

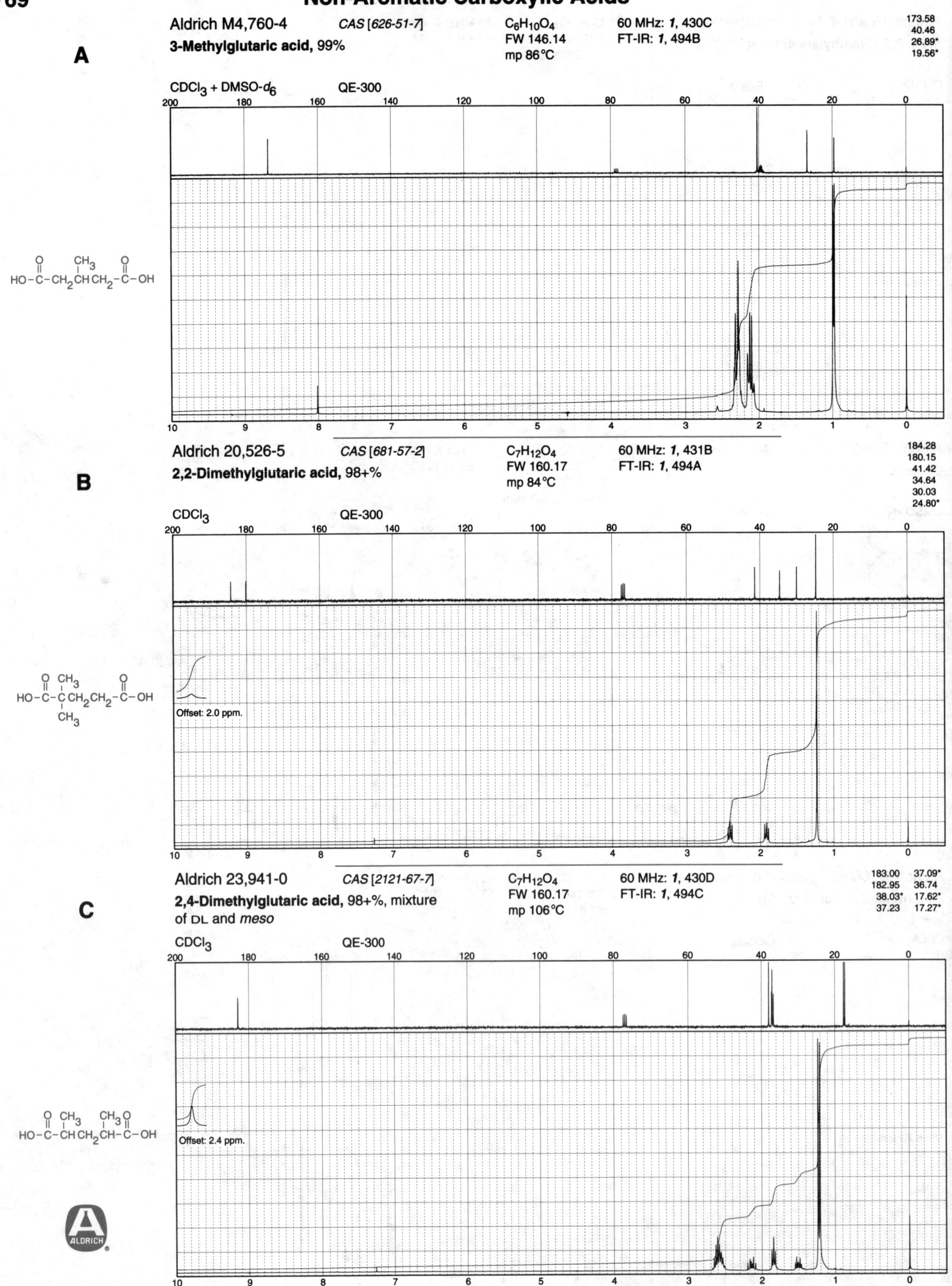

A

Aldrich M4,760-4   *CAS [626-51-7]* $C_6H_{10}O_4$ 60 MHz: *1*, 430C

3-Methylglutaric acid, 99%   FW 146.14   FT-IR: *1*, 494B   mp 86°C

173.58
40.46
26.89*
19.56*

$CDCl_3$ + DMSO-d_6 QE-300

B

Aldrich 20,526-5   *CAS [681-57-2]* $C_7H_{12}O_4$ 60 MHz: *1*, 431B

2,2-Dimethylglutaric acid, 98+%   FW 160.17   FT-IR: *1*, 494A   mp 84°C

184.28
180.15
41.42
34.64
30.03
24.80*

$CDCl_3$ QE-300

Offset: 2.0 ppm.

C

Aldrich 23,941-0   *CAS [2121-67-7]* $C_7H_{12}O_4$ 60 MHz: *1*, 430D

2,4-Dimethylglutaric acid, 98+%, mixture   FW 160.17   FT-IR: *1*, 494C

of DL and *meso*   mp 106°C

183.00   37.09*
182.95   36.74
38.03*   17.62*
37.23   17.27*

$CDCl_3$ QE-300

Offset: 2.4 ppm.

Aldrich D15,940-9 *CAS [4839-46-7]* C$_7$H$_{12}$O$_4$ 60 MHz: *1*, 431A

3,3-Dimethylglutaric acid, 98% FW 160.17 FT-IR: *1*, 494D

mp 101°C

173.09
44.72
31.69
27.33*

A

CDCl$_3$ + DMSO-*d*$_6$ QE-300

Offset: 1.9 ppm.

HO—C—CH$_2$—C—CH$_2$—C—OH (with CH$_3$ groups)

Aldrich A2,635-7 *CAS [124-04-9]* C$_6$H$_{10}$O$_4$ FT-IR: *1*, 495B

Adipic acid, 99% FW 146.14 VP-FT-IR: *3*, 582A

mp 153°C
bp 265°C (100 mm)
Fp 385°F

174.23
33.31
23.97

B

DMSO-*d*$_6$ QE-300

Offset: 2.3 ppm.

HO—C—CH$_2$CH$_2$CH$_2$CH$_2$—C—OH

Aldrich M2,740-9 *CAS [3058-01-3]* C$_7$H$_{12}$O$_4$ 60 MHz: *1*, 431C

3-Methyladipic acid, 99% FW 160.17 FT-IR: *1*, 495C

mp 101°C
bp 230°C (30 mm)

174.63
173.91
41.14
31.49
31.27
29.37*
19.25*

C

CDCl$_3$ + DMSO-*d*$_6$ QE-300

HO—C—CH$_2$CH$_2$CHCH$_2$—C—OH (with CH$_3$ group)

ALDRICH

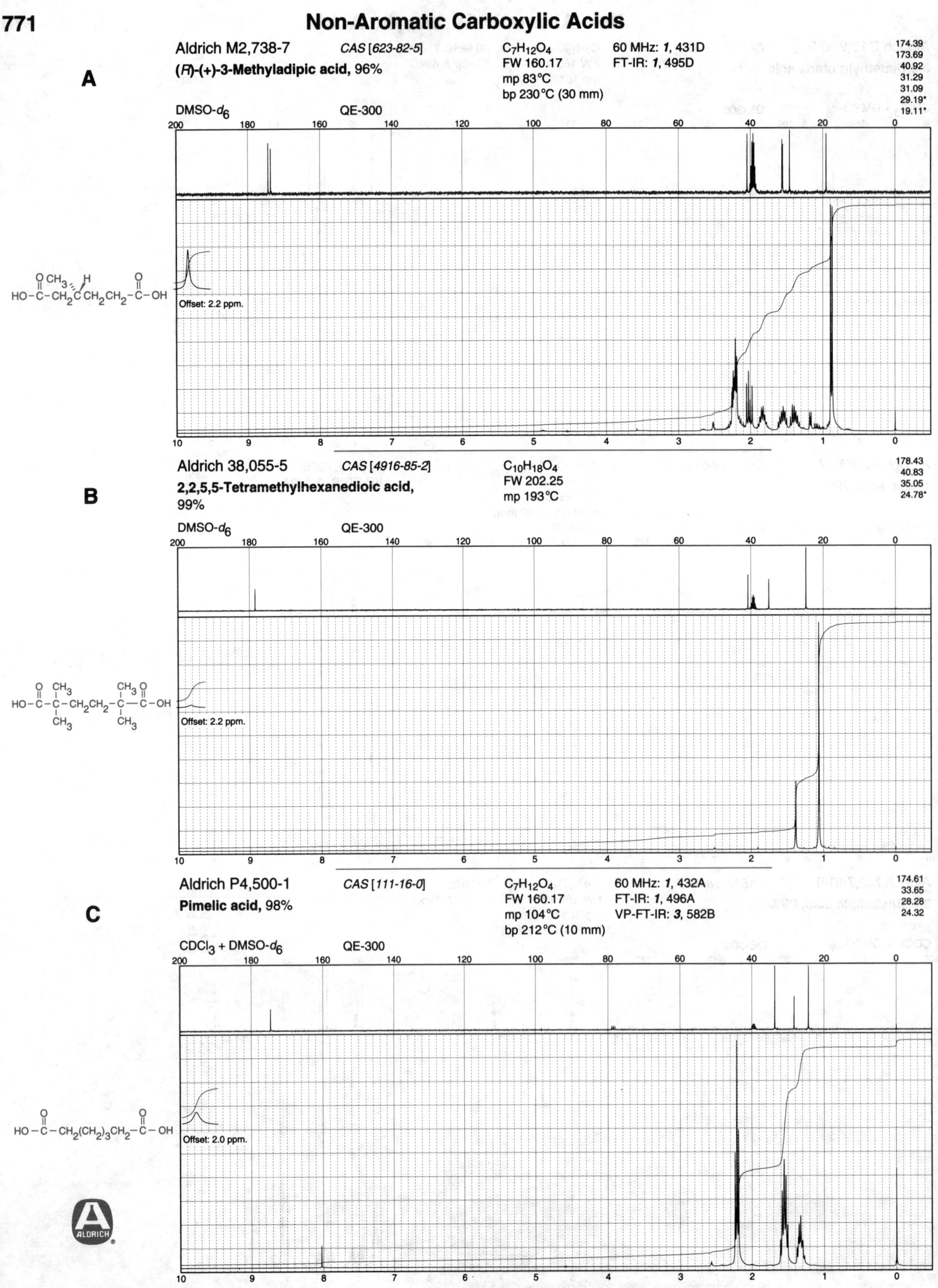

A

Aldrich M2,738-7   CAS [623-82-5] $C_7H_{12}O_4$ 60 MHz: 1, 431D

(R)-(+)-3-Methyladipic acid, 96%   FW 160.17   FT-IR: 1, 495D

mp 83°C

bp 230°C (30 mm)

174.39
173.69
40.92
31.29
31.09
29.19*
19.11*

DMSO-d_6 QE-300

Offset: 2.2 ppm.

B

Aldrich 38,055-5   CAS [4916-85-2] $C_{10}H_{18}O_4$

2,2,5,5-Tetramethylhexanedioic acid,   FW 202.25

99%   mp 193°C

178.43
40.83
35.05
24.78*

DMSO-d_6 QE-300

Offset: 2.2 ppm.

C

Aldrich P4,500-1   CAS [111-16-0] $C_7H_{12}O_4$ 60 MHz: 1, 432A

Pimelic acid, 98%   FW 160.17   FT-IR: 1, 496A

mp 104°C   VP-FT-IR: 3, 582B

bp 212°C (10 mm)

174.61
33.65
28.28
24.32

CDCl$_3$ + DMSO-d_6 QE-300

Offset: 2.0 ppm.

Non-Aromatic Carboxylic Acids

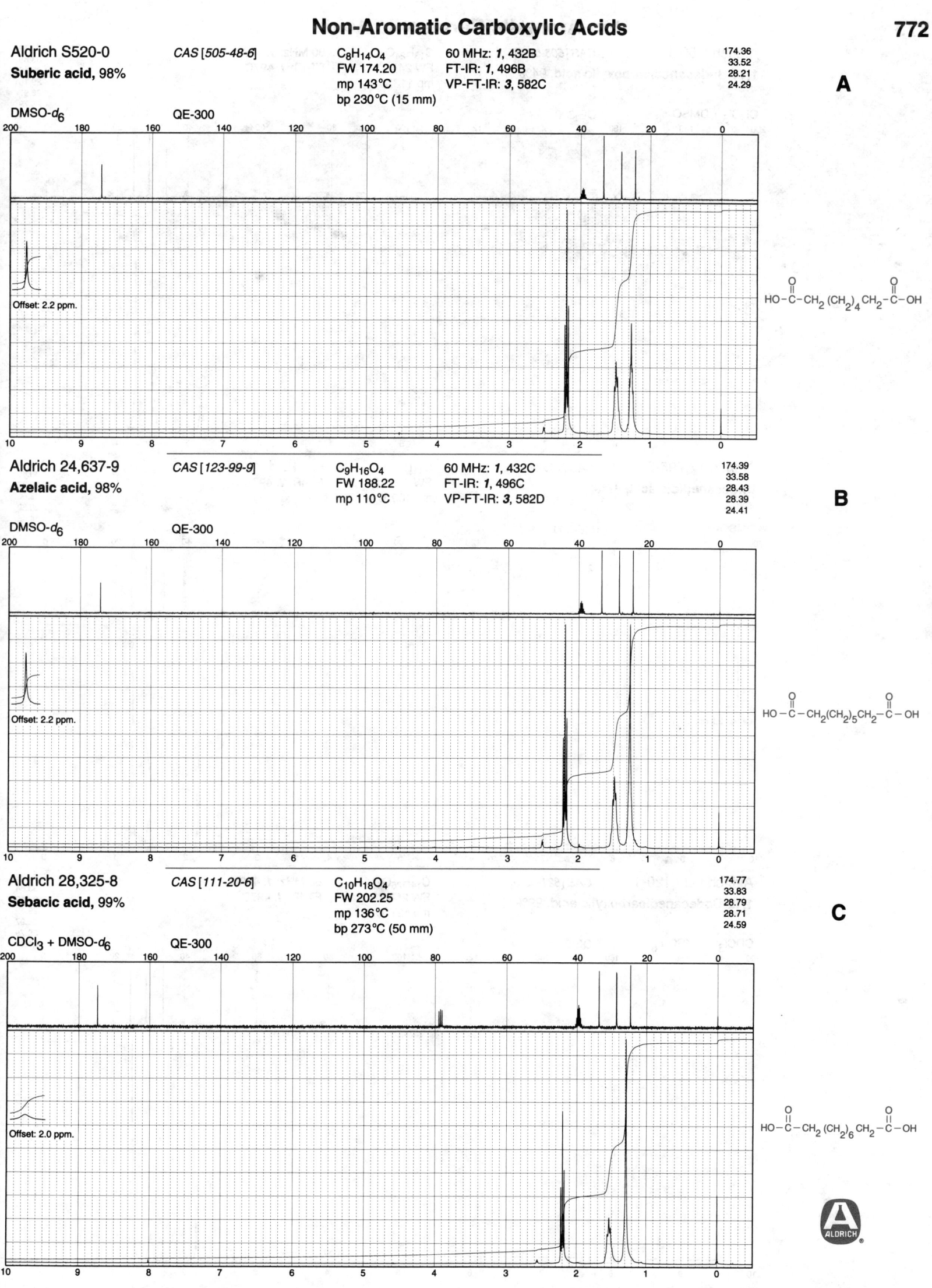

Aldrich S520-0

Suberic acid, 98%

CAS [505-48-6]

C₈H₁₄O₄
FW 174.20
mp 143°C
bp 230°C (15 mm)

60 MHz: *1*, 432B
FT-IR: *1*, 496B
VP-FT-IR: *3*, 582C

174.36
33.52
28.21
24.29

A

DMSO-*d₆*   QE-300

Offset: 2.2 ppm.

$HO-\overset{O}{\overset{\|}{C}}-CH_2\,(CH_2)_4\,CH_2-\overset{O}{\overset{\|}{C}}-OH$

Aldrich 24,637-9

Azelaic acid, 98%

CAS [123-99-9]

C₉H₁₆O₄
FW 188.22
mp 110°C

60 MHz: *1*, 432C
FT-IR: *1*, 496C
VP-FT-IR: *3*, 582D

174.39
33.58
28.43
28.39
24.41

B

DMSO-*d₆*   QE-300

Offset: 2.2 ppm.

$HO-\overset{O}{\overset{\|}{C}}-CH_2(CH_2)_5CH_2-\overset{O}{\overset{\|}{C}}-OH$

Aldrich 28,325-8

Sebacic acid, 99%

CAS [111-20-6]

C₁₀H₁₈O₄
FW 202.25
mp 136°C
bp 273°C (50 mm)

174.77
33.83
28.79
28.71
24.59

C

CDCl₃ + DMSO-*d₆*   QE-300

Offset: 2.0 ppm.

$HO-\overset{O}{\overset{\|}{C}}-CH_2\,(CH_2)_6\,CH_2-\overset{O}{\overset{\|}{C}}-OH$

ALDRICH

Non-Aromatic Carboxylic Acids

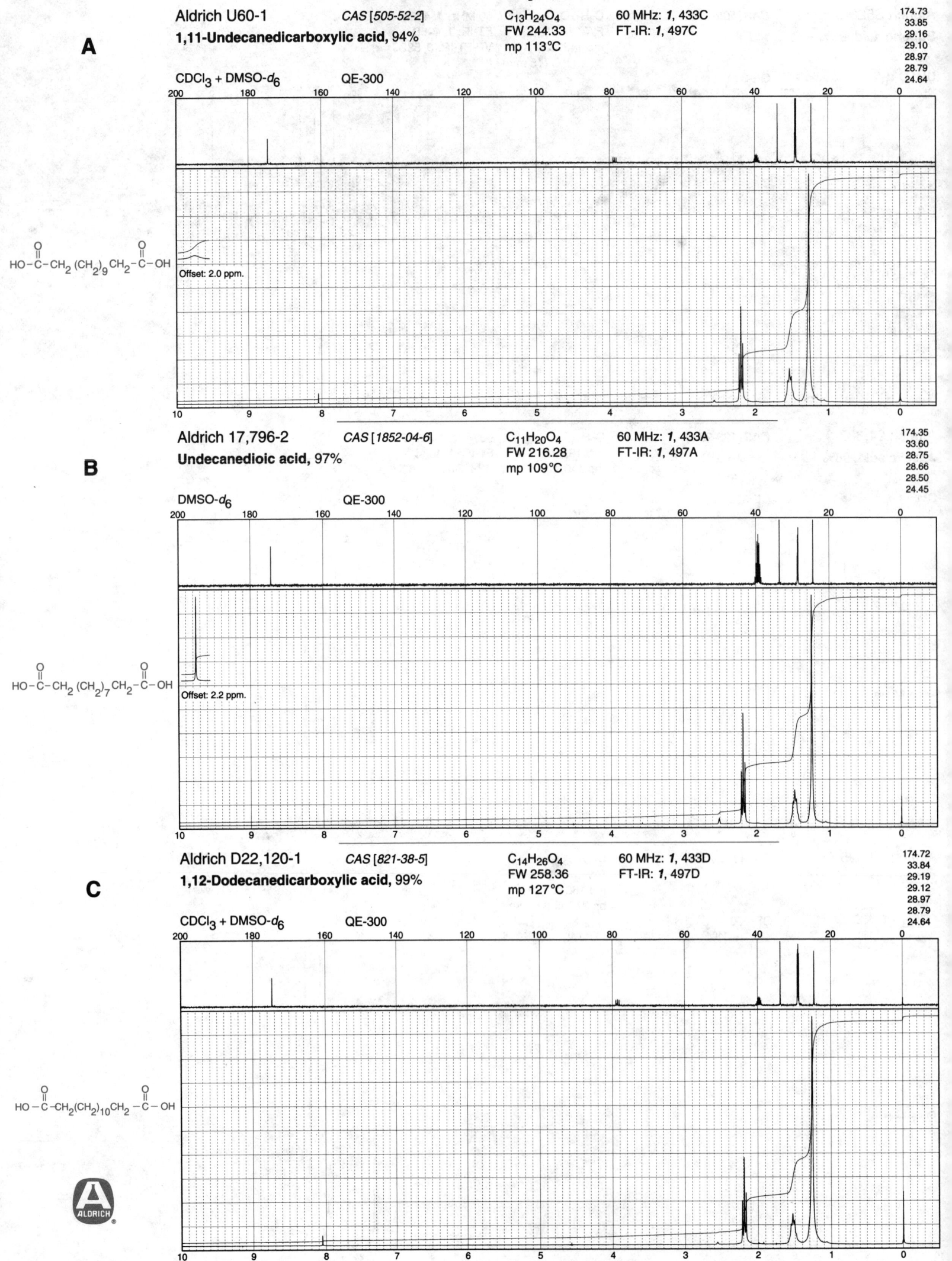

A

Aldrich U60-1   *CAS [505-52-2]*

1,11-Undecanedicarboxylic acid, 94%

$C_{13}H_{24}O_4$
FW 244.33
mp 113°C

60 MHz: *1*, 433C
FT-IR: *1*, 497C

174.73
33.85
29.16
29.10
28.97
28.79
24.64

CDCl$_3$ + DMSO-d_6 QE-300

Offset: 2.0 ppm.

B

Aldrich 17,796-2   *CAS [1852-04-6]*

Undecanedioic acid, 97%

$C_{11}H_{20}O_4$
FW 216.28
mp 109°C

60 MHz: *1*, 433A
FT-IR: *1*, 497A

174.35
33.60
28.75
28.66
28.50
24.45

DMSO-d_6 QE-300

Offset: 2.2 ppm.

C

Aldrich D22,120-1   *CAS [821-38-5]*

1,12-Dodecanedicarboxylic acid, 99%

$C_{14}H_{26}O_4$
FW 258.36
mp 127°C

60 MHz: *1*, 433D
FT-IR: *1*, 497D

174.72
33.84
29.19
29.12
28.97
28.79
24.64

CDCl$_3$ + DMSO-d_6 QE-300

ALDRICH

Non-Aromatic Carboxylic Acids

Aldrich 17,750-4 *CAS [505-54-4]*
Hexadecanedioic acid, 98%

$C_{16}H_{30}O_4$
FW 286.41
mp 125°C

60 MHz: *1*, 434A
FT-IR: *1*, 498A

174.72
33.85
29.24
29.14
28.99
28.81
24.65

A

CDCl$_3$ + DMSO-d_6 QE-300

$HO-\overset{O}{\underset{}{C}}-CH_2(CH_2)_{12}CH_2-\overset{O}{\underset{}{C}}-OH$

Aldrich 30,667-3 *CAS [505-56-6]*
Docosanedioic acid, 85%

$C_{22}H_{42}O_4$
FW 370.58
mp 122°C

174.75
33.85
29.25
29.11
28.97
28.79
24.63

B

CDCl$_3$ + DMSO-d_6 QE-300

Offset: 1.9 ppm.

$HO-\overset{O}{\underset{}{C}}-CH_2(CH_2)_{18}CH_2-\overset{O}{\underset{}{C}}-OH$

Aldrich 30,670-3 *CAS [2450-31-9]*
Tetracosanedioic acid, 80%

$C_{24}H_{46}O_4$
FW 398.63
mp 124°C

174.74 29.11
33.85   28.97
29.25   28.79
29.22   24.63

C

CDCl$_3$ + DMSO-d_6 QE-300

$HO-\overset{O}{\underset{}{C}}-CH_2(CH_2)_{20}CH_2-\overset{O}{\underset{}{C}}-OH$

A

Aldrich T5,350-3   CAS [99-14-9]   C₆H₈O₆   60 MHz: 1, 442C

Tricarballylic acid, 99%   FW 176.12   FT-IR: 1, 505B

mp 161°C

174.77
172.83
37.05*
35.05

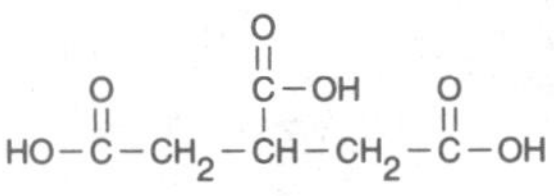

CDCl₃ + DMSO-d₆   QE-300

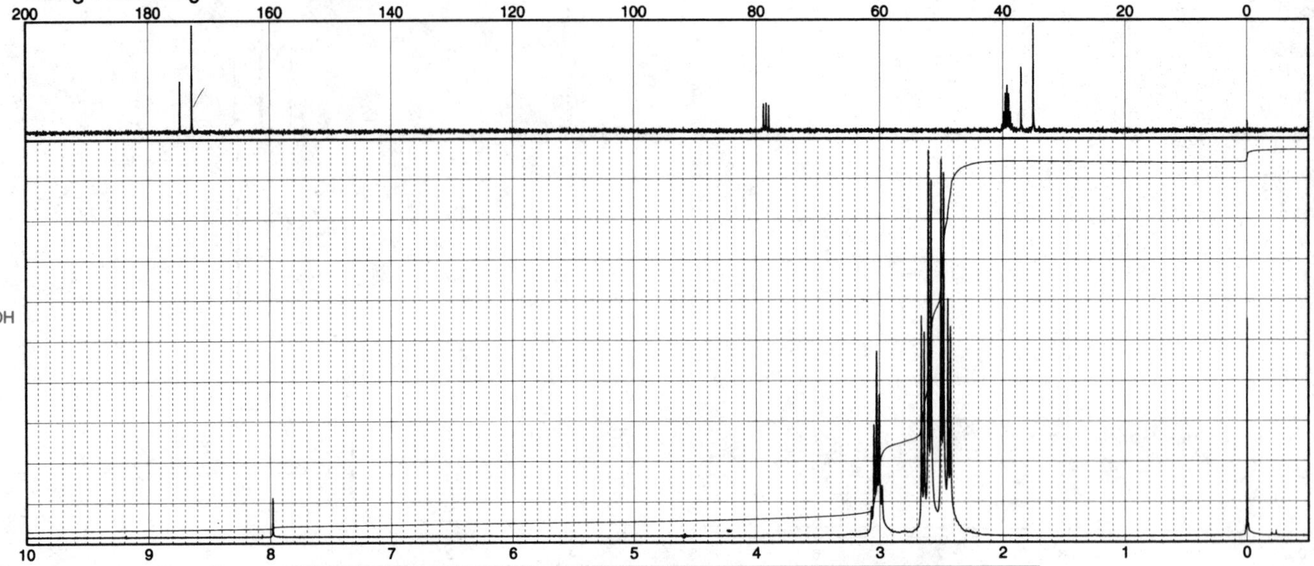

B

Aldrich M8,520-4   CAS [1590-02-9]   C₇H₁₀O₆   60 MHz: 1, 442D

β-Methyltricarballylic acid, 96%   FW 190.15   FT-IR: 1, 505C

mp 161°C d.

176.72
172.38
41.38
40.76
22.80*

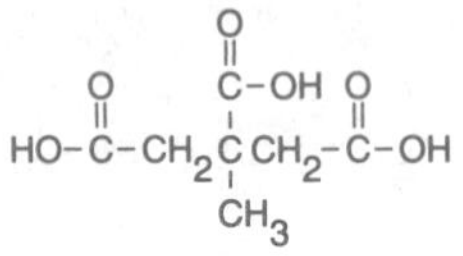

CDCl₃ + DMSO-d₆   QE-300

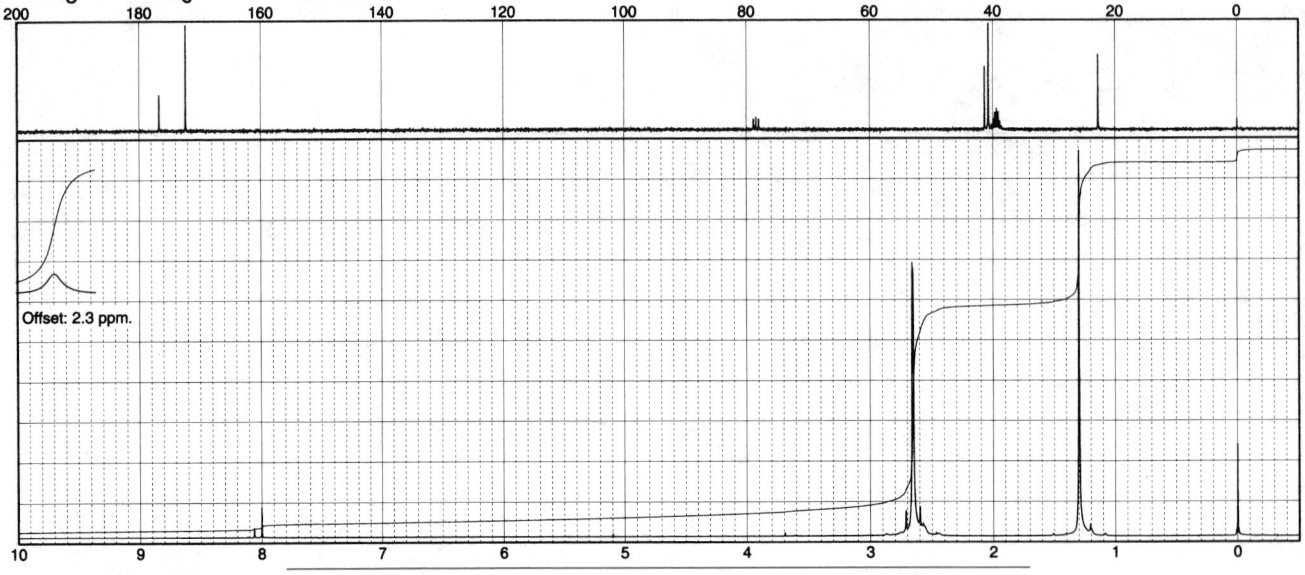

Offset: 2.3 ppm.

C

Aldrich 25,730-3   CAS [1703-58-8]   C₈H₁₀O₈   FT-IR: 1, 495A

1,2,3,4-Butanetetracarboxylic acid, 99%   FW 234.16   mp 196°C

179.19
178.62
45.22*
35.90

D₂O   QE-300

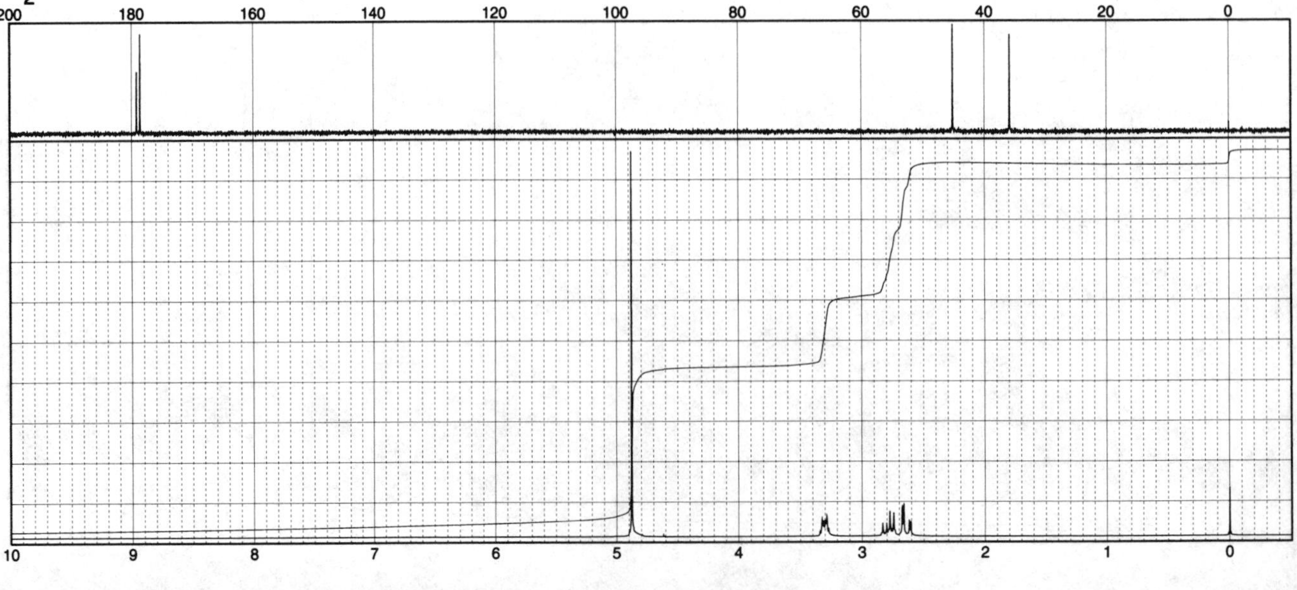

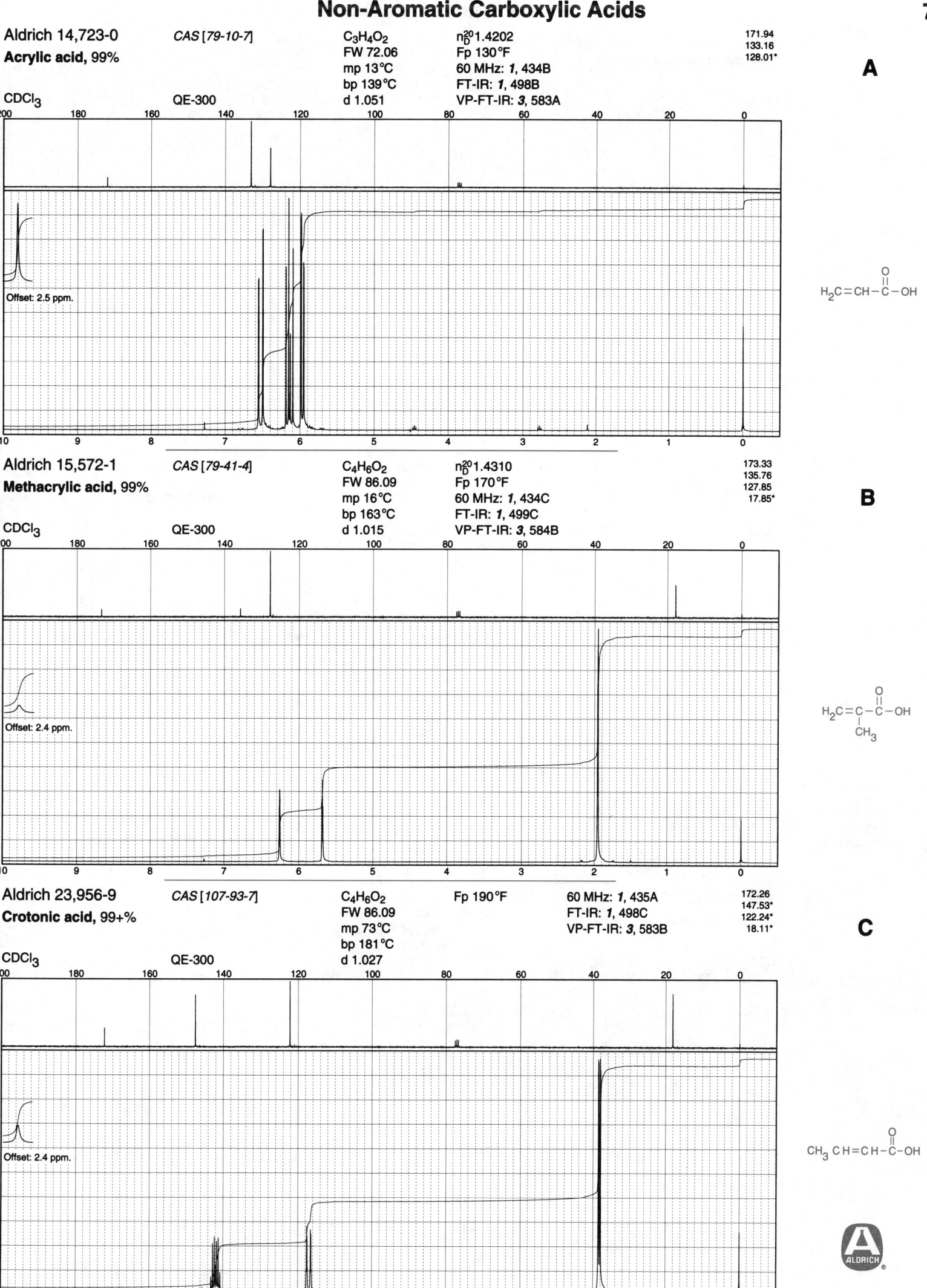

Aldrich 14,723-0
Acrylic acid, 99%

CAS [79-10-7]

C₃H₄O₂
FW 72.06
mp 13°C
bp 139°C
d 1.051

n_D^{20} 1.4202
Fp 130°F
60 MHz: *1*, 434B
FT-IR: *1*, 498B
VP-FT-IR: *3*, 583A

171.94
133.16
128.01*

A

CDCl₃   QE-300

Offset: 2.5 ppm.

H₂C=CH—C—OH (with O double bond)

Aldrich 15,572-1
Methacrylic acid, 99%

CAS [79-41-4]

C₄H₆O₂
FW 86.09
mp 16°C
bp 163°C
d 1.015

n_D^{20} 1.4310
Fp 170°F
60 MHz: *1*, 434C
FT-IR: *1*, 499C
VP-FT-IR: *3*, 584B

173.33
135.76
127.85
17.85*

B

CDCl₃   QE-300

Offset: 2.4 ppm.

H₂C=C—C—OH with CH₃ and O double bond

Aldrich 23,956-9
Crotonic acid, 99+%

CAS [107-93-7]

C₄H₆O₂
FW 86.09
mp 73°C
bp 181°C
d 1.027

Fp 190°F

60 MHz: *1*, 435A
FT-IR: *1*, 498C
VP-FT-IR: *3*, 583B

172.26
147.53*
122.24*
18.11*

C

CDCl₃   QE-300

Offset: 2.4 ppm.

CH₃CH=CH—C—OH with O double bond

ALDRICH

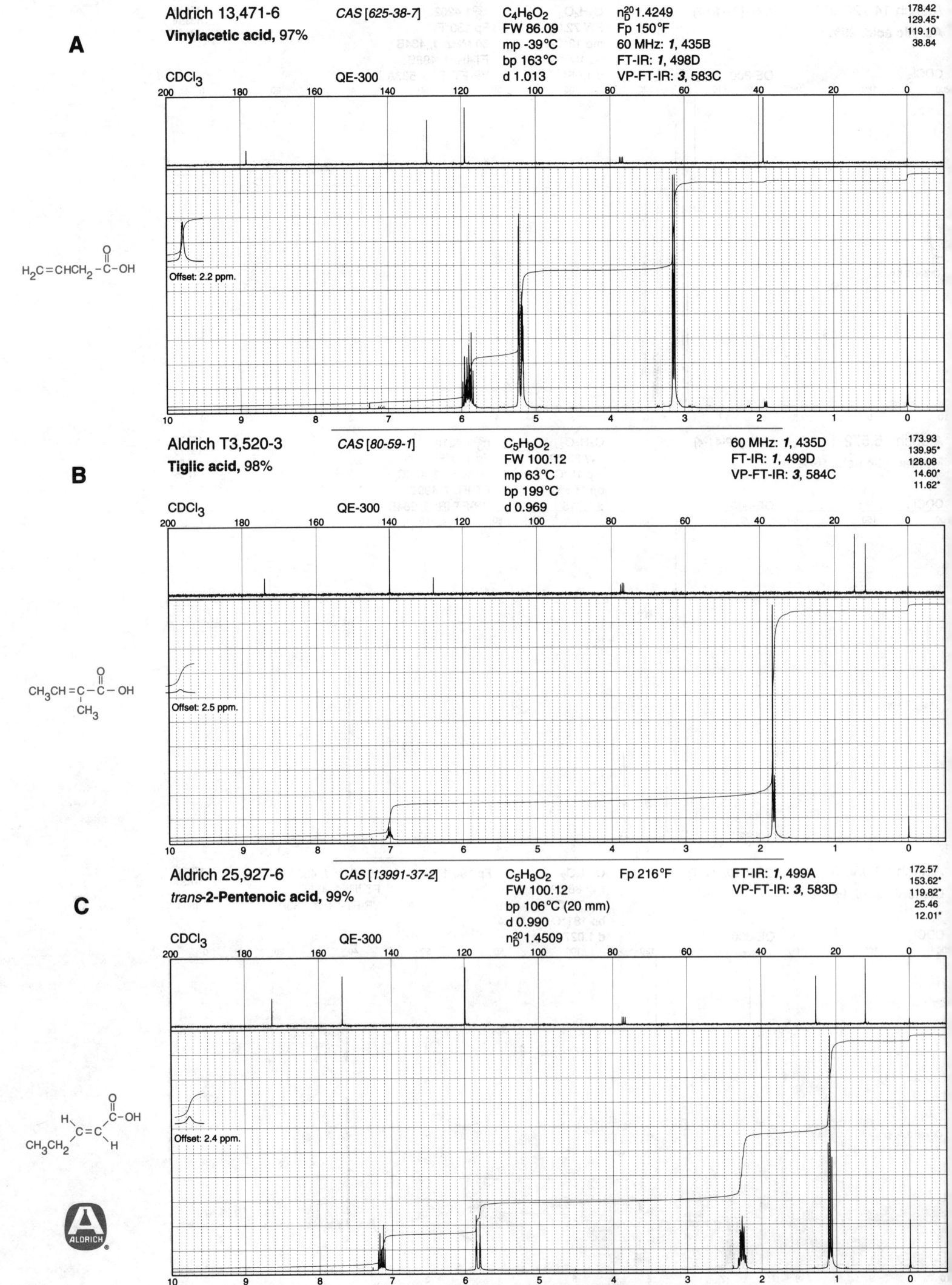

A

Aldrich 13,471-6   *CAS [625-38-7]*
Vinylacetic acid, 97%

$C_4H_6O_2$
FW 86.09
mp -39°C
bp 163°C
d 1.013

n_D^{20}1.4249
Fp 150°F
60 MHz: *1*, 435B
FT-IR: *1*, 498D
VP-FT-IR: *3*, 583C

178.42
129.45*
119.10
38.84

CDCl$_3$ QE-300

$H_2C=CHCH_2-\overset{\overset{\displaystyle O}{\|}}{C}-OH$

Offset: 2.2 ppm.

B

Aldrich T3,520-3   *CAS [80-59-1]*
Tiglic acid, 98%

$C_5H_8O_2$
FW 100.12
mp 63°C
bp 199°C
d 0.969

60 MHz: *1*, 435D
FT-IR: *1*, 499D
VP-FT-IR: *3*, 584C

173.93
139.95*
128.08
14.60*
11.62*

CDCl$_3$ QE-300

$CH_3CH=\overset{}{\underset{\underset{\displaystyle CH_3}{|}}{C}}-\overset{\overset{\displaystyle O}{\|}}{C}-OH$

Offset: 2.5 ppm.

C

Aldrich 25,927-6   *CAS [13991-37-2]*
trans-**2-Pentenoic acid, 99%**

$C_5H_8O_2$
FW 100.12
bp 106°C (20 mm)
d 0.990
n_D^{20}1.4509

Fp 216°F

FT-IR: *1*, 499A
VP-FT-IR: *3*, 583D

172.57
153.62*
119.82*
25.46
12.01*

CDCl$_3$ QE-300

CH_3CH_2 ... $\overset{\overset{\displaystyle O}{\|}}{C}-OH$

Offset: 2.4 ppm.

Non-Aromatic Carboxylic Acids

Aldrich 24,592-5
4-Pentenoic acid, 97%

CAS [591-80-0]

$C_5H_8O_2$
FW 100.12
mp -23°C
bp 84°C (12 mm)
d 0.981

n_D^{20} 1.4283
Fp 193°F
60 MHz: *1*, 435C
FT-IR: *1*, 499B
VP-FT-IR: *3*, 584A

179.80
136.24*
115.70
33.38
28.50

A

CDCl₃   QE-300

Offset: 1.8 ppm.

$H_2C=CHCH_2CH_2-\overset{\displaystyle O}{\overset{\|}{C}}-OH$

Aldrich 26,477-6
***trans*-2-Methyl-2-pentenoic acid, 99%**

CAS [16957-70-3]

$C_6H_{10}O_2$
FW 114.15
mp 27°C
bp 124°C (30 mm)
Fp 226°F

VP-FT-IR: *3*, 585A

174.17
146.68*
126.60
22.24
12.90*
11.80*

B

CDCl₃   QE-300

Offset: 2.7 ppm.

CH_3CH_2 ... CH_3 ... $C-OH$

Aldrich 30,513-8
2,2-Dimethyl-4-pentenoic acid, 95%

CAS [16386-93-9]

$C_7H_{12}O_2$
FW 128.17
bp 106°C (20 mm)
d 0.933
n_D^{20} 1.4310

Fp 191°F

VP-FT-IR: *3*, 585C

184.64
133.82*
118.18
44.40
42.17
24.55*

C

CDCl₃   QE-300

$H_2C=CHCH_2\overset{\displaystyle CH_3}{\underset{\displaystyle CH_3}{C}}-\overset{\displaystyle O}{\overset{\|}{C}}-OH$

Non-Aromatic Carboxylic Acids

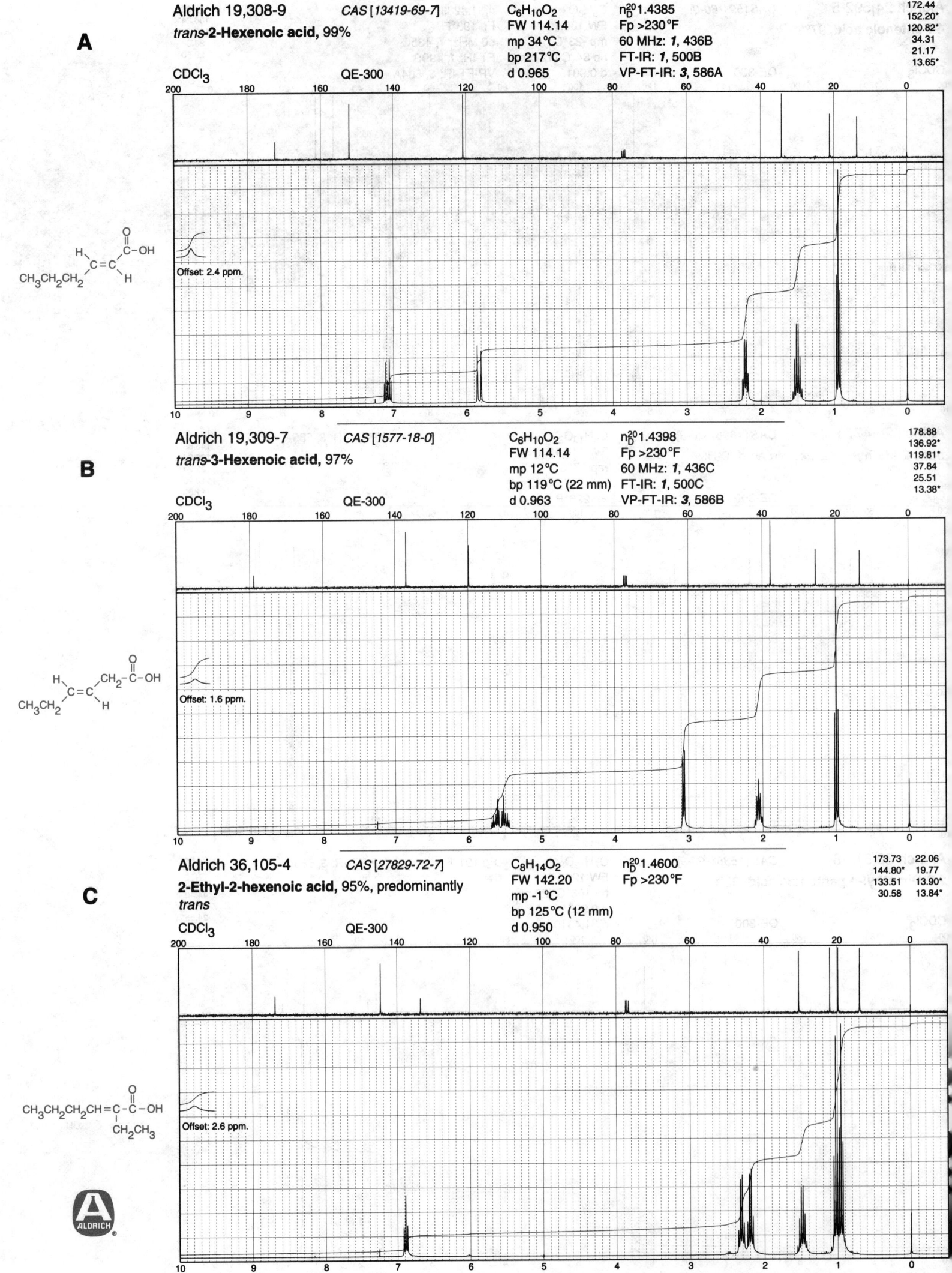

A

Aldrich 19,308-9   *CAS [13419-69-7]*   C₆H₁₀O₂   n²⁰_D 1.4385

$C_6H_{10}O_2$ n_D^{20} 1.4385

trans-2-Hexenoic acid, 99%

FW 114.14   Fp >230°F
mp 34°C   60 MHz: *1*, 436B
bp 217°C   FT-IR: *1*, 500B
d 0.965   VP-FT-IR: *3*, 586A

172.44
152.20*
120.82*
34.31
21.17
13.65*

CDCl₃   QE-300

Offset: 2.4 ppm.

B

Aldrich 19,309-7   *CAS [1577-18-0]* $C_6H_{10}O_2$ n_D^{20} 1.4398

trans-3-Hexenoic acid, 97%

FW 114.14   Fp >230°F
mp 12°C   60 MHz: *1*, 436C
bp 119°C (22 mm)   FT-IR: *1*, 500C
d 0.963   VP-FT-IR: *3*, 586B

178.88
136.92*
119.81*
37.84
25.51
13.38*

CDCl₃   QE-300

Offset: 1.6 ppm.

C

Aldrich 36,105-4   *CAS [27829-72-7]* $C_8H_{14}O_2$ n_D^{20} 1.4600

2-Ethyl-2-hexenoic acid, 95%, predominantly
trans

FW 142.20   Fp >230°F
mp -1°C
bp 125°C (12 mm)
d 0.950

173.73   22.06
144.80*   19.77
133.51   13.90*
30.58   13.84*

CDCl₃   QE-300

Offset: 2.6 ppm.

ALDRICH

Non-Aromatic Carboxylic Acids

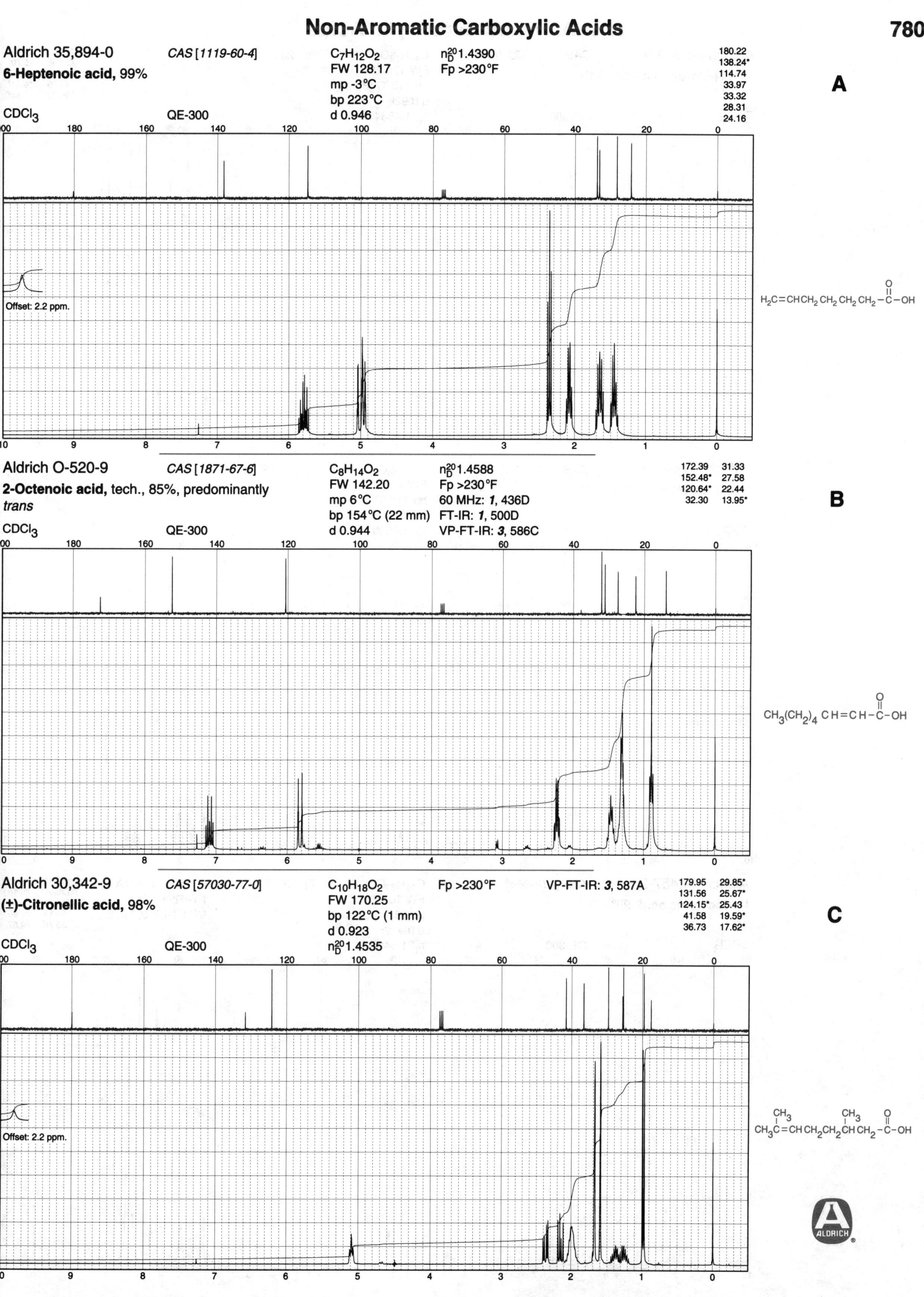

Aldrich 35,894-0   *CAS [1119-60-4]*   C₇H₁₂O₂   n²⁰_D 1.4390

6-Heptenoic acid, 99%

FW 128.17   Fp >230°F
mp -3°C
bp 223°C
d 0.946

CDCl₃   QE-300

180.22
138.24*
114.74
33.97
33.32
28.31
24.16

A

Offset: 2.2 ppm.

H₂C=CHCH₂·CH₂·CH₂·CH₂–C–OH (O)

Aldrich O-520-9   *CAS [1871-67-6]*   C₈H₁₄O₂   n²⁰_D 1.4588

2-Octenoic acid, tech., 85%, predominantly *trans*

FW 142.20   Fp >230°F
mp 6°C   60 MHz: *1*, 436D
bp 154°C (22 mm)   FT-IR: *1*, 500D
d 0.944   VP-FT-IR: *3*, 586C

CDCl₃   QE-300

172.39   31.33
152.48* 27.58
120.64* 22.44
32.30   13.95*

B

CH₃(CH₂)₄ CH=CH–C–OH (O)

Aldrich 30,342-9   *CAS [57030-77-0]*   C₁₀H₁₈O₂   Fp >230°F   VP-FT-IR: *3*, 587A

(±)-Citronellic acid, 98%

FW 170.25
bp 122°C (1 mm)
d 0.923
n²⁰_D 1.4535

CDCl₃   QE-300

179.95   29.85*
131.56   25.67*
124.15* 25.43
41.58   19.59*
36.73   17.62*

C

Offset: 2.2 ppm.

CH₃C=CHCH₂CH₂CH₂CHCH₂–C–OH (O) with CH₃ and CH₃ groups

Non-Aromatic Carboxylic Acids

A

Aldrich 34,049-9 *CAS [18951-85-4]* $C_{10}H_{18}O_2$ Fp >230°F

(*R*)-(+)-Citronellic acid, 98%

FW 170.25
bp 119°C (3 mm)
d 0.926
n_D^{20} 1.4534

180.05	29.82*
131.58	25.70*
124.12*	25.41
41.59	19.58*
36.70	17.64*

CDCl$_3$ QE-300

B

Aldrich 36,442-8 *CAS [2111-53-7]* $C_{10}H_{18}O_2$ Fp >230°F

(*S*)-(-)-Citronellic acid, 98%

FW 170.25
bp 118°C
d 0.926
n_D^{20} 1.4530

180.14	29.84*
131.53	25.70*
124.17*	25.45
41.62	19.60*
36.74	17.63*

CDCl$_3$ QE-300

Offset: 2.2 ppm.

C

Aldrich 12,467-2 *CAS [112-38-9]* $C_{11}H_{20}O_2$ Fp 300°F

Undecylenic acid, 98%

FW 184.28
bp 137°C (2 mm)
d 0.912
n_D^{20} 1.4493

60 MHz: *1*, 437A
FT-IR: *1*, 501A
VP-FT-IR: *3*, 587B

180.48	29.26
139.07*	29.18
114.11	29.05
34.14	28.90
33.78	24.67

CDCl$_3$ QE-300

Offset: 1.8 ppm.

Non-Aromatic Carboxylic Acids

Aldrich 28,673-7
Myristoleic acid, 98%

CAS [544-64-9]

$C_{14}H_{26}O_2$
FW 226.26
mp -4°C
bp 144°C
n_D^{20} 1.4562

Fp 144°F

VP-FT-IR: **3**, 587C

180.08	29.08
129.95*	27.19
129.72*	26.96
34.11	24.73
32.01	22.36
29.71	13.95*
29.15	

A

CDCl₃ QE-300

CH₃(CH₂)₃CH=CH(CH₂)₆CH₂—C(=O)—OH

Aldrich 28,692-3
Palmitoleic acid, 98%

CAS [373-49-9]

$C_{16}H_{30}O_2$
FW 254.42
mp 1°C
bp 162°C
n_D^{20} 1.4582

Fp 144°F

VP-FT-IR: **3**, 587D

180.47	29.69	27.16
129.96*	29.15	24.67
129.68*	29.07	22.68
34.12	29.04	14.11*
31.80	29.01	
29.75	27.23	

B

CDCl₃ QE-300

CH₃(CH₂)₄CH₂—CH₂(CH₂)₅CH₂—C(=O)—OH C=C

Aldrich 26,804-6
Oleic acid, 99+%

CAS [112-80-1]

$C_{18}H_{34}O_2$
FW 282.47
mp 13°C
bp 195°C (1 mm)
d 0.891

n_D^{20} 1.4595
Fp >230°F
VP-FT-IR: **3**, 588A

180.47	29.69	27.16
129.97*	29.54	24.67
129.67*	29.34	22.70
34.12	29.16	14.12*
31.92	29.06	
29.78	27.23	

C

CDCl₃ QE-300

CH₃(CH₂)₆CH₂—CH₂(CH₂)₅CH₂—C(=O)—OH C=C

ALDRICH

Non-Aromatic Carboxylic Acids

A

Aldrich E30-4

Elaidic acid, 98%

CAS [112-79-8]

$C_{18}H_{34}O_2$
FW 282.47
mp 44°C
bp 288°C (100 mm)
Fp >230°F

CDCl$_3$ QE-300

60 MHz: *1*, 437B	180.42	31.93	29.11
FT-IR: *1*, 501B	130.46*	29.68	29.05
VP-FT-IR: *3*, 588B	130.14*	29.57	28.93
	34.12	29.51	24.68
	32.62	29.34	22.70
	32.56	29.20	14.11*

B

Aldrich 28,743-1

trans-**Vaccenic acid, 98%**

CAS [693-72-1]

$C_{18}H_{34}O_2$
FW 282.47
mp 45°C
Fp >230°F

CDCl$_3$ QE-300

180.42	29.64	28.85
130.36*	29.45	24.69
130.26*	29.41	22.67
34.13	29.24	14.09*
32.61	29.12	
31.78	29.08	

C

Aldrich 28,571-4

cis-**11-Eicosenoic acid, 98%**

CAS [5561-99-9]

$C_{20}H_{38}O_2$
FW 310.53
mp 24°C
d 0.883
n$_D^{20}$ 1.4606

Fp >230°F

CDCl$_3$ QE-300

180.47	29.55	27.22
129.88*	29.49	24.68
129.78*	29.42	22.71
34.13	29.34	14.12*
31.93	29.27	
29.77	29.08	

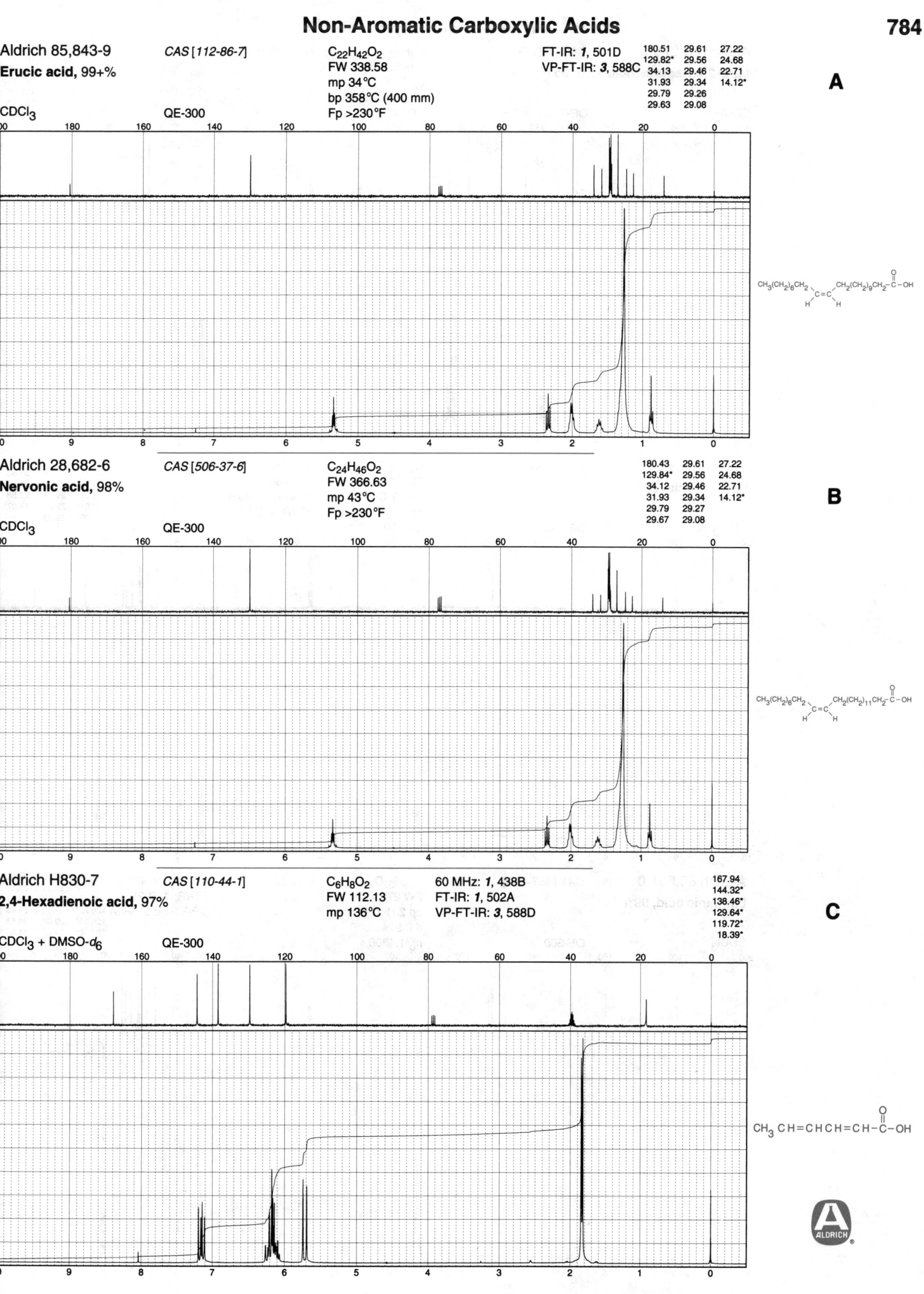

Aldrich 85,843-9 *CAS [112-86-7]* **Erucic acid, 99+%**

$C_{22}H_{42}O_2$
FW 338.58
mp 34°C
bp 358°C (400 mm)
Fp >230°F

FT-IR: *1*, 501D
VP-FT-IR: *3*, 588C

180.51	29.61	27.22
129.82*	29.56	24.68
34.13	29.46	22.71
31.93	29.34	14.12*
29.79	29.26	
29.63	29.08	

A

CDCl₃   QE-300

Aldrich 28,682-6 *CAS [506-37-6]* **Nervonic acid, 98%**

$C_{24}H_{46}O_2$
FW 366.63
mp 43°C
Fp >230°F

180.43	29.61	27.22
129.84*	29.56	24.68
34.12	29.46	22.71
31.93	29.34	14.12*
29.79	29.27	
29.67	29.08	

B

CDCl₃   QE-300

Aldrich H830-7 *CAS [110-44-1]* **2,4-Hexadienoic acid, 97%**

$C_6H_8O_2$
FW 112.13
mp 136°C

60 MHz: *1*, 438B
FT-IR: *1*, 502A
VP-FT-IR: *3*, 588D

167.94
144.32*
138.46*
129.64*
119.72*
18.39*

C

CDCl₃ + DMSO-d_6 QE-300

A

Aldrich 33,276-3   CAS *[38867-17-3]*   C₇H₁₀O₂

trans-**2,6-Heptadienoic acid**, 95%

$C_7H_{10}O_2$
FW 126.16
bp 79 °C (2 mm)
d 0.971
n_D^{20} 1.4730

Fp >230 °F

172.24*
151.23*
136.80*
121.11*
115.66
31.87
31.53

CDCl₃   QE-300

Offset: 2.7 ppm.

B

Aldrich 23,392-7   *CAS [60-33-3]*

Linoleic acid, 99%

$C_{18}H_{32}O_2$
FW 280.46
mp -5 °C
bp 230 °C (16 mm)
d 0.902

n_D^{20} 1.4697
Fp >230 °F
60 MHz: *1*, 437C
FT-IR: *1*, 501C

180.07	31.53	27.19
130.16*	29.58	25.63
129.96*	29.35	24.66
128.01*	29.14	22.58
127.84*	29.07	14.07*
34.05	29.03	

CDCl₃   QE-300

C

Aldrich 85,601-0   *CAS [463-40-1]*

Linolenic acid, 99%

$C_{18}H_{30}O_2$
FW 278.44
bp 231 °C (1 mm)
d 0.914
n_D^{20} 1.4800

Fp >230 °F

60 MHz: *1*, 437D
FT-IR: *1*, 502B
VP-FT-IR: *3*, 589A

180.47	127.06*	27.20
131.88*	34.11	25.61
130.17*	29.57	25.54
128.22*	29.14	24.65
128.19*	29.07	20.56
127.70*	29.03	14.28*

CDCl₃   QE-300

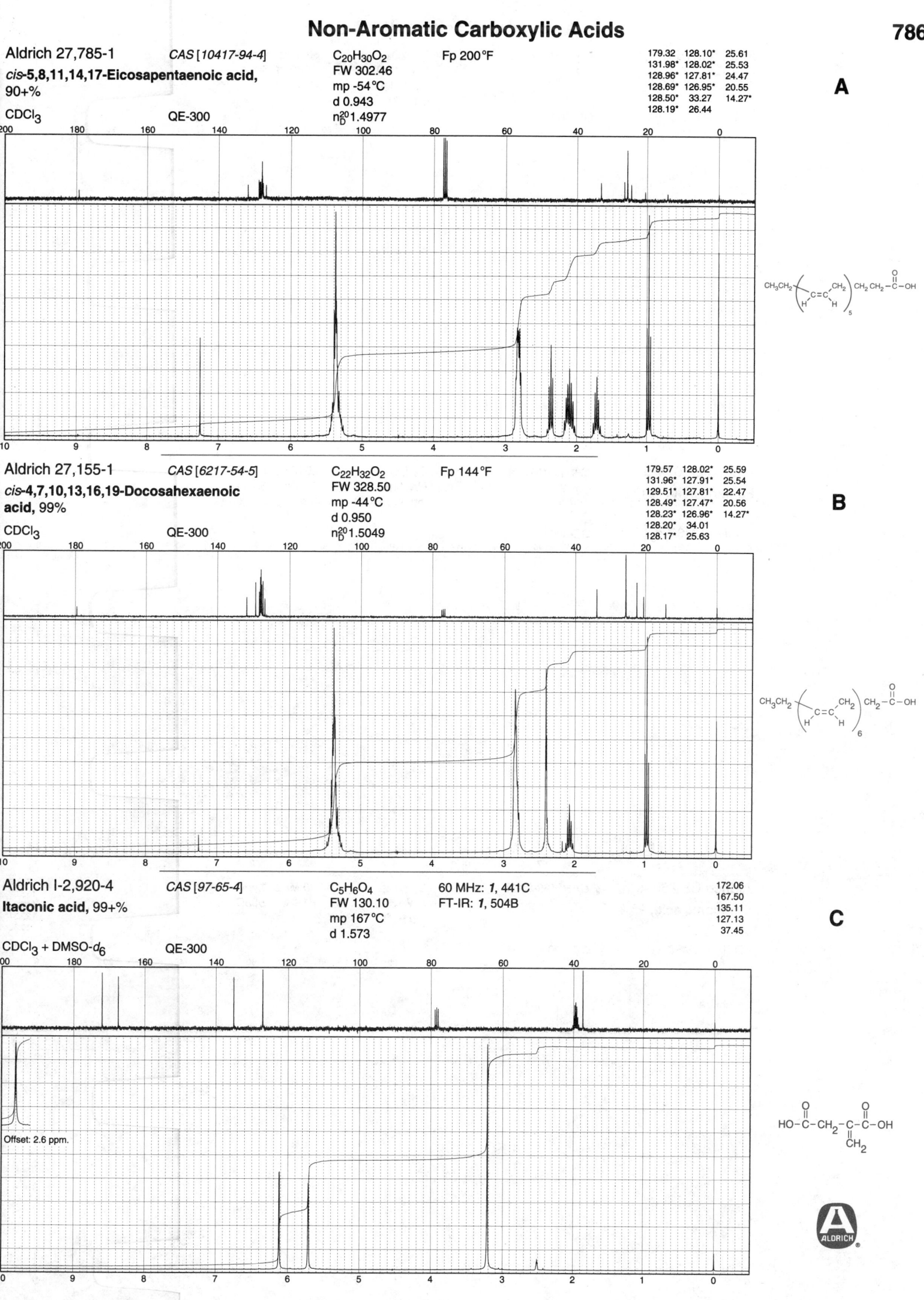

Aldrich 27,785-1   *CAS [10417-94-4]* $C_{20}H_{30}O_2$ Fp 200°F

cis-**5,8,11,14,17-Eicosapentaenoic acid,** 90+%

$CDCl_3$ QE-300

FW 302.46
mp -54°C
d 0.943
n_D^{20} 1.4977

179.32	128.10*	25.61
131.98*	128.02*	25.53
128.96*	127.81*	24.47
128.69*	126.95*	20.55
128.50*	33.27	14.27*
128.19*	26.44	

A

Aldrich 27,155-1   *CAS [6217-54-5]* $C_{22}H_{32}O_2$ Fp 144°F

cis-**4,7,10,13,16,19-Docosahexaenoic acid,** 99%

$CDCl_3$ QE-300

FW 328.50
mp -44°C
d 0.950
n_D^{20} 1.5049

179.57	128.02*	25.59
131.96*	127.91*	25.54
129.51*	127.81*	24.47
128.49*	127.47*	20.56
128.23*	126.96*	14.27*
128.20*	34.01	
128.17*	25.63	

B

Aldrich I-2,920-4   *CAS [97-65-4]* $C_5H_6O_4$ 60 MHz: *1*, 441C

Itaconic acid, 99+%

$CDCl_3$ + DMSO-d_6 QE-300

FW 130.10
mp 167°C
d 1.573

FT-IR: *1*, 504B

172.06
167.50
135.11
127.13
37.45

C

Offset: 2.6 ppm.

ALDRICH

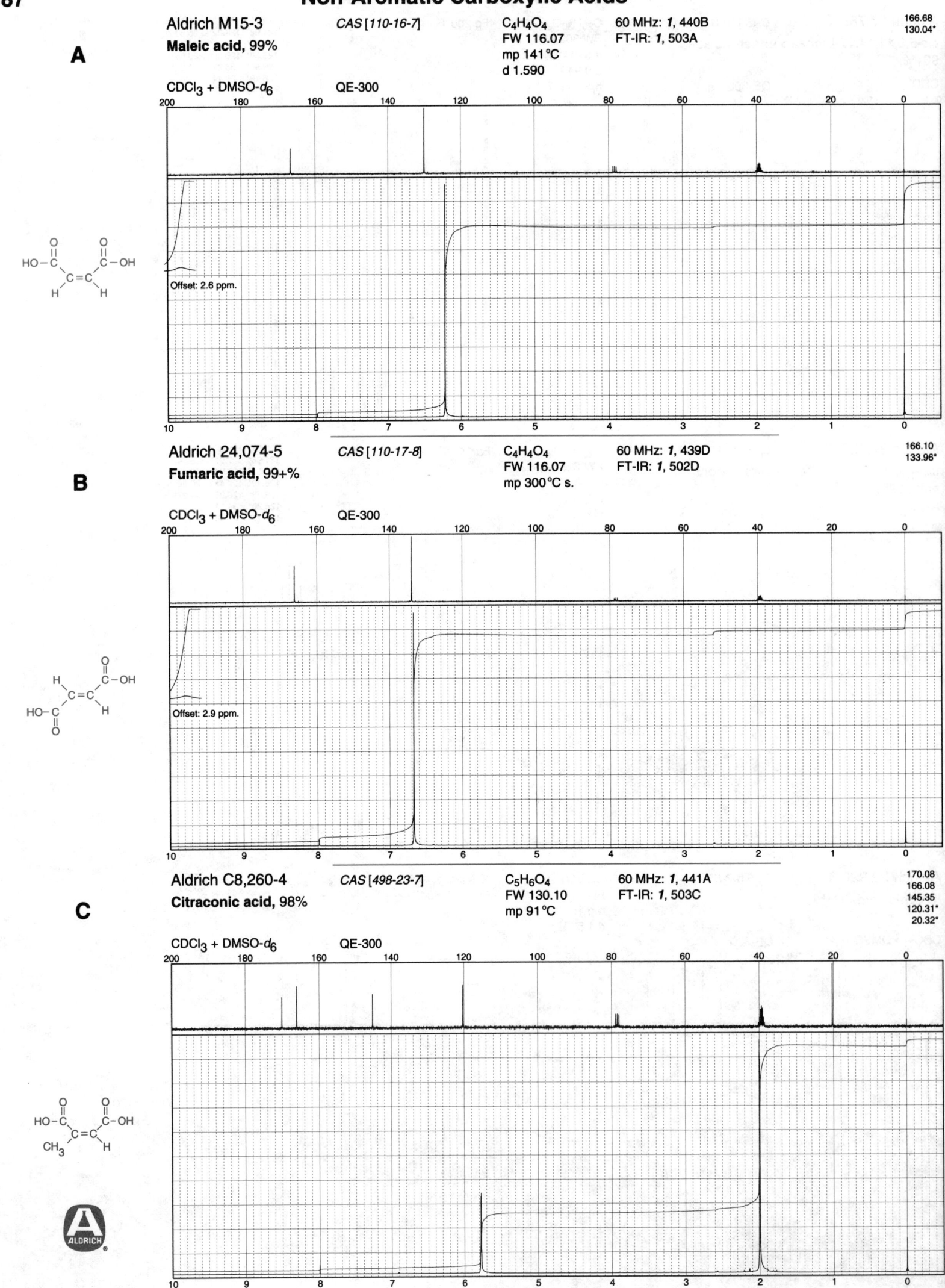

Aldrich M15-3

Maleic acid, 99%

CAS [110-16-7]

$C_4H_4O_4$
FW 116.07
mp 141°C
d 1.590

60 MHz: *1*, 440B
FT-IR: *1*, 503A

166.68
130.04*

A

CDCl$_3$ + DMSO-d_6 QE-300

Offset: 2.6 ppm.

Aldrich 24,074-5

Fumaric acid, 99+%

CAS [110-17-8]

$C_4H_4O_4$
FW 116.07
mp 300°C s.

60 MHz: *1*, 439D
FT-IR: *1*, 502D

166.10
133.96*

B

CDCl$_3$ + DMSO-d_6 QE-300

Offset: 2.9 ppm.

Aldrich C8,260-4

Citraconic acid, 98%

CAS [498-23-7]

$C_5H_6O_4$
FW 130.10
mp 91°C

60 MHz: *1*, 441A
FT-IR: *1*, 503C

170.08
166.08
145.35
120.31*
20.32*

C

CDCl$_3$ + DMSO-d_6 QE-300

ALDRICH

Non-Aromatic Carboxylic Acids

788

Aldrich 13,104-0 *CAS [498-24-8]* $C_5H_6O_4$ 60 MHz: *1*, 440D

Mesaconic acid, 99% FW 130.10 FT-IR: *1*, 503B

mp 201°C

A

168.39
167.11
142.96
126.97*
13.96*

$CDCl_3$ + DMSO-d_6 QE-300

Aldrich G260-5 *CAS [628-48-8]* $C_5H_6O_4$

trans-**Glutaconic acid** FW 130.10

60 MHz: *1*, 441B

FT-IR: *1*, 503D

B

171.47
166.77
141.10*
124.37*
36.64

DMSO-d_6 QE-300

Offset: 2.6 ppm.

Aldrich H1,785-6 *CAS [4436-74-2]* $C_6H_8O_4$ 60 MHz: *1*, 441D

trans-β-**Hydromuconic acid,** 98% FW 144.13 FT-IR: *1*, 504C

mp 196°C

C

172.52
125.82*
37.30

DMSO-d_6 QE-300

Offset: 2.4 ppm.

ALDRICH

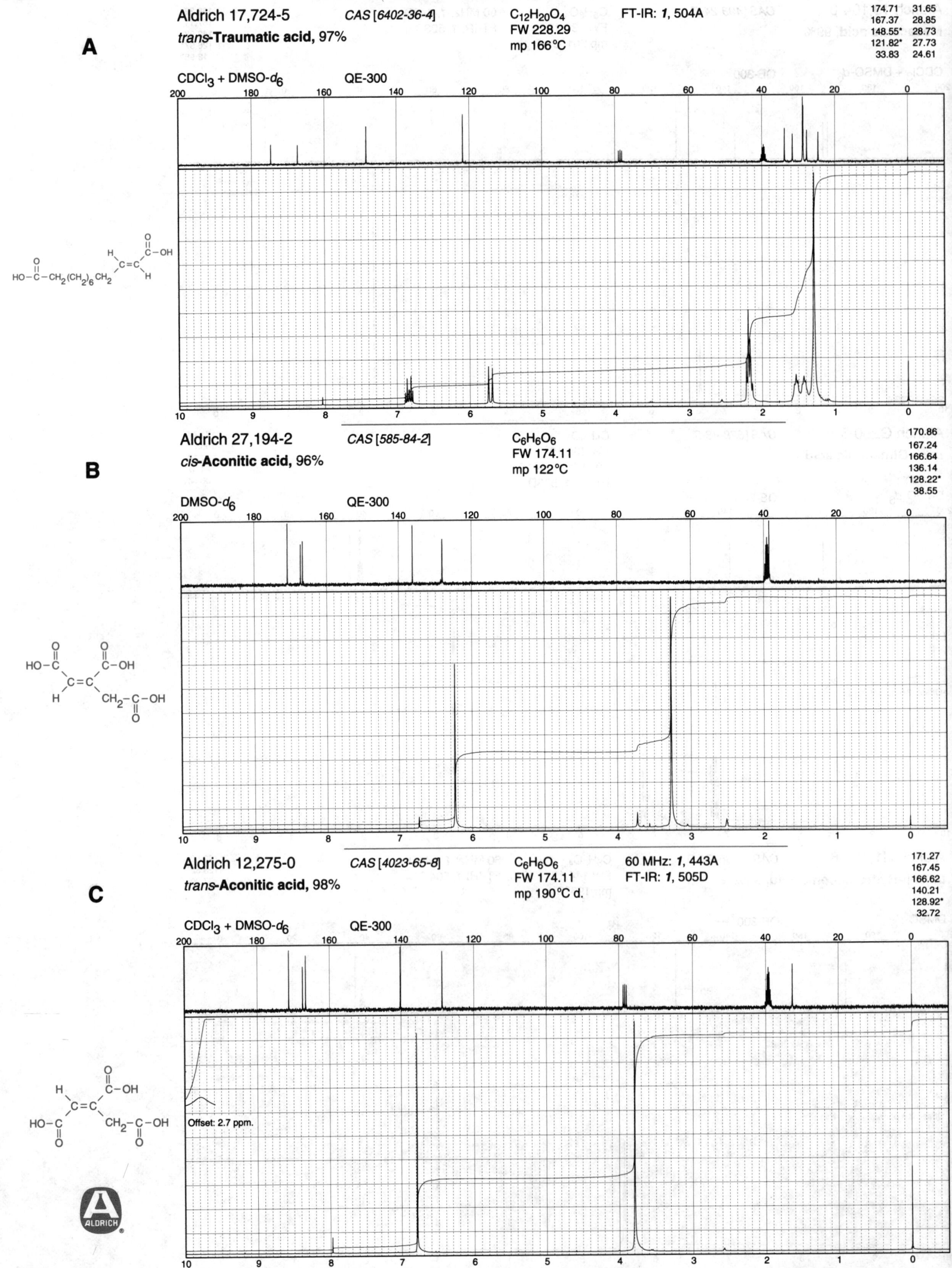

A

Aldrich 17,724-5   CAS [6402-36-4] $C_{12}H_{20}O_4$ FT-IR: *1*, 504A

trans-Traumatic acid, 97%   FW 228.29
mp 166°C

174.71 31.65
167.37 28.85
148.55* 28.73
121.82* 27.73
33.83 24.61

$CDCl_3$ + DMSO-d_6 QE-300

B

Aldrich 27,194-2   CAS [585-84-2] $C_6H_6O_6$

cis-Aconitic acid, 96%   FW 174.11
mp 122°C

170.86
167.24
166.64
136.14
128.22*
38.55

DMSO-d_6 QE-300

C

Aldrich 12,275-0   CAS [4023-65-8] $C_6H_6O_6$ 60 MHz: *1*, 443A

trans-Aconitic acid, 98%   FW 174.11   FT-IR: *1*, 505D
mp 190°C d.

171.27
167.45
166.62
140.21
128.92*
32.72

$CDCl_3$ + DMSO-d_6 QE-300

Offset: 2.7 ppm.

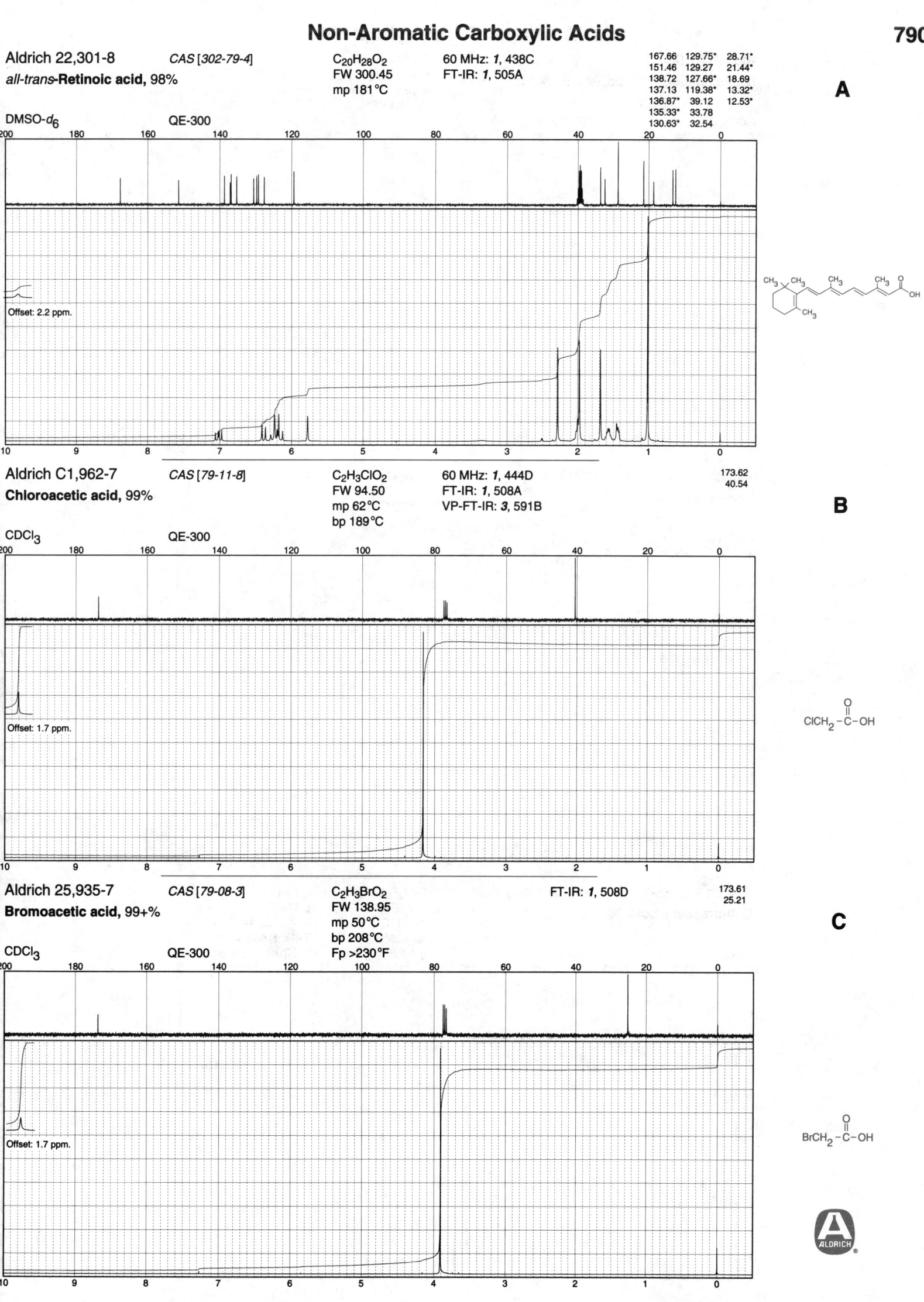

A

Aldrich 22,301-8   CAS [302-79-4] $C_{20}H_{28}O_2$ 60 MHz: *1*, 438C
all-trans-Retinoic acid, 98%   FW 300.45   FT-IR: *1*, 505A
                                              mp 181°C

167.66	129.75*	28.71*
151.46	129.27	21.44*
138.72	127.66*	18.69
137.13	119.38*	13.32*
136.87*	39.12	12.53*
135.33*	33.78	
130.63*	32.54	

DMSO-d_6 QE-300

Offset: 2.2 ppm.

B

Aldrich C1,962-7   CAS [79-11-8] $C_2H_3ClO_2$ 60 MHz: *1*, 444D
Chloroacetic acid, 99%   FW 94.50   FT-IR: *1*, 508A
                                         mp 62°C   VP-FT-IR: *3*, 591B
                                         bp 189°C

173.62
40.54

CDCl$_3$ QE-300

Offset: 1.7 ppm.

C

Aldrich 25,935-7   CAS [79-08-3] $C_2H_3BrO_2$ FT-IR: *1*, 508D
Bromoacetic acid, 99+%   FW 138.95
                                         mp 50°C
                                         bp 208°C
                                         Fp >230°F

173.61
25.21

CDCl$_3$ QE-300

Non-Aromatic Carboxylic Acids

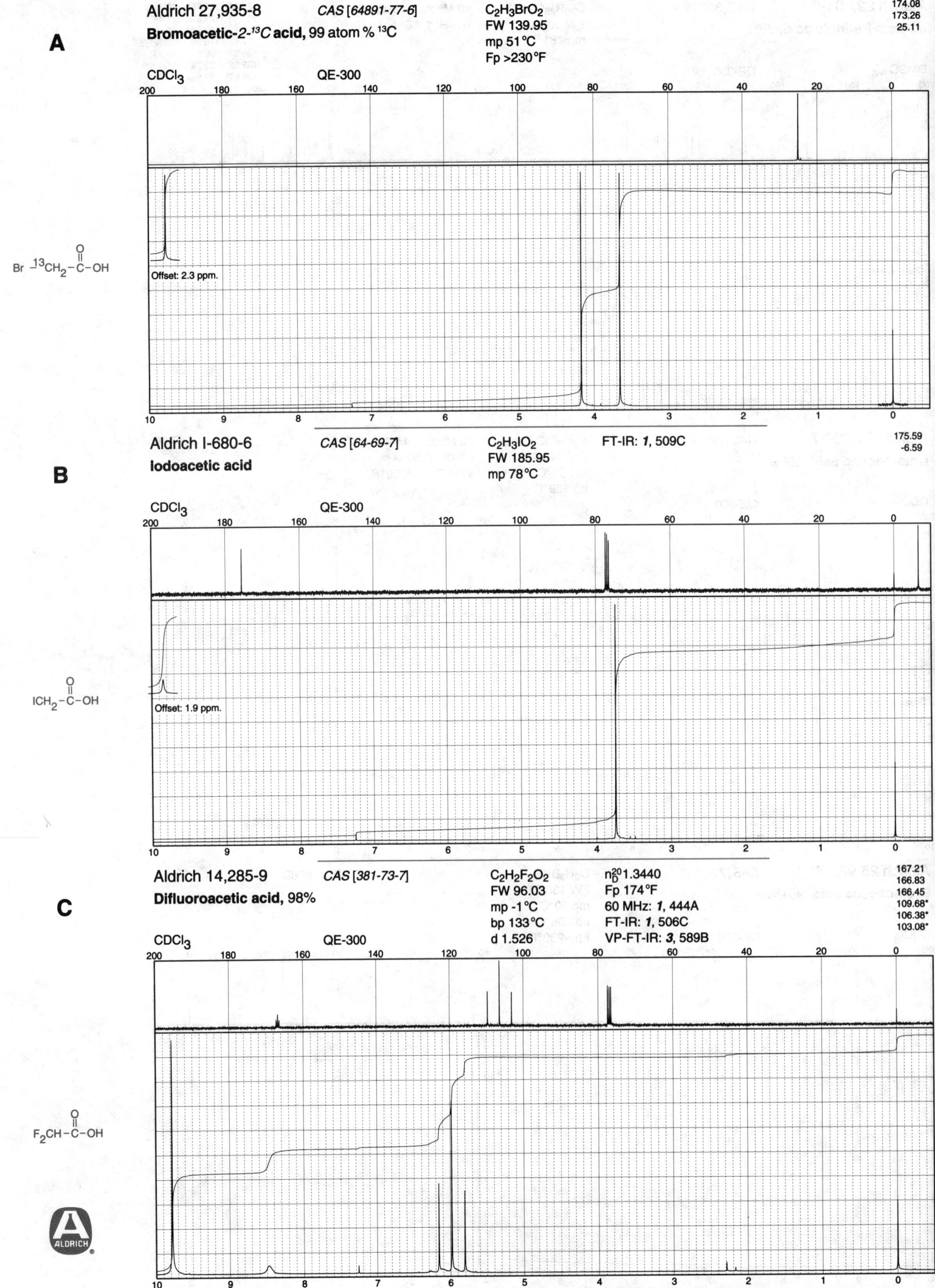

A

Aldrich 27,935-8   *CAS [64891-77-6]* $C_2H_3BrO_2$ 174.08
Bromoacetic-2-^{13}C acid, 99 atom % ^{13}C   FW 139.95   173.26
mp 51 °C   25.11
Fp >230 °F

CDCl$_3$ QE-300

Br —^{13}CH$_2$—C—OH
        ‖
        O

Offset: 2.3 ppm.

Aldrich I-680-6   *CAS [64-69-7]* $C_2H_3IO_2$ FT-IR: *1*, 509C   175.59
Iodoacetic acid   FW 185.95   -6.59
mp 78 °C

B

CDCl$_3$ QE-300

ICH$_2$—C—OH
        ‖
        O

Offset: 1.9 ppm.

Aldrich 14,285-9   *CAS [381-73-7]* $C_2H_2F_2O_2$ n_D^{20} 1.3440   167.21
Difluoroacetic acid, 98%   FW 96.03   Fp 174 °F   166.83
mp -1 °C   60 MHz: *1*, 444A   166.45
bp 133 °C   FT-IR: *1*, 506C   109.68*
d 1.526   VP-FT-IR: *3*, 589B   106.38*
103.08*

C

CDCl$_3$ QE-300

F$_2$CH—C—OH
        ‖
        O

ALDRICH

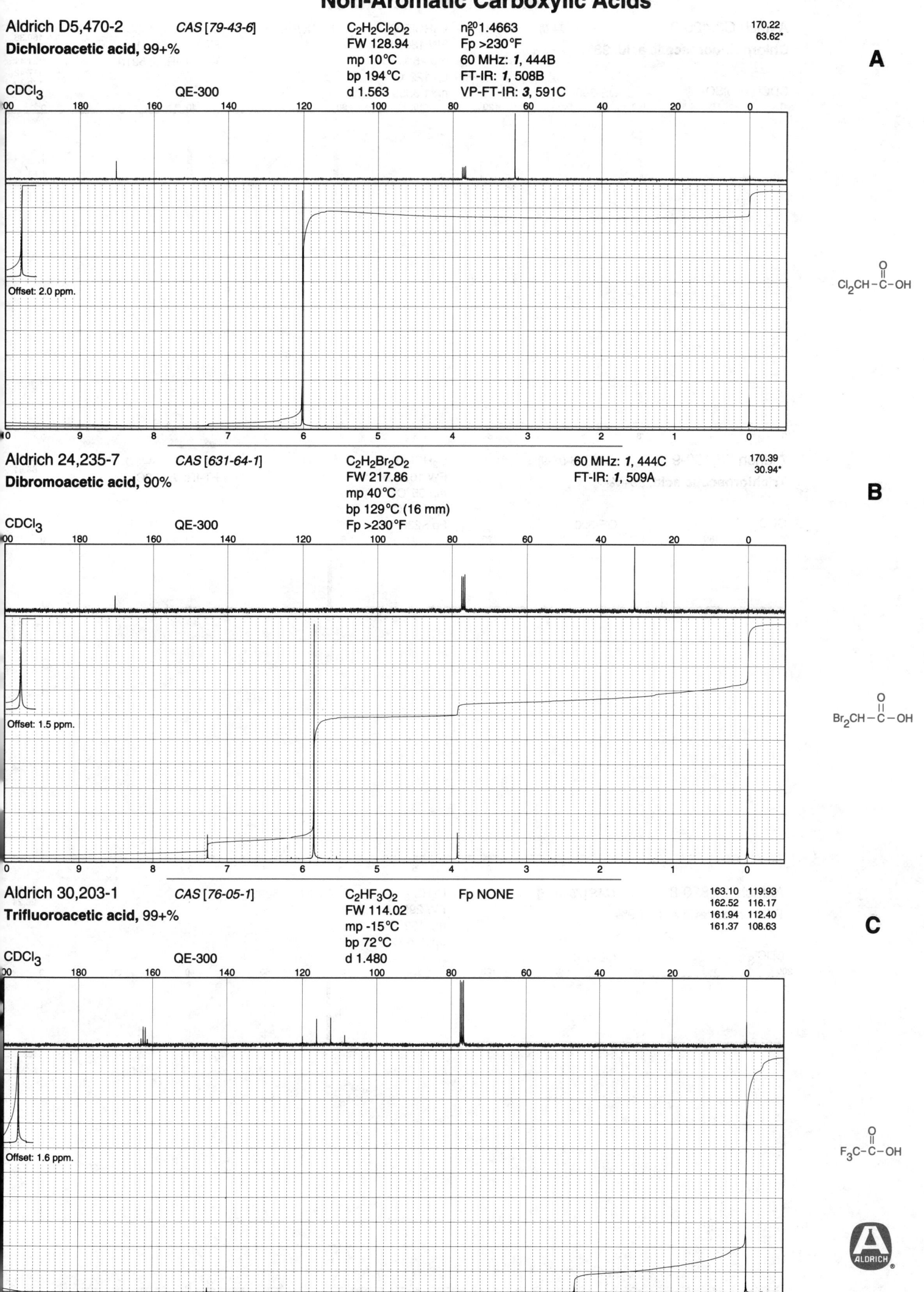

Aldrich D5,470-2 *CAS [79-43-6]*
Dichloroacetic acid, 99+%

$C_2H_2Cl_2O_2$
FW 128.94
mp 10°C
bp 194°C
d 1.563

n_D^{20} 1.4663
Fp >230°F
60 MHz: *1*, 444B
FT-IR: *1*, 508B
VP-FT-IR: *3*, 591C

170.22
63.62*

A

CDCl₃   QE-300

Offset: 2.0 ppm.

$Cl_2CH-\overset{\overset{\displaystyle O}{\|}}{C}-OH$

Aldrich 24,235-7 *CAS [631-64-1]*
Dibromoacetic acid, 90%

$C_2H_2Br_2O_2$
FW 217.86
mp 40°C
bp 129°C (16 mm)
Fp >230°F

60 MHz: *1*, 444C
FT-IR: *1*, 509A

170.39
30.94*

B

CDCl₃   QE-300

Offset: 1.5 ppm.

$Br_2CH-\overset{\overset{\displaystyle O}{\|}}{C}-OH$

Aldrich 30,203-1 *CAS [76-05-1]*
Trifluoroacetic acid, 99+%

$C_2HF_3O_2$
FW 114.02
mp -15°C
bp 72°C
d 1.480

Fp NONE

163.10 119.93
162.52 116.17
161.94 112.40
161.37 108.63

C

CDCl₃   QE-300

Offset: 1.6 ppm.

$F_3C-\overset{\overset{\displaystyle O}{\|}}{C}-OH$

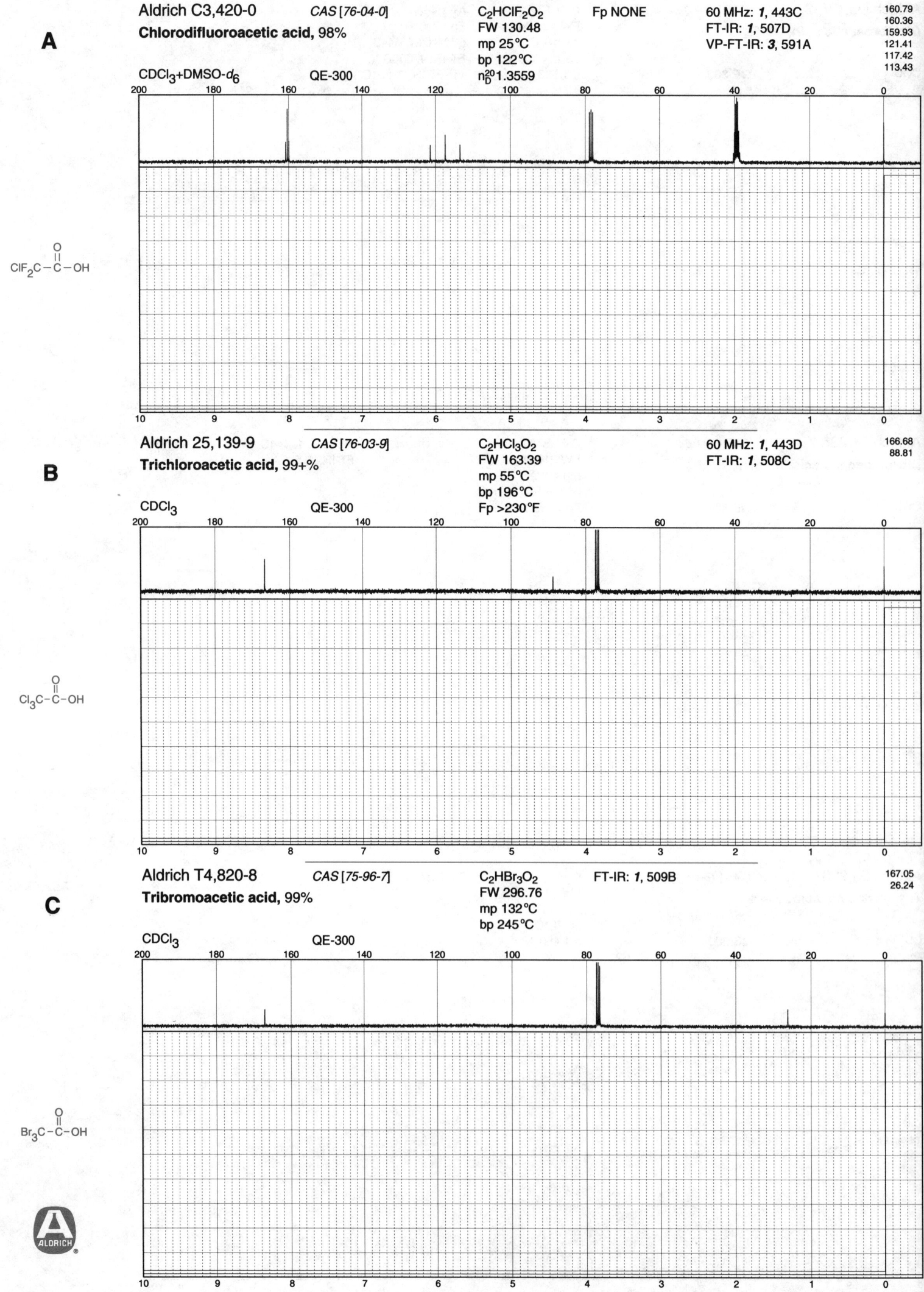

A

Aldrich C3,420-0   *CAS [76-04-0]* $C_2HClF_2O_2$ Fp NONE   60 MHz: *1*, 443C

Chlorodifluoroacetic acid, 98%   FW 130.48   FT-IR: *1*, 507D

mp 25°C   VP-FT-IR: *3*, 591A

bp 122°C

$CDCl_3$+DMSO-d_6 QE-300 n_D^{20}1.3559

160.79
160.36
159.93
121.41
117.42
113.43

$ClF_2C-\overset{\overset{\displaystyle O}{\|}}{C}-OH$

B

Aldrich 25,139-9   *CAS [76-03-9]* $C_2HCl_3O_2$ 60 MHz: *1*, 443D

Trichloroacetic acid, 99+%   FW 163.39   FT-IR: *1*, 508C

mp 55°C

bp 196°C

$CDCl_3$ QE-300   Fp >230°F

166.68
88.81

$Cl_3C-\overset{\overset{\displaystyle O}{\|}}{C}-OH$

C

Aldrich T4,820-8   *CAS [75-96-7]* $C_2HBr_3O_2$ FT-IR: *1*, 509B

Tribromoacetic acid, 99%   FW 296.76

mp 132°C

bp 245°C

$CDCl_3$ QE-300

167.05
26.24

$Br_3C-\overset{\overset{\displaystyle O}{\|}}{C}-OH$

ALDRICH

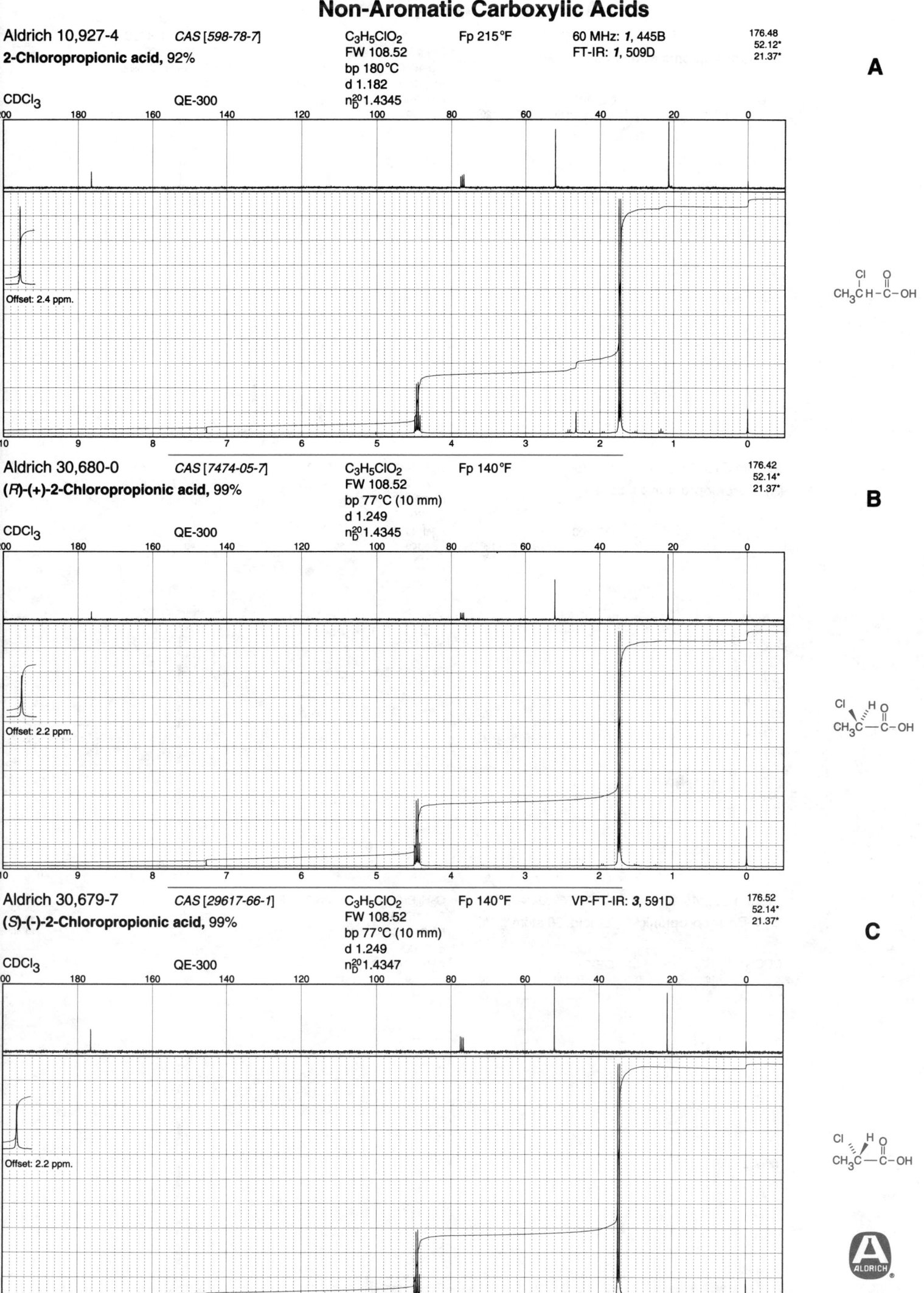

Aldrich 10,927-4 *CAS [598-78-7]* C₃H₅ClO₂ Fp 215°F 60 MHz: *1*, 445B 176.48
2-Chloropropionic acid, 92% FW 108.52 FT-IR: *1*, 509D 52.12*
bp 180°C 21.37*
CDCl₃ QE-300 d 1.182
n²⁰_D 1.4345

A

Offset: 2.4 ppm.

CH₃CHC—OH (Cl, O)

Aldrich 30,680-0 *CAS [7474-05-7]* C₃H₅ClO₂ Fp 140°F 176.42
(R)-(+)-2-Chloropropionic acid, 99% FW 108.52 52.14*
bp 77°C (10 mm) 21.37*
CDCl₃ QE-300 d 1.249
n²⁰_D 1.4345

B

Offset: 2.2 ppm.

CH₃C—C—OH (Cl, H, O)

Aldrich 30,679-7 *CAS [29617-66-1]* C₃H₅ClO₂ Fp 140°F VP-FT-IR: *3*, 591D 176.52
(S)-(-)-2-Chloropropionic acid, 99% FW 108.52 52.14*
bp 77°C (10 mm) 21.37*
CDCl₃ QE-300 d 1.249
n²⁰_D 1.4347

C

Offset: 2.2 ppm.

CH₃C—C—OH (Cl, H, O)

ALDRICH

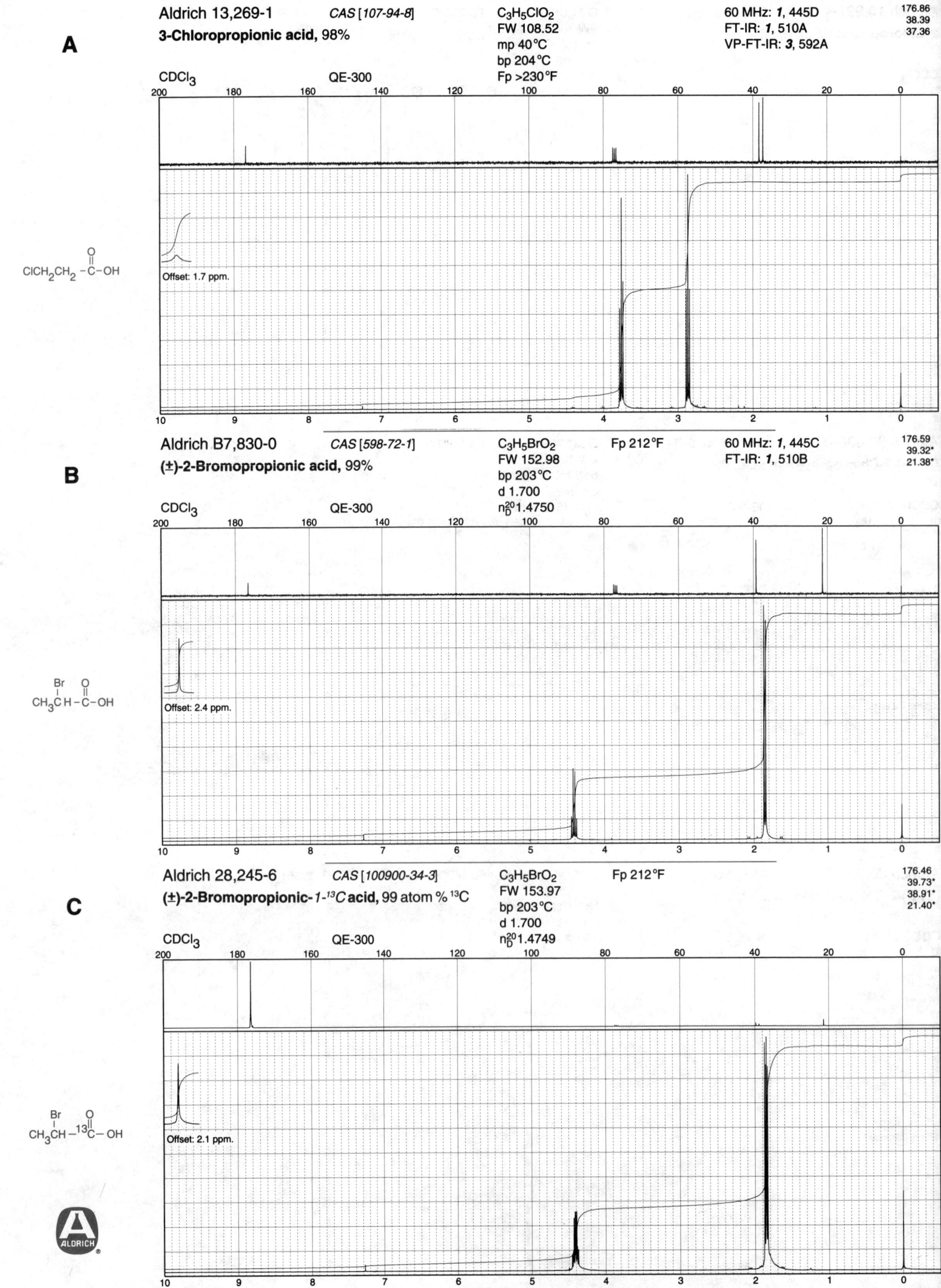

A

Aldrich 13,269-1   *CAS [107-94-8]* $C_3H_5ClO_2$ 60 MHz: *1*, 445D   176.86
                                                                                                   38.39
3-Chloropropionic acid, 98%   FW 108.52   FT-IR: *1*, 510A   37.36
                                                                mp 40°C   VP-FT-IR: *3*, 592A

CDCl₃   QE-300   bp 204°C
                                                     Fp >230°F

Offset: 1.7 ppm.

$ClCH_2CH_2-\overset{\overset{\displaystyle O}{\|}}{C}-OH$

B

Aldrich B7,830-0   *CAS [598-72-1]* $C_3H_5BrO_2$ Fp 212°F   60 MHz: *1*, 445C   176.59
                                                                                                             39.32*
(±)-2-Bromopropionic acid, 99%   FW 152.98   FT-IR: *1*, 510B   21.38*
                                                     bp 203°C
                                                     d 1.700
CDCl₃   QE-300 n_D^{20}1.4750

Offset: 2.4 ppm.

$CH_3\overset{\overset{\displaystyle Br}{|}}{C}H-\overset{\overset{\displaystyle O}{\|}}{C}-OH$

C

Aldrich 28,245-6   *CAS [100900-34-3]* $C_3H_5BrO_2$ Fp 212°F   176.46
                                                                                            39.73*
(±)-2-Bromopropionic-*1-¹³C* acid, 99 atom % ¹³C   FW 153.97   38.91*
                                                           bp 203°C   21.40*
                                                           d 1.700
CDCl₃   QE-300 n_D^{20}1.4749

Offset: 2.1 ppm.

$CH_3\overset{\overset{\displaystyle Br}{|}}{C}H-^{13}\overset{\overset{\displaystyle O}{\|}}{C}-OH$

ALDRICH

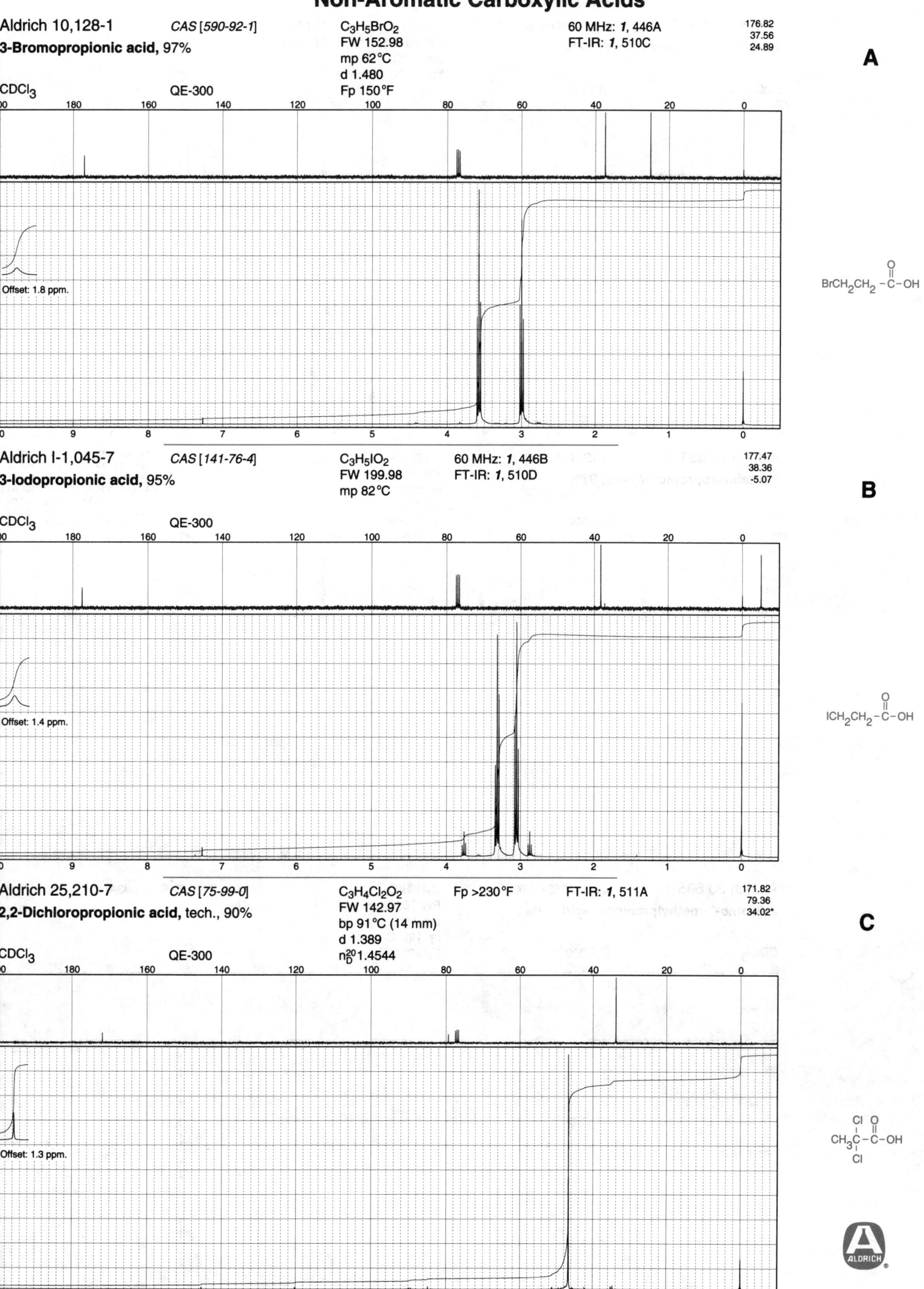

Aldrich 10,128-1 *CAS [590-92-1]*
3-Bromopropionic acid, 97%

C₃H₅BrO₂
FW 152.98
mp 62°C
d 1.480
Fp 150°F

60 MHz: *1*, 446A
FT-IR: *1*, 510C

176.82
37.56
24.89

A

CDCl₃   QE-300

Offset: 1.8 ppm.

BrCH₂CH₂–C–OH

Aldrich I-1,045-7 *CAS [141-76-4]*
3-Iodopropionic acid, 95%

C₃H₅IO₂
FW 199.98
mp 82°C

60 MHz: *1*, 446B
FT-IR: *1*, 510D

177.47
38.36
-5.07

B

CDCl₃   QE-300

Offset: 1.4 ppm.

ICH₂CH₂–C–OH

Aldrich 25,210-7 *CAS [75-99-0]*
2,2-Dichloropropionic acid, tech., 90%

C₃H₄Cl₂O₂
FW 142.97
bp 91°C (14 mm)
d 1.389
n₂₀/D 1.4544

Fp >230°F   FT-IR: *1*, 511A

171.82
79.36
34.02*

C

CDCl₃   QE-300

Offset: 1.3 ppm.

CH₃C–C–OH
Cl Cl O

ALDRICH

Non-Aromatic Carboxylic Acids

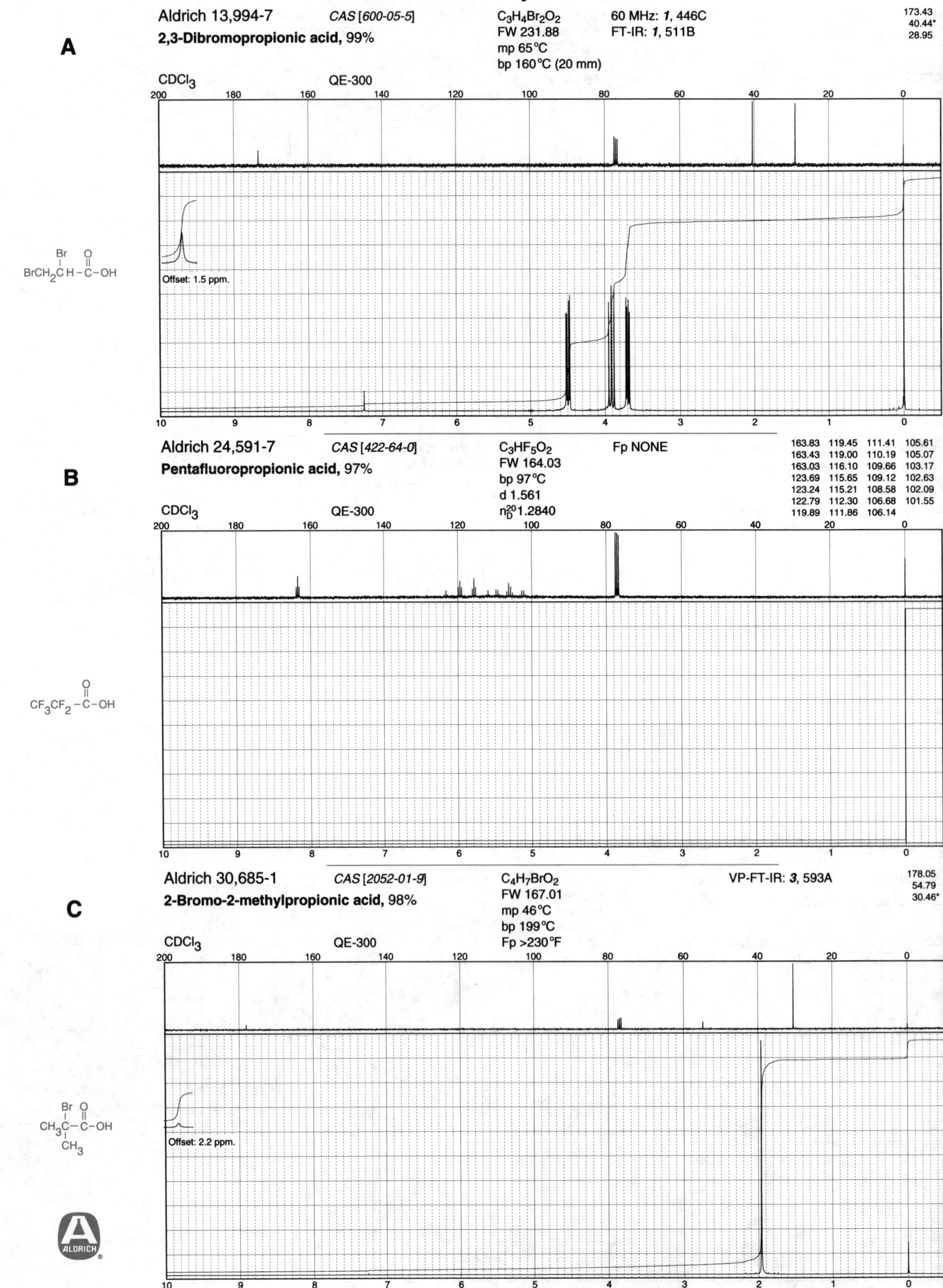

A

Aldrich 13,994-7   *CAS [600-05-5]* $C_3H_4Br_2O_2$ 60 MHz: *1*, 446C

2,3-Dibromopropionic acid, 99%   FW 231.88   FT-IR: *1*, 511B

mp 65°C

bp 160°C (20 mm)

173.43
40.44*
28.95

CDCl₃   QE-300

BrCH₂CH–C–OH (Br, O)

Offset: 1.5 ppm.

B

Aldrich 24,591-7   *CAS [422-64-0]* $C_3HF_5O_2$ Fp NONE

Pentafluoropropionic acid, 97%   FW 164.03

bp 97°C

d 1.561

n_D^{20} 1.2840

163.83	119.45	111.41	105.61
163.43	119.00	110.19	105.07
163.03	116.10	109.66	103.17
123.69	115.65	109.12	102.63
123.24	115.21	108.58	102.09
122.79	112.30	106.68	101.55
119.89	111.86	106.14	

CDCl₃   QE-300

CF₃CF₂–C–OH (O)

C

Aldrich 30,685-1   *CAS [2052-01-9]* $C_4H_7BrO_2$ VP-FT-IR: *3*, 593A

2-Bromo-2-methylpropionic acid, 98%   FW 167.01

mp 46°C

bp 199°C

Fp >230°F

178.05
54.79
30.46*

CDCl₃   QE-300

CH₃C–C–OH (Br, O)
CH₃

Offset: 2.2 ppm.

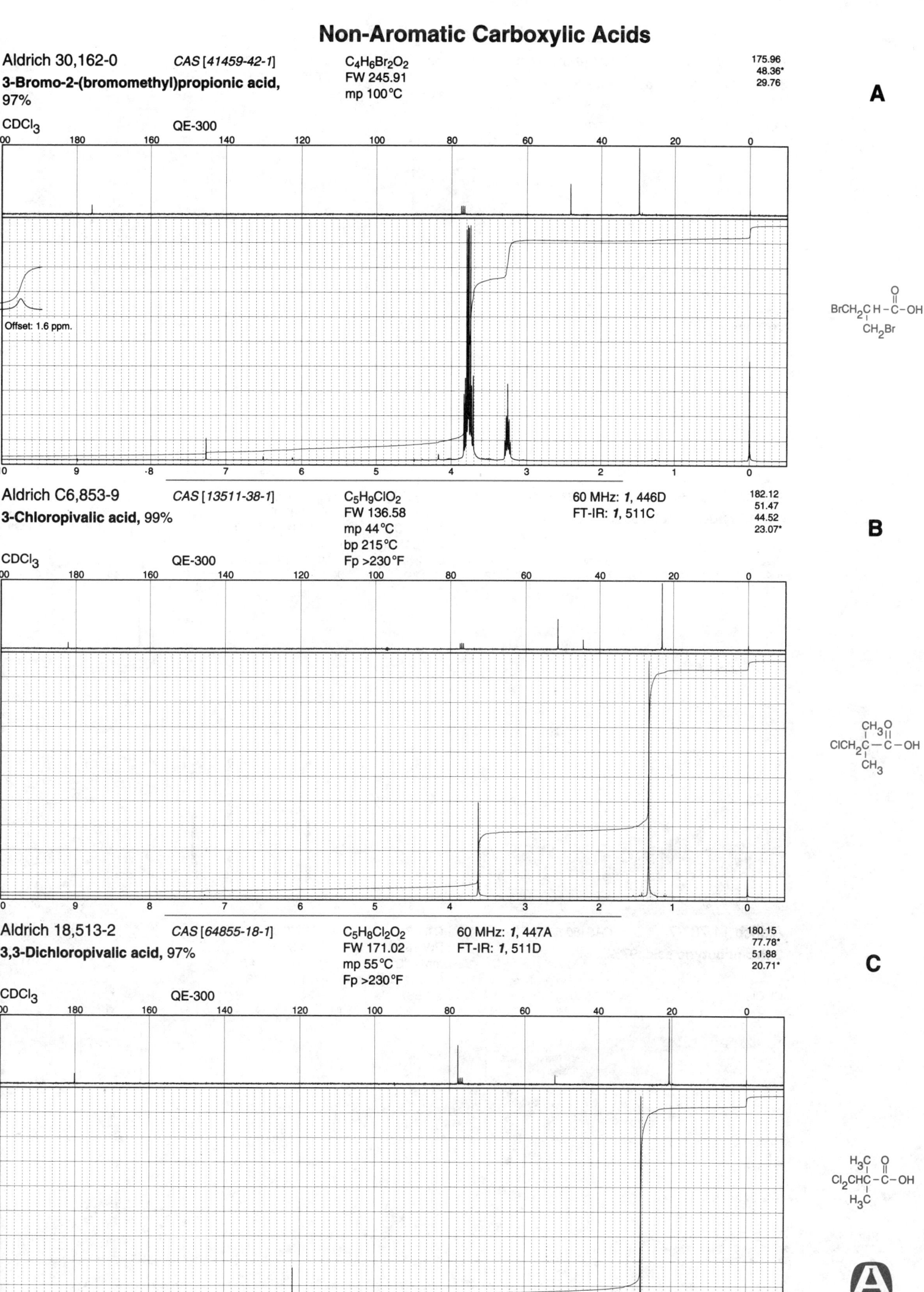

Aldrich 30,162-0   *CAS [41459-42-1]* $C_4H_6Br_2O_2$ FW 245.91   mp 100°C

3-Bromo-2-(bromomethyl)propionic acid, 97%

175.96
48.36*
29.76

A

CDCl$_3$ QE-300

Offset: 1.6 ppm.

$BrCH_2CH-C-OH$ with CH_2Br and $=O$

Aldrich C6,853-9   *CAS [13511-38-1]* $C_5H_9ClO_2$ FW 136.58   mp 44°C   bp 215°C   Fp >230°F

3-Chloropivalic acid, 99%

60 MHz: *1*, 446D
FT-IR: *1*, 511C

182.12
51.47
44.52
23.07*

B

CDCl$_3$ QE-300

$ClCH_2C-C-OH$ with two CH_3 and $=O$

Aldrich 18,513-2   *CAS [64855-18-1]* $C_5H_8Cl_2O_2$ FW 171.02   mp 55°C   Fp >230°F

3,3-Dichloropivalic acid, 97%

60 MHz: *1*, 447A
FT-IR: *1*, 511D

180.15
77.78*
51.88
20.71*

C

CDCl$_3$ QE-300

$Cl_2CHC-C-OH$ with two H_3C and $=O$

ALDRICH

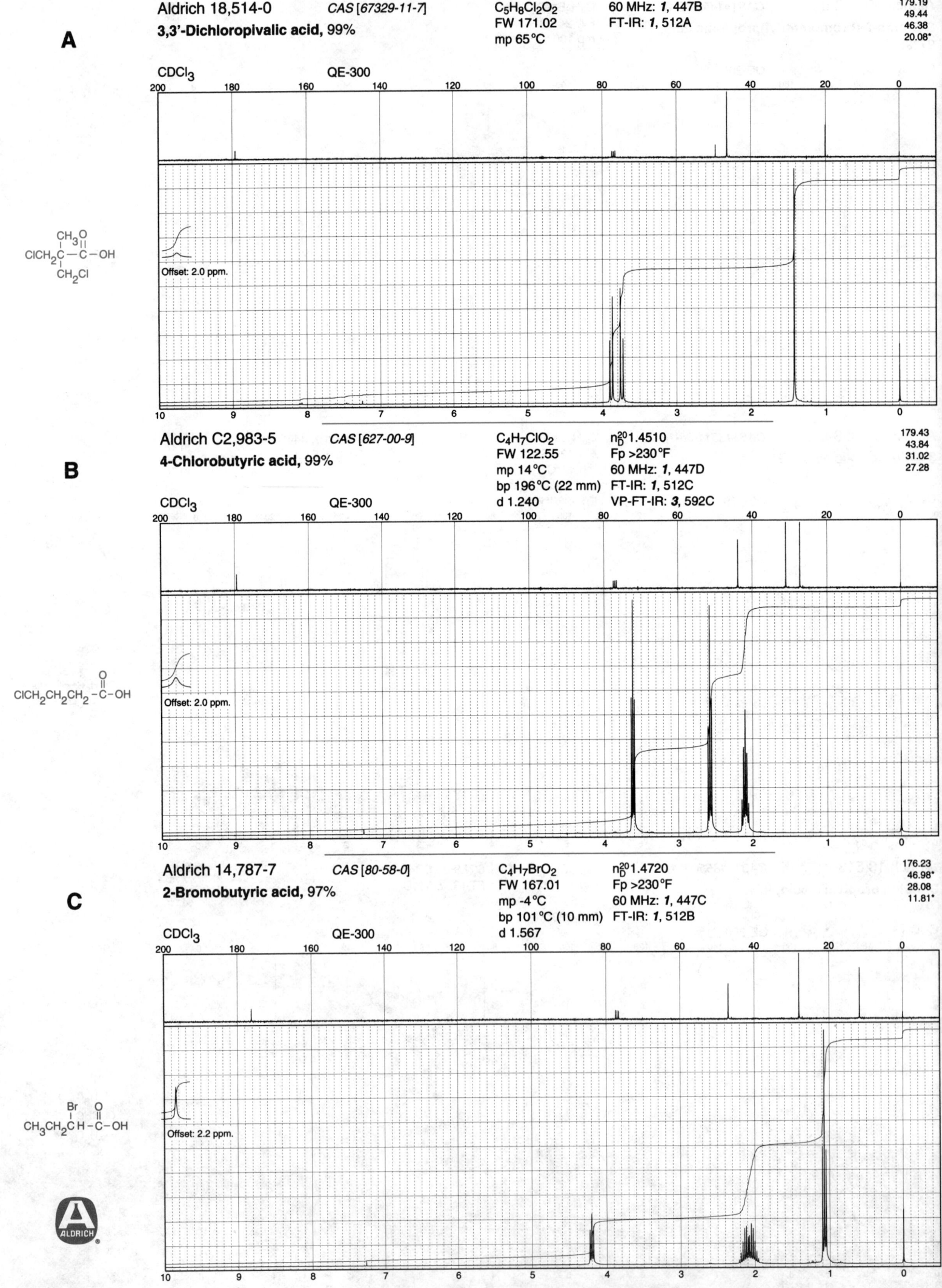

Aldrich 18,514-0   CAS [67329-11-7] $C_5H_8Cl_2O_2$ 60 MHz: *1*, 447B
3,3'-Dichloropivalic acid, 99%   FW 171.02   FT-IR: *1*, 512A
mp 65°C

179.19
49.44
46.38
20.08*

CDCl₃   QE-300

Offset: 2.0 ppm.

A

CH₃ O
ClCH₂—C—C—OH
CH₂Cl

Aldrich C2,983-5   CAS [627-00-9] $C_4H_7ClO_2$ n_D^{20}1.4510
4-Chlorobutyric acid, 99%   FW 122.55   Fp >230°F
mp 14°C   60 MHz: *1*, 447D
bp 196°C (22 mm)   FT-IR: *1*, 512C
d 1.240   VP-FT-IR: *3*, 592C

179.43
43.84
31.02
27.28

CDCl₃   QE-300

Offset: 2.0 ppm.

B

O
ClCH₂CH₂CH₂—C—OH

Aldrich 14,787-7   CAS [80-58-0] $C_4H_7BrO_2$ n_D^{20}1.4720
2-Bromobutyric acid, 97%   FW 167.01   Fp >230°F
mp -4°C   60 MHz: *1*, 447C
bp 101°C (10 mm)   FT-IR: *1*, 512B
d 1.567

176.23
46.98*
28.08
11.81*

CDCl₃   QE-300

Offset: 2.2 ppm.

C

Br O
CH₃CH₂CH—C—OH

Non-Aromatic Carboxylic Acids

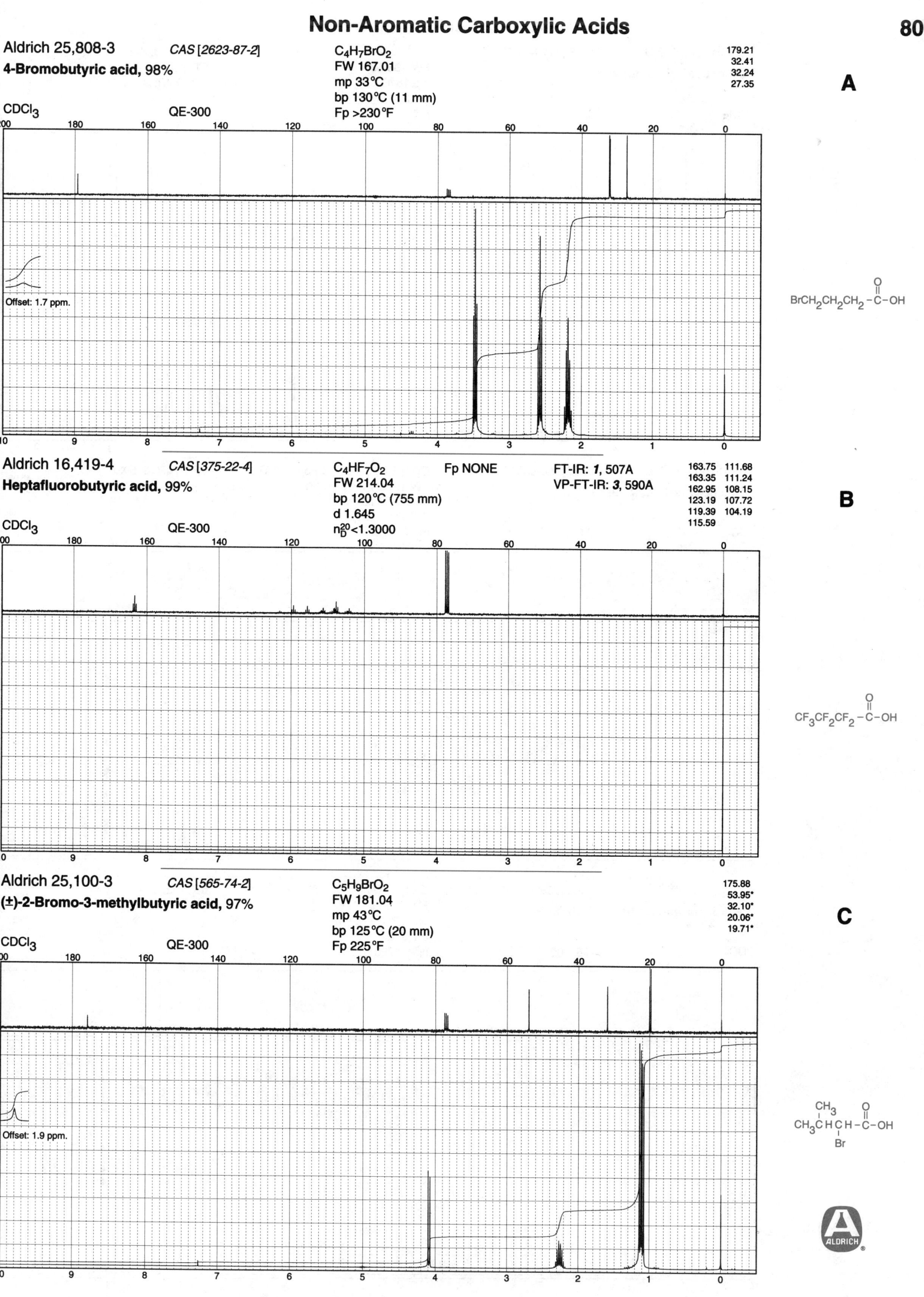

Aldrich 25,808-3 *CAS [2623-87-2]*
4-Bromobutyric acid, 98%

C₄H₇BrO₂ → $C_4H_7BrO_2$
FW 167.01
mp 33°C
bp 130°C (11 mm)
Fp >230°F

179.21
32.41
32.24
27.35

A

CDCl₃   QE-300

$BrCH_2CH_2CH_2-\overset{\displaystyle O}{\underset{\displaystyle }{C}}-OH$

Offset: 1.7 ppm.

Aldrich 16,419-4 *CAS [375-22-4]*
Heptafluorobutyric acid, 99%

$C_4HF_7O_2$
FW 214.04
bp 120°C (755 mm)
d 1.645
$n_D^{20} < 1.3000$

Fp NONE

FT-IR: *1*, 507A
VP-FT-IR: *3*, 590A

163.75 111.68
163.35 111.24
162.95 108.15
123.19 107.72
119.39 104.19
115.59

B

CDCl₃   QE-300

$CF_3CF_2CF_2-\overset{\displaystyle O}{\underset{\displaystyle }{C}}-OH$

Aldrich 25,100-3 *CAS [565-74-2]*
(±)-2-Bromo-3-methylbutyric acid, 97%

$C_5H_9BrO_2$
FW 181.04
mp 43°C
bp 125°C (20 mm)
Fp 225°F

175.88
53.95*
32.10*
20.06*
19.71*

C

CDCl₃   QE-300

$CH_3CHCH-\overset{\displaystyle O}{\underset{\displaystyle }{C}}-OH$ with CH_3 above and Br below

Offset: 1.9 ppm.

ALDRICH

Non-Aromatic Carboxylic Acids

A

Aldrich C7,290-0   *CAS [1119-46-6]*   C$_5$H$_9$ClO$_2$ Fp >230°F   60 MHz: *1*, 448A

5-Chlorovaleric acid, 98%

FW 136.58   FT-IR: *1*, 512D
mp 18°C   VP-FT-IR: *3*, 592D
d 1.342
n$_D^{20}$1.4555

179.94
44.37
33.22
31.72
21.96

CDCl$_3$ QE-300

ClCH$_2$CH$_2$CH$_2$CH$_2$–C(=O)–OH

Offset: 2.0 ppm.

B

Aldrich 26,513-6   *CAS [584-93-0]*   C$_5$H$_9$BrO$_2$ Fp >230°F   VP-FT-IR: *3*, 593B

2-Bromovaleric acid, 99%

FW 181.04
bp 134°C (25 mm)
d 1.381
n$_D^{20}$1.4709

176.36
45.08*
36.49
20.48
13.24*

CDCl$_3$ QE-300

CH$_3$CH$_2$CH$_2$CH(Br)–C(=O)–OH

Offset: 2.1 ppm.

C

Aldrich 24,283-7   *CAS [2681-83-6]*   C$_6$H$_{11}$BrO$_2$ Fp >230°F   60 MHz: *1*, 448C

(±)-2-Bromohexanoic acid, 97%

FW 195.06   FT-IR: *1*, 513A
bp 137°C (18 mm)
d 1.370
n$_D^{20}$1.4720

176.35
45.35*
34.34
29.28
21.97
13.79*

CDCl$_3$ QE-300

CH$_3$(CH$_2$)$_3$CH(Br)–C(=O)–OH

Offset: 2.1 ppm.

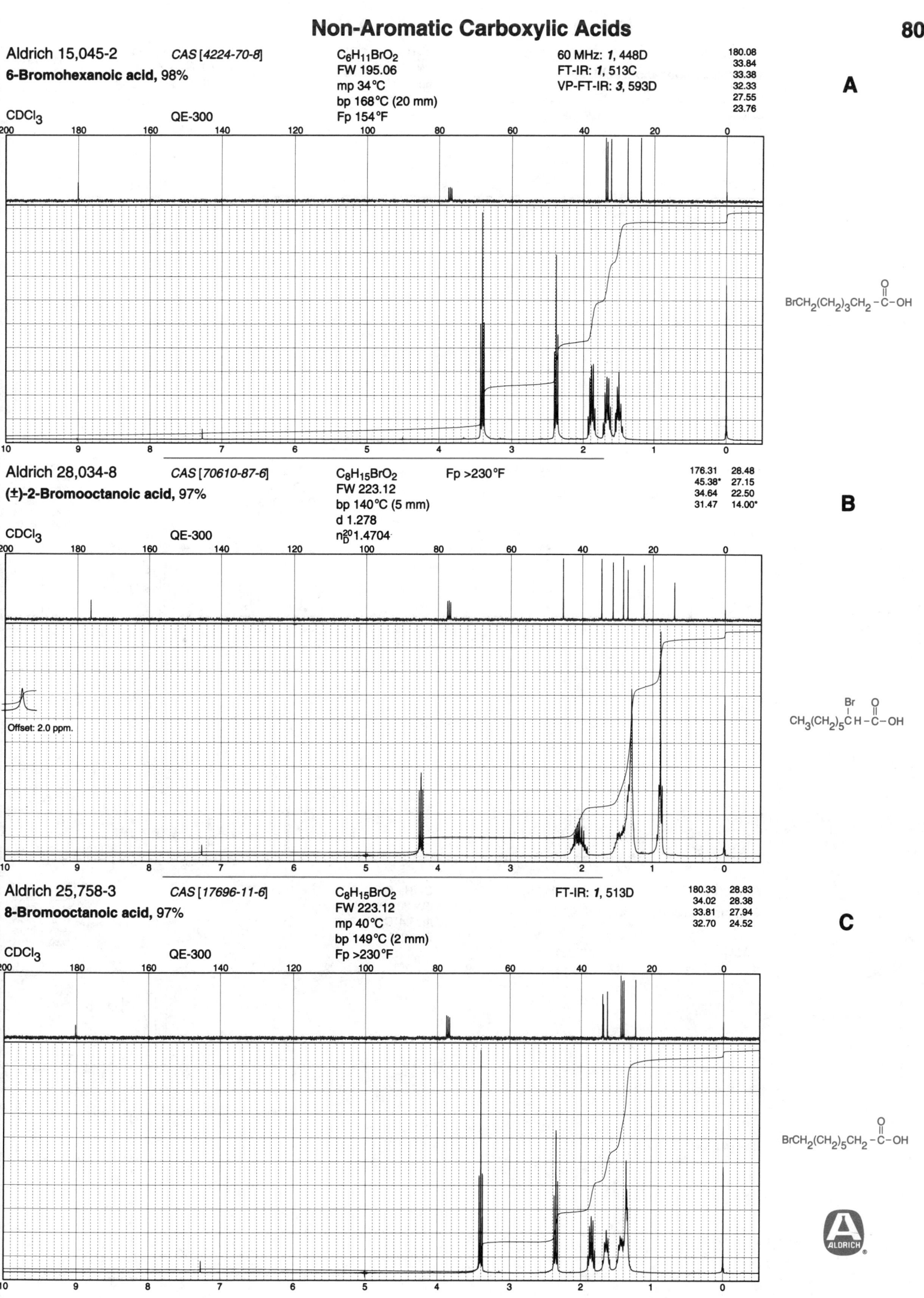

Aldrich 15,045-2 *CAS [4224-70-8]*

6-Bromohexanoic acid, 98%

C₆H₁₁BrO₂
FW 195.06
mp 34°C
bp 168°C (20 mm)
Fp 154°F

60 MHz: *1*, 448D
FT-IR: *1*, 513C
VP-FT-IR: *3*, 593D

180.08
33.84
33.38
32.33
27.55
23.76

A

CDCl₃   QE-300

$BrCH_2(CH_2)_3CH_2-\overset{\displaystyle O}{\overset{\displaystyle \|}{C}}-OH$

Aldrich 28,034-8 *CAS [70610-87-6]*

(±)-2-Bromooctanoic acid, 97%

C₈H₁₅BrO₂
FW 223.12
bp 140°C (5 mm)
d 1.278
n²⁰_D 1.4704

Fp >230°F

176.31   28.48
45.38*   27.15
34.64   22.50
31.47   14.00*

B

CDCl₃   QE-300

Offset: 2.0 ppm.

$CH_3(CH_2)_5\overset{\displaystyle Br}{\underset{\displaystyle |}{C}}H-\overset{\displaystyle O}{\overset{\displaystyle \|}{C}}-OH$

Aldrich 25,758-3 *CAS [17696-11-6]*

8-Bromooctanoic acid, 97%

C₈H₁₅BrO₂
FW 223.12
mp 40°C
bp 149°C (2 mm)
Fp >230°F

FT-IR: *1*, 513D

180.33   28.83
34.02   28.38
33.81   27.94
32.70   24.52

C

CDCl₃   QE-300

$BrCH_2(CH_2)_5CH_2-\overset{\displaystyle O}{\overset{\displaystyle \|}{C}}-OH$

Non-Aromatic Carboxylic Acids

A

Aldrich B8,280-4   *CAS [2834-05-1]*

11-Bromoundecanoic acid, 99%

$C_{11}H_{21}BrO_2$
FW 265.20
mp 50°C
bp 174°C (2 mm)
Fp >230°F

CDCl$_3$ QE-300

60 MHz: *1*, 449A
FT-IR: *1*, 514A
VP-FT-IR: *3*, 594A

180.41	29.15
34.09	29.00
33.95	28.71
32.81	28.14
29.33	24.63
29.27	

$BrCH_2(CH_2)_8CH_2-\overset{\displaystyle O}{\overset{\|}{C}}-OH$

B

Aldrich 20,099-9   *CAS [73367-80-3]*

12-Bromododecanoic acid, 97%

$C_{12}H_{23}BrO_2$
FW 279.22
mp 54°C
Fp >230°F

60 MHz: *1*, 449B
FT-IR: *1*, 514B
VP-FT-IR: *3*, 594B

180.42	29.36
34.11	29.20
33.95	29.04
32.85	28.75
29.43	28.18
29.39	24.66

CDCl$_3$ QE-300

$BrCH_2(CH_2)_9CH_2-\overset{\displaystyle O}{\overset{\|}{C}}-OH$

C

Aldrich 28,035-6   *CAS [10520-81-7]*

2-Bromotetradecanoic acid, 98%

$C_{14}H_{27}BrO_2$
FW 307.28
mp 44°C
Fp >230°F

173.46	29.50
46.39*	29.33
34.89	28.86
31.92	27.25
29.64	22.69
29.60	14.11*

CDCl$_3$ QE-300

Offset: 1.2 ppm.

$CH_3(CH_2)_{11}\overset{\displaystyle Br}{\underset{}{CH}}-\overset{\displaystyle O}{\overset{\|}{C}}-OH$

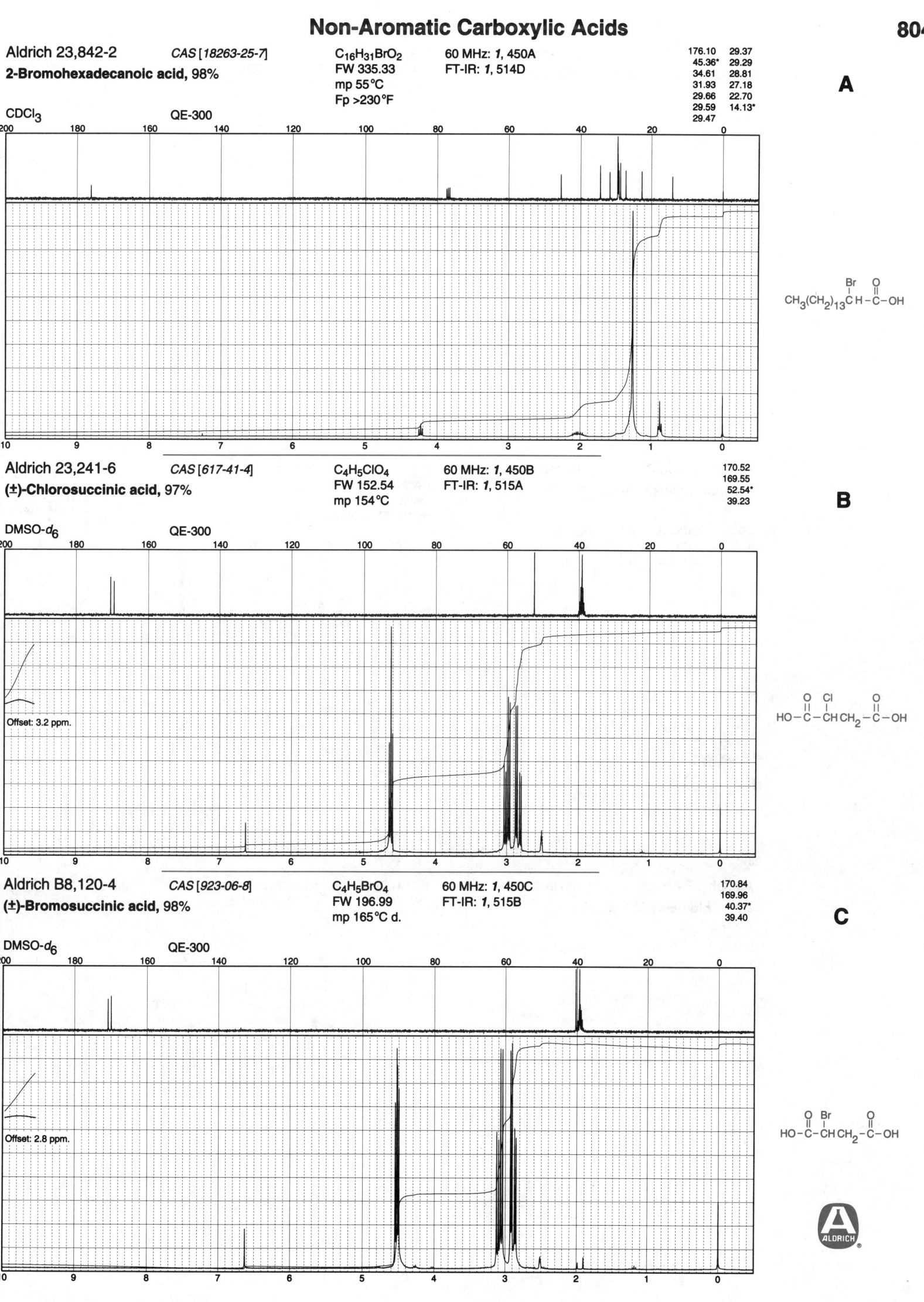

Aldrich 23,842-2   *CAS [18263-25-7]*
2-Bromohexadecanoic acid, 98%

C₁₆H₃₁BrO₂
FW 335.33
mp 55°C
Fp >230°F

60 MHz: *1*, 450A
FT-IR: *1*, 514D

176.10 29.37
45.36* 29.29
34.61   28.81
31.93   27.18
29.66   22.70
29.59   14.13*
29.47

A

CDCl₃   QE-300

CH₃(CH₂)₁₃CH–C–OH (with Br on CH and =O on C)

Aldrich 23,241-6   *CAS [617-41-4]*
(±)-Chlorosuccinic acid, 97%

C₄H₅ClO₄
FW 152.54
mp 154°C

60 MHz: *1*, 450B
FT-IR: *1*, 515A

170.52
169.55
52.54*
39.23

B

DMSO-*d₆*   QE-300

Offset: 3.2 ppm.

HO–C–CH–CH₂–C–OH (with Cl on CH, =O on both C)

Aldrich B8,120-4   *CAS [923-06-8]*
(±)-Bromosuccinic acid, 98%

C₄H₅BrO₄
FW 196.99
mp 165°C d.

60 MHz: *1*, 450C
FT-IR: *1*, 515B

170.84
169.96
40.37*
39.40

C

DMSO-*d₆*   QE-300

Offset: 2.8 ppm.

HO–C–CH–CH₂–C–OH (with Br on CH, =O on both C)

ALDRICH

Non-Aromatic Carboxylic Acids

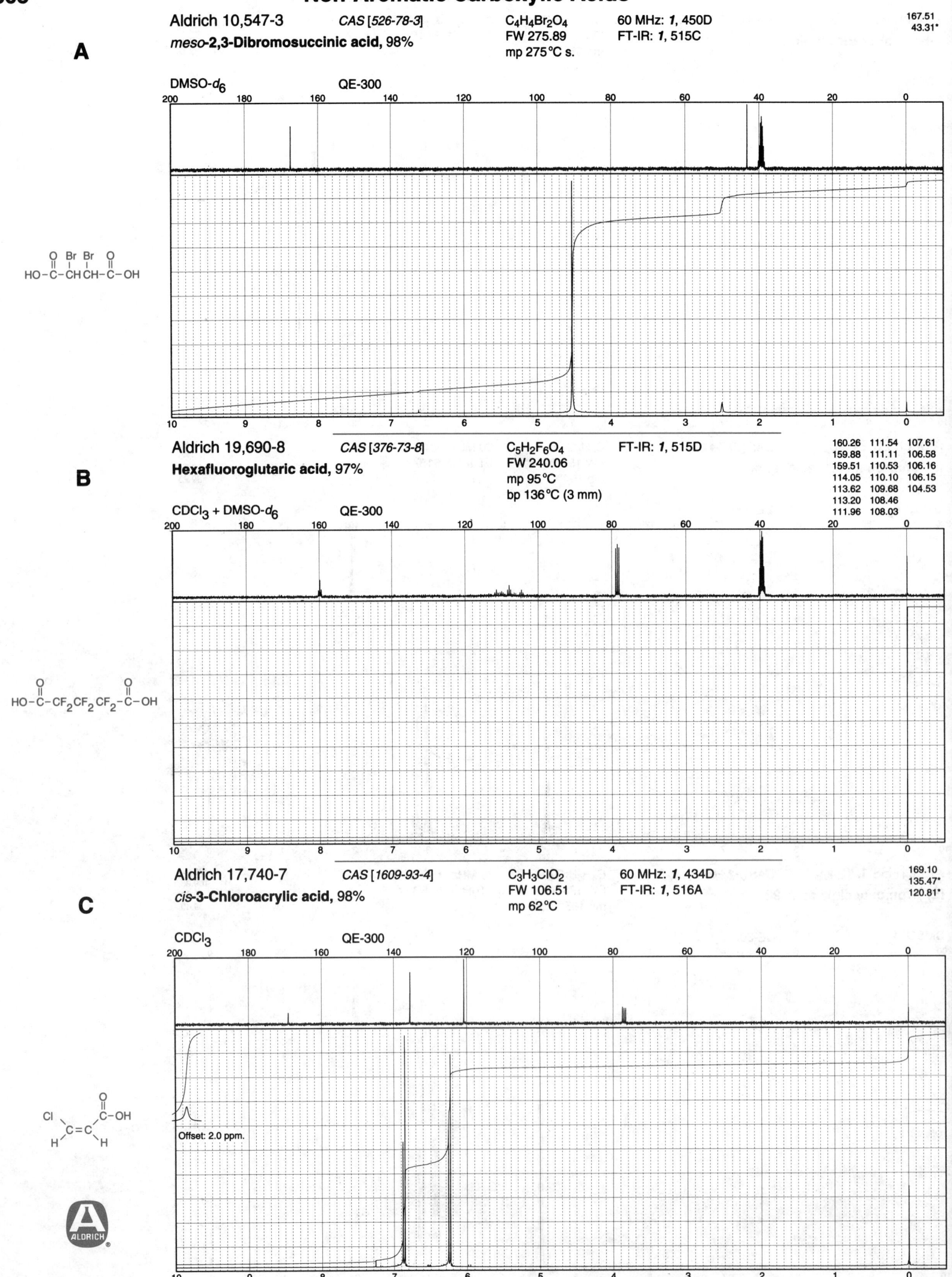

A

Aldrich 10,547-3   *CAS [526-78-3]* $C_4H_4Br_2O_4$ 60 MHz: *1*, 450D   167.51
meso-**2,3-Dibromosuccinic acid, 98%**   FW 275.89   FT-IR: *1*, 515C   43.31*
mp 275°C s.

DMSO-d_6 QE-300

B

Aldrich 19,690-8   *CAS [376-73-8]* $C_5H_2F_6O_4$ FT-IR: *1*, 515D
Hexafluoroglutaric acid, 97%   FW 240.06
mp 95°C
bp 136°C (3 mm)

160.26	111.54	107.61
159.88	111.11	106.58
159.51	110.53	106.16
114.05	110.10	106.15
113.62	109.68	104.53
113.20	108.46	
111.96	108.03	

CDCl$_3$ + DMSO-d_6 QE-300

C

Aldrich 17,740-7   *CAS [1609-93-4]* $C_3H_3ClO_2$ 60 MHz: *1*, 434D   169.10
cis-**3-Chloroacrylic acid, 98%**   FW 106.51   FT-IR: *1*, 516A   135.47*
mp 62°C   120.81*

CDCl$_3$ QE-300

Offset: 2.0 ppm.

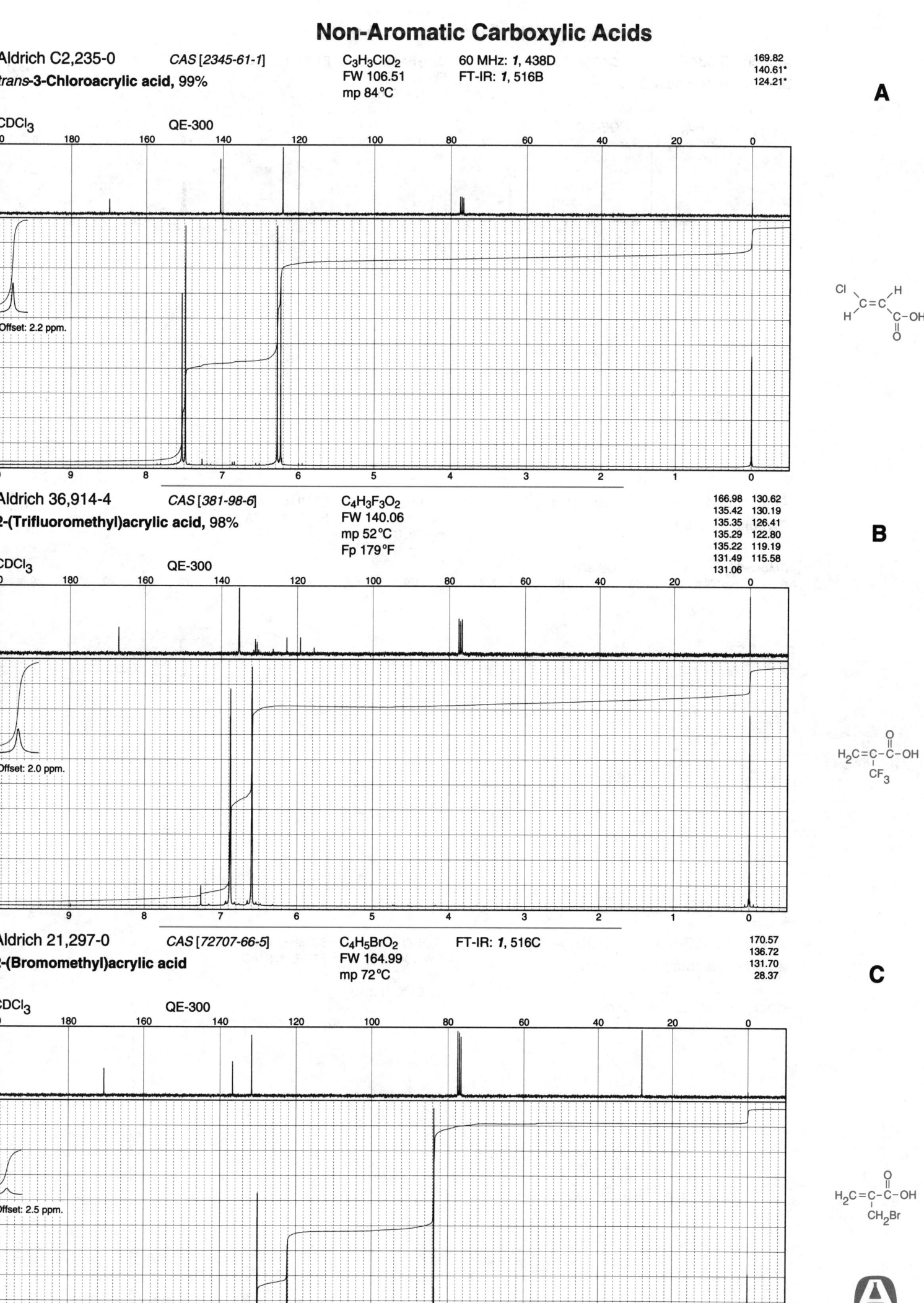

A

Aldrich C2,235-0   CAS [2345-61-1]

trans-3-Chloroacrylic acid, 99%

$C_3H_3ClO_2$
FW 106.51
mp 84°C

60 MHz: *1*, 438D
FT-IR: *1*, 516B

169.82
140.61*
124.21*

CDCl₃   QE-300

Offset: 2.2 ppm.

B

Aldrich 36,914-4   CAS [381-98-6]

2-(Trifluoromethyl)acrylic acid, 98%

$C_4H_3F_3O_2$
FW 140.06
mp 52°C
Fp 179°F

166.98   130.62
135.42   130.19
135.35   126.41
135.29   122.80
135.22   119.19
131.49   115.58
131.06

CDCl₃   QE-300

Offset: 2.0 ppm.

C

Aldrich 21,297-0   CAS [72707-66-5]

2-(Bromomethyl)acrylic acid

$C_4H_5BrO_2$
FW 164.99
mp 72°C

FT-IR: *1*, 516C

170.57
136.72
131.70
28.37

CDCl₃   QE-300

Offset: 2.5 ppm.

Non-Aromatic Carboxylic Acids

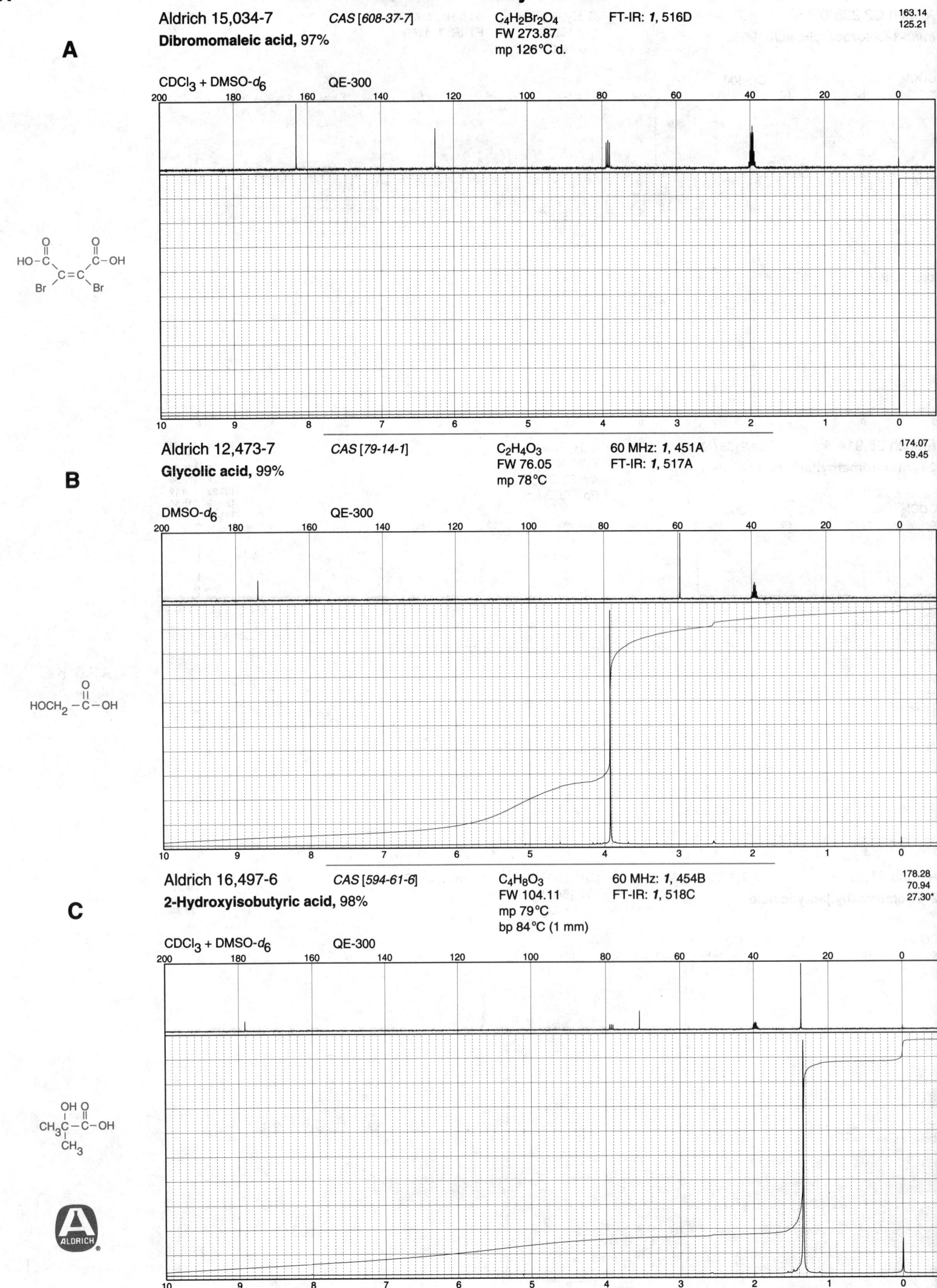

A

Aldrich 15,034-7   *CAS [608-37-7]* $C_4H_2Br_2O_4$ FT-IR: *1*, 516D   163.14
125.21
Dibromomaleic acid, 97%   FW 273.87
mp 126°C d.

CDCl$_3$ + DMSO-d_6 QE-300

B

Aldrich 12,473-7   *CAS [79-14-1]* $C_2H_4O_3$ 60 MHz: *1*, 451A   174.07
59.45
Glycolic acid, 99%   FW 76.05   FT-IR: *1*, 517A
mp 78°C

DMSO-d_6 QE-300

C

Aldrich 16,497-6   *CAS [594-61-6]* $C_4H_8O_3$ 60 MHz: *1*, 454B   178.28
70.94
2-Hydroxyisobutyric acid, 98%   FW 104.11   FT-IR: *1*, 518C   27.30*
mp 79°C
bp 84°C (1 mm)

CDCl$_3$ + DMSO-d_6 QE-300

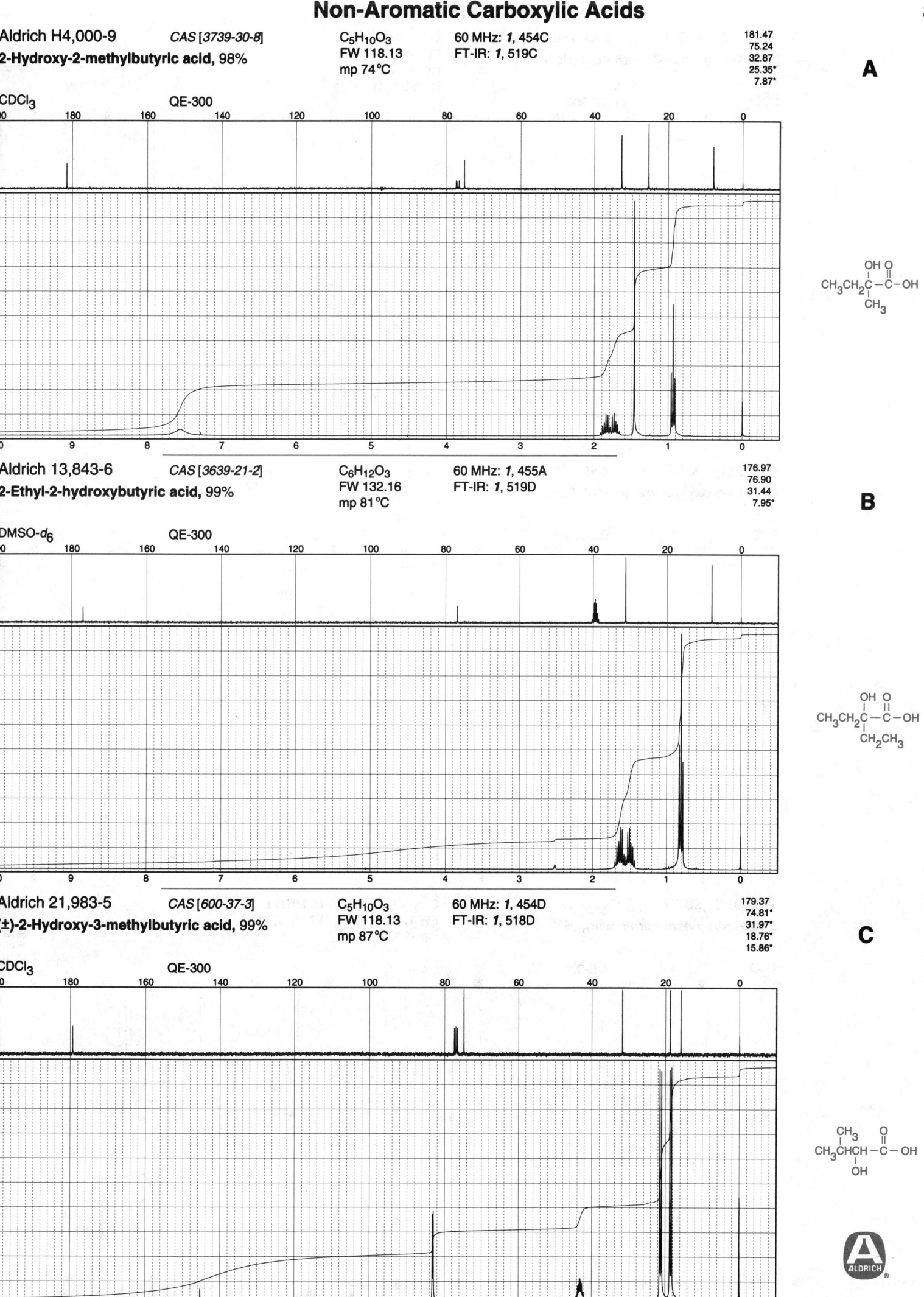

Aldrich H4,000-9   CAS [3739-30-8]
2-Hydroxy-2-methylbutyric acid, 98%

C₅H₁₀O₃
FW 118.13
mp 74°C

$C_5H_{10}O_3$
FW 118.13
mp 74°C

60 MHz: *1*, 454C
FT-IR: *1*, 519C

181.47
75.24
32.87
25.35*
7.87*

A

CDCl₃   QE-300

$CH_3CH_2C(OH)(CH_3)COOH$

Aldrich 13,843-6   CAS [3639-21-2]
2-Ethyl-2-hydroxybutyric acid, 99%

$C_6H_{12}O_3$
FW 132.16
mp 81°C

60 MHz: *1*, 455A
FT-IR: *1*, 519D

176.97
76.90
31.44
7.95*

B

DMSO-d₆   QE-300

$CH_3CH_2C(OH)(CH_2CH_3)COOH$

Aldrich 21,983-5   CAS [600-37-3]
(±)-2-Hydroxy-3-methylbutyric acid, 99%

$C_5H_{10}O_3$
FW 118.13
mp 87°C

60 MHz: *1*, 454D
FT-IR: *1*, 518D

179.37
74.81*
31.97*
18.76*
15.86*

C

CDCl₃   QE-300

$(CH_3)_2CHCH(OH)COOH$

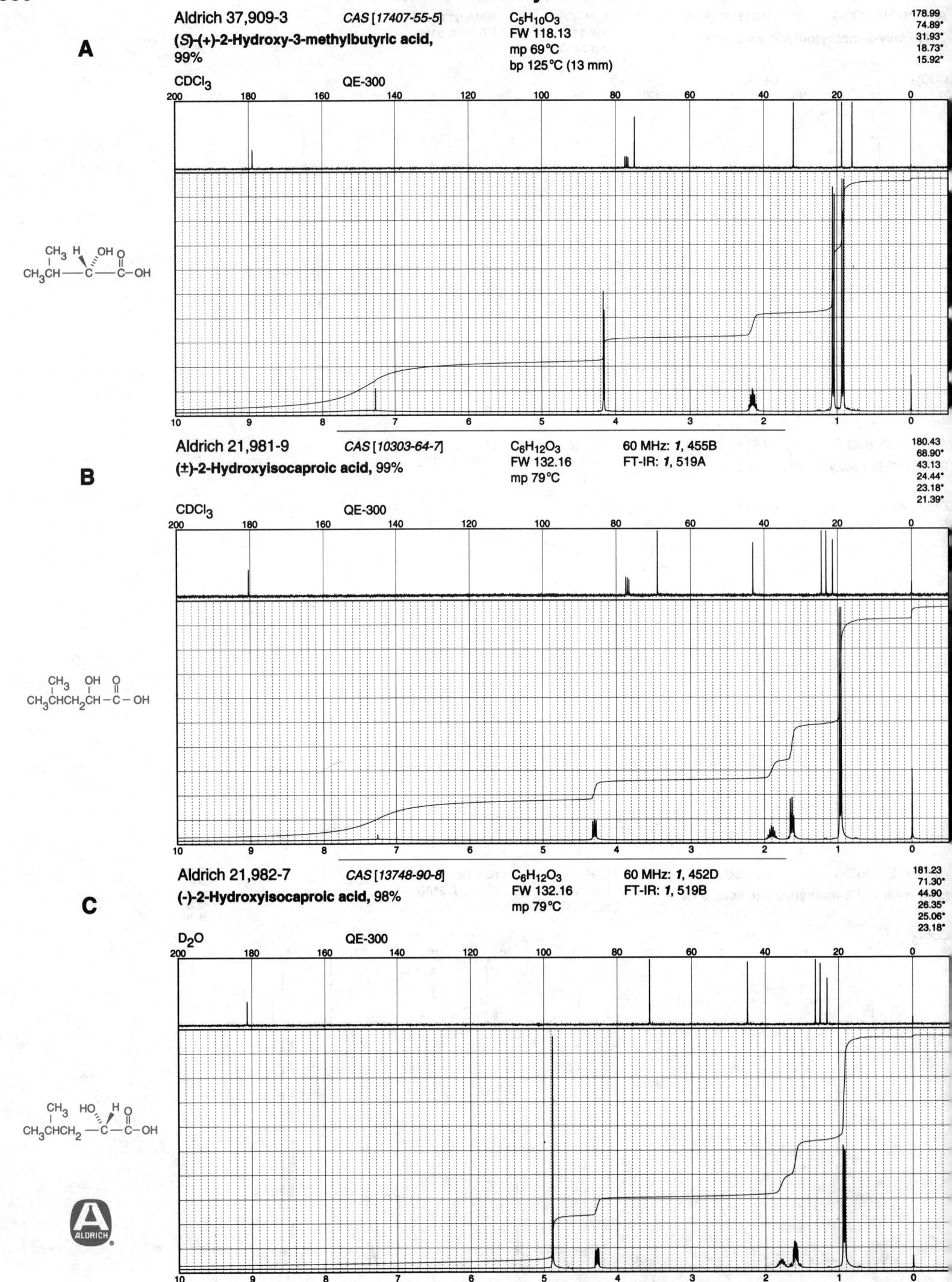

A

Aldrich 37,909-3   *CAS [17407-55-5]*

(S)-(+)-2-Hydroxy-3-methylbutyric acid, 99%

$C_5H_{10}O_3$
FW 118.13
mp 69°C
bp 125°C (13 mm)

178.99
74.89*
31.93*
18.73*
15.92*

CDCl₃   QE-300

B

Aldrich 21,981-9   *CAS [10303-64-7]*

(±)-2-Hydroxyisocaproic acid, 99%

$C_6H_{12}O_3$
FW 132.16
mp 79°C

60 MHz: *1*, 455B
FT-IR: *1*, 519A

180.43
68.90*
43.13
24.44*
23.18*
21.39*

CDCl₃   QE-300

C

Aldrich 21,982-7   *CAS [13748-90-8]*

(-)-2-Hydroxyisocaproic acid, 98%

$C_6H_{12}O_3$
FW 132.16
mp 79°C

60 MHz: *1*, 452D
FT-IR: *1*, 519B

181.23
71.30*
44.90
26.35*
25.06*
23.18*

D₂O   QE-300

Aldrich 21,980-0   *CAS [636-36-2]*

(±)-2-Hydroxycaproic acid, 99%

C₆H₁₂O₃
FW 132.16
mp 61°C
Fp >230°F

60 MHz: *1*, 452B
FT-IR: *1*, 520A

180.89
72.94*
35.77
29.15
24.43
15.85*

A

D₂O   QE-300

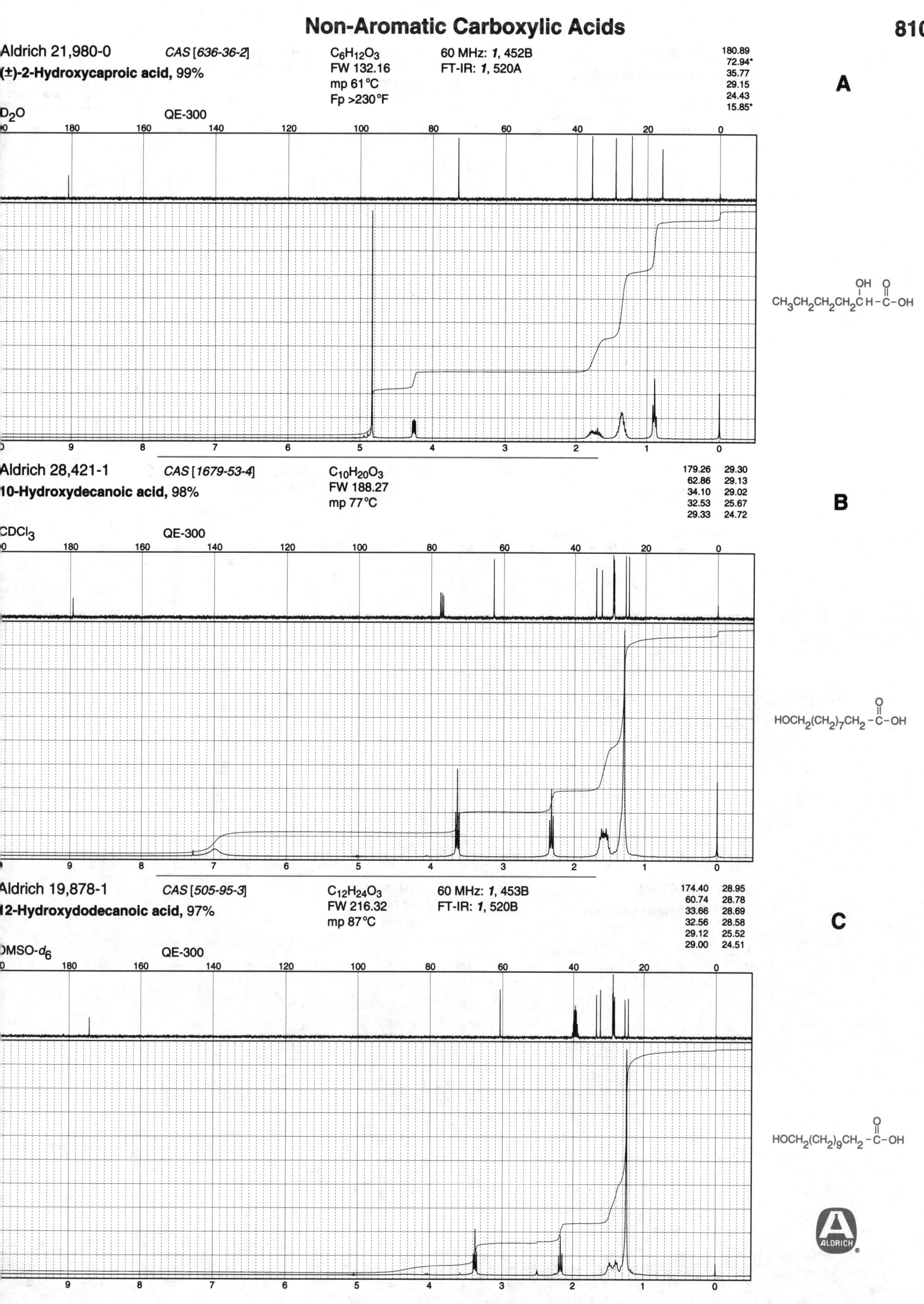

Aldrich 28,421-1   *CAS [1679-53-4]*

10-Hydroxydecanoic acid, 98%

C₁₀H₂₀O₃
FW 188.27
mp 77°C

179.26   29.30
62.86   29.13
34.10   29.02
32.53   25.67
29.33   24.72

B

CDCl₃   QE-300

Aldrich 19,878-1   *CAS [505-95-3]*

12-Hydroxydodecanoic acid, 97%

C₁₂H₂₄O₃
FW 216.32
mp 87°C

60 MHz: *1*, 453B
FT-IR: *1*, 520B

174.40   28.95
60.74   28.78
33.66   28.69
32.56   28.58
29.12   25.52
29.00   24.51

C

DMSO-d₆   QE-300

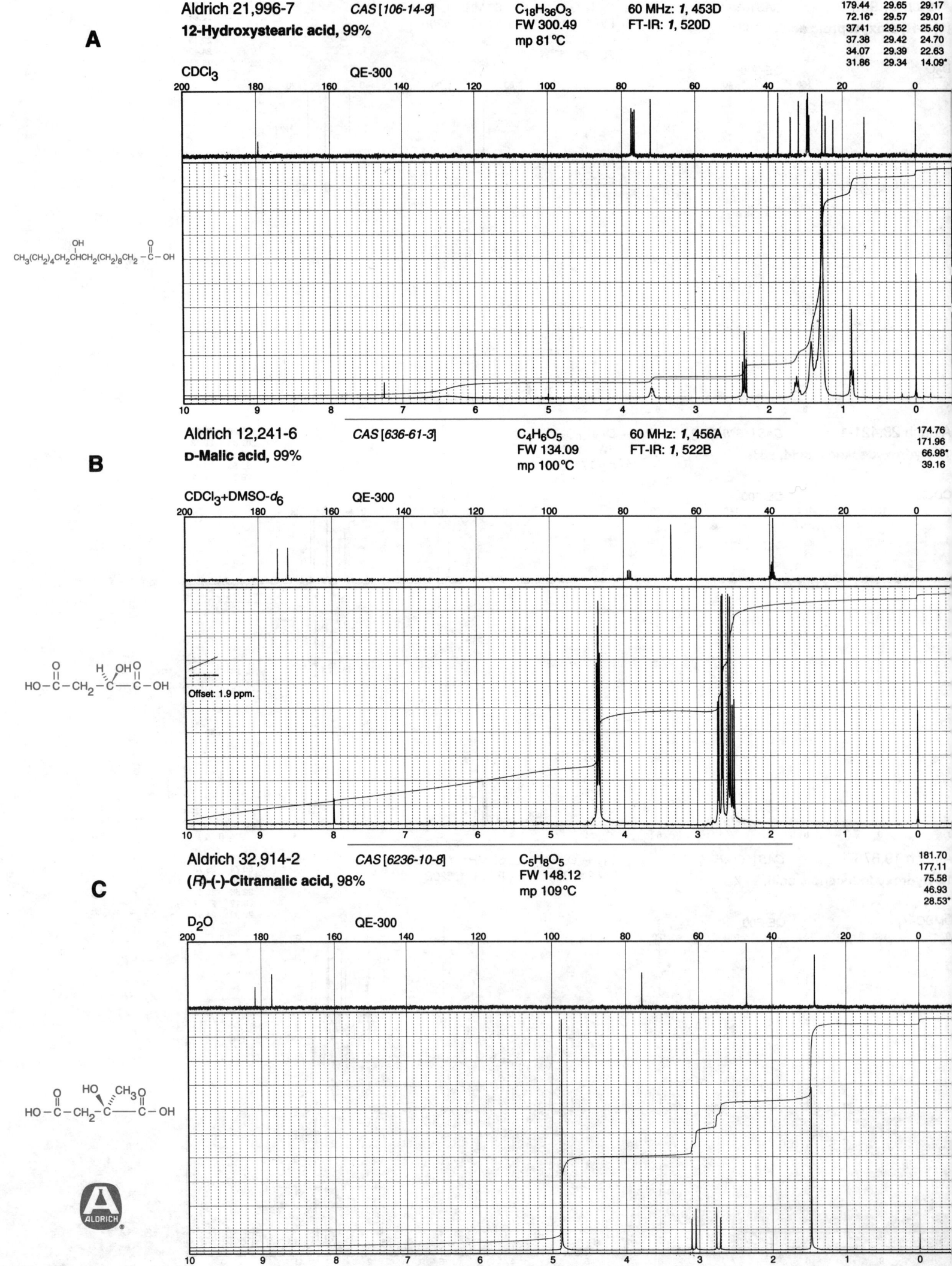

A

Aldrich 21,996-7   *CAS [106-14-9]* $C_{18}H_{36}O_3$ 60 MHz: *1*, 453D

12-Hydroxystearic acid, 99%   FW 300.49   FT-IR: *1*, 520D

mp 81°C

179.44	29.65	29.17
72.16*	29.57	29.01
37.41	29.52	25.60
37.38	29.42	24.70
34.07	29.39	22.63
31.86	29.34	14.09*

CDCl₃   QE-300

$CH_3(CH_2)_4CH_2CHCH_2(CH_2)_8CH_2-\overset{O}{\overset{\|}{C}}-OH$

B

Aldrich 12,241-6   *CAS [636-61-3]* $C_4H_6O_5$ 60 MHz: *1*, 456A

D-Malic acid, 99%   FW 134.09   FT-IR: *1*, 522B

mp 100°C

174.76
171.96
66.98*
39.16

CDCl₃+DMSO-*d₆*   QE-300

Offset: 1.9 ppm.

$HO-\overset{O}{\overset{\|}{C}}-CH_2-\overset{H}{\overset{|}{C}}-\overset{OH}{\underset{}{}}\overset{O}{\overset{\|}{C}}-OH$

C

Aldrich 32,914-2   *CAS [6236-10-8]* $C_5H_8O_5$

(*R*)-(-)-Citramalic acid, 98%   FW 148.12

mp 109°C

181.70
177.11
75.58
46.93
28.53*

D₂O   QE-300

$HO-\overset{O}{\overset{\|}{C}}-CH_2-\overset{HO}{\underset{}{\overset{|}{C}}}\overset{CH_3}{\underset{}{}}\overset{O}{\overset{\|}{C}}-OH$

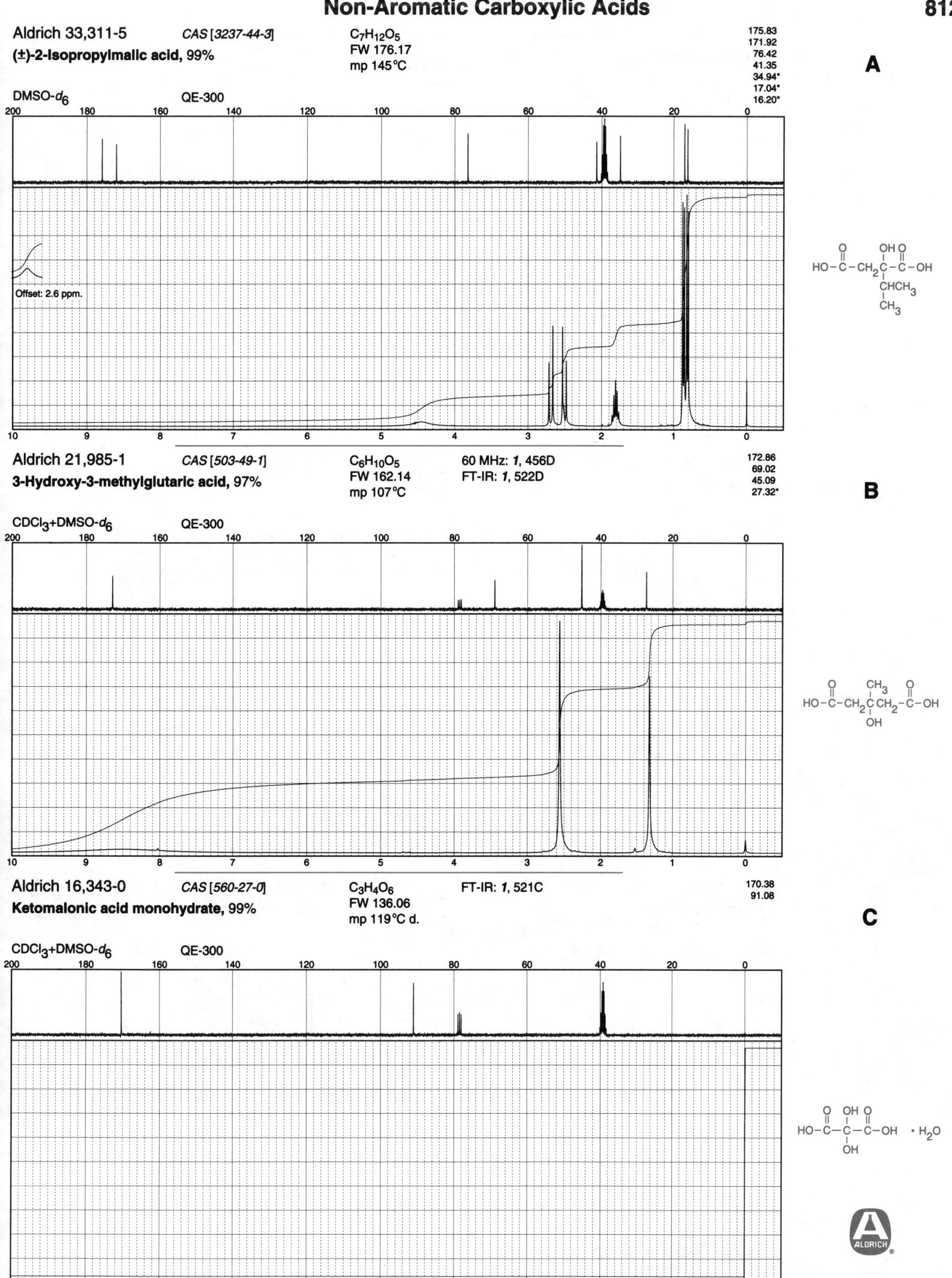

Aldrich 33,311-5   CAS [3237-44-3]   C₇H₁₂O₅
(±)-2-Isopropylmalic acid, 99%   FW 176.17
                                        mp 145°C

$C_7H_{12}O_5$
FW 176.17
mp 145°C

175.83
171.92
76.42
41.35
34.94*
17.04*
16.20*

A

DMSO-d_6 QE-300

Offset: 2.6 ppm.

Aldrich 21,985-1   CAS [503-49-1] $C_6H_{10}O_5$ 60 MHz: 1, 456D
3-Hydroxy-3-methylglutaric acid, 97%   FW 162.14   FT-IR: 1, 522D
                                        mp 107°C

172.86
69.02
45.09
27.32*

B

CDCl₃+DMSO-d_6 QE-300

Aldrich 16,343-0   CAS [560-27-0] $C_3H_4O_6$ FT-IR: 1, 521C
Ketomalonic acid monohydrate, 99%   FW 136.06
                                        mp 119°C d.

170.38
91.08

C

CDCl₃+DMSO-d_6 QE-300

Non-Aromatic Carboxylic Acids

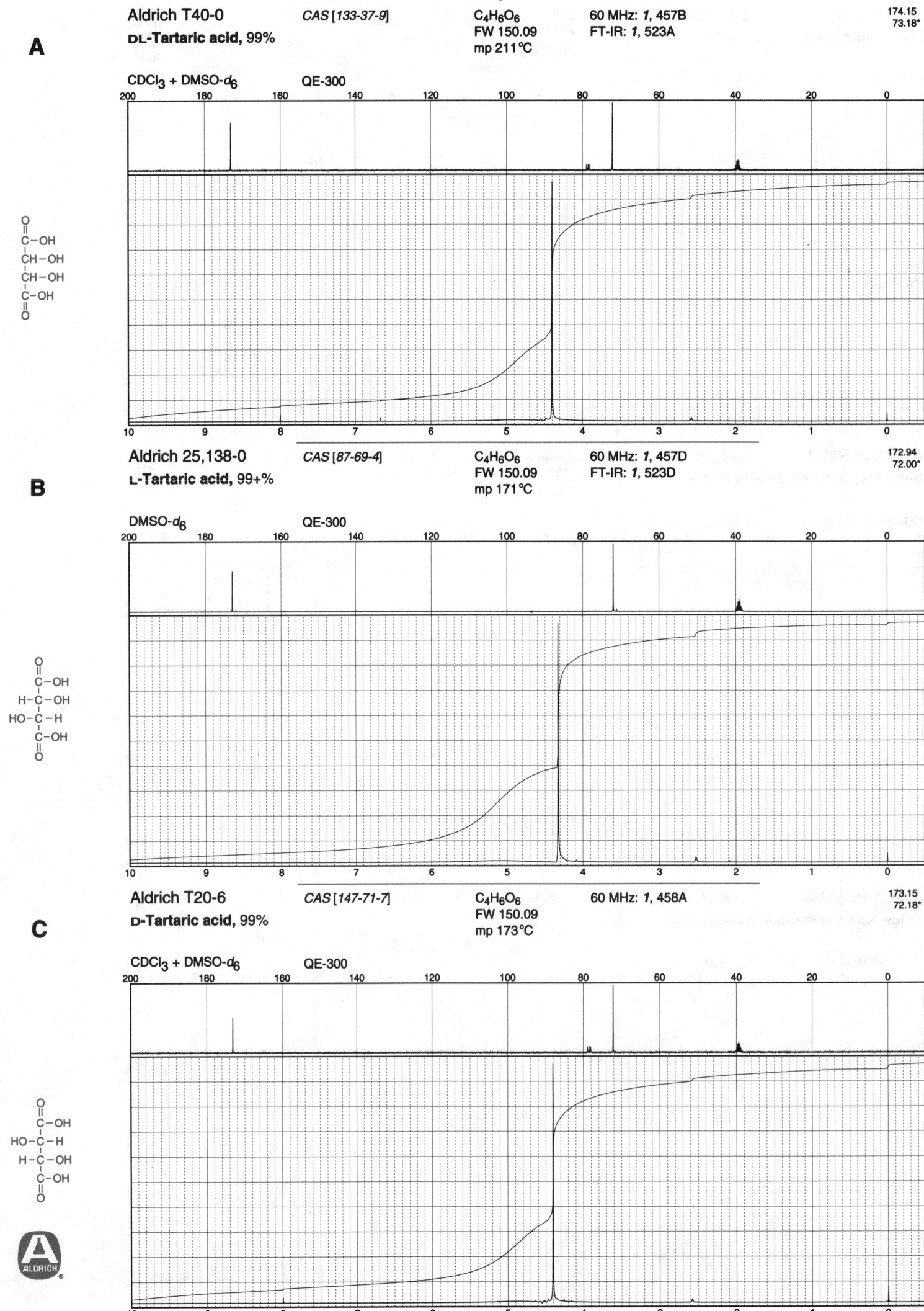

Aldrich T40-0 — CAS [133-37-9] — $C_4H_6O_6$ — FW 150.09 — mp 211°C — 60 MHz: *1*, 457B — FT-IR: *1*, 523A
DL-Tartaric acid, 99%
CDCl$_3$ + DMSO-d_6 — QE-300
174.15 73.18*

Aldrich 25,138-0 — CAS [87-69-4] — $C_4H_6O_6$ — FW 150.09 — mp 171°C — 60 MHz: *1*, 457D — FT-IR: *1*, 523D
L-Tartaric acid, 99+%
DMSO-d_6 — QE-300
172.94 72.00*

Aldrich T20-6 — CAS [147-71-7] — $C_4H_6O_6$ — FW 150.09 — mp 173°C — 60 MHz: *1*, 458A
D-Tartaric acid, 99%
CDCl$_3$ + DMSO-d_6 — QE-300
173.15 72.18*

Aldrich 25,127-5

CAS [77-92-9]

Citric acid, 99.5+%

$C_6H_8O_7$
FW 192.12
mp 153°C

60 MHz: *1*, 458D
FT-IR: *1*, 524B

174.37
171.11
72.31
42.61

A

DMSO-d_6 QE-300

Aldrich 21,783-2

CAS [666-99-9]

Agaricic acid, 97%

$C_{22}H_{40}O_7$
FW 416.56
mp 138°C

FT-IR: *1*, 506A

174.77	29.20	26.47
173.37	29.04	25.87
171.43	28.89	25.14
74.89	28.85	24.33
53.51*	28.70	22.07
40.88	28.21	13.87*
31.28	27.37	

B

DMSO-d_6 QE-300

Aldrich 13,862-2

CAS [77-95-2]

(1R,3R,4R,5R)-(-)-Quinic acid, 98%

$C_7H_{12}O_6$
FW 192.17
mp 168°C d.

60 MHz: *1*, 459C
FT-IR: *1*, 525A

175.38
74.33*
74.24
68.82*
66.51*
40.30
37.24

C

DMSO-d_6 QE-300

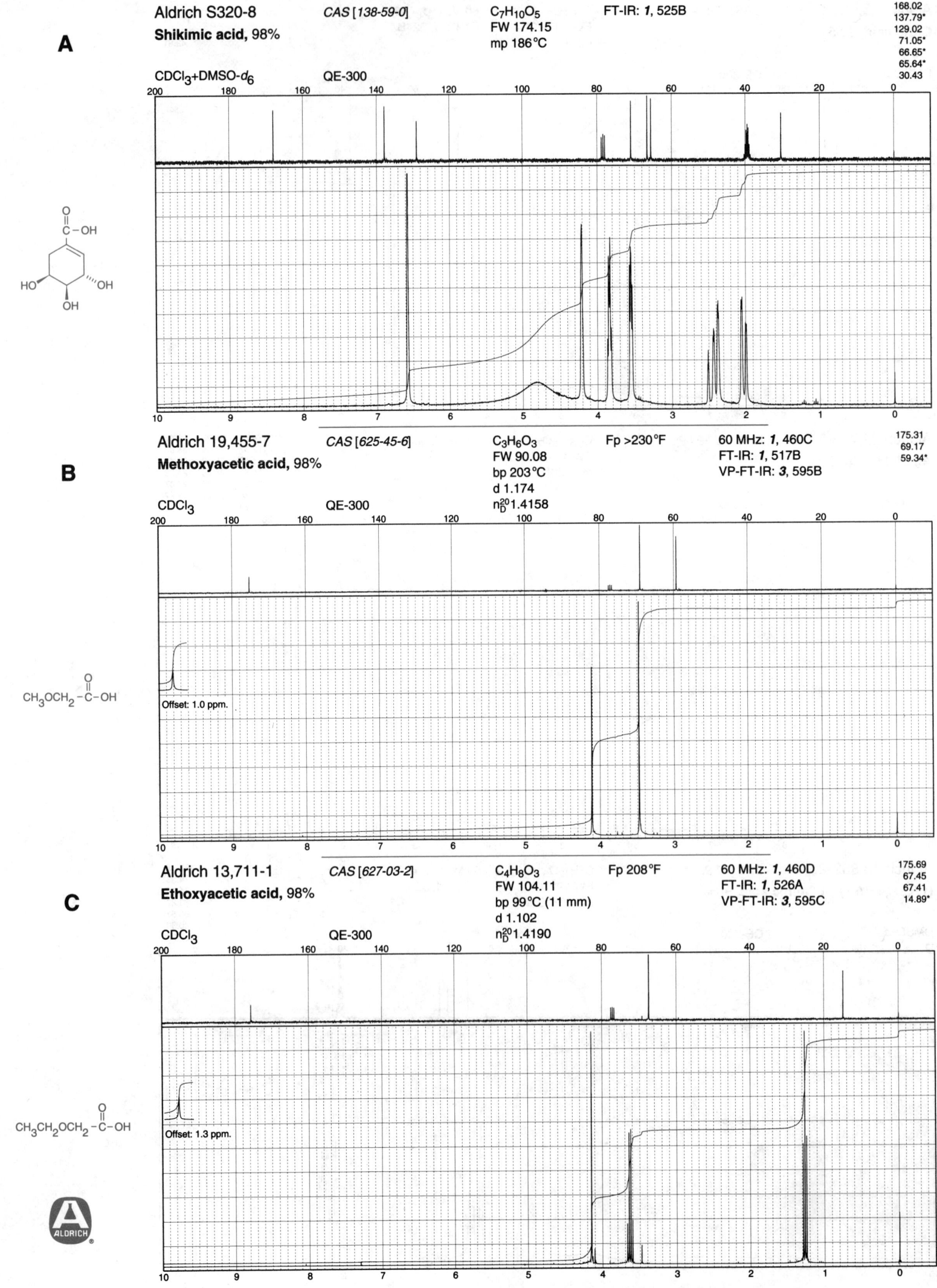

A

Aldrich S320-8
Shikimic acid, 98%
CAS [138-59-0]
$C_7H_{10}O_5$
FW 174.15
mp 186°C
FT-IR: **1**, 525B

168.02
137.79*
129.02
71.05*
66.65*
65.64*
30.43

CDCl$_3$+DMSO-d_6

QE-300

B

Aldrich 19,455-7
Methoxyacetic acid, 98%
CAS [625-45-6]
$C_3H_6O_3$
FW 90.08
bp 203°C
d 1.174
n_D^{20} 1.4158
Fp >230°F
60 MHz: **1**, 460C
FT-IR: **1**, 517B
VP-FT-IR: **3**, 595B

175.31
69.17
59.34*

CDCl$_3$

QE-300

CH$_3$OCH$_2$–C–OH

Offset: 1.0 ppm.

C

Aldrich 13,711-1
Ethoxyacetic acid, 98%
CAS [627-03-2]
$C_4H_8O_3$
FW 104.11
bp 99°C (11 mm)
d 1.102
n_D^{20} 1.4190
Fp 208°F
60 MHz: **1**, 460D
FT-IR: **1**, 526A
VP-FT-IR: **3**, 595C

175.69
67.45
67.41
14.89*

CDCl$_3$

QE-300

CH$_3$CH$_2$OCH$_2$–C–OH

Offset: 1.3 ppm.

ALDRICH

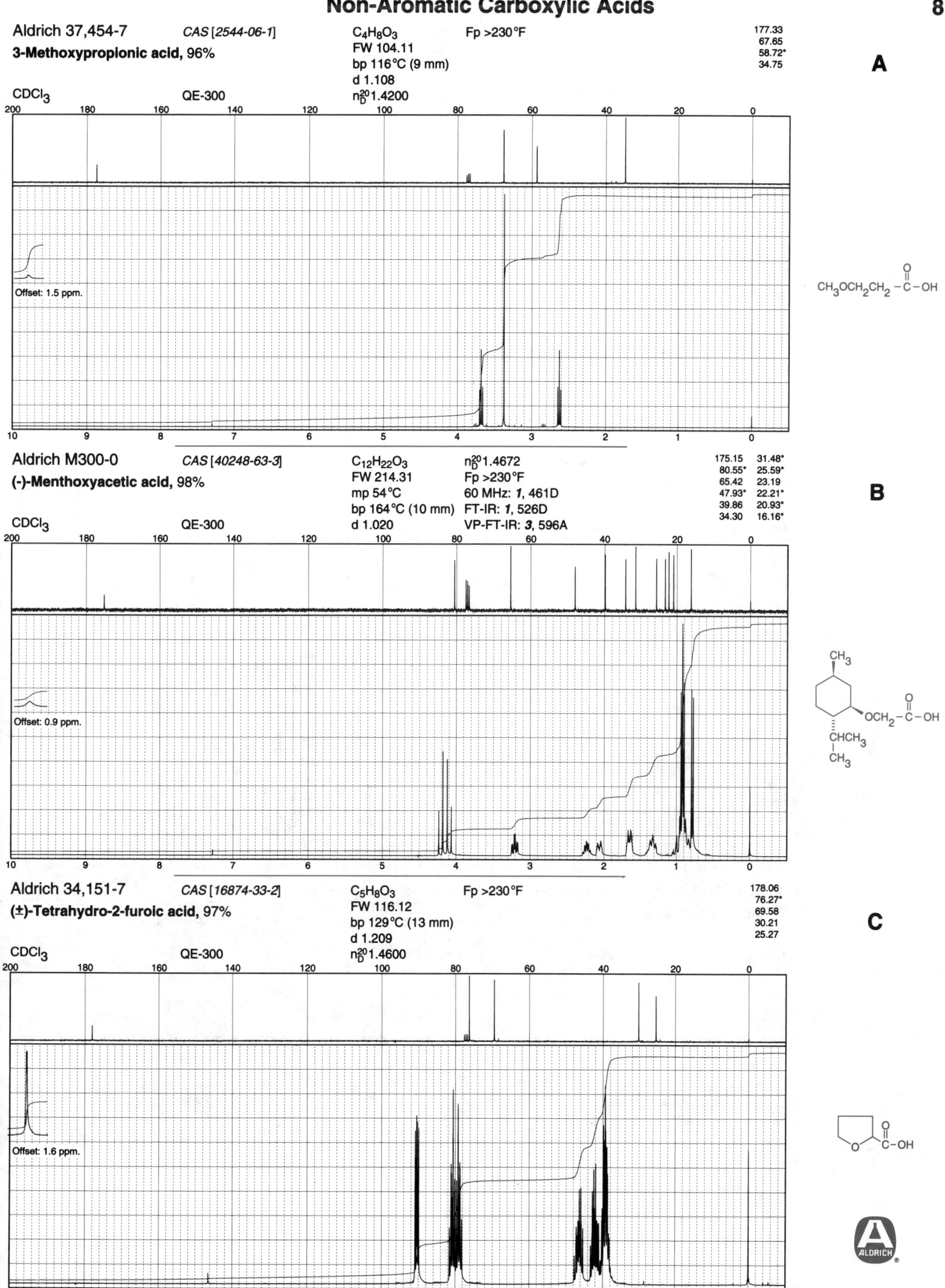

Aldrich 37,454-7 *CAS [2544-06-1]*
3-Methoxypropionic acid, 96%

$C_4H_8O_3$
FW 104.11
bp 116°C (9 mm)
d 1.108
n_D^{20}1.4200

Fp >230°F

177.33
67.65
58.72*
34.75

A

CDCl₃   QE-300

Offset: 1.5 ppm.

$CH_3OCH_2CH_2-\overset{\displaystyle O}{\overset{\|}{C}}-OH$

Aldrich M300-0 *CAS [40248-63-3]*
(-)-Menthoxyacetic acid, 98%

$C_{12}H_{22}O_3$
FW 214.31
mp 54°C
bp 164°C (10 mm)
d 1.020

n_D^{20}1.4672
Fp >230°F
60 MHz: *1*, 461D
FT-IR: *1*, 526D
VP-FT-IR: *3*, 596A

175.15   31.48*
80.55*   25.59*
65.42   23.19
47.93*   22.21*
39.86   20.93*
34.30   16.16*

B

CDCl₃   QE-300

Offset: 0.9 ppm.

Aldrich 34,151-7 *CAS [16874-33-2]*
(±)-Tetrahydro-2-furoic acid, 97%

$C_5H_8O_3$
FW 116.12
bp 129°C (13 mm)
d 1.209
n_D^{20}1.4600

Fp >230°F

178.06
76.27*
69.58
30.21
25.27

C

CDCl₃   QE-300

Offset: 1.6 ppm.

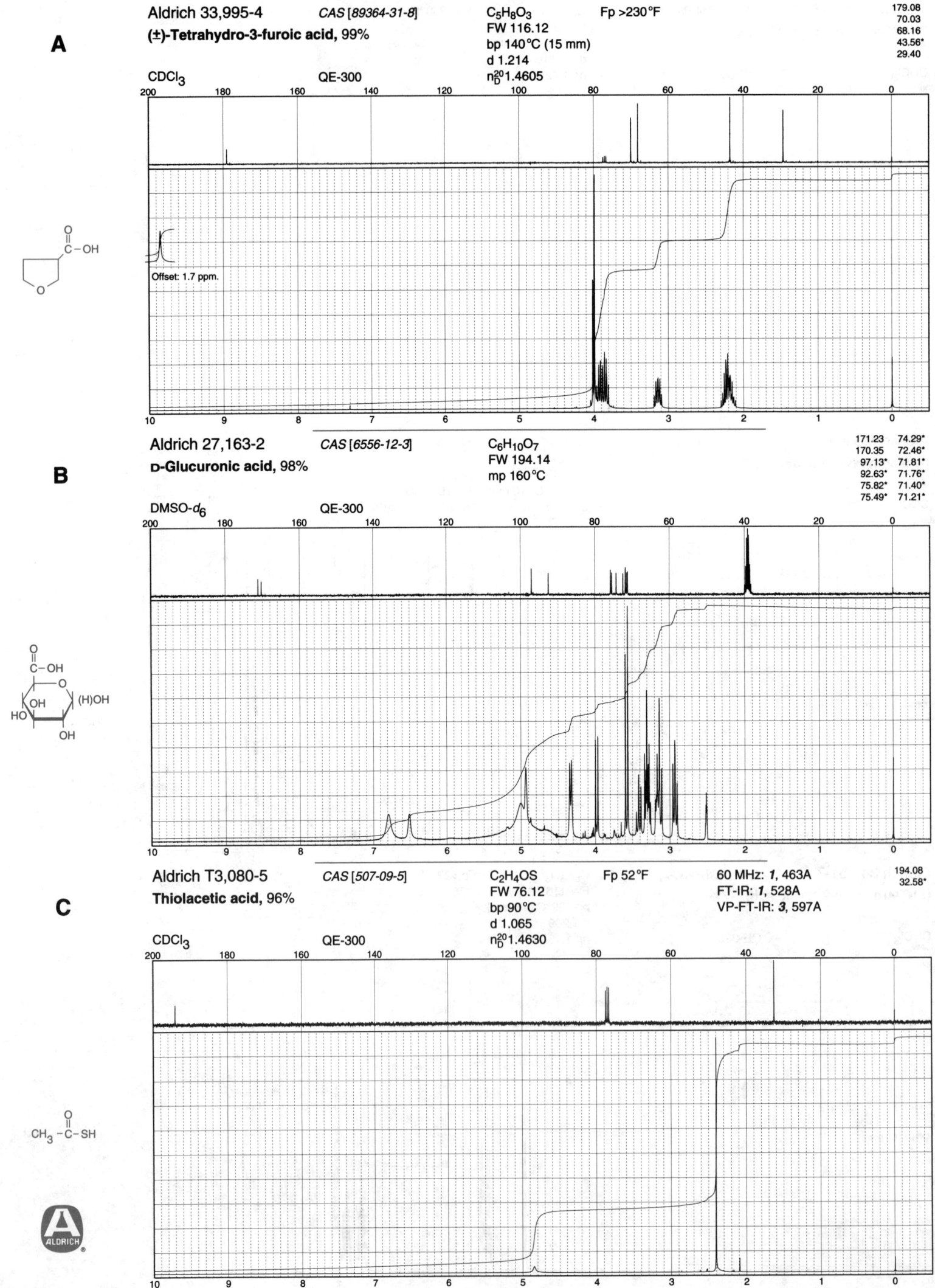

A

Aldrich 33,995-4   *CAS [89364-31-8]* $C_5H_8O_3$ Fp >230°F

(±)-**Tetrahydro-3-furoic acid, 99%**

FW 116.12
bp 140°C (15 mm)
d 1.214
n_D^{20} 1.4605

179.08
70.03
68.16
43.56*
29.40

CDCl$_3$ QE-300

Offset: 1.7 ppm.

B

Aldrich 27,163-2   *CAS [6556-12-3]* $C_6H_{10}O_7$

D-**Glucuronic acid, 98%**

FW 194.14
mp 160°C

171.23   74.29*
170.35   72.46*
97.13*   71.81*
92.63*   71.76*
75.82*   71.40*
75.49*   71.21*

DMSO-d_6 QE-300

C

Aldrich T3,080-5   *CAS [507-09-5]* C_2H_4OS Fp 52°F

Thiolacetic acid, 96%

FW 76.12
bp 90°C
d 1.065
n_D^{20} 1.4630

60 MHz: *1*, 463A
FT-IR: *1*, 528A
VP-FT-IR: *3*, 597A

194.08
32.58*

CDCl$_3$ QE-300

ALDRICH

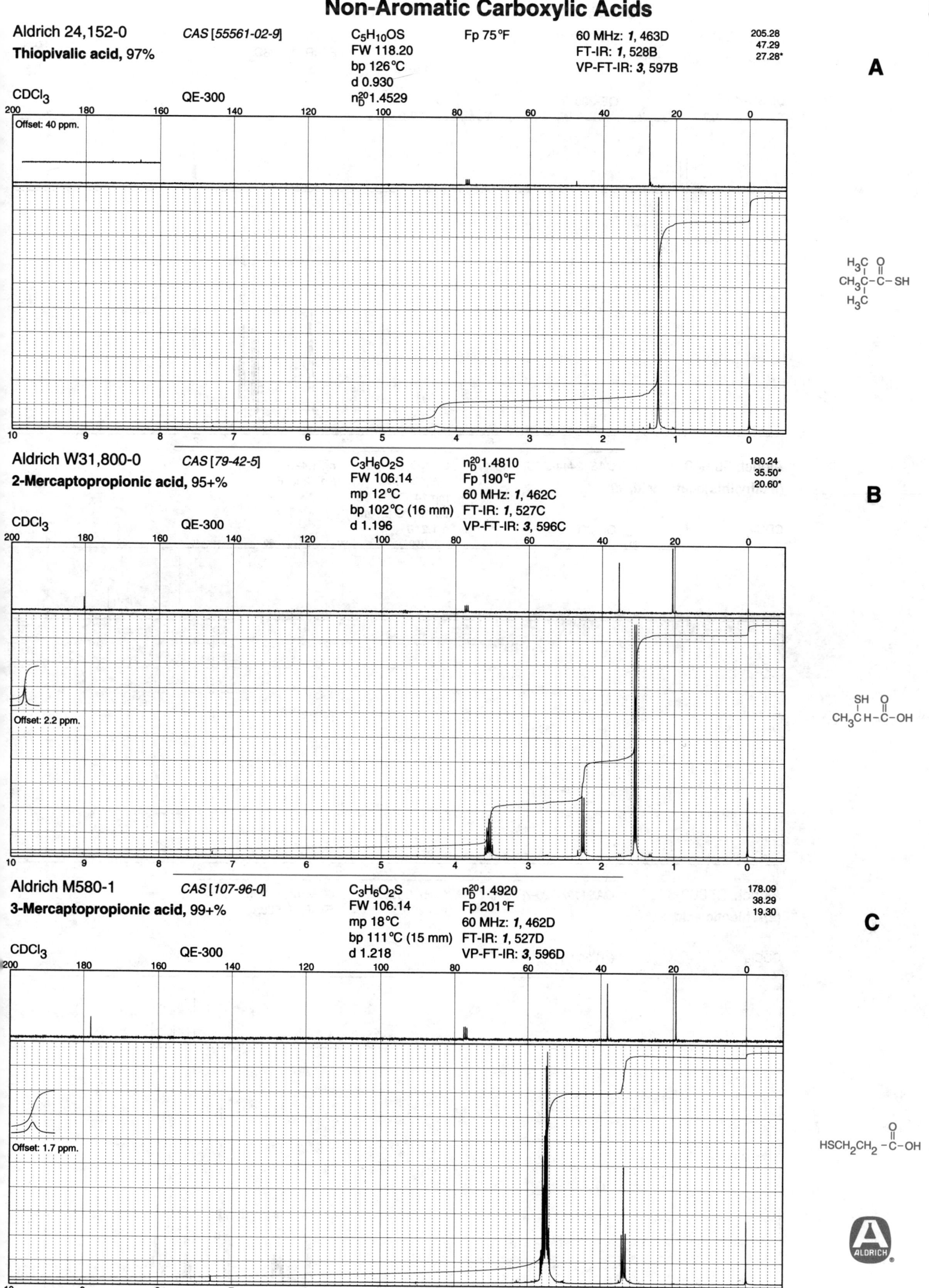

Aldrich 24,152-0 *CAS [55561-02-9]* C₅H₁₀OS Fp 75°F 60 MHz: *1*, 463D 205.28 / 47.29 / 27.28*
Thiopivalic acid, 97% FW 118.20 FT-IR: *1*, 528B
bp 126°C VP-FT-IR: *3*, 597B **A**
d 0.930
CDCl₃ QE-300 n²⁰_D 1.4529

Aldrich W31,800-0 *CAS [79-42-5]* C₃H₆O₂S n²⁰_D 1.4810 180.24 / 35.50* / 20.60*
2-Mercaptopropionic acid, 95+% FW 106.14 Fp 190°F
mp 12°C 60 MHz: *1*, 462C **B**
bp 102°C (16 mm) FT-IR: *1*, 527C
CDCl₃ QE-300 d 1.196 VP-FT-IR: *3*, 596C

Aldrich M580-1 *CAS [107-96-0]* C₃H₆O₂S n²⁰_D 1.4920 178.09 / 38.29 / 19.30
3-Mercaptopropionic acid, 99+% FW 106.14 Fp 201°F
mp 18°C 60 MHz: *1*, 462D
bp 111°C (15 mm) FT-IR: *1*, 527D **C**
CDCl₃ QE-300 d 1.218 VP-FT-IR: *3*, 596D

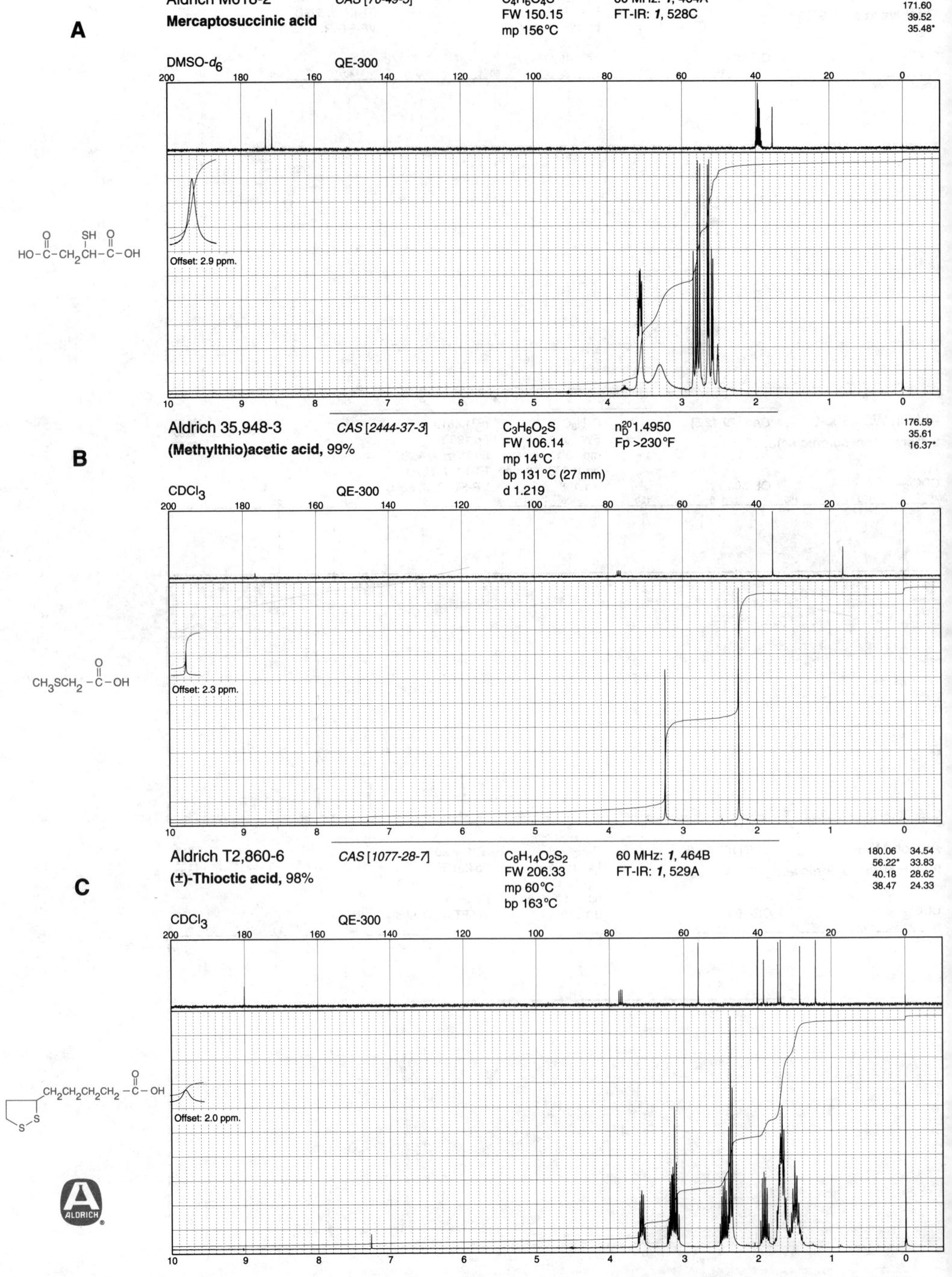

A

Aldrich M618-2 — CAS [70-49-5] — C₄H₆O₄S — 60 MHz: *1*, 464A — 173.32 171.60 39.52 35.48*
Mercaptosuccinic acid — FW 150.15 — FT-IR: *1*, 528C
mp 156°C

DMSO-*d₆* — QE-300

Offset: 2.9 ppm.

Aldrich 35,948-3 — CAS [2444-37-3] — C₃H₆O₂S — n₍D₎²⁰ 1.4950 — 176.59 35.61 16.37*
(Methylthio)acetic acid, 99% — FW 106.14 — Fp >230°F
mp 14°C
bp 131°C (27 mm)
d 1.219

B

CDCl₃ — QE-300

CH₃SCH₂–C–OH

Offset: 2.3 ppm.

Aldrich T2,860-6 — CAS [1077-28-7] — C₈H₁₄O₂S₂ — 60 MHz: *1*, 464B — 180.06 34.54 / 56.22* 33.83 / 40.18 28.62 / 38.47 24.33
(±)-Thioctic acid, 98% — FW 206.33 — FT-IR: *1*, 529A
mp 60°C
bp 163°C

C

CDCl₃ — QE-300

Offset: 2.0 ppm.

ALDRICH

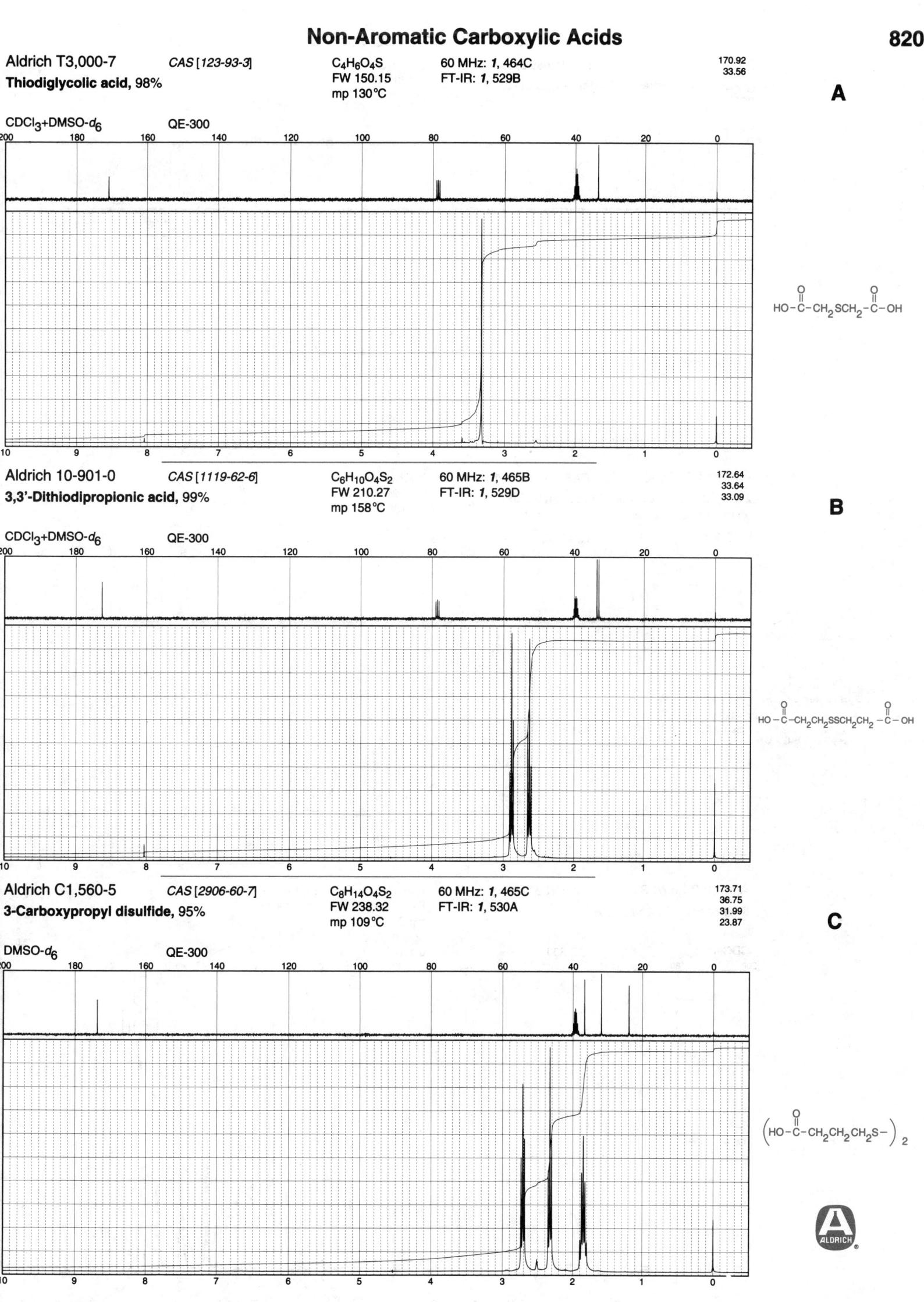

Aldrich T3,000-7   *CAS [123-93-3]* $C_4H_6O_4S$ 60 MHz: *1*, 464C   170.92
Thiodiglycolic acid, 98%   FW 150.15   FT-IR: *1*, 529B   33.56
mp 130°C

A

$CDCl_3$+DMSO-d_6 QE-300

Aldrich 10-901-0   *CAS [1119-62-6]* $C_6H_{10}O_4S_2$ 60 MHz: *1*, 465B   172.64
3,3'-Dithiodipropionic acid, 99%   FW 210.27   FT-IR: *1*, 529D   33.64
mp 158°C   33.09

B

$CDCl_3$+DMSO-d_6 QE-300

Aldrich C1,560-5   *CAS [2906-60-7]* $C_8H_{14}O_4S_2$ 60 MHz: *1*, 465C   173.71
3-Carboxypropyl disulfide, 95%   FW 238.32   FT-IR: *1*, 530A   36.75
mp 109°C   31.99
23.87

C

DMSO-d_6 QE-300

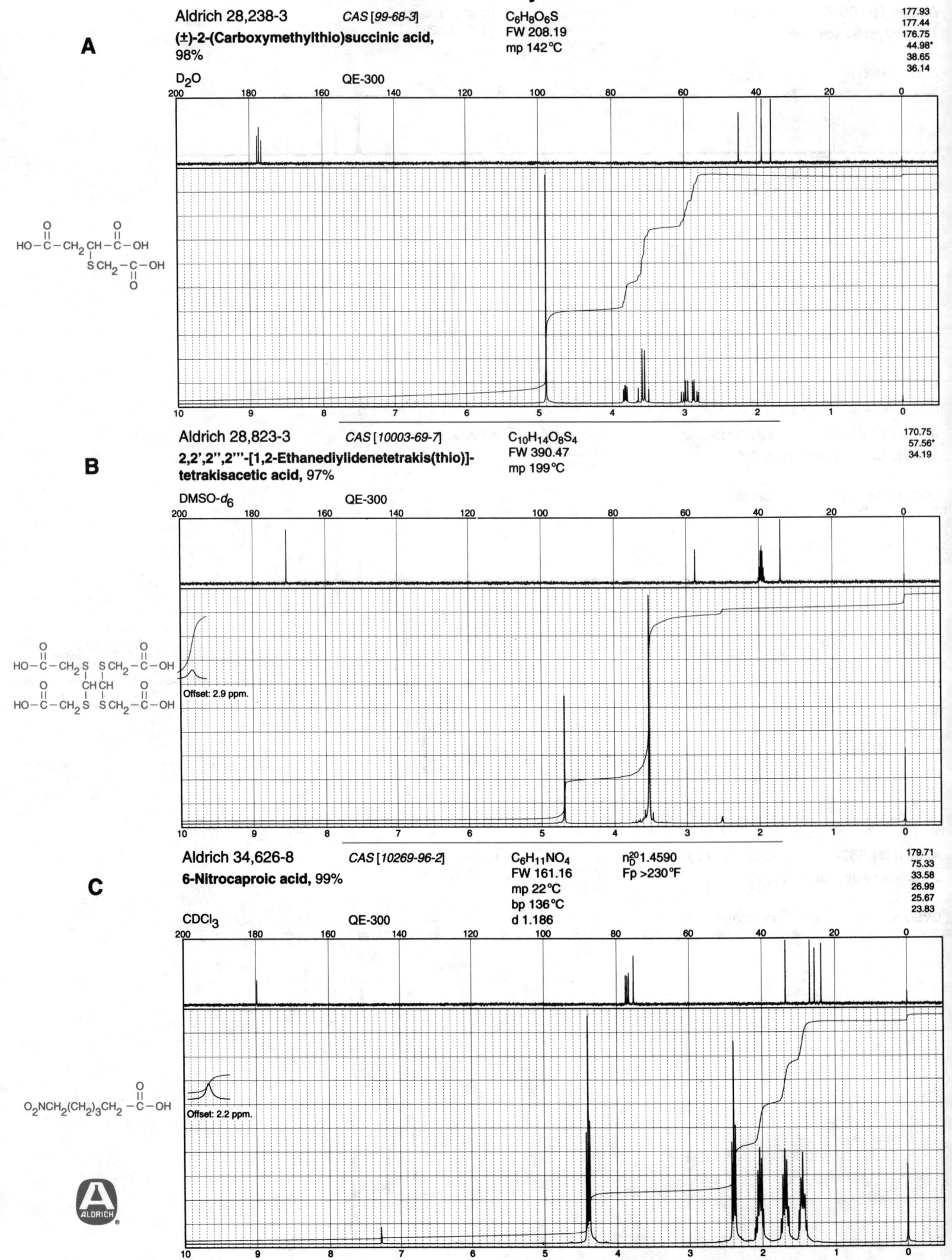

A

Aldrich 28,238-3   *CAS [99-68-3]*   C₆H₈O₆S
FW 208.19
mp 142°C

(±)-2-(Carboxymethylthio)succinic acid, 98%

177.93
177.44
176.75
44.98*
38.65
36.14

D₂O   QE-300

B

Aldrich 28,823-3   *CAS [10003-69-7]*   C₁₀H₁₄O₈S₄
FW 390.47
mp 199°C

2,2',2'',2'''-[1,2-Ethanediylidenetetrakis(thio)]-tetrakisacetic acid, 97%

170.75
57.56*
34.19

DMSO-d₆   QE-300

Offset: 2.9 ppm.

C

Aldrich 34,626-8   *CAS [10269-96-2]*   C₆H₁₁NO₄   n²⁰_D 1.4590
FW 161.16   Fp >230°F
mp 22°C
bp 136°C
d 1.186

6-Nitrocaproic acid, 99%

179.71
75.33
33.58
26.99
25.67
23.83

CDCl₃   QE-300

Offset: 2.2 ppm.

Non-Aromatic Carboxylic Acids

Aldrich 36,123-2   CAS [59085-15-3]   C$_{10}$H$_{15}$NO$_8$

Nitromethanetrispropionic acid, 97%

FW 277.23
mp 183°C

172.90
92.74
29.78
28.19

A

DMSO-d_6 QE-300

Offset: 2.5 ppm.

HO—C—CH$_2$CH$_2$ —CNO$_2$
CH$_2$CH$_2$—C—OH
CH$_2$CH$_2$—C—OH

Aldrich G1,060-1   CAS [563-96-2]   C$_2$H$_2$O$_3$

Glyoxylic acid monohydrate, 98%

FW 92.05
mp 51°C
bp 100°C
Fp >230°F

FT-IR: 1, 530B

172.17
170.15
170.02
90.23*
88.21*
86.72*

B

CDCl$_3$+DMSO-d_6 QE-300

H—C—C—OH · H$_2$O

Aldrich 28,097-6   CAS [39748-49-7]   C$_6$H$_{10}$O$_3$

(±)-3-Methyl-2-oxopentanoic acid, 98%

FW 130.15
mp 43°C
bp 79°C (12 mm)
Fp 180°F

198.93
161.08
42.23*
25.18
14.76*
11.34*

C

CDCl$_3$ QE-300

CH$_3$ O O
CH$_3$CH$_2$CH—C—C—OH

Non-Aromatic Carboxylic Acids

A

Aldrich L200-9   *CAS [123-76-2]* $C_5H_8O_3$ $n_D^{20}1.4396$ 206.87
Levulinic acid, 98%   FW 116.12   Fp 280°F   178.53
                                         mp 32°C   60 MHz: *1*, 468A   37.67
                                         bp 246°C   FT-IR: *1*, 531B   29.77*
                                                                          27.81

CDCl₃   QE-300   d 1.134

Offset: 40 ppm.

$CH_3-\overset{O}{\underset{}{C}}-CH_2CH_2-\overset{O}{\underset{}{C}}-OH$

Offset: 1.0 ppm.

B

Aldrich A1,320-4   *CAS [3128-06-1]* $C_6H_{10}O_3$ $n_D^{20}1.4451$ 208.44
4-Acetylbutyric acid, 97%   FW 130.14   Fp >230°F   179.08
                                         mp 14°C   60 MHz: *1*, 468B   42.29
                                         bp 275°C   FT-IR: *1*, 531C   32.93
                                                      VP-FT-IR: *3*, 597D 29.91*
                                                                          18.54

CDCl₃   QE-300   d 1.090

Offset: 40 ppm.

$CH_3-\overset{O}{\underset{}{C}}-CH_2CH_2CH_2-\overset{O}{\underset{}{C}}-OH$

Offset: 1.4 ppm.

C

Aldrich 34,362-5   *CAS [14112-98-2]* $C_8H_{14}O_3$ 209.23   29.88*
7-Oxooctanoic acid, 98%   FW 158.20   179.73   28.48
                                         mp 28°C   43.37   24.39
                                         bp 161°C (4 mm)   33.83   23.31
                                         Fp >230°F

CDCl₃   QE-300

Offset: 40 ppm.

$CH_3-\overset{O}{\underset{}{C}}-CH_2(CH_2)_3CH_2-\overset{O}{\underset{}{C}}-OH$

Offset: 1.8 ppm.

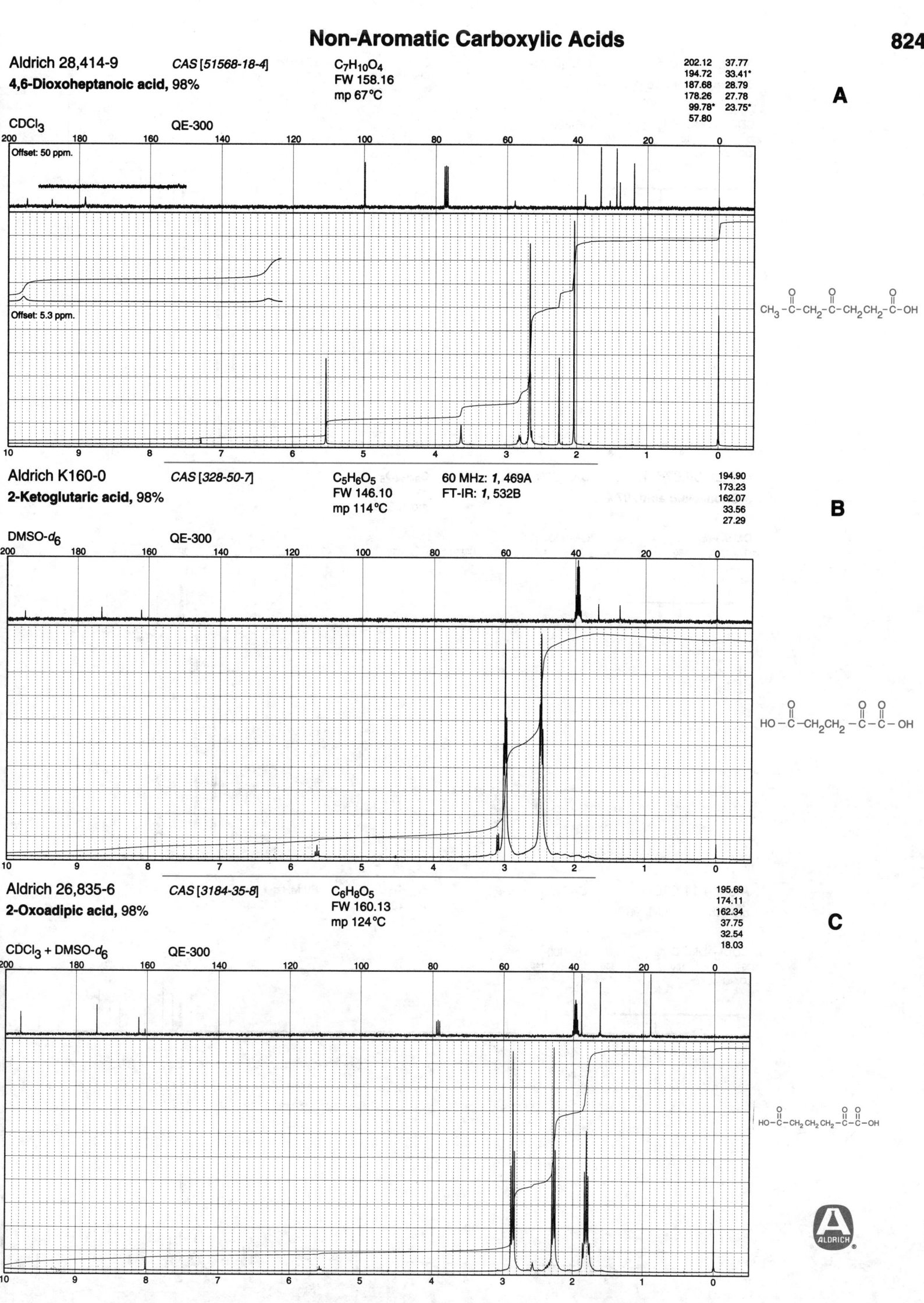

Aldrich 28,414-9   CAS [51568-18-4] $C_7H_{10}O_4$ FW 158.16   mp 67°C

4,6-Dioxoheptanoic acid, 98%

CDCl$_3$ QE-300

202.12   37.77
194.72   33.41*
187.68   28.79
178.26   27.78
99.78*   23.75*
57.80

A

Offset: 50 ppm.
Offset: 5.3 ppm.

$CH_3-C(=O)-CH_2-C(=O)-CH_2CH_2-C(=O)-OH$

Aldrich K160-0   CAS [328-50-7] $C_5H_6O_5$ FW 146.10   mp 114°C   60 MHz: 1, 469A   FT-IR: 1, 532B

2-Ketoglutaric acid, 98%

194.90
173.23
162.07
33.56
27.29

B

DMSO-d_6 QE-300

$HO-C(=O)-CH_2CH_2-C(=O)-C(=O)-OH$

Aldrich 26,835-6   CAS [3184-35-8] $C_6H_8O_5$ FW 160.13   mp 124°C

2-Oxoadipic acid, 98%

195.69
174.11
162.34
37.75
32.54
18.03

C

CDCl$_3$ + DMSO-d_6 QE-300

$HO-C(=O)-CH_2 CH_2 CH_2-C(=O)-C(=O)-OH$

A

Aldrich K350-6

4-Ketopimelic acid, 98%

CAS [502-50-1]

C$_7$H$_{10}$O$_5$
FW 174.15
mp 143°C

60 MHz: *1*, 469C
FT-IR: *1*, 532D

207.14
173.71
36.71
27.66

CDCl$_3$+DMSO-d_6 QE-300

Offset: 40 ppm.

HO–C–CH$_2$–CH$_2$–C–CH$_2$–CH$_2$–C–OH

B

Aldrich 36,976-4

5-Oxoazelaic acid, 97%

CAS [57822-06-7]

C$_9$H$_{14}$O$_5$
FW 202.21
mp 108°C

209.55
174.05
40.84
32.71
18.59

DMSO-d_6 QE-300

Offset: 40 ppm.

Offset: 2.2 ppm.

HO–C–CH$_2$–CH$_2$–CH$_2$–C–CH$_2$–CH$_2$–C–OH

C

Aldrich 11,010-8

cis-Pinonic acid, 98%

CAS [61826-55-9]

C$_{10}$H$_{16}$O$_3$
FW 184.24
mp 106°C

60 MHz: *1*, 469D
FT-IR: *1*, 533A

206.49	34.84
173.73	29.89*
53.54*	22.84
42.72	16.98*
37.61*	

CDCl$_3$+DMSO-d_6 QE-300

Offset: 40 ppm.

CH$_3$ CH$_3$

CH$_3$–C–⟨ ⟩–CH$_2$–C–OH

Aldrich 12,495-8   CAS [6003-94-7]   C₇H₄O₆   60 MHz: 1, 479C

$C_7H_4O_6$ 60 MHz: 1, 479C

Chelidonic acid monohydrate, 96%   FW 202.12   FT-IR: 1, 533C

mp 265°C d.

179.16
160.64
153.94
118.69*

A

DMSO-d_6 QE-300

HO–C ⟋ ⟍ C–OH · H₂O

Aldrich C11,660-2   CAS [1759-53-1] $C_4H_6O_2$ n_D^{20} 1.4380

181.69
12.89*
9.16

Cyclopropanecarboxylic acid, 95%   FW 86.09   Fp 161°F

mp 18°C   60 MHz: 1, 470A

bp 183°C   FT-IR: 1, 533D

d 1.088   VP-FT-IR: 3, 598B

B

CDCl₃   QE-300

Offset: 2.3 ppm.

Aldrich 20,560-5   CAS [6914-76-7] $C_5H_8O_2$ 60 MHz: 1, 470B

183.29
18.89*
18.41
17.58

1-Methylcyclopropanecarboxylic acid, 98%   FW 100.12   FT-IR: 1, 534A

mp 35°C   VP-FT-IR: 3, 598C

bp 184°C

Fp 184°F

C

CDCl₃   QE-300

Offset: 2.6 ppm.

ALDRICH

Non-Aromatic Carboxylic Acids

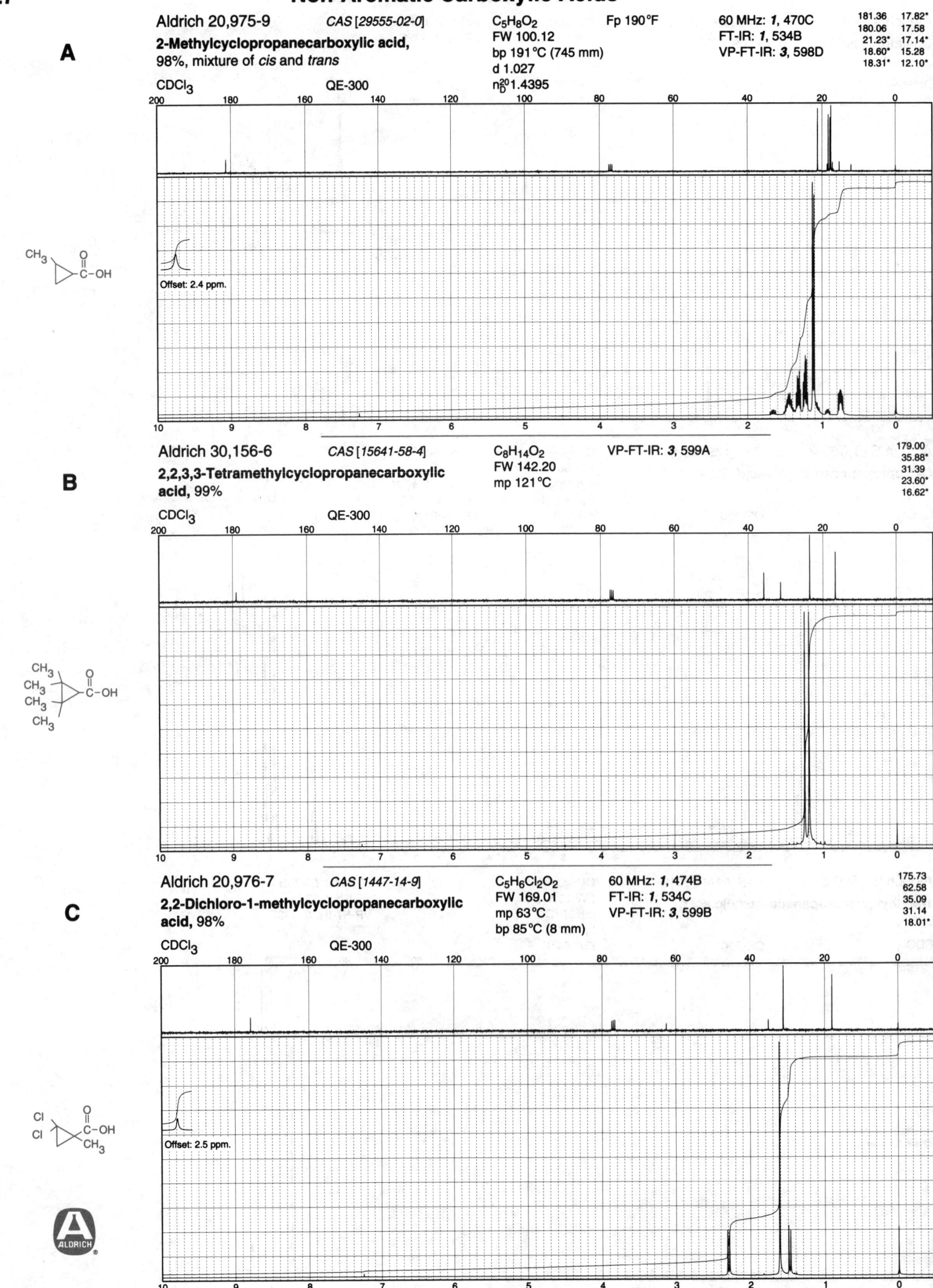

A

Aldrich 20,975-9   *CAS [29555-02-0]* $C_5H_8O_2$ Fp 190°F   60 MHz: *1*, 470C   181.36 17.82*
2-Methylcyclopropanecarboxylic acid,   FW 100.12   FT-IR: *1*, 534B   180.06 17.58
98%, mixture of *cis* and *trans*   bp 191°C (745 mm)   VP-FT-IR: *3*, 598D   21.23* 17.14*
d 1.027   18.60* 15.28
n_D^{20}1.4395   18.31* 12.10*

CDCl$_3$ QE-300

Offset: 2.4 ppm.

Aldrich 30,156-6   *CAS [15641-58-4]* $C_8H_{14}O_2$ VP-FT-IR: *3*, 599A   179.00
2,2,3,3-Tetramethylcyclopropanecarboxylic   FW 142.20   35.88*
acid, 99%   mp 121°C   31.39
23.60*
16.62*

B

CDCl$_3$ QE-300

Aldrich 20,976-7   *CAS [1447-14-9]* $C_5H_6Cl_2O_2$ 60 MHz: *1*, 474B   175.73
2,2-Dichloro-1-methylcyclopropanecarboxylic   FW 169.01   FT-IR: *1*, 534C   62.58
acid, 98%   mp 63°C   VP-FT-IR: *3*, 599B   35.09
bp 85°C (8 mm)   31.14
18.01*

C

CDCl$_3$ QE-300

Offset: 2.5 ppm.

ALDRICH

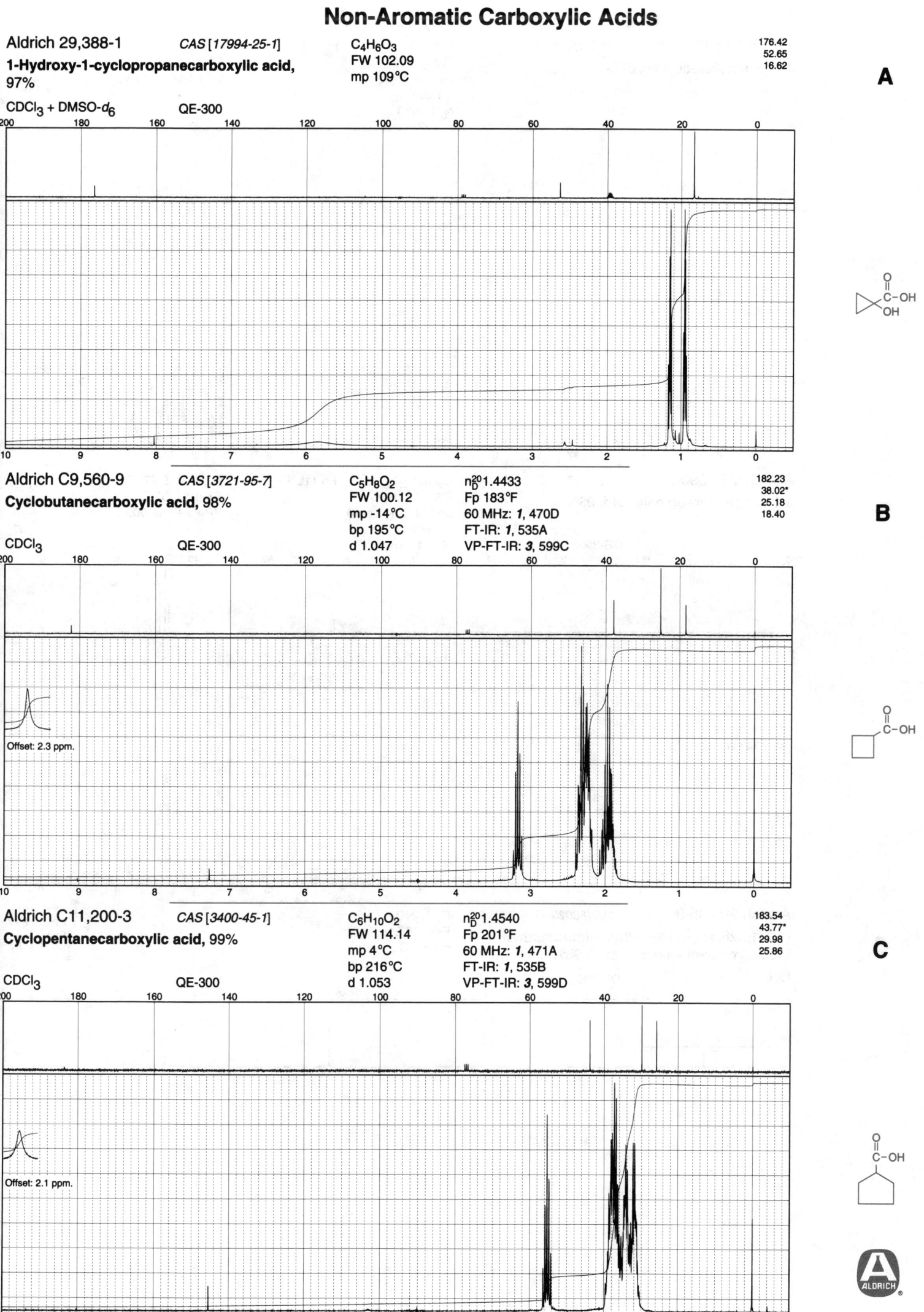

Aldrich 29,388-1   CAS [17994-25-1] $C_4H_6O_3$ 176.42 / 52.65 / 16.62

A

1-Hydroxy-1-cyclopropanecarboxylic acid, 97%   FW 102.09   mp 109°C

$CDCl_3$ + DMSO-d_6 QE-300

Aldrich C9,560-9   CAS [3721-95-7] $C_5H_8O_2$ n_D^{20}1.4433   182.23 / 38.02* / 25.18 / 18.40

B

Cyclobutanecarboxylic acid, 98%   FW 100.12   Fp 183°F
mp -14°C   60 MHz: 1, 470D
bp 195°C   FT-IR: 1, 535A

$CDCl_3$ QE-300   d 1.047   VP-FT-IR: 3, 599C

Offset: 2.3 ppm.

Aldrich C11,200-3   CAS [3400-45-1] $C_6H_{10}O_2$ n_D^{20}1.4540   183.54 / 43.77* / 29.98 / 25.86

C

Cyclopentanecarboxylic acid, 99%   FW 114.14   Fp 201°F
mp 4°C   60 MHz: 1, 471A
bp 216°C   FT-IR: 1, 535B

$CDCl_3$ QE-300   d 1.053   VP-FT-IR: 3, 599D

Offset: 2.1 ppm.

ALDRICH

Non-Aromatic Carboxylic Acids

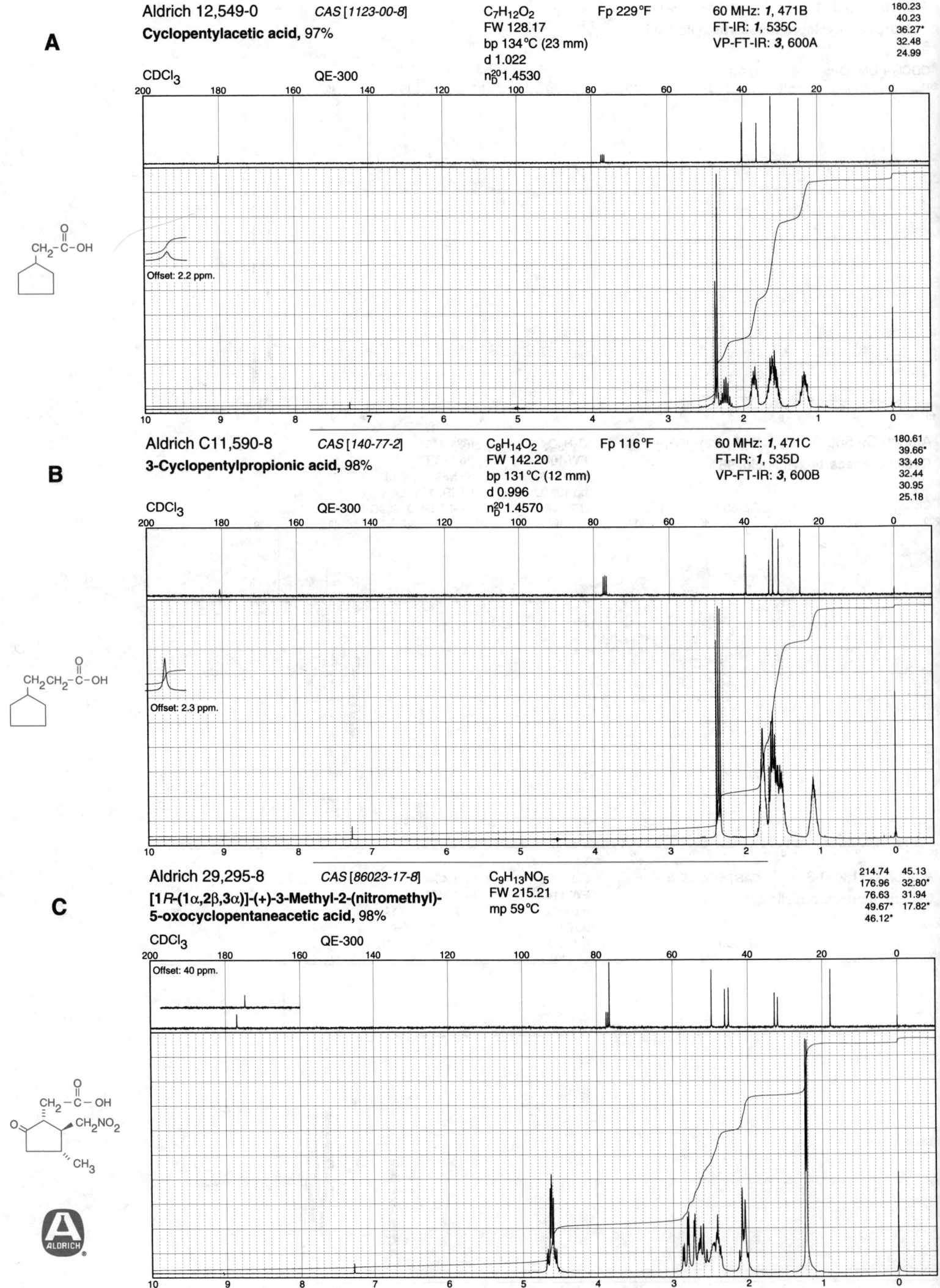

A

Aldrich 12,549-0   CAS [1123-00-8] $C_7H_{12}O_2$ Fp 229°F   60 MHz: *1*, 471B   180.23
Cyclopentylacetic acid, 97%   FW 128.17   FT-IR: *1*, 535C   40.23
bp 134°C (23 mm)   VP-FT-IR: *3*, 600A   36.27*
d 1.022   32.48
n_D^{20}1.4530   24.99

CDCl₃   QE-300

Offset: 2.2 ppm.

B

Aldrich C11,590-8   CAS [140-77-2] $C_8H_{14}O_2$ Fp 116°F   60 MHz: *1*, 471C   180.61
3-Cyclopentylpropionic acid, 98%   FW 142.20   FT-IR: *1*, 535D   39.66*
bp 131°C (12 mm)   VP-FT-IR: *3*, 600B   33.49
d 0.996   32.44
n_D^{20}1.4570   30.95
   25.18

CDCl₃   QE-300

Offset: 2.3 ppm.

C

Aldrich 29,295-8   CAS [86023-17-8] $C_9H_{13}NO_5$ 214.74   45.13
[1*R*-(1α,2β,3α)]-(+)-3-Methyl-2-(nitromethyl)-   FW 215.21   176.96   32.80*
5-oxocyclopentaneacetic acid, 98%   mp 59°C   76.63   31.94
   49.67*   17.82*
   46.12*

CDCl₃   QE-300

Offset: 40 ppm.

Aldrich C10,450-7   *CAS [5292-21-7]*

Cyclohexylacetic acid, 99%

DMSO-d_6 QE-300

$C_8H_{14}O_2$
FW 142.20
mp 32°C
bp 243°C
d 1.007

n_D^{20} 1.4630
Fp >230°F
60 MHz: *1*, 472B
FT-IR: *1*, 537A
VP-FT-IR: *3*, 602B

173.60
41.54
34.20*
32.41
25.75
25.59

A

Offset: 2.2 ppm.

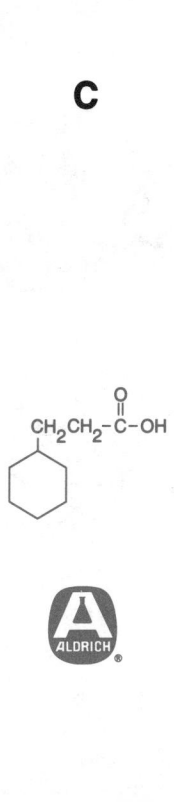

Aldrich 33,384-0   *CAS [52034-92-1]*

Dicyclohexylacetic acid, 99%

DMSO-d_6 QE-300

$C_{14}H_{24}O_2$
FW 224.35
mp 140°C

175.26   29.13
56.50*   26.15
35.68*   26.05
30.80   25.91

B

Offset: 2.1 ppm.

Aldrich 16,147-0   *CAS [701-97-3]*

Cyclohexanepropionic acid, 99%

CDCl$_3$ QE-300

$C_9H_{16}O_2$
FW 156.23
mp 16°C
bp 276°C
d 0.912

n_D^{20} 1.4636
Fp >230°F
60 MHz: *1*, 472C
FT-IR: *1*, 537B
VP-FT-IR: *3*, 602C

180.95
37.16*
32.97
32.08
31.75
26.55
26.23

C

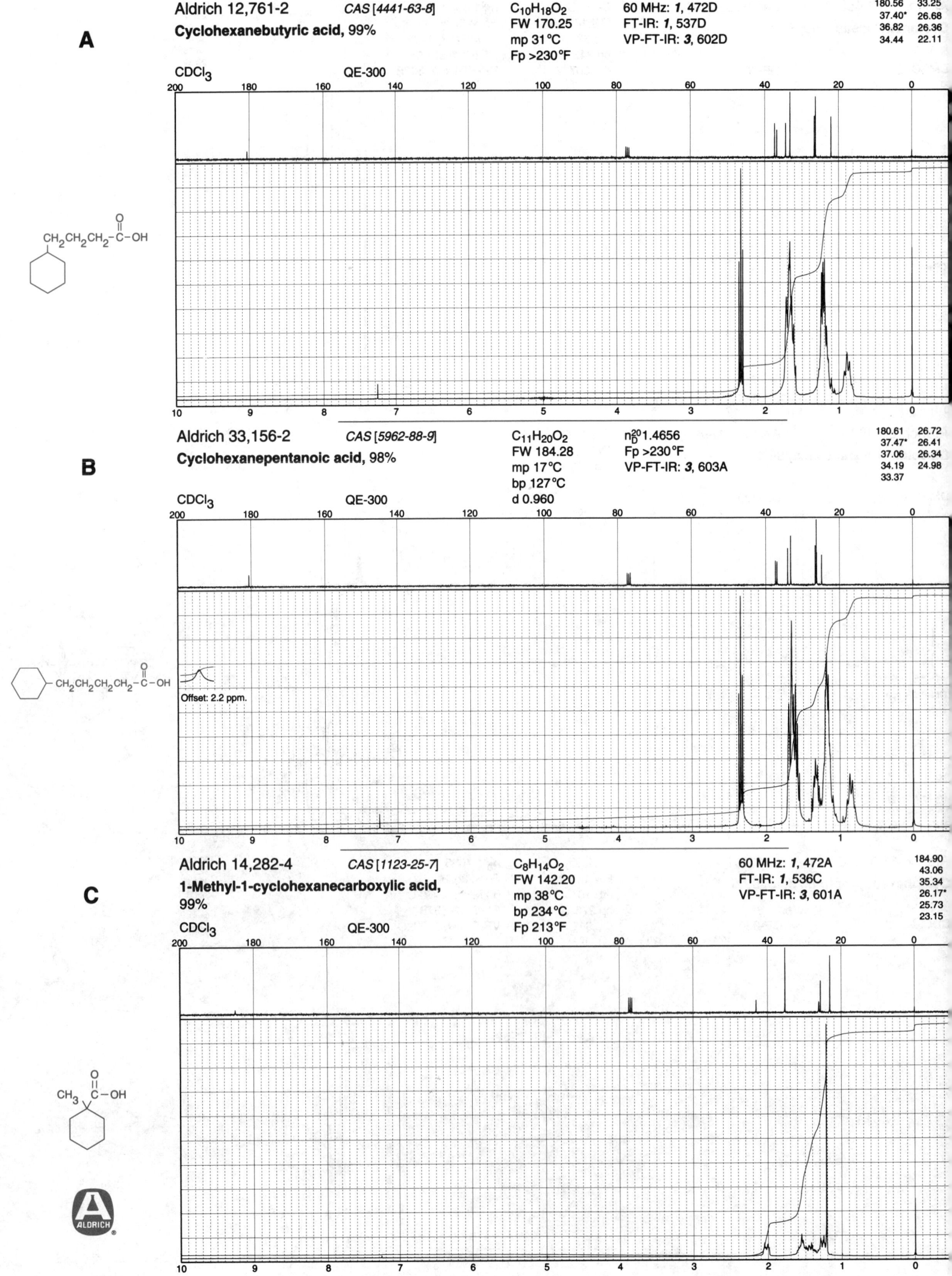

A

Aldrich 12,761-2   *CAS [4441-63-8]*

Cyclohexanebutyric acid, 99%

$C_{10}H_{18}O_2$
FW 170.25
mp 31°C
Fp >230°F

60 MHz: *1*, 472D
FT-IR: *1*, 537D
VP-FT-IR: *3*, 602D

180.56	33.25
37.40*	26.68
36.82	26.36
34.44	22.11

CDCl₃   QE-300

B

Aldrich 33,156-2   *CAS [5962-88-9]*

Cyclohexanepentanoic acid, 98%

$C_{11}H_{20}O_2$
FW 184.28
mp 17°C
bp 127°C
d 0.960

n_D^{20} 1.4656
Fp >230°F
VP-FT-IR: *3*, 603A

180.61	26.72
37.47*	26.41
37.06	26.34
34.19	24.98
33.37	

CDCl₃   QE-300

Offset: 2.2 ppm.

C

Aldrich 14,282-4   *CAS [1123-25-7]*

1-Methyl-1-cyclohexanecarboxylic acid, 99%

$C_8H_{14}O_2$
FW 142.20
mp 38°C
bp 234°C
Fp 213°F

60 MHz: *1*, 472A
FT-IR: *1*, 536C
VP-FT-IR: *3*, 601A

184.90	
43.06	
35.34	
26.17*	
25.73	
23.15	

CDCl₃   QE-300

Non-Aromatic Carboxylic Acids

Aldrich 33,060-4 CAS [56586-13-1] C$_8$H$_{14}$O$_2$ Fp 180°F VP-FT-IR: **3**, 601B

(±)-2-Methyl-1-cyclohexanecarboxylic
acid, 99%, mixture of *cis* and *trans*

CDCl$_3$ QE-300

FW 142.20
bp 242°C (746 mm)
d 1.009
n$_D^{20}$ 1.4633

183.19	31.81	23.77
182.17	31.10*	21.56
51.34*	29.87	20.65*
45.94*	25.84	15.23*
34.20	25.38	
34.09*	24.43	

A

Offset: 2.5 ppm.

Aldrich 33,061-2 CAS [13293-59-9] C$_8$H$_{14}$O$_2$ VP-FT-IR: **3**, 601C

(±)-3-Methyl-1-cyclohexanecarboxylic
acid, 98%, mixture of *cis* and *trans*

CDCl$_3$ QE-300

FW 142.20
d 0.989
n$_D^{20}$ 1.4590
Fp >230°F

182.96	34.40	25.45
182.75	33.70	22.59*
43.35*	32.04*	21.93
39.14*	28.45	21.28*
37.21	28.34*	
35.28	27.46	

B

Offset: 2.5 ppm.

Aldrich 33,062-0 CAS [4331-54-8] C$_8$H$_{14}$O$_2$ Fp >230°F VP-FT-IR: **3**, 601D

4-Methyl-1-cyclohexanecarboxylic acid,
99%, mixture of *cis* and *trans*

CDCl$_3$ QE-300

FW 142.20
bp 135°C (15 mm)
d 1.005
n$_D^{20}$ 1.4598

183.16	31.32
182.56	30.59*
43.00*	28.83
40.22*	25.91
34.16	22.45*
31.96*	21.04*

C

Offset: 2.3 ppm.

ALDRICH

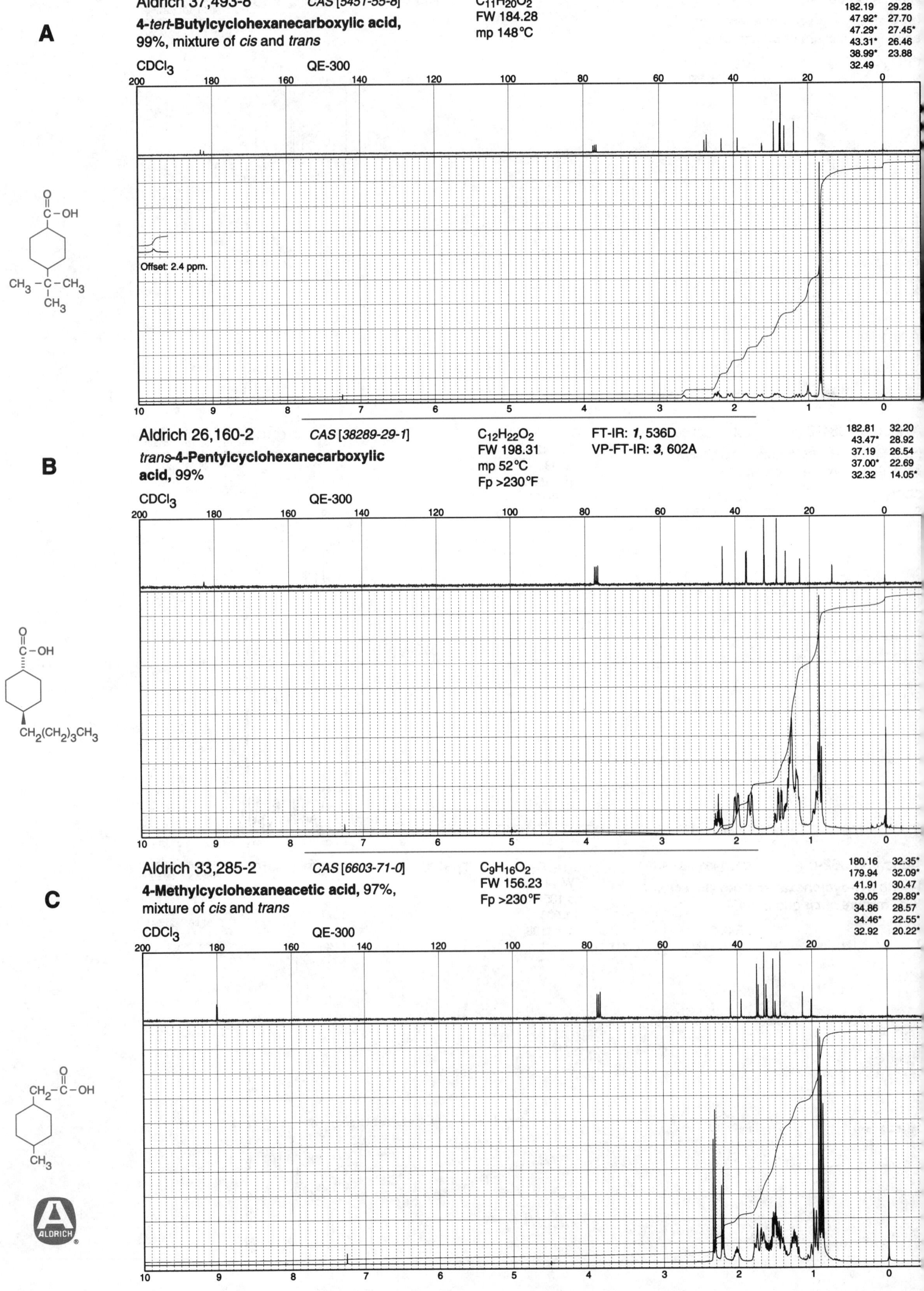

A

Aldrich 37,493-8   *CAS [5451-55-8]* $C_{11}H_{20}O_2$
FW 184.28
mp 148°C

4-*tert*-Butylcyclohexanecarboxylic acid,
99%, mixture of *cis* and *trans*

183.05	32.39
182.19	29.28
47.92*	27.70
47.29*	27.45
43.31*	26.46
38.99*	23.88
32.49	

CDCl₃   QE-300

Offset: 2.4 ppm.

B

Aldrich 26,160-2   *CAS [38289-29-1]* $C_{12}H_{22}O_2$
FW 198.31
mp 52°C
Fp >230°F

trans-4-Pentylcyclohexanecarboxylic
acid, 99%

FT-IR: *1*, 536D
VP-FT-IR: *3*, 602A

182.81	32.20
43.47*	28.92
37.19	26.54
37.00*	22.69
32.32	14.05*

CDCl₃   QE-300

C

Aldrich 33,285-2   *CAS [6603-71-0]* $C_9H_{16}O_2$
FW 156.23
Fp >230°F

4-Methylcyclohexaneacetic acid, 97%,
mixture of *cis* and *trans*

180.16	32.35*
179.94	32.09*
41.91	30.47*
39.05	29.89*
34.86	28.57
34.46*	22.55*
32.92	20.22*

CDCl₃   QE-300

ALDRICH

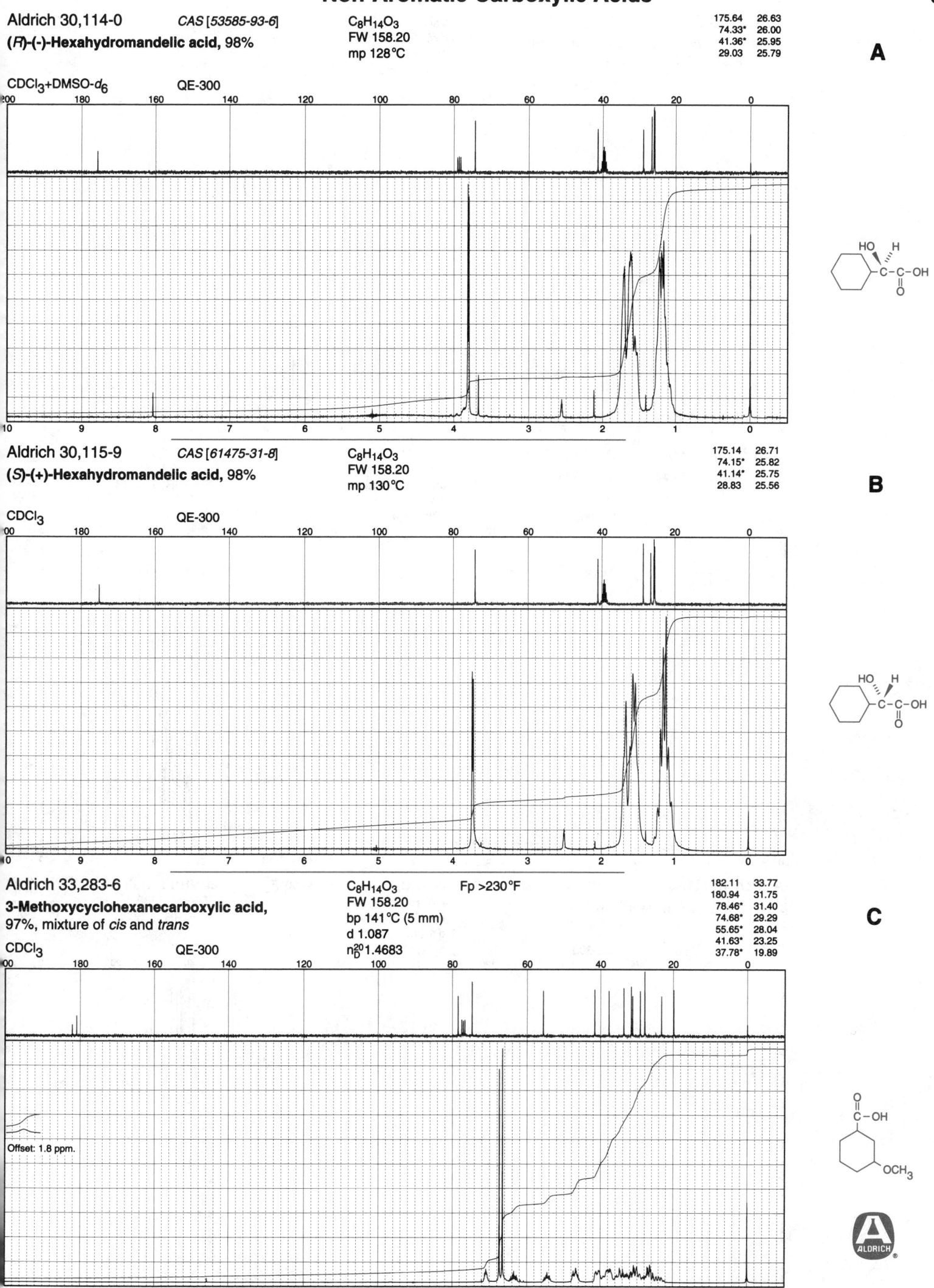

Aldrich 30,114-0 CAS [53585-93-6]

(R)-(-)-Hexahydromandelic acid, 98%

C$_8$H$_{14}$O$_3$
FW 158.20
mp 128°C

175.64 26.63
74.33* 26.00
41.36* 25.95
29.03   25.79

A

CDCl$_3$+DMSO-d_6 QE-300

Aldrich 30,115-9 CAS [61475-31-8]

(S)-(+)-Hexahydromandelic acid, 98%

C$_8$H$_{14}$O$_3$
FW 158.20
mp 130°C

175.14 26.71
74.15* 25.82
41.14* 25.75
28.83   25.56

B

CDCl$_3$ QE-300

Aldrich 33,283-6

3-Methoxycyclohexanecarboxylic acid,
97%, mixture of cis and trans

C$_8$H$_{14}$O$_3$ Fp >230°F
FW 158.20
bp 141°C (5 mm)
d 1.087
n$_D^{20}$1.4683

182.11 33.77
180.94 31.75
78.46* 31.40
74.68* 29.29
55.65* 28.04
41.63* 23.25
37.78* 19.89

C

CDCl$_3$ QE-300

Offset: 1.8 ppm.

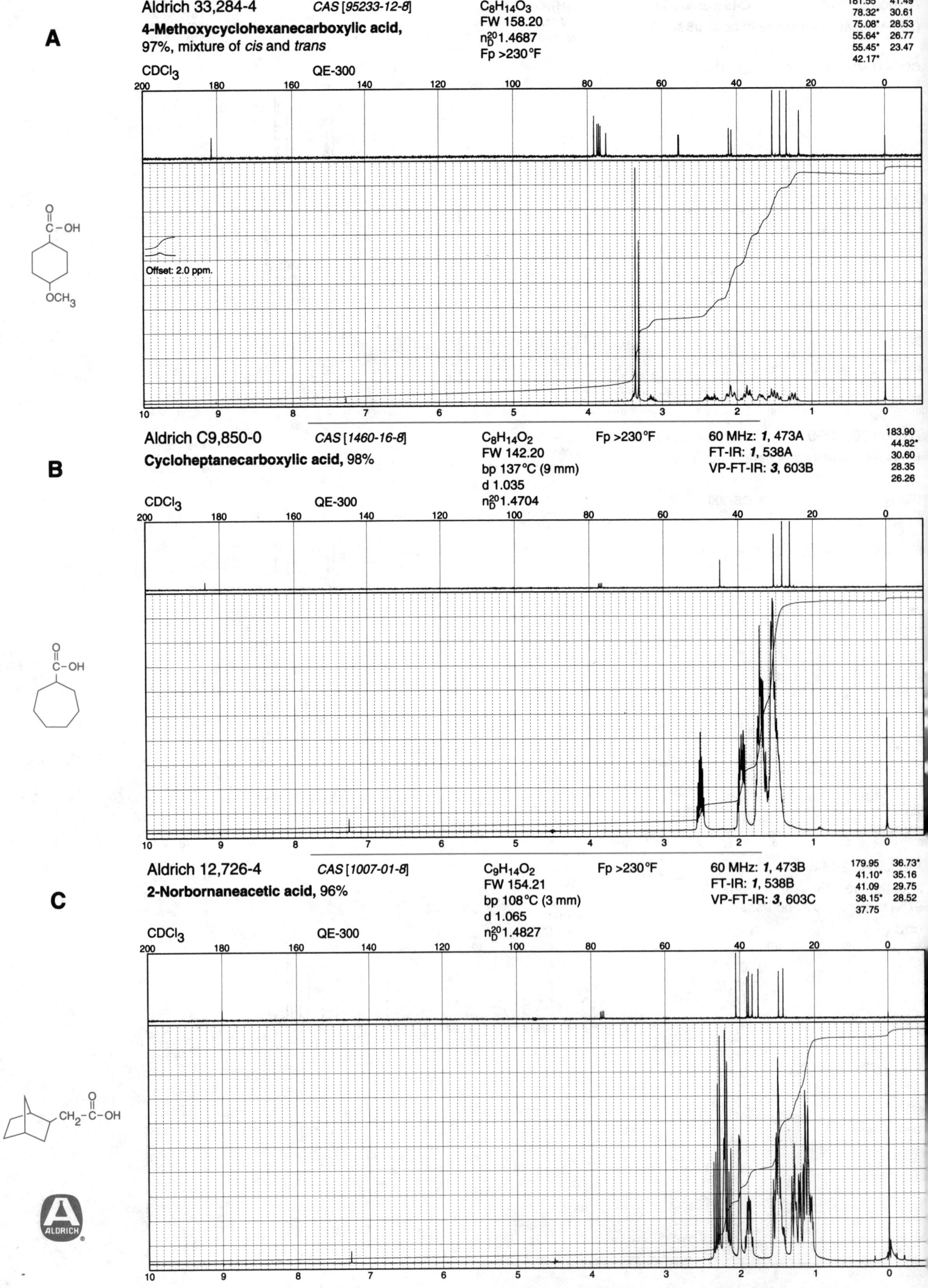

A

Aldrich 33,284-4   *CAS [95233-12-8]*

4-Methoxycyclohexanecarboxylic acid,
97%, mixture of *cis* and *trans*

$C_8H_{14}O_3$
FW 158.20
n_D^{20} 1.4687
Fp >230°F

181.55	41.49*
78.32*	30.61
75.08*	28.53
55.64*	26.77
55.45*	23.47
42.17*	

CDCl$_3$ QE-300

Offset: 2.0 ppm.

B

Aldrich C9,850-0   *CAS [1460-16-8]*

Cycloheptanecarboxylic acid, 98%

$C_8H_{14}O_2$
FW 142.20
bp 137°C (9 mm)
d 1.035
n_D^{20} 1.4704

Fp >230°F

60 MHz: *1*, 473A
FT-IR: *1*, 538A
VP-FT-IR: *3*, 603B

183.90	44.82*
30.60	
28.35	
26.26	

CDCl$_3$ QE-300

C

Aldrich 12,726-4   *CAS [1007-01-8]*

2-Norbornaneacetic acid, 96%

$C_9H_{14}O_2$
FW 154.21
bp 108°C (3 mm)
d 1.065
n_D^{20} 1.4827

Fp >230°F

60 MHz: *1*, 473B
FT-IR: *1*, 538B
VP-FT-IR: *3*, 603C

179.95	36.73*
41.10*	35.16
41.09	29.75
38.15*	28.52
37.75	

CDCl$_3$ QE-300

ALDRICH

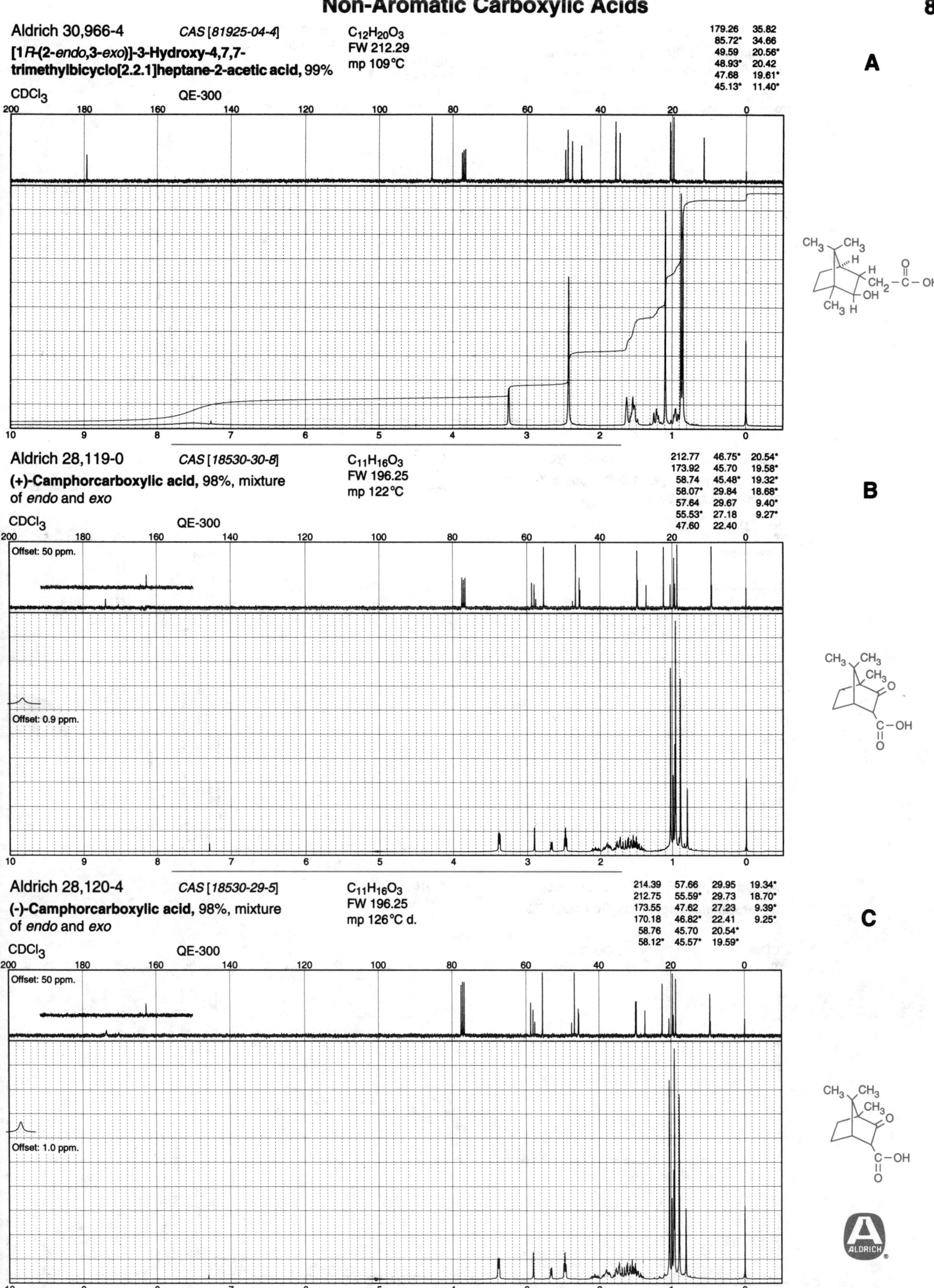

Aldrich 30,966-4   *CAS [81925-04-4]* $C_{12}H_{20}O_3$ FW 212.29   mp 109°C

[1*R*-(2-*endo*,3-*exo*)]-3-Hydroxy-4,7,7-trimethylbicyclo[2.2.1]heptane-2-acetic acid, 99%

CDCl₃   QE-300

179.26	35.82
85.72*	34.66
49.59	20.56*
48.93*	20.42
47.68	19.61*
45.13*	11.40*

A

Aldrich 28,119-0   *CAS [18530-30-8]* $C_{11}H_{16}O_3$ FW 196.25   mp 122°C

(+)-Camphorcarboxylic acid, 98%, mixture of *endo* and *exo*

CDCl₃   QE-300

212.77	46.75*	20.54*
173.92	45.70	19.58*
58.74	45.48*	19.32*
58.07*	29.84	18.68*
57.64	29.67	9.40*
55.53*	27.18	9.27*
47.60	22.40	

B

Offset: 50 ppm.

Offset: 0.9 ppm.

Aldrich 28,120-4   *CAS [18530-29-5]* $C_{11}H_{16}O_3$ FW 196.25   mp 126°C d.

(-)-Camphorcarboxylic acid, 98%, mixture of *endo* and *exo*

CDCl₃   QE-300

214.39	57.66	29.95	19.34*
212.75	55.59*	29.73	18.70*
173.55	47.62	27.23	9.39*
170.18	46.82*	22.41	9.25*
58.76	45.70	20.54*	
58.12*	45.57*	19.59*	

C

Offset: 50 ppm.

Offset: 1.0 ppm.

Non-Aromatic Carboxylic Acids

A

Aldrich 30,386-0   *CAS [18209-43-3]*   C$_9$H$_{14}$O$_2$ Fp >230 °F

cis-**Bicyclo[3.3.0]octane-2-carboxylic acid**, 98%, mixture of *endo* and *exo*

FW 154.21   VP-FT-IR: *3*, 603D
bp 106 °C
d 1.080
n$_D^{20}$1.4860

182.92	43.51*	31.90
181.19	42.97*	31.31
51.77*	35.02	29.92
48.68*	33.50	27.48
47.98*	33.46	26.33
45.64*	33.02	25.47

CDCl$_3$ QE-300

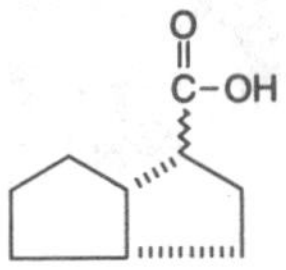

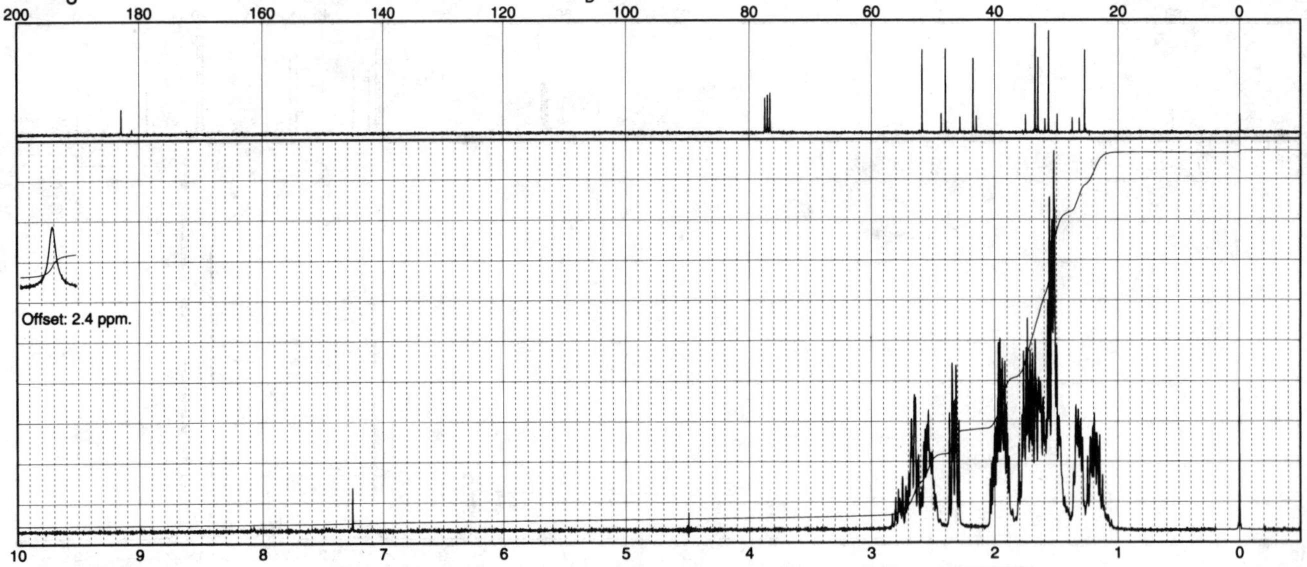

Offset: 2.4 ppm.

B

Aldrich 32,285-7   *CAS [50703-32-7]*   C$_8$H$_8$O$_3$

anti-**3-Oxotricyclo[2.2.1.0^{2,6}]heptane-7-carboxylic acid**, 97%

FW 152.15
mp 142 °C

209.38	28.64
172.43	21.26*
45.70*	19.69*
40.59*	19.35*

CDCl$_3$+DMSO-d_6 QE-300

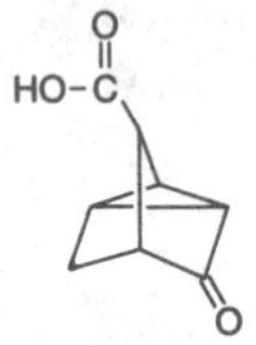

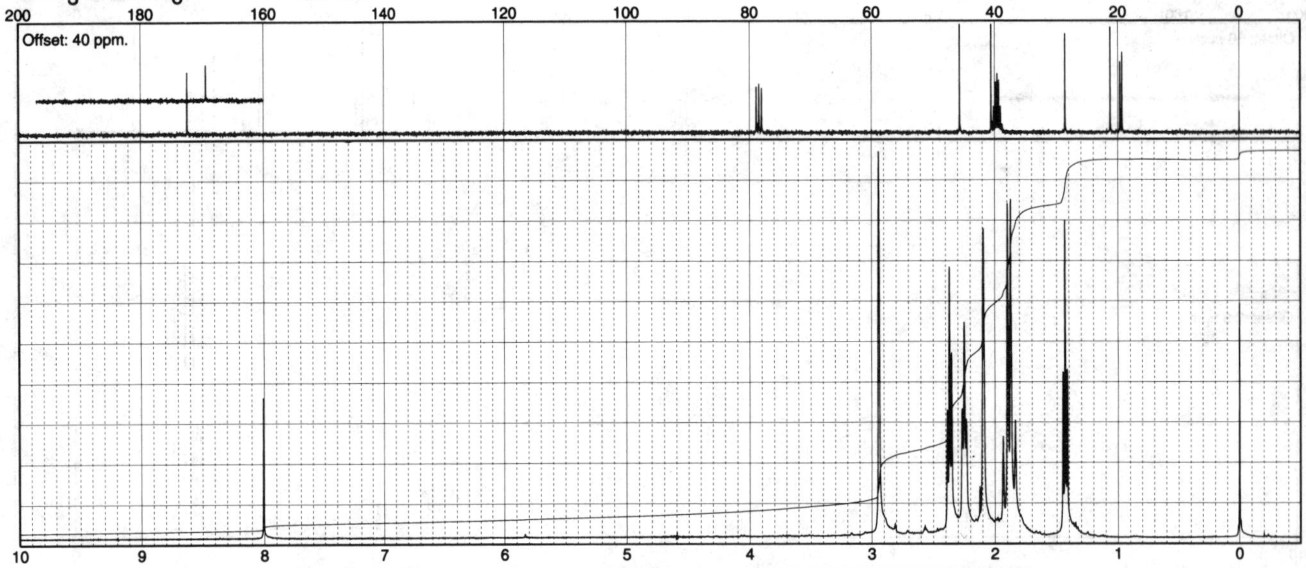

Offset: 40 ppm.

C

Aldrich 27,623-5   *CAS [16200-53-6]*   C$_{10}$H$_{14}$O$_2$

3-Noradamantanecarboxylic acid, 97%

FW 166.22
mp 106 °C

184.67
53.69
46.70
44.48*
43.61
37.48*
34.63

CDCl$_3$ QE-300

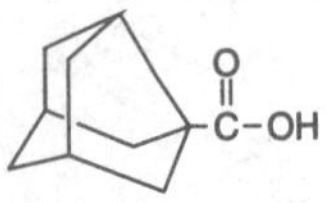

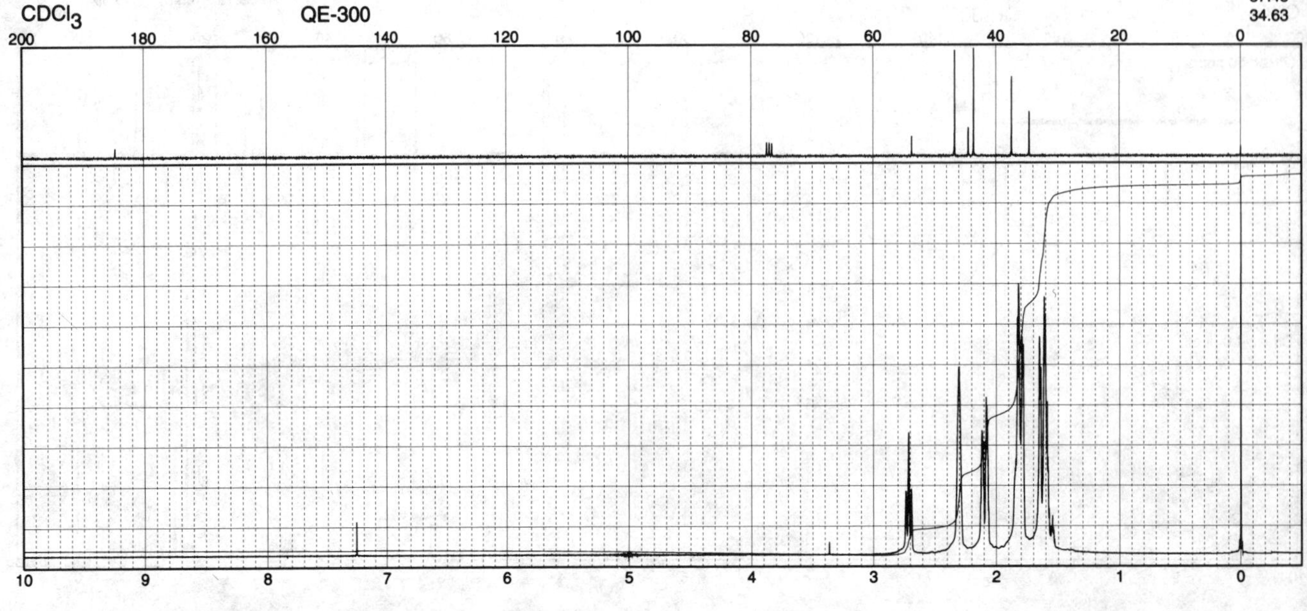

ALDRICH

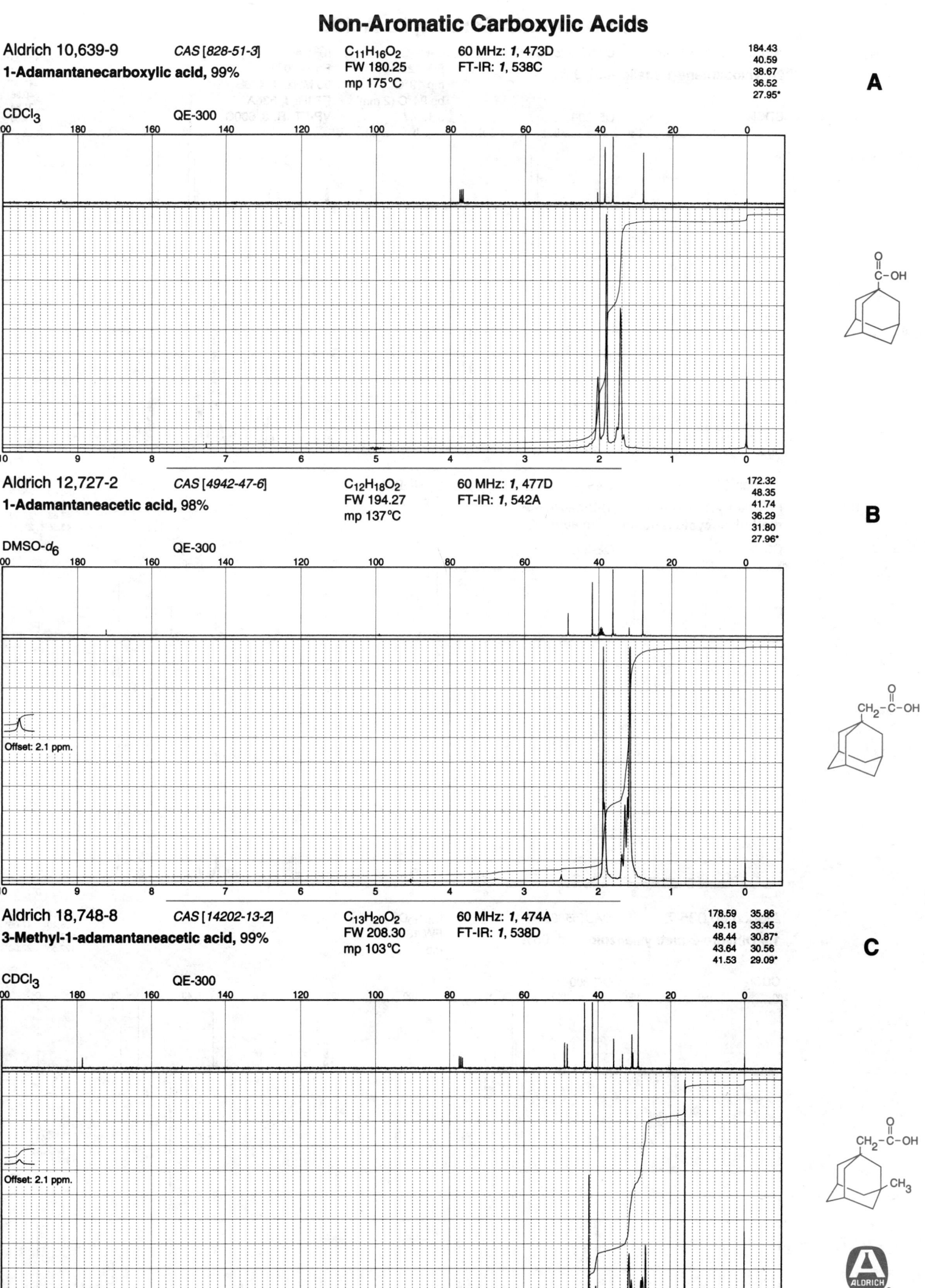

Aldrich 10,639-9   *CAS [828-51-3]*   C₁₁H₁₆O₂   60 MHz: *1*, 473D

1-Adamantanecarboxylic acid, 99%   FW 180.25   FT-IR: *1*, 538C   mp 175°C

184.43 / 40.59 / 38.67 / 36.52 / 27.95*

A

CDCl₃   QE-300

Aldrich 12,727-2   *CAS [4942-47-6]*   C₁₂H₁₈O₂   60 MHz: *1*, 477D

1-Adamantaneacetic acid, 98%   FW 194.27   FT-IR: *1*, 542A   mp 137°C

172.32 / 48.35 / 41.74 / 36.29 / 31.80 / 27.96*

B

DMSO-d₆   QE-300

Offset: 2.1 ppm.

Aldrich 18,748-8   *CAS [14202-13-2]*   C₁₃H₂₀O₂   60 MHz: *1*, 474A

3-Methyl-1-adamantaneacetic acid, 99%   FW 208.30   FT-IR: *1*, 538D   mp 103°C

178.59 / 35.86 / 49.18 / 33.45 / 48.44 / 30.87* / 43.64 / 30.56 / 41.53 / 29.09*

C

CDCl₃   QE-300

Offset: 2.1 ppm.

ALDRICH

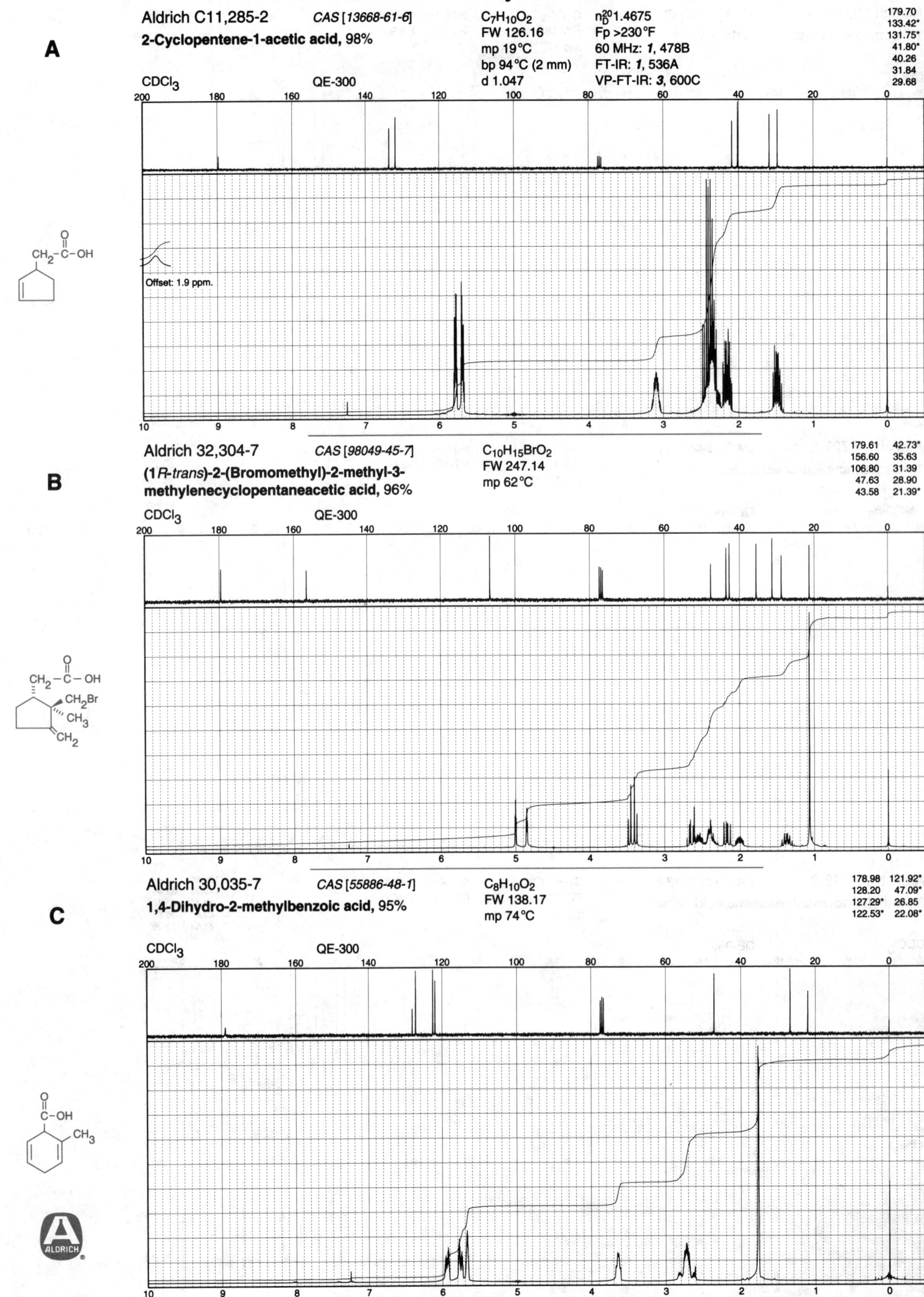

A

Aldrich C11,285-2   *CAS [13668-61-6]*

2-Cyclopentene-1-acetic acid, 98%

$C_7H_{10}O_2$
FW 126.16
mp 19 °C
bp 94 °C (2 mm)
d 1.047

n_D^{20}1.4675
Fp >230 °F
60 MHz: *1*, 478B
FT-IR: *1*, 536A
VP-FT-IR: *3*, 600C

179.70
133.42*
131.75*
41.80*
40.26
31.84
29.68

CDCl₃   QE-300

Offset: 1.9 ppm.

B

Aldrich 32,304-7   *CAS [98049-45-7]*

(1R-trans)-2-(Bromomethyl)-2-methyl-3-methylenecyclopentaneacetic acid, 96%

$C_{10}H_{15}BrO_2$
FW 247.14
mp 62 °C

179.61   42.73*
156.60   35.63
106.80   31.39
47.63   28.90
43.58   21.39*

CDCl₃   QE-300

C

Aldrich 30,035-7   *CAS [55886-48-1]*

1,4-Dihydro-2-methylbenzoic acid, 95%

$C_8H_{10}O_2$
FW 138.17
mp 74 °C

178.98   121.92*
128.20   47.09*
127.29*   26.85
122.53*   22.08*

CDCl₃   QE-300

A

Aldrich 86,216-9   *CAS [14375-45-2]*

(±)-2-*cis*,4-*trans*-Abscisic acid, 98%

$C_{15}H_{20}O_4$
FW 264.32
mp 189°C

197.13	127.47*	41.35
166.91	125.98*	24.22*
163.17	118.57*	23.14*
148.64	78.52	20.96*
136.82*	49.48	19.17*

$CDCl_3$ + DMSO-d_6 QE-300

B

Aldrich 34,341-2   *CAS [598-10-7]*

1,1-Cyclopropanedicarboxylic acid, 97%

$C_5H_6O_4$
FW 130.10
mp 135°C

171.59
27.08
16.16

DMSO-d_6 QE-300

C

Aldrich C9,580-3   *CAS [5445-51-2]*

1,1-Cyclobutanedicarboxylic acid, 99%

$C_6H_8O_4$
FW 144.13
mp 161°C

60 MHz: *1*, 474C
FT-IR: *1*, 539D

172.84
52.14
28.14
15.47

DMSO-d_6 QE-300

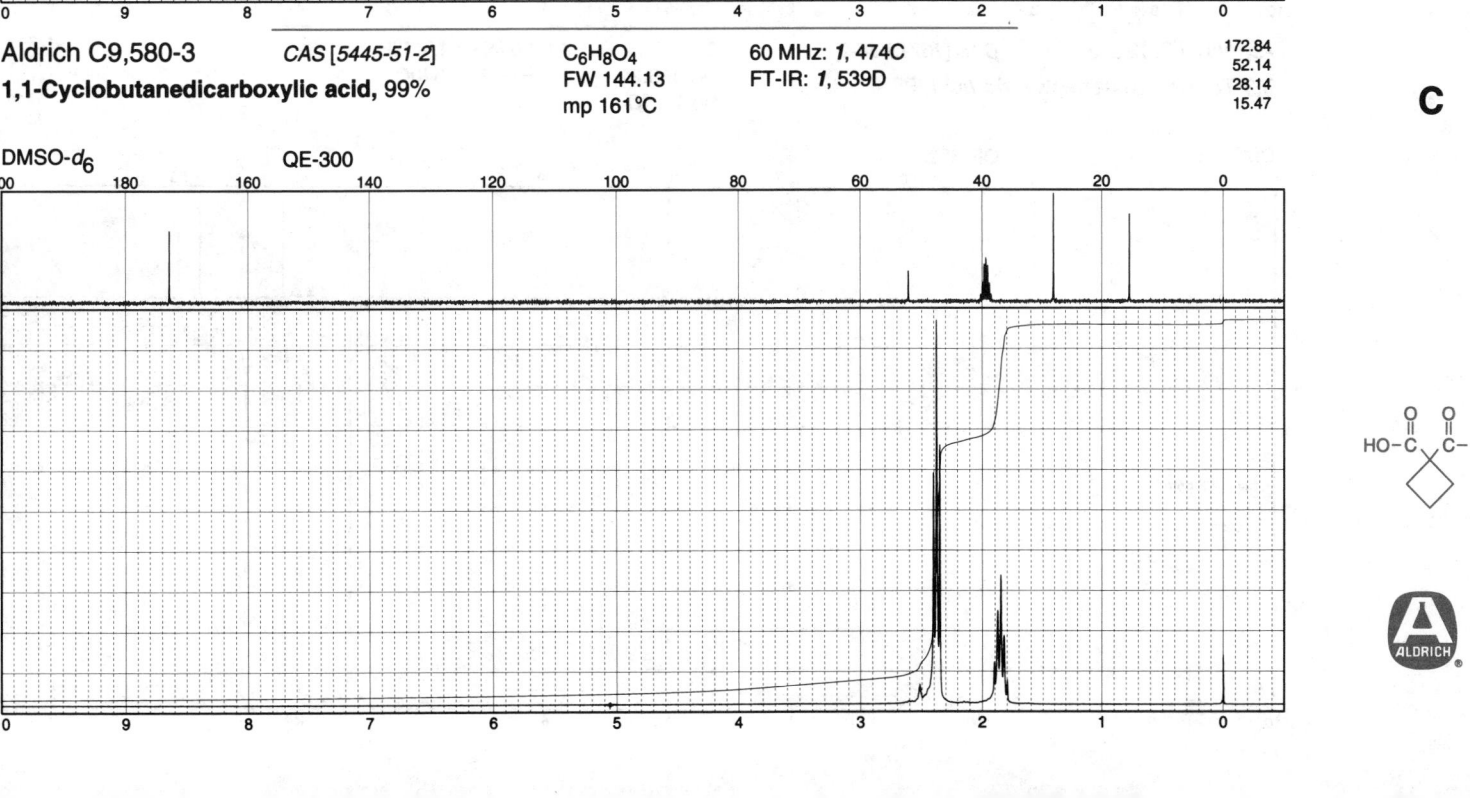

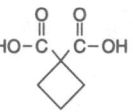

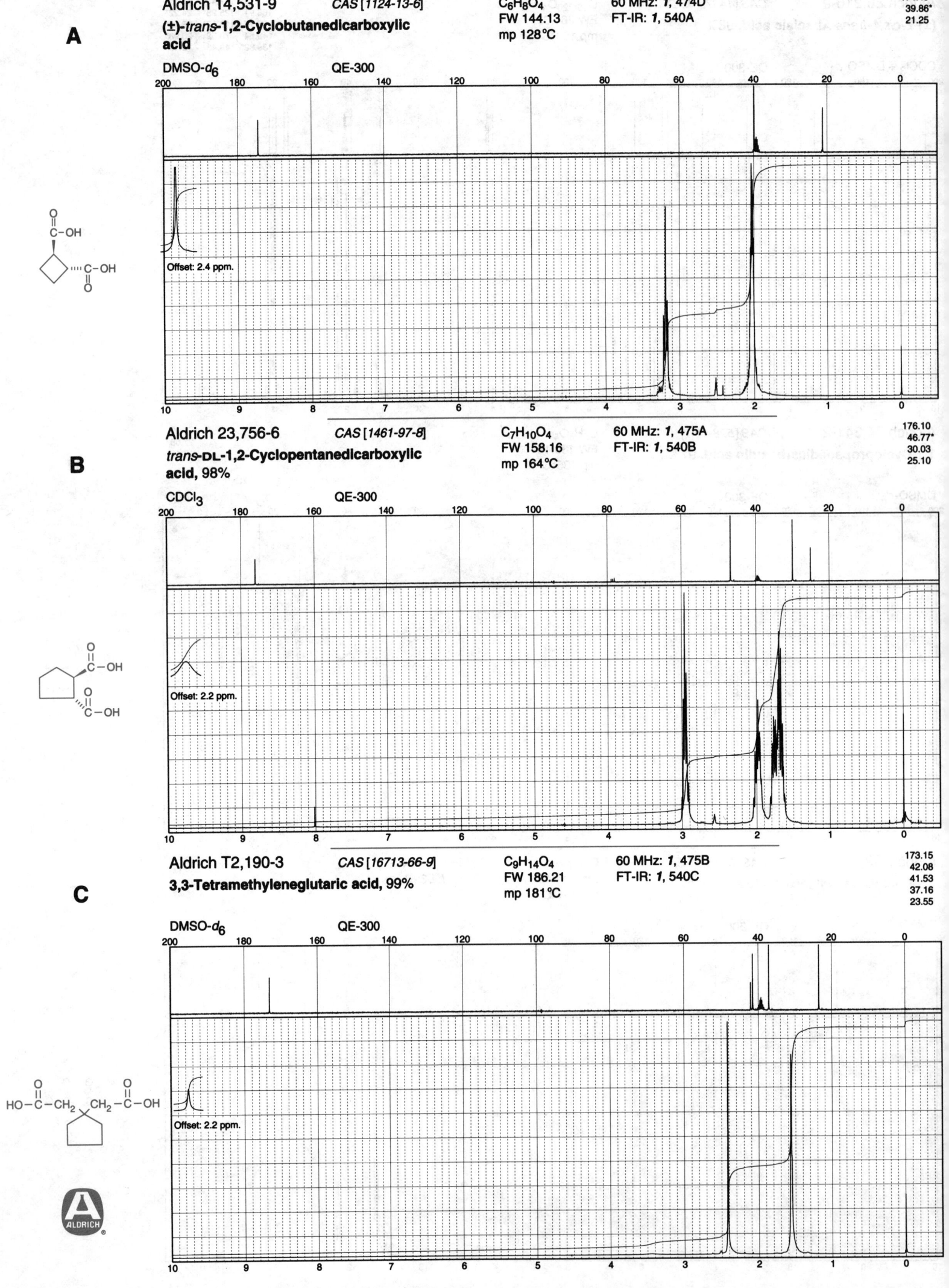

A

Aldrich 14,531-9   *CAS [1124-13-6]*

(±)-*trans*-1,2-Cyclobutanedicarboxylic acid

$C_6H_8O_4$
FW 144.13
mp 128°C

60 MHz: *1*, 474D
FT-IR: *1*, 540A

174.49
39.86*
21.25

DMSO-d_6 QE-300

Offset: 2.4 ppm.

B

Aldrich 23,756-6   *CAS [1461-97-8]*

trans-DL-1,2-Cyclopentanedicarboxylic acid, 98%

$C_7H_{10}O_4$
FW 158.16
mp 164°C

60 MHz: *1*, 475A
FT-IR: *1*, 540B

176.10
46.77*
30.03
25.10

CDCl$_3$ QE-300

Offset: 2.2 ppm.

C

Aldrich T2,190-3   *CAS [16713-66-9]*

3,3-Tetramethyleneglutaric acid, 99%

$C_9H_{14}O_4$
FW 186.21
mp 181°C

60 MHz: *1*, 475B
FT-IR: *1*, 540C

173.15
42.08
41.53
37.16
23.55

DMSO-d_6 QE-300

Offset: 2.2 ppm.

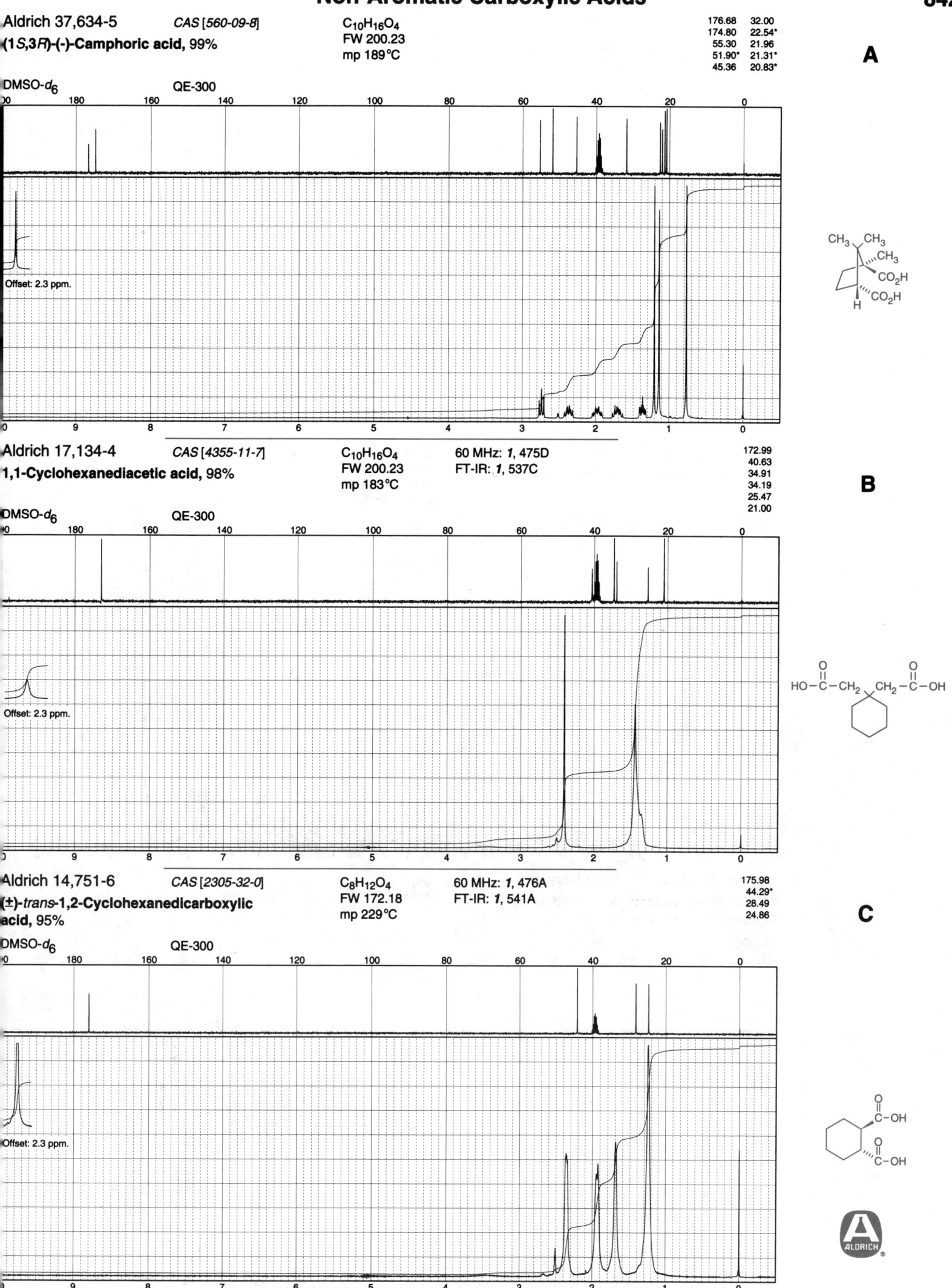

Aldrich 37,634-5   CAS [560-09-8]   C₁₀H₁₆O₄   FW 200.23   mp 189°C

(1S,3R)-(-)-Camphoric acid, 99%

DMSO-d₆   QE-300

A

176.68 32.00
174.80 22.54*
55.30   21.96
51.90* 21.31*
45.36   20.83*

Offset: 2.3 ppm.

Aldrich 17,134-4   CAS [4355-11-7]   C₁₀H₁₆O₄   FW 200.23   mp 183°C   60 MHz: 1, 475D   FT-IR: 1, 537C

1,1-Cyclohexanediacetic acid, 98%

DMSO-d₆   QE-300

B

172.99
40.63
34.91
34.19
25.47
21.00

Offset: 2.3 ppm.

Aldrich 14,751-6   CAS [2305-32-0]   C₈H₁₂O₄   FW 172.18   mp 229°C   60 MHz: 1, 476A   FT-IR: 1, 541A

(±)-trans-1,2-Cyclohexanedicarboxylic acid, 95%

DMSO-d₆   QE-300

C

175.98
44.29*
28.49
24.86

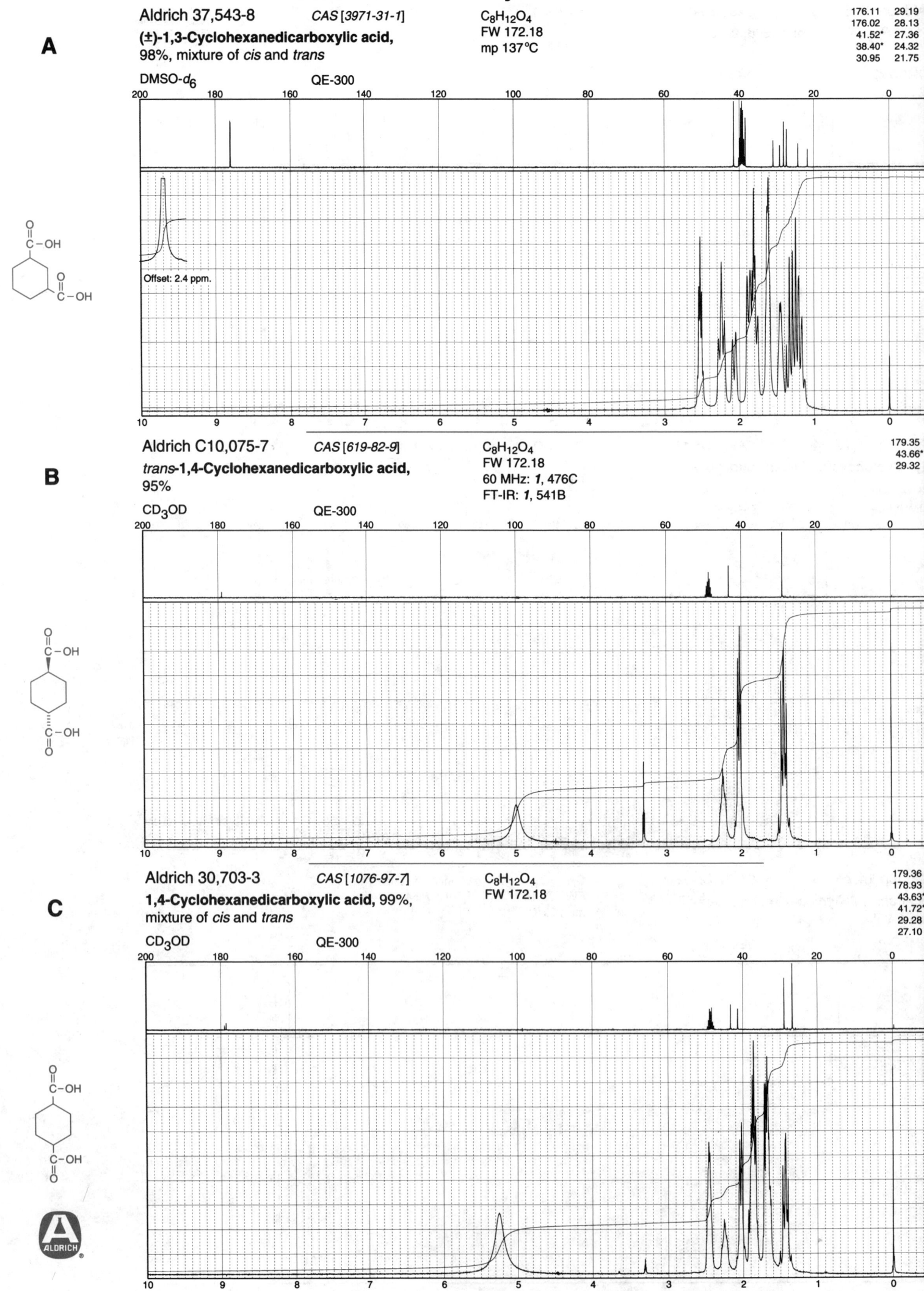

A

Aldrich 37,543-8   *CAS [3971-31-1]*   C$_8$H$_{12}$O$_4$
FW 172.18
mp 137°C

(±)-1,3-Cyclohexanedicarboxylic acid,
98%, mixture of *cis* and *trans*

176.11	29.19
176.02	28.13
41.52*	27.36
38.40*	24.32
30.95	21.75

DMSO-*d$_6$*   QE-300

Offset: 2.4 ppm.

B

Aldrich C10,075-7   *CAS [619-82-9]*   C$_8$H$_{12}$O$_4$
FW 172.18
60 MHz: *1*, 476C
FT-IR: *1*, 541B

trans-1,4-Cyclohexanedicarboxylic acid,
95%

| 179.35* |
| 43.66* |
| 29.32 |

CD$_3$OD   QE-300

C

Aldrich 30,703-3   *CAS [1076-97-7]*   C$_8$H$_{12}$O$_4$
FW 172.18

1,4-Cyclohexanedicarboxylic acid, 99%,
mixture of *cis* and *trans*

| 179.36 |
| 178.93 |
| 43.63* |
| 41.72* |
| 29.28 |
| 27.10 |

CD$_3$OD   QE-300

Aldrich 34,082-0 *CAS [39269-10-8]* $C_{12}H_{16}O_4$ 177.75
39.80
1,3-Adamantanedicarboxylic acid, 98% FW 224.26 37.56
mp 277°C 34.91
27.29*

A

DMSO-d_6 QE-300

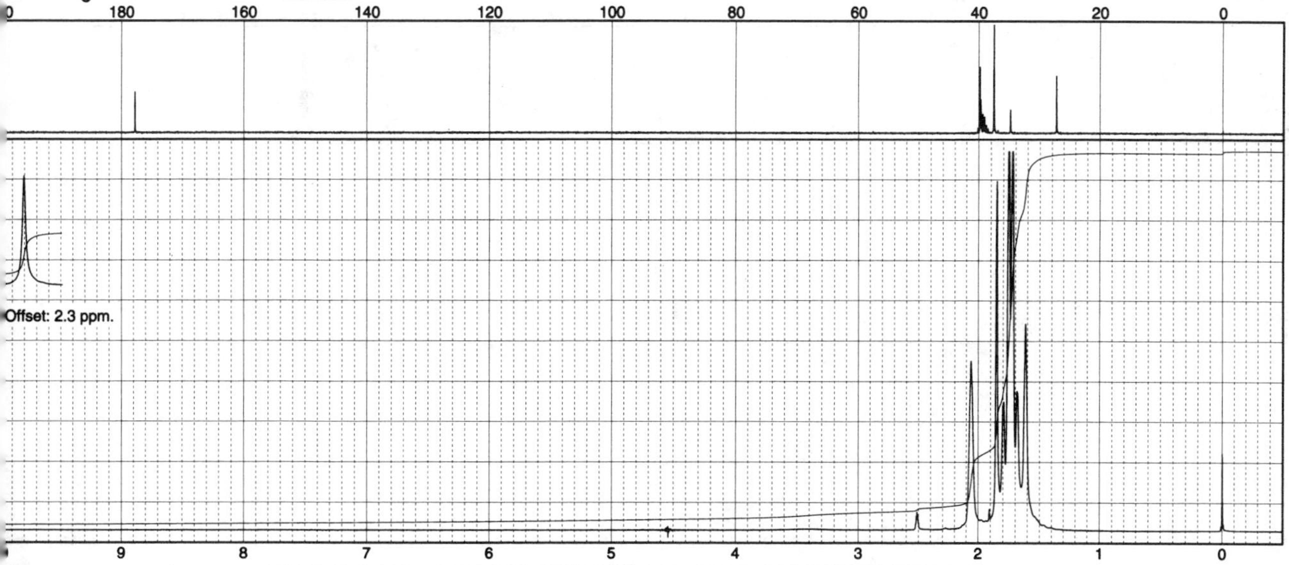

Offset: 2.3 ppm.

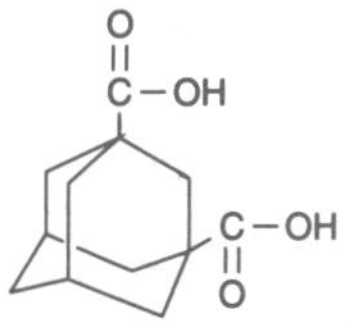

Aldrich 22,053-1 *CAS [499-02-5]* $C_6H_6O_4$ 60 MHz: *1*, 478A 170.29
129.81
3-Methylenecyclopropane-*trans*-1,2-dicarboxylic FW 142.11 FT-IR: *1*, 534D 105.05
acid, 95% mp 195°C 25.45*

B

CDCl$_3$ + DMSO-d_6 QE-300

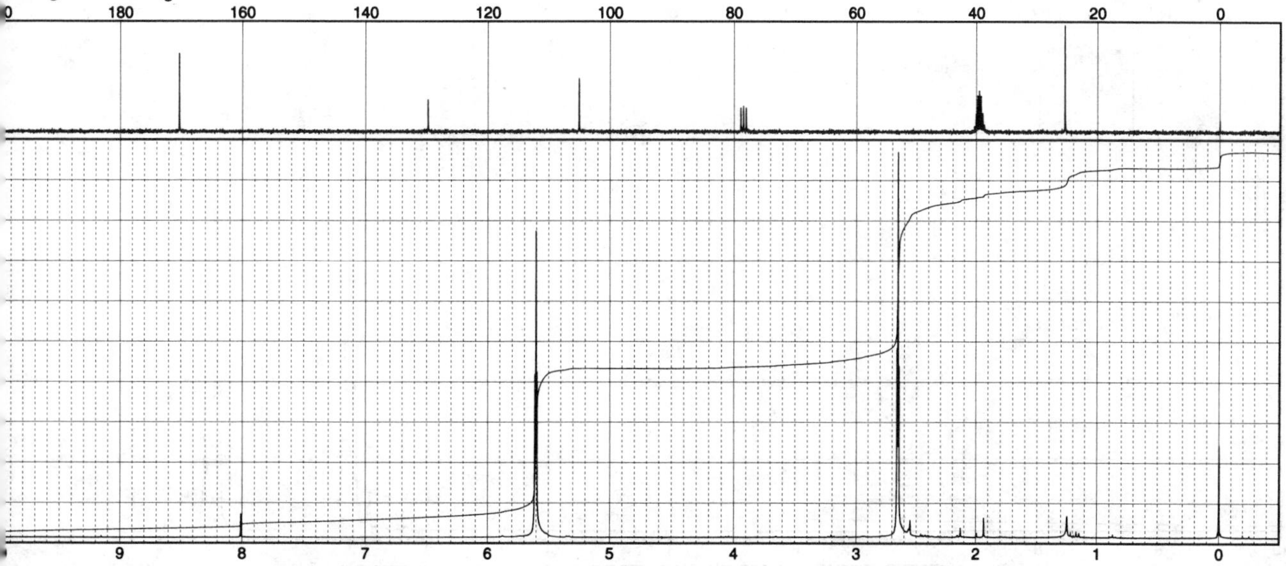

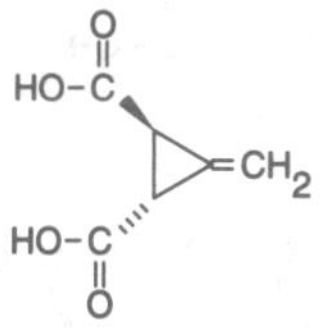

Aldrich 21,670-4 *CAS [3853-88-1]* $C_9H_{10}O_4$ FT-IR: *1*, 539A 173.48
134.42*
***cis*-5-Norbornene-*endo*-2,3-dicarboxylic** FW 182.18 48.26
acid, 98% mp 175°C d. 47.81*
45.85*

C

CDCl$_3$ + DMSO-d_6 QE-300

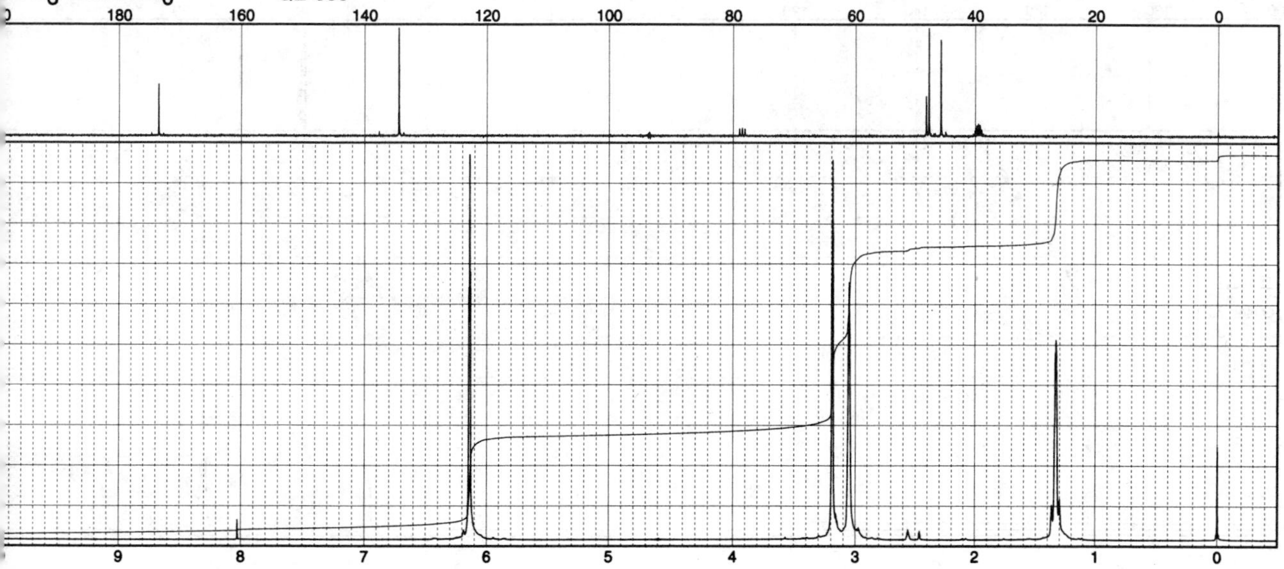

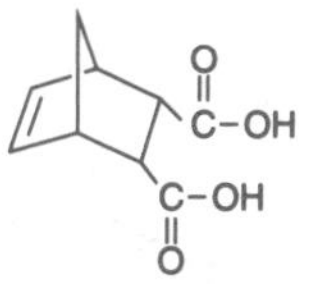

Non-Aromatic Carboxylic Acids

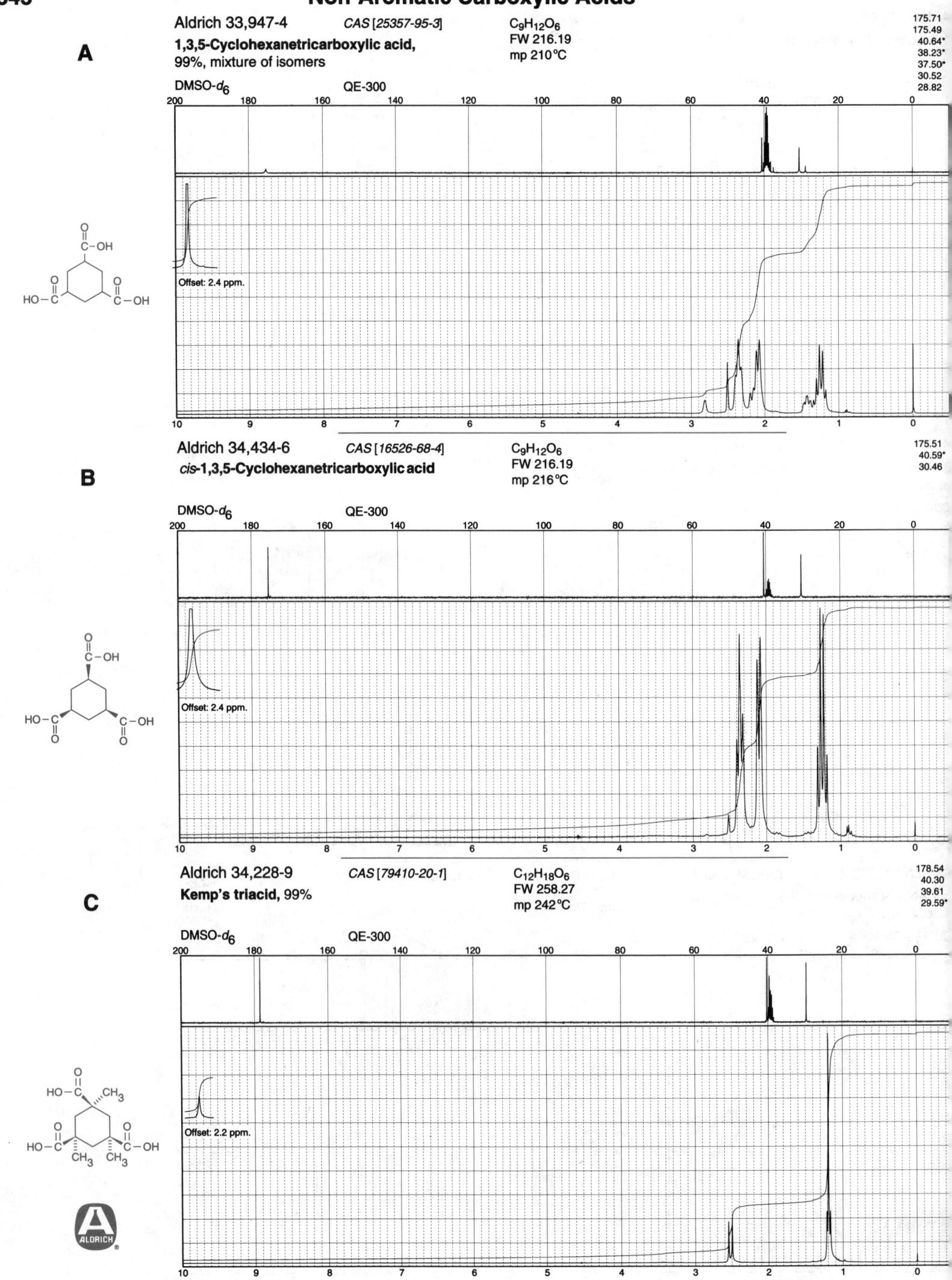

Aldrich 33,947-4   CAS [25357-95-3]   C$_9$H$_{12}$O$_6$
FW 216.19
mp 210°C

A

1,3,5-Cyclohexanetricarboxylic acid,
99%, mixture of isomers

DMSO-d_6 QE-300

175.71
175.49
40.64*
38.23*
37.50*
30.52
28.82

Offset: 2.4 ppm.

Aldrich 34,434-6   CAS [16526-68-4]   C$_9$H$_{12}$O$_6$
FW 216.19
mp 216°C

B

cis-**1,3,5-Cyclohexanetricarboxylic acid**

DMSO-d_6 QE-300

175.51
40.59*
30.46

Offset: 2.4 ppm.

Aldrich 34,228-9   CAS [79410-20-1]   C$_{12}$H$_{18}$O$_6$
FW 258.27
mp 242°C

C

Kemp's triacid, 99%

DMSO-d_6 QE-300

178.54
40.30
39.61
29.59*

Offset: 2.2 ppm.

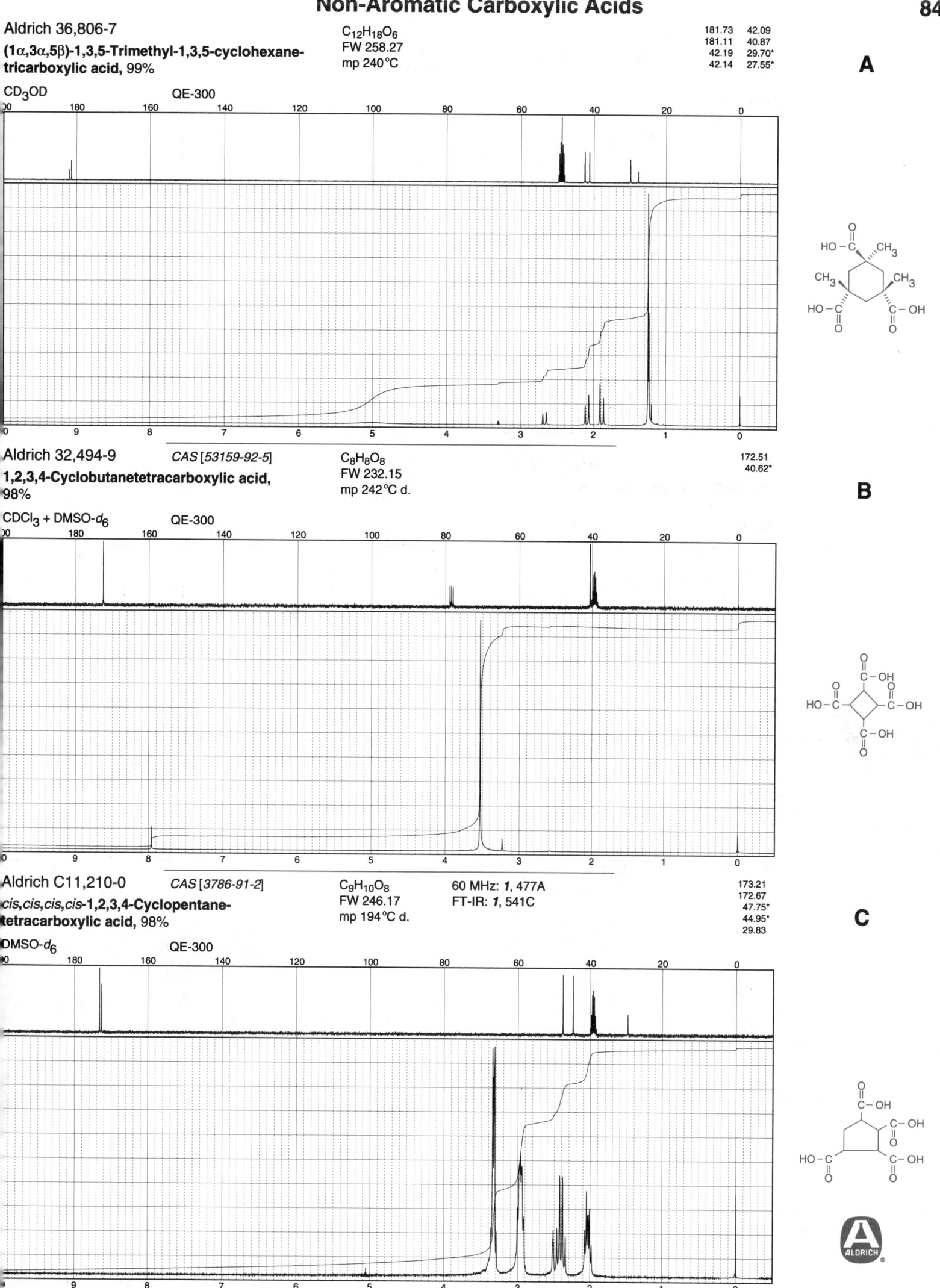

Aldrich 36,806-7

$(1\alpha,3\alpha,5\beta)$-1,3,5-Trimethyl-1,3,5-cyclohexane-tricarboxylic acid, 99%

$C_{12}H_{18}O_6$
FW 258.27
mp 240°C

181.73   42.09
181.11   40.87
42.19   29.70*
42.14   27.55*

CD$_3$OD   QE-300

A

Aldrich 32,494-9   CAS [53159-92-5]

1,2,3,4-Cyclobutanetetracarboxylic acid, 98%

$C_8H_8O_8$
FW 232.15
mp 242°C d.

172.51
40.62*

CDCl$_3$ + DMSO-d_6 QE-300

B

Aldrich C11,210-0   CAS [3786-91-2]

cis,cis,cis,cis-1,2,3,4-Cyclopentane-tetracarboxylic acid, 98%

$C_9H_{10}O_8$
FW 246.17
mp 194°C d.

60 MHz: 1, 477A
FT-IR: 1, 541C

173.21
172.67
47.75*
44.95*
29.83

DMSO-d_6 QE-300

C

A

Aldrich 36,077-5

1,2,3,4,5,6-Cyclohexanehexacarboxylic acid

$C_{12}H_{12}O_{12}$
FW 366.24
mp 216°C d.

173.50
171.48
45.83*
39.15*

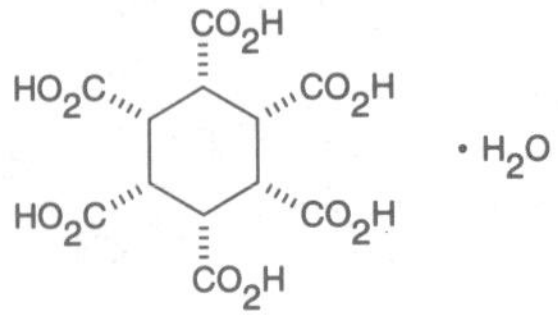

DMSO-d_6 QE-300

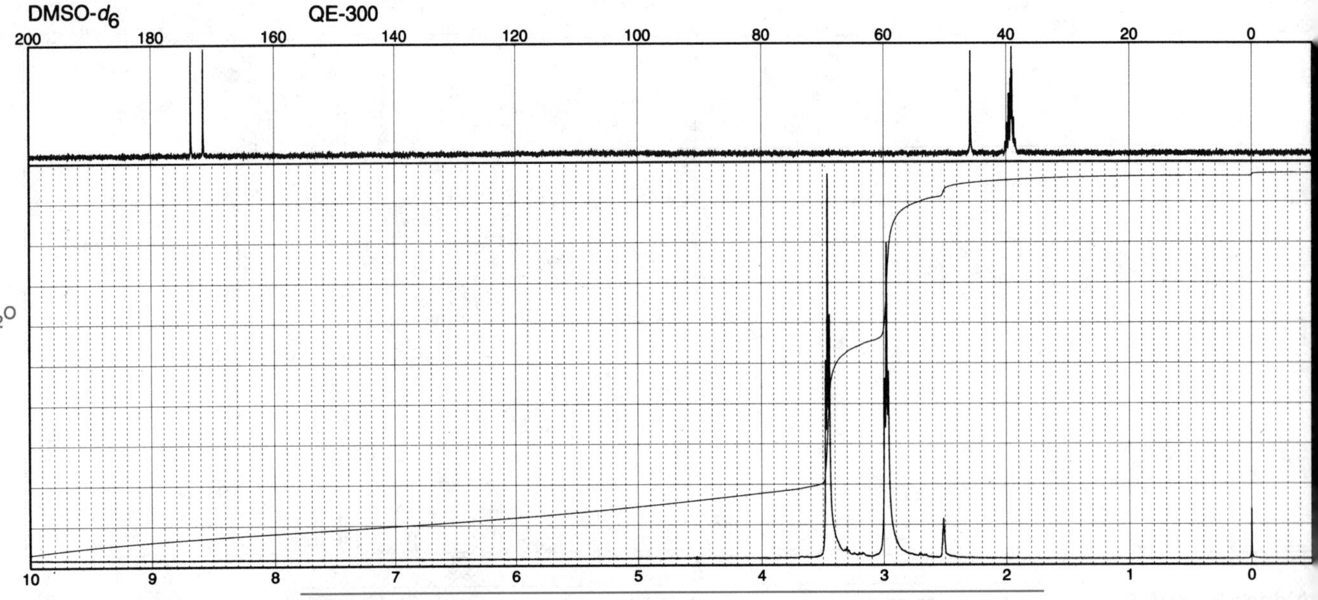

· H_2O

B

Aldrich A140-1   *CAS [5429-56-1]* $C_5H_7NO_3$ FT-IR: *1*, 542D

2-Acetamidoacrylic acid, 99%

FW 129.12
mp 186°C d.

169.17
164.91
133.07
107.69
23.70*

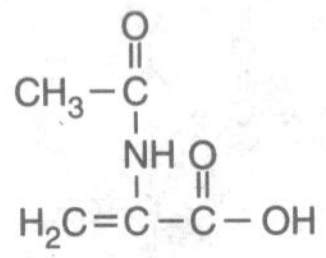

DMSO-d_6 QE-300

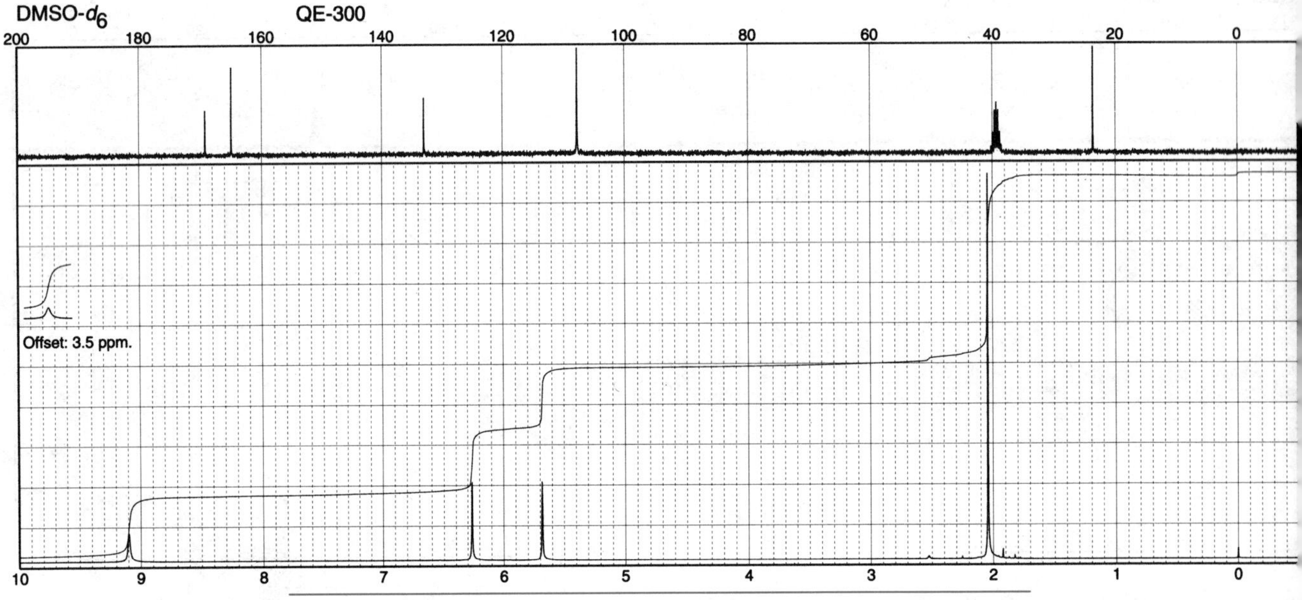

Offset: 3.5 ppm.

A

Aldrich 24,124-5   *CAS [127-09-3]*   C₂H₄O₂

$C_2H_4O_2$

FW 82.03

FT-IR: *1*, 545D

Acetic acid, sodium salt, 99+%

184.20
26.15°

B

Aldrich 34,515-6   *CAS [562-90-3]* $C_8H_{12}O_8Si$

FW 264.27

mp 113°C

bp 148°C (5 mm)

Silicon(IV) acetate, tech., 93%

167.74
22.23

C

Aldrich 33,599-1   *CAS [10534-59-5]* $C_{18}H_{39}NO_2$

FW 301.52

mp 97°C

Tetrabutylammonium acetate, 97%

183.39
60.79
25.84
21.88
15.60

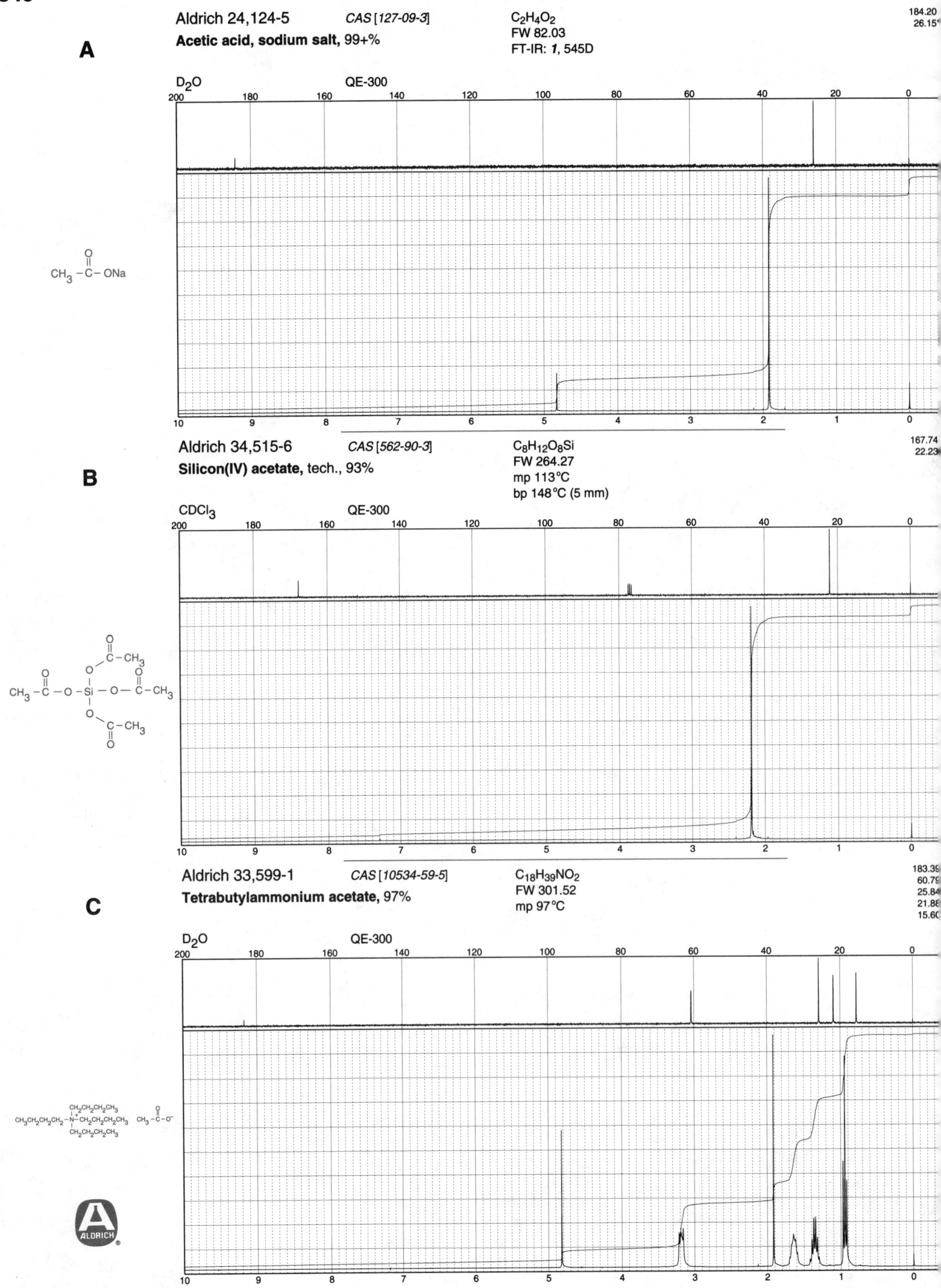

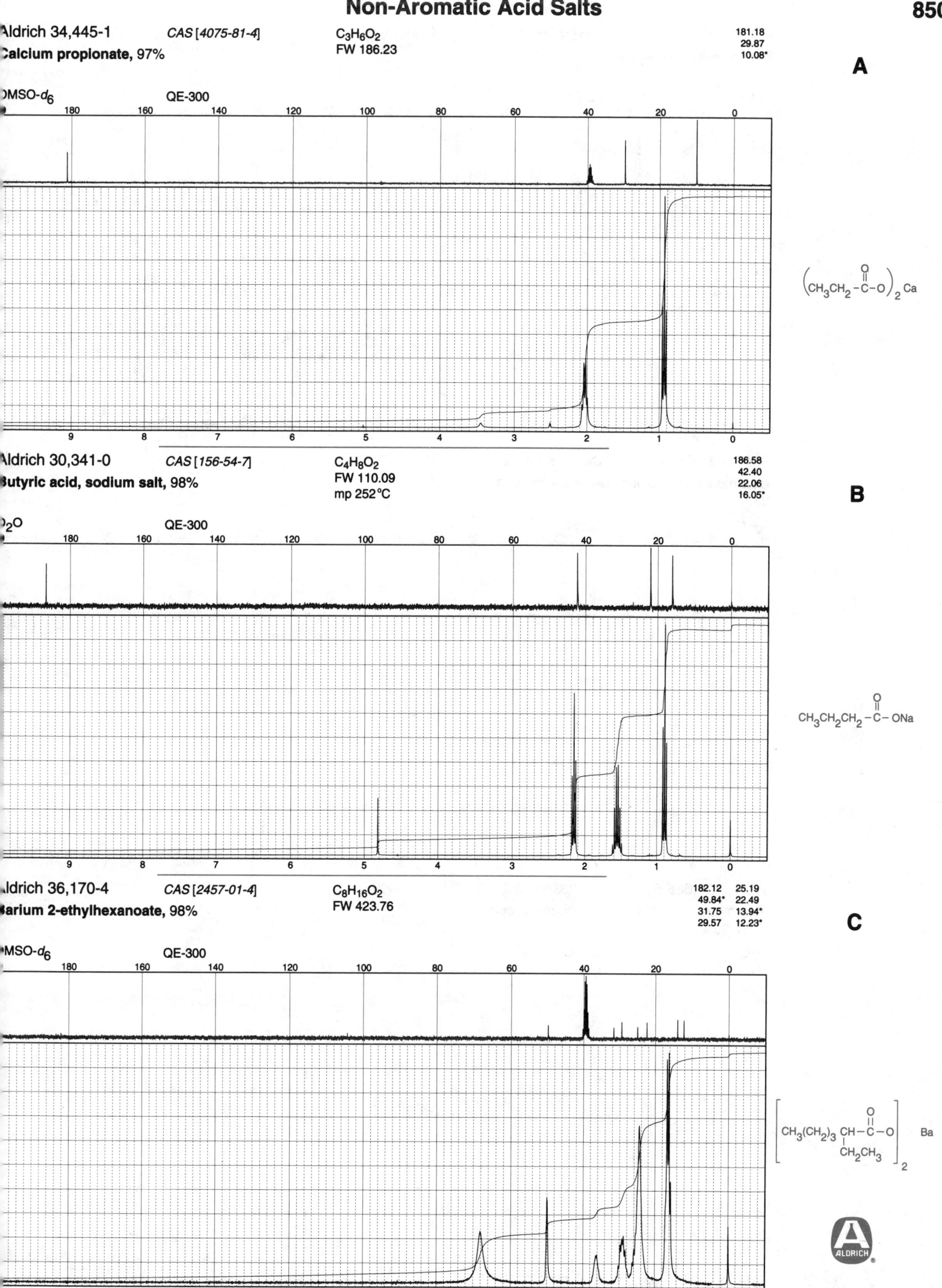

Aldrich 34,445-1   CAS [4075-81-4]   C₃H₆O₂
Calcium propionate, 97%   FW 186.23

181.18
29.87
10.08*

A

DMSO-d₆   QE-300

$\left(CH_3CH_2-\overset{\overset{\displaystyle O}{\|}}{C}-O\right)_2 Ca$

Aldrich 30,341-0   CAS [156-54-7]   C₄H₈O₂
Butyric acid, sodium salt, 98%   FW 110.09
mp 252°C

186.58
42.40
22.06
16.05*

B

D₂O   QE-300

$CH_3CH_2CH_2-\overset{\overset{\displaystyle O}{\|}}{C}-ONa$

Aldrich 36,170-4   CAS [2457-01-4]   C₈H₁₆O₂
Barium 2-ethylhexanoate, 98%   FW 423.76

182.12   25.19
49.84*   22.49
31.75   13.94*
29.57   12.23*

C

DMSO-d₆   QE-300

$\left[CH_3(CH_2)_3\ CH-\overset{\overset{\displaystyle O}{\|}}{C}-O\right]_2 Ba$
  (with CH₂CH₃ branch)

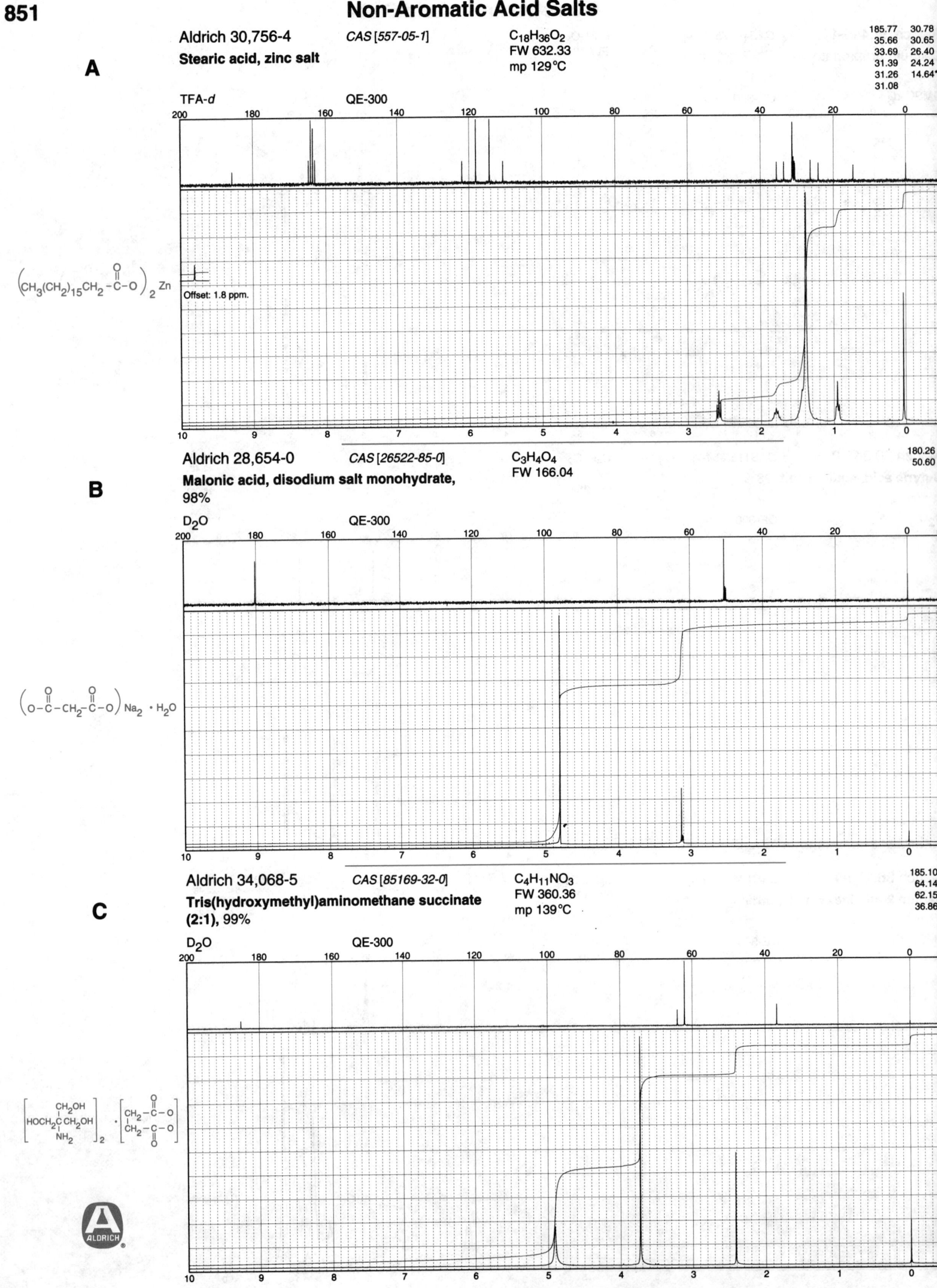

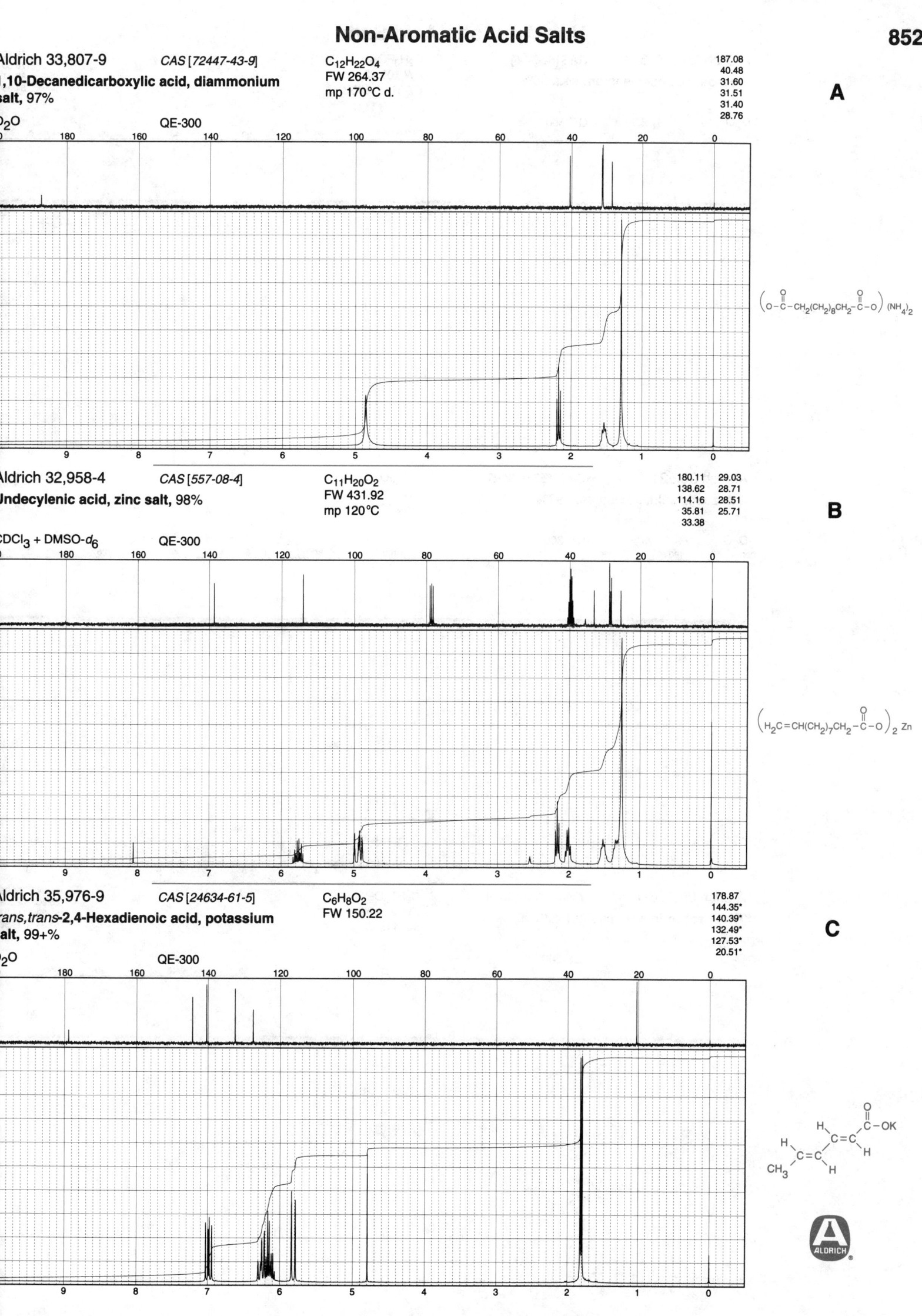

Aldrich 33,807-9 *CAS [72447-43-9]*
1,10-Decanedicarboxylic acid, diammonium salt, 97%

C₁₂H₂₂O₄
FW 264.37
mp 170°C d.

D₂O   QE-300

187.08
40.48
31.60
31.51
31.40
28.76

A

$\left(\overset{O}{\underset{\|}{\text{O}-\text{C}}}-\text{CH}_2(\text{CH}_2)_8\text{CH}_2-\overset{O}{\underset{\|}{\text{C}}}-\text{O}\right)(\text{NH}_4)_2$

Aldrich 32,958-4 *CAS [557-08-4]*
Undecylenic acid, zinc salt, 98%

C₁₁H₂₀O₂
FW 431.92
mp 120°C

CDCl₃ + DMSO-d₆   QE-300

180.11   29.03
138.62   28.71
114.16   28.51
35.81   25.71
33.38

B

$\left(\text{H}_2\text{C}{=}\text{CH}(\text{CH}_2)_7\text{CH}_2-\overset{O}{\underset{\|}{\text{C}}}-\text{O}\right)_2 \text{Zn}$

Aldrich 35,976-9 *CAS [24634-61-5]*
trans,trans-2,4-Hexadienoic acid, potassium salt, 99+%

C₆H₈O₂
FW 150.22

D₂O   QE-300

178.87
144.35*
140.39*
132.49*
127.53*
20.51*

C

A

Aldrich 25,659-5   *CAS [62-74-8]*   C₂H₃FO₂
Fluoroacetic acid, sodium salt, 98%   FW 100.02
FT-IR: *1*, 506B

179.12
178.88
83.43
81.08

D₂O   QE-300

FCH₂–C–ONa (with O double bond above C)

Aldrich 36,521-1   *CAS [65749-30-6]*   C₂H₃IO₂
Iodoacetic acid, lithium salt, 97%   FW 191.88
mp 239°C d.

179.86
4.57

B

D₂O   QE-300

ICH₂–C–OLi (with O double bond above C)

Aldrich 36,776-1   *CAS [21907-50-6]*   C₂HF₃O₂
Trifluoroacetic acid, cesium salt, 99+%   FW 245.92
mp 115°C

166.34 124.96
165.86 121.09
165.39 117.23
164.93 113.36

C

D₂O   QE-300

F₃C–C–OCs (with O double bond above C)

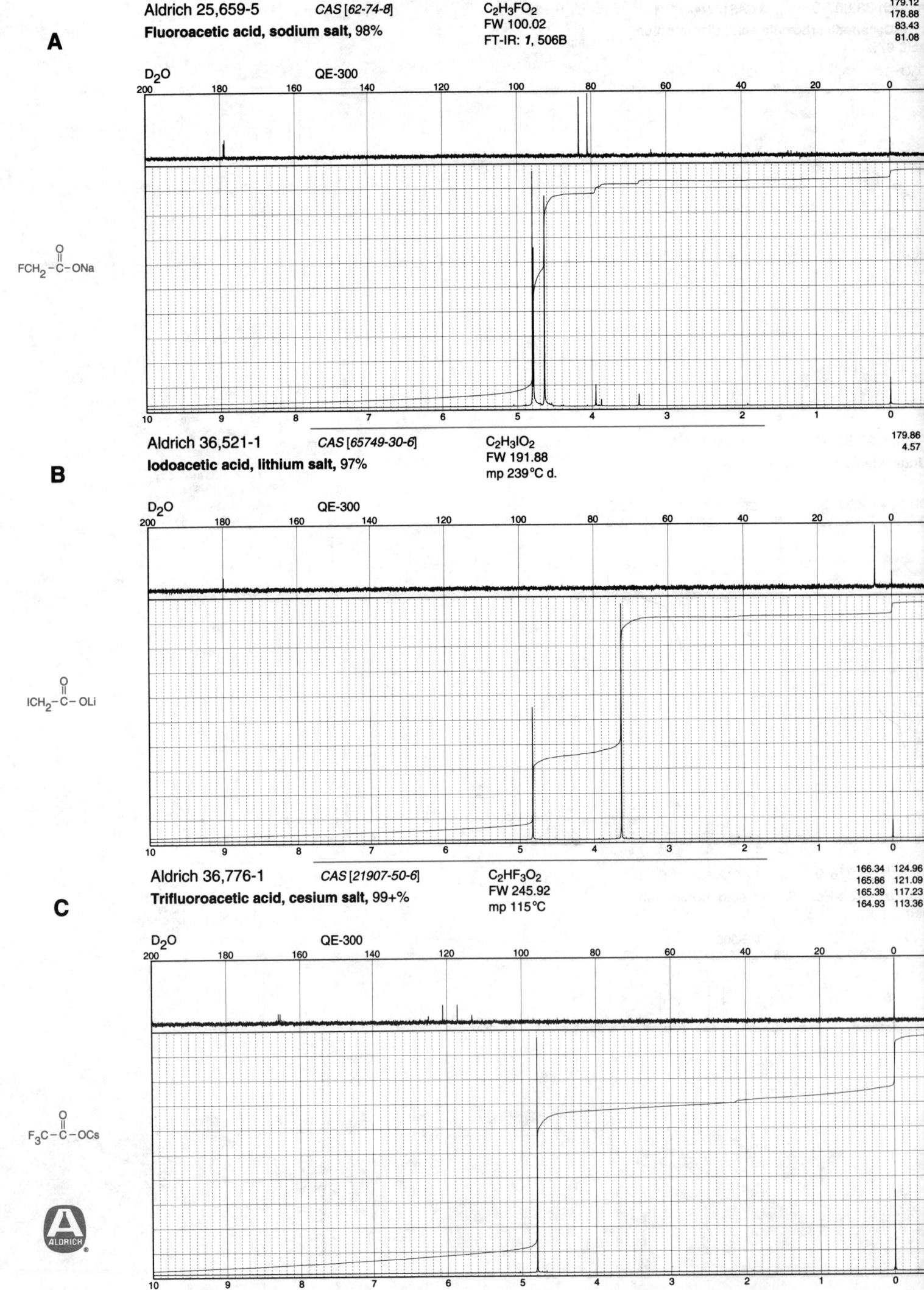

Aldrich 32,621-6   CAS [509-09-1]

Silver pentafluoropropionate, 98%

CDCl$_3$+DMSO-d_6 QE-300

C$_3$HF$_5$O$_2$
FW 270.89
mp 243°C

160.74	117.42	107.78
160.43	116.95	107.29
160.11	116.48	106.81
124.51	113.16	104.66
121.20	112.69	104.29
120.73	111.27	103.80
120.26	110.78	103.31

A

$$CF_3CF_2 - \overset{\overset{\displaystyle O}{\|}}{C} - OAg$$

Aldrich 22,011-6   CAS [19054-57-0]

(±)-2-Hydroxybutyric acid, sodium salt, 97%

D$_2$O   QE-300

C$_4$H$_8$O$_3$
FW 126.09
mp 134°C

60 MHz: *1*, 452A
FT-IR: *1*, 548D

184.28
76.17*
29.85
11.45*

B

$$CH_3CH_2\underset{\underset{\displaystyle OH}{|}}{CH} - \overset{\overset{\displaystyle O}{\|}}{C} - ONa$$

Aldrich 23,389-7   CAS [306-31-0]

(±)-3-Hydroxybutyric acid, sodium salt, 99%

D$_2$O   QE-300

C$_4$H$_8$O$_3$
FW 126.09
mp 166°C

60 MHz: *1*, 451C
FT-IR: *1*, 549A

183.17
68.41*
49.27
24.59*

C

$$CH_3\underset{\underset{\displaystyle OH}{|}}{CH}CH_2 - \overset{\overset{\displaystyle O}{\|}}{C} - ONa$$

ALDRICH

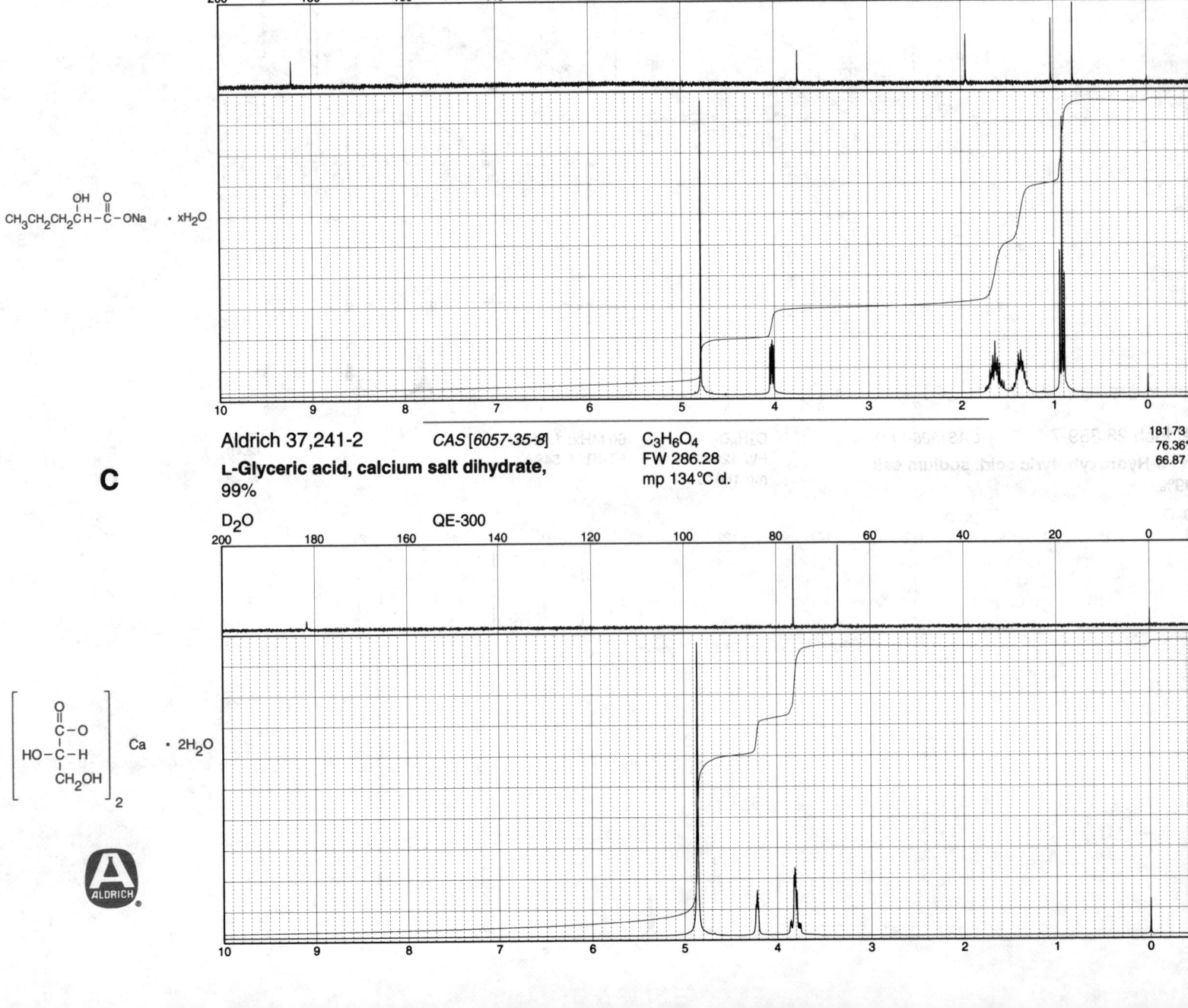

A

Aldrich H2,222-1   *CAS [502-85-2]*

4-Hydroxybutyric acid, sodium salt, 99%

$C_4H_8O_3$
FW 126.09
mp 146°C

FT-IR: *1*, 549C

185.81
64.26
36.80
31.18

D_2O QE-300

$HOCH_2CH_2CH_2-\overset{\overset{O}{\|}}{C}-ONa$

B

Aldrich 21,998-3

(±)-2-Hydroxyvaleric acid, sodium salt hydrate, 98%

$C_5H_{10}O_3$
FW 140.13
60 MHz: *1*, 452C
FT-IR: *1*, 549B

184.38
74.97*
39.02
20.63
15.96*

D_2O QE-300

$CH_3CH_2CH_2CH-\overset{\overset{O}{\|}}{C}-ONa \cdot xH_2O$
        with OH on the CH

C

Aldrich 37,241-2   *CAS [6057-35-8]*

L-Glyceric acid, calcium salt dihydrate, 99%

$C_3H_6O_4$
FW 286.28
mp 134°C d.

181.73
76.36*
66.87

D_2O QE-300

$\left[\begin{array}{c} \overset{\overset{O}{\|}}{C}-O \\ HO-C-H \\ CH_2OH \end{array} \right]_2 Ca \cdot 2H_2O$

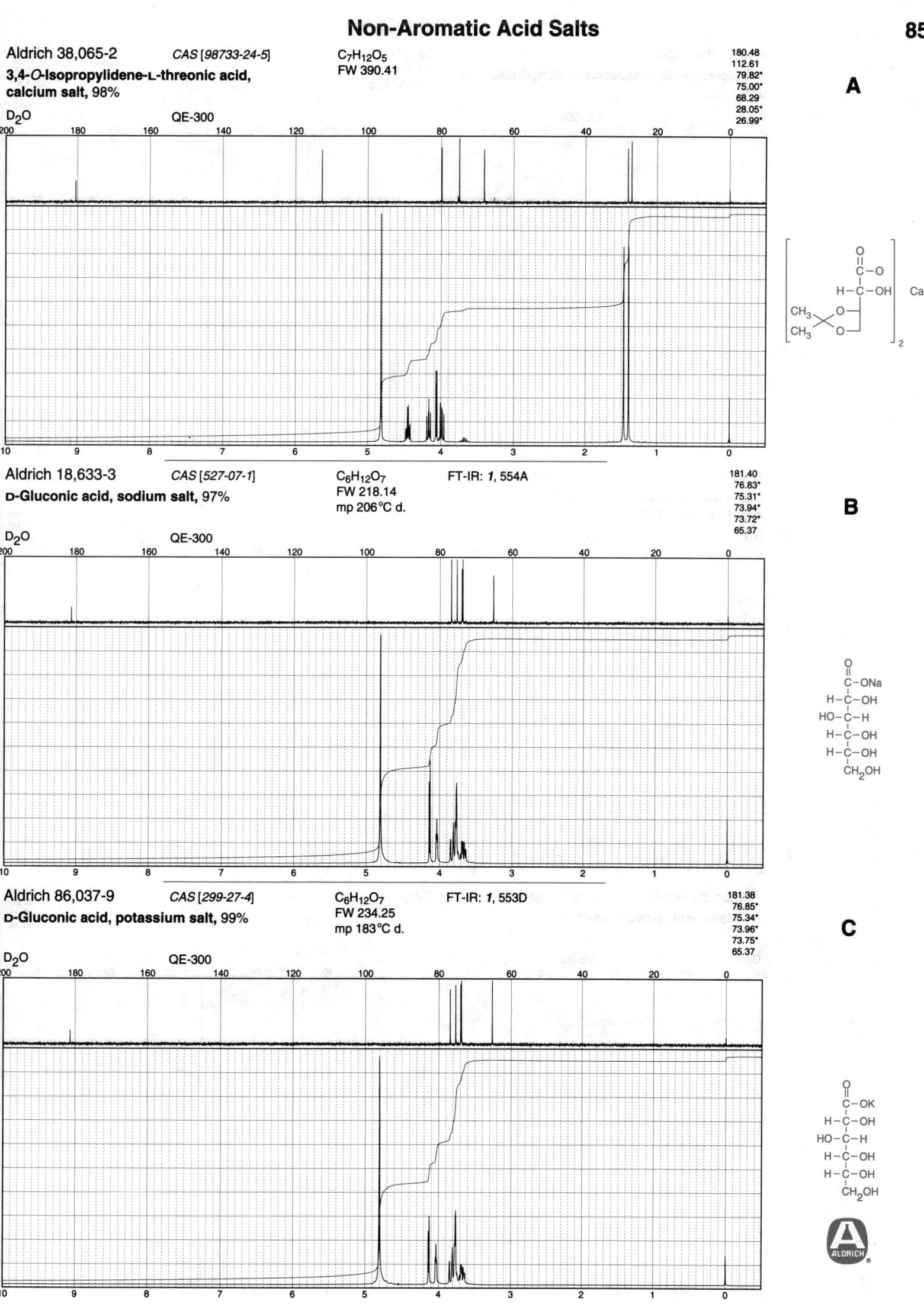

Aldrich 38,065-2   *CAS [98733-24-5]*   C₇H₁₂O₅
3,4-O-Isopropylidene-L-threonic acid,   FW 390.41
calcium salt, 98%

D₂O   QE-300

180.48
112.61
79.82*
75.00*
68.29
28.05*
26.99*

A

Aldrich 18,633-3   *CAS [527-07-1]*   C₆H₁₂O₇   FT-IR: *1*, 554A
D-Gluconic acid, sodium salt, 97%   FW 218.14
mp 206°C d.

D₂O   QE-300

181.40
76.83*
75.31*
73.94*
73.72*
65.37

B

Aldrich 86,037-9   *CAS [299-27-4]*   C₆H₁₂O₇   FT-IR: *1*, 553D
D-Gluconic acid, potassium salt, 99%   FW 234.25
mp 183°C d.

D₂O   QE-300

181.38
76.85*
75.34*
73.96*
73.75*
65.37

C

A

Aldrich 34,443-5

D-Gluconic acid, magnesium salt hydrate, 97%

$C_6H_{12}O_7$
FW 414.61
mp 200°C d.

181.25
76.82*
75.16*
73.98*
73.62*
65.42

D_2O   QE-300

B

Aldrich 30,676-2   CAS [10094-62-9]

D-glycero-D-gulo-Heptonic acid, sodium salt dihydrate, 97%

$C_7H_{14}O_8$
FW 284.20
mp 169°C

181.20
76.37*
75.15*
74.07*
73.96*
71.91*
65.53

D_2O   QE-300

C

Aldrich 30,849-8   CAS [22798-10-3]

DL-Malic acid, disodium salt, 97%

$C_4H_6O_5$
FW 178.05

183.81
182.63
73.06*
45.35

D_2O   QE-300

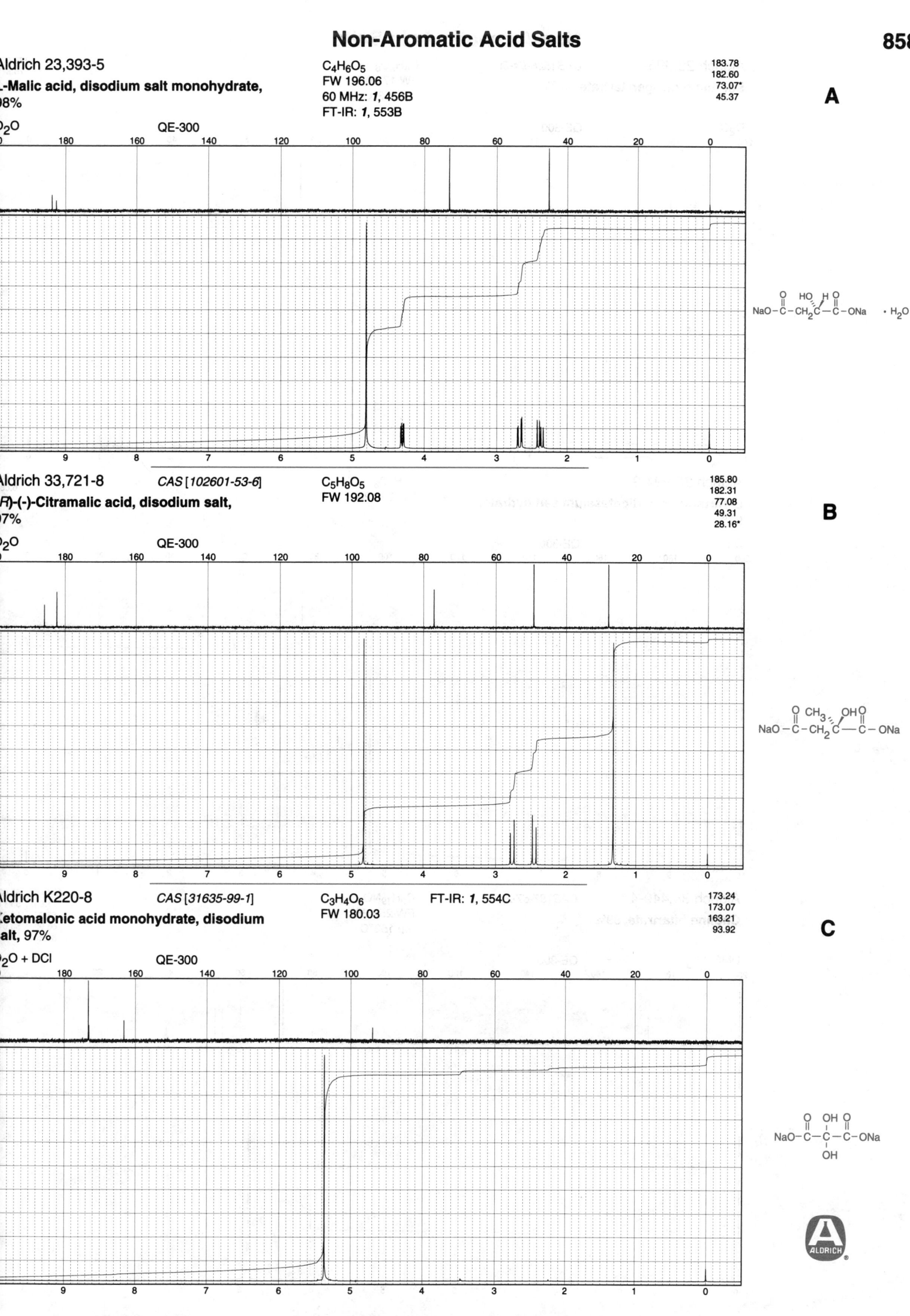

Aldrich 23,393-5

L-Malic acid, disodium salt monohydrate, 98%

C₄H₆O₅
FW 196.06
60 MHz: *1*, 456B
FT-IR: *1*, 553B

183.78
182.60
73.07*
45.37

A

₂O   QE-300

Aldrich 33,721-8   CAS [102601-53-6]

(R)-(-)-Citramalic acid, disodium salt, 97%

C₅H₈O₅
FW 192.08

185.80
182.31
77.08
49.31
28.16*

B

₂O   QE-300

Aldrich K220-8   CAS [31635-99-1]

Ketomalonic acid monohydrate, disodium salt, 97%

C₃H₄O₆
FW 180.03

FT-IR: *1*, 554C

173.24
173.07
163.21
93.92

C

₂O + DCl   QE-300

ALDRICH

A

Aldrich 28,998-1   *CAS [526-94-3]*   C4H6O6
Sodium hydrogen tartrate, 98%   FW 172.07
mp 253°C d.

179.52
75.81

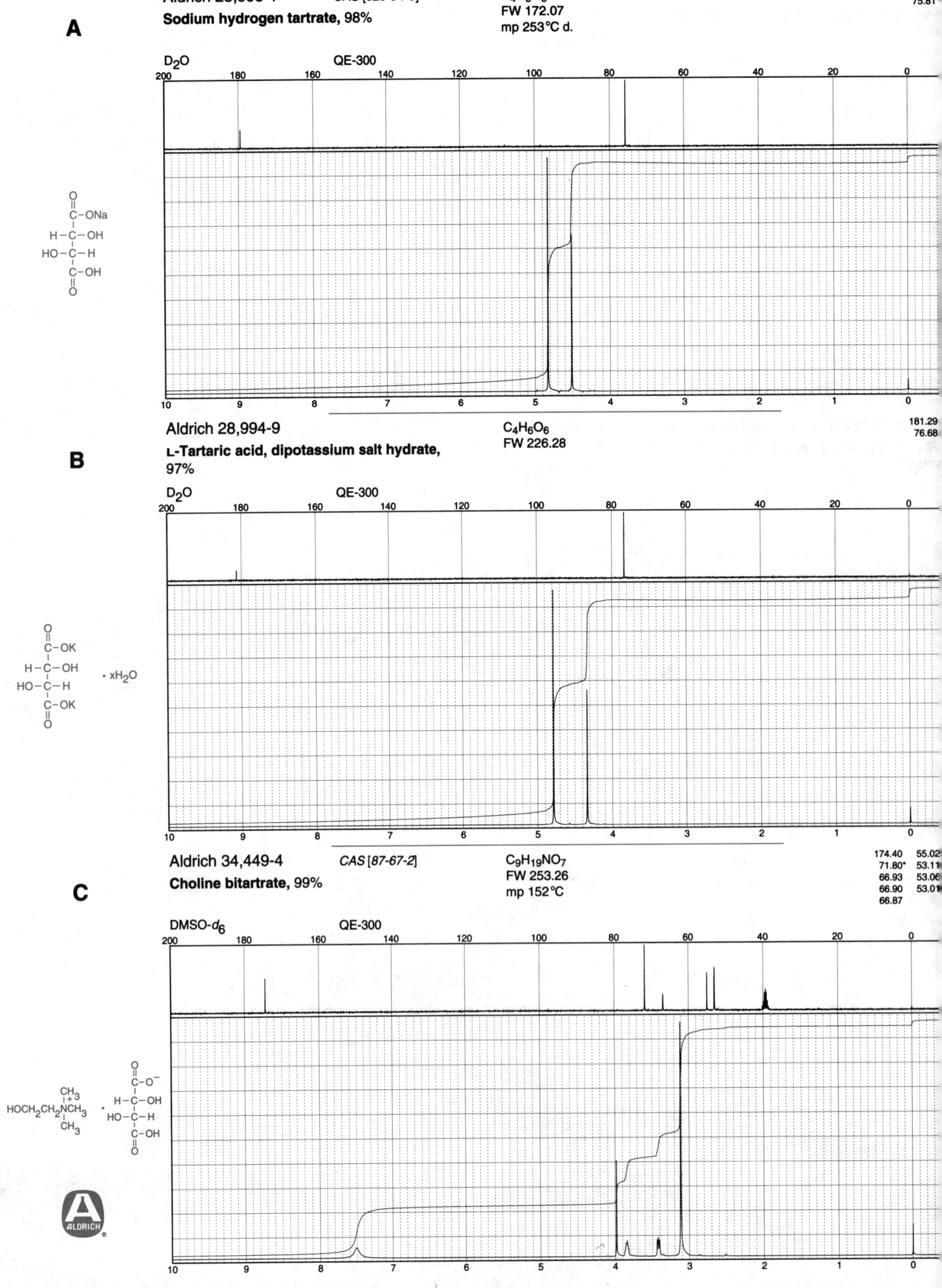

Aldrich 28,994-9   C4H6O6
FW 226.28

B

**L-Tartaric acid, dipotassium salt hydrate,
97%**

181.29
76.68

Aldrich 34,449-4   *CAS [87-67-2]*   C9H19NO7
Choline bitartrate, 99%   FW 253.26
mp 152°C

174.40   55.02
71.80*   53.11
66.93   53.06
66.90   53.01
66.87

C

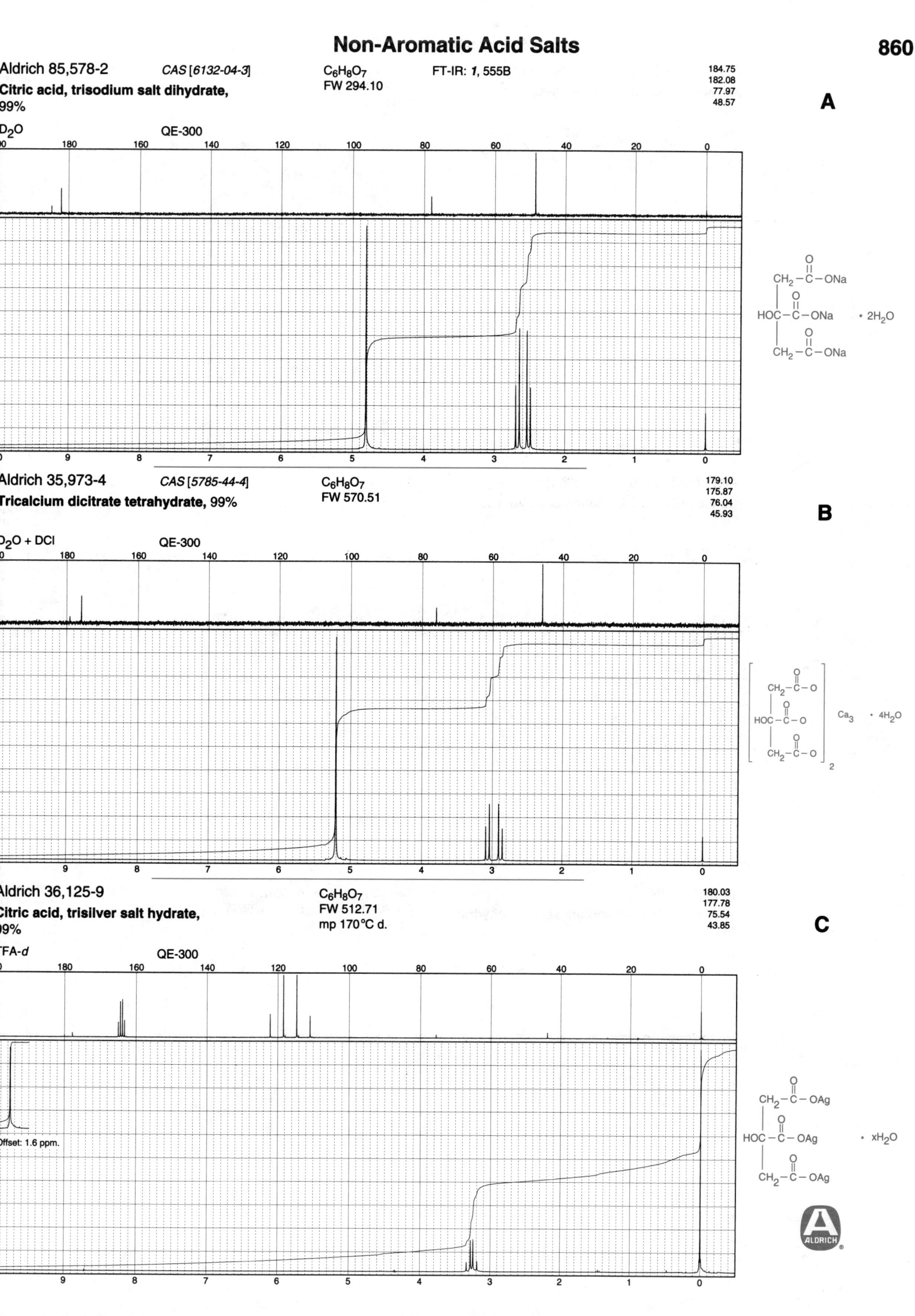

Aldrich 85,578-2   *CAS [6132-04-3]* $C_6H_8O_7$ FT-IR: *1*, 555B
Citric acid, trisodium salt dihydrate, 99%   FW 294.10

184.75
182.08
77.97
48.57

A

D_2O QE-300

Aldrich 35,973-4   *CAS [5785-44-4]* $C_6H_8O_7$
Tricalcium dicitrate tetrahydrate, 99%   FW 570.51

179.10
175.87
76.04
45.93

B

D_2O + DCl   QE-300

Aldrich 36,125-9 $C_6H_8O_7$
Citric acid, trisilver salt hydrate, 99%   FW 512.71
mp 170°C d.

180.03
177.78
75.54
43.85

C

TFA-*d*   QE-300

Offset: 1.6 ppm.

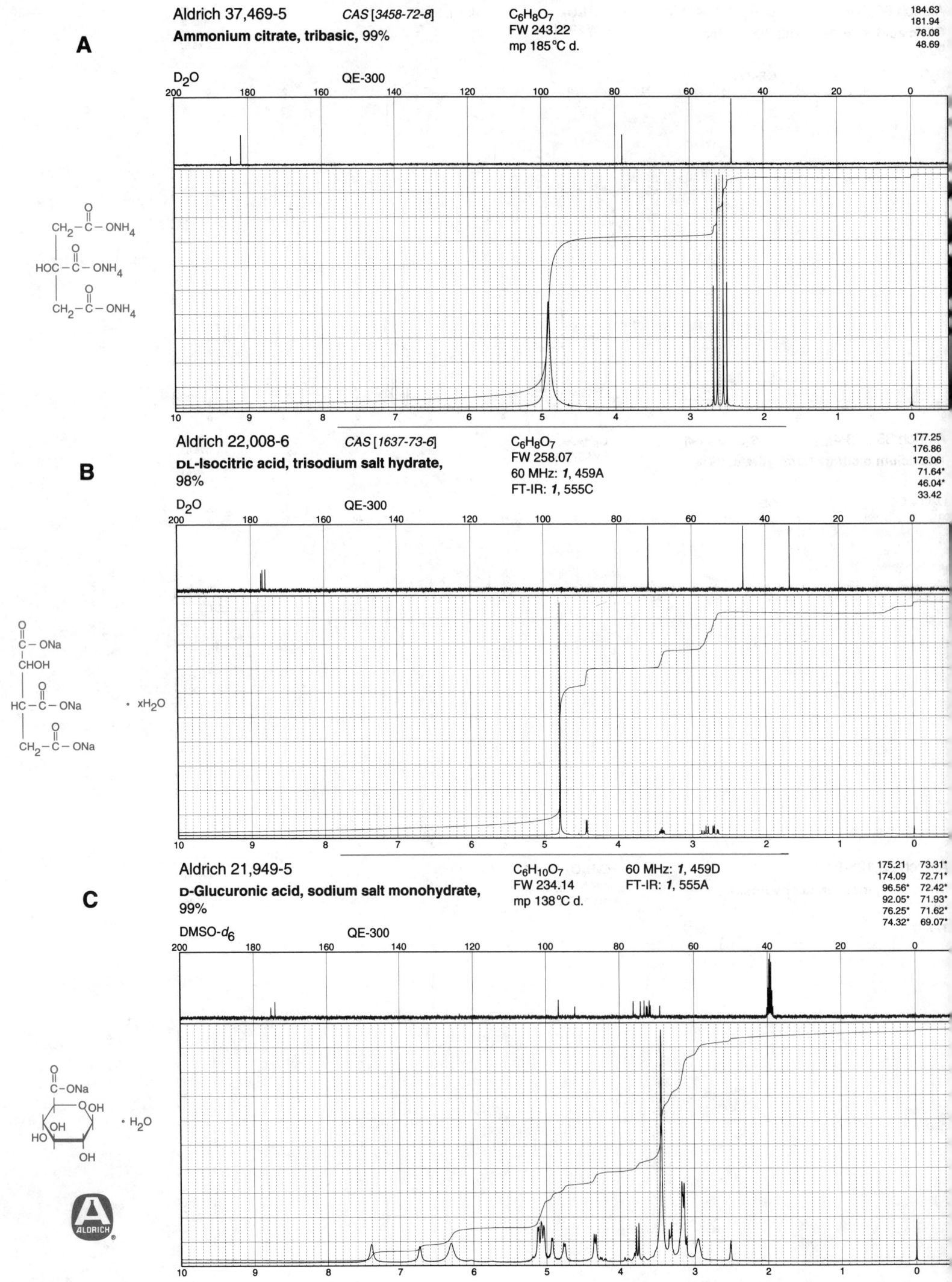

Aldrich 37,469-5   CAS [3458-72-8] $C_6H_8O_7$

Ammonium citrate, tribasic, 99%

FW 243.22
mp 185°C d.

184.63
181.94
78.08
48.69

A

D_2O QE-300

Aldrich 22,008-6   CAS [1637-73-6] $C_6H_8O_7$

DL-Isocitric acid, trisodium salt hydrate, 98%

FW 258.07
60 MHz: 1, 459A
FT-IR: 1, 555C

177.25
176.86
176.06
71.64*
46.04*
33.42

B

D_2O QE-300

• xH_2O

Aldrich 21,949-5 $C_6H_{10}O_7$ 60 MHz: 1, 459D

D-Glucuronic acid, sodium salt monohydrate, 99%

FW 234.14
mp 138°C d.
FT-IR: 1, 555A

175.21   73.31*
174.09   72.71*
96.56*   72.42*
92.05*   71.93*
76.25*   71.62*
74.32*   69.07*

C

DMSO-d_6 QE-300

• H_2O

Aldrich 17,817-9 *CAS [16698-52-5]* $C_6H_8O_3$ FT-IR: ***1***, 561C

3,4-Dihydro-2*H*-pyran-2-carboxylic acid, sodium salt, 97%

FW 150.11

mp 243°C

181.94
144.70*
104.95*
77.97*
27.97
21.35

A

D_2O QE-300

O
C–ONa
O

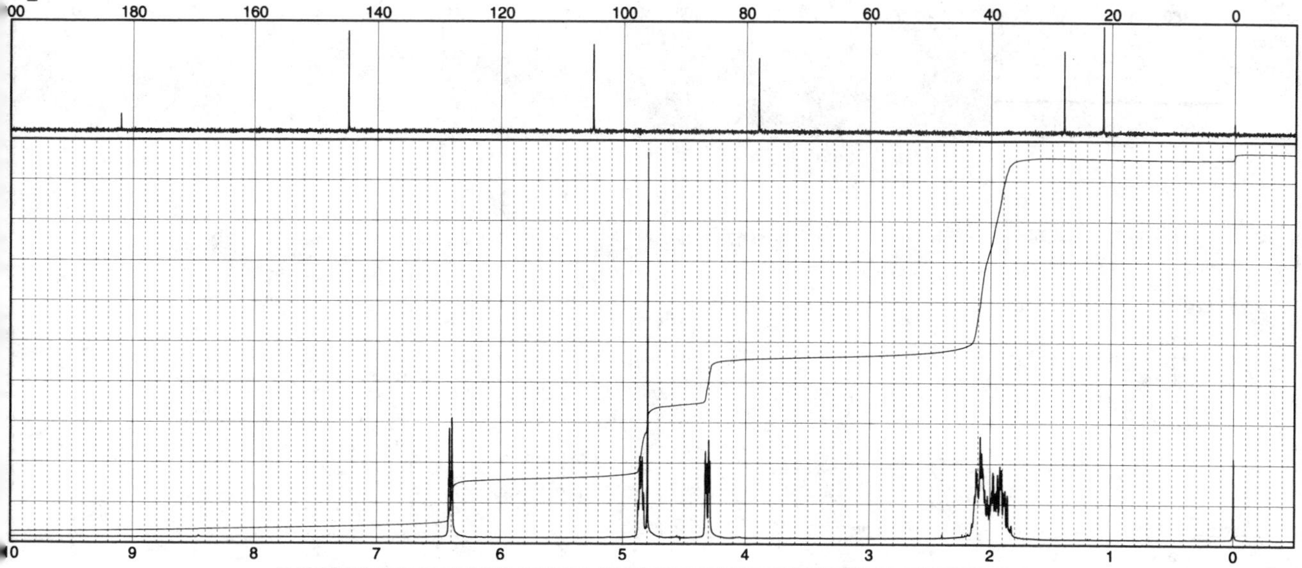

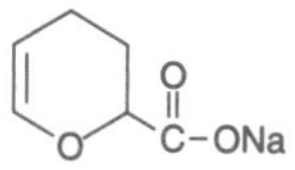

Aldrich 24,177-6 *CAS [10387-40-3]* C_2H_4OS 60 MHz: ***1***, 463B

Potassium thioacetate, 98%

FW 114.21 FT-IR: ***1***, 548C

mp 175°C

224.39
40.49*

B

D_2O QE-300

Offset: 50 ppm.

O
CH3 – C – SK

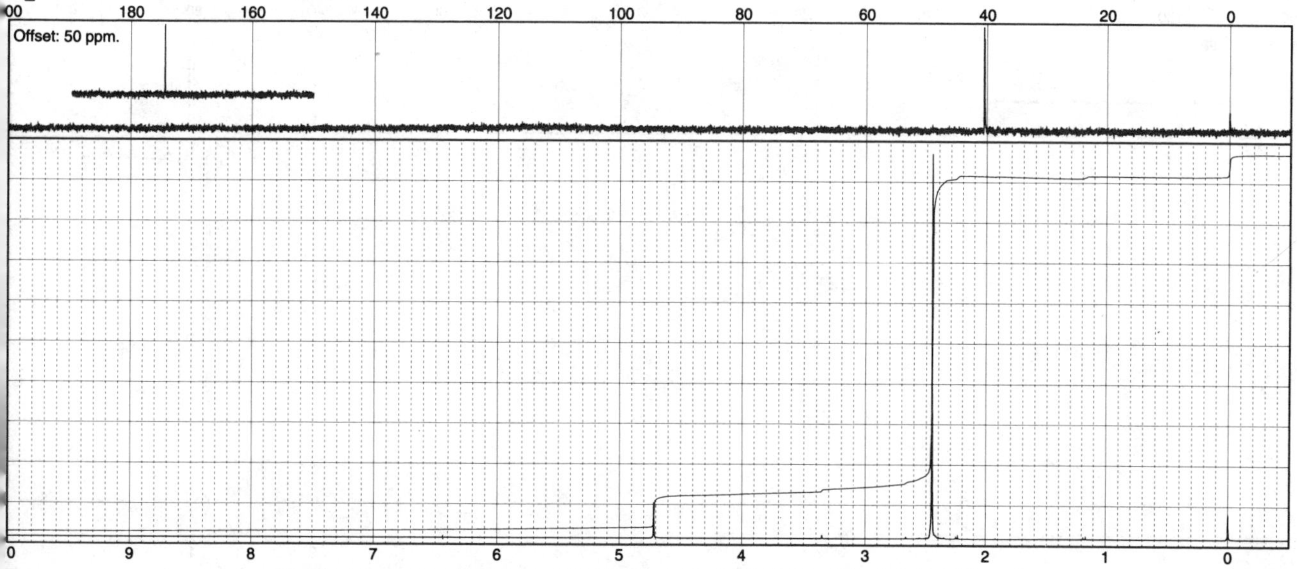

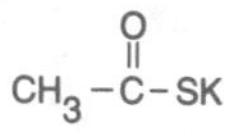

Aldrich P7,622-5 *CAS [113-24-6]* $C_3H_4O_3$ FT-IR: ***1***, 550D

Pyruvic acid, sodium salt, 98%

FW 110.04

207.85
172.69
29.29*

C

D_2O QE-300

Offset: 50 ppm.

O O
CH3 – C – C – ONa

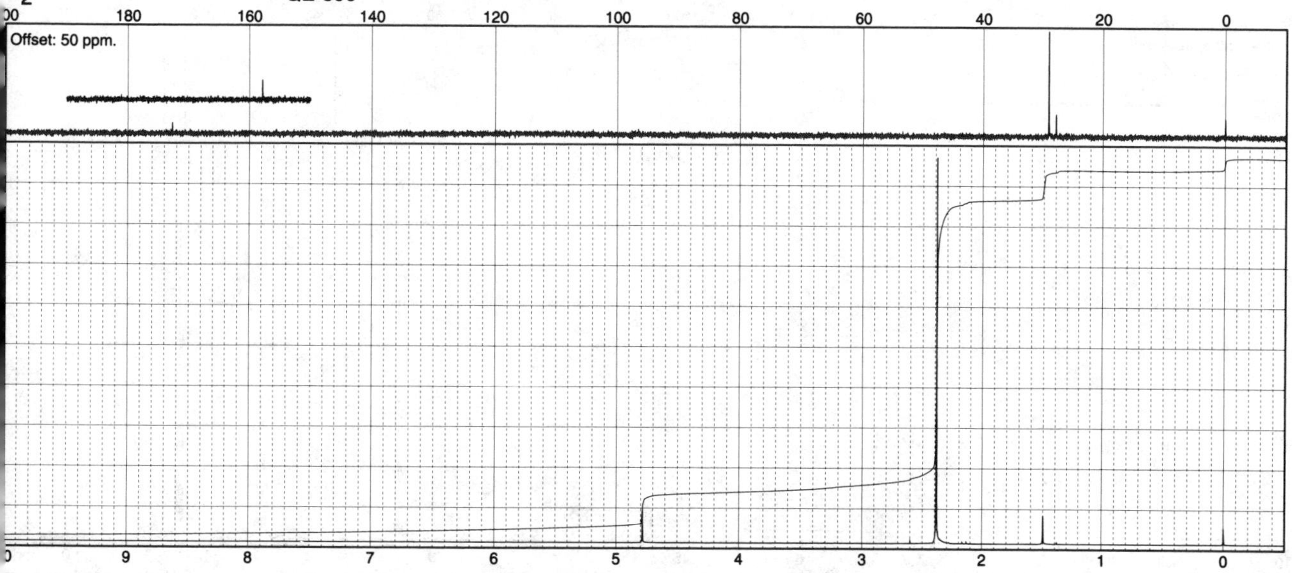

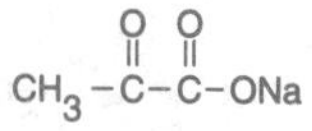

A

Aldrich 28,636-2

2-Ketobutyric acid, sodium salt monohydrate, 98%

$C_4H_6O_3$
FW 142.09
mp 210°C d.

D₂O QE-300

Offset: 50 ppm.

$CH_3CH_2-\overset{O}{\underset{||}{C}}-\overset{O}{\underset{||}{C}}-ONa$ • H_2O

211.08
173.67
35.45
9.47*

B

Aldrich 19,899-4   *CAS [3715-29-5]*

3-Methyl-2-oxobutanoic acid, sodium salt, 95%

$C_5H_8O_3$
FW 138.10
mp 229°C

60 MHz: *1*, 466B
FT-IR: *1*, 551B

D₂O QE-300

Offset: 50 ppm.

$CH_3CH-\overset{CH_3}{\underset{|}{}}\overset{O}{\underset{||}{C}}-\overset{O}{\underset{||}{C}}-ONa$

214.21
174.83
39.73*
19.20*

C

Aldrich 21,276-8   *CAS [13022-83-8]*

2-Oxopentanoic acid, sodium salt, 98%

$C_5H_8O_3$
FW 138.10
mp 236°C d.

FT-IR: *1*, 551A

D₂O QE-300

Offset: 50 ppm.

$CH_3CH_2CH_2-\overset{O}{\underset{||}{C}}-\overset{O}{\underset{||}{C}}-ONa$

210.71
173.74
43.91
19.23
15.74*

ALDRICH

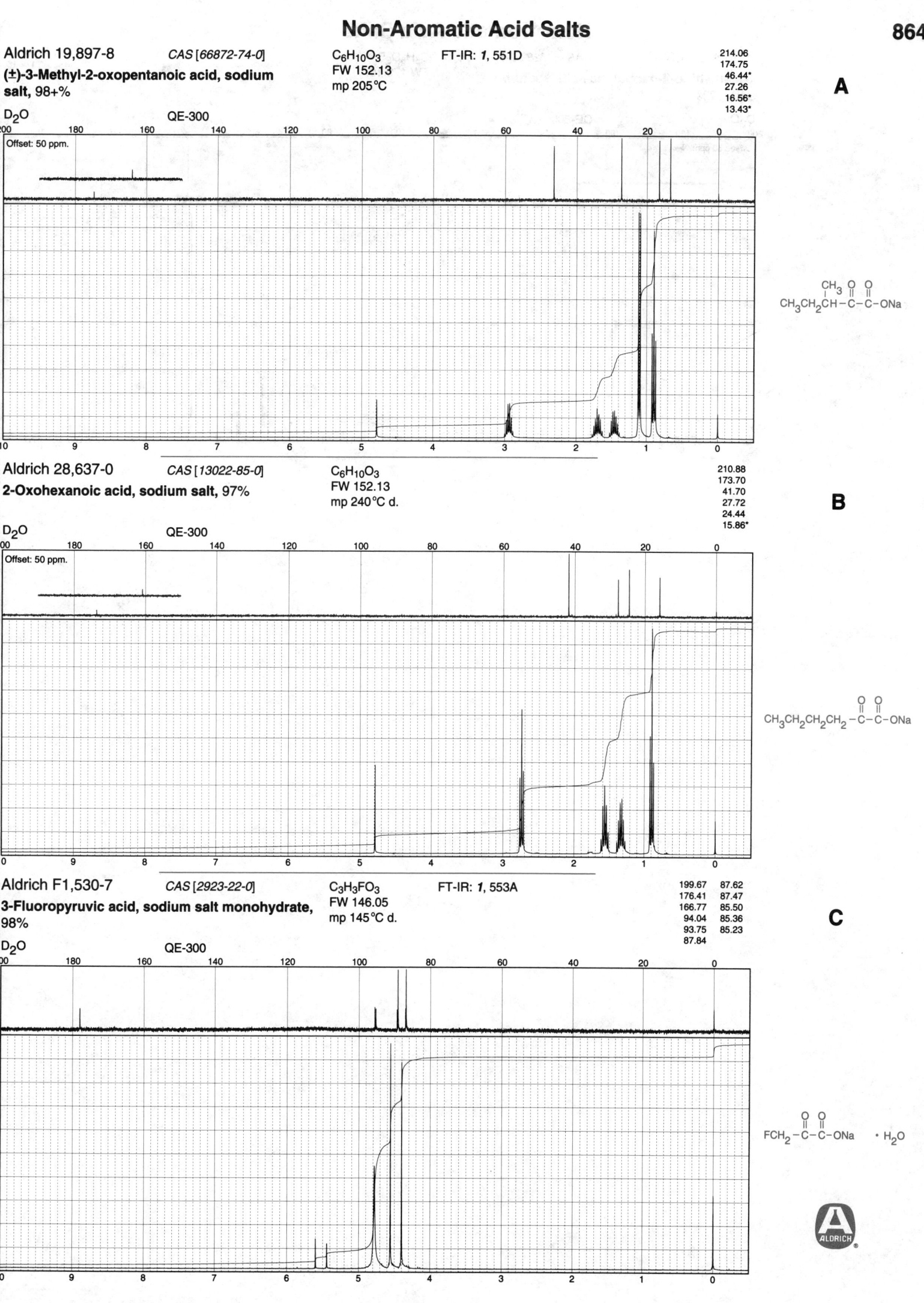

Aldrich 19,897-8   *CAS [66872-74-0]*

(±)-3-Methyl-2-oxopentanoic acid, sodium salt, 98+%

$C_6H_{10}O_3$
FW 152.13
mp 205°C

FT-IR: *1*, 551D

D_2O QE-300

Offset: 50 ppm.

214.06
174.75
46.44*
27.26
16.56*
13.43*

A

$CH_3CH_2CH-\overset{CH_3}{\underset{}{C}}-\overset{O}{\underset{}{C}}-\overset{O}{\underset{}{C}}-ONa$

Aldrich 28,637-0   *CAS [13022-85-0]*

2-Oxohexanoic acid, sodium salt, 97%

$C_6H_{10}O_3$
FW 152.13
mp 240°C d.

D_2O QE-300

Offset: 50 ppm.

210.88
173.70
41.70
27.72
24.44
15.86*

B

$CH_3CH_2CH_2CH_2-\overset{O}{\underset{}{C}}-\overset{O}{\underset{}{C}}-ONa$

Aldrich F1,530-7   *CAS [2923-22-0]*

3-Fluoropyruvic acid, sodium salt monohydrate, 98%

$C_3H_3FO_3$
FW 146.05
mp 145°C d.

FT-IR: *1*, 553A

D_2O QE-300

199.67   87.62
176.41   87.47
166.77   85.50
94.04   85.36
93.75   85.23
87.84

C

$FCH_2-\overset{O}{\underset{}{C}}-\overset{O}{\underset{}{C}}-ONa$ • H_2O

ALDRICH

Non-Aromatic Acid Salts

Aldrich 28,639-7   *CAS [51828-97-8]*   C5H8O3S
4-Methylthio-2-oxobutyric acid, sodium   FW 170.16
salt, 97%

207.42
172.16
41.56
29.51
17.11*

D2O   QE-300

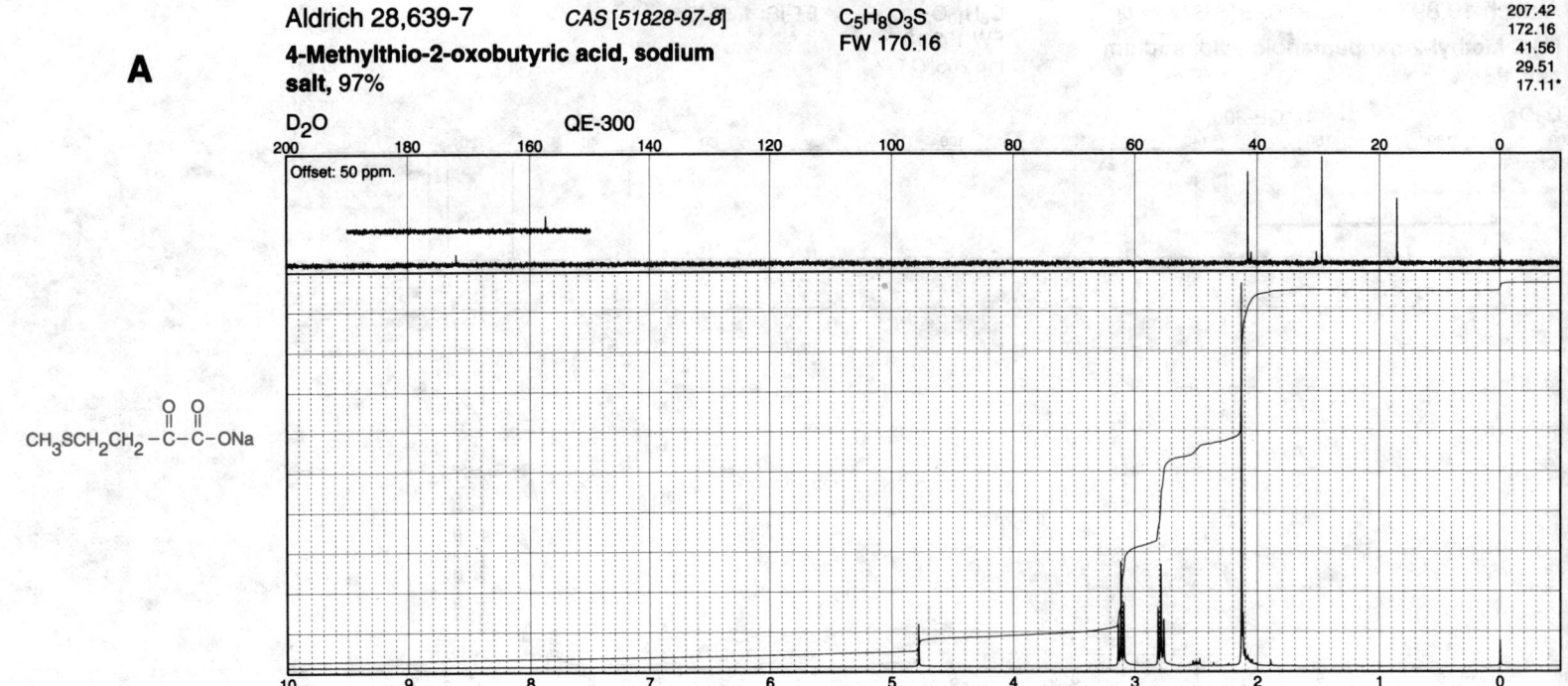

$CH_3SCH_2CH_2-\overset{\overset{O}{\|}}{C}-\overset{\overset{O}{\|}}{C}-ONa$

Aldrich 24,126-1   CAS [56-40-6] $C_2H_5NO_2$ 60 MHz: *1*, 481A

Glycine, 99+%   FW 75.07   FT-IR: *1*, 563A

mp 240°C d.

A

175.19
44.23

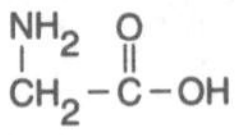

D$_2$O   QE-300

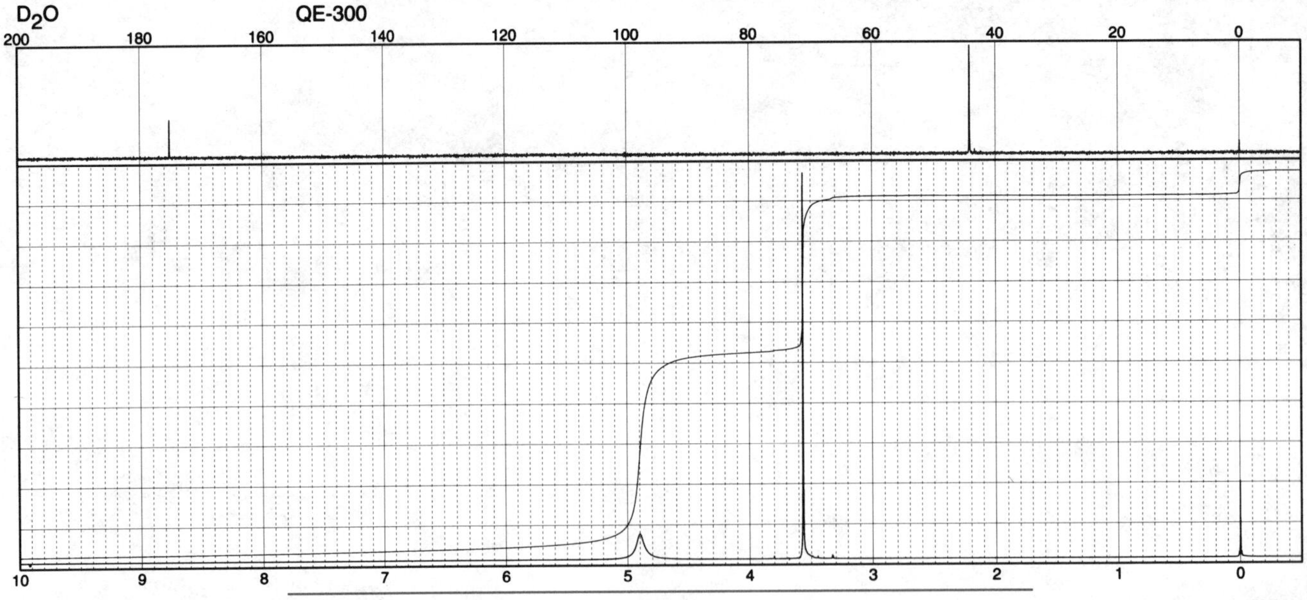

$$\underset{CH_2}{\overset{NH_2}{|}} - \underset{\overset{||}{O}}{\overset{O}{C}} - OH$$

Aldrich 33,050-7 $C_{10}H_{16}N_2O_8$

Ethylenediaminetetraacetic acid, diammonium   FW 326.31

salt hydrate, 97%   mp 218°C d.

B

173.83
60.89
54.46

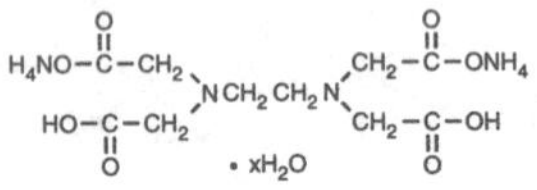

D$_2$O   QE-300

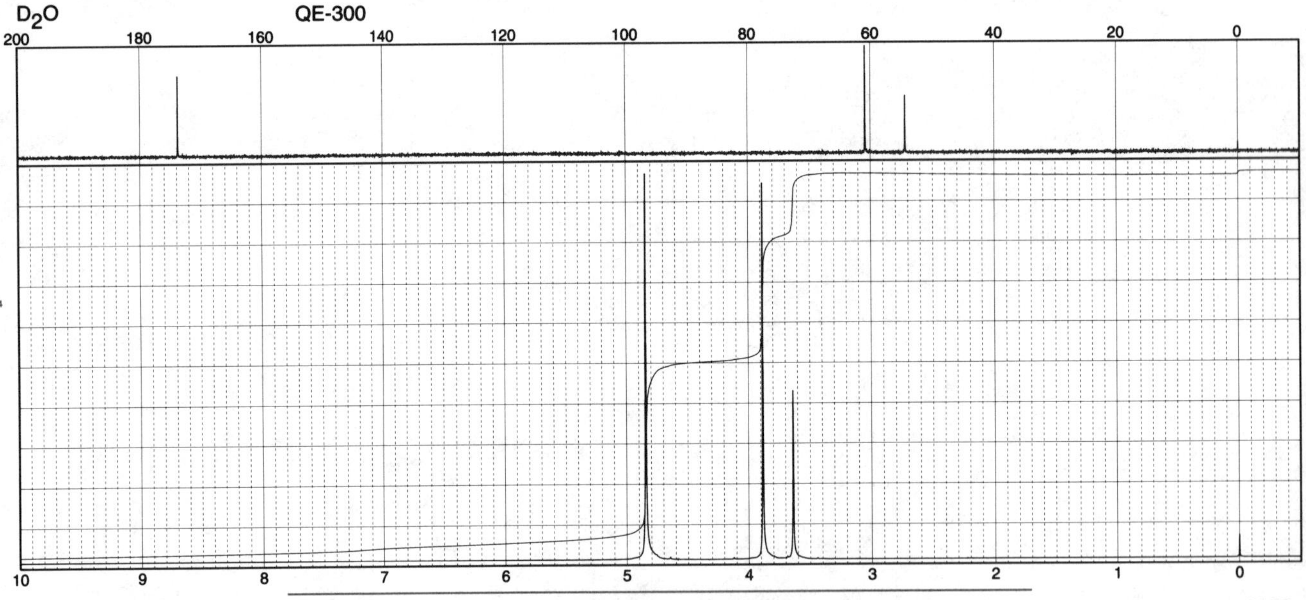

Aldrich 28,556-0   CAS [67-43-6] $C_{14}H_{23}N_3O_{10}$ FT-IR: *1*, 543C

Diethylenetriaminepentaacetic acid,   FW 393.35

98%   mp 220°C

C

181.87
181.69
61.30
60.93
54.07
53.61

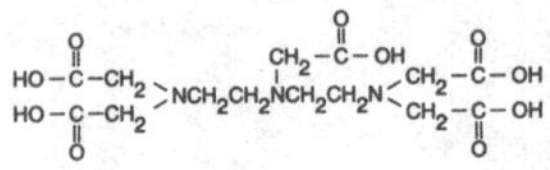

D$_2$O + NAOD   QE-300

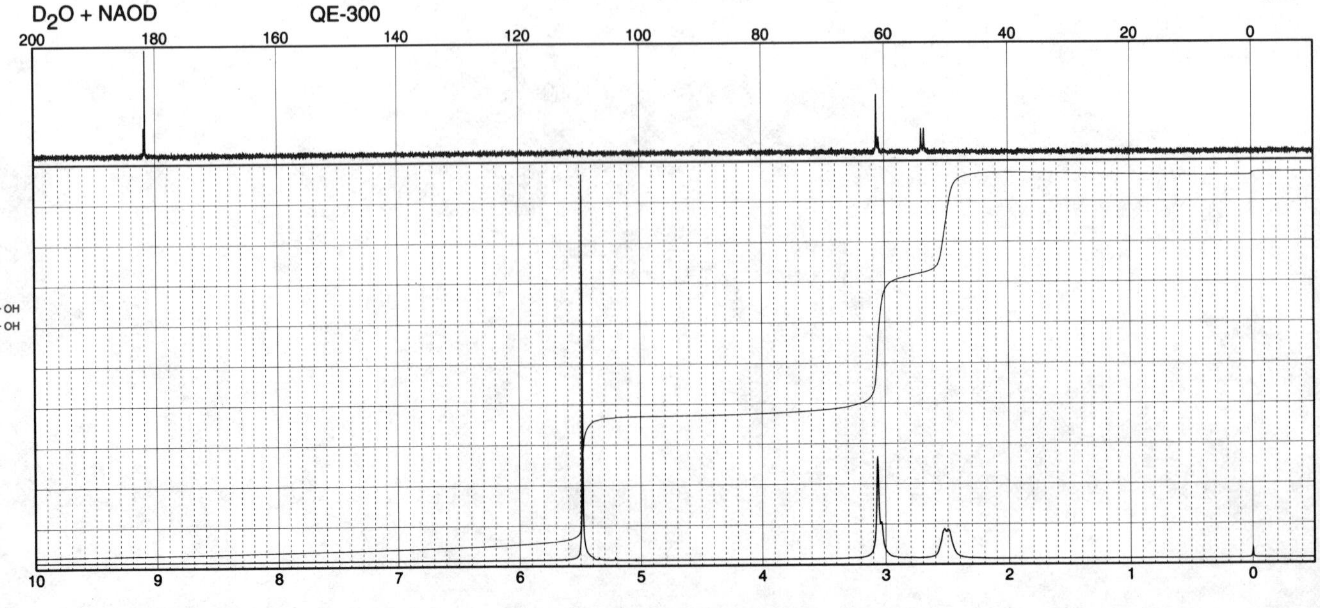

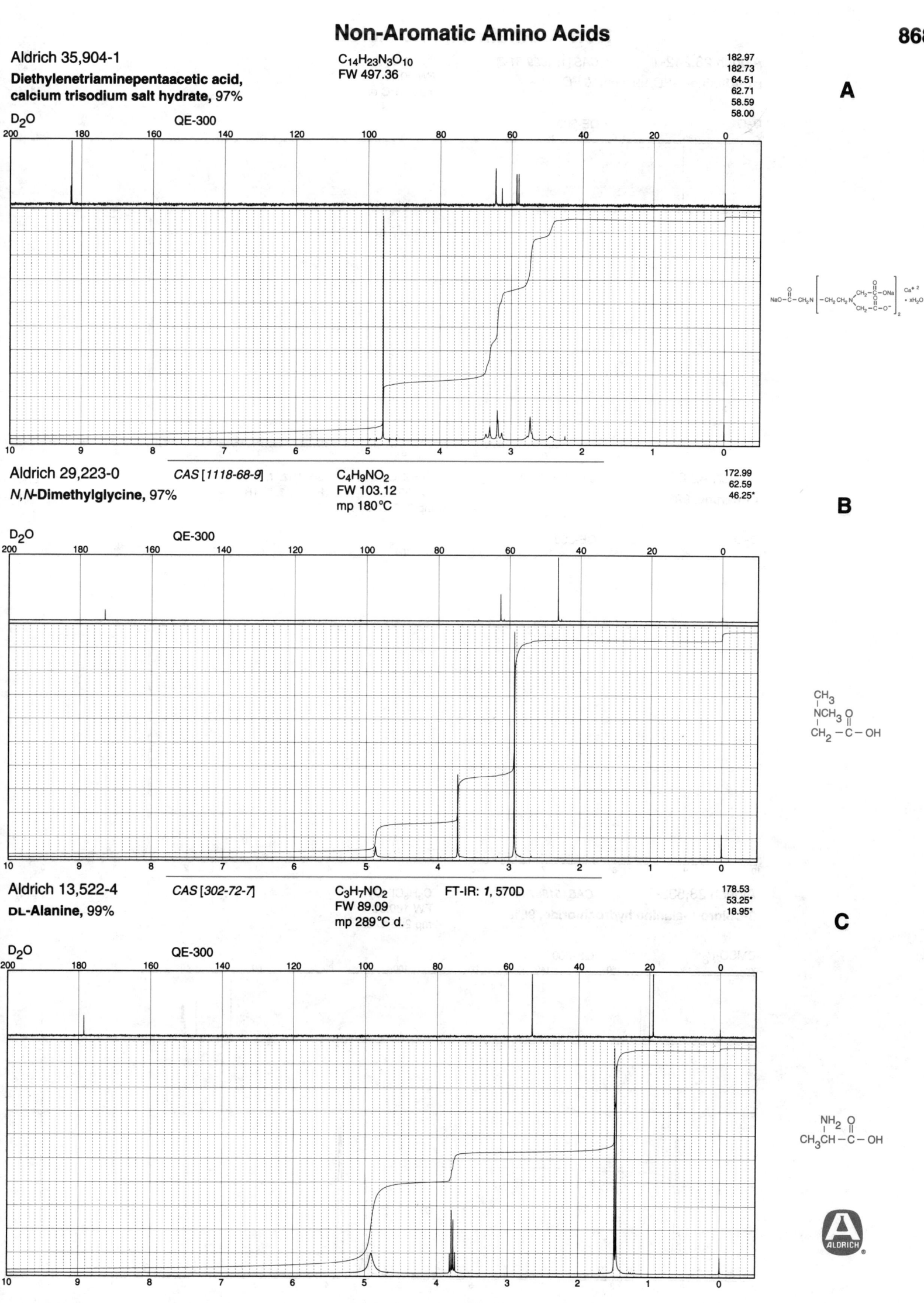

Aldrich 35,904-1

Diethylenetriaminepentaacetic acid, calcium trisodium salt hydrate, 97%

$C_{14}H_{23}N_3O_{10}$
FW 497.36

D₂O   QE-300

A

182.97
182.73
64.51
62.71
58.59
58.00

Aldrich 29,223-0   *CAS [1118-68-9]*

N,N-**Dimethylglycine, 97%**

$C_4H_9NO_2$
FW 103.12
mp 180°C

D₂O   QE-300

B

172.99
62.59
46.25*

Aldrich 13,522-4   *CAS [302-72-7]*

DL-Alanine, 99%

$C_3H_7NO_2$
FW 89.09
mp 289°C d.

FT-IR: *1*, 570D

D₂O   QE-300

C

178.53
53.25*
18.95*

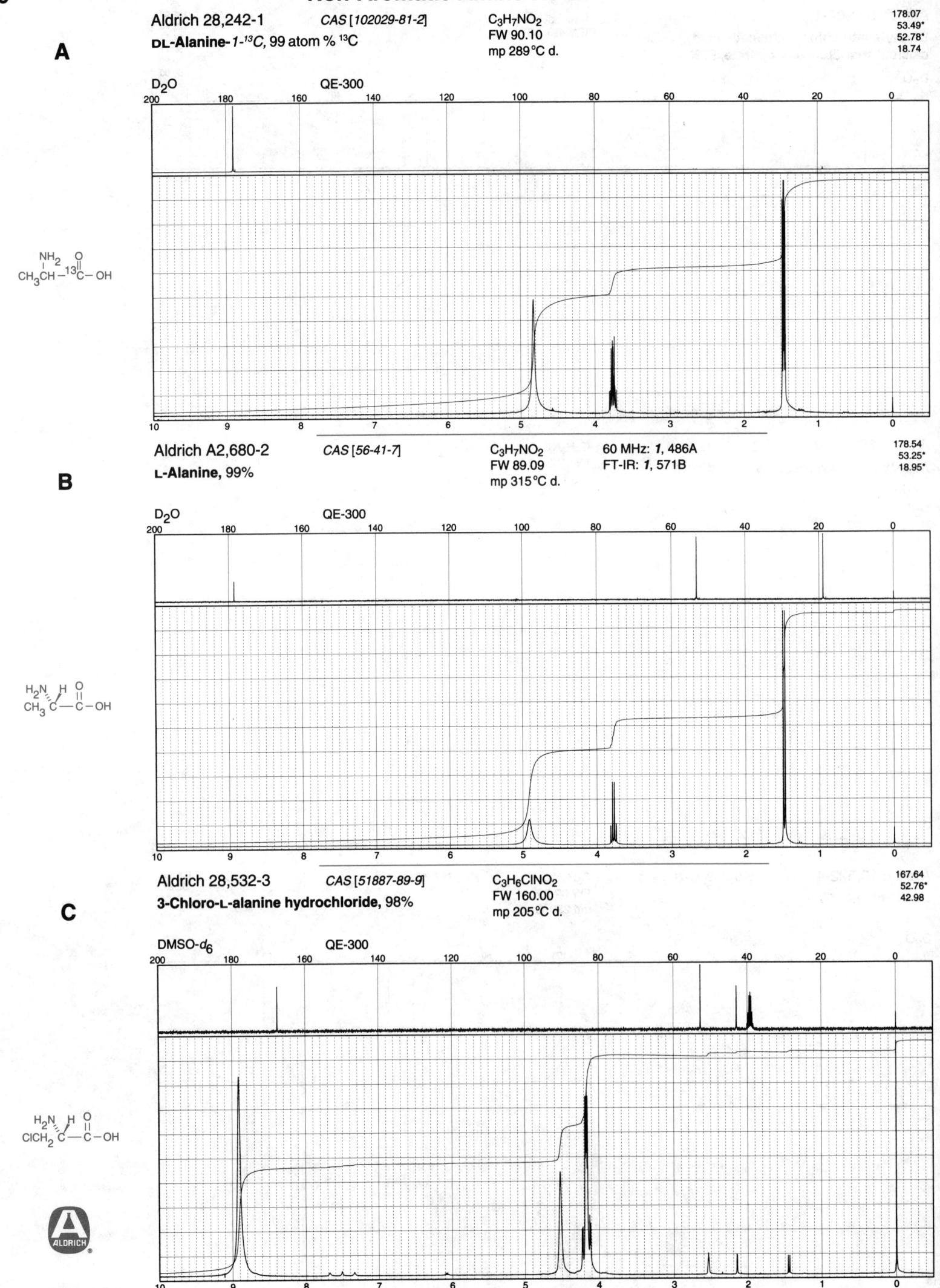

A

Aldrich 28,242-1   CAS [102029-81-2]   C₃H₇NO₂
DL-Alanine-1-¹³C, 99 atom % ¹³C   FW 90.10
                                          mp 289°C d.

178.07
53.49*
52.78*
18.74

D₂O   QE-300

$CH_3CH-^{13}C-OH$ (with NH₂ and O)

B

Aldrich A2,680-2   CAS [56-41-7]   C₃H₇NO₂   60 MHz: 1, 486A
L-Alanine, 99%   FW 89.09   FT-IR: 1, 571B
                                      mp 315°C d.

178.54
53.25*
18.95*

D₂O   QE-300

CH_3-C-OH (with H₂N, H, O)

C

Aldrich 28,532-3   CAS [51887-89-9]   C₃H₆ClNO₂
3-Chloro-L-alanine hydrochloride, 98%   FW 160.00
                                        mp 205°C d.

167.64
52.76*
42.98

DMSO-d₆   QE-300

$ClCH_2-C-OH$ (with H₂N, H, O)

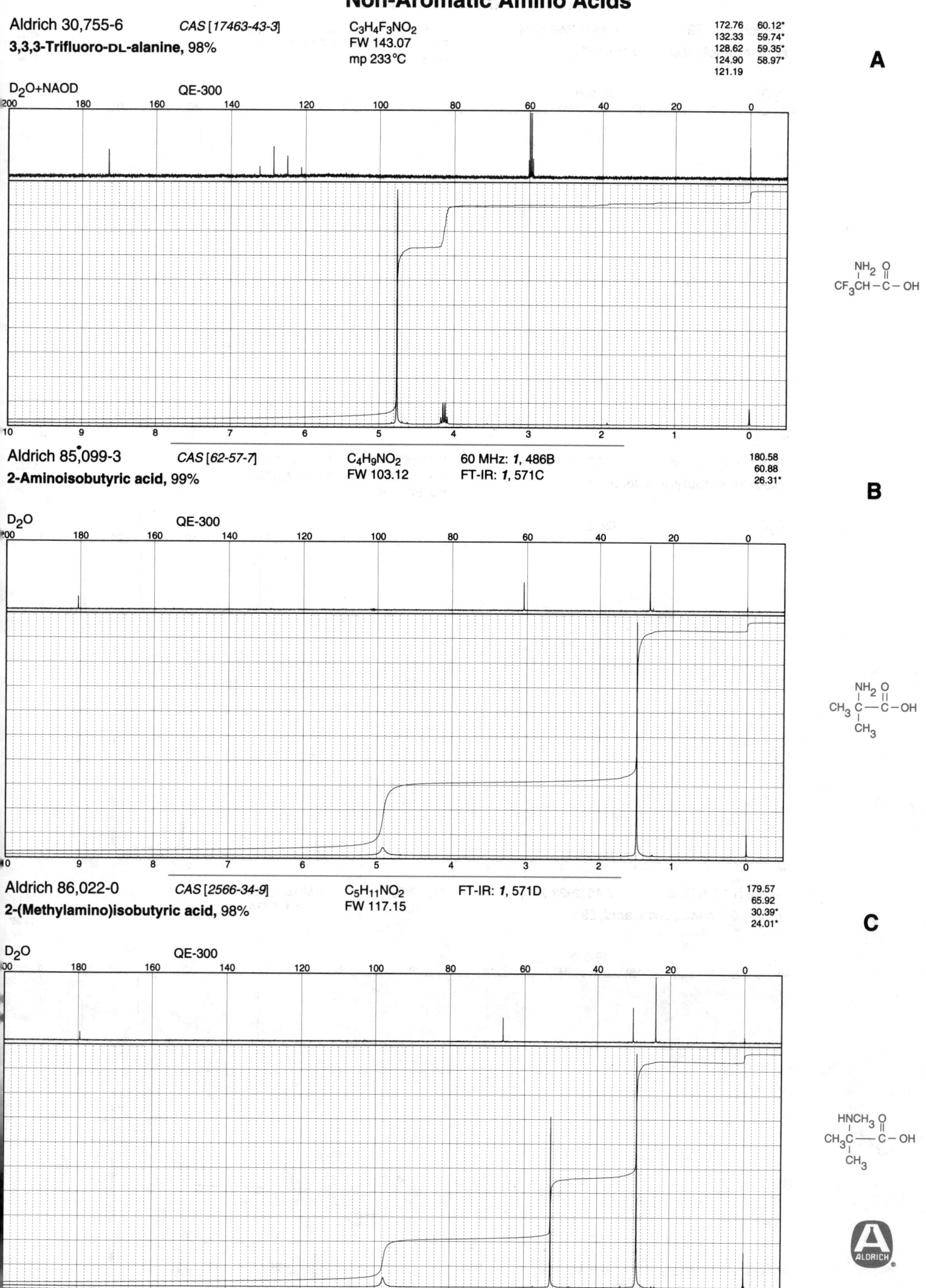

Aldrich 30,755-6   CAS [17463-43-3] $C_3H_4F_3NO_2$
3,3,3-Trifluoro-DL-alanine, 98%   FW 143.07   mp 233°C

172.76   60.12*
132.33   59.74*
128.62   59.35*
124.90   58.97*
121.19

A

D_2O+NAOD   QE-300

$CF_3CH-C-OH$ with NH_2 and O groups

Aldrich 85,099-3   CAS [62-57-7] $C_4H_9NO_2$ 60 MHz: 1, 486B
2-Aminoisobutyric acid, 99%   FW 103.12   FT-IR: 1, 571C

180.58
60.88
26.31*

B

D_2O QE-300

$CH_3-C-C-OH$ with NH_2, O, and CH_3 groups

Aldrich 86,022-0   CAS [2566-34-9] $C_5H_{11}NO_2$ FT-IR: 1, 571D
2-(Methylamino)isobutyric acid, 98%   FW 117.15

179.57
65.92
30.39*
24.01*

C

D_2O QE-300

$CH_3-C-C-OH$ with $HNCH_3$, O, and CH_3 groups

ALDRICH

Non-Aromatic Amino Acids

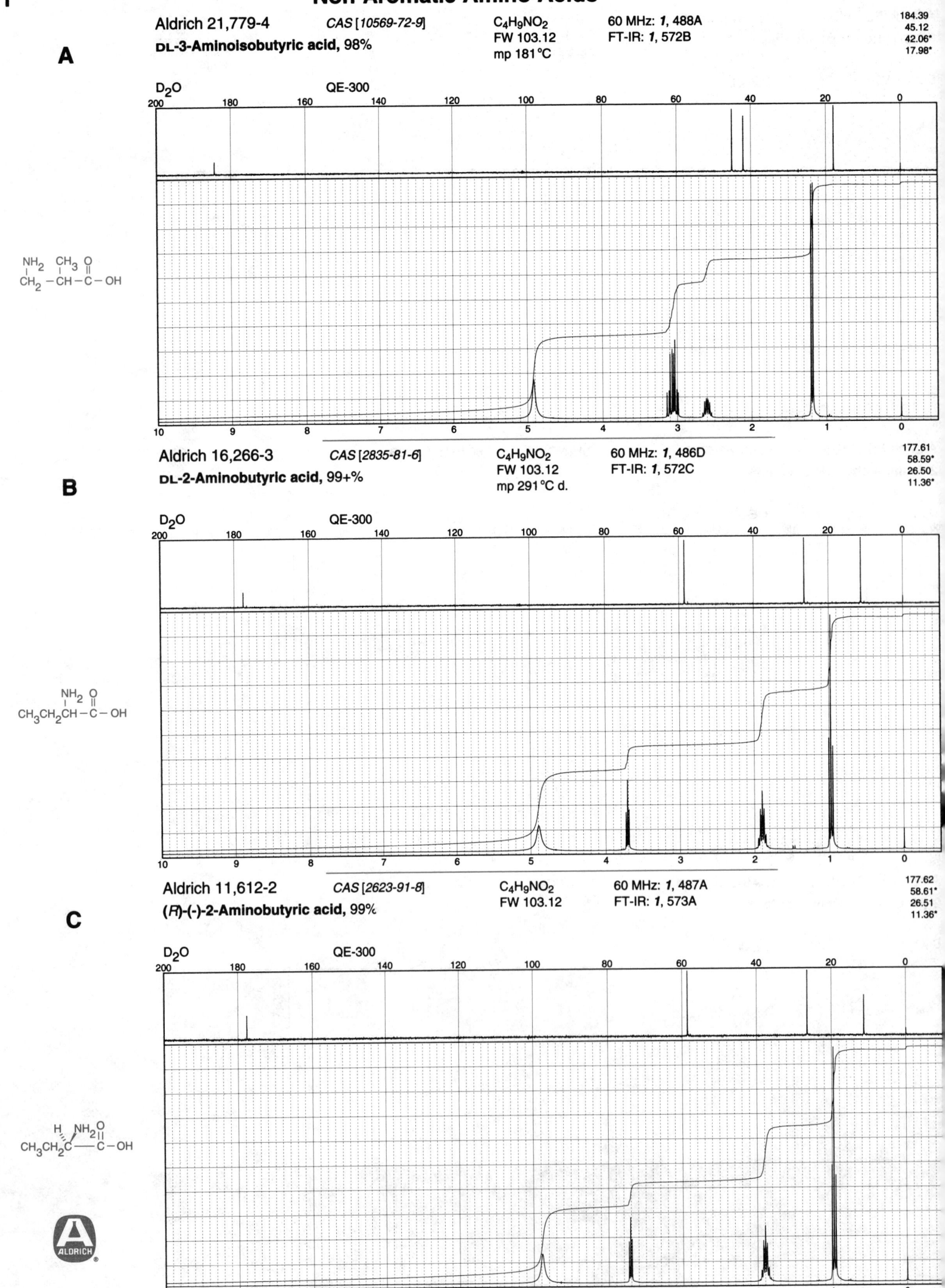

A

Aldrich 21,779-4   *CAS [10569-72-9]*   C₄H₉NO₂   60 MHz: *1*, 488A   184.39
DL-3-Aminoisobutyric acid, 98%   FW 103.12   FT-IR: *1*, 572B   45.12
mp 181 °C   42.06*
17.98*

B

Aldrich 16,266-3   *CAS [2835-81-6]*   C₄H₉NO₂   60 MHz: *1*, 486D   177.61
DL-2-Aminobutyric acid, 99+%   FW 103.12   FT-IR: *1*, 572C   58.59*
mp 291 °C d.   26.50
11.36*

C

Aldrich 11,612-2   *CAS [2623-91-8]*   C₄H₉NO₂   60 MHz: *1*, 487A   177.62
(R)-(-)-2-Aminobutyric acid, 99%   FW 103.12   FT-IR: *1*, 573A   58.61*
26.51
11.36*

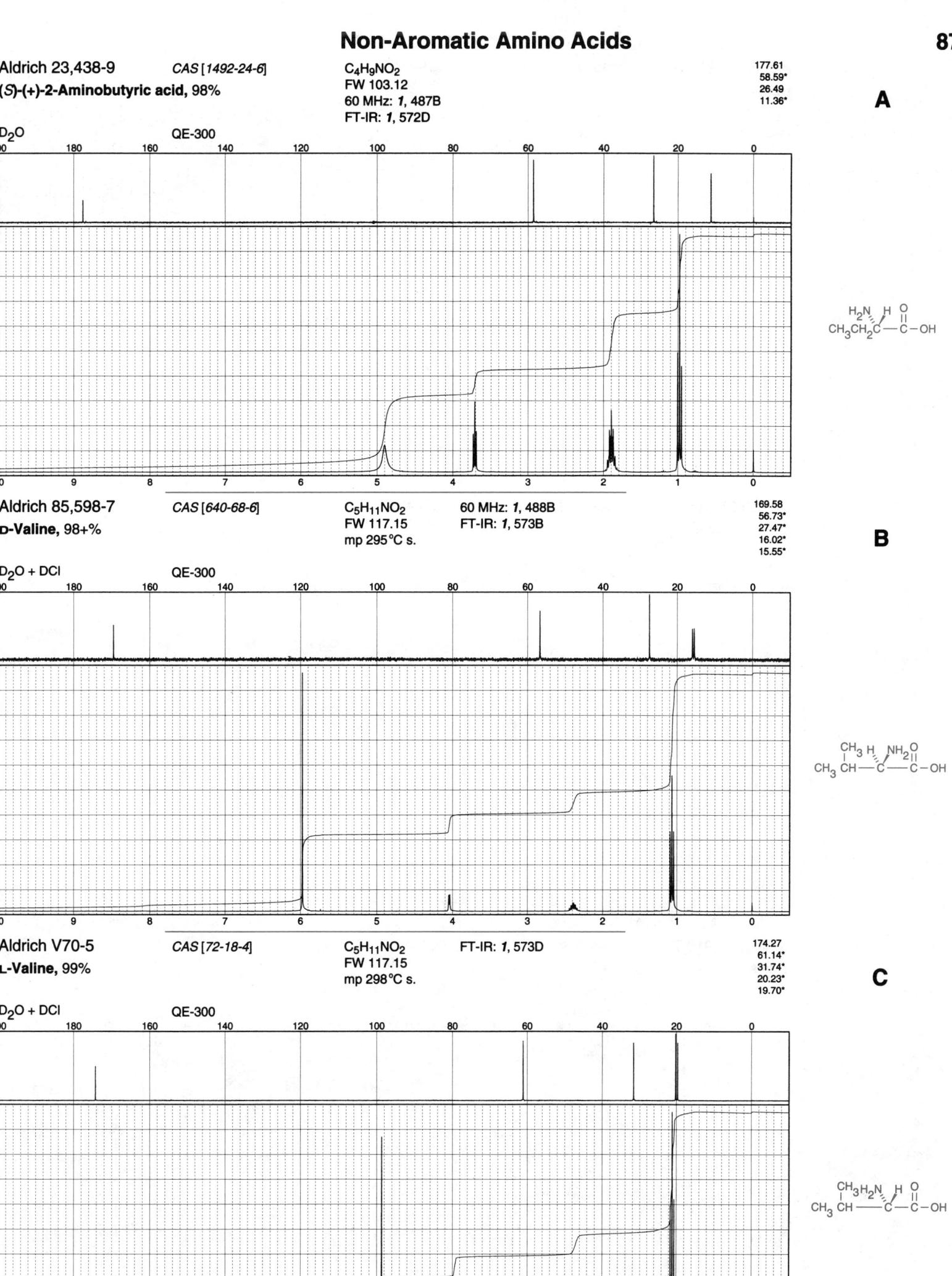

Aldrich 23,438-9   CAS [1492-24-6]

(S)-(+)-2-Aminobutyric acid, 98%

$C_4H_9NO_2$
FW 103.12
60 MHz: *1*, 487B
FT-IR: *1*, 572D

177.61
58.59*
26.49
11.36*

A

D_2O QE-300

Aldrich 85,598-7   CAS [640-68-6]

D-Valine, 98+%

$C_5H_{11}NO_2$
FW 117.15
mp 295°C s.

60 MHz: *1*, 488B
FT-IR: *1*, 573B

169.58
56.73*
27.47*
16.02*
15.55*

B

D_2O + DCl   QE-300

Aldrich V70-5   CAS [72-18-4]

L-Valine, 99%

$C_5H_{11}NO_2$
FW 117.15
mp 298°C s.

FT-IR: *1*, 573D

174.27
61.14*
31.74*
20.23*
19.70*

C

D_2O + DCl   QE-300

Non-Aromatic Amino Acids

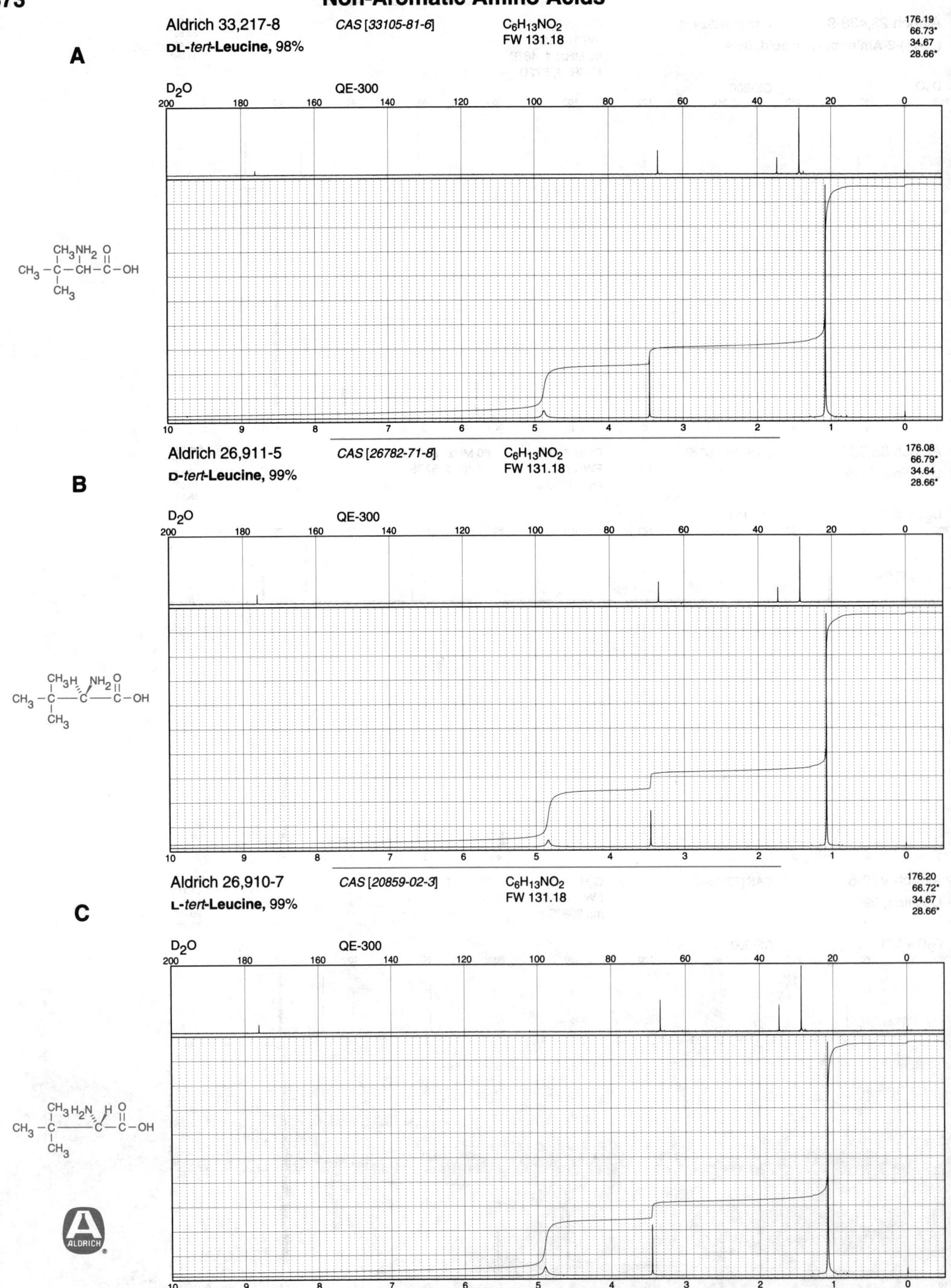

A

Aldrich 33,217-8   *CAS [33105-81-6]* $C_6H_{13}NO_2$
FW 131.18

DL-*tert*-Leucine, 98%

176.19
66.73*
34.67
28.66*

D$_2$O   QE-300

CH$_3$ NH$_2$ O
CH$_3$-C-CH-C-OH
CH$_3$

B

Aldrich 26,911-5   *CAS [26782-71-8]* $C_6H_{13}NO_2$
FW 131.18

D-*tert*-Leucine, 99%

176.08
66.79*
34.64
28.66*

D$_2$O   QE-300

CH$_3$H NH$_2$ O
CH$_3$-C-C-C-OH
CH$_3$

C

Aldrich 26,910-7   *CAS [20859-02-3]* $C_6H_{13}NO_2$
FW 131.18

L-*tert*-Leucine, 99%

176.20
66.72*
34.67
28.66*

D$_2$O   QE-300

CH$_3$ H$_2$N H O
CH$_3$-C-C-C-OH
CH$_3$

Aldrich 22,284-4   *CAS [760-78-1]*   C₅H₁₁NO₂
DL-Norvaline, 99%   FW 117.15
                                           60 MHz: *1*, 488C
                                           FT-IR: *1*, 574A

171.99
52.87*
31.80
17.92
13.11*

A

D₂O + DCl   QE-300

$CH_3CH_2CH_2CH-C-OH$ with NH_2 and O

Aldrich 85,163-9   *CAS [6600-40-4]*   C₅H₁₁NO₂   FT-IR: *1*, 574C
L-Norvaline, 99%   FW 117.15

175.71
56.00*
34.72
20.52
15.69*

B

D₂O + DCl   QE-300

$CH_3CH_2CH_2CH-C-OH$ with H_2N, H, O

Aldrich 17,344-4   *CAS [7685-44-1]*   C₅H₉NO₂   60 MHz: *1*, 490A
DL-2-Amino-4-pentenoic acid, 99%   FW 115.13   FT-IR: *1*, 574D
                                            mp 259°C

176.91
134.17*
123.29
56.80*
37.65

C

D₂O   QE-300

$H_2C=CHCH_2CH-C-OH$ with NH_2 and O

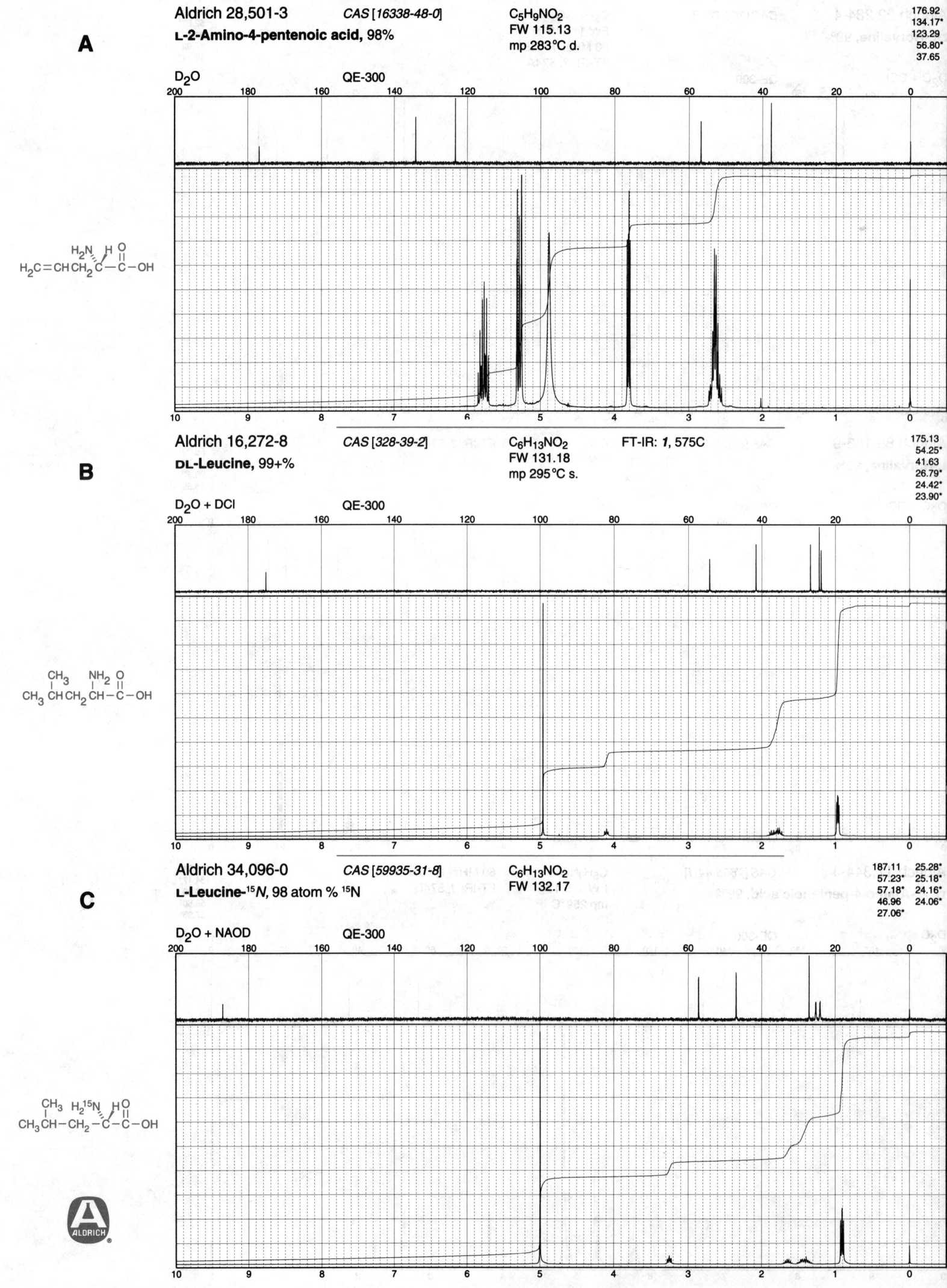

Aldrich 28,501-3   *CAS [16338-48-0]* $C_5H_9NO_2$

A

L-2-Amino-4-pentenoic acid, 98%   FW 115.13   mp 283°C d.

176.92
134.17*
123.29
56.80*
37.65

D_2O QE-300

Aldrich 16,272-8   *CAS [328-39-2]* $C_6H_{13}NO_2$ FT-IR: *1*, 575C

B

DL-Leucine, 99+%   FW 131.18   mp 295°C s.

175.13
54.25*
41.63
26.79*
24.42*
23.90*

D_2O + DCl   QE-300

Aldrich 34,096-0   *CAS [59935-31-8]* $C_6H_{13}NO_2$

C

L-Leucine-^{15}N, 98 atom % ^{15}N FW 132.17

187.11   25.28*
57.23*   25.18*
57.18*   24.16*
46.96   24.06*
27.06*

D_2O + NAOD   QE-300

A

Aldrich 21,963-0   CAS [54897-59-5] $C_3H_8N_2O_2$ 60 MHz: 1, 493D

2,3-Diaminopropionic acid monohydrochloride, 98%   FW 140.57   FT-IR: 1, 576C   mp 232°C d.

173.98
53.18*
41.67

D$_2$O   QE-300

$H_2NCH_2CH-\overset{NH_2}{\underset{}{C}}\overset{O}{\underset{}{C}}-OH$ · HCl

B

Aldrich 85,165-5   CAS [327-57-1] $C_6H_{13}NO_2$ FT-IR: 1, 576D

L-Norleucine, 99+%   FW 131.18

174.40
55.65*
31.97
28.85
24.11
15.68*

D$_2$O + DCI   QE-300

$CH_3CH_2CH_2CH_2CH_2-\overset{H_2N}{\underset{}{C}}\overset{H}{\underset{}{}}\overset{O}{\underset{}{C}}-OH$

C

Aldrich 21,770-0   CAS [644-90-6] $C_8H_{17}NO_2$ 60 MHz: 1, 490D

DL-2-Aminocaprylic acid, 99%   FW 159.23   FT-IR: 1, 577A   mp 260°C d.

175.69   30.16
56.51*   26.39
32.90   23.91
31.91   14.33*

TFA-d   QE-300

Offset: 2.0 ppm.

$CH_3(CH_2)_4CH_2CH-\overset{NH_2}{\underset{}{C}}\overset{O}{\underset{}{C}}-OH$

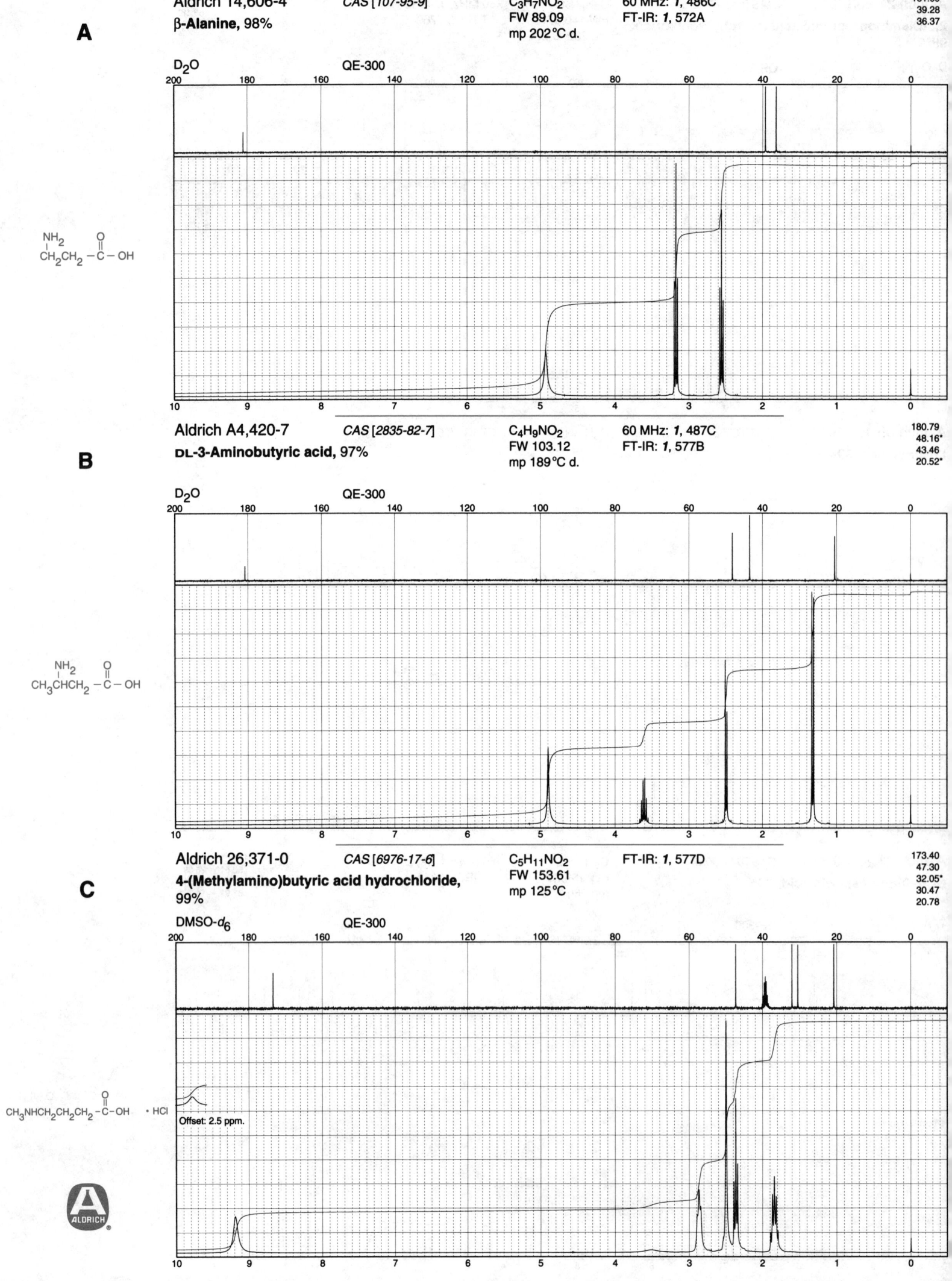

A

Aldrich 14,606-4

β-Alanine, 98%

CAS [107-95-9]

$C_3H_7NO_2$
FW 89.09
mp 202°C d.

60 MHz: *1*, 486C
FT-IR: *1*, 572A

181.09
39.28
36.37

D₂O   QE-300

B

Aldrich A4,420-7

DL-3-Aminobutyric acid, 97%

CAS [2835-82-7]

$C_4H_9NO_2$
FW 103.12
mp 189°C d.

60 MHz: *1*, 487C
FT-IR: *1*, 577B

180.79
48.16*
43.46
20.52*

D₂O   QE-300

C

Aldrich 26,371-0

**4-(Methylamino)butyric acid hydrochloride,
99%**

CAS [6976-17-6]

$C_5H_{11}NO_2$
FW 153.61
mp 125°C

FT-IR: *1*, 577D

173.40
47.30
32.05*
30.47
20.78

DMSO-d_6 QE-300

Offset: 2.5 ppm.

A

Aldrich 26,373-7   CAS [69954-66-1] $C_6H_{13}NO_2$ FT-IR: 1, 578A

4-(Dimethylamino)butyric acid hydrochloride, 98%

FW 167.64
mp 154 °C

173.25
55.53
41.68*
30.55
19.15

DMSO-d_6 QE-300

$CH_3NHCH_2CH_2CH_2-\overset{\overset{\displaystyle O}{\|}}{C}-OH$ • HCl

(with CH_3 on N)

B

Aldrich 26,365-6   CAS [6249-56-5] $C_7H_{16}ClNO_2$ FT-IR: 1, 681A

(3-Carboxypropyl)trimethylammonium chloride, 97%

FW 181.66
mp 220 °C d.

179.75   55.73*
68.31   55.67*
68.27   33.28
68.23   20.99
55.78*

D$_2$O   QE-300

$CH_3\overset{\overset{\displaystyle CH_3}{|+}}{\underset{\underset{\displaystyle CH_3}{|}}{N}}CH_2CH_2CH_2-\overset{\overset{\displaystyle O}{\|}}{C}-OH$ Cl$^-$

C

Aldrich 19,433-6   CAS [627-95-2] $C_5H_{11}NO_2$ FT-IR: 1, 578C

5-Aminovaleric acid hydrochloride, 99%

FW 153.61
mp 96 °C

181.32
42.00
36.21
29.01
24.08

D$_2$O   QE-300

$H_2NCH_2(CH_2)_2CH_2-\overset{\overset{\displaystyle O}{\|}}{C}-OH$ • HCl

ALDRICH

Non-Aromatic Amino Acids

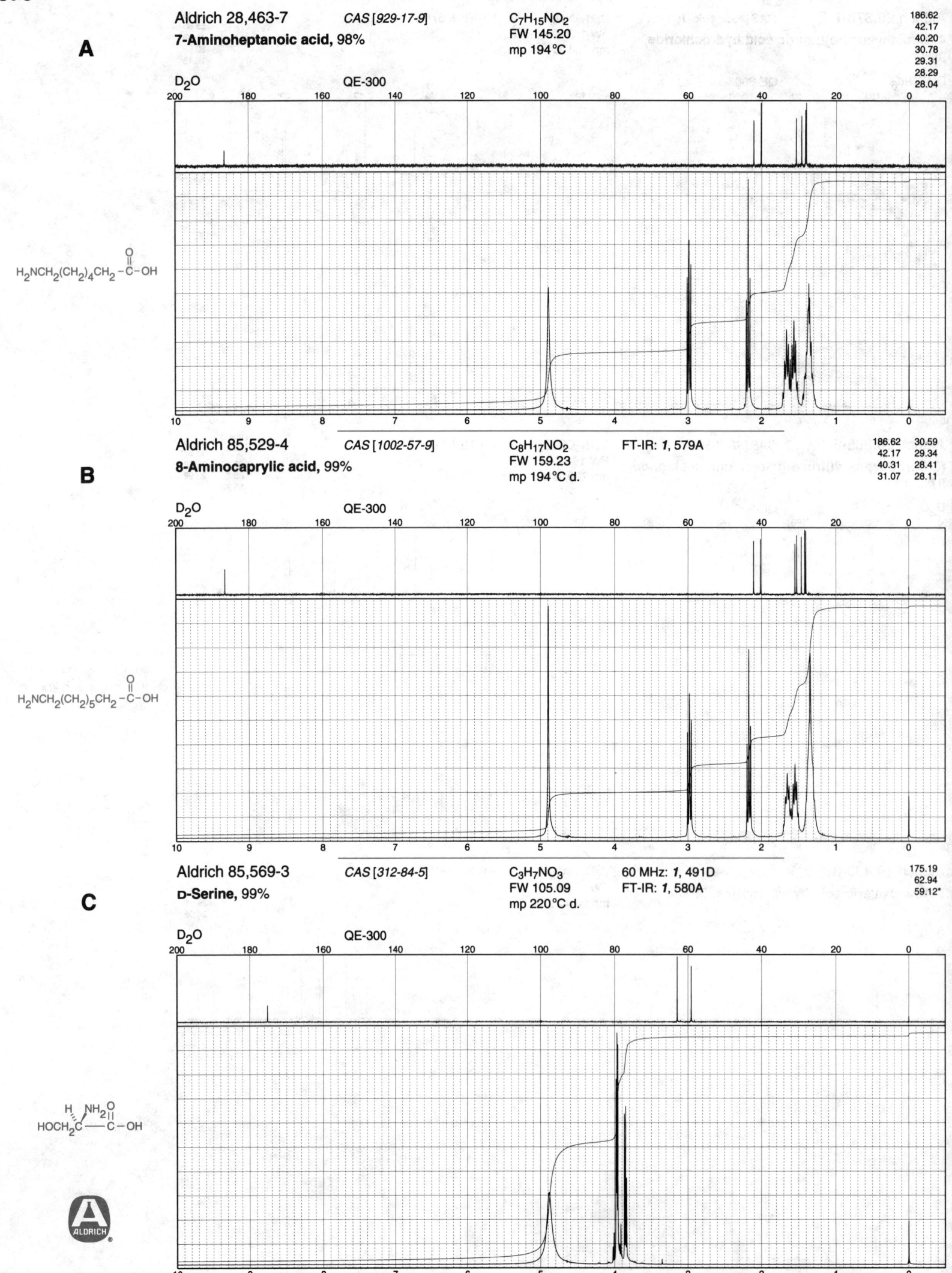

A

Aldrich 28,463-7   *CAS [929-17-9]*   C₇H₁₅NO₂

7-Aminoheptanoic acid, 98%

$C_7H_{15}NO_2$
FW 145.20
mp 194°C

186.62
42.17
40.20
30.78
29.31
28.29
28.04

D₂O   QE-300

H₂NCH₂(CH₂)₄CH₂–C(=O)–OH

B

Aldrich 85,529-4   *CAS [1002-57-9]* $C_8H_{17}NO_2$ FT-IR: *1*, 579A

8-Aminocaprylic acid, 99%

$C_8H_{17}NO_2$
FW 159.23
mp 194°C d.

186.62   30.59
42.17   29.34
40.31   28.41
31.07   28.11

D₂O   QE-300

H₂NCH₂(CH₂)₅CH₂–C(=O)–OH

C

Aldrich 85,569-3   *CAS [312-84-5]* $C_3H_7NO_3$ 60 MHz: *1*, 491D

D-Serine, 99%

$C_3H_7NO_3$
FW 105.09
mp 220°C d.

FT-IR: *1*, 580A

175.19
62.94
59.12*

D₂O   QE-300

HOCH₂C(H)(NH₂)–C(=O)–OH

ALDRICH

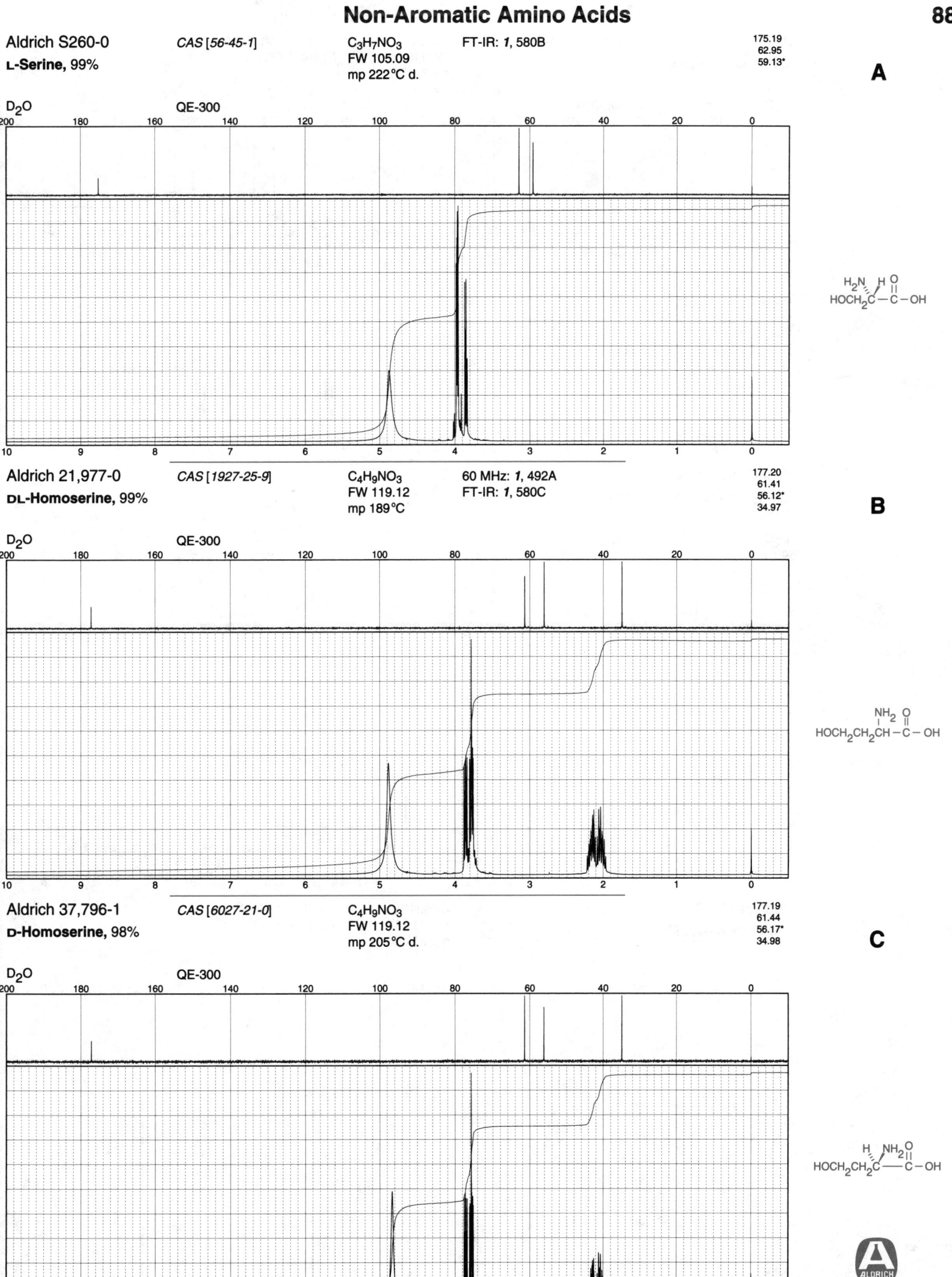

Aldrich S260-0   *CAS [56-45-1]*   C₃H₇NO₃   FT-IR: *1*, 580B
L-Serine, 99%   FW 105.09
    mp 222°C d.

175.19
62.95
59.13*

A

D₂O   QE-300

Aldrich 21,977-0   *CAS [1927-25-9]*   C₄H₉NO₃   60 MHz: *1*, 492A
DL-Homoserine, 99%   FW 119.12   FT-IR: *1*, 580C
    mp 189°C

177.20
61.41
56.12*
34.97

B

D₂O   QE-300

Aldrich 37,796-1   *CAS [6027-21-0]*   C₄H₉NO₃
D-Homoserine, 98%   FW 119.12
    mp 205°C d.

177.19
61.44
56.17*
34.98

C

D₂O   QE-300

Non-Aromatic Amino Acids

A

Aldrich 21,978-9   *CAS [672-15-1]*   C$_4$H$_9$NO$_3$ 60 MHz: *1*, 492B   177.19
L-Homoserine, 98%   FW 119.12   FT-IR: *1*, 580D   61.40
     mp 203°C d.   56.12*
     34.97

D$_2$O   QE-300

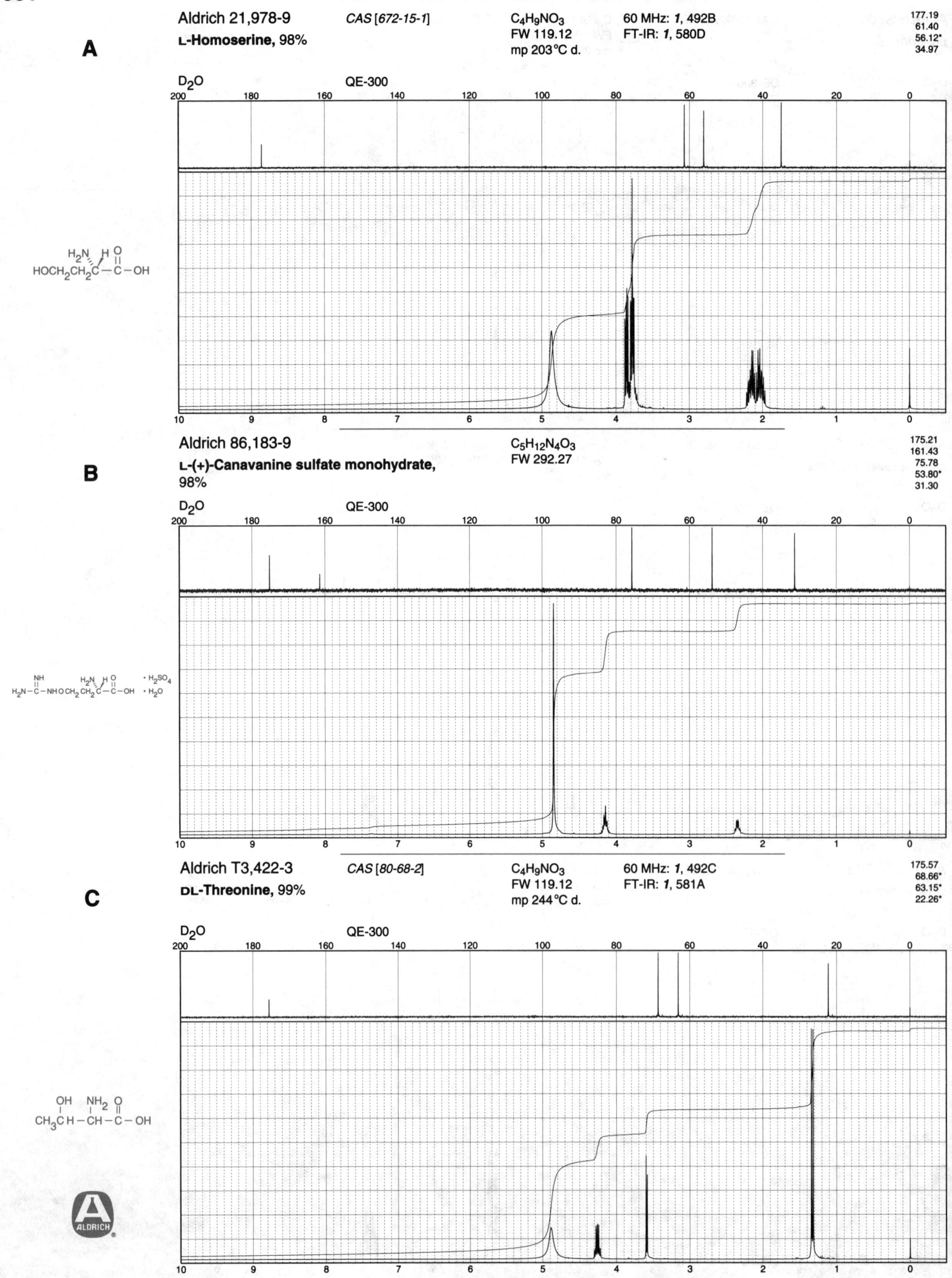

B

Aldrich 86,183-9   C$_5$H$_{12}$N$_4$O$_3$ 175.21
L-(+)-Canavanine sulfate monohydrate,   FW 292.27   161.43
98%   75.78
     53.80*
     31.30

D$_2$O   QE-300

C

Aldrich T3,422-3   *CAS [80-68-2]*   C$_4$H$_9$NO$_3$ 60 MHz: *1*, 492C   175.57
DL-Threonine, 99%   FW 119.12   FT-IR: *1*, 581A   68.66*
     mp 244°C d.   63.15*
     22.26*

D$_2$O   QE-300

Non-Aromatic Amino Acids

882

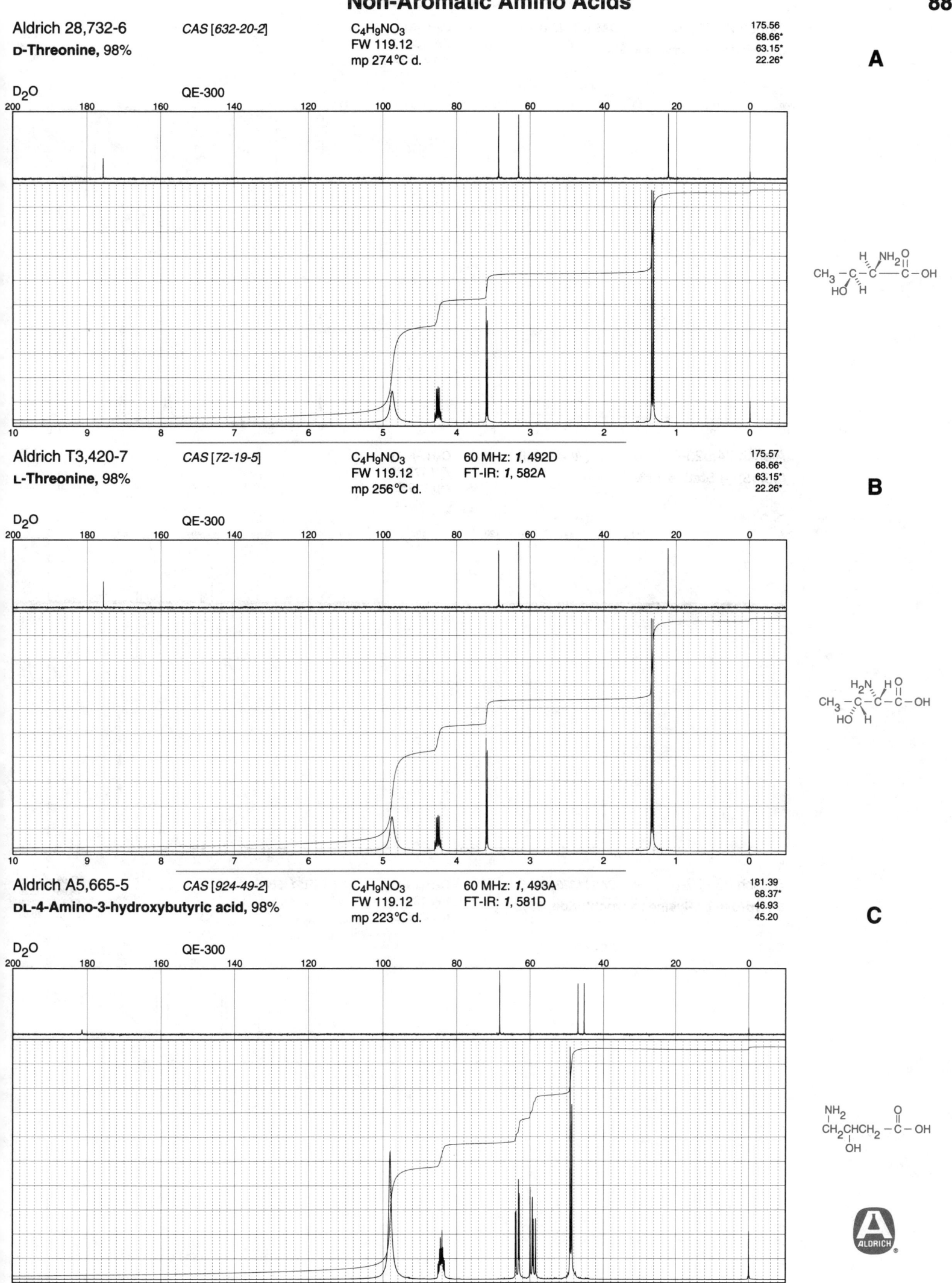

A

Aldrich 28,617-6 CAS [34042-00-7] C5H11NO3
FW 133.15
DL-3-Hydroxynorvaline, 98% mp 225°C

175.82
74.05*
61.56*
29.24
12.28*

D2O QE-300

CH3CH2CH—CH—C—OH (with OH, NH2, O substituents)

B

Aldrich 34,823-6 CAS [49642-07-1] C8H17NO3
FW 175.23
(3S,4S)-(-)-Statine, 98% mp 210°C d.

181.65 41.23
71.10* 26.70*
56.69* 24.91*
44.25   23.73*

D2O QE-300

CH3CH—CH2—CH—CH—CH2—C—OH (with CH3, H2N, OH, O substituents)

C

Aldrich 85,937-0 CAS [13204-98-3] C6H14N2O3 FT-IR: 1, 582B
FW 198.65
5-Hydroxy-DL-lysine hydrochloride, 95% mp 225°C d.

177.09 47.09
69.96* 32.36
69.86* 32.29
57.15* 29.34

D2O QE-300

H2NCH2CHCH2CH2CH—C—OH · HCl (with OH, NH2, O substituents)

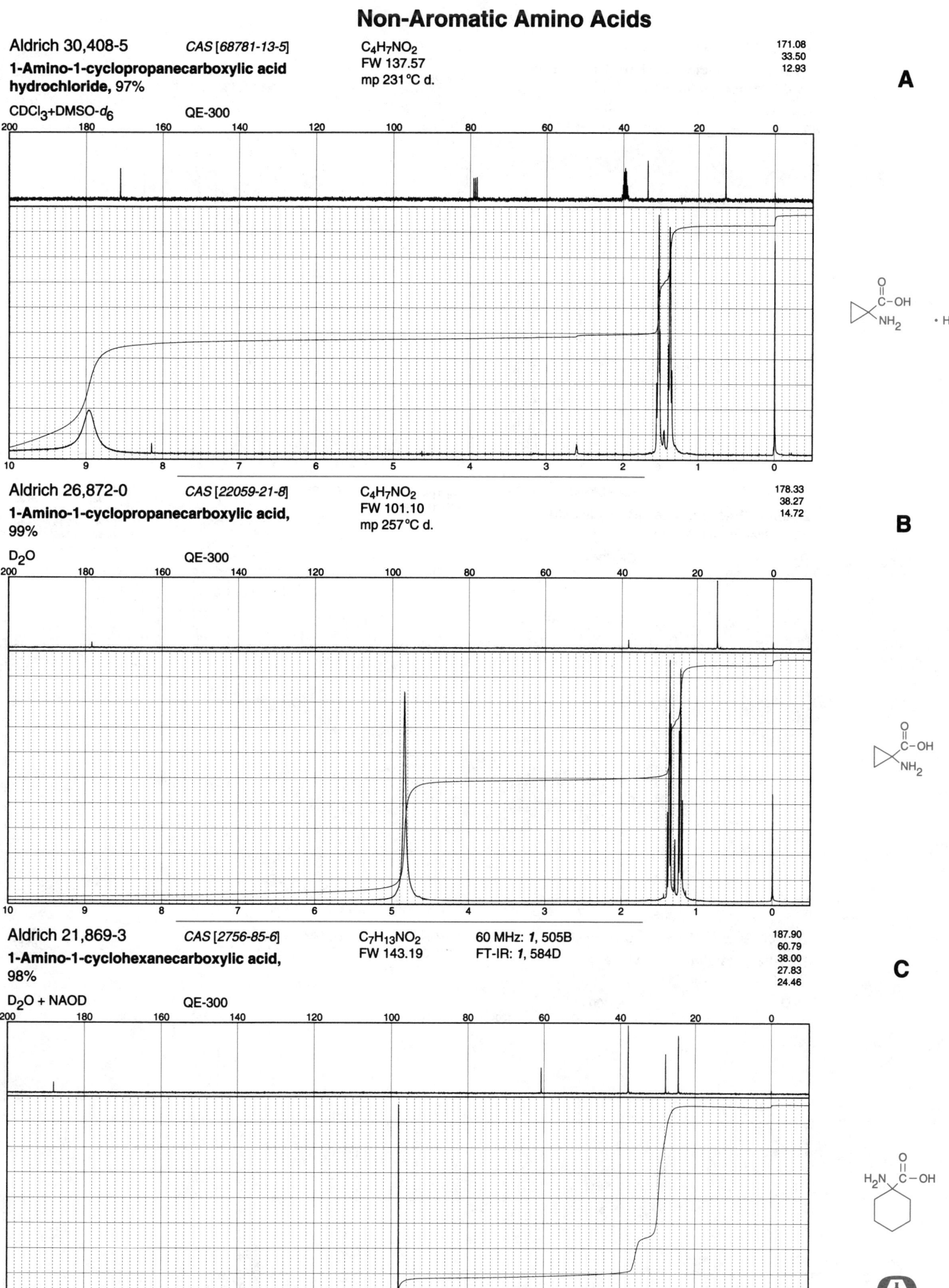

Aldrich 30,408-5   *CAS [68781-13-5]*

1-Amino-1-cyclopropanecarboxylic acid hydrochloride, 97%

$C_4H_7NO_2$
FW 137.57
mp 231°C d.

171.08
33.50
12.93

A

CDCl$_3$+DMSO-d_6 QE-300

· HCl

Aldrich 26,872-0   *CAS [22059-21-8]*

1-Amino-1-cyclopropanecarboxylic acid, 99%

$C_4H_7NO_2$
FW 101.10
mp 257°C d.

178.33
38.27
14.72

B

D$_2$O   QE-300

Aldrich 21,869-3   *CAS [2756-85-6]*

1-Amino-1-cyclohexanecarboxylic acid, 98%

$C_7H_{13}NO_2$
FW 143.19

60 MHz: *1*, 505B
FT-IR: *1*, 584D

187.90
60.79
38.00
27.83
24.46

C

D$_2$O + NAOD   QE-300

ALDRICH

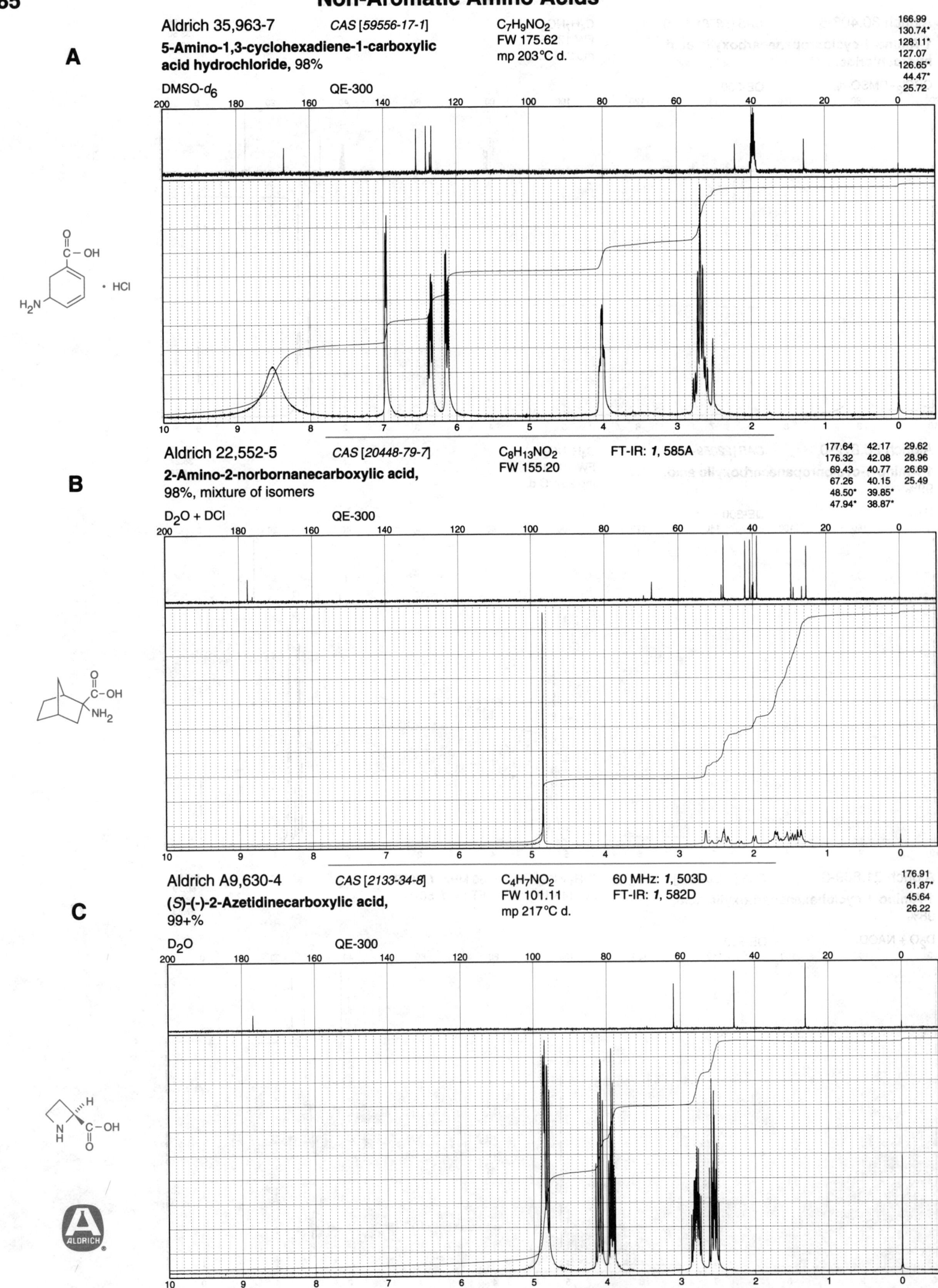

A

Aldrich 35,963-7   CAS [59556-17-1] $C_7H_9NO_2$ 166.99
                                                        130.74*
5-Amino-1,3-cyclohexadiene-1-carboxylic   FW 175.62   128.11*
acid hydrochloride, 98%   mp 203°C d.   127.07
                                                        126.65*
DMSO-d_6 QE-300   44.47*
                                                        25.72

B

Aldrich 22,552-5   CAS [20448-79-7] $C_8H_{13}NO_2$ FT-IR: 1, 585A   177.64 42.17 29.62
                                                                            176.32 42.08 28.96
2-Amino-2-norbornanecarboxylic acid,   FW 155.20   69.43   40.77 26.69
98%, mixture of isomers   67.26   40.15 25.49
                                                                            48.50* 39.85*
D_2O + DCl   QE-300   47.94* 38.87*

C

Aldrich A9,630-4   CAS [2133-34-8] $C_4H_7NO_2$ 60 MHz: 1, 503D   176.91
                                                                        61.87*
(S)-(-)-2-Azetidinecarboxylic acid,   FW 101.11   FT-IR: 1, 582D   45.64
99+%   mp 217°C d.   26.22
D_2O QE-300

Aldrich 17,182-4
DL-Proline, 99%

CAS [609-36-9]

C₅H₉NO₂
FW 115.13
mp 208°C d.

60 MHz: *1*, 504A
FT-IR: *1*, 583A

177.35
63.98*
48.83
31.77
26.55

A

D₂O   QE-300

Aldrich 85,891-9
D-Proline, 99+%

CAS [344-25-2]

C₅H₉NO₂
FW 115.13
mp 223°C d.

FT-IR: *1*, 583B

177.35
63.97*
48.83
31.77
26.55

B

D₂O   QE-300

Aldrich 13,154-7
L-Proline, 99+%

CAS [147-85-3]

C₅H₉NO₂
FW 115.13
mp 228°C d.

FT-IR: *1*, 583C

177.36
63.97*
48.83
31.78
26.56

C

D₂O   QE-300

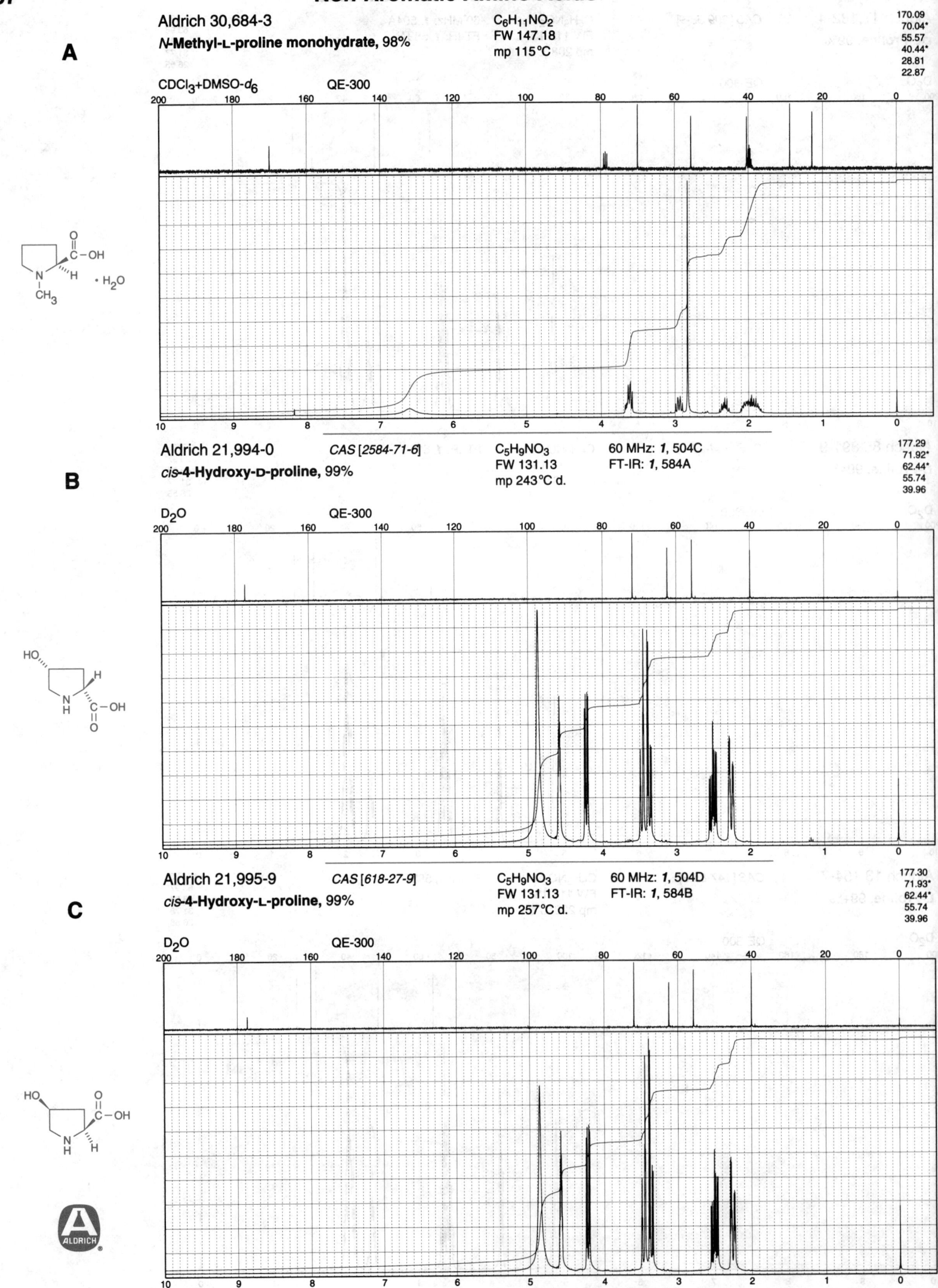

A

Aldrich 30,684-3

N-Methyl-L-proline monohydrate, 98%

C$_6$H$_{11}$NO$_2$
FW 147.18
mp 115°C

170.09
70.04*
55.57
40.44*
28.81
22.87

CDCl$_3$+DMSO-*d$_6$*   QE-300

B

Aldrich 21,994-0   *CAS [2584-71-6]*

cis-4-Hydroxy-D-proline, 99%

C$_5$H$_9$NO$_3$
FW 131.13
mp 243°C d.

60 MHz: *1*, 504C
FT-IR: *1*, 584A

177.29
71.92*
62.44*
55.74
39.96

D$_2$O   QE-300

C

Aldrich 21,995-9   *CAS [618-27-9]*

cis-4-Hydroxy-L-proline, 99%

C$_5$H$_9$NO$_3$
FW 131.13
mp 257°C d.

60 MHz: *1*, 504D
FT-IR: *1*, 584B

177.30
71.93*
62.44*
55.74
39.96

D$_2$O   QE-300

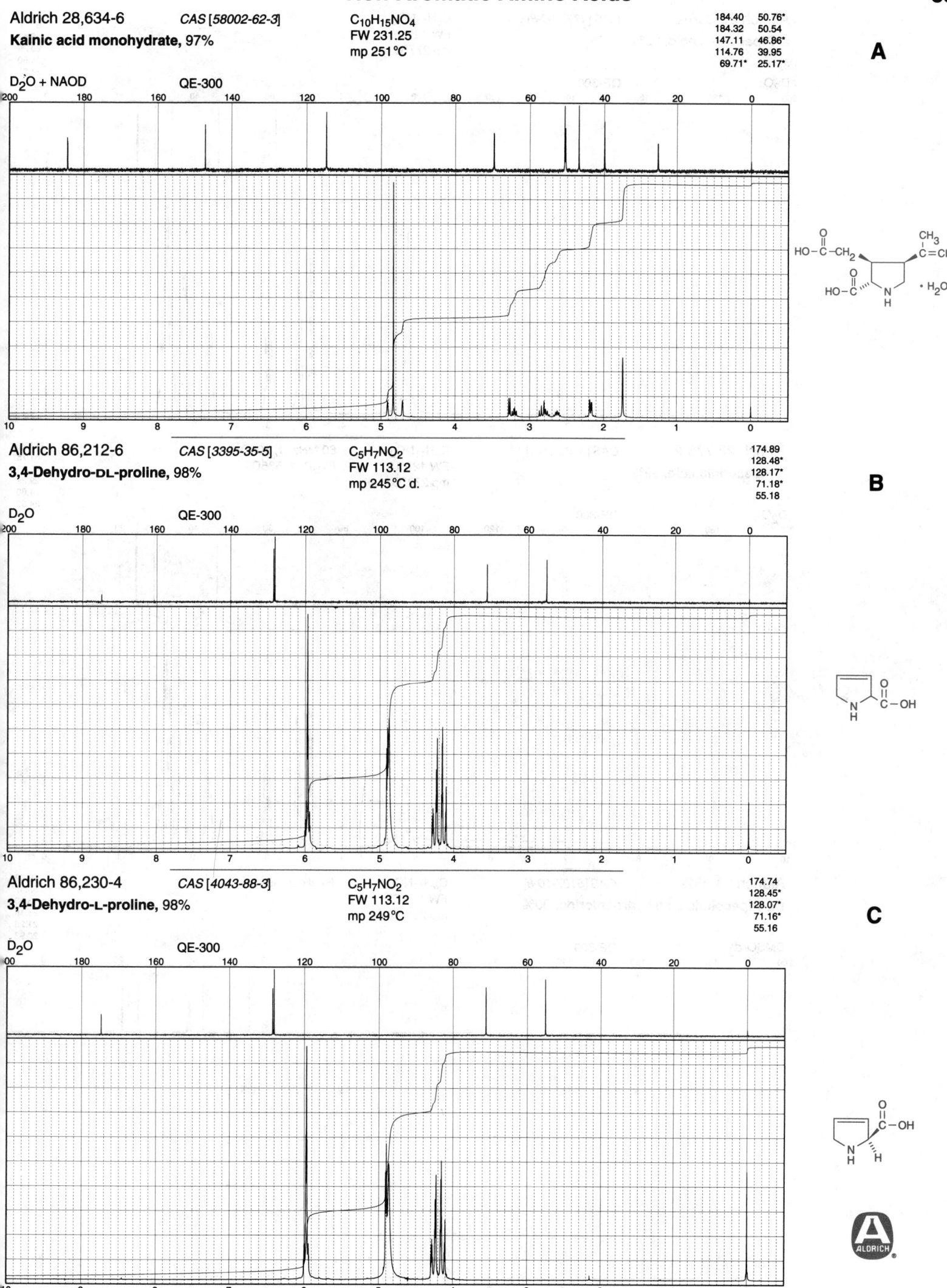

Aldrich 28,634-6   *CAS [58002-62-3]*

Kainic acid monohydrate, 97%

$C_{10}H_{15}NO_4$
FW 231.25
mp 251°C

184.40	50.76*
184.32	50.54
147.11	46.86*
114.76	39.95
69.71*	25.17*

A

D_2O + NAOD   QE-300

Aldrich 86,212-6   *CAS [3395-35-5]*

3,4-Dehydro-DL-proline, 98%

$C_5H_7NO_2$
FW 113.12
mp 245°C d.

| 174.89 |
| 128.48* |
| 128.17* |
| 71.18* |
| 55.18 |

B

D_2O QE-300

Aldrich 86,230-4   *CAS [4043-88-3]*

3,4-Dehydro-L-proline, 98%

$C_5H_7NO_2$
FW 113.12
mp 249°C

| 174.74 |
| 128.45* |
| 128.07* |
| 71.16* |
| 55.16 |

C

D_2O QE-300

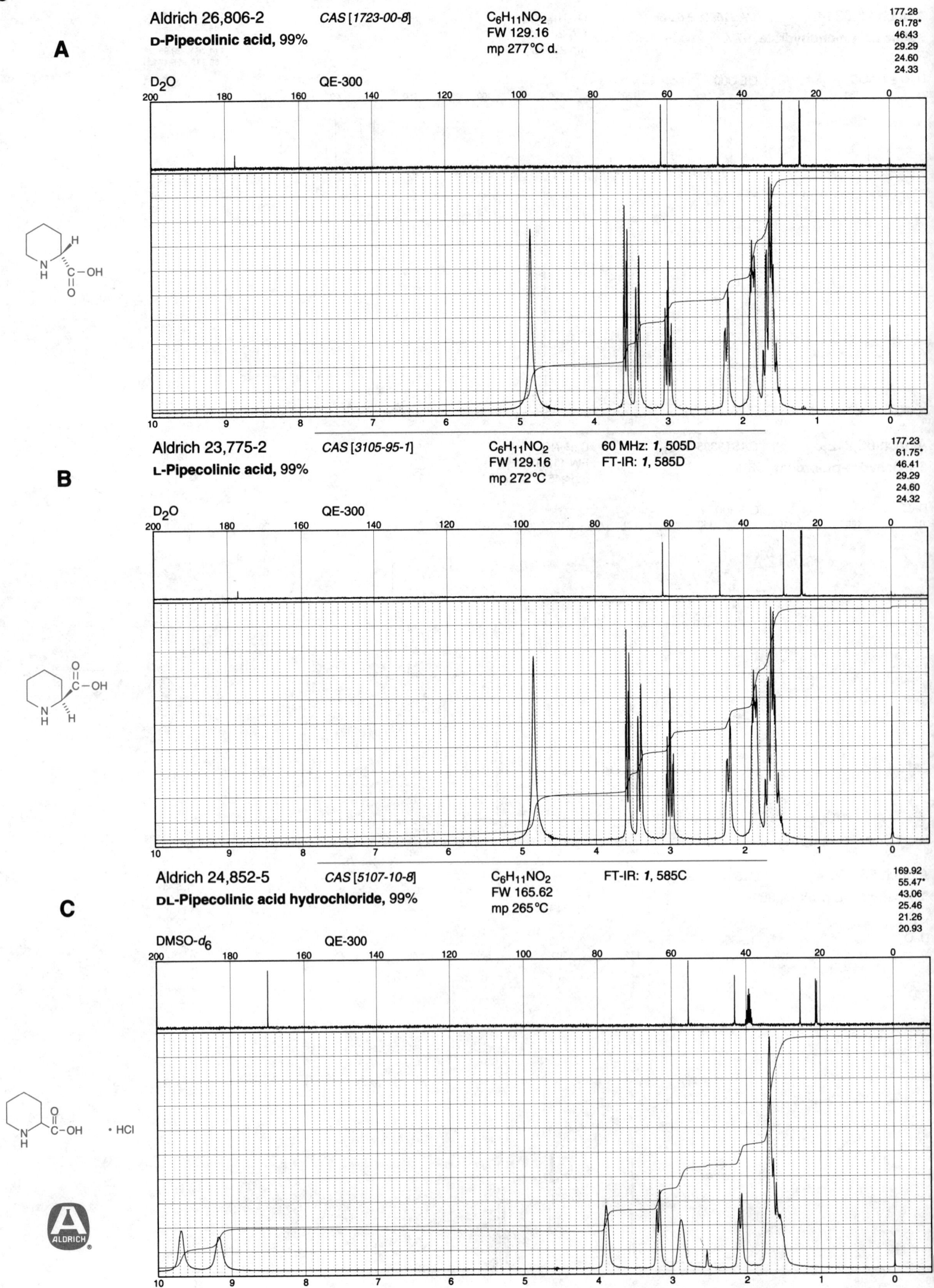

A

Aldrich 26,806-2   *CAS [1723-00-8]*   C₆H₁₁NO₂
D-Pipecolinic acid, 99%   FW 129.16
mp 277°C d.

177.28
61.78*
46.43
29.29
24.60
24.33

D₂O   QE-300

B

Aldrich 23,775-2   *CAS [3105-95-1]*   C₆H₁₁NO₂   60 MHz: *1*, 505D
L-Pipecolinic acid, 99%   FW 129.16   FT-IR: *1*, 585D
mp 272°C

177.23
61.75*
46.41
29.29
24.60
24.32

D₂O   QE-300

C

Aldrich 24,852-5   *CAS [5107-10-8]*   C₆H₁₁NO₂   FT-IR: *1*, 585C
DL-Pipecolinic acid hydrochloride, 99%   FW 165.62
mp 265°C

169.92
55.47*
43.06
25.46
21.26
20.93

DMSO-*d₆*   QE-300

Aldrich 21,167-2

Nipecotic acid, 98%

CAS [498-95-3]

$C_6H_{11}NO_2$
FW 129.16
mp 261°C d.

60 MHz: *1*, 506A
FT-IR: *1*, 586A

182.73
48.61
46.76
43.64*
28.49
23.83

A

D_2O QE-300

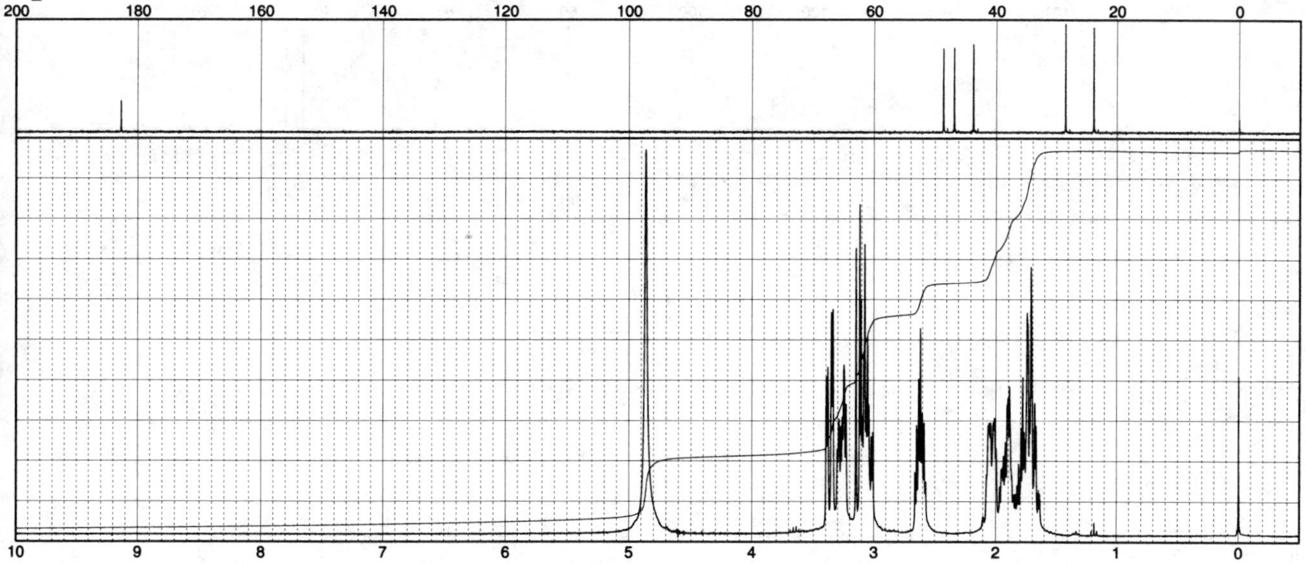

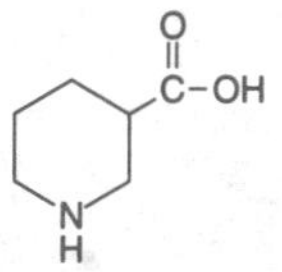

Aldrich 33,592-4

1-Piperidinepropionic acid, 99%

CAS [26371-07-3]

$C_8H_{15}NO_2$
FW 157.21
mp 108°C
bp 107°C

175.04
53.93
52.65
31.14
23.56
22.73

B

$CDCl_3$ QE-300

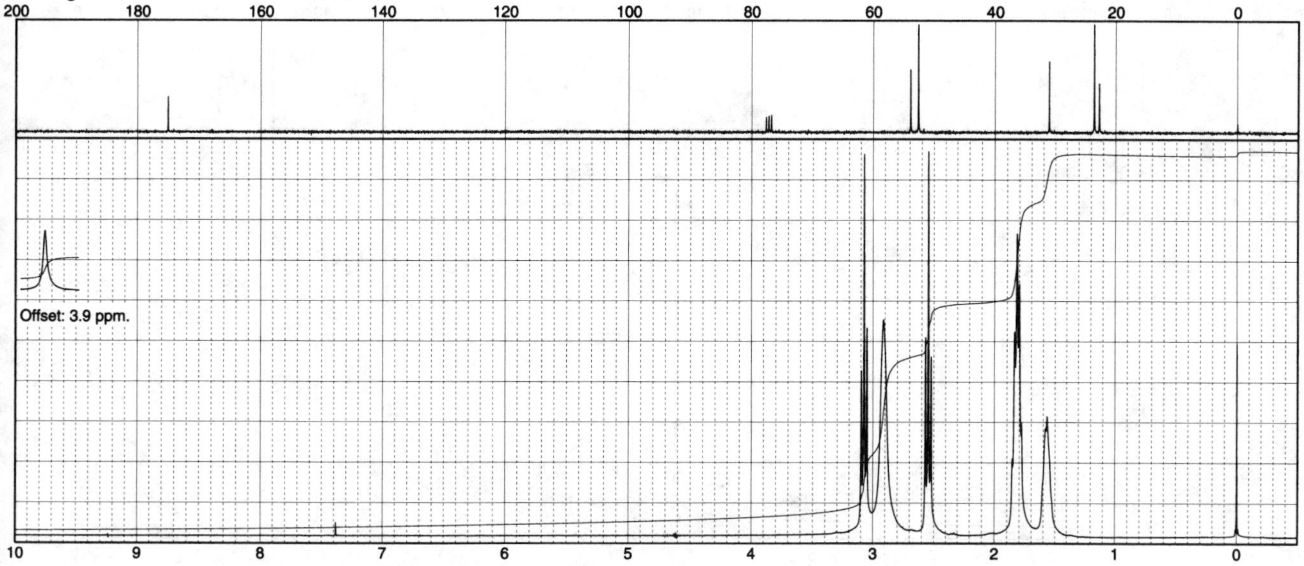

Offset: 3.9 ppm.

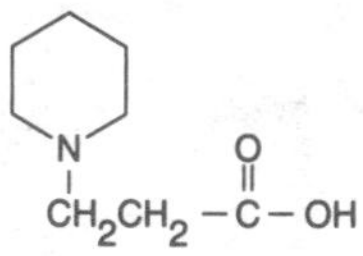

Aldrich 36,022-8

L-Lysine dihydrochloride, 99%

CAS [657-26-1]

$C_6H_{14}N_2O_2$
FW 219.11
mp 203°C

174.89
55.62*
41.95
32.14
29.14
24.30

C

D_2O QE-300

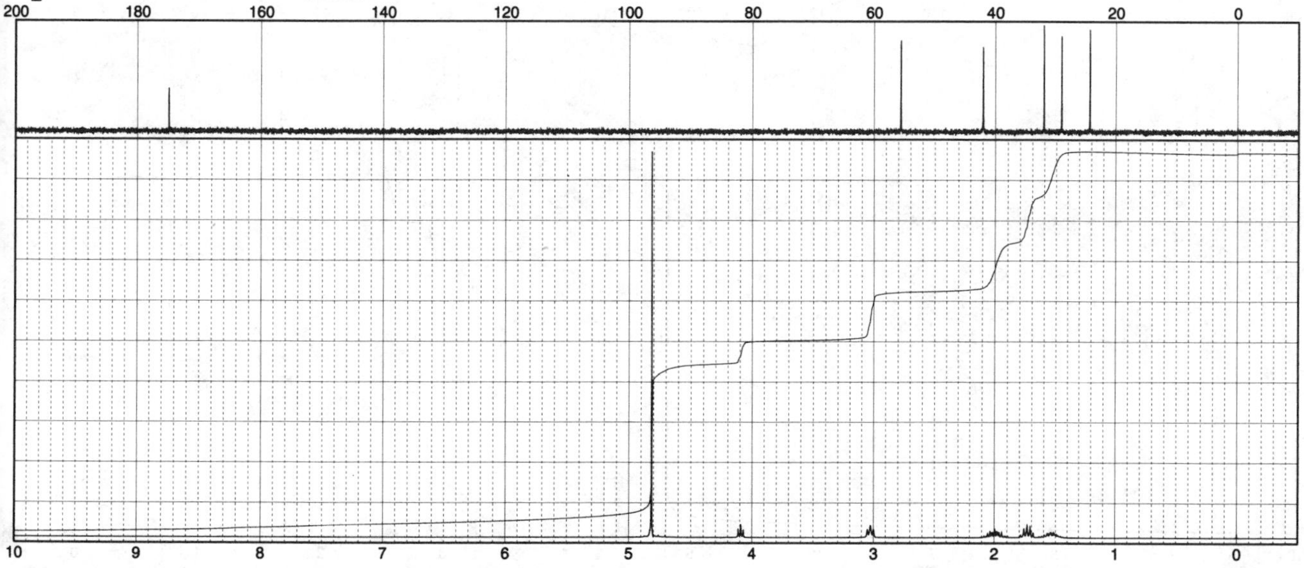

Non-Aromatic Amino Acids

A

Aldrich G279-6   CAS [19285-83-7] $C_5H_9NO_4$ FT-IR: 1, 589C

DL-Glutamic acid monohydrate, 99%   FW 165.15   mp 185°C d.

185.27
184.64
58.50*
36.82
33.98

D₂O + NAOD   QE-300

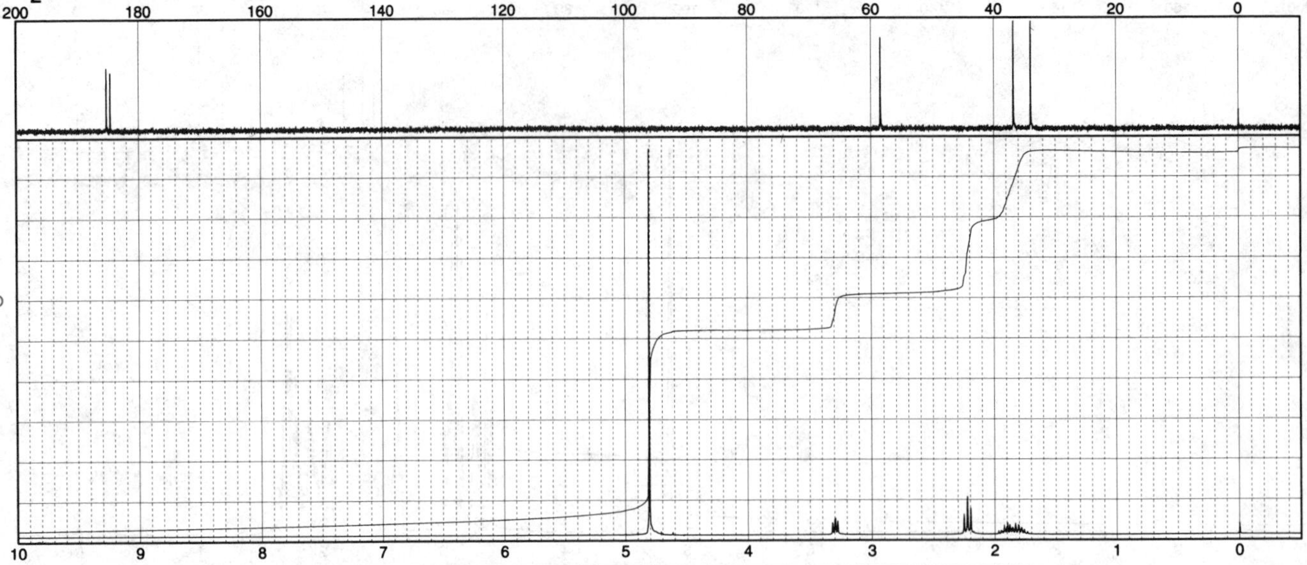

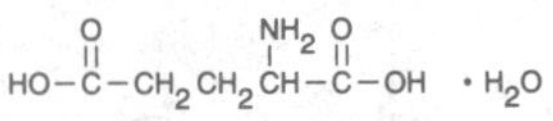

HO–C(=O)–CH₂–CH₂–CH(NH₂)–C(=O)–OH · H₂O

B

Aldrich 27,147-0   CAS [2577-62-0] $C_7H_{14}N_2O_4$

(±)-2,6-Diaminopimelic acid, 98%   FW 190.20

174.35
55.28*
32.04
23.16

D₂O + DCl   QE-300

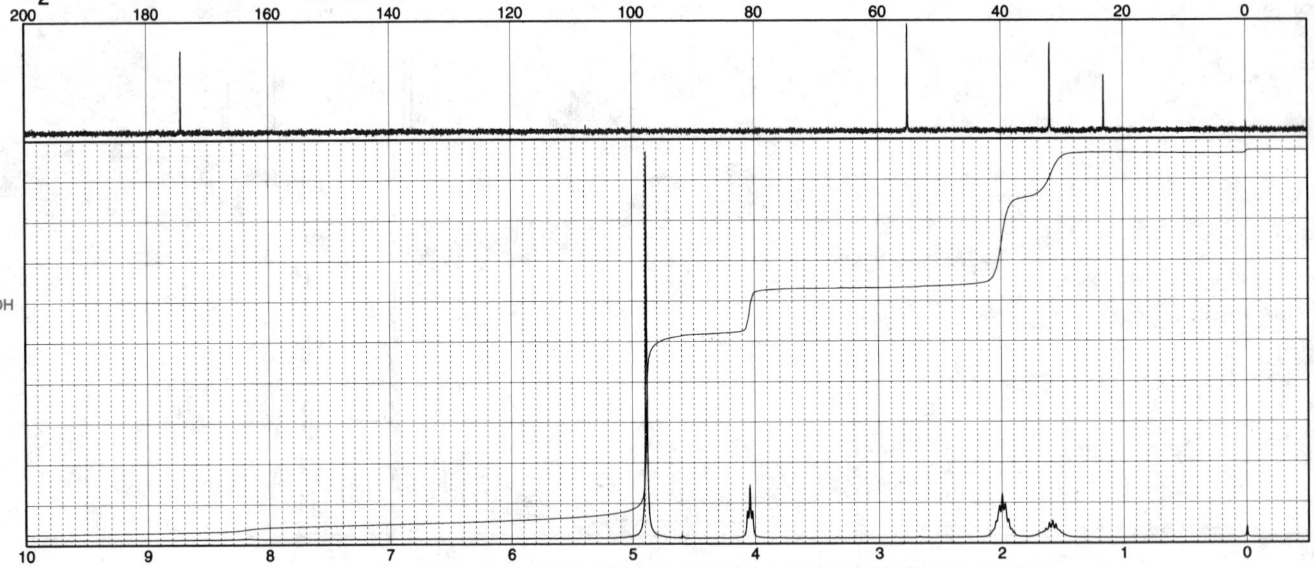

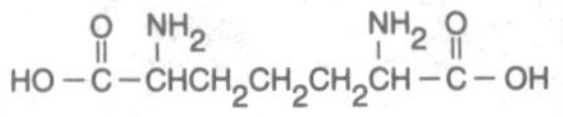

HO–C(=O)–CH(NH₂)–CH₂–CH₂–CH₂–CH(NH₂)–C(=O)–OH

C

Aldrich 86,167-7   CAS [3374-22-9] $C_3H_7NO_2S$

DL-Cysteine, 97%   FW 121.16   mp 225°C d.

183.32
62.33*
33.36

D₂O + NAOD   QE-300

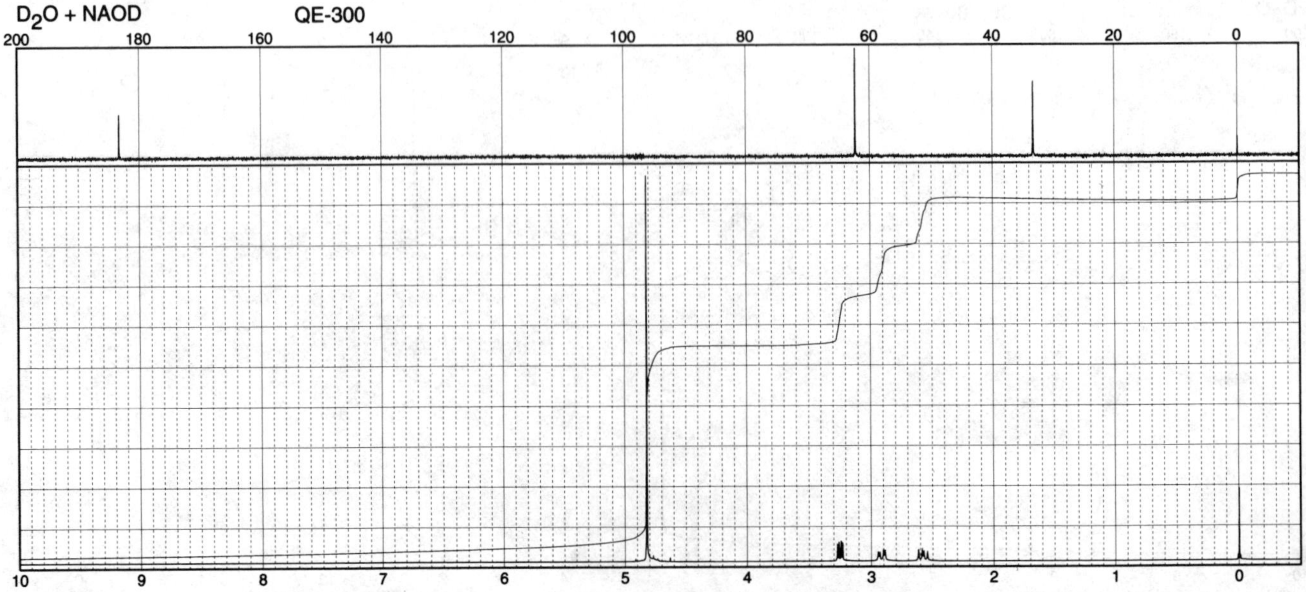

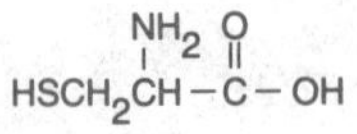

HSCH₂–CH(NH₂)–C(=O)–OH

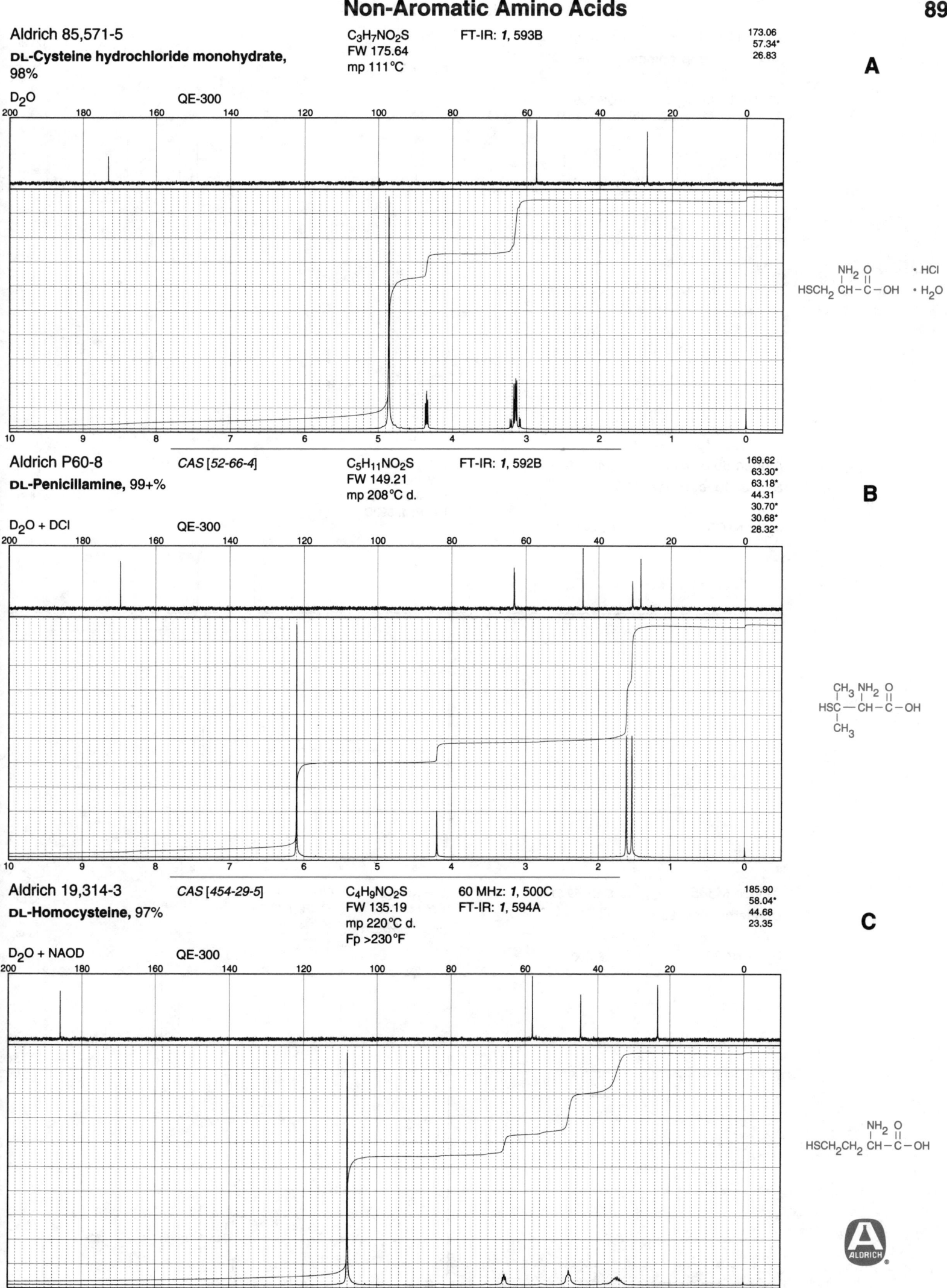

Aldrich 85,571-5

DL-Cysteine hydrochloride monohydrate, 98%

C₃H₇NO₂S
FW 175.64
mp 111°C

$C_3H_7NO_2S$
FW 175.64
mp 111°C

FT-IR: *1*, 593B

173.06
57.34*
26.83

A

D₂O   QE-300

NH_2 O • HCl
HSCH₂ CH–C–OH   • H₂O

Aldrich P60-8

DL-Penicillamine, 99+%

CAS *[52-66-4]*

$C_5H_{11}NO_2S$
FW 149.21
mp 208°C d.

FT-IR: *1*, 592B

169.62
63.30*
63.18*
44.31
30.70*
30.68*
28.32*

B

D₂O + DCl   QE-300

CH₃ NH₂ O
HSC–CH–C–OH
CH₃

Aldrich 19,314-3

DL-Homocysteine, 97%

CAS *[454-29-5]*

$C_4H_9NO_2S$
FW 135.19
mp 220°C d.
Fp >230°F

60 MHz: *1*, 500C
FT-IR: *1*, 594A

185.90
58.04*
44.68
23.35

C

D₂O + NAOD   QE-300

NH_2 O
HSCH₂CH₂ CH–C–OH

Non-Aromatic Amino Acids

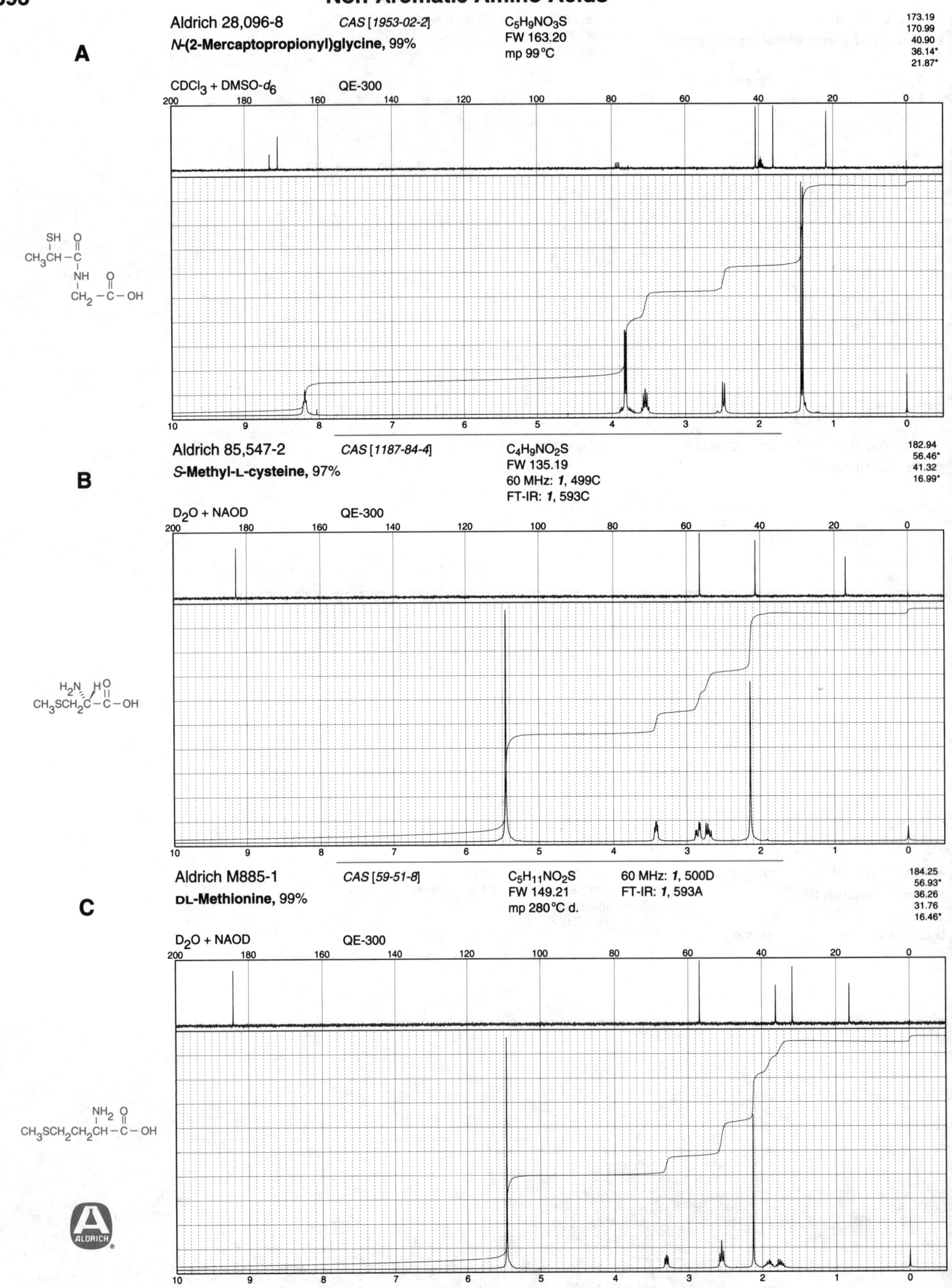

A

Aldrich 28,096-8   CAS [1953-02-2] $C_5H_9NO_3S$
N-(2-Mercaptopropionyl)glycine, 99%   FW 163.20
mp 99°C

173.19
170.99
40.90
36.14*
21.87*

$CDCl_3$ + DMSO-d_6 QE-300

B

Aldrich 85,547-2   CAS [1187-84-4] $C_4H_9NO_2S$
S-Methyl-L-cysteine, 97%   FW 135.19
60 MHz: *1*, 499C
FT-IR: *1*, 593C

182.94
56.46*
41.32
16.99*

D_2O + NAOD   QE-300

C

Aldrich M885-1   CAS [59-51-8] $C_5H_{11}NO_2S$ 60 MHz: *1*, 500D
DL-Methionine, 99%   FW 149.21   FT-IR: *1*, 593A
mp 280°C d.

184.25
56.93*
36.26
31.76
16.46*

D_2O + NAOD   QE-300

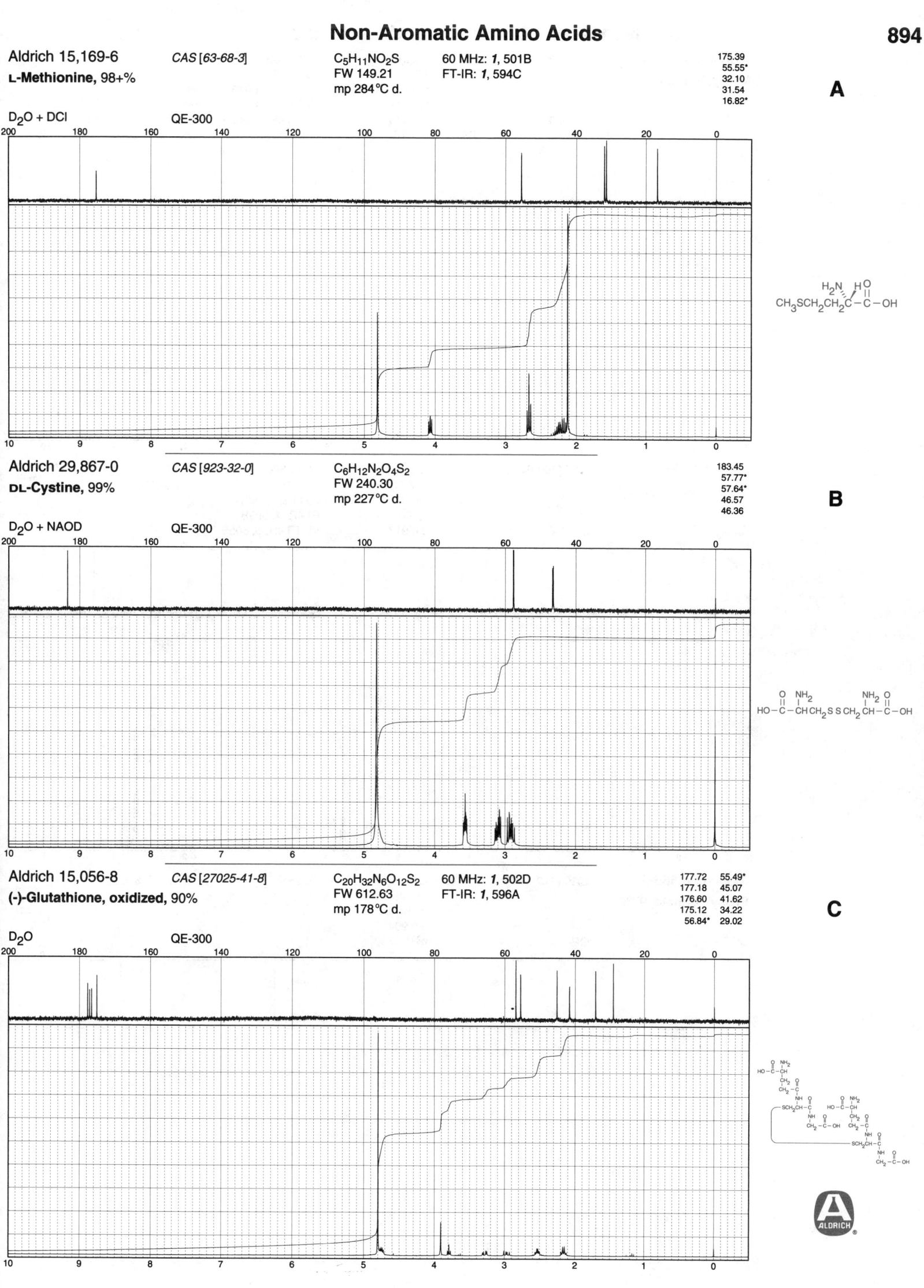

Aldrich 15,169-6

CAS [63-68-3]

L-Methionine, 98+%

C₅H₁₁NO₂S
FW 149.21
mp 284°C d.

60 MHz: *1*, 501B
FT-IR: *1*, 594C

175.39
55.55*
32.10
31.54
16.82*

A

D₂O + DCl   QE-300

Aldrich 29,867-0

CAS [923-32-0]

DL-Cystine, 99%

C₆H₁₂N₂O₄S₂
FW 240.30
mp 227°C d.

183.45
57.77*
57.64*
46.57
46.36

B

D₂O + NAOD   QE-300

Aldrich 15,056-8

CAS [27025-41-8]

(−)-Glutathione, oxidized, 90%

C₂₀H₃₂N₆O₁₂S₂
FW 612.63
mp 178°C d.

60 MHz: *1*, 502D
FT-IR: *1*, 596A

177.72   55.49*
177.18   45.07
176.60   41.62
175.12   34.22
56.84*   29.02

C

D₂O   QE-300

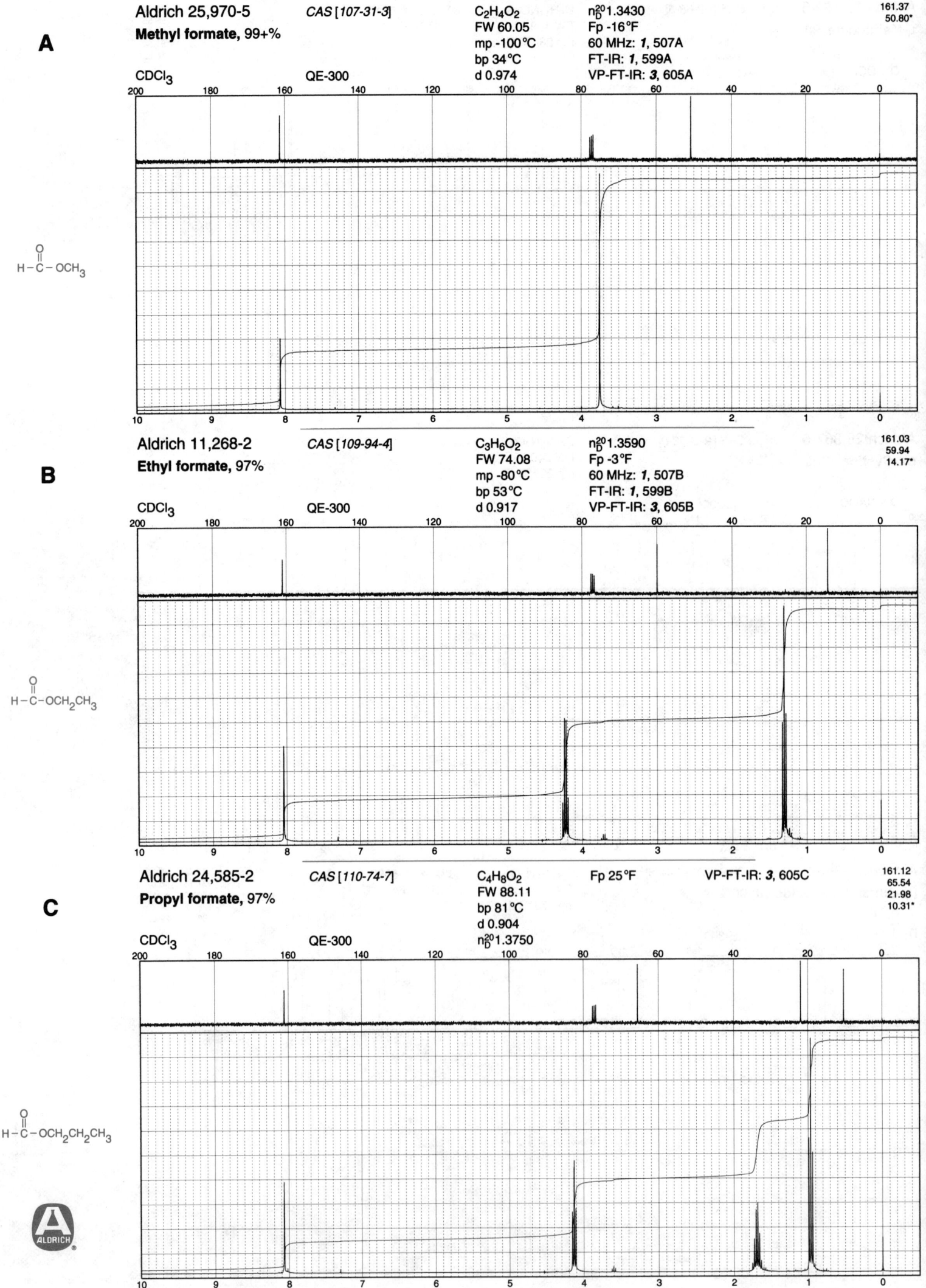

A

Aldrich 25,970-5

CAS [107-31-3]

Methyl formate, 99+%

$C_2H_4O_2$
FW 60.05
mp -100°C
bp 34°C
d 0.974

n_D^{20} 1.3430
Fp -16°F
60 MHz: *1*, 507A
FT-IR: *1*, 599A
VP-FT-IR: *3*, 605A

161.37
50.80*

CDCl₃

QE-300

B

Aldrich 11,268-2

CAS [109-94-4]

Ethyl formate, 97%

$C_3H_6O_2$
FW 74.08
mp -80°C
bp 53°C
d 0.917

n_D^{20} 1.3590
Fp -3°F
60 MHz: *1*, 507B
FT-IR: *1*, 599B
VP-FT-IR: *3*, 605B

161.03
59.94
14.17*

CDCl₃

QE-300

C

Aldrich 24,585-2

CAS [110-74-7]

Propyl formate, 97%

$C_4H_8O_2$
FW 88.11
bp 81°C
d 0.904
n_D^{20} 1.3750

Fp 25°F

VP-FT-IR: *3*, 605C

161.12
65.54
21.98
10.31*

CDCl₃

QE-300

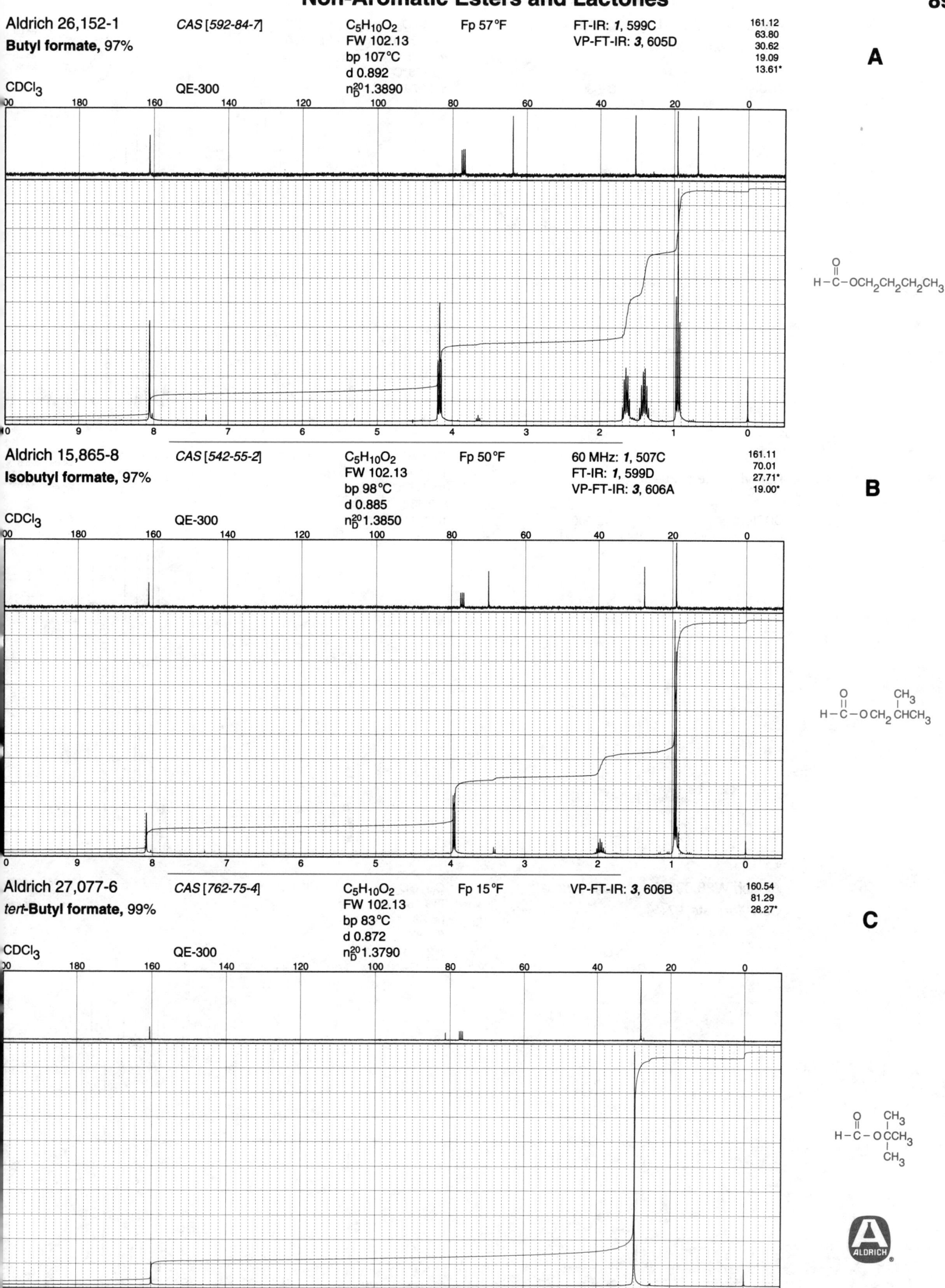

Aldrich 26,152-1   CAS [592-84-7]   C5H10O2   Fp 57°F   FT-IR: 1, 599C   161.12
Butyl formate, 97%   FW 102.13   VP-FT-IR: 3, 605D   63.80
                                        bp 107°C   30.62
                                        d 0.892   19.09
CDCl3   QE-300   n20D 1.3890   13.61*

A

Aldrich 15,865-8   CAS [542-55-2]   C5H10O2   Fp 50°F   60 MHz: 1, 507C   161.11
Isobutyl formate, 97%   FW 102.13   FT-IR: 1, 599D   70.01
                                        bp 98°C   VP-FT-IR: 3, 606A   27.71*
                                        d 0.885   19.00*
CDCl3   QE-300   n20D 1.3850

B

Aldrich 27,077-6   CAS [762-75-4]   C5H10O2   Fp 15°F   VP-FT-IR: 3, 606B   160.54
tert-Butyl formate, 99%   FW 102.13   81.29
                                        bp 83°C   28.27*
                                        d 0.872
CDCl3   QE-300   n20D 1.3790

C

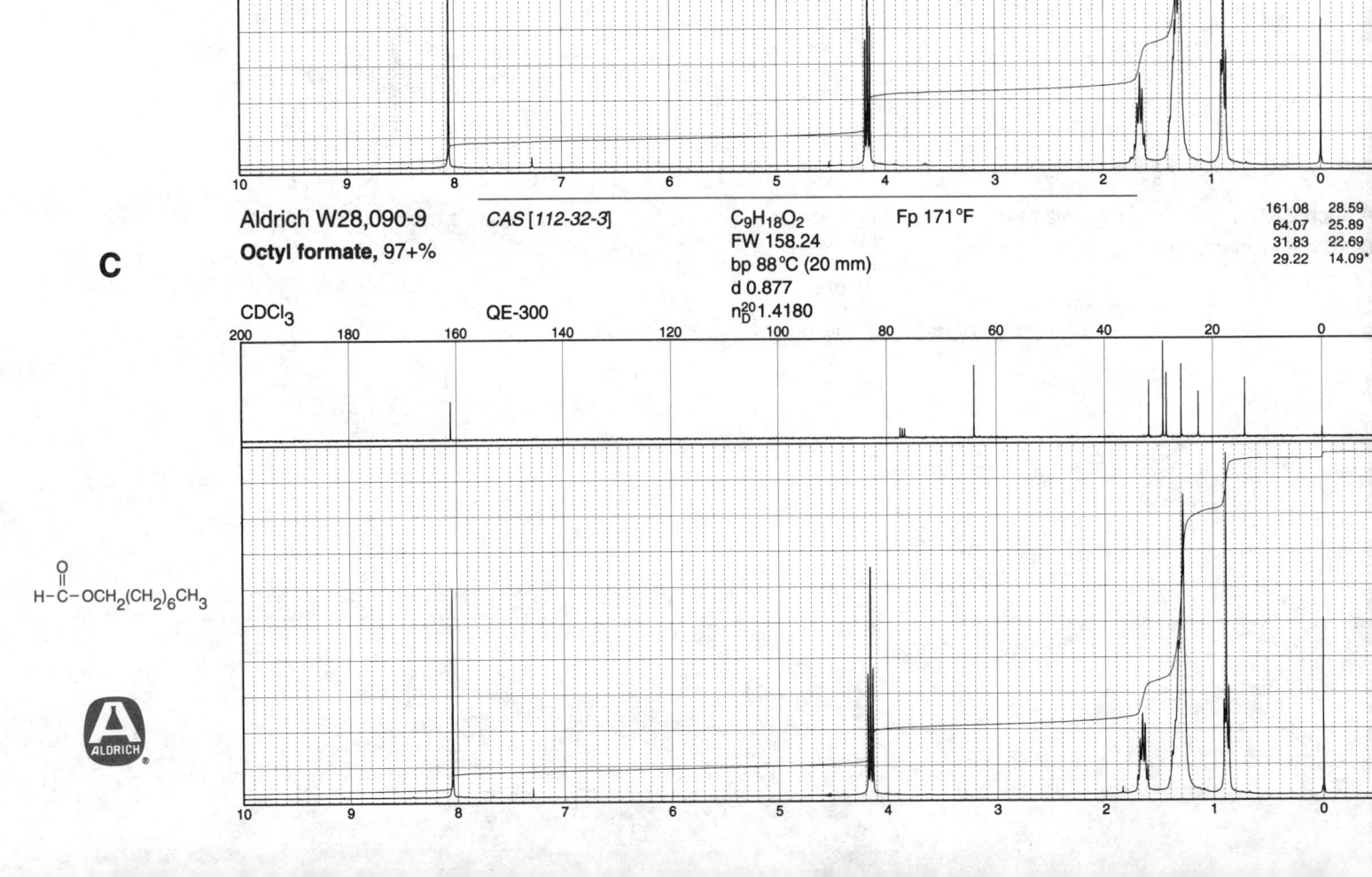

A

Aldrich W25,700-1   CAS [629-33-4] $C_7H_{14}O_2$ n_D^{20} 1.4070

Hexyl formate, 97+%   FW 130.19   Fp 118°F

mp -32°C

bp 156°C

d 0.879

CDCl$_3$ QE-300

161.02
64.09
31.43
28.59
25.56
22.55
13.94*

$H-\overset{\overset{\textstyle O}{\|}}{C}-OCH_2(CH_2)_4CH_3$

B

Aldrich W25,520-3   CAS [112-23-2] $C_8H_{16}O_2$ Fp 140°F   VP-FT-IR: 3, 607A

Heptyl formate, 99+%   FW 144.21

bp 178°C

d 0.882

CDCl$_3$ QE-300 n_D^{20} 1.4130

161.10   28.55
64.09   25.81
31.72   22.58
28.88   14.05*

$H-\overset{\overset{\textstyle O}{\|}}{C}-OCH_2(CH_2)_5CH_3$

C

Aldrich W28,090-9   CAS [112-32-3] $C_9H_{18}O_2$ Fp 171°F

Octyl formate, 97+%   FW 158.24

bp 88°C (20 mm)

d 0.877

CDCl$_3$ QE-300 n_D^{20} 1.4180

161.08   28.59
64.07   25.89
31.83   22.69
29.22   14.09*

$H-\overset{\overset{\textstyle O}{\|}}{C}-OCH_2(CH_2)_6CH_3$

ALDRICH

Aldrich 18,632-5   *CAS [79-20-9]*

Methyl acetate, 99%

$C_3H_6O_2$
FW 74.08
mp -98°C
bp 58°C
d 0.932

n_D^{20} 1.3610
Fp 15°F
60 MHz: *1*, 507D
FT-IR: *1*, 600A
VP-FT-IR: *3*, 607B

171.37
51.53*
20.63*

A

CDCl$_3$ QE-300

$CH_3 - \overset{\displaystyle O}{\overset{\|}{C}} - OCH_3$

Aldrich 15,485-7   *CAS [141-78-6]*

Ethyl acetate, 99.5+%

$C_4H_8O_2$
FW 88.11
mp -84°C
bp 77°C
d 0.902

n_D^{20} 1.3720
Fp 26°F
60 MHz: *1*, 508A
FT-IR: *1*, 600B
VP-FT-IR: *3*, 607C

170.95
60.34
20.98*
14.23*

B

CDCl$_3$ QE-300

$CH_3 - \overset{\displaystyle O}{\overset{\|}{C}} - OCH_2CH_3$

Aldrich W32,820-0   *CAS [625-60-5]*

Ethyl thioacetate, 96%

C_4H_8OS
FW 104.17
bp 116°C
d 0.979
n_D^{20} 1.4580

Fp 65°F

VP-FT-IR: *3*, 731C

195.80
30.58*
23.51
14.73*

C

CDCl$_3$ QE-300

$CH_3 - \overset{\displaystyle O}{\overset{\|}{C}} - SCH_2CH_3$

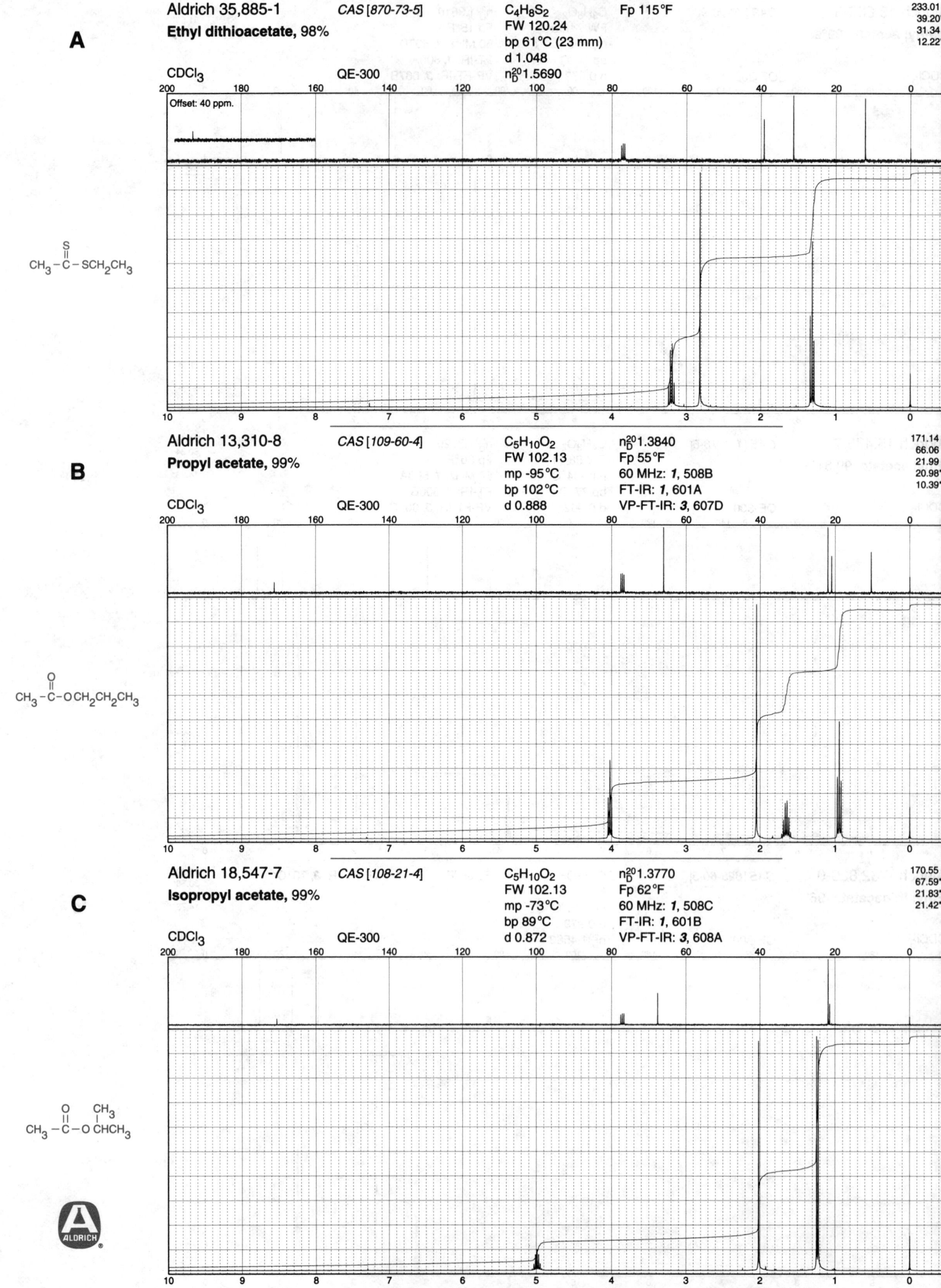

A

Aldrich 35,885-1   *CAS [870-73-5]* $C_4H_8S_2$ Fp 115°F

Ethyl dithioacetate, 98%   FW 120.24
bp 61°C (23 mm)
d 1.048

CDCl₃   QE-300 n_D^{20} 1.5690

233.01
39.20*
31.34
12.22*

Offset: 40 ppm.

S
‖
CH₃–C–SCH₂CH₃

B

Aldrich 13,310-8   *CAS [109-60-4]* $C_5H_{10}O_2$ n_D^{20} 1.3840

Propyl acetate, 99%   FW 102.13   Fp 55°F
mp -95°C   60 MHz: *1*, 508B
bp 102°C   FT-IR: *1*, 601A
d 0.888   VP-FT-IR: *3*, 607D

CDCl₃   QE-300

171.14
66.06
21.99
20.98*
10.39*

O
‖
CH₃–C–OCH₂CH₂CH₃

C

Aldrich 18,547-7   *CAS [108-21-4]* $C_5H_{10}O_2$ n_D^{20} 1.3770

Isopropyl acetate, 99%   FW 102.13   Fp 62°F
mp -73°C   60 MHz: *1*, 508C
bp 89°C   FT-IR: *1*, 601B
d 0.872   VP-FT-IR: *3*, 608A

CDCl₃   QE-300

170.55
67.59
21.83*
21.42*

O   CH₃
‖ |
CH₃–C–O CHCH₃

Aldrich 15,466-0   *CAS [123-86-4]* $C_6H_{12}O_2$ n_D^{20} 1.3940

Butyl acetate, 99.5%

FW 116.16   Fp 72°F

mp -78°C   60 MHz: *1*, 508D

bp 125°C   FT-IR: *1*, 601C

CDCl$_3$ QE-300   d 0.882   VP-FT-IR: *3*, 608B

A

171.05
64.33
30.75
20.95*
19.19
13.70*

$$CH_3-\overset{\displaystyle O}{\overset{\|}{C}}-O\,CH_2CH_2CH_2CH_3$$

Aldrich 24,259-4   *CAS [105-46-4]* $C_6H_{12}O_2$ n_D^{20} 1.3890

(±)-*sec*-Butyl acetate, 99%

FW 116.16   Fp 61°F

mp -99°C   60 MHz: *1*, 509A

bp 112°C   FT-IR: *1*, 601D

CDCl$_3$ QE-300   d 0.872   VP-FT-IR: *3*, 608C

B

170.68
72.18*
28.84
21.30*
19.44*
9.65*

$$CH_3-\overset{\displaystyle O}{\overset{\|}{C}}-O\,\overset{\displaystyle CH_3}{\overset{\displaystyle |}{CH}}CH_2CH_3$$

Aldrich 12,949-6   *CAS [110-19-0]* $C_6H_{12}O_2$ n_D^{20} 1.3900

Isobutyl acetate, 99%

FW 116.16   Fp 71°F

mp -99°C   VP-FT-IR: *3*, 608D

bp 116°C

CDCl$_3$ QE-300   d 0.868

C

171.10
70.59
27.69*
20.93*
19.08*

$$CH_3-\overset{\displaystyle O}{\overset{\|}{C}}-O\,CH_2\overset{\displaystyle CH_3}{\overset{\displaystyle |}{CH}}CH_3$$

ALDRICH

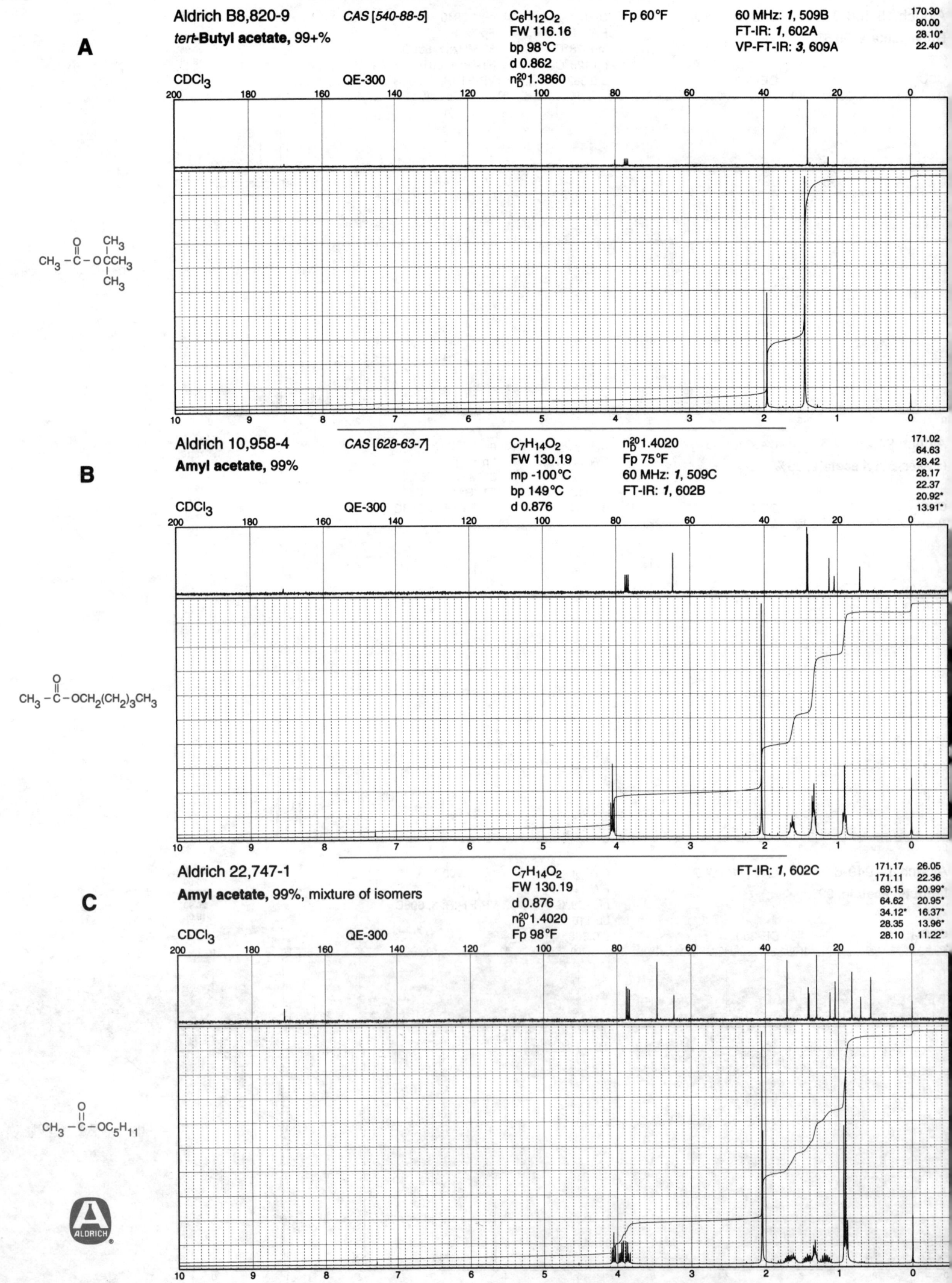

A

Aldrich B8,820-9   CAS [540-88-5]   C$_6$H$_{12}$O$_2$ Fp 60°F   60 MHz: *1*, 509B   170.30
tert-Butyl acetate, 99+%   FW 116.16   FT-IR: *1*, 602A   80.00
bp 98°C   VP-FT-IR: *3*, 609A   28.10*
d 0.862   22.40*
CDCl$_3$ QE-300   n$_D^{20}$1.3860

CH$_3$–C(=O)–O–C(CH$_3$)$_2$CH$_3$ (structure with CH$_3$ groups)

B

Aldrich 10,958-4   CAS [628-63-7]   C$_7$H$_{14}$O$_2$ n$_D^{20}$1.4020   171.02
Amyl acetate, 99%   FW 130.19   Fp 75°F   64.63
mp -100°C   60 MHz: *1*, 509C   28.42
bp 149°C   FT-IR: *1*, 602B   28.17
d 0.876   22.37
20.92*
CDCl$_3$ QE-300   13.91*

CH$_3$–C(=O)–OCH$_2$(CH$_2$)$_3$CH$_3$

C

Aldrich 22,747-1   C$_7$H$_{14}$O$_2$ FT-IR: *1*, 602C   171.17   26.05
Amyl acetate, 99%, mixture of isomers   FW 130.19   171.11   22.36
d 0.876   69.15   20.99*
n$_D^{20}$1.4020   64.62   20.95*
34.12* 16.37*
CDCl$_3$ QE-300   Fp 98°F   28.35   13.96*
28.10   11.22*

CH$_3$–C(=O)–OC$_5$H$_{11}$

Aldrich W36,440-1   *CAS [53496-15-4]*   C₇H₁₄O₂   Fp 95°F

2-Methylbutyl acetate, 94+%

FW 130.19

bp 138°C (741 mm)

d 0.876

n²⁰_D 1.4010

CDCl₃   QE-300

171.18
69.15
34.13*
26.06
20.94*
16.37*
11.22*

A

$$CH_3-\overset{\overset{\displaystyle O}{\|}}{C}-OCH_2\overset{\overset{\displaystyle CH_3}{|}}{CH}CH_2CH_3$$

Aldrich 11,267-4   *CAS [123-92-2]*   C₇H₁₄O₂   n²⁰_D 1.4000

Isoamyl acetate, 98%

FW 130.19   Fp 77°F

mp -78°C   60 MHz: *1*, 509D

bp 142°C (756 mm)   FT-IR: *1*, 602D

d 0.876   VP-FT-IR: *3*, 609C

CDCl₃   QE-300

171.15
63.12
37.31
25.05*
22.45*
21.02*

B

$$CH_3-\overset{\overset{\displaystyle O}{\|}}{C}-OCH_2CH_2\overset{\overset{\displaystyle CH_3}{|}}{CH}CH_3$$

Aldrich 10,815-4   *CAS [142-92-7]*   C₈H₁₆O₂   n²⁰_D 1.4090

Hexyl acetate, 99%

FW 144.21   Fp 99°F

mp -80°C   60 MHz: *1*, 510A

bp 169°C   FT-IR: *1*, 603A

d 0.876   VP-FT-IR: *3*, 609D

CDCl₃   QE-300

171.05   25.63
64.63   22.56
31.49   20.96*
28.65   13.97*

C

$$CH_3-\overset{\overset{\displaystyle O}{\|}}{C}-OCH_2(CH_2)_4CH_3$$

Non-Aromatic Esters and Lactones

A

Aldrich 30,797-1   *CAS [10031-87-5]*   C$_8$H$_{16}$O$_2$ Fp 127°F   VP-FT-IR: **3**, 610A

2-Ethylbutyl acetate, 98%

FW 144.21
bp 160°C (740 mm)
d 0.876
n$_D^{20}$1.4100

171.23
66.61
40.34*
23.33
20.96*
11.02*

CDCl$_3$ QE-300

$$CH_3-C(=O)-OCH_2\,CH(CH_2CH_3)CH_2CH_3$$

B

Aldrich O-550-0   *CAS [112-14-1]*   C$_{10}$H$_{20}$O$_2$ Fp 187°F   FT-IR: **1**, 603B

Octyl acetate, 99+%

FW 172.27
bp 211°C
d 0.868
n$_D^{20}$1.4180

VP-FT-IR: **3**, 610C

171.05   28.68
64.63   25.97
31.82   22.66
29.25   20.96*
29.21   14.06*

CDCl$_3$ QE-300

$$CH_3-C(=O)-OCH_2(CH_2)_6CH_3$$

C

Aldrich D22,195-3   *CAS [112-66-3]*   C$_{14}$H$_{28}$O$_2$ Fp >230°F   FT-IR: **1**, 603C

Dodecyl acetate, 97%

FW 228.38
bp 150°C (15 mm)
d 0.865
n$_D^{20}$1.4320

VP-FT-IR: **3**, 611B

171.05   29.29
64.64   28.67
31.94   25.96
29.66   22.70
29.60   20.97*
29.55   14.10*
29.36

CDCl$_3$ QE-300

$$CH_3-C(=O)-OCH_2(CH_2)_{10}CH_3$$

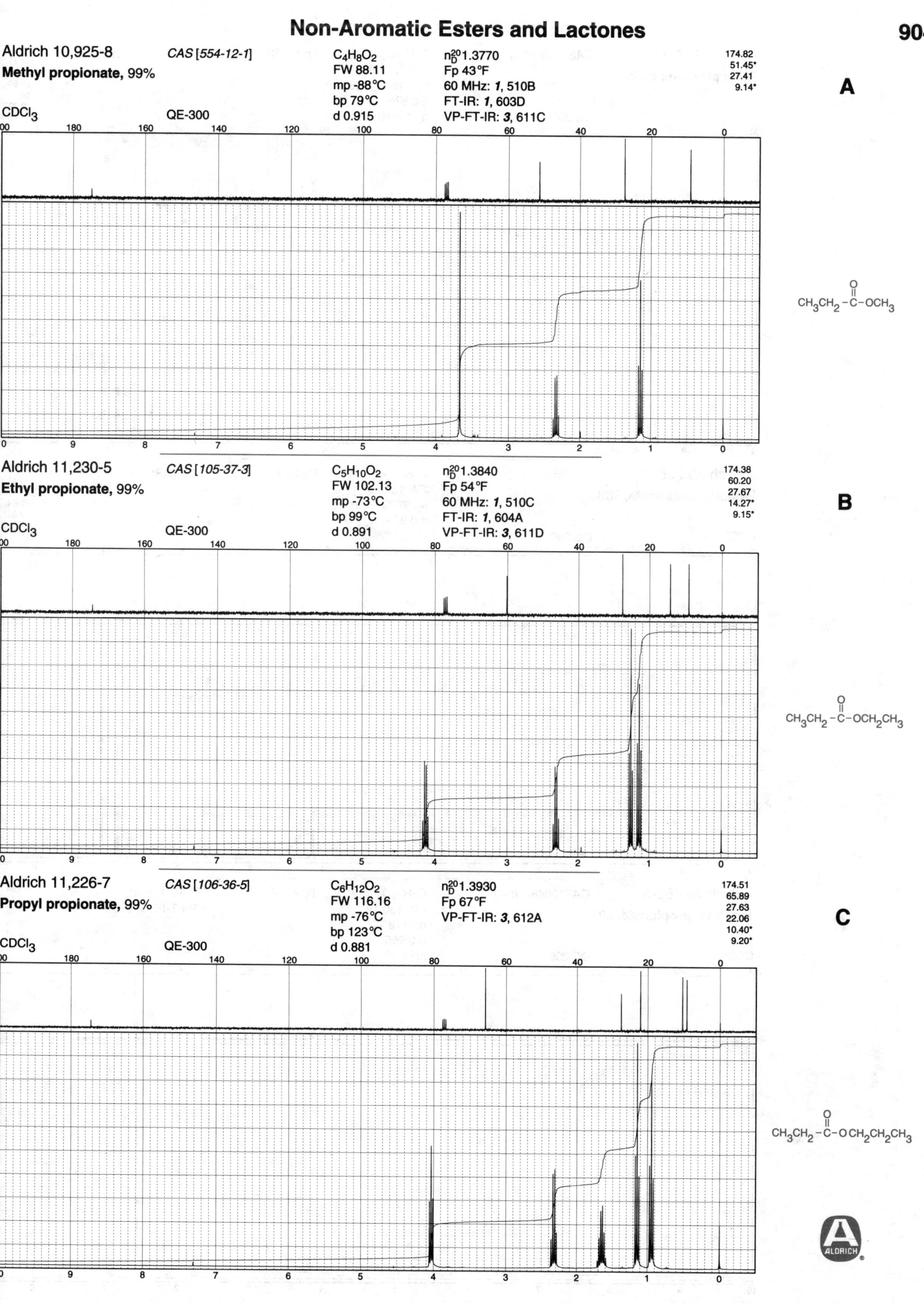

Aldrich 10,925-8 CAS [554-12-1]
Methyl propionate, 99%

C$_4$H$_8$O$_2$
FW 88.11
mp -88°C
bp 79°C
d 0.915

n$_D^{20}$1.3770
Fp 43°F
60 MHz: *1*, 510B
FT-IR: *1*, 603D
VP-FT-IR: *3*, 611C

174.82
51.45*
27.41
9.14*

A

CDCl$_3$ QE-300

CH$_3$CH$_2$–C(=O)–OCH$_3$

Aldrich 11,230-5 CAS [105-37-3]
Ethyl propionate, 99%

C$_5$H$_{10}$O$_2$
FW 102.13
mp -73°C
bp 99°C
d 0.891

n$_D^{20}$1.3840
Fp 54°F
60 MHz: *1*, 510C
FT-IR: *1*, 604A
VP-FT-IR: *3*, 611D

174.38
60.20
27.67
14.27*
9.15*

B

CDCl$_3$ QE-300

CH$_3$CH$_2$–C(=O)–OCH$_2$CH$_3$

Aldrich 11,226-7 CAS [106-36-5]
Propyl propionate, 99%

C$_6$H$_{12}$O$_2$
FW 116.16
mp -76°C
bp 123°C
d 0.881

n$_D^{20}$1.3930
Fp 67°F
VP-FT-IR: *3*, 612A

174.51
65.89
27.63
22.06
10.40*
9.20*

C

CDCl$_3$ QE-300

CH$_3$CH$_2$–C(=O)–O CH$_2$CH$_2$CH$_3$

ALDRICH

Non-Aromatic Esters and Lactones

905

A

Aldrich 30,737-8   *CAS [590-01-2]* $C_7H_{14}O_2$ Fp 101 °F   VP-FT-IR: **3**, 612B

Butyl propionate, 99%   FW 130.19
bp 145 °C (756 mm)
d 0.875

CDCl$_3$ QE-300 n_D^{20}1.4020

174.52
64.17
30.74
27.64
19.18
13.73*
9.18*

$CH_3CH_2-\overset{\overset{\displaystyle O}{\|}}{C}-OCH_2CH_2CH_2CH_3$

B

Aldrich 30,758-0   *CAS [540-42-1]* $C_7H_{14}O_2$ Fp 79 °F   VP-FT-IR: **3**, 612C

Isobutyl propionate, 98%   FW 130.19
bp 67 °C (60 mm)
d 0.869

CDCl$_3$ QE-300 n_D^{20}1.3970

174.47
70.41
27.77*
27.64
19.09*
9.21*

$CH_3CH_2-\overset{\overset{\displaystyle O}{\|}}{C}-OCH_2\overset{\overset{\displaystyle CH_3}{|}}{CH}CH_3$

C

Aldrich 25,452-5   *CAS [20487-40-5]* $C_7H_{14}O_2$ Fp 62 °F   FT-IR: **1**, 604B

tert-Butyl propionate, 99%   FW 130.19   VP-FT-IR: **3**, 612D
bp 118 °C
d 0.865

CDCl$_3$ QE-300 n_D^{20}1.3930

173.84
79.84
28.82
28.14*
9.23*

$CH_3CH_2-\overset{\overset{\displaystyle O}{\|}}{C}-O\overset{\overset{\displaystyle CH_3}{|}}{\underset{\underset{\displaystyle CH_3}{|}}{C}}CH_3$

ALDRICH

Non-Aromatic Esters and Lactones

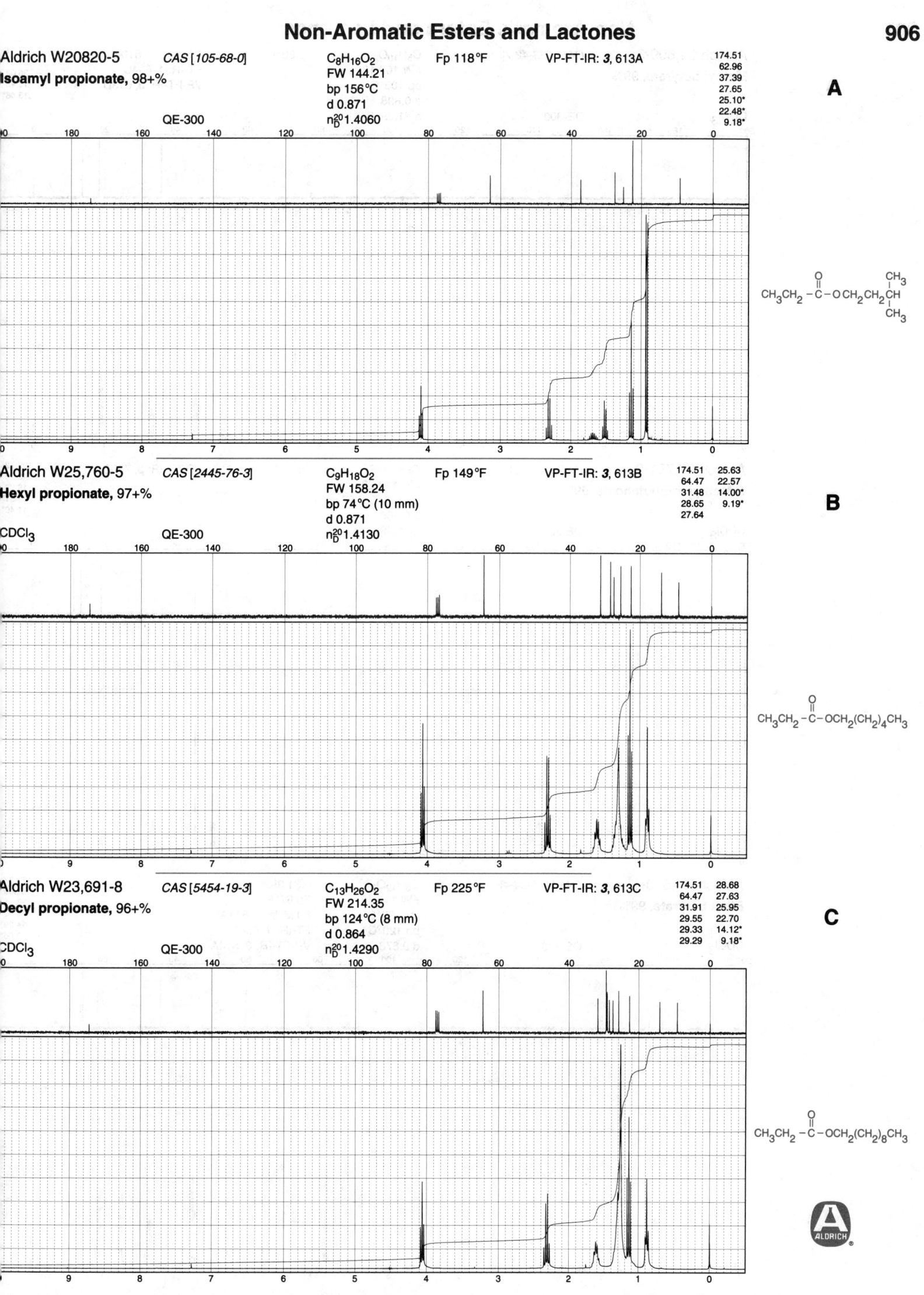

Aldrich W20820-5   CAS [105-68-0] $C_8H_{16}O_2$ Fp 118°F   VP-FT-IR: 3, 613A

Isoamyl propionate, 98+%

FW 144.21
bp 156°C
d 0.871
n_D^{20} 1.4060

QE-300

174.51
62.96
37.39
27.65
25.10*
22.48*
9.18*

A

$CH_3CH_2-\overset{\displaystyle O}{\overset{\displaystyle \|}{C}}-OCH_2CH_2CH\overset{\displaystyle CH_3}{\underset{\displaystyle CH_3}{}}$

Aldrich W25,760-5   CAS [2445-76-3] $C_9H_{18}O_2$ Fp 149°F   VP-FT-IR: 3, 613B

Hexyl propionate, 97+%

FW 158.24
bp 74°C (10 mm)
d 0.871
n_D^{20} 1.4130

CDCl₃   QE-300

174.51   25.63
64.47   22.57
31.48   14.00*
28.65   9.19*
27.64

B

$CH_3CH_2-\overset{\displaystyle O}{\overset{\displaystyle \|}{C}}-OCH_2(CH_2)_4CH_3$

Aldrich W23,691-8   CAS [5454-19-3] $C_{13}H_{26}O_2$ Fp 225°F   VP-FT-IR: 3, 613C

Decyl propionate, 96+%

FW 214.35
bp 124°C (8 mm)
d 0.864
n_D^{20} 1.4290

CDCl₃   QE-300

174.51   28.68
64.47   27.63
31.91   25.95
29.55   22.70
29.33   14.12*
29.29   9.18*

C

$CH_3CH_2-\overset{\displaystyle O}{\overset{\displaystyle \|}{C}}-OCH_2(CH_2)_8CH_3$

ALDRICH

Non-Aromatic Esters and Lactones

A

Aldrich 24,609-3   *CAS [623-42-7]* $C_5H_{10}O_2$ Fp 53°F   60 MHz: *1*, 510D   174.01
Methyl butyrate, 99%   FW 102.13   FT-IR: *1*, 604C   51.34*
bp 103°C   VP-FT-IR: *3*, 613D   36.02
d 0.898   18.49
n_D^{20}1.3860   13.68*

CDCl₃   QE-300

$CH_3CH_2CH_2-\overset{\displaystyle O}{\underset{\displaystyle \|}{C}}-OCH_3$

B

Aldrich 27,781-9   *CAS [2432-51-1]* $C_5H_{10}OS$ Fp 94°F   VP-FT-IR: *3*, 730C   199.72
S-Methyl thiobutanoate, 99%   FW 118.20   45.78
bp 143°C (757 mm)   19.23
d 0.966   13.51*
n_D^{20}1.4610   11.49*

CDCl₃   QE-300

$CH_3CH_2CH_2-\overset{\displaystyle O}{\underset{\displaystyle \|}{C}}-SCH_3$

C

Aldrich E1,570-1   *CAS [105-54-4]* $C_6H_{12}O_2$ n_D^{20}1.3920   173.57
Ethyl butyrate, 99%   FW 116.16   Fp 67°F   60.09
mp -93°C   60 MHz: *1*, 511A   36.31
bp 120°C   FT-IR: *1*, 604D   18.52
d 0.878   VP-FT-IR: *3*, 614A   14.29*
13.66*

CDCl₃   QE-300

$CH_3CH_2CH_2-\overset{\displaystyle O}{\underset{\displaystyle \|}{C}}-OCH_2CH_3$

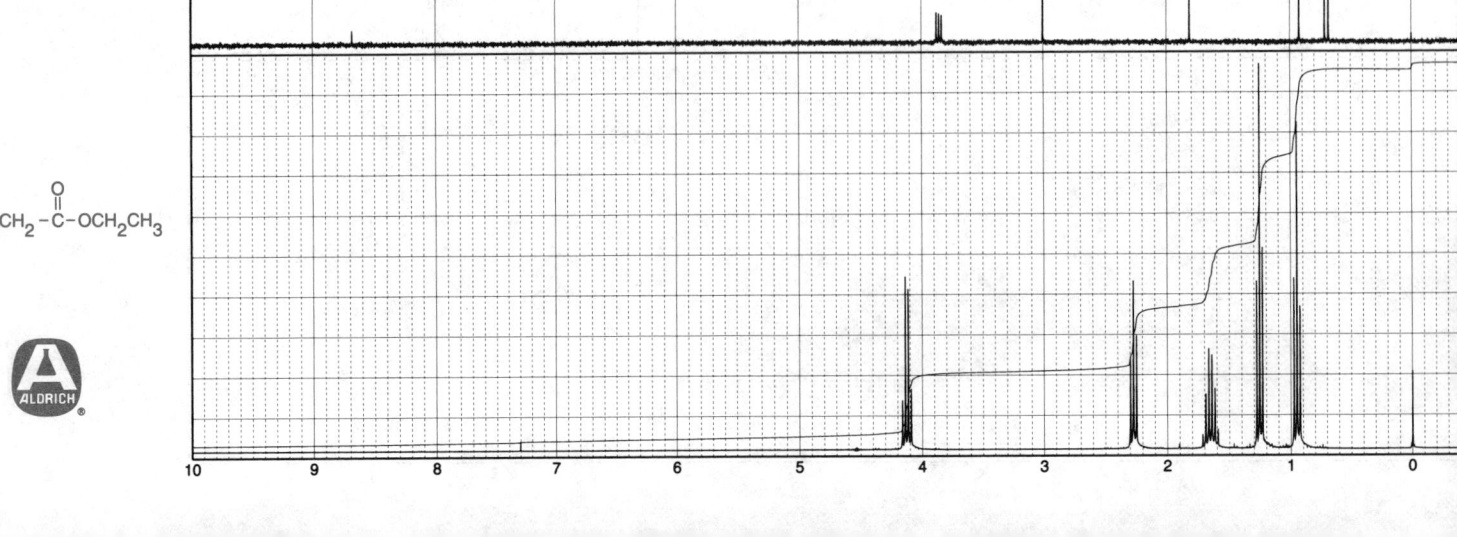

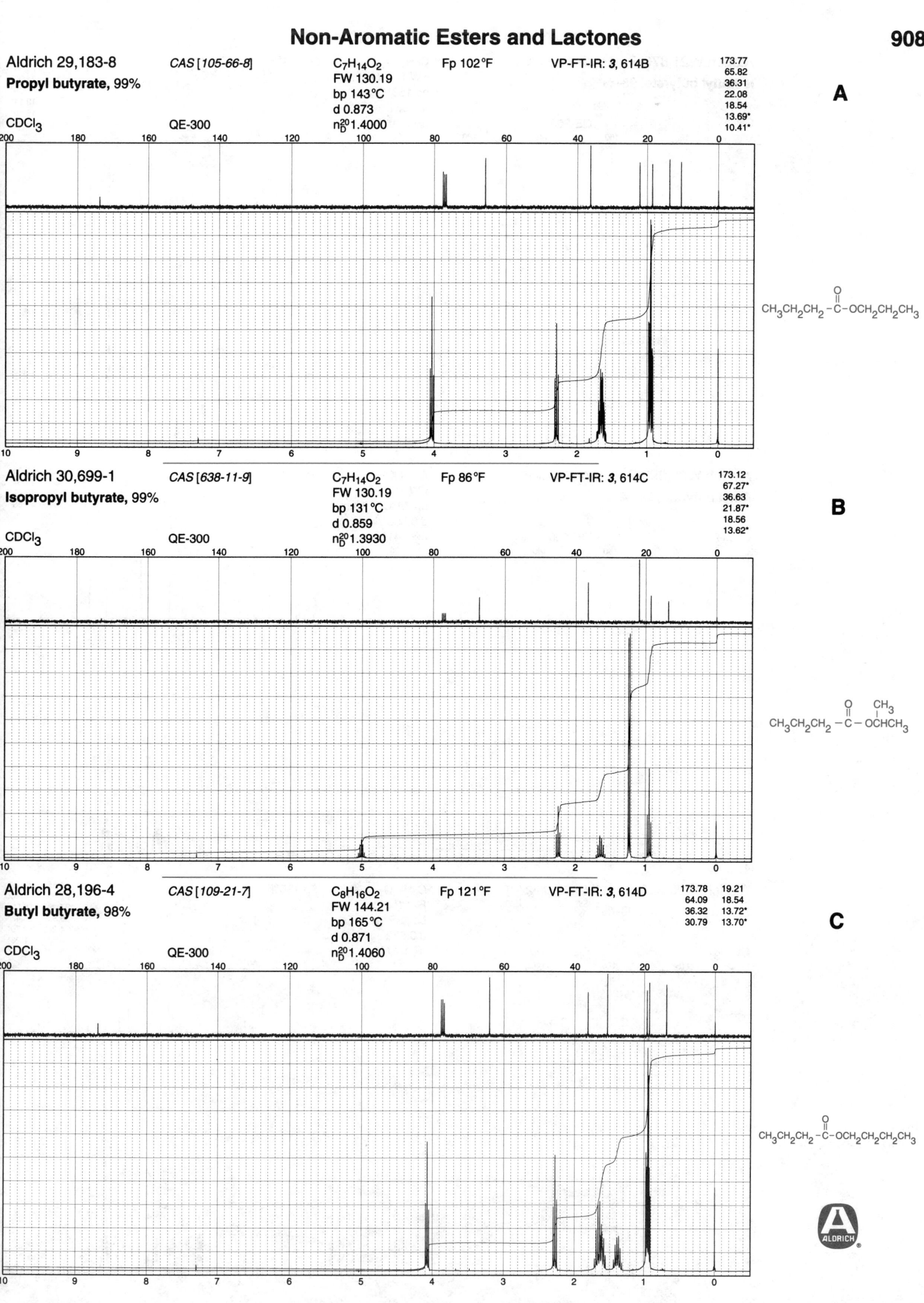

Aldrich 29,183-8
Propyl butyrate, 99%

CAS [105-66-8]

C7H14O2
FW 130.19
bp 143°C
d 0.873
n$_D^{20}$1.4000

Fp 102°F

VP-FT-IR: *3*, 614B

173.77
65.82
36.31
22.08
18.54
13.69*
10.41*

A

CDCl3   QE-300

$CH_3CH_2CH_2-C(=O)-OCH_2CH_2CH_3$

Aldrich 30,699-1
Isopropyl butyrate, 99%

CAS [638-11-9]

C7H14O2
FW 130.19
bp 131°C
d 0.859
n$_D^{20}$1.3930

Fp 86°F

VP-FT-IR: *3*, 614C

173.12
67.27*
36.63
21.87*
18.56
13.62*

B

CDCl3   QE-300

$CH_3CH_2CH_2-C(=O)-OCH(CH_3)CH_3$

Aldrich 28,196-4
Butyl butyrate, 98%

CAS [109-21-7]

C8H16O2
FW 144.21
bp 165°C
d 0.871
n$_D^{20}$1.4060

Fp 121°F

VP-FT-IR: *3*, 614D

173.78   19.21
64.09   18.54
36.32   13.72*
30.79   13.70*

C

CDCl3   QE-300

$CH_3CH_2CH_2-C(=O)-OCH_2CH_2CH_2CH_3$

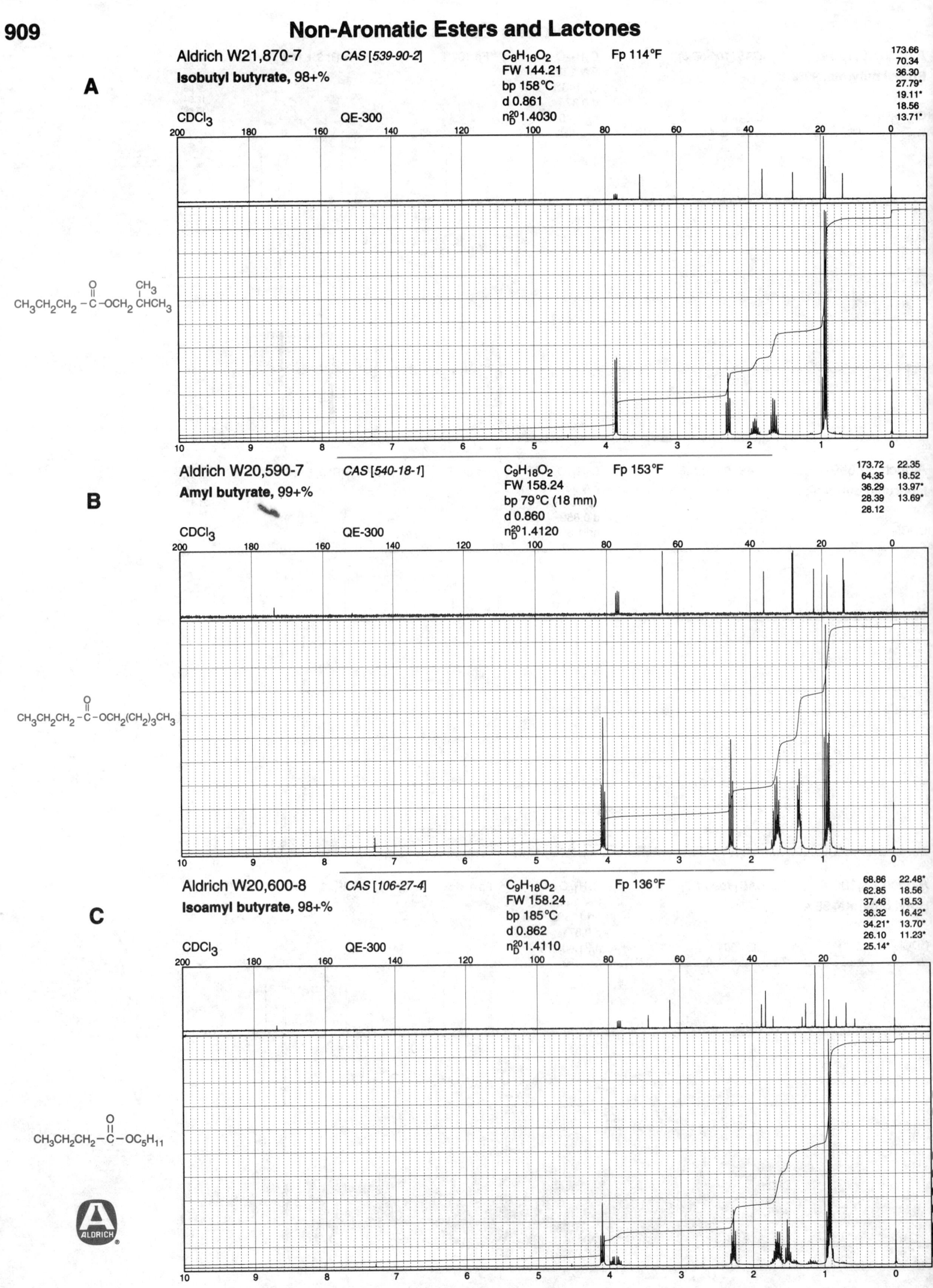

A

Aldrich W21,870-7   CAS [539-90-2]   C$_8$H$_{16}$O$_2$ Fp 114°F
Isobutyl butyrate, 98+%   FW 144.21
bp 158°C
d 0.861
n$_D^{20}$1.4030
CDCl$_3$ QE-300

173.66
70.34
36.30
27.79*
19.11*
18.56
13.71*

CH$_3$CH$_2$CH$_2$-C(=O)-OCH$_2$CHCH$_3$ (CH$_3$)

B

Aldrich W20,590-7   CAS [540-18-1]   C$_9$H$_{18}$O$_2$ Fp 153°F
Amyl butyrate, 99+%   FW 158.24
bp 79°C (18 mm)
d 0.860
n$_D^{20}$1.4120
CDCl$_3$ QE-300

173.72   22.35
64.35   18.52
36.29   13.97*
28.39   13.69*
28.12

CH$_3$CH$_2$CH$_2$-C(=O)-OCH$_2$(CH$_2$)$_3$CH$_3$

C

Aldrich W20,600-8   CAS [106-27-4]   C$_9$H$_{18}$O$_2$ Fp 136°F
Isoamyl butyrate, 98+%   FW 158.24
bp 185°C
d 0.862
n$_D^{20}$1.4110
CDCl$_3$ QE-300

68.86   22.48*
62.85   18.56
37.46   18.53
36.32   16.42*
34.21*   13.70*
26.10   11.23*
25.14*

CH$_3$CH$_2$CH$_2$-C(=O)-OC$_5$H$_{11}$

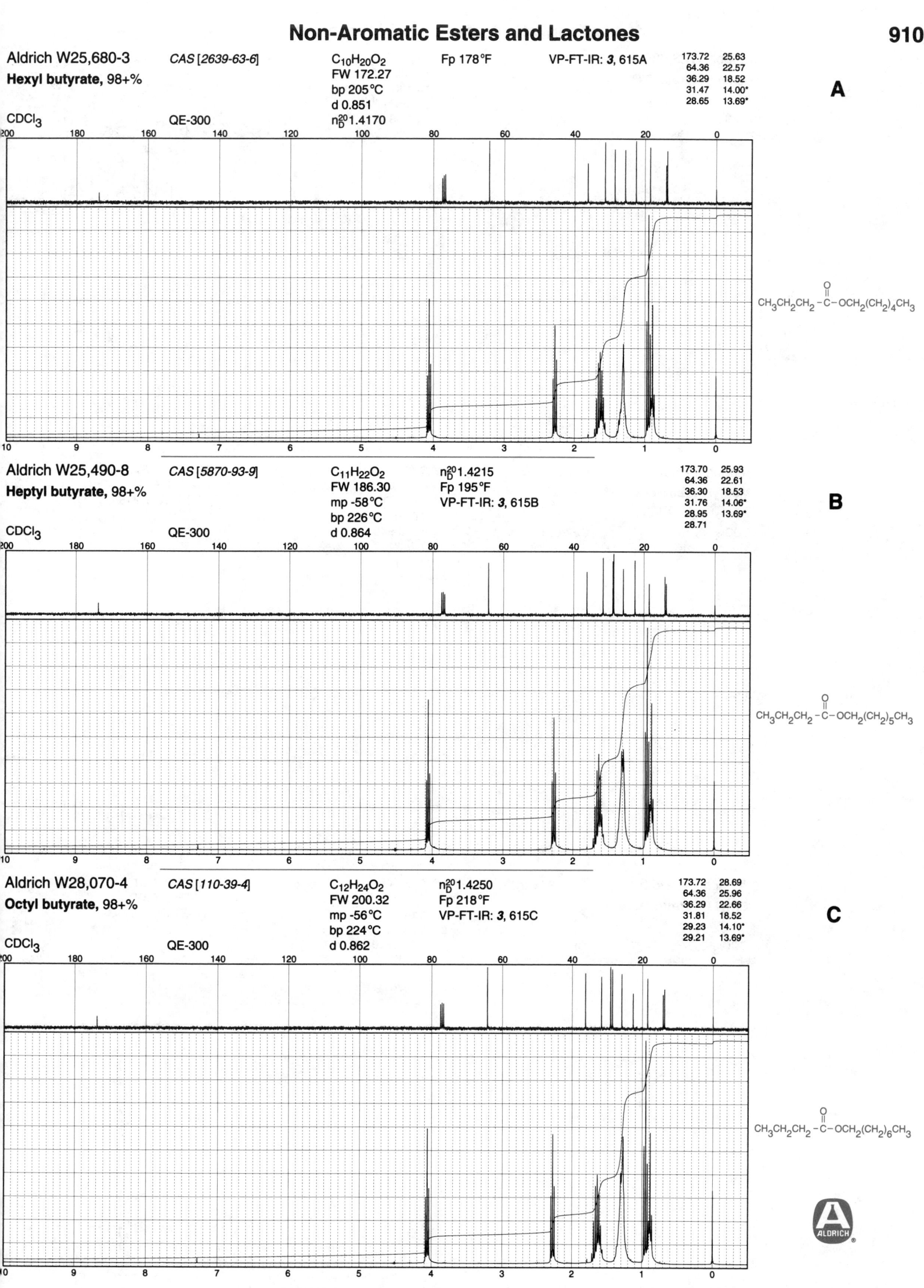

Aldrich W25,680-3   CAS [2639-63-6]   C₁₀H₂₀O₂   Fp 178°F   VP-FT-IR: **3**, 615A

Hexyl butyrate, 98+%   FW 172.27   bp 205°C   d 0.851   n₂₀ᴰ1.4170

CDCl₃   QE-300

173.72	25.63
64.36	22.57
36.29	18.52
31.47	14.00*
28.65	13.69*

A

$$CH_3CH_2CH_2-\overset{\displaystyle O}{\overset{\displaystyle \|}{C}}-OCH_2(CH_2)_4CH_3$$

Aldrich W25,490-8   CAS [5870-93-9]   C₁₁H₂₂O₂   n₂₀ᴰ1.4215

Heptyl butyrate, 98+%   FW 186.30   Fp 195°F   mp -58°C   VP-FT-IR: **3**, 615B   bp 226°C   d 0.864

CDCl₃   QE-300

173.70	25.93
64.36	22.61
36.30	18.53
31.76	14.06*
28.95	13.69*
28.71	

B

$$CH_3CH_2CH_2-\overset{\displaystyle O}{\overset{\displaystyle \|}{C}}-OCH_2(CH_2)_5CH_3$$

Aldrich W28,070-4   CAS [110-39-4]   C₁₂H₂₄O₂   n₂₀ᴰ1.4250

Octyl butyrate, 98+%   FW 200.32   Fp 218°F   mp -56°C   VP-FT-IR: **3**, 615C   bp 224°C   d 0.862

CDCl₃   QE-300

173.72	28.69
64.36	25.96
36.29	22.66
31.81	18.52
29.23	14.10*
29.21	13.69*

C

$$CH_3CH_2CH_2-\overset{\displaystyle O}{\overset{\displaystyle \|}{C}}-OCH_2(CH_2)_6CH_3$$

ALDRICH®

Non-Aromatic Esters and Lactones

A

Aldrich W23,680-2   *CAS [5454-09-1]*   C14H28O2   Fp >230°F   VP-FT-IR: *3*, 615D

Decyl butyrate, 97%

FW 228.38
bp 135°C (8 mm)
d 0.862
n$_D^{20}$1.4310

173.71	28.69
64.36	25.96
36.29	22.70
31.92	18.52
29.55	14.12*
29.33	13.69*
29.28	

CDCl3   QE-300

200 180 160 140 120 100 80 60 40 20 0

$CH_3CH_2CH_2-\overset{O}{\overset{\|}{C}}-OCH_2(CH_2)_8CH_3$

10 9 8 7 6 5 4 3 2 1 0

B

Aldrich 14,800-8   *CAS [547-63-7]*   C5H10O2   Fp 38°F   60 MHz: *1*, 511B

Methyl isobutyrate, 99%

FW 102.13
bp 90°C
d 0.891
n$_D^{20}$1.3840

FT-IR: *1*, 605A
VP-FT-IR: *3*, 616A

| 177.48 |
| 51.50* |
| 33.94* |
| 19.01* |

CDCl3   QE-300

200 180 160 140 120 100 80 60 40 20 0

$\overset{CH_3}{\underset{CH_3}{CH}}-\overset{O}{\overset{\|}{C}}-OCH_3$

10 9 8 7 6 5 4 3 2 1 0

C

Aldrich 24,608-5   *CAS [97-62-1]*   C6H12O2   Fp 57°F   60 MHz: *1*, 511C

Ethyl isobutyrate, 99%

FW 116.16
bp 113°C
d 0.869
n$_D^{20}$1.3870

FT-IR: *1*, 605B
VP-FT-IR: *3*, 616B

| 177.16 |
| 60.17 |
| 34.04* |
| 19.01* |
| 14.25* |

CDCl3   QE-300

200 180 160 140 120 100 80 60 40 20 0

$\overset{CH_3}{\underset{CH_3}{CH}}-\overset{O}{\overset{\|}{C}}-OCH_2CH_3$

10 9 8 7 6 5 4 3 2 1 0

ALDRICH

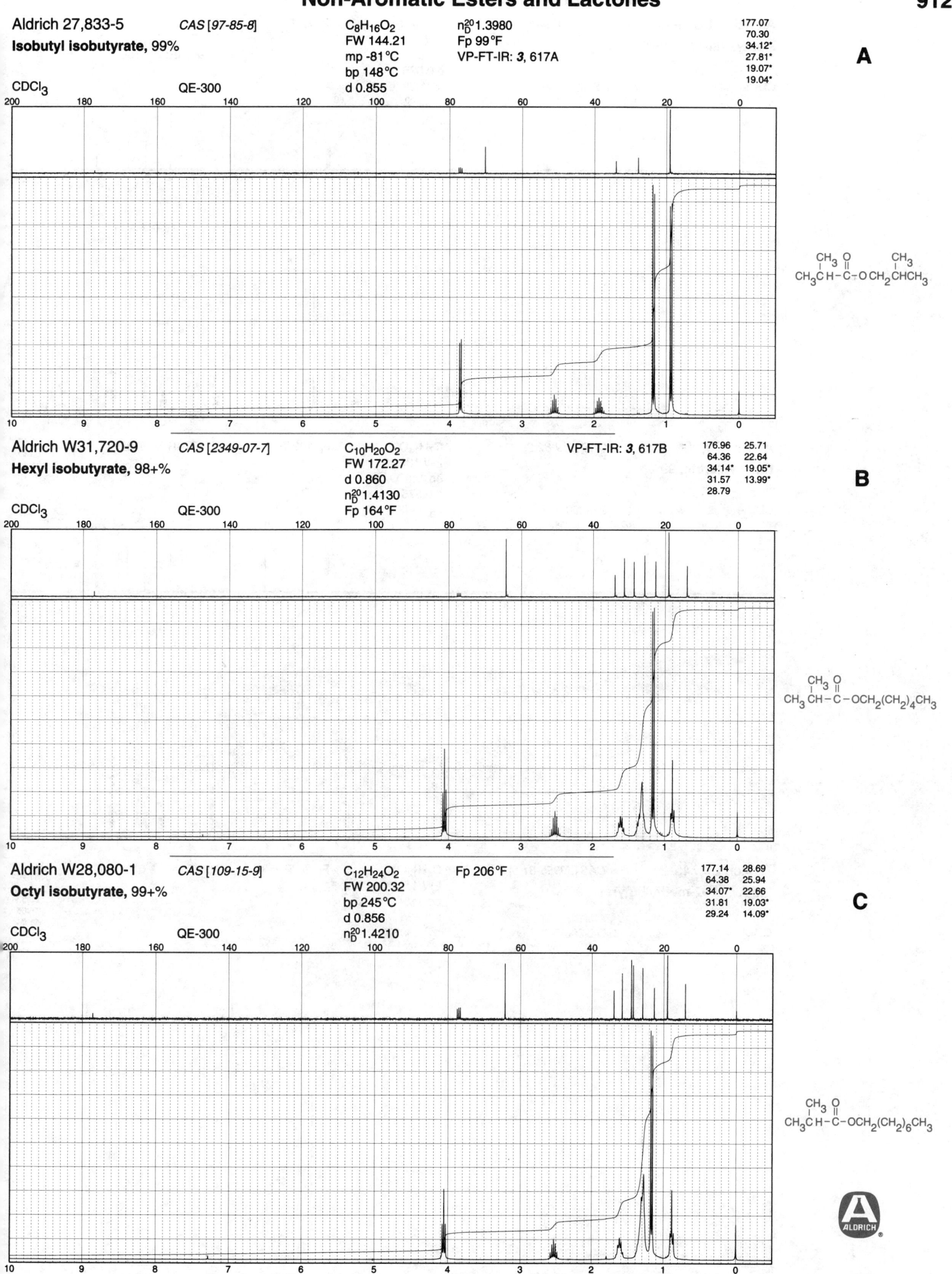

Aldrich 27,833-5   CAS [97-85-8]
Isobutyl isobutyrate, 99%

C$_8$H$_{16}$O$_2$
FW 144.21
mp -81°C
bp 148°C
d 0.855

n$_D^{20}$ 1.3980
Fp 99°F
VP-FT-IR: **3**, 617A

177.07
70.30
34.12*
27.81*
19.07*
19.04*

A

CDCl$_3$ QE-300

Aldrich W31,720-9   CAS [2349-07-7]
Hexyl isobutyrate, 98+%

C$_{10}$H$_{20}$O$_2$
FW 172.27
d 0.860
n$_D^{20}$ 1.4130
Fp 164°F

VP-FT-IR: **3**, 617B

176.96   25.71
64.36   22.64
34.14*   19.05*
31.57   13.99*
28.79

B

CDCl$_3$ QE-300

Aldrich W28,080-1   CAS [109-15-9]
Octyl isobutyrate, 99+%

C$_{12}$H$_{24}$O$_2$
FW 200.32
bp 245°C
d 0.856
n$_D^{20}$ 1.4210

Fp 206°F

177.14   28.69
64.38   25.94
34.07*   22.66
31.81   19.03*
29.24   14.09*

C

CDCl$_3$ QE-300

A

Aldrich 14,899-7   *CAS [624-24-8]* $C_6H_{12}O_2$ Fp 72°F   60 MHz: *1*, 511D

Methyl valerate, 99%   FW 116.16   FT-IR: *1*, 605C

bp 128°C   VP-FT-IR: *3*, 618C

d 0.875

CDCl$_3$ QE-300 n_D^{20}1.3970

174.18
51.36*
33.86
27.11
22.32
13.69*

$CH_3CH_2CH_2CH_2-\overset{\overset{O}{\|}}{C}-OCH_3$

B

Aldrich 29,086-6   *CAS [539-82-2]* $C_7H_{14}O_2$ Fp 102°F   VP-FT-IR: *3*, 618D

Ethyl valerate, 99%   FW 130.19

bp 145°C

d 0.875

CDCl$_3$ QE-300 n_D^{20}1.4000

173.87
60.15
34.14
27.12
22.31
14.28*
13.73*

$CH_3CH_2CH_2CH_2-\overset{\overset{O}{\|}}{C}-OCH_2CH_3$

C

Aldrich 27,745-2   *CAS [53955-81-0]* $C_6H_{12}O_2$ Fp 90°F   VP-FT-IR: *3*, 617C

Methyl (±)-2-methylbutyrate, 99%   FW 116.16

bp 115°C

d 0.885

CDCl$_3$ QE-300 n_D^{20}1.3930

177.13
51.40*
41.01*
26.85
16.65*
11.64*

$CH_3CH_2-\overset{\overset{CH_3}{|}}{C}H-\overset{\overset{O}{\|}}{C}-OCH_3$

A ALDRICH

Aldrich 30,688-6   *CAS [7452-79-1]*   C$_7$H$_{14}$O$_2$ Fp 79°F   VP-FT-IR: **3**, 617D

Ethyl (±)-2-methylbutyrate, 99%   FW 130.19
bp 133°C
d 0.869
n$_D^{20}$1.3970

CDCl$_3$ QE-300

176.64
60.02
41.13*
26.85
16.63*
14.30*
11.60*

A

CH$_3$CH$_2$CH$-$C$-$OCH$_2$CH$_3$ (with CH$_3$ and O above)

Aldrich W36,991-8   *CAS [66576-71-4]*   C$_8$H$_{16}$O$_2$ Fp 90°F

Isopropyl 2-methylbutyrate, 98+%   FW 144.21
bp 142°C (727 mm)
d 0.851
n$_D^{20}$1.3970

CDCl$_3$ QE-300

176.21   21.86*
67.09*   21.81*
41.26*   16.64*
26.84   11.59*

B

CH$_3$CH$_2$ CH$-$C$-$O CHCH$_3$ (with CH$_3$, O, CH$_3$ above)

Aldrich W33,931-8   *CAS [15706-73-7]*   C$_9$H$_{18}$O$_2$ Fp 140°F

Butyl 2-methylbutyrate, 97+%   FW 158.24
bp 175°C (730 mm)
d 0.863
n$_D^{20}$1.4090

CDCl$_3$ QE-300

176.76   19.20
63.98   16.67*
41.18*   13.72*
30.79   11.65*
26.85

C

CH$_3$CH$_2$ CH$-$C$-$OCH$_2$CH$_2$CH$_2$CH$_3$ (with CH$_3$, O above)

A

Aldrich W33,590-8   CAS [2445-78-5]

2-Methylbutyl 2-methylbutyrate, 90+%

$C_{10}H_{20}O_2$
FW 172.27
bp 71°C (11 mm)
d 0.855
n_D^{20} 1.4140

Fp 153°F

VP-FT-IR: *3*, 618A

176.75	26.08
68.75	16.68*
41.25*	16.43*
34.22*	11.65*
26.84	11.24*

CDCl₃   QE-300

$CH_3CH_2CH-C-OCH_2CHCH_2CH_3$ (CH₃ and O=C above)

B

Aldrich W34,990-9   CAS [10032-15-2]

Hexyl 2-methylbutanoate, 95+%

$C_{11}H_{22}O_2$
FW 186.30
d 0.858
n_D^{20} 1.4185
Fp 183°F

VP-FT-IR: *3*, 618B

176.75	25.62
64.26	22.56
41.17*	16.68*
31.45	13.99*
28.68	11.64*
26.84	

CDCl₃   QE-300

$CH_3CH_2CH-C-OCH_2(CH_2)_4CH_3$ (CH₃ and O=C above)

C

Aldrich 11,228-3   CAS [108-64-5]

Ethyl isovalerate, 98%

$C_7H_{14}O_2$
FW 130.19
mp -99°C
bp 132°C
d 0.868

n_D^{20} 1.3960
Fp 80°F
60 MHz: *1*, 512A
FT-IR: *1*, 605D
VP-FT-IR: *3*, 619A

173.12
60.05
43.53
25.75*
22.41*
14.30*

CDCl₃   QE-300

$CH_3CHCH_2-C-OCH_2CH_3$ (CH₃ and O=C above)

ALDRICH

Aldrich W20,850-7   CAS [659-70-1]

Isoamyl isovalerate, 98+%

$C_{10}H_{20}O_2$
FW 172.27
bp 193°C
d 0.854
n_D^{20}1.4120

Fp 152°F

VP-FT-IR: *3*, 619D

173.14	25.74*
62.76	25.10*
43.54	22.45*
37.42	22.41*

A

CDCl₃   QE-300

$$CH_3CHCH_2-C-OCH_2CH_2CHCH_3$$
with CH_3 groups and O (C=O)

Aldrich W35,060-5   CAS [2445-77-4]

2-Methylbutyl isovalerate, 98+%

$C_{10}H_{20}O_2$
FW 172.27
d 0.858
n_D^{20}1.4140
Fp 143°F

173.23	25.76*
68.81	22.43*
43.57	16.43*
34.15*	11.22*
26.07	

B

CDCl₃   QE-300

$$CH_3CHCH_2-C-OCH_2CHCH_2CH_3$$
with CH_3 groups and O (C=O)

Aldrich W28,140-9   CAS [7786-58-5]

Octyl isovalerate, 98+%

$C_{13}H_{26}O_2$
FW 214.35
bp 250°C
d 0.862
n_D^{20}1.4250

Fp >230°F

173.17	25.96
64.29	25.75*
43.54	22.65
31.80	22.41*
29.21	14.09*
28.69	

C

CDCl₃   QE-300

$$CH_3CHCH_2-C-OCH_2(CH_2)_6CH_3$$
with CH_3 group and O (C=O)

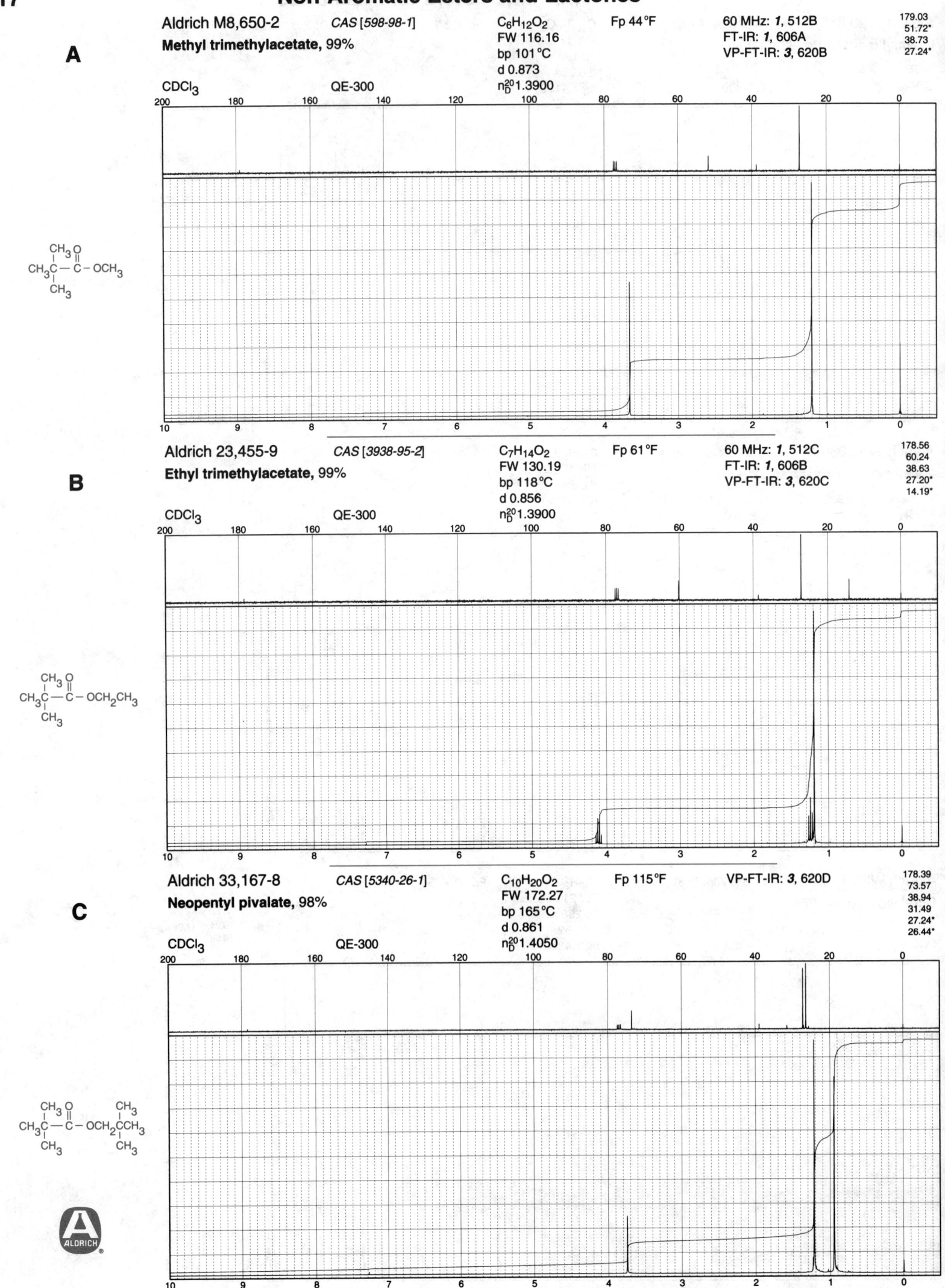

A

Aldrich M8,650-2   *CAS [598-98-1]* $C_6H_{12}O_2$ Fp 44°F   60 MHz: *1*, 512B   179.03
Methyl trimethylacetate, 99%   FW 116.16   FT-IR: *1*, 606A   51.72*
bp 101°C   VP-FT-IR: *3*, 620B   38.73
d 0.873   27.24*
n_D^{20}1.3900

CDCl$_3$ QE-300

B

Aldrich 23,455-9   *CAS [3938-95-2]* $C_7H_{14}O_2$ Fp 61°F   60 MHz: *1*, 512C   178.56
Ethyl trimethylacetate, 99%   FW 130.19   FT-IR: *1*, 606B   60.24
bp 118°C   VP-FT-IR: *3*, 620C   38.63
d 0.856   27.20*
n_D^{20}1.3900   14.19*

CDCl$_3$ QE-300

C

Aldrich 33,167-8   *CAS [5340-26-1]* $C_{10}H_{20}O_2$ Fp 115°F   VP-FT-IR: *3*, 620D   178.39
Neopentyl pivalate, 98%   FW 172.27   73.57
bp 165°C   38.94
d 0.861   31.49
n_D^{20}1.4050   27.24*
26.44*

CDCl$_3$ QE-300

Aldrich 25,994-2   CAS *[106-70-7]* $C_7H_{14}O_2$ n_D^{20}1.4050

Methyl caproate, 99%

FW 130.19   Fp 113°F

mp -71°C   FT-IR: *1*, 606C

bp 151°C   VP-FT-IR: *3*, 621A

CDCl$_3$ QE-300   d 0.885

174.18
51.35*
34.11
31.39
24.71
22.35
13.88*

A

$CH_3(CH_2)_3CH_2 - \overset{\displaystyle O}{\overset{\|}{C}} - OCH_3$

Aldrich 14,896-2   CAS *[123-66-0]* $C_8H_{16}O_2$ Fp 121°F   60 MHz: *1*, 512D

Ethyl caproate, 99+%

FW 144.21   FT-IR: *1*, 606D

bp 168°C   VP-FT-IR: *3*, 621B

d 0.873

CDCl$_3$ QE-300 n_D^{20}1.4070

173.76   24.73
60.10   22.36
34.40   14.28*
31.38   13.89*

B

$CH_3(CH_2)_3CH_2 - \overset{\displaystyle O}{\overset{\|}{C}} - OCH_2CH_3$

Aldrich W29,491-8   CAS *[626-77-7]* $C_9H_{18}O_2$ n_D^{20}1.4120

Propyl hexanoate, 98+%

FW 158.24   Fp 125°F

mp -35°C   VP-FT-IR: *3*, 621C

bp 187°C

CDCl$_3$ QE-300   d 0.867

173.88   22.36
65.80   22.07
34.37   13.91*
31.38   10.41*
24.76

C

$CH_3(CH_2)_3CH_2 - \overset{\displaystyle O}{\overset{\|}{C}} - OCH_2CH_2CH_3$

A

Aldrich W22,010-8   *CAS [626-82-4]* $C_{10}H_{20}O_2$ Fp 178°F   VP-FT-IR: **3**, 622A

Butyl hexanoate, 98+%

FW 172.27
bp 62°C (3 mm)
d 0.866
n_D^{20}1.4160

CDCl₃   QE-300

173.89	24.74
64.07	22.36
34.39	19.20
31.38	13.91*
30.77	13.72*

$CH_3(CH_2)_3CH_2-\overset{O}{\overset{\|}{C}}-OCH_2CH_2CH_2CH_3$

B

Aldrich W22,020-5   *CAS [105-79-3]* $C_{10}H_{20}O_2$ VP-FT-IR: **3**, 622B

Isobutyl hexanoate, 98+%

FW 172.27
d 0.856
n_D^{20}1.4140
Fp 169°F

CDCl₃   QE-300

173.88	24.75
70.34	22.35
34.37	19.10*
31.36	13.92*
27.75*	

$CH_3(CH_2)_3CH_2-\overset{O}{\overset{\|}{C}}-OCH_2\overset{CH_3}{\overset{|}{C}}HCH_3$

C

Aldrich W25,720-6   *CAS [6378-65-0]* $C_{12}H_{24}O_2$ n_D^{20}1.4240

Hexyl hexanoate, 97+%

FW 201.33
mp -28°C
bp 246°C
d 0.863
Fp 211°F

CDCl₃   QE-300

173.91	25.63
64.37	24.74
34.38	22.56
31.46	22.35
31.35	14.00*
28.65	13.92*

$CH_3(CH_2)_3CH_2-\overset{O}{\overset{\|}{C}}-OCH_2(CH_2)_4CH_3$

ALDRICH

Aldrich W34,880-5   CAS [28959-02-6]

Ethyl 2-methylpentanoate, 98+%

$C_8H_{16}O_2$
FW 144.21
bp 153°C
d 0.864

$n_D^{20} 1.4030$
Fp 103°F

176.84	20.44
60.03	17.06*
39.36*	14.29*
36.04	13.96*

A

CDCl₃   QE-300

$$CH_3CH_2CH_2\overset{\overset{\displaystyle CH_3}{|}}{C}H-\overset{\overset{\displaystyle O}{||}}{C}-OCH_2CH_3$$

Aldrich W27,210-8   CAS [2412-80-8]

Methyl 4-methylvalerate, 97+%

$C_7H_{14}O_2$
FW 130.19
bp 140°C
d 0.888
$n_D^{20} 1.4040$

Fp 103°F

174.41
51.43*
33.81
32.17
27.70*
22.22*

B

CDCl₃   QE-300

$$CH_3\overset{\overset{\displaystyle CH_3}{|}}{C}HCH_2CH_2-\overset{\overset{\displaystyle O}{||}}{C}-OCH_3$$

Aldrich 14,900-4   CAS [106-73-0]

Methyl enanthate, 99%

$C_8H_{16}O_2$
FW 144.21
bp 173°C
d 0.870
$n_D^{20} 1.4100$

Fp 127°F

60 MHz: *1*, 513A
FT-IR: *1*, 607A
VP-FT-IR: *3*, 623A

174.18	28.88
51.35*	24.99
34.15	22.51
31.50	13.99*

C

CDCl₃   QE-300

$$CH_3(CH_2)_4CH_2-\overset{\overset{\displaystyle O}{||}}{C}-OCH_3$$

A

Aldrich 11,236-4   CAS [106-30-9]

Ethyl heptanoate, 99%

$C_9H_{18}O_2$
FW 158.24
mp -33°C
bp 189°C
d 0.868

n_D^{20} 1.4120
Fp 151°F
VP-FT-IR: **3**, 623B

173.87	25.00
60.14	22.52
34.43	14.28*
31.50	14.02*
28.86	

CDCl₃   QE-300

$CH_3(CH_2)_4CH_2-\overset{\displaystyle O}{\overset{\|}{C}}-OCH_2CH_3$

B

Aldrich W29,480-2   CAS [7778-87-2]

Propyl heptanoate, 98+%

$C_{10}H_{20}O_2$
FW 172.27
mp -32°C
bp 208°C
d 0.869

n_D^{20} 1.4170
Fp 170°F
VP-FT-IR: **3**, 623C

173.88	25.04
65.80	22.53
34.42	22.07
31.51	14.02*
28.88	10.41*

CDCl₃   QE-300

$CH_3(CH_2)_4CH_2-\overset{\displaystyle O}{\overset{\|}{C}}-OCH_2CH_2CH_3$

C

Aldrich 26,067-3   CAS [111-11-5]

Methyl caprylate, 99%

$C_9H_{18}O_2$
FW 158.24
bp 195°C
d 0.877
n_D^{20} 1.4170

Fp 163°F

FT-IR: **1**, 607B
VP-FT-IR: **3**, 624B

174.18	28.95
51.34*	25.03
34.15	22.63
31.70	14.03*
29.17	

CDCl₃   QE-300

$CH_3(CH_2)_5CH_2-\overset{\displaystyle O}{\overset{\|}{C}}-OCH_3$

ALDRICH

Aldrich 11,232-1   *CAS [106-32-1]* $C_{10}H_{20}O_2$ n_D^{20}1.4170

Ethyl caprylate, 99+%   FW 172.27   Fp 167°F   **A**

mp -48°C   60 MHz: *1*, 513B

bp 207°C   FT-IR: *1*, 607C

CDCl_3   QE-300   d 0.878   VP-FT-IR: *3*, 624C

173.75	28.96	
60.09	25.05	
34.44	22.62	
31.71	14.28*	
29.16	14.03*	

$CH_3(CH_2)_5CH_2 - \overset{O}{\overset{\|}{C}} - OCH_2CH_3$

Aldrich 24,589-5   *CAS [1731-84-6]* $C_{10}H_{20}O_2$ Fp 184°F   60 MHz: *1*, 513C

Methyl nonanoate, 98%   FW 172.27   FT-IR: *1*, 607D   **B**

bp 214°C   VP-FT-IR: *3*, 625D

d 0.875

CDCl_3   QE-300 n_D^{20}1.4210

174.18	29.22
51.35*	29.16
34.15	25.02
31.86	22.67
29.26	14.07*

$CH_3(CH_2)_6CH_2 - \overset{O}{\overset{\|}{C}} - OCH_3$

Aldrich 11,234-8   *CAS [123-29-5]* $C_{11}H_{22}O_2$ Fp 202°F   VP-FT-IR: *3*, 626A

Ethyl nonanoate, 97%   FW 186.30   **C**

bp 119°C (23 mm)

d 0.866

CDCl_3   QE-300 n_D^{20}1.4220

173.82	29.16
60.12	25.02
34.41	22.66
31.84	14.27*
29.26	14.10*
29.18	

$CH_3(CH_2)_6CH_2 - \overset{O}{\overset{\|}{C}} - OCH_2CH_3$

Non-Aromatic Esters and Lactones

A

Aldrich W20,780-2 *CAS [7779-70-6]* C$_{14}$H$_{28}$O$_2$ Fp >230°F VP-FT-IR: **3**, 626B

Isoamyl nonanoate, 96+% FW 228.38
bp 260°C
d 0.860
n$_D^{20}$1.4290

CDCl$_3$ QE-300

173.86	29.17
62.85	25.12*
37.44	25.06
34.43	22.67
31.85	22.47*
29.27	14.09*
29.20	

CH$_3$(CH$_2$)$_6$CH$_2$—C—O—CH$_2$CH$_2$CHCH$_3$ (with O above C and CH$_3$ above CHCH$_3$)

B

Aldrich 29,903-0 *CAS [110-42-9]* C$_{11}$H$_{22}$O$_2$ Fp 202°F VP-FT-IR: **3**, 626C

Methyl decanoate, 99% FW 186.30
bp 108°C (10 mm)
d 0.873
n$_D^{20}$1.4250

CDCl$_3$ QE-300

174.19	29.29
51.35*	29.21
34.15	25.02
31.90	22.69
29.45	14.08*

CH$_3$(CH$_2$)$_7$CH$_2$—C—OCH$_3$ (with O above C)

C

Aldrich 14,897-0 *CAS [110-38-3]* C$_{12}$H$_{24}$O$_2$ Fp 216°F 60 MHz: **1**, 513D

Ethyl caprate, 99+% FW 200.32 FT-IR: **1**, 608A
bp 245°C VP-FT-IR: **3**, 626D
d 0.862
n$_D^{20}$1.4250

CDCl$_3$ QE-300

173.76	29.20
60.09	25.04
34.43	22.69
31.90	14.28*
29.45	14.08*
29.29	

CH$_3$(CH$_2$)$_7$CH$_2$—C—OCH$_2$CH$_3$ (with O above C)

ALDRICH

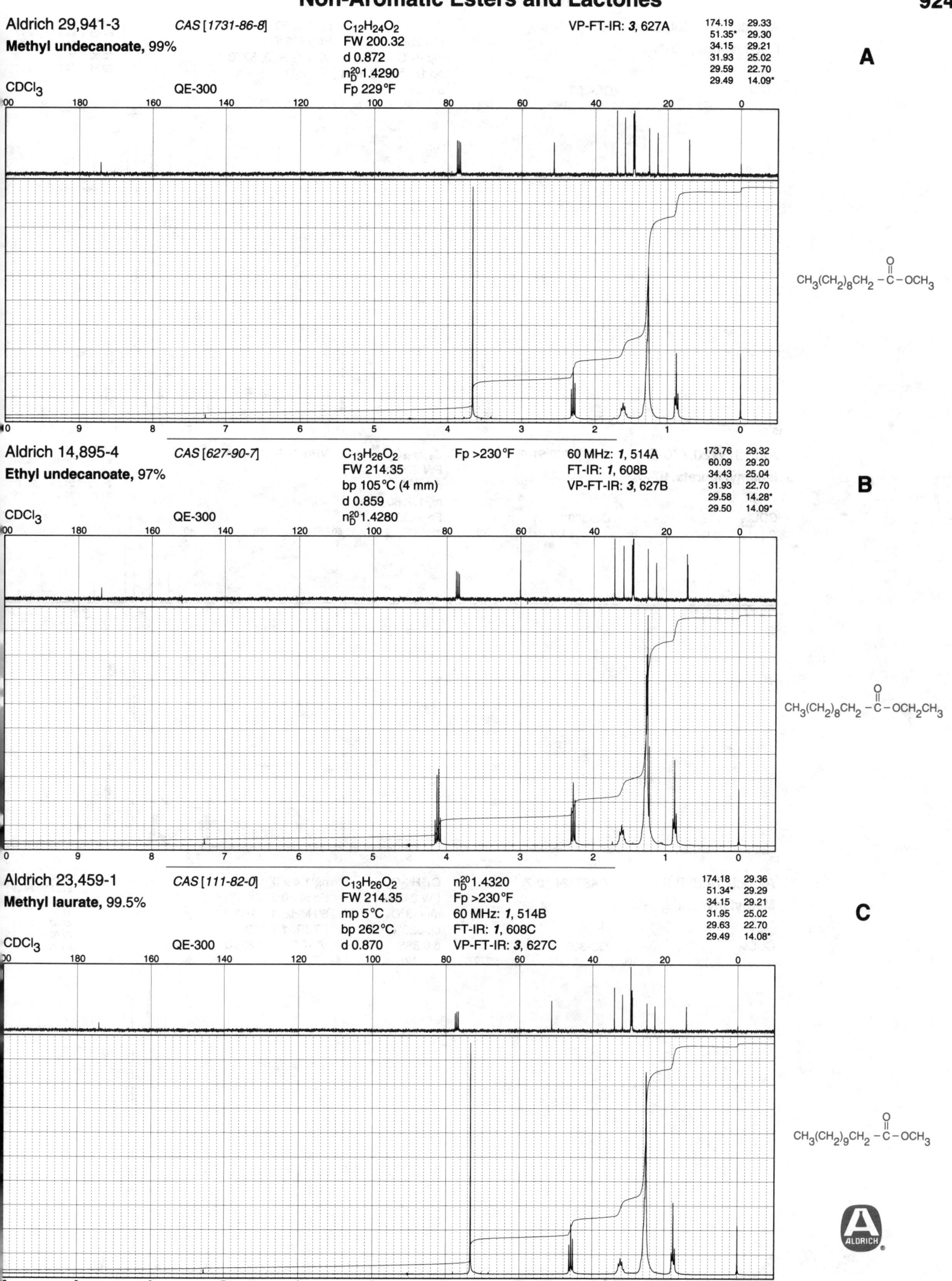

Aldrich 29,941-3   *CAS [1731-86-8]*
Methyl undecanoate, 99%

$C_{12}H_{24}O_2$
FW 200.32
d 0.872
n_D^{20} 1.4290
Fp 229°F

VP-FT-IR: *3*, 627A

174.19	29.33
51.35*	29.30
34.15	29.21
31.93	25.02
29.59	22.70
29.49	14.09*

A

CDCl₃   QE-300

$CH_3(CH_2)_8CH_2 - \overset{\overset{\displaystyle O}{\|}}{C} - OCH_3$

Aldrich 14,895-4   *CAS [627-90-7]*
Ethyl undecanoate, 97%

$C_{13}H_{26}O_2$
FW 214.35
bp 105°C (4 mm)
d 0.859
n_D^{20} 1.4280

Fp >230°F

60 MHz: *1*, 514A
FT-IR: *1*, 608B
VP-FT-IR: *3*, 627B

173.76	29.32
60.09	29.20
34.43	25.04
31.93	22.70
29.58	14.28*
29.50	14.09*

B

CDCl₃   QE-300

$CH_3(CH_2)_8CH_2 - \overset{\overset{\displaystyle O}{\|}}{C} - OCH_2CH_3$

Aldrich 23,459-1   *CAS [111-82-0]*
Methyl laurate, 99.5%

$C_{13}H_{26}O_2$
FW 214.35
mp 5°C
bp 262°C
d 0.870

n_D^{20} 1.4320
Fp >230°F
60 MHz: *1*, 514B
FT-IR: *1*, 608C
VP-FT-IR: *3*, 627C

174.18	29.36
51.34*	29.29
34.15	29.21
31.95	25.02
29.63	22.70
29.49	14.08*

C

CDCl₃   QE-300

$CH_3(CH_2)_9CH_2 - \overset{\overset{\displaystyle O}{\|}}{C} - OCH_3$

ALDRICH

A

Aldrich W22,060-4   *CAS [106-18-3]*

Butyl laurate, 99+%

$C_{16}H_{32}O_2$
FW 256.43
mp -5°C
bp 180°C (18 mm)
d 0.855

n_D^{20} 1.4350
Fp >230°F
VP-FT-IR: *3*, 627D

173.71	29.70	25.13
64.03	29.57	22.77
34.44	29.43	19.26
32.02	29.38	14.10*
30.88	29.27	13.72*

CDCl$_3$ QE-300

$CH_3(CH_2)_9CH_2-\overset{O}{\overset{\|}{C}}-OCH_2CH_2CH_2CH_3$

B

Aldrich W20,770-5   *CAS [6309-51-9]*

Isoamyl laurate, 97+%

$C_{17}H_{34}O_2$
FW 270.46
d 0.856
n_D^{20} 1.4360
Fp >230°F

VP-FT-IR: *3*, 628A

173.91	29.62	25.09
62.86	29.49	25.04*
37.39	29.35	22.70
34.43	29.29	22.47*
31.93	29.18	14.12*

CDCl$_3$ QE-300

$CH_3(CH_2)_9CH_2-\overset{O}{\overset{\|}{C}}-OCH_2CH_2\overset{CH_3}{\overset{|}{CH}}CH_3$

C

Aldrich 14,898-9   *CAS [124-10-7]*

Methyl myristate, 99%

$C_{15}H_{30}O_2$
FW 242.40
mp 18°C
bp 323°C
d 0.855

n_D^{20} 1.4360
Fp >230°F
60 MHz: *1*, 515A
FT-IR: *1*, 608D
VP-FT-IR: *3*, 628D

174.20	29.38
51.36*	29.29
34.13	29.19
31.95	25.00
29.67	22.71
29.62	14.10*
29.48	

CDCl$_3$ QE-300

$CH_3(CH_2)_{11}CH_2-\overset{O}{\overset{\|}{C}}-OCH_3$

ALDRICH

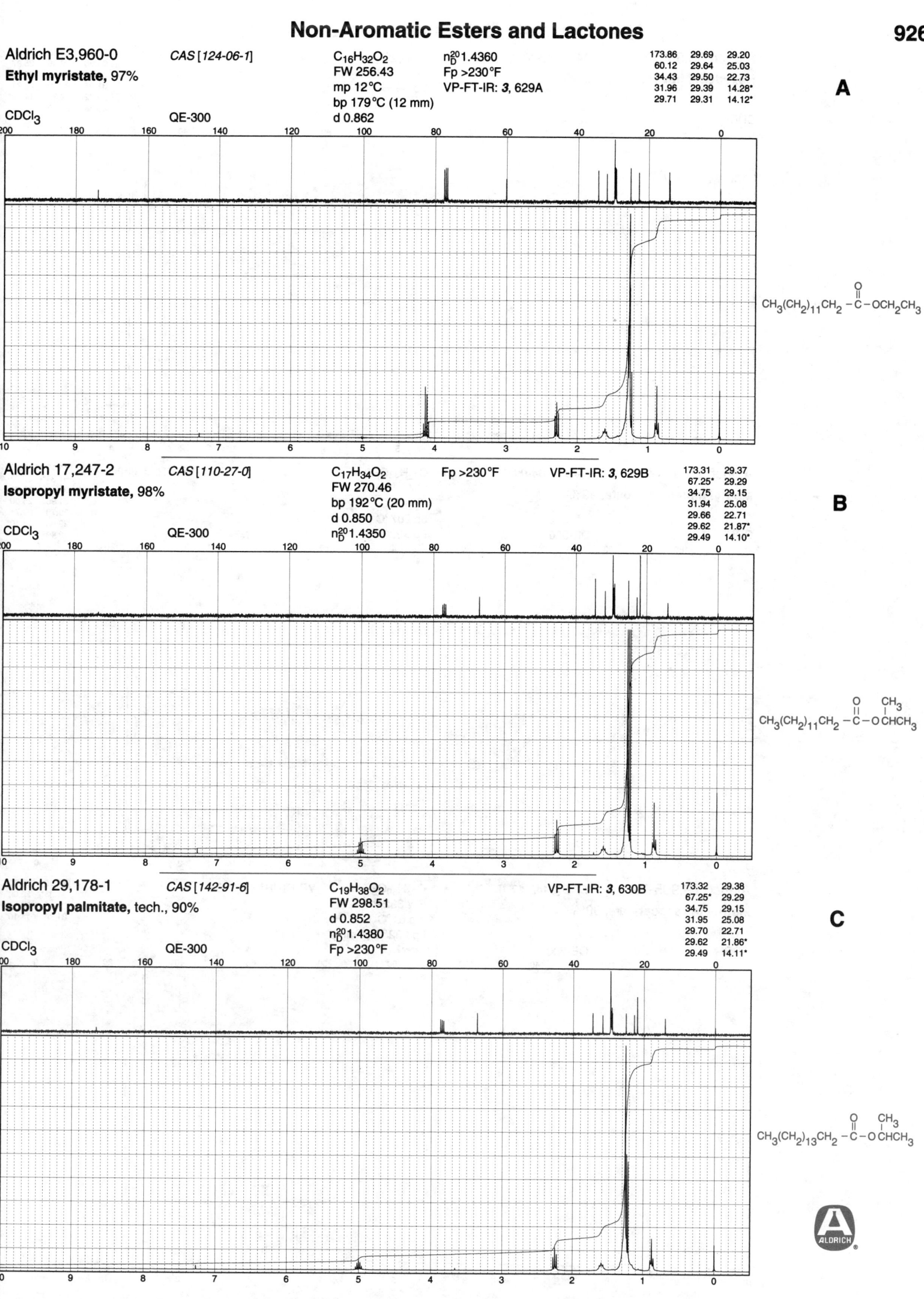

Aldrich E3,960-0
Ethyl myristate, 97%

CAS [124-06-1]

$C_{16}H_{32}O_2$
FW 256.43
mp 12°C
bp 179°C (12 mm)
d 0.862

n_D^{20} 1.4360
Fp >230°F
VP-FT-IR: *3*, 629A

173.86	29.69	29.20
60.12	29.64	25.03
34.43	29.50	22.73
31.96	29.39	14.28*
29.71	29.31	14.12*

A

CDCl₃   QE-300

$CH_3(CH_2)_{11}CH_2-\overset{\overset{\displaystyle O}{\|}}{C}-OCH_2CH_3$

Aldrich 17,247-2
Isopropyl myristate, 98%

CAS [110-27-0]

$C_{17}H_{34}O_2$
FW 270.46
bp 192°C (20 mm)
d 0.850
n_D^{20} 1.4350

Fp >230°F

VP-FT-IR: *3*, 629B

173.31	29.37
67.25*	29.29
34.75	29.15
31.94	25.08
29.66	22.71
29.62	21.87*
29.49	14.10*

B

CDCl₃   QE-300

$CH_3(CH_2)_{11}CH_2-\overset{\overset{\displaystyle O}{\|}}{C}-O\overset{\overset{\displaystyle CH_3}{|}}{C}HCH_3$

Aldrich 29,178-1
Isopropyl palmitate, tech., 90%

CAS [142-91-6]

$C_{19}H_{38}O_2$
FW 298.51
d 0.852
n_D^{20} 1.4380
Fp >230°F

VP-FT-IR: *3*, 630B

173.32	29.38
67.25*	29.29
34.75	29.15
31.95	25.08
29.70	22.71
29.62	21.86*
29.49	14.11*

C

CDCl₃   QE-300

$CH_3(CH_2)_{13}CH_2-\overset{\overset{\displaystyle O}{\|}}{C}-O\overset{\overset{\displaystyle CH_3}{|}}{C}HCH_3$

A

Aldrich 38,833-5   *CAS [540-10-3]* $C_{32}H_{64}O_2$
Hexadecyl hexadecanoate, 98%   FW 480.87
mp 56 °C

173.89	29.63	28.68
64.36	29.56	25.97
34.42	29.50	25.06
31.95	29.39	22.71
29.72	29.30	14.12*
29.68	29.19	

CDCl₃   QE-300

$CH_3(CH_2)_{13}CH_2-\overset{\overset{\displaystyle O}{\|}}{C}-OCH_2(CH_2)_{14}CH_3$

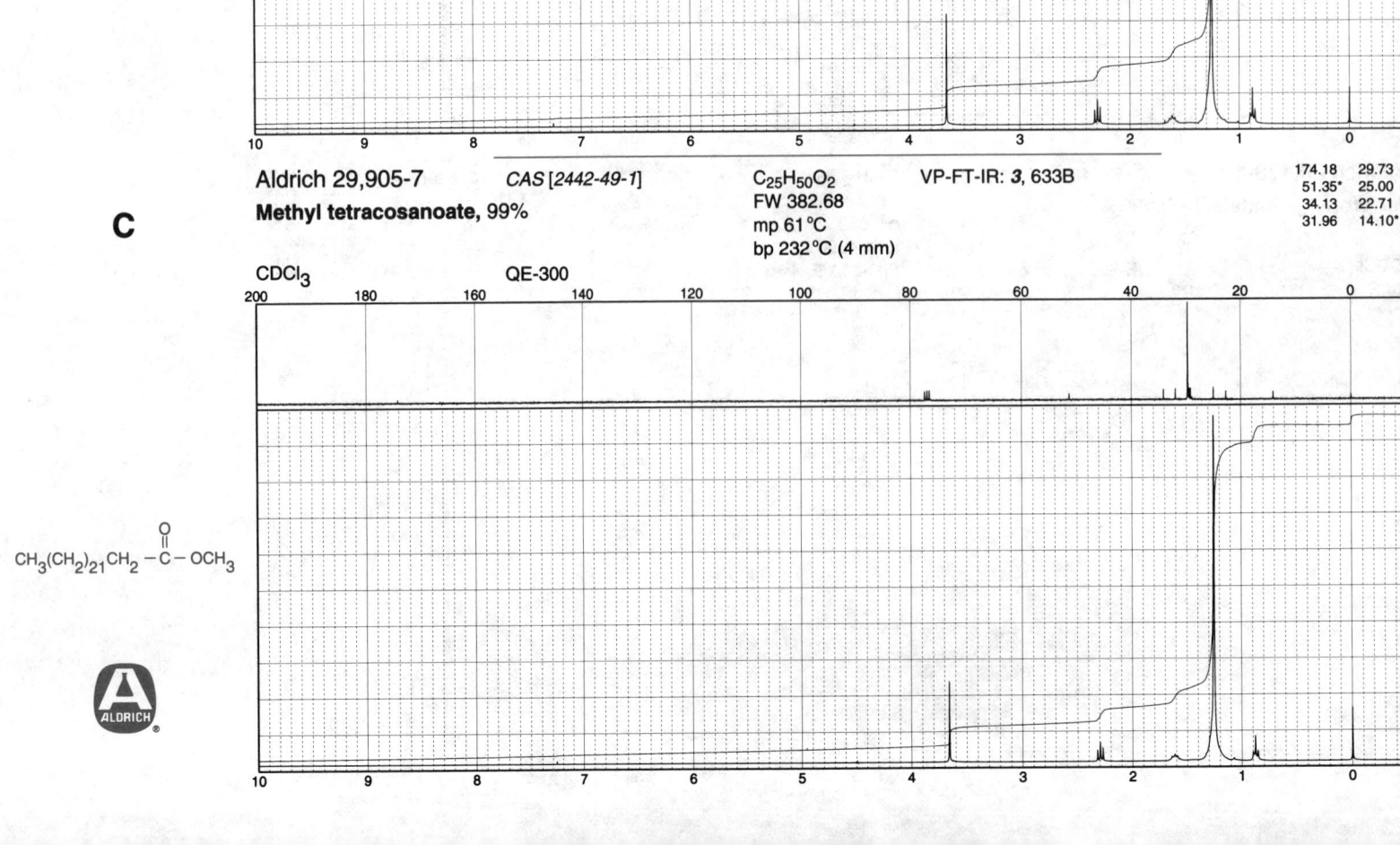

B

Aldrich 29,904-9   *CAS [6064-90-0]* $C_{22}H_{44}O_2$ VP-FT-IR: **3**, 632C
Methyl heneicosanoate, 99%   FW 340.60
mp 49 °C
bp 207 °C (4 mm)
Fp >230 °F

174.17	29.39
51.34*	29.29
34.13	29.20
31.96	25.00
29.73	22.71
29.63	14.10*
29.49	

CDCl₃   QE-300

$CH_3(CH_2)_{18}CH_2-\overset{\overset{\displaystyle O}{\|}}{C}-OCH_3$

C

Aldrich 29,905-7   *CAS [2442-49-1]* $C_{25}H_{50}O_2$ VP-FT-IR: **3**, 633B
Methyl tetracosanoate, 99%   FW 382.68
mp 61 °C
bp 232 °C (4 mm)

174.18	29.73
51.35*	25.00
34.13	22.71
31.96	14.10*

CDCl₃   QE-300

$CH_3(CH_2)_{21}CH_2-\overset{\overset{\displaystyle O}{\|}}{C}-OCH_3$

ALDRICH

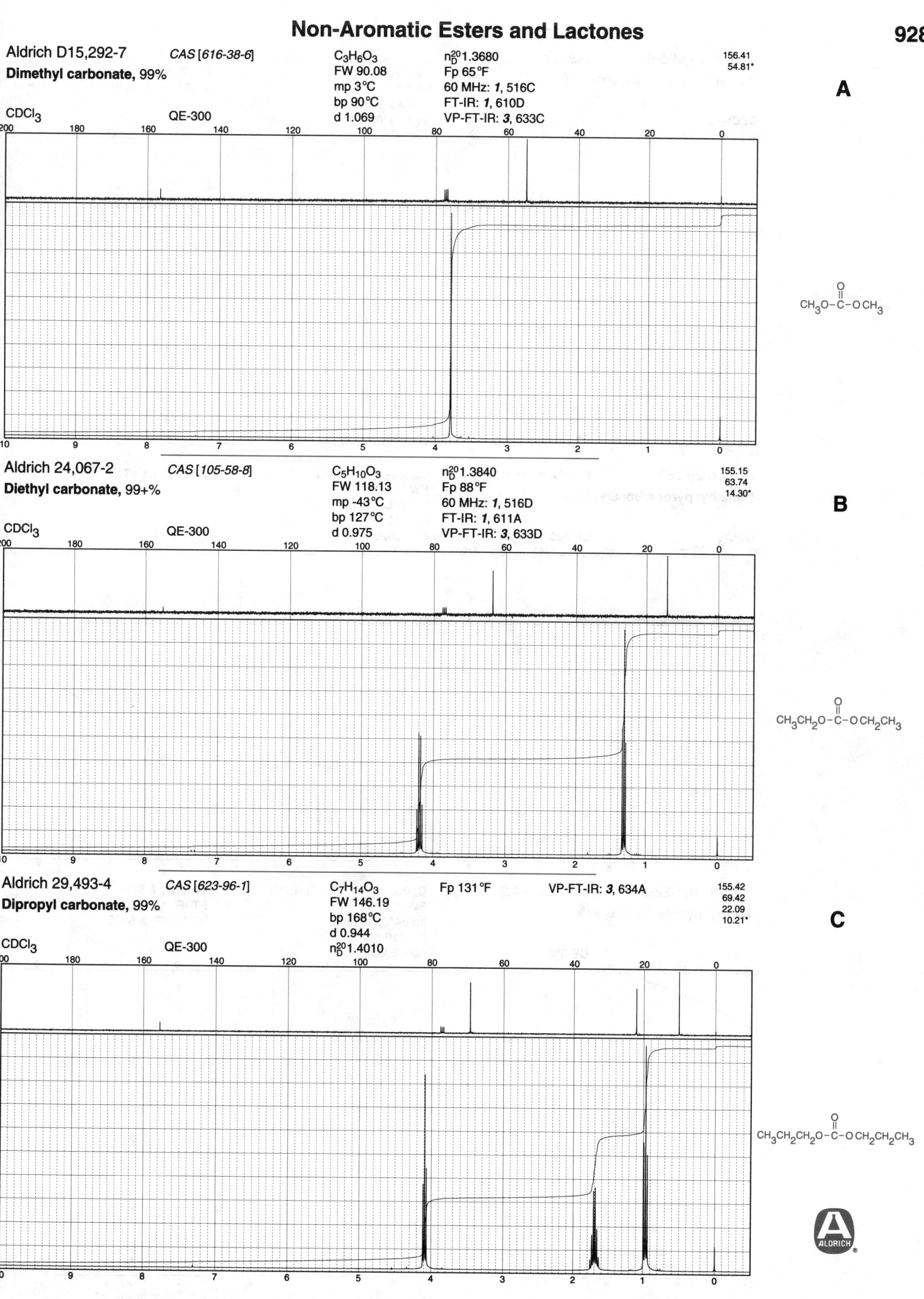

Aldrich D15,292-7   *CAS [616-38-6]*
Dimethyl carbonate, 99%

C$_3$H$_6$O$_3$
FW 90.08
mp 3°C
bp 90°C
d 1.069

n$_D^{20}$1.3680
Fp 65°F
60 MHz: *1*, 516C
FT-IR: *1*, 610D
VP-FT-IR: *3*, 633C

156.41
54.81*

A

CDCl$_3$ QE-300

Aldrich 24,067-2   *CAS [105-58-8]*
Diethyl carbonate, 99+%

C$_5$H$_{10}$O$_3$
FW 118.13
mp -43°C
bp 127°C
d 0.975

n$_D^{20}$1.3840
Fp 88°F
60 MHz: *1*, 516D
FT-IR: *1*, 611A
VP-FT-IR: *3*, 633D

155.15
63.74
14.30*

B

CDCl$_3$ QE-300

Aldrich 29,493-4   *CAS [623-96-1]*
Dipropyl carbonate, 99%

C$_7$H$_{14}$O$_3$
FW 146.19
bp 168°C
d 0.944
n$_D^{20}$1.4010

Fp 131°F   VP-FT-IR: *3*, 634A

155.42
69.42
22.09
10.21*

C

CDCl$_3$ QE-300

A

Aldrich 15,995-6   *CAS [64057-79-0]* $C_9H_{14}O_3$ Fp 163°F   60 MHz: *1*, 542D

Bis(2-methylallyl) carbonate, 98%   FW 170.21   FT-IR: *1*, 611D

bp 201°C   VP-FT-IR: *3*, 688B

d 0.943

n_D^{20}1.4370

154.96
139.38
113.46
71.11
19.31*

CDCl$_3$ QE-300

$H_2C=CCH_2O-\overset{\overset{\displaystyle O}{\|}}{C}-OCH_2C=CH_2$
  | |
  CH_3 CH_3

B

Aldrich 29,287-7   *CAS [4525-33-1]* $C_4H_6O_5$ Fp 176°F   VP-FT-IR: *3*, 634B

Dimethyl pyrocarbonate, 95%   FW 134.09

bp 46°C (5 mm)

d 1.250

n_D^{20}1.3920

149.08
56.25*

CDCl$_3$ QE-300

$CH_3O-\overset{\overset{\displaystyle O}{\|}}{C}-O-\overset{\overset{\displaystyle O}{\|}}{C}-OCH_3$

C

Aldrich 15,922-0   *CAS [1609-47-8]* $C_6H_{10}O_5$ Fp 157°F   60 MHz: *1*, 517A

Diethyl pyrocarbonate, 97%   FW 162.14   FT-IR: *1*, 611B

bp 94°C (18 mm)   VP-FT-IR: *3*, 634C

d 1.120

n_D^{20}1.3980

148.53
65.98
13.93*

CDCl$_3$ QE-300

$CH_3CH_2O-\overset{\overset{\displaystyle O}{\|}}{C}-O-\overset{\overset{\displaystyle O}{\|}}{C}-OCH_2CH_3$

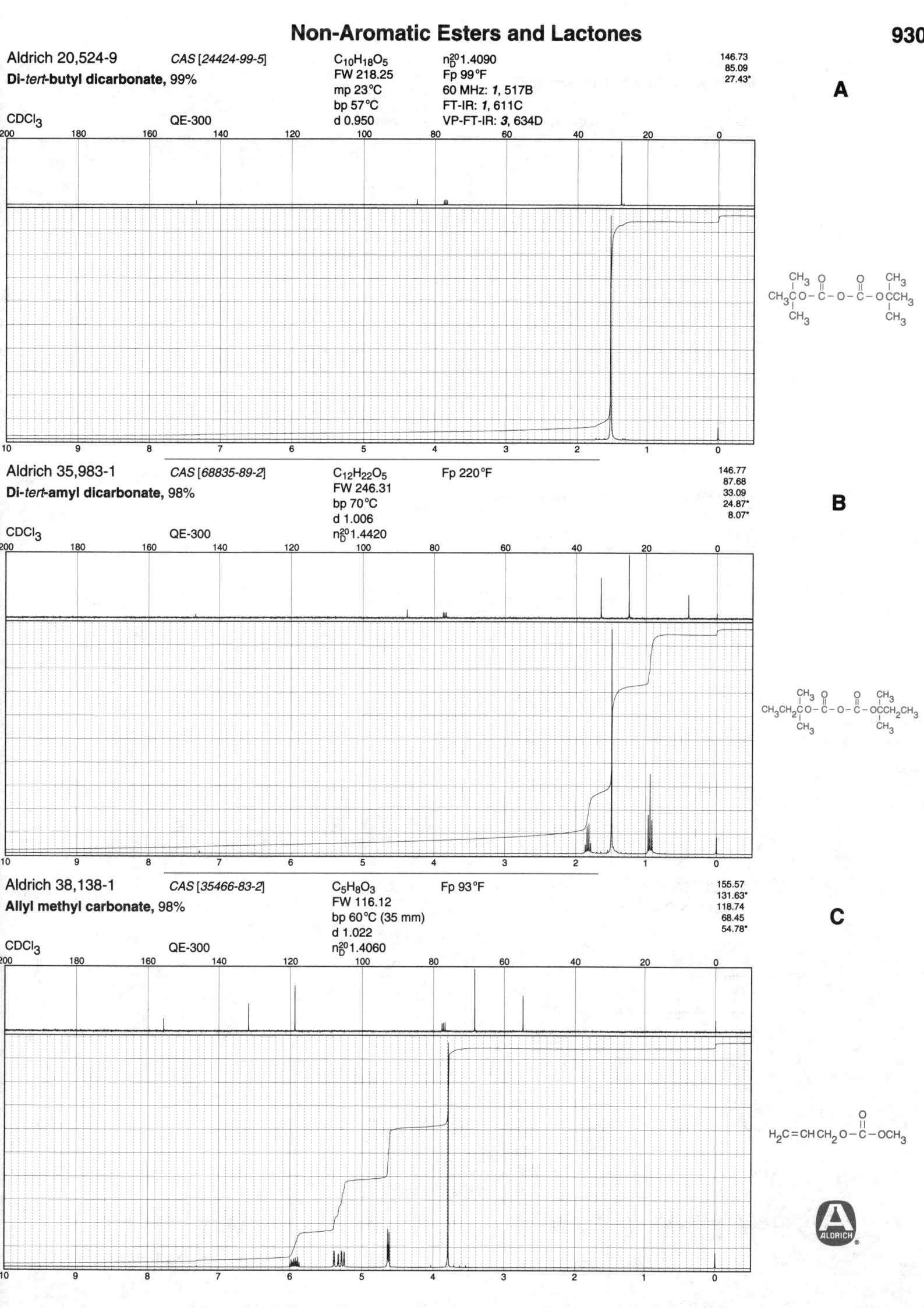

Aldrich 20,524-9   *CAS [24424-99-5]*

Di-*tert*-butyl dicarbonate, 99%

CDCl₃   QE-300

$C_{10}H_{18}O_5$
FW 218.25
mp 23°C
bp 57°C
d 0.950

n_D^{20} 1.4090
Fp 99°F
60 MHz: *1*, 517B
FT-IR: *1*, 611C
VP-FT-IR: *3*, 634D

146.73
85.09
27.43*

A

Aldrich 35,983-1   *CAS [68835-89-2]*

Di-*tert*-amyl dicarbonate, 98%

CDCl₃   QE-300

$C_{12}H_{22}O_5$
FW 246.31
bp 70°C
d 1.006
n_D^{20} 1.4420

Fp 220°F

146.77
87.68
33.09
24.87*
8.07*

B

Aldrich 38,138-1   *CAS [35466-83-2]*

Allyl methyl carbonate, 98%

CDCl₃   QE-300

$C_5H_8O_3$
FW 116.12
bp 60°C (35 mm)
d 1.022
n_D^{20} 1.4060

Fp 93°F

155.57
131.63*
118.74
68.45
54.78*

C

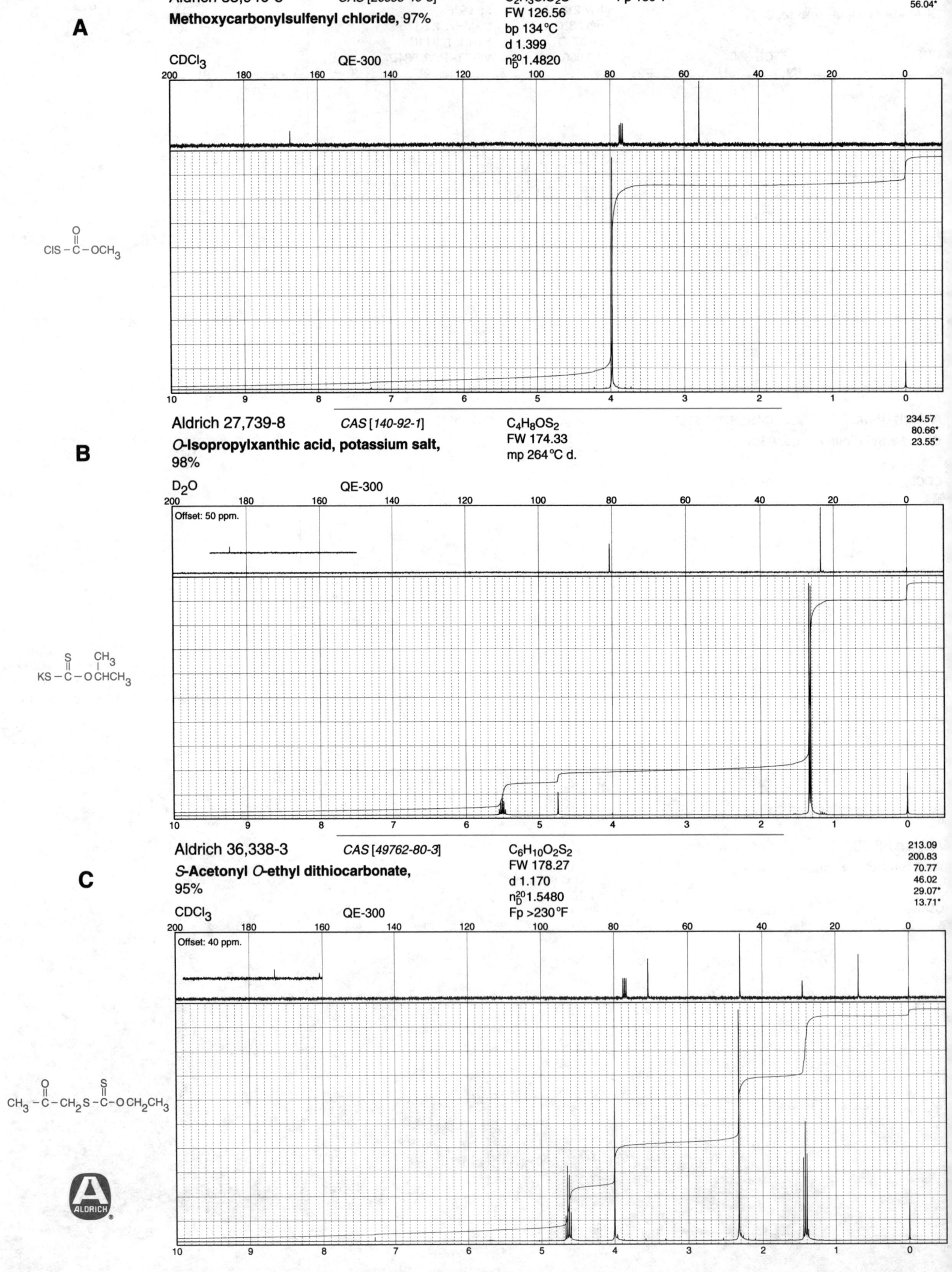

A

Aldrich 35,940-8   CAS [26555-40-8] $C_2H_3ClO_2S$ Fp 130°F

Methoxycarbonylsulfenyl chloride, 97%
FW 126.56
bp 134°C
d 1.399
n_D^{20}1.4820

167.57
56.04*

CDCl₃   QE-300

B

Aldrich 27,739-8   CAS [140-92-1] $C_4H_8OS_2$

O-Isopropylxanthic acid, potassium salt, 98%
FW 174.33
mp 264°C d.

234.57
80.66*
23.55*

D₂O   QE-300

Offset: 50 ppm.

C

Aldrich 36,338-3   CAS [49762-80-3] $C_6H_{10}O_2S_2$

S-Acetonyl O-ethyl dithiocarbonate, 95%
FW 178.27
d 1.170
n_D^{20}1.5480
Fp >230°F

213.09
200.83
70.77
46.02
29.07*
13.71*

CDCl₃   QE-300

Offset: 40 ppm.

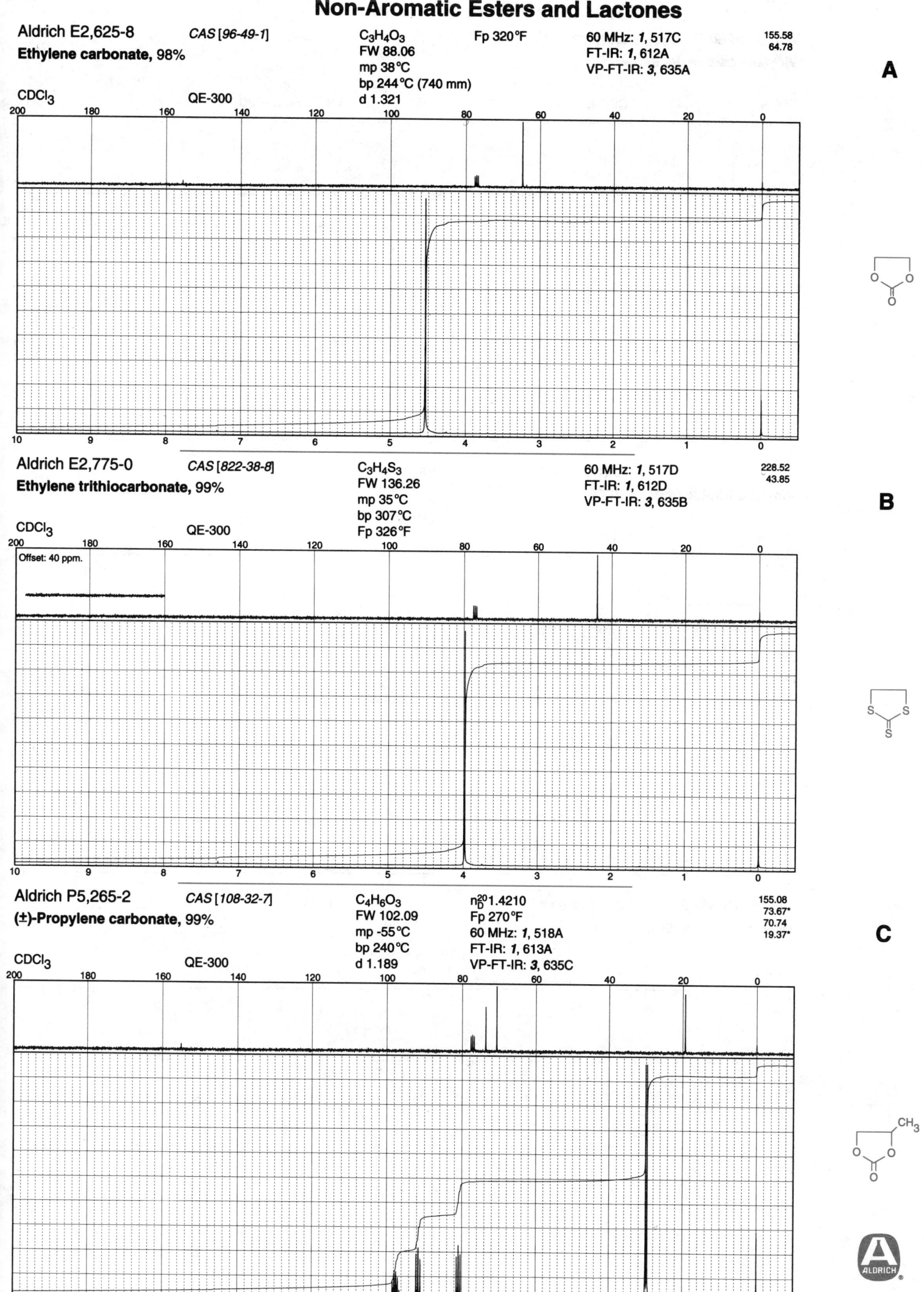

Aldrich E2,625-8   CAS [96-49-1]
Ethylene carbonate, 98%

$C_3H_4O_3$
FW 88.06
mp 38°C
bp 244°C (740 mm)
d 1.321

Fp 320°F

60 MHz: *1*, 517C
FT-IR: *1*, 612A
VP-FT-IR: *3*, 635A

155.58
64.78

A

CDCl$_3$ QE-300

Aldrich E2,775-0   CAS [822-38-8]
Ethylene trithiocarbonate, 99%

$C_3H_4S_3$
FW 136.26
mp 35°C
bp 307°C
Fp 326°F

60 MHz: *1*, 517D
FT-IR: *1*, 612D
VP-FT-IR: *3*, 635B

228.52
43.85

B

CDCl$_3$ QE-300

Offset: 40 ppm.

Aldrich P5,265-2   CAS [108-32-7]
(±)-Propylene carbonate, 99%

$C_4H_6O_3$
FW 102.09
mp -55°C
bp 240°C
d 1.189

n_D^{20}1.4210
Fp 270°F
60 MHz: *1*, 518A
FT-IR: *1*, 613A
VP-FT-IR: *3*, 635C

155.08
73.67*
70.74
19.37*

C

CDCl$_3$ QE-300

Non-Aromatic Esters and Lactones

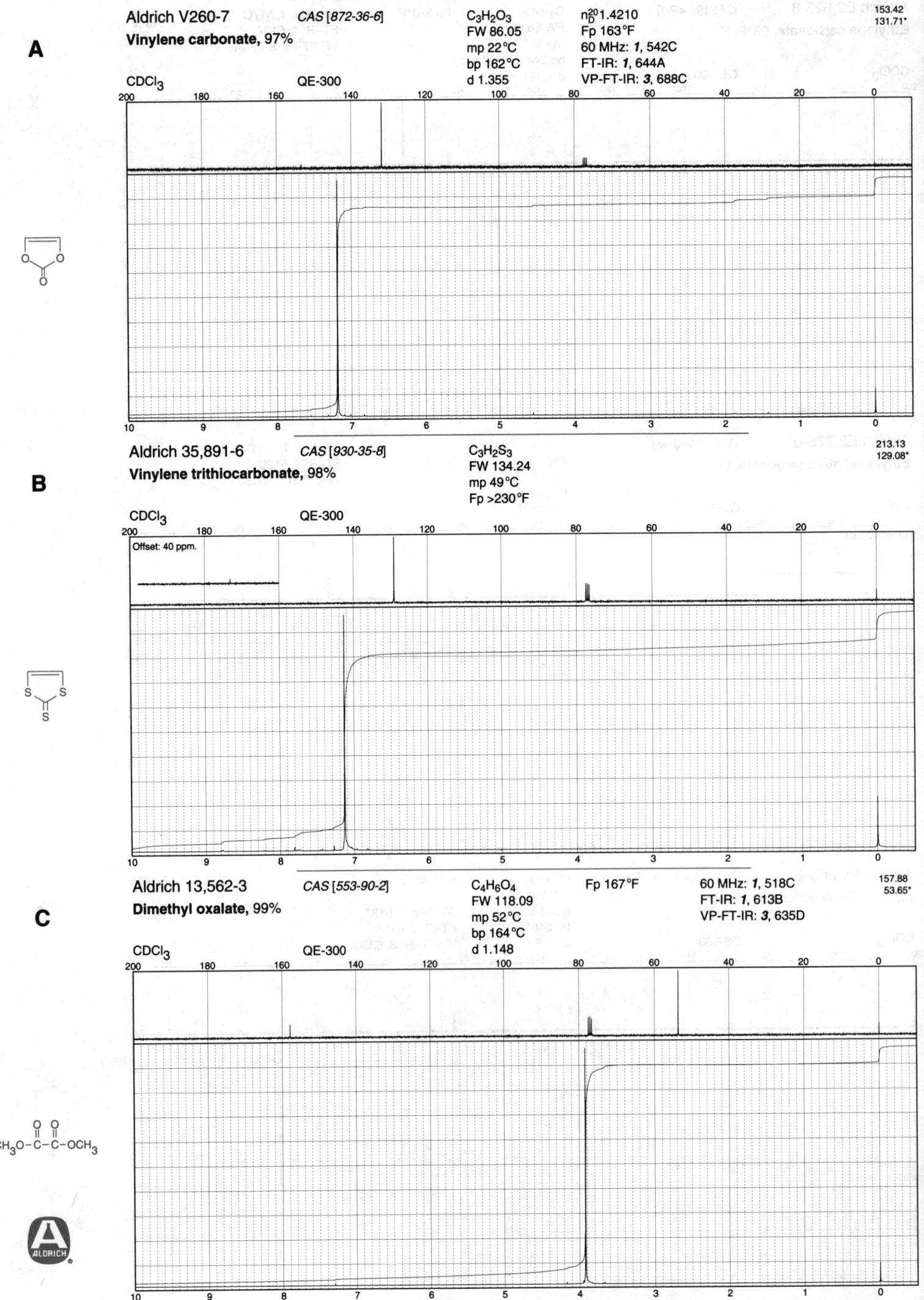

A

Aldrich V260-7 *CAS [872-36-6]* C$_3$H$_2$O$_3$ n$_D^{20}$ 1.4210 153.42 131.71*
Vinylene carbonate, 97% FW 86.05 Fp 163°F
mp 22°C 60 MHz: *1*, 542C
bp 162°C FT-IR: *1*, 644A
d 1.355 VP-FT-IR: *3*, 688C

CDCl$_3$ QE-300

B

Aldrich 35,891-6 *CAS [930-35-8]* C$_3$H$_2$S$_3$ 213.13 129.08*
Vinylene trithiocarbonate, 98% FW 134.24
mp 49°C
Fp >230°F

CDCl$_3$ QE-300

Offset: 40 ppm.

C

Aldrich 13,562-3 *CAS [553-90-2]* C$_4$H$_6$O$_4$ Fp 167°F 60 MHz: *1*, 518C 157.88 53.65*
Dimethyl oxalate, 99% FW 118.09 FT-IR: *1*, 613B
mp 52°C VP-FT-IR: *3*, 635D
bp 164°C
d 1.148

CDCl$_3$ QE-300

CH$_3$O-C-C-OCH$_3$

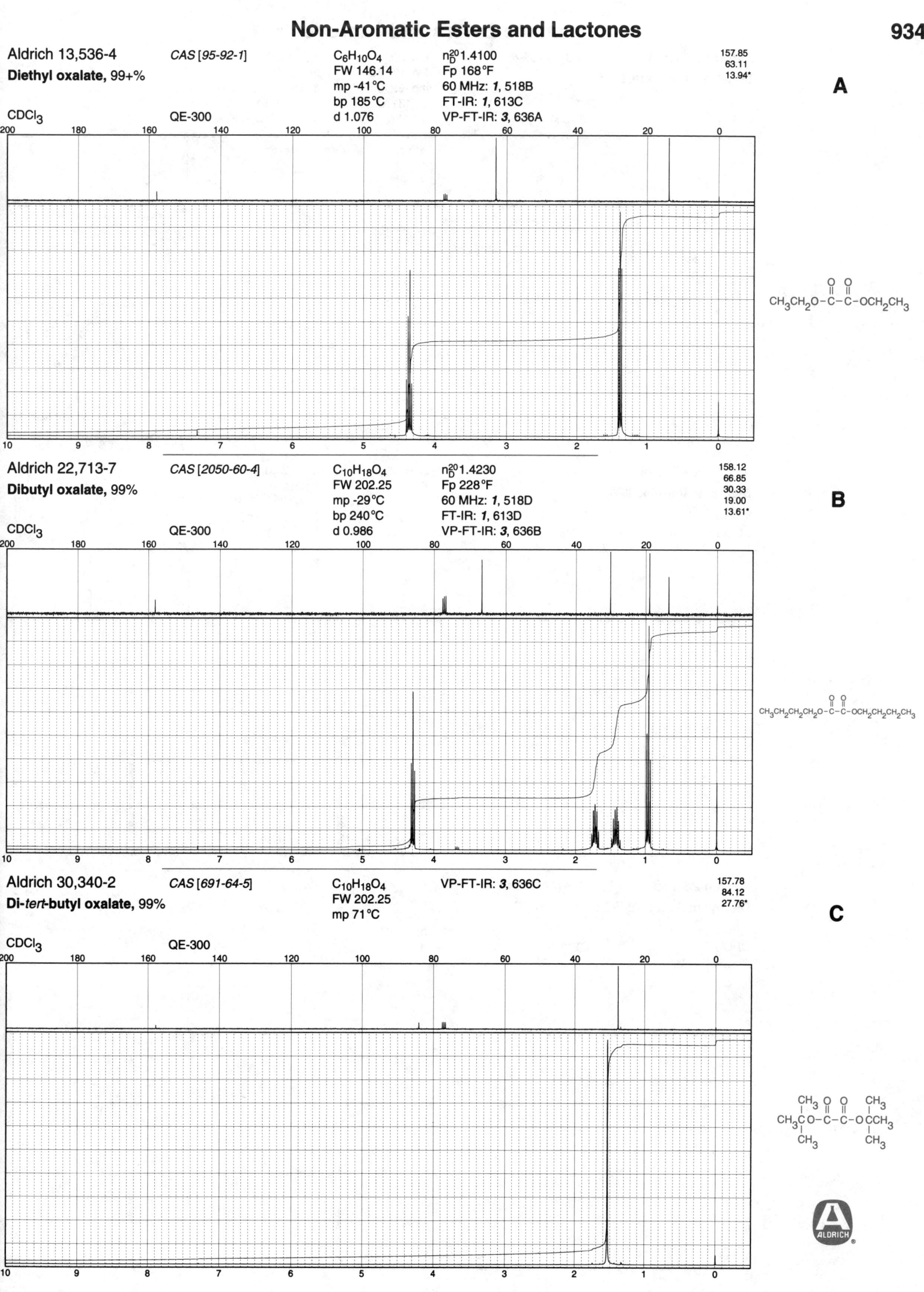

Aldrich 13,536-4   *CAS [95-92-1]* $C_6H_{10}O_4$ n_D^{20} 1.4100
Diethyl oxalate, 99+%
FW 146.14   Fp 168°F
mp -41°C   60 MHz: *1*, 518B
bp 185°C   FT-IR: *1*, 613C
d 1.076   VP-FT-IR: *3*, 636A

157.85
63.11
13.94*

A

CDCl₃   QE-300

Aldrich 22,713-7   *CAS [2050-60-4]* $C_{10}H_{18}O_4$ n_D^{20} 1.4230
Dibutyl oxalate, 99%
FW 202.25   Fp 228°F
mp -29°C   60 MHz: *1*, 518D
bp 240°C   FT-IR: *1*, 613D
d 0.986   VP-FT-IR: *3*, 636B

158.12
66.85
30.33
19.00
13.61*

B

CDCl₃   QE-300

Aldrich 30,340-2   *CAS [691-64-5]* $C_{10}H_{18}O_4$ VP-FT-IR: *3*, 636C
Di-*tert*-butyl oxalate, 99%
FW 202.25
mp 71°C

157.78
84.12
27.76*

C

CDCl₃   QE-300

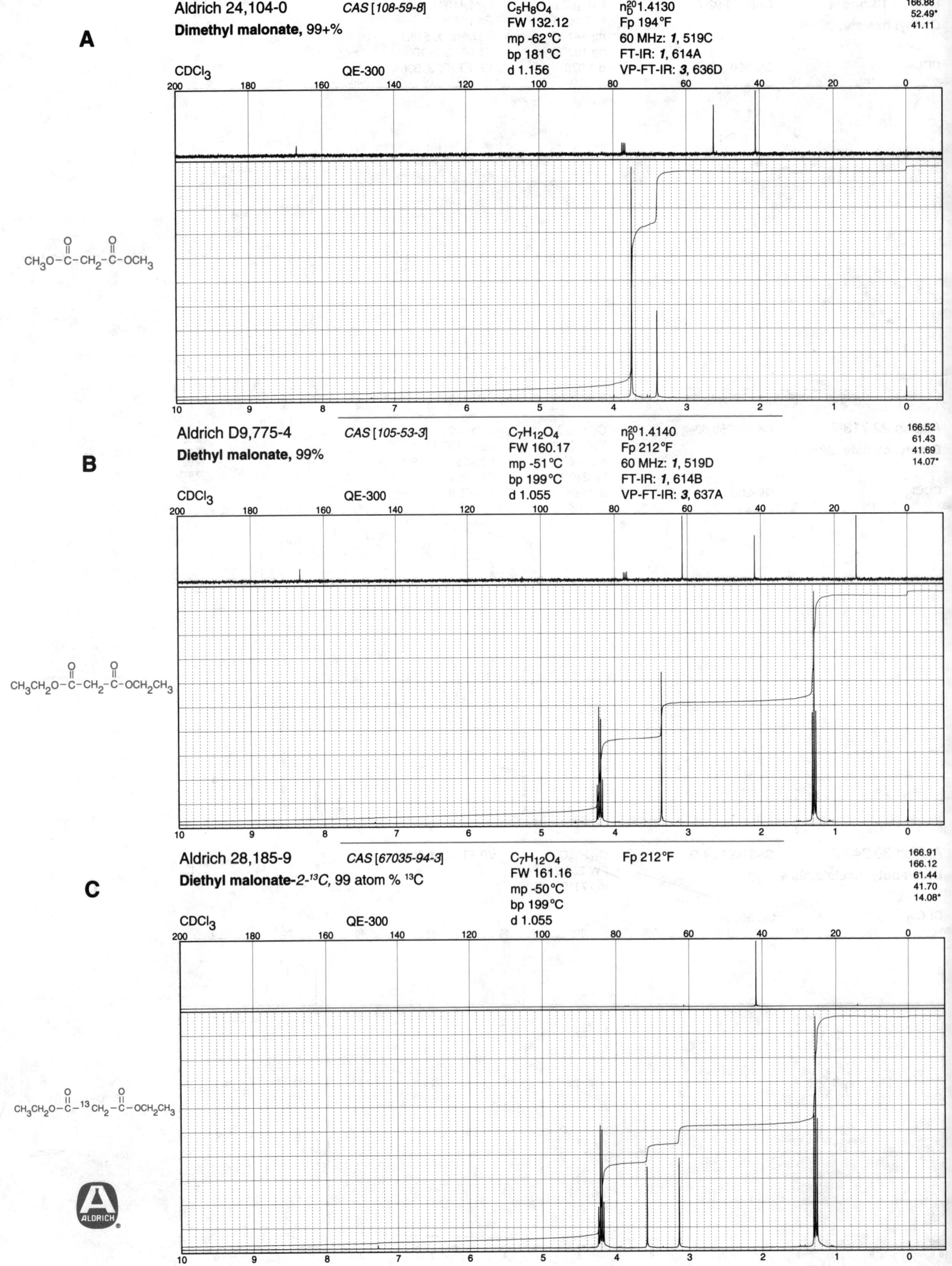

A

Aldrich 24,104-0   *CAS [108-59-8]*

Dimethyl malonate, 99+%

$C_5H_8O_4$
FW 132.12
mp -62°C
bp 181°C
d 1.156

n_D^{20} 1.4130
Fp 194°F
60 MHz: *1*, 519C
FT-IR: *1*, 614A
VP-FT-IR: *3*, 636D

166.88
52.49*
41.11

CDCl$_3$ QE-300

B

Aldrich D9,775-4   *CAS [105-53-3]*

Diethyl malonate, 99%

$C_7H_{12}O_4$
FW 160.17
mp -51°C
bp 199°C
d 1.055

n_D^{20} 1.4140
Fp 212°F
60 MHz: *1*, 519D
FT-IR: *1*, 614B
VP-FT-IR: *3*, 637A

166.52
61.43
41.69
14.07*

CDCl$_3$ QE-300

C

Aldrich 28,185-9   *CAS [67035-94-3]*

Diethyl malonate-2-^{13}C, 99 atom % ^{13}C

$C_7H_{12}O_4$
FW 161.16
mp -50°C
bp 199°C
d 1.055

Fp 212°F

166.91
166.12
61.44
41.70
14.08*

CDCl$_3$ QE-300

ALDRICH

Aldrich 36,001-5   CAS [42726-73-8]   C₈H₁₄O₄   Fp 175°F

$C_8H_{14}O_4$

tert-Butyl methyl malonate, 95%

FW 174.20
bp 80°C (11 mm)
d 1.030
n_D^{20} 1.4160

CDCl₃   QE-300

167.31
165.60
82.00
52.21*
42.69
27.97*

A

$$CH_3C O - C - CH_2 - C - OCH_3$$
(structure: tert-butyl methyl malonate)

Aldrich 30,664-9   CAS [32864-38-3]   C₉H₁₆O₄   Fp 189°F   VP-FT-IR: 3, 637B

$C_9H_{16}O_4$

tert-Butyl ethyl malonate, 95%

FW 188.22
bp 84°C (8 mm)
d 0.994
n_D^{20} 1.4160

CDCl₃   QE-300

166.90
165.73
81.89
61.23
42.97
27.92*
14.09*

B

$$CH_3CO - C - CH_2 - C - OCH_2CH_3$$
(structure: tert-butyl ethyl malonate)

Aldrich 25,448-7   CAS [541-16-2]   C₁₁H₂₀O₄ n_D^{20} 1.4190

$C_{11}H_{20}O_4$

Di-tert-butyl malonate, 99%

FW 216.28
mp -7°C
bp 111°C (22 mm)
d 0.966

Fp 192°F
FT-IR: 1, 614C
VP-FT-IR: 3, 637C

CDCl₃   QE-300

166.04
81.52
44.36
27.96*

C

$$CH_3CO - C - CH_2 - C - OCCH_3$$
(structure: di-tert-butyl malonate)

ALDRICH

Non-Aromatic Esters and Lactones

937

A

Aldrich 34,028-6   *CAS [609-02-9]* $C_6H_{10}O_4$ Fp 169°F

Dimethyl methylmalonate, 99%

FW 146.14
bp 177°C
d 1.098
n_D^{20} 1.4140

170.46
52.51*
45.82*
13.66*

CDCl$_3$ QE-300

Structure: $CH_3O-\overset{O}{\overset{\|}{C}}-\overset{}{\underset{CH_3}{CH}}-\overset{O}{\overset{\|}{C}}-OCH_3$

B

Aldrich 12,613-6   *CAS [609-08-5]* $C_8H_{14}O_4$ Fp 170°F   60 MHz: *1*, 520A

Diethyl methylmalonate, 99%

FW 174.20
bp 199°C
d 1.013
n_D^{20} 1.4130

FT-IR: *1*, 614D
VP-FT-IR: *3*, 637D

170.10
61.31
46.19*
14.07*
13.56*

CDCl$_3$ QE-300

Structure: $CH_3CH_2O-\overset{O}{\overset{\|}{C}}-\overset{}{\underset{CH_3}{CH}}-\overset{O}{\overset{\|}{C}}-OCH_2CH_3$

C

Aldrich 14,390-1   *CAS [1619-62-1]* $C_9H_{16}O_4$ Fp 160°F   FT-IR: *1*, 616A

Diethyl dimethylmalonate, 97%

FW 188.22
bp 192°C
d 0.991
n_D^{20} 1.4120

VP-FT-IR: *3*, 638A

172.83
61.16
49.83
22.75*
14.03*

CDCl$_3$ QE-300

Structure: $CH_3CH_2O-\overset{O}{\overset{\|}{C}}-\overset{CH_3}{\underset{CH_3}{C}}-\overset{O}{\overset{\|}{C}}-OCH_2CH_3$

ALDRICH

Aldrich 25,050-3 *CAS [27132-23-6]*
Dimethyl diethylmalonate, 99%

$C_9H_{16}O_4$
FW 188.23
bp 97°C (17 mm)
d 1.040
n_D^{20}1.4270

Fp 176°F

60 MHz: *1*, 520B
VP-FT-IR: *3*, 638B

172.19
58.55
52.14*
24.98
8.46*

A

CDCl$_3$ QE-300

Aldrich D9,520-4 *CAS [133-13-1]*
Diethyl ethylmalonate, 99%

$C_9H_{16}O_4$
FW 188.22
bp 76°C (5 mm)
d 1.004
n_D^{20}1.4160

Fp 191°F

FT-IR: *1*, 615A
VP-FT-IR: *3*, 638C

169.38
61.18
53.61*
22.25
14.11*
11.84*

B

CDCl$_3$ QE-300

Aldrich 15,681-7 *CAS [77-25-8]*
Diethyl diethylmalonate, 98%

$C_{11}H_{20}O_4$
FW 216.28
bp 229°C
d 0.990
n_D^{20}1.4230

Fp 202°F

60 MHz: *1*, 520C
FT-IR: *1*, 615B
VP-FT-IR: *3*, 638D

171.76
60.87
58.36
24.66
14.14*
8.35*

C

CDCl$_3$ QE-300

Non-Aromatic Esters and Lactones

A

Aldrich 22,881-8   CAS [2163-48-6]

Diethyl propylmalonate, 99%

$C_{10}H_{18}O_4$
FW 202.25
bp 222 °C
d 0.987
n_D^{20} 1.4180

Fp 197 °F

60 MHz: **1**, 520D
FT-IR: **1**, 615C
VP-FT-IR: **3**, 639A

169.51
61.20
51.87*
30.81
20.62
14.10*
13.69*

CDCl$_3$ QE-300

$CH_3CH_2O-\overset{\overset{O}{\|}}{C}-\overset{\overset{|}{CH}}{\underset{\underset{CH_2CH_2CH_3}{|}}{}}-\overset{\overset{O}{\|}}{C}-OCH_2CH_3$

B

Aldrich 11,203-8   CAS [133-08-4]

Diethyl butylmalonate, 99%

$C_{11}H_{20}O_4$
FW 216.28
bp 238 °C
d 0.983
n_D^{20} 1.4220

Fp 201 °F

60 MHz: **1**, 521A
FT-IR: **1**, 615D
VP-FT-IR: **3**, 639C

169.54   28.48
61.22   22.36
52.06*   14.10*
29.50   13.82*

CDCl$_3$ QE-300

$CH_3CH_2O-\overset{\overset{O}{\|}}{C}-\overset{\overset{|}{CH}}{\underset{\underset{CH_2CH_2CH_2CH_3}{|}}{}}-\overset{\overset{O}{\|}}{C}-OCH_2CH_3$

C

Aldrich 27,804-1   CAS [759-24-0]

Diethyl tert-butylmalonate, 96%

$C_{11}H_{20}O_4$
FW 216.28
bp 103 °C (11 mm)
d 1.014
n_D^{20} 1.4250

Fp 195 °F

VP-FT-IR: **3**, 639D

168.35
61.32*
60.78
33.60
28.05*
14.13*

CDCl$_3$ QE-300

$CH_3CH_2O-\overset{\overset{O}{\|}}{C}-\overset{\overset{|}{CH}}{\underset{\underset{CH_3-\overset{\overset{CH_3}{|}}{C}-CH_3}{|}}{}}-\overset{\overset{O}{\|}}{C}-OCH_2CH_3$

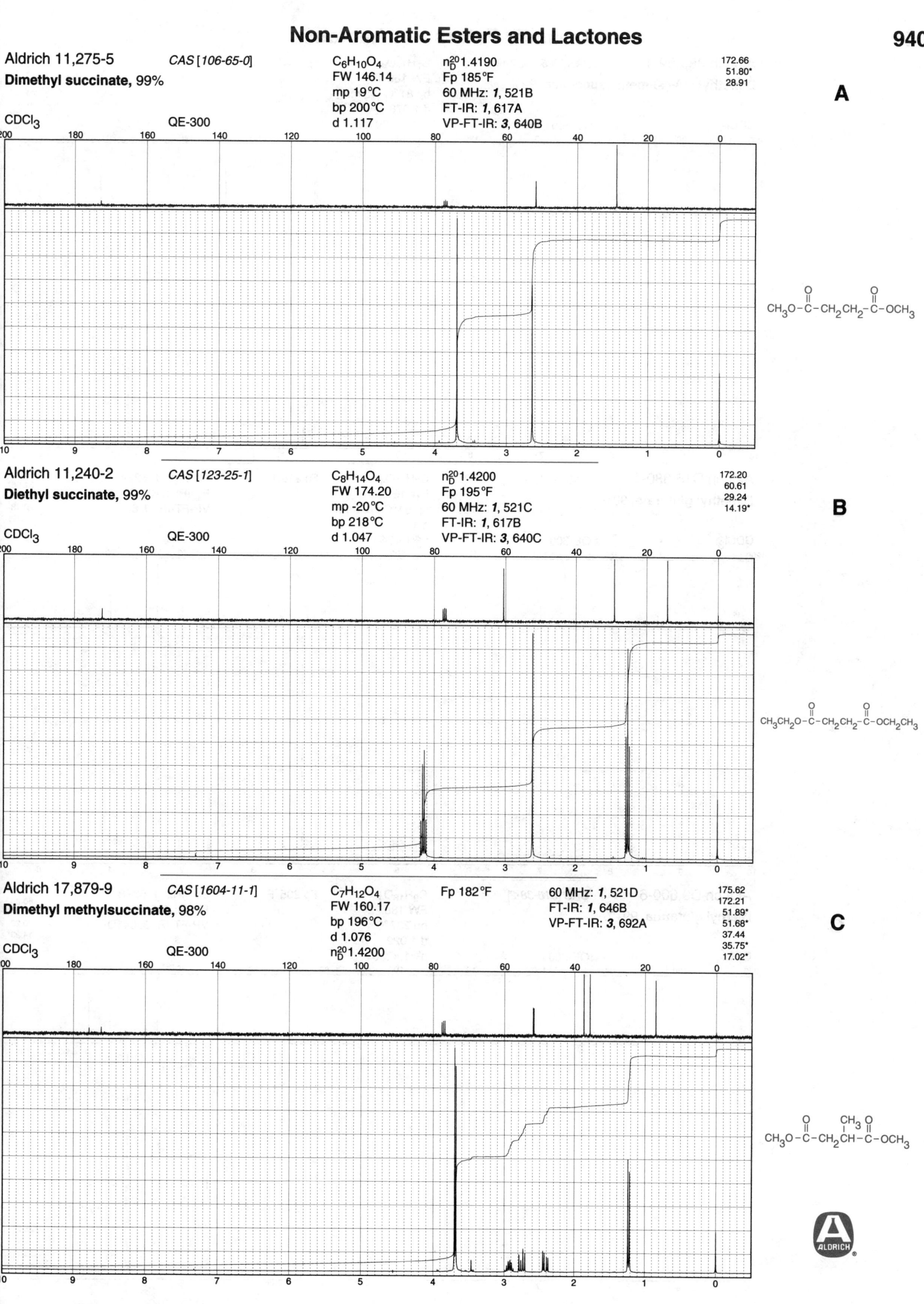

Aldrich 11,275-5   *CAS [106-65-0]* $C_6H_{10}O_4$ $n_D^{20} 1.4190$ 172.66
Dimethyl succinate, 99%   FW 146.14   Fp 185°F   51.80*
mp 19°C   60 MHz: *1*, 521B   28.91
bp 200°C   FT-IR: *1*, 617A
CDCl$_3$ QE-300   d 1.117   VP-FT-IR: *3*, 640B

A

$CH_3O-\overset{O}{\overset{\|}{C}}-CH_2CH_2-\overset{O}{\overset{\|}{C}}-OCH_3$

Aldrich 11,240-2   *CAS [123-25-1]* $C_8H_{14}O_4$ $n_D^{20} 1.4200$ 172.20
Diethyl succinate, 99%   FW 174.20   Fp 195°F   60.61
mp -20°C   60 MHz: *1*, 521C   29.24
bp 218°C   FT-IR: *1*, 617B   14.19*
CDCl$_3$ QE-300   d 1.047   VP-FT-IR: *3*, 640C

B

$CH_3CH_2O-\overset{O}{\overset{\|}{C}}-CH_2CH_2-\overset{O}{\overset{\|}{C}}-OCH_2CH_3$

Aldrich 17,879-9   *CAS [1604-11-1]* $C_7H_{12}O_4$ Fp 182°F   60 MHz: *1*, 521D   175.62
Dimethyl methylsuccinate, 98%   FW 160.17   FT-IR: *1*, 646B   172.21
bp 196°C   VP-FT-IR: *3*, 692A   51.89*
d 1.076   51.68*
CDCl$_3$ QE-300 $n_D^{20} 1.4200$ 37.44
35.75*
17.02*

C

$CH_3O-\overset{O}{\overset{\|}{C}}-CH_2\overset{CH_3}{\overset{|}{C}H}-\overset{O}{\overset{\|}{C}}-OCH_3$

Non-Aromatic Esters and Lactones

A

Aldrich 38,209-4   *CAS [22644-27-5]* $C_7H_{12}O_4$ Fp 184°F

Dimethyl (*R*)-(+)-methylsuccinate, 99% FW 160.17

bp 81°C (12 mm)

d 1.076

n_D^{20} 1.4180

CDCl$_3$ QE-300

175.62
172.20
51.90*
51.69*
37.41
35.72*
17.02*

B

Aldrich D15,880-1   *CAS [1119-40-0]* $C_7H_{12}O_4$ Fp 218°F

Dimethyl glutarate, 98% FW 160.17

bp 94°C (13 mm)

d 1.087

n_D^{20} 1.4240

CDCl$_3$ QE-300

60 MHz: *1*, 522A
FT-IR: *1*, 617C
VP-FT-IR: *3*, 641C

173.28
51.56*
33.04
20.13

C

Aldrich D9,600-6   *CAS [818-38-2]* $C_9H_{16}O_4$ Fp 205°F

Diethyl glutarate, 99+% FW 188.22

bp 237°C

d 1.022

n_D^{20} 1.4230

CDCl$_3$ QE-300

60 MHz: *1*, 522B
FT-IR: *1*, 617D
VP-FT-IR: *3*, 641D

172.85
60.33
33.36
20.22
14.23*

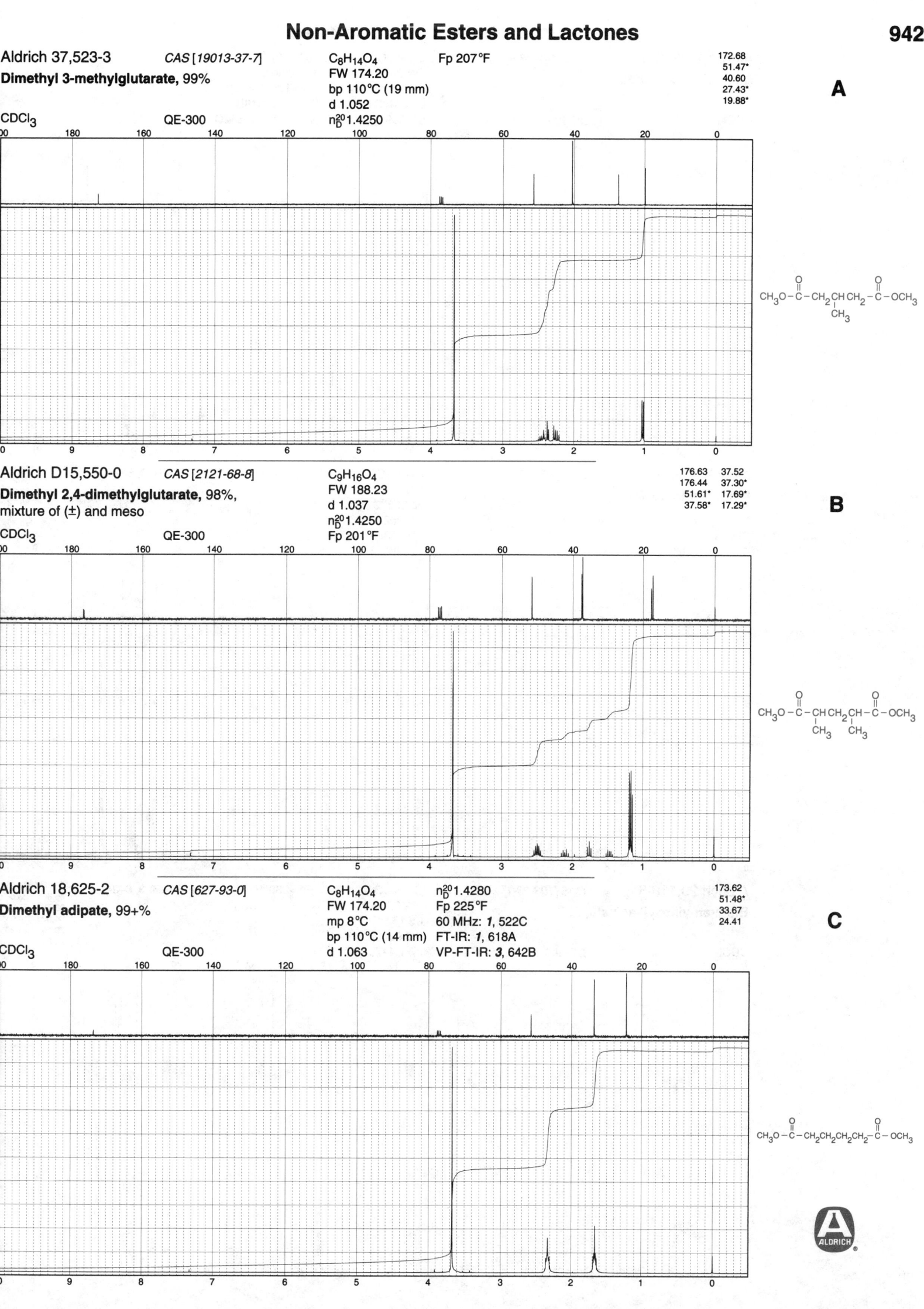

Aldrich 37,523-3   CAS [19013-37-7]   C₈H₁₄O₄   Fp 207°F

$C_8H_{14}O_4$

Dimethyl 3-methylglutarate, 99%

FW 174.20
bp 110°C (19 mm)
d 1.052
n_D^{20} 1.4250

CDCl₃   QE-300

172.68
51.47*
40.60
27.43*
19.88*

A

$CH_3O-\overset{O}{\overset{\|}{C}}-CH_2\overset{}{\underset{CH_3}{C}}HCH_2-\overset{O}{\overset{\|}{C}}-OCH_3$

Aldrich D15,550-0   CAS [2121-68-8]   C₉H₁₆O₄

$C_9H_{16}O_4$

Dimethyl 2,4-dimethylglutarate, 98%,
mixture of (±) and meso

FW 188.23
d 1.037
n_D^{20} 1.4250
Fp 201°F

CDCl₃   QE-300

176.63   37.52
176.44   37.30*
51.61*   17.69*
37.58*   17.29*

B

$CH_3O-\overset{O}{\overset{\|}{C}}-\underset{CH_3}{C}HCH_2-\underset{CH_3}{C}H-\overset{O}{\overset{\|}{C}}-OCH_3$

Aldrich 18,625-2   CAS [627-93-0]   C₈H₁₄O₄ n_D^{20} 1.4280

Dimethyl adipate, 99+%

FW 174.20   Fp 225°F
mp 8°C   60 MHz: 1, 522C
bp 110°C (14 mm)   FT-IR: 1, 618A
d 1.063   VP-FT-IR: 3, 642B

CDCl₃   QE-300

173.62
51.48*
33.67
24.41

C

$CH_3O-\overset{O}{\overset{\|}{C}}-CH_2CH_2CH_2CH_2-\overset{O}{\overset{\|}{C}}-OCH_3$

ALDRICH

A

Aldrich 24,572-0 *CAS [141-28-6]* $C_{10}H_{18}O_4$ $n_D^{20}1.4270$
Diethyl adipate, 99% FW 202.25 Fp >230°F
mp -18°C 60 MHz: *1*, 522D
bp 251°C FT-IR: *1*, 618B
d 1.009 VP-FT-IR: *3*, 642C

173.23
60.25
33.97
24.44
14.25*

CDCl$_3$ QE-300

$CH_3CH_2O-\overset{O}{\overset{\|}{C}}-CH_2CH_2CH_2CH_2-\overset{O}{\overset{\|}{C}}-OCH_2CH_3$

B

Aldrich 30,949-4 *CAS [105-99-7]* $C_{14}H_{26}O_4$ Fp >230°F VP-FT-IR: *3*, 642D
Dibutyl adipate, 96% FW 258.36
bp 305°C
d 0.962
$n_D^{20}1.4360$

173.34
64.21
33.97
30.72
24.48
19.17
13.70*

CDCl$_3$ QE-300

$CH_3CH_2CH_2CH_2O-\overset{O}{\overset{\|}{C}}-CH_2CH_2CH_2CH_2-\overset{O}{\overset{\|}{C}}-OCH_2CH_2CH_2CH_3$

C

Aldrich 29,118-8 *CAS [103-23-1]* $C_{22}H_{42}O_4$ Fp >230°F VP-FT-IR: *3*, 643A
Bis(2-ethylhexyl) adipate, 99% FW 370.58
bp 167°C (1 mm)
d 0.990
$n_D^{20}1.4470$

173.48 24.51
66.80 23.84
38.79* 22.99
34.02 14.04*
30.45 11.00*
28.96

CDCl$_3$ QE-300

$CH_3(CH_2)_3CHCH_2O-\overset{O}{\overset{\|}{C}}-CH_2CH_2CH_2CH_2-\overset{O}{\overset{\|}{C}}-OCH_2CH(CH_2)_3CH_3$

Aldrich 18,006-8

Dimethyl pimelate, 99%

CAS [1732-08-7]

$C_9H_{16}O_4$
FW 188.22
bp 122°C (11 mm)
d 1.041
n_D^{20} 1.4310

Fp >230°F

60 MHz: **1**, 523A
FT-IR: **1**, 618C
VP-FT-IR: **3**, 643D

173.92
51.44*
33.83
28.61
24.57

CDCl₃   QE-300

A

$CH_3O-\overset{O}{\overset{\|}{C}}-CH_2(CH_2)_3CH_2-\overset{O}{\overset{\|}{C}}-OCH_3$

Aldrich D9,970-6

Diethyl pimelate, 97%

CAS [2050-20-6]

$C_{11}H_{20}O_4$
FW 216.28
bp 193°C (100 mm)
d 0.994
n_D^{20} 1.4290

Fp >230°F

60 MHz: **1**, 523B
FT-IR: **1**, 618D
VP-FT-IR: **3**, 644A

173.58
60.22
34.15
28.62
24.63
14.26*

CDCl₃   QE-300

B

$CH_3CH_2O-\overset{O}{\overset{\|}{C}}-CH_2(CH_2)_3CH_2-\overset{O}{\overset{\|}{C}}-OCH_2CH_3$

Aldrich 14,901-2

Dimethyl suberate, 99%

CAS [1732-09-8]

$C_{10}H_{18}O_4$
FW 202.25
bp 268°C
d 1.014
n_D^{20} 1.4320

Fp >230°F

60 MHz: **1**, 523C
FT-IR: **1**, 619A
VP-FT-IR: **3**, 644B

174.05
51.44*
33.95
28.75
24.73

CDCl₃   QE-300

C

$CH_3O-\overset{O}{\overset{\|}{C}}-CH_2(CH_2)_4CH_2-\overset{O}{\overset{\|}{C}}-OCH_3$

A

Aldrich D10,060-9   CAS [2050-23-9] $C_{12}H_{22}O_4$ Fp >230°F   60 MHz: *1*, 523D

Diethyl suberate, 97%   FW 230.30   FT-IR: *1*, 619B

bp 282°C   VP-FT-IR: *3*, 644C

d 0.982

n_D^{20} 1.4320

173.61
60.15
34.26
28.76
24.78
14.26*

CDCl₃   QE-300

$CH_3CH_2O-\overset{O}{\underset{\|}{C}}-CH_2(CH_2)_4CH_2-\overset{O}{\underset{\|}{C}}-OCH_2CH_3$

B

Aldrich 28,063-1   CAS [16090-77-0] $C_{16}H_{30}O_4$ Fp >230°F   VP-FT-IR: *3*, 644D

Dibutyl suberate, 99%   FW 286.41

bp 176°C (4 mm)

d 0.948

n_D^{20} 1.4390

173.77   28.77
64.12   24.81
34.26   19.15
30.69   13.72*

CDCl₃   QE-300

$\left(-CH_2\,CH_2\,CH_2-\overset{O}{\underset{\|}{C}}-O\,CH_2\,CH_2\,CH_2\,CH_3\right)_2$

C

Aldrich 22,311-5   CAS [106-79-6] $C_{12}H_{22}O_4$ Fp 293°F   60 MHz: *1*, 524A

Dimethyl sebacate, 99%   FW 230.31   FT-IR: *1*, 619C

mp 30°C   VP-FT-IR: *3*, 645A

bp 158°C (10 mm)

d 0.988

174.17
51.42
34.05
29.06
24.91

CDCl₃   QE-300

$CH_3O-\overset{O}{\underset{\|}{C}}-CH_2(CH_2)_6CH_2-\overset{O}{\underset{\|}{C}}-OCH_3$

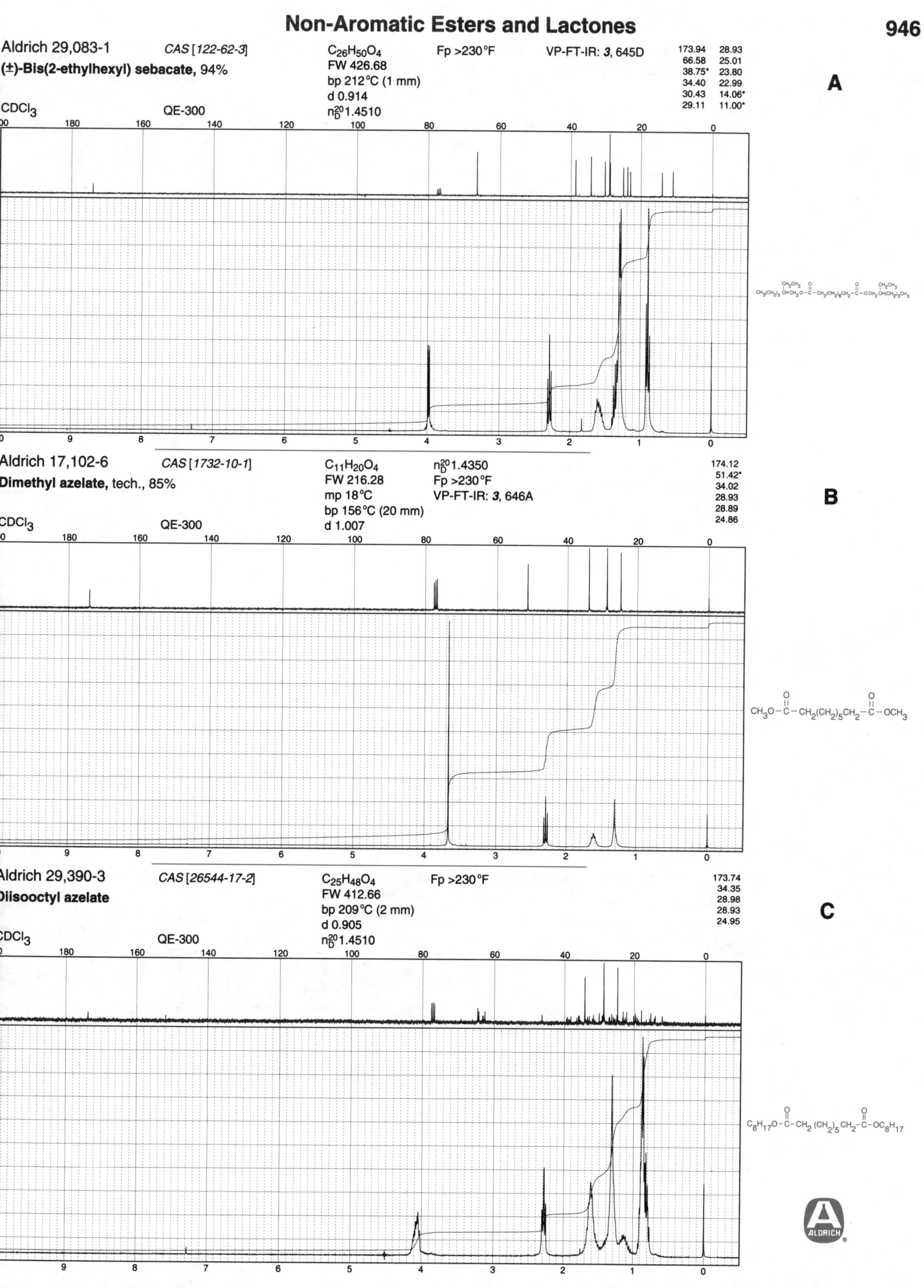

Aldrich 29,083-1   CAS [122-62-3] $C_{26}H_{50}O_4$ Fp >230°F   VP-FT-IR: **3**, 645D

(±)-Bis(2-ethylhexyl) sebacate, 94%

FW 426.68
bp 212°C (1 mm)
d 0.914
n_D^{20} 1.4510

CDCl₃   QE-300

A

173.94	28.93
66.58	25.01
38.75*	23.80
34.40	22.99
30.43	14.06*
29.11	11.00*

Aldrich 17,102-6   CAS [1732-10-1] $C_{11}H_{20}O_4$ n_D^{20} 1.4350

Dimethyl azelate, tech., 85%

FW 216.28   Fp >230°F
mp 18°C   VP-FT-IR: **3**, 646A
bp 156°C (20 mm)
d 1.007

CDCl₃   QE-300

B

| 174.12 |
| 51.42* |
| 34.02 |
| 28.93 |
| 28.89 |
| 24.86 |

Aldrich 29,390-3   CAS [26544-17-2] $C_{25}H_{48}O_4$ Fp >230°F

Diisooctyl azelate

FW 412.66
bp 209°C (2 mm)
d 0.905
n_D^{20} 1.4510

CDCl₃   QE-300

C

| 173.74 |
| 34.35 |
| 28.98 |
| 28.93 |
| 24.95 |

A

Aldrich 17,190-5   *CAS [1472-87-3]* $C_{15}H_{28}O_4$
Dimethyl brassylate, 99%   FW 272.39
mp 36°C
bp 327°C
Fp >230°F

60 MHz: *1*, 524C
FT-IR: *1*, 620A
VP-FT-IR: *3*, 647A

174.26	29.42
51.40*	29.26
34.12	29.17
29.51	24.98

CDCl$_3$ QE-300

$CH_3O-\overset{O}{\underset{\|}{C}}-CH_2(CH_2)_9CH_2-\overset{O}{\underset{\|}{C}}-OCH_3$

B

Aldrich 14,404-5   *CAS [19812-63-6]* $C_{18}H_{34}O_4$
Diethyl tetradecanedioate, 99%   FW 314.47
mp 30°C
Fp 223°F

60 MHz: *1*, 524D
FT-IR: *1*, 620B

173.81	29.26
60.11	29.14
34.38	24.99
29.55	14.26*
29.44	

CDCl$_3$ QE-300

$CH_3CH_2O-\overset{O}{\underset{\|}{C}}-CH_2(CH_2)_{10}CH_2-\overset{O}{\underset{\|}{C}}-OCH_2CH_3$

C

Aldrich W23,741-8   *CAS [626-11-9]* $C_8H_{14}O_5$
Diethyl malate, 97+%   FW 190.20
bp 123°C (12 mm)
d 1.128
Fp 185°F

173.32
170.45
67.33*
61.96
60.94
38.78
14.12*

CDCl$_3$ QE-300

$CH_3CH_2O-\overset{O}{\underset{\|}{C}}-CH_2-\overset{OH}{\underset{\|}{CH}}-\overset{O}{\underset{\|}{C}}-OCH_2CH_3$

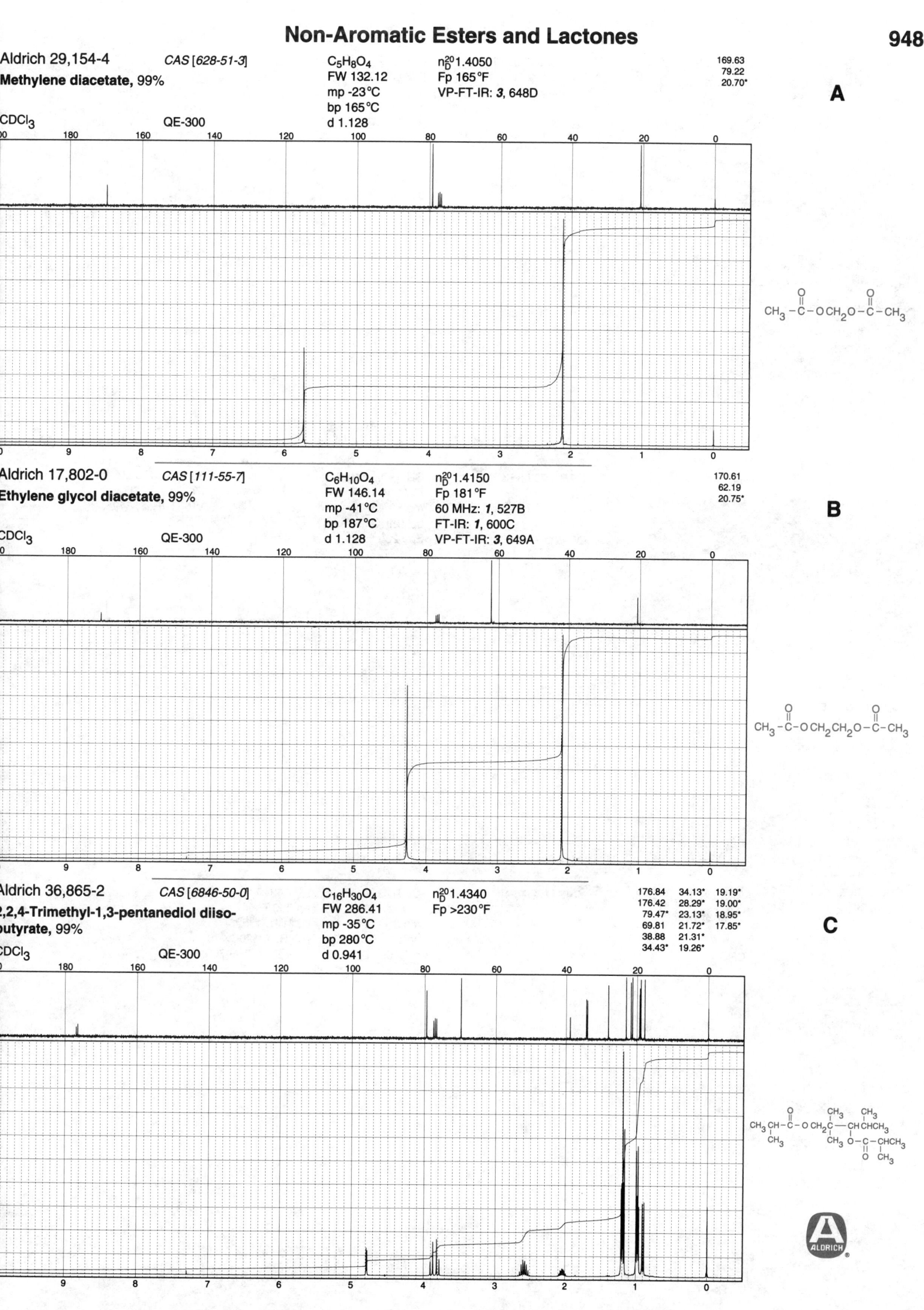

Aldrich 29,154-4 *CAS [628-51-3]*
Methylene diacetate, 99%

$C_5H_8O_4$
FW 132.12
mp -23°C
bp 165°C
d 1.128

n_D^{20} 1.4050
Fp 165°F
VP-FT-IR: *3*, 648D

169.63
79.22
20.70*

CDCl₃   QE-300   **A**

Aldrich 17,802-0 *CAS [111-55-7]*
Ethylene glycol diacetate, 99%

$C_6H_{10}O_4$
FW 146.14
mp -41°C
bp 187°C
d 1.128

n_D^{20} 1.4150
Fp 181°F
60 MHz: *1*, 527B
FT-IR: *1*, 600C
VP-FT-IR: *3*, 649A

170.61
62.19
20.75*

CDCl₃   QE-300   **B**

Aldrich 36,865-2 *CAS [6846-50-0]*
2,2,4-Trimethyl-1,3-pentanediol diiso-butyrate, 99%

$C_{16}H_{30}O_4$
FW 286.41
mp -35°C
bp 280°C
d 0.941

n_D^{20} 1.4340
Fp >230°F

176.84	34.13*	19.19*
176.42	28.29*	19.00*
79.47*	23.13*	18.95*
69.81	21.72*	17.85*
38.88	21.31*	
34.43*	19.26*	

CDCl₃   QE-300   **C**

ALDRICH

A

Aldrich 23,422-2

CAS [25395-31-7]

Diacetin, tech., 50%

$C_7H_{12}O_5$
FW 176.17
d 1.170
n_D^{20} 1.4400
Fp >230°F

171.46	72.25*	63.34	20.82*
171.19	70.05*	62.55	20.78*
171.08	69.08*	62.28	20.75*
170.80	67.61*	61.42	20.68*
170.69	65.30	60.96	
170.29	65.15		21.00*
75.04	63.71		20.87*

CDCl₃   QE-300

B

Aldrich 20,948-1

CAS [2983-35-9]

1,1,2-Triacetoxyethane, 98%

$C_8H_{12}O_6$
FW 204.18
mp 50°C
bp 128°C (11 mm)
Fp >230°F

60 MHz: *1*, 527A
FT-IR: *1*, 623D
VP-FT-IR: *3*, 649B

170.19
168.52
86.64*
62.51
20.65*
20.56*

CDCl₃   QE-300

C

Aldrich T4,370-2

CAS [102-76-1]

Triacetin, 99%

$C_9H_{14}O_6$
FW 218.21
mp 3°C
bp 258°C
d 1.155

n_D^{20} 1.4310
Fp 300°F
60 MHz: *1*, 525A
FT-IR: *1*, 620C
VP-FT-IR: *3*, 647C

170.44
170.06
69.10*
62.27
20.86*
20.66*

CDCl₃   QE-300

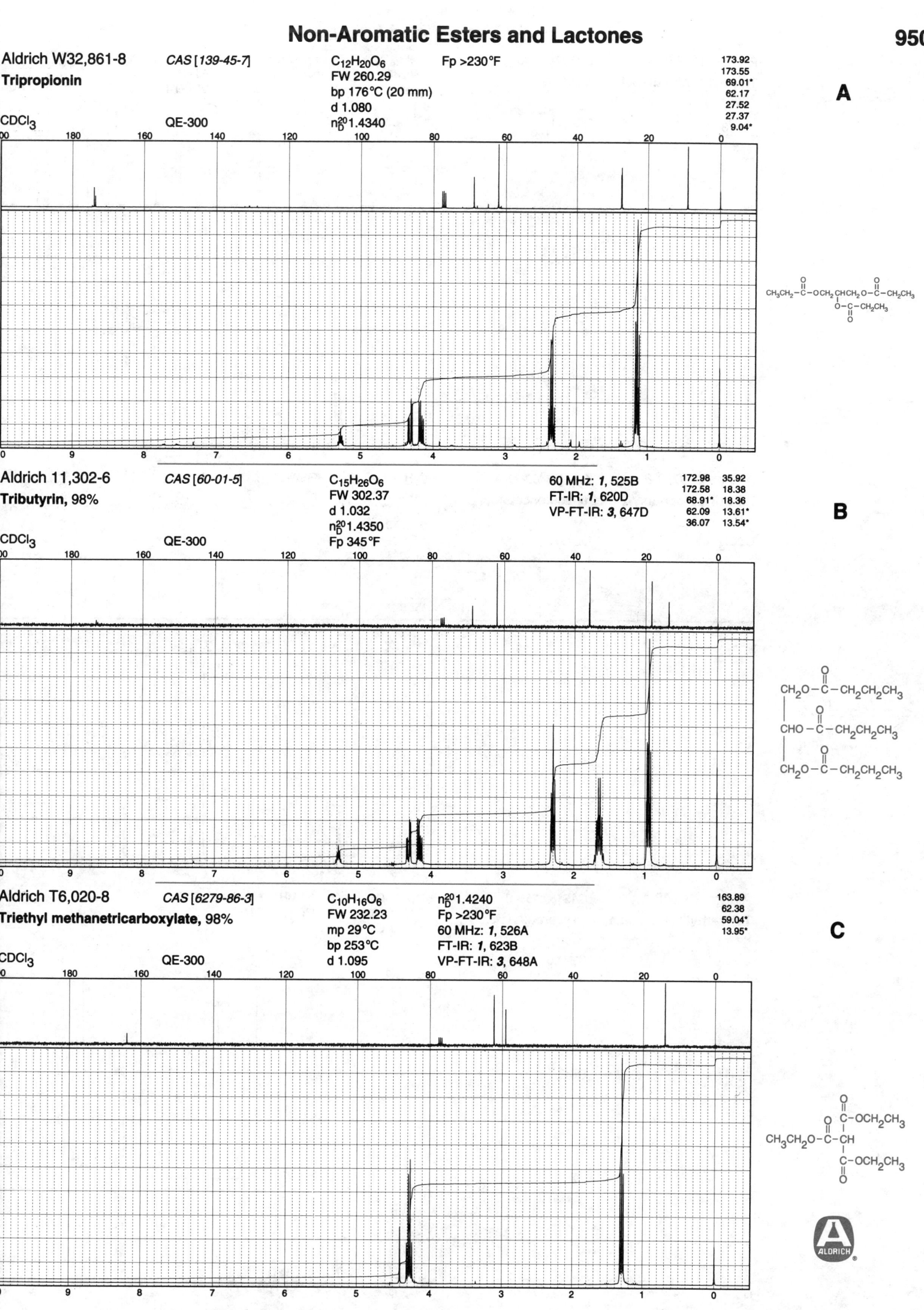

Aldrich W32,861-8   *CAS [139-45-7]*   C₁₂H₂₀O₆   Fp >230°F
Tripropionin
$C_{12}H_{20}O_6$
FW 260.29
bp 176°C (20 mm)
d 1.080
n_D^{20} 1.4340

173.92
173.55
69.01*
62.17
27.52
27.37
9.04*

A

CDCl₃   QE-300

$CH_3CH_2-C-O-CH_2CHCH_2O-C-CH_2CH_3$
$O-C-CH_2CH_3$

Aldrich 11,302-6   *CAS [60-01-5]*   C₁₅H₂₆O₆
Tributyrin, 98%
$C_{15}H_{26}O_6$
FW 302.37
d 1.032
n_D^{20} 1.4350
Fp 345°F

60 MHz: *1*, 525B
FT-IR: *1*, 620D
VP-FT-IR: *3*, 647D

172.98   35.92
172.58   18.38
68.91*   18.36
62.09   13.61*
36.07   13.54*

B

CDCl₃   QE-300

$CH_2O-C-CH_2CH_2CH_3$
$CHO-C-CH_2CH_2CH_3$
$CH_2O-C-CH_2CH_2CH_3$

Aldrich T6,020-8   *CAS [6279-86-3]*   C₁₀H₁₆O₆ n_D^{20} 1.4240
Triethyl methanetricarboxylate, 98%
$C_{10}H_{16}O_6$
FW 232.23
mp 29°C
bp 253°C
d 1.095
Fp >230°F
60 MHz: *1*, 526A
FT-IR: *1*, 623B
VP-FT-IR: *3*, 648A

163.89
62.38
59.04*
13.95*

C

CDCl₃   QE-300

$CH_3CH_2O-C-CH$
$C-OCH_2CH_3$
$C-OCH_2CH_3$

ALDRICH

Non-Aromatic Esters and Lactones

A

Aldrich T5,985-4   *CAS [7459-46-3]*

Triethyl 1,1,2-ethanetricarboxylate, 99%

C$_{11}$H$_{18}$O$_6$
FW 246.26
bp 99°C
d 1.074
n$_D^{20}$1.4290

Fp >230°F

CDCl$_3$ QE-300

60 MHz: *1*, 526B
FT-IR: *1*, 623C
VP-FT-IR: *3*, 648C

170.75	47.88*
168.36	33.18
61.75	14.13*
61.00	14.02*

B

Aldrich 22,039-6   *CAS [25409-39-6]*

(±)-Ethyl 2-acetoxy-2-methylacetoacetate, 97%

C$_9$H$_{14}$O$_5$
FW 202.21
bp 82°C (1 mm)
d 1.079
n$_D^{20}$1.4280

Fp 219°F

CDCl$_3$ QE-300

60 MHz: *1*, 526C
FT-IR: *1*, 624A
VP-FT-IR: *3*, 649C

201.13	25.76*
169.47	20.82*
167.64	19.98*
85.35	13.92*
62.21	

C

Aldrich 13,188-1   *CAS [632-56-4]*

Tetraethyl 1,1,2,2-ethanetetracarboxylate, 99%

C$_{14}$H$_{22}$O$_8$
FW 318.32
mp 76°C

60 MHz: *1*, 526D
FT-IR: *1*, 624B

167.06
62.05
51.46*
13.96*

CDCl$_3$ QE-300

Non-Aromatic Esters and Lactones

A

Aldrich 19,293-7   CAS [41498-71-9]

(±)-*exo*-2-Norbornyl formate, 97%

$C_8H_{12}O_2$	Fp 129°F
FW 140.18	
bp 66°C (16 mm)	
d 1.048	
n_D^{20} 1.4620	

FT-IR: *1*, 637D
VP-FT-IR: *3*, 672B

160.78	35.40*
77.43*	35.23
41.53*	28.10
39.53	24.25

CDCl₃   QE-300

B

Aldrich 24,651-4   CAS [622-45-7]

Cyclohexyl acetate, 99%

$C_8H_{14}O_2$	Fp 136°F
FW 142.20	
bp 173°C	
d 0.966	
n_D^{20} 1.4390	

60 MHz: *1*, 531A
FT-IR: *1*, 628B
VP-FT-IR: *3*, 655C

170.45
72.64*
31.70
25.43
23.84
21.41*

CDCl₃   QE-300

C

Aldrich 34,773-6   CAS [32210-23-4]

4-*tert*-Butylcyclohexyl acetate, 99%,
mixture of *cis* and *trans*

$C_{12}H_{22}O_2$	Fp 212°F
FW 198.31	
bp 229°C (25 mm)	
d 0.934	
n_D^{20} 1.4520	

170.58	32.09
73.65*	30.58
69.36*	27.58*
47.59*	27.45*
47.06*	25.44
32.50	21.67
32.29	21.45*

CDCl₃   QE-300

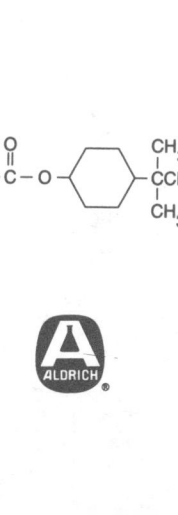

A

Aldrich 19,331-3   *CAS [21722-83-8]*

2-Cyclohexylethyl acetate, 98%

$C_{10}H_{18}O_2$
FW 170.25
bp 98°C (15 mm)
d 0.949
n_D^{20} 1.4470

Fp 178°F

60 MHz: *1*, 532A
FT-IR: *1*, 630A
VP-FT-IR: *3*, 657B

171.07	33.20
62.76	26.52
36.02	26.23
34.61*	21.02*

CDCl₃   QE-300

$CH_3-C(=O)-OCH_2CH_2-$ (cyclohexyl)

B

Aldrich W26,680-9   *CAS [16409-45-3]*

***dl*-Menthyl acetate, 97+%**

$C_{12}H_{22}O_2$
FW 198.31
bp 229°C
d 0.922
n_D^{20} 1.4450

Fp 198°F

170.55	26.34*
74.11*	23.53
47.02*	22.01*
40.95	21.30*
34.28	20.73*
31.38*	16.40*

CDCl₃   QE-300

C

Aldrich 25,024-4   *CAS [16409-45-3]*

(±)-Menthyl acetate, mixture of *cis* and *trans*

$C_{12}H_{22}O_2$
FW 198.31
d 0.919
n_D^{20} 1.4470
Fp 198°F

170.61	45.70*	31.36*	24.96	21.15*
74.09*	40.91	30.65	23.46	20.89*
71.59*	39.12	29.98	22.19*	20.74*
71.11*	35.84	29.18*	22.02*	20.69*
46.98*	35.40	27.54*	21.49*	20.42*
46.65*	34.80	26.49*	21.45*	18.95*
45.79*	34.24	26.28*	21.33*	16.36*

CDCl₃   QE-300

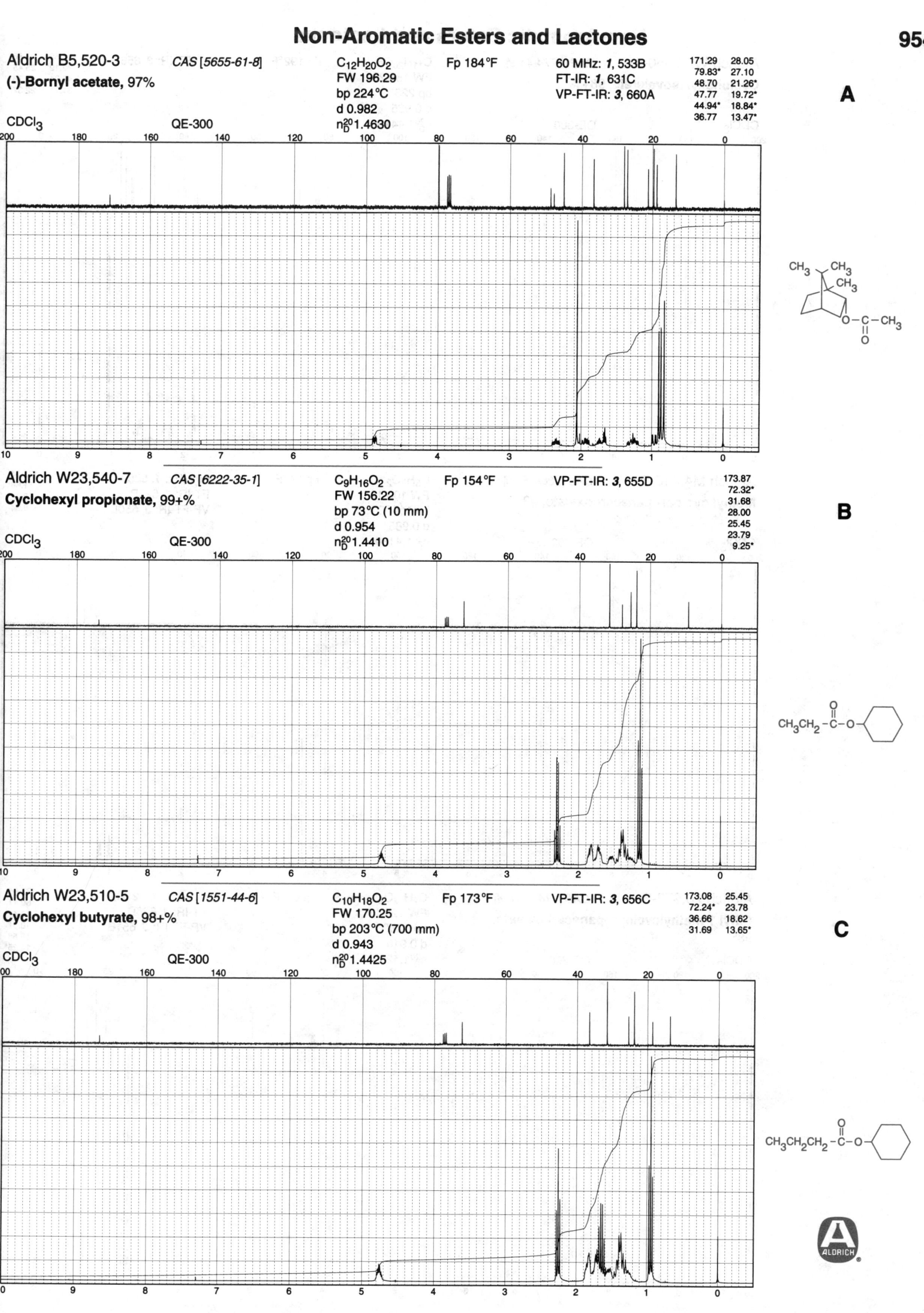

Aldrich B5,520-3   *CAS [5655-61-8]*   C₁₂H₂₀O₂   Fp 184°F   60 MHz: *1*, 533B

(-)-Bornyl acetate, 97%   FW 196.29   FT-IR: *1*, 631C

bp 224°C   VP-FT-IR: *3*, 660A

CDCl₃   QE-300   d 0.982

n²⁰_D 1.4630

A

171.29	28.05
79.83*	27.10
48.70	21.26*
47.77	19.72*
44.94*	18.84*
36.77	13.47*

Aldrich W23,540-7   *CAS [6222-35-1]*   C₉H₁₆O₂   Fp 154°F   VP-FT-IR: *3*, 655D

Cyclohexyl propionate, 99+%   FW 156.22

bp 73°C (10 mm)

CDCl₃   QE-300   d 0.954

n²⁰_D 1.4410

B

| 173.87 |
| 72.32* |
| 31.68 |
| 28.00 |
| 25.45 |
| 23.79 |
| 9.25* |

Aldrich W23,510-5   *CAS [1551-44-6]*   C₁₀H₁₈O₂   Fp 173°F   VP-FT-IR: *3*, 656C

Cyclohexyl butyrate, 98+%   FW 170.25

bp 203°C (700 mm)

CDCl₃   QE-300   d 0.943

n²⁰_D 1.4425

C

173.08	25.45
72.24*	23.78
36.66	18.62
31.69	13.65*

A

Aldrich W23,550-4   CAS [7774-44-9]   C$_{11}$H$_{20}$O$_2$ Fp 192°F   VP-FT-IR: 3, 656D

Cyclohexyl isovalerate, 98+%

FW 184.28
bp 223°C
d 0.925
n$_D^{20}$1.4410

CDCl$_3$ QE-300

172.52	25.86*
72.19*	25.45
43.91	23.78
31.71	22.37*

CH$_3$CHCH$_2$-C-O-⬡ (CH$_3$)

B

Aldrich M4,040-5   CAS [2868-37-3]   C$_5$H$_8$O$_2$ Fp 64°F   60 MHz: 1, 529A

Methyl cyclopropanecarboxylate, 98%

FW 100.12
bp 119°C
d 0.985
n$_D^{20}$1.4180

FT-IR: 1, 626B
VP-FT-IR: 3, 650D

175.34
51.66*
12.72*
8.34

CDCl$_3$ QE-300

▷-C-OCH$_3$

C

Aldrich 20,978-3   CAS [71441-76-4]   C$_7$H$_{12}$O$_2$ Fp 87°F   60 MHz: 1, 529C

Ethyl 1-methylcyclopropanecarboxylate, 98%

FW 128.17
bp 136°C
d 0.918
n$_D^{20}$1.4200

FT-IR: 1, 626D
VP-FT-IR: 3, 651B

175.96
60.41
19.42*
18.51
16.68
14.25*

CDCl$_3$ QE-300

▷-C-OCH$_2$CH$_3$ (CH$_3$)

Non-Aromatic Esters and Lactones

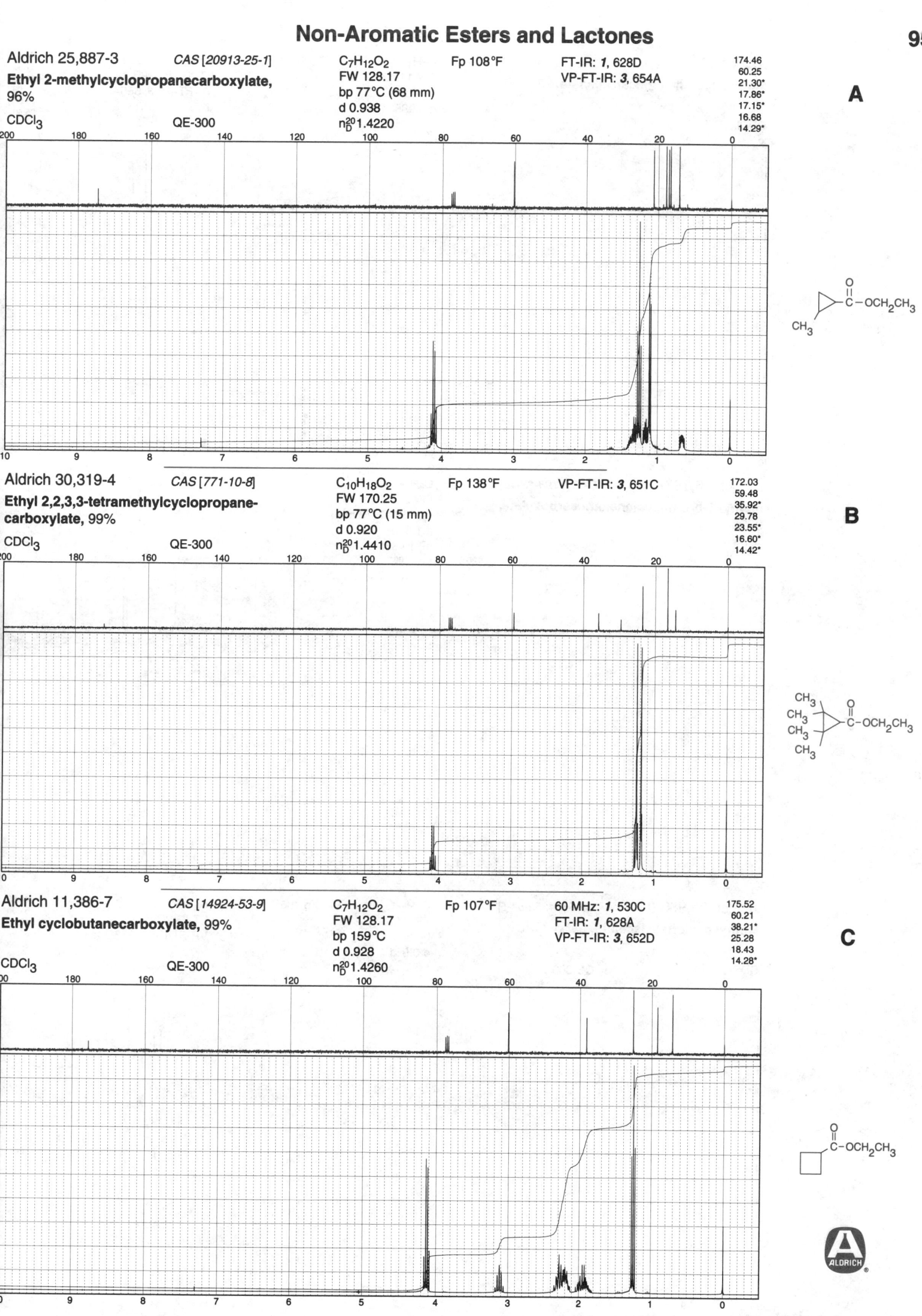

Aldrich 25,887-3   *CAS [20913-25-1]* $C_7H_{12}O_2$ Fp 108°F   FT-IR: **1**, 628D

Ethyl 2-methylcyclopropanecarboxylate,   FW 128.17   VP-FT-IR: **3**, 654A

96%   bp 77°C (68 mm)

CDCl₃   QE-300   d 0.938 n_D^{20} 1.4220

174.46
60.25
21.30*
17.86*
17.15*
16.68
14.29*

A

Aldrich 30,319-4   *CAS [771-10-8]* $C_{10}H_{18}O_2$ Fp 138°F   VP-FT-IR: **3**, 651C

Ethyl 2,2,3,3-tetramethylcyclopropane-   FW 170.25

carboxylate, 99%   bp 77°C (15 mm)

CDCl₃   QE-300   d 0.920 n_D^{20} 1.4410

172.03
59.48
35.92*
29.78
23.55*
16.60*
14.42*

B

Aldrich 11,386-7   *CAS [14924-53-9]* $C_7H_{12}O_2$ Fp 107°F   60 MHz: **1**, 530C

Ethyl cyclobutanecarboxylate, 99%   FW 128.17   FT-IR: **1**, 628A

bp 159°C   VP-FT-IR: **3**, 652D

CDCl₃   QE-300   d 0.928 n_D^{20} 1.4260

175.52
60.21
38.21*
25.28
18.43
14.28*

C

ALDRICH

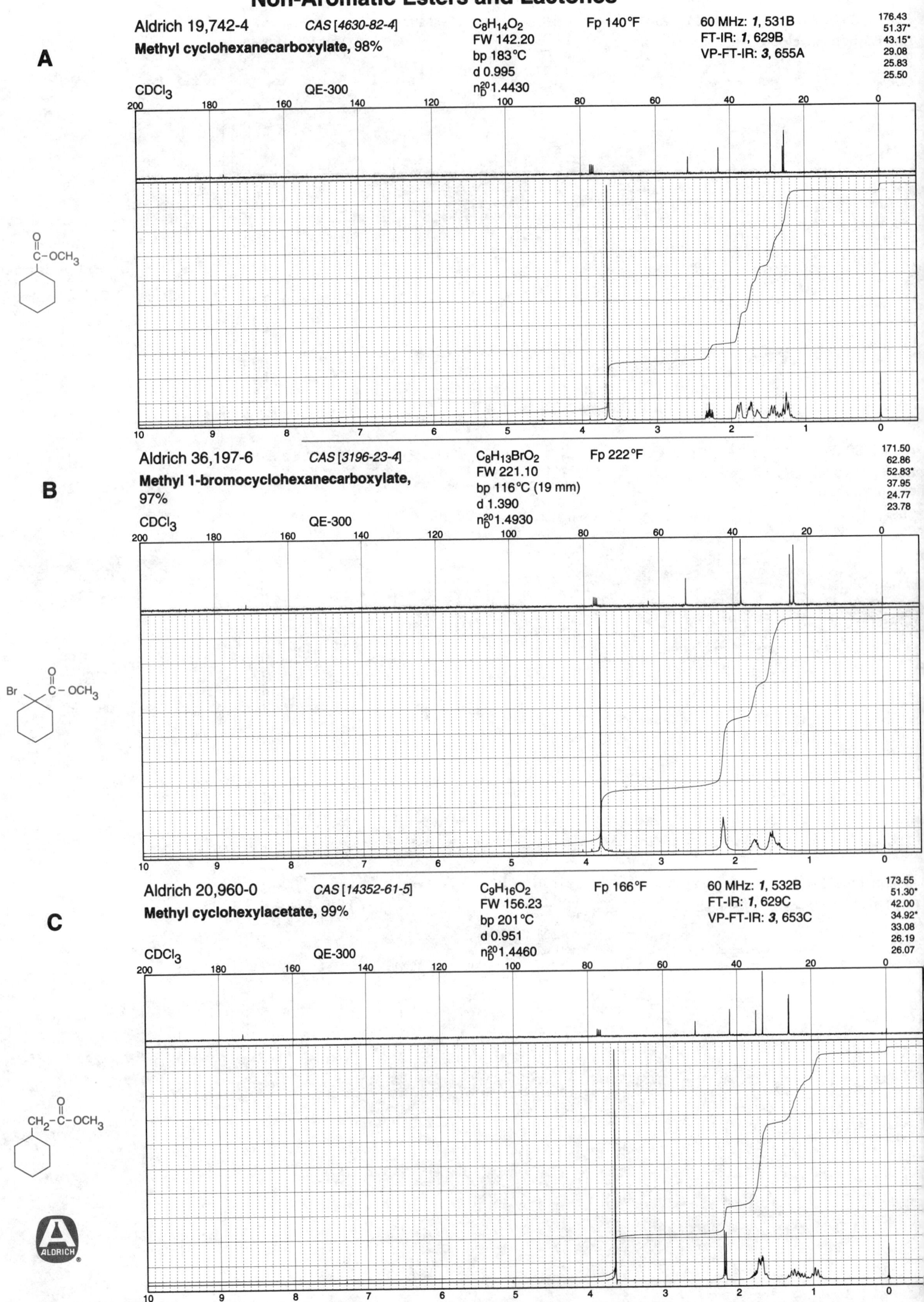

A

Aldrich 19,742-4   *CAS [4630-82-4]* $C_8H_{14}O_2$ Fp 140°F   60 MHz: *1*, 531B
Methyl cyclohexanecarboxylate, 98%   FW 142.20   FT-IR: *1*, 629B
bp 183°C   VP-FT-IR: *3*, 655A
d 0.995
n_D^{20}1.4430
CDCl$_3$ QE-300

176.43
51.37*
43.15*
29.08
25.83
25.50

B

Aldrich 36,197-6   *CAS [3196-23-4]* $C_8H_{13}BrO_2$ Fp 222°F
Methyl 1-bromocyclohexanecarboxylate, 97%   FW 221.10
bp 116°C (19 mm)
d 1.390
n_D^{20}1.4930
CDCl$_3$ QE-300

171.50
62.86
52.83*
37.95
24.77
23.78

C

Aldrich 20,960-0   *CAS [14352-61-5]* $C_9H_{16}O_2$ Fp 166°F   60 MHz: *1*, 532B
Methyl cyclohexylacetate, 99%   FW 156.23   FT-IR: *1*, 629C
bp 201°C   VP-FT-IR: *3*, 653C
d 0.951
n_D^{20}1.4460
CDCl$_3$ QE-300

173.55
51.30*
42.00
34.92*
33.08
26.19
26.07

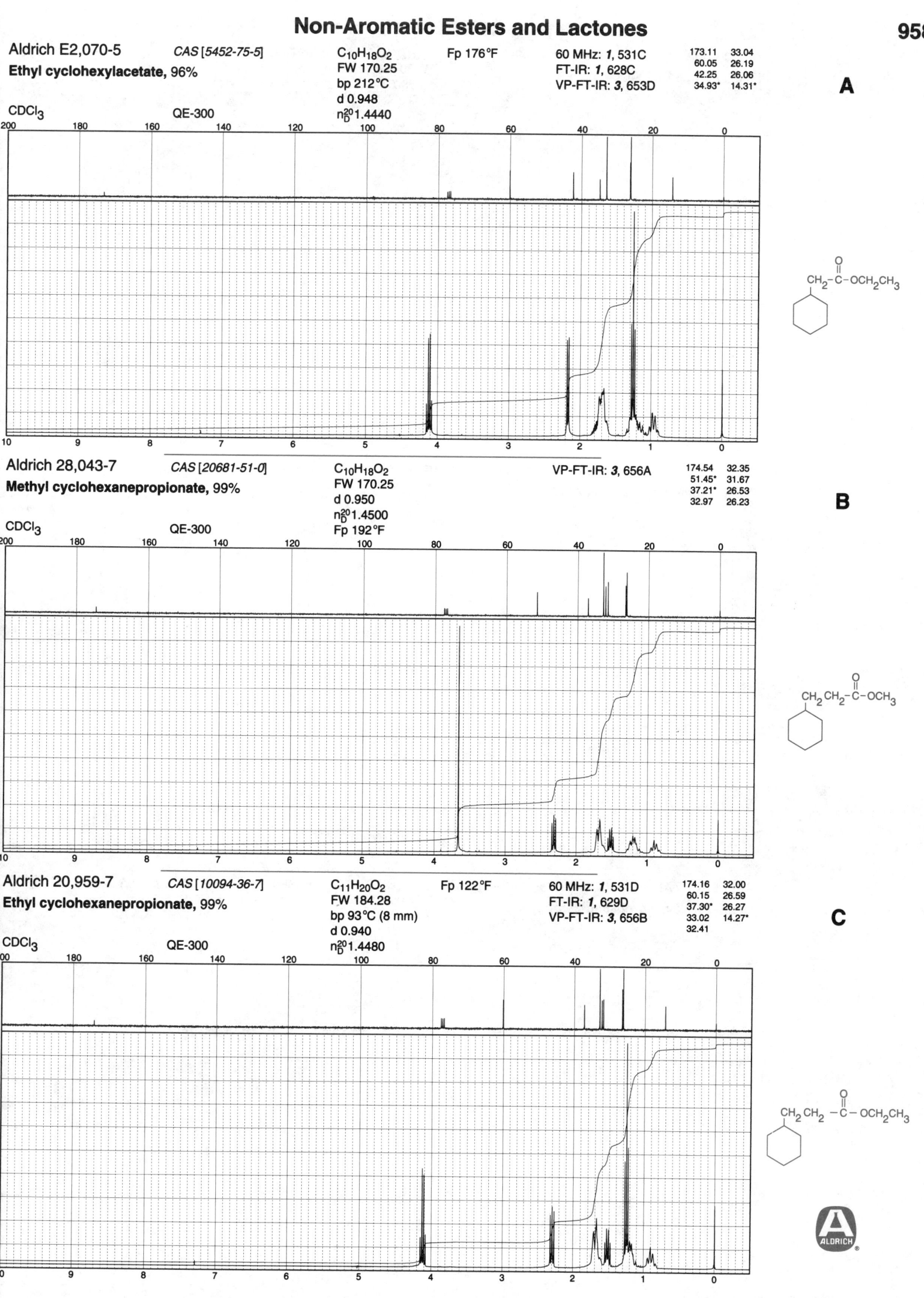

Aldrich E2,070-5   CAS [5452-75-5] $C_{10}H_{18}O_2$ Fp 176°F   60 MHz: 1, 531C
Ethyl cyclohexylacetate, 96%   FW 170.25   FT-IR: 1, 628C
bp 212°C   VP-FT-IR: 3, 653D
d 0.948
n_D^{20}1.4440

CDCl₃   QE-300

173.11   33.04
60.05   26.19
42.25   26.06
34.93*   14.31*

A

Aldrich 28,043-7   CAS [20681-51-0] $C_{10}H_{18}O_2$ VP-FT-IR: 3, 656A
Methyl cyclohexanepropionate, 99%   FW 170.25
d 0.950
n_D^{20}1.4500
Fp 192°F

CDCl₃   QE-300

174.54   32.35
51.45*   31.67
37.21*   26.53
32.97   26.23

B

Aldrich 20,959-7   CAS [10094-36-7] $C_{11}H_{20}O_2$ Fp 122°F   60 MHz: 1, 531D
Ethyl cyclohexanepropionate, 99%   FW 184.28   FT-IR: 1, 629D
bp 93°C (8 mm)   VP-FT-IR: 3, 656B
d 0.940
n_D^{20}1.4480

CDCl₃   QE-300

174.16   32.00
60.15   26.59
37.30*   26.27
33.02   14.27*
32.41

C

ALDRICH

A

Aldrich 30,635-5 CAS [34212-59-4] C₂₄H₄₂O₄

$C_{24}H_{42}O_4$
FW 394.60
mp 63°C
bp 203°C (2 mm)

VP-FT-IR: **3**, 641A

(1R)-(-)-Dimenthyl succinate, 99%

171.64	29.54
74.40*	26.27*
47.01*	23.48
40.87	22.00*
34.27	20.76*
31.38*	16.36*

CDCl₃   QE-300

B

Aldrich 30,634-7 CAS [96149-05-2]

$C_{24}H_{42}O_4$
FW 394.60
mp 63°C
bp 203°C (2 mm)

VP-FT-IR: **3**, 640D

(1S)-(+)-Dimenthyl succinate, 99%

171.65	29.54
74.41*	26.27*
47.01*	23.48
40.88	22.00*
34.27	20.76*
31.38*	16.37*

CDCl₃   QE-300

C

Aldrich 14,771-0 CAS [1559-02-0]

$C_9H_{14}O_4$
FW 186.21
bp 95°C (10 mm)
d 1.055
n²⁰_D 1.4330

Fp 188°F

60 MHz: **1**, 530A
FT-IR: **1**, 629A
VP-FT-IR: **3**, 654B

Diethyl 1,1-cyclopropanedicarboxylate, 97%

169.73
61.33
28.29
16.20
14.07*

CDCl₃   QE-300

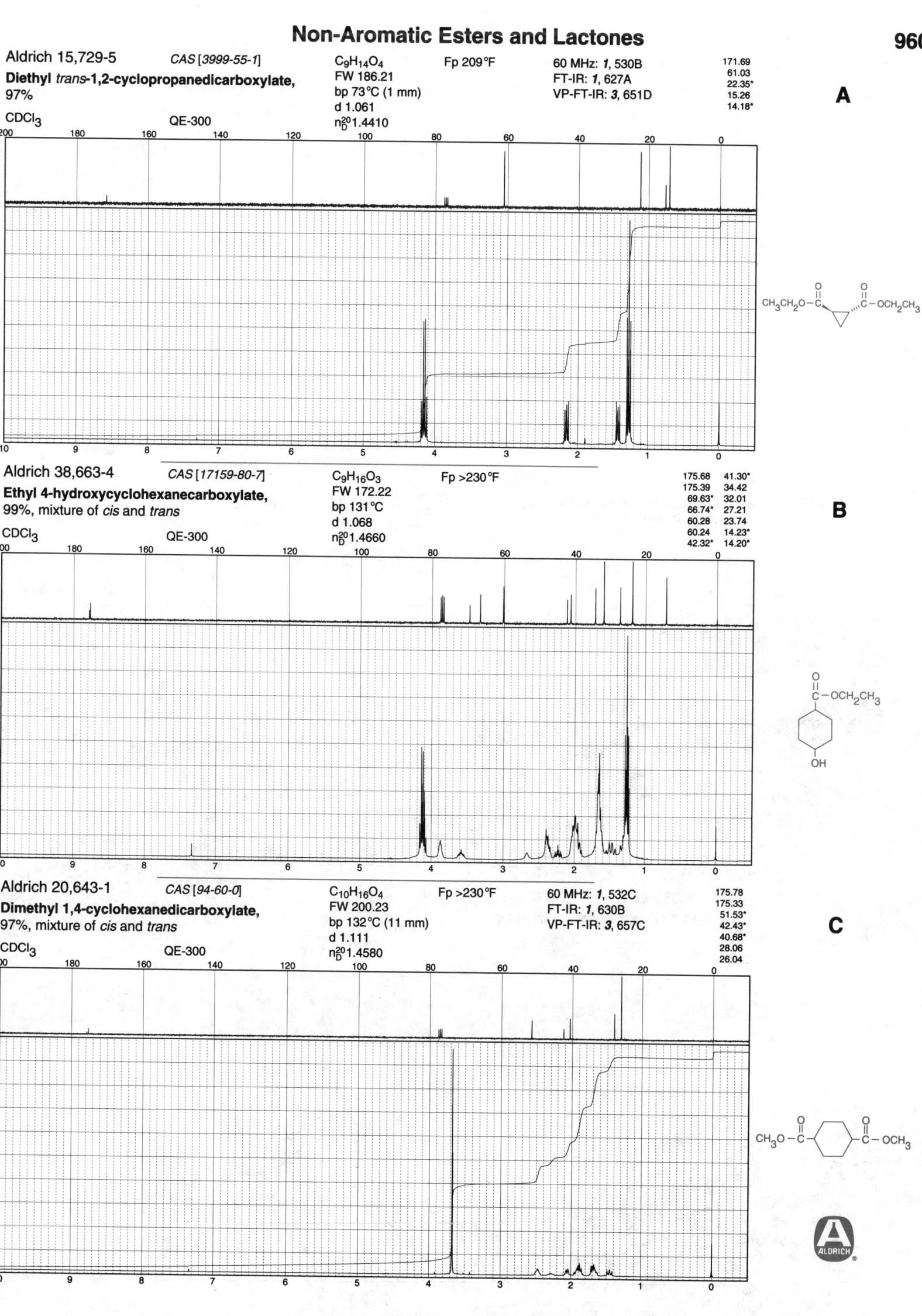

Aldrich 15,729-5   *CAS [3999-55-1]*   C$_9$H$_{14}$O$_4$ Fp 209°F   60 MHz: *1*, 530B   171.69
Diethyl *trans*-1,2-cyclopropanedicarboxylate,   FW 186.21   FT-IR: *1*, 627A   61.03
97%   bp 73°C (1 mm)   VP-FT-IR: *3*, 651D   22.35*
CDCl$_3$ QE-300   d 1.061   15.26
n$_D^{20}$1.4410   14.18*

A

Aldrich 38,663-4   *CAS [17159-80-7]*   C$_9$H$_{16}$O$_3$ Fp >230°F   175.68   41.30*
Ethyl 4-hydroxycyclohexanecarboxylate,   FW 172.22   175.39   34.42
99%, mixture of *cis* and *trans*   bp 131°C   69.63*   32.01
CDCl$_3$ QE-300   d 1.068   66.74*   27.21
n$_D^{20}$1.4660   60.28   23.74
60.24   14.23*
42.32*   14.20*

B

Aldrich 20,643-1   *CAS [94-60-0]*   C$_{10}$H$_{16}$O$_4$ Fp >230°F   60 MHz: *1*, 532C   175.78
Dimethyl 1,4-cyclohexanedicarboxylate,   FW 200.23   FT-IR: *1*, 630B   175.33
97%, mixture of *cis* and *trans*   bp 132°C (11 mm)   VP-FT-IR: *3*, 657C   51.53*
CDCl$_3$ QE-300   d 1.111   42.43*
n$_D^{20}$1.4580   40.68*
28.06
26.04

C

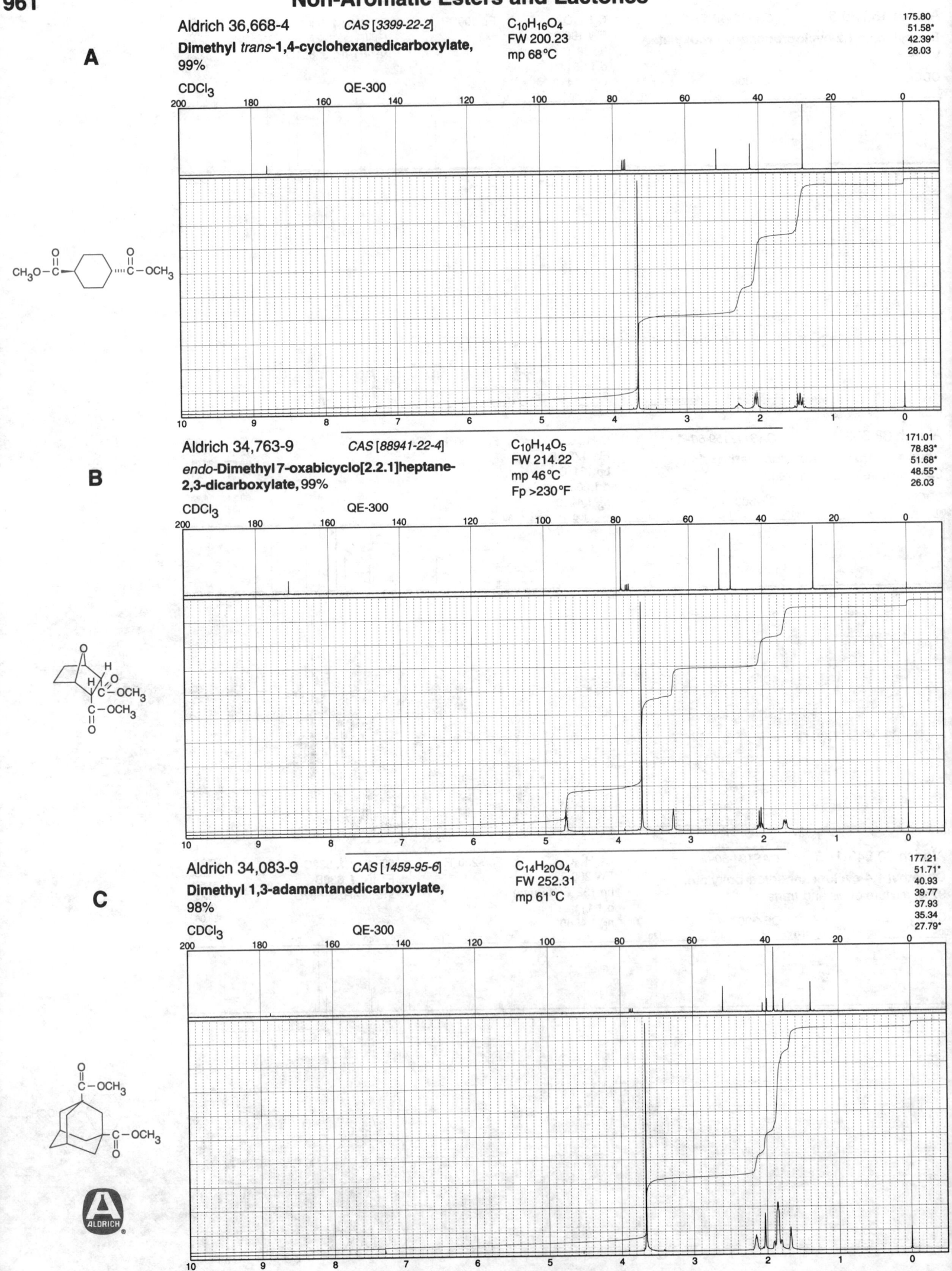

A

Aldrich 36,668-4   *CAS [3399-22-2]*

Dimethyl *trans*-1,4-cyclohexanedicarboxylate, 99%

$C_{10}H_{16}O_4$
FW 200.23
mp 68°C

175.80
51.58*
42.39*
28.03

CDCl₃   QE-300

B

Aldrich 34,763-9   *CAS [88941-22-4]*

endo-**Dimethyl 7-oxabicyclo[2.2.1]heptane-2,3-dicarboxylate, 99%**

$C_{10}H_{14}O_5$
FW 214.22
mp 46°C
Fp >230°F

171.01
78.83*
51.68*
48.55*
26.03

CDCl₃   QE-300

C

Aldrich 34,083-9   *CAS [1459-95-6]*

Dimethyl 1,3-adamantanedicarboxylate, 98%

$C_{14}H_{20}O_4$
FW 252.31
mp 61°C

177.21
51.71*
40.93
39.77
37.93
35.34
27.79*

CDCl₃   QE-300

Aldrich 36,229-8   CAS [6998-83-0] $C_{12}H_{18}O_6$ 174.37
51.85*
Trimethyl *cis,cis*-1,3,5-cyclohexane-   FW 258.27   41.73*
tricarboxylate, 99%   mp 48°C   30.46
Fp >230°F

A

CDCl_3   QE-300

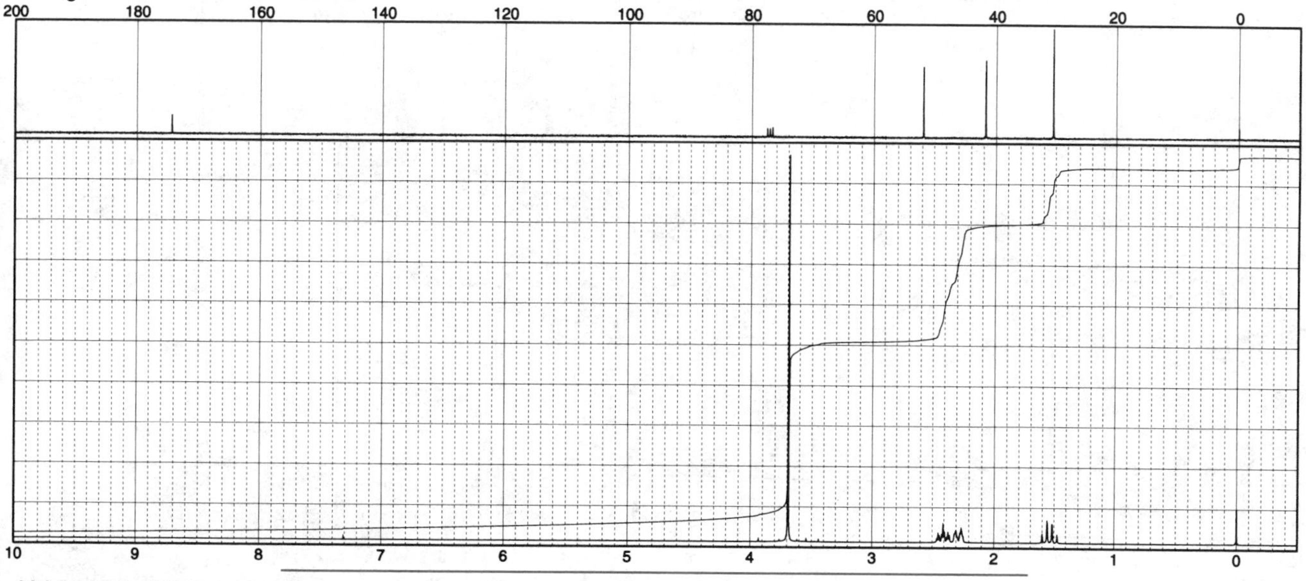

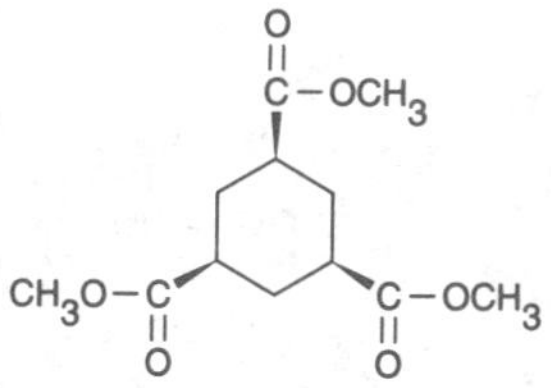

Aldrich 32,495-7   CAS [14495-41-1] $C_{12}H_{16}O_8$ 171.26
52.38*
Tetramethyl 1,2,3,4-cyclobutanetetra-   FW 288.26   40.75*
carboxylate, 98%   mp 146°C

B

CDCl_3   QE-300

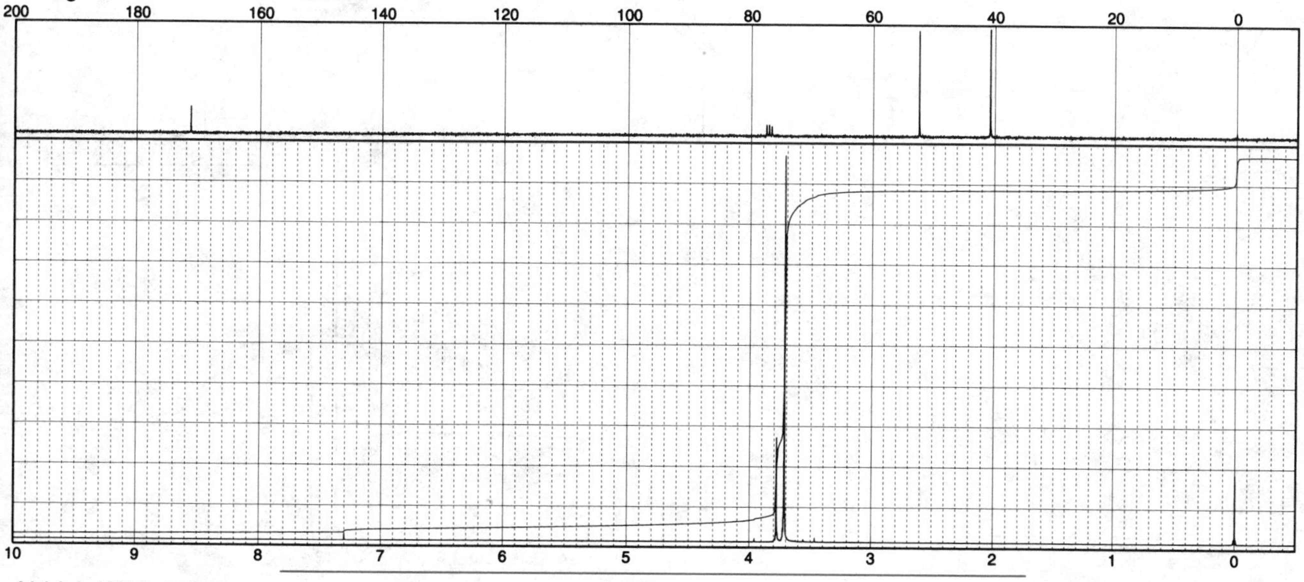

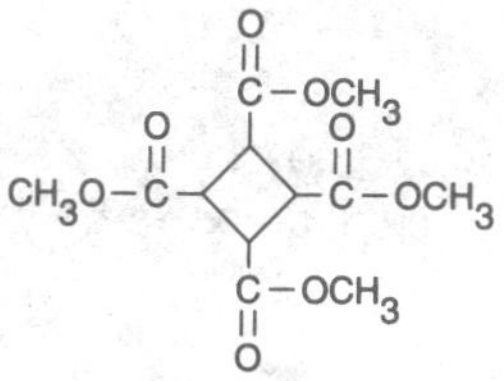

Aldrich W26,420-2   CAS [115-99-1] $C_{11}H_{18}O_2$ Fp 186°F   VP-FT-IR: *3*, 662D   160.41   40.01
140.88*   25.63*
Linalyl formate   FW 182.27   132.07   23.84*
bp 102°C (10 mm)   123.38*   22.21
d 0.914   114.29   17.60*
n_D^{20} 1.4550   83.80

C

CDCl_3   QE-300

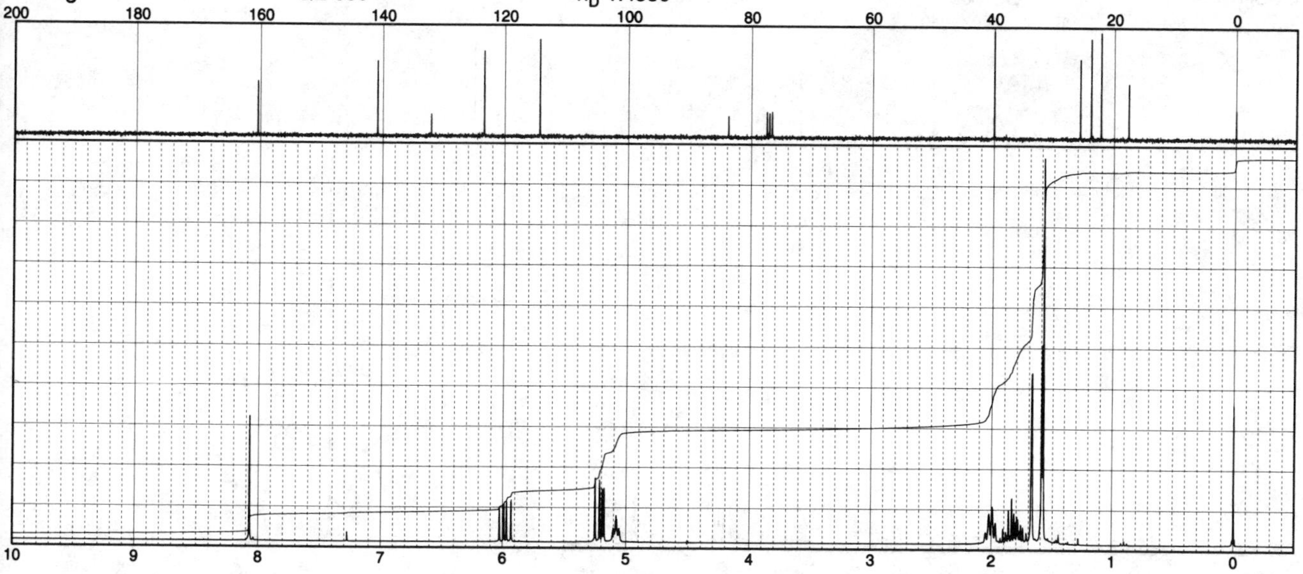

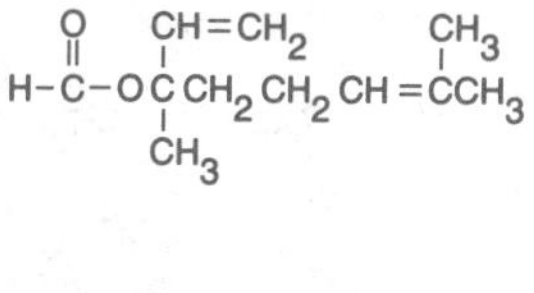

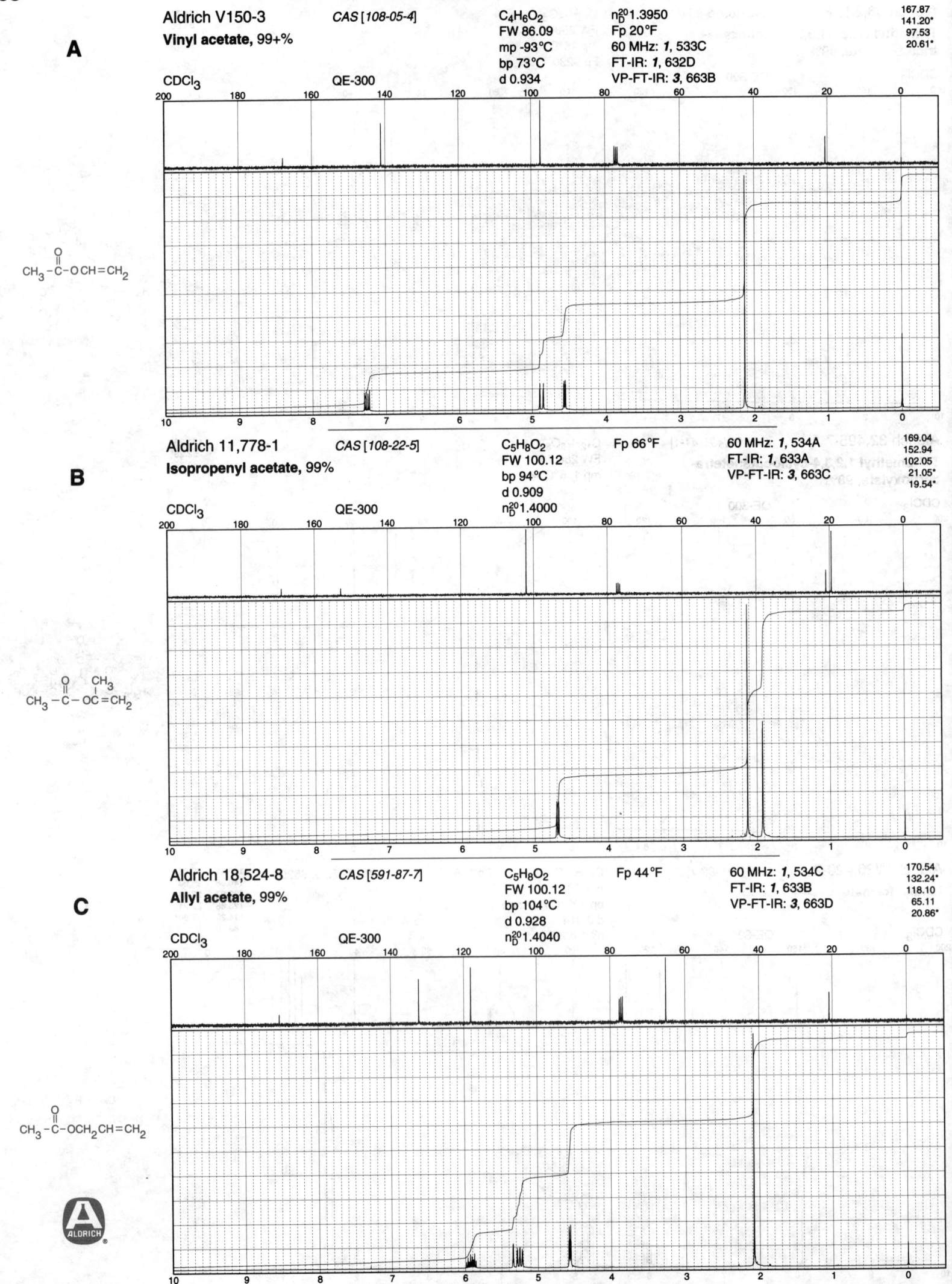

Aldrich V150-3   CAS [108-05-4]

A

Vinyl acetate, 99+%

$C_4H_6O_2$
FW 86.09
mp -93°C
bp 73°C
d 0.934

n_D^{20}1.3950
Fp 20°F
60 MHz: *1*, 533C
FT-IR: *1*, 632D
VP-FT-IR: *3*, 663B

167.87
141.20*
97.53
20.61*

CDCl₃   QE-300

$CH_3-\overset{O}{\overset{\|}{C}}-OCH=CH_2$

Aldrich 11,778-1   CAS [108-22-5]

B

Isopropenyl acetate, 99%

$C_5H_8O_2$
FW 100.12
bp 94°C
d 0.909
n_D^{20}1.4000

Fp 66°F

60 MHz: *1*, 534A
FT-IR: *1*, 633A
VP-FT-IR: *3*, 663C

169.04
152.94
102.05
21.05*
19.54*

CDCl₃   QE-300

$CH_3-\overset{O}{\overset{\|}{C}}-O\overset{CH_3}{\overset{|}{C}}=CH_2$

Aldrich 18,524-8   CAS [591-87-7]

C

Allyl acetate, 99%

$C_5H_8O_2$
FW 100.12
bp 104°C
d 0.928
n_D^{20}1.4040

Fp 44°F

60 MHz: *1*, 534C
FT-IR: *1*, 633B
VP-FT-IR: *3*, 663D

170.54
132.24*
118.10
65.11
20.86*

CDCl₃   QE-300

$CH_3-\overset{O}{\overset{\|}{C}}-OCH_2CH=CH_2$

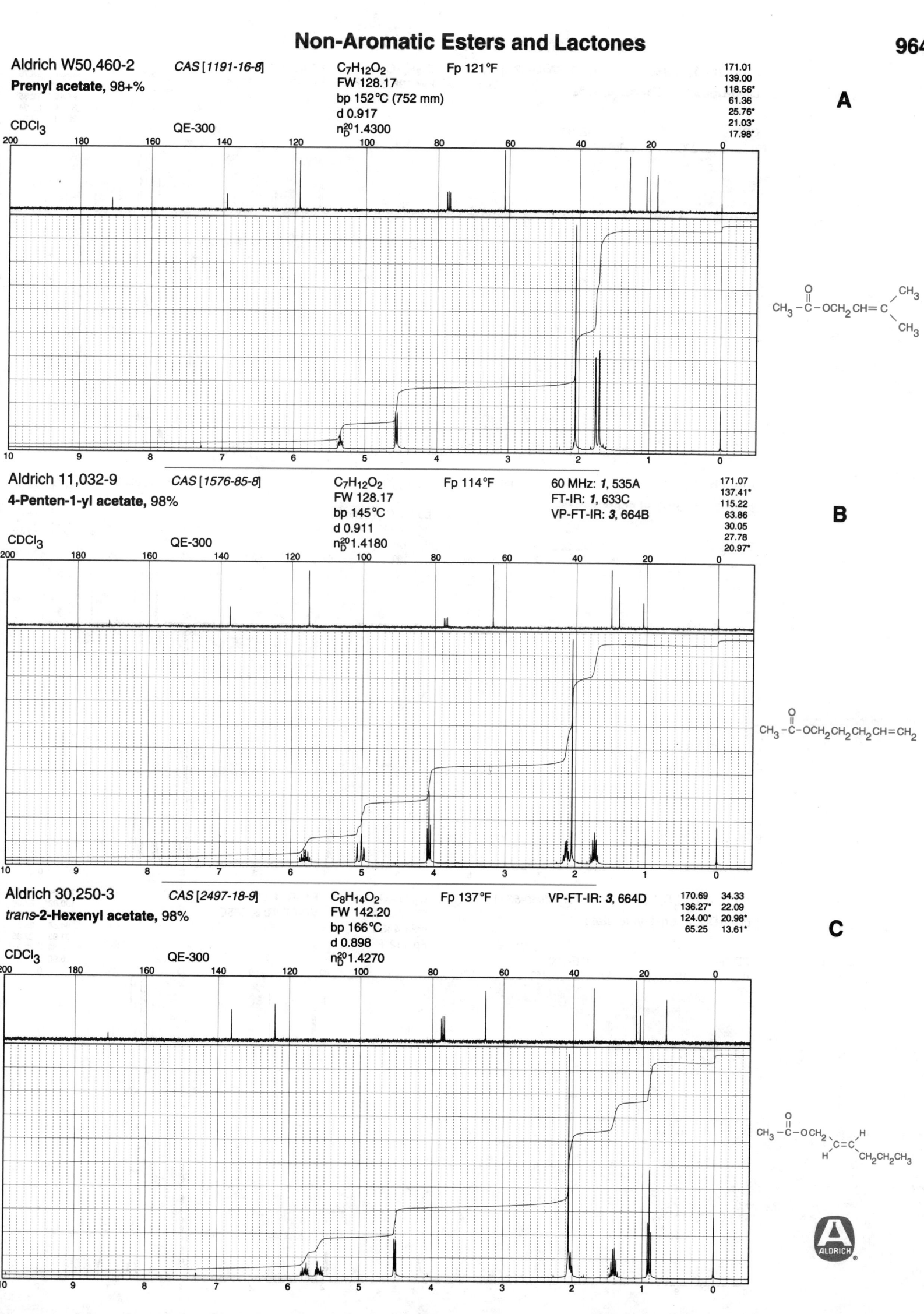

Aldrich W50,460-2   CAS [1191-16-8]   C₇H₁₂O₂   Fp 121°F
Prenyl acetate, 98+%
FW 128.17
bp 152°C (752 mm)
d 0.917
n₂₀ᴅ 1.4300

CDCl₃   QE-300

171.01
139.00
118.56*
61.36
25.76*
21.03*
17.98*

A

Aldrich 11,032-9   CAS [1576-85-8]   C₇H₁₂O₂   Fp 114°F   60 MHz: 1, 535A
4-Penten-1-yl acetate, 98%
FW 128.17   FT-IR: 1, 633C
bp 145°C   VP-FT-IR: 3, 664B
d 0.911
n₂₀ᴅ 1.4180

CDCl₃   QE-300

171.07
137.41*
115.22
63.86
30.05
27.78
20.97*

B

Aldrich 30,250-3   CAS [2497-18-9]   C₈H₁₄O₂   Fp 137°F   VP-FT-IR: 3, 664D
trans-**2-Hexenyl acetate, 98%**
FW 142.20
bp 166°C
d 0.898
n₂₀ᴅ 1.4270

CDCl₃   QE-300

170.69   34.33
136.27*   22.09
124.00*   20.98*
65.25   13.61*

C

Non-Aromatic Esters and Lactones

A

Aldrich 24,909-2   CAS [67446-07-5] $C_{12}H_{22}O_2$ FT-IR: **1**, 633D
cis-5-Decen-1-yl acetate, 92%   FW 198.31   VP-FT-IR: **3**, 665A
Fp 144°F

171.02	26.96
130.51*	26.76
128.99*	26.07
64.45	22.35
31.94	20.94*
28.26	13.97*

CDCl₃   QE-300

B

Aldrich 28,912-4   CAS [38421-90-8] $C_{12}H_{22}O_2$ VP-FT-IR: **3**, 665B
trans-5-Decen-1-yl acetate, 97%   FW 198.31
bp 95°C (14 mm)
n_D^{20}1.4390
Fp 144°F

171.07	31.80
131.03*	28.10
129.47*	25.92
64.49	22.21
32.25	20.95*
32.13	13.93*

CDCl₃   QE-300

C

Aldrich 25,982-9   CAS [14959-86-5] $C_{14}H_{26}O_2$ FT-IR: **1**, 634A
cis-7-Dodecen-1-yl acetate   FW 226.36   VP-FT-IR: **3**, 665C
n_D^{20}1.4430
Fp 142°F

171.04	28.63
130.02*	27.09
129.56*	26.94
64.58	25.86
31.98	22.35
29.62	20.96*
28.90	13.97*

CDCl₃   QE-300

Aldrich 24,906-8 *CAS [20711-10-8]* C₁₆H₃₀O₂ FT-IR: *1*, 634D

cis-11-Tetradecen-1-yl acetate, 95% FW 254.42 VP-FT-IR: *3*, 666B **A**

Fp 142°F

CDCl₃ QE-300

171.13	29.27
131.46*	28.62
129.25*	27.10
64.64	25.92
29.78	21.00*
29.52	20.52
29.42	14.40*

$CH_3-\overset{O}{\overset{\|}{C}}-OCH_2(CH_2)_8CH_2\overset{H}{\underset{H}{C=C}}CH_2CH_3$

Aldrich 21,901-7 *CAS [16974-10-0]* C₁₆H₃₀O₂ FT-IR: *1*, 635A

cis-7-Tetradecen-1-yl acetate, 96% FW 254.42 VP-FT-IR: *3*, 666C **B**

Fp 144°F

CDCl₃ QE-300

171.14	29.61	25.84
130.06*	29.00	22.67
129.53*	28.89	20.99*
64.60	28.59	14.11*
31.79	27.23	
29.73	27.09	

$CH_3-\overset{O}{\overset{\|}{C}}-OCH_2(CH_2)_4CH_2\overset{H}{\underset{H}{C=C}}CH_2(CH_2)_4CH_3$

Aldrich 24,912-2 *CAS [33189-72-9]* C₁₆H₃₀O₂ FT-IR: *1*, 635B

trans-11-Tetradecen-1-yl acetate, 97% FW 254.42 VP-FT-IR: *3*, 666D **C**

Fp 144°F

CDCl₃ QE-300

171.03	29.27
131.85*	29.17
129.33*	28.67
64.62	25.95
32.56	25.60
29.68	20.95*
29.52	14.01*

$CH_3-\overset{O}{\overset{\|}{C}}-OCH_2(CH_2)_8CH_2\overset{H}{\underset{H}{C=C}}CH_2CH_3$

A

Aldrich 25,983-7 CAS [34010-21-4] C₁₈H₃₄O₂ FT-IR: **1**, 636A
cis-11-Hexadecen-1-yl acetate, 95% FW 282.47 VP-FT-IR: **3**, 667C
n_D^{20} 1.4480
Fp 142°F

171.06	28.66
129.82*	27.22
64.63	26.94
32.00	25.94
29.78	22.36
29.52	20.97*
29.28	13.98*

CDCl₃ QE-300

$CH_3-C(=O)-O CH_2(CH_2)_8 CH_2 \quad CH_2CH_2CH_3$

B

Aldrich 22,086-8 CAS [1515-76-0] C₆H₈O₂ Fp 92°F 60 MHz: **1**, 534B
1-Acetoxy-1,3-butadiene, FW 112.13 FT-IR: **1**, 636D
mixture of *cis* and *trans* bp 61°C (40 mm) VP-FT-IR: **3**, 668A
d 0.945
n_D^{20} 1.4690

167.66	117.59
167.32	117.15
138.69*	116.01*
134.11*	113.30*
131.67*	20.61*
128.84*	

CDCl₃ QE-300

$CH_3-C(=O)-OCH=CHCH=CH_2$

C

Aldrich L280-7 CAS [115-95-7] C₁₂H₂₀O₂ Fp 194°F VP-FT-IR: **3**, 669B
Linalyl acetate, 97% FW 196.29
bp 220°C
d 0.895
n_D^{20} 1.4500

169.86	39.68
141.77*	25.67*
131.71	23.58*
123.75*	22.34
113.02	22.17*
82.82	17.56*

CDCl₃ QE-300

$CH_3-C(=O)-O C(CH=CH_2)(CH_3) CH_2 CH_2 CH=C(CH_3)CH_3$
with CH₃

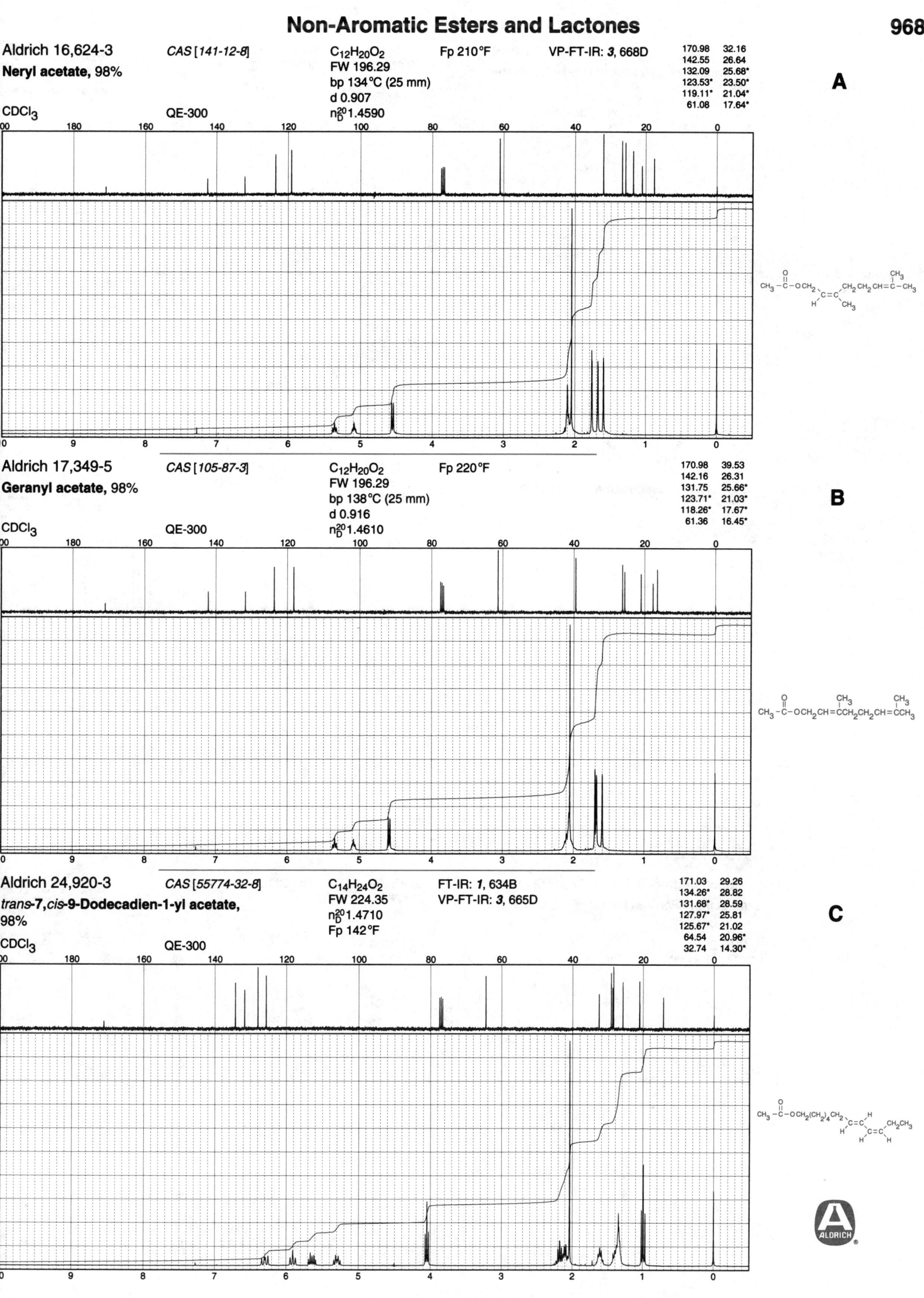

Aldrich 16,624-3 *CAS [141-12-8]* $C_{12}H_{20}O_2$ Fp 210 °F VP-FT-IR: *3*, 668D **A**
Neryl acetate, 98%
FW 196.29
bp 134 °C (25 mm)
d 0.907
$CDCl_3$ QE-300 n_D^{20} 1.4590

170.98	32.16
142.55	26.64
132.09	25.68*
123.53*	23.50*
119.11*	21.04*
61.08	17.64*

Aldrich 17,349-5 *CAS [105-87-3]* $C_{12}H_{20}O_2$ Fp 220 °F **B**
Geranyl acetate, 98%
FW 196.29
bp 138 °C (25 mm)
d 0.916
$CDCl_3$ QE-300 n_D^{20} 1.4610

170.98	39.53
142.16	26.31
131.75	25.66*
123.71*	21.03*
118.26*	17.67*
61.36	16.45*

Aldrich 24,920-3 *CAS [55774-32-8]* $C_{14}H_{24}O_2$ FT-IR: *1*, 634B **C**
trans-7,*cis*-9-Dodecadien-1-yl acetate,
98%
FW 224.35 VP-FT-IR: *3*, 665D
n_D^{20} 1.4710
Fp 142 °F
$CDCl_3$ QE-300

171.03	29.26
134.26*	28.82
131.68*	28.59
127.97*	25.81
125.67*	21.02
64.54	20.96*
32.74	14.30*

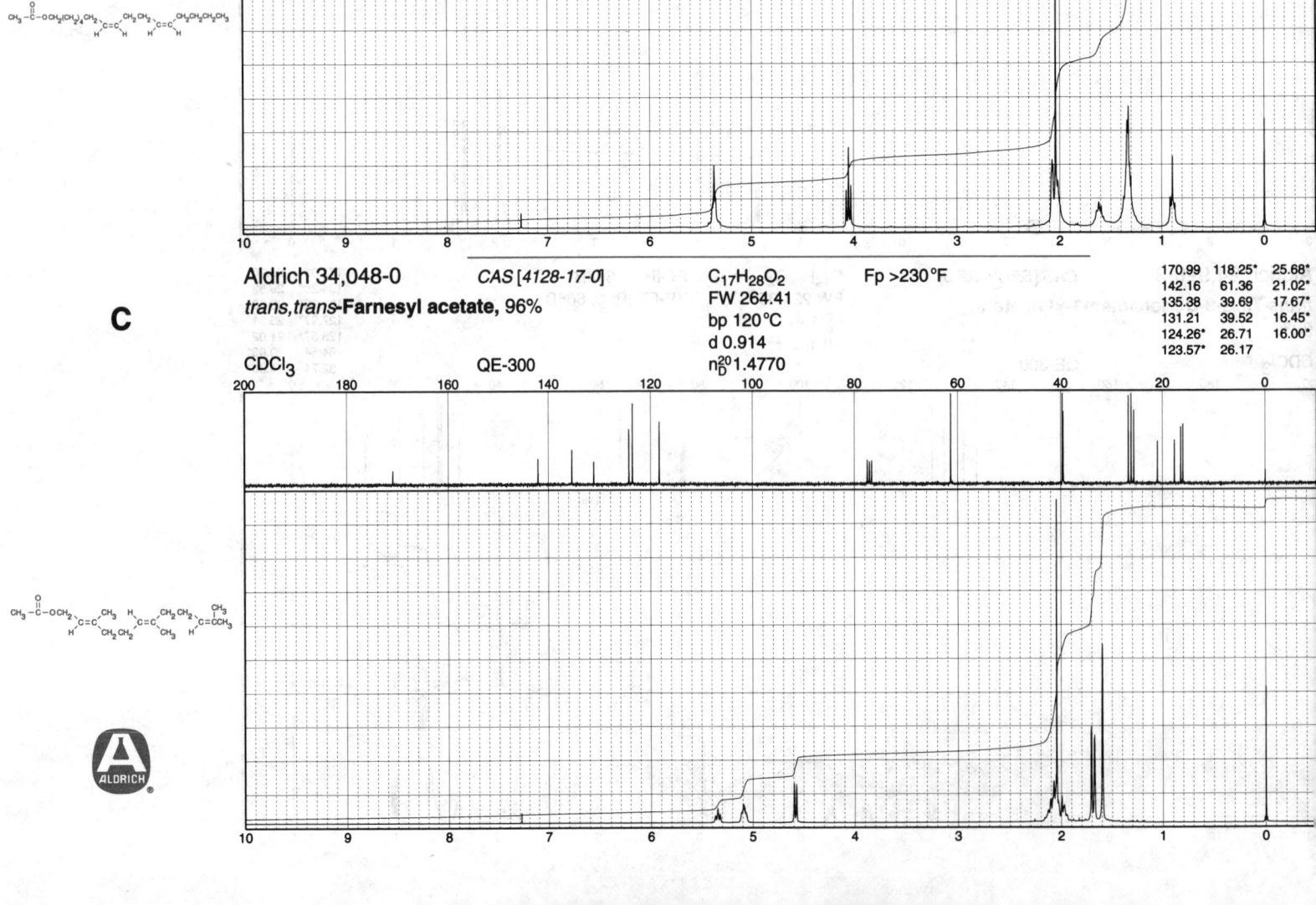

A

Aldrich 24,915-7 CAS [30562-09-5] $C_{16}H_{28}O_2$ FT-IR: *1*, 635C

cis-9,trans-11-Tetradecadien-1-yl acetate, 95% FW 252.40 VP-FT-IR: *3*, 667A
bp 86°C
Fp 142°F

CDCl$_3$ QE-300

171.19	64.62	28.58
136.07*	29.67	27.64
129.98*	29.35	25.88
128.57*	29.19	21.00*
124.62*	29.14	13.66*

B

Aldrich 24,918-1 CAS [52207-99-5] $C_{18}H_{32}O_2$ FT-IR: *1*, 636B

cis-7,cis-11-Hexadecadien-1-yl acetate, 97% FW 280.46 VP-FT-IR: *3*, 667D
Fp 144°F

CDCl$_3$ QE-300

171.18	31.92	27.13
130.30*	29.57	26.95
130.00*	28.88	25.83
129.31*	28.58	22.34
129.04*	27.42	21.00*
64.61	27.39	13.99*

C

Aldrich 34,048-0 CAS [4128-17-0] $C_{17}H_{28}O_2$ Fp >230°F

trans,trans-Farnesyl acetate, 96% FW 264.41
bp 120°C
d 0.914
n$_D^{20}$ 1.4770

CDCl$_3$ QE-300

170.99	118.25*	25.68*
142.16	61.36	21.02*
135.38	39.69	17.67*
131.21	39.52	16.45*
124.26*	26.71	16.00*
123.57*	26.17	

Non-Aromatic Esters and Lactones

A

Aldrich W33,290-9   CAS [41820-22-8]

Allyl thiopropionate, 95%

$C_6H_{10}OS$
FW 130.21
d 0.965
n_D^{20} 1.4820
Fp 119°F

VP-FT-IR: **3**, 669D

CDCl$_3$ QE-300

199.37
133.18*
117.70
37.28
31.62
9.62*

$CH_3CH_2-\overset{\overset{\displaystyle O}{\|}}{C}-S\,CH_2\,CH=CH_2$

B

Aldrich 24,652-2   CAS [2051-78-7]

Allyl butyrate, 99%

$C_7H_{12}O_2$
FW 128.17
bp 45°C (15 mm)
d 0.902
n_D^{20} 1.4140

Fp 107°F

60 MHz: **1**, 534D
FT-IR: **1**, 637B
VP-FT-IR: **3**, 670C

CDCl$_3$ QE-300

173.23
132.37*
117.94
64.86
36.15
18.46
13.66*

$CH_3CH_2CH_2-\overset{\overset{\displaystyle O}{\|}}{C}-OCH_2-CH=CH_2$

C

Aldrich W26,390-7   CAS [78-36-4]

Linalyl butyrate, 96+%

$C_{14}H_{24}O_2$
FW 224.35
bp 81°C
d 0.892
n_D^{20} 1.4500

Fp 218°F

172.37   37.30
141.98*   25.65*
131.65   23.71*
123.86*   22.40
112.91   18.57
82.58   17.55*
39.83   13.69*

CDCl$_3$ QE-300

$CH_3CH_2CH_2-\overset{\overset{\displaystyle O}{\|}}{C}-O-\underset{\underset{\displaystyle CH=CH_2}{|}}{\overset{\overset{\displaystyle CH_3}{|}}{C}}-CH_2CH_2CH=\overset{\overset{\displaystyle CH_3}{|}}{C}CH_3$

ALDRICH

Non-Aromatic Esters and Lactones

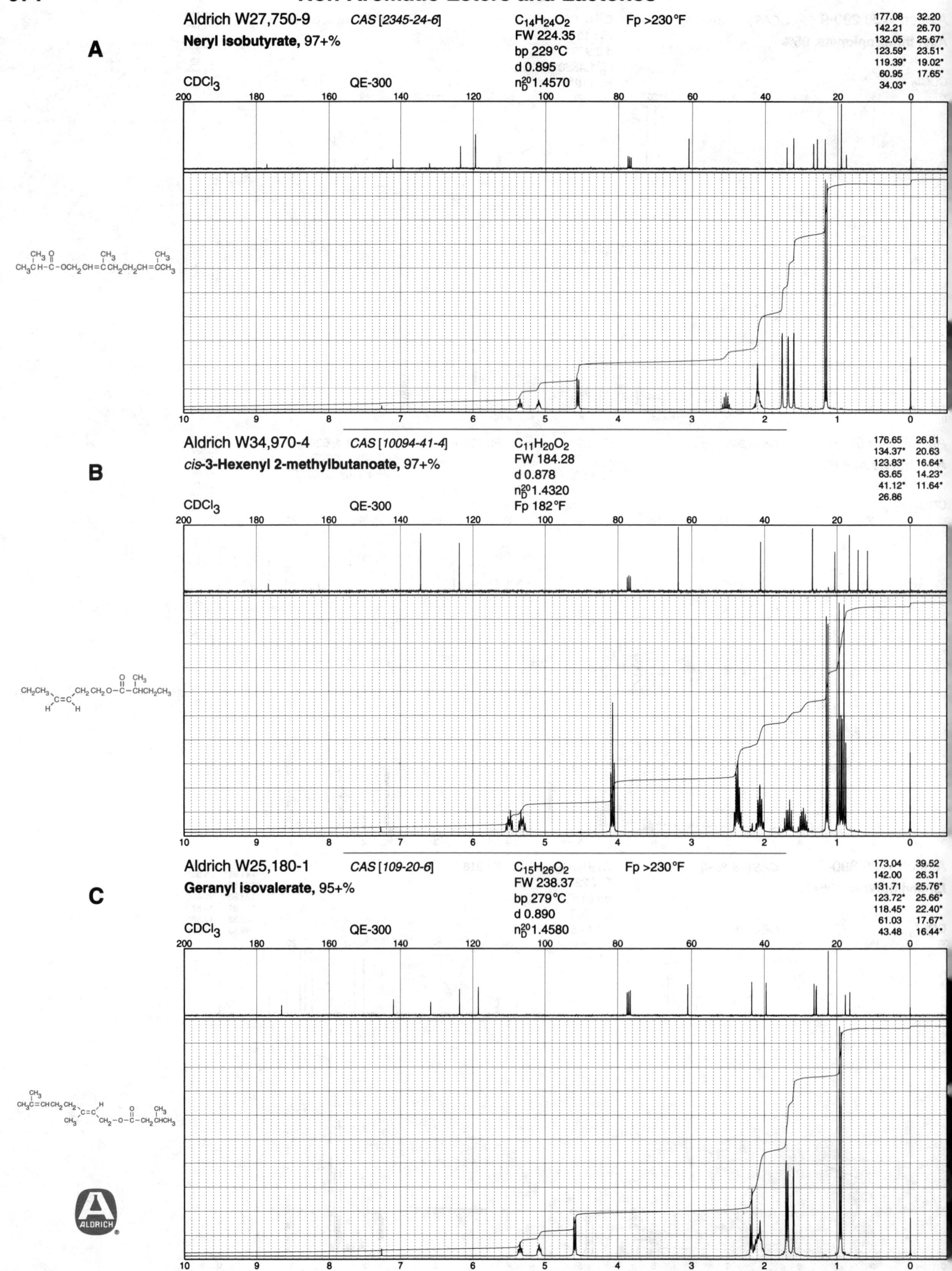

A

Aldrich W27,750-9   CAS [2345-24-6]   C₁₄H₂₄O₂   Fp >230°F

Neryl isobutyrate, 97+%

FW 224.35
bp 229°C
d 0.895
n²⁰_D 1.4570

CDCl₃   QE-300

177.08   32.20
142.21   26.70
132.05   25.67*
123.59*   23.51*
119.39*   19.02*
60.95   17.65*
34.03*

B

Aldrich W34,970-4   CAS [10094-41-4]   C₁₁H₂₀O₂

***cis*-3-Hexenyl 2-methylbutanoate, 97+%**

FW 184.28
d 0.878
n²⁰_D 1.4320
Fp 182°F

CDCl₃   QE-300

176.65   26.81
134.37*   20.63
123.83*   16.64*
63.65   14.23*
41.12*   11.64*
26.86

C

Aldrich W25,180-1   CAS [109-20-6]   C₁₅H₂₆O₂   Fp >230°F

Geranyl isovalerate, 95+%

FW 238.37
bp 279°C
d 0.890
n²⁰_D 1.4580

CDCl₃   QE-300

173.04   39.52
142.00   26.31
131.71   25.76*
123.72*   25.66*
118.45*   22.40*
61.03   17.67*
43.48   16.44*

Aldrich W20,320-3   *CAS [123-68-2]*   C$_9$H$_{16}$O$_2$ Fp 151°F

Allyl hexanoate, 98+%

FW 156.23
bp 76°C (15 mm)
d 0.887

CDCl$_3$ QE-300   n$_D^{20}$1.4240

173.40	31.32
132.34*	24.66
117.97	22.33
64.89	13.90*
34.24	

A

CH$_3$(CH$_2$)$_3$CH$_2$—C(=O)—O CH$_2$ CH=CH$_2$

Aldrich W20,310-6   *CAS [142-19-8]*   C$_{10}$H$_{18}$O$_2$

Allyl heptanoate, 97+%

FW 170.25
d 0.885
n$_D^{20}$1.4280
Fp 180°F

CDCl$_3$ QE-300

173.39	31.48
132.37*	28.84
117.97	24.95
64.89	22.50
34.29	14.02*

B

CH$_3$(CH$_2$)$_4$CH$_2$—C(=O)—OCH$_2$ CH=CH$_2$

Aldrich W20,371-8   *CAS [4230-97-1]*   C$_{11}$H$_{20}$O$_2$ Fp 203°F

Allyl octanoate, 99+%

FW 184.28
bp 88°C (6 mm)
d 0.881

CDCl$_3$ QE-300   n$_D^{20}$1.4320

173.33	29.15
132.43*	28.94
117.92	25.01
64.87	22.61
34.31	14.03*
31.69	

C

CH$_3$(CH$_2$)$_5$CH$_2$—C(=O)—O CH$_2$ CH=CH$_2$

ALDRICH

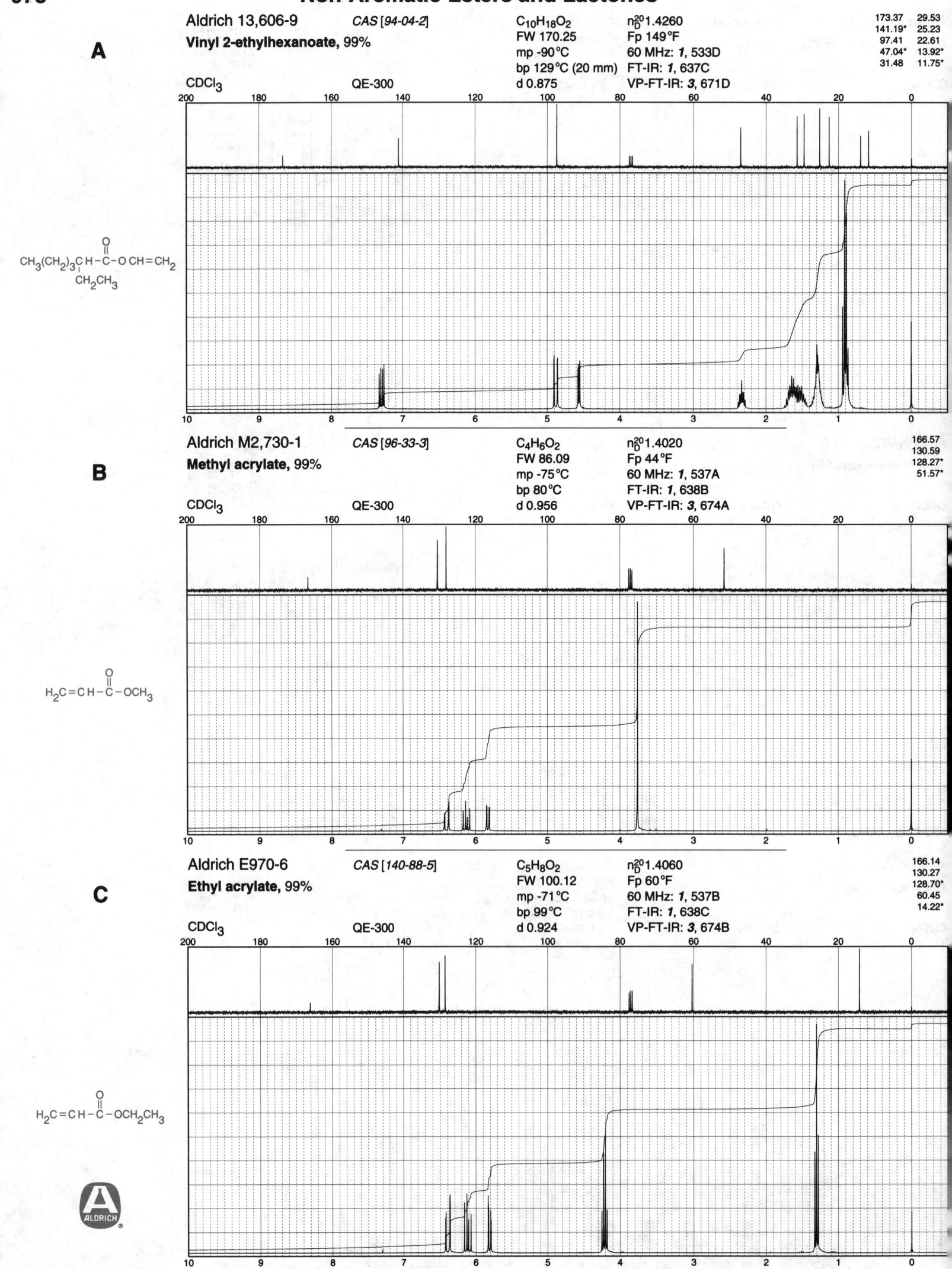

A

Aldrich 13,606-9
CAS [94-04-2]
Vinyl 2-ethylhexanoate, 99%

$C_{10}H_{18}O_2$
FW 170.25
mp -90°C
bp 129°C (20 mm)
d 0.875

n_D^{20} 1.4260
Fp 149°F
60 MHz: *1*, 533D
FT-IR: *1*, 637C
VP-FT-IR: *3*, 671D

173.37 29.53
141.19* 25.23
97.41 22.61
47.04* 13.92*
31.48 11.75*

CDCl₃ QE-300

$CH_3(CH_2)_3CH-C-O-CH=CH_2$
$\quad\quad\quad CH_2CH_3$

B

Aldrich M2,730-1
CAS [96-33-3]
Methyl acrylate, 99%

$C_4H_6O_2$
FW 86.09
mp -75°C
bp 80°C
d 0.956

n_D^{20} 1.4020
Fp 44°F
60 MHz: *1*, 537A
FT-IR: *1*, 638B
VP-FT-IR: *3*, 674A

166.57
130.59
128.27*
51.57*

CDCl₃ QE-300

$H_2C=CH-C-OCH_3$

C

Aldrich E970-6
CAS [140-88-5]
Ethyl acrylate, 99%

$C_5H_8O_2$
FW 100.12
mp -71°C
bp 99°C
d 0.924

n_D^{20} 1.4060
Fp 60°F
60 MHz: *1*, 537B
FT-IR: *1*, 638C
VP-FT-IR: *3*, 674B

166.14
130.27
128.70*
60.45
14.22*

CDCl₃ QE-300

$H_2C=CH-C-OCH_2CH_3$

A

Aldrich 23,492-3

Butyl acrylate, 99+%

CAS [141-32-2]

$C_7H_{12}O_2$
FW 128.17
bp 145°C
d 0.894
n_D^{20}1.4180

Fp 103°F

60 MHz: *1*, 541B
FT-IR: *1*, 642A
VP-FT-IR: *3*, 674C

166.23
130.23
128.69*
64.38
30.73
19.19
13.70*

CDCl₃

QE-300

$H_2C=CH-\overset{\overset{\displaystyle O}{\|}}{C}-OCH_2CH_2CH_2CH_3$

B

Aldrich 32,718-2

tert-**Butyl acrylate, 98%**

CAS [1663-39-4]

$C_7H_{12}O_2$
FW 128.17
bp 62°C (60 mm)
d 0.875
n_D^{20}1.4110

Fp 63°F

VP-FT-IR: *3*, 674D

165.48
130.32*
129.19
80.46
28.07*

CDCl₃

QE-300

$H_2C=CH-\overset{\overset{\displaystyle O}{\|}}{C}-O\overset{\overset{\displaystyle CH_3}{|}}{\underset{\underset{\displaystyle CH_3}{|}}{C}}CH_3$

C

Aldrich 29,081-4

(±)-2-Ethylhexyl acrylate, 98%

CAS [103-11-7]

$C_{11}H_{20}O_2$
FW 184.28
bp 217°C
d 0.885
n_D^{20}1.4360

Fp 175°F

VP-FT-IR: *3*, 675A

166.38   28.99
130.23   23.90
128.78* 22.99
67.00   14.03*
38.88*   11.02*
30.51

CDCl₃

QE-300

$H_2C=CH-\overset{\overset{\displaystyle O}{\|}}{C}-OCH_2\overset{\overset{\displaystyle CH_2CH_3}{|}}{CH}(CH_2)_3CH_3$

ALDRICH

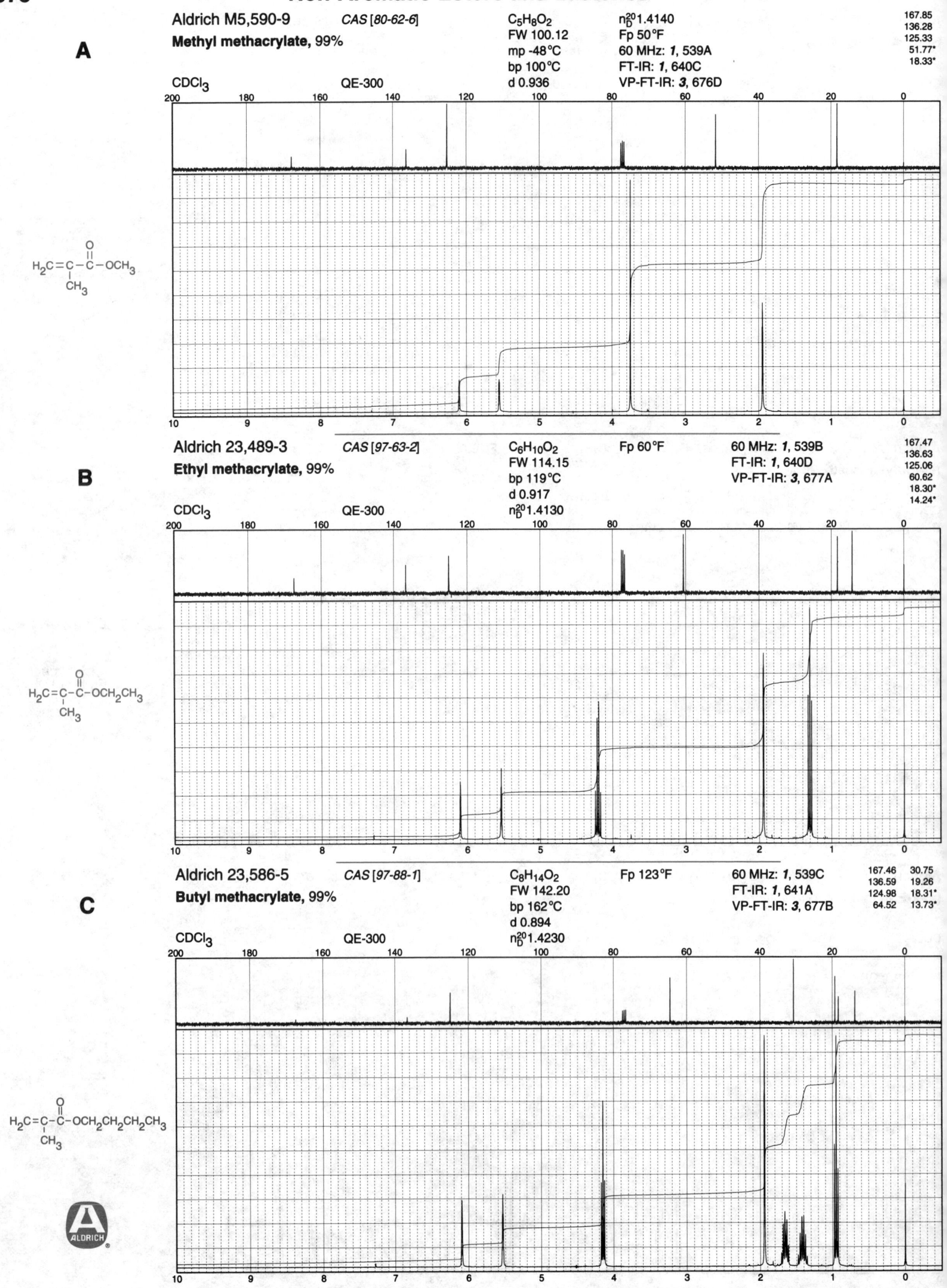

Aldrich M5,590-9   CAS [80-62-6]   C₅H₈O₂   n₂⁰ 1.4140
Methyl methacrylate, 99%   FW 100.12   Fp 50°F
mp -48°C   60 MHz: *1*, 539A
bp 100°C   FT-IR: *1*, 640C
d 0.936   VP-FT-IR: *3*, 676D

CDCl₃   QE-300

A

167.85
136.28
125.33
51.77*
18.33*

Aldrich 23,489-3   CAS [97-63-2]   C₆H₁₀O₂   Fp 60°F   60 MHz: *1*, 539B
Ethyl methacrylate, 99%   FW 114.15   FT-IR: *1*, 640D
bp 119°C   VP-FT-IR: *3*, 677A
d 0.917
n₂⁰ 1.4130

CDCl₃   QE-300

B

167.47
136.63
125.06
60.62
18.30*
14.24*

Aldrich 23,586-5   CAS [97-88-1]   C₈H₁₄O₂   Fp 123°F   60 MHz: *1*, 539C
Butyl methacrylate, 99%   FW 142.20   FT-IR: *1*, 641A
bp 162°C   VP-FT-IR: *3*, 677B
d 0.894
n₂⁰ 1.4230

CDCl₃   QE-300

C

167.46   30.75
136.59   19.26
124.98   18.31*
64.52   13.73*

Aldrich 16,991-9   *CAS [97-86-9]*   C$_8$H$_{14}$O$_2$ Fp 107°F   VP-FT-IR: *3*, 677C

Isobutyl methacrylate, 97%

FW 142.20
bp 155°C
d 0.886
n$_D^{20}$1.4200

| 167.40 |
| 136.53 |
| 125.07 |
| 70.70 |
| 27.81* |
| 19.13* |
| 18.33* |

A

CDCl$_3$ QE-300

H$_2$C=C—C—O—CH$_2$CHCH$_3$ with CH$_3$ and CH$_3$ groups

Aldrich 29,080-7   *CAS [688-84-6]*   C$_{12}$H$_{22}$O$_2$ Fp 198°F   VP-FT-IR: *3*, 678A

(±)-2-Ethylhexyl methacrylate, 98%

FW 198.31
bp 120°C (18 mm)
d 0.885
n$_D^{20}$1.4380

167.52	28.99
136.58	24.01
125.01	22.99
67.04	18.33*
38.87*	14.03*
30.60	11.07*

B

CDCl$_3$ QE-300

H$_2$C=C—C—O—CH$_2$CH(CH$_2$)$_3$CH$_3$ with CH$_3$ and CH$_2$CH$_3$ groups

---

Aldrich 29,181-1   *CAS [142-90-5]*   C$_{16}$H$_{30}$O$_2$ n$_D^{20}$1.4450

Lauryl methacrylate, 96%

FW 254.42
mp -7°C
bp 142°C (4 mm)
d 0.868

Fp >230°F
VP-FT-IR: *3*, 678B

167.50	29.68	28.69
136.65	29.61	26.04
125.02	29.57	22.73
64.85	29.39	18.32*
31.96	29.30	14.11*

C

CDCl$_3$ QE-300

H$_2$C=C—C—O—OCH$_2$(CH$_2$)$_{10}$CH$_3$ with CH$_3$ group

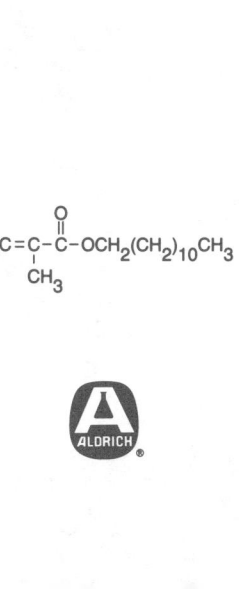

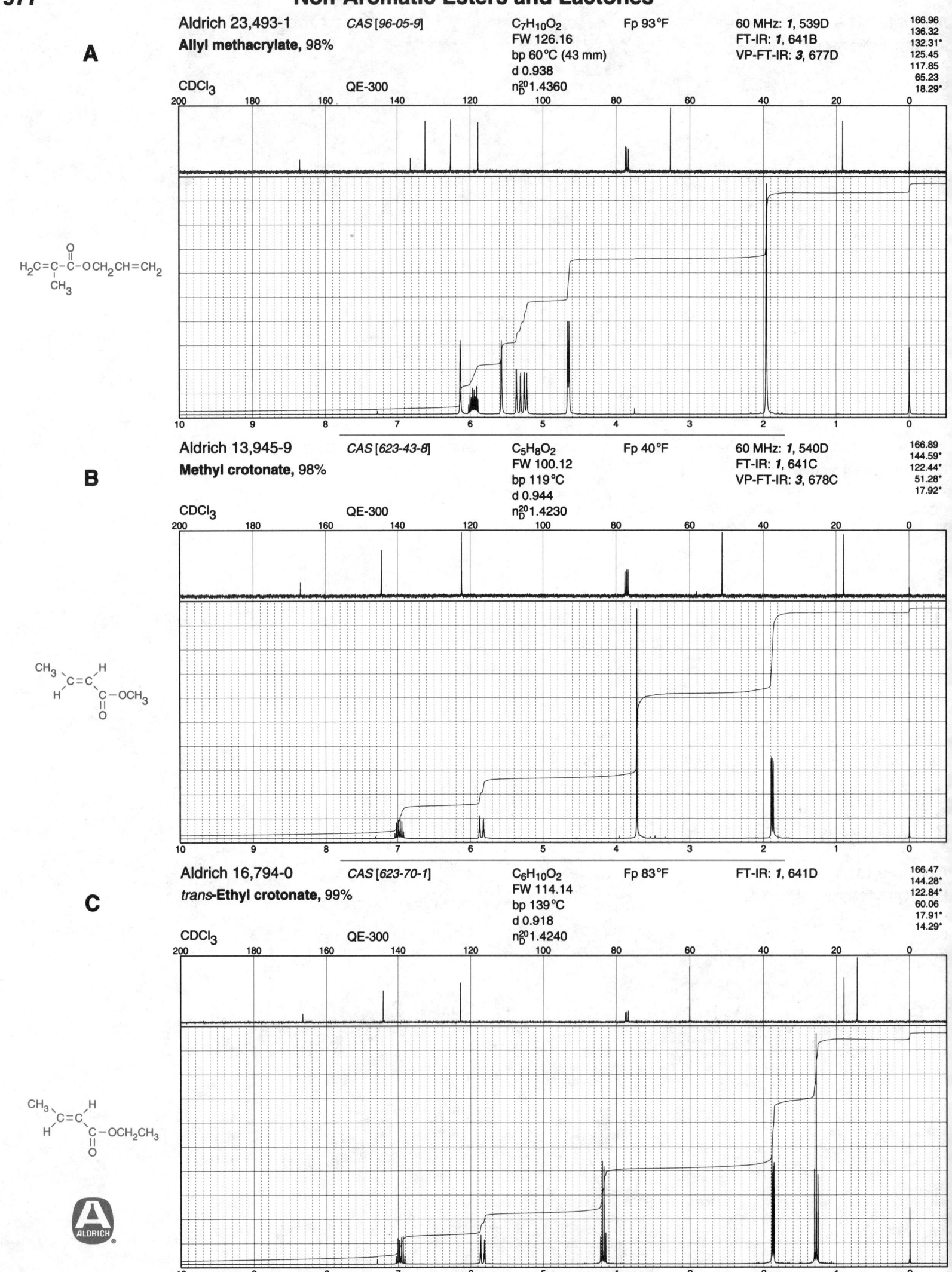

A

Aldrich 23,493-1   *CAS [96-05-9]*   C₇H₁₀O₂   Fp 93°F   60 MHz: *1*, 539D

Allyl methacrylate, 98%

$C_7H_{10}O_2$ FW 126.16   bp 60°C (43 mm)   d 0.938 n_D^{20}1.4360   FT-IR: *1*, 641B   VP-FT-IR: *3*, 677D

166.96
136.32
132.31*
125.45
117.85
65.23
18.29*

CDCl₃   QE-300

H₂C=C–C–OCH₂CH=CH₂
   | ‖
   CH₃ O

B

Aldrich 13,945-9   *CAS [623-43-8]*   Fp 40°F   60 MHz: *1*, 540D

Methyl crotonate, 98%

$C_5H_8O_2$ FW 100.12   bp 119°C   d 0.944 n_D^{20}1.4230   FT-IR: *1*, 641C   VP-FT-IR: *3*, 678C

166.89
144.59*
122.44*
51.28*
17.92*

CDCl₃   QE-300

CH₃   H
   \   /
    C=C
   /   \
  H   C–OCH₃
         ‖
         O

C

Aldrich 16,794-0   *CAS [623-70-1]*   C₆H₁₀O₂   Fp 83°F   FT-IR: *1*, 641D

***trans*-Ethyl crotonate, 99%**

$C_6H_{10}O_2$ FW 114.14   bp 139°C   d 0.918 n_D^{20}1.4240

166.47
144.28*
122.84*
60.06
17.91*
14.29*

CDCl₃   QE-300

CH₃   H
   \   /
    C=C
   /   \
  H   C–OCH₂CH₃
         ‖
         O

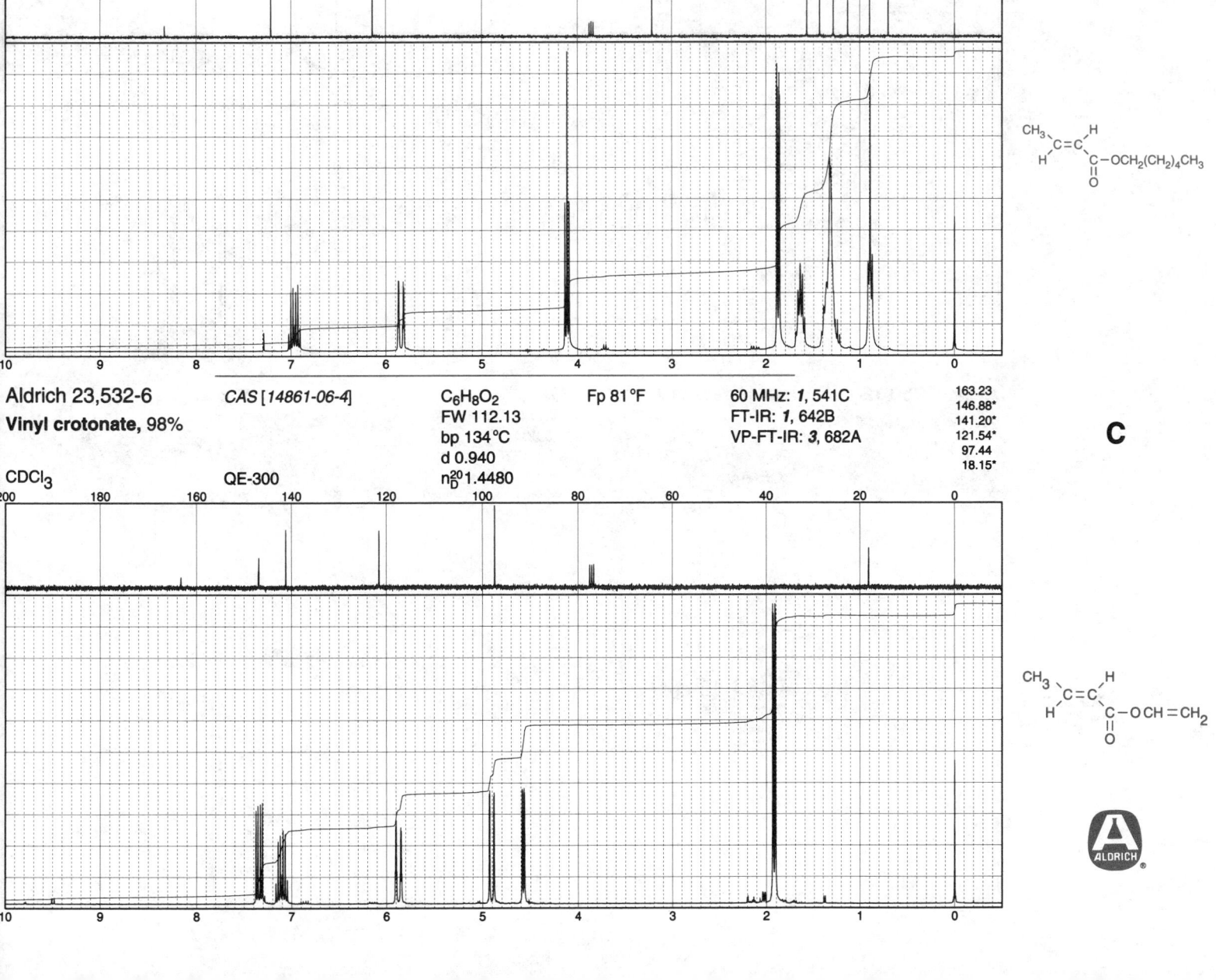

Aldrich W34,321-8   *CAS [589-66-2]*
Isobutyl *trans*-2-butenoate, 98+%

C$_8$H$_{14}$O$_2$ Fp 131°F
FW 142.19
bp 171°C
d 0.890
n$_D^{20}$1.4280

A

166.49
144.10*
122.92*
70.26
27.88*
19.13*
17.85*

CDCl$_3$ QE-300

Aldrich W33,540-1   *CAS [19089-92-0]*
Hexyl *trans*-2-butenoate, 95+%

C$_{10}$H$_{18}$O$_2$
FW 170.25
d 0.885
n$_D^{20}$1.4380
Fp 193°F

B

166.58   28.68
144.26*   25.66
122.82*   22.57
64.31   17.94*
31.49   14.01*

CDCl$_3$ QE-300

Aldrich 23,532-6   *CAS [14861-06-4]*
Vinyl crotonate, 98%

C$_6$H$_8$O$_2$ Fp 81°F   60 MHz: *1*, 541C
FW 112.13   FT-IR: *1*, 642B
bp 134°C   VP-FT-IR: *3*, 682A
d 0.940
n$_D^{20}$1.4480

C

163.23
146.88*
141.20*
121.54*
97.44
18.15*

CDCl$_3$ QE-300

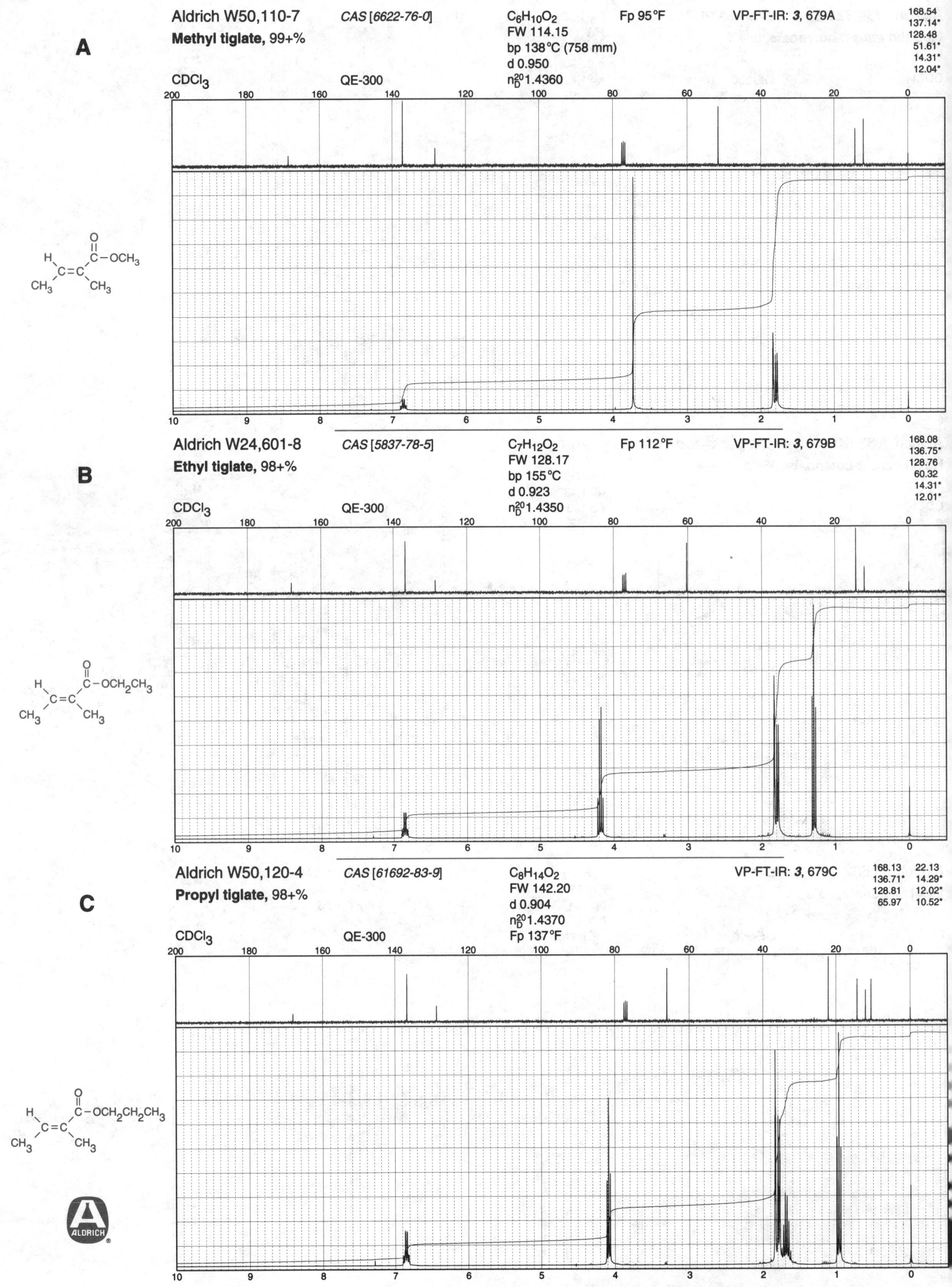

A

Aldrich W50,110-7   *CAS [6622-76-0]* $C_6H_{10}O_2$ Fp 95°F   VP-FT-IR: **3**, 679A

Methyl tiglate, 99+%

FW 114.15
bp 138°C (758 mm)
d 0.950
n_D^{20}1.4360

168.54
137.14*
128.48
51.61*
14.31*
12.04*

CDCl$_3$ QE-300

B

Aldrich W24,601-8   *CAS [5837-78-5]* $C_7H_{12}O_2$ Fp 112°F   VP-FT-IR: **3**, 679B

Ethyl tiglate, 98+%

FW 128.17
bp 155°C
d 0.923
n_D^{20}1.4350

168.08
136.75*
128.76
60.32
14.31*
12.01*

CDCl$_3$ QE-300

C

Aldrich W50,120-4   *CAS [61692-83-9]* $C_8H_{14}O_2$ VP-FT-IR: **3**, 679C

Propyl tiglate, 98+%

FW 142.20
d 0.904
n_D^{20}1.4370
Fp 137°F

168.13   22.13
136.71*   14.29*
128.81   12.02*
65.97   10.52*

CDCl$_3$ QE-300

Aldrich W32,290-3   CAS [1733-25-1]   C$_8$H$_{14}$O$_2$ VP-FT-IR: **3**, 679D

Isopropyl tiglate, 98+%

FW 142.20
d 0.896
n$_D^{20}$1.4310
Fp 123°F

167.57
136.39*
129.12
67.45*
21.91*
14.26*
12.00*

A

CDCl$_3$ QE-300

Aldrich 34,051-0   CAS [28127-58-4]   C$_9$H$_{16}$O$_2$ Fp 151°F

(±)-sec-Butyl tiglate, 97%

FW 156.23
bp 85°C (27 mm)
d 0.889
n$_D^{20}$1.4350

167.73   19.52*
136.37* 14.28*
129.11   12.05*
71.99*   9.73*
28.94

B

CDCl$_3$ QE-300

Aldrich W50,070-4   CAS [61692-84-0]   C$_9$H$_{16}$O$_2$ Fp 153°F   VP-FT-IR: **3**, 680A

Isobutyl tiglate, 98+%

FW 156.23
bp 183°C
d 0.899
n$_D^{20}$1.4380

168.09   27.87*
136.73* 19.18*
128.77   14.31*
70.49   12.03*

C

CDCl$_3$ QE-300

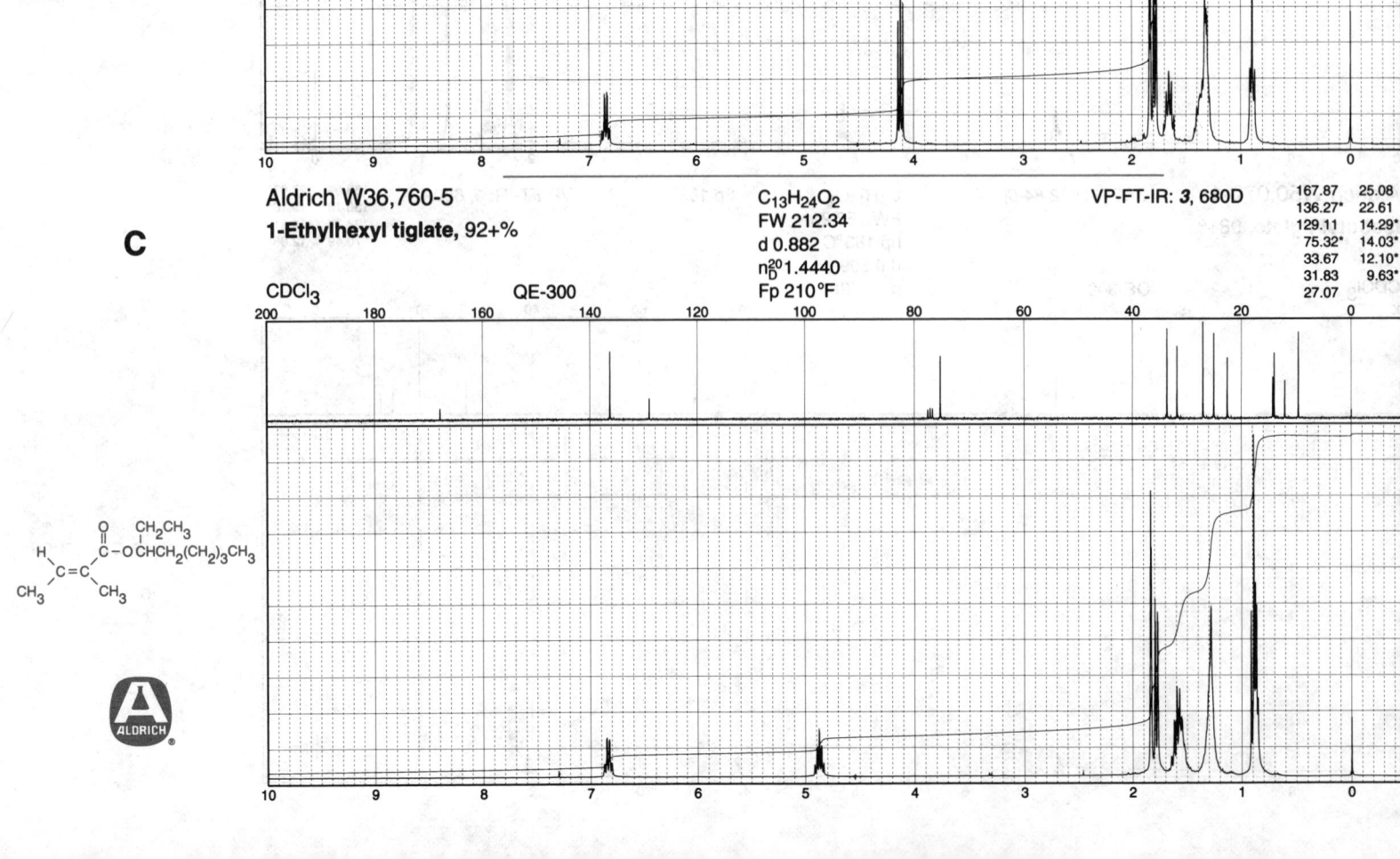

A

Aldrich W50,101-8   CAS [41519-18-0]   C$_{10}$H$_{18}$O$_2$ Fp 176°F   VP-FT-IR: **3**, 680B

Isoamyl tiglate, 97+%   FW 170.25
bp 205°C
d 0.897
n$_D^{20}$1.4410

CDCl$_3$ QE-300

168.13	25.22*
136.70*	22.51*
128.80	14.29*
63.04	12.02*
37.48	

B

Aldrich W50,090-9   CAS [16930-96-4]   C$_{11}$H$_{20}$O$_2$ Fp 209°F   VP-FT-IR: **3**, 680C

Hexyl tiglate, 97+%   FW 184.28
bp 108°C (16 mm)
d 0.894
n$_D^{20}$1.4460

CDCl$_3$ QE-300

168.14	25.73
136.69*	22.57
128.80	14.29*
64.55	14.00*
31.50	12.03*
28.71	

C

Aldrich W36,760-5   C$_{13}$H$_{24}$O$_2$ VP-FT-IR: **3**, 680D

1-Ethylhexyl tiglate, 92+%   FW 212.34
d 0.882
n$_D^{20}$1.4440
Fp 210°F

CDCl$_3$ QE-300

167.87	25.08
136.27*	22.61
129.11	14.29*
75.32*	14.03*
33.67	12.10*
31.83	9.63*
27.07	

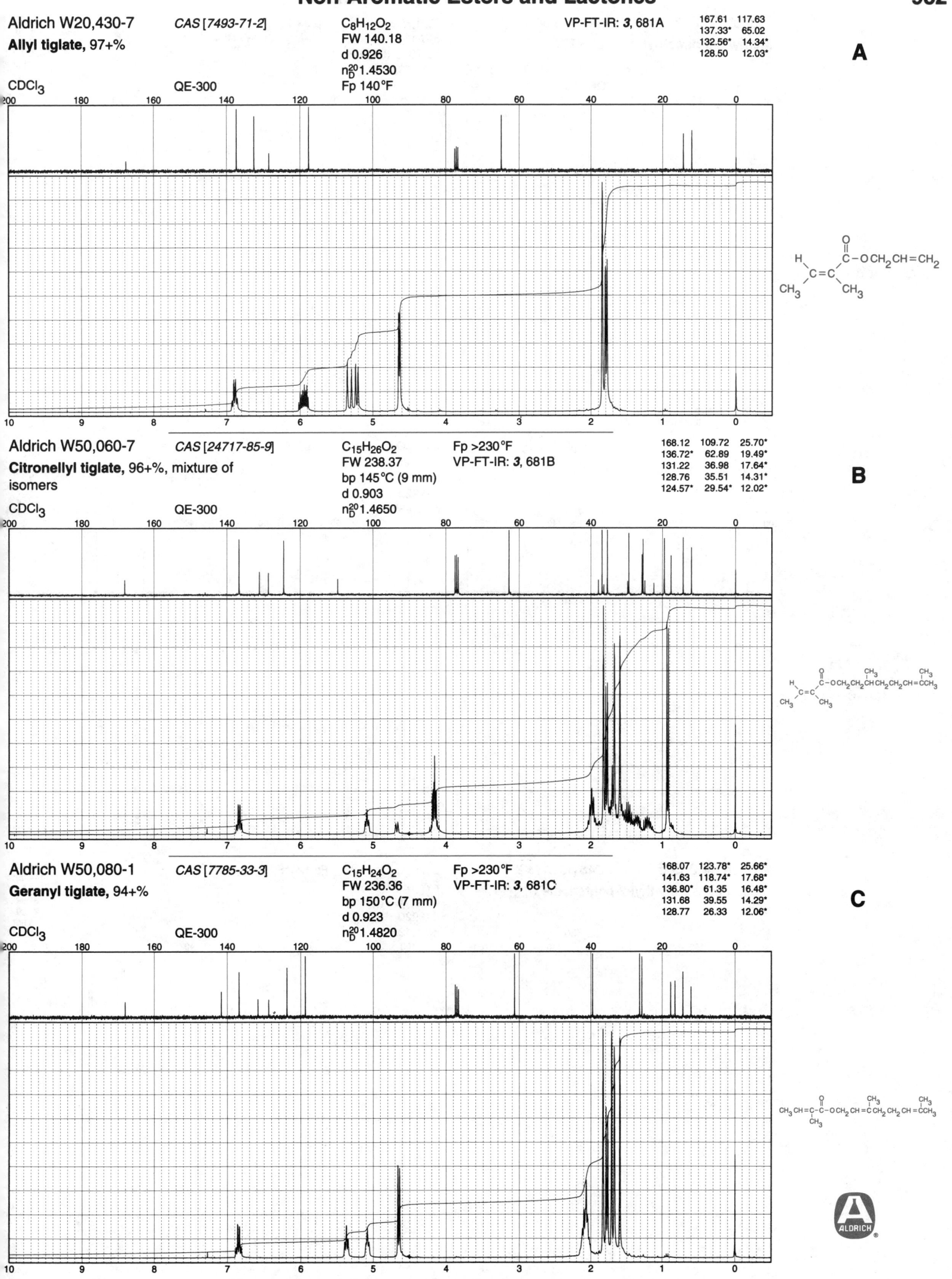

Aldrich W20,430-7   CAS [7493-71-2]   C₈H₁₂O₂   VP-FT-IR: 3, 681A
Allyl tiglate, 97+%   FW 140.18
d 0.926
n_D^{20} 1.4530
CDCl₃   QE-300   Fp 140°F

167.61	117.63
137.33*	65.02
132.56*	14.34*
128.50	12.03*

A

Aldrich W50,060-7   CAS [24717-85-9]   C₁₅H₂₆O₂   Fp >230°F
Citronellyl tiglate, 96+%, mixture of   FW 238.37   VP-FT-IR: 3, 681B
isomers   bp 145°C (9 mm)
d 0.903
CDCl₃   QE-300 n_D^{20} 1.4650

168.12	109.72	25.70*
136.72*	62.89	19.49*
131.22	36.98	17.64*
128.76	35.51	14.31*
124.57*	29.54*	12.02*

B

Aldrich W50,080-1   CAS [7785-33-3]   C₁₅H₂₄O₂   Fp >230°F
Geranyl tiglate, 94+%   FW 236.36   VP-FT-IR: 3, 681C
bp 150°C (7 mm)
d 0.923
CDCl₃   QE-300 n_D^{20} 1.4820

168.07	123.78*	25.66*
141.63	118.74*	17.68*
136.80*	61.35	16.48*
131.68	39.55	14.29*
128.77	26.33	12.06*

C

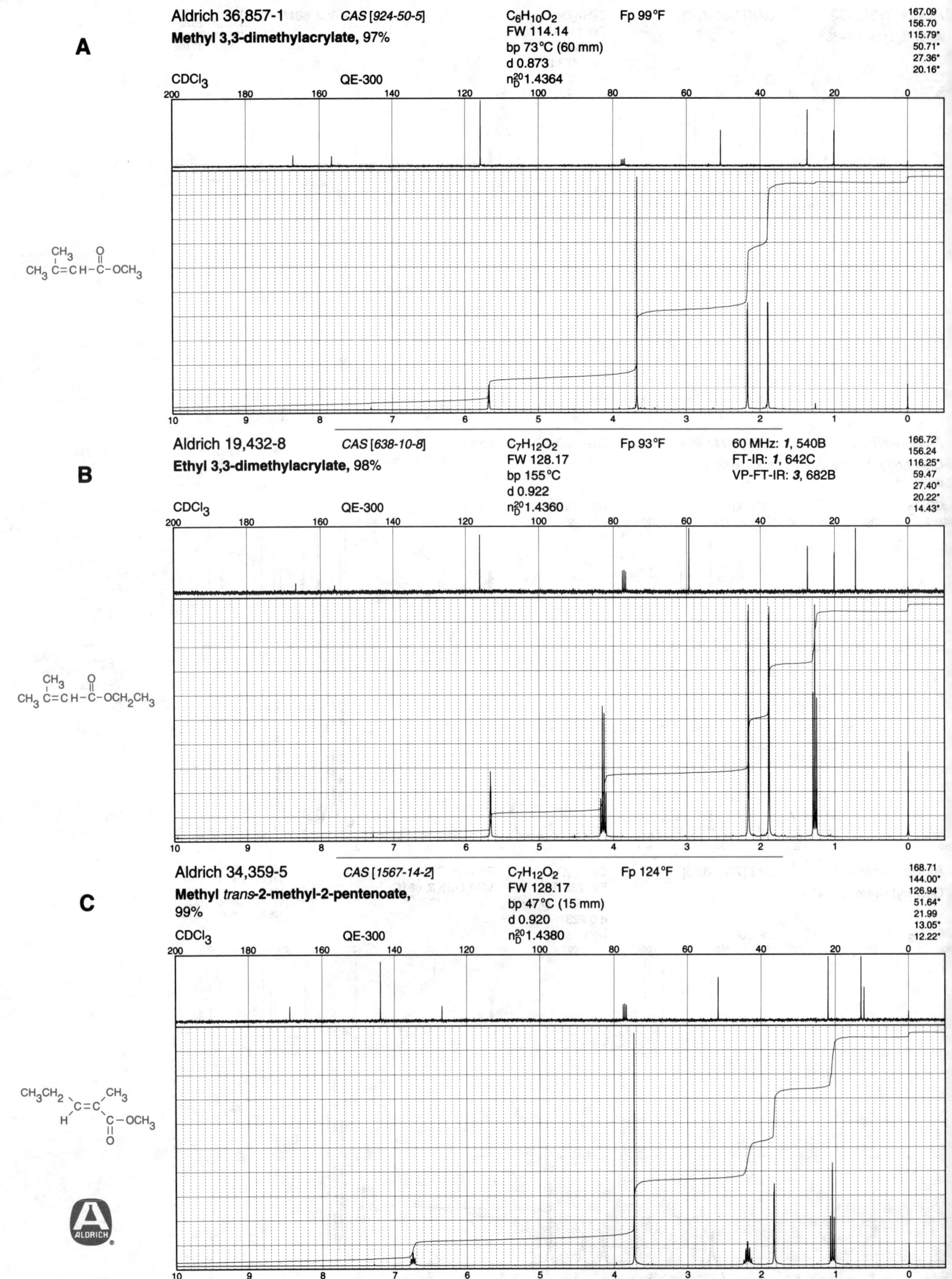

A

Aldrich 36,857-1   *CAS [924-50-5]*
Methyl 3,3-dimethylacrylate, 97%

$C_6H_{10}O_2$
FW 114.14
bp 73°C (60 mm)
d 0.873
n_D^{20} 1.4364

Fp 99°F

CDCl₃   QE-300

167.09
156.70
115.79*
50.71*
27.36*
20.16*

B

Aldrich 19,432-8   *CAS [638-10-8]*
Ethyl 3,3-dimethylacrylate, 98%

$C_7H_{12}O_2$
FW 128.17
bp 155°C
d 0.922
n_D^{20} 1.4360

Fp 93°F

60 MHz: *1*, 540B
FT-IR: *1*, 642C
VP-FT-IR: *3*, 682B

CDCl₃   QE-300

166.72
156.24
116.25*
59.47
27.40*
20.22*
14.43*

C

Aldrich 34,359-5   *CAS [1567-14-2]*
Methyl *trans*-2-methyl-2-pentenoate,
99%

$C_7H_{12}O_2$
FW 128.17
bp 47°C (15 mm)
d 0.920
n_D^{20} 1.4380

Fp 124°F

CDCl₃   QE-300

168.71
144.00*
126.94
51.64*
21.99
13.05*
12.22*

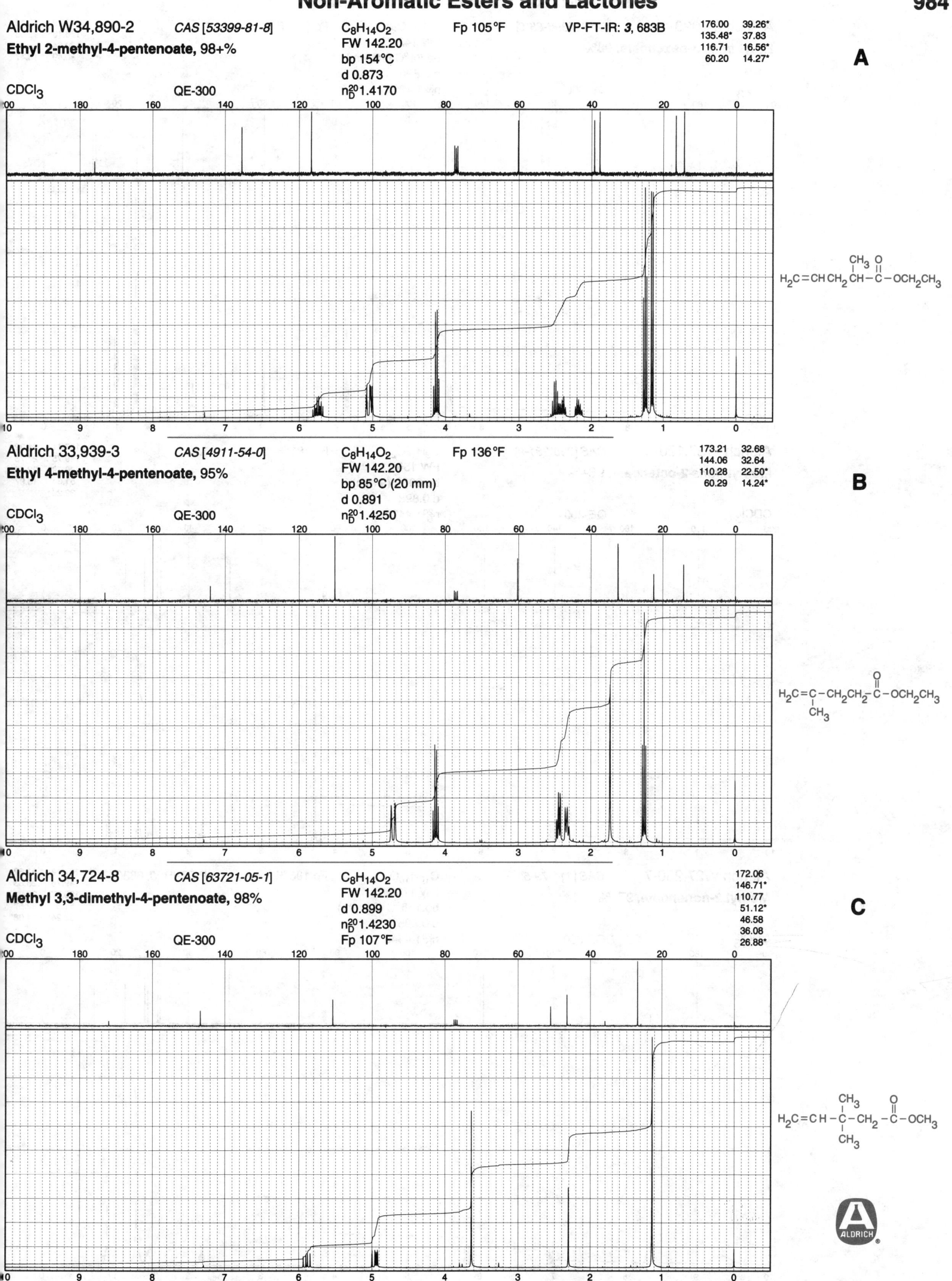

Aldrich W34,890-2   *CAS [53399-81-8]*
Ethyl 2-methyl-4-pentenoate, 98+%

$C_8H_{14}O_2$
FW 142.20
bp 154°C
d 0.873
n_D^{20} 1.4170

Fp 105°F

VP-FT-IR: *3*, 683B

176.00	39.26*
135.48*	37.83
116.71	16.56*
60.20	14.27*

CDCl₃   QE-300

A

Aldrich 33,939-3   *CAS [4911-54-0]*
Ethyl 4-methyl-4-pentenoate, 95%

$C_8H_{14}O_2$
FW 142.20
bp 85°C (20 mm)
d 0.891
n_D^{20} 1.4250

Fp 136°F

173.21	32.68
144.06	32.64
110.28	22.50*
60.29	14.24*

CDCl₃   QE-300

B

Aldrich 34,724-8   *CAS [63721-05-1]*
Methyl 3,3-dimethyl-4-pentenoate, 98%

$C_8H_{14}O_2$
FW 142.20
d 0.899
n_D^{20} 1.4230
Fp 107°F

| 172.06 |
| 146.71* |
| 110.77 |
| 51.12* |
| 46.58 |
| 36.08 |
| 26.88* |

CDCl₃   QE-300

C

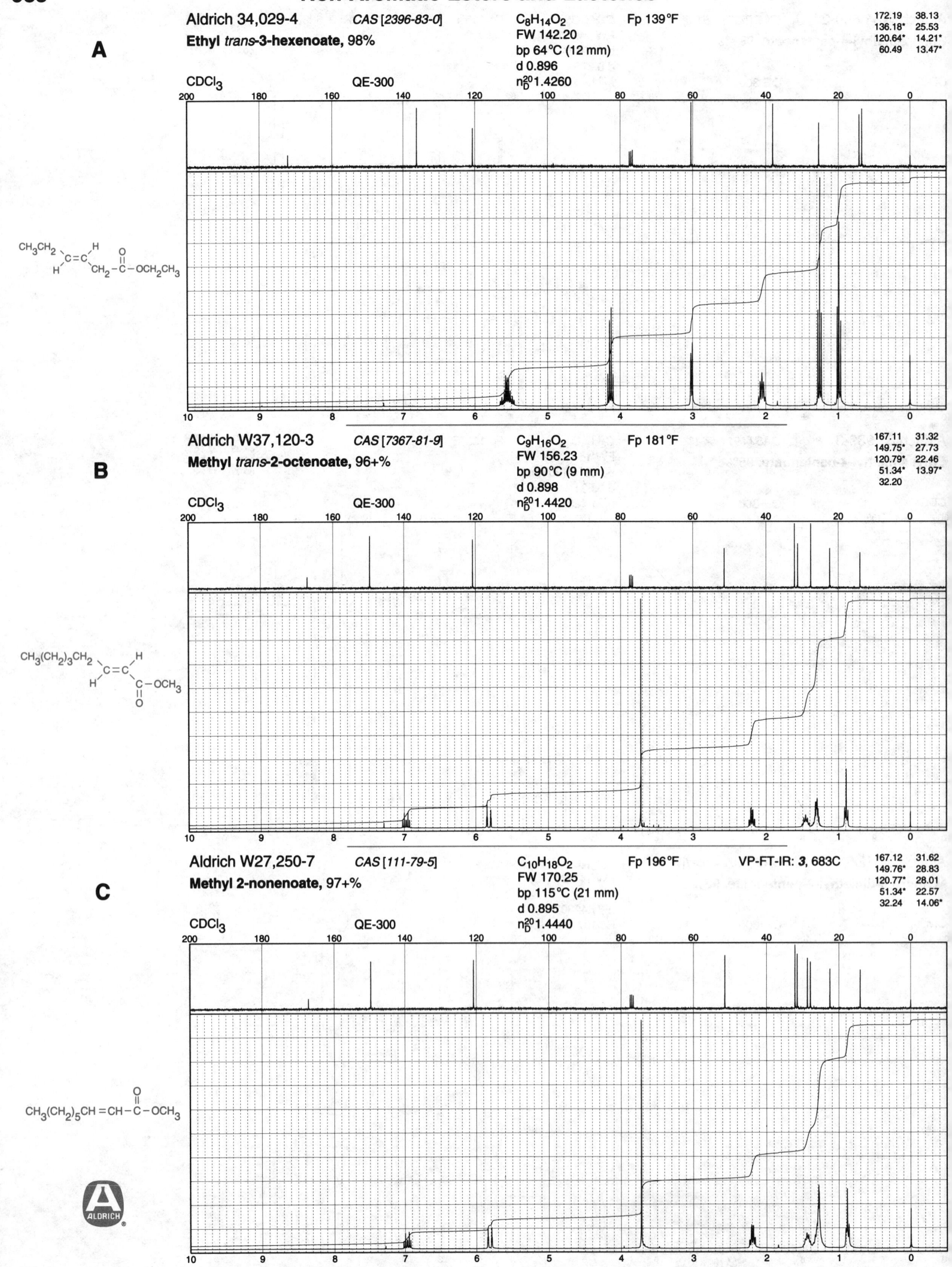

A

Aldrich 34,029-4   *CAS [2396-83-0]*   C₈H₁₄O₂   Fp 139°F

Ethyl *trans*-3-hexenoate, 98%

FW 142.20
bp 64°C (12 mm)
d 0.896
n²⁰_D 1.4260

CDCl₃   QE-300

172.19   38.13
136.18*   25.53
120.64*   14.21*
60.49   13.47*

CH₃CH₂ and H, C=C, CH₂-C(=O)-OCH₂CH₃

B

Aldrich W37,120-3   *CAS [7367-81-9]*   C₉H₁₆O₂   Fp 181°F

Methyl *trans*-2-octenoate, 96+%

FW 156.23
bp 90°C (9 mm)
d 0.898
n²⁰_D 1.4420

CDCl₃   QE-300

167.11   31.32
149.75*   27.73
120.79*   22.46
51.34*   13.97*
32.20

CH₃(CH₂)₃CH₂ and H, C=C, C(=O)-OCH₃

C

Aldrich W27,250-7   *CAS [111-79-5]*   C₁₀H₁₈O₂   Fp 196°F   VP-FT-IR: *3*, 683C

Methyl 2-nonenoate, 97+%

FW 170.25
bp 115°C (21 mm)
d 0.895
n²⁰_D 1.4440

CDCl₃   QE-300

167.12   31.62
149.76*   28.83
120.77*   28.01
51.34*   22.57
32.24   14.06*

CH₃(CH₂)₅CH=CH-C(=O)-OCH₃

ALDRICH

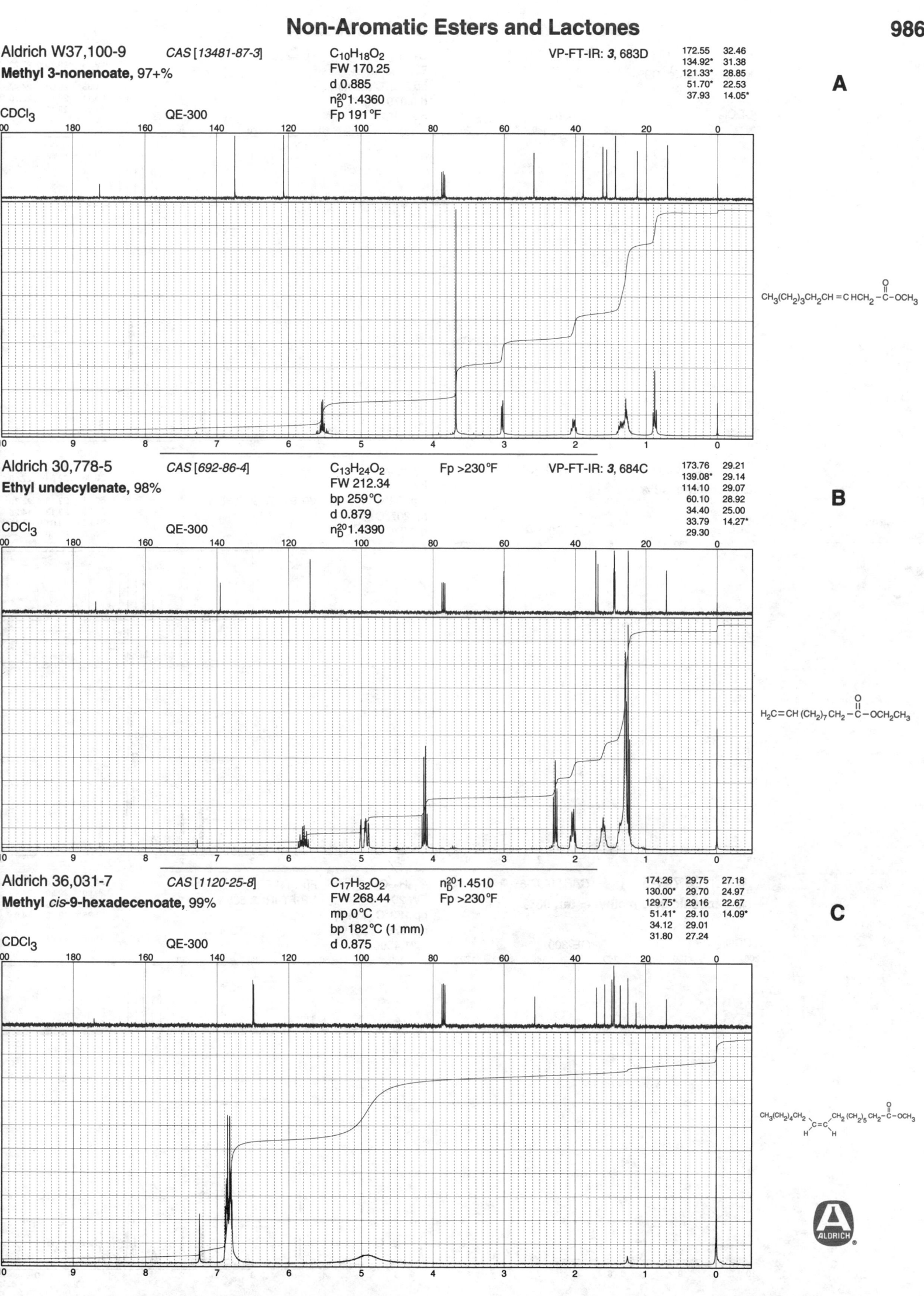

Non-Aromatic Esters and Lactones

A

Aldrich 31,111-1

CAS [112-62-9]

Methyl oleate, 99%

$C_{19}H_{36}O_2$
FW 296.50
bp 218°C (20 mm)
d 0.879
n_D^{20} 1.4520

Fp >230°F

174.18	29.79	29.11
129.95*	29.70	27.23
129.69*	29.55	27.18
51.38*	29.34	24.97
34.11	29.17	22.70
31.93	29.15	14.11*

CDCl$_3$ QE-300

CH$_3$(CH$_2$)$_6$CH$_2$ CH$_2$(CH$_2$)$_5$ C—OCH$_3$

B

Aldrich 26,801-1

CAS [111-62-6]

Ethyl oleate, 98%

$C_{20}H_{38}O_2$
FW 310.53
mp -32°C
bp 207°C
d 0.870

n_D^{20} 1.4500
Fp >230°F
VP-FT-IR: **3**, 685D

173.77	29.70	27.23
129.93*	29.54	27.18
129.70*	29.34	24.99
60.11	29.18	22.70
34.38	29.13	14.26
31.92	29.01	14.12*
29.78	28.50	

CDCl$_3$ QE-300

CH$_3$(CH$_2$)$_6$CH$_2$ CH$_2$(CH$_2$)$_5$CH$_2$ C—OCH$_2$CH$_3$

C

Aldrich 28,741-5

CAS [1937-63-9]

cis-**Vaccenic acid methyl ester, 98%**

$C_{19}H_{36}O_2$
FW 296.50
bp 163°C (1 mm)
d 0.878
n_D^{20} 1.4500

Fp 144°F
VP-FT-IR: **3**, 685A

174.16	29.78	27.24
129.89*	29.50	24.99
129.81*	29.45	22.68
51.35*	29.28	14.08*
34.13	29.19	
31.82	29.01	

CDCl$_3$ QE-300

CH$_3$(CH$_2$)$_4$CH$_2$ CH$_2$(CH$_2$)$_7$ C—OCH$_3$

ALDRICH

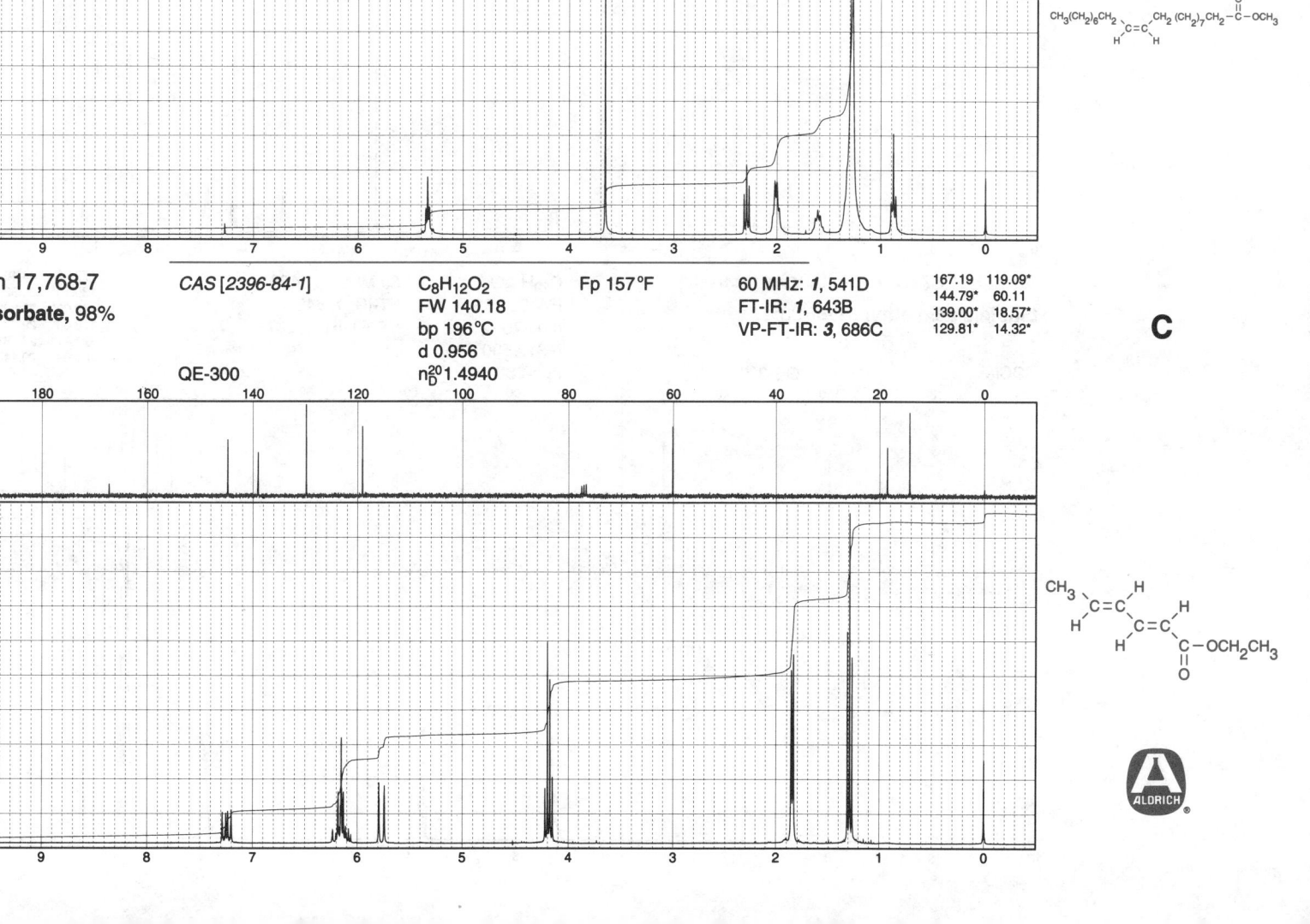

Aldrich 28,742-3   *CAS [6198-58-9]*   C$_{19}$H$_{36}$O$_2$ VP-FT-IR: *3*, 685B

trans-Vaccenic acid methyl ester, 98%

FW 296.50
bp 173 °C (3 mm)
n$_D^{20}$1.4520
Fp 144 °F

CDCl$_3$ QE-300

A

174.16	31.79	28.86
130.36*	29.66	24.99
130.28*	29.45	22.67
51.34*	29.27	14.08*
34.13	29.18	
32.61	29.14	

Aldrich 28,572-2   *CAS [2390-09-2]*   C$_{21}$H$_{40}$O$_2$ VP-FT-IR: *3*, 686A

Methyl *cis*-11-eicosenoate, 98%

FW 324.55
mp -15 °C
n$_D^{20}$1.4530
Fp >230 °F

CDCl$_3$ QE-300

B

174.18	29.79	29.18
129.87*	29.56	27.23
129.78*	29.51	24.98
51.38*	29.46	22.71
34.11	29.36	14.12*
31.94	29.29	

Aldrich 17,768-7   *CAS [2396-84-1]*   C$_8$H$_{12}$O$_2$ Fp 157 °F   60 MHz: *1*, 541D

Ethyl sorbate, 98%

FW 140.18
bp 196 °C
d 0.956
n$_D^{20}$1.4940

CDCl$_3$ QE-300

FT-IR: *1*, 643B
VP-FT-IR: *3*, 686C

C

167.19	119.09*
144.79*	60.11
139.00*	18.57*
129.81*	14.32*

A

Aldrich W31,480-3   *CAS [3025-30-7]*

Ethyl *trans*-2,*cis*-4-decadienoate, 80+%

$C_{12}H_{20}O_2$
FW 196.29
bp 71 °C
d 0.902
n_D^{20} 1.4860

Fp >230 °F

VP-FT-IR: *3*, 686D

167.22	31.40
141.54*	29.08
139.43*	28.24
126.41*	22.50
121.17*	14.33*
60.21	13.99*

CDCl₃   QE-300

B

Aldrich 10,335-7   *CAS [112-63-0]*

Methyl linoleate, 99%

$C_{19}H_{34}O_2$
FW 294.48
mp -35 °C
bp 192 °C (4 mm)
d 0.889

n_D^{20} 1.4620
Fp >230 °F
VP-FT-IR: *3*, 686B

174.18	34.09	27.20
130.14*	31.54	25.63
129.97*	29.60	24.95
128.00*	29.36	22.59
127.86*	29.17	14.07*
51.39*	29.12	

CDCl₃   QE-300

C

Aldrich 85,776-9   *CAS [544-35-4]*

Linoleic acid ethyl ester, 98%

$C_{20}H_{36}O_2$
FW 308.51
d 0.876
n_D^{20} 1.4600
Fp >230 °F

60 MHz: *1*, 542A
FT-IR: *1*, 643C
VP-FT-IR: *3*, 687A

173.71	34.40	27.23
130.16*	31.56	25.67
130.00*	29.62	25.01
128.03*	29.37	22.59
127.91*	29.18	14.27*
60.09	29.14	14.05*

CDCl₃   QE-300

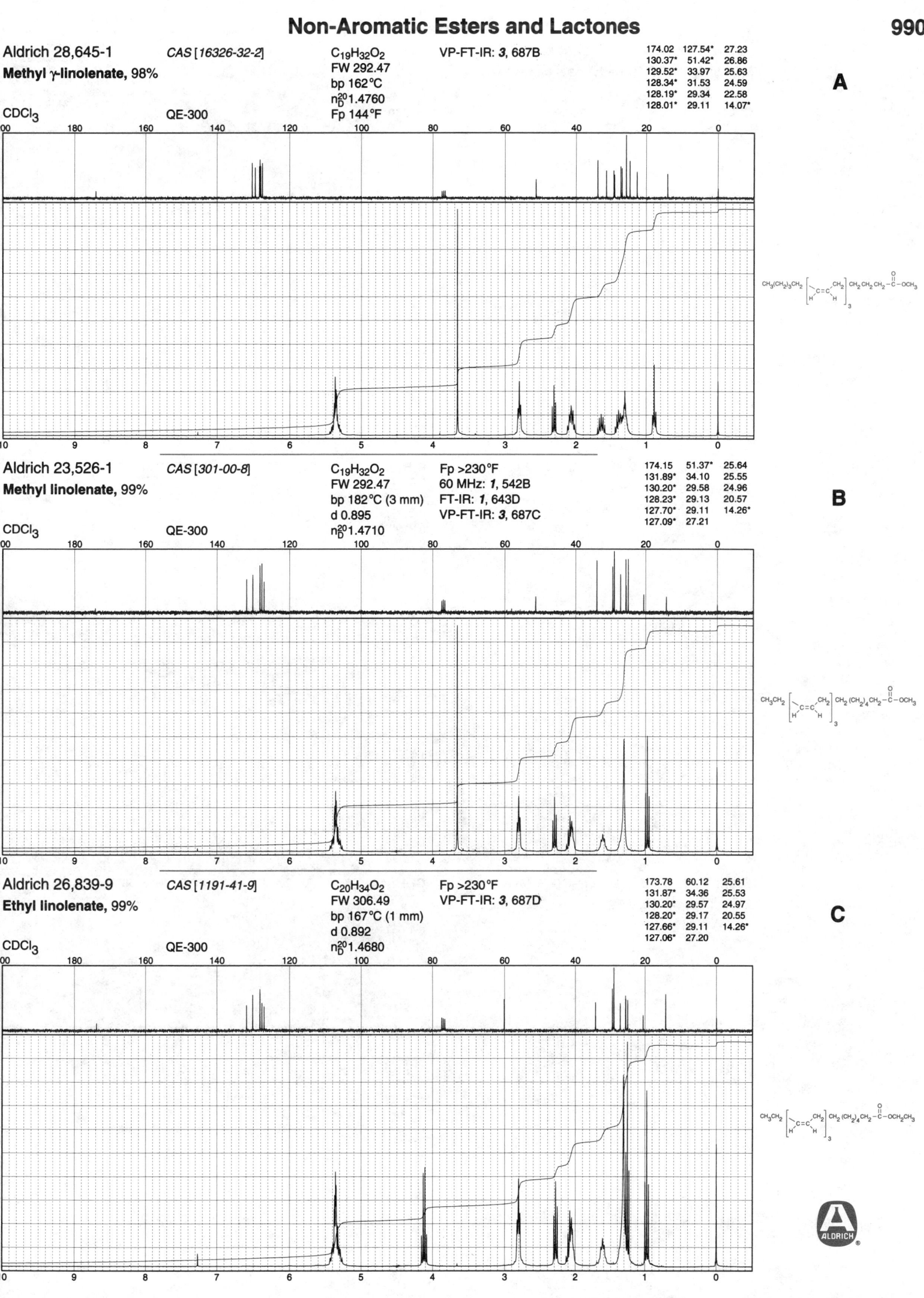

Aldrich 28,645-1   *CAS [16326-32-2]*   C₁₉H₃₂O₂
Methyl γ-linolenate, 98%   FW 292.47
                                                bp 162°C
CDCl₃   QE-300   n²⁰_D 1.4760
                                                Fp 144°F

VP-FT-IR: *3*, 687B

174.02	127.54*	27.23
130.37*	51.42*	26.86
129.52*	33.97	25.63
128.34*	31.53	24.59
128.19*	29.34	22.58
128.01*	29.11	14.07*

A

Aldrich 23,526-1   *CAS [301-00-8]*   C₁₉H₃₂O₂   Fp >230°F
Methyl linolenate, 99%   FW 292.47   60 MHz: *1*, 542B
                                              bp 182°C (3 mm)   FT-IR: *1*, 643D
CDCl₃   QE-300   d 0.895   VP-FT-IR: *3*, 687C
                                              n²⁰_D 1.4710

174.15	51.37*	25.64
131.89*	34.10	25.55
130.20*	29.58	24.96
128.23*	29.13	20.57
127.70*	29.11	14.26*
127.09*	27.21	

B

Aldrich 26,839-9   *CAS [1191-41-9]*   C₂₀H₃₄O₂   Fp >230°F
Ethyl linolenate, 99%   FW 306.49   VP-FT-IR: *3*, 687D
                                               bp 167°C (1 mm)
CDCl₃   QE-300   d 0.892
                                               n²⁰_D 1.4680

173.78	60.12	25.61
131.87*	34.36	25.53
130.20*	29.57	24.97
128.20*	29.17	20.55
127.66*	29.11	14.26*
127.06*	27.20	

C

ALDRICH

A

Aldrich 28,568-4

CAS [301-01-9]

Methyl 4,7,10,13,16,19-docosahexaenoate, 99%

CDCl$_3$ QE-300

$C_{23}H_{34}O_2$
FW 342.53
d 0.921
n$_D^{20}$1.4980
Fp 200 °F

173.45	128.06*	25.62
131.97*	128.03*	25.57
129.26*	127.82*	25.53
128.51*	126.96*	22.79
128.20*	51.54*	20.55
128.17*	34.00	14.27*

CH$_3$CH$_2$(CH=CHCH$_2$)$_6$ CH$_2$—C(=O)—OCH$_3$

B

Aldrich 28,123-9

CAS [15022-08-9]

Diallyl carbonate, 99%

CDCl$_3$ QE-300

$C_7H_{10}O_3$
FW 142.16
bp 96 °C (60 mm)
d 0.991
n$_D^{20}$1.4280

Fp 138 °F   VP-FT-IR: *3*, 688A

154.77
131.59*
118.76
68.43

H$_2$C=CHCH$_2$O—C(=O)—OCH$_2$CH=CH$_2$

C

Aldrich 10,533-3

CAS [925-16-6]

Diallyl succinate, 97%

CDCl$_3$ QE-300

$C_{10}H_{14}O_4$
FW 198.22
bp 105 °C (3 mm)
d 1.051
n$_D^{20}$1.4540

Fp >230 °F   VP-FT-IR: *3*, 688D

171.80
132.00*
118.22
65.35
29.06

H$_2$C=CHCH$_2$O—C(=O)—CH$_2$CH$_2$—C(=O)—OCH$_2$CH=CH$_2$

Aldrich 23,819-8

CAS [624-48-6]

$C_6H_8O_4$
FW 144.13
bp 205°C
d 1.152
n_D^{20} 1.4410

Fp 196°F

60 MHz: **1**, 543C
FT-IR: **1**, 644B
VP-FT-IR: **3**, 689A

165.59
129.74*
52.14*

A

Dimethyl maleate, 96%

CDCl₃

QE-300

180 160 140 120 100 80 60 40 20 0

9 8 7 6 5 4 3 2 1 0

Aldrich D9,770-3

CAS [141-05-9]

$C_8H_{12}O_4$
FW 172.18
mp -10°C
bp 225°C
d 1.064

n_D^{20} 1.4410
Fp 200°F
60 MHz: **1**, 543B
FT-IR: **1**, 644C
VP-FT-IR: **3**, 689B

165.18
129.82*
61.16
14.03*

B

Diethyl maleate, 97%

CDCl₃

QE-300

180 160 140 120 100 80 60 40 20 0

9 8 7 6 5 4 3 2 1 0

Aldrich D4,710-2

CAS [105-76-0]

$C_{12}H_{20}O_4$
FW 228.29
bp 281°C
d 0.988
n_D^{20} 1.4452

Fp >230°F

VP-FT-IR: **3**, 689D

165.26
129.73*
65.12
30.50
19.11
13.66*

C

ibutyl maleate, 98%

DCl₃

QE-300

180 160 140 120 100 80 60 40 20 0

9 8 7 6 5 4 3 2 1 0

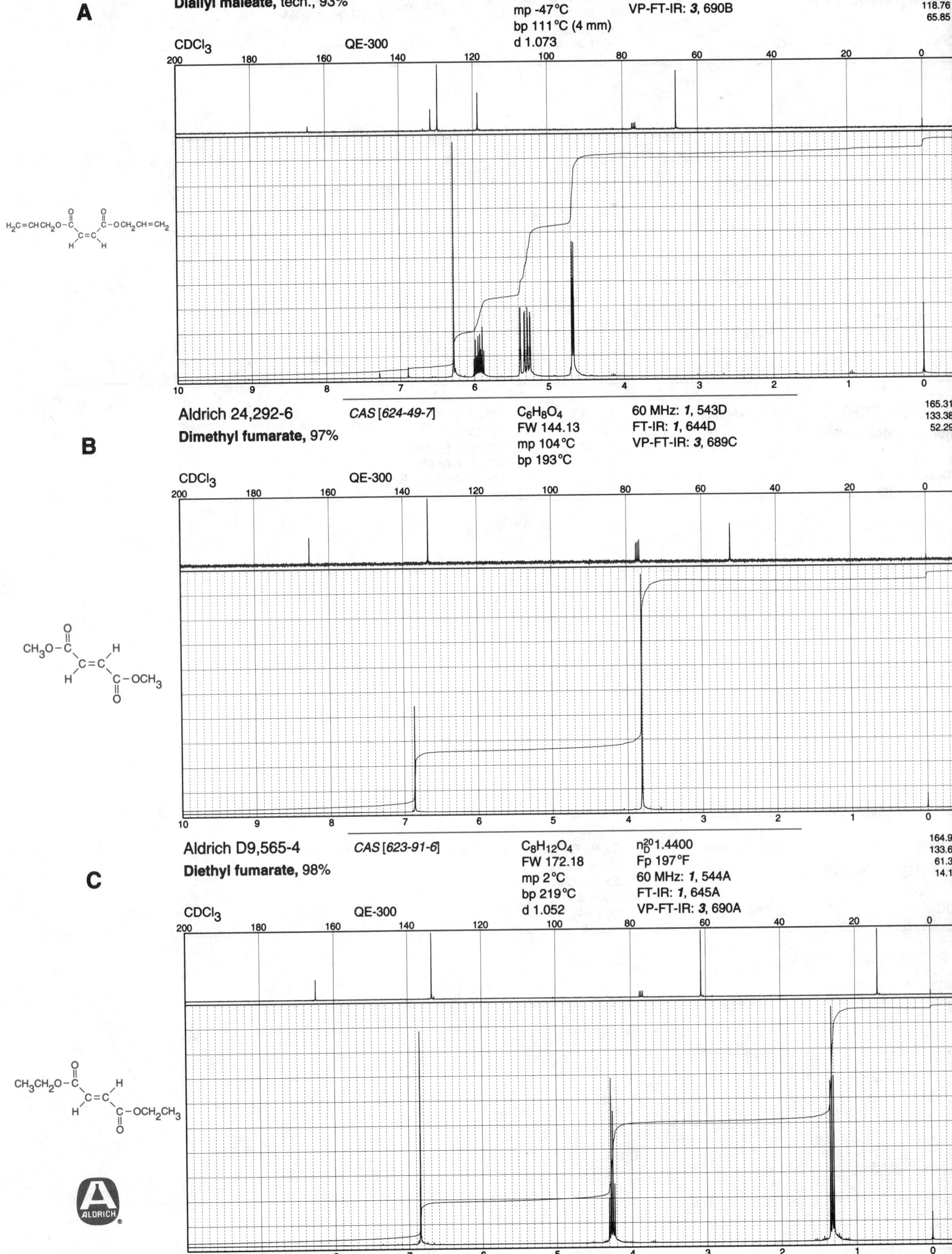

Aldrich 29,122-6   CAS [999-21-3]

Diallyl maleate, tech., 93%

$C_{10}H_{12}O_4$
FW 196.20
mp -47°C
bp 111°C (4 mm)
d 1.073

n_D^{20}1.4690
Fp >230°F
VP-FT-IR: **3**, 690B

A

CDCl$_3$ QE-300

164.65
131.65*
129.76*
118.76
65.85

Aldrich 24,292-6   CAS [624-49-7]

Dimethyl fumarate, 97%

$C_6H_8O_4$
FW 144.13
mp 104°C
bp 193°C

60 MHz: **1**, 543D
FT-IR: **1**, 644D
VP-FT-IR: **3**, 689C

B

CDCl$_3$ QE-300

165.31
133.38*
52.29*

Aldrich D9,565-4   CAS [623-91-6]

Diethyl fumarate, 98%

$C_8H_{12}O_4$
FW 172.18
mp 2°C
bp 219°C
d 1.052

n_D^{20}1.4400
Fp 197°F
60 MHz: **1**, 544A
FT-IR: **1**, 645A
VP-FT-IR: **3**, 690A

C

CDCl$_3$ QE-300

164.9(
133.62
61.3(
14.1(

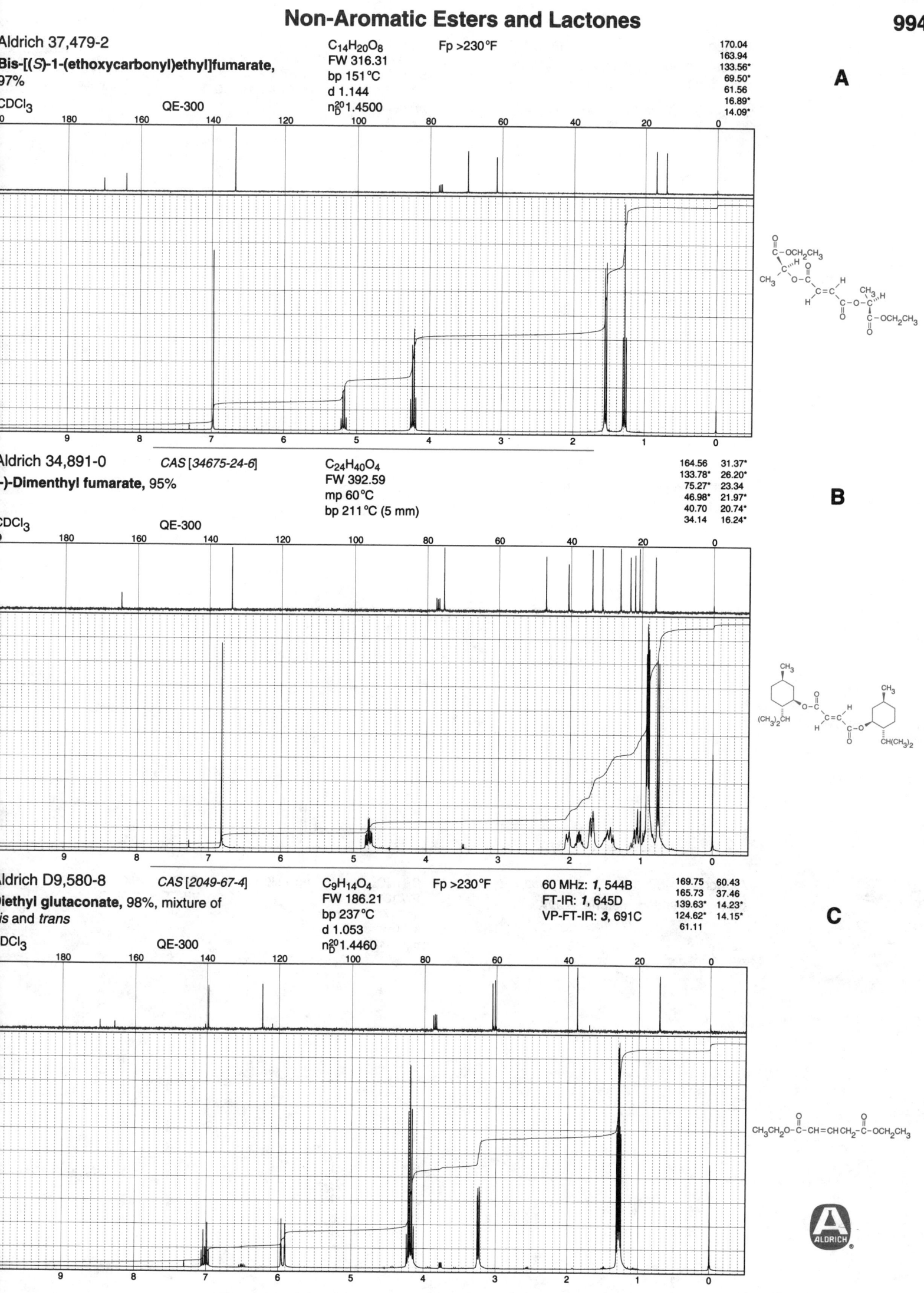

$C_{14}H_{20}O_8$
FW 316.31
bp 151 °C
d 1.144
n_D^{20} 1.4500

Fp >230 °F

170.04
163.94
133.56*
69.50*
61.56
16.89*
14.09*

A

$C_{24}H_{40}O_4$
FW 392.59
mp 60 °C
bp 211 °C (5 mm)

164.56   31.37*
133.78* 26.20*
75.27*   23.34
46.98*   21.97*
40.70   20.74*
34.14   16.24*

B

$C_9H_{14}O_4$
FW 186.21
bp 237 °C
d 1.053
n_D^{20} 1.4460

Fp >230 °F

60 MHz: 1, 544B
FT-IR: 1, 645D
VP-FT-IR: 3, 691C

169.75   60.43
165.73   37.46
139.63* 14.23*
124.62* 14.15*
61.11

C

Non-Aromatic Esters and Lactones

A

Aldrich 18,659-7   *CAS [52313-87-8]* $C_8H_{12}O_4$ Fp 208 °F

Dimethyl 3-methylglutaconate, 95%, mixture   FW 172.18   60 MHz: *1*, 544C
of *cis* and *trans*   bp 110 °C (12 mm)   FT-IR: *1*, 646A
    d 1.095   VP-FT-IR: *3*, 691D
    n_D^{20}1.4560

170.60	119.21*	50.99*
170.21	118.82*	45.65
166.41	52.07*	38.27
151.06	51.91*	25.65*
150.98	51.02*	18.91*

CDCl₃   QE-300

CH₃O–C–CH=C–CH₂–C–OCH₃ (with O, CH₃, O substituents)

B

Aldrich 25,566-1   *CAS [17041-60-0]* $C_7H_{10}O_4$ FT-IR: *1*, 645B
Dimethyl ethylidenemalonate, 98%   FW 158.16   VP-FT-IR: *3*, 691A
    d 1.111
    n_D^{20}1.4470
    Fp 207 °F

165.73
164.26
145.74*
128.95
52.28*
52.20*
15.69*

CDCl₃   QE-300

CH₃O–C–C–C–OCH₃ (with O, O, ‖CHCH₃)

C

Aldrich D9,480-1   *CAS [1462-12-0]* $C_9H_{14}O_4$ Fp >230 °F   FT-IR: *1*, 645C
Diethyl ethylidenemalonate, 99%   FW 186.21   VP-FT-IR: *3*, 691B
    bp 117 °C (17 mm)
    d 1.019
    n_D^{20}1.4420

165.34
163.89
144.54
129.80
61.14
15.47
14.13

CDCl₃   QE-300

CH₃CH₂O–C–C–C–OCH₂CH₃ (with O, O, ‖CHCH₃)

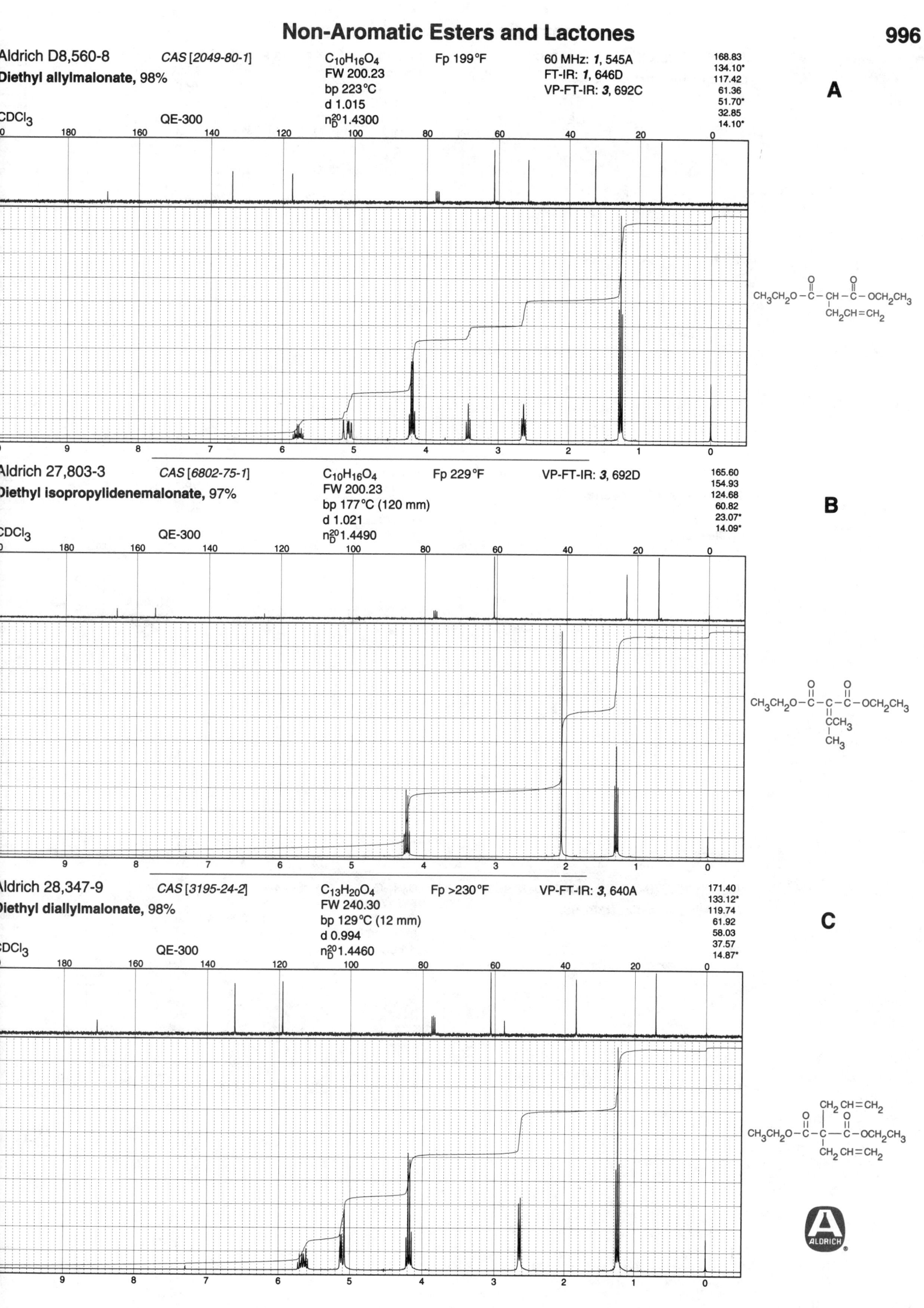

Aldrich D8,560-8   *CAS [2049-80-1]*
Diethyl allylmalonate, 98%

$C_{10}H_{16}O_4$ Fp 199°F
FW 200.23
bp 223°C
d 1.015
n_D^{20} 1.4300

60 MHz: *1*, 545A
FT-IR: *1*, 646D
VP-FT-IR: *3*, 692C

168.83
134.10*
117.42
61.36
51.70*
32.85
14.10*

A

CDCl₃   QE-300

Aldrich 27,803-3   *CAS [6802-75-1]*
Diethyl isopropylidenemalonate, 97%

$C_{10}H_{16}O_4$ Fp 229°F
FW 200.23
bp 177°C (120 mm)
d 1.021
n_D^{20} 1.4490

VP-FT-IR: *3*, 692D

165.60
154.93
124.68
60.82
23.07*
14.09*

B

CDCl₃   QE-300

Aldrich 28,347-9   *CAS [3195-24-2]*
Diethyl diallylmalonate, 98%

$C_{13}H_{20}O_4$ Fp >230°F
FW 240.30
bp 129°C (12 mm)
d 0.994
n_D^{20} 1.4460

VP-FT-IR: *3*, 640A

171.40
133.12*
119.74
61.92
58.03
37.57
14.87*

C

CDCl₃   QE-300

ALDRICH

A

Aldrich 10,953-3   *CAS [617-52-7]*

Dimethyl itaconate, 97%

C$_7$H$_{10}$O$_4$
FW 158.15
mp 39°C
bp 208°C
d 1.124

Fp 213°F

60 MHz: *1*, 544D
FT-IR: *1*, 646C
VP-FT-IR: *3*, 692B

171.05
166.54
133.68
128.48
52.11*
52.02*
37.49

CDCl$_3$ QE-300

CH$_3$O–C–CH$_2$–C–C–OCH$_3$ with O, O, CH$_2$

B

Aldrich 25,611-0   *CAS [869-29-4]*

1,1-Diacetoxy-2-propene, 98%

C$_7$H$_{10}$O$_4$
FW 158.16
bp 184°C
d 1.078
n$_D^{20}$1.4190

Fp 173°F

FT-IR: *1*, 637A
VP-FT-IR: *3*, 668B

168.51
131.29*
120.38
89.11*
20.77*

CDCl$_3$ QE-300

CH$_3$–C–O–CHCH=CH$_2$ with O–C–CH$_3$

C

Aldrich 15,679-5   *CAS [10476-95-6]*

Methallylidene diacetate, 98%

C$_8$H$_{12}$O$_4$
FW 172.18
mp -15°C
bp 191°C
d 1.039

n$_D^{20}$1.4240
Fp 182°F
VP-FT-IR: *3*, 694A

168.56
138.72
116.20
90.68
20.77
16.56

CDCl$_3$ QE-300

CH$_3$–C–O–CH–C=CH$_2$ with O, CH$_3$, O–C–CH$_3$, O

Aldrich 32,753-0   *CAS [78267-54-6]* $C_8H_{12}O_4$ Fp 189°F   VP-FT-IR: *3*, 668C

trans-**1,1-Diacetoxy-2-butene, tech.,**
80%

CDCl₃   QE-300

FW 172.18
bp 106°C (20 mm)
d 1.057
n_D^{20} 1.4290

168.65
133.12*
124.58*
89.71*
20.87*
17.59*

A

Aldrich E2,650-9   *CAS [97-90-5]* $C_{10}H_{14}O_4$ Fp >230°F   FT-IR: *1*, 642D

Ethylene glycol dimethacrylate, tech.,
90%

CDCl₃   QE-300

FW 198.22
bp 99°C (5 mm)
d 1.051
n_D^{20} 1.4540

VP-FT-IR: *3*, 682C

167.00
136.04
125.81
62.35
18.21*

B

Aldrich 26,154-8   *CAS [109-16-0]* $C_{14}H_{22}O_6$ Fp >230°F   FT-IR: *1*, 643A

Triethylene glycol dimethacrylate, 95%

CDCl₃   QE-300

FW 286.33
bp 171°C (5 mm)
d 1.092
n_D^{20} 1.4600

VP-FT-IR: *3*, 682D

167.23
136.15
125.58
70.67
69.18
63.82
18.27*

C

A

Aldrich 23,495-8   CAS [2082-81-7]

1,4-Butanediol dimethacrylate, 95%

$C_{12}H_{18}O_4$
FW 226.28
bp 133°C (4 mm)
d 1.023
n_D^{20} 1.4560

Fp >230°F

60 MHz: *1*, 538C
FT-IR: *1*, 639D
VP-FT-IR: *3*, 676B

167.28
136.34
125.27
64.15
25.44
18.29*

CDCl₃   QE-300

$$\left[\begin{array}{c} \overset{O}{\underset{CH_3}{H_2C=C-C-OCH_2CH_2-}} \end{array} \right]_2$$

B

Aldrich 24,684-0   CAS [3290-92-4]

**2-Ethyl-2-(hydroxymethyl)-1,3-propane-
diol trimethacrylate**

$C_{18}H_{26}O_6$
FW 338.40
d 1.060
n_D^{20} 1.4720
Fp >230°F

60 MHz: *1*, 540A
FT-IR: *1*, 640B

166.87   41.17
135.88   23.50
125.93   18.29*
64.29   7.50*

CDCl₃   QE-300

$$CH_2O-\overset{O}{\underset{}{C}}-\overset{CH_3}{\underset{}{C}}=CH_2$$
$$CH_3CH_2CCH_2O-\overset{O}{\underset{}{C}}-\overset{CH_3}{\underset{}{C}}=CH_2$$
$$CH_2O-\overset{O}{\underset{}{C}}-\overset{CH_3}{\underset{}{C}}=CH_2$$

C

Aldrich 33,096-5

**1,1,1-Tris(2-hydroxyethoxymethyl)propane
triacrylate,** tech.

$C_{21}H_{32}O_9$
FW 428.48
d 1.090
n_D^{20} 1.4710
Fp >230°F

165.99	130.79	71.01	69.17	64.71	63.39
165.93	130.38	70.63	69.05	64.52	63.25
165.87	128.50*	70.60	64.93	64.44	23.03
165.77	128.23*	70.54	64.89	63.71	22.89
131.20	71.22	70.41	64.84	63.65	7.53*
130.87	71.12	69.29	64.80	63.51	7.48*

CDCl₃   QE-300

$$CH_3CH_2\left(CH_2OCH_2CH_2O-\overset{O}{\underset{}{C}}-CH=CH_2 \right)_3$$

ALDRICH

Aldrich 19,457-3 *CAS [89-49-6]*
Isopulegyl acetate, 96%, mixture of isomers

CDCl$_3$ QE-300

$C_{12}H_{20}O_2$ Fp 186°F
FW 196.29
bp 113°C (8 mm)
d 0.925
n$_D^{20}$1.4560

A

170.44	73.49*	34.62	25.04
170.38	70.38*	34.14	22.43*
146.26	50.71*	31.39*	22.15*
146.19	46.86*	30.40	22.02*
111.58	40.48	28.03*	21.13*
110.50	39.31	26.65*	19.54*

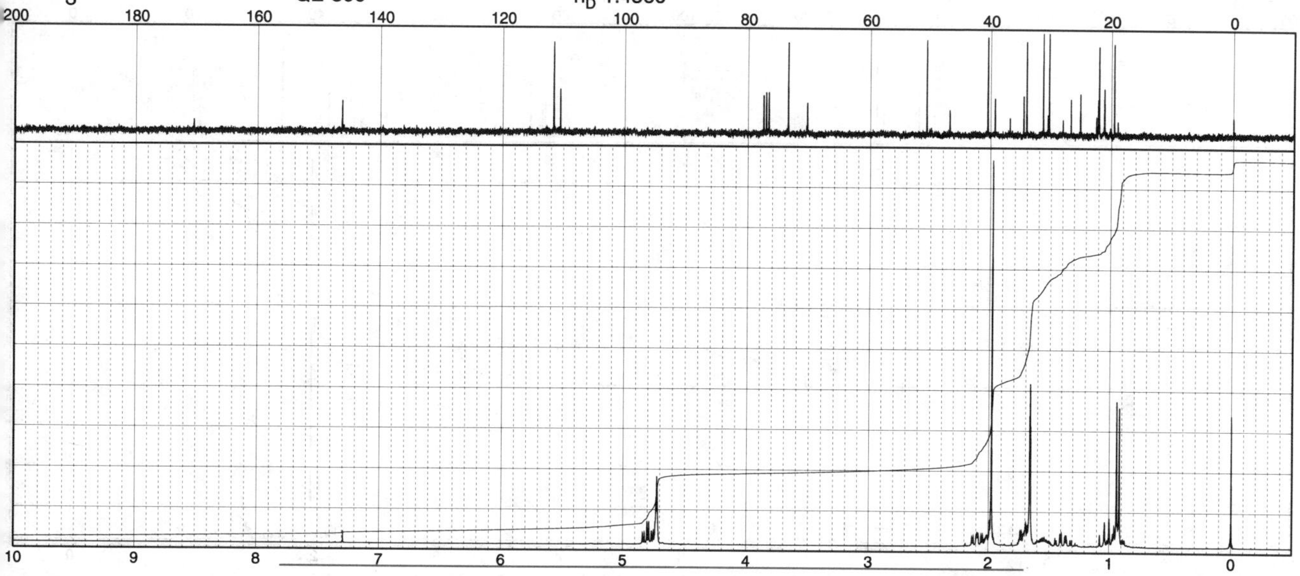

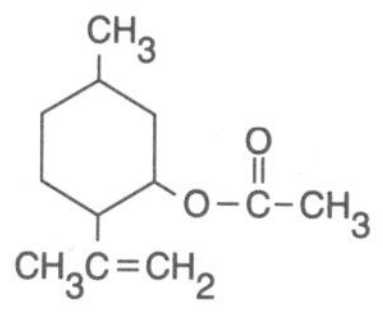

Aldrich 19,410-7 *CAS [20777-49-5]*
(-)-Dihydrocarvyl acetate, 95+%, mixture of isomers

CDCl$_3$ QE-300

$C_{12}H_{20}O_2$ Fp 194°F
FW 196.29
bp 233°C
d 0.947
n$_D^{20}$1.4590

B

170.66	78.23*	38.60*	31.24	25.88	20.83*
149.59	75.13*	37.17*	30.91	24.57*	18.22*
148.81	74.80*	36.92	30.87	21.38*	18.04*
109.44	73.47*	35.64	30.60	21.31*	16.94*
108.82	43.89*	35.03*	30.49	21.22*	11.55*
108.77	43.71*	33.27	29.16	21.16*	
108.59	39.37*	33.17	27.53	20.90*	

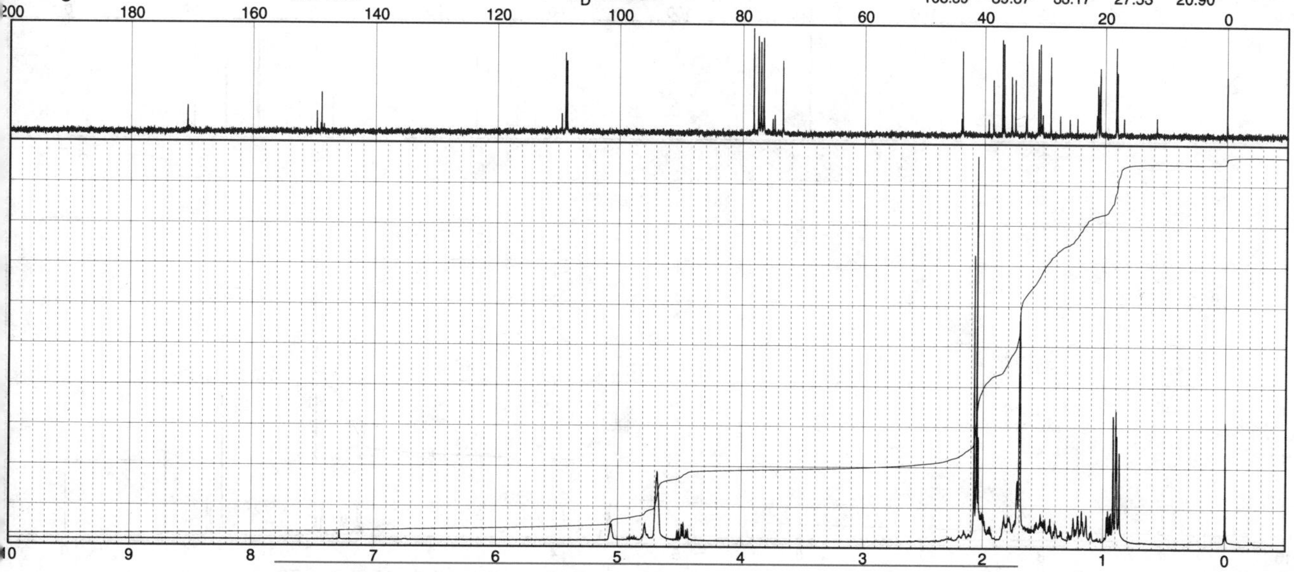

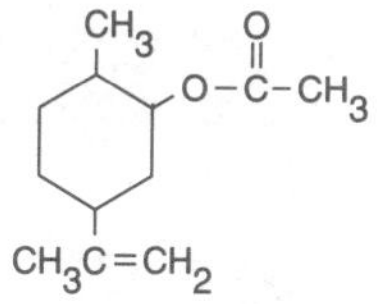

Aldrich 19,605-3 *CAS [97-42-7]*
(-)-Carvyl acetate, 98+%, mixture of isomers

CDCl$_3$ QE-300

$C_{12}H_{18}O_2$ Fp 208°F
FW 194.27
bp 78°C
d 0.976
n$_D^{20}$1.4750

C

170.78	127.79	40.33*	21.31*
170.73	125.85	35.87*	21.14*
148.65	109.31	34.06	20.80*
148.20	109.18	33.73	20.58*
132.87	73.20*	30.94	20.45*
130.96	70.62*	30.81	18.77*

ALDRICH

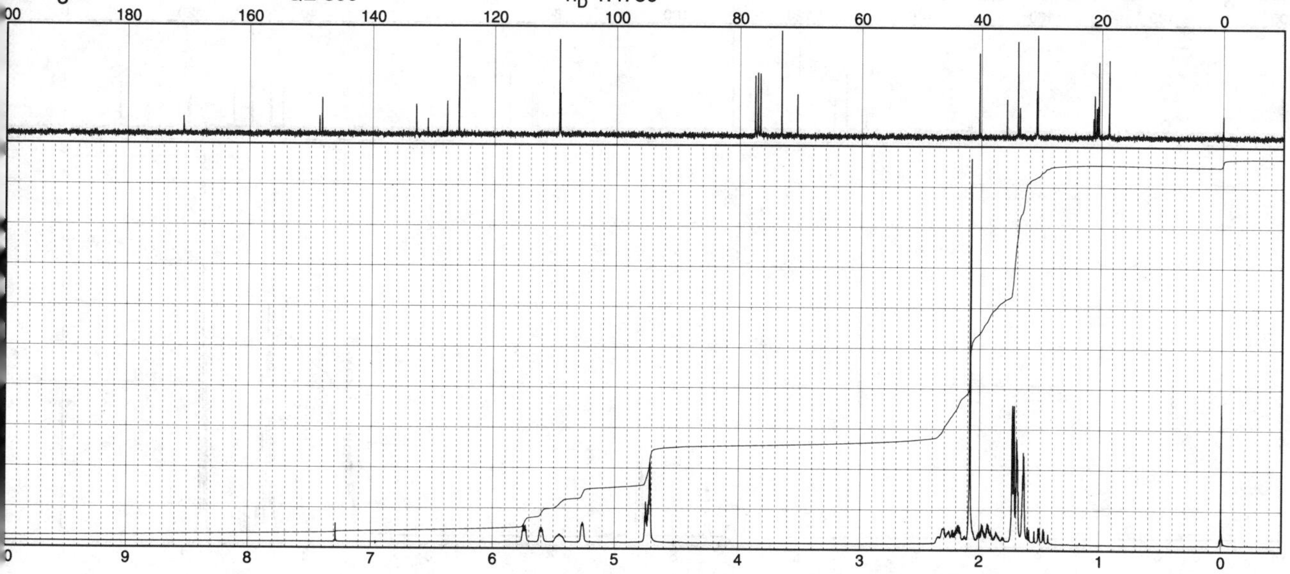

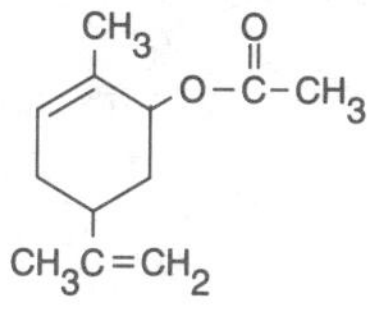

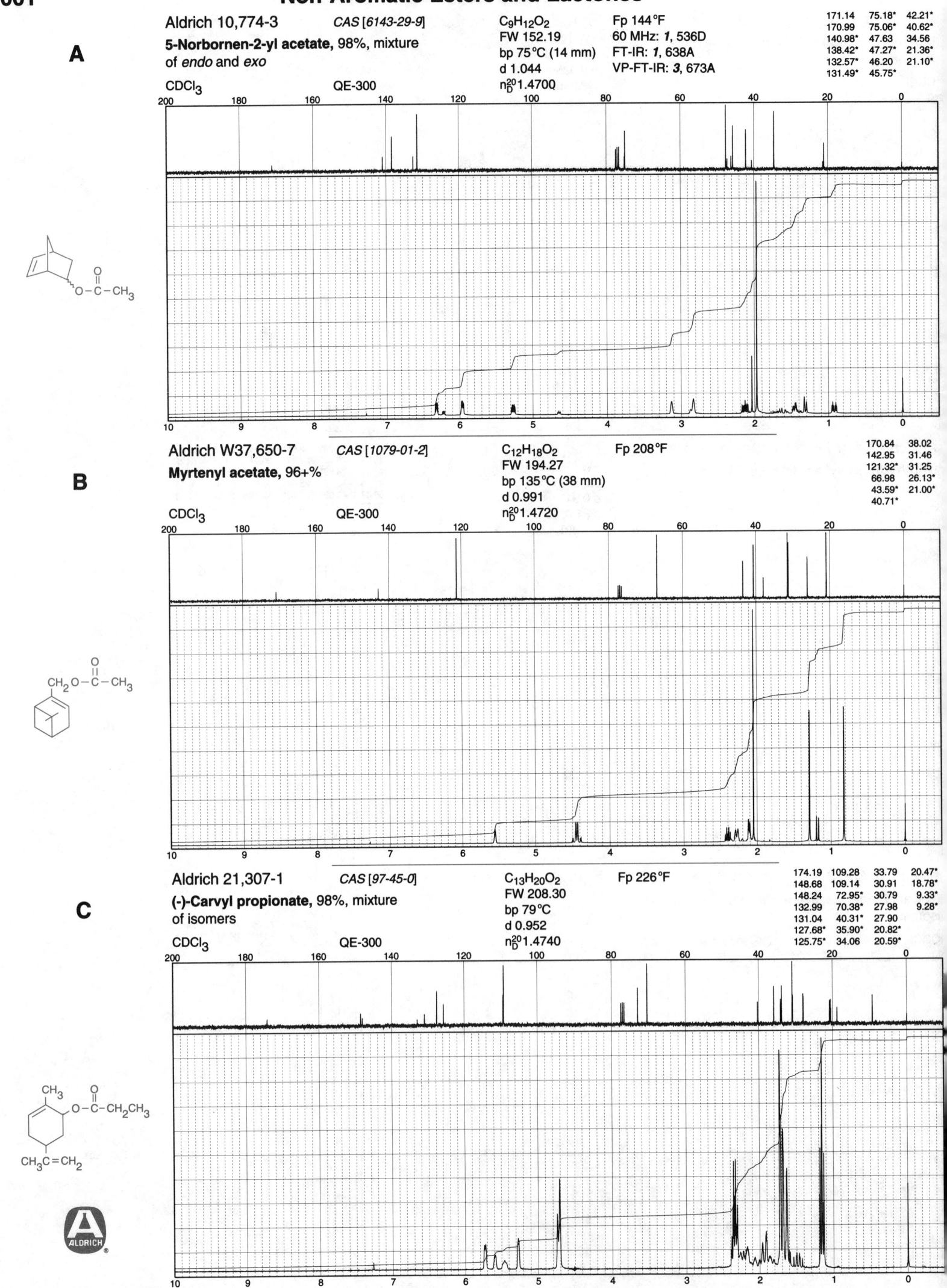

A   Aldrich 10,774-3   *CAS [6143-29-9]*

5-Norbornen-2-yl acetate, 98%, mixture of *endo* and *exo*

CDCl₃   QE-300

$C_9H_{12}O_2$
FW 152.19
bp 75°C (14 mm)
d 1.044
n_D^{20} 1.4700

Fp 144°F
60 MHz: *1*, 536D
FT-IR: *1*, 638A
VP-FT-IR: *3*, 673A

171.14	75.18*	42.21*
170.99	75.06*	40.62*
140.98*	47.63	34.56
138.42*	47.27*	21.36*
132.57*	46.20	21.10*
131.49*	45.75*	

B   Aldrich W37,650-7   *CAS [1079-01-2]*

Myrtenyl acetate, 96+%

CDCl₃   QE-300

$C_{12}H_{18}O_2$
FW 194.27
bp 135°C (38 mm)
d 0.991
n_D^{20} 1.4720

Fp 208°F

170.84	38.02
142.95	31.46
121.32*	31.25
66.98	26.13*
43.59*	21.00*
40.71*	

C   Aldrich 21,307-1   *CAS [97-45-0]*

(-)-Carvyl propionate, 98%, mixture of isomers

CDCl₃   QE-300

$C_{13}H_{20}O_2$
FW 208.30
bp 79°C
d 0.952
n_D^{20} 1.4740

Fp 226°F

174.19	109.28	33.79	20.47*
148.68	109.14	30.91	18.78*
148.24	72.95*	30.79	9.33*
132.99	70.38*	27.98	9.28*
131.04	40.31*	27.90	
127.68*	35.90*	20.82*	
125.75*	34.06	20.59*	

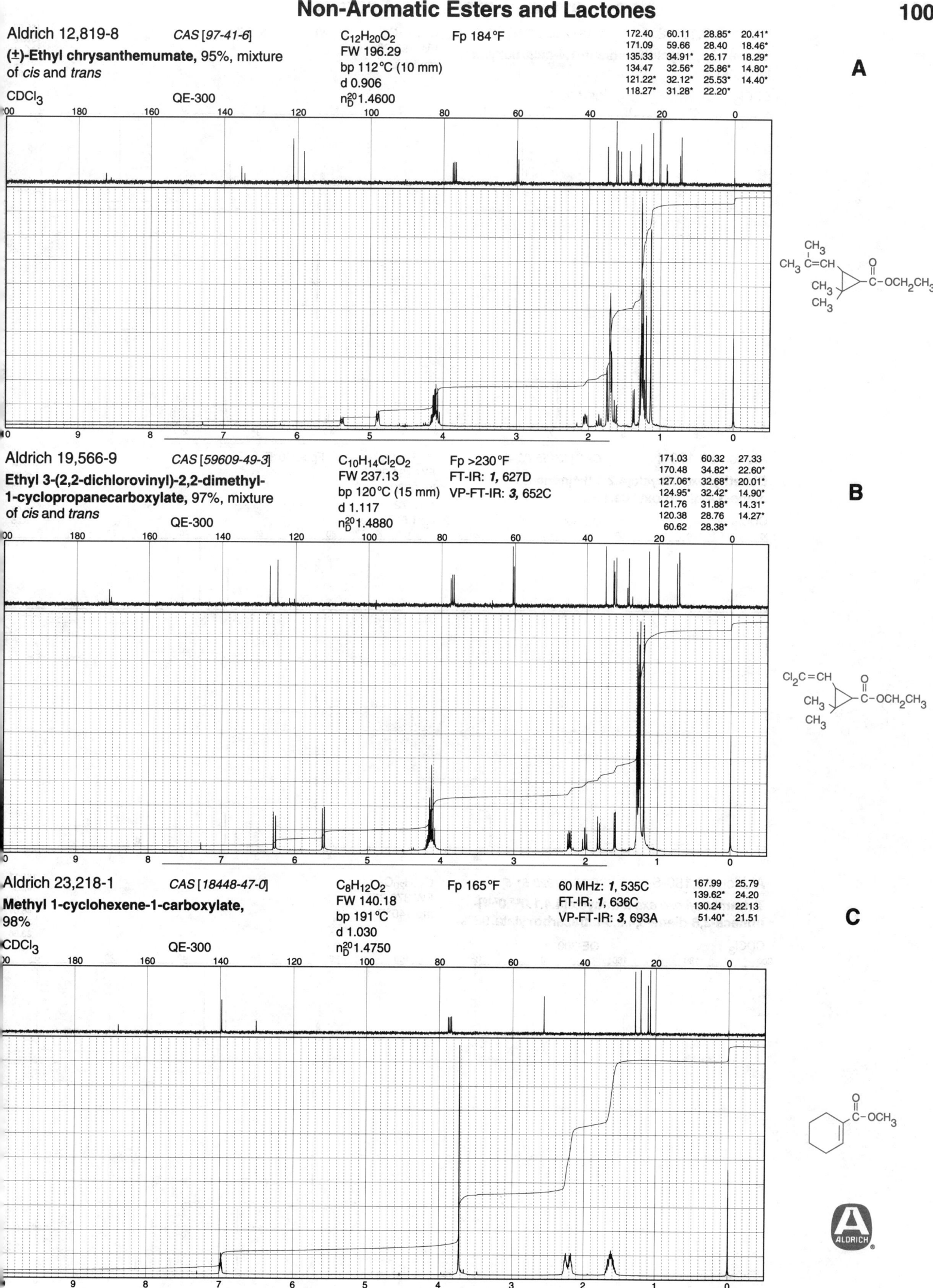

Aldrich 12,819-8   *CAS [97-41-6]*

(±)-Ethyl chrysanthemumate, 95%, mixture of *cis* and *trans*

CDCl₃   QE-300

$C_{12}H_{20}O_2$
FW 196.29
bp 112°C (10 mm)
d 0.906
n_D^{20}1.4600

Fp 184°F

172.40	60.11	28.85*	20.41*
171.09	59.66	28.40	18.46*
135.33	34.91*	26.17	18.29*
134.47	32.56*	25.86*	14.80*
121.22*	32.12*	25.53*	14.40*
118.27*	31.28*	22.20*	

A

Aldrich 19,566-9   *CAS [59609-49-3]*

Ethyl 3-(2,2-dichlorovinyl)-2,2-dimethyl-1-cyclopropanecarboxylate, 97%, mixture of *cis* and *trans*

QE-300

$C_{10}H_{14}Cl_2O_2$
FW 237.13
bp 120°C (15 mm)
d 1.117
n_D^{20}1.4880

Fp >230°F
FT-IR: *1*, 627D
VP-FT-IR: *3*, 652C

171.03	60.32	27.33
170.48	34.82*	22.60*
127.06*	32.68*	20.01*
124.95*	32.42*	14.90*
121.76	31.88*	14.31*
120.38	28.76	14.27*
60.62	28.38*	

B

Aldrich 23,218-1   *CAS [18448-47-0]*

Methyl 1-cyclohexene-1-carboxylate, 98%

CDCl₃   QE-300

$C_8H_{12}O_2$
FW 140.18
bp 191°C
d 1.030
n_D^{20}1.4750

Fp 165°F

60 MHz: *1*, 535C
FT-IR: *1*, 636C
VP-FT-IR: *3*, 693A

167.99	25.79
139.62*	24.20
130.24	22.13
51.40*	21.51

C

ALDRICH

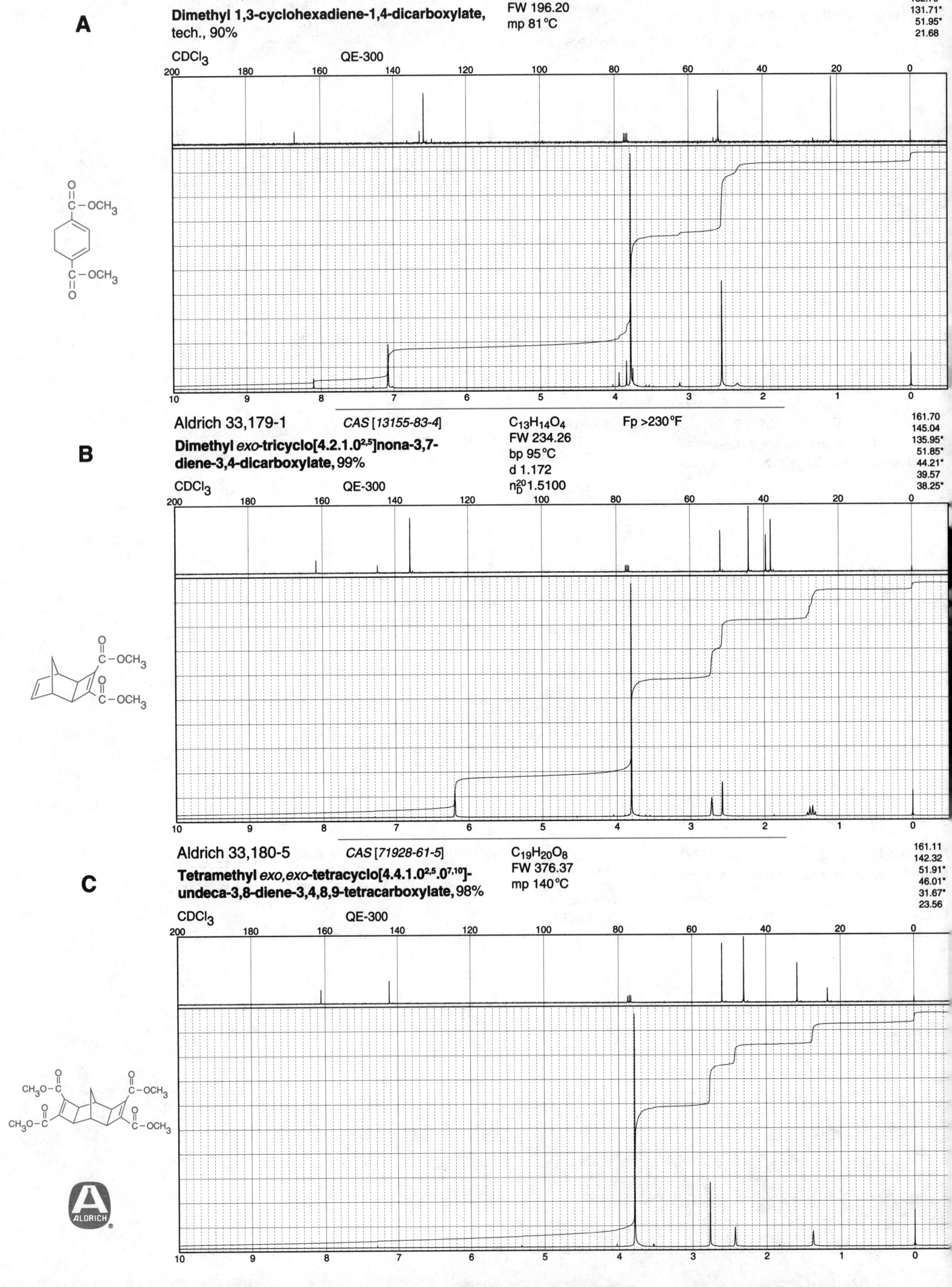

A

Aldrich 30,158-2   CAS [1659-95-6]   C₁₀H₁₂O₄   VP-FT-IR: *3*, 659B

Dimethyl 1,3-cyclohexadiene-1,4-dicarboxylate, tech., 90%

FW 196.20
mp 81°C

166.92
132.79
131.71*
51.95*
21.68

CDCl₃   QE-300

B

Aldrich 33,179-1   CAS [13155-83-4]   C₁₃H₁₄O₄   Fp >230°F

Dimethyl *exo*-tricyclo[4.2.1.0²,⁵]nona-3,7-diene-3,4-dicarboxylate, 99%

FW 234.26
bp 95°C
d 1.172
n₂₀ᴰ 1.5100

161.70
145.04
135.95*
51.85*
44.21*
39.57
38.25*

CDCl₃   QE-300

C

Aldrich 33,180-5   CAS [71928-61-5]   C₁₉H₂₀O₈

Tetramethyl *exo,exo*-tetracyclo[4.4.1.0²,⁵.0⁷,¹⁰]-undeca-3,8-diene-3,4,8,9-tetracarboxylate, 98%

FW 376.37
mp 140°C

161.11
142.32
51.91*
46.01*
31.67*
23.56

CDCl₃   QE-300

A

Aldrich 30,284-8 CAS [16691-59-1]

Pentamethyl cyclopentadiene-1,2,3,4,5-pentacarboxylate, 97%

$C_{15}H_{16}O_{10}$
FW 356.29
mp 149°C d.

170.31
118.15
53.73*

D_2O+NAOD QE-300

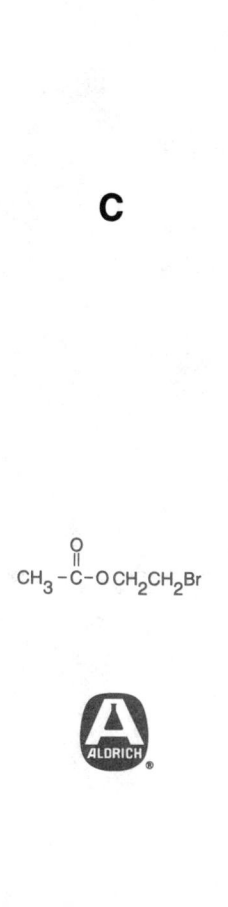

B

Aldrich 30,320-8 CAS [590-97-6]

Bromomethyl acetate, 95%

$C_3H_5BrO_2$
FW 152.98
bp 132°C (750 mm)
d 1.560
n_D^{20} 1.4470

Fp 135°F VP-FT-IR: 3, 694C

168.70
57.05
20.81*

$CDCl_3$ QE-300

$CH_3-C(=O)-O-CH_2Br$

C

Aldrich 13,768-5 CAS [927-68-4]

2-Bromoethyl acetate, 97%

$C_4H_7BrO_2$
FW 167.01
mp -14°C
bp 159°C
d 1.514

n_D^{20} 1.4540
Fp 160°F
60 MHz: 1, 546A
FT-IR: 1, 647B
VP-FT-IR: 3, 695A

170.36
63.79
28.62
20.70*

$CDCl_3$ QE-300

$CH_3-C(=O)-O-CH_2CH_2Br$

A

Aldrich 16,656-1   *CAS [628-09-1]* $C_5H_9ClO_2$ Fp 154°F   VP-FT-IR: *3*, 695B

3-Chloropropyl acetate, 98%

FW 136.58
bp 81°C (30 mm)
d 1.111
n_D^{20}1.4290

170.86
61.16
41.21
31.60
20.86*

CDCl₃   QE-300

$CH_3-\overset{\overset{\text{O}}{\|}}{C}-OCH_2CH_2CH_2Cl$

B

Aldrich C2,920-7   *CAS [6962-92-1]* $C_6H_{11}ClO_2$ Fp 148°F   FT-IR: *1*, 647C

4-Chlorobutyl acetate, 98%

FW 150.61
bp 92°C (22 mm)
d 1.072
n_D^{20}1.4340

VP-FT-IR: *3*, 695C

170.91
63.56
44.42
29.18
26.10
20.90*

CDCl₃   QE-300

$CH_3-\overset{\overset{\text{O}}{\|}}{C}-OCH_2CH_2CH_2CH_2Cl$

C

Aldrich 13,598-4   *CAS [14273-86-0]* $C_6H_{11}ClO_2$ Fp 194°F   60 MHz: *1*, 555C

Methyl 5-chlorovalerate, 97%

FW 150.61
bp 107°C (38 mm)
d 1.047
n_D^{20}1.4360

VP-FT-IR: *3*, 711A

173.48
51.55*
44.42
33.17
31.87
22.26

CDCl₃   QE-300

$ClCH_2CH_2CH_2CH_2-\overset{\overset{\text{O}}{\|}}{C}-OCH_3$

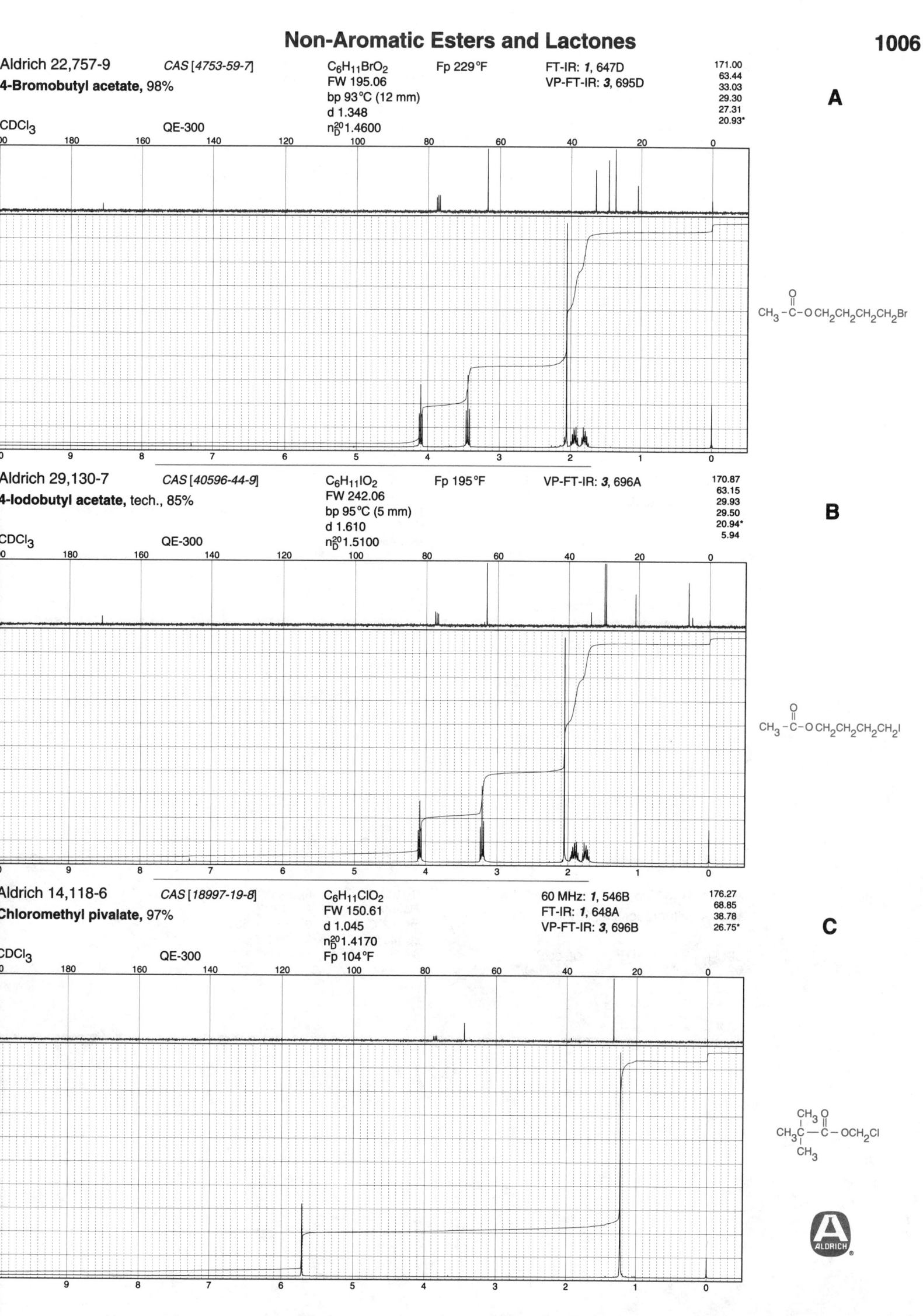

Aldrich 22,757-9   CAS [4753-59-7] $C_6H_{11}BrO_2$ Fp 229°F   FT-IR: *1*, 647D   171.00
4-Bromobutyl acetate, 98%   FW 195.06   VP-FT-IR: *3*, 695D   63.44
bp 93°C (12 mm)   33.03
d 1.348   29.30
27.31
CDCl₃   QE-300 n_D^{20}1.4600   20.93*

A

Aldrich 29,130-7   CAS [40596-44-9] $C_6H_{11}IO_2$ Fp 195°F   VP-FT-IR: *3*, 696A   170.87
4-Iodobutyl acetate, tech., 85%   FW 242.06   63.15
bp 95°C (5 mm)   29.93
d 1.610   29.50
20.94*
CDCl₃   QE-300 n_D^{20}1.5100   5.94

B

Aldrich 14,118-6   CAS [18997-19-8] $C_6H_{11}ClO_2$ 60 MHz: *1*, 546B   176.27
Chloromethyl pivalate, 97%   FW 150.61   FT-IR: *1*, 648A   68.85
d 1.045   VP-FT-IR: *3*, 696B   38.78
n_D^{20}1.4170   26.75*
Fp 104°F
CDCl₃   QE-300

C

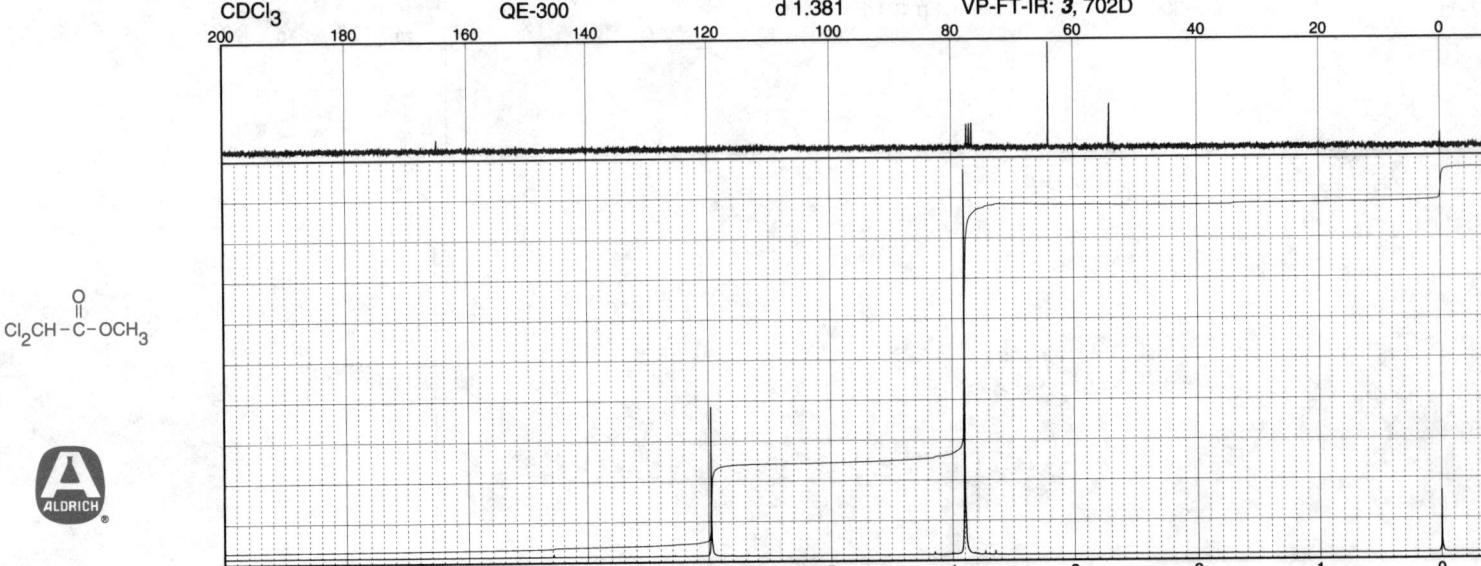

A

Aldrich 10,841-3   *CAS [96-34-4]*
Methyl chloroacetate, 99%

$C_3H_5ClO_2$
FW 108.52
mp -33°C
bp 130°C (740 mm)
d 1.238

n_D^{20} 1.4220
Fp 125°F
60 MHz: *1*, 548C
FT-IR: *1*, 650C
VP-FT-IR: *3*, 700A

167.75
53.04*
40.72

CDCl₃   QE-300

$ClCH_2-\overset{\overset{\displaystyle O}{\|}}{C}-OCH_3$

B

Aldrich 15,791-0   *CAS [96-32-2]*
Methyl bromoacetate, 97%

$C_3H_5BrO_2$
FW 152.98
bp 52°C (15 mm)
d 1.616
n_D^{20} 1.4580

Fp 145°F

60 MHz: *1*, 548D
FT-IR: *1*, 651A
VP-FT-IR: *3*, 700C

167.63
53.13*
25.51

CDCl₃   QE-300

$BrCH_2-\overset{\overset{\displaystyle O}{\|}}{C}-OCH_3$

C

Aldrich 10,840-5   *CAS [116-54-1]*
Methyl dichloroacetate, 99+%

$C_3H_4Cl_2O_2$
FW 142.97
mp -52°C
bp 143°C
d 1.381

n_D^{20} 1.4420
Fp 176°F
60 MHz: *1*, 550A
FT-IR: *1*, 652C
VP-FT-IR: *3*, 702D

164.98
64.06*
54.16*

CDCl₃   QE-300

$Cl_2CH-\overset{\overset{\displaystyle O}{\|}}{C}-OCH_3$

Non-Aromatic Esters and Lactones

Aldrich 24,983-1   *CAS [431-47-0]*   C$_3$H$_3$F$_3$O$_2$ Fp 19°F   FT-IR: *1*, 653B

Methyl trifluoroacetate, 99%   FW 128.05   VP-FT-IR: *3*, 703C

bp 43°C

d 1.273

CDCl$_3$ QE-300   n$_D^{20}$1.2900

158.98 116.66
158.42 112.88
157.85 109.12
157.29 54.30*
120.44

A

F$_3$C–C–OCH$_3$ (O)

Aldrich 24,249-7   *CAS [598-99-2]*   C$_3$H$_3$Cl$_3$O$_2$ Fp 163°F   60 MHz: *1*, 550B

Methyl trichloroacetate, 99%   FW 177.42   FT-IR: *1*, 652D

bp 153°C   VP-FT-IR: *3*, 703A

d 1.488

CDCl$_3$ QE-300   n$_D^{20}$1.4550

162.54
89.60
55.70*

B

Cl$_3$C–C–OCH$_3$ (O)

Aldrich 16,381-3   *CAS [459-72-3]*   C$_4$H$_7$FO$_2$ Fp 86°F   60 MHz: *1*, 548B

Ethyl fluoroacetate, 98%   FW 106.10   FT-IR: *1*, 650B

bp 119°C (753 mm)   VP-FT-IR: *3*, 699D

d 1.098

CDCl$_3$ QE-300   n$_D^{20}$1.3750

167.97
167.69
78.88
76.47
61.46
14.14*

C

FCH$_2$–C–OCH$_2$CH$_3$ (O)

ALDRICH

A

Aldrich E1,685-6   *CAS [105-39-5]* $C_4H_7ClO_2$ n_D^{20} 1.4210

Ethyl chloroacetate, 99%   FW 122.55   Fp 150°F

mp -26°C   FT-IR: *1*, 650D

bp 143°C   VP-FT-IR: *3*, 700B

d 1.145

167.26
62.27
40.98
14.09*

CDCl₃   QE-300

$ClCH_2-\overset{\overset{\text{O}}{\|}}{C}-OCH_2CH_3$

B

Aldrich 13,397-3   *CAS [105-36-2]* $C_4H_7BrO_2$ Fp 118°F   60 MHz: *1*, 549A

Ethyl bromoacetate, 98%   FW 167.01   FT-IR: *1*, 651B

bp 159°C   VP-FT-IR: *3*, 700D

d 1.506

n_D^{20} 1.4510

167.19
62.31
26.02
14.01*

CDCl₃   QE-300

$BrCH_2-\overset{\overset{\text{O}}{\|}}{C}-OCH_2CH_3$

C

Aldrich 28,380-0   *CAS [61898-49-5]* $C_4H_7BrO_2$ Fp 118°F

Ethyl bromoacetate-$^{13}C_2$, 98 atom % ^{13}C   FW 168.99

bp 159°C

d 1.506

n_D^{20} 1.4500

167.56
166.70
62.26
26.34
25.47
14.00*

CDCl₃   QE-300

$Br^{13}CH_2-^{13}\overset{\overset{\text{O}}{\|}}{C}-OCH_2CH_3$

ALDRICH

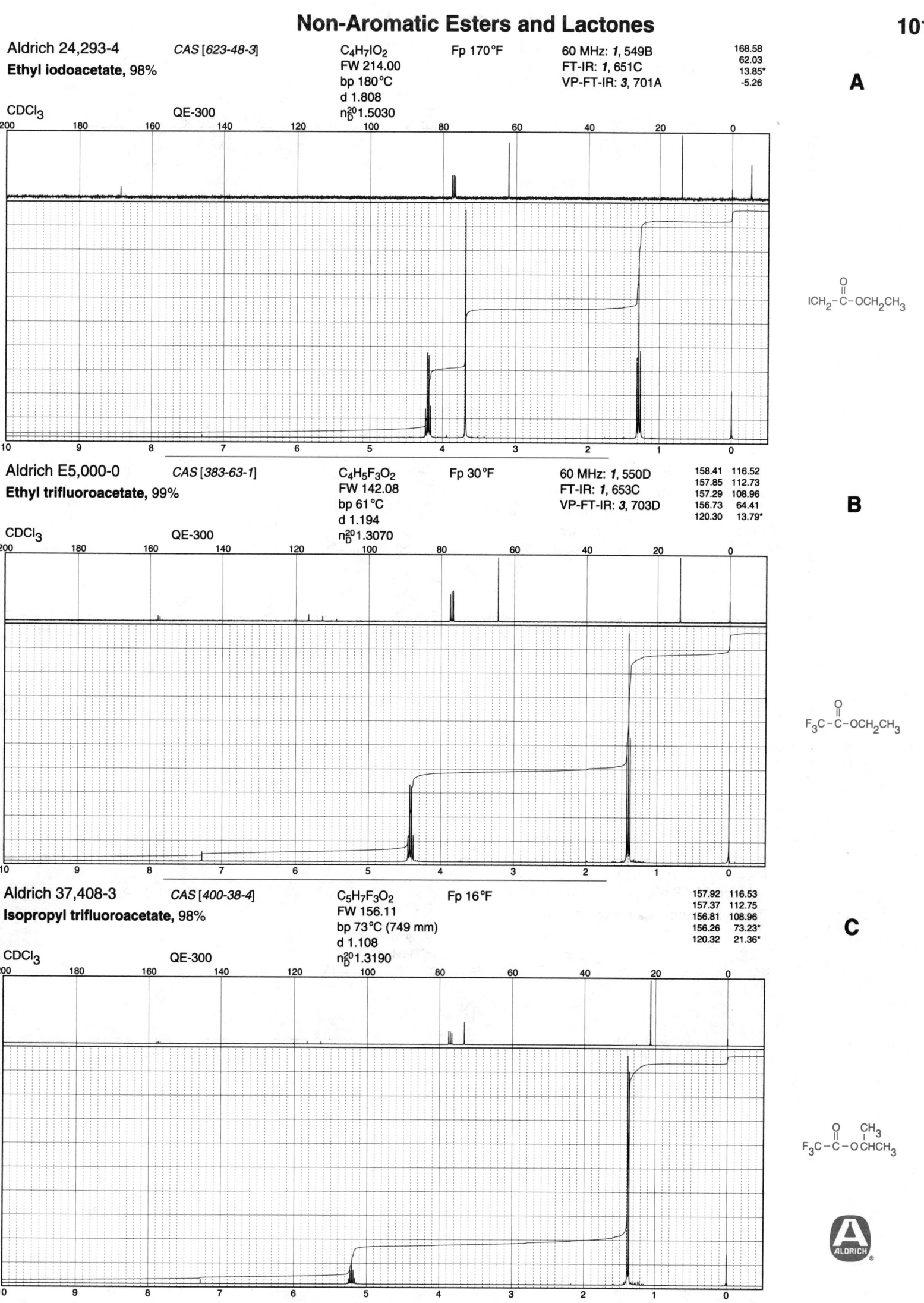

Aldrich 24,293-4   CAS [623-48-3] $C_4H_7IO_2$ Fp 170°F   60 MHz: 1, 549B

Ethyl iodoacetate, 98%   FW 214.00   FT-IR: 1, 651C   VP-FT-IR: 3, 701A

bp 180°C

d 1.808

$CDCl_3$ QE-300 n_D^{20}1.5030

168.58
62.03
13.85*
-5.26

A

$ICH_2-\overset{O}{\overset{\|}{C}}-OCH_2CH_3$

Aldrich E5,000-0   CAS [383-63-1] $C_4H_5F_3O_2$ Fp 30°F   60 MHz: 1, 550D

Ethyl trifluoroacetate, 99%   FW 142.08   FT-IR: 1, 653C

bp 61°C   VP-FT-IR: 3, 703D

d 1.194

$CDCl_3$ QE-300 n_D^{20}1.3070

158.41   116.52
157.85   112.73
157.29   108.96
156.73   64.41
120.30   13.79*

B

$F_3C-\overset{O}{\overset{\|}{C}}-OCH_2CH_3$

Aldrich 37,408-3   CAS [400-38-4] $C_5H_7F_3O_2$ Fp 16°F

Isopropyl trifluoroacetate, 98%   FW 156.11

bp 73°C (749 mm)

d 1.108

$CDCl_3$ QE-300 n_D^{20}1.3190

157.92   116.53
157.37   112.75
156.81   108.96
156.26   73.23*
120.32   21.36*

C

$F_3C-\overset{O}{\overset{\|}{C}}-O\overset{\overset{CH_3}{|}}{C}HCH_3$

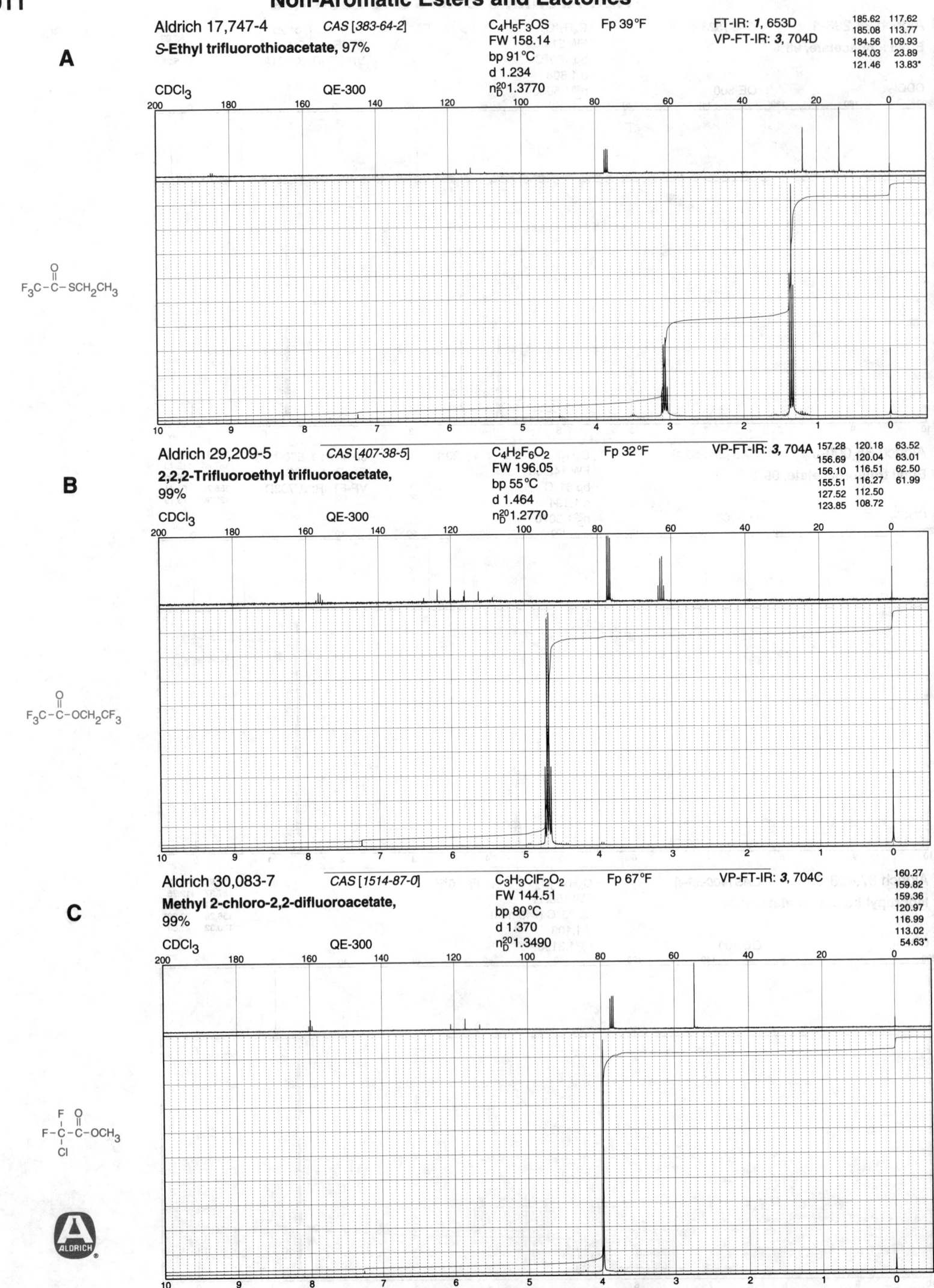

A

Aldrich 17,747-4   *CAS [383-64-2]* $C_4H_5F_3OS$ Fp 39 °F   FT-IR: *1*, 653D
S-Ethyl trifluorothioacetate, 97%   FW 158.14   VP-FT-IR: *3*, 704D
bp 91 °C
d 1.234
CDCl$_3$ QE-300   n$_D^{20}$1.3770

185.62   117.62
185.08   113.77
184.56   109.93
184.03   23.89
121.46   13.83*

$F_3C-\overset{\displaystyle O}{\overset{\|}{C}}-SCH_2CH_3$

Aldrich 29,209-5   *CAS [407-38-5]* $C_4H_2F_6O_2$ Fp 32 °F   VP-FT-IR: *3*, 704A
2,2,2-Trifluoroethyl trifluoroacetate,
99%   FW 196.05
bp 55 °C
d 1.464
CDCl$_3$ QE-300   n$_D^{20}$1.2770

B

157.28   120.18   63.52
156.69   120.04   63.01
156.10   116.51   62.50
155.51   116.27   61.99
127.52   112.50
123.85   108.72

$F_3C-\overset{\displaystyle O}{\overset{\|}{C}}-OCH_2CF_3$

Aldrich 30,083-7   *CAS [1514-87-0]* $C_3H_3ClF_2O_2$ Fp 67 °F   VP-FT-IR: *3*, 704C
Methyl 2-chloro-2,2-difluoroacetate,
99%   FW 144.51
bp 80 °C
d 1.370
CDCl$_3$ QE-300   n$_D^{20}$1.3490

C

160.27
159.82
159.36
120.97
116.99
113.02
54.63*

$F-\overset{\displaystyle F}{\underset{\displaystyle Cl}{\overset{\|}{C}}}-\overset{\displaystyle O}{\overset{\|}{C}}-OCH_3$

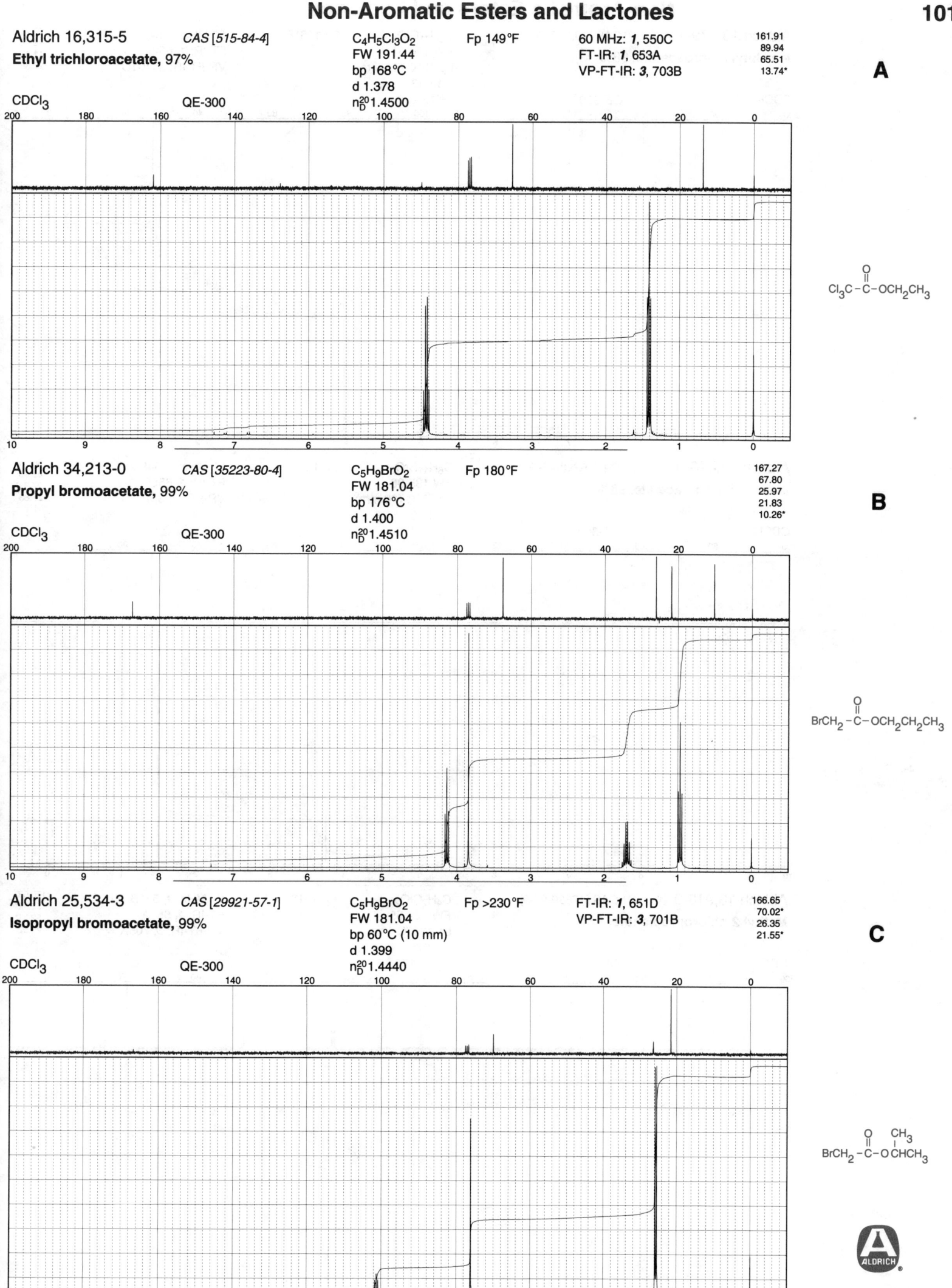

Aldrich 16,315-5   CAS [515-84-4]
Ethyl trichloroacetate, 97%

C₄H₅Cl₃O₂
FW 191.44
bp 168°C
d 1.378
n²⁰_D 1.4500

Fp 149°F

60 MHz: *1*, 550C
FT-IR: *1*, 653A
VP-FT-IR: *3*, 703B

161.91
89.94
65.51
13.74*

CDCl₃   QE-300

A

Aldrich 34,213-0   CAS [35223-80-4]
Propyl bromoacetate, 99%

C₅H₉BrO₂
FW 181.04
bp 176°C
d 1.400
n²⁰_D 1.4510

Fp 180°F

167.27
67.80
25.97
21.83
10.26*

CDCl₃   QE-300

B

Aldrich 25,534-3   CAS [29921-57-1]
Isopropyl bromoacetate, 99%

C₅H₉BrO₂
FW 181.04
bp 60°C (10 mm)
d 1.399
n²⁰_D 1.4440

Fp >230°F

FT-IR: *1*, 651D
VP-FT-IR: *3*, 701B

166.65
70.02*
26.35
21.55*

CDCl₃   QE-300

C

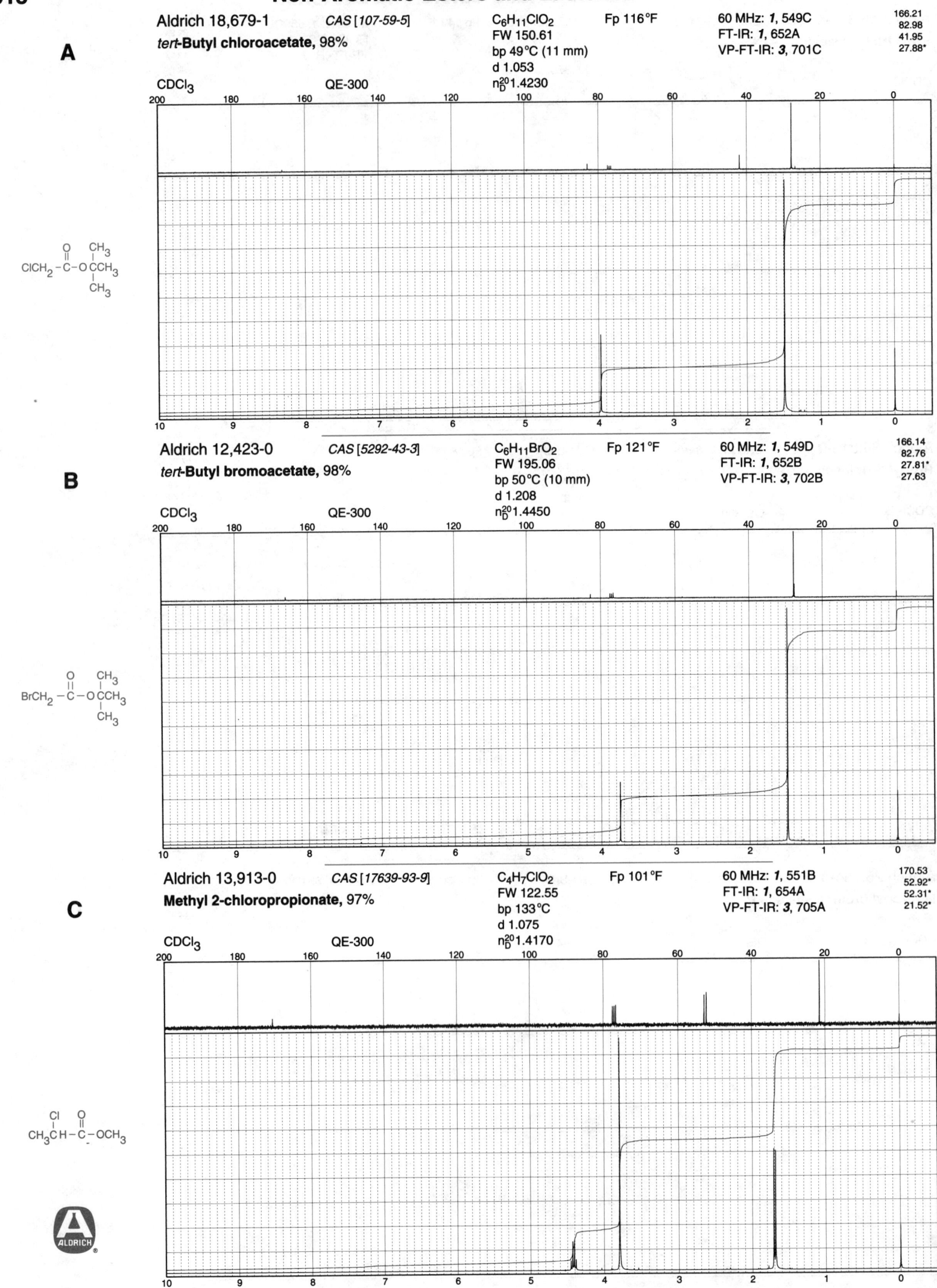

A

Aldrich 18,679-1   CAS [107-59-5]   C6H11ClO2   Fp 116°F   60 MHz: 1, 549C   166.21
tert-Butyl chloroacetate, 98%   FW 150.61   FT-IR: 1, 652A   82.98
                                            bp 49°C (11 mm)   VP-FT-IR: 3, 701C 41.95
                                            d 1.053   27.88*
CDCl3   QE-300   n20D 1.4230

B

Aldrich 12,423-0   CAS [5292-43-3]   C6H11BrO2   Fp 121°F   60 MHz: 1, 549D   166.14
tert-Butyl bromoacetate, 98%   FW 195.06   FT-IR: 1, 652B   82.76
                                            bp 50°C (10 mm)   VP-FT-IR: 3, 702B 27.81*
                                            d 1.208   27.63
CDCl3   QE-300   n20D 1.4450

C

Aldrich 13,913-0   CAS [17639-93-9]   C4H7ClO2   Fp 101°F   60 MHz: 1, 551B   170.53
Methyl 2-chloropropionate, 97%   FW 122.55   FT-IR: 1, 654A   52.92*
                                            bp 133°C   VP-FT-IR: 3, 705A 52.31*
                                            d 1.075   21.52*
CDCl3   QE-300   n20D 1.4170

Aldrich 27,775-4   *CAS [77287-29-7]*   C$_4$H$_7$ClO$_2$ Fp 99°F   VP-FT-IR: *3*, 705B   170.48

Methyl (*R*)-(+)-2-chloropropionate, 99%   FW 122.55   52.89*
bp 133°C   52.31*
d 1.152   21.52*

A

CDCl$_3$ QE-300   n$_D^{20}$1.4170

$$CH_3\overset{Cl}{\underset{}{C}}\overset{H}{\underset{}{C}}-\overset{O}{\underset{}{C}}-OCH_3$$

Aldrich 24,703-0   *CAS [73246-45-4]*   C$_4$H$_7$ClO$_2$ Fp 101°F   FT-IR: *1*, 654B   170.51

Methyl (*S*)-(-)-2-chloropropionate, 99%   FW 122.55   VP-FT-IR: *3*, 705C   52.92*
bp 81°C (110 mm)   52.32*
d 1.143   21.51*

B

CDCl$_3$ QE-300   n$_D^{20}$1.4170

$$CH_3\overset{Cl}{\underset{}{C}}\overset{H}{\underset{}{C}}-\overset{O}{\underset{}{C}}-OCH_3$$

Aldrich 16,718-5   *CAS [57885-43-5]*   C$_4$H$_7$BrO$_2$ Fp 125°F   60 MHz: *1*, 552A   170.70

Methyl (±)-2-bromopropionate, 99%   FW 167.01   FT-IR: *1*, 654D   52.97*
bp 51°C (19 mm)   VP-FT-IR: *3*, 706A   39.77*
d 1.497   21.69*

C

CDCl$_3$ QE-300   n$_D^{20}$1.4520

$$CH_3\overset{Br}{\underset{}{C}}H-\overset{O}{\underset{}{C}}-OCH_3$$

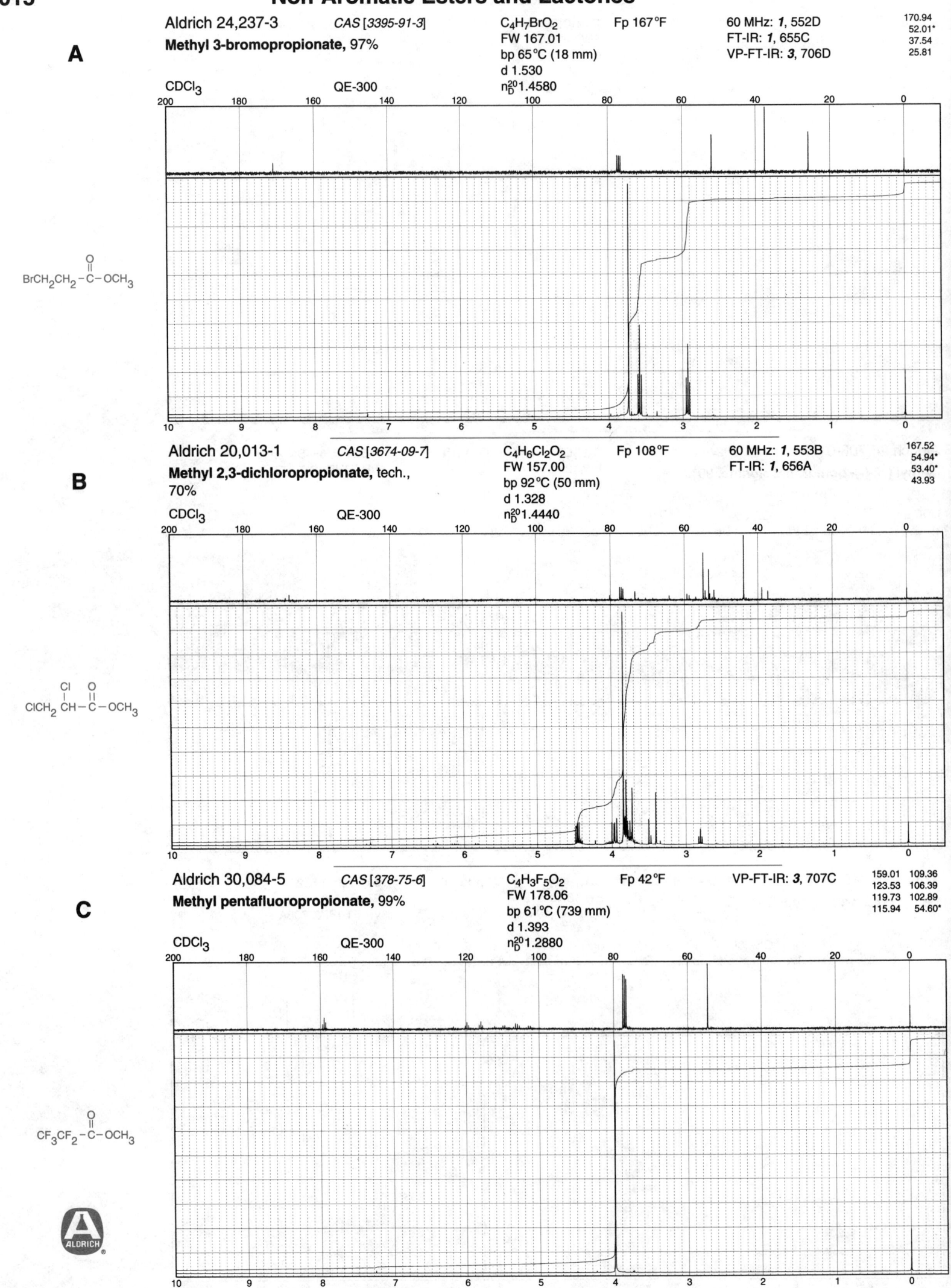

A

Aldrich 24,237-3   *CAS [3395-91-3]* $C_4H_7BrO_2$ Fp 167°F   60 MHz: *1*, 552D

Methyl 3-bromopropionate, 97%   FW 167.01   FT-IR: *1*, 655C

bp 65°C (18 mm)   VP-FT-IR: *3*, 706D

d 1.530

n_D^{20}1.4580

CDCl$_3$ QE-300

170.94
52.01*
37.54
25.81

$BrCH_2CH_2-\overset{\overset{\displaystyle O}{\|}}{C}-OCH_3$

B

Aldrich 20,013-1   *CAS [3674-09-7]* $C_4H_6Cl_2O_2$ Fp 108°F   60 MHz: *1*, 553B

Methyl 2,3-dichloropropionate, tech., 70%   FW 157.00   FT-IR: *1*, 656A

bp 92°C (50 mm)

d 1.328

n_D^{20}1.4440

CDCl$_3$ QE-300

167.52
54.94*
53.40*
43.93

$ClCH_2\overset{\overset{\displaystyle Cl}{|}}{CH}-\overset{\overset{\displaystyle O}{\|}}{C}-OCH_3$

C

Aldrich 30,084-5   *CAS [378-75-6]* $C_4H_3F_5O_2$ Fp 42°F   VP-FT-IR: *3*, 707C

Methyl pentafluoropropionate, 99%   FW 178.06

bp 61°C (739 mm)

d 1.393

n_D^{20}1.2880

CDCl$_3$ QE-300

159.01   109.36
123.53   106.39
119.73   102.89
115.94   54.60*

$CF_3CF_2-\overset{\overset{\displaystyle O}{\|}}{C}-OCH_3$

Aldrich 19,241-4   CAS [535-13-7] $C_5H_9ClO_2$ Fp 101°F   60 MHz: *1*, 551D

(±)-Ethyl 2-chloropropionate, 99%   FW 136.58   FT-IR: *1*, 654C

bp 148°C   VP-FT-IR: *3*, 705D

d 1.072

CDCl₃   QE-300 n_D^{20}1.4170

170.02
62.00
52.56*
21.49*
14.02*

A

$CH_3CH-C-OCH_2CH_3$ (with Cl and O above)

Aldrich E1,785-2   CAS [623-71-2] $C_5H_9ClO_2$ Fp 130°F   60 MHz: *1*, 552C

Ethyl 3-chloropropionate, 98%   FW 136.58   FT-IR: *1*, 655B

bp 163°C   VP-FT-IR: *3*, 706C

d 1.003

CDCl₃   QE-300 n_D^{20}1.4250

170.28
60.95
39.09
37.63
14.19*

B

$ClCH_2CH_2-C-OCH_2CH_3$ (with O above)

Aldrich E1,455-1   CAS [535-11-5] $C_5H_9BrO_2$ Fp 125°F   60 MHz: *1*, 552B

Ethyl 2-bromopropionate, 99%   FW 181.03   FT-IR: *1*, 655A

bp 158°C   VP-FT-IR: *3*, 706B

d 1.394

CDCl₃   QE-300 n_D^{20}1.4460

170.15
61.95
40.20*
21.64*
13.93*

C

$CH_3CH-C-OCH_2CH_3$ (with Br and O above)

Non-Aromatic Esters and Lactones

1017

A

Aldrich 12,816-3   CAS [539-74-2] $C_5H_9BrO_2$ Fp 175°F   60 MHz: *1*, 553A

Ethyl 3-bromopropionate, 99%   FW 181.03   FT-IR: *1*, 655D

bp 136°C (50 mm)   VP-FT-IR: *3*, 707A

d 1.412

n_D^{20} 1.4520

CDCl₃   QE-300

170.38
60.94
37.80
25.92
14.19*

$BrCH_2CH_2-C(O)-OCH_2CH_3$

B

Aldrich E2,280-5   CAS [3674-13-3] $C_5H_8Br_2O_2$ Fp 197°F   60 MHz: *1*, 553C

Ethyl 2,3-dibromopropionate, 97%   FW 259.94   FT-IR: *1*, 656B

bp 213°C (746 mm)   VP-FT-IR: *3*, 707B

d 1.788

n_D^{20} 1.5000

CDCl₃   QE-300

167.36
62.51
41.21*
29.72
13.89*

$BrCH_2CH(Br)-C(O)-OCH_2CH_3$

C

Aldrich 29,092-0   CAS [426-65-3] $C_5H_5F_5O_2$ Fp 35°F   VP-FT-IR: *3*, 708A

Ethyl pentafluoropropionate, 99%   FW 192.09

bp 76°C

d 1.299

n_D^{20} 1.3010

CDCl₃   QE-300

158.93 116.45 106.33
158.54 116.00 105.81
158.15 115.55 105.28
123.57 112.21 102.84
120.23 109.83 102.31
119.78 109.30 64.71
119.33 106.86 13.78*

$CF_3CF_2-C(O)-OCH_2CH_3$

Non-Aromatic Esters and Lactones

Aldrich E1,440-3   CAS [600-00-0] $C_6H_{11}BrO_2$ Fp 140°F   60 MHz: 1, 555A   171.57
Ethyl 2-bromoisobutyrate, 98%   FW 195.06   FT-IR: 1, 657C   61.98
bp 66°C (11 mm)   VP-FT-IR: 3, 709A   55.88
d 1.329   30.78*
CDCl₃   QE-300 n_D^{20}1.4440   13.86*

A

Aldrich 32,509-0   CAS [110556-33-7] $C_5H_9BrO_2$ Fp 165°F   VP-FT-IR: 3, 709C   173.68
Methyl (R)-(+)-3-bromo-2-methylpropionate,   FW 181.03   52.06*
97%   bp 85°C (37 mm)   42.04*
d 1.422   34.07
CDCl₃   QE-300 n_D^{20}1.4550   16.28*

B

Aldrich 32,492-2   CAS [98190-85-3] $C_5H_9BrO_2$ Fp 165°F   VP-FT-IR: 3, 709B   173.68
Methyl (S)-(-)-3-bromo-2-methylpropionate,   FW 181.03   52.07*
97%   bp 98°C (62 mm)   42.05*
d 1.422   34.08
CDCl₃   QE-300 n_D^{20}1.4550   16.29*

C

A

Aldrich M3,480-4   *CAS [3153-37-5]*   C₅H₉ClO₂   Fp 139°F   60 MHz: *1*, 554B

Methyl 4-chlorobutyrate, 98+%   FW 136.58   FT-IR: *1*, 657A

bp 176°C   VP-FT-IR: *3*, 708C

d 1.120

n_D^{20} 1.4320

CDCl₃   QE-300

172.97 / 51.64* / 44.04 / 31.00 / 27.72

$ClCH_2CH_2CH_2-\overset{O}{\overset{||}{C}}-OCH_3$

B

Aldrich 23,731-0   *CAS [69043-96-5]*   C₅H₉BrO₂   Fp 155°F   60 MHz: *1*, 553D

Methyl (±)-2-bromobutyrate, 97%   FW 181.04   FT-IR: *1*, 656C

bp 138°C (50 mm)   VP-FT-IR: *3*, 707D

d 1.573

n_D^{20} 1.4520

CDCl₃   QE-300

170.15 / 52.85* / 47.37* / 28.39 / 11.88*

$CH_3CH_2\overset{Br}{\overset{|}{C}}H-\overset{O}{\overset{||}{C}}-OCH_3$

C

Aldrich 28,229-4   *CAS [356-24-1]*   C₅H₃F₇O₂   Fp >230°F

Methyl heptafluorobutyrate, 99%   FW 228.07   VP-FT-IR: *3*, 710C

bp 81°C

d 1.472

n_D^{20} 1.2930

CDCl₃   QE-300

159.36	119.03	107.39
158.97	116.11	104.73
158.58	115.66	104.40
123.27	115.22	104.30
122.83	111.34	103.87
119.91	108.25	54.75*
119.47	107.82	

$CF_3CF_2CF_2-\overset{O}{\overset{||}{C}}-OCH_3$

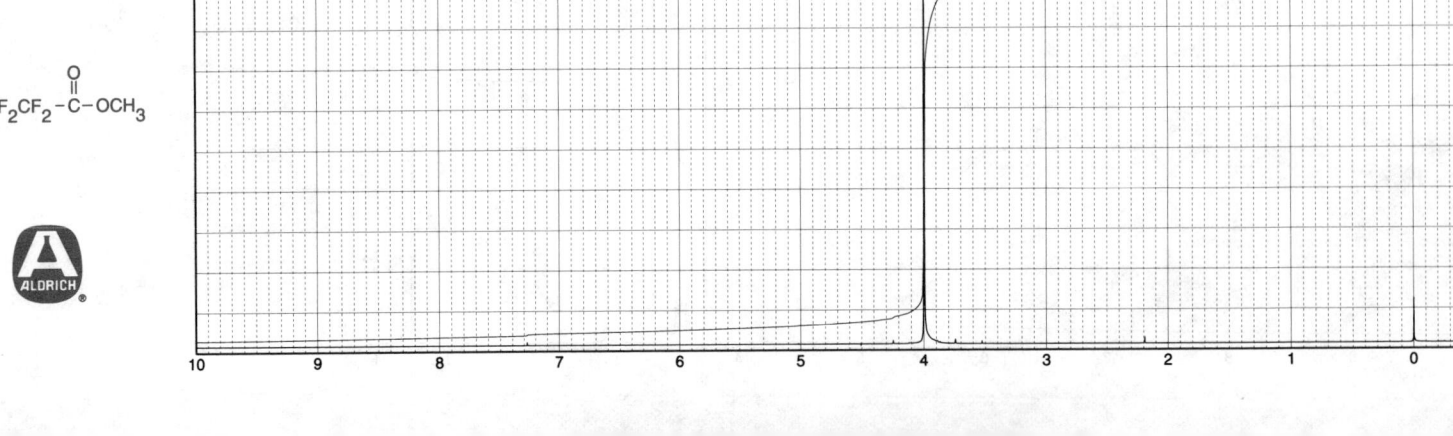

Non-Aromatic Esters and Lactones

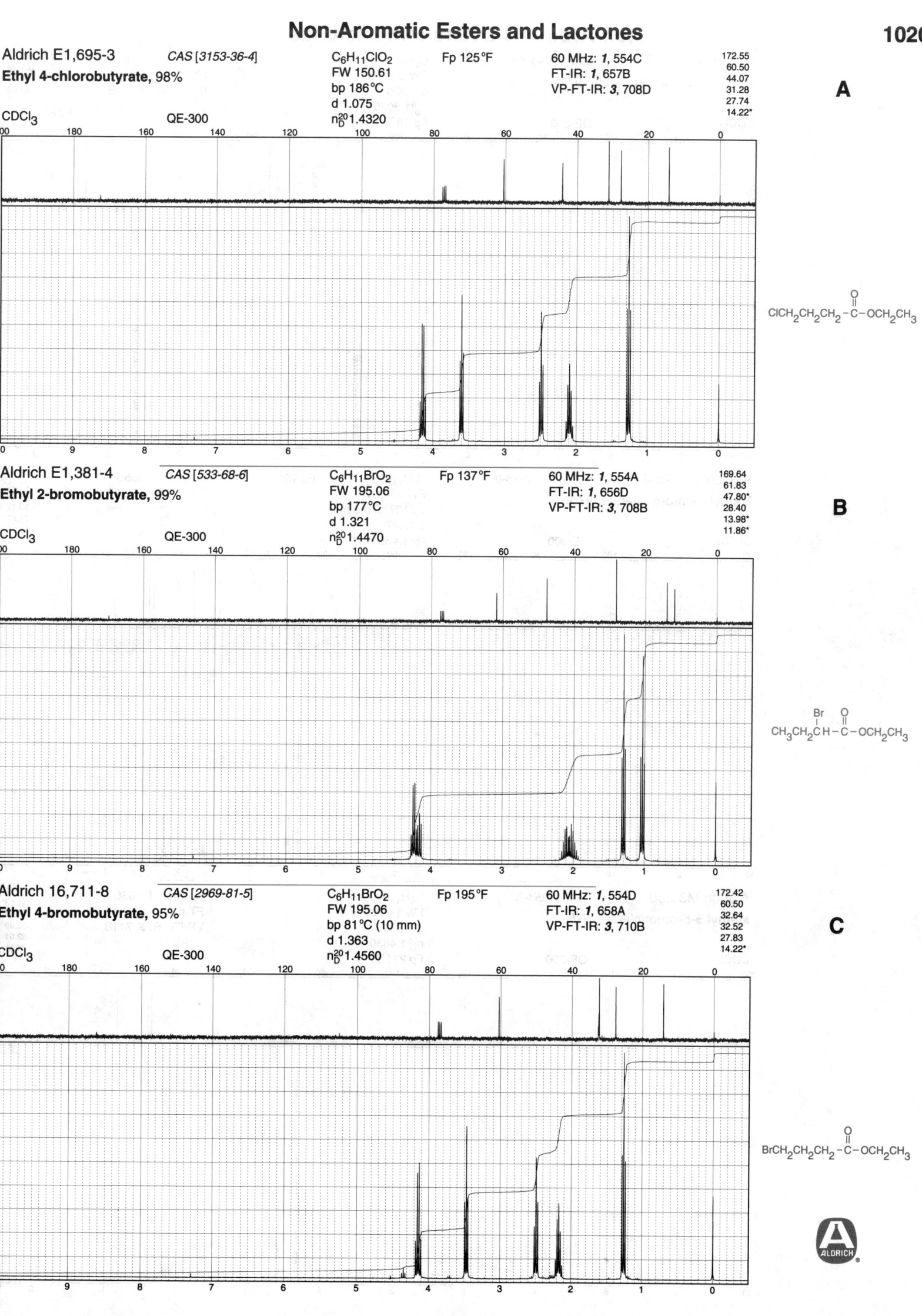

Aldrich E1,695-3   *CAS [3153-36-4]*

Ethyl 4-chlorobutyrate, 98%

$C_6H_{11}ClO_2$
FW 150.61
bp 186°C
d 1.075
n_D^{20} 1.4320

Fp 125°F

60 MHz: *1*, 554C
FT-IR: *1*, 657B
VP-FT-IR: *3*, 708D

172.55
60.50
44.07
31.28
27.74
14.22*

A

CDCl$_3$ QE-300

ClCH$_2$CH$_2$CH$_2$—C—OCH$_2$CH$_3$

Aldrich E1,381-4   *CAS [533-68-6]*

Ethyl 2-bromobutyrate, 99%

$C_6H_{11}BrO_2$
FW 195.06
bp 177°C
d 1.321
n_D^{20} 1.4470

Fp 137°F

60 MHz: *1*, 554A
FT-IR: *1*, 656D
VP-FT-IR: *3*, 708B

169.64
61.83
47.80*
28.40
13.98*
11.86*

B

CDCl$_3$ QE-300

CH$_3$CH$_2$CH—C—OCH$_2$CH$_3$

Aldrich 16,711-8   *CAS [2969-81-5]*

Ethyl 4-bromobutyrate, 95%

$C_6H_{11}BrO_2$
FW 195.06
bp 81°C (10 mm)
d 1.363
n_D^{20} 1.4560

Fp 195°F

60 MHz: *1*, 554D
FT-IR: *1*, 658A
VP-FT-IR: *3*, 710B

172.42
60.50
32.64
32.52
27.83
14.22*

C

CDCl$_3$ QE-300

BrCH$_2$CH$_2$CH$_2$—C—OCH$_2$CH$_3$

ALDRICH

A

Aldrich 17,514-5   *CAS [356-27-4]* $C_6H_5F_7O_2$ 60 MHz: *1*, 551A
Ethyl heptafluorobutyrate, 97%   FW 242.09   FT-IR: *1*, 650A
    d 1.396   VP-FT-IR: *3*, 699B
    n_D^{20}1.3020
    Fp 58°F

159.10 111.50
158.71 107.98
158.33 104.46
123.53 65.03
119.73 13.80*
115.93

CDCl₃   QE-300

$CF_3CF_2CF_2-\overset{\displaystyle O}{\overset{\|}{C}}-O\,CH_2CH_3$

B

Aldrich 13,598-4   *CAS [14273-86-0]* $C_6H_{11}ClO_2$ Fp 194°F   60 MHz: *1*, 555C
Methyl 5-chlorovalerate, 97%   FW 150.61   VP-FT-IR: *3*, 711A
    bp 107°C (38 mm)
    d 1.047
    n_D^{20}1.4360

173.48
51.55*
44.42
33.17
31.87
22.26

CDCl₃   QE-300

$ClCH_2CH_2CH_2CH_2-\overset{\displaystyle O}{\overset{\|}{C}}-OCH_3$

C

Aldrich M3,260-7   *CAS [5454-83-1]* $C_6H_{11}BrO_2$ 60 MHz: *1*, 555D
Methyl 5-bromovalerate, 97%   FW 195.06   FT-IR: *1*, 658C
    d 1.363   VP-FT-IR: *3*, 711B
    n_D^{20}1.4630
    Fp 211°F

173.39
51.53*
33.05
32.90
32.01
23.51

CDCl₃   QE-300

$BrCH_2CH_2CH_2CH_2-\overset{\displaystyle O}{\overset{\|}{C}}-OCH_3$

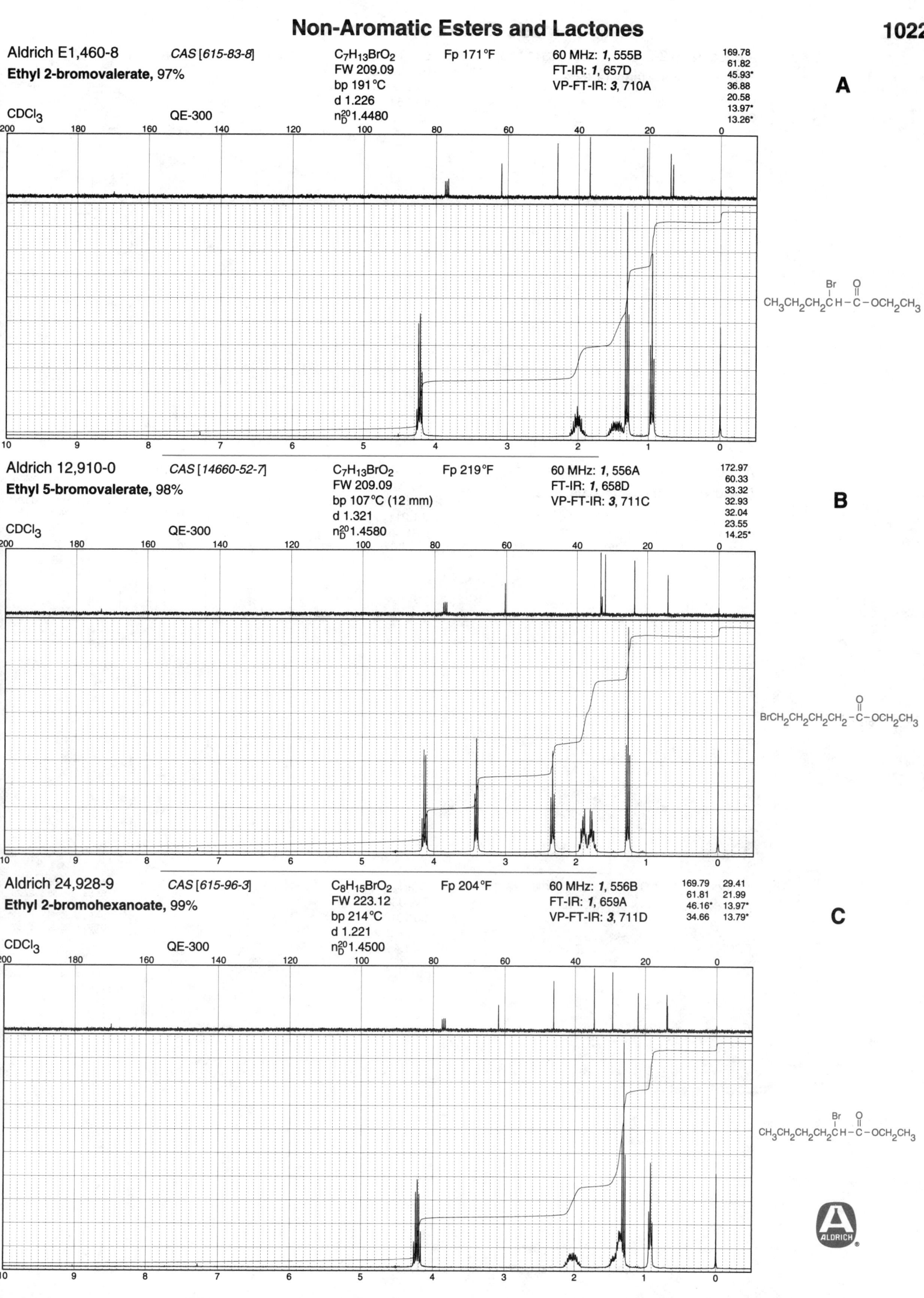

Aldrich E1,460-8   CAS [615-83-8]

Ethyl 2-bromovalerate, 97%

C$_7$H$_{13}$BrO$_2$ Fp 171°F
FW 209.09
bp 191°C
d 1.226
n$_D^{20}$1.4480

60 MHz: *1*, 555B
FT-IR: *1*, 657D
VP-FT-IR: *3*, 710A

169.78
61.82
45.93*
36.88
20.58
13.97*
13.26*

A

CDCl$_3$ QE-300

CH$_3$CH$_2$CH$_2$CH—C—OCH$_2$CH$_3$ (Br, O)

Aldrich 12,910-0   CAS [14660-52-7]

Ethyl 5-bromovalerate, 98%

C$_7$H$_{13}$BrO$_2$ Fp 219°F
FW 209.09
bp 107°C (12 mm)
d 1.321
n$_D^{20}$1.4580

60 MHz: *1*, 556A
FT-IR: *1*, 658D
VP-FT-IR: *3*, 711C

172.97
60.33
33.32
32.93
32.04
23.55
14.25*

B

CDCl$_3$ QE-300

BrCH$_2$CH$_2$CH$_2$CH$_2$—C—OCH$_2$CH$_3$ (O)

Aldrich 24,928-9   CAS [615-96-3]

Ethyl 2-bromohexanoate, 99%

C$_8$H$_{15}$BrO$_2$ Fp 204°F
FW 223.12
bp 214°C
d 1.221
n$_D^{20}$1.4500

60 MHz: *1*, 556B
FT-IR: *1*, 659A
VP-FT-IR: *3*, 711D

169.79   29.41
61.81   21.99
46.16*   13.97*
34.66   13.79*

C

CDCl$_3$ QE-300

CH$_3$CH$_2$CH$_2$CH$_2$CH—C—OCH$_2$CH$_3$ (Br, O)

ALDRICH

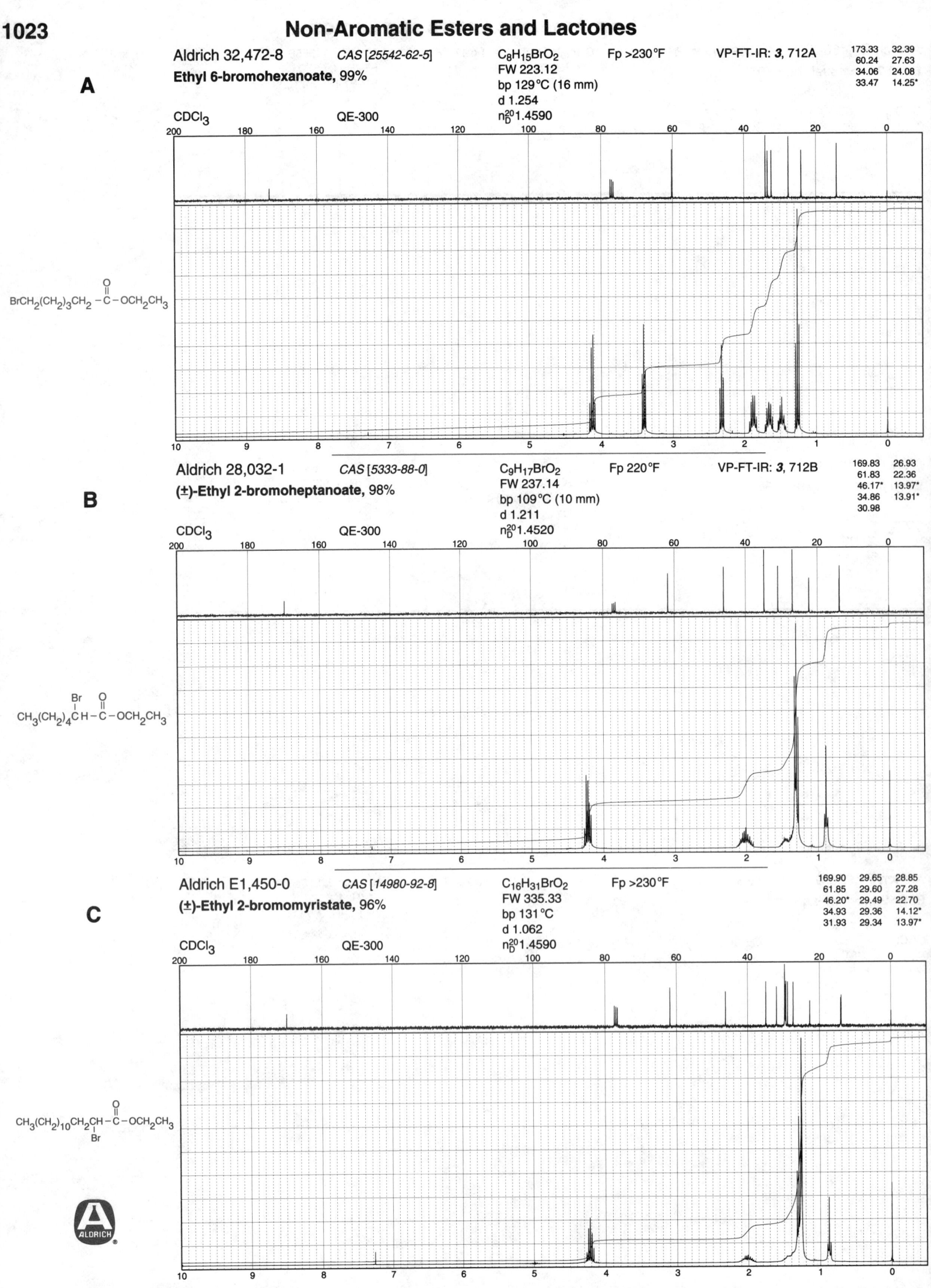

A

Aldrich 32,472-8   *CAS [25542-62-5]* $C_8H_{15}BrO_2$ Fp >230°F   VP-FT-IR: **3**, 712A

Ethyl 6-bromohexanoate, 99%

FW 223.12
bp 129°C (16 mm)
d 1.254
n_D^{20}1.4590

CDCl₃   QE-300

173.33	32.39
60.24	27.63
34.06	24.08
33.47	14.25*

$BrCH_2(CH_2)_3CH_2-C(=O)-OCH_2CH_3$

B

Aldrich 28,032-1   *CAS [5333-88-0]* $C_9H_{17}BrO_2$ Fp 220°F   VP-FT-IR: **3**, 712B

(±)-Ethyl 2-bromoheptanoate, 98%

FW 237.14
bp 109°C (10 mm)
d 1.211
n_D^{20}1.4520

CDCl₃   QE-300

169.83	26.93
61.83	22.36
46.17*	13.97*
34.86	13.91*
30.98	

$CH_3(CH_2)_4CH(Br)-C(=O)-OCH_2CH_3$

C

Aldrich E1,450-0   *CAS [14980-92-8]* $C_{16}H_{31}BrO_2$ Fp >230°F

(±)-Ethyl 2-bromomyristate, 96%

FW 335.33
bp 131°C
d 1.062
n_D^{20}1.4590

CDCl₃   QE-300

169.90	29.65	28.85
61.85	29.60	27.28
46.20*	29.49	22.70
34.93	29.36	14.12*
31.93	29.34	13.97*

$CH_3(CH_2)_{10}CH_2CH(Br)-C(=O)-OCH_2CH_3$

ALDRICH

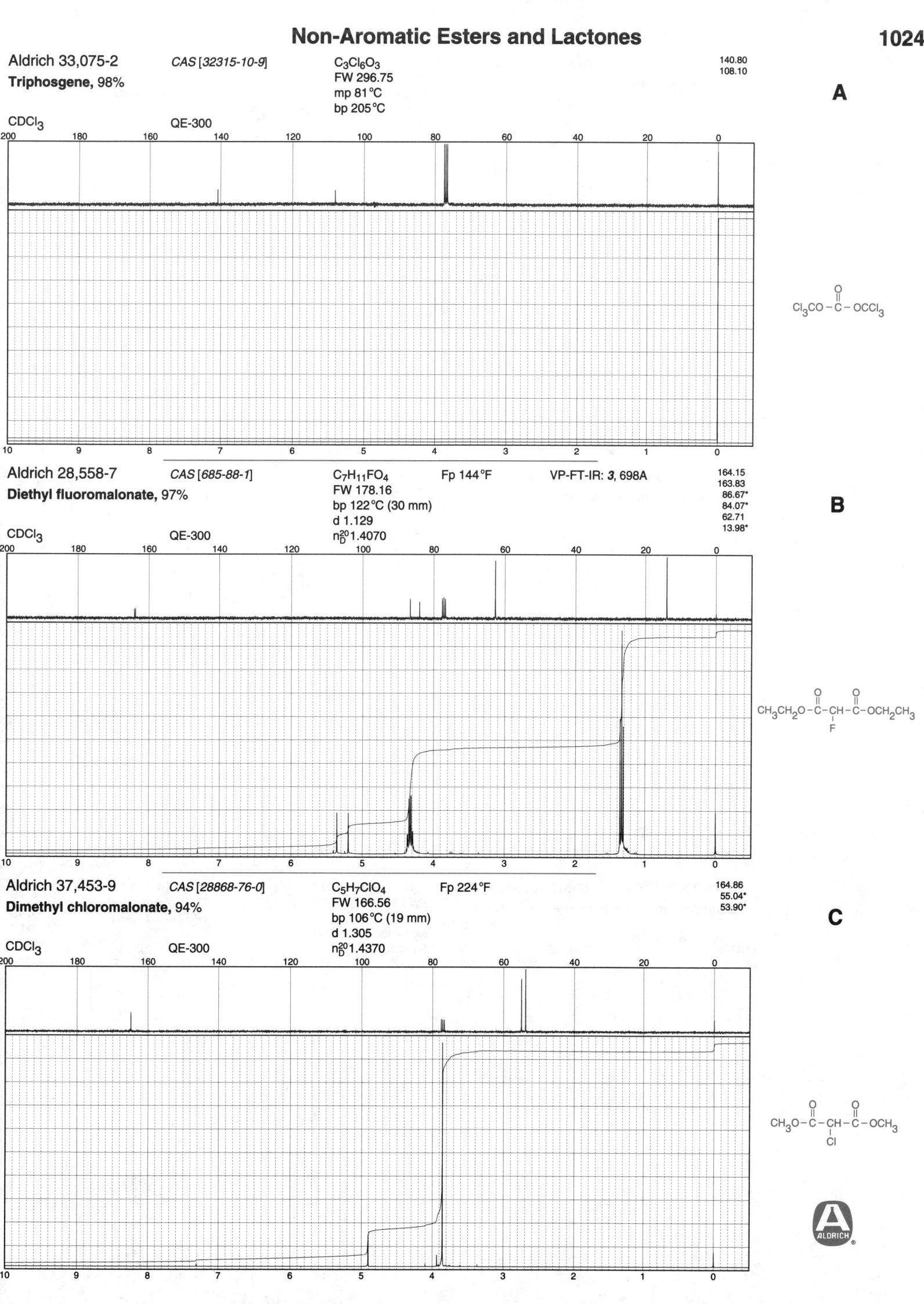

Aldrich 33,075-2
Triphosgene, 98%
CAS [32315-10-9]
C3Cl6O3
FW 296.75
mp 81°C
bp 205°C

140.80
108.10

A

CDCl3   QE-300

Cl3CO—C(=O)—OCCl3

Aldrich 28,558-7
Diethyl fluoromalonate, 97%
CAS [685-88-1]
C7H11FO4
FW 178.16
bp 122°C (30 mm)
d 1.129
n$_D^{20}$1.4070
Fp 144°F
VP-FT-IR: **3**, 698A

164.15
163.83
86.67*
84.07*
62.71
13.98*

B

CDCl3   QE-300

CH3CH2O—C(=O)—CHF—C(=O)—OCH2CH3

Aldrich 37,453-9
Dimethyl chloromalonate, 94%
CAS [28868-76-0]
C5H7ClO4
FW 166.56
bp 106°C (19 mm)
d 1.305
n$_D^{20}$1.4370
Fp 224°F

164.86
55.04*
53.90*

C

CDCl3   QE-300

CH3O—C(=O)—CHCl—C(=O)—OCH3

ALDRICH

Non-Aromatic Esters and Lactones

A

Aldrich D9,160-8   CAS [14064-10-9]

Diethyl chloromalonate, 98%

$C_7H_{11}ClO_4$
FW 194.61
d 1.204
n_D^{20} 1.4310
Fp 218°F

60 MHz: *1*, 547A
FT-IR: *1*, 649B
VP-FT-IR: *3*, 698B

164.45
63.13
55.48*
13.92*

CDCl$_3$ QE-300

$CH_3CH_2O-\overset{O}{\overset{\|}{C}}-\underset{Cl}{CH}-\overset{O}{\overset{\|}{C}}-OCH_2CH_3$

B

Aldrich 35,989-0   CAS [868-26-8]

Dimethyl bromomalonate, tech., 90%

$C_5H_7BrO_4$
FW 211.02
bp 107°C (11 mm)
d 1.601
n_D^{20} 1.4620

Fp >230°F

164.93
53.89*
41.59*

CDCl$_3$ QE-300

$CH_3O-\overset{O}{\overset{\|}{C}}-\underset{Br}{CH}-\overset{O}{\overset{\|}{C}}-OCH_3$

C

Aldrich 13,944-0   CAS [29263-94-3]

Diethyl 2-bromo-2-methylmalonate, 98%

$C_8H_{13}BrO_4$
FW 253.10
d 1.325
n_D^{20} 1.4490
Fp >230°F

60 MHz: *1*, 547C
FT-IR: *1*, 649D
VP-FT-IR: *3*, 698D

167.39
63.01
56.68
26.75*
13.84*

CDCl$_3$ QE-300

$CH_3CH_2O-\overset{O}{\overset{\|}{C}}-\underset{\underset{CH_3}{|}}{\overset{\overset{Br}{|}}{C}}-\overset{O}{\overset{\|}{C}}-OCH_2CH_3$

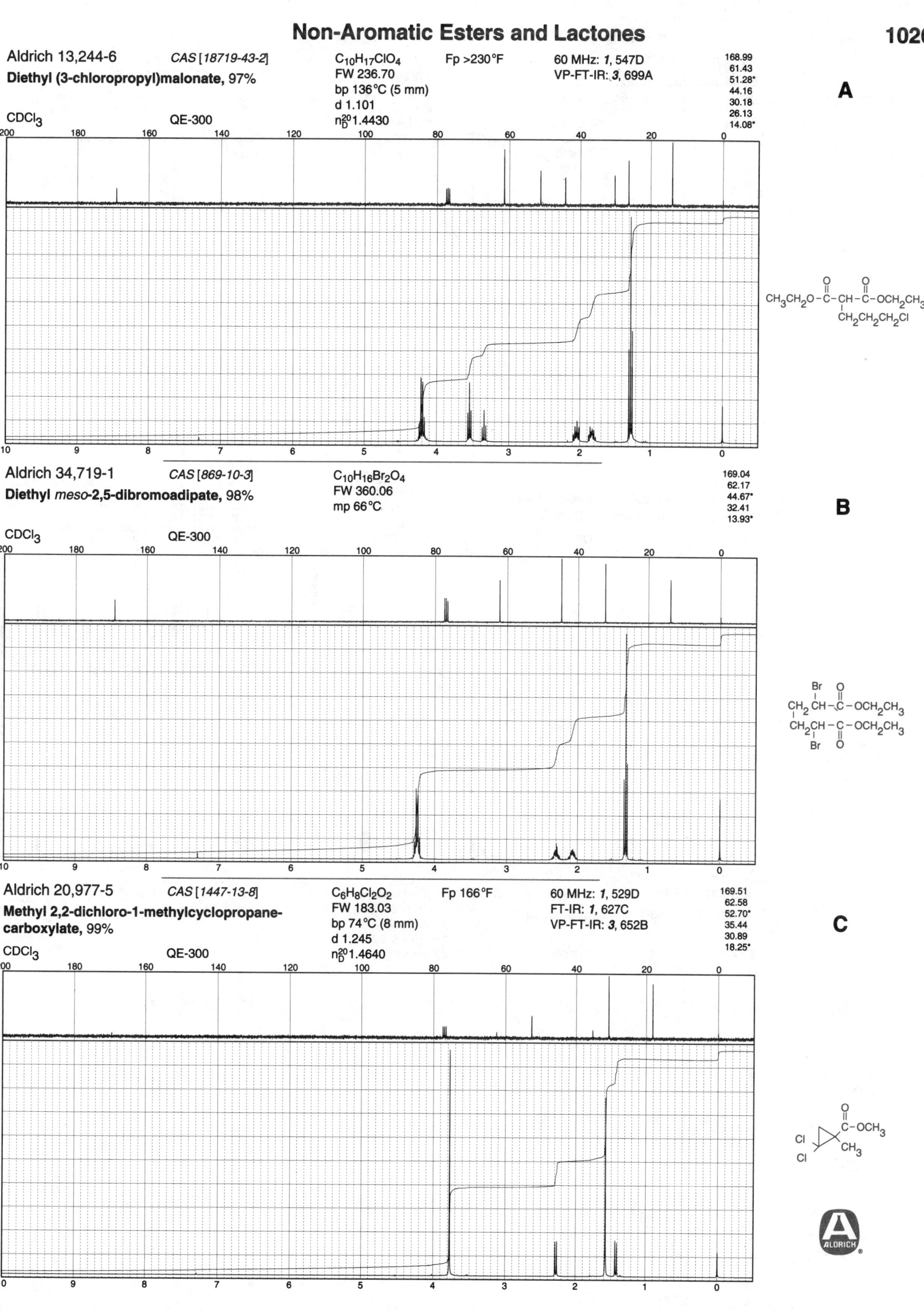

Aldrich 13,244-6   *CAS [18719-43-2]*
Diethyl (3-chloropropyl)malonate, 97%

$C_{10}H_{17}ClO_4$
FW 236.70
bp 136°C (5 mm)
d 1.101
n_D^{20} 1.4430

Fp >230°F

60 MHz: *1*, 547D
VP-FT-IR: *3*, 699A

168.99
61.43
51.28*
44.16
30.18
26.13
14.08*

A

CDCl₃   QE-300

Aldrich 34,719-1   *CAS [869-10-3]*
Diethyl *meso*-2,5-dibromoadipate, 98%

$C_{10}H_{16}Br_2O_4$
FW 360.06
mp 66°C

169.04
62.17
44.67*
32.41
13.93*

B

CDCl₃   QE-300

Aldrich 20,977-5   *CAS [1447-13-8]*
Methyl 2,2-dichloro-1-methylcyclopropane-carboxylate, 99%

$C_6H_8Cl_2O_2$
FW 183.03
bp 74°C (8 mm)
d 1.245
n_D^{20} 1.4640

Fp 166°F

60 MHz: *1*, 529D
FT-IR: *1*, 627C
VP-FT-IR: *3*, 652B

169.51
62.58
52.70*
35.44
30.89
18.25*

C

CDCl₃   QE-300

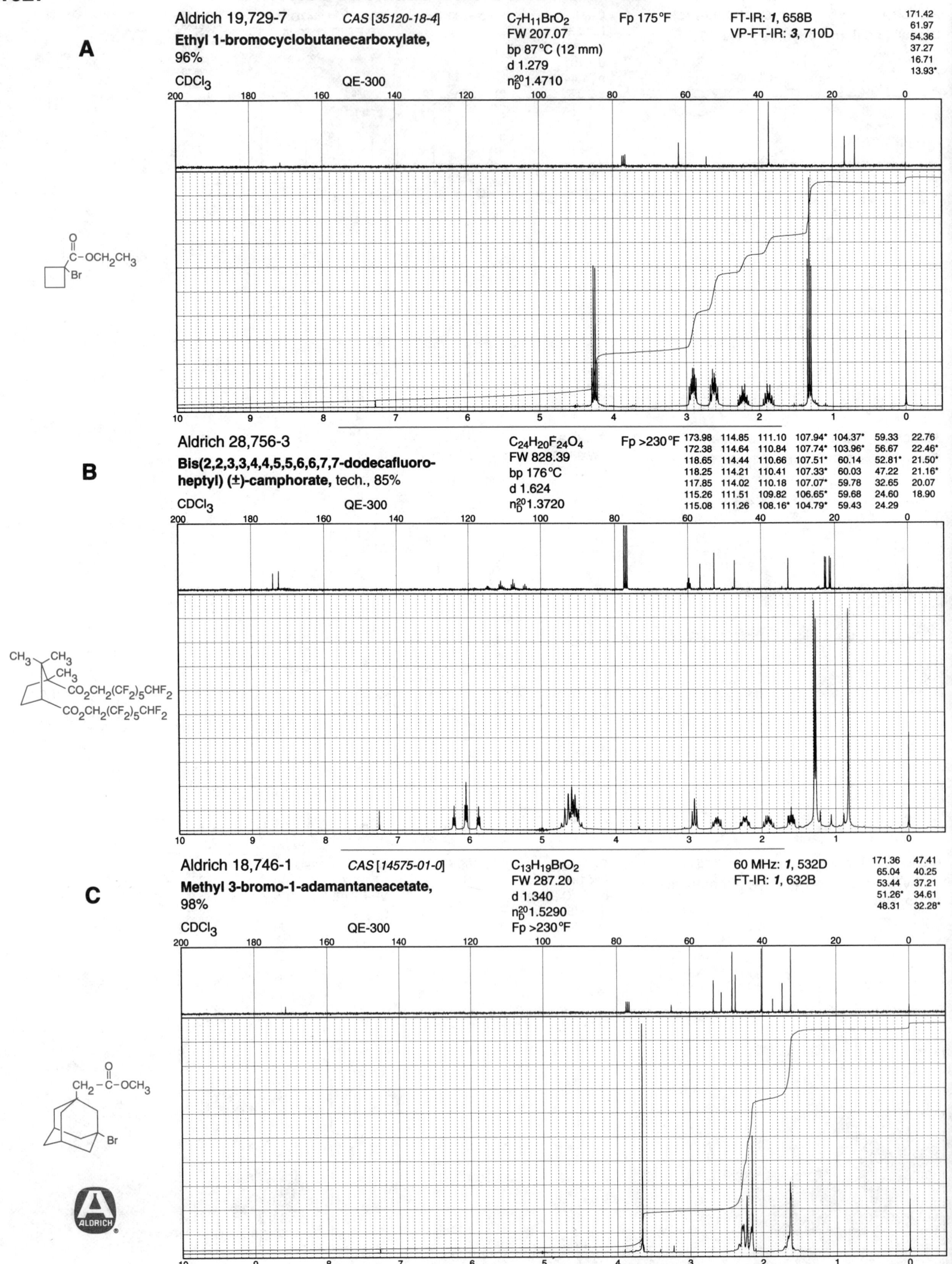

A

Aldrich 19,729-7   CAS [35120-18-4] $C_7H_{11}BrO_2$ Fp 175°F   FT-IR: *1*, 658B

Ethyl 1-bromocyclobutanecarboxylate, 96%

FW 207.07
bp 87°C (12 mm)
d 1.279
n_D^{20} 1.4710

VP-FT-IR: *3*, 710D

171.42
61.97
54.36
37.27
16.71
13.93*

CDCl₃   QE-300

B

Aldrich 28,756-3

Bis(2,2,3,3,4,4,5,5,6,6,7,7-dodecafluoro-heptyl) (±)-camphorate, tech., 85%

$C_{24}H_{20}F_{24}O_4$
FW 828.39
bp 176°C
d 1.624
n_D^{20} 1.3720

Fp >230°F

CDCl₃   QE-300

173.98	114.85	111.10	107.94*	104.37*	59.33	22.76
172.38	114.64	110.84	107.74*	103.96*	56.67	22.46*
118.65	114.44	110.66	107.51*	60.14	52.81*	21.50*
118.25	114.21	110.41	107.33*	60.03	47.22	21.16*
117.85	114.02	110.18	107.07*	59.78	32.65	20.07
115.26	111.51	109.82	106.65*	59.68	24.60	18.90
115.08	111.26	108.16*	104.79*	59.43	24.29	

C

Aldrich 18,746-1   CAS [14575-01-0] $C_{13}H_{19}BrO_2$ 60 MHz: *1*, 532D

Methyl 3-bromo-1-adamantaneacetate, 98%

FW 287.20
d 1.340
n_D^{20} 1.5290
Fp >230°F

FT-IR: *1*, 632B

171.36   47.41
65.04   40.25
53.44   37.21
51.26*   34.61
48.31   32.28*

CDCl₃   QE-300

Aldrich 18,749-6   CAS [16668-48-7] $C_{16}H_{22}Cl_2O_4$ 60 MHz: 1, 533A

Dimethyl α,α'-dichloro-1,3-adamantane-diacetate

FW 349.25
mp 93°C

FT-IR: 1, 632C

168.73	37.71
66.79*	37.19
52.43*	37.06
39.45	35.47
37.80	27.97*

A

CDCl₃   QE-300

Aldrich 29,077-7   CAS [692-72-8]

2-Chloroallyl acetate, 97%

$C_5H_7ClO_2$
FW 134.56
bp 145°C
d 1.121
n_D^{20} 1.4370

Fp 119°F

169.97
135.89
114.87
65.98
20.71*

B

CDCl₃   QE-300

Aldrich 29,772-0   CAS [407-47-6]

2,2,2-Trifluoroethyl acrylate, 99%

$C_5H_5F_3O_2$
FW 154.09
bp 46°C (125 mm)
d 1.216
n_D^{20} 1.3500

Fp 54°F   VP-FT-IR: 3, 696C

164.35	117.60
132.99	61.23
128.61	60.74
126.82*	60.26
124.94	59.77
121.27	

C

CDCl₃   QE-300

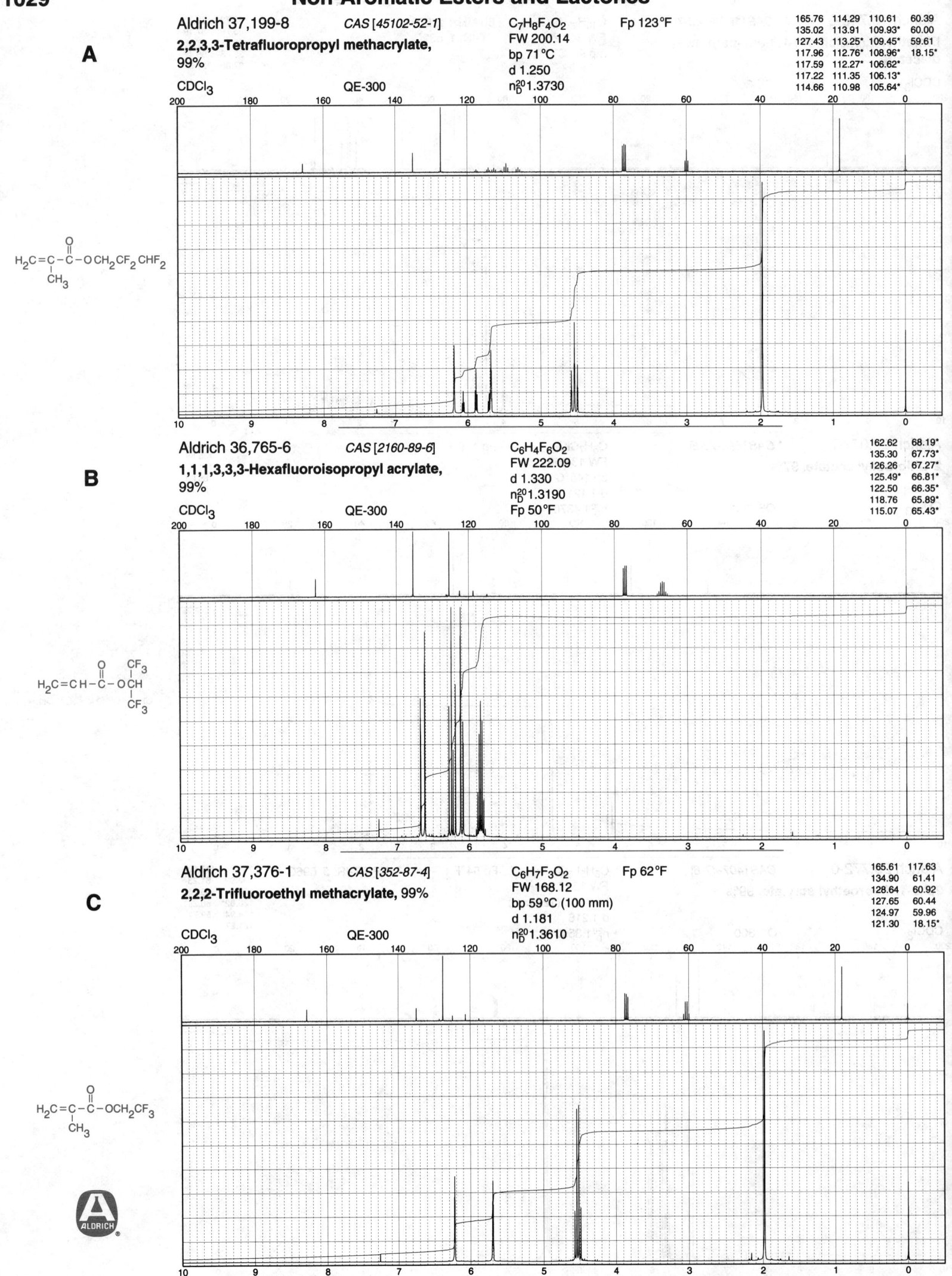

A

Aldrich 37,199-8   *CAS [45102-52-1]*

2,2,3,3-Tetrafluoropropyl methacrylate, 99%

CDCl$_3$ QE-300

$C_7H_8F_4O_2$
FW 200.14
bp 71°C
d 1.250
n$_D^{20}$1.3730

Fp 123°F

165.76	114.29	110.61	60.39
135.02	113.91	109.93*	60.00
127.43	113.25*	109.45*	59.61
117.96	112.76*	108.96*	18.15*
117.59	112.27*	106.62*	
117.22	111.35	106.13*	
114.66	110.98	105.64*	

H$_2$C=C–C–OCH$_2$CF$_2$CHF$_2$
         | ‖
        CH$_3$ O

B

Aldrich 36,765-6   *CAS [2160-89-6]*

1,1,1,3,3,3-Hexafluoroisopropyl acrylate, 99%

CDCl$_3$ QE-300

$C_6H_4F_6O_2$
FW 222.09
d 1.330
n$_D^{20}$1.3190
Fp 50°F

162.62	68.19*
135.30	67.73*
126.26	67.27*
125.49*	66.81*
122.50	66.35*
118.76	65.89*
115.07	65.43*

H$_2$C=CH–C–OCH
         ‖ |
         O   CF$_3$ / CF$_3$

C

Aldrich 37,376-1   *CAS [352-87-4]*

2,2,2-Trifluoroethyl methacrylate, 99%

CDCl$_3$ QE-300

$C_6H_7F_3O_2$
FW 168.12
bp 59°C (100 mm)
d 1.181
n$_D^{20}$1.3610

Fp 62°F

165.61	117.63
134.90	61.41
128.64	60.92
127.65	60.44
124.97	59.96
121.30	18.15*

H$_2$C=C–C–OCH$_2$CF$_3$
         | ‖
        CH$_3$ O

ALDRICH

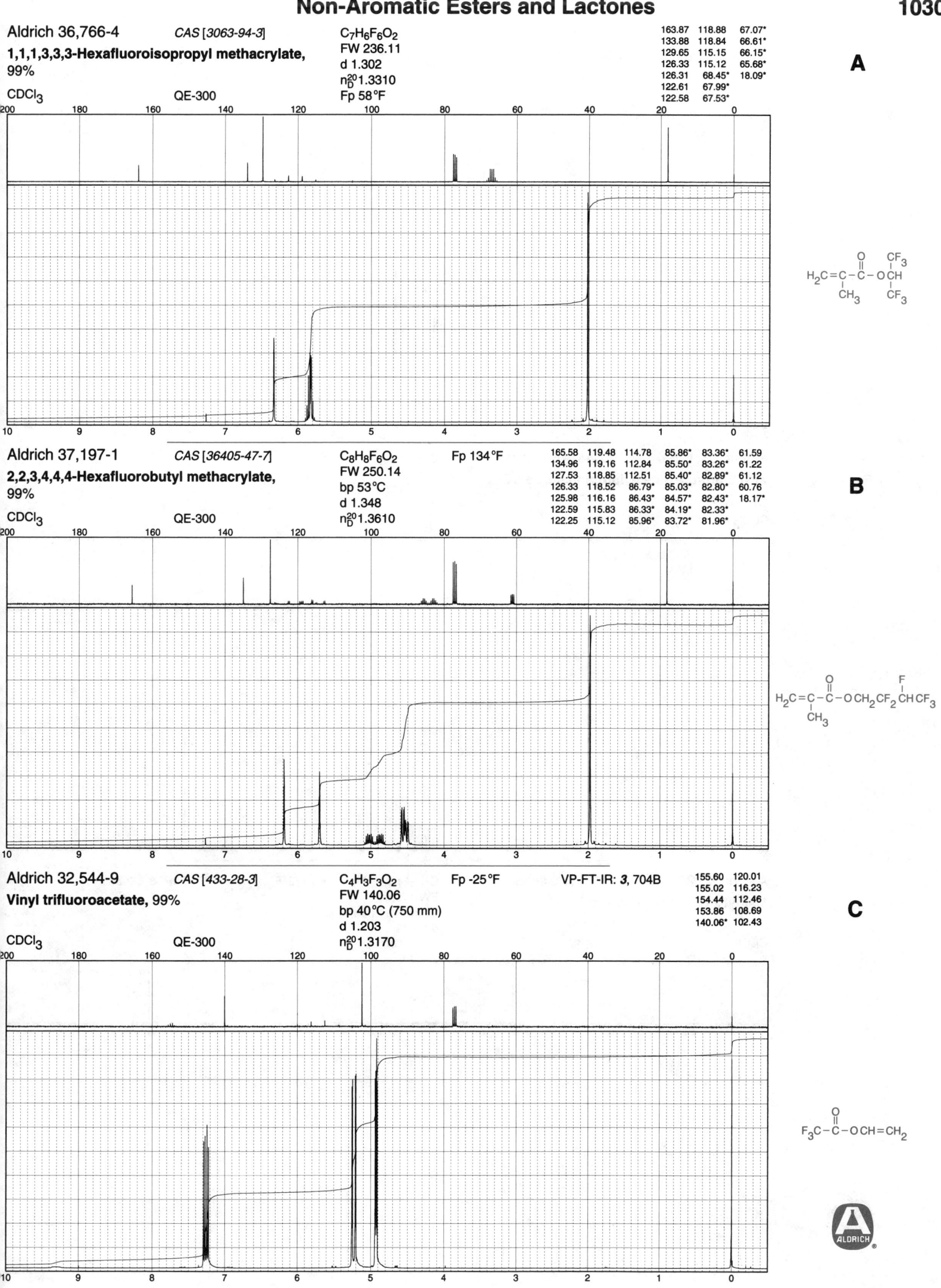

Aldrich 36,766-4   *CAS [3063-94-3]* $C_7H_6F_6O_2$
FW 236.11

1,1,1,3,3,3-Hexafluoroisopropyl methacrylate, 99%

d 1.302
n_D^{20} 1.3310
Fp 58 °F

CDCl$_3$ QE-300

163.87	118.88	67.07*
133.88	118.84	66.61*
129.65	115.15	66.15*
126.33	115.12	65.68*
126.31		68.45*	18.09*
122.61		67.99*
122.58		67.53*

A

Aldrich 37,197-1   *CAS [36405-47-7]* $C_8H_8F_6O_2$ Fp 134 °F
FW 250.14

2,2,3,4,4,4-Hexafluorobutyl methacrylate, 99%

bp 53 °C
d 1.348
n_D^{20} 1.3610

CDCl$_3$ QE-300

165.58	119.48	114.78	85.86*	83.36*	61.59
134.96	119.16	112.84	85.50*	83.26*	61.22
127.53	118.85	112.51	85.40*	82.89*	61.12
126.33	118.52	86.79*	85.03*	82.80*	60.76
125.98	116.16	86.43*	84.57*	82.43*	18.17*
122.59	115.83	86.33*	84.19*	82.33*
122.25	115.12	85.96*	83.72*	81.96*

B

Aldrich 32,544-9   *CAS [433-28-3]* $C_4H_3F_3O_2$ Fp -25 °F   VP-FT-IR: *3*, 704B
FW 140.06

Vinyl trifluoroacetate, 99%

bp 40 °C (750 mm)
d 1.203
n_D^{20} 1.3170

CDCl$_3$ QE-300

155.60	120.01
155.02	116.23
154.44	112.46
153.86	108.69
140.06*	102.43

C

ALDRICH

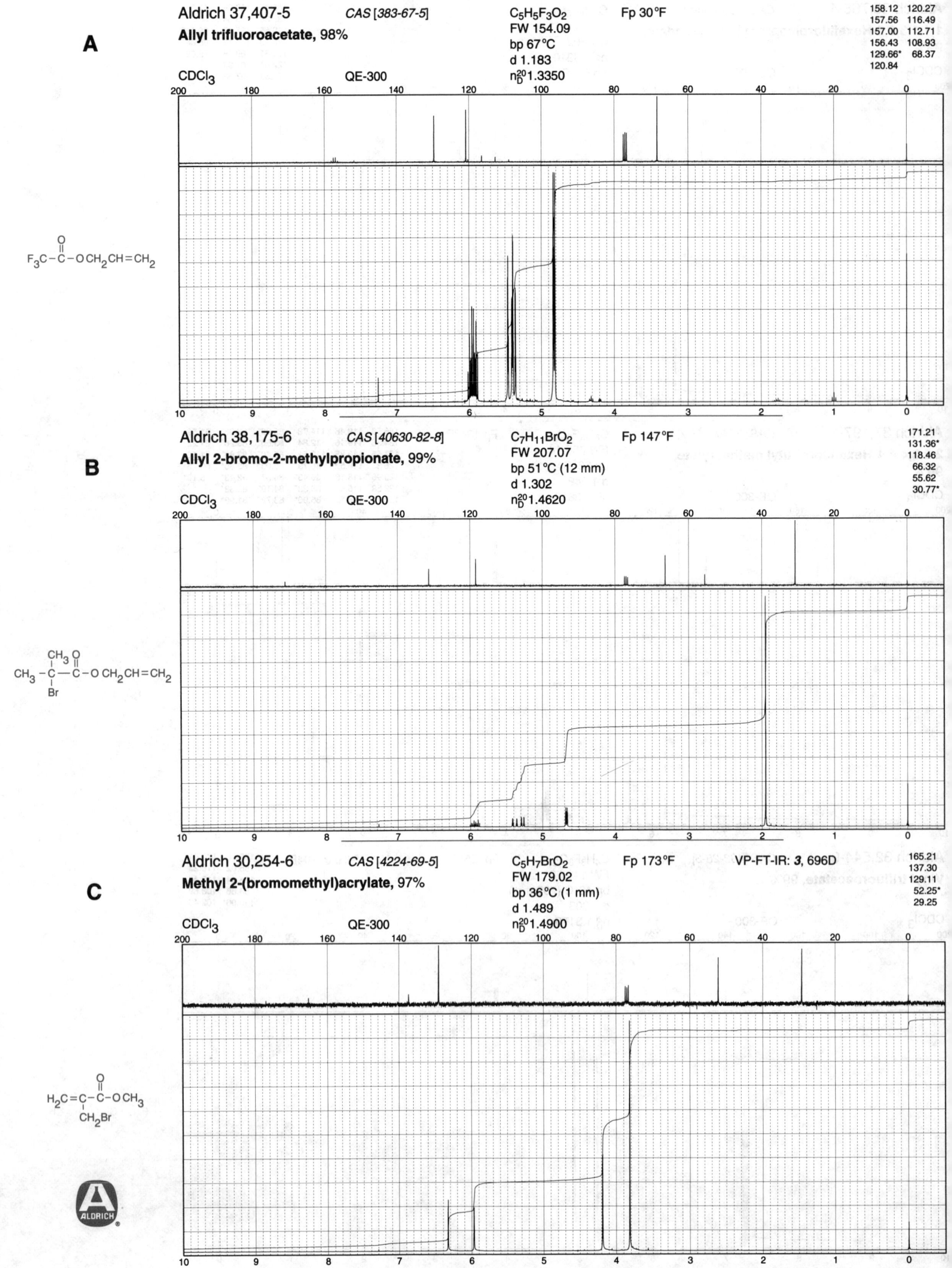

A

Aldrich 37,407-5   *CAS [383-67-5]*   C$_5$H$_5$F$_3$O$_2$ Fp 30°F
Allyl trifluoroacetate, 98%   FW 154.09
bp 67°C
d 1.183
n$_D^{20}$1.3350

CDCl$_3$ QE-300

158.12 120.27
157.56 116.49
157.00 112.71
156.43 108.93
129.66* 68.37
120.84

F$_3$C–C–O CH$_2$CH=CH$_2$

B

Aldrich 38,175-6   *CAS [40630-82-8]*   C$_7$H$_{11}$BrO$_2$ Fp 147°F
Allyl 2-bromo-2-methylpropionate, 99%   FW 207.07
bp 51°C (12 mm)
d 1.302
n$_D^{20}$1.4620

CDCl$_3$ QE-300

171.21
131.36*
118.46
66.32
55.62
30.77*

CH$_3$–C–C–O CH$_2$CH=CH$_2$

C

Aldrich 30,254-6   *CAS [4224-69-5]*   C$_5$H$_7$BrO$_2$ Fp 173°F   VP-FT-IR: *3*, 696D
Methyl 2-(bromomethyl)acrylate, 97%   FW 179.02
bp 36°C (1 mm)
d 1.489
n$_D^{20}$1.4900

CDCl$_3$ QE-300

165.21
137.30
129.11
52.25*
29.25

H$_2$C=C–C–O CH$_3$
CH$_2$Br

Non-Aromatic Esters and Lactones

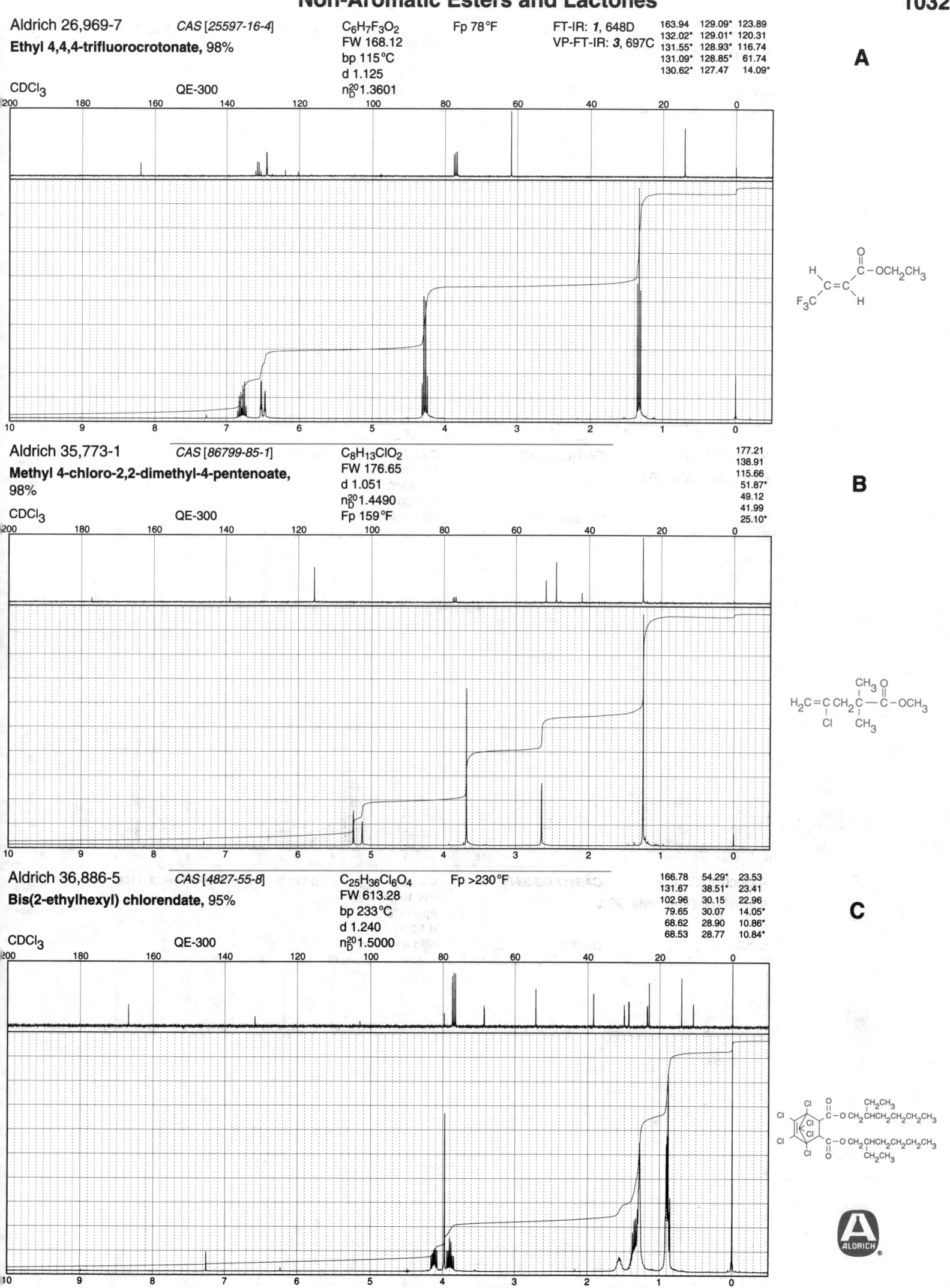

Aldrich 26,969-7 *CAS [25597-16-4]* C₆H₇F₃O₂ Fp 78°F
Ethyl 4,4,4-trifluorocrotonate, 98%
FW 168.12
bp 115°C
d 1.125
nᴰ²⁰1.3601
CDCl₃   QE-300

FT-IR: *1*, 648D
VP-FT-IR: *3*, 697C

163.94 129.09* 123.89
132.02* 129.01* 120.31
131.55* 128.93* 116.74
131.09* 128.85* 61.74
130.62* 127.47 14.09*

A

Aldrich 35,773-1 *CAS [86799-85-1]* C₈H₁₃ClO₂
Methyl 4-chloro-2,2-dimethyl-4-pentenoate, 98%
FW 176.65
d 1.051
nᴰ²⁰1.4490
Fp 159°F
CDCl₃   QE-300

177.21
138.91
115.66
51.87*
49.12
41.99
25.10*

B

Aldrich 36,886-5 *CAS [4827-55-8]* C₂₅H₃₆Cl₆O₄ Fp >230°F
Bis(2-ethylhexyl) chlorendate, 95%
FW 613.28
bp 233°C
d 1.240
nᴰ²⁰1.5000
CDCl₃   QE-300

166.78 54.29* 23.53
131.67 38.51* 23.41
102.96 30.15 22.96
79.65 30.07 14.05*
68.62 28.90 10.86*
68.53 28.77 10.84*

C

A

Aldrich 32,526-0   CAS [96-35-5]   C3H6O3   Fp 153°F   VP-FT-IR: 3, 721A

173.85
60.51
52.23*

Methyl glycolate, 98%

FW 90.08
bp 150°C
d 1.167
n_D^20 1.4170

CDCl3   QE-300

HOCH2 — C(=O) — OCH3

B

Aldrich 36,484-3   CAS [623-50-7]   C4H8O3   Fp 143°F

173.38
61.45
60.65
14.17*

Ethyl glycolate, 98%

FW 104.11
bp 159°C
d 1.100
n_D^20 1.4190

CDCl3   QE-300

HOCH2 — C(=O) — OCH2CH3

C

Aldrich 27,776-2   CAS [17392-83-5]   C4H8O3   Fp 121°F   VP-FT-IR: 3, 712D

176.10
66.79*
52.49*
20.34*

Methyl (R)-(+)-lactate, 98%

FW 104.11
bp 145°C
d 1.090
n_D^20 1.4130

CDCl3   QE-300

CH3, H
   C
HO   C — OCH3
      ‖
      O

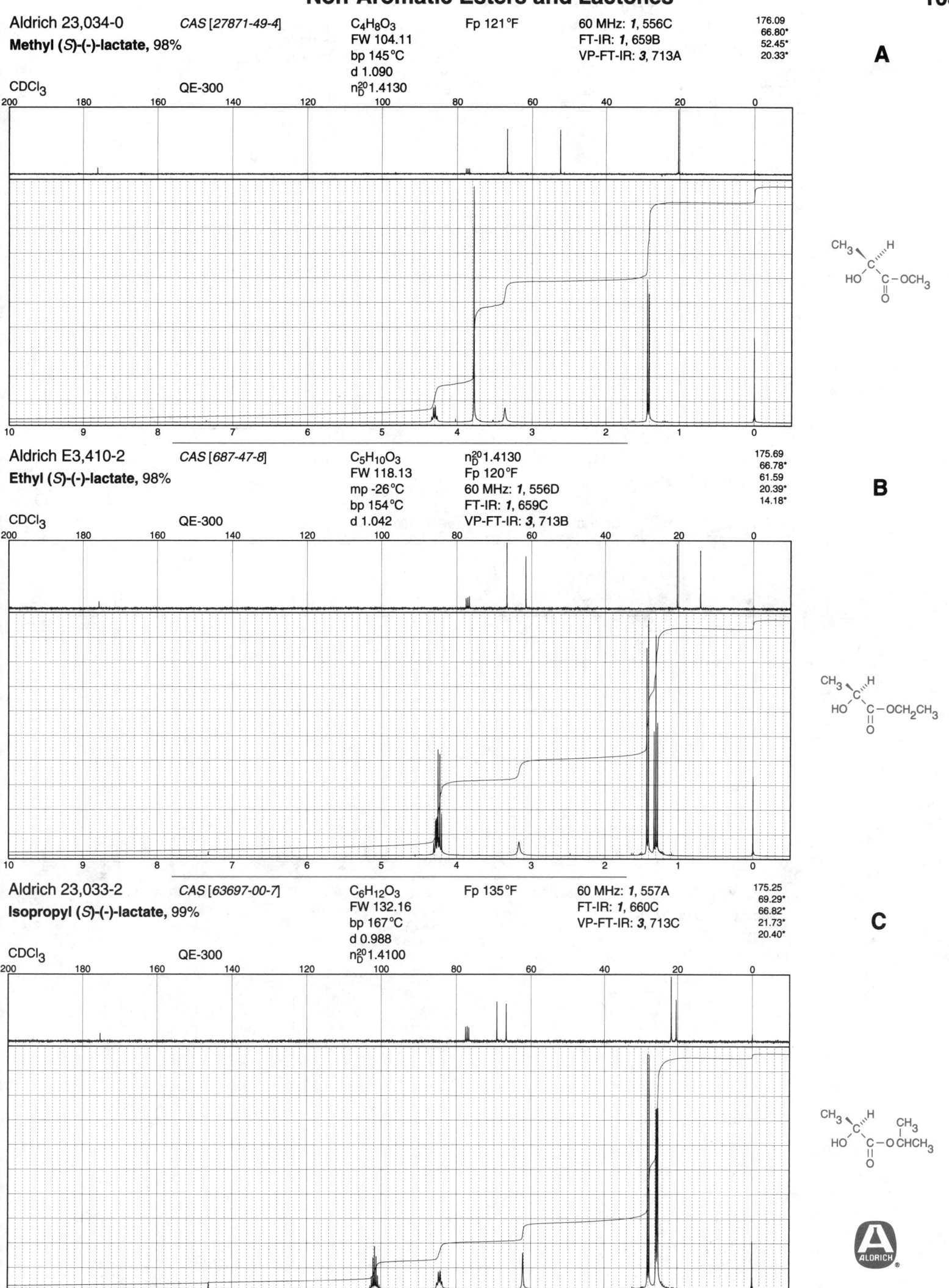

Aldrich 23,034-0 CAS [27871-49-4] C4H8O3 Fp 121°F 60 MHz: *1*, 556C 176.09
Methyl (S)-(-)-lactate, 98% FW 104.11 FT-IR: *1*, 659B 66.80*
bp 145°C VP-FT-IR: *3*, 713A 52.45*
d 1.090 20.33*
CDCl₃ QE-300 n²⁰_D 1.4130

A

Aldrich E3,410-2 CAS [687-47-8] C5H10O3 n²⁰_D 1.4130 175.69
Ethyl (S)-(-)-lactate, 98% FW 118.13 Fp 120°F 66.78*
mp -26°C 60 MHz: *1*, 556D 61.59
bp 154°C FT-IR: *1*, 659C 20.39*
d 1.042 VP-FT-IR: *3*, 713B 14.18*
CDCl₃ QE-300

B

Aldrich 23,033-2 CAS [63697-00-7] C6H12O3 Fp 135°F 60 MHz: *1*, 557A 175.25
Isopropyl (S)-(-)-lactate, 99% FW 132.16 FT-IR: *1*, 660C 69.29*
bp 167°C VP-FT-IR: *3*, 713C 66.82*
d 0.988 21.73*
CDCl₃ QE-300 n²⁰_D 1.4100 20.40*

C

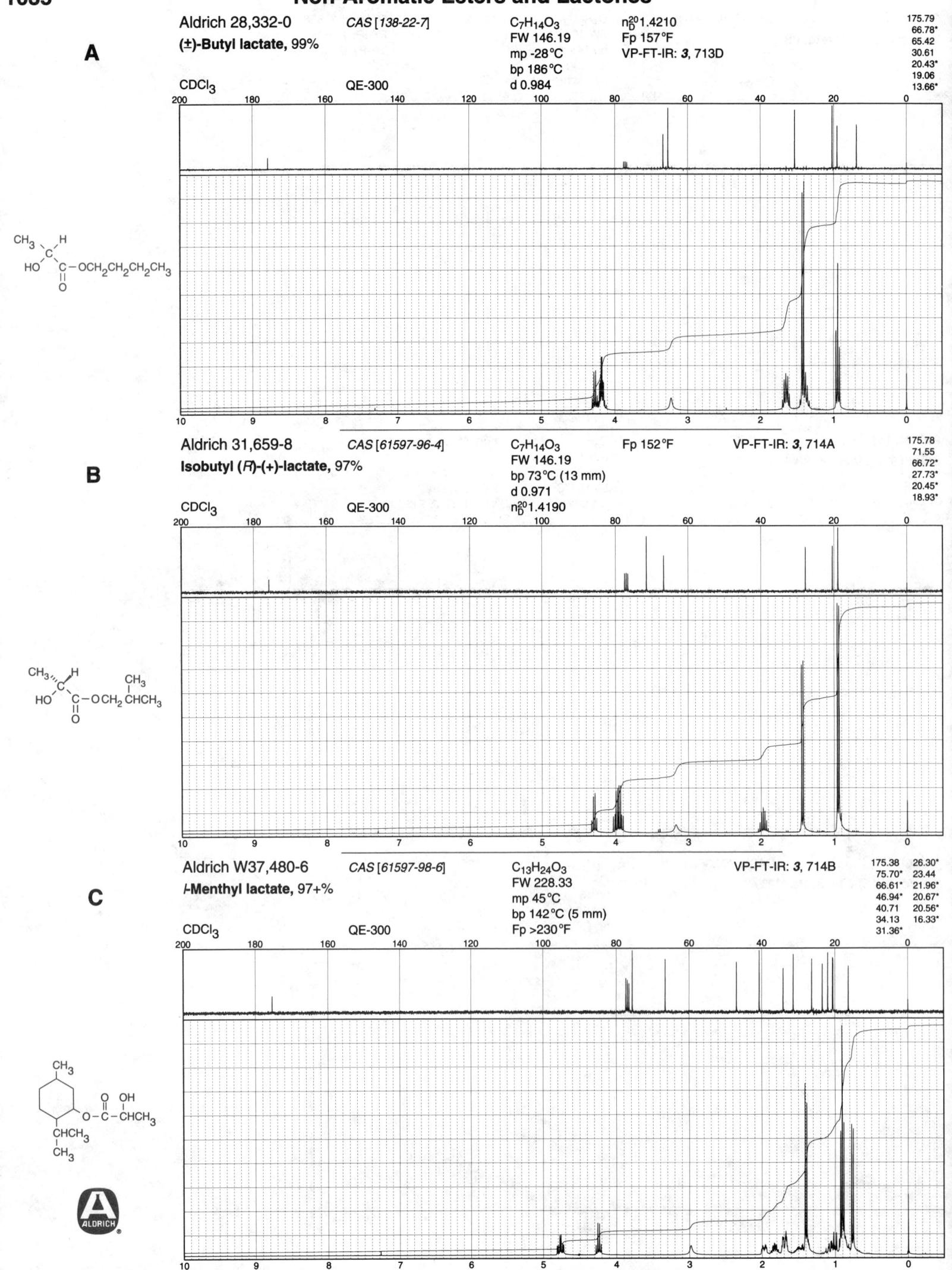

A Aldrich 28,332-0 CAS [138-22-7] C₇H₁₄O₃ nD²⁰ 1.4210 175.79
(±)-Butyl lactate, 99% FW 146.19 Fp 157°F 66.78*
mp -28°C VP-FT-IR: 3, 713D 65.42
bp 186°C 30.61
d 0.984 20.43*
19.06
13.66*

CDCl₃ QE-300

B Aldrich 31,659-8 CAS [61597-96-4] C₇H₁₄O₃ Fp 152°F VP-FT-IR: 3, 714A 175.78
Isobutyl (R)-(+)-lactate, 97% FW 146.19 71.55
bp 73°C (13 mm) 66.72*
d 0.971 27.73*
nD²⁰ 1.4190 20.45*
18.93*

CDCl₃ QE-300

C Aldrich W37,480-6 CAS [61597-98-6] C₁₃H₂₄O₃ VP-FT-IR: 3, 714B 175.38 26.30*
l-Menthyl lactate, 97+% FW 228.33 75.70* 23.44
mp 45°C 66.61* 21.96*
bp 142°C (5 mm) 46.94* 20.67*
Fp >230°F 40.71 20.56*
34.13 16.33*
31.36*

CDCl₃ QE-300

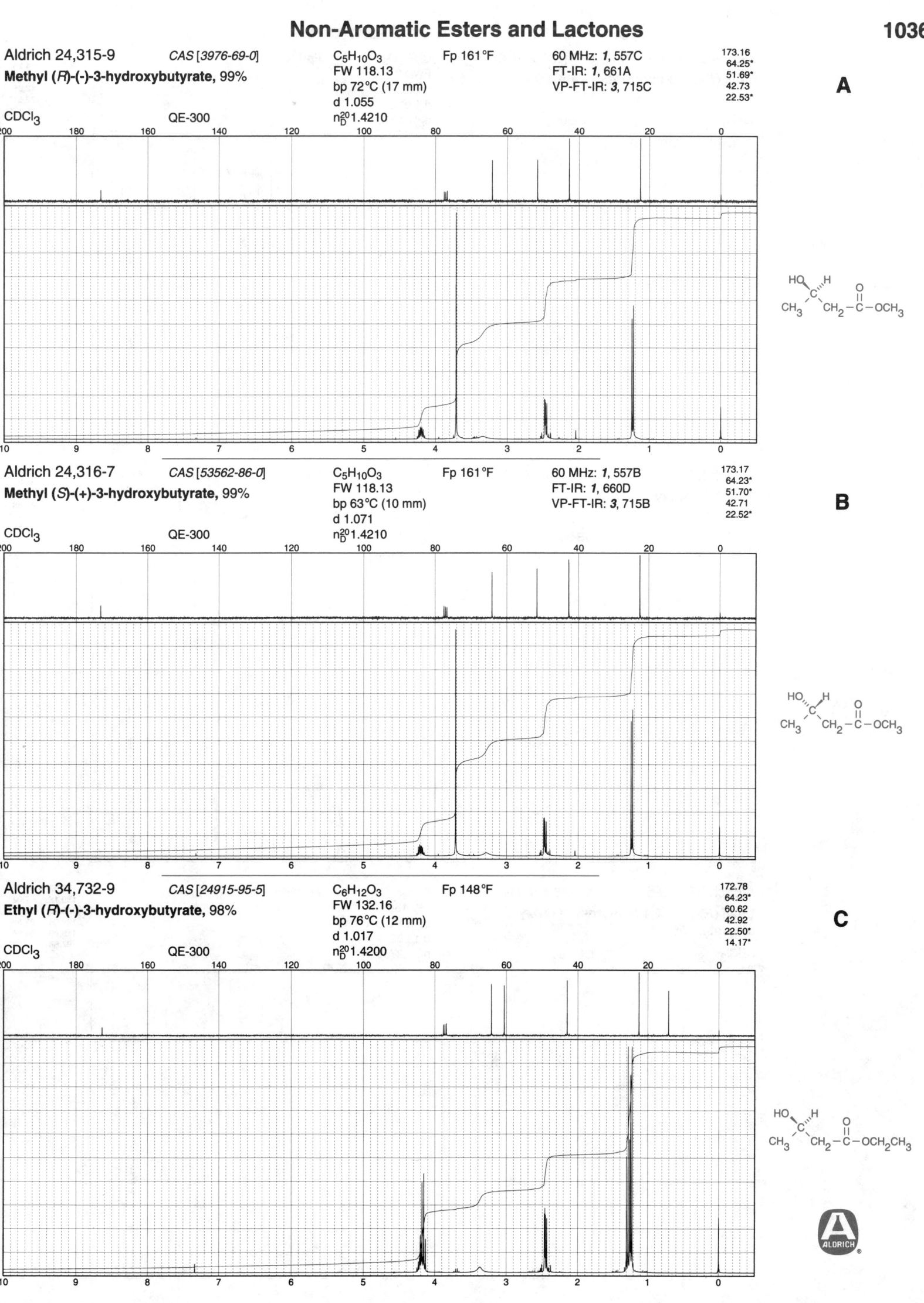

Aldrich 24,315-9   *CAS [3976-69-0]*

Methyl (R)-(-)-3-hydroxybutyrate, 99%

$C_5H_{10}O_3$ Fp 161°F

FW 118.13

bp 72°C (17 mm)

d 1.055

n_D^{20} 1.4210

CDCl₃   QE-300

60 MHz: *1*, 557C

FT-IR: *1*, 661A

VP-FT-IR: *3*, 715C

173.16
64.25*
51.69*
42.73
22.53*

A

Aldrich 24,316-7   *CAS [53562-86-0]*

Methyl (S)-(+)-3-hydroxybutyrate, 99%

$C_5H_{10}O_3$ Fp 161°F

FW 118.13

bp 63°C (10 mm)

d 1.071

n_D^{20} 1.4210

CDCl₃   QE-300

60 MHz: *1*, 557B

FT-IR: *1*, 660D

VP-FT-IR: *3*, 715B

173.17
64.23*
51.70*
42.71
22.52*

B

Aldrich 34,732-9   *CAS [24915-95-5]*

Ethyl (R)-(-)-3-hydroxybutyrate, 98%

$C_6H_{12}O_3$ Fp 148°F

FW 132.16

bp 76°C (12 mm)

d 1.017

n_D^{20} 1.4200

CDCl₃   QE-300

172.78
64.23*
60.62
42.92
22.50*
14.17*

C

Non-Aromatic Esters and Lactones

A

Aldrich 37,470-9   *CAS [56816-01-4]* $C_6H_{12}O_3$ Fp 148°F

Ethyl (S)-(+)-3-hydroxybutyrate, 99%   FW 132.16
bp 181°C
d 1.012
n_D^{20}1.4210

172.76
64.25*
60.62
42.96
22.51*
14.17*

CDCl$_3$ QE-300

B

Aldrich E3,060-3   *CAS [5405-41-4]* $C_6H_{12}O_3$ Fp 148°F   60 MHz: *1*, 557D

Ethyl 3-hydroxybutyrate, 98+%   FW 132.16   FT-IR: *1*, 661C
bp 170°C   VP-FT-IR: *3*, 716A
d 1.017
n_D^{20}1.4200

172.79
64.22*
60.60
42.88
22.47*
14.16*

CDCl$_3$ QE-300

C

Aldrich 37,679-5   *CAS [85571-85-3]* $C_6H_9F_3O_3$ Fp 186°F

Ethyl (R)-(+)-4,4,4-trifluoro-3-hydroxy-butyrate, 99%   FW 186.13
bp 89°C (21 mm)
d 1.259
n_D^{20}1.3750

170.89   67.45*
130.12   67.02*
126.40   66.59*
122.68   61.69
118.96   34.96
67.88*   14.03*

CDCl$_3$ QE-300

Aldrich E3,120-0   *CAS [80-55-7]*

Ethyl 2-hydroxyisobutyrate, 98%

$C_6H_{12}O_3$
FW 132.16
bp 150°C
d 0.965
n_D^{20}1.4080

Fp 112°F

FT-IR: *1*, 662C
VP-FT-IR: *3*, 716D

177.41
71.90
61.72
27.17*
14.15*

CDCl$_3$ QE-300

A

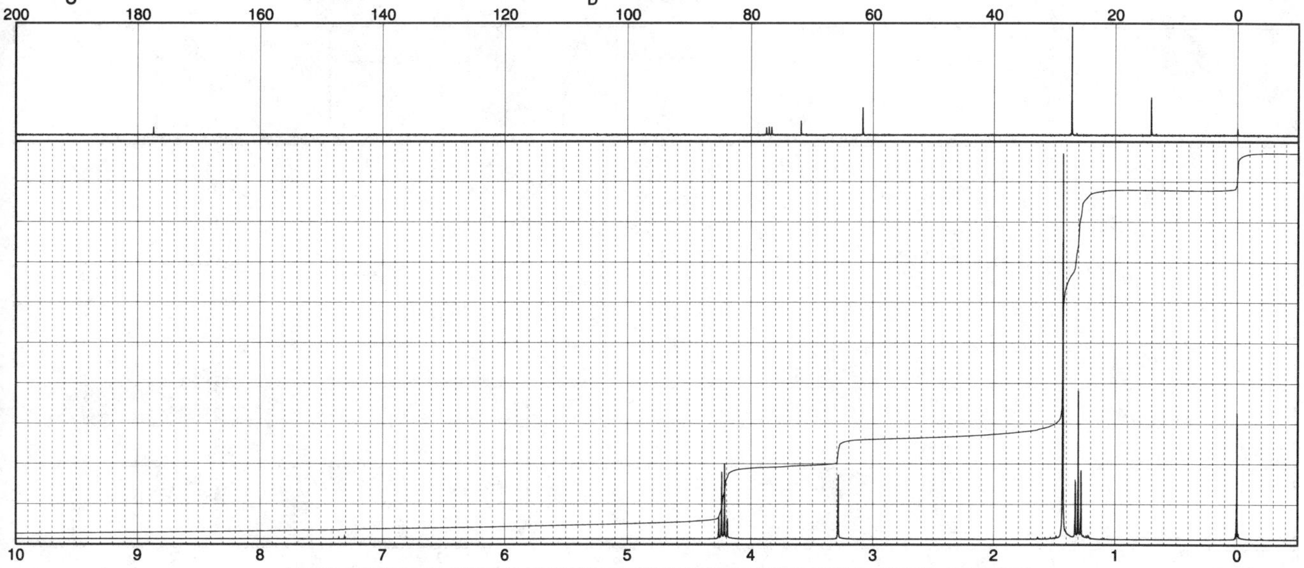

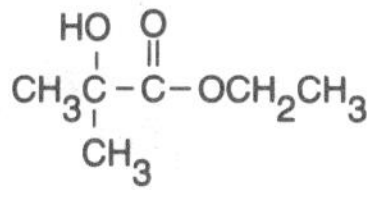

Aldrich 38,234-5   *CAS [36293-63-7]*

***tert*-Butyl-2-hydroxyisobutyrate, 99%**

$C_8H_{16}O_3$
FW 160.22
d 0.924
n_D^{20}1.4040
Fp 121°F

176.78
81.94
71.89
27.87*
27.12*

CDCl$_3$ QE-300

B

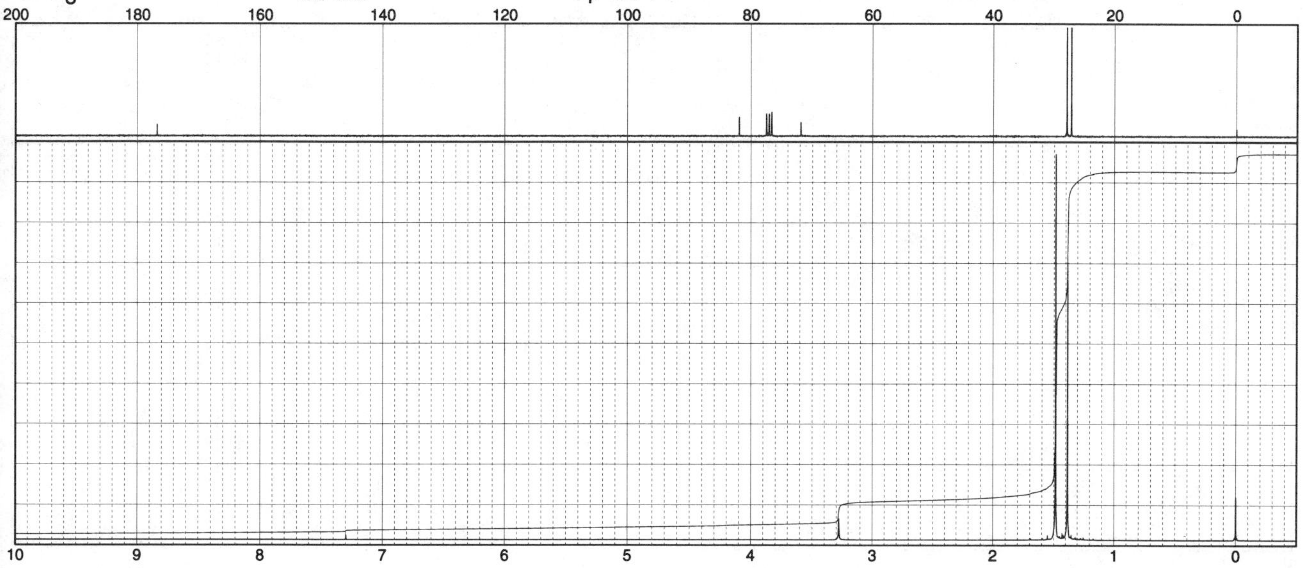

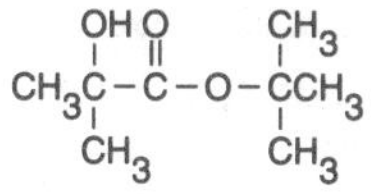

Aldrich 27,014-8   *CAS [72657-23-9]*

Methyl (*R*)-(-)-3-hydroxy-2-methylpropionate, 99%

$C_5H_{10}O_3$
FW 118.13
bp 77°C (12 mm)
d 1.066
n_D^{20}1.4250

Fp 178°F

VP-FT-IR: *3*, 715A

176.07
64.49
51.83*
41.82*
13.45*

CDCl$_3$ QE-300

C

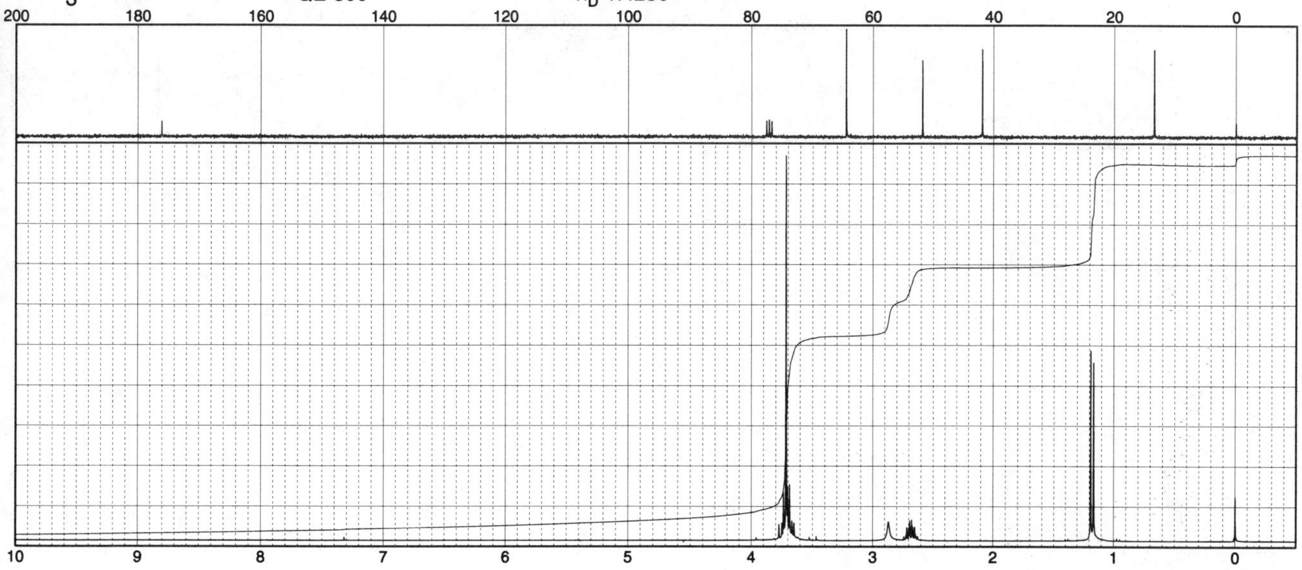

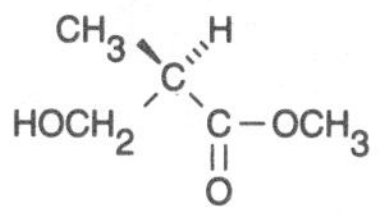

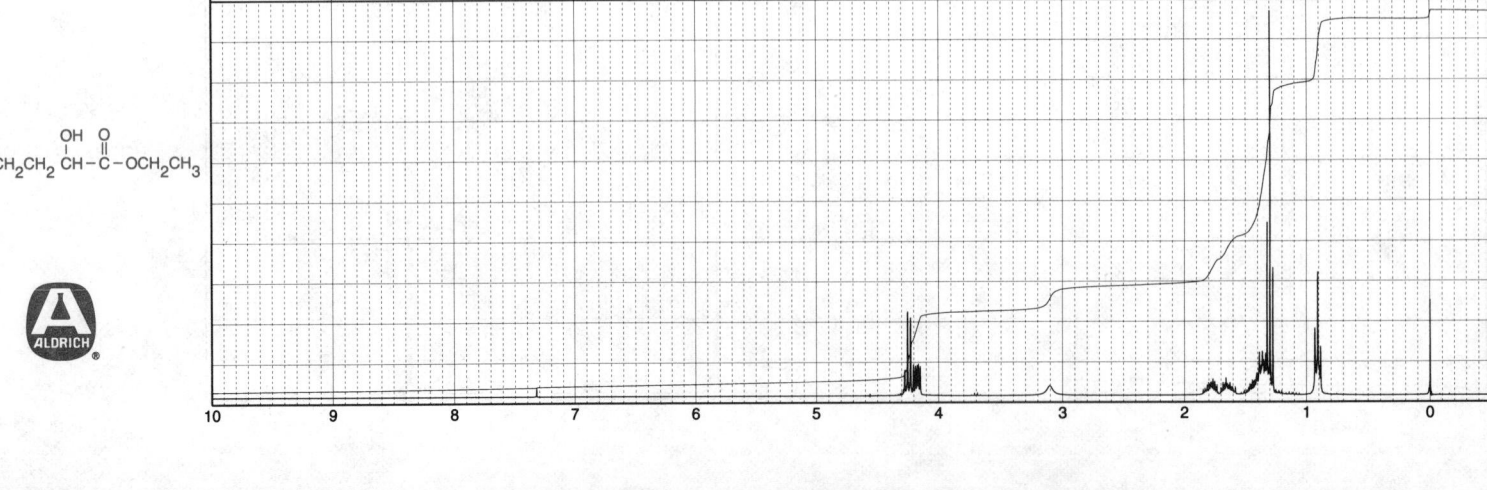

A

Aldrich 27,012-1   *CAS [80657-57-4]*
Methyl (S)-(+)-3-hydroxy-2-methylpropionate,
99%

$C_5H_{10}O_3$
FW 118.13
bp 74°C (10 mm)
d 1.071
n_D^{20}1.4250

Fp 178°F

VP-FT-IR: *3*, 714D

176.07
64.49
51.83*
41.81*
13.45*

CDCl₃   QE-300

200 180 160 140 120 100 80 60 40 20 0

B

Aldrich 22,745-5   *CAS [14002-80-3]*
Methyl 2,2-dimethyl-3-hydroxypropionate,
98%

$C_6H_{12}O_3$
FW 132.16
bp 178°C (740 mm)
d 1.036
n_D^{20}1.4280

Fp 169°F

60 MHz: *1*, 558A
FT-IR: *1*, 661D
VP-FT-IR: *3*, 716B

177.97
69.60
51.94*
44.26
22.06*

CDCl₃   QE-300

200 180 160 140 120 100 80 60 40 20 0

C

Aldrich 23,258-0   *CAS [124439-28-7]*
Ethyl (±)-2-hydroxycaproate, 99%

$C_8H_{16}O_3$
FW 160.22
bp 195°C
d 0.967
n_D^{20}1.4240

Fp 179°F

60 MHz: *1*, 558C
FT-IR: *1*, 662B
VP-FT-IR: *3*, 717B

175.40 26.93
70.48* 22.45
61.51   14.22*
34.13   13.94*

CDCl₃   QE-300

200 180 160 140 120 100 80 60 40 20 0

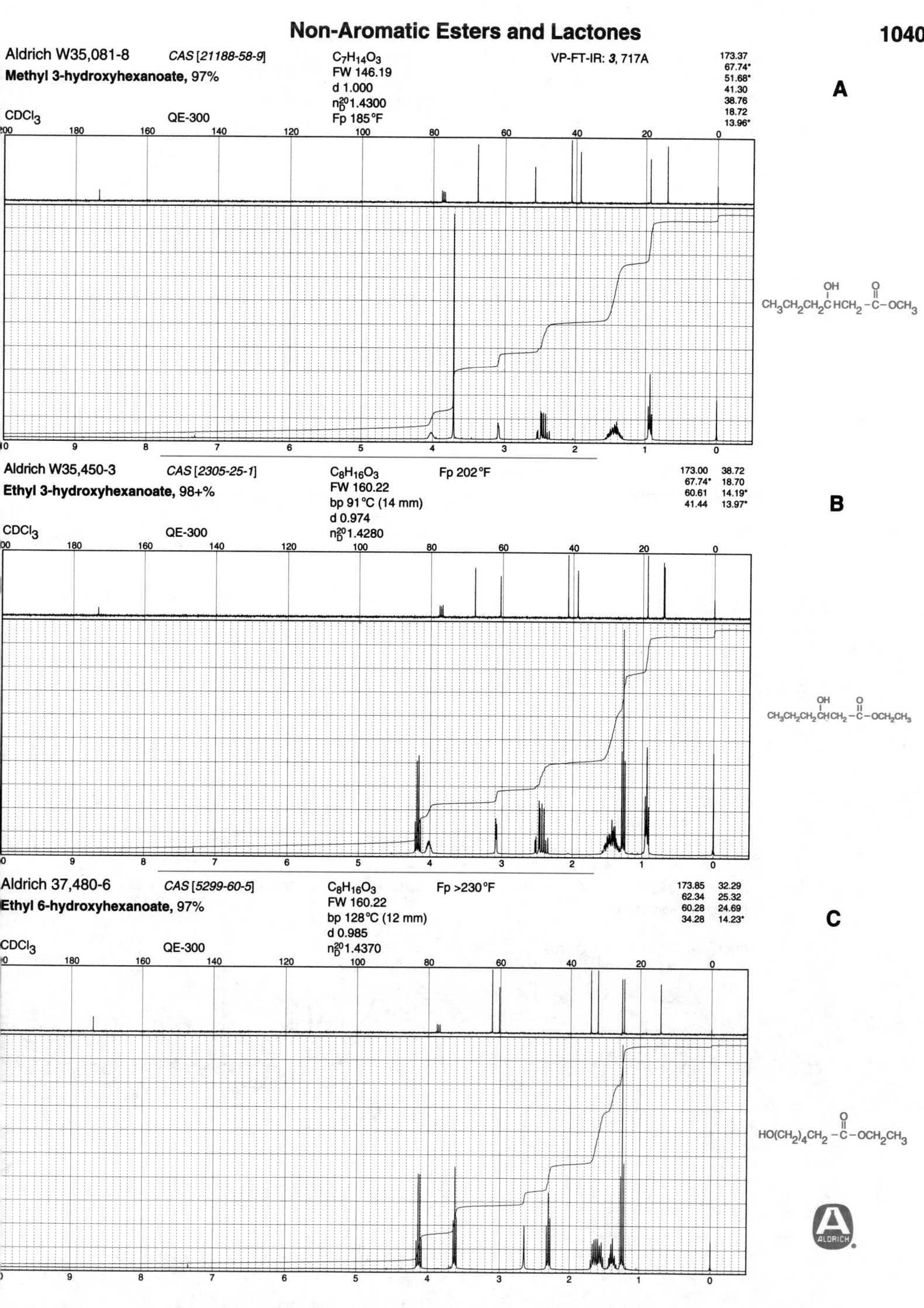

Aldrich W35,081-8   *CAS [21188-58-9]*
Methyl 3-hydroxyhexanoate, 97%

C₇H₁₄O₃
FW 146.19
d 1.000
n²⁰_D 1.4300
Fp 185°F

VP-FT-IR: *3*, 717A

173.37
67.74*
51.68*
41.30
38.76
18.72
13.96*

A

CDCl₃   QE-300

Aldrich W35,450-3   *CAS [2305-25-1]*
Ethyl 3-hydroxyhexanoate, 98+%

C₈H₁₆O₃
FW 160.22
bp 91°C (14 mm)
d 0.974
n²⁰_D 1.4280

Fp 202°F

173.00   38.72
67.74*   18.70
60.61   14.19*
41.44   13.97*

B

CDCl₃   QE-300

Aldrich 37,480-6   *CAS [5299-60-5]*
Ethyl 6-hydroxyhexanoate, 97%

C₈H₁₆O₃
FW 160.22
bp 128°C (12 mm)
d 0.985
n²⁰_D 1.4370

Fp >230°F

173.85   32.29
62.34   25.32
60.28   24.69
34.28   14.23*

C

CDCl₃   QE-300

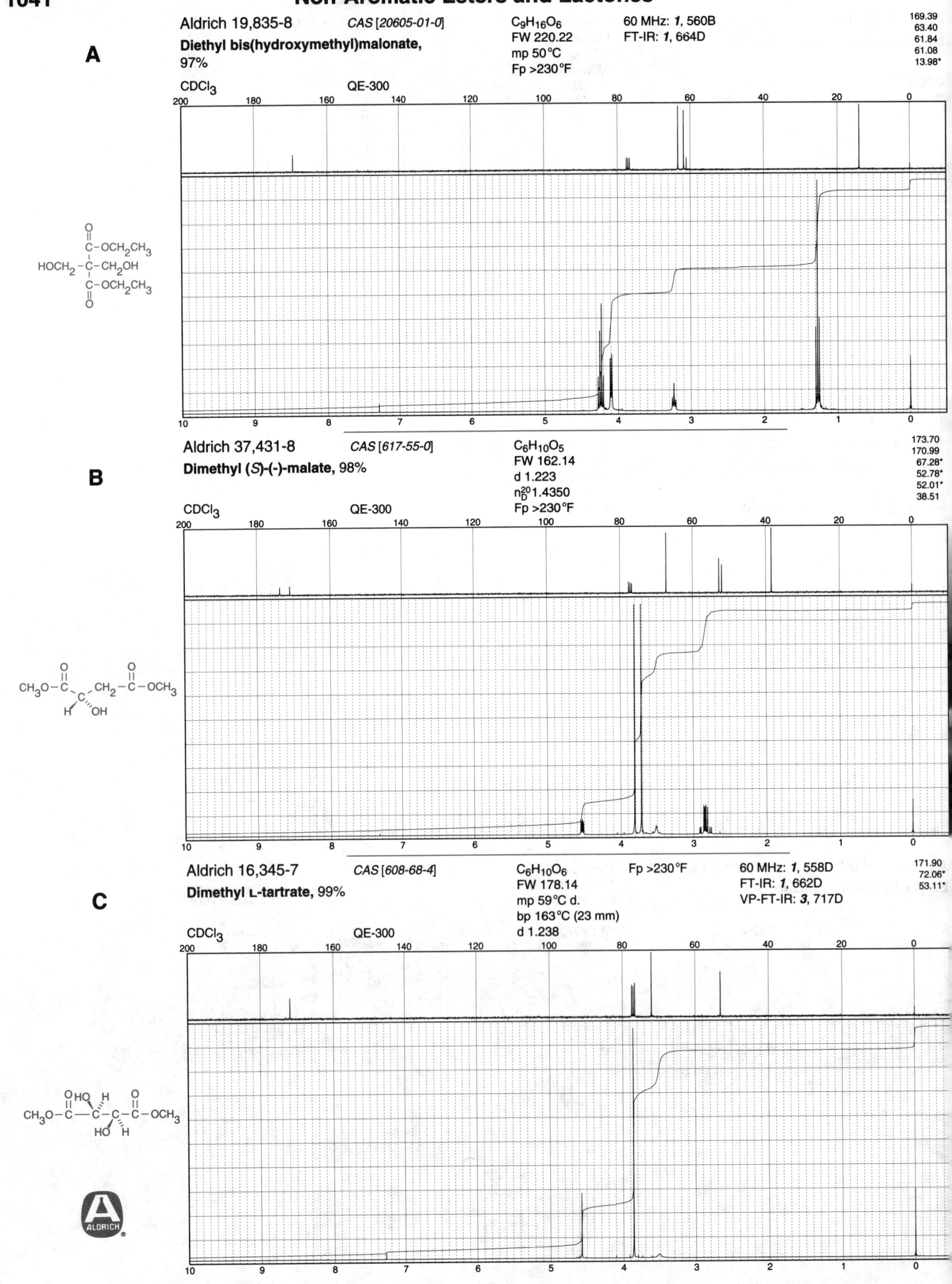

A

Aldrich 19,835-8   CAS [20605-01-0]   C₉H₁₆O₆   60 MHz: *1*, 560B

Diethyl bis(hydroxymethyl)malonate,   FW 220.22   FT-IR: *1*, 664D

97%   mp 50°C

Fp >230°F

CDCl₃   QE-300

169.39
63.40
61.84
61.08
13.98*

B

Aldrich 37,431-8   CAS [617-55-0]   C₆H₁₀O₅

Dimethyl (S)-(-)-malate, 98%   FW 162.14

d 1.223

n₂₀ᴰ 1.4350

Fp >230°F

CDCl₃   QE-300

173.70
170.99
67.28*
52.78*
52.01*
38.51

C

Aldrich 16,345-7   CAS [608-68-4]   C₆H₁₀O₆   Fp >230°F   60 MHz: *1*, 558D

Dimethyl L-tartrate, 99%   FW 178.14   FT-IR: *1*, 662D

mp 59°C d.   VP-FT-IR: *3*, 717D

bp 163°C (23 mm)

d 1.238

CDCl₃   QE-300

171.90
72.06*
53.11*

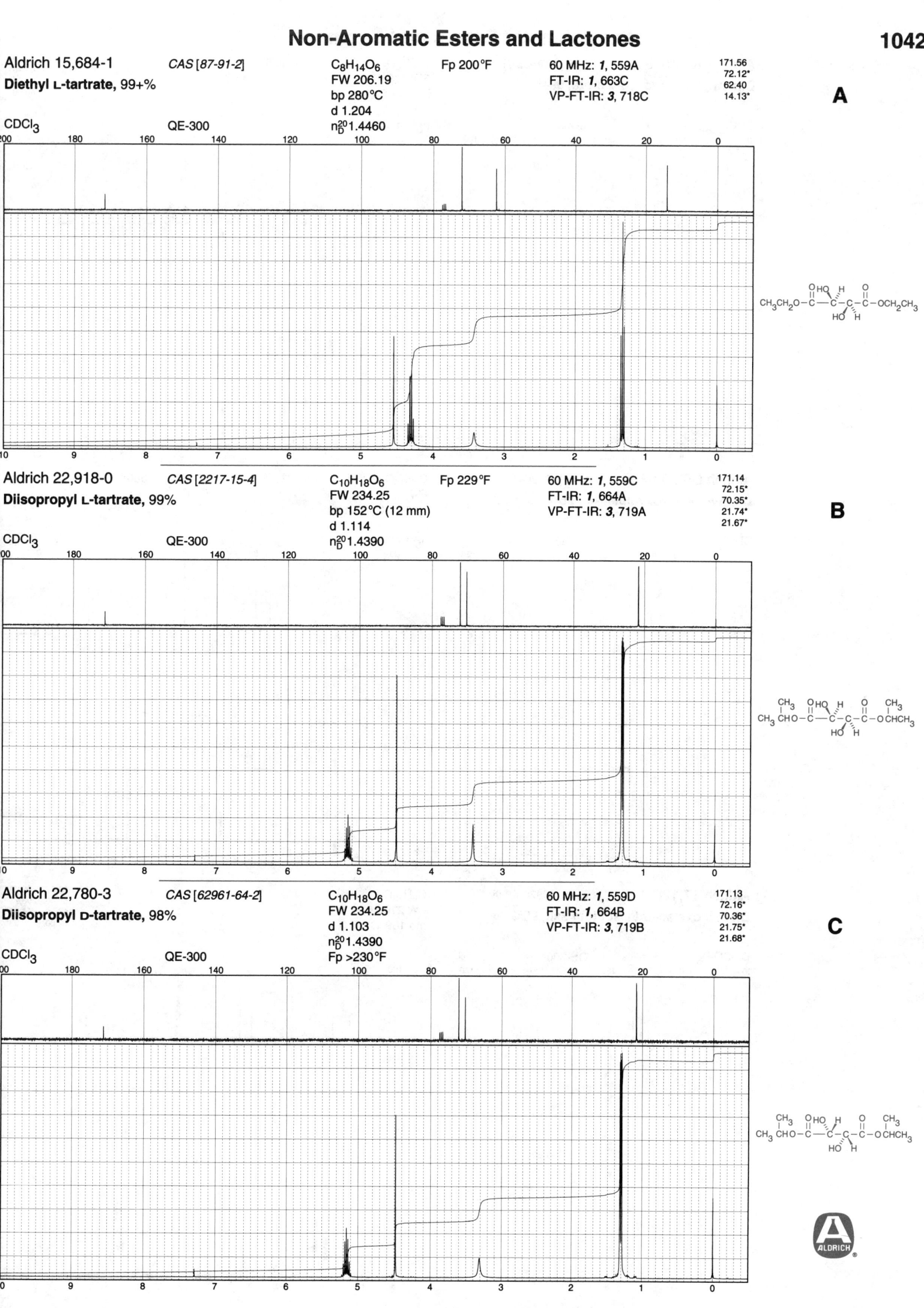

Aldrich 15,684-1
Diethyl L-tartrate, 99+%

CAS [87-91-2]

C$_8$H$_{14}$O$_6$
FW 206.19
bp 280°C
d 1.204
n$_D^{20}$1.4460

Fp 200°F

60 MHz: *1*, 559A
FT-IR: *1*, 663C
VP-FT-IR: *3*, 718C

171.56
72.12*
62.40
14.13*

A

CDCl$_3$ QE-300

CH$_3$CH$_2$O—C—C—C—C—OCH$_2$CH$_3$

Aldrich 22,918-0
Diisopropyl L-tartrate, 99%

CAS [2217-15-4]

C$_{10}$H$_{18}$O$_6$
FW 234.25
bp 152°C (12 mm)
d 1.114
n$_D^{20}$1.4390

Fp 229°F

60 MHz: *1*, 559C
FT-IR: *1*, 664A
VP-FT-IR: *3*, 719A

171.14
72.15*
70.35*
21.74*
21.67*

B

CDCl$_3$ QE-300

CH$_3$CHO—C—C—C—C—O CHCH$_3$

Aldrich 22,780-3
Diisopropyl D-tartrate, 98%

CAS [62961-64-2]

C$_{10}$H$_{18}$O$_6$
FW 234.25
d 1.103
n$_D^{20}$1.4390
Fp >230°F

60 MHz: *1*, 559D
FT-IR: *1*, 664B
VP-FT-IR: *3*, 719B

171.13
72.16*
70.36*
21.75*
21.68*

C

CDCl$_3$ QE-300

CH$_3$CHO—C—C—C—C—O CHCH$_3$

ALDRICH

A

Aldrich 16,697-9   *CAS [7250-55-7]* $C_7H_{12}O_5$ Fp >230°F

Dimethyl 3-hydroxyglutarate, 99+%

FW 176.17
bp 140°C (8 mm)
d 1.192
n_D^{20} 1.4420

CDCl₃   QE-300

172.09
64.70*
51.80*
40.58

CH₃O—C—CH₂CHCH₂—C—OCH₃
        ‖ | ‖
        O   OH   O

B

Aldrich D9,700-2   *CAS [32328-03-3]* $C_9H_{16}O_5$ Fp >230°F   60 MHz: *1*, 560A

Diethyl 3-hydroxyglutarate, 95%

FW 204.22
bp 157°C (23 mm)
d 1.103
n_D^{20} 1.4390

FT-IR: *1*, 663B
VP-FT-IR: *3*, 718B

171.71
64.75*
60.74
40.75
14.16*

CDCl₃   QE-300

CH₃CH₂O—C—CH₂CHCH₂—C—OCH₂CH₃
        ‖ | ‖
        O   OH   O

C

Aldrich 17,126-3   *CAS [40876-98-0]* $C_8H_{12}O_5$

Diethyl oxalacetate, sodium salt, 95%

FW 210.16
mp 189°C

169.57   60.02
168.40   56.73
167.30   14.63*
82.80*   13.93*

DMSO-d₆   QE-300

CH₃CH₂O—C—C—C=CH—C—OCH₂CH₃
        ‖   ‖   ‖
        O   O⁻   O

Na⁺

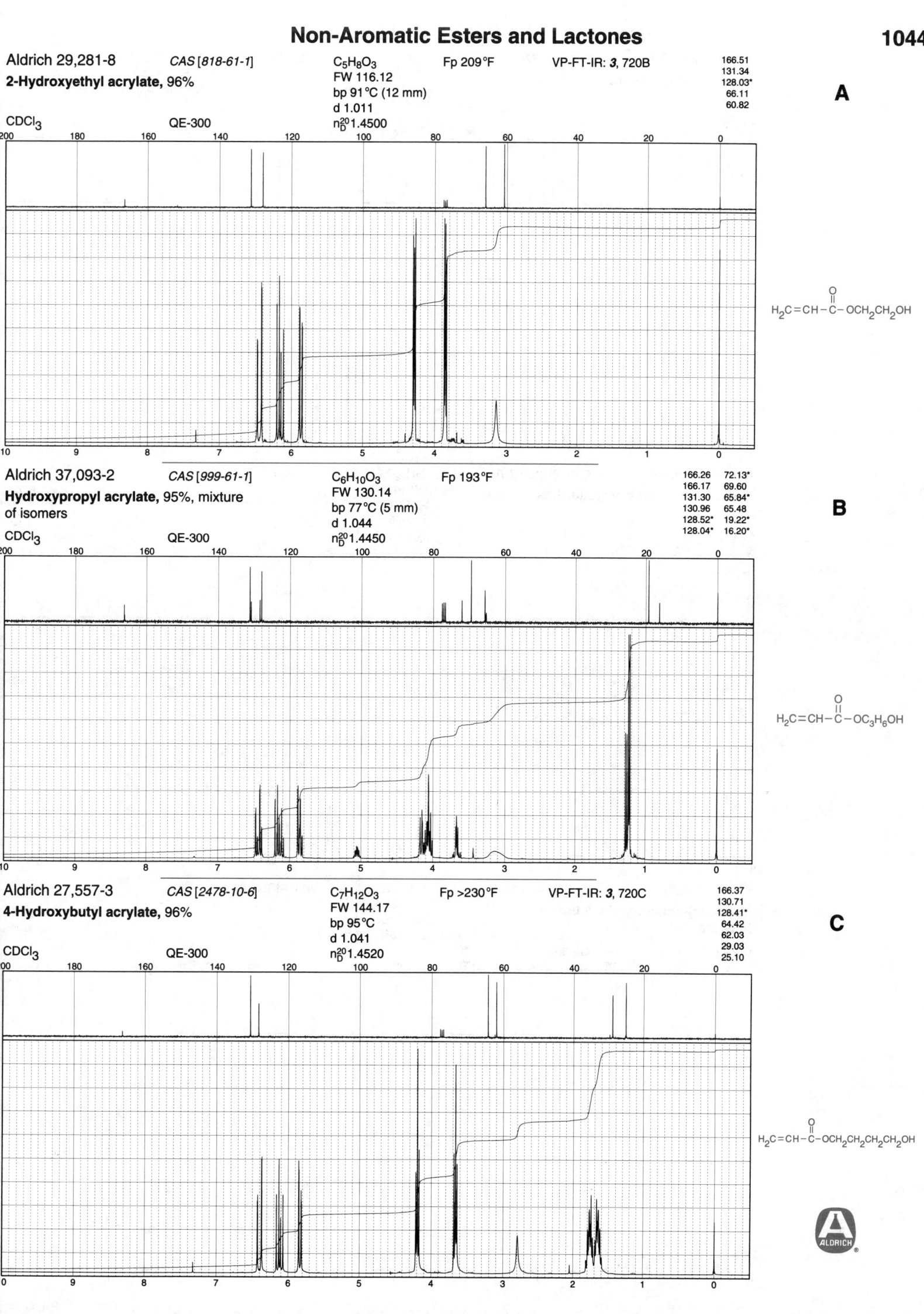

Aldrich 29,281-8   CAS [818-61-1]   C₅H₈O₃   Fp 209°F   VP-FT-IR: *3*, 720B

2-Hydroxyethyl acrylate, 96%

FW 116.12
bp 91°C (12 mm)
d 1.011
n²⁰_D 1.4500

CDCl₃   QE-300

166.51
131.34
128.03*
66.11
60.82

A

H₂C=CH−C−OCH₂CH₂OH

Aldrich 37,093-2   CAS [999-61-1]   C₆H₁₀O₃   Fp 193°F

Hydroxypropyl acrylate, 95%, mixture of isomers

FW 130.14
bp 77°C (5 mm)
d 1.044
n²⁰_D 1.4450

CDCl₃   QE-300

166.26   72.13*
166.17   69.60
131.30   65.84*
130.96   65.48
128.52*   19.22*
128.04*   16.20*

B

H₂C=CH−C−OC₃H₆OH

Aldrich 27,557-3   CAS [2478-10-6]   C₇H₁₂O₃   Fp >230°F   VP-FT-IR: *3*, 720C

4-Hydroxybutyl acrylate, 96%

FW 144.17
bp 95°C
d 1.041
n²⁰_D 1.4520

CDCl₃   QE-300

166.37
130.71
128.41*
64.42
62.03
29.03
25.10

C

H₂C=CH−C−OCH₂CH₂CH₂CH₂OH

A

Aldrich 12,863-5   *CAS [868-77-9]*

2-Hydroxyethyl methacrylate, 97%

$C_6H_{10}O_3$
FW 130.14
bp 67°C (3 mm)
d 1.034
n_D^{20} 1.4520

Fp 207°F

60 MHz: *1*, 560C
FT-IR: *1*, 665B
VP-FT-IR: *3*, 720D

167.75
136.04
125.97
66.33
61.05
18.27*

CDCl₃   QE-300

$H_2C=C-C-OCH_2CH_2OH$
   | ||
   CH₃ O

B

Aldrich 26,854-2   *CAS [27813-02-1]*

Hydroxypropyl methacrylate, 97%

$C_7H_{12}O_3$
FW 144.17
bp 57°C
d 1.066
n_D^{20} 1.4470

Fp 206°F

167.44   69.78
167.39   66.00*
136.36   65.77
135.99   19.25*
125.95   18.31*
125.67   16.24*
72.31*

CDCl₃   QE-300

$H_2C=C-C-OC_3H_6OH$
   | ||
   CH₃ O

C

Aldrich 28,835-7   *CAS [19487-29-7]*

Methyl (±)-3-hydroxy-4,4,4-trichloro-butyrate, 99%

$C_5H_7Cl_3O_3$
FW 221.47
mp 64°C
bp 136°C (13 mm)

VP-FT-IR: *3*, 715D

171.10
102.12
79.27*
52.39*
36.96

CDCl₃   QE-300

          OH   O
          | ||
$Cl_3C CHCH_2-C-OCH_3$

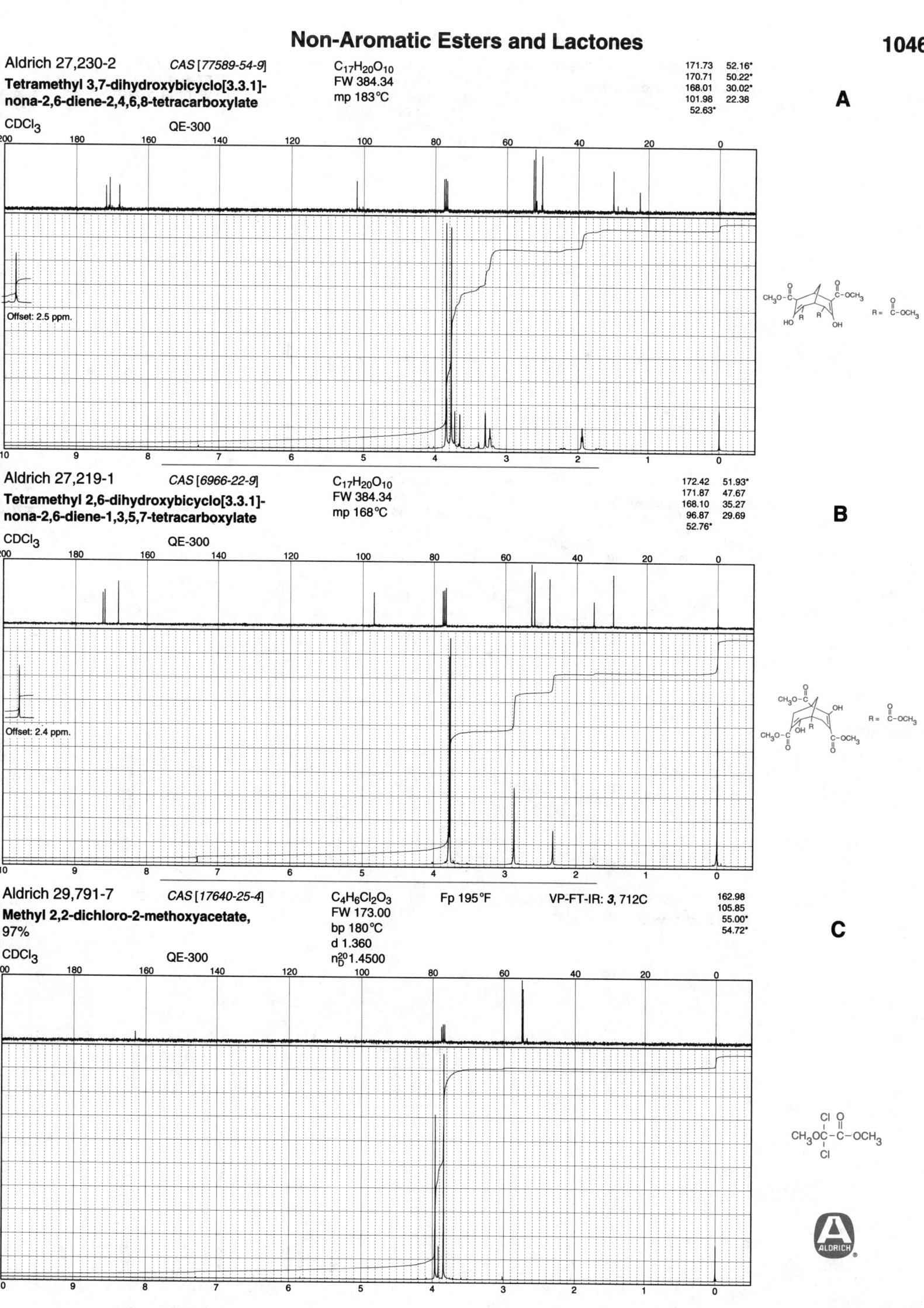

Aldrich 27,230-2   CAS [77589-54-9]

Tetramethyl 3,7-dihydroxybicyclo[3.3.1]-nona-2,6-diene-2,4,6,8-tetracarboxylate

$C_{17}H_{20}O_{10}$
FW 384.34
mp 183°C

171.73   52.16*
170.71   50.22*
168.01   30.02*
101.98   22.38
52.63*

A

CDCl$_3$ QE-300

Offset: 2.5 ppm.

Aldrich 27,219-1   CAS [6966-22-9]

Tetramethyl 2,6-dihydroxybicyclo[3.3.1]-nona-2,6-diene-1,3,5,7-tetracarboxylate

$C_{17}H_{20}O_{10}$
FW 384.34
mp 168°C

172.42   51.93*
171.87   47.67
168.10   35.27
96.87   29.69
52.76*

B

CDCl$_3$ QE-300

Offset: 2.4 ppm.

Aldrich 29,791-7   CAS [17640-25-4]

Methyl 2,2-dichloro-2-methoxyacetate, 97%

$C_4H_6Cl_2O_3$
FW 173.00
bp 180°C
d 1.360
n$_D^{20}$1.4500

Fp 195°F   VP-FT-IR: *3*, 712C

162.98
105.85
55.00*
54.72*

C

CDCl$_3$ QE-300

A

Aldrich 23,259-9   *CAS [3938-96-3]*   C$_5$H$_{10}$O$_3$ Fp 115°F   60 MHz: *1*, 561A

Ethyl methoxyacetate, 98%   FW 118.13

bp 45°C (9 mm)   FT-IR: *1*, 666B

d 1.007   VP-FT-IR: *3*, 721D

n$_D^{20}$1.3990

170.17
69.87
60.84
59.30*
14.21*

CDCl$_3$ QE-300

CH$_3$OCH$_2$ – C(=O) – OCH$_2$CH$_3$

B

Aldrich 37,417-2   *CAS [817-95-8]*   C$_6$H$_{12}$O$_3$ Fp 119°F

Ethyl ethoxyacetate, 98%   FW 132.16

bp 156°C

d 0.975

n$_D^{20}$1.4020

170.47
68.12
67.15
60.77
15.00*
14.22*

CDCl$_3$ QE-300

CH$_3$CH$_2$OCH$_2$ – C(=O) – OCH$_2$CH$_3$

C

Aldrich 17,828-4   *CAS [763-69-9]*   C$_7$H$_{14}$O$_3$ Fp 126°F   60 MHz: *1*, 561B

Ethyl 3-ethoxypropionate, 99+%   FW 146.19

bp 166°C (755 mm)

d 0.949

n$_D^{20}$1.4050

171.60
66.34
65.84
60.42
35.23
15.09*
14.20*

CDCl$_3$ QE-300

CH$_3$CH$_2$OCH$_2$CH$_2$ – C(=O) – OCH$_2$CH$_3$

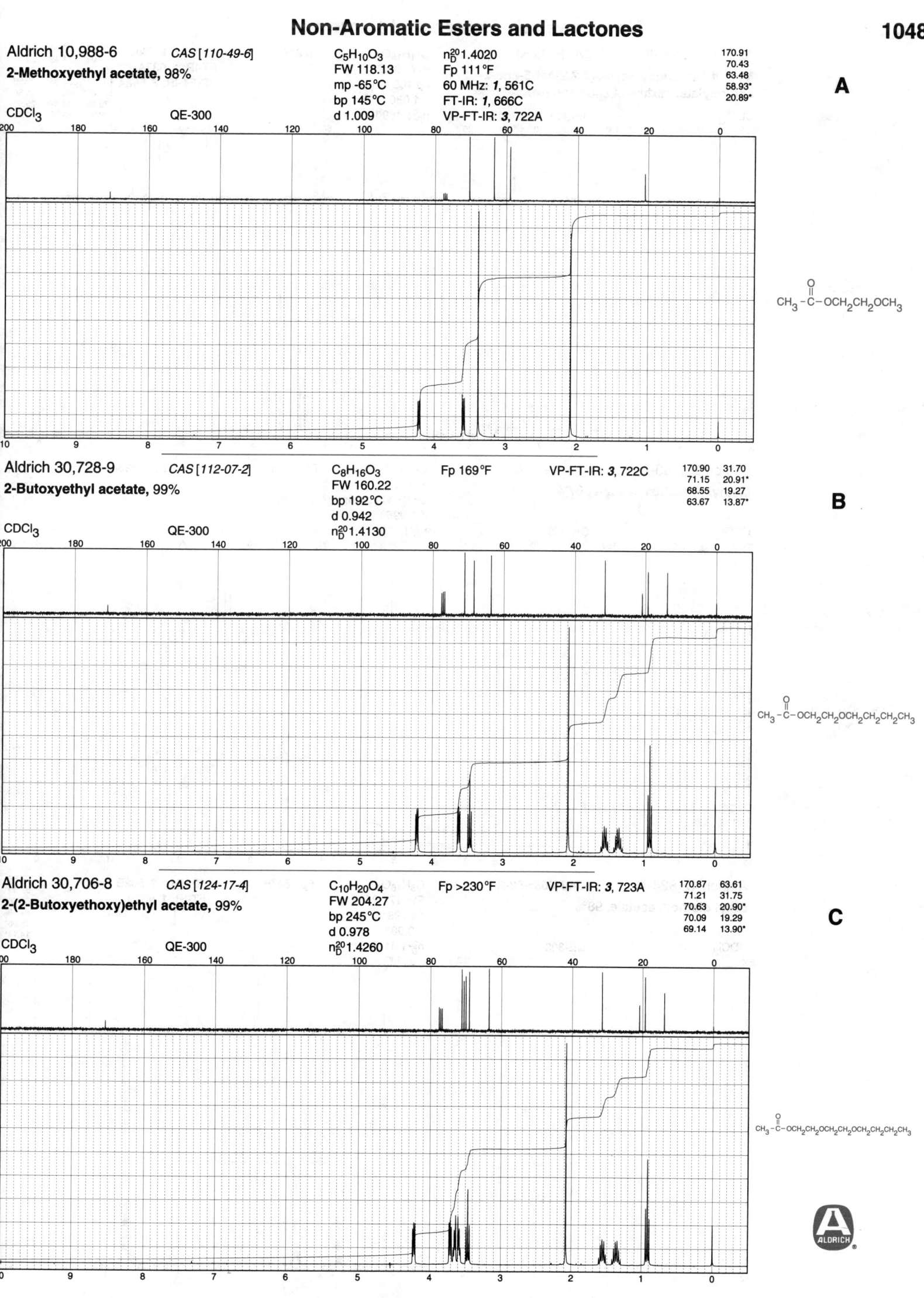

Aldrich 10,988-6   *CAS [110-49-6]*

2-Methoxyethyl acetate, 98%

$C_5H_{10}O_3$
FW 118.13
mp -65°C
bp 145°C
d 1.009

n_D^{20}1.4020
Fp 111°F
60 MHz: *1*, 561C
FT-IR: *1*, 666C
VP-FT-IR: *3*, 722A

170.91
70.43
63.48
58.93*
20.89*

A

CDCl₃   QE-300

$CH_3-C-OCH_2CH_2OCH_3$

Aldrich 30,728-9   *CAS [112-07-2]*

2-Butoxyethyl acetate, 99%

$C_8H_{16}O_3$
FW 160.22
bp 192°C
d 0.942
n_D^{20}1.4130

Fp 169°F

VP-FT-IR: *3*, 722C

170.90   31.70
71.15   20.91*
68.55   19.27
63.67   13.87*

B

CDCl₃   QE-300

$CH_3-C-OCH_2CH_2OCH_2CH_2CH_2CH_3$

Aldrich 30,706-8   *CAS [124-17-4]*

2-(2-Butoxyethoxy)ethyl acetate, 99%

$C_{10}H_{20}O_4$
FW 204.27
bp 245°C
d 0.978
n_D^{20}1.4260

Fp >230°F

VP-FT-IR: *3*, 723A

170.87   63.61
71.21   31.75
70.63   20.90*
70.09   19.29
69.14   13.90*

C

CDCl₃   QE-300

$CH_3-C-OCH_2CH_2OCH_2CH_2OCH_2CH_2CH_2CH_3$

A

Aldrich 21,760-3   *CAS [5259-50-7]*

Methyl 1-methoxybicyclo[2.2.2]oct-5-ene-2-carboxylate, mixture of *endo* and *exo*

CDCl$_3$ QE-300

$C_{11}H_{16}O_3$ Fp 218°F
FW 196.25
bp 105°C (17 mm)
d 1.086
n$_D^{20}$1.4890

60 MHz: *1*, 536C
FT-IR: *1*, 632A
VP-FT-IR: *3*, 662A

174.92	51.61*	30.91
134.09*	51.49*	29.51*
133.79*	51.30*	29.35*
132.61*	50.86*	28.56
132.40*	46.41*	25.56
79.28	44.57*	25.51
78.97	33.00	24.85

B

Aldrich 29,463-2   *CAS [89-91-8]*

Methyl dimethoxyacetate, 97%

CDCl$_3$ QE-300

$C_5H_{10}O_4$ Fp 146°F
FW 134.13
bp 67°C (18 mm)
d 1.096
n$_D^{20}$1.4050

VP-FT-IR: *3*, 723D

167.46
98.85*
53.89*
52.37*

C

Aldrich 12,824-4   *CAS [6065-82-3]*

Ethyl diethoxyacetate, 98%

CDCl$_3$ QE-300

$C_8H_{16}O_4$ Fp 164°F
FW 176.21
bp 199°C
d 0.985
n$_D^{20}$1.4100

60 MHz: *1*, 562B
FT-IR: *1*, 667B
VP-FT-IR: *3*, 724A

167.58
97.49*
62.36
61.36
15.06*
14.14*

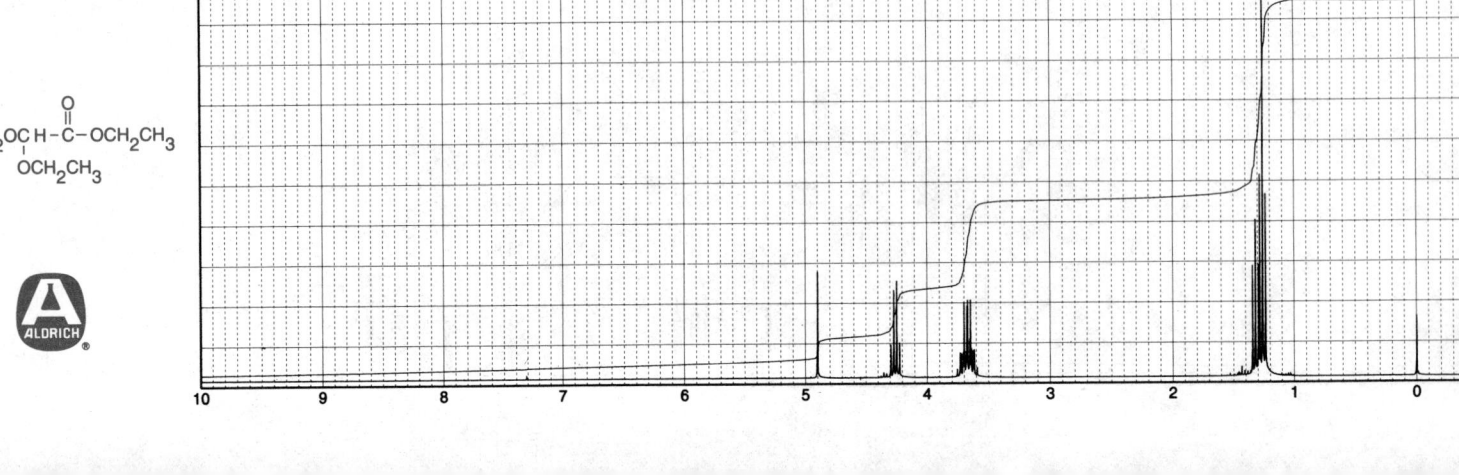

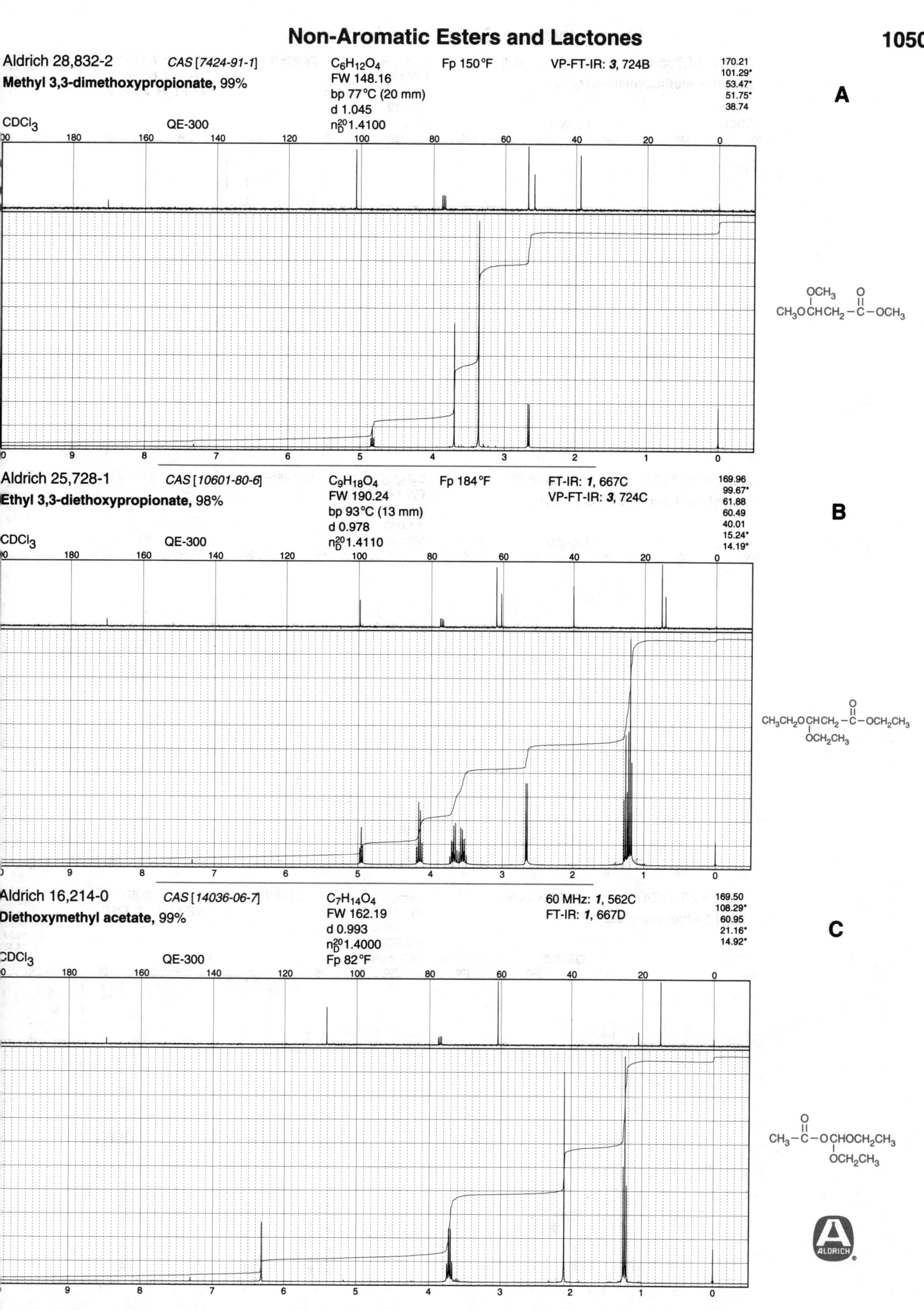

Aldrich 28,832-2   *CAS [7424-91-1]*   C₆H₁₂O₄   Fp 150°F   VP-FT-IR: *3*, 724B $C_6H_{12}O_4$
Methyl 3,3-dimethoxypropionate, 99%   FW 148.16   bp 77°C (20 mm)   d 1.045 n_D^{20}1.4100

170.21
101.29*
53.47*
51.75*
38.74

A

CDCl₃   QE-300

Aldrich 25,728-1   *CAS [10601-80-6]*   C₉H₁₈O₄   Fp 184°F   FT-IR: *1*, 667C   VP-FT-IR: *3*, 724C
Ethyl 3,3-diethoxypropionate, 98%   FW 190.24   bp 93°C (13 mm)   d 0.978 n_D^{20}1.4110

169.96
99.67*
61.88
60.49
40.01
15.24*
14.19*

B

CDCl₃   QE-300

Aldrich 16,214-0   *CAS [14036-06-7]*   C₇H₁₄O₄   60 MHz: *1*, 562C   FT-IR: *1*, 667D
Diethoxymethyl acetate, 99%   FW 162.19   d 0.993 n_D^{20}1.4000   Fp 82°F

169.50
108.29*
60.95
21.16*
14.92*

C

CDCl₃   QE-300

A

Aldrich 24,785-5   *CAS [5018-30-4]*   C6H10O5   Fp 226°F   60 MHz: *1*, 563A

Dimethyl methoxymalonate, 98%   FW 162.14   FT-IR: *1*, 668B

bp 215°C (745 mm)   VP-FT-IR: *3*, 725B

d 1.172

n_D^20 1.4230

166.64
80.24*
58.73*
52.92*

CDCl3   QE-300

CH3O–C–CH–C–OCH3
　　║　　｜　　║
　　O　OCH3　O

B

Aldrich 36,237-9   *CAS [5788-17-0]*   C5H8O3   Fp 145°F

Methyl *trans*-3-methoxyacrylate, 99%   FW 116.12

bp 56°C (18 mm)

d 1.080

n_D^20 1.4510

167.97
163.18*
95.75*
57.29*
51.03*

CDCl3   QE-300

CH3OCH=CH–C–OCH3
　　　　　　║
　　　　　　O

C

Aldrich 25,012-0   *CAS [1001-26-9]*   C7H12O3   Fp 169°F   FT-IR: *1*, 665D

Ethyl 3-ethoxyacrylate, 98%   FW 144.17   VP-FT-IR: *3*, 721B

bp 196°C

d 0.998

n_D^20 1.4460

167.83
162.23*
96.40*
66.61
59.69
14.43*
14.38

CDCl3   QE-300

CH3CH2OCH=CH–C–OCH2CH3
　　　　　　　║
　　　　　　　O

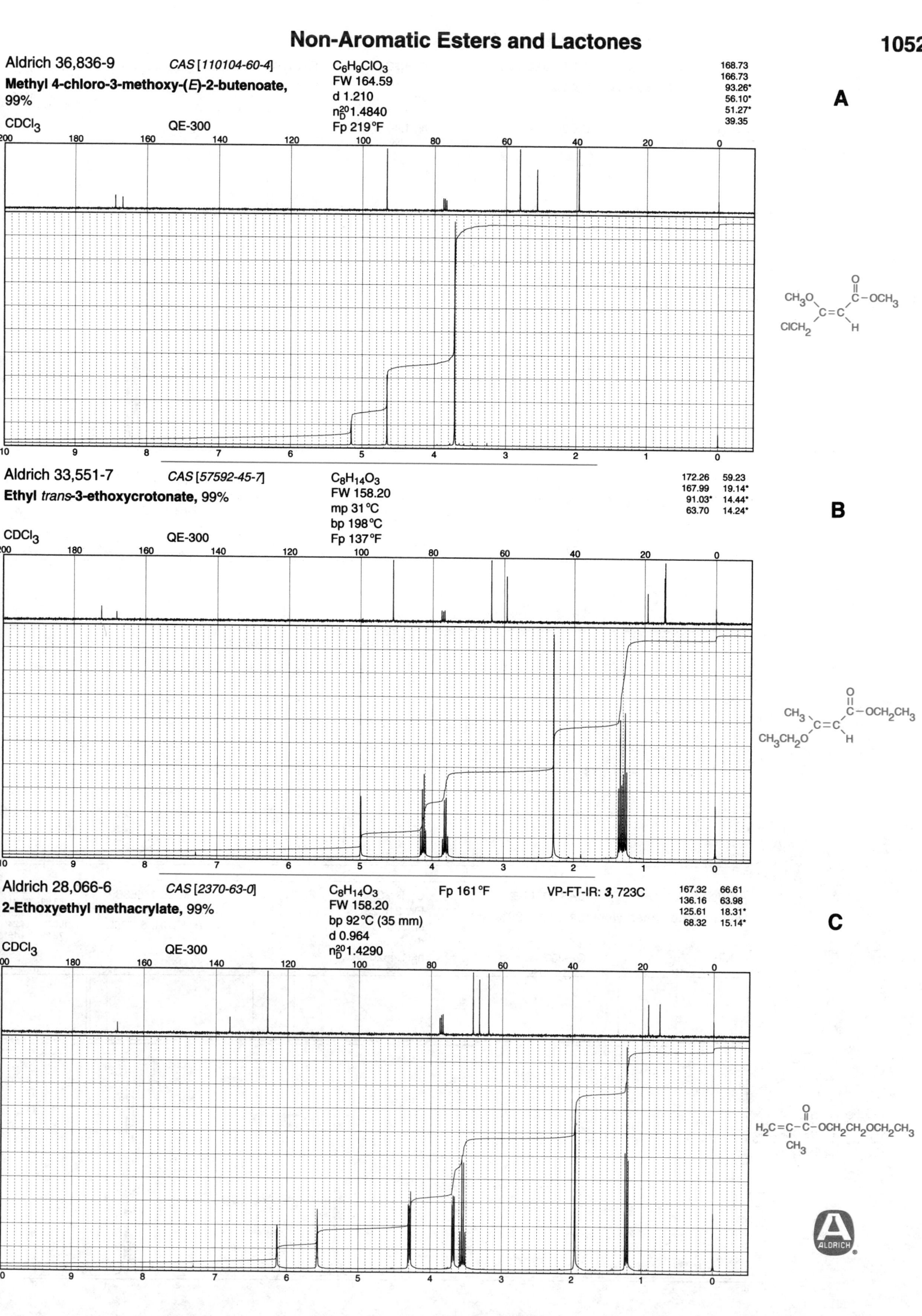

Aldrich 36,836-9   CAS [110104-60-4]   C₆H₉ClO₃
Methyl 4-chloro-3-methoxy-(E)-2-butenoate,
99%

FW 164.59
d 1.210
n²⁰_D 1.4840
Fp 219°F

168.73
166.73
93.26*
56.10*
51.27*
39.35

A

CDCl₃   QE-300

Aldrich 33,551-7   CAS [57592-45-7]   C₈H₁₄O₃
Ethyl trans-3-ethoxycrotonate, 99%

FW 158.20
mp 31°C
bp 198°C
Fp 137°F

172.26   59.23
167.99   19.14*
91.03*   14.44*
63.70   14.24*

B

CDCl₃   QE-300

Aldrich 28,066-6   CAS [2370-63-0]   C₈H₁₄O₃   Fp 161°F   VP-FT-IR: 3, 723C
2-Ethoxyethyl methacrylate, 99%

FW 158.20
bp 92°C (35 mm)
d 0.964
n²⁰_D 1.4290

167.32   66.61
136.16   63.98
125.61   18.31*
68.32   15.14*

C

CDCl₃   QE-300

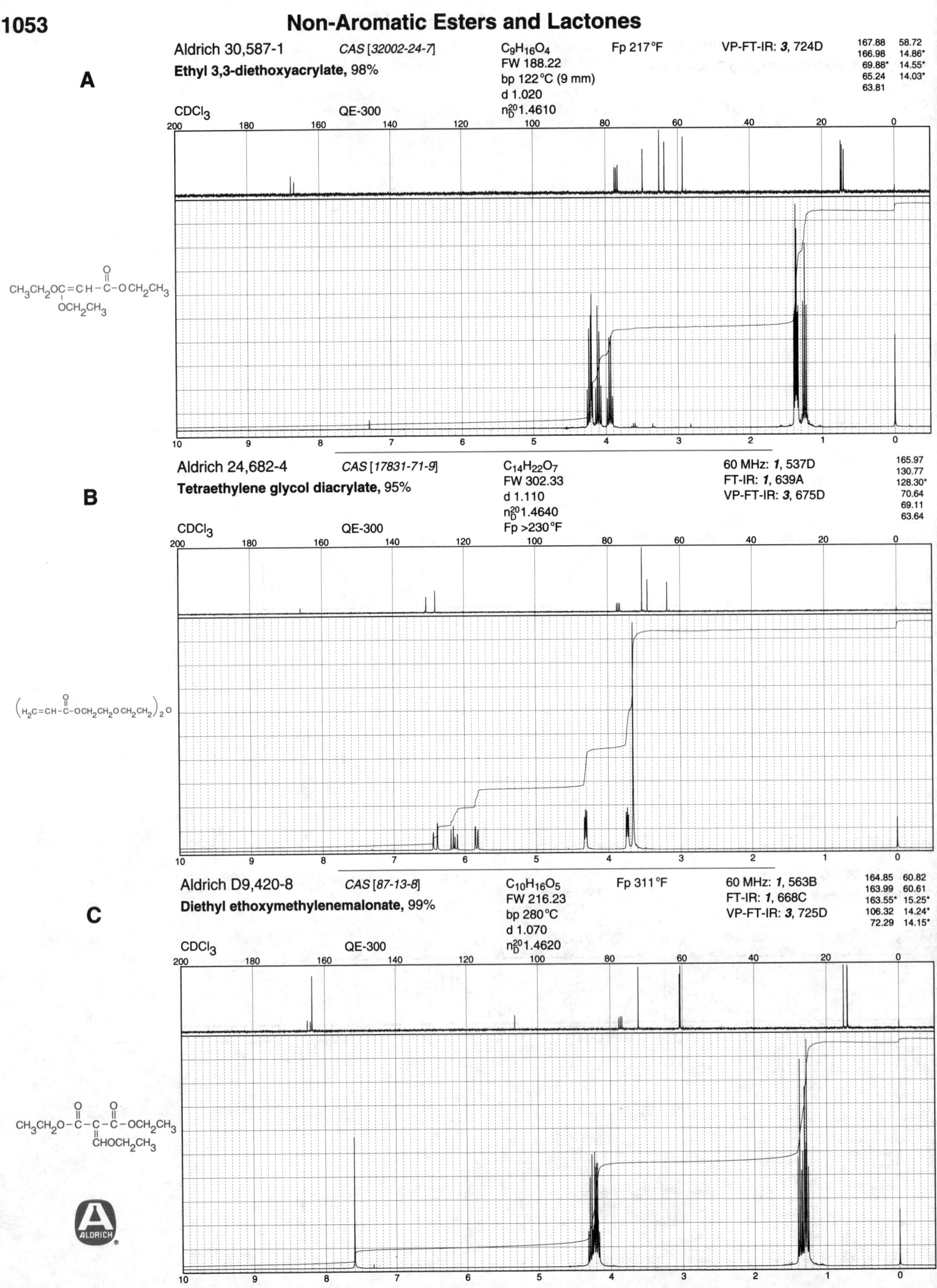

A

Aldrich 30,587-1   *CAS [32002-24-7]* $C_9H_{16}O_4$ Fp 217°F   VP-FT-IR: **3**, 724D

Ethyl 3,3-diethoxyacrylate, 98%

FW 188.22
bp 122°C (9 mm)
d 1.020
n_D^{20}1.4610

167.88   58.72
166.98   14.86*
69.88*   14.55*
65.24   14.03*
63.81

CDCl$_3$ QE-300

$CH_3CH_2OC=CH-C-OCH_2CH_3$
$\quad\quad | \quad\quad\quad\quad \overset{\parallel}{O}$
$\quad OCH_2CH_3$

B

Aldrich 24,682-4   *CAS [17831-71-9]* $C_{14}H_{22}O_7$ 60 MHz: **1**, 537D

Tetraethylene glycol diacrylate, 95%

FW 302.33
d 1.110
n_D^{20}1.4640
Fp >230°F

FT-IR: **1**, 639A
VP-FT-IR: **3**, 675D

165.97
130.77
128.30*
70.64
69.11
63.64

CDCl$_3$ QE-300

$(H_2C=CH-C-OCH_2CH_2OCH_2CH_2)_2O$
$\quad\quad\quad \overset{\parallel}{O}$

C

Aldrich D9,420-8   *CAS [87-13-8]* $C_{10}H_{16}O_5$ Fp 311°F   60 MHz: **1**, 563B

Diethyl ethoxymethylenemalonate, 99%

FW 216.23
bp 280°C
d 1.070
n_D^{20}1.4620

FT-IR: **1**, 668C
VP-FT-IR: **3**, 725D

164.85   60.82
163.99   60.61
163.55*   15.25*
106.32   14.24*
72.29   14.15*

CDCl$_3$ QE-300

$CH_3CH_2O-\overset{\overset{\textstyle O}{\parallel}}{C}-\overset{}{C}-\overset{\overset{\textstyle O}{\parallel}}{C}-OCH_2CH_3$
$\quad\quad\quad\quad\quad || $
$\quad\quad\quad\quad CHOCH_2CH_3$

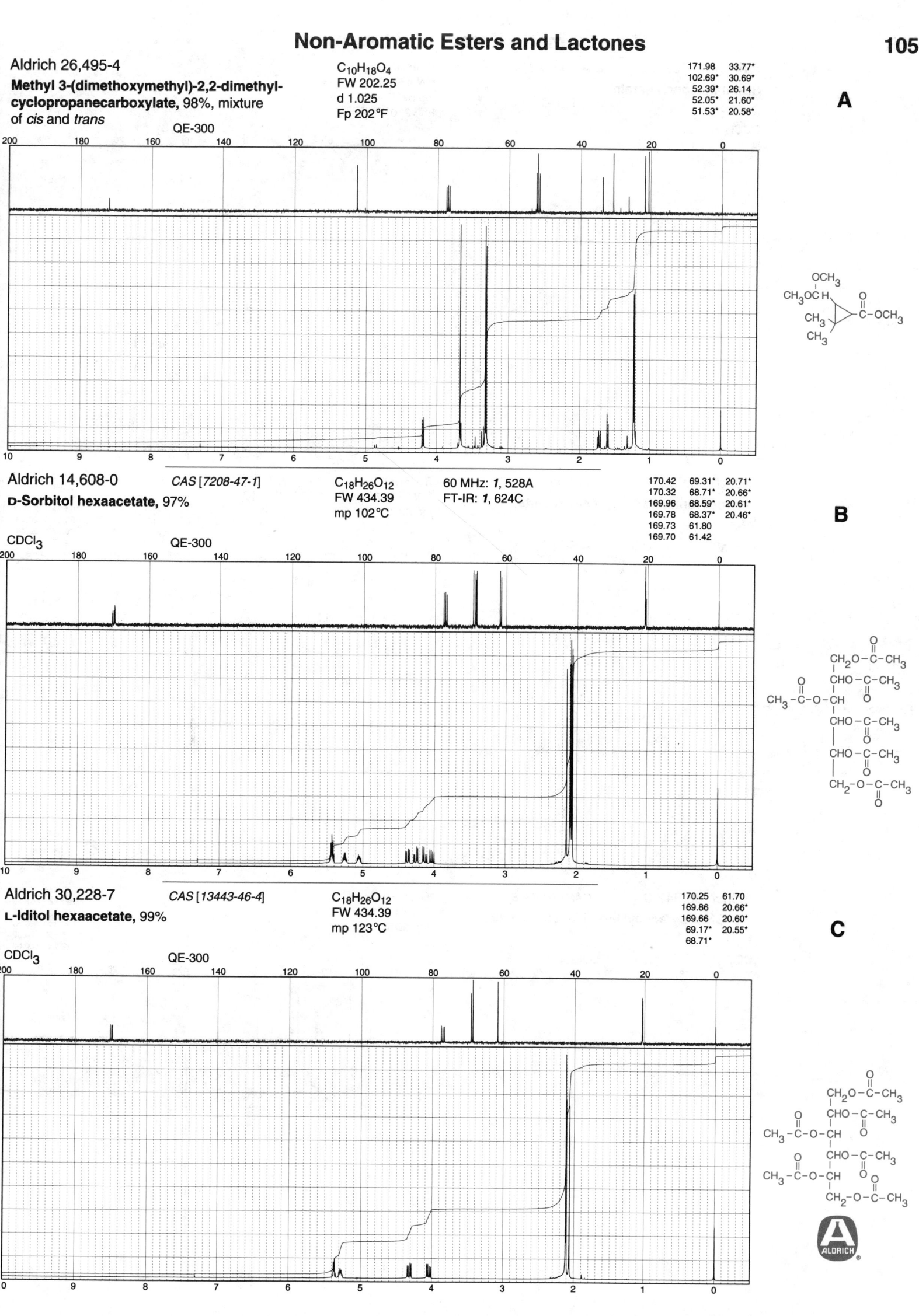

Aldrich 26,495-4

Methyl 3-(dimethoxymethyl)-2,2-dimethyl-cyclopropanecarboxylate, 98%, mixture of *cis* and *trans*

QE-300

$C_{10}H_{18}O_4$
FW 202.25
d 1.025
Fp 202°F

171.98	33.77*
102.69*	30.69*
52.39*	26.14
52.05*	21.60*
51.53*	20.58*

A

Aldrich 14,608-0   *CAS [7208-47-1]*

D-Sorbitol hexaacetate, 97%

CDCl₃   QE-300

$C_{18}H_{26}O_{12}$
FW 434.39
mp 102°C

60 MHz: *1*, 528A
FT-IR: *1*, 624C

170.42	69.31*	20.71*
170.32	68.71*	20.66*
169.96	68.59*	20.61*
169.78	68.37*	20.46*
169.73	61.80	
169.70	61.42	

B

Aldrich 30,228-7   *CAS [13443-46-4]*

L-Iditol hexaacetate, 99%

CDCl₃   QE-300

$C_{18}H_{26}O_{12}$
FW 434.39
mp 123°C

170.25	61.70
169.86	20.66*
169.66	20.60*
69.17*	20.55*
68.71*	

C

A

Aldrich 31,822-1   *CAS [1338-41-6]* $C_{24}H_{46}O_6$
FW 430.63
mp 57°C
Fp >230°F

Sorbitan monostearate

CDCl$_3$ QE-300

$CH_2O-C-CH_2(CH_2)_{15}CH_3$

HO-C-H

OH

OH

B

Aldrich 15,902-6   *CAS [13035-61-5]* $C_{13}H_{18}O_9$ 60 MHz: *1*, 528D
FW 318.28   FT-IR: *1*, 624D
mp 82°C

β-D-Ribofuranose 1,2,3,5-tetraacetate, 98%

CDCl$_3$ QE-300

170.38	70.57*
169.64	63.66
169.37	21.01*
168.94	20.70*
98.21*	20.46*
79.34*	20.44*
74.17*	

$CH_3-C-OCH_2$ $O-C-CH_3$

CH_3-C-O $O-C-CH_3$

C

Aldrich 36,842-3   *CAS [572-09-8]* $C_{14}H_{19}BrO_9$
FW 411.21
mp 90°C

α-D-Glucopyranosyl bromide tetraacetate, 97%

CDCl$_3$ QE-300

170.37	70.14*
169.72	67.15*
169.67	60.93
169.35	20.63*
86.56*	20.60*
72.13*	20.53*
70.57*	

CH_2OR

RO OR Br $R = -C-CH_3$

OR

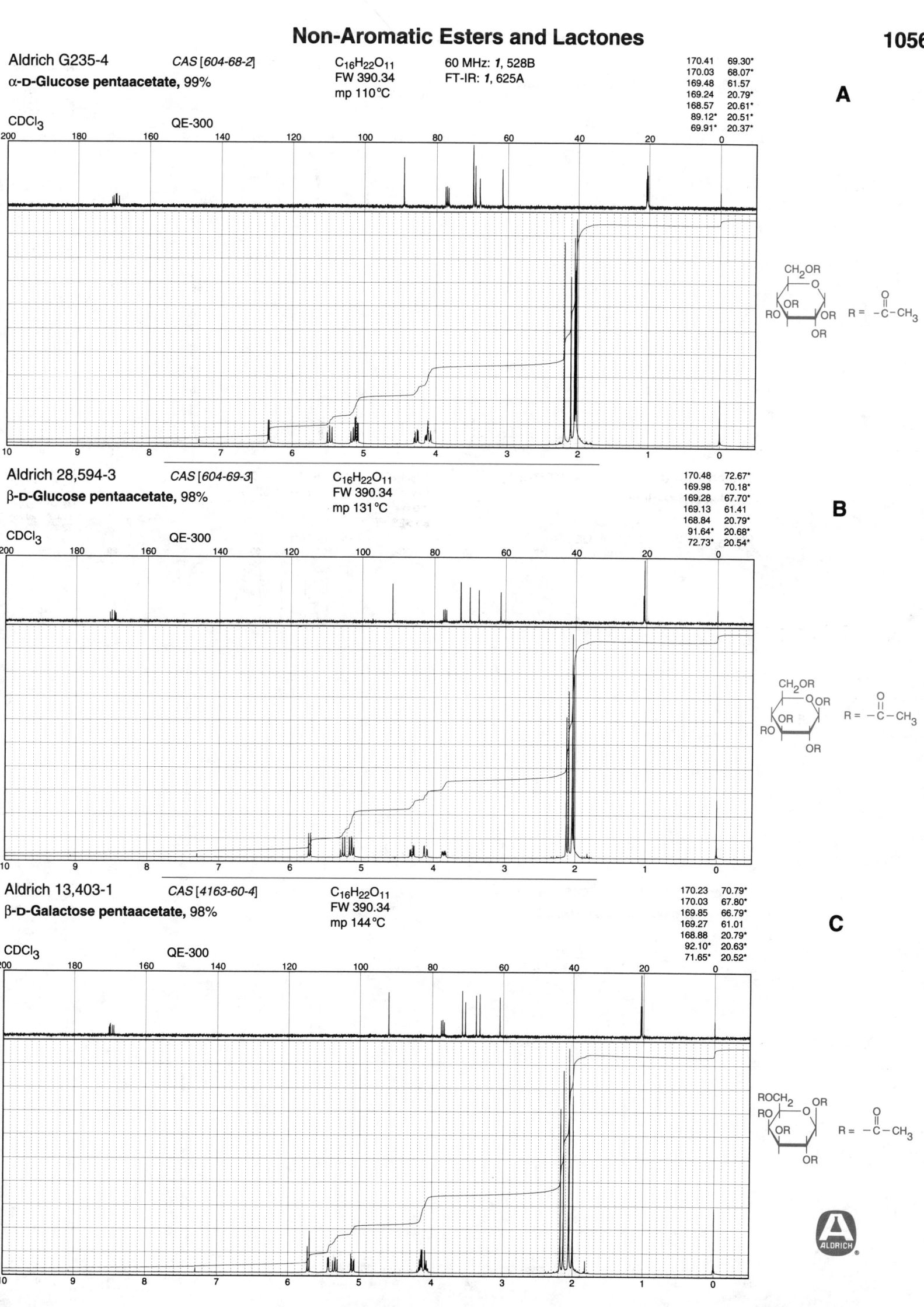

Aldrich G235-4   *CAS [604-68-2]*

α-D-Glucose pentaacetate, 99%

$C_{16}H_{22}O_{11}$
FW 390.34
mp 110°C

60 MHz: *1*, 528B
FT-IR: *1*, 625A

170.41	69.30*
170.03	68.07*
169.48	61.57
169.24	20.79*
168.57	20.61*
89.12*	20.51*
69.91*	20.37*

A

CDCl₃   QE-300

R = $-\overset{O}{\overset{\|}{C}}-CH_3$

Aldrich 28,594-3   *CAS [604-69-3]*

β-D-Glucose pentaacetate, 98%

$C_{16}H_{22}O_{11}$
FW 390.34
mp 131°C

170.48	72.67*
169.98	70.18*
169.28	67.70*
169.13	61.41
168.84	20.79*
91.64*	20.68*
72.73*	20.54*

B

CDCl₃   QE-300

R = $-\overset{O}{\overset{\|}{C}}-CH_3$

Aldrich 13,403-1   *CAS [4163-60-4]*

β-D-Galactose pentaacetate, 98%

$C_{16}H_{22}O_{11}$
FW 390.34
mp 144°C

170.23	70.79*
170.03	67.80*
169.85	66.79*
169.27	61.01
168.88	20.79*
92.10*	20.63*
71.65*	20.52*

C

CDCl₃   QE-300

R = $-\overset{O}{\overset{\|}{C}}-CH_3$

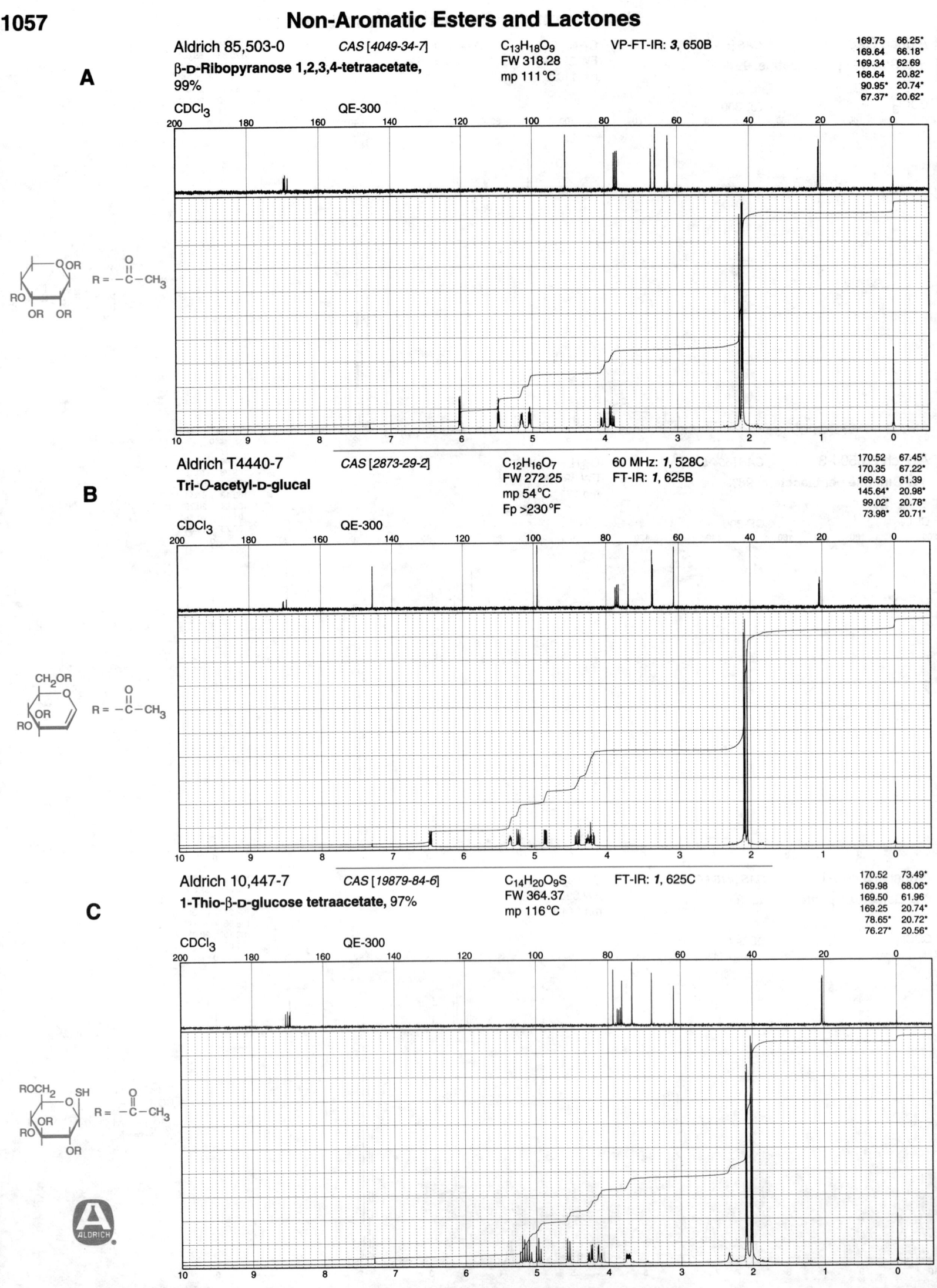

A

Aldrich 85,503-0   CAS [4049-34-7] $C_{13}H_{18}O_9$ VP-FT-IR: **3**, 650B

β-D-Ribopyranose 1,2,3,4-tetraacetate, 99%

FW 318.28
mp 111°C

169.75	66.25*
169.64	66.18*
169.34	62.69
168.64	20.82*
90.95*	20.74*
67.37*	20.62*

CDCl₃   QE-300

$R = -\overset{O}{\underset{\parallel}{C}}-CH_3$

B

Aldrich T4440-7   CAS [2873-29-2] $C_{12}H_{16}O_7$ 60 MHz: **1**, 528C

Tri-O-acetyl-D-glucal

FW 272.25
mp 54°C
Fp >230°F

FT-IR: **1**, 625B

170.52	67.45*
170.35	67.22*
169.53	61.39
145.64*	20.98*
99.02*	20.78*
73.98*	20.71*

CDCl₃   QE-300

$R = -\overset{O}{\underset{\parallel}{C}}-CH_3$

C

Aldrich 10,447-7   CAS [19879-84-6] $C_{14}H_{20}O_9S$ FT-IR: **1**, 625C

1-Thio-β-D-glucose tetraacetate, 97%

FW 364.37
mp 116°C

170.52	73.49*
169.98	68.06*
169.50	61.96
169.25	20.74*
78.65*	20.72*
76.27*	20.56*

CDCl₃   QE-300

$R = -\overset{O}{\underset{\parallel}{C}}-CH_3$

Aldrich 31,024-7   *CAS [18968-05-3]*

1,3,4,6-Tetra-*O*-acetyl-β-D-mannopyranose, 98%

CDCl₃   QE-300

$C_{14}H_{20}O_{10}$
FW 348.31
mp 161 °C

170.72	68.31*
170.08	65.20*
169.52	61.98
168.53	20.87*
91.65*	20.79*
73.01*	20.74*
72.82*	20.66*

A

ROCH₂ ... OR / OR HO / RO $R = -\overset{O}{\underset{}{C}}-CH_3$

Aldrich 36,168-2   *CAS [5019-24-9]*

Methyl-α-D-mannopyranoside 2,3,4,6-tetra-acetate, 99%

CDCl₃   QE-300

$C_{15}H_{22}O_{10}$
FW 362.33
mp 65 °C

170.49	68.44*
169.90	66.24*
169.74	62.55
169.62	55.26*
98.60*	20.83*
69.54*	20.65*
69.08*	

B

CH₂OR ... O / OR RO / RO OCH₃ $R = -\overset{O}{\underset{}{C}}-CH_3$

Aldrich 33,229-1   *CAS [34819-86-8]*

3,4-Di-*O*-acetyl-6-deoxy-L-glucal, 98%

CDCl₃   QE-300

$C_{10}H_{14}O_5$
FW 214.22
bp 69 °C
d 1.116
n_D^{20} 1.4540

Fp >230 °F

170.53	71.79*
169.79	68.27*
145.91*	21.06*
98.74*	20.88*
72.47*	16.55*

C

RO ... O / CH₃ / OR $R = -\overset{O}{\underset{}{C}}-CH_3$

ALDRICH

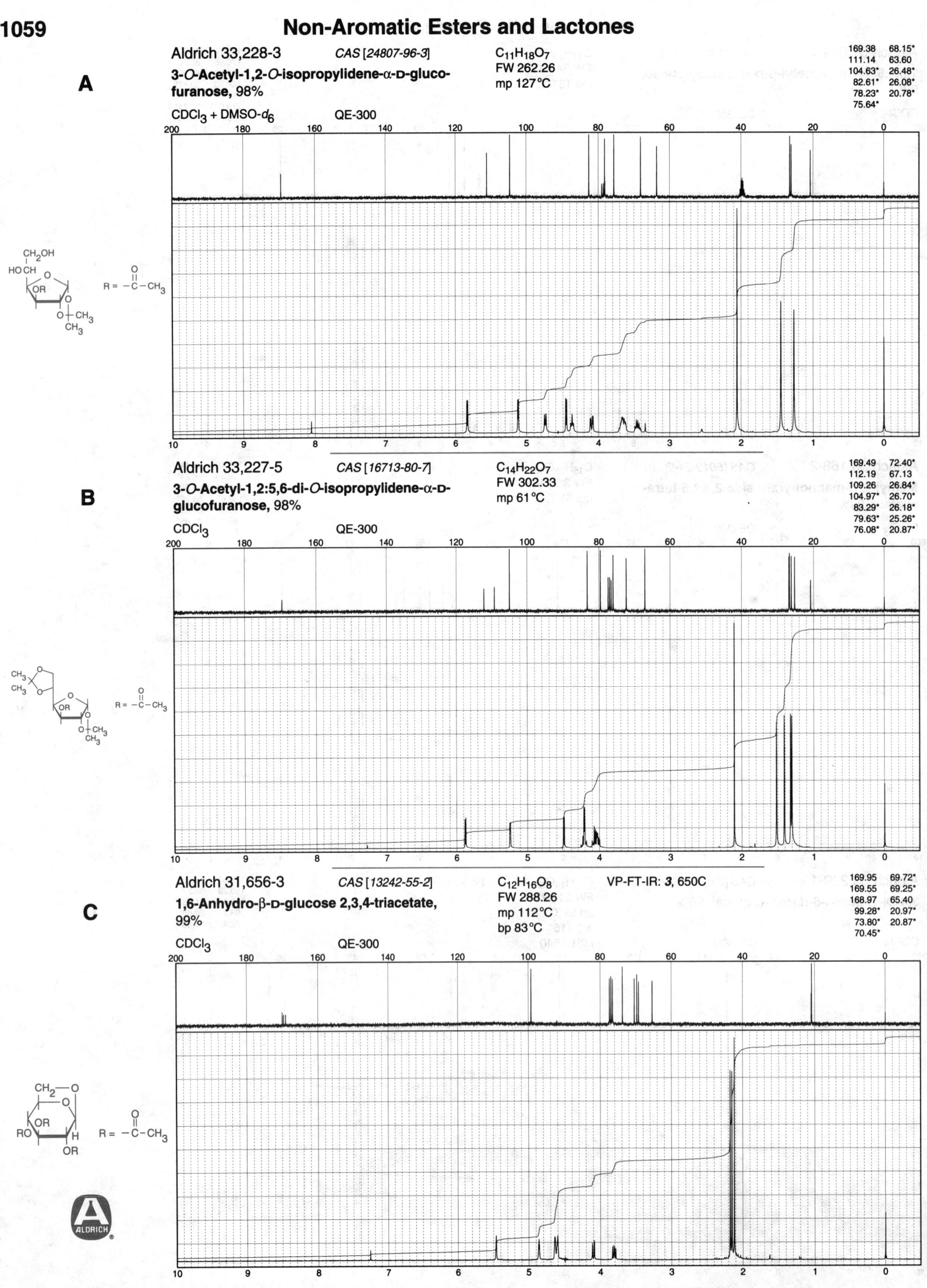

A Aldrich 33,228-3   *CAS [24807-96-3]* $C_{11}H_{18}O_7$

3-*O*-Acetyl-1,2-*O*-isopropylidene-α-D-gluco-furanose, 98%

FW 262.26
mp 127°C

169.38   68.15*
111.14   63.60
104.63* 26.48*
82.61*   26.08*
78.23*   20.78*
75.64*

CDCl₃ + DMSO-d₆   QE-300

R = −C(=O)−CH₃

B Aldrich 33,227-5   *CAS [16713-80-7]* $C_{14}H_{22}O_7$

3-*O*-Acetyl-1,2:5,6-di-*O*-isopropylidene-α-D-glucofuranose, 98%

FW 302.33
mp 61°C

169.49   72.40*
112.19   67.13
109.26   26.84*
104.97* 26.70*
83.29*   26.18*
79.63*   25.26*
76.08*   20.87*

CDCl₃   QE-300

R = −C(=O)−CH₃

C Aldrich 31,656-3   *CAS [13242-55-2]* $C_{12}H_{16}O_8$ VP-FT-IR: *3*, 650C

1,6-Anhydro-β-D-glucose 2,3,4-triacetate, 99%

FW 288.26
mp 112°C
bp 83°C

169.95   69.72*
169.55   69.25*
168.97   65.40
99.28*   20.97*
73.80*   20.87*
70.45*

CDCl₃   QE-300

R = −C(=O)−CH₃

Non-Aromatic Esters and Lactones

1060

Aldrich 25,260-3 CAS [126-14-7] $C_{28}H_{38}O_{19}$ FT-IR: *1*, 625D
Sucrose octaacetate, 99% FW 678.60 mp 88 °C

CDCl$_3$ QE-300

170.65	169.48	70.29*	61.77
170.44	104.01	69.64*	20.69*
170.07	89.95*	68.52*	20.63*
170.00	79.14*	68.23*	20.56*
169.86	75.72*	63.64	
169.63	75.02*	62.88	

A

Aldrich 32,193-1 CAS [79409-46-4] $C_{27}H_{36}O_{15}$
Secologanin ethylene acetal tetraacetate, 98% FW 600.58 mp 124 °C

CDCl$_3$ QE-300

170.54	120.30	70.57*	32.47
170.14	111.19	68.12*	26.95*
169.27	103.25*	64.72	20.74*
168.92	96.23*	64.64	20.59*
166.94	95.82*	61.67	20.20*
150.43*	72.50*	51.23*	
133.18*	72.14*	43.15*	

B

Aldrich 33,262-3 CAS [23739-88-0] $C_{84}H_{112}O_{56}$
Triacetyl-β-cyclodextrin FW 2017.79 mp 205 °C

CDCl$_3$ QE-300

170.63	70.80*
170.34	70.37*
169.33	69.54*
96.68*	62.45
76.69*	20.77*

C

ALDRICH

A

Aldrich E3,430-7   CAS [623-51-8] $C_4H_8O_2S$ Fp 118°F   60 MHz: 1, 564B

Ethyl 2-mercaptoacetate, 97%   FW 120.17   FT-IR: 1, 669D

bp 54°C (12 mm)   VP-FT-IR: 3, 728A

d 1.096

n_D^{20} 1.4570

170.78
61.64
26.54
14.09*

CDCl₃   QE-300

$HSCH_2-\overset{O}{\overset{\|}{C}}-OCH_2CH_3$

B

Aldrich 38,149-7   CAS [10193-96-1] $C_{12}H_{20}O_6S_3$ Fp >230°F

2-Ethyl-2-(hydroxymethyl)-1,3-propanediol   FW 356.48

tris(2-mercaptoacetate)   bp 245°C (1 mm)

d 1.280

n_D^{20} 1.5290

170.43
64.59
41.15
26.35
22.92
7.35*

CDCl₃   QE-300

$CH_3CH_2-\overset{\overset{\displaystyle CH_2O-\overset{O}{\overset{\|}{C}}-CH_2SH}{|}}{\underset{\underset{\displaystyle CH_2O-\overset{O}{\overset{\|}{C}}-CH_2SH}{|}}{C}}-CH_2O-\overset{O}{\overset{\|}{C}}-CH_2SH$

C

Aldrich W32,790-5   CAS [19788-49-9] $C_5H_{10}O_2S$ Fp 137°F   VP-FT-IR: 3, 728C

Ethyl 2-mercaptopropionate, 96%   FW 134.20

bp 41°C (1 mm)

d 1.031

n_D^{20} 1.4490

173.50
61.30
35.68*
21.11*
14.05*

CDCl₃   QE-300

$CH_3\overset{SH}{\overset{|}{C}}H-\overset{O}{\overset{\|}{C}}-OCH_2CH_3$

ALDRICH

Non-Aromatic Esters and Lactones

Aldrich 10,898-7   CAS [2935-90-2]   C₄H₈O₂S   Fp 140°F   60 MHz: 1, 564D

$C_4H_8O_2S$
FW 120.17
bp 55°C (14 mm)
d 1.085
n_D^{20} 1.4640

Methyl 3-mercaptopropionate, 98%

CDCl₃   QE-300

60 MHz: 1, 564D
FT-IR: 1, 670A
VP-FT-IR: 3, 728D

177.93
51.73*
38.26
19.72

A

$HSCH_2CH_2-\overset{\overset{\textstyle O}{\|}}{C}-OCH_3$

Aldrich 30,710-6   CAS [5466-06-8]   C₅H₁₀O₂S   Fp 163°F   VP-FT-IR: 3, 729A

$C_5H_{10}O_2S$
FW 134.20
bp 76°C (10 mm)
d 1.039
n_D^{20} 1.4570

Ethyl 3-mercaptopropionate, 99%

CDCl₃   QE-300

171.47
60.62
38.52
19.77
14.22*

B

$HSCH_2CH_2-\overset{\overset{\textstyle O}{\|}}{C}-OCH_2CH_3$

Aldrich 38,145-4   CAS [16215-21-7]   C₇H₁₄O₂S   Fp 200°F

$C_7H_{14}O_2S$
FW 162.25
bp 101°C (12 mm)
d 0.999
n_D^{20} 1.4570

Butyl 3-mercaptopropionate, 98%

CDCl₃   QE-300

171.57
64.54
38.52
30.64
19.80
19.13
13.69*

C

$HSCH_2CH_2-\overset{\overset{\textstyle O}{\|}}{C}-OCH_2CH_2CH_2CH_3$

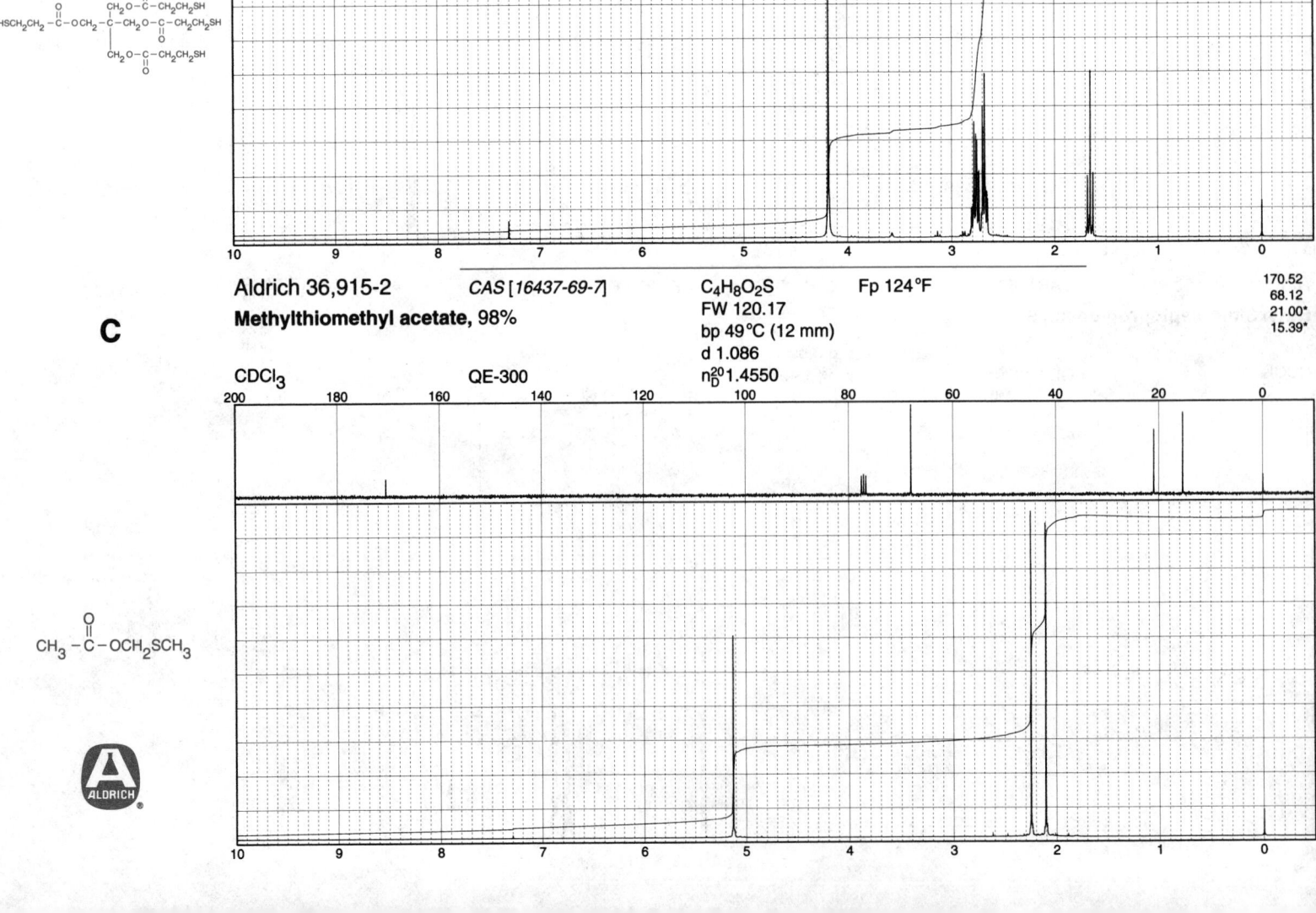

A

Aldrich 38,148-9   *CAS [33007-83-9]* $C_{15}H_{26}O_6S_3$ Fp 205°F

2-Ethyl-2-(hydroxymethyl)-1,3-propanediol tris(3-mercaptopropionate)

FW 398.56
bp 220°C
d 1.210
$n_D^{20}1.5180$

CDCl$_3$ QE-300

171.10
63.75
40.66
38.29
22.88
19.64
7.34*

B

Aldrich 38,146-2   *CAS [7575-23-7]* $C_{17}H_{28}O_8S_4$ Fp >230°F

Pentaerythritol tetrakis(3-mercapto-propionate)

FW 488.66
bp 275°C (1 mm)
d 1.280
$n_D^{20}1.5310$

CDCl$_3$ QE-300

170.90
62.09
42.00
38.17
19.57

C

Aldrich 36,915-2   *CAS [16437-69-7]* $C_4H_8O_2S$ Fp 124°F

Methylthiomethyl acetate, 98%

FW 120.17
bp 49°C (12 mm)
d 1.086
$n_D^{20}1.4550$

CDCl$_3$ QE-300

170.52
68.12
21.00*
15.39*

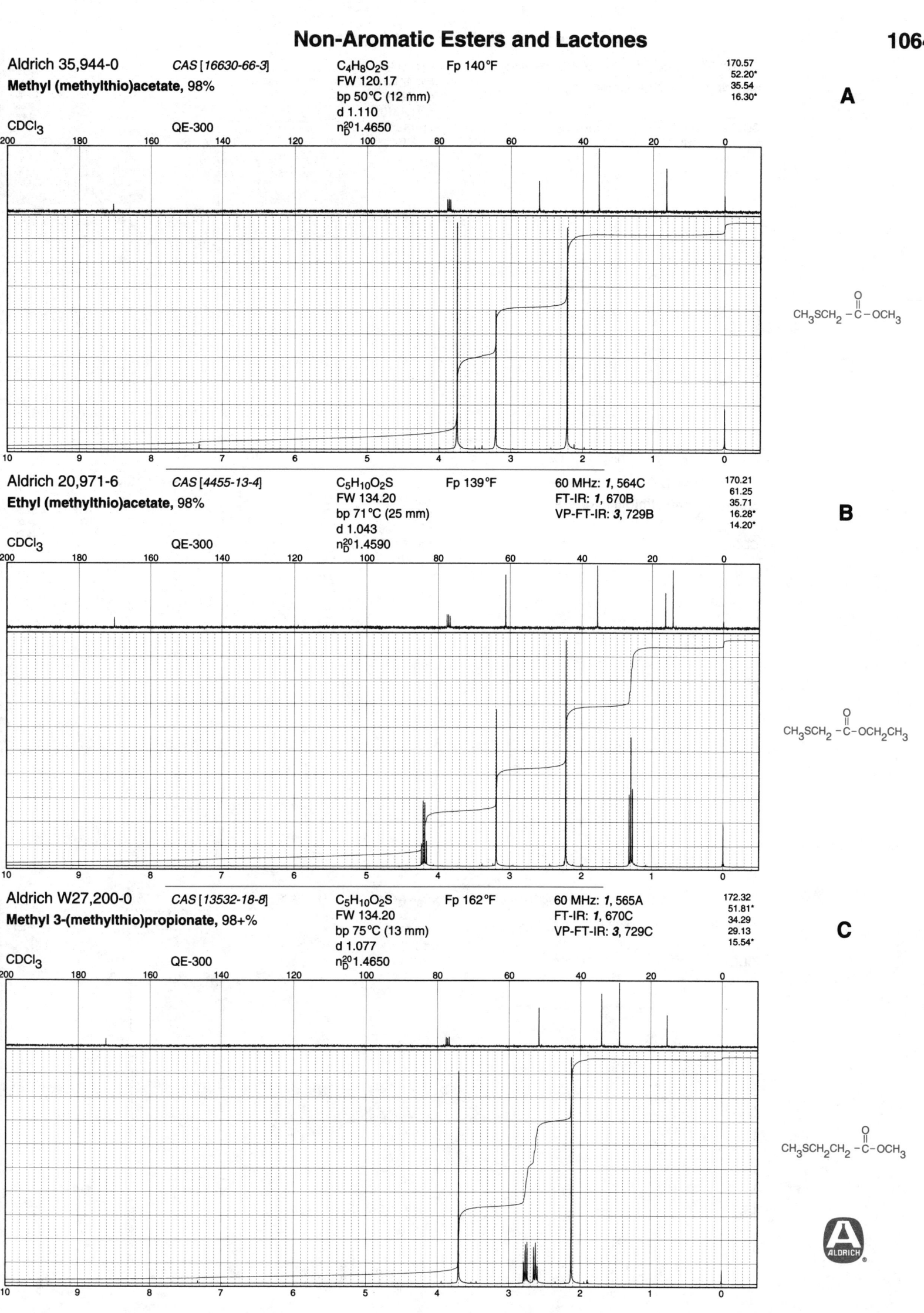

Aldrich 35,944-0   *CAS [16630-66-3]* $C_4H_8O_2S$ Fp 140°F

Methyl (methylthio)acetate, 98%

FW 120.17
bp 50°C (12 mm)
d 1.110
n_D^{20} 1.4650

CDCl$_3$ QE-300

170.57
52.20*
35.54
16.30*

A

$CH_3SCH_2-\overset{\displaystyle O}{\overset{\|}{C}}-OCH_3$

Aldrich 20,971-6   *CAS [4455-13-4]* $C_5H_{10}O_2S$ Fp 139°F

Ethyl (methylthio)acetate, 98%

FW 134.20
bp 71°C (25 mm)
d 1.043
n_D^{20} 1.4590

CDCl$_3$ QE-300

60 MHz: *1*, 564C
FT-IR: *1*, 670B
VP-FT-IR: *3*, 729B

170.21
61.25
35.71
16.28*
14.20*

B

$CH_3SCH_2-\overset{\displaystyle O}{\overset{\|}{C}}-OCH_2CH_3$

Aldrich W27,200-0   *CAS [13532-18-8]* $C_5H_{10}O_2S$ Fp 162°F

Methyl 3-(methylthio)propionate, 98+%

FW 134.20
bp 75°C (13 mm)
d 1.077
n_D^{20} 1.4650

CDCl$_3$ QE-300

60 MHz: *1*, 565A
FT-IR: *1*, 670C
VP-FT-IR: *3*, 729C

172.32
51.81*
34.29
29.13
15.54*

C

$CH_3SCH_2CH_2-\overset{\displaystyle O}{\overset{\|}{C}}-OCH_3$

A

Aldrich 31,967-8   *CAS [51534-66-8]* $C_6H_{12}O_2S$ VP-FT-IR: *3*, 730D

Methyl 2-(methylthio)butyrate, 98+%
FW 148.22
d 1.028
n_D^{20} 1.4610
Fp 148°F

172.71
52.03*
48.91*
24.00
13.65*
11.88*

CDCl₃   QE-300

CH₃S O
  | ||
CH₃CH₂CH–C–OCH₃

B

Aldrich W34,120-7   *CAS [53053-51-3]* $C_6H_{12}O_2S$ VP-FT-IR: *3*, 731A

Methyl 4-(methylthio)butyrate, 97+%
FW 148.22
d 1.036
n_D^{20} 1.4660
Fp 183°F

173.47
51.54*
33.43
32.66
24.09
15.26*

CDCl₃   QE-300

            O
            ||
CH₃SCH₂CH₂CH₂–C–OCH₃

C

Aldrich 22,634-3   *CAS [20461-95-4]* $C_8H_{16}O_2S_2$ Fp 77°F   60 MHz: *1*, 565C

Ethyl bis(ethylthio)acetate, 99%
FW 208.34
bp 70°C
d 1.071
n_D^{20} 1.4970

FT-IR: *1*, 671A
VP-FT-IR: *3*, 731B

169.35
61.87
50.34*
25.07
14.20*
14.08*

CDCl₃   QE-300

                 O
                 ||
CH₃CH₂SCH–C–OCH₂CH₃
       |
      SCH₂CH₃

ALDRICH

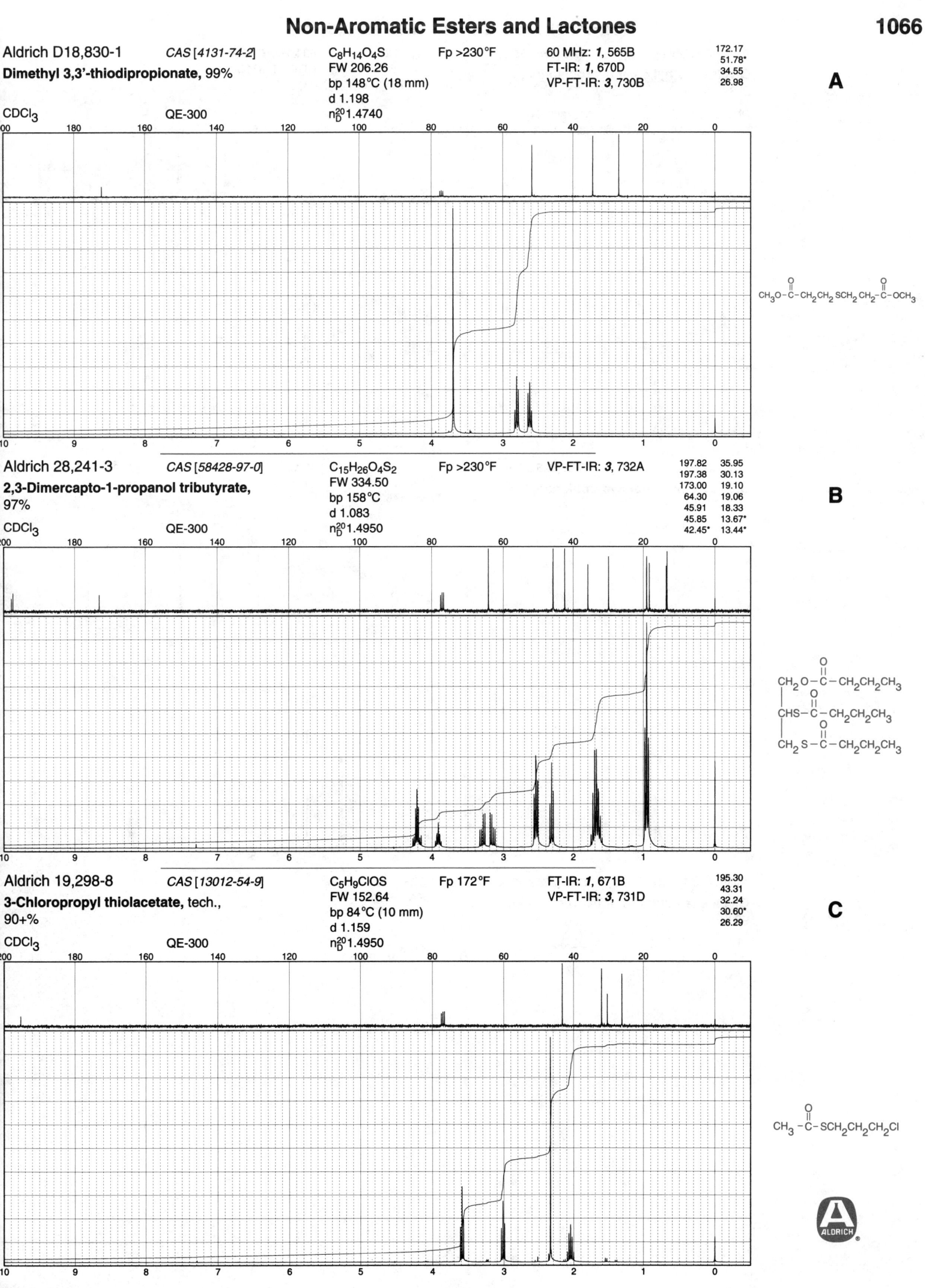

Aldrich D18,830-1   *CAS [4131-74-2]*
Dimethyl 3,3'-thiodipropionate, 99%

C₈H₁₄O₄S
FW 206.26
bp 148°C (18 mm)
d 1.198
n²⁰_D 1.4740

Fp >230°F

60 MHz: *1*, 565B
FT-IR: *1*, 670D
VP-FT-IR: *3*, 730B

172.17
51.78*
34.55
26.98

A

CDCl₃   QE-300

$CH_3O-\overset{O}{\overset{\|}{C}}-CH_2CH_2SCH_2CH_2-\overset{O}{\overset{\|}{C}}-OCH_3$

Aldrich 28,241-3   *CAS [58428-97-0]*
2,3-Dimercapto-1-propanol tributyrate, 97%

C₁₅H₂₆O₄S₂
FW 334.50
bp 158°C
d 1.083
n²⁰_D 1.4950

Fp >230°F

VP-FT-IR: *3*, 732A

197.82   35.95
197.38   30.13
173.00   19.10
64.30   19.06
45.91   18.33
45.85   13.67*
42.45*   13.44*

B

CDCl₃   QE-300

$CH_2O-\overset{O}{\overset{\|}{C}}-CH_2CH_2CH_3$
$CHS-\overset{O}{\overset{\|}{C}}-CH_2CH_2CH_3$
$CH_2S-\overset{O}{\overset{\|}{C}}-CH_2CH_2CH_3$

Aldrich 19,298-8   *CAS [13012-54-9]*
3-Chloropropyl thiolacetate, tech., 90+%

C₅H₉ClOS
FW 152.64
bp 84°C (10 mm)
d 1.159
n²⁰_D 1.4950

Fp 172°F

FT-IR: *1*, 671B
VP-FT-IR: *3*, 731D

195.30
43.31
32.24
30.60*
26.29

C

CDCl₃   QE-300

$CH_3-\overset{O}{\overset{\|}{C}}-SCH_2CH_2CH_2Cl$

ALDRICH

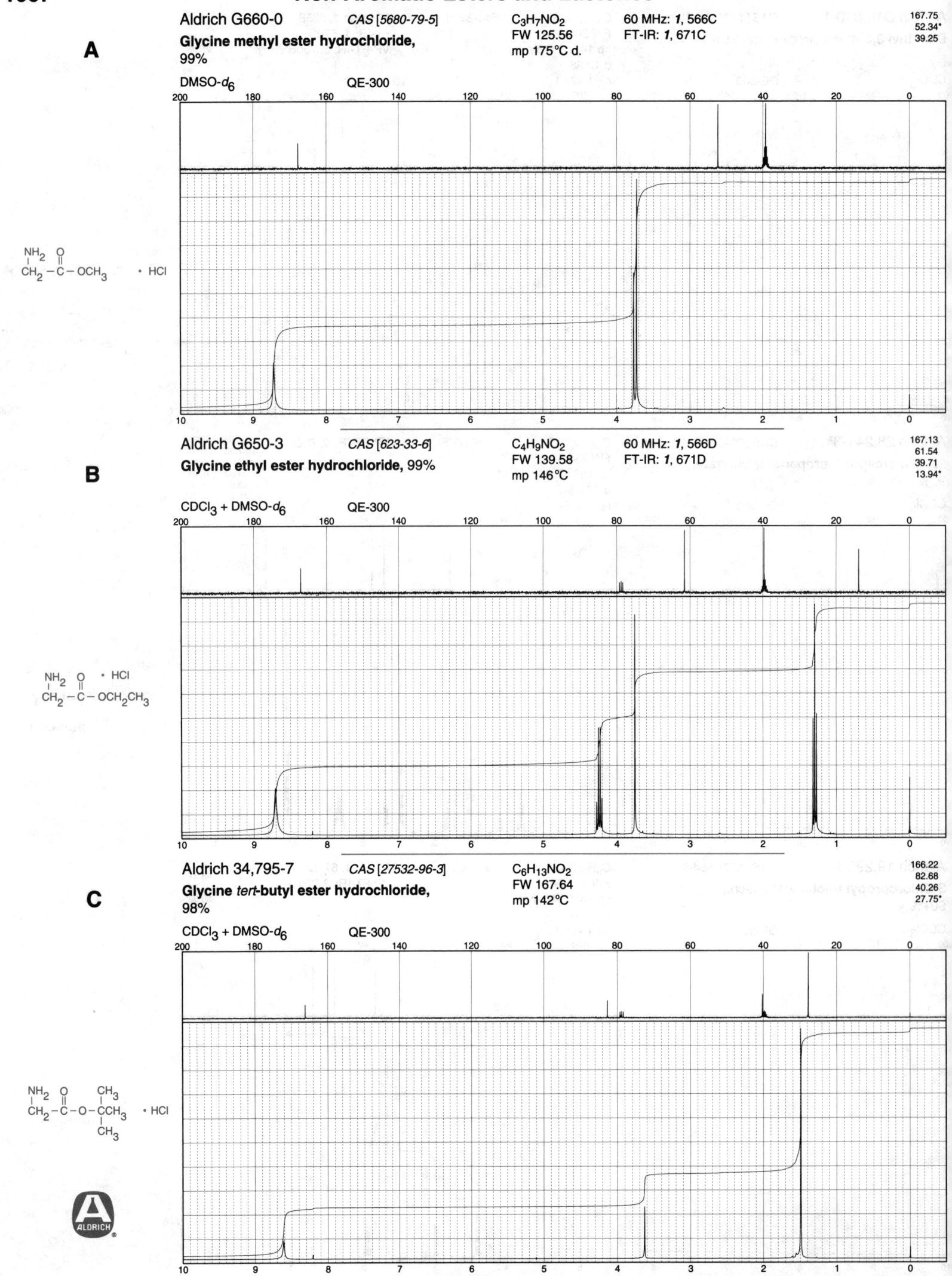

A

Aldrich G660-0 CAS [5680-79-5] C₃H₇NO₂ 60 MHz: *1*, 566C 167.75
Glycine methyl ester hydrochloride, FW 125.56 FT-IR: *1*, 671C 52.34*
99% mp 175°C d. 39.25

DMSO-d₆ QE-300

$$\underset{CH_2}{\overset{NH_2}{|}} - \underset{}{\overset{O}{\underset{\parallel}{C}}} - OCH_3 \quad \cdot \; HCl$$

B

Aldrich G650-3 CAS [623-33-6] C₄H₉NO₂ 60 MHz: *1*, 566D 167.13
Glycine ethyl ester hydrochloride, 99% FW 139.58 FT-IR: *1*, 671D 61.54
mp 146°C 39.71
13.94*

CDCl₃ + DMSO-d₆ QE-300

$$\underset{CH_2}{\overset{NH_2}{|}} - \underset{}{\overset{O}{\underset{\parallel}{C}}} - OCH_2CH_3 \quad \cdot \; HCl$$

C

Aldrich 34,795-7 CAS [27532-96-3] C₆H₁₃NO₂ 166.22
Glycine *tert*-butyl ester hydrochloride, FW 167.64 82.68
98% mp 142°C 40.26
27.75*

CDCl₃ + DMSO-d₆ QE-300

$$\underset{CH_2}{\overset{NH_2}{|}} - \underset{}{\overset{O}{\underset{\parallel}{C}}} - O - \underset{\underset{CH_3}{|}}{\overset{CH_3}{\overset{|}{C}}}CH_3 \quad \cdot \; HCl$$

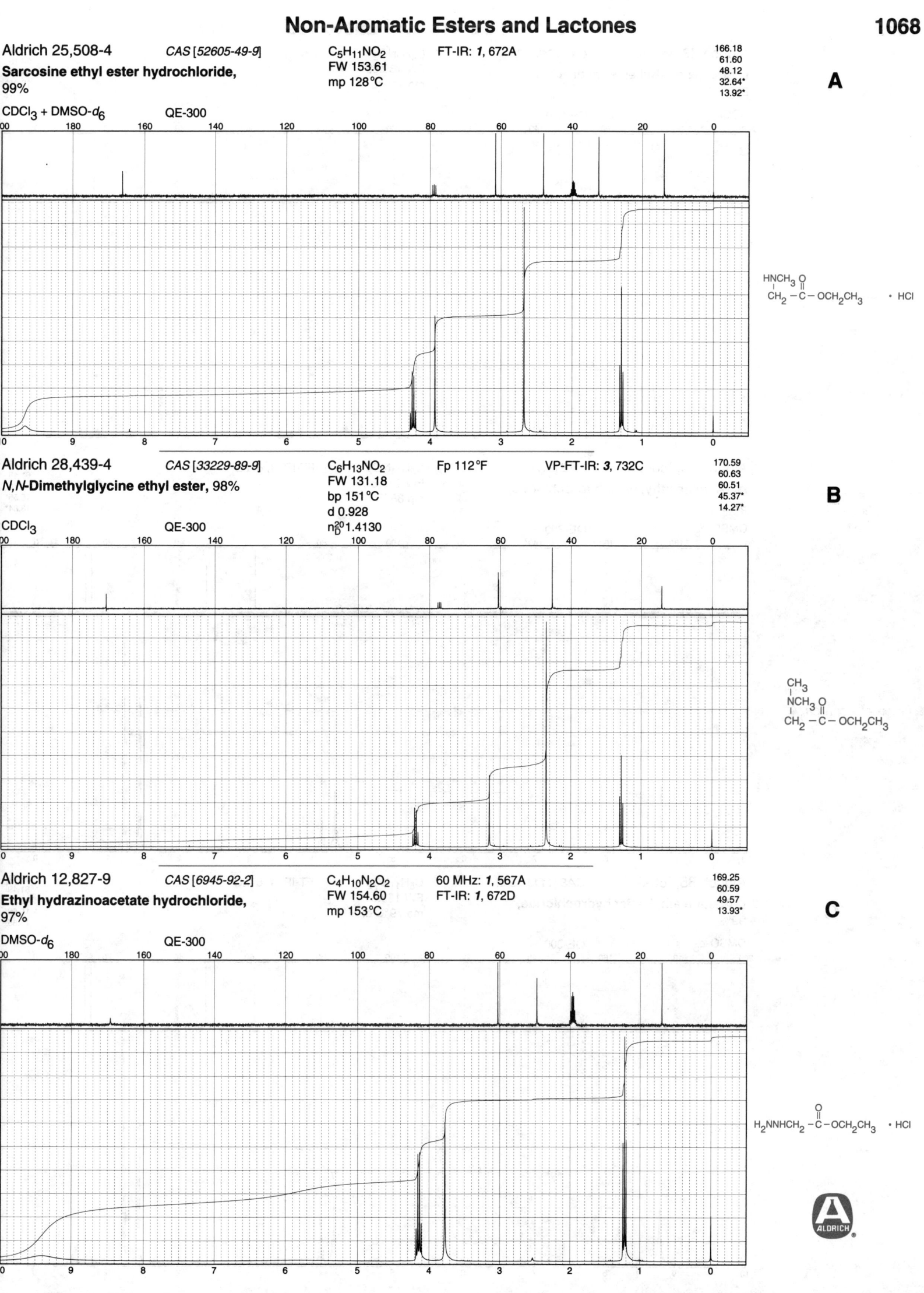

Aldrich 25,508-4   *CAS [52605-49-9]*   C₅H₁₁NO₂   FT-IR: *1*, 672A

$C_5H_{11}NO_2$
FW 153.61
mp 128°C

Sarcosine ethyl ester hydrochloride, 99%

166.18
61.60
48.12
32.64*
13.92*

A

CDCl₃ + DMSO-*d₆*   QE-300

Aldrich 28,439-4   *CAS [33229-89-9]* $C_6H_{13}NO_2$ Fp 112°F   VP-FT-IR: *3*, 732C

N,N-Dimethylglycine ethyl ester, 98%

FW 131.18
bp 151°C
d 0.928
n_D^{20} 1.4130

170.59
60.63
60.51
45.37*
14.27*

B

CDCl₃   QE-300

Aldrich 12,827-9   *CAS [6945-92-2]* $C_4H_{10}N_2O_2$ 60 MHz: *1*, 567A

Ethyl hydrazinoacetate hydrochloride, 97%

FW 154.60
mp 153°C
FT-IR: *1*, 672D

169.25
60.59
49.57
13.93*

C

DMSO-*d₆*   QE-300

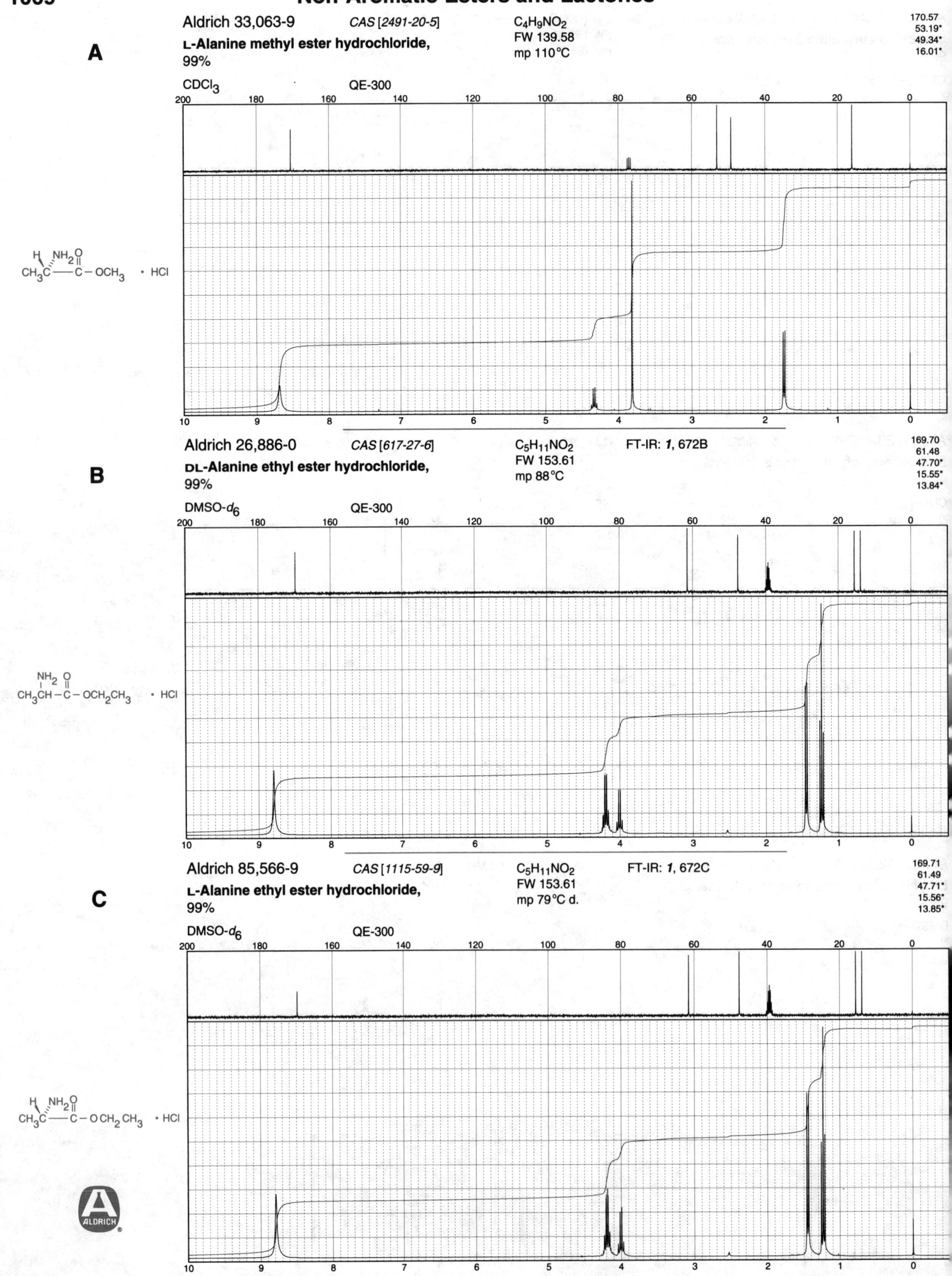

A

Aldrich 33,063-9   *CAS [2491-20-5]* $C_4H_9NO_2$
FW 139.58
mp 110°C

L-Alanine methyl ester hydrochloride, 99%

170.57
53.19*
49.34*
16.01*

$CDCl_3$ QE-300

B

Aldrich 26,886-0   *CAS [617-27-6]* $C_5H_{11}NO_2$
FW 153.61
mp 88°C   FT-IR: *1*, 672B

DL-Alanine ethyl ester hydrochloride, 99%

169.70
61.48
47.70*
15.55*
13.84*

DMSO-d_6 QE-300

C

Aldrich 85,566-9   *CAS [1115-59-9]* $C_5H_{11}NO_2$
FW 153.61
mp 79°C d.   FT-IR: *1*, 672C

L-Alanine ethyl ester hydrochloride, 99%

169.71
61.49
47.71*
15.56*
13.85*

DMSO-d_6 QE-300

Aldrich 30,614-2   CAS [4244-84-2]

β-Alanine ethyl ester hydrochloride, 98%

$C_5H_{11}NO_2$
FW 153.61
mp 71°C

171.19
61.27
35.73
31.43
14.09*

A

CDCl₃   QE-300

$$NH_2$$
$$CH_2CH_2-C-OCH_2CH_3 \cdot HCl$$

Aldrich 24,559-3   CAS [3853-06-3]

Methyl 3-(dimethylamino)propionate, 99%

$C_6H_{13}NO_2$
FW 131.18
bp 153°C
d 0.917
n_D^{20}1.4180

Fp 124°F

60 MHz: *1*, 567D
FT-IR: *1*, 673D
VP-FT-IR: *3*, 733B

172.82
54.78
51.58*
45.27*
32.72

B

CDCl₃   QE-300

$$CH_3 \quad\quad O$$
$$CH_3NCH_2CH_2-C-OCH_3$$

Aldrich E1,055-6   CAS [5303-65-1]

Ethyl 3-aminobutyrate, tech., 90%

$C_6H_{13}NO_2$
FW 131.18
bp 61°C (13 mm)
d 0.894
n_D^{20}1.4241

Fp 108°F

60 MHz: *1*, 567B
FT-IR: *1*, 673B
VP-FT-IR: *3*, 733A

172.35
60.25
44.40
44.05*
23.63*
14.24*

C

CDCl₃   QE-300

$$O$$
$$CH_3CHCH_2-C-OCH_2CH_3$$
$$NH_2$$

A

Aldrich 22,784-6   *CAS [13013-02-0]*

Methyl 4-nitrobutyrate, 97%

$C_5H_9NO_4$
FW 147.13
bp 108°C (09 mm)
d 1.149
n_D^{20} 1.4380

FT-IR: *1*, 782C
VP-FT-IR: *3*, 736A

172.31
74.35
51.89*
30.26
22.39

CDCl$_3$ QE-300

$O_2NCH_2CH_2CH_2-\overset{\overset{\displaystyle O}{\|}}{C}-OCH_3$

B

Aldrich 86,027-1   *CAS [6306-52-1]*

L-Valine methyl ester hydrochloride, 99%

$C_6H_{13}NO_2$
FW 167.64
mp 172°C

FT-IR: *1*, 674A

168.83
58.62*
52.85*
29.85*
18.48*
18.28*

CDCl$_3$ QE-300

$CH_3CH-\overset{\overset{\displaystyle CH_3}{|}}{\underset{}{C}}-\overset{\overset{\displaystyle NH_2}{|}}{\underset{}{C}}-\overset{\overset{\displaystyle O}{\|}}{C}-OCH_3 \cdot HCl$

C

Aldrich 22,069-8   *CAS [17609-47-1]*

L-Valine ethyl ester hydrochloride

$C_7H_{15}NO_2$
FW 181.66
mp 104°C

60 MHz: *1*, 568B
FT-IR: *1*, 674B

168.39
61.36
57.27*
29.18*
18.48*
17.42*
13.92*

DMSO-d_6 QE-300

$CH_3CH-\overset{\overset{\displaystyle CH_3}{|}}{\underset{}{C}}-\overset{\overset{\displaystyle NH_2}{|}}{\underset{}{C}}-\overset{\overset{\displaystyle O}{\|}}{C}-OCH_2CH_3 \cdot HCl$

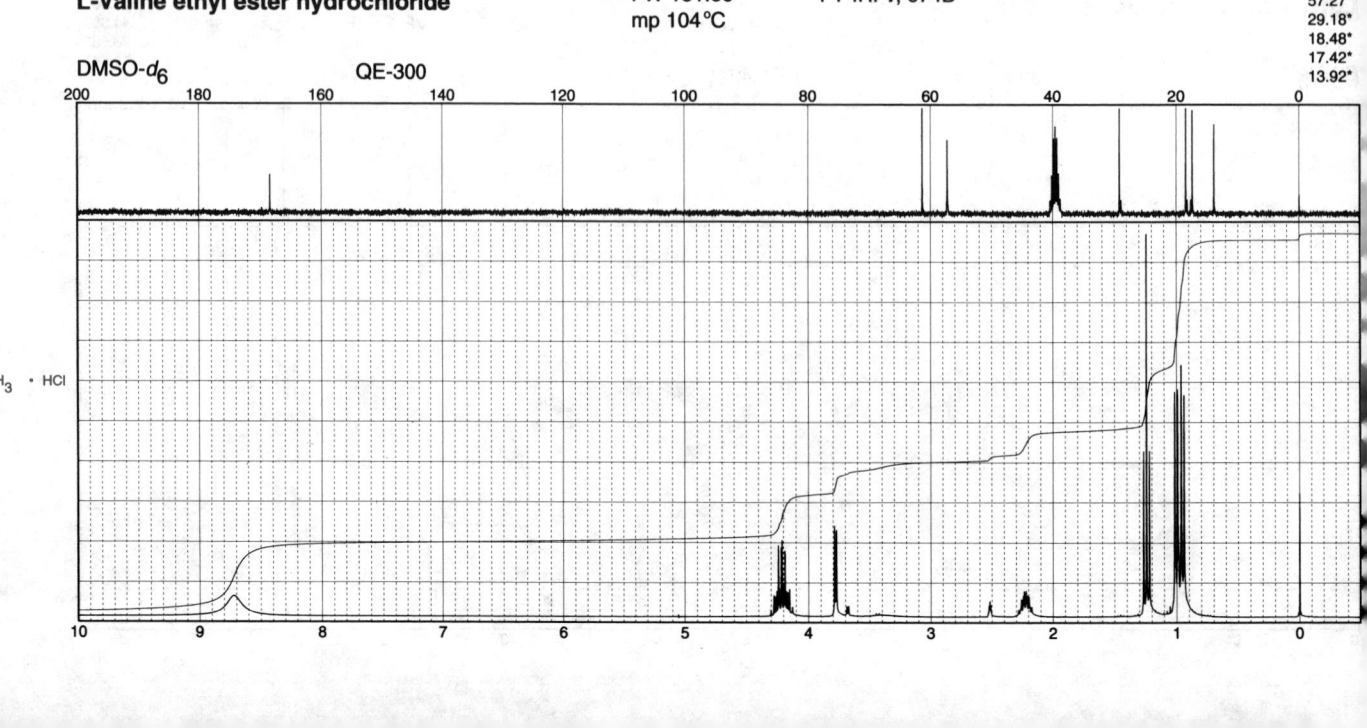

ALDRICH

Non-Aromatic Esters and Lactones

A

Aldrich L100-2   *CAS [7517-19-3]*

L-Leucine methyl ester hydrochloride, 98%

$C_7H_{15}NO_2$
FW 181.66
mp 149 °C d.

60 MHz: *1*, 568C
FT-IR: *1*, 674C

170.06
52.52*
50.74*
39.21
23.87*
22.19*
21.91*

CDCl$_3$ + DMSO-d_6 QE-300

$CH_3\ CH_2\ CH_2\!-\!\underset{CH_3}{\overset{H}{C}}\!\!-\!\!\underset{NH_2}{\overset{}{C}}\!-\!\overset{O}{\overset{\|}{C}}\!-\!OCH_3$ · HCl

B

Aldrich 39,120-4   *CAS [32559-18-5]*

Methyl pipecolinate hydrochloride, 97%

$C_7H_{13}NO_2$
FW 179.65
mp 205 °C d.

168.91
55.29*
52.64*
43.13
25.38
21.06
20.89

DMSO-d_6 QE-300

· HCl

C

Aldrich 19,971-2   *CAS [14205-39-1]*

Methyl 3-aminocrotonate, 97%

$C_5H_9NO_2$
FW 115.13
mp 82 °C

60 MHz: *1*, 569A
FT-IR: *1*, 675A
VP-FT-IR: *3*, 733C

170.52
159.92
83.58*
50.05*
22.28*

CDCl$_3$ QE-300

$CH_3\underset{NH_2}{\overset{}{C}}\!=\!CH\!-\!\overset{O}{\overset{\|}{C}}\!-\!OCH_3$

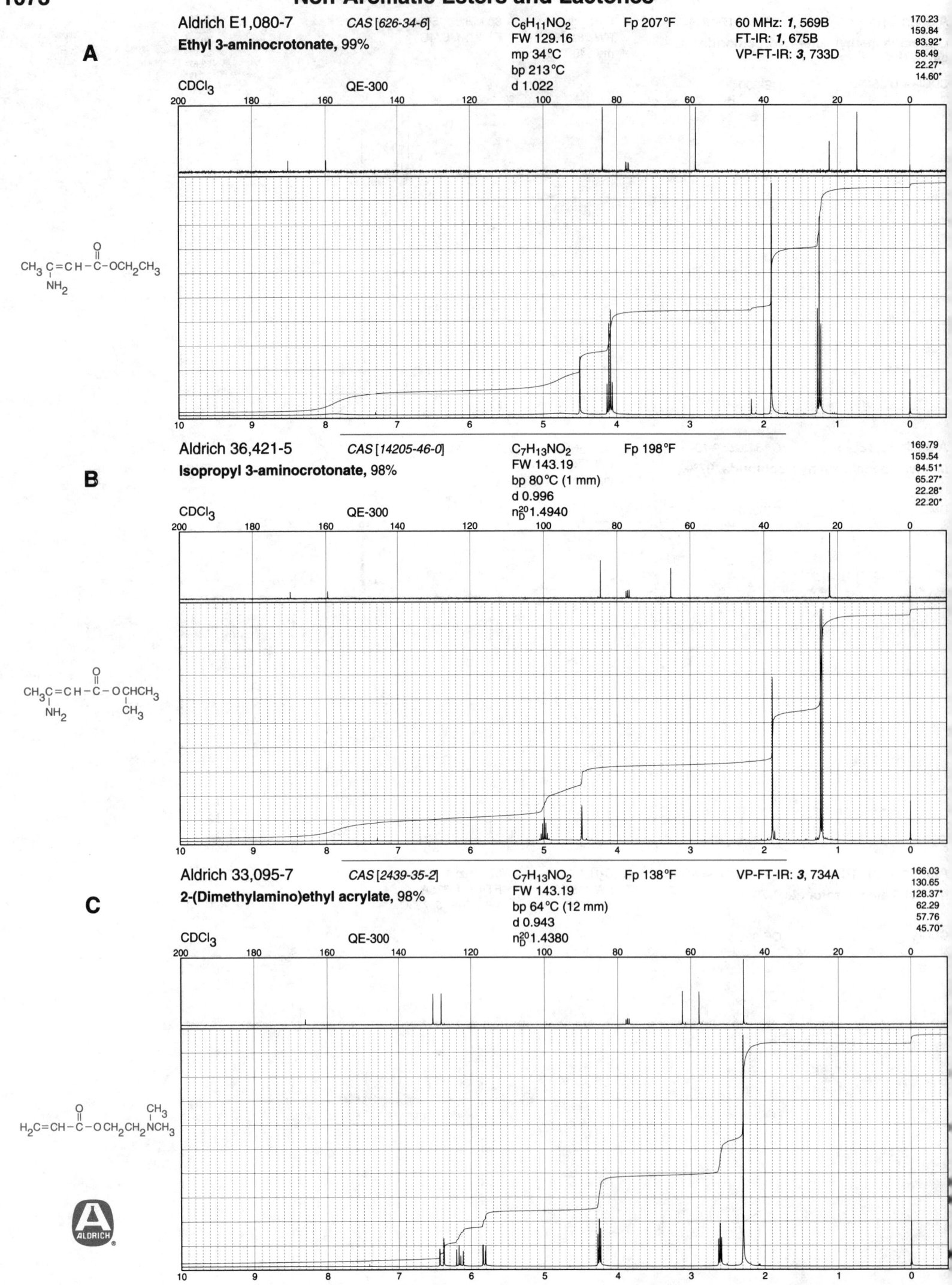

A

Aldrich E1,080-7   CAS [626-34-6]   C₆H₁₁NO₂   Fp 207°F   60 MHz: *1*, 569B
Ethyl 3-aminocrotonate, 99%   FW 129.16   FT-IR: *1*, 675B
                                   mp 34°C   VP-FT-IR: *3*, 733D
                                   bp 213°C
CDCl₃   QE-300   d 1.022

170.23
159.84
83.92*
58.49
22.27*
14.60*

B

Aldrich 36,421-5   CAS [14205-46-0]   C₇H₁₃NO₂   Fp 198°F
Isopropyl 3-aminocrotonate, 98%   FW 143.19
                                      bp 80°C (1 mm)
                                      d 0.996
CDCl₃   QE-300   n₂₀ᴰ 1.4940

169.79
159.54
84.51*
65.27*
22.28*
22.20*

C

Aldrich 33,095-7   CAS [2439-35-2]   C₇H₁₃NO₂   Fp 138°F   VP-FT-IR: *3*, 734A
2-(Dimethylamino)ethyl acrylate, 98%   FW 143.19
                                           bp 64°C (12 mm)
                                           d 0.943
CDCl₃   QE-300   n₂₀ᴰ 1.4380

166.03
130.65
128.37*
62.29
57.76
45.70*

Non-Aromatic Esters and Lactones

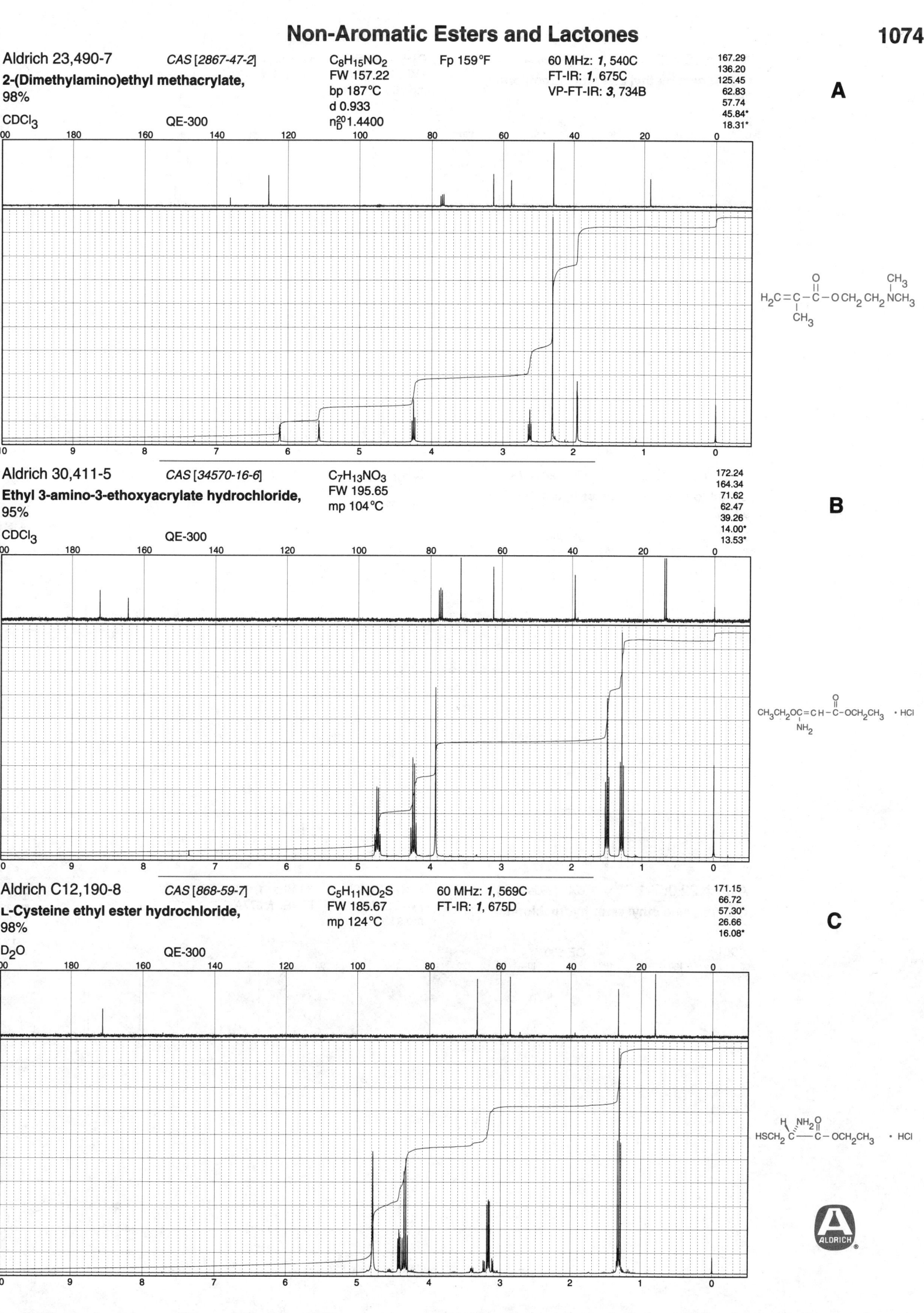

Aldrich 23,490-7   *CAS [2867-47-2]*

2-(Dimethylamino)ethyl methacrylate, 98%

C₈H₁₅NO₂
FW 157.22
bp 187°C
d 0.933
n²⁰_D 1.4400

Fp 159°F

60 MHz: *1*, 540C
FT-IR: *1*, 675C
VP-FT-IR: *3*, 734B

167.29
136.20
125.45
62.83
57.74
45.84*
18.31*

A

CDCl₃   QE-300

Aldrich 30,411-5   *CAS [34570-16-6]*

Ethyl 3-amino-3-ethoxyacrylate hydrochloride, 95%

C₇H₁₃NO₃
FW 195.65
mp 104°C

172.24
164.34
71.62
62.47
39.26*
14.00*
13.53*

B

CDCl₃   QE-300

Aldrich C12,190-8   *CAS [868-59-7]*

L-Cysteine ethyl ester hydrochloride, 98%

C₅H₁₁NO₂S
FW 185.67
mp 124°C

60 MHz: *1*, 569C
FT-IR: *1*, 675D

171.15
66.72
57.30*
26.66
16.08*

C

D₂O   QE-300

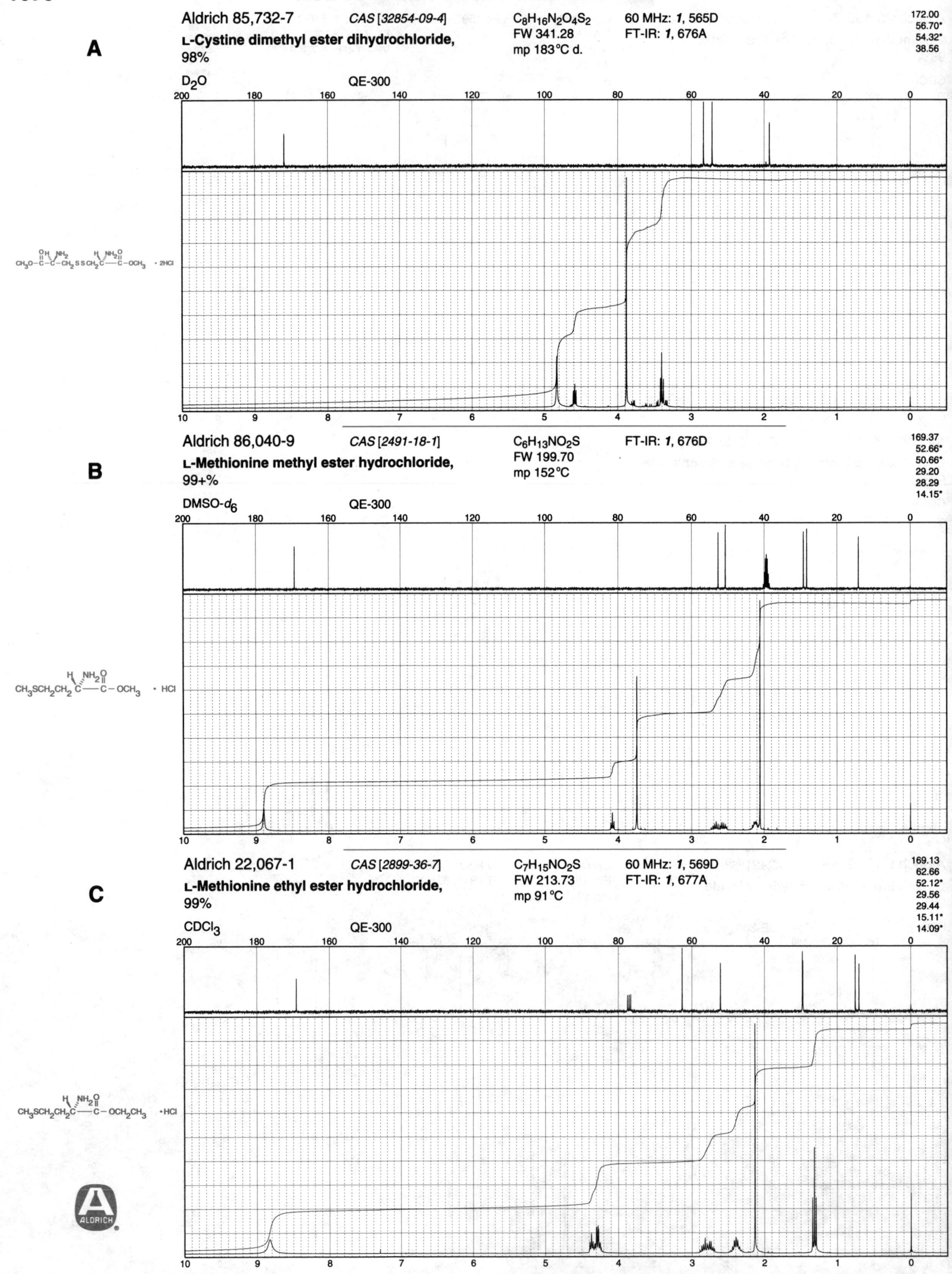

A

Aldrich 85,732-7   *CAS [32854-09-4]* $C_8H_{16}N_2O_4S_2$ 60 MHz: *1*, 565D

L-Cystine dimethyl ester dihydrochloride, 98%

FW 341.28
mp 183°C d.

FT-IR: *1*, 676A

172.00
56.70*
54.32*
38.56

D_2O QE-300

B

Aldrich 86,040-9   *CAS [2491-18-1]* $C_6H_{13}NO_2S$ FT-IR: *1*, 676D

L-Methionine methyl ester hydrochloride, 99+%

FW 199.70
mp 152°C

169.37
52.66*
50.66*
29.20
28.29
14.15*

DMSO-d_6 QE-300

C

Aldrich 22,067-1   *CAS [2899-36-7]* $C_7H_{15}NO_2S$ 60 MHz: *1*, 569D

L-Methionine ethyl ester hydrochloride, 99%

FW 213.73
mp 91°C

FT-IR: *1*, 677A

169.13
62.66
52.12*
29.56
29.44
15.11*
14.09*

$CDCl_3$ QE-300

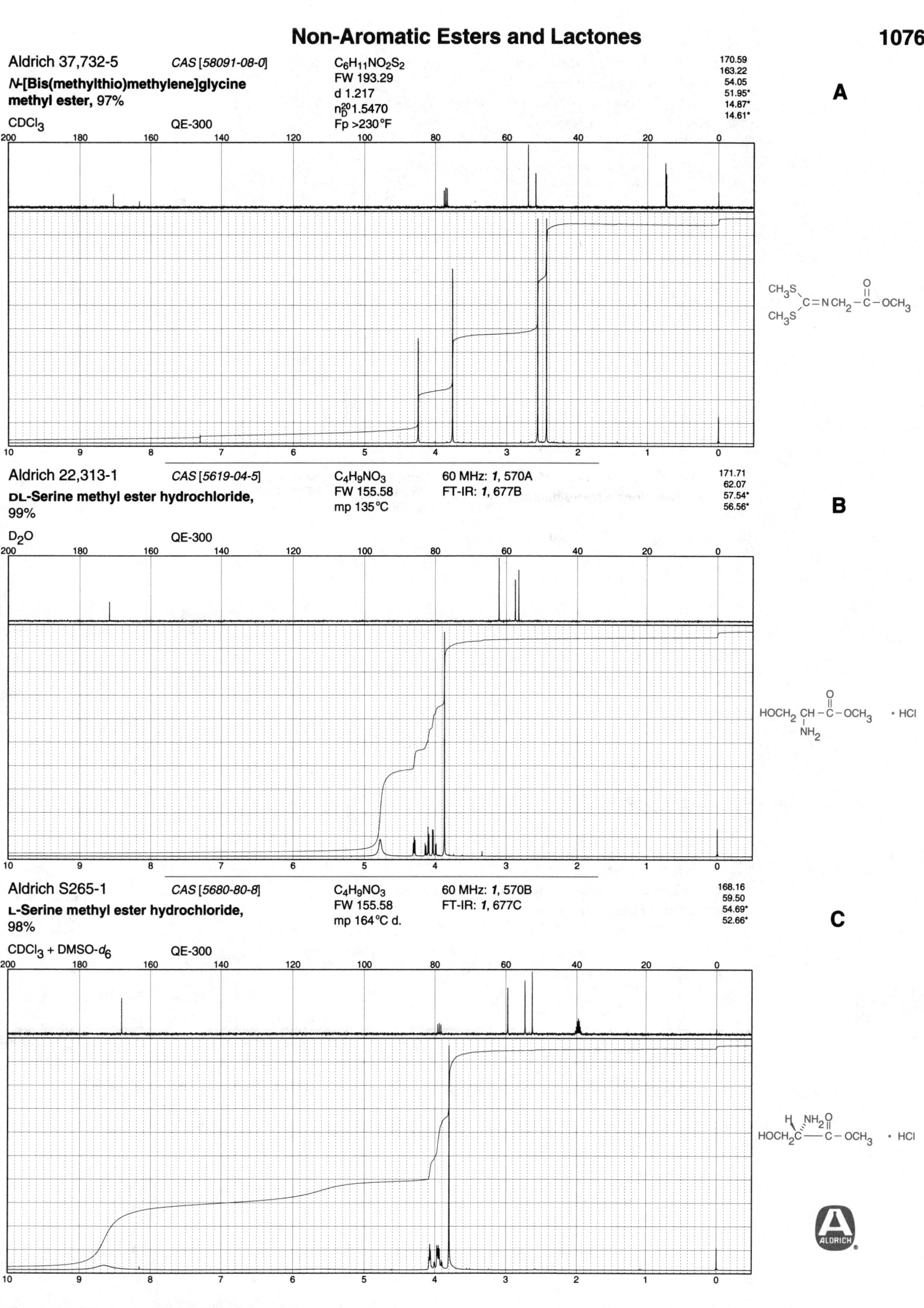

Aldrich 37,732-5   CAS [58091-08-0] $C_6H_{11}NO_2S_2$
N-[Bis(methylthio)methylene]glycine methyl ester, 97%

FW 193.29
d 1.217
n_D^{20} 1.5470
Fp >230°F

170.59
163.22
54.05
51.95*
14.87*
14.61*

A

CDCl₃   QE-300

Aldrich 22,313-1   CAS [5619-04-5]
DL-Serine methyl ester hydrochloride, 99%

$C_4H_9NO_3$
FW 155.58
mp 135°C

60 MHz: *1*, 570A
FT-IR: *1*, 677B

171.71
62.07
57.54*
56.56*

B

D₂O   QE-300

Aldrich S265-1   CAS [5680-80-8]
L-Serine methyl ester hydrochloride, 98%

$C_4H_9NO_3$
FW 155.58
mp 164°C d.

60 MHz: *1*, 570B
FT-IR: *1*, 677C

168.16
59.50
54.69*
52.66*

C

CDCl₃ + DMSO-d₆   QE-300

A

Aldrich 22,312-3   *CAS [26348-61-8]* $C_5H_{11}NO_3$ 60 MHz: *1*, 570C   167.66
L-Serine ethyl ester hydrochloride, 99%   FW 169.61   FT-IR: *1*, 677D   61.70
mp 131°C   59.52
54.69*
13.92*

$CDCl_3$ + DMSO-d_6 QE-300

$HOCH_2C—C—OCH_2CH_3$ · HCl

B

Aldrich D8,758-9   *CAS [13433-00-6]* $C_7H_{13}NO_4$ 60 MHz: *1*, 570D   163.48
Diethyl aminomalonate hydrochloride,   FW 211.65   FT-IR: *1*, 678A   62.83
98%   mp 166°C d.   54.91*
13.78*

$CDCl_3$ + DMSO-d_6 QE-300

$CH_3CH_2O—C—CH—C—OCH_2CH_3$ · HCl

C

Aldrich 30,934-6   *CAS [1118-89-4]* $C_9H_{17}NO_4$ 177.07   32.48
L-Glutamic acid diethyl ester hydrochloride,   FW 239.70   172.34   27.66
97%   mp 109°C   66.51   16.18*
64.87   16.06*
54.96*

D_2O QE-300

$CH_3CH_2O—C—CH_2CH_2—C—OCH_2CH_3$ · HCl

A

Aldrich 13,535-6   CAS [60-31-1]   C₇H₁₆ClNO₂   FT-IR: 1, 678C

Acetylcholine chloride, 98%

$C_7H_{16}ClNO_2$
FW 181.66
mp 148°C

169.96
64.58
58.02
54.11*
21.01*

CDCl₃   QE-300

$CH_3-\overset{O}{\overset{\|}{C}}-OCH_2CH_2\overset{CH_3}{\overset{|}{\underset{CH_3}{\overset{+}{N}}}}CH_3 \cdot Cl^-$

B

Aldrich A2,230-0   CAS [1866-15-5]   C₇H₁₆INOS   60 MHz: 1, 571C

S-Acetylthiocholine iodide, 98%

$C_7H_{16}INOS$
FW 289.18
mp 207°C

FT-IR: 1, 679C

193.84
64.00
52.54*
52.50*
52.46*
30.44*
21.86

CDCl₃ + DMSO-d₆   QE-300

$CH_3-\overset{O}{\overset{\|}{C}}-SCH_2CH_2\overset{CH_3}{\overset{|}{\underset{CH_3}{\overset{+}{N}}}}CH_3 \cdot I^-$

C

Aldrich A1,800-1   CAS [62-51-1]   C₈H₁₈ClNO₂   60 MHz: 1, 571D

Acetyl-β-methylcholine chloride, 98%

$C_8H_{18}ClNO_2$
FW 195.69
mp 172°C

FT-IR: 1, 679D

169.75
68.56
65.47*
54.19*
21.36*
18.84*

CDCl₃   QE-300

$CH_3-\overset{O}{\overset{\|}{C}}-OCHCH_2-\overset{CH_3}{\overset{|}{\underset{CH_3}{\overset{+}{N}}}}CH_3 \quad Cl^-$
with CH₃ on the OCH carbon

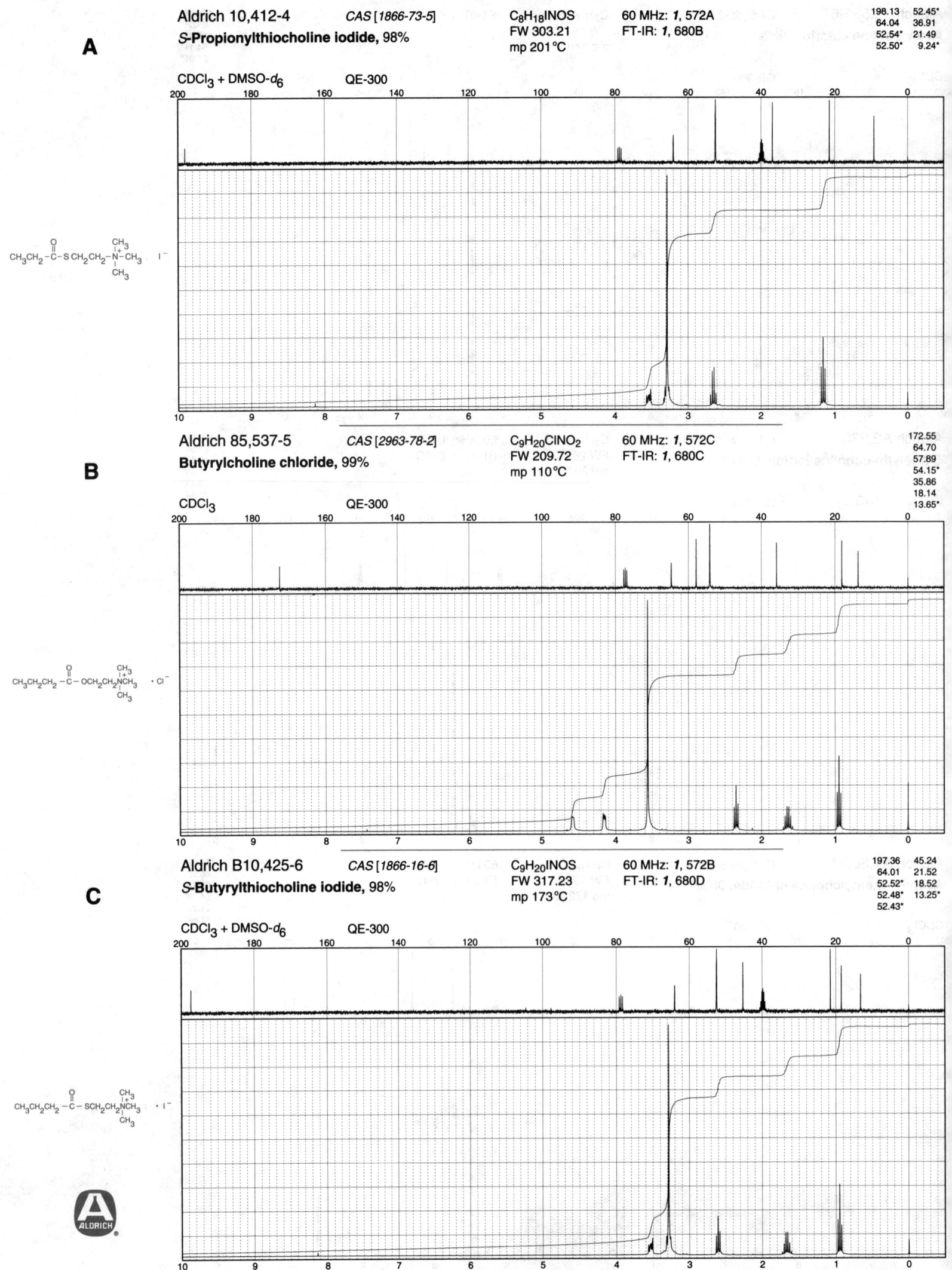

Aldrich 10,412-4   *CAS [1866-73-5]* $C_8H_{18}INOS$ 60 MHz: *1*, 572A   198.13   52.45*
S-Propionylthiocholine iodide, 98%   FW 303.21   FT-IR: *1*, 680B   64.04   36.91
mp 201 °C   52.54*   21.49
52.50*   9.24*

CDCl$_3$ + DMSO-d_6 QE-300

B

Aldrich 85,537-5   *CAS [2963-78-2]* $C_9H_{20}ClNO_2$ 60 MHz: *1*, 572C   172.55
Butyrylcholine chloride, 99%   FW 209.72   FT-IR: *1*, 680C   64.70
mp 110 °C   57.89
54.15*
35.86
18.14
13.65*

CDCl$_3$ QE-300

C

Aldrich B10,425-6   *CAS [1866-16-6]* $C_9H_{20}INOS$ 60 MHz: *1*, 572B   197.36   45.24
S-Butyrylthiocholine iodide, 98%   FW 317.23   FT-IR: *1*, 680D   64.01   21.52
mp 173 °C   52.52*   18.52
52.48*   13.25*
52.43*

CDCl$_3$ + DMSO-d_6 QE-300

Non-Aromatic Esters and Lactones

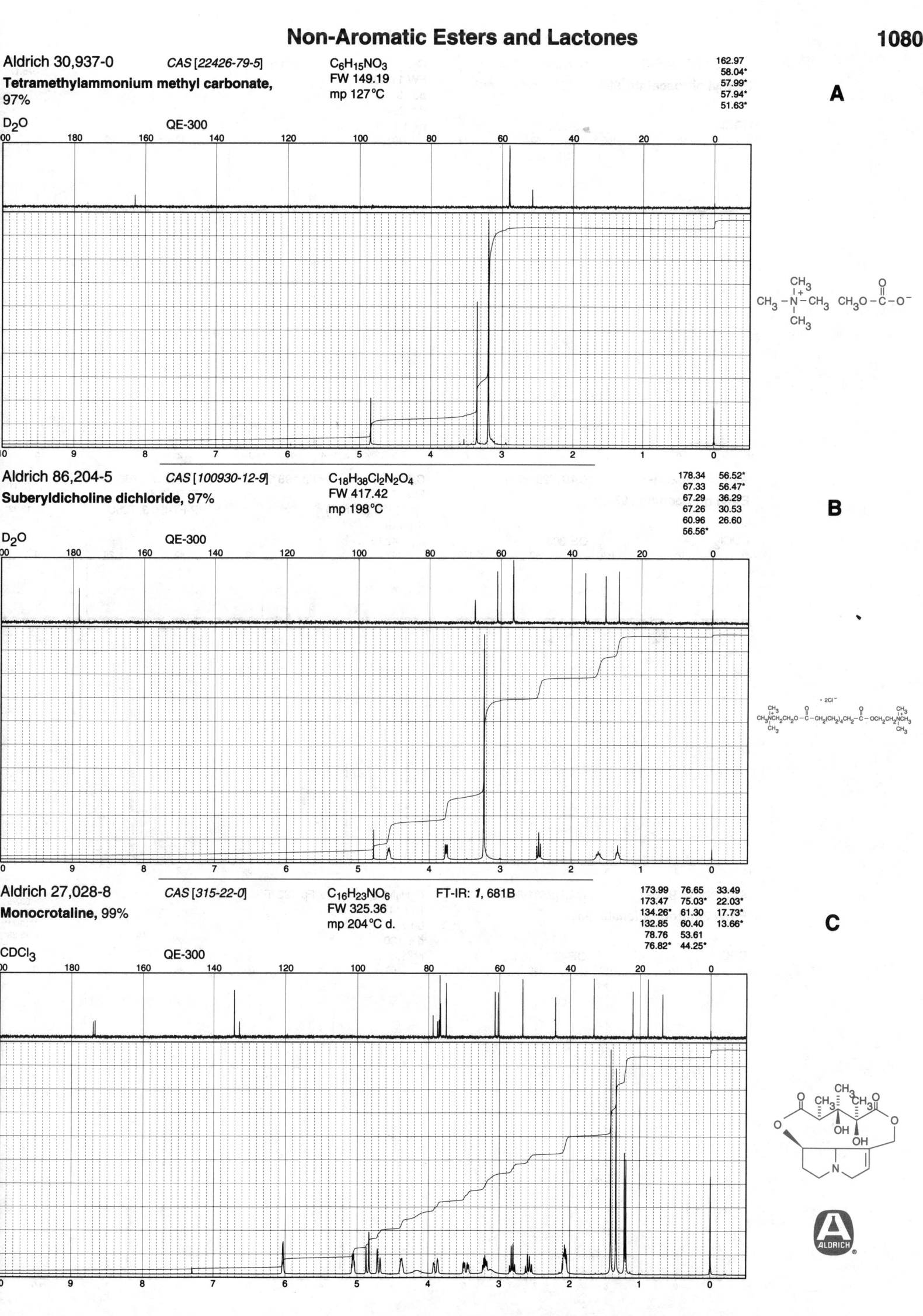

Aldrich 30,937-0   *CAS [22426-79-5]* $C_6H_{15}NO_3$
Tetramethylammonium methyl carbonate, 97%   FW 149.19   mp 127°C

162.97	
58.04*	
57.99*	
57.94*	
51.63*	

A

D_2O QE-300

Aldrich 86,204-5   *CAS [100930-12-9]* $C_{18}H_{38}Cl_2N_2O_4$
Suberyldicholine dichloride, 97%   FW 417.42   mp 198°C

178.34	56.52*
67.33	56.47*
67.29	36.29
67.26	30.53
60.96	26.60
56.56*	

B

D_2O QE-300

Aldrich 27,028-8   *CAS [315-22-0]* $C_{16}H_{23}NO_6$ FT-IR: *1*, 681B
Monocrotaline, 99%   FW 325.36   mp 204°C d.

173.99	76.65	33.49
173.47	75.03*	22.03*
134.26*	61.30	17.73*
132.85	60.40	13.66*
78.76	53.61	
76.82*	44.25*	

C

$CDCl_3$ QE-300

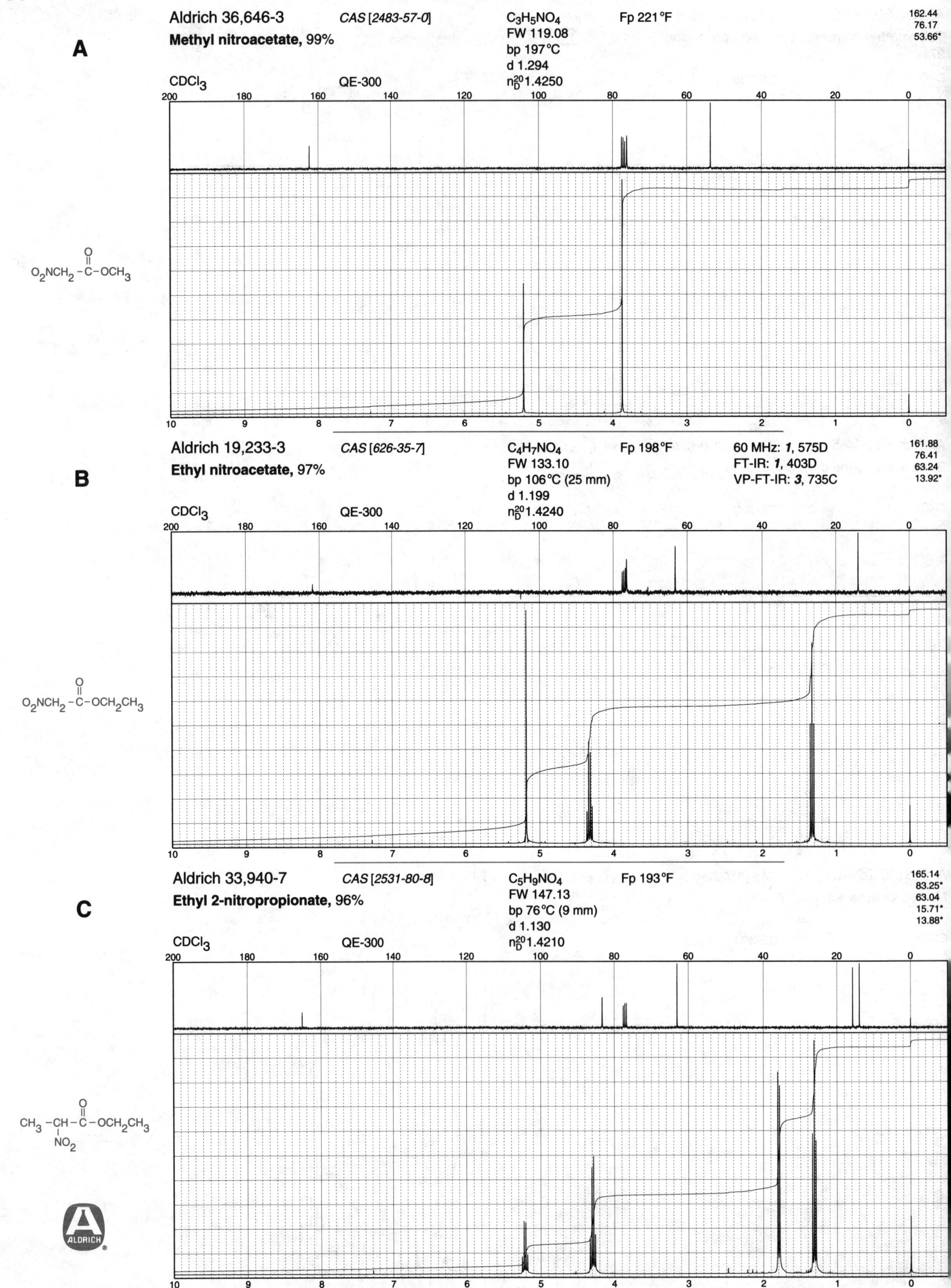

A

Aldrich 36,646-3   *CAS [2483-57-0]*   C₃H₅NO₄   Fp 221°F

Methyl nitroacetate, 99%

$C_3H_5NO_4$
FW 119.08
bp 197°C
d 1.294
n_D^{20} 1.4250

162.44
76.17
53.66*

CDCl₃   QE-300

$O_2NCH_2-\overset{\displaystyle O}{\overset{\|}{C}}-OCH_3$

B

Aldrich 19,233-3   *CAS [626-35-7]*   C₄H₇NO₄   Fp 198°F

Ethyl nitroacetate, 97%

$C_4H_7NO_4$
FW 133.10
bp 106°C (25 mm)
d 1.199
n_D^{20} 1.4240

60 MHz: *1*, 575D
FT-IR: *1*, 403D
VP-FT-IR: *3*, 735C

161.88
76.41
63.24
13.92*

CDCl₃   QE-300

$O_2NCH_2-\overset{\displaystyle O}{\overset{\|}{C}}-OCH_2CH_3$

C

Aldrich 33,940-7   *CAS [2531-80-8]*   C₅H₉NO₄   Fp 193°F

Ethyl 2-nitropropionate, 96%

$C_5H_9NO_4$
FW 147.13
bp 76°C (9 mm)
d 1.130
n_D^{20} 1.4210

165.14
83.25*
63.04
15.71*
13.88*

CDCl₃   QE-300

$CH_3-\underset{\underset{\displaystyle NO_2}{|}}{CH}-\overset{\displaystyle O}{\overset{\|}{C}}-OCH_2CH_3$

Non-Aromatic Esters and Lactones

Aldrich 31,687-3   CAS [20497-95-4]   C4H7NO4   Fp >230°F   VP-FT-IR: 3, 735D

Methyl 3-nitropropionate, 98%

FW 133.10
bp 68°C (1 mm)
d 1.249
nD20 1.4340

169.97
69.73
52.39*
30.85

A

CDCl3   QE-300

$O_2NCH_2CH_2-\overset{\overset{\displaystyle O}{\|}}{C}-OCH_3$

Aldrich 37,364-8   CAS [2531-81-9]   C6H11NO4   Fp 186°F

Ethyl 2-nitrobutyrate, 96%

FW 161.16
bp 83°C (8 mm)
d 1.096
nD20 1.4230

164.47
89.43*
62.91
23.93
13.91*
10.17*

B

CDCl3   QE-300

$CH_3CH_2\underset{\underset{\displaystyle NO_2}{|}}{CH}-\overset{\overset{\displaystyle O}{\|}}{C}-OCH_2CH_3$

Aldrich 38,848-3   CAS [16507-02-1]   C7H13NO4   Fp >230°F

Methyl 4-methyl-4-nitropentanoate, 96%

FW 175.18
bp 79°C (1 mm)
d 1.114
nD20 1.4410

172.53
87.26
51.88*
35.32
29.05
25.73*

C

CDCl3   QE-300

$CH_3-\underset{\underset{\displaystyle NO_2}{|}}{\overset{\overset{\displaystyle CH_3}{|}}{C}}-CH_2CH_2-\overset{\overset{\displaystyle O}{\|}}{C}-OCH_3$

1082

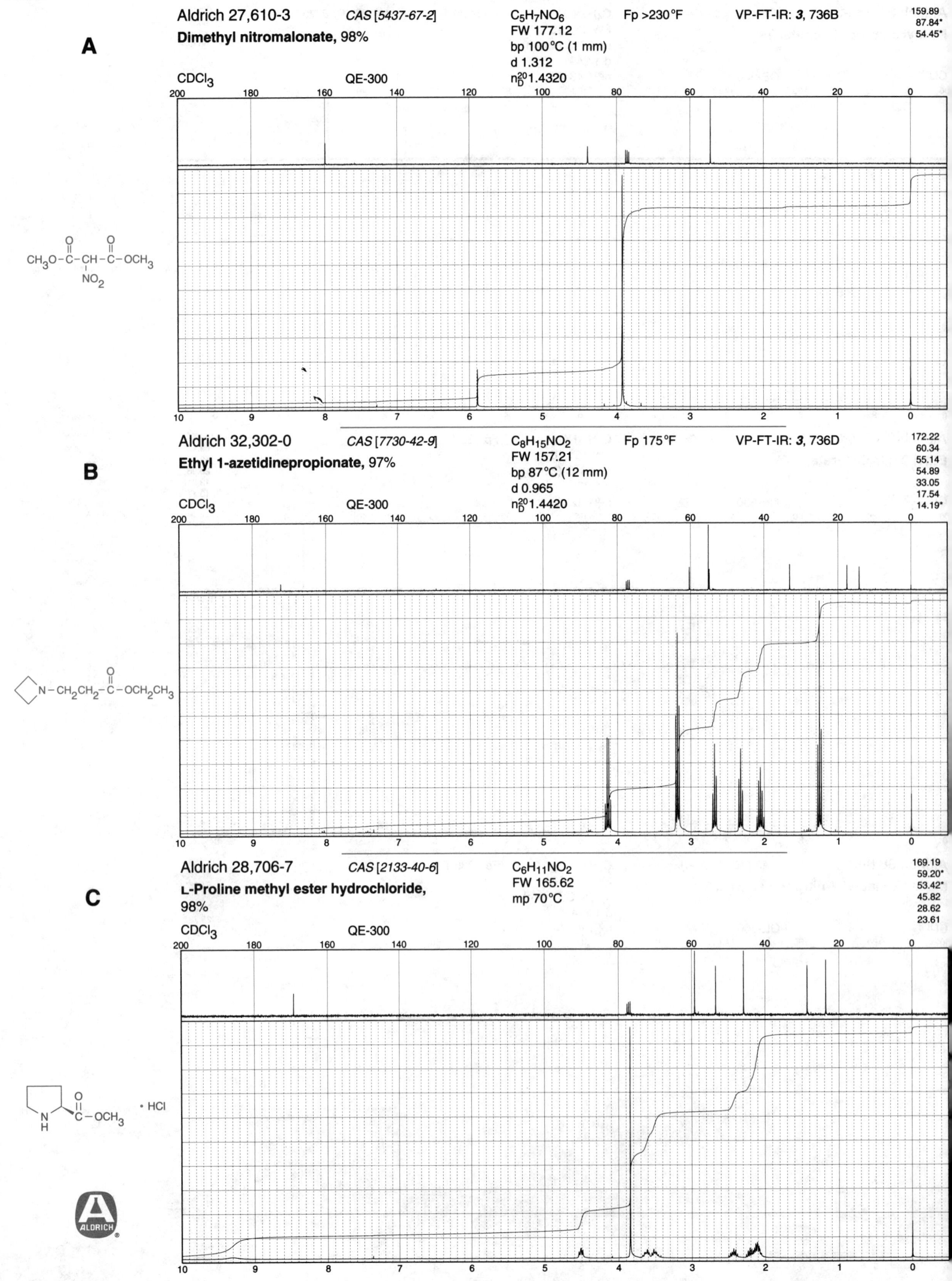

A

Aldrich 27,610-3   CAS [5437-67-2] $C_5H_7NO_6$ Fp >230°F   VP-FT-IR: **3**, 736B   159.89 / 87.84* / 54.45*

Dimethyl nitromalonate, 98%   FW 177.12   bp 100°C (1 mm)   d 1.312 n_D^{20} 1.4320

$CDCl_3$ QE-300

$CH_3O-\overset{O}{\overset{\|}{C}}-\overset{}{\underset{\underset{NO_2}{|}}{CH}}-\overset{O}{\overset{\|}{C}}-OCH_3$

B

Aldrich 32,302-0   CAS [7730-42-9] $C_8H_{15}NO_2$ Fp 175°F   VP-FT-IR: **3**, 736D   172.22 / 60.34 / 55.14 / 54.89 / 33.05 / 17.54 / 14.19*

Ethyl 1-azetidinepropionate, 97%   FW 157.21   bp 87°C (12 mm)   d 0.965 n_D^{20} 1.4420

$CDCl_3$ QE-300

$\square N-CH_2CH_2-\overset{O}{\overset{\|}{C}}-OCH_2CH_3$

C

Aldrich 28,706-7   CAS [2133-40-6] $C_6H_{11}NO_2$ 169.19 / 59.20* / 53.42* / 45.82 / 28.62 / 23.61

L-Proline methyl ester hydrochloride, 98%   FW 165.62   mp 70°C

$CDCl_3$ QE-300

$\underset{H}{\overset{}{N}}\overset{O}{\overset{\|}{C}}-OCH_3$ · HCl

ALDRICH

Aldrich 26,375-3 *CAS [22041-19-6]*
Ethyl 1-pyrrolidineacetate, 97%

CDCl₃　　　　QE-300

$C_8H_{15}NO_2$
FW 157.22
bp 60°C (2 mm)
d 0.982
n_D^{20} 1.4470

Fp 176°F

FT-IR: *1*, 673A
VP-FT-IR: *3*, 732D

170.74
60.49
57.08
53.99
23.76
14.25*

A

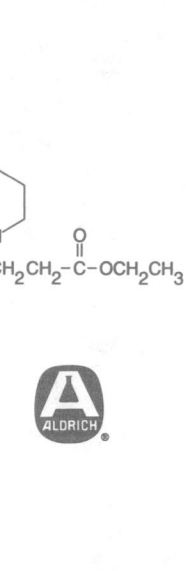

Aldrich 37,939-5 *CAS [23853-10-3]*
Ethyl 1-piperidineacetate, 98%

CDCl₃　　　　QE-300

$C_9H_{17}NO_2$
FW 171.24
bp 110°C (25 mm)
d 0.984
n_D^{20} 1.4530

Fp 196°F

170.55
60.41
60.38
54.29
25.82
23.91
14.27*

B

Aldrich 19,601-0 *CAS [19653-33-9]*
Ethyl 1-piperidinepropionate, 97%

CDCl₃　　　　QE-300

$C_{10}H_{19}NO_2$
FW 185.27
bp 218°C
d 0.927
n_D^{20} 1.4540

Fp 190°F

60 MHz: *1*, 573C
FT-IR: *1*, 682A
VP-FT-IR: *3*, 737A

172.71　32.34
60.29　25.98
54.32　24.31
54.26　14.23*

C

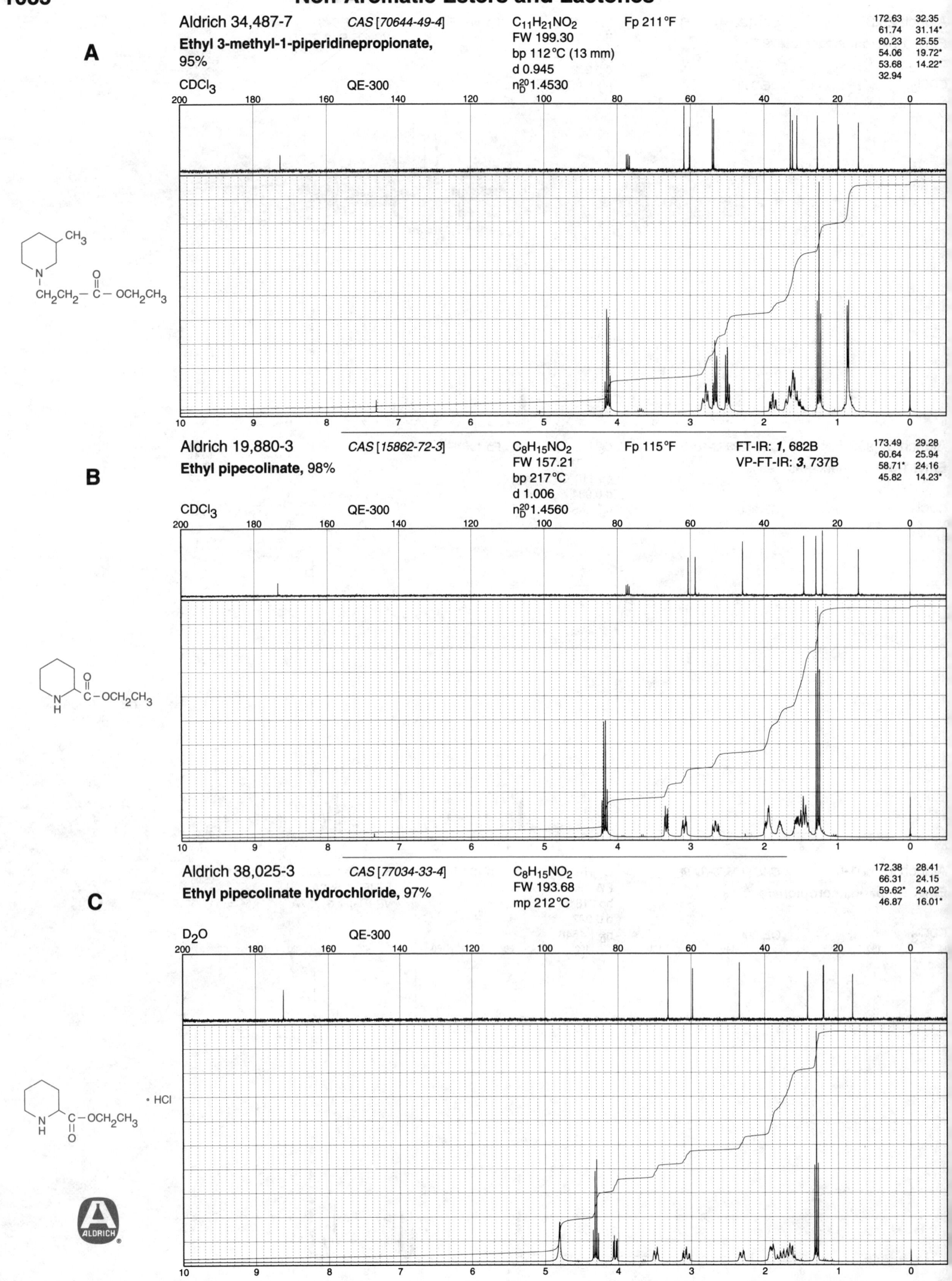

A

Aldrich 34,487-7   CAS [70644-49-4] $C_{11}H_{21}NO_2$ Fp 211°F

Ethyl 3-methyl-1-piperidinepropionate,
95%

FW 199.30
bp 112°C (13 mm)
d 0.945
n_D^{20} 1.4530

$CDCl_3$ QE-300

172.63	32.35
61.74	31.14*
60.23	25.55
54.06	19.72*
53.68	14.22*
32.94	

B

Aldrich 19,880-3   CAS [15862-72-3] $C_8H_{15}NO_2$ Fp 115°F   FT-IR: **1**, 682B

Ethyl pipecolinate, 98%

FW 157.21
bp 217°C
d 1.006
n_D^{20} 1.4560

VP-FT-IR: **3**, 737B

$CDCl_3$ QE-300

173.49	29.28
60.64	25.94
58.71*	24.16
45.82	14.23*

C

Aldrich 38,025-3   CAS [77034-33-4] $C_8H_{15}NO_2$

Ethyl pipecolinate hydrochloride, 97%

FW 193.68
mp 212°C

D_2O QE-300

172.38	28.41
66.31	24.15
59.62*	24.02
46.87	16.01*

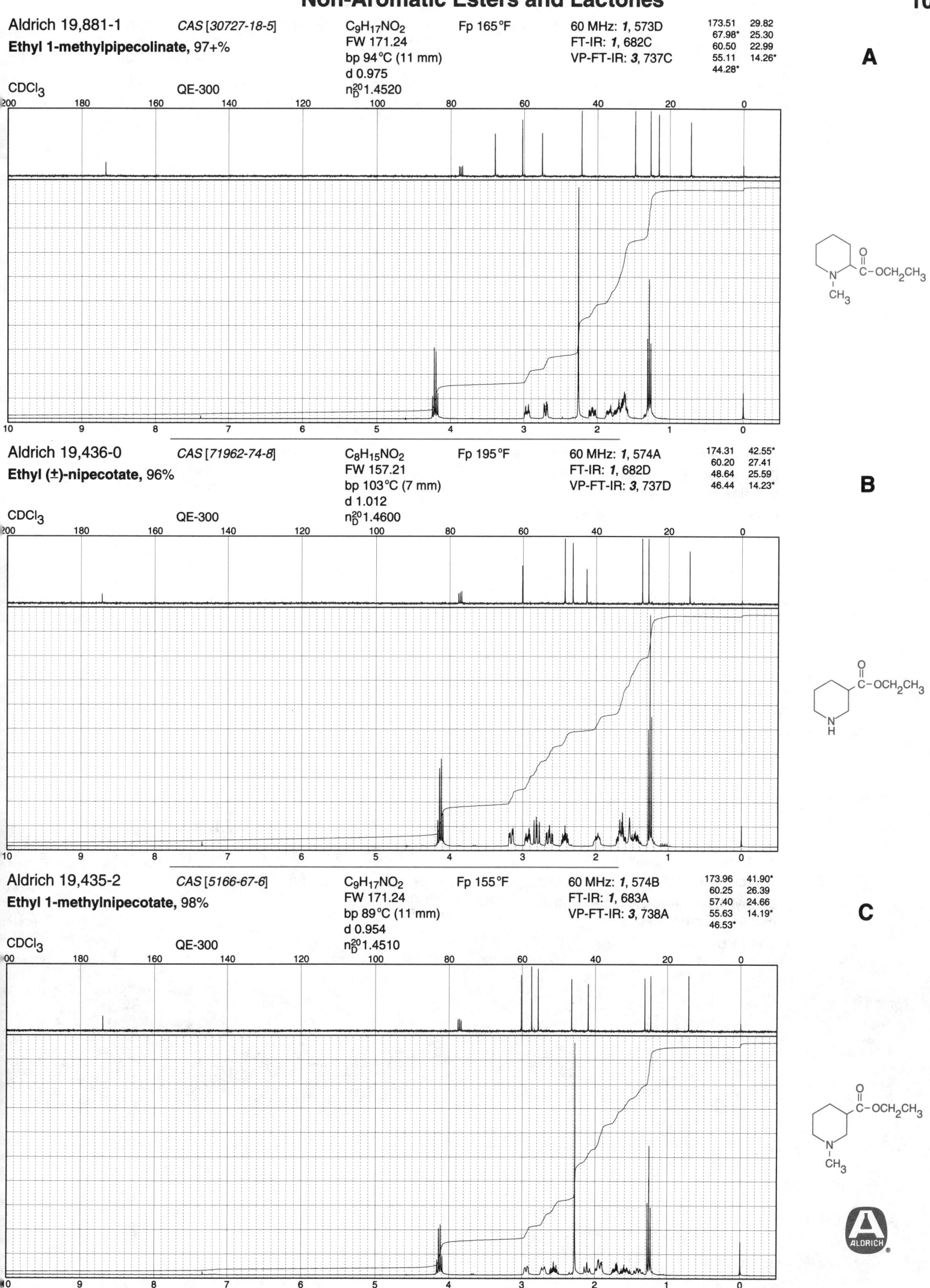

Aldrich 19,881-1 *CAS [30727-18-5]*
Ethyl 1-methylpipecolinate, 97+%

$C_9H_{17}NO_2$
FW 171.24
bp 94°C (11 mm)
d 0.975
n_D^{20} 1.4520

Fp 165°F

60 MHz: *1*, 573D
FT-IR: *1*, 682C
VP-FT-IR: *3*, 737C

173.51 29.82
67.98* 25.30
60.50 22.99
55.11 14.26*
44.28*

A

CDCl₃   QE-300

Aldrich 19,436-0 *CAS [71962-74-8]*
Ethyl (±)-nipecotate, 96%

$C_8H_{15}NO_2$
FW 157.21
bp 103°C (7 mm)
d 1.012
n_D^{20} 1.4600

Fp 195°F

60 MHz: *1*, 574A
FT-IR: *1*, 682D
VP-FT-IR: *3*, 737D

174.31 42.55*
60.20 27.41
48.64 25.59
46.44 14.23*

B

CDCl₃   QE-300

Aldrich 19,435-2 *CAS [5166-67-6]*
Ethyl 1-methylnipecotate, 98%

$C_9H_{17}NO_2$
FW 171.24
bp 89°C (11 mm)
d 0.954
n_D^{20} 1.4510

Fp 155°F

60 MHz: *1*, 574B
FT-IR: *1*, 683A
VP-FT-IR: *3*, 738A

173.96 41.90*
60.25 26.39
57.40 24.66
55.63 14.19*
46.53*

C

CDCl₃   QE-300

ALDRICH

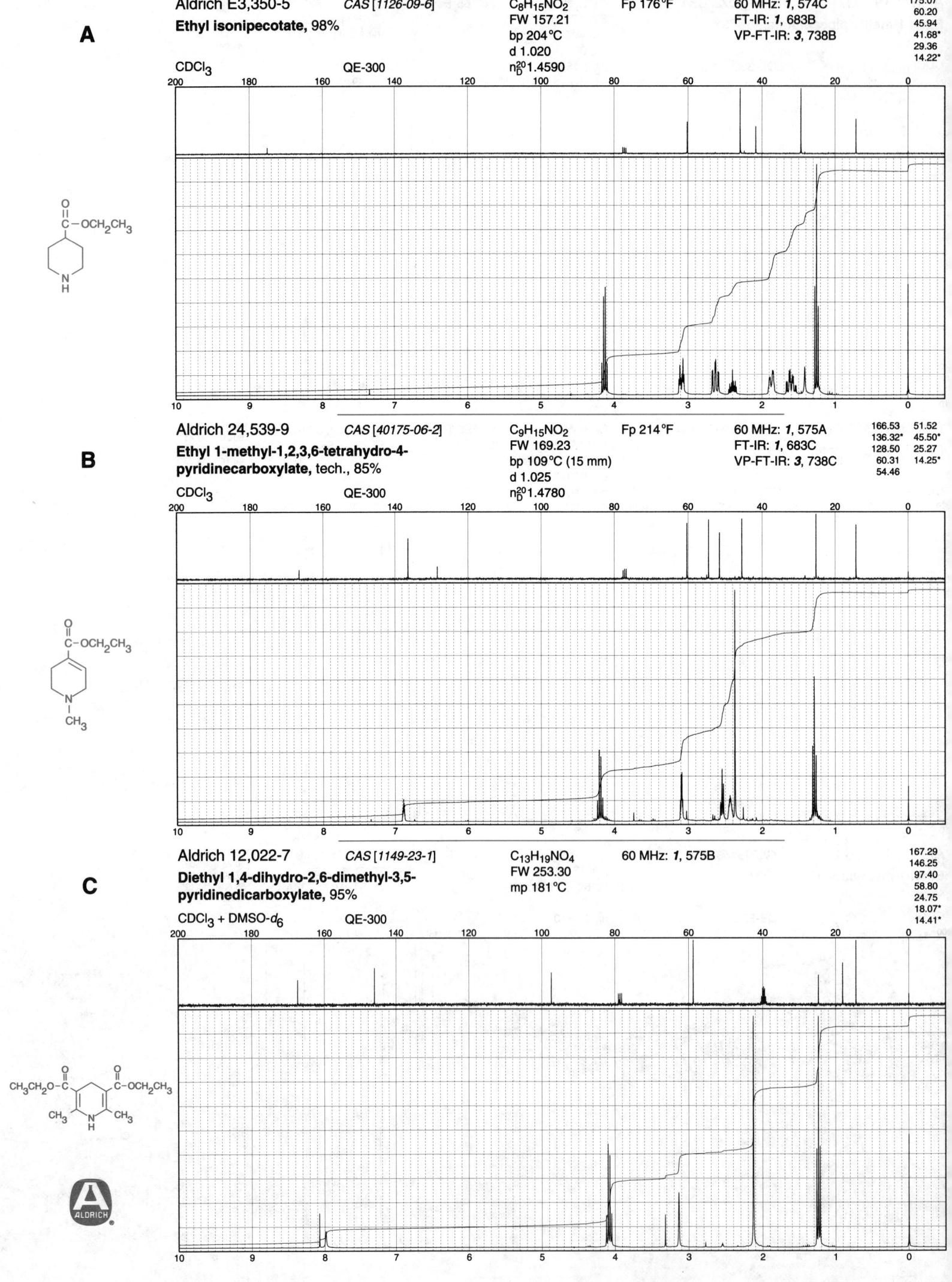

A

Aldrich E3,350-5   *CAS [1126-09-6]* $C_8H_{15}NO_2$ Fp 176°F   60 MHz: *1*, 574C

Ethyl isonipecotate, 98%   FW 157.21   FT-IR: *1*, 683B

bp 204°C   VP-FT-IR: *3*, 738B

CDCl₃   QE-300   d 1.020

n_D^{20} 1.4590

175.07
60.20
45.94
41.68*
29.36
14.22*

B

Aldrich 24,539-9   *CAS [40175-06-2]* $C_9H_{15}NO_2$ Fp 214°F   60 MHz: *1*, 575A

Ethyl 1-methyl-1,2,3,6-tetrahydro-4-   FW 169.23   FT-IR: *1*, 683C

pyridinecarboxylate, tech., 85%   bp 109°C (15 mm)   VP-FT-IR: *3*, 738C

CDCl₃   QE-300   d 1.025

n_D^{20} 1.4780

166.53   51.52
136.32*   45.50*
128.50   25.27
60.31   14.25*
54.46

C

Aldrich 12,022-7   *CAS [1149-23-1]* $C_{13}H_{19}NO_4$ 60 MHz: *1*, 575B

Diethyl 1,4-dihydro-2,6-dimethyl-3,5-   FW 253.30

pyridinedicarboxylate, 95%   mp 181°C

CDCl₃ + DMSO-d_6 QE-300

167.29
146.25
97.40
58.80
24.75
18.07*
14.41*

Non-Aromatic Esters and Lactones

A

Aldrich 13,703-0   CAS [632-93-9]
Diethyl 1,4-dihydro-2,4,6-trimethyl-3,5-pyridinedicarboxylate, 99%
$C_{14}H_{21}NO_4$
FW 267.33
mp 131°C
60 MHz: *1*, 575C
FT-IR: *1*, 681D

167.93	28.50*
144.63	22.25*
104.38	19.34*
59.57	14.44*

CDCl$_3$ QE-300

B

Aldrich 33,812-5   CAS [60456-26-0]
(R)-(-)-Glycidyl butyrate, 96%
$C_7H_{12}O_3$
FW 144.17
bp 90°C (19 mm)
d 1.018
n_D^{20} 1.4280
Fp 185°F

| 173.26 |
| 64.73 |
| 49.38* |
| 44.63 |
| 35.89 |
| 18.36 |
| 13.64* |

CDCl$_3$ QE-300

C

Aldrich 29,065-3   CAS [106-90-1]
Glycidyl acrylate, tech., 90%
$C_6H_8O_3$
FW 128.13
bp 115°C (78 mm)
d 1.099
n_D^{20} 1.4490
Fp 169°F
VP-FT-IR: *3*, 738D

| 165.71 |
| 131.48 |
| 127.86* |
| 65.09 |
| 49.32* |
| 44.63 |

CDCl$_3$ QE-300

A

Aldrich 15,123-8   *CAS [106-91-2]*

Glycidyl methacrylate, 97%

$C_7H_{10}O_3$
FW 142.15
bp 189°C
d 1.042
n_D^{20}1.4490

Fp 169°F

60 MHz: *1*, 563C
FT-IR: *1*, 668D
VP-FT-IR: *3*, 726C

166.93
135.82
126.13
65.17
49.38*
44.59
18.28*

CDCl3   QE-300

$H_2C=C-C-OCH_2CH-CH_2$
CH_3

B

Aldrich W30,550-2   *CAS [637-64-9]*

Tetrahydrofurfuryl acetate, 97+%

$C_7H_{12}O_3$
FW 144.17
bp 195°C (753 mm)
d 1.061
n_D^{20}1.4370

Fp 184°F

VP-FT-IR: *3*, 726D

170.92
76.45*
68.40
66.55
27.96
25.65
20.91*

CDCl3   QE-300

$CH_3-C-O-CH_2$

C

Aldrich W30,580-4   *CAS [637-65-0]*

dl-Tetrahydrofurfuryl propionate, 97+%

$C_8H_{14}O_3$
FW 158.20
bp 207°C
d 1.040
n_D^{20}1.4380

Fp 198°F

174.35   27.99
76.53*   27.45
68.41   25.67
66.39   9.09*

CDCl3   QE-300

$CH_3CH_2-C-O-CH_2$

ALDRICH

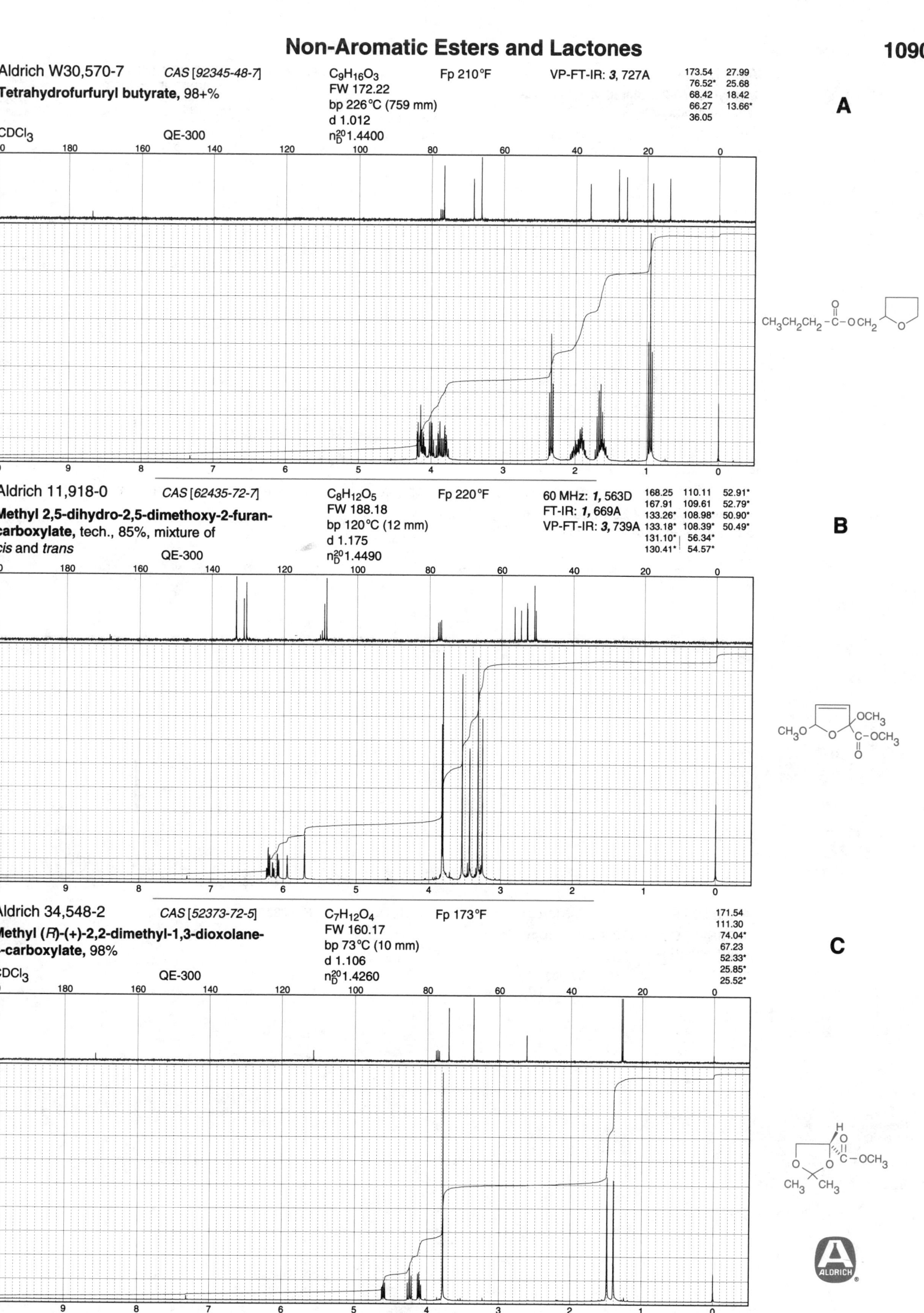

Aldrich W30,570-7   CAS [92345-48-7]   C₉H₁₆O₃   Fp 210°F   VP-FT-IR: 3, 727A
Tetrahydrofurfuryl butyrate, 98+%   FW 172.22
bp 226°C (759 mm)
d 1.012
CDCl₃   QE-300   n₂₀ᴰ 1.4400

173.54 27.99
76.52* 25.68
68.42 18.42
66.27 13.66*
36.05

A

Aldrich 11,918-0   CAS [62435-72-7]   C₈H₁₂O₅   Fp 220°F   60 MHz: 1, 563D
Methyl 2,5-dihydro-2,5-dimethoxy-2-furan-   FW 188.18   FT-IR: 1, 669A
carboxylate, tech., 85%, mixture of   bp 120°C (12 mm)   VP-FT-IR: 3, 739A
cis and *trans*   d 1.175
QE-300   n₂₀ᴰ 1.4490

168.25 110.11 52.91*
167.91 109.61 52.79*
133.26* 108.98* 50.90*
133.18* 108.39* 50.49*
131.10* 56.34*
130.41* 54.57*

B

Aldrich 34,548-2   CAS [52373-72-5]   C₇H₁₂O₄   Fp 173°F
Methyl (R)-(+)-2,2-dimethyl-1,3-dioxolane-   FW 160.17
4-carboxylate, 98%   bp 73°C (10 mm)
d 1.106
CDCl₃   QE-300   n₂₀ᴰ 1.4260

171.54
111.30
74.04*
67.23
52.33*
25.85*
25.52*

C

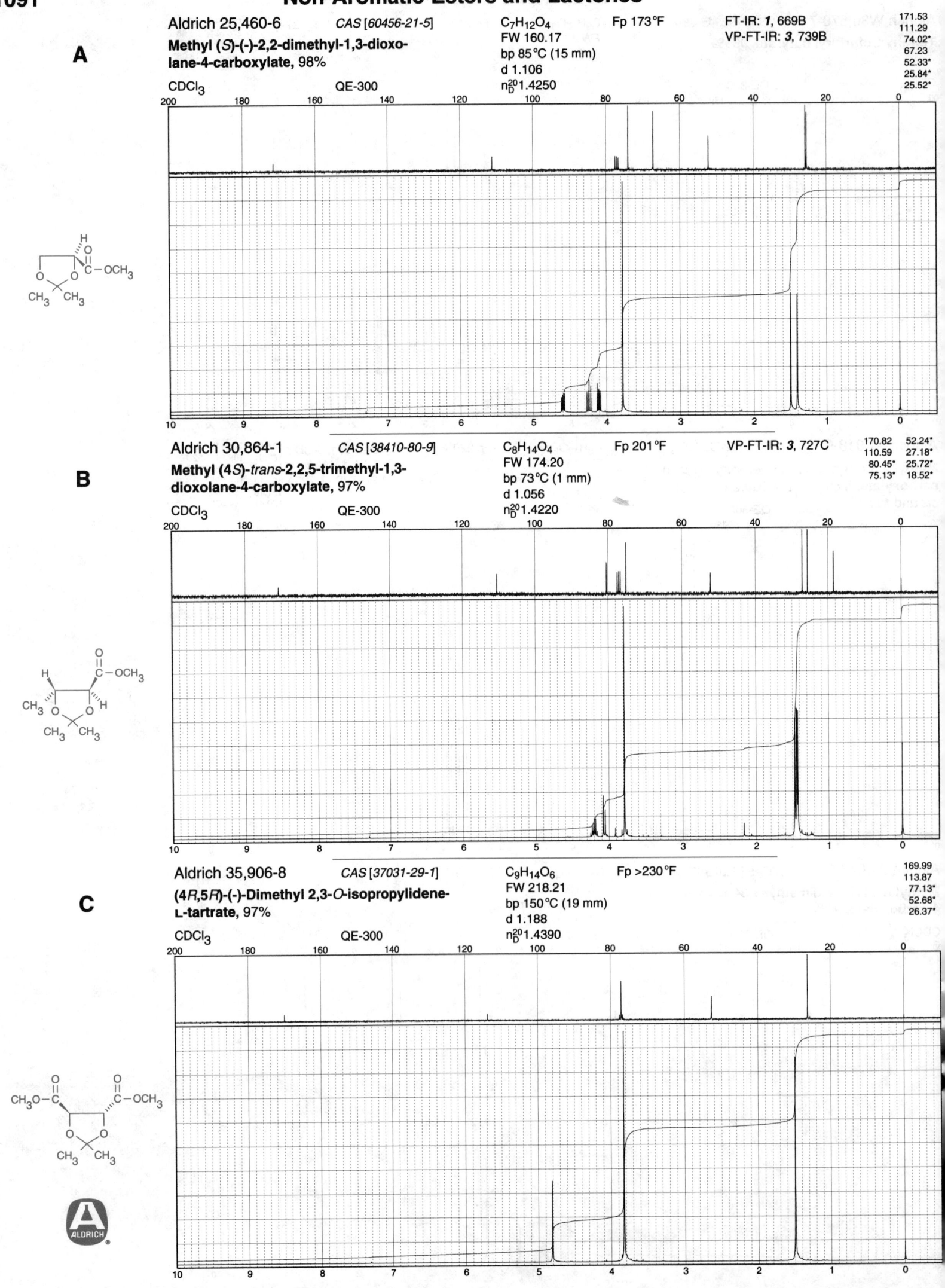

A

Aldrich 25,460-6   *CAS [60456-21-5]*

Methyl (S)-(-)-2,2-dimethyl-1,3-dioxo-lane-4-carboxylate, 98%

CDCl₃   QE-300

$C_7H_{12}O_4$ Fp 173°F   FT-IR: *1*, 669B
FW 160.17   VP-FT-IR: *3*, 739B
bp 85°C (15 mm)
d 1.106
n_D^{20}1.4250

171.53
111.29
74.02*
67.23
52.33*
25.84*
25.52*

B

Aldrich 30,864-1   *CAS [38410-80-9]*

Methyl (4S)-*trans*-2,2,5-trimethyl-1,3-dioxolane-4-carboxylate, 97%

CDCl₃   QE-300

$C_8H_{14}O_4$ Fp 201°F   VP-FT-IR: *3*, 727C
FW 174.20
bp 73°C (1 mm)
d 1.056
n_D^{20}1.4220

170.82   52.24*
110.59   27.18*
80.45*   25.72*
75.13*   18.52*

C

Aldrich 35,906-8   *CAS [37031-29-1]*

(4R,5R)-(-)-Dimethyl 2,3-*O*-isopropylidene-L-tartrate, 97%

CDCl₃   QE-300

$C_9H_{14}O_6$ Fp >230°F
FW 218.21
bp 150°C (19 mm)
d 1.188
n_D^{20}1.4390

169.99
113.87
77.13*
52.68*
26.37*

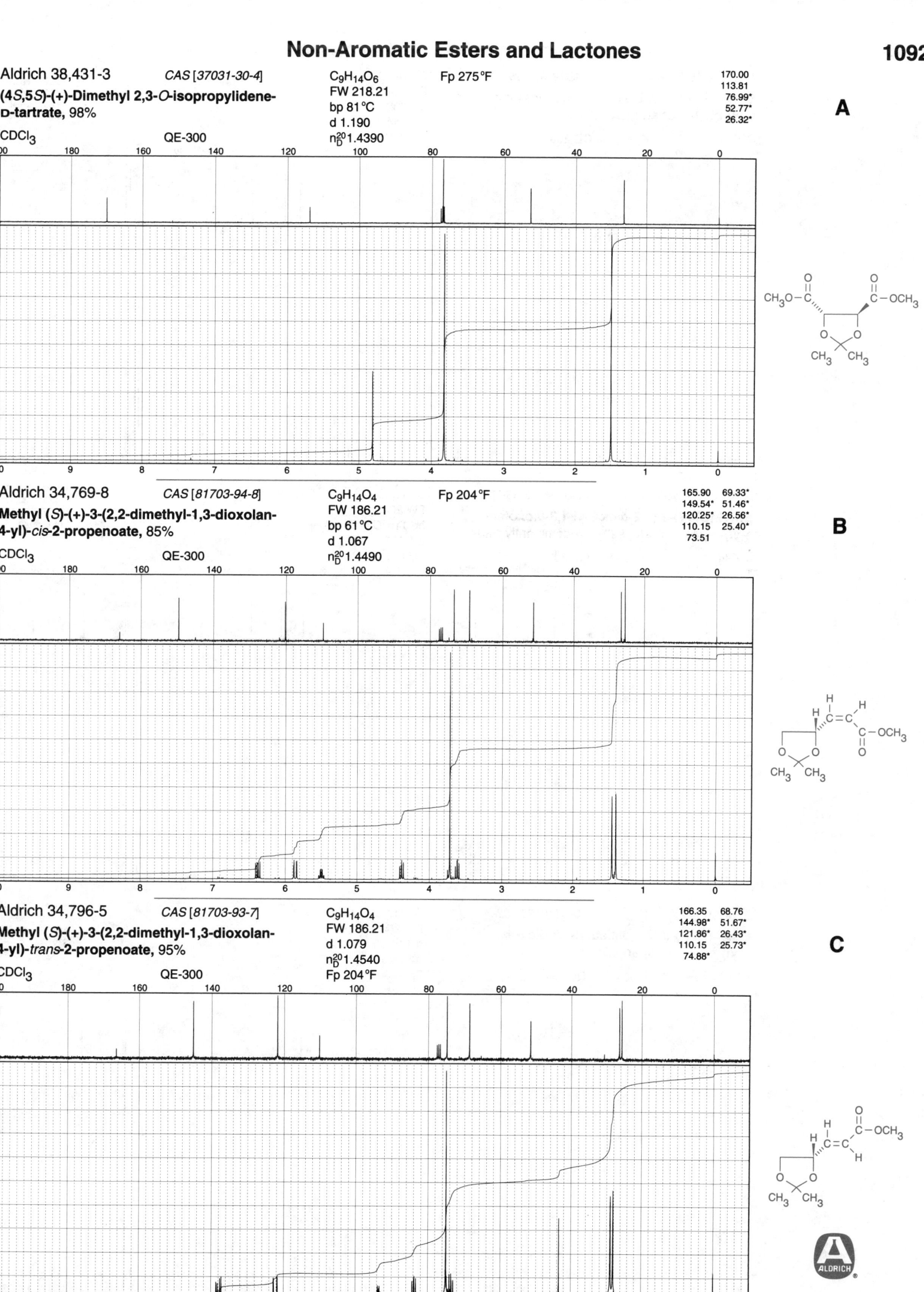

Aldrich 38,431-3   *CAS [37031-30-4]*

(4S,5S)-(+)-Dimethyl 2,3-O-isopropylidene-D-tartrate, 98%

CDCl₃   QE-300

C₉H₁₄O₆
FW 218.21
bp 81 °C
d 1.190
n²⁰_D 1.4390

Fp 275 °F

170.00
113.81
76.99*
52.77*
26.32*

A

Aldrich 34,769-8   *CAS [81703-94-8]*

Methyl (S)-(+)-3-(2,2-dimethyl-1,3-dioxolan-4-yl)-cis-2-propenoate, 85%

CDCl₃   QE-300

C₉H₁₄O₄
FW 186.21
bp 61 °C
d 1.067
n²⁰_D 1.4490

Fp 204 °F

165.90   69.33*
149.54*   51.46*
120.25*   26.56*
110.15   25.40*
73.51

B

Aldrich 34,796-5   *CAS [81703-93-7]*

Methyl (S)-(+)-3-(2,2-dimethyl-1,3-dioxolan-4-yl)-trans-2-propenoate, 95%

CDCl₃   QE-300

C₉H₁₄O₄
FW 186.21
d 1.079
n²⁰_D 1.4540
Fp 204 °F

166.35   68.76
144.98*   51.67*
121.86*   26.43*
110.15   25.73*
74.88*

C

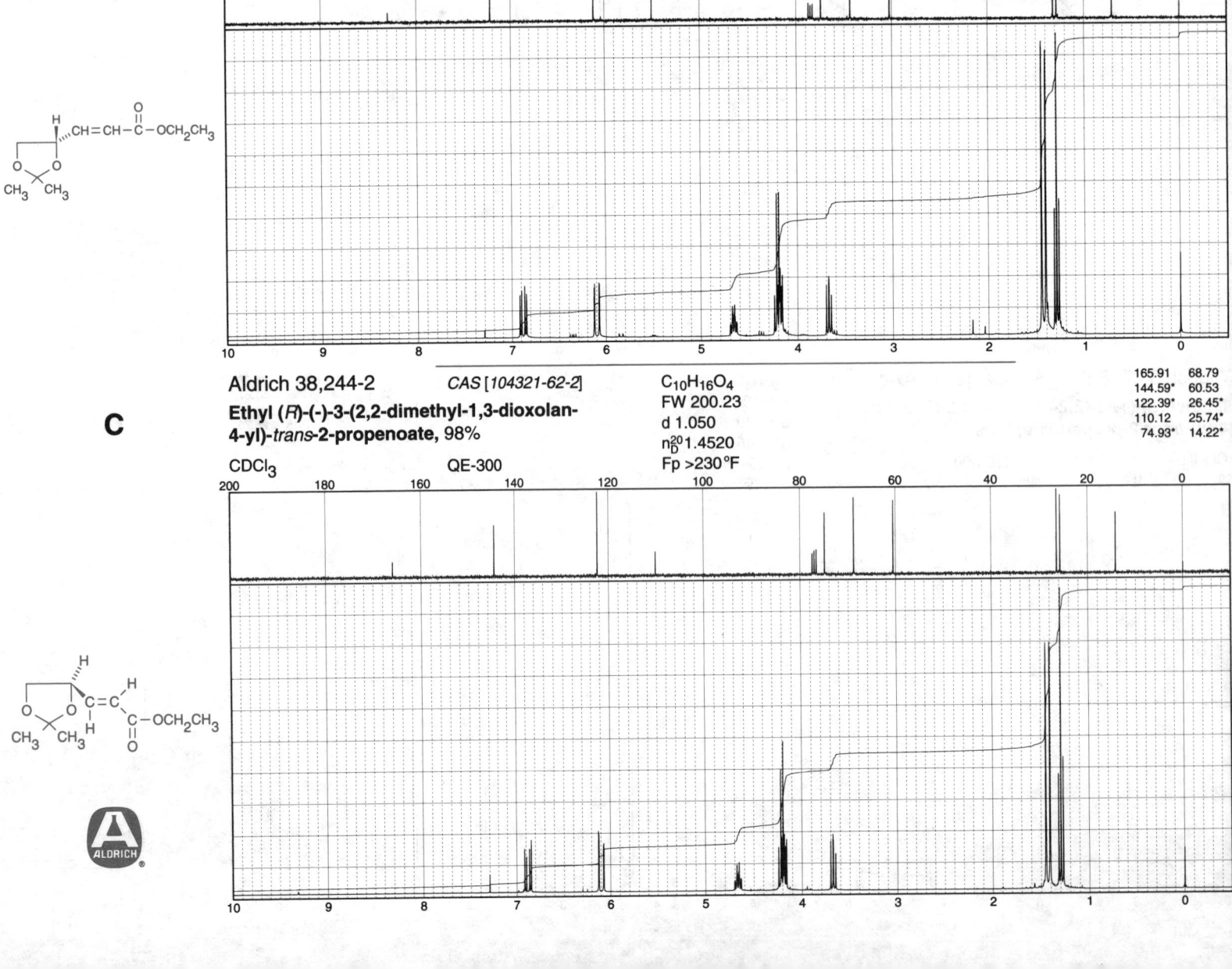

A

Aldrich 36,962-4   *CAS [91926-90-8]*

Ethyl (*S*)-(+)-3-(2,2-dimethyl-1,3-dioxolan-4-yl)-*cis*-2-propenoate, 99%

$C_{10}H_{16}O_4$
FW 200.23
d 1.038
n_D^{20}1.4470
Fp 207°F

CDCl₃   QE-300

165.55	69.38
149.22*	60.39
120.72*	26.56*
109.63	25.40*
73.53*	14.18*

B

Aldrich 34,681-0   *CAS [64520-58-7]*

Ethyl (*S*)-(+)-3-(2,2-dimethyl-1,3-dioxolan-4-yl)-2-propenoate, 99%, predominantly *trans*

$C_{10}H_{16}O_4$ Fp >230°F
FW 200.23
bp 116°C (12 mm)
d 1.035
n_D^{20}1.4505

CDCl₃   QE-300

165.90	68.78
144.58*	60.52
122.37*	26.45*
110.11	25.73*
74.92*	14.21*

C

Aldrich 38,244-2   *CAS [104321-62-2]*

Ethyl (*R*)-(−)-3-(2,2-dimethyl-1,3-dioxolan-4-yl)-*trans*-2-propenoate, 98%

$C_{10}H_{16}O_4$
FW 200.23
d 1.050
n_D^{20}1.4520
Fp >230°F

CDCl₃   QE-300

165.91	68.79
144.59*	60.53
122.39*	26.45*
110.12	25.74*
74.93*	14.22*

ALDRICH

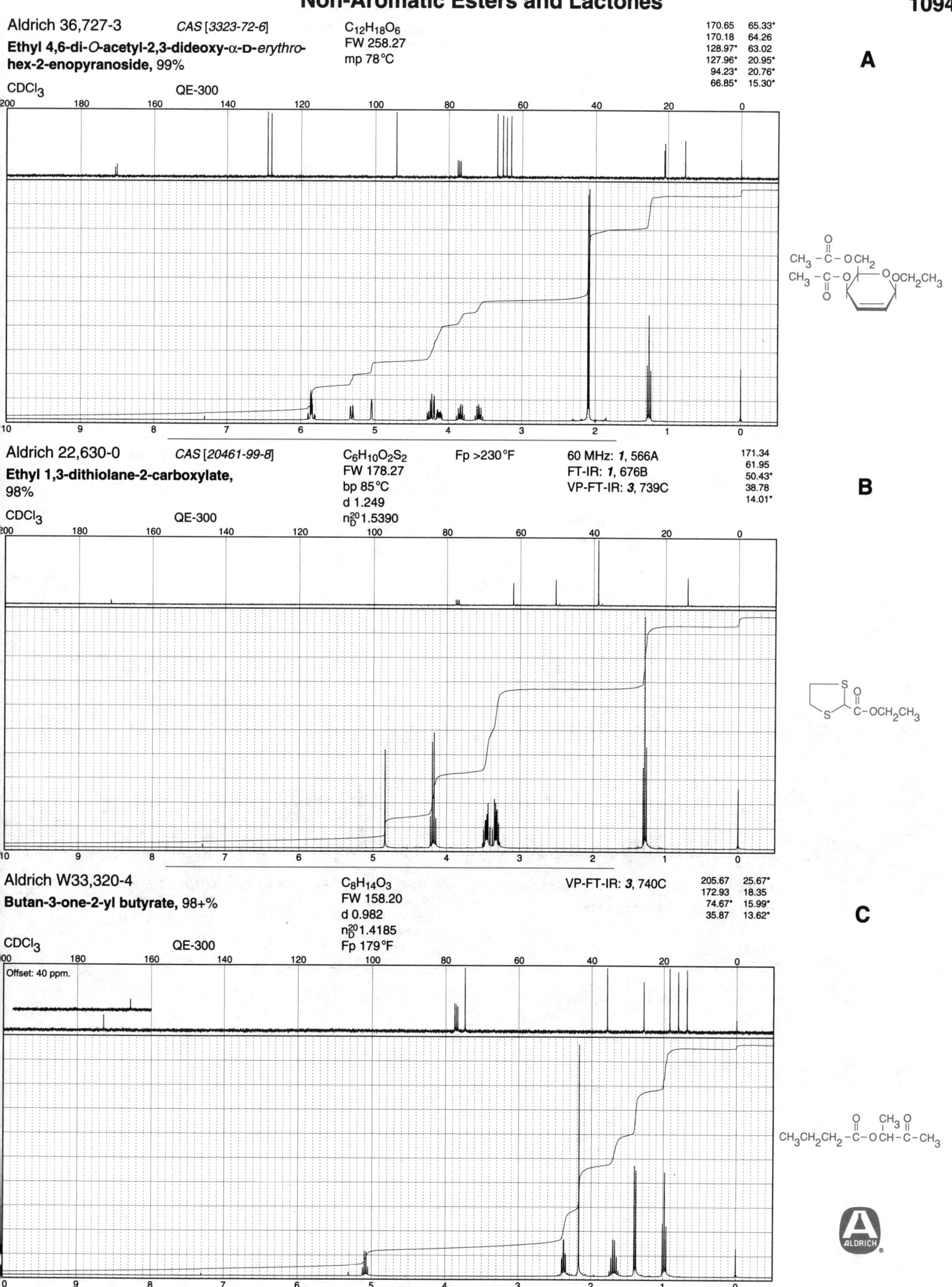

Aldrich 36,727-3   *CAS [3323-72-6]*

Ethyl 4,6-di-O-acetyl-2,3-dideoxy-α-D-erythro-hex-2-enopyranoside, 99%

CDCl₃   QE-300

$C_{12}H_{18}O_6$
FW 258.27
mp 78 °C

170.65	65.33*
170.18	64.26
128.97*	63.02
127.96*	20.95*
94.23*	20.76*
66.85*	15.30*

A

Aldrich 22,630-0   *CAS [20461-99-8]*

Ethyl 1,3-dithiolane-2-carboxylate, 98%

CDCl₃   QE-300

$C_6H_{10}O_2S_2$
FW 178.27
bp 85 °C
d 1.249
n_D^{20} 1.5390

Fp >230 °F

60 MHz: *1*, 566A
FT-IR: *1*, 676B
VP-FT-IR: *3*, 739C

171.34
61.95
50.43*
38.78
14.01*

B

Aldrich W33,320-4

Butan-3-one-2-yl butyrate, 98+%

CDCl₃   QE-300

$C_8H_{14}O_3$
FW 158.20
d 0.982
n_D^{20} 1.4185
Fp 179 °F

VP-FT-IR: *3*, 740C

205.67	25.67*
172.93	18.35
74.67*	15.99*
35.87	13.62*

C

Offset: 40 ppm.

Non-Aromatic Esters and Lactones

A

Aldrich M7,980-8   *CAS [600-22-6]*   C$_4$H$_6$O$_3$ Fp 103°F   60 MHz: *1*, 576A
Methyl pyruvate, 98%   FW 102.09   FT-IR: *1*, 684A
                                                 bp 136°C   VP-FT-IR: *3*, 740B
                                                 d 1.130
                                                 n$_D^{20}$1.4060

191.54
161.14
52.99*
26.70*

CDCl$_3$ QE-300

CH$_3$-C-C-OCH$_3$ (with two =O groups)

B

Aldrich E4,780-8   *CAS [617-35-6]*   C$_5$H$_8$O$_3$ Fp 114°F   60 MHz: *1*, 576B
Ethyl pyruvate, 98%   FW 116.12   FT-IR: *1*, 684B
                                                 bp 144°C   VP-FT-IR: *3*, 740D
                                                 d 1.060
                                                 n$_D^{20}$1.4050

192.04
160.86
62.49
26.65*
14.01*

CDCl$_3$ QE-300

CH$_3$-C-C-OCH$_2$CH$_3$ (with two =O groups)

C

Aldrich 21,845-6   *CAS [20201-24-5]*   C$_7$H$_{12}$O$_3$ Fp 110°F   60 MHz: *1*, 576C
Ethyl 3-methyl-2-oxobutyrate, 97%   FW 144.17   FT-IR: *1*, 684C
                                                 bp 62°C (11 mm)   VP-FT-IR: *3*, 741B
                                                 d 0.989
                                                 n$_D^{20}$1.4110

198.19
161.81
62.20
37.01*
17.16*
14.04*

CDCl$_3$ QE-300

CH$_3$CH-C-C-OCH$_2$CH$_3$ (with CH$_3$ and two =O groups)

ALDRICH

A

Aldrich 19,209-0   *CAS [13984-57-1]*

Ethyl 4-acetylbutyrate, 98%

$C_8H_{14}O_3$
FW 158.20
bp 222°C
d 0.989
n_D^{20} 1.4270

Fp 157°F

60 MHz: *1*, 578D
FT-IR: *1*, 686C
VP-FT-IR: *3*, 742C

207.90	33.24
173.07	29.92*
60.31	18.90
42.46	14.23*

$CDCl_3$ QE-300

Offset: 40 ppm.

$CH_3-\overset{O}{\overset{\|}{C}}-CH_2CH_2CH_2-\overset{O}{\overset{\|}{C}}-OCH_2CH_3$

B

Aldrich D9,900-5   *CAS [6317-49-3]*

Diethyl 4-oxopimelate, 99%

$C_{11}H_{18}O_5$
FW 230.26
d 1.073
n_D^{20} 1.4410
Fp >230°F

60 MHz: *1*, 578C
FT-IR: *1*, 686D

| 206.94 |
| 172.60 |
| 60.59 |
| 37.08 |
| 27.98 |
| 14.16* |

$CDCl_3$ QE-300

Offset: 40 ppm.

$CH_2-\overset{O}{\overset{\|}{C}}-OCH_2CH_3$
CH_2
$C=O$
CH_2
$CH_2-\overset{O}{\overset{\|}{C}}-OCH_2CH_3$

C

Aldrich 16,526-3   *CAS [607-97-6]*

Ethyl 2-ethylacetoacetate, 95%

$C_8H_{14}O_3$
FW 158.20
bp 189°C (743 mm)
d 0.981
n_D^{20} 1.4220

Fp >230°F

203.25	28.81*
169.83	21.63
61.42*	14.13*
61.25	11.90*

$CDCl_3$ QE-300

Offset: 50 ppm.

$CH_3-\overset{O}{\overset{\|}{C}}-CH-\overset{O}{\overset{\|}{C}}-OCH_2CH_3$
$\qquad\quad|$
$\qquad\quad CH_2CH_3$

ALDRICH

A

Aldrich 36,075-9 *CAS* [*7779-75-1*]
Isobutyl acetoacetate, 98%

$C_8H_{14}O_3$
FW 158.20
bp 100°C (22 mm)
d 0.980
n_D^{20} 1.4240
Fp 173°F

CDCl$_3$ QE-300

200.33
167.09
71.42
50.09
30.08*
27.69*
19.00*

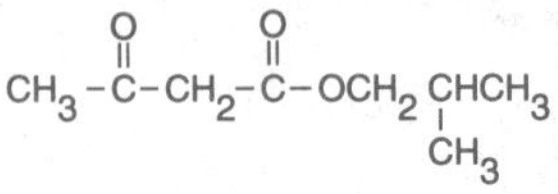

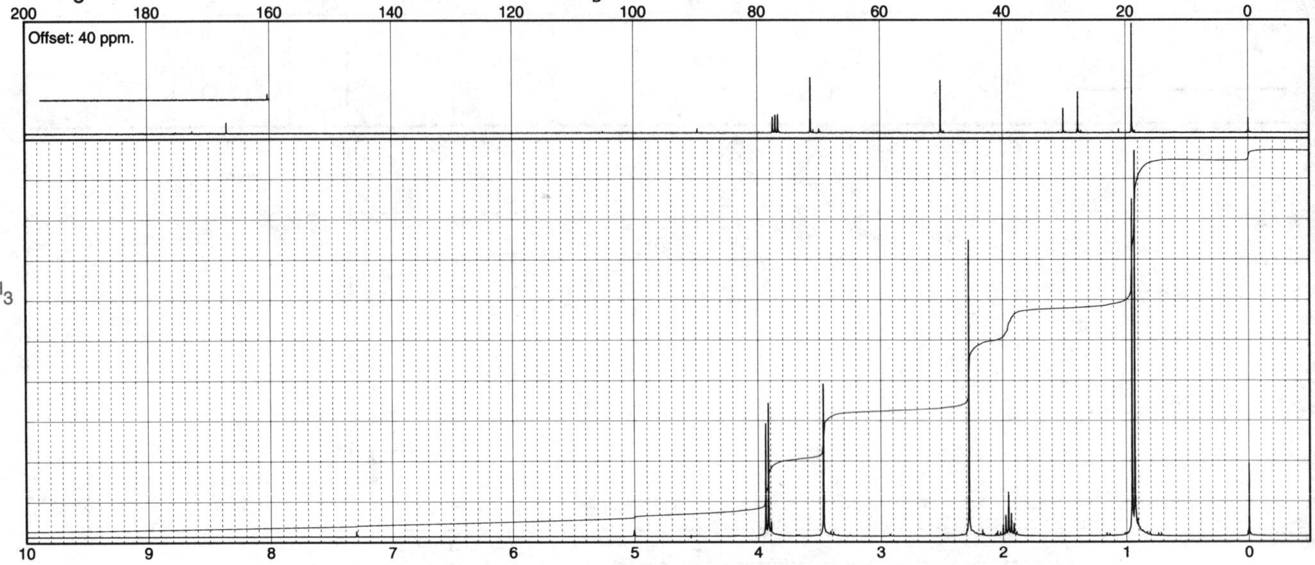
Offset: 40 ppm.

B

Aldrich B8,860-8 *CAS* [*1694-31-1*]
***tert*-Butyl acetoacetate**, 98%

$C_8H_{14}O_3$
FW 158.20
d 0.954
n_D^{20} 1.4200
Fp 141°F

60 MHz: ***1***, 579B
FT-IR: ***1***, 688C

CDCl$_3$ QE-300

201.02
166.31
81.86
51.47
30.00*
27.94*

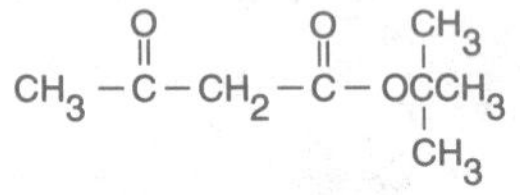

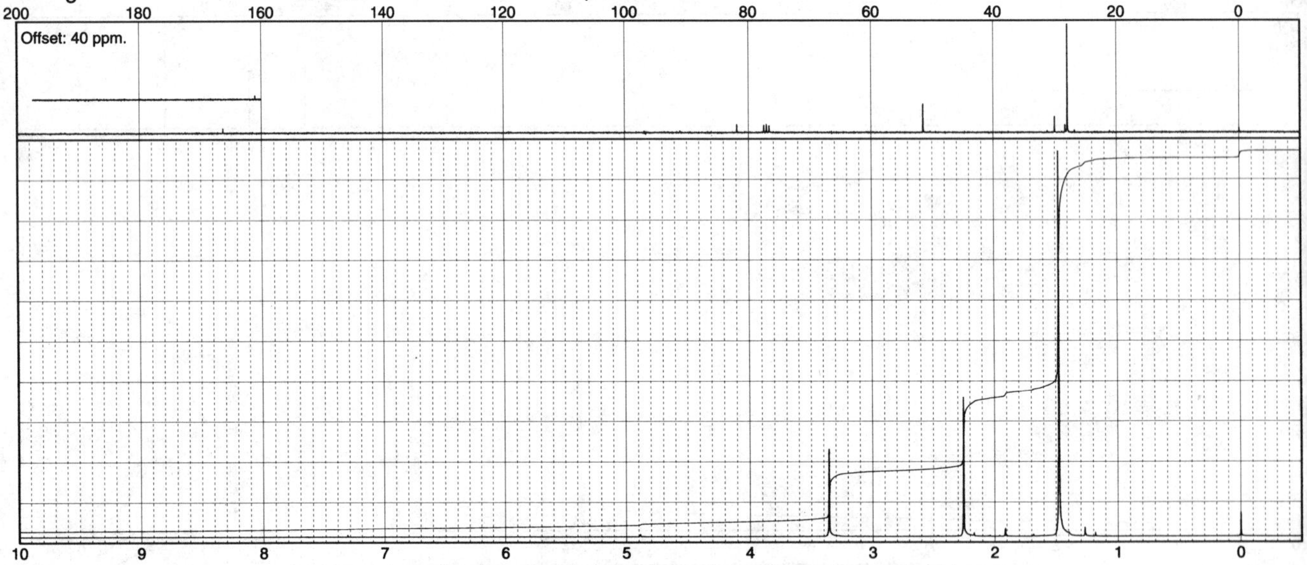
Offset: 40 ppm.

C

Aldrich 32,363-2 *CAS* [*15925-47-0*]
***S-tert*-Butyl acetothioacetate**, 99%

$C_8H_{14}O_2S$
FW 174.26
bp 98°C
d 0.994
n_D^{20} 1.4860
Fp 182°F

CDCl$_3$ QE-300

200.05
192.45
59.11
48.97
30.12*
29.56*

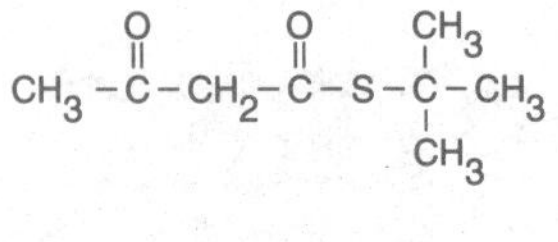

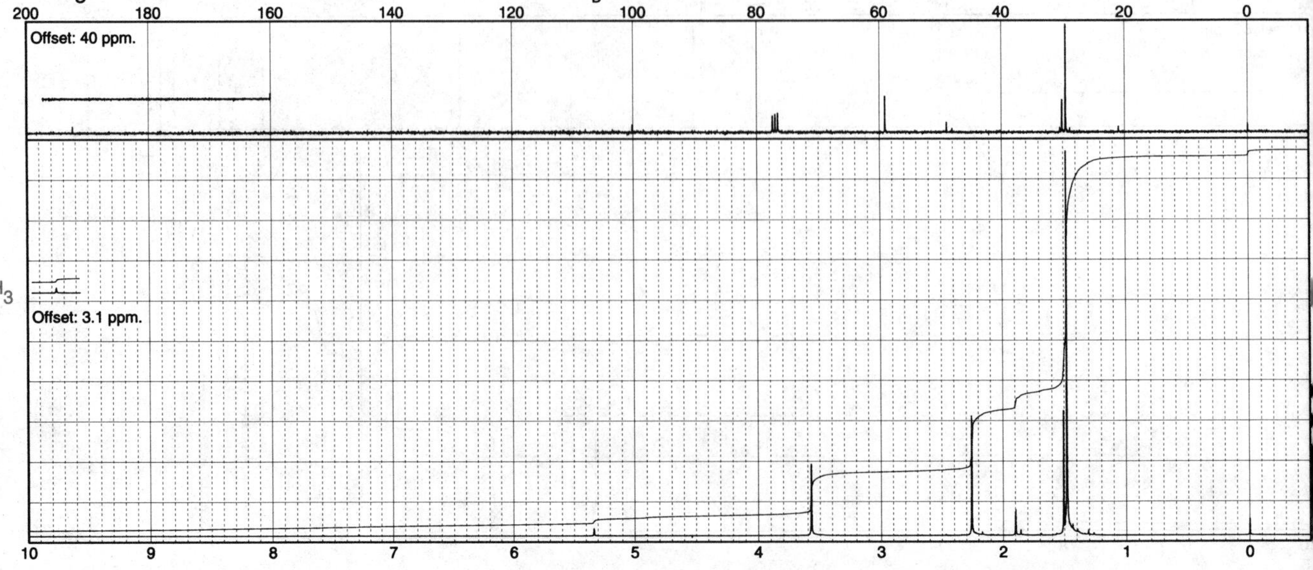
Offset: 40 ppm.
Offset: 3.1 ppm.

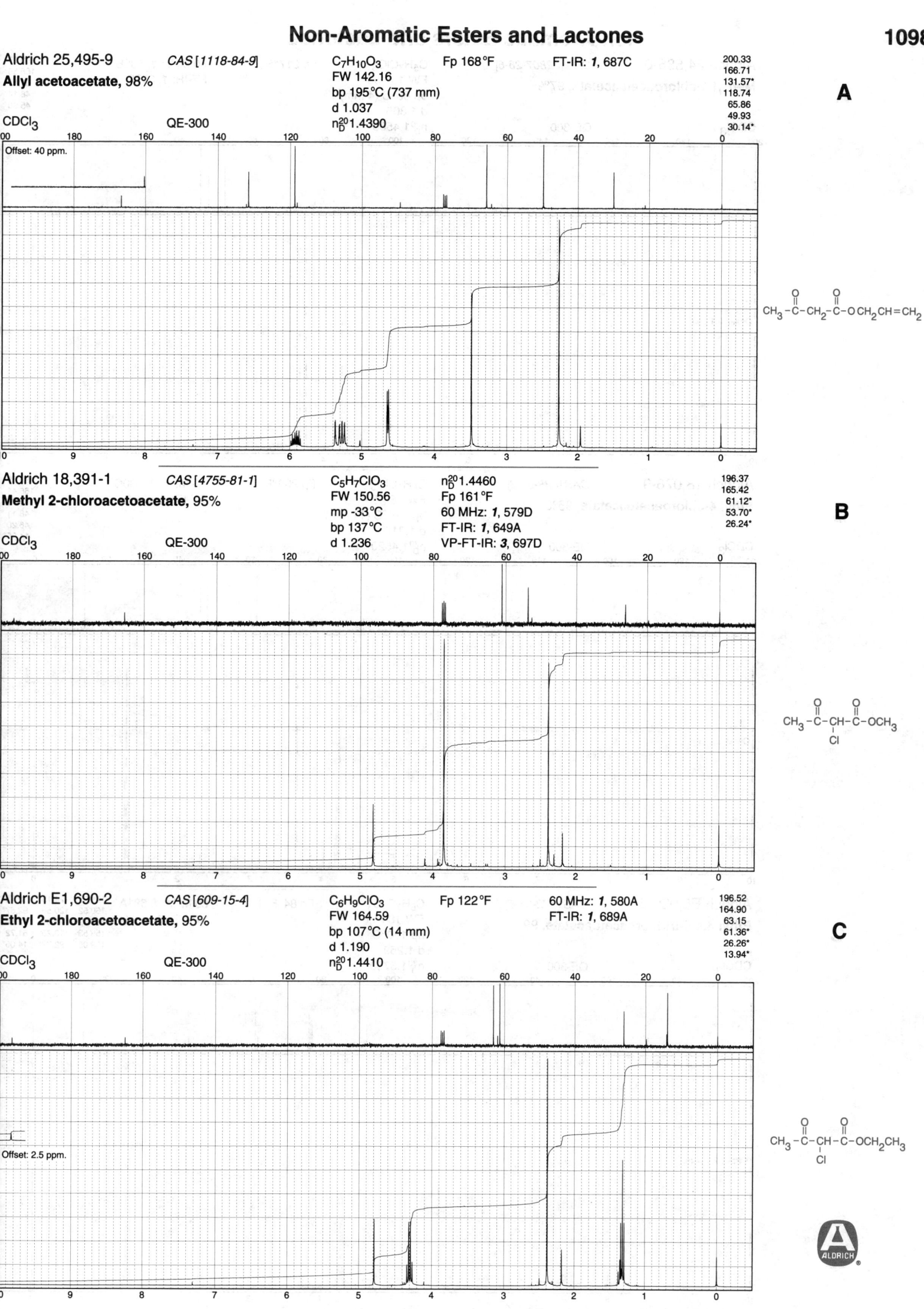

Aldrich 25,495-9
CAS [1118-84-9]
Allyl acetoacetate, 98%

C₇H₁₀O₃
FW 142.16
bp 195°C (737 mm)
d 1.037
n²⁰_D 1.4390

Fp 168°F

FT-IR: *1*, 687C

CDCl₃
QE-300

200.33
166.71
131.57*
118.74
65.86
49.93
30.14*

A

Offset: 40 ppm.

$CH_3-\overset{O}{\overset{||}{C}}-CH_2-\overset{O}{\overset{||}{C}}-O\,CH_2\,CH=CH_2$

Aldrich 18,391-1
CAS [4755-81-1]
Methyl 2-chloroacetoacetate, 95%

C₅H₇ClO₃
FW 150.56
mp -33°C
bp 137°C
d 1.236

n²⁰_D 1.4460
Fp 161°F
60 MHz: *1*, 579D
FT-IR: *1*, 649A
VP-FT-IR: *3*, 697D

196.37
165.42
61.12*
53.70*
26.24*

B

CDCl₃
QE-300

$CH_3-\overset{O}{\overset{||}{C}}-\underset{\underset{Cl}{|}}{C}H-\overset{O}{\overset{||}{C}}-OCH_3$

Aldrich E1,690-2
CAS [609-15-4]
Ethyl 2-chloroacetoacetate, 95%

C₆H₉ClO₃
FW 164.59
bp 107°C (14 mm)
d 1.190
n²⁰_D 1.4410

Fp 122°F

60 MHz: *1*, 580A
FT-IR: *1*, 689A

196.52
164.90
63.15
61.36*
26.26*
13.94*

C

CDCl₃
QE-300

Offset: 2.5 ppm.

$CH_3-\overset{O}{\overset{||}{C}}-\underset{\underset{Cl}{|}}{C}H-\overset{O}{\overset{||}{C}}-OCH_2CH_3$

ALDRICH

Non-Aromatic Esters and Lactones

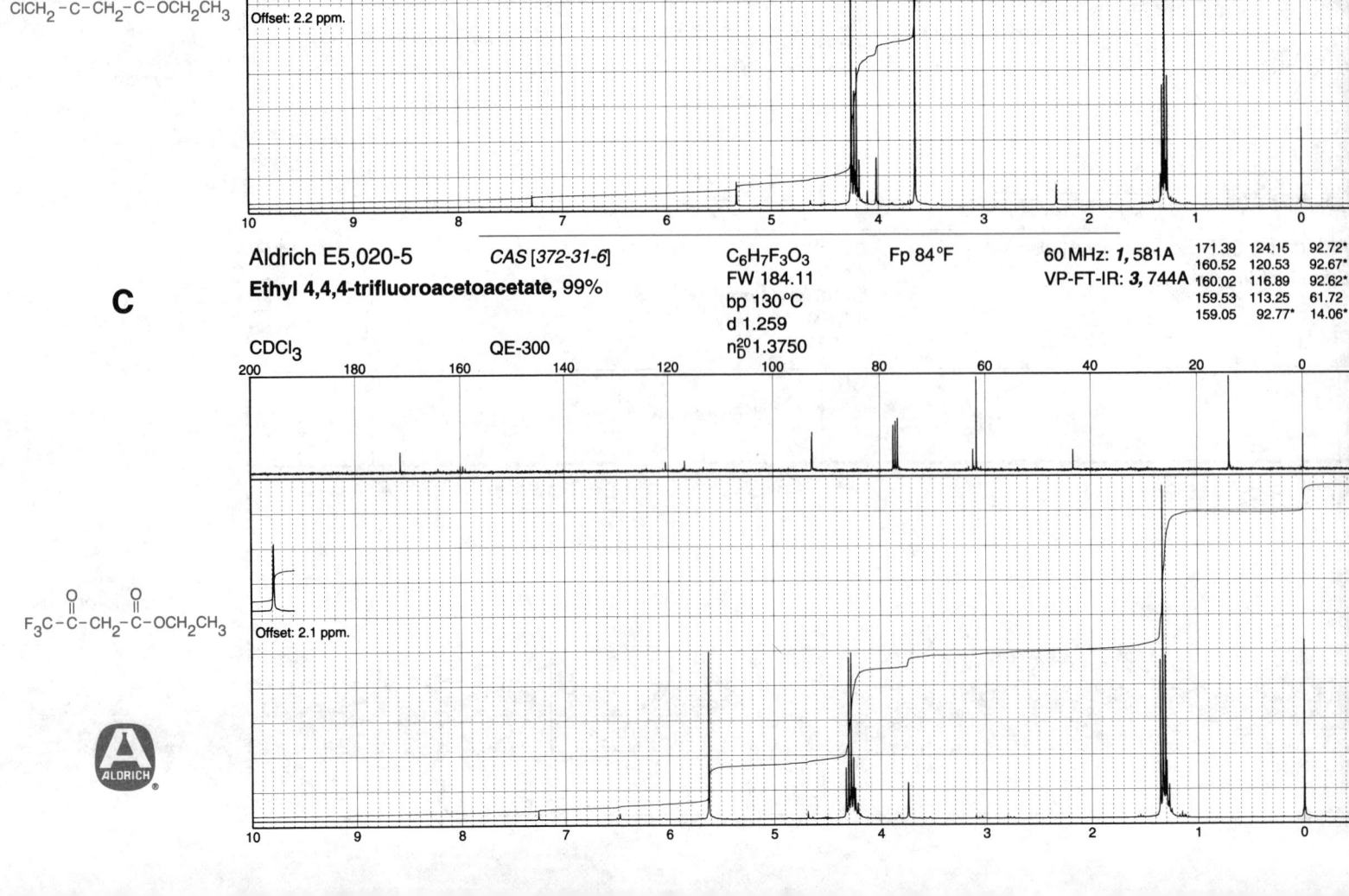

A

Aldrich 24,586-0   CAS [32807-28-6]   C₅H₇ClO₃   Fp 217°F   60 MHz: 1, 580B

$C_5H_7ClO_3$

Methyl 4-chloroacetoacetate, 97%   FW 150.56   FT-IR: 1, 687D
bp 85°C (4 mm)
d 1.305
n_D^{20}1.4560

195.27
166.86
52.63*
48.14
45.90

CDCl₃   QE-300

$ClCH_2-C-CH_2-C-OCH_3$

Offset: 2.1 ppm.

B

Aldrich 18,076-9   CAS [638-07-3]   C₆H₉ClO₃   Fp 206°F   60 MHz: 1, 580C

$C_6H_9ClO_3$

Ethyl 4-chloroacetoacetate, 95%   FW 164.59   FT-IR: 1, 688A
bp 115°C (14 mm)
d 1.218
n_D^{20}1.4520

195.39
166.41
61.76
48.11
46.20
14.05*

CDCl₃   QE-300

$ClCH_2-C-CH_2-C-OCH_2CH_3$

Offset: 2.2 ppm.

C

Aldrich E5,020-5   CAS [372-31-6]   C₆H₇F₃O₃   Fp 84°F   60 MHz: 1, 581A

$C_6H_7F_3O_3$

Ethyl 4,4,4-trifluoroacetoacetate, 99%   FW 184.11   VP-FT-IR: 3, 744A
bp 130°C
d 1.259
n_D^{20}1.3750

171.39	124.15	92.72*
160.52	120.53	92.67*
160.02	116.89	92.62*
159.53	113.25	61.72
159.05	92.77*	14.06*

CDCl₃   QE-300

$F_3C-C-CH_2-C-OCH_2CH_3$

Offset: 2.1 ppm.

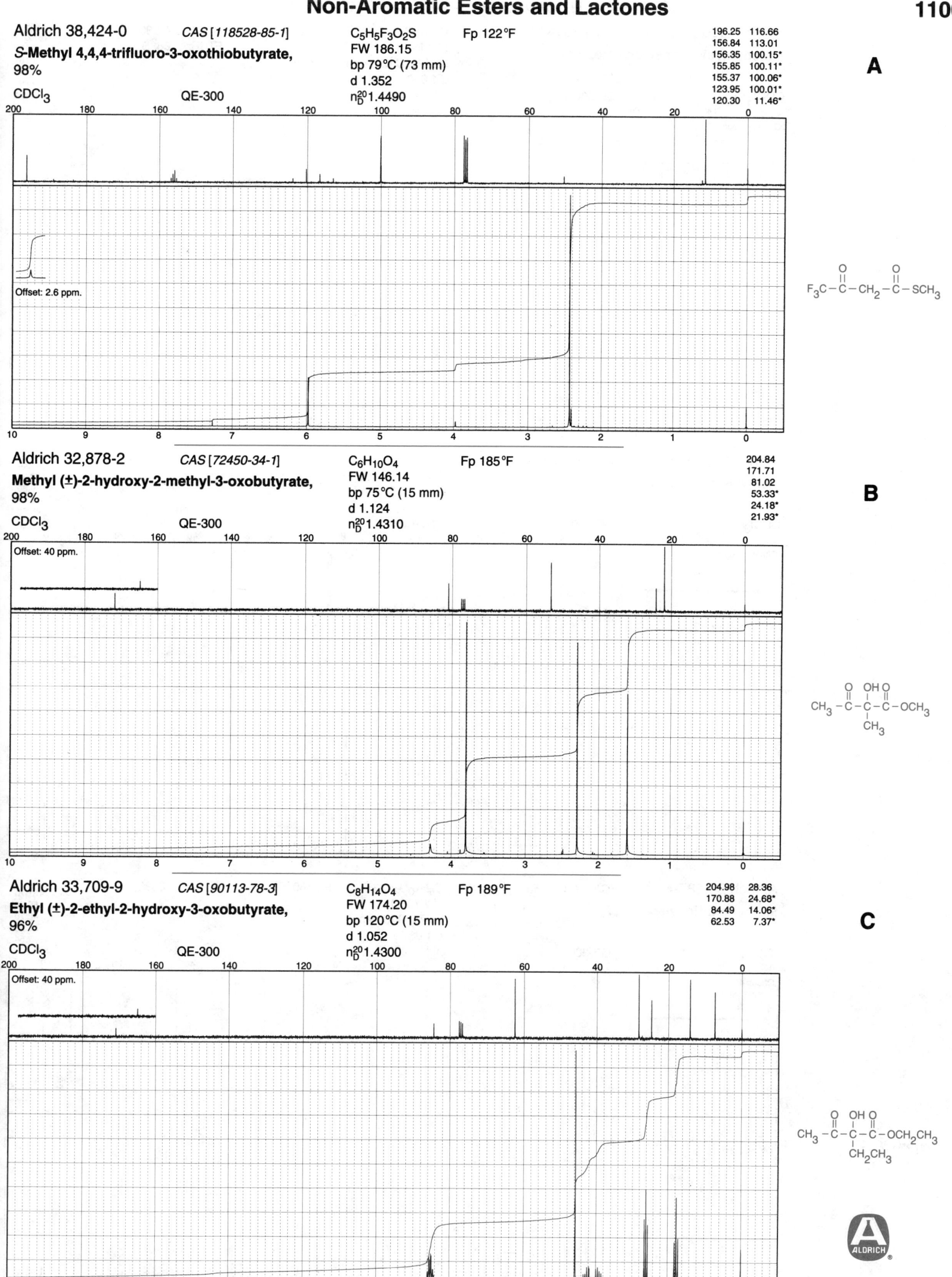

Aldrich 38,424-0   CAS [118528-85-1]   C5H5F3O2S   Fp 122°F

S-Methyl 4,4,4-trifluoro-3-oxothiobutyrate, 98%

FW 186.15
bp 79°C (73 mm)
d 1.352
n20D 1.4490

CDCl3   QE-300

196.25 116.66
156.84 113.01
156.35 100.15*
155.85 100.11*
155.37 100.06*
123.95 100.01*
120.30   11.46*

A

Offset: 2.6 ppm.

Aldrich 32,878-2   CAS [72450-34-1]   C6H10O4   Fp 185°F

Methyl (±)-2-hydroxy-2-methyl-3-oxobutyrate, 98%

FW 146.14
bp 75°C (15 mm)
d 1.124
n20D 1.4310

CDCl3   QE-300

204.84
171.71
81.02
53.33*
24.18*
21.93*

B

Offset: 40 ppm.

Aldrich 33,709-9   CAS [90113-78-3]   C8H14O4   Fp 189°F

Ethyl (±)-2-ethyl-2-hydroxy-3-oxobutyrate, 96%

FW 174.20
bp 120°C (15 mm)
d 1.052
n20D 1.4300

CDCl3   QE-300

204.98   28.36
170.88   24.68*
 84.49   14.06*
 62.53   7.37*

C

Offset: 40 ppm.

A

Aldrich 23,082-0   CAS [22502-03-0] $C_7H_{12}O_4$ Fp 218°F   60 MHz: **1**, 580D

2-Methoxyethyl acetoacetate, 97%   FW 160.17   FT-IR: **1**, 688B

bp 120°C (20 mm)

d 1.090

n_D^{20} 1.4340

CDCl$_3$ QE-300

Offset: 40 ppm.

200.36
167.07
70.17
64.18
58.88*
49.89
30.05*

$CH_3-\overset{O}{\overset{\|}{C}}-CH_2-\overset{O}{\overset{\|}{C}}-OCH_2CH_2OCH_3$

B

Aldrich 28,150-6   CAS [41051-15-4] $C_6H_{10}O_4$ Fp 193°F

Methyl 4-methoxyacetoacetate, 97%   FW 146.14

bp 89°C (8 mm)

d 1.129

n_D^{20} 1.4310

CDCl$_3$ QE-300

Offset: 40 ppm.

201.51
167.44
77.32
59.39*
52.41*
45.53

$CH_3OCH_2-\overset{O}{\overset{\|}{C}}-CH_2-\overset{O}{\overset{\|}{C}}-OCH_3$

C

Aldrich 25,587-4   CAS [30414-53-0] $C_6H_{10}O_3$ Fp 160°F   FT-IR: **1**, 689C

Methyl propionylacetate, 98%   FW 130.15

bp 74°C (5 mm)

d 1.037

n_D^{20} 1.4220

CDCl$_3$ QE-300

Offset: 40 ppm.

Offset: 2.2 ppm.

203.16
167.70
52.28*
48.70
36.33
7.54*

$CH_3CH_2-\overset{O}{\overset{\|}{C}}-CH_2-\overset{O}{\overset{\|}{C}}-OCH_3$

Aldrich 24,716-2   CAS [4949-44-4]   C₇H₁₂O₃   Fp 172°F   60 MHz: 1, 581C

Ethyl propionylacetate, 97%

$C_7H_{12}O_3$
FW 144.17
bp 84°C (12 mm)
d 1.012
n_D^{20} 1.4220

FT-IR: 1, 689D

203.28
167.26
61.30
49.01
36.29
14.11*
7.55*

A

CDCl₃   QE-300

Offset: 40 ppm.

Offset: 2.3 ppm.

$CH_3CH_2-C(=O)-CH_2-C(=O)-OCH_2CH_3$

Aldrich 12,262-9   CAS [539-88-8]

Ethyl levulinate, 99%

$C_7H_{12}O_3$
FW 144.17
bp 94°C (18 mm)
d 1.016
n_D^{20} 1.4220

Fp 195°F

60 MHz: 1, 577B
FT-IR: 1, 685B
VP-FT-IR: 3, 741D

206.62
172.67
60.58
37.94
29.86*
28.02
14.17*

B

CDCl₃   QE-300

Offset: 40 ppm.

$CH_3-C(=O)-CH_2CH_2-C(=O)-OCH_2CH_3$

Aldrich 17,953-1   CAS [2052-15-5]

Butyl levulinate, 98%

$C_9H_{16}O_3$
FW 172.22
bp 107°C (5 mm)
d 0.974
n_D^{20} 1.4270

Fp 197°F

60 MHz: 1, 577C
FT-IR: 1, 685C

206.55   29.85*
172.73   28.01
64.49   19.12
37.95   13.70*
30.64

C

CDCl₃   QE-300

Offset: 40 ppm.

$CH_3-C(=O)-CH_2CH_2-C(=O)-OCH_2CH_2CH_2CH_3$

ALDRICH

Non-Aromatic Esters and Lactones

A

Aldrich 28,506-4 *CAS [79416-27-6]*

Methyl 5-aminolevulinate hydrochloride, 98%

$C_6H_{11}NO_3$
FW 181.62
mp 124°C

206.73
178.02
55.23*
49.87
37.17
30.12

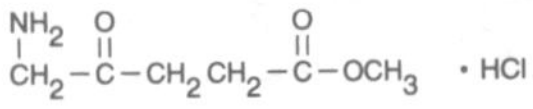

D_2O QE-300

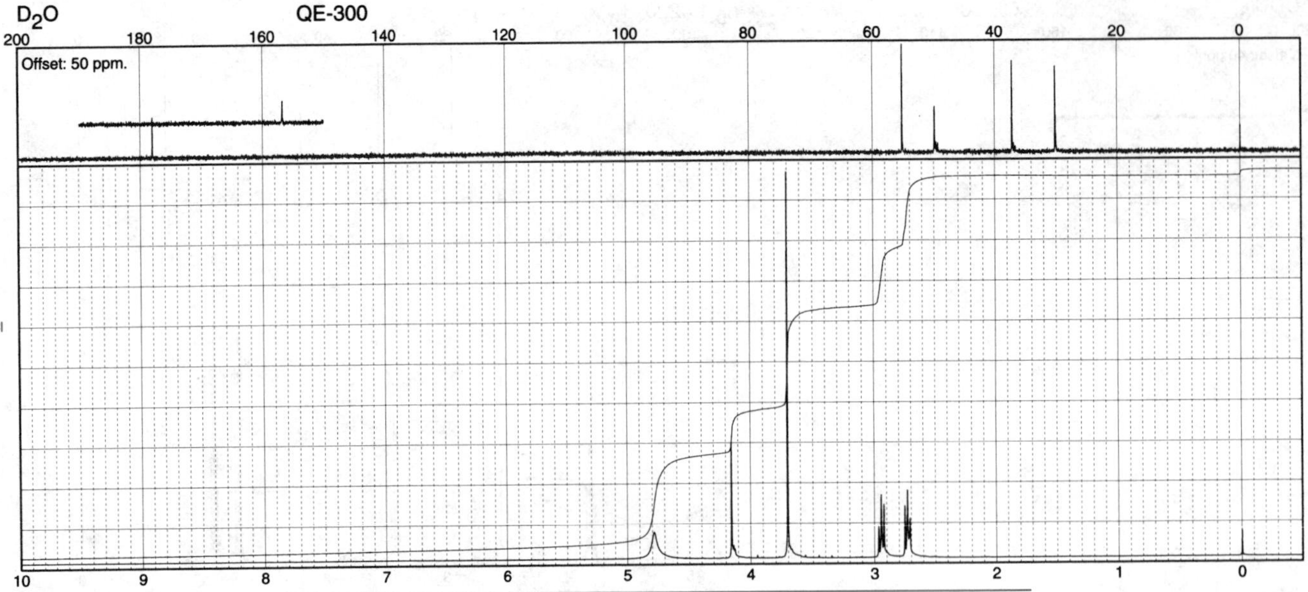

Offset: 50 ppm.

B

Aldrich E3,320-3 *CAS [7152-15-0]*

Ethyl isobutyrylacetate, 95%

$C_8H_{14}O_3$
FW 158.20
mp -9°C
bp 173°C
d 0.980

n_D^{20} 1.4250
Fp 128°F
60 MHz: ***1***, 581B
FT-IR: ***1***, 689B
VP-FT-IR: ***3***, 744B

206.53
167.37
61.27
47.12
41.20*
17.92*
14.11*

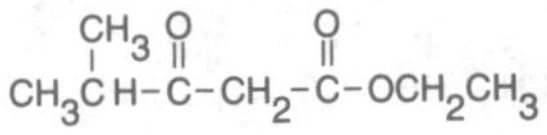

$CDCl_3$ QE-300

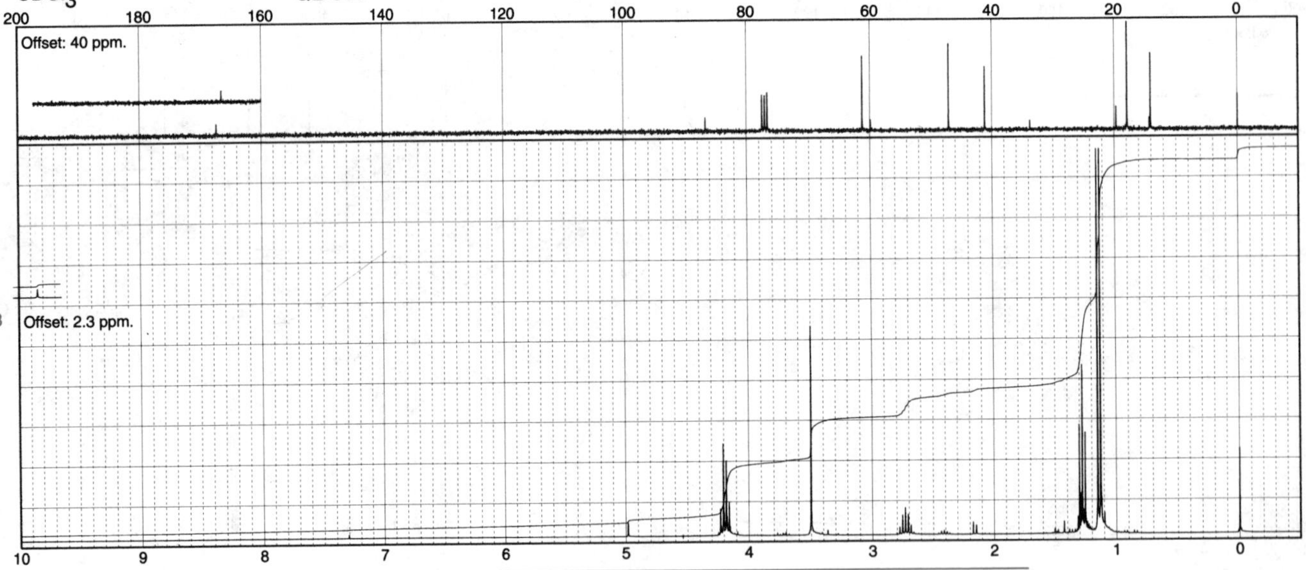

Offset: 40 ppm.
Offset: 2.3 ppm.

C

Aldrich 26,240-4 *CAS [55107-14-7]*

Methyl 4,4-dimethyl-3-oxopentanoate, 99%

$C_8H_{14}O_3$
FW 158.20
bp 69°C (13 mm)
d 0.990
n_D^{20} 1.4310

Fp 171°F

207.89 44.76
168.14 43.69
85.18* 27.43*
52.20* 26.06*
51.08*

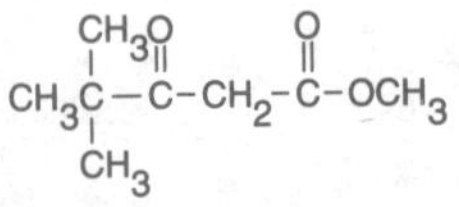

$CDCl_3$ QE-300

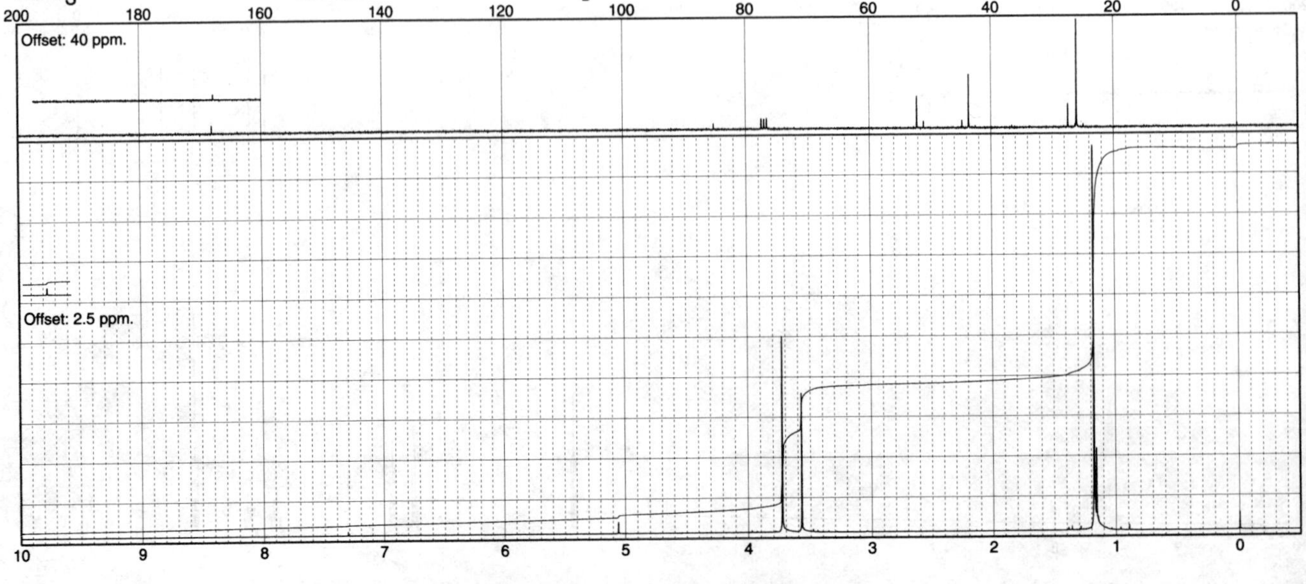

Offset: 40 ppm.
Offset: 2.5 ppm.

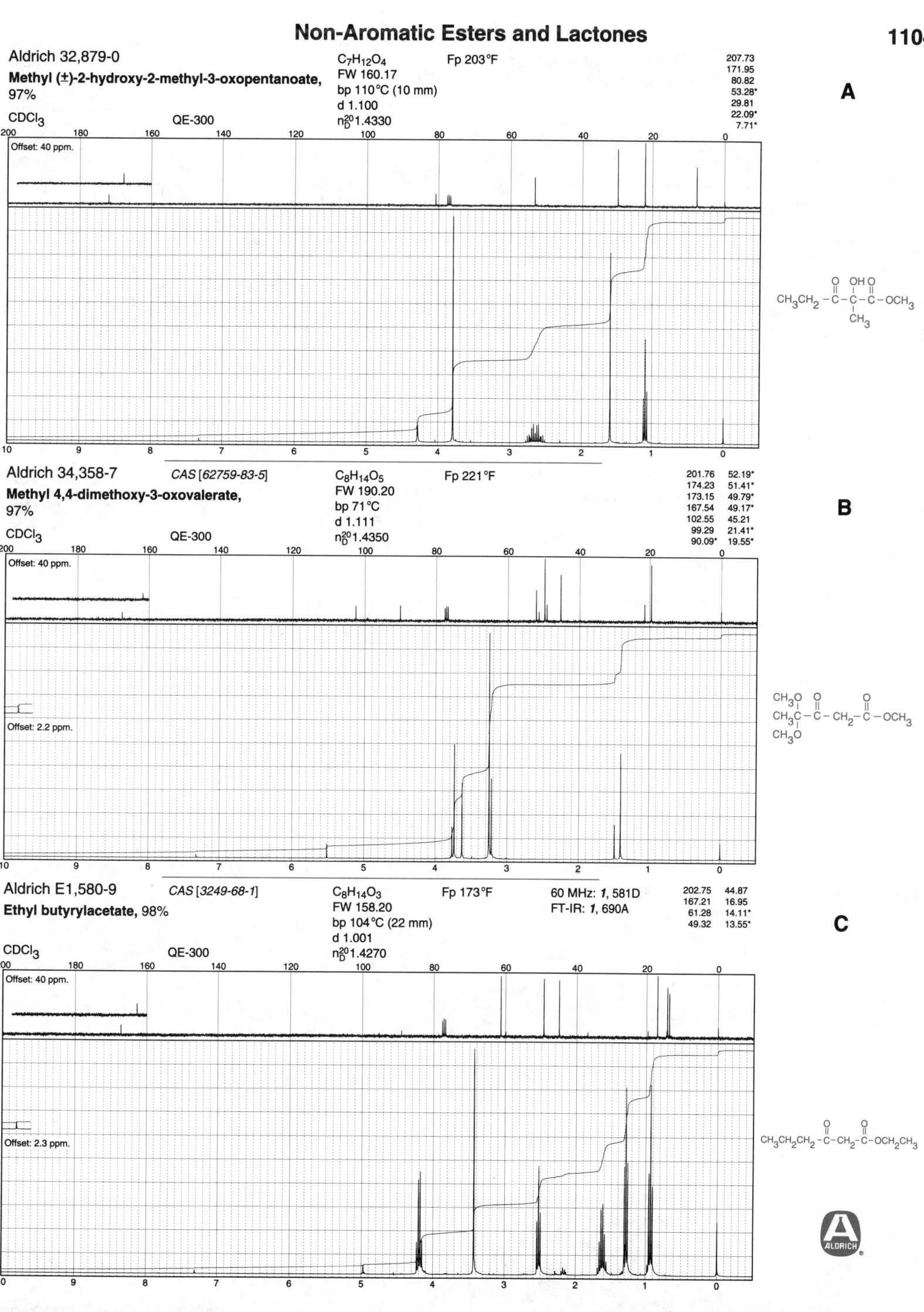

Aldrich 32,879-0

Methyl (±)-2-hydroxy-2-methyl-3-oxopentanoate, 97%

CDCl$_3$ QE-300

$C_7H_{12}O_4$ Fp 203°F
FW 160.17
bp 110°C (10 mm)
d 1.100
n_D^{20} 1.4330

207.73
171.95
80.82
53.28*
29.81
22.09*
7.71*

A

Offset: 40 ppm.

CH$_3$CH$_2$—C—C—C—OCH$_3$ (O / OH O / CH$_3$)

Aldrich 34,358-7   CAS [62759-83-5]

Methyl 4,4-dimethoxy-3-oxovalerate, 97%

CDCl$_3$ QE-300

$C_8H_{14}O_5$ Fp 221°F
FW 190.20
bp 71°C
d 1.111
n_D^{20} 1.4350

201.76   52.19*
174.23   51.41*
173.15   49.79*
167.54   49.17*
102.55   45.21
99.29   21.41*
90.09*   19.55*

B

Offset: 40 ppm.

Offset: 2.2 ppm.

CH$_3$O O O
CH$_3$—C—C—CH$_2$—C—OCH$_3$
CH$_3$O

Aldrich E1,580-9   CAS [3249-68-1]

Ethyl butyrylacetate, 98%

CDCl$_3$ QE-300

$C_8H_{14}O_3$ Fp 173°F
FW 158.20
bp 104°C (22 mm)
d 1.001
n_D^{20} 1.4270

60 MHz: 1, 581D
FT-IR: 1, 690A

202.75   44.87
167.21   16.95
61.28   14.11*
49.32   13.55*

C

Offset: 2.3 ppm.

CH$_3$CH$_2$CH$_2$—C—CH$_2$—C—OCH$_2$CH$_3$ (O / O)

ALDRICH

Non-Aromatic Esters and Lactones

A

Aldrich 34,570-9 | CAS [2380-20-3]
Methyl 5-oxooctadecanoate, 99%

$C_{19}H_{36}O_3$
FW 312.50
mp 59°C
Fp >230°F

210.25	31.93	29.37
173.55	29.68	29.26
51.50*	29.66	23.87
42.88	29.63	22.70
41.44	29.49	18.91
33.07	29.43	14.12*

CDCl$_3$ QE-300

Offset: 40 ppm.

$CH_3(CH_2)_{11}CH_2-C-CH_2CH_2CH_2-C-OCH_3$

B

Aldrich 30,517-0 | CAS [2380-22-5]
Methyl 7-oxooctadecanoate, 98%

$C_{19}H_{36}O_3$
FW 312.50
mp 51°C
Fp >230°F

210.97	31.92	28.72
173.92	29.62	24.72
51.40*	29.49	23.92
42.87	29.43	23.41
42.43	29.34	22.69
33.86	29.29	14.09*

CDCl$_3$ QE-300

Offset: 40 ppm.

$CH_3(CH_2)_9CH_2-C-CH_2(CH_2)_3CH_2-C-OCH_3$

C

Aldrich 30,335-6 | CAS [2380-27-0]
Methyl 12-oxooctadecanoate, 98%

$C_{19}H_{36}O_3$
FW 312.50
mp 47°C
bp 179°C
Fp >230°F

211.30	31.63	28.97
174.13	29.38	24.97
51.33*	29.28	23.90
42.81	29.22	22.50
34.11	29.15	13.99*

CDCl$_3$ QE-300

Offset: 40 ppm.

$CH_3(CH_2)_4CH_2-C-CH_2(CH_2)_8CH_2-C-OCH_3$

ALDRICH

Non-Aromatic Esters and Lactones

Aldrich 24,962-9   CAS [603-69-0]   C₈H₁₂O₄   Fp 185°F   FT-IR: 1, 690C

Ethyl diacetoacetate, 97%

$C_8H_{12}O_4$
FW 172.18
bp 105°C (17 mm)
d 1.104
n_D^{20} 1.4700

CDCl₃   QE-300

196.54
167.12
108.67
60.68
25.96*
14.25*

A

Aldrich 29,617-1   CAS [13984-53-7]   C₉H₁₄O₄   Fp >230°F

Methyl 4-acetyl-5-oxohexanoate, 98%

$C_9H_{14}O_4$
FW 186.21
bp 182°C (48 mm)
d 1.066
n_D^{20} 1.4590

CDCl₃   QE-300

203.62   34.59
191.25   31.34
172.91   29.29*
172.85   23.03
108.63   22.86*
66.99*   22.84
51.70*

B

Aldrich 30,160-4   CAS [2832-10-2]   C₁₀H₁₆O₄   Fp >230°F

Ethyl 4-acetyl-5-oxohexanoate, 98%

$C_{10}H_{16}O_4$
FW 200.23
bp 154°C (19 mm)
d 1.067
n_D^{20} 1.4570

CDCl₃   QE-300

203.56   31.65
191.20   29.25*
172.43   23.07
108.71   22.94
67.13*   22.88*
60.57   14.21*
34.90

C

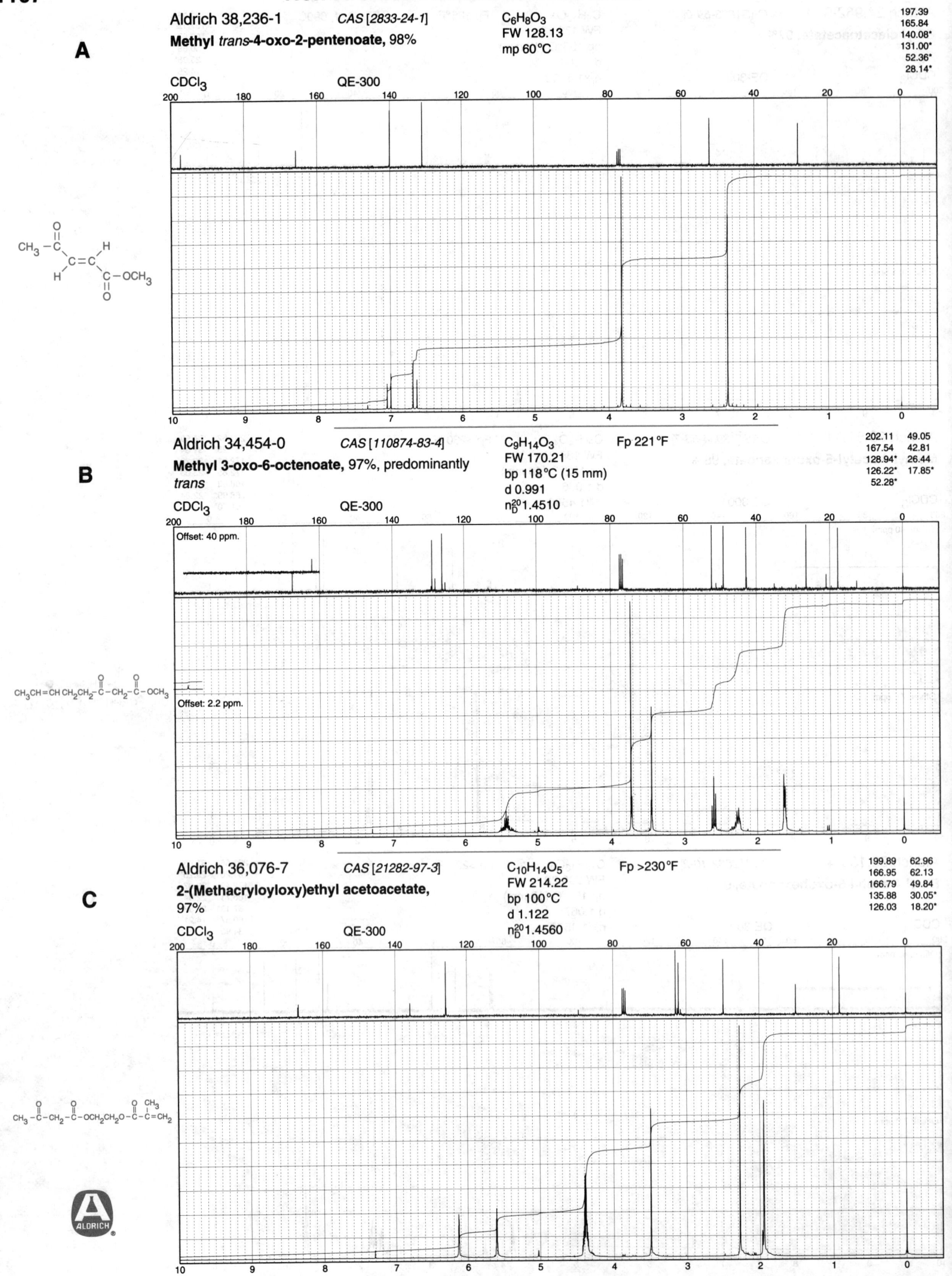

A

Aldrich 38,236-1   *CAS [2833-24-1]* $C_6H_8O_3$
Methyl *trans*-4-oxo-2-pentenoate, 98%   FW 128.13
mp 60°C

197.39
165.84
140.08*
131.00*
52.36*
28.14*

CDCl₃   QE-300

B

Aldrich 34,454-0   *CAS [110874-83-4]* $C_9H_{14}O_3$ Fp 221°F
Methyl 3-oxo-6-octenoate, 97%, predominantly   FW 170.21
trans   bp 118°C (15 mm)
d 0.991
n_D^20 1.4510

202.11   49.05
167.54   42.81
128.94*   26.44
126.22*   17.85*
52.28*

CDCl₃   QE-300

Offset: 40 ppm.

Offset: 2.2 ppm.

C

Aldrich 36,076-7   *CAS [21282-97-3]* $C_{10}H_{14}O_5$ Fp >230°F
2-(Methacryloyloxy)ethyl acetoacetate,   FW 214.22
97%   bp 100°C
d 1.122
n_D^20 1.4560

199.89   62.96
166.95   62.13
166.79   49.84
135.88   30.05*
126.03   18.20*

CDCl₃   QE-300

Aldrich D9,740-1   CAS [609-09-6] $C_7H_{10}O_5$ Fp >230°F   60 MHz: *1*, 577D   177.96

Diethyl ketomalonate, 98%   FW 174.15   FT-IR: *1*, 685D   160.01

bp 209°C   VP-FT-IR: *3*, 742A   63.40

d 1.142   13.91*

n_D^{20}1.4150

CDCl$_3$ QE-300

A

$CH_3CH_2O-\overset{O}{\underset{}{C}}-\overset{O}{\underset{}{C}}-\overset{O}{\underset{}{C}}-OCH_2CH_3$

Aldrich 13,776-6   CAS [5965-53-7] $C_9H_{14}O_5$ Fp >230°F   60 MHz: *1*, 578A   189.65   61.67

Diethyl oxalpropionate, 97%   FW 202.21   FT-IR: *1*, 686A   169.70   48.49*

bp 138°C (23 mm)   160.45   13.97*

d 1.073   62.75   11.78*

n_D^{20}1.4320

CDCl$_3$ QE-300

B

Offset: 2.7 ppm.

$CH_3CH_2O-\overset{O}{\underset{}{C}}-\overset{CH_3}{\underset{}{CH}}-\overset{O}{\underset{}{C}}-\overset{O}{\underset{}{C}}-OCH_2CH_3$

Aldrich D13,800-2   CAS [1830-54-2] $C_7H_{10}O_5$ Fp >230°F   60 MHz: *1*, 582C   195.22

Dimethyl 1,3-acetonedicarboxylate, 97%   FW 174.15   FT-IR: *1*, 691B   167.10

bp 150°C (25 mm)   52.47*

d 1.185   48.68

n_D^{20}1.4440

CDCl$_3$ QE-300

C

Offset: 2.2 ppm.

$CH_3O-\overset{O}{\underset{}{C}}-CH_2-\overset{O}{\underset{}{C}}-CH_2-\overset{O}{\underset{}{C}}-OCH_3$

ALDRICH

Non-Aromatic Esters and Lactones

A

Aldrich 16,512-3   *CAS [105-50-0]* $C_9H_{14}O_5$ Fp 187°F   60 MHz: *1*, 582D
Diethyl 1,3-acetonedicarboxylate, 97%   FW 202.21   FT-IR: *1*, 691C
bp 250°C
d 1.113
n_D^{20} 1.4400

195.45
166.67
61.53
48.98
14.06*

CDCl₃   QE-300

CH₃CH₂O-C-CH₂-C-CH₂-C-OCH₂CH₃

Offset: 2.3 ppm.

B

Aldrich 27,081-4   *CAS [28009-80-5]* $C_{13}H_{22}O_5$
Di-*tert*-butyl 1,3-acetonedicarboxylate,   FW 258.32
98%   mp 59°C

196.25
165.90
82.19
50.21
27.94*

CDCl₃   QE-300

CH₃C-O-C-CH₂-C-CH₂-C-O-CCH₃

C

Aldrich D16,760-6   *CAS [5457-44-3]* $C_8H_{12}O_5$ Fp >230°F
Dimethyl 3-oxoadipate, 97%   FW 188.18
bp 150°C (18 mm)
d 1.169
n_D^{20} 1.4450

200.95   51.85*
172.80   48.97
167.41   37.40
52.36*   27.68

CDCl₃   QE-300

Offset: 40 ppm.

CH₃O-C-CH₂CH₂-C-CH₂-C-OCH₃

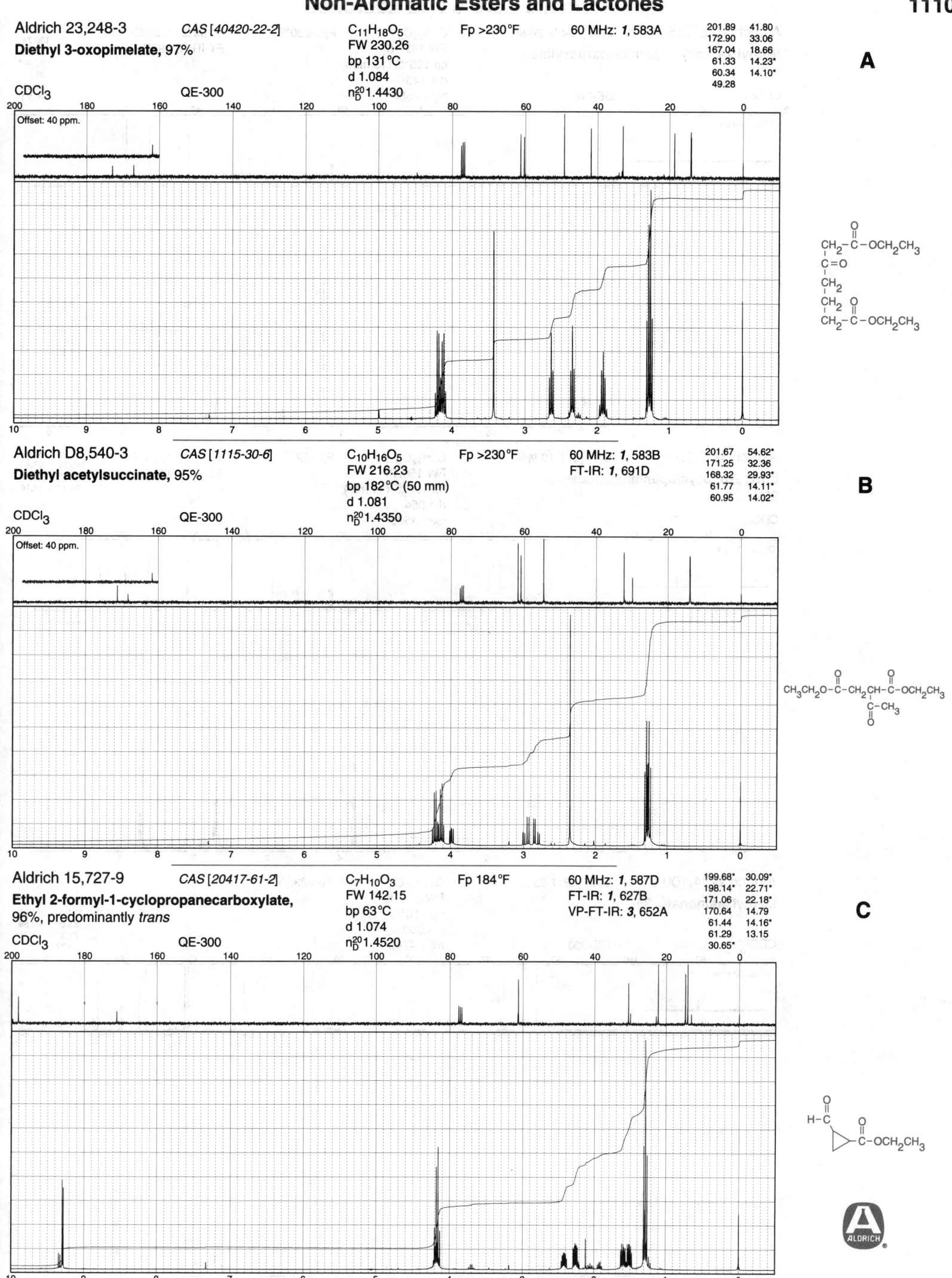

Aldrich 23,248-3   *CAS [40420-22-2]* $C_{11}H_{18}O_5$ Fp >230°F   60 MHz: *1*, 583A
Diethyl 3-oxopimelate, 97%
FW 230.26
bp 131°C
d 1.084
n_D^{20} 1.4430

201.89	41.80
172.90	33.06
167.04	18.66
61.33	14.23*
60.34	14.10*
49.28	

CDCl₃   QE-300   Offset: 40 ppm.

A

Aldrich D8,540-3   *CAS [1115-30-6]* $C_{10}H_{16}O_5$ Fp >230°F   60 MHz: *1*, 583B
Diethyl acetylsuccinate, 95%
FW 216.23   FT-IR: *1*, 691D
bp 182°C (50 mm)
d 1.081
n_D^{20} 1.4350

201.67	54.62*
171.25	32.36
168.32	29.93*
61.77	14.11*
60.95	14.02*

CDCl₃   QE-300   Offset: 40 ppm.

B

Aldrich 15,727-9   *CAS [20417-61-2]* $C_7H_{10}O_3$ Fp 184°F   60 MHz: *1*, 587D
Ethyl 2-formyl-1-cyclopropanecarboxylate,
96%, predominantly *trans*
FW 142.15   FT-IR: *1*, 627B
bp 63°C   VP-FT-IR: *3*, 652A
d 1.074
n_D^{20} 1.4520

199.68*	30.09*
198.14*	22.71*
171.06	22.18*
170.64	14.79
61.44	14.16*
61.29	13.15
30.65*	

CDCl₃   QE-300

C

ALDRICH

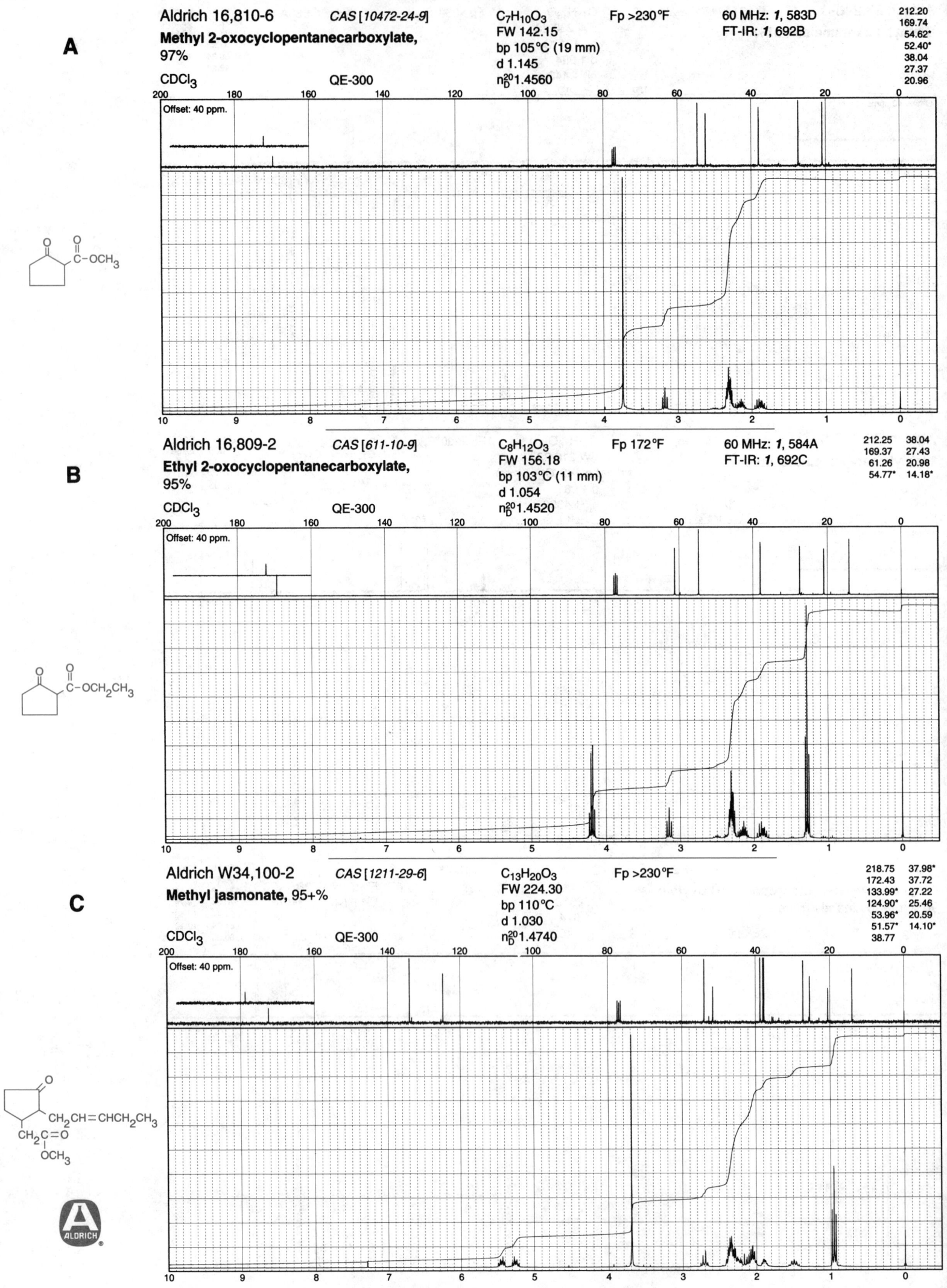

A

Aldrich 16,810-6   CAS [10472-24-9]   C$_7$H$_{10}$O$_3$ Fp >230°F   60 MHz: **1**, 583D   212.20
Methyl 2-oxocyclopentanecarboxylate,   FW 142.15   FT-IR: **1**, 692B   169.74
97%   bp 105°C (19 mm)   54.62*
    d 1.145   52.40*
CDCl$_3$ QE-300   n$_D^{20}$1.4560   38.04
    27.37
    20.96

Offset: 40 ppm.

B

Aldrich 16,809-2   CAS [611-10-9]   C$_8$H$_{12}$O$_3$ Fp 172°F   60 MHz: **1**, 584A   212.25   38.04
Ethyl 2-oxocyclopentanecarboxylate,   FW 156.18   FT-IR: **1**, 692C   169.37   27.43
95%   bp 103°C (11 mm)   61.26   20.98
    d 1.054   54.77*   14.18*
CDCl$_3$ QE-300   n$_D^{20}$1.4520

Offset: 40 ppm.

C

Aldrich W34,100-2   CAS [1211-29-6]   C$_{13}$H$_{20}$O$_3$ Fp >230°F   218.75   37.98*
Methyl jasmonate, 95+%   FW 224.30   172.43   37.72
    bp 110°C   133.99*   27.22
    d 1.030   124.90*   25.46
CDCl$_3$ QE-300   n$_D^{20}$1.4740   53.96*   20.59
    51.57*   14.10*
    38.77

Offset: 40 ppm.

ALDRICH

Aldrich E2,000-4 *CAS [1655-07-8]*
Ethyl 2-cyclohexanonecarboxylate, 95%

CDCl₃ QE-300

C₉H₁₄O₃ Fp 185°F
FW 170.21
bp 106°C (11 mm)
d 1.064
n²⁰_D 1.4770

60 MHz: *1*, 584B
FT-IR: *1*, 692D

206.19	60.12	23.31
172.71	57.22*	22.42
171.93	41.55	21.97
169.93	29.98	14.33*
97.72	29.10	14.17*
61.04	27.13	

A

Offset: 40 ppm.
Offset: 2.4 ppm.

Aldrich 32,062-5 *CAS [17159-79-4]*
Ethyl 4-oxocyclohexanecarboxylate, 97%

CDCl₃ QE-300

C₉H₁₄O₃ Fp >230°F
FW 170.21
bp 151°C (40 mm)
d 1.068
n²⁰_D 1.4590

VP-FT-IR: *3*, 744C

209.88
174.05
60.67
40.69*
39.71
28.53
14.22*

B

Offset: 40 ppm.

Aldrich E3,570-2 *CAS [13537-82-1]*
Ethyl 4-methyl-2-cyclohexanone-1-carboxylate,
contains 10% of the methyl ester

CDCl₃ QE-300

C₁₀H₁₆O₃ Fp >230°F
FW 184.24
bp 130°C (32 mm)
d 1.045
n²⁰_D 1.4730

206.44	61.28	49.87	30.07	14.31*
205.73	60.94	48.62	28.60	14.17*
172.64	60.86	37.20	28.27*	14.12*
171.51	60.13	34.73*	27.73	
170.10	56.82*	33.87*	22.07	
169.83	55.66*	32.34	21.46*	
97.23	50.64*	30.55	21.28*	

C

Offset: 40 ppm.
Offset: 2.4 ppm.

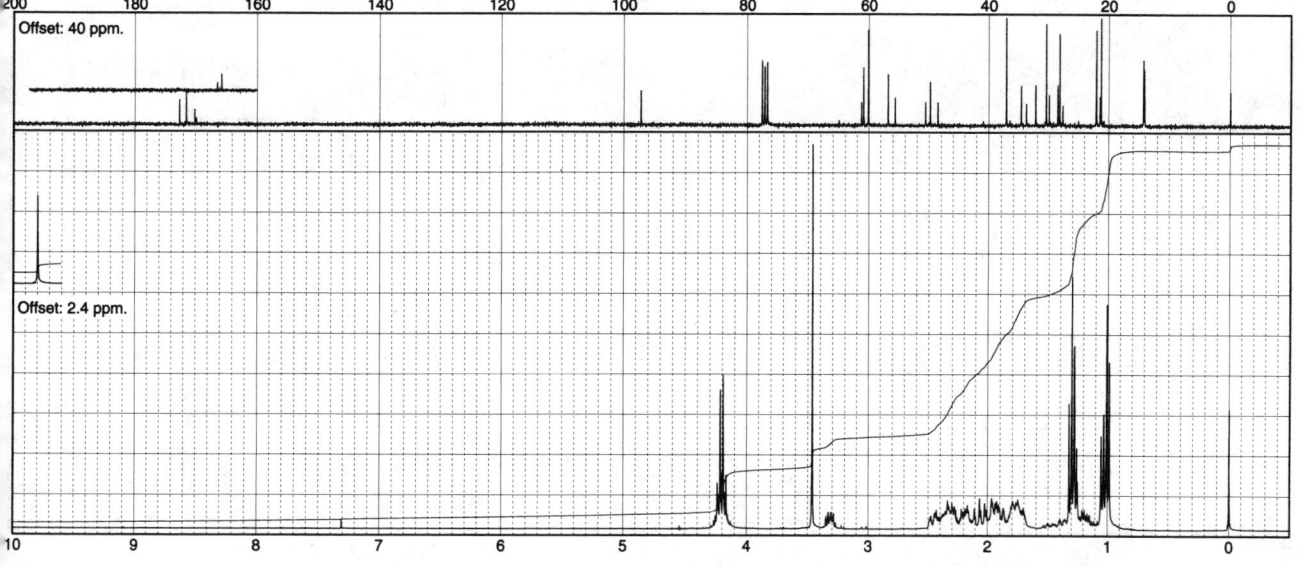
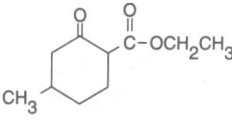

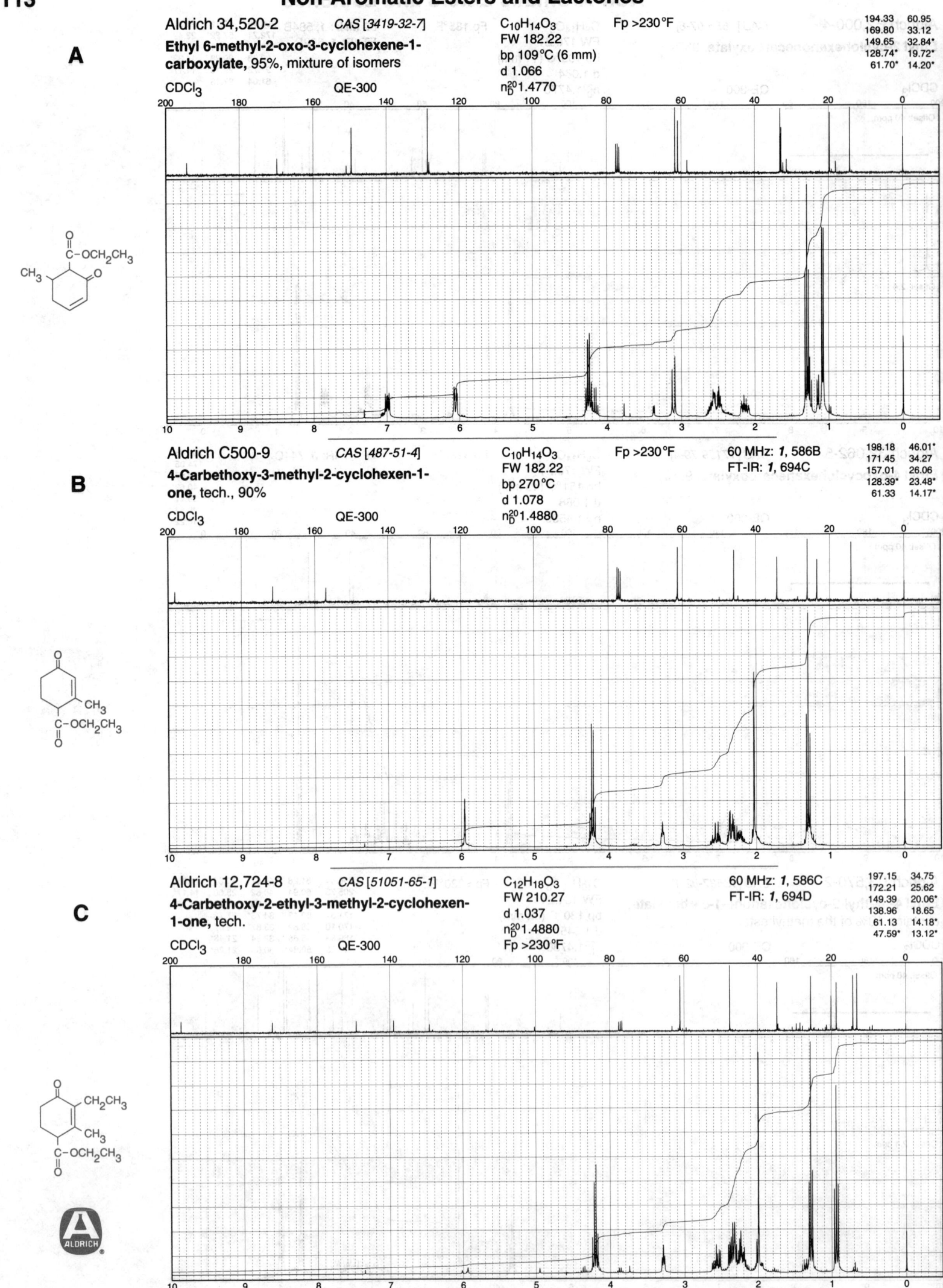

A

Aldrich 34,520-2   CAS [3419-32-7]

Ethyl 6-methyl-2-oxo-3-cyclohexene-1-carboxylate, 95%, mixture of isomers

CDCl₃   QE-300

$C_{10}H_{14}O_3$
FW 182.22
bp 109°C (6 mm)
d 1.066
n_D^{20}1.4770

Fp >230°F

194.33 60.95
169.80 33.12
149.65 32.84*
128.74* 19.72*
61.70* 14.20*

B

Aldrich C500-9   CAS [487-51-4]

4-Carbethoxy-3-methyl-2-cyclohexen-1-one, tech., 90%

CDCl₃   QE-300

$C_{10}H_{14}O_3$
FW 182.22
bp 270°C
d 1.078
n_D^{20}1.4880

Fp >230°F

60 MHz: 1, 586B
FT-IR: 1, 694C

198.18 46.01*
171.45 34.27
157.01 26.06
128.39* 23.48*
61.33 14.17*

C

Aldrich 12,724-8   CAS [51051-65-1]

4-Carbethoxy-2-ethyl-3-methyl-2-cyclohexen-1-one, tech.

CDCl₃   QE-300

$C_{12}H_{18}O_3$
FW 210.27
d 1.037
n_D^{20}1.4880
Fp >230°F

60 MHz: 1, 586C
FT-IR: 1, 694D

197.15 34.75
172.21 25.62
149.39 20.06*
138.96 18.65
61.13 14.18*
47.59* 13.12*

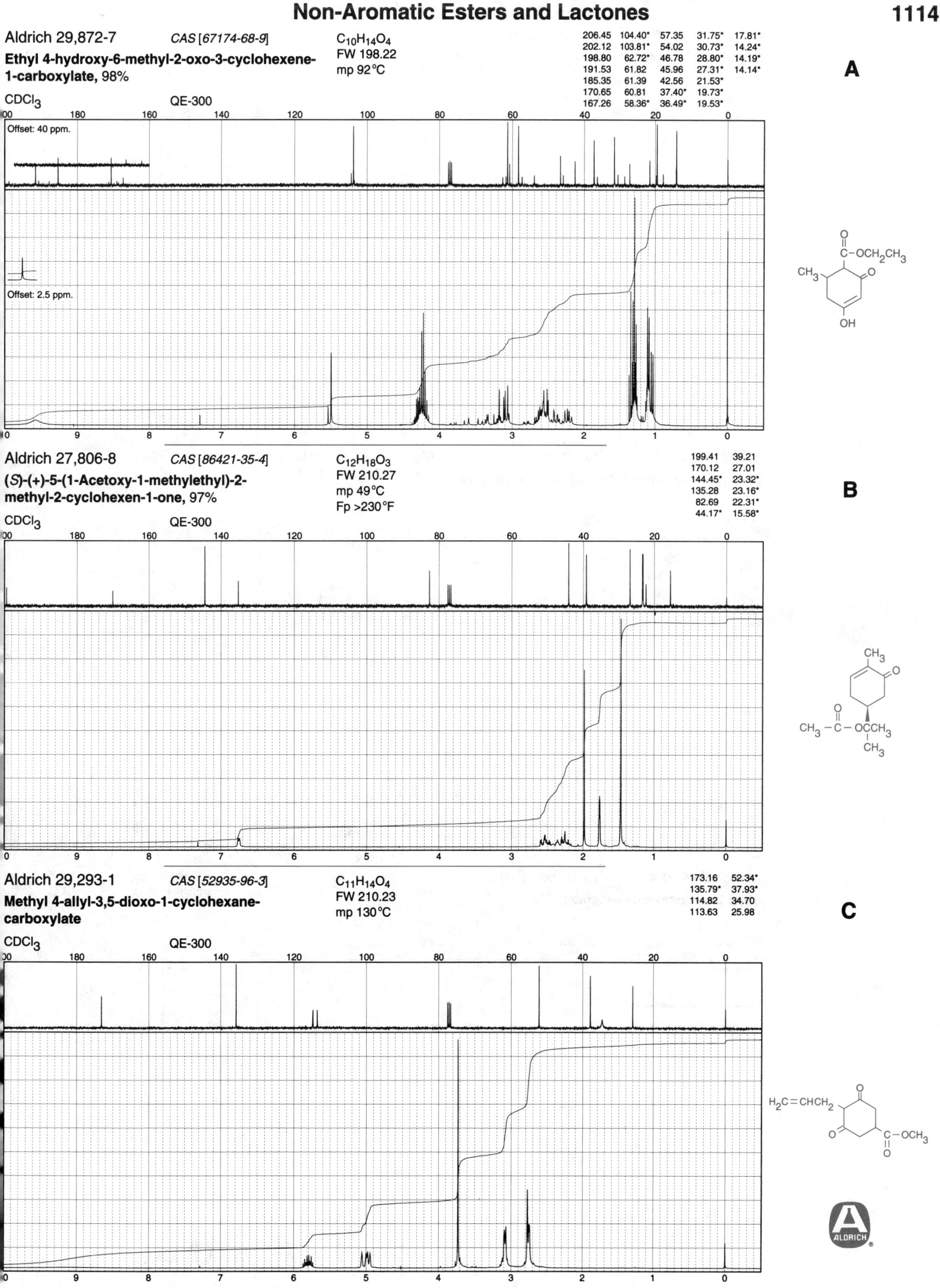

Aldrich 29,872-7 *CAS [67174-68-9]*

Ethyl 4-hydroxy-6-methyl-2-oxo-3-cyclohexene-1-carboxylate, 98%

CDCl₃   QE-300

$C_{10}H_{14}O_4$
FW 198.22
mp 92 °C

206.45	104.40*	57.35	31.75*	17.81*
202.12	103.81*	54.02	30.73*	14.24*
198.80	62.72*	46.78	28.80*	14.19*
191.53	61.82	45.96	27.31*	14.14*
185.35	61.39	42.56	21.53*	
170.65	60.81	37.40*	19.73*	
167.26	58.36*	36.49*	19.53*	

A

Offset: 40 ppm.

Offset: 2.5 ppm.

Aldrich 27,806-8 *CAS [86421-35-4]*

(S)-(+)-5-(1-Acetoxy-1-methylethyl)-2-methyl-2-cyclohexen-1-one, 97%

CDCl₃   QE-300

$C_{12}H_{18}O_3$
FW 210.27
mp 49 °C
Fp >230 °F

199.41	39.21
170.12	27.01
144.45*	23.32*
135.28	23.16*
82.69	22.31*
44.17*	15.58*

B

Aldrich 29,293-1 *CAS [52935-96-3]*

Methyl 4-allyl-3,5-dioxo-1-cyclohexane-carboxylate

CDCl₃   QE-300

$C_{11}H_{14}O_4$
FW 210.23
mp 130 °C

173.16	52.34*
135.79*	37.93*
114.82	34.70
113.63	25.98

C

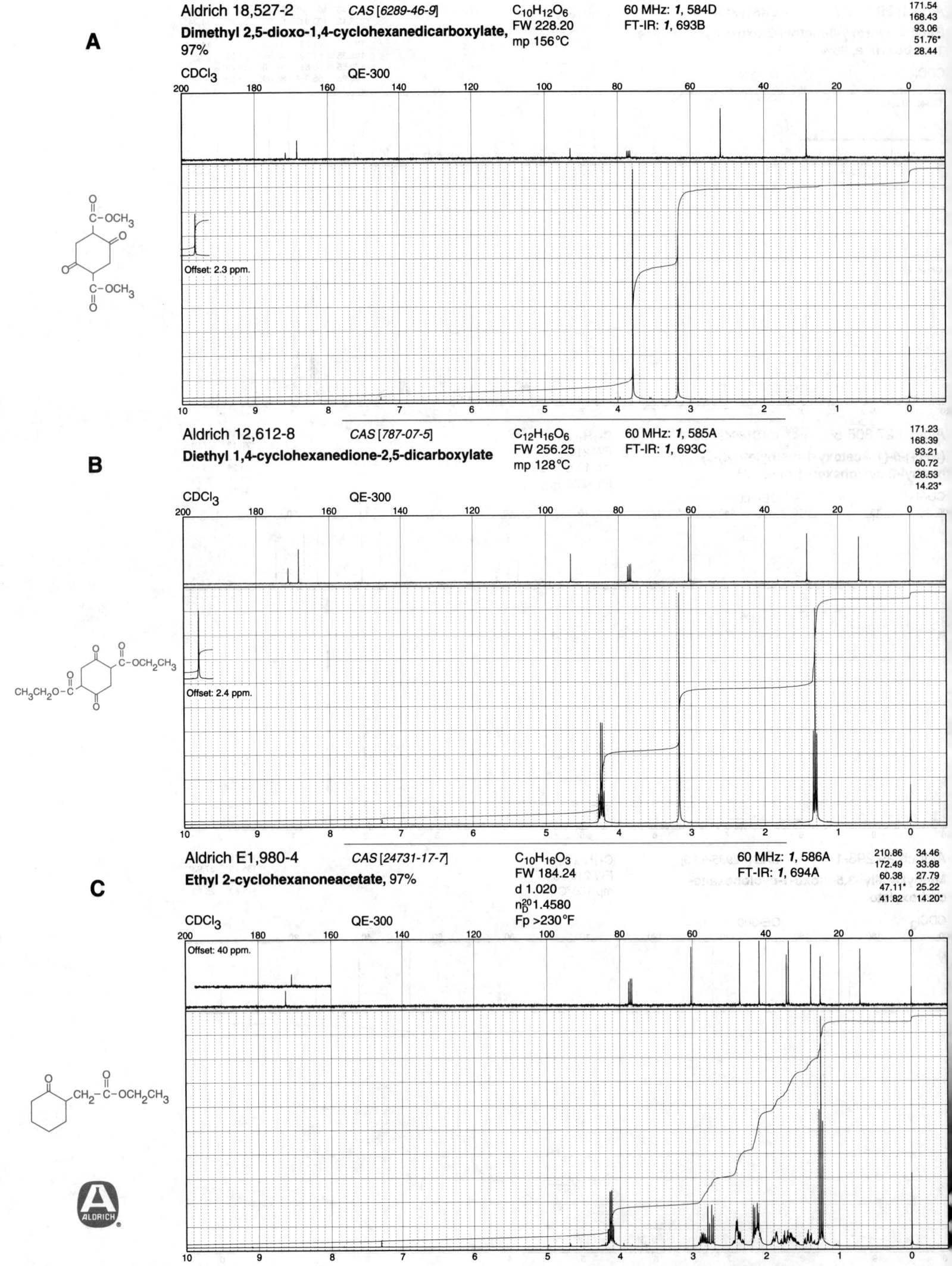

A

Aldrich 18,527-2   CAS [6289-46-9] $C_{10}H_{12}O_6$ 60 MHz: *1*, 584D

Dimethyl 2,5-dioxo-1,4-cyclohexanedicarboxylate,   FW 228.20   FT-IR: *1*, 693B

97%   mp 156°C

171.54
168.43
93.06
51.76*
28.44

CDCl$_3$ QE-300

Offset: 2.3 ppm.

B

Aldrich 12,612-8   CAS [787-07-5] $C_{12}H_{16}O_6$ 60 MHz: *1*, 585A

Diethyl 1,4-cyclohexanedione-2,5-dicarboxylate   FW 256.25   FT-IR: *1*, 693C

mp 128°C

171.23
168.39
93.21
60.72
28.53
14.23*

CDCl$_3$ QE-300

Offset: 2.4 ppm.

C

Aldrich E1,980-4   CAS [24731-17-7] $C_{10}H_{16}O_3$ 60 MHz: *1*, 586A

Ethyl 2-cyclohexanoneacetate, 97%   FW 184.24   FT-IR: *1*, 694A

d 1.020
n$_D^{20}$1.4580
Fp >230°F

210.86   34.46
172.49   33.88
60.38   27.79
47.11*   25.22
41.82   14.20*

CDCl$_3$ QE-300

Offset: 40 ppm.

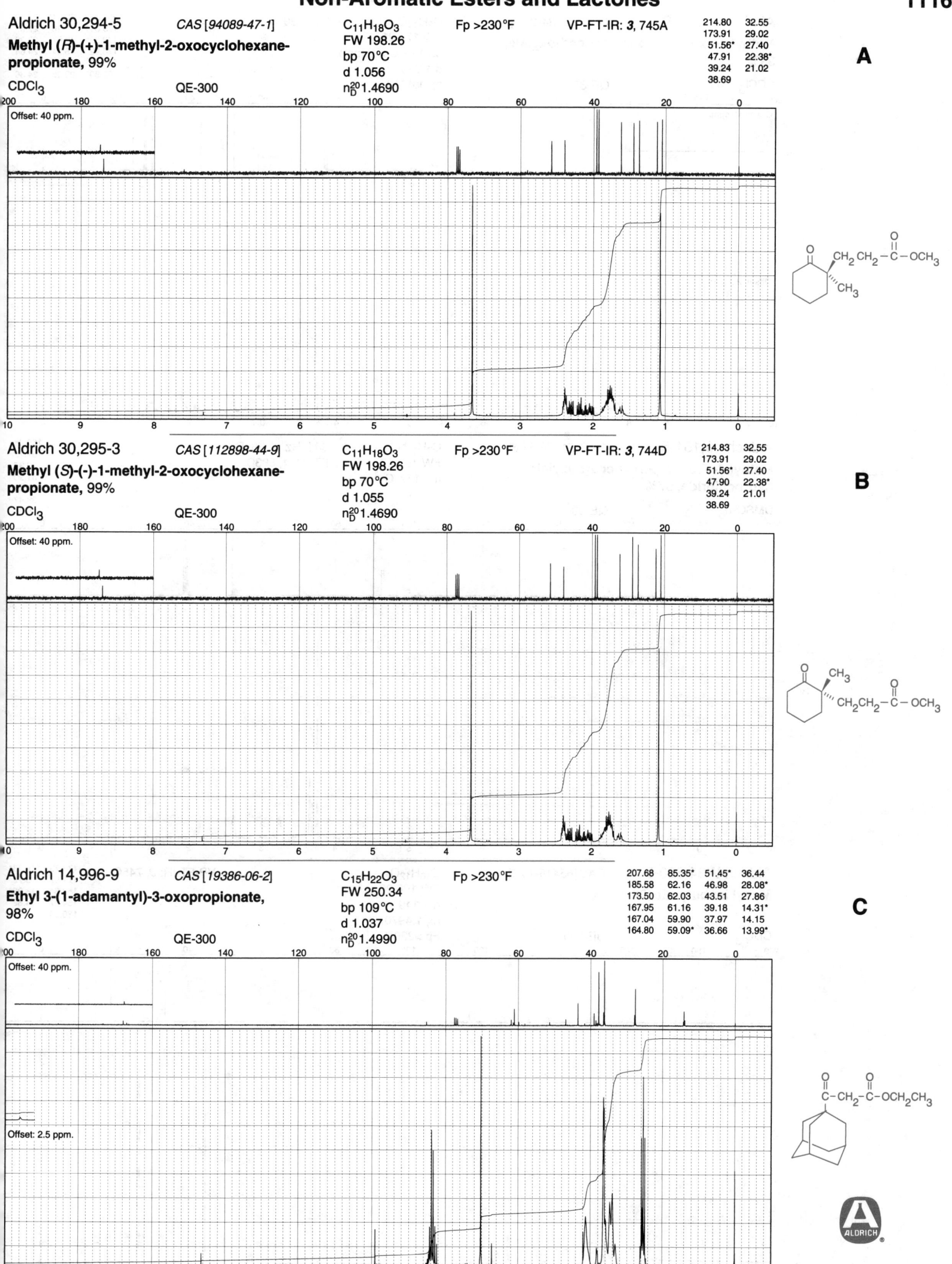

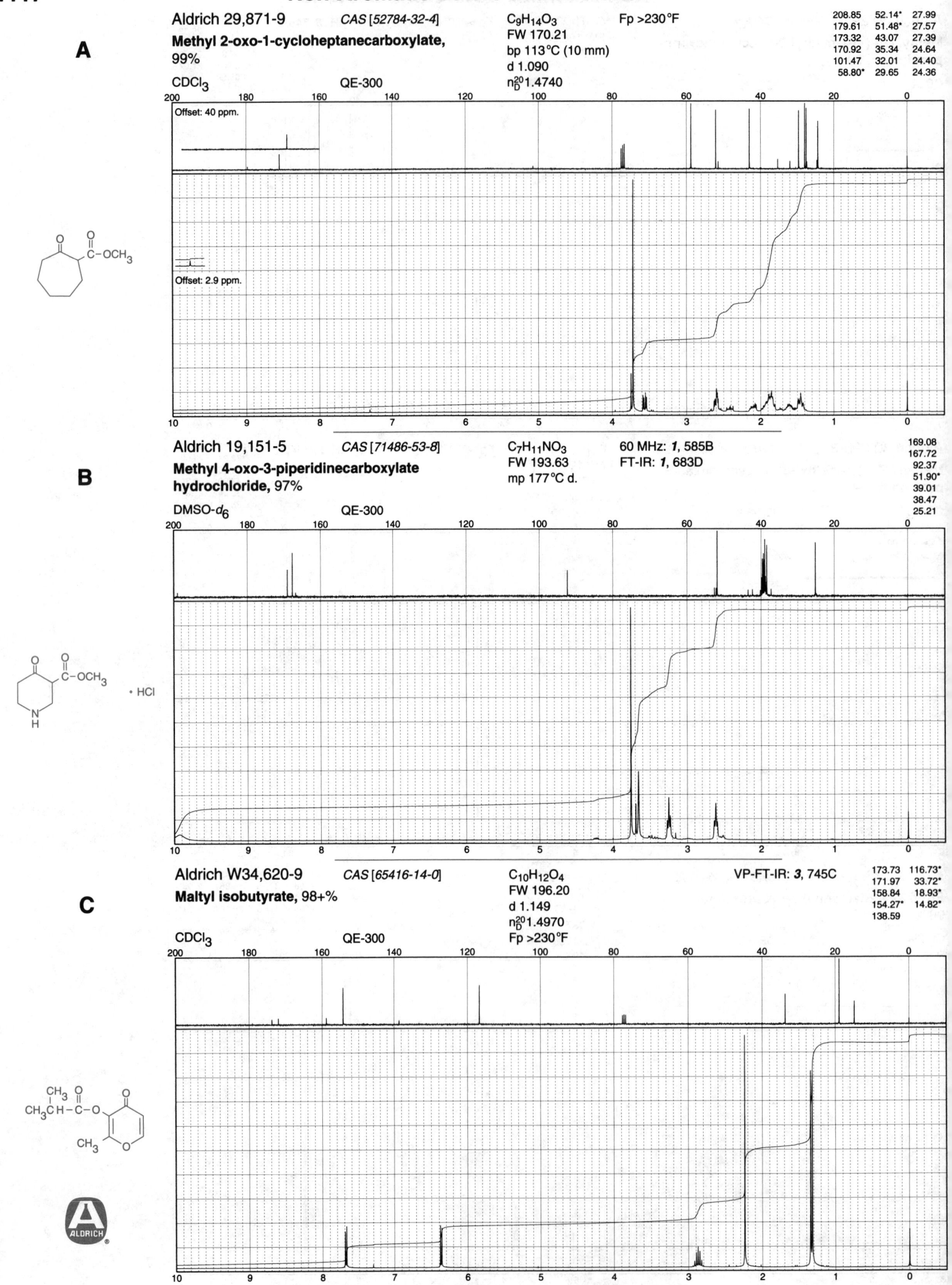

A

Aldrich 29,871-9   *CAS [52784-32-4]*   C$_9$H$_{14}$O$_3$ Fp >230°F

Methyl 2-oxo-1-cycloheptanecarboxylate,
99%

FW 170.21
bp 113°C (10 mm)
d 1.090
n$_D^{20}$1.4740

208.85	52.14*	27.99
179.61	51.48*	27.57
173.32	43.07	27.39
170.92	35.34	24.64
101.47	32.01	24.40
58.80*	29.65	24.36

CDCl$_3$ QE-300

Offset: 40 ppm.

Offset: 2.9 ppm.

B

Aldrich 19,151-5   *CAS [71486-53-8]*   C$_7$H$_{11}$NO$_3$ 60 MHz: *1*, 585B

Methyl 4-oxo-3-piperidinecarboxylate
hydrochloride, 97%

FW 193.63
mp 177°C d.

FT-IR: *1*, 683D

169.08
167.72
92.37
51.90*
39.01
38.47
25.21

DMSO-d$_6$ QE-300

· HCl

C

Aldrich W34,620-9   *CAS [65416-14-0]*   C$_{10}$H$_{12}$O$_4$ VP-FT-IR: *3*, 745C

Maltyl isobutyrate, 98+%

FW 196.20
d 1.149
n$_D^{20}$1.4970
Fp >230°F

173.73	116.73*
171.97	33.72*
158.84	18.93*
154.27*	14.82*
138.59	

CDCl$_3$ QE-300

A

Aldrich 18,605-8 CAS [532-34-3]

Butopyronoxyl, 85+%

C$_{12}$H$_{18}$O$_4$
FW 226.27
d 1.054
n$_D^{20}$1.4770
Fp >230°F

60 MHz: *1*, 599C
FT-IR: *1*, 693D

193.66	47.61
162.05	30.38
157.20	25.78*
107.24*	19.06
82.51	13.65*
66.20	

CDCl$_3$ QE-300

B

Aldrich 36,089-9 CAS [6148-64-7]

Ethyl malonate, potassium salt, 98%

C$_5$H$_8$O$_4$
FW 170.21
mp 200°C d.

176.78
174.25
64.79
47.52
16.06*

D$_2$O QE-300

C

Aldrich M8,110-1 CAS [3878-55-5]

***mono*-Methyl succinate, 95%**

C$_5$H$_8$O$_4$
FW 132.12
mp 58°C
bp 151°C (20 mm)

60 MHz: *1*, 588A
FT-IR: *1*, 695B
VP-FT-IR: *3*, 745D

173.34
172.37
51.29*
28.74
28.65

CDCl$_3$ + DMSO-d_6 QE-300

Offset: 2.2 ppm.

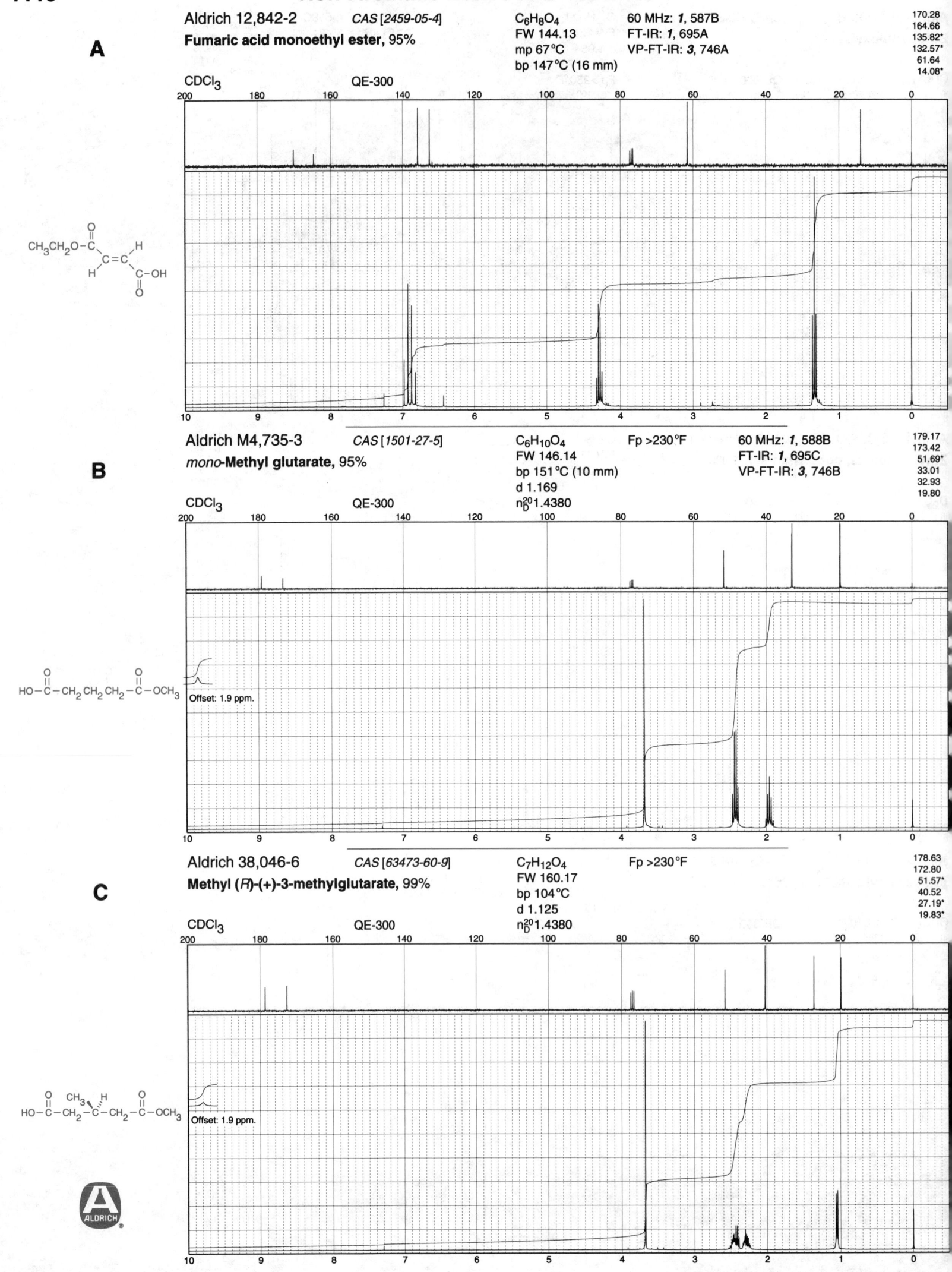

A

Aldrich 12,842-2   CAS [2459-05-4] $C_6H_8O_4$ 60 MHz: *1*, 587B   170.28
Fumaric acid monoethyl ester, 95%   FW 144.13   FT-IR: *1*, 695A   164.66
    mp 67°C   VP-FT-IR: *3*, 746A   135.82*
    bp 147°C (16 mm)   132.57*
            61.64
            14.08*

CDCl₃   QE-300

B

Aldrich M4,735-3   CAS [1501-27-5] $C_6H_{10}O_4$ Fp >230°F   60 MHz: *1*, 588B   179.17
mono-**Methyl glutarate, 95%**   FW 146.14   FT-IR: *1*, 695C   173.42
    bp 151°C (10 mm)   VP-FT-IR: *3*, 746B   51.69*
    d 1.169   33.01
    n_D^{20}1.4380   32.93
            19.80

CDCl₃   QE-300

Offset: 1.9 ppm.

C

Aldrich 38,046-6   CAS [63473-60-9] $C_7H_{12}O_4$ Fp >230°F   178.63
Methyl (*R*)-(+)-3-methylglutarate, 99%   FW 160.17   172.80
    bp 104°C   51.57*
    d 1.125   40.52
    n_D^{20}1.4380   27.19*
            19.83*

CDCl₃   QE-300

Offset: 1.9 ppm.

ALDRICH

Non-Aromatic Esters and Lactones

Aldrich 85,826-9   *CAS [1499-55-4]* $C_6H_{11}NO_4$ FT-IR: *1*, 678B

L-Glutamic acid 5-methyl ester, 99%   FW 161.16   mp 182°C d.

178.04
176.67
56.75*
55.17*
32.51
28.28

A

D_2O QE-300

CH₃O−C−CH₂CH₂−C−C−OH · HCl

Aldrich A2,640-3   *CAS [627-91-8]* $C_7H_{12}O_4$ n_D^{20} 1.4400

Adipic acid monomethyl ester, 99%   FW 160.17   Fp >230°F   mp 9°C   60 MHz: *1*, 588C   bp 162°C (10 mm)   FT-IR: *1*, 695D   d 1.081   VP-FT-IR: *3*, 746C

179.58
173.81
51.61*
33.65
24.27
24.06

B

$CDCl_3$ QE-300

HO−C−CH₂CH₂CH₂CH₂−C−OCH₃

Aldrich 12,276-9   *CAS [626-86-8]* $C_8H_{14}O_4$ n_D^{20} 1.4390

Adipic acid monoethyl ester, 98%   FW 174.20   Fp >230°F   mp 29°C   60 MHz: *1*, 588D   bp 180°C (18 mm)   FT-IR: *1*, 696A   d 0.980   VP-FT-IR: *3*, 746D

179.54   33.67
173.39   24.30
60.41   24.07
33.92   14.22*

C

$CDCl_3$ QE-300

HO−C−CH₂CH₂CH₂CH₂−C−OCH₂CH₃

ALDRICH

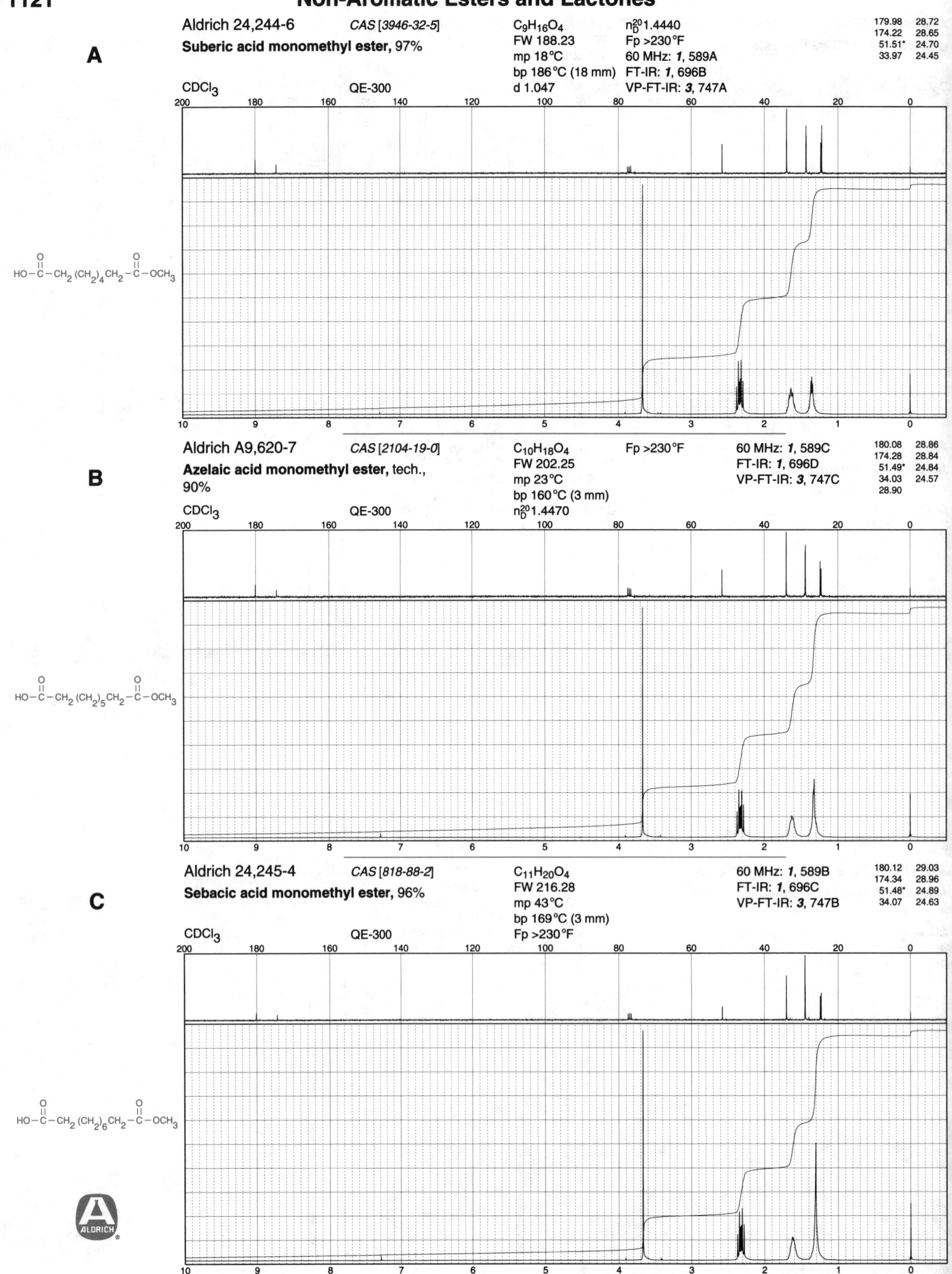

A

Aldrich 24,244-6   CAS [3946-32-5]

Suberic acid monomethyl ester, 97%

$C_9H_{16}O_4$
FW 188.23
mp 18°C
bp 186°C (18 mm)
d 1.047

n_D^{20} 1.4440
Fp >230°F
60 MHz: **1**, 589A
FT-IR: **1**, 696B
VP-FT-IR: **3**, 747A

179.98   28.72
174.22   28.65
51.51*   24.70
33.97   24.45

CDCl₃   QE-300

$HO-\overset{O}{\underset{}{C}}-CH_2\ (CH_2)_4\ CH_2-\overset{O}{\underset{}{C}}-OCH_3$

B

Aldrich A9,620-7   CAS [2104-19-0]

Azelaic acid monomethyl ester, tech., 90%

$C_{10}H_{18}O_4$
FW 202.25
mp 23°C
bp 160°C (3 mm)
n_D^{20} 1.4470

Fp >230°F

60 MHz: **1**, 589C
FT-IR: **1**, 696D
VP-FT-IR: **3**, 747C

180.08   28.86
174.28   28.84
51.49*   24.84
34.03   24.57
28.90

CDCl₃   QE-300

$HO-\overset{O}{\underset{}{C}}-CH_2\ (CH_2)_5\ CH_2-\overset{O}{\underset{}{C}}-OCH_3$

C

Aldrich 24,245-4   CAS [818-88-2]

Sebacic acid monomethyl ester, 96%

$C_{11}H_{20}O_4$
FW 216.28
mp 43°C
bp 169°C (3 mm)
Fp >230°F

60 MHz: **1**, 589B
FT-IR: **1**, 696C
VP-FT-IR: **3**, 747B

180.12   29.03
174.34   28.96
51.48*   24.89
34.07   24.63

CDCl₃   QE-300

$HO-\overset{O}{\underset{}{C}}-CH_2\ (CH_2)_6\ CH_2-\overset{O}{\underset{}{C}}-OCH_3$

Aldrich 30,234-1 *CAS [13831-30-6]* $C_4H_6O_4$ VP-FT-IR: *3*, 747D
Acetoxyacetic acid, 99%
FW 118.09
mp 68°C
bp 142°C (12 mm)

173.32
170.62
60.27
20.37*

A

CDCl$_3$ QE-300

Offset: 1.1 ppm.

$CH_3-\overset{O}{\overset{\|}{C}}-O\,CH_2-\overset{O}{\overset{\|}{C}}-OH$

Aldrich 36,278-6 *CAS [6034-46-4]* $C_5H_8O_4$ Fp >230°F
(S)-(-)-2-Acetoxypropionic acid, 99%
FW 132.12
bp 116°C (2 mm)
d 1.183
n_D^{20} 1.4230

176.53
170.66
68.24*
20.57*
16.77*

B

CDCl$_3$ QE-300

Offset: 1.9 ppm.

$CH_3-\overset{O}{\overset{\|}{C}}-O$
$CH_3-\overset{|}{\underset{H}{C}}-\overset{O}{\overset{\|}{C}}-OH$

Aldrich 17,302-9 *CAS [6326-83-6]* $C_5H_6O_4S_3$ 60 MHz: *1*, 543A
Bis(carboxymethyl) trithiocarbonate, FW 226.29 FT-IR: *1*, 612C
98% mp 174°C

222.22
167.89
38.71

C

DMSO-d_6 QE-300

Offset: 40 ppm.

$HO-\overset{O}{\overset{\|}{C}}-CH_2\,S-\overset{S}{\overset{\|}{C}}-S\,CH_2-\overset{O}{\overset{\|}{C}}-OH$

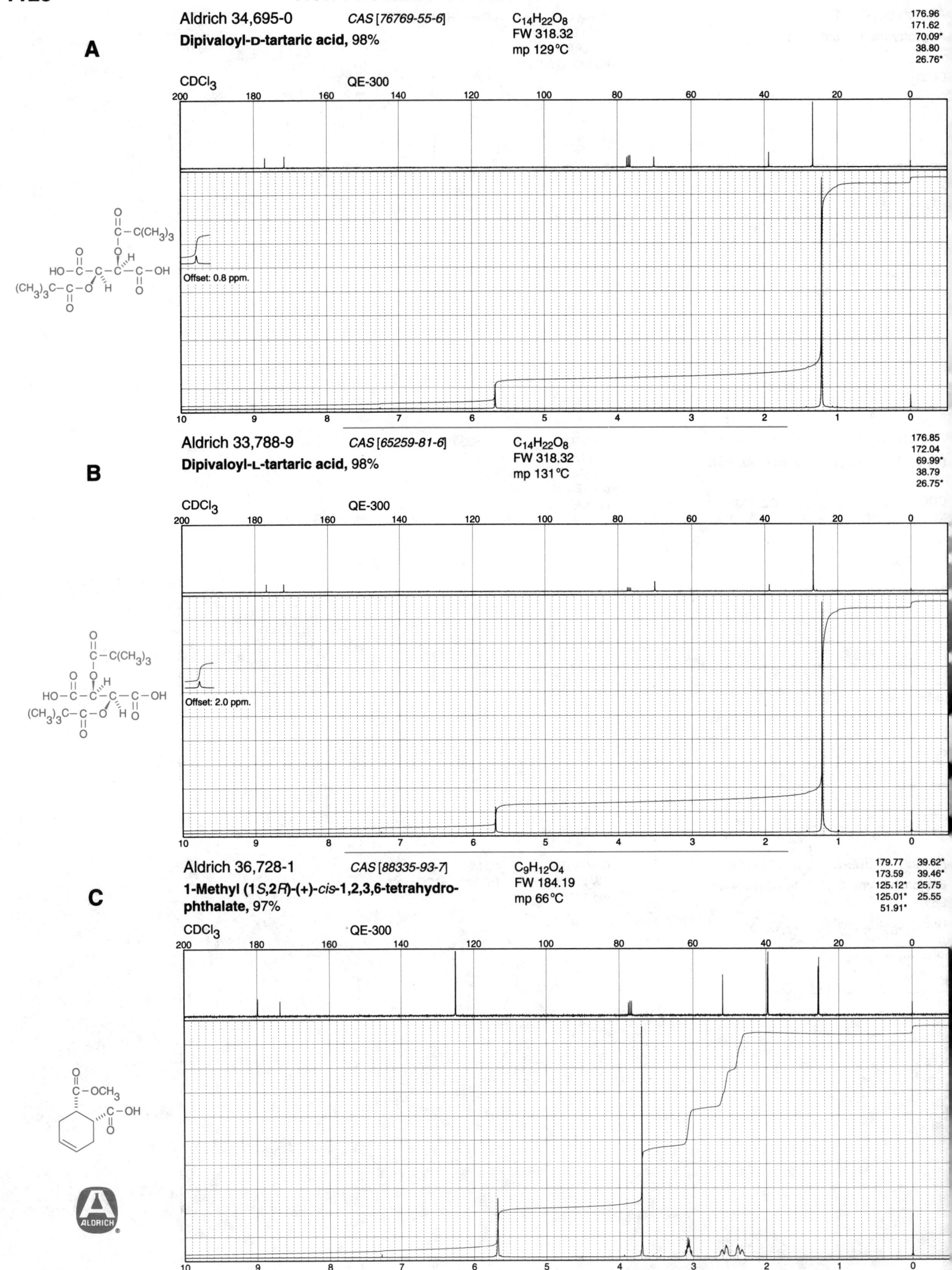

A

Aldrich 34,695-0   *CAS [76769-55-6]* $C_{14}H_{22}O_8$
Dipivaloyl-D-tartaric acid, 98%   FW 318.32
mp 129°C

176.96
171.62
70.09*
38.80
26.76*

CDCl$_3$ QE-300

Offset: 0.8 ppm.

B

Aldrich 33,788-9   *CAS [65259-81-6]* $C_{14}H_{22}O_8$
Dipivaloyl-L-tartaric acid, 98%   FW 318.32
mp 131°C

176.85
172.04
69.99*
38.79
26.75*

CDCl$_3$ QE-300

Offset: 2.0 ppm.

C

Aldrich 36,728-1   *CAS [88335-93-7]* $C_9H_{12}O_4$
1-Methyl (1S,2R)-(+)-*cis*-1,2,3,6-tetrahydro-   FW 184.19
phthalate, 97%   mp 66°C

179.77   39.62*
173.59   39.46*
125.12*   25.75
125.01*   25.55
51.91*

CDCl$_3$ QE-300

Aldrich 34,756-6 CAS [96185-91-0] $C_{10}H_{12}O_4$

mono-Methyl *cis*-5-norbornene-*endo*-2,3-dicarboxylate, 98%

FW 196.20
mp 104°C

178.62	48.75
172.79	48.23*
135.49*	48.04*
134.23*	46.54*
51.47*	46.05*

A

CDCl₃ QE-300

Aldrich 21,912-6 CAS [36536-46-6] $C_4H_6O_2$ n_D^{20} 1.4110

(±)-β-Butyrolactone, 98+%

FW 86.09
mp -44°C
bp 72°C (29 mm)
d 1.056

Fp 140°F
60 MHz: *1*, 589D
FT-IR: *1*, 697A
VP-FT-IR: *3*, 748A

| 168.20 |
| 68.03* |
| 44.30 |
| 20.60* |

B

CDCl₃ QE-300

Aldrich 30,205-8 CAS [674-82-8] $C_4H_4O_2$ Fp 94°F

Diketene

FW 84.07
bp 70°C (100 mm)
d 1.090
n_D^{20} 1.4390

| 165.13 |
| 147.65 |
| 86.99 |
| 42.38 |

C

CDCl₃ QE-300

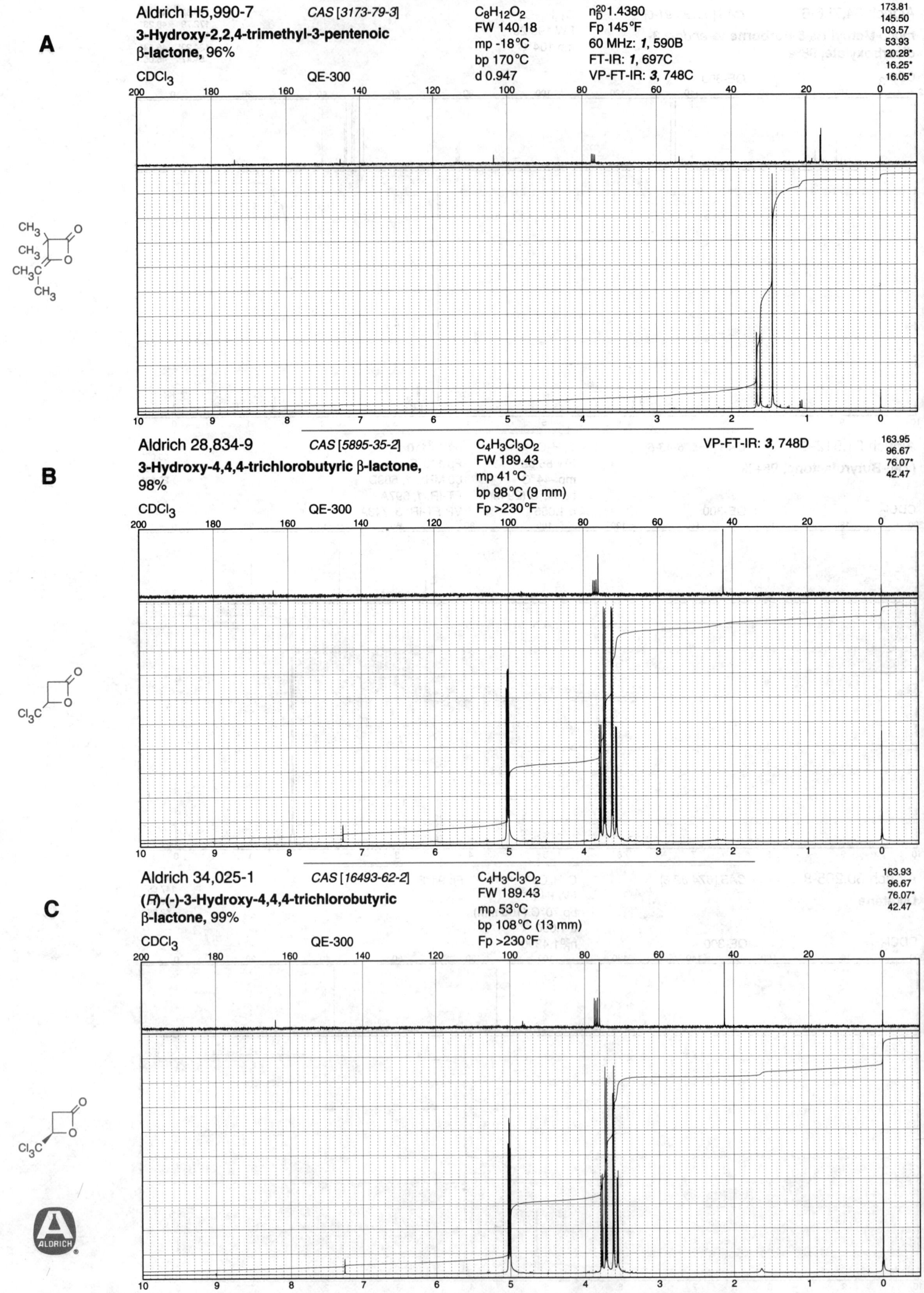

A

Aldrich H5,990-7   *CAS [3173-79-3]*

3-Hydroxy-2,2,4-trimethyl-3-pentenoic β-lactone, 96%

CDCl₃   QE-300

$C_8H_{12}O_2$
FW 140.18
mp -18°C
bp 170°C
d 0.947

n_D^{20} 1.4380
Fp 145°F
60 MHz: *1*, 590B
FT-IR: *1*, 697C
VP-FT-IR: *3*, 748C

173.81
145.50
103.57
53.93
20.28*
16.25*
16.05*

B

Aldrich 28,834-9   *CAS [5895-35-2]*

3-Hydroxy-4,4,4-trichlorobutyric β-lactone, 98%

CDCl₃   QE-300

$C_4H_3Cl_3O_2$
FW 189.43
mp 41°C
bp 98°C (9 mm)
Fp >230°F

VP-FT-IR: *3*, 748D

163.95
96.67
76.07*
42.47

C

Aldrich 34,025-1   *CAS [16493-62-2]*

(R)-(-)-3-Hydroxy-4,4,4-trichlorobutyric β-lactone, 99%

CDCl₃   QE-300

$C_4H_3Cl_3O_2$
FW 189.43
mp 53°C
bp 108°C (13 mm)
Fp >230°F

163.93
96.67
76.07*
42.47

A

Aldrich 34,027-8   CAS [16493-63-3]

(S)-(+)-3-Hydroxy-4,4,4-trichlorobutyric
β-lactone, 99%

CDCl₃   QE-300

C₄H₃Cl₃O₂
FW 189.43
mp 53°C
bp 103°C (12 mm)
Fp >230°F

163.94
96.67
76.07*
42.47

B

Aldrich 34,023-5   CAS [93239-42-0]

(R)-(+)-3-Hydroxy-3-methyl-4,4,4-tri-
chlorobutyric β-lactone, 98%

CDCl₃   QE-300

C₅H₅Cl₃O₂
FW 203.45
mp 44°C
bp 120°C
Fp >230°F

164.09
101.80
81.77
47.74
20.90*

C

Aldrich 34,024-3   CAS [93206-60-1]

(S)-(-)-3-Hydroxy-3-methyl-4,4,4-tri-
chlorobutyric β-lactone, 98%

CDCl₃   QE-300

C₅H₅Cl₃O₂
FW 203.45
mp 44°C
bp 120°C
Fp >230°F

164.10
101.80
81.77
47.74
20.90*

Non-Aromatic Esters and Lactones

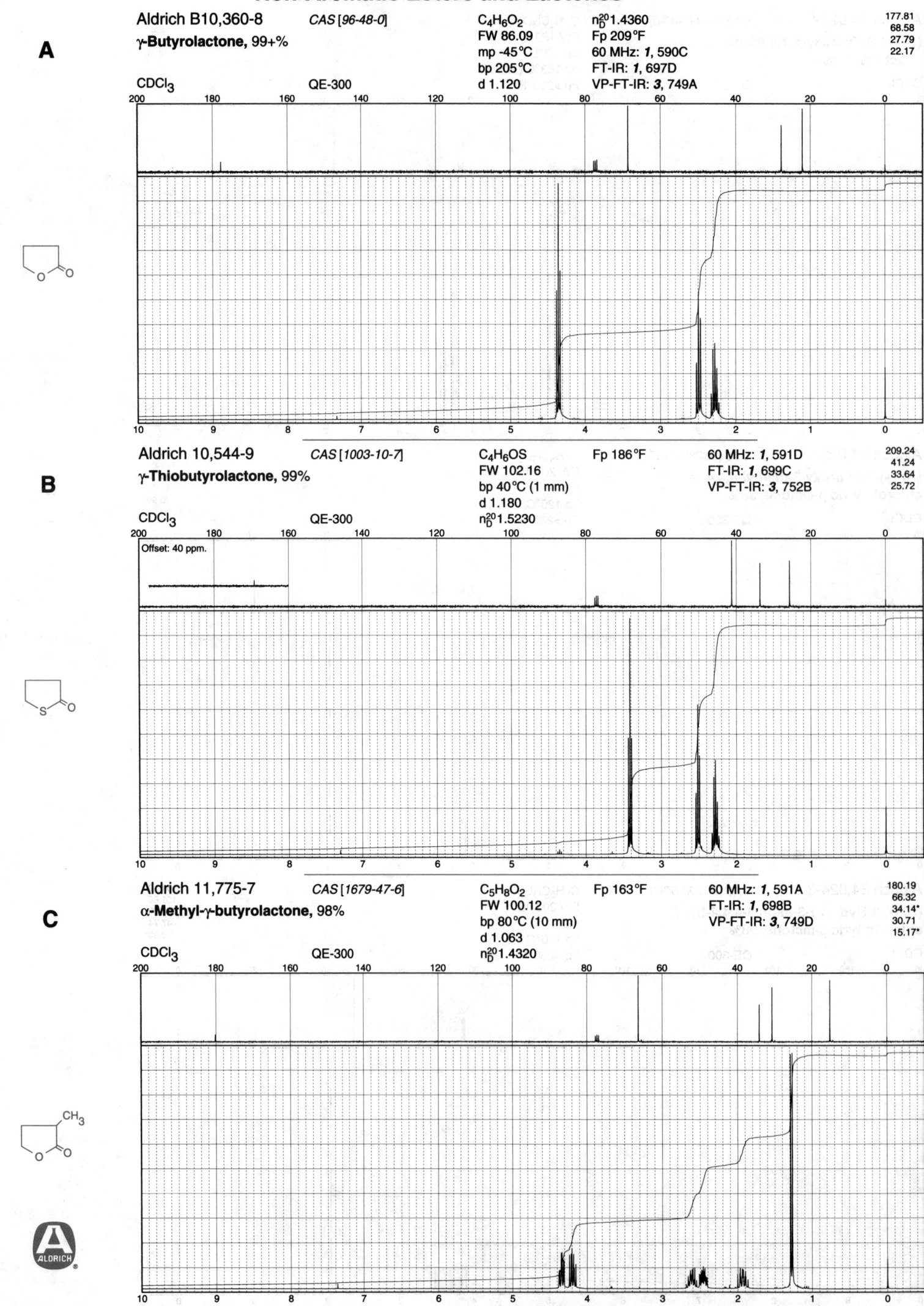

A

Aldrich B10,360-8 CAS [96-48-0]

γ-Butyrolactone, 99+%

C₄H₆O₂
FW 86.09
mp -45°C
bp 205°C
d 1.120

n_D^{20} 1.4360
Fp 209°F
60 MHz: *1*, 590C
FT-IR: *1*, 697D
VP-FT-IR: *3*, 749A

177.81
68.58
27.79
22.17

CDCl₃   QE-300

B

Aldrich 10,544-9 CAS [1003-10-7]

γ-Thiobutyrolactone, 99%

C₄H₆OS
FW 102.16
bp 40°C (1 mm)
d 1.180

Fp 186°F

60 MHz: *1*, 591D
FT-IR: *1*, 699C
VP-FT-IR: *3*, 752B

209.24
41.24
33.64
25.72

CDCl₃   QE-300 n_D^{20} 1.5230

Offset: 40 ppm.

C

Aldrich 11,775-7 CAS [1679-47-6]

α-Methyl-γ-butyrolactone, 98%

C₅H₈O₂
FW 100.12
bp 80°C (10 mm)
d 1.063

Fp 163°F

60 MHz: *1*, 591A
FT-IR: *1*, 698B
VP-FT-IR: *3*, 749D

180.19
66.32
34.14*
30.71
15.17*

CDCl₃   QE-300 n_D^{20} 1.4320

Aldrich 33,371-9   CAS [64190-48-3]   C$_5$H$_8$O$_2$ Fp 193°F   VP-FT-IR: 3, 750A

(S)-(-)-β-Methyl-γ-butyrolactone, 97%

177.27
74.69
36.09
30.36*
17.88*

A

CDCl$_3$ QE-300   n$_D^{20}$1.4320

Aldrich V40-3   CAS [108-29-2]   C$_5$H$_8$O$_2$ n$_D^{20}$1.4330

γ-Valerolactone, 99%   FW 100.12   Fp 178°F

177.24
77.28*
29.68
29.09
21.04*

B

mp -31°C   60 MHz: 1, 590D
bp 208°C   FT-IR: 1, 698A

CDCl$_3$ QE-300   d 1.057   VP-FT-IR: 3, 749C

Aldrich 30,383-6   CAS [695-06-7]   C$_6$H$_{10}$O$_2$ Fp 209°F   VP-FT-IR: 3, 750C

γ-Caprolactone, 98%   FW 114.14
bp 219°C
d 1.027

177.12
82.13*
28.83
28.51
27.48
9.38*

C

CDCl$_3$ QE-300   n$_D^{20}$1.4390

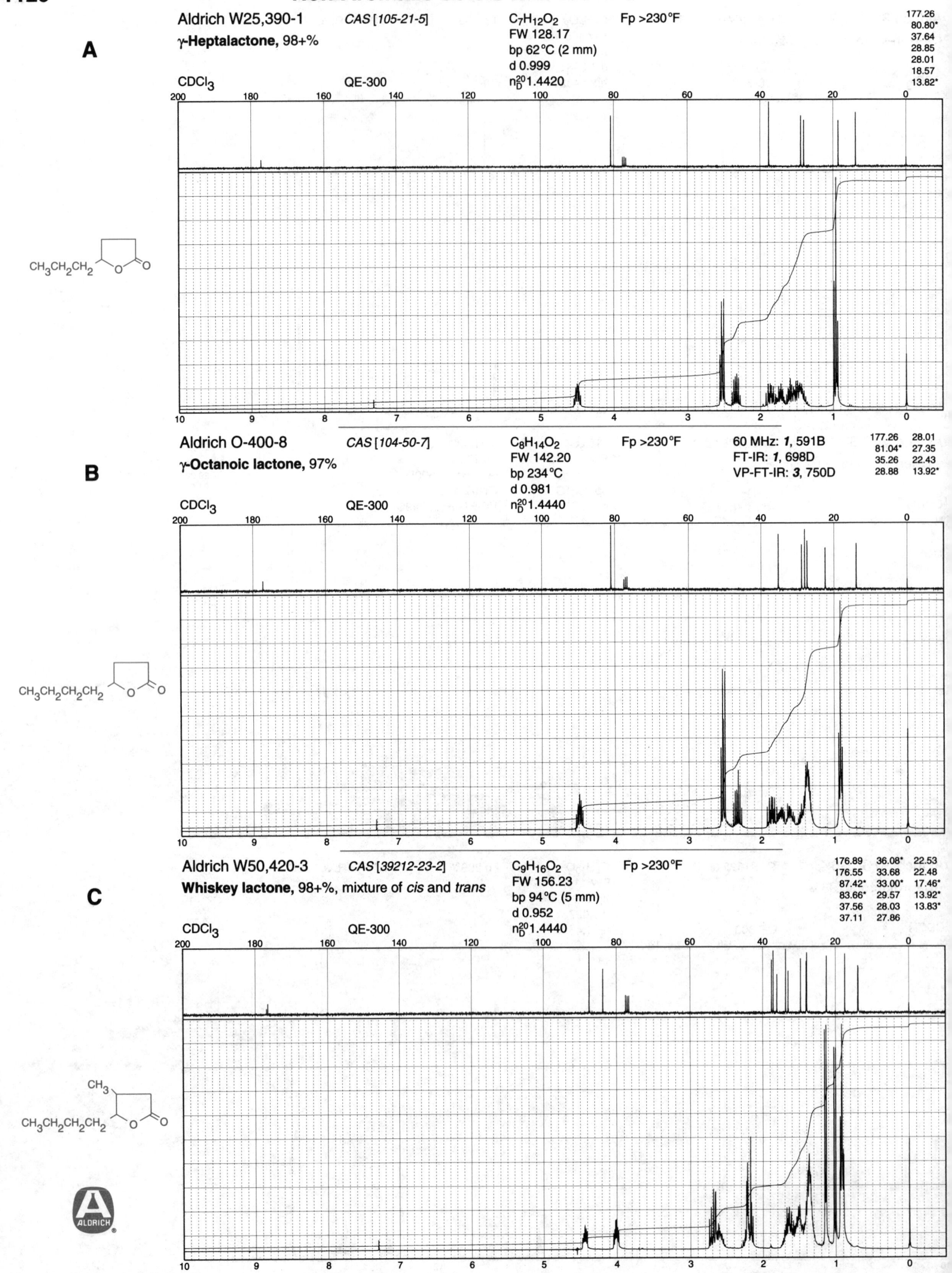

A

Aldrich W25,390-1 *CAS [105-21-5]* C$_7$H$_{12}$O$_2$ Fp >230°F

γ-**Heptalactone**, 98+%

FW 128.17
bp 62°C (2 mm)
d 0.999
n$_D^{20}$1.4420

CDCl$_3$ QE-300

177.26
80.80*
37.64
28.85
28.01
18.57
13.82*

CH$_3$CH$_2$CH$_2$

B

Aldrich O-400-8 *CAS [104-50-7]* C$_8$H$_{14}$O$_2$ Fp >230°F

γ-**Octanoic lactone**, 97%

FW 142.20
bp 234°C
d 0.981
n$_D^{20}$1.4440

60 MHz: *1*, 591B
FT-IR: *1*, 698D
VP-FT-IR: *3*, 750D

177.26 28.01
81.04* 27.35
35.26 22.43
28.88 13.92*

CDCl$_3$ QE-300

CH$_3$CH$_2$CH$_2$CH$_2$

C

Aldrich W50,420-3 *CAS [39212-23-2]* C$_9$H$_{16}$O$_2$ Fp >230°F

Whiskey lactone, 98+%, mixture of *cis* and *trans*

FW 156.23
bp 94°C (5 mm)
d 0.952
n$_D^{20}$1.4440

176.89 36.08* 22.53
176.55 33.68 22.48
87.42* 33.00* 17.46*
83.66* 29.57 13.92*
37.56 28.03 13.83*
37.11 27.86

CDCl$_3$ QE-300

CH$_3$

CH$_3$CH$_2$CH$_2$CH$_2$

Aldrich 29,237-0   *CAS [104-61-0]*

γ-Nonanoic lactone, 97%

C₉H₁₆O₂	Fp >230°F
FW 156.23	
bp 122°C (6 mm)	
d 0.976	
n²⁰_D 1.4470	

$C_9H_{16}O_2$ Fp >230°F
FW 156.23
bp 122°C (6 mm)
d 0.976
n_D^{20} 1.4470

VP-FT-IR: *3*, 751B

177.25	28.01
81.06*	24.91
35.54	22.50
31.51	13.96*
28.87	

A

CDCl₃   QE-300

$CH_3(CH_2)_3CH_2$ — (lactone ring)

Aldrich D80-4   *CAS [706-14-9]*

γ-Decanolactone, 99%

$C_{10}H_{18}O_2$
FW 170.25
d 0.948
n_D^{20} 1.4490
Fp >230°F

60 MHz: *1*, 591C
FT-IR: *1*, 699A
VP-FT-IR: *3*, 751C

177.23	28.86
81.04*	28.01
35.58	25.20
31.65	22.53
29.00	14.04*

B

CDCl₃   QE-300

$CH_3(CH_2)_4CH_2$ — (lactone ring)

Aldrich U80-6   *CAS [104-67-6]*

Undecanoic γ-lactone, 99%

$C_{11}H_{20}O_2$ Fp >230°F
FW 184.28
bp 165°C (13 mm)
d 0.949
n_D^{20} 1.4500

VP-FT-IR: *3*, 751D

177.23	28.86
81.05*	28.02
35.58	25.24
31.73	22.62
29.30	14.07*
29.13	

C

CDCl₃   QE-300

$CH_3(CH_2)_5CH_2$ — (lactone ring)

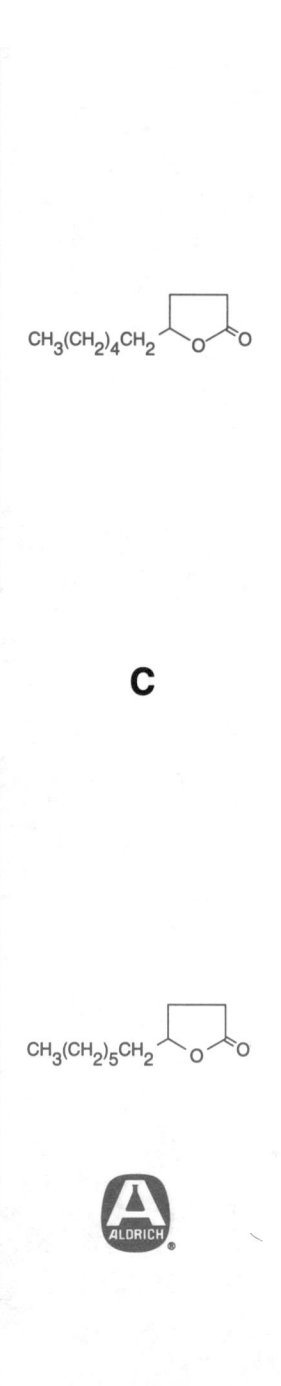

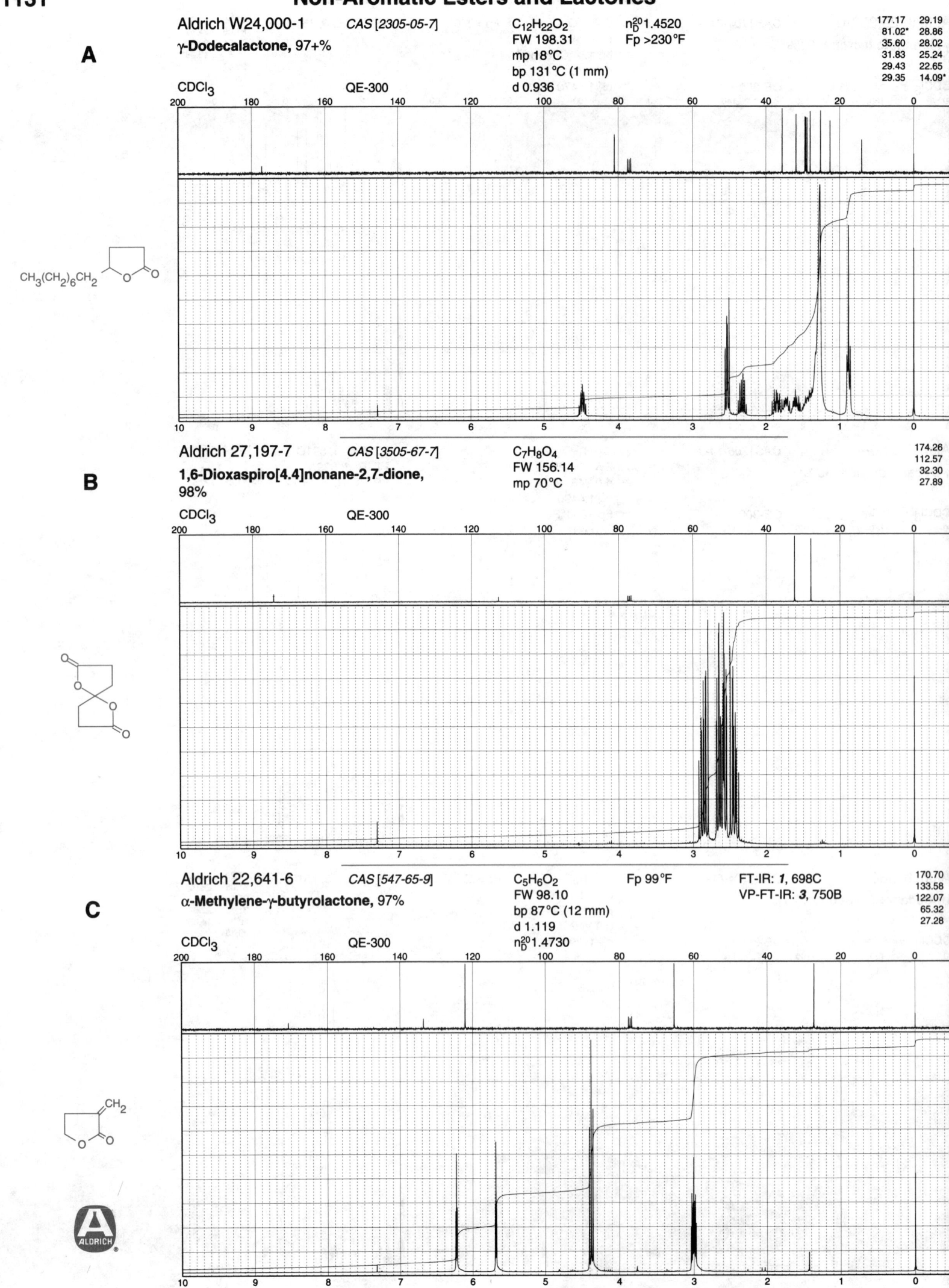

A

Aldrich W24,000-1   CAS [2305-05-7] $C_{12}H_{22}O_2$ $n_D^{20}1.4520$

γ-Dodecalactone, 97+%   FW 198.31   Fp >230°F

mp 18°C

bp 131°C (1 mm)

CDCl$_3$ QE-300   d 0.936

177.17 29.19
81.02* 28.86
35.60   28.02
31.83   25.24
29.43   22.65
29.35   14.09*

$CH_3(CH_2)_6CH_2$

B

Aldrich 27,197-7   CAS [3505-67-7] $C_7H_8O_4$

1,6-Dioxaspiro[4.4]nonane-2,7-dione, 98%   FW 156.14

mp 70°C

CDCl$_3$ QE-300

174.26
112.57
32.30
27.89

C

Aldrich 22,641-6   CAS [547-65-9] $C_5H_6O_2$ Fp 99°F   FT-IR: 1, 698C

α-Methylene-γ-butyrolactone, 97%   FW 98.10   VP-FT-IR: 3, 750B

bp 87°C (12 mm)

d 1.119

CDCl$_3$ QE-300 $n_D^{20}1.4730$

170.70
133.58
122.07
65.32
27.28

CH$_2$

Aldrich 27,300-7   CAS [10008-73-8] $C_5H_6O_2$ Fp 170°F

γ-**Methylene**-γ-**butyrolactone, 99%**

FW 98.10
bp 80°C (17 mm)
d 1.096
n_D^{20} 1.4660

174.94
155.74
88.56
27.98
25.08

A

CDCl$_3$ QE-300

Aldrich 27,992-7   CAS [54483-22-6] $C_7H_8O_2$

cis-(+)-3,3a,6,6a-**Tetrahydro-2H-cyclopenta-**
[b]**furan-2-one, 99%**

FW 124.14
mp 47°C
Fp >230°F

176.68
131.26*
129.62*
83.03*
45.56*
39.51
33.27

B

CDCl$_3$ QE-300

Aldrich B5,960-8   CAS [5061-21-2] $C_4H_5BrO_2$ Fp >230°F

α-**Bromo**-γ-**butyrolactone, 97%**

FW 164.99
bp 138°C (6 mm)
d 1.990
n_D^{20} 1.5080

60 MHz: 1, 592A
FT-IR: 1, 699D
VP-FT-IR: 3, 752D

173.10
67.04
37.70*
33.61

C

CDCl$_3$ QE-300

ALDRICH

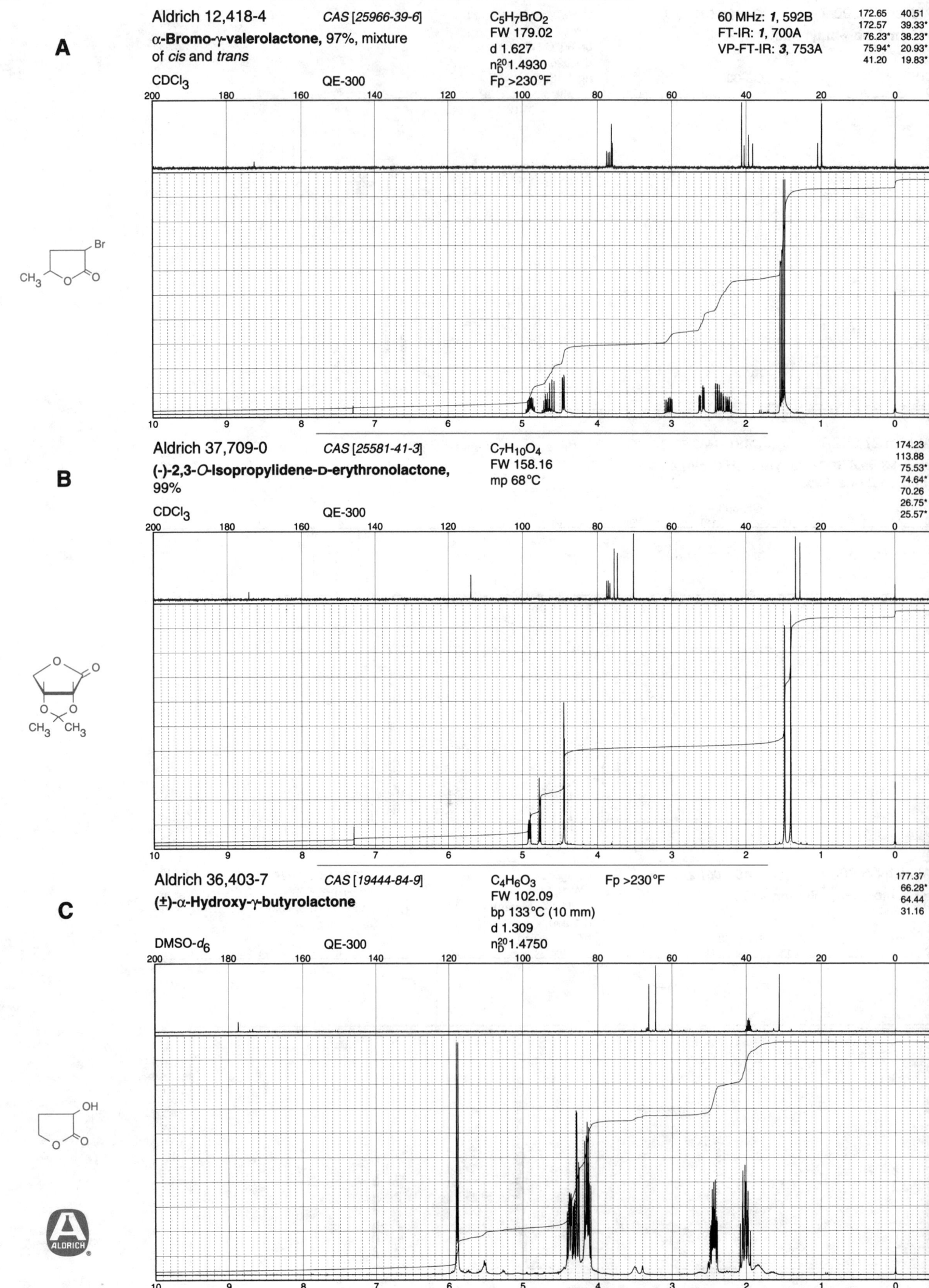

A

Aldrich 12,418-4   CAS [25966-39-6]   C$_5$H$_7$BrO$_2$ 60 MHz: **1**, 592B   172.65   40.51
α-**Bromo**-γ-**valerolactone**, 97%, mixture   FW 179.02   FT-IR: **1**, 700A   172.57   39.33*
of *cis* and *trans*   d 1.627   VP-FT-IR: **3**, 753A   76.23*   38.23*
                              n$_D^{20}$1.4930   75.94*   20.93*
CDCl$_3$ QE-300   Fp >230°F   41.20   19.83*

B

Aldrich 37,709-0   CAS [25581-41-3]   C$_7$H$_{10}$O$_4$ 174.23
(-)-**2,3**-*O*-**Isopropylidene**-D-**erythronolactone**,   FW 158.16   113.88
99%   mp 68°C   75.53*
                              74.64*
CDCl$_3$ QE-300   70.26
                              26.75*
                              25.57*

C

Aldrich 36,403-7   CAS [19444-84-9]   C$_4$H$_6$O$_3$ Fp >230°F   177.37
(±)-α-**Hydroxy**-γ-**butyrolactone**   FW 102.09   66.28*
                              bp 133°C (10 mm)   64.44
                              d 1.309   31.16
DMSO-*d$_6$*   QE-300   n$_D^{20}$1.4750

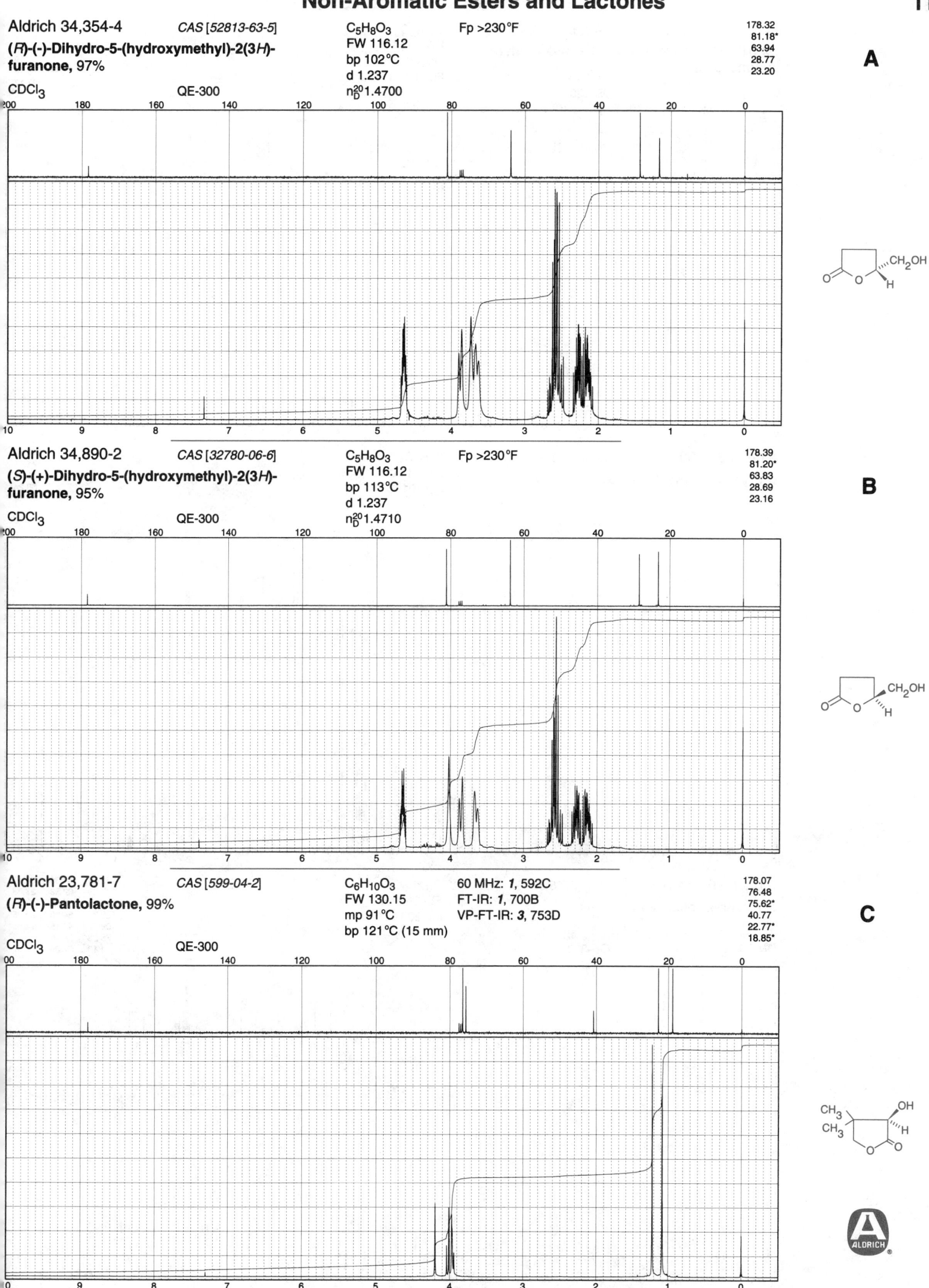

Aldrich 34,354-4   CAS [52813-63-5]   C₅H₈O₃   Fp >230°F

$C_5H_8O_3$
FW 116.12
bp 102°C
d 1.237
n_D^{20} 1.4700

(R)-(-)-Dihydro-5-(hydroxymethyl)-2(3H)-furanone, 97%

CDCl₃   QE-300

178.32
81.18*
63.94
28.77
23.20

A

Aldrich 34,890-2   CAS [32780-06-6]   C₅H₈O₃   Fp >230°F

$C_5H_8O_3$
FW 116.12
bp 113°C
d 1.237
n_D^{20} 1.4710

(S)-(+)-Dihydro-5-(hydroxymethyl)-2(3H)-furanone, 95%

CDCl₃   QE-300

178.39
81.20*
63.83
28.69
23.16

B

Aldrich 23,781-7   CAS [599-04-2]   C₆H₁₀O₃   60 MHz: 1, 592C

$C_6H_{10}O_3$
FW 130.15
mp 91°C
bp 121°C (15 mm)

FT-IR: 1, 700B
VP-FT-IR: 3, 753D

(R)-(-)-Pantolactone, 99%

CDCl₃   QE-300

178.07
76.48
75.62*
40.77
22.77*
18.85*

C

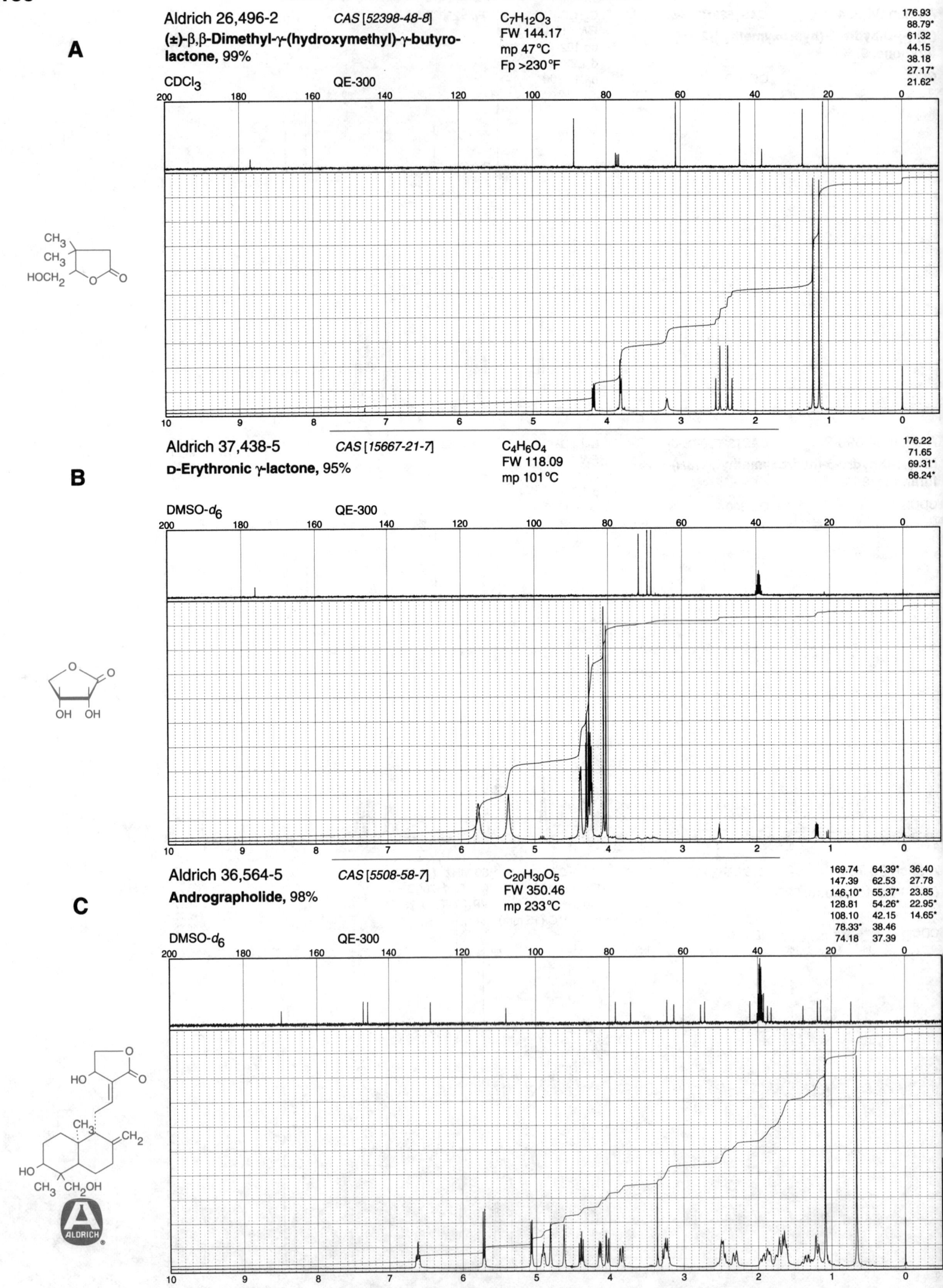

A

Aldrich 26,496-2   CAS [52398-48-8]   C$_7$H$_{12}$O$_3$
FW 144.17
(±)-β,β-Dimethyl-γ-(hydroxymethyl)-γ-butyro-   mp 47°C
lactone, 99%   Fp >230°F

176.93
88.79*
61.32
44.15
38.18
27.17*
21.62*

CDCl$_3$ QE-300

B

Aldrich 37,438-5   CAS [15667-21-7]   C$_4$H$_6$O$_4$
FW 118.09
D-Erythronic γ-lactone, 95%   mp 101°C

176.22
71.65
69.31*
68.24*

DMSO-d_6 QE-300

C

Aldrich 36,564-5   CAS [5508-58-7]   C$_{20}$H$_{30}$O$_5$
FW 350.46
Andrographolide, 98%   mp 233°C

169.74	64.39*	36.40
147.39	62.53	27.78
146,10*	55.37*	23.85
128.81	54.26*	22.95*
108.10	42.15	14.65*
78.33*	38.46	
74.18	37.39	

DMSO-d_6 QE-300

Non-Aromatic Esters and Lactones

Aldrich 85,729-7   CAS [5336-08-3]   C5H8O5   60 MHz: 1, 595B   181.29
D-(+)-Ribonic γ-lactone, 98%   FW 148.11   FT-IR: 1, 705D   89.53*
mp 86°C   72.30*
71.77*
63.35

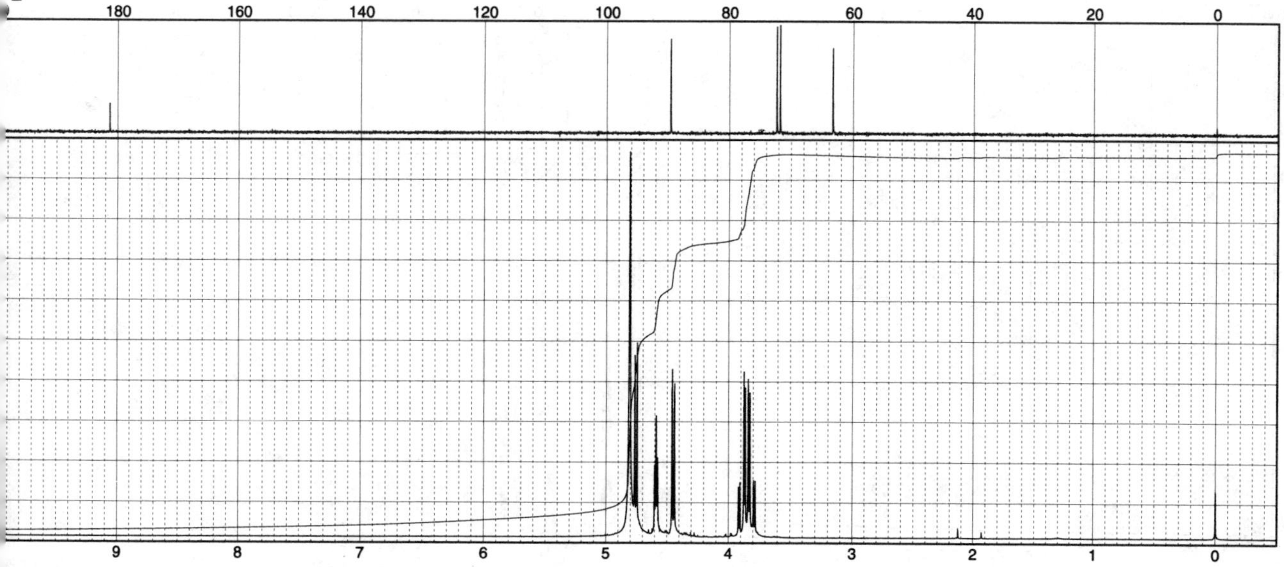

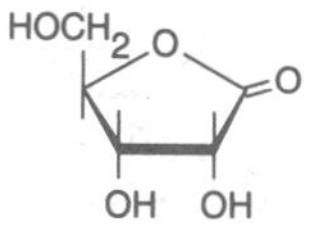

A

D2O   QE-300

Aldrich 37,369-9   CAS [30725-00-9]   C8H12O5   174.14 74.90*
2,3-O-Isopropylidene-D-ribonic γ-lactone,   FW 188.18   111.50 60.27
99%   mp 137°C   82.13* 26.44*
77.99* 25.03*

B

DMSO-d6   QE-300

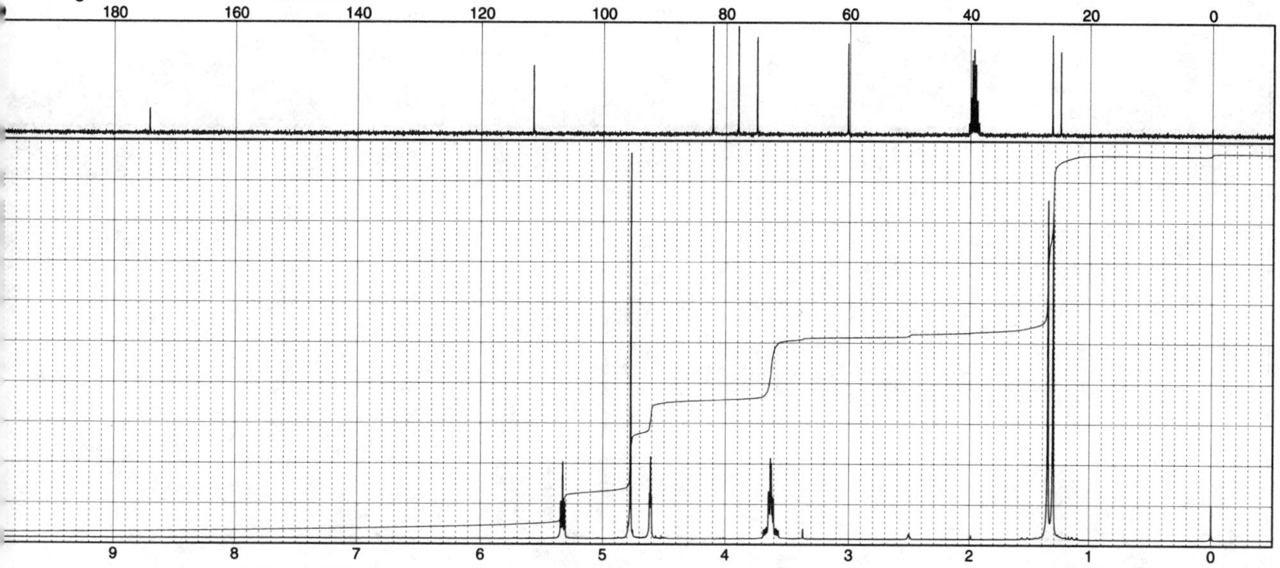

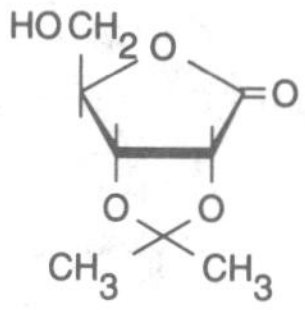

Aldrich 38,907-2   CAS [15384-37-9]   C5H8O5   180.03
D-Xylonic γ-lactone, 96%   FW 148.11   83.23*
mp 103°C   75.99*
75.02*
61.76

C

D2O   QE-300

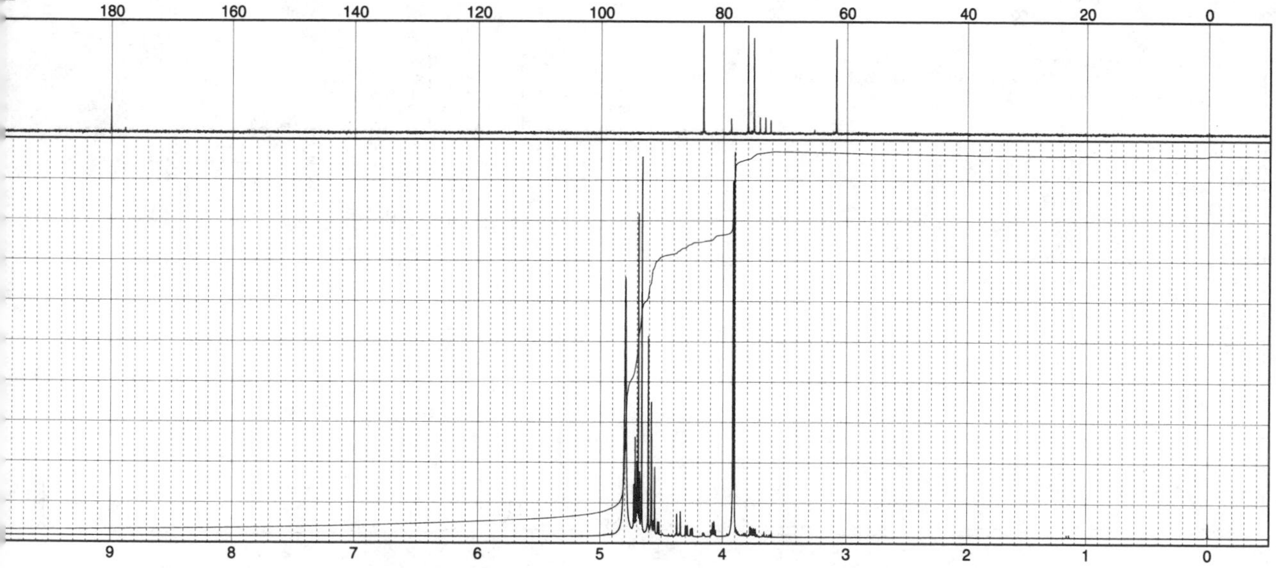

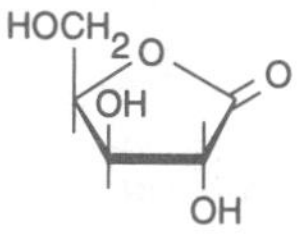

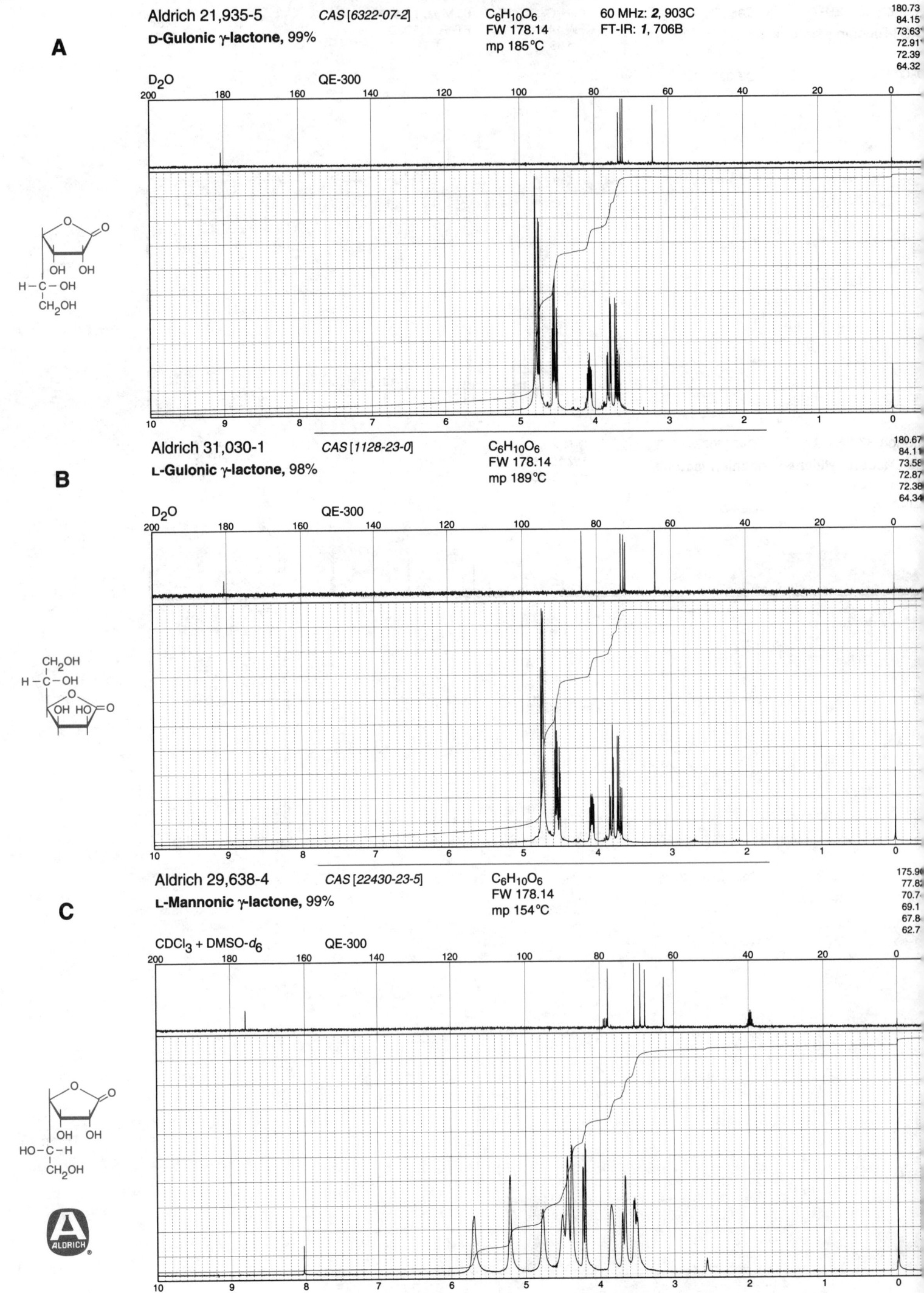

A

Aldrich 21,935-5   CAS [6322-07-2]   C₆H₁₀O₆   60 MHz: *2*, 903C
D-Gulonic γ-lactone, 99%   FW 178.14   FT-IR: *1*, 706B
mp 185°C

180.73
84.15
73.63
72.91
72.39
64.32

D₂O   QE-300

B

Aldrich 31,030-1   CAS [1128-23-0]   C₆H₁₀O₆
L-Gulonic γ-lactone, 98%   FW 178.14
mp 189°C

180.67
84.11
73.58
72.87
72.38
64.34

D₂O   QE-300

C

Aldrich 29,638-4   CAS [22430-23-5]   C₆H₁₀O₆
L-Mannonic γ-lactone, 99%   FW 178.14
mp 154°C

175.9
77.8
70.7
69.1
67.8
62.7

CDCl₃ + DMSO-d₆   QE-300

Aldrich 20,666-0   CAS [22404-04-2] $C_7H_{12}O_7$ 60 MHz: **1**, 593C
ɒ-glycero-ʟ-manno-**Heptonic γ-lactone**   FW 208.17   FT-IR: **1**, 707B
   mp 150 °C

ᴅMSO-d_6 QE-300

A

176.00
77.21*
70.70*
69.68*
69.00*
65.23*
61.95

Aldrich 12,564-4   CAS [79703-26-7] $C_7H_{12}O_7$ 60 MHz: **1**, 595C
α-ᴅ-**Glucoheptonic γ-lactone, 99%**   FW 208.17   FT-IR: **1**, 702D
   mp 154 °C

ᴅ₂O   QE-300

B

180.74
83.35*
73.80*
73.47*
73.33*
73.06*
64.71

Aldrich 20,664-4   CAS [6968-62-3] $C_8H_{14}O_8$ 60 MHz: **2**, 903D
,β-**Glucooctanoic γ-lactone**   FW 238.20   FT-IR: **1**, 706D
   mp 188 °C d.

ᴅMSO-d_6 QE-300

C

176.44
85.86*
71.70*
70.63*
69.65*
68.28*
62.85

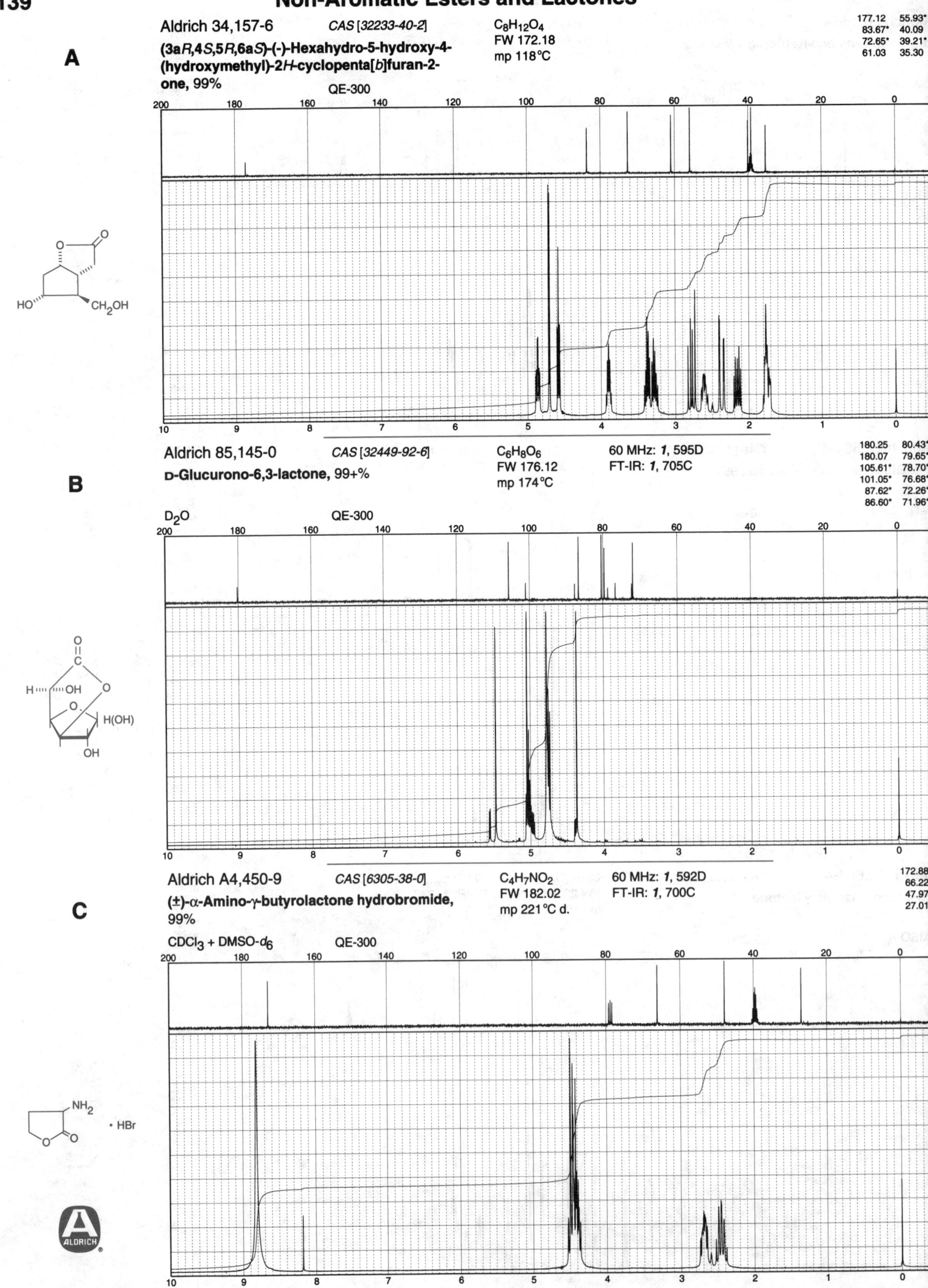

A

Aldrich 34,157-6 CAS [32233-40-2] C₈H₁₂O₄ FW 172.18 mp 118°C

(3aR,4S,5R,6aS)-(-)-Hexahydro-5-hydroxy-4-(hydroxymethyl)-2H-cyclopenta[b]furan-2-one, 99%

QE-300

177.12	55.93*
83.67*	40.09
72.65*	39.21*
61.03	35.30

B

Aldrich 85,145-0 CAS [32449-92-6] C₆H₈O₆ FW 176.12 mp 174°C 60 MHz: 1, 595D FT-IR: 1, 705C

D-Glucurono-6,3-lactone, 99+%

D₂O   QE-300

180.25	80.43*
180.07	79.65*
105.61*	78.70*
101.05*	76.68*
87.62*	72.26*
86.60*	71.96*

C

Aldrich A4,450-9 CAS [6305-38-0] C₄H₇NO₂ FW 182.02 mp 221°C d. 60 MHz: 1, 592D FT-IR: 1, 700C

(±)-α-Amino-γ-butyrolactone hydrobromide, 99%

CDCl₃ + DMSO-d₆   QE-300

| 172.88 |
| 66.22 |
| 47.97 |
| 27.01 |

ALDRICH

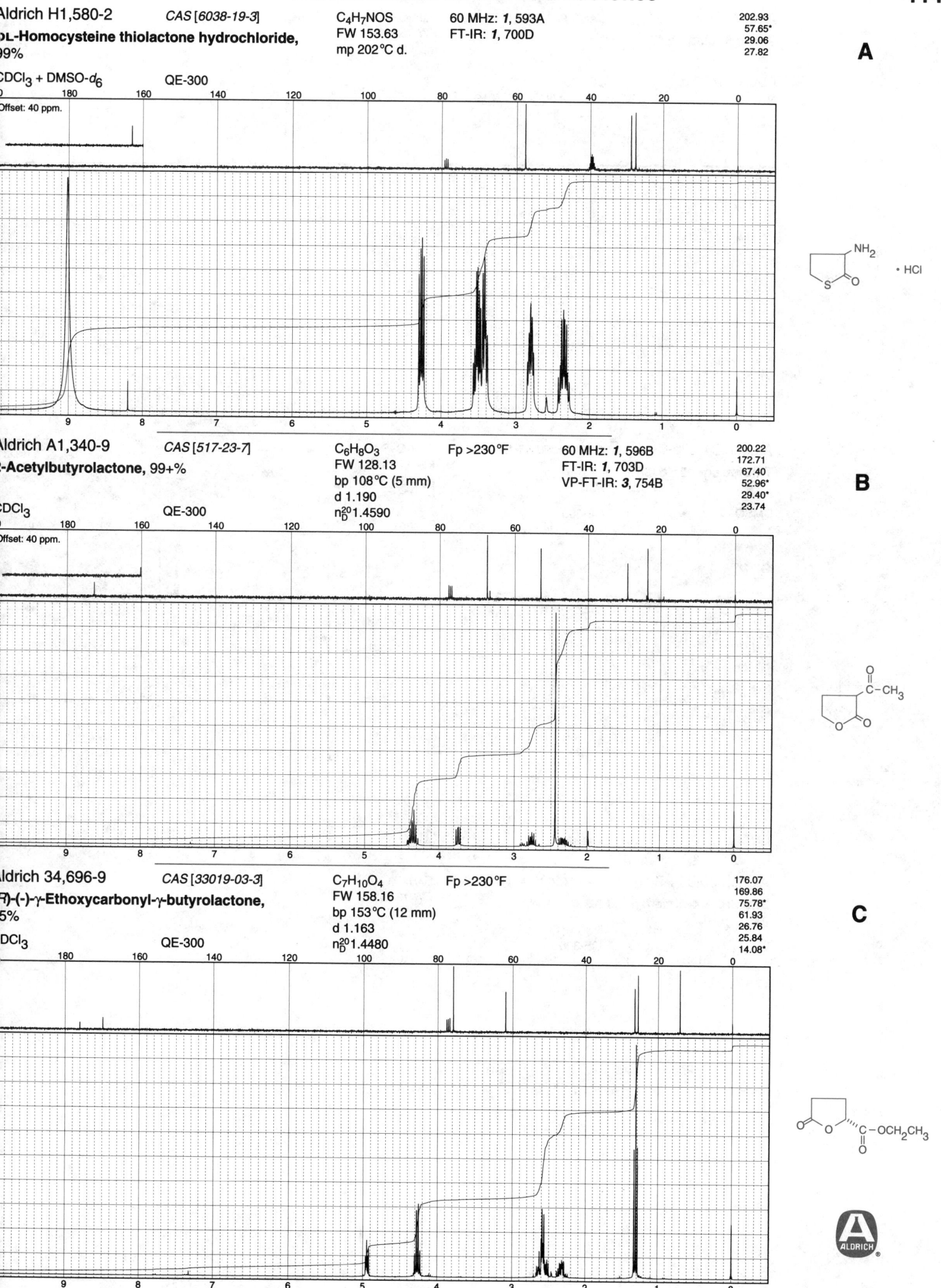

Aldrich H1,580-2   CAS [6038-19-3]   C₄H₇NOS   60 MHz: 1, 593A   202.93
                                                FW 153.63   FT-IR: 1, 700D   57.65*
DL-Homocysteine thiolactone hydrochloride,   mp 202°C d.   29.06
99%   27.82

CDCl₃ + DMSO-d₆   QE-300
Offset: 40 ppm.

Aldrich A1,340-9   CAS [517-23-7]   C₆H₈O₃   Fp >230°F   60 MHz: 1, 596B   200.22
                                                FW 128.13   FT-IR: 1, 703D   172.71
α-Acetylbutyrolactone, 99+%   bp 108°C (5 mm)   VP-FT-IR: 3, 754B 67.40
                                                d 1.190   52.96*
                                                n₂₀D 1.4590   29.40*
CDCl₃   QE-300   23.74
Offset: 40 ppm.

Aldrich 34,696-9   CAS [33019-03-3]   C₇H₁₀O₄   Fp >230°F   176.07
                                                FW 158.16   169.86
(R)-(-)-γ-Ethoxycarbonyl-γ-butyrolactone,   bp 153°C (12 mm)   75.78*
5%   d 1.163   61.93
                                                n₂₀D 1.4480   26.76
CDCl₃   QE-300   25.84
                                                                                  14.08*

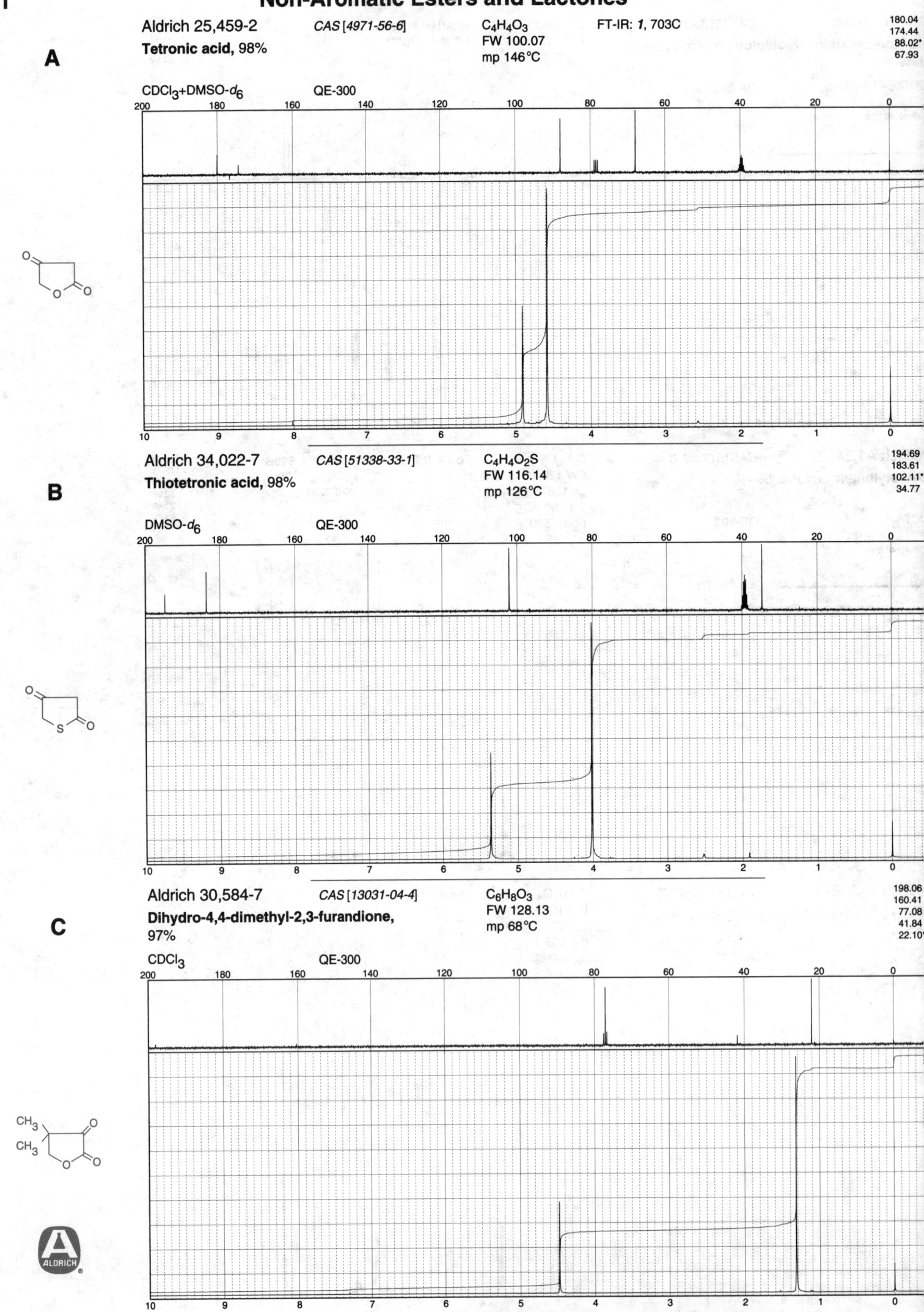

A

Aldrich 25,459-2 *CAS [4971-56-6]* C₄H₄O₃ FT-IR: *1*, 703C
Tetronic acid, 98% FW 100.07
mp 146°C

180.04
174.44
88.02*
67.93

CDCl₃+DMSO-*d₆* QE-300

B

Aldrich 34,022-7 *CAS [51338-33-1]* C₄H₄O₂S
Thiotetronic acid, 98% FW 116.14
mp 126°C

194.69
183.61
102.11*
34.77

DMSO-*d₆* QE-300

C

Aldrich 30,584-7 *CAS [13031-04-4]* C₆H₈O₃
Dihydro-4,4-dimethyl-2,3-furandione, FW 128.13
97% mp 68°C

198.06
160.41
77.08
41.84
22.10*

CDCl₃ QE-300

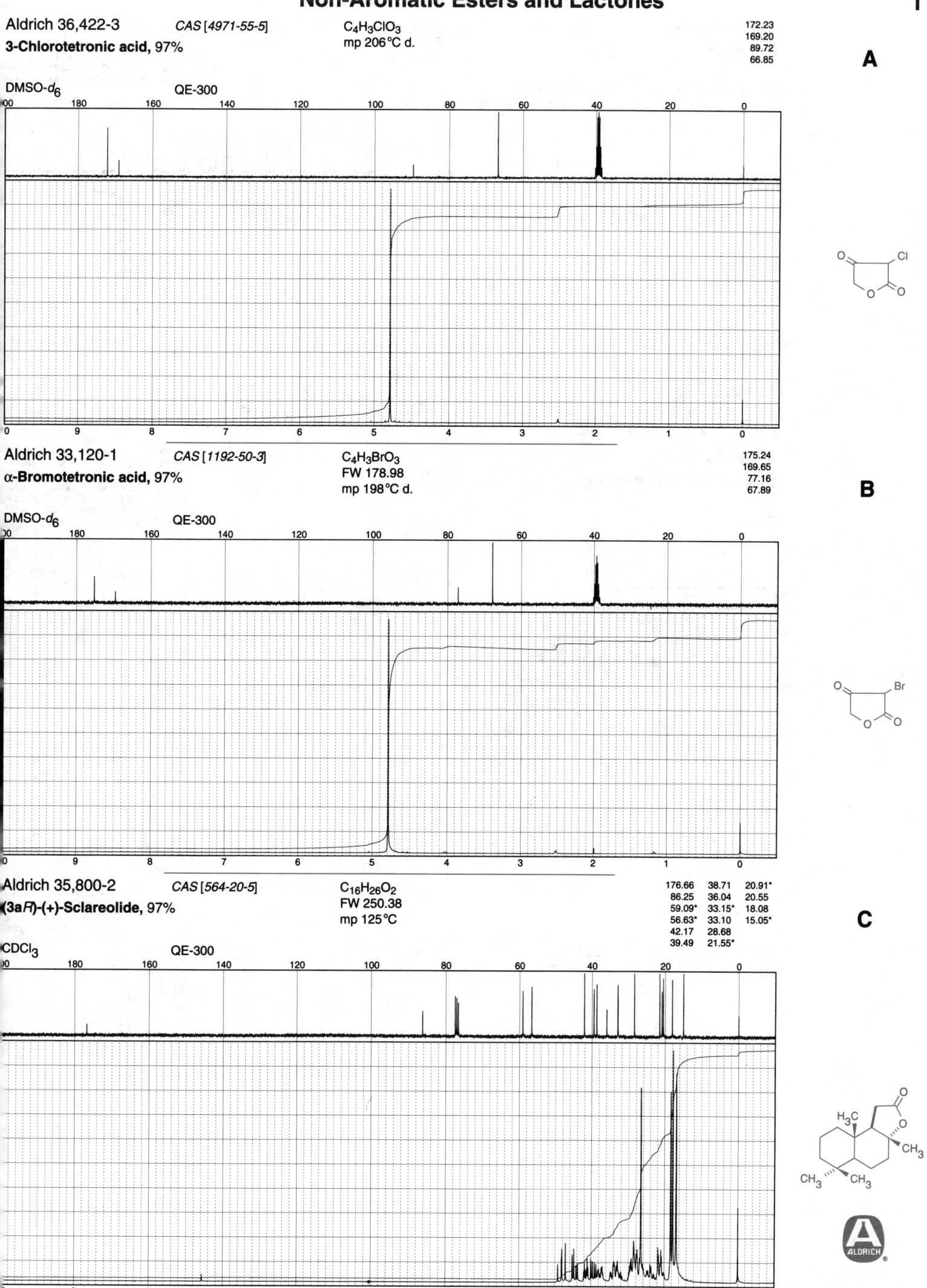

Aldrich 36,422-3   CAS [4971-55-5] $C_4H_3ClO_3$ mp 206°C d.

3-Chlorotetronic acid, 97%

172.23
169.20
89.72
66.85

A

DMSO-d_6 QE-300

Aldrich 33,120-1   CAS [1192-50-3] $C_4H_3BrO_3$ FW 178.98   mp 198°C d.

α-Bromotetronic acid, 97%

175.24
169.65
77.16
67.89

B

DMSO-d_6 QE-300

Aldrich 35,800-2   CAS [564-20-5] $C_{16}H_{26}O_2$ FW 250.38   mp 125°C

(3aR)-(+)-Sclareolide, 97%

176.66	38.71	20.91*
86.25	36.04	20.55
59.09*	33.15*	18.08
56.63*	33.10	15.05*
42.17	28.68	
39.49	21.55*	

C

CDCl$_3$ QE-300

ALDRICH

A

Aldrich P4,470-6   CAS [124-87-8] $C_{30}H_{34}O_{13}$ FT-IR: 1, 701B
Picrotoxin   FW 602.59
mp 201 °C

174.52	85.25	72.89	49.96*	29.36*
174.24	84.56	72.00	47.92*	28.10*
169.45	80.17*	67.83	47.50	22.81*
169.34	79.83*	61.37*	46.29	16.36*
139.89	77.33*	52.11*	43.71	15.81*
111.28	76.98*	50.64*	42.84	

$CDCl_3$ + DMSO-d_6 QE-300

B

Aldrich 85,619-3 $C_{37}H_{67}NO_{13}$ FT-IR: 1, 703B
(-)-Erythromycin hydrate, 98%   FW 733.98
mp 139 °C

221.49	76.98*	49.47*	35.07	18.33*
175.76	75.01	45.06*	28.84	16.26*
103.29*	74.79	44.95*	26.93*	15.89*
96.36*	72.66	40.29*	21.52*	11.98*
83.64*	71.04*	39.55*	21.40*	10.69*
80.07*	68.94*	38.60	21.22	9.16*
78.07*	65.59*	38.12*	18.67*	

$CDCl_3$ QE-300

Offset: 40 ppm.

C

Aldrich 22,308-5   CAS [481-06-1] $C_{15}H_{18}O_3$ FT-IR: 1, 709B
Santonin, 99%   FW 246.29
mp 173 °C

186.12	125.80*	37.91
177.42	81.34*	25.14*
154.81*	53.59*	23.08
150.98	41.37	12.49*
128.60	40.97*	10.88*

$CDCl_3$ QE-300

Aldrich 31,047-6   CAS [53558-93-3]   C₅H₆O₄   VP-FT-IR: **3**, 754C

$C_5H_6O_4$
FW 130.10
mp 72°C
bp 166°C

(R)-(-)-5-Oxo-2-tetrahydrofurancarboxylic acid, 98%

DMSO-d_6 QE-300

A

176.19
171.41
75.44*
26.73
25.56

Aldrich 30,146-9   CAS [21461-84-7]

(S)-(+)-5-Oxo-2-tetrahydrofurancarboxylic acid, 98%

$C_5H_6O_4$
FW 130.10
mp 72°C
bp 153°C

CDCl₃ + DMSO-d_6 QE-300

B

176.28
171.49
75.50*
26.75
25.57

Aldrich 22,293-3   CAS [389-36-6]

D-Saccharic 1,4-lactone monohydrate, 98%

$C_6H_8O_7$
FW 210.13
mp 92°C

60 MHz: **1**, 593D
FT-IR: **1**, 706C

DMSO-d_6 QE-300

C

175.38
172.03
78.82*
73.17*
71.85*
68.71*

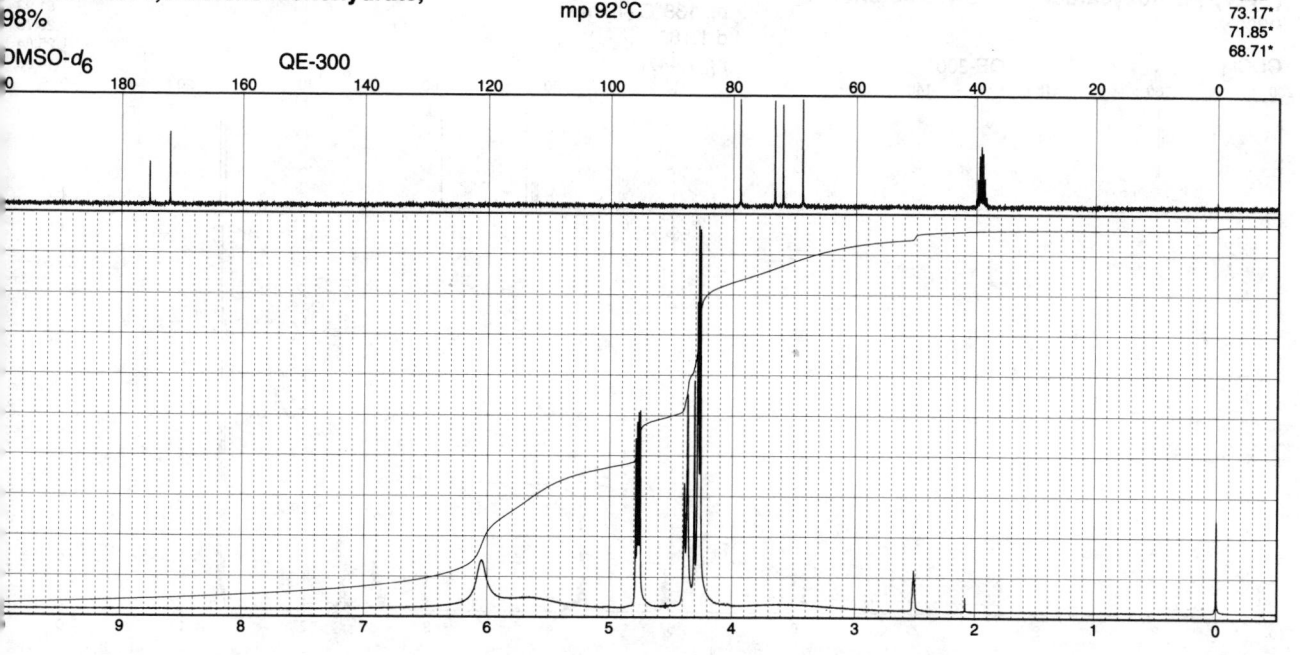

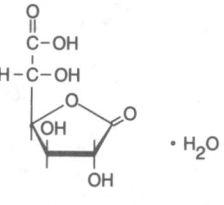

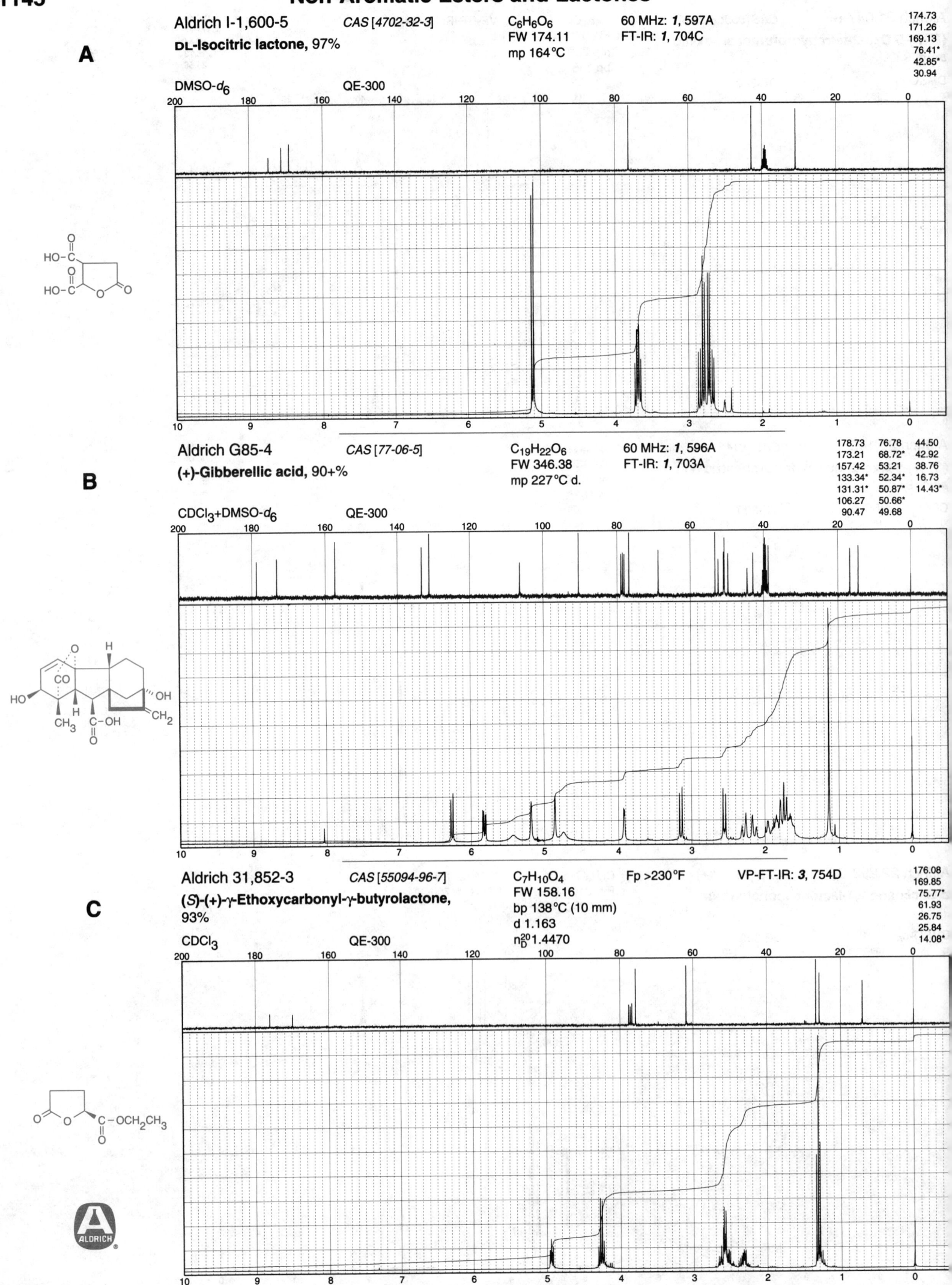

A

Aldrich I-1,600-5   CAS [4702-32-3] $C_6H_6O_6$ 60 MHz: **1**, 597A
DL-Isocitric lactone, 97%   FW 174.11   FT-IR: **1**, 704C
mp 164°C

174.73
171.26
169.13
76.41*
42.85*
30.94

DMSO-d_6 QE-300

B

Aldrich G85-4   CAS [77-06-5] $C_{19}H_{22}O_6$ 60 MHz: **1**, 596A
(+)-Gibberellic acid, 90+%   FW 346.38   FT-IR: **1**, 703A
mp 227°C d.

178.73 76.78 44.50
173.21 68.72* 42.92
157.42 53.21 38.76
133.34* 52.34* 16.73
131.31* 50.87* 14.43*
106.27 50.66*
90.47 49.68

CDCl$_3$+DMSO-d_6 QE-300

C

Aldrich 31,852-3   CAS [55094-96-7] $C_7H_{10}O_4$ Fp >230°F   VP-FT-IR: **3**, 754D
(S)-(+)-γ-Ethoxycarbonyl-γ-butyrolactone, 93%   FW 158.16
bp 138°C (10 mm)
d 1.163
n$_D^{20}$1.4470

CDCl$_3$ QE-300

176.08
169.85
75.77*
61.93
26.75
25.84
14.08*

Aldrich 37,636-1   CAS [102096-60-6] $C_9H_{12}O_4$ n_D^{20} 1.4600

(R)-(+)-α-Acryloyloxy-β,β-dimethyl-γ-butyro-lactone, 95%

FW 184.19   Fp >230°F
mp 8°C
bp 84°C
d 1.125

172.25	75.10*
164.72	40.33
132.77	22.96*
126.98*	19.85*
	76.17

A

CDCl₃   QE-300

Aldrich 29,594-9   CAS [4423-79-4] $C_8H_{12}O_3$ VP-FT-IR: **3**, 755C

1,4-Dioxaspiro[4.5]decan-2-one, 95%

FW 156.18
mp 34°C
bp 70°C
Fp 221°F

171.53
113.50
63.28
35.33
24.42
23.00

B

CDCl₃   QE-300

Aldrich 28,375-4   CAS [497-23-4] $C_4H_4O_2$ n_D^{20} 1.4690

2(5H)-Furanone, 98%

FW 84.07   Fp 214°F
mp 5°C   VP-FT-IR: **3**, 749B
bp 87°C (12 mm)
d 1.185

173.81
153.40*
121.21*
72.30

C

CDCl₃   QE-300

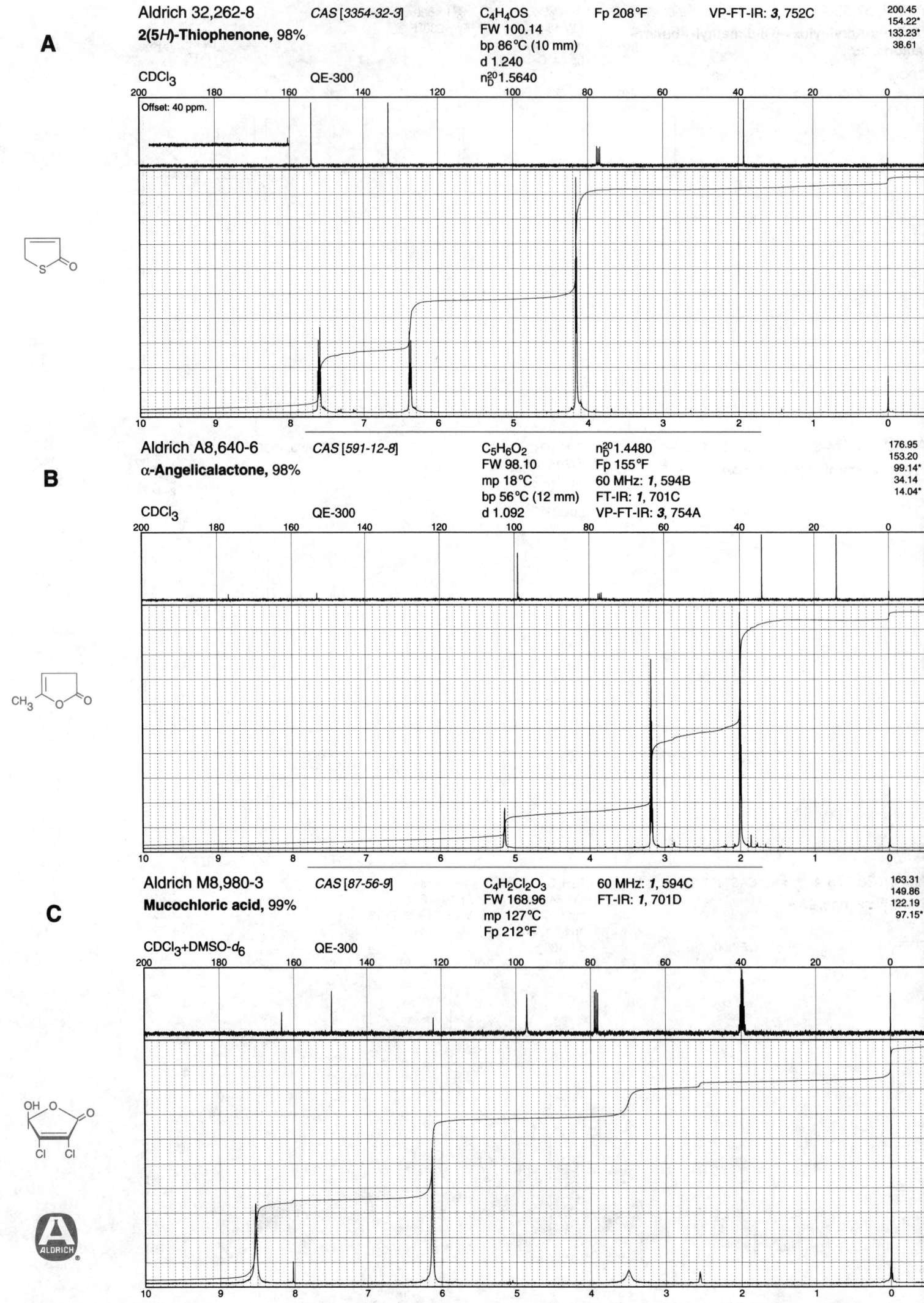

A

Aldrich 32,262-8   *CAS [3354-32-3]* C_4H_4OS Fp 208°F   VP-FT-IR: *3*, 752C   200.45 / 154.22* / 133.23* / 38.61

2(5H)-Thiophenone, 98%   FW 100.14 / bp 86°C (10 mm) / d 1.240 / n_D^{20} 1.5640

CDCl₃   QE-300

Offset: 40 ppm.

B

Aldrich A8,640-6   *CAS [591-12-8]* $C_5H_6O_2$ n_D^{20} 1.4480   176.95 / 153.20 / 99.14* / 34.14 / 14.04*

α-Angelicalactone, 98%   FW 98.10   Fp 155°F

mp 18°C   60 MHz: *1*, 594B

bp 56°C (12 mm)   FT-IR: *1*, 701C

d 1.092   VP-FT-IR: *3*, 754A

CDCl₃   QE-300

CH₃

C

Aldrich M8,980-3   *CAS [87-56-9]* $C_4H_2Cl_2O_3$ 60 MHz: *1*, 594C   163.31 / 149.86 / 122.19 / 97.15*

Mucochloric acid, 99%   FW 168.96   FT-IR: *1*, 701D

mp 127°C

Fp 212°F

CDCl₃+DMSO-*d₆*   QE-300

OH

Cl   Cl

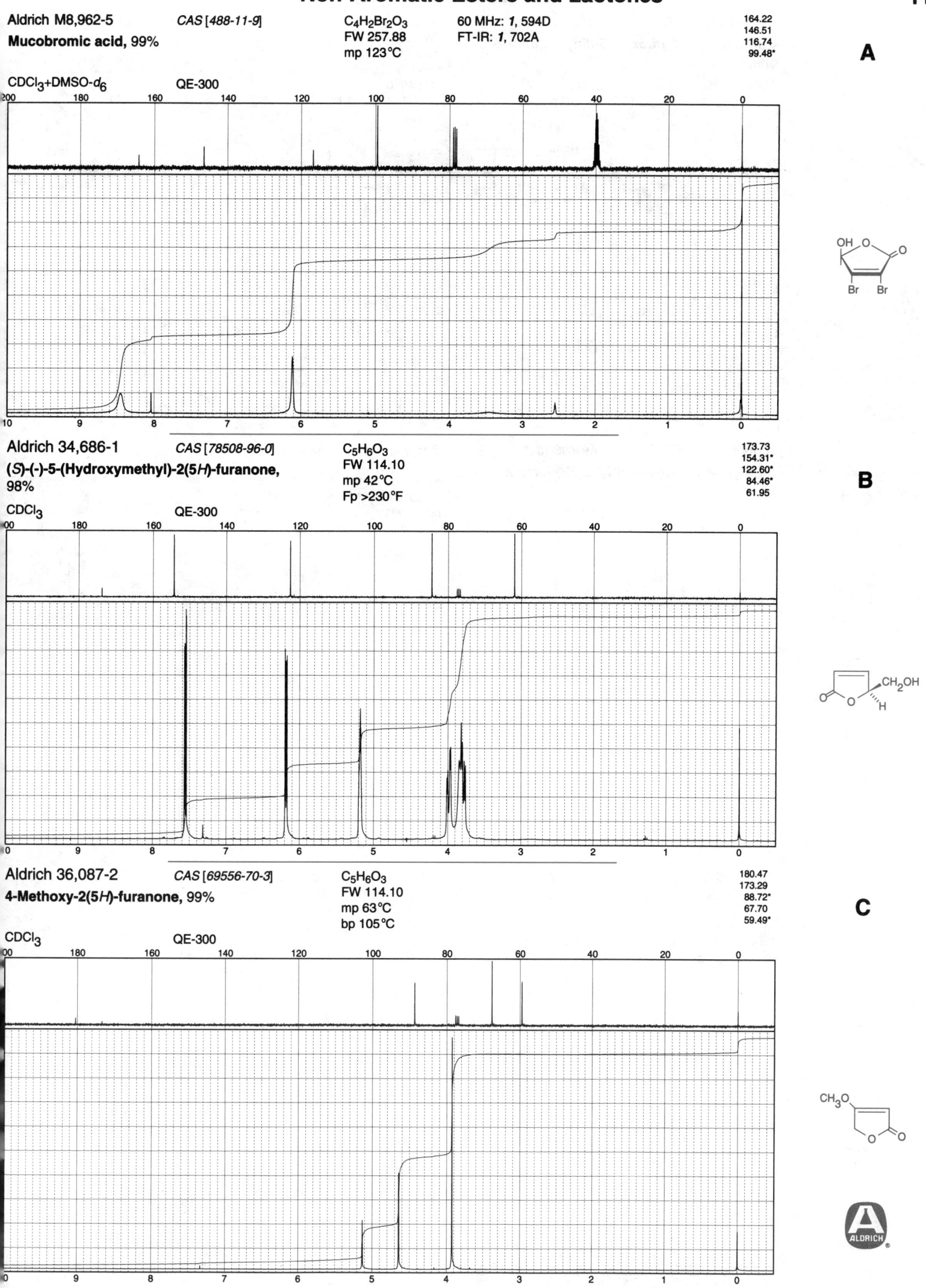

Aldrich M8,962-5

Mucobromic acid, 99%

CAS [488-11-9]

C₄H₂Br₂O₃
FW 257.88
mp 123°C

60 MHz: *1*, 594D
FT-IR: *1*, 702A

164.22
146.51
116.74
99.48*

A

CDCl₃+DMSO-d₆   QE-300

Aldrich 34,686-1   CAS [78508-96-0]

(S)-(-)-5-(Hydroxymethyl)-2(5H)-furanone, 98%

C₅H₆O₃
FW 114.10
mp 42°C
Fp >230°F

173.73
154.31*
122.60*
84.46*
61.95

B

CDCl₃   QE-300

Aldrich 36,087-2   CAS [69556-70-3]

4-Methoxy-2(5H)-furanone, 99%

C₅H₆O₃
FW 114.10
mp 63°C
bp 105°C

180.47
173.29
88.72*
67.70
59.49*

C

CDCl₃   QE-300

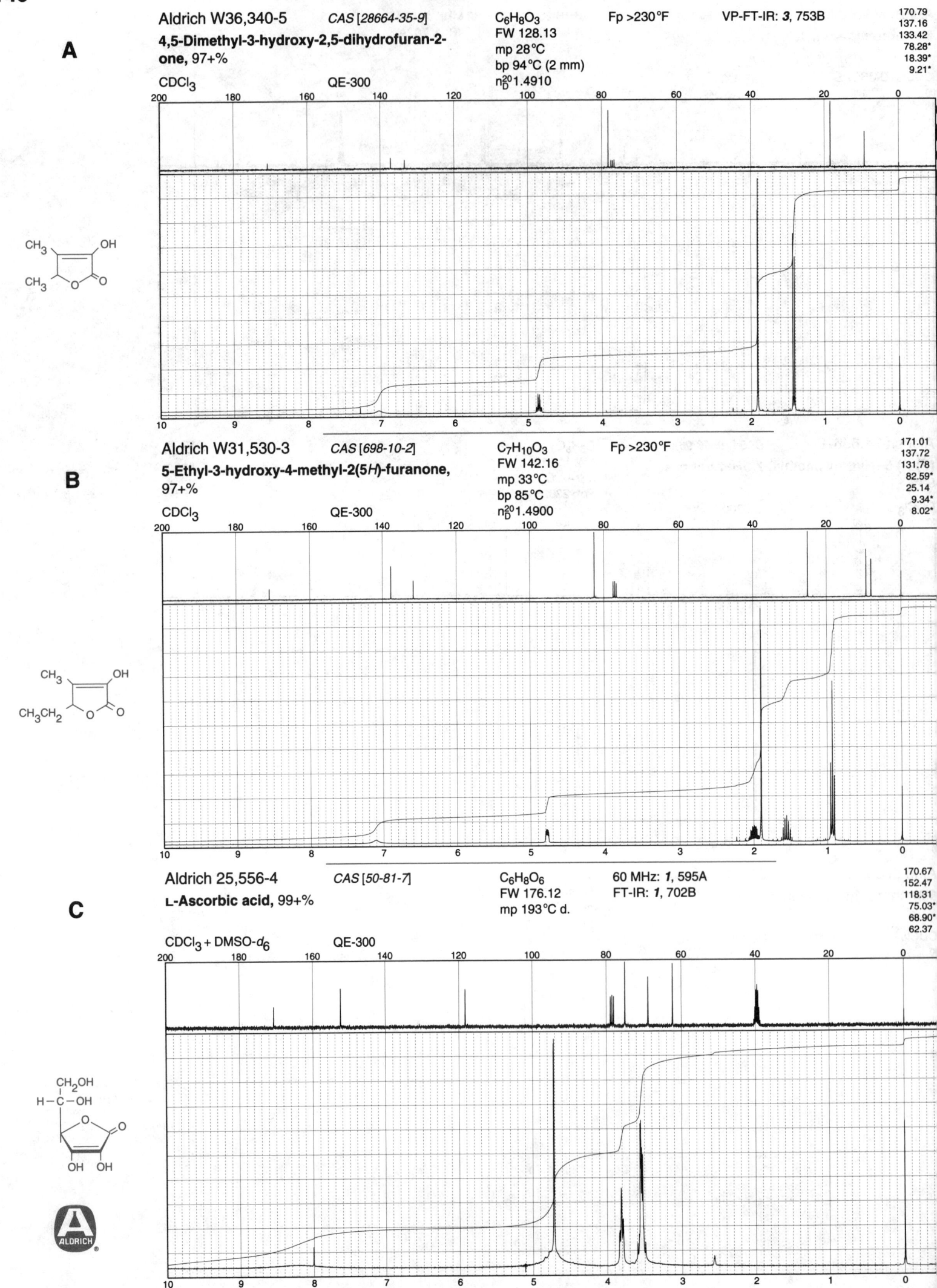

A

Aldrich W36,340-5   *CAS [28664-35-9]* $C_6H_8O_3$ Fp >230°F   VP-FT-IR: *3*, 753B

4,5-Dimethyl-3-hydroxy-2,5-dihydrofuran-2-one, 97+%

FW 128.13
mp 28°C
bp 94°C (2 mm)
n_D^{20}1.4910

CDCl$_3$ QE-300

170.79
137.16
133.42
78.28*
18.39*
9.21*

B

Aldrich W31,530-3   *CAS [698-10-2]* $C_7H_{10}O_3$ Fp >230°F

5-Ethyl-3-hydroxy-4-methyl-2(5H)-furanone, 97+%

FW 142.16
mp 33°C
bp 85°C
n_D^{20}1.4900

CDCl$_3$ QE-300

171.01
137.72
131.78
82.59*
25.14
9.34*
8.02*

C

Aldrich 25,556-4   *CAS [50-81-7]* $C_6H_8O_6$ 60 MHz: *1*, 595A

L-Ascorbic acid, 99+%

FW 176.12
mp 193°C d.

FT-IR: *1*, 702B

CDCl$_3$ + DMSO-d_6 QE-300

170.67
152.47
118.31
75.03*
68.90*
62.37

Aldrich 35,964-5 *CAS [5743-28-2]*

L-Ascorbic acid, calcium salt dihydrate,
99%

D₂O   QE-300

$C_6H_8O_6$		179.95
FW 426.35		178.35
mp 166°C d.		115.85
		81.24*
		72.49*
		65.38

A

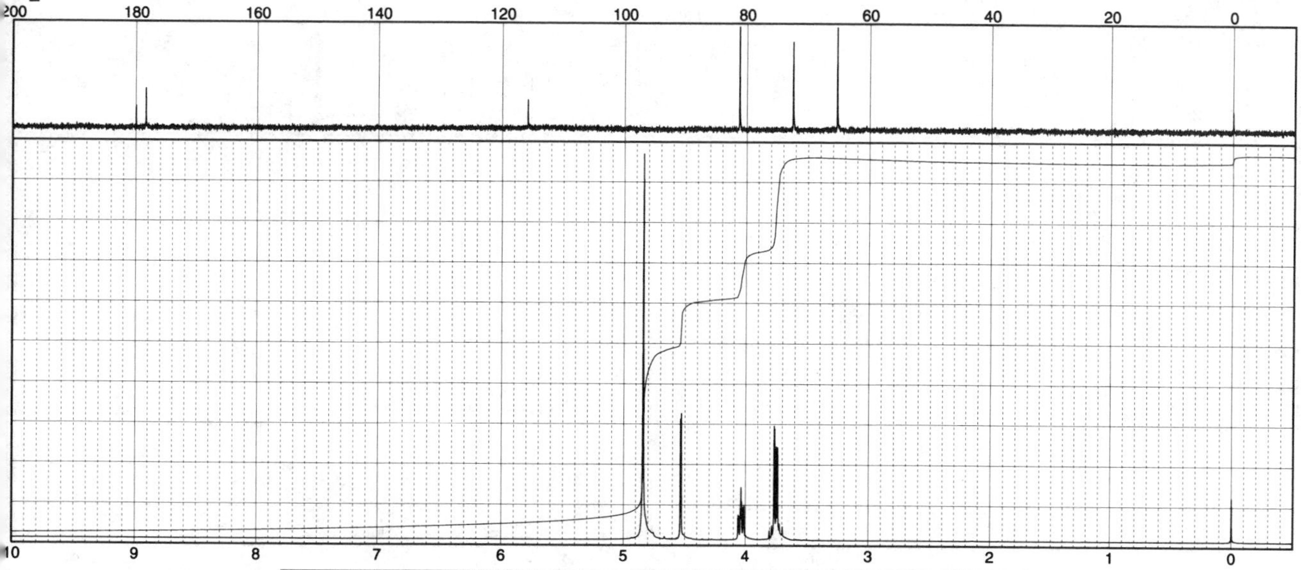

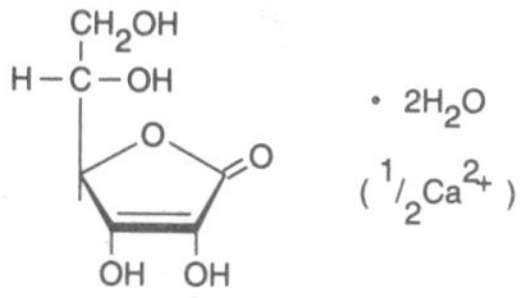

Aldrich 85,606-1 *CAS [89-65-6]*

D-Isoascorbic acid, 98%

CDCl₃ + DMSO-d_6 QE-300

$C_6H_8O_6$ FT-IR: *1*, 707A	170.31
FW 176.12	152.22
mp 171°C d.	118.46
	76.92*
	70.87*
	62.04

B

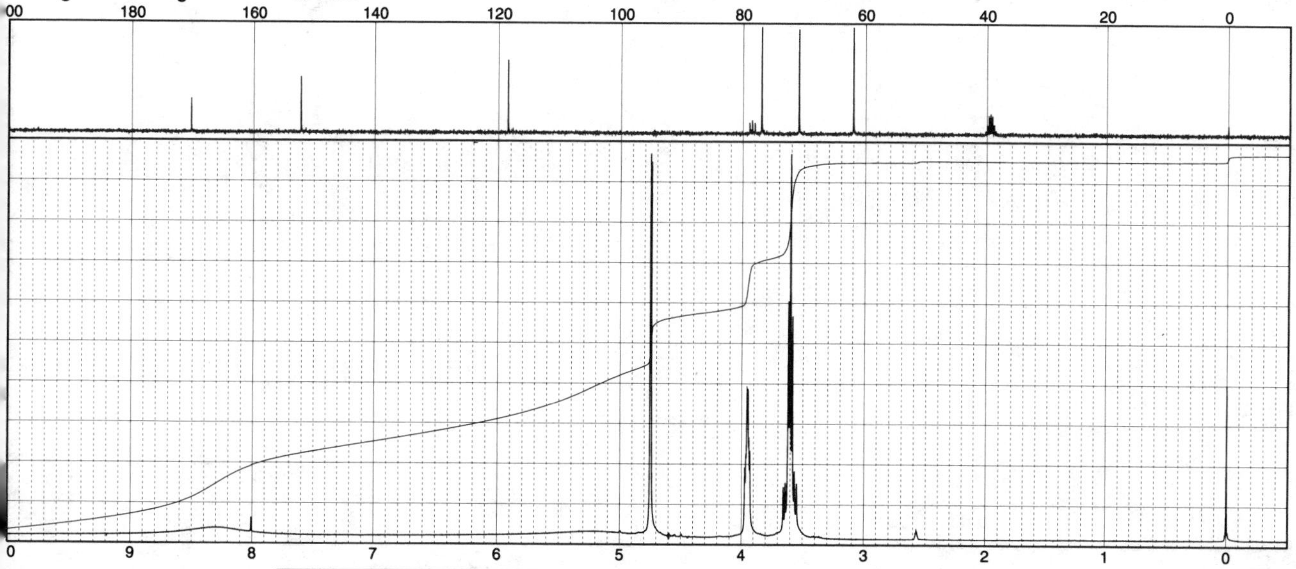

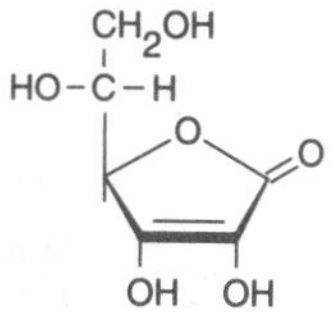

Aldrich 30,136-1 *CAS [15042-01-0]*

5,6-Isopropylidene-L-ascorbic acid, 98%

CDCl₃ + DMSO-d_6 QE-300

$C_9H_{12}O_6$	170.00	73.99*
FW 216.19	151.58	64.98
mp 210°C d.	118.54	25.89*
	109.14	25.47*
	74.57*	

C

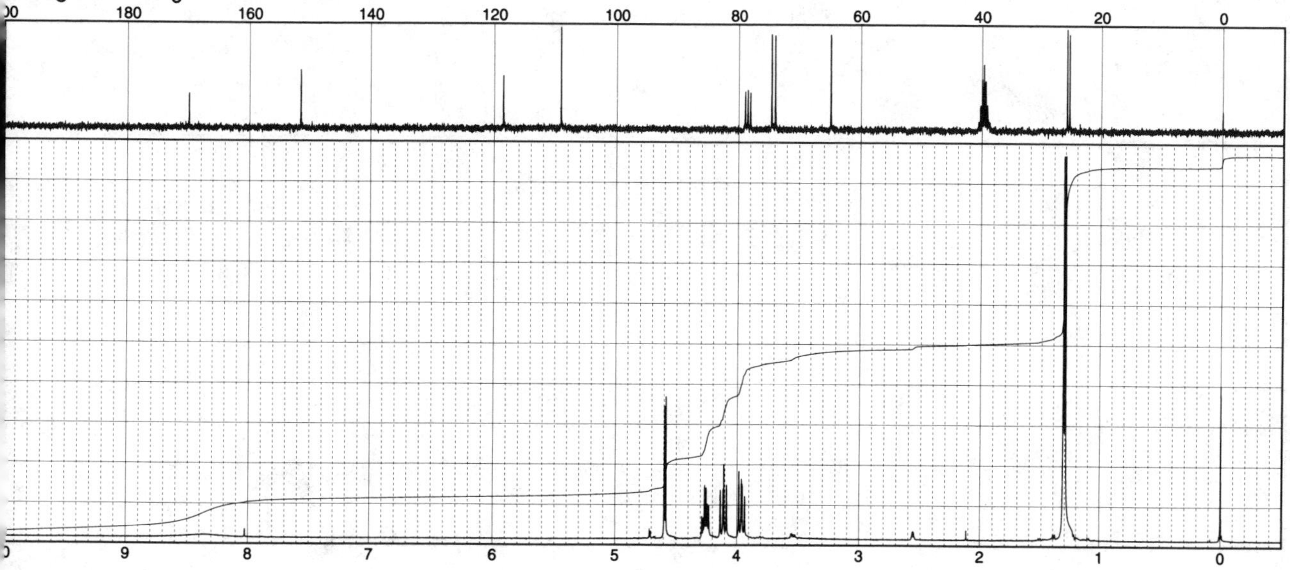

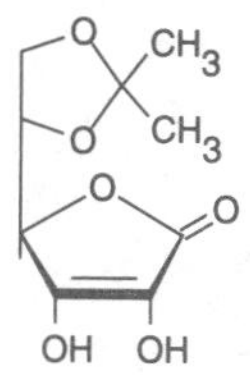

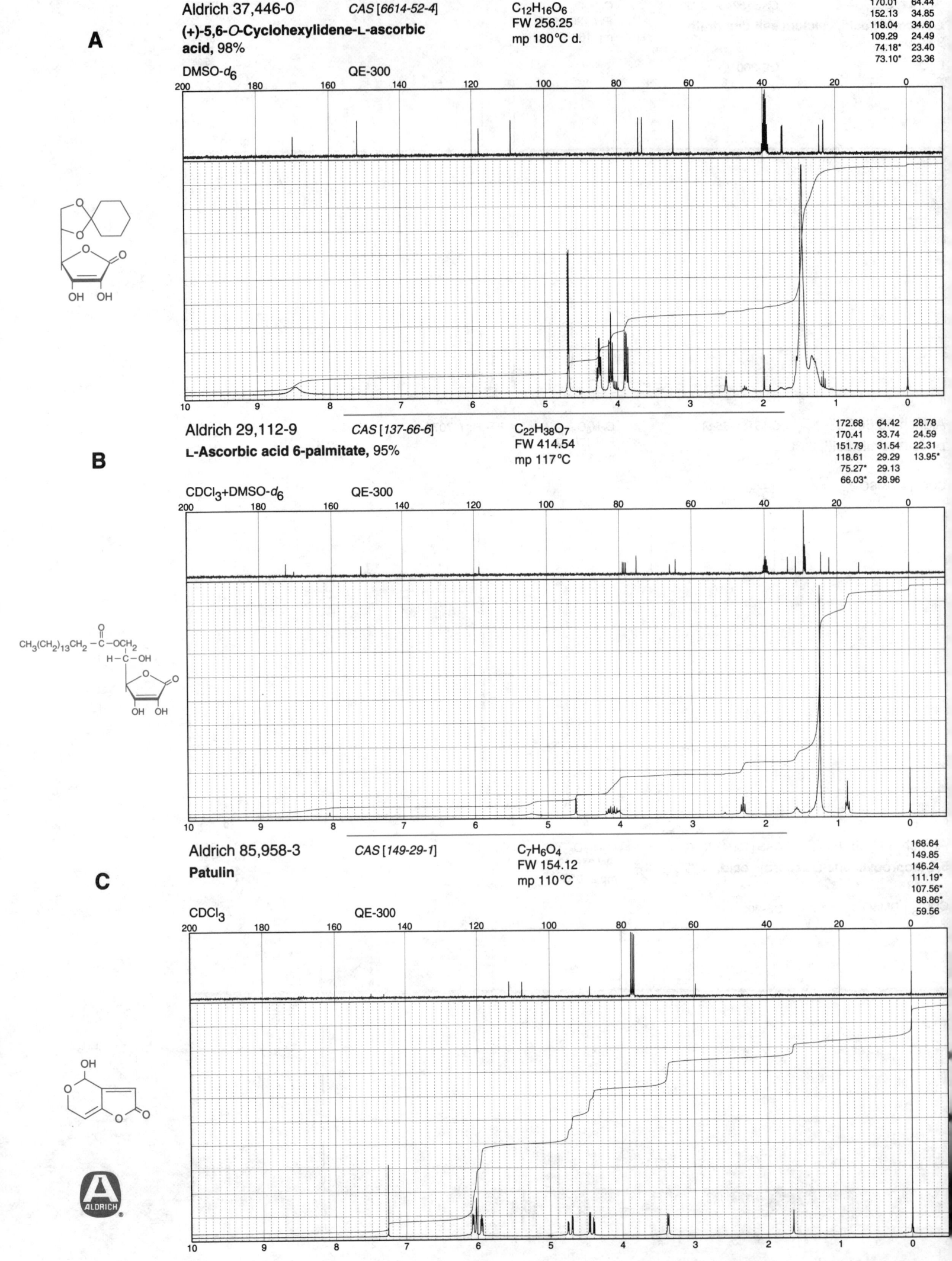

A

Aldrich 37,446-0   *CAS [6614-52-4]*   C$_{12}$H$_{16}$O$_6$
FW 256.25
(+)-5,6-O-Cyclohexylidene-L-ascorbic acid, 98%   mp 180°C d.

170.01	64.44	
152.13	34.85	
118.04	34.60	
109.29	24.49	
74.18*	23.40	
73.10*	23.36	

DMSO-d$_6$ QE-300

B

Aldrich 29,112-9   *CAS [137-66-6]*   C$_{22}$H$_{38}$O$_7$
FW 414.54
L-Ascorbic acid 6-palmitate, 95%   mp 117°C

172.68	64.42	28.78
170.41	33.74	24.59
151.79	31.54	22.31
118.61	29.29	13.95*
75.27*	29.13	
66.03*	28.96	

CDCl$_3$+DMSO-d$_6$ QE-300

C

Aldrich 85,958-3   *CAS [149-29-1]*   C$_7$H$_6$O$_4$
FW 154.12
Patulin   mp 110°C

168.64	
149.85	
146.24	
111.19*	
107.56*	
88.86*	
59.56	

CDCl$_3$ QE-300

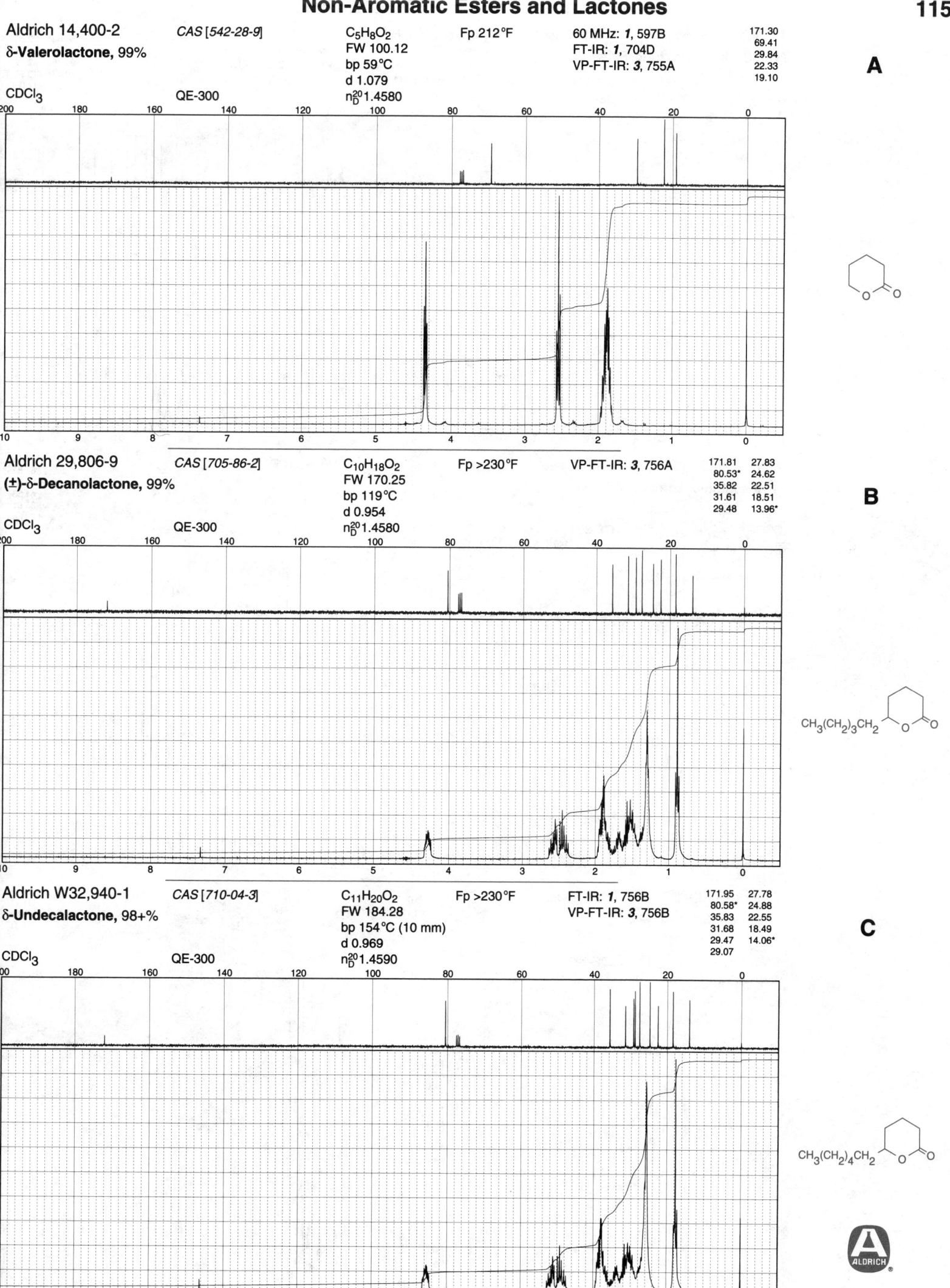

Aldrich 14,400-2

δ-Valerolactone, 99%

CAS [542-28-9]

$C_5H_8O_2$
FW 100.12
bp 59 °C
d 1.079
n_D^{20} 1.4580

Fp 212 °F

60 MHz: *1*, 597B
FT-IR: *1*, 704D
VP-FT-IR: *3*, 755A

171.30
69.41
29.84
22.33
19.10

A

CDCl₃   QE-300

Aldrich 29,806-9

(±)-δ-Decanolactone, 99%

CAS [705-86-2]

$C_{10}H_{18}O_2$
FW 170.25
bp 119 °C
d 0.954
n_D^{20} 1.4580

Fp >230 °F

VP-FT-IR: *3*, 756A

171.81   27.83
80.53*   24.62
35.82   22.51
31.61   18.51
29.48   13.96*

B

CDCl₃   QE-300

$CH_3(CH_2)_3CH_2$

Aldrich W32,940-1

δ-Undecalactone, 98+%

CAS [710-04-3]

$C_{11}H_{20}O_2$
FW 184.28
bp 154 °C (10 mm)
d 0.969
n_D^{20} 1.4590

Fp >230 °F

FT-IR: *1*, 756B
VP-FT-IR: *3*, 756B

171.95   27.78
80.58*   24.88
35.83   22.55
31.68   18.49
29.47   14.06*
29.07

C

CDCl₃   QE-300

$CH_3(CH_2)_4CH_2$

ALDRICH

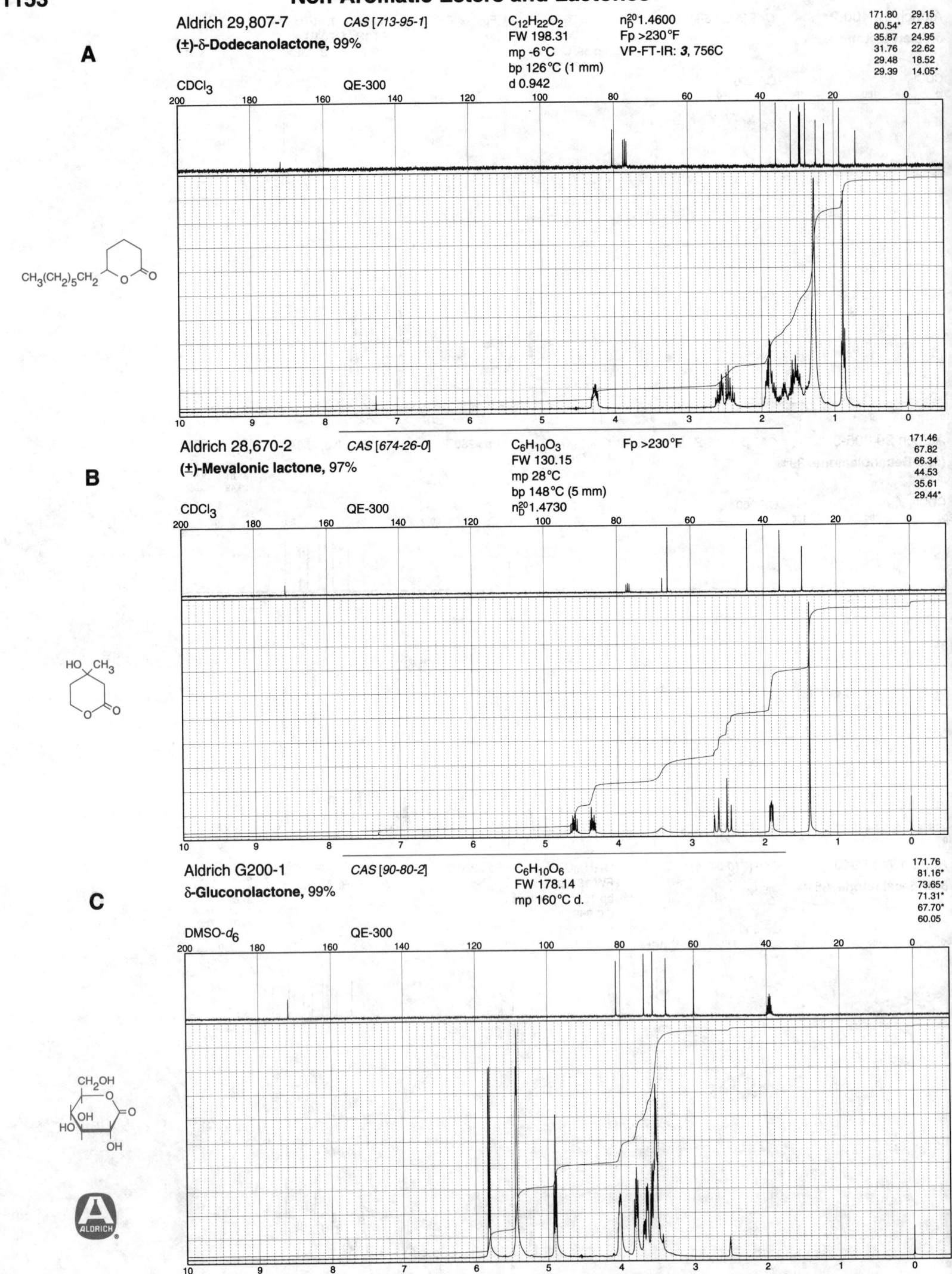

A

Aldrich 29,807-7   *CAS [713-95-1]* $C_{12}H_{22}O_2$ $n_D^{20}1.4600$

(±)-δ-Dodecanolactone, 99%

FW 198.31   Fp >230°F
mp -6°C   VP-FT-IR: *3*, 756C
bp 126°C (1 mm)
d 0.942

171.80   29.15
80.54*   27.83
35.87   24.95
31.76   22.62
29.48   18.52
29.39   14.05*

CDCl₃   QE-300

CH₃(CH₂)₅CH₂ (lactone structure)

B

Aldrich 28,670-2   *CAS [674-26-0]* $C_6H_{10}O_3$ Fp >230°F

(±)-Mevalonic lactone, 97%

FW 130.15
mp 28°C
bp 148°C (5 mm)
$n_D^{20}1.4730$

171.46
67.82
66.34
44.53
35.61
29.44*

CDCl₃   QE-300

HO CH₃ (lactone structure)

C

Aldrich G200-1   *CAS [90-80-2]* $C_6H_{10}O_6$

δ-Gluconolactone, 99%

FW 178.14
mp 160°C d.

171.76
81.16*
73.65*
71.31*
67.70*
60.05

DMSO-d_6 QE-300

CH₂OH (structure) OH, HO, OH

ALDRICH

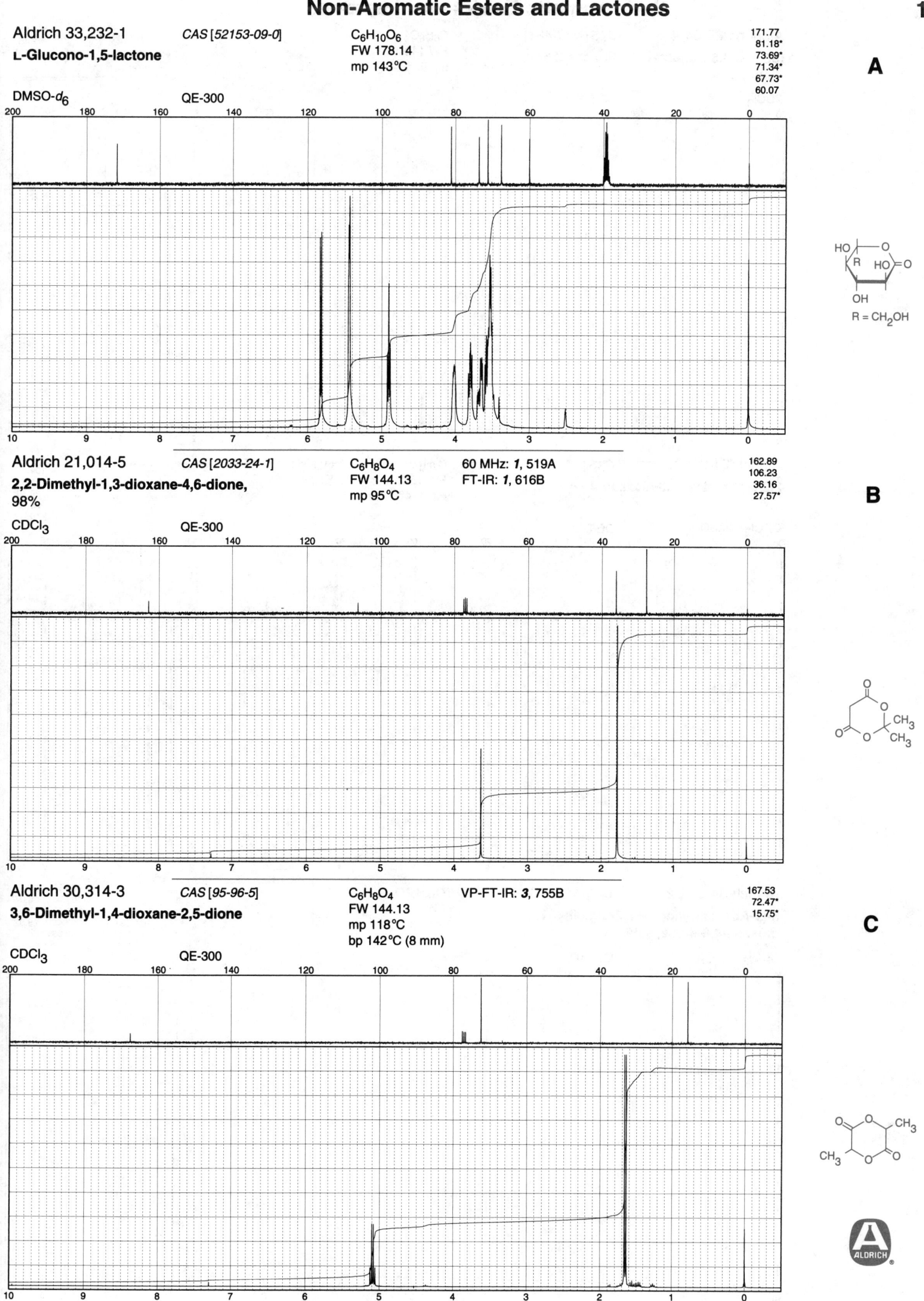

Aldrich 33,232-1
L-Glucono-1,5-lactone

CAS [52153-09-0]

$C_6H_{10}O_6$
FW 178.14
mp 143°C

171.77
81.18*
73.69*
71.34*
67.73*
60.07

A

DMSO-d_6 QE-300

R = CH_2OH

Aldrich 21,014-5
2,2-Dimethyl-1,3-dioxane-4,6-dione, 98%

CAS [2033-24-1]

$C_6H_8O_4$
FW 144.13
mp 95°C

60 MHz: *1*, 519A
FT-IR: *1*, 616B

162.89
106.23
36.16
27.57*

B

$CDCl_3$ QE-300

CH_3
CH_3

Aldrich 30,314-3
3,6-Dimethyl-1,4-dioxane-2,5-dione

CAS [95-96-5]

$C_6H_8O_4$
FW 144.13
mp 118°C
bp 142°C (8 mm)

VP-FT-IR: *3*, 755B

167.53
72.47*
15.75*

C

$CDCl_3$ QE-300

CH_3
CH_3

ALDRICH

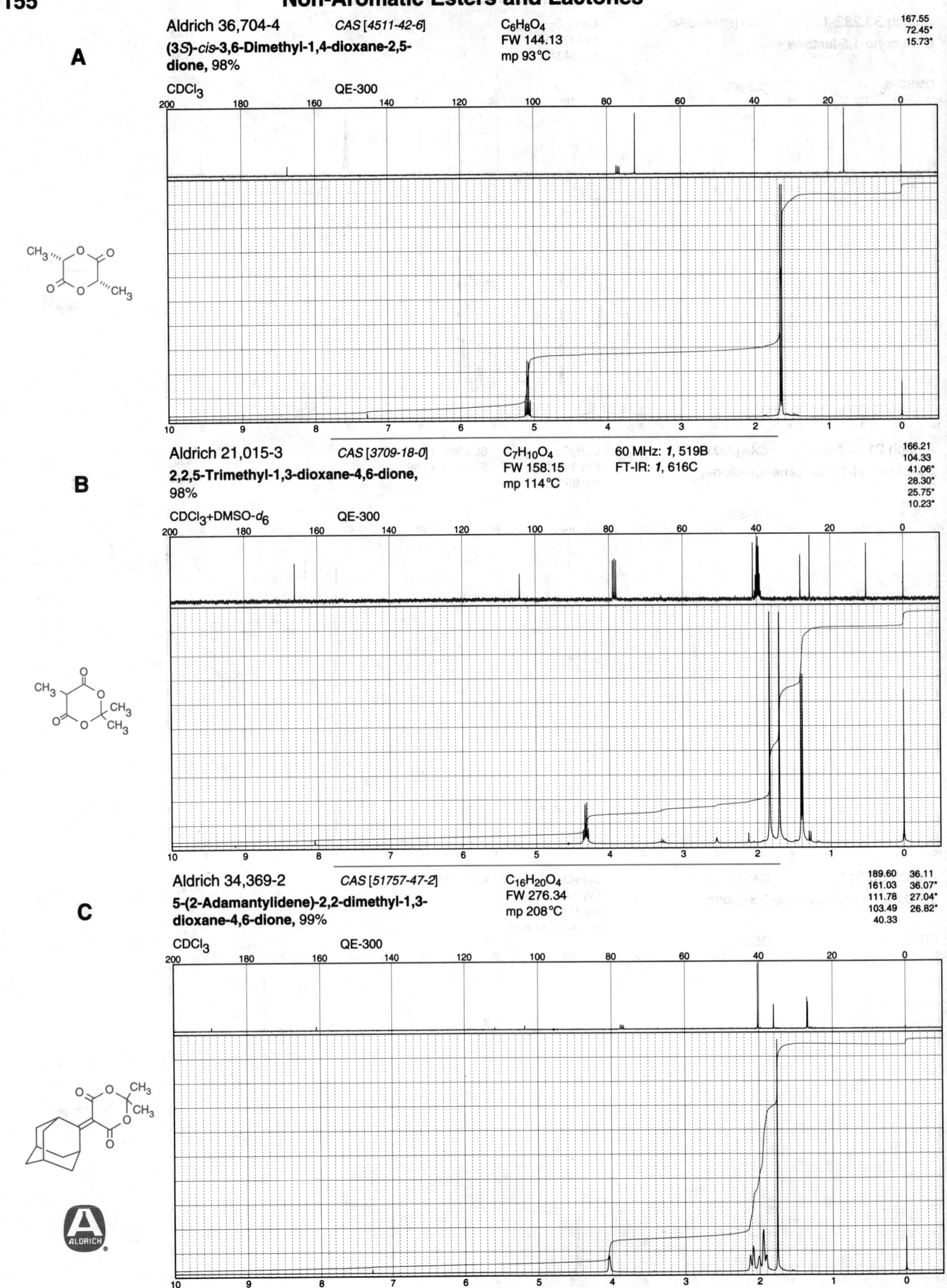

A

Aldrich 36,704-4   *CAS [4511-42-6]*   C₆H₈O₄   167.55
                                                                          72.45*
(3S)-cis-3,6-Dimethyl-1,4-dioxane-2,5-   FW 144.13   15.73*
dione, 98%   mp 93°C

CDCl₃   QE-300

B

Aldrich 21,015-3   *CAS [3709-18-0]*   C₇H₁₀O₄   60 MHz: *1*, 519B   166.21
                                                                                              104.33
2,2,5-Trimethyl-1,3-dioxane-4,6-dione,   FW 158.15   FT-IR: *1*, 616C   41.06*
98%   mp 114°C   28.30*
                                                                                25.75*
                                                                                10.23*

CDCl₃+DMSO-*d*₆   QE-300

C

Aldrich 34,369-2   *CAS [51757-47-2]*   C₁₆H₂₀O₄   189.60   36.11
                                                                                        161.03   36.07*
5-(2-Adamantylidene)-2,2-dimethyl-1,3-   FW 276.34   111.78   27.04*
dioxane-4,6-dione, 99%   mp 208°C   103.49   26.82*
                                                          40.33

CDCl₃   QE-300

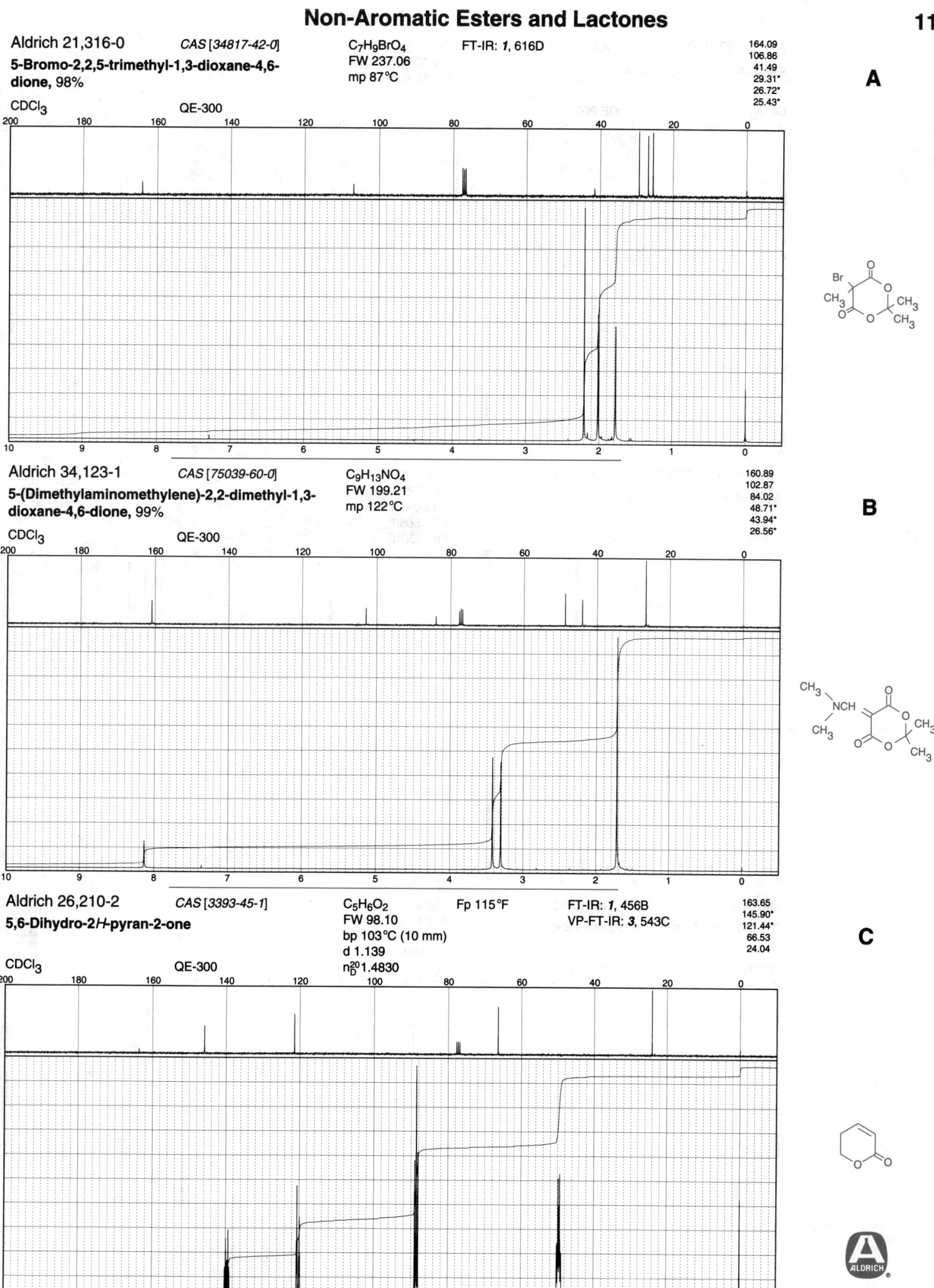

Aldrich 21,316-0   CAS [34817-42-0]   C₇H₉BrO₄   FT-IR: 1, 616D

5-Bromo-2,2,5-trimethyl-1,3-dioxane-4,6-dione, 98%

FW 237.06
mp 87 °C

164.09
106.86
41.49
29.31*
26.72*
25.43*

CDCl₃   QE-300

A

Aldrich 34,123-1   CAS [75039-60-0]   C₉H₁₃NO₄

5-(Dimethylaminomethylene)-2,2-dimethyl-1,3-dioxane-4,6-dione, 99%

FW 199.21
mp 122 °C

160.89
102.87
84.02
48.71*
43.94*
26.56*

CDCl₃   QE-300

B

Aldrich 26,210-2   CAS [3393-45-1]   C₅H₆O₂   Fp 115 °F   FT-IR: 1, 456B

5,6-Dihydro-2H-pyran-2-one

FW 98.10
bp 103 °C (10 mm)
d 1.139
n²⁰_D 1.4830

VP-FT-IR: 3, 543C

163.65
145.90*
121.44*
66.53
24.04

CDCl₃   QE-300

C

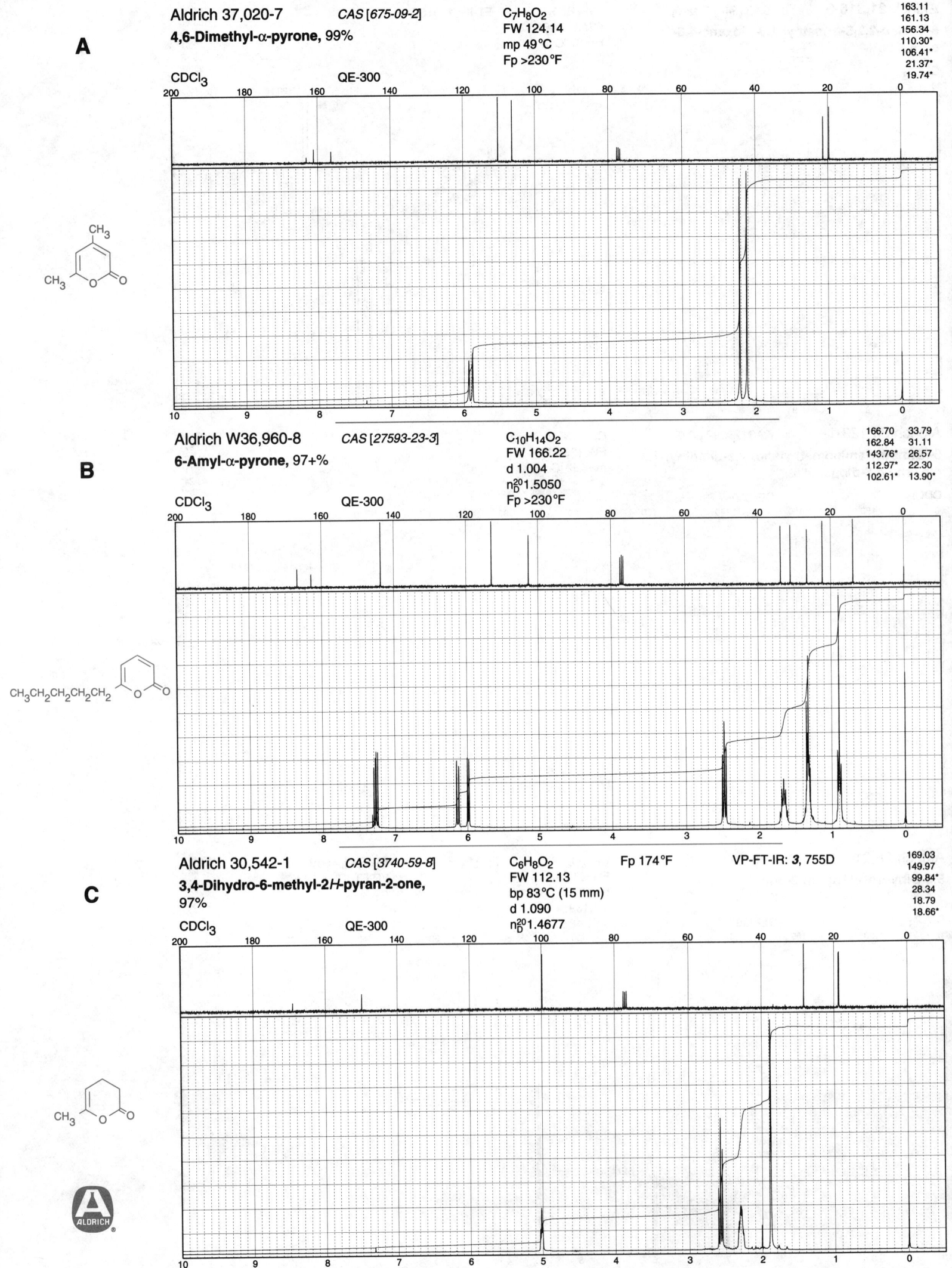

A Aldrich 37,020-7   *CAS [675-09-2]*

4,6-Dimethyl-α-pyrone, 99%

$C_7H_8O_2$
FW 124.14
mp 49°C
Fp >230°F

163.11
161.13
156.34
110.30*
106.41*
21.37*
19.74*

CDCl$_3$ QE-300

B Aldrich W36,960-8   *CAS [27593-23-3]*

6-Amyl-α-pyrone, 97+%

$C_{10}H_{14}O_2$
FW 166.22
d 1.004
n_D^{20} 1.5050
Fp >230°F

166.70 33.79
162.84 31.11
143.76* 26.57
112.97* 22.30
102.61* 13.90*

CDCl$_3$ QE-300

CH$_3$CH$_2$CH$_2$CH$_2$CH$_2$

C Aldrich 30,542-1   *CAS [3740-59-8]*   Fp 174°F   VP-FT-IR: *3*, 755D

3,4-Dihydro-6-methyl-2H-pyran-2-one, 97%

$C_6H_8O_2$
FW 112.13
bp 83°C (15 mm)
d 1.090
n_D^{20} 1.4677

169.03
149.97
99.84*
28.34
18.79
18.66*

CDCl$_3$ QE-300

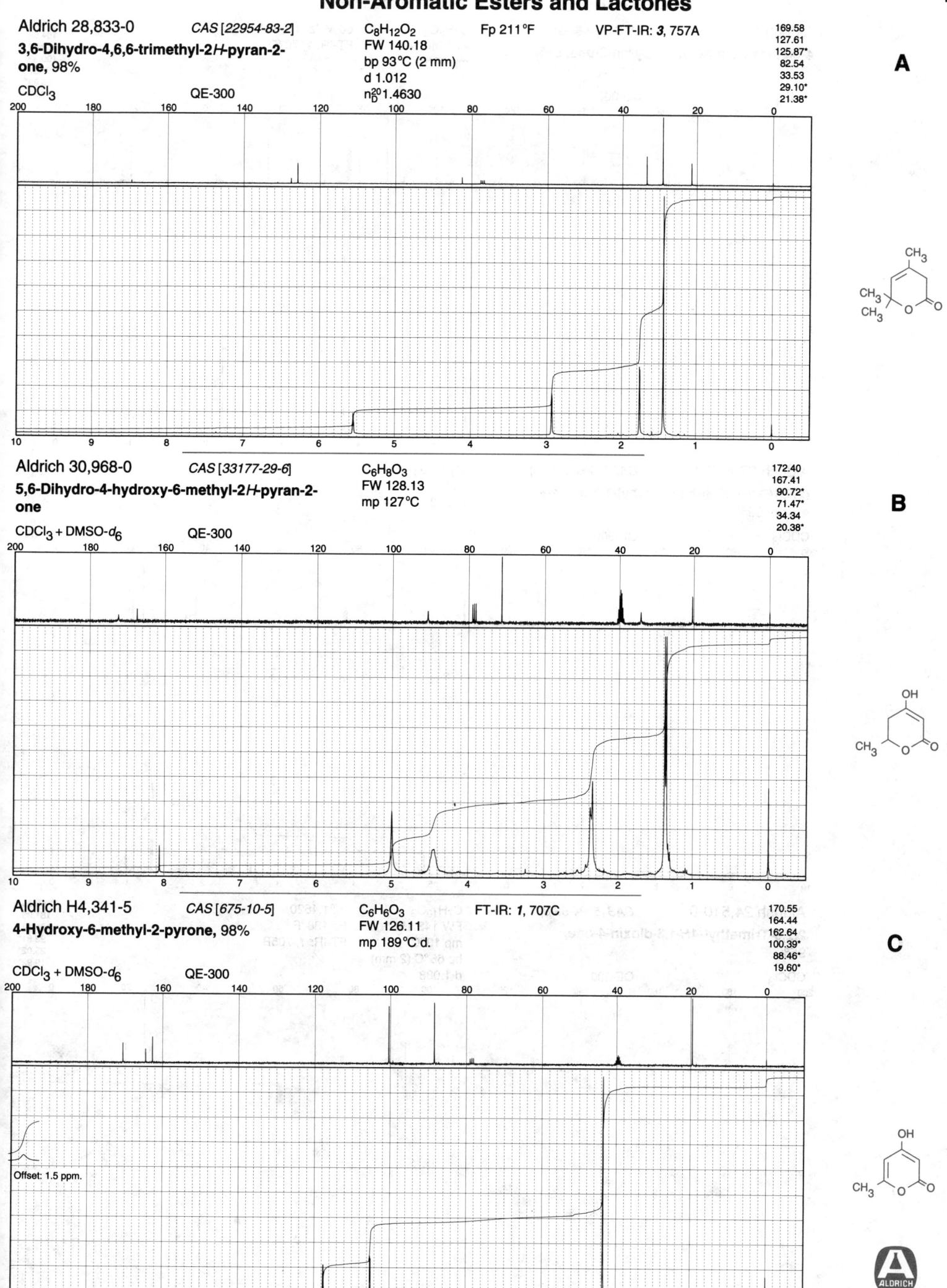

Aldrich 28,833-0   CAS [22954-83-2]

3,6-Dihydro-4,6,6-trimethyl-2H-pyran-2-one, 98%

CDCl₃   QE-300

$C_8H_{12}O_2$
FW 140.18
bp 93°C (2 mm)
d 1.012
n_D^{20} 1.4630

Fp 211°F

VP-FT-IR: *3*, 757A

169.58
127.61
125.87*
82.54
33.53
29.10*
21.38*

A

Aldrich 30,968-0   CAS [33177-29-6]

5,6-Dihydro-4-hydroxy-6-methyl-2H-pyran-2-one

CDCl₃ + DMSO-d_6 QE-300

$C_6H_8O_3$
FW 128.13
mp 127°C

172.40
167.41
90.72*
71.47*
34.34
20.38*

B

Aldrich H4,341-5   CAS [675-10-5]

4-Hydroxy-6-methyl-2-pyrone, 98%

CDCl₃ + DMSO-d_6 QE-300

$C_6H_6O_3$
FW 126.11
mp 189°C d.

FT-IR: *1*, 707C

170.55
164.44
162.64
100.39*
88.46*
19.60*

C

Offset: 1.5 ppm.

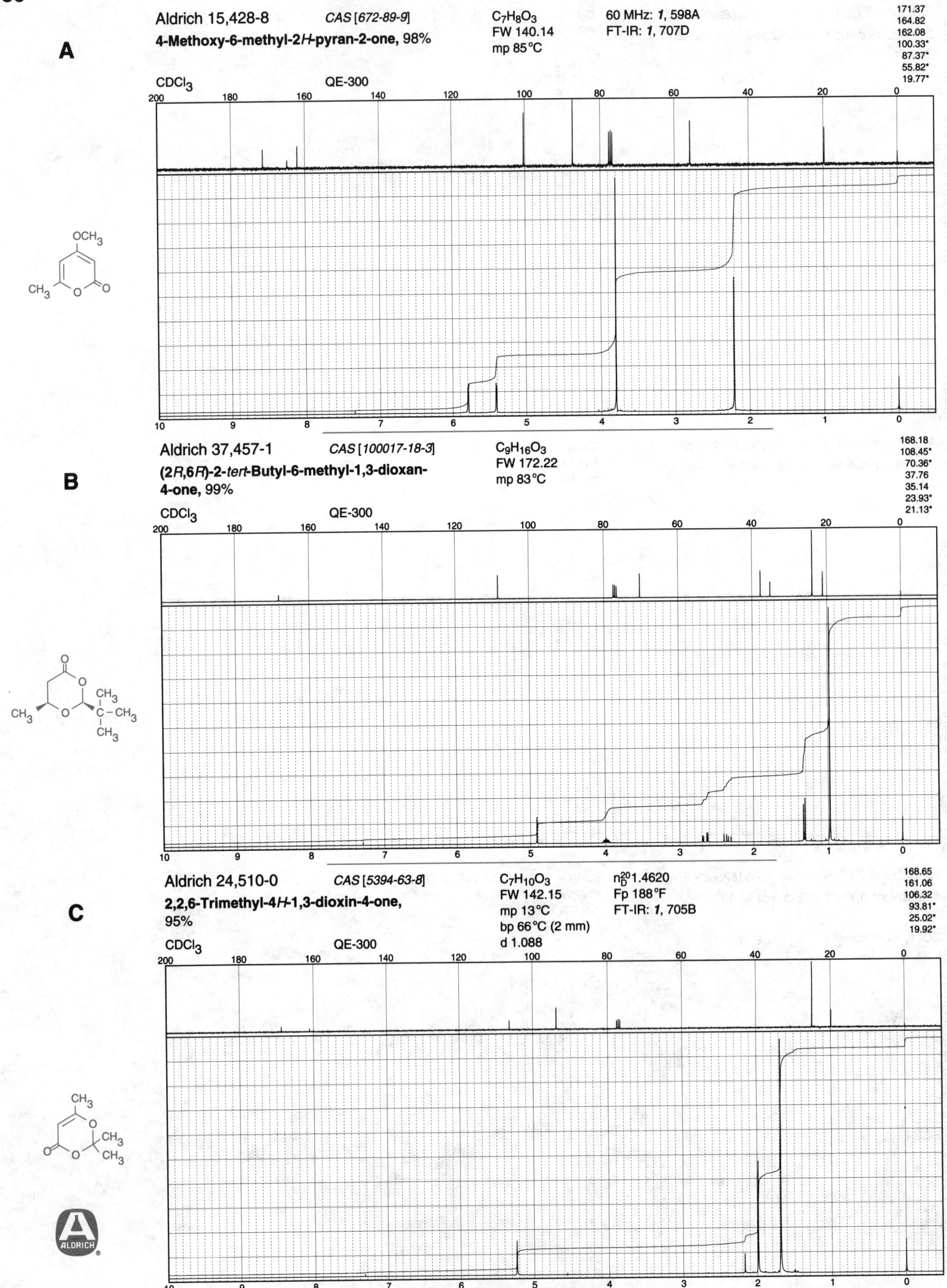

A

Aldrich 15,428-8   CAS [672-89-9]

4-Methoxy-6-methyl-2H-pyran-2-one, 98%

C$_7$H$_8$O$_3$
FW 140.14
mp 85°C

60 MHz: *1*, 598A
FT-IR: *1*, 707D

171.37
164.82
162.08
100.33*
87.37*
55.82*
19.77*

CDCl$_3$ QE-300

B

Aldrich 37,457-1   CAS [100017-18-3]

(2R,6R)-2-tert-Butyl-6-methyl-1,3-dioxan-4-one, 99%

C$_9$H$_{16}$O$_3$
FW 172.22
mp 83°C

168.18
108.45*
70.36*
37.76
35.14
23.93*
21.13*

CDCl$_3$ QE-300

C

Aldrich 24,510-0   CAS [5394-63-8]

2,2,6-Trimethyl-4H-1,3-dioxin-4-one, 95%

C$_7$H$_{10}$O$_3$
FW 142.15
mp 13°C
bp 66°C (2 mm)
d 1.088

n_D^{20}1.4620
Fp 188°F
FT-IR: *1*, 705B

168.65
161.06
106.32
93.81*
25.02*
19.92*

CDCl$_3$ QE-300

Aldrich C8,540-9   CAS [500-05-0]

Coumalic acid, 97%

C6H4O4
FW 140.10
mp 204°C d.
bp 218°C (120 mm)

60 MHz: *1*, 598B
FT-IR: *1*, 708A

164.17
159.82
158.76*
142.42*
114.50*
112.05

A

DMSO-*d*6   QE-300

Aldrich 27,369-4   CAS [27593-22-2]

2-Oxo-6-pentyl-2*H*-pyran-3-carboxylic acid, 97%

C11H14O4
FW 210.23
mp 123°C

173.14   34.27
165.44   31.10
162.68   26.54
151.44*   22.22
112.50   13.75*
105.36*

B

CDCl3   QE-300

Aldrich 26,143-2   CAS [6018-41-3]

Methyl coumalate, 98%

C7H6O4
FW 154.12
mp 66°C
bp 179°C (60 mm)

FT-IR: *1*, 708B
VP-FT-IR: *3*, 757B

163.34
159.70
158.12*
141.58*
115.23*
111.92
52.44*

C

CDCl3   QE-300

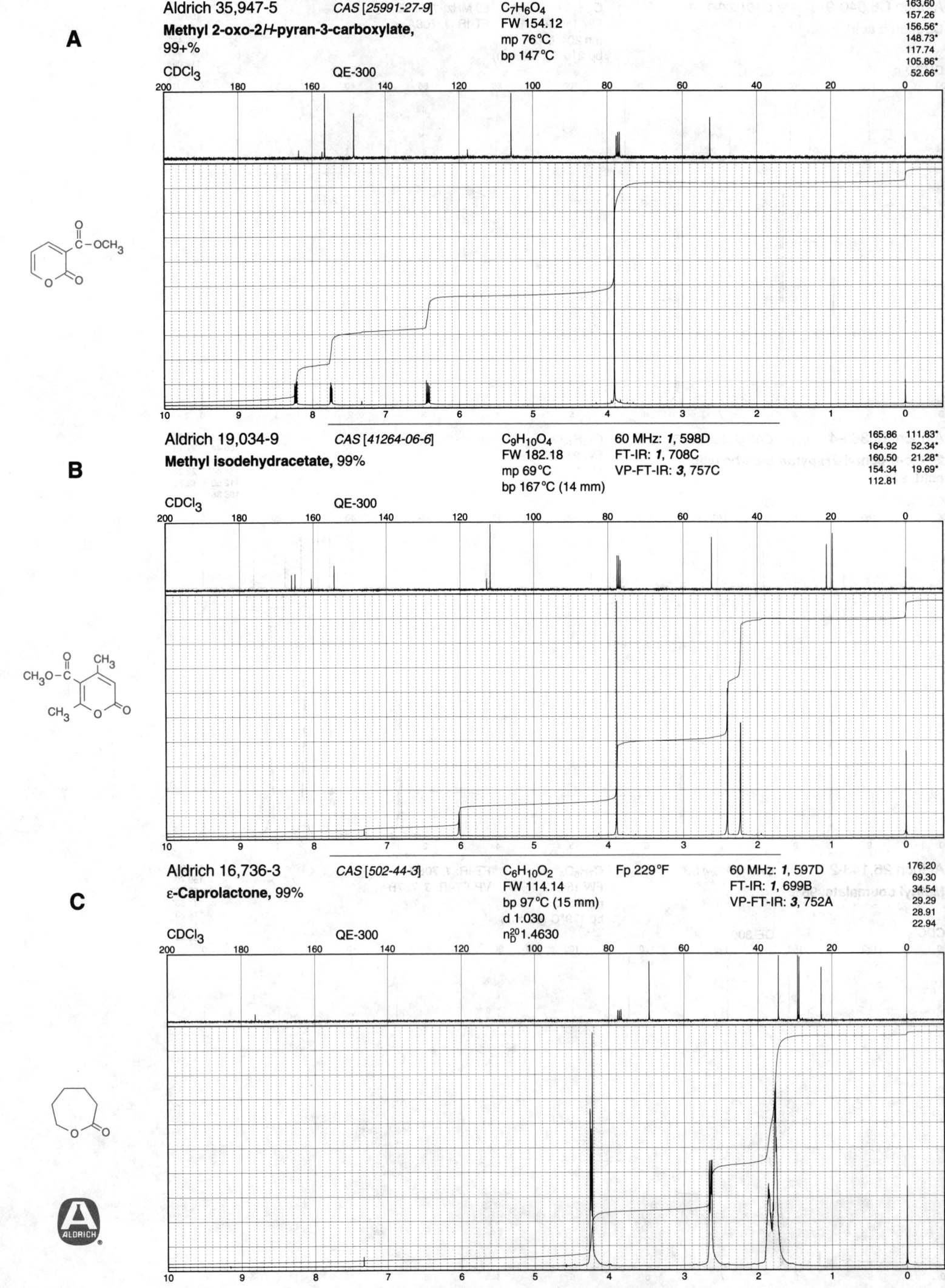

A

Aldrich 35,947-5   CAS [25991-27-9] $C_7H_6O_4$ 163.60
Methyl 2-oxo-2H-pyran-3-carboxylate,   FW 154.12   157.26
99+%   mp 76°C   156.56*
    bp 147°C   148.73*
        117.74
CDCl₃   QE-300   105.86*
        52.66*

B

Aldrich 19,034-9   CAS [41264-06-6] $C_9H_{10}O_4$ 60 MHz: 1, 598D   165.86   111.83*
Methyl isodehydracetate, 99%   FW 182.18   FT-IR: 1, 708C   164.92   52.34*
    mp 69°C   VP-FT-IR: 3, 757C   160.50   21.28*
    bp 167°C (14 mm)   154.34   19.69*
            112.81

CDCl₃   QE-300

C

Aldrich 16,736-3   CAS [502-44-3] $C_6H_{10}O_2$ Fp 229°F   60 MHz: 1, 597D   176.20
ε-Caprolactone, 99%   FW 114.14   FT-IR: 1, 699B   69.30
    bp 97°C (15 mm)   VP-FT-IR: 3, 752A   34.54
    d 1.030   29.29
    n²⁰_D 1.4630   28.91
                22.94

CDCl₃   QE-300

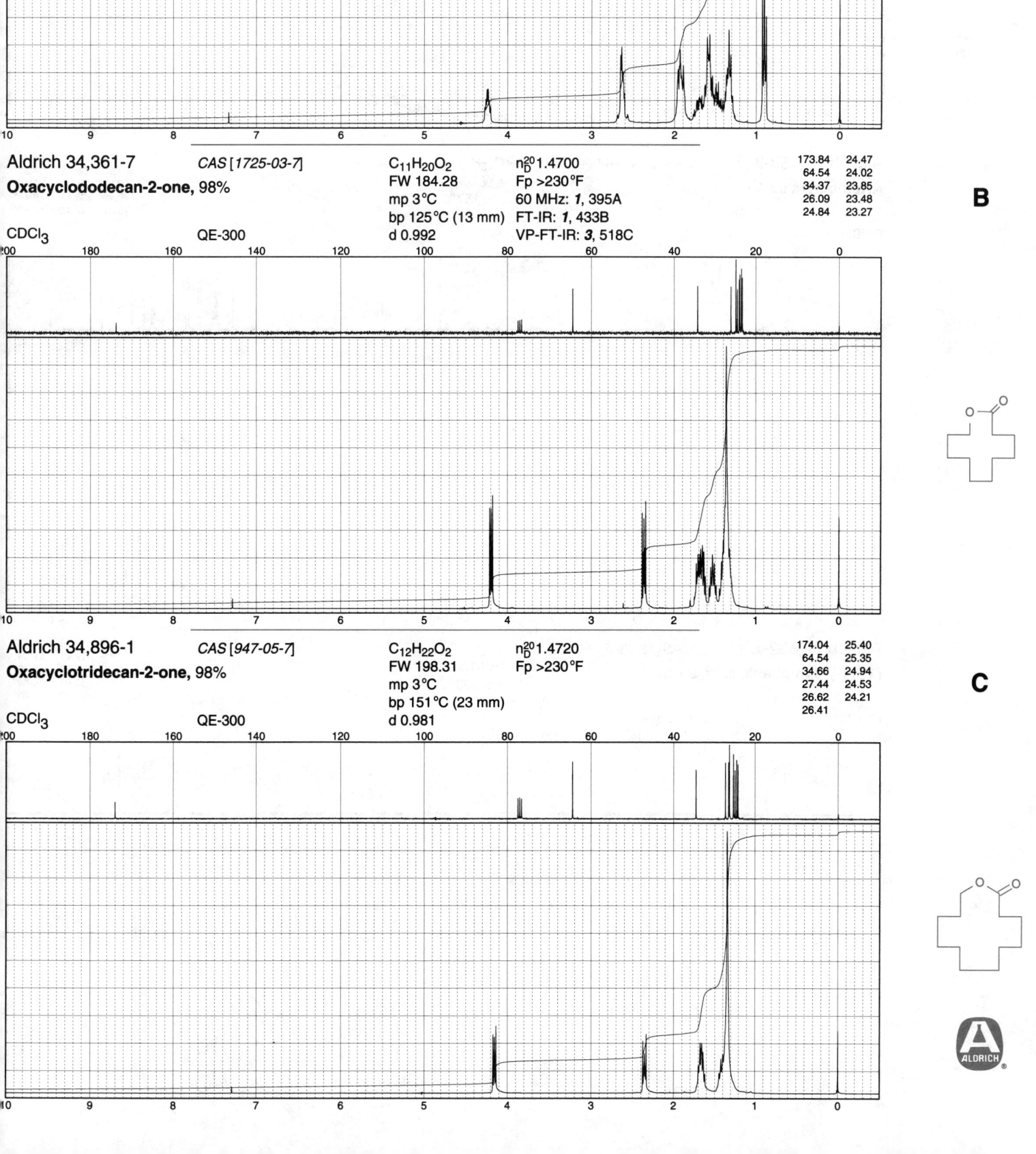

Aldrich W36,130-5   *CAS [5579-78-2]*   C$_{10}$H$_{18}$O$_2$ Fp >230°F

ε-**Decalactone**, 99+%

FW 170.25
bp 117°C (3 mm)
d 0.976

CDCl$_3$ QE-300   n$_D^{20}$1.4610

A

175.72	28.30
80.51*	27.56
36.12	23.10
34.95	22.48
34.60	13.96*

CH$_3$CH$_2$CH$_2$CH$_2$

Aldrich 34,361-7   *CAS [1725-03-7]*   C$_{11}$H$_{20}$O$_2$ n$_D^{20}$1.4700

Oxacyclododecan-2-one, 98%

FW 184.28
mp 3°C
bp 125°C (13 mm)
d 0.992

Fp >230°F
60 MHz: *1*, 395A
FT-IR: *1*, 433B
VP-FT-IR: *3*, 518C

CDCl$_3$ QE-300

B

173.84	24.47
64.54	24.02
34.37	23.85
26.09	23.48
24.84	23.27

Aldrich 34,896-1   *CAS [947-05-7]*   C$_{12}$H$_{22}$O$_2$ n$_D^{20}$1.4720

Oxacyclotridecan-2-one, 98%

FW 198.31
mp 3°C
bp 151°C (23 mm)
d 0.981

Fp >230°F

CDCl$_3$ QE-300

C

174.04	25.40
64.54	25.35
34.66	24.94
27.44	24.53
26.62	24.21
26.41	

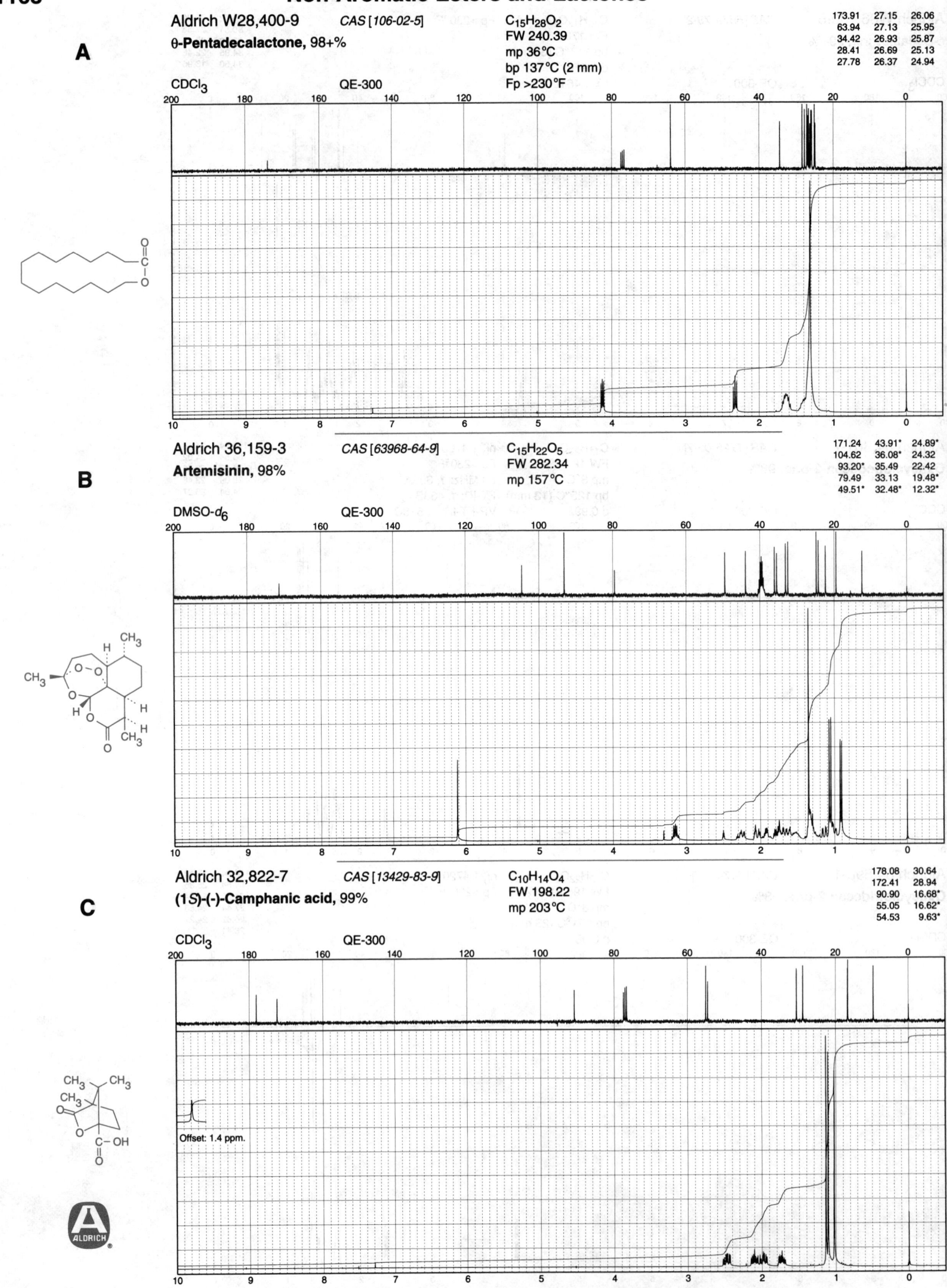

A

Aldrich W28,400-9   CAS [106-02-5]

θ-**Pentadecalactone, 98+%**

$C_{15}H_{28}O_2$
FW 240.39
mp 36°C
bp 137°C (2 mm)
Fp >230°F

173.91	27.15	26.06
63.94	27.13	25.95
34.42	26.93	25.87
28.41	26.69	25.13
27.78	26.37	24.94

CDCl$_3$ QE-300

B

Aldrich 36,159-3   CAS [63968-64-9]

Artemisinin, 98%

$C_{15}H_{22}O_5$
FW 282.34
mp 157°C

171.24	43.91*	24.89*
104.62	36.08*	24.32
93.20*	35.49	22.42
79.49	33.13	19.48*
49.51*	32.48*	12.32*

DMSO-d_6 QE-300

C

Aldrich 32,822-7   CAS [13429-83-9]

(1S)-(-)-**Camphanic acid, 99%**

$C_{10}H_{14}O_4$
FW 198.22
mp 203°C

178.08	30.64
172.41	28.94
90.90	16.68*
55.05	16.62*
54.53	9.63*

CDCl$_3$ QE-300

Offset: 1.4 ppm.

A

Aldrich 22,616-5

(1S)-(-)-Camphanic acid monohydrate,
99%

CDCl$_3$ + DMSO-d_6 QE-300

C$_{10}$H$_{14}$O$_4$ FT-IR: *1*, 526C
FW 216.24
mp 124°C

177.79	30.24
168.58	28.64
90.87	16.54*
54.30	16.45*
53.31	9.52*

B

Aldrich 31,814-0

Retrorsine, 97%

CAS [480-54-6]

CDCl$_3$ + DMSO-d_6 QE-300

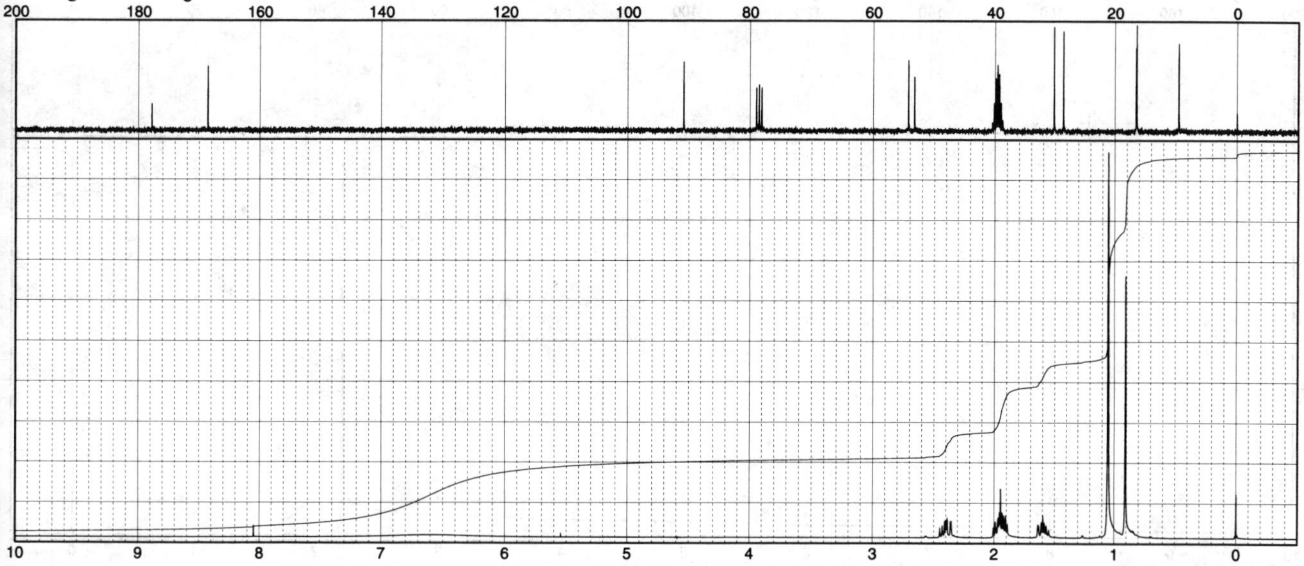

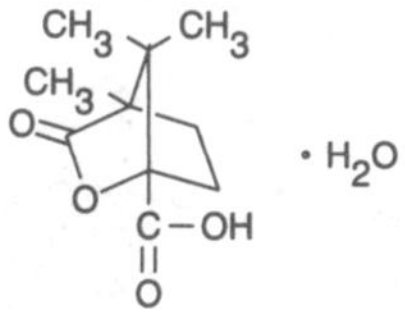

C$_{18}$H$_{25}$NO$_6$
FW 351.40
mp 210°C

174.43	77.13*	37.95
166.83	74.63*	35.39*
135.30*	66.90	34.39
132.95*	62.59	14.73*
131.93	59.31	11.19*
81.23	52.60	

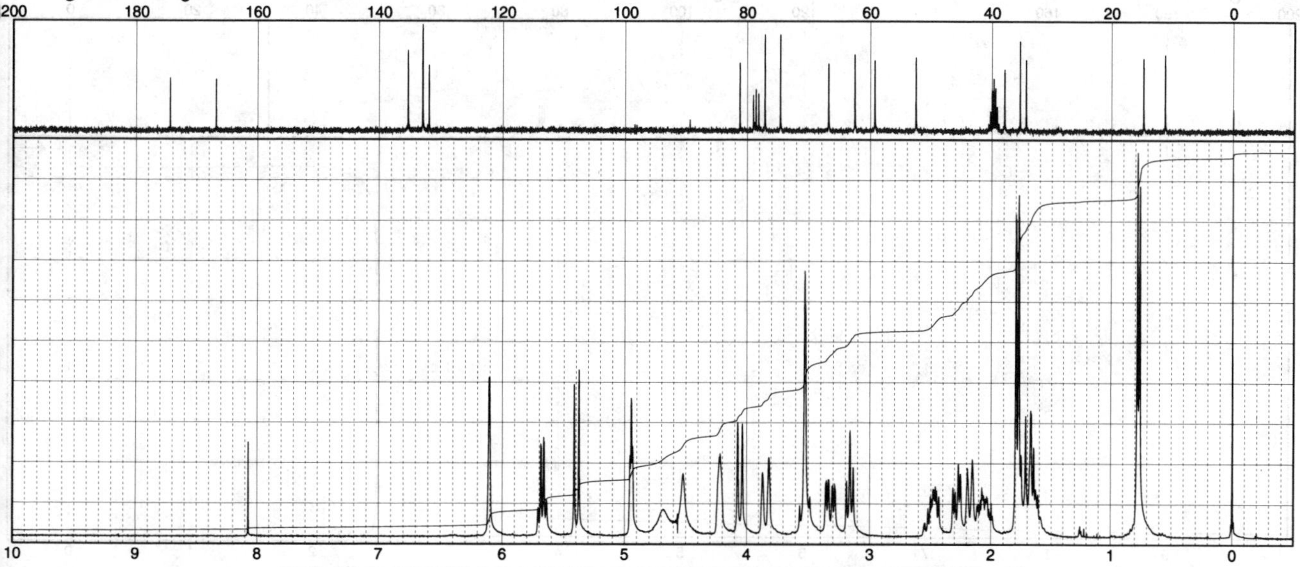

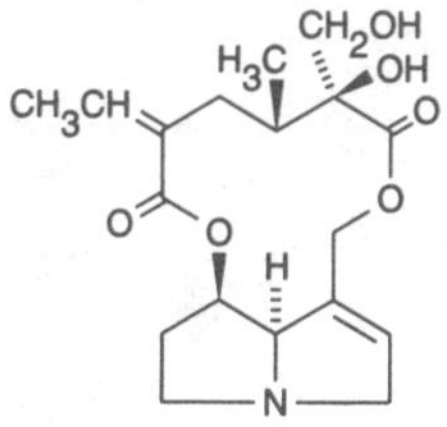

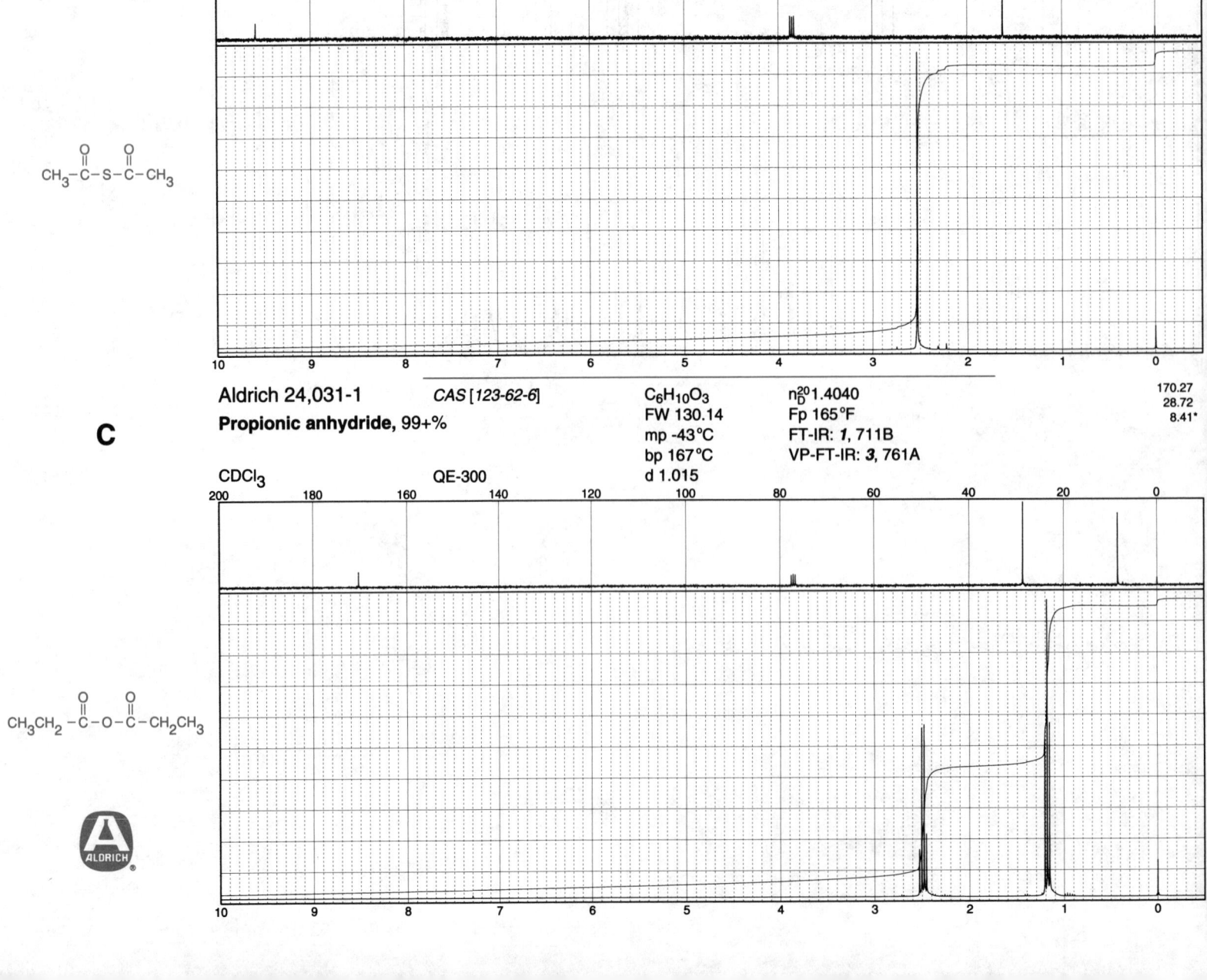

A

Aldrich 24,284-5   *CAS [108-24-7]*

Acetic anhydride, 98%

$C_4H_6O_3$
FW 102.09
mp -73°C
bp 139°C
d 1.080

n_D^{20} 1.3900
Fp 130°F
60 MHz: *1*, 601A
FT-IR: *1*, 711A
VP-FT-IR: *3*, 760C

166.40
22.14*

CDCl₃   QE-300

$CH_3-\overset{O}{\underset{||}{C}}-O-\overset{O}{\underset{||}{C}}-CH_3$

B

Aldrich A2,220-3   *CAS [3232-39-1]*

Acetyl sulfide

$C_4H_6O_2S$
FW 118.15
bp 120°C
d 1.117
n_D^{20} 1.4790

Fp 80°F   60 MHz: *1*, 601B
VP-FT-IR: *3*, 760D

191.69
32.60*

CDCl₃   QE-300

$CH_3-\overset{O}{\underset{||}{C}}-S-\overset{O}{\underset{||}{C}}-CH_3$

C

Aldrich 24,031-1   *CAS [123-62-6]*

Propionic anhydride, 99+%

$C_6H_{10}O_3$
FW 130.14
mp -43°C
bp 167°C
d 1.015

n_D^{20} 1.4040
Fp 165°F
FT-IR: *1*, 711B
VP-FT-IR: *3*, 761A

170.27
28.72
8.41*

CDCl₃   QE-300

$CH_3CH_2-\overset{O}{\underset{||}{C}}-O-\overset{O}{\underset{||}{C}}-CH_2CH_3$

ALDRICH

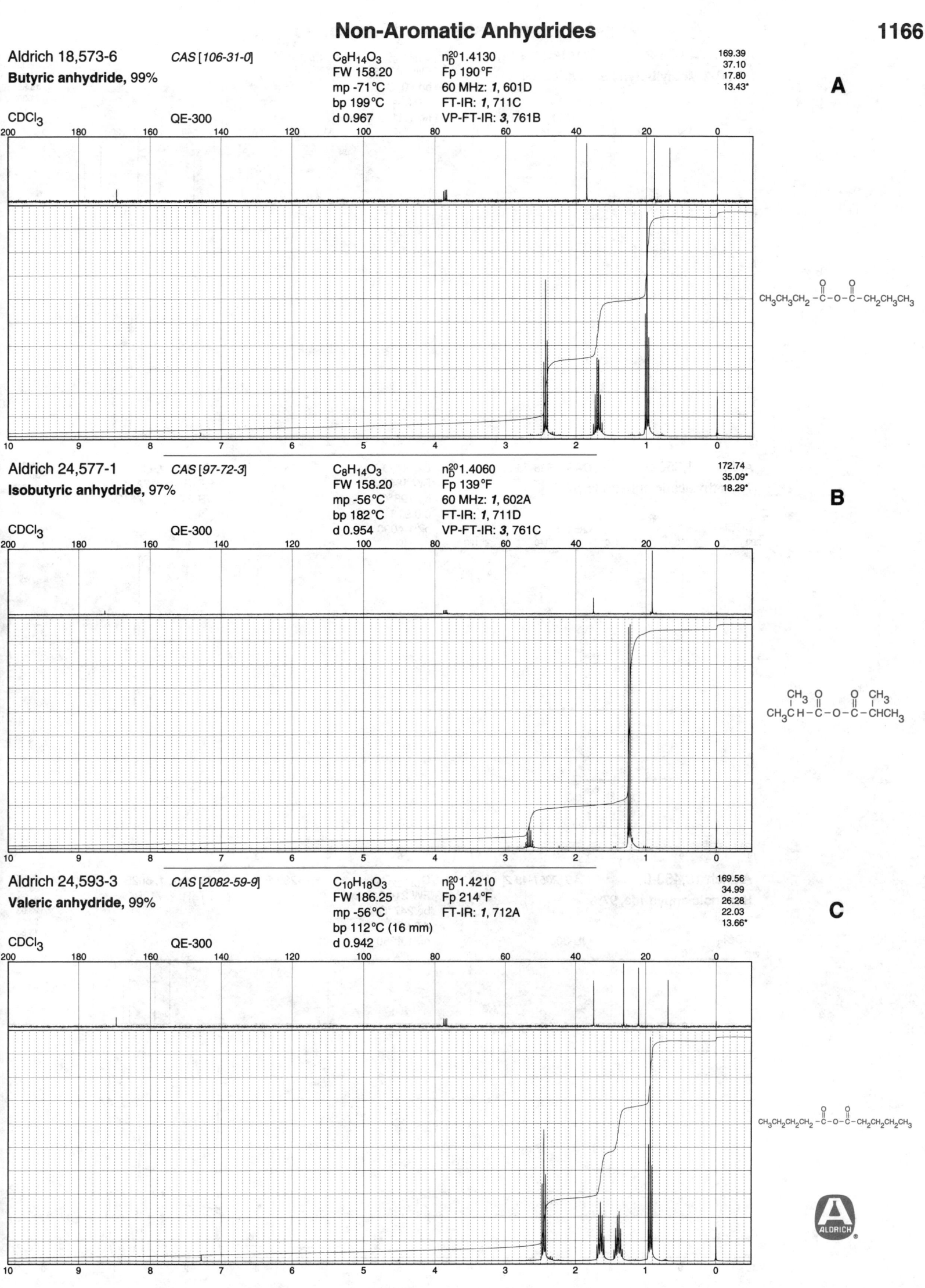

Aldrich 18,573-6
Butyric anhydride, 99%

CAS [106-31-0]

$C_8H_{14}O_3$
FW 158.20
mp -71°C
bp 199°C

n_D^{20} 1.4130
Fp 190°F
60 MHz: *1*, 601D
FT-IR: *1*, 711C
VP-FT-IR: *3*, 761B

169.39
37.10
17.80
13.43*

A

CDCl₃   QE-300   d 0.967

$CH_3CH_2CH_2-\overset{O}{\overset{\|}{C}}-O-\overset{O}{\overset{\|}{C}}-CH_2CH_2CH_3$

Aldrich 24,577-1
Isobutyric anhydride, 97%

CAS [97-72-3]

$C_8H_{14}O_3$
FW 158.20
mp -56°C
bp 182°C

n_D^{20} 1.4060
Fp 139°F
60 MHz: *1*, 602A
FT-IR: *1*, 711D
VP-FT-IR: *3*, 761C

172.74
35.09*
18.29*

B

CDCl₃   QE-300   d 0.954

$\underset{CH_3CH}{\overset{CH_3}{}}-\overset{O}{\overset{\|}{C}}-O-\overset{O}{\overset{\|}{C}}-\underset{CHCH_3}{\overset{CH_3}{}}$

Aldrich 24,593-3
Valeric anhydride, 99%

CAS [2082-59-9]

$C_{10}H_{18}O_3$
FW 186.25
mp -56°C
bp 112°C (16 mm)

n_D^{20} 1.4210
Fp 214°F
FT-IR: *1*, 712A

169.56
34.99
26.28
22.03
13.66*

C

CDCl₃   QE-300   d 0.942

$CH_3CH_2CH_2CH_2-\overset{O}{\overset{\|}{C}}-O-\overset{O}{\overset{\|}{C}}-CH_2CH_2CH_2CH_3$

ALDRICH

Non-Aromatic Esters and Lactones

A

Aldrich 34,856-2   *CAS [84131-91-9]*   C₁₀H₁₈O₃   Fp 194°F

(S)-(+)-2-Methylbutyric anhydride, 94%

$C_{10}H_{18}O_3$
FW 186.25
bp 60°C
d 0.934
n_D^{20} 1.4180

172.39
42.03*
26.21
15.85*
11.36*

CDCl₃   QE-300

B

Aldrich 14,350-2   *CAS [1538-75-6]*

Trimethylacetic anhydride, 99%

$C_{10}H_{18}O_3$
FW 186.25
bp 193°C
d 0.918
n_D^{20} 1.4090

Fp 135°F

60 MHz: *1*, 601C
FT-IR: *1*, 713A
VP-FT-IR: *3*, 761D

173.91
40.18
26.50*

CDCl₃   QE-300

C

Aldrich 19,453-0   *CAS [2051-49-2]*

Hexanoic anhydride, 97%

$C_{12}H_{22}O_3$
FW 214.31
bp 247°C
d 0.926
n_D^{20} 1.4280

Fp >230°F

60 MHz: *1*, 602B
FT-IR: *1*, 712B

169.57
35.25
31.03
23.93
22.28
13.85*

CDCl₃   QE-300

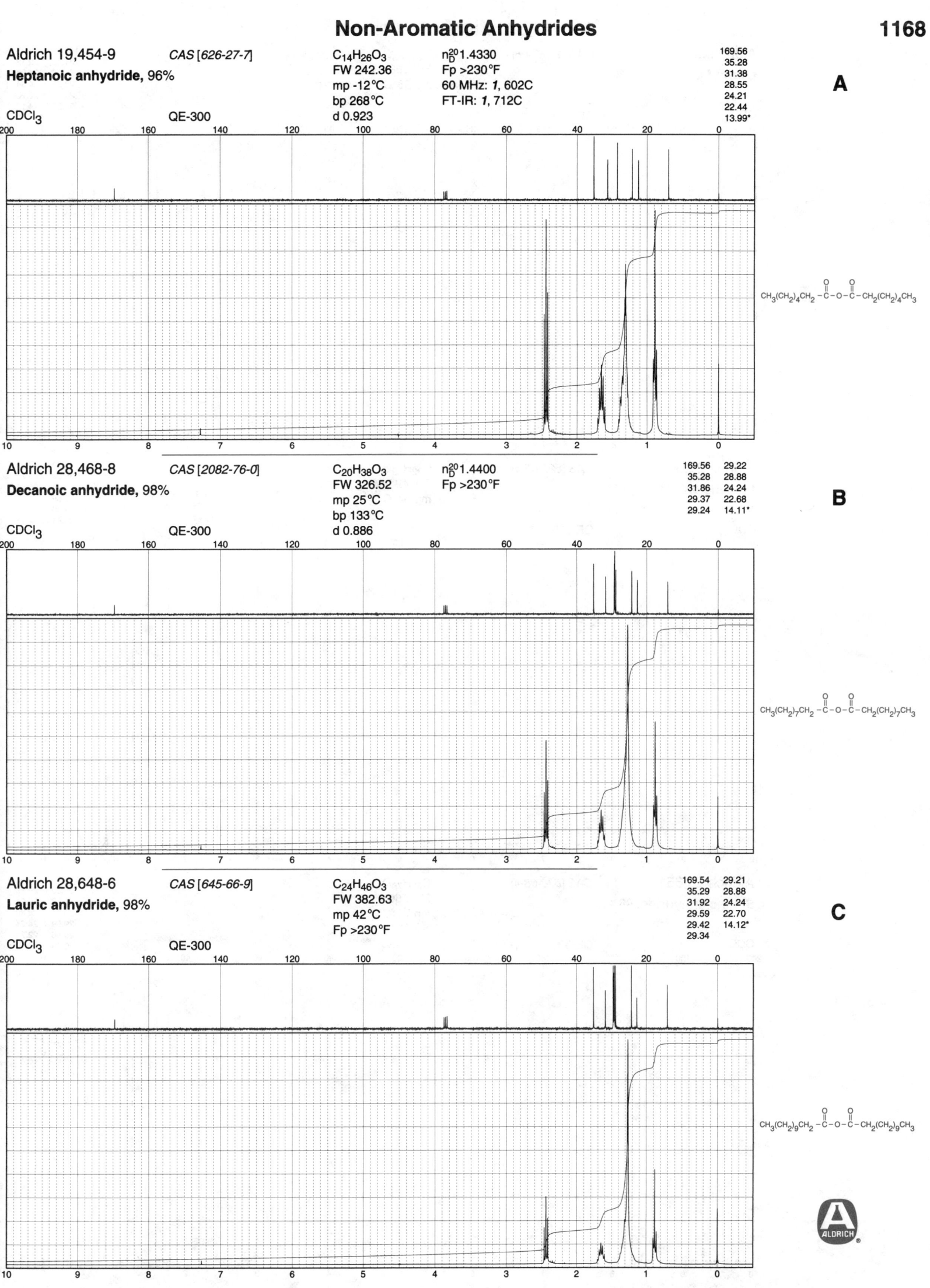

Aldrich 19,454-9   CAS [626-27-7]
Heptanoic anhydride, 96%

$C_{14}H_{26}O_3$
FW 242.36
mp -12°C
bp 268°C
d 0.923

n_D^{20} 1.4330
Fp >230°F
60 MHz: *1*, 602C
FT-IR: *1*, 712C

169.56
35.28
31.38
28.55
24.21
22.44
13.99*

A

CDCl₃   QE-300

$CH_3(CH_2)_4CH_2-C-O-C-CH_2(CH_2)_4CH_3$

Aldrich 28,468-8   CAS [2082-76-0]
Decanoic anhydride, 98%

$C_{20}H_{38}O_3$
FW 326.52
mp 25°C
bp 133°C
d 0.886

n_D^{20} 1.4400
Fp >230°F

169.56   29.22
35.28   28.88
31.86   24.24
29.37   22.68
29.24   14.11*

B

CDCl₃   QE-300

$CH_3(CH_2)_7CH_2-C-O-C-CH_2(CH_2)_7CH_3$

Aldrich 28,648-6   CAS [645-66-9]
Lauric anhydride, 98%

$C_{24}H_{46}O_3$
FW 382.63
mp 42°C
Fp >230°F

169.54   29.21
35.29   28.88
31.92   24.24
29.59   22.70
29.42   14.12*
29.34

C

CDCl₃   QE-300

$CH_3(CH_2)_9CH_2-C-O-C-CH_2(CH_2)_9CH_3$

Non-Aromatic Anhydrides

A

Aldrich 29,078-5

Lauroyl peroxide, 97%

CAS [105-74-8]

$C_{24}H_{46}O_4$
FW 398.63
mp 56°C

169.19	29.35
31.92	29.12
30.02	28.94
29.59	24.84
29.58	22.70
29.37	14.12*

CDCl₃   QE-300

$CH_3(CH_2)_9CH_2-\overset{O}{\overset{\|}{C}}-O-O-\overset{O}{\overset{\|}{C}}-CH_2(CH_2)_9CH_3$

B

Aldrich 28,650-8

Palmitic anhydride, 97%

CAS [623-65-4]

$C_{32}H_{62}O_3$
FW 494.85
mp 64°C

169.53	29.39
35.29	29.22
31.95	28.89
29.69	24.25
29.59	22.71
29.41	14.13*

CDCl₃   QE-300

$CH_3(CH_2)_{13}CH_2-\overset{O}{\overset{\|}{C}}-O-\overset{O}{\overset{\|}{C}}-CH_2(CH_2)_{13}CH_3$

C

Aldrich 28,651-6

Stearic anhydride, 98%

CAS [638-08-4]

$C_{36}H_{70}O_3$
FW 550.96
mp 71°C

169.53	29.42
35.29	29.39
31.95	29.21
29.71	28.89
29.69	24.24
29.66	22.71
29.59	14.13*

CDCl₃   QE-300

$CH_3(CH_2)_{15}CH_2-\overset{O}{\overset{\|}{C}}-O-\overset{O}{\overset{\|}{C}}-CH_2(CH_2)_{15}CH_3$

ALDRICH

Non-Aromatic Anhydrides

Aldrich 28,647-8 CAS [55726-23-3] $C_{44}H_{86}O_3$ FW 663.18 mp 82°C
Docosanoic anhydride, 98%

169.55 29.38
35.29 29.20
31.93 28.88
29.71 24.23
29.59 22.70
29.41 14.13*

A

CDCl₃ QE-300

$CH_3(CH_2)_{19}CH_2-C(=O)-O-C(=O)-CH_2(CH_2)_{19}CH_3$

Aldrich 13,097-4 CAS [78957-07-0] $C_8H_{10}O_3$ FW 154.17 bp 248°C d 1.040 n_D^{20} 1.4740 Fp 231°F
Crotonic anhydride, 95%, predominantly *trans*

60 MHz: 1, 603B
FT-IR: 1, 712D

161.79
149.35*
121.92*
18.34*

B

CDCl₃ QE-300

$CH_3CH=CH-C(=O)-O-C(=O)-CH=CHCH_3$

Aldrich 27,668-5 CAS [760-93-0] $C_8H_{10}O_3$ FW 154.17 bp 87°C (13 mm) d 1.035 n_D^{20} 1.4530 Fp 184°F
Methacrylic anhydride, 94%

163.01
135.67
129.03
17.92*

C

CDCl₃ QE-300

$H_2C=C(CH_3)-C(=O)-O-C(=O)-C(CH_3)=CH_2$

ALDRICH

Non-Aromatic Anhydrides

Aldrich 28,415-7
CAS [24909-72-6]

Oleic anhydride, 98%

A

CDCl₃ QE-300

$C_{36}H_{66}O_3$
FW 546.93
mp 23°C
bp 208°C (11 mm)
Fp >230°F

169.49	29.66	27.22
129.98*	29.54	27.15
129.61*	29.33	24.21
35.26	29.11	22.69
31.92	29.03	14.12*
29.77	28.85	

$CH_3(CH_2)_7CH=CH(CH_2)_6CH_2-C$
$CH_3(CH_2)_7CH=CH(CH_2)_6CH_2-C$

Aldrich 28,649-4
CAS [24909-68-0]

Linoleic anhydride, 98%

B

CDCl₃ QE-300

$C_{36}H_{62}O_3$
FW 542.89
mp -1°C
d 0.901
n_D^{20} 1.4760

Fp >230°F

169.49	31.53	27.20
130.14*	29.56	27.17
129.89*	29.35	25.63
128.04*	29.11	24.20
127.83*	29.03	22.58
35.25	28.84	14.08*

$CH_3(CH_2)_4(CH=CHCH_2)_2(CH_2)_5CH_2-C$
$CH_3(CH_2)_4(CH=CHCH_2)_2(CH_2)_5CH_2-C$

Aldrich 21,516-3
CAS [541-88-8]

Chloroacetic anhydride, 97%

C

CDCl₃ QE-300

$C_4H_4Cl_2O_3$
FW 170.98
mp 58°C
bp 122°C (20 mm)

60 MHz: *1*, 602D
FT-IR: *1*, 713B

161.88
40.98

$ClCH_2-C-O-C-CH_2Cl$

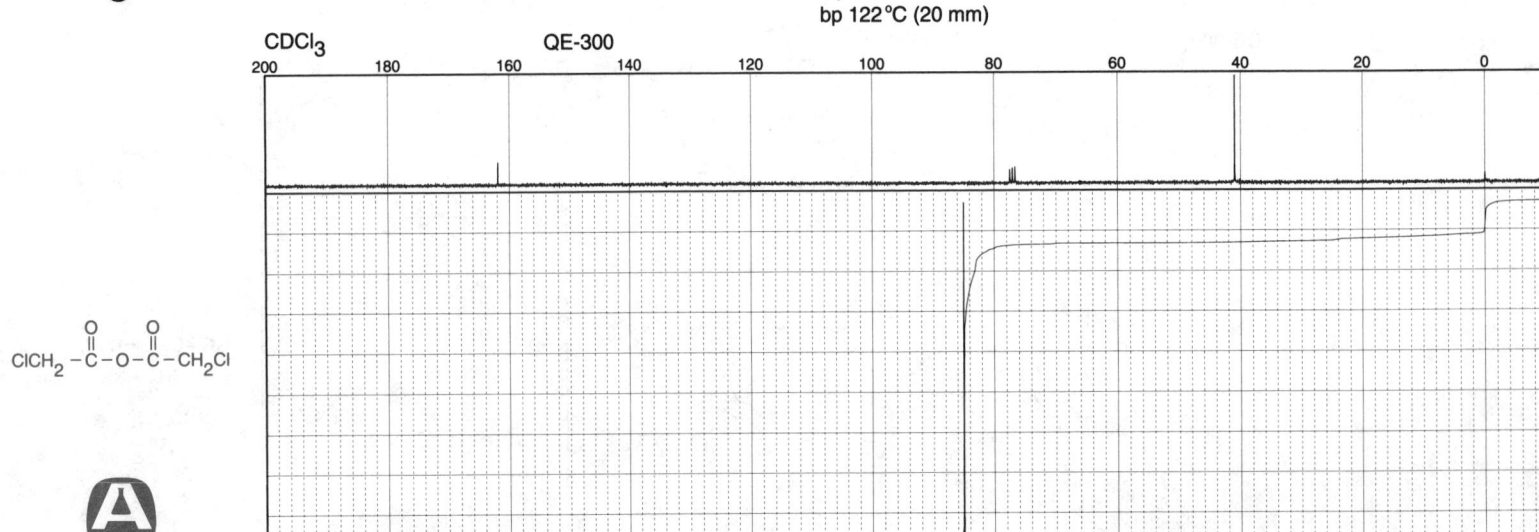

Non-Aromatic Anhydrides

Aldrich 28,426-2 *CAS [54907-61-8]*
Iodoacetic anhydride

$C_4H_4I_2O_3$
FW 353.88
mp 50°C
Fp >230°F

162.95
-6.64

A

CDCl₃ · · · QE-300

$ICH_2-\overset{\displaystyle O}{\overset{\|}{C}}-O-\overset{\displaystyle O}{\overset{\|}{C}}-CH_2I$

Aldrich 10,623-2 *CAS [407-25-0]*
Trifluoroacetic anhydride, 99+%

$C_4F_6O_3$
FW 210.03
mp -65°C
bp 40°C
d 1.487

n_D^{20}<1.3000
Fp NONE
FT-IR: *1*, 714B
VP-FT-IR: *3*, 762B

150.77 118.98
150.13 115.20
149.49 111.41
148.85 107.63

B

CDCl₃ · · · QE-300

$F_3C-\overset{\displaystyle O}{\overset{\|}{C}}-O-\overset{\displaystyle O}{\overset{\|}{C}}-CF_3$

Aldrich 25,206-9 *CAS [2834-23-3]*
Chlorodifluoroacetic anhydride, 98%

$C_4Cl_2F_4O_3$
FW 242.94
bp 97°C
d 1.395
n_D^{20}1.3470

Fp NONE

FT-IR: *1*, 714A
VP-FT-IR: *3*, 762A

151.90
151.38
150.86
119.48
115.49
111.51

C

CDCl₃ · · · QE-300

$CF_2Cl-\overset{\displaystyle O}{\overset{\|}{C}}-O-\overset{\displaystyle O}{\overset{\|}{C}}-CF_2Cl$

A

Aldrich 25,239-5   *CAS [4124-31-6]*   C$_4$Cl$_6$O$_3$ Fp NONE   FT-IR: *1*, 713D   153.93 / 87.83

Trichloroacetic anhydride, 99%

FW 308.76
bp 140°C (60 mm)
d 1.690
n_D^{20}1.4840

CDCl$_3$ QE-300

Cl$_3$C–C(=O)–O–C(=O)–CCl$_3$

B

Aldrich 25,238-7   *CAS [356-42-3]*   C$_6$F$_{10}$O$_3$ Fp NONE

Pentafluoropropionic anhydride, 99%

FW 310.05
bp 70°C (735 mm)
d 1.571
n_D^{20}<1.3000

151.46	119.12	111.09	105.16
151.01	118.68	109.80	104.62
150.57	115.77	109.26	102.71
123.37	115.33	108.71	102.16
122.92	114.89	108.17	101.61
122.48	111.97	106.26	101.07
119.56	111.53	105.71	

CDCl$_3$ QE-300

CF$_3$CF$_2$–C(=O)–O–C(=O)–CF$_2$CF$_3$

C

Aldrich 15,739-2   *CAS [336-59-4]*   C$_8$F$_{14}$O$_3$ Fp NONE   FT-IR: *1*, 714D

Heptafluorobutyric anhydride, 98%

FW 410.06
bp 109°C
d 1.653
n_D^{20}<1.3000

VP-FT-IR: *3*, 762D

150.77
122.89
119.08
115.27
110.68
107.11
103.54

CDCl$_3$ QE-300

CF$_3$CF$_2$CF$_2$–C(=O)–O–C(=O)–CF$_2$CF$_2$CF$_3$

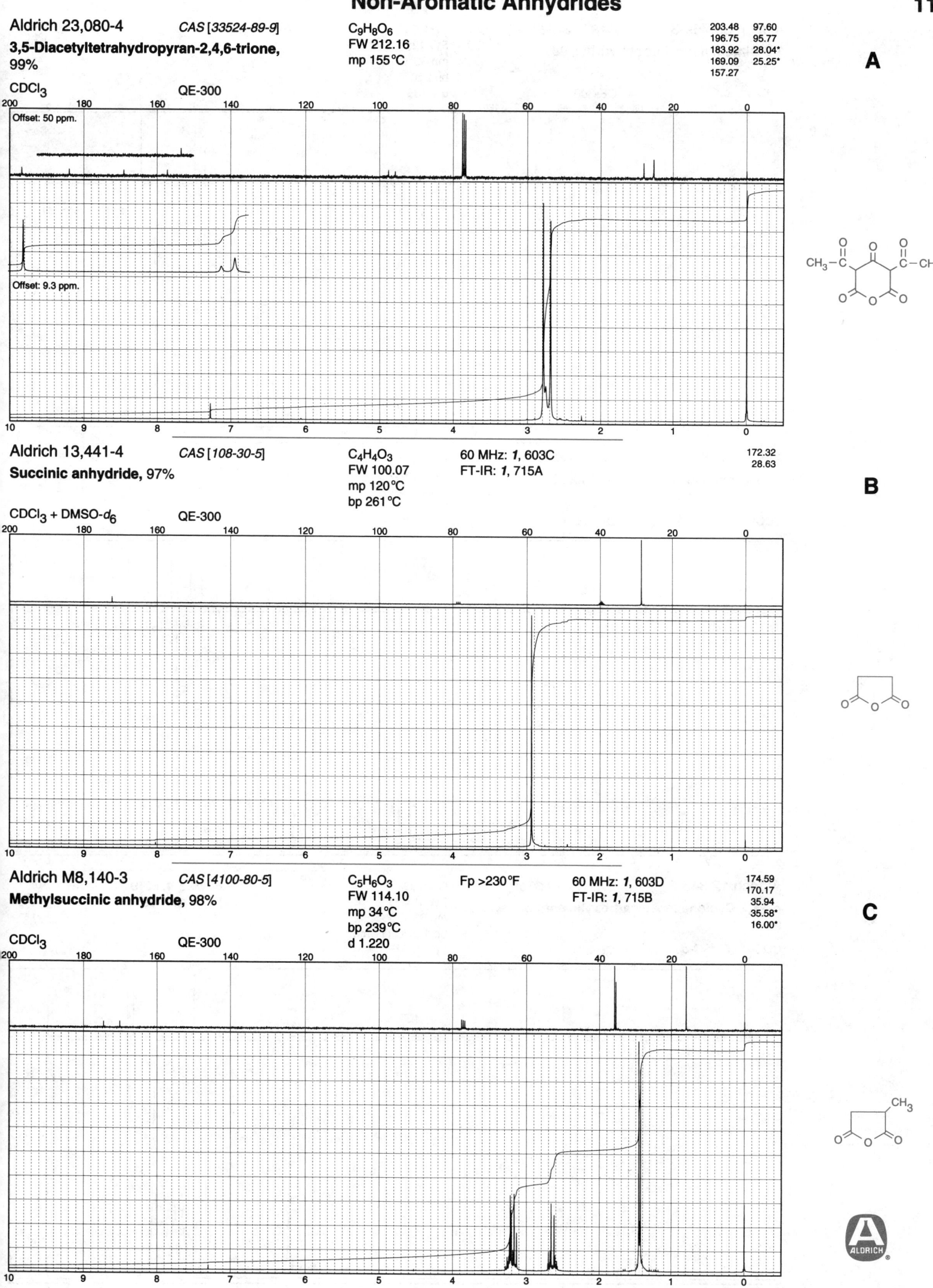

Aldrich 23,080-4   CAS [33524-89-9] $C_9H_8O_6$ FW 212.16   mp 155°C

3,5-Diacetyltetrahydropyran-2,4,6-trione, 99%

203.48   97.60
196.75   95.77
183.92   28.04*
169.09   25.25*
157.27

A

CDCl$_3$ QE-300

Offset: 50 ppm.

Offset: 9.3 ppm.

Aldrich 13,441-4   CAS [108-30-5] $C_4H_4O_3$ FW 100.07   mp 120°C   bp 261°C   60 MHz: 1, 603C   FT-IR: 1, 715A

Succinic anhydride, 97%

172.32
28.63

B

CDCl$_3$ + DMSO-d_6 QE-300

Aldrich M8,140-3   CAS [4100-80-5] $C_5H_6O_3$ FW 114.10   mp 34°C   bp 239°C   d 1.220   Fp >230°F   60 MHz: 1, 603D   FT-IR: 1, 715B

Methylsuccinic anhydride, 98%

174.59
170.17
35.94
35.58*
16.00*

C

CDCl$_3$ QE-300

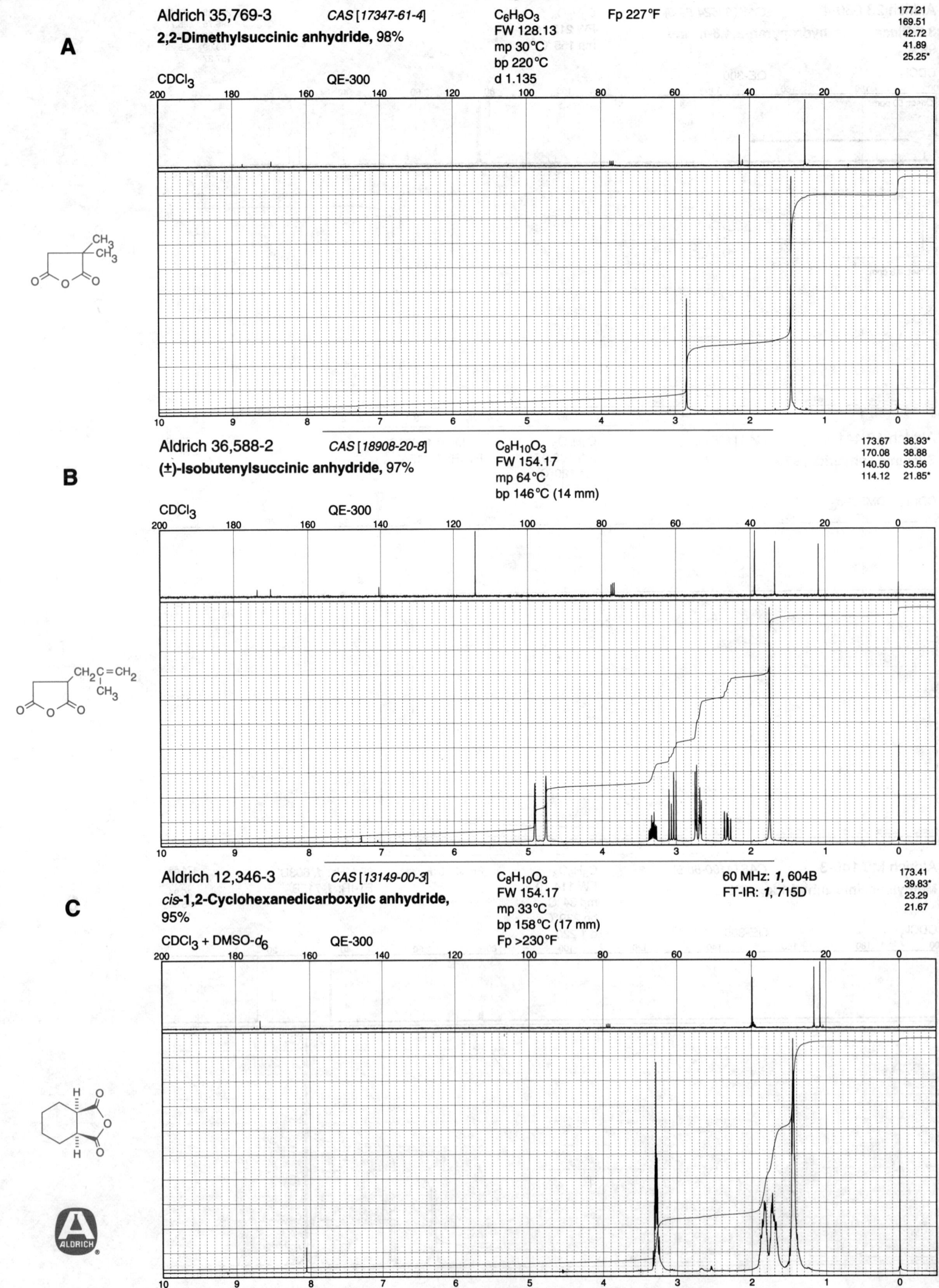

A

Aldrich 35,769-3   *CAS [17347-61-4]*   C$_6$H$_8$O$_3$ Fp 227°F

2,2-Dimethylsuccinic anhydride, 98%

FW 128.13
mp 30°C
bp 220°C
d 1.135

177.21
169.51
42.72
41.89
25.25*

CDCl$_3$ QE-300

B

Aldrich 36,588-2   *CAS [18908-20-8]*   C$_8$H$_{10}$O$_3$

(±)-Isobutenylsuccinic anhydride, 97%

FW 154.17
mp 64°C
bp 146°C (14 mm)

173.67   38.93*
170.08   38.88
140.50   33.56
114.12   21.85*

CDCl$_3$ QE-300

C

Aldrich 12,346-3   *CAS [13149-00-3]*   C$_8$H$_{10}$O$_3$

cis-1,2-Cyclohexanedicarboxylic anhydride, 95%

FW 154.17
mp 33°C
bp 158°C (17 mm)
Fp >230°F

60 MHz: *1*, 604B
FT-IR: *1*, 715D

173.41
39.83*
23.29
21.67

CDCl$_3$ + DMSO-*d$_6$*   QE-300

A

Aldrich 14,829-6 *CAS [14166-21-3]*

(±)-trans-1,2-Cyclohexanedicarboxylic anhydride

$C_8H_{10}O_3$
FW 154.17
mp 146°C

60 MHz: *1*, 604C
FT-IR: *1*, 716A

171.03
45.60*
24.52
24.44

CDCl$_3$ + DMSO-d_6 QE-300

Offset: 2.1 ppm.

B

Aldrich 14,993-4 *CAS [19438-60-9]*

(±)-Hexahydro-4-methylphthalic anhydride, 97%, mixture of *cis* and *trans*

$C_9H_{12}O_3$
FW 168.19
d 1.162
n_D^{20} 1.4770
Fp >230°F

60 MHz: *1*, 605A
FT-IR: *1*, 716B

173.15	40.27*	29.21*
172.98	39.92*	28.23*
172.88	34.48	26.01
172.67	30.60	22.00*
41.01*	30.37	21.60*
40.82*	29.46	21.28

CDCl$_3$ QE-300

C

Aldrich 25,992-6 *CAS [2170-03-8]*

Itaconic anhydride, 95%

$C_5H_4O_3$
FW 112.09
mp 71°C
bp 115°C (12 mm)

FT-IR: *1*, 716D

170.16
166.59
132.68
125.99
34.62

CD$_3$CN   QE-300

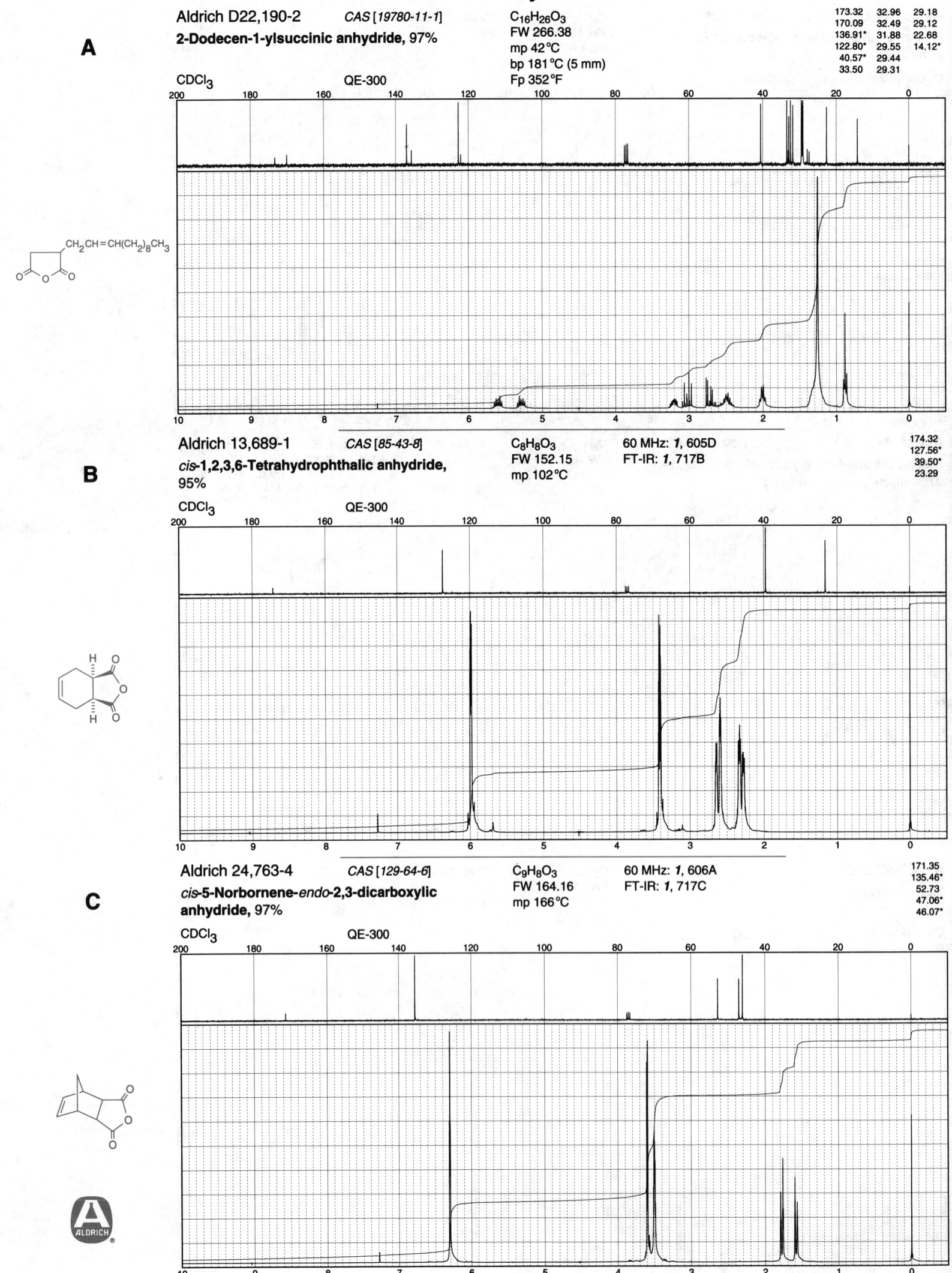

A

Aldrich D22,190-2   *CAS [19780-11-1]* $C_{16}H_{26}O_3$
2-Dodecen-1-ylsuccinic anhydride, 97%   FW 266.38
mp 42°C
bp 181°C (5 mm)
Fp 352°F

173.32	32.96	29.18
170.09	32.49	29.12
136.91*	31.88	22.68
122.80*	29.55	14.12*
40.57*	29.44	
33.50	29.31	

CDCl₃   QE-300

CH₂CH=CH(CH₂)₈CH₃

B

Aldrich 13,689-1   *CAS [85-43-8]* $C_8H_8O_3$ 60 MHz: *1*, 605D
cis-1,2,3,6-Tetrahydrophthalic anhydride, 95%   FW 152.15   FT-IR: *1*, 717B
mp 102°C

174.32
127.56*
39.50*
23.29

CDCl₃   QE-300

C

Aldrich 24,763-4   *CAS [129-64-6]* $C_9H_8O_3$ 60 MHz: *1*, 606A
cis-5-Norbornene-*endo*-2,3-dicarboxylic anhydride, 97%   FW 164.16   FT-IR: *1*, 717C
mp 166°C

171.35
135.46*
52.73
47.06*
46.07*

CDCl₃   QE-300

ALDRICH

Aldrich 10,904-5   *CAS [826-62-0]* $C_9H_8O_3$

5-Norbornene-2,3-dicarboxylic anhydride

171.34
135.46*
52.73
47.06*
46.06*

A

CDCl$_3$ QE-300

Aldrich 10,914-2   *CAS [24327-08-0]* $C_{10}H_{10}O_3$ 60 MHz: *1*, 606B

endo-**Bicyclo[2.2.2]oct-5-ene-2,3-dicarboxylic anhydride**

FW 178.19   FT-IR: *1*, 717D
mp 146 °C

173.29
132.81*
44.62*
31.34*
22.62

B

CDCl$_3$ + DMSO-d_6 QE-300

Aldrich 23,543-1   *CAS [25134-21-8]* $C_{10}H_{10}O_3$

Methyl-5-norbornene-2,3-dicarboxylic anhydride

FW 178.19
d 1.232
n_D^{20} 1.5060
Fp >230 °F

172.00	127.12*	50.38*	47.00*
171.69	58.40	49.31*	46.48*
148.19	52.78	49.22*	46.04*
142.87	51.59*	48.28*	43.52
137.95*	51.36*	48.03*	17.24*
135.48*	51.27*	47.55*	16.43*
130.36*	50.69*	47.11*	15.18*

C

CDCl$_3$ QE-300

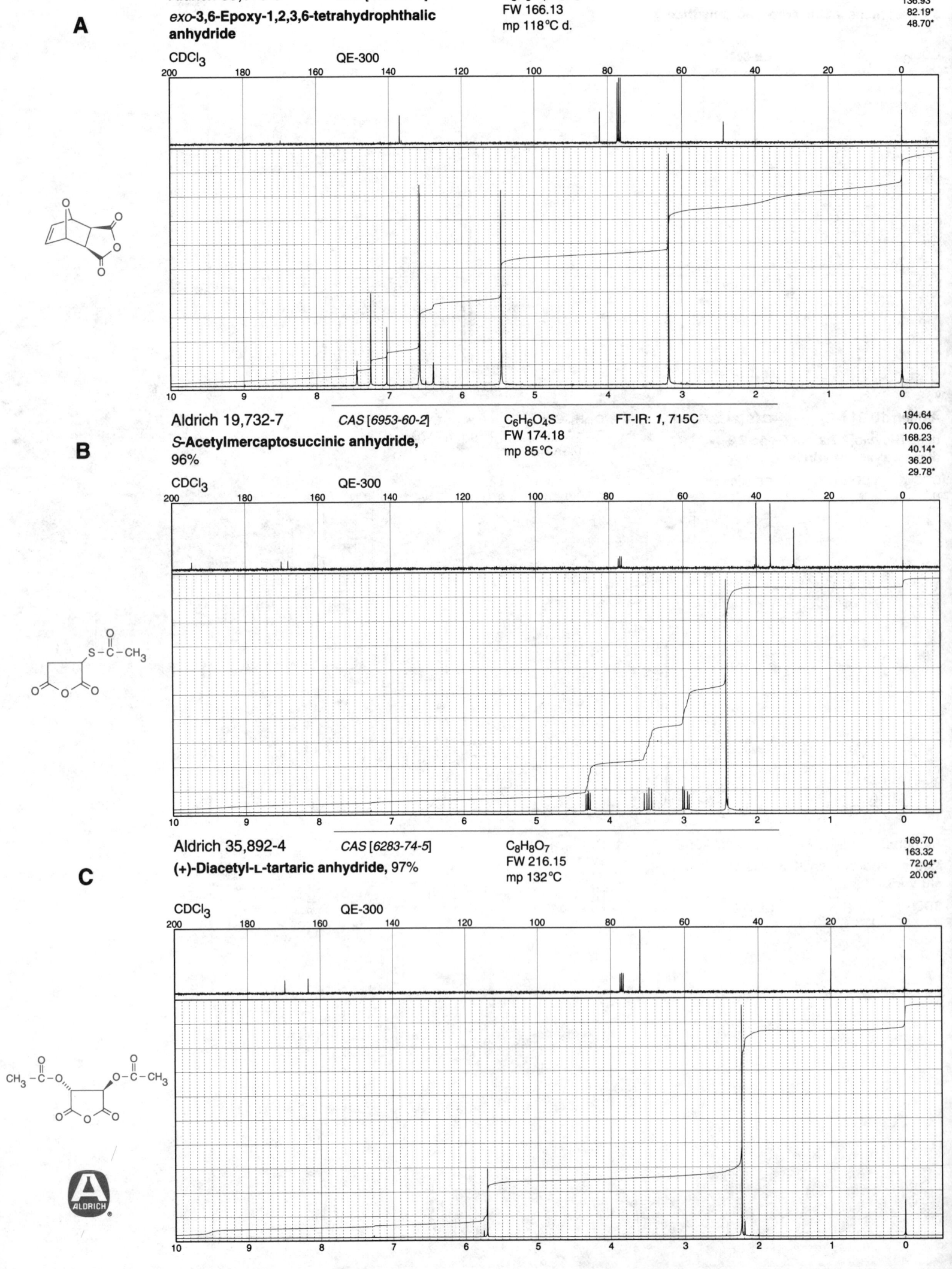

A

Aldrich 33,073-6 *CAS [6118-51-0]* C8H6O4 FW 166.13 mp 118°C d.

169.80
136.93*
82.19*
48.70*

exo-3,6-Epoxy-1,2,3,6-tetrahydrophthalic anhydride

CDCl3 QE-300

B

Aldrich 19,732-7 *CAS [6953-60-2]* C6H6O4S FW 174.18 mp 85°C FT-IR: *1*, 715C

194.64
170.06
168.23
40.14*
36.20
29.78*

S-Acetylmercaptosuccinic anhydride, 96%

CDCl3 QE-300

C

Aldrich 35,892-4 *CAS [6283-74-5]* C8H8O7 FW 216.15 mp 132°C

169.70
163.32
72.04*
20.06*

(+)-Diacetyl-L-tartaric anhydride, 97%

CDCl3 QE-300

ALDRICH

Aldrich 16,133-0   *CAS [4415-87-6]* $C_8H_4O_6$ FW 196.11   170.36 / 41.43*

1,2,3,4-Cyclobutanetetracarboxylic dianhydride, 97%

DMSO-d_6 QE-300

A

FT-IR: *1*, 716C

Aldrich 17,858-6   *CAS [1719-83-1]* $C_{12}H_8O_6$ FW 248.19   172.21 / 132.39* / 42.83* / 33.22*

Bicyclo[2.2.2]oct-7-ene-2,3,5,6-tetra-carboxylic dianhydride

DMSO-d_6 QE-300

B

Aldrich M18-8   *CAS [108-31-6]* $C_4H_2O_3$ FW 98.06   mp 55°C   bp 200°C   Fp 218°F   60 MHz: *1*, 607A   FT-IR: *1*, 718C   164.23 / 136.55*

Maleic anhydride

CDCl$_3$ QE-300

C

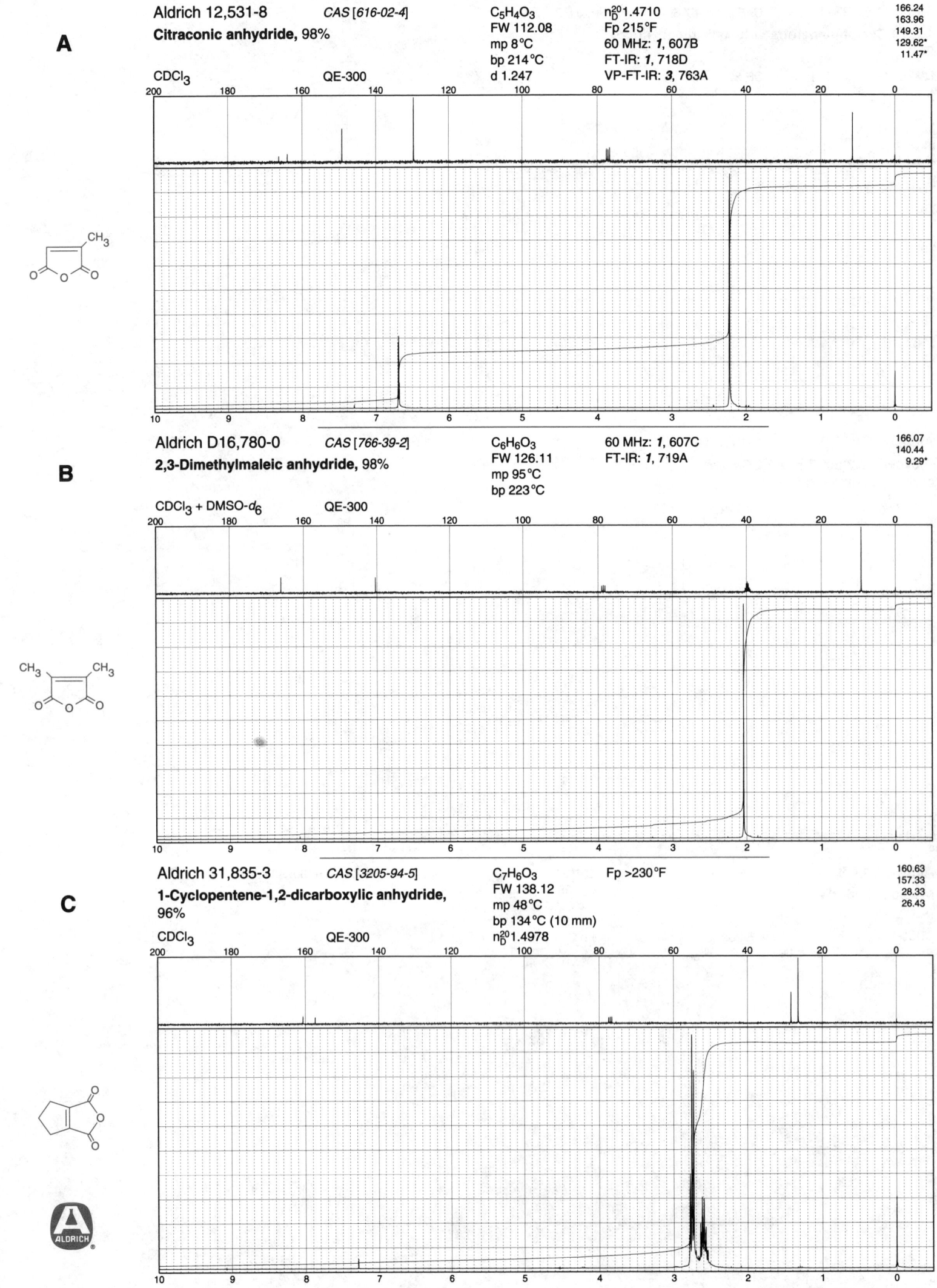

Aldrich 12,531-8   CAS [616-02-4]

Citraconic anhydride, 98%

$C_5H_4O_3$
FW 112.08
mp 8°C
bp 214°C
d 1.247

n_D^{20} 1.4710
Fp 215°F
60 MHz: **1**, 607B
FT-IR: **1**, 718D
VP-FT-IR: **3**, 763A

166.24
163.96
149.31
129.62*
11.47*

CDCl₃   QE-300

Aldrich D16,780-0   CAS [766-39-2]

2,3-Dimethylmaleic anhydride, 98%

$C_6H_6O_3$
FW 126.11
mp 95°C
bp 223°C

60 MHz: **1**, 607C
FT-IR: **1**, 719A

166.07
140.44
9.29*

CDCl₃ + DMSO-d_6 QE-300

Aldrich 31,835-3   CAS [3205-94-5]

1-Cyclopentene-1,2-dicarboxylic anhydride, 96%

$C_7H_6O_3$
FW 138.12
mp 48°C
bp 134°C (10 mm)
n_D^{20} 1.4978

Fp >230°F

160.63
157.33
28.33
26.43

CDCl₃   QE-300

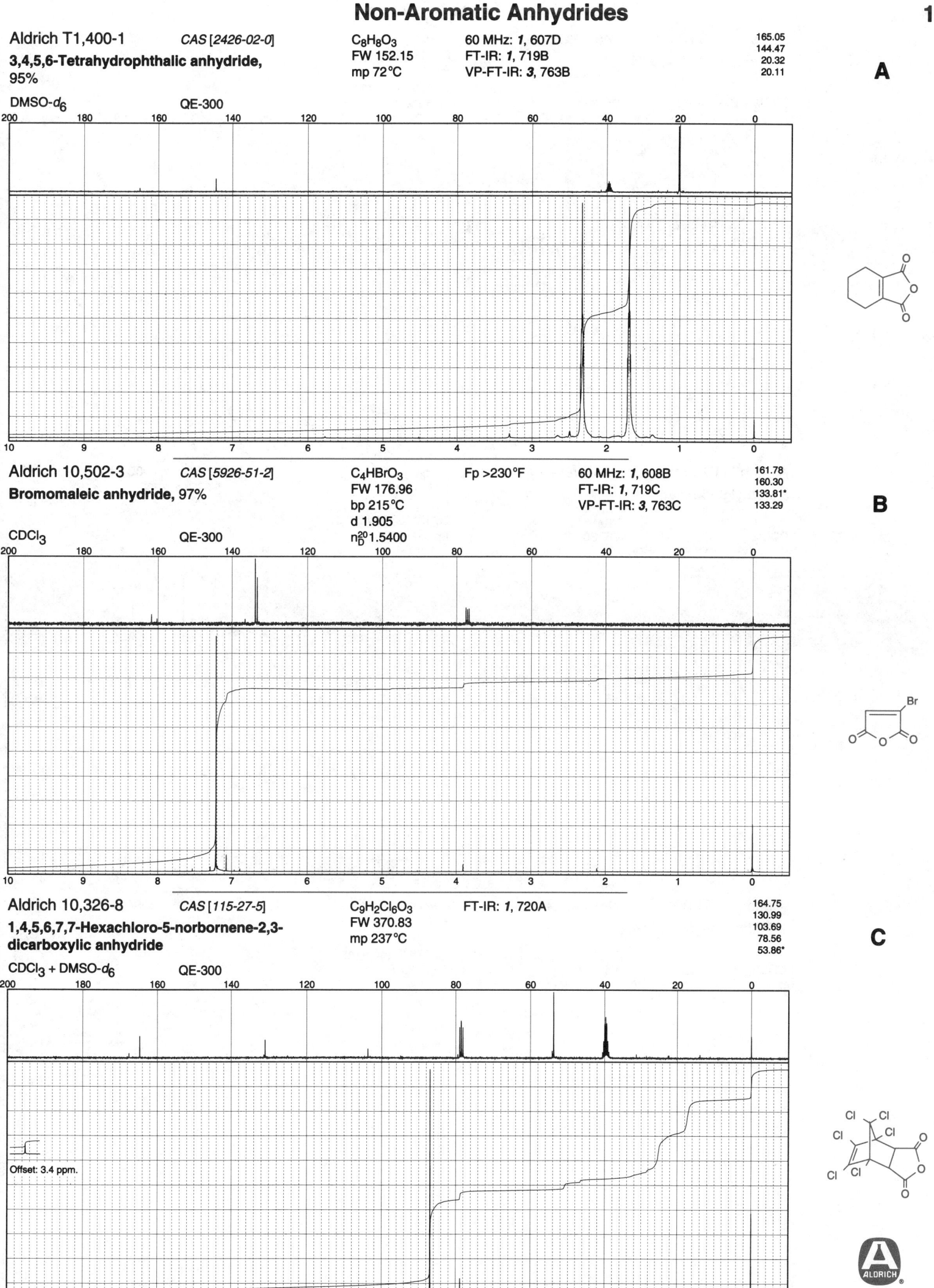

Aldrich T1,400-1 CAS [2426-02-0]
3,4,5,6-Tetrahydrophthalic anhydride, 95%

C₈H₈O₃
FW 152.15
mp 72°C

60 MHz: 1, 607D
FT-IR: 1, 719B
VP-FT-IR: 3, 763B

165.05
144.47
20.32
20.11

A

DMSO-d₆   QE-300

Aldrich 10,502-3 CAS [5926-51-2]
Bromomaleic anhydride, 97%

C₄HBrO₃
FW 176.96
bp 215°C
d 1.905
n²⁰_D 1.5400

Fp >230°F

60 MHz: 1, 608B
FT-IR: 1, 719C
VP-FT-IR: 3, 763C

161.78
160.30
133.81*
133.29

B

CDCl₃   QE-300

Aldrich 10,326-8 CAS [115-27-5]
1,4,5,6,7,7-Hexachloro-5-norbornene-2,3-dicarboxylic anhydride

C₉H₂Cl₆O₃
FW 370.83
mp 237°C

FT-IR: 1, 720A

164.75
130.99
103.69
78.56
53.86*

C

CDCl₃ + DMSO-d₆   QE-300

Offset: 3.4 ppm.

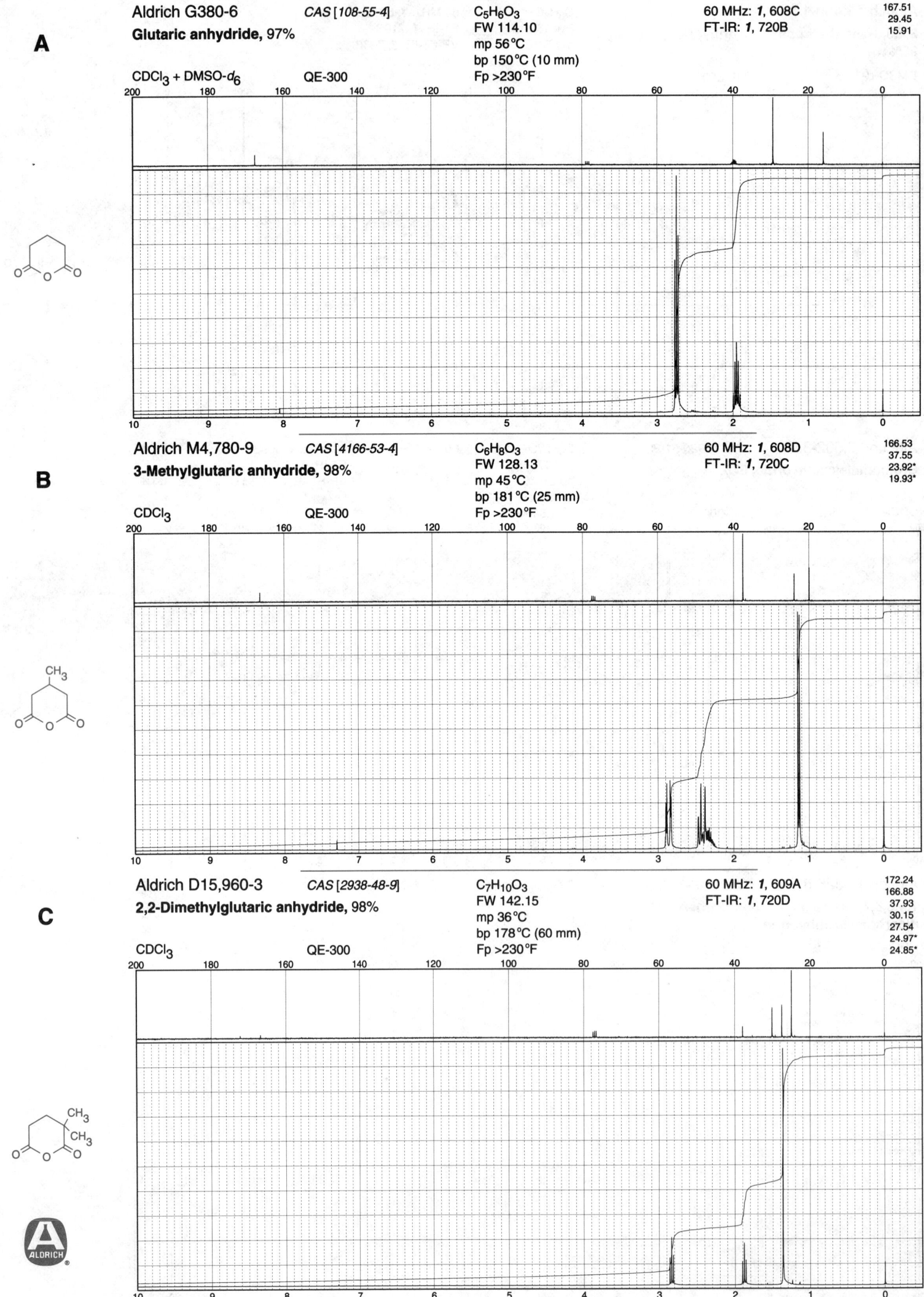

A

Aldrich G380-6 *CAS [108-55-4]*
Glutaric anhydride, 97%

$C_5H_6O_3$
FW 114.10
mp 56°C
bp 150°C (10 mm)
Fp >230°F

CDCl$_3$ + DMSO-d_6 QE-300

60 MHz: *1*, 608C
FT-IR: *1*, 720B

167.51
29.45
15.91

B

Aldrich M4,780-9 *CAS [4166-53-4]*
3-Methylglutaric anhydride, 98%

$C_6H_8O_3$
FW 128.13
mp 45°C
bp 181°C (25 mm)
Fp >230°F

CDCl$_3$ QE-300

60 MHz: *1*, 608D
FT-IR: *1*, 720C

166.53
37.55
23.92*
19.93*

C

Aldrich D15,960-3 *CAS [2938-48-9]*
2,2-Dimethylglutaric anhydride, 98%

$C_7H_{10}O_3$
FW 142.15
mp 36°C
bp 178°C (60 mm)
Fp >230°F

CDCl$_3$ QE-300

60 MHz: *1*, 609A
FT-IR: *1*, 720D

172.24
166.88
37.93
30.15
27.54
24.97*
24.85*

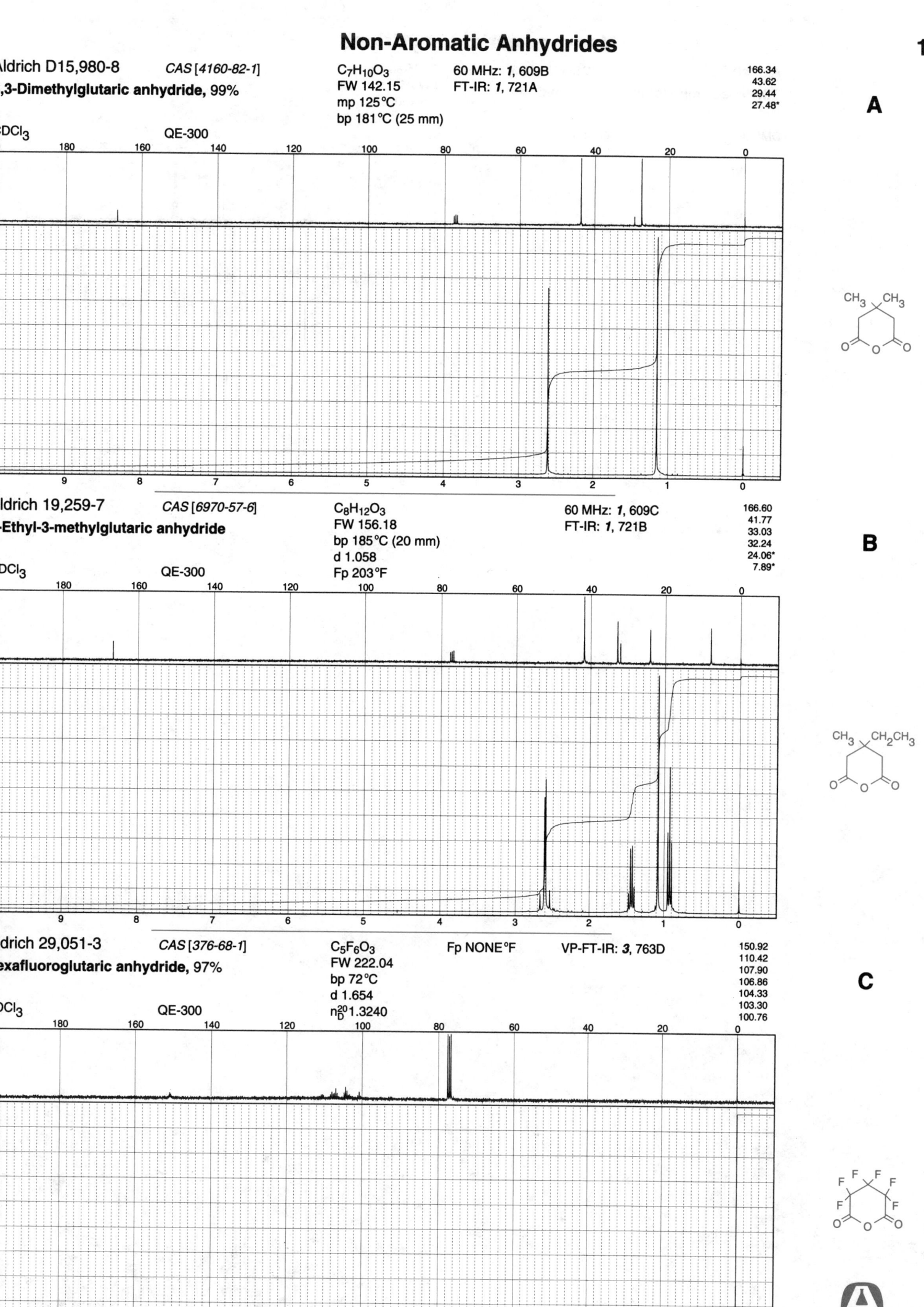

Aldrich D15,980-8   *CAS [4160-82-1]*

3,3-Dimethylglutaric anhydride, 99%

$C_7H_{10}O_3$
FW 142.15
mp 125°C
bp 181°C (25 mm)

60 MHz: *1*, 609B
FT-IR: *1*, 721A

166.34
43.62
29.44
27.48*

A

CDCl₃   QE-300

Aldrich 19,259-7   *CAS [6970-57-6]*

3-Ethyl-3-methylglutaric anhydride

$C_8H_{12}O_3$
FW 156.18
bp 185°C (20 mm)
d 1.058
Fp 203°F

60 MHz: *1*, 609C
FT-IR: *1*, 721B

166.60
41.77
33.03
32.24
24.06*
7.89*

B

CDCl₃   QE-300

Aldrich 29,051-3   *CAS [376-68-1]*

Hexafluoroglutaric anhydride, 97%

$C_5F_6O_3$
FW 222.04
bp 72°C
d 1.654
n_D^{20} 1.3240

Fp NONE°F   VP-FT-IR: *3*, 763D

150.92
110.42
107.90
106.86
104.33
103.30
100.76

C

CDCl₃   QE-300

Non-Aromatic Anhydrides

Aldrich 33,204-6   *CAS [23911-25-3]*   C$_{10}$H$_{12}$N$_2$O$_6$
Ethylenediaminetetraacetic dianhydride   FW 256.22
mp 190°C d.

165.58
52.14
51.08

A

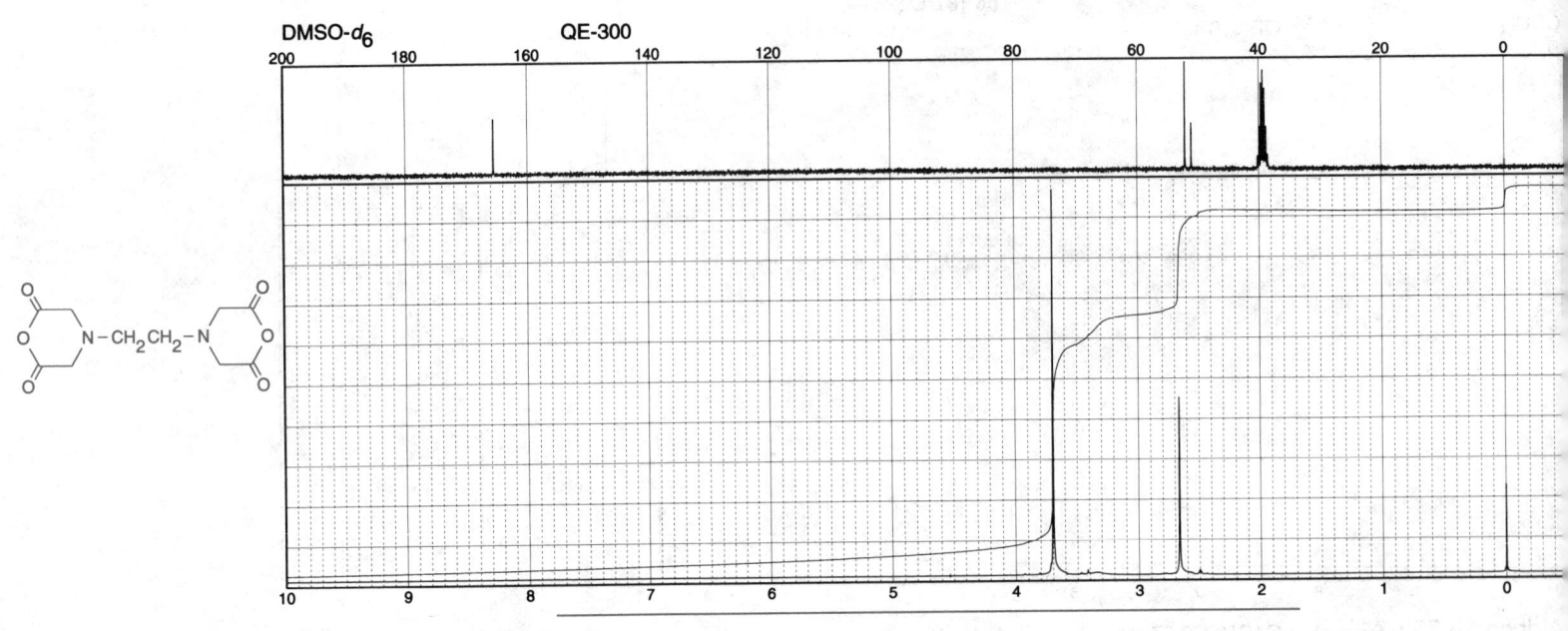

DMSO-d_6 QE-300

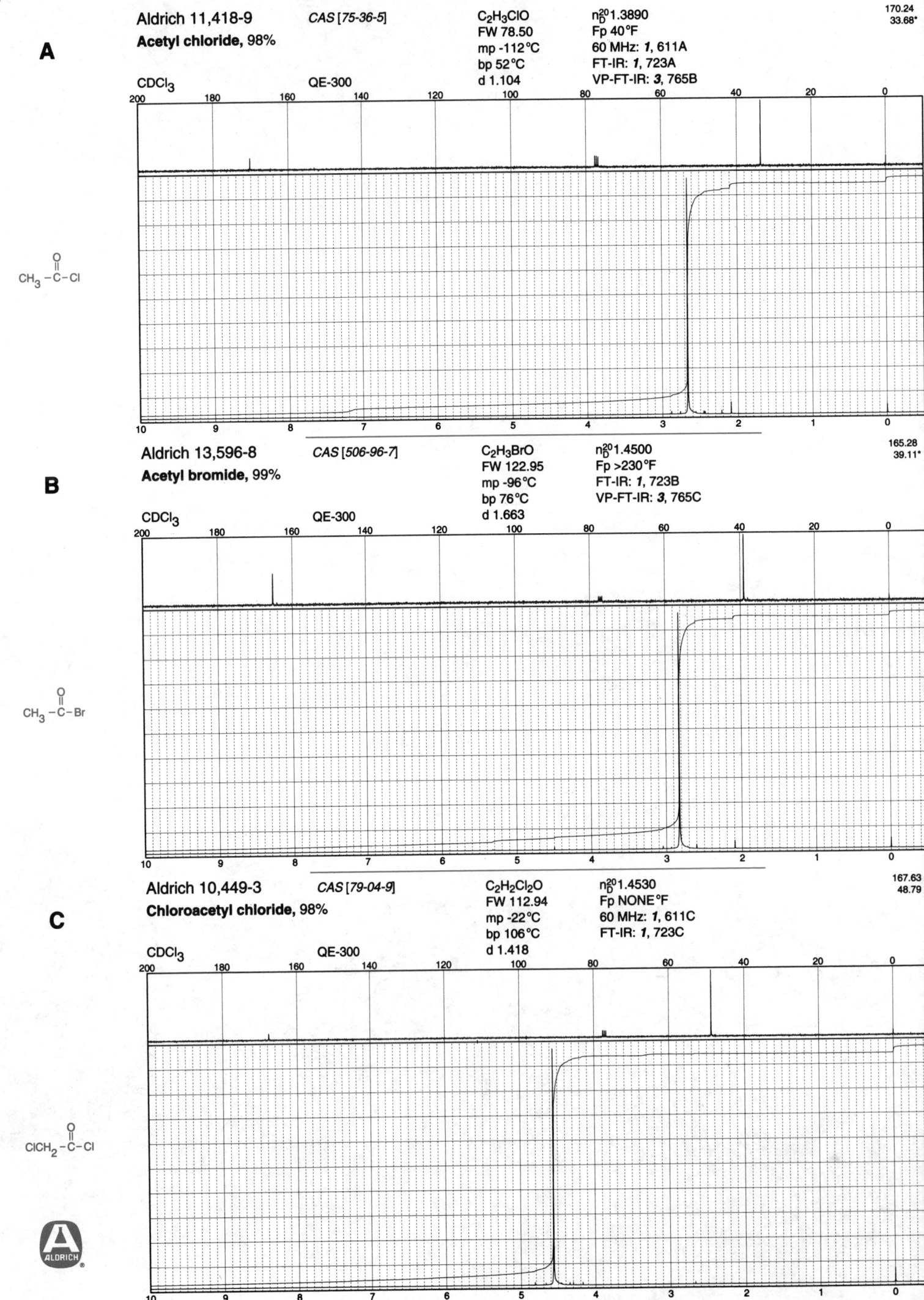

A

Aldrich 11,418-9   *CAS [75-36-5]*   C₂H₃ClO n_D^{20}1.3890   170.24
33.68*

Acetyl chloride, 98%

FW 78.50   Fp 40°F
mp -112°C   60 MHz: *1*, 611A
bp 52°C   FT-IR: *1*, 723A
d 1.104   VP-FT-IR: *3*, 765B

CDCl₃   QE-300

B

Aldrich 13,596-8   *CAS [506-96-7]*   C₂H₃BrO n_D^{20}1.4500   165.28
39.11*

Acetyl bromide, 99%

FW 122.95   Fp >230°F
mp -96°C   FT-IR: *1*, 723B
bp 76°C   VP-FT-IR: *3*, 765C
d 1.663

CDCl₃   QE-300

C

Aldrich 10,449-3   *CAS [79-04-9]*   C₂H₂Cl₂O n_D^{20}1.4530   167.63
48.79

Chloroacetyl chloride, 98%

FW 112.94   Fp NONE°F
mp -22°C   60 MHz: *1*, 611C
bp 106°C   FT-IR: *1*, 723C
d 1.418

CDCl₃   QE-300

Aldrich 20,955-4   CAS [22118-09-8] C_2H_2BrClO Fp NONE°F   60 MHz: 1, 611B   166.27
Bromoacetyl chloride, 95%   FW 157.40   FT-IR: 1, 723D   34.20
bp 128°C   VP-FT-IR: 3, 765D
d 1.908

A

$CDCl_3$ QE-300 n_D^{20} 1.4960

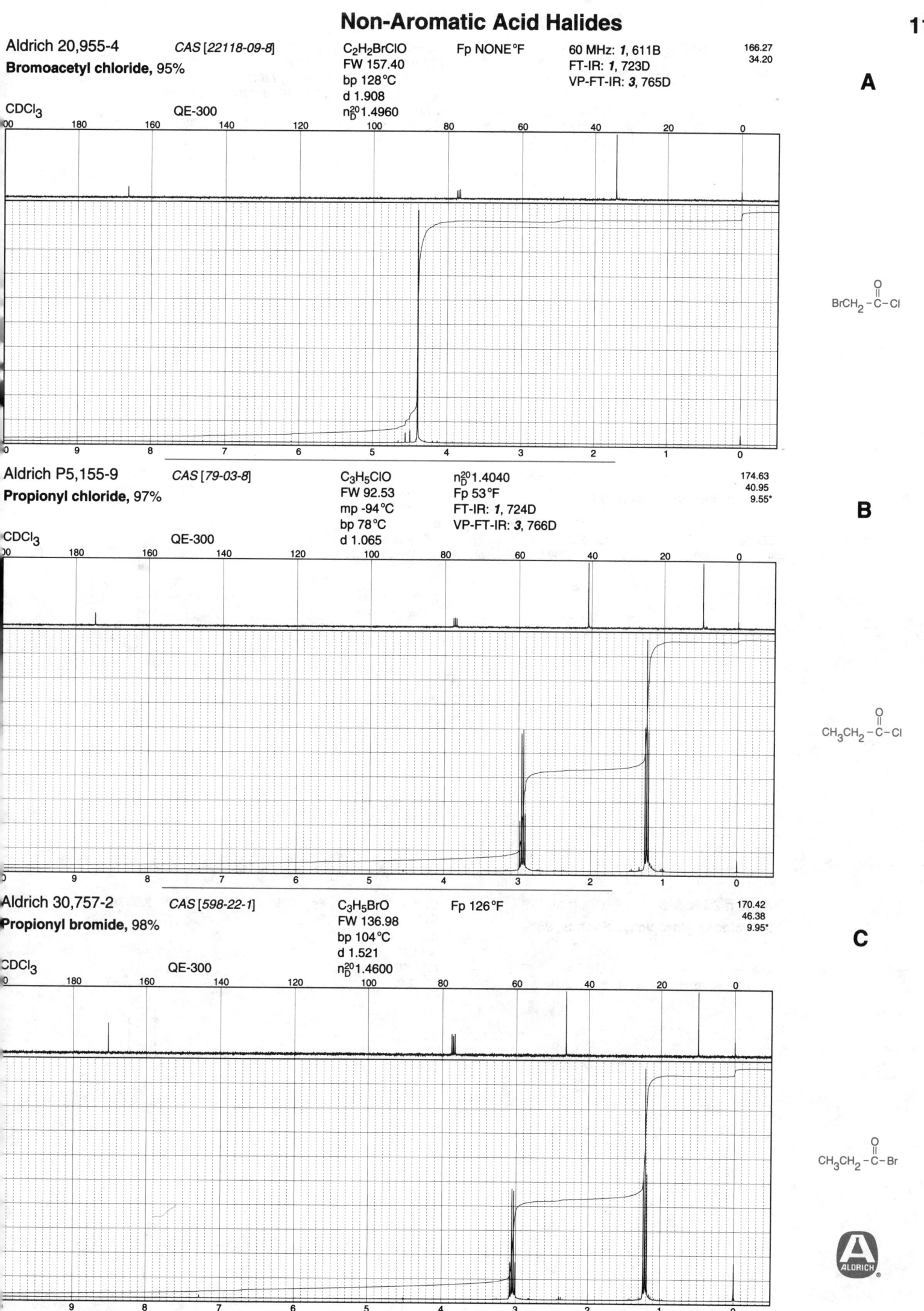

Aldrich P5,155-9   CAS [79-03-8] C_3H_5ClO n_D^{20} 1.4040   174.63
Propionyl chloride, 97%   FW 92.53   Fp 53°F   40.95
mp -94°C   FT-IR: 1, 724D   9.55*
bp 78°C   VP-FT-IR: 3, 766D

B

$CDCl_3$ QE-300   d 1.065

$CH_3CH_2 - \overset{O}{\underset{||}{C}} - Cl$

Aldrich 30,757-2   CAS [598-22-1] C_3H_5BrO Fp 126°F   170.42
Propionyl bromide, 98%   FW 136.98   46.38
bp 104°C   9.95*
d 1.521

C

$CDCl_3$ QE-300 n_D^{20} 1.4600

$CH_3CH_2 - \overset{O}{\underset{||}{C}} - Br$

ALDRICH

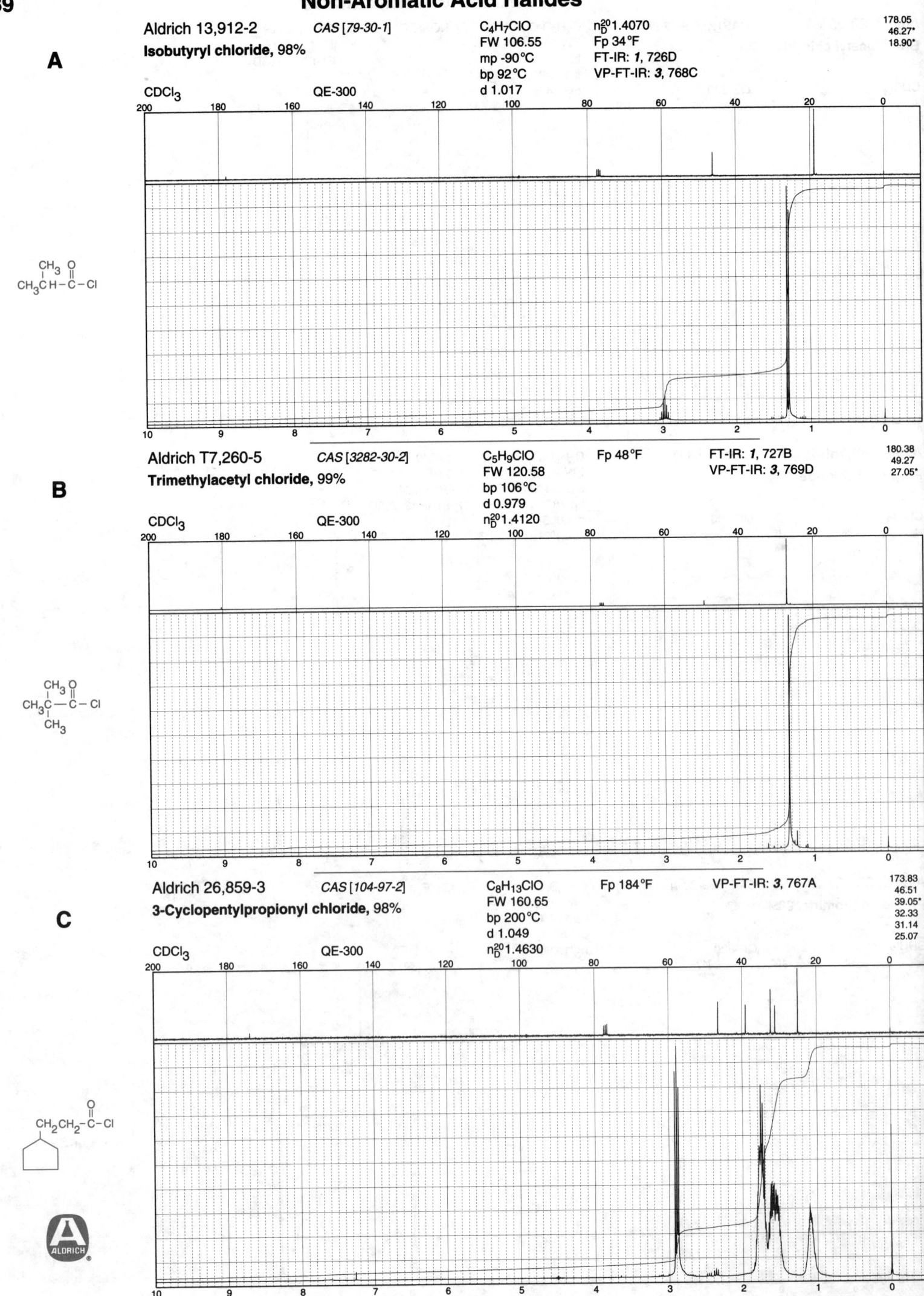

A Aldrich 13,912-2 *CAS [79-30-1]* C₄H₇ClO n₂₀ 1.4070 178.05
Isobutyryl chloride, 98% FW 106.55 Fp 34°F 46.27*
mp -90°C FT-IR: *1*, 726D 18.90*
bp 92°C VP-FT-IR: *3*, 768C
d 1.017

CDCl₃ QE-300

CH_3 O
$CH_3CH-C-Cl$

B Aldrich T7,260-5 *CAS [3282-30-2]* C₅H₉ClO Fp 48°F FT-IR: *1*, 727B 180.38
Trimethylacetyl chloride, 99% FW 120.58 VP-FT-IR: *3*, 769D 49.27
bp 106°C 27.05*
d 0.979
n₂₀ 1.4120

CDCl₃ QE-300

CH_3 O
$CH_3C-C-Cl$
CH_3

C Aldrich 26,859-3 *CAS [104-97-2]* C₈H₁₃ClO Fp 184°F VP-FT-IR: *3*, 767A 173.83
3-Cyclopentylpropionyl chloride, 98% FW 160.65 46.51
bp 200°C 39.05*
d 1.049 32.33
n₂₀ 1.4630 31.14
25.07

CDCl₃ QE-300

CH_2CH_2-C-Cl
O

Non-Aromatic Acid Halides

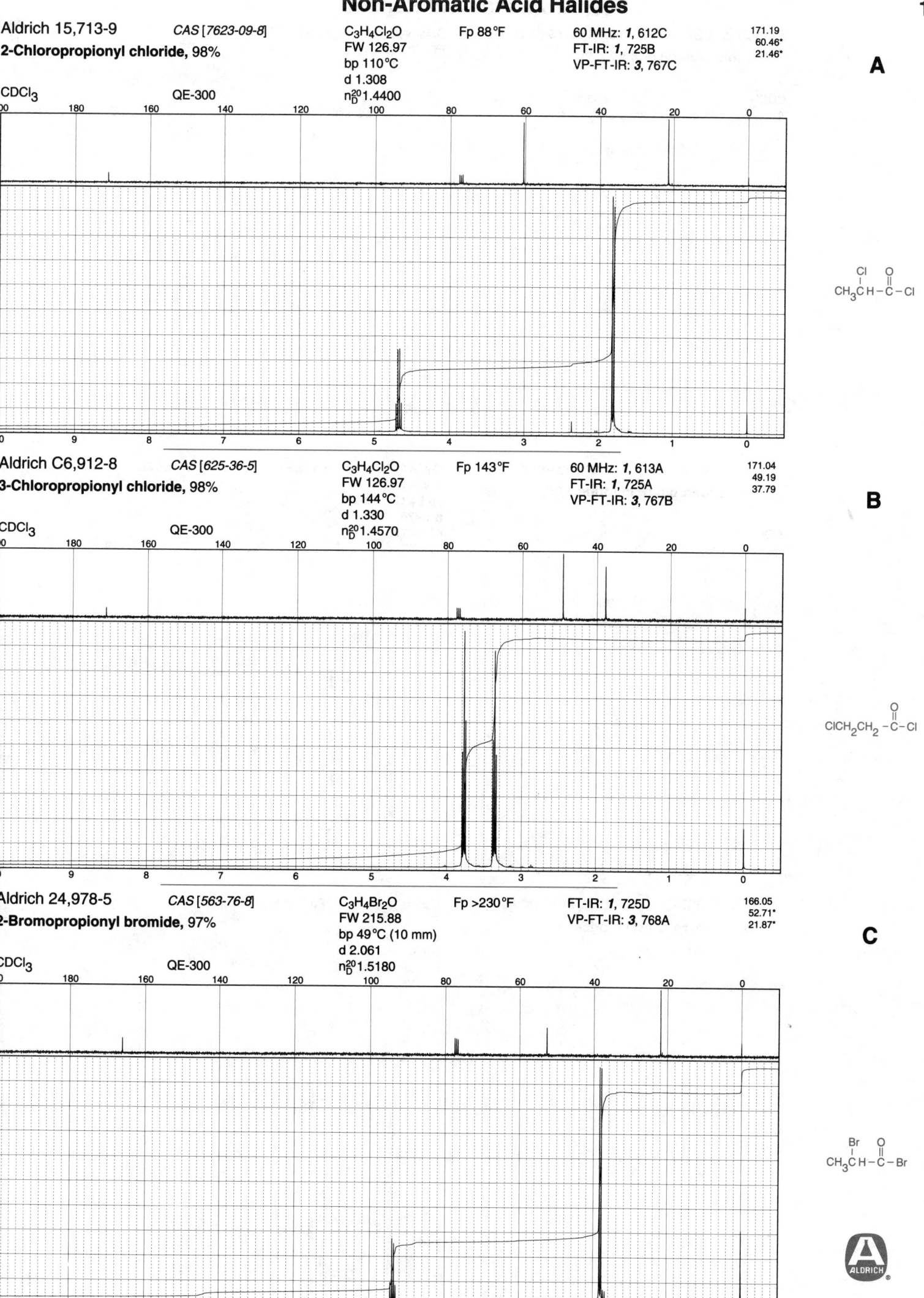

Aldrich 15,713-9   *CAS [7623-09-8]* $C_3H_4Cl_2O$ Fp 88°F   60 MHz: *1*, 612C   171.19

2-Chloropropionyl chloride, 98%   FW 126.97   FT-IR: *1*, 725B   60.46*

bp 110°C   VP-FT-IR: *3*, 767C   21.46*

d 1.308

$CDCl_3$ QE-300 n_D^{20} 1.4400

A

$CH_3CH–C–Cl$ (with Cl on CH and O on C)

Aldrich C6,912-8   *CAS [625-36-5]* $C_3H_4Cl_2O$ Fp 143°F   60 MHz: *1*, 613A   171.04

3-Chloropropionyl chloride, 98%   FW 126.97   FT-IR: *1*, 725A   49.19

bp 144°C   VP-FT-IR: *3*, 767B   37.79

d 1.330

$CDCl_3$ QE-300 n_D^{20} 1.4570

B

$ClCH_2CH_2–C–Cl$ (with O on C)

Aldrich 24,978-5   *CAS [563-76-8]* $C_3H_4Br_2O$ Fp >230°F   FT-IR: *1*, 725D   166.05

2-Bromopropionyl bromide, 97%   FW 215.88   VP-FT-IR: *3*, 768A   52.71*

bp 49°C (10 mm)   21.87*

d 2.061

$CDCl_3$ QE-300 n_D^{20} 1.5180

C

$CH_3CH–C–Br$ (with Br on CH and O on C)

1190

Non-Aromatic Acid Halides

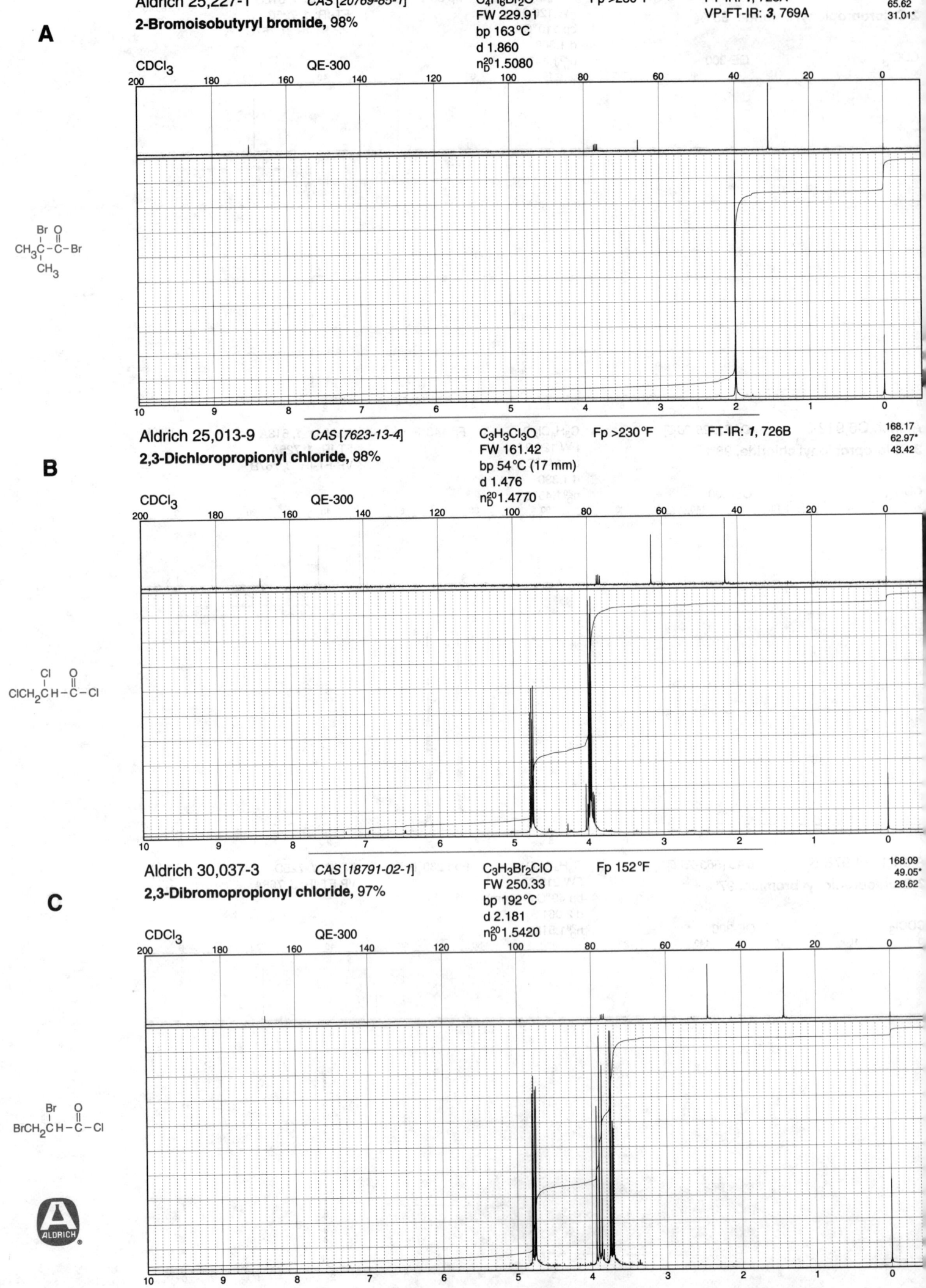

A

Aldrich 25,227-1   CAS [20769-85-1]   C$_4$H$_6$Br$_2$O   Fp >230°F   FT-IR: *1*, 728A

2-Bromoisobutyryl bromide, 98%   FW 229.91   VP-FT-IR: *3*, 769A

bp 163°C
d 1.860
n$_D^{20}$1.5080

170.20
65.62
31.01*

CDCl$_3$ QE-300

$CH_3-\overset{\overset{Br}{|}}{\underset{\underset{CH_3}{|}}{C}}-\overset{\overset{O}{\|}}{C}-Br$

B

Aldrich 25,013-9   CAS [7623-13-4]   C$_3$H$_3$Cl$_3$O   Fp >230°F   FT-IR: *1*, 726B

2,3-Dichloropropionyl chloride, 98%   FW 161.42

bp 54°C (17 mm)
d 1.476
n$_D^{20}$1.4770

168.17
62.97*
43.42

CDCl$_3$ QE-300

$ClCH_2-\overset{\overset{Cl}{|}}{C}H-\overset{\overset{O}{\|}}{C}-Cl$

C

Aldrich 30,037-3   CAS [18791-02-1]   C$_3$H$_3$Br$_2$ClO   Fp 152°F

2,3-Dibromopropionyl chloride, 97%   FW 250.33

bp 192°C
d 2.181
n$_D^{20}$1.5420

168.09
49.05*
28.62

CDCl$_3$ QE-300

$BrCH_2-\overset{\overset{Br}{|}}{C}H-\overset{\overset{O}{\|}}{C}-Cl$

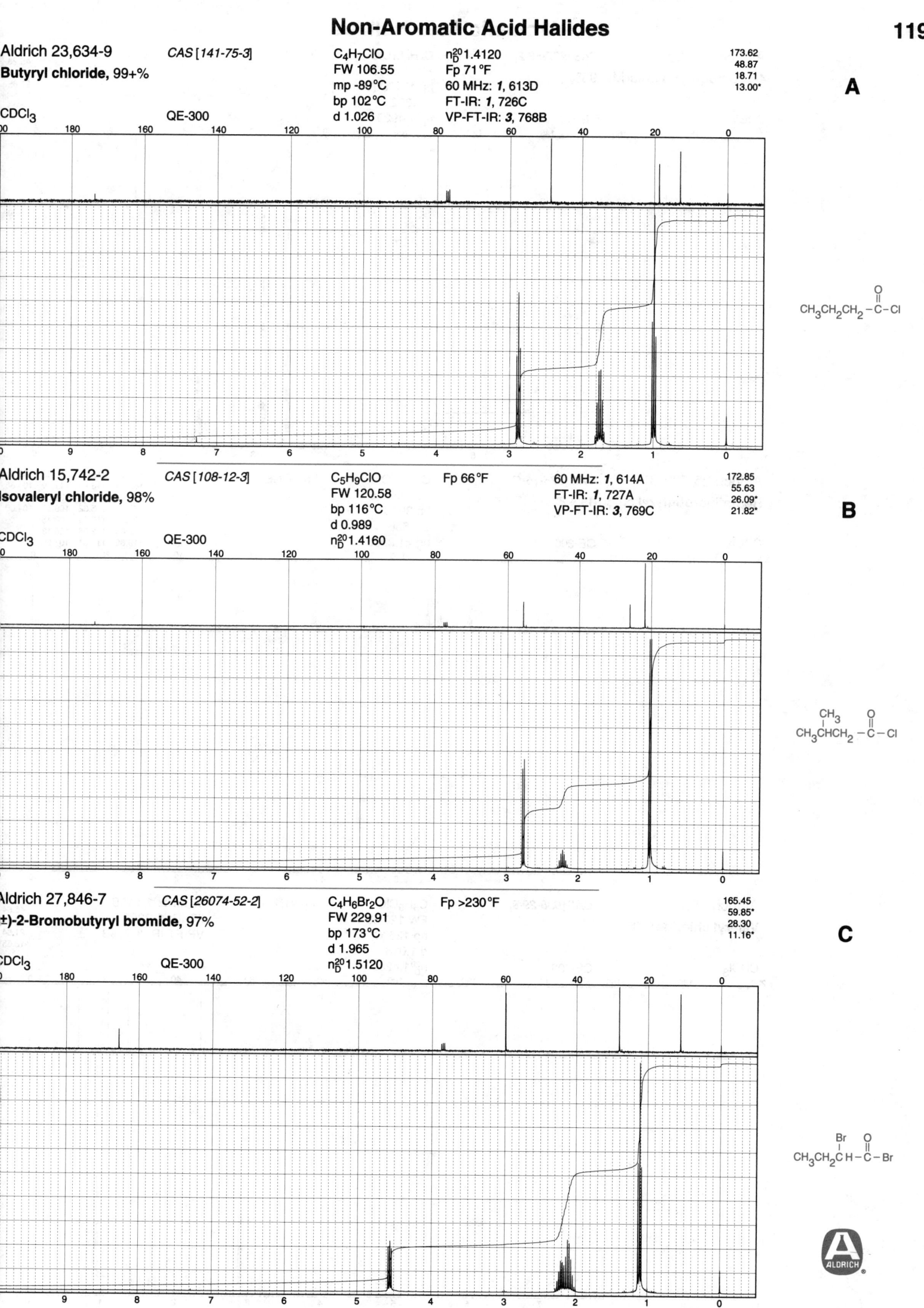

Aldrich 23,634-9   *CAS [141-75-3]* C_4H_7ClO n_D^{20} 1.4120   173.62
Butyryl chloride, 99+%   FW 106.55   Fp 71°F   48.87
mp -89°C   60 MHz: *1*, 613D   18.71
bp 102°C   FT-IR: *1*, 726C   13.00*
CDCl₃   QE-300   d 1.026   VP-FT-IR: *3*, 768B

A

$CH_3CH_2CH_2-\overset{\overset{\displaystyle O}{\|}}{C}-Cl$

Aldrich 15,742-2   *CAS [108-12-3]* C_5H_9ClO Fp 66°F   60 MHz: *1*, 614A   172.85
Isovaleryl chloride, 98%   FW 120.58   FT-IR: *1*, 727A   55.63
bp 116°C   VP-FT-IR: *3*, 769C   26.09*
d 0.989   21.82*
CDCl₃   QE-300 n_D^{20} 1.4160

B

$CH_3CHCH_2-\overset{\overset{\displaystyle O}{\|}}{C}-Cl$
    $\overset{\displaystyle CH_3}{|}$ above

Aldrich 27,846-7   *CAS [26074-52-2]* $C_4H_6Br_2O$ Fp >230°F   165.45
(±)-2-Bromobutyryl bromide, 97%   FW 229.91   59.85*
bp 173°C   28.30
d 1.965   11.16*
CDCl₃   QE-300 n_D^{20} 1.5120

C

$CH_3CH_2CH-\overset{\overset{\displaystyle O}{\|}}{C}-Br$
        $\overset{\displaystyle Br}{|}$ above

1193

Non-Aromatic Acid Halides

A

Aldrich 25,193-3   *CAS [927-58-2]* C_4H_6BrClO Fp 196°F   FT-IR: *1*, 727D   173.01
4-Bromobutyryl chloride, 95%   FW 185.46   45.23
                                              bp 101°C (37 mm)   31.31
                                                d 1.602   27.65

CDCl₃   QE-300 n_D^{20} 1.4920

$BrCH_2CH_2CH_2 - \overset{\displaystyle O}{\overset{\|}{C}} - Cl$

B

Aldrich 25,792-3   *CAS [375-16-6]* C_4ClF_7O Fp NONE°F

162.53	119.43	111.62	105.01
162.05	118.99	111.19	104.86
161.56	116.05	109.00	104.43
123.67	115.62	108.57	104.00
123.23	115.18	108.46	
122.80	112.06	108.13	
119.86	111.81	107.60	

Heptafluorobutyryl chloride, 98%   FW 232.49
                        bp 39°C
                        d 1.556

CDCl₃   QE-300 n_D^{20} <1.3000

$CF_3CF_2CF_2 - \overset{\displaystyle O}{\overset{\|}{C}} - Cl$

C

Aldrich 15,714-7   *CAS [638-29-9]* C_5H_9ClO Fp 91°F   60 MHz: *1*, 615B   173.76
Valeryl chloride, 98%   FW 120.58   FT-IR: *1*, 728C   46.84
                                             bp 126°C   VP-FT-IR: *3*, 769B   27.08
                                               d 1.016   21.64
                                                    13.55°

CDCl₃   QE-300 n_D^{20} 1.4200

$CH_3CH_2CH_2CH_2 - \overset{\displaystyle O}{\overset{\|}{C}} - Cl$

ALDRICH

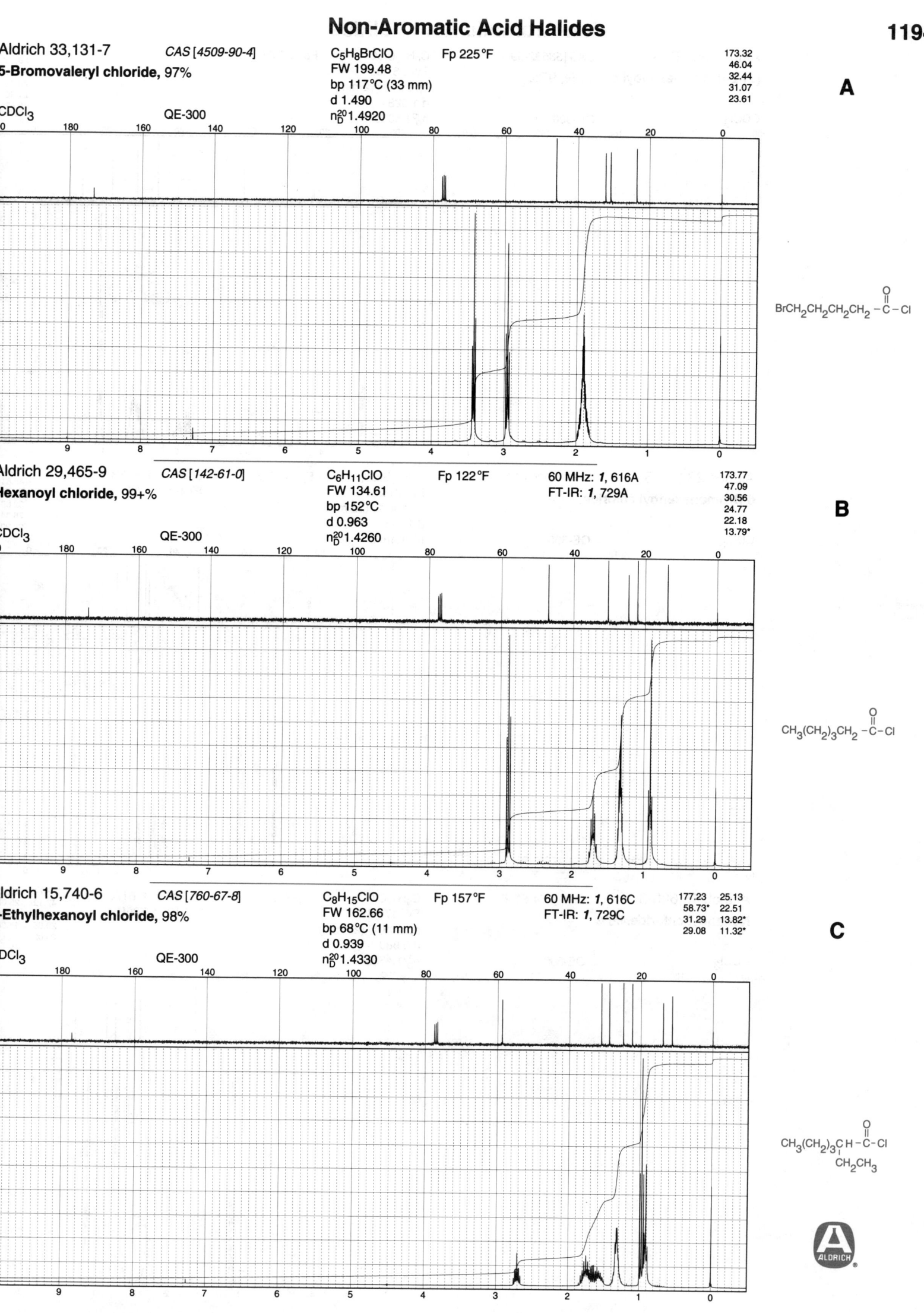

Aldrich 33,131-7   CAS [4509-90-4]   C₅H₈BrClO   Fp 225°F
5-Bromovaleryl chloride, 97%
C_5H_8BrClO
FW 199.48
bp 117°C (33 mm)
d 1.490
n_D^{20} 1.4920

CDCl₃   QE-300

173.32
46.04
32.44
31.07
23.61

A

$BrCH_2CH_2CH_2CH_2 - \overset{\displaystyle O}{\overset{\|}{C}} - Cl$

Aldrich 29,465-9   CAS [142-61-0]   C₆H₁₁ClO   Fp 122°F   60 MHz: 1, 616A   FT-IR: 1, 729A
Hexanoyl chloride, 99+%
$C_6H_{11}ClO$
FW 134.61
bp 152°C
d 0.963
n_D^{20} 1.4260

CDCl₃   QE-300

173.77
47.09
30.56
24.77
22.18
13.79*

B

$CH_3(CH_2)_3CH_2 - \overset{\displaystyle O}{\overset{\|}{C}} - Cl$

Aldrich 15,740-6   CAS [760-67-8]   C₈H₁₅ClO   Fp 157°F   60 MHz: 1, 616C   FT-IR: 1, 729C
2-Ethylhexanoyl chloride, 98%
$C_8H_{15}ClO$
FW 162.66
bp 68°C (11 mm)
d 0.939
n_D^{20} 1.4330

CDCl₃   QE-300

177.23   25.13
58.73*   22.51
31.29   13.82*
29.08   11.32*

C

$CH_3(CH_2)_3\underset{\underset{\displaystyle CH_2CH_3}{|}}{CH} - \overset{\displaystyle O}{\overset{\|}{C}} - Cl$

ALDRICH

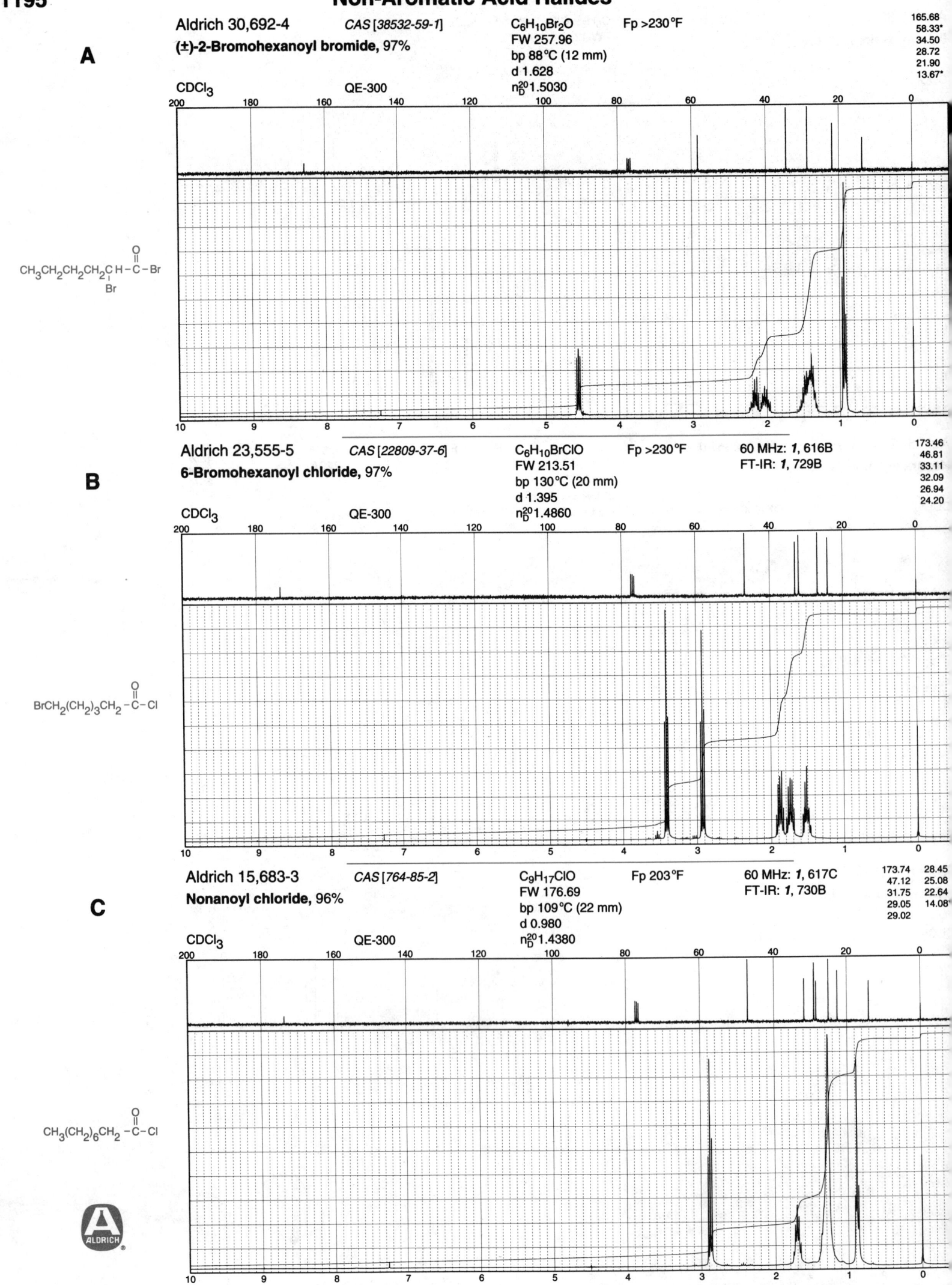

A

Aldrich 30,692-4   CAS [38532-59-1]   C$_6$H$_{10}$Br$_2$O   Fp >230°F

(±)-2-Bromohexanoyl bromide, 97%

FW 257.96
bp 88°C (12 mm)
d 1.628
n$_D^{20}$1.5030

165.68
58.33*
34.50
28.72
21.90
13.67*

CDCl$_3$ QE-300

CH$_3$CH$_2$CH$_2$CH$_2$CH–C–Br
                      | ||
                      Br   O

B

Aldrich 23,555-5   CAS [22809-37-6]   C$_6$H$_{10}$BrClO   Fp >230°F   60 MHz: 1, 616B

6-Bromohexanoyl chloride, 97%

FW 213.51
bp 130°C (20 mm)
d 1.395
n$_D^{20}$1.4860

FT-IR: 1, 729B

173.46
46.81
33.11
32.09
26.94
24.20

CDCl$_3$ QE-300

BrCH$_2$(CH$_2$)$_3$CH$_2$–C–Cl
                         ||
                         O

C

Aldrich 15,683-3   CAS [764-85-2]   C$_9$H$_{17}$ClO   Fp 203°F   60 MHz: 1, 617C

Nonanoyl chloride, 96%

FW 176.69
bp 109°C (22 mm)
d 0.980
n$_D^{20}$1.4380

FT-IR: 1, 730B

173.74   28.45
47.12   25.08
31.75   22.64
29.05   14.08*
29.02

CDCl$_3$ QE-300

CH$_3$(CH$_2$)$_6$CH$_2$–C–Cl
                      ||
                      O

Non-Aromatic Acid Halides

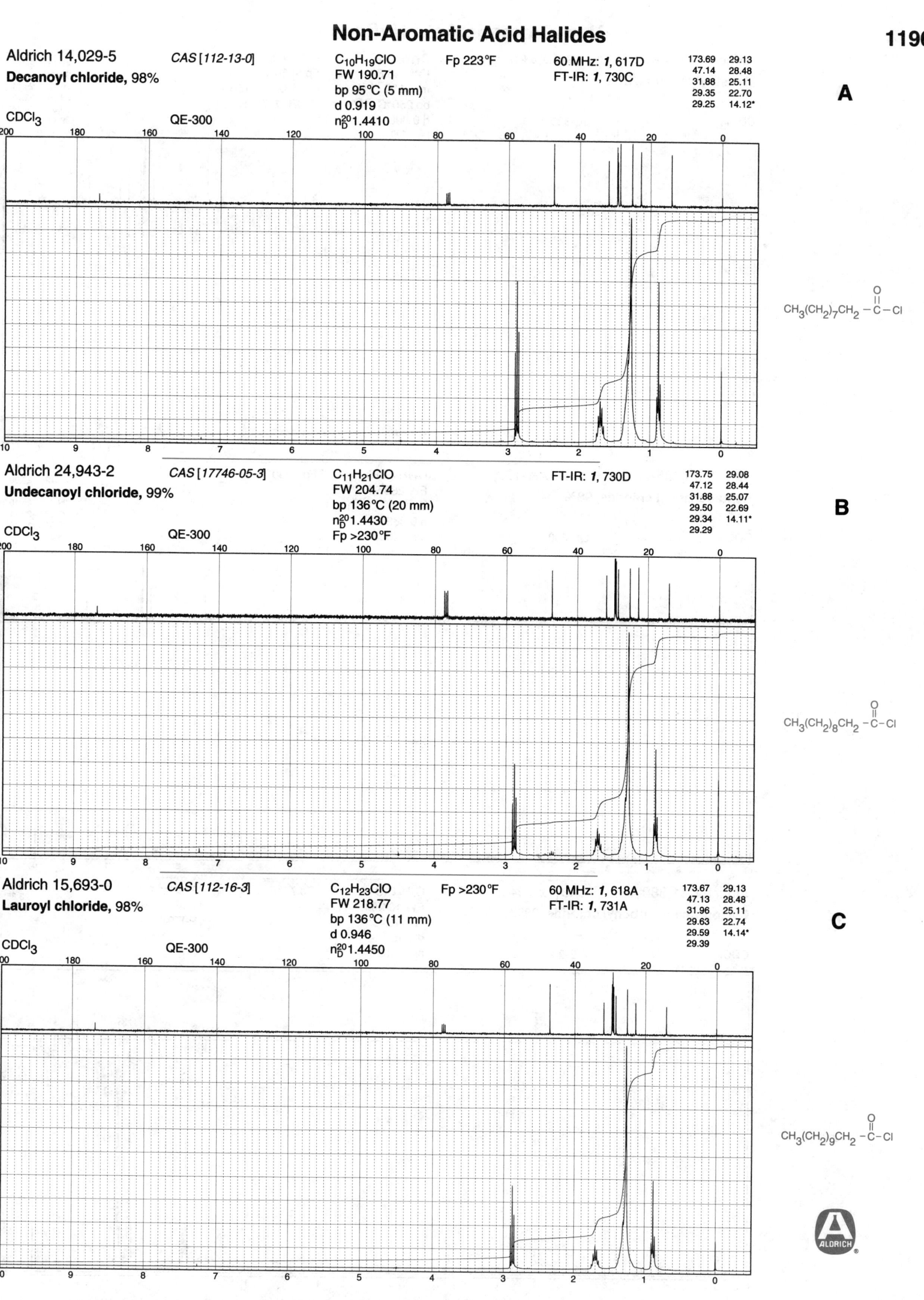

Aldrich 14,029-5
Decanoyl chloride, 98%

CAS [112-13-0]

C₁₀H₁₉ClO
FW 190.71
bp 95°C (5 mm)
d 0.919
n_D^{20} 1.4410

Fp 223°F

60 MHz: *1*, 617D
FT-IR: *1*, 730C

173.69	29.13
47.14	28.48
31.88	25.11
29.35	22.70
29.25	14.12*

A

CDCl₃   QE-300

Aldrich 24,943-2
Undecanoyl chloride, 99%

CAS [17746-05-3]

C₁₁H₂₁ClO
FW 204.74
bp 136°C (20 mm)
n_D^{20} 1.4430
Fp >230°F

FT-IR: *1*, 730D

173.75	29.08
47.12	28.44
31.88	25.07
29.50	22.69
29.34	14.11*
29.29	

B

CDCl₃   QE-300

Aldrich 15,693-0
Lauroyl chloride, 98%

CAS [112-16-3]

C₁₂H₂₃ClO
FW 218.77
bp 136°C (11 mm)
d 0.946
n_D^{20} 1.4450

Fp >230°F

60 MHz: *1*, 618A
FT-IR: *1*, 731A

173.67	29.13
47.13	28.48
31.96	25.11
29.63	22.74
29.59	14.14*
29.39	

C

CDCl₃   QE-300

A

Aldrich 18,520-5   *CAS [112-64-1]*

Myristoyl chloride, 97%

$C_{14}H_{27}ClO$
FW 246.82
mp -1°C
bp 250°C (100 mm)
d 0.908

$n_D^{20} 1.4490$
Fp >230°F
60 MHz: *1*, 618B
FT-IR: *1*, 731B

173.65	29.39
47.13	29.13
31.98	28.48
29.72	25.11
29.68	22.75
29.59	14.14*
29.41	

CDCl$_3$ QE-300

$CH_3(CH_2)_{11}CH_2-\overset{\displaystyle O}{\underset{\displaystyle }{C}}-Cl$

B

Aldrich 30,665-7   *CAS [40480-10-2]*

Heptadecanoyl chloride, 98%

$C_{17}H_{33}ClO$
FW 288.91
bp 176°C (4 mm)
d 0.883
$n_D^{20} 1.4530$

Fp >230°F

173.66	29.35
47.12	29.09
31.95	28.46
29.70	25.10
29.62	22.71
29.54	14.10*
29.38	

CDCl$_3$ QE-300

$CH_3(CH_2)_{14}CH_2-\overset{\displaystyle O}{\underset{\displaystyle }{C}}-Cl$

C

Aldrich C11,680-1   *CAS [4023-34-1]*

Cyclopropanecarbonyl chloride, 98%

C_4H_5ClO
FW 104.54
bp 119°C
d 1.152
$n_D^{20} 1.4520$

Fp 74°F

60 MHz: *1*, 627A
FT-IR: *1*, 744B
VP-FT-IR: *3*, 774C

175.21	
23.85*	
12.44	

CDCl$_3$ QE-300

Aldrich 30,318-6   *CAS [24303-61-5]*

2,2,3,3-Tetramethyl-1-cyclopropanecarbonyl chloride

$C_8H_{13}ClO$
FW 160.65
mp 40°C
Fp 151°F

169.39
48.12*
37.58
23.23*
16.78*

A

CDCl$_3$ QE-300

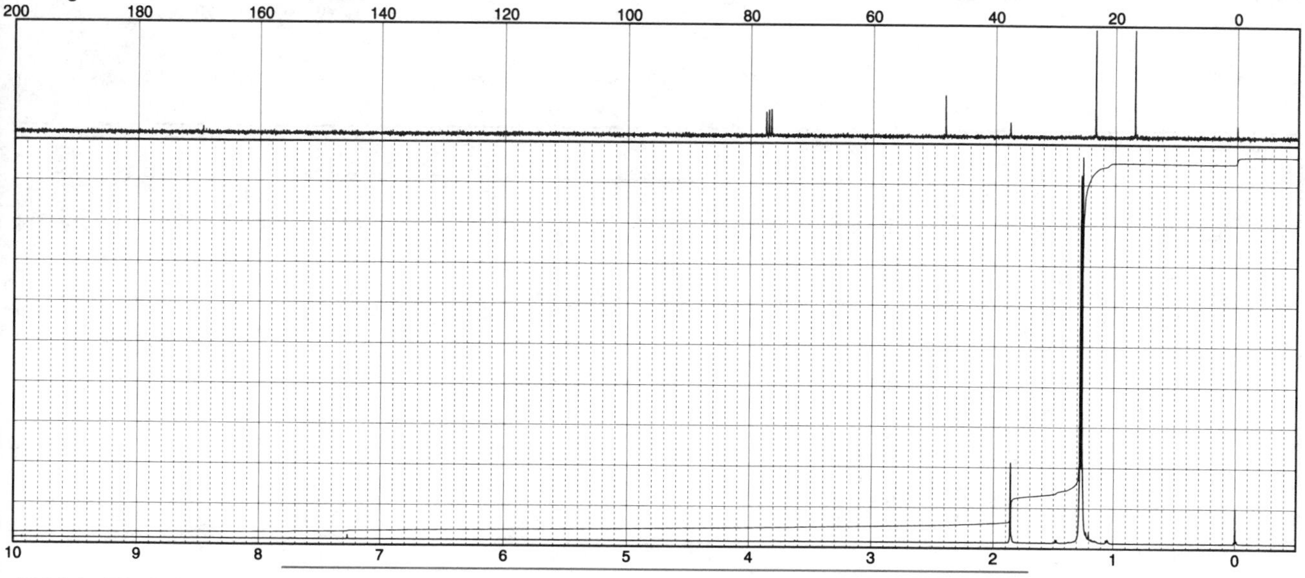

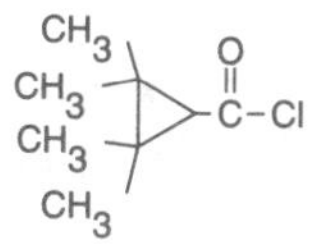

Aldrich C9,570-6   *CAS [5006-22-4]*

Cyclobutanecarbonyl chloride, 98%

C_5H_7ClO
FW 118.56
bp 60°C (50 mm)
n$_D^{20}$1.4550
Fp 99°F

FT-IR: *1*, 744C

175.54
49.07*
26.14
17.27

B

CDCl$_3$ QE-300

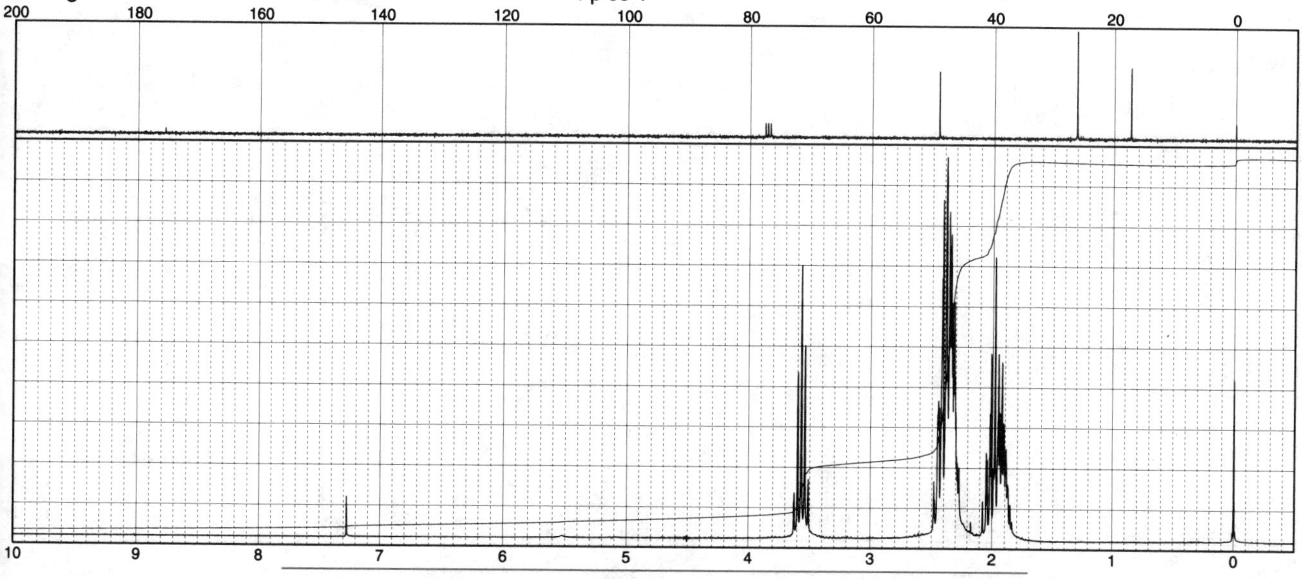

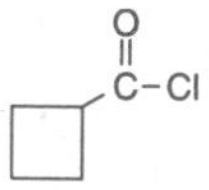

Aldrich 32,831-6   *CAS [4524-93-0]*

Cyclopentanecarbonyl chloride, 98%

C_6H_9ClO
FW 132.59
bp 162°C
d 1.091
n$_D^{20}$1.4622

Fp 140°F

176.96
55.90*
30.19
25.59

C

CDCl$_3$ QE-300

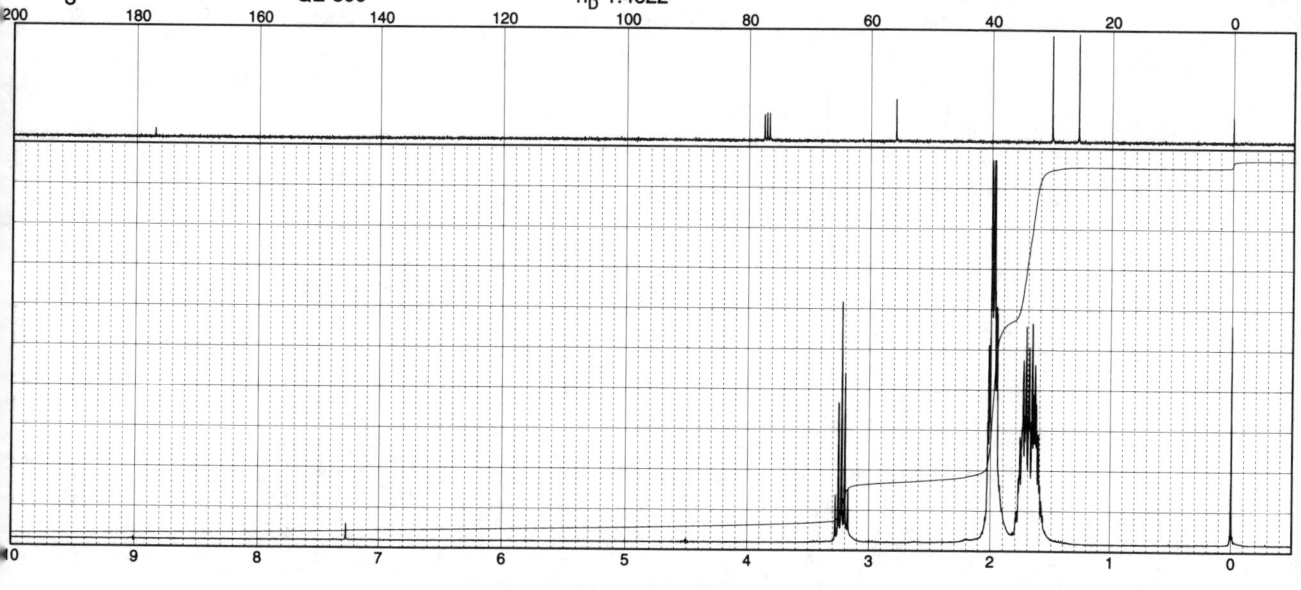

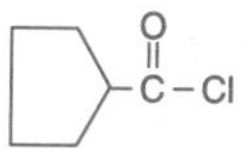

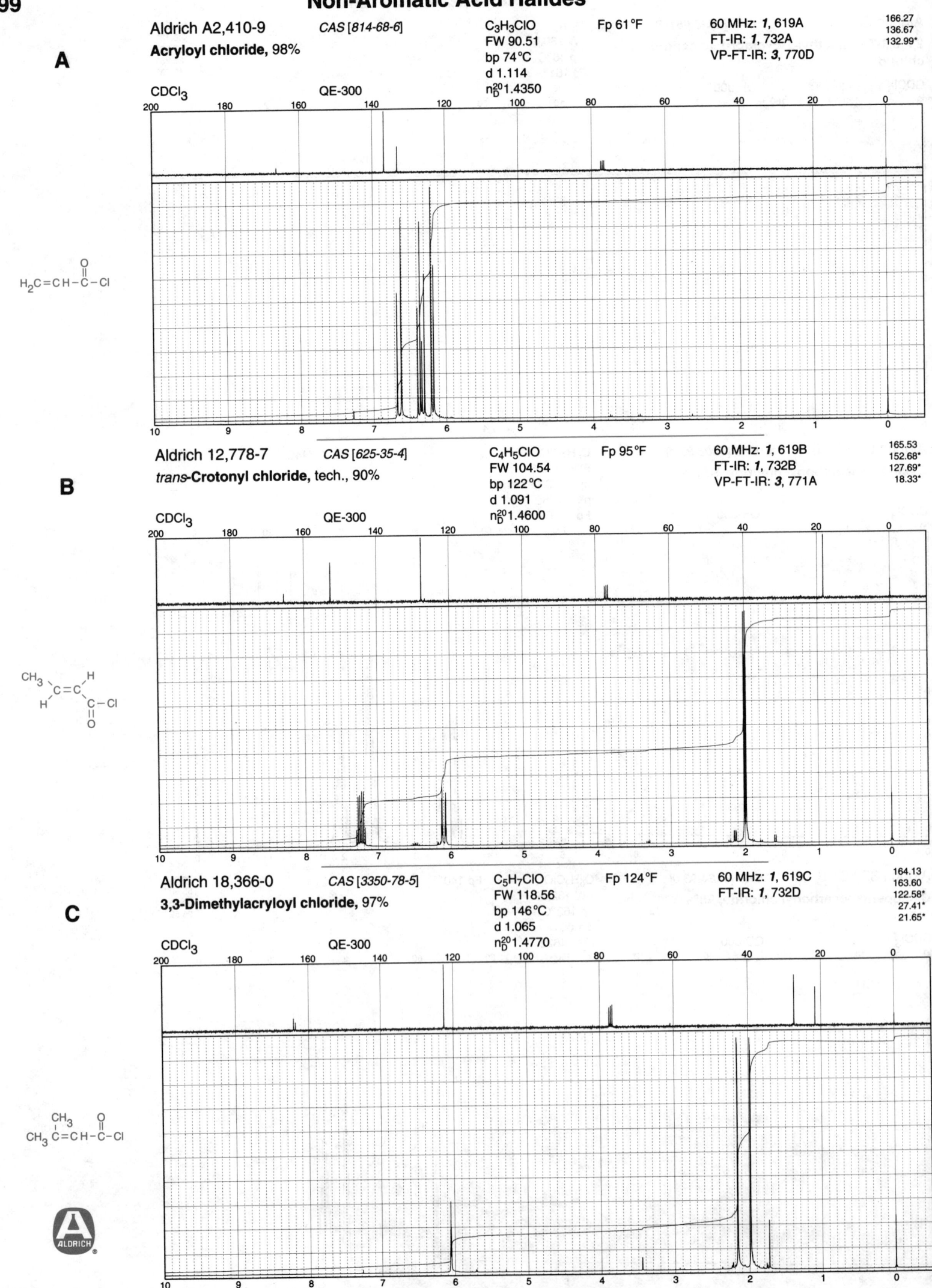

A

Aldrich A2,410-9   *CAS [814-68-6]*   C₃H₃ClO   Fp 61 °F   60 MHz: *1*, 619A   166.27
Acryloyl chloride, 98%   FW 90.51   FT-IR: *1*, 732A   136.67
                                          bp 74 °C   VP-FT-IR: *3*, 770D 132.99*
                                          d 1.114
                                          n²⁰_D 1.4350

CDCl₃   QE-300

$H_2C=CH-\overset{O}{\overset{\|}{C}}-Cl$

B

Aldrich 12,778-7   *CAS [625-35-4]*   C₄H₅ClO   Fp 95 °F   60 MHz: *1*, 619B   165.53
***trans*-Crotonyl chloride, tech., 90%** FW 104.54   FT-IR: *1*, 732B   152.68*
                                          bp 122 °C   VP-FT-IR: *3*, 771A 127.69*
                                          d 1.091   18.33*
                                          n²⁰_D 1.4600

CDCl₃   QE-300

C

Aldrich 18,366-0   *CAS [3350-78-5]*   C₅H₇ClO   Fp 124 °F   60 MHz: *1*, 619C   164.13
3,3-Dimethylacryloyl chloride, 97%   FW 118.56   FT-IR: *1*, 732D   163.60
                                         bp 146 °C   122.58*
                                         d 1.065   27.41*
                                         n²⁰_D 1.4770   21.65*

CDCl₃   QE-300

Non-Aromatic Acid Halides

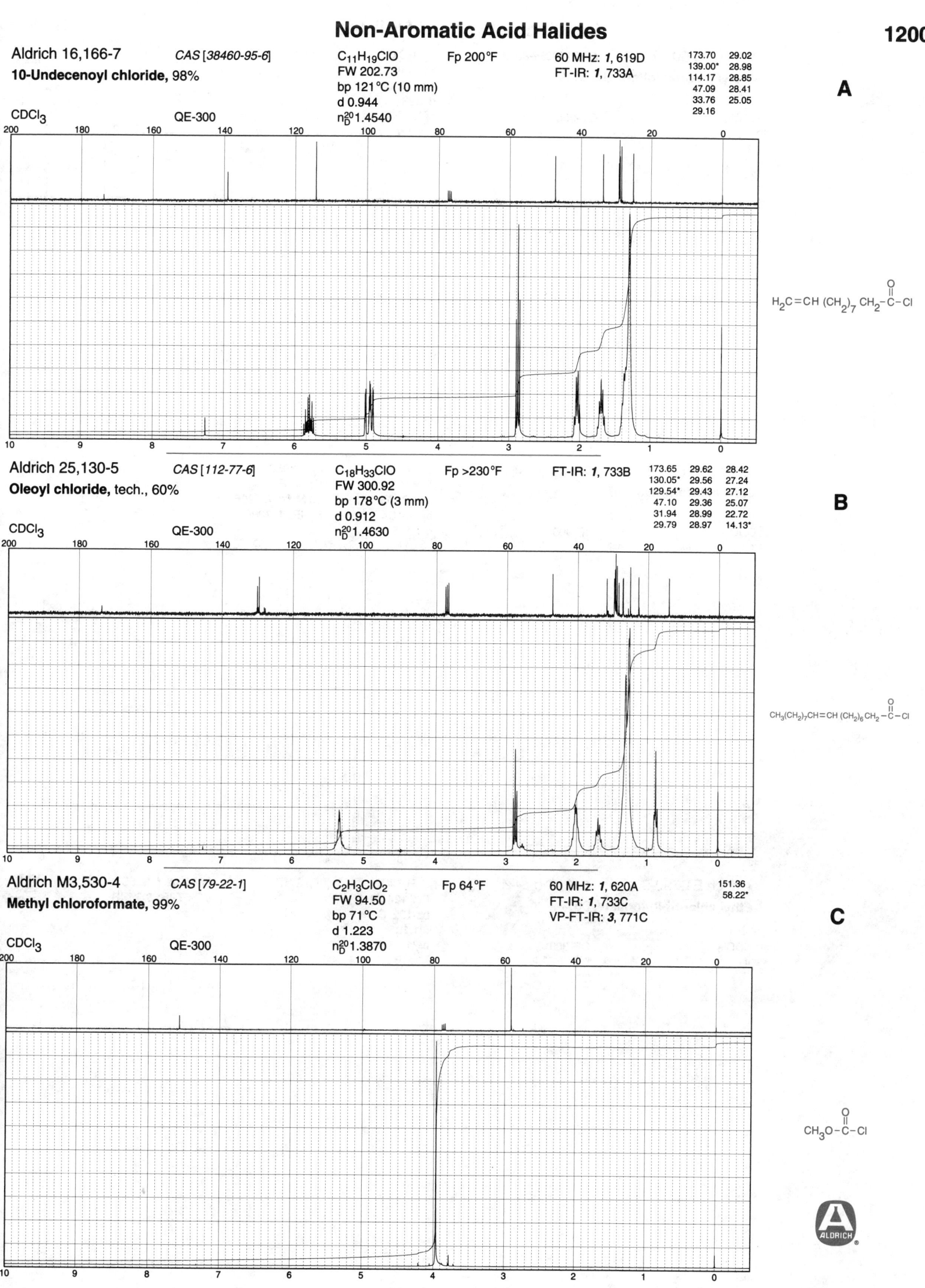

Aldrich 16,166-7   *CAS [38460-95-6]*

10-Undecenoyl chloride, 98%

$C_{11}H_{19}ClO$
FW 202.73
bp 121°C (10 mm)
d 0.944
$n_D^{20}1.4540$

Fp 200°F

60 MHz: *1*, 619D
FT-IR: *1*, 733A

173.70	29.02
139.00*	28.98
114.17	28.85
47.09	28.41
33.76	25.05
29.16	

CDCl$_3$ QE-300

A

$H_2C{=}CH\,(CH_2)_7\,CH_2{-}\underset{\underset{O}{\parallel}}{C}{-}Cl$

Aldrich 25,130-5   *CAS [112-77-6]*

Oleoyl chloride, tech., 60%

$C_{18}H_{33}ClO$
FW 300.92
bp 178°C (3 mm)
d 0.912
$n_D^{20}1.4630$

Fp >230°F

FT-IR: *1*, 733B

173.65	29.62	28.42
130.05*	29.56	27.24
129.54*	29.43	27.12
47.10	29.36	25.07
31.94	28.99	22.72
29.79	28.97	14.13*

CDCl$_3$ QE-300

B

$CH_3(CH_2)_7CH{=}CH\,(CH_2)_6\,CH_2{-}\underset{\underset{O}{\parallel}}{C}{-}Cl$

Aldrich M3,530-4   *CAS [79-22-1]*

Methyl chloroformate, 99%

$C_2H_3ClO_2$
FW 94.50
bp 71°C
d 1.223
$n_D^{20}1.3870$

Fp 64°F

60 MHz: *1*, 620A
FT-IR: *1*, 733C
VP-FT-IR: *3*, 771C

151.36	
58.22*	

CDCl$_3$ QE-300

C

$CH_3O{-}\underset{\underset{O}{\parallel}}{C}{-}Cl$

A

Aldrich 36,501-7   *CAS [18369-83-0]*   C2H3ClOS   Fp 88°F   166.29
Methyl chlorothiolformate, 97%   FW 110.56   16.27*
bp 110°C
d 1.288
n_D^{20}1.4900

CDCl3   QE-300

$CH_3S-\overset{O}{\underset{|}{C}}-Cl$

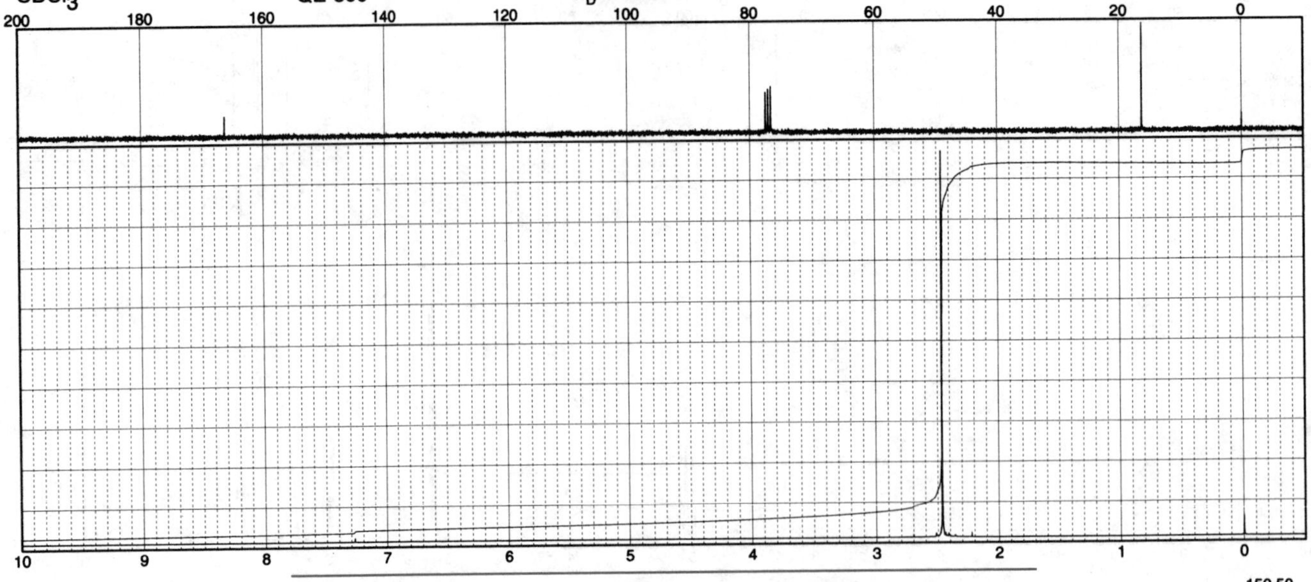

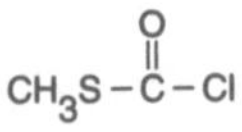

B

Aldrich 18,589-2   *CAS [541-41-3]*   C3H5ClO2 n_D^{20}1.3950   150.50
Ethyl chloroformate, 97%   FW 108.52   Fp 36°F   68.35
mp -81°C   60 MHz: *1*, 620B   13.97*
bp 93°C   FT-IR: *1*, 733D
d 1.135

CDCl3   QE-300

$CH_3CH_2O-\overset{O}{\underset{|}{C}}-Cl$

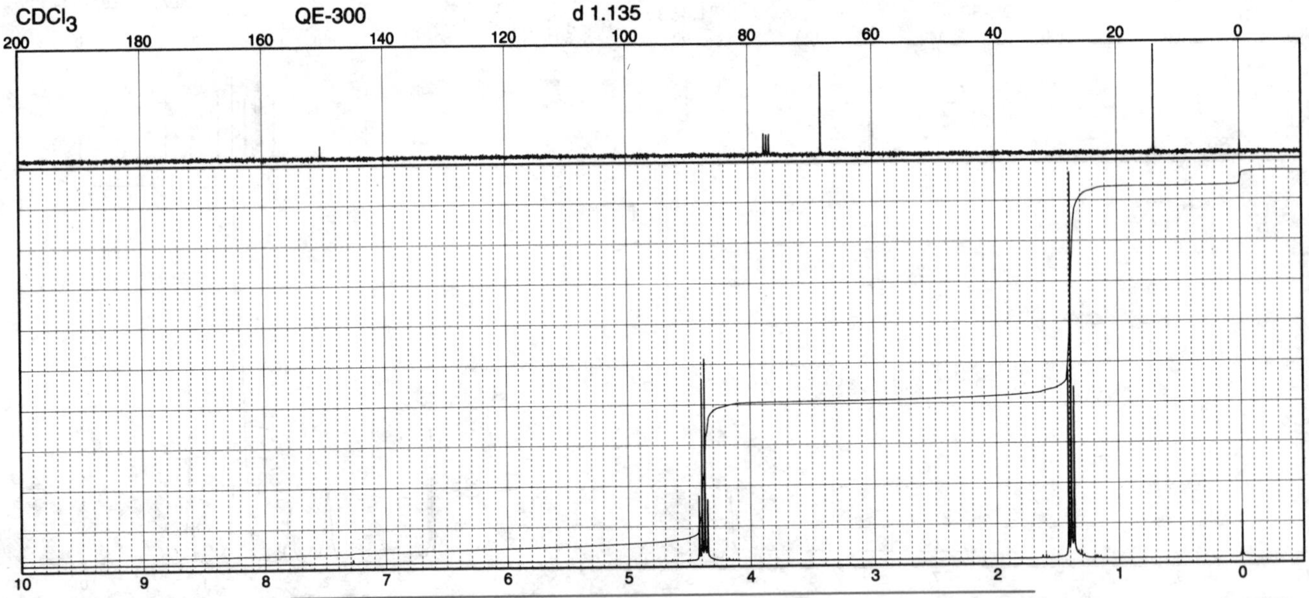

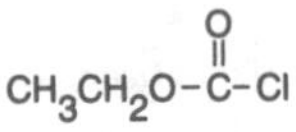

C

Aldrich E1,790-9   *CAS [2941-64-2]*   C3H5ClOS   Fp 87°F   60 MHz: *1*, 626C   165.72
Ethyl chlorothiolformate, 98%   FW 124.59   VP-FT-IR: *3*, 774B   28.48
bp 132°C   14.00*
d 1.195
n_D^{20}1.4820

CDCl3   QE-300

$CH_3CH_2S-\overset{O}{\underset{|}{C}}-Cl$

ALDRICH

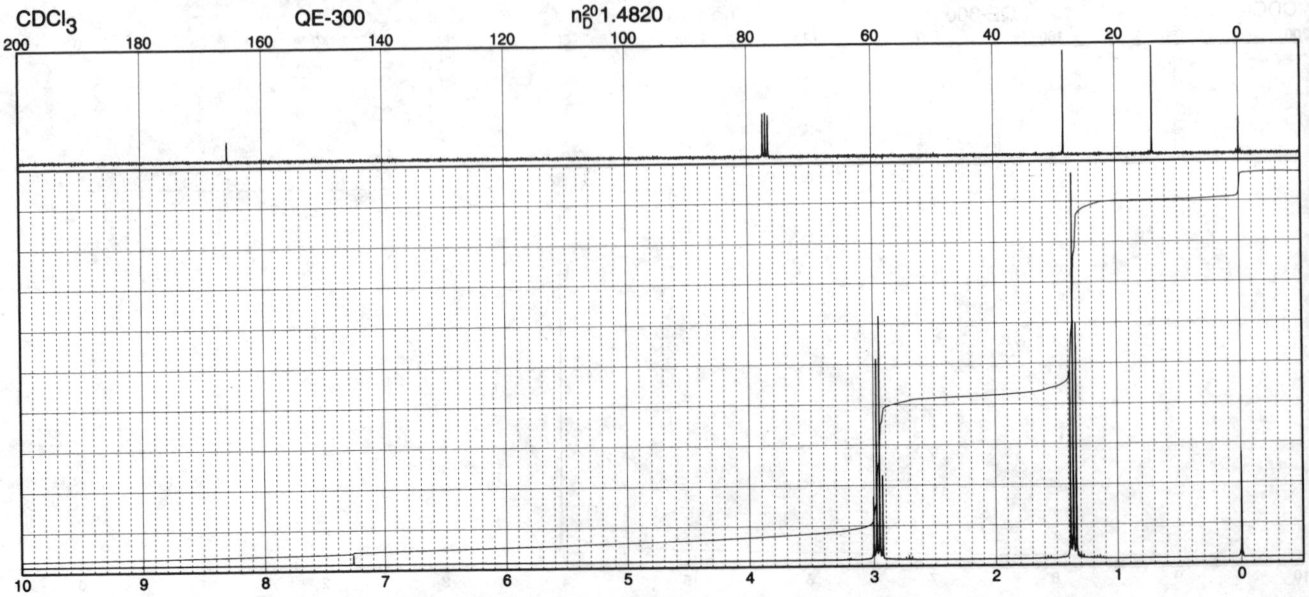

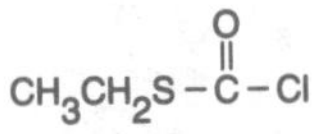

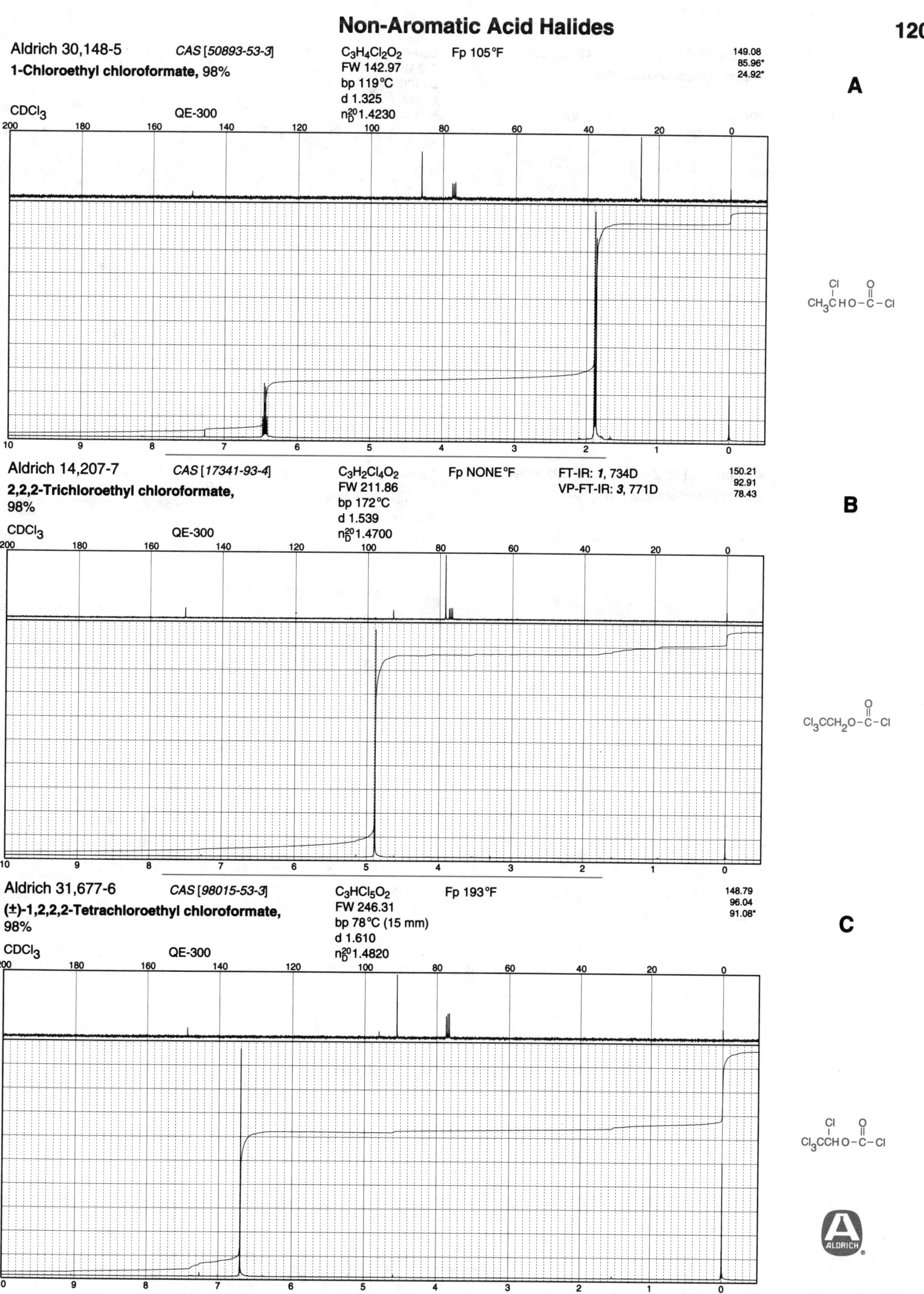

Aldrich 30,148-5   CAS [50893-53-3] $C_3H_4Cl_2O_2$ Fp 105°F   149.08
1-Chloroethyl chloroformate, 98%   FW 142.97   85.96*
                            bp 119°C   24.92*
                            d 1.325
CDCl₃   QE-300 n_D^{20} 1.4230

A

Aldrich 14,207-7   CAS [17341-93-4] $C_3H_2Cl_4O_2$ Fp NONE°F   FT-IR: *1*, 734D   150.21
2,2,2-Trichloroethyl chloroformate,   FW 211.86   VP-FT-IR: *3*, 771D   92.91
98%   bp 172°C   78.43
                            d 1.539
CDCl₃   QE-300 n_D^{20} 1.4700

B

Aldrich 31,677-6   CAS [98015-53-3] $C_3HCl_5O_2$ Fp 193°F   148.79
(±)-1,2,2,2-Tetrachloroethyl chloroformate,   FW 246.31   96.04
98%   bp 78°C (15 mm)   91.08*
                            d 1.610
CDCl₃   QE-300 n_D^{20} 1.4820

C

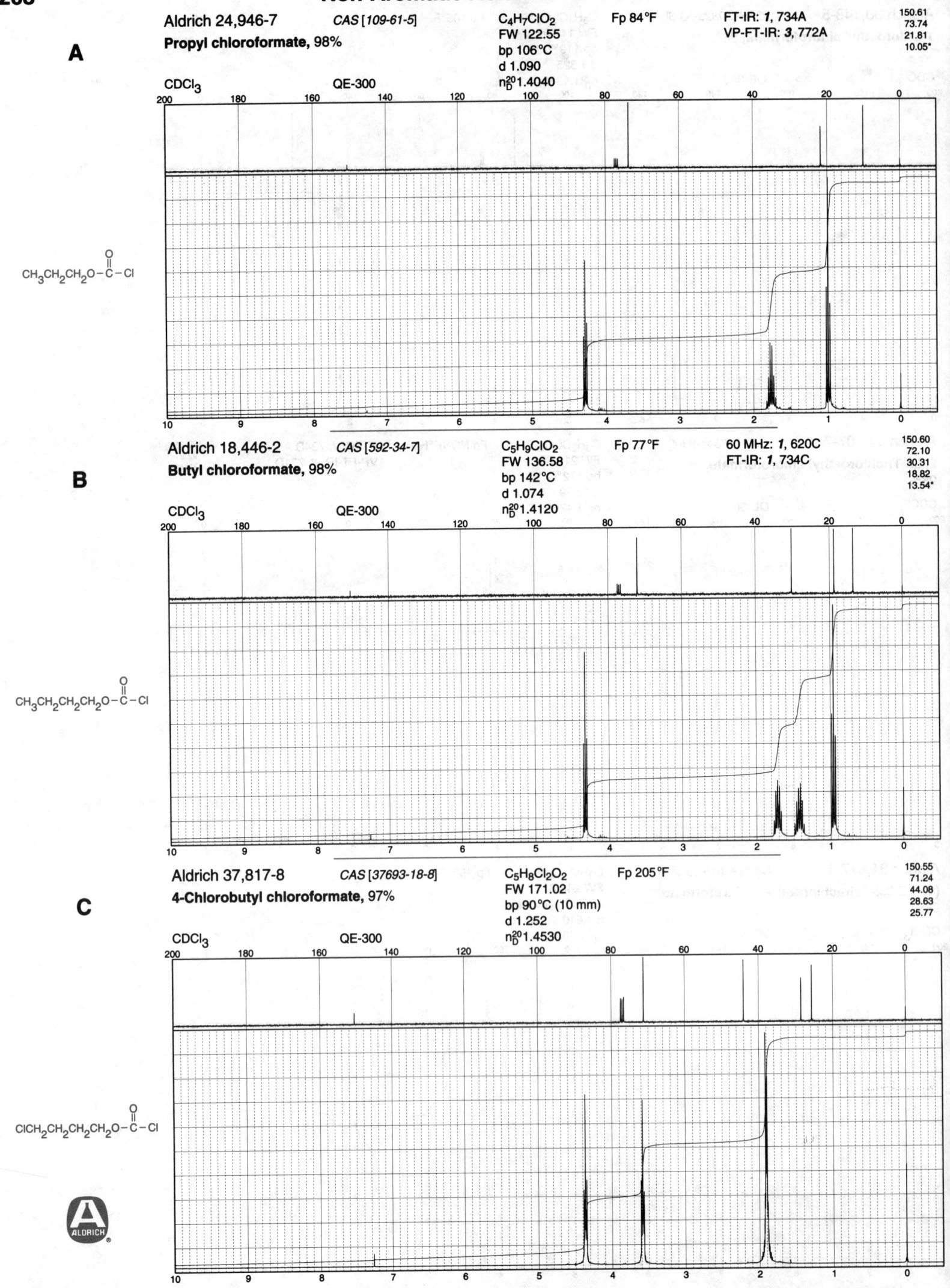

A

Aldrich 24,946-7 *CAS [109-61-5]* C$_4$H$_7$ClO$_2$ Fp 84°F FT-IR: *1*, 734A 150.61
Propyl chloroformate, 98% FW 122.55 VP-FT-IR: *3*, 772A 73.74
bp 106°C 21.81
d 1.090 10.05*
n$_D^{20}$1.4040

CDCl$_3$ QE-300

CH$_3$CH$_2$CH$_2$O–C(=O)–Cl

B

Aldrich 18,446-2 *CAS [592-34-7]* C$_5$H$_9$ClO$_2$ Fp 77°F 60 MHz: *1*, 620C 150.60
Butyl chloroformate, 98% FW 136.58 FT-IR: *1*, 734C 72.10
bp 142°C 30.31
d 1.074 18.82
n$_D^{20}$1.4120 13.54*

CDCl$_3$ QE-300

CH$_3$CH$_2$CH$_2$CH$_2$O–C(=O)–Cl

C

Aldrich 37,817-8 *CAS [37693-18-8]* C$_5$H$_8$Cl$_2$O$_2$ Fp 205°F 150.55
4-Chlorobutyl chloroformate, 97% FW 171.02 71.24
bp 90°C (10 mm) 44.08
d 1.252 28.63
n$_D^{20}$1.4530 25.77

CDCl$_3$ QE-300

ClCH$_2$CH$_2$CH$_2$CH$_2$O–C(=O)–Cl

ALDRICH

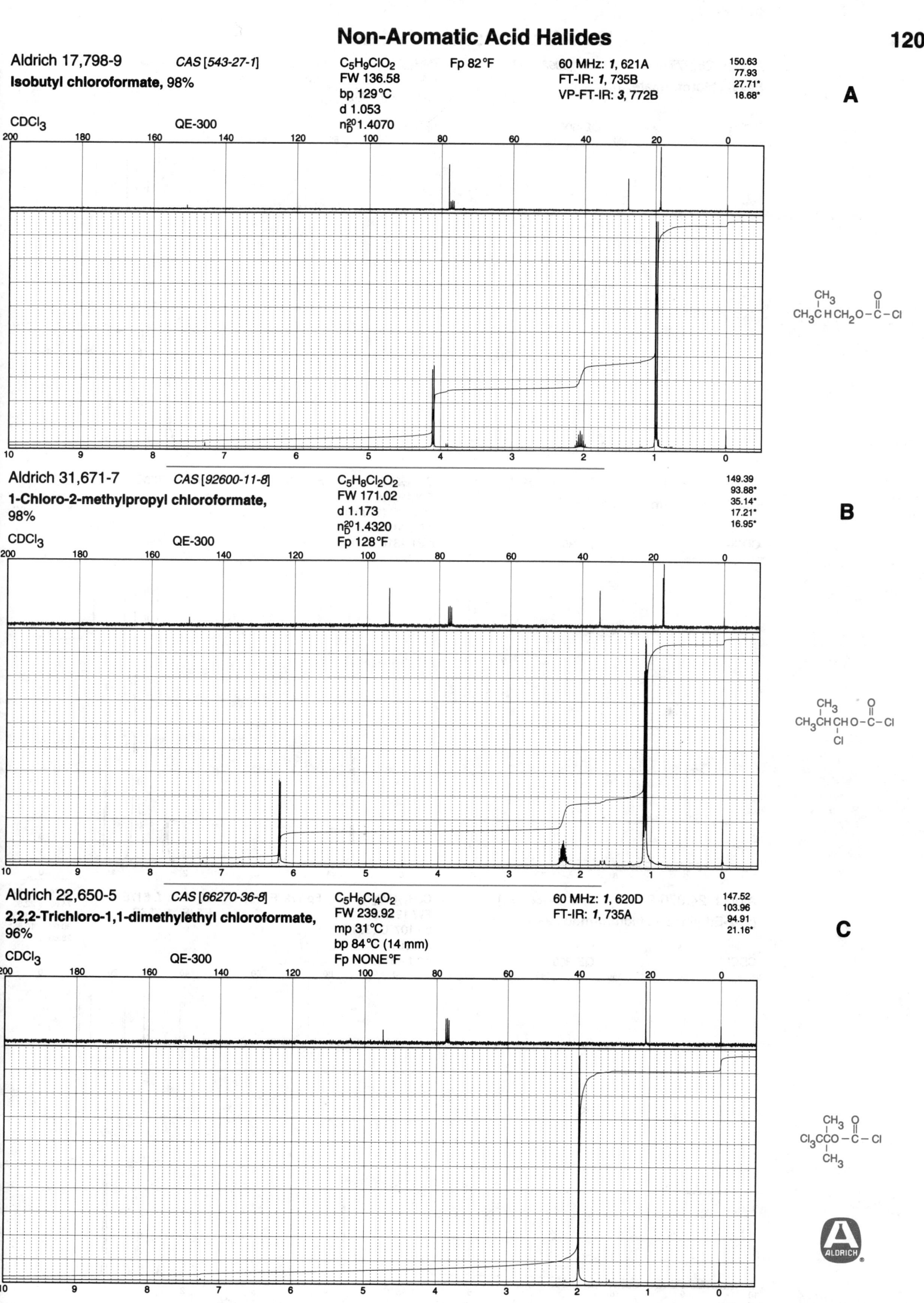

Aldrich 17,798-9   CAS [543-27-1]   C₅H₉ClO₂   Fp 82°F   60 MHz: 1, 621A

Isobutyl chloroformate, 98%

FW 136.58
bp 129°C
d 1.053
n²⁰_D 1.4070

FT-IR: 1, 735B
VP-FT-IR: 3, 772B

150.63
77.93
27.71*
18.68*

A

CDCl₃   QE-300

Aldrich 31,671-7   CAS [92600-11-8]   C₅H₈Cl₂O₂

1-Chloro-2-methylpropyl chloroformate, 98%

FW 171.02
d 1.173
n²⁰_D 1.4320
Fp 128°F

149.39
93.88*
35.14*
17.21*
16.95*

B

CDCl₃   QE-300

Aldrich 22,650-5   CAS [66270-36-8]   C₅H₆Cl₄O₂   60 MHz: 1, 620D

2,2,2-Trichloro-1,1-dimethylethyl chloroformate, 96%

FW 239.92
mp 31°C
bp 84°C (14 mm)
Fp NONE°F

FT-IR: 1, 735A

147.52
103.96
94.91
21.16*

C

CDCl₃   QE-300

Non-Aromatic Acid Halides

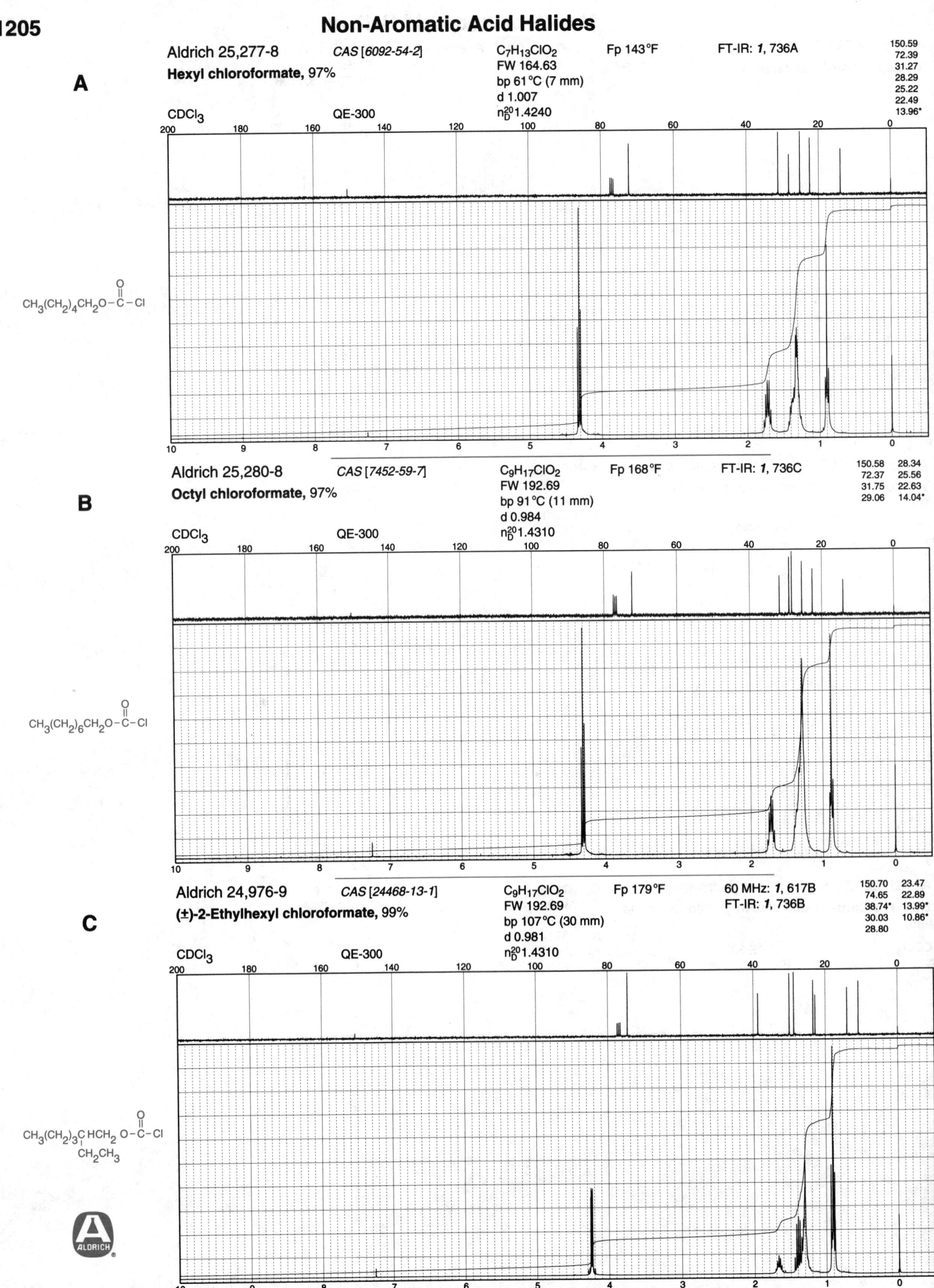

A

Aldrich 25,277-8 | *CAS [6092-54-2]* | C₇H₁₃ClO₂ | Fp 143°F | FT-IR: *1*, 736A

Hexyl chloroformate, 97%

$C_7H_{13}ClO_2$
FW 164.63
bp 61°C (7 mm)
d 1.007
n_D^{20} 1.4240

150.59
72.39
31.27
28.29
25.22
22.49
13.96*

CDCl₃   QE-300

$CH_3(CH_2)_4CH_2O-\overset{\overset{\displaystyle O}{\|}}{C}-Cl$

B

Aldrich 25,280-8 | *CAS [7452-59-7]* | C₉H₁₇ClO₂ | Fp 168°F | FT-IR: *1*, 736C

Octyl chloroformate, 97%

$C_9H_{17}ClO_2$
FW 192.69
bp 91°C (11 mm)
d 0.984
n_D^{20} 1.4310

150.58 28.34
72.37 25.56
31.75 22.63
29.06 14.04*

CDCl₃   QE-300

$CH_3(CH_2)_6CH_2O-\overset{\overset{\displaystyle O}{\|}}{C}-Cl$

C

Aldrich 24,976-9 | *CAS [24468-13-1]* | C₉H₁₇ClO₂ | Fp 179°F | 60 MHz: *1*, 617B

(±)-2-Ethylhexyl chloroformate, 99% | | | | FT-IR: *1*, 736B

$C_9H_{17}ClO_2$
FW 192.69
bp 107°C (30 mm)
d 0.981
n_D^{20} 1.4310

150.70 23.47
74.65 22.89
38.74* 13.99*
30.03 10.86*
28.80

CDCl₃   QE-300

$CH_3(CH_2)_3\underset{\underset{\displaystyle CH_2CH_3}{|}}{C}HCH_2O-\overset{\overset{\displaystyle O}{\|}}{C}-Cl$

ALDRICH

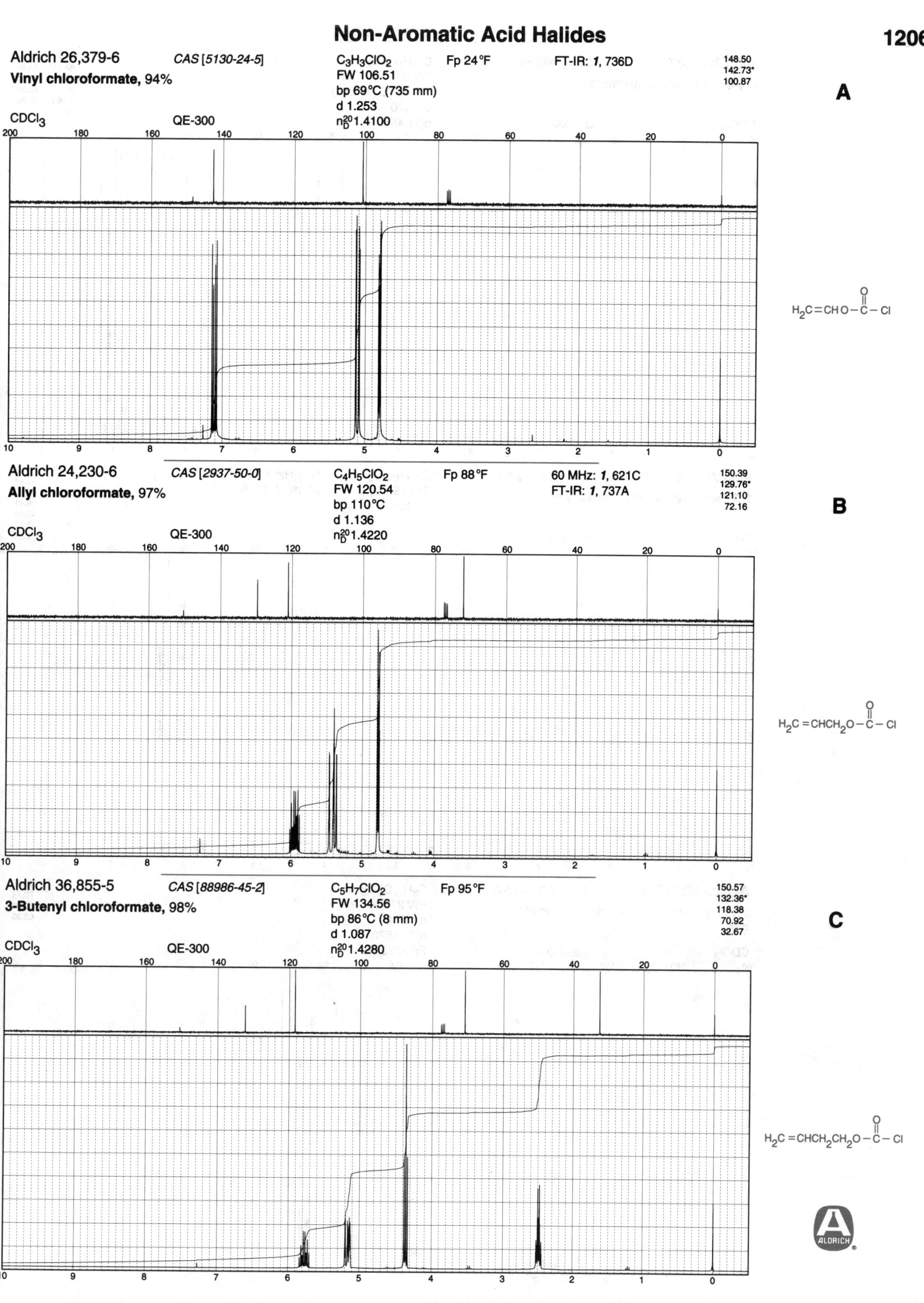

Aldrich 26,379-6   *CAS [5130-24-5]*   C₃H₃ClO₂   Fp 24°F   FT-IR: *1*, 736D

Vinyl chloroformate, 94%

$C_3H_3ClO_2$
FW 106.51
bp 69°C (735 mm)
d 1.253
$n_D^{20} 1.4100$

148.50
142.73*
100.87

A

CDCl₃   QE-300

Aldrich 24,230-6   *CAS [2937-50-0]*   C₄H₅ClO₂   Fp 88°F   60 MHz: *1*, 621C

Allyl chloroformate, 97%

$C_4H_5ClO_2$
FW 120.54
bp 110°C
d 1.136
$n_D^{20} 1.4220$

FT-IR: *1*, 737A

150.39
129.76*
121.10
72.16

B

CDCl₃   QE-300

Aldrich 36,855-5   *CAS [88986-45-2]*   C₅H₇ClO₂   Fp 95°F

3-Butenyl chloroformate, 98%

$C_5H_7ClO_2$
FW 134.56
bp 86°C (8 mm)
d 1.087
$n_D^{20} 1.4280$

150.57
132.36*
118.38
70.92
32.67

C

CDCl₃   QE-300

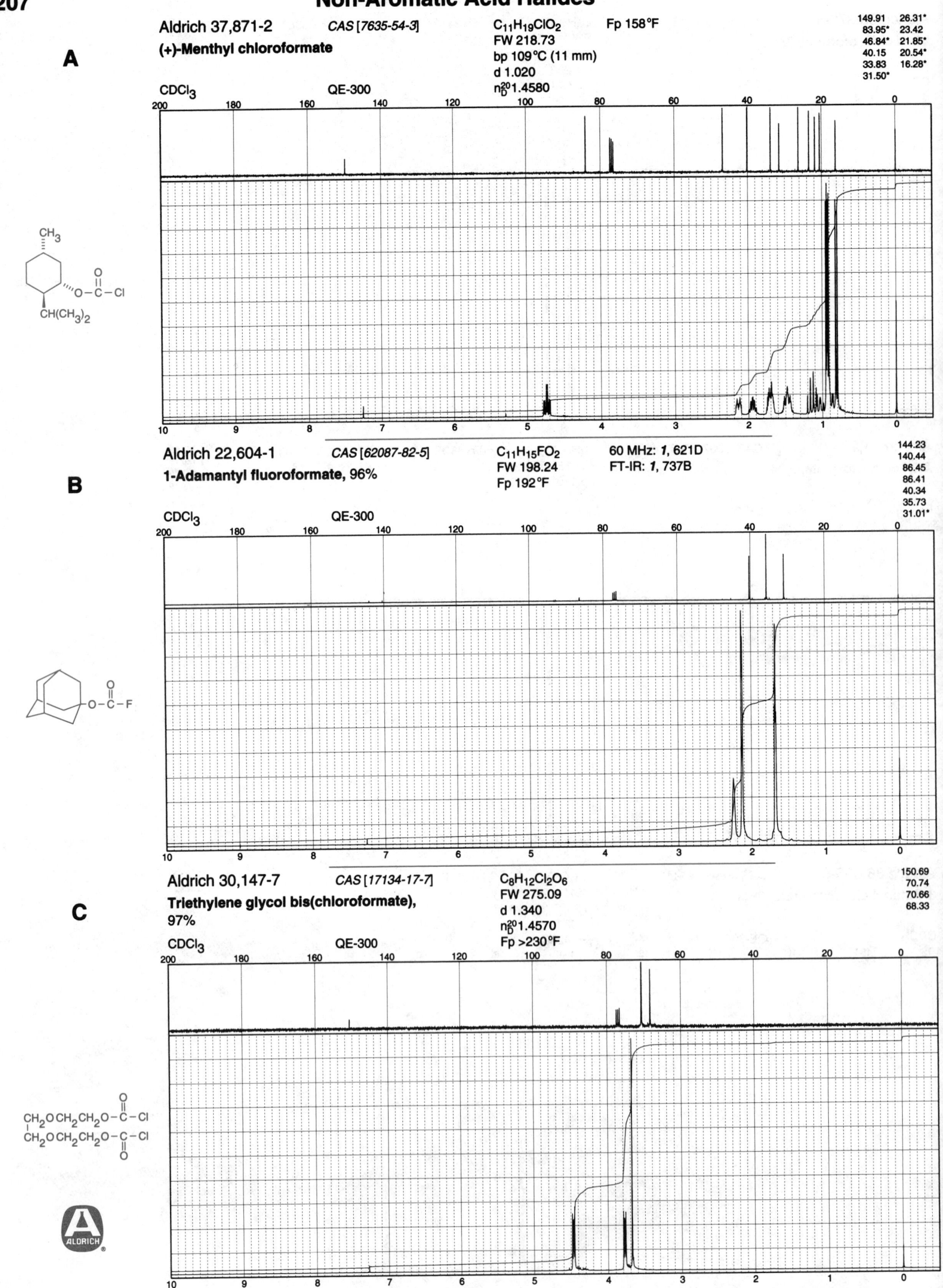

A

Aldrich 37,871-2   *CAS [7635-54-3]* $C_{11}H_{19}ClO_2$ Fp 158°F

(+)-Menthyl chloroformate

FW 218.73
bp 109°C (11 mm)
d 1.020
n_D^{20} 1.4580

149.91   26.31*
83.95*   23.42
46.84*   21.85*
40.15   20.54*
33.83   16.28*
31.50*

CDCl$_3$ QE-300

B

Aldrich 22,604-1   *CAS [62087-82-5]* $C_{11}H_{15}FO_2$ 60 MHz: *1*, 621D

1-Adamantyl fluoroformate, 96%

FW 198.24   FT-IR: *1*, 737B
Fp 192°F

144.23
140.44
86.45
86.41
40.34
35.73
31.01*

CDCl$_3$ QE-300

C

Aldrich 30,147-7   *CAS [17134-17-7]* $C_8H_{12}Cl_2O_6$

Triethylene glycol bis(chloroformate), 97%

FW 275.09
d 1.340
n_D^{20} 1.4570
Fp >230°F

150.69
70.74
70.66
68.33

CDCl$_3$ QE-300

Non-Aromatic Acid Halides

Aldrich 15,144-0 *CAS [5781-53-3]* C₃H₃ClO₃   Fp 116°F   60 MHz: *1*, 622B

Methyl oxalyl chloride, 96%

$C_3H_3ClO_3$
FW 122.51
bp 119°C
d 1.332
n_D^{20} 1.4190

60 MHz: *1*, 622B
FT-IR: *1*, 737D
VP-FT-IR: *3*, 772C

160.87
156.12
55.05*

A

CDCl₃   QE-300

$$CH_3O-\overset{O}{\underset{\|}{C}}-\overset{O}{\underset{\|}{C}}-Cl$$

Aldrich E4,310-1 *CAS [4755-77-5]*

Ethyl oxalyl chloride, 98%

$C_4H_5ClO_3$
FW 136.53
bp 135°C
d 1.222
n_D^{20} 1.4160

Fp 107°F

60 MHz: *1*, 622C
FT-IR: *1*, 738A
VP-FT-IR: *3*, 772D

161.15
155.65
65.13
13.86*

B

CDCl₃   QE-300

$$CH_3CH_2O-\overset{O}{\underset{\|}{C}}-\overset{O}{\underset{\|}{C}}-Cl$$

Aldrich 16,401-1 *CAS [37517-81-0]*

Methyl malonyl chloride, 95%

$C_4H_5ClO_3$
FW 136.53
bp 58°C (12 mm)
d 1.273
n_D^{20} 1.4320

Fp 176°F

166.32
164.25
53.08*
51.87

C

CDCl₃   QE-300

$$CH_3O-\overset{O}{\underset{\|}{C}}-CH_2-\overset{O}{\underset{\|}{C}}-Cl$$

ALDRICH

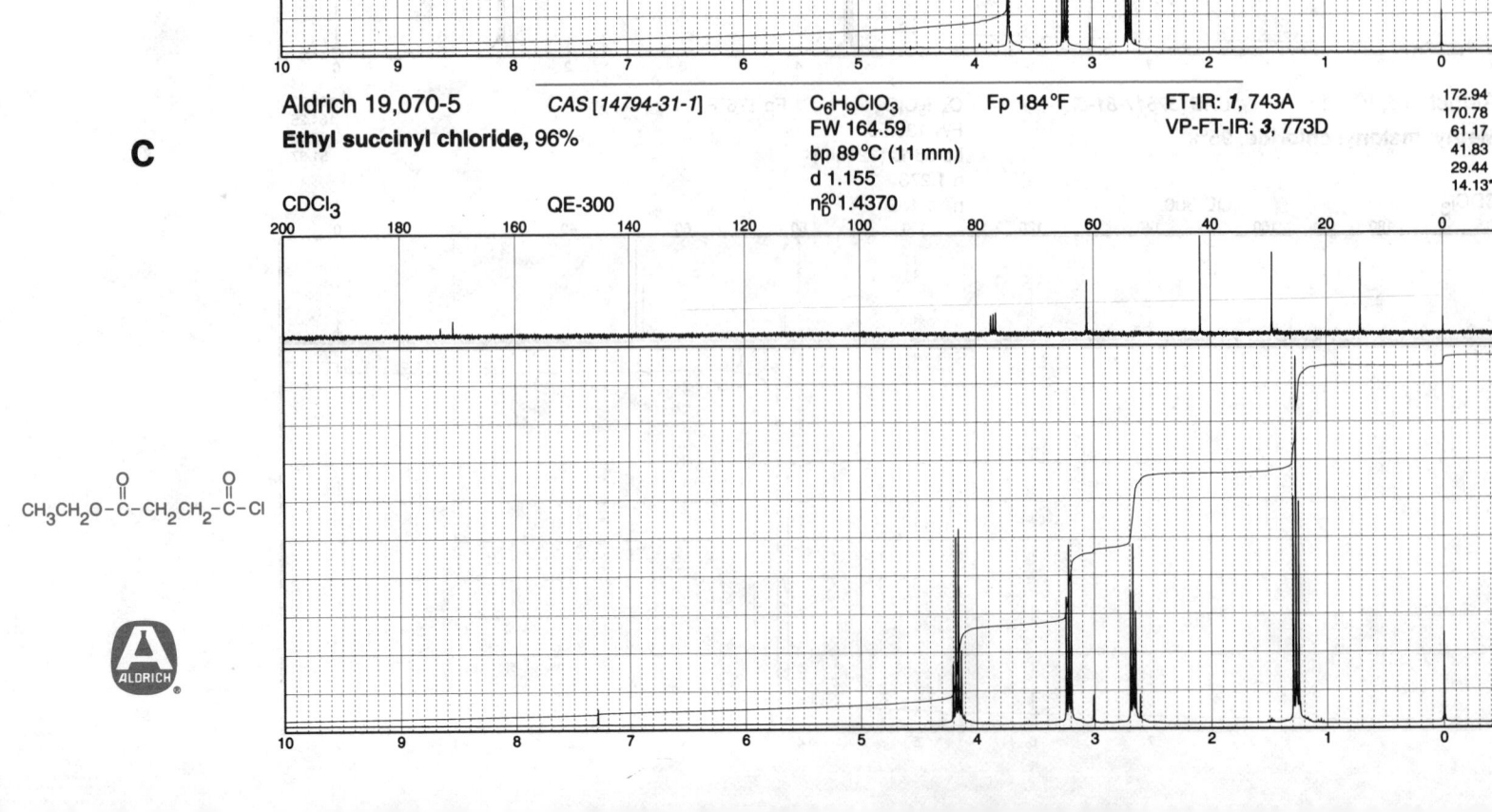

A

Aldrich 33,550-9   *CAS [36239-09-5]* $C_5H_7ClO_3$ Fp 149°F
Ethyl malonyl chloride, 97%   FW 150.56
bp 61°C (10 mm)
d 1.182
n_D^{20}1.4280

CDCl$_3$ QE-300

166.38
163.76
62.36
52.19
13.97*

$CH_3CH_2O-\overset{O}{\underset{\|}{C}}-CH_2-\overset{O}{\underset{\|}{C}}-Cl$

B

Aldrich C1,104-9   *CAS [1490-25-1]* $C_5H_7ClO_3$ Fp 165°F   60 MHz: *1*, 623A
3-Carbomethoxypropionyl chloride, 97%   FW 150.56   FT-IR: *1*, 738C
bp 62°C (3 mm)
d 1.223
n_D^{20}1.4400

CDCl$_3$ QE-300

173.02
171.34
52.17*
41.79
29.14

$CH_3O-\overset{O}{\underset{\|}{C}}-CH_2CH_2-\overset{O}{\underset{\|}{C}}-Cl$

C

Aldrich 19,070-5   *CAS [14794-31-1]* $C_6H_9ClO_3$ Fp 184°F   FT-IR: *1*, 743A
Ethyl succinyl chloride, 96%   FW 164.59   VP-FT-IR: *3*, 773D
bp 89°C (11 mm)
d 1.155
n_D^{20}1.4370

CDCl$_3$ QE-300

172.94
170.78
61.17
41.83
29.44
14.13*

$CH_3CH_2O-\overset{O}{\underset{\|}{C}}-CH_2CH_2-\overset{O}{\underset{\|}{C}}-Cl$

Non-Aromatic Acid Halides

Aldrich M3,535-5 *CAS [1501-26-4]* $C_6H_9ClO_3$ Fp 180°F 60 MHz: *1*, 623B

Methyl 4-(chloroformyl)butyrate, 98% FW 164.59 FT-IR: *1*, 738D

bp 110°C (17 mm)

d 1.191

CDCl$_3$ QE-300 n_D^{20} 1.4460

173.34
172.71
51.76*
46.01
32.09
20.21

A

$CH_3O-\overset{O}{\overset{\|}{C}}-CH_2CH_2CH_2-\overset{O}{\overset{\|}{C}}-Cl$

Aldrich 38,556-5 *CAS [41624-92-4]* $C_9H_{15}ClO_3$

Methyl suberyl chloride, 96% FW 206.67

d 1.456

n_D^{20} 1.4500

CDCl$_3$ QE-300 Fp >230°F

173.89 28.54
173.61 28.04
51.46* 24.84
46.95 24.57
33.84

B

$CH_3O-\overset{O}{\overset{\|}{C}}-CH_2(CH_2)_4 CH_2-\overset{O}{\overset{\|}{C}}-Cl$

Aldrich 30,853-6 *CAS [14113-02-1]* $C_{10}H_{17}ClO_3$ Fp >230°F

Ethyl 7-(chloroformyl)heptanoate, 98% FW 220.70

bp 146°C (12 mm)

d 1.045

CDCl$_3$ QE-300 n_D^{20} 1.4480

173.64 28.53
173.49 28.05
60.22 24.84
46.96 24.60
34.12 14.25*

C

$Cl-\overset{O}{\overset{\|}{C}}-CH_2(CH_2)_4CH_2-\overset{O}{\overset{\|}{C}}-OCH_2CH_3$

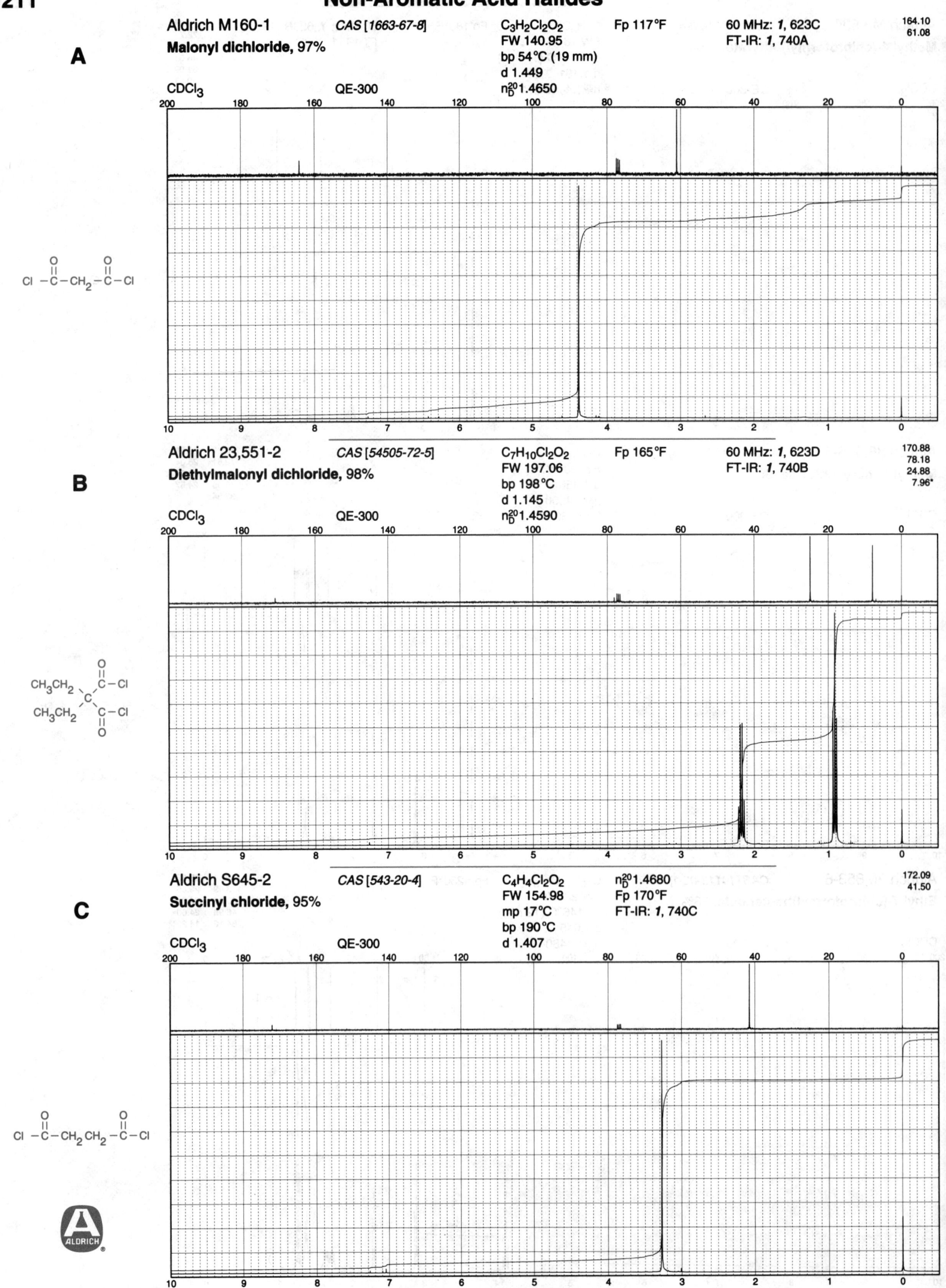

A

Aldrich M160-1

Malonyl dichloride, 97%

CAS [1663-67-8]

$C_3H_2Cl_2O_2$
FW 140.95
bp 54°C (19 mm)
d 1.449
n_D^{20}1.4650

Fp 117°F

60 MHz: *1*, 623C
FT-IR: *1*, 740A

164.10
61.08

CDCl₃ QE-300

B

Aldrich 23,551-2

Diethylmalonyl dichloride, 98%

CAS [54505-72-5]

$C_7H_{10}Cl_2O_2$
FW 197.06
bp 198°C
d 1.145
n_D^{20}1.4590

Fp 165°F

60 MHz: *1*, 623D
FT-IR: *1*, 740B

170.88
78.18
24.88
7.96*

CDCl₃ QE-300

C

Aldrich S645-2

Succinyl chloride, 95%

CAS [543-20-4]

$C_4H_4Cl_2O_2$
FW 154.98
mp 17°C
bp 190°C
d 1.407

n_D^{20}1.4680
Fp 170°F
FT-IR: *1*, 740C

172.09
41.50

CDCl₃ QE-300

Aldrich G460-8   *CAS [2873-74-7]* $C_5H_6Cl_2O_2$ Fp 224°F   60 MHz: *1*, 624A   173.07
45.04
20.32

Glutaryl dichloride, 97%   FW 169.01   FT-IR: *1*, 740D

bp 217°C

d 1.324

CDCl₃   QE-300 n_D^{20} 1.4720

A

$$Cl-\overset{\overset{\displaystyle O}{\|}}{C}-CH_2\,CH_2\,CH_2-\overset{\overset{\displaystyle O}{\|}}{C}-Cl$$

Aldrich 23,552-0   *CAS [44987-62-4]* $C_7H_{10}Cl_2O_2$ Fp >230°F   60 MHz: *1*, 624C   173.31
172.50
53.60
44.54
30.73
29.63*
18.66*

3-Methyladipoyl chloride, 97%   FW 197.06   FT-IR: *1*, 741B

bp 118°C (10 mm)

d 1.217

CDCl₃   QE-300 n_D^{20} 1.4720

B

$$Cl-\overset{\overset{\displaystyle O}{\|}}{C}-CH_2\overset{\overset{\displaystyle CH_3}{|}}{CH}CH_2CH_2-\overset{\overset{\displaystyle O}{\|}}{C}-Cl$$

Aldrich 30,666-5   *CAS [4834-98-4]* $C_{12}H_{20}Cl_2O_2$ Fp >230°F   173.65
47.07
29.15
28.98
28.38
25.05

Dodecanedioyl dichloride, 98%   FW 267.20

bp 140°C

d 1.069

CDCl₃   QE-300 n_D^{20} 1.4680

C

$$Cl-\overset{\overset{\displaystyle O}{\|}}{C}-CH_2\,(CH_2)_8\,CH_2-\overset{\overset{\displaystyle O}{\|}}{C}-Cl$$

ALDRICH®

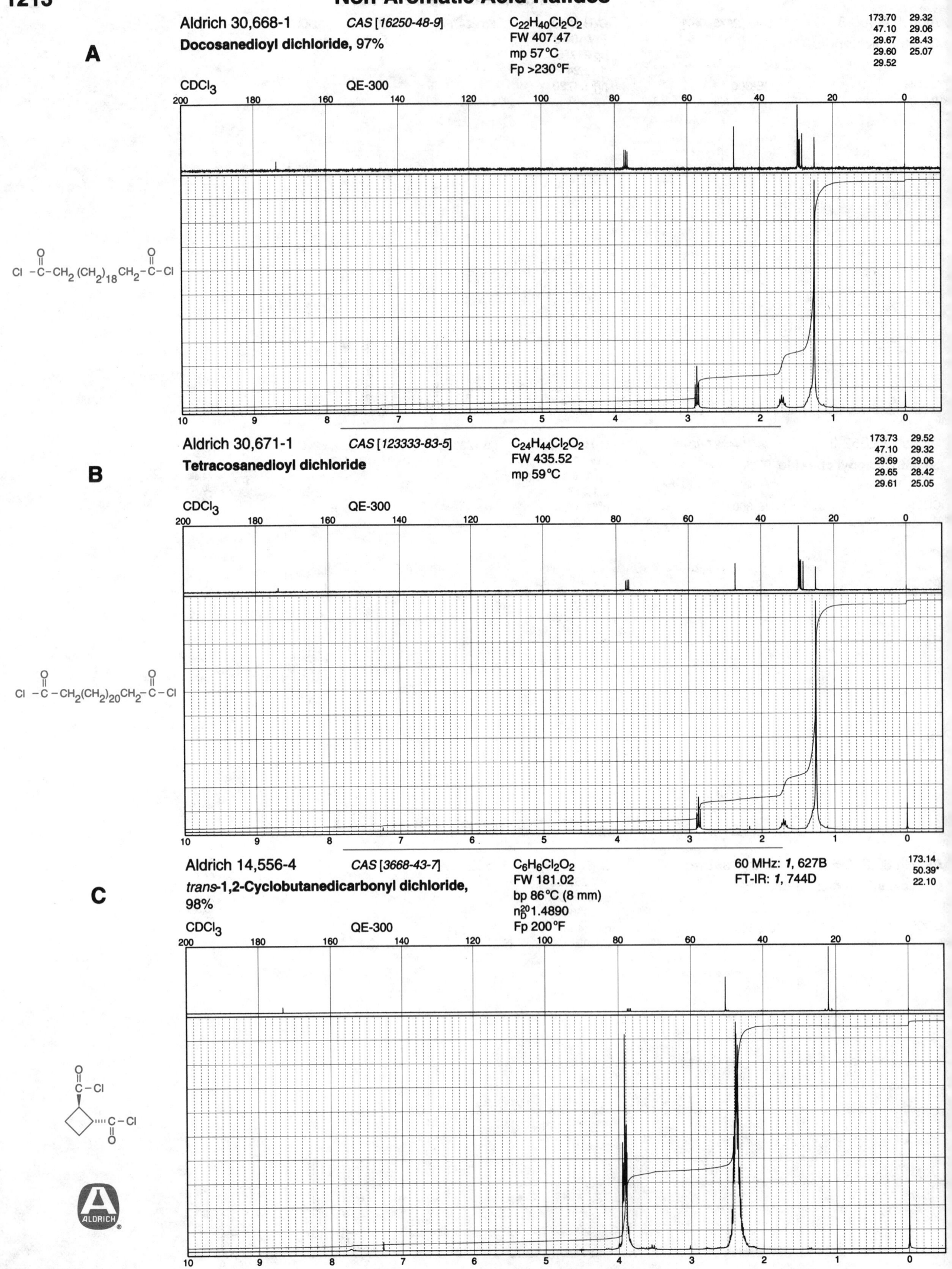

A

Aldrich 30,668-1   *CAS [16250-48-9]* $C_{22}H_{40}Cl_2O_2$

Docosanedioyl dichloride, 97%   FW 407.47

mp 57°C

Fp >230°F

173.70	29.32
47.10	29.06
29.67	28.43
29.60	25.07
29.52	

CDCl$_3$ QE-300

Cl—C(=O)—CH$_2$(CH$_2$)$_{18}$CH$_2$—C(=O)—Cl

B

Aldrich 30,671-1   *CAS [123333-83-5]* $C_{24}H_{44}Cl_2O_2$

Tetracosanedioyl dichloride   FW 435.52

mp 59°C

173.73	29.52
47.10	29.32
29.69	29.06
29.65	28.42
29.61	25.05

CDCl$_3$ QE-300

Cl—C(=O)—CH$_2$(CH$_2$)$_{20}$CH$_2$—C(=O)—Cl

C

Aldrich 14,556-4   *CAS [3668-43-7]* $C_6H_6Cl_2O_2$

trans-1,2-Cyclobutanedicarbonyl dichloride,   FW 181.02

98%

bp 86°C (8 mm)

n$_D^{20}$1.4890

Fp 200°F

60 MHz: *1*, 627B

FT-IR: *1*, 744D

173.14	
50.39*	
22.10	

CDCl$_3$ QE-300

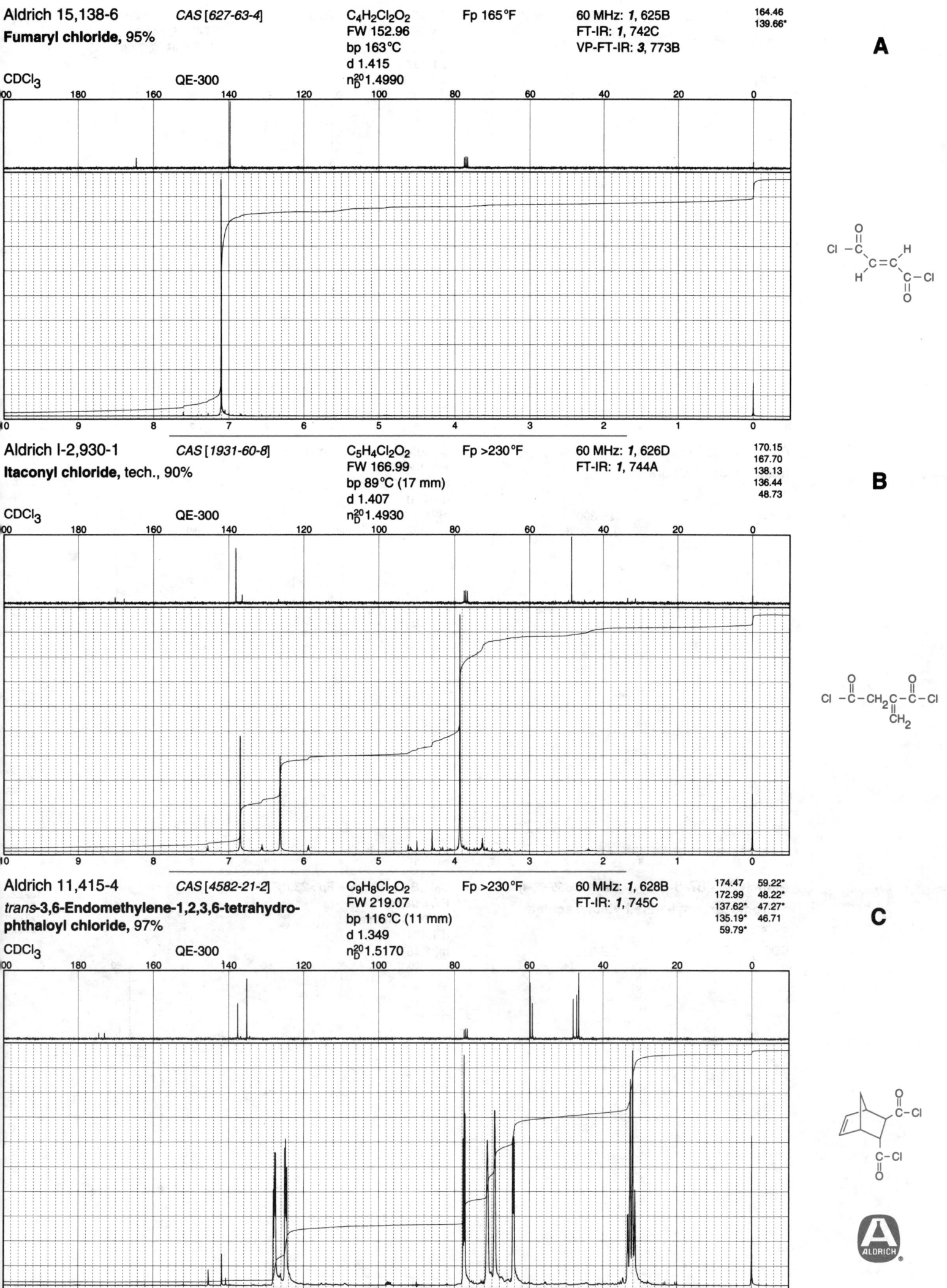

A

Aldrich 15,138-6   CAS [627-63-4]   C₄H₂Cl₂O₂   Fp 165°F   60 MHz: *1*, 625B   164.46
Fumaryl chloride, 95%   FW 152.96   FT-IR: *1*, 742C   139.66*
                                         bp 163°C   VP-FT-IR: *3*, 773B
                                         d 1.415
CDCl₃   QE-300   n²⁰_D 1.4990

B

Aldrich I-2,930-1   CAS [1931-60-8]   C₅H₄Cl₂O₂   Fp >230°F   60 MHz: *1*, 626D   170.15
Itaconyl chloride, tech., 90%   FW 166.99   FT-IR: *1*, 744A   167.70
                                         bp 89°C (17 mm)   138.13
                                         d 1.407   136.44
CDCl₃   QE-300   n²⁰_D 1.4930   48.73

C

Aldrich 11,415-4   CAS [4582-21-2]   C₉H₈Cl₂O₂   Fp >230°F   60 MHz: *1*, 628B   174.47   59.22*
***trans*-3,6-Endomethylene-1,2,3,6-tetrahydro-**   FW 219.07   FT-IR: *1*, 745C   172.99   48.22*
phthaloyl chloride, 97%   bp 116°C (11 mm)   137.62*   47.27*
                                         d 1.349   135.19*   46.71
CDCl₃   QE-300   n²⁰_D 1.5170   59.79*

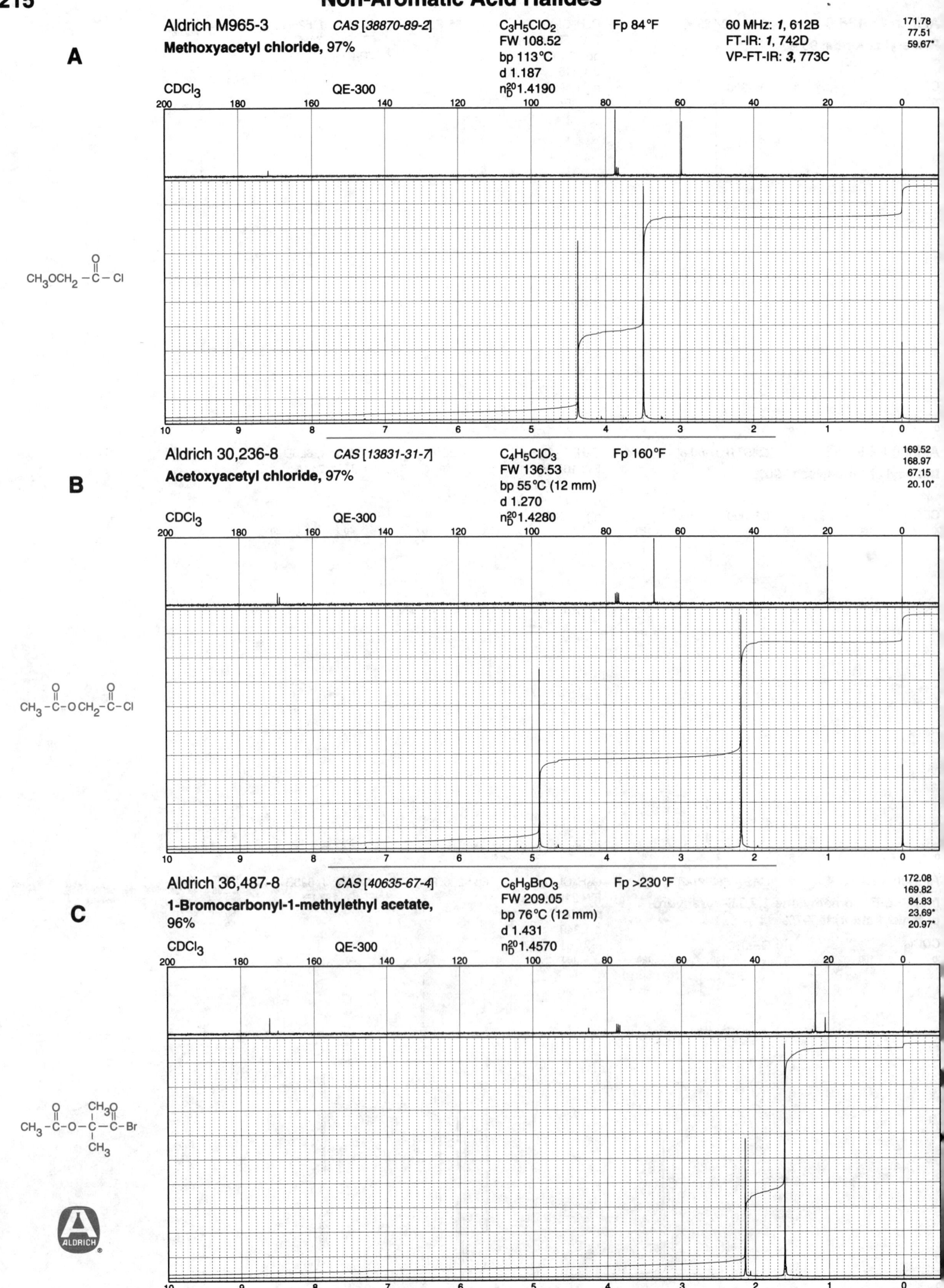

A

Aldrich M965-3   *CAS [38870-89-2]*   C₃H₅ClO₂   Fp 84°F   60 MHz: *1*, 612B   171.78
Methoxyacetyl chloride, 97%   FW 108.52   FT-IR: *1*, 742D   77.51
bp 113°C   VP-FT-IR: *3*, 773C   59.67*
d 1.187
CDCl₃   QE-300   n²⁰_D 1.4190

CH₃OCH₂—C(=O)—Cl

B

Aldrich 30,236-8   *CAS [13831-31-7]*   C₄H₅ClO₃   Fp 160°F   169.52
Acetoxyacetyl chloride, 97%   FW 136.53   168.97
bp 55°C (12 mm)   67.15
d 1.270   20.10*
CDCl₃   QE-300   n²⁰_D 1.4280

CH₃—C(=O)—O—CH₂—C(=O)—Cl

C

Aldrich 36,487-8   *CAS [40635-67-4]*   C₆H₉BrO₃   Fp >230°F   172.08
1-Bromocarbonyl-1-methylethyl acetate,   FW 209.05   169.82
96%   bp 76°C (12 mm)   84.83
d 1.431   23.69*
CDCl₃   QE-300   n²⁰_D 1.4570   20.97*

CH₃—C(=O)—O—C(CH₃)(CH₃)—C(=O)—Br

Aldrich 32,617-8   CAS [40635-66-3] $C_6H_9ClO_3$ Fp 155°F   174.82
                                        FW 164.59   169.87
1-Chlorocarbonyl-1-methylethyl acetate,   bp 56°C (6 mm)   83.09
95%   d 1.136   24.01*
                                        n_D^{20} 1.4280   20.82*

A

CDCl$_3$ QE-300

Aldrich 37,815-1   CAS [21062-20-4] $C_4H_4Cl_2O_3$ Fp >230°F   170.95
                                        FW 170.98   75.07
Diglycolyl chloride, 95%   bp 86°C (2 mm)
                                        d 1.439
                                        n_D^{20} 1.4730

B

CDCl$_3$ QE-300

Aldrich 22,617-3   CAS [39637-74-6] $C_{10}H_{13}ClO_3$ 60 MHz: 1, 628C   176.50   31.48
                                        FW 216.67   FT-IR: 1, 745D   170.88   28.79
(1S)-(-)-Camphanic chloride, 98%   mp 72°C   94.86   16.69*
                                                                                  55.54   16.61*
                                                                                  55.41   9.61*

C

CDCl$_3$ QE-300

Non-Aromatic Acid Halides

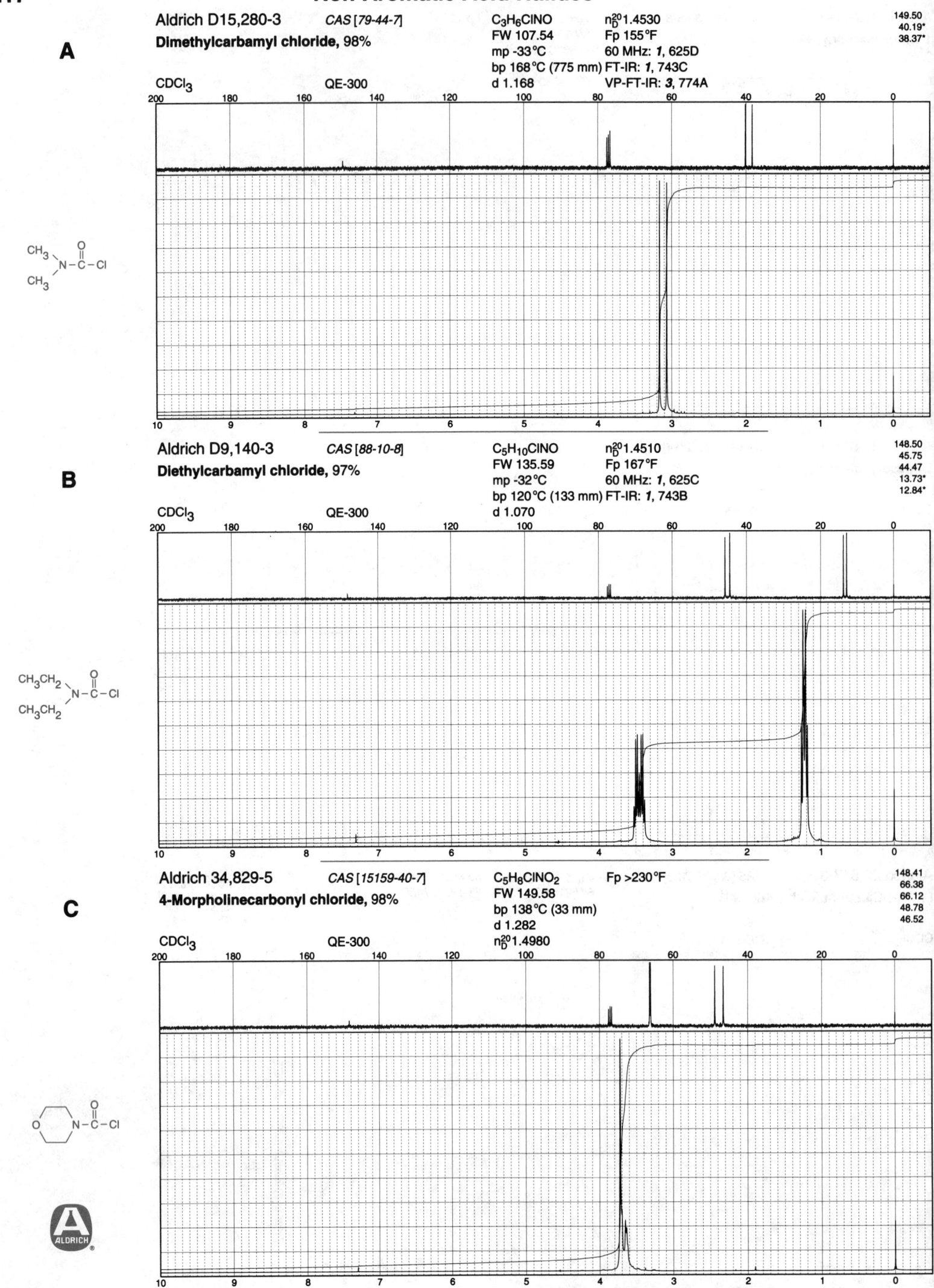

A

Aldrich D15,280-3   CAS [79-44-7]

Dimethylcarbamyl chloride, 98%

C_3H_6ClNO
FW 107.54
mp -33°C
bp 168°C (775 mm)
d 1.168

n_D^{20} 1.4530
Fp 155°F
60 MHz: *1*, 625D
FT-IR: *1*, 743C
VP-FT-IR: *3*, 774A

149.50
40.19*
38.37*

CDCl$_3$ QE-300

B

Aldrich D9,140-3   CAS [88-10-8]

Diethylcarbamyl chloride, 97%

$C_5H_{10}ClNO$
FW 135.59
mp -32°C
bp 120°C (133 mm)
d 1.070

n_D^{20} 1.4510
Fp 167°F
60 MHz: *1*, 625C
FT-IR: *1*, 743B

148.50
45.75
44.47
13.73*
12.84*

CDCl$_3$ QE-300

C

Aldrich 34,829-5   CAS [15159-40-7]

4-Morpholinecarbonyl chloride, 98%

$C_5H_8ClNO_2$
FW 149.58
bp 138°C (33 mm)
d 1.282
n_D^{20} 1.4980

Fp >230°F

148.41
66.38
66.12
48.78
46.52

CDCl$_3$ QE-300

ALDRICH

Aldrich 13,589-5   *CAS [16420-13-6]*

Dimethylthiocarbamoyl chloride, 97%

C₃H₆ClNS
FW 123.61
mp 43°C
bp 93°C
Fp 209°F

60 MHz: *1*, 626A
FT-IR: *1*, 743D

175.17
45.64*
45.00*

A

CDCl₃   QE-300

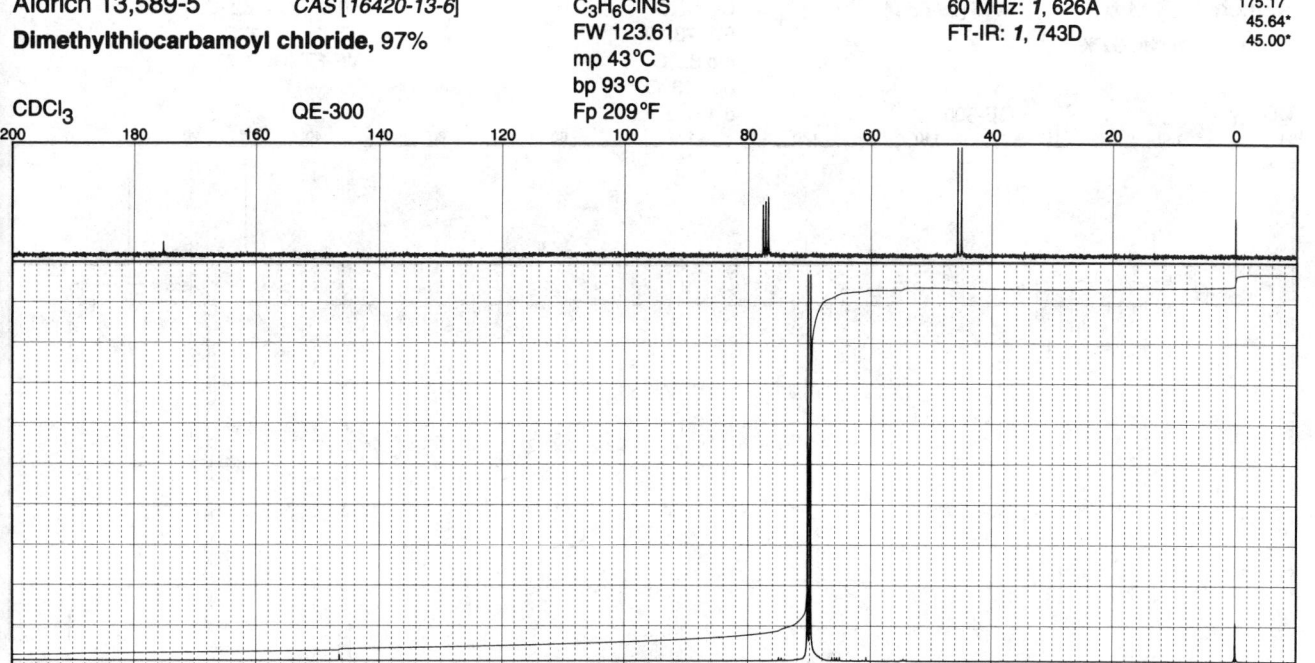

Non-Aromatic Amides

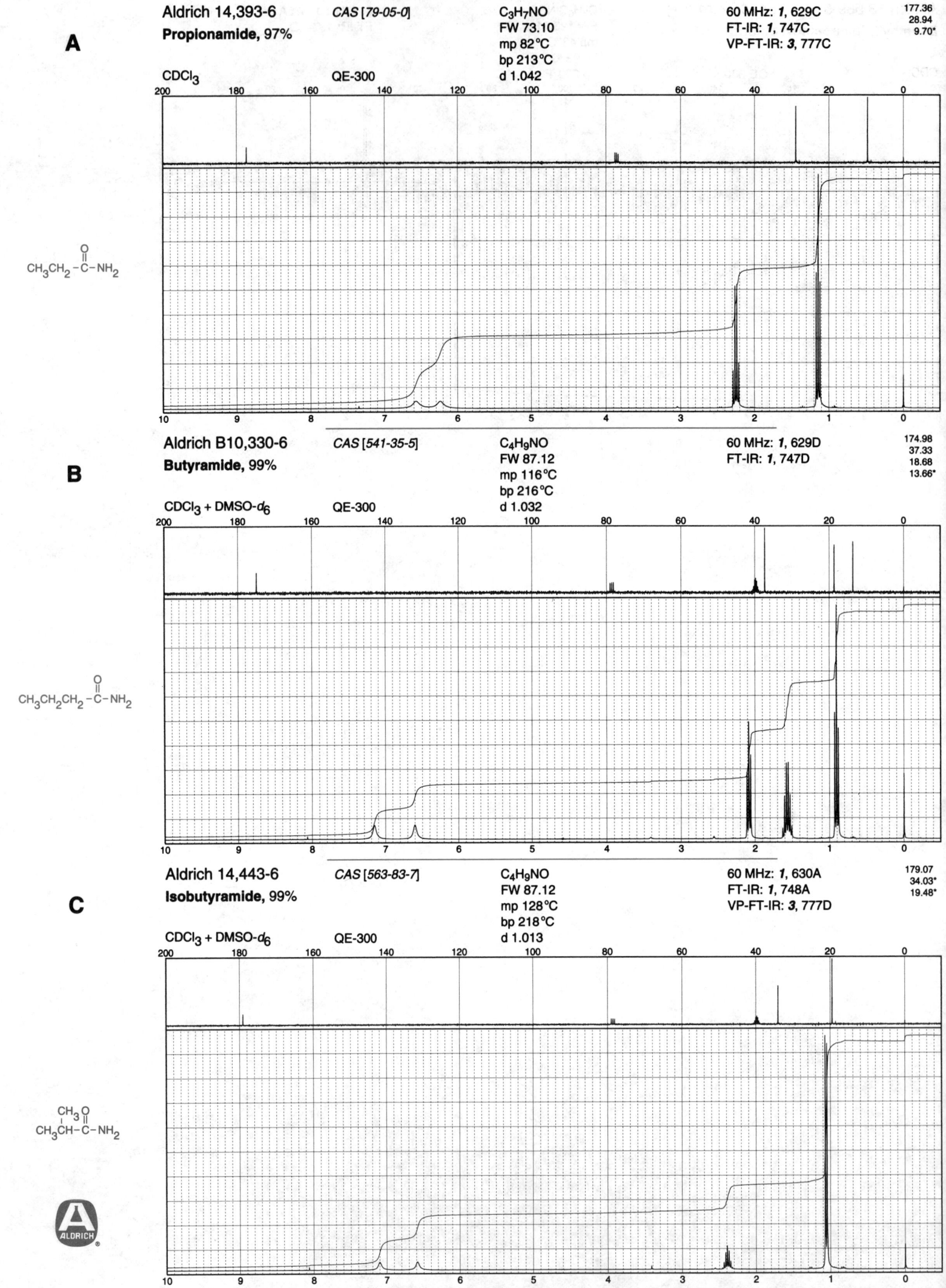

A

Aldrich 14,393-6
Propionamide, 97%

CAS [79-05-0]

C₃H₇NO
FW 73.10
mp 82°C
bp 213°C
d 1.042

60 MHz: *1*, 629C
FT-IR: *1*, 747C
VP-FT-IR: *3*, 777C

177.36
28.94
9.70*

CDCl₃   QE-300

CH₃CH₂–C–NH₂

B

Aldrich B10,330-6
Butyramide, 99%

CAS [541-35-5]

C₄H₉NO
FW 87.12
mp 116°C
bp 216°C
d 1.032

60 MHz: *1*, 629D
FT-IR: *1*, 747D

174.98
37.33
18.68
13.66*

CDCl₃ + DMSO-d₆   QE-300

CH₃CH₂CH₂–C–NH₂

C

Aldrich 14,443-6
Isobutyramide, 99%

CAS [563-83-7]

C₄H₉NO
FW 87.12
mp 128°C
bp 218°C
d 1.013

60 MHz: *1*, 630A
FT-IR: *1*, 748A
VP-FT-IR: *3*, 777D

179.07
34.03*
19.48*

CDCl₃ + DMSO-d₆   QE-300

CH₃CH–C–NH₂
CH₃

Aldrich T7,160-9   *CAS [754-10-9]*   C₅H₁₁NO   60 MHz: *1*, 630B

Trimethylacetamide, 98%   FW 101.15   FT-IR: *1*, 748B

mp 156°C
bp 212°C

$C_5H_{11}NO$

FW 101.15

180.52
38.00
27.48*

A

CDCl₃ + DMSO-*d₆*   QE-300

H₃C   O
| ‖
CH₃–C–C–NH₂
|
H₃C

Aldrich 29,339-3   *CAS [628-02-4]*   C₆H₁₃NO   VP-FT-IR: *3*, 778A

Hexanoamide, 98%   FW 115.18

mp 101°C

176.40
35.97
31.42
25.28
22.39
13.90*

B

CDCl₃   QE-300

CH₃(CH₂)₃CH₂–C–NH₂
            ‖
            O

Aldrich 22,729-3   *CAS [1122-56-1]*   C₇H₁₃NO   FT-IR: *1*, 748D

Cyclohexanecarboxamide, 97%   FW 127.19

mp 187°C

178.02
43.94*
29.27
25.59
25.44

C

CDCl₃ + DMSO-*d₆*   QE-300

O
‖
C–NH₂

ALDRICH®

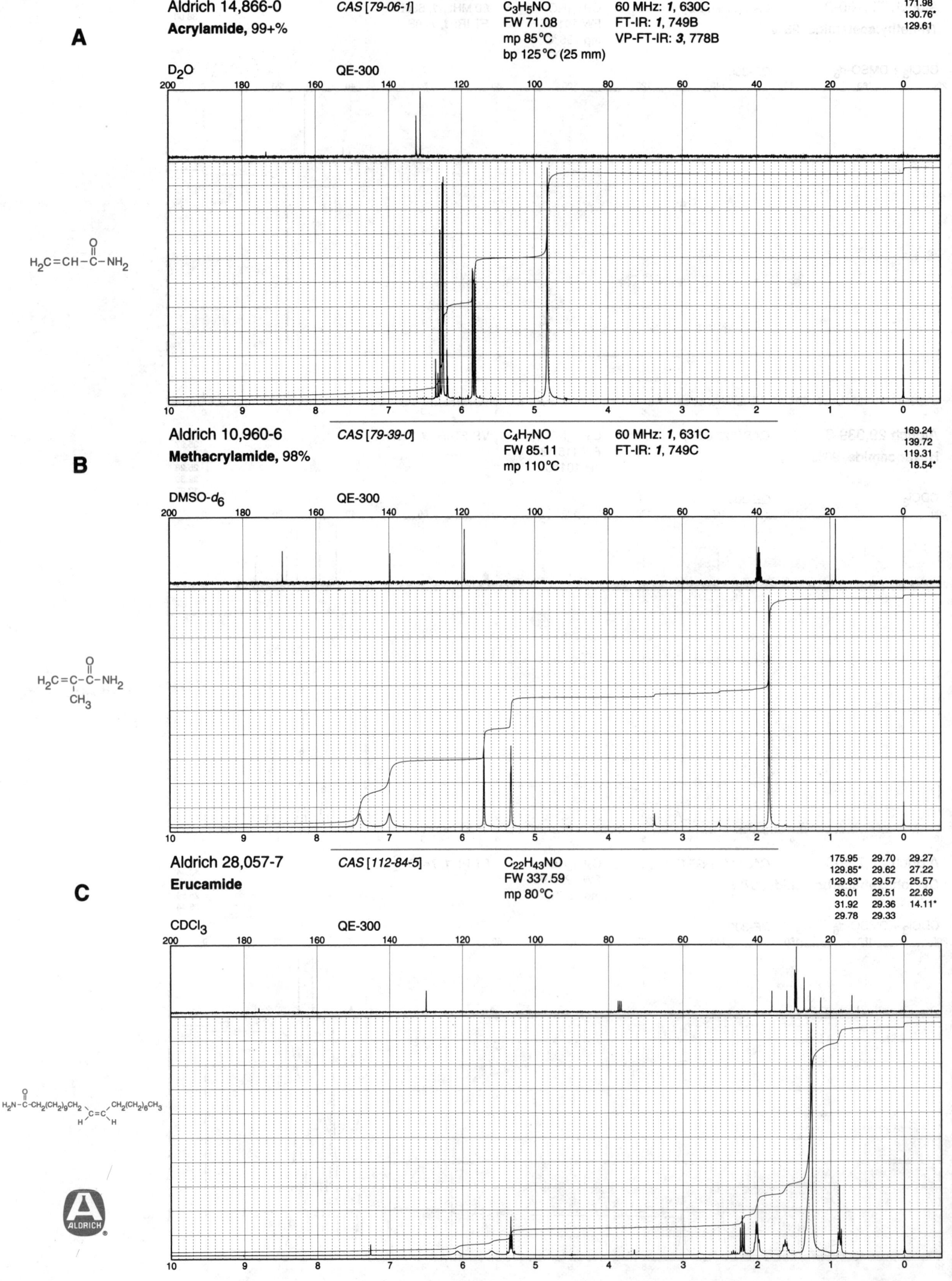

A

Aldrich 14,866-0

CAS [79-06-1]

Acrylamide, 99+%

C₃H₅NO
C_3H_5NO
FW 71.08
mp 85°C
bp 125°C (25 mm)

60 MHz: *1*, 630C
FT-IR: *1*, 749B
VP-FT-IR: *3*, 778B

171.98
130.76*
129.61

D₂O   QE-300

H₂C=CH−C−NH₂

B

Aldrich 10,960-6

CAS [79-39-0]

Methacrylamide, 98%

C_4H_7NO
FW 85.11
mp 110°C

60 MHz: *1*, 631C
FT-IR: *1*, 749C

169.24
139.72
119.31
18.54*

DMSO-*d₆*   QE-300

H₂C=C−C−NH₂
        |
        CH₃

C

Aldrich 28,057-7

CAS [112-84-5]

Erucamide

$C_{22}H_{43}NO$
FW 337.59
mp 80°C

175.95	29.70	29.27
129.85*	29.62	27.22
129.83*	29.57	25.57
36.01	29.57	22.69
31.92	29.51	14.11*
29.78	29.36	
	29.33	

CDCl₃   QE-300

H₂N−C−CH₂(CH₂)₉CH₂   CH₂(CH₂)₆CH₃
                      \\   /
                       C=C
                      /   \\
                     H   H

ALDRICH

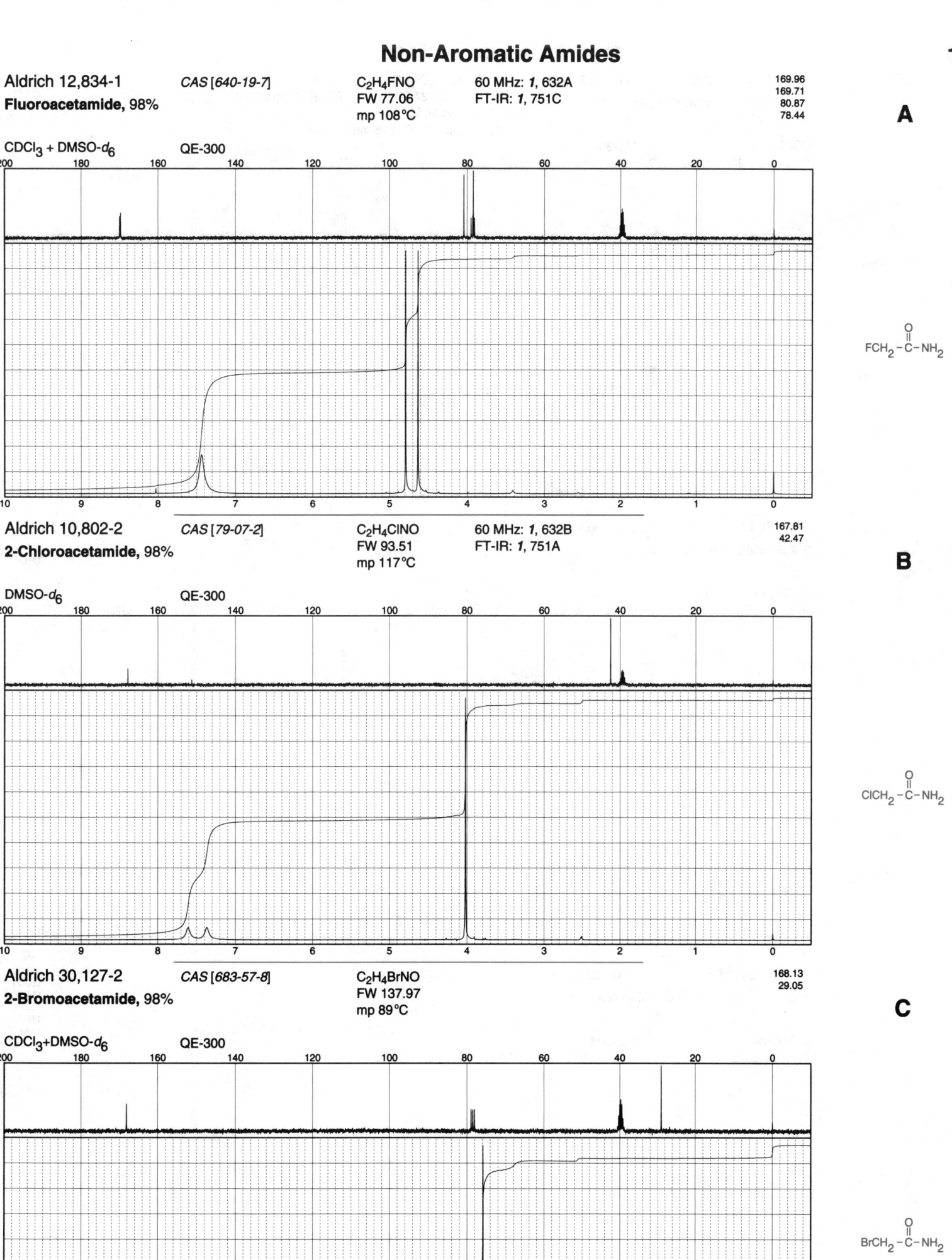

Aldrich 12,834-1   CAS [640-19-7]   C₂H₄FNO   60 MHz: *1*, 632A
Fluoroacetamide, 98%   FW 77.06   FT-IR: *1*, 751C   mp 108°C

C_2H_4FNO

FW 77.06

mp 108°C

60 MHz: *1*, 632A

FT-IR: *1*, 751C

169.96
169.71
80.87
78.44

CDCl₃ + DMSO-*d*₆   QE-300

$FCH_2-\overset{O}{\underset{||}{C}}-NH_2$

Aldrich 10,802-2   CAS [79-07-2]   C₂H₄ClNO   60 MHz: *1*, 632B
2-Chloroacetamide, 98%   FW 93.51   FT-IR: *1*, 751A   mp 117°C

167.81
42.47

B

DMSO-*d*₆   QE-300

$ClCH_2-\overset{O}{\underset{||}{C}}-NH_2$

Aldrich 30,127-2   CAS [683-57-8]   C₂H₄BrNO
2-Bromoacetamide, 98%   FW 137.97   mp 89°C

168.13
29.05

C

CDCl₃+DMSO-*d*₆   QE-300

$BrCH_2-\overset{O}{\underset{||}{C}}-NH_2$

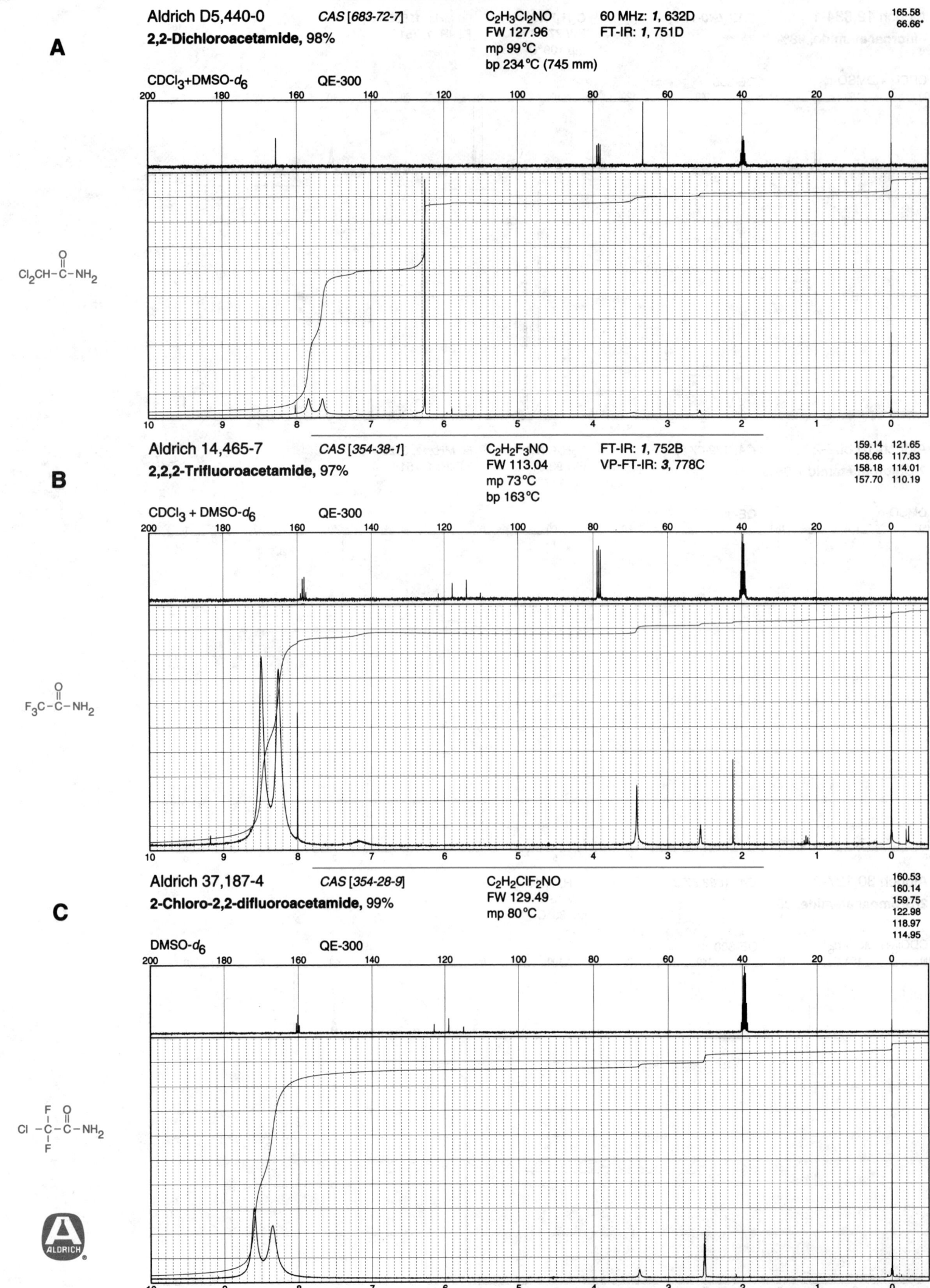

A

Aldrich D5,440-0   *CAS [683-72-7]*   C₂H₃Cl₂NO   60 MHz: *1*, 632D

2,2-Dichloroacetamide, 98%   FW 127.96   FT-IR: *1*, 751D

mp 99°C

bp 234°C (745 mm)

165.58
66.66*

CDCl₃+DMSO-*d₆*   QE-300

Cl₂CH–C–NH₂

B

Aldrich 14,465-7   *CAS [354-38-1]*   C₂H₂F₃NO   FT-IR: *1*, 752B

2,2,2-Trifluoroacetamide, 97%   FW 113.04   VP-FT-IR: *3*, 778C

mp 73°C

bp 163°C

159.14 121.65
158.66 117.83
158.18 114.01
157.70 110.19

CDCl₃ + DMSO-*d₆*   QE-300

F₃C–C–NH₂

C

Aldrich 37,187-4   *CAS [354-28-9]*   C₂H₂ClF₂NO

2-Chloro-2,2-difluoroacetamide, 99%   FW 129.49

mp 80°C

160.53
160.14
159.75
122.98
118.97
114.95

DMSO-*d₆*   QE-300

Cl –C–C–NH₂

ALDRICH

Non-Aromatic Amides

Aldrich 21,734-4   CAS [594-65-0]   C₂H₂Cl₃NO   FT-IR: *1*, 752A   163.07 / 93.04

2,2,2-Trichloroacetamide, 99%   FW 162.40   mp 142°C   bp 239°C

A

CDCl₃ + DMSO-*d₆*   QE-300

$Cl_3C-\overset{\displaystyle O}{\overset{\displaystyle \|}{C}}-NH_2$

Aldrich 19,239-2   CAS [27816-36-0]   C₃H₆ClNO   60 MHz: *1*, 633A   173.02 / 55.05* / 22.48*

(±)-2-Chloropropionamide, 98%   FW 107.54   FT-IR: *1*, 752C   mp 79°C

B

CDCl₃   QE-300

$CH_3\overset{\displaystyle Cl}{\underset{\displaystyle }{C}}H-\overset{\displaystyle O}{\overset{\displaystyle \|}{C}}-NH_2$

Aldrich 30,859-5   CAS [5875-25-2]   C₃H₆BrNO   171.13 / 43.51* / 21.80*

(±)-2-Bromopropionamide, 99%   FW 152.00   mp 124°C

C

CDCl₃ + DMSO-*d₆*   QE-300

$CH_3\underset{\displaystyle Br}{C}H-\overset{\displaystyle O}{\overset{\displaystyle \|}{C}}-NH_2$

ALDRICH

Non-Aromatic Amides

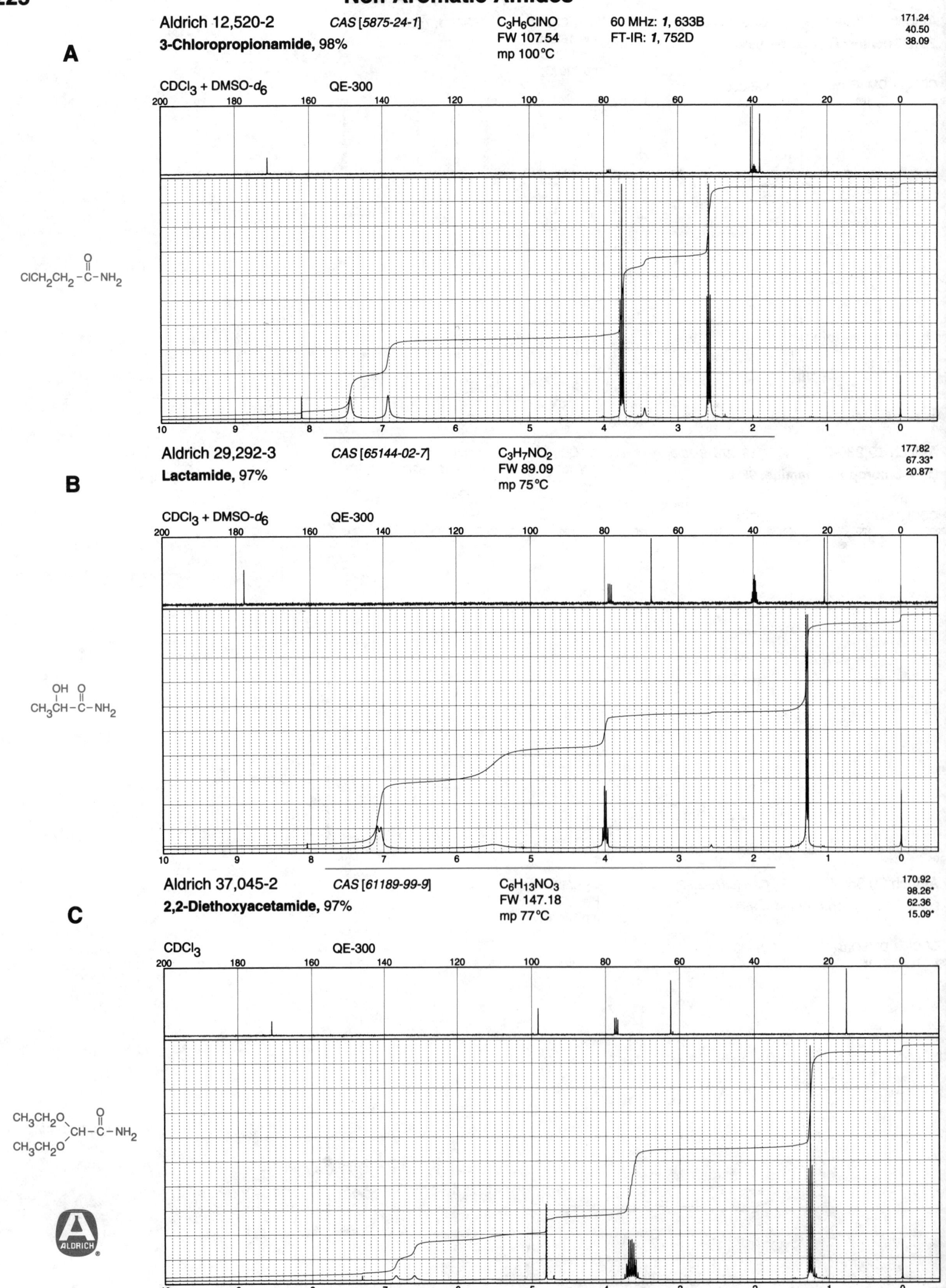

Aldrich 12,520-2   *CAS [5875-24-1]* C_3H_6ClNO 60 MHz: *1*, 633B

3-Chloropropionamide, 98%   FW 107.54   FT-IR: *1*, 752D

mp 100°C

171.24
40.50
38.09

A

$CDCl_3 + DMSO\text{-}d_6$ QE-300

$ClCH_2CH_2-\overset{\overset{\text{O}}{\|}}{C}-NH_2$

Aldrich 29,292-3   *CAS [65144-02-7]* $C_3H_7NO_2$

Lactamide, 97%   FW 89.09

mp 75°C

177.82
67.33*
20.87*

B

$CDCl_3 + DMSO\text{-}d_6$ QE-300

$CH_3\overset{\overset{\text{OH}}{|}}{C}H-\overset{\overset{\text{O}}{\|}}{C}-NH_2$

Aldrich 37,045-2   *CAS [61189-99-9]* $C_6H_{13}NO_3$

2,2-Diethoxyacetamide, 97%   FW 147.18

mp 77°C

170.92
98.26*
62.36
15.09*

C

$CDCl_3$ QE-300

$\begin{array}{l} CH_3CH_2O \\ CH_3CH_2O \end{array}\!\!\!\overset{\overset{\text{O}}{\|}}{\underset{}{CH-C-NH_2}}$

ALDRICH

Non-Aromatic Amides

Aldrich T2,880-0   *CAS [3206-73-3]*

(±)-6-Thioctic amide, 98%

$C_8H_{15}NOS_2$
FW 205.34
mp 128°C

60 MHz: *1*, 634B
FT-IR: *1*, 753B

173.96	34.80
56.02*	34.05
39.80	28.28
37.98	24.76

A

DMSO-*d*$_6$ QE-300

Aldrich 28,642-7   *CAS [10466-61-2]*

L-Leucinamide hydrochloride, 99%

$C_6H_{14}N_2O$
FW 166.65
mp 255°C

170.96
51.29*
40.11
23.85*
22.49*
22.26*

B

CDCl$_3$ + DMSO-*d*$_6$ QE-300

Aldrich 37,579-9   *CAS [65414-74-6]*

L-Serinamide hydrochloride, 98%

$C_3H_8N_2O_2$
FW 140.57
mp 187°C

168.40
60.02
54.15*

C

DMSO-*d*$_6$ QE-300

Non-Aromatic Amides

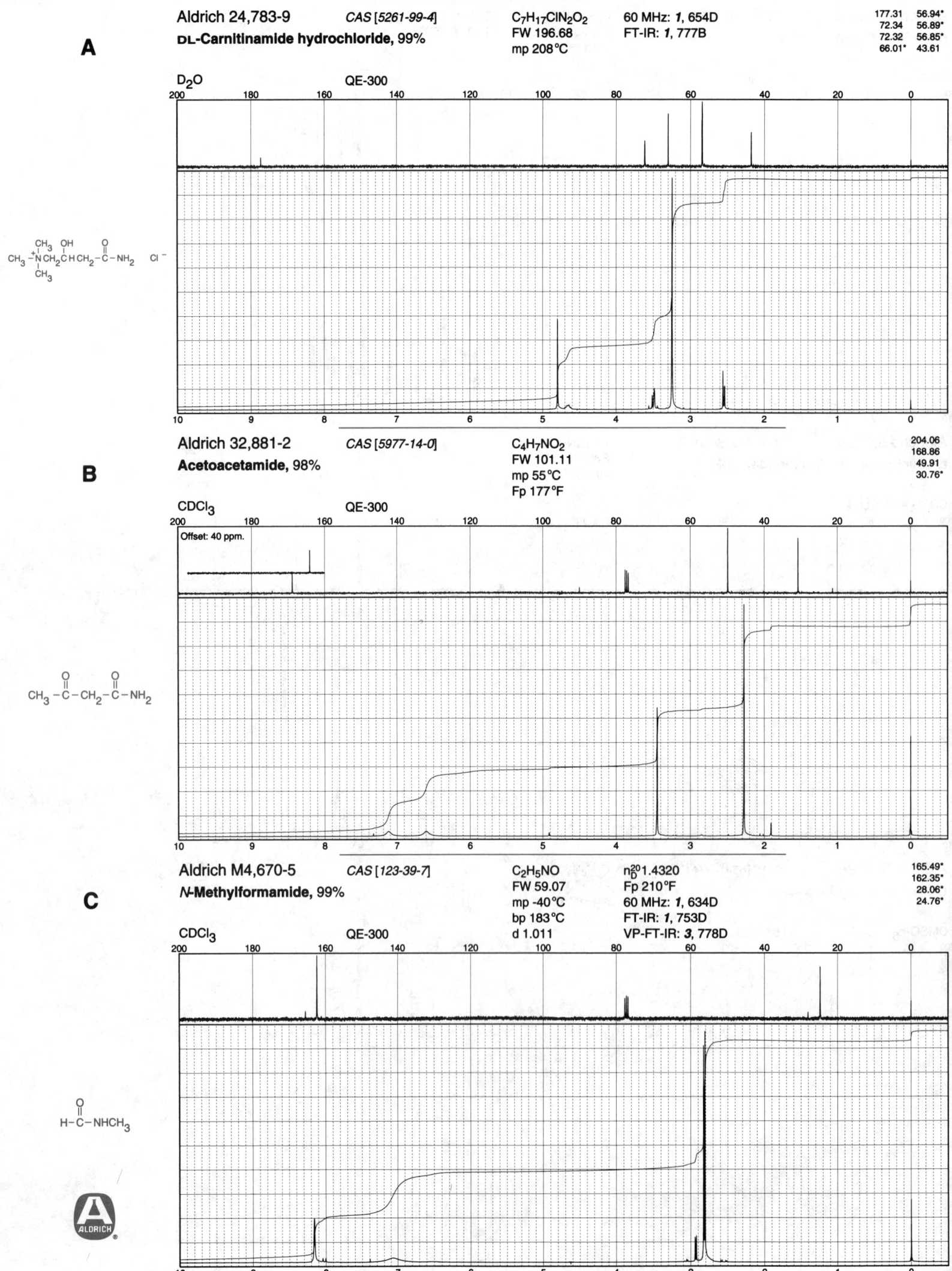

A

Aldrich 24,783-9   *CAS [5261-99-4]*   C₇H₁₇ClN₂O₂   60 MHz: *1*, 654D
DL-Carnitinamide hydrochloride, 99%   FW 196.68   FT-IR: *1*, 777B
mp 208°C

177.31 56.94*
72.34   56.89*
72.32   56.85*
66.01* 43.61

D₂O   QE-300

B

Aldrich 32,881-2   *CAS [5977-14-0]*   C₄H₇NO₂
Acetoacetamide, 98%   FW 101.11
mp 55°C
Fp 177°F

204.06
168.86
49.91
30.76*

CDCl₃   QE-300
Offset: 40 ppm.

C

Aldrich M4,670-5   *CAS [123-39-7]*   C₂H₅NO   n²⁰_D 1.4320
***N*-Methylformamide**, 99%   FW 59.07   Fp 210°F
mp -40°C   60 MHz: *1*, 634D
bp 183°C   FT-IR: *1*, 753D
d 1.011   VP-FT-IR: *3*, 778D

165.49*
162.35*
28.06*
24.76*

CDCl₃   QE-300

A

Aldrich 11,747-1   *CAS [2425-74-3]*

N-tert-Butylformamide, 98%

C$_5$H$_{11}$NO
FW 101.15
mp 16°C
bp 202°C
d 0.903

n$_D^{20}$ 1.4330
Fp 203°F
FT-IR: *1*, 754A
VP-FT-IR: *3*, 779A

163.15*
160.67*
51.21
50.31
30.81*
28.90*

CDCl$_3$ QE-300

B

Aldrich 12,068-5   *CAS [766-93-8]*

N-Cyclohexylformamide, 99%

C$_7$H$_{13}$NO
FW 127.19
mp 39°C
bp 113°C (700 mm)
Fp >230°F

60 MHz: *1*, 635B
FT-IR: *1*, 754B
VP-FT-IR: *3*, 779B

163.77*   32.92
160.56*   25.43
51.20*   25.01
47.07*   24.79
34.59

CDCl$_3$ QE-300

C

Aldrich 15,973-5   *CAS [4774-33-8]*

N,N',N''-Methylidynetrisformamide, 97%

C$_4$H$_7$N$_3$O$_3$
FW 145.12
mp 171°C d.

60 MHz: *1*, 635A
FT-IR: *1*, 754C

163.47*
160.47*
160.28*
64.42*
60.45*
56.36*

DMSO-d_6 QE-300

Non-Aromatic Amides

A

Aldrich M2,630-5   *CAS [79-16-3]*   C₃H₇NO n_D^{20}1.4330

N-Methylacetamide, 99+%

C_3H_7NO
FW 73.10
mp 27°C
bp 205°C
d 0.957

Fp 227°F
60 MHz: *1*, 635C
FT-IR: *1*, 754D
VP-FT-IR: *3*, 779C

171.27
26.26*
22.87*

CDCl₃   QE-300

$$CH_3-\overset{O}{\overset{\|}{C}}-NHCH_3$$

B

Aldrich 14,740-0   *CAS [625-50-3]*   C₄H₉NO   Fp 224°F   VP-FT-IR: *3*, 779D

N-Ethylacetamide, 99%

C_4H_9NO
FW 87.12
bp 91°C (8 mm)
d 0.924
n_D^{20}1.4330

170.35
34.41
23.13*
14.71*

CDCl₃   QE-300

$$CH_3-\overset{O}{\overset{\|}{C}}-NHCH_2CH_3$$

C

Aldrich 13,710-3   *CAS [880-52-4]*   C₁₂H₁₉NO   60 MHz: *1*, 636C

N-(1-Adamantyl)acetamide, 99%

$C_{12}H_{19}NO$
FW 193.29
mp 149°C

168.83
50.81
41.14
36.23
29.04*
23.90*

CDCl₃+DMSO-*d₆*   QE-300

Aldrich 27,796-7   CAS [26389-62-8]   C₇H₁₃NO   VP-FT-IR: 3, 780A

N-Isopropylcyclopropanecarboxamide, 98%

FW 127.19
mp 92°C

$C_7H_{13}NO$

172.63
41.31*
22.81*
14.65*
6.80

A

CDCl₃   QE-300

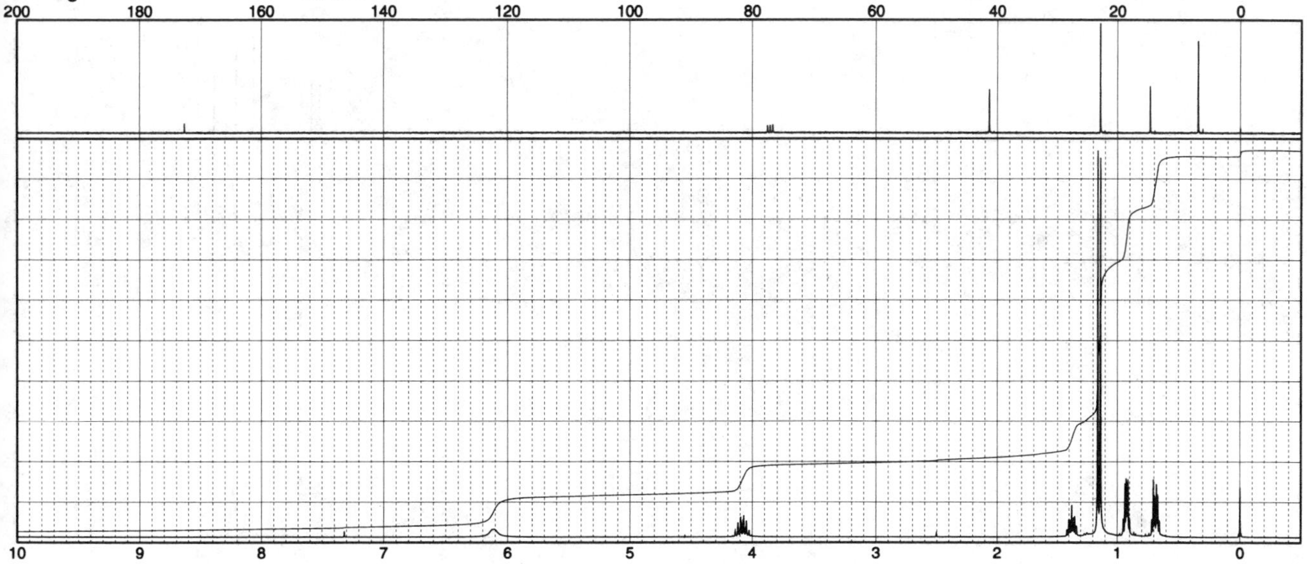

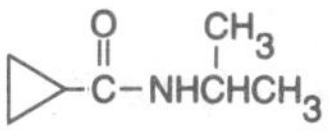

Aldrich 30,321-6   CAS [55265-53-7] $C_8H_{15}NO$

2,2,3,3-Tetramethylcyclopropanecarboxamide, 98%

FW 141.22
mp 93°C

174.54
36.83*
28.55
23.67*
16.76*

B

CDCl₃   QE-300

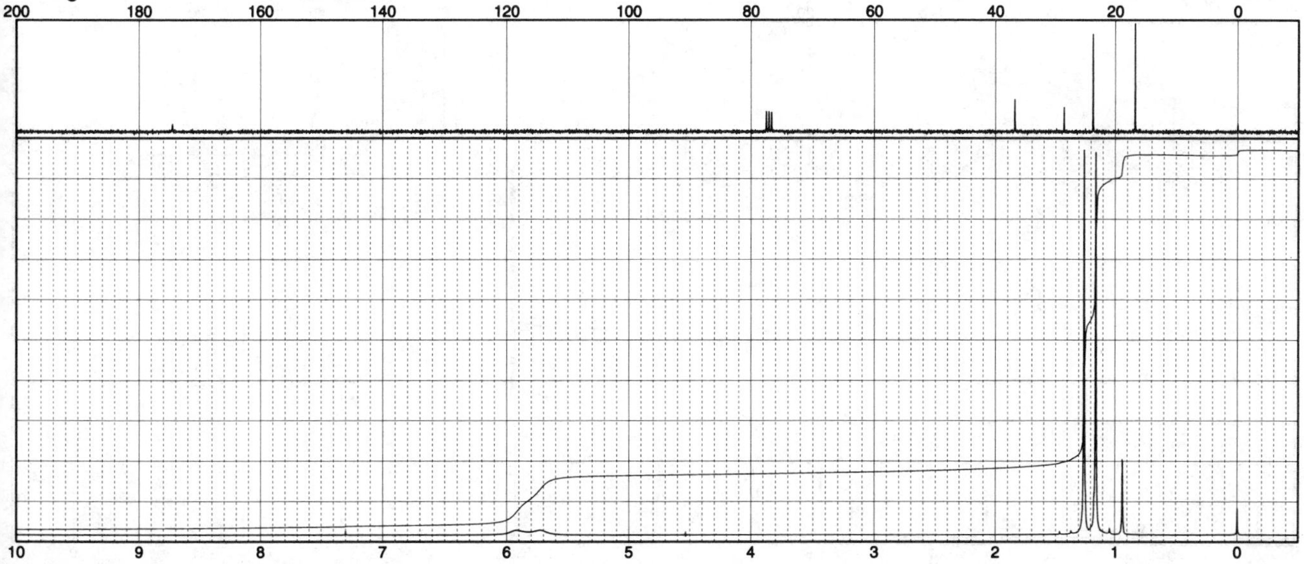

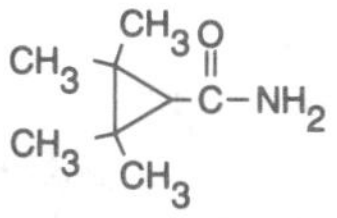

Aldrich 14,832-6   CAS [110-26-9] $C_7H_{10}N_2O_2$ 60 MHz: 1, 645A

N,N'-Methylenebisacrylamide, 99+%

FW 154.17
mp 300°C

FT-IR: 1, 749D

164.79
131.28*
125.76
43.27

C

DMSO-d_6 QE-300

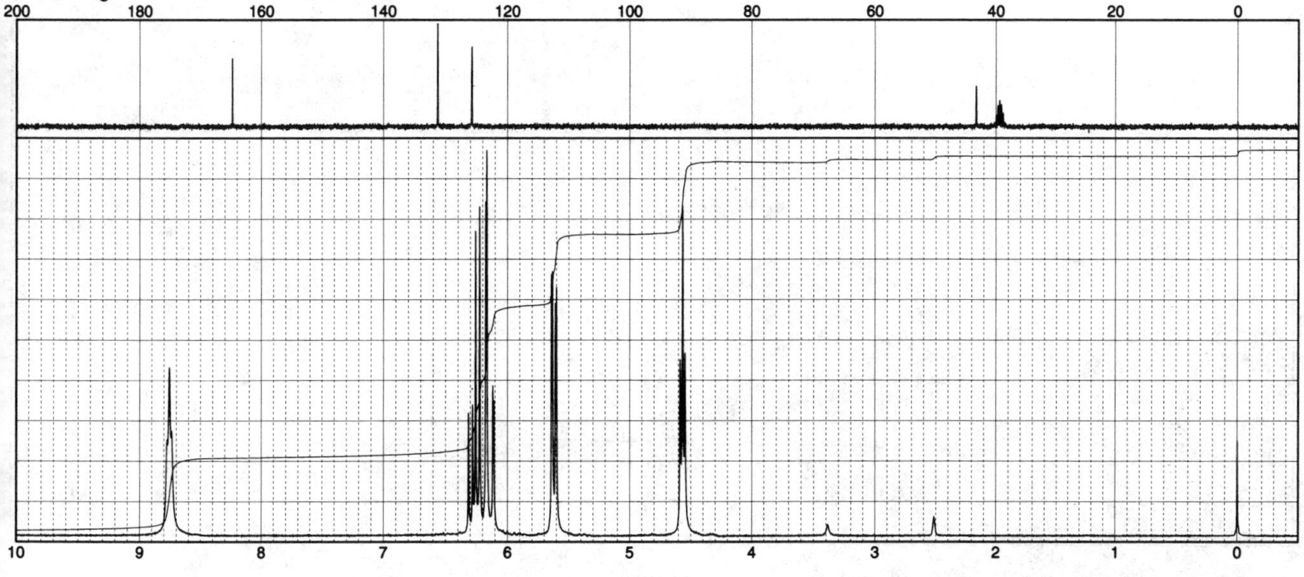

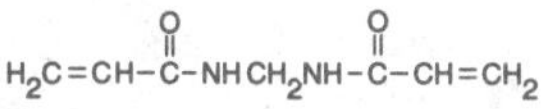

A

Aldrich 22,423-5   *CAS [3073-59-4]* $C_{10}H_{20}N_2O_2$ 60 MHz: *1*, 637C

N,N'-Hexamethylenebisacetamide, 98%   FW 200.28   FT-IR: *1*, 755A

mp 129°C

169.30
38.69
29.16
26.26
22.64*

CDCl₃+DMSO-d_6 QE-300

$CH_3-\overset{O}{\overset{\|}{C}}-NHCH_2(CH_2)_4CH_2NH-\overset{O}{\overset{\|}{C}}-CH_3$

B

Aldrich 36,495-9   *CAS [13880-05-2]* $C_7H_{13}NO_4$

N-[Tris(hydroxymethyl)methyl]acrylamide, 93%   FW 175.18

mp 140°C

165.37
132.16*
125.00
62.42
60.51

DMSO-d_6 QE-300

$H_2C=CH\overset{O}{\overset{\|}{C}}-NH-\overset{CH_2OH}{\underset{CH_2OH}{\overset{|}{\underset{|}{C}}}}-CH_2OH$

C

Aldrich 24,209-8   *CAS [815-06-5]* $C_3H_4F_3NO$ 60 MHz: *1*, 636A

N-Methyl-2,2,2-trifluoroacetamide, 98%   FW 127.07   FT-IR: *1*, 755C

mp 52°C   VP-FT-IR: *3*, 780B

bp 157°C

Fp 165°F

158.90   117.88
158.41   114.08
157.92   110.28
157.43   26.39*
121.68

CDCl₃   QE-300

$F_3C-\overset{O}{\overset{\|}{C}}-NHCH_3$

ALDRICH

A

Aldrich 22,639-4 CAS [685-27-8] C$_5$H$_3$F$_6$NO$_2$ Fp 108°F FT-IR: **1**, 755D

N-Methylbis(trifluoroacetamide), 98% FW 223.08 VP-FT-IR: **3**, 780C

bp 121°C
d 1.547
n$_D^{20}$1.3460

CDCl$_3$ QE-300

160.26 117.14
159.71 113.34
159.16 109.54
158.61 32.09*
120.95

F$_3$C–C(=O)–N(CH$_3$)–C(=O)–CF$_3$

B

Aldrich 14,750-8 CAS [1477-57-2] C$_{12}$H$_{20}$Cl$_4$N$_2$O$_2$ 60 MHz: **1**, 633C

ertilysin™, 97% FW 366.12 FT-IR: **1**, 753A

mp 126°C

DMSO-d$_6$ QE-300

164.27
67.87*
40.22
29.45
29.39
27.03

Cl$_2$CH–C(=O)–NHCH$_2$(CH$_2$)$_6$CH$_2$NH–C(=O)–CHCl$_2$

C

Aldrich D595-0 CAS [625-77-4] C$_4$H$_7$NO$_2$ 60 MHz: **1**, 636B

acetamide, 97% FW 101.11 FT-IR: **1**, 756A

mp 76°C
bp 223°C

CDCl$_3$ QE-300

172.24
24.87*

CH$_3$–C(=O)–NH–C(=O)–CH$_3$

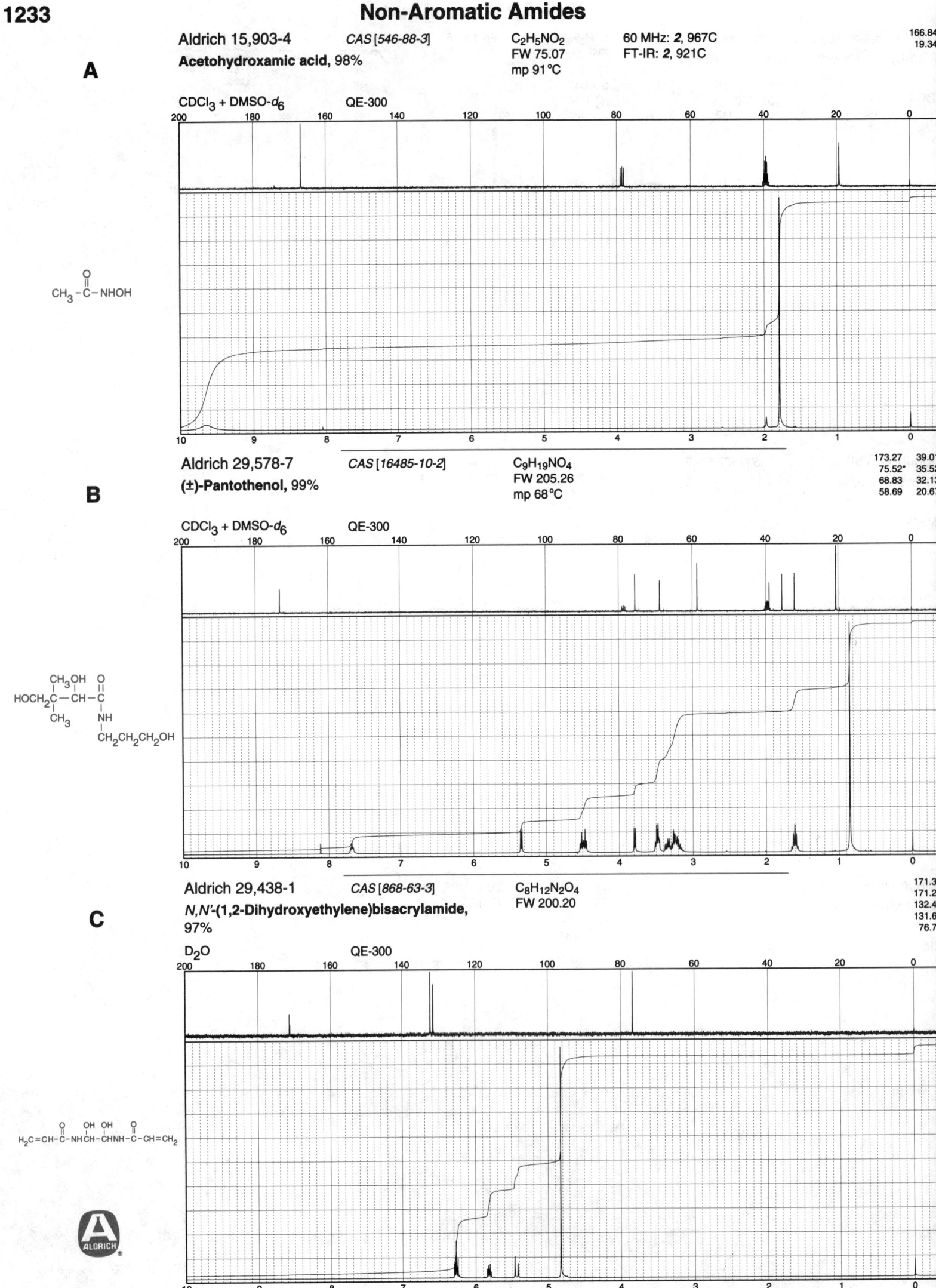

A

Aldrich 15,903-4   *CAS [546-88-3]* $C_2H_5NO_2$ 60 MHz: **2**, 967C   166.84

Acetohydroxamic acid, 98%   FW 75.07   FT-IR: **2**, 921C   19.34

mp 91°C

$CDCl_3$ + DMSO-d_6 QE-300

B

Aldrich 29,578-7   *CAS [16485-10-2]* $C_9H_{19}NO_4$ 173.27   39.01

(±)-Pantothenol, 99%   FW 205.26   75.52*   35.52

mp 68°C   68.83   32.13

58.69   20.67

$CDCl_3$ + DMSO-d_6 QE-300

C

Aldrich 29,438-1   *CAS [868-63-3]* $C_8H_{12}N_2O_4$ 171.3

***N,N'*-(1,2-Dihydroxyethylene)bisacrylamide,**   FW 200.20   171.2

97%   132.4

131.6

76.7

D_2O QE-300

Aldrich 15,686-8   CAS [58477-85-3]

(+)-N,N'-Diallyltartardiamide, 99+%

C₁₀H₁₆N₂O₄
FW 228.25
mp 187°C

60 MHz: 1, 631D
FT-IR: 1, 750B

172.36
135.72*
115.46
73.15*
41.28

A

DMSO-d₆   QE-300

Aldrich 24,409-0   CAS [1001-53-2]

N-Acetylethylenediamine, 96%

C₄H₁₀N₂O
FW 102.14
mp 51°C
bp 128°C (3 mm)
Fp >230°F

60 MHz: 1, 637B
FT-IR: 1, 756C
VP-FT-IR: 3, 780D

170.65
42.45
41.48
23.13*

B

CDCl₃   QE-300

$CH_3-C(=O)-NHCH_2CH_2NH_2$

Aldrich 19,131-0   CAS [7355-58-0]

N-(2-Chloroethyl)acetamide

C₄H₈ClNO
FW 121.57
d 1.204
n²⁰_D 1.4790
Fp >230°F

60 MHz: 1, 636D
FT-IR: 1, 756D

171.48
43.29
41.61
22.69*

C

CDCl₃   QE-300

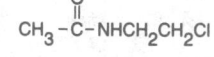

$CH_3-C(=O)-NHCH_2CH_2Cl$

ALDRICH

A

Aldrich 29,087-4   CAS [1187-58-2]   C$_4$H$_9$NO   n$_D^{20}$1.4360   174.92
                                        FW 87.12   Fp 222°F   29.50
N-Methylpropionamide, 98%   mp -43°C   VP-FT-IR: 3, 781A 26.19*
                                        bp 146°C (90 mm)   9.99*
                                        d 0.931

CDCl$_3$ QE-300

CH$_3$CH$_2$-C(=O)-NHCH$_3$

B

Aldrich 36,148-8   CAS [81778-06-5]   C$_5$H$_{10}$ClNO$_2$ 177.45
                                        FW 151.59   55.01
3-Chloro-N-hydroxy-2,2-dimethylpropionamide, mp 143°C d.   46.01
97%   24.97*

D$_2$O   QE-300

ClCH$_2$-C(CH$_3$)(CH$_3$)-C(=O)-NHOH

C

Aldrich 32,470-1   CAS [42222-06-0]   C$_8$H$_{15}$NO$_2$ 204.86
                                        FW 157.21   164.59
N-tert-Butylacetoacetamide, 99%   mp 50°C   51.29
                                        bp 73°C   51.01
                                        Fp >230°F   30.90*
                                                               28.63*

CDCl$_3$ QE-300

Offset: 40 ppm.

CH$_3$-C(=O)-CH$_2$-C(=O)-NH-C(CH$_3$)(CH$_3$)(CH$_3$)

A

Aldrich 22,234-8   CAS [2873-97-4]
Diacetone acrylamide, 99%

C$_9$H$_{15}$NO$_2$
FW 169.23
mp 55°C
Fp >230°F

60 MHz: *1*, 644D
FT-IR: *1*, 750C

207.92   52.29
165.17   51.10
131.91* 31.71*
125.49   27.40*

CDCl$_3$ QE-300

Offset: 40 ppm.

$H_2C=CH-C(O)-NHC(CH_3)(CH_3)CH_2-C(O)-CH_3$

B

Aldrich 36,334-0   CAS [1190-73-4]
N-Acetylcysteamine, 97%

C$_4$H$_9$NOS
FW 119.19
mp 7°C
bp 139°C (7 mm)
d 1.121

n$_D^{20}$ 1.5110
Fp >230°F

170.58
42.60
24.41
23.12*

CDCl$_3$ QE-300

$CH_3-C(O)-NHCH_2CH_2SH$

C

Aldrich 17,830-6   CAS [1420-88-8]
N,S-Diacetylcysteamine, 98%

C$_6$H$_{11}$NO$_2$S
FW 161.22
mp 33°C
bp 182°C (15 mm)
Fp >230°F

60 MHz: *1*, 638A

195.82
170.36
39.50
30.55*
28.85
23.06*

CDCl$_3$ QE-300

$CH_3-C(O)-S-CH_2CH_2-N(H)-C(O)-CH_3$

ALDRICH

A

Aldrich 28,316-9   CAS [867-81-2] $C_9H_{17}NO_5$
FW 241.22
mp 175°C

(+)-Pantothenic acid, sodium salt, 98%

182.98	39.47
177.60	38.92
78.59*	23.30*
71.18	22.00*
41.35	

D_2O QE-300

B

Aldrich 30,525-1   CAS [777-33-3] $C_6H_4F_3NO_4$
FW 211.10
mp 141°C

(S)-(-)-2-(Trifluoroacetamido)succinic anhydride, 97%

170.06	121.16
168.58	117.36
157.69	113.55
157.19	109.75
156.69	49.27*
156.19	34.27

$CDCl_3$+DMSO-d_6 QE-300

Offset: 0.3 ppm.

C

Aldrich 85,134-5   CAS [14215-68-0] $C_8H_{15}NO_6$ FT-IR: 1, 757C
FW 221.21
mp 161°C

N-Acetyl-D-galactosamine

169.57	67.34*
90.85*	60.62
70.24*	50.23*
68.22*	22.62*

DMSO-d_6 QE-300

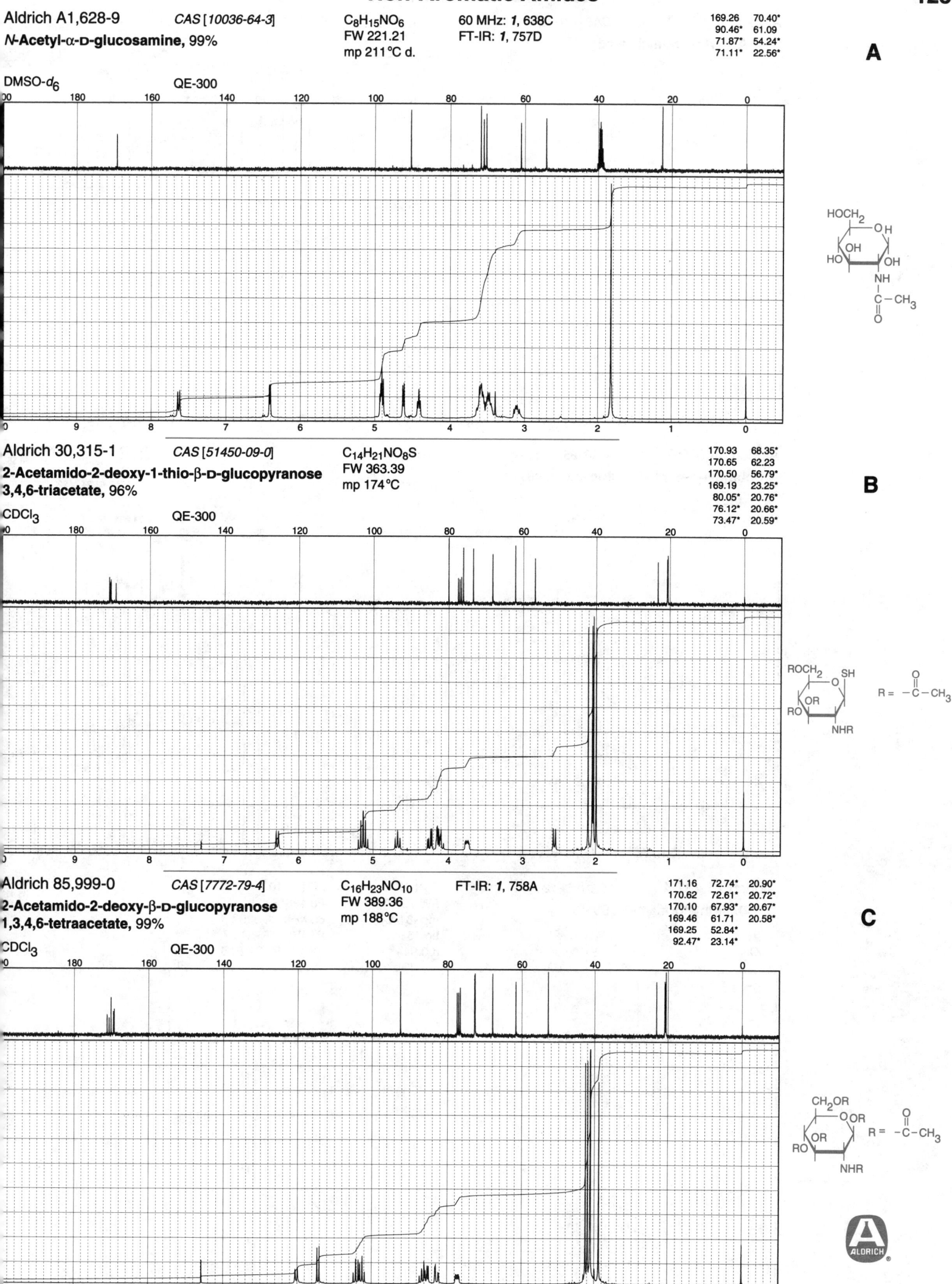

Aldrich A1,628-9   *CAS [10036-64-3]*   C₈H₁₅NO₆   60 MHz: *1*, 638C

N-Acetyl-α-ᴅ-glucosamine, 99%   FW 221.21   FT-IR: *1*, 757D
mp 211°C d.

169.26	70.40*
90.46*	61.09
71.87*	54.24*
71.11*	22.56*

A

DMSO-*d₆*   QE-300

Aldrich 30,315-1   *CAS [51450-09-0]*   C₁₄H₂₁NO₈S   FW 363.39
mp 174°C

**2-Acetamido-2-deoxy-1-thio-β-ᴅ-glucopyranose
3,4,6-triacetate, 96%**

170.93	68.35*
170.65	62.23
170.50	56.79*
169.19	23.25*
80.05*	20.76*
76.12*	20.66*
73.47*	20.59*

B

CDCl₃   QE-300

Aldrich 85,999-0   *CAS [7772-79-4]*   C₁₆H₂₃NO₁₀   FT-IR: *1*, 758A
FW 389.36
mp 188°C

**2-Acetamido-2-deoxy-β-ᴅ-glucopyranose
1,3,4,6-tetraacetate, 99%**

171.16	72.74*	20.90*
170.62	72.61*	20.72*
170.10	67.93*	20.67*
169.46	61.71	20.58*
169.25	52.84*	
92.47*	23.14*	

C

CDCl₃   QE-300

A

Aldrich 85,565-0   CAS [131-48-6]   C$_{11}$H$_{19}$NO$_9$ FT-IR: *1*, 758C

(-)-*N*-Acetylneuraminic acid   FW 309.28   mp 185°C d.

177.66	69.60*
176.32	66.06
98.23	54.98*
73.30*	41.77
73.06*	25.00*
71.16*	

D$_2$O   QE-300

B

Aldrich 34,389-7   CAS [85316-98-9]   C$_{15}$H$_{31}$NO$_6$

N-Octanoyl-*N*-methyl-D-glucamine, 98%   FW 321.42   mp 86°C

172.50	70.72*	36.60*	28.57
172.29	69.77*	33.43*	28.55
72.42*	69.25*	32.60	24.87
71.52*	63.24	32.10	24.53
71.42*	63.20	31.17	22.01
71.33*	51.81	28.80	13.87*
71.23*	50.74	28.71	

DMSO-d$_6$ QE-300

CH$_3$(CH$_2$)$_5$CH$_2$—C—NCH$_3$

C

Aldrich 15,481-4   CAS [68-12-2]   C$_3$H$_7$NO   n$_D^{20}$1.4310

N,N-Dimethylformamide, 99+%   FW 73.10   Fp 136°F

mp -61°C   60 MHz: *1*, 639A

bp 153°C   FT-IR: *1*, 758D

d 0.944   VP-FT-IR: *3*, 781B

162.42*
36.41*
31.35*

CDCl$_3$ QE-300

H—C—NCH$_3$ with CH$_3$

Aldrich 37,368-0   *CAS [29584-42-7]* $C_3H_7NO_4S$ 162.19*
35.71*
30.69*

Sulfur trioxide *N,N*-dimethylformamide complex, 97%   FW 153.16
mp 157°C

A

DMSO-d_6 QE-300

Offset: 4.2 ppm.

Aldrich 18,631-7   *CAS [617-84-5]* $C_5H_{11}NO$ Fp 141°F   60 MHz: *1*, 639B   162.08*
41.83
36.58
14.91*
12.78*

***N,N*-Diethylformamide, 99%**   FW 101.15
bp 177°C
d 0.908
n_D^{20}1.4340   FT-IR: *1*, 759A
VP-FT-IR: *3*, 781D

B

CDCl$_3$ QE-300

Aldrich 22,625-4   *CAS [2700-30-3]* $C_7H_{15}NO$ n_D^{20}1.4370   161.48*
46.59*
43.85*
23.41*
20.18*

***N,N*-Diisopropylformamide, 98%**   FW 129.20
mp 11°C
bp 196°C
d 0.890   Fp 167°F
60 MHz: *1*, 639C
FT-IR: *1*, 759B
VP-FT-IR: *3*, 782A

C

CDCl$_3$ QE-300

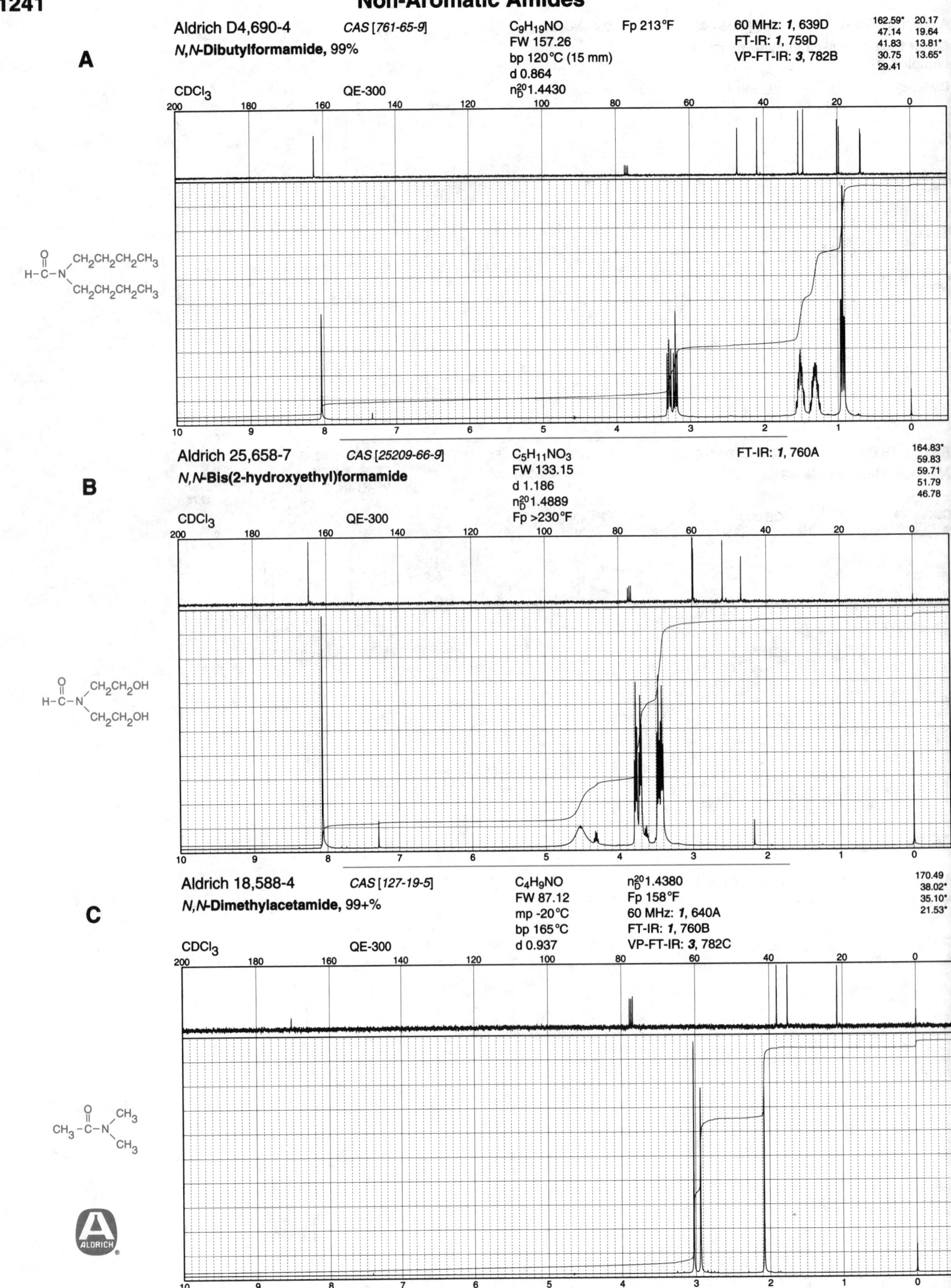

A

Aldrich D4,690-4 CAS [761-65-9]

N,N-Dibutylformamide, 99%

C$_9$H$_{19}$NO
FW 157.26
bp 120°C (15 mm)
d 0.864
n$_D^{20}$1.4430

Fp 213°F

60 MHz: *1*, 639D
FT-IR: *1*, 759D
VP-FT-IR: *3*, 782B

162.59*
47.14
41.83
30.75
29.41

20.17
19.64
13.81*
13.65*

CDCl$_3$ QE-300

B

Aldrich 25,658-7 CAS [25209-66-9]

N,N-Bis(2-hydroxyethyl)formamide

C$_5$H$_{11}$NO$_3$
FW 133.15
d 1.186
n$_D^{20}$1.4889
Fp >230°F

FT-IR: *1*, 760A

164.83*
59.83
59.71
51.79
46.78

CDCl$_3$ QE-300

C

Aldrich 18,588-4 CAS [127-19-5]

N,N-Dimethylacetamide, 99+%

C$_4$H$_9$NO
FW 87.12
mp -20°C
bp 165°C
d 0.937

n$_D^{20}$1.4380
Fp 158°F
60 MHz: *1*, 640A
FT-IR: *1*, 760B
VP-FT-IR: *3*, 782C

170.49
38.02*
35.10*
21.53*

CDCl$_3$ QE-300

Aldrich 13,752-9   CAS [685-91-6]   C₆H₁₃NO   Fp 159°F   60 MHz: *1*, 643B

$C_6H_{13}NO$ Fp 159°F

N,N-Diethylacetamide, 97%   FW 115.18   FT-IR: *1*, 760C
bp 184°C   VP-FT-IR: *3*, 782D
d 0.925
CDCl₃   QE-300   n²⁰_D 1.4400

169.58
42.86
39.95
21.44*
14.18*
13.10*

A

Aldrich 34,909-7   CAS [759-22-8]   C₈H₁₇NO   Fp 172°F

$C_8H_{17}NO$ Fp 172°F

N,N-Diisopropylacetamide, 99%   FW 143.23
bp 196°C
d 0.891
CDCl₃   QE-300   n²⁰_D 1.4410

169.45
49.29*
45.45*
23.96*
20.96*
20.63*

B

Aldrich 25,513-0   CAS [3195-78-6]   C₅H₉NO   Fp 138°F   FT-IR: *1*, 759C

C_5H_9NO Fp 138°F

N-Methyl-N-vinylacetamide, 98%   FW 99.13   VP-FT-IR: *3*, 783A
bp 70°C (25 mm)
d 0.959
CDCl₃   QE-300   n²⁰_D 1.4830

169.24   32.12*
134.41*   28.35*
132.39*   22.67*
93.52   21.85*
93.23

C

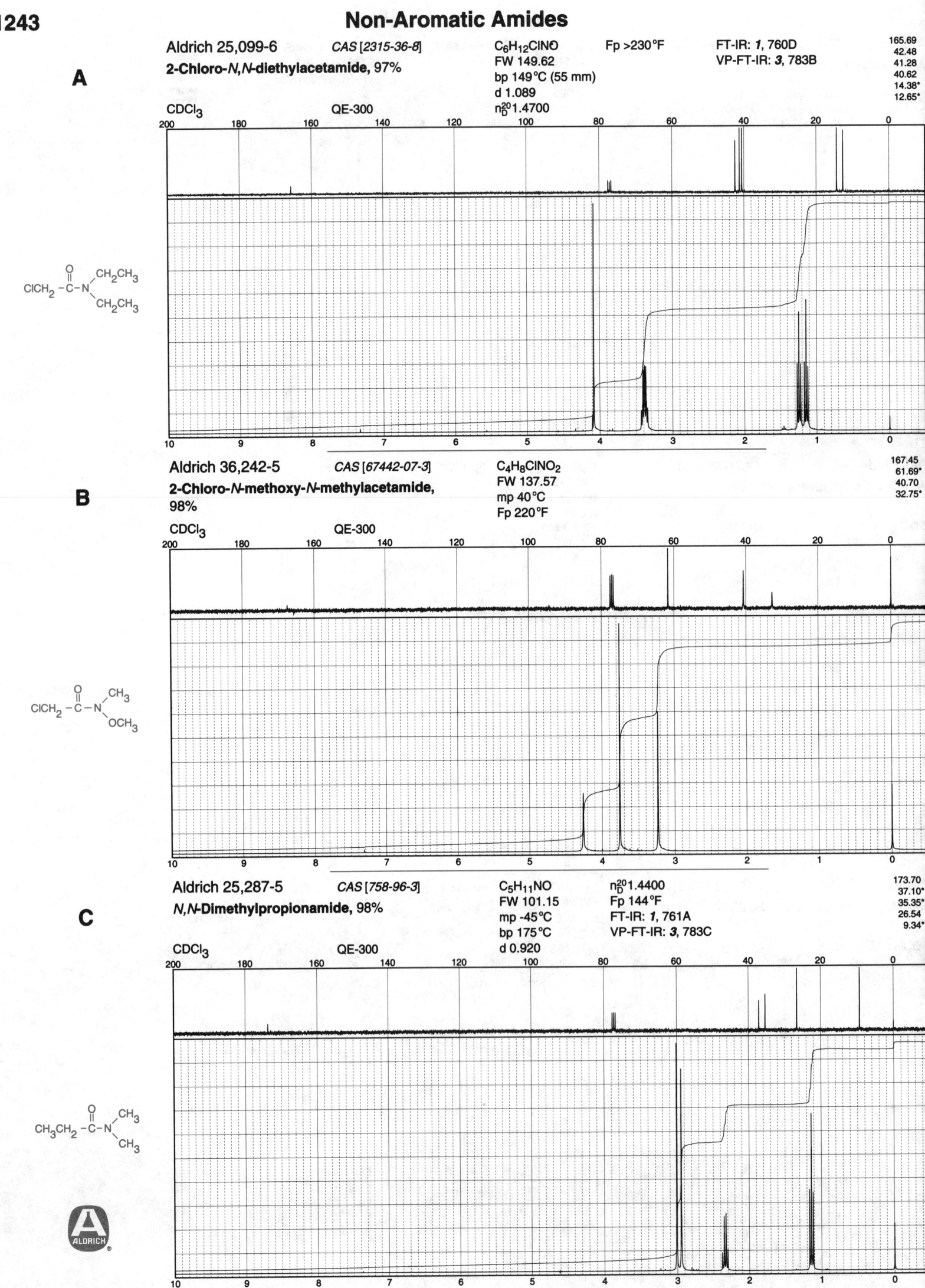

A

Aldrich 25,099-6   CAS [2315-36-8] $C_6H_{12}ClNO$ Fp >230°F   FT-IR: *1*, 760D

2-Chloro-N,N-diethylacetamide, 97%   FW 149.62   VP-FT-IR: *3*, 783B

bp 149°C (55 mm)

d 1.089

n_D^{20} 1.4700

CDCl$_3$ QE-300

165.69
42.48
41.28
40.62
14.38*
12.65*

ClCH$_2$—C—N(CH$_2$CH$_3$)(CH$_2$CH$_3$)

B

Aldrich 36,242-5   CAS [67442-07-3] $C_4H_8ClNO_2$

2-Chloro-N-methoxy-N-methylacetamide, 98%   FW 137.57

mp 40°C

Fp 220°F

CDCl$_3$ QE-300

167.45
61.69*
40.70
32.75*

ClCH$_2$—C—N(CH$_3$)(OCH$_3$)

C

Aldrich 25,287-5   CAS [758-96-3] $C_5H_{11}NO$ n_D^{20} 1.4400

N,N-Dimethylpropionamide, 98%   FW 101.15   Fp 144°F

mp -45°C   FT-IR: *1*, 761A

bp 175°C   VP-FT-IR: *3*, 783C

d 0.920

CDCl$_3$ QE-300

173.70
37.10*
35.35*
26.54
9.34*

CH$_3$CH$_2$—C—N(CH$_3$)(CH$_3$)

A

Aldrich 30,257-0   CAS [1114-51-8]

N,N-Diethylpropionamide, 99%

$C_7H_{15}NO$ Fp 163°F   FT-IR: **1**, 27C

FW 129.20   VP-FT-IR: **3**, 783D

bp 77°C (12 mm)

d 0.897

n_D^{20} 1.4420

CDCl$_3$ QE-300

172.80
41.83
40.05
26.26
14.31*
13.12*
9.63*

$CH_3CH_2-\overset{\overset{\displaystyle O}{\|}}{C}-N\overset{\displaystyle CH_2CH_3}{\underset{\displaystyle CH_2CH_3}{<}}$

B

Aldrich 34,715-9   CAS [21678-37-5]

N,N,2-Trimethylpropionamide, 99%

$C_6H_{13}NO$ Fp 154°F

FW 115.18

bp 176°C (744 mm)

d 0.892

n_D^{20} 1.4393

CDCl$_3$ QE-300

176.88
37.05*
35.54*
30.17*
19.25*

$CH_3CH-\overset{\overset{\displaystyle O}{\|}}{C}-N\overset{\displaystyle CH_3}{\underset{\displaystyle CH_3}{<}}$
$\quad\overset{\displaystyle |}{CH_3}$

C

Aldrich 27,413-5   CAS [2680-03-7]

N,N-Dimethylacrylamide, 99%

C_5H_9NO Fp 161°F   VP-FT-IR: **3**, 784B

FW 99.13

bp 81°C (20 mm)

d 0.962

n_D^{20} 1.4730

CDCl$_3$ QE-300

166.43
127.70*
127.32
37.31*
35.55*

$H_2C=CH-\overset{\overset{\displaystyle O}{\|}}{C}-N\overset{\displaystyle CH_3}{\underset{\displaystyle CH_3}{<}}$

A

Aldrich 19,419-0   CAS [3352-87-2]   C₁₆H₃₃NO   Fp >230°F   60 MHz: *1*, 640B

$C_{16}H_{33}NO$

N,N-Diethyldodecanamide, 98%

FW 255.45
bp 167°C (2 mm)
d 0.847
n_D^{20} 1.4540

FT-IR: *1*, 761B
VP-FT-IR: *3*, 784A

172.24	29.35
41.98	25.55
40.03	22.69
33.20	14.44*
31.93	14.09*
29.64	13.14*
29.57	

CDCl₃   QE-300

$CH_3(CH_2)_9CH_2 - \overset{\overset{\displaystyle O}{\|}}{C} - N \overset{CH_2CH_3}{\underset{CH_2CH_3}{<}}$

B

Aldrich 32,929-0   CAS [5810-11-7]   C₆H₁₀ClNO₂   Fp >230°F

$C_6H_{10}ClNO_2$

2-Chloro-N,N-dimethylacetoacetamide, 96%

FW 163.61
bp 95°C
d 1.203
n_D^{20} 1.4840

198.50
164.88
59.71*
37.78*
36.15*
26.80*

CDCl₃   QE-300

$CH_3 - \overset{\overset{\displaystyle O}{\|}}{C} - \underset{\underset{\displaystyle Cl}{|}}{CH} - \overset{\overset{\displaystyle O}{\|}}{C} - N \overset{CH_3}{\underset{CH_3}{<}}$

C

Aldrich N810-5   CAS [4138-26-5]   C₆H₁₂N₂O   60 MHz: *1*, 641A

$C_6H_{12}N_2O$

Nipecotamide, 95%

FW 128.18
mp 105°C

FT-IR: *1*, 762A

177.53
50.48
47.42
44.56*
29.21
26.80

DMSO-d₆   QE-300

ALDRICH

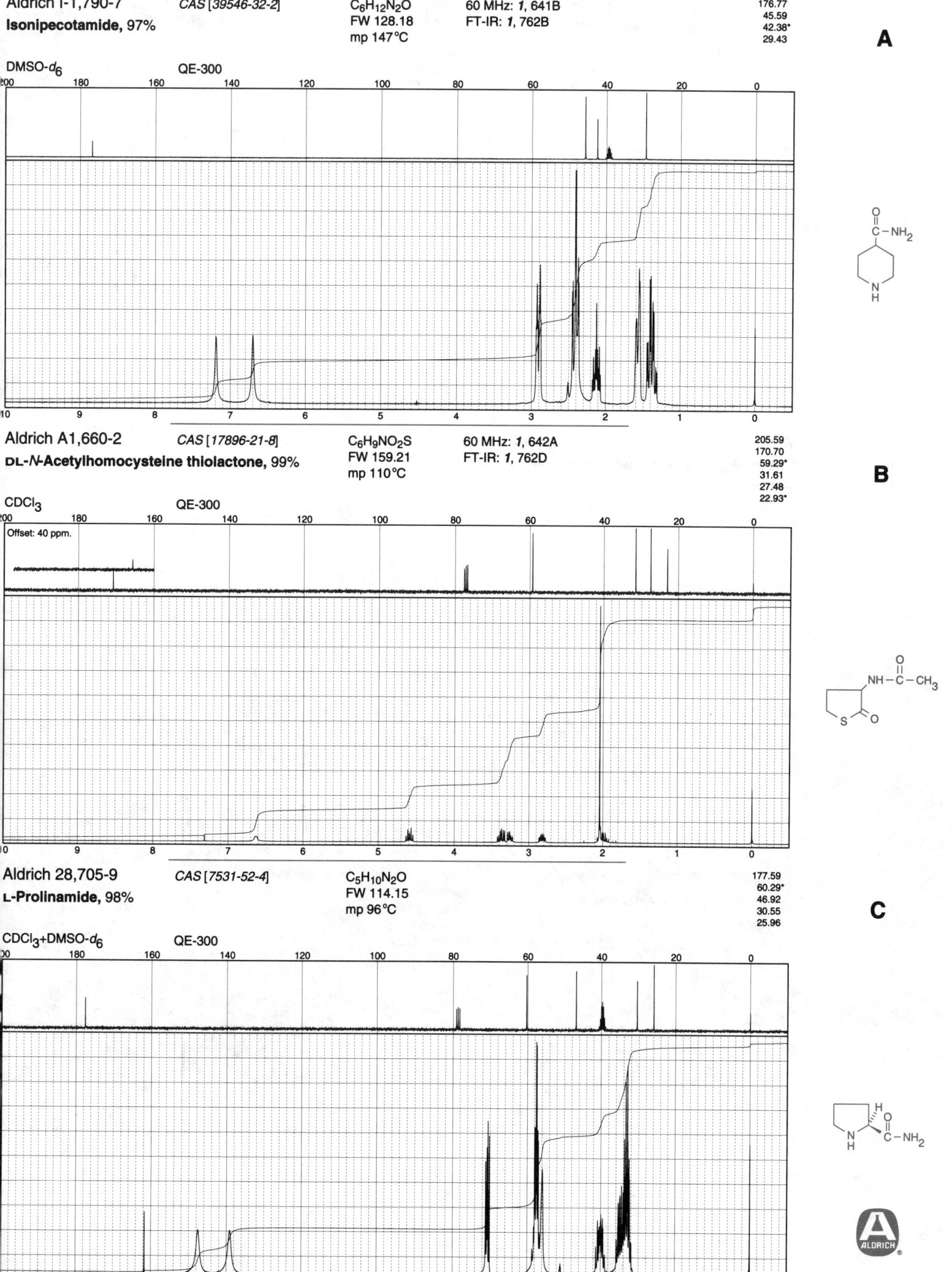

Aldrich I-1,790-7
Isonipecotamide, 97%

CAS [39546-32-2]

C6H12N2O
FW 128.18
mp 147°C

60 MHz: *1*, 641B
FT-IR: *1*, 762B

176.77
45.59
42.38*
29.43

A

DMSO-*d6*   QE-300

Aldrich A1,660-2
DL-*N*-Acetylhomocysteine thiolactone, 99%

CAS [17896-21-8]

C6H9NO2S
FW 159.21
mp 110°C

60 MHz: *1*, 642A
FT-IR: *1*, 762D

205.59
170.70
59.29*
31.61
27.48
22.93*

B

CDCl3   QE-300

Offset: 40 ppm.

Aldrich 28,705-9
L-Prolinamide, 98%

CAS [7531-52-4]

C5H10N2O
FW 114.15
mp 96°C

177.59
60.29*
46.92
30.55
25.96

C

CDCl3+DMSO-*d6*   QE-300

ALDRICH

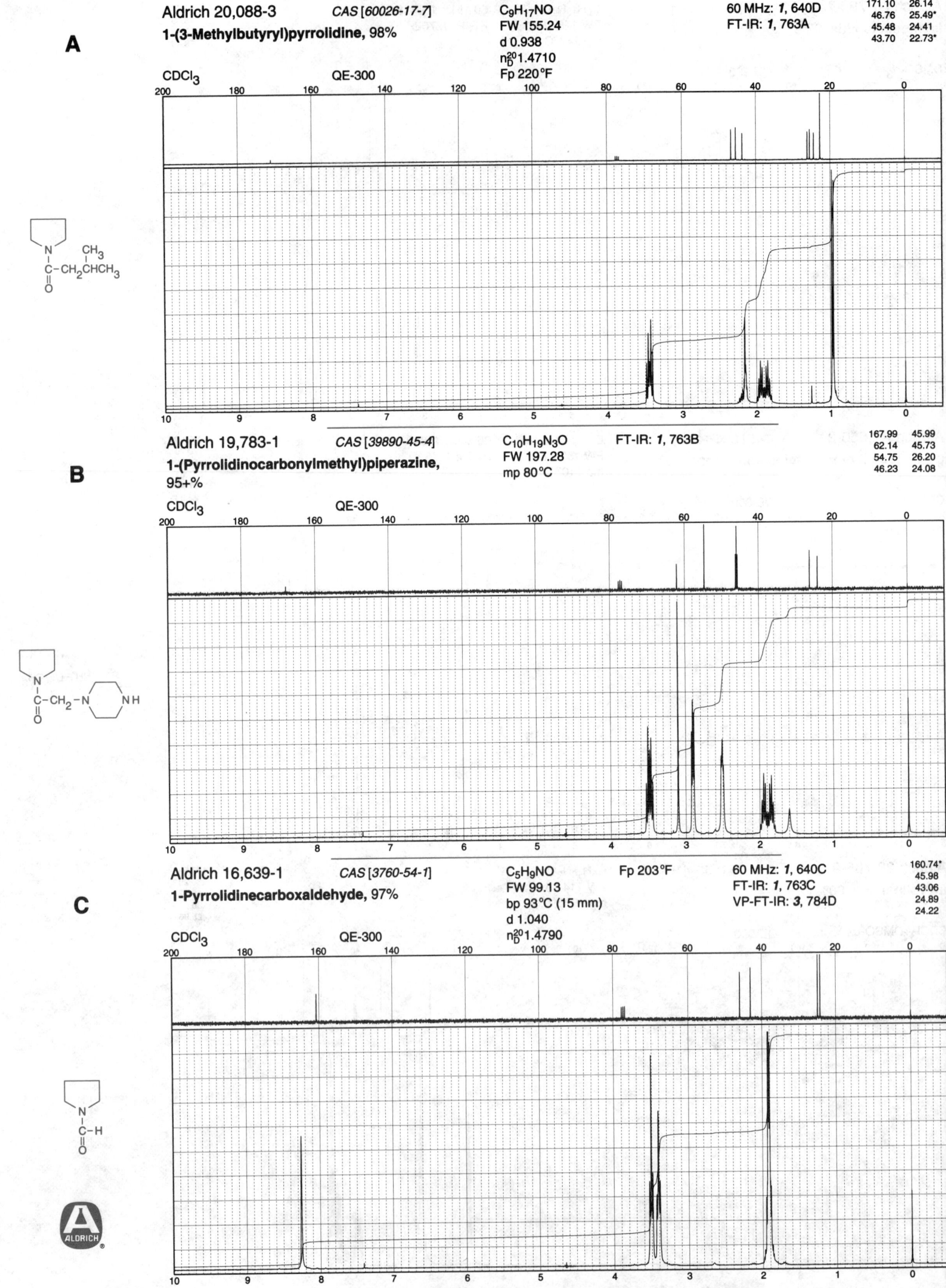

A

Aldrich 20,088-3 *CAS [60026-17-7]*
1-(3-Methylbutyryl)pyrrolidine, 98%

$C_9H_{17}NO$
FW 155.24
d 0.938
n_D^{20}1.4710
Fp 220°F

60 MHz: *1*, 640D
FT-IR: *1*, 763A

171.10 26.14
46.76   25.49*
45.48   24.41
43.70   22.73*

CDCl$_3$ QE-300

B

Aldrich 19,783-1 *CAS [39890-45-4]*
1-(Pyrrolidinocarbonylmethyl)piperazine, 95+%

$C_{10}H_{19}N_3O$
FW 197.28
mp 80°C

FT-IR: *1*, 763B

167.99 45.99
62.14   45.73
54.75   26.20
46.23   24.08

CDCl$_3$ QE-300

C

Aldrich 16,639-1 *CAS [3760-54-1]*
1-Pyrrolidinecarboxaldehyde, 97%

C_5H_9NO
FW 99.13
bp 93°C (15 mm)
d 1.040
n_D^{20}1.4790

Fp 203°F

60 MHz: *1*, 640C
FT-IR: *1*, 763C
VP-FT-IR: *3*, 784D

160.74*
45.98
43.06
24.89
24.22

CDCl$_3$ QE-300

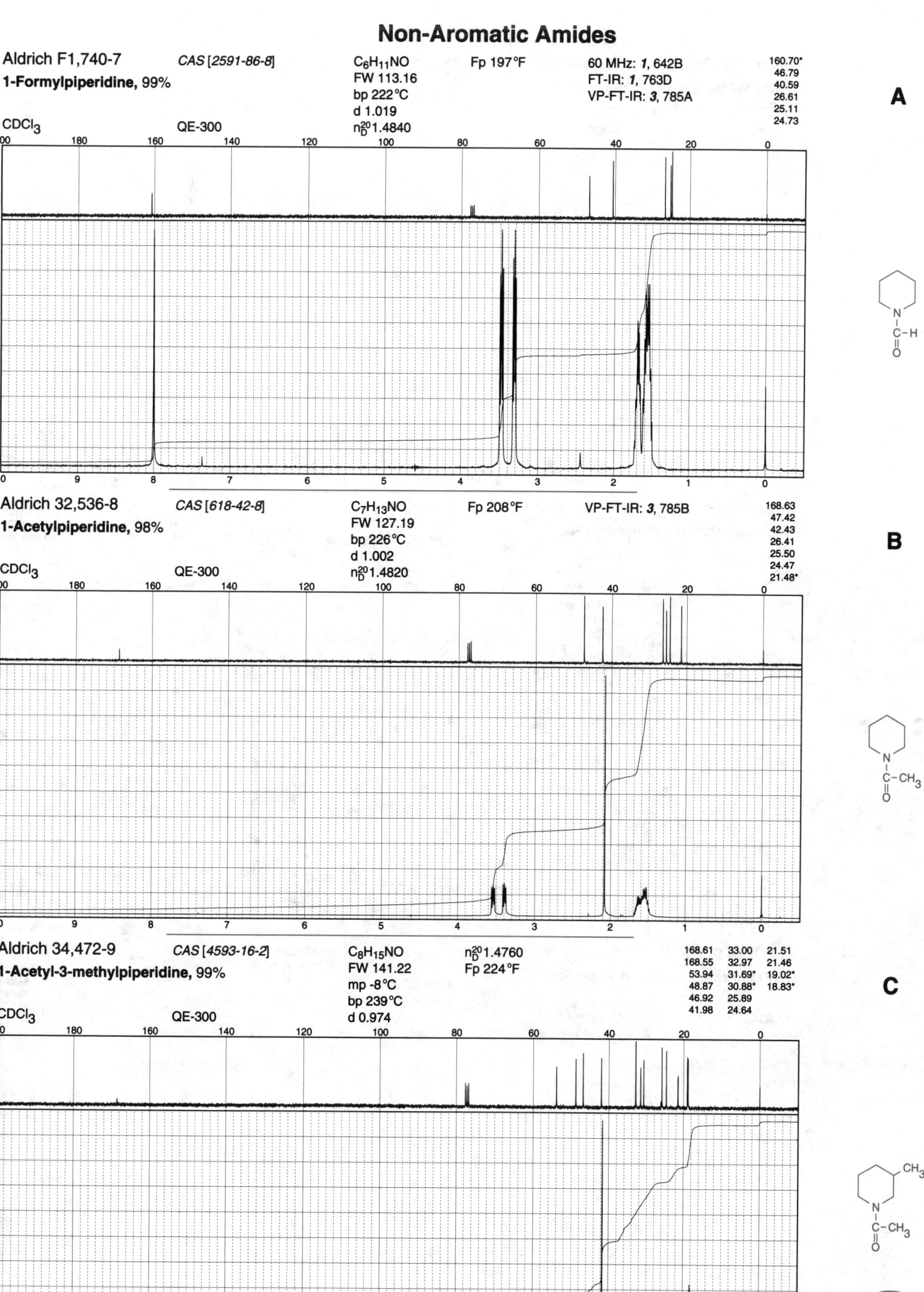

Aldrich F1,740-7   CAS [2591-86-8]
1-Formylpiperidine, 99%

C$_6$H$_{11}$NO   Fp 197°F   60 MHz: *1*, 642B
FW 113.16   FT-IR: *1*, 763D
bp 222°C   VP-FT-IR: *3*, 785A
d 1.019
n$_D^{20}$1.4840

CDCl$_3$ QE-300

160.70*
46.79
40.59
26.61
25.11
24.73

Aldrich 32,536-8   CAS [618-42-8]
1-Acetylpiperidine, 98%

C$_7$H$_{13}$NO   Fp 208°F   VP-FT-IR: *3*, 785B
FW 127.19
bp 226°C
d 1.002
n$_D^{20}$1.4820

CDCl$_3$ QE-300

168.63
47.42
42.43
26.41
25.50
24.47
21.48*

Aldrich 34,472-9   CAS [4593-16-2]
1-Acetyl-3-methylpiperidine, 99%

C$_8$H$_{15}$NO   n$_D^{20}$1.4760
FW 141.22   Fp 224°F
mp -8°C
bp 239°C
d 0.974

CDCl$_3$ QE-300

168.61	33.00	21.51
168.55	32.97	21.46
53.94	31.69*	19.02*
48.87	30.88*	18.83*
46.92	25.89	
41.98	24.64	

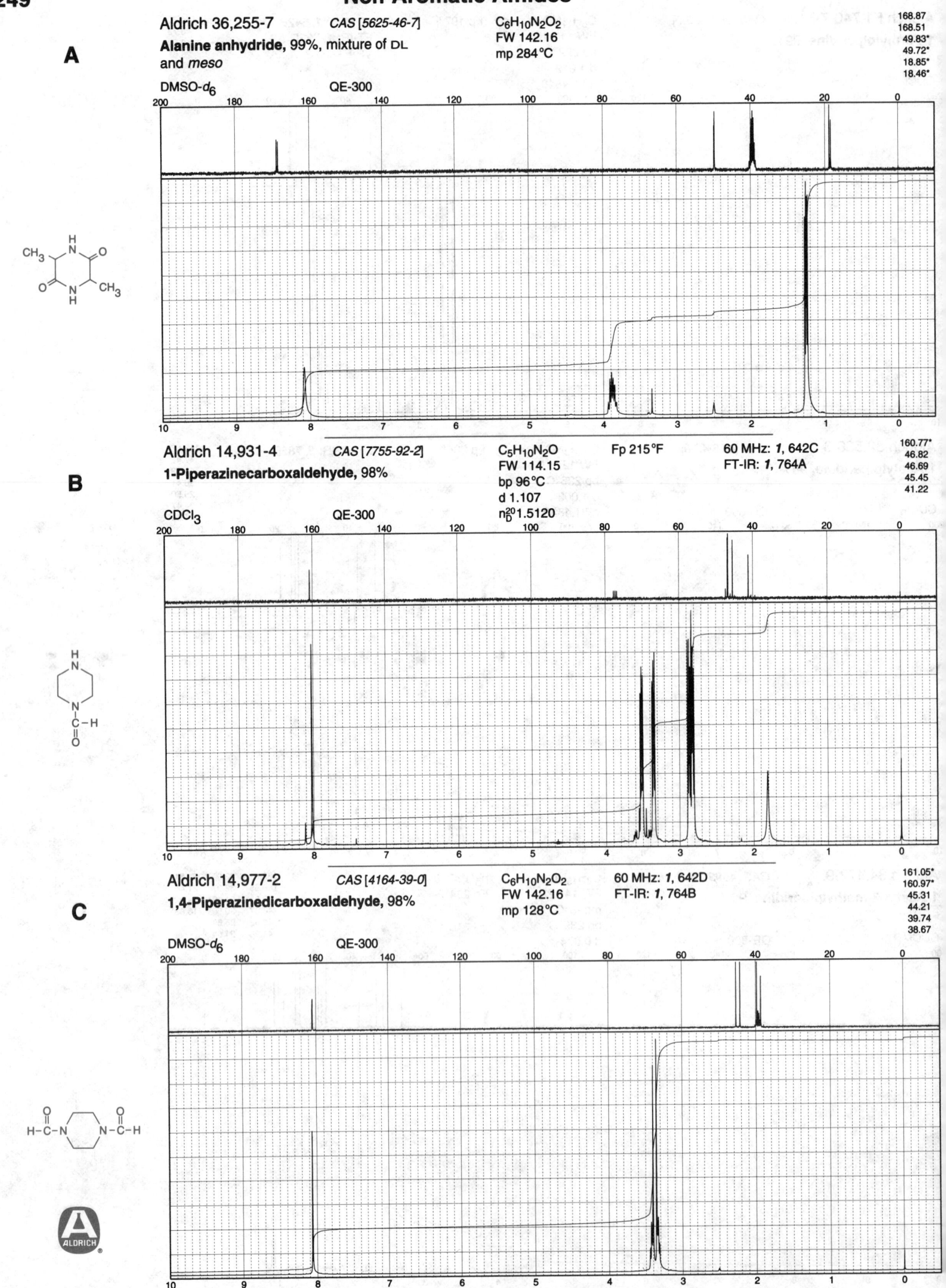

A

Aldrich 36,255-7   *CAS [5625-46-7]*   C₆H₁₀N₂O₂

$C_6H_{10}N_2O_2$

Alanine anhydride, 99%, mixture of DL
and *meso*

FW 142.16
mp 284°C

168.87
168.51
49.83*
49.72*
18.85*
18.46*

DMSO-*d₆*   QE-300

Aldrich 14,931-4   *CAS [7755-92-2]* $C_5H_{10}N_2O$ Fp 215°F   60 MHz: *1*, 642C

1-Piperazinecarboxaldehyde, 98%

FW 114.15
bp 96°C
d 1.107
n_D^{20} 1.5120

FT-IR: *1*, 764A

160.77*
46.82
46.69
45.45
41.22

B

CDCl₃   QE-300

Aldrich 14,977-2   *CAS [4164-39-0]* $C_6H_{10}N_2O_2$ 60 MHz: *1*, 642D

1,4-Piperazinedicarboxaldehyde, 98%

FW 142.16
mp 128°C

FT-IR: *1*, 764B

161.05*
160.97*
45.31
44.21
39.74
38.67

C

DMSO-*d₆*   QE-300

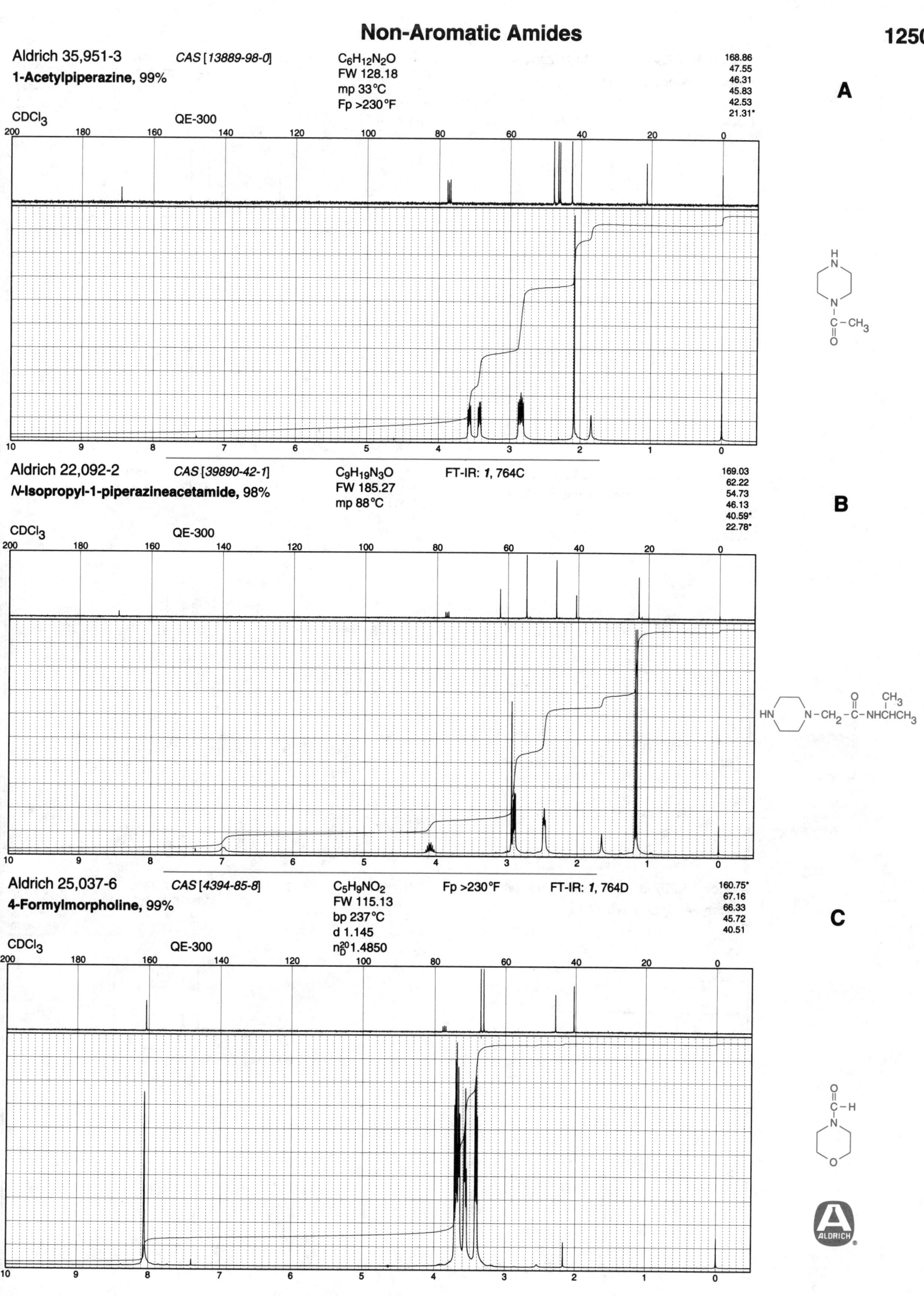

Aldrich 35,951-3 — CAS [13889-98-0]
1-Acetylpiperazine, 99%
$C_6H_{12}N_2O$
FW 128.18
mp 33°C
Fp >230°F

CDCl₃ QE-300

168.86
47.55
46.31
45.83
42.53
21.31*

A

Aldrich 22,092-2 — CAS [39890-42-1]
N-Isopropyl-1-piperazineacetamide, 98%
$C_9H_{19}N_3O$
FW 185.27
mp 88°C
FT-IR: 1, 764C

CDCl₃ QE-300

169.03
62.22
54.73
46.13
40.59*
22.78*

B

Aldrich 25,037-6 — CAS [4394-85-8]
4-Formylmorpholine, 99%
$C_5H_9NO_2$
FW 115.13
bp 237°C
d 1.145
n²⁰_D 1.4850
Fp >230°F
FT-IR: 1, 764D

CDCl₃ QE-300

160.75*
67.16
66.33
45.72
40.51

C

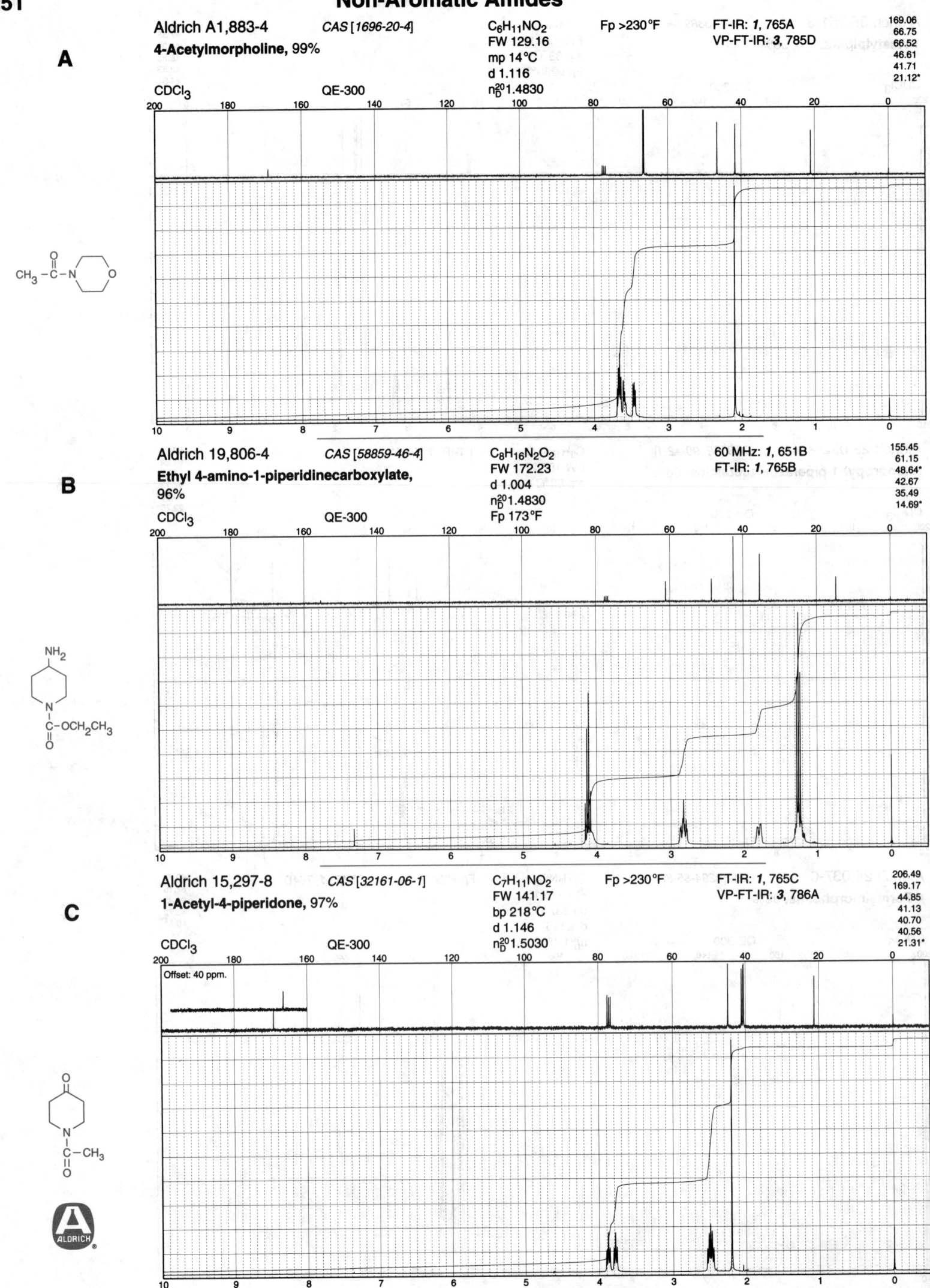

A

Aldrich A1,883-4   CAS [1696-20-4]

4-Acetylmorpholine, 99%

CDCl₃   QE-300

$C_6H_{11}NO_2$
FW 129.16
mp 14°C
d 1.116
n_D^{20}1.4830

Fp >230°F

FT-IR: *1*, 765A
VP-FT-IR: *3*, 785D

169.06
66.75
66.52
46.61
41.71
21.12*

B

Aldrich 19,806-4   CAS [58859-46-4]

Ethyl 4-amino-1-piperidinecarboxylate, 96%

CDCl₃   QE-300

$C_8H_{16}N_2O_2$
FW 172.23
d 1.004
n_D^{20}1.4830
Fp 173°F

60 MHz: *1*, 651B
FT-IR: *1*, 765B

155.45
61.15
48.64*
42.67
35.49
14.69*

C

Aldrich 15,297-8   CAS [32161-06-1]

1-Acetyl-4-piperidone, 97%

CDCl₃   QE-300

$C_7H_{11}NO_2$
FW 141.17
bp 218°C
d 1.146
n_D^{20}1.5030

Fp >230°F

FT-IR: *1*, 765C
VP-FT-IR: *3*, 786A

206.49
169.17
44.85
41.13
40.70
40.56
21.31*

Offset: 40 ppm.

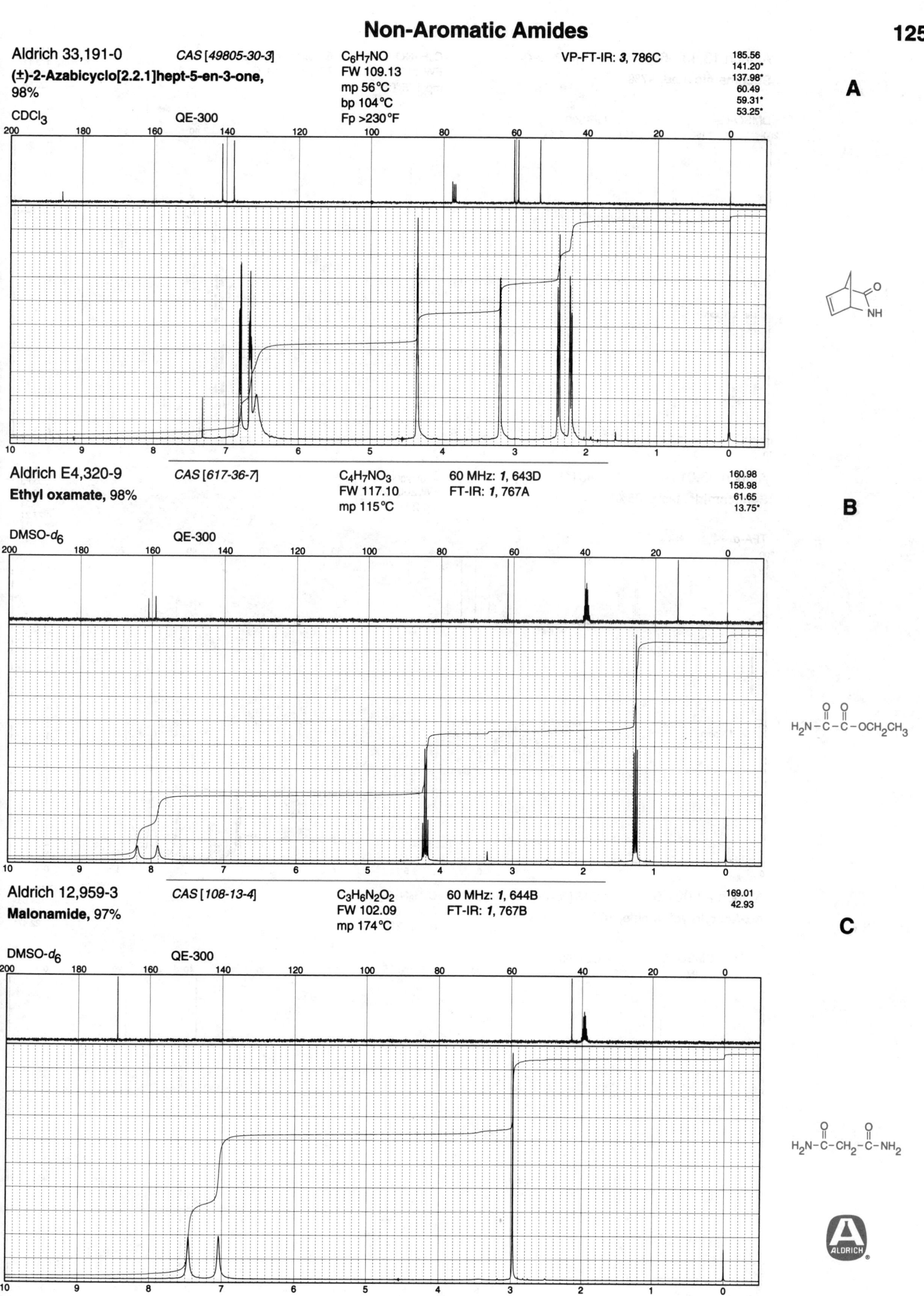

Aldrich 33,191-0   CAS [49805-30-3]   C$_6$H$_7$NO   VP-FT-IR: *3*, 786C

(±)-2-Azabicyclo[2.2.1]hept-5-en-3-one,
98%

FW 109.13
mp 56°C
bp 104°C
Fp >230°F

185.56
141.20*
137.98*
60.49
59.31*
53.25*

A

CDCl$_3$ QE-300

Aldrich E4,320-9   CAS [617-36-7]   C$_4$H$_7$NO$_3$ 60 MHz: *1*, 643D

Ethyl oxamate, 98%

FW 117.10
mp 115°C

FT-IR: *1*, 767A

160.98
158.98
61.65
13.75*

B

DMSO-d_6 QE-300

Aldrich 12,959-3   CAS [108-13-4]   C$_3$H$_6$N$_2$O$_2$ 60 MHz: *1*, 644B

Malonamide, 97%

FW 102.09
mp 174°C

FT-IR: *1*, 767B

169.01
42.93

C

DMSO-d_6 QE-300

A

Aldrich 13,437-6   *CAS [638-32-4]*   C$_4$H$_7$NO$_3$ 60 MHz: *1*, 644C

Succinamic acid, 97%

FW 117.10   FT-IR: *1*, 767C

mp 155°C

173.83
173.11
29.58
28.88

DMSO-*d$_6$*   QE-300

Offset: 2.3 ppm.

H$_2$N–C–CH$_2$CH$_2$–C–OH

B

Aldrich 36,011-2   *CAS [1740-54-1]*   C$_{10}$H$_{20}$N$_2$O$_2$

Sebacamide, tech., 85%

FW 200.28

mp 201°C

185.71
35.88
30.41
30.34
27.21

TFA-*d*   QE-300

Offset: 1.9 ppm.

H$_2$N–C–CH$_2$(CH$_2$)$_6$CH$_2$–C–NH$_2$

C

Aldrich 28,021-6   *CAS [2620-63-5]*   C$_4$H$_8$N$_2$O$_2$

Nα-Acetylglycinamide, 97%

FW 116.12

mp 142°C

171.48
169.88
42.03
22.47*

CDCl$_3$ + DMSO-*d$_6$*   QE-300

CH$_3$–C–NHCH$_2$–C–NH$_2$

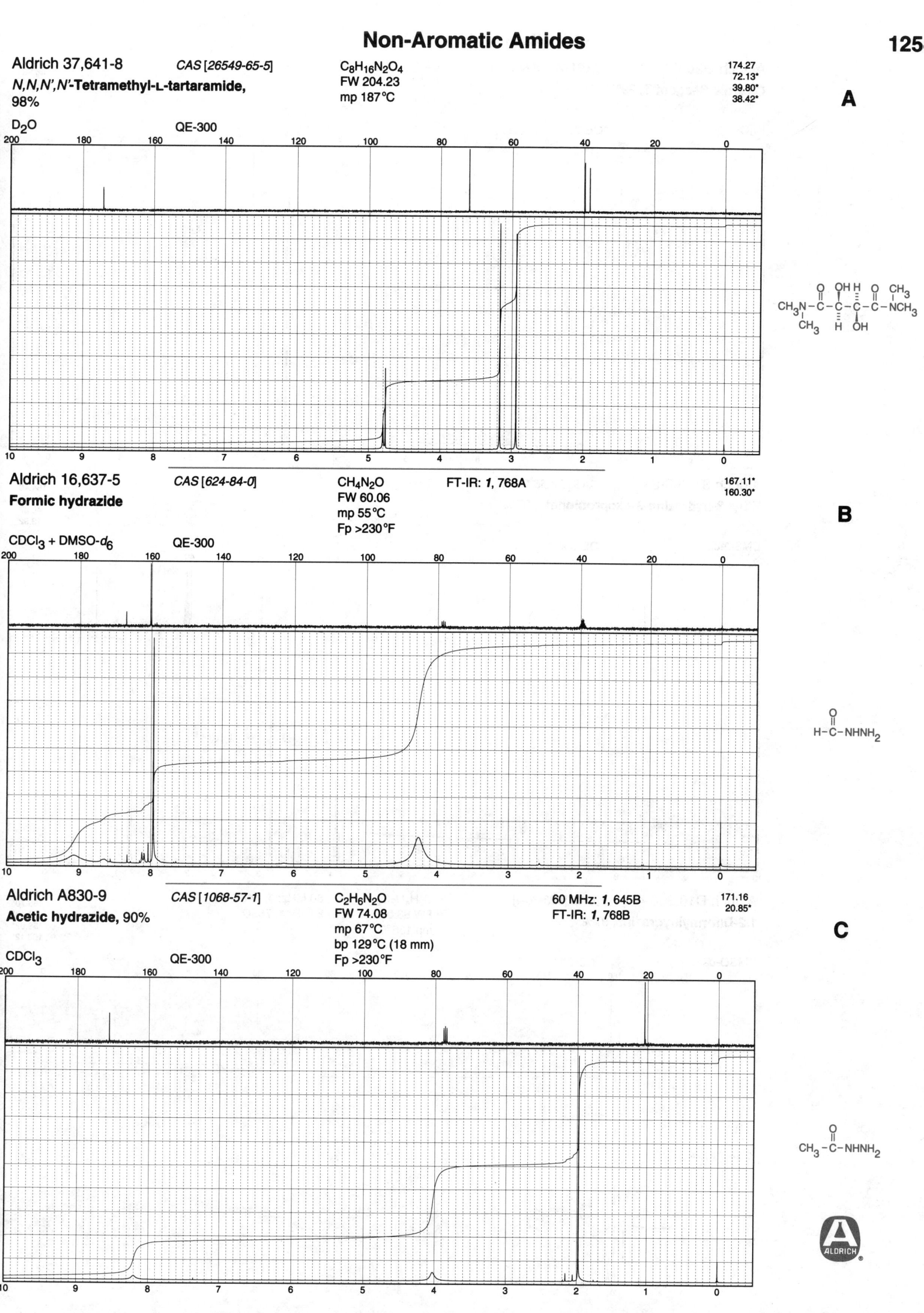

Aldrich 37,641-8   CAS [26549-65-5] $C_8H_{16}N_2O_4$ 174.27

N,N,N',N'-Tetramethyl-L-tartaramide,   FW 204.23   72.13*

98%   mp 187°C   39.80*   38.42*

A

D_2O QE-300

Aldrich 16,637-5   CAS [624-84-0] CH_4N_2O FT-IR: 1, 768A   167.11*

Formic hydrazide   FW 60.06   160.30*

mp 55°C

Fp >230°F

B

$CDCl_3$ + DMSO-d_6 QE-300

Aldrich A830-9   CAS [1068-57-1] $C_2H_6N_2O$ 60 MHz: 1, 645B   171.16

Acetic hydrazide, 90%   FW 74.08   FT-IR: 1, 768B   20.85*

mp 67°C

bp 129°C (18 mm)

Fp >230°F

C

$CDCl_3$ QE-300

A

Aldrich G90-0
Girard's Reagent T, 99%

CAS [123-46-6]

$C_5H_{14}ClN_3O$
FW 167.64
mp 190°C d.

FT-IR: 1, 768C

165.61
66.98
57.10*

D₂O   QE-300

$$CH_3 \atop CH_3-\overset{\overset{\displaystyle +}{|}}{\underset{\underset{\displaystyle CH_3}{|}}{N}}-CH_2-\overset{\overset{\displaystyle O}{\|}}{C}-NHNH_2 \quad Cl^-$$

Aldrich 37,890-9
Ethyl 3-hydrazino-3-oxopropionate, 97%

CAS [30866-24-1]

$C_5H_{10}N_2O_3$
FW 146.15
mp 74°C

167.61
164.46
60.38
40.57
13.92*

B

DMSO-d₆   QE-300

$$H_2NNH-\overset{\overset{\displaystyle O}{\|}}{C}-CH_2-\overset{\overset{\displaystyle O}{\|}}{C}-OCH_2CH_3$$

Aldrich D10,300-4
1,2-Diformylhydrazine, 97%

CAS [628-36-4]

$C_2H_4N_2O_2$
FW 88.07
mp 156°C

60 MHz: 1, 646C
FT-IR: 1, 768D

166.91*
166.70*
161.43*
158.92*
158.12*

C

DMSO-d₆   QE-300

Offset: 0.3 ppm.

$$H-\overset{\overset{\displaystyle O}{\|}}{C}-NHNH-\overset{\overset{\displaystyle O}{\|}}{C}-H$$

ALDRICH

Non-Aromatic Amides

Aldrich D840-2   CAS [3148-73-0]   C₄H₈N₂O₂   60 MHz: 1, 646D

1,2-Diacetylhydrazine, 98%

$C_4H_8N_2O_2$
FW 116.12
mp 139°C
bp 209°C (15 mm)

60 MHz: 1, 646D
FT-IR: 1, 769A

167.90
20.31*

A

DMSO-d_6 QE-300

$CH_3-\overset{\displaystyle O}{\overset{\|}{C}}-NHNH-\overset{\displaystyle O}{\overset{\|}{C}}-CH_3$

Aldrich S550-2   CAS [4146-43-4]

Succinic dihydrazide, 96%

$C_4H_{10}N_4O_2$
FW 146.15
mp 171°C

60 MHz: 1, 645C
FT-IR: 1, 770B

170.58
28.81

B

DMSO-d_6 QE-300

$H_2NNH-\overset{\displaystyle O}{\overset{\|}{C}}-CH_2CH_2-\overset{\displaystyle O}{\overset{\|}{C}}-NHNH_2$

Aldrich 17,794-6   CAS [1596-84-5]

Succinic 2,2-dimethylhydrazide, 99%

$C_6H_{12}N_2O_3$
FW 160.17
mp 163°C

60 MHz: 1, 645D
FT-IR: 1, 771C

173.91	46.27*
173.61	28.91
173.17	28.53
168.24	28.26
47.47*	26.48

C

DMSO-d_6 QE-300

$HO-\overset{\displaystyle O}{\overset{\|}{C}}-CH_2CH_2-\overset{\displaystyle O}{\overset{\|}{C}}-NHN\overset{\displaystyle CH_3}{\underset{\displaystyle CH_3}{<}}$

ALDRICH

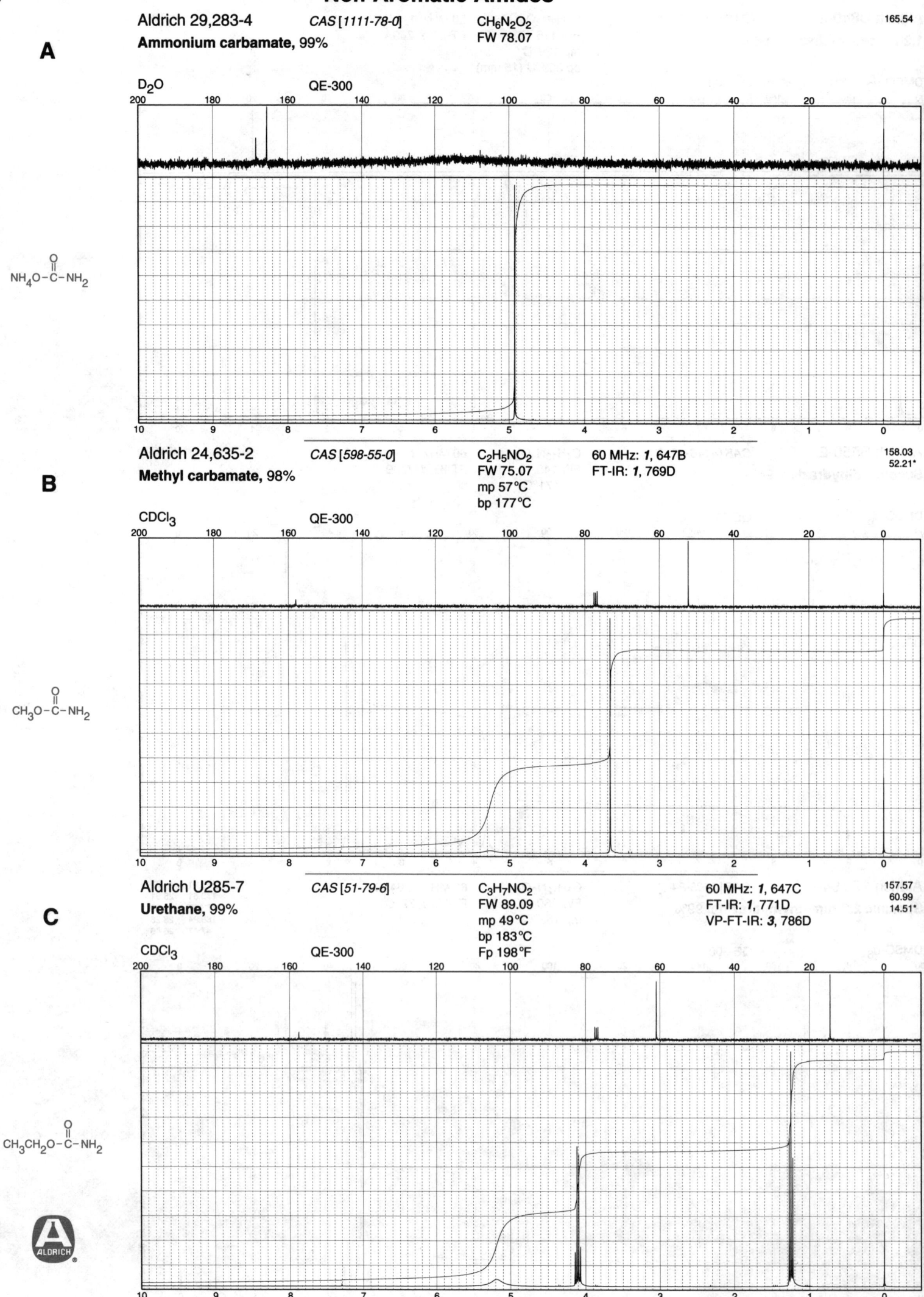

A

Aldrich 29,283-4   *CAS [1111-78-0]*   CH₆N₂O₂   165.54
Ammonium carbamate, 99%   FW 78.07

$CH_6N_2O_2$
FW 78.07

D₂O   QE-300

$NH_4O-\overset{\overset{\displaystyle O}{\|}}{C}-NH_2$

B

Aldrich 24,635-2   *CAS [598-55-0]*   C₂H₅NO₂   60 MHz: *1*, 647B   158.03
Methyl carbamate, 98%   FW 75.07   FT-IR: *1*, 769D   52.21*
mp 57°C
bp 177°C

CDCl₃   QE-300

$CH_3O-\overset{\overset{\displaystyle O}{\|}}{C}-NH_2$

C

Aldrich U285-7   *CAS [51-79-6]*   C₃H₇NO₂   60 MHz: *1*, 647C   157.57
Urethane, 99%   FW 89.09   FT-IR: *1*, 771D   60.99
mp 49°C   VP-FT-IR: *3*, 786D   14.51*
bp 183°C
Fp 198°F

CDCl₃   QE-300

$CH_3CH_2O-\overset{\overset{\displaystyle O}{\|}}{C}-NH_2$

ALDRICH

A

Aldrich 14,209-3   CAS [13698-16-3]

N,N-Dichlorourethane, tech., 90%

C3H5Cl2NO2   Fp 206°F   60 MHz: 1, 648C   157.58
FW 157.98   FT-IR: 1, 772A   67.04
bp 56°C (15 mm)   14.17*
d 1.349
n_D^{20} 1.4600

CDCl3   QE-300

$CH_3CH_2O-\overset{\overset{\displaystyle O}{\|}}{C}-N\overset{\displaystyle Cl}{\underset{\displaystyle Cl}{}}$

B

Aldrich 11,947-4   CAS [589-41-3]

N-Hydroxyurethane

C3H7NO3   60 MHz: 1, 648D   159.65
FW 105.09   FT-IR: 1, 772B   62.25
bp 115°C (3 mm)   14.37*
n_D^{20} 1.4450

CDCl3   QE-300

$CH_3CH_2O-\overset{\overset{\displaystyle O}{\|}}{C}-NHOH$

C

Aldrich 36,928-4

Ethyl N-methoxy-N-methylcarbamate, 98%

C5H11NO3   Fp 127°F   157.31
FW 133.15   62.06
bp 153°C   61.57*
d 1.022   35.59*
n_D^{20} 1.4120   14.60*

CDCl3   QE-300

$CH_3O-\overset{\displaystyle CH_3}{\underset{}{N}}-\overset{\overset{\displaystyle O}{\|}}{C}-OCH_2CH_3$

ALDRICH

A

Aldrich E5,122-8   CAS [623-78-9] $C_5H_{11}NO_2$ Fp 168 °F   60 MHz: *1*, 649A   156.62
60.53
N-Ethylurethane, 99%   FW 117.15   FT-IR: *1*, 772C   35.76
bp 85 °C (20 mm)   VP-FT-IR: *3*, 787A   15.29*
d 0.981   14.68*
n_D^{20}1.4210

CDCl$_3$ QE-300

$CH_3CH_2O-\overset{\overset{\displaystyle O}{\|}}{C}-NHCH_2CH_3$

B

Aldrich T6,030-5   CAS [3206-31-3] $C_9H_{15}NO_6$ Fp >230 °F   60 MHz: *1*, 648B   150.26
64.29
Triethyl nitrilotricarboxylate, 97%   FW 233.22   FT-IR: *1*, 772D   13.89*
bp 147 °C (12 mm)
d 1.135
n_D^{20}1.4280

CDCl$_3$ QE-300

$CH_3CH_2O-\overset{\overset{\displaystyle O}{\|}}{C}-N-\overset{\overset{\displaystyle O}{\|}}{C}-OCH_2CH_3$
$\overset{|}{\underset{\underset{\displaystyle O}{\|}}{C}}-OCH_2CH_3$

C

Aldrich B9,080-7   CAS [592-35-8] $C_5H_{11}NO_2$ 157.77
64.98
Butyl carbamate, 98%   FW 117.15   31.01
mp 54 °C   19.06
Fp 228 °F   13.73*

CDCl$_3$ QE-300

$CH_3CH_2CH_2CH_2O-\overset{\overset{\displaystyle O}{\|}}{C}-NH_2$

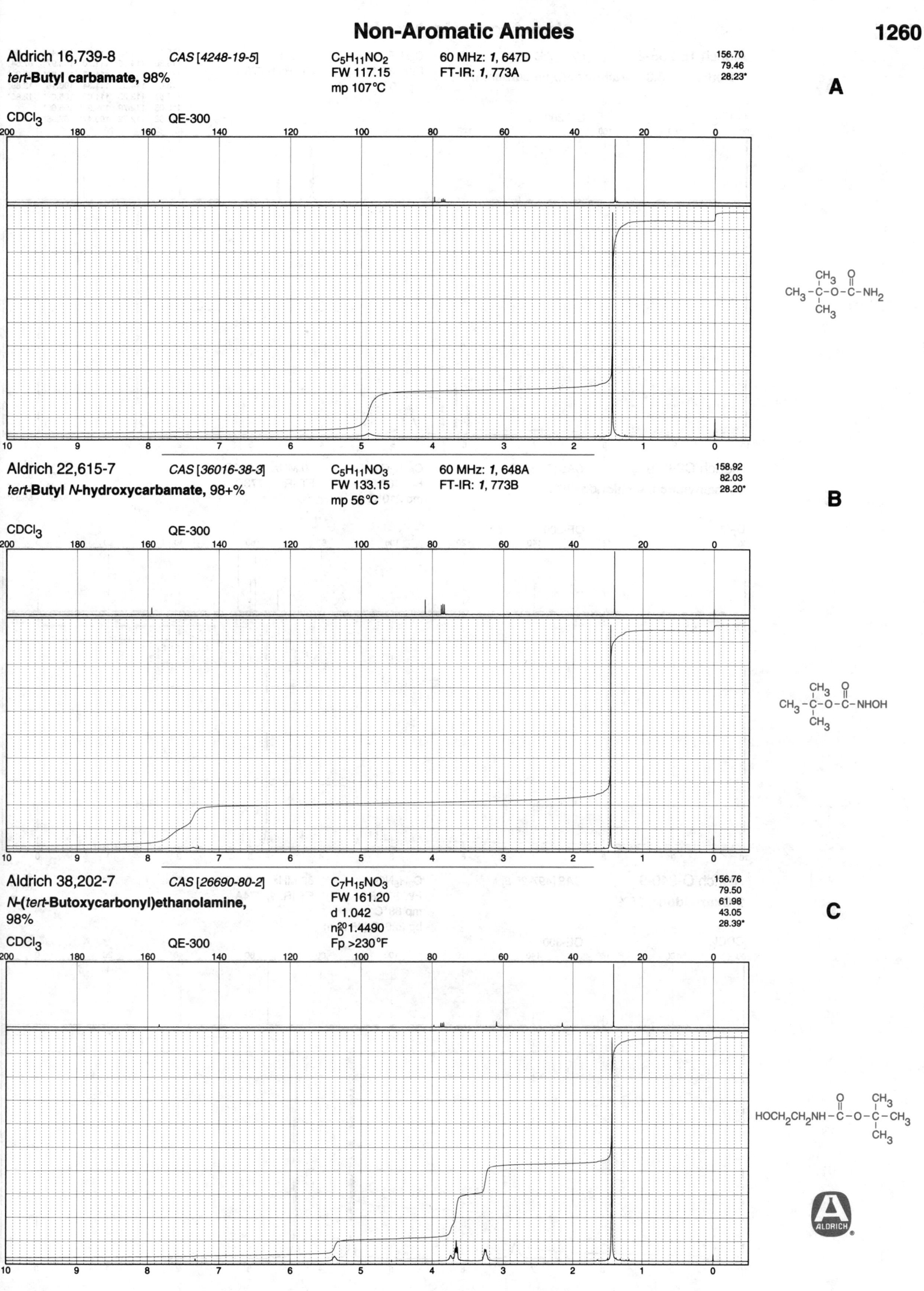

Aldrich 16,739-8   CAS [4248-19-5]   C$_5$H$_{11}$NO$_2$ 60 MHz: *1*, 647D   156.70
tert-Butyl carbamate, 98%   FW 117.15   FT-IR: *1*, 773A   79.46
mp 107°C   28.23*

A

CDCl$_3$ QE-300

Aldrich 22,615-7   CAS [36016-38-3]   C$_5$H$_{11}$NO$_3$ 60 MHz: *1*, 648A   158.92
tert-**Butyl N-hydroxycarbamate**, 98+%   FW 133.15   FT-IR: *1*, 773B   82.03
mp 56°C   28.20*

B

CDCl$_3$ QE-300

Aldrich 38,202-7   CAS [26690-80-2]   C$_7$H$_{15}$NO$_3$ 156.76
N-(*tert*-**Butoxycarbonyl)ethanolamine**,   FW 161.20   79.50
98%   d 1.042   61.98
n$_D^{20}$1.4490   43.05
28.39*

CDCl$_3$ QE-300   Fp >230°F

C

A

Aldrich 15,238-2   *CAS [756-48-9]* $C_5H_7F_4NO_2$ 60 MHz: *1*, 649B

1-Methyl-2,2,3,3-tetrafluoropropyl carbamate   FW 189.11   FT-IR: *1*, 773C

mp 60°C

155.15	114.68	112.29	109.41*	67.64*
118.36	114.66	111.70	108.97*	67.54*
118.04	114.36	111.37	106.60*	67.22*
118.01	114.32	111.34	106.16*	12.89*
117.67	113.23	111.00	106.09*	12.86*
115.05	112.79	109.91*	105.66*	
115.02	112.72	109.47*	67.96*	

CDCl₃   QE-300

F₂CHCF₂CHO–C–NH₂ (CH₃, O)

B

Aldrich C240-9   *CAS [51-83-2]* $C_6H_{15}ClN_2O_2$ 60 MHz: *1*, 649C

Carbamylcholine chloride, 99%   FW 182.65   FT-IR: *1*, 773D

mp 210°C d.

160.67	61.34
67.67	56.62*
67.63	56.57*
67.59	56.52*

D₂O   QE-300

CH₃–⁺N CH₂CH₂O–C–NH₂ Cl⁻ (CH₃, CH₃, O)

C

Aldrich O-940-9   *CAS [497-25-6]* $C_3H_5NO_2$ 60 MHz: *1*, 649D

2-Oxazolidone, 98%   FW 87.08   FT-IR: *1*, 774A

mp 88°C

bp 220°C (48 mm)

161.13
65.14
40.79

CDCl₃   QE-300

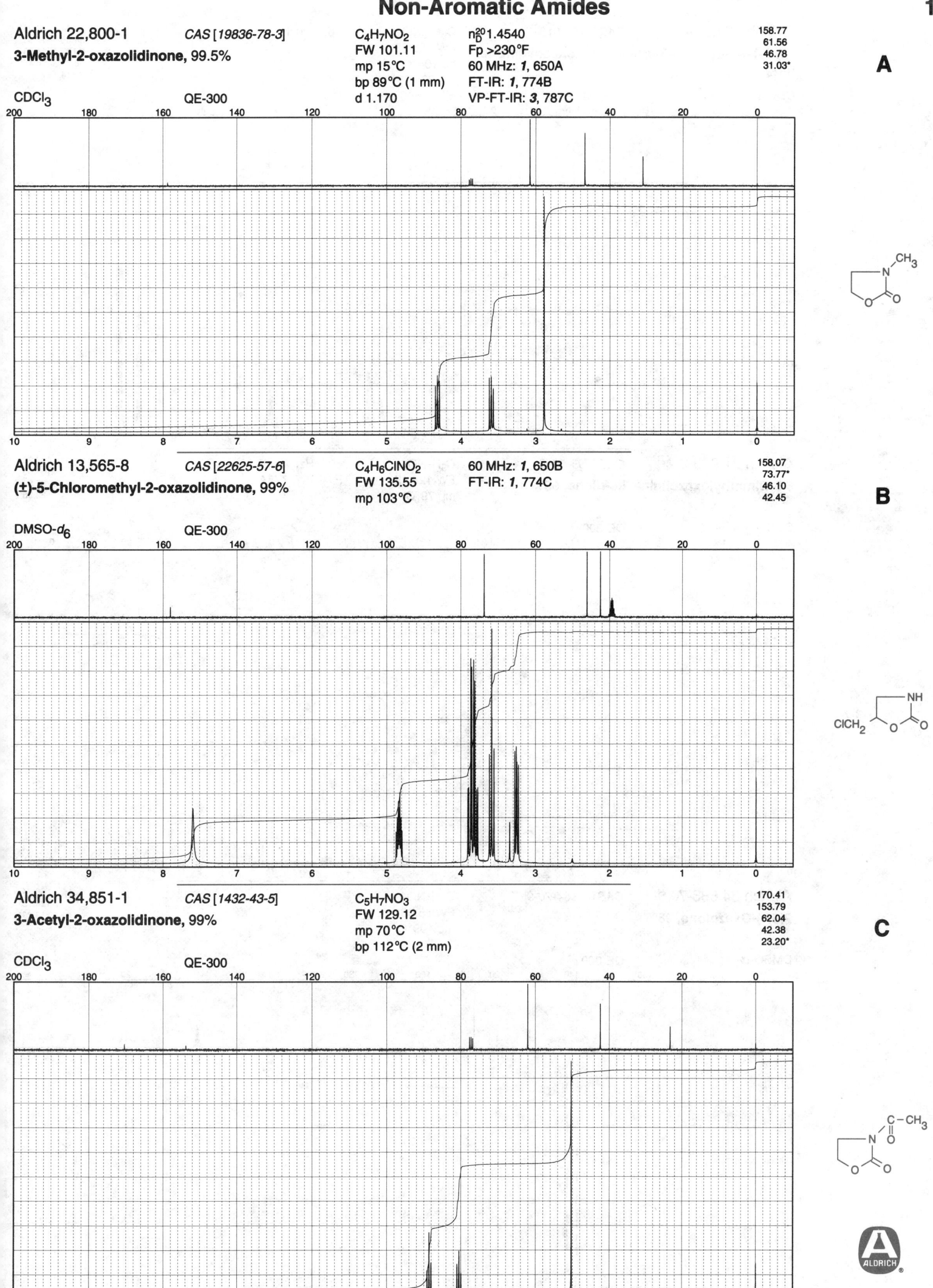

Aldrich 22,800-1   *CAS [19836-78-3]* $C_4H_7NO_2$ n_D^{20} 1.4540   158.77
3-Methyl-2-oxazolidinone, 99.5%   FW 101.11   Fp >230°F   61.56
mp 15°C   60 MHz: *1*, 650A   46.78
bp 89°C (1 mm)   FT-IR: *1*, 774B   31.03*
d 1.170   VP-FT-IR: *3*, 787C

A

CDCl$_3$ QE-300

Aldrich 13,565-8   *CAS [22625-57-6]* $C_4H_6ClNO_2$ 60 MHz: *1*, 650B   158.07
(±)-5-Chloromethyl-2-oxazolidinone, 99%   FW 135.55   FT-IR: *1*, 774C   73.77*
mp 103°C   46.10
42.45

B

DMSO-d_6 QE-300

Aldrich 34,851-1   *CAS [1432-43-5]* $C_5H_7NO_3$ 170.41
3-Acetyl-2-oxazolidinone, 99%   FW 129.12   153.79
mp 70°C   62.04
bp 112°C (2 mm)   42.38
23.20*

C

CDCl$_3$ QE-300

ALDRICH

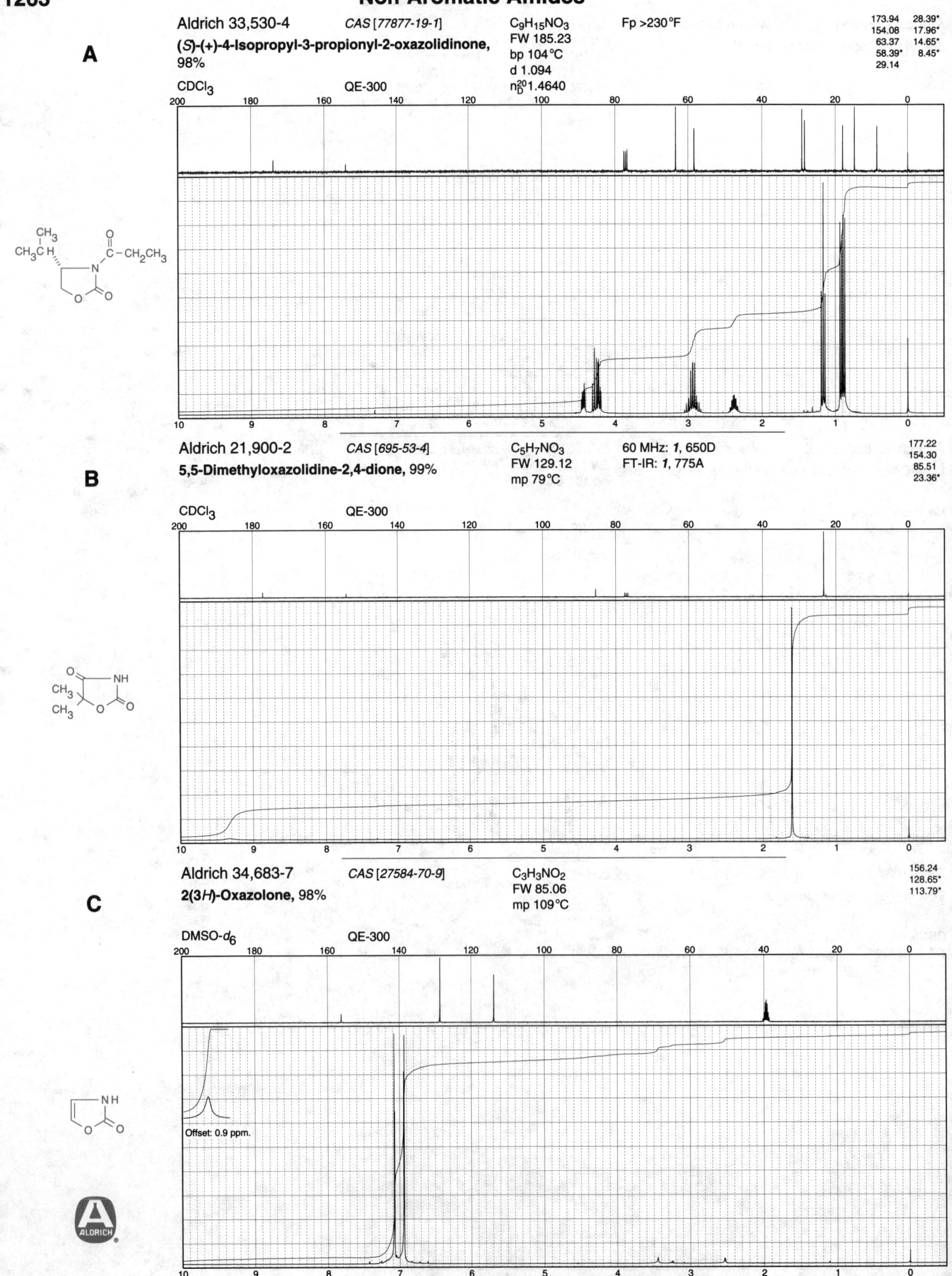

A

Aldrich 33,530-4   CAS [77877-19-1] $C_9H_{15}NO_3$ Fp >230°F
(S)-(+)-4-Isopropyl-3-propionyl-2-oxazolidinone, 98%
FW 185.23
bp 104°C
d 1.094
n_D^{20} 1.4640

CDCl₃   QE-300

173.94   28.39*
154.08   17.96*
63.37   14.65*
58.39*   8.45*
29.14

B

Aldrich 21,900-2   CAS [695-53-4] $C_5H_7NO_3$ 60 MHz: 1, 650D
5,5-Dimethyloxazolidine-2,4-dione, 99%
FW 129.12   FT-IR: 1, 775A
mp 79°C

CDCl₃   QE-300

177.22
154.30
85.51
23.36

C

Aldrich 34,683-7   CAS [27584-70-9] $C_3H_3NO_2$
2(3H)-Oxazolone, 98%
FW 85.06
mp 109°C

DMSO-d₆   QE-300

Offset: 0.9 ppm.

156.24
128.65*
113.79*

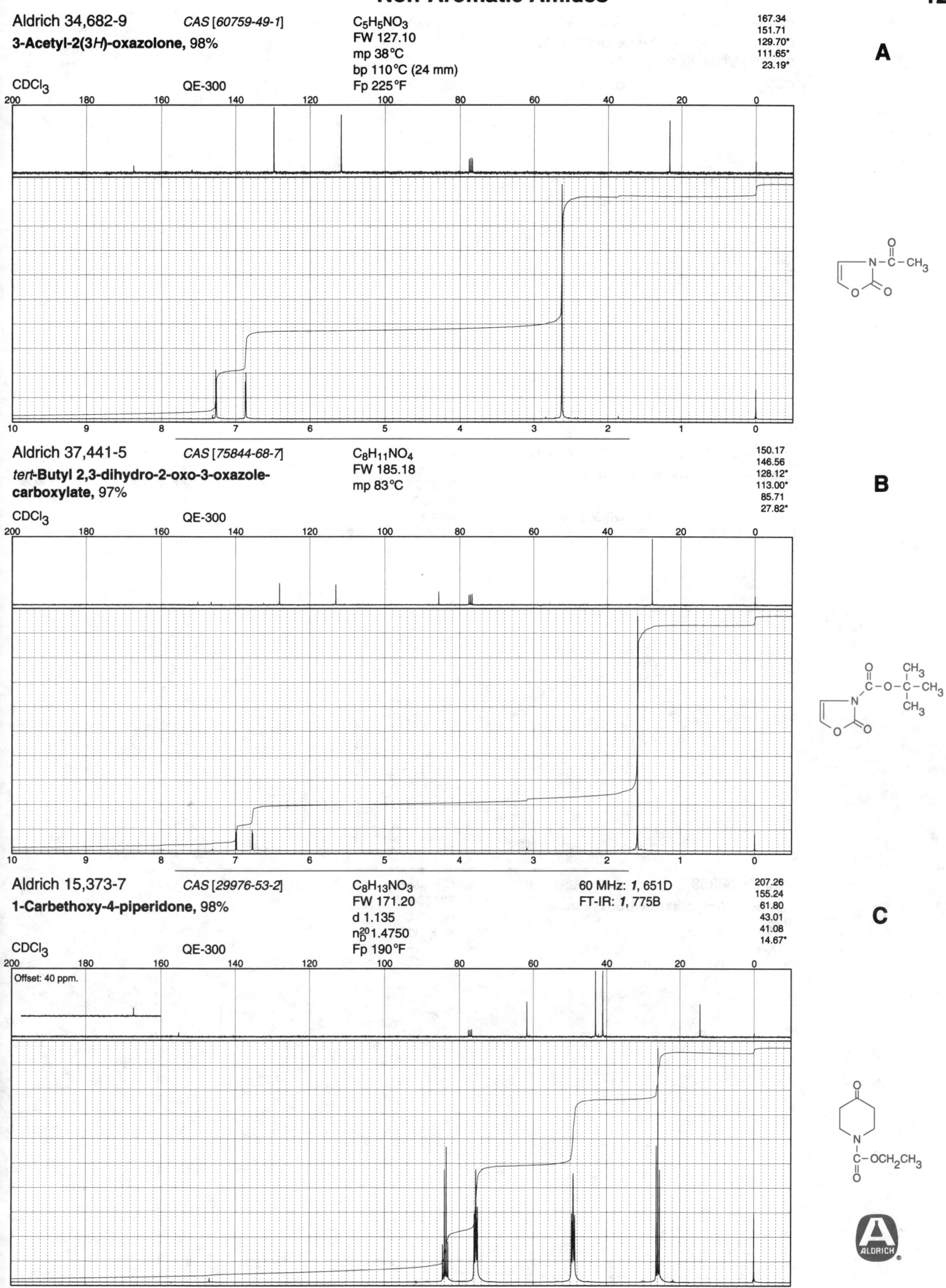

A

Aldrich 34,682-9   CAS [60759-49-1]
3-Acetyl-2(3H)-oxazolone, 98%

$C_5H_5NO_3$
FW 127.10
mp 38°C
bp 110°C (24 mm)
Fp 225°F

167.34
151.71
129.70*
111.65*
23.19*

CDCl₃   QE-300

B

Aldrich 37,441-5   CAS [75844-68-7]
***tert*-Butyl 2,3-dihydro-2-oxo-3-oxazole-carboxylate, 97%**

$C_8H_{11}NO_4$
FW 185.18
mp 83°C

150.17
146.56
128.12*
113.00*
85.71
27.82*

CDCl₃   QE-300

C

Aldrich 15,373-7   CAS [29976-53-2]
1-Carbethoxy-4-piperidone, 98%

$C_8H_{13}NO_3$
FW 171.20
d 1.135
n_D^{20} 1.4750
Fp 190°F

60 MHz: *1*, 651D
FT-IR: *1*, 775B

207.26
155.24
61.80
43.01
41.08
14.67*

CDCl₃   QE-300

Offset: 40 ppm.

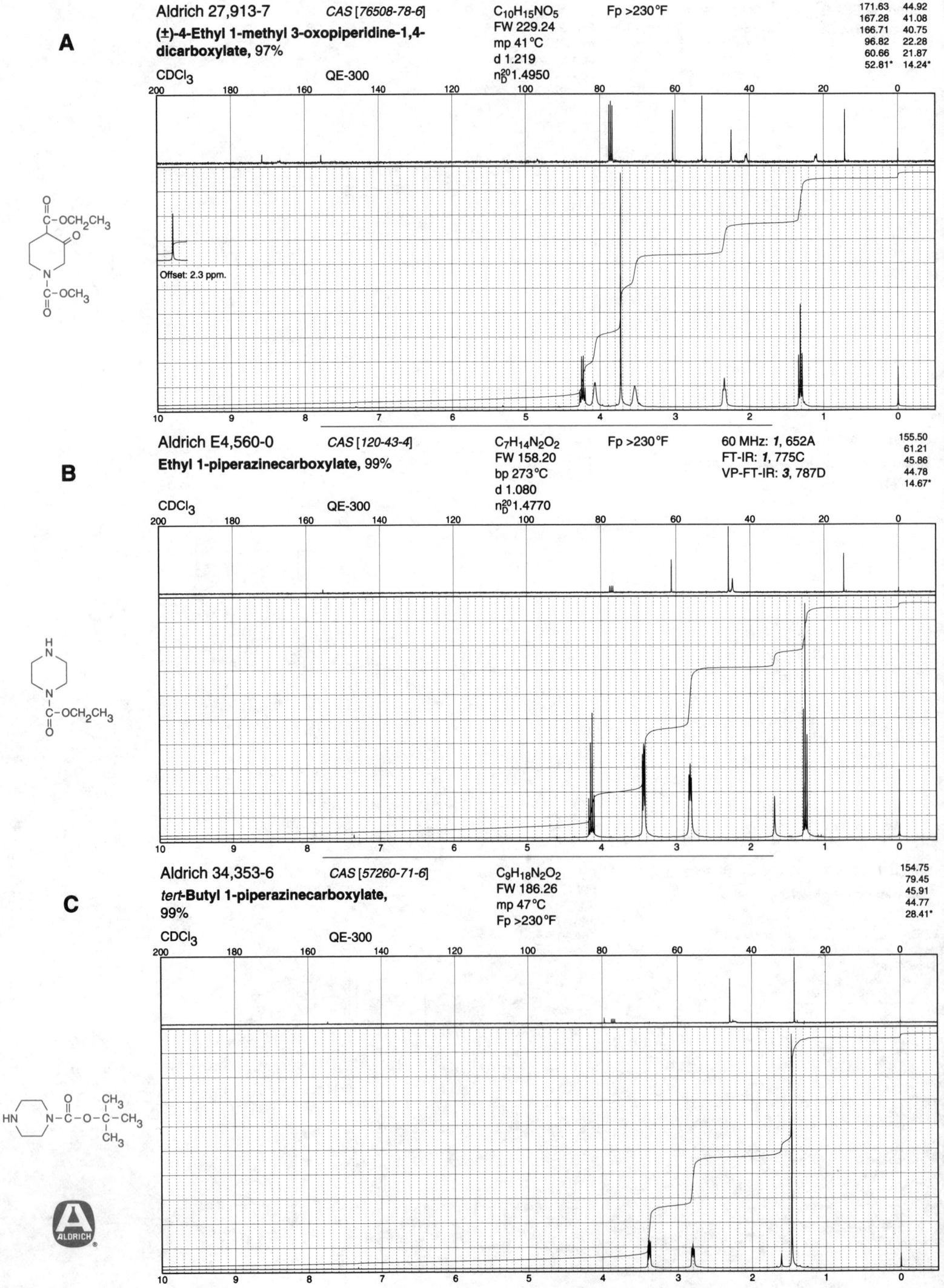

A

Aldrich 27,913-7   CAS [76508-78-6] $C_{10}H_{15}NO_5$ Fp >230°F

(±)-4-Ethyl 1-methyl 3-oxopiperidine-1,4-dicarboxylate, 97%

FW 229.24
mp 41°C
d 1.219
n_D^{20} 1.4950

171.63	44.92
167.28	41.08
166.71	40.75
96.82	22.28
60.66	21.87
52.81*	14.24*

CDCl₃   QE-300

Offset: 2.3 ppm.

B

Aldrich E4,560-0   CAS [120-43-4] $C_7H_{14}N_2O_2$ Fp >230°F

Ethyl 1-piperazinecarboxylate, 99%

FW 158.20
bp 273°C
d 1.080
n_D^{20} 1.4770

60 MHz: *1*, 652A
FT-IR: *1*, 775C
VP-FT-IR: *3*, 787D

155.50
61.21
45.86
44.78
14.67*

CDCl₃   QE-300

C

Aldrich 34,353-6   CAS [57260-71-6] $C_9H_{18}N_2O_2$

tert-Butyl 1-piperazinecarboxylate, 99%

FW 186.26
mp 47°C
Fp >230°F

154.75
79.45
45.91
44.77
28.41*

CDCl₃   QE-300

ALDRICH

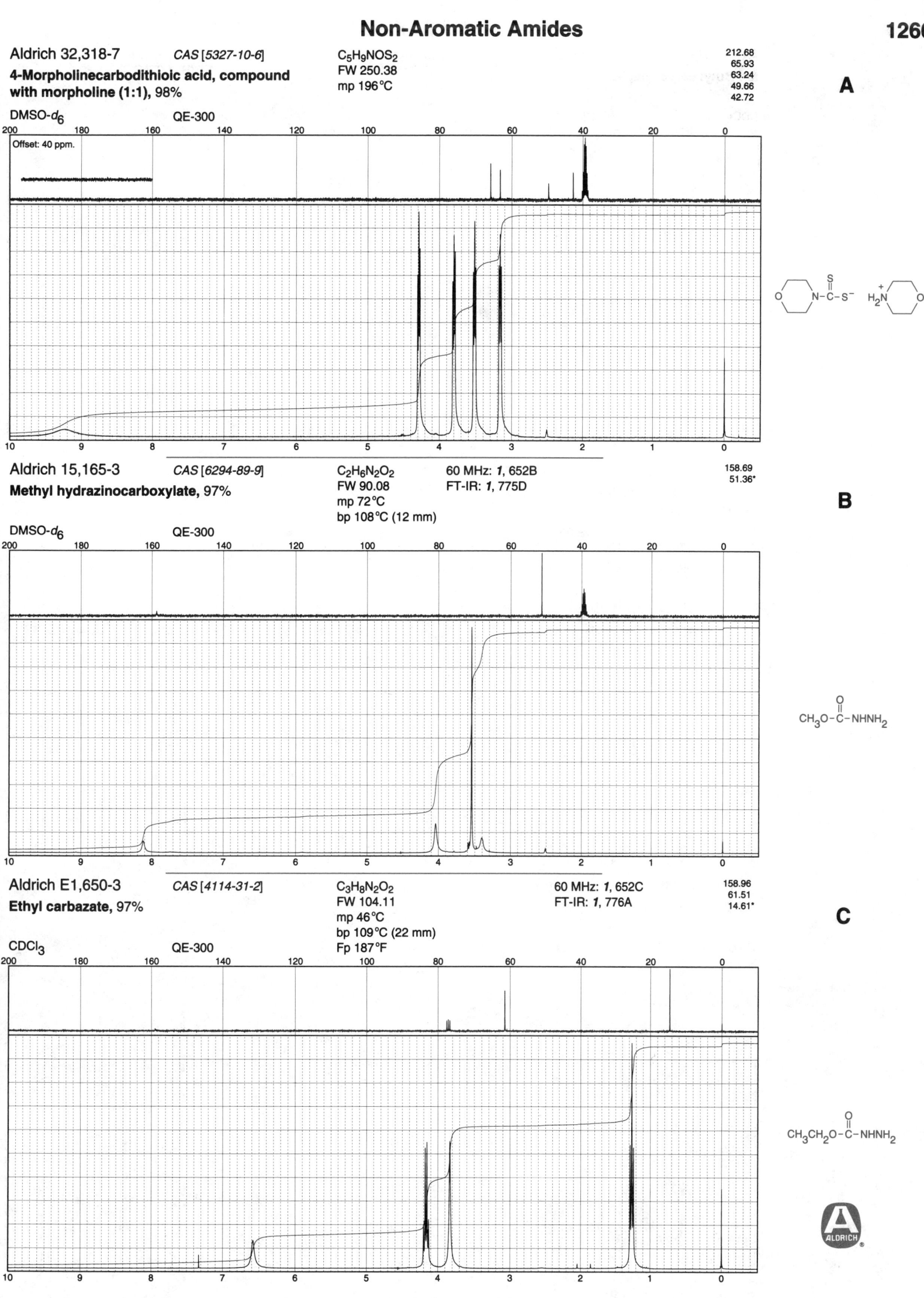

Aldrich 32,318-7   CAS [5327-10-6]
**4-Morpholinecarbodithioic acid, compound
with morpholine (1:1), 98%**

C$_5$H$_9$NOS$_2$
FW 250.38
mp 196°C

212.68
65.93
63.24
49.66
42.72

A

DMSO-d_6 QE-300

Offset: 40 ppm.

200 180 160 140 120 100 80 60 40 20 0
10 9 8 7 6 5 4 3 2 1 0

Aldrich 15,165-3   CAS [6294-89-9]
Methyl hydrazinocarboxylate, 97%

C$_2$H$_6$N$_2$O$_2$
FW 90.08
mp 72°C
bp 108°C (12 mm)

60 MHz: **1**, 652B
FT-IR: **1**, 775D

158.69
51.36*

B

DMSO-d_6 QE-300

200 180 160 140 120 100 80 60 40 20 0
10 9 8 7 6 5 4 3 2 1 0

CH$_3$O–C–NHNH$_2$

Aldrich E1,650-3   CAS [4114-31-2]
Ethyl carbazate, 97%

C$_3$H$_8$N$_2$O$_2$
FW 104.11
mp 46°C
bp 109°C (22 mm)
Fp 187°F

60 MHz: **1**, 652C
FT-IR: **1**, 776A

158.96
61.51
14.61*

C

CDCl$_3$ QE-300

200 180 160 140 120 100 80 60 40 20 0
10 9 8 7 6 5 4 3 2 1 0

CH$_3$CH$_2$O–C–NHNH$_2$

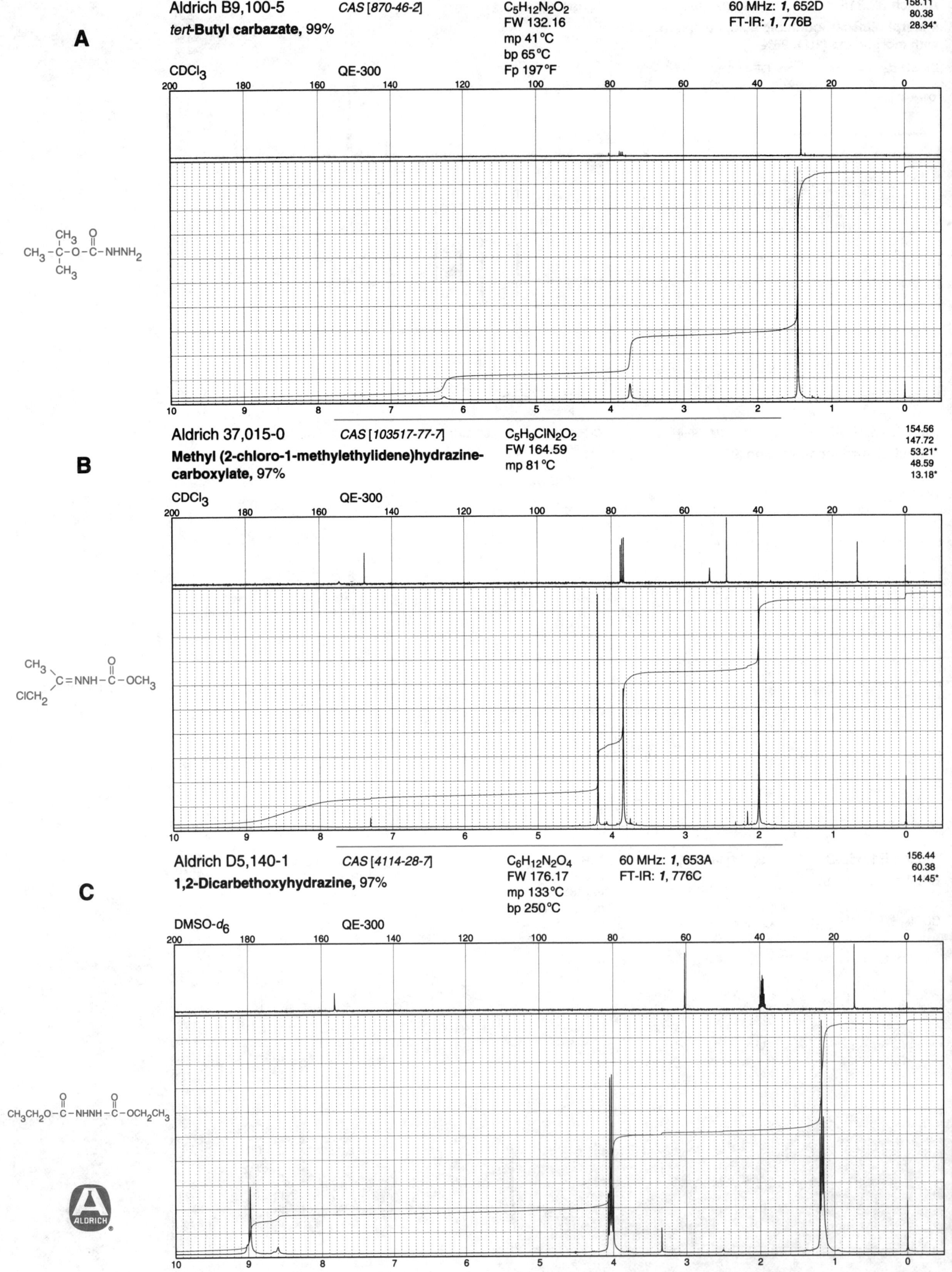

A Aldrich B9,100-5   CAS [870-46-2] $C_5H_{12}N_2O_2$
tert-Butyl carbazate, 99%   FW 132.16
mp 41°C
bp 65°C
Fp 197°F

60 MHz: *1*, 652D
FT-IR: *1*, 776B

158.11
80.38
28.34*

CDCl₃   QE-300

B Aldrich 37,015-0   CAS [103517-77-7] $C_5H_9ClN_2O_2$
Methyl (2-chloro-1-methylethylidene)hydrazine-   FW 164.59
carboxylate, 97%   mp 81°C

154.56
147.72
53.21*
48.59
13.18*

CDCl₃   QE-300

C Aldrich D5,140-1   CAS [4114-28-7] $C_6H_{12}N_2O_4$
1,2-Dicarbethoxyhydrazine, 97%   FW 176.17
mp 133°C
bp 250°C

60 MHz: *1*, 653A
FT-IR: *1*, 776C

156.44
60.38
14.45*

DMSO-*d*₆   QE-300

Non-Aromatic Amides

1268

A

Aldrich 13,453-8 *CAS [4530-20-5]* C₇H₁₃NO₄ 60 MHz: *1*, 654B

$C_7H_{13}NO_4$
FW 175.18
mp 88°C

N-(*tert*-Butoxycarbonyl)glycine, 98%

60 MHz: *1*, 654B
FT-IR: *1*, 777A

174.75	80.38
173.93	43.34
157.28	42.18
155.95	28.25*
81.77	

CDCl₃ QE-300

Offset: 2.1 ppm.

B

Aldrich 35,972-6 *CAS [13734-41-3]*

N-(*tert*-Butoxycarbonyl)-L-valine, 99%

$C_{10}H_{19}NO_4$
FW 217.27
mp 80°C

173.34	29.48*
155.63	28.13*
77.85	19.06*
59.00*	18.06*

DMSO-d₆ QE-300

Offset: 2.7 ppm.

C

Aldrich 35,965-3 *CAS [13139-16-7]*

N-(*tert*-Butoxycarbonyl)-L-isoleucine, 98%

$C_{11}H_{21}NO_4$
FW 231.30
mp 71°C

173.33	28.13*
155.51	24.71
77.86	15.49*
57.99*	11.15*
36.01*	

DMSO-d₆ QE-300

Offset: 2.6 ppm.

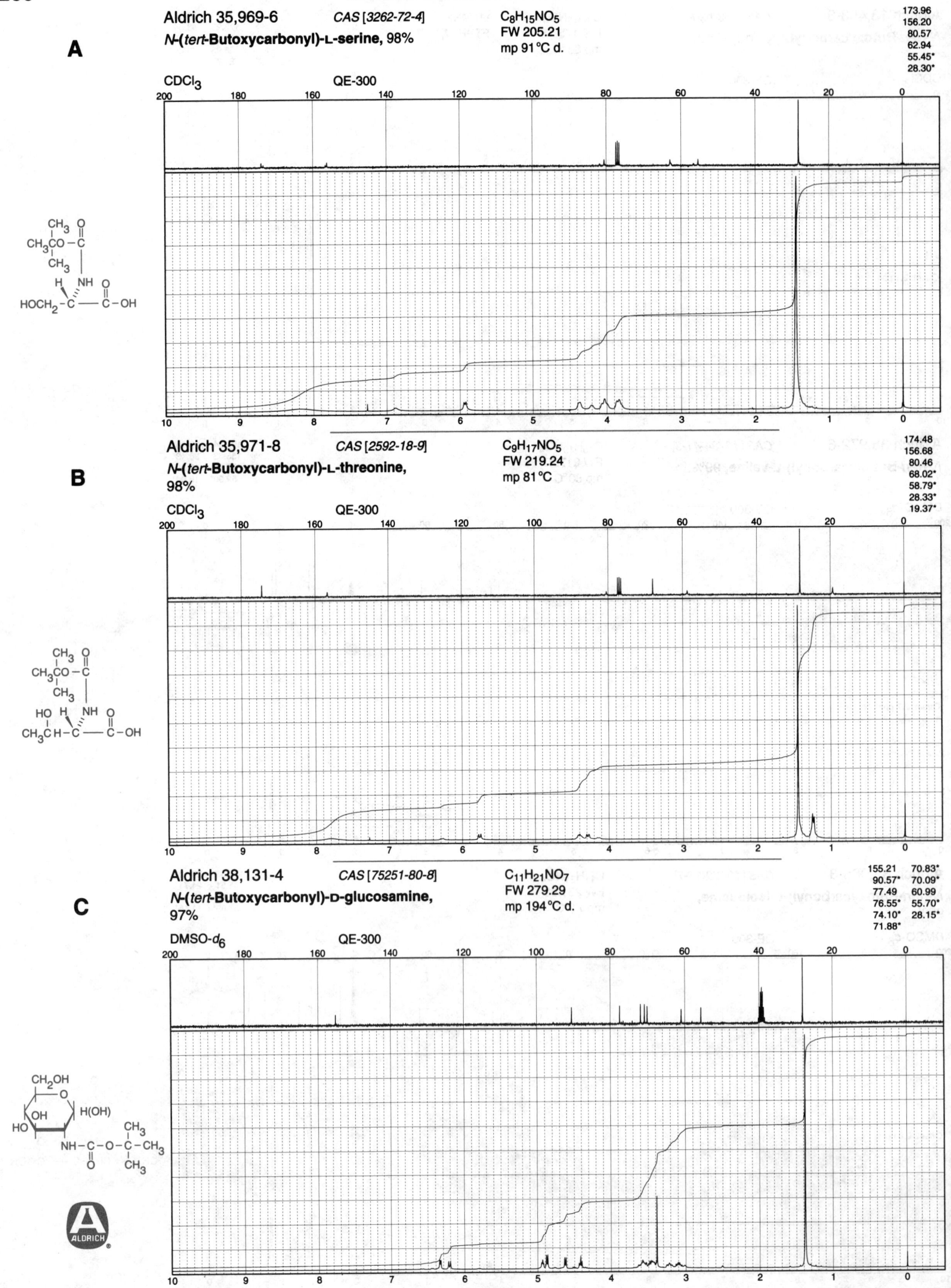

A

Aldrich 35,969-6 *CAS [3262-72-4]* $C_8H_{15}NO_5$ FW 205.21 mp 91°C d.

N-(tert-Butoxycarbonyl)-L-serine, 98%

173.96
156.20
80.57
62.94
55.45*
28.30*

CDCl₃ QE-300

B

Aldrich 35,971-8 *CAS [2592-18-9]* $C_9H_{17}NO_5$ FW 219.24 mp 81°C

N-(tert-Butoxycarbonyl)-L-threonine, 98%

174.48
156.68
80.46
68.02*
58.79*
28.33*
19.37*

CDCl₃ QE-300

C

Aldrich 38,131-4 *CAS [75251-80-8]* $C_{11}H_{21}NO_7$ FW 279.29 mp 194°C d.

N-(tert-Butoxycarbonyl)-D-glucosamine, 97%

155.21 70.83*
90.57* 70.09*
77.49 60.99
76.55* 55.70*
74.10* 28.15*
71.88*

DMSO-d₆ QE-300

ALDRICH

A

Aldrich 34,636-5   *CAS [61315-59-1]*

$C_{10}H_{19}NO_4S$
FW 430.66
mp 142°C

N-(*tert*-Butoxycarbonyl)-D-methionine,
dicyclohexylammonium salt, 98%

CDCl$_3$ QE-300

175.71	30.34
155.35	29.06
78.58	28.45*
55.16*	25.17
52.51*	24.77
33.74	15.46*

B

Aldrich 35,968-8   *CAS [13734-28-6]*

$C_{11}H_{22}N_2O_4$
FW 246.31
mp 207°C d.

*N*α-(*tert*-Butoxycarbonyl)-L-lysine, 99%

CD$_3$OD   QE-300

179.22	33.70
157.56	28.79*
80.03	28.23
56.72*	23.42
40.54	

C

Aldrich 35,966-1   *CAS [2418-95-3]*

$C_{11}H_{22}N_2O_4$
FW 246.31
mp 250°C d.

*N*ε-(*tert*-Butoxycarbonyl)-L-lysine, 97%

D$_2$O + NAOD   QE-300

184.80	37.08
159.10	31.41
81.23	30.30*
58.20*	24.89
42.49	

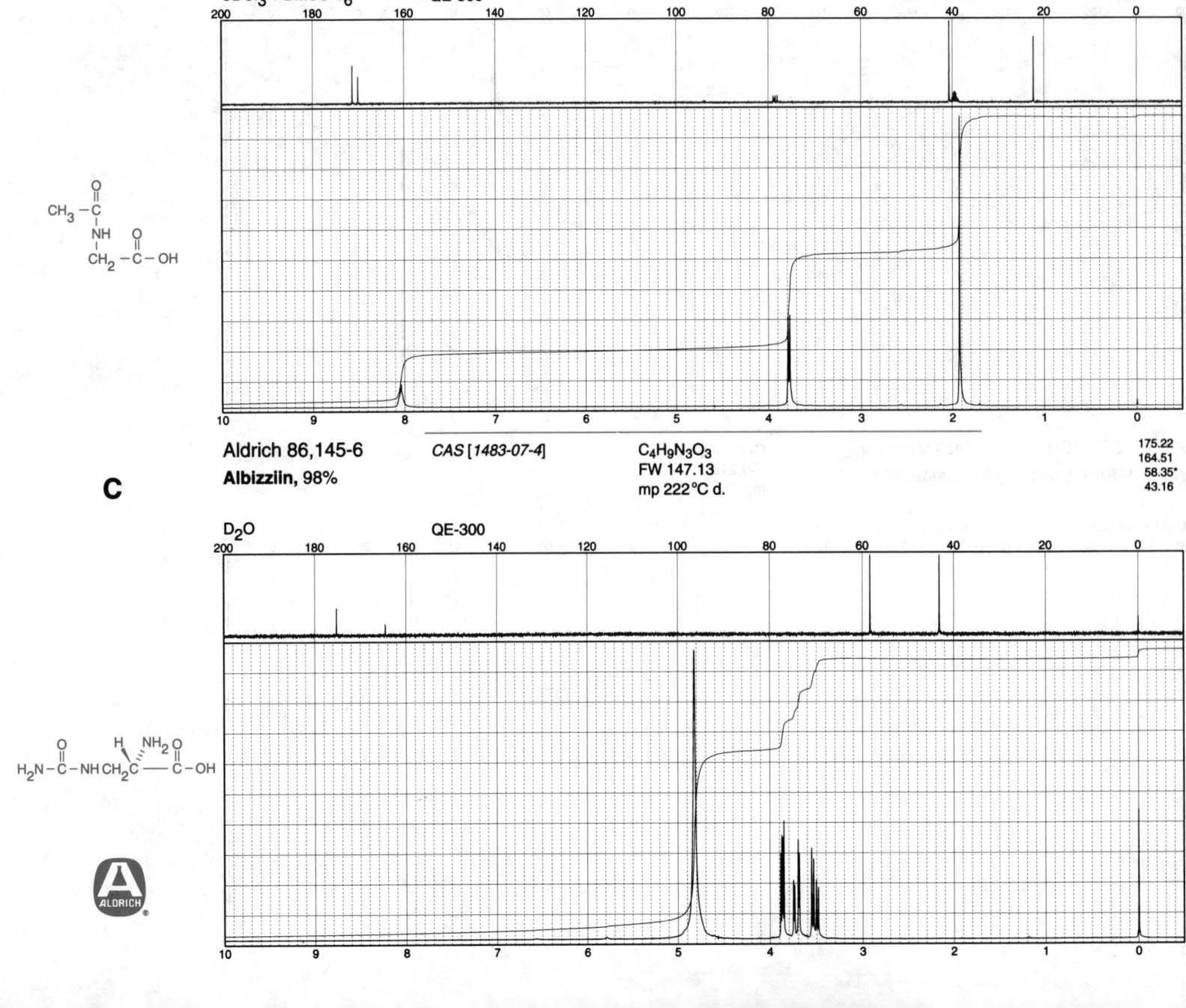

A Aldrich 33,203-8   *CAS [51779-32-9]*   C$_{10}$H$_{19}$NO$_4$
Di-*tert*-butyl iminodicarboxylate, 95%   FW 217.27
mp 91 °C

149.78
81.92
28.03*

CDCl$_3$ QE-300

B Aldrich A1,630-0   *CAS [543-24-8]*   C$_4$H$_7$NO$_3$ 60 MHz: *1*, 655A
***N*-Acetylglycine, 99%**   FW 117.10   FT-IR: *1*, 777C
mp 208 °C

171.31
170.04
40.74
22.30*

CDCl$_3$ + DMSO-*d$_6$*   QE-300

C Aldrich 86,145-6   *CAS [1483-07-4]*   C$_4$H$_9$N$_3$O$_3$
Albizziin, 98%   FW 147.13
mp 222 °C d.

175.22
164.51
58.35*
43.16

D$_2$O   QE-300

A

Aldrich 23,329-3 *CAS [3103-38-6]* C$_6$H$_9$NO$_4$ FT-IR: *1*, 777D

N-(Acetoacetyl)glycine, 95+% FW 159.14
mp 105°C

CDCl$_3$ + DMSO-d$_6$ QE-300

Offset: 40 ppm.

202.38
170.95
166.45
50.94
40.85
29.82*

CH$_3$-C(=O)-CH$_2$-C(=O)-NHCH$_2$-C(=O)-OH

B

Aldrich 30,268-6 *CAS [77402-03-0]* C$_7$H$_{11}$NO$_4$

Methyl 2-acrylamido-2-methoxyacetate, 98% FW 173.17
mp 74°C

CDCl$_3$ QE-300

168.48
165.76
130.03*
128.33
78.40*
56.71*
52.88*

H$_2$C=CH-C(=O)-NHCH-C(=O)-OCH$_3$
           |
          OCH$_3$

C

Aldrich 26,050-9 *CAS [4387-85-3]* C$_8$H$_{10}$N$_2$O$_4$

2,2-Bis(acrylamido)acetic acid, 96% FW 198.18
FT-IR: *1*, 778A

DMSO-d$_6$ QE-300

169.40
164.26
130.73*
126.42
55.93*

H$_2$C=CH-C(=O)-NH
                \
                 CH-C(=O)-OH
                /
H$_2$C=CH-C(=O)-NH

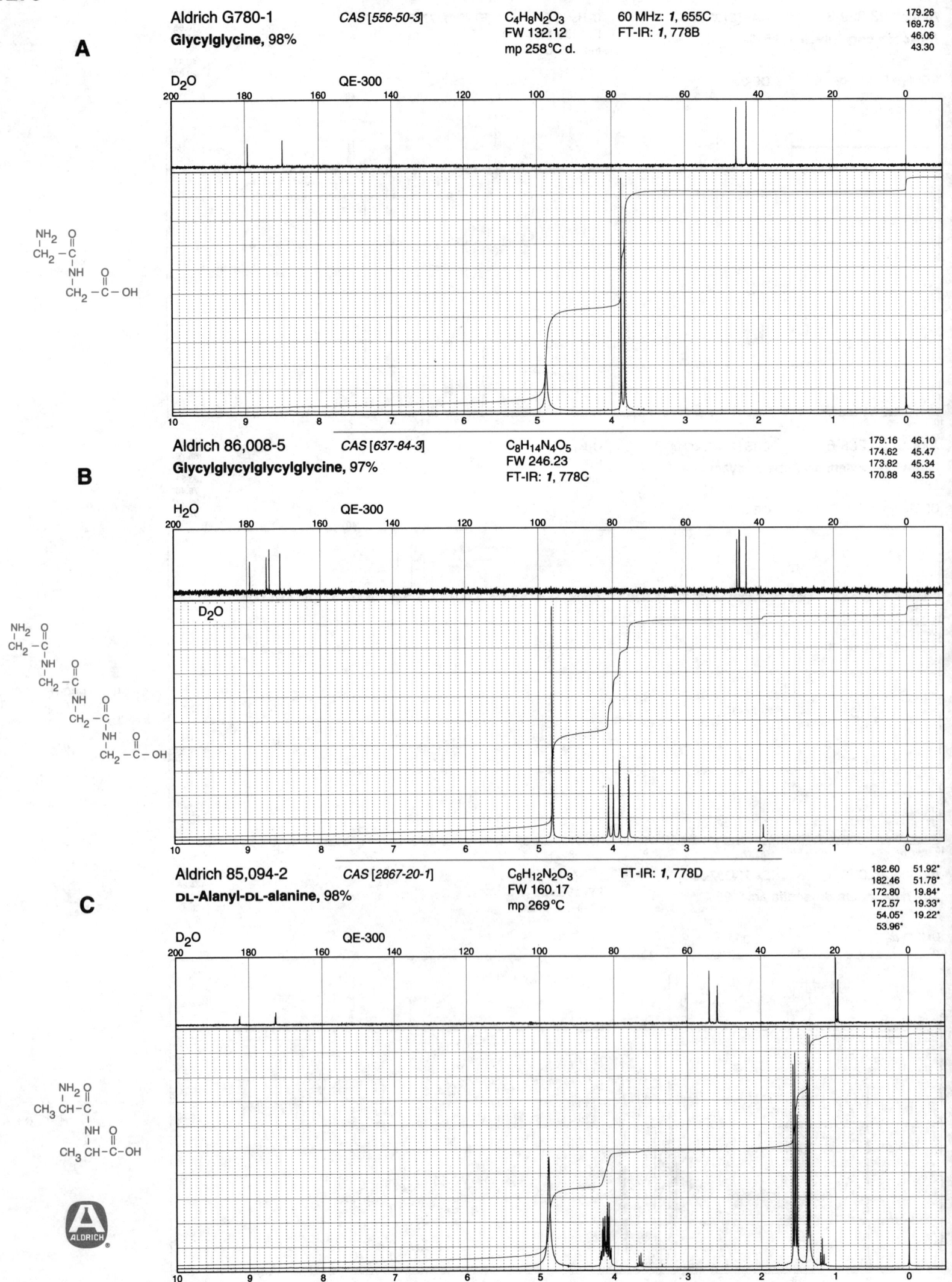

A

Aldrich G780-1
Glycylglycine, 98%

CAS [556-50-3]

$C_4H_8N_2O_3$
FW 132.12
mp 258°C d.

60 MHz: *1*, 655C
FT-IR: *1*, 778B

179.26
169.78
46.06
43.30

D₂O QE-300

B

Aldrich 86,008-5
Glycylglycylglycylglycine, 97%

CAS [637-84-3]

$C_8H_{14}N_4O_5$
FW 246.23
FT-IR: *1*, 778C

179.16   46.10
174.62   45.47
173.82   45.34
170.88   43.55

H₂O QE-300

D₂O

C

Aldrich 85,094-2
DL-Alanyl-DL-alanine, 98%

CAS [2867-20-1]

$C_6H_{12}N_2O_3$
FW 160.17
mp 269°C

FT-IR: *1*, 778D

182.60   51.92*
182.46   51.78*
172.80   19.84*
172.57   19.33*
54.05*   19.22*
53.96*

D₂O QE-300

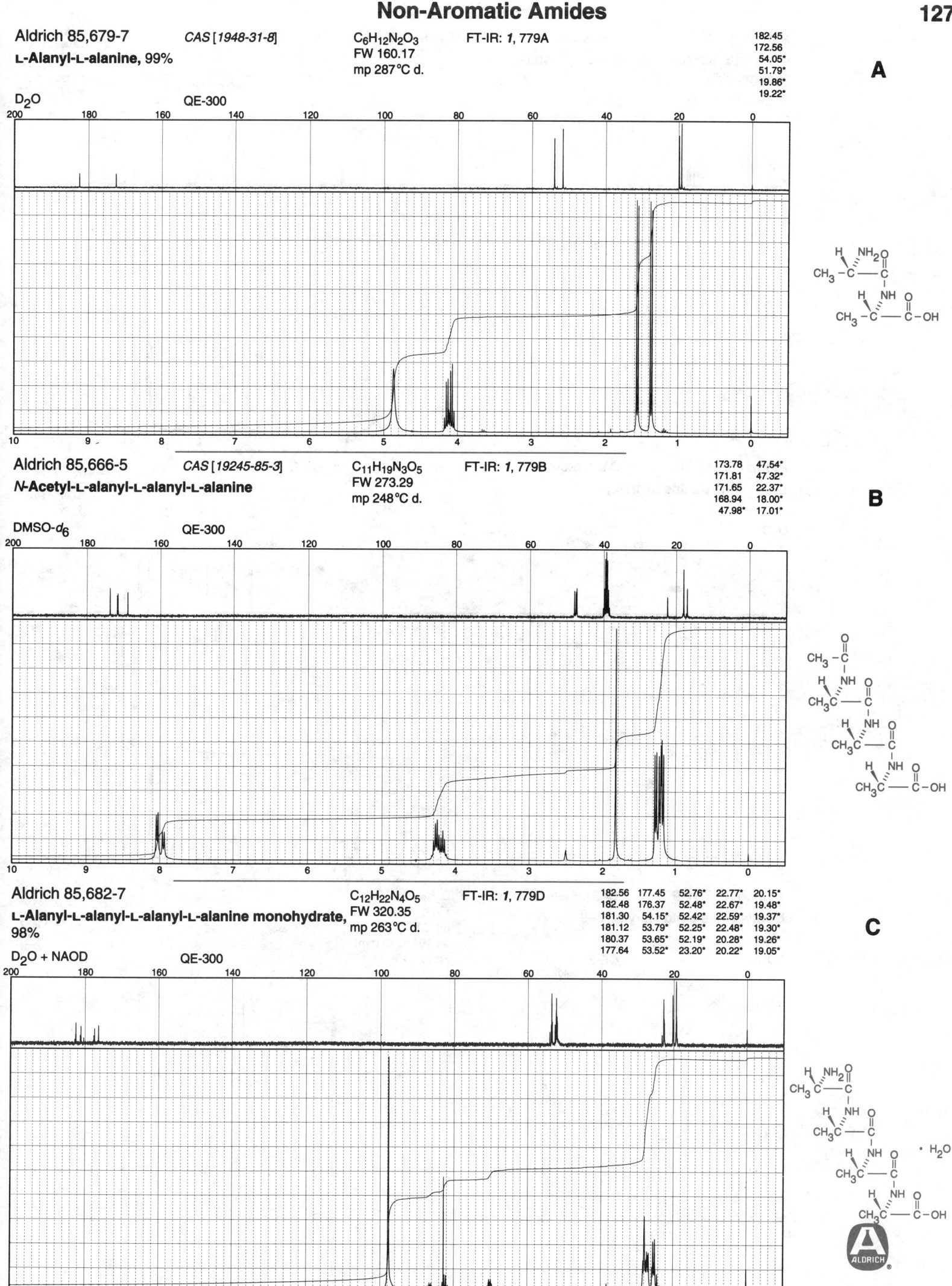

Aldrich 85,679-7
L-Alanyl-L-alanine, 99%

CAS [1948-31-8]

$C_6H_{12}N_2O_3$
FW 160.17
mp 287°C d.

FT-IR: *1*, 779A

182.45
172.56
54.05*
51.79*
19.86*
19.22*

A

D_2O QE-300

Aldrich 85,666-5
***N*-Acetyl-L-alanyl-L-alanyl-L-alanine**

CAS [19245-85-3]

$C_{11}H_{19}N_3O_5$
FW 273.29
mp 248°C d.

FT-IR: *1*, 779B

173.78	47.54*
171.81	47.32*
171.65	22.37*
168.94	18.00*
47.98*	17.01*

B

DMSO-d_6 QE-300

Aldrich 85,682-7
L-Alanyl-L-alanyl-L-alanyl-L-alanine monohydrate, 98%

$C_{12}H_{22}N_4O_5$
FW 320.35
mp 263°C d.

FT-IR: *1*, 779D

182.56	177.45	52.76*	22.77*	20.15*
182.48	176.37	52.48*	22.67*	19.48*
181.30	54.15*	52.42*	22.59*	19.37*
181.12	53.79*	52.25*	22.48*	19.30*
180.37	53.65*	52.19*	20.28*	19.26*
177.64	53.52*	23.20*	20.22*	19.05*

C

D_2O + NAOD   QE-300

Non-Aromatic Amides

A

Aldrich 85,926-5   *CAS [41036-27-5]*   C₁₀H₁₉N₃O₄   FT-IR: *1, 780A*

L-Alanyl-L-alanyl-L-alanine methyl ester acetate, 99%

C$_{10}$H$_{19}$N$_3$O$_4$
FW 305.33
mp 153°C

174.00	48.00*
173.06	47.59*
172.82	22.44*
172.17	19.67*
51.77*	18.31*
49.49*	17.00*

CDCl₃ + DMSO-*d*₆   QE-300

B

Aldrich 85,125-6   *CAS [7298-84-2]*   C₉H₁₈N₂O₃   FT-IR: *1, 780B*

L-Leucyl-L-alanine hydrate, 99%

C$_9$H$_{18}$N$_2$O$_3$
FW 202.25
mp 260°C

182.07	26.55*
172.17	24.60*
54.65*	23.85*
54.07*	20.03*
42.56	

D₂O   QE-300

C

Aldrich 31,751-9   *CAS [35356-70-8]*   C₆H₉NO₃

Methyl 2-acetamidoacrylate, 99%

C$_6$H$_9$NO$_3$
FW 143.14
mp 51°C
bp 104°C (8 mm)
Fp >230°F

168.84
164.53
130.92
108.67
52.94*
24.61*

CDCl₃   QE-300

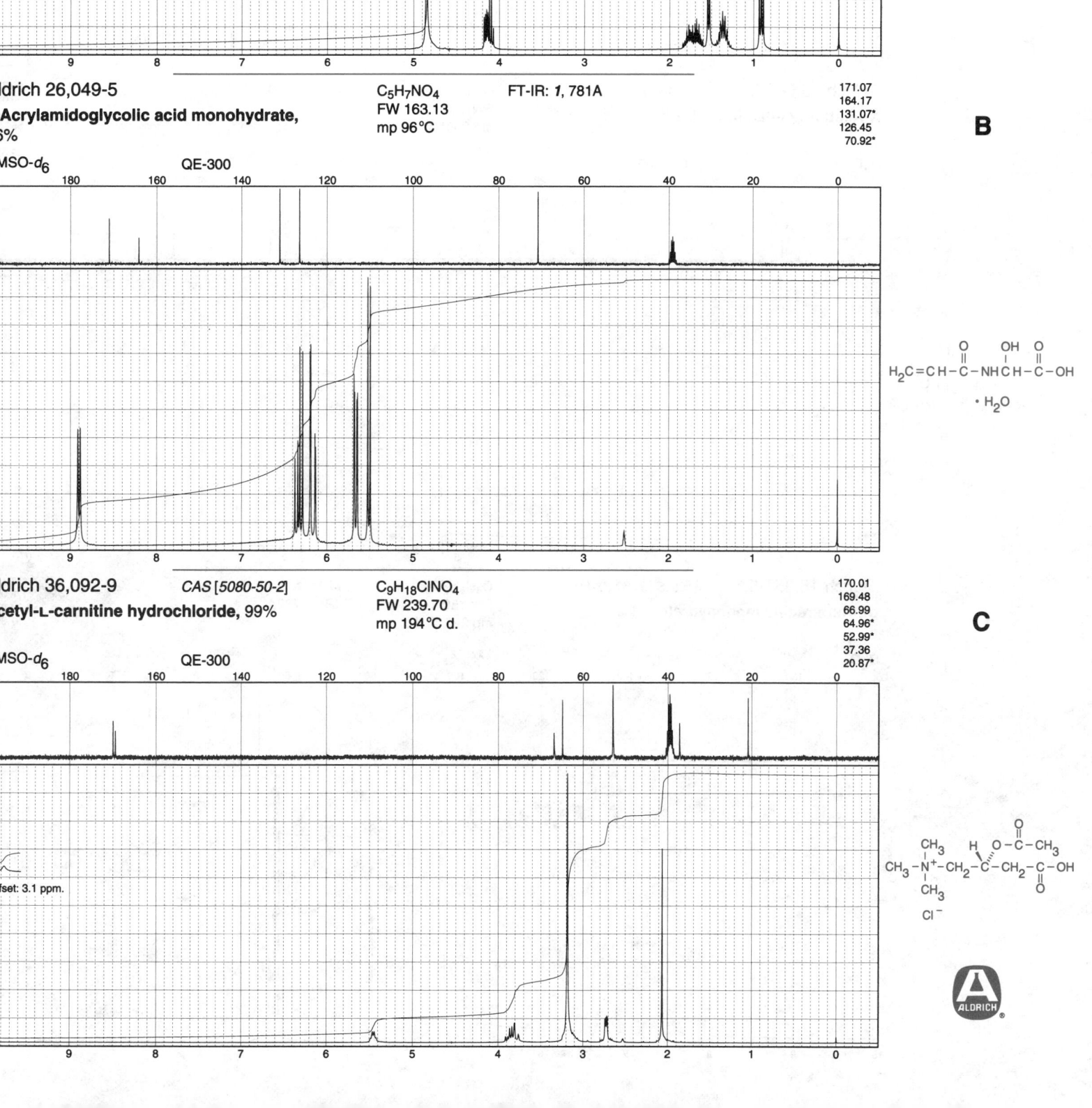

Aldrich 85,001-2   *CAS [2325-18-0]*   C₈H₁₆N₂O₃   FT-IR: *1*, 780C

DL-Alanyl-DL-norvaline
FW 188.23
mp 241°C

182.05	36.27
173.19	21.58
58.04*	19.56*
52.01*	15.71*

A

D₂O   QE-300

Aldrich 26,049-5

2-Acrylamidoglycolic acid monohydrate, 96%

C₅H₇NO₄   FT-IR: *1*, 781A
FW 163.13
mp 96°C

171.07
164.17
131.07*
126.45
70.92*

B

DMSO-*d₆*   QE-300

Aldrich 36,092-9   *CAS [5080-50-2]*   C₉H₁₈ClNO₄

Acetyl-L-carnitine hydrochloride, 99%

FW 239.70
mp 194°C d.

170.01
169.48
66.99
64.96*
52.99*
37.36
20.87*

C

DMSO-*d₆*   QE-300

Offset: 3.1 ppm.

A

Aldrich 36,093-7   *CAS [18877-64-0]*   C$_{23}$H$_{46}$ClNO$_4$
Palmitoyl-L carnitine hydrochloride, mp 170°C
95%

CD$_3$OD   QE-300

174.03	35.08	30.12
172.25	32.98	25.66
69.45	30.68	23.64
66.21*	30.50	14.39*
54.63*	30.37	
37.91	30.32	

B

Aldrich 85,564-2   *CAS [1188-37-0]*   C$_7$H$_{11}$NO$_5$ FT-IR: *1*, 781D
N-Acetyl-L-glutamic acid, 99%   FW 189.17
mp 200°C

D$_2$O   QE-300

180.85
180.06
176.77
56.59*
33.69
29.61
24.71*

C

Aldrich 15,357-5   *CAS [3130-87-8]*   C$_4$H$_8$N$_2$O$_3$ 60 MHz: *1*, 656B
DL-Asparagine monohydrate, 98%   FW 150.14   FT-IR: *1*, 782A
mp 220°C d.

177.15
176.09
54.04*
37.30

D$_2$O   QE-300

ALDRICH

Aldrich A9,300-3 *CAS [70-47-3]* C₄H₈N₂O₃

L-Asparagine, 99%

FW 132.12
mp 235°C d.

177.15
176.07
54.05*
37.29

A

D₂O   QE-300

H₂N–C–CH₂–C–C–OH (with O, H, NH₂, O groups)

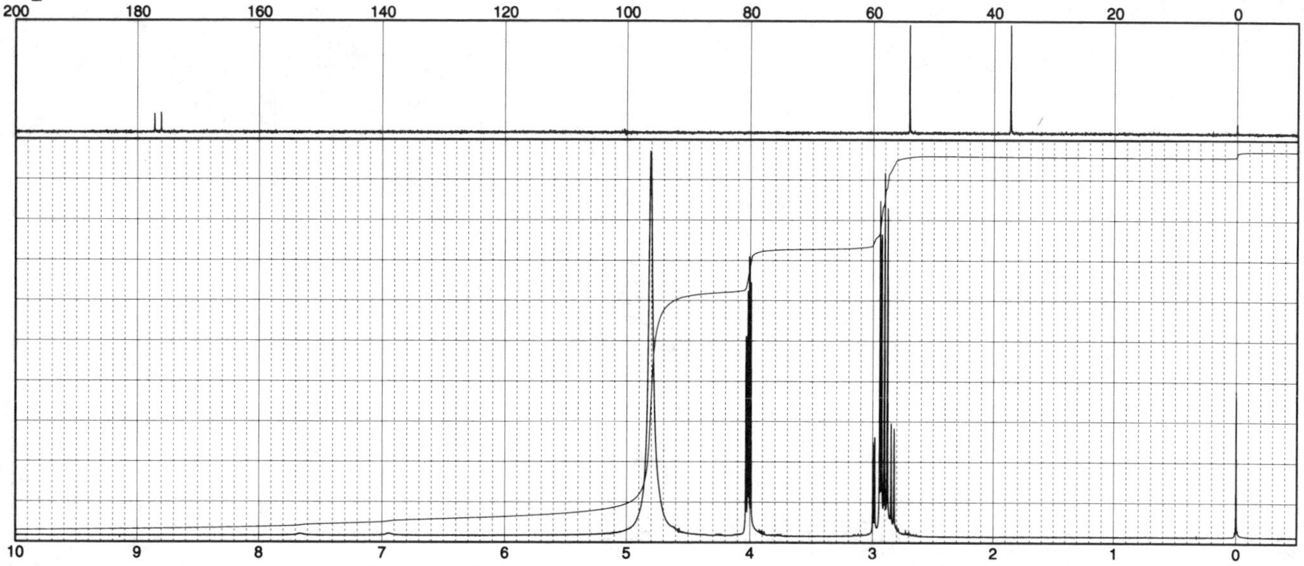

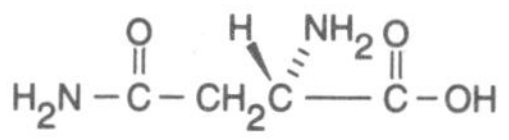

Aldrich G320-2 *CAS [56-85-9]* C₅H₁₀N₂O₃   FT-IR: *1*, 782D

L-Glutamine, 99%

FW 146.15
mp 185°C d.

180.30
176.71
56.91*
33.63
28.99

B

D₂O   QE-300

H₂N–C–CH₂CH₂–C–C–OH (with O, H, NH₂, O groups)

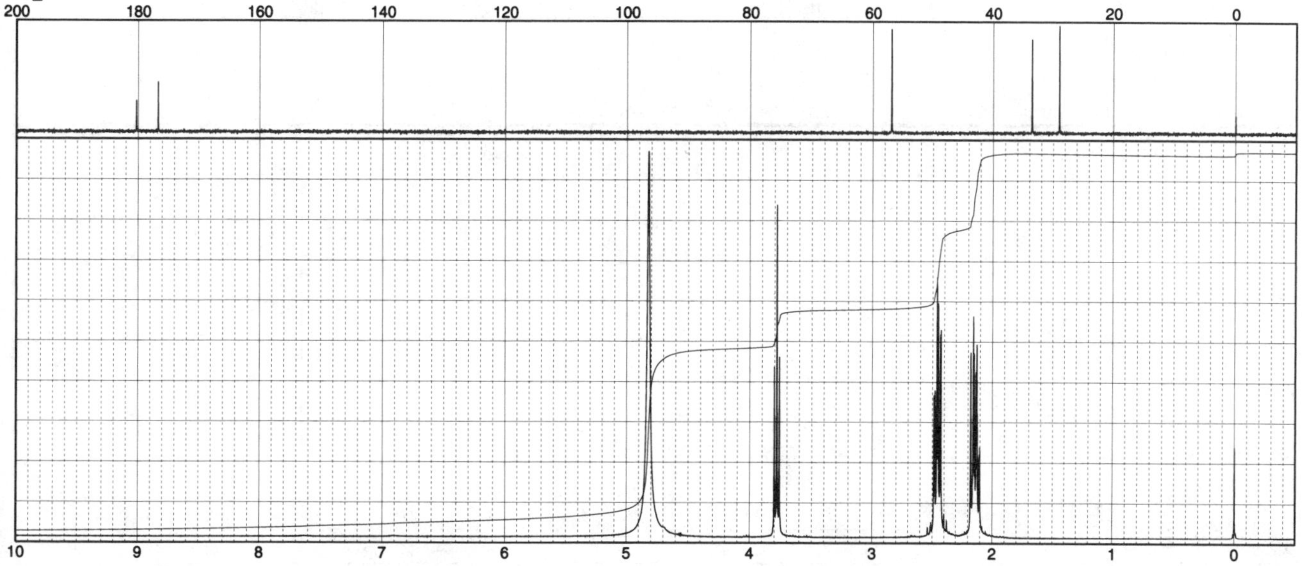

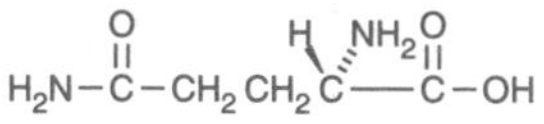

Aldrich 85,925-7 *CAS [81013-00-5]* C₉H₁₉N₃O₂   FT-IR: *1*, 783B

Nα-Acetyl-L-lysine-N-methylamide monohydrate

FW 219.29
mp 175°C

172.13   31.28
169.07   25.40*
52.50*   22.70
40.49   22.42*
31.69

C

DMSO-d₆   QE-300

CH₃–C...NH...C–NHCH₃ • H₂O
H₂NCH₂CH₂CH₂CH₂C

ALDRICH

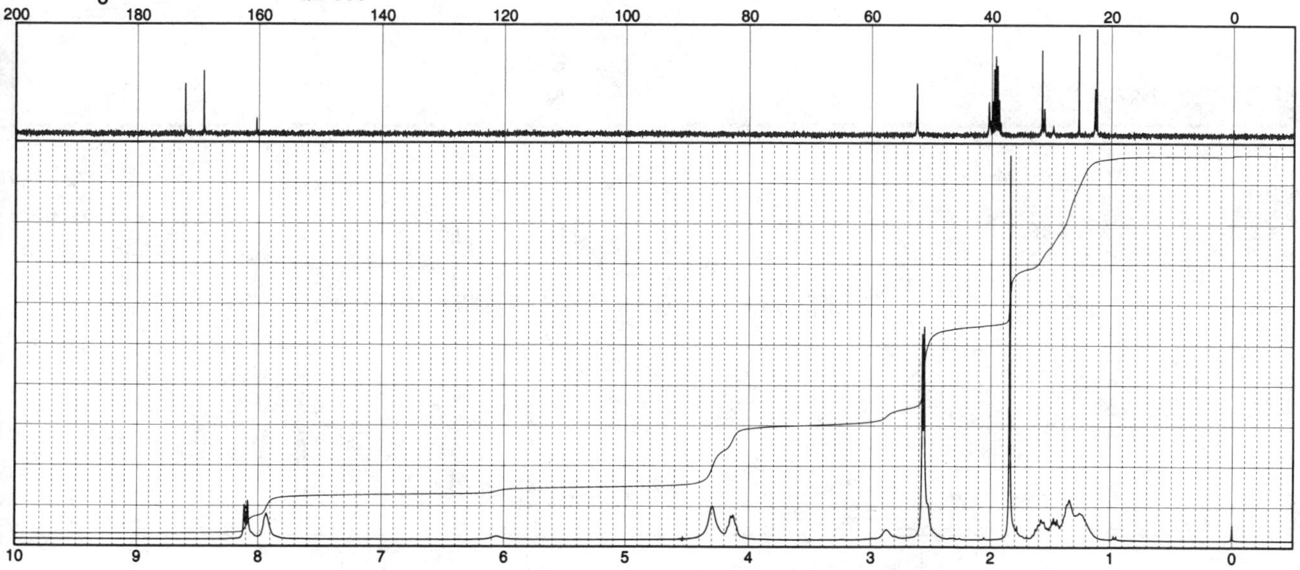

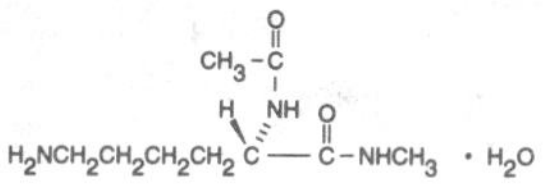

A

Aldrich 11,579-7   CAS [692-04-6]   C₈H₁₆N₂O₃   60 MHz: *1*, 657C

$N\varepsilon$-Acetyl-L-lysine, 99%

FW 188.23   FT-IR: *1*, 783C

mp 250°C d.

177.45	32.86
176.63	30.73
57.44*	24.66*
41.78	24.55

D₂O   QE-300

B

Aldrich 85,909-5   CAS [20911-93-7]   C₉H₁₈N₂O₃   FT-IR: *1*, 784A

$N\alpha$-Acetyl-L-lysine methyl ester hydrochloride, 98%

FW 238.72

mp 111°C

172.51	30.03
169.41	26.28
51.81*	22.22*
51.66*	22.12
38.10	

DMSO-d_6 QE-300

C

Aldrich C8,370-8   CAS [372-75-8]   C₆H₁₃N₃O₃   60 MHz: *1*, 658B

L-Citrulline, 98%

FW 175.19   FT-IR: *1*, 784B

mp 221°C d.

177.35
164.25
57.28*
42.03
30.54
27.79

D₂O   QE-300

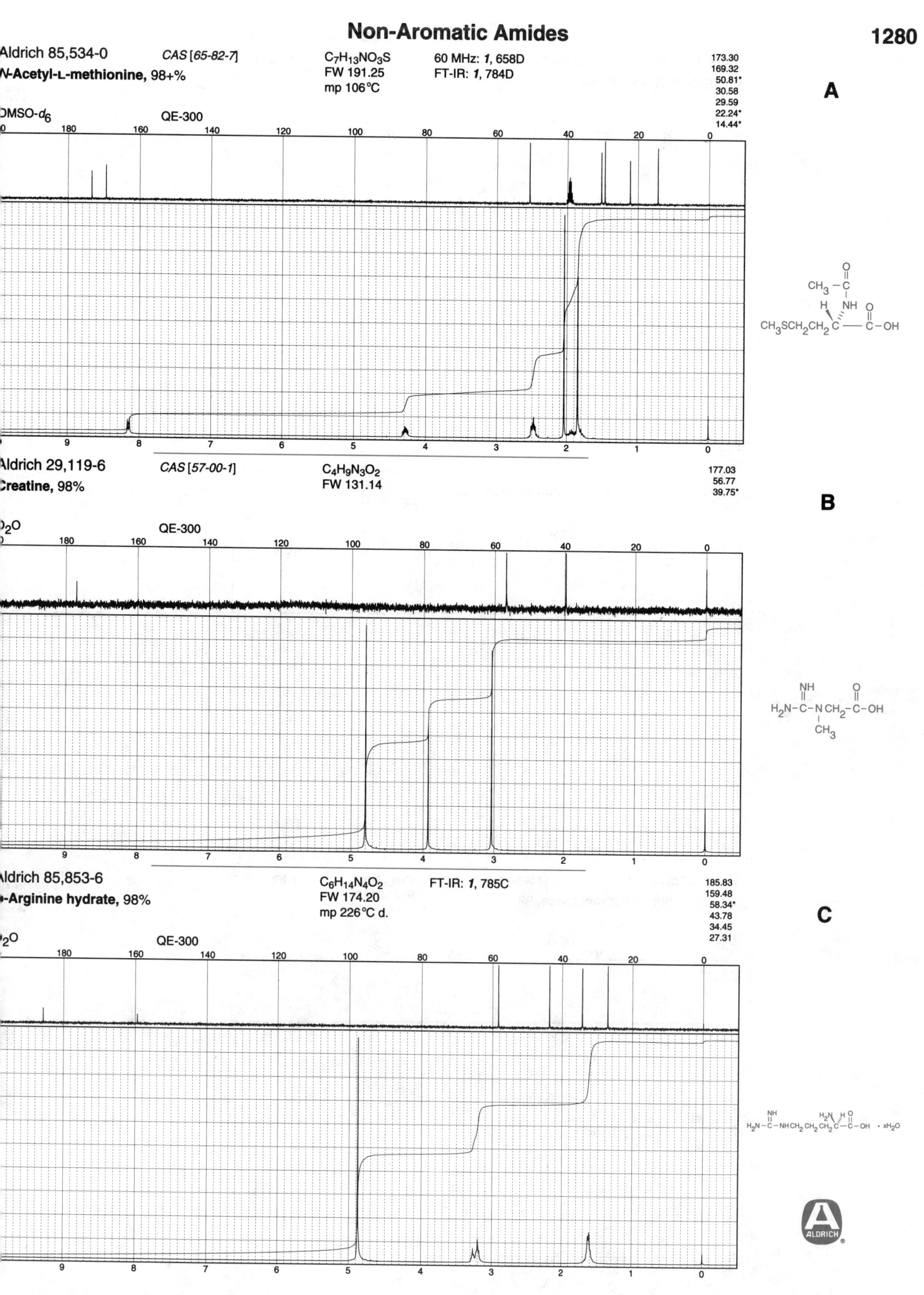

A

Aldrich 85,534-0   *CAS [65-82-7]*
N-Acetyl-L-methionine, 98+%

$C_7H_{13}NO_3S$
FW 191.25
mp 106°C

60 MHz: *1*, 658D
FT-IR: *1*, 784D

DMSO-d_6 QE-300

173.30
169.32
50.81*
30.58
29.59
22.24*
14.44*

B

Aldrich 29,119-6   *CAS [57-00-1]*
Creatine, 98%

$C_4H_9N_3O_2$
FW 131.14

177.03
56.77
39.75*

D_2O QE-300

C

Aldrich 85,853-6
Arginine hydrate, 98%

$C_6H_{14}N_4O_2$
FW 174.20
mp 226°C d.

FT-IR: *1*, 785C

185.83
159.48
58.34*
43.78
34.45
27.31

D_2O QE-300

A

Aldrich 85,567-7   *CAS [627-75-8]*   C_6H_{14}N_4O_2   60 MHz: *1*, 495D

D-Arginine hydrochloride   FW 210.67   FT-IR: *1*, 785D

mp 217°C

177.13
159.46
57.09*
43.32
30.35
26.72

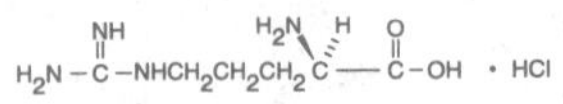

D_2O   QE-300

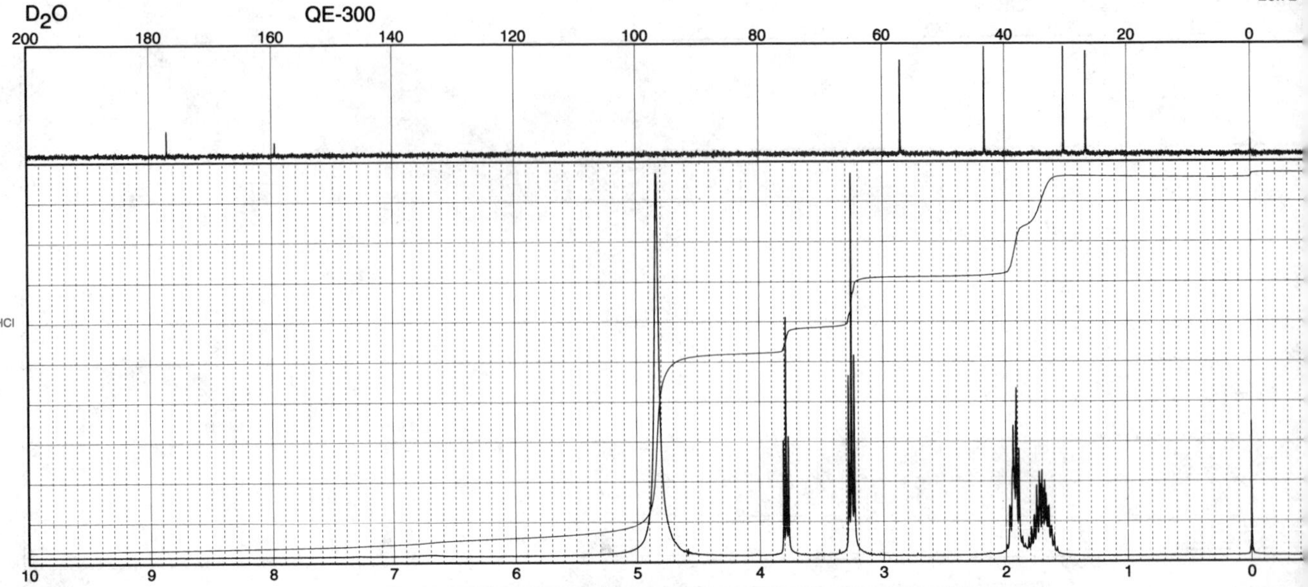

B

Aldrich 13,846-0   C_6H_{14}N_4O_2

DL-Arginine hydrochloride monohydrate,   FW 228.68

99%   mp 129°C

176.92
159.53
57.14*
43.34
30.34
26.73

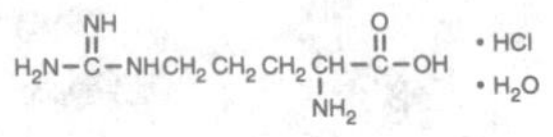

D_2O   QE-300

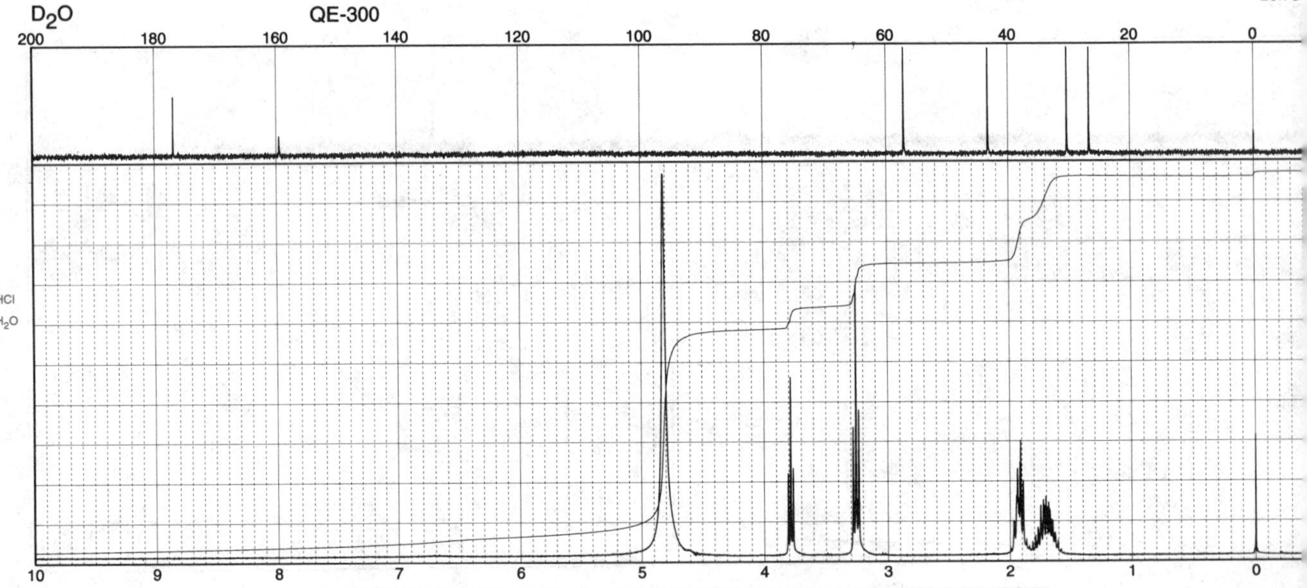

C

Aldrich 15,711-2   *CAS [1483-01-8]*   C_7H_{16}N_4O_2   FT-IR: *1*, 786C

L-Homoarginine hydrochloride, 99+%   FW 224.69

mp 214°C

177.44
159.41
57.38
43.54
32.76
30.36
24.35

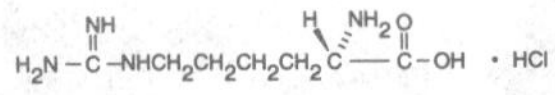

D_2O   QE-300

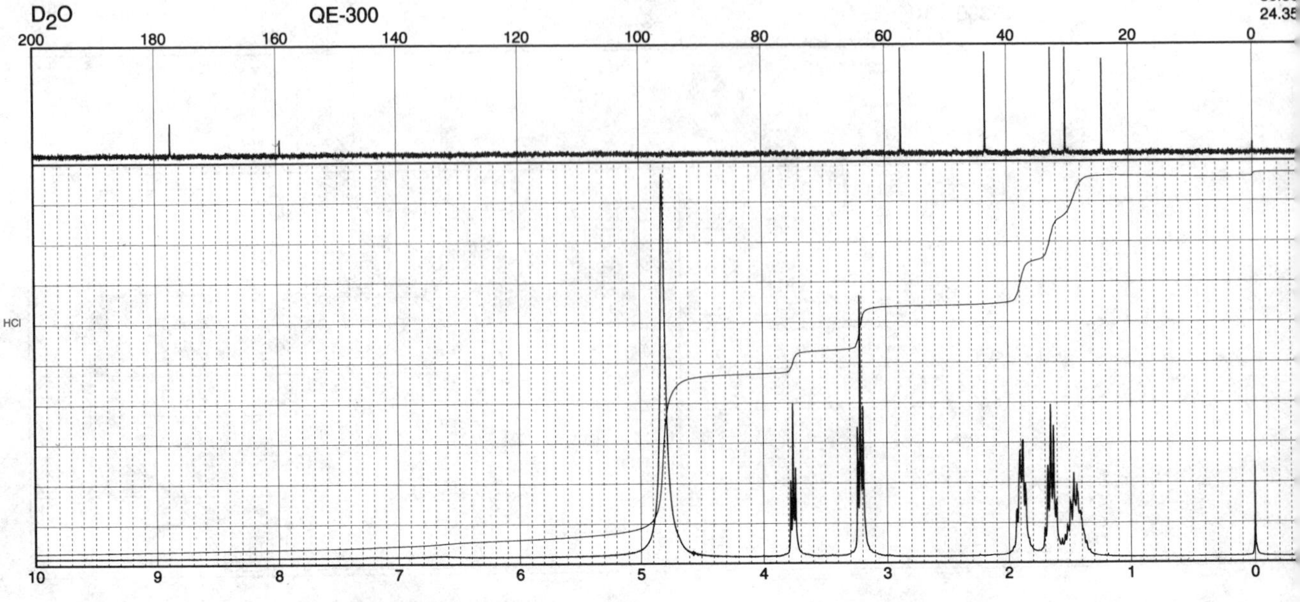

Non-Aromatic Amides

Aldrich 28,393-2

Nopaline monohydrate, 98%

$C_{11}H_{20}N_4O_6$
FW 322.33
mp 183°C d.

176.78	40.42
173.67	29.28
171.54	25.60
156.64	25.05
64.01*	24.67
57.02*	

A

DMSO-d_6 QE-300

Aldrich 13,806-1   *CAS [616-91-1]*

N-Acetyl-L-cysteine, 98%

$C_5H_9NO_3S$
FW 163.20
mp 110°C

60 MHz: *1*, 659D
FT-IR: *1*, 786D

171.55
169.27
54.22*
25.49
22.25*

B

DMSO-d_6 QE-300

Aldrich A1,900-8   *CAS [59-53-0]*

N-Acetyl-DL-penicillamine, 98%

$C_7H_{13}NO_3S$
FW 191.25
mp 189°C d.

60 MHz: *1*, 660A
FT-IR: *1*, 787A

171.10
169.27
61.24*
44.99
29.81*
29.40*
22.22*

C

DMSO-d_6 QE-300

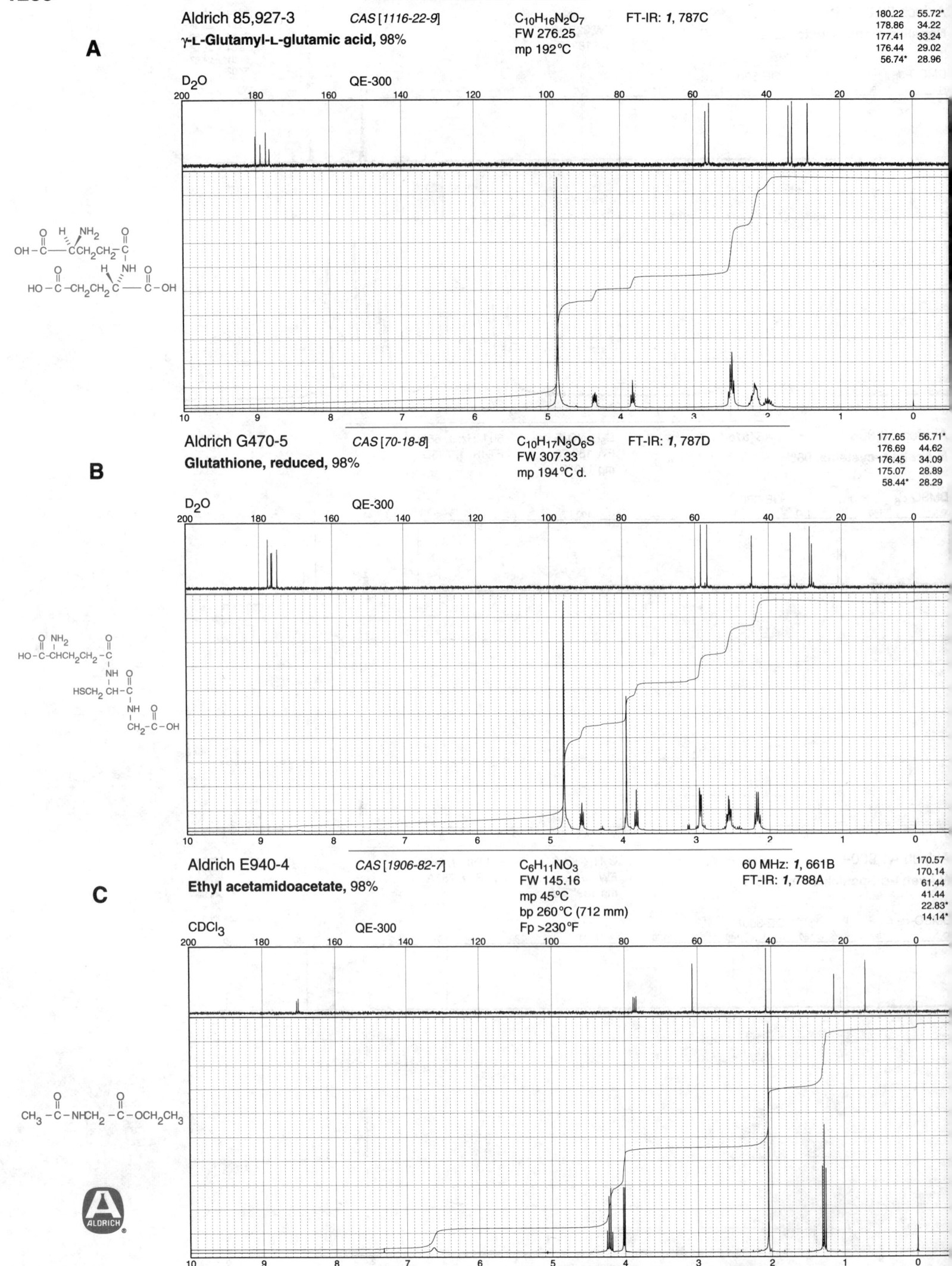

Aldrich 85,927-3   *CAS [1116-22-9]* $C_{10}H_{16}N_2O_7$ FT-IR: *1*, 787C

γ-L-Glutamyl-L-glutamic acid, 98%   FW 276.25   mp 192°C

A

D₂O   QE-300

180.22   55.72*
178.86   34.22
177.41   33.24
176.44   29.02
56.74*   28.96

Aldrich G470-5   *CAS [70-18-8]* $C_{10}H_{17}N_3O_6S$ FT-IR: *1*, 787D

Glutathione, reduced, 98%   FW 307.33   mp 194°C d.

B

D₂O   QE-300

177.65   56.71*
176.69   44.62
176.45   34.09
175.07   28.89
58.44*   28.29

Aldrich E940-4   *CAS [1906-82-7]* $C_6H_{11}NO_3$ 60 MHz: *1*, 661B

Ethyl acetamidoacetate, 98%   FW 145.16   FT-IR: *1*, 788A
mp 45°C
bp 260°C (712 mm)
Fp >230°F

C

CDCl₃   QE-300

170.57
170.14
61.44
41.44
22.83*
14.14*

Non-Aromatic Amides

1284

A

Aldrich D9,560-3　　CAS [6326-44-9]
Diethyl formamidomalonate, 97%

C₈H₁₃NO₅
FW 203.19
mp 54°C
bp 175°C (11 mm)
Fp >230°F

60 MHz: *1*, 660D
FT-IR: *1*, 788B

165.99
160.67*
62.79
55.05*
13.97*

CDCl₃　QE-300

B

Aldrich D8,460-1　　CAS [1068-90-2]
Diethyl acetamidomalonate, 98%

C₉H₁₅NO₅
FW 217.22
mp 97°C
bp 185°C (20 mm)

60 MHz: *1*, 661A
FT-IR: *1*, 788C

169.78
166.44
62.52
56.48*
22.67*
13.99*

CDCl₃　QE-300

C

Aldrich 85,959-1　　CAS [2001-95-8]
(+)-Valinomycin, 98%

C₅₄H₉₀N₆O₁₈
FW 1111.36
mp 188°C

170.99 57.82* 18.68*
170.13 57.56* 18.54*
169.85 29.83* 17.84*
168.67 29.73* 16.80*
77.86* 29.24* 16.69*
69.34* 18.82*

DMSO-d₆　QE-300

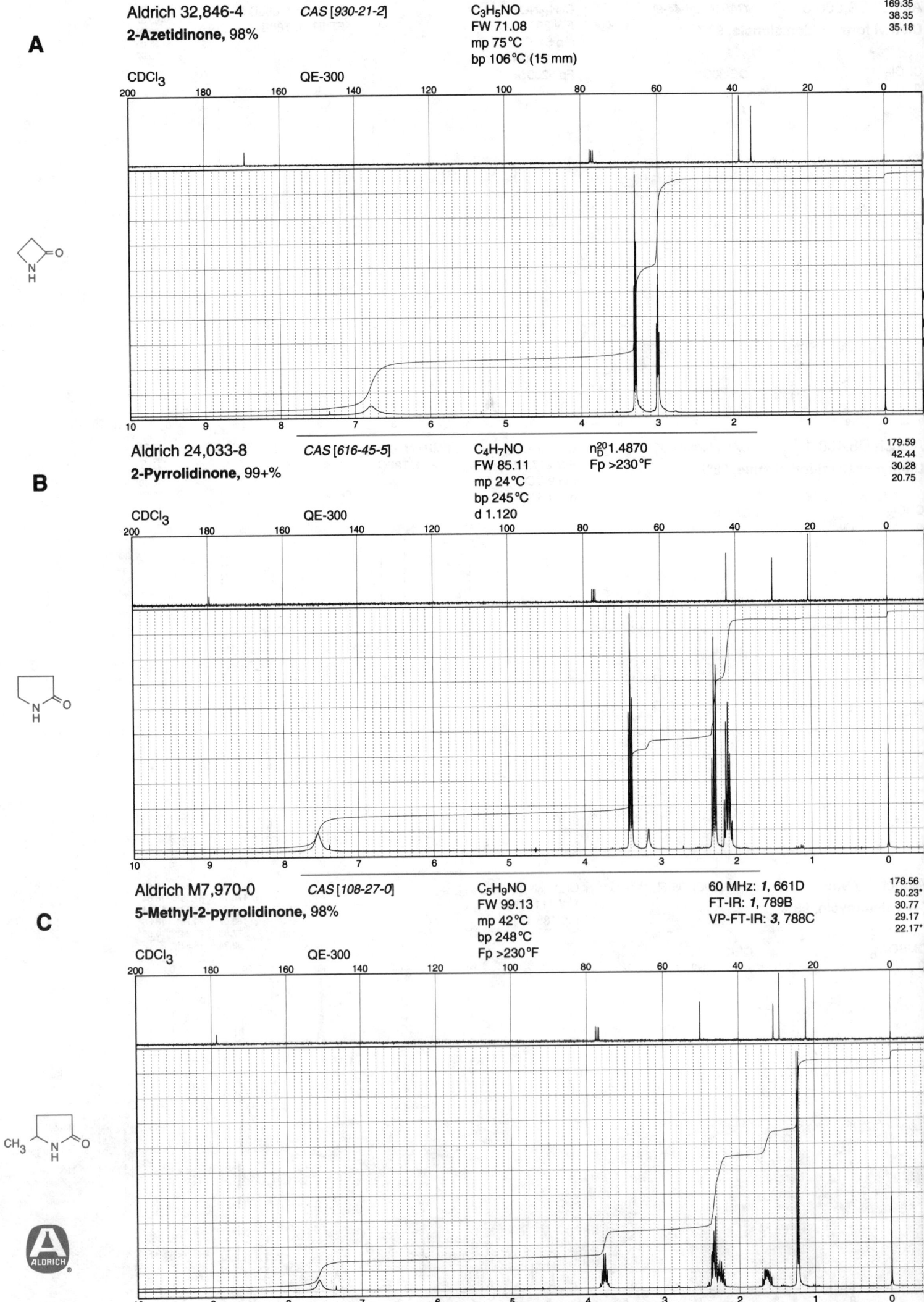

A

Aldrich 32,846-4   *CAS [930-21-2]* C_3H_5NO
2-Azetidinone, 98%   FW 71.08
                                         mp 75°C
                                         bp 106°C (15 mm)

169.35
38.35
35.18

CDCl$_3$ QE-300

B

Aldrich 24,033-8   *CAS [616-45-5]* C_4H_7NO $n_D^{20}1.4870$
2-Pyrrolidinone, 99+%   FW 85.11   Fp >230°F
                                         mp 24°C
                                         bp 245°C
                                         d 1.120

179.59
42.44
30.28
20.75

CDCl$_3$ QE-300

C

Aldrich M7,970-0   *CAS [108-27-0]* C_5H_9NO 60 MHz: *1*, 661D
5-Methyl-2-pyrrolidinone, 98%   FW 99.13   FT-IR: *1*, 789B
                                         mp 42°C   VP-FT-IR: *3*, 788C
                                         bp 248°C
                                         Fp >230°F

178.56
50.23*
30.77
29.17
22.17*

CDCl$_3$ QE-300

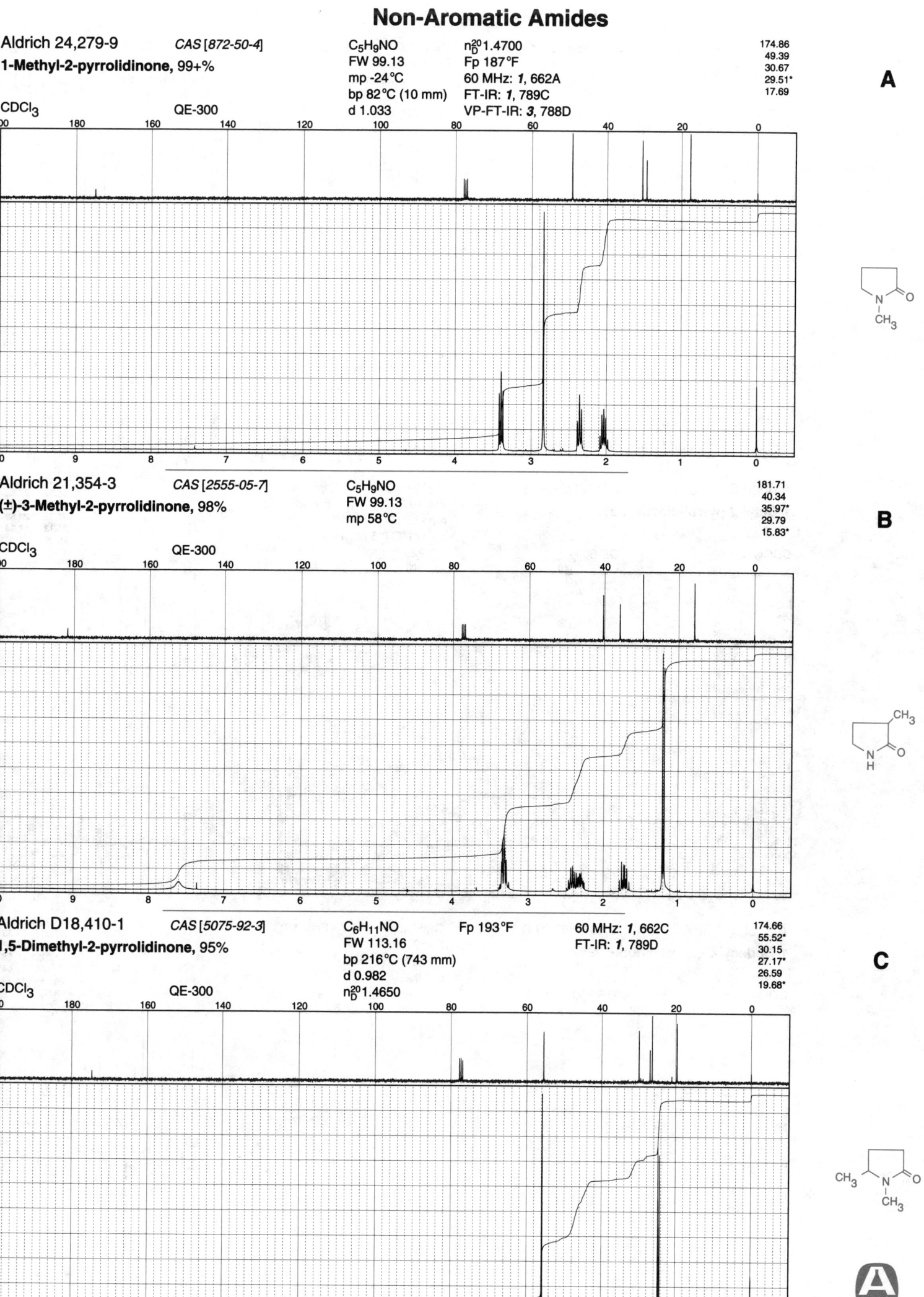

Aldrich 24,279-9   *CAS [872-50-4]*

1-Methyl-2-pyrrolidinone, 99+%

C$_5$H$_9$NO
FW 99.13
mp -24°C
bp 82°C (10 mm)
d 1.033

n$_D^{20}$1.4700
Fp 187°F
60 MHz: *1*, 662A
FT-IR: *1*, 789C
VP-FT-IR: *3*, 788D

174.86
49.39
30.67
29.51*
17.69

A

CDCl$_3$ QE-300

Aldrich 21,354-3   *CAS [2555-05-7]*

(±)-3-Methyl-2-pyrrolidinone, 98%

C$_5$H$_9$NO
FW 99.13
mp 58°C

181.71
40.34
35.97*
29.79
15.83*

B

CDCl$_3$ QE-300

Aldrich D18,410-1   *CAS [5075-92-3]*

1,5-Dimethyl-2-pyrrolidinone, 95%

C$_6$H$_{11}$NO
FW 113.16
bp 216°C (743 mm)
d 0.982
n$_D^{20}$1.4650

Fp 193°F

60 MHz: *1*, 662C
FT-IR: *1*, 789D

174.66
55.52*
30.15
27.17*
26.59
19.68*

C

CDCl$_3$ QE-300

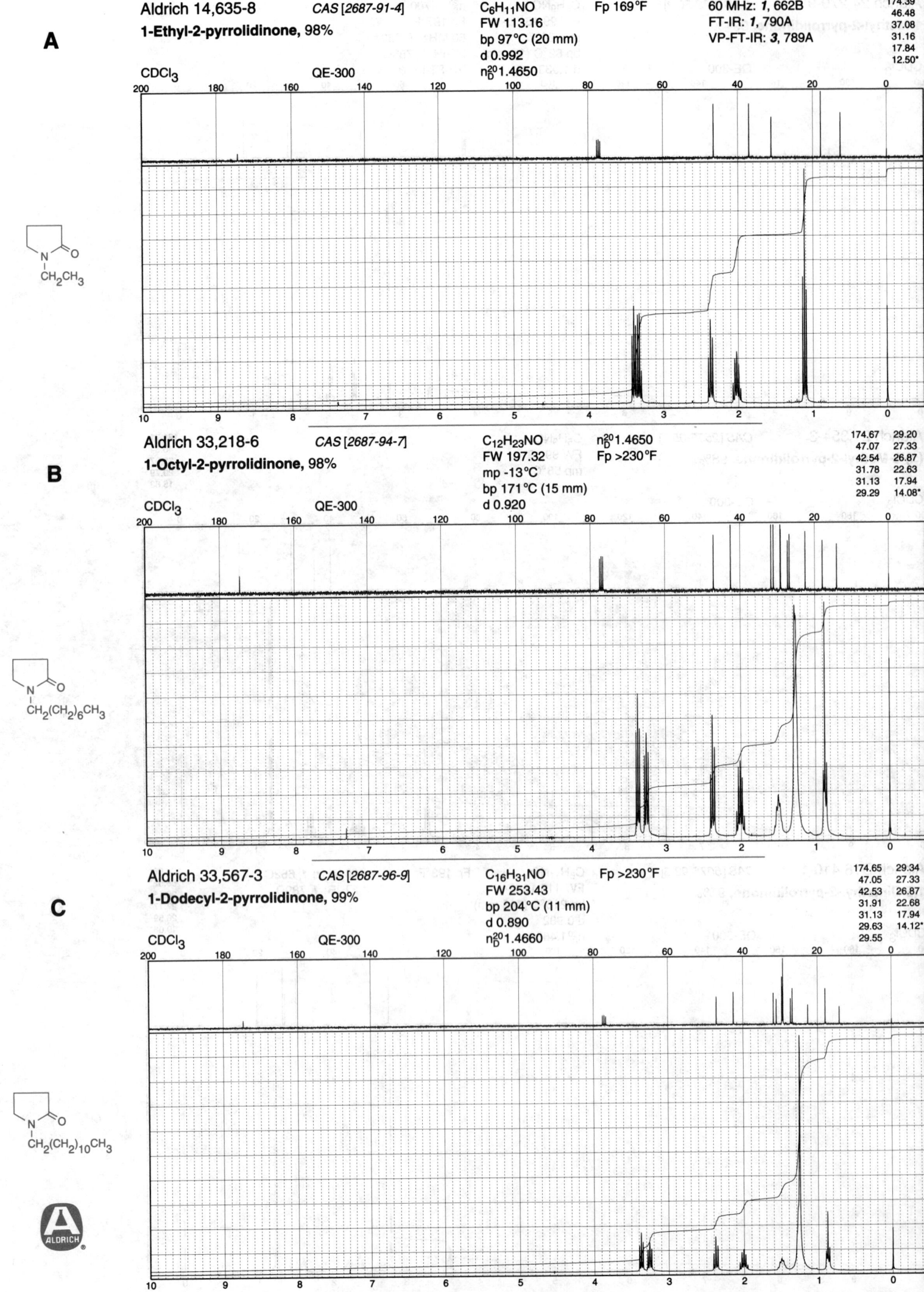

A

Aldrich 14,635-8   CAS [2687-91-4]   C₆H₁₁NO   Fp 169°F
1-Ethyl-2-pyrrolidinone, 98%   FW 113.16   FT-IR: *1*, 790A
bp 97°C (20 mm)   60 MHz: *1*, 662B
VP-FT-IR: *3*, 789A
d 0.992
n$_D^{20}$1.4650

174.39
46.48
37.08
31.16
17.84
12.50*

CDCl₃   QE-300

B

Aldrich 33,218-6   CAS [2687-94-7]   C₁₂H₂₃NO   n$_D^{20}$1.4650
1-Octyl-2-pyrrolidinone, 98%   FW 197.32   Fp >230°F
mp -13°C
bp 171°C (15 mm)
d 0.920

174.67   29.20
47.07   27.33
42.54   26.87
31.78   22.63
31.13   17.94
29.29   14.08*

CDCl₃   QE-300

C

Aldrich 33,567-3   CAS [2687-96-9]   C₁₆H₃₁NO   Fp >230°F
1-Dodecyl-2-pyrrolidinone, 99%   FW 253.43
bp 204°C (11 mm)
d 0.890
n$_D^{20}$1.4660

174.65   29.34
47.05   27.33
42.53   26.87
31.91   22.68
31.13   17.94
29.63   14.12*
29.55

CDCl₃   QE-300

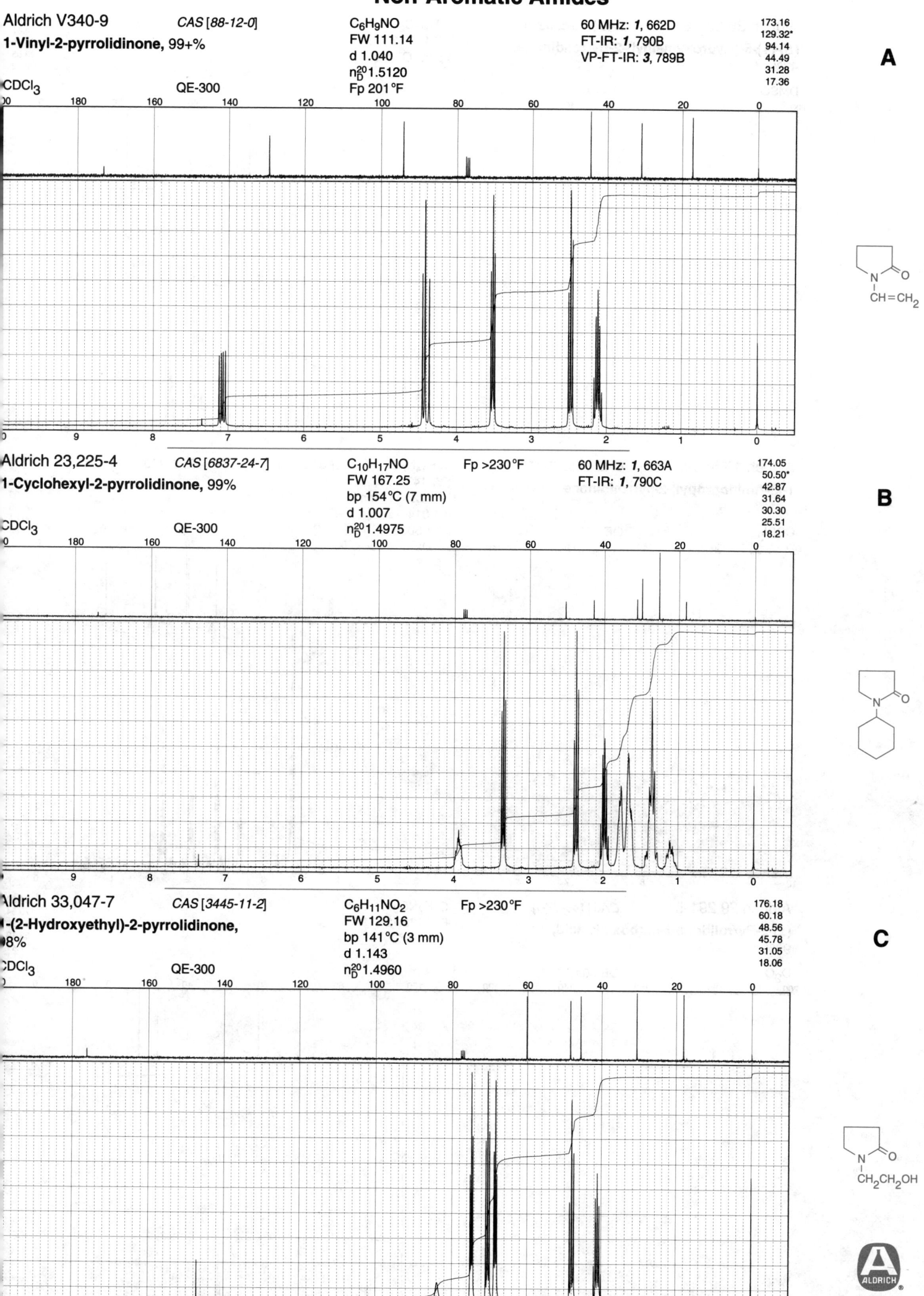

A

Aldrich V340-9   *CAS [88-12-0]*
1-Vinyl-2-pyrrolidinone, 99+%

CDCl₃   QE-300

C₆H₉NO
FW 111.14
d 1.040
n²⁰_D 1.5120
Fp 201°F

60 MHz: *1*, 662D
FT-IR: *1*, 790B
VP-FT-IR: *3*, 789B

173.16
129.32*
94.14
44.49
31.28
17.36

B

Aldrich 23,225-4   *CAS [6837-24-7]*
1-Cyclohexyl-2-pyrrolidinone, 99%

CDCl₃   QE-300

C₁₀H₁₇NO
FW 167.25
bp 154°C (7 mm)
d 1.007
n²⁰_D 1.4975

Fp >230°F

60 MHz: *1*, 663A
FT-IR: *1*, 790C

174.05
50.50*
42.87
31.64
30.30
25.51
18.21

C

Aldrich 33,047-7   *CAS [3445-11-2]*
1-(2-Hydroxyethyl)-2-pyrrolidinone, 98%

CDCl₃   QE-300

C₆H₁₁NO₂
FW 129.16
bp 141°C (3 mm)
d 1.143
n²⁰_D 1.4960

Fp >230°F

176.18
60.18
48.56
45.78
31.05
18.06

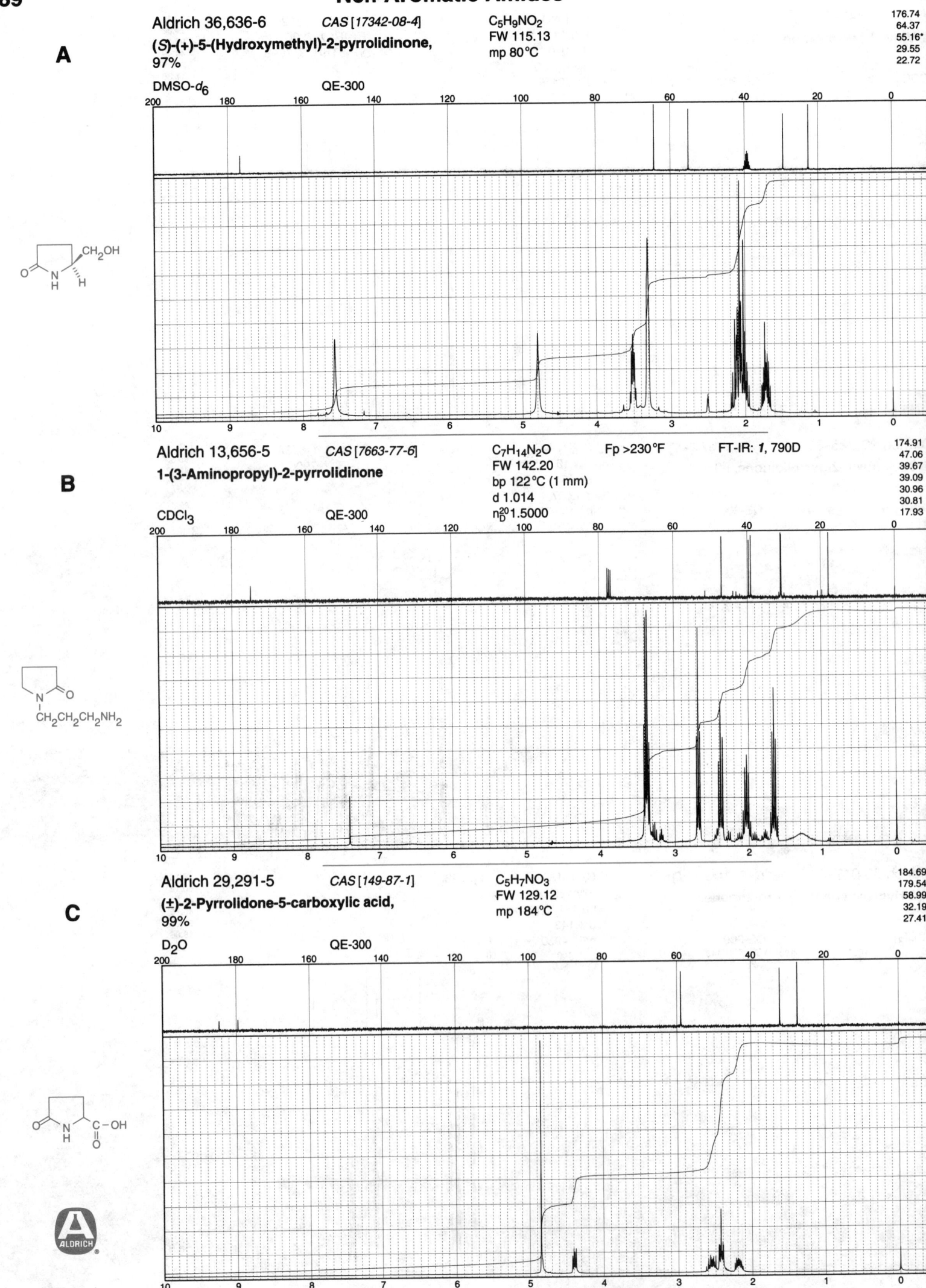

A

Aldrich 36,636-6   CAS [17342-08-4] $C_5H_9NO_2$

(S)-(+)-5-(Hydroxymethyl)-2-pyrrolidinone, 97%

FW 115.13
mp 80°C

176.74
64.37
55.16*
29.55
22.72

DMSO-d_6 QE-300

B

Aldrich 13,656-5   CAS [7663-77-6] $C_7H_{14}N_2O$ Fp >230°F   FT-IR: 1, 790D

1-(3-Aminopropyl)-2-pyrrolidinone

FW 142.20
bp 122°C (1 mm)
d 1.014
n_D^{20}1.5000

174.91
47.06
39.67
39.09
30.96
30.81
17.93

CDCl$_3$ QE-300

C

Aldrich 29,291-5   CAS [149-87-1] $C_5H_7NO_3$

(±)-2-Pyrrolidone-5-carboxylic acid, 99%

FW 129.12
mp 184°C

184.69
179.54
58.99*
32.19
27.41

D$_2$O   QE-300

A

Aldrich P7,520-2   CAS [98-79-3]

(S)-(-)-2-Pyrrolidone-5-carboxylic acid, 97%

$C_5H_7NO_3$
FW 129.12
mp 162°C

60 MHz: *1*, 663C
FT-IR: *1*, 791A

176.87
174.20
54.67*
28.95
24.50

DMSO-d_6 QE-300

Offset: 3.0 ppm.

B

Aldrich 32,756-5   CAS [2761-13-9]

1,1'-Ethylenebis(5-oxo-3-pyrrolidine-carboxylic acid), 98%

$C_{12}H_{16}N_2O_6$
FW 284.27
mp 247°C

174.12
174.09
172.21
48.06
38.67
35.43*
33.28

DMSO-d_6 QE-300

Offset: 2.8 ppm.

C

Aldrich 19,633-9   CAS [59776-88-4]

Methyl 2-oxo-1-pyrrolidineacetate, 97%

$C_7H_{11}NO_3$
FW 157.17
d 1.131
n_D^{20} 1.4720
Fp >230°F

60 MHz: *1*, 663D
FT-IR: *1*, 791B

175.46
169.10
52.09*
47.67
43.96
30.27
17.97

CDCl$_3$ QE-300

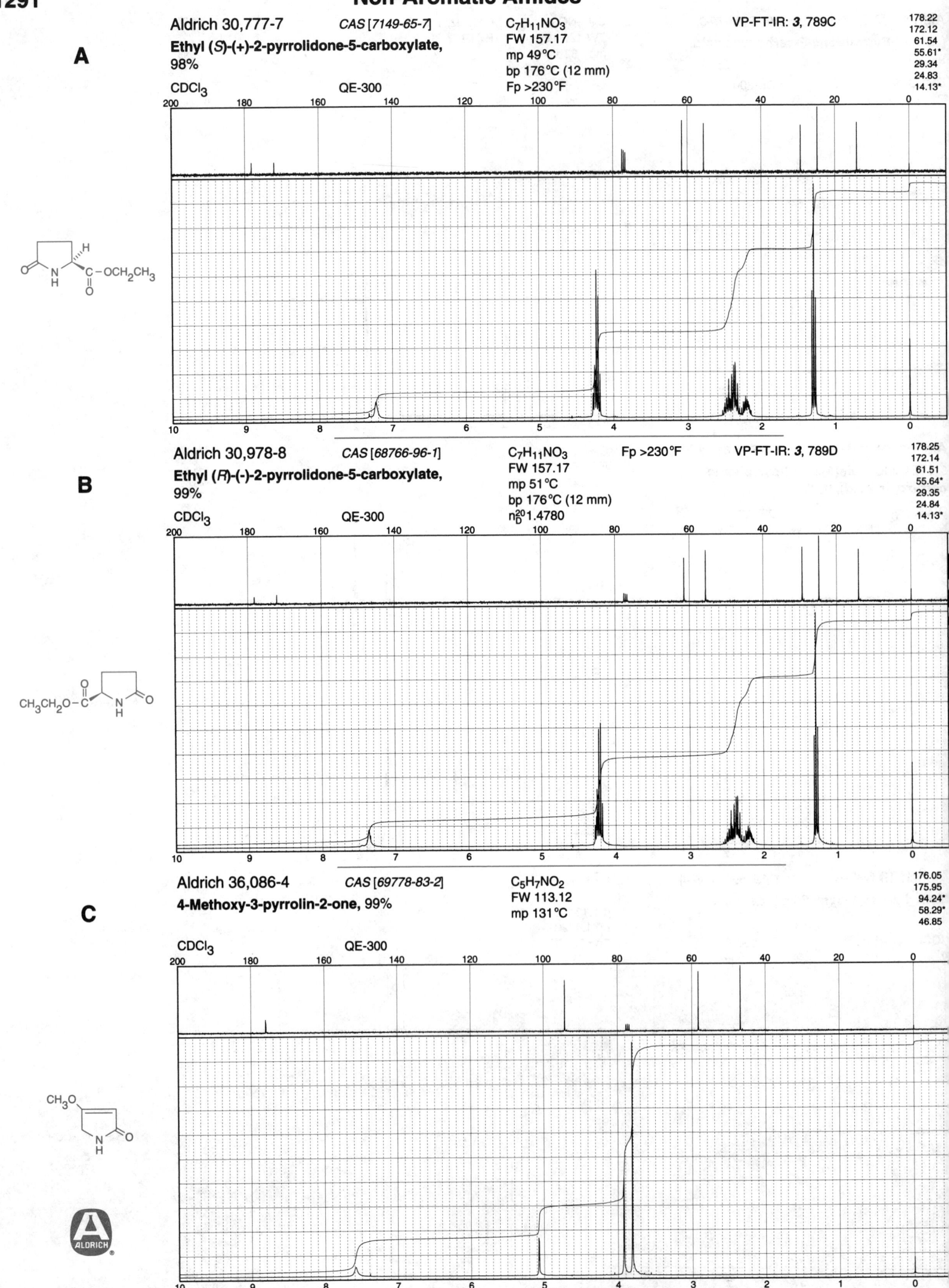

A

Aldrich 30,777-7   *CAS [7149-65-7]*   C₇H₁₁NO₃   VP-FT-IR: *3*, 789C

$C_7H_{11}NO_3$
FW 157.17
mp 49 °C
bp 176 °C (12 mm)
Fp >230 °F

Ethyl (*S*)-(+)-2-pyrrolidone-5-carboxylate, 98%

CDCl₃   QE-300

178.22
172.12
61.54
55.61*
29.34
24.83
14.13*

B

Aldrich 30,978-8   *CAS [68766-96-1]*   C₇H₁₁NO₃   Fp >230 °F   VP-FT-IR: *3*, 789D

$C_7H_{11}NO_3$
FW 157.17
mp 51 °C
bp 176 °C (12 mm)
n_D^{20} 1.4780

Ethyl (*R*)-(−)-2-pyrrolidone-5-carboxylate, 99%

CDCl₃   QE-300

178.25
172.14
61.51
55.64*
29.35
24.84
14.13*

C

Aldrich 36,086-4   *CAS [69778-83-2]*   C₅H₇NO₂

4-Methoxy-3-pyrrolin-2-one, 99%

$C_5H_7NO_2$
FW 113.12
mp 131 °C

CDCl₃   QE-300

176.05
175.95
94.24*
58.29*
46.85

Non-Aromatic Amides

Aldrich 33,759-5   CAS [119838-38-9]   C₁₃H₂₄N₂O₃   170.53 39.59

A

(S)-(-)-1-(tert-Butoxycarbonyl)-2-tert-butyl-3-methyl-4-imidazolidinone, 99%

Aldrich 33,759-5   CAS [119838-38-9]

$C_{13}H_{24}N_2O_3$
FW 256.35
mp 69 °C

170.53 39.59
154.51 31.50*
82.55* 28.29*
81.19 26.01*
49.86

CDCl₃   QE-300

Aldrich 29,888-3   CAS [17016-83-0]

(4S)-(-)-4-Isopropyl-2-oxazolidinone, 99%

$C_6H_{11}NO_2$
FW 129.16
mp 74 °C

160.77
68.63
58.43*
32.71*
17.96*
17.65*

CDCl₃   QE-300

B

Aldrich V20-9   CAS [675-20-7]

δ-Valerolactam, 99%

C_5H_9NO
FW 99.13
mp 39 °C
bp 256 °C
Fp >230 °F

60 MHz: 1, 664C
FT-IR: 1, 791D

172.71
42.12
31.50
22.30
20.90

CDCl₃   QE-300

C

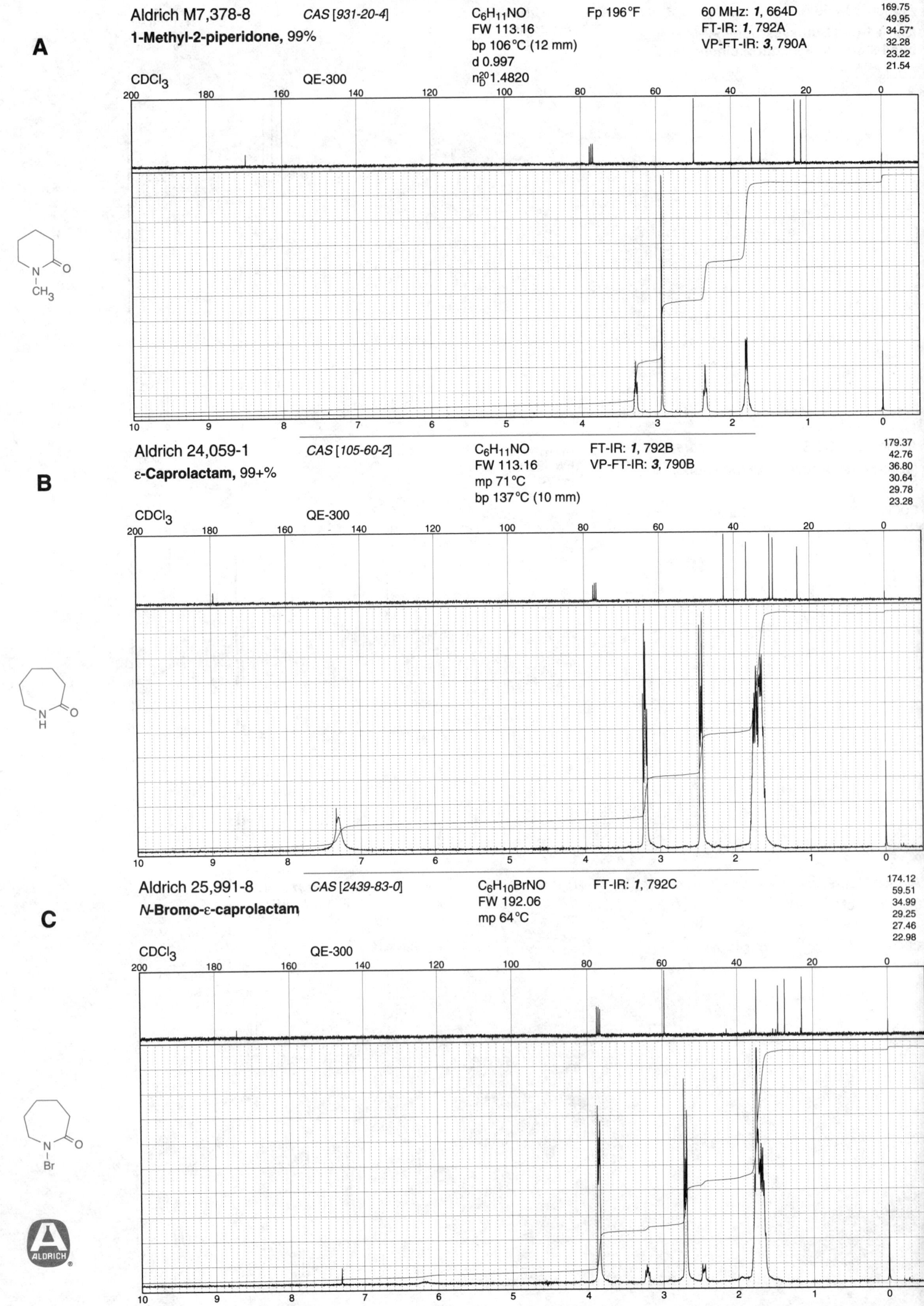

A

Aldrich M7,378-8

1-Methyl-2-piperidone, 99%

CAS [931-20-4]

C_6H_11NO
FW 113.16
bp 106°C (12 mm)
d 0.997
n_D^{20} 1.4820

Fp 196°F

60 MHz: *1*, 664D
FT-IR: *1*, 792A
VP-FT-IR: *3*, 790A

169.75
49.95
34.57*
32.28
23.22
21.54

CDCl_3   QE-300

B

Aldrich 24,059-1

ε-Caprolactam, 99+%

CAS [105-60-2]

C_6H_11NO
FW 113.16
mp 71°C
bp 137°C (10 mm)

FT-IR: *1*, 792B
VP-FT-IR: *3*, 790B

179.37
42.76
36.80
30.64
29.78
23.28

CDCl_3   QE-300

C

Aldrich 25,991-8

N-Bromo-ε-caprolactam

CAS [2439-83-0]

C_6H_10BrNO
FW 192.06
mp 64°C

FT-IR: *1*, 792C

174.12
59.51
34.99
29.25
27.46
22.98

CDCl_3   QE-300

A

Aldrich 22,476-6   *CAS [2556-73-2]*

N-Methylcaprolactam, 99%

C$_7$H$_{13}$NO
FW 127.19
bp 107°C (6 mm)
d 0.991
n$_D^{20}$ 1.4840

Fp 218°F

60 MHz: *1*, 665C
FT-IR: *1*, 792D

175.77
51.34
36.95
35.71*
29.89
27.68
23.38

CDCl$_3$ QE-300

B

Aldrich 30,064-0   *CAS [17929-90-7]*

DL-α-Amino-ε-caprolactam, 97%

C$_6$H$_{12}$N$_2$O
FW 128.18
mp 76°C
bp 170°C (13 mm)

179.73
53.86*
41.75
33.90
29.04
28.47

CDCl$_3$ QE-300

C

Aldrich 26,358-3   *CAS [28957-33-7]*

D-α-Amino-ε-caprolactam, 97%

C$_6$H$_{12}$N$_2$O
FW 128.18
mp 58°C

FT-IR: *1*, 793A

179.61
53.91*
41.83
33.96
29.10
28.48

CDCl$_3$ QE-300

ALDRICH®

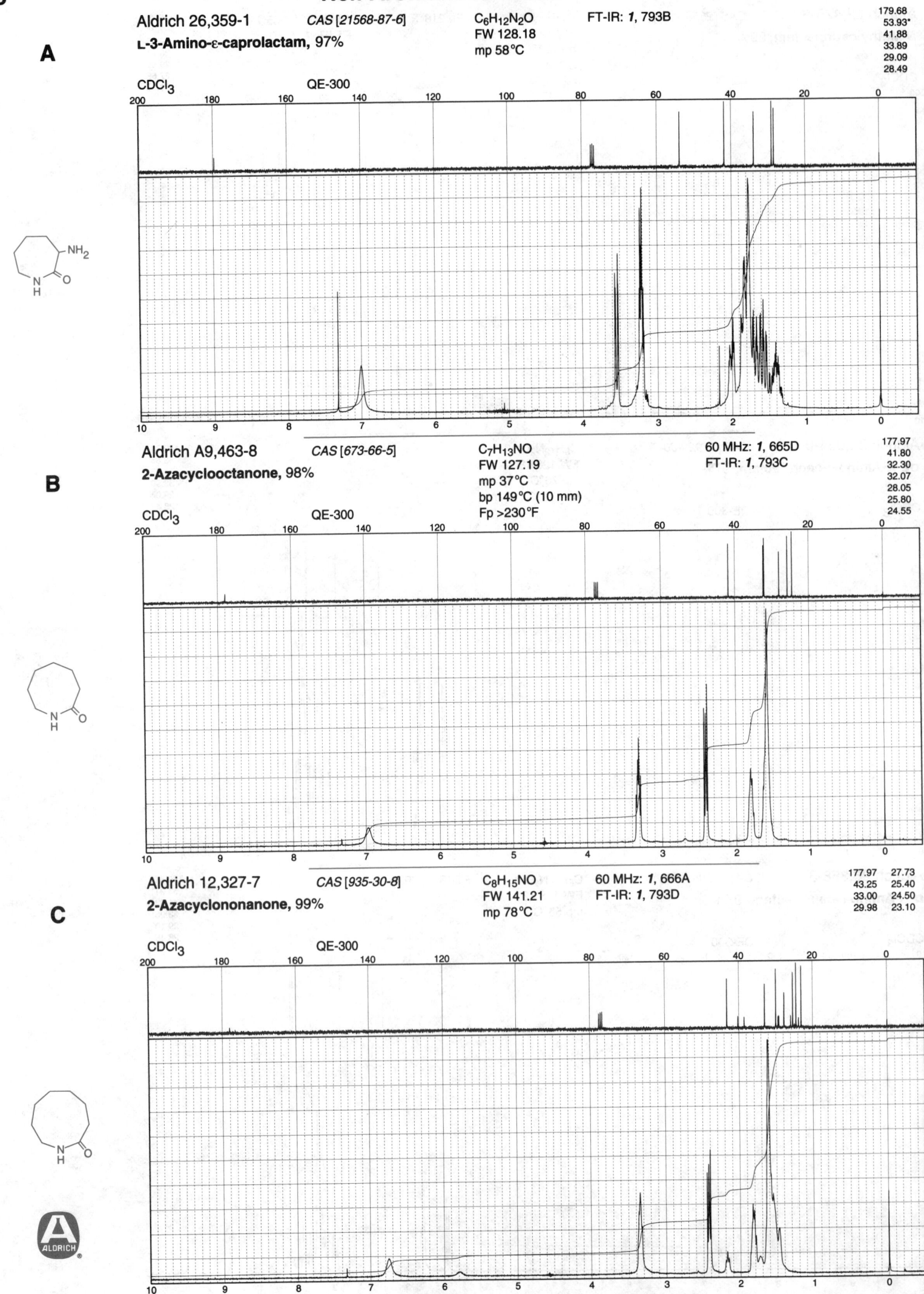

A

Aldrich 26,359-1   *CAS [21568-87-6]* $C_6H_{12}N_2O$ FT-IR: *1*, 793B

L-3-Amino-ε-caprolactam, 97%

FW 128.18
mp 58°C

179.68
53.93*
41.88
33.89
29.09
28.49

CDCl₃   QE-300

B

Aldrich A9,463-8   *CAS [673-66-5]* $C_7H_{13}NO$ 60 MHz: *1*, 665D

2-Azacyclooctanone, 98%

FW 127.19
mp 37°C
bp 149°C (10 mm)
Fp >230°F

FT-IR: *1*, 793C

177.97
41.80
32.30
32.07
28.05
25.80
24.55

CDCl₃   QE-300

C

Aldrich 12,327-7   *CAS [935-30-8]* $C_8H_{15}NO$ 60 MHz: *1*, 666A

2-Azacyclononanone, 99%

FW 141.21
mp 78°C

FT-IR: *1*, 793D

177.97   27.73
43.25   25.40
33.00   24.50
29.98   23.10

CDCl₃   QE-300

Aldrich A9,465-4   CAS [947-04-6]   C₁₂H₂₃NO   60 MHz: *1*, 666B
2-Azacyclotridecanone, 98%   FW 197.32   FT-IR: *1*, 794A
    mp 152°C

173.48	26.30
39.04	25.83
36.85	25.23
28.32	25.04
26.82	24.78
26.42	24.03

A

CDCl₃   QE-300

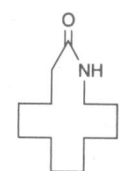

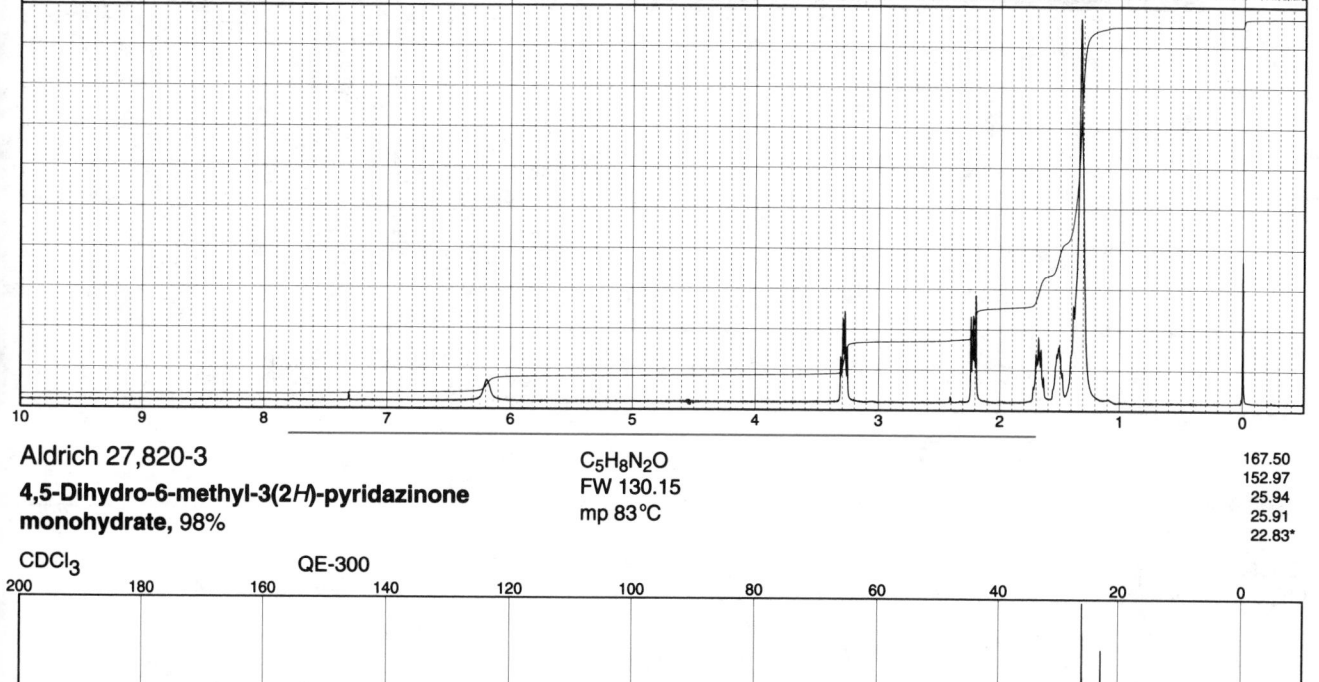

Aldrich 27,820-3   C₅H₈N₂O
4,5-Dihydro-6-methyl-3(2H)-pyridazinone   FW 130.15
monohydrate, 98%   mp 83°C

167.50
152.97
25.94
25.91
22.83*

B

CDCl₃   QE-300

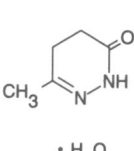

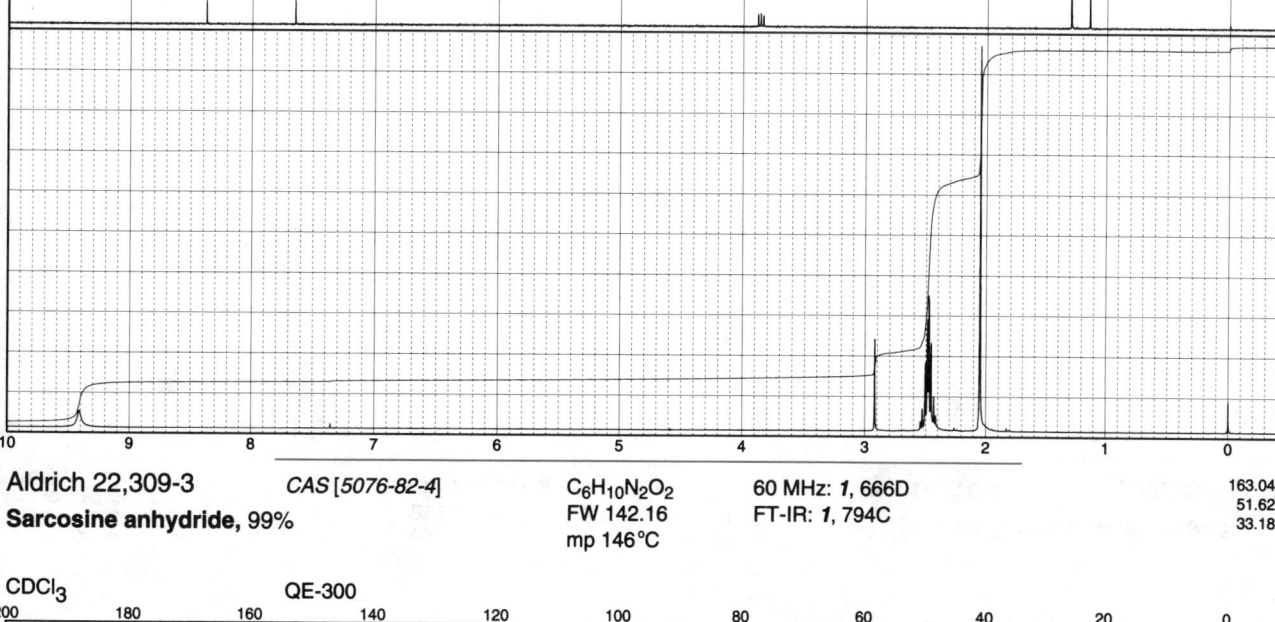

Aldrich 22,309-3   CAS [5076-82-4]   C₆H₁₀N₂O₂   60 MHz: *1*, 666D
Sarcosine anhydride, 99%   FW 142.16   FT-IR: *1*, 794C
    mp 146°C

163.04
51.62
33.18*

C

CDCl₃   QE-300

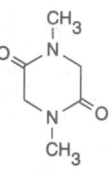

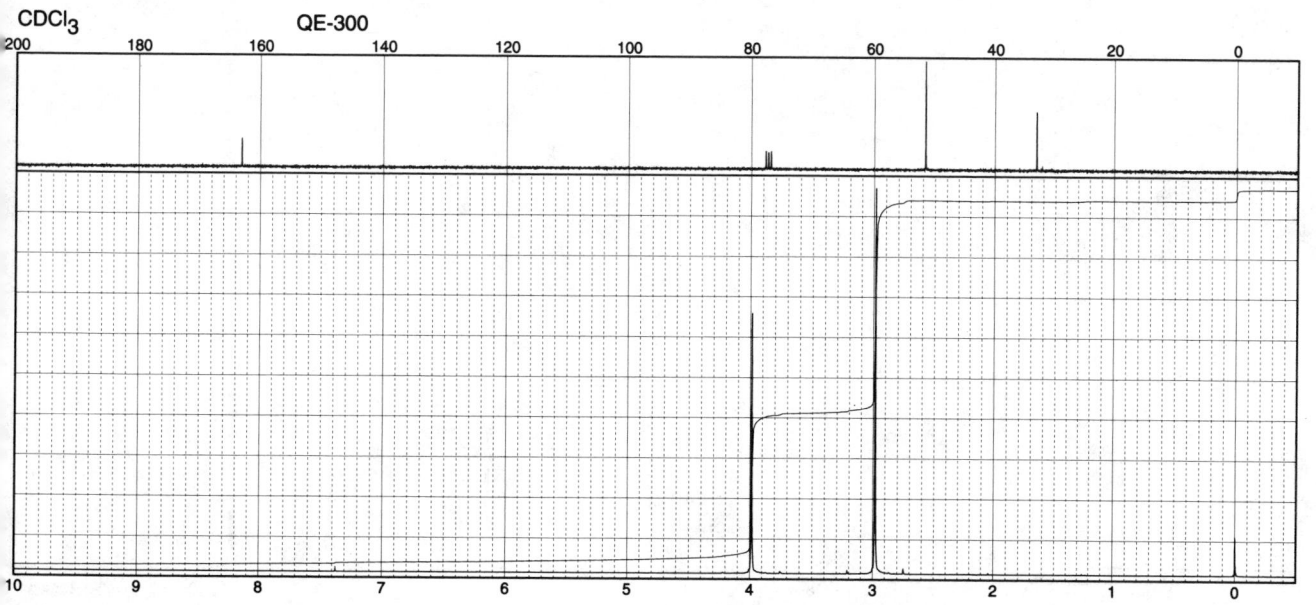

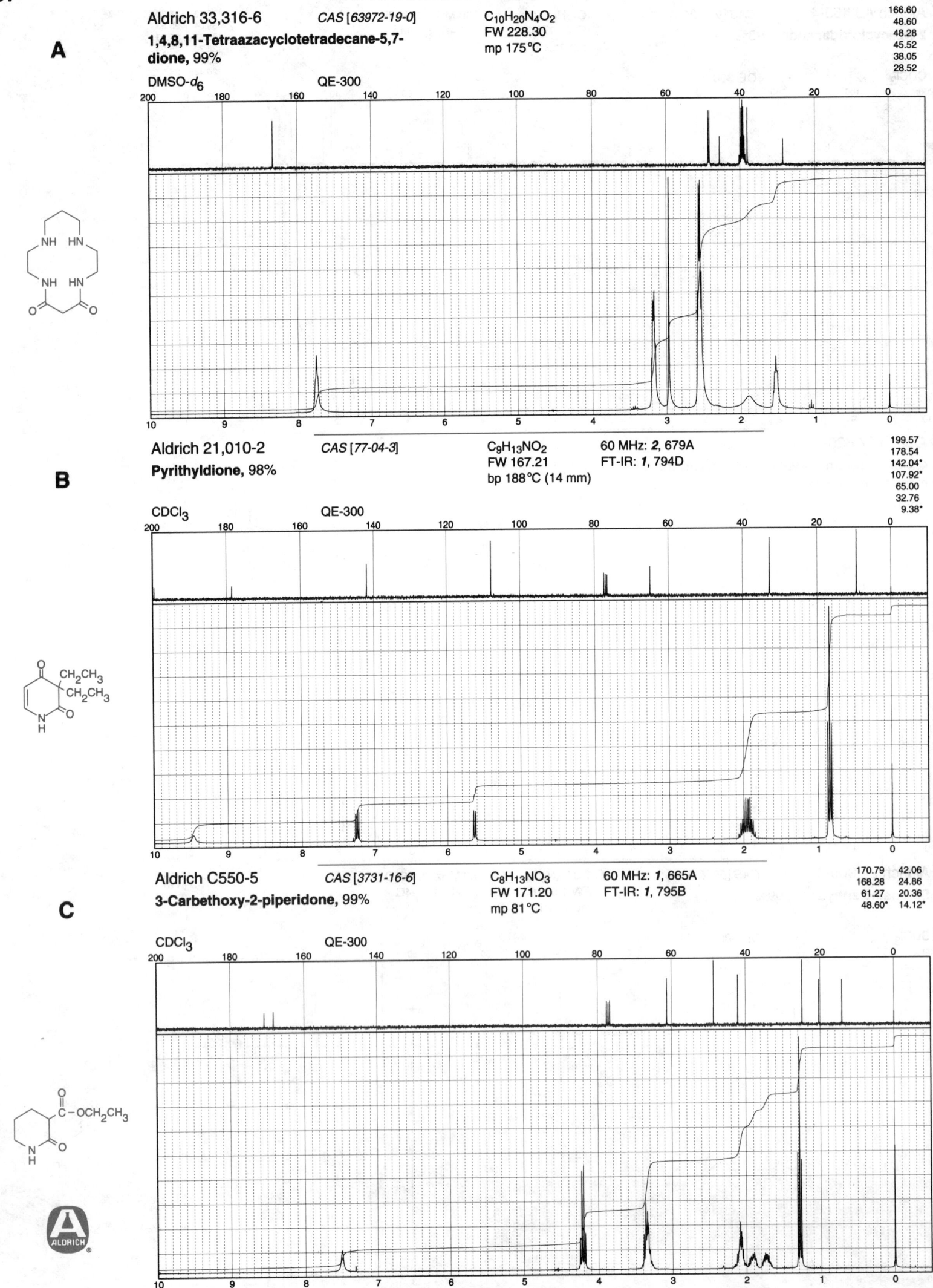

A

Aldrich 33,316-6   *CAS [63972-19-0]*

1,4,8,11-Tetraazacyclotetradecane-5,7-dione, 99%

$C_{10}H_{20}N_4O_2$
FW 228.30
mp 175°C

166.60
48.60
48.28
45.52
38.05
28.52

DMSO-d_6 QE-300

B

Aldrich 21,010-2   *CAS [77-04-3]*

Pyrithyldione, 98%

$C_9H_{13}NO_2$
FW 167.21
bp 188°C (14 mm)

60 MHz: *2*, 679A
FT-IR: *1*, 794D

199.57
178.54
142.04*
107.92*
65.00
32.76
9.38*

CDCl₃   QE-300

C

Aldrich C550-5   *CAS [3731-16-6]*

3-Carbethoxy-2-piperidone, 99%

$C_8H_{13}NO_3$
FW 171.20
mp 81°C

60 MHz: *1*, 665A
FT-IR: *1*, 795B

170.79   42.06
168.28   24.86
61.27   20.36
48.60*   14.12*

CDCl₃   QE-300

Aldrich 28,301-0   CAS [1888-91-1] $C_8H_{13}NO_2$ Fp >230°F

N-Acetylcaprolactam, 99%

FW 155.20
bp 135°C (26 mm)
d 1.094
n_D^{20} 1.4890

CDCl₃   QE-300

177.63	29.18
172.93	28.52
42.98	27.42*
39.69	23.74

A

Aldrich 29,964-2   CAS [10333-11-6] $C_9H_{13}NO$

3,4,5,6,7,8-Hexahydro-2(1H)-quinolinone, 97%

FW 151.21
mp 145°C

CDCl₃   QE-300

171.62	26.14
128.36	25.83
109.43	22.73
30.71	22.22
27.84	

B

Aldrich 33,423-5   CAS [98203-44-2] $C_{10}H_{17}NO_2$ Fp 137°F

(+)-3-Isopropyl-7a-methyltetrahydro-pyrrolo[2,1-b]oxazol-5(6H)-one, 95%

FW 183.25
bp 70°C
d 1.027
n_D^{20} 1.4730

CDCl₃   QE-300

178.80	33.45*
99.83	32.96
70.89	24.94*
61.62*	20.52*
33.93	19.01*

C

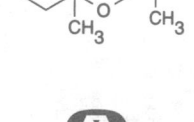

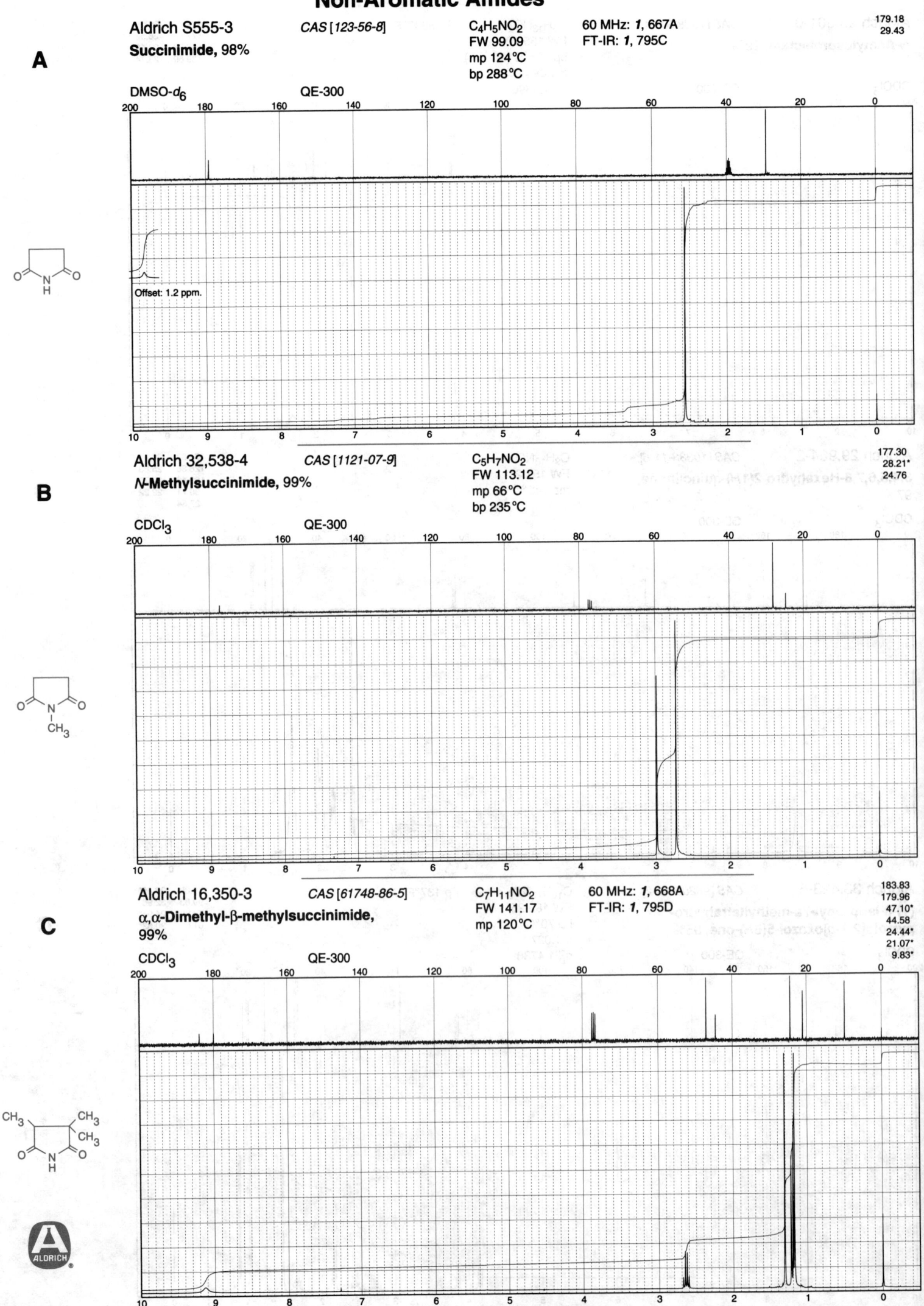

A

Aldrich S555-3   *CAS [123-56-8]* $C_4H_5NO_2$ 60 MHz: *1*, 667A

Succinimide, 98%   FW 99.09   FT-IR: *1*, 795C

mp 124°C

bp 288°C

DMSO-d_6 QE-300

Offset: 1.2 ppm.

B

Aldrich 32,538-4   *CAS [1121-07-9]* $C_5H_7NO_2$

N-Methylsuccinimide, 99%   FW 113.12

mp 66°C

bp 235°C

CDCl$_3$ QE-300

C

Aldrich 16,350-3   *CAS [61748-86-5]* $C_7H_{11}NO_2$ 60 MHz: *1*, 668A   183.83
179.96
47.10*
44.58
24.44*
21.07*
9.83*

α,α-Dimethyl-β-methylsuccinimide, 99%   FW 141.17   FT-IR: *1*, 795D

mp 120°C

CDCl$_3$ QE-300

ALDRICH

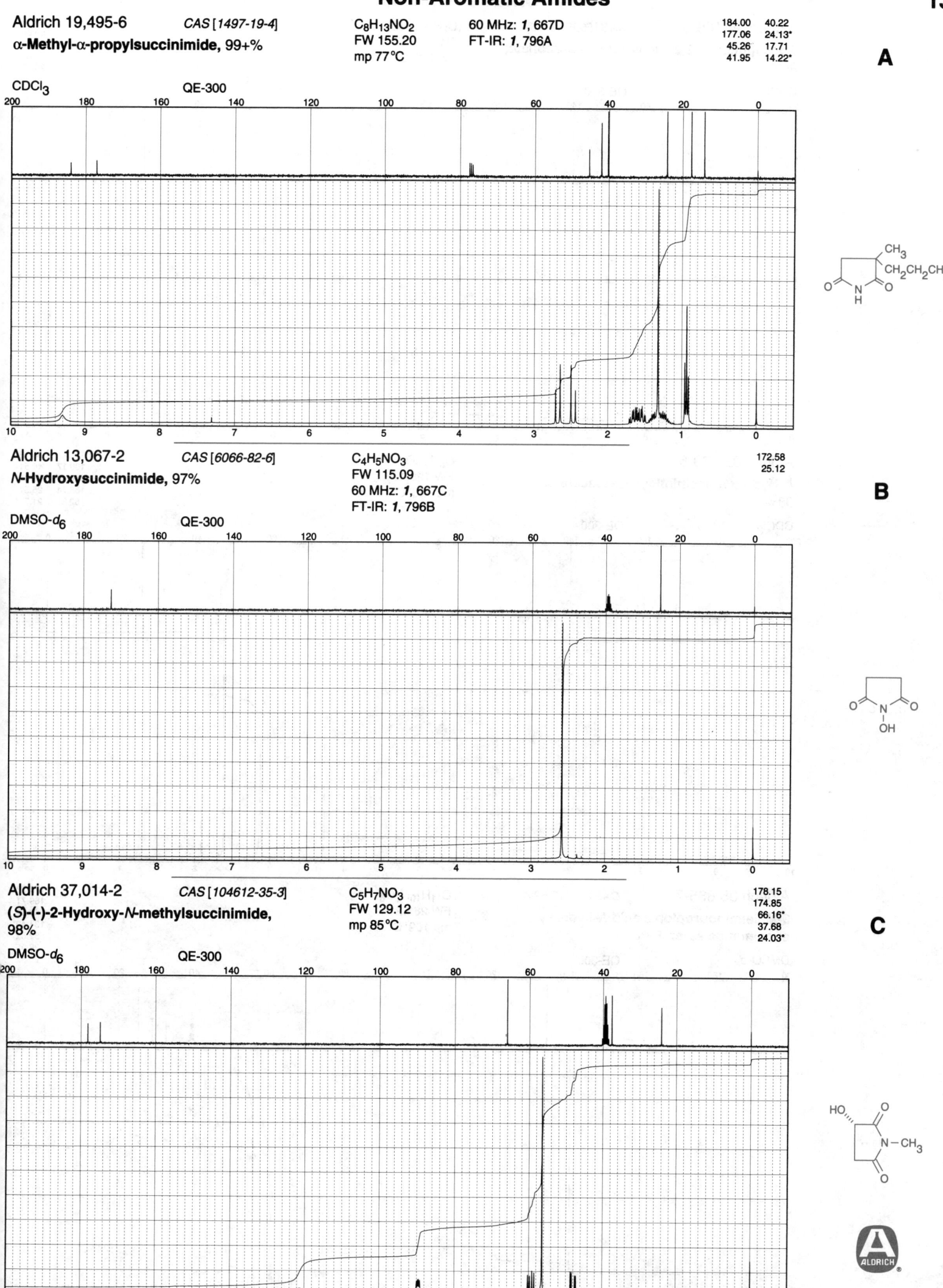

Aldrich 19,495-6   *CAS [1497-19-4]*

α-Methyl-α-propylsuccinimide, 99+%

$C_8H_{13}NO_2$
FW 155.20
mp 77°C

60 MHz: *1*, 667D
FT-IR: *1*, 796A

184.00	40.22
177.06	24.13*
45.26	17.71
41.95	14.22*

A

CDCl₃   QE-300

Aldrich 13,067-2   *CAS [6066-82-6]*

N-Hydroxysuccinimide, 97%

$C_4H_5NO_3$
FW 115.09
60 MHz: *1*, 667C
FT-IR: *1*, 796B

172.58
25.12

B

DMSO-d_6 QE-300

Aldrich 37,014-2   *CAS [104612-35-3]*

(S)-(-)-2-Hydroxy-N-methylsuccinimide, 98%

$C_5H_7NO_3$
FW 129.12
mp 85°C

178.15
174.85
66.16*
37.68
24.03*

C

DMSO-d_6 QE-300

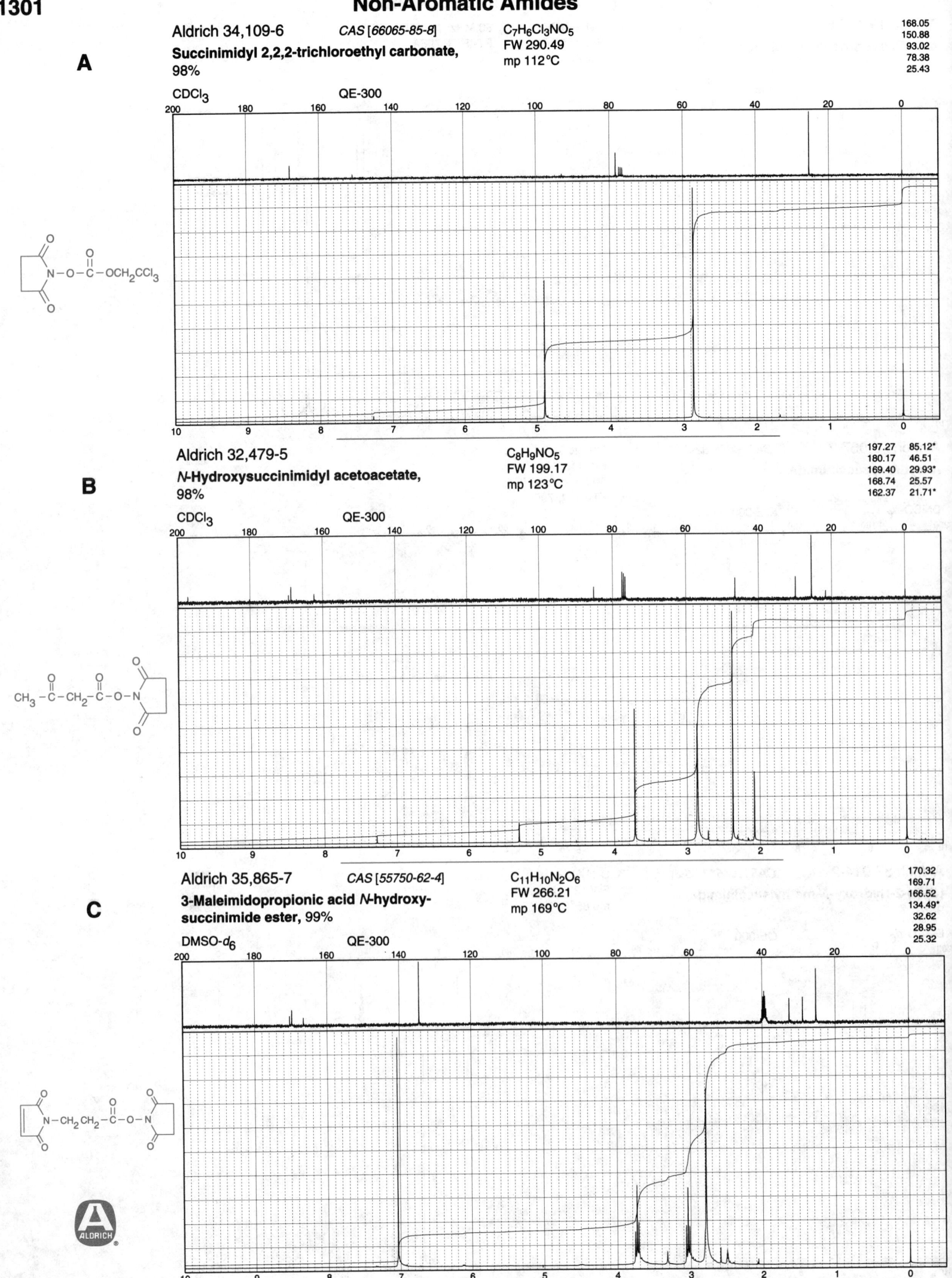

A

Aldrich 34,109-6   *CAS [66065-85-8]*   C₇H₆Cl₃NO₅
Succinimidyl 2,2,2-trichloroethyl carbonate, 98%   FW 290.49
mp 112°C

168.05
150.88
93.02
78.38
25.43

CDCl₃   QE-300

B

Aldrich 32,479-5   C₈H₉NO₅
N-Hydroxysuccinimidyl acetoacetate, 98%   FW 199.17
mp 123°C

197.27   85.12*
180.17   46.51
169.40   29.93*
168.74   25.57
162.37   21.71*

CDCl₃   QE-300

C

Aldrich 35,865-7   *CAS [55750-62-4]*   C₁₁H₁₀N₂O₆
3-Maleimidopropionic acid N-hydroxy-succinimide ester, 99%   FW 266.21
mp 169°C

170.32
169.71
166.52
134.49*
32.62
28.95
25.32

DMSO-d₆   QE-300

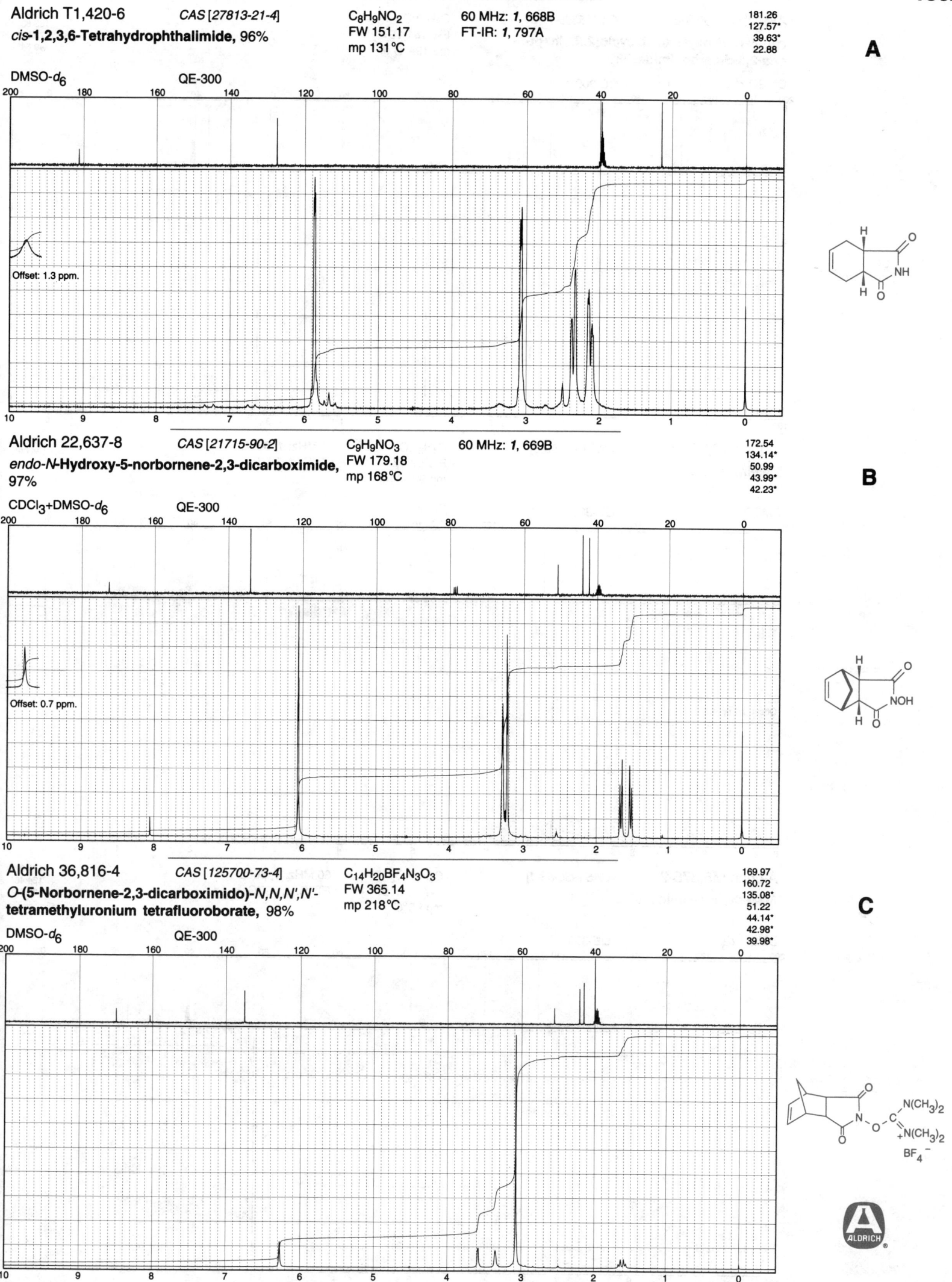

Aldrich T1,420-6 *CAS [27813-21-4]*
$C_8H_9NO_2$
60 MHz: *1*, 668B
181.26
127.57*
39.63*
22.88

cis-1,2,3,6-Tetrahydrophthalimide, 96%
FW 151.17
mp 131 °C
FT-IR: *1*, 797A

A

DMSO-d_6 QE-300

Offset: 1.3 ppm.

Aldrich 22,637-8 *CAS [21715-90-2]*
$C_9H_9NO_3$
60 MHz: *1*, 669B
172.54
134.14*
50.99
43.99*
42.23*

endo-*N*-**Hydroxy-5-norbornene-2,3-dicarboximide,**
97%
FW 179.18
mp 168 °C

B

CDCl$_3$+DMSO-d_6 QE-300

Offset: 0.7 ppm.

Aldrich 36,816-4 *CAS [125700-73-4]*
$C_{14}H_{20}BF_4N_3O_3$
169.97
160.72
135.08*
51.22
44.14*
42.98*
39.98*

O-(**5-Norbornene-2,3-dicarboximido)-***N,N,N',N'*-
tetramethyluronium tetrafluoroborate, 98%
FW 365.14
mp 218 °C

C

DMSO-d_6 QE-300

Non-Aromatic Amides

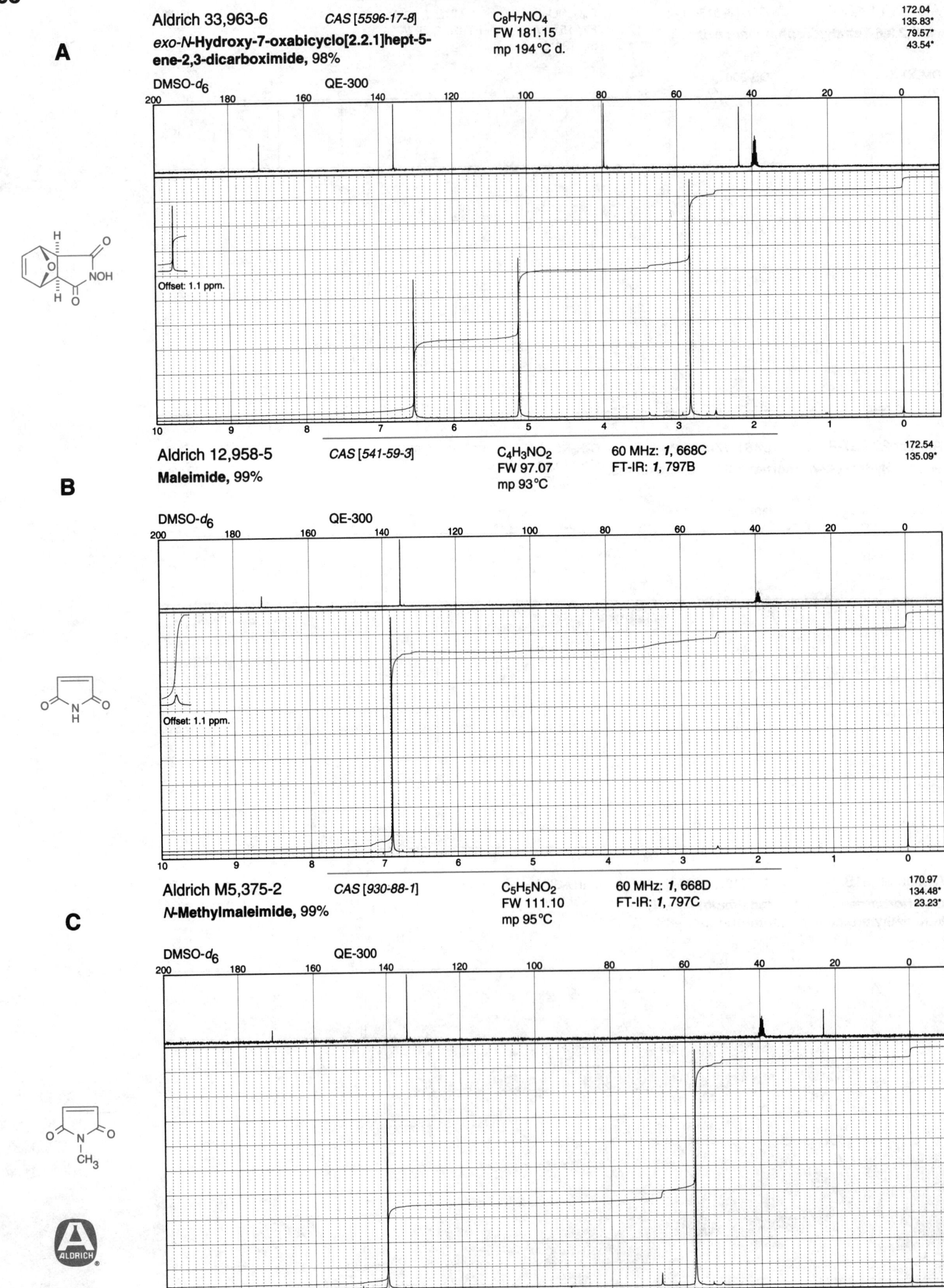

A

Aldrich 33,963-6   *CAS [5596-17-8]*   C₈H₇NO₄
FW 181.15
mp 194°C d.

172.04
135.83*
79.57*
43.54*

exo-N-Hydroxy-7-oxabicyclo[2.2.1]hept-5-
ene-2,3-dicarboximide, 98%

DMSO-*d*₆   QE-300

Offset: 1.1 ppm.

B

Aldrich 12,958-5   *CAS [541-59-3]*   C₄H₃NO₂
FW 97.07
mp 93°C

60 MHz: *1*, 668C
FT-IR: *1*, 797B

172.54
135.09*

Maleimide, 99%

DMSO-*d*₆   QE-300

Offset: 1.1 ppm.

C

Aldrich M5,375-2   *CAS [930-88-1]*   C₅H₅NO₂
FW 111.10
mp 95°C

60 MHz: *1*, 668D
FT-IR: *1*, 797C

170.97
134.48*
23.23*

N-Methylmaleimide, 99%

DMSO-*d*₆   QE-300

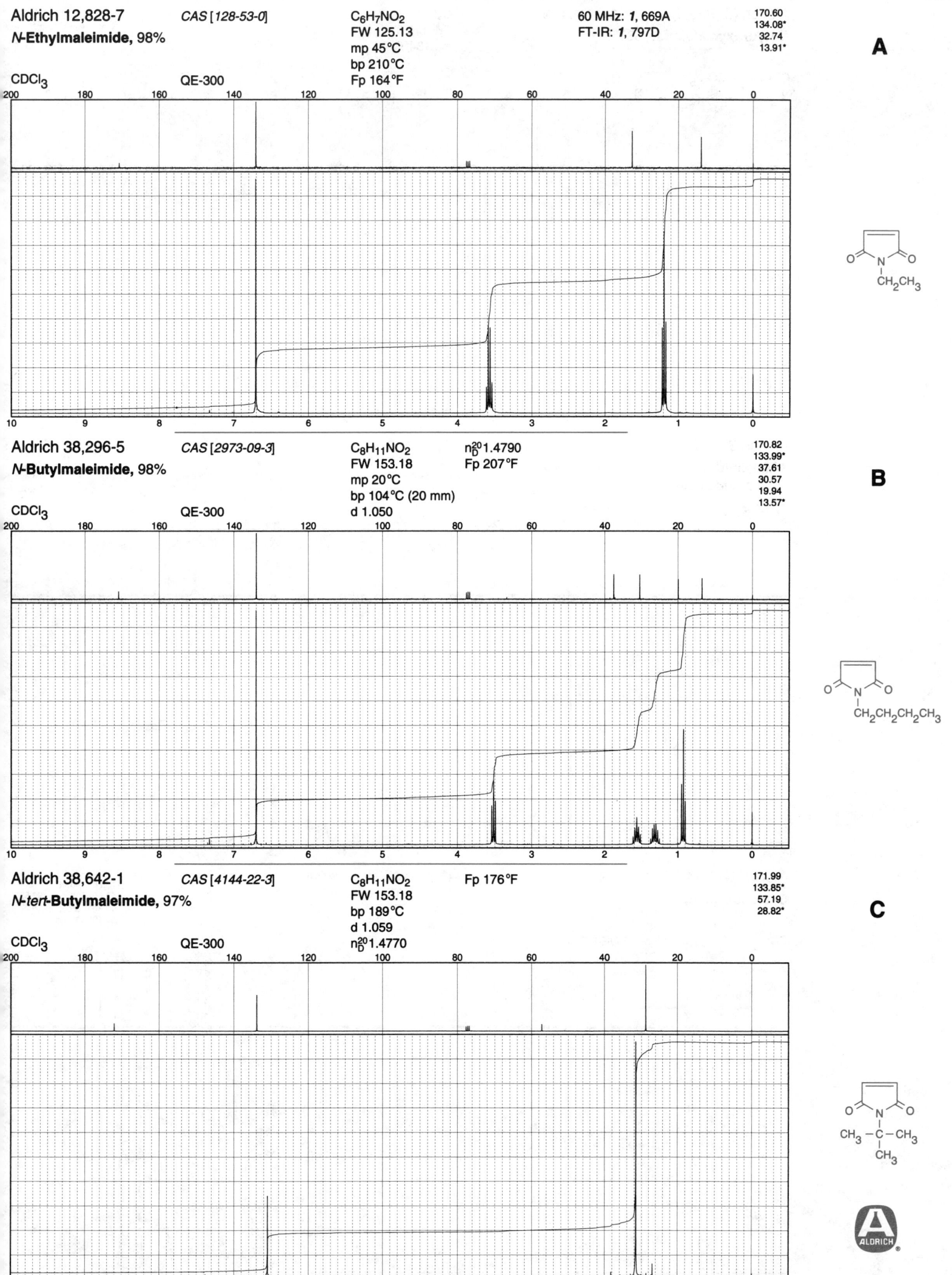

Aldrich 12,828-7   *CAS [128-53-0]* $C_6H_7NO_2$ 60 MHz: *1*, 669A

N-Ethylmaleimide, 98%   FW 125.13   FT-IR: *1*, 797D
mp 45°C
bp 210°C
Fp 164°F

CDCl₃   QE-300

170.60
134.08*
32.74
13.91*

A

Aldrich 38,296-5   *CAS [2973-09-3]* $C_8H_{11}NO_2$ n_D^{20}1.4790

N-Butylmaleimide, 98%   FW 153.18   Fp 207°F
mp 20°C
bp 104°C (20 mm)

CDCl₃   QE-300   d 1.050

170.82
133.99*
37.61
30.57
19.94
13.57*

B

Aldrich 38,642-1   *CAS [4144-22-3]* $C_8H_{11}NO_2$ Fp 176°F

N-tert-Butylmaleimide, 97%   FW 153.18
bp 189°C
d 1.059

CDCl₃   QE-300 n_D^{20}1.4770

171.99
133.85*
57.19
28.82*

C

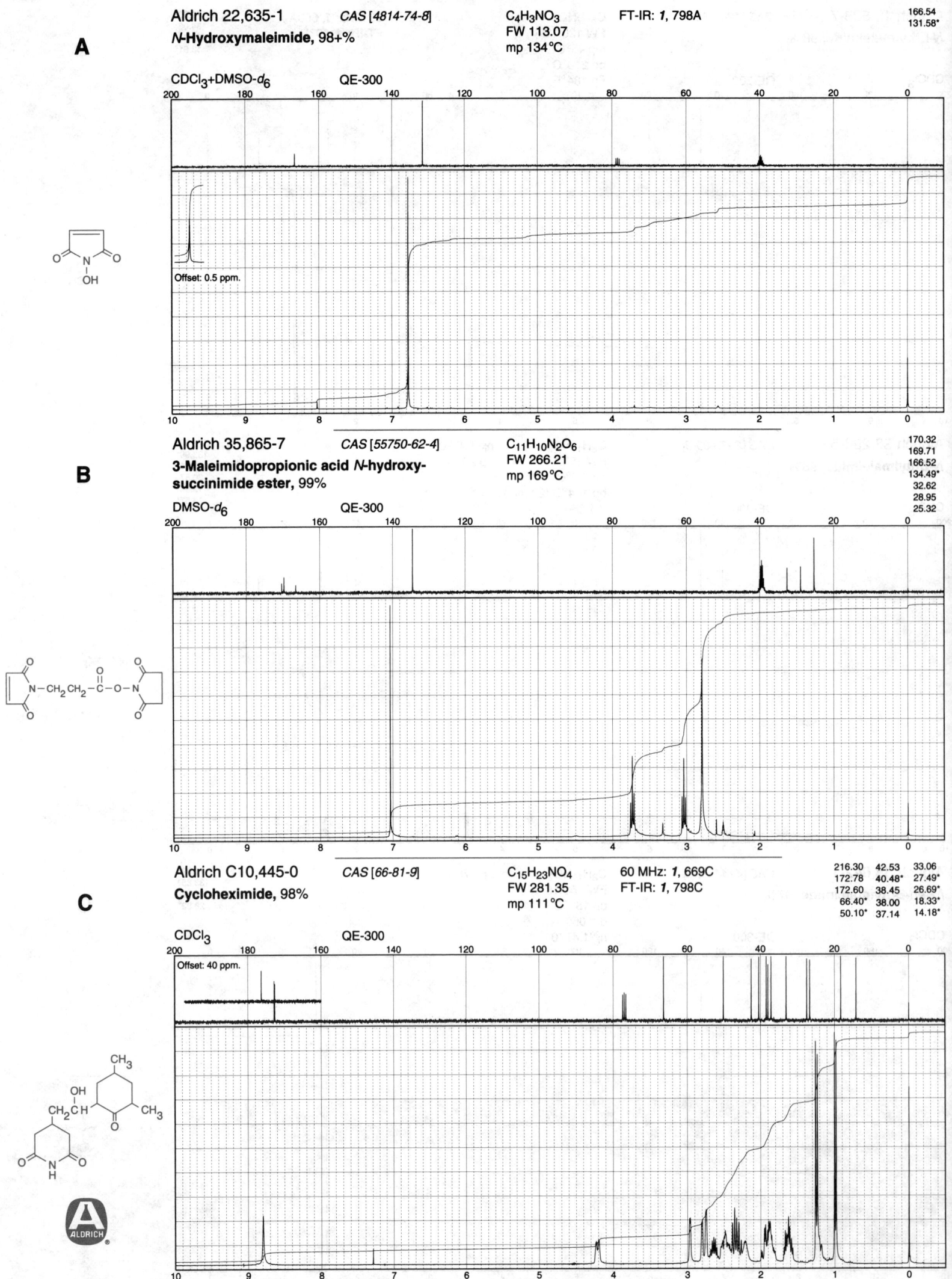

A

Aldrich 22,635-1   *CAS [4814-74-8]*   C$_4$H$_3$NO$_3$ FT-IR: *1*, 798A   166.54
131.58*

N-Hydroxymaleimide, 98+%   FW 113.07
mp 134°C

CDCl$_3$+DMSO-*d$_6$*   QE-300

Offset: 0.5 ppm.

B

Aldrich 35,865-7   *CAS [55750-62-4]*   C$_{11}$H$_{10}$N$_2$O$_6$ 170.32
169.71

3-Maleimidopropionic acid N-hydroxy-   FW 266.21   166.52
134.49*

succinimide ester, 99%   mp 169°C   32.62
28.95
25.32

DMSO-*d$_6$*   QE-300

C

Aldrich C10,445-0   *CAS [66-81-9]*   C$_{15}$H$_{23}$NO$_4$ 60 MHz: *1*, 669C   216.30 42.53 33.06
172.78 40.48* 27.49*

Cycloheximide, 98%   FW 281.35   FT-IR: *1*, 798C   172.60 38.45 26.69*
66.40* 38.00 18.33*

mp 111°C   50.10* 37.14 14.18*

CDCl$_3$ QE-300

Offset: 40 ppm.

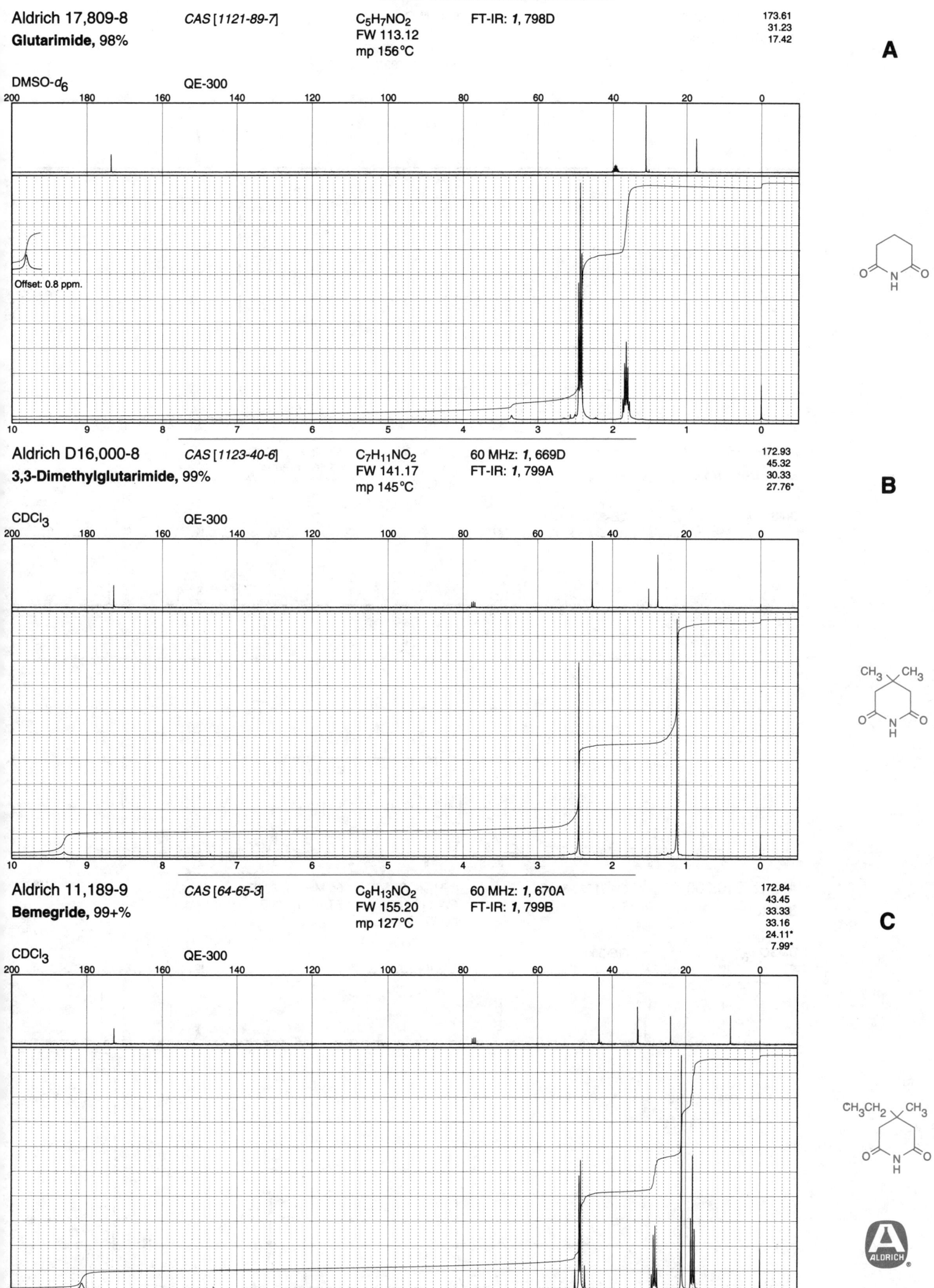

Aldrich 17,809-8   *CAS [1121-89-7]*   C$_5$H$_7$NO$_2$ FT-IR: *1*, 798D   173.61
31.23
17.42

Glutarimide, 98%   FW 113.12
mp 156°C

A

DMSO-*d$_6$*   QE-300

Offset: 0.8 ppm.

Aldrich D16,000-8   *CAS [1123-40-6]*   C$_7$H$_{11}$NO$_2$ 60 MHz: *1*, 669D   172.93
45.32
30.33
27.76*

3,3-Dimethylglutarimide, 99%   FW 141.17   FT-IR: *1*, 799A
mp 145°C

B

CDCl$_3$ QE-300

Aldrich 11,189-9   *CAS [64-65-3]*   C$_8$H$_{13}$NO$_2$ 60 MHz: *1*, 670A   172.84
43.45
33.33
33.16
24.11*
7.99*

Bemegride, 99+%   FW 155.20   FT-IR: *1*, 799B
mp 127°C

C

CDCl$_3$ QE-300

ALDRICH

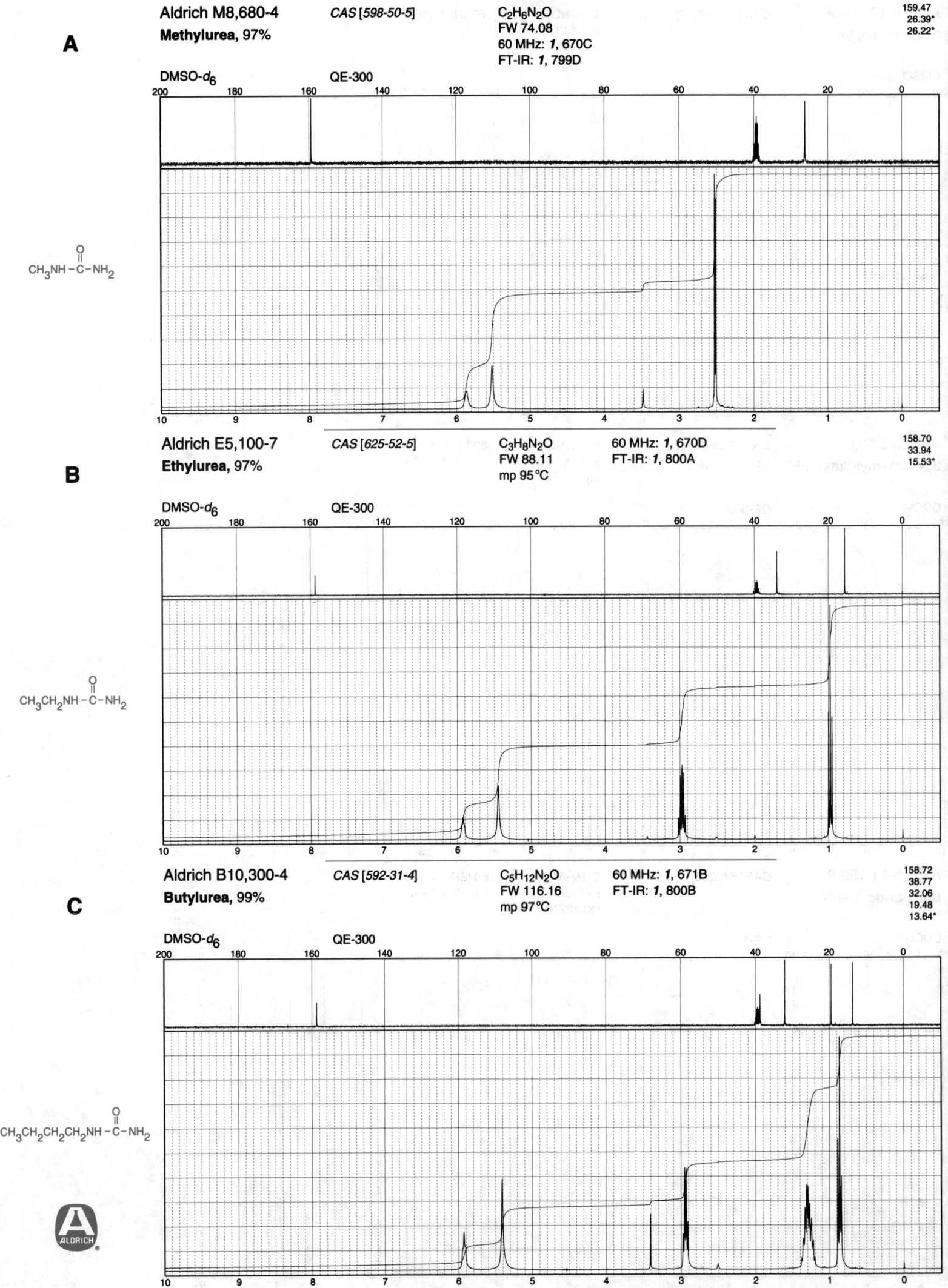

Aldrich M8,680-4
Methylurea, 97%

CAS [598-50-5]

C₂H₆N₂O
FW 74.08
60 MHz: *1,* 670C
FT-IR: *1,* 799D

159.47
26.39*
26.22*

DMSO-d₆ QE-300

CH₃NH–C–NH₂ (O)

Aldrich E5,100-7
Ethylurea, 97%

CAS [625-52-5]

C₃H₈N₂O
FW 88.11
mp 95°C

60 MHz: *1,* 670D
FT-IR: *1,* 800A

158.70
33.94
15.53*

DMSO-d₆ QE-300

CH₃CH₂NH–C–NH₂ (O)

Aldrich B10,300-4
Butylurea, 99%

CAS [592-31-4]

C₅H₁₂N₂O
FW 116.16
mp 97°C

60 MHz: *1,* 671B
FT-IR: *1,* 800B

158.72
38.77
32.06
19.48
13.64*

DMSO-d₆ QE-300

CH₃CH₂CH₂CH₂NH–C–NH₂ (O)

ALDRICH

Aldrich 14,014-7   *CAS [13072-69-0]* $C_{11}H_{18}N_2O$ FT-IR: *1*, 800C

N-(1-Adamantyl)urea, 98%   FW 194.28
mp 250°C d.

157.61
49.06
41.78
36.03
28.83*

A

DMSO-d_6 QE-300

Aldrich A3,680-8   *CAS [557-11-9]* $C_4H_8N_2O$ 60 MHz: *1*, 671C
FW 100.12   FT-IR: *1*, 800D

Allylurea, 95%   mp 85°C

158.51
136.64*
114.08
41.50

B

DMSO-d_6 QE-300

Aldrich 21,729-8   *CAS [1801-72-5]* $C_7H_{12}N_2O$ 60 MHz: *1*, 674C
FW 140.19   FT-IR: *1*, 801A

1,3-Diallylurea, 98%   mp 92°C

159.08
135.63*
114.99
42.58

C

CDCl$_3$ QE-300

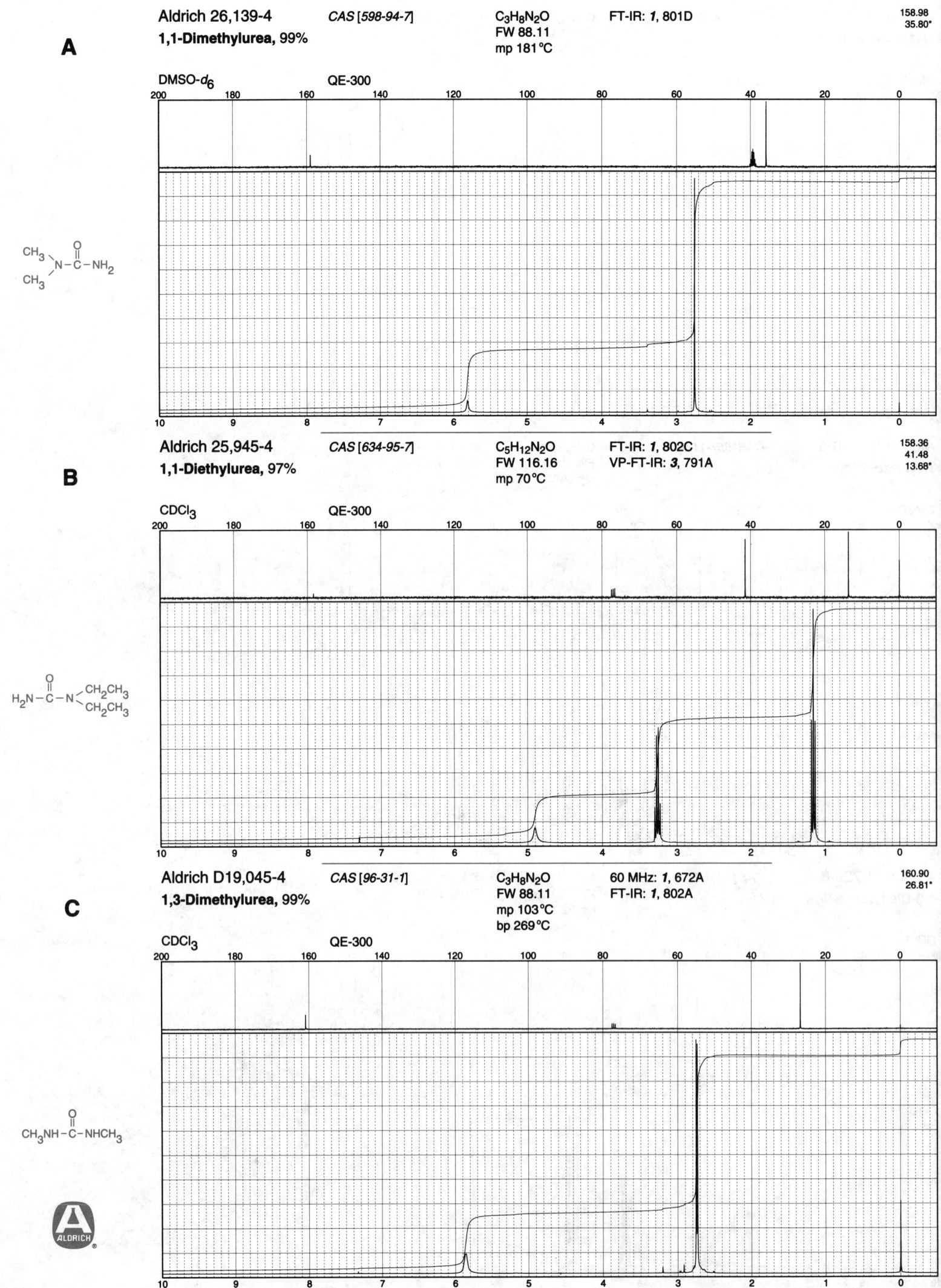

A

Aldrich 26,139-4   *CAS [598-94-7]*   C₃H₈N₂O   FT-IR: *1*, 801D
1,1-Dimethylurea, 99%   FW 88.11
                                                mp 181°C

158.98
35.80*

DMSO-*d₆*   QE-300

B

Aldrich 25,945-4   *CAS [634-95-7]*   C₅H₁₂N₂O   FT-IR: *1*, 802C
1,1-Diethylurea, 97%   FW 116.16   VP-FT-IR: *3*, 791A
                                                mp 70°C

158.36
41.48
13.68*

CDCl₃   QE-300

C

Aldrich D19,045-4   *CAS [96-31-1]*   C₃H₈N₂O   60 MHz: *1*, 672A
1,3-Dimethylurea, 99%   FW 88.11   FT-IR: *1*, 802A
                                                mp 103°C
                                                bp 269°C

160.90
26.81*

CDCl₃   QE-300

A

Aldrich D10,108-7   *CAS [623-76-7]* $C_5H_{12}N_2O$ 60 MHz: *1*, 672B   159.34
1,3-Diethylurea, 97%   FW 116.16   FT-IR: *1*, 802B   34.85
mp 113°C   15.58*

CDCl$_3$ QE-300

$CH_3CH_2NH-\overset{\overset{\text{O}}{\|}}{C}-NHCH_2CH_3$

B

Aldrich D8,080-0   *CAS [2387-23-7]* $C_{13}H_{24}N_2O$ 60 MHz: *1*, 672C   158.97
1,3-Dicyclohexylurea, 98%   FW 224.35   FT-IR: *1*, 802D   54.29*
mp 233°C   34.05
26.36
25.96

TFA-*d*   QE-300

Offset: 1.6 ppm.

C

Aldrich 26,374-5   *CAS [6257-10-9]* $C_{14}H_{26}N_2O$ n_D^{20}1.4980   150.62   33.67
N,N'-Dicyclohexyl-O-methylisourea, 95%   FW 238.38   Fp >230°F   52.30*   25.73
mp 33°C   FT-IR: *1*, 803A   51.75*   25.29
bp 163°C (11 mm)   50.11*   25.09
d 0.990   34.44   24.49

DMSO-*d*$_6$ QE-300

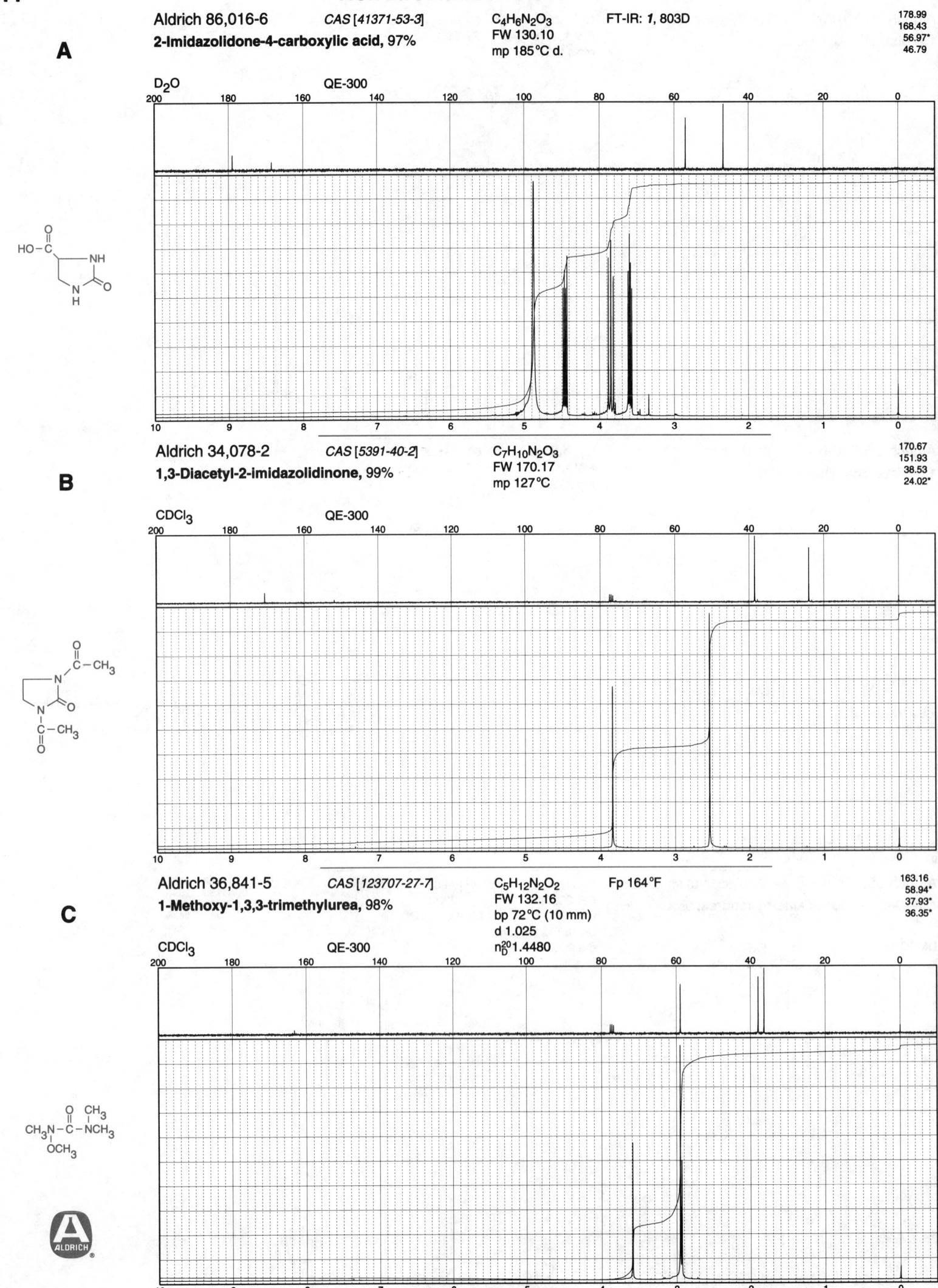

A

Aldrich 86,016-6   *CAS [41371-53-3]*   C₄H₆N₂O₃   FT-IR: *1*, 803D

$C_4H_6N_2O_3$
FW 130.10
mp 185°C d.

2-Imidazolidone-4-carboxylic acid, 97%

178.99
168.43
56.97*
46.79

D₂O   QE-300

B

Aldrich 34,078-2   *CAS [5391-40-2]* $C_7H_{10}N_2O_3$
FW 170.17
mp 127°C

1,3-Diacetyl-2-imidazolidinone, 99%

170.67
151.93
38.53
24.02*

CDCl₃   QE-300

C

Aldrich 36,841-5   *CAS [123707-27-7]* $C_5H_{12}N_2O_2$
FW 132.16
bp 72°C (10 mm)
d 1.025
n_D^{20}1.4480

1-Methoxy-1,3,3-trimethylurea, 98%   Fp 164°F

163.16
58.94*
37.93*
36.35*

CDCl₃   QE-300

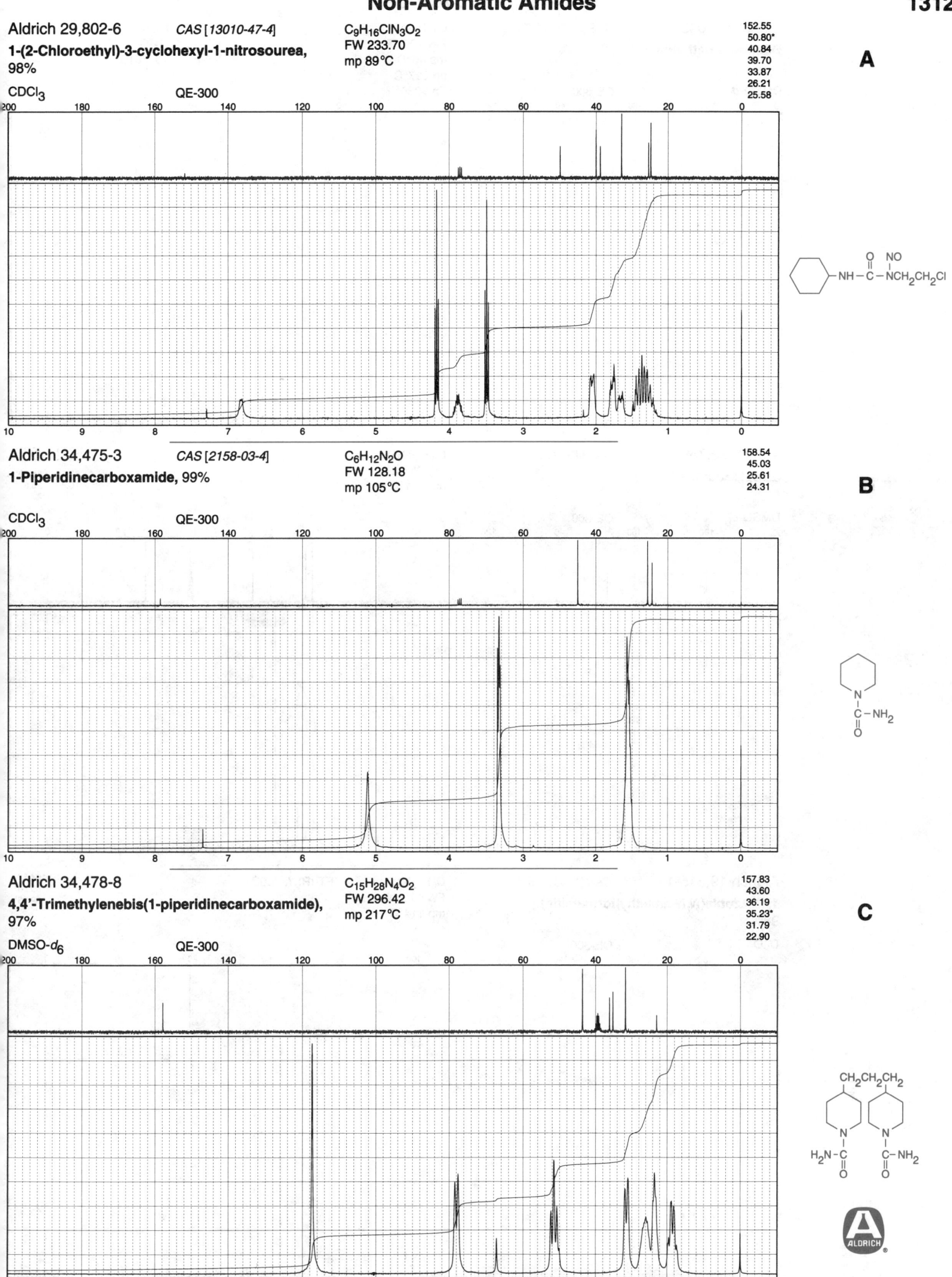

Aldrich 29,802-6 *CAS [13010-47-4]* C₉H₁₆ClN₃O₂ $C_9H_{16}ClN_3O_2$
FW 233.70
mp 89°C

1-(2-Chloroethyl)-3-cyclohexyl-1-nitrosourea, 98%

CDCl₃   QE-300

152.55
50.80*
40.84
39.70
33.87
26.21
25.58

A

Aldrich 34,475-3 *CAS [2158-03-4]* $C_6H_{12}N_2O$
FW 128.18
mp 105°C

1-Piperidinecarboxamide, 99%

CDCl₃   QE-300

158.54
45.03
25.61
24.31

B

Aldrich 34,478-8 $C_{15}H_{28}N_4O_2$
FW 296.42
mp 217°C

4,4'-Trimethylenebis(1-piperidinecarboxamide), 97%

DMSO-d_6 QE-300

157.83
43.60
36.19
35.23*
31.79
22.90

C

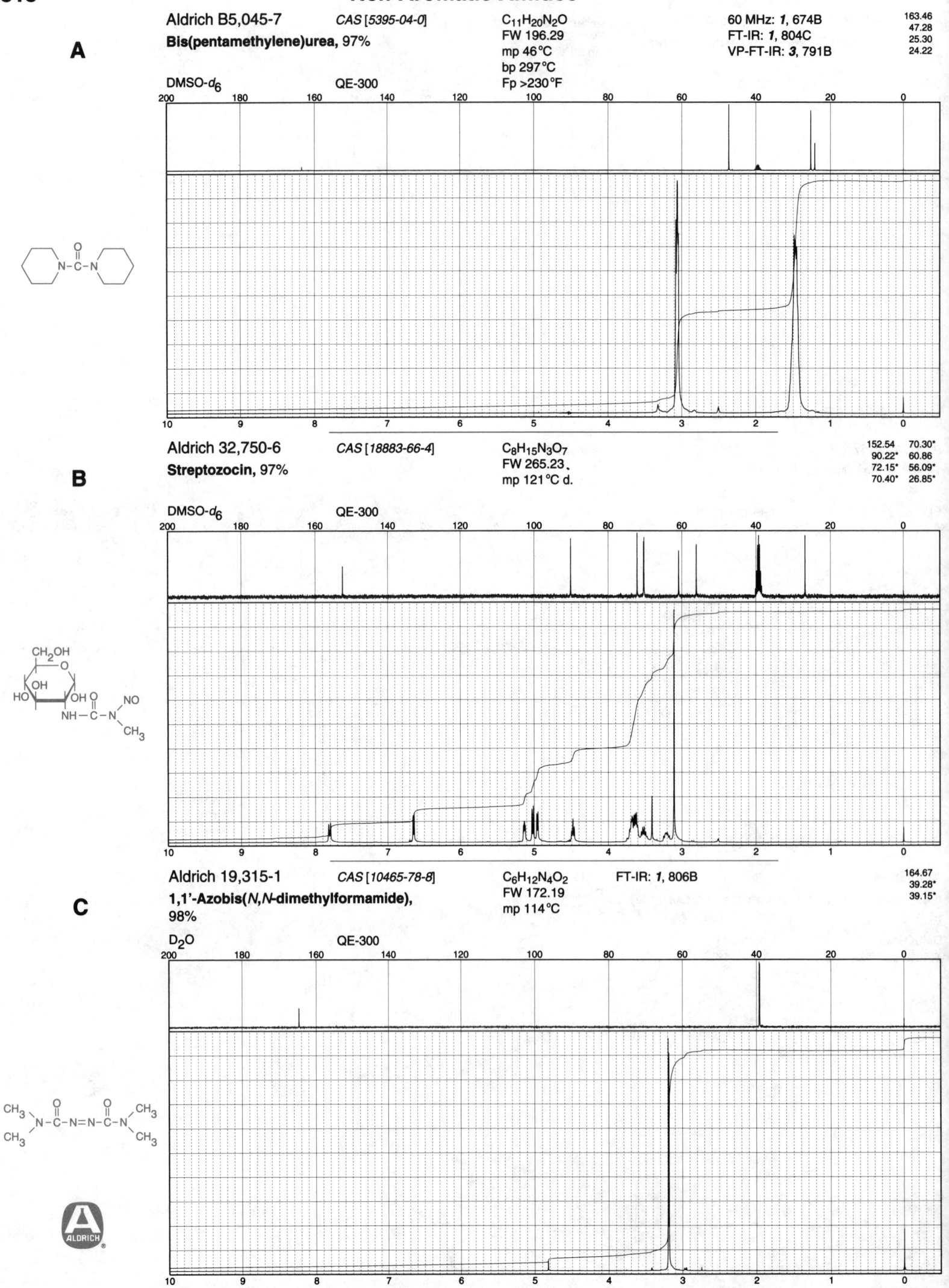

A

Aldrich B5,045-7   *CAS [5395-04-0]* $C_{11}H_{20}N_2O$ 60 MHz: *1*, 674B   163.46
Bis(pentamethylene)urea, 97%   FW 196.29   FT-IR: *1*, 804C   47.28
mp 46°C   VP-FT-IR: *3*, 791B   25.30
bp 297°C   24.22
DMSO-d_6 QE-300   Fp >230°F

B

Aldrich 32,750-6   *CAS [18883-66-4]* $C_8H_{15}N_3O_7$ 152.54   70.30*
Streptozocin, 97%   FW 265.23   90.22*   60.86
mp 121°C d.   72.15*   56.09*
70.40*   26.85*
DMSO-d_6 QE-300

C

Aldrich 19,315-1   *CAS [10465-78-8]* $C_6H_{12}N_4O_2$ FT-IR: *1*, 806B   164.67
1,1'-Azobis(*N,N*-dimethylformamide),   FW 172.19   39.28*
98%   mp 114°C   39.15*
D$_2$O   QE-300

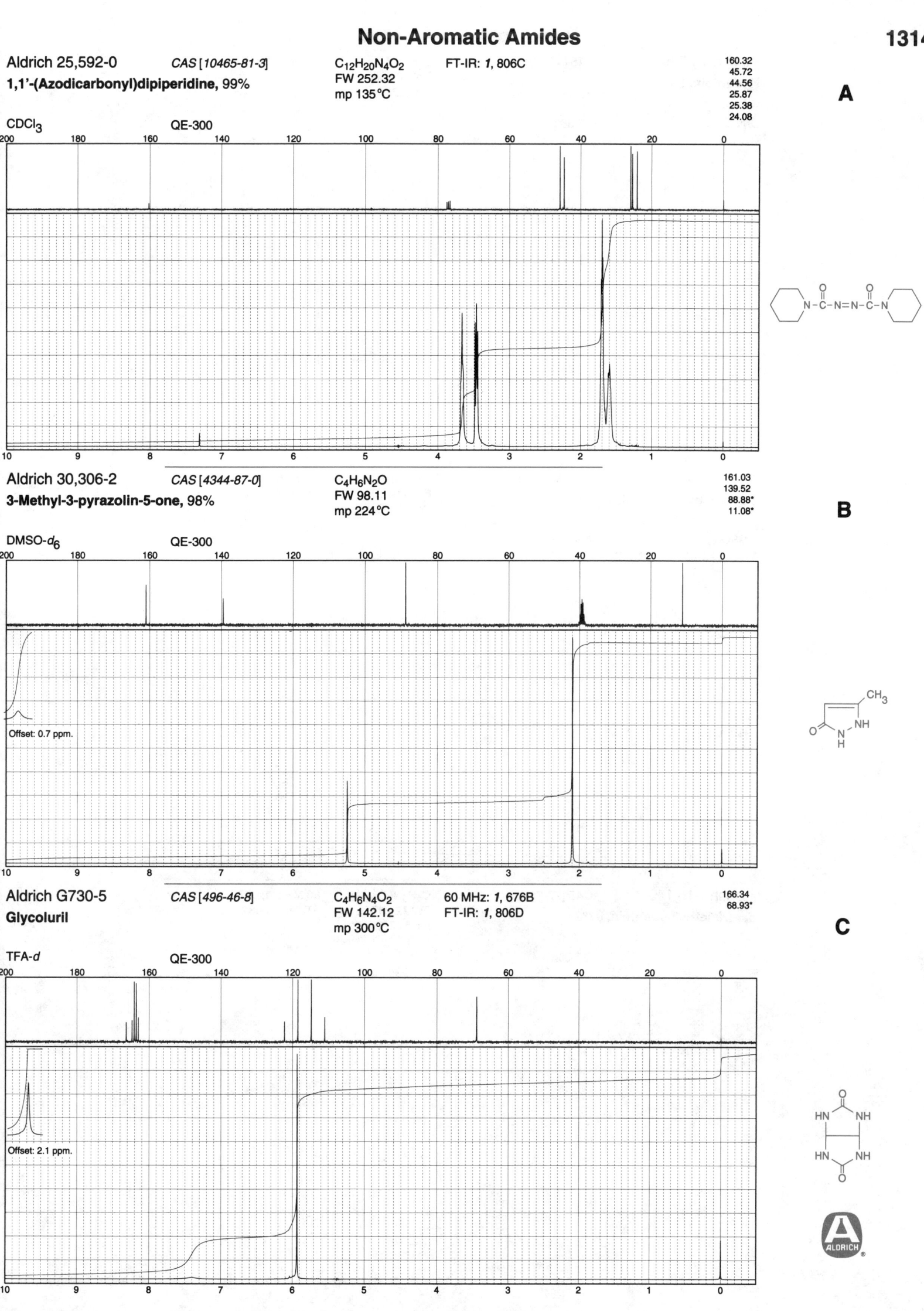

Aldrich 25,592-0   CAS [10465-81-3] $C_{12}H_{20}N_4O_2$ FT-IR: 1, 806C

1,1'-(Azodicarbonyl)dipiperidine, 99%

FW 252.32
mp 135°C

160.32
45.72
44.56
25.87
25.38
24.08

A

CDCl₃   QE-300

Aldrich 30,306-2   CAS [4344-87-0] $C_4H_6N_2O$

3-Methyl-3-pyrazolin-5-one, 98%

FW 98.11
mp 224°C

161.03
139.52
88.88*
11.08*

B

DMSO-d_6 QE-300

Offset: 0.7 ppm.

Aldrich G730-5   CAS [496-46-8] $C_4H_6N_4O_2$ 60 MHz: 1, 676B

Glycoluril

FW 142.12
mp 300°C   FT-IR: 1, 806D

166.34
68.93*

C

TFA-d QE-300

Offset: 2.1 ppm.

ALDRICH

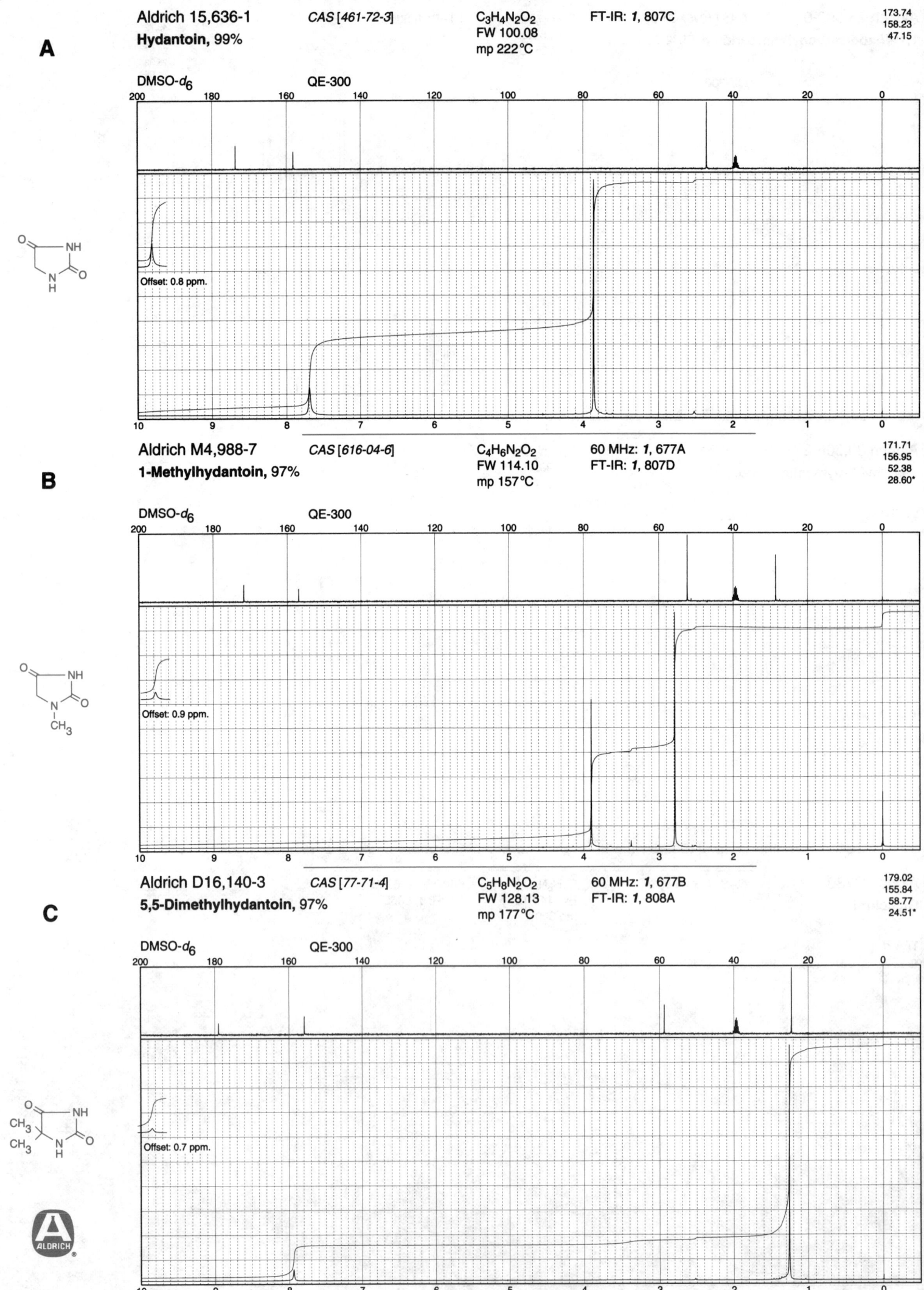

A

Aldrich 15,636-1
Hydantoin, 99%
CAS [461-72-3]
C₃H₄N₂O₂
FW 100.08
mp 222°C
FT-IR: *1*, 807C

173.74
158.23
47.15

DMSO-*d₆*   QE-300

Offset: 0.8 ppm.

B

Aldrich M4,988-7
1-Methylhydantoin, 97%
CAS [616-04-6]
C₄H₆N₂O₂
FW 114.10
mp 157°C
60 MHz: *1*, 677A
FT-IR: *1*, 807D

171.71
156.95
52.38
28.60*

DMSO-*d₆*   QE-300

Offset: 0.9 ppm.

C

Aldrich D16,140-3
5,5-Dimethylhydantoin, 97%
CAS [77-71-4]
C₅H₈N₂O₂
FW 128.13
mp 177°C
60 MHz: *1*, 677B
FT-IR: *1*, 808A

179.02
155.84
58.77
24.51*

DMSO-*d₆*   QE-300

Offset: 0.7 ppm.

ALDRICH

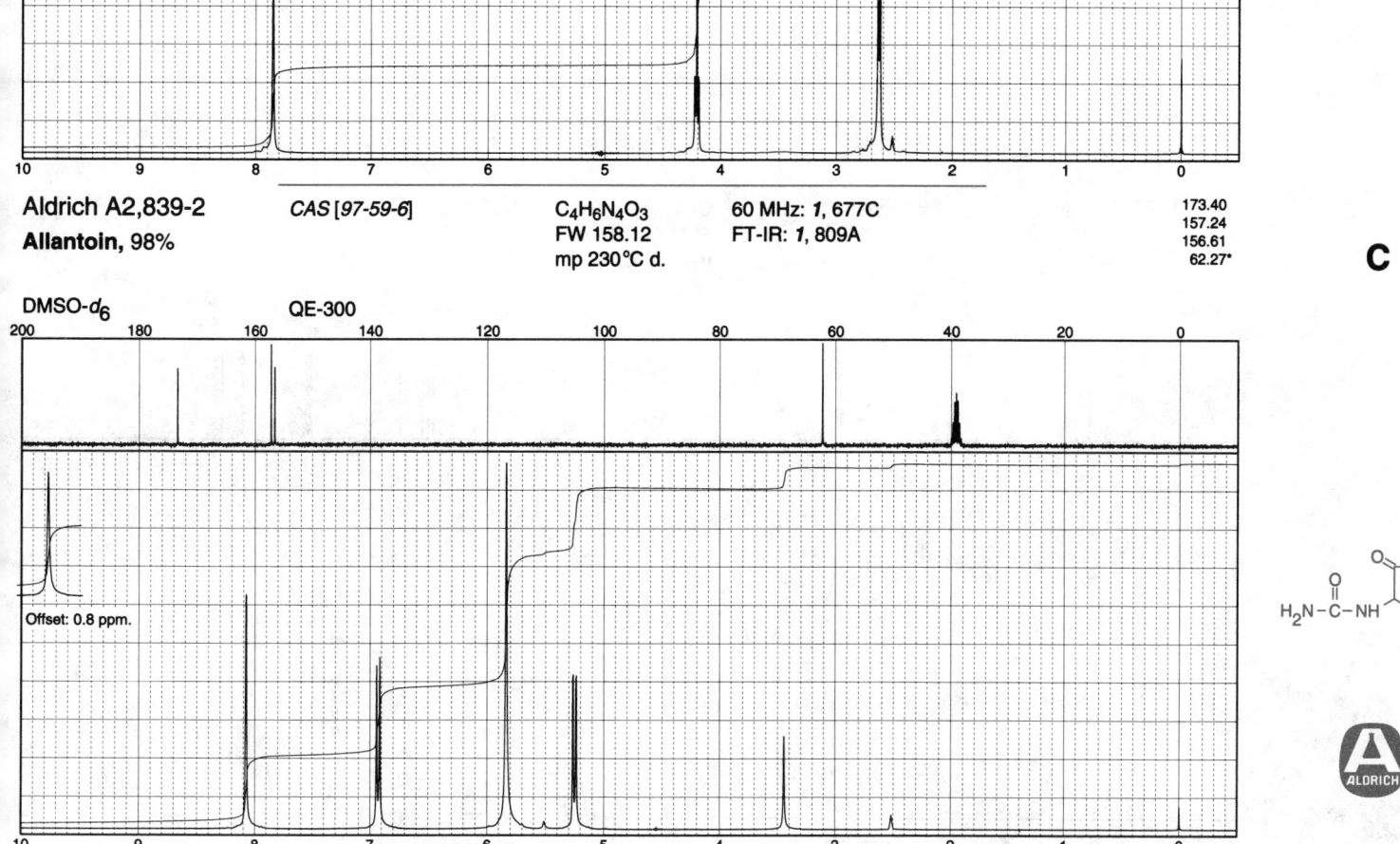

Aldrich 36,825-3   *CAS [15871-54-2]*

7,7,9,9-Tetramethyl-1,3,8-triazaspiro[4.5]-decane-2,4-dione, 98%

$C_{11}H_{19}N_3O_2$
mp 300 °C

181.56
161.01
63.50
58.03
42.97
33.10*
27.81*

A

D$_2$O + DCl   QE-300

Aldrich 85,062-4   *CAS [5427-26-9]*

5-Hydantoinacetic acid, 98%

$C_5H_6N_2O_4$
FW 158.11
mp 215 °C d.

FT-IR: *1*, 808D

175.28
170.93
157.53
54.25*
35.30

B

DMSO-*d$_6$*   QE-300

Offset: 2.7 ppm.

Aldrich A2,839-2   *CAS [97-59-6]*

Allantoin, 98%

$C_4H_6N_4O_3$
FW 158.12
mp 230 °C d.

60 MHz: *1*, 677C
FT-IR: *1*, 809A

173.40
157.24
156.61
62.27*

C

DMSO-*d$_6$*   QE-300

Offset: 0.8 ppm.

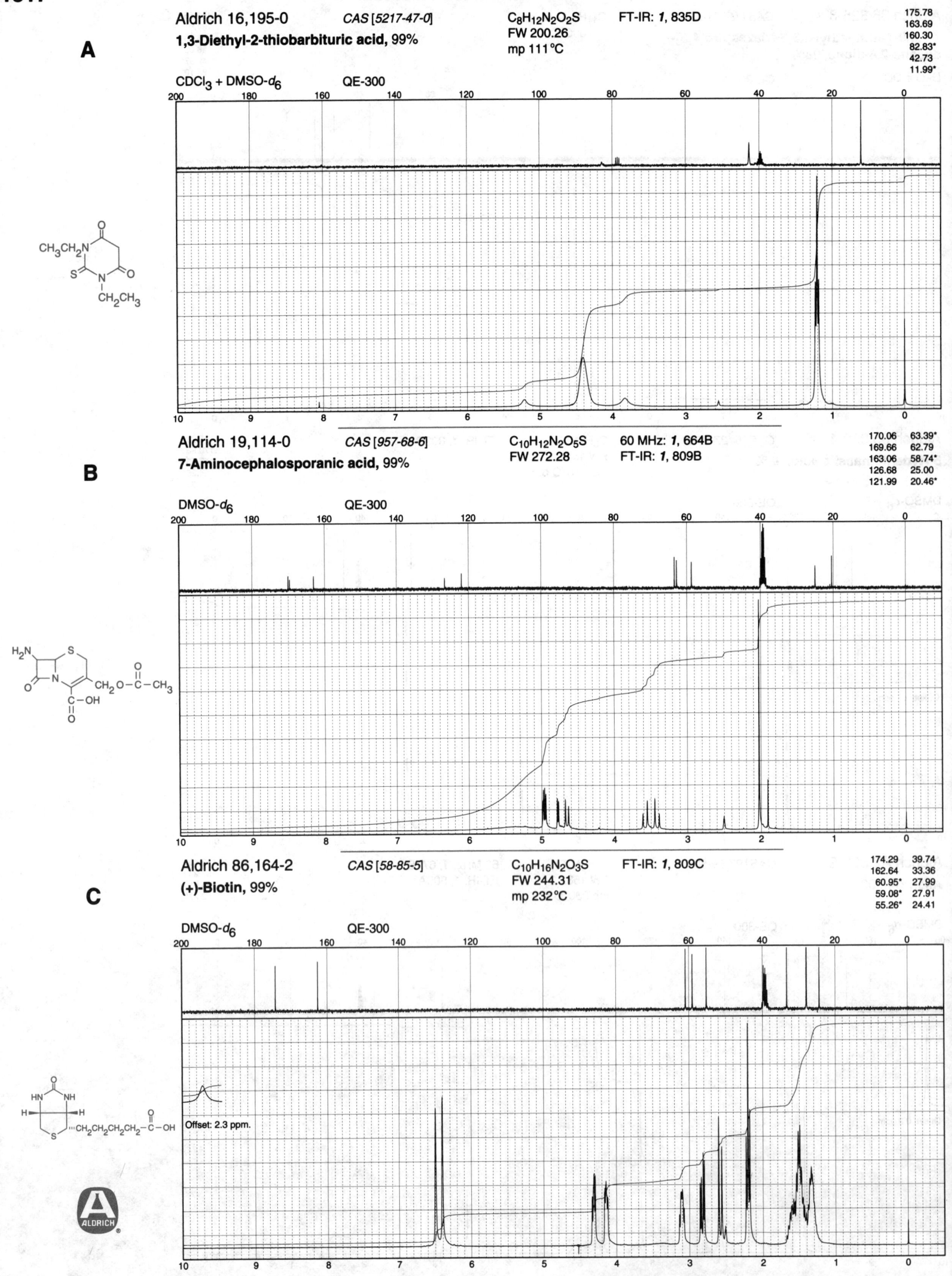

A

Aldrich 16,195-0   *CAS [5217-47-0]*   C₈H₁₂N₂O₂S   FT-IR: *1*, 835D
1,3-Diethyl-2-thiobarbituric acid, 99%   FW 200.26
mp 111°C

175.78
163.69
160.30
82.83*
42.73
11.99*

CDCl₃ + DMSO-d₆   QE-300

B

Aldrich 19,114-0   *CAS [957-68-6]*   C₁₀H₁₂N₂O₅S   60 MHz: *1*, 664B
7-Aminocephalosporanic acid, 99%   FW 272.28   FT-IR: *1*, 809B

170.06   63.39*
169.66   62.79
163.06   58.74*
126.68   25.00
121.99   20.46*

DMSO-d₆   QE-300

C

Aldrich 86,164-2   *CAS [58-85-5]*   C₁₀H₁₆N₂O₃S   FT-IR: *1*, 809C
(+)-Biotin, 99%   FW 244.31
mp 232°C

174.29   39.74
162.64   33.36
60.95*   27.99
59.08*   27.91
55.26*   24.41

DMSO-d₆   QE-300

Offset: 2.3 ppm.

ALDRICH®

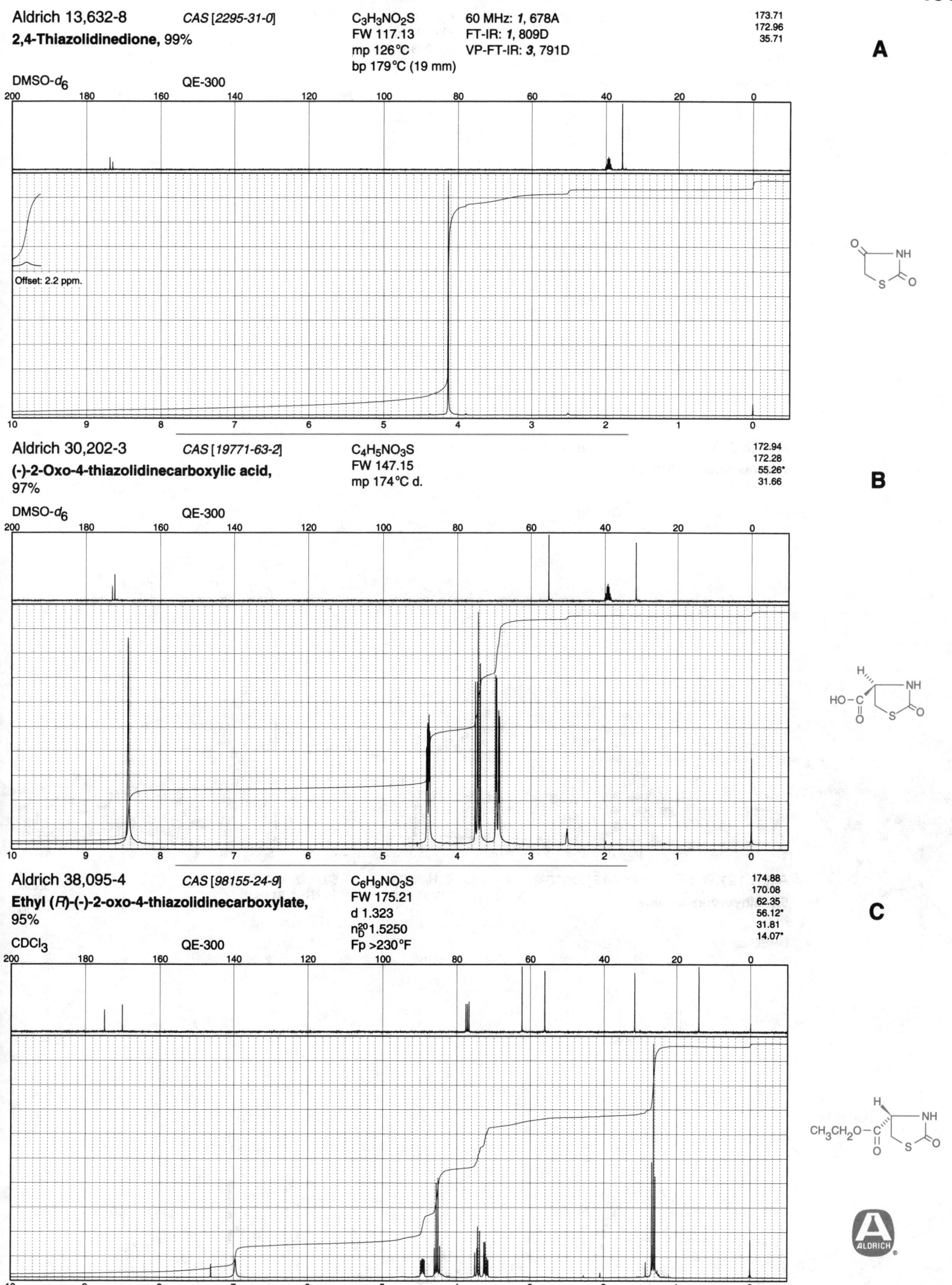

Aldrich 13,632-8   CAS [2295-31-0]
2,4-Thiazolidinedione, 99%

$C_3H_3NO_2S$
FW 117.13
mp 126°C
bp 179°C (19 mm)

60 MHz: *1*, 678A
FT-IR: *1*, 809D
VP-FT-IR: *3*, 791D

173.71
172.96
35.71

A

DMSO-d_6 QE-300

Offset: 2.2 ppm.

Aldrich 30,202-3   CAS [19771-63-2]
(-)-2-Oxo-4-thiazolidinecarboxylic acid, 97%

$C_4H_5NO_3S$
FW 147.15
mp 174°C d.

172.94
172.28
55.26*
31.66

B

DMSO-d_6 QE-300

Aldrich 38,095-4   CAS [98155-24-9]
Ethyl (*R*)-(-)-2-oxo-4-thiazolidinecarboxylate, 95%

$C_6H_9NO_3S$
FW 175.21
d 1.323
n_D^{20}1.5250
Fp >230°F

174.88
170.08
62.35
56.12*
31.81
14.07*

C

CDCl$_3$ QE-300

A

Aldrich U260-1

Urazole, 97%

CAS [3232-84-6]

$C_2H_3N_3O_2$
FW 101.07
mp 243°C

FT-IR: *1*, 810C

DMSO-d_6 QE-300

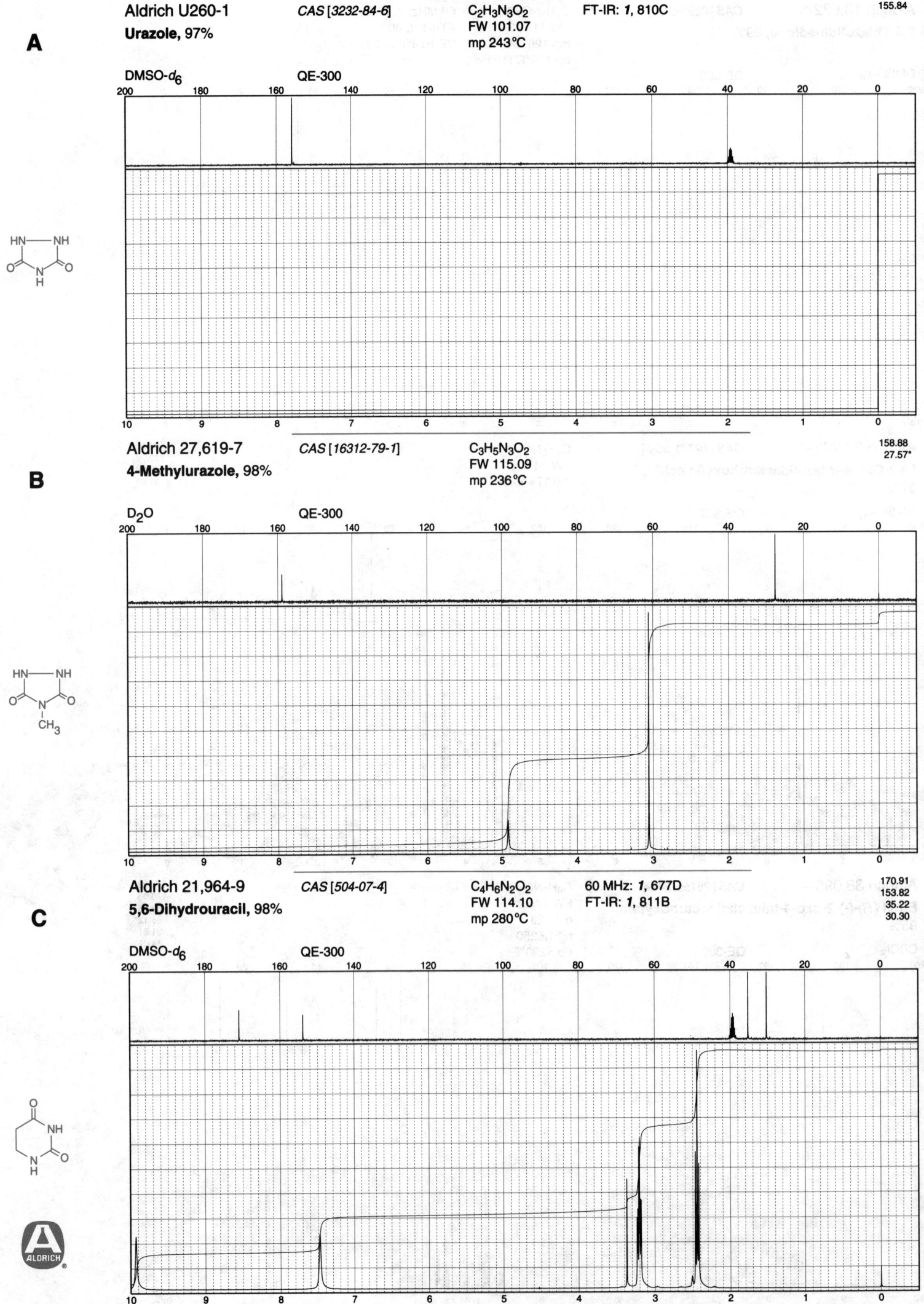

Aldrich 27,619-7

4-Methylurazole, 98%

CAS [16312-79-1]

$C_3H_5N_3O_2$
FW 115.09
mp 236°C

158.88
27.57*

B

D$_2$O   QE-300

Aldrich 21,964-9

5,6-Dihydrouracil, 98%

CAS [504-07-4]

$C_4H_6N_2O_2$
FW 114.10
mp 280°C

60 MHz: *1*, 677D
FT-IR: *1*, 811B

170.91
153.82
35.22
30.30

C

DMSO-d_6 QE-300

Non-Aromatic Amides

1320

Aldrich B20-8   *CAS [67-52-7]* $C_4H_4N_2O_3$

Barbituric acid, 98%

FW 128.09

60 MHz: *2*, 693A

167.61
151.46
39.22

A

DMSO-d_6 QE-300

200   180   160   140   120   100   80   60   40   20   0

Offset: 1.3 ppm.

10   9   8   7   6   5   4   3   2   1   0

Aldrich 31,800-0   *CAS [769-42-6]* $C_6H_8N_2O_3$

1,3-Dimethylbarbituric acid, 98%

FW 156.14

mp 125°C

164.74
151.83
39.39
28.43*

B

CDCl$_3$ QE-300

200   180   160   140   120   100   80   60   40   20   0

10   9   8   7   6   5   4   3   2   1   0

Aldrich 30,855-2   *CAS [511-67-1]* $C_4H_2Br_2N_2O_3$

5,5-Dibromobarbituric acid, 97%

FW 285.89

mp 241°C

162.97
148.46
48.07

C

CDCl$_3$ + DMSO-d_6 QE-300

200   180   160   140   120   100   80   60   40   20   0

Offset: 2.1 ppm.

10   9   8   7   6   5   4   3   2   1   0

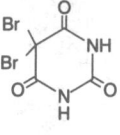

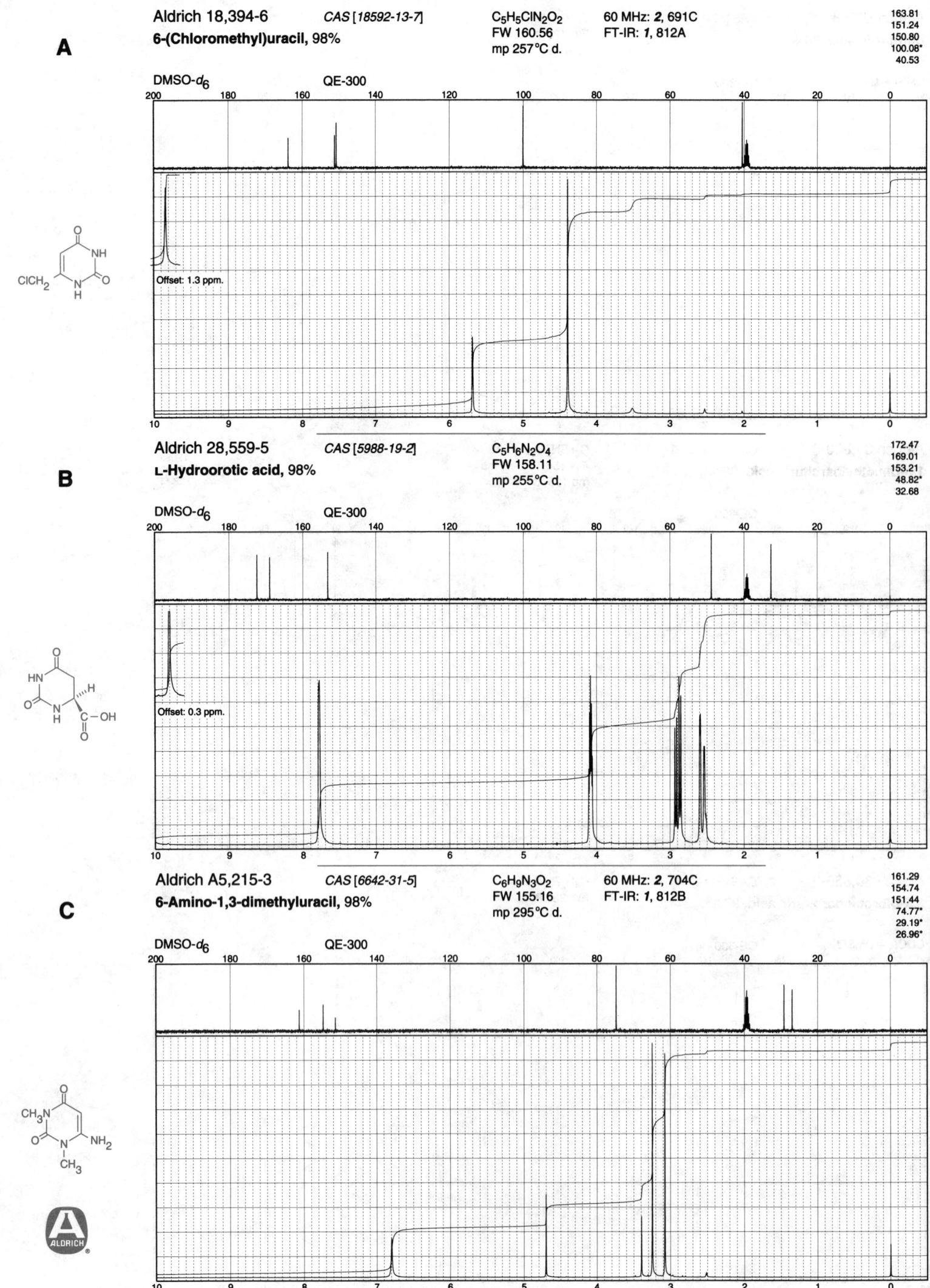

A

Aldrich 18,394-6   *CAS [18592-13-7]* $C_5H_5ClN_2O_2$ 60 MHz: *2*, 691C

6-(Chloromethyl)uracil, 98%   FW 160.56   FT-IR: *1*, 812A

mp 257°C d.

163.81
151.24
150.80
100.08*
40.53

DMSO-d_6 QE-300

Offset: 1.3 ppm.

B

Aldrich 28,559-5   *CAS [5988-19-2]* $C_5H_6N_2O_4$

L-Hydroorotic acid, 98%   FW 158.11

mp 255°C d.

172.47
169.01
153.21
48.82*
32.68

DMSO-d_6 QE-300

Offset: 0.3 ppm.

C

Aldrich A5,215-3   *CAS [6642-31-5]* $C_6H_9N_3O_2$ 60 MHz: *2*, 704C

6-Amino-1,3-dimethyluracil, 98%   FW 155.16   FT-IR: *1*, 812B

mp 295°C d.

161.29
154.74
151.44
74.77*
29.19*
26.96*

DMSO-d_6 QE-300

ALDRICH

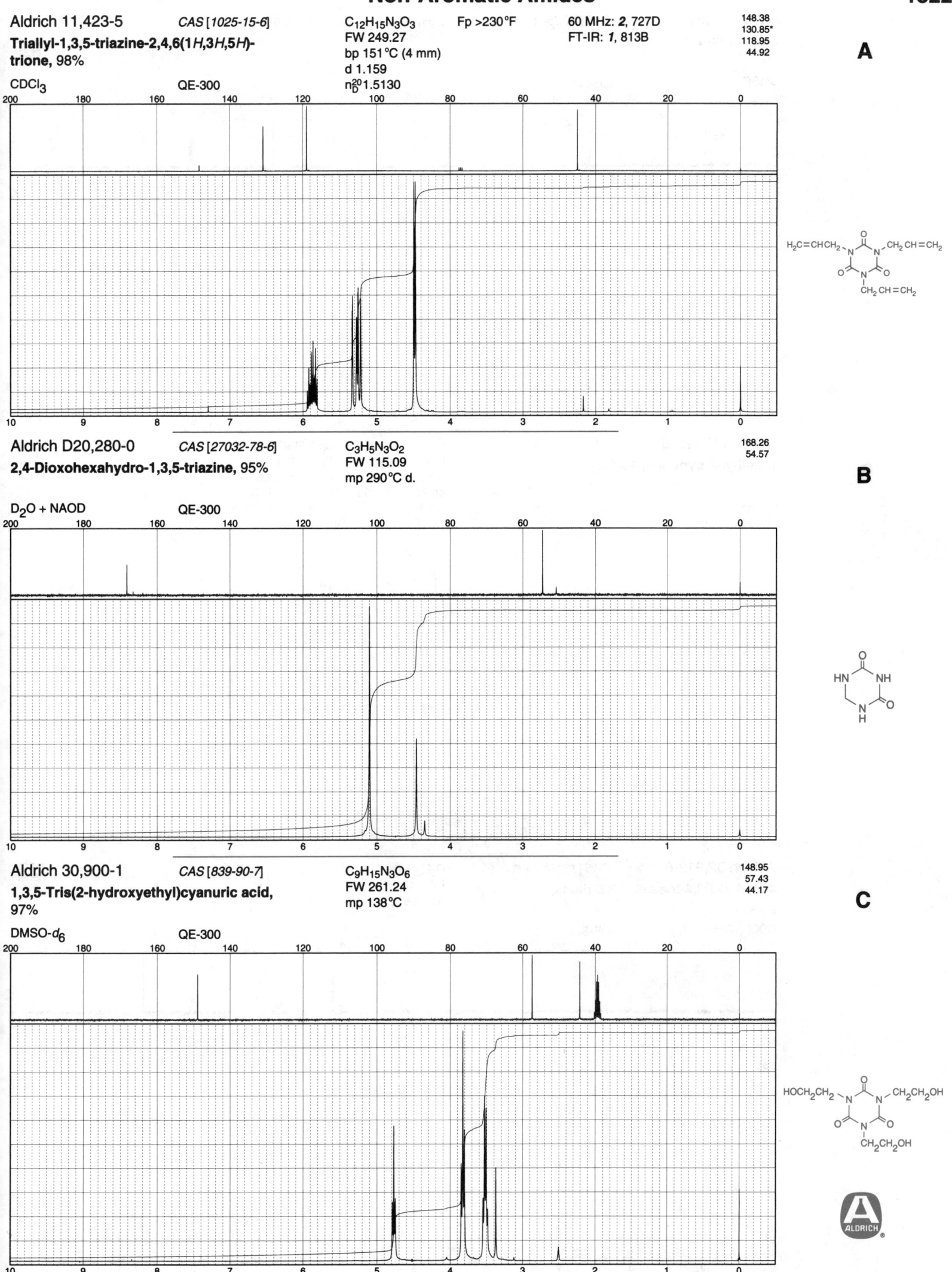

Aldrich 11,423-5 CAS [1025-15-6] $C_{12}H_{15}N_3O_3$ Fp >230°F 60 MHz: **2**, 727D 148.38
130.85*
Triallyl-1,3,5-triazine-2,4,6(1H,3H,5H)-trione, 98% FW 249.27 FT-IR: **1**, 813B 118.95
44.92

bp 151°C (4 mm) **A**

CDCl₃ QE-300 d 1.159
n_D^{20} 1.5130

Aldrich D20,280-0 CAS [27032-78-6] $C_3H_5N_3O_2$ 168.26
54.57
2,4-Dioxohexahydro-1,3,5-triazine, 95% FW 115.09
mp 290°C d.

B

D₂O + NAOD QE-300

Aldrich 30,900-1 CAS [839-90-7] $C_9H_{15}N_3O_6$ 148.95
57.43
1,3,5-Tris(2-hydroxyethyl)cyanuric acid, 97% FW 261.24 44.17
mp 138°C

C

DMSO-d_6 QE-300

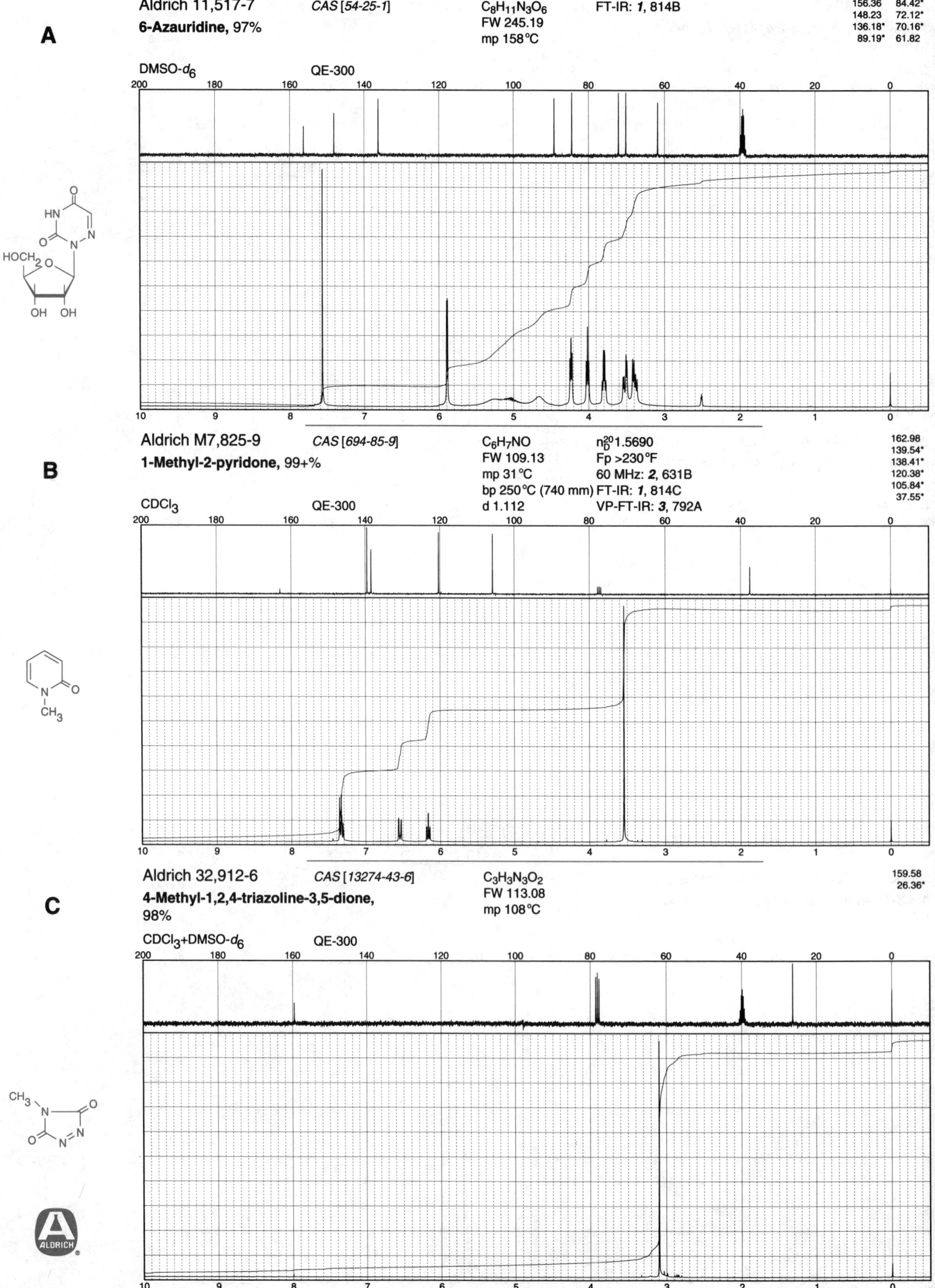

A

Aldrich 11,517-7

6-Azauridine, 97%

CAS [54-25-1]

$C_8H_{11}N_3O_6$
FW 245.19
mp 158°C

FT-IR: *1*, 814B

156.36   84.42*
148.23   72.12*
136.18*   70.16*
89.19*   61.82

DMSO-d_6 QE-300

B

Aldrich M7,825-9

1-Methyl-2-pyridone, 99+%

CAS [694-85-9]

C_6H_7NO
FW 109.13
mp 31°C
bp 250°C (740 mm)
d 1.112

n_D^{20} 1.5690
Fp >230°F
60 MHz: *2*, 631B
FT-IR: *1*, 814C
VP-FT-IR: *3*, 792A

162.98
139.54*
138.41*
120.38*
105.84*
37.55*

CDCl$_3$ QE-300

C

Aldrich 32,912-6

4-Methyl-1,2,4-triazoline-3,5-dione,
98%

CAS [13274-43-6]

$C_3H_3N_3O_2$
FW 113.08
mp 108°C

159.58
26.36*

CDCl$_3$+DMSO-d_6 QE-300

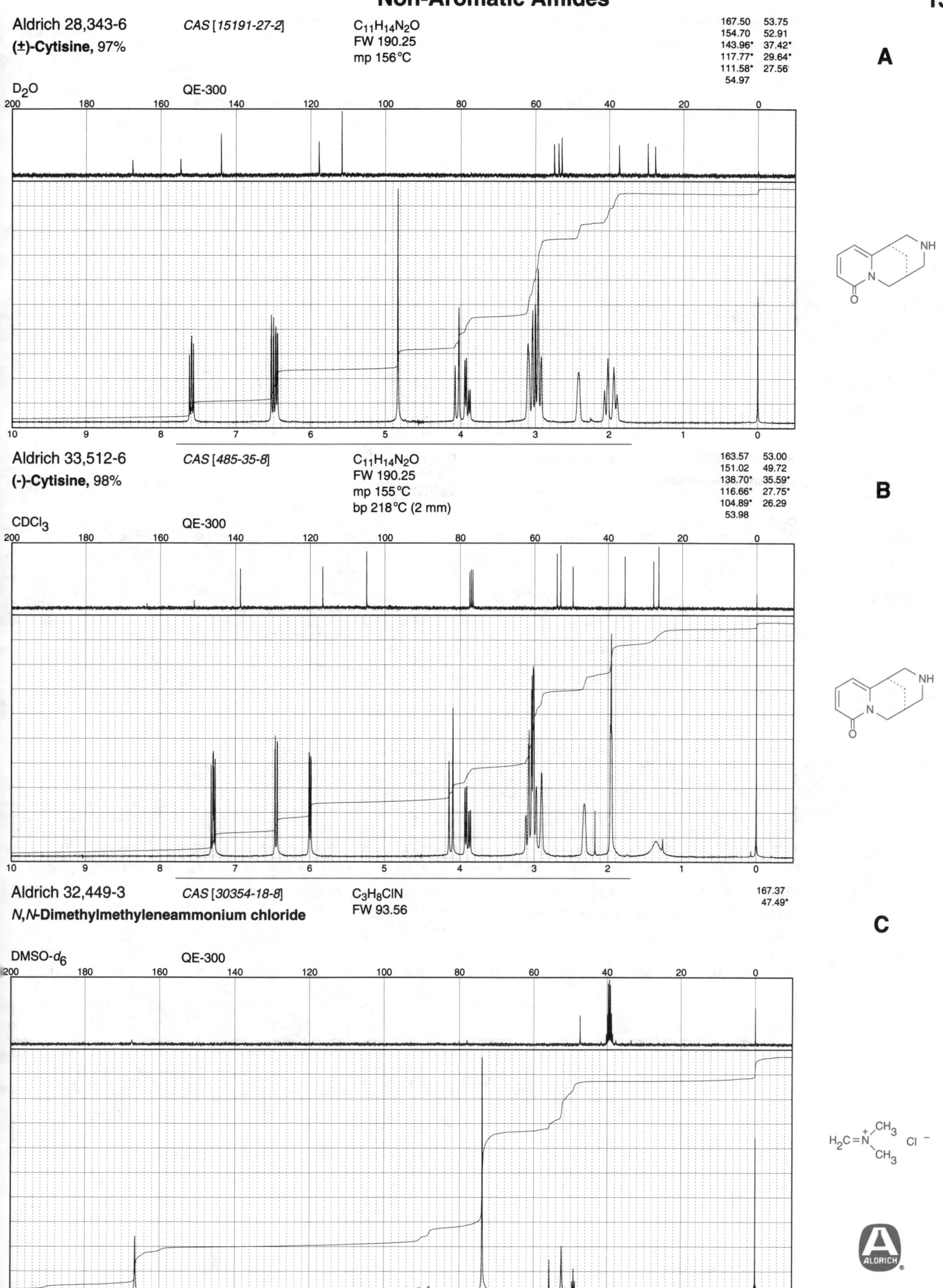

Aldrich 28,343-6

(±)-Cytisine, 97%

CAS [15191-27-2]

$C_{11}H_{14}N_2O$
FW 190.25
mp 156°C

167.50	53.75
154.70	52.91
143.96*	37.42*
117.77*	29.64*
111.58*	27.56*
54.97	

A

D_2O QE-300

Aldrich 33,512-6

(−)-Cytisine, 98%

CAS [485-35-8]

$C_{11}H_{14}N_2O$
FW 190.25
mp 155°C
bp 218°C (2 mm)

163.57	53.00
151.02	49.72
138.70*	35.59*
116.66*	27.75*
104.89*	26.29
53.98	

B

$CDCl_3$ QE-300

Aldrich 32,449-3

N,N-Dimethylmethyleneammonium chloride

CAS [30354-18-8]

C_3H_8ClN
FW 93.56

167.37
47.49*

C

DMSO-d_6 QE-300

Non-Aromatic Amides

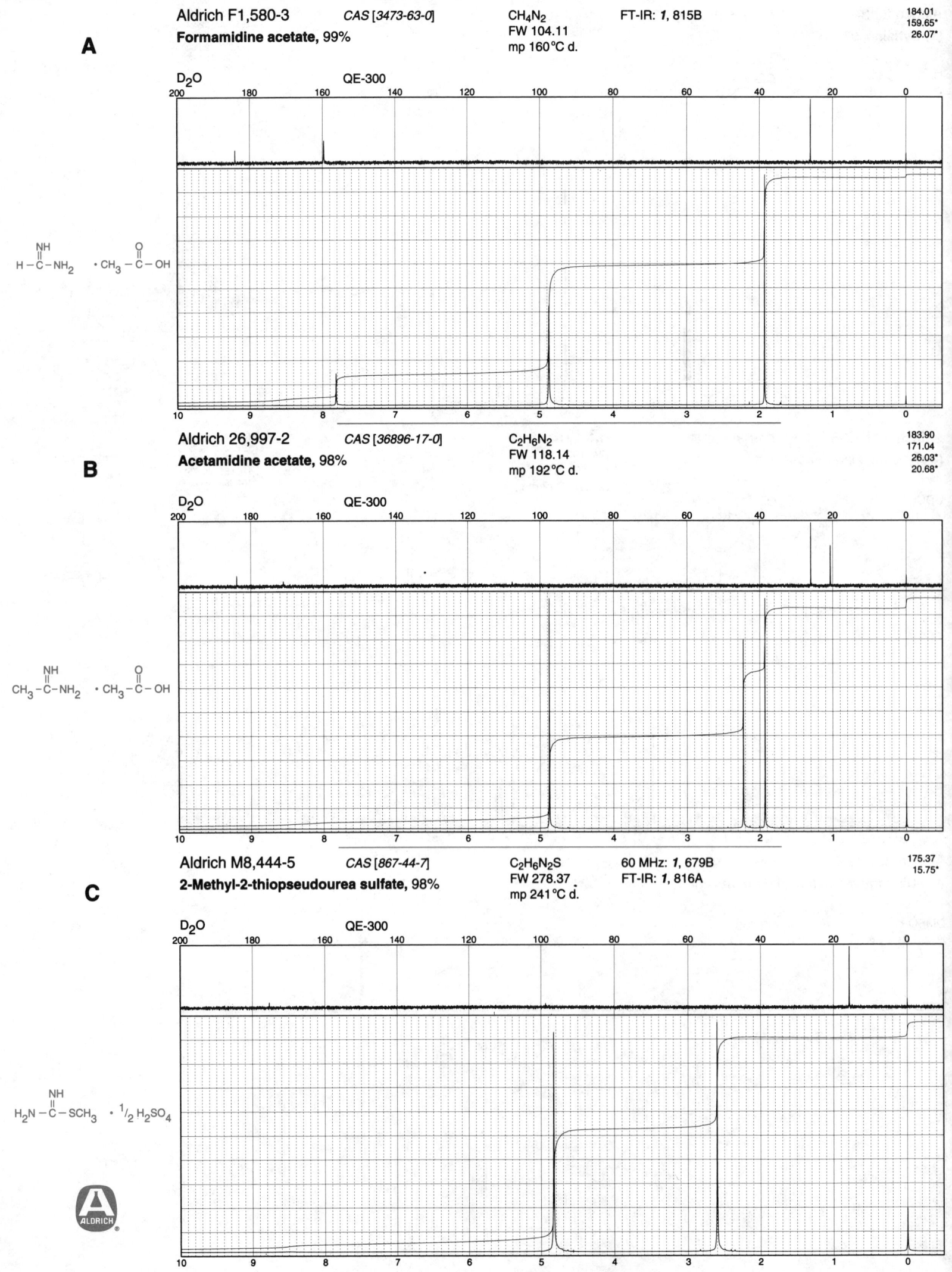

A

Aldrich F1,580-3   *CAS [3473-63-0]*   CH₄N₂   FT-IR: *1*, 815B

Formamidine acetate, 99%   FW 104.11
mp 160°C d.

184.01
159.65*
26.07*

D₂O   QE-300

H–C–NH₂ · CH₃–C–OH (NH, O)

Aldrich 26,997-2   *CAS [36896-17-0]*   C₂H₆N₂
Acetamidine acetate, 98%   FW 118.14
mp 192°C d.

183.90
171.04
26.03*
20.68*

B

D₂O   QE-300

CH₃–C–NH₂ · CH₃–C–OH (NH, O)

Aldrich M8,444-5   *CAS [867-44-7]*   C₂H₆N₂S   60 MHz: *1*, 679B
2-Methyl-2-thiopseudourea sulfate, 98%   FW 278.37   FT-IR: *1*, 816A
mp 241°C d.

175.37
15.75*

C

H₂N–C–SCH₃ · ½ H₂SO₄ (NH)

D₂O   QE-300

Aldrich 29,906-5   *CAS [23314-06-9]* $C_7H_{16}N_2$ Fp 94°F   VP-FT-IR: **3**, 792B

N'-*tert*-Butyl-N,N-dimethylformamidine,
97%

FW 128.22
bp 139°C
d 0.815
n_D^{20}1.4450

150.84*
52.79
36.96*
31.38*

A

CDCl$_3$ QE-300

Aldrich M5,370-1   *CAS [29427-58-5]* $C_2H_6N_2O$ 60 MHz: **1**, 679C

O-Methylisourea hydrogen sulfate, 99%

FW 172.16
mp 119°C

FT-IR: **1**, 816D

162.24
57.42*

B

DMSO-*d*$_6$ QE-300

Aldrich 34,624-1   *CAS [16694-44-3]* C_4H_9NO

Isopropyl formimidate hydrochloride

FW 123.58
mp 108°C d.
Fp >230°F

169.50*
82.32*
21.74*

C

DMSO-*d*$_6$ QE-300

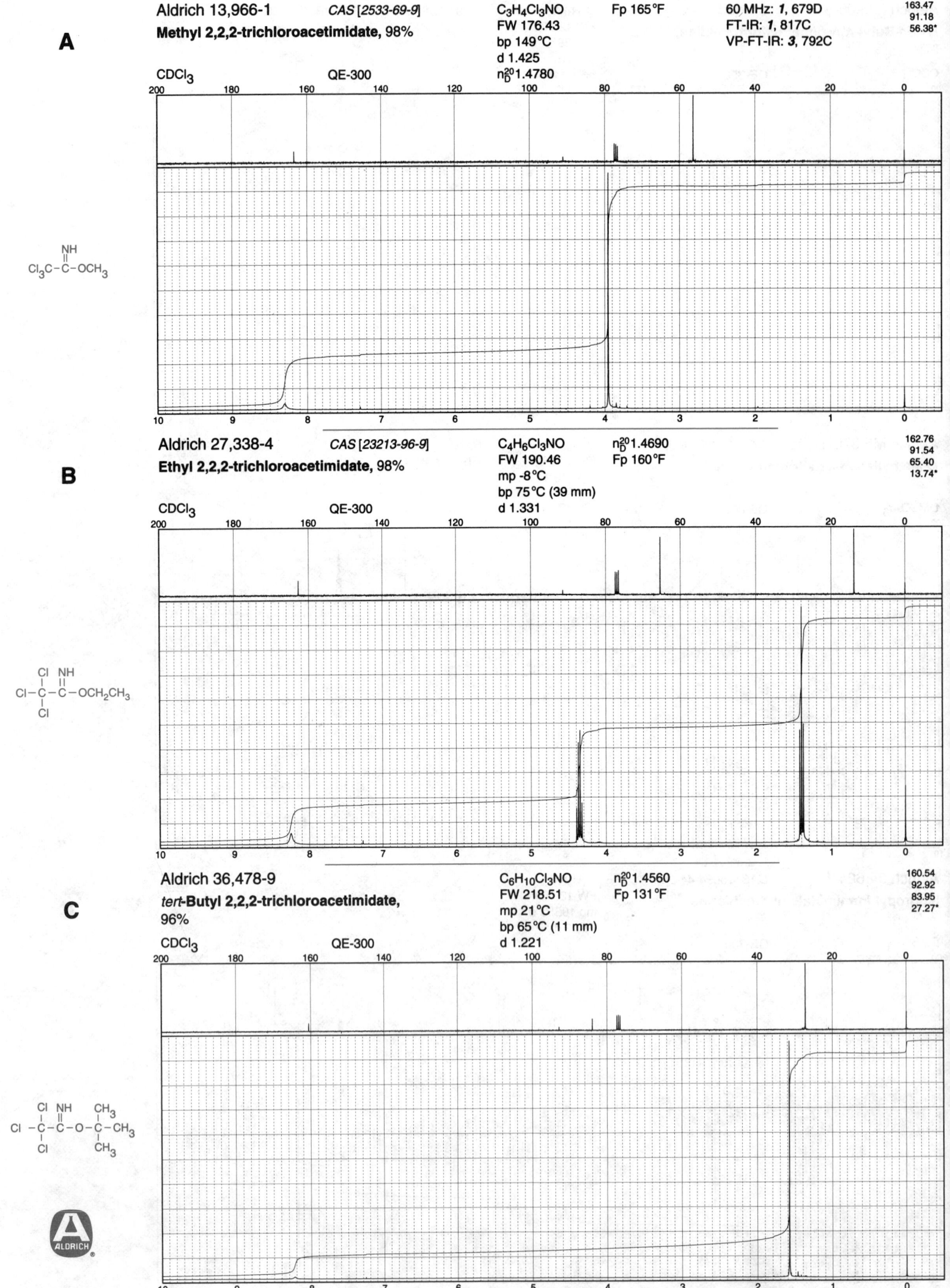

A

Aldrich 13,966-1   *CAS [2533-69-9]*   C3H4Cl3NO   Fp 165°F   60 MHz: *1*, 679D   163.47
Methyl 2,2,2-trichloroacetimidate, 98%   FW 176.43   FT-IR: *1*, 817C   91.18
bp 149°C   VP-FT-IR: *3*, 792C   56.38*
d 1.425
n_D^{20} 1.4780

CDCl3   QE-300

$Cl_3C-\overset{\overset{NH}{\|}}{C}-OCH_3$

B

Aldrich 27,338-4   *CAS [23213-96-9]*   C4H6Cl3NO n_D^{20} 1.4690   162.76
Ethyl 2,2,2-trichloroacetimidate, 98%   FW 190.46   Fp 160°F   91.18
mp -8°C   65.40
bp 75°C (39 mm)   13.74*
d 1.331

CDCl3   QE-300

$Cl-\overset{\overset{Cl}{|}}{\underset{\underset{Cl}{|}}{C}}-\overset{\overset{NH}{\|}}{C}-OCH_2CH_3$

C

Aldrich 36,478-9   C6H10Cl3NO n_D^{20} 1.4560   160.54
tert-**Butyl 2,2,2-trichloroacetimidate,**   FW 218.51   Fp 131°F   92.92
96%   mp 21°C   83.95
bp 65°C (11 mm)   27.27*
d 1.221

CDCl3   QE-300

$Cl-\overset{\overset{Cl}{|}}{\underset{\underset{Cl}{|}}{C}}-\overset{\overset{NH}{\|}}{C}-O-\overset{\overset{CH_3}{|}}{\underset{\underset{CH_3}{|}}{C}}-CH_3$

ALDRICH

Non-Aromatic Amides

Aldrich 17,830-6   CAS [58537-94-3]   C9H18N2O2
Dimethyl pimelimidate dihydrochloride   FW 259.18
mp 122°C d.

A

D2O   QE-300

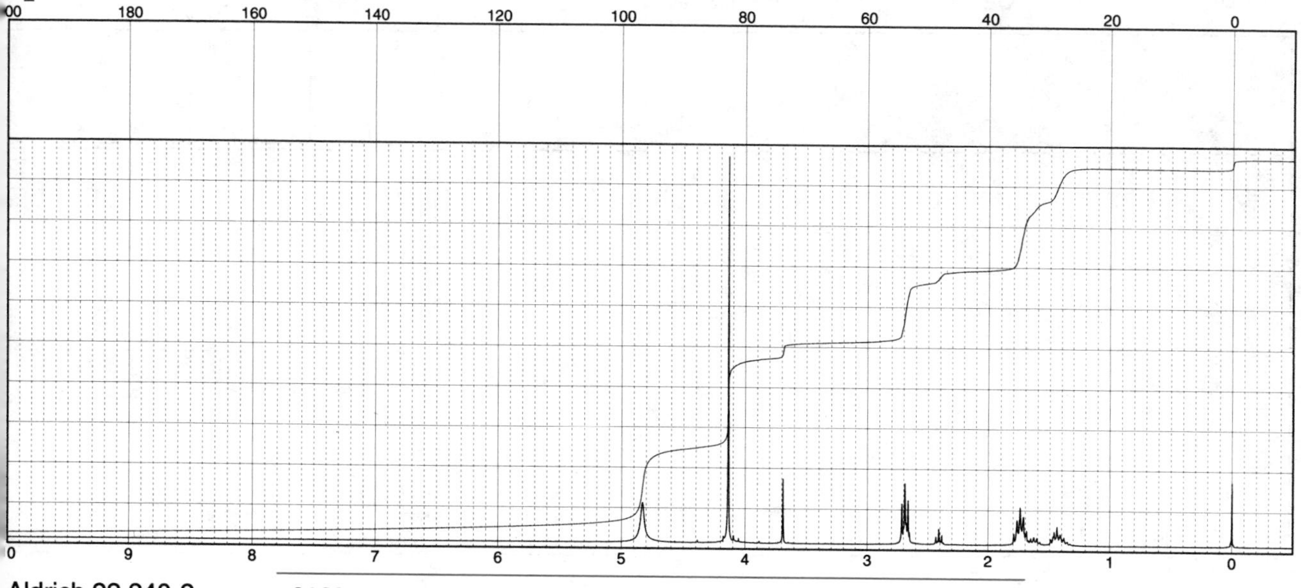

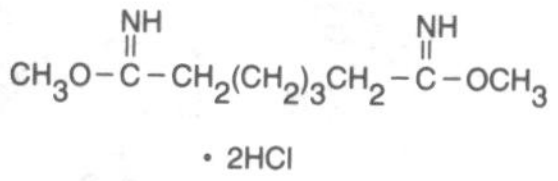

$CH_3O-\overset{NH}{\overset{\|}{C}}-CH_2(CH_2)_3CH_2-\overset{NH}{\overset{\|}{C}}-OCH_3$

· 2HCl

Aldrich 22,240-2   CAS [22661-87-6]   C2H7N3
1-Methylguanidine hydrochloride, 98%   FW 109.56
60 MHz: 1, 681A
FT-IR: 1, 818D

B

157.69
27.57*

DMSO-d6   QE-300

$H_2N-\overset{NH}{\overset{\|}{C}}-NHCH_3$ · HCl

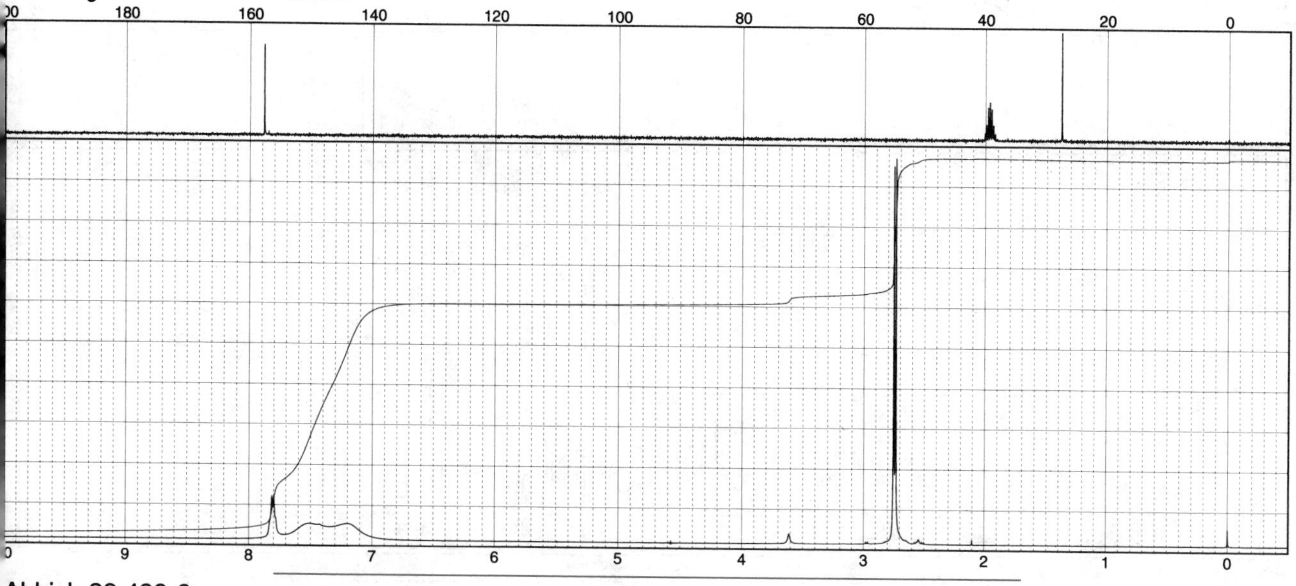

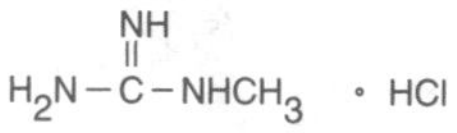

Aldrich 29,489-6   CAS [19341-54-9]   C3H9N3
1-Ethylguanidine hydrochloride, 99%   FW 123.59

C

156.80
35.51
14.09*

DMSO-d6   QE-300

$H_2N-\overset{NH}{\overset{\|}{C}}-NHCH_2CH_3$ · HCl

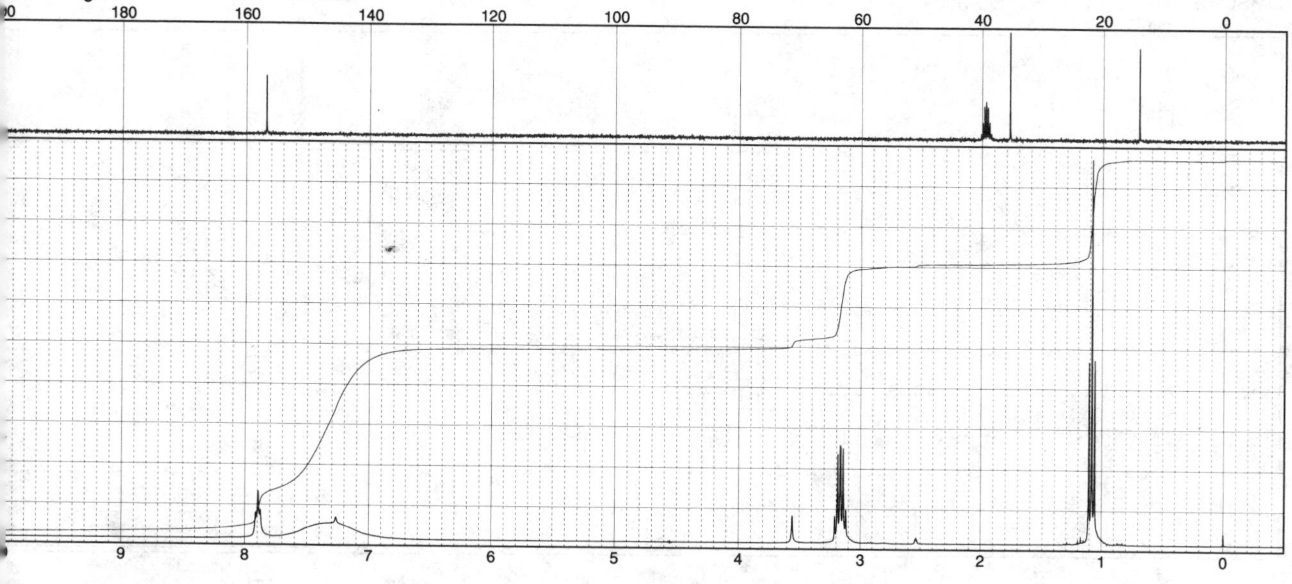

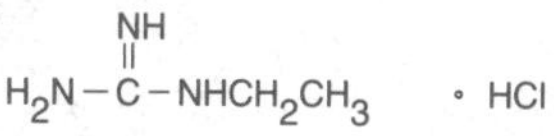

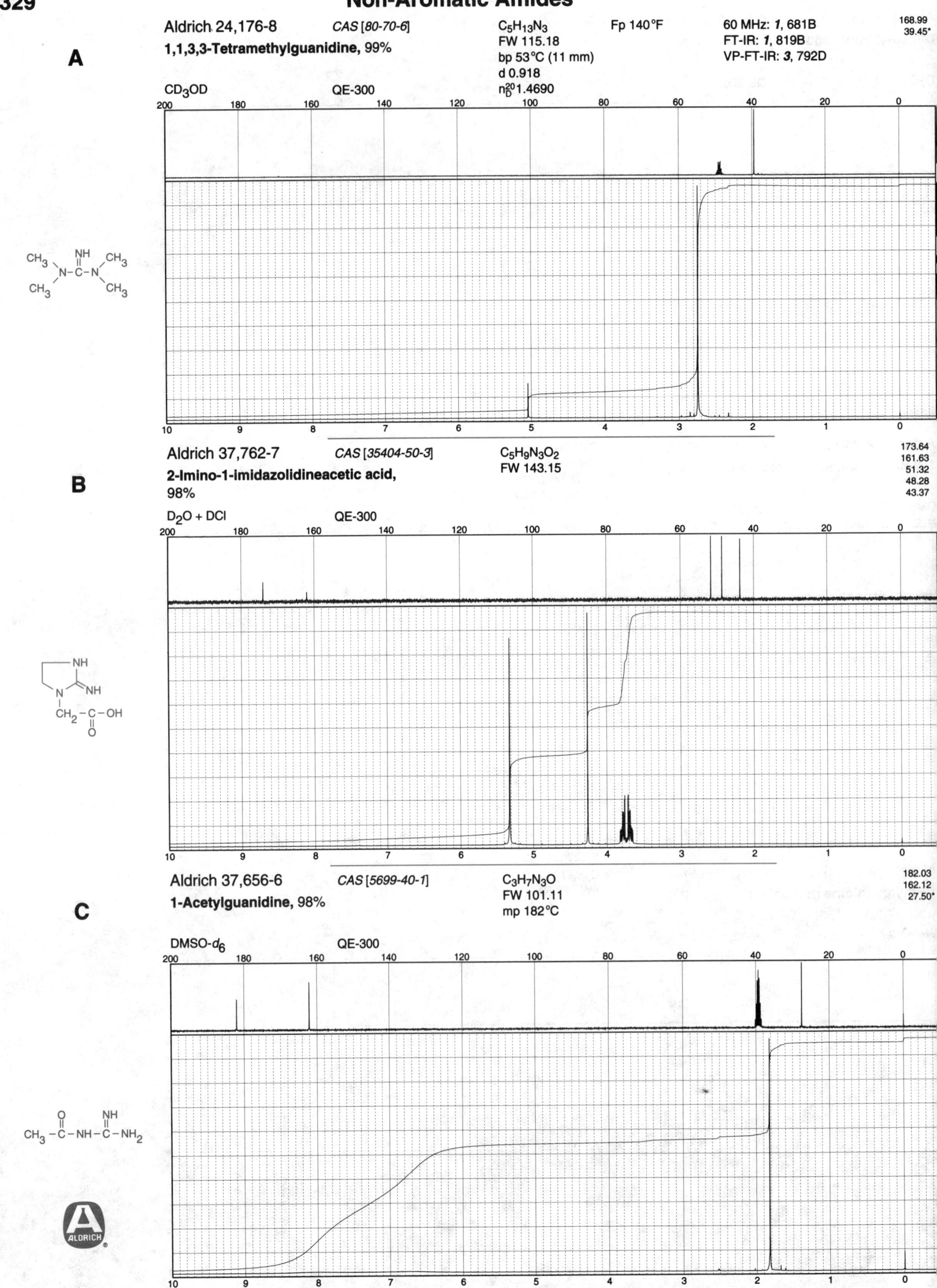

A

Aldrich 24,176-8   *CAS [80-70-6]*   C$_5$H$_{13}$N$_3$ Fp 140°F   60 MHz: *1*, 681B   168.99
1,1,3,3-Tetramethylguanidine, 99%   FW 115.18   FT-IR: *1*, 819B   39.45*
bp 53°C (11 mm)   VP-FT-IR: *3*, 792D
d 0.918
n$_D^{20}$1.4690

CD$_3$OD   QE-300

B

Aldrich 37,762-7   *CAS [35404-50-3]*   C$_5$H$_9$N$_3$O$_2$ 173.64
2-Imino-1-imidazolidineacetic acid,   FW 143.15   161.63
98%   51.32
48.28
43.37

D$_2$O + DCl   QE-300

C

Aldrich 37,656-6   *CAS [5699-40-1]*   C$_3$H$_7$N$_3$O   182.03
1-Acetylguanidine, 98%   FW 101.11   162.12
mp 182°C   27.50*

DMSO-d_6 QE-300

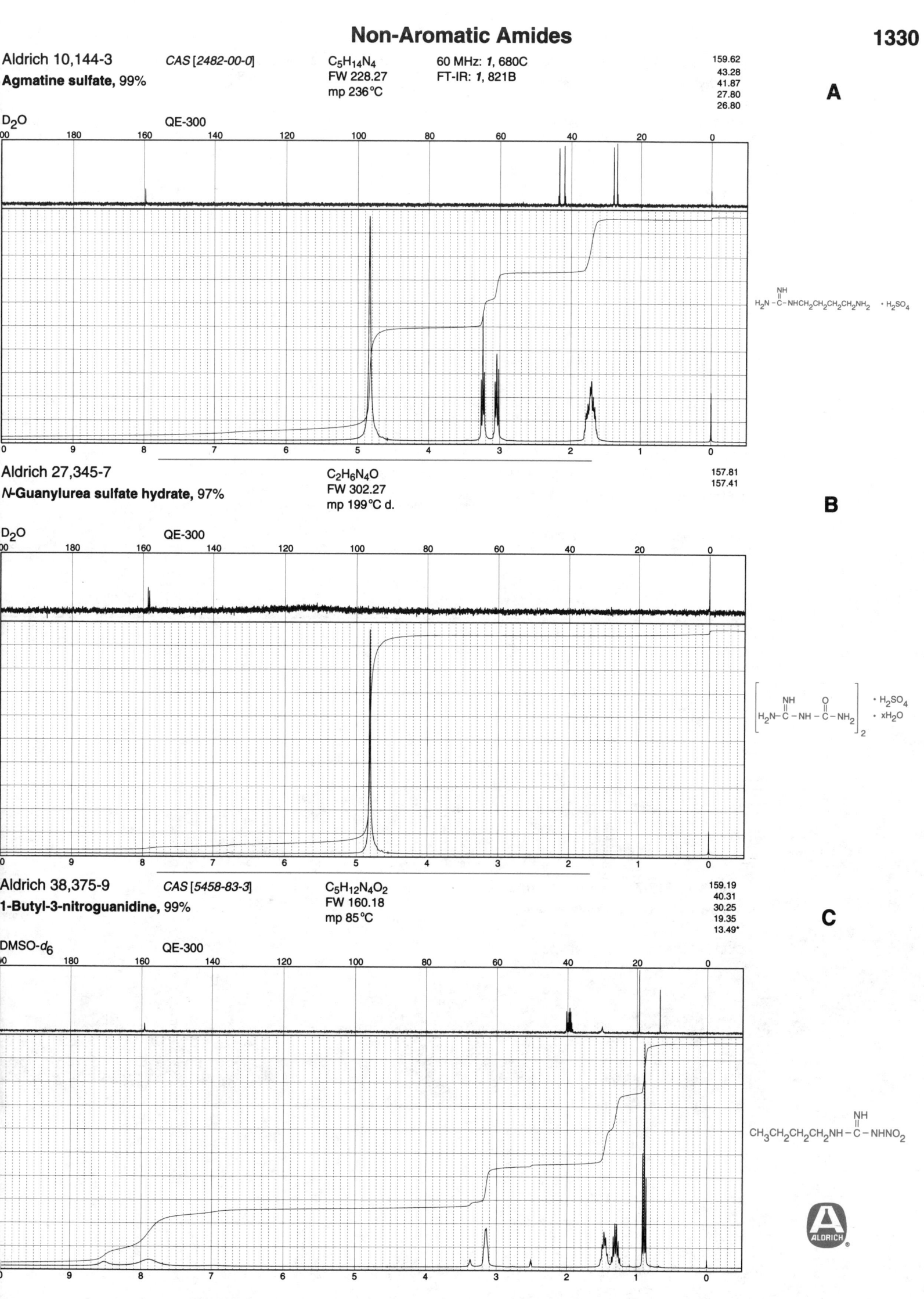

Aldrich 10,144-3   *CAS [2482-00-0]*   C₅H₁₄N₄   60 MHz: *1*, 680C

$C_5H_{14}N_4$ FW 228.27   mp 236°C

FT-IR: *1*, 821B

Agmatine sulfate, 99%

159.62
43.28
41.87
27.80
26.80

A

D₂O   QE-300

Aldrich 27,345-7

N-Guanylurea sulfate hydrate, 97%

$C_2H_6N_4O$ FW 302.27   mp 199°C d.

157.81
157.41

B

D₂O   QE-300

Aldrich 38,375-9   *CAS [5458-83-3]* $C_5H_{12}N_4O_2$ FW 160.18   mp 85°C

159.19
40.31
30.25
19.35
13.49*

C

1-Butyl-3-nitroguanidine, 99%

DMSO-d_6 QE-300

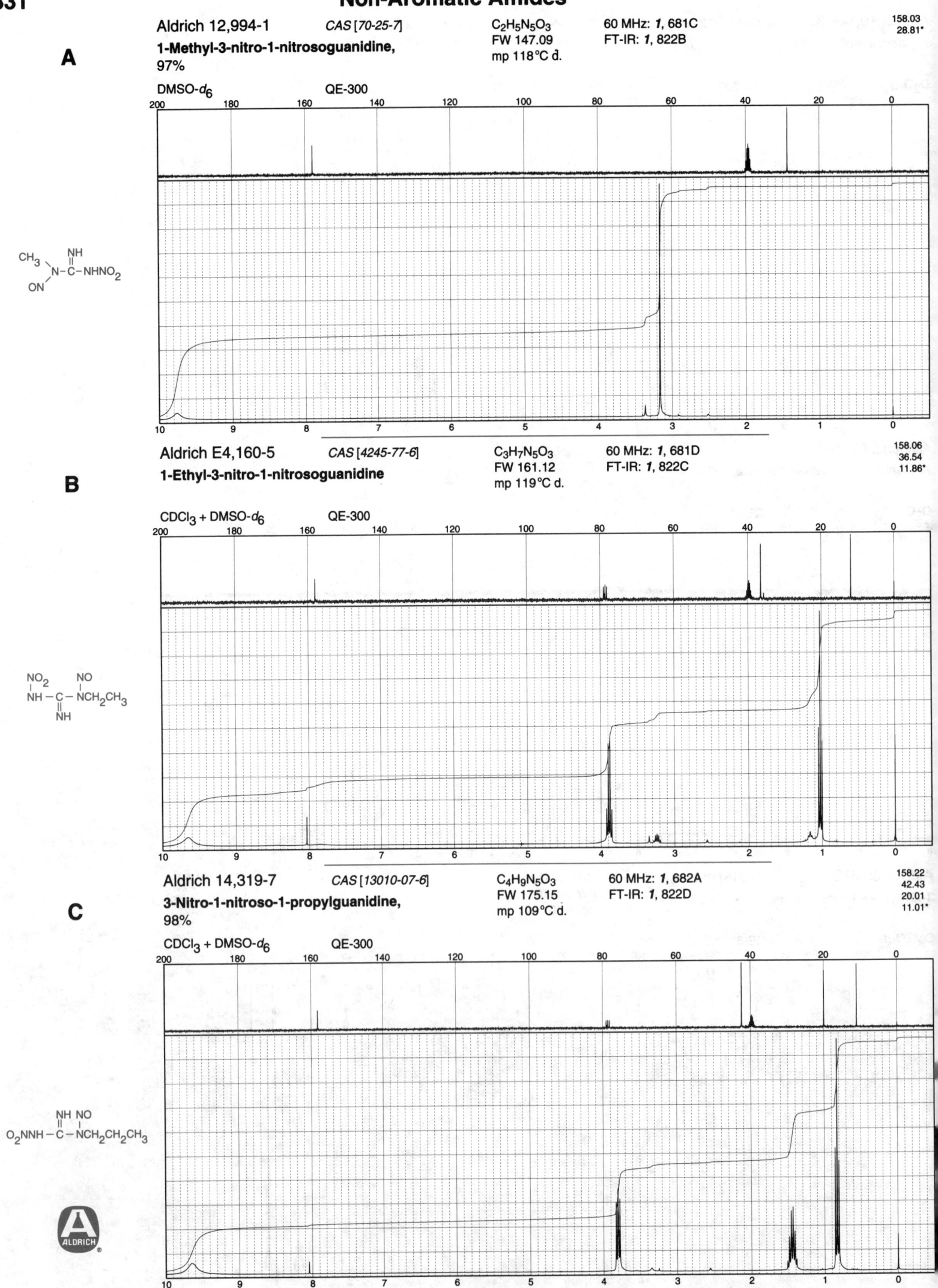

A

Aldrich 12,994-1   *CAS [70-25-7]*   C₂H₅N₅O₃   60 MHz: *1*, 681C   158.03
28.81*
1-Methyl-3-nitro-1-nitrosoguanidine,   FW 147.09   FT-IR: *1*, 822B
97%   mp 118°C d.

DMSO-*d₆*   QE-300

B

Aldrich E4,160-5   *CAS [4245-77-6]*   C₃H₇N₅O₃   60 MHz: *1*, 681D   158.06
36.54
1-Ethyl-3-nitro-1-nitrosoguanidine   FW 161.12   FT-IR: *1*, 822C   11.86*
mp 119°C d.

CDCl₃ + DMSO-*d₆*   QE-300

C

Aldrich 14,319-7   *CAS [13010-07-6]*   C₄H₉N₅O₃   60 MHz: *1*, 682A   158.22
42.43
3-Nitro-1-nitroso-1-propylguanidine,   FW 175.15   FT-IR: *1*, 822D   20.01
98%   mp 109°C d.   11.01*

CDCl₃ + DMSO-*d₆*   QE-300

Aldrich 14,223-9   *CAS [13010-08-7]* $C_5H_{11}N_5O_3$ 60 MHz: *1*, 682B

1-Butyl-3-nitro-1-nitrosoguanidine,
97%

FW 189.18   FT-IR: *1*, 823A
mp 120°C d.

157.88
41.00
28.36
19.44
13.33*

A

DMSO-d_6 QE-300

$CH_3CH_2CH_2CH_2$ —N—C—NHNO$_2$ (with NH above C=, ON below N)

Aldrich 13,744-8   *CAS [1120-64-5]* C_4H_7NO Fp 68°F   60 MHz: *1*, 682C

2-Methyl-2-oxazoline, 98%

FW 85.11   FT-IR: *1*, 823C
bp 110°C   VP-FT-IR: *3*, 793A
d 1.005
n_D^{20} 1.4340

165.35
67.38
54.59
13.79*

B

CDCl$_3$ QE-300

(oxazoline ring with N, O, CH_3)

Aldrich 13,745-6   *CAS [10431-98-8]* C_5H_9NO n_D^{20} 1.4370

2-Ethyl-2-oxazoline, 99+%

FW 99.13   Fp 85°F
mp -62°C   60 MHz: *1*, 682D
bp 128°C   FT-IR: *1*, 823D
d 0.982   VP-FT-IR: *3*, 793B

169.46
67.23
54.40
21.40
10.31*

C

CDCl$_3$ QE-300

(oxazoline ring with N, O, CH_2CH_3)

ALDRICH

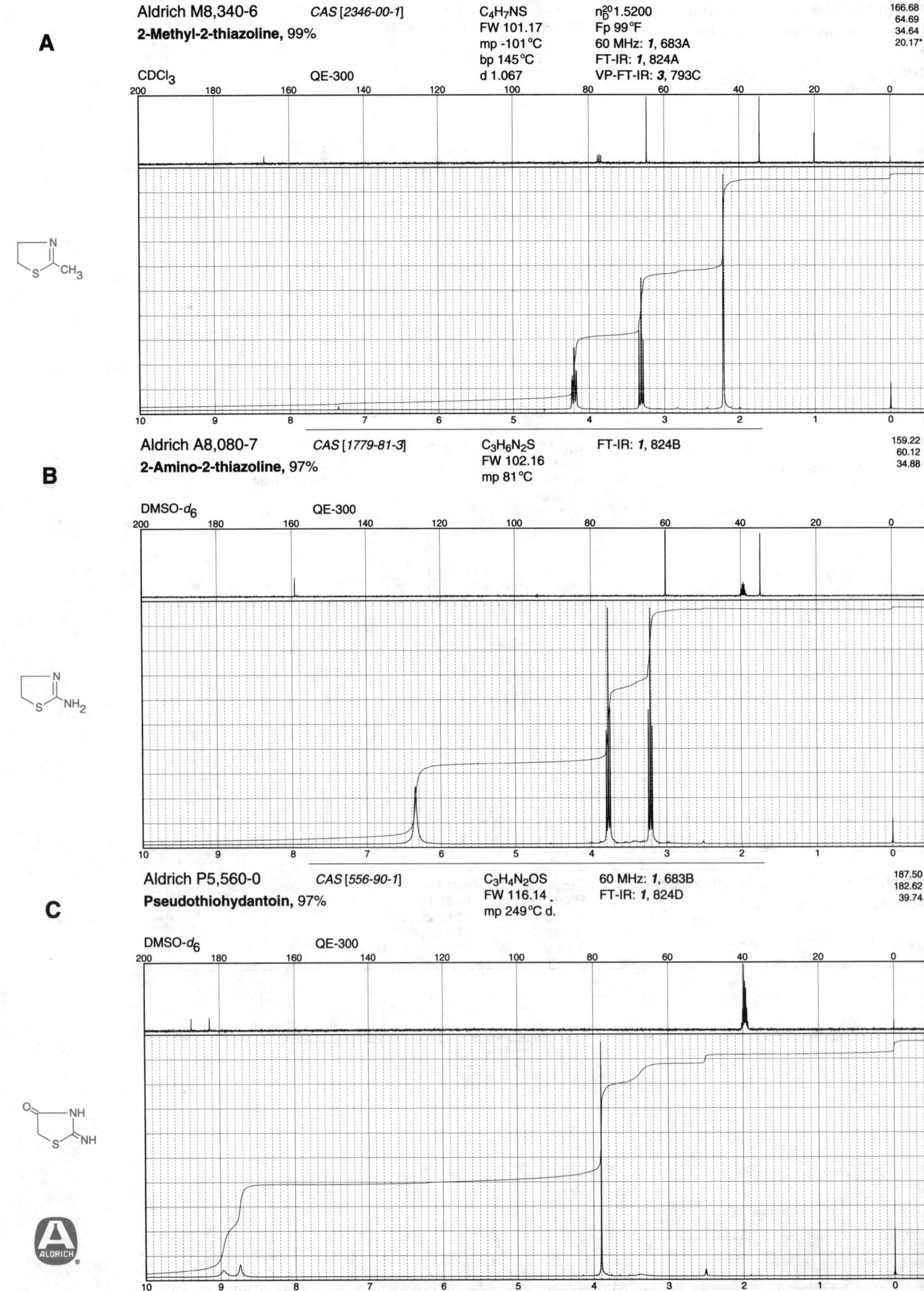

A

Aldrich M8,340-6   CAS [2346-00-1]

2-Methyl-2-thiazoline, 99%

C_4H_7NS
FW 101.17
mp -101°C
bp 145°C
d 1.067

n_D^{20}1.5200
Fp 99°F
60 MHz: **1**, 683A
FT-IR: **1**, 824A
VP-FT-IR: **3**, 793C

166.68
64.69
34.64
20.17*

$CDCl_3$ QE-300

B

Aldrich A8,080-7   CAS [1779-81-3]

2-Amino-2-thiazoline, 97%

$C_3H_6N_2S$
FW 102.16
mp 81°C

FT-IR: **1**, 824B

159.22
60.12
34.88

DMSO-d_6 QE-300

C

Aldrich P5,560-0   CAS [556-90-1]

Pseudothiohydantoin, 97%

$C_3H_4N_2OS$
FW 116.14
mp 249°C d.

60 MHz: **1**, 683B
FT-IR: **1**, 824D

187.50
182.62
39.74

DMSO-d_6 QE-300

Aldrich A9,580-4 *CAS [2525-16-8]*

1-Aza-2-methoxy-1-cycloheptene, 99%

CDCl$_3$ QE-300

C$_7$H$_{13}$NO
FW 127.19
bp 53°C (18 mm)
d 0.887
n$_D^{20}$1.4630

Fp 124°F

60 MHz: *1*, 683D
FT-IR: *1*, 825D

169.65
52.44*
48.72
32.11
31.31
28.00
23.52

A

Aldrich 13,658-1 *CAS [3001-72-7]*

1,5-Diazabicyclo[4.3.0]non-5-ene, 98%

CDCl$_3$ QE-300

C$_7$H$_{12}$N$_2$
FW 124.19
bp 97°C (7 mm)
d 1.005
n$_D^{20}$1.5190

Fp 202°F

60 MHz: *1*, 684A
FT-IR: *1*, 826A
VP-FT-IR: *3*, 794A

160.58
51.42
44.04
43.08
31.47
20.86
19.69

B

Aldrich 13,900-9 *CAS [6674-22-2]*

1,8-Diazabicyclo[5.4.0]undec-7-ene, 98%

CDCl$_3$ QE-300

C$_9$H$_{16}$N$_2$
FW 152.24
bp 82°C
d 1.018
n$_D^{20}$1.5220

Fp >230°F

60 MHz: *1*, 684B
FT-IR: *1*, 826B
VP-FT-IR: *3*, 794B

161.41 29.76
52.80 28.53
48.34 26.01
44.24 22.50
37.40

C

ALDRICH®

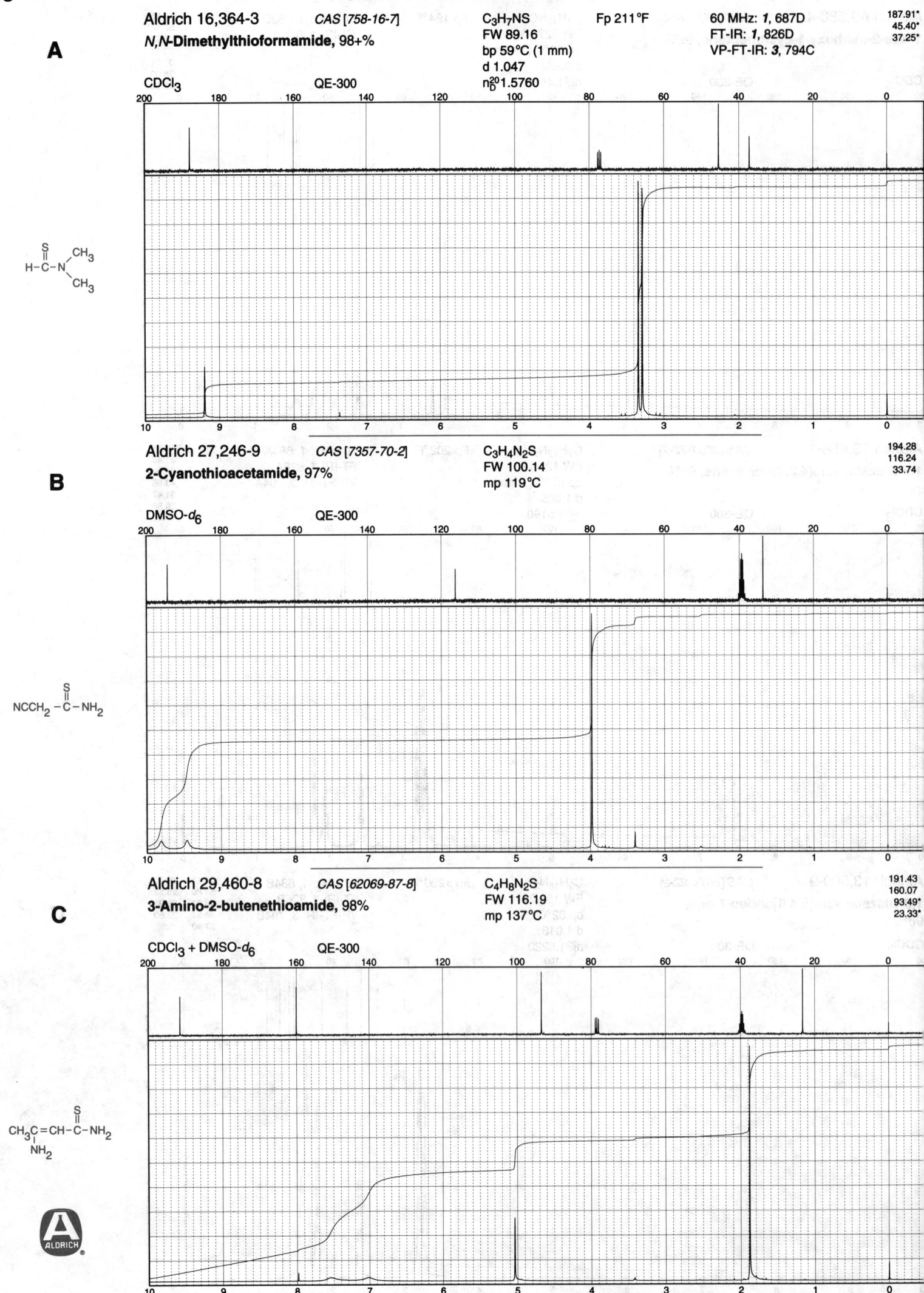

A

Aldrich 16,364-3   *CAS [758-16-7]*   C₃H₇NS   Fp 211°F   60 MHz: *1*, 687D   187.91*
45.40*
N,N-Dimethylthioformamide, 98+%   FW 89.16   FT-IR: *1*, 826D   37.25*
bp 59°C (1 mm)   VP-FT-IR: *3*, 794C
d 1.047
CDCl₃   QE-300   n₀²⁰1.5760

B

Aldrich 27,246-9   *CAS [7357-70-2]*   C₃H₄N₂S   194.28
116.24
2-Cyanothioacetamide, 97%   FW 100.14   33.74
mp 119°C

DMSO-*d*₆   QE-300

C

Aldrich 29,460-8   *CAS [62069-87-8]*   C₄H₈N₂S   191.43
160.07
3-Amino-2-butenethioamide, 98%   FW 116.19   93.49*
mp 137°C   23.33*

CDCl₃ + DMSO-*d*₆   QE-300

ALDRICH

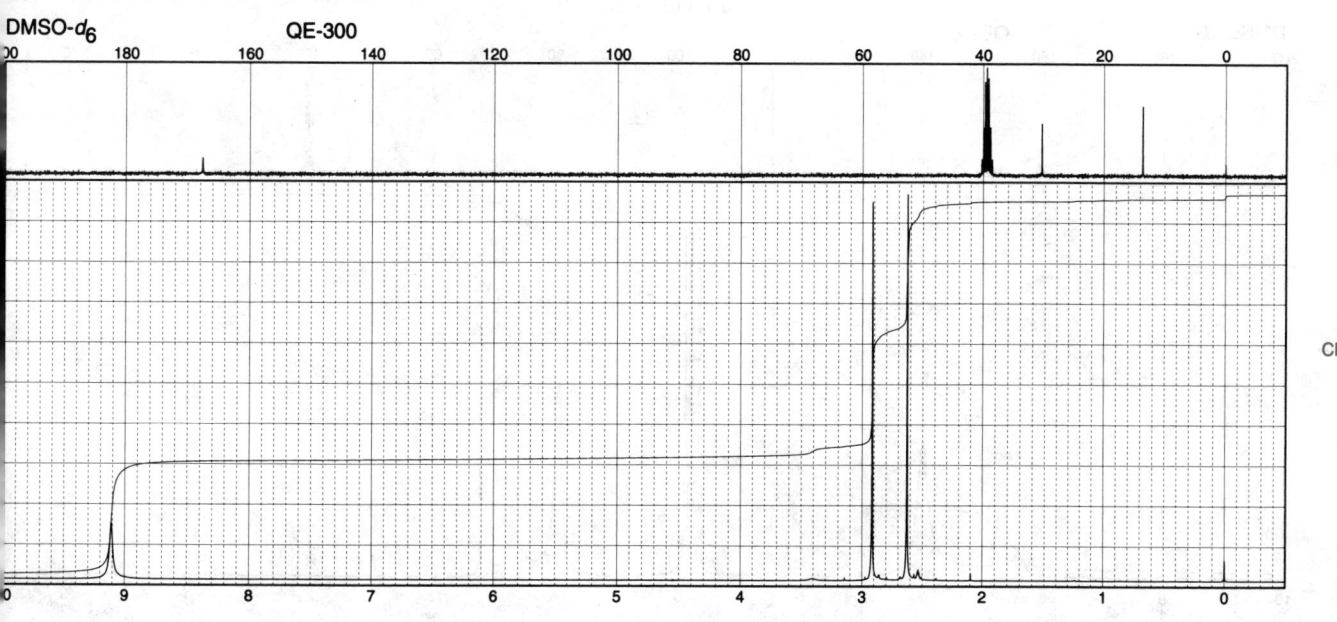

Aldrich M8,460-7   *CAS [598-52-7]*   C₂H₆N₂S   60 MHz: *1*, 684D

1-Methyl-2-thiourea, 97%   FW 90.15   FT-IR: *1*, 827C

mp 120°C

181.14
180.92
34.03*
32.25*

A

D₂O   QE-300

$CH_3NH-\overset{\overset{\displaystyle S}{\|}}{C}-NH_2$

Aldrich 25,253-0   *CAS [625-53-6]*   C₃H₈N₂S   FT-IR: *1*, 827D

1-Ethyl-2-thiourea, 99%   FW 104.17

mp 112°C

182.62
179.76
38.56
37.51
14.38*
13.82*

B

DMSO-d₆   QE-300

$CH_3CH_2NH-\overset{\overset{\displaystyle S}{\|}}{C}-NH_2$

Aldrich 39,196-4   *CAS [41306-45-0]*   C₃H₈N₂S

1,2-Dimethyl-2-thiopseudourea hydriodide,   FW 232.09
98%   mp 135°C

167.62
30.55*
13.63*

C

DMSO-d₆   QE-300

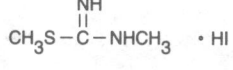

$CH_3S-\overset{\overset{\displaystyle NH}{\|}}{C}-NHCH_3 \cdot HI$

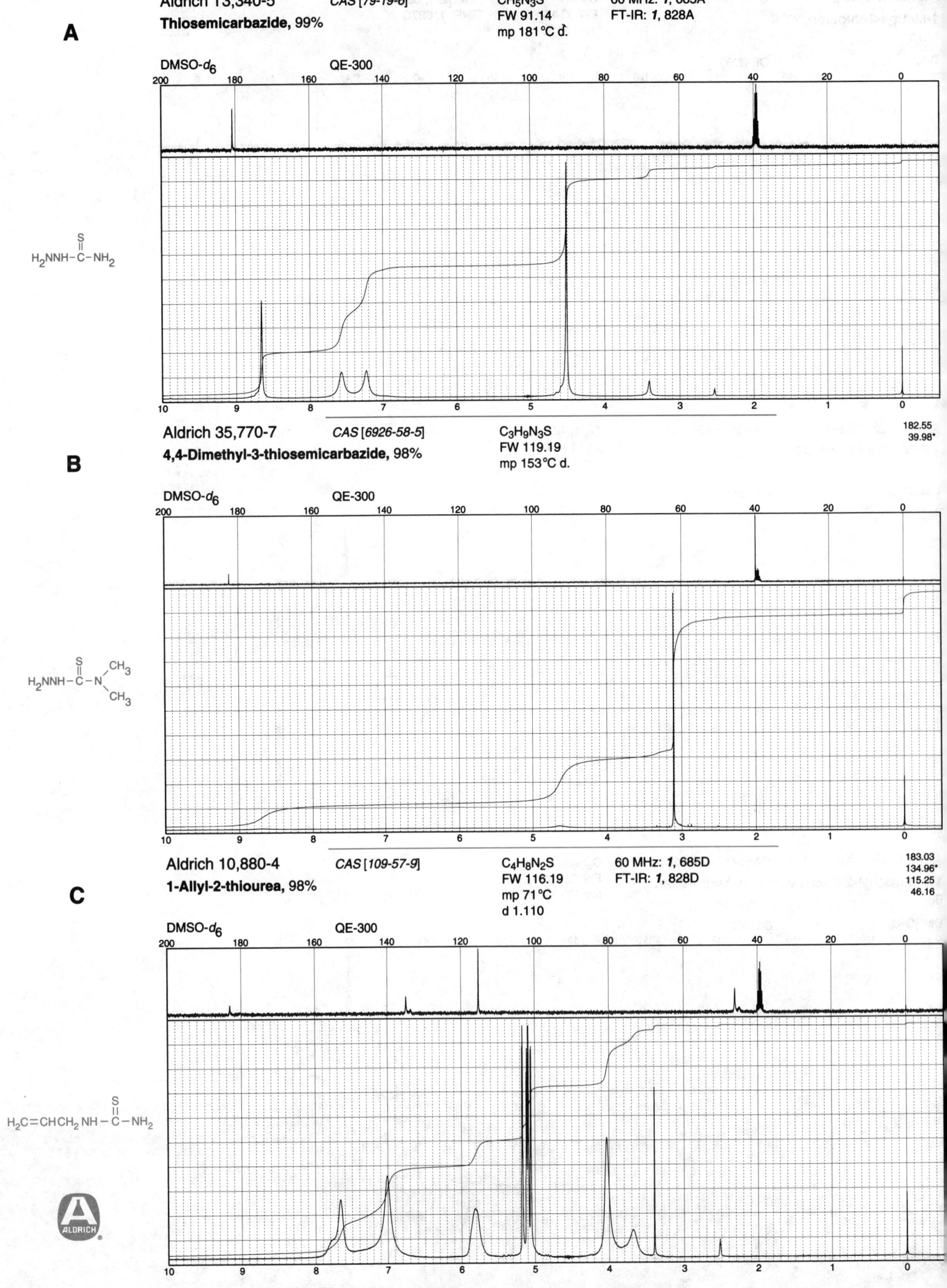

A

Aldrich T3,340-5   *CAS [79-19-6]* CH_5N_3S 60 MHz: *1*, 685A
Thiosemicarbazide, 99%   FW 91.14   FT-IR: *1*, 828A
mp 181°C d.

DMSO-d_6 QE-300

$H_2NNH-\overset{\overset{\displaystyle S}{\|}}{C}-NH_2$

182.55
39.98*

B

Aldrich 35,770-7   *CAS [6926-58-5]* $C_3H_9N_3S$
4,4-Dimethyl-3-thiosemicarbazide, 98%   FW 119.19
mp 153°C d.

DMSO-d_6 QE-300

$H_2NNH-\overset{\overset{\displaystyle S}{\|}}{C}-N\overset{\displaystyle CH_3}{\underset{\displaystyle CH_3}{}}$

Aldrich 10,880-4   *CAS [109-57-9]* $C_4H_8N_2S$ 60 MHz: *1*, 685D
1-Allyl-2-thiourea, 98%   FW 116.19   FT-IR: *1*, 828D
mp 71°C
d 1.110

183.03
134.96*
115.25
46.16

C

DMSO-d_6 QE-300

$H_2C=CHCH_2NH-\overset{\overset{\displaystyle S}{\|}}{C}-NH_2$

Aldrich A3,590-9 *CAS [3766-55-0]* C$_4$H$_9$N$_3$S FT-IR: *1*, 829A

181.05
135.31
114.99*
45.06

4-Allyl-3-thiosemicarbazide, 97% FW 131.20 mp 95°C

A

DMSO-d_6 QE-300

$$H_2NNH-\overset{\displaystyle S}{\overset{\|}{C}}-NHCH_2CH=CH_2$$

Aldrich 27,888-2 *CAS [5117-16-8]* C$_3$H$_8$N$_2$Se

177.46
45.25*
38.17*

1,1-Dimethyl-2-selenourea, 97% FW 151.07 mp 173°C

B

DMSO-d_6 QE-300

$$\begin{matrix} CH_3 \\ \quad \diagdown \\ \qquad N-\overset{\displaystyle Se}{\overset{\|}{C}}-NH_2 \\ \quad \diagup \\ CH_3 \end{matrix}$$

Aldrich D18,870-0 *CAS [534-13-4]* C$_3$H$_8$N$_2$S

182.47
30.97*

1,3-Dimethyl-2-thiourea, 99% FW 104.18
60 MHz: *1*, 686A
FT-IR: *1*, 829B

C

CDCl$_3$ QE-300

$$CH_3NH-\overset{\displaystyle S}{\overset{\|}{C}}-NHCH_3$$

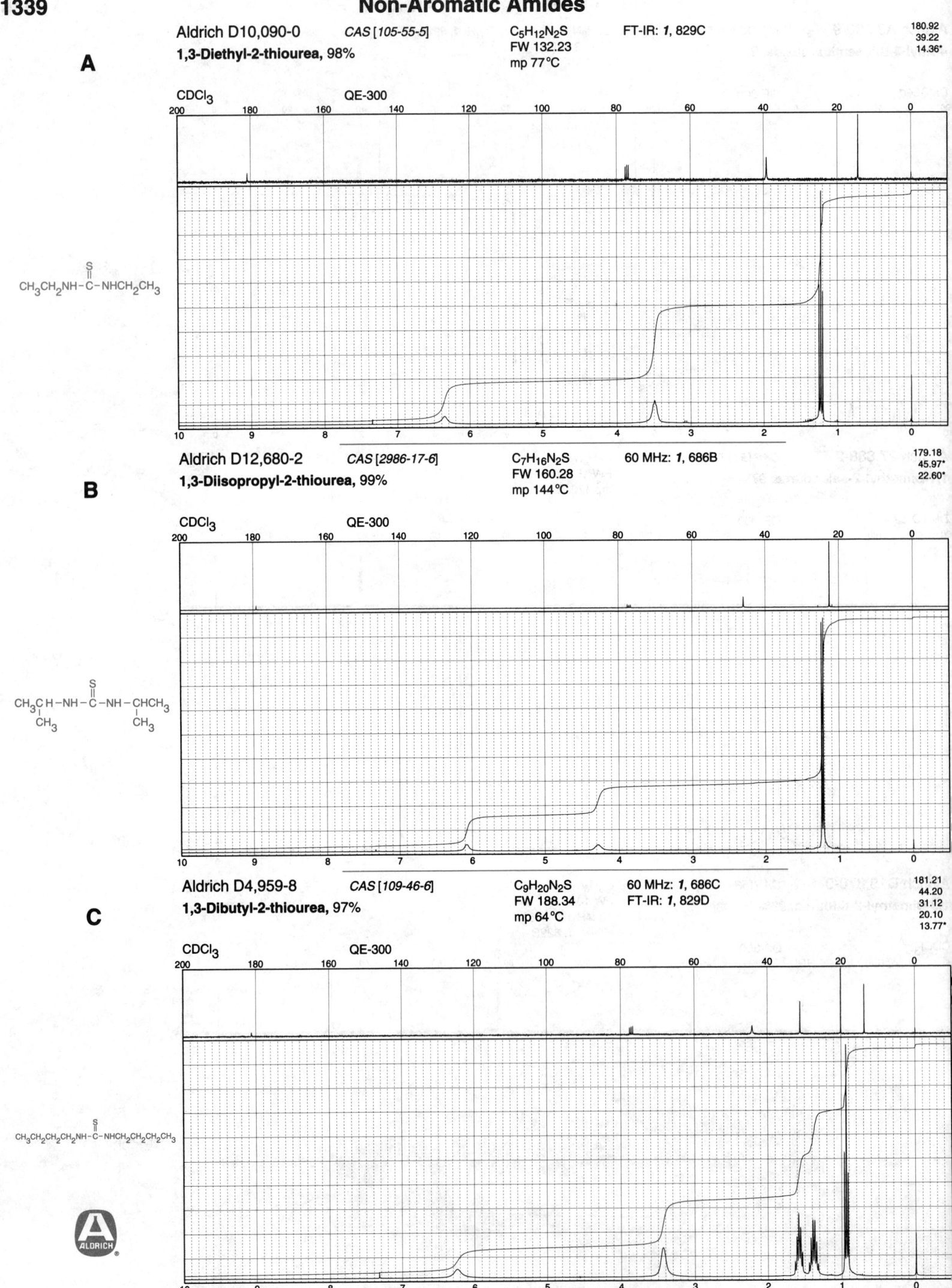

A

Aldrich D10,090-0   *CAS [105-55-5]*   C₅H₁₂N₂S   FT-IR: *1*, 829C

1,3-Diethyl-2-thiourea, 98%   FW 132.23
mp 77°C

180.92
39.22
14.36*

CDCl₃   QE-300

CH₃CH₂NH–C–NHCH₂CH₃ (with S double bonded to C)

B

Aldrich D12,680-2   *CAS [2986-17-6]*   C₇H₁₆N₂S   60 MHz: *1*, 686B

1,3-Diisopropyl-2-thiourea, 99%   FW 160.28
mp 144°C

179.18
45.97*
22.60*

CDCl₃   QE-300

CH₃CH–NH–C–NH–CHCH₃ (with S double bonded to C; CH₃ groups below)

C

Aldrich D4,959-8   *CAS [109-46-6]*   C₉H₂₀N₂S   60 MHz: *1*, 686C

1,3-Dibutyl-2-thiourea, 97%   FW 188.34   FT-IR: *1*, 829D
mp 64°C

181.21
44.20
31.12
20.10
13.77*

CDCl₃   QE-300

CH₃CH₂CH₂CH₂NH–C–NHCH₂CH₂CH₂CH₃ (with S double bonded to C)

ALDRICH

Aldrich 19,046-2   *CAS [21018-38-2]* $C_6H_{12}N_2S$ FT-IR: *1*, 801B

1-Methallyl-3-methyl-2-thiourea, 98%

FW 144.24
mp 61°C

A

182.60
141.20
111.67
49.91
31.14*
20.24*

$CDCl_3$ QE-300

$H_2C=C\ CH_2\ NH-C-NHCH_3$ with CH_3 and S

Aldrich I-50-4   *CAS [96-45-7]* $C_3H_6N_2S$ 60 MHz: *1*, 687B

2-Imidazolidinethione, 98%

FW 102.16
mp 199°C
FT-IR: *1*, 830B

B

183.19
43.94

DMSO-d_6 QE-300

Aldrich 12,202-5   *CAS [10574-66-0]* $C_5H_7NO_2S$ 60 MHz: *1*, 693A

3-Ethyl-2-thioxo-4-oxazolidinone, 97%

FW 145.18
mp 40°C
Fp >230°F

C

190.48
170.79
69.82
38.10
12.03*

$CDCl_3$ QE-300

ALDRICH

Non-Aromatic Amides

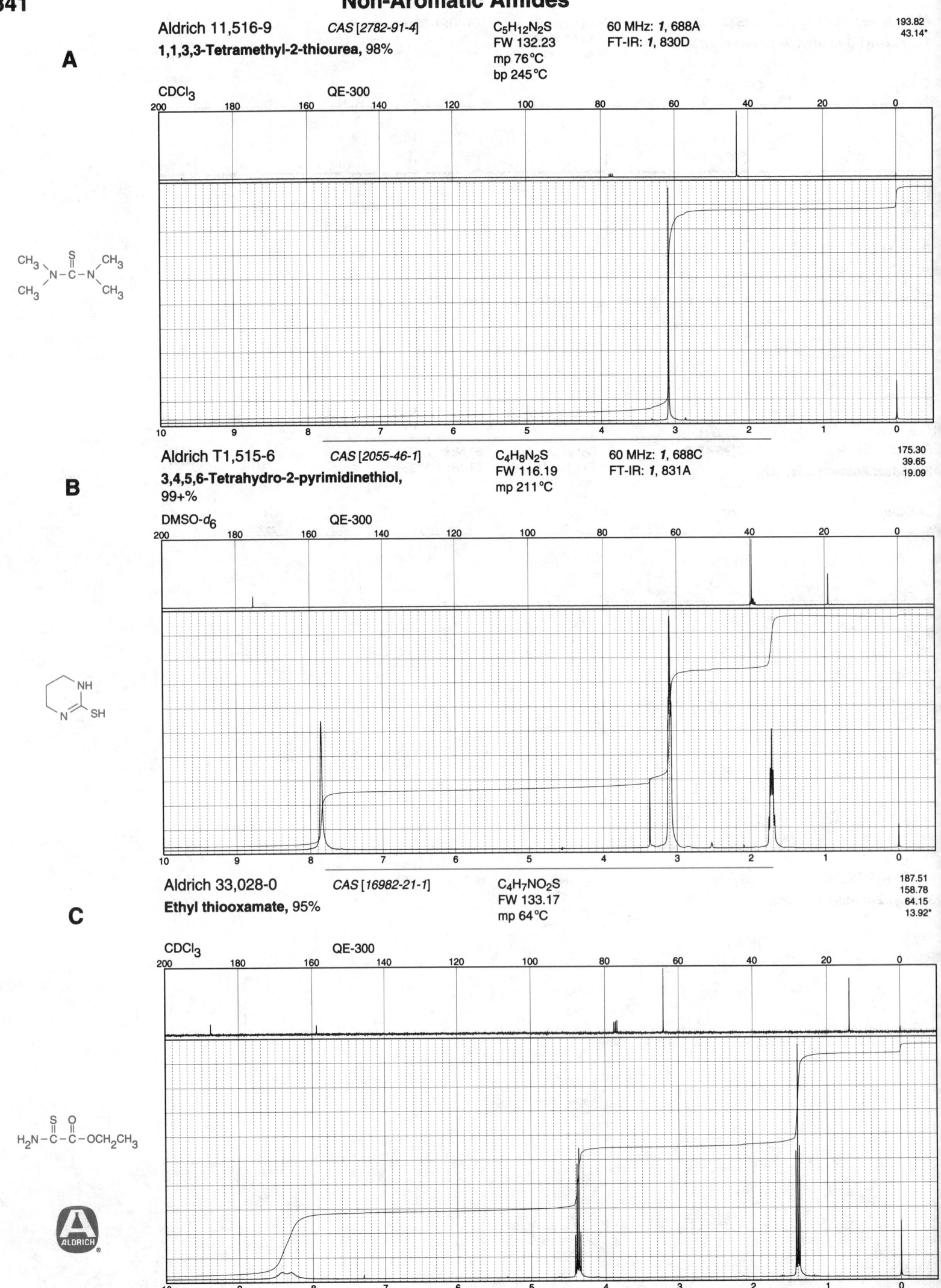

1341

A

Aldrich 11,516-9   *CAS [2782-91-4]*   C₅H₁₂N₂S   60 MHz: *1*, 688A   193.82 43.14*
1,1,3,3-Tetramethyl-2-thiourea, 98%   FW 132.23   FT-IR: *1*, 830D
mp 76°C
bp 245°C

CDCl₃   QE-300

B

Aldrich T1,515-6   *CAS [2055-46-1]*   C₄H₈N₂S   60 MHz: *1*, 688C   175.30 39.65 19.09
3,4,5,6-Tetrahydro-2-pyrimidinethiol, 99+%   FW 116.19   FT-IR: *1*, 831A
mp 211°C

DMSO-*d₆*   QE-300

C

Aldrich 33,028-0   *CAS [16982-21-1]*   C₄H₇NO₂S   187.51 158.78 64.15 13.92*
Ethyl thiooxamate, 95%   FW 133.17
mp 64°C

CDCl₃   QE-300

Aldrich A3,280-2   *CAS [105-81-7]*   C₆H₁₂N₂OS
1-Allyl-3-(2-hydroxyethyl)-2-thiourea, 97%
FW 160.24
mp 78 °C

$C_6H_{12}N_2OS$
FW 160.24
mp 78 °C

182.59
134.73*
115.28
60.05
46.44
46.20

A

CDCl₃ + DMSO-*d₆*   QE-300

H₂C=CHCH₂-NH-C(=S)-NHCH₂CH₂OH

Aldrich A2,285-8   *CAS [591-08-2]*   C₃H₆N₂OS   60 MHz: *1*, 688D
1-Acetyl-2-thiourea, 99%
FW 118.16   FT-IR: *1*, 831B
mp 167 °C

181.44
171.74
23.66*

B

DMSO-*d₆*   QE-300

Offset: 1.3 ppm.

CH₃-C(=O)-NH-C(=S)-NH₂

Aldrich 24,318-3   *CAS [5442-32-0]*   C₅H₁₂N₄S₂   60 MHz: *1*, 689A
S,S'-(1,3-Propanediyl)bis(isothiouronium bromide), 98%
FW 354.14   FT-IR: *1*, 831C
mp 200 °C

169.32
28.56
28.37

C

DMSO-*d₆*   QE-300

H₂N-C(=NH)-SCH₂CH₂CH₂S-C(=NH)-NH₂

• 2HBr

ALDRICH

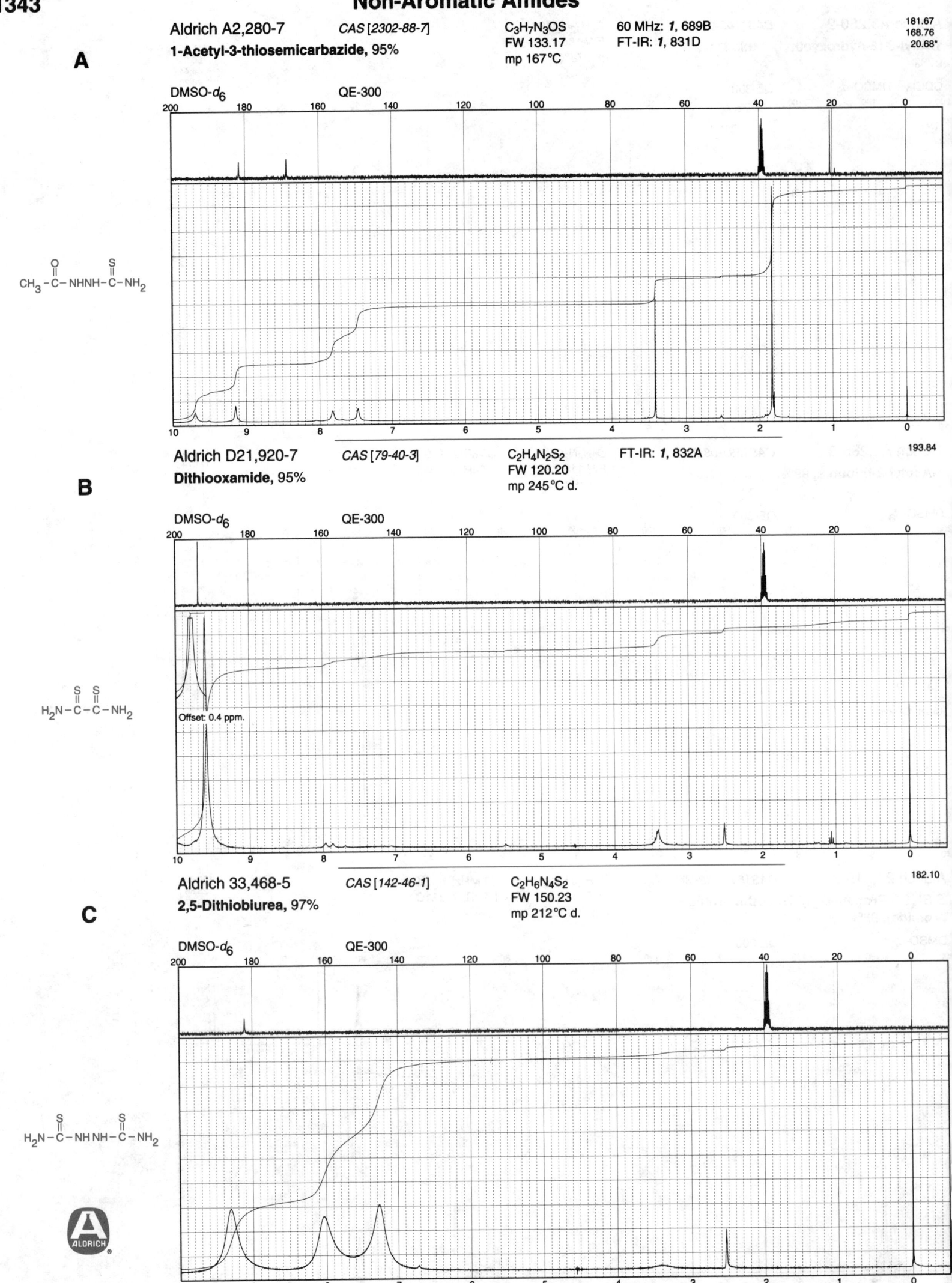

A

Aldrich A2,280-7   CAS [2302-88-7]   C₃H₇N₃OS   60 MHz: 1, 689B   181.67
1-Acetyl-3-thiosemicarbazide, 95%   FW 133.17   FT-IR: 1, 831D   168.76
mp 167°C   20.68*

DMSO-*d*₆   QE-300

CH₃—C—NHNH—C—NH₂
      ‖   ‖
      O   S

193.84

B

Aldrich D21,920-7   CAS [79-40-3]   C₂H₄N₂S₂   FT-IR: 1, 832A
Dithiooxamide, 95%   FW 120.20
mp 245°C d.

DMSO-*d*₆   QE-300

Offset: 0.4 ppm.

H₂N—C—C—NH₂
    ‖   ‖
    S   S

182.10

C

Aldrich 33,468-5   CAS [142-46-1]   C₂H₆N₄S₂
2,5-Dithiobiurea, 97%   FW 150.23
mp 212°C d.

DMSO-*d*₆   QE-300

H₂N—C—NH NH—C—NH₂
    ‖   ‖
    S   S

A

Aldrich 85,778-5   *CAS [94-37-1]*

Dicyclopentamethylenethiuram disulfide, 98%

CDCl$_3$ QE-300

C$_{12}$H$_{20}$N$_2$S$_4$
FW 320.56
60 MHz: *1*, 690A
FT-IR: *1*, 832C

192.45
55.70
52.61
26.42
25.85
24.17
24.02

B

Aldrich T1,160-6   *CAS [97-77-8]*

Tetraethylthiuram disulfide, 97%

CDCl$_3$ QE-300

C$_{10}$H$_{20}$N$_2$S$_4$
FW 296.54
mp 72°C

60 MHz: *1*, 689D
FT-IR: *1*, 833B

192.50
51.98
47.57
13.48*
11.44*

C

Aldrich 32,971-1   *CAS [137-30-4]*

Dimethyldithiocarbamic acid, zinc salt, 97%

DMSO-*d$_6$*   QE-300

C$_3$H$_7$NS$_2$
FW 305.80
mp 251°C

203.46
44.50*

Offset: 40 ppm.

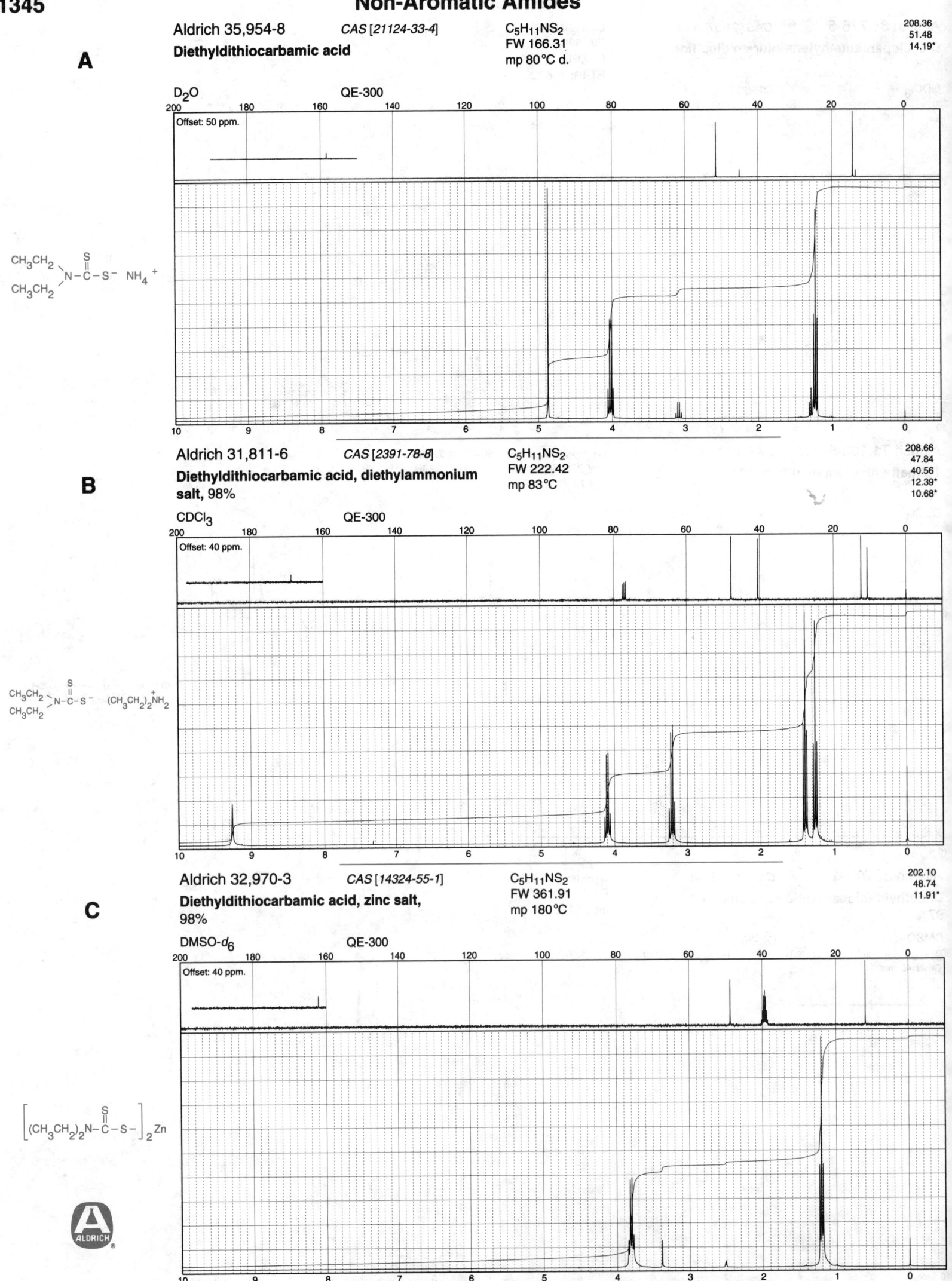

A

Aldrich 35,954-8   *CAS [21124-33-4]* $C_5H_{11}NS_2$
Diethyldithiocarbamic acid   FW 166.31
mp 80°C d.

208.36
51.48
14.19*

D$_2$O   QE-300

Offset: 50 ppm.

CH$_3$CH$_2$
          N–C–S⁻ NH$_4$$^+$
CH$_3$CH$_2$ ‖
             S

B

Aldrich 31,811-6   *CAS [2391-78-8]* $C_5H_{11}NS_2$
Diethyldithiocarbamic acid, diethylammonium   FW 222.42
salt, 98%   mp 83°C

208.66
47.84
40.56
12.39*
10.68*

CDCl$_3$ QE-300

Offset: 40 ppm.

CH$_3$CH$_2$ S
          N–C–S⁻ (CH$_3$CH$_2$)$_2$NH$_2$$^+$
CH$_3$CH$_2$ ‖

C

Aldrich 32,970-3   *CAS [14324-55-1]* $C_5H_{11}NS_2$
Diethyldithiocarbamic acid, zinc salt,   FW 361.91
98%   mp 180°C

202.10
48.74
11.91*

DMSO-d_6 QE-300

Offset: 40 ppm.

$\left[(CH_3CH_2)_2N-\overset{\overset{S}{\|}}{C}-S- \right]_2 Zn$

ALDRICH

A

Aldrich 36,298-0 *CAS [27268-57-1]* $C_5H_{10}N_2S_2$ FW 162.28 mp 117°C

Methyl 2-(1-methylethylidene)hydrazine-carbodithioate, 96%

200.54
152.75
25.15*
17.69*
16.94*

CDCl₃ QE-300

Offset: 40 ppm.

B

Aldrich T2,828-2 *CAS [7203-96-5]* $C_6H_{11}NS$ FW 129.23 mp 109°C 60 MHz: *1*, 691B FT-IR: *1*, 834D

ω-Thiocaprolactam, 98%

207.56
45.52
44.61
29.39
27.68
23.85

DMSO-*d₆* QE-300

Offset: 40 ppm.

Offset: 0.4 ppm.

C

Aldrich 34,540-7 *CAS [104499-08-3]* $C_6H_{11}NOS$ FW 145.23 mp 52°C

(S)-(-)-4-Isopropyl-2-oxazolidinethione, 99%

189.66
73.51
62.46*
32.19*
18.04*
17.87*

CDCl₃ QE-300

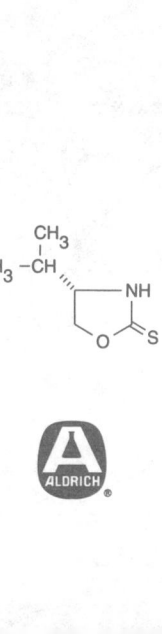

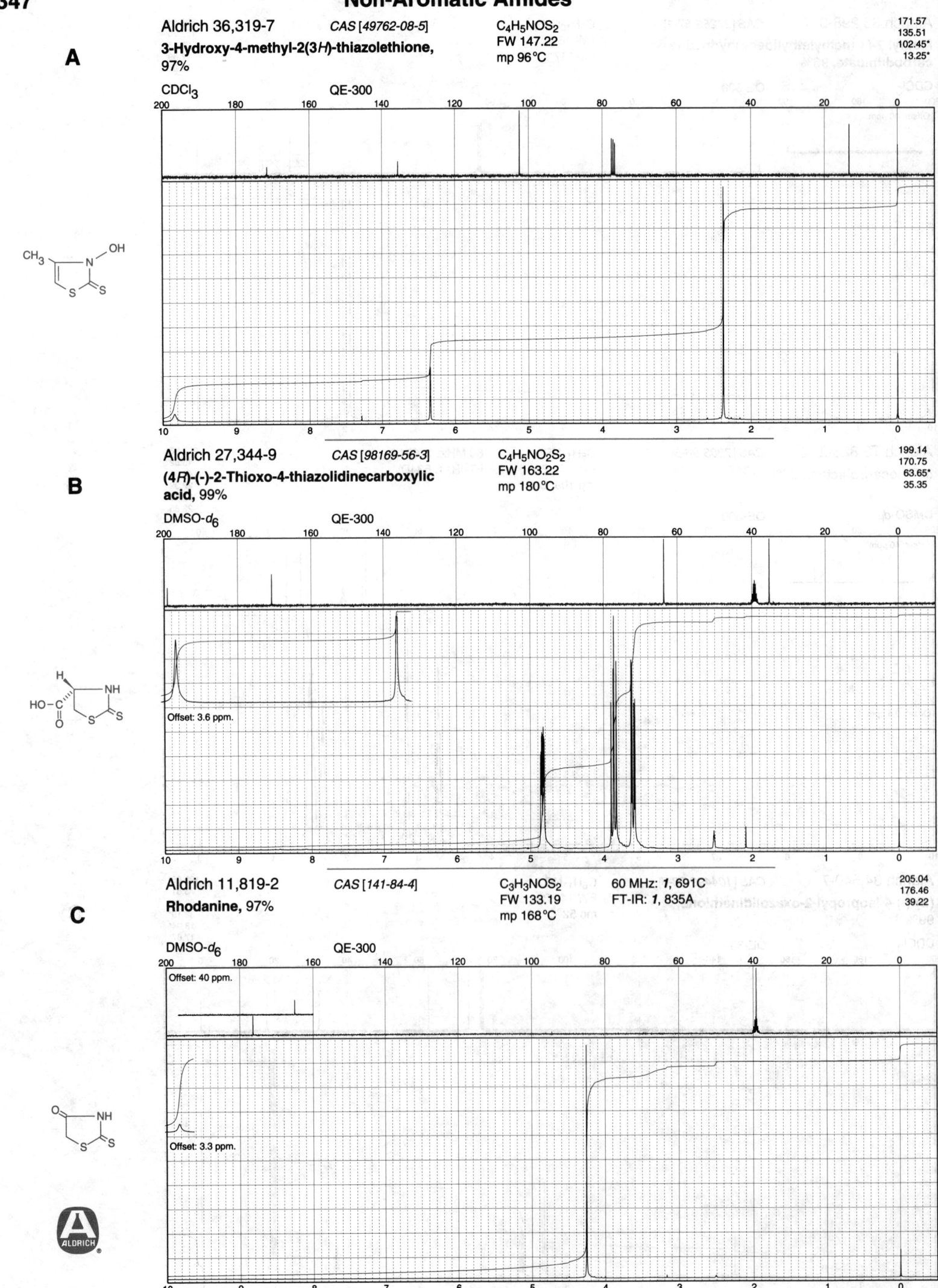

A

Aldrich 36,319-7   *CAS [49762-08-5]* $C_4H_5NOS_2$
3-Hydroxy-4-methyl-2(3H)-thiazolethione,
97%

FW 147.22
mp 96°C

171.57
135.51
102.45*
13.25*

$CDCl_3$ QE-300

B

Aldrich 27,344-9   *CAS [98169-56-3]* $C_4H_5NO_2S_2$
(4R)-(-)-2-Thioxo-4-thiazolidinecarboxylic
acid, 99%

FW 163.22
mp 180°C

199.14
170.75
63.65*
35.35

DMSO-d_6 QE-300

Offset: 3.6 ppm.

C

Aldrich 11,819-2   *CAS [141-84-4]* $C_3H_3NOS_2$ 60 MHz: *1*, 691C
Rhodanine, 97%

FW 133.19   FT-IR: *1*, 835A
mp 168°C

205.04
176.46
39.22

DMSO-d_6 QE-300

Offset: 40 ppm.

Offset: 3.3 ppm.

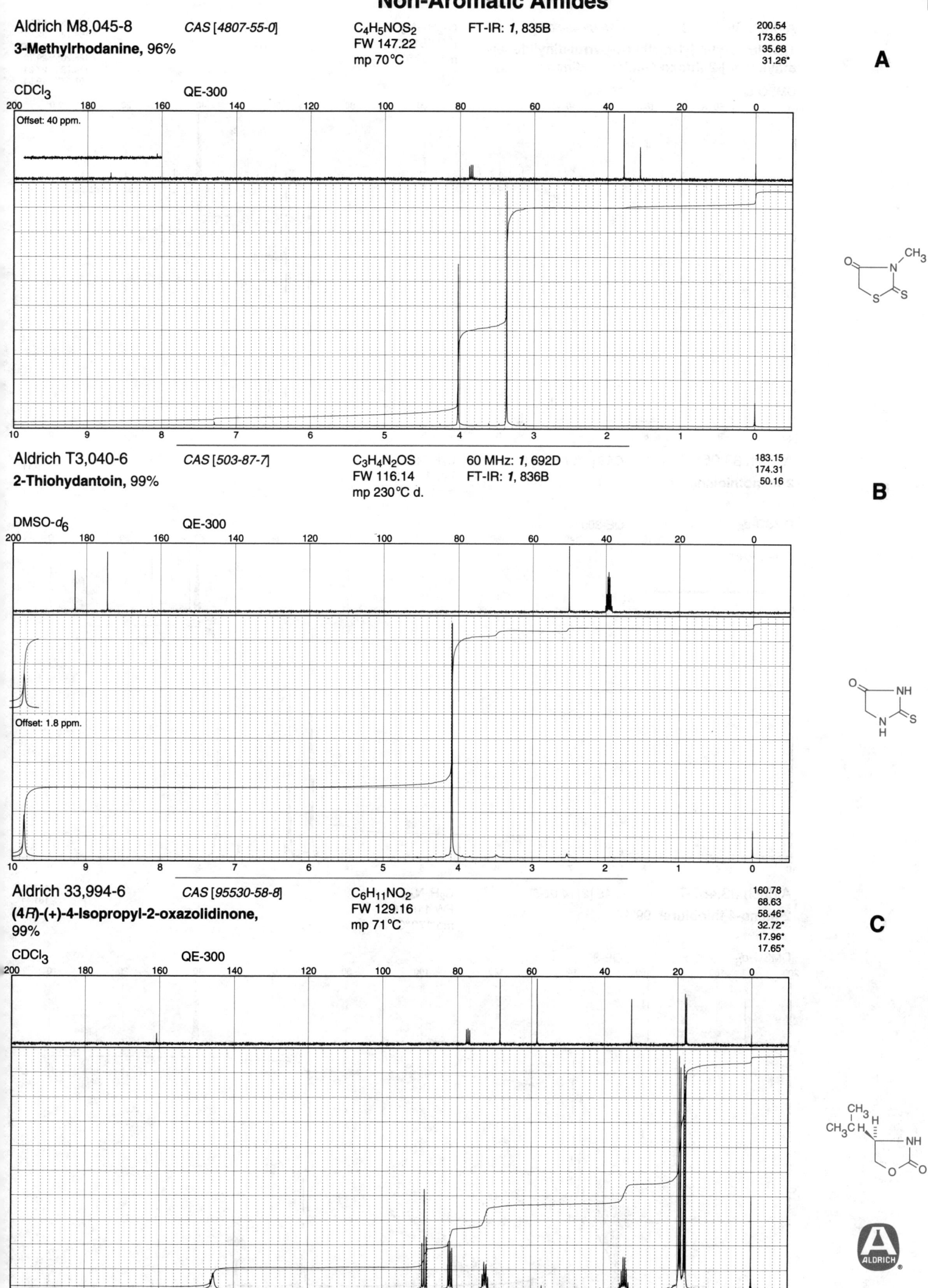

Aldrich M8,045-8   CAS [4807-55-0]   C₄H₅NOS₂   FT-IR: 1, 835B

3-Methylrhodanine, 96%   FW 147.22
mp 70 °C

200.54
173.65
35.68
31.26*

A

CDCl₃   QE-300
Offset: 40 ppm.

Aldrich T3,040-6   CAS [503-87-7]   C₃H₄N₂OS   60 MHz: 1, 692D

2-Thiohydantoin, 99%   FW 116.14   FT-IR: 1, 836B
mp 230 °C d.

183.15
174.31
50.16

B

DMSO-d₆   QE-300
Offset: 1.8 ppm.

Aldrich 33,994-6   CAS [95530-58-8]   C₆H₁₁NO₂

(4R)-(+)-4-Isopropyl-2-oxazolidinone,   FW 129.16
99%   mp 71 °C

160.78
68.63
58.46*
32.72*
17.96*
17.65*

C

CDCl₃   QE-300

1349

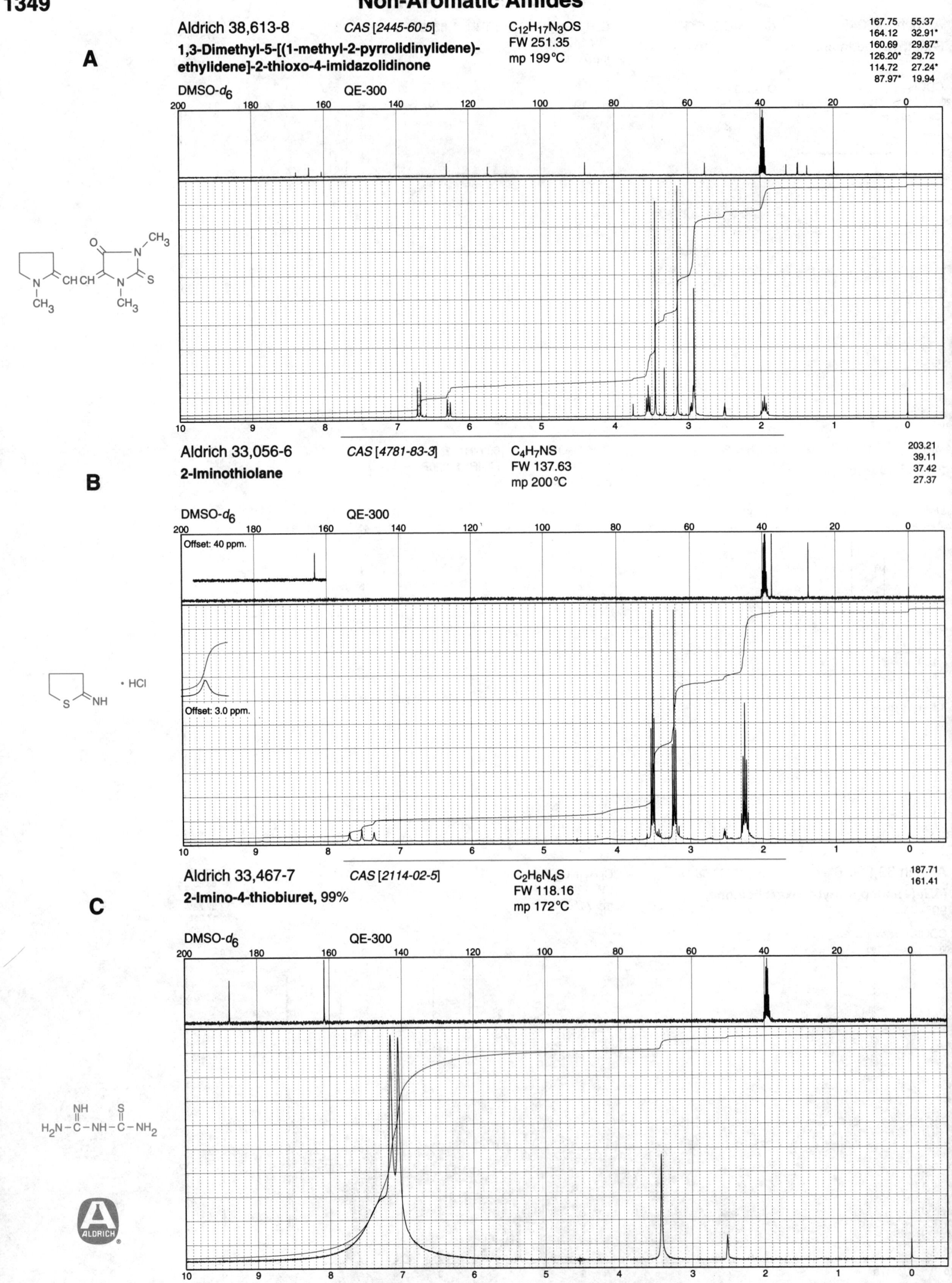

A

Aldrich 38,613-8   *CAS [2445-60-5]*

1,3-Dimethyl-5-[(1-methyl-2-pyrrolidinylidene)-ethylidene]-2-thioxo-4-imidazolidinone

$C_{12}H_{17}N_3OS$
FW 251.35
mp 199°C

167.75	55.37
164.12	32.91*
160.69	29.87*
126.20*	29.72
114.72	27.24*
87.97*	19.94

DMSO-d_6 QE-300

B

Aldrich 33,056-6   *CAS [4781-83-3]*

2-Iminothiolane

C_4H_7NS
FW 137.63
mp 200°C

203.21
39.11
37.42
27.37

DMSO-d_6 QE-300

Offset: 40 ppm.

Offset: 3.0 ppm.

• HCl

C

Aldrich 33,467-7   *CAS [2114-02-5]*

2-Imino-4-thiobiuret, 99%

$C_2H_6N_4S$
FW 118.16
mp 172°C

187.71
161.41

DMSO-d_6 QE-300

Non-Aromatic Nitriles and Cumulated Double Bonds

A

Aldrich 15,460-1   *CAS [75-05-8]* C_2H_3N $n_D^{20}1.3440$ 116.53
Acetonitrile, 99.5+%   FW 41.05   Fp 42°F   1.81*
                                            mp -48°C   60 MHz: *1*, 695A
                                            bp 82°C   FT-IR: *1*, 837A
CDCl₃   QE-300   d 0.786   VP-FT-IR: *3*, 795A

CH_3CN

B

Aldrich 18,559-0   *CAS [107-12-0]* C_3H_5N $n_D^{20}1.3660$ 120.71
Propionitrile, 99%   FW 55.08   Fp 43°F   10.89
                                            mp -93°C   60 MHz: *1*, 695C   10.51*
                                            bp 97°C   FT-IR: *1*, 837B
CDCl₃   QE-300   d 0.772   VP-FT-IR: *3*, 795B

CH_3CH_2CN

C

Aldrich B10,380-2   *CAS [109-74-0]* C_4H_7N $n_D^{20}1.3840$ 119.75
Butyronitrile, 98%   FW 69.11   Fp 62°F   19.19
                                            mp -112°C   60 MHz: *1*, 695D   19.02
                                            bp 116°C   13.29*
CDCl₃   QE-300   d 0.794

$CH_3CH_2CH_2CN$

ALDRICH

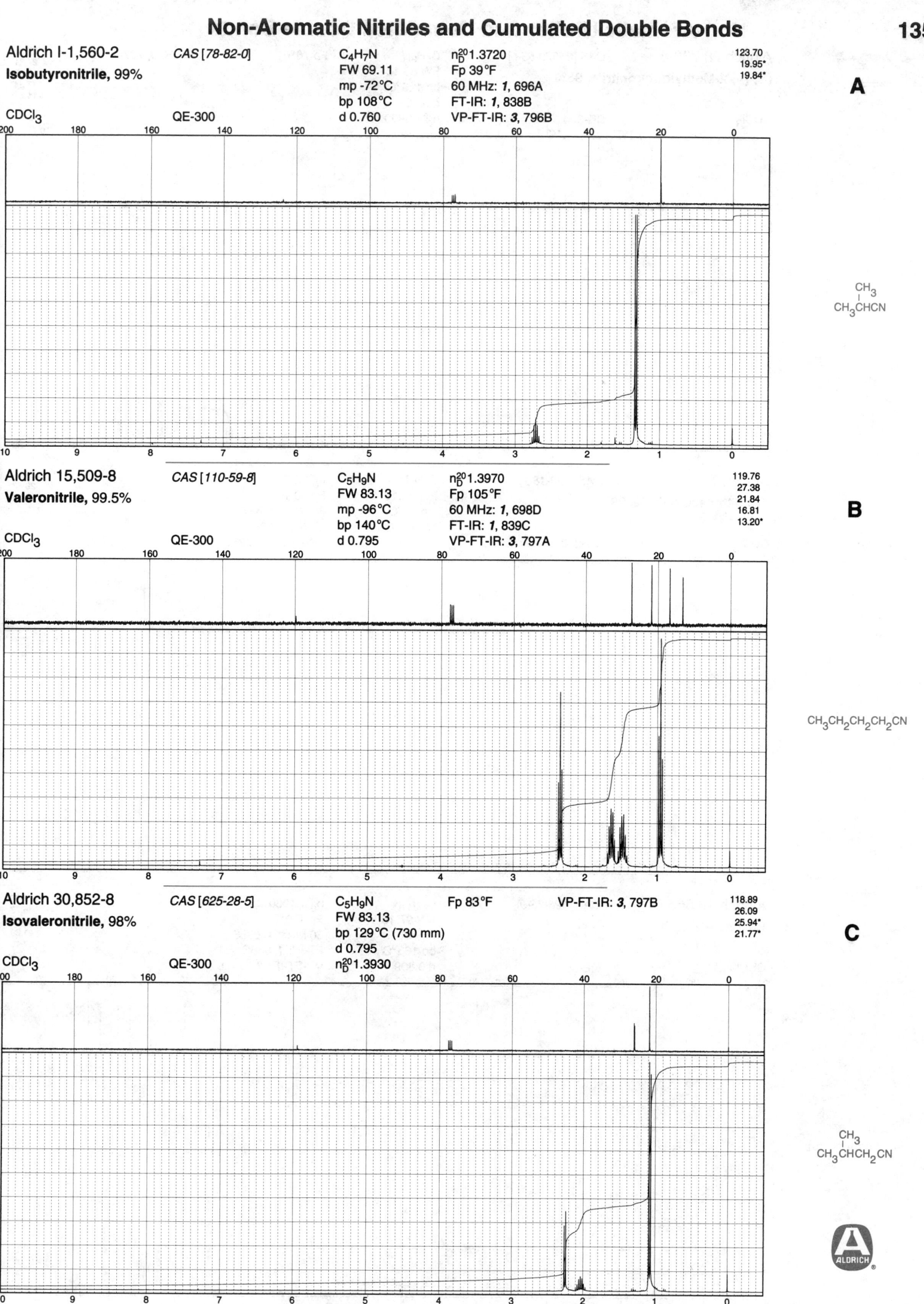

Aldrich I-1,560-2
Isobutyronitrile, 99%

CAS [78-82-0]

C$_4$H$_7$N
FW 69.11
mp -72°C
bp 108°C
d 0.760

CDCl$_3$

QE-300

n$_D^{20}$1.3720
Fp 39°F
60 MHz: *1*, 696A
FT-IR: *1*, 838B
VP-FT-IR: *3*, 796B

123.70
19.95*
19.84*

A

Aldrich 15,509-8
Valeronitrile, 99.5%

CAS [110-59-8]

C$_5$H$_9$N
FW 83.13
mp -96°C
bp 140°C
d 0.795

CDCl$_3$

QE-300

n$_D^{20}$1.3970
Fp 105°F
60 MHz: *1*, 698D
FT-IR: *1*, 839C
VP-FT-IR: *3*, 797A

119.76
27.38
21.84
16.81
13.20*

B

Aldrich 30,852-8
Isovaleronitrile, 98%

CAS [625-28-5]

C$_5$H$_9$N
FW 83.13
bp 129°C (730 mm)
d 0.795
n$_D^{20}$1.3930

Fp 83°F

VP-FT-IR: *3*, 797B

118.89
26.09
25.94*
21.77*

CDCl$_3$

QE-300

C

Non-Aromatic Nitriles and Cumulated Double Bonds

A

Aldrich 28,730-8   CAS [25570-03-0]   C$_5$H$_9$N   Fp 76°F   VP-FT-IR: **3**, 797C

(S)-(+)-2-Methylbutyronitrile, 98%   FW 83.13
bp 126°C
d 0.786
n$_D^{20}$1.3900

122.90
27.31
27.09*
17.64*
11.41*

CDCl$_3$ QE-300

200 180 160 140 120 100 80 60 40 20 0

H CN
| |
CH$_3$CH$_2$—C—CH$_3$

10 9 8 7 6 5 4 3 2 1 0

B

Aldrich T7,200-1   CAS [630-18-2]   C$_5$H$_9$N   n$_D^{20}$1.3770

Trimethylacetonitrile, 98%   FW 83.13   Fp 40°F
mp 16°C   60 MHz: **1**, 697A
bp 106°C   FT-IR: **1**, 839A
d 0.752   VP-FT-IR: **3**, 796C

125.81
28.42*
28.09

CDCl$_3$ QE-300

200 180 160 140 120 100 80 60 40 20 0

CH$_3$
|
CH$_3$—C—CN
|
CH$_3$

10 9 8 7 6 5 4 3 2 1 0

C

Aldrich 16,665-0   CAS [628-73-9]   C$_6$H$_{11}$N   n$_D^{20}$1.4060

Hexanenitrile, 98%   FW 97.16   Fp 110°F
mp -80°C   60 MHz: **1**, 696B
bp 163°C   FT-IR: **1**, 838C
d 0.809   VP-FT-IR: **3**, 797D

119.85
30.77
25.09
21.89
17.10
13.76*

CDCl$_3$ QE-300

200 180 160 140 120 100 80 60 40 20 0

CH$_3$CH$_2$CH$_2$CH$_2$CH$_2$CN

10 9 8 7 6 5 4 3 2 1 0

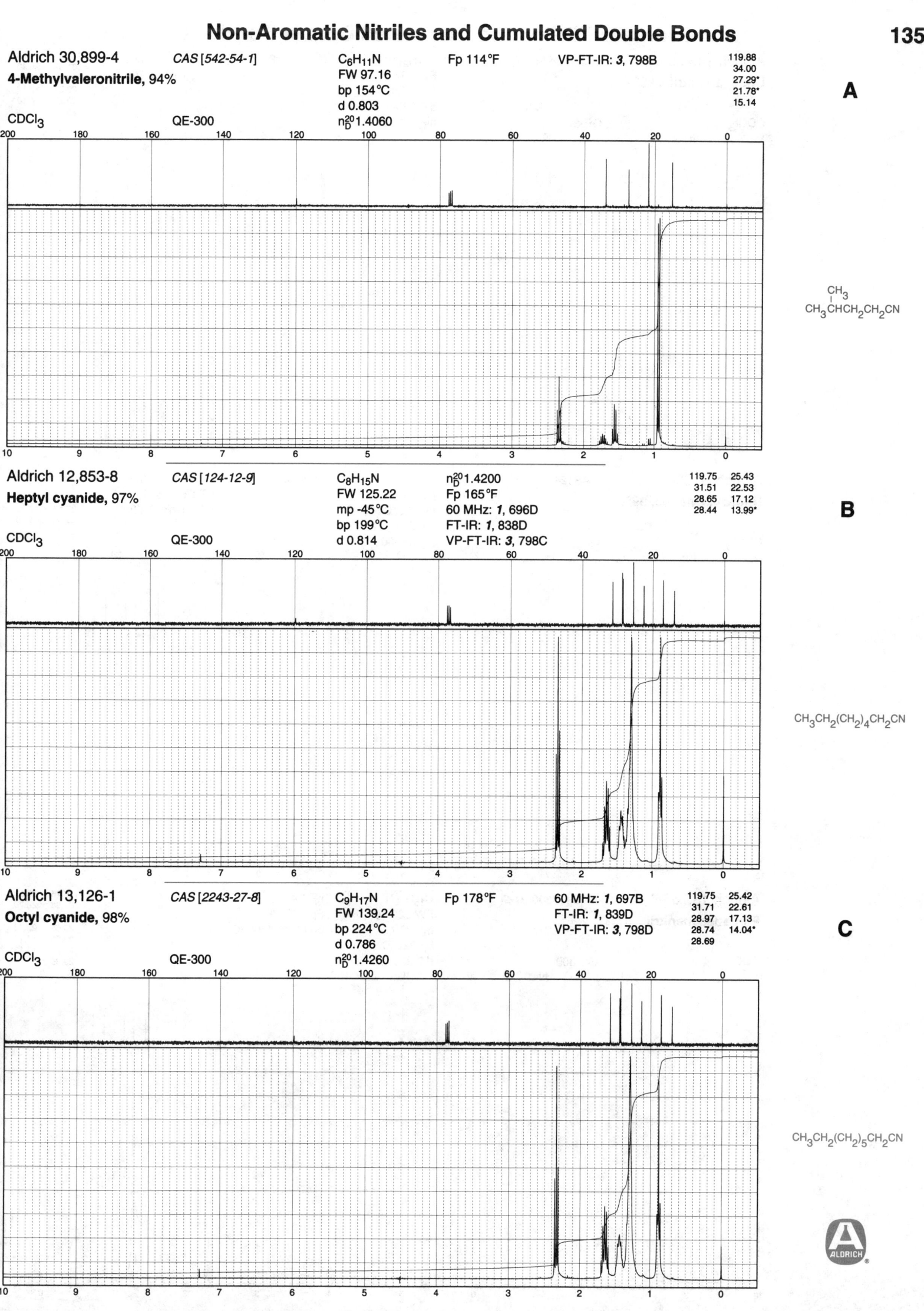

Aldrich 30,899-4
4-Methylvaleronitrile, 94%

CAS [542-54-1]

$C_6H_{11}N$
FW 97.16
bp 154°C
d 0.803
n_D^{20} 1.4060

Fp 114°F

VP-FT-IR: *3*, 798B

119.88
34.00
27.29*
21.78*
15.14

A

CDCl₃   QE-300

$$CH_3CHCH_2CH_2CN$$ with CH_3

Aldrich 12,853-8
Heptyl cyanide, 97%

CAS [124-12-9]

$C_8H_{15}N$
FW 125.22
mp -45°C
bp 199°C
d 0.814

n_D^{20} 1.4200
Fp 165°F
60 MHz: *1*, 696D
FT-IR: *1*, 838D
VP-FT-IR: *3*, 798C

119.75   25.43
31.51   22.53
28.65   17.12
28.44   13.99*

B

CDCl₃   QE-300

$$CH_3CH_2(CH_2)_4CH_2CN$$

Aldrich 13,126-1
Octyl cyanide, 98%

CAS [2243-27-8]

$C_9H_{17}N$
FW 139.24
bp 224°C
d 0.786
n_D^{20} 1.4260

Fp 178°F

60 MHz: *1*, 697B
FT-IR: *1*, 839D
VP-FT-IR: *3*, 798D

119.75   25.42
31.71   22.61
28.97   17.13
28.74   14.04*
28.69

C

CDCl₃   QE-300

$$CH_3CH_2(CH_2)_5CH_2CN$$

ALDRICH

Non-Aromatic Nitriles and Cumulated Double Bonds

A

Aldrich 14,456-8   *CAS [2244-07-7]* $C_{11}H_{21}N$ Fp >230°F   60 MHz: *1*, 697C

119.80	28.68
31.87	25.40
29.47	22.68
29.32	17.13
29.27	14.10*
28.78	

Undecanenitrile, 99%   FW 167.30   VP-FT-IR: *3*, 799A

bp 253°C

d 0.823

CDCl₃   QE-300 $n_D^{20}1.4330$

$CH_3(CH_2)_8CH_2CN$

B

Aldrich U160-5   *CAS [2437-25-4]* $C_{12}H_{23}N$ Fp >230°F   60 MHz: *1*, 697D

119.79	28.70
31.91	25.44
29.57	22.69
29.53	17.13
29.32	14.10*
28.79	

Undecyl cyanide, 99%   FW 181.32   FT-IR: *1*, 840A

bp 198°C (100 mm)   VP-FT-IR: *3*, 799B

d 0.827

CDCl₃   QE-300 $n_D^{20}1.4360$

$CH_3(CH_2)_9CH_2CN$

C

Aldrich 13,763-4   *CAS [18300-91-9]* $C_{15}H_{29}N$ $n_D^{20}1.4420$

119.76	28.81
31.96	28.70
29.69	25.43
29.65	22.73
29.55	17.12
29.40	14.13*
29.35	

Pentadecanenitrile, 97%   FW 223.40   Fp >230°F

mp 22°C   60 MHz: *1*, 698B

bp 322°C

CDCl₃   QE-300   d 0.825

$CH_3(CH_2)_{12}CH_2CN$

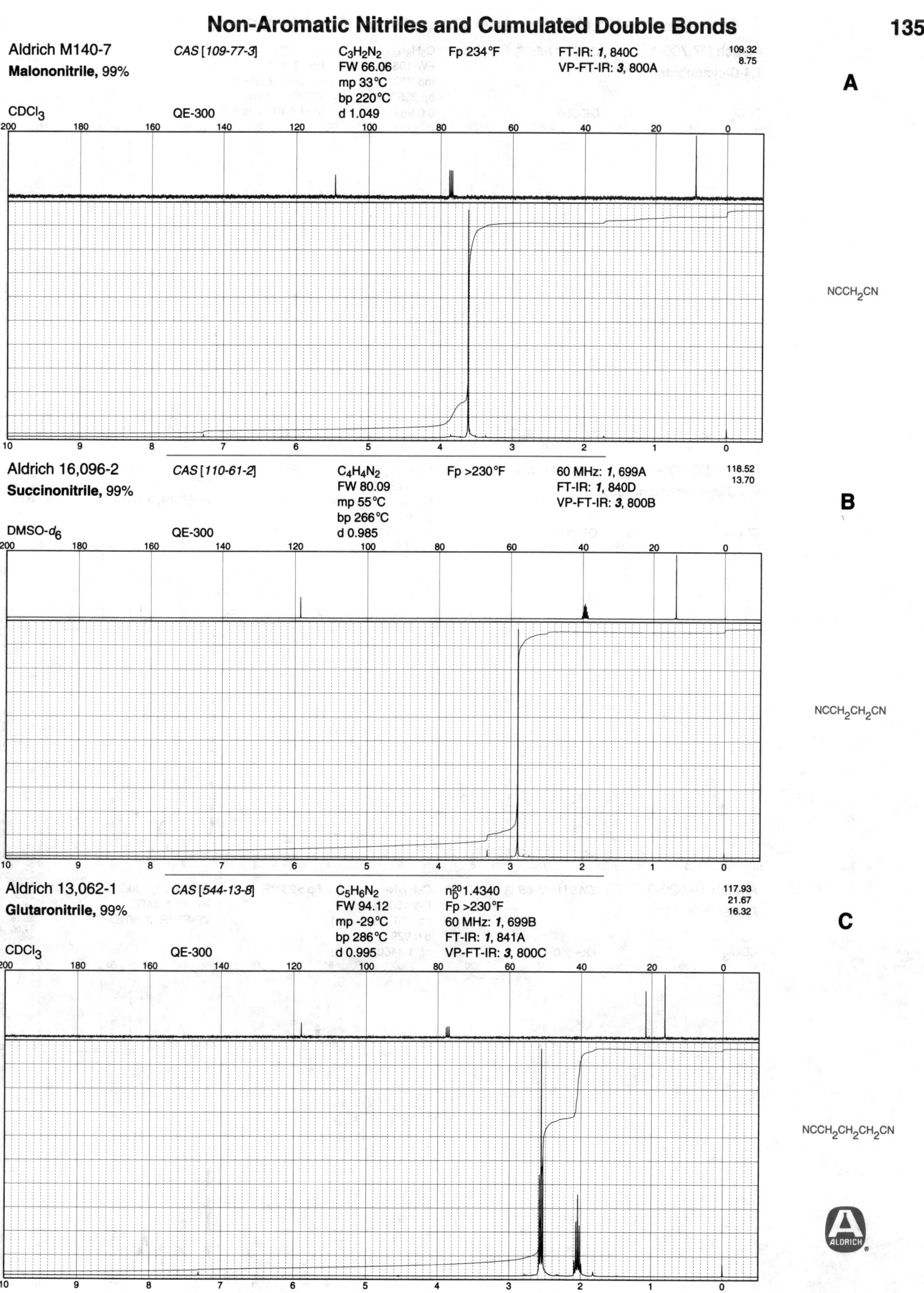

Aldrich M140-7   *CAS [109-77-3]* $C_3H_2N_2$ Fp 234°F   FT-IR: *1*, 840C   109.32   8.75

Malononitrile, 99%   FW 66.06   VP-FT-IR: *3*, 800A

mp 33°C
bp 220°C

CDCl₃   QE-300   d 1.049

A

NCCH₂CN

Aldrich 16,096-2   *CAS [110-61-2]* $C_4H_4N_2$ Fp >230°F   60 MHz: *1*, 699A   118.52   13.70

Succinonitrile, 99%   FW 80.09   FT-IR: *1*, 840D

mp 55°C   VP-FT-IR: *3*, 800B
bp 266°C

DMSO-d_6 QE-300   d 0.985

B

NCCH₂CH₂CN

Aldrich 13,062-1   *CAS [544-13-8]* $C_5H_6N_2$ n_D^{20} 1.4340   117.93   21.67   16.32

Glutaronitrile, 99%   FW 94.12   Fp >230°F

mp -29°C   60 MHz: *1*, 699B
bp 286°C   FT-IR: *1*, 841A

CDCl₃   QE-300   d 0.995   VP-FT-IR: *3*, 800C

C

NCCH₂CH₂CH₂CN

A

Aldrich D7,700-1   *CAS [111-69-3]* $C_6H_8N_2$ $n_D^{20}1.4380$ 118.87
24.27
1,4-Dicyanobutane, 99%   FW 108.14   Fp >230°F   16.64
mp 2°C   60 MHz: *1*, 699C
bp 295°C   FT-IR: *1*, 841C
d 0.951   VP-FT-IR: *3*, 800D

CDCl₃   QE-300

$NCCH_2CH_2CH_2CH_2CN$

B

Aldrich D7,900-4   *CAS [646-20-8]* $C_7H_{10}N_2$ Fp >230°F   60 MHz: *1*, 700A   119.30
27.68
1,5-Dicyanopentane, 98%   FW 122.17   FT-IR: *1*, 841D   24.65
bp 176°C (14 mm)   VP-FT-IR: *3*, 801A   16.95
d 0.951
CDCl₃   QE-300 $n_D^{20}1.4410$

$NCCH_2(CH_2)_3CH_2CN$

C

Aldrich 19,020-9   *CAS [1675-69-0]* $C_9H_{14}N_2$ Fp >230°F   60 MHz: *1*, 700C   119.67
28.30
Azelanitrile, 95%   FW 150.23   FT-IR: *1*, 842B   27.94
bp 176°C (11 mm)   VP-FT-IR: *3*, 801D   25.16
d 0.929   17.06
CDCl₃   QE-300 $n_D^{20}1.4460$

$NCCH_2(CH_2)_5CH_2CN$

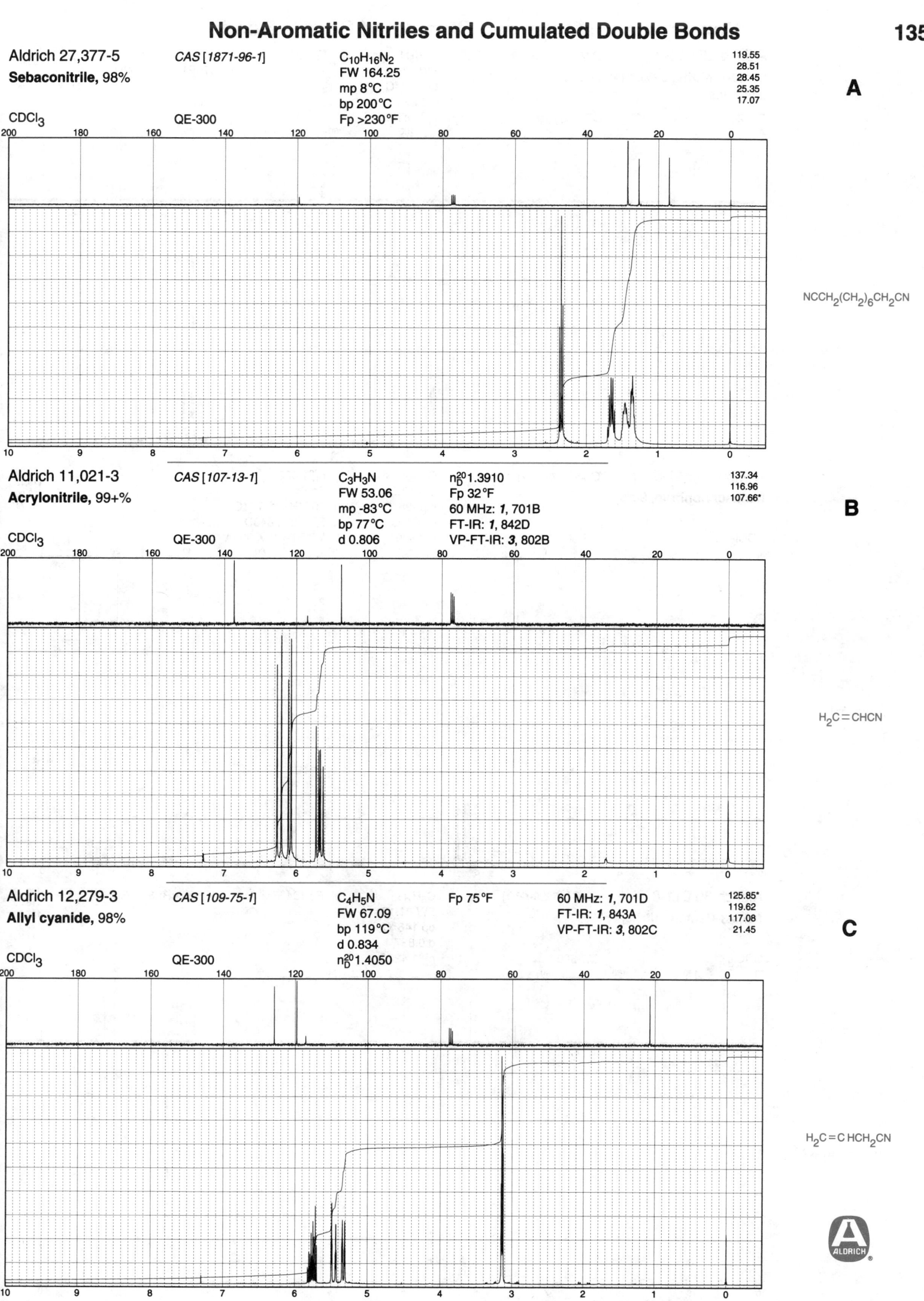

Aldrich 27,377-5   CAS [1871-96-1]   C₁₀H₁₆N₂
Sebaconitrile, 98%   FW 164.25
                                        mp 8°C
                                        bp 200°C
                                        Fp >230°F

CDCl₃   QE-300

119.55
28.51
28.45
25.35
17.07

A

NCCH₂(CH₂)₆CH₂CN

Aldrich 11,021-3   CAS [107-13-1]   C₃H₃N   n²⁰_D 1.3910
Acrylonitrile, 99+%   FW 53.06   Fp 32°F
                                        mp -83°C   60 MHz: *1*, 701B
                                        bp 77°C   FT-IR: *1*, 842D
                                        d 0.806   VP-FT-IR: *3*, 802B

CDCl₃   QE-300

137.34
116.96
107.66*

B

H₂C=CHCN

Aldrich 12,279-3   CAS [109-75-1]   C₄H₅N   Fp 75°F   60 MHz: *1*, 701D
Allyl cyanide, 98%   FW 67.09   FT-IR: *1*, 843A
                                        bp 119°C   VP-FT-IR: *3*, 802C
                                        d 0.834
CDCl₃   QE-300   n²⁰_D 1.4050

125.85*
119.62
117.08
21.45

C

H₂C=CHCH₂CN

A

Aldrich 25,252-2   CAS [4786-20-3]   C₄H₅N   Fp 68°F   FT-IR: 1, 843C

C_4H_5N
FW 67.09
bp 121°C
d 0.824
n_D^{20}1.4190

VP-FT-IR: 3, 802D

151.23*	101.05*
150.04*	100.67*
117.39	19.11*
115.86	17.48*

Crotononitrile, 99%, mixture of *cis* and *trans*

CDCl₃   QE-300

$CH_3\,CH{=}CHCN$

B

Aldrich 19,541-3   CAS [126-98-7]

Methacrylonitrile, 99%

C_4H_5N
FW 67.09
mp -36°C
bp 91°C
d 0.800

n_D^{20}1.4000
Fp 54°F
60 MHz: 1, 701C
FT-IR: 1, 843D
VP-FT-IR: 3, 803A

131.18
119.12
118.00
20.71*

CDCl₃   QE-300

$$\begin{array}{c} CH_3 \\ | \\ H_2C{=}CCN \end{array}$$

C

Aldrich 30,611-8   CAS [16529-66-1]

trans-3-Pentenenitrile, 95%

C_5H_7N
FW 81.12
bp 146°C
d 0.837
n_D^{20}1.4220

Fp 104°F   VP-FT-IR: 3, 803C

130.92*
118.39*
117.80
20.35
17.56*

CDCl₃   QE-300

ALDRICH

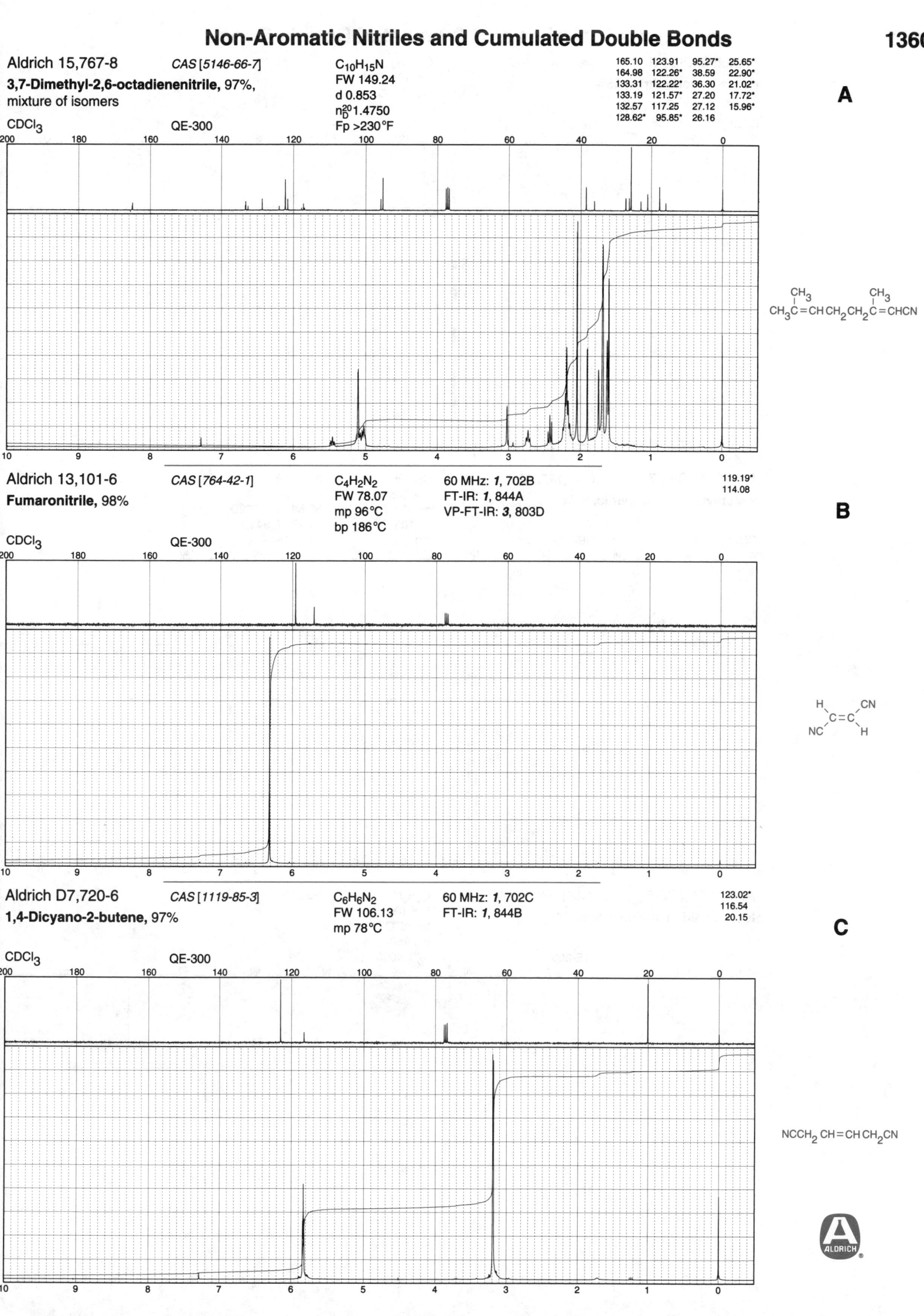

Aldrich 15,767-8   *CAS [5146-66-7]*   C₁₀H₁₅N

A

3,7-Dimethyl-2,6-octadienenitrile, 97%,
mixture of isomers

CDCl₃   QE-300

$C_{10}H_{15}N$
FW 149.24
d 0.853
n_D^{20} 1.4750
Fp >230°F

165.10	123.91	95.27*	25.65*
164.98	122.26*	38.59	22.90*
133.31	122.22*	36.30	21.02*
133.19	121.57*	27.20	17.72*
132.57	117.25	27.12	15.96*
128.62*	95.85*	26.16	

Aldrich 13,101-6   *CAS [764-42-1]*   C₄H₂N₂

B

Fumaronitrile, 98%

CDCl₃   QE-300

$C_4H_2N_2$
FW 78.07
mp 96°C
bp 186°C

60 MHz: *1*, 702B
FT-IR: *1*, 844A
VP-FT-IR: *3*, 803D

119.19*
114.08

Aldrich D7,720-6   *CAS [1119-85-3]*   C₆H₆N₂

C

1,4-Dicyano-2-butene, 97%

CDCl₃   QE-300

$C_6H_6N_2$
FW 106.13
mp 78°C

60 MHz: *1*, 702C
FT-IR: *1*, 844B

123.02*
116.54
20.15

1361

A

Aldrich 34,894-5   *CAS [1557-59-1]*   C$_6$H$_4$N$_2$
cis,cis-Muconitrile, 98%   FW 104.11
mp 130°C

142.61*
114.65
106.10*

CDCl$_3$ QE-300

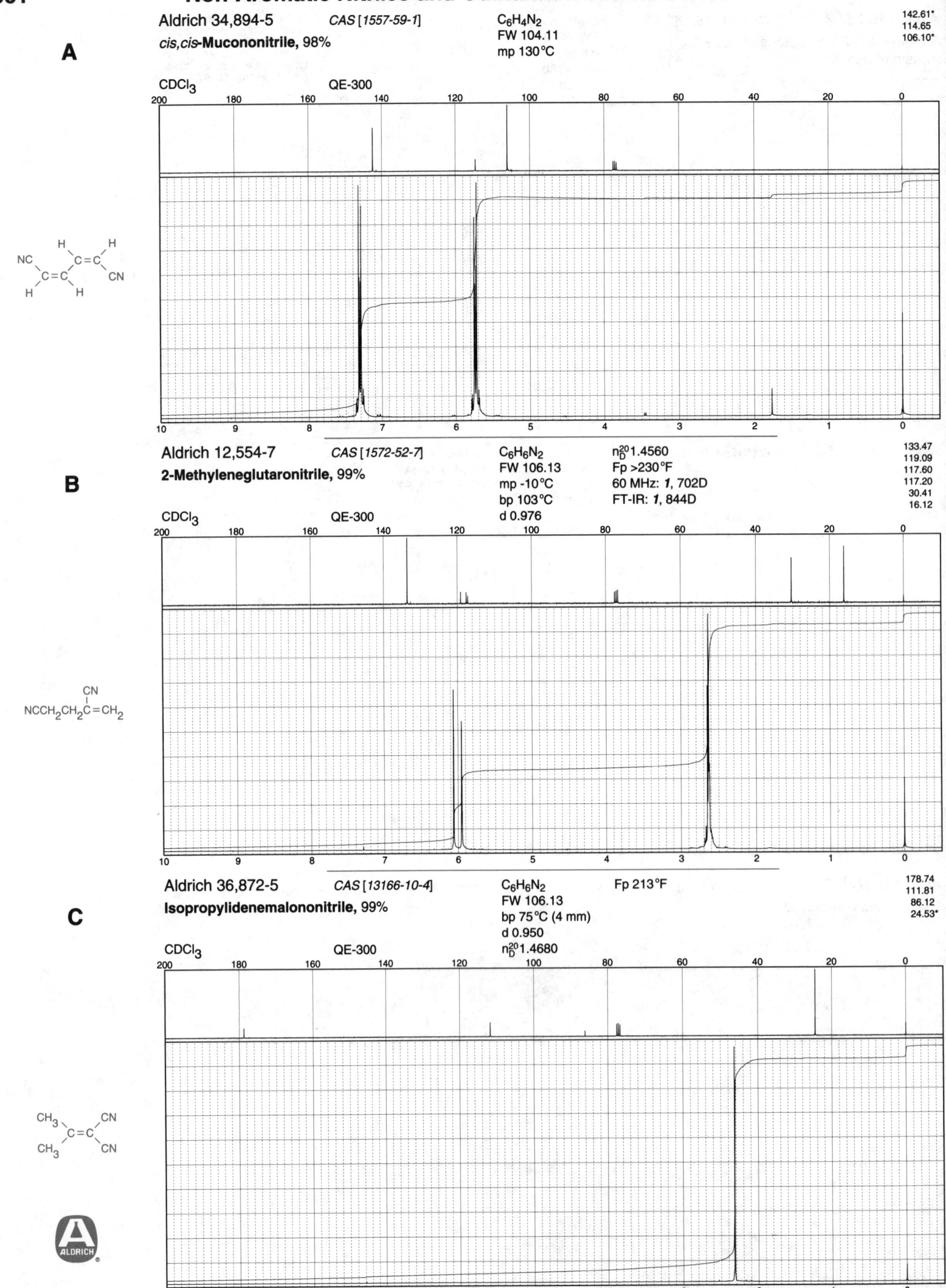

Aldrich 12,554-7   *CAS [1572-52-7]*   C$_6$H$_6$N$_2$ n$_D^{20}$1.4560
2-Methyleneglutaronitrile, 99%   FW 106.13   Fp >230°F
mp -10°C   60 MHz: *1*, 702D
bp 103°C   FT-IR: *1*, 844D
d 0.976

133.47
119.09
117.60
117.20
30.41
16.12

B

CDCl$_3$ QE-300

NCCH$_2$CH$_2$C(CN)=CH$_2$

Aldrich 36,872-5   *CAS [13166-10-4]*   C$_6$H$_6$N$_2$ Fp 213°F
Isopropylidenemalononitrile, 99%   FW 106.13
bp 75°C (4 mm)
d 0.950
n$_D^{20}$1.4680

178.74
111.81
86.12
24.53*

C

CDCl$_3$ QE-300

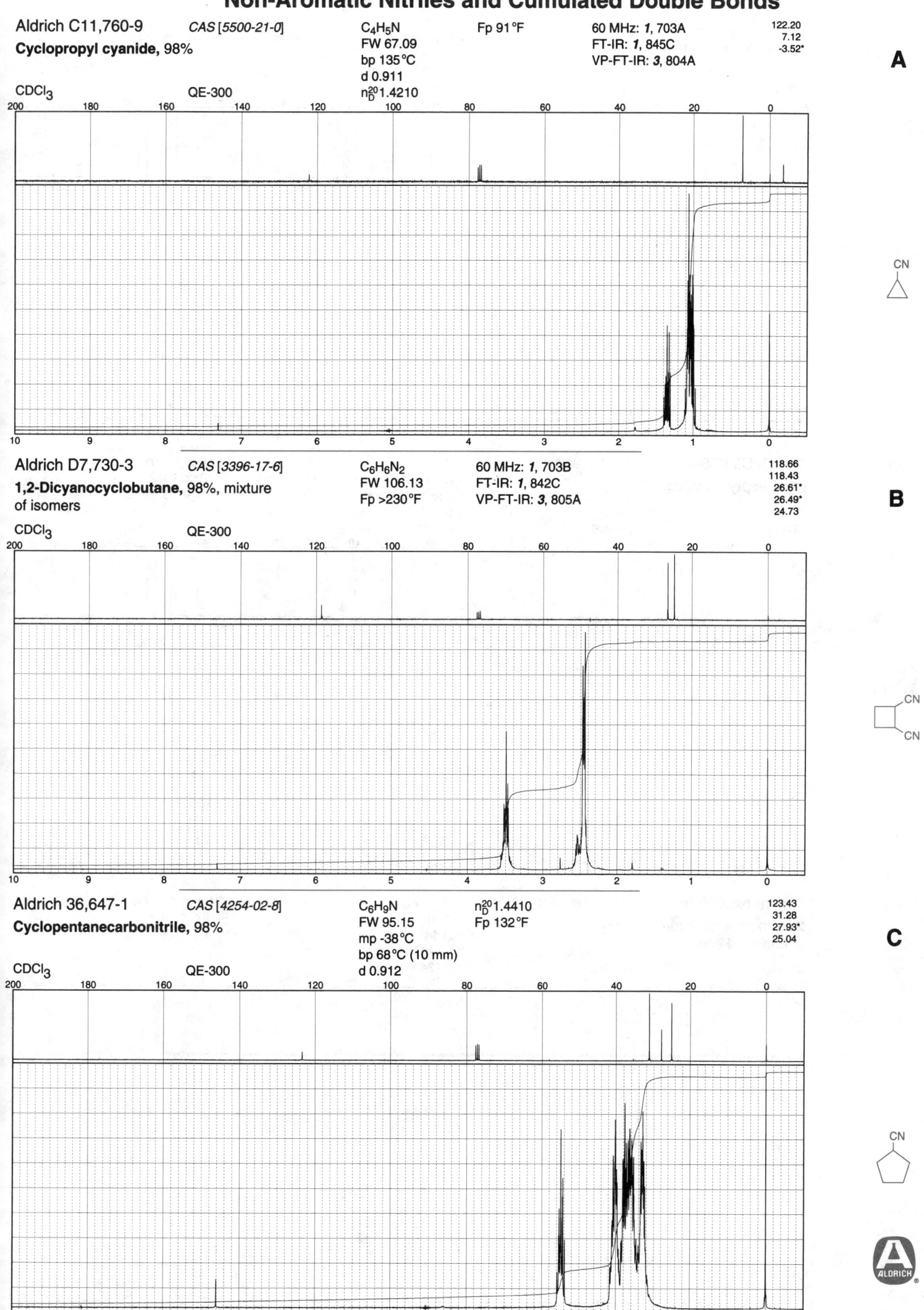

A

Aldrich C11,760-9   CAS [5500-21-0]

Cyclopropyl cyanide, 98%

C_4H_5N
FW 67.09
bp 135°C
d 0.911
n_D^{20} 1.4210

Fp 91°F

60 MHz: *1*, 703A
FT-IR: *1*, 845C
VP-FT-IR: *3*, 804A

122.20
7.12
-3.52*

$CDCl_3$ QE-300

B

Aldrich D7,730-3   CAS [3396-17-6]

1,2-Dicyanocyclobutane, 98%, mixture of isomers

$C_6H_6N_2$
FW 106.13
Fp >230°F

60 MHz: *1*, 703B
FT-IR: *1*, 842C
VP-FT-IR: *3*, 805A

118.66
118.43
26.61*
26.49*
24.73

$CDCl_3$ QE-300

C

Aldrich 36,647-1   CAS [4254-02-8]

Cyclopentanecarbonitrile, 98%

C_6H_9N
FW 95.15
mp -38°C
bp 68°C (10 mm)
d 0.912

n_D^{20} 1.4410
Fp 132°F

123.43
31.28
27.93*
25.04

$CDCl_3$ QE-300

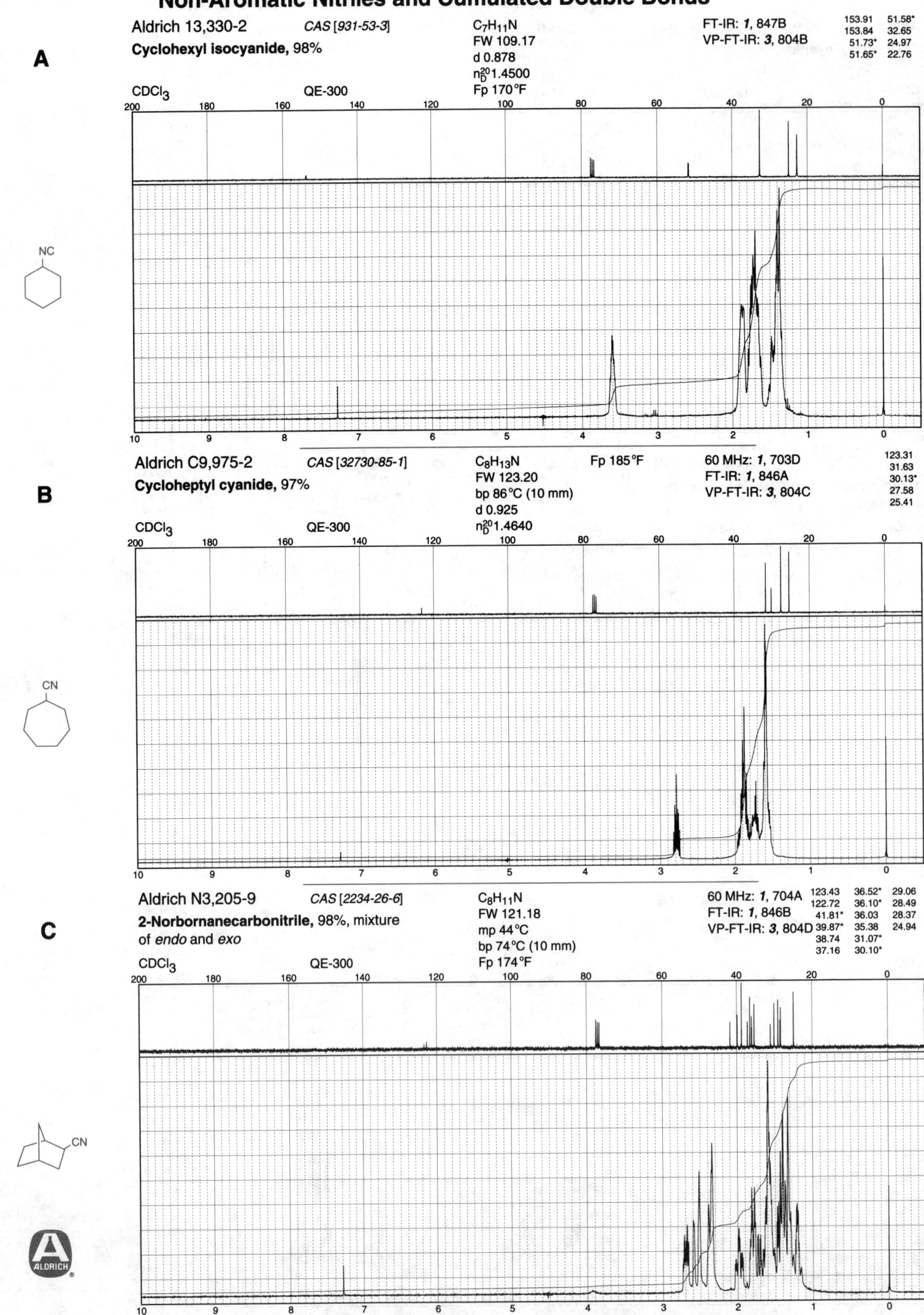

Aldrich 13,330-2   *CAS [931-53-3]*   C$_7$H$_{11}$N   FT-IR: *1*, 847B

Cyclohexyl isocyanide, 98%   FW 109.17   VP-FT-IR: *3*, 804B

d 0.878
n$_D^{20}$1.4500
Fp 170°F

153.91 51.58*
153.84 32.65
51.73* 24.97
51.65* 22.76

A

CDCl$_3$ QE-300

Aldrich C9,975-2   *CAS [32730-85-1]*   C$_8$H$_{13}$N   Fp 185°F   60 MHz: *1*, 703D

Cycloheptyl cyanide, 97%   FW 123.20   FT-IR: *1*, 846A

bp 86°C (10 mm)   VP-FT-IR: *3*, 804C
d 0.925
n$_D^{20}$1.4640

123.31
31.63
30.13*
27.58
25.41

B

CDCl$_3$ QE-300

Aldrich N3,205-9   *CAS [2234-26-6]*   C$_8$H$_{11}$N   60 MHz: *1*, 704A

2-Norbornanecarbonitrile, 98%, mixture   FW 121.18   FT-IR: *1*, 846B
of *endo* and *exo*   mp 44°C   VP-FT-IR: *3*, 804D
bp 74°C (10 mm)
Fp 174°F

123.43 36.52* 29.06
122.72 36.10* 28.49
41.81* 36.03 28.37
39.87* 35.38 24.94
38.74   31.07*
37.16   30.10*

C

CDCl$_3$ QE-300

ALDRICH

Aldrich 13,805-3   *CAS [23074-42-2]*

1-Adamantanecarbonitrile, 97%

$C_{11}H_{15}N$
FW 161.25
mp 195°C

60 MHz: *1*, 704B
FT-IR: *1*, 846C

125.12
39.95
35.76
30.20
27.13*

A

CDCl$_3$ QE-300

Aldrich 12,163-0   *CAS [22734-04-9]*

1-Cyclopenteneacetonitrile, 90%

C_7H_9N
FW 107.16
bp 124°C (100 mm)
d 0.951
n$_D^{20}$ 1.4670

Fp 157°F

60 MHz: *1*, 704C
FT-IR: *1*, 846D
VP-FT-IR: *3*, 805B

132.47
129.48*
117.53
34.73
32.48
23.49
19.68

B

CDCl$_3$ QE-300

Aldrich C10,340-3   *CAS [6975-71-9]*

1-Cyclohexenylacetonitrile, 92%

$C_8H_{11}N$
FW 121.18
bp 144°C (90 mm)
d 0.947
n$_D^{20}$ 1.4780

Fp 183°F

60 MHz: *1*, 705A
FT-IR: *1*, 847A
VP-FT-IR: *3*, 805D

127.09   25.77
126.26*   25.11
117.74   22.45
28.03   21.76

C

CDCl$_3$ QE-300

A

Aldrich 15,057-6   *CAS [95-11-4]*

5-Norbornene-2-carbonitrile, 98%, mixture of isomers

C_8H_9N
FW 119.17
mp 13°C
bp 84°C (10 mm)

n_D^{20} 1.4880
Fp 150°F
60 MHz: *1*, 705B
FT-IR: *1*, 847C
VP-FT-IR: *3*, 806B

138.79*	123.04	42.38*
138.07*	48.51	41.82*
134.00*	47.45*	32.48
132.69*	47.16	32.21
123.55	45.75*	27.20*

CDCl₃   QE-300

B

Aldrich 25,744-3   *CAS [503-20-8]*

Fluoroacetonitrile, 99%

C_2H_2FN
FW 59.04
bp 80°C
d 1.061
n_D^{20} 1.3330

Fp 7°F

FT-IR: *1*, 847D
VP-FT-IR: *3*, 806C

114.10
113.76
68.14
65.77

CDCl₃   QE-300

FCH₂CN

C

Aldrich C1,965-1   *CAS [107-14-2]*

Chloroacetonitrile, 99%

C_2H_2ClN
FW 75.50
bp 125°C
d 1.193
n_D^{20} 1.4220

Fp 118°F

60 MHz: *1*, 705D
FT-IR: *1*, 848A
VP-FT-IR: *3*, 806D

114.73
24.79

CDCl₃   QE-300

ClCH₂CN

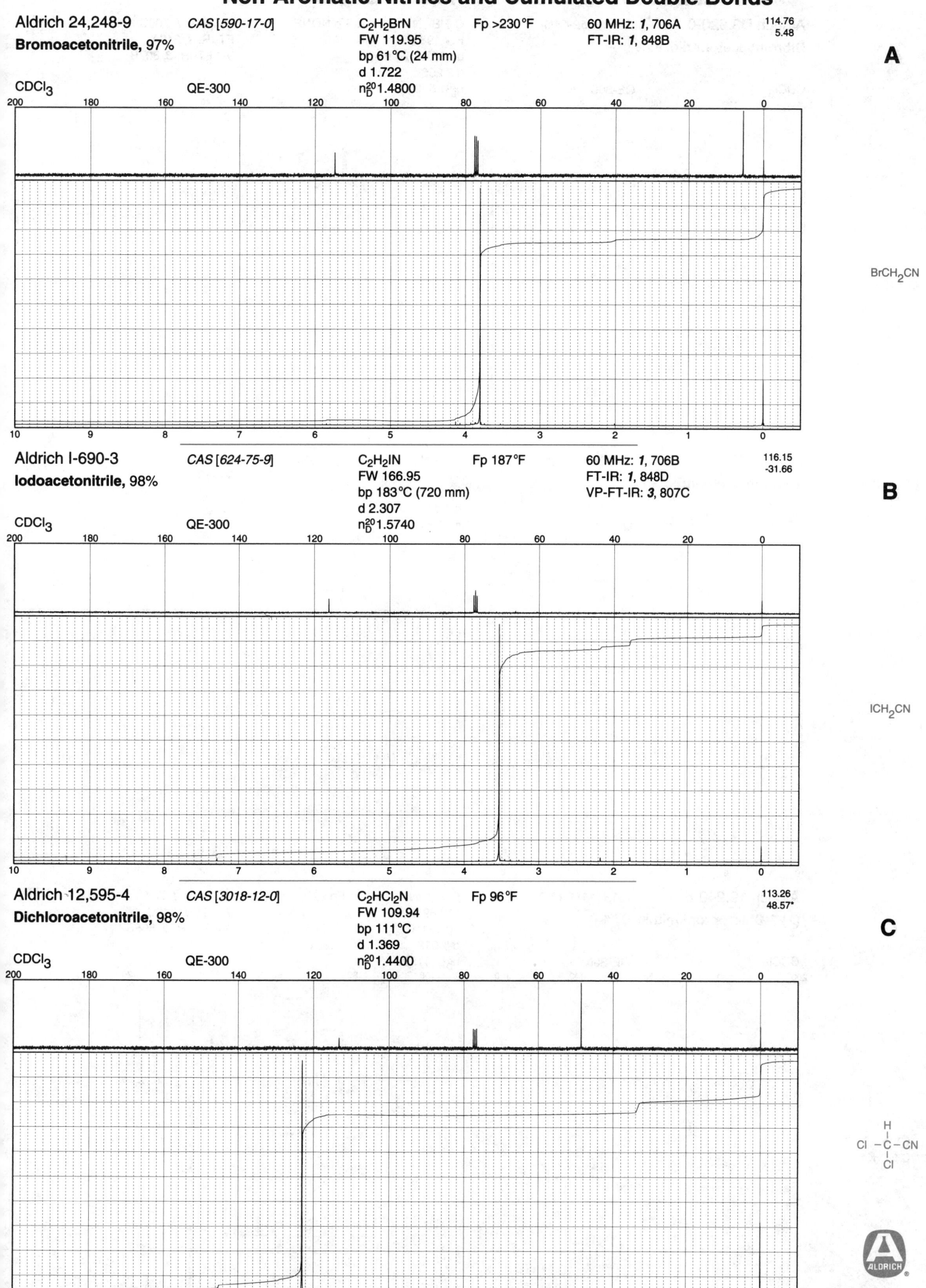

Aldrich 24,248-9

CAS [590-17-0]

C_2H_2BrN
FW 119.95
bp 61 °C (24 mm)
d 1.722
n_D^{20} 1.4800

Fp >230 °F

60 MHz: *1*, 706A
FT-IR: *1*, 848B

114.76
5.48

A

Bromoacetonitrile, 97%

$CDCl_3$

QE-300

$BrCH_2CN$

Aldrich I-690-3

CAS [624-75-9]

C_2H_2IN
FW 166.95
bp 183 °C (720 mm)
d 2.307
n_D^{20} 1.5740

Fp 187 °F

60 MHz: *1*, 706B
FT-IR: *1*, 848D
VP-FT-IR: *3*, 807C

116.15
-31.66

B

Iodoacetonitrile, 98%

$CDCl_3$

QE-300

ICH_2CN

Aldrich 12,595-4

CAS [3018-12-0]

C_2HCl_2N
FW 109.94
bp 111 °C
d 1.369
n_D^{20} 1.4400

Fp 96 °F

113.26
48.57*

C

Dichloroacetonitrile, 98%

$CDCl_3$

QE-300

A

Aldrich D3,820-0   *CAS [3252-43-5]* C_2HBr_2N Fp NONE   60 MHz: *1*, 706C   113.87
Dibromoacetonitrile, 95%   FW 198.86   FT-IR: *1*, 849A   5.35*
                                             bp 68°C (24 mm)   VP-FT-IR: *3*, 807B
                                             d 2.296
CDCl₃   QE-300 $n_D^{20}1.5390$

Br
|
Br CHCN

B

Aldrich T5,380-5   *CAS [545-06-2]* C_2Cl_3N Fp NONE   FT-IR: *1*, 848C   113.10
Trichloroacetonitrile, 98%   FW 144.39   VP-FT-IR: *3*, 807A   70.07
                                          bp 84°C
                                          d 1.440
CDCl₃   QE-300 $n_D^{20}1.4400$

Cl
|
Cl —C— CN
|
Cl

C

Aldrich 19,240-6   *CAS [1617-17-0]* C_3H_4ClN Fp 92°F   60 MHz: *1*, 707A   117.76
(±)-2-Chloropropionitrile, 97%   FW 89.53   FT-IR: *1*, 849C   37.73*
                                          bp 121°C   VP-FT-IR: *3*, 808B   23.43*
                                          d 1.012
CDCl₃   QE-300 $n_D^{20}1.4130$

Cl
|
CH₃CHCN

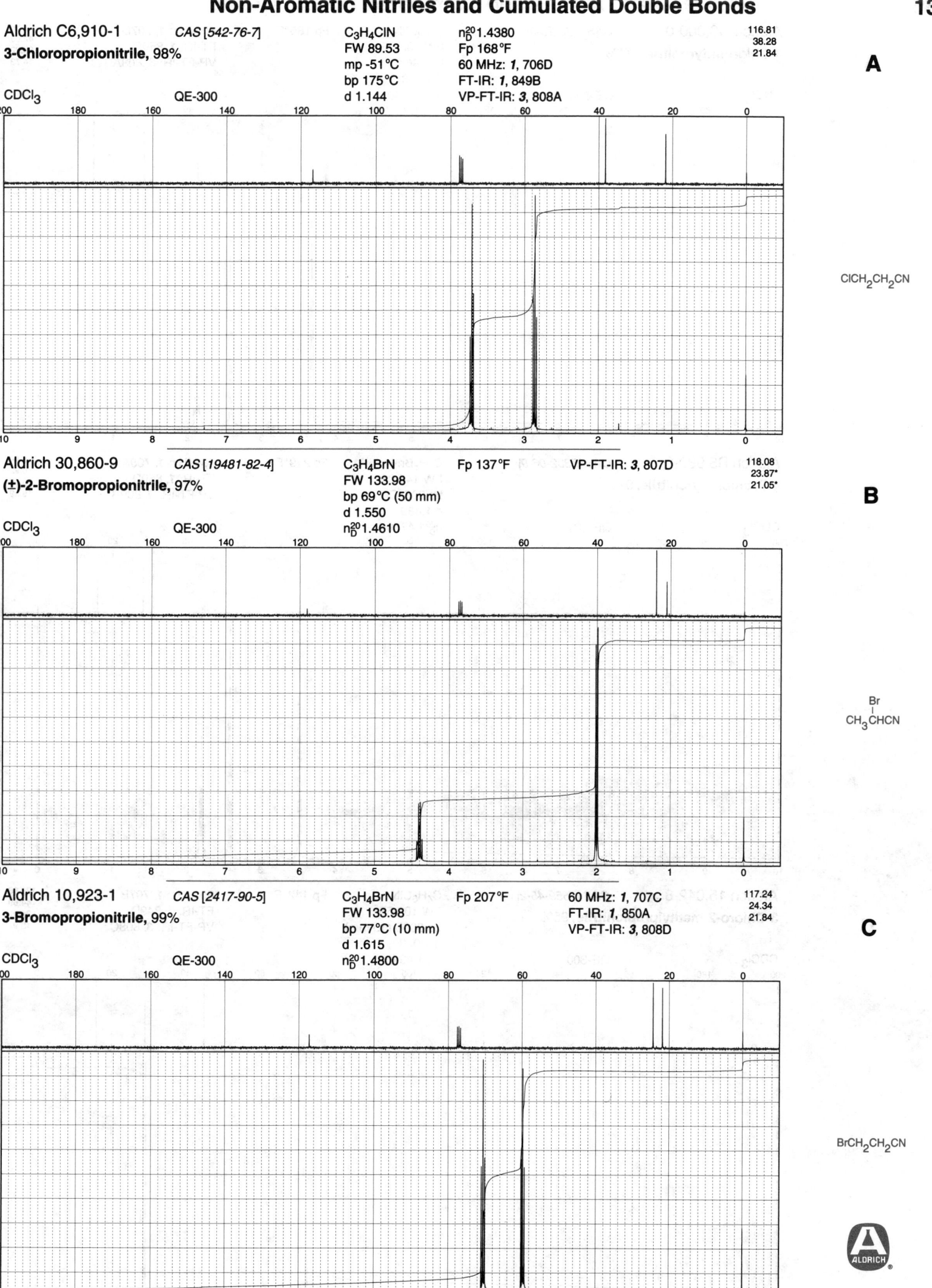

Aldrich C6,910-1 CAS [542-76-7] C₃H₄ClN n²⁰_D 1.4380 116.81
3-Chloropropionitrile, 98% FW 89.53 Fp 168°F 38.28
mp -51°C 60 MHz: 1, 706D 21.84
bp 175°C FT-IR: 1, 849B
CDCl₃ QE-300 d 1.144 VP-FT-IR: 3, 808A

A

ClCH₂CH₂CN

Aldrich 30,860-9 CAS [19481-82-4] C₃H₄BrN Fp 137°F VP-FT-IR: 3, 807D 118.08
(±)-2-Bromopropionitrile, 97% FW 133.98 23.87*
bp 69°C (50 mm) 21.05*
d 1.550
CDCl₃ QE-300 n²⁰_D 1.4610

B

Br
|
CH₃CHCN

Aldrich 10,923-1 CAS [2417-90-5] C₃H₄BrN Fp 207°F 60 MHz: 1, 707C 117.24
3-Bromopropionitrile, 99% FW 133.98 24.34
bp 77°C (10 mm) FT-IR: 1, 850A 21.84
d 1.615 VP-FT-IR: 3, 808D
CDCl₃ QE-300 n²⁰_D 1.4800

C

BrCH₂CH₂CN

A

Aldrich C3,000-0   *CAS [628-20-6]*

4-Chlorobutyronitrile, 97%

C_4H_6ClN
FW 103.55
bp 196°C
d 1.158
n_D^{20} 1.4440

Fp 185°F

60 MHz: *1*, 707D
FT-IR: *1*, 850C
VP-FT-IR: *3*, 809A

118.58
42.69
28.22
14.73

CDCl$_3$ QE-300

$ClCH_2CH_2CH_2CN$

B

Aldrich B5,980-2   *CAS [5332-06-9]*

4-Bromobutyronitrile, 97%

C_4H_6BrN
FW 148.01
bp 205°C
d 1.489
n_D^{20} 1.4780

Fp 219°F

60 MHz: *1*, 708A
FT-IR: *1*, 850D
VP-FT-IR: *3*, 809B

118.44
30.92
28.15
15.99

CDCl$_3$ QE-300

$BrCH_2CH_2CH_2CN$

C

Aldrich 15,042-8   *CAS [7659-45-2]*

3-Chloro-2-methylpropionitrile, 95%

C_4H_6ClN
FW 103.55
bp 60°C (15 mm)
d 1.078
n_D^{20} 1.4340

Fp 142°F

60 MHz: *1*, 707B
FT-IR: *1*, 849D
VP-FT-IR: *3*, 808C

120.08
44.89
28.92*
15.97*

CDCl$_3$ QE-300

$\overset{\displaystyle CH_3}{\underset{\displaystyle |}{ClCH_2CHCN}}$

ALDRICH

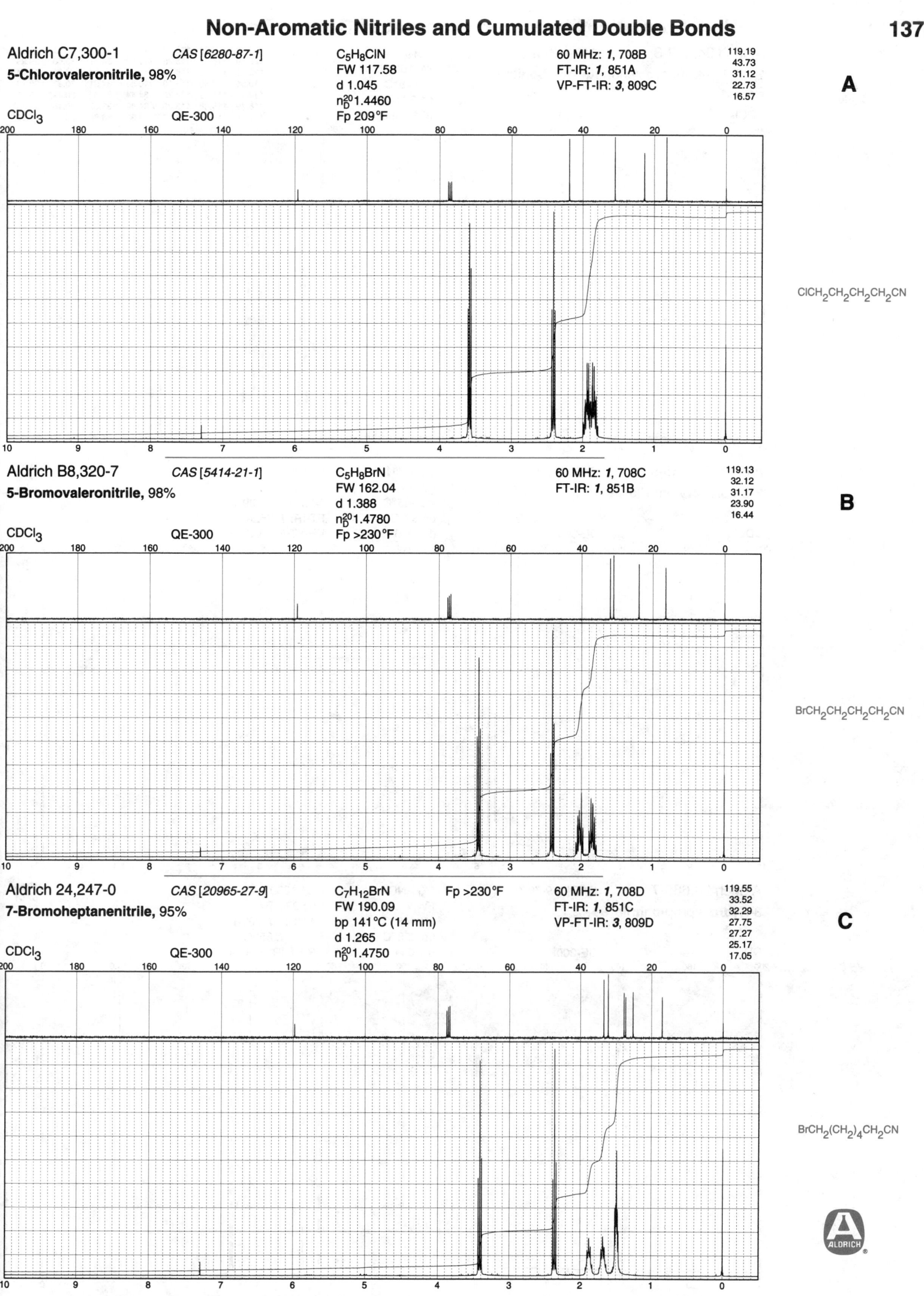

Aldrich C7,300-1   CAS [6280-87-1]   C5H8ClN   60 MHz: *1*, 708B   119.19
5-Chlorovaleronitrile, 98%   FW 117.58   FT-IR: *1*, 851A   43.73
d 1.045   VP-FT-IR: *3*, 809C   31.12
n_D^20 1.4460   22.73
CDCl3   QE-300   Fp 209°F   16.57

A

ClCH2CH2CH2CH2CN

Aldrich B8,320-7   CAS [5414-21-1]   C5H8BrN   60 MHz: *1*, 708C   119.13
5-Bromovaleronitrile, 98%   FW 162.04   FT-IR: *1*, 851B   32.12
d 1.388   31.17
n_D^20 1.4780   23.90
CDCl3   QE-300   Fp >230°F   16.44

B

BrCH2CH2CH2CH2CN

Aldrich 24,247-0   CAS [20965-27-9]   C7H12BrN   Fp >230°F   60 MHz: *1*, 708D   119.55
7-Bromoheptanenitrile, 95%   FW 190.09   FT-IR: *1*, 851C   33.52
bp 141°C (14 mm)   VP-FT-IR: *3*, 809D   32.29
d 1.265   27.75
n_D^20 1.4750   27.27
CDCl3   QE-300   25.17
17.05

C

BrCH2(CH2)4CH2CN

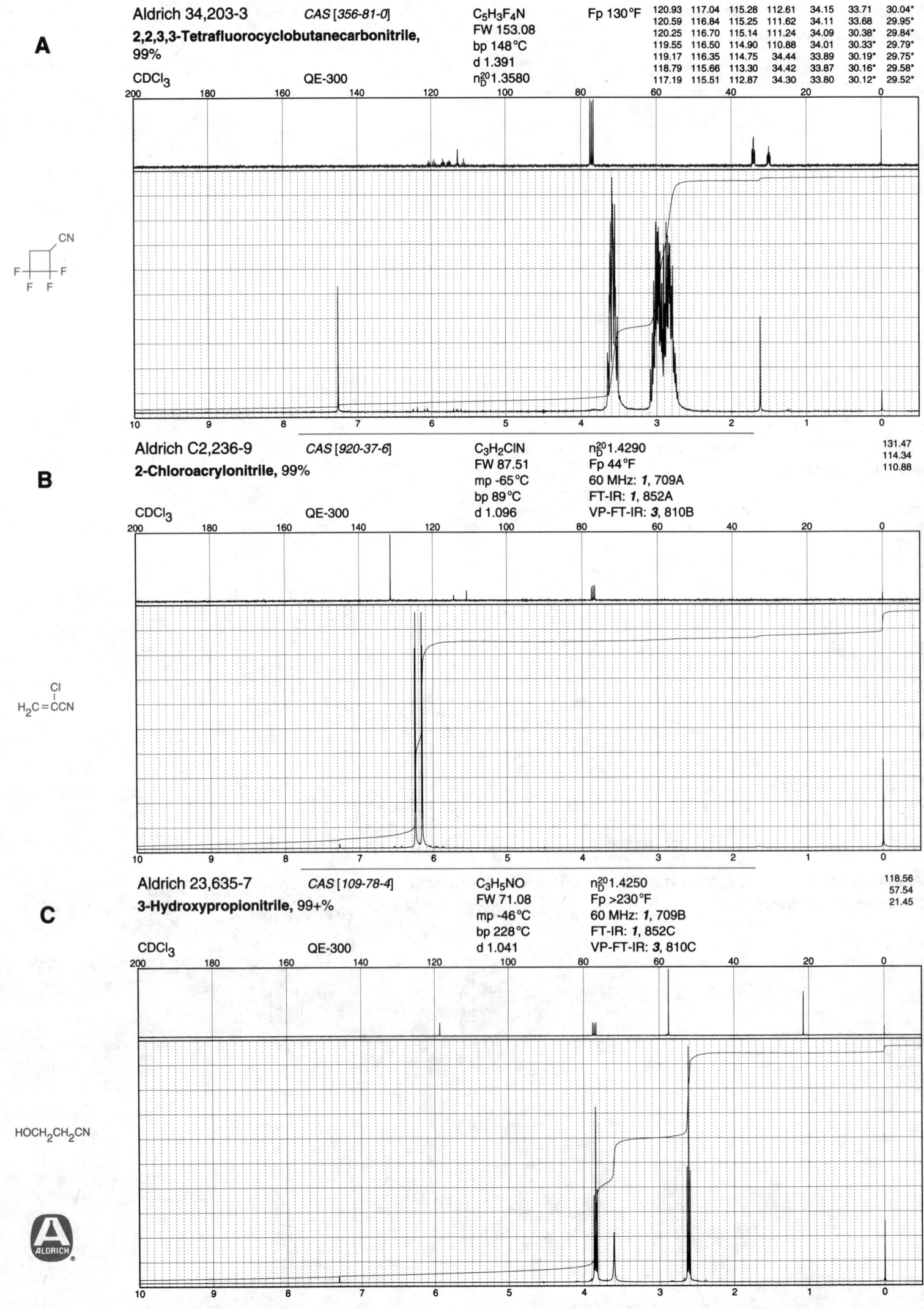

A

Aldrich 34,203-3   *CAS [356-81-0]*   C$_5$H$_3$F$_4$N   Fp 130°F
2,2,3,3-Tetrafluorocyclobutanecarbonitrile, 99%
FW 153.08
bp 148°C
d 1.391
n$_D^{20}$1.3580

CDCl$_3$ QE-300

120.93	117.04	115.28	112.61	34.15	33.71	30.04*
120.59	116.84	115.25	111.62	34.11	33.68	29.95*
120.25	116.70	115.14	111.24	34.09	30.38*	29.84*
119.55	116.50	114.90	110.88	34.01	30.33*	29.79*
119.17	116.35	114.75	34.44	33.89	30.19*	29.75*
118.79	115.66	113.30	34.42	33.87	30.16*	29.58*
117.19	115.51	112.87	34.30	33.80	30.12*	29.52*

B

Aldrich C2,236-9   *CAS [920-37-6]*   C$_3$H$_2$ClN   n$_D^{20}$1.4290
2-Chloroacrylonitrile, 99%
FW 87.51   Fp 44°F
mp -65°C   60 MHz: *1*, 709A
bp 89°C   FT-IR: *1*, 852A
d 1.096   VP-FT-IR: *3*, 810B

CDCl$_3$ QE-300

131.47
114.34
110.88

C

Aldrich 23,635-7   *CAS [109-78-4]*   C$_3$H$_5$NO   n$_D^{20}$1.4250
3-Hydroxypropionitrile, 99+%
FW 71.08   Fp >230°F
mp -46°C   60 MHz: *1*, 709B
bp 228°C   FT-IR: *1*, 852C
d 1.041   VP-FT-IR: *3*, 810C

CDCl$_3$ QE-300

118.56
57.54
21.45

Aldrich M900-9   CAS [1738-36-9] C_3H_5NO Fp 89°F   60 MHz: **1**, 709D   115.88
Methoxyacetonitrile, 99+%   FW 71.08   FT-IR: **1**, 853A   59.05*
bp 119°C (731 mm)   VP-FT-IR: **3**, 810D   57.86
d 0.956

CDCl$_3$ QE-300 n_D^{20} 1.3810

A

CH_3OCH_2CN

Aldrich 16,082-2   CAS [35120-10-6] C_3H_5NS Fp 154°F   60 MHz: **1**, 695B   116.34
(Methylthio)acetonitrile, 99%   FW 87.14   FT-IR: **1**, 853B   19.15
bp 62°C (15 mm)   VP-FT-IR: **3**, 811A   15.88*
d 1.039

CDCl$_3$ QE-300 n_D^{20} 1.4830

B

CH_3SCH_2CN

Aldrich 11,762-5   CAS [110-67-8] C_4H_7NO Fp 142°F   VP-FT-IR: **3**, 811B   117.97
3-Methoxypropionitrile, 98%   FW 85.11   67.01
bp 165°C   58.86*
d 0.937   18.71

CDCl$_3$ QE-300 n_D^{20} 1.4030

C

$CH_3OCH_2CH_2CN$

ALDRICH

1373

A

Aldrich E780-0   *CAS [2141-62-0]* C_5H_9NO Fp 147°F   FT-IR: *1*, 853C

3-Ethoxypropionitrile, 99%   FW 99.14   VP-FT-IR: *3*, 811C

bp 172°C
d 0.911
n_D^{20}1.4060

CDCl₃   QE-300

118.01
66.69
65.01
18.92
14.97*

$CH_3CH_2OCH_2CH_2CN$

B

Aldrich 37,861-5   *CAS [2032-34-0]* $C_7H_{13}NO_2$ Fp 183°F

3,3-Diethoxypropionitrile, 95%   FW 143.19

bp 92°C (11 mm)
d 0.954
n_D^{20}1.4150

CDCl₃   QE-300

116.39
98.04*
62.79
24.01
15.09*

CH_3CH_2O
       CHCH₂CN
CH_3CH_2O

C

Aldrich C9,120-4   *CAS [1656-48-0]* $C_6H_8N_2O$ Fp >230°F   60 MHz: *1*, 710A

2-Cyanoethyl ether, 98%   FW 124.14   FT-IR: *1*, 853D

bp 111°C   VP-FT-IR: *3*, 811D
d 1.043
n_D^{20}1.4400

CDCl₃   QE-300

117.69
65.75
18.80

$NCCH_2CH_2OCH_2CH_2CN$

ALDRICH

Aldrich 36,236-0 CAS *[60838-50-8]*

3-Methoxyacrylonitrile, 98%, mixture of isomers

CDCl₃   QE-300

C_4H_5NO
FW 83.09
d 0.990
n_D^{20} 1.4550
Fp 169°F

165.23*	74.84*
164.32*	74.27*
117.40	61.77*
115.27	57.66*

A

$CH_3OCH=CHCN$

Aldrich 26,369-9 CAS *[61310-53-0]*

3-Ethoxyacrylonitrile, 97%, mixture of *cis* and *trans*

CDCl₃   QE-300

C_5H_7NO
FW 97.12
bp 91°C (19 mm)
d 0.944
n_D^{20} 1.4540

Fp 179°F

FT-IR: *1*, 854A
VP-FT-IR: *3*, 812A

164.47*	73.93*
163.20*	70.84
117.71	67.35
115.54	15.15*
75.05*	14.30*

B

$CH_3CH_2OCH=CHCN$

Aldrich E620-0 CAS *[123-06-8]*

Ethoxymethylenemalononitrile, 98%

CDCl₃   QE-300

$C_6H_6N_2O$
FW 122.13
mp 65°C
bp 160°C (12 mm)

60 MHz: *1*, 710C
FT-IR: *1*, 854B
VP-FT-IR: *3*, 812B

| 173.99* |
| 112.06 |
| 109.86 |
| 75.05 |
| 67.05 |
| 15.14* |

C

$$CH_3CH_2OCH=C\overset{\displaystyle CN}{\underset{\displaystyle CN}{|}}$$

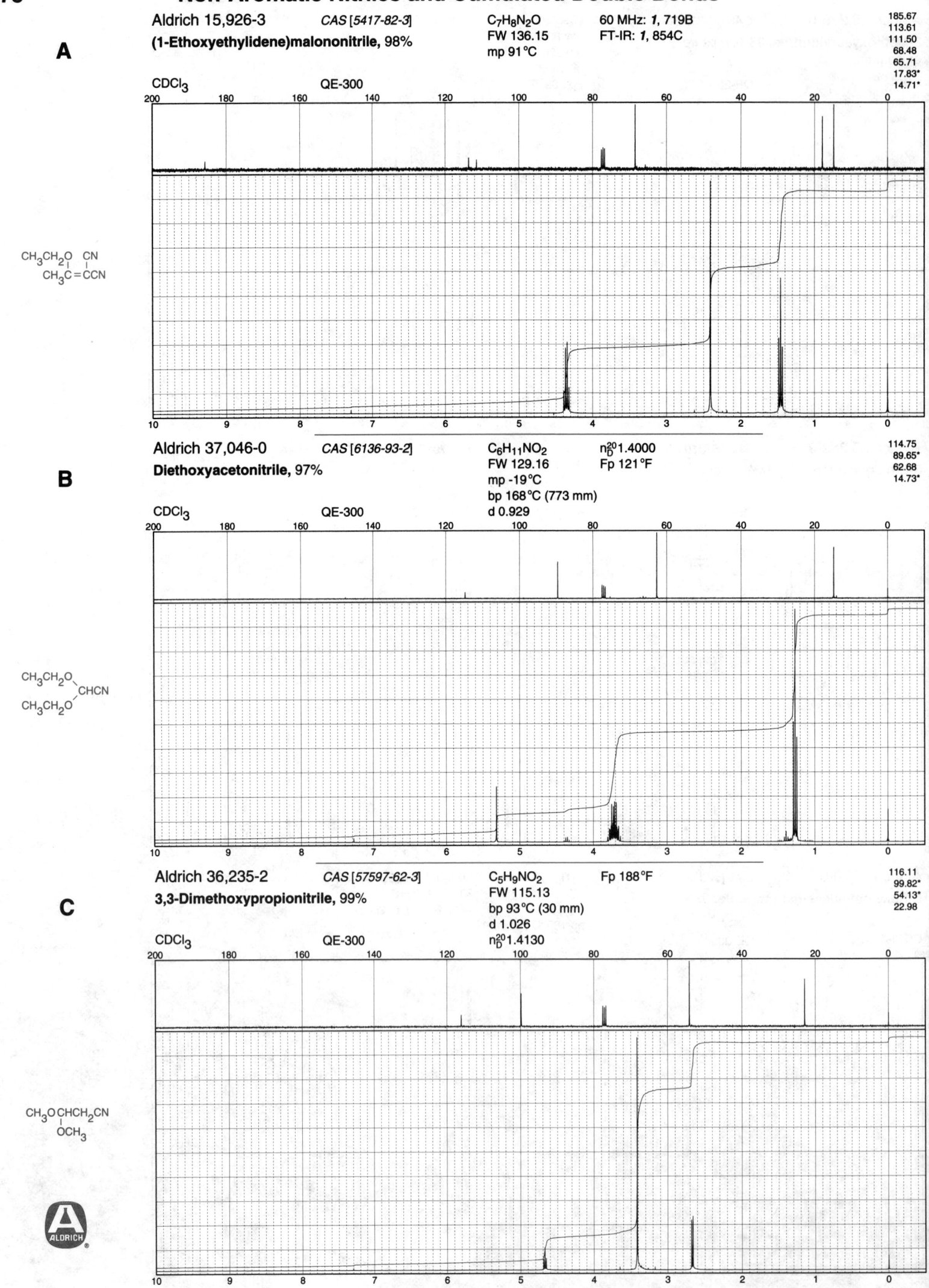

A

Aldrich 15,926-3   *CAS [5417-82-3]* $C_7H_8N_2O$ 60 MHz: *1*, 719B

(1-Ethoxyethylidene)malononitrile, 98%   FW 136.15   FT-IR: *1*, 854C

mp 91°C

185.67
113.61
111.50
68.48
65.71
17.83*
14.71*

CDCl₃   QE-300

B

Aldrich 37,046-0   *CAS [6136-93-2]* $C_6H_{11}NO_2$ n_D^{20}1.4000

Diethoxyacetonitrile, 97%   FW 129.16   Fp 121°F

mp -19°C

bp 168°C (773 mm)

d 0.929

114.75
89.65*
62.68
14.73*

CDCl₃   QE-300

C

Aldrich 36,235-2   *CAS [57597-62-3]* $C_5H_9NO_2$ Fp 188°F

3,3-Dimethoxypropionitrile, 99%   FW 115.13

bp 93°C (30 mm)

d 1.026

n_D^{20}1.4130

116.11
99.82*
54.13*
22.98

CDCl₃   QE-300

Aldrich 15,958-1 *CAS [14618-78-1]*
3-Cyanopropionaldehyde dimethyl acetal, 98%
CDCl₃   QE-300

C₆H₁₁NO₂   Fp 193°F
FW 129.16
bp 91°C (14 mm)
d 0.992
n²⁰_D 1.4190

119.37
102.67*
53.78*
28.53
12.42

A

NCCH₂CH₂CH HOCH₃ (OCH₃)

Aldrich 16,517-4 *CAS [18381-45-8]*
3-Cyanopropionaldehyde diethyl acetal, 98%
CDCl₃   QE-300

C₈H₁₅NO₂   Fp 195°F
FW 157.21
bp 105°C (10 mm)
d 0.937
n²⁰_D 1.4190

60 MHz: *1*, 710B
FT-IR: *1*, 855A
VP-FT-IR: *3*, 812C

119.48
100.84*
62.34
29.64
15.25*
12.50

B

NCCH₂CH₂CHOCH₂CH₃ (OCH₂CH₃)

Aldrich 31,119-7 *CAS [3189-43-3]*
Tetracyanoethylene oxide
ACETONE-d₆   QE-300

C₆N₄O
FW 144.09

108.87
47.84

C

NC, CN / NC, CN (O)

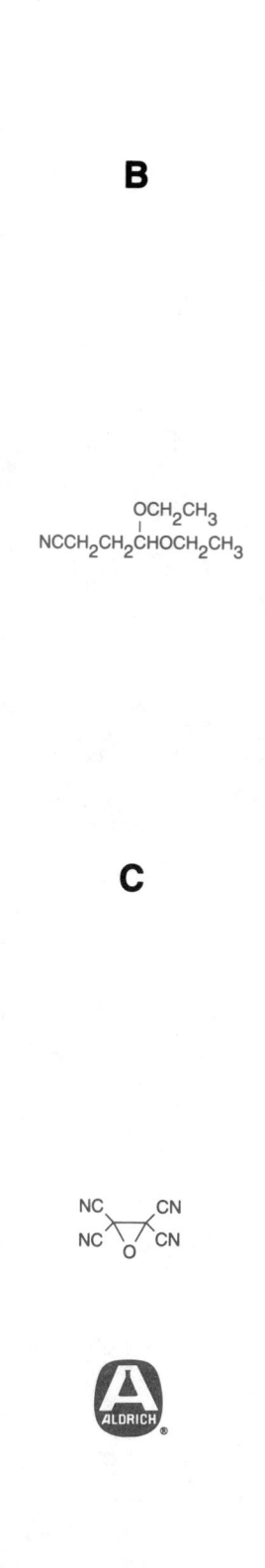

118.94

Aldrich 17,832-2   CAS [1934-75-4] C_2HN_3
FW 89.03
FT-IR: *1*, 855C

Sodium dicyanamide, 96%

A

DMSO-d_6 QE-300

$NaN(CN)_2$

Aldrich 21,380-2   CAS [1467-79-4] $C_3H_6N_2$ Fp 137°F   60 MHz: *1*, 721D   119.32
                                       FW 70.09   FT-IR: *1*, 855D   40.49*
Dimethylcyanamide, 99%   bp 162°C   VP-FT-IR: *3*, 812D
                                       d 0.867
                                       n_D^{20}1.4100

B

CDCl$_3$ QE-300

$$\begin{matrix} CH_3 \\ \quad\quad N-CN \\ CH_3 \end{matrix}$$

Aldrich 21,381-0   CAS [617-83-4] $C_5H_{10}N_2$ Fp 157°F   60 MHz: *1*, 722A   117.29
                                      FW 98.15   FT-IR: *1*, 856A   45.91
Diethylcyanamide, 99%   bp 187°C   VP-FT-IR: *3*, 813A   12.88*
                                      d 0.846
                                      n_D^{20}1.4230

C

CDCl$_3$ QE-300

$$\begin{matrix} CH_3CH_2 \\ \quad\quad\quad N-CN \\ CH_3CH_2 \end{matrix}$$

Aldrich 21,382-9   CAS [3085-76-5]   C₇H₁₄N₂   Fp 174°F   60 MHz: *1*, 729C   114.81
Diisopropylcyanamide, 97+%   FW 126.20   FT-IR: *1*, 856B   50.96*
  bp 94°C (25 mm)   VP-FT-IR: *3*, 813B   21.41*
CDCl₃   QE-300   d 0.839
  n²⁰_D 1.4270

A

Aldrich 21,853-7   CAS [1530-88-7]   C₅H₈N₂   Fp 225°F   60 MHz: *1*, 722B   117.82
1-Pyrrolidinecarbonitrile, 97%   FW 96.13   FT-IR: *1*, 856C   50.50
  bp 76°C (1 mm)   VP-FT-IR: *3*, 813C   25.67
CDCl₃   QE-300   d 0.954
  n²⁰_D 1.4690

B

Aldrich 21,383-7   CAS [1530-87-6]   C₆H₁₀N₂   Fp 207°F   60 MHz: *1*, 722C   118.56
1-Piperidinecarbonitrile, 99%   FW 110.16   FT-IR: *1*, 856D   50.20
  bp 102°C (10 mm)   VP-FT-IR: *3*, 813D   24.59
CDCl₃   QE-300   d 0.951   23.02
  n²⁰_D 1.4700

C

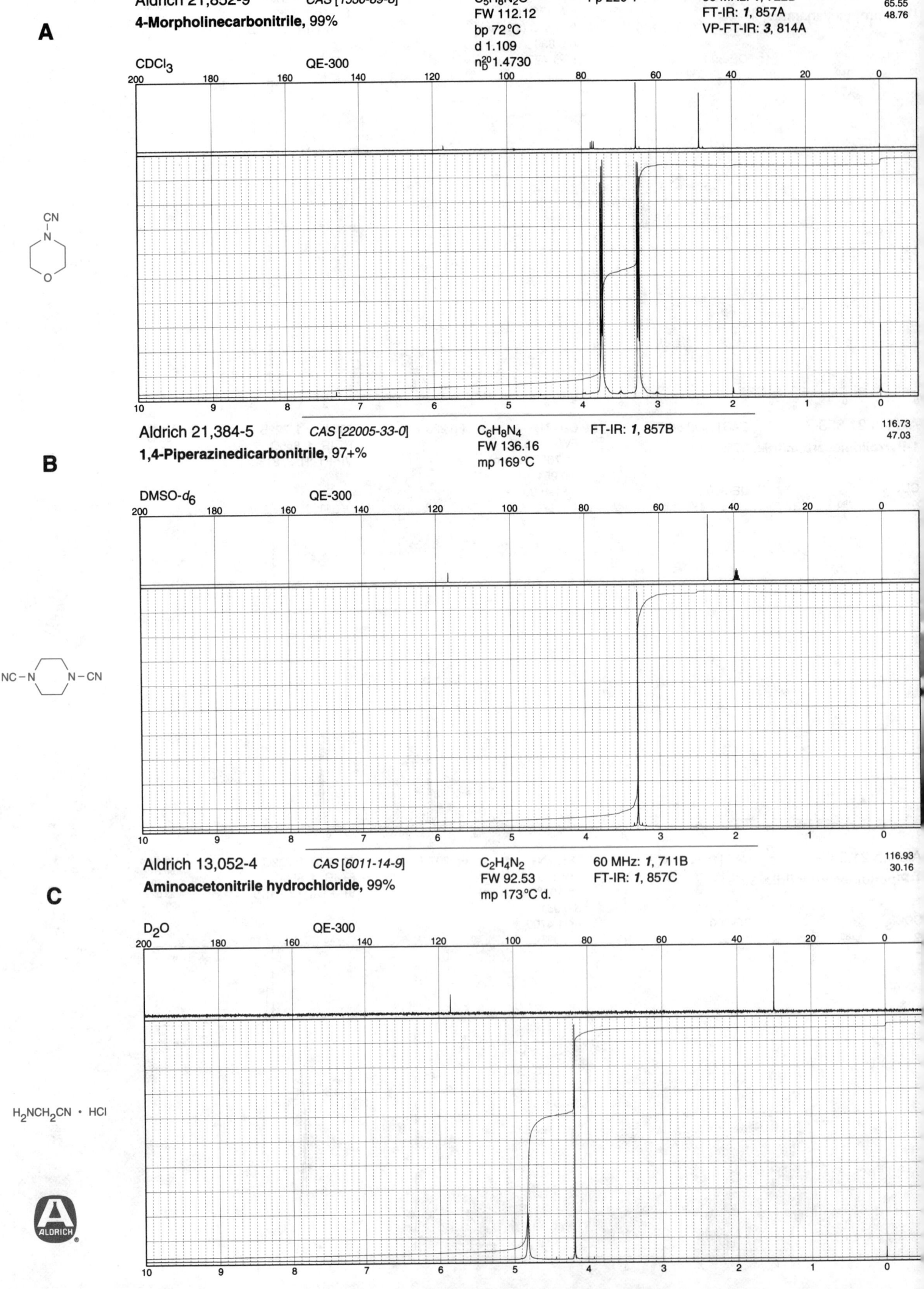

A

Aldrich 21,852-9   *CAS [1530-89-8]*   C5H8N2O   Fp 220°F   60 MHz: *1*, 722D   117.19
65.55
48.76
4-Morpholinecarbonitrile, 99%   FW 112.12   FT-IR: *1*, 857A
bp 72°C   VP-FT-IR: *3*, 814A
d 1.109
n_D^{20}1.4730

CDCl3   QE-300

B

Aldrich 21,384-5   *CAS [22005-33-0]*   C6H8N4   FT-IR: *1*, 857B   116.73
47.03
1,4-Piperazinedicarbonitrile, 97+%   FW 136.16
mp 169°C

DMSO-d_6 QE-300

C

Aldrich 13,052-4   *CAS [6011-14-9]*   C2H4N2   60 MHz: *1*, 711B   116.93
30.16
Aminoacetonitrile hydrochloride, 99%   FW 92.53   FT-IR: *1*, 857C
mp 173°C d.

D2O   QE-300

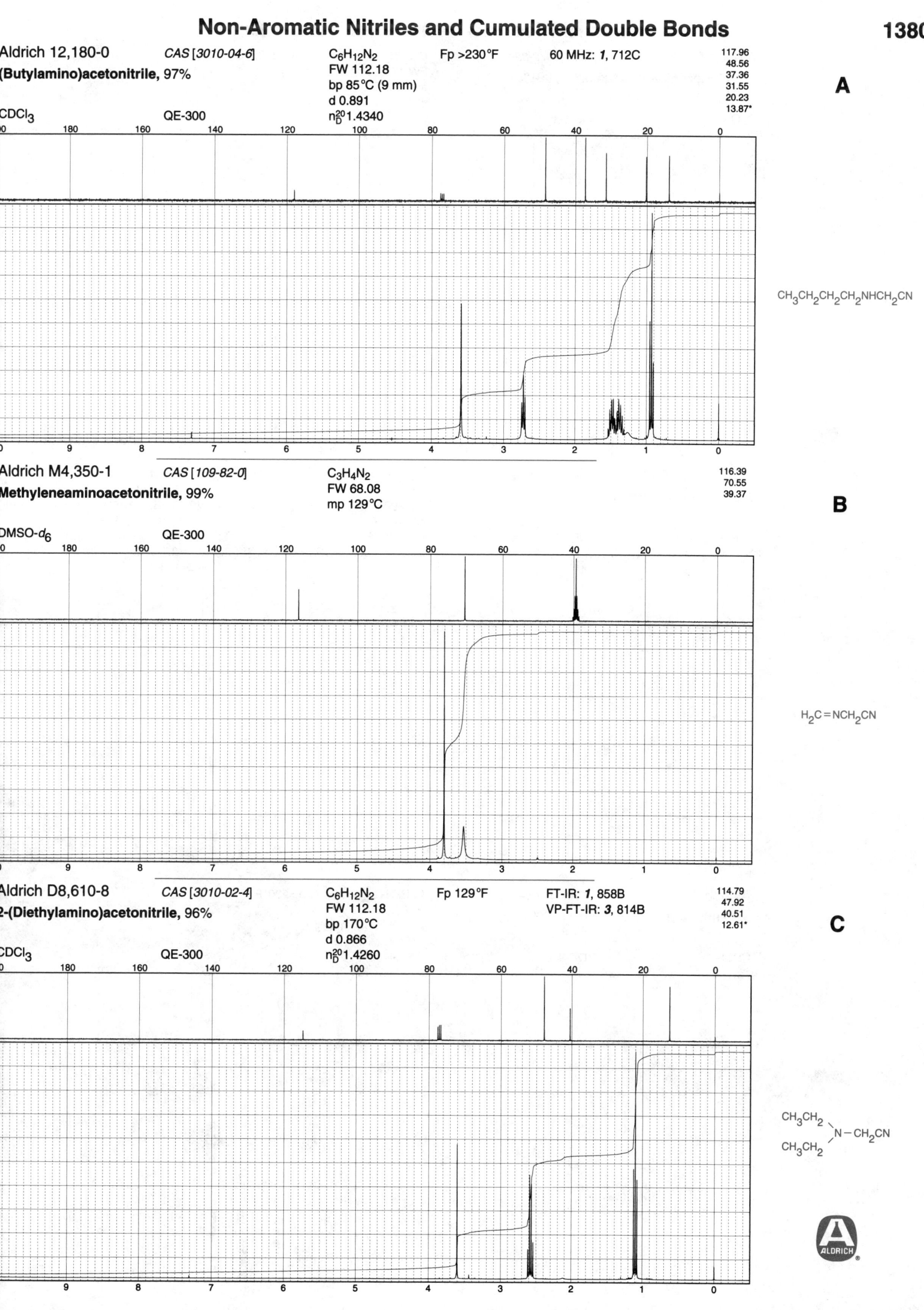

Aldrich 12,180-0   CAS [3010-04-6]   C$_6$H$_{12}$N$_2$ Fp >230°F   60 MHz: 1, 712C   117.96
(Butylamino)acetonitrile, 97%   FW 112.18   48.56
bp 85°C (9 mm)   37.36
d 0.891   31.55
n$_D^{20}$1.4340   20.23
13.87*

A

CDCl$_3$ QE-300

CH$_3$CH$_2$CH$_2$CH$_2$NHCH$_2$CN

Aldrich M4,350-1   CAS [109-82-0]   C$_3$H$_4$N$_2$ 116.39
Methyleneaminoacetonitrile, 99%   FW 68.08   70.55
mp 129°C   39.37

B

DMSO-d$_6$ QE-300

H$_2$C=NCH$_2$CN

Aldrich D8,610-8   CAS [3010-02-4]   C$_6$H$_{12}$N$_2$ Fp 129°F   FT-IR: 1, 858B   114.79
2-(Diethylamino)acetonitrile, 96%   FW 112.18   VP-FT-IR: 3, 814B   47.92
bp 170°C   40.51
d 0.866   12.61*
n$_D^{20}$1.4260

C

CDCl$_3$ QE-300

CH$_3$CH$_2$
CH$_3$CH$_2$ N—CH$_2$CN

ALDRICH

A

Aldrich A7,642-7   *CAS [2079-89-2]*   C₃H₆N₂   FT-IR: *1*, 858D

3-Aminopropionitrile fumarate, 99+%   FW 128.13
mp 177°C d.

177.28
138.11*
120.46
38.02
18.60

D₂O   QE-300

H₂NCH₂CH₂CN · ¹/₂ (C₄H₄O₄)

Aldrich M2,760-3   *CAS [693-05-0]*   C₄H₈N₂   60 MHz: *1*, 712D

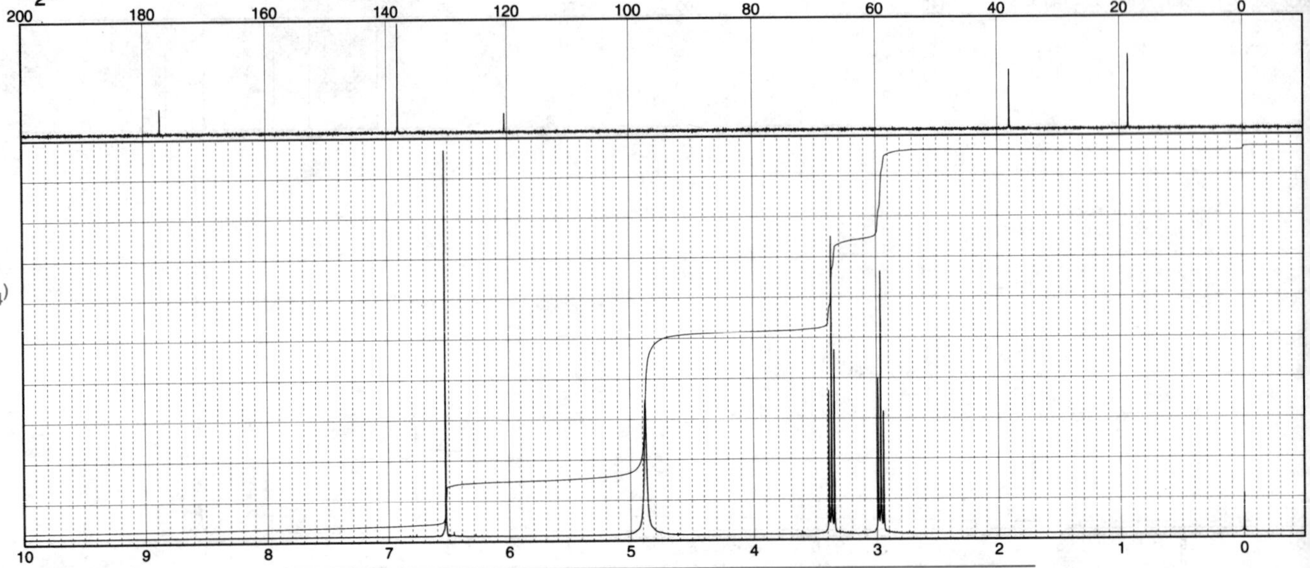

B

N-Methyl-β-alaninenitrile, 98%   FW 84.12
d 0.899
n$_D^{20}$1.4310
Fp >230°F

FT-IR: *1*, 859B
VP-FT-IR: *3*, 815B

118.78
47.00
35.72*
18.31

CDCl₃   QE-300

CH₃NHCH₂CH₂CN

Aldrich 31,730-6   *CAS [111-94-4]*   C₆H₉N₃   Fp >230°F

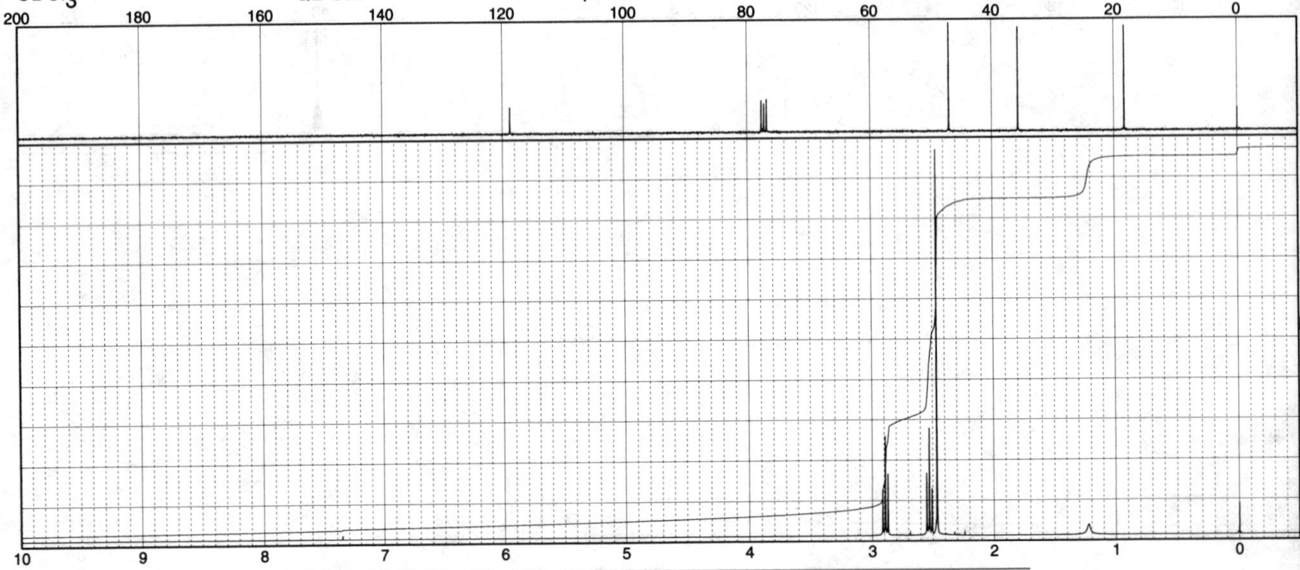

C

3,3'-Iminodipropionitrile, tech., 90%   FW 123.16
bp 205°C (25 mm)
d 1.020
n$_D^{20}$1.4700

118.73
44.41
18.81

CDCl₃   QE-300

HN⟨CH₂CH₂CN / CH₂CH₂CN⟩

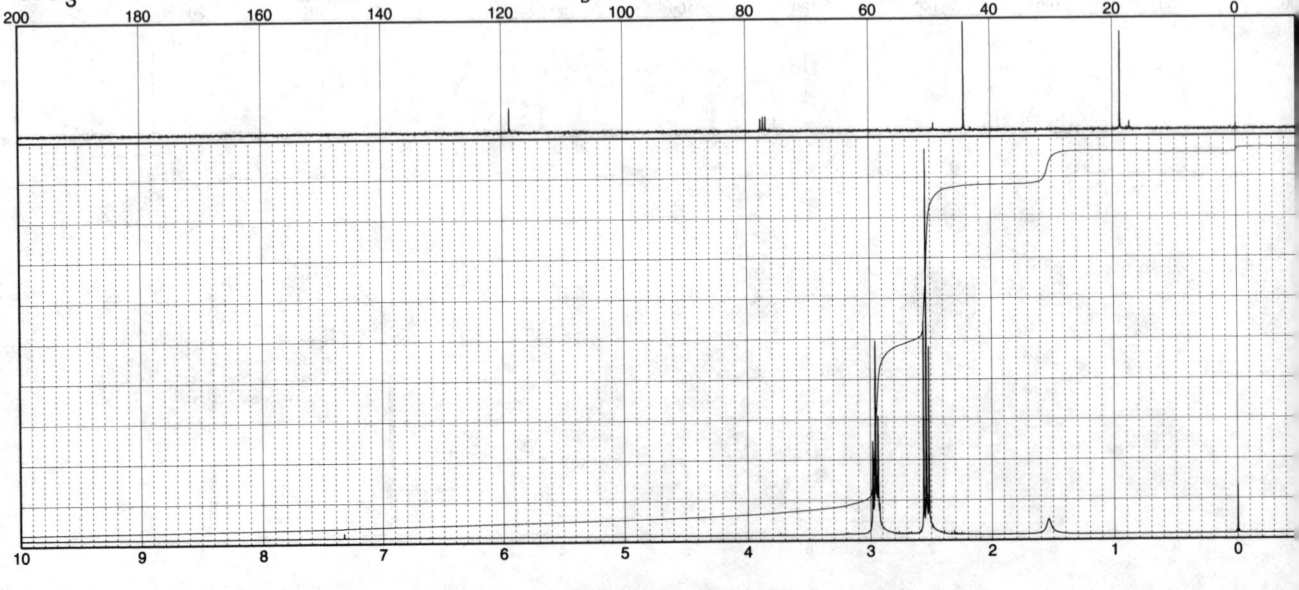

Aldrich 23,448-6   *CAS [1738-25-6]* $C_5H_{10}N_2$ n_D^{20} 1.4260

3-(Dimethylamino)propionitrile, 98%

FW 98.15   Fp 145°F
mp -43°C   60 MHz: *1*, 713A
bp 171°C (750 mm)   FT-IR: *1*, 859C
d 0.870   VP-FT-IR: *3*, 815C

118.75
54.56
44.90*
16.24

A

CDCl₃   QE-300

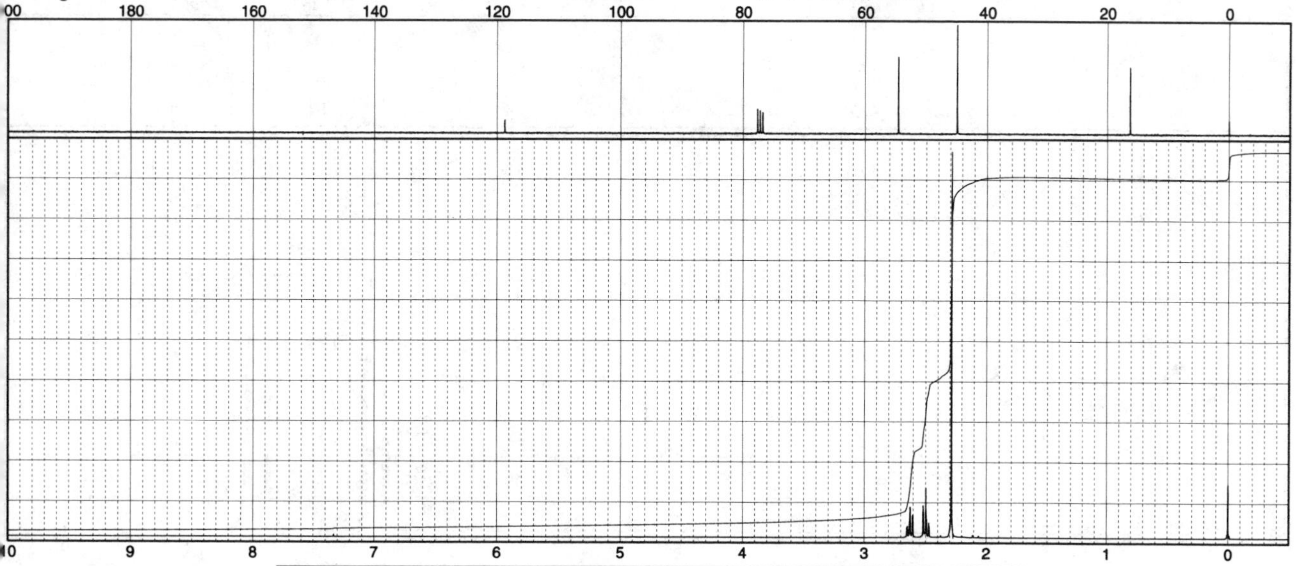

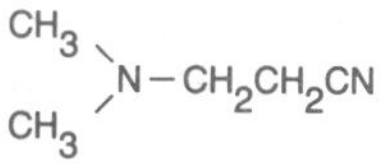

Aldrich 19,602-9   *CAS [3088-41-3]* $C_8H_{14}N_2$ Fp 217°F

1-Piperidinepropionitrile, 99%

FW 138.21   60 MHz: *1*, 713B
bp 111°C (16 mm)   FT-IR: *1*, 859A
d 0.933   VP-FT-IR: *3*, 815A
n_D^{20} 1.4700

119.01
54.12
53.94
25.83
24.08
15.66

B

CDCl₃   QE-300

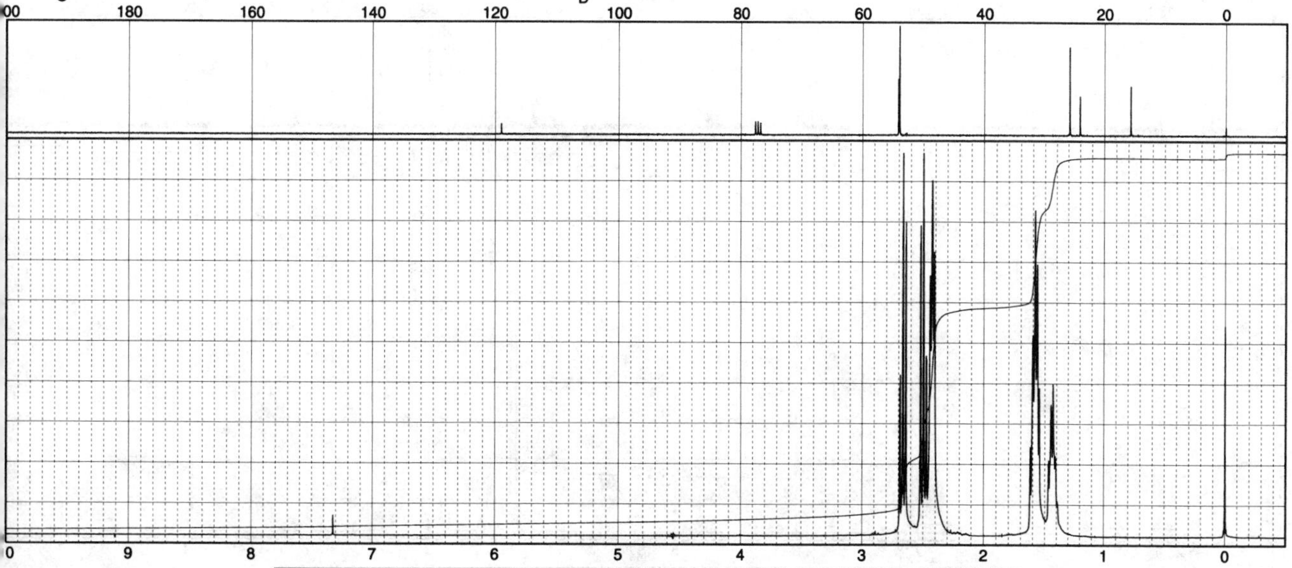

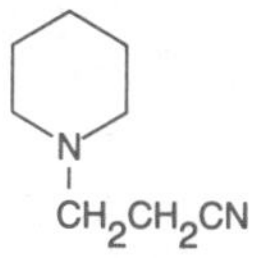

Aldrich 32,512-0   *CAS [4542-47-6]* $C_7H_{12}N_2O$ n_D^{20} 1.4720

4-Morpholinepropionitrile, 98%

FW 140.19   Fp >230°F
mp 21°C   VP-FT-IR: *3*, 816A
bp 121°C (2 mm)
d 1.037

118.73
66.61
53.50
52.94
15.58

C

CDCl₃   QE-300

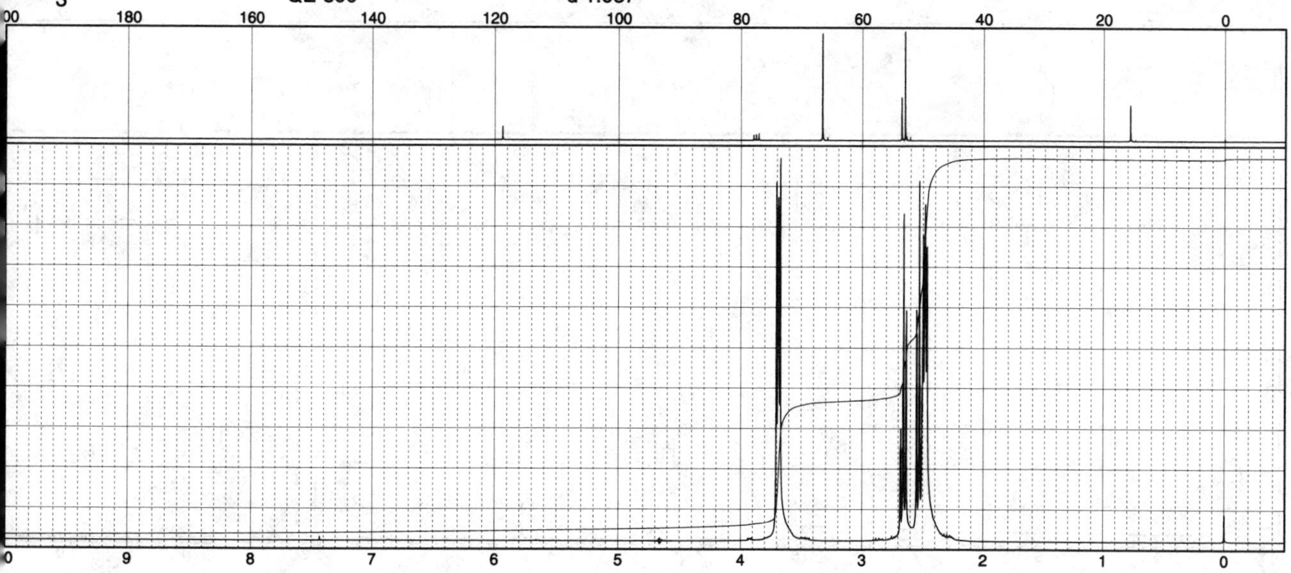

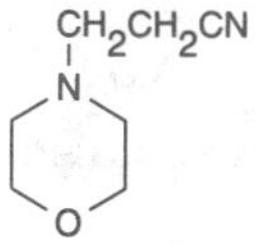

A

Aldrich 25,968-3   *CAS [2432-74-8]* $C_6H_{12}N_2$ Fp >230°F   FT-IR: *1*, 860A

6-Aminocapronitrile, 99%   FW 112.18   VP-FT-IR: *3*, 816B

121.12
42.19
33.01
27.02
26.32
17.23

CD₃OD   QE-300   bp 128°C (27 mm)
d 0.907
n_D^{20}1.4480

$H_2NCH_2CH_2CH_2CH_2CH_2CN$

B

Aldrich 14,770-2   *CAS [35520-41-3]* $C_5H_8N_2$ Fp >230°F   60 MHz: *1*, 714A

trans-3-Dimethylaminoacrylonitrile,   FW 96.13   FT-IR: *1*, 860C
95%   bp 78°C
d 0.878

154.40*
122.25
59.96*
39.84*

CDCl₃   QE-300 n_D^{20}1.5330

$(CH_3)_2N$ and H, H and CN (structure)

C

Aldrich 11,764-1   *CAS [1118-61-2]* $C_4H_6N_2$

3-Aminocrotononitrile, 96%, mixture   FW 82.11
of *cis* and *trans*   60 MHz: *1*, 714C
FT-IR: *1*, 861A

162.12   63.80*
161.52   61.85*
121.91   21.08*
119.85   19.42*

CDCl₃   QE-300

CH_3 $C=CHCN$ with NH_2 (structure)

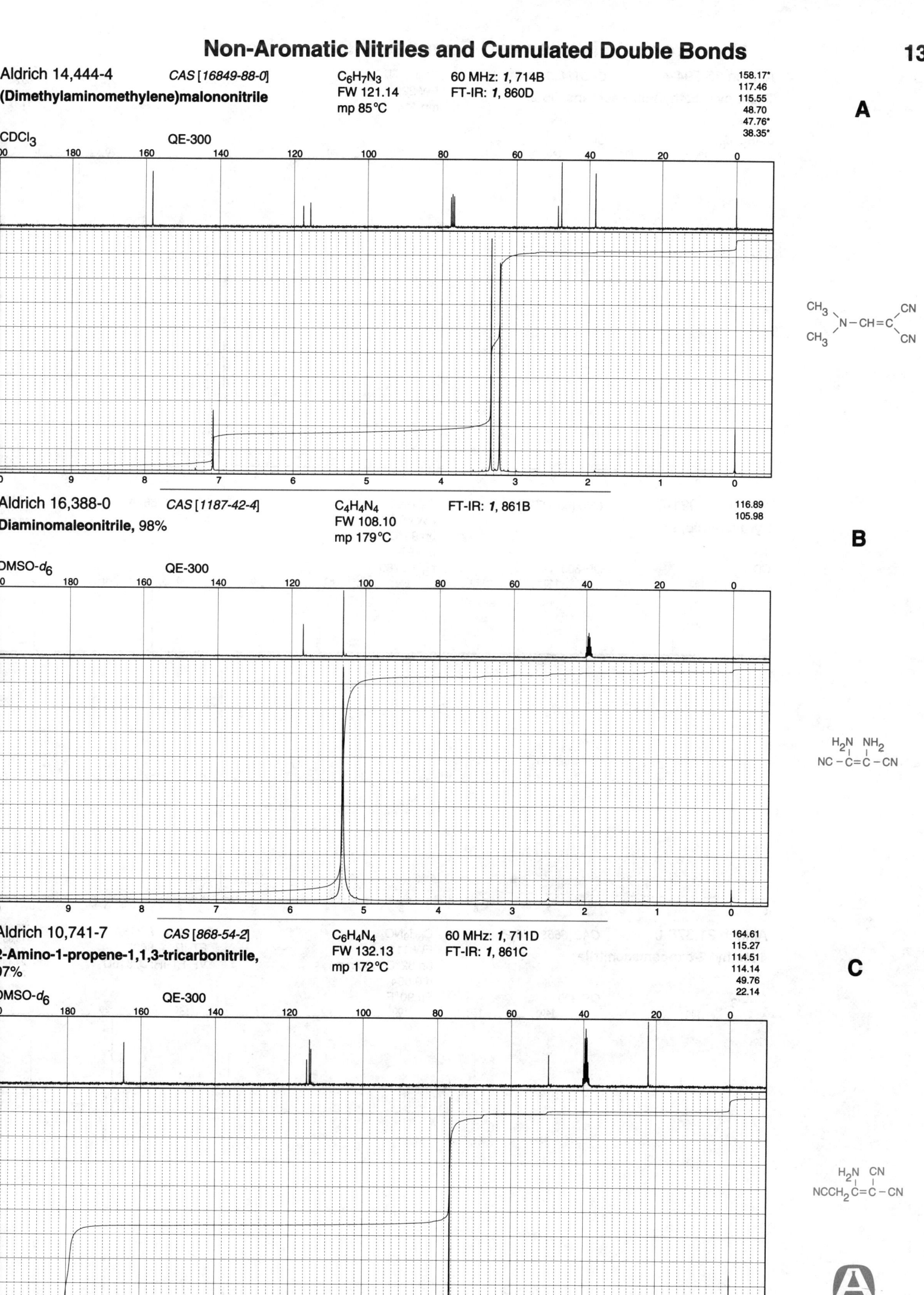

Aldrich 14,444-4 CAS [16849-88-0] C6H7N3 60 MHz: 1, 714B 158.17*
(Dimethylaminomethylene)malononitrile FW 121.14 FT-IR: 1, 860D 117.46
mp 85°C 115.55
48.70
47.76*
38.35*

A

CDCl3 QE-300

Aldrich 16,388-0 CAS [1187-42-4] C4H4N4 FT-IR: 1, 861B 116.89
Diaminomaleonitrile, 98% FW 108.10 105.98
mp 179°C

B

DMSO-d6 QE-300

Aldrich 10,741-7 CAS [868-54-2] C6H4N4 60 MHz: 1, 711D 164.61
2-Amino-1-propene-1,1,3-tricarbonitrile, FW 132.13 FT-IR: 1, 861C 115.27
97% mp 172°C 114.51
114.14
49.76
22.14

C

DMSO-d6 QE-300

Non-Aromatic Nitriles and Cumulated Double Bonds

A

Aldrich 35,794-4   *CAS [1466-48-4]* $C_{10}H_{12}N_4O_2$ 119.28
Tris(2-cyanoethyl)nitromethane, 95%   FW 220.23   90.84
mp 114°C   29.20
11.47

DMSO-d_6 QE-300

```
CH₂CH₂CN
O₂N-C-CH₂CH₂CN
CH₂CH₂CN
```

$O_2N-\overset{CH_2CH_2CN}{\underset{CH_2CH_2CN}{C}}-CH_2CH_2CN$

B

Aldrich 26,921-2   *CAS [631-57-2]* C_3H_3NO Fp 58°F   FT-IR: *1*, 862A   173.47
Pyruvonitrile, 95%   FW 69.06   VP-FT-IR: *3*, 816C   113.51
bp 93°C   32.07*
d 0.974
n_D^{20}1.3780

CDCl$_3$ QE-300

$CH_3-\overset{O}{\overset{\|}{C}}-CN$

C

Aldrich 21,378-0   *CAS [66582-16-9]* C_6H_9NO 60 MHz: *1*, 715A   176.66
4-Methyl-2-oxopentanenitrile   FW 111.14   FT-IR: *1*, 862B   113.38
bp 82°C (100 mm)   VP-FT-IR: *3*, 816D   53.56
d 0.894   24.46*
Fp 90°F   22.10*

CDCl$_3$ QE-300

$CH_3\overset{CH_3}{\underset{}{CH}}CH_2-\overset{O}{\overset{\|}{C}}-CN$

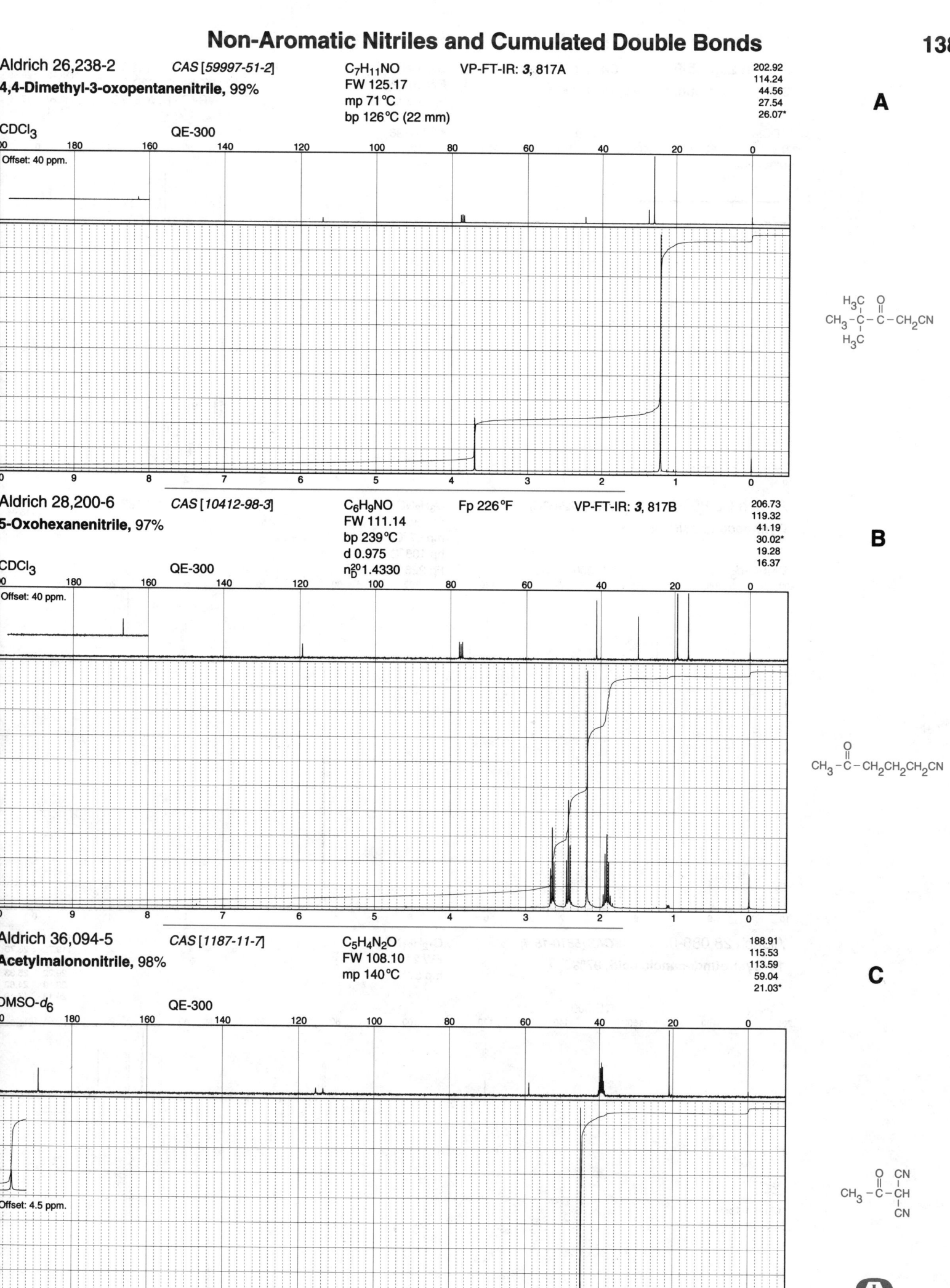

Aldrich 26,238-2   *CAS [59997-51-2]* $C_7H_{11}NO$ VP-FT-IR: **3**, 817A

4,4-Dimethyl-3-oxopentanenitrile, 99%

FW 125.17
mp 71°C
bp 126°C (22 mm)

202.92
114.24
44.56
27.54
26.07*

A

CDCl₃   QE-300

Offset: 40 ppm.

Aldrich 28,200-6   *CAS [10412-98-3]* C_6H_9NO Fp 226°F   VP-FT-IR: **3**, 817B

5-Oxohexanenitrile, 97%

FW 111.14
bp 239°C
d 0.975
n_D^{20} 1.4330

206.73
119.32
41.19
30.02*
19.28
16.37

B

CDCl₃   QE-300

Offset: 40 ppm.

Aldrich 36,094-5   *CAS [1187-11-7]* $C_5H_4N_2O$

Acetylmalononitrile, 98%

FW 108.10
mp 140°C

188.91
115.53
113.59
59.04
21.03*

C

DMSO-d_6 QE-300

Offset: 4.5 ppm.

A

Aldrich 20,645-8   *CAS [4594-78-9]* $C_9H_{13}NO$ Fp >230°F   60 MHz: *1*, 715D

2-Oxo-1-cyclohexanepropionitrile   FW 151.21   FT-IR: *1*, 862D   211.71 27.97
bp 140°C (10 mm)   VP-FT-IR: *3*, 817C   119.73 25.58
d 1.006   48.84* 25.12
n_D^{20}1.4736   42.20 15.19
34.21

CDCl$_3$ QE-300

Offset: 40 ppm.

B

Aldrich C8,850-5   *CAS [372-09-8]* $C_3H_3NO_2$ 60 MHz: *1*, 716B   165.58
Cyanoacetic acid, 99%   FW 85.06   115.44
mp 67°C   24.57
bp 108°C
Fp 226°F

DMSO-d_6 QE-300

NCCH$_2$—C—OH

C

Aldrich 28,086-0   *CAS [5810-18-4]* $C_{12}H_{21}NO_2$ 180.14 28.97
11-Cyanoundecanoic acid, 97%   FW 211.31   119.79 28.70
mp 57°C   34.07 28.62
29.22 25.33
29.19 24.62
29.13 17.10

CDCl$_3$ QE-300

NCCH$_2$(CH$_2$)$_8$CH$_2$—C—OH

Offset: 1.5 ppm.

ALDRICH

Aldrich 31,981-3   *CAS [17640-15-2]* $C_3H_3NO_2$ Fp 79 °F   144.72
Methyl cyanoformate, 99%   FW 85.06   109.21
                                                     bp 101 °C   55.02*
                                                     d 1.072
CDCl$_3$ QE-300 n_D^{20}1.3740

A

$$NC-\overset{\displaystyle O}{\overset{\|}{C}}-OCH_3$$

Aldrich E1,885-9   *CAS [623-49-4]* $C_4H_5NO_2$ Fp 76 °F   60 MHz: *1*, 717A   144.26
Ethyl cyanoformate, 99%   FW 99.09   FT-IR: *1*, 864B   109.42
                                                   bp 116 °C   VP-FT-IR: *3*, 817D   65.27
                                                   d 1.003   13.72*
CDCl$_3$ QE-300 n_D^{20}1.3820

B

$$NC-\overset{\displaystyle O}{\overset{\|}{C}}-OCH_2CH_3$$

Aldrich 36,095-3   *CAS [7790-01-4]* $C_6H_6N_2O_2$ 167.62
1,1-Dicyanoethyl acetate, 98%   FW 138.13   112.53
                                                    mp 70 °C   57.61
                                                                             25.83*
                                                                             20.00*
CDCl$_3$ QE-300

C

$$CH_3-\overset{\displaystyle O}{\overset{\|}{C}}-O-\overset{\displaystyle CN}{\underset{\displaystyle CN}{C}}-CH_3$$

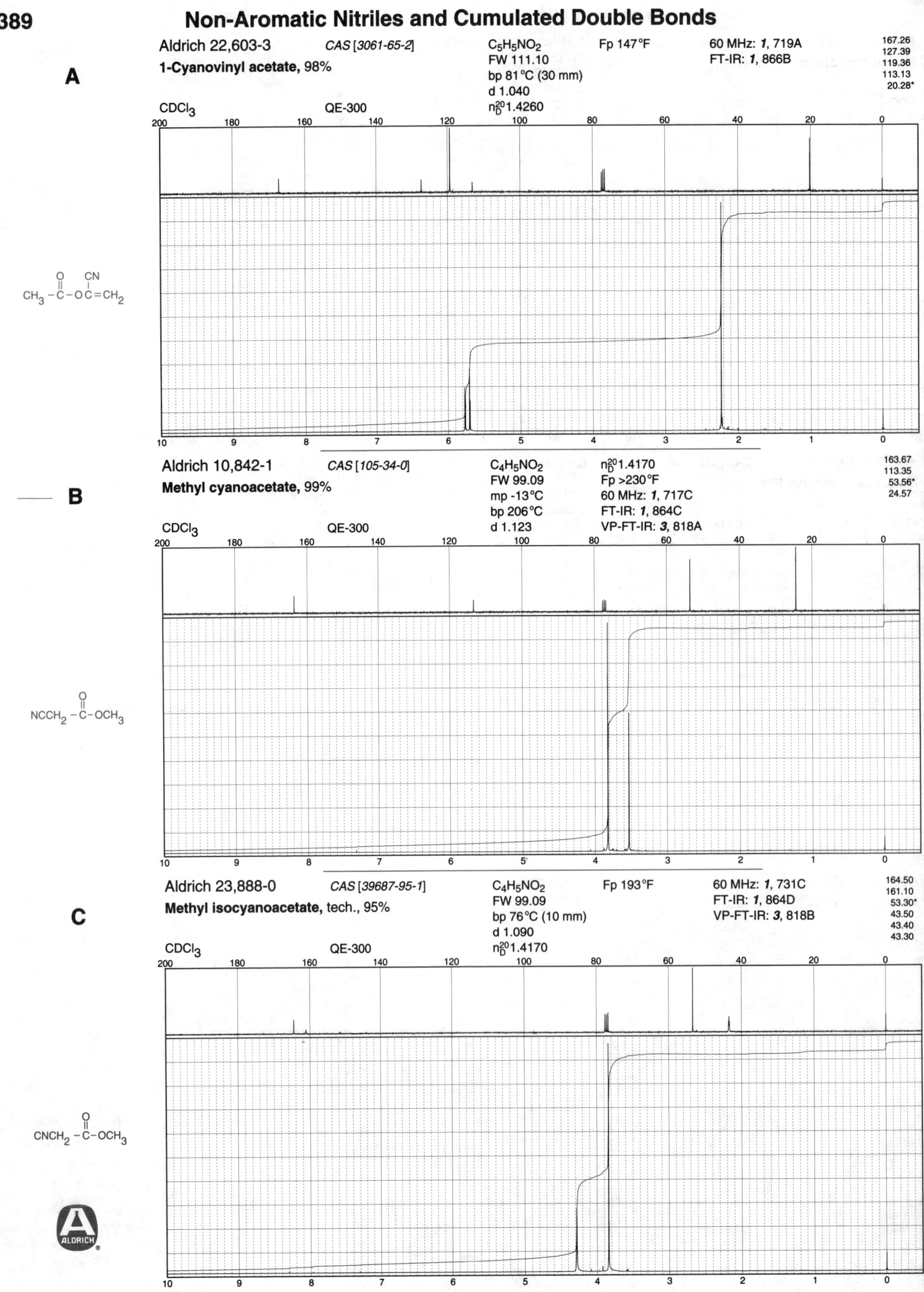

Aldrich 22,603-3   *CAS [3061-65-2]*   C₅H₅NO₂   Fp 147°F   60 MHz: *1*, 719A   167.26
1-Cyanovinyl acetate, 98%   FW 111.10   FT-IR: *1*, 866B   127.39
                              bp 81°C (30 mm)   119.36
                              d 1.040   113.13
CDCl₃   QE-300   n²⁰_D 1.4260   20.28*

Aldrich 10,842-1   *CAS [105-34-0]*   C₄H₅NO₂   n²⁰_D 1.4170   163.67
Methyl cyanoacetate, 99%   FW 99.09   Fp >230°F   113.35
                              mp -13°C   60 MHz: *1*, 717C   53.56*
                              bp 206°C   FT-IR: *1*, 864C   24.57
CDCl₃   QE-300   d 1.123   VP-FT-IR: *3*, 818A

Aldrich 23,888-0   *CAS [39687-95-1]*   C₄H₅NO₂   Fp 193°F   60 MHz: *1*, 731C   164.50
Methyl isocyanoacetate, tech., 95%   FW 99.09   FT-IR: *1*, 864D   161.10
                              bp 76°C (10 mm)   VP-FT-IR: *3*, 818B   53.30*
                              d 1.090   43.50
CDCl₃   QE-300   n²⁰_D 1.4170   43.40
                                                                                       43.30

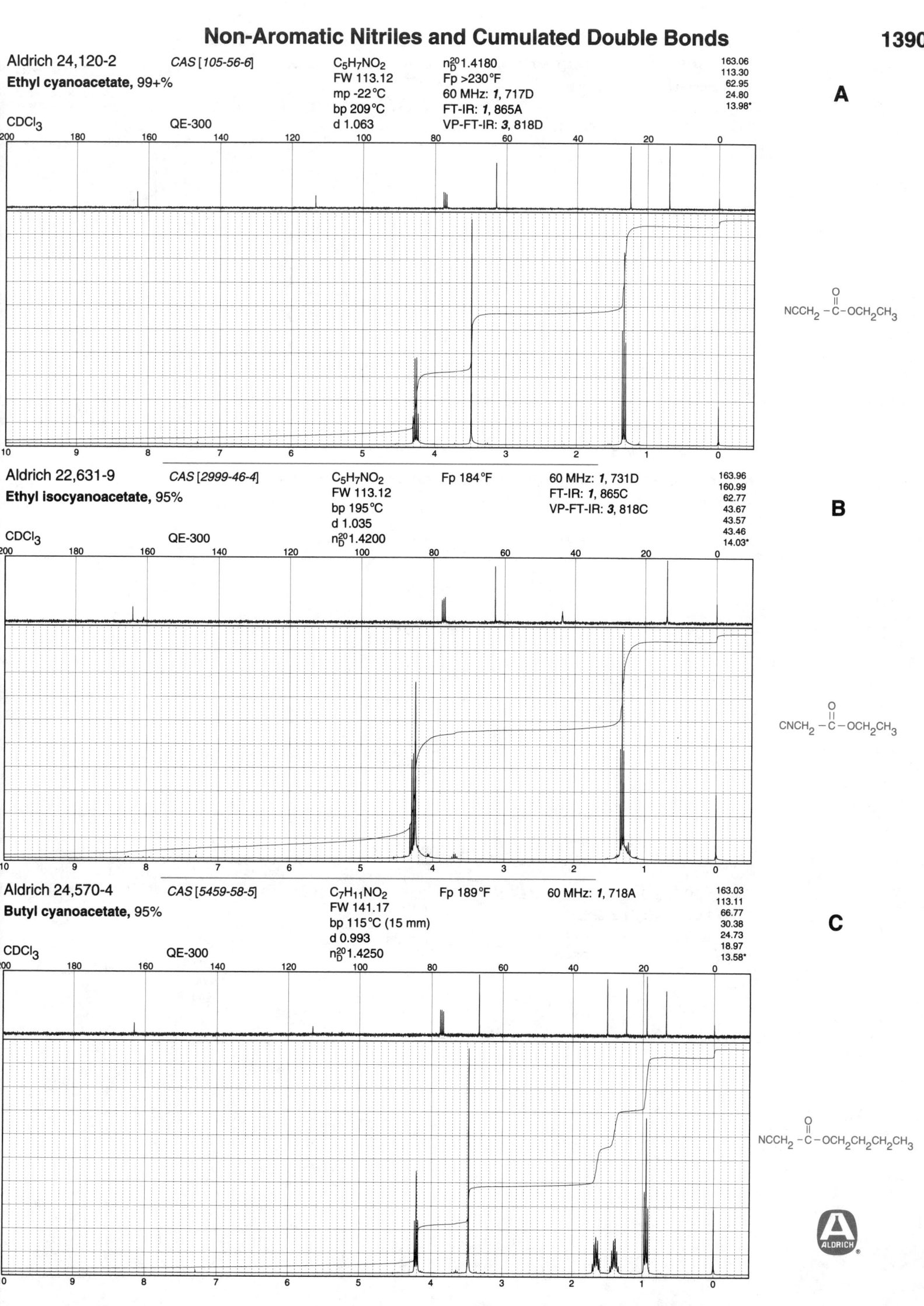

Aldrich 24,120-2 CAS [105-56-6] $C_5H_7NO_2$ n_D^{20}1.4180
Ethyl cyanoacetate, 99+% FW 113.12 Fp >230°F
mp -22°C 60 MHz: **1**, 717D
CDCl₃ QE-300 bp 209°C FT-IR: **1**, 865A
d 1.063 VP-FT-IR: **3**, 818D

163.06
113.30
62.95
24.80
13.98*

A

$NCCH_2-\overset{O}{\overset{\|}{C}}-OCH_2CH_3$

Aldrich 22,631-9 CAS [2999-46-4] $C_5H_7NO_2$ Fp 184°F 60 MHz: **1**, 731D
Ethyl isocyanoacetate, 95% FW 113.12 FT-IR: **1**, 865C
bp 195°C VP-FT-IR: **3**, 818C
CDCl₃ QE-300 d 1.035
n_D^{20}1.4200

163.96
160.99
62.77
43.67
43.57
43.46
14.03*

B

$CNCH_2-\overset{O}{\overset{\|}{C}}-OCH_2CH_3$

Aldrich 24,570-4 CAS [5459-58-5] $C_7H_{11}NO_2$ Fp 189°F 60 MHz: **1**, 718A
Butyl cyanoacetate, 95% FW 141.17
bp 115°C (15 mm)
CDCl₃ QE-300 d 0.993
n_D^{20}1.4250

163.03
113.11
66.77
30.38
24.73
18.97
13.58*

C

$NCCH_2-\overset{O}{\overset{\|}{C}}-OCH_2CH_2CH_2CH_3$

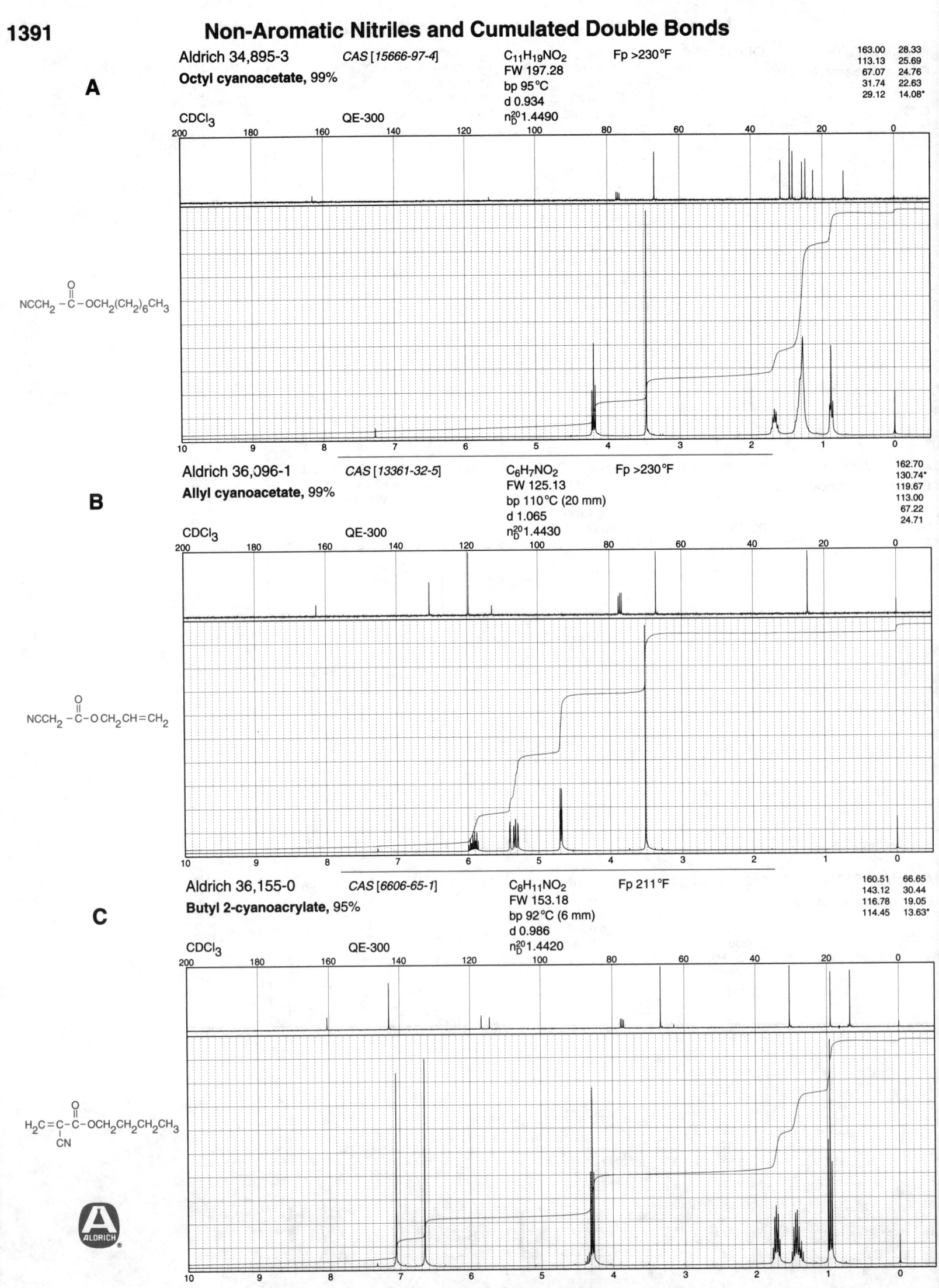

A

Aldrich 34,895-3   CAS [15666-97-4] $C_{11}H_{19}NO_2$ Fp >230°F

Octyl cyanoacetate, 99%

FW 197.28
bp 95°C
d 0.934
n_D^{20} 1.4490

CDCl₃   QE-300

163.00	28.33
113.13	25.69
67.07	24.76
31.74	22.63
29.12	14.08*

$NCCH_2-\overset{\overset{\displaystyle O}{\|}}{C}-OCH_2(CH_2)_6CH_3$

B

Aldrich 36,096-1   CAS [13361-32-5] $C_6H_7NO_2$ Fp >230°F

Allyl cyanoacetate, 99%

FW 125.13
bp 110°C (20 mm)
d 1.065
n_D^{20} 1.4430

CDCl₃   QE-300

| 162.70 |
| 130.74* |
| 119.67 |
| 113.00 |
| 67.22 |
| 24.71 |

$NCCH_2-\overset{\overset{\displaystyle O}{\|}}{C}-OCH_2CH=CH_2$

C

Aldrich 36,155-0   CAS [6606-65-1] $C_8H_{11}NO_2$ Fp 211°F

Butyl 2-cyanoacrylate, 95%

FW 153.18
bp 92°C (6 mm)
d 0.986
n_D^{20} 1.4420

CDCl₃   QE-300

160.51	66.65
143.12	30.44
116.78	19.05
114.45	13.63*

$H_2C=\overset{}{C}-\overset{\overset{\displaystyle O}{\|}}{C}-OCH_2CH_2CH_2CH_3$
$\quad\quad|$
$\quad\;\,CN$

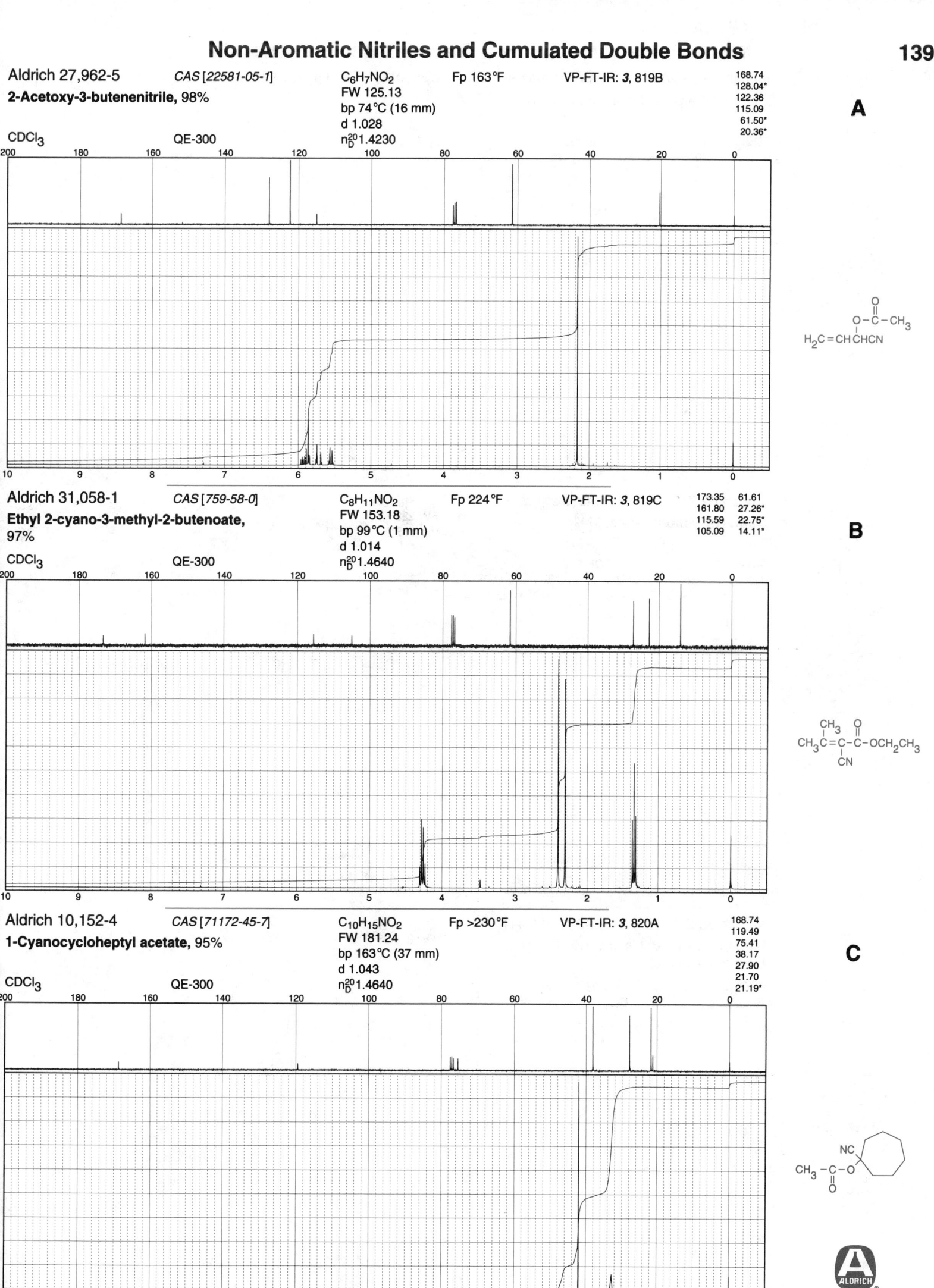

Aldrich 27,962-5   CAS [22581-05-1]   C₆H₇NO₂   Fp 163°F   VP-FT-IR: 3, 819B
2-Acetoxy-3-butenenitrile, 98%
FW 125.13
bp 74°C (16 mm)
d 1.028
n_D^{20} 1.4230
CDCl₃   QE-300

168.74
128.04*
122.36
115.09
61.50*
20.36*

A

Aldrich 31,058-1   CAS [759-58-0]   C₈H₁₁NO₂   Fp 224°F   VP-FT-IR: 3, 819C
Ethyl 2-cyano-3-methyl-2-butenoate, 97%
FW 153.18
bp 99°C (1 mm)
d 1.014
n_D^{20} 1.4640
CDCl₃   QE-300

173.35   61.61
161.80   27.26*
115.59   22.75*
105.09   14.11*

B

Aldrich 10,152-4   CAS [71172-45-7]   C₁₀H₁₅NO₂   Fp >230°F   VP-FT-IR: 3, 820A
1-Cyanocycloheptyl acetate, 95%
FW 181.24
bp 163°C (37 mm)
d 1.043
n_D^{20} 1.4640
CDCl₃   QE-300

168.74
119.49
75.41
38.17
27.90
21.70
21.19*

C

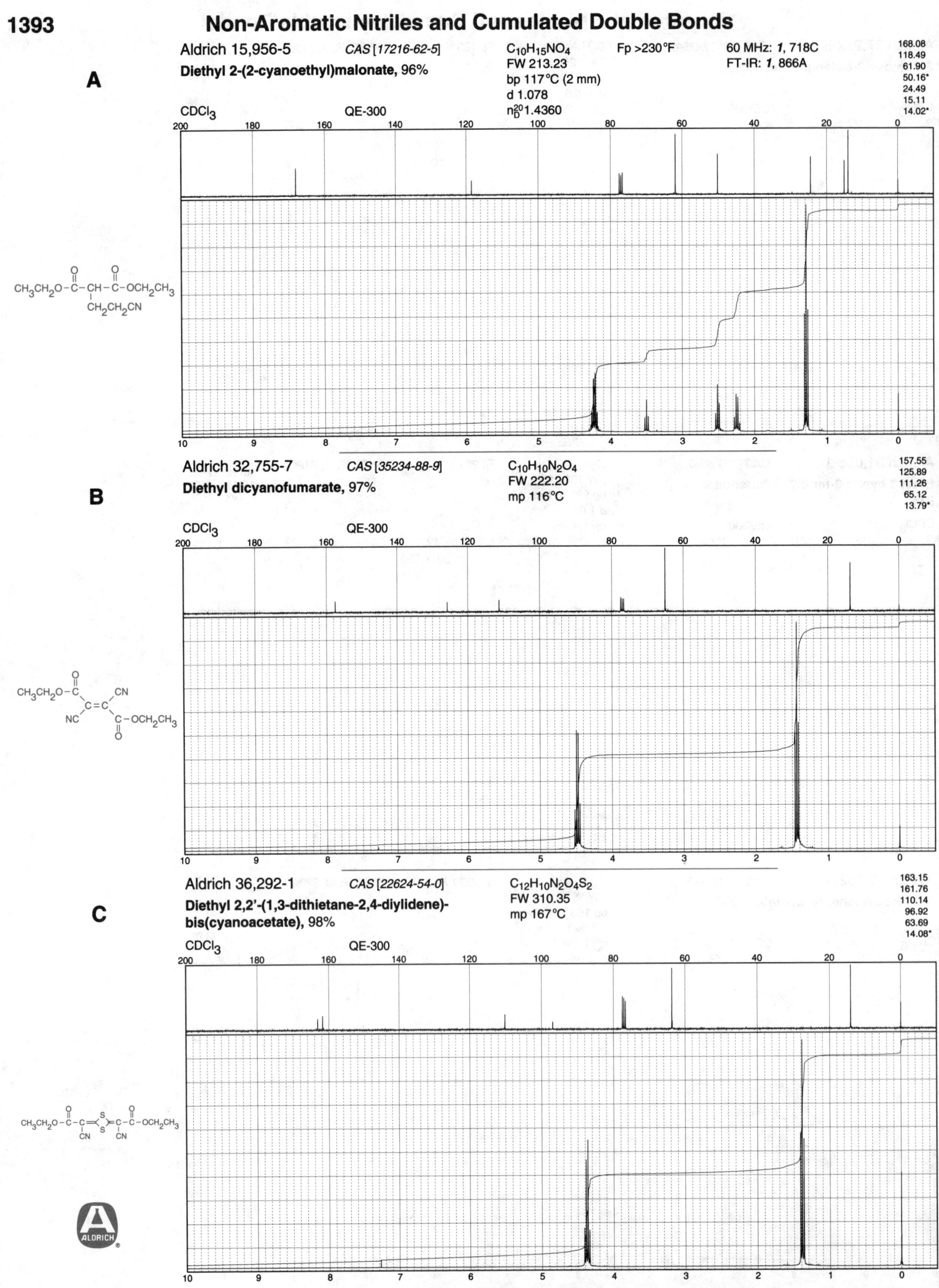

A

Aldrich 15,956-5   CAS [17216-62-5] $C_{10}H_{15}NO_4$ Fp >230°F   60 MHz: *1*, 718C

Diethyl 2-(2-cyanoethyl)malonate, 96%   FW 213.23

bp 117°C (2 mm)   FT-IR: *1*, 866A

d 1.078

n_D^{20} 1.4360

CDCl$_3$ QE-300

168.08
118.49
61.90
50.16*
24.49
15.11
14.02*

B

Aldrich 32,755-7   CAS [35234-88-9] $C_{10}H_{10}N_2O_4$

Diethyl dicyanofumarate, 97%   FW 222.20

mp 116°C

CDCl$_3$ QE-300

157.55
125.89
111.26
65.12
13.79*

C

Aldrich 36,292-1   CAS [22624-54-0] $C_{12}H_{10}N_2O_4S_2$

Diethyl 2,2'-(1,3-dithietane-2,4-diylidene)-   FW 310.35
bis(cyanoacetate), 98%   mp 167°C

CDCl$_3$ QE-300

163.15
161.76
110.14
96.92
63.69
14.08*

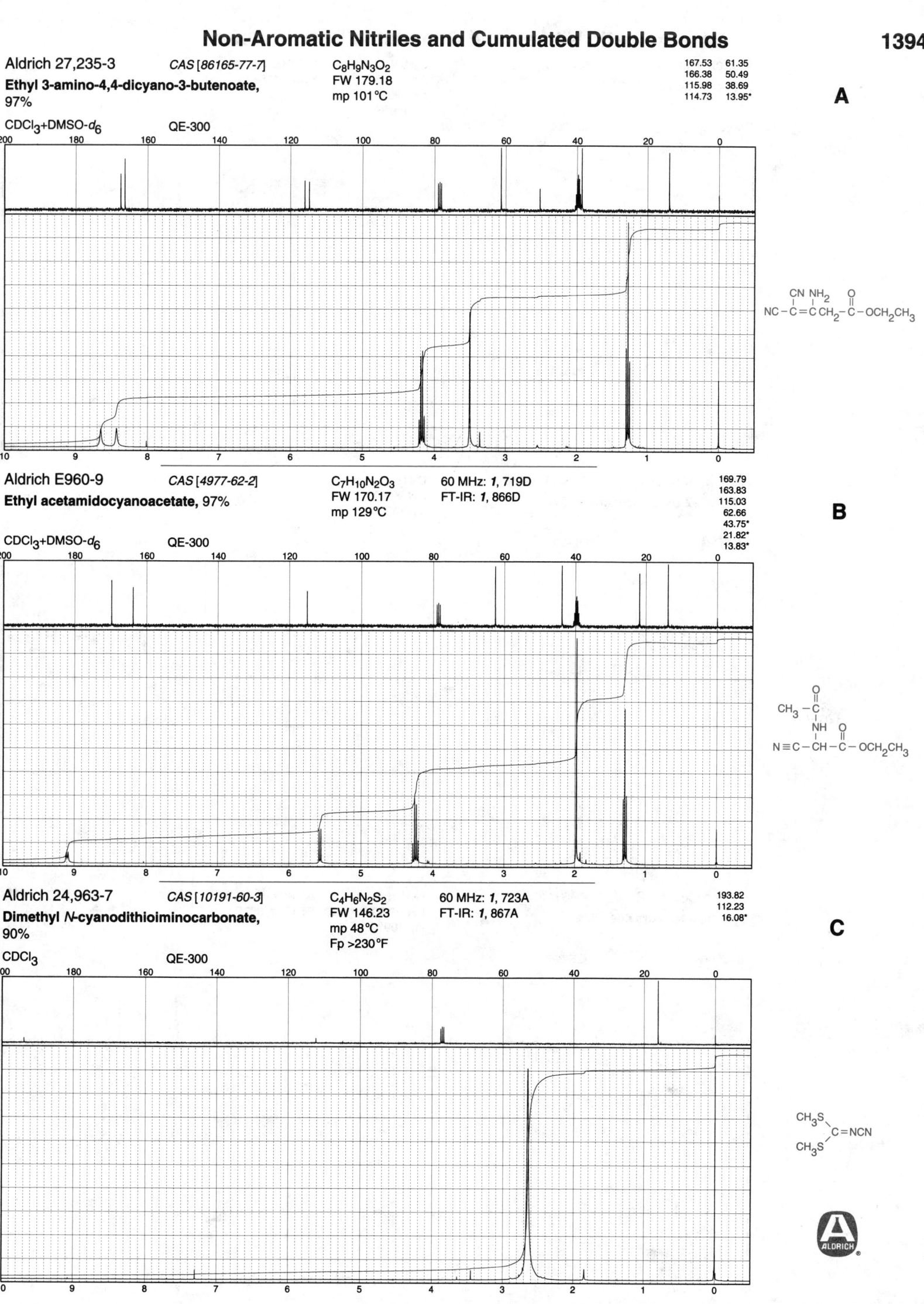

167.53   61.35
166.38   50.49
115.98   38.69
114.73   13.95*

A

NC–C=C CH2–C–OCH2CH3

169.79
163.83
115.03
62.66
43.75*
21.82*
13.83*

B

CH3–C
NH
N≡C–CH–C–OCH2CH3

193.82
112.23
16.08*

C

CH3S
C=NCN
CH3S

ALDRICH

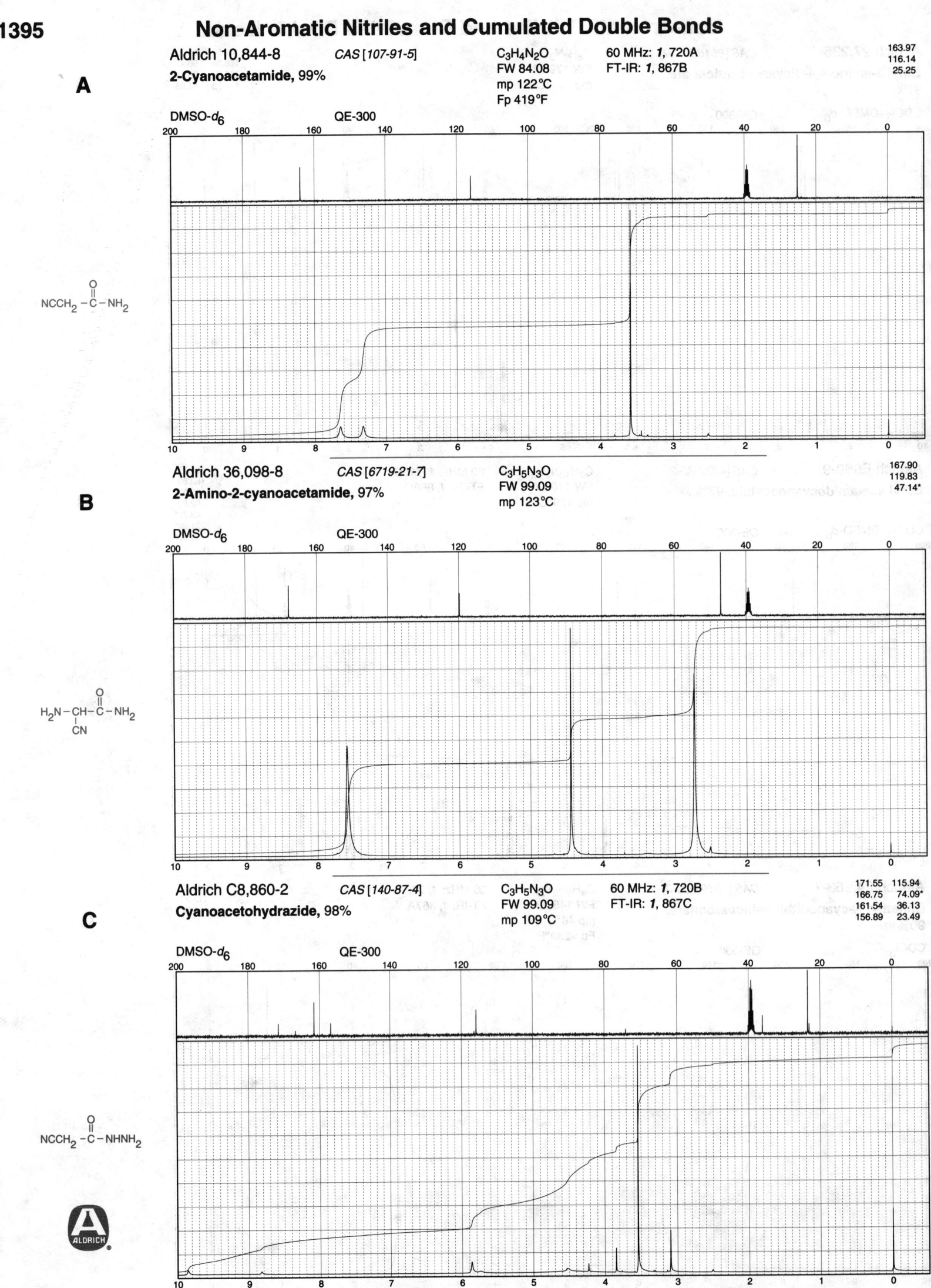

A

Aldrich 10,844-8   CAS [107-91-5]   C₃H₄N₂O   60 MHz: *1*, 720A   163.97
116.14
2-Cyanoacetamide, 99%   FW 84.08   FT-IR: *1*, 867B   25.25
mp 122°C
Fp 419°F

DMSO-*d₆*   QE-300

NCCH₂−C(=O)−NH₂

B

Aldrich 36,098-8   CAS [6719-21-7]   C₃H₅N₃O   167.90
119.83
2-Amino-2-cyanoacetamide, 97%   FW 99.09   47.14*
mp 123°C

DMSO-*d₆*   QE-300

H₂N−CH(CN)−C(=O)−NH₂

C

Aldrich C8,860-2   CAS [140-87-4]   C₃H₅N₃O   60 MHz: *1*, 720B   171.55 115.94
166.75 74.09*
Cyanoacetohydrazide, 98%   FW 99.09   FT-IR: *1*, 867C   161.54 36.13
mp 109°C   156.89 23.49

DMSO-*d₆*   QE-300

NCCH₂−C(=O)−NHNH₂

ALDRICH

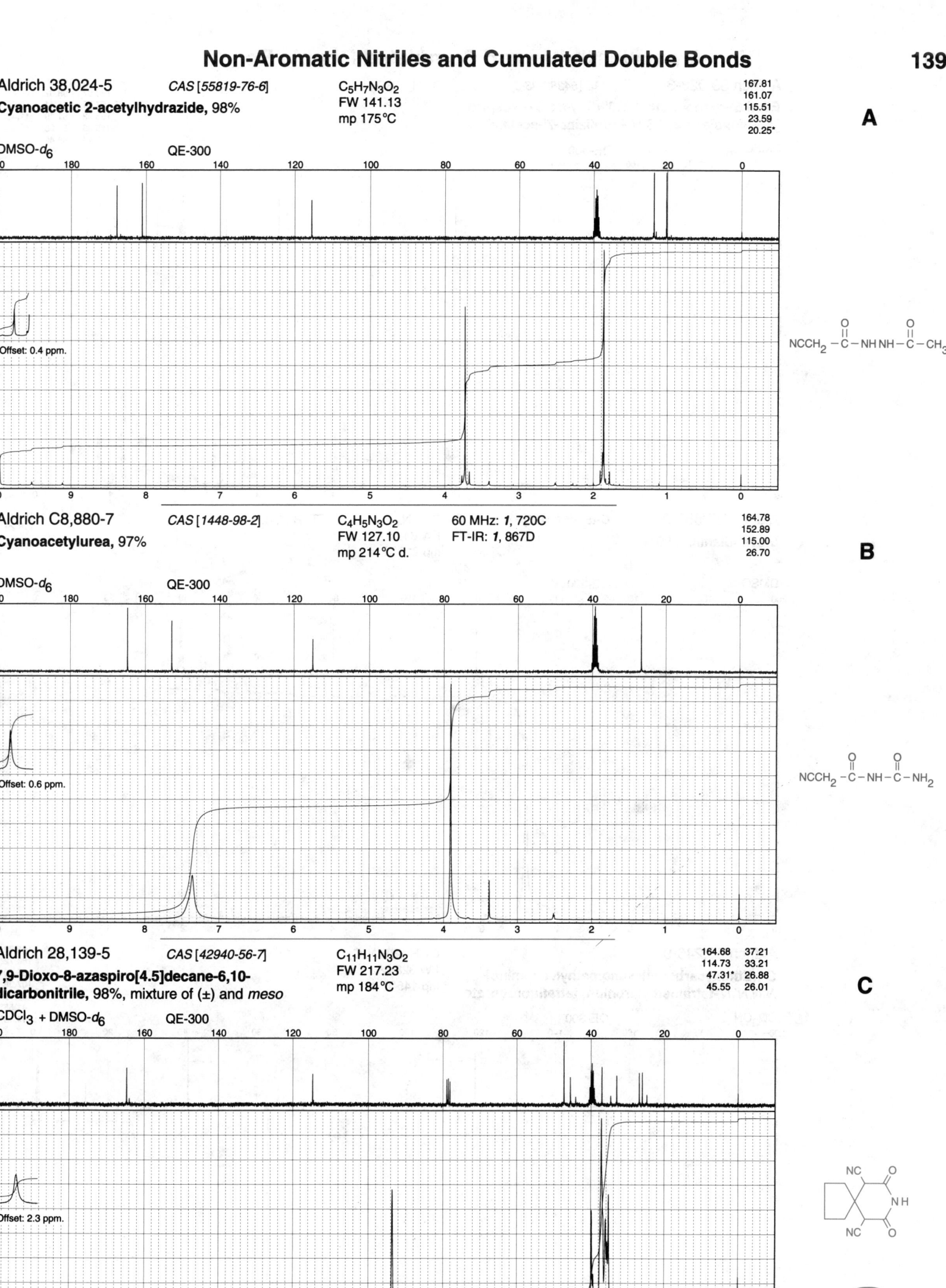

Aldrich 38,024-5   CAS [55819-76-6]   C₅H₇N₃O₂   167.81 / 161.07 / 115.51 / 23.59 / 20.25*

Cyanoacetic 2-acetylhydrazide, 98%   FW 141.13   mp 175°C   **A**

DMSO-d₆   QE-300

Offset: 0.4 ppm.

NCCH₂—C(=O)—NHNH—C(=O)—CH₃

Aldrich C8,880-7   CAS [1448-98-2]   C₄H₅N₃O₂   60 MHz: 1, 720C   164.78 / 152.89 / 115.00 / 26.70

Cyanoacetylurea, 97%   FW 127.10   FT-IR: 1, 867D   mp 214°C d.   **B**

DMSO-d₆   QE-300

Offset: 0.6 ppm.

NCCH₂—C(=O)—NH—C(=O)—NH₂

Aldrich 28,139-5   CAS [42940-56-7]   C₁₁H₁₁N₃O₂   164.68 / 114.73 / 47.31* / 45.55   37.21 / 33.21 / 26.88 / 26.01

7,9-Dioxo-8-azaspiro[4.5]decane-6,10-dicarbonitrile, 98%, mixture of (±) and *meso*   FW 217.23   mp 184°C   **C**

CDCl₃ + DMSO-d₆   QE-300

Offset: 2.3 ppm.

Non-Aromatic Nitriles and Cumulated Double Bonds

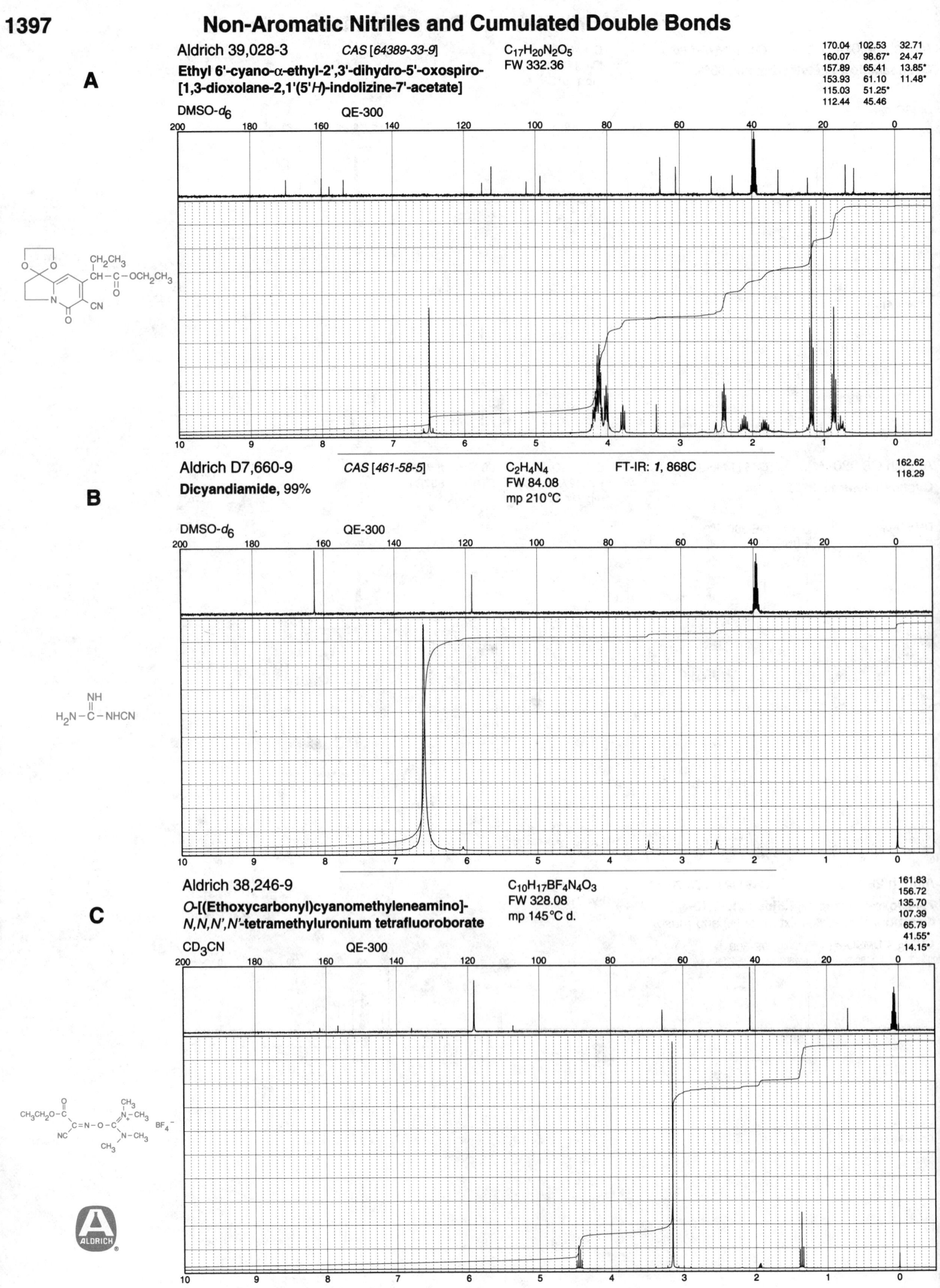

A

Aldrich 39,028-3   *CAS [64389-33-9]*   C₁₇H₂₀N₂O₅
FW 332.36

**Ethyl 6'-cyano-α-ethyl-2',3'-dihydro-5'-oxospiro-
[1,3-dioxolane-2,1'(5'H)-indolizine-7'-acetate]**

DMSO-*d₆*   QE-300

170.04	102.53	32.71
160.07	98.67*	24.47
157.89	65.41	13.85*
153.93	61.10	11.48*
115.03	51.25*	
112.44	45.46	

B

Aldrich D7,660-9   *CAS [461-58-5]*   C₂H₄N₄   FT-IR: *1*, 868C
FW 84.08
mp 210°C

Dicyandiamide, 99%

DMSO-*d₆*   QE-300

162.62
118.29

C

Aldrich 38,246-9   C₁₀H₁₇BF₄N₄O₃
FW 328.08
mp 145°C d.

O-[(Ethoxycarbonyl)cyanomethyleneamino]-
N,N,N',N'-tetramethyluronium tetrafluoroborate

CD₃CN   QE-300

161.83
156.72
135.70
107.39
65.79
41.55*
14.15*

Aldrich C8,900-5 *CAS [6629-04-5]* C$_6$H$_8$N$_2$O$_3$ 60 MHz: *1*, 721C

N-Cyanoacetylurethane, 98%

FW 156.14 FT-IR: *1*, 868D

mp 168°C

164.07
151.52
115.14
61.44
27.70
14.02*

A

DMSO-d$_6$ QE-300

Offset: 1.2 ppm.

NCCH$_2$ -C-NH- C-OCH$_2$CH$_3$

Aldrich 19,291-0 *CAS [5447-66-5]* C$_8$H$_{10}$N$_4$O$_2$ 60 MHz: *1*, 721B

2,4-Dicyano-3-methylglutaramide, 99%,
mixture of isomers

FW 194.19 FT-IR: *1*, 869A

mp 160°C

165.19	116.26	33.68*
164.93	115.90	32.78*
164.87	41.86*	15.39*
164.74	41.59*	13.98*
116.85	41.39*	13.94*
116.61	40.61*	

B

DMSO-d$_6$ QE-300

H$_2$N-C-CHCHCH-C-NH$_2$
       | |
       CN   CN
       |
       CH$_3$

Aldrich 21,388-8 *CAS [67944-71-2]* C$_3$H$_5$N$_3$S 60 MHz: *1*, 723B

1-Cyano-3-methylisothiourea, sodium
salt, 94+%

FW 137.14 FT-IR: *1*, 869B

mp 290°C d.

190.63
187.37
122.80
122.52
31.41*
29.64*

C

DMSO-d$_6$ QE-300

SNa
|
CH$_3$NHC=NCN

ALDRICH®

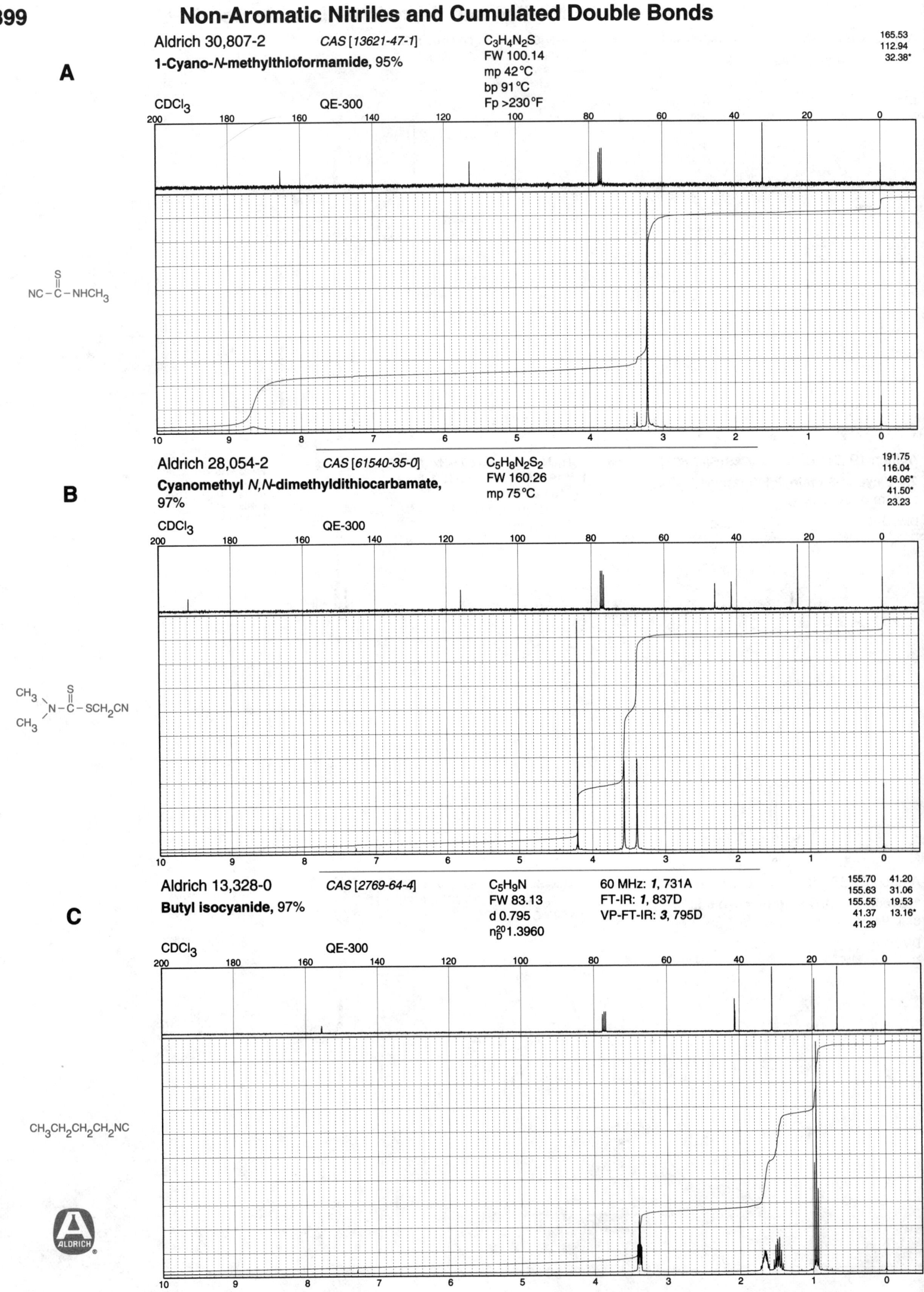

A

Aldrich 30,807-2   CAS [13621-47-1]

1-Cyano-N-methylthioformamide, 95%

$C_3H_4N_2S$
FW 100.14
mp 42°C
bp 91°C
Fp >230°F

CDCl$_3$ QE-300

165.53
112.94
32.38*

NC–C(=S)–NHCH$_3$

B

Aldrich 28,054-2   CAS [61540-35-0]

Cyanomethyl N,N-dimethyldithiocarbamate, 97%

$C_5H_8N_2S_2$
FW 160.26
mp 75°C

CDCl$_3$ QE-300

191.75
116.04
46.06*
41.50*
23.23

(CH$_3$)$_2$N–C(=S)–SCH$_2$CN

C

Aldrich 13,328-0   CAS [2769-64-4]

Butyl isocyanide, 97%

C_5H_9N
FW 83.13
d 0.795
n$_D^{20}$1.3960

60 MHz: 1, 731A
FT-IR: 1, 837D
VP-FT-IR: 3, 795D

155.70   41.20
155.63   31.06
155.55   19.53
41.37   13.16*
41.29

CDCl$_3$ QE-300

CH$_3$CH$_2$CH$_2$CH$_2$NC

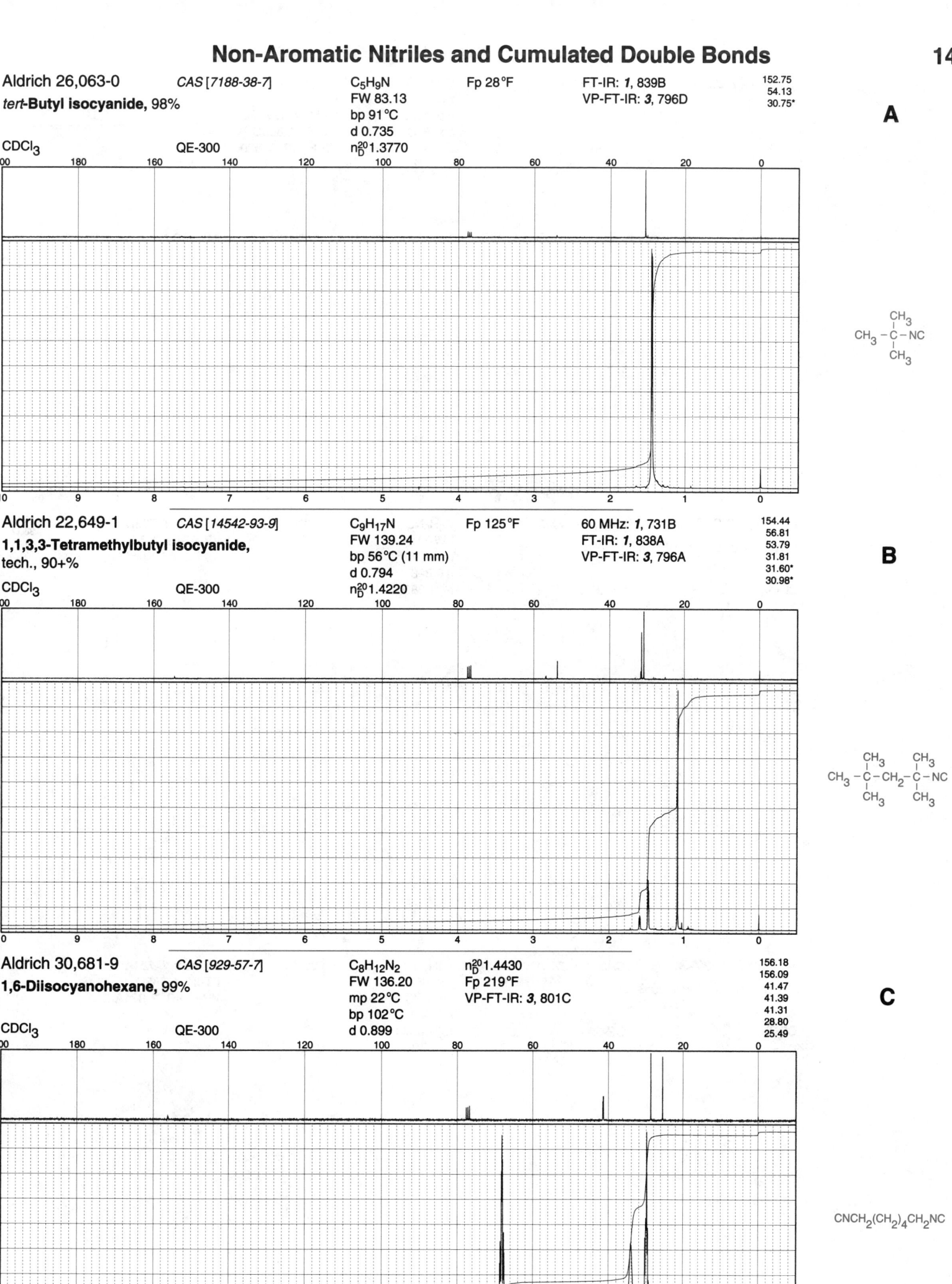

Aldrich 26,063-0 *CAS [7188-38-7]* C₅H₉N Fp 28°F FT-IR: *1*, 839B 152.75
54.13
30.75*

tert-Butyl isocyanide, 98% FW 83.13 VP-FT-IR: *3*, 796D

bp 91°C

d 0.735

CDCl₃ QE-300 n²⁰_D 1.3770

A

Aldrich 22,649-1 *CAS [14542-93-9]* C₉H₁₇N Fp 125°F 60 MHz: *1*, 731B 154.44
56.81
53.79
31.81
31.60*
30.98*

1,1,3,3-Tetramethylbutyl isocyanide, FW 139.24 FT-IR: *1*, 838A

tech., 90+% bp 56°C (11 mm) VP-FT-IR: *3*, 796A

d 0.794

CDCl₃ QE-300 n²⁰_D 1.4220

B

Aldrich 30,681-9 *CAS [929-57-7]* C₈H₁₂N₂ n²⁰_D 1.4430 156.18
156.09
41.47
41.39
41.31
28.80
25.49

1,6-Diisocyanohexane, 99% FW 136.20 Fp 219°F

mp 22°C VP-FT-IR: *3*, 801C

bp 102°C

CDCl₃ QE-300 d 0.899

C

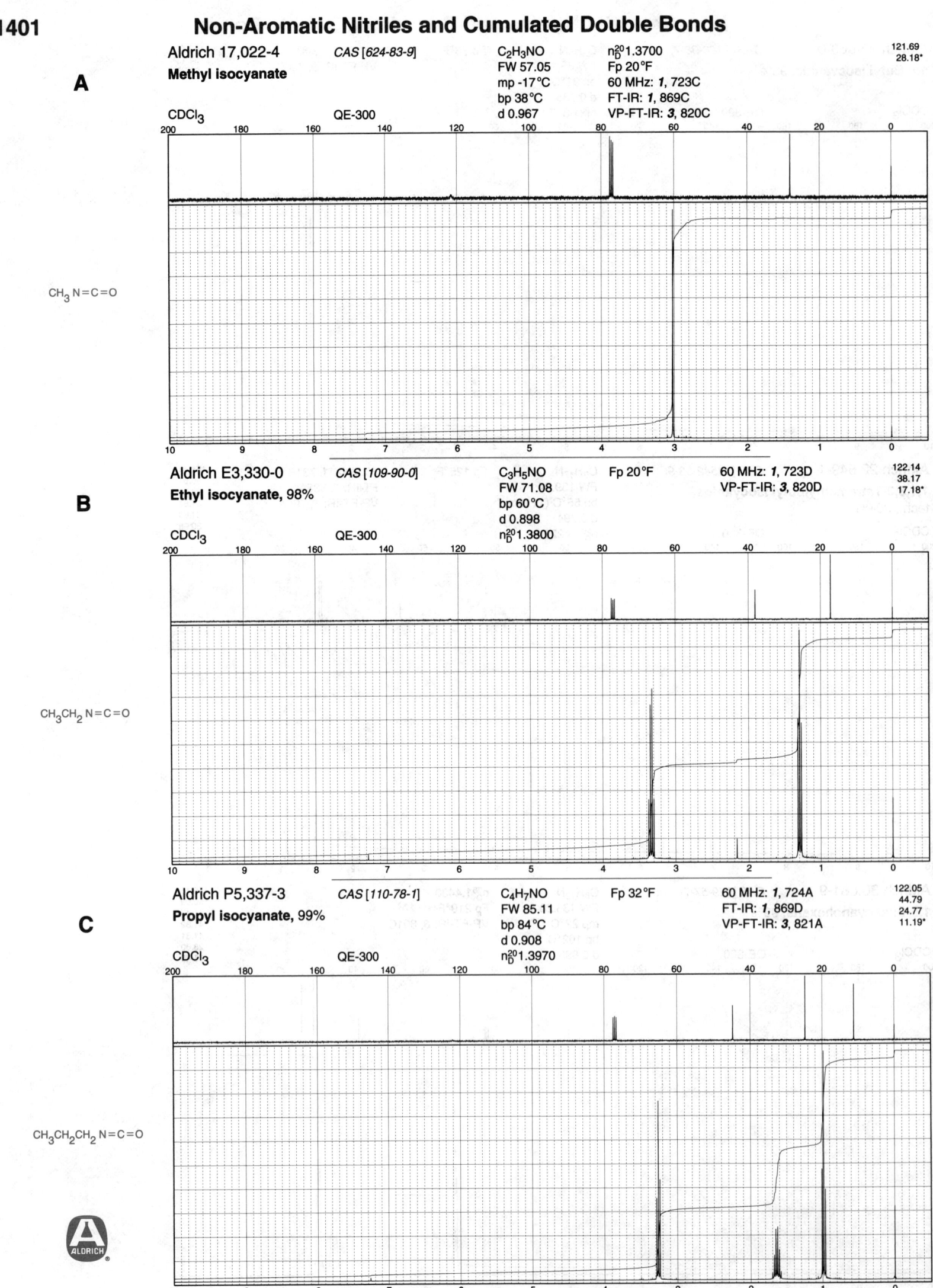

Aldrich 17,022-4   *CAS [624-83-9]* C_2H_3NO n_D^{20} 1.3700   121.69
A   **Methyl isocyanate**   FW 57.05   Fp 20°F   28.18*
     mp -17°C   60 MHz: *1*, 723C
     CDCl₃   QE-300   bp 38°C   FT-IR: *1*, 869C
     d 0.967   VP-FT-IR: *3*, 820C

$CH_3 N=C=O$

Aldrich E3,330-0   *CAS [109-90-0]* C_3H_5NO Fp 20°F   60 MHz: *1*, 723D   122.14
B   **Ethyl isocyanate,** 98%   FW 71.08   VP-FT-IR: *3*, 820D   38.17
     bp 60°C   17.18*
     d 0.898
     CDCl₃   QE-300 n_D^{20} 1.3800

$CH_3CH_2 N=C=O$

Aldrich P5,337-3   *CAS [110-78-1]* C_4H_7NO Fp 32°F   60 MHz: *1*, 724A   122.05
C   **Propyl isocyanate,** 99%   FW 85.11   FT-IR: *1*, 869D   44.79
     bp 84°C   VP-FT-IR: *3*, 821A   24.77
     d 0.908   11.19*
     CDCl₃   QE-300 n_D^{20} 1.3970

$CH_3CH_2CH_2 N=C=O$

Aldrich 14,107-0
Isopropyl isocyanate, 98+%

CAS [1795-48-8]

C₄H₇NO
FW 85.11
bp 75°C
d 0.866
n₂₀ᴰ 1.3820

Fp 27°F

60 MHz: *1*, 724B
FT-IR: *1*, 870A
VP-FT-IR: *3*, 821B

122.40
47.29*
25.07*

A

CDCl₃   QE-300

$$CH_3CHN=C=O$$
with CH₃ above

Aldrich B9,573-6
Butyl isocyanate, 98%

CAS [111-36-4]

C₅H₉NO
FW 99.13
bp 115°C
d 0.880
n₂₀ᴰ 1.4060

Fp 64°F

60 MHz: *1*, 724C
FT-IR: *1*, 870B
VP-FT-IR: *3*, 821C

121.92
42.70
33.34
19.74
13.41*

B

CDCl₃   QE-300

$$CH_3CH_2CH_2CH_2N=C=O$$

Aldrich 14,445-2
tert-**Butyl isocyanate, 97%**

CAS [1609-86-5]

C₅H₉NO
FW 99.13
bp 86°C
d 0.868
n₂₀ᴰ 1.3860

Fp 24°F

60 MHz: *1*, 724D
FT-IR: *1*, 870C
VP-FT-IR: *3*, 821D

122.35
55.52
31.90*

C

CDCl₃   QE-300

$$CH_3-C-N=C=O$$
with CH₃ above and CH₃ below

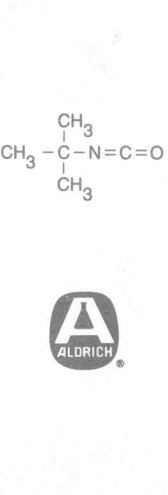

Non-Aromatic Nitriles and Cumulated Double Bonds

A

Aldrich 32,974-6   *CAS [3158-26-7]* $C_9H_{17}NO$ Fp 160 °F   VP-FT-IR: *3*, 822A

Octyl isocyanate, 97%

FW 155.24
bp 202 °C
d 0.880
n_D^{20} 1.4320

121.99	28.97
43.03	26.59
31.80	22.67
31.37	14.08*
29.18	

CDCl$_3$ QE-300

$CH_3(CH_2)_6CH_2N=C=O$

B

Aldrich C10,519-8   *CAS [3173-53-3]* $C_7H_{11}NO$ Fp 120 °F   60 MHz: *1*, 726A

Cyclohexyl isocyanate, 98%

FW 125.17
bp 169 °C
d 0.980
n_D^{20} 1.4550

VP-FT-IR: *3*, 822C

| 122.41 |
| 53.37* |
| 34.82 |
| 25.26 |
| 23.82 |

CDCl$_3$ QE-300

N=C=O

C

Aldrich 37,506-3   *CAS [4411-25-0]* $C_{11}H_{15}NO$

1-Adamantyl isocyanate, 98%

FW 177.25
mp 145 °C

| 122.76 |
| 55.67 |
| 45.25 |
| 35.75 |
| 29.69* |

CDCl$_3$ QE-300

NCO

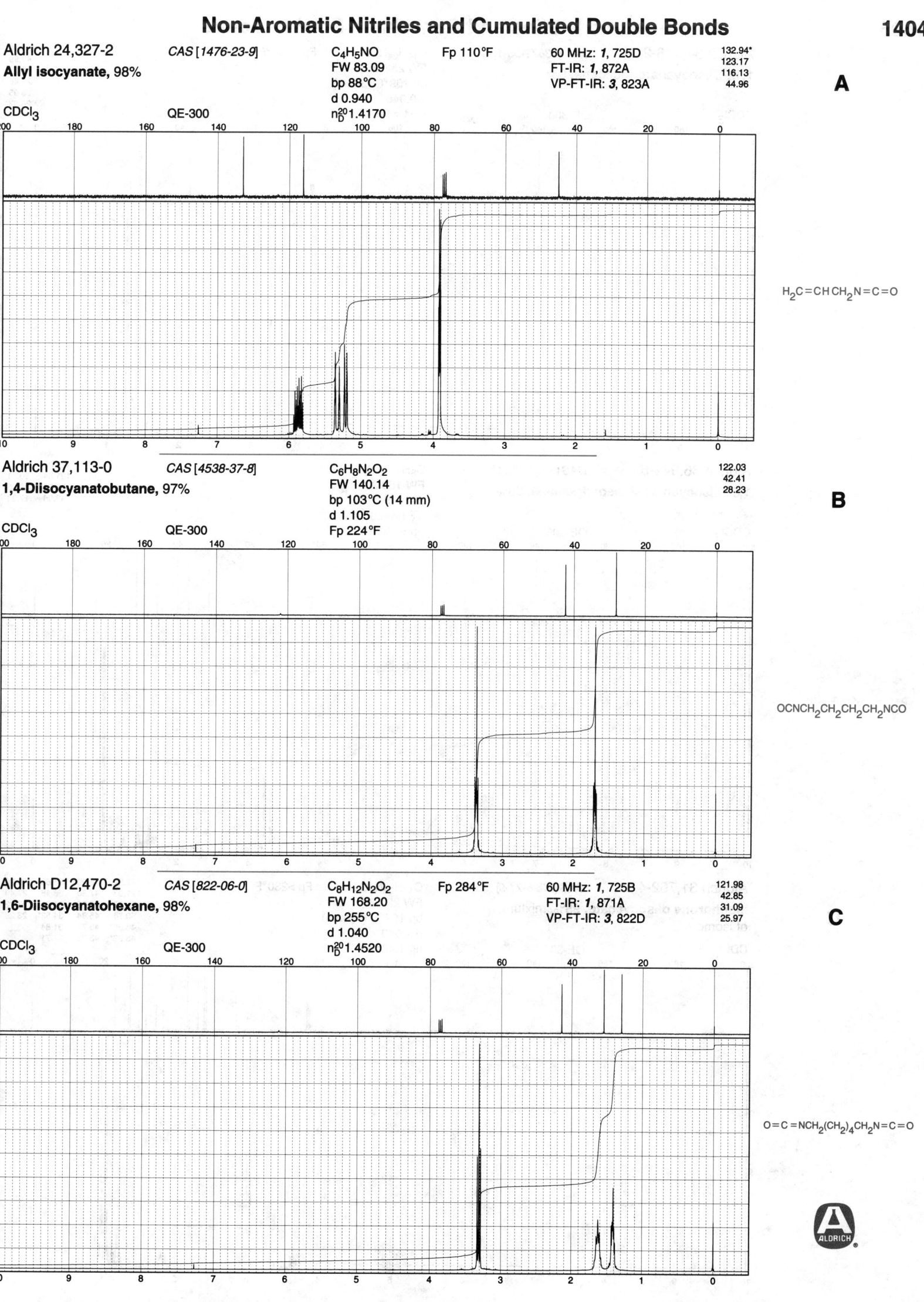

Aldrich 24,327-2
Allyl isocyanate, 98%

CAS [1476-23-9]

C₄H₅NO
FW 83.09
bp 88°C
d 0.940
n²⁰_D 1.4170

Fp 110°F

60 MHz: *1*, 725D
FT-IR: *1*, 872A
VP-FT-IR: *3*, 823A

132.94*
123.17
116.13
44.96

A

CDCl₃   QE-300

H₂C=CHCH₂N=C=O

Aldrich 37,113-0
1,4-Diisocyanatobutane, 97%

CAS [4538-37-8]

C₆H₈N₂O₂
FW 140.14
bp 103°C (14 mm)
d 1.105
Fp 224°F

122.03
42.41
28.23

B

CDCl₃   QE-300

OCNCH₂CH₂CH₂CH₂NCO

Aldrich D12,470-2
1,6-Diisocyanatohexane, 98%

CAS [822-06-0]

C₈H₁₂N₂O₂
FW 168.20
bp 255°C
d 1.040
n²⁰_D 1.4520

Fp 284°F

60 MHz: *1*, 725B
FT-IR: *1*, 871A
VP-FT-IR: *3*, 822D

121.98
42.85
31.09
25.97

C

CDCl₃   QE-300

O=C=NCH₂(CH₂)₄CH₂N=C=O

ALDRICH®

A

Aldrich 34,176-2   *CAS [13879-35-1]*

1,12-Diisocyanatododecane, 95%

$C_{14}H_{24}N_2O_2$
FW 252.36
bp 169°C (3 mm)
d 0.940
n_D^{20} 1.4590

Fp >230°F

121.90
42.99
31.31
29.46
28.95
26.53

CDCl₃   QE-300

OCNCH₂(CH₂)₁₀CH₂NCO

B

Aldrich 36,944-6   *CAS [34813-62-2]*

1,5-Diisocyanato-2-methylpentane, 98%

$C_8H_{12}N_2O_2$
FW 168.20
d 1.049
n_D^{20} 1.4550
Fp >230°F

122.05	34.23*
121.84	30.71
48.73	28.65
43.08	17.33*

CDCl₃   QE-300

O=C=N CH₂CH₂CH₂CH CH₂N=C=O
                    |
                   CH₃

C

Aldrich 31,762-4   *CAS [4098-71-9]*

Isophorone diisocyanate, 98%, mixture of isomers

$C_{12}H_{18}N_2O_2$
FW 222.29
bp 159°C (15 mm)
d 1.049
n_D^{20} 1.4840

Fp >230°F

122.82	48.00	36.63	29.80*
121.93	47.80	36.46	27.49*
56.77	46.23	34.65*	27.02*
50.74	45.84	34.53*	23.29*
48.56*	43.77	31.84	
48.33*	43.32	31.71	

CDCl₃   QE-300

CH₃   CH₂N=C=O
  |
  (cyclohexane ring)
       CH₃
       CH₃
O=C=N

Aldrich 26,936-0   CAS [7517-76-2]   C₈H₁₀N₂O₂
$C_8H_{10}N_2O_2$
FW 166.18
mp 61 °C

trans-1,4-Cyclohexylene diisocyanate, 97%

122.35
51.76*
31.45

A

CDCl₃   QE-300

Aldrich 25,076-7   CAS [30121-98-3] C_2Cl_3NO Fp NONE   FT-IR: 1, 871C   150.46
Trichloromethyl isocyanate   FW 160.39   VP-FT-IR: 3, 823C   137.17
bp 121 °C
d 1.564

B

CDCl₃   QE-300 n_D^{20} 1.4770

$Cl_3CN=C=O$

Aldrich 24,164-4   CAS [1943-83-5] C_3H_4ClNO Fp 134 °F   60 MHz: 1, 725C   124.23
2-Chloroethyl isocyanate, 95%   FW 105.52   FT-IR: 1, 871D   44.94
bp 135 °C   VP-FT-IR: 3, 823D   44.43
d 1.237

C

CDCl₃   QE-300 n_D^{20} 1.4490

$ClCH_2CH_2N=C=O$

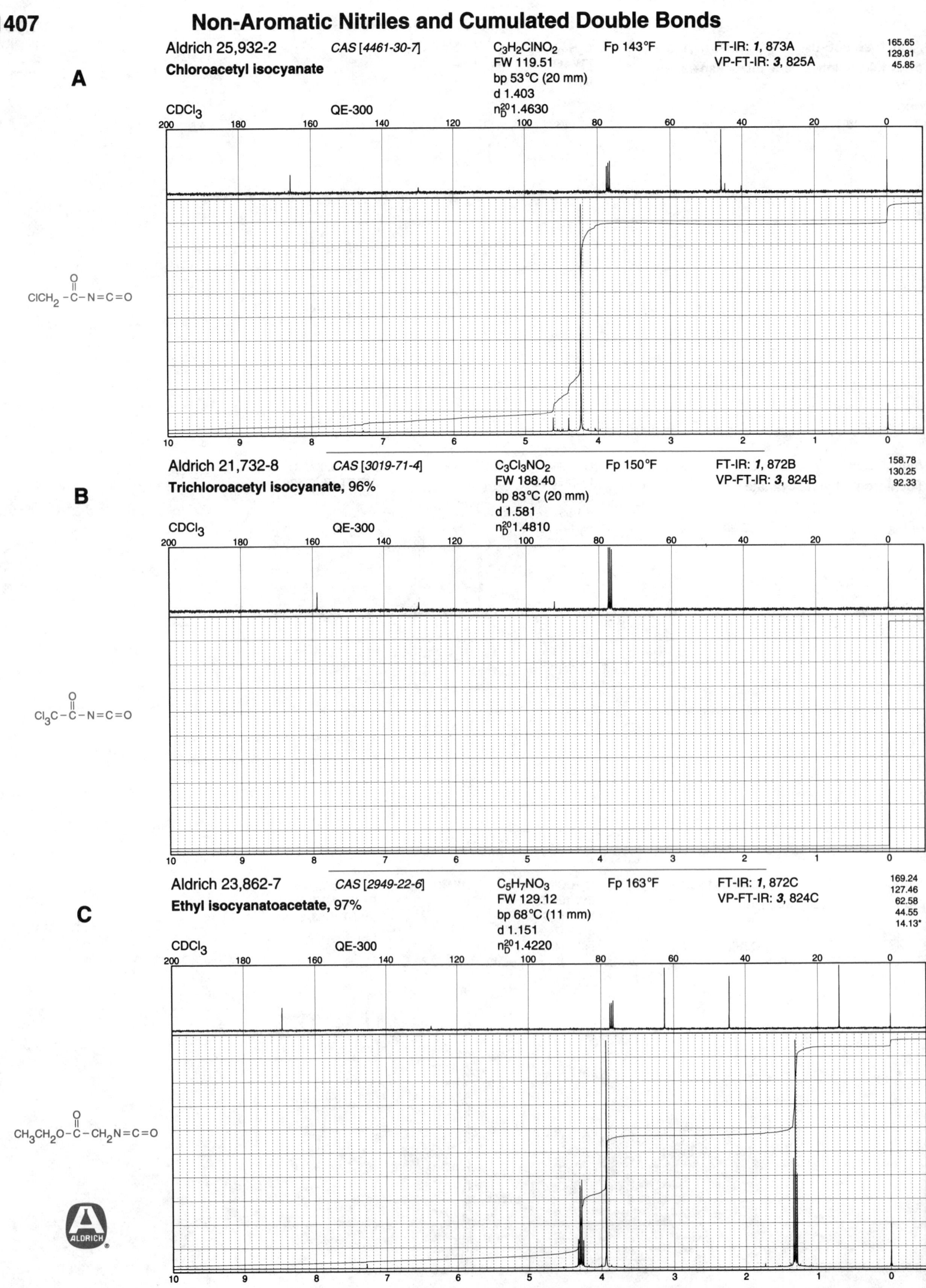

A

Aldrich 25,932-2   *CAS [4461-30-7]*   C₃H₂ClNO₂   Fp 143°F   FT-IR: *1*, 873A   165.65
Chloroacetyl isocyanate   FW 119.51   VP-FT-IR: *3*, 825A   129.81
bp 53°C (20 mm)   45.85
d 1.403
CDCl₃   QE-300   n²⁰_D 1.4630

ClCH₂–C(=O)–N=C=O

B

Aldrich 21,732-8   *CAS [3019-71-4]*   C₃Cl₃NO₂   Fp 150°F   FT-IR: *1*, 872B   158.78
Trichloroacetyl isocyanate, 96%   FW 188.40   VP-FT-IR: *3*, 824B   130.25
bp 83°C (20 mm)   92.33
d 1.581
CDCl₃   QE-300   n²⁰_D 1.4810

Cl₃C–C(=O)–N=C=O

C

Aldrich 23,862-7   *CAS [2949-22-6]*   C₅H₇NO₃   Fp 163°F   FT-IR: *1*, 872C   169.24
Ethyl isocyanatoacetate, 97%   FW 129.12   VP-FT-IR: *3*, 824C   127.46
bp 68°C (11 mm)   62.58
d 1.151   44.55
CDCl₃   QE-300   n²⁰_D 1.4220   14.13*

CH₃CH₂O–C(=O)–CH₂N=C=O

Aldrich 30,822-6 *CAS [19617-43-7]* C$_4$H$_5$NO$_3$ Fp 96 °F VP-FT-IR: *3*, 825B 149.24 / 130.01 / 65.07 / 13.95*

Ethoxycarbonyl isocyanate

FW 115.09
bp 25 °C (10 mm)
d 1.115
n$_D^{20}$ 1.4080

CDCl$_3$ QE-300

A

$$CH_3CH_2O-\overset{\overset{\displaystyle O}{\|}}{C}-N=C=O$$

Aldrich 14,658-7 *CAS [556-64-9]* C$_2$H$_3$NS n$_D^{20}$ 1.4680 113.21 / 16.32*

Methyl thiocyanate, 99%

FW 73.12 Fp 101 °F
mp -5 °C 60 MHz: *1*, 726B
bp 131 °C FT-IR: *1*, 873C
d 1.068 VP-FT-IR: *3*, 825D

CDCl$_3$ QE-300

B

CH$_3$SCN

Aldrich 30,895-1 *CAS [542-90-5]* C$_3$H$_5$NS Fp 109 °F VP-FT-IR: *3*, 826A 111.99 / 28.58 / 15.40*

Ethyl thiocyanate, 98%

FW 87.14
bp 145 °C
d 1.012
n$_D^{20}$ 1.4620

CDCl$_3$ QE-300

C

CH$_3$CH$_2$SCN

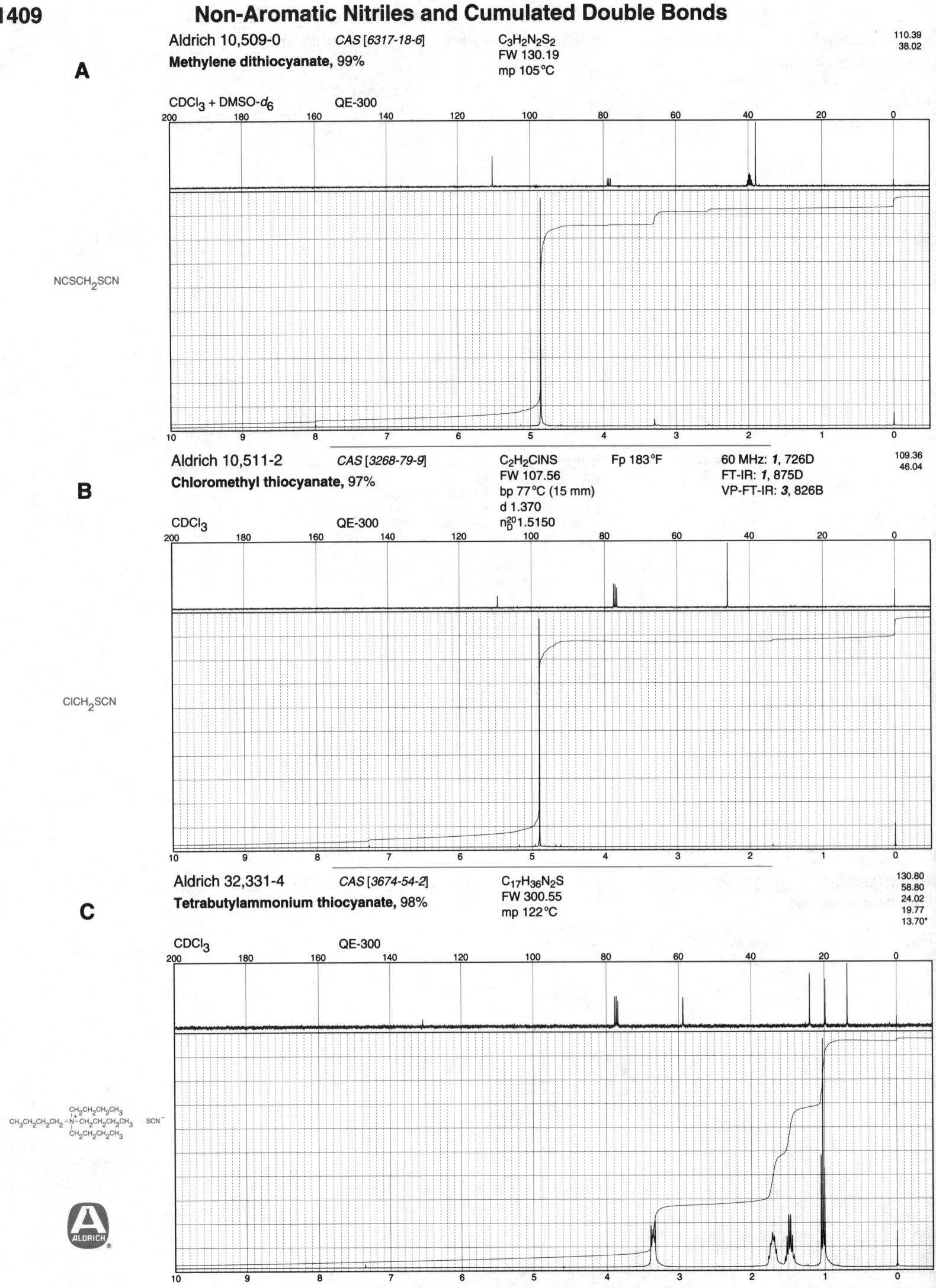

A

Aldrich 10,509-0   *CAS [6317-18-6]* $C_3H_2N_2S_2$
Methylene dithiocyanate, 99%   FW 130.19
mp 105°C

110.39
38.02

$CDCl_3$ + DMSO-d_6 QE-300

$NCSCH_2SCN$

B

Aldrich 10,511-2   *CAS [3268-79-9]* C_2H_2ClNS Fp 183°F   60 MHz: *1,* 726D
Chloromethyl thiocyanate, 97%   FW 107.56   FT-IR: *1,* 875D
bp 77°C (15 mm)   VP-FT-IR: *3,* 826B
d 1.370
n_D^{20}1.5150

109.36
46.04

$CDCl_3$ QE-300

$ClCH_2SCN$

C

Aldrich 32,331-4   *CAS [3674-54-2]* $C_{17}H_{36}N_2S$
Tetrabutylammonium thiocyanate, 98%   FW 300.55
mp 122°C

130.80
58.80
24.02
19.77
13.70*

$CDCl_3$ QE-300

$CH_3CH_2CH_2CH_2$—N$^+$—$CH_2CH_2CH_2CH_3$ SCN^-
with $CH_2CH_2CH_2CH_3$ groups

Aldrich 11,277-1   *CAS [556-61-6]*   C$_2$H$_3$NS   Fp 90°F   60 MHz: *1*, 727A   128.97
30.35*

Methyl isothiocyanate, 97%   FW 73.12   FT-IR: *1*, 873D
mp 32°C   VP-FT-IR: *3*, 826C

A

CDCl$_3$ QE-300   bp 118°C
d 1.069

CH$_3$N=C=S

Aldrich E3,390-4   *CAS [542-85-8]*   C$_3$H$_5$NS   n$_D^{20}$1.5120   129.57
40.19
15.59*

Ethyl isothiocyanate, 97%   FW 87.14   Fp 90°F

B

mp -6°C   60 MHz: *1*, 727B
bp 131°C   FT-IR: *1*, 874A

CDCl$_3$ QE-300   d 0.995   VP-FT-IR: *3*, 826D

CH$_3$CH$_2$N=C=S

Aldrich 25,394-4   *CAS [628-30-8]*   C$_4$H$_7$NS   Fp 5°F   FT-IR: *1*, 874B   129.50
46.76
23.58
11.26*

Propyl isothiocyanate, 98%   FW 101.17   VP-FT-IR: *3*, 827A

C

bp 153°C

CDCl$_3$ QE-300   d 0.978
n$_D^{20}$1.5040

CH$_3$CH$_2$CH$_2$N=C=S

ALDRICH

A

Aldrich 25,379-0   *CAS [592-82-5]*
Butyl isothiocyanate, 99%

C$_5$H$_9$NS
FW 115.20
bp 71°C (35 mm)
d 0.955
n$_D^{20}$1.5000

Fp 151°F

60 MHz: *1*, 727C
FT-IR: *1*, 874C
VP-FT-IR: *3*, 827B

129.32
44.74
31.89
19.77
13.28*

CDCl$_3$ QE-300

CH$_3$CH$_2$CH$_2$CH$_2$N=C=S

B

Aldrich 25,185-2   *CAS [590-42-1]*
***tert*-Butyl isothiocyanate, 99%**

C$_5$H$_9$NS
FW 115.20
mp 11°C
bp 31°C (10 mm)
d 0.908

n$_D^{20}$1.4800
Fp 101°F
60 MHz: *1*, 727D
VP-FT-IR: *3*, 827C

129.79
58.27
30.69*

CDCl$_3$ QE-300

$$CH_3-\underset{\underset{CH_3}{|}}{\overset{\overset{CH_3}{|}}{C}}-N=C=S$$

C

Aldrich C10,540-6   *CAS [1122-82-3]*
Cyclohexyl isothiocyanate, 98%

C$_7$H$_{11}$NS
FW 141.24
bp 219°C
d 0.996
n$_D^{20}$1.5380

Fp 204°F

60 MHz: *1*, 728C
FT-IR: *1*, 875B
VP-FT-IR: *3*, 828A

129.73
55.40*
33.21
25.06
23.24

CDCl$_3$ QE-300

N=C=S

ALDRICH

Aldrich 19,612-6   *CAS [16182-04-0]*   C₄H₅NO₂S   Fp 123°F   FT-IR: *1*, 873B

Ethoxycarbonyl isothiocyanate, 98%   FW 131.15   VP-FT-IR: *3*, 825C

bp 56°C (18 mm)

d 1.112

CDCl₃   QE-300   n₂₀D 1.5000

A

149.35
147.11
65.17
13.91*

$CH_3CH_2O-\overset{\overset{\text{O}}{\|}}{C}-N=C=S$

Aldrich 33,858-3   *CAS [14152-97-7]*   C₁₅H₁₉NO₉S

2,3,4,6-Tetra-*O*-acetyl-β-ᴅ-glucopyranosyl   FW 389.38

isothiocyanate   mp 115°C

CDCl₃   QE-300

B

170.48   72.45*
170.02   71.83*
169.13   67.61*
168.96   61.47
144.20   20.69*
83.45*   20.52*
74.02*

$R = -\overset{\overset{\text{O}}{\|}}{C}-CH_3$

Aldrich D12,540-7   *CAS [693-13-0]*   C₇H₁₄N₂   Fp 93°F   60 MHz: *1*, 729A

1,3-Diisopropylcarbodiimide, 99%   FW 126.20   FT-IR: *1*, 876A

bp 147°C   VP-FT-IR: *3*, 828D

d 0.806

CDCl₃   QE-300   n₂₀D 1.4330

C

140.19
48.95*
24.61*

$CH_3CHN=C=NCHCH_3$
with CH_3 groups

A

Aldrich 23,556-3   CAS [691-24-7] $C_9H_{18}N_2$ Fp 95°F   60 MHz: **1**, 729D   139.14
1,3-Di-*tert*-butylcarbodiimide, 99%   FW 154.26   FT-IR: **1**, 876B   54.85
                                            bp 49°C (12 mm)   VP-FT-IR: **3**, 829A   31.31*
                                            d 0.800
                                            n_D^{20}1.4280

CDCl$_3$ QE-300

$$CH_3-\underset{\underset{CH_3}{|}}{\overset{\overset{CH_3}{|}}{C}}-N=C=N-\underset{\underset{CH_3}{|}}{\overset{\overset{CH_3}{|}}{C}}-CH_3$$

B

Aldrich D8,000-2   CAS [538-75-0] $C_{13}H_{22}N_2$ 60 MHz: **1**, 730A   139.75
1,3-Dicyclohexylcarbodiimide, 99%   FW 206.33   FT-IR: **1**, 876C   55.72*
                                          mp 35°C   VP-FT-IR: **3**, 829B   34.92
                                          bp 123°C (6 mm)   25.46
                                          Fp >230°F   24.70

CDCl$_3$ QE-300

$$\text{cyclohexyl}-N=C=N-\text{cyclohexyl}$$

C

Aldrich 16,534-4   CAS [22572-40-3] $C_9H_{20}IN_3$ 139.31
1-(3-Dimethylaminopropyl)-3-ethylcarbodiimide   FW 297.18   64.96
methiodide   mp 98°C   54.10*
                                                                     43.27
                                                                     41.46
                                                                     25.13
                                                                     16.93*

CDCl$_3$ QE-300

$$CH_3CH_2N=C=NCH_2CH_2CH_2\overset{\overset{CH_3}{|}}{\underset{\underset{CH_3}{|}}{N^+}}CH_3 \quad I^-$$

A

Aldrich 22,640-8

O-Methyl-*N,N'*-diisopropylisourea, 99%

CAS [54648-79-2]

C₈H₁₈N₂O
FW 158.25
bp 51°C
d 0.871
n²⁰_D 1.4360

Fp 96°F

60 MHz: **2**, 969B
FT-IR: **1**, 878B
VP-FT-IR: **3**, 830A

151.03
51.81*
44.48*
42.71*
24.71*
23.42*

DMSO-*d*₆   QE-300

```
        CH3          CH3
CH3CHNHC=NCHCH3
             OCH3
```

B

Aldrich D9,000-8

Diethyl azodicarboxylate

CAS [1972-28-7]

C₆H₁₀N₂O₄
FW 174.16
bp 106°C (13 mm)
d 1.106
n²⁰_D 1.4200

Fp >230°F

60 MHz: **1**, 653C
FT-IR: **1**, 776D
VP-FT-IR: **3**, 788A

160.28
65.48
14.05*

CDCl₃   QE-300

```
         O          O
          ‖          ‖
CH3CH2O-C-N=N-C-OCH2CH3
```

C

Aldrich 22,554-1

Diisopropyl azodicarboxylate, 95%

CAS [2446-83-5]

C₈H₁₄N₂O₄
FW 202.21
bp 75°C
d 1.027
n²⁰_D 1.4200

Fp 223°F

60 MHz: **1**, 653D
FT-IR: **1**, 878C

159.90
74.32*
21.58*

CDCl₃   QE-300

```
 CH3   O         O   CH3
  |     ‖          ‖    |
CH3CH-O-C-N=N-C-O-CHCH3
```

Non-Aromatic Nitriles and Cumulated Double Bonds

A

Aldrich 13,599-2   *CAS [870-50-8]* $C_{10}H_{18}N_2O_4$ 60 MHz: *1*, 654A

Di-*tert*-butyl azodicarboxylate, 98%   FW 230.26
mp 91 °C

158.73
87.48
27.12*

DMSO-d_6 QE-300

$$CH_3-\overset{\underset{\displaystyle CH_3}{|}}{\underset{\underset{\displaystyle CH_3}{|}}{C}}-O-\overset{\overset{\displaystyle O}{||}}{C}-N=N-\overset{\overset{\displaystyle O}{||}}{C}-O-\overset{\underset{\displaystyle CH_3}{|}}{\underset{\underset{\displaystyle CH_3}{|}}{C}}-CH_3$$

B

Aldrich 29,153-6   *CAS [38857-88-4]* $C_6H_4Cl_6N_2O_4$

Bis(2,2,2-trichloroethyl) azodicarboxylate, 98%   FW 380.83
mp 110 °C

158.33
93.05
76.84

CDCl$_3$ QE-300

$$Cl_3CCH_2O-\overset{\overset{\displaystyle O}{||}}{C}-N=N-\overset{\overset{\displaystyle O}{||}}{C}-OCH_2CCl_3$$

C

Aldrich 38,021-0   *CAS [2094-98-6]* $C_{14}H_{20}N_4$

1,1'-Azobis(cyclohexanecarbonitrile), 98%   FW 244.34
mp 117 °C

118.25
74.33
33.85
24.34
21.96

CDCl$_3$ QE-300

Aldrich 27,621-9

CAS [24886-73-5]

1-Azidoadamantane, 97%

$C_{10}H_{15}N_3$
FW 177.25
mp 81°C
Fp 185°F

58.97
41.51
35.91
29.81*

A

CDCl$_3$

QE-300

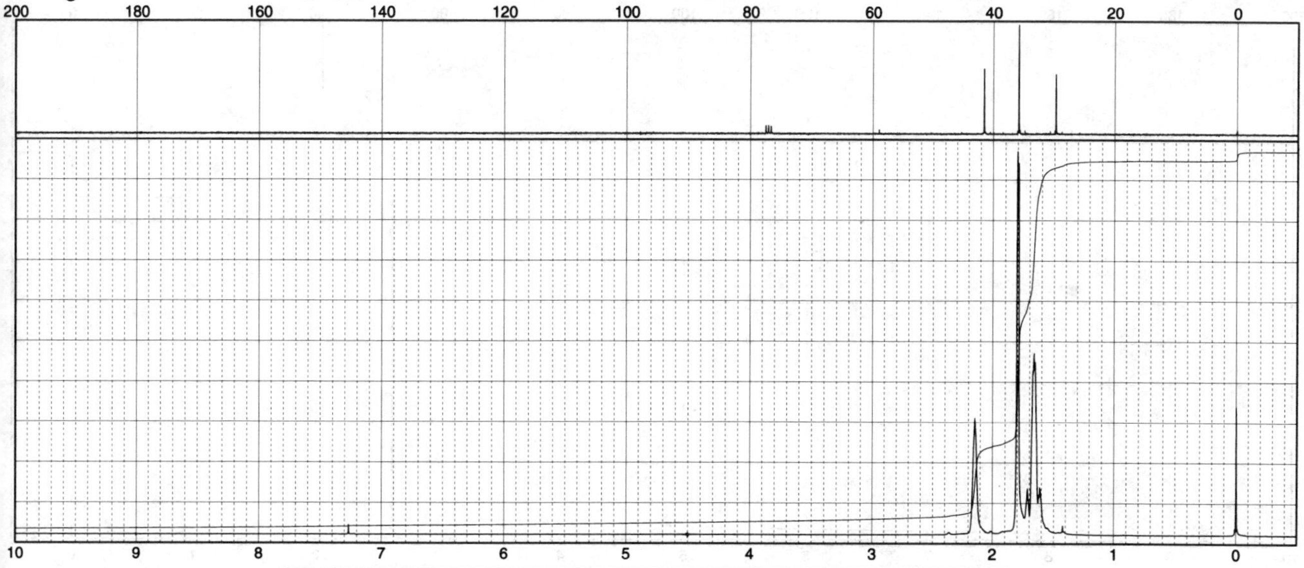

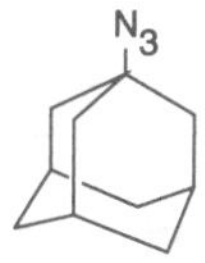

Aldrich 27,315-5

CAS [627-70-3]

Acetone azine, 98%

$C_6H_{12}N_2$
FW 112.18
mp -63°C
bp 133°C
d 0.842

n_D^{20} 1.4540
Fp 88°F

159.62
25.07*
17.84*

B

CDCl$_3$

QE-300

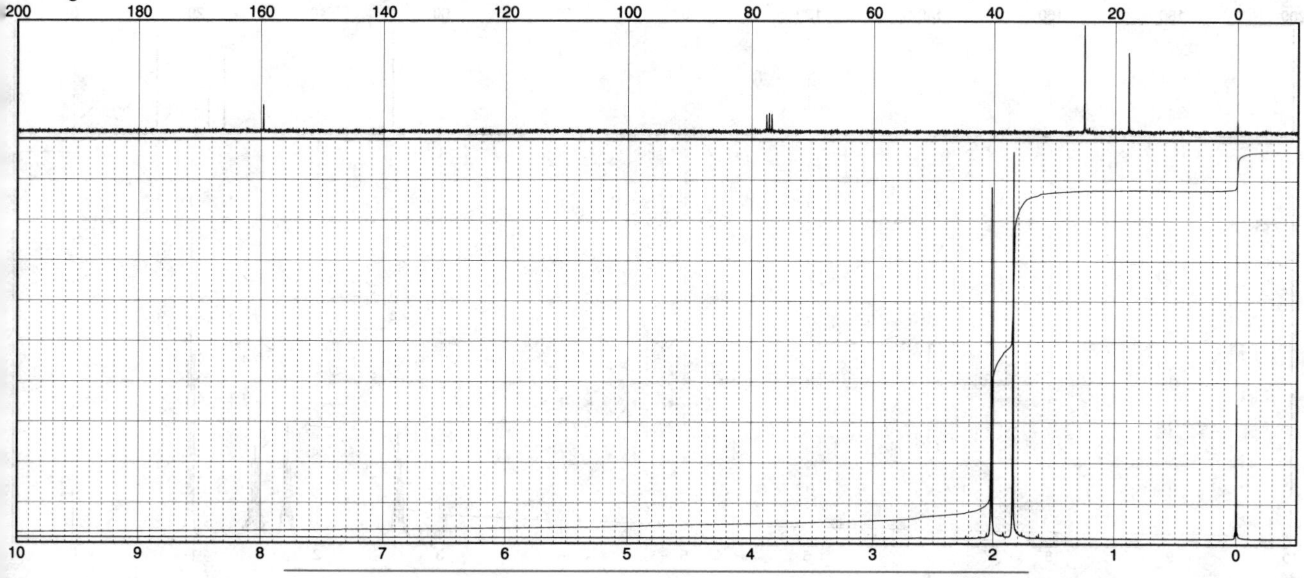

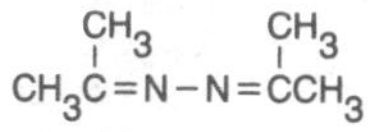

$$CH_3\underset{\underset{CH_3}{|}}{C}=N-N=\underset{\underset{CH_3}{|}}{C}CH_3$$

A

Aldrich 15,493-8   *CAS [67-68-5]*

Methyl sulfoxide, 99.9%

C_2H_6OS
FW 78.13
mp 18°C
bp 189°C
d 1.101

n_D^{20} 1.4790
Fp 203°F
60 MHz: *2*, 781A
FT-IR: *1*, 879A
VP-FT-IR: *3*, 831A

40.98*

CDCl$_3$ QE-300

$CH_3-\overset{\overset{O}{\|}}{S}-CH_3$

B

Aldrich B10,240-7   *CAS [2168-93-6]*

Butyl sulfoxide, 96%

$C_8H_{18}OS$
FW 162.30
mp 33°C
bp 250°C
d 0.832

n_D^{20} 1.4670
Fp 250°F
60 MHz: *2*, 781B
FT-IR: *1*, 879B

52.14
24.61
22.08
13.70*

CDCl$_3$ QE-300

$CH_3CH_2CH_2CH_2-\overset{\overset{O}{\|}}{S}-CH_2CH_2CH_2CH_3$

C

Aldrich T2,240-3   *CAS [1600-44-8]*

Tetramethylene sulfoxide, 96%

C_4H_8OS
FW 104.17
d 1.158
n_D^{20} 1.5200
Fp >230°F

60 MHz: *2*, 781C
FT-IR: *1*, 879C

54.47
25.43

CDCl$_3$ QE-300

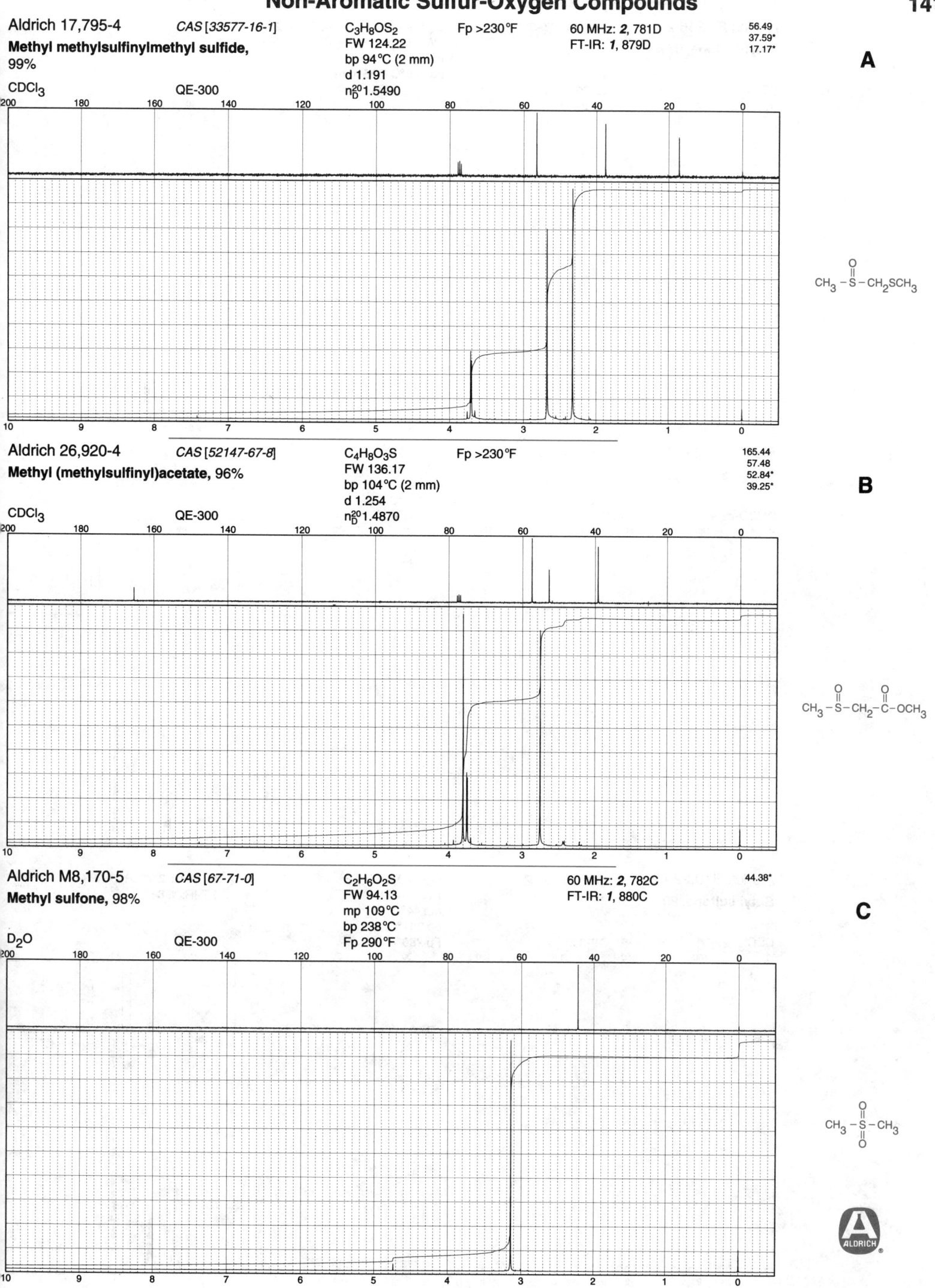

Aldrich 17,795-4   *CAS [33577-16-1]*
Methyl methylsulfinylmethyl sulfide,
99%

$C_3H_8OS_2$
FW 124.22
bp 94°C (2 mm)
d 1.191
n_D^{20} 1.5490

Fp >230°F

60 MHz: **2**, 781D
FT-IR: **1**, 879D

56.49
37.59*
17.17*

A

CDCl₃   QE-300

$CH_3-\overset{O}{\overset{\|}{S}}-CH_2SCH_3$

Aldrich 26,920-4   *CAS [52147-67-8]*
Methyl (methylsulfinyl)acetate, 96%

$C_4H_8O_3S$
FW 136.17
bp 104°C (2 mm)
d 1.254
n_D^{20} 1.4870

Fp >230°F

165.44
57.48
52.84*
39.25*

B

CDCl₃   QE-300

$CH_3-\overset{O}{\overset{\|}{S}}-CH_2-\overset{O}{\overset{\|}{C}}-OCH_3$

Aldrich M8,170-5   *CAS [67-71-0]*
Methyl sulfone, 98%

$C_2H_6O_2S$
FW 94.13
mp 109°C
bp 238°C
Fp 290°F

60 MHz: **2**, 782C
FT-IR: **1**, 880C

44.38*

C

D₂O   QE-300

$CH_3-\overset{O}{\underset{O}{\overset{\|}{\underset{\|}{S}}}}-CH_3$

A

Aldrich 24,246-2   *CAS [597-35-3]*   C$_4$H$_{10}$O$_2$S   60 MHz: *2*, 782D   46.16
Ethyl sulfone, 97%   FW 122.19   FT-IR: *1*, 880D   6.56*
                                                      mp 74°C
                                                      bp 246°C (755 mm)

CDCl$_3$ QE-300

CH$_3$CH$_2$—S—CH$_2$CH$_3$ (with O above and O below S)

B

Aldrich P5,429-9   *CAS [598-03-8]*   C$_6$H$_{14}$O$_2$S   60 MHz: *2*, 783A   53.20
Propyl sulfone, 99%   FW 150.24   FT-IR: *1*, 881B   15.12
                                                      mp 29°C   VP-FT-IR: *3*, 831B   12.81*
                                                      bp 270°C
                                                      Fp 260°F

DMSO-*d*$_6$ QE-300

CH$_3$CH$_2$CH$_2$—S—CH$_2$CH$_2$CH$_3$ (with O above and O below S)

C

Aldrich B10,220-2   *CAS [598-04-9]*   C$_8$H$_{18}$O$_2$S   60 MHz: *2*, 783B   52.48
Butyl sulfone, 99%   FW 178.29   FT-IR: *1*, 881C   23.95
                                                      mp 44°C   21.79
                                                      bp 291°C   13.55*
                                                      Fp 290°F

CDCl$_3$ QE-300

CH$_3$CH$_2$CH$_2$CH$_2$—S—CH$_2$CH$_2$CH$_2$CH$_3$ (with O above and O below S)

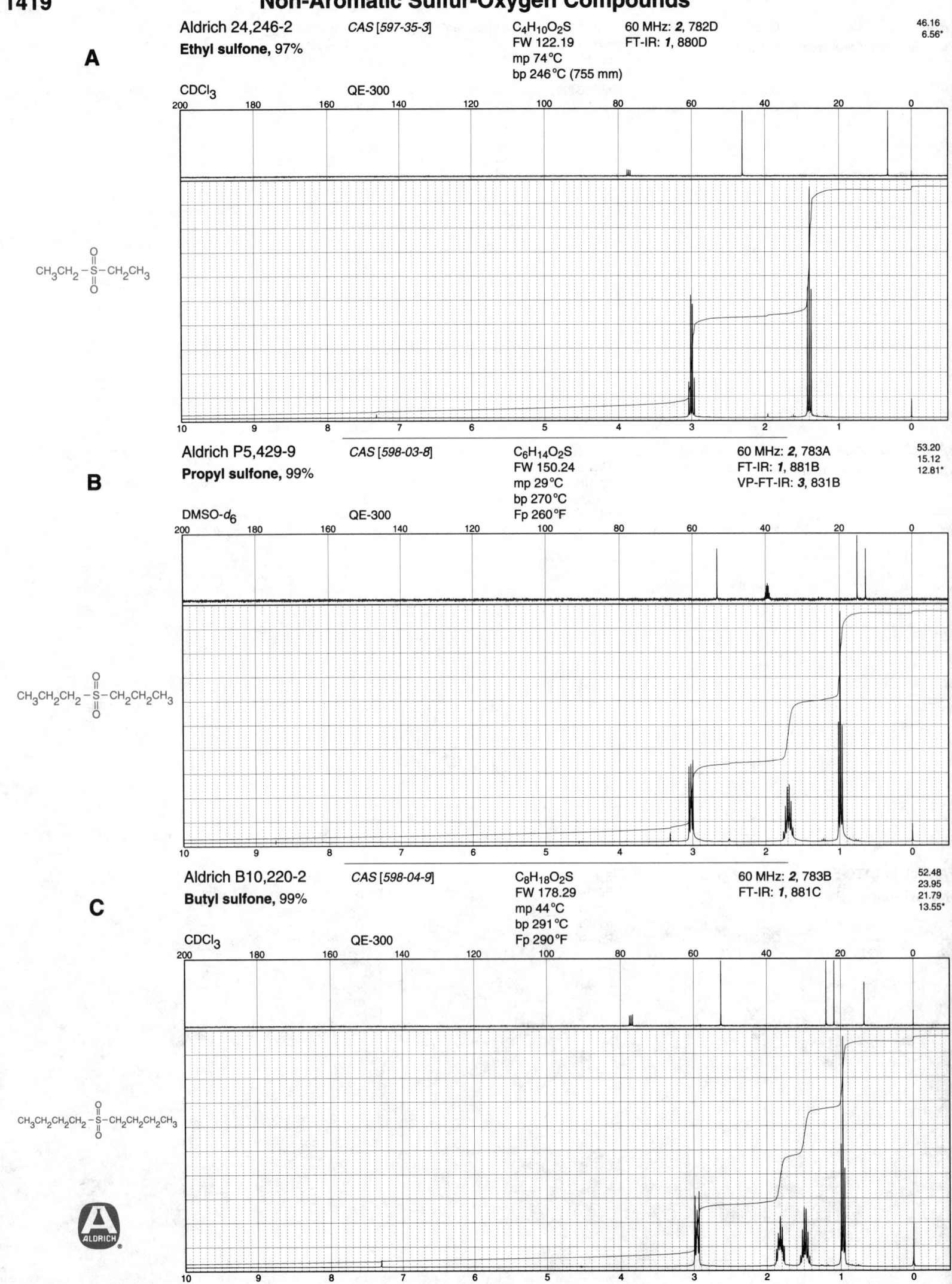

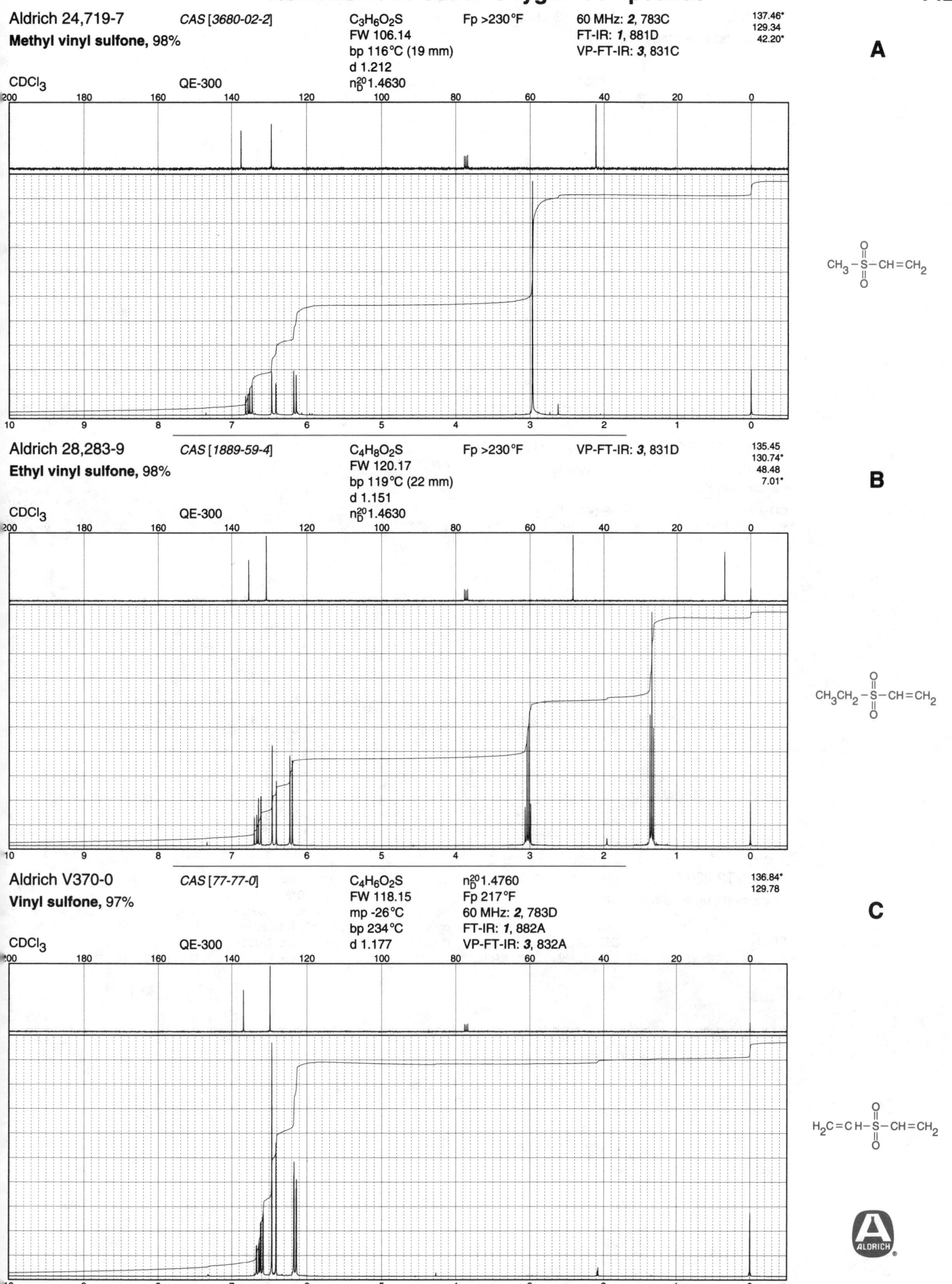

Aldrich 24,719-7   CAS [3680-02-2]   C₃H₆O₂S   Fp >230°F   60 MHz: 2, 783C   137.46*
Methyl vinyl sulfone, 98%   FW 106.14   FT-IR: 1, 881D   129.34
bp 116°C (19 mm)   VP-FT-IR: 3, 831C   42.20*
CDCl₃   QE-300   d 1.212
n²⁰_D 1.4630

A

$CH_3-S-CH=CH_2$ (with O above and O below S)

Aldrich 28,283-9   CAS [1889-59-4]   C₄H₈O₂S   Fp >230°F   VP-FT-IR: 3, 831D   135.45
Ethyl vinyl sulfone, 98%   FW 120.17   130.74*
bp 119°C (22 mm)   48.48
CDCl₃   QE-300   d 1.151   7.01*
n²⁰_D 1.4630

B

$CH_3CH_2-S-CH=CH_2$ (with O above and O below S)

Aldrich V370-0   CAS [77-77-0]   C₄H₆O₂S   n²⁰_D 1.4760   136.84*
Vinyl sulfone, 97%   FW 118.15   Fp 217°F   129.78
mp -26°C   60 MHz: 2, 783D
bp 234°C   FT-IR: 1, 882A
CDCl₃   QE-300   d 1.177   VP-FT-IR: 3, 832A

C

$H_2C=CH-S-CH=CH_2$ (with O above and O below S)

ALDRICH

A

Aldrich 36,354-5   *CAS [15205-66-0]* $C_3H_8O_3S$
2-(Methylsulfonyl)ethanol, 98%   FW 124.16
mp 32°C
bp 149°C
Fp >230°F

56.96
56.38
42.72*

$CH_3 - \overset{O}{\underset{O}{S}} - CH_2CH_2OH$

CDCl$_3$ QE-300

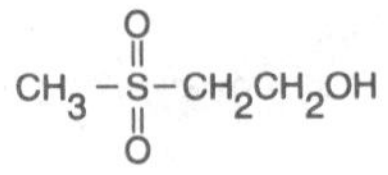

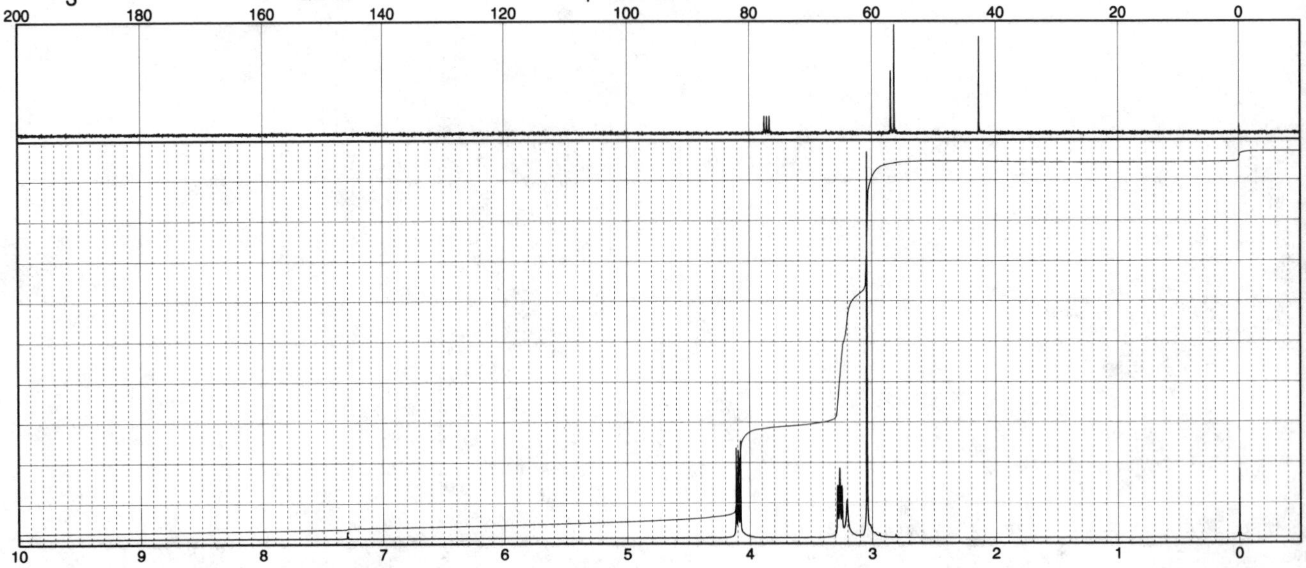

B

Aldrich 36,789-3   *CAS [57903-15-8]* $C_8H_{11}NO_7S$
2-(Methylsulfonyl)ethyl succinimidyl   FW 265.24
carbonate, 98%   mp 117°C

170.51
152.10
65.33
53.55
42.65*
26.29

CD$_3$CN   QE-300

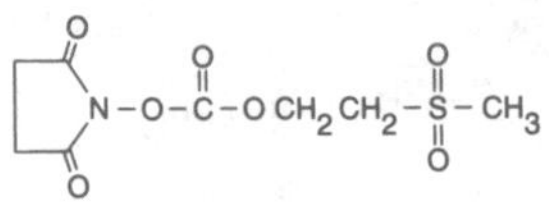

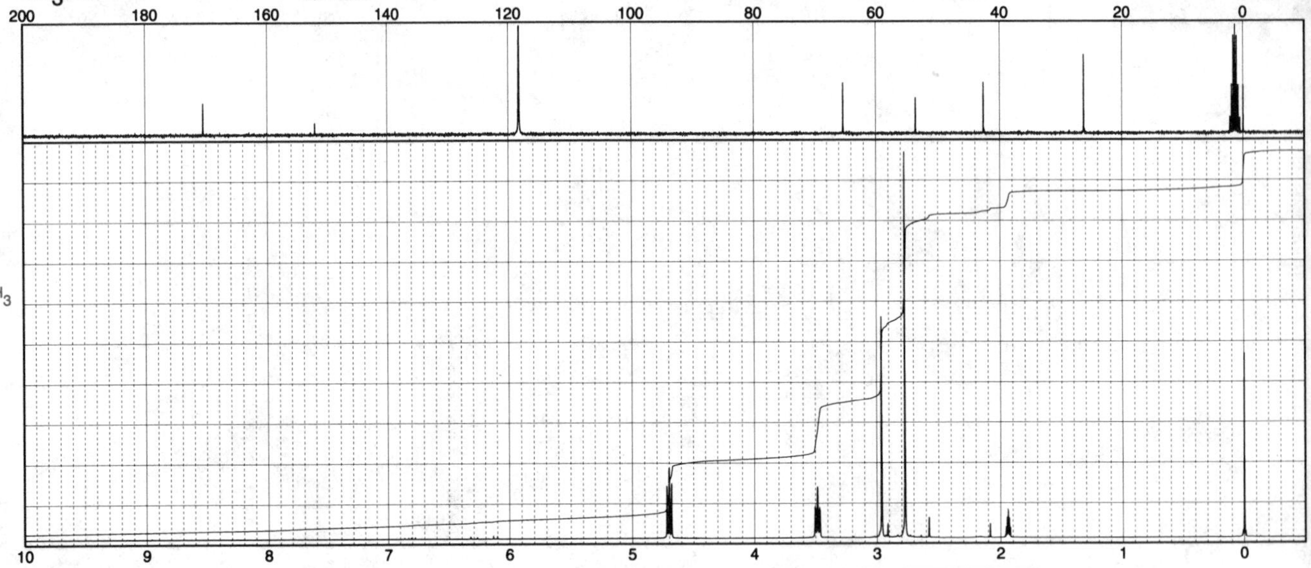

C

Aldrich T2,220-9   *CAS [126-33-0]* $C_4H_8O_2S$ n_D^{20} 1.4840
Tetramethylene sulfone, 99%   FW 120.17   Fp 330°F
mp 27°C   60 MHz: **2**, 784C
bp 285°C   FT-IR: **1**, 882C
d 1.261   VP-FT-IR: **3**, 832B

51.12
22.72

CDCl$_3$ QE-300

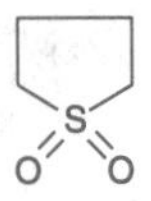

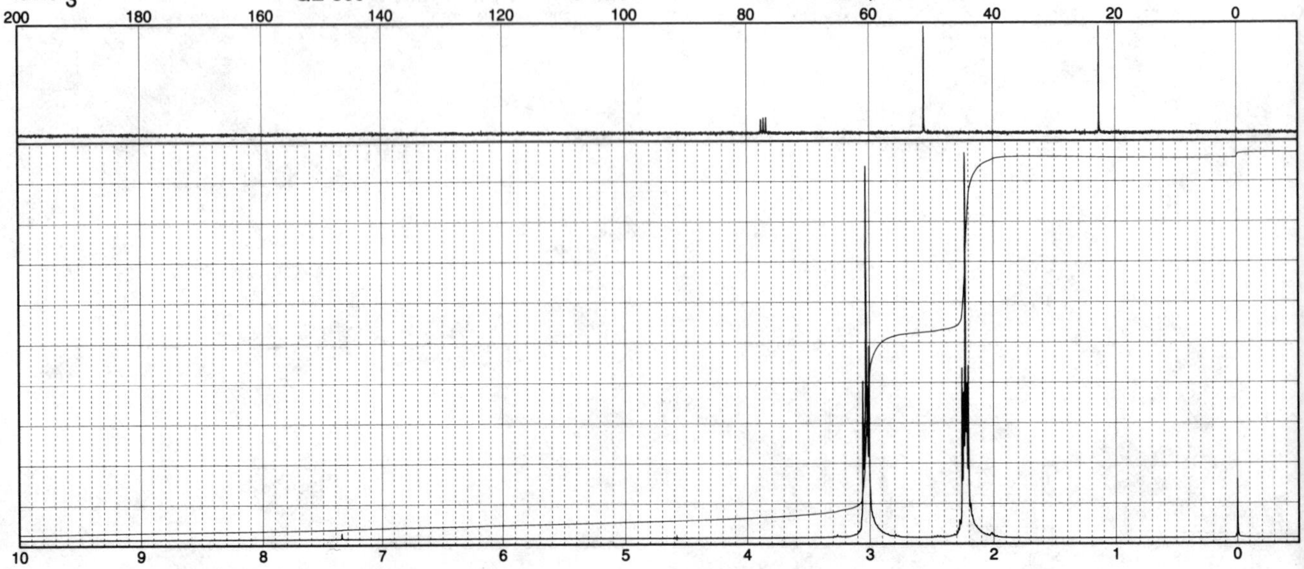

ALDRICH

Non-Aromatic Sulfur-Oxygen Compounds

1422

Aldrich M8,165-9   *CAS [872-93-5]* $C_5H_{10}O_2S$ Fp >230°F   60 MHz: *2*, 784D   58.23
52.64
31.61*
30.94
19.30*

3-Methylsulfolane, 99%   FW 134.20   FT-IR: *1*, 883A
bp 276°C   VP-FT-IR: *3*, 832C
d 1.191
n_D^{20}1.4770

CDCl₃   QE-300

A

Aldrich D18,640-6   *CAS [1003-78-7]* $C_6H_{12}O_2S$ n_D^{20}1.4730   58.53*
57.96
39.82
29.01*
19.45*
12.14*

2,4-Dimethylsulfolane, 95%   FW 148.22   Fp 294°F
mp -3°C   60 MHz: *2*, 785A
bp 281°C   FT-IR: *1*, 883B
d 1.136

CDCl₃   QE-300

B

Aldrich 16,219-1   *CAS [49592-61-2]* $C_4H_7ClO_3S$ 60 MHz: *2*, 785D   73.15*
58.15*
57.06
56.58

4-Chlorotetrahydrothiophene-3-ol-1,1-   FW 170.61   FT-IR: *1*, 883C
dioxide, 98%   mp 163°C

DMSO-d_6 QE-300

C

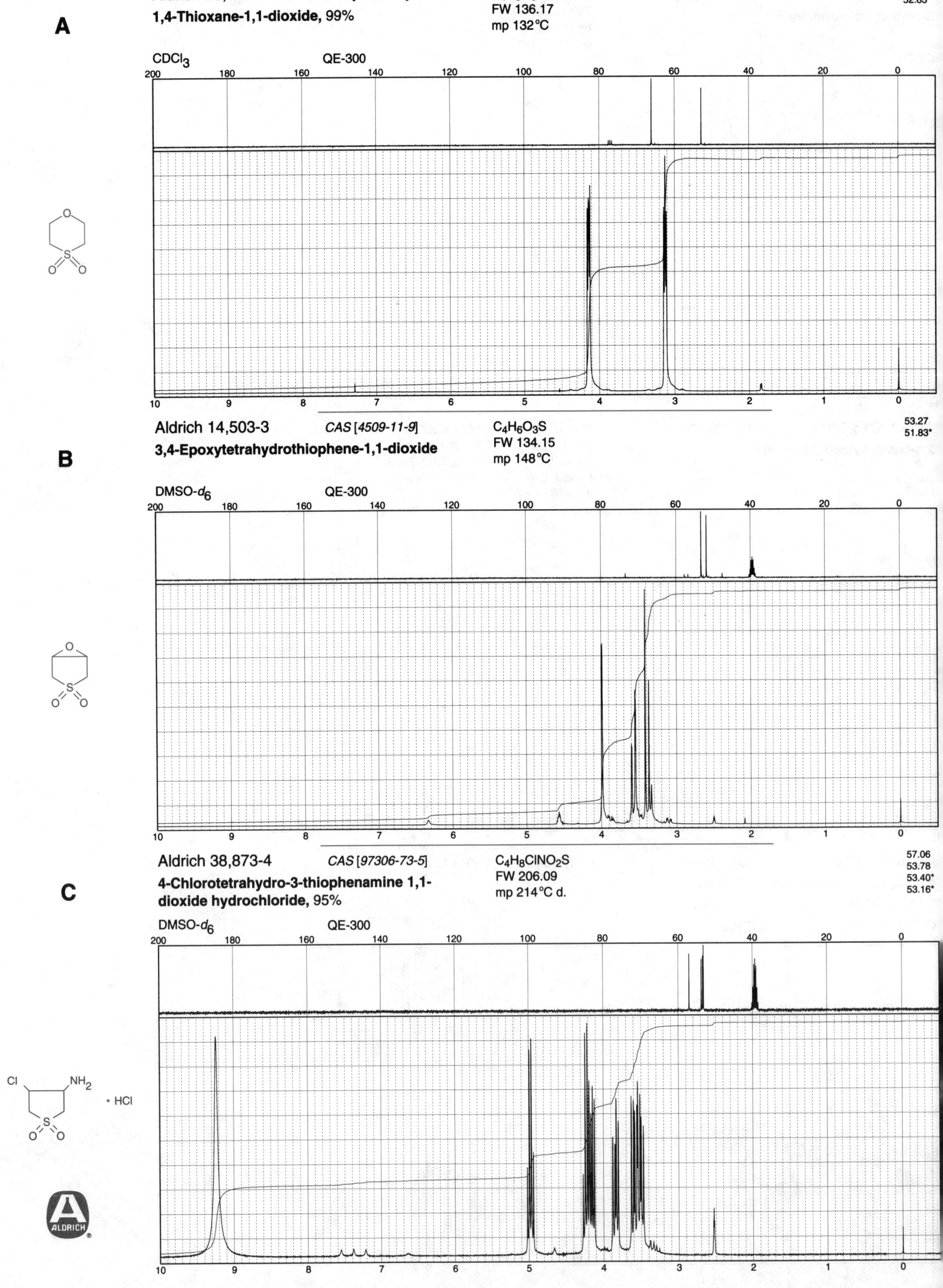

A

Aldrich 28,319-3   CAS [107-61-9] $C_4H_8O_3S$
1,4-Thioxane-1,1-dioxide, 99%   FW 136.17   mp 132°C

66.14
52.83

CDCl$_3$ QE-300

B

Aldrich 14,503-3   CAS [4509-11-9] $C_4H_6O_3S$
3,4-Epoxytetrahydrothiophene-1,1-dioxide   FW 134.15   mp 148°C

53.27
51.83*

DMSO-d_6 QE-300

C

Aldrich 38,873-4   CAS [97306-73-5] $C_4H_8ClNO_2S$
4-Chlorotetrahydro-3-thiophenamine 1,1-dioxide hydrochloride, 95%   FW 206.09   mp 214°C d.

57.06
53.78
53.40*
53.16*

DMSO-d_6 QE-300

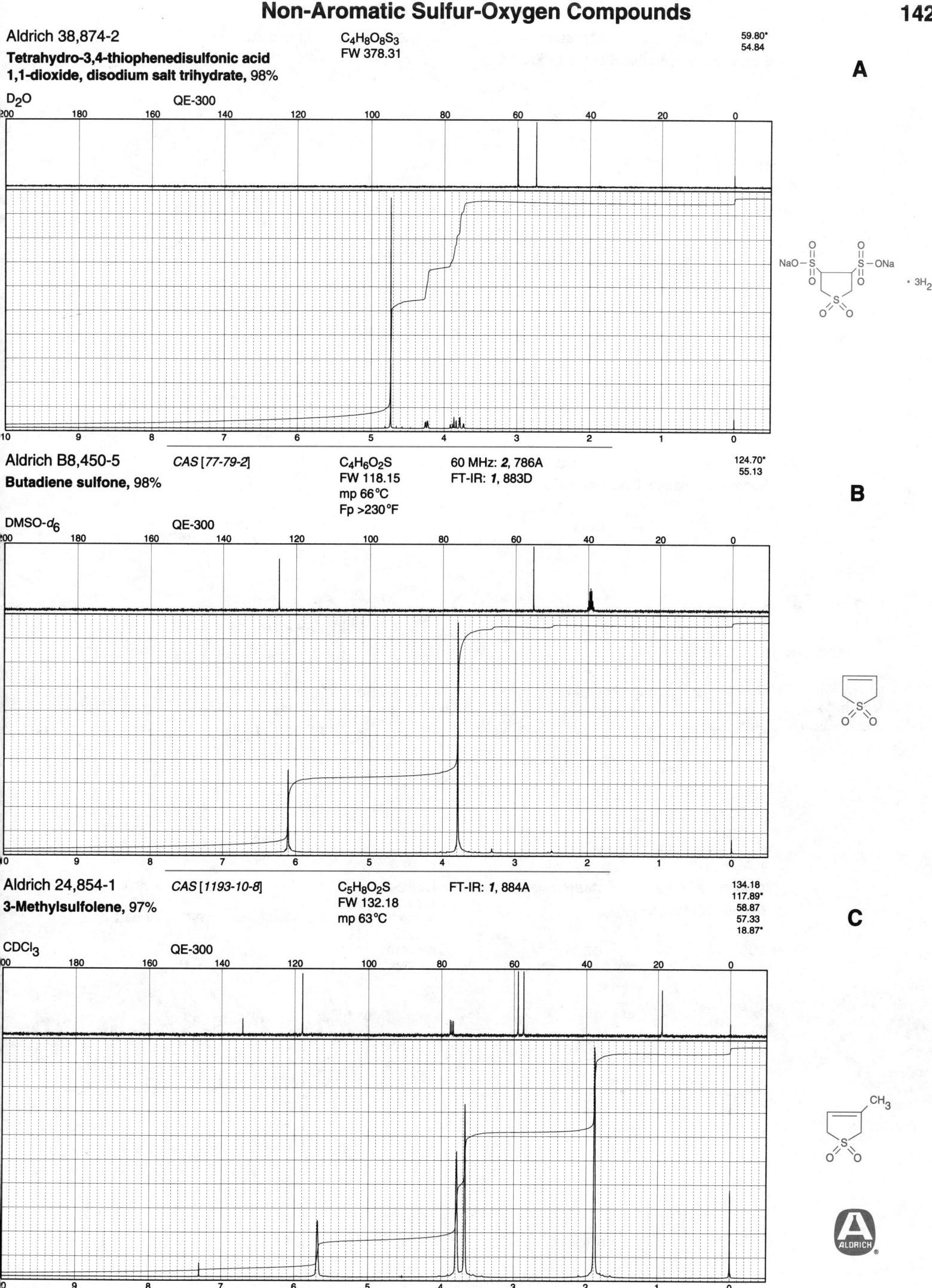

Aldrich 38,874-2

Tetrahydro-3,4-thiophenedisulfonic acid 1,1-dioxide, disodium salt trihydrate, 98%

C₄H₈O₈S₃
FW 378.31

59.80*
54.84

A

D₂O QE-300

Aldrich B8,450-5 CAS [77-79-2]

Butadiene sulfone, 98%

C₄H₆O₂S
FW 118.15
mp 66°C
Fp >230°F

60 MHz: 2, 786A
FT-IR: 1, 883D

124.70*
55.13

B

DMSO-d₆ QE-300

Aldrich 24,854-1 CAS [1193-10-8]

3-Methylsulfolene, 97%

C₅H₈O₂S
FW 132.18
mp 63°C

FT-IR: 1, 884A

134.18
117.89*
58.87
57.33
18.87*

C

CDCl₃ QE-300

Non-Aromatic Sulfur-Oxygen Compounds

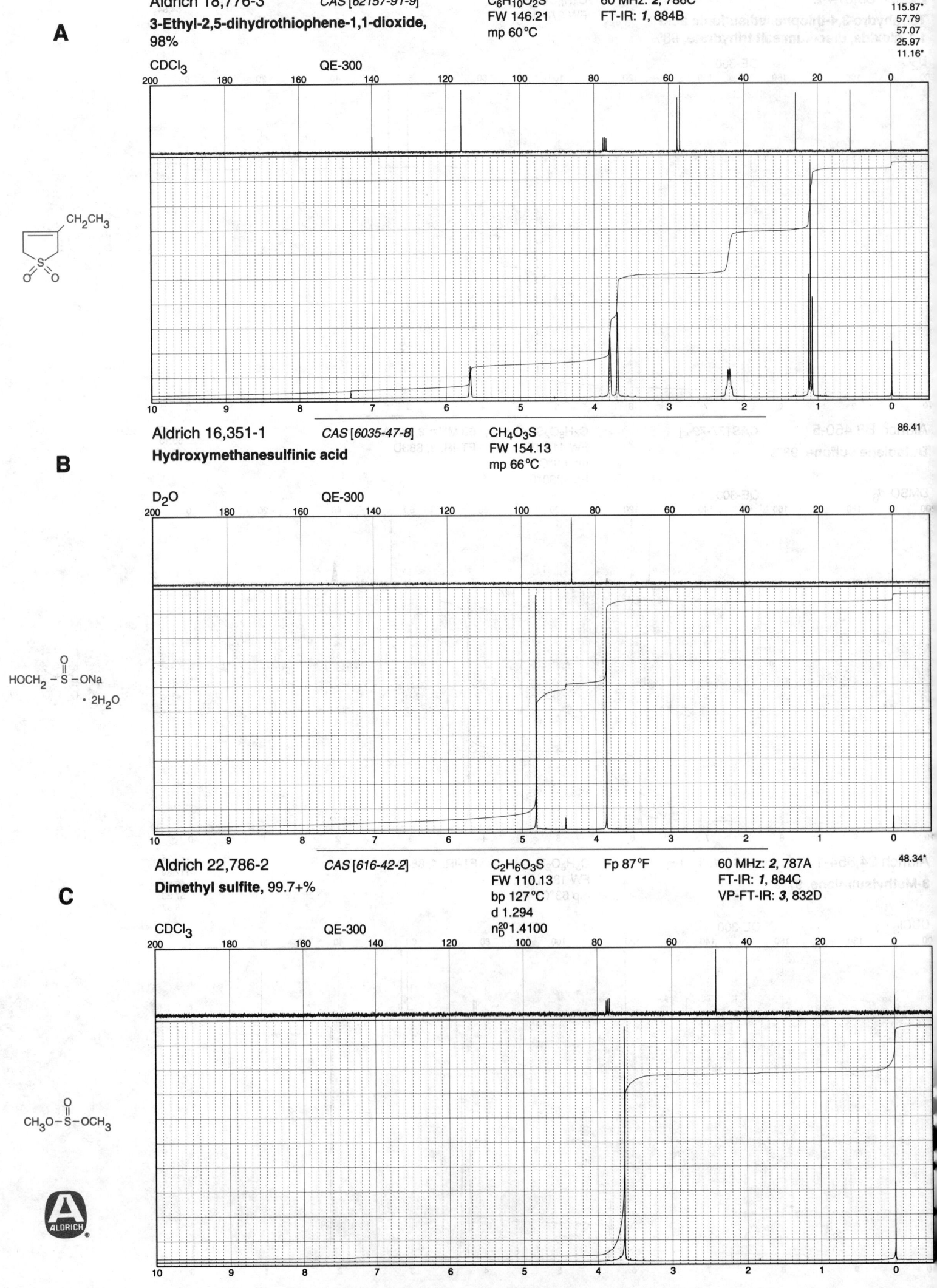

A

Aldrich 18,776-3   *CAS [62157-91-9]* $C_6H_{10}O_2S$ 60 MHz: **2**, 786C

3-Ethyl-2,5-dihydrothiophene-1,1-dioxide, 98%   FW 146.21   FT-IR: **1**, 884B

mp 60°C

139.90
115.87*
57.79
57.07
25.97
11.16*

CDCl₃   QE-300

86.41

Aldrich 16,351-1   *CAS [6035-47-8]* CH_4O_3S

Hydroxymethanesulfinic acid   FW 154.13

mp 66°C

B

D₂O   QE-300

48.34*

Aldrich 22,786-2   *CAS [616-42-2]* $C_2H_6O_3S$ Fp 87°F   60 MHz: **2**, 787A

Dimethyl sulfite, 99.7+%   FW 110.13   FT-IR: **1**, 884C

bp 127°C   VP-FT-IR: **3**, 832D

d 1.294

n_D^{20} 1.4100

C

CDCl₃   QE-300

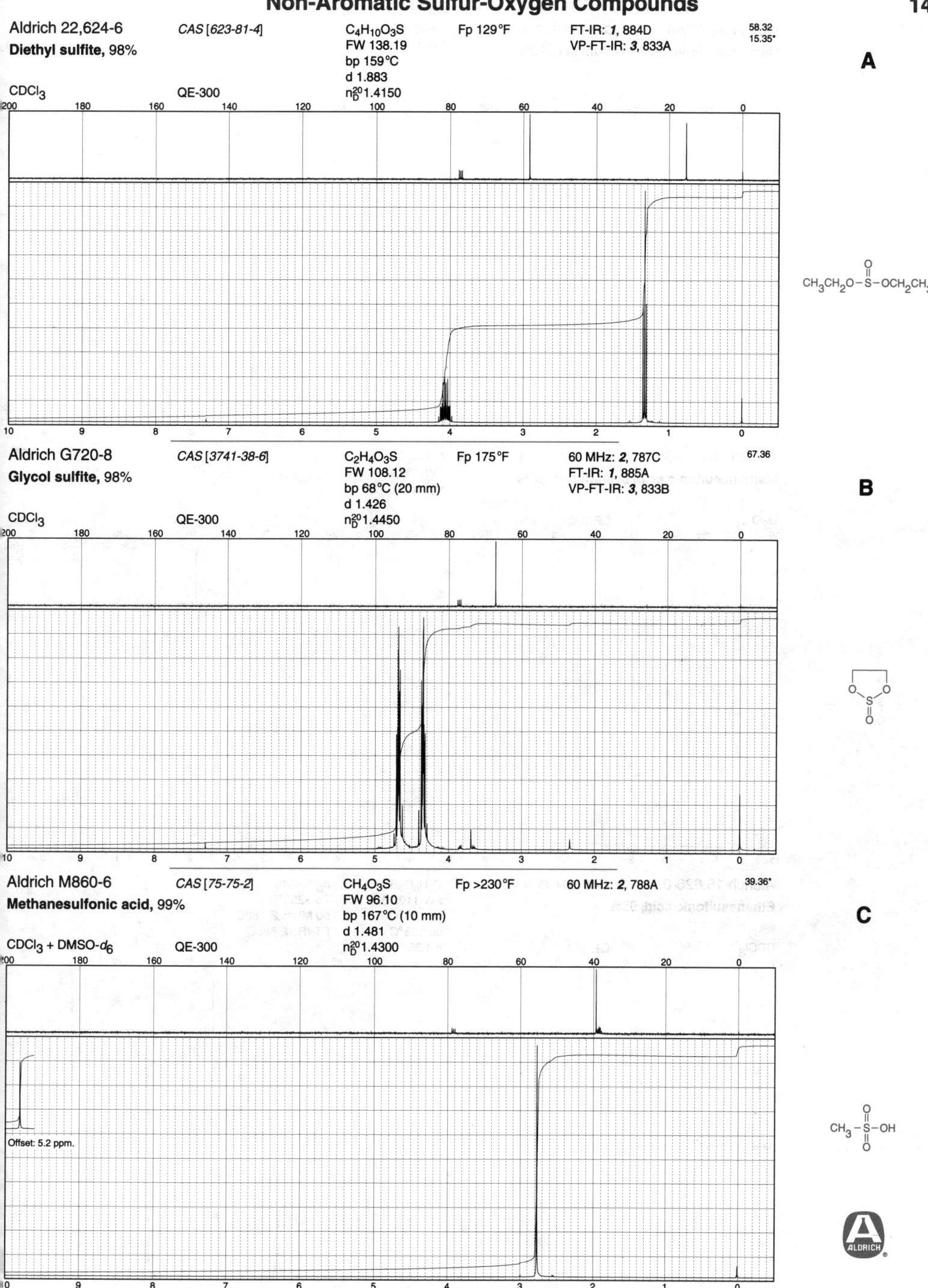

Aldrich 22,624-6
Diethyl sulfite, 98%

CAS [623-81-4]

$C_4H_{10}O_3S$
FW 138.19
bp 159°C
d 1.883
n_D^{20} 1.4150

Fp 129°F

FT-IR: *1*, 884D
VP-FT-IR: *3*, 833A

58.32
15.35*

A

$CDCl_3$

QE-300

$CH_3CH_2O-\overset{\displaystyle O}{\underset{\displaystyle }{S}}-OCH_2CH_3$

Aldrich G720-8
Glycol sulfite, 98%

CAS [3741-38-6]

$C_2H_4O_3S$
FW 108.12
bp 68°C (20 mm)
d 1.426
n_D^{20} 1.4450

Fp 175°F

60 MHz: *2*, 787C
FT-IR: *1*, 885A
VP-FT-IR: *3*, 833B

67.36

B

$CDCl_3$

QE-300

Aldrich M860-6
Methanesulfonic acid, 99%

CAS [75-75-2]

CH_4O_3S
FW 96.10
bp 167°C (10 mm)
d 1.481
n_D^{20} 1.4300

Fp >230°F

60 MHz: *2*, 788A

39.36*

C

$CDCl_3$ + DMSO-d_6

QE-300

Offset: 5.2 ppm.

$CH_3-\overset{\displaystyle O}{\underset{\displaystyle O}{S}}-OH$

ALDRICH

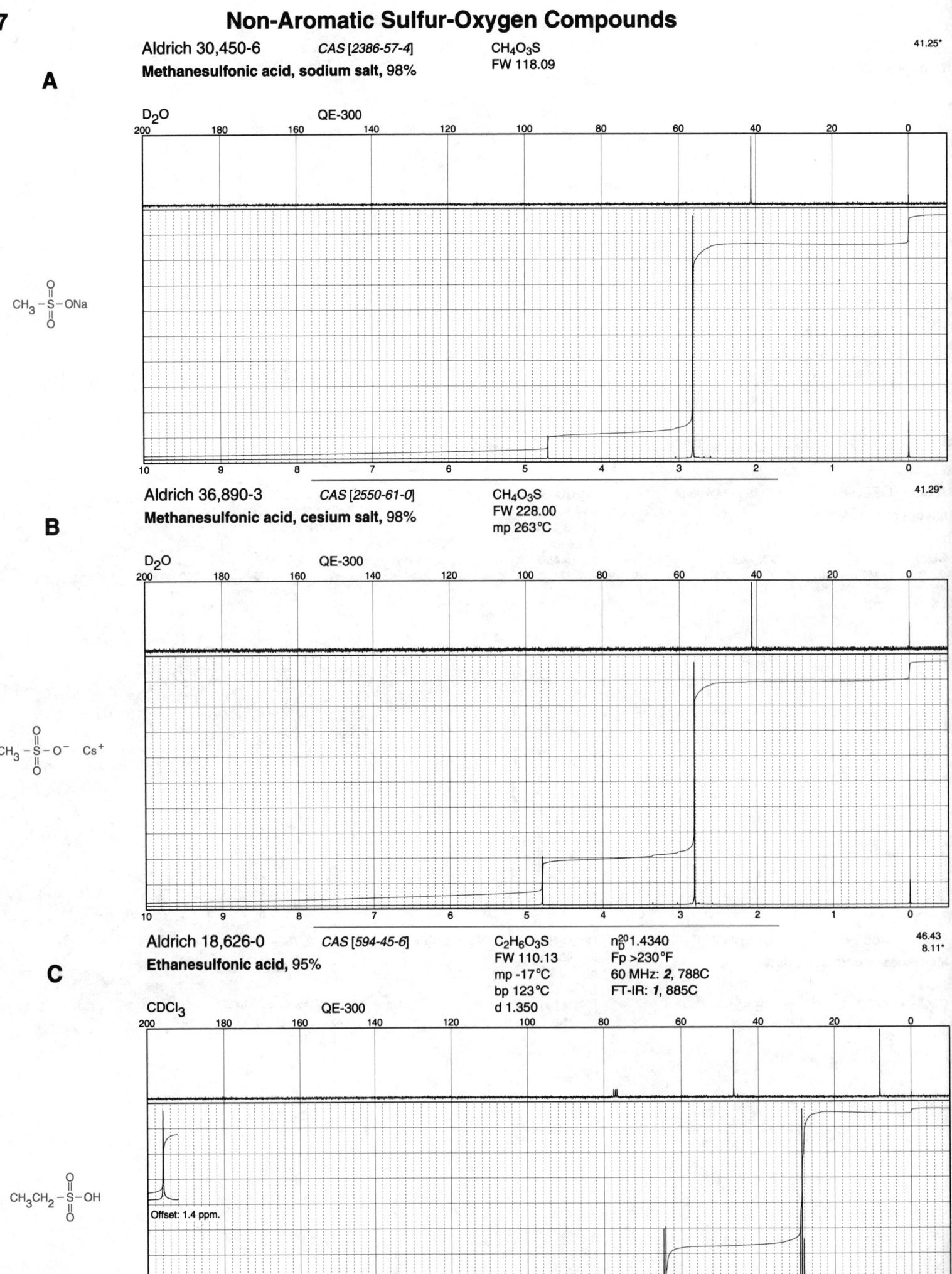

A

Aldrich 30,450-6   *CAS [2386-57-4]* CH_4O_3S
Methanesulfonic acid, sodium salt, 98%   FW 118.09

41.25*

$CH_3-\overset{\overset{\displaystyle O}{\|}}{\underset{\underset{\displaystyle O}{\|}}{S}}-ONa$

D₂O   QE-300

B

Aldrich 36,890-3   *CAS [2550-61-0]* CH_4O_3S
Methanesulfonic acid, cesium salt, 98%   FW 228.00
mp 263°C

41.29*

$CH_3-\overset{\overset{\displaystyle O}{\|}}{\underset{\underset{\displaystyle O}{\|}}{S}}-O^-$ Cs^+

D₂O   QE-300

C

Aldrich 18,626-0   *CAS [594-45-6]* $C_2H_6O_3S$ n_D^{20} 1.4340
Ethanesulfonic acid, 95%   FW 110.13   Fp >230°F
mp -17°C   60 MHz: **2**, 788C
bp 123°C   FT-IR: **1**, 885C
d 1.350

46.43
8.11*

$CH_3CH_2-\overset{\overset{\displaystyle O}{\|}}{\underset{\underset{\displaystyle O}{\|}}{S}}-OH$

CDCl₃   QE-300

Offset: 1.4 ppm.

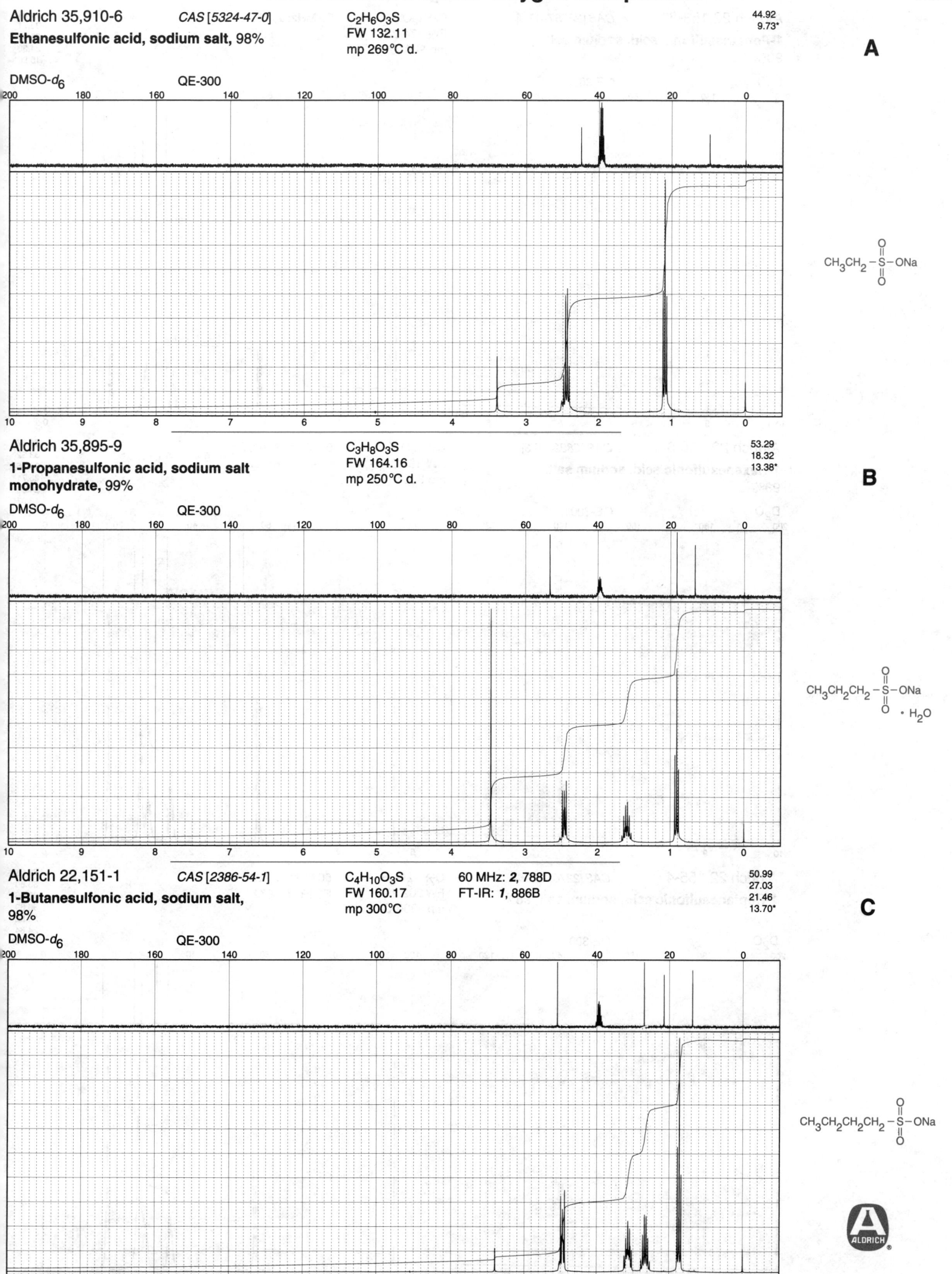

Aldrich 35,910-6   *CAS [5324-47-0]*   C₂H₆O₃S
FW 132.11
Ethanesulfonic acid, sodium salt, 98%   mp 269°C d.

44.92
9.73*

A

DMSO-*d*₆   QE-300

$CH_3CH_2 - \overset{\overset{\displaystyle O}{\|}}{\underset{\underset{\displaystyle O}{\|}}{S}} - ONa$

Aldrich 35,895-9   C₃H₈O₃S
FW 164.16
1-Propanesulfonic acid, sodium salt   mp 250°C d.
monohydrate, 99%

53.29
18.32
13.38*

B

DMSO-*d*₆   QE-300

$CH_3CH_2CH_2 - \overset{\overset{\displaystyle O}{\|}}{\underset{\underset{\displaystyle O}{\|}}{S}} - ONa$
$\cdot H_2O$

Aldrich 22,151-1   *CAS [2386-54-1]*   C₄H₁₀O₃S   60 MHz: **2**, 788D
FW 160.17   FT-IR: **1**, 886B
1-Butanesulfonic acid, sodium salt,   mp 300°C
98%

50.99
27.03
21.46
13.70*

C

DMSO-*d*₆   QE-300

$CH_3CH_2CH_2CH_2 - \overset{\overset{\displaystyle O}{\|}}{\underset{\underset{\displaystyle O}{\|}}{S}} - ONa$

ALDRICH

A

Aldrich 22,153-8   *CAS [22767-49-3]*   C$_5$H$_{12}$O$_3$S   60 MHz: *2*, 789A

1-Pentanesulfonic acid, sodium salt, 98%   FW 174.20   FT-IR: *1*, 886C   mp 300°C

53.84
32.71
26.43
24.30
15.89*

D$_2$O   QE-300

CH$_3$(CH$_2$)$_3$CH$_2$—S—ONa (with O double bonds)

B

Aldrich 22,154-6   *CAS [2832-45-3]*   C$_6$H$_{14}$O$_3$S   60 MHz: *2*, 789B

1-Hexanesulfonic acid, sodium salt, 98%   FW 188.22   FT-IR: *1*, 886D   mp 300°C

53.87
33.32
30.15
26.71
24.50
16.07*

D$_2$O   QE-300

CH$_3$(CH$_2$)$_4$CH$_2$—S—ONa (with O double bonds)

C

Aldrich 22,155-4   *CAS [22767-50-6]*   C$_7$H$_{16}$O$_3$S   60 MHz: *2*, 789C

1-Heptanesulfonic acid, sodium salt, 98%   FW 202.25   FT-IR: *1*, 887A   mp >300°C

53.85
33.86
30.95
30.69
26.82
24.84
16.26*

D$_2$O   QE-300

CH$_3$(CH$_2$)$_5$CH$_2$—S—ONa • xH$_2$O (with O double bonds)

ALDRICH

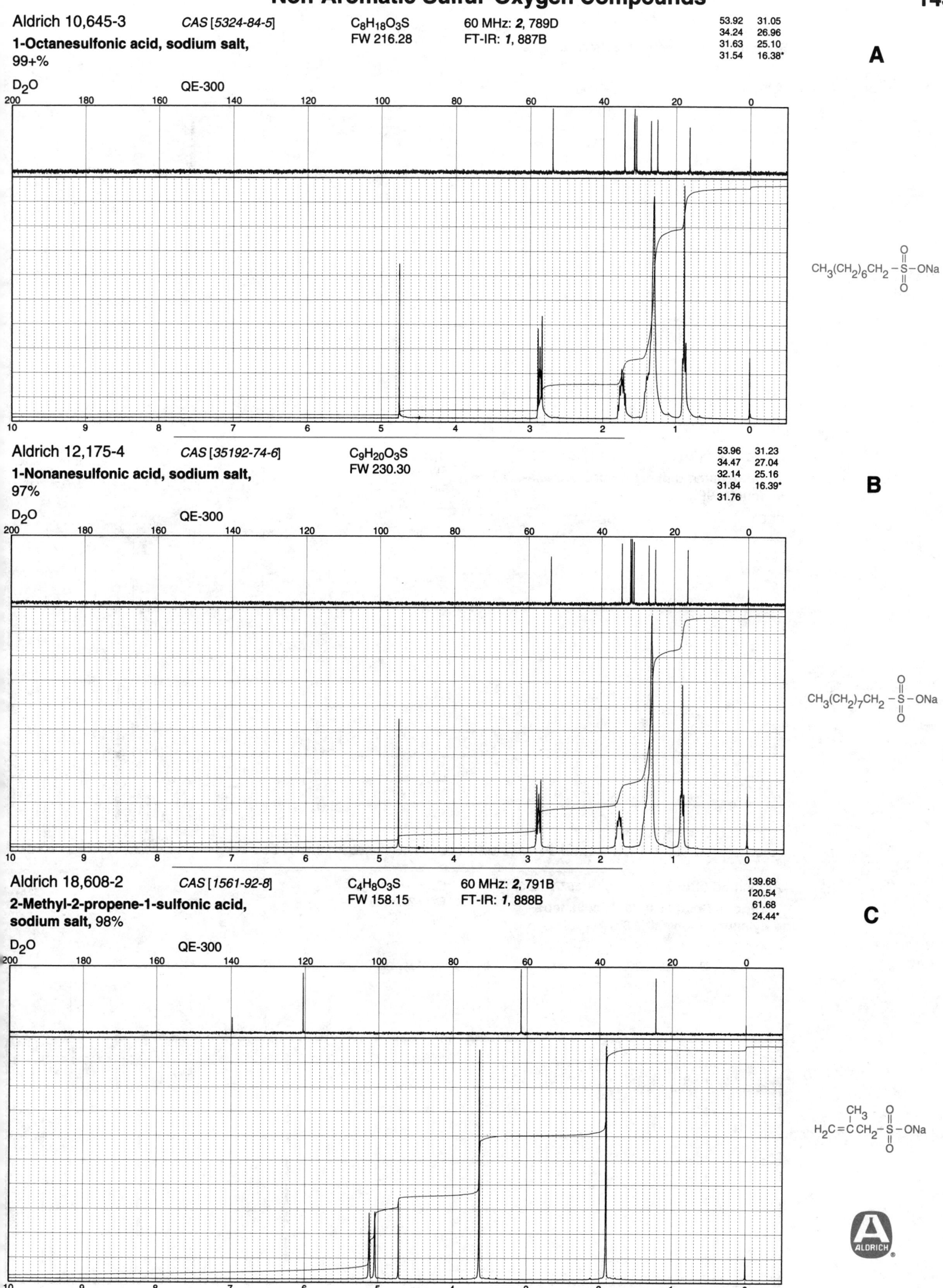

Aldrich 10,645-3   *CAS [5324-84-5]* $C_8H_{18}O_3S$ 60 MHz: *2*, 789D   FT-IR: *1*, 887B

1-Octanesulfonic acid, sodium salt, 99+%

FW 216.28

53.92	31.05
34.24	26.96
31.63	25.10
31.54	16.38*

A

D₂O   QE-300

$CH_3(CH_2)_6CH_2 - \overset{\overset{O}{\|}}{\underset{\underset{O}{\|}}{S}} - ONa$

Aldrich 12,175-4   *CAS [35192-74-6]* $C_9H_{20}O_3S$ FW 230.30

1-Nonanesulfonic acid, sodium salt, 97%

53.96	31.23
34.47	27.04
32.14	25.16
31.84	16.39*
31.76	

B

D₂O   QE-300

$CH_3(CH_2)_7CH_2 - \overset{\overset{O}{\|}}{\underset{\underset{O}{\|}}{S}} - ONa$

Aldrich 18,608-2   *CAS [1561-92-8]* $C_4H_8O_3S$ 60 MHz: *2*, 791B

2-Methyl-2-propene-1-sulfonic acid, sodium salt, 98%

FW 158.15   FT-IR: *1*, 888B

139.68
120.54
61.68
24.44*

C

D₂O   QE-300

$H_2C=\overset{\overset{CH_3}{|}}{C}CH_2 - \overset{\overset{O}{\|}}{\underset{\underset{O}{\|}}{S}} - ONa$

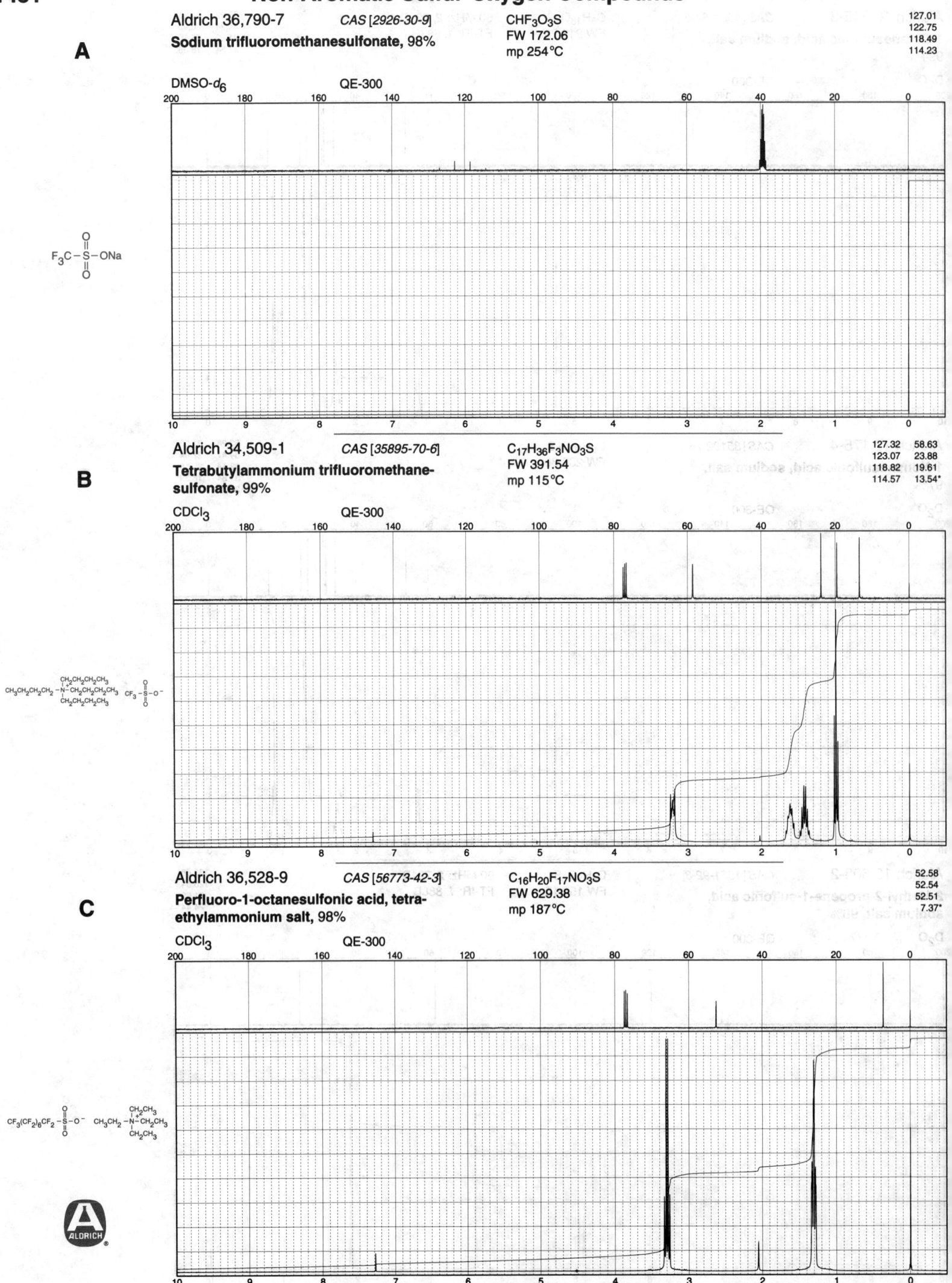

A

Aldrich 36,790-7   *CAS [2926-30-9]*   CHF₃O₃S
Sodium trifluoromethanesulfonate, 98%   FW 172.06
mp 254°C

127.01
122.75
118.49
114.23

DMSO-*d*₆   QE-300

$F_3C-S-ONa$

B

Aldrich 34,509-1   *CAS [35895-70-6]*   C₁₇H₃₆F₃NO₃S
Tetrabutylammonium trifluoromethane-   FW 391.54
sulfonate, 99%   mp 115°C

127.32 58.63
123.07 23.88
118.82 19.61
114.57 13.54*

CDCl₃   QE-300

C

Aldrich 36,528-9   *CAS [56773-42-3]*   C₁₆H₂₀F₁₇NO₃S
Perfluoro-1-octanesulfonic acid, tetra-   FW 629.38
ethylammonium salt, 98%   mp 187°C

52.58
52.54
52.51
7.37*

CDCl₃   QE-300

Aldrich 22,007-8 *CAS [1562-00-1]* $C_2H_6O_4S$ 60 MHz: *2*, 796C 59.75
55.61

Isethionic acid, sodium salt, 98% FW 148.11 FT-IR: *1*, 889A
mp 193°C

A

D$_2$O QE-300

HOCH$_2$CH$_2$–S–ONa (with two =O on S)

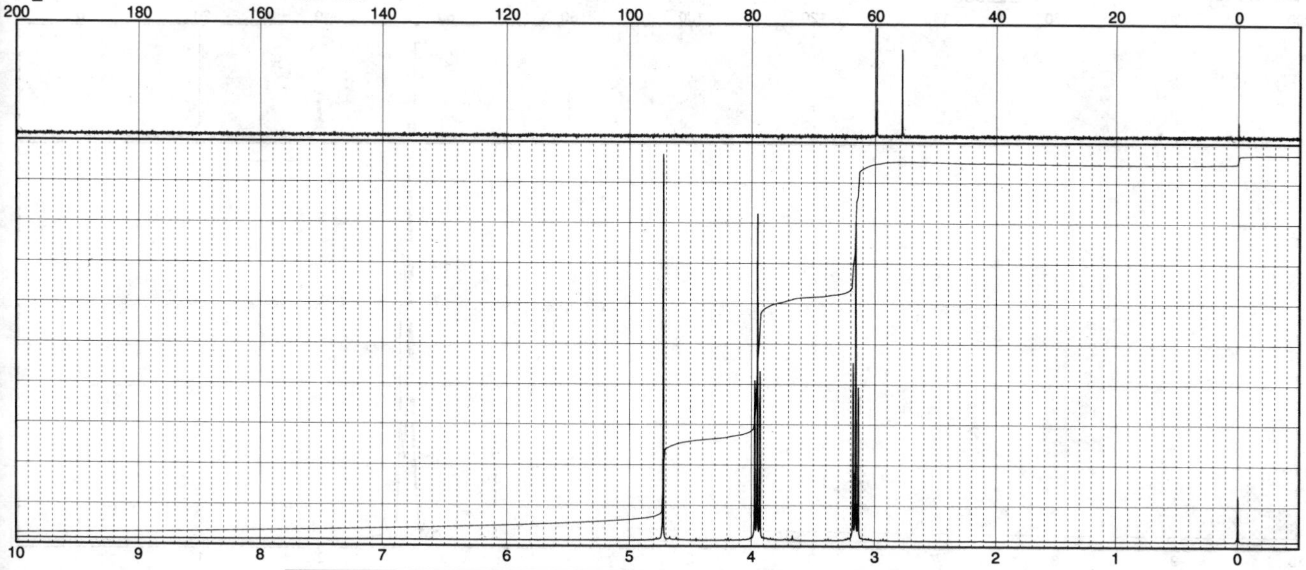

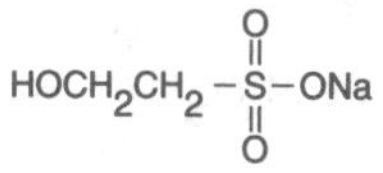

Aldrich 38,980-3 *CAS [3542-44-7]* $C_3H_8O_4S$ FT-IR: *1*, 895D 62.97
50.74

3-Hydroxy-1-propanesulfonic acid, sodium FW 162.15 29.78
salt, tech., 80% mp 260°C d.

B

D$_2$O QE-300

HOCH$_2$CH$_2$CH$_2$–S–ONa (with two =O on S)

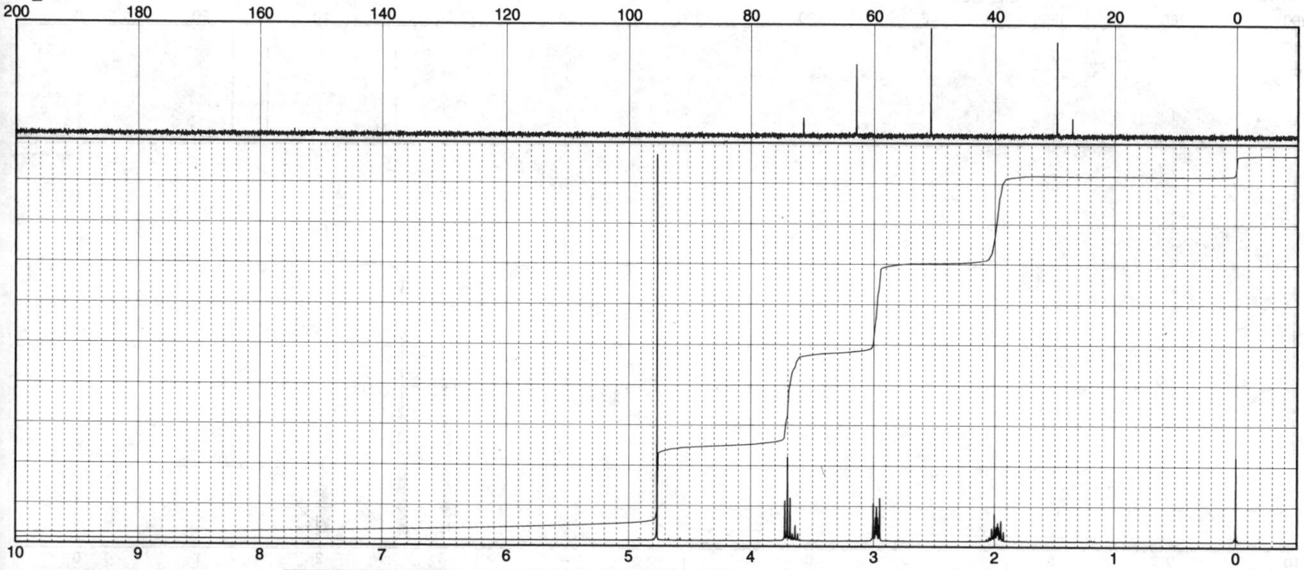

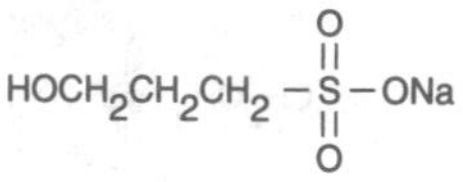

Aldrich 32,533-3 *CAS [143218-48-8]* $C_3H_7ClO_4S$ 69.96*
56.95

(±)-3-Chloro-2-hydroxy-1-propanesulfonic FW 196.59 51.18
acid, sodium salt hydrate, 95% mp 256°C d.

C

D$_2$O QE-300

OH
ClCH$_2$C HCH$_2$SO$_3$Na · xH$_2$O

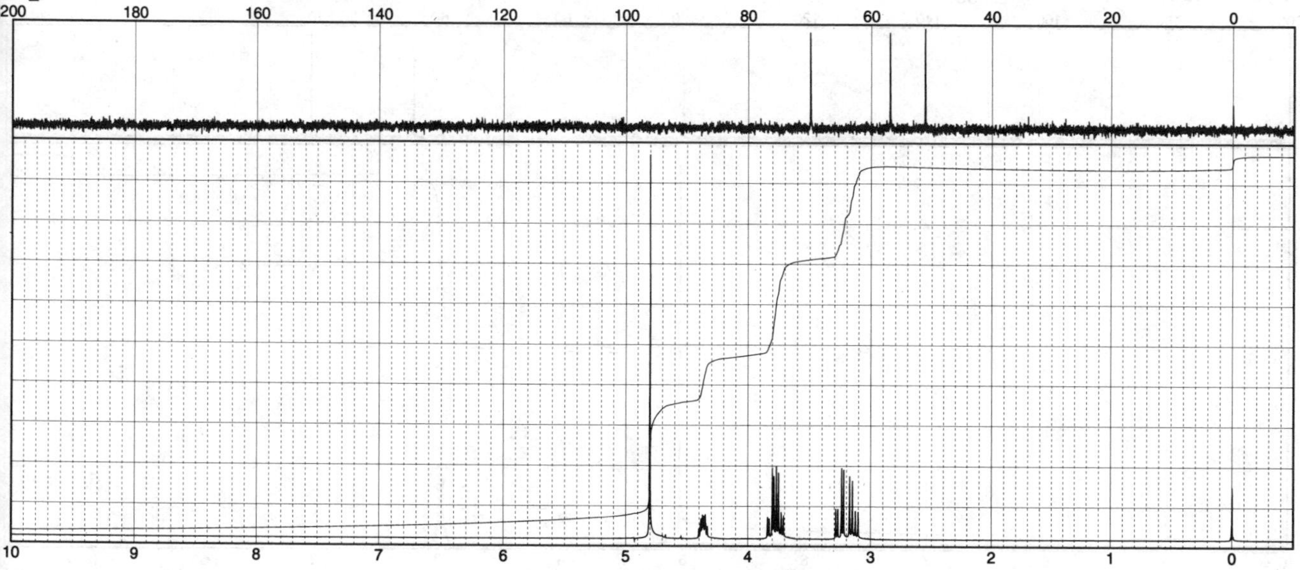

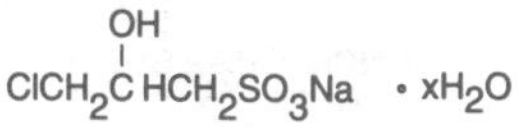

A

Aldrich 26,836-4   *CAS [19767-45-4]* $C_2H_6O_3S_2$
2-Mercaptoethanesulfonic acid, sodium salt, 98%
FW 164.18

55.00
19.56

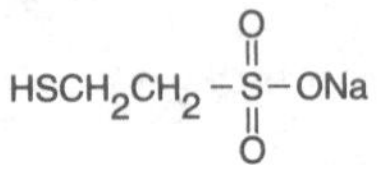

DMSO-d_6 QE-300

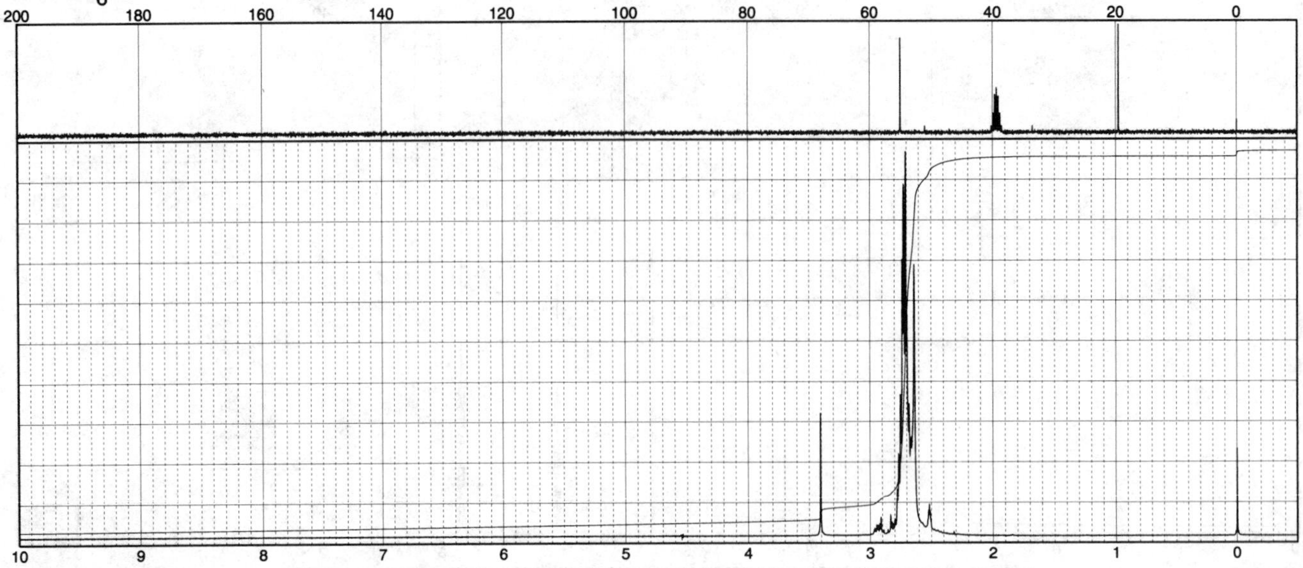

HSCH$_2$CH$_2$—S(=O)(=O)—ONa

B

Aldrich 25,168-2   *CAS [17636-10-1]* $C_3H_8O_3S_2$
3-Mercapto-1-propanesulfonic acid, sodium salt, tech., 90%
FW 178.21

52.21
31.13
25.32

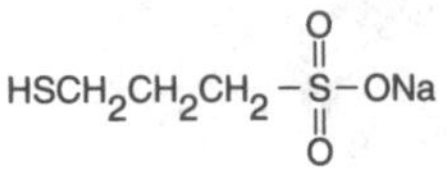

D$_2$O   QE-300

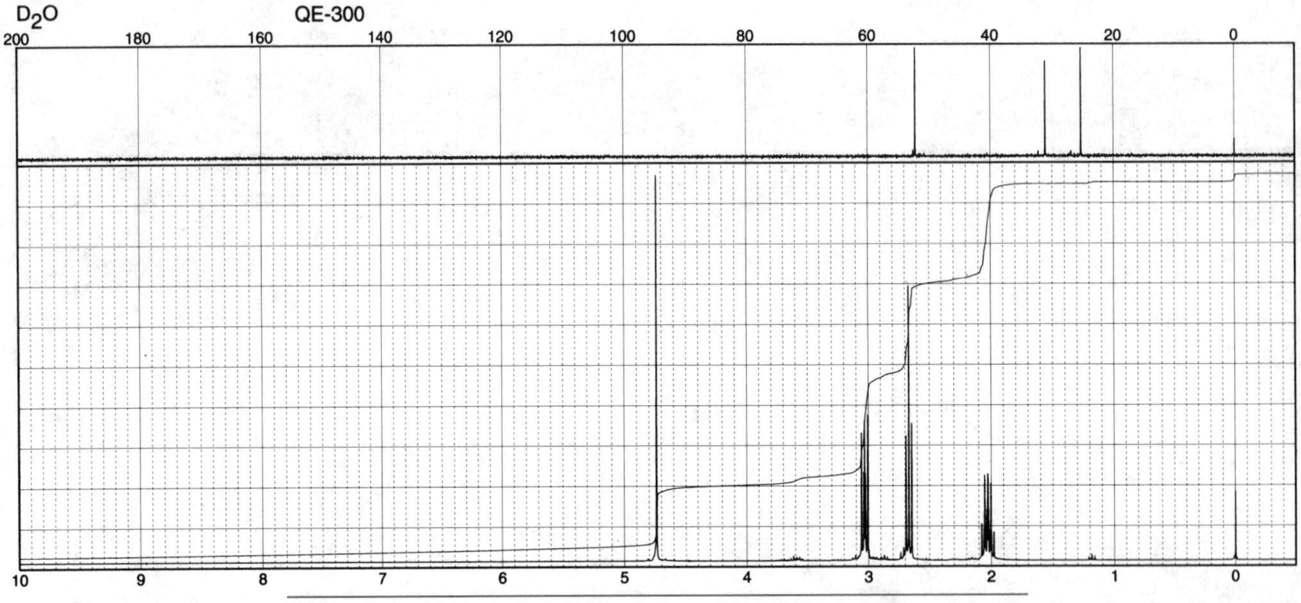

HSCH$_2$CH$_2$CH$_2$—S(=O)(=O)—ONa

C

Aldrich 15,224-2   *CAS [107-35-7]* $C_2H_7NO_3S$ 60 MHz: *2*, 792A
Taurine, 99%
FW 125.15   FT-IR: *1*, 890A
mp 300°C

55.90
39.15

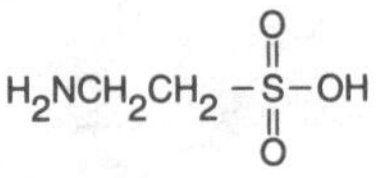

D$_2$O + NAOD   QE-300

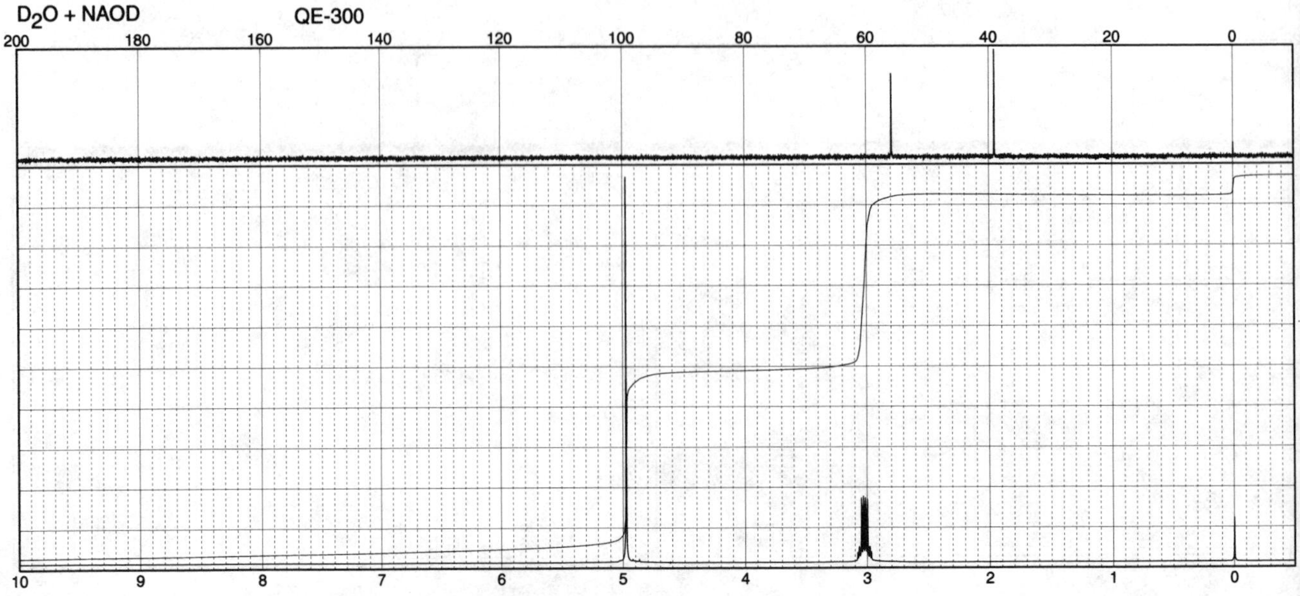

H$_2$NCH$_2$CH$_2$—S(=O)(=O)—OH

Aldrich A7,612-5   CAS [81028-90-2]   C₃H₉NO₃S   FT-IR: 1, 889C

**3-Amino-1-propanesulfonic acid,
sodium salt dihydrate**

FW 197.19
mp 148°C d.

49.06
40.76
28.95

A

DMSO-d₆   QE-300

H₂NCH₂CH₂CH₂—S(=O)(=O)—ONa • 2H₂O

Aldrich 33,596-7   CAS [14933-08-5]   C₁₇H₃₇NO₃S

**Dodecyldimethyl(3-sulfopropyl)ammonium
hydroxide, inner salt, 98%**

FW 335.55
mp 246°C

64.32	29.60	26.37
63.20	29.51	22.74
50.73*	29.48	22.68
47.89	29.32	19.48
31.89	29.25	14.13*

B

CDCl₃   QE-300

CH₃(CH₂)₁₀CH₂—N⁺(CH₃)(CH₃)—CH₂CH₂CH₂—S(=O)(=O)—O⁻

Aldrich 34,016-2   CAS [68399-79-1]   C₇H₁₇NO₅S

**3-[(1,1-Dimethyl-2-hydroxyethyl)amino]-2-
hydroxy-1-propanesulfonic acid, 98%**

FW 227.28
mp 221°C d.

67.25
66.62*
63.13
57.62
48.51
22.77*
22.75*

C

D₂O   QE-300

HOCH₂C(CH₃)(CH₃)—NH—CH₂CH(OH)CH₂—S(=O)(=O)—OH

ALDRICH

Non-Aromatic Sulfur-Oxygen Compounds

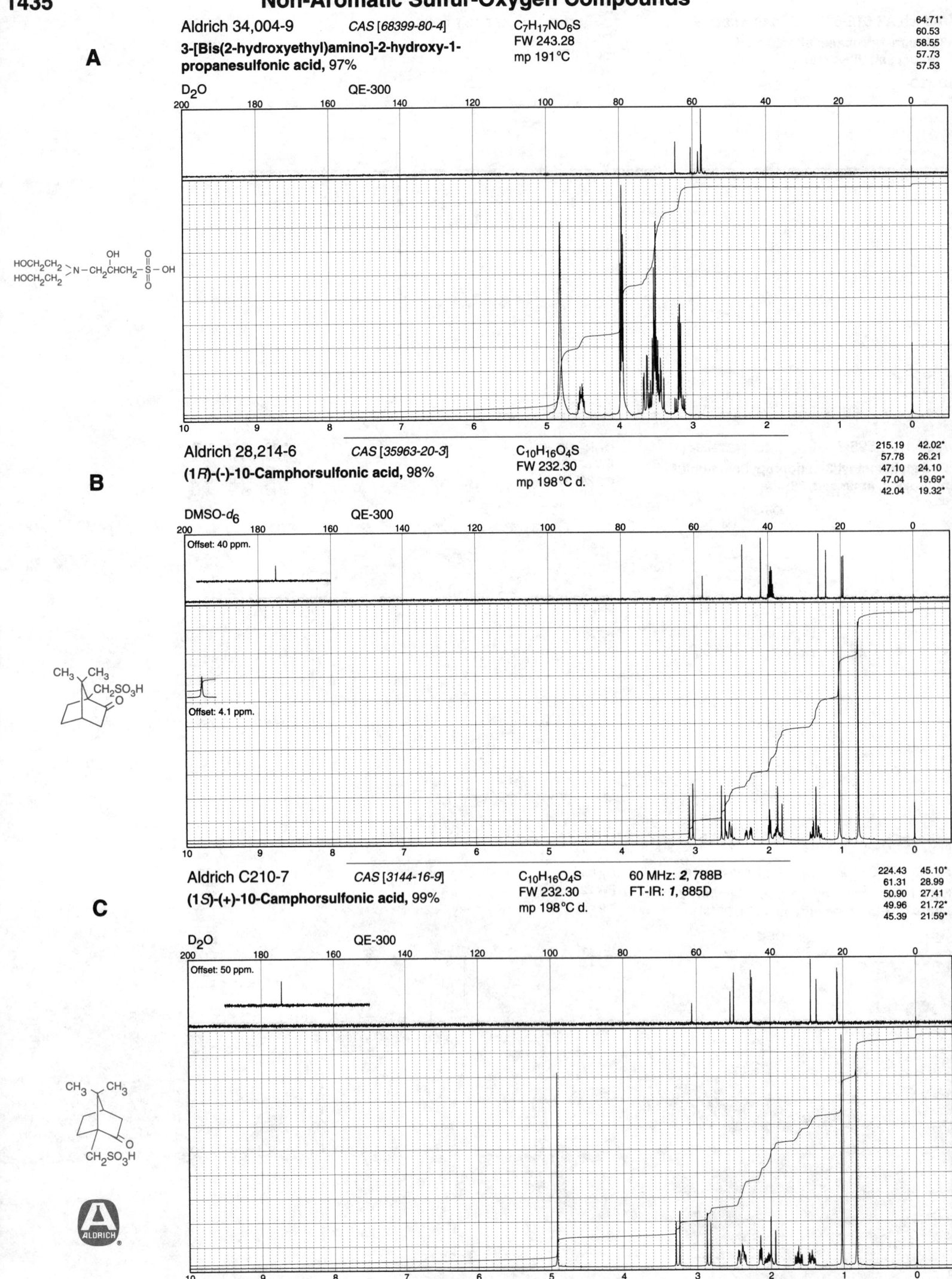

A

Aldrich 34,004-9   *CAS [68399-80-4]*   C₇H₁₇NO₆S
FW 243.28
mp 191°C

3-[Bis(2-hydroxyethyl)amino]-2-hydroxy-1-propanesulfonic acid, 97%

$C_7H_{17}NO_6S$
FW 243.28
mp 191°C

64.71*
60.53
58.55
57.73
57.53

D₂O   QE-300

HOCH₂CH₂ ⟩N—CH₂CHCH₂S—OH

B

Aldrich 28,214-6   *CAS [35963-20-3]*   C₁₀H₁₆O₄S
(1R)-(-)-10-Camphorsulfonic acid, 98%
FW 232.30
mp 198°C d.

215.19   42.02*
57.78   26.21
47.10   24.10
47.04   19.69*
42.04   19.32*

DMSO-d₆   QE-300

Offset: 40 ppm.

Offset: 4.1 ppm.

CH₃ CH₃
CH₂SO₃H
O

C

Aldrich C210-7   *CAS [3144-16-9]*   C₁₀H₁₆O₄S   60 MHz: **2**, 788B
(1S)-(+)-10-Camphorsulfonic acid, 99%   FW 232.30   FT-IR: **1**, 885D
mp 198°C d.

224.43   45.10*
61.31   28.99
50.90   27.41
49.96   21.72*
45.39   21.59*

D₂O   QE-300

Offset: 50 ppm.

CH₃ CH₃
O
CH₂SO₃H

ALDRICH

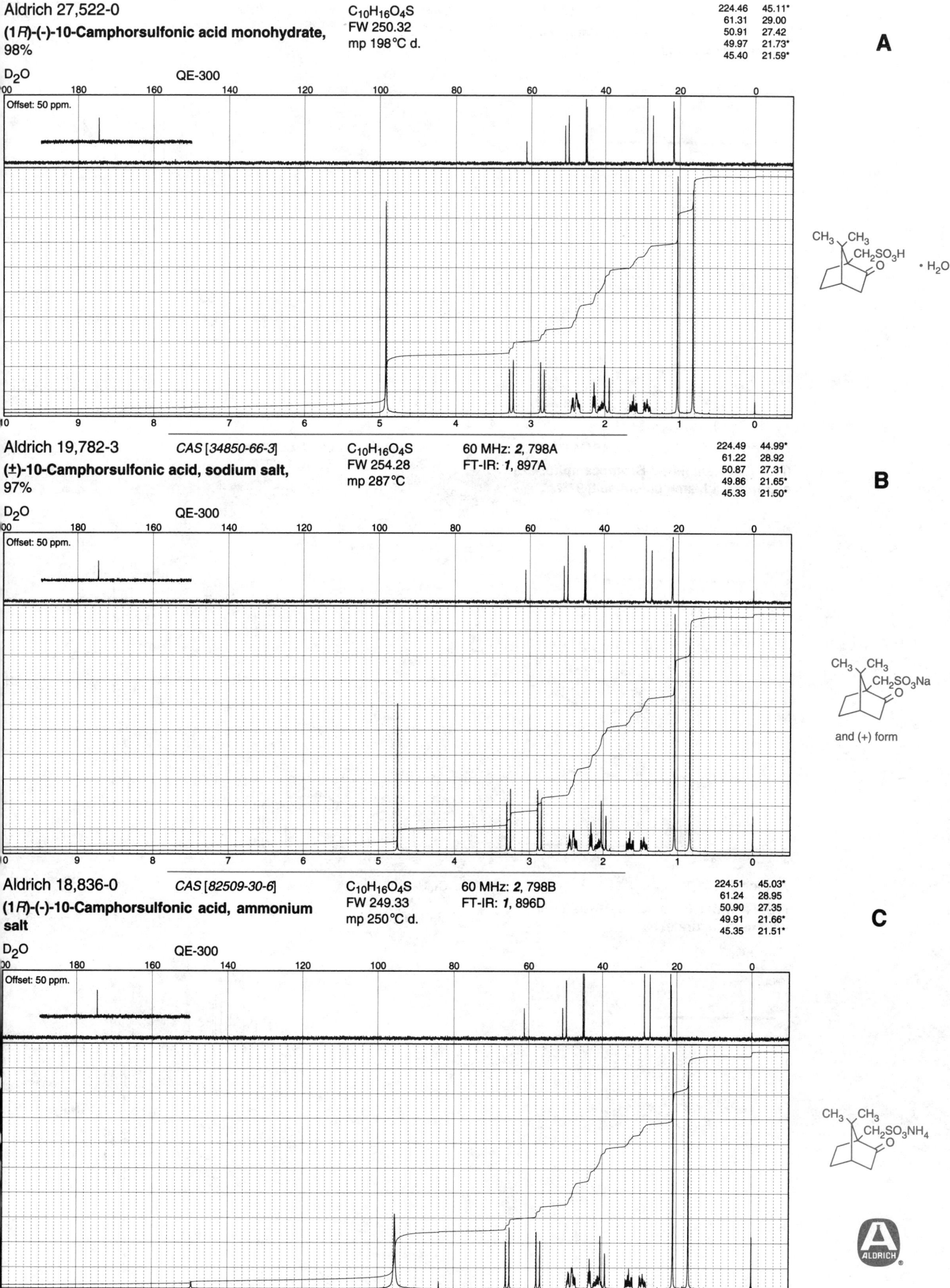

Aldrich 27,522-0

(1R)-(-)-10-Camphorsulfonic acid monohydrate, 98%

C₁₀H₁₆O₄S
FW 250.32
mp 198°C d.

224.46	45.11*
61.31	29.00
50.91	27.42
49.97	21.73*
45.40	21.59*

A

D₂O   QE-300
Offset: 50 ppm.

Aldrich 19,782-3   CAS [34850-66-3]

(±)-10-Camphorsulfonic acid, sodium salt, 97%

C₁₀H₁₆O₄S
FW 254.28
mp 287°C

60 MHz: 2, 798A
FT-IR: 1, 897A

224.49	44.99*
61.22	28.92
50.87	27.31
49.86	21.65*
45.33	21.50*

B

D₂O   QE-300
Offset: 50 ppm.

and (+) form

Aldrich 18,836-0   CAS [82509-30-6]

(1R)-(-)-10-Camphorsulfonic acid, ammonium salt

C₁₀H₁₆O₄S
FW 249.33
mp 250°C d.

60 MHz: 2, 798B
FT-IR: 1, 896D

224.51	45.03*
61.24	28.95
50.90	27.35
49.91	21.66*
45.35	21.51*

C

D₂O   QE-300
Offset: 50 ppm.

ALDRICH

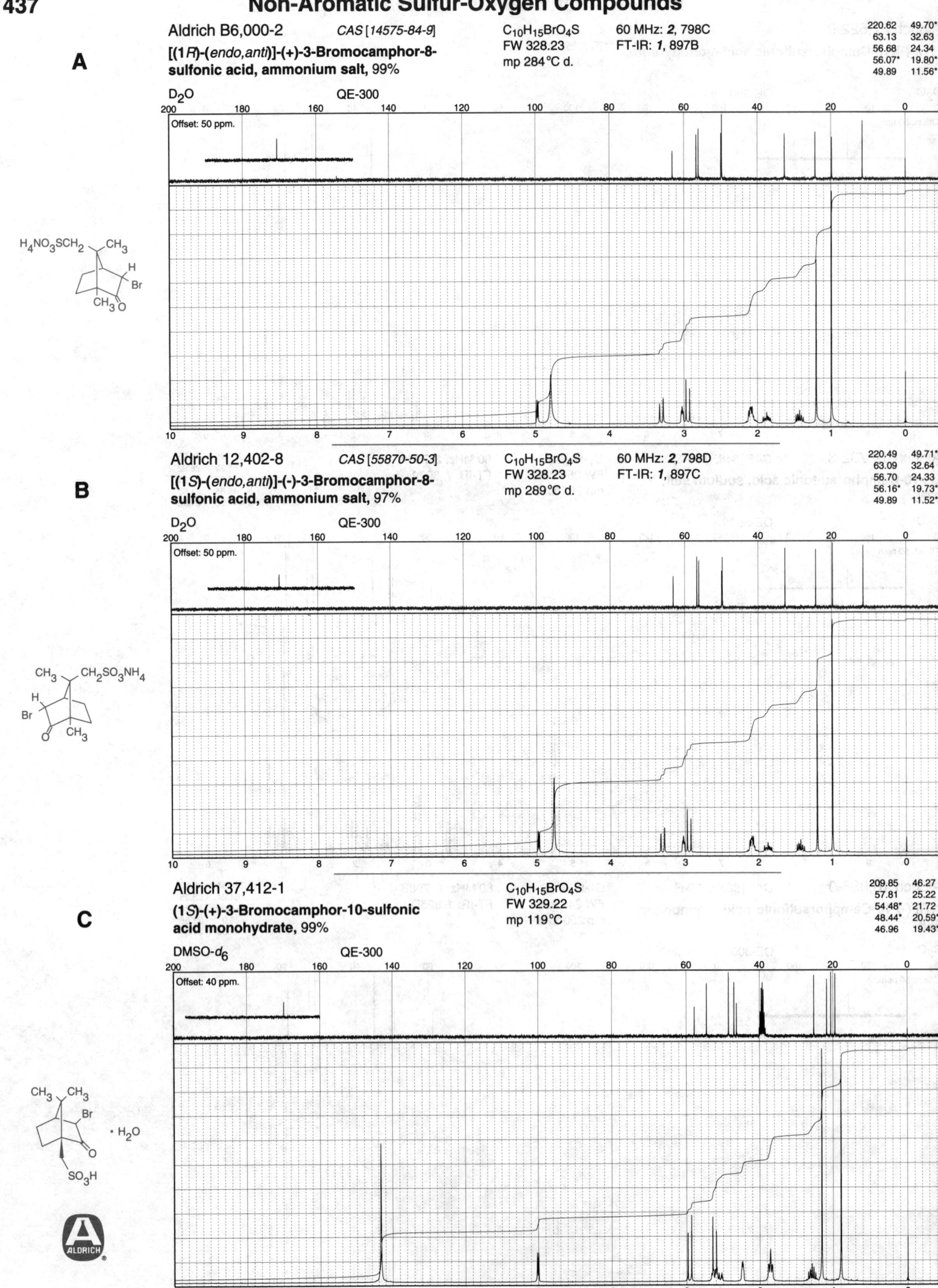

A

Aldrich B6,000-2   CAS [14575-84-9]

[(1*R*)-(*endo,anti*)]-(+)-3-Bromocamphor-8-
sulfonic acid, ammonium salt, 99%

C₁₀H₁₅BrO₄S
FW 328.23
mp 284°C d.

60 MHz: **2**, 798C
FT-IR: **1**, 897B

220.62	49.70*
63.13	32.63
56.68	24.34
56.07*	19.80*
49.89	11.56*

D₂O   QE-300

Offset: 50 ppm.

H₄NO₃SCH₂ ─ CH₃

B

Aldrich 12,402-8   CAS [55870-50-3]

[(1*S*)-(*endo,anti*)]-(-)-3-Bromocamphor-8-
sulfonic acid, ammonium salt, 97%

C₁₀H₁₅BrO₄S
FW 328.23
mp 289°C d.

60 MHz: **2**, 798D
FT-IR: **1**, 897C

220.49	49.71*
63.09	32.64
56.70	24.33
56.16*	19.73*
49.89	11.52*

D₂O   QE-300

Offset: 50 ppm.

CH₃ ─ CH₂SO₃NH₄

C

Aldrich 37,412-1

(1*S*)-(+)-3-Bromocamphor-10-sulfonic
acid monohydrate, 99%

C₁₀H₁₅BrO₄S
FW 329.22
mp 119°C

209.85	46.27
57.81	25.22
54.48*	21.72
48.44*	20.59*
46.96	19.43*

DMSO-*d*₆   QE-300

Offset: 40 ppm.

CH₃ CH₃
         Br
              · H₂O
         O
SO₃H

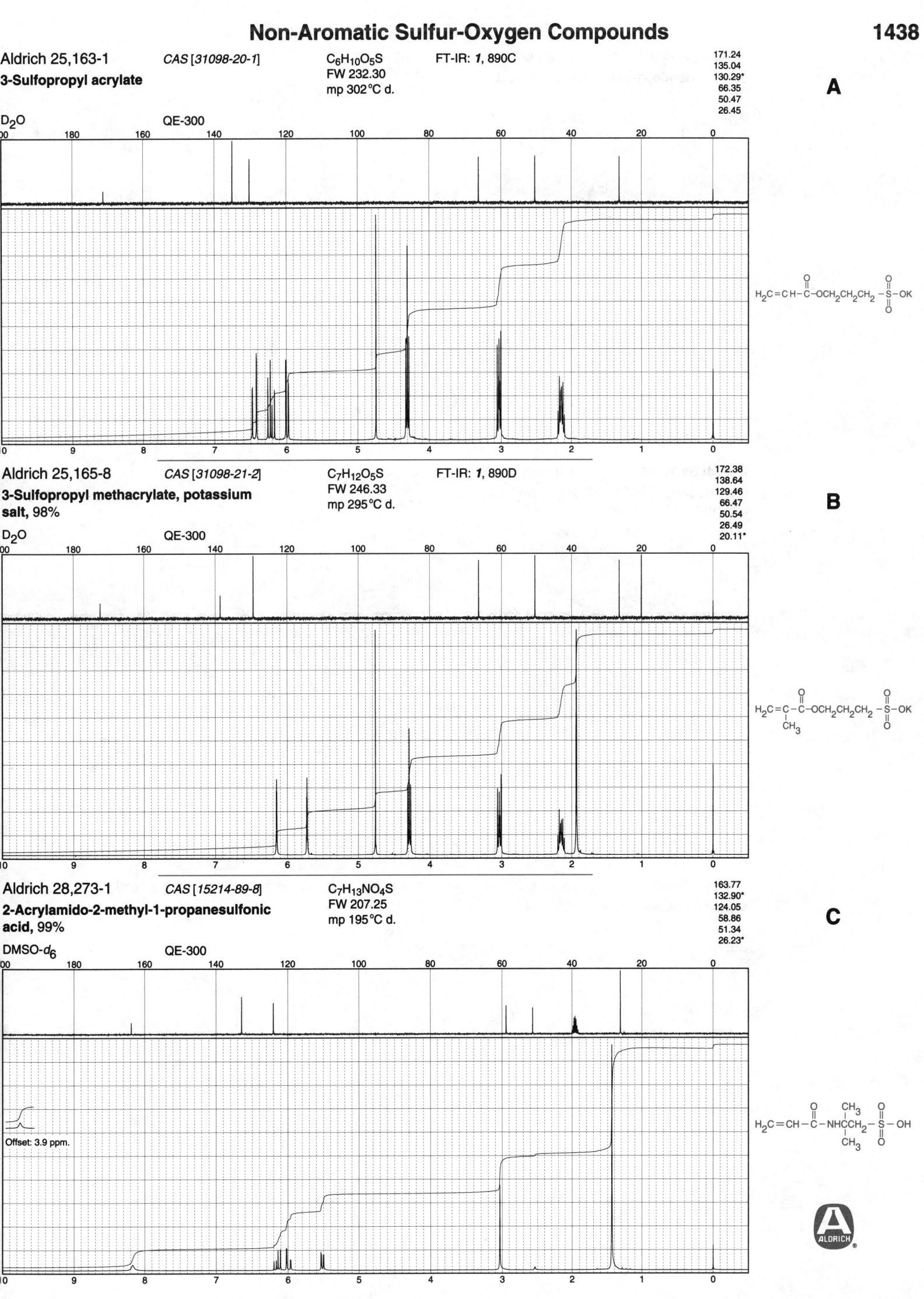

A

Aldrich 25,170-4   *CAS [21668-81-5]* $C_4H_{10}N_2O_3S_2$

3-(Amidinothio)-1-propanesulfonic acid, 97%

FW 198.26
mp 255°C

173.66
51.73
31.99
26.66

D_2O QE-300

$H_2N-\overset{\overset{NH}{\|}}{C}-SCH_2CH_2CH_2-\overset{\overset{O}{\|}}{\underset{\underset{O}{\|}}{S}}-OH$

B

Aldrich 34,185-1   *CAS [106627-54-7]* $C_4H_5NO_6S$

(±)-1-Hydroxy-2,5-dioxo-3-pyrrolidine-sulfonic acid, monosodium salt, 97%

FW 217.13
mp 250°C d.

176.57
172.30
59.17*
32.28

D_2O QE-300

C

Aldrich 12,992-5   *CAS [66-27-3]* $C_2H_6O_3S$ Fp 220°F   60 MHz: **2**, 799D

Methyl methanesulfonate, 99%

FW 110.13
bp 203°C
d 1.300
n_D^{20} 1.4140

FT-IR: **1**, 898A
VP-FT-IR: **3**, 833C

55.76*
36.44*

$CDCl_3$ QE-300

$CH_3-\overset{\overset{O}{\|}}{\underset{\underset{O}{\|}}{S}}-OCH_3$

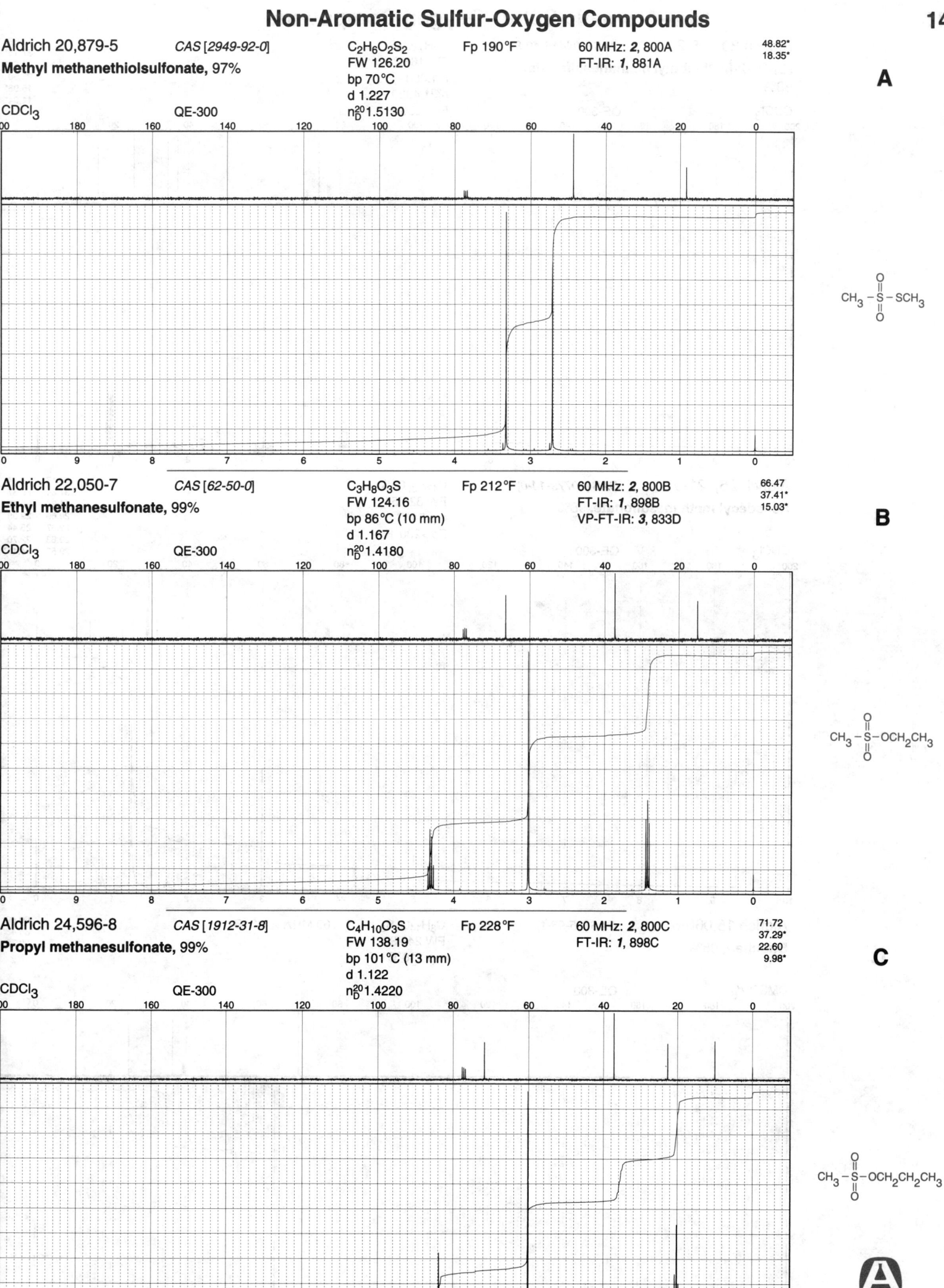

Aldrich 20,879-5 *CAS [2949-92-0]* C$_2$H$_6$O$_2$S$_2$ Fp 190°F 60 MHz: **2**, 800A 48.82*
Methyl methanethiolsulfonate, 97% FW 126.20 FT-IR: **1**, 881A 18.35*
bp 70°C
d 1.227
CDCl$_3$ QE-300 n$_D^{20}$1.5130

A

CH$_3$–S–SCH$_3$ (with O above and O below S)

Aldrich 22,050-7 *CAS [62-50-0]* C$_3$H$_8$O$_3$S Fp 212°F 60 MHz: **2**, 800B 66.47
Ethyl methanesulfonate, 99% FW 124.16 FT-IR: **1**, 898B 37.41*
bp 86°C (10 mm) VP-FT-IR: **3**, 833D 15.03*
d 1.167
CDCl$_3$ QE-300 n$_D^{20}$1.4180

B

CH$_3$–S–OCH$_2$CH$_3$ (with O above and O below S)

Aldrich 24,596-8 *CAS [1912-31-8]* C$_4$H$_{10}$O$_3$S Fp 228°F 60 MHz: **2**, 800C 71.72
Propyl methanesulfonate, 99% FW 138.19 FT-IR: **1**, 898C 37.29*
bp 101°C (13 mm) 22.60
d 1.122 9.98*
CDCl$_3$ QE-300 n$_D^{20}$1.4220

C

CH$_3$–S–OCH$_2$CH$_2$CH$_3$ (with O above and O below S)

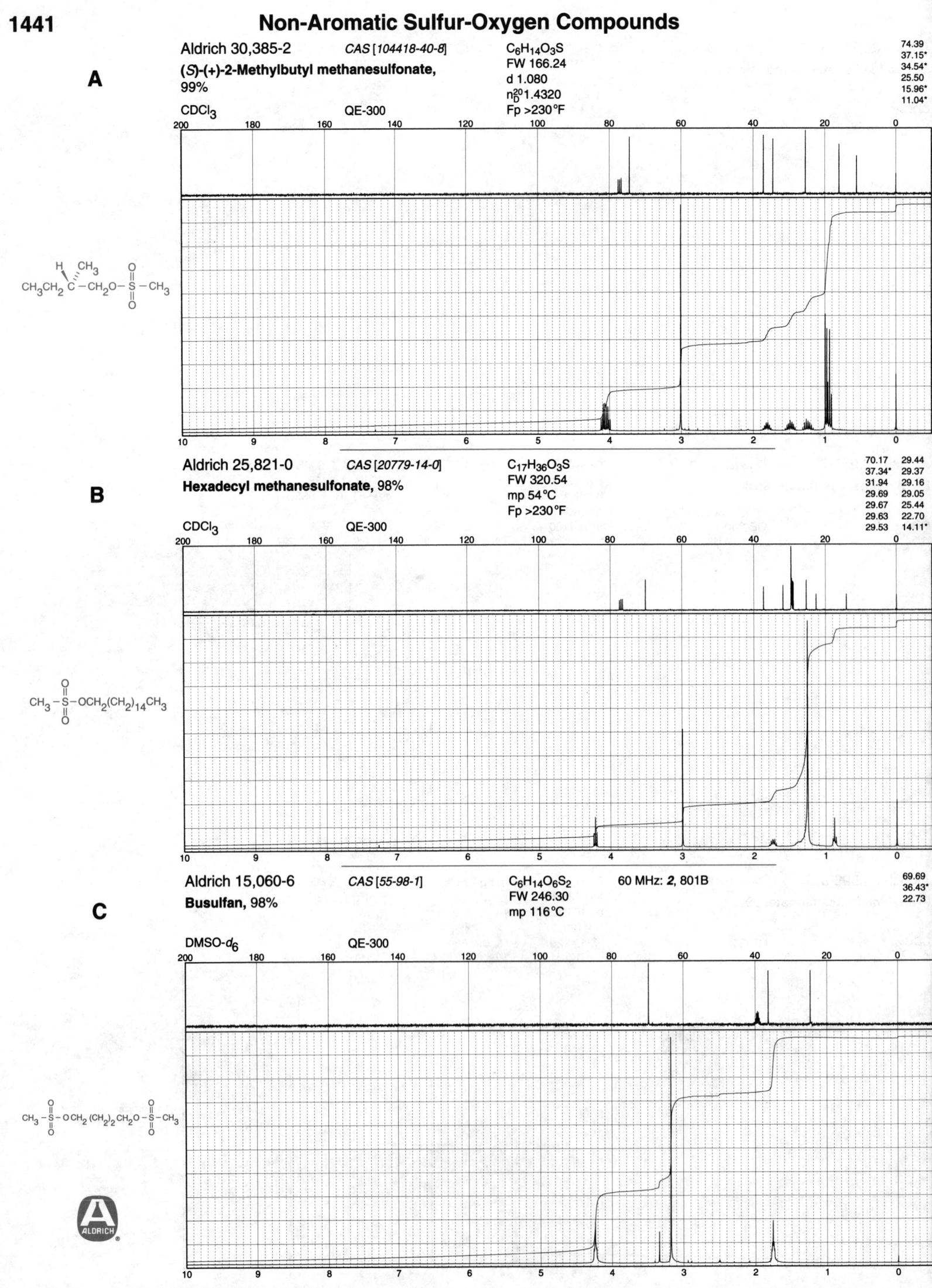

A

Aldrich 30,385-2   *CAS [104418-40-8]* $C_6H_{14}O_3S$

(*S*)-(+)-2-Methylbutyl methanesulfonate, 99%

FW 166.24
d 1.080
n_D^{20} 1.4320
Fp >230°F

74.39
37.15*
34.54*
25.50
15.96*
11.04*

CDCl$_3$ QE-300

B

Aldrich 25,821-0   *CAS [20779-14-0]* $C_{17}H_{36}O_3S$

Hexadecyl methanesulfonate, 98%

FW 320.54
mp 54°C
Fp >230°F

70.17	29.44
37.34*	29.37
31.94	29.16
29.69	29.05
29.67	25.44
29.63	22.70
29.53	14.11*

CDCl$_3$ QE-300

C

Aldrich 15,060-6   *CAS [55-98-1]* $C_6H_{14}O_6S_2$ 60 MHz: *2*, 801B

Busulfan, 98%

FW 246.30
mp 116°C

69.69
36.43*
22.73

DMSO-*d$_6$*   QE-300

A

Aldrich C4,170-3   *CAS [3570-58-9]*

2-Chloroethyl methanesulfonate, 99+%

$C_3H_7ClO_3S$
FW 158.60
bp 126°C (9 mm)
d 1.390
n_D^{20} 1.4560

Fp >230°F

60 MHz: *2*, 801C
FT-IR: *1*, 899C

68.85
41.33
37.73*

CDCl₃   QE-300

$CH_3-\overset{O}{\underset{O}{S}}-OCH_2CH_2Cl$

B

Aldrich 37,522-5

2-(2-Chloroethoxy)ethyl methanesulfonate, 97%

$C_5H_{11}ClO_4S$
FW 202.66
d 1.326
n_D^{20} 1.4610
Fp >230°F

71.29
69.17
68.89
42.82
37.65*

CDCl₃   QE-300

$CH_3-\overset{O}{\underset{O}{S}}-OCH_2CH_2OCH_2CH_2Cl$

C

Aldrich 33,230-5   *CAS [5450-26-0]*

1,2:5,6-Di-*O*-isopropylidene-3-*O*-(methylsulfonyl)-α-D-glucofuranose

$C_{13}H_{22}O_8S$
FW 338.38
mp 80°C

112.63	67.58
109.56	37.99*
105.17*	26.94*
83.71*	26.64*
82.68*	26.17*
79.75*	25.20*
72.05*	

CDCl₃   QE-300

$R = -\overset{O}{\underset{O}{S}}-CH_3$

A

Aldrich 26,456-3   CAS [58742-64-6]

Ethyl L-2-[(methylsulfonyl)oxy]propionate,
tech., 90%

$C_6H_{12}O_5S$
FW 196.22
d 1.213
n_D^{20} 1.4320
Fp >230°F

VP-FT-IR: **3**, 834A

169.43
74.30*
62.04
39.11*
18.37*
14.05*

CDCl$_3$ QE-300

B

Aldrich 16,248-3   CAS [333-27-7]

Methyl trifluoromethanesulfonate, 99+%

$C_2H_3F_3O_3S$
FW 164.10
bp 97°C
d 1.450
n_D^{20} 1.3260

Fp 101°F

60 MHz: **2**, 800D
FT-IR: **1**, 898D
VP-FT-IR: **3**, 834B

125.13
120.90
116.67
112.44
61.74*

CDCl$_3$ QE-300

C

Aldrich 24,653-0   CAS [425-75-2]

Ethyl trifluoromethanesulfonate, 99%

$C_3H_5F_3O_3S$
FW 178.13
bp 115°C
d 1.374
n_D^{20} 1.3360

Fp 96°F

60 MHz: **2**, 801A
FT-IR: **1**, 899A

125.08
120.85
116.63
112.40
74.19
15.25*

CDCl$_3$ QE-300

Aldrich 37,463-6 *CAS [84028-88-6]* C$_6$H$_9$F$_3$O$_5$S Fp 174°F

Ethyl L-2-[(trifluoromethylsulfonyl)oxy]propionate, 98%

FW 250.20
bp 34°C
d 1.342
n$_D^{20}$ 1.3740

CDCl$_3$ QE-300

167.46	80.24*
124.94	62.84
120.71	18.11*
116.49	13.96*
112.27	

A

Aldrich 31,025-5 *CAS [92051-23-5]* C$_{15}$H$_{19}$F$_3$O$_{12}$S

1,3,4,6-Tetra-*O*-acetyl-2-*O*-trifluoro-methanesulfonyl-β-D-mannopyranose, 99%

FW 480.37
mp 121°C

CDCl$_3$ QE-300

170.50	116.36	64.72*
169.78	112.13	61.74
169.08	89.12*	20.64*
167.89	81.35*	20.52*
124.82	73.53*	20.42*
120.58	69.66*	20.40*

B

Aldrich P5,070-6 *CAS [1120-71-4]* C$_3$H$_6$O$_3$S Fp >230°F

1,3-Propane sultone, 98%

FW 122.14
mp 32°C
bp 180°C (30 mm)
d 1.392

60 MHz: *2*, 801D
FT-IR: *1*, 899D
VP-FT-IR: *3*, 834D

69.07
44.11
23.64

CDCl$_3$ QE-300

C

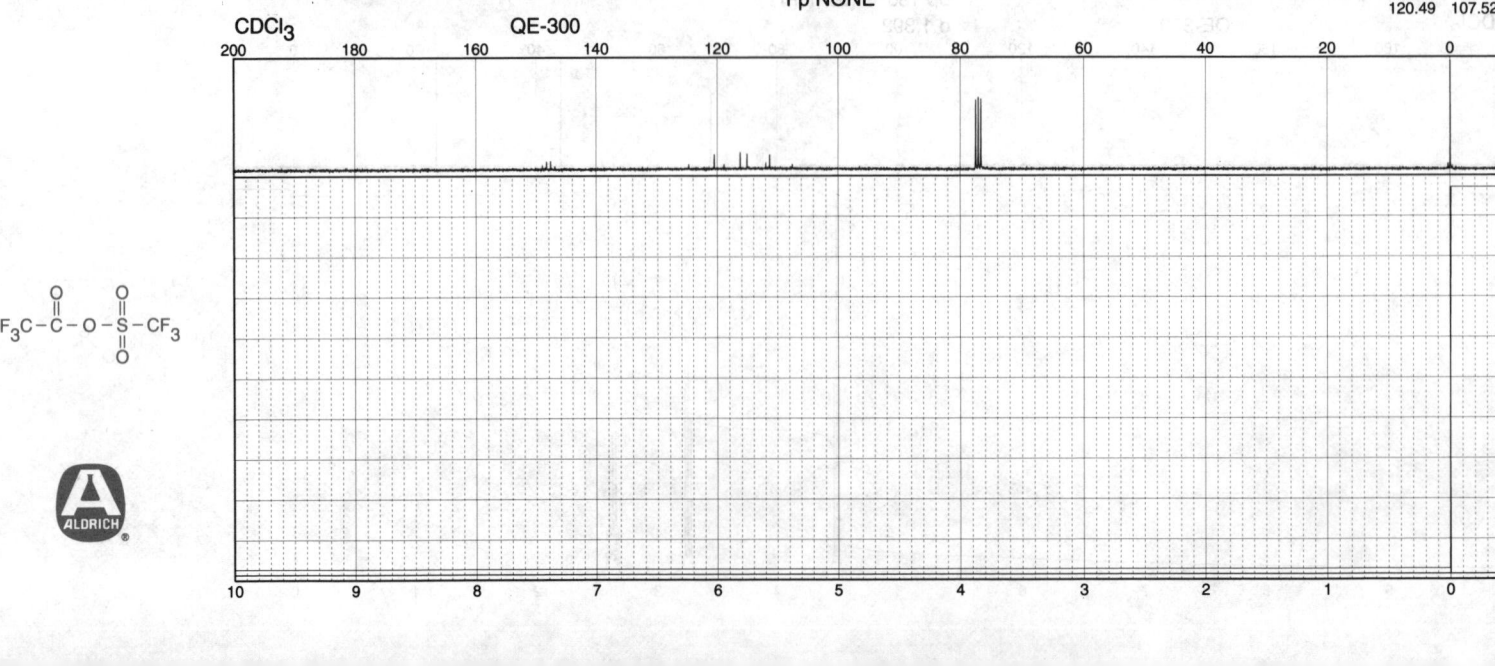

A

Aldrich B8,550-1   *CAS [1633-83-6]*   C₄H₈O₃S

1,4-Butane sultone, 99+%

FW 136.17
d 1.331
n_D^{20} 1.4640
Fp >230°F

60 MHz: *2*, 802A
FT-IR: *1*, 900A

74.13
48.25
23.52
22.82

CDCl₃   QE-300

B

Aldrich 26,919-0   *CAS [7143-01-3]*   C₂H₆O₅S₂

Methanesulfonic anhydride, 97%

FW 174.20
mp 70°C
bp 125°C (4 mm)

41.35*

CDCl₃   QE-300

CH₃–S(=O)(=O)–O–S(=O)(=O)–CH₃

C

Aldrich 30,868-4   *CAS [68602-57-3]*   C₃F₆O₄S

Trifluoroacetic acid, anhydride with
trifluoromethanesulfonic acid, tech., 90%

FW 246.09
bp 63°C
Fp NONE

149.01	118.88
148.35	116.23
147.69	115.10
147.02	111.96
124.75	111.31
120.49	107.52

CDCl₃   QE-300

F₃C–C(=O)–O–S(=O)(=O)–CF₃

ALDRICH

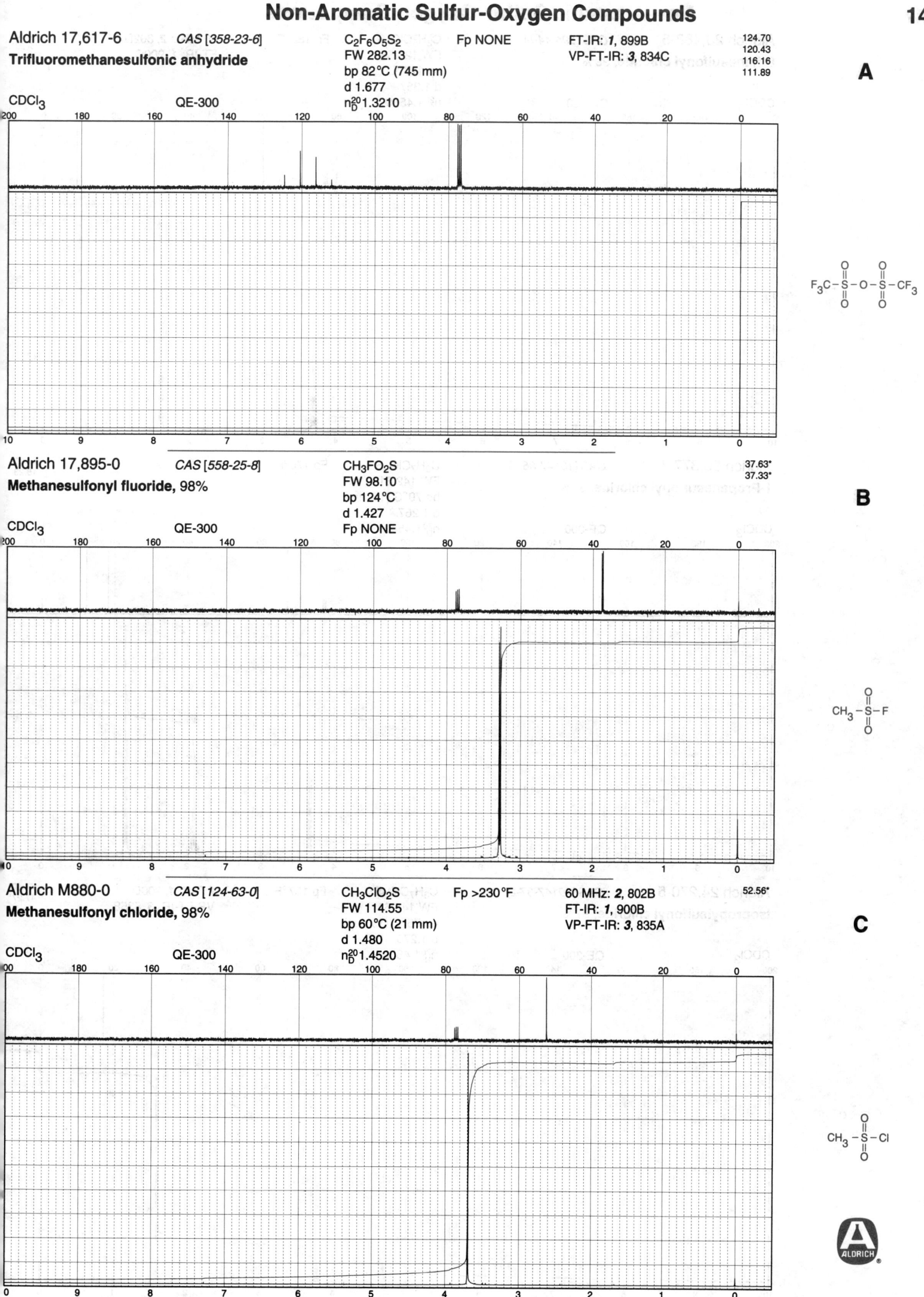

Aldrich 17,617-6   *CAS [358-23-6]* $C_2F_6O_5S_2$ Fp NONE   FT-IR: *1*, 899B
Trifluoromethanesulfonic anhydride   FW 282.13   VP-FT-IR: *3*, 834C
bp 82°C (745 mm)
d 1.677
CDCl$_3$ QE-300   n$_D^{20}$1.3210

124.70
120.43
116.16
111.89

A

Aldrich 17,895-0   *CAS [558-25-8]* CH_3FO_2S
Methanesulfonyl fluoride, 98%   FW 98.10
bp 124°C
d 1.427
CDCl$_3$ QE-300   Fp NONE

37.63*
37.33*

B

Aldrich M880-0   *CAS [124-63-0]* CH_3ClO_2S Fp >230°F   60 MHz: *2*, 802B
Methanesulfonyl chloride, 98%   FW 114.55   FT-IR: *1*, 900B
bp 60°C (21 mm)   VP-FT-IR: *3*, 835A
d 1.480
CDCl$_3$ QE-300   n$_D^{20}$1.4520

52.56*

C

A

Aldrich 23,488-5   *CAS [594-44-5]* $C_2H_5ClO_2S$ Fp 182°F   60 MHz: *2*, 802D   60.13
Ethanesulfonyl chloride, 99%   FW 128.58   FT-IR: *1*, 900C   9.23*
bp 177°C
d 1.357

CDCl$_3$ QE-300 n_D^{20}1.4530

$CH_3CH_2-\overset{\overset{O}{\|}}{\underset{\underset{O}{\|}}{S}}-Cl$

B

Aldrich 30,377-1   *CAS [10147-36-1]* $C_3H_7ClO_2S$ Fp 176°F   67.01
1-Propanesulfonyl chloride, 97%   FW 142.61   18.23
bp 79°C (15 mm)   12.26*
d 1.267

CDCl$_3$ QE-300 n_D^{20}1.4540

$CH_3CH_2CH_2-\overset{\overset{O}{\|}}{\underset{\underset{O}{\|}}{S}}-Cl$

C

Aldrich 24,270-5   *CAS [10147-37-2]* $C_3H_7ClO_2S$ Fp 187°F   FT-IR: *1*, 900D   67.40*
Isopropylsulfonyl chloride, 97%   FW 142.61   VP-FT-IR: *3*, 835B   17.27*
bp 75°C (19 mm)
d 1.270

CDCl$_3$ QE-300 n_D^{20}1.4530

$CH_3CH\overset{\overset{\displaystyle CH_3}{\,}}{\underset{\,}{}}-\overset{\overset{O}{\|}}{\underset{\underset{O}{\|}}{S}}-Cl$

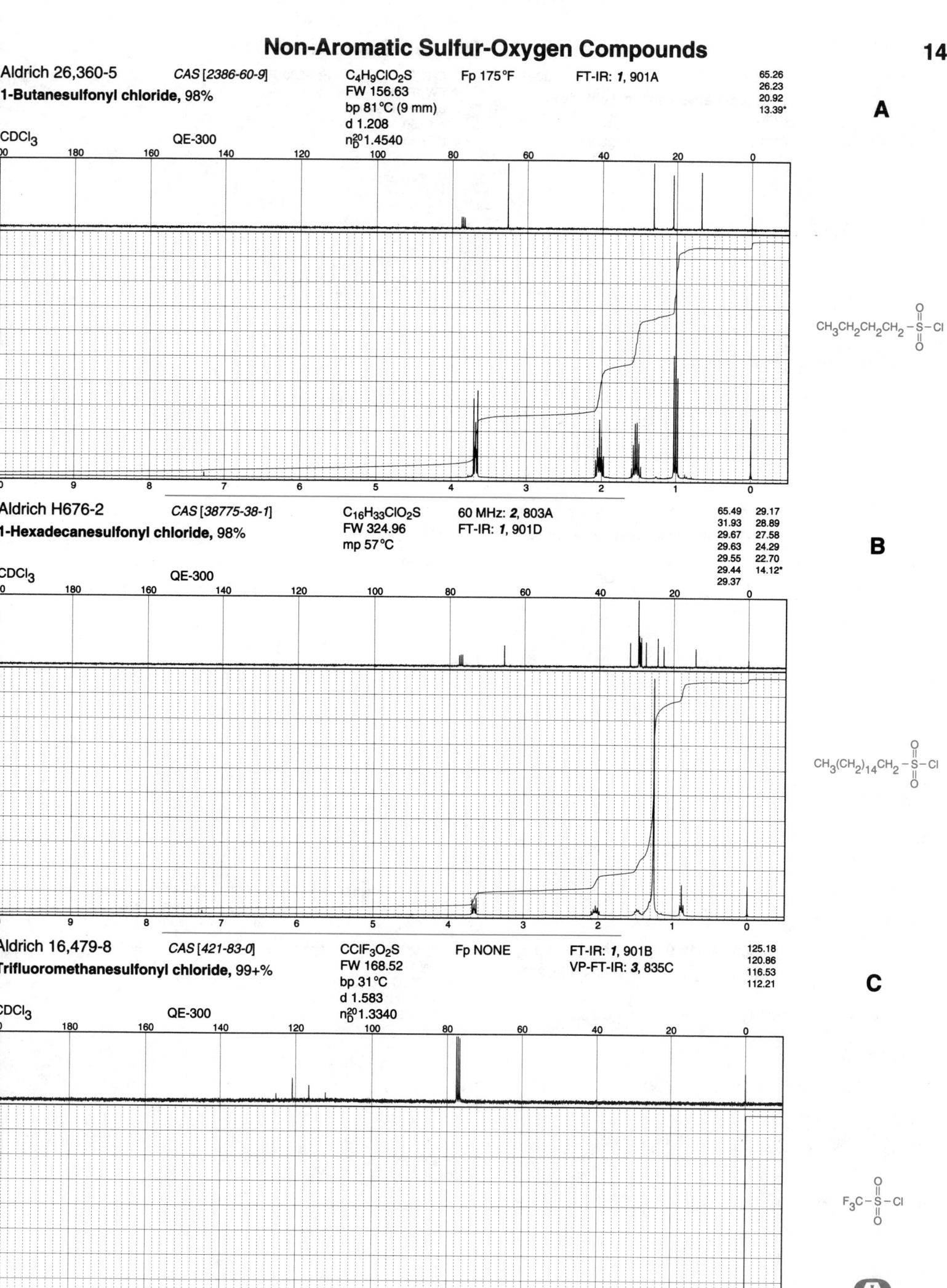

A

Aldrich 26,360-5   *CAS [2386-60-9]* $C_4H_9ClO_2S$ Fp 175°F   FT-IR: *1*, 901A

1-Butanesulfonyl chloride, 98%

FW 156.63
bp 81°C (9 mm)
d 1.208
n_D^{20} 1.4540

CDCl₃   QE-300

65.26
26.23
20.92
13.39*

$CH_3CH_2CH_2CH_2-\overset{\displaystyle O}{\underset{\displaystyle O}{S}}-Cl$

B

Aldrich H676-2   *CAS [38775-38-1]* $C_{16}H_{33}ClO_2S$ 60 MHz: *2*, 803A

1-Hexadecanesulfonyl chloride, 98%

FW 324.96
mp 57°C

FT-IR: *1*, 901D

CDCl₃   QE-300

65.49   29.17
31.93   28.89
29.67   27.58
29.63   24.29
29.55   22.70
29.44   14.12*
29.37

$CH_3(CH_2)_{14}CH_2-\overset{\displaystyle O}{\underset{\displaystyle O}{S}}-Cl$

C

Aldrich 16,479-8   *CAS [421-83-0]* $CClF_3O_2S$ Fp NONE   FT-IR: *1*, 901B

Trifluoromethanesulfonyl chloride, 99+%

FW 168.52
bp 31°C
d 1.583
n_D^{20} 1.3340

VP-FT-IR: *3*, 835C

CDCl₃   QE-300

125.18
120.86
116.53
112.21

$F_3C-\overset{\displaystyle O}{\underset{\displaystyle O}{S}}-Cl$

ALDRICH

A

Aldrich 30,232-5   *CAS [1622-32-8]* $C_2H_4Cl_2O_2S$ Fp >230°F

2-Chloro-1-ethanesulfonyl chloride, 97%

FW 163.02
bp 85°C (15 mm)
d 1.560

CDCl$_3$ QE-300 n_D^{20}1.4930

65.63
35.41

$ClCH_2CH_2-\overset{\displaystyle O}{\underset{\displaystyle O}{S}}-Cl$

B

Aldrich 32,478-7   *CAS [1648-99-3]* $C_2H_2ClF_3O_2S$ Fp 162°F

2,2,2-Trifluoroethanesulfonyl chloride, 99%

FW 182.55
bp 66°C (45 mm)
d 1.651

CDCl$_3$ QE-300 n_D^{20}1.3880

125.56   65.50
121.87   65.05
118.17   64.59
114.48   64.14

$CF_3CH_2-\overset{\displaystyle O}{\underset{\displaystyle O}{S}}-Cl$

C

Aldrich 12,519-9   *CAS [1633-82-5]* $C_3H_6Cl_2O_2S$ Fp >230°F

3-Chloropropanesulfonyl chloride, 99%

FW 177.05
bp 70°C
d 1.456

CDCl$_3$ QE-300 n_D^{20}1.4890

60 MHz: *2*, 803B
FT-IR: *1*, 902A

62.41
41.50
27.37

$ClCH_2CH_2CH_2-\overset{\displaystyle O}{\underset{\displaystyle O}{S}}-Cl$

ALDRICH

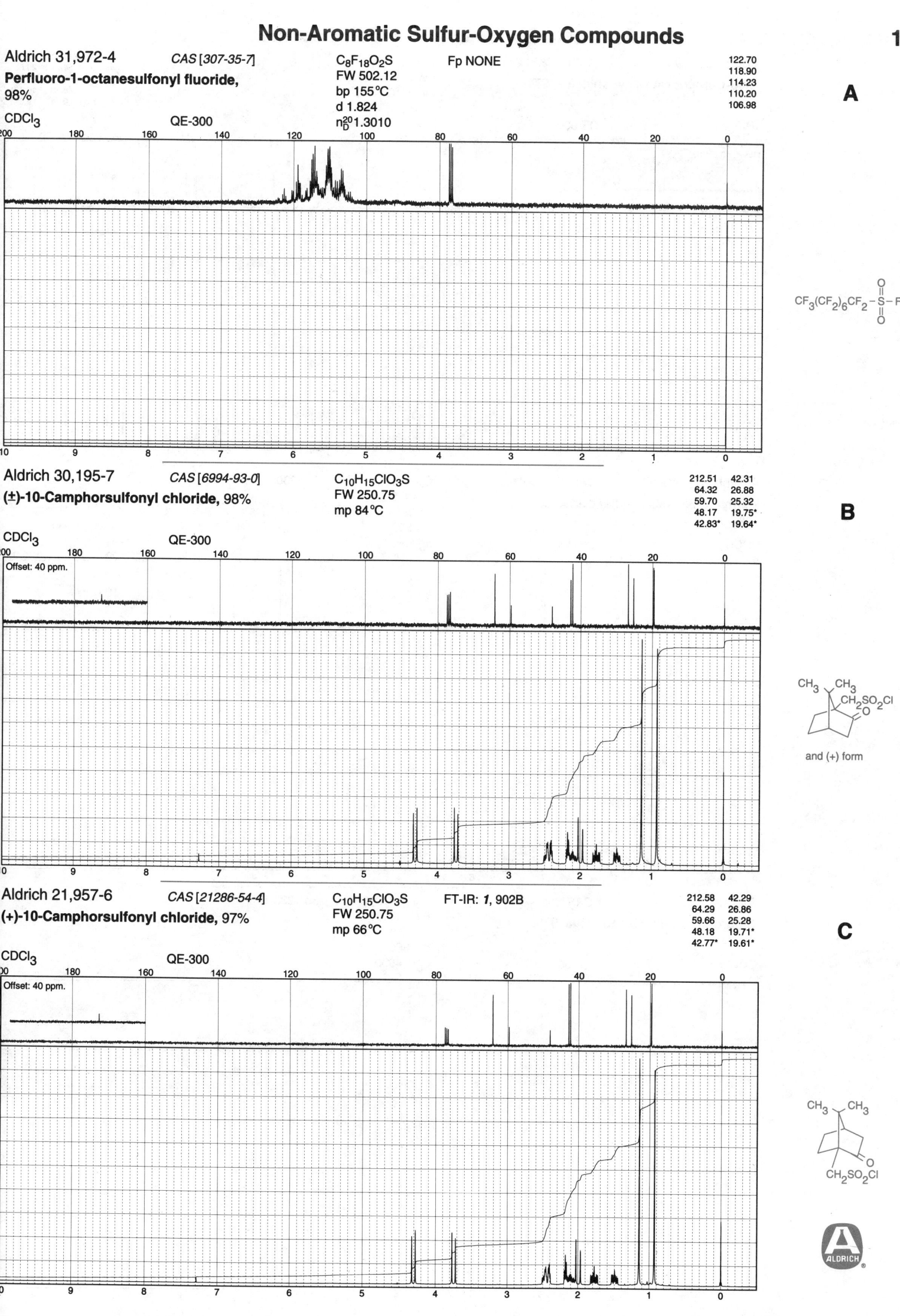

Aldrich 31,972-4   *CAS [307-35-7]* $C_8F_{18}O_2S$ Fp NONE

Perfluoro-1-octanesulfonyl fluoride, 98%

FW 502.12
bp 155°C
d 1.824
n_D^{20} 1.3010

CDCl$_3$ QE-300

122.70
118.90
114.23
110.20
106.98

A

$CF_3(CF_2)_6CF_2-\overset{\overset{\displaystyle O}{\|}}{\underset{\underset{\displaystyle O}{\|}}{S}}-F$

Aldrich 30,195-7   *CAS [6994-93-0]* $C_{10}H_{15}ClO_3S$

(±)-10-Camphorsulfonyl chloride, 98%

FW 250.75
mp 84°C

CDCl$_3$ QE-300
Offset: 40 ppm.

212.51 42.31
64.32   26.88
59.70   25.32
48.17   19.75*
42.83* 19.64*

B

and (+) form

Aldrich 21,957-6   *CAS [21286-54-4]* $C_{10}H_{15}ClO_3S$ FT-IR: *1*, 902B

(+)-10-Camphorsulfonyl chloride, 97%

FW 250.75
mp 66°C

CDCl$_3$ QE-300
Offset: 40 ppm.

212.58 42.29
64.29   26.86
59.66   25.28
48.18   19.71*
42.77* 19.61*

C

A

Aldrich 30,194-9   *CAS [39262-22-1]*   C$_{10}$H$_{15}$ClO$_3$S
FW 250.75
mp 67°C

(-)-10-Camphorsulfonyl chloride, 97%

212.65	42.29
64.24	26.85
59.65	25.26
48.18	19.72*
42.74*	19.62*

CDCl$_3$ QE-300

Offset: 40 ppm.

200 180 160 140 120 100 80 60 40 20 0

10 9 8 7 6 5 4 3 2 1 0

B

Aldrich 37,049-5   *CAS [56146-83-9]*   C$_3$H$_5$ClO$_4$S
FW 172.59
bp 98°C (8 mm)
d 1.505
n$_D^{20}$ 1.4670

Fp >230°F

160.58
66.84
53.96*

Methyl (chlorosulfonyl)acetate, tech., 90%

CDCl$_3$ QE-300

200 180 160 140 120 100 80 60 40 20 0

10 9 8 7 6 5 4 3 2 1 0

Cl —S—CH$_2$—C—OCH$_3$

C

Aldrich 25,977-2   *CAS [4025-77-8]*   C$_2$H$_2$Cl$_2$O$_3$S
FW 177.01
bp 72°C (1 mm)
d 1.669
n$_D^{20}$ 1.4930

Fp >230°F

161.02
73.49

Chlorosulfonylacetyl chloride, 97%

CDCl$_3$ QE-300

200 180 160 140 120 100 80 60 40 20 0

10 9 8 7 6 5 4 3 2 1 0

Cl —C—CH$_2$—S—Cl

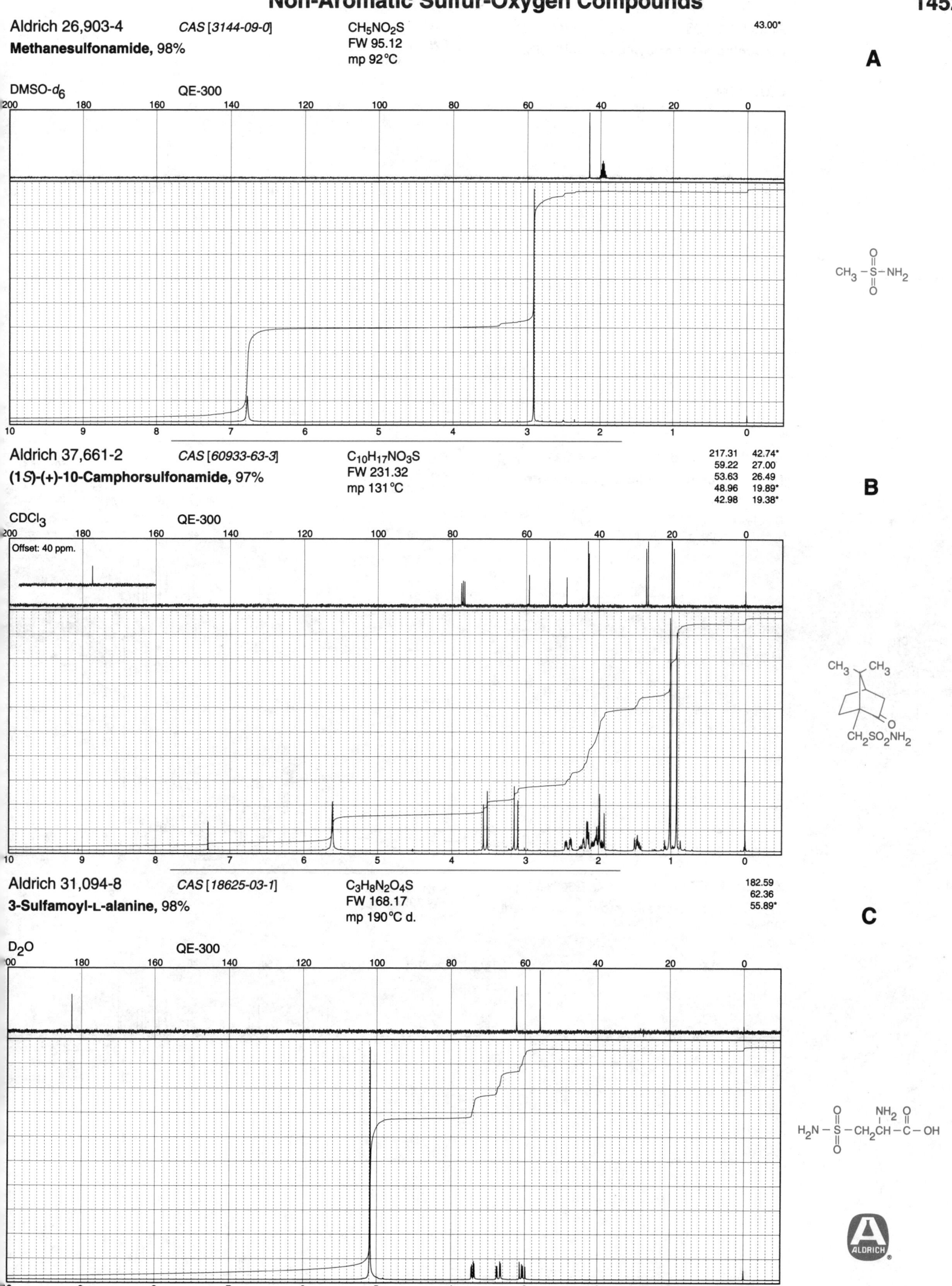

Aldrich 26,903-4   *CAS [3144-09-0]*   CH₅NO₂S   43.00*
Methanesulfonamide, 98%   FW 95.12
mp 92°C

CH_3NO_2S
FW 95.12
mp 92°C

A

DMSO-*d*₆   QE-300

Aldrich 37,661-2   *CAS [60933-63-3]*   C₁₀H₁₇NO₃S   217.31 42.74*
(1S)-(+)-10-Camphorsulfonamide, 97%   FW 231.32   59.22 27.00
mp 131°C   53.63 26.49
48.96 19.89*
42.98 19.38*

B

CDCl₃   QE-300

Offset: 40 ppm.

Aldrich 31,094-8   *CAS [18625-03-1]*   C₃H₈N₂O₄S   182.59
3-Sulfamoyl-L-alanine, 98%   FW 168.17   62.36
mp 190°C d.   55.89*

C

D₂O   QE-300

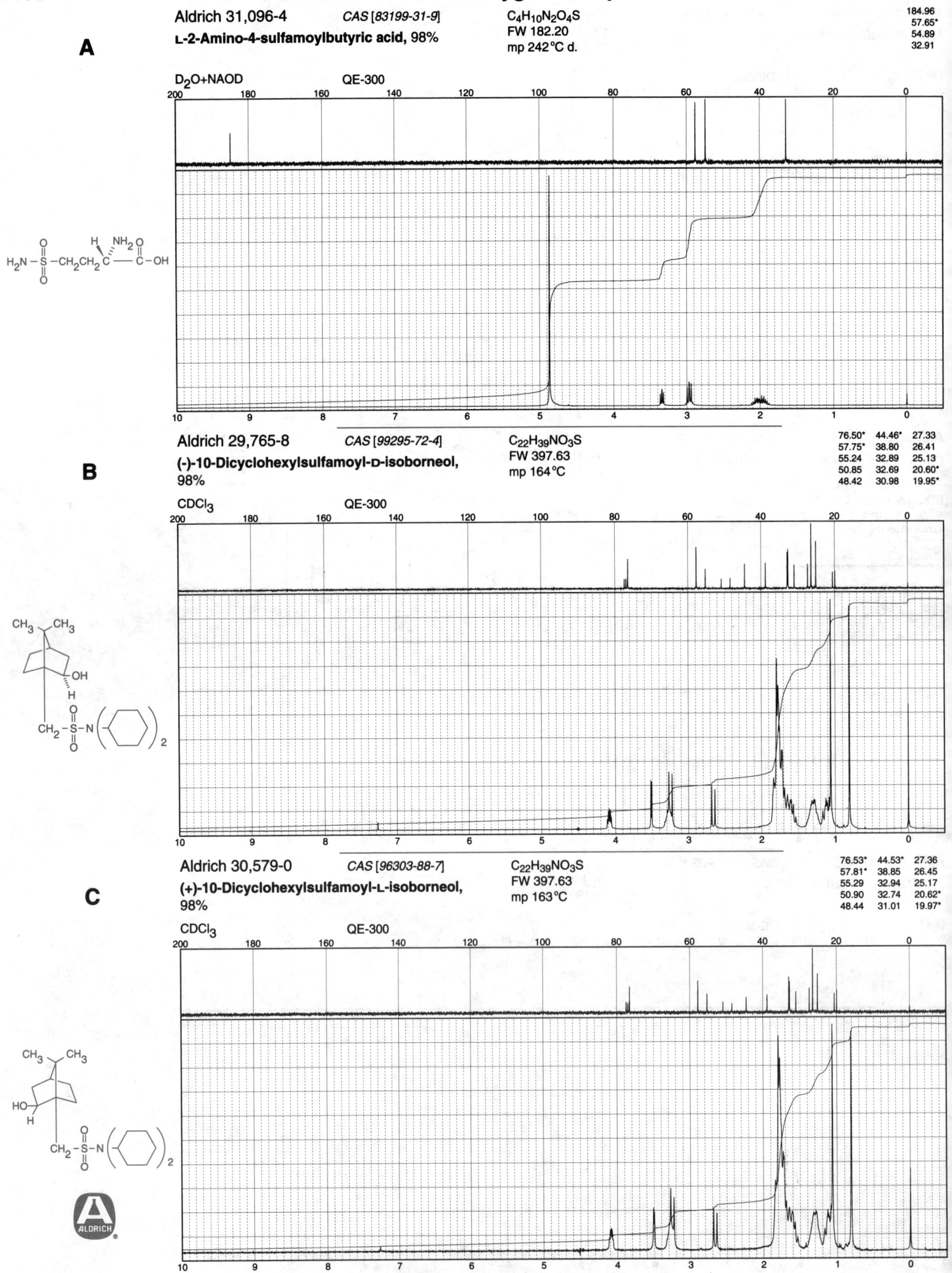

A

Aldrich 31,096-4   *CAS [83199-31-9]*   C₄H₁₀N₂O₄S
FW 182.20
mp 242°C d.

L-2-Amino-4-sulfamoylbutyric acid, 98%

184.96
57.65*
54.89
32.91

D₂O+NAOD   QE-300

B

Aldrich 29,765-8   *CAS [99295-72-4]*   C₂₂H₃₉NO₃S
FW 397.63
mp 164°C

(-)-10-Dicyclohexylsulfamoyl-D-isoborneol, 98%

76.50*	44.46*	27.33
57.75*	38.80	26.41
55.24	32.89	25.13
50.85	32.69	20.60*
48.42	30.98	19.95*

CDCl₃   QE-300

C

Aldrich 30,579-0   *CAS [96303-88-7]*   C₂₂H₃₉NO₃S
FW 397.63
mp 163°C

(+)-10-Dicyclohexylsulfamoyl-L-isoborneol, 98%

76.53*	44.53*	27.36
57.81*	38.85	26.45
55.29	32.94	25.17
50.90	32.74	20.62*
48.44	31.01	19.97*

CDCl₃   QE-300

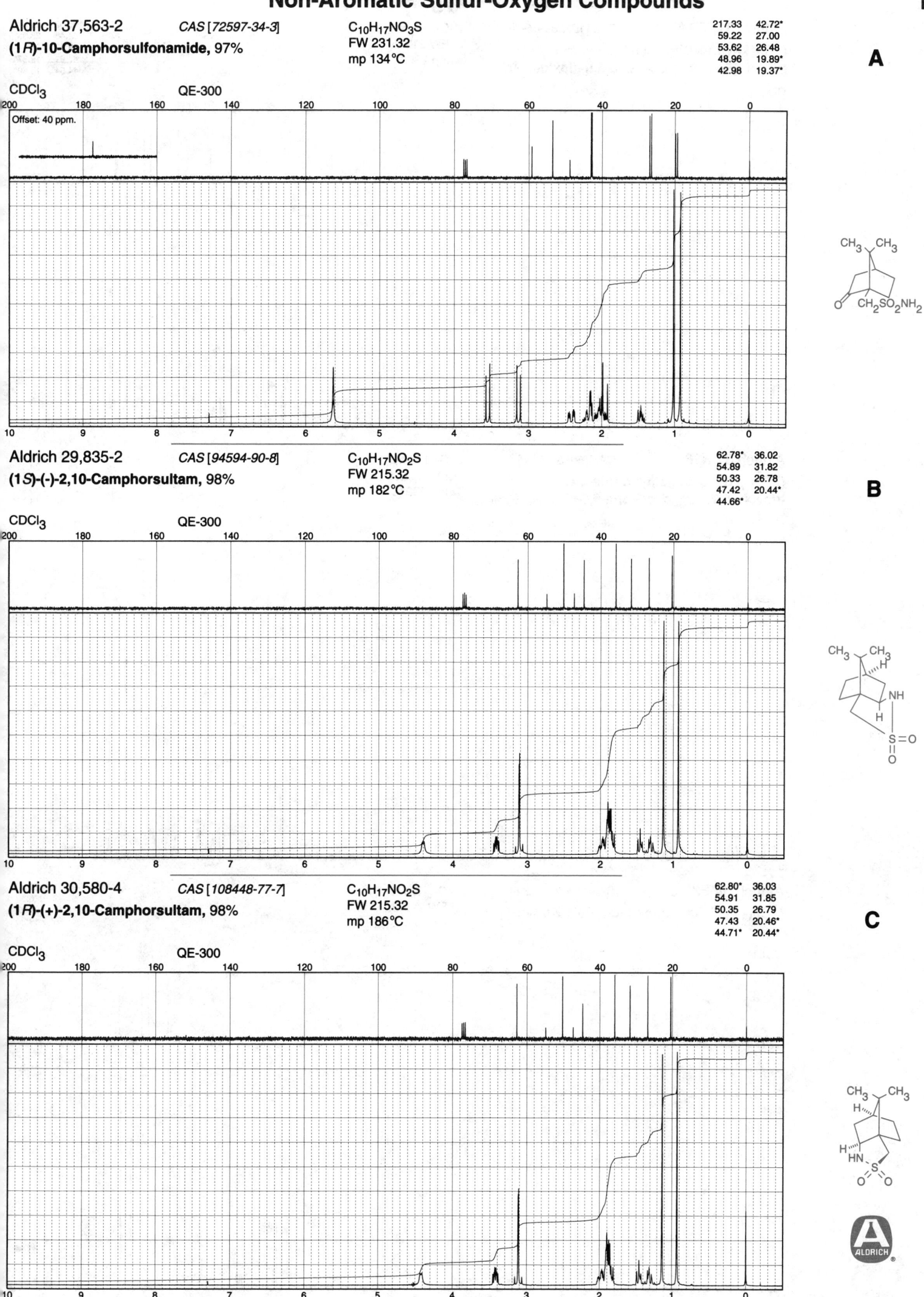

Aldrich 37,563-2   *CAS [72597-34-3]*

(1R)-10-Camphorsulfonamide, 97%

$C_{10}H_{17}NO_3S$
FW 231.32
mp 134 °C

217.33	42.72*
59.22	27.00
53.62	26.48
48.96	19.89*
42.98	19.37*

A

CDCl₃   QE-300

Offset: 40 ppm.

Aldrich 29,835-2   *CAS [94594-90-8]*

(1S)-(-)-2,10-Camphorsultam, 98%

$C_{10}H_{17}NO_2S$
FW 215.32
mp 182 °C

62.78*	36.02
54.89	31.82
50.33	26.78
47.42	20.44*
44.66*	

B

CDCl₃   QE-300

Aldrich 30,580-4   *CAS [108448-77-7]*

(1R)-(+)-2,10-Camphorsultam, 98%

$C_{10}H_{17}NO_2S$
FW 215.32
mp 186 °C

62.80*	36.03
54.91	31.85
50.35	26.79
47.43	20.46*
44.71*	20.44*

C

CDCl₃   QE-300

ALDRICH

Non-Aromatic Sulfur-Oxygen Compounds

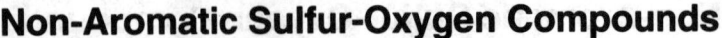

A

Aldrich 31,971-6 *CAS* [*107869-45-4*]

(7*R*)-(+)-10,10-Dimethyl-5-thia-4-aza-tricyclo[5.2.1.0^3,7]dec-3-ene-5,5-dioxide, 99%

$C_{10}H_{15}NO_2S$
FW 213.30
mp 230°C

195.53	35.83
64.48	28.33
49.37	26.54
47.92	19.37*
44.54*	18.89*

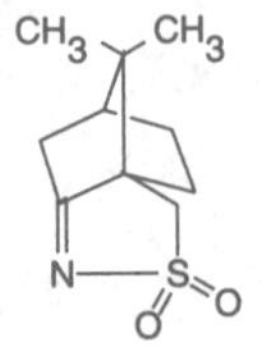

$CDCl_3$ QE-300

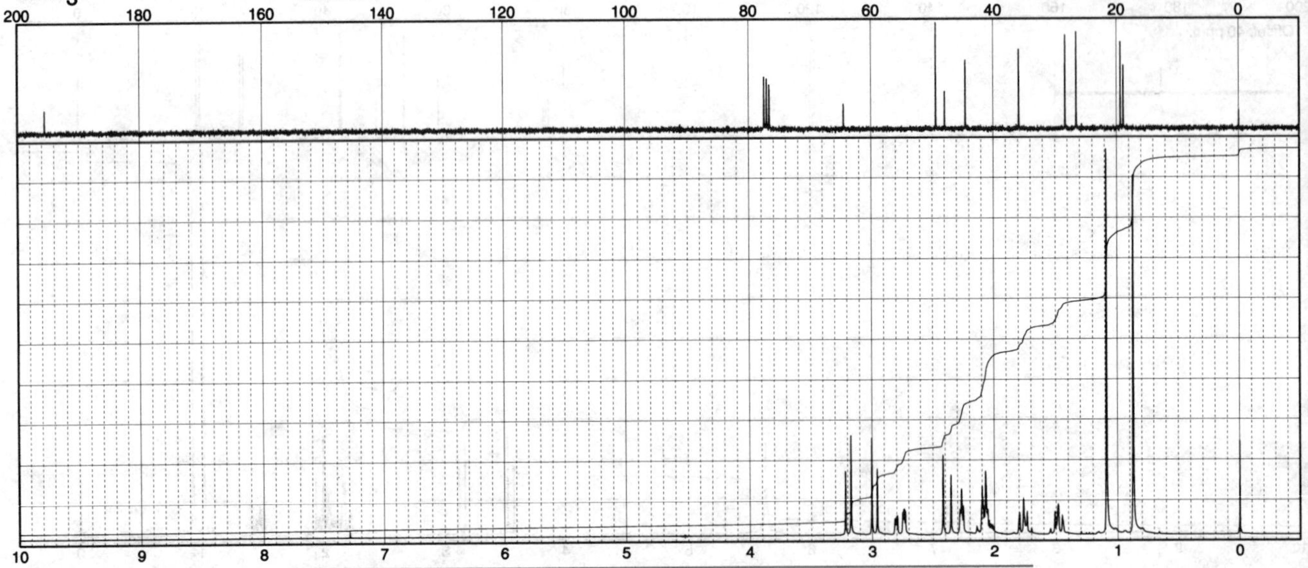

B

Aldrich 31,836-1 *CAS* [*60886-80-8*]

(7*S*)-(-)-10,10-Dimethyl-5-thia-4-aza-tricyclo[5.2.1.0^3,7]dec-3-ene-5,5-dioxide, 99%

$C_{10}H_{15}NO_2S$
FW 213.30
mp 229°C

195.53	35.83
64.49	28.33
49.37	26.54
47.92	19.37*
44.54*	18.89*

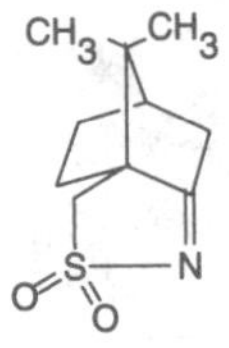

$CDCl_3$ QE-300

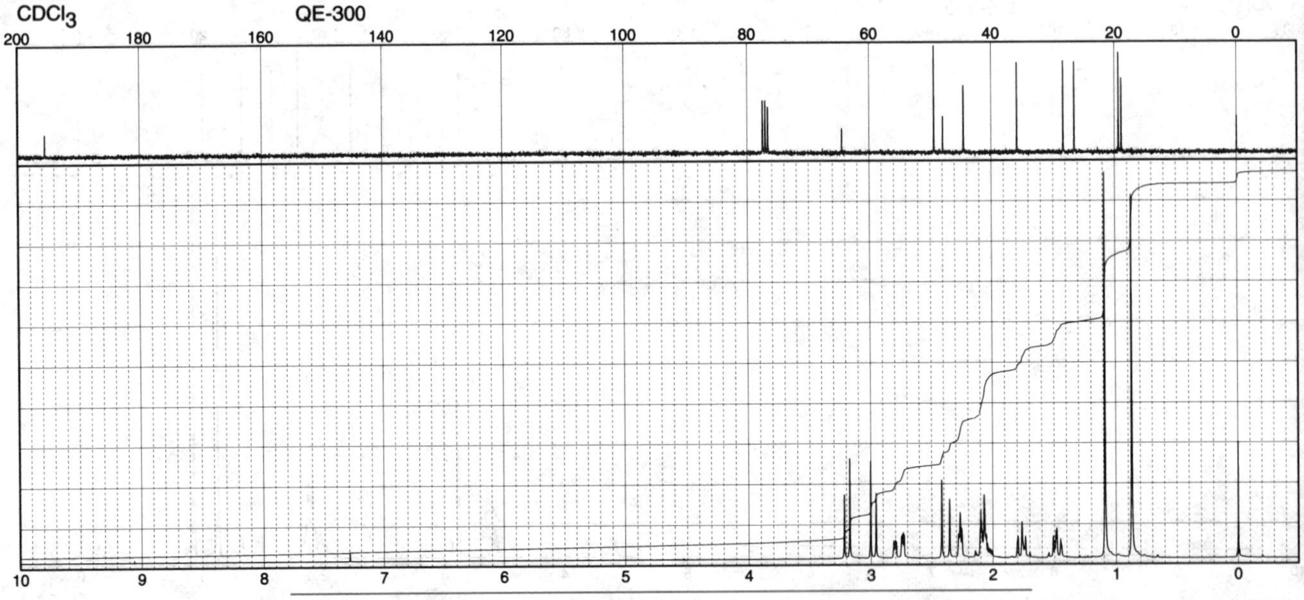

C

Aldrich 34,535-0 *CAS* [*104322-63-6*]

(1*S*)-(+)-(10-Camphorsulfonyl)oxaziridine

$C_{10}H_{15}NO_3S$
FW 229.30
mp 173°C

98.61	28.22
53.92	26.39
48.17	20.37*
45.66*	19.30*
33.49	

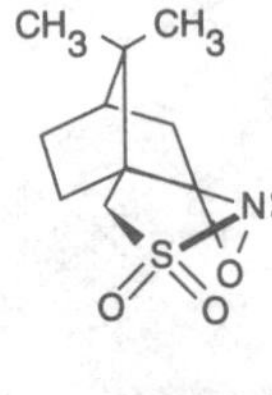

$CDCl_3$ QE-300

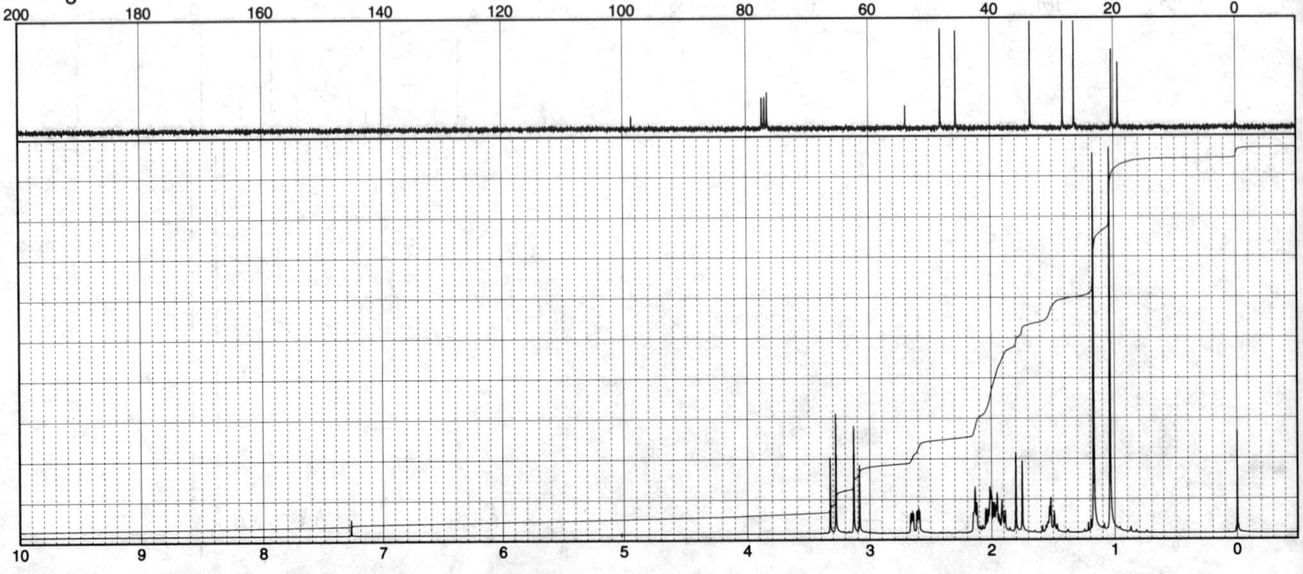

Aldrich 30,359-3   *CAS [2181-44-4]* $C_4H_{12}O_4S_2$
FW 188.27
mp 93°C

Trimethylsulfonium methyl sulfate, 98%

58.12*
29.42*

A

D$_2$O   QE-300

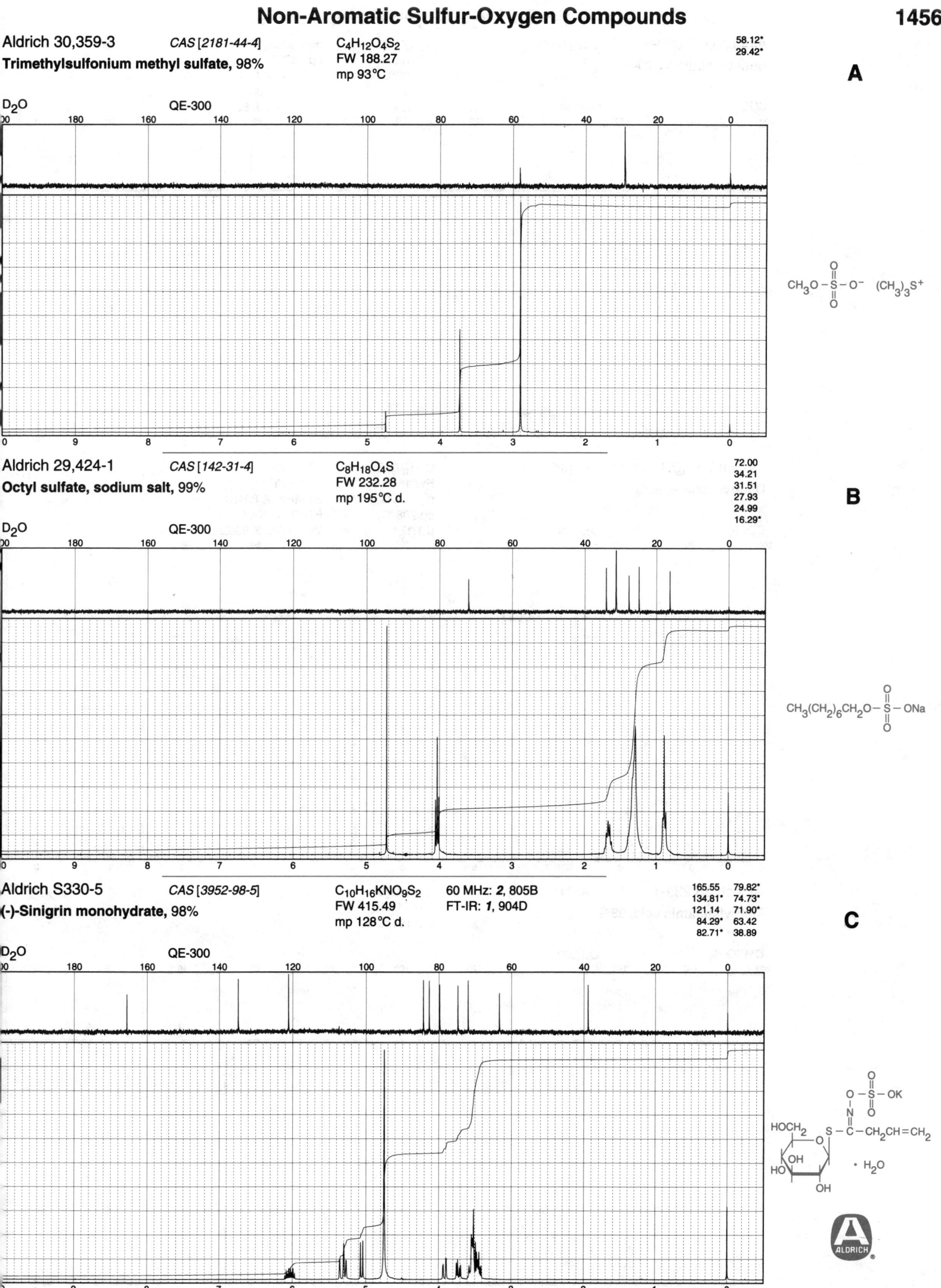

$CH_3O-\overset{\displaystyle O}{\underset{\displaystyle O}{S}}-O^-$ $(CH_3)_3S^+$

Aldrich 29,424-1   *CAS [142-31-4]* $C_8H_{18}O_4S$
FW 232.28
mp 195°C d.

Octyl sulfate, sodium salt, 99%

72.00
34.21
31.51
27.93
24.99
16.29*

B

D$_2$O   QE-300

$CH_3(CH_2)_6CH_2O-\overset{\displaystyle O}{\underset{\displaystyle O}{S}}-ONa$

Aldrich S330-5   *CAS [3952-98-5]* $C_{10}H_{16}KNO_9S_2$ 60 MHz: *2*, 805B
FW 415.49   FT-IR: *1*, 904D
mp 128°C d.

(-)-Sinigrin monohydrate, 98%

165.55   79.82*
134.81* 74.73*
121.14   71.90*
84.29*   63.42
82.71*   38.89

C

D$_2$O   QE-300

· H$_2$O

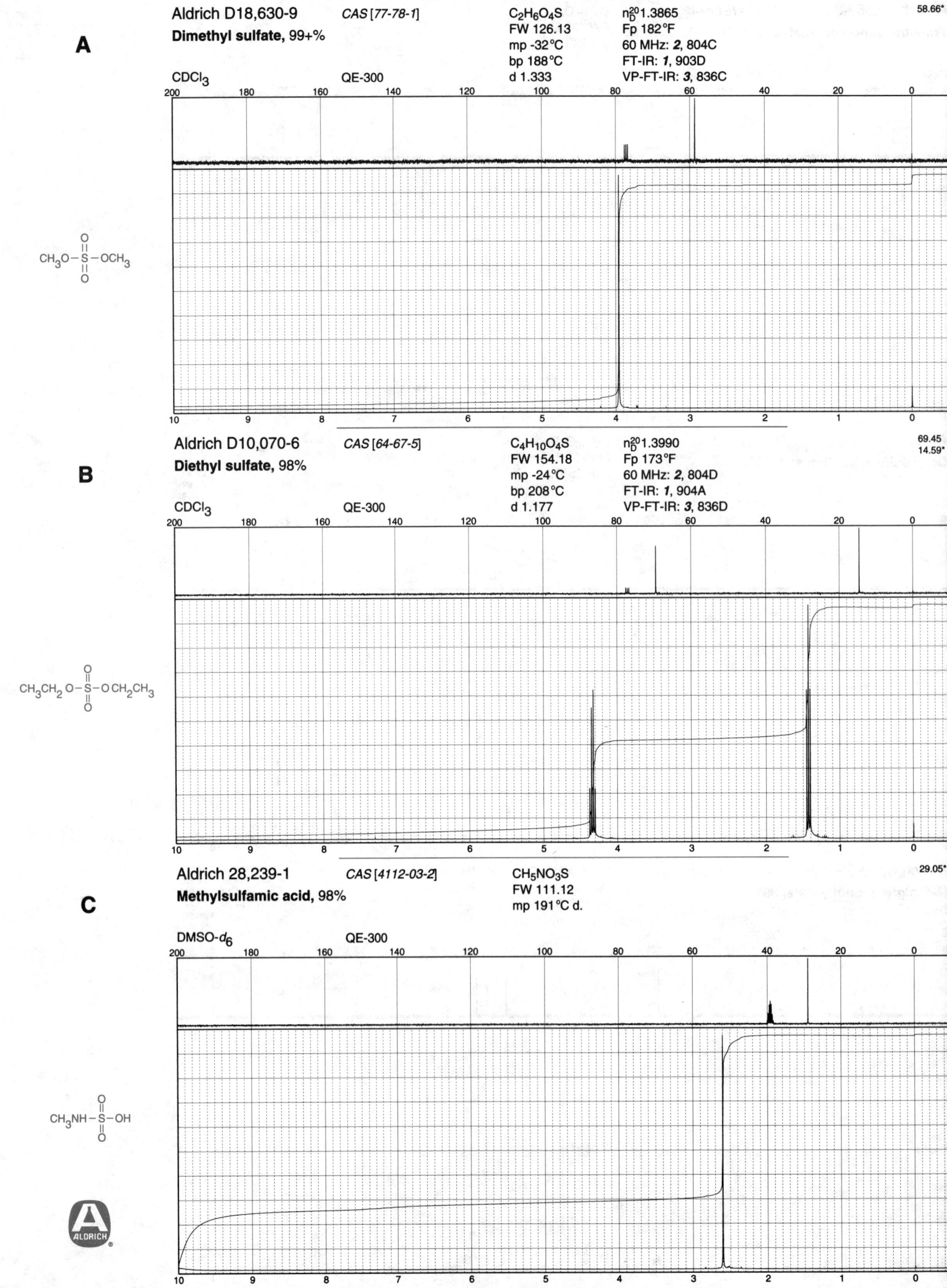

A

Aldrich D18,630-9   *CAS [77-78-1]* $C_2H_6O_4S$ $n_D^{20} 1.3865$ 58.66*

Dimethyl sulfate, 99+%

FW 126.13   Fp 182°F

mp -32°C   60 MHz: *2*, 804C

bp 188°C   FT-IR: *1*, 903D

d 1.333   VP-FT-IR: *3*, 836C

CDCl$_3$ QE-300

$CH_3O-\overset{\displaystyle O}{\underset{\displaystyle O}{S}}-OCH_3$

B

Aldrich D10,070-6   *CAS [64-67-5]* $C_4H_{10}O_4S$ $n_D^{20} 1.3990$ 69.45 / 14.59*

Diethyl sulfate, 98%

FW 154.18   Fp 173°F

mp -24°C   60 MHz: *2*, 804D

bp 208°C   FT-IR: *1*, 904A

d 1.177   VP-FT-IR: *3*, 836D

CDCl$_3$ QE-300

$CH_3CH_2O-\overset{\displaystyle O}{\underset{\displaystyle O}{S}}-OCH_2CH_3$

C

Aldrich 28,239-1   *CAS [4112-03-2]* CH_5NO_3S 29.05*

Methylsulfamic acid, 98%

FW 111.12

mp 191°C d.

DMSO-d_6 QE-300

$CH_3NH-\overset{\displaystyle O}{\underset{\displaystyle O}{S}}-OH$

Aldrich 29,002-5   CAS [39085-61-5]   C$_4$H$_{11}$NO$_3$S
Butylsulfamic acid, 98%   FW 153.20
mp 189°C d.

43.67
27.77
19.27
13.39*

A

DMSO-d_6 QE-300

$CH_3CH_2CH_2CH_2NH-\overset{\overset{O}{\|}}{\underset{\underset{O}{\|}}{S}}-OH$

Aldrich D18,625-2   CAS [13360-57-1]   C$_2$H$_6$ClNO$_2$S   Fp 202°F   60 MHz: **2**, 803C   39.56*
Dimethylsulfamoyl chloride, 99%   FW 143.59   FT-IR: **1**, 902C
bp 114°C (75 mm)   VP-FT-IR: **3**, 835D
d 1.337
n$_D^{20}$1.4520

B

CDCl$_3$ QE-300

$(CH_3)_2N-\overset{\overset{O}{\|}}{\underset{\underset{O}{\|}}{S}}-Cl$

Aldrich 36,548-3   CAS [29684-56-8]   C$_8$H$_{18}$N$_2$O$_4$S
(Methoxycarbonylsulfamoyl)triethylammonium   FW 238.31
hydroxide, inner salt, 97%   mp 78°C

158.08
53.18*
50.52
9.36*

C

CDCl$_3$ QE-300

$CH_3O-\overset{\overset{O}{\|}}{C}-N-\overset{\overset{O}{\|}}{\underset{\underset{O}{\|}}{S}}-\overset{CH_2CH_3}{\underset{CH_2CH_3}{N^+-CH_2CH_3}}$

Non-Aromatic Phosphorus Compounds

A

Aldrich 36,271-9   *CAS [819-19-2]*   C$_8$H$_{19}$P   31.68*
FW 146.22   31.51*
Di-*tert*-butylphosphine, 98%   d 0.790   29.90
Fp -1°F   29.69

CDCl$_3$ QE-300

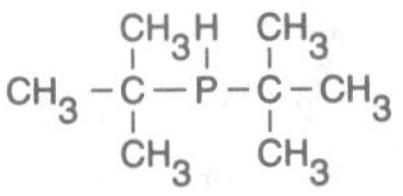

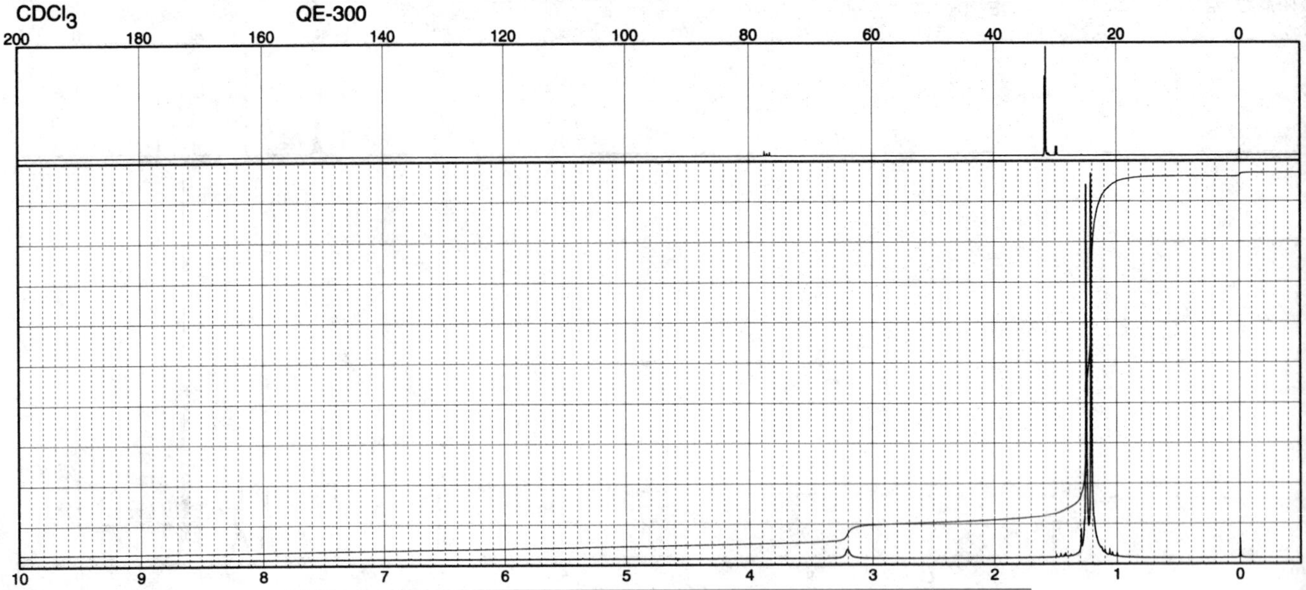

Aldrich 32,332-2   *CAS [594-09-2]*   C$_3$H$_9$P   Fp -22°F   16.34*
FW 76.08   16.18*
Trimethylphosphine, 97%   mp -43°C
bp 39°C

B

CDCl$_3$ QE-300   d 0.735

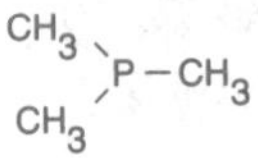

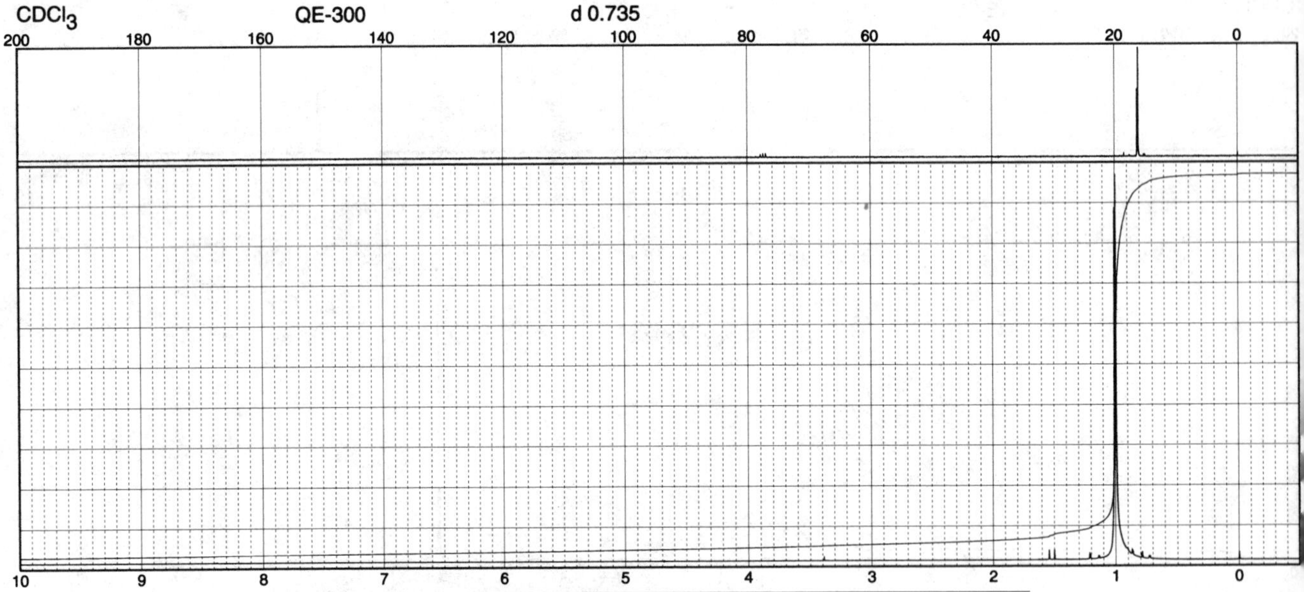

Aldrich 33,697-1   *CAS [2234-97-1]*   C$_9$H$_{21}$P   Fp 144°F   29.85
FW 160.24   29.69
Tripropylphosphine, 97%   bp 73°C (12 mm)   19.49
d 0.801   19.31
n$_D^{20}$ 1.4584   16.24*
16.09*

C

CDCl$_3$ QE-300

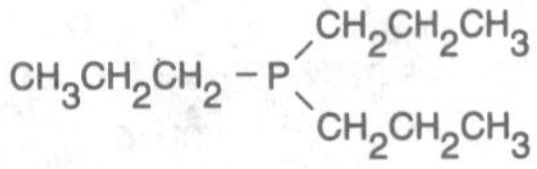

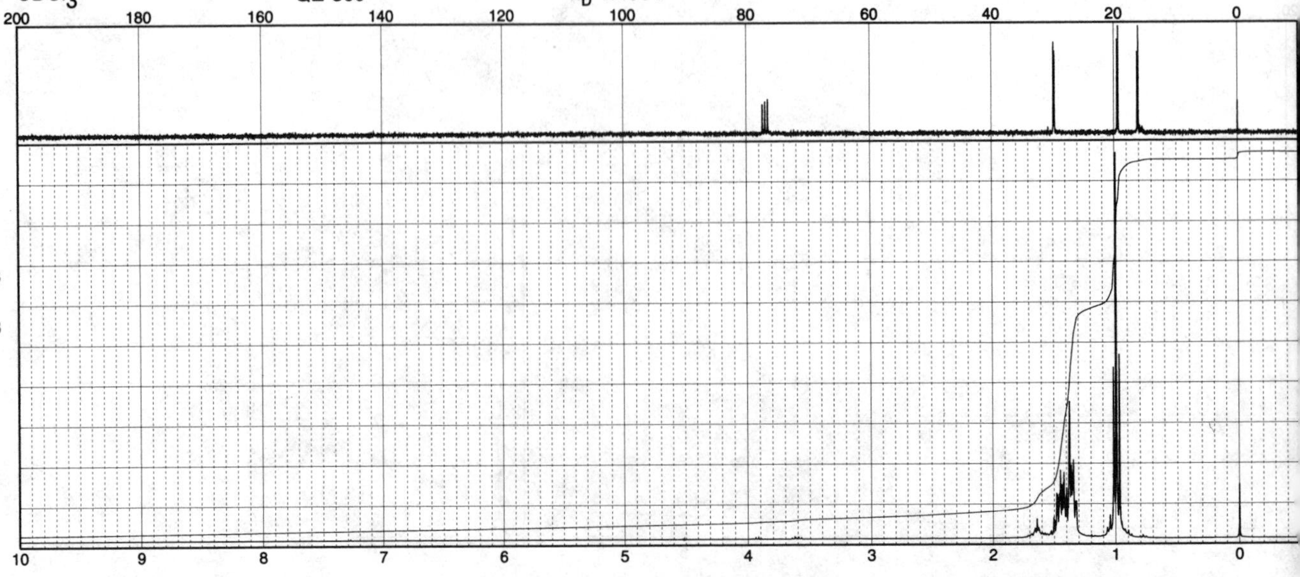

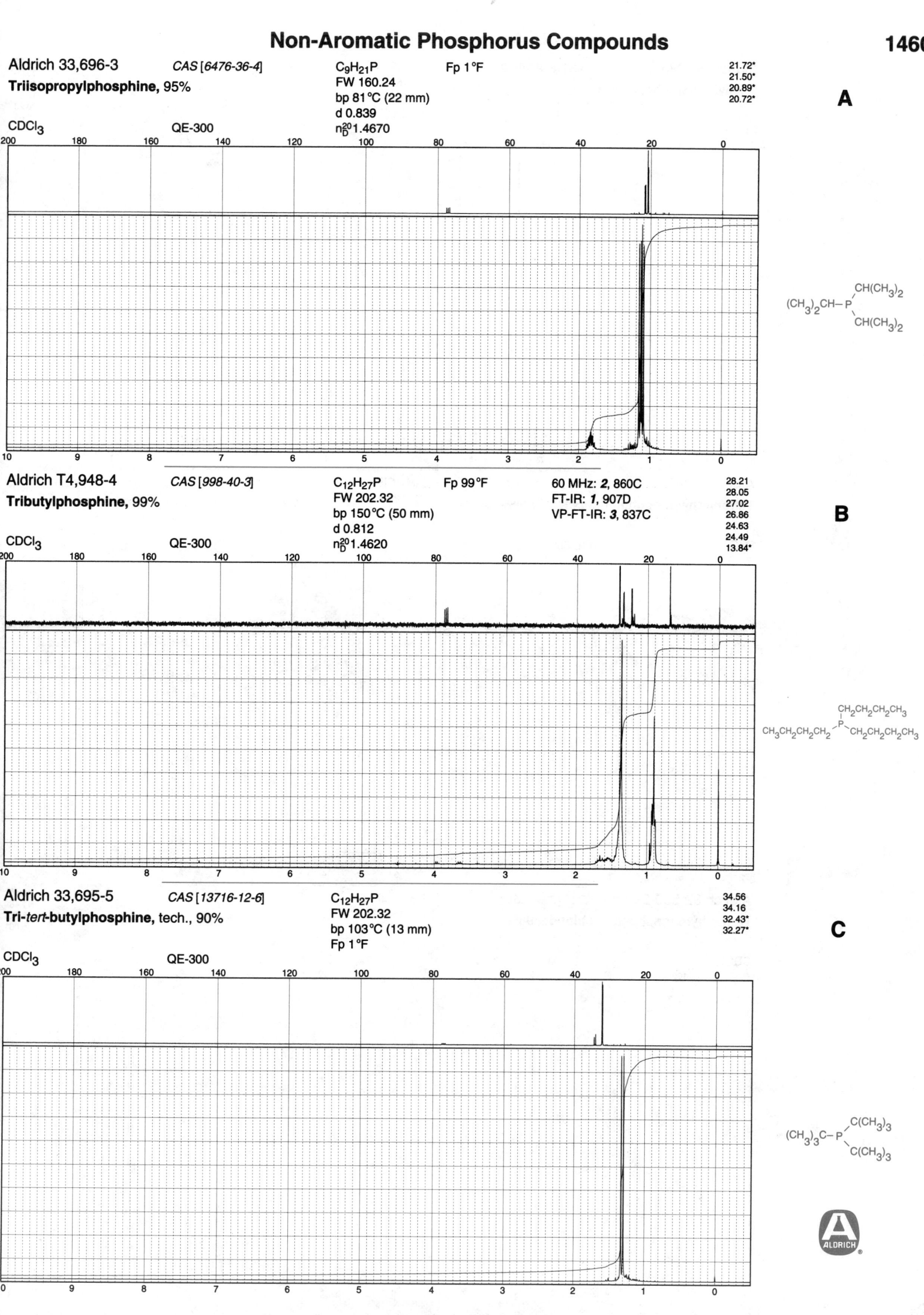

Aldrich 33,696-3   *CAS [6476-36-4]*   C₉H₂₁P   Fp 1 °F

Triisopropylphosphine, 95%

CDCl₃   QE-300

$C_9H_{21}P$
FW 160.24
bp 81 °C (22 mm)
d 0.839
n_D^{20} 1.4670

21.72*
21.50*
20.89*
20.72*

A

$(CH_3)_2CH-P$ ⟨ $CH(CH_3)_2$ / $CH(CH_3)_2$ ⟩

Aldrich T4,948-4   *CAS [998-40-3]*

Tributylphosphine, 99%

CDCl₃   QE-300

$C_{12}H_{27}P$
FW 202.32
bp 150 °C (50 mm)
d 0.812
n_D^{20} 1.4620

Fp 99 °F

60 MHz: **2**, 860C
FT-IR: **1**, 907D
VP-FT-IR: **3**, 837C

28.21
28.05
27.02
26.86
24.63
24.49
13.84*

B

$CH_3CH_2CH_2CH_2-P$ ⟨ $CH_2CH_2CH_2CH_3$ / $CH_2CH_2CH_2CH_3$ ⟩

Aldrich 33,695-5   *CAS [13716-12-6]*

Tri-*tert*-butylphosphine, tech., 90%

CDCl₃   QE-300

$C_{12}H_{27}P$
FW 202.32
bp 103 °C (13 mm)
Fp 1 °F

34.56
34.16
32.43*
32.27*

C

$(CH_3)_3C-P$ ⟨ $C(CH_3)_3$ / $C(CH_3)_3$ ⟩

1461

A

Aldrich 33,693-9   *CAS [16523-89-0]*   C₉H₁₅P
Triallylphosphine, 98%

$C_9H_{15}P$
FW 154.19
bp 70°C (10 mm)
Fp 203°F

133.17*
133.09*
116.66
116.55
30.23
30.01

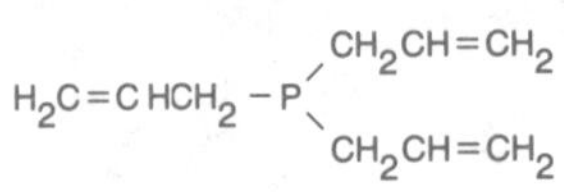

CDCl₃   QE-300

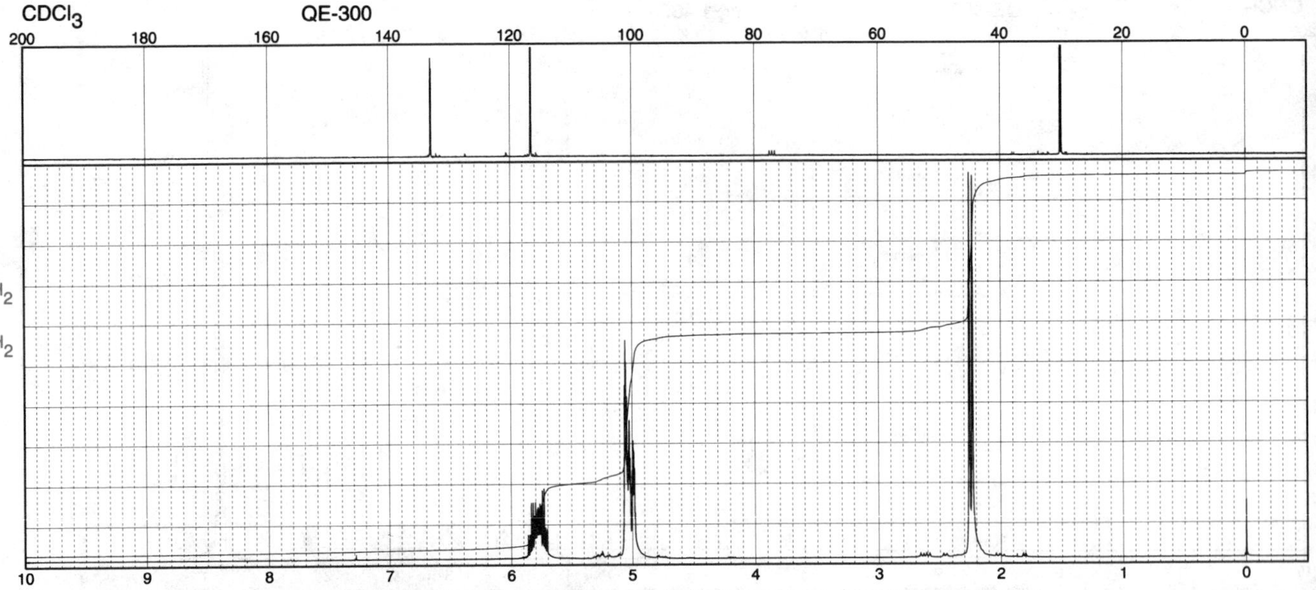

H₂C=CHCH₂—P—CH₂CH=CH₂ / CH₂CH=CH₂

B

Aldrich 28,827-6   *CAS [1941-19-1]*   C₄H₁₂ClP
Tetramethylphosphonium chloride, 98%   FW 126.57

12.16*
11.41*

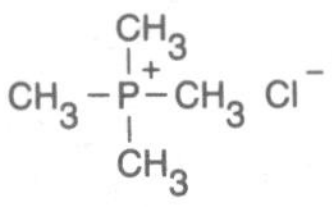

D₂O   QE-300

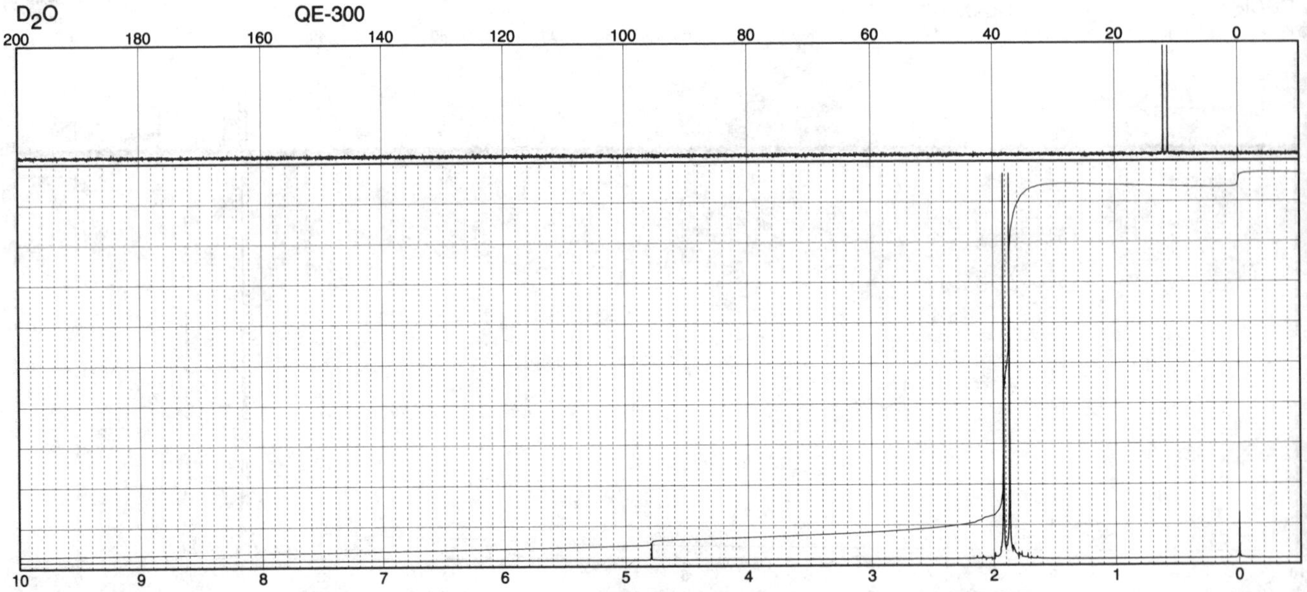

CH₃—P⁺(CH₃)(CH₃)CH₃ Cl⁻

C

Aldrich 32,539-2   *CAS [7368-65-2]*   C₈H₂₀ClP
Tetraethylphosphonium chloride, 98%   FW 182.68

12.10
11.46
6.07*
6.00*

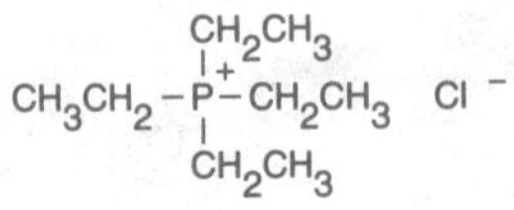

CDCl₃   QE-300

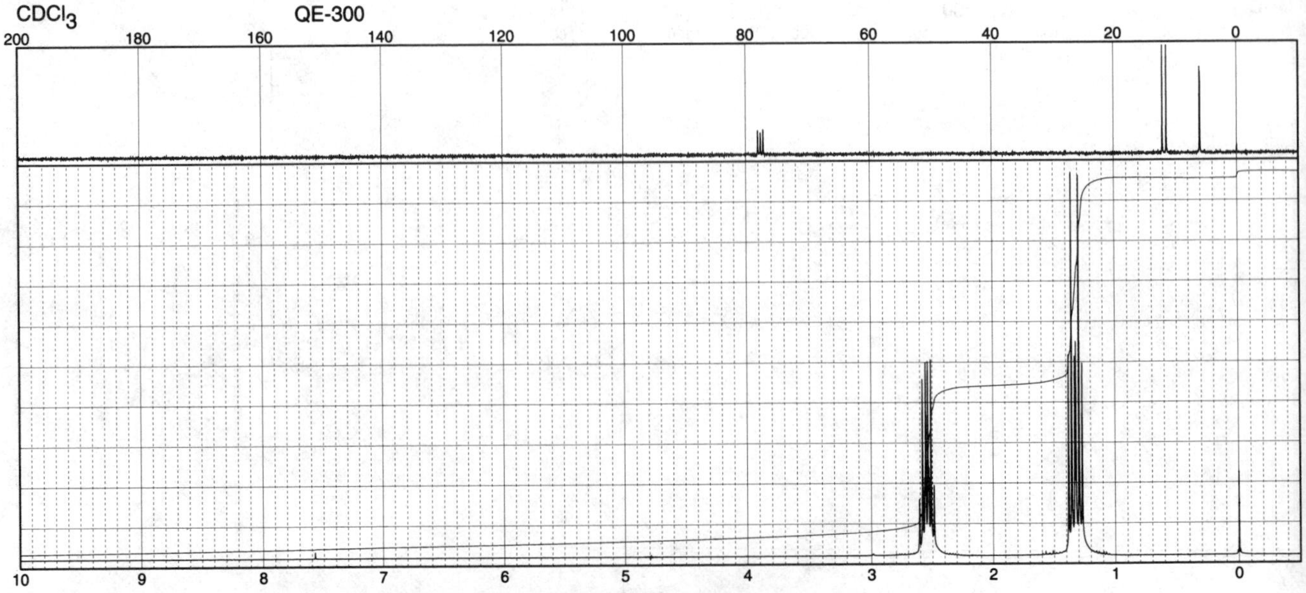

CH₃CH₂—P⁺(CH₂CH₃)(CH₂CH₃)CH₂CH₃ Cl⁻

ALDRICH®

Aldrich 33,365-4   *CAS [4317-07-1]*   C$_8$H$_{20}$BrP

Tetraethylphosphonium bromide, 97%

FW 227.13
mp 334 °C

12.31
11.66
6.14*
6.07*

A

CDCl$_3$ QE-300

$$CH_3CH_2 - \overset{\overset{\displaystyle CH_2CH_3}{|}}{\underset{\underset{\displaystyle CH_2CH_3}{|}}{\overset{+}{P}}} - CH_2CH_3 \quad\quad Br^-$$

Aldrich 32,540-6   *CAS [4317-06-0]*   C$_8$H$_{20}$IP

Tetraethylphosphonium iodide, 97%

FW 274.13
mp 302 °C

11.23
10.58
5.55*
5.48*

B

CDCl$_3$ + DMSO-d_6 QE-300

$$CH_3CH_2 - \overset{\overset{\displaystyle CH_2CH_3}{|}}{\underset{\underset{\displaystyle CH_2CH_3}{|}}{\overset{+}{P}}} - CH_2CH_3 \quad\quad I^-$$

Aldrich 14,480-0   *CAS [2304-30-5]*   C$_{16}$H$_{36}$ClP

Tetrabutylphosphonium chloride, 96%

FW 294.89
mp 82 °C

60 MHz: *2*, 861C
FT-IR: *1*, 909A

24.12
23.90
23.83
19.34
18.71
13.56*

C

CDCl$_3$ QE-300

$$CH_3CH_2CH_2CH_2 - \overset{\overset{\displaystyle CH_2CH_2CH_2CH_3}{|}}{\underset{\underset{\displaystyle CH_2CH_2CH_2CH_3}{|}}{\overset{+}{P}}} - CH_2CH_2CH_2CH_3 \quad\quad Cl^-$$

A

Aldrich 17,276-6   *CAS [814-29-9]*   C$_{12}$H$_{27}$OP   60 MHz: *2*, 860D

Tributylphosphine oxide, 95%   FW 218.32   FT-IR: *1*, 908A

mp 67°C
bp 150°C (1 mm)

28.14
27.28
24.42
24.23
23.83
23.78
13.65*

CDCl$_3$ QE-300

CH$_3$CH$_2$CH$_2$CH$_2$—P—CH$_2$CH$_2$CH$_2$CH$_3$
                         ‖
                         O
          |
   CH$_2$CH$_2$CH$_2$CH$_3$

B

Aldrich 22,330-1   *CAS [78-50-2]*   C$_{24}$H$_{51}$OP   60 MHz: *2*, 861A

Trioctylphosphine oxide, 99%   FW 386.65   FT-IR: *1*, 908B

mp 52°C
bp 202°C (2 mm)
Fp >230°F

31.82   27.58
31.32   22.64
31.14   21.76
29.13   21.72
29.08   14.09*
28.44

CDCl$_3$ QE-300

CH$_3$(CH$_2$)$_6$CH$_2$—P—CH$_2$(CH$_2$)$_6$CH$_3$
                         ‖
                         O
          |
   CH$_2$(CH$_2$)$_6$CH$_3$

C

Aldrich 33,683-1   *CAS [686-69-1]*   C$_4$H$_{10}$ClP   Fp 66°F

Chlorodiethylphosphine, 95%   FW 124.55

bp 134°C
d 1.023
n$_D^{20}$ 1.4750

27.09
26.71
8.41*
8.23*

CDCl$_3$ QE-300

CH$_3$CH$_2$
          \
           P—Cl
          /
CH$_3$CH$_2$

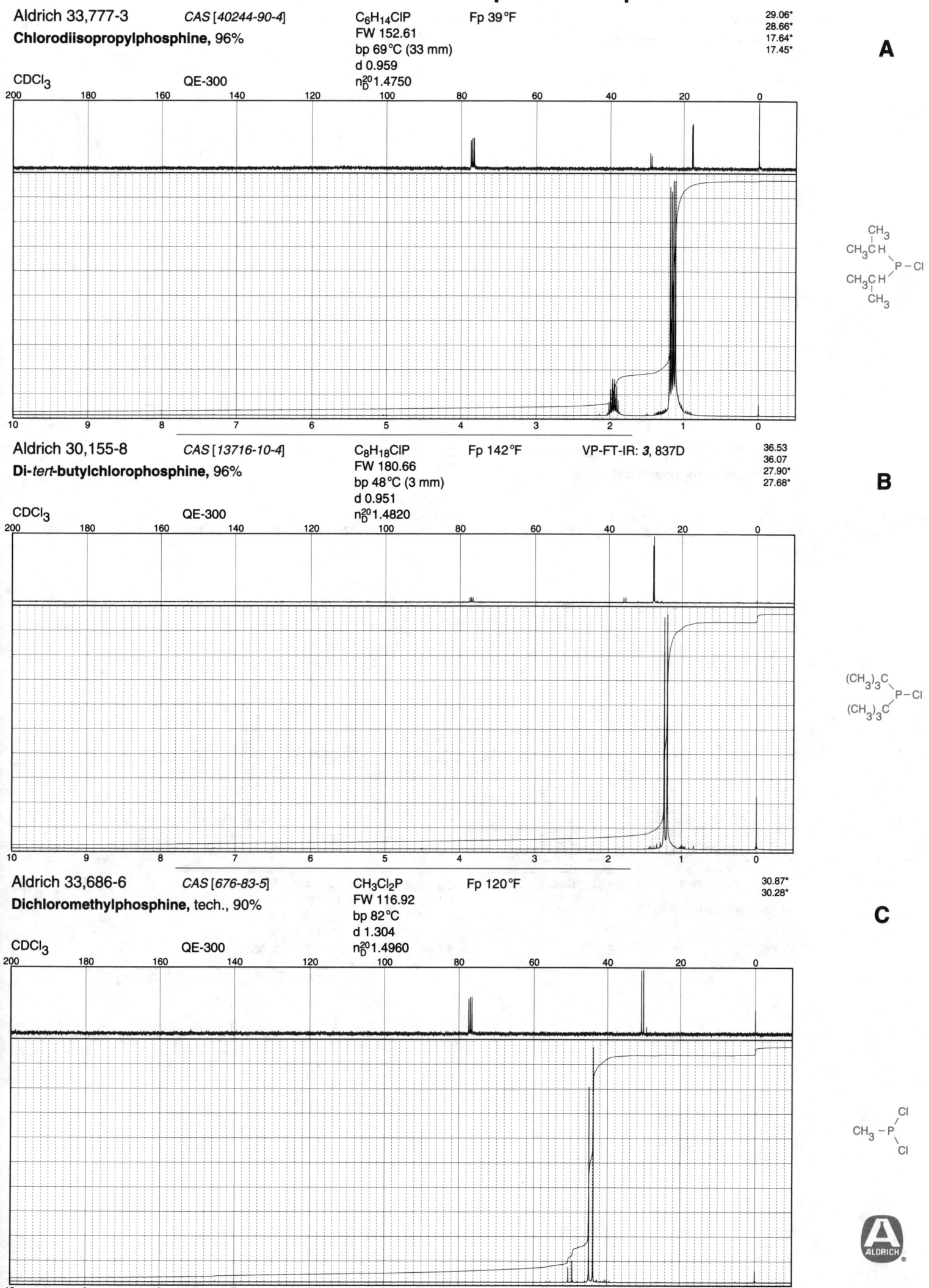

$C_6H_{14}ClP$
FW 152.61
bp 69°C (33 mm)
d 0.959
n_D^{20} 1.4750

Fp 39°F

29.06*
28.66*
17.64*
17.45*

A

CDCl₃   QE-300

200 180 160 140 120 100 80 60 40 20 0

10 9 8 7 6 5 4 3 2 1 0

$C_8H_{18}ClP$
FW 180.66
bp 48°C (3 mm)
d 0.951
n_D^{20} 1.4820

Fp 142°F

VP-FT-IR: **3**, 837D

36.53
36.07
27.90*
27.68*

B

CDCl₃   QE-300

200 180 160 140 120 100 80 60 40 20 0

10 9 8 7 6 5 4 3 2 1 0

CH_3Cl_2P
FW 116.92
bp 82°C
d 1.304
n_D^{20} 1.4960

Fp 120°F

30.87*
30.28*

C

CDCl₃   QE-300

200 180 160 140 120 100 80 60 40 20 0

10 9 8 7 6 5 4 3 2 1 0

ALDRICH®

Non-Aromatic Phosphorus Compounds

A

Aldrich 33,685-8   CAS [1498-40-4]   C₂H₅Cl₂P   Fp 91°F

$C_2H_5Cl_2P$

Dichloroethylphosphine, 95%

FW 130.94
bp 112°C
d 1.260
n_D^{20}1.4950

36.13
35.56
6.99*
6.82*

CDCl₃   QE-300

CH_3CH_2-P (with Cl, Cl)

B

Aldrich 26,192-0   CAS [28240-69-9]   C₂H₄Cl₄P₂   Fp NONE

1,2-Bis(dichlorophosphino)ethane, 97%

$C_2H_4Cl_4P_2$
FW 231.81
bp 70°C (1 mm)
d 1.536
n_D^{20}1.5880

34.99
34.88
34.34
34.22

CDCl₃   QE-300

$Cl_2P-CH_2CH_2-PCl_2$

C

Aldrich 24,090-7   CAS [121-45-9]   C₃H₉O₃P n_D^{20}1.4080

Trimethyl phosphite, 99+%

FW 124.08
mp -78°C
bp 112°C
d 1.052

Fp 82°F
60 MHz: 2, 862B
FT-IR: 1, 910A
VP-FT-IR: 3, 838B

49.10*
48.96*

CDCl₃   QE-300

$CH_3O-P(OCH_3)OCH_3$

Aldrich T6,780-6   CAS [116-17-6]

Triisopropyl phosphite, 95%

$C_9H_{21}O_3P$
FW 208.24
bp 64°C (11 mm)
d 0.844
n_D^{20}1.4110

Fp 154°F

60 MHz: **2**, 862D
FT-IR: **1**, 910C
VP-FT-IR: **3**, 838D

66.12*
65.94*
24.64*
24.59*

A

CDCl₃   QE-300

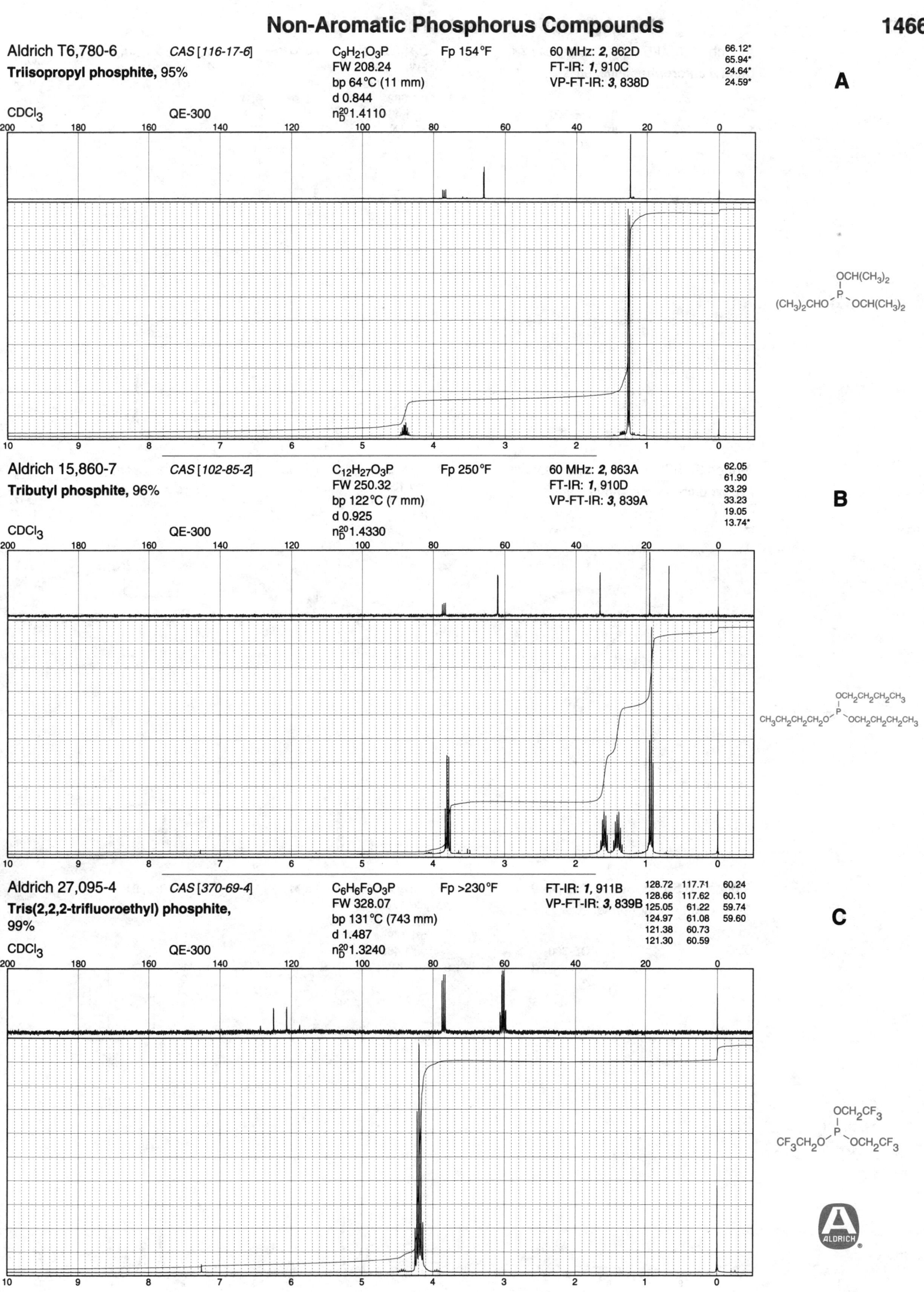

Aldrich 15,860-7   CAS [102-85-2]

Tributyl phosphite, 96%

$C_{12}H_{27}O_3P$
FW 250.32
bp 122°C (7 mm)
d 0.925
n_D^{20}1.4330

Fp 250°F

60 MHz: **2**, 863A
FT-IR: **1**, 910D
VP-FT-IR: **3**, 839A

62.05
61.90
33.29
33.23
19.05
13.74*

B

CDCl₃   QE-300

Aldrich 27,095-4   CAS [370-69-4]

Tris(2,2,2-trifluoroethyl) phosphite, 99%

$C_6H_6F_9O_3P$
FW 328.07
bp 131°C (743 mm)
d 1.487
n_D^{20}1.3240

Fp >230°F

FT-IR: **1**, 911B
VP-FT-IR: **3**, 839B

128.72	117.71	60.24
128.66	117.62	60.10
125.05	61.22	59.74
124.97	61.08	59.60
121.38	60.73	
121.30	60.59	

C

CDCl₃   QE-300

A

Aldrich 25,550-5   *CAS [589-57-1]* $C_4H_{10}ClO_2P$ Fp 34°F   60 MHz: **2**, 864A

Diethyl chlorophosphite, 98%   FW 156.55   FT-IR: **1**, 911D

bp 57°C (30 mm)   VP-FT-IR: **3**, 839C

d 1.082

n_D^{20} 1.4340

CDCl₃   QE-300

61.23
61.17
16.34*
16.28*

CH_3CH_2O
CH_3CH_2O — P—Cl

B

Aldrich 30,570-7   *CAS [13040-68-1]* $C_3H_7Cl_2OP$ Fp 146°F   VP-FT-IR: **3**, 844D

Propyl dichlorophosphite, 99%   FW 160.97

bp 40°C (13 mm)

d 1.228

n_D^{20} 1.4670

CDCl₃   QE-300

70.11
69.98
23.20
23.17
10.12*

$CH_3CH_2CH_2O$ — P — Cl, Cl

C

Aldrich 30,569-3   *CAS [10496-13-6]* $C_4H_9Cl_2OP$ Fp 71°F   VP-FT-IR: **3**, 840A

Butyl dichlorophosphite, 98%   FW 175.00

bp 161°C (750 mm)

d 1.179

n_D^{20} 1.4680

CDCl₃   QE-300

68.39
68.26
31.78
31.74
18.79
13.53*

$CH_3CH_2CH_2CH_2O$ — P — Cl, Cl

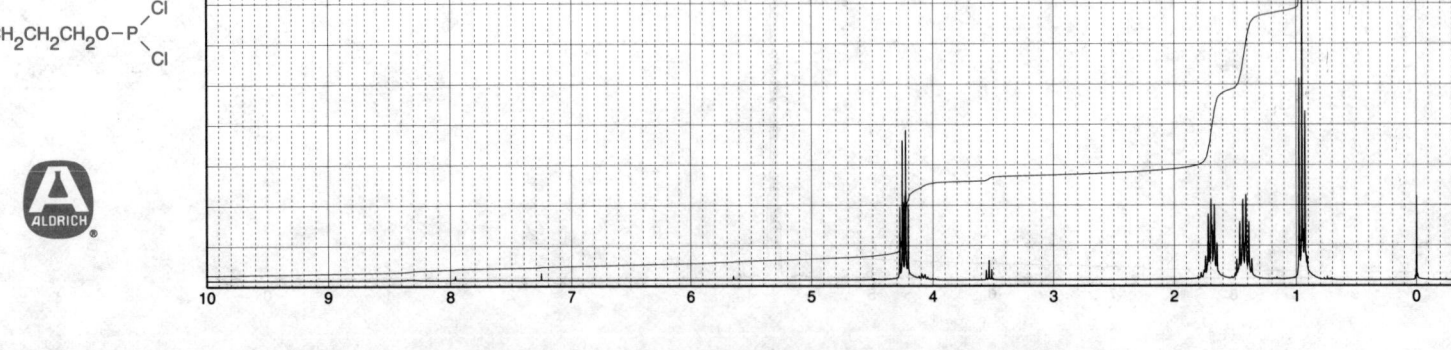

Aldrich 21,954-1   *CAS [60010-51-7]*

2,2,2-Trichloroethyl phosphorodichloridite, 97%

CDCl₃   QE-300

C₂H₂Cl₅OP	Fp NONE°F	60 MHz: *2*, 864B	94.11
FW 250.28		FT-IR: *1*, 912B	94.07
bp 42°C			76.30
d 1.606			76.15
n²⁰_D 1.5210			

A

CCl_3CH_2O-P Cl Cl

Aldrich 25,822-9   *CAS [39177-74-7]*

2,2,2-Trichloro-1,1-dimethylethyl dichloro-phosphite, 97%

CDCl₃   QE-300

C₄H₆Cl₅OP	Fp >230°F	FT-IR: *1*, 912C	105.07
FW 278.33			93.53
bp 118°C (16 mm)			24.48*
d 1.528			24.35*
n²⁰_D 1.5230			

B

CH₃ Cl
CCl_3C-O-P
CH₃ Cl

Aldrich 36,596-3   *CAS [117924-33-1]*

Di-*tert*-butyl diethylphosphoramidite, tech., 93%

CDCl₃   QE-300

C₁₂H₂₈NO₂P	Fp 173°F	74.41	30.94*
FW 249.34		74.27	30.83*
bp 40°C		37.62	14.89*
d 0.896		37.34	14.83*
n²⁰_D 1.4334			

C

CH₃CH₂ OC(CH₃)₃
N—P
CH₃CH₂ OC(CH₃)₃

Non-Aromatic Phosphorus Compounds

A

Aldrich 26,252-8   *CAS [86030-43-5]*   C$_7$H$_{17}$ClNOP   Fp 163°F   VP-FT-IR: *3*, 844A

N,N-Diisopropylmethylphosphonamidic chloride, 95%

FW 197.65
bp 35°C
d 1.018
n$_D^{20}$1.4680

CDCl$_3$ QE-300

53.08*	
52.83*	
45.85*	
45.69*	
23.93*	
23.38*	

B

Aldrich 30,230-9   *CAS [89992-70-1]*   C$_9$H$_{18}$ClN$_2$OP   Fp >230°F

2-Cyanoethyl N,N-diisopropylchloro-phosphoramidite

FW 236.68
bp 104°C
d 1.061
n$_D^{20}$1.4780

CDCl$_3$ QE-300

116.85	23.99*
60.56	23.35*
60.30	23.21*
46.27*	19.95
46.10*	19.86
24.03*	

C

Aldrich 30,599-5   *CAS [102691-36-1]*   C$_{15}$H$_{32}$N$_3$OP   Fp 141°F

2-Cyanoethyl N,N,N',N'-tetraisopropyl-phosphorodiamidite, 97%

FW 301.42
bp 100°C
d 0.949
n$_D^{20}$1.4700

CDCl$_3$ QE-300

117.91	24.50*
59.48	23.90*
59.15	23.83*
44.68*	20.65
44.51*	20.53
24.61*	

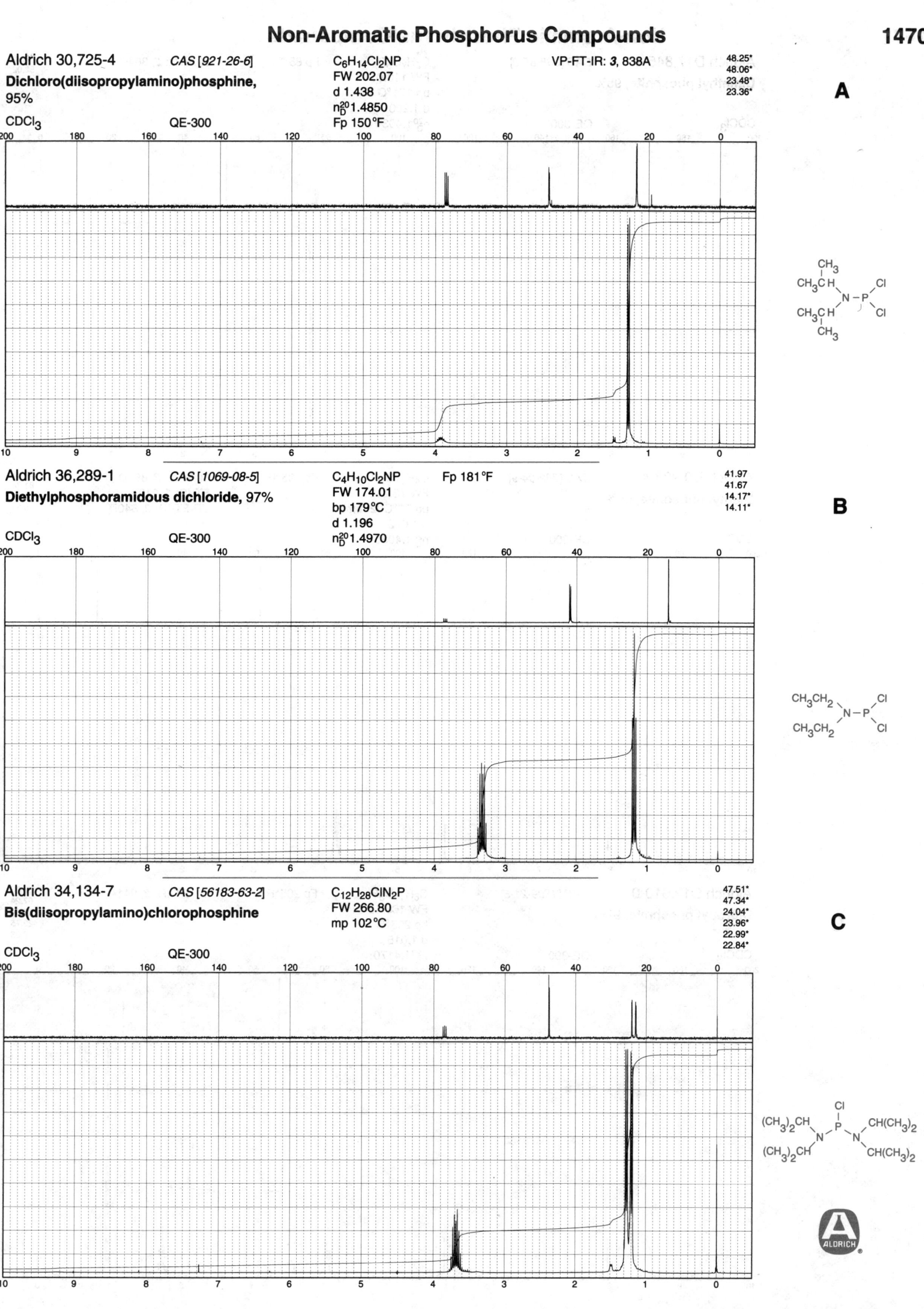

Aldrich 30,725-4   CAS [921-26-6]
Dichloro(diisopropylamino)phosphine, 95%
CDCl₃   QE-300

$C_6H_{14}Cl_2NP$
FW 202.07
d 1.438
n_D^{20} 1.4850
Fp 150°F

VP-FT-IR: **3**, 838A

48.25*
48.06*
23.48*
23.36*

A

Aldrich 36,289-1   CAS [1069-08-5]
Diethylphosphoramidous dichloride, 97%
CDCl₃   QE-300

$C_4H_{10}Cl_2NP$
FW 174.01
bp 179°C
d 1.196
n_D^{20} 1.4970

Fp 181°F

41.97
41.67
14.17*
14.11*

B

Aldrich 34,134-7   CAS [56183-63-2]
Bis(diisopropylamino)chlorophosphine
CDCl₃   QE-300

$C_{12}H_{28}ClN_2P$
FW 266.80
mp 102°C

47.51*
47.34*
24.04*
23.96*
22.99*
22.84*

C

Non-Aromatic Phosphorus Compounds

A

Aldrich D17,845-4   *CAS [868-85-9]* $C_2H_7O_3P$ Fp 85°F   60 MHz: *2*, 864C   52.10*
Dimethyl phosphite, 98%   FW 110.05   52.02*
bp 171°C
d 1.200
n_D^{20}1.4030

CDCl₃   QE-300

$CH_3O-\overset{\overset{O}{\|}}{\underset{H}{P}}-OCH_3$

B

Aldrich D9,923-4   *CAS [762-04-9]* $C_4H_{11}O_3P$ Fp 195°F   60 MHz: *2*, 864D   61.81
Diethyl phosphite, 98%   FW 138.10   FT-IR: *1*, 913A   61.74
bp 51°C (2 mm)   VP-FT-IR: *3*, 840D   16.39*
d 1.072   16.31*
n_D^{20}1.4080

CDCl₃   QE-300

$CH_3CH_2O-\overset{\overset{O}{\|}}{\underset{H}{P}}-OCH_2CH_3$

C

Aldrich D12,610-0   *CAS [1809-21-8]* $C_6H_{15}O_3P$ Fp 205°F   VP-FT-IR: *3*, 841A   67.32
Dipropyl phosphite, 97%   FW 166.16   67.24
bp 203°C   23.84
d 1.018   23.75
n_D^{20}1.4170   10.08*

CDCl₃   QE-300

$CH_3CH_2CH_2O-\overset{\overset{O}{\|}}{\underset{H}{P}}-OCH_2CH_2CH_3$

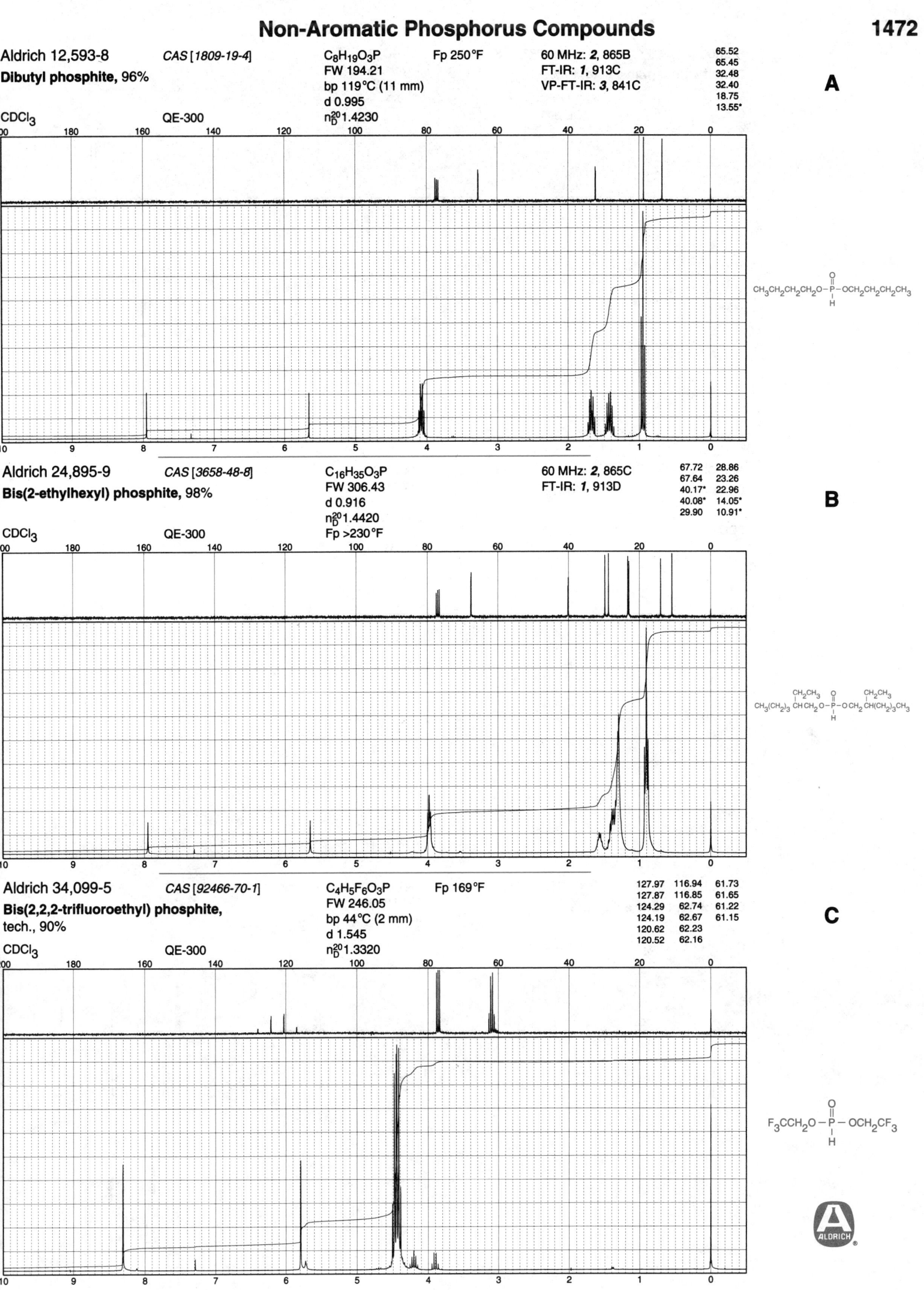

A

Aldrich 12,593-8
Dibutyl phosphite, 96%

CAS [1809-19-4]

$C_8H_{19}O_3P$
FW 194.21
bp 119°C (11 mm)
d 0.995
n_D^{20} 1.4230

Fp 250°F

60 MHz: *2*, 865B
FT-IR: *1*, 913C
VP-FT-IR: *3*, 841C

65.52
65.45
32.48
32.40
18.75
13.55*

$CDCl_3$

QE-300

$CH_3CH_2CH_2CH_2O-\overset{\overset{O}{\|}}{\underset{\underset{H}{\|}}{P}}-OCH_2CH_2CH_2CH_3$

B

Aldrich 24,895-9
Bis(2-ethylhexyl) phosphite, 98%

CAS [3658-48-8]

$C_{16}H_{35}O_3P$
FW 306.43
d 0.916
n_D^{20} 1.4420
Fp >230°F

60 MHz: *2*, 865C
FT-IR: *1*, 913D

67.72   28.86
67.64   23.26
40.17* 22.96
40.08* 14.05*
29.90   10.91*

$CDCl_3$

QE-300

$CH_3(CH_2)_3\underset{\underset{\overset{|}{CH_2CH_3}}{\overset{\overset{CH_2CH_3}{|}}{CH}}}{}CH_2O-\overset{\overset{O}{\|}}{\underset{\underset{H}{\|}}{P}}-OCH_2\underset{\underset{\overset{|}{CH_2CH_3}}{}}{\overset{\overset{CH_2CH_3}{|}}{CH}}(CH_2)_3CH_3$

C

Aldrich 34,099-5
Bis(2,2,2-trifluoroethyl) phosphite,
tech., 90%

CAS [92466-70-1]

$C_4H_5F_6O_3P$
FW 246.05
bp 44°C (2 mm)
d 1.545
n_D^{20} 1.3320

Fp 169°F

127.97   116.94   61.73
127.87   116.85   61.65
124.29   62.74   61.22
124.19   62.67   61.15
120.62   62.23
120.52   62.16

$CDCl_3$

QE-300

$F_3CCH_2O-\overset{\overset{O}{\|}}{\underset{\underset{H}{\|}}{P}}-OCH_2CF_3$

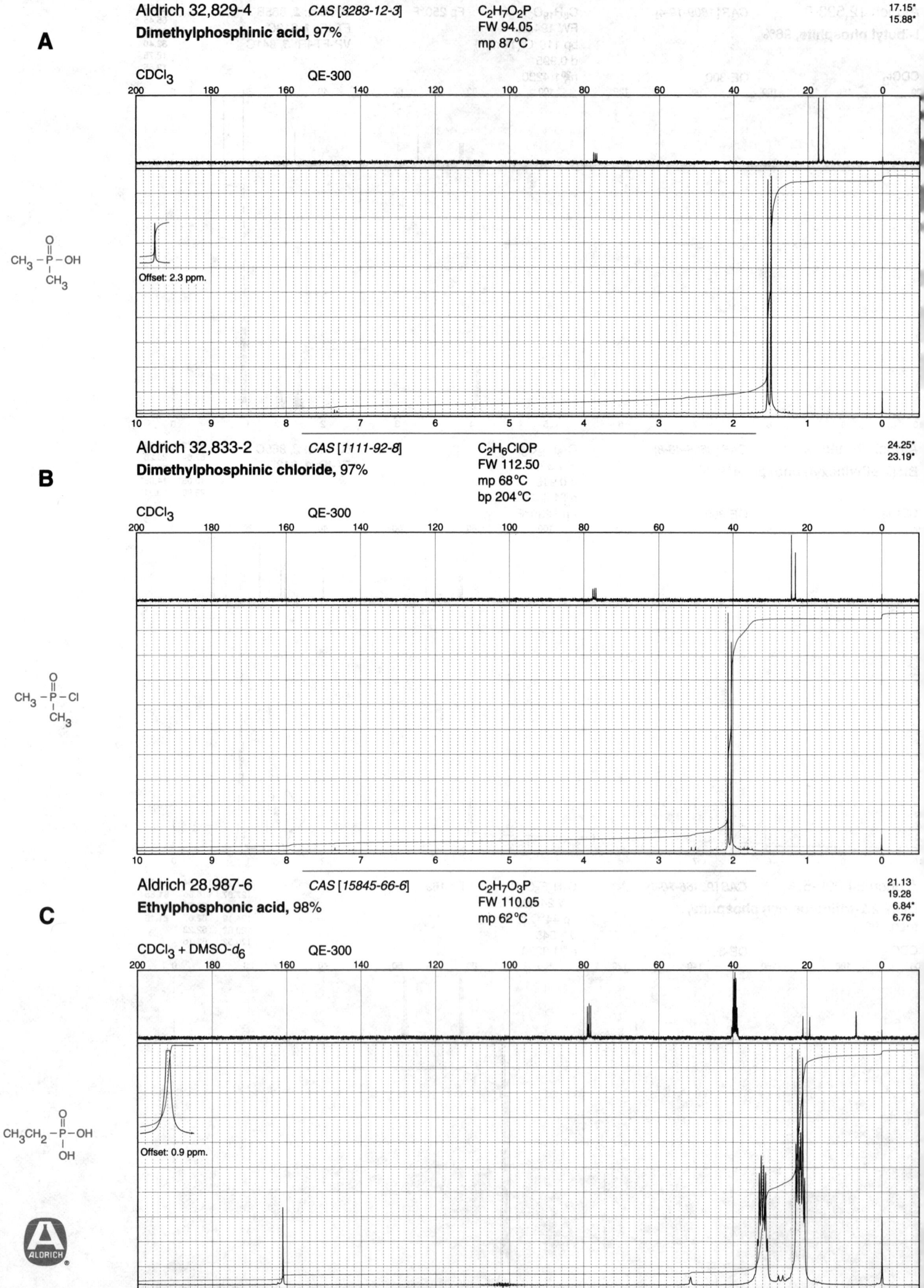

Aldrich 32,829-4   *CAS [3283-12-3]*   C₂H₇O₂P
Dimethylphosphinic acid, 97%   FW 94.05
mp 87°C

17.15*
15.88*

CDCl₃   QE-300

Offset: 2.3 ppm.

$CH_3 - \overset{\overset{O}{\|}}{\underset{\underset{CH_3}{|}}{P}} - OH$

Aldrich 32,833-2   *CAS [1111-92-8]*   C₂H₆ClOP
Dimethylphosphinic chloride, 97%   FW 112.50
mp 68°C
bp 204°C

24.25*
23.19*

CDCl₃   QE-300

$CH_3 - \overset{\overset{O}{\|}}{\underset{\underset{CH_3}{|}}{P}} - Cl$

Aldrich 28,987-6   *CAS [15845-66-6]*   C₂H₇O₃P
Ethylphosphonic acid, 98%   FW 110.05
mp 62°C

21.13
19.28
6.84*
6.76*

CDCl₃ + DMSO-d₆   QE-300

Offset: 0.9 ppm.

$CH_3CH_2 - \overset{\overset{O}{\|}}{\underset{\underset{OH}{|}}{P}} - OH$

ALDRICH

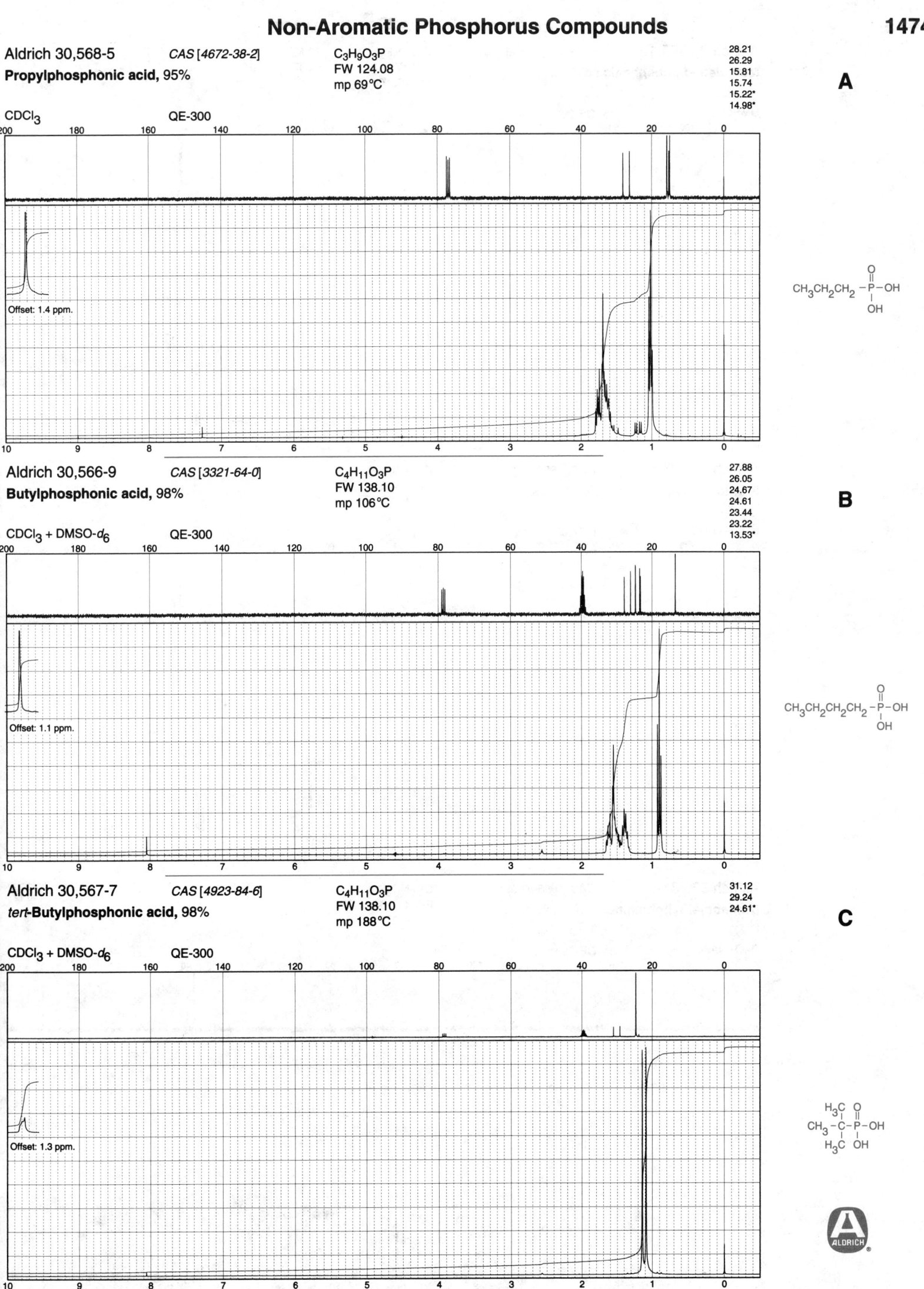

Aldrich 30,568-5   *CAS [4672-38-2]* $C_3H_9O_3P$
Propylphosphonic acid, 95%   FW 124.08
mp 69°C

28.21
26.29
15.81
15.74
15.22*
14.98*

A

CDCl₃   QE-300

Offset: 1.4 ppm.

$CH_3CH_2CH_2-\overset{\displaystyle O}{\underset{\displaystyle OH}{P}}-OH$

Aldrich 30,566-9   *CAS [3321-64-0]* $C_4H_{11}O_3P$
Butylphosphonic acid, 98%   FW 138.10
mp 106°C

27.88
26.05
24.67
24.61
23.44
23.22
13.53*

B

CDCl₃ + DMSO-d₆   QE-300

Offset: 1.1 ppm.

$CH_3CH_2CH_2CH_2-\overset{\displaystyle O}{\underset{\displaystyle OH}{P}}-OH$

Aldrich 30,567-7   *CAS [4923-84-6]* $C_4H_{11}O_3P$
tert-**Butylphosphonic acid, 98%**   FW 138.10
mp 188°C

31.12
29.24
24.61*

C

CDCl₃ + DMSO-d₆   QE-300

Offset: 1.3 ppm.

$\begin{array}{c} H_3C \\ CH_3-C-\overset{\displaystyle O}{\underset{\displaystyle OH}{P}}-OH \\ H_3C \end{array}$

ALDRICH

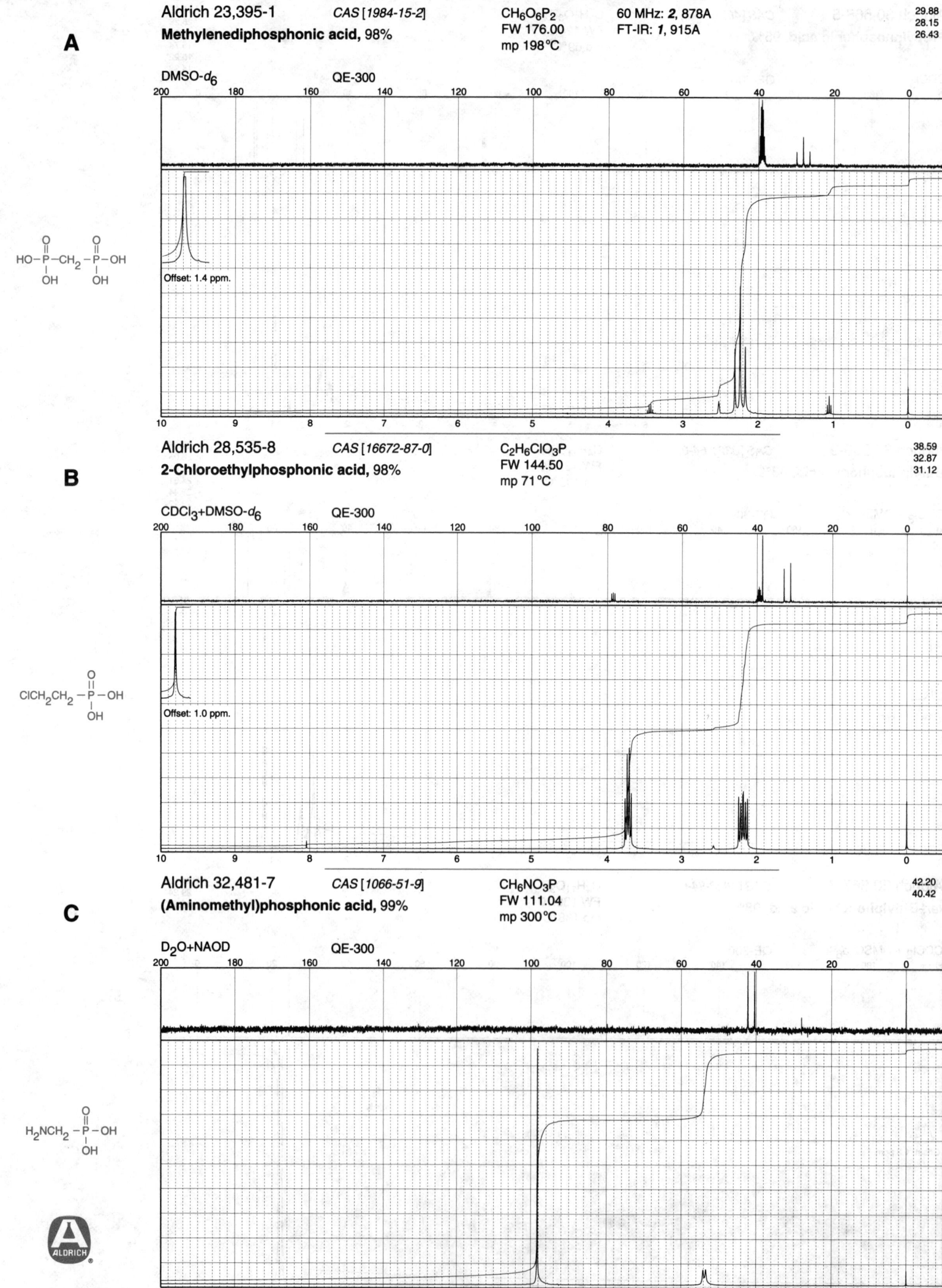

A

Aldrich 23,395-1   CAS [1984-15-2] $CH_6O_6P_2$ 60 MHz: *2*, 878A

Methylenediphosphonic acid, 98%   FW 176.00   FT-IR: *1*, 915A

mp 198°C

29.88
28.15
26.43

DMSO-d_6 QE-300

Offset: 1.4 ppm.

B

Aldrich 28,535-8   CAS [16672-87-0] $C_2H_6ClO_3P$

2-Chloroethylphosphonic acid, 98%   FW 144.50

mp 71°C

38.59
32.87
31.12

$CDCl_3$+DMSO-d_6 QE-300

Offset: 1.0 ppm.

C

Aldrich 32,481-7   CAS [1066-51-9] CH_6NO_3P

(Aminomethyl)phosphonic acid, 99%   FW 111.04

mp 300°C

42.20
40.42

D_2O+NAOD   QE-300

Aldrich 36,754-0   *CAS [60687-36-7]*   C₂H₈NO₃P
FW 125.07

$C_2H_8NO_3P$

(*R*)-(-)-(1-Aminoethyl)phosphonic acid, 99%

FW 125.07
mp 290°C d.

47.68*
45.67*
15.91*
15.88*

A

D₂O + DCl   QE-300

Aldrich 36,755-9   *CAS [66068-76-6]* $C_2H_8NO_3P$
FW 125.07
mp 290°C d.

(*S*)-(+)-(1-Aminoethyl)phosphonic acid, 99%

47.78*
45.79*
15.99*
15.95*

B

D₂O + DCl   QE-300

Aldrich 26,866-6   *CAS [6323-97-3]* $C_2H_8NO_3P$
FW 125.07
mp 284°C

(±)-1-Aminoethylphosphonic acid, 96%

48.58*
46.66*
16.56*
16.53*

C

D₂O   QE-300

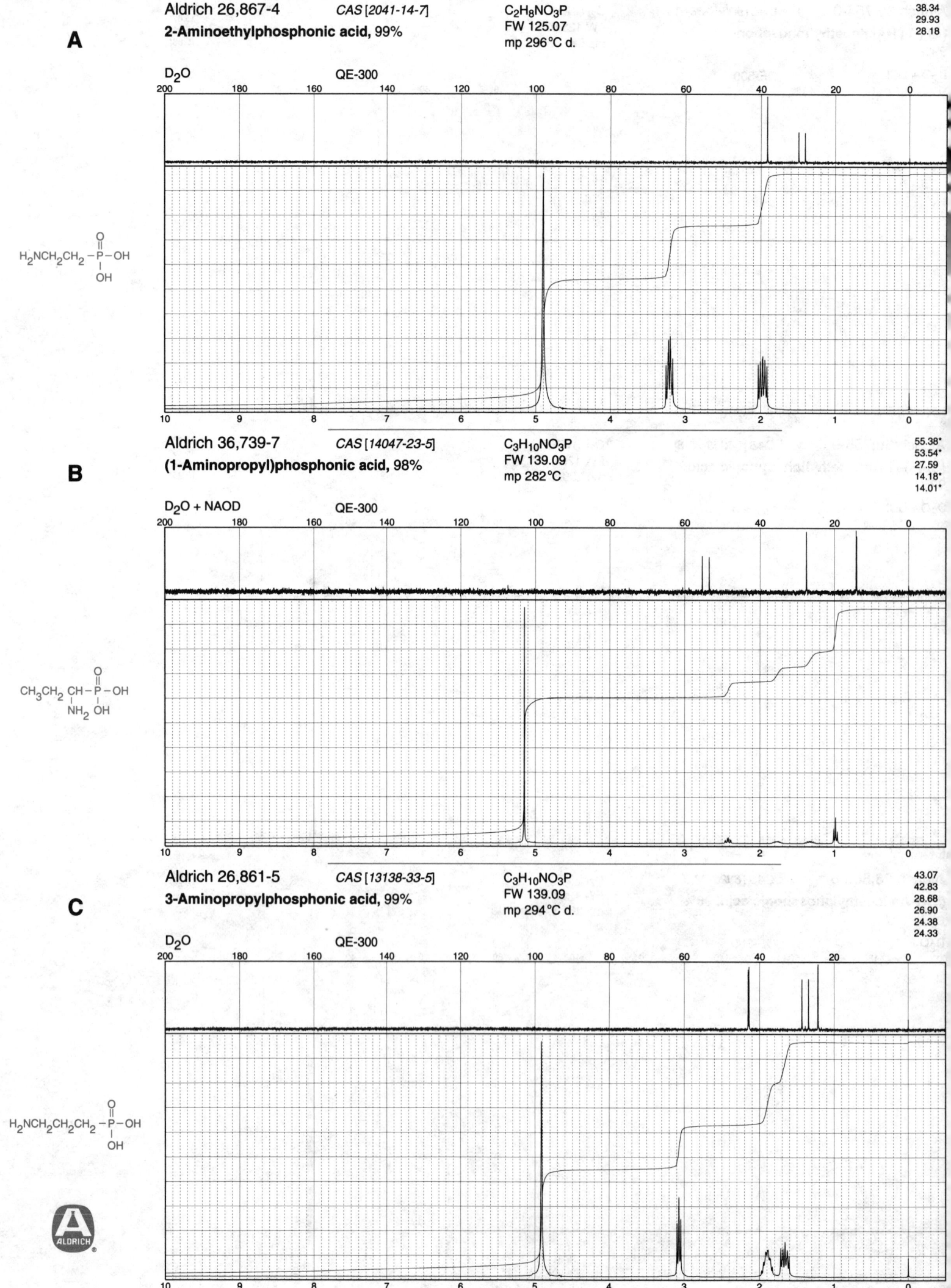

A

Aldrich 26,867-4   *CAS [2041-14-7]* $C_2H_8NO_3P$
FW 125.07
mp 296°C d.

2-Aminoethylphosphonic acid, 99%

38.34
29.93
28.18

D_2O QE-300

$H_2NCH_2CH_2-\overset{\overset{O}{\|}}{\underset{\underset{OH}{|}}{P}}-OH$

B

Aldrich 36,739-7   *CAS [14047-23-5]* $C_3H_{10}NO_3P$
FW 139.09
mp 282°C

(1-Aminopropyl)phosphonic acid, 98%

55.38*
53.54*
27.59
14.18*
14.01*

D_2O + NAOD   QE-300

$CH_3CH_2-\underset{\underset{NH_2}{|}}{CH}-\overset{\overset{O}{\|}}{\underset{\underset{OH}{|}}{P}}-OH$

C

Aldrich 26,861-5   *CAS [13138-33-5]* $C_3H_{10}NO_3P$
FW 139.09
mp 294°C d.

3-Aminopropylphosphonic acid, 99%

43.07
42.83
28.68
26.90
24.38
24.33

D_2O QE-300

$H_2NCH_2CH_2CH_2-\overset{\overset{O}{\|}}{\underset{\underset{OH}{|}}{P}}-OH$

ALDRICH

Non-Aromatic Phosphorus Compounds

Aldrich 36,736-2   *CAS [13138-36-8]*   C$_4$H$_{12}$NO$_3$P
FW 153.12
mp 287°C d.

(±)-(1-Aminobutyl)phosphonic acid, 98%

52.75*
50.86*
33.32
33.30
21.88
21.76
15.78*

A

D$_2$O   QE-300

CH$_3$CH$_2$CH$_2$-CH-P-OH with O, NH$_2$, OH

Aldrich 28,427-0   *CAS [4408-78-0]*   C$_2$H$_5$O$_5$P
FW 140.03
mp 145°C

Phosphonoacetic acid, 99%

168.02
167.94
37.03
35.34

B

CDCl$_3$ + DMSO-d_6 QE-300

Offset: 1.7 ppm.

HO-P-CH$_2$-C-OH with O, OH, O

Aldrich 22,855-9   *CAS [5962-42-5]*   C$_3$H$_7$O$_5$P
FW 154.06
FT-IR: 1, 918D

2-Carboxyethylphosphonic acid, tech., 94%

173.54
173.30
27.68
27.64
23.72
21.88

C

DMSO-d_6 QE-300

Offset: 1.5 ppm.

HO-C-CH$_2$CH$_2$-P-OH with O, O, OH

A

Aldrich 36,738-9   *CAS [20263-06-3]* $C_3H_8NO_5P$
DL-2-Amino-3-phosphonopropionic acid,   FW 169.08
98%   mp 228°C

186.14
185.91
56.00*
55.94*
39.11
37.42

D_2O+NAOD   QE-300

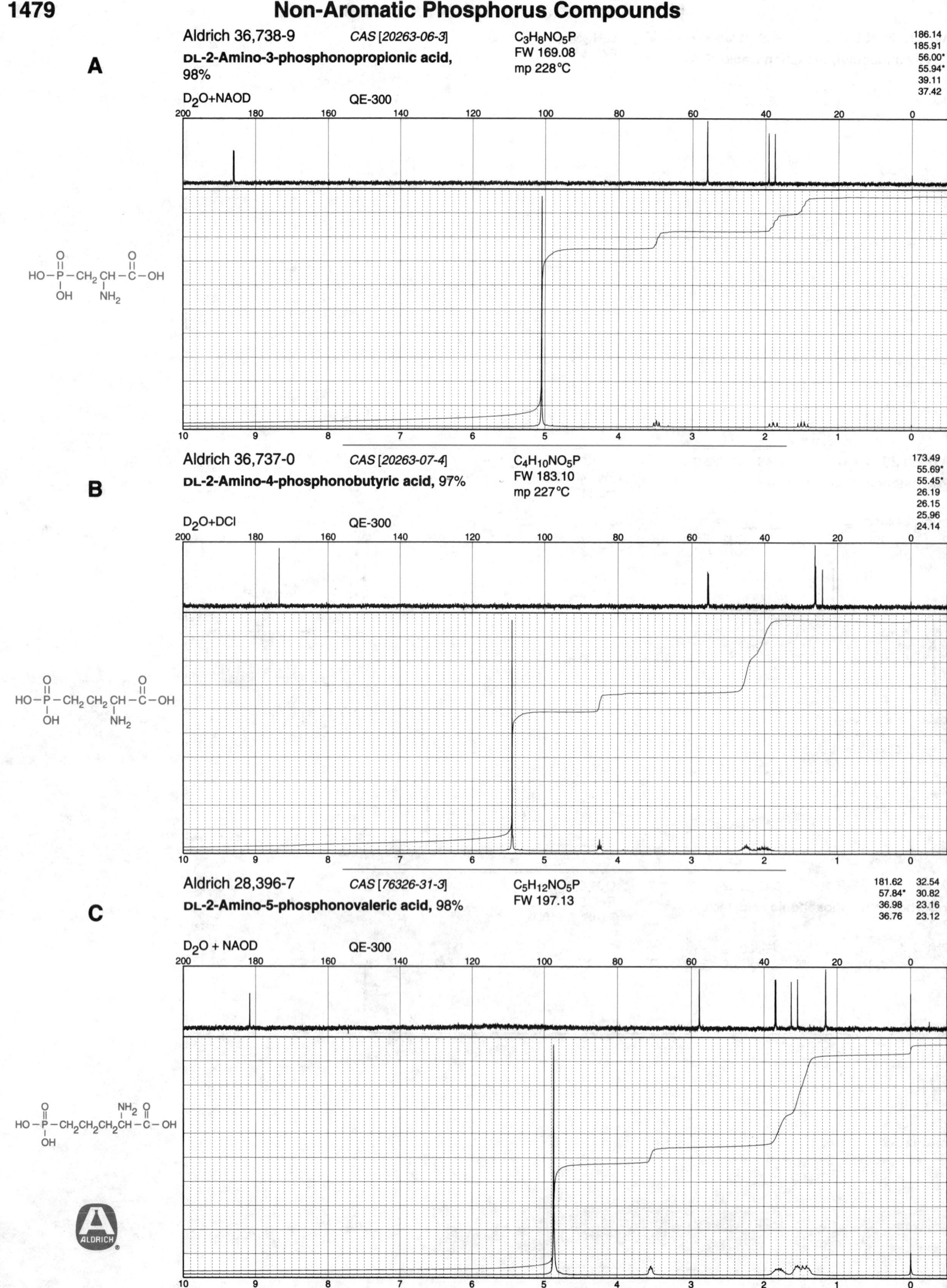

B

Aldrich 36,737-0   *CAS [20263-07-4]* $C_4H_{10}NO_5P$
DL-2-Amino-4-phosphonobutyric acid, 97%   FW 183.10
mp 227°C

173.49
55.69*
55.45*
26.19
26.15
25.96
24.14

D_2O+DCl   QE-300

C

Aldrich 28,396-7   *CAS [76326-31-3]* $C_5H_{12}NO_5P$
DL-2-Amino-5-phosphonovaleric acid, 98%   FW 197.13

181.62   32.54
57.84*   30.82
36.98   23.16
36.76   23.12

D_2O + NAOD   QE-300

A

Aldrich 33,775-7   CAS [1071-83-6] $C_3H_8NO_5P$
FW 169.08
mp 230°C d.

N-(Phosphonomethyl)glycine, 96%

174.21
53.74
53.66
48.95
47.22

D_2O + NAOD   QE-300

HO–P–CH₂ structure:
$$HO-\underset{HO}{\overset{O}{P}}-CH_2-NH-CH_2-\underset{}{\overset{O}{C}}-OH$$

B

Aldrich 38,656-1   CAS [1832-53-7] $C_3H_9O_3P$
FW 124.08
d 1.172
n_D^{20} 1.4260
Fp >230°F

Ethyl methylphosphonate, 98%

61.17
61.08
16.33*
16.24*
12.60*
10.66*

$CDCl_3$ QE-300

Offset: 2.4 ppm.

$$CH_3-\underset{OH}{\overset{O}{P}}-OCH_2CH_3$$

C

Aldrich D16,910-2   CAS [756-79-6] $C_3H_9O_3P$ Fp 156°F   60 MHz: **2**, 866D
FW 124.08   FT-IR: **1**, 914D
bp 181°C   VP-FT-IR: **3**, 841D
d 1.145
n_D^{20} 1.4130

Dimethyl methylphosphonate, 97%

52.23*
52.14*
10.83*
8.92*

$CDCl_3$ QE-300

$$CH_3-\underset{OCH_3}{\overset{O}{P}}-OCH_3$$

A

Aldrich 26,811-9   *CAS [683-08-9]*   C$_5$H$_{13}$O$_3$P   Fp 168°F   VP-FT-IR: *3*, 842A

61.48
61.40
16.46*
16.38*
12.18*
10.27*

Diethyl methylphosphonate, 97%

FW 152.13
bp 194°C
d 1.041
n$_D^{20}$1.4140

CDCl$_3$ QE-300

Aldrich 35,918-1   *CAS [1660-94-2]*   C$_9$H$_{22}$O$_6$P$_2$ Fp >230°F

62.57   23.63
62.54   16.39*
62.50   16.35*
27.25   16.32*
25.44

B

Tetraethyl methylenediphosphonate, 97%

FW 288.22
bp 173°C (11 mm)
d 1.160
n$_D^{20}$1.4400

CDCl$_3$ QE-300

Aldrich 11,613-0   *CAS [682-30-4]*   C$_6$H$_{13}$O$_3$P   Fp >230°F   60 MHz: *2*, 867D

135.33   61.84
135.31   61.77
127.23*   16.40*
124.79*   16.32*

C

Diethyl vinylphosphonate, 97%

FW 164.14
bp 202°C
d 1.068
n$_D^{20}$1.4290

FT-IR: *1*, 916B
VP-FT-IR: *3*, 842D

CDCl$_3$ QE-300

Aldrich 30,563-4   CAS [2799-58-8] $C_5H_{11}Cl_2O_3P$ Fp >230°F   VP-FT-IR: 3, 842B

Bis(2-chloroethyl) methylphosphonate,
95%

CDCl₃   QE-300

FW 221.02
bp 125°C (3 mm)
d 1.344
n_D^{20} 1.4700

65.24
65.16
42.93
42.84
12.19*
10.27*

A

$$CH_3-\overset{\overset{O}{\|}}{\underset{\underset{OCH_2CH_2Cl}{|}}{P}}-OCH_2CH_2Cl$$

Aldrich 26,361-3   CAS [3167-63-3] $C_5H_{12}ClO_3P$ Fp 188°F   FT-IR: 1, 915B

Diethyl chloromethylphosphonate, 97%

CDCl₃   QE-300

FW 186.58
bp 110°C (10 mm)
d 1.200
n_D^{20} 1.4400

63.43
63.34
34.44
32.32
16.46*
16.38*

B

$$ClCH_2-\overset{\overset{O}{\|}}{\underset{\underset{OCH_2CH_3}{|}}{P}}-OCH_2CH_3$$

Aldrich 37,452-0   CAS [866-23-9] $C_5H_{10}Cl_3O_3P$ Fp >230°F

Diethyl (trichloromethyl)phosphonate,
97%

CDCl₃   QE-300

FW 255.47
bp 131°C (14 mm)
d 1.362
n_D^{20} 1.4630

90.07
87.46
67.09
66.99
16.43*
16.36*

C

$$Cl-\overset{\overset{Cl}{|}}{\underset{\underset{Cl}{|}}{C}}-\overset{\overset{O}{\|}}{\underset{\underset{OCH_2CH_3}{|}}{P}}-OCH_2CH_3$$

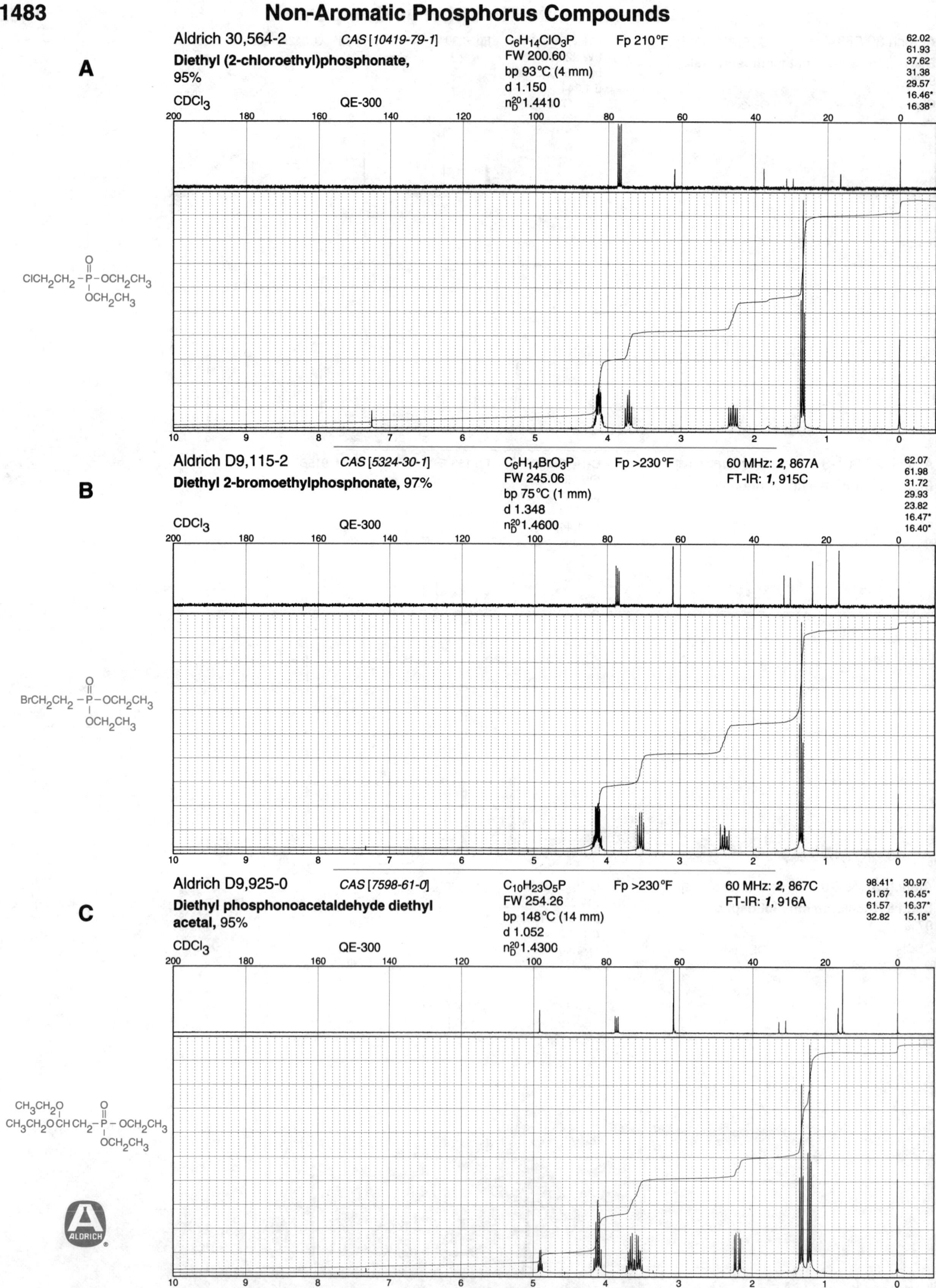

A

Aldrich 30,564-2   *CAS [10419-79-1]* $C_6H_{14}ClO_3P$ Fp 210°F

Diethyl (2-chloroethyl)phosphonate, 95%

FW 200.60
bp 93°C (4 mm)
d 1.150
n_D^{20} 1.4410

CDCl₃   QE-300

62.02
61.93
37.62
31.38
29.57
16.46*
16.38*

ClCH₂CH₂–P(=O)(OCH₂CH₃)OCH₂CH₃

B

Aldrich D9,115-2   *CAS [5324-30-1]* $C_6H_{14}BrO_3P$ Fp >230°F   60 MHz: **2**, 867A

Diethyl 2-bromoethylphosphonate, 97%

FW 245.06
bp 75°C (1 mm)
d 1.348
n_D^{20} 1.4600

FT-IR: **1**, 915C

CDCl₃   QE-300

62.07
61.98
31.72
29.93
23.82
16.47*
16.40*

BrCH₂CH₂–P(=O)(OCH₂CH₃)OCH₂CH₃

C

Aldrich D9,925-0   *CAS [7598-61-0]* $C_{10}H_{23}O_5P$ Fp >230°F   60 MHz: **2**, 867C

Diethyl phosphonoacetaldehyde diethyl acetal, 95%

FW 254.26
bp 148°C (14 mm)
d 1.052
n_D^{20} 1.4300

FT-IR: **1**, 916A

CDCl₃   QE-300

98.41*   30.97
61.67   16.45*
61.57   16.37*
32.82   15.18*

CH₃CH₂O / CH₃CH₂O CH–CH₂–P(=O)(OCH₂CH₃)OCH₂CH₃

Aldrich 36,666-8   *CAS [28460-01-7]* $C_6H_{15}O_3PS$ Fp 215°F

Diethyl (methylthiomethyl)phosphonate,
96%

FW 198.22
bp 120°C (10 mm)
d 1.130
n_D^{20} 1.4650

CDCl₃   QE-300

62.58	17.36*
62.49	17.34*
28.43	16.52*
26.44	16.45*

A

$CH_3SCH_2-P(=O)(OCH_2CH_3)(OCH_2CH_3)$

Aldrich 15,653-1   *CAS [54091-78-0]* $C_7H_{17}O_3PS$ 60 MHz: **2**, 867B

Diethyl (ethylthiomethyl)phosphonate

FW 212.25
d 1.097
n_D^{20} 1.4640
Fp >230°F

FT-IR: **1**, 915D

CDCl₃   QE-300

62.62	23.87
62.53	16.53*
27.49	16.45*
27.45	14.17*
25.86	

B

$CH_3CH_2SCH_2-P(=O)(OCH_2CH_3)(OCH_2CH_3)$

Aldrich 37,416-4   *CAS [51868-96-3]* $C_9H_{20}NO_3P$ Fp 217°F

Diethyl (pyrrolidinomethyl)phosphonate,
97%

FW 221.24
bp 133°C (10 mm)
d 1.064
n_D^{20} 1.4530

CDCl₃   QE-300

61.93	50.45
61.84	23.78
56.11	16.53*
55.96	16.46*
52.64	

C

pyrrolidine-$N-CH_2-P(=O)(OCH_2CH_3)(OCH_2CH_3)$

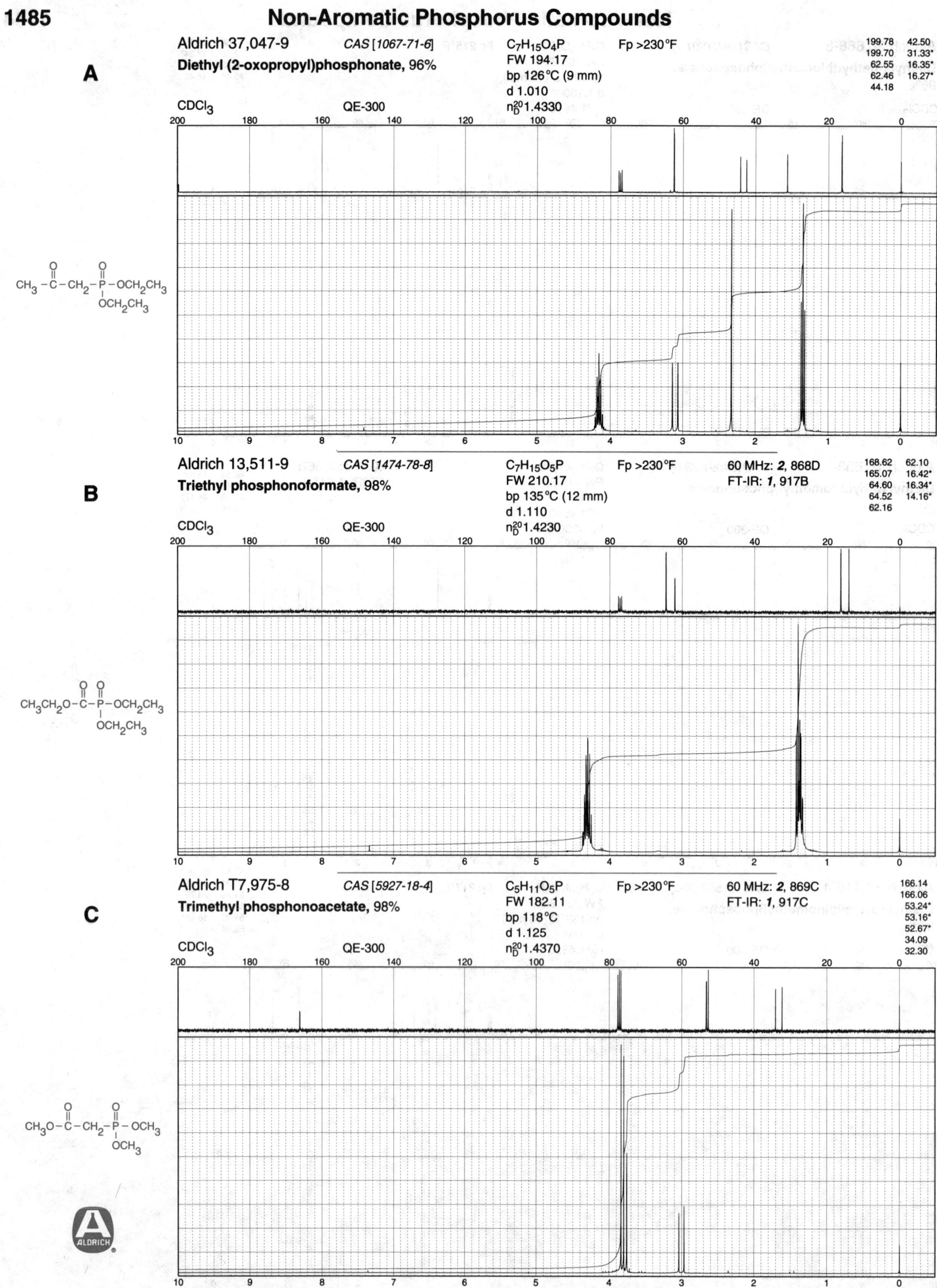

A

Aldrich 37,047-9   *CAS [1067-71-6]* $C_7H_{15}O_4P$ Fp >230°F
Diethyl (2-oxopropyl)phosphonate, 96%   FW 194.17
bp 126°C (9 mm)
d 1.010
n_D^{20} 1.4330

CDCl₃   QE-300

199.78   42.50
199.70   31.33*
62.55   16.35*
62.46   16.27*
44.18

$CH_3-\overset{O}{\overset{\|}{C}}-CH_2-\overset{O}{\overset{\|}{P}}-OCH_2CH_3$
$\overset{|}{OCH_2CH_3}$

B

Aldrich 13,511-9   *CAS [1474-78-8]* $C_7H_{15}O_5P$ Fp >230°F   60 MHz: **2**, 868D
Triethyl phosphonoformate, 98%   FW 210.17   FT-IR: **1**, 917B
bp 135°C (12 mm)
d 1.110
n_D^{20} 1.4230

CDCl₃   QE-300

168.62   62.10
165.07   16.42*
64.60   16.34*
64.52   14.16*
62.16

$CH_3CH_2O-\overset{O}{\overset{\|}{C}}-\overset{O}{\overset{\|}{P}}-OCH_2CH_3$
$\overset{|}{OCH_2CH_3}$

C

Aldrich T7,975-8   *CAS [5927-18-4]* $C_5H_{11}O_5P$ Fp >230°F   60 MHz: **2**, 869C
Trimethyl phosphonoacetate, 98%   FW 182.11   FT-IR: **1**, 917C
bp 118°C
d 1.125
n_D^{20} 1.4370

CDCl₃   QE-300

166.14
166.06
53.24*
53.16*
52.67*
34.09
32.30

$CH_3O-\overset{O}{\overset{\|}{C}}-CH_2-\overset{O}{\overset{\|}{P}}-OCH_3$
$\overset{|}{OCH_3}$

Aldrich 36,526-2   *CAS [311-46-6]* $C_6H_{13}O_5P$ Fp 210°F

Ethyl dimethylphosphonoacetate, 98%

FW 196.14
bp 134°C (10 mm)
d 1.188
n_D^{20} 1.4350

165.60	53.09*
165.52	34.36
61.65	32.57
53.18*	14.08*

A

CDCl₃   QE-300

$CH_3CH_2O-\overset{O}{\overset{\|}{C}}-CH_2-\overset{O}{\overset{\|}{P}}-OCH_3$
                          OCH_3

Aldrich 15,876-3   *CAS [1067-74-9]* $C_7H_{15}O_5P$ Fp >230°F

Methyl diethylphosphonoacetate, 97%

FW 210.17
bp 129°C (9 mm)
d 1.145
n_D^{20} 1.4340

60 MHz: **2**, 869D
FT-IR: **1**, 917D
VP-FT-IR: **3**, 843A

166.26	35.02
166.18	33.24
62.75	16.38*
62.67	16.30*
52.54*	

B

CDCl₃   QE-300

$CH_3O-\overset{O}{\overset{\|}{C}}-CH_2-\overset{O}{\overset{\|}{P}}-OCH_2CH_3$
                          OCH_2CH_3

Aldrich T6,130-1   *CAS [867-13-0]* $C_8H_{17}O_5P$ Fp >230°F

Triethyl phosphonoacetate, 99%

FW 224.19
bp 144°C (9 mm)
d 1.130
n_D^{20} 1.4310

60 MHz: **2**, 870A
FT-IR: **1**, 918A

165.77	35.27
165.69	33.49
62.68	16.38*
62.59	16.29*
61.51	14.10*

C

CDCl₃   QE-300

$CH_3CH_2O-\overset{O}{\overset{\|}{C}}-CH_2-\overset{O}{\overset{\|}{P}}-OCH_2CH_3$
                          OCH_2CH_3

Non-Aromatic Phosphorus Compounds

1487

A

Aldrich 29,320-2   CAS [82426-28-6]   C$_8$H$_{17}$O$_5$P   Fp >230°F
Triethyl phosphonoacetate-2-^{13}C, 99 atom % ^{13}C   FW 225.19
bp 144°C (9 mm)
d 1.130
n$_D^{20}$1.4310

CDCl$_3$ QE-300

166.15	61.49
166.07	35.33
165.37	33.55
165.29	16.37*
62.68	16.29*
62.59	14.11*

CH$_3$CH$_2$O–C–^{13}CH$_2$–P–OCH$_2$CH$_3$ (O, OCH$_2$CH$_3$)

B

Aldrich 28,384-3   CAS [100940-60-1]   C$_8$H$_{17}$O$_5$P   Fp >230°F
Triethyl phosphonoacetate-^{13}C$_2$, 99 atom % ^{13}C   FW 226.18
bp 144°C
d 1.130
n$_D^{20}$1.4310

CDCl$_3$ QE-300

166.15	35.74
166.07	34.96
165.37	33.96
165.29	33.18
62.67	16.38*
62.59	16.30*
61.47	14.10*

CH$_3$CH$_2$O–^{13}C–^{13}CH$_2$–P–OCH$_2$CH$_3$ (O, OCH$_2$CH$_3$)

C

Aldrich 34,833-3   CAS [27784-76-5]   C$_{10}$H$_{21}$O$_5$P   Fp >230°F
***tert*-Butyl diethylphosphonoacetate,**   FW 252.25
95%   bp 102°C (1 mm)
d 1.074
n$_D^{20}$1.4310

CDCl$_3$ QE-300

164.78	34.70
81.91	27.91*
62.46	16.39*
62.38	16.30*
36.47	

CH$_3$–C(CH$_3$)$_2$–O–C–CH$_2$–P–OCH$_2$CH$_3$ (O, OCH$_2$CH$_3$)

ALDRICH

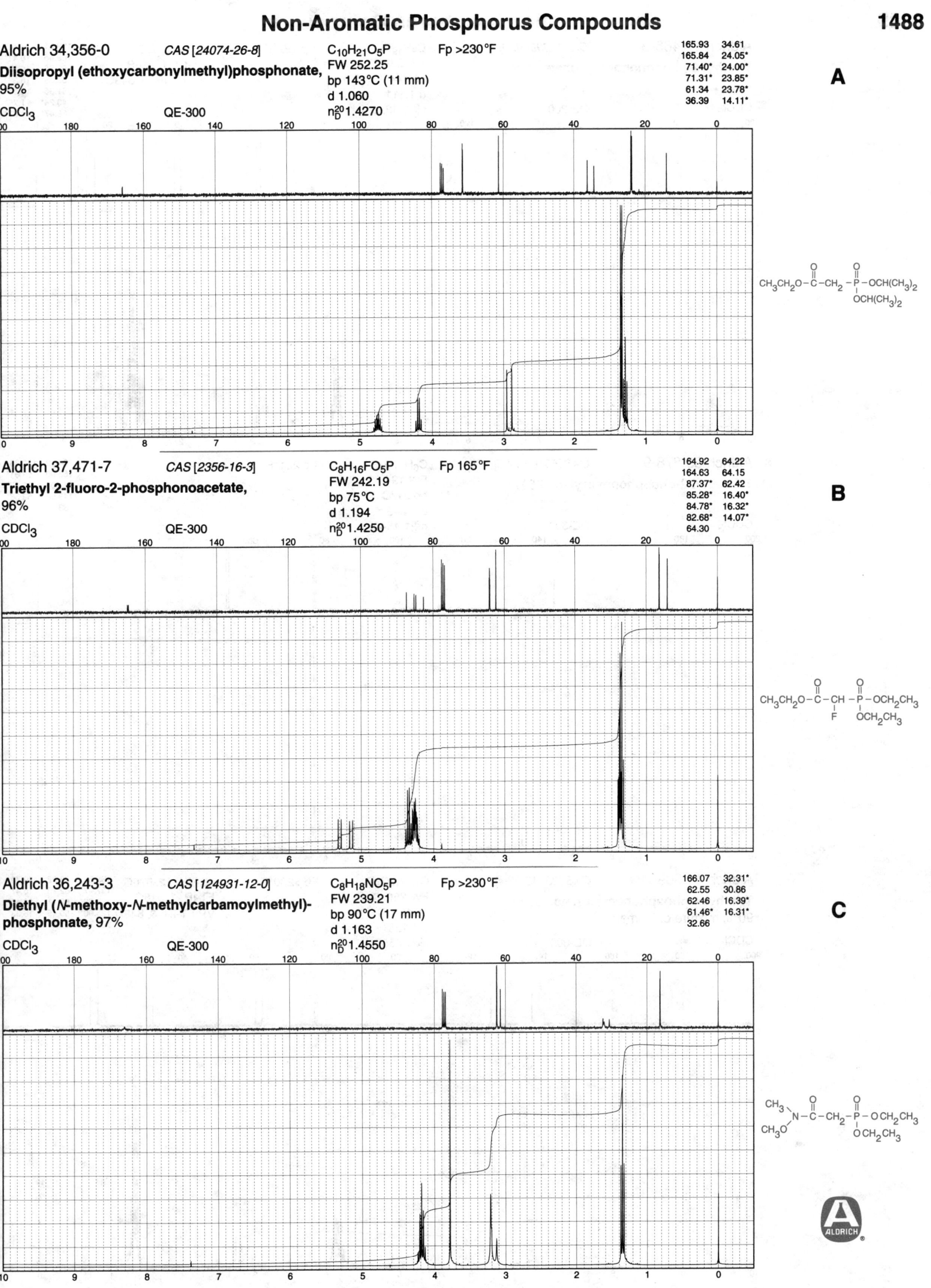

Aldrich 34,356-0   CAS [24074-26-8]   C₁₀H₂₁O₅P   Fp >230°F

Diisopropyl (ethoxycarbonylmethyl)phosphonate, 95%

FW 252.25
bp 143°C (11 mm)
d 1.060
nD²⁰ 1.4270

CDCl₃   QE-300

165.93	34.61
165.84	24.05*
71.40*	24.00*
71.31*	23.85*
61.34	23.78*
36.39	14.11*

A

Aldrich 37,471-7   CAS [2356-16-3]   C₈H₁₆FO₅P   Fp 165°F

Triethyl 2-fluoro-2-phosphonoacetate, 96%

FW 242.19
bp 75°C
d 1.194
nD²⁰ 1.4250

CDCl₃   QE-300

164.92	64.22
164.63	64.15
87.37*	62.42
85.28*	16.40*
84.78*	16.32*
82.68*	14.07*
64.30	

B

Aldrich 36,243-3   CAS [124931-12-0]   C₈H₁₈NO₅P   Fp >230°F

Diethyl (N-methoxy-N-methylcarbamoylmethyl)-phosphonate, 97%

FW 239.21
bp 90°C (17 mm)
d 1.163
nD²⁰ 1.4550

CDCl₃   QE-300

166.07	32.31*
62.55	30.86
62.46	16.39*
61.46*	16.31*
32.66	

C

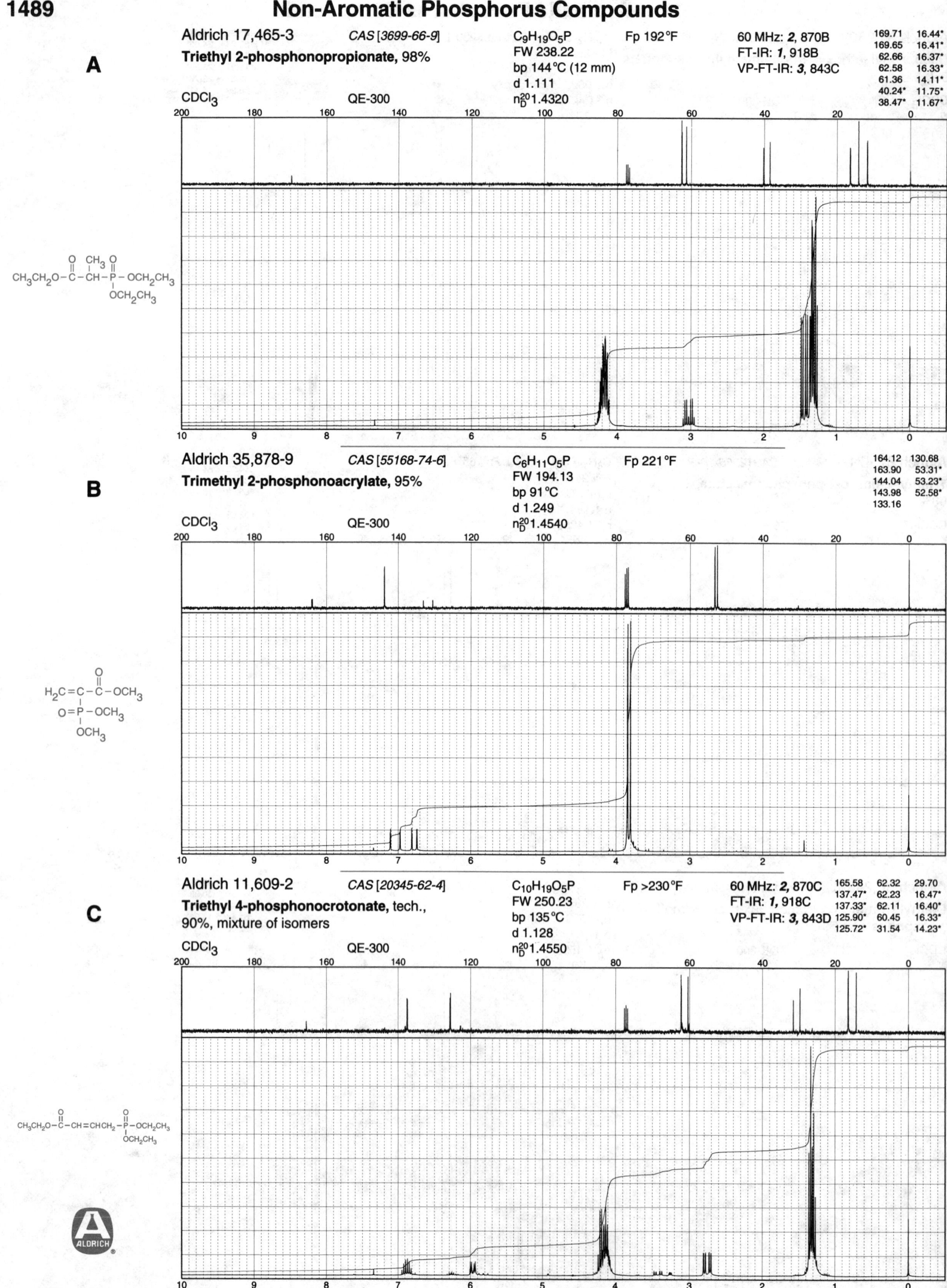

A

Aldrich 17,465-3   *CAS [3699-66-9]* $C_9H_{19}O_5P$ Fp 192 °F   60 MHz: *2*, 870B

Triethyl 2-phosphonopropionate, 98%   FW 238.22   FT-IR: *1*, 918B
VP-FT-IR: *3*, 843C

bp 144 °C (12 mm)
d 1.111
n_D^{20} 1.4320

CDCl$_3$ QE-300

169.71	16.44*
169.65	16.41*
62.66	16.37*
62.58	16.33*
61.36	14.11*
40.24*	11.75*
38.47*	11.67*

B

Aldrich 35,878-9   *CAS [55168-74-6]* $C_6H_{11}O_5P$ Fp 221 °F

Trimethyl 2-phosphonoacrylate, 95%   FW 194.13
bp 91 °C
d 1.249
n_D^{20} 1.4540

CDCl$_3$ QE-300

164.12	130.68
163.90	53.31*
144.04	53.23*
143.98	52.58*
133.16	

C

Aldrich 11,609-2   *CAS [20345-62-4]* $C_{10}H_{19}O_5P$ Fp >230 °F   60 MHz: *2*, 870C

Triethyl 4-phosphonocrotonate, tech.,   FW 250.23   FT-IR: *1*, 918C
90%, mixture of isomers   bp 135 °C   VP-FT-IR: *3*, 843D
d 1.128
n_D^{20} 1.4550

CDCl$_3$ QE-300

165.58	62.32	29.70
137.47*	62.23	16.47*
137.33*	62.11	16.40*
125.90*	60.45	16.33*
125.72*	31.54	14.23*

A

Aldrich 33,895-8   CAS [41891-54-7]

Triethyl 3-methyl-4-phosphonocrotonate,
tech., 80%, mixture of *cis* and *trans*

CDCl₃   QE-300

$C_{11}H_{21}O_5P$
FW 264.26
d 1.089
n_D^{20} 1.4600
Fp >230°F

166.02	118.96*	39.43	16.44*
165.96	118.81*	37.64	16.37*
165.90	62.25	32.14	16.29*
149.57	62.16	30.37	14.27*
149.42	62.04	26.15*	
120.05*	61.95	20.03*	
119.89*	59.71	20.00*	

$$CH_3CH_2O-\overset{O}{\overset{\|}{C}}-CH=\underset{CH_3}{C}CH_2-\overset{O}{\overset{\|}{P}}-OCH_2CH_3$$
$$OCH_2CH_3$$

B

Aldrich 27,425-9   CAS [88738-78-7]

Bis(2,2,2-trifluoroethyl) (methoxycarbonyl-methyl)phosphonate, 95%

CDCl₃   QE-300

$C_7H_9F_6O_5P$
FW 318.11
d 1.504
n_D^{20} 1.3700
Fp >230°F

VP-FT-IR: **3**, 843B

165.23	120.68	62.53
165.18	117.13	62.46
128.13	117.06	62.03
128.02	63.54	61.95
124.45	63.47	53.04*
124.35	63.04	34.76
120.79	62.97	32.85

$$CH_3O-\overset{O}{\overset{\|}{C}}-CH_2-\overset{O}{\overset{\|}{P}}-OCH_2CF_3$$
$$OCH_2CF_3$$

C

Aldrich 24,673-5   CAS [2942-58-7]

Diethyl cyanophosphonate, 93%

CDCl₃   QE-300

$C_5H_{10}NO_3P$
FW 163.11
bp 105°C (19 mm)
d 1.075
n_D^{20} 1.4010

Fp 177°F

60 MHz: **2**, 870D
FT-IR: **1**, 919A

115.22
112.36
66.10
66.02
15.95*
15.87*

$$CH_3CH_2O-\overset{O}{\overset{\|}{P}}-OCH_2CH_3$$
$$CN$$

A

Aldrich D9,170-5   *CAS [2537-48-6]* $C_6H_{12}NO_3P$ Fp >230°F   60 MHz: *2*, 871A   112.85 17.41

Diethyl cyanomethylphosphonate, 98%   FW 177.14   FT-IR: *1*, 919B   112.70 16.36*

bp 102°C   63.91 16.28*

d 1.095   63.82 15.50

CDCl$_3$ QE-300 n_D^{20} 1.4320

$CH_3CH_2O-\overset{\displaystyle O}{\underset{\displaystyle CH_2CN}{P}}-OCH_2CH_3$

B

Aldrich 35,986-6   *CAS [41003-94-5]* $C_6H_{12}NO_3P$ Fp >230°F   161.21

Diethyl (isocyanomethyl)phosphonate,   FW 177.14   64.01

97%   bp 127°C (2 mm)   63.92

d 1.105   38.68

36.62

CDCl$_3$ QE-300 n_D^{20} 1.4330   16.41*

16.33*

$CNCH_2-\overset{\displaystyle O}{\underset{\displaystyle OCH_2CH_3}{P}}-OCH_2CH_3$

C

Aldrich 22,805-2   *CAS [676-97-1]* CH_3Cl_2OP Fp >230°F   60 MHz: *2*, 873A   30.81*

Methylphosphonic dichloride, 98%   FW 132.91   FT-IR: *1*, 922C   29.41*

mp 36°C

bp 163°C

CDCl$_3$ QE-300   d 1.468

$CH_3-\overset{\displaystyle O}{\underset{\displaystyle Cl}{P}}-Cl$

Aldrich 27,596-4   *CAS [1066-50-8]*   C₂H₅Cl₂OP   Fp >230°F

Ethylphosphonic dichloride, 98%

$C_2H_5Cl_2OP$

FW 146.94

bp 72°C (12 mm)

d 1.376

n_D^{20}1.4650

CDCl₃   QE-300

A

37.20
35.88
7.39*
7.28*

$$CH_3CH_2-\overset{\overset{\textstyle O}{\|}}{\underset{\underset{\textstyle Cl}{|}}{P}}-Cl$$

Aldrich 30,558-8   *CAS [2302-80-9]*   C₄H₉Cl₂OP   Fp >230°F   VP-FT-IR: *3*, 845B

Butylphosphonic dichloride, 98%

$C_4H_9Cl_2OP$

FW 175.00

bp 97°C (15 mm)

d 1.241

n_D^{20}1.4660

CDCl₃   QE-300

B

43.41
42.12
24.96
24.87
22.97
22.67
13.39*

$$CH_3CH_2CH_2CH_2-\overset{\overset{\textstyle O}{\|}}{\underset{\underset{\textstyle Cl}{|}}{P}}-Cl$$

Aldrich 30,562-6   *CAS [928-64-3]*   C₆H₁₃Cl₂OP   Fp >230°F

Hexylphosphonic dichloride, tech., 90%

$C_6H_{13}Cl_2OP$

FW 203.05

bp 118°C (16 mm)

d 1.185

n_D^{20}1.4650

CDCl₃   QE-300

C

43.68   22.94
42.40   22.86
31.05   22.26
29.40   13.90*
29.12

$$CH_3(CH_2)_4CH_2-\overset{\overset{\textstyle O}{\|}}{\underset{\underset{\textstyle Cl}{|}}{P}}-Cl$$

ALDRICH

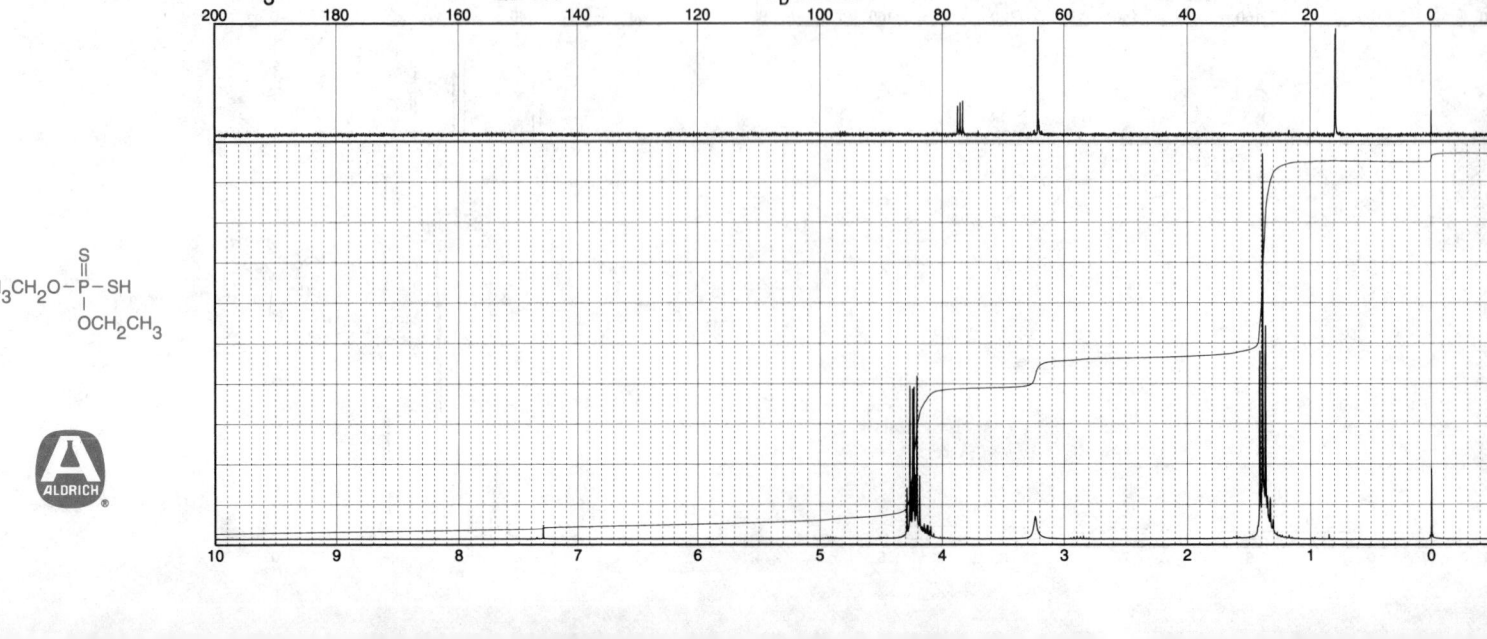

A

Aldrich 30,556-1   *CAS [1498-47-1]* $C_3H_5Cl_2OP$ Fp >230°F   VP-FT-IR: **3**, 845C

Allylphosphonic dichloride

FW 158.95
bp 76°C (16 mm)
d 1.334

CDCl₃   QE-300 n_D^{20}1.4870

124.61
124.35
124.11*
123.91*
48.25
46.95

$H_2C=CHCH_2-\overset{\overset{O}{\|}}{\underset{\underset{Cl}{|}}{P}}-Cl$

B

Aldrich C5,625-5   *CAS [1983-26-2]* CH_2Cl_3OP Fp >230°F   60 MHz: **2**, 874A

Chloromethylphosphonic dichloride, 97%

FW 167.36
bp 80°C (10 mm)
d 1.638

CDCl₃   QE-300 n_D^{20}1.4980

FT-IR: **1**, 923D

46.41
44.86

$ClCH_2-\overset{\overset{O}{\|}}{\underset{\underset{Cl}{|}}{P}}-Cl$

C

Aldrich D9,360-0   *CAS [298-06-6]* $C_4H_{11}O_2PS_2$ Fp 181°F   60 MHz: **2**, 875C

Diethyl dithiophosphate, tech., 90%

FW 186.23
bp 60°C (1 mm)
d 1.111

CDCl₃   QE-300 n_D^{20}1.5120

FT-IR: **1**, 925B

64.35
64.27
15.88*
15.77*

$CH_3CH_2O-\overset{\overset{S}{\|}}{\underset{\underset{OCH_2CH_3}{|}}{P}}-SH$

ALDRICH

Aldrich 34,722-1   CAS [3338-24-7] $C_4H_{11}O_2PS_2$
Diethyl dithiophosphate   FW 208.22
d 1.220
n_D^{20} 1.4720
Fp NONE

65.73
65.64
18.19*
18.08*

A

D_2O QE-300

$$CH_3CH_2O-\overset{\overset{\displaystyle S}{\|}}{\underset{\underset{\displaystyle CH_3CH_2O}{|}}{P}}-SNa$$

Aldrich D15,295-1   CAS [2524-03-0] $C_2H_6ClO_2PS$ Fp 221°F   60 MHz: 2, 874D   55.70*
Dimethyl chlorothiophosphate, 97%   FW 160.56   FT-IR: 1, 924D   55.61*
bp 67°C (16 mm)   VP-FT-IR: 3, 846A
d 1.322
n_D^{20} 1.4820

B

$CDCl_3$ QE-300

$$CH_3O-\overset{\overset{\displaystyle S}{\|}}{\underset{\underset{\displaystyle OCH_3}{|}}{P}}-Cl$$

Aldrich D9,168-3   CAS [2524-04-1] $C_4H_{10}ClO_2PS$ Fp >230°F   FT-IR: 1, 924C   66.14
Diethyl chlorothiophosphate, 97%   FW 188.61   VP-FT-IR: 3, 845D   66.05
bp 45°C (3 mm)   15.68*
d 1.200   15.57*
n_D^{20} 1.4720

C

$CDCl_3$ QE-300

$$CH_3CH_2O-\overset{\overset{\displaystyle S}{\|}}{\underset{\underset{\displaystyle OCH_2CH_3}{|}}{P}}-Cl$$

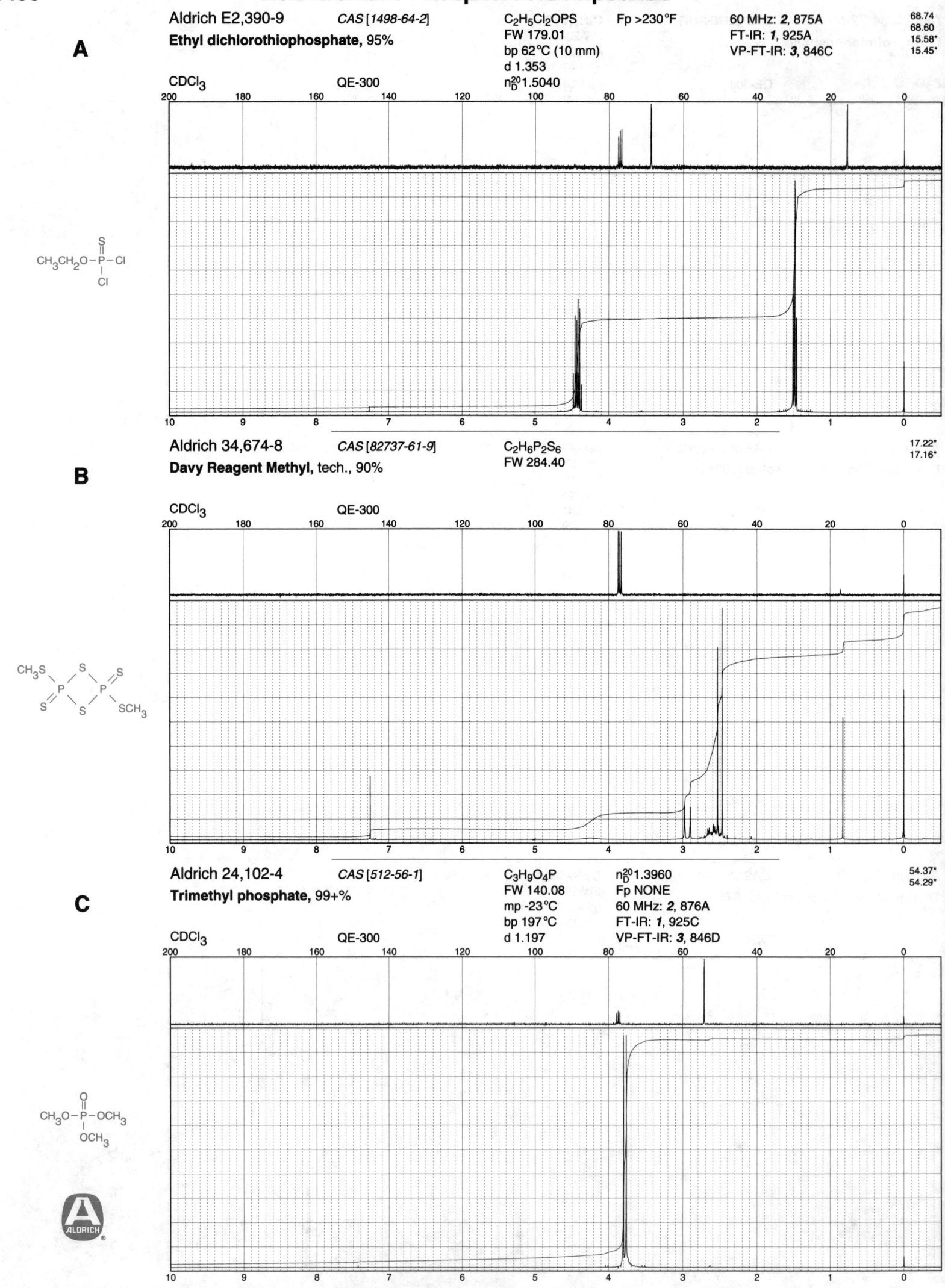

Aldrich E2,390-9   *CAS [1498-64-2]* $C_2H_5Cl_2OPS$ Fp >230°F   60 MHz: *2*, 875A   68.74 / 68.60
Ethyl dichlorothiophosphate, 95%   FW 179.01   FT-IR: *1*, 925A   15.58*
bp 62°C (10 mm)   VP-FT-IR: *3*, 846C   15.45*
d 1.353
n_D^{20} 1.5040

A

CDCl₃   QE-300

$CH_3CH_2O-\overset{\overset{S}{\|}}{\underset{\underset{Cl}{|}}{P}}-Cl$

Aldrich 34,674-8   *CAS [82737-61-9]* $C_2H_6P_2S_6$ 17.22* / 17.16*
Davy Reagent Methyl, tech., 90%   FW 284.40

B

CDCl₃   QE-300

$CH_3S-P(S)(S)-P(S)(S)-SCH_3$

Aldrich 24,102-4   *CAS [512-56-1]* $C_3H_9O_4P$ n_D^{20} 1.3960   54.37* / 54.29*
Trimethyl phosphate, 99+%   FW 140.08   Fp NONE
mp -23°C   60 MHz: *2*, 876A
bp 197°C   FT-IR: *1*, 925C
d 1.197   VP-FT-IR: *3*, 846D

C

CDCl₃   QE-300

$CH_3O-\overset{\overset{O}{\|}}{\underset{\underset{OCH_3}{|}}{P}}-OCH_3$

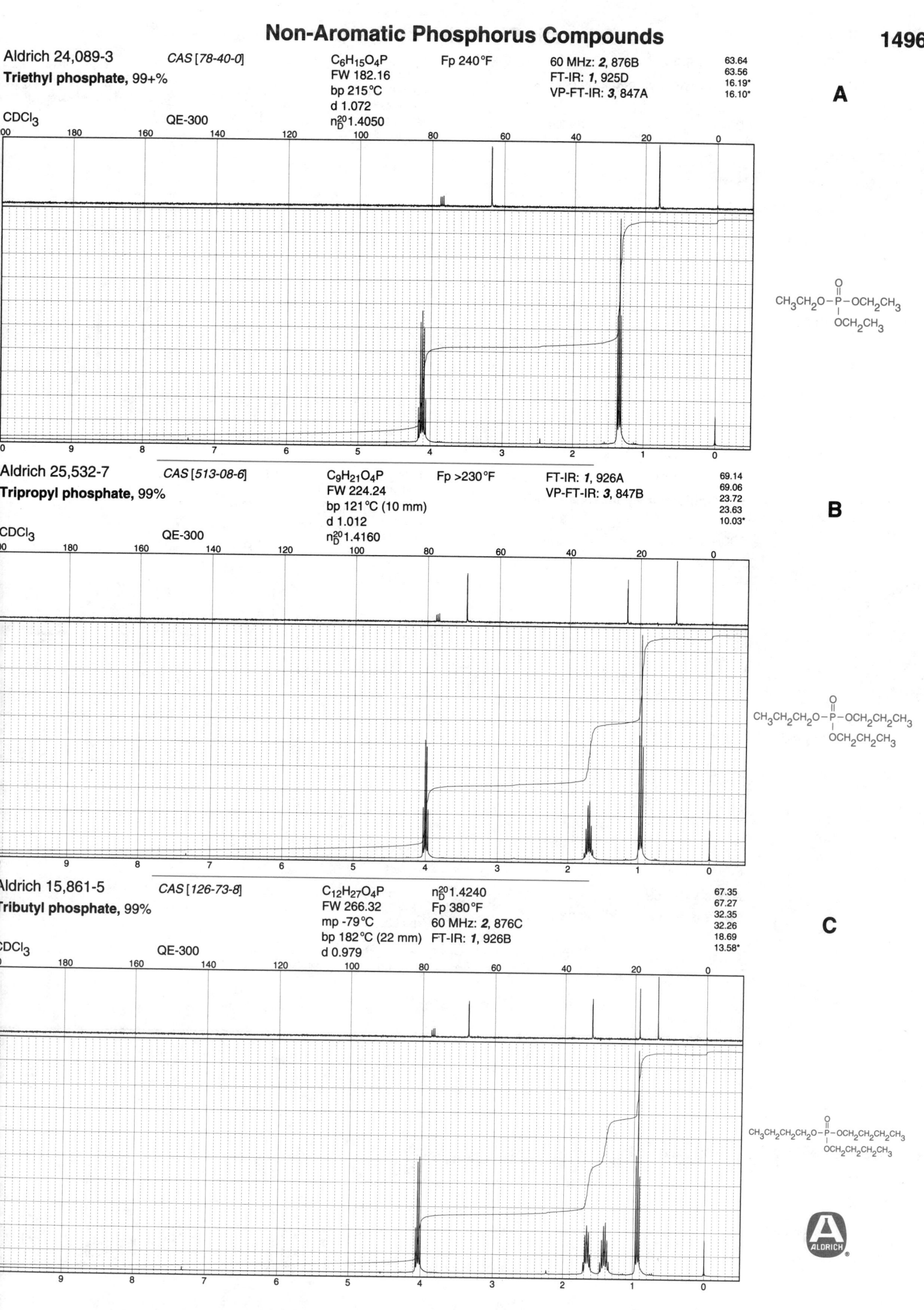

Aldrich 24,089-3   *CAS [78-40-0]* $C_6H_{15}O_4P$ Fp 240°F   60 MHz: **2**, 876B   63.64
Triethyl phosphate, 99+%   FW 182.16   FT-IR: **1**, 925D   63.56
bp 215°C   VP-FT-IR: **3**, 847A   16.19*
CDCl₃   QE-300   d 1.072   16.10*
n_D^{20} 1.4050

A

$CH_3CH_2O-\overset{\overset{O}{\|}}{\underset{\underset{OCH_2CH_3}{|}}{P}}-OCH_2CH_3$

Aldrich 25,532-7   *CAS [513-08-6]* $C_9H_{21}O_4P$ Fp >230°F   FT-IR: **1**, 926A   69.14
Tripropyl phosphate, 99%   FW 224.24   VP-FT-IR: **3**, 847B   69.06
bp 121°C (10 mm)   23.72
CDCl₃   QE-300   d 1.012   23.63
n_D^{20} 1.4160   10.03*

B

$CH_3CH_2CH_2O-\overset{\overset{O}{\|}}{\underset{\underset{OCH_2CH_2CH_3}{|}}{P}}-OCH_2CH_2CH_3$

Aldrich 15,861-5   *CAS [126-73-8]* $C_{12}H_{27}O_4P$ n_D^{20} 1.4240   67.35
Tributyl phosphate, 99%   FW 266.32   Fp 380°F   67.27
mp -79°C   60 MHz: **2**, 876C   32.35
bp 182°C (22 mm)   FT-IR: **1**, 926B   32.26
CDCl₃   QE-300   d 0.979   18.69
13.58*

C

$CH_3CH_2CH_2CH_2O-\overset{\overset{O}{\|}}{\underset{\underset{OCH_2CH_2CH_2CH_3}{|}}{P}}-OCH_2CH_2CH_2CH_3$

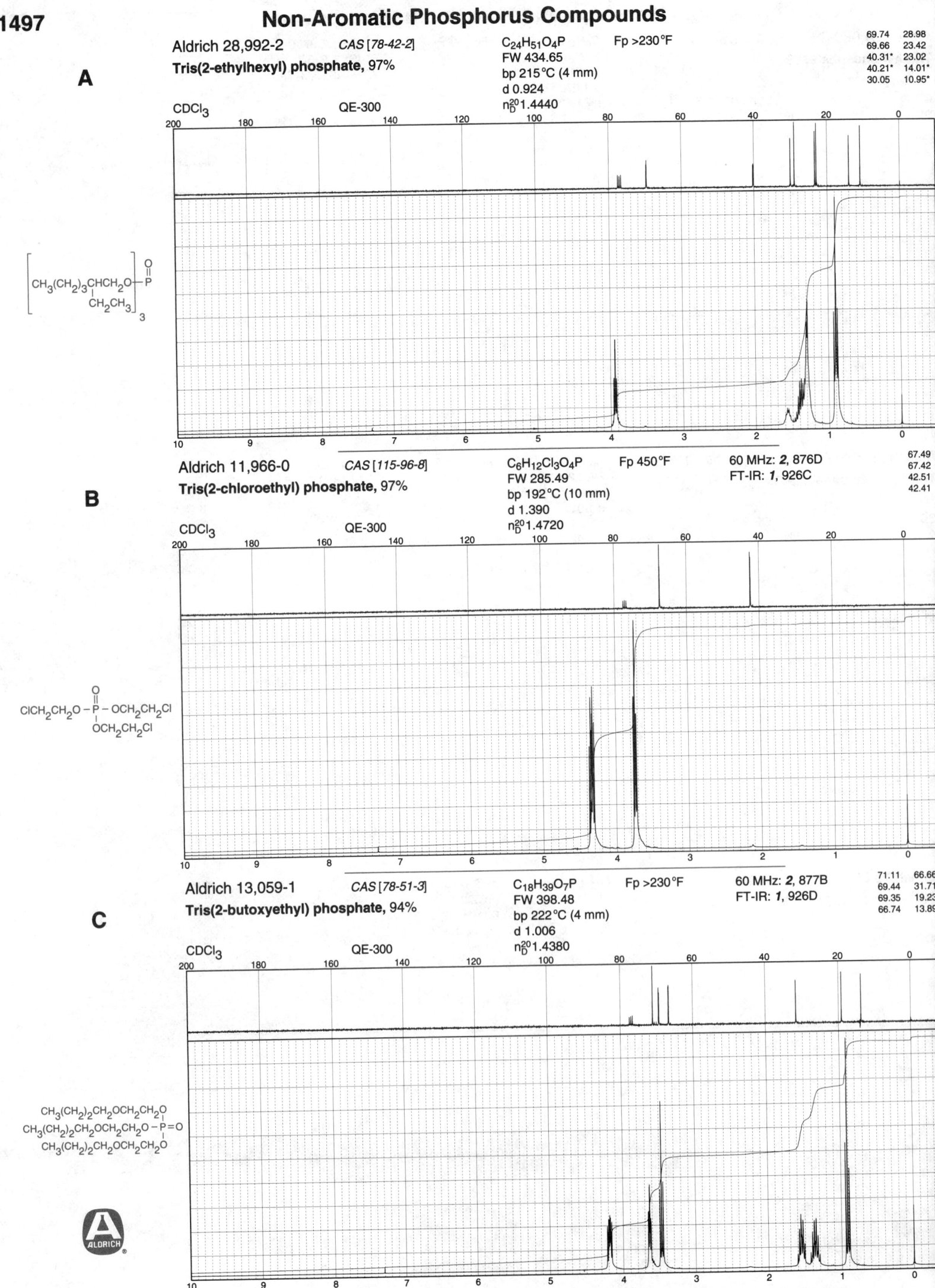

A

Aldrich 28,992-2   *CAS [78-42-2]* $C_{24}H_{51}O_4P$ Fp >230°F
Tris(2-ethylhexyl) phosphate, 97%   FW 434.65
bp 215°C (4 mm)
d 0.924
n_D^{20} 1.4440

69.74 28.98
69.66 23.42
40.31* 23.02
40.21* 14.01*
30.05 10.95*

CDCl₃   QE-300

$\left[CH_3(CH_2)_3CHCH_2O-\underset{\underset{CH_2CH_3}{|}}{\overset{\overset{O}{||}}{P}} \right]_3$

Aldrich 11,966-0   *CAS [115-96-8]* $C_6H_{12}Cl_3O_4P$ Fp 450°F   60 MHz: *2*, 876D
Tris(2-chloroethyl) phosphate, 97%   FW 285.49   FT-IR: *1*, 926C
bp 192°C (10 mm)
d 1.390
n_D^{20} 1.4720

67.49
67.42
42.51
42.41

B

CDCl₃   QE-300

$ClCH_2CH_2O-\underset{\underset{OCH_2CH_2Cl}{|}}{\overset{\overset{O}{||}}{P}}-OCH_2CH_2Cl$

Aldrich 13,059-1   *CAS [78-51-3]* $C_{18}H_{39}O_7P$ Fp >230°F   60 MHz: *2*, 877B
Tris(2-butoxyethyl) phosphate, 94%   FW 398.48   FT-IR: *1*, 926D
bp 222°C (4 mm)
d 1.006
n_D^{20} 1.4380

71.11 66.66
69.44 31.71
69.35 19.23
66.74 13.89

C

CDCl₃   QE-300

$CH_3(CH_2)_2CH_2OCH_2CH_2O$
$CH_3(CH_2)_2CH_2OCH_2CH_2O-P=O$
$CH_3(CH_2)_2CH_2OCH_2CH_2O$

ALDRICH

Non-Aromatic Phosphorus Compounds

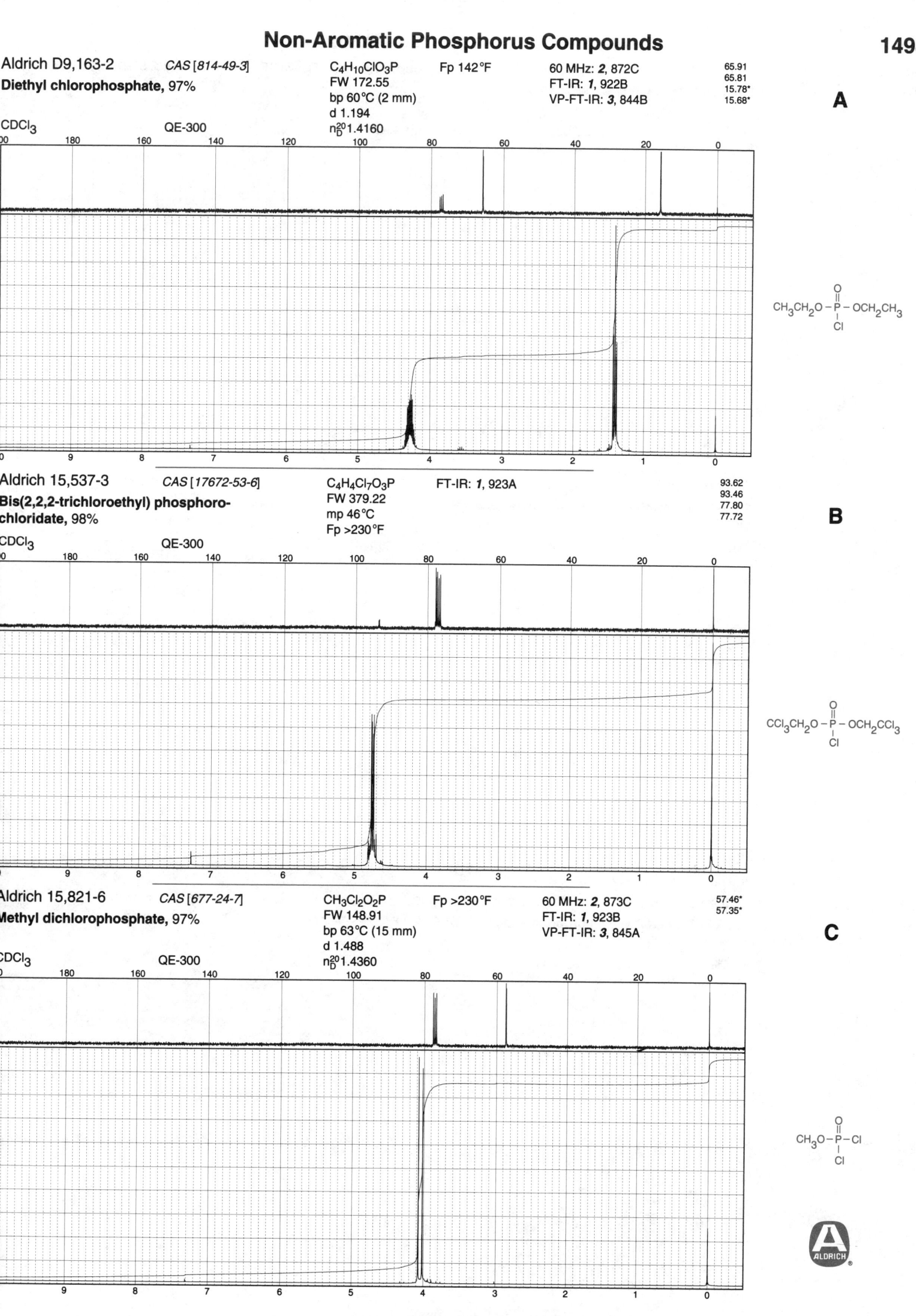

Aldrich D9,163-2   *CAS [814-49-3]* $C_4H_{10}ClO_3P$ Fp 142°F   60 MHz: *2*, 872C   65.91
Diethyl chlorophosphate, 97%   FW 172.55   FT-IR: *1*, 922B   65.81
    bp 60°C (2 mm)   VP-FT-IR: *3*, 844B   15.78*
    d 1.194   15.68*
CDCl$_3$ QE-300 n_D^{20}1.4160

A

$CH_3CH_2O-\overset{\overset{O}{\|}}{\underset{\underset{Cl}{\|}}{P}}-OCH_2CH_3$

Aldrich 15,537-3   *CAS [17672-53-6]* $C_4H_4Cl_7O_3P$ FT-IR: *1*, 923A   93.62
Bis(2,2,2-trichloroethyl) phosphoro-   FW 379.22   93.46
chloridate, 98%   mp 46°C   77.80
    Fp >230°F   77.72
CDCl$_3$ QE-300

B

$CCl_3CH_2O-\overset{\overset{O}{\|}}{\underset{\underset{Cl}{\|}}{P}}-OCH_2CCl_3$

Aldrich 15,821-6   *CAS [677-24-7]* $CH_3Cl_2O_2P$ Fp >230°F   60 MHz: *2*, 873C   57.46*
Methyl dichlorophosphate, 97%   FW 148.91   FT-IR: *1*, 923B   57.35*
    bp 63°C (15 mm)   VP-FT-IR: *3*, 845A
    d 1.488
CDCl$_3$ QE-300 n_D^{20}1.4360

C

$CH_3O-\overset{\overset{O}{\|}}{\underset{\underset{Cl}{\|}}{P}}-Cl$

Non-Aromatic Phosphorus Compounds

A

Aldrich E2,370-4   *CAS [1498-51-7]*   C₂H₅Cl₂O₂P   Fp >230°F   60 MHz: *2*, 873D

$C_2H_5Cl_2O_2P$

Ethyl dichlorophosphate, 96%

FW 162.94
bp 63°C (10 mm)
d 1.373
n_D^{20} 1.4340

FT-IR: *1*, 923C

68.70
68.58
15.61*
15.49*

CDCl₃   QE-300

$CH_3CH_2O-\overset{\overset{O}{\|}}{\underset{\underset{Cl}{|}}{P}}-Cl$

B

Aldrich 24,216-0   *CAS [18868-46-7]*   C₂H₂Cl₅O₂P   60 MHz: *2*, 874B

$C_2H_2Cl_5O_2P$

2,2,2-Trichloroethyl phosphorodichloridate, 98%

FW 266.28
mp 29°C
bp 57°C
Fp >230°F

FT-IR: *1*, 924A

93.08
92.90
78.31
78.21

CDCl₃   QE-300

$CCl_3CH_2O-\overset{\overset{O}{\|}}{\underset{\underset{Cl}{|}}{P}}-Cl$

C

Aldrich 24,428-7   *CAS [53676-22-5]*   C₂H₂Br₃Cl₂O₂P   60 MHz: *2*, 874C

$C_2H_2Br_3Cl_2O_2P$

2,2,2-Tribromoethyl dichlorophosphate, 97%

FW 399.64
mp 35°C
bp 91°C
Fp >230°F

FT-IR: *1*, 924B

80.79
80.69
32.35
32.16

CDCl₃   QE-300

$CBr_3CH_2O-\overset{\overset{O}{\|}}{\underset{\underset{Cl}{|}}{P}}-Cl$

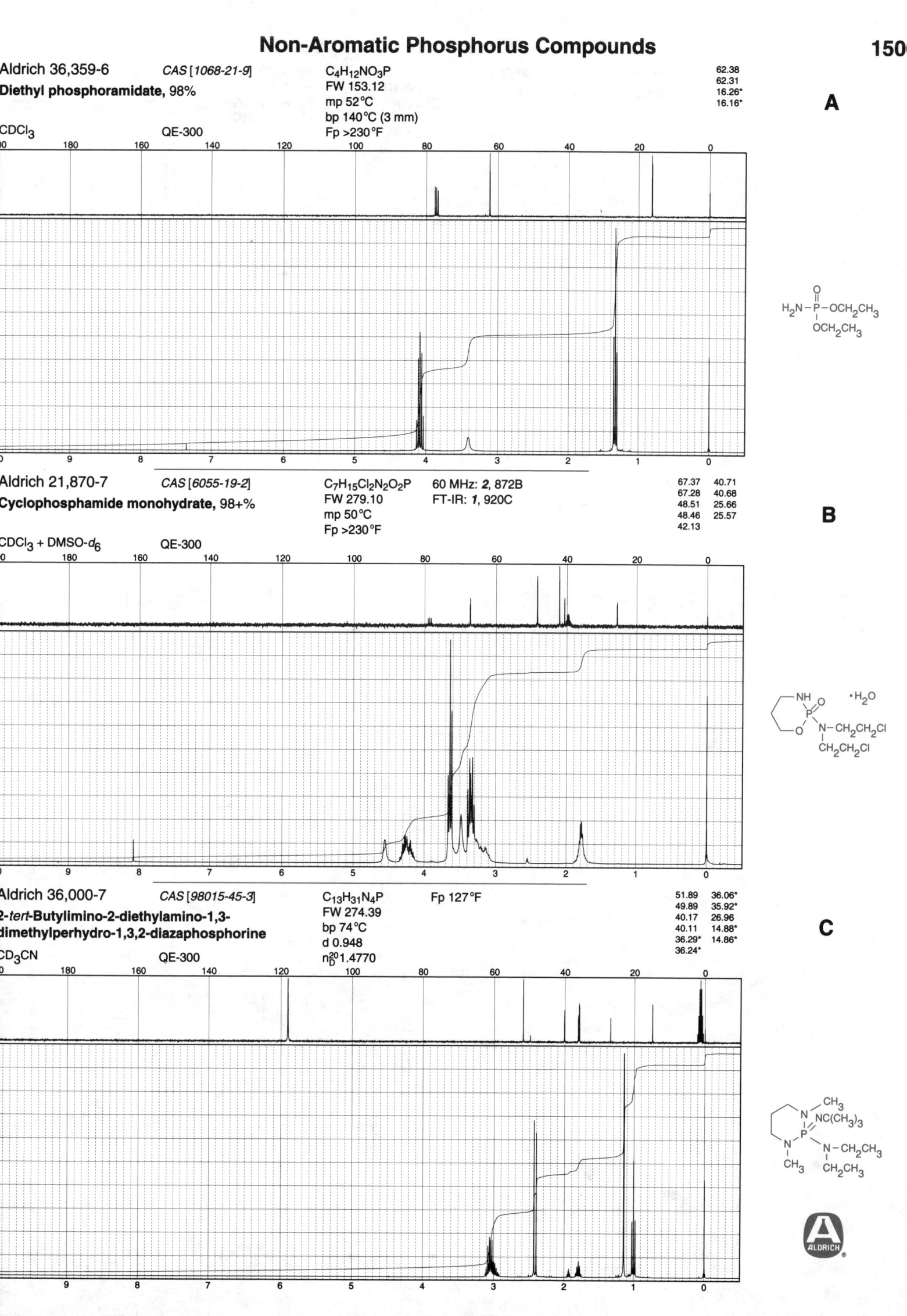

Aldrich 36,359-6 *CAS [1068-21-9]*

Diethyl phosphoramidate, 98%

C$_4$H$_{12}$NO$_3$P
FW 153.12
mp 52°C
bp 140°C (3 mm)
Fp >230°F

62.38
62.31
16.26*
16.16*

A

CDCl$_3$ QE-300

Aldrich 21,870-7 *CAS [6055-19-2]*

Cyclophosphamide monohydrate, 98+%

C$_7$H$_{15}$Cl$_2$N$_2$O$_2$P   60 MHz: **2**, 872B
FW 279.10   FT-IR: **1**, 920C
mp 50°C
Fp >230°F

67.37   40.71
67.28   40.68
48.51   25.66
48.46   25.57
42.13

B

CDCl$_3$ + DMSO-d_6 QE-300

Aldrich 36,000-7 *CAS [98015-45-3]*

2-*tert*-Butylimino-2-diethylamino-1,3-dimethylperhydro-1,3,2-diazaphosphorine

C$_{13}$H$_{31}$N$_4$P   Fp 127°F
FW 274.39
bp 74°C
d 0.948
n$_D^{20}$ 1.4770

51.89   36.06*
49.89   35.92*
40.17   26.96
40.11   14.88*
36.29*   14.86*
36.24*

C

CD$_3$CN   QE-300

Non-Aromatic Phosphorus Compounds

A

Aldrich H1,160-2   CAS [680-31-9] $C_6H_{18}N_3OP$ n_D^{20} 1.4580   36.86*
                                                                                    36.81*

Hexamethylphosphoramide, 99%   FW 179.20   Fp 222°F

mp 7°C   60 MHz: **2**, 872A

bp 231°C (740 mm) FT-IR: **1**, 921C

d 1.030

CDCl₃   QE-300

$(CH_3)_2N-\underset{\underset{N(CH_3)_2}{|}}{\overset{\overset{O}{\|}}{P}}-N(CH_3)_2$

B

Aldrich 21,625-9   CAS [4441-17-2] $C_{15}H_{30}N_3OP$ 60 MHz: **2**, 871D   45.78
                                                                                          45.76
Tripiperidinophosphine oxide, 98%   FW 299.40   FT-IR: **1**, 921D   26.54
                                                                                          26.47
mp 41°C   24.84

Fp >230°F

CDCl₃   QE-300

C

Aldrich 25,129-1 $C_3H_9O_6P$ FT-IR: **1**, 927B   77.68*
                                                            77.62*
Glycerol 2-phosphate, disodium salt   FW 216.04   65.19
hydrate, 98%   mp 103°C   65.14

D₂O   QE-300

$\underset{HOCH_2}{\overset{HOCH_2}{\diagdown}}CH-O-\underset{\underset{ONa}{|}}{\overset{\overset{O}{\|}}{P}}-ONa \quad \cdot xH_2O$

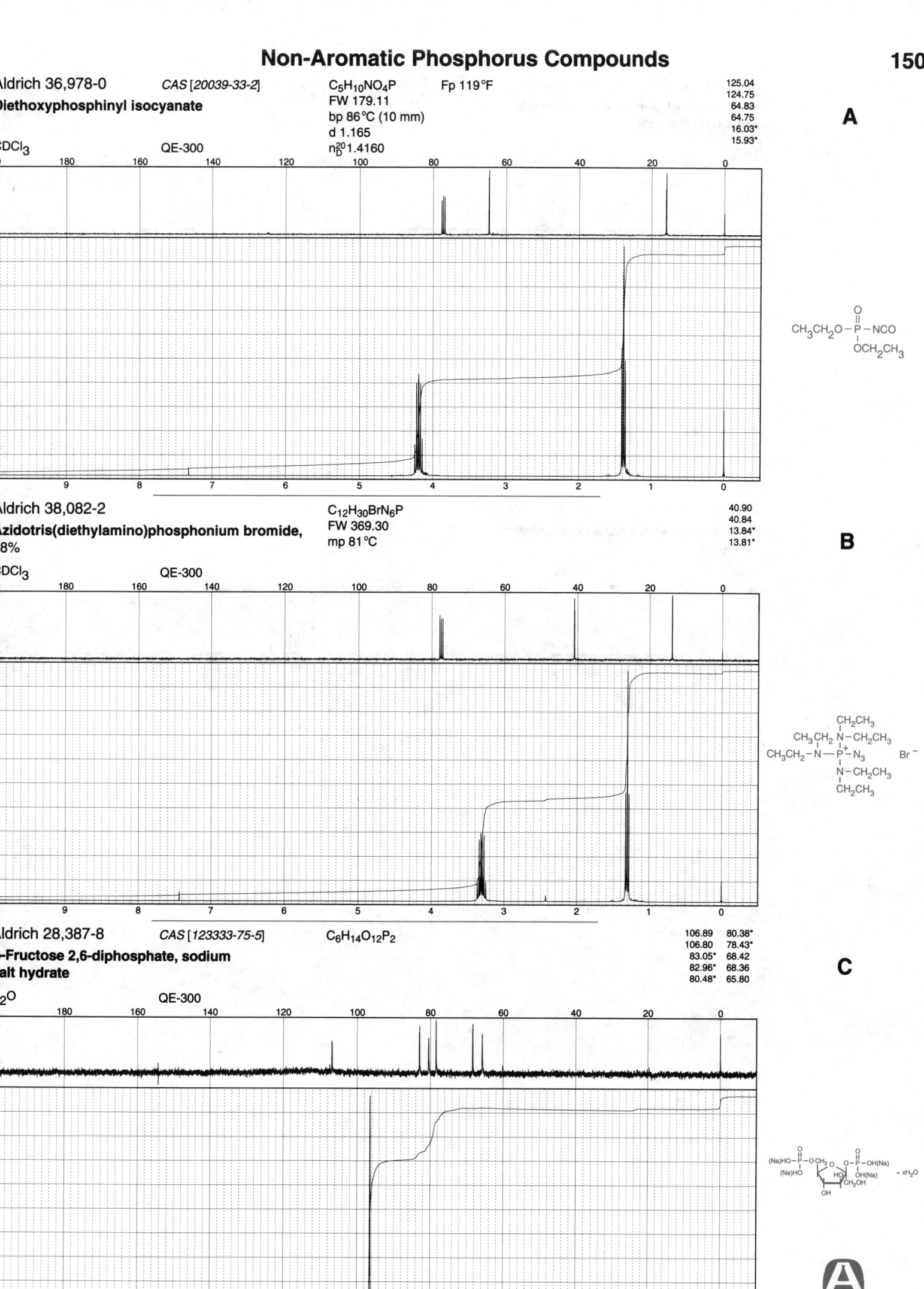

Aldrich 36,978-0   *CAS [20039-33-2]*   C₅H₁₀NO₄P   Fp 119°F
Diethoxyphosphinyl isocyanate   FW 179.11
bp 86°C (10 mm)
d 1.165
CDCl₃   QE-300 n_D^{20} 1.4160

125.04
124.75
64.83
64.75
16.03*
15.93*

A

$CH_3CH_2O-\overset{\overset{O}{\|}}{\underset{\underset{OCH_2CH_3}{|}}{P}}-NCO$

Aldrich 38,082-2   C₁₂H₃₀BrN₆P
Azidotris(diethylamino)phosphonium bromide,   FW 369.30
98%   mp 81°C
CDCl₃   QE-300

40.90
40.84
13.84*
13.81*

B

$CH_3CH_2-N-\overset{\overset{CH_2CH_3}{|}}{\underset{\underset{CH_2CH_3}{|}}{\underset{N-CH_2CH_3}{|}}{\overset{N-CH_2CH_3}{|}}}-N_3$ Br⁻

Aldrich 28,387-8   *CAS [123333-75-5]*   C₆H₁₄O₁₂P₂
ᴅ-Fructose 2,6-diphosphate, sodium
salt hydrate
D₂O   QE-300

106.89   80.38*
106.80   78.43*
83.05*   68.42
82.96*   68.36
80.48*   65.80

C

$(Na)HO-\overset{O}{\underset{(Na)HO}{\overset{\|}{P}}}-OCH_2O\quad\overset{O}{\underset{OH(Na)}{\overset{\|}{P}}}-OH(Na)$ • xH₂O

A

Aldrich 85,887-0   *CAS [5996-14-5]* $C_6H_{13}O_9P$ 60 MHz: **2**, 878C

α-D-Glucose-1-phosphate, dipotassium salt dihydrate, 98%

FW 372.36   FT-IR: **1**, 929B

mp 209°C d.

96.39*	74.94*
96.33*	74.79*
75.94*	72.54*
75.03*	63.50

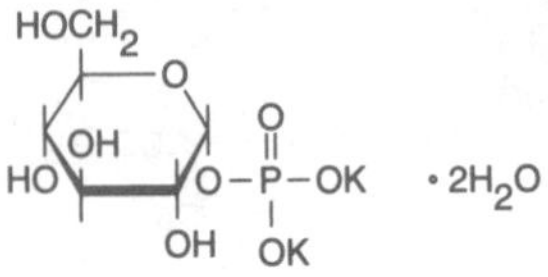

D₂O   QE-300

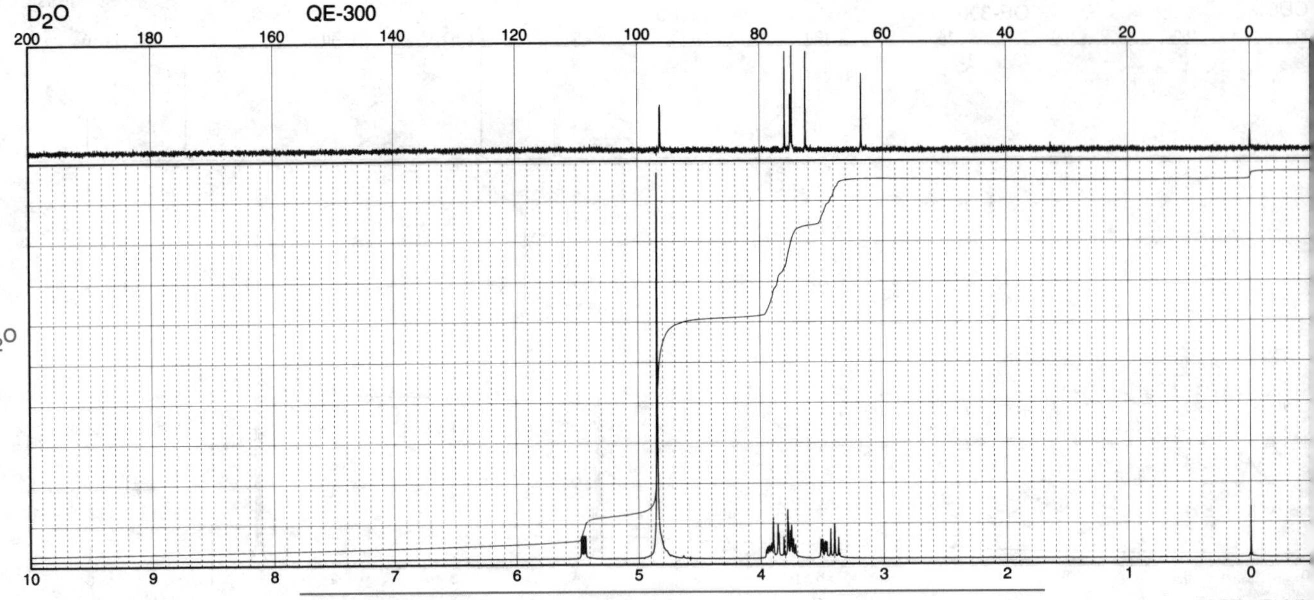

B

Aldrich 28,597-8   *CAS [54010-71-8]* $C_6H_{13}O_9P$

β-D-Glucose-6-phosphate, monosodium salt, 98%

FW 282.12

mp 204°C d.

98.79*	74.24*
94.98*	73.38*
78.27*	73.28*
77.70*	71.93*
77.60*	66.73
76.90*	66.67
75.35*	66.65

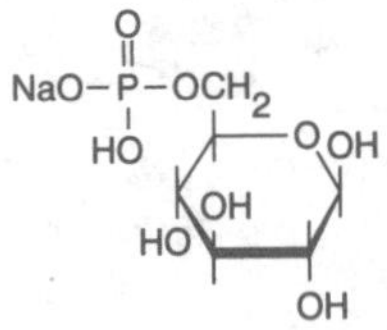

D₂O   QE-300

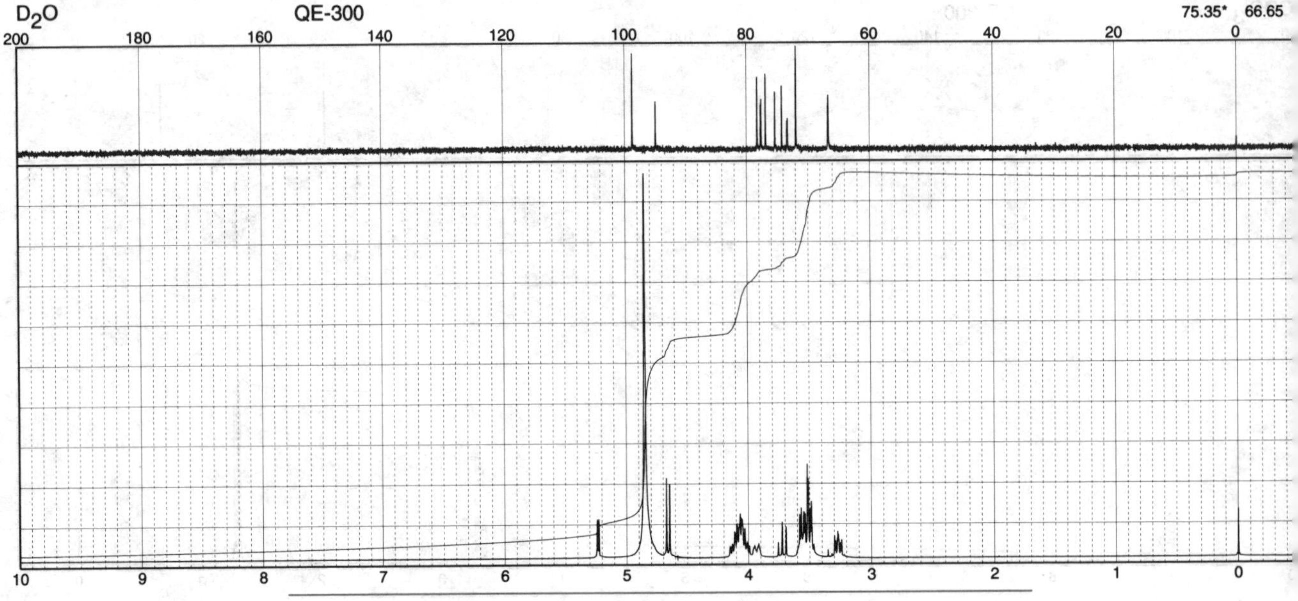

ALPHABETICAL
INDEX
VOLUME 1